W9-BQL-317

The Encyclopedia of Seeds

Science, Technology and Uses

This book is dedicated to the memory of Christine E. Bewley. We will miss her.

The Encyclopedia of Seeds

Science, Technology and Uses

Edited by

Michael Black

King's College, London, UK

J. Derek Bewley

University of Guelph, Ontario, Canada

Peter Halmer

Germain's Technology Group, Norfolk, UK

CABI is a trading name of CAB International

CABI Head Office
Nosworthy Way
Wallingford
Oxfordshire OX10 8DE
UK
Tel: +44 (0)1491 832111
Fax: +44 (0)1491 833508
E-mail: cabi@cabi.org
Website: www.cabi.org

CABI North American Office
875 Massachusetts Avenue
7th Floor
Cambridge, MA 02139
USA
Tel: +1 617 395 4056
Fax: +1 617 354 6875
E-mail: cabi-nao@cabi.org

A catalogue record for this book is available from the British Library, London, UK.

A catalogue record for this book is available from the Library of Congress, Washington, DC.

ISBN-10: 0–85199–723–6
ISBN-13: 978–0–85199–723–0

Typeset by Columns Design Ltd., Reading RG4 7DH, UK
Printed and bound in the UK by Cromwell Press, Trowbridge

Contributors

Arias, Ingrid, Becker Underwood, Harwood Industrial Estate, Harwood Road, Littlehampton, West Sussex BN17 7AU, UK

Babu, Nirmal, Indian Institute of Spices Research, Calicut, 673012, India

Bassel, George W., Department of Molecular and Cellular Biology, University of Guelph, Ontario, Canada, N1G 2W1

Benech-Arnold, Roberta, IFEVA, Facultad de Agronomia, Universidad de Buenos Aires, Buenos Aires, Argentina

Bethke, Paul, University of California at Berkeley, Dept of Plant and Microbial Biology, 311A Koshland Hall, Berkeley, CA 94720-3102, USA

Bewley, J. Derek, Department of Molecular and Cellular Biology, University of Guelph, Ontario, Canada, N1G 2W1

Biddle, Anthony, Processors and Growers Research Organisation, Great North Road, Thornhaugh, Peterborough PE8 6HJ, UK

Black, Michael, King's College, Franklin Wilkins Building, 150 Stamford Street, London SE1 9NH, UK

Bradbeer, J.W., 85 Court Lane, London, SE21 7EF, UK

Bradford, Kent J., University of California, Department of Vegetable Crops, One Shields Avenue, Davis, CA 95616-8631, USA

Buitink, Julia, UMR Physiologie Moleculaire des Semences, 16 Bd Lavoisier, F49045 Angers, France

Bullock, James M., NERC Centre for Ecology and Hydrology, CEH Dorset, Winfrith Technology Centre, Dorchester, Dorset DT2 8ZD, UK

Burbulis, Natalija, Department Plant Agriculture, University of Guelph, Guelph, Ontario, Canada, N1G 2W1

Burris, Joe, Burris Consulting, 1707 Burnett Ave, Ames, IA 500010, USA

Chaudhury, Abdul, CSIRO Plant Industry, GPO Box 1600, Canberra, ACT 2601, Australia

Cockerell, Valerie, Scottish Agricultural Science Agency (SASA), East Craigs, Edinburgh EH 12 8NJ, UK

Cohn, Marc A., Department of Plant Pathology and Crop Physiology, 302 Life Sciences Building, Louisiana State University, Baton Rouge, LA 70803, USA

Côme, Daniel, Université Pierre et Marie Curie, Laboratoire de Physiologie Végétale Appliquée, site d'Ivry, Batiment B, 2ème étage, 3 rue Galilée, 94200 Ivry sur Seine, France

Copeland, Lawrence O., Michigan State University, Dept of Crop & Soil Sciences, 278 Plant & Soil Sciences Bld., East Lansing, MI 48824-1325, USA

Corbineau, Françoise, Université Pierre et Marie Curie, Laboratoire de Physiologie Végétale Appliquée, site d'Ivry, Batiment B, 2ème étage, 3 rue Galilée, 94200 Ivry sur Seine, France

Costa, J.M., Instituto Superior da Agronomia, Dep. Botânica e Engenharia Biológica, Universidade Técnica de Lisboa, Tapada da Ajuda, 1349-017 Lisboa, Portugal

Counce, Paul, University of Arkansas, Rice Research and Extension Center, Dept of Crops, Soil & Environmental Science, PO Box 1168, Stuttgart, AR 72160, USA

de la Vega, Abelardo, Advanta Semillas S.A.I.C., Ruta 33, Km 636, C.C. 559, 2600 – Venado Tuerto, Santa Fe, Argentina

de Milliano, Walter A.J., African Centre for Crop Improvement, Plant Pathology, UKZNP, Private Bag X01, Scottsville, 3209, KwaZulu-Natal, South Africa

de Rooij, Petra, Syngenta Seeds B.V., Westeinde 62, PO Box 2, 1600 AA Enkhuizen, The Netherlands

Devlin, P., Dept of Biological Sciences, Royal Holloway College, Egham, Surrey TW20 0EX, UK

Dickie, John, Royal Botanic Gardens Kew, Wakehurst Place, Ardingly, West Sussex RH17 6TN, UK

Don, Ronald, Scottish Agricultural Science Agency (SASA), East Craigs, Edinburgh EH 12 8NJ, UK

Downey, Keith, Agriculture and Agri-Food Canada, Saskatoon Research Centre, 107 Science Place, Saskatoon, Canada, S7N 0X2

Egli, Dennis B., Department of Agronomy, University of Kentucky, Lexington, KY 40546-0091, USA

Evers, Anthony, Talybont, Albert Street, Markyate, Herts AL3 8HY, UK

Falk, Duane E., Department of Plant Agriculture, University of Guelph, Guelph, Ontario, Canada, N1G 2W1

Finch-Savage, William, Warwick-HRI, Warwick University, Wellesbourne, Warwick, CV35 9EF, UK

Foley, Michael E., USDA-ARS, Biosciences Research Lab, Fargo, ND 58105-5674, USA

Footitt, Steven, Crop Performance and Improvement Division, Rothamsted Research, Harpenden, Herts AL5 2JQ, UK

Gong, Xuemei, Department of Molecular and Cellular Biology, University of Guelph, Guelph, Ontario, Canada, N1G 2W1

Gooding, Mike J., Dept of Agriculture, University of Reading, Earley Gate, Whiteknights Road, Reading RG6 6AR, UK

Gosling, Peter, Alice Holt Forestry Research Station, Wrecclesham, Farnham, Surrey GU10 4LU, UK

Gutterman, Y., Jacob Blaustein Institute for Desert Research, Ben Gurion University of the Negev, Sede Boker Campus, 84990, Israel

Habstritt, Charles H., 101A Owen Hall, University of Minnesota, 2900 University Avenue, Crookston, MN 56716-5001, USA

Hall, Antonio J., University of Buenos Aires, IFEVA, Agricultural Plant Physiology and Ecological Research Institute, Facultad de Agronomia, Av San Martin 4453, 1417 Buenos Aires, Argentina

Halmer, Peter, GTG – Germain's Technology Group, Hansa Road, Hardwick Industrial Estate, King's Lynn, Norfolk PE30 4LG, UK

Hampton, John G., New Zealand Seed Technology Institute, PO Box 84, Lincoln University, Canterbury, New Zealand,

Heuvelink, Ep, Wageningen University, HPC, Marijkeweg 22, 6709 PG Wageningen, The Netherlands

Hilhorst, Henk, Laboratory of Plant Physiology, Department of Plant Sciences, Wageningen University, Arboretumlaan 4, NL-6703 BD Wageningen, The Netherlands

Hodges, Rick, Natural Resources Institute, Chatham Maritime, Kent ME4 4TB, UK

Holdsworth, Michael, University of Nottingham, Dept of Crop Science, Sutton Bonington Campus, Loughborough, Leics LE12 5RD, UK

Houghton, Peter J., Department of Pharmacy, King's College London, Franklin Wilkins Building, 150 Stamford Street, London SE1 6NN, UK

Jack, Anne M., Department of Agronomy, University of Kentucky, College of Agriculture, 1405 Veterans Drive, Lexington, KY 40546-0312, USA

Jaggard, Keith, IACR Brooms Barn, Higham, Bury St Edmunds, Suffolk IP28 6NP, UK

James, Martha G., 2152 Molecular Biology Building, Iowa State University, Ames, IA 50011, USA

Jones, Russell L., University of California at Berkeley, Dept of Plant & Microbial Biology, 311A Koshland Hall, Berkeley, CA 94720-3102, USA

Jordan, David L., Department of Crop Science, College of Agriculture and Life Science, North Carolina State University, Williams Hall 4124, PO Box 762, Raleigh, NC 27695-7620, USA

Kermode, Allison R., Department of Biological Sciences, Simon Fraser University, Burnaby BC, Canada, V5A 1S6

Kinney, Anthony J., Research Leader, Dupont Experiment Station, Crop Genetics, PO Box 80402, Wilmington, DE 19880-0402, USA

Kott, Laima S., Department of Plant Agriculture, University of Guelph, Guelph, Ontario, Canada, N1G 2W1

le Buanec, Bernard, Secretary General, ISF, Chemin du Reposir 7, 1260 Nyon, Switzerland

Leprince, Olivier, UMR Physiologie Moleculaire des Semences, 16 Bd Lavoisier, F49045 Angers, France

Linington, Simon H., Millennium Seed Bank Project, Seed Conservation Department, Royal Botanic Gardens Kew, Wakehurst Place, Ardingly, West Sussex RH17 6TN, UK

Loeffler, Tim, Seminis Vegetable Seeds Inc., 2700 Camino del Sol, Oxnard, CA 93030-7967, USA.

Lycett, Grantley, School of Biosciences, University of Nottingham, Sutton Bonington Campus, Loughborough, Leics LE12 5RD, UK

Miller, Robert D., Department of Agronomy, University of Kentucky, College of Agriculture, 1405 Veterans Drive, Lexington, KY 40546-0312, USA

Morris, Peter, School of Life Sciences, Heriot-Watt University, Riccarton, Edinburgh, EH14 4AS, UK

Mullen, Rob, Department of Molecular and Cellular Biology, University of Guelph, Guelph, Ontario, Canada, N1G 2W1

Murray, James Ross, School of Agronomy and Horticulture, The University of Queensland, Gatton Campus, Gatton, Qld 4343, Australia

Negbi, Moshe, 11 Yehuda Hanasi, Tel Aviv 69200, Israel

Nesbitt, Mark, Centre for Economic Botany, Royal Botanic Gardens Kew, Richmond, Surrey TW9 3AE, UK

Offler, Christina E., Department of Biological Sciences, University of Newcastle, Newcastle, NSW 2308, Australia

Oliver, Melvin J., USDA-ARS Plant Genetics Research Unit, 205 Curtis Hall, University of Missouri, Columbia, Columbia, MO 65211, USA

Olsen, Odd-Arne, Pioneer Hi-Bred International Inc, 7300 NW 62nd Avenue, Johnston, IA 50131-1004, USA

Palmer, G., Institute of Brewing and Distilling Technology, School of Life Sciences, Heriot-Watt University, Edinburgh EH14 4AS, UK

Patrick, John W., Department of Biological Sciences, University of Newcastle, Newcastle, NSW 2308, Australia

Peter, K.V., Vice Chancellor, Kerala Agricultural University, KAU-PO, Velanikkara, Thrissur, Kerala State, 680656, India

Peumans, Willy J., Ghent University, Department of Molecular Biotechnology, Coupure Links 653, 9000 Gent, Belgium

Polito, V.S., University of California, Department of Pomology, 1035 Wickson Hall, One Shields Avenue, Davis, CA 95616-8683, USA

Powell, Alison, Department of Agriculture and Forestry, University of Aberdeen, Aberdeen, AB24 5UA, UK

Probert, Robin, Millennium Seed Bank Project, Royal Botanic Gardens Kew, Wakehurst Place, Ardingly, West Sussex RH17 6TN, UK

Quiros, Carlos F., University of California, Department of Vegetable Crops, One Shields Avenue, Davis, CA 95616, USA

Reid, Alexandra J., Agriculture and Agri-Food Canada/Agriculture et Agroalimentaire Canada, 1391 Sandford St, London, Ontario, Canada, N5V 4T3

Roberts, J., School of Biosciences, University of Nottingham, Sutton Bonington Campus, Loughborough, Leics LE12 5RD, UK

Rodriguez, M.V., IFEVA, Facultad de Agronomia, Universidad de Buenos Aires, Buenos Aires, Argentina

Rodriguez-Sotres, Rogelio, Departamento de Bioquimica, Facultad de Quimica, UNAM, Ave. Universidad y Copilco, Mexico 04510 DF, Mexico

Ryder, Edward J., U.S. Agricultural Research Station, 1636 East Alisal Street, Salinas, CA 93905, USA

Sakai, Hajime, DuPont Agriculture and Nutrition, DTP200, 1 Innovation Way, Newark, DE 19711, USA

Sanchez, R.A., IFEVA, Facultad de Agronomia, Universidad de Buenos Aires, Buenos Aires, Argentina

Scudamore, Keith, KAS Mycotoxins, 6 Fern Drive, Taplow, Maidenhead, Berkshire SL6 0JS, UK

Shewry, Peter, Rothamsted Research, Harpenden, Hertfordshire AL5 2JQ, UK

Smith, Brian, Horticulture Research International, Wellesbourne, Warwick CV35 9EF, UK

Smith, C. Wayne, Cotton Improvement Lab., Dept of Soil and Crop Sci. Texas A&M University, College Station, TX 77843-2474, USA.

Spears, Jan, Department of Crop Science, College of Agriculture and Life Science, North Carolina State University, Williams Hall 4124, PO Box 762, Raleigh, NC 27695-7620, USA

Spoelstra, Patrick, Syngenta Seeds B.V., Westeinde 62, PO Box 2, 1600 AA Enkhuizen, The Netherlands

Stuppy, Wolfgang, Royal Botanic Gardens Kew, Wakehurst Place, Ardingly, West Sussex RH17 6TN, UK

TeKrony, Dennis, Department of Agronomy, University of Kentucky College of Agriculture, Plant Sciences Building, 1405 Veterans Drive, Lexington, Kentucky 40546-0312, USA

Thaxton, Peggy, Delta Research and Extension Center, 82 Stoneville Road, PO Box 197, Stoneville, MS 38776, USA

Thompson, Ken, Department of Animal and Plant Sciences, University of Sheffield, Sheffield S10 2TN, UK

Throne-Holst, Mimmi, Department of Biological Sciences, University of Newcastle, Newcastle, NSW 2308, Australia

Triplett, B., USDA-ARS, Southern Regional Research Center, 1100 Robert E Lee Blvd, New Orleans, LA 70124, USA

Tripp, Robert, Overseas Development Institute, 111 Westminster Bridge Road, London SE1 7JD, UK

Van Damme, Els, Ghent University, Department of Molecular Biotechnology, Coupure Links 653, 9000 Gent, Belgium

van Kester, Wim, Syngenta Seeds B.V., Westeinde 62, PO Box 2, 1600 AA Enkhuizen, The Netherlands

Vazquez Ramos, Jorge, Departamento de Bioquimica, Facultad de Quimica, UNAM, Ave. Universidad y Copilco, Mexico 04510 DF, Mexico

von Aderkas, Patrick, Department of Biology, PO Box 3020 Stn CSC, University of Victoria, Victoria, BC, Canada, V8W 3N5

Votava, Eric J., Manager Plant Breeding, Johnny's Selected Seeds, 955 Benton Avenue, Winslow, ME 04901, USA

Voysest-Voysest, Oswaldo, 1225 Bushnell Street, Beloit, WI 53511, USA

Walters, Christina, Plant Germplasm Preservation, USDA-ARS, 1111 South Mason St, Ft. Collins, CO 80521-4500, USA

Watkinson, Jonathan I., Dept of Plant Physiology, 103 Plant Molecular Biology Building, Old Glade Road, Virginia Polytechnic Institute and State Univ, Blacksburg, VA 24061-0327, USA

Way, Michael, Seed Conservation Department, Royal Botanic Gardens Kew, Wakehurst Place, Ardingly, Haywards Heath, West Sussex RH17 6TN, UK

Welbaum, Greg E., Department of Horticulture, Virginia Polytechnic Institute and State University, Blacksburg, VA 24061-0327, USA

Whalley, W. Richard, Silsoe Research Institute, Wrest Park, Silsoe, Bedford, MK45 4HS, UK

Whipps, John M., Warwick-HRI, Warwick University, Wellesbourne, Warwick, CV35 9EF, UK

Williams, Paul M., Becker Underwood, Harwood Industrial Estate, Harwood Road, Littlehampton, West Sussex BN17 7AU, UK

Wilson, Karl A., Department of Biological Sciences, SUNY, Binghamton, NY 13902-6000, USA

Guide to Author's Initials

AAP	Powell, Alison A.
AB	Biddle, Anthony
AC	Chaudhury, Abdul
AE	Evers, Anthony
AJ	Jack, Anne
AJdlV	de la Vega, Abelardo J.
AJH	Hall, Antonio
AJK	Kinney, Anthony J.
AJR	Reid, Alexandra J.
ARK	Kermode, Allison R.
BAT	Triplett, Barbara A.
BlB	le Buanec, Bernard
BS	Smith, Brian
CEO	Offler, Christina E.
CFQ	Quiros, Carlos
CHH	Habstritt, Charles
CWS	Smith, C. Wayne
CW	Walters, Christina
DBE	Egli, Dennis B.
DC	Côme, Daniel
DF	Falk, Duane E.
DLJ	Jordan, David L.
DTK	TeKrony, Dennis
EH	Heuvelink, Ep
EJR	Ryder, Edward J.
EJV	Votava, Eric
EVD	van Damme, Els
FC	Corbineau, Francoise
GEW	Welbaum, Greg
GHP	Palmer, Geoffrey H.
GL	Lycett, Grantley
GWB	Bassel, George W.
HS	Sakai, Hajime
HWMH	Hilhorst, Henk W.M.H.
IA	Arias, Ingrid
JB	Buitink, Julia
JD	Dickie, John
JDB	Bewley, J. Derek
JS	Spears, Jan F.
JGH	Hampton, John G.
JIW	Watkinson, Jonathan I.
JMB	Bullock, James
JMC	Costa, J.M.
JMW	Whipps, John
JR	Roberts, Jerry
JRM	Murray, James Ross
JSB	Burris, Joe S.
JVR	Vazquez Ramos, Jorge
JWB	Bradbeer, J. William
JWP	Patrick, John W.
KAS	Scudamore, Keith A.
KAW	Wilson, Karl A.
KD	Downey, Keith
KJ	Jaggard, Keith
KJB	Bradford, Kent J.
KNB	Babu, K. Nirmal
KT	Thompson, Ken
KVP	Peter, Kuruppacharil V.
LOC	Copeland, Lawrence O.
LSK	Kott, Laima
MAC	Cohn, Marc
MB	Black, Michael
MEF	Foley, Michael E.
MGJ	James, Martha G.
MH	Holdsworth, Michael
MJG	Gooding, Mike
MJO	Oliver, Melvin J.
MN	Nesbitt, Mark
MoN	Negbi, Moshe
MT-H	Throne-Holst, Mimmi
MVR	Rodriguez, M.V.
MW	Way, Michael
NB	Burbulis, Natalija
OAO	Olsen, Odd-Arne
OL	Leprince, Olivier
OV-V	Voysest-Voysest, Oswaldo
PB	Bethke, Paul
PC	Counce, Paul
PD	Devlin, Paul
PDR	de Rooij, Petra
PG	Gosling, Peter
PH	Halmer, Peter
PJH	Houghton, Peter J.
PM	Morris, Peter
PMW	Williams, Paul
PRS	Shewry, Peter R.
PS	Spoelstra, Patrick
PT	Thaxton, Peggy
PvA	Von Aderkas, Patrick
RAS	Sanchez, Rudolfo A.
RD	Don, Ronald
RDM	Miller, Robert
RH	Hodges, Rick
RJ	Jones, Russell
RJP	Probert, Robin J.
RLB-A	Benech-Arnold, Roberto
RRS	Rodriguez-Sotres, Rogelio
RT	Tripp, Robert
RTM	Mullen, Robert T.
SF	Footitt, Steven
SHL	Linington, Simon H.
TL	Loeffler, Timothy

VC	Cockerell, Valerie
VP	Polito, Victor S.
WAJdeM	de Milliano, Walter A.J.
WFS	Finch-Savage, William
WJP	Peumans, Willy J.
WRW	Whalley, Richard
WS	Stuppy, Wolfgang
WvK	van Kester, Wim
XG	Gong, Xuemei
YG	Gutterman, Yitzchak

Preface

Our goal in producing this encyclopedia is to provide comprehensive coverage of seeds – for the first time together in a single volume. The reader will find here all the basics of seed biology, as well as the background information necessary for understanding them: seed structure, composition, development, viability, germination, dormancy, hormones and metabolism, seedling emergence, ecology and evolution and domestication. Interwoven with this are accounts of the technological principles of seed and grain production, drying, processing, storage and testing, of treatments used to enhance seed performance and to control diseases and pests, along with the uses of seeds to propagate the major and minor food, spice, fodder, forage and industrial crops in worldwide agriculture and horticulture, and the systems, laws and regulations that govern the seed trade. The third major strand of information concerns the uses of seeds as food – as grain sources of carbohydrates, proteins, oils and fats for humans and animals – as well as beverages, spices, flavourings and food additives, pharmaceuticals and poisons, and their uses as industrial sources of chemicals and fibres, along with aspects of cultural practices and social and scientific history.

The encyclopedia aims to address the needs of a diverse readership. Seeds are the focus of attention of a multiplicity of scientists, technologists and other professionals, of scholars and researchers in academia and commerce, and of those who work in the seed, food and crop protection industries and in government and the law. We therefore expect that our readers will approach the subject of seeds from many disciplinary directions: as biologists, agriculturalists and horticulturalists, agronomists, breeders and geneticists, economists, salespeople and marketers, conservationists, ethnobotanists, nutritionists, pharmacologists, teachers and legislators, and commercial producers of varieties and crop protection agents – and that does not exhaust the list! We have tried to anticipate and serve the needs of all of these and of those who might seek information *ad hoc*. Indeed, we hope that many will experience the joy of serendipity, discovering new topics by browsing or through accidents of alphabetical adjacency.

How the encyclopedia was constructed

The original and central idea of this book was to provide up-to-date and comprehensive coverage of seed science and technology for those working in institutes and research organizations and the seed industry, as a compendium of different topics or terms, authored by a set of international experts, and arranged alphabetically. The size of each entry would depend on the complexity and detail required to describe it – from simple definitions to short descriptive articles with references to specialist literature for further reading, bearing in mind the wide and varied potential readership.

Development of the book began in earnest in 2002, when the Editors produced a provisional compilation of contents to serve as a first draft template. Then began the process of gathering expert contributors, each commissioned to write on a thematically linked group of topics. Authors were recruited to the cause, in many cases selecting colleagues to assist them, so that altogether 110 people, in addition to the editors, have had a hand in writing this book. Some years after its formation in embryo, elaborated but very much in line with its guiding concept, *The Encyclopedia of Seeds* finally emerges into the light – perhaps a little larger than originally conceived and rather later too.

Although the choice of much subject matter was straightforward, sometimes we faced difficult editorial decisions about where to set the boundaries. Broadly speaking, the plant life cycle is covered here from the development of reproductive structures through to the establishment of developed seedlings, although the important role that seeds play in delivering protection against some pests and diseases that afflict mature crops is also included. But, for example, only certain aspects of genetics and breeding, of agriculture and horticulture, and of food science and industrial processing are touched on. Also, though facets of molecular biology, biochemistry, cell biology, anatomy and morphology, pathology, botany and ecology are directly dealt with, readers are assumed either to have a basic grounding in these subjects or to know how to gain that knowledge from other books and the Internet.

As the contributed articles started to be submitted in 2003 – usually in the form of extended essays – the Editors began the task of arranging and titling the information in a logical and coherent way, bearing in mind the alphabetical arrangement of the book and anticipating the requirements of readers to look up a topic directly in the body of the encyclopedia without necessarily using the Index. Quite often a submitted piece had to be split into several smaller articles and short definitions had to be created. Conversely, it was sometimes necessary to meld sections of one contributor's article with those of another or to prune overlapping submissions. Frequently gaps and omissions became apparent, and additions and elaborations had to be made – mostly written by the Editors themselves, who worked to achieve a uniformity and consistency of presentation throughout all the articles. As much illustrative picture material as possible was included: the contributors provided some and the Editors have gathered more, including some specially prepared items, such as the new maps in the Crop Atlas Appendix. In this way the architecture of the book slowly grew, and its full form only became apparent when the entire manuscript was consolidated and cross-referenced in mid-2005.

Clearly, it is to be expected that as knowledge of seeds continues to move forward, some of the information here will become dated; however, this Encyclopedia should still provide a foundation upon which an understanding of new discoveries can be built.

How to use the encyclopedia

The encyclopedia contains articles on major topics (ranging in size from just a few paragraphs to several pages), definitions and short entries directing the reader to more detailed information. The initials at the end of an article indicate authorship, with full names listed at the front of the book. In some cases the Editors have constructed an article from several different contributions, and where appropriate the authors of individual sections are listed. Cross referencing is used throughout, by putting in **bold**, on first mention, a topic that is also covered in a separate article or definition elsewhere in the volume, or by the referral, '**see:**.......'. The reader can therefore work his or her way through the many facets of a particular topic by moving from entry to entry guided by these cross-references. The reader can also navigate the book using the network diagram on the following page, which groups selected key articles by title and shows some of the conceptual linkages between these clusters. Information on many subjects that do not have special entries devoted to them can be traced by reference to the Index. Plant species whose seeds receive mention are also in the Index of Species, which contains common names and binomial nomenclature.

Many of the articles conclude with suggested reading matter. Each collection is not intended to be a comprehensive list of references but only a guide to where further information may be found.

Acknowledgements

We express our gratitude to all the contributors for providing us with articles, and for their patience when faced with requests for editorial changes. Without them, this venture could never have come to fruition.

We also thank the following for providing illustrations:

- Mike Amphlett (CABI) who specially photographed many seeds;
- Steve Jones and Cathy Carlin (NIAB Cambridge) for images from the NIAB *Seed Identification Handbook – Agriculture, Horticulture and Weeds*.*
- Mark Nesbitt, Wolfgang Stuppy and Elly Vaes of Royal Botanic Gardens Kew, UK for providing and photographing seed material;
- and all those, too numerous to mention here, whose names are shown with the individual images.

Finally, we thank the editorial and publication team at CABI who have supported us throughout, including Rebecca Stubbs for assistance and advice in the planning and early stages of the project, Jenny Thorp, our copy editor, who attentively read and carefully questioned and corrected the manuscript from cover to cover, and Tracy Ehrlich, Claire Parfitt and Jenny Dunhill for their assistance and patience during the production stages.

Accomplishing this encyclopedia has taken us on a strenuous journey of discovery, guided through thousands of editorial hours by our original plan that we refined as we went along. Any errors, omissions and flaws in structure and content are therefore our responsibility.

Michael Black
J. Derek Bewley
Peter Halmer

Spring 2006

The editors and publishers are grateful to Germain's Technology Group, King's Lynn, UK, www.germains.com, for funding the provision of the Colour Plates in this Volume.

Network diagram of major Encyclopedia contents

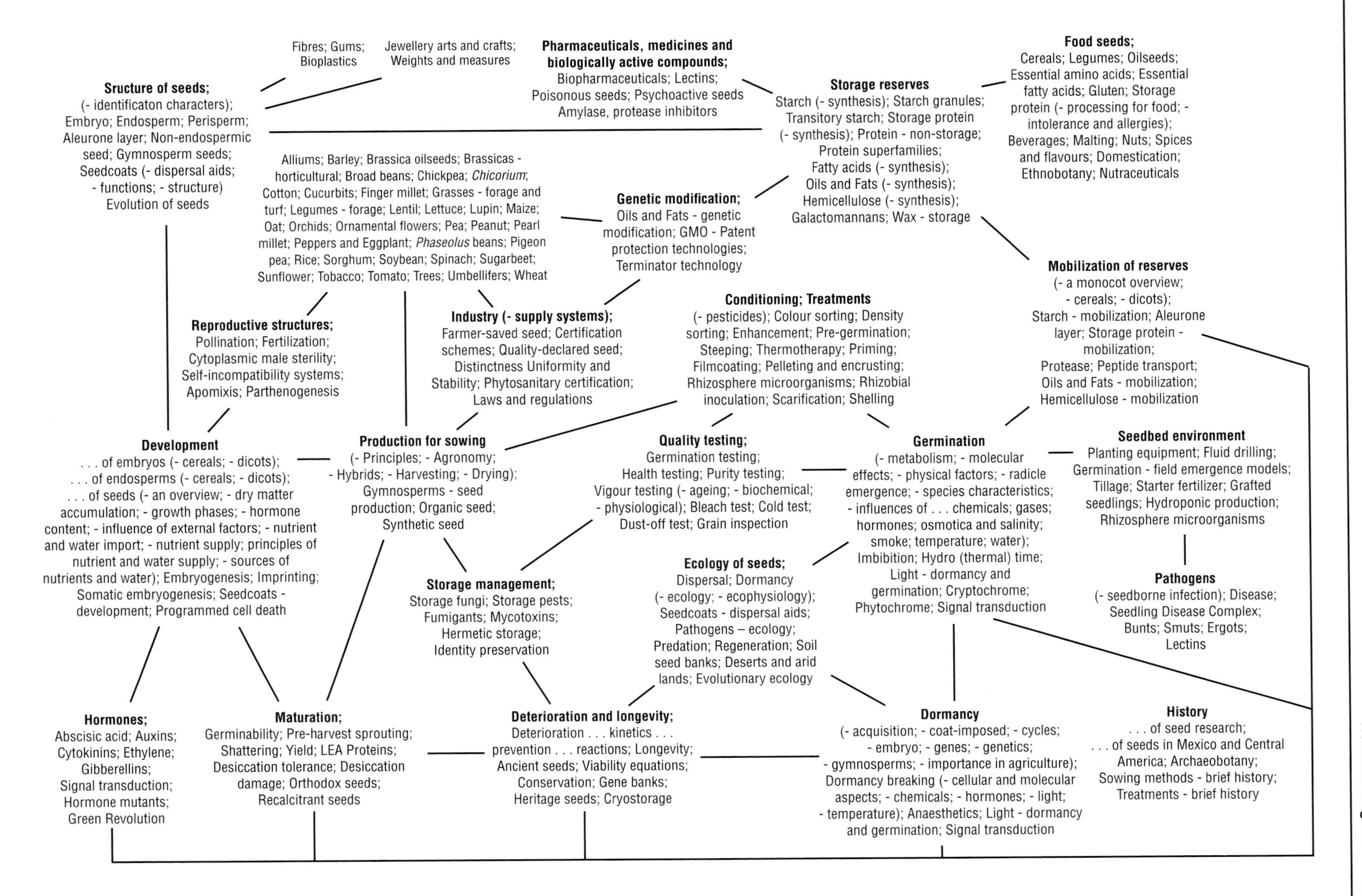

A

ABA
See: Abscisic acid

ABA genes
Genes involved in the biosynthetic pathway of **abscisic acid** (ABA). Three of these genes have been described and identified: *ABA1*, encoding for zeaxanthin epoxidase, *ABA2*, encoding for xanthoxin oxidase, and *ABA3*, encoding for molybdenum cofactor sulfurase. **(See: Dormancy – genes)**

ABI genes
Genes involved in **abscisic acid** (ABA) signalling in seeds and seedlings and therefore in sensitivity to ABA. Five of these genes have been described: *ABI1*, *ABI2*, *ABI3*, *ABI4*, *ABI5*. Of these, *ABI3* is considered to be a developmental gene. **Mutants** lacking these genes all show a reduced dormancy phenotype. **(See: Dormancy – genes; Signal transduction)**

ABI proteins
Gene products of the *ABI* genes, identified as serine/threonine phosphatase 2C (ABI1, ABI2), B3 domain protein with B1 and B2 domain (ABI3), APETALA2 domain protein (ABI4) and a basic **leucine zipper protein** (ABI5). **(See: Abscisic acid; Dormancy – genes; Signal transduction)**

Abnormal seedlings
Seedlings that are judged, in a germination test, not to show the capacity for continued development into satisfactory plants when grown in good quality soil and under favourable conditions of moisture, temperature and light; as distinguished from **normal seedlings**. Seedlings may be regarded as abnormal because they are damaged, deformed, decayed or show other defects. Criteria for abnormal seedlings are set out in the ISTA Testing Rules. **(See: Germination testing)**

Don, R. (2003) *ISTA Handbook on Seedling Evaluation*, 3rd edn, ISTA, Basserdorf, Switzerland.

ABRE
See: **Signal transduction – some terminology**

Abscisic acid
Abscisic acid (ABA) was independently discovered in two ways. Firstly a compound that accelerated organ abscission was named abscisin and secondly a compound that promoted bud dormancy was named dormin. When it was found that the compounds were the same one, it was renamed abscisic acid, usually abbreviated to ABA. Paradoxically, it is now recognized that ABA is not primarily involved in either of these processes!

As well as in flowering plants, ABA is present in lower plants, algae, cyanobacteria and fungi, though a similar compound, lunularic acid, has also been found in algae and liverworts. It is found mainly in seeds, where it promotes **maturation** and **dormancy**, the synthesis of certain **storage proteins** and inhibits **germination**, and in leaves, roots, dormant buds, potato tubers and ripening fruit (**see: Storage reserves synthesis – regulation**). Increases in ABA content in plants have long been observed to result from dehydration, low and high temperature, salt and flooding or anoxia.

The structure of ABA is shown in Fig. A.1. In order to be biologically active, ABA must have a pentadienoic side chain with two double bonds in the 2-*cis*, 4-*trans* configuration and a carboxyl group. The all-*trans* form does show biological activity due to racemization in the presence of UV light. The double bond in the cyclohexane ring and the keto group are also essential. ABA has one asymmetric carbon atom and can therefore exist as *S*-(+)-ABA or *R*-(–)-ABA. Only the former is naturally occurring.

Biosynthesis in fungi occurs through the relatively simple direct pathway. This involves synthesis from isopentenyl diphosphate via farnesyl diphosphate. However, in flowering plants, an indirect pathway operates that is blocked by inhibitors of carotenoid biosynthesis such a norflurazon and fluridone. Synthesis of the 15-carbon ABA molecule is by cleavage of a 40-carbon carotenoid precursor. The first step in the synthesis is the terpenoid pathway that synthesizes geranylgeranyl diphosphate. Two of these units are then joined by phytoene synthase to form the 40-carbon compound, phytoene. A series of desaturations then gives rise to lycopene which cyclizes to produce β-carotene (Fig. A.2) which is then

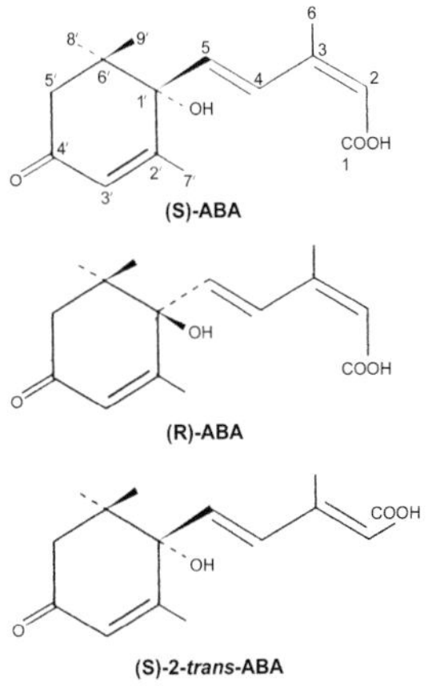

Fig. A.1. The structures of naturally-occurring ABA [(S)-ABA], and synthetic (R)-ABA and *trans*-ABA. Modified from Crozier, A., Kamiya, Y., Bishop, G. and Yokota, T. (2000) Biosynthesis of hormones and elicitor molecules. In: Buchanan, B.B., Gruissem, W. and Jones, R.L. (eds) *Biochemistry and Molecular Biology of Plants.* American Society of Plant Physiologists, Rockville, MD, USA.

oxidized to antheraxanthin. At this point the biosynthetic pathway branches: each branch gives rise to a compound where the double bond nearest but one to the cyclohexane ring is in the *cis* configuration. This is then cleaved to produce the 15-carbon xanthoxin. Whereas all of the carotenoid metabolism occurs within plastids, the final two steps occur in the cytoplasm. Xanthoxin is oxidized successively to ABA aldehyde and finally to ABA (Fig. A.3).

The enzymes involved in these transformations have not all been characterized but a series of biosynthetic **mutants** of maize, tomato, *Arabidopsis* and other species are known to be blocked at different steps in the pathway (Table A.1, and see Figs A.2 and A.3). These mutants are assisting greatly in the characterization of the enzymes.

Mutants blocked in conversions of zeaxanthin to other xanthophylls generally show symptoms characteristic of ABA deficiency and not of carotenoid deficiency. Therefore, this is taken to be the true start point of the biosynthesis of ABA. The gene encoding zeaxanthin epoxidase (ZEP), which converts zeaxanthin to antheraxanthin has been cloned and is regulated by diurnal rhythms and, in roots, by water stress. Changes in ZEP in seeds mirror the changes in ABA content. The gene for 9-*cis*-epoxycarotenoid dioxygenase (NCED) which cleaves the xanthophyll molecule to produce xanthoxin, has also been cloned and characterized. NCED appears to be induced by daily changes in light level, unlike ZEP that obeys a true diurnal rhythm. NCED is induced by drought in leaves and particularly strongly in roots.

Degradation of ABA is mainly by oxidation to 8′-hydroxy ABA, probably by a membrane-bound P450 mono-oxygenase, and then to phaseic acid (Fig. A.4). 8′-Hydroxy ABA retains some activity and will stimulate long chain fatty acid

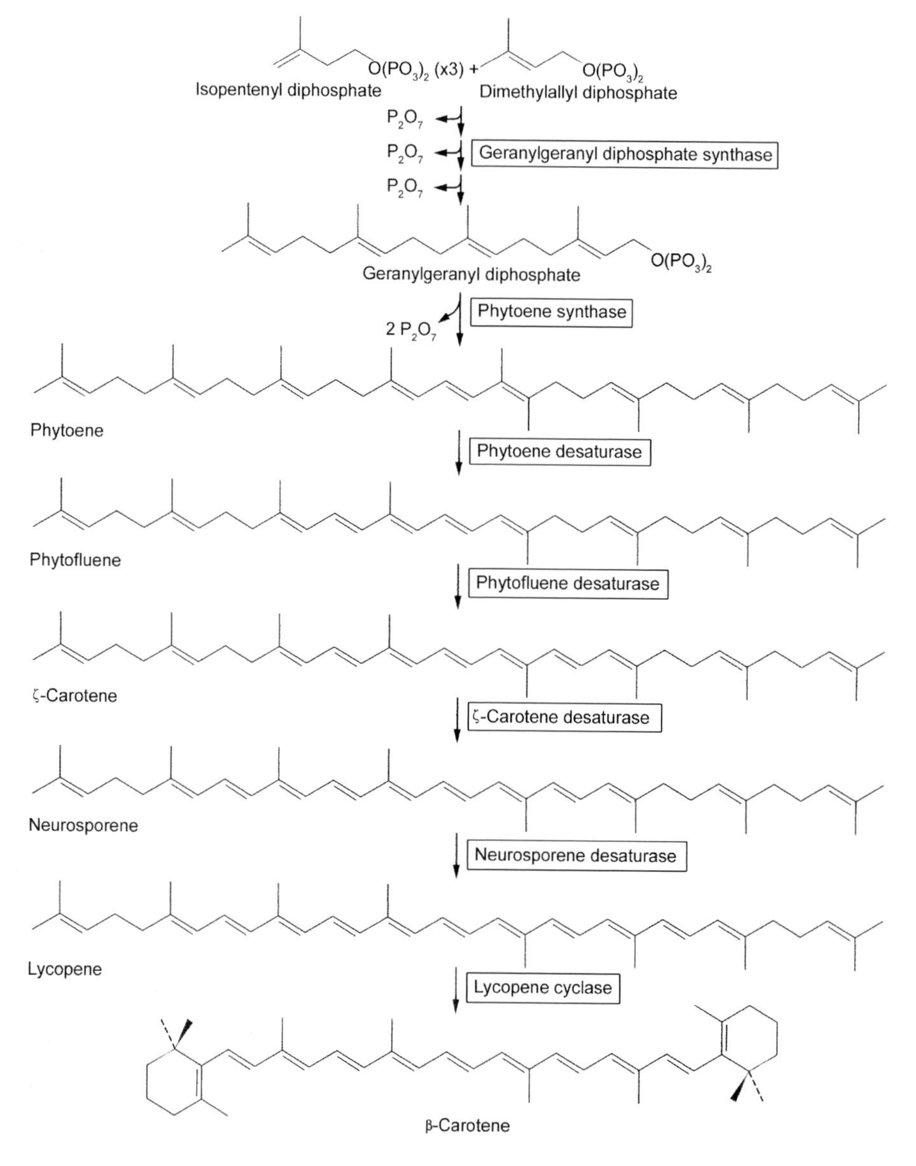

Fig. A.2. The early stages of ABA biosynthesis. This part of the pathway leads to ABA but also to carotenoid pigments and other compounds. Modified from Crozier, A., Kamiya, Y., Bishop, G. and Yokota, T. (2000) Biosynthesis of hormones and elicitor molecules. In: Buchanan, B.B., Gruissem, W. and Jones, R.L. (eds) *Biochemistry and Molecular Biology of Plants*. American Society of Plant Physiologists, Rockville, MD, USA.

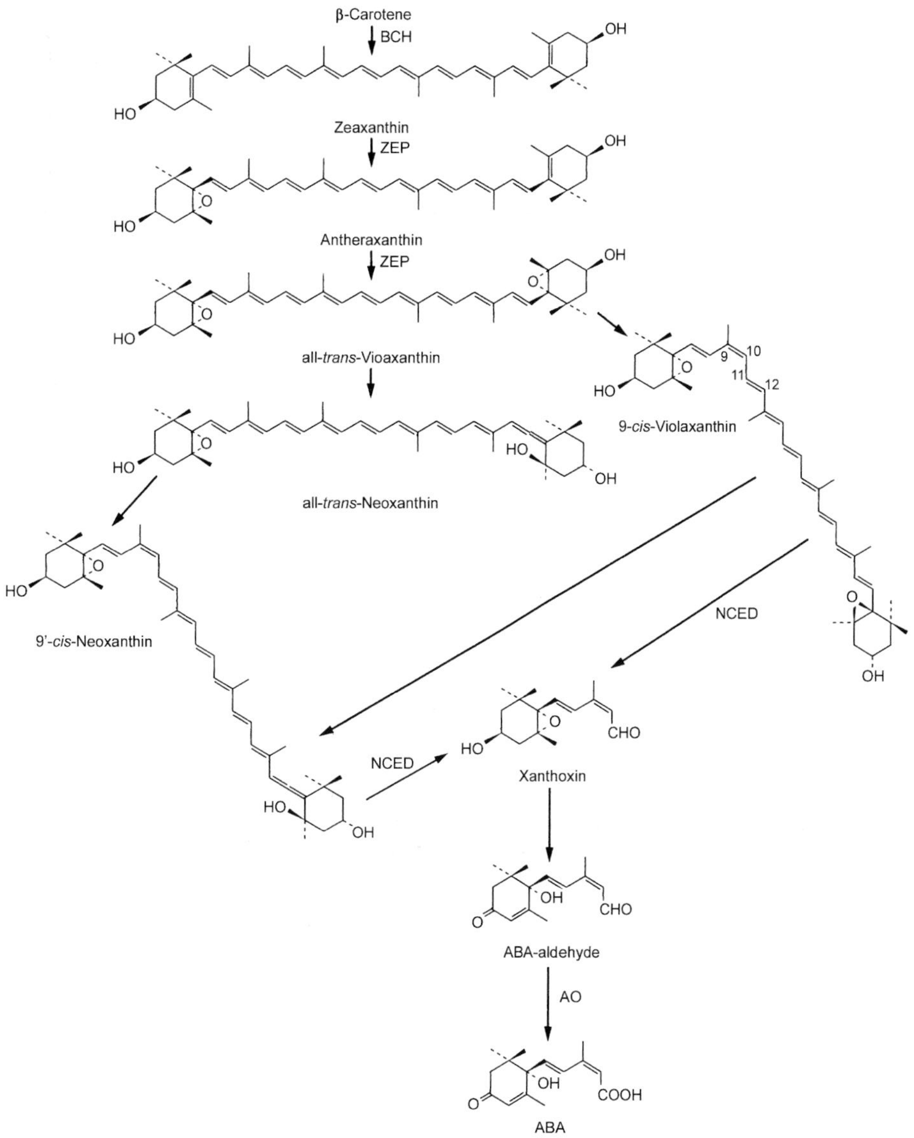

Fig. A.3. The later stages of ABA biosynthesis. Zeaxanthin is the start of the pathway leading to ABA proper. Enzymes involved are: AO, aldehyde oxidase; BCH, β-carotene hydroxylase; ZEP, zeaxanthin epoxidase; NCED, 9-*cis*-epoxycarotenoid dioxygenase; enzymes involved in other steps have not been fully characterized. Modified from Crozier, A., Kamiya, Y., Bishop, G. and Yokota, T. (2000) Biosynthesis of hormones and elicitor molecules. In: Buchanan, B.B., Gruissem, W. and Jones, R.L. (eds) *Biochemistry and Molecular Biology of Plants.* American Society of Plant Physiologists, Rockville, MD, USA.

biosynthesis in embryos of *Brassica* spp. 7′-Hydroxy ABA inhibits wheat embryo germination. Hence these metabolites must be degraded further and reduction of phaseic acid to the main catabolite, dihydrophaseic acid (DPA) or a more minor catabolite, *epi*-DPA completes the process. There is also a hydroxylation to 9′-hydroxyl ABA which is converted to neophaseic acid. Other more minor metabolites are sometimes formed and the importance of these varies from species to species. The oxidation to phaseic acid seems to be induced by ABA itself. High 8′-hydroxylase activity occurs in many tissues, including developing and germinating seeds.

Formation of the glucose ester of ABA or its metabolites seems to be another way of permanently inactivating ABA. The ester accumulates in the vacuole and does not seem to be converted back to ABA even under conditions of water stress, which strongly induce ABA production. However, when ABA glucosyl ester is applied, it is readily hydrolysed to ABA, which is then able to enter the cells. Concentrations of ABA glucosyl ester increase to a peak 18 hours after imbibition of non-dormant lettuce seeds but not in dormant ones, suggesting that the formation of glucosyl ester is a means of overcoming the germination inhibitory effect of ABA in some seeds.

Addition of glucosyl residues at the C1′ or C4′ positions gives rise to ABA-O-glucosides. It is not clear whether these are storage compounds that can be hydrolysed back to ABA or more permanent products.

Commercial preparations behave differently from natural plant products. When ABA glucosyl ester is applied to plants, it

Table A.1. ABA biosynthetic mutants and the steps blocked.

Mutant	Species	Step affected
vp12	*Zea mays* (maize)	Geranylgeranyl diphosphate synthase
vp2, vp5	*Zea mays*	Phytoene desaturase
vp9	*Zea mays*	ζ-Carotene desaturase
vp7	*Zea mays*	Lycopene cyclase
aba2	*Nicotiana plumbaginifolia* (sweet scented tobacco)	Zeaxanthin epoxidase (ZEP)
aba1	*Arabidopsis thaliana*	Zeaxanthin epoxidase
vp14	*Zea mays*	9-*cis*-Epoxycarotenoid dioxygenase (NCED)
notabilis	*Lycopersicon esculentum* (tomato)	9-*cis*-Epoxycarotenoid dioxygenase
aba2	*Arabidopsis thaliana*	Xanthoxin oxidase
aba3	*Arabidopsis thaliana*	ABA-aldehyde oxidase (molybdenum cofactor)
nar2a	*Hordeum vulgare* (barley)	ABA-aldehyde oxidase (molybdenum cofactor)
flacca	*Lycopersicon esculentum*	ABA-aldehyde oxidase (molybdenum cofactor)
aba1	*Nicotiana plumbaginifolia*	ABA-aldehyde oxidase (molybdenum cofactor)
sitiens	*Lycopersicon esculentum*	ABA-aldehyde oxidase
droopy	*Solanum phureja* (Andean potato)	ABA-aldehyde oxidase

Note: The *vp* mutants cause **vivipary** in immature maize grains. The *aba* and *sitiens* mutants affect **dormancy** and **germination**. **(See: Dormancy – genes)**

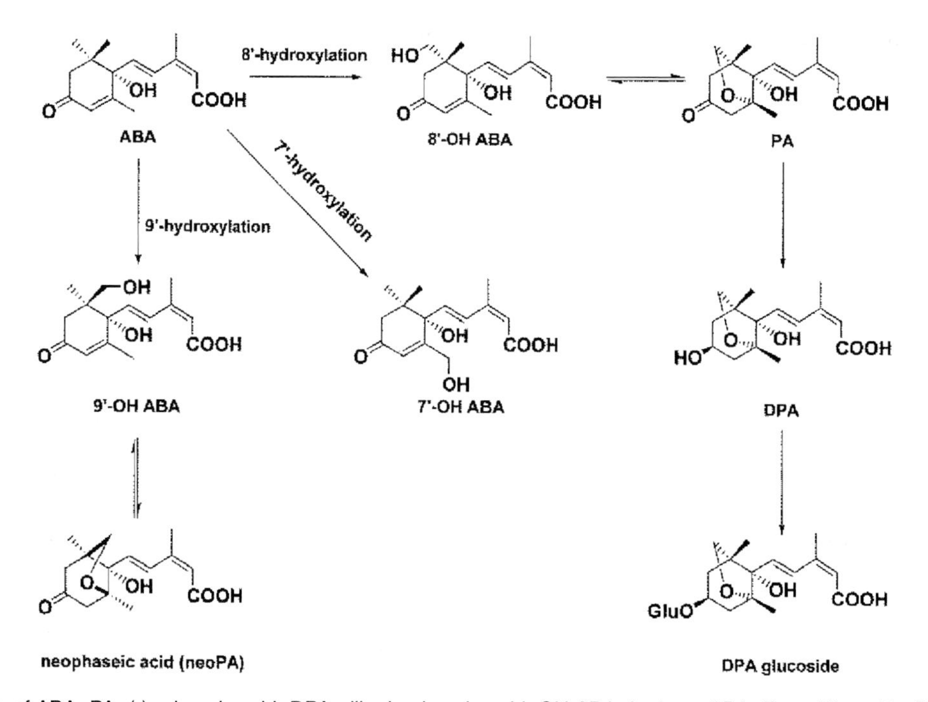

Fig. A.4. Breakdown of ABA. PA, (-)- phaseic acid; DPA, dihydrophaseic acid; OH ABA, hydroxy ABA. From Zhou, R., Cutler, A.J., Ambrose, S.J., Galka, M.M., Nelson, K.M., Squires, T.M., Loewen, M.K., Jadhav, A.S., Ross, A.R.S., Taylor, D.C. and Abrams, S.R. (2003) A new abscisic acid catabolic pathway. *Plant Physiology* 134, 361–369. Copyrighted by the American Society of Plant Biologists and reprinted with permission.

is unable to enter the cells but it is readily hydrolysed to ABA, which is then able to enter. Endogenous ABA glucosyl ester is confined to the vacuole, where it seems to be metabolically inert. Commercial preparations of ABA are a racemic mixture of the natural identical *S*-(+) form and the artificial *R*-(–) form. The latter is not readily oxidized and therefore it is catabolized more slowly than natural ABA. It is converted mainly to the glucosyl ester with only a small amount being oxidized to 7′-hydroxy ABA. **(See: Development of seeds – hormone content; Signal transduction – hormones)** (GL, JR)

Buchanan, B.B., Gruissem, W. and Jones, R.L. (eds) *Biochemistry and Molecular Biology of Plants.* American Society of Plant Physiologists, Rockville, MD, USA, pp. 850–929.

Crozier, A., Kamiya, Y., Bishop, G. and Yokota, T. (2000).Biosynthesis of hormones and elicitor molecules. In: Buchanan, B.B., Gruissem, W. and Jones, R.L. (eds) *Biochemistry and Molecular Biology of Plants.* American Society of Plant Physiologists, Rockville, MD, USA, pp. 850–929.

Finkelstein, R.R. and Rock, C.D. (2002) Abscisic acid biosynthesis and response. In: Somerville, C.R. and Meyerowitz, E.M. (eds)

The Arabidopsis Book. American Society of Plant Biologists, Rockville, MD, USA.

Nambara, E. and Marion-Poll, A. (2003) ABA action and interactions in seeds. *Trends in Plant Science* 8, 213–217.

Taylor, I.B., Burbidge, A. and Thompson, A.J. (2000) Control of abscisic acid synthesis. *Journal of Experimental Botany* 51, 1563–1574.

Zeevart, J.A.D. (1999) Abscisic acid metabolism and its regulation. In: Hooykaas, P.J.J., Hall, M.A. and Libbenga, K.R. (eds) *Biochemistry and Molecular Biology of Plant Hormones*. Elsevier, Amsterdam, The Netherlands, pp. 189–207.

Acampylotropous

Type of ovule. The same as campylotropous. (**See:** Fig. R.3 in **Reproductive structures, 1. Female**)

Accelerated ageing

The rapid loss of germination percentage in seedlots stored for a brief period under stressful conditions. The term was coined in a series of presentations in the 1960s that culminated in a 1973 report of a test developed to predict seed **vigour** and **longevity**. The test distinguishes seedlots that have high germination percentages in the laboratory but varied field emergence and uses the assumption that stress tolerance is an indicator of seed vigour. In accelerated **ageing** tests, reduction in germination is measured after seeds are exposed to humid (100% or 75% RH) and warm (45 or 30°C) conditions for a few days or weeks. In a similar test called **controlled deterioration**, seed water contents are adjusted to 18–22% (fresh weight basis) at 10°C before a brief exposure to 45°C. Independently, researchers testing parameters for accelerated ageing and the **viability equations** showed that the kinetics of deterioration under extreme warm, moist conditions correlate with the kinetics of deterioration under ambient conditions and this led to the widely-held but still disputed hypothesis (**see: deterioration reactions**) that the mechanism of seed deterioration is similar across broad ranges of moisture and temperature. The term 'accelerated ageing' may refer specifically to a stress test using extreme moist, warm conditions or to the general concept of using stress to alter seed viability. (**See: Vigour testing – ageing**) (CW)

Delouche, J.C. and Baskin, C.C. (1973) Accelerated aging techniques for predicting the relative storability of seedlots. *Seed Science and Technology* 1, 427–452.

Powell, A.A. and Matthews, S. (1981) Evaluation of controlled deterioration, a new vigour test for small seeded vegetables. *Seed Science and Technology* 9, 633–640.

Accessory fruits

Those derived from simple or compound ovaries and some non-ovarian tissues. Examples, and species, are burrs (*Xanthium*, cocklebur), coenocarpium (*Ananas*, pineapple), diclesium (*Mirabilis*, Four-o'-clock), glans (*Quercus*, oak), hip (*Rosa*, rose), pome (*Malus*, apple), pseudocarp (*Fragaria*, strawberry), synocium (*Ficus*, fig), tryma (*Carya*, pecan), winged nut (*Carpinus*, hornbeam). (**See** Fig. F.11 in **Fruit – types**)

Achene

A one-seeded, dry, indehiscent fruit (does not split at maturity) with the seed attached to the fruit wall at one point only. It is derived from a one-loculed superior ovary. Seedcoat is insignificant compared with the fruit coat (**pericarp**), e.g. *Helianthus annuus*, sunflower. (**See: Inflorescence; Reproductive structures, 1. Female**)

Acorn

The fruit of oak (*Quercus* spp.). (**See: Glans**)

Active ingredient (ai)

For agrochemicals, such as those used in pesticide seed treatments: the chemical in the formulation that has the specific biological effect on the target organism – that is, to prevent, destroy, repel or mitigate the 'pest(s)' or, in the case of a plant regulator, alter the rate of growth, maturation or other plant behaviour through its physiological action. Active ingredients are given common and full chemical name(s) that comply with accepted guidelines (**see: Treatment – pesticides, 1. Introduction**). For many non-synthetic, natural biopesticides, however, the active ingredients might not be fully chemically characterized. The same point applies to biologically active ingredient component(s) in complex plant extracts – including those from seeds (**see: Pharmaceuticals, medicines and biologically active compounds**).

(AI is also an abbreviation for **amylase inhibitors**.) (PH)

Adenosine triphosphate

See: ATP

Adlay

See: Job's tears

Adzuki bean

1. Worldwide importance and distribution

The warm-season legume, adzuki bean (*Vigna angularis* syn. *Phaseolus angularis* [Willd.]) also called azuki, is one of the 12 most important **grain legumes** in the world. It is produced in temperate regions of east Asia, northeast Asia being the most important, particularly China, Japan, South Korea and Taiwan, but also in India, Thailand, the Philippines, New Zealand, Australia, USA and Argentina. Adzuki is Japan's and Korea's second most important legume after the **soybean**. (**See: Legumes**)

2. Origin

The Far East is the probable centre of origin. The earliest reported site of adzuki use and production is in North Korea (approximately 3000 years ago). Cultivation of adzuki in Japan is thought to have started later than in China and Korea, around 1000 years ago. It has been postulated that adzuki of Japan came along two paths from Southeast Asia; one path through China, across the Korean peninsula, and the other via Taiwan to Okinawa. The adzuki bean is not found in the wild.

3. Seed characteristics

Adzuki seeds are approx. 5 mm long, subcylindrical with subtruncated ends, characterized by a smooth coat, a strongly defined cotyledonary ridge and an elevated, countersunk micropyle. The hilum is white and approximately 2.4–3.3 mm long and 0.6–0.8 mm wide. A thin caruncle encompasses the hilum cushion. Seed colours range from solid maroon to solid black, blue-black, green, ash grey, green-grey, brown-red, reddish-brown (Colour Plate 1A), straw yellow, white and various mottled combinations.

4. Nutritional value

Nutritional components of dry and processed adzuki seeds are shown in Table A.2.

Adzuki protein is composed of approximately 47% **albumins** and 37% **globulins**. The **essential amino acid** contents are superior or comparable to those in soy protein. **Oil** content is about 1% (one of the lowest among **grain legumes**), composed of 75% unsaturated (linoleic, oleic and linolenic) and 25% saturated **fatty acids** (stearic, palmitic and canaubic). The beans contain less soluble and insoluble (in the seedcoat) dietary fibre, and ash than common (*Phaseolus*) bean. The **starch** is 35% **amylose**. Among the flatulogenic factors, adzuki bean is reported to have 0.3–0.5% raffinose and 2.8–4.1% stachyose, less than in common bean.

5. Uses

In China, Korea and Japan adzuki is traditionally served in a sweetened form on festive days. Around 70% is principally used to make a sweetened paste called *an* in Japanese, composed of nearly equal parts of adzuki paste and sugar, or *yokan* (with agar). Adzuki *an*, either in a smooth, whole-bean or intermediate form is used in numerous east Asian food and Western-style desserts, approximately 30% of it by the Japanese and Korean ice cream industries. The sweetened paste is also used as the base for a soft drink served hot. A white-seeded adzuki is used to make high quality, white, sweetened paste for speciality Japanese bakery. Adzuki is also consumed in Japan as candied or boiled seeds. The beans are also used as a component in sweet soups, mixed with rice, sprouted, made into flour or used as a substitute for coffee. In North America and Europe adzuki bean is valued for its relatively short cooking time as well as its low fat and high protein and natural sugar nutritional profile. Adzuki is one of the preferred seeds in Asia for bean sprouts: the sprouted beans are particularly tasty with a nutty flavour. The young tender pods may be harvested for use as snap beans.

In addition to the use as a food product, adzuki beans are also used for making facial cream and shampoos, and medicines, and the plants as a nitrogen-fixing green manure, livestock feed and a soil stabilizer.

Table A.2. Nutritional content of dry and processed adzuki beans.

Component /100 g fresh weight	Dry seeds	Processed seeds
Protein (g)	20.3–25.3	4.4–26.2
Fat (g)	0.6–2.2	0.4–1.0
Carbohydrate (g)	54.4–57.4	22.3–65.2
Fibre (g)	4.3–5.7	0.9–1.9
Ca (mg)	75–253	13–60
P (mg)	350	80–220
Fe (mg)	5.4–7.6	80–220
Na (mg)	1	1–90
K (mg)	1500	60–460
Vitamin A (IU)	15	0
Vitamin B1 (mg)	0.45–0.57	0.01–0.15
Vitamin B2 (mg)	0.16–0.18	0.03–0.06
Niacin (μg)	2.2–3.2	0–0.8

Lumpkin, T.A. and McClary, D.C. (1994) *Azuki Bean: Botany, Production and Uses.* CAB International, Wallingford, UK.

6. Market classes and economics

Red is normally the preferred colour of beans for consumption. Japan, the main consumer, has two distinct market classes: the regular type (>4.2 mm length) and a large *dainagon* type (>4.8 mm length). Here, four grades are used to classify adzuki seeds based on percentages of moisture, foreign material and damaged/immature seeds. Scarlet *dainagon* seeds are used for high quality confection products and customarily command a price 40–90% higher than standard grade adzuki. China uses a three-grade classification incorporating percentage of moisture, foreign material and damaged, immature seeds, as well as **testa** colour and smell. Grade one requires a minimum of 96% adzuki seeds with a diameter of more than 2 mm. This grade commands a higher price and is almost exclusively exported while grades two and three are sold on the domestic market.

The international trade is divided into four product categories: raw adzuki and various forms of *an* (see section 5). World prices for adzuki are largely determined by supply and demand in the Japanese market that consumes more than 100,000 t annually.

7. Plant types

Adzuki shows **hypogeal** seedling growth. Plants exhibit both **indeterminate** and **determinate** flowering habits. Maturity time and environmental conditions during the growing season are reported to influence the degree of indeterminacy. Early-maturing cultivars are semi-determinate. Vining types are grown in China, India, Taiwan and Thailand; bush or erect plant types are grown in both northern Japan and countries in the western hemisphere. (OV-V)

Lumpkin, T.A. and McClary, D.C. (1994) *Azuki Bean: Botany, Production and Uses.* CAB International, Wallingford, UK.

Aerial seeding

The broadcast sowing of seed from aircraft, such as for rice (**water seeding**), some grasses (**see: Grasses – forage and turf**) or for remediation or reclamation of land areas. Distinct from **air seeding**.

Aerobic respiration

Respiration that occurs in the presence of sufficient oxygen to allow for the oxidation of organic molecules (e.g. sugars, fatty acids, amino acids) to CO_2 and H_2O, and the incorporation of the released free energy into a readily usable currency of metabolic energy, adenosine triphosphate (**ATP**).

Aflatoxin

A toxin produced by fungal infection with *Aspergillus* spp. especially of groundnut (peanut) and some nuts. (**See: Mycotoxins**)

African Seed Trade Association (AFSTA)

Established in 2000 and based in Nairobi, Kenya, this organization's objectives include promoting the use of improved-quality seed, strengthening communication between African and world seed industries, developing contacts with the official African governmental organizations, helping to establish national seed trade associations, and promoting regulatory harmonization throughout Africa to facilitate movement of seed, and training seed personnel. www.afsta.org (**See: Industry**) (BlB)

Afterripening

The progressive loss of **dormancy** in mature dry seed. Environmental factors such as increased oxygen and temperature accelerate the rate at which seed afterripen. Increasing moisture content reduces this phenomenon. The mechanistic basis of this process is uncharacterized. (**See: Dormancy breaking – temperature**)

Agamospermy

See: Parthenogenesis

Ageing

Generally referred to as the process that seeds undergo when stored in less than optimal conditions. This results in a variety of symptoms including: (i) reduced **viability** leading to poor or sporadic germination; (ii) reduced seedling **vigour** (including abnormal seedling growth); or even (iii) complete loss of **germinability**. **Accelerated ageing** is achieved experimentally by placing seeds in conditions that promote ageing, e.g. high temperature and humidity. (**See: Deterioration; Vigour testing – ageing**)

Agrobacterium tumefaciens

A soil-borne pathogen that naturally transfers a segment of DNA (called T-DNA) from its large tumour-inducing **(Ti) plasmid** through the plant membranes and incorporates it into the genomic DNA of plant cells adjacent to a wound site. *Agrobacterium* strains used in plant transformation have their Ti plasmid disarmed and their T-DNA replaced with multi-cloning sites into which genes of interest as well as dominant selection markers can be integrated. (**See: Transformation**)

AI (ai)

See: Active ingredient

Air-screen cleaner

A basic piece of **conditioning** equipment for cleaning seed for sowing, utilizing a combination of air flow and oscillating screens. Seed is separated on the basis of size, width, thickness, specific gravity and resistance to air flow. **See: Conditioning, II. Cleaning**

Air seeding

The operating principle underlying a type of pneumatic seeder used for drill planting, by which seed is conveyed through one or more dividing heads to a number of outlets located symmetrically around it, as an alternative to traditional mass-flow-meter and gravity-drop-delivery systems. (**See: Planting equipment – placement in soil**) Distinct from **aerial seeding**.

Ajowan

The seed spice, ajowan, ajwain or Bishop's weed (*Trachyspermum ammi* Apiaceae) (**see: Spices and flavours**, Table S.12), is an herbaceous annual up to 90 cm tall. The 'seeds' are grey-brown, ovoid, 2 mm-long mericarps, smelling of thymol. In India ajowan seeds are used as spice and for flavouring pickles, biscuits, confectionery and beverages.

1. Constituents

The characteristic odour and taste is due to the **essential oil** (2–4% fresh weight). This is distilled from the seeds and contains a large proportion of thymol (Table A.3). The oil is colourless or brownish yellow possessing a characteristic odour of thymol and a sharp burning taste.

Twenty-seven compounds are identified in the seed oil of which thymol is present in greatest quantity. Additional constituents are other phenols such as carvacrol. (**See:** Fig. S.44 in **Spices and flavours**)

2. Uses

The use of ajowan is confined almost to central Asia and northern India. It is particularly popular in savoury Indian recipes like savoury pastries, snacks and breads, but also in the Arab world and Ethiopia. Ajowan has long been used in indigenous medicines. The essential oil is a highly valued ingredient in the formulation of Unani and Ayurvedic medicines and is employed as an antiseptic, aromatic and carminative. Thymol, which is crystallized from the oil, is sold in Indian markets as *ajowan ka-phool* or *sat-ajowan*. It is also used in the perfume industry. (KVP, NB)

Guenther, E. (1978) *The Essential Oils*, Vol. 2. Robert E. Krieger Publishing Company, New York, USA.

Peter, K.V. (ed.) (2001) *Handbook of Herbs and Spices*. Woodhead Publishing Company, Cambridge, UK.

Weiss, E.A. (2002) *Spice Crops*. CAB International, Wallingford, UK.

Albumins

See: Osborne fractions; Storage protein

Table A.3. Main constituents of ajowan essential oil.

Constituents	Percentage (wt)
Phenolics	**(% in phenolics)**
Thymol	87.75
Carvacrol	11.17
Non-phenolics	**(% in non-phenolics)**
α-Thujene	0.27
α-Pinene	0.28
β-Pinene	2.38
Para-cymene	60.78
Limonene	8.36
γ-Terpinene	22.26
Linalool	0.27
Camphor	0.28
β-Terpineol	1.35
Borneol	0.49

Battacharya, A.K., Kaul, P.N. and Rajeswara Rao, B.R. (1998) Essential oil composition of ajwain (*Trachyspermum ammi* L. Sprague) seeds produced in Andhra Pradesh. *Indian Perfumer* 42, 65–67.

Aleurone layer

One to several layers of cells that cover the surface of the **endosperm** in cereal, grass and some endospermic **legume** seeds, immediately below the **pericarp**, or fruit coat. In maize and wheat, the aleurone layer is only one cell layer thick, but in rice it is one to several cell layers, and in barley it is three cell layers. Aleurone layer and endosperm cells have a common lineage and thus are related, although they differ morphologically, with aleurone layer cells assuming a characteristic cuboidal shape (**see: Development of endosperms – cereals**). Aleurone layer and endosperm cell identities can switch, although aleurone layer identity typically is specified and maintained by positional cues at the endosperm surface throughout seed development. Like endosperm cells, the nuclei of aleurone layer cells have three sets of chromosomes, two sets from the female parent and one set from the male parent.

The aleurone layer serves as a storage site for many compounds. The cytoplasm of aleurone layer cells accumulates numerous 'aleurone grains', which are small vacuoles that contain a mix of **phytin**, **storage protein** and **oil**, or a protein–starch mix. Furthermore, aleurone layer cells often are pigmented, containing anthocyanins that are responsible for the colourful grains of maize.

The primary function of the aleurone layer is as a digestive tissue. Towards the end of seed maturation, aleurone layer cells are rendered tolerant to desiccation, which allows them to survive as the seed undergoes **maturation drying**. Following germination, aleurone layer cells are stimulated by **gibberellic acid** to secrete a variety of hydrolytic enzymes that break down cell walls, **starch** and storage proteins in the starchy endosperm, making simple sugars and amino acids available for uptake by the growing seedling. As enzymes are released from the aleurone layer to mobilize stored reserves in the starchy endosperm, it undergoes **programmed cell death**.

(MGJ)

(**See: Cereals; Endosperm; Mobilization of reserves – cereals; Mobilization of reserves – a monocot overview; Starch – mobilization; Xenia**)

Becraft, P.W. and Asuncion-Crabb, Y. (2000) Positional cues specify and maintain aleurone cell fate in maize endosperm development. *Development* 127, 4039–4048.

Olsen, O.-A. (2001) Endosperm development: cellularization and cell fate specification. *Annual Review of Plant Physiology and Plant Molecular Biology* 52, 233–267.

Alfalfa

Medicago sativa L. is an important and nutritious forage crop, rich in minerals, vitamins and proteins, but its seed is not used extensively other than for replanting. Alfalfa seed sprouts (seedlings) are used widely in salads.

Alfalfa is grown in almost every state in the USA and is called 'The Queen of the Forages'. As a perennial **cool-season legume** it can survive temperature extremes of −25 to over 50°C. It becomes quiescent under severe drought and usually will resume growth when conditions improve. Alfalfa (European name: lucerne) most likely originated in the Middle East, probably near Iran; it is grown in one form or the other on nearly all the continents and represents one of the oldest forage crops.

Seeds are small, 1–2 mm in length and about half as wide and they are slightly kidney shaped. At maturity they have a residual **endosperm**, with a few layers of thickened **hemicellulose** cell walls and an outer **aleurone layer**. Impermeable or hard seeds are common, and are problematical for uniform germination; **scarification** of the coat may be necessary. **See: Legumes – forage**

For information on the particular contributions of the seed to seed science **see: Research seed species – contributions to seed science**.

(JDB)

A-line (Male-sterile line)

See: Cytoplasmic male sterility

Alkaloids

Chemically heterogeneous organic, nitrogen-containing compounds generally of plant origin that are physiologically active and in many cases poisonous. They occur in many seeds. (**See: Pharmaceuticals and pharmacologically active compounds**)

Alleles (Allelic gene)

Alleles are alternative forms of a **gene**, a single allele of which is inherited from each parent. They are located on corresponding loci of homologous chromosomes. Both parental alleles are similar, but their differences are defined by specific and unique DNA sequences. Alleles encode alternative forms of a single trait, which can result in different phenotypes. There are dominant and recessive alleles. A dominant allele (or character) expresses itself to the exclusion of its contrasting recessive alternative. A heterozygous organism has one, several or many genes with unlike alleles.

Allelopathy

Allelopathy occurs when plant-derived chemicals (allelochemicals) inhibit germination, growth or development of other plants. These leach into the soil from living plant parts (including seeds) or from decomposing plant material, are exuded from roots, or are emitted from plants as volatiles. For example, fresh or dead leaves or rainwater dripping from chamise (*Adenostoma fasciculatum*), a shrub of the Californian chaparral, inhibit germination of a large number of species growing in this habitat. While best studied in Californian chaparral, similar studies have suggested allelopathic effects on germination in a wide range of species and ecosystems around the world, e.g. *Eucalyptus* species in Australian mallee, the heather (*Erica scoparia*) in Iberian heaths, crowberry (*Empetrum hermaphroditum*) in boreal forests, and creosote bush (*Larrea tridentate*) in the Mojave desert. Sometimes species inhibit germination of their own seeds by allelopathy (autotoxicity).

While in some cases allelopathy is thought to be an evolved strategy which allows plants to inhibit other species, autotoxicity shows that it is often difficult to explain this effect in evolutionary terms. It may be more a side effect of plant biochemical processes. However, it is an important factor in determining the species composition and vegetation structure of plant communities. Autotoxic inhibition of germination has been shown to cause declines in weedy species in abandoned arable fields in the USA, which allows other species to colonize. Allelochemicals can cause large bare zones around some species (e.g. chamise).

Many plant-derived compounds have been identified as having allelopathic qualities, including **phenolics**, terpenoids, alkaloids, polyacetylenes, **fatty acids** and steroids. These are generally secondary plant metabolites (end products of biochemical pathways). Leachates from vegetation litter inhibiting germination of *Nicotiana attenuata* in the western USA have been reported to include **abscisic acid** and several terpenes. Allelopathic leachates are usually a mixture of many potentially toxic substances. For example, germination-inhibiting leachates of *Erica scoparia* leaves contain seven phenolic compounds, and similarly active leachates of *Eucalyptus baxteri* litter contain phenolic glycosides, tannins and genistic, caffeic, gallic and ellagic acids. **(See: Germination – influences of chemicals)**

Because plants interact in many ways, it can be difficult to isolate allelopathic effects and to demonstrate them conclusively. Thus, there is controversy about the extent and importance of allelopathy in nature. (JMB)

Baskin, C.C. and Baskin, J.M. (2001) *Seeds: Ecology, Biogeography of Dormancy and Germination.* Academic Press, San Diego, USA.

Krock, B., Schmidt, S., Hertweck, C. and Baldwin, I.T. (2002) Vegetation-derived abscisic acid and four terpenes enforce dormancy in seeds of the post-fire annual, *Nicotiana attenuata*. *Seed Science Research* 12, 239–252.

Inderjit, Dakshini, K.M.M. and Foy, C.L. (1999) *Principles and Practices in Plant Ecology: Allelochemical Interactions.* CRC Press, London.

Allergen

A substance, often a protein, recognised by the immune system of some individuals who respond by showing adverse (allergic) reactions of different degrees of severity. Seeds of many species including some **cereals**, **legumes** and **nuts** contain allergens. **See: Storage proteins – intolerance and allergies**

Alliums

Alliums are now placed in their own family, the Alliaceae. There are around 300 species known, with seven of significant commercial importance (Table A.4). Shallots and multiplier onions, once classified as separate species, are now regarded as conspecific with bulb onions.

1. History, uses, distribution, economic importance

Alliums are versatile and potent flavouring ingredients, accepted by almost all traditions and cultures. All alliums have long been valued for their medicinal properties and they may have significant health benefits. Their flavour chemistry is based on the release after cutting or crushing of the enzyme, alliinase, which reacts with precursor chemicals (cysteine sulphoxides) to produce an array of breakdown products within the cells that are different for each species, resulting in a signature flavour and odour.

Onions are grown wherever plants are farmed, exhibiting a great diversity in form including flesh and skin colour, shape, dry matter content and pungency or sweetness. They can be used cooked or raw and are dehydrated, pickled, canned or frozen, either alone or as a major constituent of many pre-prepared foods. Onion is the second most important horticultural crop after tomatoes, with current production being around 44 million t, accounting for around 10% of all the world's vegetable production. The species adapt to a wide range of environments and because of their storage characteristics and durability for shipping, including as dehydrated products, onions and garlic have always been traded widely. Bulb formation is largely genetically controlled by a daylength response but forms have been selected that allow production from tropical areas to near the Arctic Circle.

Onions are also grown from 'sets'. Sets are small bulbs that have been initially grown from seed at very high density that limit bulb size to 15–20 mm in diameter. They are then harvested and stored for later planting. Their purpose is to get earlier maturity, or for cropping in short growing seasons. They are easier to plant and faster to establish than seed although costs per hectare are higher.

Leek is an important crop in Europe, partly due to its winter hardiness, and also in Asia and the Middle East where young plants are used as flavourings and in salads. Varieties differ in maturity period, height, thickness, shape, colour and degree of bulbing.

Garlic is normally sterile and has to be propagated vegetatively: 'hard neck' types produce a scape that either has completely sterile flowers, or bulbils, or a mixture of bulbils and flowers; 'soft neck' types never produce a flower stalk.

Table A.4. Classification of cultivated alliums.

Species	Subdivision	Horticultural name	Edible part
A. cepa	*cepa*	Bulb onion	Bulb, leaf, leaf base
	ascalonicum	Shallot	Bulb
	aggregatum	Potato or multiplier onion	Bulb
	proliferum	Tree onion	Topset bulb on scape
A. fistulosum		Japanese bunching onion (Welsh onion)	Pseudostem, leaf, leaf base
A. sativum		Garlic	Cloves
A. ampeloprasum	*porrum*	Leek	Pseudostem
	aegyptiacum	Kurrat	Leaves
A. schoenoprasum		Chives	Leaf and leaf base
A. chinense		Rakkyo	Bulbs, leaf base
A. tuberosum		Chinese chives	Leaf, flower scape, flower

The bulb onion is not found as a wild species (this may have become extinct): it may be a hybrid from ancient cultivation. Wild Allium species are found widely in Europe, Asia, North America and North Africa and are generally adapted to temperate environments with low or erratic supplies of water, although some species can occupy a wide range of ecological niches.

Various Alliums were being grown in India, China and Egypt by around 2000–3000 BC and it is likely that they were first cultivated in Central Asia well over 5000 years ago. Ancient Egyptian inscriptions show how much was spent on onions and garlic during the construction of the pyramids, and onions, garlic and leeks are mentioned several times in the Old Testament (Numbers XI:5). The Egyptians and Romans dipped cereal seeds in onion brine before sowing, apparently to protect seeds against diseases. Leeks featured in many recipes in Apicius' 1st century cookbook and the Romans are thought to have spread Alliums throughout Europe. The Spanish and the first pilgrims introduced onions and garlic to the Americas, although many wild forms occur indigenously in North America; the name Chicago was derived from the native American name for 'the place which smells of onions'.

2. Genetics and breeding

The basic chromosome number in the cultivated Alliums is eight: bulb onion, bunching onion and garlic are diploid with 16 chromosomes; leek and Chinese chives are tetraploid with 32. The genomes are very large, with leek having over 140 times the genome size of *Arabidopsis*. Compared to some crops, Allium genetics are relatively poorly understood, partly because of their large genome. However, genetic maps are now becoming available and DNA marker-assisted selection can be used in the rapid identification of male sterile lines in onions. GM technology has recently been reported with the initial aim of introducing herbicide resistance, and modifying the alliinase enzyme that initiates the flavour reaction.

In European and North America onion production hybrids now dominate commercial production. At least some of the advantage lies purely with seed companies protecting their breeding lines, rather than in performance gain *per se*. Onion was used as the original model for the wide-scale development of F_1 hybrids using the system of **cytoplasmic male sterility** that was first identified in the USA in 1925 (**see: Production for sowing, III. Hybrids**). Similar systems are used for bunching onions and chives. In leeks, the genetics is complicated by tetrasomic inheritance (recombination among four sets of chromosomes); male sterility is used in seed production, but the sterile lines have to be maintained vegetatively. Open-pollinated varieties are still widely grown in many regions, bred either by mass selection or more sophisticated forms of recurrent half-**sib** or S1 family selection.

Breeding objectives vary with local need but concentrate on extended storage life, **bolting** resistance, maturity date, skin and bulb qualities, colour and pungency. Progress in improving pest and disease resistance has been relatively limited. Onions bred specifically for dehydration are selected for high dry matter content.

3. Development

Onion is a biennial that produces a large bulb in the first year of growth. Each cultivar has a critical daylength for bulb induction although temperature also has some influence. After **vernalization**, rapid re-growth of foliage in spring is quickly followed by the production of several tubular flower stalks each producing a single, more or less spherical, umbel (see Fig. I.1). Flowers, depending on species, range from mainly white in onion to deep purple and blue in some ornamental wild relatives. For flower initiation the plants or bulbs need to vernalize at temperatures and for times according to the cultivar (15–20°C for tropical types, around 7°C in the UK). After flower initiation the subsequent growth of the scape depends on temperature and daylength; relatively cool temperatures and long days are optimal. High temperatures combined with long days can result in a competition phase with the plant trying to re-initiate bulbing that can result in flower abortion. Self-pollination is discouraged but not totally prevented by **protandry** (the development of male organs before the female ones); the style only becomes receptive 1–2 days after the anthers shed pollen.

Seeds are borne in a pseudo-umbel at the apex of a floral stem or scape. In onion the scape is hollow, swollen in the middle and typically around 1 m tall. Depending on variety and bulb size each plant may produce up to ten or more scapes but three to four is more common. Leeks produce much taller and sturdier solid scapes, sometimes reaching 1.8 m. The fruit is a three-lobed, three-celled capsule each containing up to six black seeds. Most of the seed is oily **endosperm** with a curved cylindrical **embryo**, about 5–6 mm long, embedded in it.

4. Production

In onion, both seed-to-seed and bulb-to-seed methods are used for seed production (**see: Production for sowing, I. Principles, 1. Starting material**) The former involves sowing in mid/late summer, overwintering as small immature plants and then flowering the following year – a method that does not allow bulb inspection and **roguing**. The **seed-to-seed** method is normally used for salad onions and leeks and the longer, but more flexible, **bulb-to-seed** method for bulb onions and the **root-to-seed** method for some leeks.

Bulb production for seeding is essentially the same as for commercial bulb production although seeding rates may be higher to produce smaller bulbs, normally known as mother bulbs. Often the mother bulbs for seed production are grown in the usual area for commercial bulb production, they are then lifted, stored and shipped to regions where the climate is more optimal for seed production, warm dry summers being best. Low humidity reduces the risk of disease and sunny weather during flowering is needed for good pollinating insect activity. In Europe the preferred areas for seed production are central France and northeast Italy.

Mother bulbs are normally planted in rows 70–100 cm apart. Seed yield increases with plant density but closer spacing increases risk of disease. For production of hybrids the usual ratio of sterile to fertile parent is 1:3 or 1:4, although usually they are planted in multiple row plots (often 2:6) for easier inspection and handling at harvest. Onions are insect-pollinated and the minimum recommended **isolation distance** (1000 m) should be strictly followed.

Flower synchronization (**nicking**), so that sterile and fertile lines flower together, is vital for the production of hybrid seed to guarantee pollination of the sterile line and to minimize contamination from other pollen sources. Little pollination occurs without insect (honeybee) activity and hives are introduced when around 25% of the umbels have open flowers. It is vital in the production of hybrid seed that the male sterile line produces sufficient nectar to attract insects. Irrigation is important to maximize yield: there can be a large decline in **water potential** from the scape to the umbel and water stress aggravated by root disease can lead to death of developing seeds at the early stages of endosperm development.

Good weed, pest and disease control is critical to successful seed production: Alliums compete poorly with weeds. Seed crops are affected by much the same range of pests and diseases that affect the bulb crop. Onion thrip (*Thrips tabaci*) is the major pest and this can invade the umbel and result in poor seed development. It is more serious in warmer, drier climates and chemical control is not easy. Downy mildew (*Peronospora destructor*), first appearing on the foliage, can be serious in damper climates and where growth is luxuriant. Neck rot (*Botrytis allii*) can be a major problem in storing bulbs in preparation for a seed crop. The symptoms usually appear after apparently healthy bulbs are placed in store. The disease causes a black mass of **sclerotia** below the dry outer skin of the bulb and a grey mould may develop on the bulb surface. The disease is latent in live green tissue and symptomless. However, the disease is seedborne and can result in severe crop loss.

Seed maturation is quite rapid, with around 600 **day-degrees** above 10°C needed to pass from anthesis to shedding ripe seed. **See: Colour Plate 6A.** Seed is harvested either by hand or machine. To minimize seed losses for hand harvesting, the crop needs to be harvested when around 25% of the umbels show ripe seed as the capsules dehisce. The capsules **shatter** easily and precise timing of harvest relies on local experience. Umbels are cut with around 15 cm of scape attached. Machine harvesting is better a few days earlier when ripe seed is just being seen. Seed moisture content at harvest is likely to be around 30% and further drying is necessary, traditionally done in the open on large tarpaulins but now with forced air circulation. Threshing is easy as soon as the seed is properly dry although onion seed is easily damaged. For cleaning, air-screen cleaners are mainly used. Some difficult seed lots are cleaned using water separation, the good seeds sinking and the poor seeds and pedicels floating off. After this, the seed should be rapidly re-dried by centrifuging and forced air to less than 12% moisture.

Seed yields are very variable; from an open-pollinated cultivar in ideal conditions 2000 kg/ha is possible but 1000 kg/ha, or less, is more realistic. Hybrid yields inevitably are less and can be as low as 50–100 kg/ha. Seed weight is around 250–350 seeds per gram.

5. Quality, vigour and dormancy

At high storage temperatures and humidities onion seeds lose viability quicker than many vegetable seeds and this can be a serious problem in local seed production in tropical areas. Seed for commercial storage is dried to about 6.3% moisture-content and sealed in moisture-proof containers. Under these conditions high-quality seed will remain in good condition for at least 3 years.

There is no strong evidence for **dormancy** in onion seed but leeks sometimes show dormancy at temperatures above 20°C. This appears to occur when seed heads develop at cool temperatures of approximately 15°C and then the seed matures, or is artificially dried, at higher temperatures.

6. Sowing

In northwest Europe onion crops are established either by direct sowing or by planting prepared sets; transplants are rarely if ever used but in some parts of the world transplanting is common. In the UK the majority of leeks are direct sown although in the Netherlands, Belgium and France transplanting is the normal method of establishment, partly because transplants can be planted deeper to get a longer pseudostem in line with local consumer preferences.

From direct sowing a high standard of husbandry and seedbed preparation is needed to get good germination and establishment, as all *Allium* species are slow to germinate and develop. Leeks are especially slow even compared to onion. Onion seed can imbibe water to the point of radicle emergence even in quite dry soil but needs abundant moisture for germination. Therefore irrigation is essential for sowing at times when seedbeds are dry. A firm **seedbed** is always desired, with the intention of placing the seed on a moist firm underlying soil layer that ensures good capillary conductivity for water.

Precise control of plant populations is necessary because bulb size and hence the premium marketable fraction of the crop is largely determined by plant density. Seeds are normally sown about 1.5–2 cm deep and a large proportion of the seed direct-sown in the UK is **pelleted** before sowing for precision drilling. Emergence in the field can be calculated from the laboratory germination figure: in ideal conditions 90% of the laboratory germination will emerge but around 70% is considered average.

After germination, the **cotyledon** appears as a loop just above the soil surface (the loop or crook stage): an epigeal mode of seedling establishment (**see: Epigeal**, Fig. E.8c). The primary root only develops to 8–10 cm and all subsequent roots are adventitious, and are readily colonized by **arbuscular mycorhizzal** fungi, which assist with nutrient absorption.

7. Treatments

Pesticide coating of onion and leek seed is now standard practice to control seedling **damping off** fungi and to provide some protection against onion fly, bean seed fly and onion thrips. The major use of seed treatment is in the control of the seedborne, onion neck rot caused by *Botrytis allii*. Fungicide seed treatments are routinely used to control onion smut in northern onion growing states of the USA. The cost–benefit ratio of a relatively small cost for seed treatment, which reduced losses to low levels, is enormous, and onion seeds are now routinely coated with appropriate fungicides. Insecticide seed treatments are additionally used in some areas, for instance to control the onion fly maggot (*Delia antiqua*) and bean-seed fly (*D. platura*).

Seed inoculants and some biological control agents are

currently being investigated. Mycorhizzal fungi preparations have, at an experimental level, also shown beneficial effects on early seedling growth and establishment but neither is widely used commercially.

Priming seeds of leek and onion is now fairly normal practice in some regions, such as western Europe, and can improve time to 90% germination by 5–6 days at 15°C, although the shelf life of the seed is limited. The gains of rapid emergence and quicker establishment often makes other aspects of crop management such as post-emergence weed control much easier, but are often not significant in total crop yield at harvest. Seeds are often sown in **filmcoated, encrusted** or **pelleted** forms. (**See: Colour Plate 11A**)

(BS)

Brewster, J.L. (1994) *Onions and Other Vegetable Alliums*. Crop Production Science in Horticulture 3. CAB International, Wallingford, UK.

Kelly, A.F. and George, R.A.T. (1998) Alliaceae: Vegetable Crops. In: *Encyclopaedia of Seed Production of World Crops*. John Wiley, Chichester, UK, pp. 153–159.

Allogamy

Cross-fertilization: the transfer of pollen to the stigma from the anther of plants of different clones or lines, and the resulting union of the egg and pollen gametes to form a zygote. (**See: Fertilization; Reproductive structures**)

Allo(poly)ploid

A polyploid formed from the union of two separate chromosome sets and their subsequent doubling. An organism produced by hybridization of two species followed by chromosome doubling. (**See: Polyploid**)

Allspice

Allspice (*Pimenta officinalis*, Myrtaceae) is the dried, unripe berries of the small tropical tree (**see: Spices and flavours,** Table S.12). The 5–8 mm fruits consist of two one-seeded loculi: the seeds form the spice element (Fig. A.5). The fruits have a mixed spicy aroma, combining cloves with cinnamon, juniper, nutmeg and pepper – hence the name, allspice.

Fig. A.5. Allspice fruits and seeds (image by Mike Amphlett, CABI). (Scale = mm)

Allspice contains 2–5% **essential oil** of which the major constituent (65–85%) is eugenol (Fig. S.44 in **Spices and flavours**); eugenol methylether, 1,8-cineol and α-phellandrene are also present.

The spice is used in pickles and condiments and especially in meat dishes. Its consumption in Europe is almost confined to Britain but it does feature in some Dutch and Scandinavian dishes. Meat cured in allspice, then roasted on a 'boucan', was common in parts of the Caribbean. Sailors who relied on this as a major source of animal protein came to be called 'buccaneers'. Allspice does not feature to any great degree in Asian cooking but it is used in south-east Europe, the eastern Mediterranean and East Africa. It is utilized in toiletries and has several medical applications. The eugenol in allspice is mildly analgesic.

(KVP, NB)

Peter, K.V. (ed.) (2001) *Handbook of Herbs and Spices*. Woodhead Publishing, CRC Press, UK.

Purseglove, J.W., Brown, E.G., Green, C.L. and Robbins, S.R.J. (1981) *Spices*. Vol. 1. Tropical Agriculture Series, Longman, New York, USA.

Weiss, E.A. (2002) *Spice Crops*. CAB International, Wallingford, UK.

Almond

The almond (*Prunus dulcis*, Rosaceae), of temperate climates, is commonly termed a nut though it is not a true nut in the botanical sense (**see: Nut**) since the shell is derived only from the inner part of the **pericarp**, the **endocarp** (Colour Plate 4A). The inner, edible embryo (kernel) is about 2 cm long and 1.3 cm at its widest. The fruit, borne on the tree that can reach 12 m in height, is a **drupe**, similar to its close relative, the peach (*Prunus persica*), although the **mesocarp** becomes a dry, leathery hull at maturity, rather than the succulent tissue that characterizes the peach. Peach and almond hybridize; the hybrid is used as a rootstock. Both species derive from an ancestral *Prunus* species in central Asia which spread southwest to evolve as the modern almond and eastward into China where the peach evolved. The almond has been under cultivation for at least 5000 years and is now the most important tree-nut crop. From 1999–2003, the world's leading producers of almonds were the USA (41% of world production), Spain (15%) and Italy (6%). (**See: Crop Atlas Appendix, Map 4**)

Almonds are rich in **oil** (triacylglycerol) (**see: Nut**, Table N.2) and have the highest concentration among the nuts of the antioxidant vitamin E (α-tocopherol) which is thought to confer beneficial effects on human health.

The nuts have numerous uses in baking, cooking, confectionery and as snack food. The oil is used for culinary, cosmetic and pharmaceutical purposes.

A variety of *P. dulcis*, *P. amara*, is the bitter almond which contains the bitter, cyanogenic glycoside, amygdalin. The bitterness deters the consumption of sufficient quantity to cause death by cyanide poisoning. (**See: Poisonous seeds**)

(VP)

Vaughan, J.G. and Geissler, C.A. (1997) *The New Oxford Book of Food Plants*. Oxford University Press, Oxford, pp. 30–31.

Amadori product

Complex sugar-protein compounds that are present in seeds that show deterioration changes. They consist of the non-cyclized isomer of a glycosylamine (R-C(=O)CH_2-NH-R′), where R is usually part of a reducing sugar and R′ is usually a protein, formed from the dehydration and rearrangement of a **Schiff's base**. A Schiff's base is formed from the reaction of an amine group (NH_2) with an aldehyde (HC=O) (e.g. the carbonyl group in the non-cyclized form of glucose). Amadori products may also form directly from reactive carbonyl by-products of **peroxidation**. Amadori products are intermediates of the **Maillard Reaction** and are degraded into a variety of molecules depending on the moisture, temperature and pH of the reaction mixture in reactions that lead to non-enzymatic oxidation of biological molecules. (**See: Deterioration and longevity**) (CW)

Amaranthus

See: **Pseudocereals**

American Seed Trade Association

See: **ASTA**

Amino acids

See: **Protein and amino acids**

Amphicarpic

The production by a plant of fruit of two kinds, either in form or in time of ripening. This phenomenon occurs in only about 100 species. Several species, for example, produce flowers both above and below ground (**See: Deserts and arid lands, 3. Seed dispersal strategies and mechanisms**), which require different pollination strategies.

Amphidiploid

Alternative name for an allotetraploid: a plant that has a diploid set of chromosomes from each parent. An allotetraploid (amphipolyploid) that appears to be a normal diploid. Examples include horticultural *Brassica* spp. (**See: Ploidy; Polyploid**)

Amphiphilic/amphipathic

Characterizing molecules containing both hydrophilic (water-attracting) and hydrophobic (water-repelling) domains and therefore exhibiting a physico-chemical affinity for both aqueous and non-aqueous phases. The hydrophilic group forms an interface with water and the hydrophobic group is sequestered in a water-insoluble interior, such as in fatty acids. Typically phospholipids are amphipaths. (**See: Dehydrins; Desiccation tolerance – protection by stabilization of macromolecules; Oleosins; Partitioning of molecules**)

Amphitropous

Type of ovule. (**See:** Fig. R.3 in **Reproductive structures, 1. Female**)

Amylase inhibitors

Several types of structurally unrelated seed proteins have been identified that inhibit the activity of insect and/or mammalian α-amylases. Most of these inhibitors belong to the so-called cereal inhibitor (super)family, which is exclusively found in **cereals** and comprises trypsin inhibitors, amylase inhibitors (AI), bifunctional trypsin/AI and other proteins without any known biological activity. Endosperms of cereals like wheat, barley, finger millet (ragi) and others contain a complex mixture of related AI that all belong to the cereal inhibitor superfamily. Some of these AI occur as monomers, whereas others are homodimers and heterotetramers. In barley, the monomeric and dimeric amylase inhibitors consist of polypeptides of 132 amino acid residues whereas the tetrameric AI is built up of one CMa chain (119 amino acids), one CMb chain (124 amino acids) and two CMd chains (146 amino acids). The bifunctional inhibitor from finger millet (RBI; RATI) is a monomer of a single 122-amino acid residue polypeptide. All members of the family of trypsin/α-amylase inhibitors from cereals share a fairly high sequence similarity (95–40% identity) and have a similar folded structure that is stabilized by five intra-chain disulfide bridges. Though all of the AI inhibit mammalian and insect α-amylases, there are pronounced differences between their specific activities and substrate specificities.

In addition to cereal trypsin/α-amylase inhibitors, some cereals (e.g. rice, wheat and barley) contain bifunctional α-amylase/subtilisin I inhibitors that belong to the Kunitz-type inhibitors (**see: Protease inhibitors**).

A third type of AI occurs in kidney, French or common bean (*Phaseolus vulgaris*) seeds and consists of a single protomer that shares high sequence similarity with the bean **lectin (phytohaemagglutinin**, PHA) but unlike the lectin it is post-translationally cleaved into two smaller polypeptides of 76 (α-chain) and 137 (β-chain) amino acid residues, respectively. The bean AI is a potent inhibitor of both mammalian and insect α-amylases. Studies with transgenic peas have demonstrated that the bean AI may be involved in plant/seed defence against insects. Feeding trials with rats indicated that the bean AI blocks the degradation of starch in the digestive system of rats and at a high dose causes them severe health problems.

Three other seed AI have been identified. Seeds of the panicoid cereal *Coix lachryma-jobi* (Job's tears) contain a bifunctional AI/endochitinase that inhibits insect α-amylases and exhibits endochitinase activity. A 95-amino acid residue protein (Ragi I-2) occurs in the endosperm of finger millet that inhibits mammalian α-amylase and shares sequence similarity with a non-specific phospholipid transfer protein. A very small AI has been isolated from the endosperm of *Amaranthus hypochondriacus*. This 32-amino acid residue inhibitor of insect α-amylases contains three intra-chain disulfide bonds and shows some similarity with antimicrobial peptides from *Mirabilis jalapa*, the squash family of protease inhibitors and the knottins. (EVD, WJP)

Carbonero, P. and Garcia-Olmedo, F. (1999) A multigene family of trypsin/α-amylase inhibitors from cereals. In: Shewry, P.R. and Casey, R. (eds) *Seed Proteins*. Kluwer Academic, Dordrecht, The Netherlands, pp. 617–633.

Grossi de Sa, M.F., Mirkov, T.E., Ishimoto, M., Colucci, G., Bateman, K.S. and Chrispeels, M.J. (1997) Molecular characterization of a bean alpha-amylase inhibitor that inhibits the alpha-amylase of the Mexican bean weevil *Zabrotes subfasciatus*. *Planta* 203, 295–303.

Amyloid

Xyloglucan structural component of primary cell walls, and in some seeds also a storage carbohydrate, which is stained blue-black with iodine/potassium iodide solution. (**See: Hemicellulose; Xyloglucan**)

Amylopectin

Together with **amylose** this is a component of **starch**. It is the branched glucan polymer that accounts for **starch granule** crystallinity. Amylopectin is composed of approximately 10^4–10^5 glucose monomers joined via α-(1→4) linkages to form linear chains and α-(1→6) linkages to form branches. The distributions of linear chains and branch linkages are non-random, such that the average chain length is 20–30 glucosyl units and the median length is about 12 units. Amylopectin branch frequency is about 5%, with regions of high branch frequency alternating with areas nearly devoid of branches. This cluster organization allows for the crystalline packing of linear amylopectin chains into closely associated double helices termed 'crystalline lamellae', which alternate with 'amorphous lamellae'. (MGJ)

Amyloplast

A type of **plastid** in which **starch** is synthesized and stored. In cereal **endosperms** and the **cotyledons** of **dicots** the bounding membrane structures lose their integrity during **maturation drying**. (**See: Starch granule**)

Amylose

Together with **amylopectin** this is a component of **starch**. It is essentially a linear polymer of glucose units (monomers) coupled by α-(1→4) linkages. Some amylose molecules may also have a few long-chain branches. Typical amylose molecules contain approximately 10^2–10^4 glucosyl monomers. Amylose comprises 25–30% of starch but is not necessary for the formation of **starch granules**, as evidenced by plant species with mutations that result in a loss of amylose (i.e. *waxy* mutations): such granules are composed almost exclusively of amylopectin. Within granules, amylose chains may form double helices with one another or with long amylopectin chains, or may be complexed with lipids. High-amylose maize starches (50–70% amylose) have useful functionalities, including production of firm and opaque gels. (MGJ)

Anacampylotropous

Alternative name for campylotropous, a type of ovule. (**See:** Fig. R.4 in **Reproductive structures, 1. Female**)

Anaerobic respiration

Respiration which occurs under conditions of low (hypoxia) or no (anoxia) oxygen. This results in fermentative metabolism whereby the pyruvic acid produced as a result of glycolysis is generally converted to alcohol (ethanol).

Anaesthetics

Breaking of dormancy can be achieved by exposure of seeds to low molecular weight amphipathic (**amphiphilic**) molecules such as primary alcohols, monocarboxylic acids and anaesthetics such as chloroform and halothane. This observation led to the so-called 'anaesthetic hypothesis' which related the dormancy-breaking activity of these compounds to their lipophilicity, and resulting changes in the fluidity of **membranes** within the cells of the seed. A more recent suggestion is that the action of small amphipathic molecules influences the head-group spacing of the **phospholipid** components of the membrane, either increasing or decreasing it. This is turn facilitates the binding and activation of a peripheral membrane protein (PMP) which is a component of a **signal transduction** pathway that is essential for the completion of germination.

There are seeds of some species, however, for which small molecules, including anaesthetics, inhibit germination. In these cases the head-group spacing is proposed to be altered to the extent that it does not permit the binding of a PMP. (**See: Dormancy breaking – chemicals**) (JDB)

Hallett, B.P. and Bewley, J.D. (2002) Membranes and seed dormancy: beyond the anaesthetic hypothesis. *Seed Science Research* 12, 69–82.

Analytical purity testing

See: Purity testing

Anamorph

The asexual ('imperfect state') stage of the life cycle of fungi. In a fungus that has a life stage where it produces sexual spores, and another stage where it reproduces by other means, the stage that produces sexual spores is called the **teleomorph** (or perfect stage), and the asexual life stage is called the anamorph (or imperfect stage). Because these are often visually quite distinct, the teleomorph and the anamorph of the same fungus have different names. Members of the Fungi Imperfecti are only known by their anamorphic name. (**See: Diseases; Pathogens**)

Anatropous

The most common type of ovule in the **angiosperms** and more than 200 families are exclusively anatropous. Atropous (**orthotropous**) is a different ovule type. (**See:** Fig. R.3 in **Reproductive structures, 1. Female**)

Ancient seeds

Over the last two centuries there have been numerous claims, mainly in the popular press, that 3000-year-old cereal seeds recovered from Egyptian tombs had retained their **viability**. The likelihood of cereal seeds remaining viable for such periods under the relatively dry (~ 20% RH) but warm (~ 25°C) conditions inside a tomb can be tested using the seed **viability equations**. Even the most optimistic prediction suggests that seeds would be completely dead after a few decades, supporting the fact that all rigorously controlled scientific attempts to germinate such seeds have consistently failed (Table A.5). However, the viability equations do predict such extreme **longevity** if seeds are held under very dry and cold conditions, such as those used in seed **gene banks**.

Seeds that possess **physical dormancy** such that the internal tissues will remain very dry, protected by the impermeable seedcoat, are thought to be amongst the longest-lived under natural conditions. Seeds of the sacred lotus (*Nelumbo nucifera*) are of this kind and one recovered from an ancient lake bed in China holds the record for the most ancient,

Table A.5. Claims for extended longevity of seeds.

Species	Location	Age	Status of seed	Comments
Stored in dry conditions				
Barley	Tomb of King Tutankhamun	*ca.* 3350 years	Non-viable	Extensively carbonized
Wheat	Various ancient Egyptian tombs	3000 years or more	Not known	Age and source of grains never authenticated
Wheat	Thebes	4000–5000 years	Non-viable	Some cell fine structure conserved,
	Feyum	6400 years	Non-viable	although degrades upon imbibition
Canna compacta	Santa Rosa de Tastil, Argentina	*ca.* 600 years	Viable	Enclosed in a nutshell forming part of rattle. Shell dated at 600 years and seed probably of same age
Albizia julibrissin	China to British Museum	200 years	Viable in 1940	Germination started accidentally. Viable seedlings produced
Cassia multijuga	Museum of Natural History, Paris	158 years	Viable	Wholly authenticated history from collection to sowing
Date palm	Herod's Palace, Masada, Israel	2000 years	Viable	One seed germinated. Other seeds at site radio dated
Buried in soil or water				
Arctic lupin (*Lupinus articus*)	Miller Creek, Yukon	>10,000 years	Viable	Invalid. Age deduced indirectly from geological data. No direct dating evidence
Indian or sacred lotus (*Nelumbo nucifera*)	Kemigawa, near Tokyo	3000 years	Viable	Seeds found on submerged boat radio-dated at 3000 years. No direct measurements of age of seed, which could have settled in sediments after shedding from modern plants
	Xipaozi, Laoning Province, China	1280 years	Viable	Seeds from ^{14}C-dated fruits found in dried lake bed. The same lake bed has also yielded viable seeds dated at 200–500 years old.[a]
Chenopodium album *Spergula arvenis*	Denmark and Sweden, archaeological digs	>1700 years	Viable	No direct dating of seeds. Could be modern seeds dispersed into archaeological digs

From Bewley, J.D and Black, M. (1994) *Seeds Physiology of Development and Germination*, 2nd edn. Plenum Press, NY, USA.
[a] Shen-Miller, J. (2002) Sacred lotus, the long-living fruits of *China Antique*. *Seed Science Research* 12, 131–143.

authenticated (by radiocarbon dating), viable seed with an age of about 1200 years.

Considerable longevity has also been reported in seeds that do not possess impermeable seedcoats. For example, seeds of *Verbascum blattaria* recovered from the famous buried seed experiment begun by W.J. Beal at Michigan State University in 1879 were still viable (46% germination) after 120 years of burial in sandy soil. The next test is due in 2019. Viable seeds of a range of species estimated to be between 180 and 200 years old have also been recovered from adobe walls of historic buildings in California and Mexico. (RJP, SHL)

Shen-Miller, J. (2002) Sacred lotus, the long-lived fruits of *China Antique*. *Seed Science Research* 12, 131–143.

Telewski, F.W. and Zeevaart, J.A.D. (2002) The 120-year period for Dr. Beal's seed viability experiment. *American Journal of Botany* 89, 1285–1288.

Androgenesis

The production of plants by regeneration from microspores (pollen grains) in culture. Regeneration in many species, e.g. barley, maize, is via **embryogenesis** in which **haploid** embryos are produced.

Anemochory

Dispersal of seeds, fruits and other dispersal units (commonly termed diaspores) of a plant by wind.

Aneuploid

Polyploid plant having a chromosome number that is not an exact multiple of the haploid number, with either fewer or more than the normal number in the cell. Contrast **amphipolyploid**. (**See: Polyploid**)

Angiosperms

(Greek: *angeion* = vessel, container; *sperma* = seed) A division of the Spermatophyta producing their ovules and seeds within closed megasporophylls (carpels) in contrast to **gymnosperms** where the ovules and seeds on the megasporophylls lie 'naked' and are openly exposed to the environment. Angiosperms are also unique in displaying **double fertilization** which gives rise to a diploid embryo and a triploid storage tissue (**endosperm**). In English, the angiosperms are often also called 'flowering plants' which, however, is not strictly correct since the reproductive organs of gymnosperms are also borne in structures that fulfil the criteria of the definition of a flower. (**See: Reproductive structures**)

Anhydrobiote

Organism or organ that survives essentially complete dehydration. Dry anhydrobiotes remain viable for extensive periods when kept under cold and low relative humidity conditions. They exhibit very little or no metabolic activity depending on their water content. Upon rehydration, they resume metabolism. Examples of anhydrobiotes are found in numerous invertebrates, lower (algae, lichens, mosses) and higher plants (seeds, pollens, resurrection plants) and fungi (yeast). (**See:** entries on **Desiccation tolerance**)

Anise

Anise (aniseed) (*Pimpinella anisum* L., Apiaceae) is one of the oldest of medicines and seed spices originating in the Mediterranean region (**see: Spices and flavours,** Table S.12). The plant is an erect aromatic herb, around 50 cm tall. The 'seed' (fruit) is a short, hairy, brownish **cremocarp** 3–5 mm, splitting into 2–5 ribbed, 3–5-mm-long **mericarps** at maturity, each containing a single seed (Fig. A.6). Whole or ground, these comprise the spice. The 'seed' has a characteristic sweet smell and pleasant aromatic taste.

A colourless to pale yellow, intensely sweet, mild **essential oil** is obtained from the 'seeds' through distillation (about 4% of seed weight). The main constituents of anise seed oil are over 90% *trans* anethole, the main flavouring compound, and methyl chavicol (Table A.6) (**see**: Fig. S.44 in **Spices and flavours**). It also has a small quantity of *cis* anethole, which has toxic properties.

The crushed seed or powder is used to flavour curries, sweets, confectionery and baked goods. It is important in various alcoholic drinks or beverages (anisette, arrak, pastis, ouzo). Anise seeds are considered stomachic, carminative, diaphoretic, diuretic, antispasmodic, antiseptic and stimulative. They flavour cough medicines and cough syrups. (KVP, NB)

Fig. A.6. Anise 'seeds' (image by Mike Amphlett, CABI). (Scale = mm)

Table A.6. Composition of anise essential oil.

Compound	Percentage (wt)
trans-Anethole	94.7
cis-Anethole	0.4
Anisaldehyde	0.6
Anisyl alcohol	0.4
Linalol	0.1
Limonene	0.1
Methyl chavicol	2.1
α-Pinene	trace
β-Pinene	trace

Guenther, E. (1978) *The Essential Oils*, Vol. 2. Robert E. Krieger Publishing, New York, USA.
Peter, K.V. (ed.) (2001) *Handbook of Herbs and Spices*. Woodhead Publishing, CRC Press, UK.
Weiss, E.A. (2002) *Spice Crops*. CAB International, Wallingford, UK.

Anisogamy

The production of and union between two **gametes** of unequal size. In the case of seed-bearing plants (Spermatophytes), between the smaller gametic male nucleus from the male gametophyte, the pollen grain, and the larger female gamete, the egg cell (ovum) of the female gametophyte. They fuse to produce the **zygote**. (**See: Fertilization; Reproductive structures**)

Anisoploid

A mixture of diploid, triploid and tetraploid plants obtained from seed harvested from a mixture of diploid and tetraploid parent plants. (**See: Ploidy; Polyploid**)

Annatto

Annatto or achiote (*Bixa orellana*, Bixaceae) seeds are used both as a **spice** and a dye. The species is native to South America but the current leading producers are Brazil, Peru, Kenya: significant production also occurs in the Philippines, Ecuador, Guatemala, Ivory Coast, India and Spain. The plant is a shrub which bears fruit as red capsules containing many red, tetrahedral seeds, each 3–5 mm across the base (Colour Plate 7A).

The seeds are used primarily as a food colorant. The red colour of the seeds is caused by several apocarotenoids (oxidative products of carotenoids) that they contain, the main one being bixin, as well as by some carotenoids. These compounds may make up to 7% fresh weight of the seed. The distinctive, flowery aroma of annatto is brought about by ishwarane, a sesquiterpene hydrocarbon.

An orange-yellow paste is produced by soaking crushed seeds in water that is later allowed to evaporate. Paste produced in South America is exported to North America and Europe to be used as a colouring in many foods such as soups, stews, butter, cheese, margarine, ice cream, popcorn and other yellow or orange foodstuffs. In South America and the Caribbean seeds are fried in fat to colour it before being used for cooking. The ancient Aztecs and Maya added annatto to the chocolate drunk by the priests and rulers. (**See: Cacao; History of seeds in Mexico and Central America**)

Indigenous South American rainforest peoples have used annatto seeds as body paint and as a fabric dye. It was also used by the Maya to colour foods, for body paints and in arts, crafts and murals.

Annatto is also used currently as a natural textile dye and to colour paints, varnish, cosmetics and soap. (MB)

Anthesis

The period during which a flower becomes open and functional. During the **development of seeds,** experimentally it is desirable to time the events that are occurring with respect to a starting point. Since it is virtually impossible to record when **fertilization**, resulting in **zygote** and **endosperm** cell formation, is completed, which is the starting point of development, a more tractable marker is used. Hence time of seed development is frequently recorded as 'time (days, weeks) after anthesis' or, by the same parameter, 'time after flowering'.

Anthracnose

Plant diseases, some caused by seedborne **pathogens**, that are characterized by limited black lesions in many parts of the plant, often sunken in fleshy tissues, and in some cases post-emergence seedling damping-off, caused by certain imperfect fungi that produce conidia in acervuli. The name is derived from the Greek 'anthrax', meaning coal or charcoal. Seed is a major source of infection of *Colletotrichum* (or *Gloeosporium*) *lindemuthianum* anthracnose of common bean and pigeon pea (**See: Legumes**), where the pathogen can survive for long periods, either as dormant mycelium within the seedcoat (evident as brown spots) or as spores between or within cotyledons. Similarly, *Colletotrichum gossypii* in cotton appears to be primarily seed-transmitted by internal infection within bolls, and *C. linicola* in flax by resting hyphae in the outer seedcoat layers. *Colletotrichum kahawae*, though not seedborne or seed-transmitted, directly destroys the young, expanding coffee berries and causes premature shedding. Anthracnose is controlled by combination of fungicide seed treatment and, in some crop species, by the use of field-resistant cultivars. (PH)

Anti-nutritional agent

A component of a seed that is regarded as compromising the nutrition of humans and domestic animals. **Phytin** is one example, which binds ions such as calcium within the gut, and can lead to reduced bone strength in chickens. The term might also be applied to **allergens** that affect digestibility of foods, to toxins such as **protease inhibitors**, **lectins** and tannins. (**See: Bean – common; Phytohaemagglutinin; Soybean**)

Antioxidants

Diverse molecules that serve as a defence system against oxidative damage by confining and neutralizing **reactive oxygen species** (ROS), other **free radicals**, or catalysts to oxidizing reactions. They are important in seeds as they protect against certain deteriorative changes. Antioxidants quench or scavenge ROS by donating electrons, and this process lowers the toxicity of ROS. The enzyme **superoxide dismutase** (SOD) catalyses the reduction of superoxide into hydrogen peroxide (H_2O_2) producing triplet oxygen (a less reactive form of O_2) as a by-product. Catalases and peroxidases convert hydrogen peroxide into water and triplet oxygen. The sulfhydryl groups of glutathione (GSH) molecules, a tripeptide composed of γ-glutamic acid, cysteine and glycine, are oxidized in the presence of hydrogen peroxide to form gluthathione disulfide (GSSG) and reconverted to GSH through the activity of glutathione reductase and the cofactor NADPH. Plant phenols (e.g. tocopherol) and carotenoids (e.g. β-carotene) are the most important lipid-soluble antioxidants and can scavenge ROS (hydroxyl radical or singlet oxygen, respectively) or terminate auto-oxidation chain reactions by reacting with lipid-peroxyl radicals. Donation of one electron creates a relatively stable radical because the remaining unpaired electron is de-localized within conjugated unsaturated bonds. The radical form of tocopherol recycles to the antioxidant form by abstracting an electron from ascorbate, which also may scavenge ROS directly. The resulting ascorbyl radical, which is relatively unreactive, converts to ascorbate and oxidized products that are eventually degraded. Phenolic antioxidants (e.g. catechin from green tea and butylated hydroxytoluene [BHT] commonly used in laboratory assays) also suppress non-enzymatic browning reactions such as the **Maillard reaction**. Other hydroxy radical scavengers include mannitol and dimethylsulphoxide (DMSO), molecules that also have cryoprotectant activity. Phytates and other metal chelators sequester transition metal ions that catalyse oxidizing reactions and therefore have important prophylactic activity. Decrease in antioxidant concentrations during oxidative stress demonstrates that the defence system is working to limit oxidative damage. (**See:** articles on **Deterioration; Desiccation tolerance – protection against oxidative damage**) (CW)

Halliwell, B. and Gutteridge, J.M.C. (1999) *Free Radicals in Biology and Medicine*, 3rd edn. Oxford University Press, Oxford, UK.

Antiraphe

The side of the ovule or seed that lies opposite the **raphe**. (**See: Reproductive structures, 1. Female**)

AOSA

See: Association of Official Seed Analysts

AOSCA

See: Association of Official Seed Certifying Agencies

Apiaceae (Umbelliferae)

See: Umbellifers

Apogamy

The development of an embryo without fertilization, generally from antipodal or synergidal cells of the **embryo sac**. Such embryos are **haploid**. (**See: Apomixis; Parthenogenesis; Reproductive structures, 1. Female; Zygotic embryo**)

Apomixis

The phenomenon of **embryogenesis** without the formation of gametes, or fertilization, i.e. the ability of plants to reproduce asexually through seeds (agamospermy) – and, in the widest sense, by vegetative reproduction from somatic tissues of the mother plant as well. Plants grown from these seeds therefore contain only the maternal complement of genes (are diploid maternal), and are identical to the plant on which the seeds developed. Three different types of apomictic processes are known: diplospory, apospory and adventitious embryony (Fig. A.7). There may also be mixed apomixis involving more than one type. The former two types are known as gametophytic apomixis and the latter as sporophytic apomixis.

(a) *Diplosporous apomixis* is caused by unreduced **embryo sac** formation, i.e. without reduction division (meiosis), through modified development of the **megaspore**, by mitosis directly, or by interrupted meiosis. Examples include *Taraxacum* (dandelion) and some members of the Poaceae (Kentucky Blue Grass).

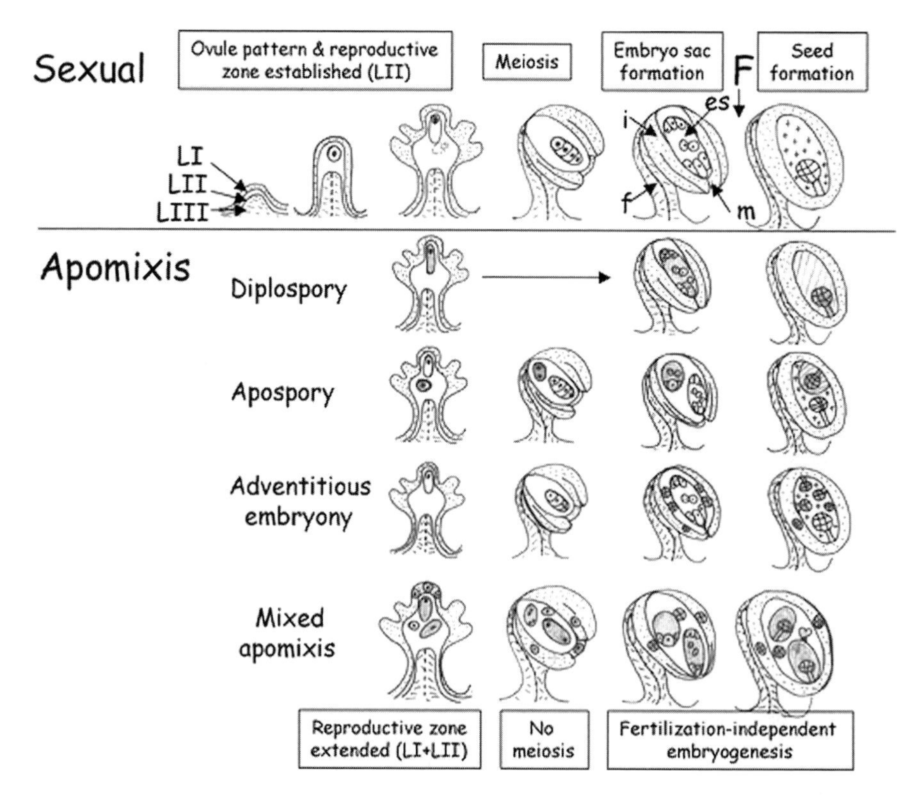

Fig. A.7. Sexual and apomictic reproduction in the ovule of flowering plants. The diagram summarizes ovule development beginning with the formation of the primordium made up of the LI, LII and LIII cell layers. The events of sexual reproduction are displayed in the top panel above the line. Apomictic processes are shown below the line and cells or structures involved in apomixis are shaded in grey while those pertaining to sexual reproduction remain white. Abbreviations: es, embryo sac; f, funiculus; F, double fertilization; i, integuments; m, micropyle. From: Koltunow, A.M. and Grossniklaus, U. (2003) Apomixis: A developmental perspective. *Annual Review of Plant Biology* 54, 547–574. Reprinted with permission, from *Annual Reviews of Plant Biology*, Volume 54 copyright 2003 by Annual Reviews. www.annualreviews.org

(b) *Apospory* is when cells of the **nucellus** (which is composed of somatic cells, not the megaspore) develop into an unreduced embryo sac (i.e. no meiotic divisions occur); the nuclei are of the same **ploidy** (2n) as the female parent. The normal sexual embryo sac often remains intact also. It is present in some members of the panicoid grasses within the Poaceae, for example millet.

The formation of the endosperm in (a) and (b) may be autonomous (fertilization-independent) or fertilization-induced (pseudogamous).

(c) *Adventitious embryony* is often induced from nucellar or integument cells without formation of an embryo sac. This is a rare type of apomixis, but occurs in *Citrus* species.

The formation of the endosperm occurs without **fertilization** of the diploid polar (endosperm mother) cell. (**See: Reproductive structures**)

Apomixis is found in about 400 species in several plant families, including Chinese chive (*Allium tuberosum* and *A. ramosum*) and several wild *Pennisetum* species. Apomictic genes from the wild plant species, *P. squamulatum*, have been transferred by backcrossing to cultivated pearl millet (*P. glaucum*). Some genetic data suggest that only a few genes control apomixis, and attempts have been made to find mutants that form full or partial seeds without fertilization. Mutations of three *FIS* (fertilization-independent seed) genes (*FIS1*, *FIS2* and *FIS3*) result in autonomous seed development in *Arabidopsis*, and mutations of *FIE* (fertilization-independent endosperm) genes result in defective endosperm development, thus decoupling the formation of these seed parts from fertilization. However, embryo and endosperm development is incomplete; embryos with *fis1* and *fis2* mutations do not develop beyond the globular stage (**see: Development of embryos – dicots**), endosperm nuclei are diploid and do not develop beyond the point of cellularization (**see: Development of endosperms – dicots**). In *fis3* mutants, neither the endosperm nor the embryo reaches these stages of development. Thus *FIS* are not the genes, or the only genes, controlling embryo apomixis, although mutations to, or down-regulation of them result in this taking place.

Apomixis is considered to benefit agriculture through its capacity to produce offspring by means of vegetative propagation, which are genetically identical to the parent plant. This could help maintain the genetics of F_1 hybrids, with hybrid vigour (**heterosis**) fully sustained in subsequent generations. In *Tripsacum dactyloides*, Eastern gamagrass, a chromosome arm has been identified which confers apomixis; attempts have been made to transfer this chromosomal region from *Tripsacum* to maize for breeding purposes. Apomixis has applications also for crops that cannot be propagated by grafting, and for the production of disease-free progeny. (**See: Development of seeds – an overview**) (O-AO, HS, XG)

Spillane, C.A., Steimer, A. and Grossniklaus, U. (2001) Apomixis in agriculture: the quest for clonal seeds. *Sexual Plant Reproduction* 14, 179–187.

Apricot

The fruit of the apricot (*Prunus armeniaca*) is a drupe similar to peach or almond. The kernel (embryo) within the stony endocarp is used for the manufacture of a flour or paste employed as a marzipan substitute (called persipan) in baking and biscuit making. The kernels also are the basis of some alcoholic liqueurs. Embryos contain 50-60% oil which can be used for cosmetic or pharmaceutical purposes. Also present is a cyanogenic glycoside (**see: Pharmaceuticals and pharmacologically active compounds**) named laetrile, sometimes referred to as a vitamin B17. Laetrile has attracted attention as a possible treatment for cancer but it has no official medical approval. (MB)

APSA

APSA, the Asia and Pacific Seed Association, has members including National Seed Associations, government research and seed enterprises, private seed companies, regional research centres and other national and regional organizations concerned with seed throughout the region. Its aim is to improve the production and trade of quality seed and planting material of agricultural and horticultural crops, by making representations to governments, disseminating information and organizing training and cultivar testing programmes in the Asia-Pacific region. Each year APSA organizes the Asian Seed Conference. www.apsaseed.com (**See: Industry**) (BIB)

Arabidopsis

Arabidopsis thaliana (wall cress, mouse-ear cress, thale cress), a member of the mustard family (Brassicaceae), is a small (15–20 cm flowering stem height), annual rosette plant, and is a persistent weed in many habitats. It has many ecotypes, but those most commonly used in research are Landsberg, Wassilewskija and Colombia, although others such as Cvi (Cape Verde Islands) are popular. It is the major plant species for molecular genetic studies. The importance of *Arabidopsis* in plant research comes from its relatively small genome on five chromosomes, which has been completely sequenced, its rapid 'cycling', i.e. growth from seed to subsequent seed production takes about 6–8 weeks, its size (large numbers of plants can be grown in a relatively small area), its ready and frequent self-pollination, its copious production of small seeds (about 5000 per plant) in siliques (siliquae) which generally do not readily **shatter** before seed maturation, and the comparative ease with which large numbers of mutants can be amassed. There are approximately 26,000 nuclear, 80 chloroplast and 60 mitochondrial genes encoding proteins in this species; the nuclear genome contains about 120 Mb (million base pairs) of DNA; the amount of DNA in the genome of rice, which also has been sequenced, is about three times larger, of maize about 20 times larger, and of wheat 120 times larger. Genetic and physical maps have been created for all five of the *Arabidopsis* chromosomes. (JDB, MB)

For information on the particular contributions of the seed to seed science **see: Research seed species – contributions to seed science.**

Somerville, C.R. and Meyerowitz, E.M. (eds) (2002) *The Arabidopsis Book*. American Society of Plant Biologists, Rockville, MD, USA (doi/10.1199/tab.0009, www.aspb.org/publications/arabidopsis/)

Arabinogalactans

A polymeric carbohydrate (**hemicellulose** polysaccharide) most commonly present as a non-storage component of the **pectin** matrix of cell walls. The highly branched Type I arabinogalactans (AGs) are composed of an α-1,4-linked galactose backbone with unit α-1,3-arabinose side chains. Type II arabinogalactans constitute a broad group of short galactose-containing chains connected to each other and are associated with structural proteins, the arabinogalactan proteins, AGPs (proteoglycans, 5% protein, 95% carbohydrate) whose precise function is unknown. The latter occur in large amounts in larch tree wood (from which they are harvested and sold as **nutraceuticals** and therapeutic agents), and their presence in wheat flour influences its dough-mixing characteristics. In cotyledons of some legumes, e.g. *Lupinus angustifolius*, arabinogalactans (along with arabinans and sometimes galactoxyloglucans) are present in thickened **cell walls** as a storage component which is mobilized following germination. (**See: Hemicellulose; Hemicellulose – mobilization**) (JDB)

Arabinoxylans

A **hemicellulose** polysaccharide which is a major component of the **cell walls** of the starchy **endosperm** of some cereal grains, e.g. wheat (85%) and of the **aleurone layer** of others, e.g. barley (71%), containing a β-1,4-xylose-linked backbone with unit α-1,2-arabinose side chains. They are not regarded as storage forms of hemicellulose. The amount of arabinoxylans synthesized during grain development is influenced by the environment. Extractable from the cell walls with hot water they (usually with co-extracted mixed linkage **β-glucans**) form solutions of high viscosity which are problematical in the brewing industry because they interfere with filtration. Their presence in poultry feed and poor digestibility increases the viscosity of faeces which sticks to eggs and imposes added washing expenses. In the wheat and rye bread- and biscuit-making industries, use has been made of the water retention capacity of arabinoxylans and their value as a source of dietary fibre. Claims that they have immune-system boosting, anti-tumour properties are strongly disputed. (JDB)

Arbuscular mycorrhiza

The term mycorrhiza, which literally means 'fungus root', describes the symbiotic association between plant roots and fungi, and the fungi capable of this relationship. Arbuscular mycorrhiza (AM) is the most widespread type, being geographically ubiquitous and forming symbiotic associations with a wide range of plant species. They are estimated to be present in about 90% of vascular plants (in contrast to **rhizobia** which have a host range limited to the **legumes**). The symbiosis develops in the plant roots where the fungus colonizes cells of the cortex to form distinct structures called arbuscules, which are sites of nutrient transfer.

The AM fungus receives carbon from the plant, and the plant, in turn, may receive a range of benefits, including improved plant growth and fitness (through increased mineral nutrition, principally phosphate), improved water relations and protection from pathogens. The growth of hyphae from the mycorrhizal root increases the volume of soil from which nutrients and water can be absorbed, thus extending the root system. The complex inter- and extra-cellular relationship between plant roots and AM fungi requires a continuous exchange of signals to ensure the proper development of the symbiosis. Plant **hormones** and certain flavonoids in root exudates, such as biochanin, may play a role in the regulation of the symbiosis, and many similarities between the signalling in the rhizobial and the AM symbiosis have been found. Several aspects of sexual reproduction can be influenced by AM, including the timing of reproductive events and number of flowers, fruits and seeds. Seed quality can also be strongly influenced by mycorrhizal infection, resulting in variation in seedling vigour and resultant competitive ability.

1. Inoculation and potential uses

The beneficial effects of AM on plant growth and health have prompted an increased interest in the use of mycorrhizas, exploiting this association as one of the most useful biological means of assuring high quality plant production with minimal input of chemicals.

Mycorrhizal inoculation is most beneficial where indigenous mycorrhizal fungi are absent or scarce, such as disinfected soils – normally used in commercial horticulture and forestry nurseries – inert substrates, micropropagation techniques, artificial landscapes, amenity and urban settings. However, the successful use of AM fungi can only be achieved under certain conditions: benefits will only be obtained by a careful selection of compatible host/AM/substratum combinations.

A variety of mycorrhizal inoculants are commercially available – some being blends of AM and ectomycorrhizas. AM fungi are obligate biotrophs unable to grow in pot cultures; hence inoculum has to be produced on living roots, and includes spores, mycelia and mycorrhizal root fragments. These are normally incorporated into a carrier, such as sand, soil, **peat**, clay, **vermiculite** and other substrates. The development of seed**coating** or liquid delivery systems has been limited mainly by the propagule size and most commercial inoculants are granular formulations.

The most common application method places the inoculum below the seed or seedling before planting, which ensures close proximity with the developing root for effective colonization. The development of an efficient AM at the nursery and seedling stage is essential to produce high quality seedlings and the general rule is that the earlier the inoculant is applied, the greater the benefit. AM inoculation of turf grasses at seeding is becoming a common practice on golf courses and football pitches, where inoculants are applied at 20 g/m^2 with calibrated granule applicators. Many commercial horticulture and ornamental crops, such as tomatoes, aubergines, peppers, leeks, asparagus and Gerbera, are also being routinely inoculated at seeding time, for example in Japan and Thailand (**see: Rhizosphere microorganisms**).

Also, all **orchids** have a requirement for mycorrhiza at germination (Basidiomycetes mainly of genus *Rhizoctonia*); though *in vitro* germination of some species is possible without a fungal symbiont, cultivation requires infection for continued development of the protocorm. (IA)

Azcon-Aguilar, C. and Barea, J.M. (1997) Applying mycorrhiza biotechnology to horticulture: significance and potentials. *Scientia Horticulture* 68, 1–24.

Koide, R.T. (2000) Mycorrhizal symbiosis and plant reproduction. In: Kapulnick, Y. and Douds, D.D., Jr (eds) *Arbuscular Mycorrhizas: Physiology and Function.* Kluwer Academic Press, Dordrecht, The Netherlands and Boston, USA, pp. 19–46.

Lovato, P., Schuepp, H., Trouvelot, A. and Gianinazzi, S. (1995) Application of arbuscular mycorrhizal fungi (AMF) in orchard and ornamental plants. In: Varma, A., Hock, B. (eds) *Mycorrhiza: Structure, Function, Molecular Biology and Biotechnology.* Springer, Berlin, pp. 443–467.

Archaeobotany

Archaeobotany (or palaeoethnobotany) is the study of plant remains from archaeological sites, with the aim of understanding past human diet, food gathering and cultivation, and environmental change. The term encompasses both macroremains (seeds and wood/charcoal) and microremains (pollen and phytoliths). Most archaeobotanists work on seed remains, including in modern forensic science, here broadly defined to include all kinds of propagules.

1. History

Interest in archaeobotany started in the late 19th century, with the discovery of desiccated plant remains in ancient Egyptian tombs, and Oswald Heer's classic 1865 report on Neolithic plant-remains from Swiss lake villages. Until the 1960s, archaeobotany was usually a part-time occupation of botanists and agronomists. Recovery of plant remains by excavators of archaeological sites was piecemeal and depended on the discovery of obvious deposits, such as burnt storerooms containing jars or silos of seeds. With exceptions, such as the work of the pioneering Danish archaeobotanist Hans Helbaek (1907–1981), reports were of variable quality. In the 1960s archaeologists developed a stronger interest in economic and environmental aspects of ancient societies, and the development of flotation techniques allowed far more reliable recovery of plant remains. Archaeobotany is now a discipline in its own

Table A.7. Seed remains found in jars at the Late Bronze Age shipwreck (1300 BC) at Ulu Burun, off the southwest coast of Turkey.

Fig	*Ficus carica*
Grape	*Vitis vinifera*
Pomegranate	*Punica granatum*
Olive	*Olea europaea*
Almond	*Amygdalus communis*
Pine	*Pinus pinea*
Coriander	*Coriandrum sativum*
Black cumin	*Nigella sativa*
Sumac	*Rhus coriaria*

Note the concentration of fruits and spices, rather than the cereals and pulses typical of daily subsistence.

right, and an integral part of many archaeological projects. The work of archaeobotanists is now mainly carried out within academic or commercial archaeological organizations, although the cross-disciplinary work continues to require strong botanical skills.

2. Preservation and recovery

Most seeds either germinate and establish seedlings, or are consumed by animals or microorganisms. Uncharred seeds and other plant parts only survive from antiquity under unusual conditions. One such case is in hyper-arid areas, such as the deserts of North Africa or the American southwest. At dry sites such as Qasr Ibrim, in southern Egypt, vast quantities of plant remains survive, ranging from food debris to baskets and paper documents. Permanently waterlogged sites, such as the Viking levels at the city of York in northern England, Windover in the wetlands of Florida, or the cargoes of shipwrecks, also preserve a wide range of plant materials (Table A.7), although with some loss of soft tissues such as **endosperm**. Comparable preservation occurs in mineral-rich deposits such as latrines, where carbonates and phosphates replace plant tissues leading to the survival of mineralized plant remains.

Except in these unusual cases, by far the most widespread form of preservation is by charring in fires. Plant materials that fall into ash, or which are in heaps or jars, will char and turn black at temperatures between about 150 and 400°C. Charring preserves the shape of seeds remarkably well, and also fine features such as seed anatomy and **seedcoat** cell patterns. Wood and tubers also char well, but light seeds and plant parts, such as leaves, tend to burn to ash. In addition to favouring more solid plant parts, charring will also disproportionately favour plants used for fuel. For example, cereal chaff and straw is often used to fuel fast fires or as kindling, so is likely to be charred. Spices and medicinal plants will be carefully hoarded and are less likely to be burnt. Charred seeds can also derive from a less direct route – burning of animal dung, which contains undigested seeds from the animals' forage and fodder. In general, charred plant remains are dominated by food and fuelplants, and other uses of plants will be more difficult to trace in the archaeological record. Unlike seeds in **soil seed banks** or pollen in lake beds, archaeological seeds cannot be considered a fully representative sample of plants used at or growing near a settlement.

Fragments of DNA have been successfully extracted from **ancient seeds**, but appear to be less degraded in desiccated material than in charred. It is often claimed that ancient seeds, particularly grain from Egyptian tombs ('mummy wheat') can be germinated. However, with the exception of the sacred lotus, *Nelumbo nucifera* (*ca.* 1200 years old), there are no documented cases of germination of truly **ancient seeds**. The degree of fragmentation of DNA and other chemicals within the seed rules this out. **(See: Longevity; Seed banks)**

3. Identification

Most seed material can be identified using gross morphology, by comparison to a seed reference collection of modern, identified material. Seed identification manuals have a limited role, as they often depend on characters that are lost in waterlogged or charred material, such as colour or appendages. Scanning electron microscopy is much used for observation of seedcoat patterns. Published identifications vary in quality, depending on the researcher's experience and access to reference collections. In particular, species-level identifications are problematic unless published in sufficient detail as to explain how the other candidate species were excluded. **(See: Seedcoats – structure; Structure of seeds – identification characters of seeds)**

4. Forensic science

Although less commonly used than entomology and palynology, seed identification is an important tool in specific instances. Seeds are used as trace and contact evidence, when their presence may indicate that two or more objects or people have been in contact, and in search and location enquiries for missing people. It is rare that seeds can be demonstrated to come from a single location, but it is often the case that they derive from plants with very specific habitat preferences, which can be highly informative in a local context. For example, presence of seeds on a body, from plants absent from the scene of crime, may help point to prior hiding places. Seed identification is also important in the investigation of human and animal stomach contents, both in cases of accidental or deliberate poisoning, and in characterizing meal contents and time of death. The comminuted nature of stomach contents means that plant anatomical skills are usually necessary for identification of seedcoat fragments. In 2003 a plant anatomist at Kew Gardens was able to identify fragments of the seedcoat of the toxic calabar bean (*Physostigma venenosum*) in the intestines of an unidentified body, recovered from the River Thames, London. Calabar bean's use as an ordeal plant in West Africa suggested a motive for the murder. **(See: Pharmaceuticals and pharmacologically active compounds; Poisonous seeds)**

The Tyrolean Iceman, dating to the Neolithic period (*ca.* 5300 years old), is a well-known example of the application of forensic skills to an archaeological case. Minute quantities of the Iceman's colon contents were examined through transmitting and scanning electron microscopes. The main component was bran of **einkorn** wheat (*Triticum monococcum*), with small quantities of muscle fibres (perhaps from ibex meat), pollen, and the eggs of an intestinal parasite, whipworm (*Trichuris trichiura*). The small size of the bran particles suggests they had been finely ground, probably for preparation of bread rather than gruel.

5. Key issues in archaeobotany

(a) *Hunter-gatherers.* Plant remains are poorly preserved at archaeological sites more than 20,000 years old. From the Late Upper Palaeolithic (Old World) or Palaeoindian (New World) (*ca.* 20,000–10,000 years ago) onwards, archaeobotany has proved informative about the role of plants in the pre-agrarian, foraging societies that subsisted on wild animals and plants. Grinding stones and mortars are abundant in some areas, and it is likely that seeds would have been processed for human consumption using these tools and fire, for example in roasting nuts or extracting oil.

Archaeobotany has demonstrated that seeds were a major resource in ancient foraging societies: the energy costs of processing seeds to food are offset by the seeds' abundance, storability and nutritional quality (e.g. **storage protein, oils**). The evidence also shows that a very diverse range of species was consumed: for example, seeds of over 150 species of edible plant were recovered from the Epipalaeolithic site of Abu Hureyra in Syria; tubers, edible greens and other plant parts would also have been eaten. This conforms with ethnographic evidence that hunter-gatherer diets are generally high in protein and fibre and low in saturated fat, with high levels of micronutrients. Ethnographic evidence suggests that perhaps 65% of energy would have been derived from plant foods. **(See: Ethnobotany)**

(b) *Domestication*. Archaeologists have devoted much attention to crop domestication and the origins of agriculture, because of farming's central role in the evolution of complex, literate civilizations. Domesticated plants can be distinguished by a range of adaptations to cultivation, most notably the loss of the capacity to disperse seed without human intervention. The wild ancestors of seed crops also typically have smaller seeds. Many claims for early finds of domesticated plants have proved over-optimistic. The relatively new technique of accelerator radiocarbon dating has allowed individual seeds to be dated, demonstrating that many archaeological plant remains are more recent than previously thought. Dates obtained before the mid-1980s must be treated with caution. **(See: Domestication)** In particular, re-dating of plant remains suggests that agriculture in the Americas is significantly younger than once thought. Claims for domestication as early as 10,000 years ago for potato and 7000 years for maize and bean are not supported by direct dating of early plant remains.

Agriculture evolved independently in at least six areas. In the Fertile Crescent of the Near East, **wheat, barley, lentils** and **peas** were domesticated about 10,000 years ago, eventually spreading as crops through much of the temperate Old World. The beginning of farming in the Fertile Crescent may have been triggered by climate change at the end of the last Ice Age, leading to increased populations that could not be supported by foraging for wild foods. In the rest of the world, domestication occurred later, without obvious climatic triggers, and is still poorly documented by archaeological remains. In Southeast Asia, **rice** was probably first domesticated in the Yangtze valley *ca.* 8000 years ago. The third independent centre of domestication in the Old World is thought to be sub-Saharan African, where **sorghum** and **pearl millet** were domesticated by 4000 years ago.

In the New World, agriculture also appears to have started in three centres. Archaeobotanical evidence from Mexico shows that the classic Mesoamerican group of crops, **maize, squash** and **common beans** were domesticated at different times, with maize appearing by 6000 years ago and beans 4000 years ago. In the Andean highlands of Peru, incomplete evidence shows that common bean (domesticated independently from that in Mesoamerica), **lima bean**, potato and **quinoa** were domesticated by 5000 years ago. In eastern North America, research in the last two decades has shown that a range of small-seeded crops, such as goosefoot (*Chenopodium berlandieri*) and marsh elder (*Iva annua*), were domesticated by 4000 years ago, prior to the arrival in the region of Mesoamerican crops such as maize. **(See: Cereals; Legumes; Pseudocereals)**

(c) *Agriculture*. Archaeobotany has proved to be a powerful tool for identifying the range of crops grown in past societies and, more interestingly, in identifying crop husbandry regimes. When combined with studies of field and settlement distribution, archaeobotany can be used to identify economic changes in farming that link to wider socio-political changes. Identification of cultivation regimes, e.g. fallowing, manuring, and irrigation, depends on accurate identification of ancient weed seeds, and their ecological interpretation. Ethnographic work in current-day traditional farming settlements has given insights into the taphonomy of archaeological plant remains, i.e. the processes that lead to the incorporation of plant remains into the archaeological record. The resulting studies of crop-processing and fuel use have proved essential in interpreting ancient plant remains, and are also of interest to a wider user group of ethnobotanists and agronomists.

(d) *Foods*. There has been increasing interest in studying the consumption of plant foods, in addition to the aspects of production discussed above. Food is a difficult subject for archaeology because, by their very nature, plant-based foodstuffs are consumed and do not enter the archaeological record. Rare exceptions include desiccated coprolites (preserved faeces) from dry areas such as the American southwest, and the stomach contents of the Iron Age bog bodies of northern Europe. Prehistoric coprolites from the Lower Pecos region of Texas and New Mexico contain a wide range of seed types, dominated by prickly pear (*Opuntia ficus-indica*) and wild grass seeds. Ritual food deposits, such as beer residues and bread loaves, made from **emmer** (*Triticum dicoccum*) and **barley** (*Hordeum vulgare*) have been found in ancient Egyptian tombs. Studies of consumption therefore depend on integrating evidence from archaeobotany with that for food-processing tools, cooking installations such as hearths and ovens, with evidence from ethnography. (MN)

Hastorf, C. (1999) Recent research and innovations in paleoethnobotany. *Journal of Archaeological Research* 7, 55–103.

Jacomet, S. and Kreuz, A. (1999) *Archäobotanik: Aufgaben, Methoden und Ergebnisse vegetations- und agrargeschichtlicher Forschung*. Eugen Ulmer, Stuttgart.

Pearsall, D.M. (2000) *Paleoethnobotany. A Handbook of Procedures*, 2nd edn. Academic Press, San Diego, CA, USA.

Smith, B.D. (1998) *The Emergence of Agriculture*. Scientific American Library, New York, USA.

Archegonia

Female sexual organs producing and containing the female **gametes**; fully developed in bryophytes and pteridophytes (mosses and ferns) in the broadest sense, only rudimentary in gymnosperms. True archegonia are absent from angiosperms (with the three-celled egg apparatus as the homologue – egg cell flanked by two synergids). **(See: Reproductive structures, 1. Female)**

Aril

Pulpy structure that grows from some part of the **ovule** or **funiculus** after fertilization and covers part of, or the whole seed. Some authors on occasion distinguish so-called localized arils or 'arillodes' that develop from some part of the ovule (e.g. **exostome, raphe, chalaza**) from 'true', funicular arils.

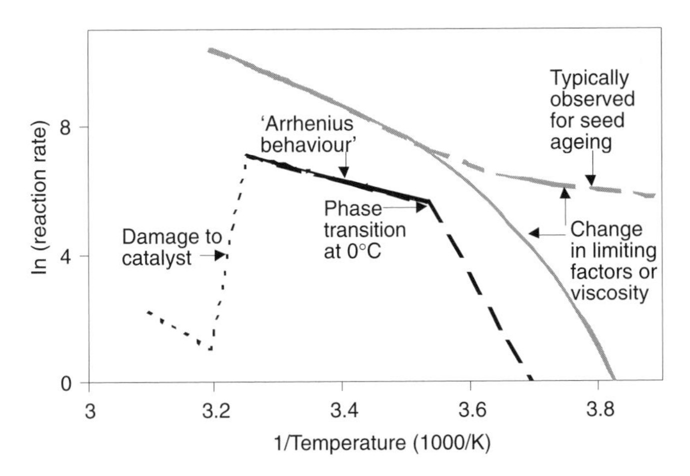

Fig. A.8. Arrhenius plots demonstrating possible effects of temperature on reaction kinetics. Typically, the ordinate of Arrhenius plots is the natural logarithm of the reaction rate and the abscissa is the reciprocal of temperature (in Kelvin). The units of the abscissa are multiplied by 1000 in this graph to avoid decimals (i.e. 1/298 x 1000 = 0.00336 x 1000 = 3.36). The solid black curve describes the linear relationship typically known as 'Arrhenius behaviour', and for most reactions is apparent for a relatively narrow temperature range. A sudden break in Arrhenius behaviour with loss of activity at high temperatures (black dotted line) suggests that the enzyme catalysing the reaction was denatured. This type of change in temperature coefficient has never been reported for seed ageing, perhaps suggesting that the deteriorative reactions are not enzyme-mediated. A break in Arrhenius behaviour with a steeper slope at low temperature (black dashed line) is usually interpreted as a structural or phase change (in membranes or proteins) that increases the activation energy of the reaction. More typically, Arrhenius plots are curvilinear (solid grey curve), suggesting a progressive change in structure, viscosity, or equilibrium coefficient of the reaction. (**See: Deterioration kinetics**)

If the aril is of a double origin, arising from both the funiculus and the testa, it is called a 'complex' aril. (**See: Seedcoats – dispersal aids**)

Arrhenius plot

The graphical depiction of the effect of temperature on reaction kinetics sometimes used to analyse temperature-dependent changes in seeds such as deterioration. It is based on the equation $k = A \exp(-E_a/RT)$ where k is the rate of the reaction at temperature T, R is the ideal gas constant, A relates to the collision frequency of reacting molecules and $(-E_a/RT)$ is the Boltzmann distribution describing the distribution of energy in molecules and E_a is the amount of energy needed to effect the reaction (also known as the activation energy or temperature coefficient). The pre-exponential factor A and the activation energy E_a can be calculated from the intercept (lnA) and slope $(-E_a/R)$ of the Arrhenius plot where the natural logarithm of k is the ordinate and 1/T (in Kelvin) is the abscissa. If the relationship is linear, the reaction is said to exhibit Arrhenius behaviour, and the reaction rate at any temperature can be predicted. The value of E_a is typically between 30 and 110 kJ/mol for simple reactions and between 40 and 75 kJ/mol if diffusion of substrates is the rate-limiting step. Catalysts lower E_a so that reactions are energetically feasible at lower temperatures. The E_a calculated for the effect of temperature on seed ageing according to **Harrington's Thumb Rules** is about 85 kJ/mol.

Biological reactions frequently do not show Arrhenius behaviour (i.e. Arrhenius plots are not linear), and the reasons for this may be important to understanding how reactions proceed across wide temperature ranges. The predominant reason for curvilinear Arrhenius plots is that the reaction requires several steps with different temperature coefficients (E_a). Alternatively, with wide temperature ranges, the convention of treating the pre-exponential factor A (related to frequency of collisions) may cause errors since this parameter is a function of the square root of temperature and the equilibrium coefficient (K), which is also dependent on temperature. In biology, non-linear Arrhenius behaviour is most commonly attributed to a phase change (also called a change of state), which affects the assumptions of the Boltzmann distribution or the properties of the catalyst. Departure from linear behaviour has been used to suggest enzyme denaturation, membrane **phase transitions**, or **glass** transitions. The latter explanation has received recent attention in the seed deterioration literature (deterioration kinetics) following discovery of glassy states in dried seeds. (CW)

Artificial seed

See: **Synthetic seed**

Arugula

See: **Brassica – horticultural**

Asia and Pacific Seed Association

See: **APSA**

Aspiration

A conditioning process, in essence using the principle of **winnowing** to separate seed according to resistance to air flow (for example, in modern air-machines and aspirators and more ancient 'winnowing fans'). Separation depends on a combination of how a seed's density, shape and surface texture affects its terminal velocity in a flowing current of air. (**See: Conditioning, II. Cleaning**)

In a different sense, aspiration is a key operating principle to convey seed by airflow in some designs of mechanized seed sowers. (**See: Planting equipment – metering and delivery**)

ASSINSEL

The International Association of Plant Breeders for the Protection of Intellectual Property, which in 2002 merged with FIS to become the International Seed Federation. (**See: ISF**)

Association of Official Seed Analysts (AOSA)

The association of official seed testing laboratories across the USA and Canada, whose activities include the establishment of standardized rules and procedures for seed **germination testing**. The membership mainly comprises the official state, federal and university seed laboratories in the USA and Canada, together with allied laboratories in government agencies and institutions outside these countries and individual affiliate, associate and honorary members.

The primary functions of the AOSA are:

- to establish the AOSA Rules for Testing Seeds, which are generally adopted in North America;
- to contribute to the refinement and modification of the rules and procedures for seed testing;
- to ensure that testing procedures are standardized between analysts and between laboratories; and
- to influence and assist in enforcement of appropriate seed legislation at state and federal levels.

Specialized committees conduct research and propose new or revised testing procedures whose adoption is decided at the annual AOSA meeting. Many individuals within the member laboratories acquire AOSA Certified Seed Analyst status through extensive training followed by a mandatory certification testing process. www.aosaseed.com (**See**, for comparison: **International Seed Testing Association (ISTA)**) (RD)

Association of Official Seed Certifying Agencies (AOSCA)

The organization composed mostly of certification agencies from the USA and Canada, equivalent to certification agencies in other countries affiliated to the **OECD Seed Schemes**. www.aosca.org (**See: Certification schemes**)

ASTA

The membership of the American Seed Trade Association (ASTA) consists of about 850 companies involved in seed production and distribution, plant breeding and related industries in North America. Through its Corn & Sorghum, Farm Seed, Lawn Seed, Soybean, and Vegetable & Flower, Seed Divisions, the Association represents industry interests, with focuses placed on regulatory and legislative matters, new technologies, and communications to members and the public about research developments, science, policy and environmental issues. www.amseed.com (**See: Industry**)

Asteraceae

The plant family that contains **sunflower**, **lettuce**, *Cichorium* (**endive** and **chicory**) and **safflower** from amongst the major world crops, and marigolds and chrysanthemums from the cultivated **ornamental flowers**.

ATP, adenosine triphosphate

A ribonucleoside triphosphate in which three phosphate (phosphoryl) groups are linked to the ribose moiety of adenosine. These links contain considerable chemical potential energy, and by donating the phosphate groups ATP participates in many biosynthetic reactions. ATP is formed from adenosine diphosphate (ADP) during oxidative phosphorylation in mitochondria, or less efficiently by substrate oxidation. (**See: Germination – metabolism; Respiration**) (JDB)

Atropous

Type of ovule, synonymous with **orthotropous**.

Aubergine

See: Pepper and Eggplant (aubergine)

Auto(poly)ploid

Polyploid plants arising from the duplication of **genomes** of a single species.

Autogamy

Pollination of the stigma by pollen produced on anthers within the same flower or another flower on the same plant or within the same clone, resulting in self-fertilization. (**See: Fertilization; Self-pollination and fertilization**)

Autophagic vacuoles

Autophagy represents a mechanism by which the seedling can recover metabolites from the cytosol of senescing **cotyledon** parenchyma cells after germination. This is accomplished by membrane-bound vacuoles containing cytoplasmic material, present within lytic vacuoles that are formed by coalescence of **protein bodies** during post-germinative storage protein mobilization. In the parenchyma cells of mung bean (*Vigna radiata*) cotyledons they appear to be formed by invagination of the lytic vacuole membrane, with subsequent pinching off of the invagination and its included cytosol into the interior of the lytic vacuole. Autophagic vacuoles may contain ribosomes and membranous organelles (e.g. Golgi bodies and mitochondria), which are digested by the acid hydrolases in the lytic vacuole. (**See: Cells and cell contents; Storage protein – mobilization**) (KAW)

Herman, E.M., Baumgartner, B. and Chrispeels, M.J. (1981) Uptake and apparent digestion of cytoplasmic organelles by protein bodies (protein storage vacuoles) in mung bean cotyledons. *European Journal of Cell Biology* 24, 226–235.

Toyooka, K., Okamoto, T. and Minamikawa, T. (2001) Cotyledon cells of *Vigna mungo* seedlings use at least two distinct autophagic machineries for degradation of starch granules and cellular components. *The Journal of Cell Biology* 154, 973–982.

Auxins

Their name derived from the Greek word 'auxein' meaning 'to increase', auxins comprise a group of **hormones** of which indole-3-acetic acid (IAA) is the best documented. IAA has been isolated from a wide range of taxonomic groups in the plant kingdom including mosses, liverworts and some green algae. There is evidence that auxins play key roles throughout the life cycle of a plant from the establishment of cell and organ polarity (e.g. in **embryogenesis**) to the regulation of cell division and expansion. IAA has been shown to contribute to **embryo** development (**see: Development of embryos – dicots**), orientation of roots and shoots, emergence of lateral roots and buds, vascular differentiation, fruit development and the shedding of plant organs.

The biosynthesis of IAA is thought to occur via a number of routes *in vivo*, dependent on the type of tissue and the species. The best documented pathway commences with the amino acid tryptophan which is converted to indole-3-pyruvic acid (IPA) (Fig. A.9). Other proposed routes are via the intermediate tryptamine: in *Arabidopsis* indole-3-acetonitrile is highly abundant and may be a biosynthetic component. There is evidence from radiolabelling studies that a tryptophan-independent pathway also exists in plants with indole itself or other derivatives acting as the precursor.

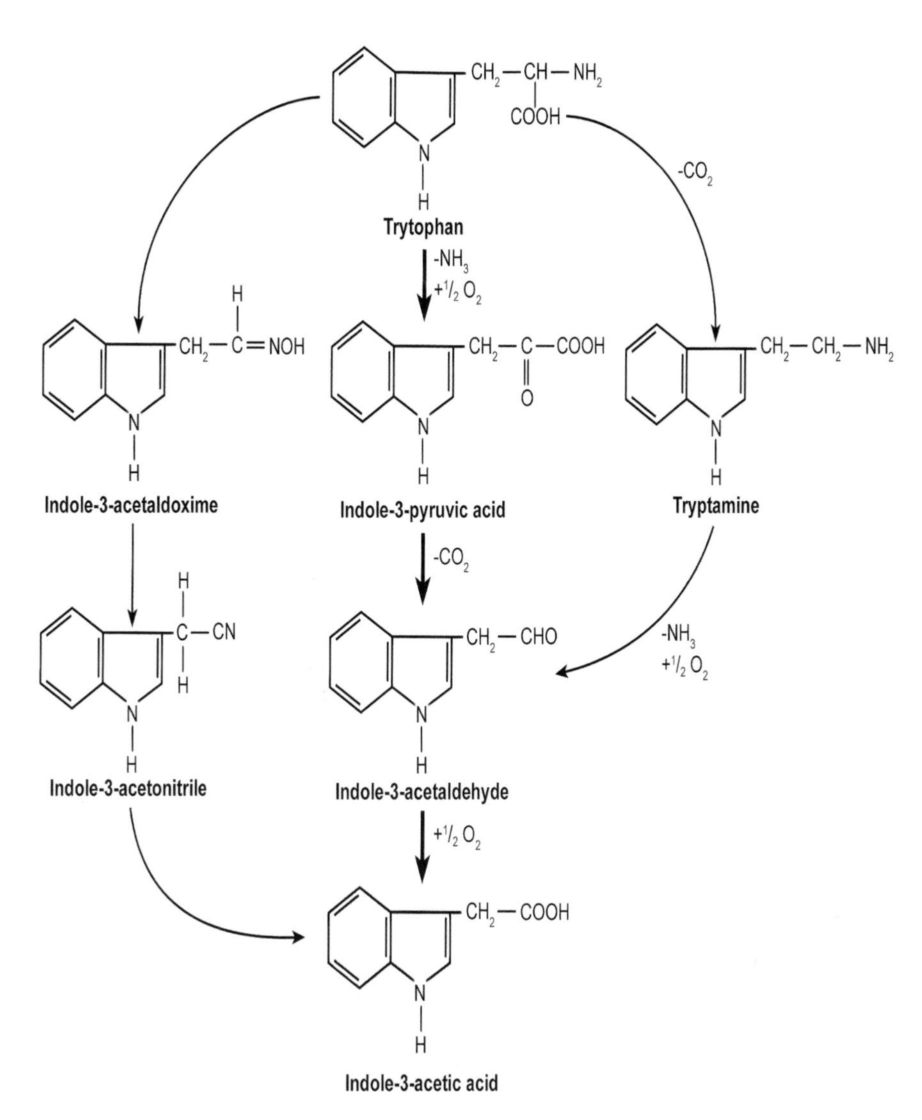

Fig. A.9. Biosynthetic pathway of indole acetic acid from the amino acid tryptophan. Modified from Srivastava, L.M. (2002) Auxins. In: *Plant Growth and Development. Hormones and Environment*, pp. 155–169.

IAA is most abundant in young tissues including fruits and immature seeds. These tissues may also represent sites of greatest synthesis; however, until the key enzymes that regulate production have been identified this cannot be confirmed.

Concentrations of IAA seem to be developmentally regulated and conjugation of IAA may provide a mechanism for controlling the amount of 'free' auxin in plant cells. Most commonly IAA is conjugated to sugars, sugar alcohols and amino acids such as aspartic acid, glutamine and alanine. In seeds of *Arabidopsis* and legumes the principle conjugate is IAA amide. In developing maize kernels the vast majority of the IAA that accumulates is in the esterified form and after germination this is thought to be hydrolysed in the seedlings to increase the amount of 'free' auxin.

IAA is transported through the plant in a polar fashion from the shoot to the root. This phenomenon is important and if disrupted, by the application of chemicals or because of the mutation of a transport protein, the phenotypic consequences are profound. Such plants may exhibit aberrant embryo development, fail to develop vasculature or be unable to respond to environmental signals such as light or gravity. Polar transport takes place in parenchyma cells and involves specific IAA influx and efflux carriers. The latter have been shown to be targeted asymmetrically and in roots are localized to the basal ends of cells.

IAA can be metabolized to inactive derivatives either by decarboxylation of the side chain or oxidation of the indole ring. Evidence to support the existence of the former pathway originates largely from *in vitro* studies but the latter has been documented in many plants including seeds of *Brassica* and *Ribes*.

Although no mutants of IAA biosynthesis have been isolated, **mutants** have been identified where IAA homeostasis has been disrupted and plants overproduce the auxin and/or their production of conjugates is attenuated. The phenotype of such mutants is seriously affected and may exhibit abnormal root development, epinasty and reduced fertility.

In seeds of the legume family and in pine a chlorinated derivative of IAA, 4-chloro-indole-3-acetic acid has been isolated. This compound is an order of magnitude more active than IAA in auxin bioassays, however its physiological role has not yet been determined. Another IAA-derivative that may function as an auxin *in vivo* is indole-3-butyric acid. In species such as tomato, tobacco and maize a non-indole compound, phenylacetic acid, has been identified that has auxin-like activity.

A number of synthetic compounds have been identified that have auxin-like activity including the growth regulators dichlorophenoxy acetic acid (2,4-D), 4-chloro-2 methyl-phenoxyacetic acid (MCPA) and naphthalene acetic acid (1-NAA). At low concentrations these compounds have potent abilities as auxin analogues whilst at high doses they can have herbicidal activity. (**See: Development of seeds – hormones content**) (GL, JR)

Bartel, B. (1997) Auxin biosynthesis. *Annual Reviews of Plant Physiology and Plant Molecular Biology* 48, 51–66.

Normanly, J. and Bartel, B. (1999) Redundancy as a way of life: IAA metabolism. *Current Opinions in Plant Biology* 2, 207–213.

Srivastava, L.M. (2002) *Plant Growth and Development. Hormones and Environment*. Academic Press, New York, USA, pp. 155–169.

Awn

The bristle or beard present at the end of the **lemma** of barley, oats, wheat and other grains and grasses, or any similar bristle-like appendage. Its length (or presence) varies between species, or even between cultivars of the same species. Awned cultivars of wheat and other cereals assimilate more carbon through photosynthesis than awnless ones, and contribute to the gain in weight by the grains during **development**, particularly in dryer conditions. (**See: Barley – cultivation**, Fig. B.1 for picture of awns, and Figs D.30, D.31 in **Domestication**)

Various mechanical techniques are used to remove awns and other **chaff** from seed that is to be planted (**See: Conditioning I. Precleaning; Grasses – forage and turf; Planting equipment – agronomic requirements; Threshing**).

(JDB)

Evans, L.T. and Rawson, H.M. (1970) Photosynthesis and respiration by the flag leaf and components of the ear during grain development in wheat. *Australian Journal of Biological Science* 23, 245–254.

Axis

See: Embryonic axis

B

Bacillus thuringiensis (Bt)

Bacillus thuringiensis (commonly abbreviated to 'Bt') is a Gram-positive soil-inhabiting bacteria which produces abundant amounts of insecticidal protein during sporulation, commonly designated as 'Cry' proteins (Table B.1), in the form of a crystal or crystal-complex inclusions. Formulations of Bt microbial insecticides, derived from killed bacteria, have been sold for many years in several countries for a wide range of purposes, including as grain protectants by treating bulk surfaces of grain such as cereals, soybean and sunflower against some **storage pests**. The proteins are consumed, solubilized and activated by proteinases in the alkaline conditions of the midgut of Lepidopteran larvae, causing cell lysis in the epithelial cells and perforation of the stomach. Other types are toxic to Dipteran and Coleopteran insects in similar ways.

Insect-resistant **genetically modified** crop varieties have recently been commercialized in which transgenically introduced Cry toxin genes produce insecticidal proteins – notably marketed in the USA on cotton (for control of tobacco budworm and cotton bollworm) and maize (primarily for control of European corn borer, and also of corn earworm and Southwestern corn borer) and potato (for control of Colorado potato beetle). (PH)

Table B.1. Genes encoding Cry proteins produced by different strains of Bt.

Cry gene designation	Toxic to these insect orders
CryIA(a), CryIA(b), CryIA(c)	Lepidoptera
Cry1B, Cry1C, Cry1D	Lepidoptera
CryII	Lepidoptera, Diptera
CryIII	Coleoptera
CryIV	Diptera
CryV	Lepidoptera, Coleoptera

The designator *Cry* is not to be confused with the notation for *cryptochrome* genes.

Baked beans

A cooked form of navy bean (common bean, *Phaseolus vulgaris*) popular in some countries. (**See: Bean – common**)

Bambara groundnut

Seeds of this leguminous plant (*Voandzeia subterranea* syn. *Vigna subterranea*) are an important food in parts of Africa, particularly in semi-arid regions that are subject to intermittent rains: it is third in importance to groundnut (**peanut**) and **cowpea**. Forms of *Voandzeia* occurred in the wild in the Sudan zones before the introduction of the groundnut. The Arabs and Fulani call the groundnut by the names guerte, gerti, or gertere, respectively, names preceding Bambara. It was carried to America in the slave trade, but has not become as popular there as the peanut, which is more nutritious. Often cultivated in mixed-cropping systems, this annual plant has a similar growth habit to peanut with the seed-bearing pods carried on gynophores on or just below the ground surface. In contrast with peanut, after germination of Bambara groundnut the cotyledons remain below the soil (**hypogeal seedling establishment**).

The wrinkled pods are about 2 cm long, each containing one or two seeds. These are spherical-ovoid, smooth, hard when dried, red, brown-black, cream or white, with an average length of about 11 mm. The content of **storage protein** is 8–20% fw, carbohydrate (mostly **starch**) 51–57% and **oil** (triacylglycerol) 3.5–12%. Seeds are relatively rich in calcium, iron and potassium.

The seeds may be eaten raw when immature, sometimes after being boiled in the pod. Mature seeds are too hard to eat raw but after roasting or boiling, they are sweet and palatable, often eaten mixed with maize grains. Seeds (sometimes after roasting) are ground into flour from which cake or porridge is made. They are also used medicinally. After seed collection the remaining plants may be used as forage. (**See: Legumes**)

(MB)

Heller, J., Begemann, F. and Mushonga, J. (eds) (1997) Bambara groundnut. *Vigna subterranea* (L.) Verdc. *Proceedings of the Workshop on Conservation and Improvement of Bambara Groundnut*, 1995, IPGRI, Rome, Italy.

Linneman, A.R. and Azam-Ali, S. (1993) Bambara groundnut, *Vigna subterranea* (L.) Verdc. In: Williams, J.T. (ed.) *Underutilized Crops, Series 2 Vegetables and Pulses*. Chapman & Hall, London, UK.

Banding, band drilling

The practice, common amongst vegetable farmers, of growing crops in closely spaced rows, with a wider spacing between the bands of multiple rows, sometimes combined with mulching under plastic sheets. This technique is adopted in some carrot production systems, for example, to produce very uniform roots.

Banding (or 'band application') also describes the practice of applying subsurface solid granular or water-dissolved fertilizers, along with pH-buffering compounds, or agrochemicals, in strips close above, below, or alongside the seed in the row before, during or after planting, to correct soil nutrient deficiencies and promote early seedling growth. (**See: Planting equipment – agronomic requirements, Planter Components; Starter fertilizer**)

Barley

1. Importance and economics

Barley is one of the most widely grown crops in the world, cultivated from above the Arctic Circle in the north, through the tropics down to the southern reaches of Argentina and Chile in the southern hemisphere. In terms of production quantity and area of cultivation barley is the world's fourth most important cereal grain (**see: Cereals,** Tables C.8 and C.9; **Crop Atlas Appendix, Maps 1, 8, 21**). Barley is second only to maize in tonnage of feed grain produced worldwide and is the most commonly used feed grain in the shorter season regions of the higher latitudes. Major exporters in 2002 were Australia, Ukraine, Russian Federation, Germany and France at a total value of approximately US$2.5 billion. Major importers were Belgium, China, Saudi Arabia, Spain and Japan (FAO data). **Hulled** two- and six-row barleys dominate world production (mainly for animal feed and brewing), but naked barley (always six-row) is important in east Asia, where barley is an important food crop. Hulless barley is becoming increasingly popular as weight-for-weight the nutrient content is higher than in hulled types, which have more fibre.

2. Origin

Barley (*Hordeum vulgare*) originated in the Near and Middle East ('Fertile Crescent') about 10,000 years ago, though there is evidence that its use was preceded by that of wild barley (*H. spontaneum*). The wild *H. spontaneum* is only known to occur in a two-rowed form so the first **domestication** was as a two-row, hulled form. The six-rowed type of *H. vulgare* most likely originated through mutation in a cultivated crop probably about 9000 years ago. It is likely that there were multiple domestication events for barley with major potential sites being the Near East and Tibet/China. There are two different loci in cultivated barley associated with the tough rachis trait (*bt1* and *bt2*) that distinguishes domesticated barley from wild barley which has a brittle, disarticulating **rachis** which remains attached to the grain. Naked grain barley appeared after the six-rowed type. In naked barley the silica-rich hull (**lemma** and **palea**) is more easily removed from the grain thus enhancing grain use and digestibility.

3. Grain structure and composition

Most barley grains have a hull of adherent palea and lemma, which are removable only with difficulty (Colour Plate 8A). In the rarer naked forms, the hull is readily lost during threshing. The hull amounts to about 13% of the grain (by weight) on average, the proportion ranging from 7 to 25% according to type, variety, grain size and latitude where the barley is grown. Winter barleys have more hull than spring types: six-row (12.5%) more than two-row (10.4%). Large, heavy grains have proportionately less hull than small, lightweight grains. The lemma may be terminated by an awn or a hood, or may be awnless; the **awns** are generally broken off during the threshing operation.

Grains are generally larger and more pointed than **wheat**, they have a ventral crease, which is shallower than those of wheat and **rye**, and its presence is obscured by the adherent palea and lemma. Weight ranges from about 35–80 mg per grain depending on factors such as position on ear, climate, cultivation conditions, cultivar, etc. The **aleurone layer** tissue has two to four (mostly three) cell layers, each cell being about 30 μm in diameter. Blue colour may be present due to anthocyanidin pigmentation. In the starchy **endosperm** two populations of **starch** granules are present in most types, though in some **mutants**, exploited for their chemically different starch, only one population of granules may be present. (**See: Cereals,** section 4)

Whole grains contain about 10% fw **storage protein**. Major proteins are the low-lysine **prolamins**, hordein, and hordenin in lesser amounts. Carbohydrate is about 64% fw (62% **starch**), and **oil** (triacylglycerol), 2% fw. Mineral ions are calcium (50 mg/100 g), iron (6 mg/100 g), potassium (560 mg/100 g) and sodium (4 mg/100 g). Vitamins B1, B2, E and folate are present.

Typical of cereals, grains are relatively rich in **phytate**. Concentrations of **β-glucans** and **arabinoxylans** are also relatively high at 3–8%. Some of these properties affect the nutritional and 'industrial' value of the grains, such as the low lysine content, the β-glucan (disadvantageous in livestock feed and for malting) and the phytate (deleterious effects in animal feed). For these reasons, barley mutants have been generated with higher lysine (*hyproly*) by enrichment of the glutelin component at the expense of the zein, low phytate and low glucans. (**See: Cereals – composition and nutritional quality**)

4. Uses and processing

Various types of barley have been developed under cultivation to suit particular end uses. The majority of barley grain (approx. 75%) is used for livestock feed. **Malting** utilizes about 15%, and 5% is consumed directly as human food (soups, stews, etc.). Barley is particularly suited to the production of malt because the hull covers the developing embryo and protect it from damage during the malting process. Two-rowed barley is most widely used in the malting and brewing industry where the uniform kernel size gives very uniform malt and subsequent efficient performance in the brewery. (**See: Malting – barley**) Unlike wheat flour, barley flour cannot be used to make 'risen' bread as the chemical and physical properties of the **gluten** do not support the leavening process, but it is used in various other prepared dishes. Hulless (naked) barley, with its higher energy and protein concentrations, is used in animal feed and in some human food products.

Prior to their utilization the grains may be processed to remove the hull, by abrasive milling or pearling technology, involving grinding by carborundum stones. The products are pot barley (used in soups, stews and broths) and the more refined pearl barley which can be ground into a flour that may be utilized in sauces, soups, pancakes and baby foods. Since the bran is not easily removed from the ventral crease of the grain the flour has a speckled appearance. The 'fines' from milling which contain bran, embryo (germ) material and some broken starchy endosperm are used for animal feed. Hulless barley can be roller milled like wheat. Barley is the major cereal used in the production of beer. (**See: Storage protein – processing for food**) For information on the particular contributions of the seed to seed science **see: Research seed species – contributions to seed science**.

(AE, MB, DF, MN)

Briggs, D.E. (1978) *Barley*. Chapman and Hall, London.

Briggs, D.E. (1998) *Malts and Malting*. Blackie, London.

MacGregor, A.W. and Bhatty, R.S. (eds) (1993) *Barley: Chemistry and Technology*. American Association of Cereal Chemists, Inc. St Paul, MN, USA and CAB International, Wallingford, UK.

Shewry, P.R. (ed.) (1992) *Barley: Genetics, Biochemistry, Molecular Biology and Biotechnology*. CAB International, Wallingford, UK.

Barley – cultivation

Barley is a cool-season grass with worldwide distribution, belonging to the *Hordeum* genus in the family Poaceae and tribe Triticeae. All cultivated barley, both two-rowed and six-rowed forms, is designated as *Hordeum vulgare* L., having 2n = 2x = 14 chromosomes with the so-called 'H' genome; although 2n = 4x = 28 chromosome tetraploids have been produced, they generally exhibit high levels of sterility and low yield. There are approximately 30 other species of *Hordeum*, most of them perennial and diploid, with some tetraploid and hexaploid. The majority of wild species are self-pollinated but some can cross-pollinate, including *H. vulgare* and *H. spontaneum* which, as they cross readily and are fully interfertile, should be considered to be the same botanical species, with two subspecies – one cultivated with a non-brittle **rachis** and occurring in both two- and six-rowed forms, the other with a brittle rachis and occurring in only the two-rowed form. Although the two-rowed and six-rowed forms are completely interfertile, they represent two different gene pools due to the complex of traits that affect their different morphological development types. The six-rowed type typically has fewer tillers, wider leaves, fewer nodes in the spike and smaller kernels compared to the two-rowed type. Hooded barley is used as whole-plant silage and for dried hay because the softer hoods do not cause the same problems to animals chewing on it as the sharp (and often rough) **awns** present on most barleys. For the geographic origins, evolution, history, world distribution and economic importance **see: Barley**.

1. Genetics and breeding

Barley is highly **self-pollinated**, with 95–99% selfing under most situations. Many two-rowed barleys flower while the **spike** is still entirely encased in the leaf sheath (boot), thus preventing the possibility of outcrossing (chasmogamy/ cleistogamy). By comparison, six-rowed barleys often have more physiologically developed spikes at the time of anthesis and flowering, with greater anther protrusion and a higher percentage of flowers that open widely. Winter barley varieties (see next section) tend to have somewhat higher outcrossing rates than spring barleys, because they flower under cooler conditions with greater diurnal temperature fluctuations, which seem to lead to greater anther protrusion from the flowers at the time of **anthesis**. Stress can cause some male sterility, which may then lead to higher levels of outcrossing than usual (and also often leads to infection of the spikes with **ergot** [*Claviceps purpurea*] and subsequent contamination of the grain with **mycotoxins**).

Most breeding methods that are commonly used in self-pollinated crops are applied to barley cultivar development. Barley is readily adapted to **single-seed descent** procedures and as many as five generations per year can be produced under the proper conditions. Although normally highly self-pollinated, over 50 different recessive gene loci have been identified for genetic male sterility, a trait that has become widely used in breeding programmes as a means of emasculation in crossing and recurrent selection schemes, and in some hybridization systems. **Cytoplasmic male sterility** has also been found, along with **nuclear fertility restoration** factors.

Barley has long been used as a model species for self-pollinated, diploid genetic studies. Its cytological, genetic and chromosome maps have been well documented and integrated – long before molecular techniques have made this possible in many other species – due to the availability of a considerable number of **mutants** with varied morphological traits, of reasonably fertile trisomics (**aneuploids** that have one extra chromosome), of easily induced translocations, and to the large, reasonably distinct chromosomes. This utility of barley has led to several comparative mapping projects around the world, and the development of molecular marker-assisted selection for a number of characteristics of scientific and economic importance. Furthermore, the development of efficient haploid induction procedures (*H. bulbosum* pollination and anther/microspore culture) has been used to produce double haploid populations, which are now commonly used in molecular studies of genetic structure and function.

2. Development

Barley is grown in a wide array of environments and climates, which requires varieties adapted to completing the normal growth cycle. Barley occurs as both 'spring' and 'winter' types; the former occur as daylength-neutral and, in most varieties, as long-daylength types (that require a minimum of 14 h of light before initiation of reproductive development); most winter types (that require **vernalization**) are daylength-sensitive too. However, there is a wide range of daylengths required to initiate reproductive growth, with effects that vary both quantitatively and in 'strength'. Barley may thus be characterized as a facultative long-day plant with a quantitative response to increasing photoperiod. True winter barleys are grown in much of Europe, Japan, southern and eastern China, the mid-western and northwestern USA and southern Canada, and some in the higher altitude regions of the Middle East. A considerable amount of spring barley varieties is grown as winter annual crops in some Mediterranean climates, such as in southern and Eastern Europe, southern China, northern Africa, India, Australia and New Zealand, South America, Mexico and the southern USA. (**See: Crop Atlas Appendix, Map 8**)

The barley **spike** is **indeterminate** and has three spikelets at each node, arranged alternately along the spike rachis. Each spikelet has a single **floret** and a vestigal rachilla and is subtended by a pair of small glumes. In wild barley and cultivated two-rowed barley only the central spikelets are fertile; thus, the spike viewed from the apex appears to have two rows of kernels on opposite sides of the rachis. In six-rowed barley, however, all three spikelets at each node have fertile florets, giving the appearance of three rows of **kernels** on each side of the spike; the lateral kernels may be considerably smaller than the central kernel, however, and generally are asymmetrically arranged, due to a twist at the basal attachment point. Ears of both types are shown in Fig. B.1. Each fertile floret consists of a **lemma** and **palea**

Fig. B.1. Ears of two- and six-rowed barley (right and left, respectively). Note the small, sterile florets in the two-rowed type, and the long awns extending from the lemma in both types.

enclosing three anthers and a single ovary with a branched **stigma**; sterile lateral florets are reduced in size, have no ovary but may have anthers.

The pattern of grain development is typical of cereals: fertilization usually takes place within 1 h of pollination; initial cell divisions in both **embryo** and **endosperm** (3n) usually begin within 15–20 h afterwards. The embryo grows and differentiates rapidly to become functional and capable of germination in 7–10 days from fertilization if dissected out and placed on a suitable culture medium, and becomes mature at 15–20 days after fertilization. Cell division in the endosperm is rapid and initially without cell wall formation, which starts to form at the outside of the developing grain (which later becomes the **aleurone layer**) and proceeds towards the interior of the endosperm, accompanied by the accumulation at maturity of starch granules in **amyloplasts** embedded in a **storage protein** matrix. (**See: Development of embryos – cereals; Development of endosperms – cereals**) After **physiological maturity** is reached, the seed begins to dehydrate and the embryo becomes dormant. **Embryo dormancy** generally lasts for several weeks and functions to prevent **preharvest sprouting**, but is an undesirable trait in malting varieties (see below).

3. Production

Most barley seed is matured on plants standing in the field, where it is then cut and threshed in a single operation with a combine harvester. Some combines are modified to strip only the spikes off the standing straw rather than cutting the straw and processing both the straw and the grain through the threshing machine. The crop may be swathed by cutting and **windrowing** in the field prior to **threshing** in regions with cool, wet harvesting conditions, in situations with substantial weed populations, or straw which does not dry down at maturity, or where barley is to be followed by another crop or in short season areas where frost or snow is a threat at the end of the cropping season.

Barley is usually dried to 14–18% moisture standing in the field prior to harvest and can usually be stored with no additional postharvest drying. If harvested at 15–20% moisture, aeration alone is generally sufficient to bring the grain moisture down to 14% or less, which is the target for safe, long-term **storage management**: to avoid growth of **storage fungi**, with its associated odour, processing and palatability problems, as well as heating and the initiation of germination in the bin, which significantly reduces quality, particularly for end use as malt.

Seed quality in grain destined for **malting** is critical and complex. Grain should be fully mature and not stressed at any point in development, well-filled with a high test-weight and moderate protein content (9–12%, depending on the region, genotype and processing system), bright (indicating no preharvest rain or weathering) and have no preharvest sprouting. The grain should be handled gently so that it is not skinned or broken during the threshing operation.

4. Quality, vigour and dormancy

Barley generally produces a very vigorous, uniform germination and early seedling growth because it has been selected for these properties over many generations of use in malting and brewing. If the seed is physically sound and has been properly matured and handled under good conditions, there is seldom a problem with vigour in barley.

Some varieties of barley have significant pre- and post-harvest dormancies. This is generally not a problem as any genotypes that deviate significantly from the 'normal' will have been strongly selected against in the breeding process. Some preharvest dormancy is desired in most situations to prevent sprouting in the spikes in the event of significant or untimely rainfall prior to harvest. Most barleys have adequate levels of this type of dormancy. However, some malting barley varieties that have been selected for very high activities of α-amylase may have less dormancy than most feed barleys; they can have significant problems with preharvest sprouting under some field conditions, and may need special handling in these cases. Too much dormancy in malting barley can also be a problem; certain weather conditions, especially during final ripening can accentuate this condition, and some varieties may require several weeks or months of postharvest maturation before they germinate well in the malthouse. Heat treatment (40°C for 2–3 weeks) upon intake in the storage facility can reduce this type of dormancy and give rapid, vigorous, uniform germination in the malthouse shortly after harvest. (**See:** entries under **Malting**)

For a discussion of the genetics of barley dormancy, **see: Pre-harvest sprouting – genetics, 3. Barley**.

5. Sowing

Barley seed is generally sown at a depth of 2–5 cm below the soil surface with a grain drill. The depth is not as critical as with some other grain crops, but in very dry conditions, sowing the seed deeper to place it in moist soil may more than offset the disadvantages of deep planting, up to about 10 cm. Some landraces have very long **coleoptiles** and can emerge from greater planting depths; however, many semi-dwarf cultivars have short coleoptiles and must not be planted too deeply as they may fail completely to emerge, resulting in complete crop loss. Barley is more sensitive to soil structure and general conditions at planting than other cereals, particularly to lack of aeration around the germinating seed and young roots, though it can be grown in minimum **tillage** or no-till systems in lighter, sandier soils. (**See: Seedbed environment**)

Barley (and most other small cereal grains) are generally sown in narrow rows (15–18 cm) to enable the crop to close the space between them rapidly and reduce weed competition, and conserve moisture in some environments. Target plant populations vary greatly, depending on environment, end use, harvesting methods, and cultivar. Lower populations are generally used in drier conditions or in regions with long vegetative development periods (lower latitudes), especially with winter barley; higher populations are used in shorter season areas where tillering may be limited or in highly productive environments, such as irrigated deserts with ample moisture combined with high levels of radiation. Early sowing is generally considered to be desirable as it lengthens the early part of the growing season where most of the yield potential is developed, helps the plant to mature earlier and avoid terminal stresses that occur in most barley-growing environments.

6. Treatments

A number of different protective fungicides are used as seed treatments to combat surface-borne fungi such as net blotch, scald, seedling blight, root rot and leaf stripe, and with systemic fungicide to treat for bunt, loose smut and other seedborne pathogens. In some countries systemic insecticidal seed treatments (neonicotinoids, such as imidacloprid) are used to prevent infection with the early-autumn barley yellow dwarf virus (BYDV), which is transmitted by aphids. (**See: Treatments – pesticides**) Pyrethroids are used to control soilborne wheat bulb fly and wireworm.

Nutrient treatments include formulations of manganese, for sowing in deficient areas. (DEF)

Foster, E. and Prentice, N. (1987) Barley. In: Olson, R.A. and Frey, K.J. (eds) *Nutritional Quality of Cereal Grains. Genetic and Agronomic Improvement*. American Society of Agronomy, Madison, Wisconsin, USA, pp. 337–396.

Rasmusson, D.C. (ed.) (1985) *Barley*. American Society of Agronomy Monograph No. 26. Madison, Wisconsin, USA.

Shewry, P.R. (ed.) (1992) *Barley: Genetics, Biochemistry, Molecular Biology and Biotechnology*. CAB International, Wallingford, UK.

Simmonds, N.W. (ed.) (1976) *Evolution of Crop Plants*. Longman, New York, USA.

Slafer, G.A., Molina-Cano, J.L., Savin, R., Araus, J.L. and Romagosa, I. (eds) (2002) *Barley Science*. Haworth Press, New York, USA.

Barley – malting

For information on the use of barley in malting **see** entries under **Malting**.

Base-temperature/-potential

Two allied mathematical concepts used in germination and field emergence models, wherein **germination rate** (the reciprocal of time to complete germination up to the optimum temperature) increases in proportion to the difference between the ambient temperature or **water potential** and a base threshold, at which it is theoretically infinitely slow. Base values for a given percentile of the seed population are calculated by mathematical extrapolation. The theoretical base temperature may or may not be identical to the minimum temperature, however. (**See: Germination – influences of temperature; Germination – field emergence models**)

Basic seed

An official term used in the **OECD Seed Scheme** for seed that has been produced by, or under the responsibility of, the breeder and intended for the production of **Certified Seed**. Called 'basic seed' because it is the basis for certified seed, its production is the last stage that the breeder normally supervises closely. Pre-basic seed is seed material at any generation between the parental material and basic seed. A term equivalent to **'Foundation Seed'** used formally in North America, and the informal term **'Elite seed'** used in some countries. (**See: Production for Sowing, I. Principles**)

BCA

See: Biological control agent

Bean

The generally edible seed, sometimes in the pod, of many plants, most commonly but not exclusively **legumes**.

Included are:

- Borlotto, black turtle, bush, canellini, common, flageolet, French, green, haricot, kidney, Mexican, navy, pinto, Romano, shelly, snap, string, wax. *Phaseolus vulgaris* – **see: Bean – common**.
- Broad, faba (fava), field, horse, tick, Windsor. *Vicia faba* – **see: Broad bean**.
- Lima, butter. *Phaseolus lunatus* – **see: Lima bean**.
- Scarlet runner. *Phaseolus coccineus* – **see: Scarlet runner bean**.
- Asparagus, long horn, yard long. All *V. unguinculata sesquipedalis*.
- Garbanzo, chickpea, *Cicer arietinum* – **see: Chickpea**.
- Adzuki, azuki. *Vigna angularis* – **see: Adzuki bean**.
- Mung. *Phaseolus aureus* – **see: Mung bean**.
- Winged, four angled, goa, manila (also known as asparagus pea or bean). *Psophocarpus tetragonolobus* – **see: Winged bean**.
- Tepary, Texas. *Phaseolus acutifolius* – **see: Tepary bean**.
- Hyacinth, Indian, Egyptian. Bonavista, dolichos. *Dolichos lablab* syn. *Lablab purpureus* – **see: Hyacinth bean**.
- Locust bean, *Ceratonia siliqua* – **see: Carob**
- Cluster bean, *Cyamopsis tetragonoloba* – **see: Guar**
- **Jack bean**, *Canavalia ensiformis*
- **Swordbean**, *Canavalia gladiata*
- **Mescal bean**, *Sophora secundifolia*
- Soybean, *Glycine max* – **see: Soybean**

Cultivated non-leguminous bean crops include:

- **Cacao**, *Theobroma* spp.
- **Castor**, *Ricinus communis*
- **Coffee**, *Coffea* spp.

Bean – common

1. World importance and distribution

The warm-season common bean (*Phaseolus vulgaris* L.), the most important **pulse**, is the third most important food legume worldwide, superseded only by **soybean** and **peanut**. There are many different named types of common bean, indicated in Table B.4. Around 14 million ha are planted, concentrated in tropical and subtropical Latin America, followed by sub-Saharan Africa. Brazil and Mexico are the largest producers and consumers. It is estimated that worldwide around 18 million t of dry beans are produced annually. Rwanda and Burundi have the highest yearly per capita consumption (>40 kg) (see section 9, below). (**See: Crop Atlas Appendix, Maps 3, 19**)

2. Origin

It is only since the late 19th century that scientists have accepted a New World origin for the common bean. This type evolved from a wild-growing vine distributed in the highlands of Middle America and the southern Andes into a widely-grown crop of at least seven growth habits and a great array of seed characteristics (Colour Plate 1B) among other differences in important traits. There is evidence that the **domestication** of beans occurred at least 2000–3000 years ago (**see: Legumes**). Early samples to reach Europe in the 16th century were probably the red, kidney-shaped type, to be called 'kidney beans'. The Aztec name *ayecotl* was corrupted by the French to 'haricot'. In England today haricot beans are the small, white, dried bean.

3. Races

The most apparent changes undergone under domestication include: the appearance of indeterminate and determinate upright bush growth habits; gigantism of leaf, pod and seed characteristics; suppression of explosive pod dehiscence (**shattering**); loss of seed **dormancy**; appearance of a vast variety of seed sizes, shapes and colours (Colour Plate 1B); and selection for insensitivity to photoperiod (**See: Domestication**). Probably because different wild bean populations participated in the initial domestication in different regions, plus the self-pollinating nature of the species, six distinct groups of related cultivated populations of common beans evolved, commonly known as races, a term used to denote a group of related **landraces** within the Middle American and Andean components of the species. The criteria used for race identification were: shape and size of leaflet, leaf hairiness, length of the internodes, number of nodes to flower, shape and size of bracteoles, inflorescence, petal standard, pod beak origin, days to maturity, seed size and shape, growth habit, characteristics of phaseolin seed protein, and allozymes. Table B.2 presents some characteristics of the races of cultivated common bean.

4. Seed characteristics

Seed size varies from 15 to 100 g/100 seeds, and shape can be cylindrical, kidney, rhombohedric, oval, or round. Testa colour shows different shades and tones of white, cream, yellow, brown, pink, red, purple and black and combinations of these

Table B.2. Some characteristics of the races of cultivated common bean (*Phaseolus vulgaris* L.).

Race	Size[a]	Shape	Bracteole	Growth habit[b]	Phaseolin[c]	Allozyme[d]	Distribution
			Middle America				
Mesoamerica	small	cylindrical, kidney, oval	large cordate	I, II, III, IV	S, Sb, B	Me^{98} or Diap-2^{105}	Lowlands of Latin America
Durango	medium	rhombohedric	small ovate	III	S, Sd	Me^{102}	Semi-arid highlands of Mexico
Jalisco	medium	oval, round, cylindrical	medium lanceolate	IV	S	Me^{100} or Mdh-2^{102}	Humid highlands of Mexico
			Andean South America				
Nueva Granada	medium & large	cylindrical, kidney	small lanceolate or triangular	I, II, III	T	$Rbsc^{100}$ or Me^{100}	Intermediate altitude of Andes (<2700 masl)[e]
Chile	medium	oval, round	small triangular	III	C, H	Mdh-1^{100}	Southern Andes
Peru	medium & large	oval, round	large lanceolate or cordate	IV	T, C, H	Mdh-1^{100}	Andean highlands (>2000 masl)

[a]Small: <25 g/100 seeds. Medium: 25–45 g/100 seeds. Large: >45 g/100 seeds.
[b]Type I = determinate, erect; Type II = indeterminate, erect; Type III = indeterminate, prostrate; Type IV = indeterminate, climber.
[c]Phaseolin is the major **storage protein** in bean seeds. Two major and several minor phaseolin types have been identified on the basis of their banding patterns in one-dimensional SDS polyacrylamide gels. The major types present in cultivated germplasm are S (for cv. Sanilac) and T (for cv. Tendergreen) and other minor types Sb, Sd, B, C, H.
[d]Me^{98}, Me^{100}, Me^{102} = allele loci of the malic enzyme (ME); Mdh-1^{100}, Mdh-2^{102} = allele loci of malate dehydrogenase (MHD); Diap-2^{105} = allele locus of diaphorase (DIAP); $Rbsc^{100}$ = allele locus of ribulose bisphosphate carboxylase (Rbsc).
[e]masl: metres above sea level.

colours in different patterns: spotting, striping, speckling, mottling, etc. (Colour Plate 1B). Days to seed maturity can vary from 60 to 270 days after sowing.

5. Nutritional quality

The nutrients for the human and animal diet that the bean plant provides will be outlined throughout various stages of its growth: as young tender leaves, as pods with immature seeds enclosed, as immature green seeds or as mature dry seeds (Table B.3). Although the dry grain appears as a more important source of calories, protein, thiamin and niacin than the rest of the edible parts of the bean plant, since in practice dry seeds are consumed in a cooked state at about 70% moisture content, on a per-serving basis they have only a slight nutrient advantage over leaves, green pods and green shelled grains.

The nutritional importance of beans in the diet is determined by the percentage of recommended dietary allowance (RDA) for nutrients that is satisfied by beans at the particular level of consumption. Assuming a range of consumption of 30 g/day (low) to 130 g/day (high), dry beans can provide 13–57% RDA of protein, 5–22% RDA of calcium, 4–20% RDA of certain B vitamins and 5–23% of daily energy requirements. So in countries where bean consumption per capita is especially high (e.g. Rwanda, Burundi) dry beans provide more than one-half of the dietary protein, up to one-quarter of the energy requirements and over 100% of the RDA for iron. At moderate consumption levels (e.g. Brazil, Kenya) dry beans provide over one-quarter of the protein needs and around 60% of the RDA for iron. At low intake levels dry beans occupy a supplementary role but nevertheless still remain an important source of iron.

Dry beans have a **storage protein** content ranging from 19 to 29%. The nutritional value of seed protein is also determined by the amino acid composition (protein quality), digestibility and any **antinutritional** factors present. Bean proteins are relatively rich in some **essential amino acids**, particularly lysine, threonine, isoleucine, phenylalanine and valine, but deficient in the sulphur-containing amino acids, methionine and cystine. Digestibility among different market classes is very variable and total biological utilization of the legume protein is relatively low with digestibility less than 76%. Antinutritional compounds which must be removed during processing may impair protein digestibility.

The total carbohydrates of dry beans range from 24 to 68%, mostly **starch** (24–56%). There is a small percentage of soluble sugars, the flatulogenic **oligosaccharides** (raffinose, stachyose, verbascose and ajugose) accounting for 31–76% of the total. Beans have a high content of non-starch polysaccharides with reported hypocholesterolemic and hypoglycaemic effects, i.e. 'soluble fibre' (actually a better source than the cereals). Water-insoluble fibre consists mainly of **cellulose**, but there are also some **hemicelluloses** and lignin. **Cotyledon** cell walls contain higher amounts of **pectin** than of cellulose while the **seedcoats** are primarily composed of cellulose and small amounts of lignin.

Beans are a good source of several mineral ions (calcium, iron, phosphorus, potassium, zinc and magnesium). Dry beans provide several water-soluble vitamins such as thiamin, riboflavin, niacin and folic acid but little ascorbic acid.

Lipid content is low (1–2%) consisting of neutral oils (60% of the total), phospholipids (24–35%) and glycolipids (1%). Fatty acid composition is variable. Linolenic acid is the predominant unsaturated fatty acid whereas palmitic acid is the main saturated one.

Anti-nutritionals. Beans contain lectins, tannins and phytin, regarded as antinutrients because they limit absorption of nutrients. The procyanidin and tannins are localized in the bean seedcoat and also in the cotyledons. The tannins (**polyphenols**) bind proteins and precipitate iron in food or in the gut but are also flavour components. Lectins bind some **storage proteins** and some enzymes. Some lectin genes encode α-amylase inhibitors. (**See: Amylase inhibitors; Lectins; Phytohaemagglutinin (PHA)**). **Phytin** sequesters important mineral ions reducing their availability from the diet.

6. Culinary quality

This refers to the aggregate of properties bearing on the preferences and requirements of consumers and processers for the dry and cooked beans. The following characteristics of dry seed influence consumer acceptability:

- Seed size, shape and colour;
- Cooking time;
- Broth appearance;
- Storability.

Table B.3. Nutrient composition of 100 g of various components of the bean plant on a fresh-weight and dry-weight basis.

Food	Food energy (calories)	Water (%)	Protein (g)	Lipid (g)	Carbohydrate (g)	Ca (mg)	Fe (mg)	Thiamin (mg)	Niacin (mg)
Fresh weight									
Leaves (raw)	36	87	3.6	0.4	6.6	274	9.2	0.18	1.3
Green beans (raw)	36	89	2.5	0.2	7.9	43	1.4	0.08	0.5
Immature seeds (cooked)	50	88	1.7	3.2	5.5	42	0.8	–	–
Dry beans (raw)	336	12	21.7	1.5	60.9	120	8.2	0.37	2.4
Dry weight									
Leaves (raw)	272	–	27.3	0.3	50.0	2075	69.7	1.36	9.8
Green beans (raw)	318	–	22.1	1.8	69.9	380	12.4	0.70	4.4
Immature seeds (cooked)	417	–	14.2	26.7	45.8	350	6.6	–	–
Dry beans (raw)	383	–	24.7	1.7	69.4	136	9.4	0.42	2.7

Source: FAO/HEW, 1968.

Table B.4. Most commonly used bean market classes and alternative names in important producing regions.

International name	Latin America	Africa	West Asia
		White-seeded	
Fabes			
Canellini	Alubia		Horos (Turkey)
White kidney			Selanic (Turkey)
Large marrow	Caballero	Gros Blanc	Seker (Turkey)
Great Northern			Dermanson (Turkey)/Safed (Iran)
Small white (haricot)	Panamito/Arroz		
Navy			
		Cream-coloured	
Cranberry/Borlotto		Speckled Sugar (South Africa)	Barbunya (Turkey)/Chitti or Talash (Iran)
Pinto	Hallados	Mwitemania (Kenya)	
Carioca			
		Yellow grain	
Peruano	Canario/Mayocoba		
		Pink	
Pink			
Light red kidney			
		Red	
Dark red kidney			
Red pinto	Nima/Calima	Rose Coco (Kenya)	
Red marrow	Radical		
Red Mexican		Red Haricot (Kenya)	
Small red	Rojo pequeño	Massai Red (Kenya)	
		Black seeded	
Black turtle soup	Negro/Preto	Ikinimba	

Table B.5. Principal seed characteristics of the international bean market classes and their cultivated area.

Market class	Size (g/100 seeds)	Shape	Colour	Cultivated area (10^3 ha)
		Colour group: white		
Fabes	100	Oblong, slightly flat	White	na
Canellini	55–65	Cylindrical	Glossy milk white	250
White kidney	45–50	Kidney-shaped	White	na
Large marrow	>60	Round	White	na
Great Northern	34–40	Flat, somewhat round	White	700
Small white	15–17	Somewhat flat	White	250
Navy	19–22	Rather round	White	na
		Colour group: cream		
Cranberry/Borlotto	40–45	Oblong	Variegated red on cream	800
Pinto	35–45	Somewhat ovate-flat	Tan and brown mottled	800
Carioca	23–25	Somewhat flat	Cream with brown stripes	2000
		Colour group: yellow		
Peruano	35–38	Elongated, plump	Light sulphur yellow	150
		Colour group: pink		
Pink	30–35	Somewhat ovate-flat	Light pink	20
Light red kidney	50–55	Kidney-shaped	Pink	300
		Colour group: red		
Dark red kidney	50–55	Kidney-shaped	Dark red	500
Red pinto	40–45	Oblong	Variegated white on red	1500
Red marrow	>60	Round	Red	50
Red Mexican	30–35	Somewhat ovate-flat	Dark red	30
Small red	22–25	Elongated	Red, shiny or opaque	250
		Colour group: black		
Black turtle soup	19–25	Elongated	Black, opaque	3500

na, Not available.

Size, shape and colour. The wide range of seed characteristics has been formalized in the bean world into distinct commercial or market classes (Tables B.2, B.4 and B.5).

Cooking time. This has genetic and environmental components and thus is influenced by season, site, and various possible interactions of these effects. Two strongly influential traits are a hard seedcoat and the hard-to-cook defect, both with different mechanisms of inheritance and environmental situations under which they develop (**see: Hard seeds**). Hard seedcoat, 'hardshell', is a condition of seed **dormancy** where the seedcoat is poorly permeable to water; the beans therefore require a longer cooking period. The incidence of hard coats increases as seed moisture content decreases under conditions of relatively high temperature and low relative humidity. Hardness can be overcome by various treatments such as an acid soak, scarification, radiation and blanching or naturally under high **relative humidity**.

Seeds with the hard-to-cook defect imbibe water but do not soften sufficiently during cooking. The defect is irreversible and develops during storage under high temperatures (above 21°C) and especially under high relative humidity when seeds have increased respiration. Beans at below 10% seed moisture content were stored at 25°C for 2 years without adverse effects whereas storage at above 13% moisture content at 25°C for 6 months resulted in reduced cookability. High moisture beans also darkened more rapidly during storage.

It has been postulated that the development of the hard-to-cook defect involves enzymatic and non-enzymatic processes that result in a reduced cell separation rate in the cotyledons. A high relative humidity and temperature during storage are thought to permit restricted metabolism that allows an excess of divalent cations (Mg^{2+} and Ca^{2+}) to accumulate from the hydrolysis of **phytin** by phytase. These ions might then combine with pectin in the middle lamella and form unsoluble pectates. Normal seed softening during cooking occurs through the conversion of insoluble Ca^{2+}/Mg^{2+} pectate in the middle lamella to soluble Na^{+}/K^{+} pectate. Attempts to control enzymatic degradation by phytase with heat treatment revealed that a non-enzymatic process is also involved in the hard-to-cook defect. While heat treatments significantly decreased the development of the hard-to-cook defect, the defect was also found to be correlated with the 'lignification' of protein in the cotyledons apparently involving phenolic compounds.

Bean broth. This, and eating quality, are characteristics which influence consumer acceptability of determined cultivars. The nature and variation for these traits are not well known.

7. Uses

The different types of dry bean seeds are the basis of a wide range of cooked dishes in many parts of the world and for many populations they are the major source of dietary protein. Beans are used in soups, stews, and are served in many other forms (**see: History of seeds in Mexico and Central America**). The baked beans that are popular in several countries are navy beans. In addition to the dry seed harvest there is a substantial production and consumption of immature pods of snap, green or stringless bean and green-shelled seeds: in some African countries tender leaves are also harvested for human consumption. In Peru and Bolivia a particular type of bean known as *ñuñas* or popping beans (the bean counterpart of popcorn), which are thought to be ancient pre-ceramic landraces, are consumed roasted. Heated with a little oil or lime, the kernel bursts out of its seedcoat which opens like a small butterfly spreading its wings. The resulting product is soft and tastes somewhat like roasted peanuts.

Snap bean comprises a group of common bean that has been selected for succulent pods (pericarp) with reduced fibre which when immature are consumed as green vegetables. Cultivars with fibre in the pod suture, which has to be removed manually before cooking, are the 'string beans'. The stringless trait was discovered in 1879 and was incorporated into improved varieties in Latin America, the most important bean-producing and consuming region.

The cultivars harvested for green shelled bean are often large-seeded, cream-mottled, red-mottled, pink-mottled or white-mottled. The distinguishing characteristic of such cultivars is that the pods change colour (turn red or purple, with or without stripes) when fresh seed is ready to be harvested for consumption.

8. Market classes

Dry beans are primarily characterized by the great diversity of seed types. A range of colours and colour patterns, varying degrees of glossiness, and several seed shapes and sizes exist: each of these – even subtle differences – determine influence on the acceptability of a specific variety, sometimes between regions of the same country. To facilitate the trade of these beans a commercial classification system has been developed. There are at least 20 market classes recognized worldwide: market classes of local importance outnumber the international ones. The most important dry bean international market classes are shown in Table B.4 and the important characteristics of these seed classes and their area of cultivation are shown in Table B.5.

Snap bean market classes are determined on the basis of pod shape (flat, cylindrical, or oval), colour (dark green, yellow, or purple) and length (or sieve size). Names such as 'Green Bean', 'Wax Bean' or 'Romano' describe commercial classes.

9. Marketing and economics

The FAO estimates the 2002 production of dry edible beans at about 55 million t grown in 70 million ha at an average yield of 791 kg/ha. The same source gives the top five bean exporters as Myanmar (Burma), China, the USA, Argentina and Canada. Unfortunately, however, many species such as **lima beans**, **adzuki beans**, mung beans, and even **broad beans**, **cowpeas** and **chickpeas**, and other minor species of *Phaseolus*, are included in the **dry beans** statistics, distorting the figures, particularly of the Asian countries (Myanmar, China, India, Pakistan). The amounts and values of the main dry bean imports and exports and the countries involved, according to the FAO, are shown in Table B.6. The actual main exporter of dry common beans is, however, the USA which exports more than one-fifth of its supply annually. The leading export varieties in 2000 were pinto (18% of dry bean value) to Mexico, Haiti and the Dominican Republic, navy (18%), to the UK, Canada and Italy, and Great Northern (13%) to Algeria, France and Turkey. Argentina is a net exporter of beans. White beans, mainly the market class Alubia

Table B.6. Dry beans: main exporter and importer countries and crop values (2002).

Dry bean exports			Dry bean imports		
Country	Tonnes (1000)	Value (million US$)	Country	Tonnes (1000)	Value (million US$)
Myanmar	1034	281	India	164	56
China	640	258	USA	136	77
USA	332	182	Japan	135	86
Argentina	265	129	Brazil	130	51
Canada	251	251	Mexico	127	58
Australia	53	23	Italy	98	61
Turkey	45	32	Pakistan	55	15
Netherlands	29	28	France	54	38

FAO STAT

(70,000–80,000 t/year) are sent mostly to Spain, Italy, France, and similar quantities of black beans to Brazil, Venezuela, Mexico and Germany. Canada, another important/exporter, sells navy beans, red kidney and cranberry. Exports from China and Myanmar appear to surpass those from the USA but, as mentioned above, much of the quantities listed as dry beans are actually not the common bean. Myanmar produces large quantities of some species which are preferred by India, Indonesia, Bangladesh, Pakistan and the Philippines.

Imports by the USA have consistently accounted for only 4–6% of domestic consumption over the past 20 years. Japan is an important bean importer but the figures shown in Table B.6 include a substantial amount of adzuki beans sold by China and Myanmar; the actual main importers of common beans are Mexico and Brazil. Most European countries are bean importers, Italy predominating.

10. Plant types

Common bean is polymorphic for growth habit, seed characteristics and days to maturity, among other traits. The main growth habits are indicated in Table B.2.

Growth habits are characterized by the following features: the type of terminal bud on the main stem (terminal = **determinate** growth, growth of the main stem terminated by the flower; axial = **indeterminate** growth, flower produced on a lateral shoot and the main stem continues to grow) and number of branches at flowering, stem strength, angle formed by lateral branches and main stem, number of nodes on the main stem, internode length, climbing ability and fruiting patterns. (**See: Legumes; *Phaseolus* beans**) (OV-V)

Broughton, W.J., Hernandez, G., Blair, M., Beebe, S., Gepts, P. and Vanderleyden, J. (2003) Beans (*Phaseolus* spp.) – model food legumes. *Plant and Soil* 252, 55–128.

Gepts, P. (1998) Origin and evolution of common bean: past events and recent trends. *HortScience* 33, 1125–1130.

Schoonhoven, A., van and Voysest, O. (eds) (1991) *Common Beans: Research for Crop Improvement.* CAB International, Wallingford, UK and Tropical Center for International Agriculture (CIAT), Cali, Colombia.

Voysest-Voysest, O. (2000) *Mejoramiento Genético del Frijol (*Phaseolus vulgaris *L.). Legado de Variedades de América Latina 1930–1999.* Centro Internacional de Agricultura Tropical, Cali, Colombia.

Bed planting

A method of planting in which the seed is planted on slightly raised areas between **furrows** with two or more seed rows sometimes planted on each bed. Bed planting can be integrated with reduced or zero tillage practices, such as in reduced-cost wheat cultivation systems. (**See: Tillage; Seedbed**)

Beet

See: Sugarbeet

Berry

In most modern textbooks this is simply defined as an indehiscent fruit in which all layers of the fruit coat (**pericarp**) become fleshy, e.g. *Persea americana* (avocado), *Vitis vinifera* (grape), *Actinidia sinensis* (kiwi), *Lycopersicon esculentum* (tomato). Sometimes also bacca, the Latin word for berry. (**See: Inflorescence**)

Betel nut

This masticatory with mild stimulant properties consists of leaves of *Piper betel* containing shredded seeds of the areca nut (*Areca catechu*). The active principle is the seed alkaloid arcoline (**see: Pharmaceuticals and pharmacologically active compounds**).

Beverages

Seeds furnish some of the world's most important beverages. Prominent among them are **cacao** (chocolate), **coffee**, **cola** and malted and unmalted cereal grains that provide beers and spirits. Others are of great local importance, such as **coconut** milk and **guarana**. Speciality drinks made from seeds are soymilk, almond milk, **rice** products and similar concoctions. Several types of seeds, especially the **spices**, are used to flavour beverages such as milk and yoghurt, particularly in Asiatic countries, or are used for distilled spirits, for example **caraway** for kummel.

For further details see the entries in bold above and those connected with **malting**. (MB)

Biliprotein

A protein with a bilin **chromophore** attached. A bilin is a biological pigment belonging to a series of yellow, green, red, or brown non-metallic open-chain **tetrapyrrole** compounds so named because of their presence in the bile pigments of mammals. Examples of biliproteins in seeds include **phytochrome** where the bilin chromophore is phytochromobilin.

Bin-run seed

A seed-trade term used in North America to describe seed taken directly from the previous crop and planted by the farmer to produce the next year's crop, often without any significant quality testing or verification. Also known as **'Farmer-Saved' seed**, and sometimes as 'Common' seed. (See: **Industry – supply systems**)

Biofungicides

In seed technology: microorganisms that can be inoculated in soil or onto seeds, before or at sowing time, as **biological control agents** to control fungal pathogens (such as certain species of *Alternaria*, *Aspergillus*, *Fusarium*, *Phytophthora*, *Pythium*, *Rhizoctonia* or *Sclerotinia*). (**See: Rhizosphere microorganisms**)

Biolistics

The 'Gene Gun' method (also known as microprojectile bombardment or biolistics) is based on bombarding tissues with microscopic, DNA-coated tungsten or gold particles. This is the most frequently used procedure for transient transformation of tissues and is also used for the stable transformation of monocot species. (**See: Transformation**)

Biological control agent (BCA)

A general term for any living organism used to control the action of another harmful or undesirable organism. The BCAs of most interest in seed technology control seed- or soil-borne antagonistic symbionts, such as the fungal and bacterial plant **pathogens** which attack or kill the seedling. (**See: Rhizosphere microorganisms**)

Biopesticides

In seed technology: microorganisms (notably **biofungicides**) that can be inoculated in soil or onto seeds before or at sowing time as **biological control agents**, that have a disease or pest biocontrol activity; sometimes also called 'bioprotectants'. (**See: Pesticides; Rhizosphere microorganisms**)

Biopharmaceuticals

The production of **recombinant proteins** in plants, including their seeds (termed 'molecular farming', or

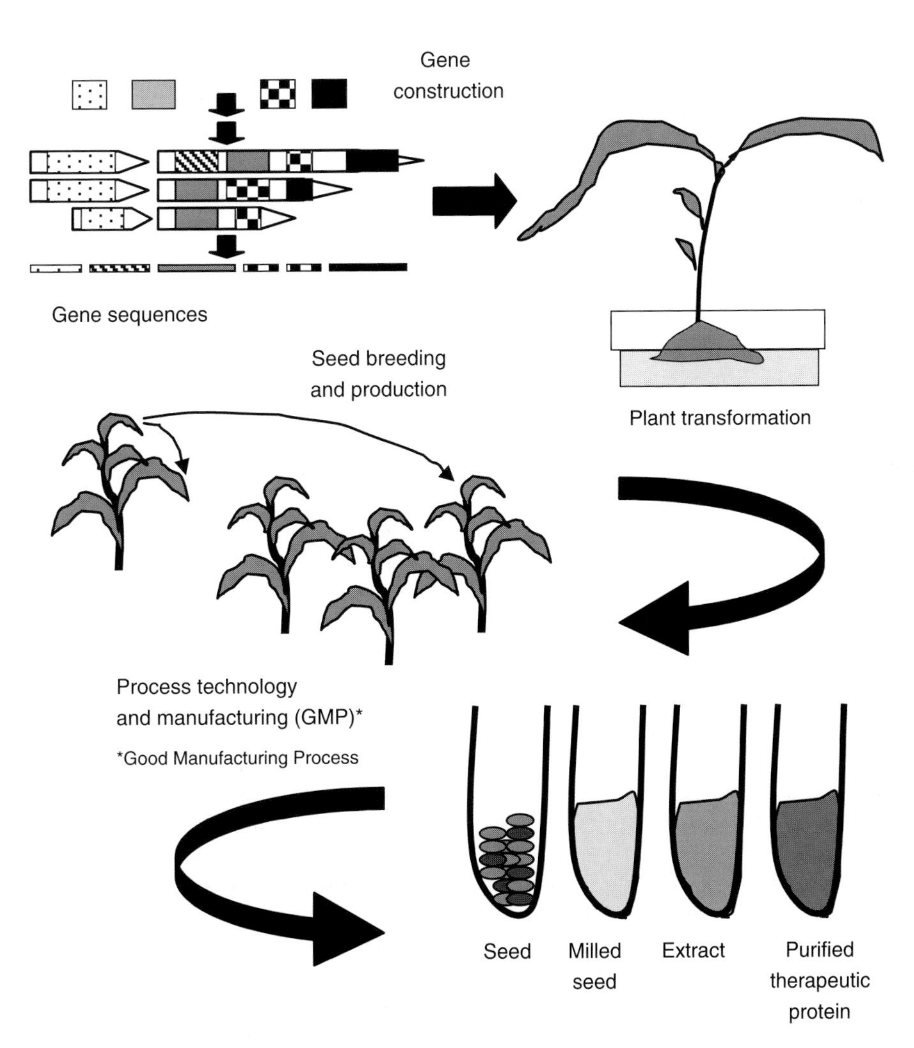

Fig. B.2. A summary of the major steps involved in the manufacture of biopharmaceuticals in plants.

sometimes 'molecular pharming') has many potential advantages for generating biopharmaceuticals relevant to clinical medicine (Fig. B.2). The potential for using plants as a production system for recombinant protein pharmaceuticals was established by Barta *et al.* in the mid-1980s with the successful expression of a human growth factor fusion protein. A crucial advance arrived soon after with the successful expression of functional antibodies in plants by Hiatt *et al.* This was a significant breakthrough for it showed that plants have the potential to produce complex mammalian proteins of medical importance. While the general eukaryotic protein synthesis pathway is conserved between plants and animals, the post-translational modifications carried out by plants and animals are not identical. There are minor differences in the structure of complex glycans in glycoproteins (**see: Storage protein–synthesis**) such as the presence of the plant-specific residues α1,3-fucose and β1,2-xylose. However, studies using mice given a recombinant IgG isolated from plants showed that, while there were some differences in the glycan groups, neither the antibody nor the glycans caused adverse immunogenic reactions. A number of vaccines have been introduced, for example into rice seeds, as has the ferritin gene, leading to enhanced accumulation of iron and zinc, and also the genes for human lactoferrin and lysozyme.

Plants are more economical producers of biopharmaceuticals than industrial facilities using fermentation or bioreactor systems. The technology is already available for harvesting and processing plants and plant products on a large scale. Plants can be directed to target proteins into intracellular compartments in which they are more stable, or even directed to express them within certain compartments. Transgenic plants can also produce seeds rich in a recombinant protein for its long-term storage. Seeds can also be used to minimize the costs associated with the purification of the pharmaceutical proteins produced in plants.

One successful example is the targeting of pharmaceutical proteins to seed oil bodies, e.g. hirudin, an anticoagulant first isolated from the leech *Hirudo medicinalis.* An **oleosin**-hirudin fusion protein has been targeted to **oil bodies** of *Brassica napus* seeds and purified by flotation centrifugation for commercial production in Canada. Proteins expressed in such a manner can accumulate to commercially acceptable concentrations. (**See: Transformation**) (AJR)

Barta, A., Sommergruber, K., Thompson, D., Hartmuth, K., Matzke, M. and Matzke, A. (1986) The expression of a nopaline syntase-human growth hormone chimaeric gene in transformed and sunflower callus tissue. *Plant Molecular Biology* 6, 347–357.

Boothe, J.G., Saponja, J.A. and Parmenter, D.L. (1997) Molecular farming in plants: oilseeds as vehicles for production of pharmaceutical proteins. *Drug Development Research* 42, 172–181.

Chargelegue, D., Vine, N., van Dolleweerd, C., Drake, P.M. and Ma, J. (2000) A murine monoclonal antibody produced in transgenic plants with plant-specific glycans is not immunogenic in mice. *Transgenic Research* 9, 187–194.

Fischer, R., Stoger, E., Schillberg, S., Christou, P. and Twyman, R.M. (2004) Plant-based production of biopharmaceuticals. *Current Opinion in Plant Biology* 7, 152–158.

Hiatt, A., Cafferkey, R. and Bowdish, K. (1989) Production of antibodies in transgenic plants. *Nature* 342, 76–78.

Ma, J.K-C., Chikwamba, R., Sparrow, P., Fischer, R., Mahoney, R. and Twyman, R.M. (2005) Plant-derived pharmaceuticals – the road forward. *Trends in Plant Science* 10, 580-585.

Takagi, H., Saito, S., Yang, L., Nagasaka, S., Nishizawa., N. and Takaiwa, F. (2005) Oral immunotherapy against a pollen allergy using a seed-based peptide vaccine. *Plant Biotechnology Journal* 3 (5), 521-533.

doi: 10.1111/j.1467-7652.2005.00143.x

Bioplastics

There is considerable potential for increasing the yield of biopolymers in crops, which will create new value-added markets for agriculture and provide a low-cost basis for the manufacture of renewable resource-based polymers and chemicals for industry.

Polyhydroxyalkanoate (PHA) biopolymers are a broad family of microbial storage polymers, which accumulate as granular inclusions in a wide range of bacteria. PHA formation has also been achieved in developing seeds. The ability to modify the material properties of PHAs by controlling their monomeric composition, molecular weight, and final physical form as either a plastic resin or an amorphous latex suspension has attracted significant industrial interest. In plants, each PHA monomer needs to be synthesized by diverting a normal pathway of plant metabolism. PHAs can be divided into two groups according to the length of their side chains. Those with short side chains, such as poly-3-hydroxybutyrate (PHB), a homopolymer of *R*-3-hydroxybutyric acid units, are crystalline thermoplastics, whereas PHAs with medium-length side chains, such as polyhydroxyoctanoic or polyhydroxydecanoic acid, are more elastomeric (flexible). Plant **peroxisomes** and plastids are attractive organelles in which to engineer medium chain length PHA formation in that they possess metabolic pathways that proceed through medium chain length 3-hydroxyacyl intermediates.

Medium chain length PHA formation has been achieved in developing seeds of *Arabidopsis* (Fig. B.3). For example, expression of PHA synthase (*phaC*) directed by the seed-specific napin (a storage protein) **promoter** yielded 0.006% PHA per unit of dry cell weight within *Arabidopsis* seeds, whereas higher yields of 0.02% PHA per unit dry cell weight have been obtained when *phaC* was expressed under the constitutive CaMV 35S promoter. Attempts to improve polymer yields in developing seeds have included the expression of *phaC* in mutant *Arabidopsis* plants containing reduced *sn*-1,2-diacylglycerol acyltransferase (DGAT) activity and co-expression of *phaC* with a plastid-targeted acyl-acyl carrier protein thioesterase (FatB3) from the Mexican shrub *Cuphea hookeriana*. **Mutant** seeds containing reduced DGAT activity and engineered with a peroxisomally targeted PhaC yielded tenfold more PHA than the 0.006% PHA per unit of dry cell weight produced in seeds containing a peroxisomally targeted PhaC. Likewise, an 18-fold increase in polymer yield was observed in seeds expressing **plastid**-targeted FatB3 and peroxisomally targeted PhaC compared to seeds containing only a peroxisomally targeted PhaC. In both cases, the increased flux of carbon for PHA synthesis was likely due to the plant's ability to channel to the peroxisomes, for degradation, excess fatty acids that are not incorporated into **oils**.

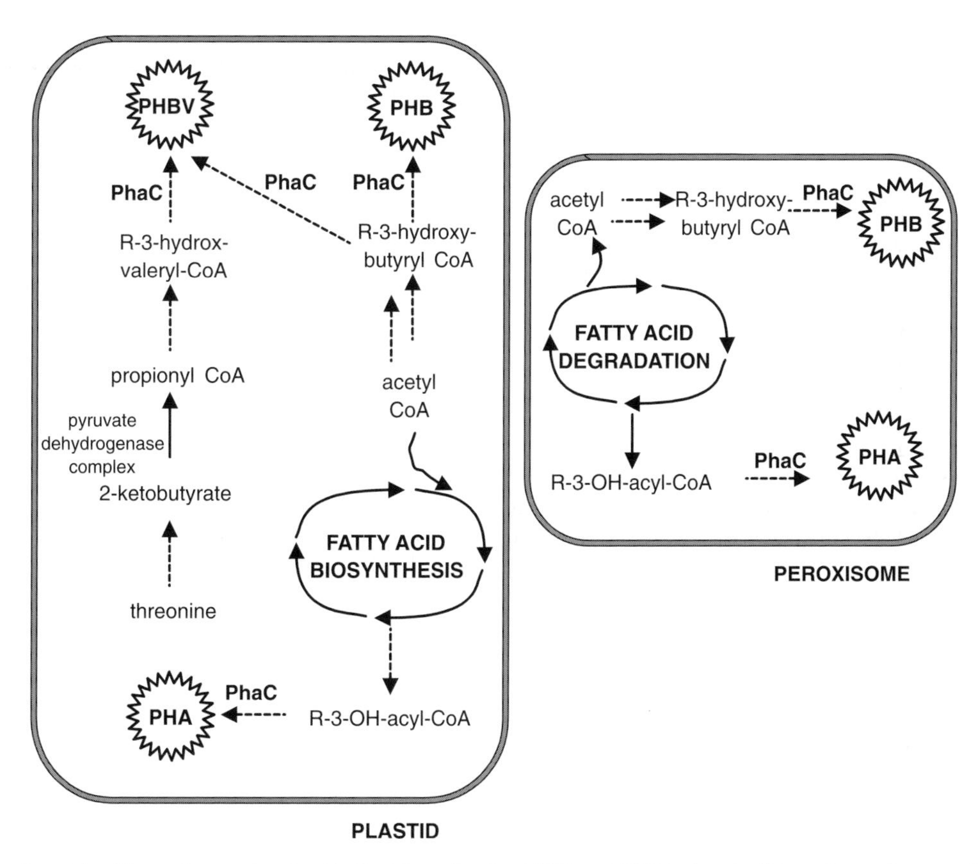

Fig. B.3. A general schematic of the pathways involved in the biosynthesis of PHA and PHB in the plastids and peroxisomes of seeds.

Several reports have described plastid-targeted PHB formation (Fig. B.3) within *Brassica* seeds upon expression of *Ralstonia eutropha* (a chemoautotrophic bacterium) genes encoding plastid-targeted thiolase, reductase and synthase. Multigene transformation vectors have been constructed in which each gene is individually expressed under the control of the seed-specific *Lesquerella* hydroxylase promoter; PHB synthesis has been achieved at 7.7% of the fresh seed weight. Plants transformed with cassettes expressing three plastid-targeted proteins for PHB synthesis, i.e. PhaA, PhaB (other enzymes in the PHB synthetic pathway, a ketothiolase and CoA reductase, respectively) and PhaC, produced this biopolymer at 0.02–7.7% of the fresh seed weight, whereas PHB-producing plants transformed with cassettes expressing plastid-targeted BktB (an alternative ketothiolase to PhA), PhaB and PhaC contained PHB at 0.02–6.3% of the fresh seed weight. (**See: Transformation**) (AJR)

Houmiel, K.L., Slater, S., Broyles, D., Casagrande, L., Colburn, S., Gonzalez, K., Mitsky, T.A., Reiser, S.E., Shah, D., Taylor, N.B., Tran, M., Valentin, H.F. and Gruys, K.J. (1999) Poly(β-hydroxybutyrate) production in oilseed leucoplasts of *Brassica napus*. *Planta* 209, 547–550.

Poirier, Y., Ventre, G. and Caldelari, D. (1999) Increased flow of fatty acids toward β-oxidation in developing seeds of *Arabidopsis* deficient in diacylglycerol acyltransferase activity or synthesizing medium-chain-length fatty acids. *Plant Physiology* 121, 1359–1366.

Valentin, H.E., Broyles, D.L., Casagrande, L.A., Colburn, S.M., Creely, W.L., DeLaquil, P.A., Felton, H.M., Gonzalez, K.A., Houmiel, K.L., Lutke, K., Mehadeo, D.A., Mitsky, T.A., Padgette, S.R., Reiser, S.E., Slater, S., Stark, D.M., Stock, R.T., Stone, D.A., Taylor, N.B., Thorne, G.M., Tran, M. and Gruys, K. (1999) PHA production, from bacteria to plants. *International Journal of Macromolecules* 25, 303–306.

Bioreactor

Mass production of cells *in vitro* is achieved by growing them in liquid culture in large fermenter vessels (bioreactors). The highly controlled conditions of bioreactors permit multiplication at larger volumes than are typical of research/laboratory-scale systems. Bioreactors have inlets and outlets for gases and liquids, ports for regulation of specific processes, impellers or other engineered methods of controlling liquid movement, and heat regulation. The major application is large-scale production of bacteria and bacterial proteins. Bioreactors are used to multiply plant cells *in vitro*, either to optimize the initial stages of **somatic embryogenesis**, or to induce further development in cultured cells, or more uncommonly, to produce secondary metabolites from cultured plant cells.

(Bioreactor technology is also used in large-scale seed osmopriming and hydropriming. **See: Priming – technology, 2. Techniques and terminologies**.) (PvA)

Biotroph

An organism entirely dependant upon another living organism (e.g. a seed or fruit) as a source of nutrients; organisms that are parasitic in nature, cause minimal damage to the host, have a narrow host range and usually cannot be grown in pure culture. (**See: Pathogen – seedborne infection**)

Black cumin

See: **Nigella**

Black gram

The black gram (or urd) (*Phaseolus mungo* syn. *Vigna mungo*) is an especially valued pulse in the Indian sub-continent and it has lesser use in other parts of Asia, the Caribbean and the Americas. It is thought to have originated in India (where wild relatives can be found) and evidence exists for its cultivation there 3000–4000 years ago. The plant spread to south and east Asia and later to America and Australia. The main cultivation of black gram is in India where 500,000–700,000 t are produced annually on approx. 2 million ha. The plant is an annual, 20–90 cm high, bearing slender, hairy, 4–7-cm-long pods, each with 4–10 seeds. These are about 4 mm long, rounded cuboid, weigh 15–40 mg and are generally black (though grey, green and brown forms exist) with a white, concave **hilum**. Seeds contain 20–24% fw **storage protein**, 1–2% **oil** (triacylglycerol) and 55–65% carbohydrate (mainly **starch**): they are relatively rich in iron, potassium and vitamin B. A trypsin inhibitor is present. (**See: Protease inhibitors**)

The immature pods can be eaten as a vegetable. The main harvest products are the **dry beans**, which are cooked, principally in India, as *dhal*. A flour is made from the seeds which is used for preparing soups, vegetables, porridge, biscuits, bread and also as a kind of soap. The seeds have several medicinal uses, e.g. to reduce inflammation. The plants can be used as green manure or forage. (**See: Legumes**)

(MB)

Swaminathan, M.S. and Jain. H.K. (1975) Food legumes in Indian agriculture. In: Milner, N. (ed.) *Nutritional Improvement of Food Legumes by Breeding*. John Wiley, New York, USA, pp. 69–82.

Black pepper

Black pepper (*Piper nigrum*, Piperaceae), the world's most important spice, is the dried fruit of the perennial woody, climbing vine (Colour Plate 3A) (**see: Spices and flavours,** Table S.12 and Fig. S.44).

India maintains the largest collection of black pepper germplasm. Over 15 varieties of improved black pepper are available.

Black pepper is produced from whole, unripe, but fully developed single-seeded fruits, 3–6 mm in diameter. White pepper is the dried seed after the removal of pericarp. It has a less pungent, mellower flavour and is low in fibre and high in starch. The vast bulk of world production still uses traditional sun-drying. Blanching of the berry clusters in boiling water for about 10 min prior to sun-drying is a common practice in some countries. Modern drying involves heating the berries in a hot airflow at about 80°C, for two periods of 4½ h with a 6-h gap in between. White pepper is prepared from fully ripe, decorticated and dried pepper fruit by loosening the pericarp in running water followed by drying. Black pepper is about one-third and white pepper is one-quarter by weight of the original green pepper. Green pepper is made from mature, green berries either dehydrated or preserved in vinegar, brine or other liquids and is mainly used as a table spice. Pink pepper is prepared by pickling the ripe red berries in brine.

Table B.7. Major constituents of essential oil of black pepper var. Panniyur-1.

Compound	Composition (%)
α-Pinene	5.28
Sabinene	8.50
β-Pinene	11.08
Myrcene	2.23
α-Phellandrene	0.68
δ-3-Carene	2.82
Limonene	21.06
Linalool	0.22
Terpinen-4-ol	0.19
α-Terpinyl acetate	0.86
α-Cubenene/-δ-elemene	3.25
α-Copaene	0.82
β-Caryophyllene	21.59
α-Amorphene	1.51
β-Bisabolene+α-Bisolene	4.25
Caryophyllene alcohol	0.12
Caryophyllene oxide	0.90
α-Cadinol	1.51

Adapted from Narayanan, C.S. (2000) In: Ravindran, P.N. (ed.) *Black Pepper Piper nigrum*. Harwood Academic, The Netherlands.

1. Chemical constituents

Black pepper is evaluated on the basis of appearance, pungency level and the aroma and flavour quality: these depend on volatile **essential oil** and **oleoresins** and the microbiological contamination levels. The aroma is contributed by the essential oil in the fruits, while the pungency is due to non-steam-volatile **alkaloids**, principally piperine. Black pepper has 10 to 15% **oleoresin**, which contains about 3% essential oil. The oleoresin is extracted by solvent extraction, a method superior to steam distillation as it preserves important heat-sensitive components. Essential oil consists predominantly of monoterpenes – sabinene, α-pinene, β-pinene, limonene and 1,8-cineol. Sesquiterpenes make up 20% of the essential oil – β-caryophyllene, humulene, β-bisabolone and caryophyllene oxide (Table B.7, also **see: Spices and flavours,** Fig. S.44). **Starch** is a major constituent of black pepper contributing to 35–45% of its weight.

The pungency level (piperine 3–9%) and the content and aroma/flavour character of the volatile oil are primarily determined by the intrinsic characteristics of the variety or cultivar grown and the stage of maturity of the berries. The volatile oil content of immature green pepper reaches a maximum at about 4½ months after fruit setting and then diminishes while the piperine content continues to increase.

2. Uses

Black pepper has multiple uses in the processed food industry, in the kitchen, and on the dining table as a spice, in beverages, perfumery and allied industries, in traditional medicine, even in beauty care. It is an indispensable ingredient in cooking, occupying a proud place in the cuisines of both east and west. Pepper oleoresin is mainly used for preserving, seasoning and flavouring meat products.

Pepper is one of the most important drugs in Indian systems of medicine – *Ayurveda*, *Unani* and *Sidha*. Pepper is pungent and acrid, hot, rubefacient, carminative, dry corrosive, antihelminthic and germicidal. It promotes salivation, helps digestion, and is thought to cure coughs, dyspnoea, cardiac diseases, colic, worms, diabetes, piles, epilepsy and almost all diseases caused by the disorders of *vata* and *pitta*. Pepper is prescribed in many ailments. Pepper has insecticidal and/or insect repellent and ovicidal activities to different insects and other harmful pests. The major alkaloid piperine is more toxic than pyrethrin to houseflies.

Besides these 'modern' uses of pepper the seed has featured historically for other purposes, for example in many countries as currency. And pepper seeds were so valuable that in AD 408 3000 pounds of them were part of the ransom, accompanying gold and silver, demanded by Alaric the Goth for sparing Rome when he besieged the city. (KVP, NB)

Govindarajan, V.S. (1977) Pepper – chemistry, technology and quality evaluation. *Critical Reviews in Food Science and Nutrition* 9, 115–225.

Purseglove, J.W., Brown, E.G., Green, C.L. and Robbins, S.R.J. (1981) *Spices*. Vol. 1. Tropical Agriculture Series, Longman, New York, USA.

Peter, K.V. (ed.) (2001) *Handbook of Herbs and Spices*. Woodhead Publishing, CRC Press, Boca Raton, FL, USA.

Ravindran, P.N. (ed.) (2000) *Black Pepper*. Piper nigrum. Harwood Academic, The Netherlands.

Weiss, E.A. (2002) *Spice Crops*. CAB International, Wallingford, UK.

Blackleg

Withering **diseases** of plants, such as those caused in canola seedling roots by seedborne *Leptosphaeria maculans* (**see: Brassica oilseeds – cultivation**) and in sugarbeet seedling roots by seedborne *Phoma* or by soilborne *Aphanomyces* in warm soils, which are controllable by seed-applied fungicides. (**See: Pathogens; Treatments – pesticides**)

Bleach test

Hot alkaline bleach is used for Class Determination in official **grain inspection** protocols, notably the determination of pigmented wheat and sorghum varieties that may contaminate 'white' variety seed lots. The test (immersion in sodium hypochlorite and concentrated sodium hydroxide for a few minutes, followed by rinsing) causes the **pericarp** constituents in tannin sorghum varieties (proanthocyanidins) to oxidize and form black pigments on the surface of caryopsis; likewise, the 'red wheat' class turns dark red, compared to the straw yellow colour formed in 'white wheat'. A similar procedure is used for determining sorghum and wheat kernels that are to be classified as 'germ-damaged' as a result of heat injury.

In an unrelated technique, brief soaking in dilute bleach (1% sodium hypochlorite) is used as a simple on-site test in the field to reveal seedcoat damage in soybean, to help adjust the combining operation throughout the harvest day and to fine-tune handling processes during subsequent **conditioning**. Seeds considered to be damaged take up solution readily and appear to swell to two to three times their original size or develop a large blister or break apart, while slightly damaged seeds become only wrinkled and undamaged seeds appear to be hardly imbibed.

Short treatments with bleach are also used to eradicate some seedborne **diseases**. (PH)

Blights

The common name of a variety of diseases characterized by their potential for catastrophic epidemic spread that cause general killing of the plant host, typically after widespread blackening of leaves and shoots, caused by fungal or bacterial **pathogens**, but also by intensive infestation by aphids.

Head blights of grasses and cereals are characterized by prematurely bleached or rotting spikelets that usually contain sterile or shrivelled seeds, in nearly all instances caused by pathogenic fungi in the genus *Fusarium*, such as *F. culmorum*, culm rot in wheat.

Kernel blight ('black point') of cereals, such as caused in wheat and barley by *Cochliobolus sativus*, is characterized by the embryo end of kernels becoming dark-brown to black; also the ability to germinate may be decreased.

Seedling blights are one expression of the **seedling disease complex** that is caused by several soilborne fungal pathogens affecting a wide range of plants, such as *Alternaria* in brassicas, or bacterial blight pathogens (for example in rice by seedborne *Xanthomonas oryzae*). (PH)

Blind seed

The effect of a **disease**, for example affecting ryegrass, or the action of a parasitic pest in the field, where the seed is fully developed and does not rot, yet is incapable of germinating. *Gloeotina*, for example, affects grasses, rye and barley in this way. (**See: Embryoless seed**)

In the unrelated 'blind plant' condition, the primary shoot **meristem** stops growing early in young plant development, sometimes within 2 weeks of sowing, such as in some cultivated tomato varieties, resulting in a seedling with no or very few true leaves, and to a check on normal growth. Mechanical injury during seed handling can lead to a similar 'bald head' symptom in common beans (**see: Phaseolus beans – cultivation**). (PH)

Blotter testing

A method of germination and **health testing**, approved for many species, whereby seeds are placed on top of wet absorbent paper sheets or pads and incubated as required before assessment. (**See: Germination testing, 1. Substrates**)

Blue eye

A spoilage condition that can develop in maize grain stored at too high moisture contents, in which the fungus *Penicillium oxalicum* fruits below the **pericarp** in the embryo (Colour Plate 10A). (**See: Diseases**)

Boerner divider

One type of equipment widely used in seed laboratories to reduce the size of the composite sample of seed lots for subsequent **quality testing** and **health testing**, based on seed falling under gravity and on to a series of baffles, which ultimately directs the seed into two spouts. (**See: Sampling, 3. Sample reduction**)

Table B.8. Horticultural *Brassica* species.

	Common name	Part of plant eaten
Brassica oleracea var. *capitata*	Cabbage	Leaves of terminal bud
B. oleracea var. *botrytis*	Cauliflower	Inflorescence
B. oleracea var. *gemmifera*	Brussels sprouts	Axillary buds
B. oleracea var. *gongylodes*	Kohl rabi, cabbage turnip, stem turnip	Swollen stem
B. oleracea var. *italica*	Broccoli, calabrese, sprouting broccoli,	Inflorescence
B. rapa var. *rapa*; syn *B. campestris* var. *rapa*, *B. campestris* ssp. *rapifera*	Turnip	Swollen hypocotyl and root, young leaves
B. rapa ssp. *chinensis*; syn *B. campestris* ssp. *chinensis*, *B. rapa* var. *chinensis*, *B. chinensis*	Chinese mustard, pak choi, bok choy, choysum	Leaves
B. rapa ssp. *pekinensis*; syn *B. pekinensis*	Chinese cabbage, Pe-tsai	Leaves
B. napus var. *napobrassica*	Swede, rutabaga, Swedish turnip	Swollen hypocotyl and root
Raphanus sativus var. *sativus*	Radish, small radish, turnip radish	Root
R. sativus var. *caudatus*	Rat-tail radish, mougri	Root, leaves, young seed pod
R. sativus var. *niger*	Chinese radish, Japanese radish, Oriental radish	Root
Crambe maritima	Seakale	Etiolated sprouts
Eruca sativa	Rocket, arugula, rucola	Leaves

Boll

The seed pod or capsule of certain plants, such as **cotton**, within which the seed and **lint** develop, and also the capsule-like fruit of **linseed/flax**.

Bolting

Production of flower (later seed) stalks which usually occurs after a response to photoperiod, temperature or **vernalization**. In biennial crops, it is the undesired formation of premature seed heads in the first year – in common idiom, 'running to seed' or 'gone to seed' – leading to partial losses in harvest yield or quality of the edible plant part, such as the root or bulb (for example, beet and onion), stem (celery) or leaf (lettuce, spinach and cabbages). Premature seed formation and scattering before harvest can also lead to the build-up in the soil of long-lived weed populations that affect succeeding crop years and can be difficult to control, such as 'weed beet' in sugarbeet fields. Bolting may be the consequence of exposing young plants to prolonged cool temperatures, when they become vernalized. In cultivated crops, it may also arise from the sowing of a **variety** in an inappropriate growing region, or delays in harvesting in over-hot conditions; it may also be the result of contamination by pollen from wild individuals during seed production. In bolting-susceptible crop plants, particular care must be taken about the time of direct sowing in the field, or in the acclimatization of young plants during the process of transplanting (**see: Hardening**). Bolting-resistance can be an important trait in breeding programmes, but care must be taken with its expression so that seed production efficiency itself is not impaired. (PH)

Bound water

See: Water binding

Brassica – horticultural

1. World distribution, origins, uses and economic importance

The horticultural crops within the family Brassicaceae fall into three *Brassica* species, *B. oleracea*, *B. rapa* and *B. napus*, and the three species *Raphanus sativus*, *Crambe maritima* and *Eruca sativa* (Table B.8).

The five varieties of *B. oleracea*, commonly referred to as the cole crops, include vegetables widely grown in temperate regions in which different plant parts have been modified into storage organs for consumption by man. Globally the two most important cole crops are cabbage and cauliflower: cabbage is grown on five continents and in more than 90 countries (3 Mha in 2000), whereas there is a much smaller production area of cauliflower (0.8 Mha worldwide), mostly in India and China. Brussels sprout production is concentrated in west and mid-Europe, Japan and North America, especially California. The large-rooted varieties of radish (*Raphanus sativus*), which are grown throughout Asia, have far greater importance than the small-rooted European type: Japan alone produces 30 times more radish by weight than does the whole of Europe. Traditionally both Chinese cabbage and mustard are important crops in China, although the distinction between them is often vague and, confusingly, there can be overlapping use of the same Chinese name. Seakale (*Crambe maritima*) originated from a wild plant of the northwest European coast and the Black Sea region and, although not now cultivated widely, there has been commercial production in northern France since 1988. Rocket (*Eruca sativa*) has also been known since ancient times, and is now grown as a minor leaf salad crop.

Many of the vegetables in Western Europe and the USA are sold through supermarkets that demand a continuity of supply of high quality, uniform produce. To achieve the specified uniformity of size for mechanized once-over harvesting in calabrese, cabbage, Brussels sprouts and cauliflower, growers prepare a sequence of plantings. This method of production has consequences for both cultivar selection and seed quality.

B. oleracea crops are thought to have originated from wild cabbage or kale, native to the coasts of northwest Europe and the Mediterranean. Cultivars spread through the Mediterranean and North Atlantic, leading to a range of specialized forms adapted to local climates. Cabbage, one of the oldest recognized vegetables, was probably in general use 2000–2500 BC. Cauliflower originated in the eastern

Mediterranean and was described in Spain by an Arab botanist in the 12th century. Broccoli is thought to have been introduced from the Levant/Cyprus to Italy, whereas kohl rabi was developed in northern Europe in the 15th century. The Brussels sprouts grown today probably first appeared in Belgium in 1750. True turnips originate in cooler parts of Europe, probably from biennial oilseed forms of *B. rapa* coming from southwest Asia in pre-classical times. It is likely that Chinese cabbage arose from the hybridization of Chinese mustard and turnips: the primary oriental form is a loose-leafed plant, first mentioned in the 5th century. Swedes (rutabaga) (*B. napus*), first recorded in Europe in 1620, are thought to be a cross between *B. oleracea* and *B. rapa*, that may have occurred several times by spontaneous hybridization where turnips and kale were being grown together. There is no one species identified as the source of cultivated radish, and the crop most likely originated both in the eastern Mediterranean and in Asia, through a cross between *R. landra* and *R. maritimus*, or from a differentiated form of *R. raphanistratum*.

2. Classification and taxonomy

B. oleracea (n = 9) is a modified amphidiploid of a cross between two five-chromosome species, followed by the loss of one chromosome pair. *B. rapa* (n = 10) includes the true turnip and some oriental vegetable species. There are many other salad vegetables within the species including *B. rapa* ssp. *narinosa* or *Taatsai* and *B. rapa* ssp. *nipposinica*, where var. *Mizuna* is recognized as *Mizuna* greens. It is likely that few genes separate some of the subspecies, particularly the oriental forms. *B. napus* (n = 19) is an **amphidiploid** of *B. rapa* (n = 10) and *B. oleracea* (n = 9). Of the remaining three species, the radish *Raphanus sativus* (n = 9) includes three varieties that are horticultural crops, all of which intercross freely with each other and related wild species.

3. Genetics/breeding

A common feature within the breeding programmes of the horticultural brassicas is the extent of **F_1** hybrid production, the uniformity of hybrids in size and maturity being a requirement of production for the main outlets of the crop. This is possible by taking advantage of the sporophytic **incompatibility (self-incompatibility, SI)** system that prevents self-fertilization in most brassica crops. SI has been exploited in China and Japan where almost all the vegetable brassicas are hybrids and systems for the production of single, double and three-way crosses of Chinese cabbage have been described. (**See: Production for sowing, III. Hybrids, Crosses**) Similarly in *B. oleracea*, open-pollinated cultivars have been largely replaced by F_1 hybrids in Europe, Japan and North America, particularly in specialist and widely grown crops. In Japan, hybrid breeding of the large-rooted radish varieties has been a feature since the 1960s. In contrast, there has been little investment in F_1 cultivars of radish in Europe until recently. This can be attributed to the low seed costs, since SI was available and the **bud pollination** needed to produce the inbred lines was easy. Hybrid varieties are found least in the low-value swede where F_1 breeding is being focused on the oilseed forms.

Problems with using the SI system of F_1 production include the breakdown of the SI of the parental lines, leading to **sib** (selfed, inbred) production, and the difficulty of generating the parental inbreds. In *B. oleracea*, the intense selection for uniformity within the crop has led to an increase in compatibility within Brussels sprouts and kohl rabi. At the extreme, summer cabbage has been placed under such intense pressure that the SI system has largely been lost and there is little or no inbreeding depression following self-fertilization. Thus, selection for SI is important when identifying the parental lines of potential hybrids. Production of the SI inbred lines themselves is laborious when bud pollination has to be used to achieve self-fertilization. The most practical alternative to hand pollination has proved to be application of high (3–6%) concentrations of CO_2 to break down the SI of the parent.

Cytoplasmic male sterility (CMS) provides an alternative to SI systems of hybrid production and has been found in Japanese and Chinese types of radish. Although this can be difficult to manipulate in commercial production, the overall costs of CMS production are less than for SI and a large range of F_1 and other hybrids are being bred by CMS in Japan and China. The Ogura CMS has also been bred into other radish types, including the European small-rooted type and F_1 cultivars are starting to appear. In addition, CMS lines of *B. rapa* are being produced by introgression from radish, and restorer systems are being perfected. Male sterility is also known in broccoli, cauliflower, cabbage, turnip and swede.

An important advance in brassica breeding has been the identification of a source of resistance to club root (*Plasmodiophora brassicae*) in the European turnip, ECD04. This resistance has been used extensively in improving varieties of oriental types of *B. rapa* and ECD04 has been hybridized with *B. oleracea* to produce a range of club root resistance in the resultant swede.

Future breeding objectives within this group of crops would include the introduction of club root resistance to the oriental *B. rapa* types and possibly to *B. oleracea*, enhancement of flavour and texture of culinary swede, and improvements in the quality, rather than yield of *B. oleracea* crops. In radish emphasis is likely to be placed on exploiting the diversity of *Raphanus* in breeding for disease resistance.

4. Seed structure

The round seeds (1–5 mg, yellow to black) of Brassicaceae have two cotyledons, are rich in oil and also contain glucosinolates. There is no endosperm. The seedcoat is made up of four layers, the epidermis, sub-epidermis, palisade and parenchyma. (**See: Seedcoats – development**, for a detailed account in *Arabidopsis*) Species of *Brassica* produce little or no mucilage when the seedcoat is moistened, unlike other Brassicaceae seeds. Seedling establishment is **epigeal**, revealing wide, deeply notched cotyledons above ground. (**See: Brassica oilseeds**)

5. Seed development and maturation

Brassica species show **indeterminate** growth and produce their flowers in the form of a raceme over an extended period. In broccoli, cabbage and Brussels sprouts there can be a 20–30 day difference in the timing of flowering within the central raceme, although in the shorter lateral racemes, flowering takes place over a shorter time. Thus there is considerable variation in the physiological stage of maturity of seeds both within and between racemes.

During brassica seed development there is a rapid increase in fresh weight and a sigmoid increase in dry weight due to an accumulation of **storage reserves** in the form of oil (54% and 70% in rapeseed and cabbage, respectively) and protein (20% and 30%) (**see: Oils and fats; Oilseeds – major; Storage protein**). The end of reserve deposition usually occurs at approximately 50% seed moisture content and it is at this point that connections to the mother plant are cut. Both **germination rate** and **viability** of seeds increase as **physiological maturity** is reached. However, these vary within a population of seeds as a result of the variability in the stage of development arising from the range of flowering times. Seed quality may also show a decline after physiological maturity as a result of deterioration. This results in complex decisions regarding when to harvest a seed crop, particularly due to the predisposition of the **siliquae** of Brassicaceae to dehisce if harvest is delayed (see next section).

6. Seed production

Seed production of brassica crops is limited to fields with a minimum 5-year gap between seed crops and at least 2 years' exclusion of any *Brassica* species. In addition careful attention has to be paid to the exclusion of cruciferous weeds and isolation from other brassicas, and crops cannot be grown on land infected with *Plasmodiophora brassicae*.

In most *B. oleracea* seed crops, plants are produced in beds and transplanted at the 5–7 leaf stage, although in kohl rabi, seeds are direct drilled and a **seed-to-seed production** system is used. In all other brassicas, the root-to-seed (swede, turnip, radish) or head-to-seed (Chinese cabbage, Chinese mustard) methods are used for the production of Basic or **Breeder's seed**, which allows for **roguing** at the time of transplanting; the seed-to-seed method is used for later stages of seed production. (**See: Production for sowing, 1. starting material**)

All brassica crops are dependent on **vernalization** to stimulate flowering, according to the crop and cultivar: sowing, or transplanting has to be timed for optimum vernalization. Flowering is also encouraged in cabbage by making an incision across the head of compact types to reduce mechanical impedance to the growing point from the leaves. Lateral growth is encouraged in Brussels sprouts by removal of the terminal point after the final roguing, and in swede and turnip the removal of the top 10 cm of the terminal shoot when flowering shoots are 30–40 cm high encourages growth from secondary inflorescences and reduces the range of the maturity period.

Cross pollination occurs in all species and is achieved mainly by natural populations of bees, although supplementary hives may be introduced. Blowflies are effective pollinators when crops are grown in confined spaces such as polythene tunnels.

In all brassica seed crops, pod shatter is a problem (**see: Shattering; Brassica oilseeds – cultivation**) and harvesting has to be timed in order to reduce the loss of seeds. In *B. oleracea*, the drying of the plant and the development of an orange brown colour indicates that the majority of the seeds are approaching maturity; at this stage the seeds in the oldest pods will be relatively firm when pressed between the thumb and forefinger. Swede and turnip are judged to be mature when the stalk turns straw-coloured, the seeds are light brown and firm; in Chinese cabbage and mustard, light brown pods indicate maturity. The 'Green Seed' problem refers to the existence of immature seed in the harvest, which retain relatively high amounts of chlorophyll.

The seed crop is then cut and left in **windrows** to mature further until the seeds separate easily from the pods. The crop is picked up from the windrows by combine and fed to the thresher. However, in small-scale production, plants may be cut by hand to windrows or sheets to contain the seeds before extraction with stationary threshers. The approach to harvest differs only in radish, where the pods do not shatter easily. In this case it is best to harvest in dry conditions when the pods are brittle and seed extraction is easier. The crop is harvested using a combine with a roller attachment that is adjusted to crack, not crush, the pods. In all species, a slow cylinder speed of the thresher is essential due to the susceptibility of the seed to cracking.

Seeds are dried after harvest, before further processing, to below 10% moisture content for storage in sacks or 8% for bulk storage. When the moisture content is above 18%, the air temperature for drying should not exceed 27°C; when below 18% moisture content, drying temperature should not exceed 38°C. Subsequent cleaning of the seed is achieved using an **air-screen cleaner**, light and shrivelled seed are removed using a gravity separator or possibly a spiral separator (**see: Separators**), which also can remove split seeds. Seed moisture content is then reduced further to around 5% if seeds are to be stored in moisture-proof containers.

Further selection of brassica seeds to improve seed quality is possible using the technique of chlorophyll fluorescence (CF). (**See: Colour sorting**) The amount of chlorophyll within a seed is directly related to the stage of maturity of the seeds, with seeds having the lowest CF being the most mature and having the highest germination capacity. Even seed samples that are visually similar in terms of colour reveal differences in CF that reflect their maturity and germination. Chorophyll fluorescence therefore offers a method for the evaluation of maturity and germination that is both rapid and non-destructive. Fluorescence can also be used to identify germinable and non-germinable brassica seeds following the leakage of sinapine on to the seed surface. High amounts of **leakage** indicate non-germinable seeds.

Seed selection by weight is also common for brassica seeds that are to be used for precision sowing (**see: Planting equipment**) and production of **transplants** in modules. These are referred to as 'precision seeds'. Precision seeds have a uniform size and high germination and are sold by number.

7. Seed quality

The characteristics of germination and vigour are particularly important in the horticultural brassicas, in which the establishment phase has a significant impact on the timely production of uniform, high-quality vegetable produce. Where crops are directly sown to a stand in the field, reliable and uniform establishment ensures that the desired spacing of plants is achieved and, subsequently, there is little variation in plant size at maturity. In addition, rapid, uniform establishment in the field results in the production of uniform seedlings giving a uniform crop that be mechanically harvested. Alternatively, many horticultural brassicas are established from transplants grown in glasshouses in modular trays (see next section). Reliable germination and emergence is again essential to avoid financial losses due to empty modules and rapid, uniform emergence will produce transplants of even size.

The minimum germination requirements for horticultural brassicas in the UK are 70% for cauliflower and radish, 75% for cabbage, calabrese, Chinese cabbage, Brussels sprouts, kohl rabi and sprouting broccoli and 80% for swede. The requirements and expense of the methods employed to establish these crops demand germination levels well above these minimum levels and high vigour. Differences in seed vigour may lead to differences in emergence: increasingly seeds are being tested for vigour, with the **controlled deterioration** test, which predicts both emergence and storage potential, being the most appropriate test identified for use with small-seeded vegetables.

Other aspects of seed quality are purity and seed health. All brassica crops should have a minimum analytical purity of 97% with a maximum of 1% (by weight) of seeds of other plant species. The only standard for seed health in brassicas in the UK is that in **Basic Seed** there should be no infection with *Phoma lingam*, the pathogen that causes canker.

8. Sowing

Seeds are usually sown in the field by precision seeders, to achieve accurate spacing, with a single plant at each station. Increasingly though, brassica seedlings are produced in modular cell trays in glasshouses before transplanting (e.g. over 90% of *B. oleracea* crops in the UK); seeds are sown using automated **vacuum precision seeders** that place a single seed into each cell, and automatic transplanters place the seedlings into the field at the desired spacing. One reason for the move to **transplant** production is to avoid problems of emergence associated with variable field conditions, confounded by variable seed quality.

9. Treatments

Fungicide treatments for horticultural brassica crops include the protectants, iprodione and thiram, and the systemics, carbendazim and metalaxyl-M. Thiram, metalaxyl and carbendazim all protect the seed from infection by the soil-borne fungi (*Pythium* and *Phoma*) that cause seedling **damping-off** diseases (**see: Treatments – pesticides, fungicides**). In addition the systemics protect the cotyledons of brassicas against attack by species of *Peronospora* that cause downy mildew of the foliage. Development of the *Alternaria* diseases at later stages of crop growth leads to stem and pod blight, and infection by *A. brassicae*, *A. brassicicola* and *A. raphani* can result in production of shrivelled seeds or the death of pod stalks before seed formation. Problems due to infection during seed production can be limited by spray treatments of iprodione.

These fungicides are sometimes and in some countries applied in conjunction with an insecticide and acaricide, such as chlorpyrifos, or the neonicotinoid insecticide imidacloprid, used against peach-potato aphid in broccoli, calabrese, cabbage, cauliflower and Brussels sprouts, and chlorpyrifos is applied against cabbage root fly.

All these crop protection seed treatments are most commonly applied in a polymer **filmcoating**. Other seed treatments such as **priming** and **pelleting** are not generally used for horticultural brassicas. However, seeds may be surface sterilized before pesticide coating and this is not always indicated on the label. Some seed may be treated by hot water or other means to remove seedborne bacteria such as *Xanthomonas* spp. following **health testing**. (AAP)

For information on the particular contributions of the seed to seed science **see: Research seed species – contributions to seed science**; and ***Arabidopsis***.

George, R.A.T. (1999) *Vegetable Seed Production*. CAB International, Wallingford, UK.

Jalink, H., van der Schoor, R., Frandas, A., van Pijlen, J.G. and Bino, R.J. (1998) Chlorophyll fluorescence of *Brassica oleracea* seeds as a non-destructive marker for seed maturity and seed performance. *Seed Science Research* 8, 437–443.

Maude, R.B. (1996) *Seed Borne Diseases and Their Control: Principles and Practice*. CAB International, Wallingford, UK.

Powell, A.A. and Matthews, S. (1984) Application of the controlled deterioration test to detect seedlots of Brussels sprouts with low potential for storage under commercial conditions. *Seed Science and Technology* 12, 649–657.

Smartt, J. and Simmonds, N.W. (1995) *Evolution of Crop Plants*. Longman, London.

Still, D.W. and Bradford, K.J. (1998) Using hydrotime and ABA time models to quantify seed quality of Brassicas during development. *Journal of the American Horticultural Society* 123, 692–699.

Whitehead, R. (2003) *The UK Pesticide Guide 2003*. CAB International, Wallingford and British Crop Protection Council, Farnham, UK.

Brassica - oilseeds

1. The crop

The brassica oilseed crops include the species *Brassica napus*, *B. rapa* (syn. *B. campestris*) and *B. juncea*, respectively called oilseed rape, turnip rape and oilseed mustard. Collectively they constitute the world's third most important source of edible oil after palm and soy. These species are cool-season crops best adapted to the extremities of the temperate zones or at higher elevations, or as winter crops in the sub-tropics. The main producing countries are China, India, Canada, the European Union and Australia, although commercial production occurs in over 14 other producing nations. In 2002 total world exports reached 7.7 million t at a value of US$1.9 billion, the principal exporter being Canada. (**See: Oilseeds – major**, Table O.1 (**See: Crop Atlas Appendix, Map 16**))

Initially Brassica oilseeds contained two important anti-nutritional factors. The oil contained 20 to 50% of the long chain monounsaturated **fatty acid** erucic (C22:1) and the seed and meal contained substantial amounts of the goitergenic sulphur compounds called glucosinolates. (**See: Oils and fats**) Selection and breeding in Canada resulted in the elimination of both undesirable factors creating a highly nutritious seed, oil and meal. This new quality seed and its products were given the trademarked name 'canola™' which is defined as seed, oil and meal from *B. napus*, *B. rapa* and *B.*

juncea that contain less than 2% of the total fatty acids as erucic acid and less than 30 μmol/g of aliphatic glucosinolates in the moisture-free meal. Only canola-quality crops are now grown for edible vegetable oil in North America, Europe and Australia with conversion to canola quality underway in China and the Indian subcontinent. (See: **Oils and fats – genetic modification**) *(Editors' note: The author of this article, Keith Downey, is known in Canada as the Father of Canola, for he was instrumental in bringing about the major changes in its nutritional quality and removal of undesirable compounds, e.g. glucosinolates and erucic acid, to make it acceptable for consumption.)* (**See: Oilseeds – major,** Table O.2, for composition)

The most widely grown species, *B. napus*, has both a spring (annual) and winter (biennial) form. In northern Europe and China the winter form predominates while the spring form is dominant in Canada and Australia. The principal species on the Indian subcontinent is *B. juncea*. The importance of *B. rapa* has diminished over the years and now occupies only limited hectarage in Canada, India, Sweden, Finland and China. A fourth species, *B. carinata* Braun called carinata or Ethiopian mustard, now grown to a very limited extent in northeast Africa, may become an important oilseed.

All four species are closely related to one another and to the cole vegetables, such as cabbage, broccoli, turnips, etc. (Fig. B.4) (**see: Brassica – horticultural**). The three species with the highest chromosome number, *B. napus*, *B. juncea* and *B. carinata* are **amphidiploids** derived from crosses between the diploid species, *B. nigra*, *B. rapa* and *B. oleracea*.

2. Origin

The tribe Brassicae appears to have originated near the Himalayan region with *B. rapa* probably one of the earliest crop plants selected by humans. *B. rapa* has the widest distribution with secondary centres of origin in Europe, Russia, Central Asia and the Near East. In contrast, *B. napus* is considered to have evolved more recently in the Mediterranean region where both putative species, *B. rapa* and *B. oleracea* were present. It is thought that *B. juncea* arose at several different locations where both putative species *B. nigra* and *B. rapa* would have been abundant, resulting in centres of diversity in China, eastern India and the Caucasus.

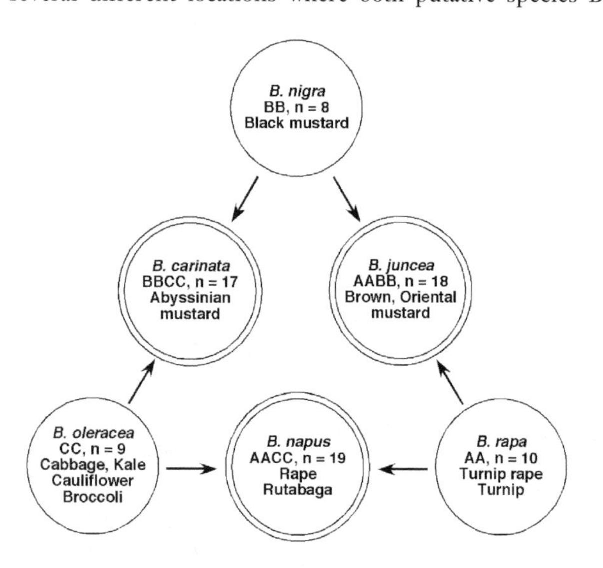

Fig. B.4. Genome and chromosome relationships of some *Brassica* species.

3. Fruit and seed

Following pollination by wind or insect the petals are shed and the ovary elongates into a bivalved (bilocular) fruit (a **silique**), 3–12 cm long, separated by a false septum with a single row of seeds in each loculus. The number of seeds per silique varies between 15 and 40 depending on the species and growing conditions. Seeds account for 15–35% of the dry matter at harvest. (**See: Brassica oilseeds – cultivation**)

The round seed is small (2–3 g/1000 in *B. rapa* and 4–5 g/1000 for *B. napus* and *B. juncea*). Seed colour can be black, reddish-brown or yellow and, depending on the species, the seedcoat exhibits slight (*B. napus*), to marked (*B. rapa*), to very marked (*B. juncea*) reticulation (Colour Plates 12A, 12B). Seeds of all species show a distinct radicle ridge, flat **hilum** and raised **chalaza**. The **seedcoat** consists of an epidermal layer which may be slightly mucilaginous, a palisade layer of thick walled columnar-shaped cells that are normally pigmented and a layer of crushed parenchyma. All that is left of the **endosperm** is a single layer of **aleurone layer** cells containing minute **protein bodies** and **oil bodies** and a thin layer of flattened parenchyma. The immature green **embryo** gradually turns to a uniform bright yellow at maturity.

The embryo parenchyma tissue shows little differentiation except for procambium bundles and palisade cells on the inner sides of the **cotyledons**. Cell inclusions are oil bodies and protein bodies of up to 20 μm in diameter as well as myrosin bodies that compartmentalize the myrosinase enzyme that breaks down the anti-nutritional glucosinolates when cells are broken and moisture is present.

The cotyledons of the brassica seeds are conduplicate, the outer folded over the smaller inner cotyledon with the **hypocotyl–radicle** axis located at the base and between the two cotyledons. The seed oil is primarily located in oil bodies within the cotyledon cells, with a very small amount present in the aleurone layer. Oil contents of over 50% of the dry weight of the seed have been reported but 43 to 46% is normally obtained.

4. Uses

Through conventional, mutagenic and transgenic means, breeders have succeeded in altering the composition of canola oil to specific requirements, namely, high oleic, high lauric, high erucic, low linolenic, and low saturated fatty acids.

Canola oil is used as a salad and cooking oil and in margarine manufacture. Methyl esters of the oil are used extensively as biodiesel in Europe. Oils high in erucic acid are produced under contract for industrial uses as a slip agent in plastic manufacture and as a lubricant. The nutritionally well-balanced seed protein makes up about 40% of the oil-free meal. The meal contains almost the entire complex of amino acids essential in human and animal diets as well as high amounts of calcium, iron and vitamins (**see: Essential amino acids; Storage protein**). Globally, canola/rapeseed meal is the second most widely used protein ingredient in animal feeds after soybean meal. Canola meal is generally traded at a lower price than soybean meal since the protein content is lower. In addition, since the hull is not removed during processing, canola meal contains more fibre, resulting in a lower biological energy content than soybean meal. Plant breeders have reduced the fibre somewhat by developing thinner, yellow seedcoated varieties. The major genetic improvement in the Brassica meals has been the reduction in glucosinolate content to very low concentrations. Removing the goitergenicity and feed intake depressing effects of the glucosinolates has resulted in its widespread use in all animal feeds. (KD)

(See: Brassica oilseeds – cultivation)

Downey, R.K. and Robbelen, G. (1989) Brassica species. In: Robbelen, G., Downey, R.K. and Ashri, A. (eds) *Oil Crops of the World*. McGraw-Hill, New York, USA, pp. 339–362.

Eskin, N.A.M., McDonald, B.E., Przybylski, R., Malcolmson, L.J., Scarth, R., Mag, T., Ward, K. and Adolph, D. (1996) Canola. In: Hui, Y.H. (ed.) *Bailey's Industrial Oil & Fat Products*, Vol. 2, 5th edn. John Wiley, New York, USA, pp. 1–95.

Vaughan, J.G. (1970) *The Structure and Utilization of Oil Seeds*. Chapman and Hall, London, UK.

Brassica oilseeds – cultivation

1. Genetics and breeding

Breeding of oilseed rape/canola is based primarily on developing cultivars that have high seed yields. Targeted agronomic traits include: (i) seedling vigour for rapid plant establishment; (ii) a leaf area index that exploits the greatest photosynthetic potential; (iii) a flowering and maturation period that is appropriate for the specific geographic region; (iv) lodging resistance that prevents loss of seed to rodents and premature pod **shattering**; (v) an ideal plant height for efficient harvesting; and (vi) resistances to diseases and insect pests that prevent direct seed losses by physical damage and secondary infections (major **pathogens** are *Phoma lingam*, *Sclerotinia sclerotiorum*, *Plasmodiophora brassicae*, *Verticillium dahliae* and *Alternaria brassicae*). Light relations in the crop canopy play an extremely important role in the final seed yield and traits that improve them are valued, such as apetalous, upright pods, and larger pods with more seeds.

Pedigree breeding is the most widely utilized method for breeding cultivars of the amphidiploid species *Brassica napus* and *B. juncea*, which exhibit out-crossing up to 30% or more. Alternatively, the **double haploid** method allows direct field selection of agronomic and quality traits within the first field year, using haploid plant embryos generated *in vitro* from young pollen grains (microspores) of F_1 plants to produce an array of homozygous plants, each carrying a unique combination of traits derived from the parent plants. The self-incompatible diploid species, *B. rapa*, is bred on a population basis through recurrent selection. In this species 'synthetic' varieties can be produced by mixing two parental populations in the field, whereby the resultant out-crossing produces seedlots composed of ~50% hybrid seed.

Hybrids can out-yield standard cultivars by 10–40%, and have been extensively developed. The **cytoplasmic male sterility** (CMS) system is naturally occurring and widely used. **Self-incompatibility** (SI) alleles introduced into *B. napus* from SI diploid species (*B. rapa* or *B. oleracea*) are used to control pollination in double-cross hybrid production, for example, using a total of four different SI-alleles (Fig. B.5).

Several other production methods of hybrid seed production have been explored. The Nuclear Male Sterility (NMS) system confers male sterility on one parent of the hybrid. In the genetically modified version of the NMS system, the pollen parent can be removed from the field by herbicide application, since the inserted male sterility construct is also linked to a herbicide resistance gene.

The already narrow genetic background of oilseed rape became even narrower during the development of 'canola'-quality rapeseed (**see: Brassica oilseeds**), because the traits that reduced erucic acid and glucosinolates (cultivars denoted as 'double zero') were derived from a small number of genetic sources. In order to expand the genetic base, and to capture specific traits that are unavailable in *B. napus*, breeders are making interspecific crosses with related *Brassica* species. Furthermore, in an effort to generate greater genetic variability or capture new traits, breeders are utilizing protoplast fusion, ionizing irradiation *in vitro* and chemical seed treatments to induce novel mutations, as well as introducing natural or artificially generated genetic constructs through **genetic modification**.

Canola is easily amenable to biotechnological manipulation and therefore is at the leading edge of genetic transformation research. To date by far the most crop hectarage of the genetically engineered canola cultivars consist of herbicide tolerance (to glyphosate or glufosinate), particularly in North America. Other genetically modified canola varieties in development exhibit increased lauric acid, increased oleic acid, nuclear male sterility for hybrid production, resistance to sulphonyl urea herbicides, resistance to insect feeding using *Bacillus thuringiensis* toxin genes, elevation of specific saturated **fatty acids**, and many others. (**See: Brassica oilseeds; Oils and fats – genetic modification; Oilseeds – major**)

2. Development and production

During the first weeks of seed development inside the young pod (a **siliqua**), the seedcoat expands to almost full size. At this stage the seed is somewhat translucent and the embryo develops rapidly within the seedcoat, filling the space previously occupied by fluid. Seed weight increases and about

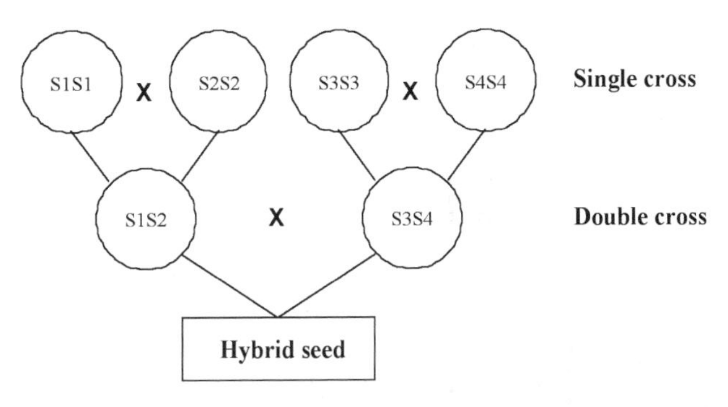

Fig. B.5. Double cross crossing scheme for production of hybrids using four self-incompatibility alleles (S1, S2, S3, S4). These crosses are made by mixing the seed and growing the two components together in the field. The method produces vast amounts of hybrid seed since every plant is both a pollen donor and a seed producer.

35–45 days after pollination seed filling is complete, by which time **cotyledons** are filled with oil and protein reserves. At maturity the **embryo**, which fills the entire seed, is bright yellow with seed moisture ranging between ~20 and 10%. For further information on seed structure **see: Brassica oilseeds. (See: Brassica – horticultural)**

Seed filling is followed by maturation and ripening which is characterized by colour changes. As the seed matures, the seedcoats begin to change from green to yellow to brown, while the silique also changes from green to a straw colour. Because the inflorescence is formed from multiple **indeterminate** racemes, by the time that the lower pods on the main raceme have elongated and turned green, most of the leaves on the plant have died, and the silique walls and green stem become the major photosynthetic source for continued seed maturation. Usually the siliques formed first at the base of the primary racemes are the longest and contain the largest seeds; but pollination, seed filling and maturation continues on lateral racemes as long as adequate photosynthate is available or until autumn frost interrupts development. At 40 to 60 days after first flowering, the seeds in the lower pods have ripened fully and changed to their final seedcoat colour, either black or mixed, depending on species. *B. rapa* flowers earlier and therefore is generally a smaller plant than *B. napus*, thus severely restricting its yield potential.

Seed yield is the result of the following factors: number of plants/unit area, the number of pods/plant, the number of seeds/pod and the mean individual seed weight. Each component is strongly affected by environmental factors such as moisture, temperature, soil texture/structure, nitrogen availability, weed competition and insect or disease damage.

Each cultivar requires a minimum number of leaves before initiating flowering (6–12), and is bred to flower and mature either earlier or later in the season; **vernalization** or photoperiod requirements may delay flowering and produce plants with more leaves and greater photosynthetic capacity.

Up to 50% of flowers do not produce pods to maturity due to heavy losses on secondary racemes towards the end of flowering. Although the potential number of pods per plant and seeds per pod is set at flowering, the final number is not established until later in the growing season: seed-filling places large demands on soil moisture and nutrients; and seed abortion, or reduction in seed weight, can be caused by any adverse environmental factor that interferes with plant function during this period.

Mature pod formation is directly related to the amount of radiation intercepted by the canopy during and shortly after flowering. A leaf canopy that intercepts 90% of sunlight radiation is ideal for optimal yield. The number of surviving seeds that undergo rapid growth is also related to the amount of radiation intercepted by the pod.

Harvesting methods such as direct combining or swathing can critically affect final yield. Canola seed pods dehisce (by **shattering**) very easily when ripe, due to disturbance of the canopy by wind or harvesting machinery. Typical main crop losses vary between 8 and 12% of the potential yield, but reductions can be considerably more in poor weather conditions. Seed production crops are generally swathed (**windrowed**) before the pods are fully mature to allow even and rapid drying for up to 2 weeks in the field, to achieve more uniform pod maturity and easy combinability of the crop, and avoid losses by pod shattering. Field seed samples must be taken to monitor seed colour changes before swathing, which is done for optimum yield when the seeds contain about 30–35% moisture and the seed colour has changed (such as from green to brown in black-seeded varieties). However, in regions where dry harvest climates prevail, producers prefer to combine the crop directly, for which purpose short or medium-height cultivars are most efficient.

Green seed results when fields are harvested too early, when the crop has suffered a sudden frost before maturity, effectively halting further development, or when plants are exposed to high heat and drought near maturity. Due to the high chlorophyll content, green seed increases crushing costs associated with its removal from the oil.

3. Quality and vigour

The highest seed grade (for example, 'Canada Certified No. 1') must have a minimum germination of 90% when tested under standardized procedures (on moist blotter papers at 25°C/15°C day/night, with 8 h light, assessed after 7 days). Seedling **vigour** is assessed by measuring seedling fresh weight after 7 days at 20°C.

Secondary dormancy, developing in imbibed seed in particular circumstances, is a potential limiting factor for establishment of susceptible cultivars. The phenomenon, its genetic background and relation to darkness, water stress, low temperatures and low oxygen availability, are still poorly understood.

4. Sowing and treatment

Poor establishment of oilseed rape/canola is a long-recognized problem – due not so much to low plant populations, since the crop can compensate well, but due to patchiness or complete emergence failures. Normally canola is sown at a depth of 12–25 mm into a **seedbed**, which ideally is level, uniform, well packed, firm, warm, aerated and moist throughout its depth. Compacted, dry and waterlogged soils, frost-heave, soilborne fungi, insect and other pests can all reduce establishment. Surface crusting on clay soils that prevent seedling emergence often occurs after heavy rains or with excessive packing of wet clays soon after planting. Seed size, vigour, age and chemical composition all have effects, when combined with these environmental factors.

Ideally the seeding rate aims to establish plant populations of 80–180 plants/m^2; however, with poor germination, a stand of 40 plants/m^2 would still be adequate, as plants increase branching to fill in the available space. Higher seeding rates produce plants with fewer branches that mature earlier, but also increase plant lodging.

Seeding date varies with many factors: number of frost-free days, soil temperature and moisture, soil texture and slope, and the specific canola variety to be planted. Germination proceeeds in soil temperatures above 10°C. Although seedlings can withstand −8°C, early planting is desirable, especially where a killing frost could affect seed filling and maturity. Late seeding will normally increase seed oil content, but protein content is highest in early-sown canola. Winter rape/canola is planted in early autumn (late August) in order to allow for adequate plant establishment before winter.

In Canada, 'fall (autumn) seeding' of spring canola in areas where there are freezing temperatures throughout the winter has shown a gain of 12 to 14 days in harvest maturity, thereby avoiding some diseases and pests. Theoretically, seeds can take advantage of optimal soil temperature and moisture conditions in the early spring by germinating immediately, while normal spring planting requires waiting for drier conditions in order to prepare the field. Because the crop is capable of a substantial degree of compensatory growth, not all the seeds need to survive the winter. Optimization of fall seeding includes use of herbicide-resistant varieties and direct planting into stubble. Ideally fall seeding is done just prior to the onset of soil freezing. An imbibition-resistant **coating** has also been developed to increase the safety margin in the face of the likely onset of the freezing conditions. However, fall seeding of spring canola varieties is generally not recommended in the Canadian western provinces due to inconsistent results and a high risk of crop failure.

5. Treatment

Canola seedlings are extremely susceptible to pest and disease damage during the first weeks after emergence, including blackleg, *Alternaria*, *Phoma*, *Rhizoctonia* and flea beetles. Cold unfavourable weather conditions together with a poorly packed seedbed can result in a complex of symptoms during establishment including seed decay, pre- and post-emergence seedling blight or rots primarily caused by *Rhizoctonia solani* (root rot), *Fusarium* sp. (foot rot) and *Pythium* sp. (seedling **damping-off**). Maximizing conditions favouring rapid germination and seedling establishment by shallow seeding into warm, moist soil with optimum fertilization will largely prevent serious damage to a canola crop by soil fungi. No single fungicide product is effective against all three fungi.

The major insect pest attacking canola shortly after germination is the flea beetle (*Phyllotretra* spp.). If more than 50% of the leaf area has been removed, seedlings are weakened, wilt and die, resulting in serious loss of yield.

Various seed treatment combinations of fungicides (such as thiram, carboxin and iprodione) and insecticides (lindane [currently being deregistered], imidacloprid, thiamethoxam, beta-cyfluthrin), according to their registration status in different countries, are widely applied as pre-planting treatments by filmcoating or encrusting to protect against fungal and insect damage. Some chemical treatments include an initial fertilizer. Granular chemical treatments (carbofuran, or organophosphates) added to the furrow at planting time also provide protection.

In Canada, a microbial seed inoculant is registered to replace the application of phosphate to the soil, based on spores of a naturally occurring soil fungus (*Penicillium bilaji*) that increases the availability of phosphate to the crop.

(LSK, NB)

Canola Council of Canada. *Canola Growers Manual.* www.canola-council.org

Downey, R.K. and Robbelen, G. (1989) Brassica species. In: Robbelen, G., Downey, R.K. and Ashri, A. (eds) *Oil Crops of the World.* McGraw-Hill, New York, USA, pp. 339–362.

Freidt, W. and Luhs, W.W. (1999) Breeding of rapeseed (*Brassica napus*) for modified seed quality – synergy of conventional and modern approaches. *10th International Rapeseed Congress.* Canberra, Australia.

Gomez-Campo, C. (ed.) (1999) *Biology of* Brassica *Coenospecies.* Elsevier, Amsterdam, The Netherlands.

Kott, L.S. (1995) Production of mutants using the rapeseed doubled haploid system. In: *Induced Mutations and Molecular Techniques for Crop Improvement. Proceedings of an International Symposium on the use of Induced Mutations and Molecular Techniques for Crop Improvement.* IAEA/FAO of United Nations, Vienna, 19–23 June 1995.

Kott, L.S. (1995) Hybrid production systems based on self-incompatibility in oilseed *Brassica*. *9th International Rapeseed Congress.* Cambridge, UK, 4–7 July 1995.

Mendham, N.J. (1995) Physiological basis of seed yield and quality in oilseed rape. *9th International Rapeseed Congress.* Cambridge, UK, 4–7 July 1995.

Brassinosteroid

Originally isolated from *Brassica napus* pollen, the brassinosteroid (BR) brassinolide (BL) was the first plant steroid to be shown to have hormonal properties, in that it can regulate stem elongation. Castasterone (CS) is a brassinosteroid found in tobacco and tomato. Since then, BRs have been shown to exist throughout the plant kingdom in nearly all tissues analysed, being particularly abundant in pollen and immature seeds. BRs are essential for elongation in stems, and may be important for seed germination (**see: Germination – influences of hormones**), xylem differentiation and pollen tube growth. Brassinosteroids can modulate the effects of light on plant development and deficient mutants of *Arabidopsis* and tomato when grown in the dark lack an apical hook, exhibit leaf expansion and express photosynthetic genes.

The starting point for synthesis of BL is campesterol that can be converted via capestanol into castasterone by two parallel pathways. The early C-6 oxidation route is thought to occur in tobacco, rice and *Arabidopsis*, while the late C-6 pathway maybe more prevalent in pea and tomato (Fig. B.6). Both BL and CS have brassinosteroid activity but in some species such as tobacco and tomato only the latter compound has been isolated. A number of brassinosteroid-deficient mutants have been identified in *Arabidopsis*, pea and tomato and their cloned wild-type genes encode enzymes that catalyse key steps in the biosynthetic pathway. All mutants are dwarf, and growth can be restored to that of the wild type by application of BL or CS.

Brassinosteroids can be metabolized to a spectrum of compounds including hydroxyl and glucosyl derivatives. Fatty acid conjugates have been isolated from lily and may represent a strategy for storing BRs. In mammals, steroids may be inactivated by sulphonation and in *Brassica napus* a gene encoding an enzyme that specifically catalyses the o-sulphonation at 22OH of BRs has been isolated. (GL, JR)

(See: Signal transduction – hormones)

Bishop, G.J. and Yokota, T. (2001) Plant steroid hormones, brassinosteroids: current highlights of molecular aspects on their synthesis/metabolism, transport, perception and response. *Plant and Cell Physiology* 49, 114–120.

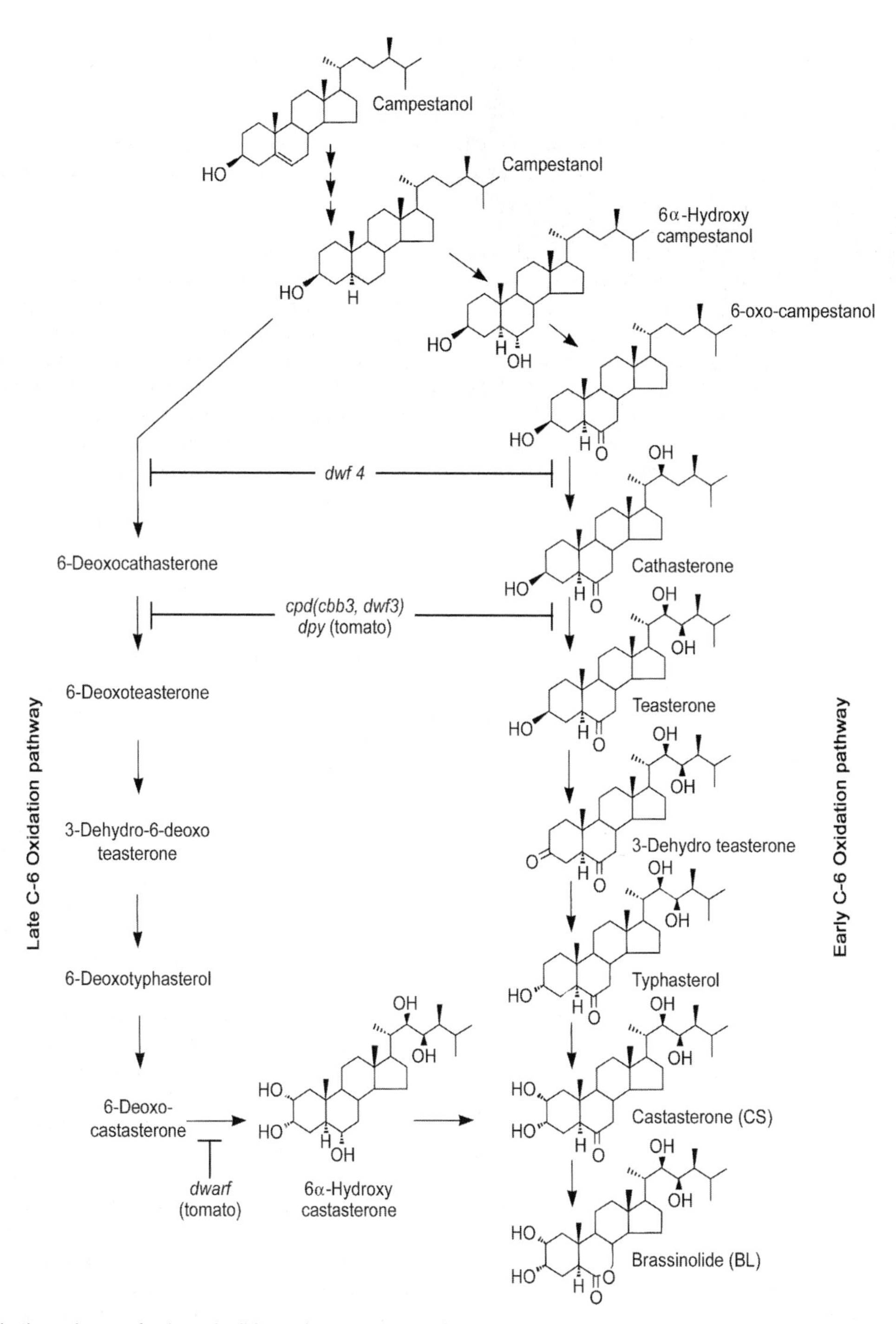

Fig. B.6. Biosynthetic pathways for brassinolide and castasterone. Stages in the pathways that have been identified as being blocked in *Arabidopsis* or tomato mutants are identified. Modified from Srivastava, L.M. (2002) Brassinosteroids. In: *Plant Growth and Development. Hormones and Environment.* Academic Press, New York, USA, pp. 205–215.

Clouse, S.D. and Sasse, J.M. (1998) Brassinosteroids: essential regulators of plant growth and development. *Annual Reviews of Plant Physiology and Plant Molecular Biology* 49, 427–451.

Srivastava, L.M. (2002) *Plant Growth and Development. Hormones and Environment.* Academic Press, New York, USA, pp. 205–215.

Brazil nut

The Brazil nut (*Bertholletia excelsa*, Lecythidaceae), of warm climates, is not a true nut according to the botanical definition (**see: Nut**) since the hard, outer shell is not the pericarp (fruit coat) but the **testa**; the 'nut' in this case is therefore a seed. Most Brazil nuts are grown in the wild in the forests of the Amazon river valley but plantations are now being established. From 1999–2003, the world's leading producers were Bolivia (51% of world production), Brazil (41%) and Côte d'Ivoire (7%). The only other producer is Peru with less than 1%.

Nut-bearing fruits (on trees which may reach 50 m high) are woody capsules (nearly 2 kg in weight and 10–12 cm in

Fig. B.7. Brazil nut. (A) a cut fruit, with seeds replaced inside and (B) kernels (embryos), with the seedcoats removed. For dimensions see text. (A, image by Dr Carlos E. Velazco, kindly supplied through Resources Brazil. B, image by Mike Amphlett, CABI.)

diameter) containing 10–25 seeds measuring 3.5–5 cm by 2 cm, triangular in cross section, and weighing 4–10 g, the Brazil nuts of commerce (Colour Plate 4B; Figs B.7A, B.7B). The agouti, a rodent with chisel-like incisors, can open the capsule and serves to disperse the seeds. The hard, bony testa encloses the 3–4-cm long kernel (embryo) whose **hypocotyl** is greatly swollen as the repository of the storage reserves: the cotyledons are much reduced and are visible only on very close inspection. The hypocotyl contains nearly 70% **oil** (triacylglycerol) and 12–15% **storage protein** among which are those rich in S-containing amino acids (**see: Nut,** Table N.2), but also allergens (**see: Storage protein – intolerance and allergies**). The nuts are relatively rich in selenium.

The Brazil nut is of high value in the confectionery and baking trades. The oil is extracted for minor culinary use.

(VP, MB)

Breeder seed

An official term used in North America for seed (or vegetative propagating material) of a cultivar that is increased on an initial small-scale by or on behalf of the breeding institute or company and used as the source of producing **Foundation seed**, and thence **Registered** and **Certified** seed. A term equivalent to 'Pre-Basic Seed' in the **OECD Seed Schemes**. (**See: Production for sowing, I. Principles**)

Broad bean

1. Worldwide importance and distribution

Broad bean or fava (faba) bean (also known as horse bean, field bean, tick bean or Windsor bean) (*Vicia faba* L.) is a cool-season legume popular in the Middle East, Europe, China and in the highlands of South America. Production is concentrated in nine major agroecological regions namely the Mediterranean, the Nile valley, Ethiopia, Central Asia, East Asia, Oceania, northern Europe, Latin America and North America. (**See: Legumes; Crop Atlas Appendix, Map 3**)

2. Origin

The origin of these beans is unclear but the best information places it in the Mediterranean area and Central Asia. There is some genetic and archaeological evidence for the co-cultivation of broad bean with the early wheats, emmer and einkorn, in the Near East. The earliest remains are from about 8500 years ago (Neolithic period) and there are widespread remains in the Mediterranean and central Europe dating from about 5000 years ago: it was a popular food in ancient civilizations (Egyptian, Greek, Roman). It is thought to have been introduced into China about 2000 years ago. The wild progenitor and the exact biological origin remain unknown.

3. Seed types

The varieties can be divided into four main groups based on seed size (Table B.9).

Seeds are variable in size and shape, but usually are nearly round, and coloured white, green, brown, purple, black or buff (Colour Plate 1C, D).

About 30% in a population of fava beans are cross-fertilizing, the main insect pollinators being bumblebees. Subspecies *paucijuga*, mainly grown in Central Asia, is mostly self-pollinating.

Table B.9. *Vicia faba* types.

Common name	Subspecies	Seed characteristics	Use
Beck, tick or pigeon bean	*Vicia faba* var. *minor*	Small, rounded (1 cm long)	feed
Horse bean	*V. faba* var. *equina*	Medium sized (1.5 cm)	feed
Broad bean	*V. faba* var. *major*	Large broad flat (2.5 cm) 8 seeds/pod	food
Windsor bean	*V. faba* var. *major*	Large broad flat (2.5 cm) 4 seeds/pod	food
—	*V. faba* var. *paucijuga*	Small, rounded (1 cm long)	feed

4. Nutritional quality

The protein content is relatively high, in the range 20–40% fresh weight (fw). Legumin (predominantly) and vicilin are the globulins present (**see: Storage protein**). Except for tryptophan and methionine, the contents of the **essential amino acids** are relatively high, especially lysine, whose concentration is twice as high as that in cereal grains. Total carbohydrate, most of which is **starch**, can reach as much as 60% fw; sucrose is also present together with approx. 5% fw of the flatulogenic, **raffinose-series oligosaccharides**, mainly stachyose and verbascose (sometimes considered as antinutritionals). **Oil** (triacylglycerol) is approx. 1.3% fw, fibre 6.8% and minerals 3% (Ca, Fe, P, Na and K). The vitamin content is higher that that of rice and wheat. The B complex and ascorbic acid predominate.

The beans contain several **antinutritional** factors. The protease (e.g. trypsin) inhibitors are, however, at much lower concentration than in, for example, **soybean** (**see: Protease inhibitors**). Where these beans are eaten regularly as a major component of the diet, a paralytic condition known as favism can occur. (Even inhalation of pollen may incite this severe haemolytic anaemia in susceptible individuals.) This disease develops only in humans that have a congenital deficiency of the enzyme glucose-6-phosphate dehydrogenase in the red blood cells and is induced by the alkaloidal glycosides, vicine and convicine and their hydrolytic derivatives divicine and isouramil in the alimentary canal (**see: Pharmaceuticals and pharmacologically active compounds**). The concentration of haemagglutinins (**lectins**) (**see: Phytohaemagglutinin**) is higher in fava bean than in other legumes. These substances are destroyed by heat during food preparation. Other detrimental factors include cyanogens, **phytin** and tannins.

5. Uses

To be used as a vegetable, beans are picked green, when they have reached full size, but before the pods dry. They are also used as a dry bean for food and livestock feed. In India and in the Andean regions of Bolivia, Ecuador and Peru, broad beans are eaten roasted. Fava bean is one of the most important winter crops for human consumption in the Middle East, Mediterranean region, China and Ethiopia, where it is a common breakfast food. The *V. faba* var. *minor* is the beck, tick or pigeon bean, much consumed as food in the Arabic world. The *V. faba* var. *equina* is mainly used for animal forage, especially for horses. The *V. faba* var. *major* also known as broad bean, Windsor or straight bean is mostly used for human consumption. The Indian varieties of the *V. faba* var. *paucijuga* are generally dried and eaten as **pulses**. The straw from bean plants is also used for brick making and as a fuel in parts of Sudan and Ethiopia.

6. World production

In 2002 the worldwide production of green and dried beans was about 5.3 million t, dry beans accounting for about 80%. The major producer was China followed by Algeria, Morocco, Egypt and Ethiopia. The export trade in dry and green beans in that year was valued at approx. US$164 million. The major exporters were Australia, China, France, Mexico, Spain and the UK (FAOSTAT). (OV-V)

Dawkins, T.C.K., Heath, M.C. and Lockwood, G. (eds) (1984) *Vicia faba:* agronomy, physiology and breeding. In: *World Crops, Production, Utilization and Description*, Vol. 10. Martinus Nijhoff, The Hague, The Netherlands.

Filippetti, A. and Ricciardi, L. (1993) Faba bean, *Vicia faba* L. In: Kalloo, G. and Bergh, B.O. (eds) *Genetic Improvement of Vegetable Crops*. Pergamon Press, Oxford, UK, pp. 353–385.

Hebblethwaite, P.D. (1983) *Faba Bean (*Vicia faba *L.)*. Butterworth-Heinemann, London, UK.

Muehlbauer, F.J. and Kaiser, W.J. (eds) (1994) *Expanding the Production and Use of Cool-season Food Legumes*. Kluwer Academic, Dordrecht, The Netherlands.

Summerfield, R.J. (ed.) (1988) *World Crops: Cool Season Food Legumes. A Global Perspective of the Problems and Prospects for Crop Improvement in Pea, Lentil, Faba bean and Chickpea*. Kluwer Academic, Dordrecht, The Netherlands.

Broad beans – cultivation

See: Broad bean for information on economic importance, origins, seeds, uses, etc. Faba (fava) beans (*Vicia faba*) can be grown as a dry harvest crop or as green vegetable crop. The seed types vary but mainly are elongated and flattened with a smooth seedcoat. Colours vary from green to brown but as the seeds age in store, they darken to a deep brown. Varieties have been developed with very low tannin content, making the seed more of a grey colour. Parent plants of tannin-free beans have white flowers whereas normally the varieties have a range of flower colours from light pink to blue and mixtures of white and blue petals.

The faba bean seed contains a high level of protein (legumin and vicilin), around 25–30%. (**See: Storage protein**)

1. Breeding

Vicia faba can be grown as a dry-harvested crop or for picking at the immature stage when the seed is eaten as a fresh vegetable. Varieties have also been selected for winter hardiness and have been developed for autumn or spring planting. Beans favour the cooler temperate areas of the world, as they are susceptible to moisture stress and high temperatures. Varieties have been selected for yield and seed size but also because the plants are mainly single stemmed, some degree of determinacy of height and earliness of harvesting is desirable.

Beans are susceptible to fungal **pathogens** especially *Peronospora viciae* which causes downy mildew, and European breeding efforts have concentrated on this aspect. Some effort has also been made to develop varieties with a white flower concomitantly reducing the amount of tannin in the seed. This compound has an antinutritive effect in some animals and therefore tannin-free beans can be used at a higher inclusion rate in animal feeding stuffs. Winter bean breeding is currently focused in the UK where this crop is a popular legume break crop. Spring varieties are bred mainly in Europe, although an increasing area is grown in Australia and China and variety development is taking place in these countries.

2. Development

The seed size of beans varies, from 800 g to 400 g per 1000 seed; the number of seeds per pod can range from two to eight – usually a stable character for a variety. However, the number of fertile flowers per plant is affected by environmental and climatic conditions. Although many flowers are self-fertile, pod set is increased in crops where pollinating insects, especially honey bees, are active. The number of flowers produced in a raceme at each reproductive node always exceeds the number of pods: a high proportion of young pods often abscise during development. Abscission of flowers and pods is dependent on a number of factors, but light deficiency seems to be the most important for flower and pod set, which occurs when the plants are growing densely and there is competition for light by the leaf canopy.

3. Production

Vicia faba seed is harvested when mature but before the seed moisture level falls below 14%, to avoid damage to the seed during the harvesting operation. However, in many seasons, particularly in Europe, the relatively late harvesting date can mean that seed moisture levels are above this and therefore in most years, seed drying is essential. Field beans have a large seed size and generally a thick testa. Bulk drying, which allows the diffusion of moisture from the centre of the seed to its surface, from where it can be removed with ventilating air, is a slow process compared to that with cereals.

For storage, the seed must be dried to 14%. Where heated air is used for forced air-ventilated drying, the temperature should not exceed 38°C or the germination of the seed can be affected or testa splitting can occur. Varieties with a very thin testa are more susceptible to this type of damage, and this can be exacerbated during any seed handling operations.

Bean seed is not often affected by dormancy and even the varieties of winter beans do not need a period of chilling before germination takes place.

4. Sowing and treatments

Winter beans are sown in late autumn to uncultivated stubbles and then inverted by ploughing. In the UK, this practice is now changing to a more direct seeding operation using wide-spaced drill coulters attached to deeply penetrating tines. Spring planting is usually made in previously ploughed and lightly cultivated soil.

The cotyledons remain below the ground after germination, and are degraded and consumed as a source of reserves to support autotropic growth: a hypogeal mode of seedling establishment (**see: Hypogeal**, Fig. H.9a).

Seed can be host to several seedborne pathogens which may affect emergence and establishment. The most common fungal pathogen is *Ascochyta fabae*, which causes a leaf and pod spot. Seed with significant levels of seed infection should be discarded although some control of the disease can be provided by a seed treatment with a systemic fungicide such as thiabendazole or fludioxinil. Seed may also be infested with stem and bulb nematode (*Ditylenchus dipsaci*), which causes severe growth damage to the plant and remains as a free-living nematode in the soil for many years. Seedborne viruses such as broad-bean stain virus and bean yellow mosaic virus do not usually affect seedling establishment although they remain as a risk to the developing crop. Seed treatments are not usually applied to field bean seed as the high tannin levels in the testa are fungistats and give some protection to the seedling from soil-borne fungi such as *Pythium ultimum*. (AB)

Dawkins, T.C.K., Heath, M.C. and Lockwood, G. (eds) (1984) *Vicia faba*: agronomy, physiology and breeding. In: *World Crops, Production, Utilization and Description*, Vol. 10. Martinus Nijhoff, The Hague, The Netherlands.

Filippetti, A. and Ricciardi, L. (1993). Faba bean, *Vicia faba* L. In: Kalloo, G. and Bergh, B.O. (eds) *Genetic Improvement of Vegetable Crops*. Pergamon Press, Oxford, UK, pp. 353–385.

Hebblethwaite, P.D. (1983) *Faba Bean (*Vicia faba L.*)*. Butterworth-Heinemann, London, UK.

Muehlbauer, F.J. and Kaiser, W.J. (eds) (1994) *Expanding the Production and Use of Cool-season Food Legumes*. Kluwer Academic, Dordrecht, The Netherlands.

Summerfield, R.J. (ed.) (1988) *World Crops: Cool Season Food Legumes. A Global Perspective of the Problems and Prospects for Crop Improvement in Pea, Lentil, Faba bean and Chickpea*. Kluwer Academic, Dordrecht, The Netherlands.

Broadcast seeding

The planting pattern resulting from the random scattering of seeds over the soil surface. For forage crops, for example, seed is spread uniformly over a firm, prepared seedbed; then it may be pressed into the seedbed surface with a corrugated roller. (**See: Planting equipment – agronomic requirements,** Fig. P.15.)

Broccoli

See: Brassica – horticultural

Brussels sprouts

See: Brassica – horticultural

Bucket elevators

Conveying equipment (sometimes known as 'legs') widely used in seed-conditioning plants to move seed to overhead storage or surge bins, designed to cause minimal mechanical damage to seed. (**See: Conveyors**)

Buckwheat

See: Pseudocereals

Bud pollination

A method, also known as bud selfing, used to maintain inbred self-incompatible parental lines in hybrid seed production based on **self-incompatibility systems (SI)**, notably in some vegetable **brassicas**. Pollen is placed by hand on an immature stigma that has not yet developed the incompatibility reaction, so avoiding the barrier to self-fertilization. The bud pollination process, however, tends to select for self-fertility over time, leading to the unwanted production of **sibs** (self-pollinating plants) in the eventual hybrid cross, which impairs varietal quality.

Bulb-to-seed production

The production of seed from bulbs, usually for biennial crops such as onion. The equivalent root-to-seed method is used for crops such as table beet, sugarbeet and carrot. (**See: Production for sowing, 1. Principles**)

Bumper mill

Also called Timothy bumper mill. Seed conditioning equipment specially designed to remove contaminant seeds from timothy grass (*Phleum pratense*), consisting of two slightly inclined decks that are knocked to impart an 'uphill' motion to the seeds to be separated. (**See: Conditioning, II. Cleaning; Pathogens,** *Tilletia*)

Bunts

Common bunt is a seedborne (or, rarely, soil-borne) fungal **disease** of cereal and grass inflorescences, affecting the quality of harvested cereal grain, which is caused by *Tilletia* spp. Spores adhere to the seed surface and then emerge during booting (the stage of growth of the cereal plant between stem elongation and ear emergence). Infected kernels are usually entirely transformed into 'stinking **smut**' – grey-brown bunt-balls about the same shape and size as normal kernels containing masses of fungus spores – a foul-smelling, dark-brown powdery dust, also called pepperbrand. In wheat affected with common bunt, for example, the ears are smaller with spreading **glumes** which tend to gape open, exposing the bunt balls, and stand more erect than healthy heads. Many bunt balls are crushed at harvest and the grain can be coloured grey by the masses of spores released; when infection is high, a dark smoke-like cloud can be released behind the harvester. The fungus imparts a characteristic fetid taste and smell of decaying fish (due to trimethylamine) to flour made from bunted wheat kernels, whose market value is discounted for this reason. Bunt infection is thus a key quality criterion during **grain inspection**. Effective seed treatments to control the seedborne **inoculum** have been available for the last 100 years: a wide range of contact and systemic fungicides are now used (**see: fungicides; Treatments – pesticides**), so that the disease is now generally at a very low level in most countries. Control measures also involve resistant varieties and crop rotation. (PH)

Butenolide

The butenolide 3-methyl-2H-furo[2,3-*c*]pyran-2-one is a compound present in plant- and cellulose-derived smoke which promotes germination, and is active at very low concentrations (< 1 ppb, 10^{-9} M). (**See: Germination – influences of smoke**)

C

C4 photosynthesis

C4 photosynthesis is thus named because the first products of carbon dioxide fixation are C4 organic acids. The carbon dioxide is later released for fixation by the C3 photosynthetic pathway where the first product is a C3 compound. Most plants fix carbon dioxide only by C3 photosynthesis. A key enzyme (RuBP carboxylase) in the conversion of carbon from CO_2 to sugar during C3 photosynthesis is relatively inefficient, and due to its activity up to half of the fixed carbon can be subsequently lost by photorespiration in C3 plants. Plants conducting only C3 photosynthesis have a low water use efficiency, i.e. a relatively large amount of water is lost from the plant per unit weight of dry matter gained.

The C4 type of photosynthesis is exhibited in a limited number of plants, but includes the tropical grasses such as **maize**, sugarcane, **sorghum** and **millet**. In these species the leaves have a distinctive anatomy that is related to a more efficient action of RuBP carboxylase, such that there is little or no photorespiration, and a higher water use efficiency.

(JDB, MB)

Cabbage

See: **Brassica – horticultural**

Cacao

Cacao seeds (*Theobroma cacao* L., Sterculiaceae) are the raw materials for cocoa, drinking and eating chocolate and cocoa butter.

1. The plant

The tropical, evergreen cacao trees are about 8 m high when fully grown, producing flowers directly on the woody main stem and branches. The trees enjoy relatively high temperatures (31–32°C average high, 18–21°C average low), plentiful rainfall and shaded conditions. The fruits, generally referred to as pods, are more or less ellipsoidal, 10–32 cm long, smooth or warty, some with five to ten furrows, white, red or green when unripe, and yellow, green or purple when ripe (Colour Plate 7B). Though originating in Central/South America the species is now cultivated in Africa, Melanesia and Southeast Asia.

There are three varieties. The source of what is considered the highest quality of processed cacao is 'criollo', of difficult cultivation and limited to Mexico, its neighbours and Venezuela. Now the major type, 'forastero' is grown widely in plantations of Africa, Asia and Brazil: within forastero is the sub-type, amelonado. 'Trinitario' yields a good quality product with less difficult cultivation: it is a cross between the previous two.

2. Origins and history

Theobroma cacao is a plant of the New World, of Central and South America. The seeds were used by the Mayans and Aztecs to make a beverage consumed by the priests and upper echelons of society (**see: Historical aspects of seeds in Mexico and Central America**). Archaeological studies at the Mayan site at Colha in northern Belize have revealed the presence of cacao residues in ancient ceramic vessels indicating that the preclassic Maya consumed chocolate as early as 600 BC. Such was the special place of cacao that it was thought of as food of the gods, a sentiment reflected in Linnaeus's choice of the generic name *Theobroma*. The specific name is derived from the Olmec and Mayan word *kawkaw*. These peoples called the beverage *cacahuatl*, and the Aztecs, *xocoatl*, which has been corrupted to give 'chocolate'.

In 1528, cacao was brought back to Spain by Cortés, the conqueror of the Aztecs, and over the subsequent years the chocolate beverage was modified from the Aztec original by the omission of chillis and **annatto** and their replacement with sugar and cinnamon, in which form it soon became extremely popular. Cacao plantations in Trinidad and Hispaniola were established by the Spaniards and in the late 17th century cacao was introduced by France to islands in the Caribbean. Somewhat later, the English and the Dutch set up plantations in their possessions in that region. In the 19th century cacao cultivation spread thoughout Africa, starting in Principe in 1822, and culminating in Nigeria in 1874 and Ghana in 1879. Cacao from Trinidad was planted in Sri Lanka firstly in 1834 from where it was later introduced into Singapore, Fiji, Samoa, Tanzania and Madagascar. It was established in Java in 1880. In the 18th and 19th centuries, techniques for treating cacao products developed in England, France and Switzerland and modern chocolate evolved.

3. Production

Cacao is cultivated in countries that lie between latitudes 20° north or south of the equator. West Africa is the major region of production followed by Southeast Asia (Table C.1). (**See: Crop Atlas Appendix, Map 20**) In 2002 world exports of beans were valued at close to US$4 billion, major exporters being, in descending order, Côte d'Ivoire, Indonesia, Ghana, Nigeria and Cameroon. The main importers, again in descending order, were The Netherlands, USA, Germany, France, UK and Belgium (FAOSTAT).

4. Seeds

Cacao seeds ('beans') ripen 5–6 months after pollination. Pods contain five rows of 20–60 whitish seeds lying in a mucilaginous pulp, each 2–4 cm long and 1–1.2 cm wide,

Table C.1. Cacao bean production (2003) (FAOSTAT).

Country	Production (thousands of tonnes)
Brazil	170
Cameroon	125
Côte d'Ivoire	1225
Ghana	475
Indonesia	426
Nigeria	380
World	3257

weighing approximately 1 g (Colour Plate 7C). The embryo (referred to in the cacao industry as the 'nib') consists of two irregularly folded, whitish-purple cotyledons and a relatively small axis and is surrounded by a firm seed coat.

The seeds are classed as **recalcitrant** (i.e. with respect to viability they are incapable of withstanding substantial water loss or cool temperatures) and have no dormancy.

5. Seed composition

The beans have been reported to contain approximately 13% **storage protein**, 48% **oil** (fat), and 35% total carbohydrate (fresh weight basis). The oil **fatty acids** include approximately 26% palmitic and shorter fatty acids (i.e. those with less than 16 C atoms), 34% stearic, 37% oleic and 2% linoleic acids (**see: Oils and fats; Oilseeds, major**). The high content of saturated fatty acids renders the extracted triacylglycerol solid at normal temperatures. B vitamins are present, with Vitamin C, **phenolic acids**, polyphenols and the methylxanthine alkaloids, theobromine (about 2.5%) and caffeine (about 0.8%) (**see: Pharmaceuticals and pharmacologically active compounds**). The bean composition is altered substantially during processing (see below). Theobromine is present in the seed coat also, together with about 13% protein, 3% triacylglycerol, 17% crude fibre, 9% tannins and 6% pentosans (all % coat fresh weight).

The edible pulp surrounding the seeds contains 8–13% glucose, nearly 1% sucrose, tartaric acid, some protein and mineral elements.

6. Processing

For the manufacture of cocoa, chocolate and cocoa butter, seeds and pulp are removed from the pod and together set to ferment in containers for up to 6–7 days, depending on the type of bean. Several microorganisms, mainly yeasts, participate in fermentation, producing alcohol, lactic and acetic acids. During this process the beans are turned for aeration which enhances bacterial activity and encourages a rise in temperature to about 45°C. The combination of acetic acid and the high temperature kills the seeds and provokes many chemical changes within them as well as in the pulp. In the seeds, these include breakdown of the embryo cell walls, the production of numerous volatile compounds and the development of chocolate colour and flavour. The beans are then dried either in the sun or artificially in containers to about 7.5% water content, and cleaned mechanically prior to sorting, grading (Colour Plate 7D) and export to processing plants. Here, the beans are roasted at 100–120°C for 45–70 min which releases water and acids and encourages further flavour development. The seed coats ('shells') are removed (**winnowing**) and can be used in animal feed, as a mulch, extracted to yield theobromine or pressed to give a lower-quality cocoa butter. The embryos ('nibs') are milled into the cocoa liquor which is later pressed to extract a quantity of cocoa butter, leaving the cocoa presscake which is pulverized into cocoa powder. Treatments such as alkalization occur at points during this process. Chocolate is made from cocoa liquor by the addition of cocoa butter and other ingredients such as milk and sugar depending on the manufacturer's requirements. White chocolate is made from the cocoa butter (Fig. C.1).

During processing numerous chemical changes occur in the beans. The final cocoa contains over 300 volatile compounds among which are the aliphatic esters, polyphenols, pyrazines, theobromine, aromatic carbonyls and diketopiperazines that contribute the unique flavour. Also present are caffeine, tyramine, dopamine, trigonelline, amino acids, proteins, tannins, fats and phospholipids.

7. Other species related to cacao

Three other species of *Theobroma*, *T. bicolor*, *T. grandiflorum* and *T. speciosum* are the sources of beverages and low-quality chocolate. The first, cultivated from southern Mexico to northern Brazil and Bolivia, produces beans called *pataxte*

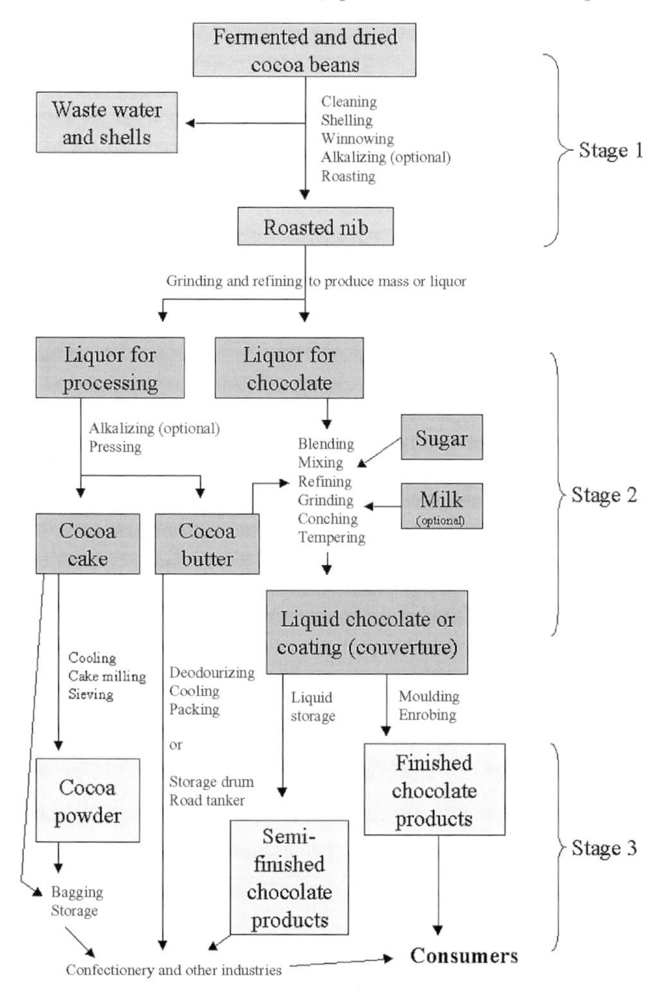

Fig. C.1. Processing of cacao (cocoa) beans. (With thanks to the International Cocoa Organization, www.icco.org/menuqa.htm)

from which a drink may be made. The second, *capuacu* from Brazil, is also the source of a beverage, made from the mucilage surrounding the seeds. And in Amazonia seeds of the last are used for chocolate and the pulp as a food. (MB)

Coe, S.D. and Coe, M.D. (1996) *The True History of Chocolate*. Thames and Hudson, London, UK.

Dand, R. (1993) *The International Cocoa Trade*. Woodhead Publishing, Cambridge, UK.

Kennedy, A.J. (1995) Cacao, *Theobroma cacao* (Sterculiaceae). In: Smartt, J. and Simmonds, N.W. (eds) *Evolution of Crop Plants*. Longman, London, pp. 472–475.

Wood, G.A.R. and Lass, R.A. (1985) *Cocoa*, 4th edn. Longman, London, UK.

Young, A.M. (1994) *The Chocolate Tree. A Natural History of Cacao*. Smithsonian Institution, Washington, DC, USA.

Calabrese

See: **Brassica – horticultural**

Calcineurin-B

A calcium- or calmodulin-binding protein generally associated with a calcineurin-A protein that on activation has phosphatase activity. Participates in signal transduction pathways. (See: **Mobilization of reserves – cereals**)

Callus

A mass of uniform plant cells that are the result of multiplication of parenchymatous (relatively unspecialized) cells in response to wounding or biochemical induction. Callus appears on wounds of many plants, sealing the injured surface and providing a mass of cells from which shoots may regenerate, as is the case in coppiced trees (those regenerated from young tree stumps). Callus was first controllably induced *in vitro* by R.J. Gautheret and P. Nobécourt in France and P.R. White in the USA, all three of whom independently came upon the callus-inducing power of **auxin** in 1938/39. Callus can be further induced to differentiate into other cell types (e.g. xylem vessels), organs, or **somatic embryos**. As a relatively uniform tissue type, callus is experimentally suited for many types of plant molecular biological and physiological studies. (PvA)

Gautheret, R.J. (1938) Sur le repiquage des cultures de tissu cambial de *Salix caprea*. *Comptes Rendues de L'academie de Sciences, Paris, Series III* 206, 125–127.

Nobécourt, P. (1938) Sur les proliferations sontanées de fragments de tubercules de carotte et leur culture sur milieu synthétique. *Bulletin de la Société Botanique Française* 85, 1–7.

White, P.R. (1939) Potentially unlimited growth of excised tomato root-tips in a liquid medium. *Plant Physiology* 9, 585–600.

Calmodulin

A calcium-binding protein that participates in signal transduction pathways.

(See: **Signal transduction – an overview**)

Campylotropous

Campylotropous (also acampylotropous or anacampylotropous) ovules or seeds occur (predominantly or as an exception) in more than 65 families belonging to both **dicotyledons** and **monocotyledons**. The ovules have a curved longitudinal axis resulting in a curved embryo because the nucellus and integument(s) develop more extensively on the antiraphal (raphe) side, which becomes (up to ten times) longer than the raphal side. Rarely, the converse development takes place in which the raphe of the anatropous ovule enlarges more than the **antiraphe** (some legumes and Vitaceae, grapes); such seeds are called 'obcampylotropous'. The primary advantage of campylotropy is that the embryo can reach up to twice the length of the seed and therefore give rise to a larger, more successfully competing seedling. (**See:** Fig. R.4 in **Reproductive structures, 1. Female**)

Canola™

See: **Brassica oilseeds; Brassica oilseeds – cultivation; Oilseeds – major**

Capping

See: **Crusting**

Capsule

A uni- or multi-celled, multi-seeded dehiscent or indehiscent fruit. (**See: Deserts and arid lands, 3. Seed dispersal strategies and mechanisms; Fruit – types; Orchids**)

Caraway

Caraway (*Carum carvi*, Apiaceae) 'seeds' (actually fruits) are used as a spice and flavour. (**See: Spices and flavours,** Table S.12) The plant is a glabrous herb, biennial but cultivated as an annual: there are also true annual forms. The dried fruits of commerce are generally separate, one-seeded, yellow-brown, tapered and ridged **mericarps**, each 4–6 mm long (Fig. C.2).

Seeds of the biennial contain 3–7% essential oil obtained by steam distillation: annual caraway has slightly less. The oil contains mainly two monoterpenes, D-carvone and D-limonene, and traces of pinene, camphor, caveol and others. The characteristic flavour and the biological properties of caraway are mainly due to D-carvone. (**See:** Fig. S.44 in **Spices and flavours**)

The seed has a distinct warm, slightly sweet, very sharp, somewhat acrid but pleasant taste with a pleasant aroma: hence

Fig. C.2. Caraway 'seeds' (image by Mike Amphlett, CABI).

it is used as culinary spice, in flavouring food, breads, meat products, salads and sauces. It is especially popular in some European countries (e.g. The Netherlands, Germany) and, after **coating** with sugar, has been a popular confectionery known for sweetening the breath (once called 'comfits' in England). Caraway oil is used for flavouring liquors, in perfumery and in soaps. In the pharmaceutical industry it is mostly employed as a carminative and stomachic. Carvone isolated from caraway seed is used for the treatment of cancer. The essential oil is spasmolytic, anti-microbial, and applied for dyspeptic complaints such as mild gastro-intestinal spasm, bloating and fullness and helps to alleviate bowel spasm: it has antibacterial and larvicidal properties. Caraway julep was taken for the relief of nervous indigestion, flatulence and was sometimes given in cases of hysteria. A combination of caraway with the other carminative herbs, anise and fennel, has been shown to be helpful in dealing with conditions of flatulence, especially in children. Its actions are analgesic, anaesthetic, anodyne, anti-anxiety, antibacterial, anti-parasitic, antiseptic, diuretic, mildly expectorant, fungicidal, muscle relaxant, soporific, stimulant, tonic, urinary antiseptic. Caraway strengthens the urinary organs, soothes irritation and expels stones. It is also well known that this oil can help with menstrual cramps, is good for the skin and decreases bruising, as well as increasing the appetite and relieving dyspepsia. Laryngitis, bronchitis and coughs are easy targets for the soothing vapours of caraway. This oil has traditionally been used as a remedy for colds and can also promote milk secretion. (KV, NB)

Guenther, E. (1978) *The Essential Oils*, Vol. 2. Robert E. Krieger Publishing, New York, USA.

Peter, K.V. (ed.) (2001) *Handbook of Herbs and Spices*. Woodhead Publishing, CRC Press, UK.

Toxopeus, H. and Lubberts, J.H. (1999) *Carum carvi* L. In: de Guzman, C.C. and Siemonsma, J.S. (eds) *Plant Resources of South-East Asia* No. 13. *Spices*. Backhuys Publishers, Leiden, The Netherlands, pp. 91–94.

Weiss, E.A. (2002) *Spice Crops*. CAB International, Wallingford, UK.

Carbohydrates

Carbohydrates are the class of compounds with the empirical formula $(CH_2O)_n$ in which n is 3 or greater. They include monomeric sugars (monosaccharides), oligosaccharides (their oligomers, usually composed of two to five monomeric sugars), and polysaccharides (polymers of many sugar monomers). Oligosaccharides and polysaccharides comprise two, several or many monomeric sugars covalently linked to each other by glycosidic bonds. Glycosidic bonds may have one of two orientations, termed α or β, and the glycosidic link may be formed from a carbon atom of one monomer through an oxygen atom to a particular carbon atom of the next monomer, depending on the compound: both greatly affect the properties of the resulting oligosaccharide or polysaccharide. Sugars may also be attached, singly or as groups, to other polymers, e.g. proteins, to form glycoproteins, and lipids, to form glycolipids.

Monosaccharides are single sugar molecules containing three to seven carbon atoms. Sugars fall into groups depending on the number of C atoms they contain: trioses (three-carbon sugars), pentoses (five-carbon sugars, e.g. arabinose, xylose and ribose) and hexoses (six-carbon sugars, including the common glucose and fructose, galactose and mannose, the sugar acids glucuronic acid and galacturonic acid, and the deoxysugars rhamnose and fucose). Free four-carbon (tetrose) and seven-carbon (heptose) sugars are rare. Many monosaccharides occur in modified (or derivatized) forms within cells, such as sugar phosphates (e.g. glucose-6-phosphate), amino sugars (e.g. N-acetyl glucosamine), sugar alcohols (e.g. *myo*-inositol and other cyclitols), deoxy sugars (e.g. deoxyribose, rhamnose and fucose) and sugar acids (e.g. glucuronic and galacturonic acid), where they serve as widespread and important metabolic products and intermediates, or as a component of polymeric structures. Sugars also may be attached to nucleotides as nucleotide diphosphate sugars during important synthetic reactions, e.g. ADPglucose in starch synthesis. Another terminology that is used to distinguish these sugars is to classify them as furanoses (sugars with a five-membered ring, four carbons and one oxygen), or as pyranoses (five carbons and one oxygen composing a six-membered ring). The ring structure that a sugar adopts is not defined by the number of its carbons, e.g. among the hexoses fructose is a furanose (two carbons external to the ring), whereas glucose is a pyranose (with one ring-external carbon). Monosaccharide molecules may have an aldehyde group (an aldose, e.g. glucose, ribose) or a ketone group (a ketose, e.g. fructose, ribulose), which form the glycosidic links.

The most common of the **oligosaccharides** is sucrose, which contains two glycosidic-linked sugars, glucose and fructose (a disaccharide). Other disaccharides include maltose, **galactinol** and cellobiose. Oligosaccharides with three sugars are trisaccharides, e.g. **raffinose**. Some oligosaccharides serve storage functions in cells (e.g. sucrose, raffinose), whereas others accumulate as breakdown products from polysaccharides (e.g. maltose from starch, cellobiose from cellulose).

Polysaccharides are polymers of monomeric sugars (a few hundred to thousands) covalently linked to form a long chain. The main 'backbone' chain is often branched due to the attachment of short (in **hemicellulose**) or long (**starch**) side chains. Starches consist of two major forms, straight-chain **amylose** and branched **amylopectin**.

Homopolymers contain one type of monomeric sugar, for example starch and cellulose, which both consist entirely of glucose molecules but with different glycosidic linkages (α or β, respectively). Heteropolymers consist of more than one type of sugar, e.g. **galactomannans**, which have a (mannan) backbone of linked mannose monomers, many of which may have single galactose side chains attached to them. Some polysaccharides have more than one sugar in their backbone, e.g. glucose and mannose in glucomannans and glucogalactomannans. Most isolated (oligo/poly)saccharide molecules have a **reducing sugar** group at one end and **non-reducing sugar** group(s) at the other(s), except that in some molecules (such as sucrose, the **raffinose series oligosaccharides** and **fructans**) all terminal monosaccharides are non-reducing (that is, they are all glycosidically linked through their reducing groups with their adjacent monosaccharides).

(**See** also: **Pharmaceuticals and pharmacologically active compounds**, **3. Glycosides**, and **4. Phenolics, volatile oils, and polysaccharides**) (JDB)

Cardamom - large

A seed spice commonly known as the large, black or Nepal cardamom (*Amomum subulatum*, syn. *Cardamomum subulatum*, Zingiberaceae), the plant is a tall, perennial shade-loving herb having a subterranean rhizome with aerial leafy shoots and spikes. The large cardamom of commerce is the grey-brown, red-brown dried capsule, approximately 25 mm long, containing 40–50 seeds (Colour Plate 3B, and **see: Spices and flavours,** Table S.12).

Approximately 14 cultivars are in popular use. They are propagated through rhizome cuttings as well as seeds and harvested only once a year. Main production areas are Nepal and Sikkim. The individual capsules are separated by hand and dried. Retention of the reddish colour of the capsules is a positive index of quality and sometimes fresh capsules are treated with dilute HCl (0.025%) to improve colour.

The fruit on average has 70% seeds and 30% skin and dry recovery of 30–35%. The chemical components vary with variety, region and age of the product. Volatile oil is the principal component responsible for the typical odour and can be obtained by steam distillation of crushed seeds. The major constituent of this essential oil is 1,8-cineole (65–80%) (Table C.2, and see **Spices and flavours,** Fig. S.44).

Large cardamom is used for flavouring various vegetables and meat preparations, food, confectionery, hot or sweet pickles and in beverages. Seed and powder are essential ingredients in mixed spice preparations and masala mixtures. The seeds are used as a substitute for small cardamom to sweeten food. The **essential oil** and **oleoresin** are used to flavour beverages, liquors and biscuits. The seeds also find application in the treatment of a wide range of medical conditions (e.g. indigestion, vomiting, biliousness, abdominal pains and rectal diseases, and infections of the teeth and gums). Aromatic oil from the seeds is applied to the eyes to allay inflammation. (**See: Cardamom – small**) (KVP, NB)

Anonymous (1985) *The Wealth of India*, Vol. I. Publications and Informations Directorate, New Delhi, pp. 227–229.

Guenther, E. (1978) *The Essential Oils*, Vol. 2. Robert E Krieger Publishing, New York, USA.

Table C.2. Composition of large cardamom essential oil.

Components	Percentage % (wt)
α-Pinene	1.1
β-Pinene	2.7
Myrcene	0.3
Limonene	2.9
1,8-Cineole	72.7
γ-Terpinene	0.4
p-Cymene	0.3
Terpinolene	tr
Terpinen-4-ol	4.7
δ-Terpineol	1.0
α-Terpineol	13.3
(E)-Nerolidol	0.5

Atta-ur-Rahman, M.I., Choudhary, A., Farooq, A., Ahmed, A., Demirci, B., Demirci, F. and Baser, K.H.C. (2000) Antifungal activities and essential oil constituents of some spices from Pakistan. *Journal of the Chemical Society of Pakistan* 22, 60–65.

Madhusoodanan, K.J. and Rao, Y.S. (2001) Cardamom (large). In: Peter, K.V. (ed.) *Handbook of Herbs and Spices*, Woodhead Publishing Co., Cambridge, UK, pp. 134–142.

Purseglove, J.W., Brown, E.G., Green, C.L. and Robbins S.R.J. (1981) *Spices*, Vol. 1. Tropical Agricuture Series, Longman, New York, USA.

Weiss, E.A. (2002) *Spice Crops*. CAB International, Wallingford, UK.

Cardamom - small

Small (or green) cardamom (*Elettaria cardamomum*, Zingiberaceae), 'The Queen of Spices', is the dried fruit of the rhizomatous, herbaceous perennial (**see: Spices and flavours,** Table S.12). The ovoid fruit is 1–2 cm long (Colour Plate 3C) containing approximately 15, 3 mm-long, hard, dark brown, sculptured seeds: the seeds constitute the spice element. Cardamom is naturally cross-pollinated and consists of three morphologically distinct types namely, Malabar, Mysore and Vazhukka. It is propagated both by seeds and by suckers.

The crop comes to harvest from the third year onwards. Capsules are harvested 3 months after flowering. The freshly picked green cardamoms are cleaned and cured using a flue-heating system, sometimes preceded by soaking in sodium carbonate to preserve the green colour. The green curing procedures used in India, Guatemala and Sri Lanka are basically similar but with some local modifications. In India, a proportion of the crop is bleached after sun-drying by exposing the spice to fumes from burning sulphur. These bleached cardamoms are particularly favoured in Iran.

The quality of cardamoms is assessed on the basis of their appearance (small olive-green capsules), the aroma/flavour character and the volatile oil content. The appearance of the spice and its organoleptic quality are regarded as closely related by many consumers. The stage of maturity of the fruits

Table C.3. Components of small cardamom essential oil.

Components	Total oil (%)
α-Pinene	1.5
β-Pinene	0.2
Sabinene	2.8
Myrcene	1.6
α-Phellandrene	0.2
Limonene	11.6
1,8-Cineole	36.3
γ-Terpinene	0.7
p-Cymene	0.1
Terpinolene	0.5
Linalool	3.0
Linalyl acetate	2.5
Terpinen-4-ol	0.9
α-Terpineol	2.6
α-Terpinyl acetate	31.3
Citronellol	0.3
Nerol	0.5
Geraniol	0.5
Methyl eugenols	0.2
Trans-nerolidol	2.7

Govindarajan, V.S., Narasimhan, S., Raghuveer, K.G. and Lewis, Y.S. (1982) Cardamom – Production Technology, Chemistry and Quality. *CRC Critical Reviews in Food Science and Technology* 16, 229–326.

at harvest influences the quality of the final product. Harvesting should be undertaken when the fruits are fully developed but still green in order to produce the dried, green form of the spice. Decorticated cardamom seed generally commands a disproportionately lower price. Seeds are classified by weight into 'prime' and 'light' seed, the former fetching a better price.

1. Constituents

Cardamom **oleoresin** is obtained by solvent extraction. Commercial oleoresins have been offered on the market with volatile oil contents ranging between 52 and 58%. The oleoresin is used for food flavouring.

The most significant component of cardamom as a spice is the volatile essential oil (3–8% fresh weight), extracted by steam distillation and described as sweet, aromatic, spicy and camphory (Table C.3, and **see: Spices and flavours,** Fig. S.44).

2. Uses and potential

Cardamom is used as an aromatic stimulant and flavouring agent in cooking. Cardamom oil is used in the food, perfumery, liquor and pharmaceutical industries as a flavour and a carminative. The oil is gaining increasing use in perfumery, with a trend to spicy tones modifying the dominant lavender group perfumes for women. Cardamom is used in indigenous medicine to treat urinary complaints, piles and jaundice, and for reducing body fat. The seeds are aromatic, acrid, sweet, cooling and warming; a digestive stimulant, carminative, aphrodisiac, digestive, stomachic, diuretic, cardiotonic, abortifacient, elexeteric, expectorant and tonic. They are used for asthma, bronchitis, haemorrhoids, strangury, renal and vesical calculi, halitosis, cardiac disorders, anorexia, dyspepsia, and several other medical conditions. Cardamom oil is used in several pharmaceutical preparations to check nausea, vomiting, headache, hepatic colic and as a cardiac stimulant, and an aid to digestion. (**See: Cardamom – large**) (KVP, NB)

Anonymous (1985) *The Wealth of India*, Vol. I. Publications and Informations Directorate, New Delhi, pp. 227–229.

Guenther, E. (1978) *The Essential Oils*, Vol. 2. Robert E. Krieger Publishing, New York.

Peter, K.V. (ed.) (2001) *Handbook of Herbs and Spices*. Woodhead Publishing, CRC Press, UK.

Purseglove, J.W., Brown, E.G., Green, C.L. and Robbins, S.R.J. (1981) *Spices*, Vol. 1. Tropical Agriculture Series, Longman, New York.

Ravindran, P.N. and Madhusoodanan, K.J. (2002) *Cardamom – The genus* Elettaria. Taylor and Francis, London.

Weiss, E.A. (2002) *Spice Crops*. CABI Publishing, Wallingford, UK.

Windholz, M. (1983) *The Merck Index: An Encyclopedia of Chemicals, Drugs and Biologicals*, 10th edn. Merck and Co., USA.

Carob

1. World importance and distribution

Carob (algarroba, locust bean, St John's bread) (*Ceratonia siliqua*) is a warm-climate legume seed produced mostly in the Mediterranean countries – Cyprus, Malta, Spain (the Atlantic coast and Majorca), Italy (Sicily, Sardinia, Adriatic coast), Portugal, Egypt and Turkey. It also occurs in Tunisia, Israel, Australia, South Africa, Mexico, Brazil and Chile and some states of the USA. (**See: Legumes**)

2. Origin

Carob originated in the Mediterranean region and western part of Asia. Charred remains of carob wood in Israel have been dated to 4000 BC and it is thought to have grown wild in Syria and the southern coast of Turkey in historic times. The pods were probably the proverbial 'locusts' eaten in the wilderness by St John the Baptist. The Greeks introduced it to Greece and Italy and the Arabs into Morocco and Spain. The carob was taken by Spanish missionaries to Mexico, southern California and South America, and by the British to South Africa, India and Australia.

3. Seed characteristics

There are up to 15 seeds inside the pod surrounded by a saccharine pulp but when the pod is fully ripe and dry, the seeds become loose. The ovate to oblong, flattish, hard seeds are about 9 mm long, 7 mm wide and 4 mm thick with hard, smooth, glossy, brown coats (Colour Plate 2E). Unlike most legume seed types they have an abundant and hard **endosperm**, amounting to 42–46% of the seed weight (Colour Plate 2F). The seeds are extremely uniform in weight and are thought to have been the original gauge for the 'carat' used by jewellers (**see: Weights and measures**).

4. Nutritional value

The seeds are rich in protein (approx. 52% fresh weight, fw), with about 8% **oil** and about 20% carbohydrate. The major proportion of the latter is **galactomannan** which makes up 30–40% of the endosperm. Carob pods (i.e. minus seeds) contain about 55% sugars (40% sucrose and the remainder, glucose, fructose and maltose), 8–10% storage protein and approx. 0.6% **oil**. Vitamins A and B and several important minerals (e.g. potassium and calcium) are present. The pods contain up to 1.5% fw tannins which can affect digestibility of protein. Carob tannins bind to (and thereby inactivate) toxins and inhibit growth of bacteria.

5. Uses

Food for both human and animal consumption is obtained from the seeds, the seed pods, and pulp.

The seeds are used to make locust bean gum, sometimes known as ceratonia gum or carob bean gum. This edible **gum** (galactomannan from the endosperm), called in the trade 'Tragasol', is a substitute for Gum Tragacanth (*Astragalus* sp.). The gum is an important commercial stabilizer and thickener in bakery goods, ice cream, salad dressings, sauces, cheese, salami, bologna, canned meat and fish, jelly, mustard, and other food products: it is used to prevent sugar crystallization and also as an egg substitute. The gum is also much employed in the manufacture of cosmetic face-packs, pharmaceutical products, detergents, paint, ink, shoe polish, adhesives, sizing for textiles, photographic paper, insecticides and match heads and in tanning. In rubber latex production, the gum is added to cause the solid to rise to the surface. It is also used for bonding paper pulp and thickening silkscreen pastes, and some derivatives are added to drilling mud. The seeds are ground to produce a protein-rich (up to 60%) flour that contains no starch or sugar and is ideal for diabetics.

In some countries roasted seeds are used a substitute for coffee or to mix with coffee. The sweet pulp of the pod, containing up to 50% sugars, can be eaten both green or dried, raw or ground into powder and used as a sweetener. Pods can be chewed as a sweetmeat. A fibre-rich flour made from the pods is utilized in breakfast foods or in confections. The extracted sugars of the pod (sucrose, glucose, fructose and maltose) can be utilized to make industrial alcohol.

The pods furnish a chocolate substitute after roasting and processing to a cocoa-like flour: this can be added to milk as a drink, or combined with wheat flour to make bread. This carob powder is free of the allergenic and addictive effects of caffeine and theobromine in **cocoa**. It contains much less fat than cocoa (7% vs 23%) and more sugars (ca. 45% vs 5%), and carob 'chocolate' is advertised as a dietary alternative to that made from cocoa. Carob flour is rich in **pectin**, and has no oxalic acid, which interferes with absorption of calcium. It contains magnesium, calcium, iron, phosphorus, potassium, manganese, barium, copper, nickel and the vitamins A, B and D.

Pods and seeds serve as cattle feed but because of the relatively high tannin content they should be only a minor proportion of the feed.

Carob also has some medicinal uses including the treatment of coughs and diarrohea.

6. Plant type

The carob is a slow-growing, multi-stemmed, medium-sized, warm-climate evergreen tree which reaches 15–17 m in height. The trees bear either male, female or hermaphrodite inflorescences (racemes). The pod is light to dark brown, 10–30 cm long, 1–2.2 cm wide, tough and fibrous, filled with soft, semi-translucent, pale-brown pulp, scant or plentiful. The production of fruit begins around the age of 15 years. A mature tree can produce over 180 kg of pods and seeds annually.

7. Marketing and economics

In 2002, nearly 214,000 t of carob pods were produced (FAOSTAT). About 35,000 t of seeds were used for gum production. Main exporters were Spain, Morocco and Portugal (approx. 36,000 t, 17,000 t and 12,000 t, respectively): total world exports were valued at just over US$55 million. Main importers were Italy (47,000 t) and Spain (19,000 t).

(OV-V)

Carlson, W.A. (1986) The carob: evaluation of trees, pods and kernels. *International Tree Crops Journal* 3, 281–290.

Tous, J.I. and Ferguson, L. (1996) Mediterranean fruits. In: Janick, J. (ed.) *Progress in New Crops*. ASHS Press, Alexandria, Virginia, USA, pp. 416–430.

www.ipgri.cgiar.org/publications/pdf/347.pdf

Carpel

The fertile leaf (**megasporophyll**) of the Spermatophyta (seed plants) within the flower that produces the **ovules**. There are several carpel (gynoecial) types, depending upon the extent to which they are fused. They include: apocarpous: separate carpels; semicarpous: ovaries partly fused, stigmas and styles separate; syncarpous: stigmas, styles and ovaries fused; unicarpellous: solitary free carpel. (**See: Reproductive structures, 3. Female**)

Carrot

See: Apiaceae (Umbelliferae)

Caruncle

A localized **aril**, the result of proliferation of the **seedcoat** (testa) near the hilum, e.g. in castor bean (*Ricinus communis*). Used synonymously with the term **strophiole**. (**See: Seedcoat – dispersal aids,** Colour Plate 15B)

Caryopsis

A term traditionally applied to the fruit of the grasses (Poaceae, [Gramineae]). It is a small indehiscent fruit having a single seed with such a thin, closely adherent **pericarp** (fruit coat) that a single unit, the grain, is formed. Similar to the **achene**, but the pericarp is not readily distinguishable from the seedcoat. (**See: Cereals; Fruit – type**)

Case-hardening

The situation in **conifers** where rapid drying causes insufficient expansion of cone scales, making it difficult to extract seeds. Moisture is trapped inside because the outer layers dry too quickly and 'collapse', or alternatively if cones are tightly packed in bags or containers. In seed processing, **pre-curing** of cones eases extraction: remoistening fruits, followed by a slower drying process. (**See: Tree seeds**)

Cashew

Cashew nuts (*Anacardium occidentale*, Anacardiaceae) are produced on a medium-sized, warm-climate tree native to South and Central America, probably originating in what is now northeast Brazil. It was introduced into India and Africa by the Portuguese. In 2003, approximately 2 million t were produced, the world's leading producers being Vietnam (32% of world production), India (23%) and Nigeria (9%) (FAOSTAT). (**See: Crop Atlas Appendix, Map 4**)

The 'nut'of commerce is the embryo (kernel) from the seed of a **drupe**, i.e. a fruit in which the inner tissue of the **pericarp**, the endocarp, forms a hard covering (shell) of the single seed. As the drupe develops, the stalk and receptacle on which it is borne swell to form a red or yellow, fleshy body called the 'apple': when mature the drupe, 2–4 cm long, lies at the apex of the 8–9 cm 'apple' (Colour Plate 4C, D). The shell around the 'nut' contains an oily liquid which has severe skin irritant and toxic components. Great caution must be exercised in extracting the kernel from the shell by hand and therefore the majority of drupes are processed by steaming and heating them in oil to remove the embryos. The kernels contain about 17% fresh weight **storage protein** and 48% **oil** of which the principal **fatty acid** is the monounsaturated oleic acid (**see: Nut,** Table N.2). Cashew nuts are consumed as snack food and are also used in confectionery, bakery and in certain food recipes. The 'apple' is used mainly locally for wine, fruit juices and jams.

(MB)

Castor bean

1. The crop

Also called castor oil plant, castor bean (*Ricinus communis*, 2n = 20, Euphorbiaceae) is a minor oil crop but because of its **fatty acid** composition the oil has important industrial applications. Castor seed oil is unique in that it contains some 88% of the total fatty acid content as ricinoleate (12-hydroxy-oleate). However, the presence of the highly toxic protein, ricin, and the allergenic storage proteins (**see: Storage proteins – intolerance and allergies**) are serious obstacles to growing and processing castor. (**See: Oils and fats – genetic modification; Oilseeds – major**)

India is the largest producer and exporter of castor accounting for 85% of world exports. Other significant producers are China and Brazil, with smaller amounts being produced by the former USSR and Thailand.

The centre of origin of the castor plant is thought to be eastern Africa, most likely Ethiopia.

2. Fruit and seed

After fertilization, the pistillate flowers develop into globular, usually spiny fruit with three carpels, each containing a seed. At maturity wild-type plants may split along the dorsal suture of the three carpels scattering the seeds. Commercial cultivars, however, retain their seed for weeks, even after frost, with little loss.

Castor seeds are large, weighing 3.0 to 3.5 g, laterally flattened with an oval shape and a prominent **caruncle** at the base. The shiny, brittle **seedcoat** makes up about 25% of the seed weight. Mottled coloured seeds are the most common (**see: Seedcoats – dispersal aids**, Colour Plate 15B), but white, brown, black and red seeds also occur. The pure white **endosperm** fills the seed enclosing the paper thin, fragile **cotyledons** of the **embryo**. A small hypocotyl and radicle are present at the caruncle end of the seed. The epidermal cells of the testa are pigmented, have a thick cuticle and show characteristic pitting. Spongy parenchyma, with some inner layers pigmented, lie below the epidermis. A 300–350 µm wide layer of curved, pitted fibres is a conspicuous feature of the testa. Vascular tissue that has developed from the outer epidermis of the inner **integument**, is present in the inner part of the testa. The parenchyma tissue of the endosperm contains **oil bodies** and **protein bodies** with **crystalloids** and **globoids** (**see: Storage protein – synthesis**). The histological features of the delicate embryo are essentially the same as the endosperm except for a developed vascular system.

Following germination in the soil, the cotyledons emerge and become photosynthetic - an epigeal mode of seedling establishment (**see: Epigeal**, Fig. E.8b).

3. Uses

Castor oil and the bio-based products derived from it are used for lubricants, paints, coatings, plastics and anti-fungals. The medicinal properties of the oil, particularly as a laxative, are also of long standing.

Although castor seed meal has a high **storage protein** content (20% in undecorticated seeds, 45% in decorticated seeds) it is unsuited as a feed due to the presence of the highly toxic protein **ricin**, and the alkaloid ricinine. Heat treatment can remove these toxins but the meal is almost always used as a fertilizer due to its high nitrogen content. (**See: Poisonous seeds**) (KD)

For information on the particular contributions of the seed to seed science **see: Research seed species – contributions to seed science.**

Weiss, E.A. (2000) *Oil Crops*. Blackwell Science, Oxford, UK.

Atsmon, D. (1989) Castor. In: Robbelen, G., Downey, R.K. and Ashri, A. (eds) *Oil Crops of the World*. McGraw-Hill, New York, USA, pp. 438–447.

Vaughan, J.G. (1970) *The Structure and Utilization of Oil Seeds*. Chapman and Hall, London, UK.

Cauliflower

See: Brassica – horticultural

Cavitation

In seed pathology, a disorder affecting pea seed development, also called **hollow heart**. It has an alternative meaning with respect to the movement of water in the xylem: cavitation occurs when air is introduced into the water column (also called embolism).

cDNA

Complementary DNA (cDNA) is a DNA molecule copied from a messenger RNA (mRNA) molecule by the enzyme reverse transcriptase; it therefore lacks the introns (non-coding regions) present in the gene for that message. Sequencing of a cDNA to obtain the order of its bases permits the amino acid sequence of the encoded protein to be determined. Expression of cDNAs in recombinant (bacterial) cells can be used to produce large quantities of their encoded proteins *in vitro*. (**See: Nucleic acids; Recombinant protein; Transcriptomics**)

Celery (celeriac)

See: Umbellifers

Cell cycle

Cell proliferation requires cell division; to accomplish this, cells must first increase in size, duplicate their chromosomes, and separate them exactly between the original and daughter cell. The cell division cycle is the fundamental means by which all living eukaryotic organisms increase their cell count during growth, and incorporates a period of DNA synthesis, mitosis (nuclear division) and **cytokinesis** (cell division), interspersed with Gaps, or rest periods, as illustrated in Fig. C.3. Cells generally spend the longest time of the cycle in the Gap 1 (G1) phase. In this state the cells are diploid (with a 2C DNA content), metabolically active and initially undergo growth; they are in the period of interphase when they do not

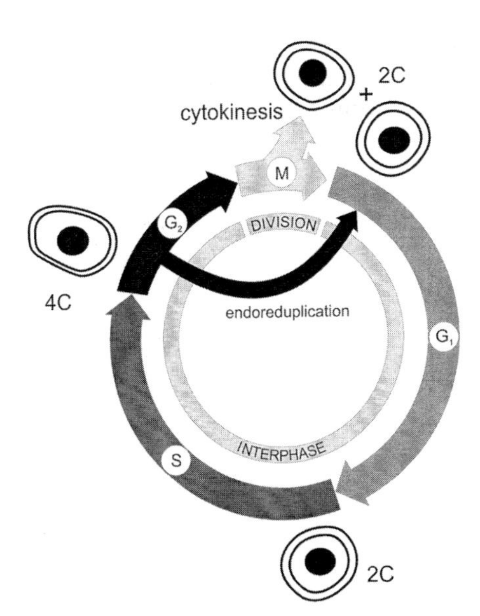

Fig. C.3. Phases of the eukaryotic cell cycle. During the Gap 1 (G1, 2C) and Gap 2 (G2, 4C) phases the cells can grow, in the early and late interphase stages of the cycle. DNA replication is confined to the part of the interphase called the S phase. Mitosis, the division of DNA and chromosomes (which condense and become visible), occurs during the M phase, followed immediately by cell division (cytokinesis). Endoreduplication of the DNA occurs without cell division, and the cells progressively increase from 4C to 8C, 16C, 32C, etc.

Based on Alberts, B., Bray, D., Lewis, J., Raff, M., Roberts, K. and Watson, J.D. (1994) *Molecular Biology of the Cell* 3rd edn. Garland Publishing, New York, USA, pp. 863–870.

replicate their DNA. Some cells exit G1 to enter a quiescent stage of the cell cycle called G0, where they remain metabolically active but no longer proliferate. For cell division to occur, duplication of DNA must occur to a transient 4C state, and this is initiated during the S (synthesis) phase of the cell cycle, the late interphase state. When DNA synthesis is completed, and prior to mitotic division of the cell, there is a second interval, Gap 2 (G2) which gives the cell time for growth and protein synthesis before division. **Mitosis** (M, from prophase to anaphase), the period of nuclear (DNA and chromosomal) division, followed by cellular division (cytokinesis), is then completed and two diploid daughter cells are formed, each with DNA in the 2C state. These cells then grow during the subsequent G1 phase as the cycle is repeated (**see: Germination**).

The state of cells in the embryos of mature dry seeds is frequently G1 or G0, i.e. DNA is in the 2C state, with cells containing DNA in the 4C state and higher in some species (Table C.4). The cells of living endosperms are triploid, and some have higher than the expected 3C nuclear DNA content due to **endoreduplication** of the DNA during development. Endoreduplication occurs when nuclei pass from the DNA duplication phase (S) to G2, and then directly to G1, omitting the mitotic (M) and cell division stages in the cycle. In the endosperms of mature cereal grains, endoreduplication during development may result in cells with DNA states from 3 to 96C, although higher values have been reported. These cells cease to undergo mitotic divisions early during endosperm development, and are committed to storage reserve synthesis

Table C.4. Nuclear DNA content of selected species of mature dry seeds expressed as C values.

	Nuclear DNA (%)									
Species and tissue	1C	2C	3C	4C	6C	8C	12C	16C	32C	64C
Lettuce (*Lactuca sativa*)										
radicle tip		100								
cotyledon		100								
endosperm			100							
Pepper (*Capsicum annuum*)										
embryo		100								
radicle tip		100								
endosperm			91		9					
French (common) bean (*Phaseolus vulgaris)*										
embryo		57		41		2				
radicle tip		34		55		11				
cotyledon				18		9		14	36	23
Spinach (*Spinacea oleracea*)										
embryo		61		32		7				
radicle tip		47		29		24				
Black, Austrian pine (*Pinus nigra*)										
embryo		100								
radicle tip		100								
megagametophyte	100									
Tomato (*Lycopersicon esculentum*)										
embryo		96		4						
radicle tip		89		11						
endosperm					75		25			

Data from Bino, R.J., Lanteri, S., Verhoeven, H.A. and Kraak, H.L. (1993) Flow cytometric determination of nuclear replication stages in seed tissues. *Annals of Botany* 72, 181–187.

(see: **Endosperm development – cereals**). Thus the onset of endoreduplication marks the end of the cell division phase. Endoreduplication can also occur in developing cotyledons of dicots, e.g. of legumes such as pea, some cultivars of which have DNA states of at least 64C. (**See: Priming – physiological mechanisms; Germination – metabolism**)

(JDB with input from Dr E. Śliwińska)

Cell tray

A container widely used in some horticultural and ornamental crop production systems for growing seedling **transplant** plugs (seeds sown in, or seedlings transplanted into, small volumes of growth media, e.g. soil, peat mixes, vermiculite). Various tray sizes are used with different numbers of cells in which the plugs are contained; these optimize floor space efficiency. Typical commercially available trays have about 70 to 200 cells (or more), each containing about 45 to 25 ml (or less) of the growing medium.

For plug production, one or more seeds are placed per cell, depending on the species and production system. Automated seeding equipment feeds the trays on to the line, fills the cells with growing medium, places seeds into them (for example, using vacuum nozzles), covers them to an appropriate depth, and waters them. Planting depth is an important consideration since large seeds tend to migrate to the surface during watering if planted too shallowly. A rule of thumb is to plant seeds at least twice as deep as their diameter. When multiple seedlings develop per plug, they may then be thinned to a single healthy and vigorous plant. (GEW)

Cell wall

Surrounding the cells of plants, the wall mechanically ensures their structural integrity. Primary cell walls are formed by growing cells and are composed of **cellulose, hemicelluloses, pectins** and proteins. Secondary cell walls may form following the completion of cell growth; the deposition of **lignin** and more cellulose to the inside of the primary wall adds to their rigidity and strength. Lignified secondary cell walls are present in seeds only within the **seedcoat** and in embryonic conducting tissue (xylem). In some seed tissues the cell wall is composed of hemicellulose and is a store of carbohydrate, although it can be, e.g. in cereal **endosperms**, that the hemicellulose-containing primary cell walls are structural. (JDB)

(See: Amyloid; Arabinogalactans; Arabinoxylans; Expansin; Extensin; Galactomannan; Glucan; Glucomannan; Mannans; Xyloglucan)

Carpita, N. and McCann, M. (2000) The cell wall. In: Buchanan, B.B., Gruissem, W. and Jones, R.L. (eds) *Biochemistry and Molecular Biology of Plants*. American Society of Plant Physiologists, Rockville, MD, USA, pp. 52–108.

Cells and cell components

Plant cells (Fig. C.4) are surrounded by a primary **cell wall** composed of cellulose fibres linked by hemicelluloses, and embedded in a pectin matrix. The pectin-rich middle lamella forms the interface between the primary cell walls of neighbouring cells. Intercellular connections occur through microscopic gaps in the wall, called plasmodesmata. Secondary cell walls are laid down to the inside of the primary wall, and may be composed of lignin, e.g. in seedcoats, or storage hemicelluloses such as galactomannans in endosperms.

To the inside of the cell wall is the plasma membrane (plasmalemma) which surrounds the cytoplasm (cytosol, protoplasm) in which lie many cellular inclusions, and where numerous metabolic reactions take place. The plasma membrane has many important functions. While water passes freely through the membrane, the passage of molecules from small ions to large proteins is restricted. Special transporters (with specificity for particular molecules or types of molecules), which are integral proteins, must be present to facilitate their movement in and out of the cell. Also present in the plasma membrane are enzymes, e.g. cellulose synthase, receptor proteins for externally derived signals, e.g. **hormones**, and trans-membrane proton transporters which set up electrochemical gradients across the membrane.

The nucleus is the site of most of the cell's genetic material, in chromatin (or in distinct chromosomes during cell division), made up of DNA complexed with protein. It is the control centre of the cell. The nucleus is surrounded by a membrane, the nuclear envelope, and the chromatin and

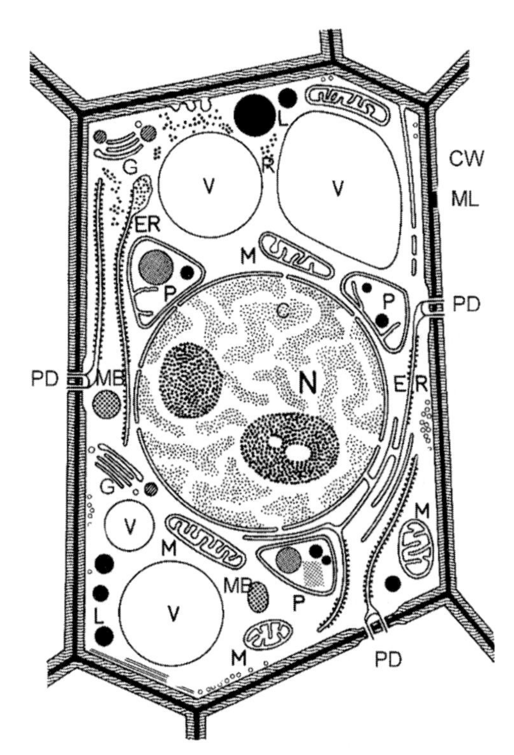

Fig. C.4. Stylized diagram of a seed cell to show its major components. CW: cell wall; ER: endoplasmic reticulum (the attached dots in certain regions represent the association of ribosomes in the rough ER; the regions without ribosomes are smooth ER); G: Golgi apparatus; L: lipid (oil) body; M: mitochondrion; MB: microbody; ML: middle lamella; N: nucleus, containing dense nucleoli and dispersed DNA as chromatin (c); P: plastid; PD: plasmodesmata; R: ribosomes, assembled as polyribosomes in the cytoplasm (free polyribosomes); V: vacuole. Not shown are structures present in some storage cells, such as protein bodies (protein storage vacuoles) and amyloplasts. Based on Mohr, H. and Schopfer, P. (1995) *Plant Physiology*. Springer, Berlin, pp. 21–37.

nucleolus (the site of ribosomal RNA synthesis) are contained within the nuclear matrix, the nucleoplasm.

The organelles within a cell play vital metabolic roles. **Mitochondria** house the respiratory machinery that generates **ATP** by way of the citric acid cycle (tricarboxylic acid or Krebs' cycle) and associated electron transport chain (**aerobic respiration**). Plastids of several types may be present. Proplastids are precursors of several plastid types, which include chloroplasts, the site of photosynthesis, which are present in some developing seeds but do not contribute effectively to their carbon economy. **Fatty acids**, a component of lipids (**oils**, fats) are synthesized in plastids (no specific name given) in seeds (this occurs in chloroplasts in leaves), and **starch** is synthesized and stored in **amyloplasts**.

Microbodies (glyoxysomes) participate in lipid mobilization in germinated seeds, wherein the fatty acids are broken down and converted to a precursor of sugars. These microbodies are converted to peroxisomes in greening cotyledons where they play a role in photorespiration.

An extensive and versatile network of membranes, the endoplasmic reticulum, ER (endomembrane system) is present throughout the cytoplasm of a cell. In some regions it is associated with the protein synthesizing complex, the polyribosomes (polysomes), to form rough ER, and in other regions it is devoid of these (smooth ER). It participates in many cellular functions, including the synthesis, processing and sorting of proteins destined for secretion and storage, and in the modification of fatty acids and synthesis of lipids. The **oil** (fat, lipid) **body** or oleosome is formed directly from the ER, as are some protein bodies. ER has conveniently been classified into three types: nuclear envelope, smooth ER, and rough ER (to which are attached ribosomes, the site of protein synthesis), but additional domains are now recognized for the binding to other cell components, including mitochondria and the plasma membrane.

The Golgi apparatus (Golgi body, trans-Golgi network) is a distinct, stacked membrane system which is vital for the transport of molecules within and from a cell. *De novo* synthesized proteins can be directed to the vacuole for storage or the plasma membrane for secretion, for example. The Golgi apparatus is also the site of synthesis for cell wall hemicelluloses and pectin.

The vacuole is a fluid-filled compartment which is separated from the cytoplasm by a tonoplast membrane, which is similar to the plasma membrane. In the compact cells of seeds there are initially many small vacuoles which, following germination, may coalesce into one or a few large vacuoles. They are the site of storage of many molecules, including organic acids, amino acids, **phenolics**, sugars, enzymes, storage proteins and **phytin**. During seed development, as storage proteins and phytin are sequestered within a large vacuole, it fragments to become many small **protein bodies** (protein storage vacuoles, PSVs). (JDB)

Cellulose

An insoluble (homo)polymer of glucose which is a vital structural component of many **cell walls**; for exceptions **see: Hemicellulose.** It exists in microfibrils, which are paracrystalline assemblies of 36 parallel cellulose chains of β-1,4-linked glucose units, 2–3 μm long. Microfibrils are embedded in a **pectin** matrix and held in place in the cell wall by cross-linking hemicelluloses, **xyloglucans** (dicots), and glucuronoarabinoxylans or mixed linkage **glucans** (monocots).

Cellobiose is a β-1,4-linked dimer of glucose, and cellotriose a trimer.

Cereals

1. Introduction

Cereals are food plants belonging to the grass family, Poaceae (previously Gramineae), and are primarily cultivated for their grains (strictly **caryopses**). Cereal grains provide about half the energy consumed by humans worldwide; more if animal feed is taken into account. **See: Crop Atlas Appendix, Map 25**. The grains are starchy, dry and relatively low in oils, allowing for easy storage. The term **millet** is used for small-seeded cereals. **Pseudocereals** are food plants with similarly starchy seeds but which belong to other plant families.

2. Types

Of the approx. 10,000 species of grass, perhaps 50 are cultivated as cereals, and of these fewer than 12 qualify as major crops (Table C.5). The grass family is divided by taxonomists into 40 tribes; within each tribe, genera often share morphological and ecological characteristics and food properties. In general, the Triticeae, Aveneae and Oryzeae account for the temperate cereals, which fail to thrive at high temperatures, while the other tribes contain tropical cereals (with the exception of **foxtail** and **proso** millet). Temperate cereals can be grown in the tropics, at high altitudes.

Selection and breeding has led to the production of a great number of cultivars of most of the cereals shown in Table C.5, to suit regional or local needs, of different composition and yield characteristics, disease resistance, etc.

3. Origins

Cereals were among the World's first crops (Table C.5): indeed, it is arguable that their use sustained the development of the various civilizations. In the Old World, **wheat** and **barley** originated and were domesticated by Neolithic peoples about 10,000 years ago in the 'Fertile Crescent' of the Near and Middle East. From these sites, the cereals spread throughout the Mediterranean and Eurasia, and shortly after, the other temperate cereals, oat and rye, came into use. Botany and archaeology place the origins of **rice** in China or India and there is evidence for domestication of the cereal in parts of China, India and Southeast Asia at times between about 6000 and 8500 years ago, although there are claims for this occurring up to 14,000 years ago in Korea. **Sorghum** and some of the millets originated in Africa where there is evidence of their use about 5000 years ago. **Maize** is thought to have originated in Mesoamerica 7000–10,000 years ago and there is evidence dating from 8000 years ago for its cultivation in ancient Central and South American civilizations. (**See: History of seeds in Mexico and Central America**)

4. Basic grain anatomy

The compound inflorescences of most grasses are made up of a number of **spikelets**, each consisting of one to several florets. A **floret** consists of two bracts (**palea** and **lemma**, the

Table C.5. The major cereals.

Tribe	Cereal	Domestication centre	Domestication time (est.)
Triticeae	**Wheat** (*Triticum* spp.)		
	Barley (*Hordeum* spp.)	Near East	10,000–11,000 ya
	Rye (*Secale cereale*)		
	Triticale	NA	NA
Oryzeae	**Rice** (*Oryza sativa*)	South China	8000 ya
	Red rice (*O. glaberrima*)	Africa (tropical)	?
	Wild rice (*Zizania palustris*)	North America	?
Aveneae	**Oat** (*Avena sativa*)	Near East (?)	3000–4000 ya?
Eragrostideae	**Finger millet** (*Eleusine coracana*)	East Africa	5000 ya (?)
	Teff (*Eragrostis tef*)	Ethiopia	?
Paniceae	**Foxtail millet** (*Setaria italica*)	Central Asia	7000 ya
	Proso millet (*Panicum mileaceum*)	Central Asia	7000 ya
	Pearl millet (*Pennisetum glaucum*)	West Africa	4000 ya (?)
	White fonio (*Digitaria exilis*)	West Africa	?
	Black fonio (*D. iburua*)	West Africa	?
Andropogoneae	**Maize** (*Zea mays*)	South Mexico	8000 ya
	Sorghum (*Sorghum bicolor*)	African savannah	?
	Job's tears (adlay) (*Coix lacryma-jobi*)	India?	?

Note: Cereals in bold lettering have separate entries. There are several other relatively minor or 'local' millets mentioned under **millets**. NA; not applicable, since this is a recent synthetic hybrid. ya: years ago.

latter in some cases extended into an **awn**), an ovary with two stigmas and three (in some cases, six) stamens. The base of the spikelet is subtended by other bracts, the **glumes**. Some grasses, such as maize and wild rice, have unisexual flowers located on different parts of the plant. After pollination and fertilization the ovary develops into the fruit enclosing the fertilized ovule, the seed. The fruit of a grass is a caryopsis: the single seed accounts for the greater part of the entire fruit when mature. The seed comprises the **embryonic axis, scutellum, endosperm, nucellus** and testa or **seedcoat**, and it is surrounded by the fruit coat or **pericarp**. The basic structural form of cereal caryopses is surprisingly consistent, to the extent that a generalized cereal grain can be described (Fig. C.5).

The embryonic axis has the potential to grow into a plant of the next (filial) generation. It is connected to and couched in the shield-like scutellum, which lies between it and the endosperm (**see: Embryo**, Fig. E.2). The scutellum behaves as a secretory and absorptive organ, serving the nutritional requirements of the embryonic axis. When germination occurs, exchange of water and solutes between the scutellum and starchy endosperm is extremely rapid, with secretion of hormones and enzymes and absorption of solubilized nutrients occurring across this boundary following germination. (**See: Mobilization of reserves – a monocot overview**)

The embryonic axis is well supplied with conducting tissues of a simple type and some are also present in the scutellum. The endosperm is of the same generation as the embryo and it comprises two clearly distinguished components: starchy endosperm and **aleurone layer** (sometimes referred to solely as aleurone). Starchy endosperm forms a central mass consisting of cells packed with nutrients, predominantly **starch** but also **storage protein**, that can be mobilized to support growth of the embryonic axis following germination. In all cereals there is an inverse gradient involving these two components, the protein percentage per unit mass of endosperm tissue increasing towards the periphery. Cell size also diminishes towards the outside and this is accompanied by increasing cell wall thickness. The walls of the starchy endosperm of wheat are composed mainly of arabinoxylans, while in barley and oats (1→3) and (1→4) β-D glucans predominate. Cellulose contributes little to cereal endosperm walls except in the case of rice. In most cases the

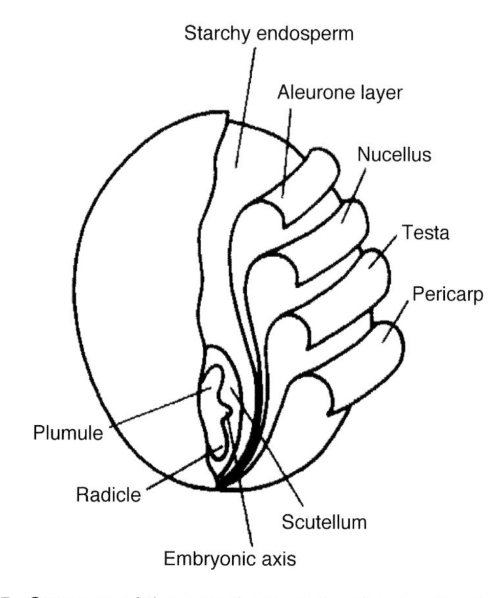

Fig. C.5. Structure of the cereal caryopsis, showing the tissue types that are present.

starchy endosperm cells are dead at seed maturity as the vital contents (membranes, nucleus, etc.) become disrupted by the accumulating reserves.

Surrounding the starchy endosperm cells, the aleurone layer consists of thick-walled cells with dense contents and prominent nuclei. Wheat, rye, oats, maize, rice and sorghum have an aleurone layer that is one cell thick (in the region adjacent to the conducting tissue the number of cell layers may be greater; in rice it can be to up to six) and in barley it is three cells thick. **Aleurone layer** cells contain no starch but they have high protein content, they are rich in oil, and contain **phytin**. When the seed germinates, hydrolytic enzymes responsible for solubilizing the reserves are synthesized in the aleurone layer cells of most cereal species. The balance between aleurone layer and scutellum in this role varies among species. A further function of some aleurone layer cells is the transfer of metabolites into the starchy endosperm during grain maturation. Though the starchy endosperm is non-living at maturity, the aleurone layer cells remain alive until well after the seed has germinated and is involved in reserve mobilization. (**See: Cereals – composition and nutritional quality**)

The nucellar epidemis, all that remains of the nucellus, surrounds the endosperm and embryo. In most cereals a cuticle is present on its outer surface. The nucellus, like all the tissues lying outside it, belongs to the current or parental generation.

The outermost tissue of the seed proper is the testa or seedcoat. It may consist of one or two cellular layers. In some varieties of sorghum a testa is absent altogether. Where two layers are present the long axes of their elongated cells lie at approximately 90° to each other. Frequently the cells of the testa accumulate corky substances during grain ripening and this may confer colour on the grain (e.g. red) and it reduces the permeability of the testa. A cuticle, thicker than that of the nucellar epidermis, is typical, and this also plays a role in regulating water and gaseous exchange.

The pericarp, or fruit coat, consists of several complete and incomplete layers. In all mature cereal grains, most cells in all pericarp tissues are dry and empty. With the seedcoat(s) (testa and nucellar epidermis) it protects the endosperm and embryo within against physical damage and fungal and insect attack. In certain types of sorghum some cells containing **starch granules** persist to maturity in the pericarp, but in most species both the granules and the cells in which they occurred during development are digested before maturity.

The outer epidermis has a cuticle that controls water relations in developing grains but this generally becomes leaky on drying. Hairs (**trichomes**) are present at the non-embryo end of wheat, rye, barley, **triticale** and oats. They are collectively known as the 'brush' and they have a high silicon content.

A summary of the presence or absence of individual covering layers showing the variation among species is given in Table C.6.

In some cereal grains, such as oats, barley and rice, the lemma and palea are usually closely adherent and are thus not removed by threshing; in other cereals (millets, oats) the lemma and palea are not adherent but are very rigid and do not fall off during threshing. These **hulls** (husks) are therefore part of the grain as traded, and their additional contribution to grain mass has to be borne in mind when comparing the relative proportions of nutrients in different species (Table C.7).

Cereals are attractive and important crops, particularly because of their large endosperm with its energy-rich storage reserves. The dry seed can be easily stored for long periods. The cereals have a naturally high fruit set so that yield potentials are favourable. Another important feature is that cereals are non-**shattering**, i.e. the mature fruits are not dispersed from the plants. This presumably was a tendency of the early domesticated types that was increasingly selected, together with the potential towards high yields. (**See: Domestication; Yield potential**)

5. Production and economic importance

Three species accounted for 85% of the world's cereals in 2001: maize, rice and wheat (Table C.8). Since 1960, the

Table C.6. Tissue layers present in grains of different cereal caryopses.

	Fruit coats							
Cereal	Epidermis	Hypodermis	Intermediate layer	Cross cell layer[a]	Tube cell layer	Testa[a]	Nucellar epidermis	Aleurone layer[a]
Barley	+	+		2	+	1	+	2–4[d]
Maize	+	+		+	+	[b]	[b]	1
Proso millet	+	+		+	+	[b]	[b]	1
Oat	+	+				1	+	1
Rice	+	+		+	+	1	+	1–3[d]
Rye	+	+	+	+	+	2	+	1
Sorghum	+	+		+[c]	+	0	+	1
Triticale	+	+	+	+	+	2	+	1
Wheat	+	+	+	+	+	2	+	1

[a]Numbers indicate cell layers.
[b]A cuticular skin persists to maturity.
[c]Incomplete layer.
[d]In barley the variation is varietal, while for rice it occurs within a single grain. In rice, aleurone layers are more abundant in lowland than upland rice and generally less in *Indica* than *Japonica* type, but the number can increase in *Japonica* if night temperatures are high during histodifferentiation. (**See: Development of seeds – an overview**)

Table C.7. Grain weights and typical proportions (%) of grain parts in some cereals.

Cereal	Grain weight (mg)	Hull	Pericarp and testa	Aleurone layer	Starchy endosperm	Embryo	
						Embryonic axis	Scutellum
Naked grains							
Wheat	27–50	–	8.5	6.7	82	1.3	1.5
Maize	150–600	–	6.0	2.7	77.8	1.5	12.0
Rye	15–40	–	10.0		86.5	1.8	1.7
Sorghum	8–50	–	7.9		82.3	9.8	
Proso millet	n/a	16.0	3.0	6.0	70.0	5.0	
Hulled grains							
Rice	n/a	20.0	4.8 (6.0)		72.7 (90.9)	1.0 (1.2)	1.5 (1.9)
Barley	32–36	13.0	2.9 (3.3)	4.8 (5.5)	76.3 (87.0)	1.7 (1.9)	1.3 (1.5)
Oats	n/a	25.0	9.0 (12.0)		63.2 (84.0)	1.2 (1.6)	1.6 (2.1)

Where cells are merged the number relates to the total value for the two tissues. Values for individual tissues were not found.
Values in parenthesis are proportions of grain excluding hull.
Based on information in Pomeranz, Y. (1998) Chemical composition of kernel structure. In: Pomeranz, Y. (ed.) *Wheat Chemistry and Technology*. American Association of Cereal Chemists, St Paul, MN, USA.

Table C.8. World production of cereals in 2001.

Cereal	Area harvested (million ha)	Yield (hg/ha)	World production	World production (%)	Export trade	Animal feed	Main countries of cultivation, with annual production
Barley	56.2	25.7	144.1	6.8	27.4	90.1	Russian Federation (19.5), Germany (13.5), Canada (10.9), Ukraine (10.2)
Fonio	0.355	7.2	0.257	0.01	–	–	Guinea (0.13), Mali (0.02)
Maize	139.1	44.2	614.5	29.2	81.6	407.0	USA (241.5), China (114.3), Brazil (41.4), Mexico (20.1)
Millets	37.0	7.9	29.1	1.4	0.249	2.6	India (11.4), Nigeria (5.5), Niger (2.4), China (2.0)
Oats	13.2	20.6	27.2	1.3	3.3	19.7	Russian Federation (7.7), Canada (2.7), USA (1.7)
Rice, paddy	151.2	39.5	597.8	28.4	41.2	10.2	China (179.3), India (139.7), Indonesia (50.5), Bangladesh (36.3)
Rye	9.9	23.6	23.3	1.1	1.6	10.9	Russian Federation (6.6), Germany (5.1), Poland (4.9)
Sorghum	44.2	13.5	59.5	2.8	7.0	29.4	USA (13.1), India (7.8), Nigeria (7.1)
Triticale	2.9	35.9	10.4	0.5	–	–	Germany (3.4), Poland (2.7), France (1.1)
Wheat	214.8	27.5	590.5	28.0	134.0	102.4	China (93.9), India (69.7), USA (53.3), Russian Federation (47.0)
Other	–	–	10.2	0.5	–	–	–
TOTAL	676.6	–	2106.9	–	–	–	–

Source: FAOSTAT (http://apps.fao.org/). Figures are in millions of tonnes unless otherwise stated.

'Green Revolution' combination of plant breeding and higher inputs of irrigation and fertilizers has enabled world production to keep pace with population; overall production of these three cereals has nearly tripled (from 640 million t in 1961 to 1800 million t in 2001), with only a 20% increase in the area cultivated. Production of these major cereals is linked to complex patterns of government subsidies and world trade. Minor cereals, such as the African millets, are being rapidly replaced by the major cereals, particularly maize, in tropical countries. However, minor cereals are now recognized as being of local importance, particularly because of their adaptation to poor, dry soils, and plant-breeding efforts are increasing in this area.

Production and consumption of cereals within individual countries and continents are not matched and movements of cereal grains contribute significantly to world trade. In 2001 the value of the world trade in cereals was approximately US$ 40 billion. Trade overall, and in the case of individual countries, is influenced by commodity price, which can be influenced by supply and demand. Annual surpluses of cereals might be stored for following seasons because harvests are subject to variation due to climatic conditions and decisions by growers as to the area planted. Also, trading is highly sophisticated and importing nations source their supplies according to supply and price at the time of need. Consequently, trading patterns at any time give only a poor indication of events before and after the time of reporting. Table C.9 reflects the position for those cereals that contribute in a major way to world trade during the 2002/3 season. (**See: Crop Atlas Appendix, Maps 1, 5, 6, 7, 8, 9, 10, 11, 12, 21, 23, 24**).

6. Grain quality

'Quality' in the general sense means 'suitability for a particular purpose'. As applied to cereals, the criteria of quality may be described in terms of: (i) *yield* (of grain for the grower; of high value components for the primary processor, and of final product for the end user); (ii) *ease of processing*; nature of the resulting end product: uniformity, appearance, chemical composition, and in the case of food and feed products: wholesomeness and palatability.

These criteria of quality are dependent upon the variety of cereal grown and upon environment, climate, soil and manure or fertilizer treatment. Manner of harvesting and storage also exert a profound influence on grain quality. Grain exposed to wet conditions before harvest may exhibit signs of germination being completed, such as high content of hydrolytic enzymes or, in extreme cases, sprouting of shoot and roots from the embryo. High enzyme production is undesirable, particularly for breadmaking. The enzyme of most concern is α-amylase, because conversion of starch to sugars in the dough gives rise to heavy, sticky bread, which is difficult to slice. (**See: Pre-harvest sprouting**)

Drying the grain before or during storage can also cause deterioration. Safe drying temperature declines at higher moisture contents. Use of excessive temperatures reduces germinability and denatures proteins important for subsequent processing. Some of the hazards encountered during storage are discussed in **Cereals – storage** and **Storage Management**.

Because of the many uses to which cereals are put, there are many different perceptions of quality, for example, in the case of wheat, a baker and his customers are likely to be interested in baking properties while a stock feeder will be concerned with nutritional properties. To a minority of end users who demand organically grown crops the manner of crop production and storage is the major quality requirement. Others may impose equally strong conditions for exclusion of products from genetically modified plants.

For marketing purposes it has been customary for many years to grade cereal samples according to certain criteria but the number of these is, of necessity, restricted and incapable of covering requirements for all end-uses. Most grading systems employ a system of segregation by class as well as by grade. Distinction of wheat by class may be based on grain colour, endosperm texture and growing season, as in the USA grading system for common wheat (classes are defined as hard red spring, soft white, etc. and must include only cultivars accepted within the respective categories), or grain length, as in the USA rice grading system. In general, achievement of a grade depends upon conformity to the description of the class, freedom from contamination by damaged, shrivelled, sprouted and diseased grains, fruits (seeds) of non-conforming plants, soil and other foreign matter. Within a class, grades may be defined by protein content and quality, and bulk density. Although fairly consistent, the grade limits may be changed from year to year, according to the quality of the harvest.

Competition among cereals merchants as well as exporting countries has led to a move away from rigid classification, towards greater flexibility aimed at meeting precise requirements of purchasers. For many end-uses, measurable criteria cannot fully define requirements so specification of variety required or to be excluded is specified in purchase orders.

7. Primary processing

For humans and most other animals to digest the nutritional parts of cereals, grains are best initially ground. Often, a further cooking stage is required before digestion can be achieved. Most types of primary processing involve reduction in particle size. Some include chemical treatment (e.g. maize).

Table C.9. Weights of major cereal grains traded in 2002/3 season (tonnes × 1000).

Wheat		Maize		Rice (2002 calendar year)		Barley		Sorghum	
Total World Trade 107,800		Total World Trade 77,960		Total World Trade 27,880		Total World Trade 16,090		Total World Trade 5830	
				Exporters					
USA	22,970	USA	41,000	Thailand	7,245	EU	5,000	USA	4,900
EU	16,000	China	15,240	India	6,650	Russia	3,200	Argentina	600
Russia	12,620	Argentina	12,500	USA	3,295	Ukraine	2,300	Sudan	190
Australia	10,950	Brazil	3,200	Vietnam	3,245	Australia	2,200		
Canada	9,390	Hungary	1,500	China	1,960	Turkey	700		
				Importers					
EU	12,000	Japan	16,500	Indonesia	3,500	Saudi Arabia	6,000	Mexico	3,400
Brazil	6,630	S Korea	8,790	Nigeria	1,820	China	1,790	Japan	1,500
Egypt	6,300	Mexico	5,500	Philippines	1,250	Japan	1,300	EU	400
Algeria	6,000	Egypt	5,000	Iraq	1,180	EU	700		
Japan	5,580	Taiwan	4,800	Saudi Arabia	940	Jordan	500		

Values ex *World Grain* sourcing from International Grains Council London, and United States, Department of Agriculture's Foreign Agricultural Service. Washington, D.C.

Cereal grains are harvested and may be stored, after which they are subjected to cleaning and processing. The products may be further processed; for example, wheat flour may be processed into bread or other baked products, but in this Encyclopedia we confine consideration to the first or primary processing and this usually involves treatments described as milling. Although the single term is applied, it does not necessarily involve the same treatment or series of treatments. For example, wheat is milled into a fine powder while rice milling results in whole endosperms with the outer tissues removed. Milling carried out in the presence of more water than can be absorbed by the grains is described as wet milling. Dry milling may involve the addition of water to grain to improve its behaviour during grinding but the amount of water added is well below that which can be absorbed by the grain tissues. One type of primary processing that involves water but no milling is **malting**, in which water is added to initiate processes associated with germination.

Because there are many different processes described as milling, it is best to consider the different types of milling separately. (AE, MN)

(**See** the individual entries for the following cereals: **Barley, Einkorn, Emmer, Finger Millet, Foxtail Millet, Job's Tears (Adlay), Maize, Oats, Pearl Millet, Proso or Common Millet, Rice, Rye, Spelt, Teff, Triticale, Wheat, Wild rice; See: Cereals – composition and nutritional quality; Cereals – storage**)

Campbell, G.M., Webb, C. and McKee, S.L. (eds) (1997) *Cereals: Novel Uses and Processes.* Kluwer Academic, Dordrecht, The Netherlands.

Dendy, D.A.V. and Dobraszczyk, B.J. (eds) (2000) *Cereals and Cereal Products: Chemistry and Technology.* C.H.I.P.S., Texas, USA.

Grubben, G.J.H. and Soetjipto Partohardjono (eds) (1996) *Plant Resources of South-East Asia No 10 Cereals.* Backhuys Publishers, Leiden, The Netherlands.

Kent, N.L. and Evers, A.D. (1994)*Technology of Cereals* 4th edn. Pergamon Press, Oxford, UK.

Nesbitt, M. (2005) Grains. In: Prance, G. and Nesbitt, M. (eds) *Cultural History of Plants.* Routledge, London, UK, pp. 45–60.

Owens, G. (ed.) (2001) *Cereals Processing Technology.* C.H.I.P.S., Texas, USA.

Vaughan, J.G. and Geissler, C.A. (1997) *The New Oxford Book of Food Plants.* Oxford University Press, Oxford, UK.

Zohary, D. and Hopf, M. (1993) *Domestication of Plants in the Old World, the Origin and Spread of Cultivated Plants in West Asia, Europe, and the Nile Valley*, 2nd edn. Clarendon Press, Oxford, UK.

Cereals – composition and nutritional quality

1. Composition and quality

Composition of grains in individual species of cereals can vary and also values for the same constituent are to some extent dependent on the analytical method used. Hence there are no absolute values for composition. The values for the main constituents of grains thus have only a comparative significance. Table C.10 is provided for this purpose.

The main contribution that all cereals make to diet is the carbohydrate **starch**, which can account for more than 70% of total dry mass of grains. It is a major source of energy. For human nutrition it is now widely recognized that, where choice is available, cereal products, vegetables (including potatoes) fruit and pulses, should provide approximately 55–60% of a healthy diet, and complex carbohydrates (starch) should contribute 40–50% of the total kilojoules consumed.

Different chemical components of cereal grains are distributed differentially. Most of the protein, for example, is in the **endosperm**, but some is in the **embryo** and the **aleurone layer** tissue. Moreover, the parts differ as to the types of protein they contain. Prolamins, glutelins and albumins are in most cases found in the starchy endosperm, and globulins in the embryo and aleurone layer (**see: Osborne fractions; Storage proteins**). Since these proteins differ in their nutritional value the nutritional qualities of the different parts must also differ. The prolamins, which in most cases are the dominant endosperm proteins, are deficient in lysine, threonine and tryptophan and some other **essential amino acids** (see below). Although the aleurone layer and embryo are proportionately rich in proteins, especially globulins, compared with the starchy endosperm, these tissues have little impact on overall nutritional properties of the grain since in most cereals prepared for human consumption the aleurone layer and embryo account only for a small fraction of grain mass and in any case are usually removed during grain

Table C.10. Approximate analysis of cereal grains (and products if data for grains are unavailable).

Cereal grain or product	Moisture	Protein	Oil	Carbohydrate*	Fibre†
Barley – whole grain	11.7	10.6	2.1	64.0	14.8[a]
– pearl, raw	10.6	7.9	1.7	83.6	5.9[b]
Foxtail Millet – flour	13.3	5.8	1.7	75.4	
Maize – whole	15.5	11 0	5 0	65 0	
Oats – oatmeal raw	8.9	12.4	8.7	72.8	6.3[a] 6.8[b]
Rice – brown	13.9	6.7	2.8	81.3	3.8[a] 1.9[b]
– white	11.4	7.3	3.6	85.8	2.7[a] 0.4[b]
Rye flour – whole	15.0	8.2	2.0	75.9	11.7[b]
Sorghum	15.5	11.2	3.7	74.1	2.6[c]
Triticale	10.5	13.05	2.09	72.13	2.6[c]
Wheat – wholemeal	14.0	12.7	2.2	63.9	8.6[a] 9.0[b]

*Mostly starch.
†Superscript letters indicate different methods of determination.
Values are per cent fresh weight (in the cases of whole grains).

processing. The maize embryo, on the other hand, accounts for 10–11% of the grain and so its nutritional contribution, say in whole-grain feed, is more significant. Starch is almost completely confined to the endosperm while the embryos are relatively rich in soluble carbohydrates (sugars and **oligosaccharides**). The aleurone layer tissue and starchy endosperm contain a proportion of **oil** but in most cases the bulk of it resides in the embryo (**scutellum**): that of maize is an example of an embryo that is particularly rich in oil (**see: Maize**).

Fibre is a nutritional factor whose importance in the human diet has been emphasized. Adequate fibre in the diet ensures that food and food waste pass through the digestive tract with sufficient speed to prevent diseases that result from excessive residence times. The term fibre covers many compounds but most are complex carbohydrates (e.g. **glucans**, **galactomannans**). Because fibre includes all parts of the diet that are not digested, its measurement is more difficult than that of a single compound. In consequence there are several definitions of fibre, some of which relate only to the method of analysis. Irrespective of measurement method, whole grain products have higher fibre content than other foods. Wholemeal bread, for example, contains more dietary fibre than dried prunes weight for weight and $1\frac{1}{2}$ times as much as cooked **pulses** such as beans and lentils.

In cereals, fibre is concentrated in the **pericarp**, seedcoats and endosperm cell walls. The pericarp in mature grains comprises mainly the lignified walls of empty cells. In oats and barley mixed linkage β-glucans predominate, while in wheat and other cereals **arabinoxylans** are the major component in endosperm cell walls. Both types have the ability to bind large quantities of water, helping to maintain a well hydrated faecal mass that travels easily through the bowels. Because of the high fibre content of brans, nutritionists recommend that a proportion of the cereals part of the diet should be consumed as products made from unrefined meal.

Cereal grains contain relatively high amounts of **phytate**, which is a storage form of phosphorus and various mineral ions, to be made available after germination to the growing seedling. Any available phosphate groups in phytate in animal feed can bind free ions, such as iron, calcium or zinc, forming salts that are largely excreted. Cereal-based diets can thus provoke mineral depletion and deficiency. One approach to this problem, so far as animal diets are concerned, is to add phytase to the food preparation. Another approach is the isolation of low-phytic acid **mutants**, using them to breed first-generation low-phytate hybrids, cultivars and lines, for example of maize, barley and rice. Phytic acid in the seeds can thus be reduced by 50–95%. Transgenic technology could also be applied to elevate the grain phytase content.

Cereals are a rich source of B vitamins. These, and other vitamins, are not distributed uniformly in the grain, neither is the distribution consistent for all vitamins. The distribution of the main vitamins present among the major components of the wheat grain is shown in Fig. C.6. The aleurone layer, although only contributing 19% of total grain dry mass has between 30 and 82% of these vitamins.

The importance to the diet of the vitamins present in cereal products and the relationships that their contributions bear to recommended daily requirements for some of them are shown in Table C.11.

Cereals proteins are important components of the diet, but cereals are not high protein foods. Storage proteins, which account for the majority of this nutrient in cereals, do not have as good a balance of amino acids as some other foods in that some essential amino acids are deficient. The first limiting amino acid in most cereals is lysine. Attempts to increase the proportion of lysine have been made, particularly in maize, which have met with some success. Oat and rice are richer in globulin and therefore their protein is less deficient in lysine. The **gluten** proteins of certain cereals, which are responsible for the viscoelastic properties of dough made from the flour (**see: Storage proteins – processing for food**), in certain people can provoke an inflammatory condition of the gastrointestinal tract known as coeliac disease. The condition is a reaction towards specific amino acid sequences in the prolamin protein, the major component of gluten. Because wheat, rye and barley are closely related their prolamins (gliadin, secalin and hordein, respectively) share the features that induce coeliac disease and therefore sensitive individuals must avoid food products from all three genera (**see: Storage proteins – intolerance and allergies**).

Some proteins present in many cereals have inhibitory effects on certain insect and mammalian enzymes, particularly amylases and proteolytic enzymes (**see: Amylase inhibitors; Protease inhibitors**). Wheat and barley grains contain a family of small proteins, the **thionins** which, though cytotoxic, are harmless when taken by mouth. But because they show resistance to proteolytic attack they reduce the overall digestibility of the grain protein.

Cereals make a useful contribution to mineral requirements in the human diet (but see above discussion of phytate). The concentration of some minerals in a selection of cereal products is shown in Table C.12.

2. Supplementation of human nutrition

Because cereals are such an important staple, many governments decree that nutrients that are critical in their respective populations should be added as supplements. The purpose may be to bring a refined product up to the nutritional value of the raw product, as is the case when nutrients equivalent to those present in the grain parts that are excluded from white flour are added to the flour. Alternatively the cereal product may provide a vehicle for conveying nutrients, not necessarily present in the unrefined grain, to consumers who might otherwise suffer deficiencies. An example of this is the addition of calcium, as chalk, to flours. In addition, manufacturers may make addition of nutrients as a supplement, such as when folic acid is added to breakfast cereals and other products. The traditional use of lime in *tortilla* preparation adds calcium and releases niacin from a bound form (**see: Maize**).

3. Animal nutrition

Although, as with human consumers of cereals, birds and mammals reared on farms benefit from fibre and minor nutritional components, the main benefit gained by animals consuming cereal grains lies in the energy they acquire from the digestion of **starch**. The ease with which this is achieved varies with the animal and cereal species concerned.

Differences in granular form and **amylose/amylopectin**

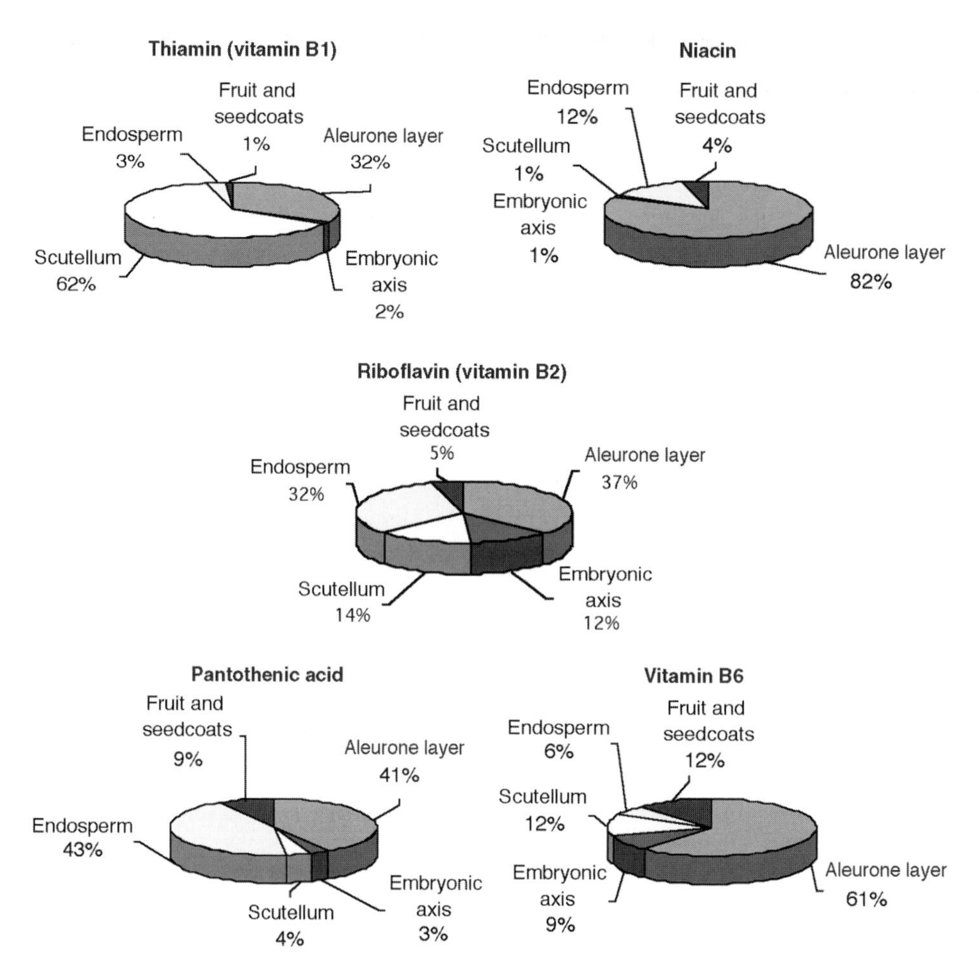

Fig. C.6. Distribution of B vitamins among the tissues of wheat grain.

ratio exist among starches from different cereals, but the main differences in starch digestibility result from variation in the proteins and non-starch polysaccharides (NSPs) that surround the granules and the relative ease with which animals can cope with them. It is the soluble components of NSPs that provide the greatest obstacle to digestion by chickens. Rice, sorghum and maize are more easily digested than wheat, triticale and particularly barley and rye. Enzymes such as glucanases and xylanase can bring about improvements when added to the feed.

Cereals are fed to farm animals as raw whole grains and in several processed forms. Poultry that lives long enough to develop crop and gizzard sufficiently to grind the grain can release energy from the starch in the endosperm. Sheep chew efficiently and can do the same. For other farm stock, grains require some processing for digestion to proceed. The simplest treatment is rolling, in which grains are subjected to pressure as they pass through a narrow gap between smooth-surfaced rollers. This serves to rupture the seedcoats and fruit coats and the **hull**, where present, thus increasing accessibility by digestive enzymes to the endosperm. Accessibility is further increased by reducing particle size by grinding. The effects of these treatments and the others described below are not consistent in all animal species and undesirable effects may accompany increased digestibility in some animals. Heating grains at high moisture content also increases digestibility since it gelatinizes some or all starch granules. Heating of whole grains may be by steaming, micronization or extrusion or it may occur to ground stocks during the pelleting (extrusion) process.

4. Fish nutrition

Freshwater fish production by aquaculture more than doubled in the 10 years up to 2000 and the methods of rearing have become more intensive. Both these factors have increased the demand for fish feed. About 85% of aquaculture production occurs in Asia, with China by far the biggest single producer.

All cereals are used in fish feed, and even in traditional methods of culture whole grains or milled fractions are scattered to increase the natural food sources present in ponds. For carnivorous fish species, the diet needs a very high protein content. The supply of fishmeal, which traditionally provided the ideal food, is inadequate to meet increased demand and alternative cereal-based foods have been developed. High protein concentrates (gluten) from wheat and maize are both used but the wheaten product offers the advantage of binding the food constituents as well as supplying the protein. Combined binding and nutritional properties (energy generating in this case) can also be provided by starch that has been steamed. Starch can contribute significantly to the diet of

Table C.11. Comparison between some cereal products with the food that has the highest vitamin content, for some B vitamins and with the recommended daily requirements of the respective vitamins.

Vitamin	Cereal product	Concentration (mg/100 g[a])	Food with highest concentration	Concentration (mg/100 g[a])	Recommended daily intake (mg)
Thiamin (B1)	Wheatgerm	1.5	Dried yeast	13.0	0.8–1.1
Niacin	Bran breakfast cereal:	49.0	Yeast extract	110.0	12–20
	Unprocessed bran:	29.6			
	Wheatgerm:	5.8			
Pyridoxin (B6)	Unprocessed bran:	1.38	Dried yeast	2.0	0.9–1.9
	Bran breakfast cereal:	0.83			
	Wholemeal flour:	0.50			
	Rye flour:	0.35			
Folic acid	Wheatgerm:	330 μg	Dried yeast	4000 μg	200 μg
	Unprocessed bran:	260 μg			
	Bran breakfast cereal:	100 μg			
Biotin	Rice bran:	60 μg	Dried yeast	200 μg	100–200 μg
	Rolled oats:	20 μg			
	Unprocessed bran:	14 μg			
	Wholemeal bread:	6 μg			
Pantothenic acid	Unprocessed bran:	2.4	Dried yeast	11.0	4–7
	Wheatgerm:	1.7			
	Bran breakfast cereal:	1.7			
Vitamin E	Wheatgerm oil:	133.0	Wheatgerm oil	133.0	8–10
	Maize oil:	11.2			
	Wheatgerm:	11.0			
Vitamin K	Wholemeal flour:	17 μg	Turnip greens	650 μg	70–140 μg

[a]Except where μg /100g indicated.

Table C.12. Selected mineral contents of some cereal-derived foods (values per 100 g).

Cereal product	K (mg)	Ca (mg)	Fe (mg)	Zn (mg)	Se (μg)
Barley, pearled, boiled	92	7	1	0.7	t
Maize flour	61	15	1.4	0.3	0.5
Rye flour, 100% extraction rate	435	34	2.9	3	1.9
Wheat bran	1340	63	12.6	9.1	2
Wheat germ	844	59	8.3	12.9	4.6
Wheat flour, white	195	21	1.4	0.8	8
Wheat, wholemeal	407	36	3.3	2.1	5
Oat bran	550	56	5.3	3	2
Oatmeal	484	102	6.9	2.7	2
Rice, brown, boiled	99	4	0.5	0.7	2.3
Rice, white, polished, boiled	11	3	0.1	0.9	t

Values are approximate and variable among samples.
t, trace.

non-carnivorous fish species. It, together with other grain components or whole grains, can increase the supply of natural food sources present in ponds. (AE, MN)

Athar, N., Spriggs, T.W. and Liu, P. (1999) *The Concise New Zealand Food Composition Tables*, 4th edn. New Zealand Institute for Crop & Food Research, Palmerston North, NZ.

Bewley, J.D. and Black, M. (1994) *Seeds: Physiology of Development and Germination*, 2nd edn. Plenum Press, London, UK.

Black, J.L. (ed.) (1999) Premium grains for livestock. *Australian Journal of Agricultural Research* 50, 629–908.

Holland, B., Welch, A.A., Unwin, I.D., Buss, D.H., Paul, A.A. and Southgate, D.A.T. (eds) (1991) *McCance and Widdowson's The Composition of Foods*, 5th edn. Ministry of Agriculture, Fisheries and Food and The Royal Society of Chemistry, Cambridge, UK.

Monro, J. and Humphrey-Taylor, V. (1994) *Composition of New Zealand Foods, 5. Bread and Flour*. New Zealand Institute for Crop & Food Research, Palmerston North, NZ.

Shewry, P.R. and Halford, N.G. (2002) Cereal seed storage proteins: structures, properties and role in grain utilization. *Journal of Experimental Botany* 53, 947–958.

Cereals - reserve mobilization

See: Cereals; Cereals – composition and nutritional quality; Malting; Mobilization of reserves – a monocot overview; Mobilization of reserves – cereals

Cereals - storage

1. Conditions

Compared with soft fruits and indeed most other natural products, cereal grains can be stored for long periods without deterioration. Acceptable storage time depends not only on storage conditions but also on the intended use for the cereal. For example, wheat of suitably low moisture content, experimentally stored at ambient temperature and moisture for 18 years in oxygen-depleted conditions, retained its baking properties. However, germinability was drastically reduced (**see: Longevity; Viability**). For this to be maintained, oxygen-depleted storage at 5°C was necessary. Such long periods of storage are unusual but cereals are routinely stored for the period between harvests (approximately 11–12 months) and often longer in a wide variety of purpose-built and multi-use structures. Storage in outside heaps is suitable only for short periods and then only where dry conditions are assured for the duration of the storage. (**See: Storage management**)

For successful storage it is important to avoid excessive temperatures, high moisture, predation by insects and other invertebrates, and access by birds and mammals such as rodents. Although sealing and insulating stores can deliver some independence of ambient conditions, the ease with which suitable conditions can be achieved varies according to storage location. For this reason periods of storage vary according to location. In general, longer periods of storage are possible in dry, colder conditions than in hot, humid conditions. A major factor contributing to this variation is the effect of temperature on 'water activity'. (**See: Deterioration reactions**)

Water is present in grain in different states, thus: (i) Strongly-bound water, in chemical union with grain components (it is not easily removed even at high temperatures – in cereals it accounts for 5–6% of the total moisture content); (ii) water bound by molecular attraction. This can be removed by use of high temperature; and (iii) water that is free or loosely bound by capillary action. Such water raises grain moisture content above 13–14%. (**See: Waterbinding**)

In a grain bulk the water that is not strongly bound is continually moving between grain and the air surrounding the grains. At some stage, if the temperature and relative humidity around the grain bulk remain constant, a state of equilibrium is reached in which flux between grain and surrounding air ceases. **Relative humidity** (RH) is the ratio between the amount of water vapour present in the air and the amount that the air is capable of holding at the same temperature. It is expressed as a percentage. **Equilibrium relative humidity** (ERH) is the relative humidity value for air in equilibrium with the grain it surrounds. At lower temperatures more water is bound to the grain substrate and consequently the ERH of grain at a given moisture content is lower at a low temperature than at a higher temperature.

ERH, or the related value water activity, is a major factor in determining the presence or absence of living organisms in spaces within and around the grain bulk. Water activity (a_w) is numerically expressed as one-hundredth of the RH value. Moulds cannot develop below 68%RH (0.68 a_w). The majority of stored grain insects can breed and develop at ERH 60% (0.6 a_w), although mites and psocids (booklice) require higher values.

Because the composition of cereal grains varies, the amount of bound water present in them varies. Thus the relationship between moisture content and ERH (a_w) at a given temperature is specific to each species. At 25°C a RH of 75% relates to a moisture content of 13.4% for oats and 15.3% for sorghum. Other cereals fall within this range. Endosperm hardness also influences the relationship and hard wheats are associated with lower moisture contents than soft wheats for the same ERH values. An ERH of 65% (0.65 a_w) is considered safe for long-term storage of cereals.

Spoilage agents may be a problem even when the overall moisture content in the store is below the safe level. This can result from air movements leading to **moisture migration**. Warm air moving to a cooler area will give up moisture to grains, thus remaining in equilibrium with them. Unless temperature gradients are extreme the exchanges occur in the vapour phase; nevertheless, variations in moisture content up to 10% within a store are possible. In the presence of oxygen, respiration increases, producing more heat and water (**see: Storage management**). If the moisture content is allowed to rise to 30% a succession of progressively heat-tolerant micro-organisms develops. Above 40°C mesophilic organisms give way to thermophiles. Thermophilic fungi die at 60°C and the process is kept going by spore-forming bacteria and thermophilic yeasts up to 70°C.

2. Storage fungi

The principal pathogens involved are various *Aspergillus* and *Penicillium* species. The predominance of species is linked to the conditions in which the grain is stored and changes as the fungi develop. Both genera produce **mycotoxins**, including aflatoxin from *A. flavus*. They also produce large numbers of spores, which can cause respiratory problems if the contaminated grain is fed to animals. (**See: Dust**)

Storage fungi can cause considerable losses of yield and quality in stored grain of all small grain cereals. Extensive growth results in substantial heating, 'caking', and even charring if heat is great.

3. Insects and arachnids

Primary invertebrate pests are capable of attacking whole grains while secondary pests feed on grains already attacked. All arachnid pests belong to the order Acarina (mites), which includes primary and secondary pests. Primary insect pests include grain beetles, grain borers, weevils and Angoumois grain moth. Secondary pests include several moth and beetle species. In regions where termites occur they are a problem and may penetrate into sealed underground stores. (**See: Storage pests**)

4. Vertebrate pests

Rodents and birds are the main vertebrate pests in grain stores. In addition to consuming grain they introduce contaminants through their excretions, which contain pathogenic bacteria. The best control measure is denial of access.

5. Pesticides

Pesticides have played a part in controlling unwanted organisms in grain stores for many years. Both liquid and **fumigant** types are used. The widely used fumigant, methyl bromide, is an ozone-depleting substance and is therefore being phased out, at least in developed countries. Phosphine is the most widely used alternative but others are in use and are being researched. In sealed stores, replacement of air surrounding grains with carbon dioxide can be effective through exclusion of the oxygen required by common pests. (**See: Storage management; Treatments – pesticides**)

Small residues of pesticides used in store and, to a lesser extent in the field, are a matter of concern and strict limits are set in many countries. However, testing for the minute amounts involved (0.01 mg/kg in some cases) is expensive.

(AE, MN)

Gooding, M.J. and Davies, W.P. (1997) *Wheat Production and Utilization*. CAB International, Wallingford, UK.

Kent, N.L. and Evers, A.D. (1994) *Technology of Cereals*, 4th edn. Pergamon Press, Oxford, UK.

Certification schemes

Seed certification, narrowly defined, is the assurance that seed is of a specified **variety** and is sufficiently pure. Certification schemes verify the source of the seed used for commercial seed multiplication, inspect seed multiplication fields and conditioning facilities, and carry out post-harvest tests to guarantee that various aspects of character, quality, pedigree and varietal quality of the end product can be assured. (**See: Production for Sowing, I. Principles.**) ('Certification' in this sense should not be confused with the same term used in the **grain inspection** of lots traded for food or industrial processing, or within **phytosanitary certification** systems aimed at stopping the spread of insects and pathogens.)

Some seed certification schemes are restricted to only a few quality aspects. For example, the **OECD Seed Scheme** is concerned only with varietal aspects. Certification may be complemented by tests carried out to ensure that only material of acceptable physical, physiological and health quality is eventually released to growers. For example, EC (European Commission) schemes specify additional standards, including: percentage analytical purity and germination, and limits on the content of moisture, disease and seeds of other crop and weed species. (**See: Quality testing**) Other quality aspects, that may be important to the buyer but which are less likely to be dealt with in seed certification schemes are seed size, density and weight characteristics and seed **vigour**. According to the wide range of schemes operating in different countries, certification may or may not be done by an official certification agency.

It is possible to have a certification system without having a seed law, and also to have a seed law without having certification; however, most commonly, both go together. The **Laws and regulations** governing seeds lay down the type of scheme (for example, set minimum standards, truth-in-labelling, etc.), and also give details of any standards, control measures and any penalties if the law is broken. Whatever the situation, it is important not to confuse certification with legislation.

Certification schemes can be compulsory or voluntary. In other words, it is possible to have schemes in which all, or only some, of the seed is certified. Both systems have their advantages and disadvantages, as listed in Table C.13. (**See: Quality-declared seed**)

1. Certification authorities

A Certification Agency or Authority is a body established to control the production of seed and ensure its quality. The control is achieved in two ways:

- controlling the multiplication and processing of seed in such a way that mechanical and genetic contamination is prevented/minimized;
- setting standards and checking each seed stock against these standards.

If seed is produced under these conditions and reaches such standards it is certified by the authority, and labels and seals attached to containers indicate this. The certification body may be part of a Ministry or Department of Agriculture, or it may be a para-statal organization such as a National Seed Corporation, or it could be an independent body made up of representatives of government, breeders, farmers, seed processors and merchants.

2. Operation of a scheme

Controls are exercised in two stages. Before-harvest controls are a pre-control plot check, and an inspection of the growing

Table C.13. A comparison of the advantages and disadvantages of compulsory and voluntary seed certification schemes.

	Compulsory Certification
Advantages	All seed on the market is certified, and standards can be easily maintained. The quality of the seed is guaranteed to the farmer.
Disadvantages	The input made by the Certifying Authority is costly and time consuming. Because of the amount of seed needed, inevitably some will be produced by inefficient growers and processors and may be of borderline quality.
	Voluntary Certification
Advantages	Usually only the very best farmers produce seed and the quality is very high.
Disadvantages	Does not eliminate poor quality uncertified seed from the market. Can be confusing to farmers to have some seed certified and some uncertified.

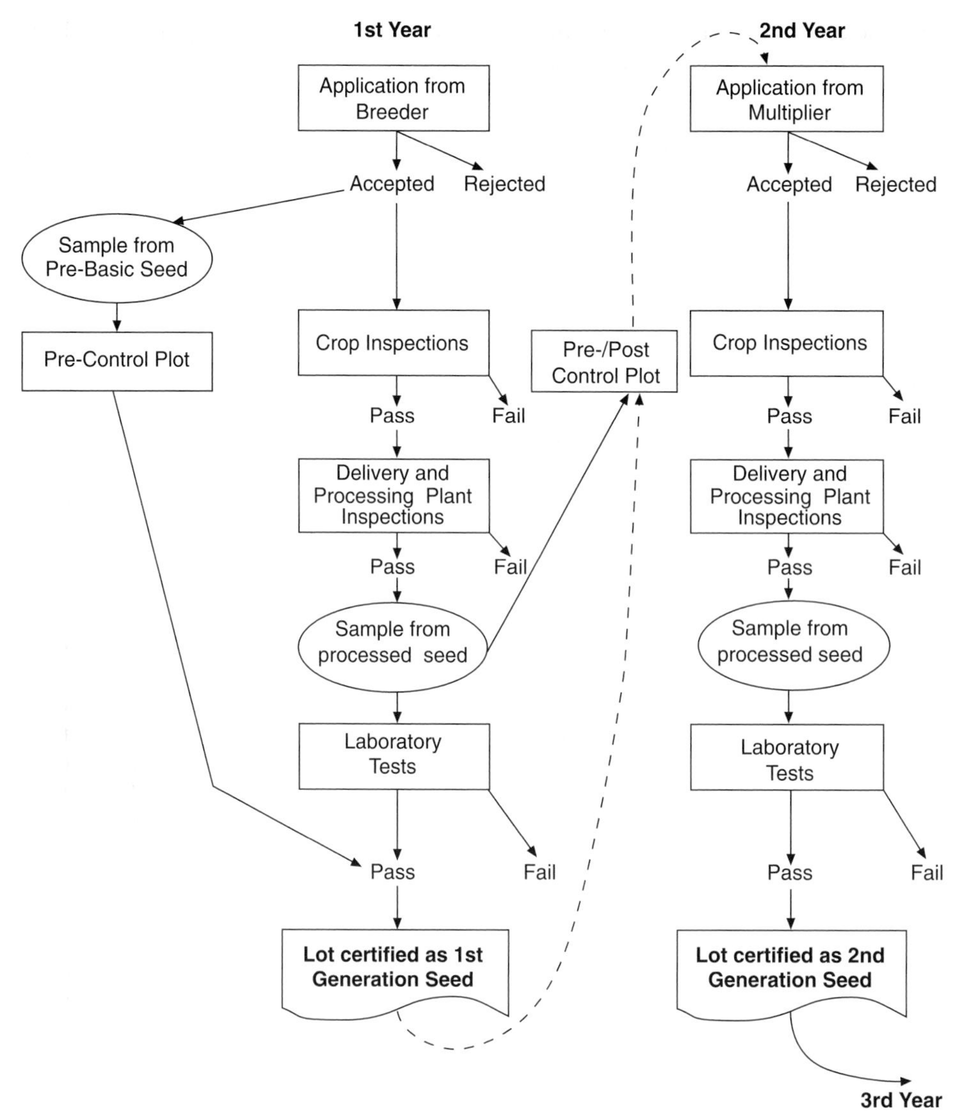

Fig. C.7. A diagrammatic representation of general certification procedures involved in producing Second Generation Certified Seed from Basic Seed. This scheme accords to a certain extent with all the different types of seed certification schemes operated worldwide.

Table C.14. Colour-coding system in the OECD certification scheme. (A different categorization is used for forest reproductive material.)

Category of Seed	Colour of Label
Breeder's seed	Violet
Pre-Basic Seed	White with diagonal violet stripe
Basic seed	White
Certified seed of the 1st Generation	Blue
Certified seed of the 2nd and 3rd Generations	Red

Table C.15. Comparison of terminology for the classes of seed used as successive stages in the production of Certified Seed.

USA (**AOSCA**)	EU and other countries (**OECD Seed Scheme**)
Breeder seed	Pre-Basic Seed
Foundation Seed	Basic Seed
Registered Seed	Certified Seed, 1st generation
Certified Seed	Certified Seed, 2nd generation

crop in the field. After-harvest controls are: inspection of deliveries to seed processing plants; sampling and testing of processed seed; and post-control plot tests and/or laboratory tests for varietal (cultivar) purity. The steps involved in the production of certified seed are outlined in Fig. C.7, and elaborated upon in sections (a)–(e) below.

(a) *Pre-control plot check*. All multiplication-category seed lots are checked in control field plots that are readily accessible to the inspection staff. Certain quality parameters, such as varietal purity and some seedborne diseases, are easier to assess in the field than in a seed laboratory. In many species, varietal differences are difficult or impossible to see in seed samples because they are only visible at certain stages of growth; for example, differences in height, earliness, growth habit or colour may be transient and only visible for a few days. If the inspection visit does not coincide with the optimum time for seeing such differences, the crop inspector might miss them. Control plots can be inspected on a day-today basis, thus ensuring detection of all varietal problems and also seedborne diseases in relation to seed standards, such as loose smut in wheat.

(b) *Inspection of the growing crop in the field*, usually at flowering when varietal differences are clearest, to examine whether it is suitable to produce the standard of seed required – inspecting isolation standards, general crop condition, health, and presence of other varieties, weeds and other crop species.

(c) *Inspection of deliveries to seed processing plants* to ensure that the harvested crop arrives intact at the designated processing plant and uncontaminated with impurities.

(d) *Sampling and testing processed seed* usually for analytical purity, 'other seed' determination and germination, and sometimes seed health and moisture content. (**See: Health testing; Purity testing**)

(e) *Post-control plots* grown to assess final-generation seedlots against the varietal and disease standards required for sale.

3. Labelling

One feature common to all certification schemes is that each bag of seed carries a label issued by the Certification Authority. In most schemes the label is colour-coded (Table C.14) according to the generation of the seed, to make it is possible to tell at a glance its certification category. In addition, labels usually carry information on the species, variety, seed lot reference number, weight and date of seed production. (RD)

Certified Seed

Seed produced on a large scale for general sale to farmers for sowing commercial crops, so as also to meet the standards set by the local official seed certification agency. Usually the plant breeder produces only a small quantity of seed of any new variety or of an older variety that is being maintained, including parental lines in the case of a hybrid variety. Depending upon the popularity of a variety and the inherent biological multiplication rate of the species, the seed from the breeder then has to be multiplied in order to obtain enough for farmers' needs. Production is usually carried out by specialized seed growers, and is managed to maintain **variety** (cultivar) identity and purity, under the auspices of **certification schemes** (Table C.15).

Multiplication takes several generations for most major species, through successive steps. (**See: Production for sowing, I. Principles**) Accordingly, in the official terminology of the **OECD Seed Scheme**: Certified Seed is the first or second generation progeny of **Basic Seed**. In the terminology of North American seed certification programmes, it is the progeny of **Registered** or **Foundation** seed. (RD)

CGIAR

See: Consultative Group on International Agricultural Research

Chaff

Chaff describes the **glumes, lemmas, paleas** and light plant-tissue fragments ordinarily broken during threshing and ordinarily removed during seed cleaning. A number of cultivated pasture and amenity grass seeds are termed chaffy because of their fluffy appearance, which is due to various appendages that remain more securely attached to or intermixed with the seed, such as large **awns**, sterile **spikelets**, surface hairs and bristles. In the mass, chaffy seeds are light and bulky, and do not flow freely because the appendages readily become entangled. They cannot be gathered in the field with conventional equipment or sown uniformly through traditional seeders, which adds considerably to the costs of their harvesting, cleaning, testing, storage, transport and use. (PH)

(See: Conditioning, II. Cleaning)

Chalaza

A topographic term to describe the region at the base of the ovule (the opposite end from the micropyle) where the integument(s) are inserted. The funiculus (funicule), chalaza and raphe form a continuous tissue without any sharp delimitations. (**See: Structure of seeds**, Fig. S.61; **Seedcoat – structure**)

Charles-Edwards Model

This model describes the number of developing vegetative or reproductive sinks (N_G : number of **floret** primordia or seeds per unit area) produced by a plant or a plant community as a function of daily photosynthesis (∇_F), the proportion of daily photosynthesis that is partitioned to the developing sink (η_G), and the minimum assimilate flux required for continued growth of an individual sink (a_G) (Eq. 1). This model was developed by D.A. Charles-Edwards,

$$N_G = \eta_G \nabla_F / a_G \quad \text{[Eq. 1]}$$

to provide a conceptual framework to describe the ordered development of plants – the production of branches, tillers, fruits and seeds. The model describes the number of these sinks as a direct function of photosynthesis (∇_F) during the critical period of development when these structures are initiated. Since supply of assimilate to the developing structure is probably the determining factor, photosynthesis (∇_F) must be adjusted for the proportion (η_G) that is partitioned to the developing sink. There is a wealth of published data supporting a direct relationship between photosynthesis and the number of fruits, seeds, tillers or branches for all agronomic crops.

A major contribution of this model was to identify a role for the characteristics of the developing sink in determining the number of plant parts produced. Charles-Edwards hypothesized that there is a stage in the early development of a vegetative or reproductive sink when a minimum flux of assimilate (a_G) is required for continued development. If the supply drops below this minimum, development will cease and the sink will abort. This minimum flux is a characteristic of the sink. The model predicts that plants with the same level of photosynthesis and partitioning will produce different numbers of seeds if there are differences in a_G.

Photosynthesis and partitioning are easily understood, but a_G is less well defined. It is not clear if it is a flux or a concentration of assimilate, or how or where it should be measured. The specific mechanism(s) that causes a developing sink to abort when subjected to an inadequate assimilate supply is unknown. Seed growth rate (rate of dry matter accumulation by an individual seed, mg/seed/day) (**see: Development of seeds – influence of external factors**), a characteristic of the seed, has been used successfully as a proxy for a_G. However, the inability to precisely define a_G does not negate its importance in understanding the factors that determine branch, tiller, fruit and seed numbers in plants.

The model accurately describes the many factors known to affect fruit and seed **number**. Changes in photosynthesis (∇_F) created by shade, extra light, drought stress, CO_2 enrichment, and plant density treatments during the critical period when fruit and seed number are determined always cause corresponding changes in the number of reproductive sinks.

The proportion of assimilate that is partitioned to reproductive growth is an important factor determining fruit and seed number and yield. Much of the historical increase in yield in some crops is attributed to changes in partitioning, which, although important, is difficult to measure and the mechanisms regulating it are not well understood.

Including the characteristics of the sink (fruit or seed) in the model explains variation in fruit and seed numbers not related to photosynthesis or partitioning. The a_G term explains the widely observed inverse relationship between genetic differences in seed size and seed number. Increasing seed size results in a reduction in seed number because the increase in a_G with a constant assimilate supply (η_G and ∇_F) must be offset by a change in N_G. Thus, the model provides a mechanistic explanation for the classical example of yield-component compensation, an increase in seed size does not increase yield because it is offset by a change in seed number. The characteristics of the sink (a_G) also explain the large differences in seed numbers between species (e.g. rice [*Oryza sativa*] typically produces nearly ten times as many seeds per unit area as maize [*Zea mays*]) that cannot be ascribed to photosynthesis or partitioning but can be explained by differences in seed size.

The Charles-Edwards model provides a reasonable explanation of many of the factors that influence fruit and seed number, but it has some weaknesses. For example, it does not include time, which may play an important role in determining fruit and seed number. The Charles-Edwards model describes the final product, seed number at maturity, but it does not account for the dynamic processes of pollination, fertilization, seed development and abortion that determine how many seeds will survive until maturity (**see: Potential seed number**).

The model also does not account for variation in seed number that is a result of poor pollination or fertilization, e.g. stress-mediated reductions in pollen viability or **protandry**. Such stress limitations are not common, but they can have devastating results when they occur. Implicit in the model is the assumption that the number of flowers does not limit seed number, an assumption that holds for the many plant species which have high levels of flower and fruit abortion. The model does not provide a role for stored carbohydrate in determining seed number, nor does it allow the size of the sink to influence photosynthetic activity. Seed number in the model is driven by the supply of current assimilate, but there is no provision for a direct effect of mineral nutrients, such as N, or hormones. Mineral nutrients, however, could affect seed number indirectly by affecting photosynthesis (∇_F). Although the information needed to discount these limitations is not necessarily complete, it seems likely that none of them seriously restricts the usefulness of the model.

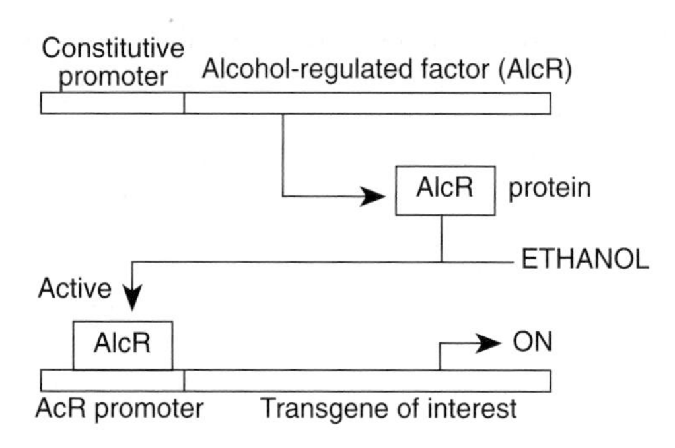

Fig. C.8. The ethanol-inducible promoter system: a model of chemically inducible promoter systems. One gene is introduced into a plant that produces a protein with which the inducing chemical interacts (ethanol with the AlcR protein). The activated AlCR then binds to the promoter region of the second introduced gene causing it to be expressed.

The Charles-Edwards model was intended to apply to all plant species. Such broad coverage is possible because the characteristics of seed growth (**see: Development of seeds – influence of external factors**) and the mechanisms regulating seed number are generally consistent across species. The Charles-Edwards model, which extends earlier conceptual models developed by A.R. Sheldrake and W.G. Duncan, accurately characterizes most of the factors that are known to affect seed number, and greatly improves our understanding of the effect of these factors on crop productivity. (**See: Development of seeds – nutrient supply; Yield**) (DBE)

Charles-Edwards, D.A. (1982) *Physiological Determinants of Crop Growth*. Academic Press Australia, North Ryde, NSW, Australia.

Charles-Edwards, D.A. (1984) On the ordered development of plants. 1. An hypothesis. *Annals of Botany* 53, 699–707.

Charles-Edwards, D.A., Doley, D. and Rimmington, G.M. (1986) *Modelling Plant Growth and Development*. Academic Press Australia, North Ryde, NSW, Australia.

Egli, D.B. (1998) *Seed Biology and the Yield of Grain Crops*. CAB International, Wallingford, UK.

Check-row planting

A square-grid pattern resulting from the accurate and indexed placement (and subsequent covering) of individual or groups of seed, resulting in plants equidistant apart and aligned in perpendicular rows. (**See: Planting equipment – agronomic requirements**, Fig. P.15.)

Chemically inducible gene promoter

These are gene promoters that regulate transcription in response to a specific external chemical stimulus. The term chemically inducible promoter is not strictly accurate as chemical induction of gene expression in general requires two genes: one that encodes a protein transcription factor with which the chemical interacts and a second gene whose activity is controlled by the binding of the transcription factor in the presence or absence of the chemical inducer (Fig. C.8). A more precise term would be a chemically inducible promoter system.

Such promoter systems offer the ability to activate a particular trait on demand and only when needed. This attribute makes such promoter systems particularly useful in plant biotechnology. Chemically inducible promoter systems, to be of use, must have no residual activity in the absence of the inducing chemical and must respond to a chemical that is not normally generated within the plant or encountered within its environment. In general, most chemically inducible promoter systems that have been developed for use in plant biotechnology have come from non-plant sources including bacteria, fungi and animals. One example is an ethanol-inducible system that utilizes a fungal (*Aspergillus*) alcohol-regulated transcription factor AlcR (see Fig. C.18). Here the gene for an alcohol-regulated (alcohol-activated) protein (produced under the control of a constitutive promoter) is introduced into a plant, and in the presence of ethanol this protein can bind to the specific alcohol-sensitive promoter to induce the second introduced gene of interest. There are two chemically inducible promoter systems that have been developed from plant sources: one that is based on a benzothiadazole-inducible pathogen-related promoter (from genes that respond to pathogen attack) and a second that is a safener-inducible promoter that responds to proprietary chemicals (**safeners**) that are used in herbicide mixes to reduce chemical damage during application. (MJO)
(See: Patent protection technologies)

Padidam, M. (2003) Chemically regulated gene expression in plants. *Current Opinion in Plant Biology* 6, 169–177.

Chestnuts

There are several species of the temperate-climate chestnut (*Castanea* spp. [Fagaceae]): *C. crenata*, Japanese chestnut; *C. dentata*, American chestnut; *C. mollissima*, Chinese chestnut; *C. sativa*, European, Spanish or sweet chestnut. The Chinese chestnut (mature trees 12–16 m high) is the most important in terms of commercial production and the European or sweet (32 m), second. In 2003, just over 1 million t of chestnuts were produced, the world's leading producers being China (71% of world production), Korea (Republic) (7%), and Italy and Turkey (5% each). **(See: Crop Atlas Appendix, Map 4)** Chestnuts are true nuts in the botanical sense, i.e. single-seeded, indehiscent fruits having a hard, dry, fibrous or woody pericarp (**see: Nut**). In most *Castanea* species there are three flowers in each compact inflorescence, each flower producing one nut: the cluster of three nuts is surrounded by spiny involucral bracts which open at maturity to expose the nuts (Colour Plate 4E). Nuts weigh 10–25 g , measuring 2–4 cm wide and 1.5–2.5 cm long. Chestnuts differ from other nuts in being high in moisture content, and in starch, and low in oil (**see: Nut**, Table N.2). They are eaten boiled, roasted, pureed, in *glacé* form or cooked in various dishes. (VP, MB)

Chickling pea

This species (*Lathyrus sativus* L.), also known as grass pea or Indian vetch, is a minor **cool-season legume (pulse)** cultivated in the Indian subcontinent, the Near and Middle East and southwest Asia: it also occurs in North America, temperate South America, North and East Africa, France and Spain. It is a low-value pulse that is eaten, for example, in times of famine in some parts of the world. Its presumed centre of origin is southwest and central Asia: archaeological remains, 7000–8000 years old, have been found in the Near East. The growth habit is similar to that of the pea, *Pisum sativum*, i.e. a climbing or straggling herbaceous winter annual with stems from 0.6 m to several metres long. The curved, flattish oblong pods are 2.5–4.0 cm long, each with 3–5 white, grey-brown or yellowy, spotted or mottled, 4–7 mm long, angled, wedge-shaped seeds.

Seeds contain 18–35% fw **storage protein** (more than half of which is globulin, and about one third albumin, some prolamin and glutelin, **see: Osborne fractions**), 0.6% **oil**, 58% carbohydrate (about 35% **starch**) and nearly 4% **phytin**. A non-protein amino acid, β-N-oxalylamino-L-alanine (BOAA), is present: this is a neurotoxin that can cause the paralytic condition, **lathyrism (see: Pharmaceuticals and pharmacologically active compounds)**.

The seeds are boiled and consumed as a pulse, e.g. in *dhal*. Flour made from the seeds can be used for chapattis, in curries, and in paste balls such as *bhajis*. Seeds can be used for making local beverages. They can be mixed with oil and salts to provide a feed for poultry and livestock. Seeds also have traditional medicinal uses. (MB)
(See: Legumes)

Campbell, C.G. (1997) *Grass pea.* Lathyrus sativus *L. Promoting the Conservation and Use of Underutilized and Neglected Crops.* 18. International Plant Genetic Resources, Rome.

Smartt, J., Kaul, A., Wolde Amlak Araya, Rahman, M.M. and Kearney, J. (1994) Grasspea (*Lathyrus sativus* L.) as a potentially safe food legume crop. In: Muehlbauer, F.J. and Kaiser, W.J. (eds) *Expanding the Production and Use of Cool Season Food Legumes.* Kluwer Academic, Dordrecht, The Netherlands, pp. 144–155.

Chickpea

1. Worldwide importance and distribution

Chickpea (*Cicer arietinum*), a **cool-season legume**, ranks as the 15th most important crop in area planted annually in the world. It is the second most important **pulse** crop in the world after dry beans (mostly *Phaseolus*) and makes up 15% of the world pulse production. Around 9 million ha are planted in 33 countries in the world, 90% in semi-arid tropical countries. Asia is by far the most important producing region with India leading the list, having 75% of the total chickpea area in the world. **(See: Crop Atlas Appendix, Map 3)** Its prominence in the Indian subcontinent is reflected in the alternative name, Bengal gram. **(See: Legumes)**

2. Origin

Cicer arietinum has never been found in the wild but closely related taxa have. The most recently discovered, *Cicer reticulatum*, has been postulated to be a wild progenitor. The area of present-day southeastern Turkey and adjoining Syria is the most probable centre of origin of chickpea. It is thought to have spread throughout the Mediterranean by about 6000 years ago and to have reached India by about 4000 years ago.

3. Seed characteristics and chickpea types

There are two common types of chickpeas: *microcarpum*, those with small (120–300 g/1000 seeds), yellow, green, light-brown or even black, angular seeds, known as *desi* types, and *macrocarpum*, the larger (260–600 g/1000 seeds), more rounded or brain-shaped types which are normally beige/buff with a zero-tannin seedcoat, known as *kabuli* types, sometimes called garbanzo beans (Colour Plate 2A). *Desi* chickpea plants usually are shorter, higher yielding, earlier maturing and more resistant to disease, frost and insect damage than *kabuli*. The *desi* types are mostly found in the Indian subcontinent and Iraq, Ethiopia and parts of Central America, whereas the *kabuli* types occur throughout southern Europe, western Asia, the Nile Valley, North Africa and North and South America. Most of world's chickpea production is *desi*; only about 10 to 15% is *kabuli*.

Chickpeas are grown as a winter crop in many parts of the world. Preferred day and night temperatures are 21–29°C and 18–21°C, respectively. *Kabuli* types are more sensitive to cold than the *desi* chickpeas which may endure soil temperatures even as low as 5°C, while *kabuli* types will not germinate in soils colder than 10°C. Chickpea plants are relatively drought-tolerant. The long taproots allow them to obtain water at a greater depth than can other pulse crops.

4. Nutritional quality

Dried chickpea contains about 20% **storage protein**, 61% carbohydrates (mostly **starch**) and 5% **oil**. It is a relatively rich source of lecithin and potassium and provides small quantities of vitamins A, B and C. *Desi* and *kabuli* types differ qualitatively and quantitatively in their dietary fibre, *kabuli* containing higher amounts, particularly **cellulose** and **hemicellulose**.

The content of sulphur-containing amino acids, methionine and cystine are limiting and threonine and valine are below satisfactory dietary levels: the content of tryptophan is satisfactory. Nonetheless, the protein of chickpeas on average is of higher nutritive value than that of most grain legumes.

Of the storage proteins, the major proportion is globulin, which is low in sulphur amino acids; albumins, glutelins and prolamins are also present (**see: Osborne fractions**).

Starch is the principal carbohydrate (48% fresh weight, fw); soluble sugars (6% fw) are also present. Chickpeas can provide a significant amount of oil in the human diet. The hypocholesterolaemic effect of chickpea may be due to the high content of **essential fatty acid** in the seeds, particularly linoleic and linolenic acids. Crude fibre is an important constituent, at a concentration directly related to the proportion of seedcoat (higher in *desi* than in *kabuli* cultivars). Chickpeas are rich in minerals and vitamins. Calcium and iron are important nutrients but since the seedcoat contributes about 70% of the total calcium most of it is lost when the coats are discarded in seed processing.

Various **antinutritional** factors are present, including **phytin**, trypsin and chymotrypsin inhibitors, **amylase inhibitors**, **oligosaccharides** and polyphenols. But chickpeas offer less of a problem than other grain legumes (**soybean, broad beans, pea, lentil**) as far as **protease inhibitors** are concerned. A pancreatic amylase inhibitor occurs in most legume seeds, but appears to be lower in chickpeas than in other food legumes.

Chickpeas provoke more flatulence due to the higher content of the flatulogenic, raffinose-series **oligosaccharides**, **raffinose**, stachyose and verbascose. Starch and **hemicellulose** in chickpeas may also contribute towards flatulence.

5. Uses

In the main growing areas around 85% of the domestic supplies of chickpea are chiefly used for human consumption and about 15% are used for animal feed except in North and Central America where this latter share is above 40%.

(a) *Mature seeds.* The majority of the chickpeas are consumed as whole ripe seeds. In India the large, white-seeded *kabuli* types are mainly used whole. In many countries whole chickpeas are a popular constituent of salads, soups, stews, etc.

(b) *Fresh (immature) seeds.* Fresh seeds are well liked in most countries that cultivate chickpeas. In India, Pakistan, Turkey and Ethiopia plants bearing filled pods are sold and are sometimes placed in a fire and the parched seeds eaten as a snack. Unripe, green chickpeas are also sold like shelled peas. These, too, may be eaten as snack food.

(c) *Puffed chickpeas.* Puffing of chickpeas, when the volume increases by 20–40%, is obtained by subjecting the grains to high temperatures for a short time: this is a common practice in the Indian subcontinent and many African countries.

(d) *Split seeds.* Chickpeas are consumed in India mainly as *dhal*, i.e. the split, dehusked seed, consisting of the cotyledons only. This *dhal* is prepared by splitting large or medium seeds in a mill. To facilitate the removal of the husk the seeds, after cleaning, are sprinkled with water and left overnight. Drying, milling and sieving then produces a uniform product. Uniformity in seed size is a very important variable for a quality product. Treatment with turmeric powder produces a yellow colour attractive to the consumers but with the exception of the *kabuli* types most Indian chickpeas are naturally yellow.

(e) *Sprouted seeds.* These are consumed in soups or as a vegetable with sauce.

(f) *Flour and meal.* Chickpea meal or flour (*besan*) is prepared in India from split seeds or from broken remains after the preparation of *dhal*. Ground chickpea is used as *hummus*, particularly in the Middle East and eastern Mediterranean, where it may also be fried to make *falafel*.

(g) *Medicinal uses.* The seeds have wide medicinal applications, particularly on the Indian sub continent. They are used as an appetizer, as an anthelmintic and to treat bronchitis, leprosy, skin diseases, blood disorders, throat problems and biliousness.

6. Postharvest technology, eating and nutritional quality

Dehulling, splitting, grinding, puffing, parching and toasting are some of the common primary processes to which chickpeas are subjected, to improve their appearance, texture, culinary properties and palatability. Processing technologies have evolved from largely traditional home-scale techniques to partially mechanized versions. The evaluation of quality requires a definition of the parameters necessary for the utilization of chickpeas as food. Because of the varied usages and local preferences, however, no standard quality parameters have been developed but certain generalizations may be made.

Before any processing operations it is often necessary to remove the seedcoat. Uniformity of seed size is essential for this mechanical procedure: hence seeds are graded by sieves before milling. With regard to the puffing of chickpea, cultivars are preferred that have been grown in specified agroclimatic regions which give superior products with good aroma. Puff volume is greater in grains with low husk content, e.g. 12–14% husk content.

With respect to cooking, quality parameters other than size are important, such as appearance of the food, taste, and texture in the mouth. The physical hardness (texture) of the seed is an important determinant of the eating quality. At least three factors affect seed hardness: (i) variability of available nutrients and moisture during the maturation phase; (ii) weather conditions, particularly temperature; and (iii) genetic constitution of the plant.

Hydration capacity of the seeds directly relates to cooking time as it determines the extent to which the seeds absorb water on soaking. The rate of penetration of water is affected by the permeability of the **seedcoat** (in turn affected by its chemical make-up, including tannins and lignin) and the **cotyledons**, by the hardness of the seed and by chemical composition of the cell walls, the seed size, the type of water used in the cooking process, by altitude, and by preparatory operations (soaking overnight, for example).

7. Market classes of chickpeas

Seed size, including uniformity, are important variables for the marketing of chickpeas. Seeds vary from about 120 mg per seed for some *desi* types to 600 mg for the larger *kabuli* types. In some cases colour and shape are more important. In foods that require *kabuli* types size is important, and for export a seed 'diameter' of 8 mm is optimum, corresponding to a weight of about 420–450 mg per seed. In some countries (e.g. Morocco, Spain and Mexico) there are official commercial grades according to seed size. In India the quality depends mainly on the cultivar: (i) *Desi* gram: brown, angular and small seeds; (ii) *Kabuli* gram: also called 'Punjab'; white or yellowish seeds; and (iii) Green gram: rather big, angular, green seeds.

Although only relatively limited amounts of chickpeas are traded on world markets (about 15% of production in 2001), India is the main buyer: the country is also a major producer. Other important producers and exporters are Australia, Mexico and Canada. According to FAO statistics, in 2001 exports averaged nearly 1 million t, mainly by Australia, Turkey, Canada, Mexico and Iran. The major importing countries were India, Pakistan, Spain, Bangladesh and Saudi Arabia. Trade figures for chickpea during 2001 are shown in Table C.16. (OV-V)

Table C.16. Major countries exporting and importing chickpea.

Country	Imports (tonnes)	Value (1000 US$)	Country	Exports (tonnes)	Value (1000 US$)
India	516,819	191,895	Australia	266,519	81,279
Pakistan	106,123	37,392	Mexico	207,093	128,000
Algeria	70,496	50,520	Turkey	153,953	75,288
Spain	68,734	48,475	Canada	149,212	53,606
Bangladesh	37,500	14,000	Iran	123,522	58,367
Saudi Arabia	25,384	7,938	USA	29,613	14,008
Italy	22,743	14,772			
Jordan	21,942	10,755			

FAO STAT 2001.

Saxena, M.C. and Singh, K.B. (eds) (1987) *The Chickpea*. CAB International, Wallingford, UK, and The International Center for Agricultural Research in the Dry Areas, Aleppo, Syria.
Summerfield, R.J. (ed.) (1988) *World Crops: Cool Season Food Legumes. A Global Perspective of the Problems and Prospects for Crop Improvement in Pea, Lentil, Faba Bean and Chickpea*. Kluwer Academic, Dordrecht, The Netherlands.
van der Maesen, L.J.G. (1972) Cicer *L. A monograph of the genus with special reference to the chickpea (*Cicer arietinum*), its ecology and cultivation*. Veenman and Zonen, Wageningen, The Netherlands.

Chickpea – cultivation

See: **Chickpea** for information on economic importance, origins, seed, types, uses, etc.

Chickpeas are a cool-season crop, usually grown as a winter (India, the Middle East) or spring crop (Australia and South and Central and North America). *Desi* types are less sensitive to cold than *kabuli* types and will tolerate lower night temperatures.

Most of the breeding work is concerned with *kabuli* types. Several varieties have been developed for the Australian and North American production areas. Most of the crop is grown without irrigation although some varieties are being selected to suit irrigated areas.

Plants are erect and free-standing, 30–70 cm tall with frond-like leaves and a number of woody stems. Pods are formed either singly or in pairs and contain two seeds.

Chickpeas prefer a well drained, neutral to alkaline soil. The seed is very fragile with a thin seedcoat which is susceptible to mechanical damage. Emergence is slow, taking between 20 and 30 days. Delayed seedling emergence causes reduced plant growth, a shortened flowering period and low yield. Chickpeas require high temperatures (above 15°C) for effective fertilization of flowers and pod set. Plant density will depend on the growing area but around 35 plants per m^2 is optimum.

Chickpeas are harvested mechanically when seed moisture is around 13%. The *kabuli*-type seeds are large and combine harvester settings should be altered to slow the drum speed and increase the space between the concave and the drum. Most *kabuli* types are grown as a premium crop for human consumption where appearance is important. Chickpeas will tend to darken if they are left unharvested and begin to weather.

Most seed is inoculated with **rhizobia** and generally initially treated with a fungicide seed treatment such as thiram to protect against soil borne **damping-off** diseases. (AB)

Saxena, M.C. and Singh, K.B. (eds) (1987) *The Chickpea*. CAB International, Wallingford, UK, and The International Center for Agricultural Research in the Dry Areas, Aleppo, Syria.
Summerfield, R.J. and Roberts, E.H. (eds) (1985) *Grain Legume Crops*. William Collins, London, UK.

Chicory

See: *Cichorium* **(endive and chicory)**

Chilling injury

Commonly observed mainly in **recalcitrant seeds**, or seedlings of numerous species of tropical or subtropical origin, this corresponds to biological disorders occurring at low, non-freezing temperatures (generally between 0 and 15°C). As in other plant organs, chilling injury in seeds and seedlings is associated with various biochemical and metabolic alterations, including changes in membrane composition and properties, decrease in ATPase activity, impairment of respiratory activity and protein synthesis, **free-radical** accumulation and lipid **peroxidation**. The disorders appear first as browning resulting from oxidation of phenolic compounds (**see: Phenolics**), and then as necroses of parts or all the tissues of the seed or the seedling. A feature of this phenomenon is its reversibility following short exposure to low temperature, but if the cellular damage is critical it quickly becomes lethal, mainly in the radicle prior to germination, or if the root of the seedling is involved. The noxious effect of chilling can also be less discernible and even difficult to detect. It can also become apparent only a long time after the application of cold or after the organs are returned to warmer temperatures. If the period at a chilling temperature is not too prolonged, or if the temperature is not too low, the damage may not be visible but it can cause reduced or abnormal growth of the seedlings or the young plants. The sensitivity of seeds to chilling is related to their water content. **Orthodox seeds** of tropical origin tolerate very low temperatures, even that of liquid nitrogen, when they are sufficiently dry. On the other hand, tropical recalcitrant seeds, which must be maintained hydrated for their survival, are always very sensitive to chilling.

Table C.17 gives the approximate limiting temperatures under which seeds or seedlings of some species show chilling injury, in comparison with the thermal minima for germination. Seeds of various species (*Cedrela odorata*, *Gossypium hirsutum*, *Symphonia globulifera*) do not germinate, and they lose their **viability** below the limit temperature of their sensitivity to chilling. Those of other species (*Agathis moorei*, *Hevea brasiliensis*, *Hopea odorata*, *Mangifera indica*, *Shorea roxburghii*) are able to germinate at temperatures lower than those which induce chilling disorders, but their seedlings do not survive. It is difficult, however, to determine accurately the limit temperature for injury because the time at which the disorders become apparent largely varies depending on the species, the stage of germination or seedling growth, and of course the temperature applied. Moreover, all the seed or seedling organs are not equally sensitive to cold. Depending on the species the damage appears first in the **radicle** (or the young roots), the **plumule** (or the young shoot) or the **cotyledons**. For example, seeds of *Hevea brasiliensis* are capable of germinating at 10°C and those of *Mangifera indica* at 5°C, but the roots of the seedlings are rapidly damaged at these temperatures. The **epicotyl** is less sensitive, but the seedlings cannot survive because of lack of a vigorous root system.

The development of injury usually increases with decreasing temperature and with exposure time. A few minutes or few hours at a harmful temperature are sometimes enough to induce irreversible damage. Thus, 30 min to 1 h at 5°C at the beginning of **imbibition** of cotton seeds is sufficient to reduce by 25% the growth of their roots. Only 2 min at 5°C results in browning of the cotyledons of soybean seeds. Young mango plants obtained from seeds stored hydrated for some months at 12°C do not show symptoms of chilling injury, but they are more puny and produce fewer leaves than those from freshly collected seeds; the longer the seed storage the more marked is this phenomenon.

The deleterious effects of cold may vary with the germination stage or with previous treatments applied to the seeds or the seedlings. The greatest sensitivity to low temperatures generally occurs during the first hours of the imbibition phase. Leakage of solutes from chilled seeds suggests that this results from impairment of membrane reorganization during the initial fast water uptake (**see: Germination – physical factors**). Such damage is often reduced when seeds are imbibed at higher temperatures before the cold treatment. Cotton seeds, for example, show two periods of maximum sensitivity to chilling, one at the very beginning of imbibition and the other about 1 day later. But they become able to tolerate chilling temperatures when they have been previously imbibed for a few hours at 31°C, and this chilling tolerance is maintained if seeds are dried back before the cold treatment. The chilling sensitivity of maize seeds appears at the second day of imbibition and is maximal at the third day. This sensitivity to chilling is also reduced by a previous imbibition of seeds at high temperature. In soybean seeds, the deleterious effect of imbibition at 1°C is maximal after 4 days and is reduced after a pre-osmoconditioning (**see: Priming**), the beneficial effect of

Table C.17. Examples of approximate thermal minima for germination, and temperatures at which symptoms of chilling injury appear in imbibed seeds or in seedlings.

Species	Thermal minimum for germination (°C)	Limit temperature for chilling injury (°C)
Agathis moorei (Moore kauri)	10	15–17
Cedrela odorata (West Indian cedar)	15	10
Dryobalanops aromatica (mahoborn teak)	5	5
Gossypium hirsutum (cotton)	10–12	10
Hevea brasiliensis (para rubber-plant)	10	15–16
Hopea odorata (merawan)	5	12
Mangifera indica (mango)	5	12
Shorea roxburghii	5	12–15
Symphonia globulifera (chewstick)	15	15

Compilation from various publications.

the latter on chilling resistance increasing during the osmotic treatment, being maximal after about 3 days.

The sensitivity of seeds or seedlings to chilling adds a difficulty to the introduction of plants from warm areas to temperate zones because the relatively low spring temperatures do not allow normal germination of seeds or growth of seedlings. In the case of maize, this problem has been solved to a large degree by breeding, but the chilling sensitivity has not been completely eliminated. It is the reason why a test for chilling resistance, called the '**cold test**', has been developed for this species, and likewise for soybean and cotton. Chilling injury is also an obstacle to the storage of recalcitrant seeds from tropical or subtropical species. These seeds must be maintained hydrated, but they then germinate. For avoiding or reducing their germination it would be necessary to store them at low temperature, but this is not possible because of chilling injury. There is no satisfactory solution for prolonged storage of these seeds. (DC, FC)

Côme, D. and Corbineau, F. (1992) Les semences et le froid. In: Côme, D. (ed.) *Les Végétaux et le Froid*. Hermann, Paris, France, pp. 401–461.

Patterson, B.D. and Graham, D. (1979) Adaptation to chilling; survival, germination, respiration and protoplasmic dynamics. In: Lyons, J.M., Graham, D. and Raison, J.K. (eds) *Low Temperature Stress in Crop Plants. The Role of the Membrane*. Academic Press, New York, USA, pp. 25–35.

Salveit, M.E., Jr and Morris, L.L. (1990) Overview of chilling injury of horticultural crops. In: Wang, C.Y. (ed.) *Chilling Injury of Horticultural Crops*. CRC Press, Boca Raton, FL, USA, pp. 3–15.

Chimeric gene

A gene construct whose components have different origins. For example, any part of the gene – the 5′ upstream region, 5′ untranslated region, coding region and 3′ downstream region – may come from a different origin. Often chimeric genes contain the coding region of a **reporter gene**. (**See: Transformation; Transgene**)

Chitting

A term used synonymously with shooting or sprouting: informally applied to describe the state in which seeds show the first sign of completion of germination, usually evidenced by the visible emergence of the radicle from the seed. The term is also used informally in seed technology to describe **enhancement** treatments designed to bring seeds to this stage, such as in **fluid drilling** and **pre-germination**.

Chlorophyll sorting technique

A conditioning technique based on the use of laser-induced fluorescence to quantitatively detect residual amounts of chlorophyll remaining in **seedcoats** (which occurs when seeds are harvested before **physiological maturity**) as a basis for the **colour sorting** of seeds one at a time, and hence upgrade the quality of the seedlot as a whole. (**See: Brassica – horticultural; Conditioning, II. Cleaning, 7. Other sorting principles**)

Jalink, H., van der Schoor, R., Frandas, A., van Pijlen, J.G. and Bino, R.J. (1998) Chlorophyll fluorescence of *Brassica oleracea* seeds as a non-destructive marker for seed maturity and seed performance. *Seed Science Research* 8, 437-443.

Chocolate

See: Cacao

Chromatin

The complex of DNA and proteins (histones) that comprise the chromosomes of eukaryotes. (**See: Cell cycle**)

Chromophore

A chemical group that when attached to a protein or other molecule confers an ability to absorb light energy on the conjugated molecule. The chromophore itself has a particular colour due to absorbance of a certain range of light wavelengths. The wavelengths absorbed may be modified slightly by its binding to an associated molecule. Both the **phytochrome** and **cryptochrome** plant photoreceptors in seeds contain chromophores (as do the chlorophylls in green-plant photosynthetic systems).

Chromoprotein

A protein that can absorb light in the visible region of the spectrum due to attachment of a coloured **chromophore**. The wavelengths that a chromoprotein absorbs are dependent on the properties of the specific chromophore. (**See: Cryptochrome; Phytochrome**)

Chromosome number

The single-copy or haploid chromosome number of the gametes (i.e. the number in the pollen or egg cells), designated by the symbol *n*. Hence the somatic chromosome number (that present in non-reproductive tissues) of the resultant plant progeny produced from the zygote, the product of fusion of the gametes, is designated by 2n (diploid). Variability can evolve or be created by breeding, resulting in changes in chromosome number, commonly in multiples of the complete chromosome set (designated by the symbol x) that is basic to the species (euploidy), but also by the addition or deletion of specific chromosomes (aneuploidy). (**See: Ploidy; Polyploids**)

Cichorium (endive and chicory)

Endive, *Cichorium endivia* L. and chicory, *C. intybus* L., are grown primarily for use in salads. The principal producers are Italy, France, Spain, the USA and the Netherlands. These are the two principal species out of nine in the genus, which is in the family Asteraceae. Chicory and endive are quite closely related; substantial interspecific crossing occurs when *C. endivia* is the male parent.

(a) *Endive* is similar to lettuce in many respects, the main differences being in top growth appearance and in reproductive structure. The rosette occurs in two forms, both relatively prostrate. One consists of narrow leaves with highly frilled margins (frisée), in which the outer leaves are dark green and bitter, and the inner leaves are yellowish or whitish and milder; the other is a broad leaf type, known as escarole. Endive is a self-pollinated biennial. The seed stalk forms on plants that have been **vernalized**, with flower heads comprising 16–19 florets contained in involucres.

(b) *Chicory* is of two types. The witloof type (in which leaves are forced in darkness from stored roots) is mostly grown in France, Belgium and the Netherlands. The non-forced head types are grown primarily in Italy and France. Both the wild and some cultivated types have dark green, long, strap-shaped leaves, forming a broad rosette on a compressed stem; another main cultivated outdoor type is radicchio, which has round or elongated dark or light green heads that may contain anthocyanin. Chicory is also a biennial and requires **vernalization**. The seedstalk, flowers and **achenes** are similar to those of endive. The flowers are usually blue, and are about 3.5 cm in diameter. Each flower head contains 11–28 florets. All forms of chicory are cross-pollinated; selfing is prevented by sporophytic **self-incompatibility**.

1. Origins

Chicory probably originated in the Mediterranean Basin and is now widely distributed over Europe, North Africa and portions of Asia. The ancient Egyptians, Greeks and Romans used the green leaves as a salad vegetable and the roots for medicinal purposes. It was harvested in the wild before the various specialized forms were developed. The first mention of cultivation was in 1616, in Germany. Later it was cultivated in England.

In 1775, two French physicians discovered that chicory roots could be dried, roasted and ground for use as a substitute or a flavouring for coffee. It was later noted that discarded roots sprouted in the dark, forming elongated and loose white leaves (hence the name ‘witloof’, or Belgian endive). When larger roots were used and forced under a cover of sand or soil, small tight heads (chicons) form. Witloof later became popular in France and The Netherlands.

Italy is the home of the greatest variety of non-forced chicory types. Outdoor heading types may be spherical or elongated with different combinations of red, white or green leaves. The rosette type is similar in appearance to the wild form, open with elongated leaves, and green or red midribs, that continue to form until the reproductive phase. The roots may be dried, roasted and ground as a coffee substitute, and their high fructan **carbohydrate** content enables them to be used for the industrial production of fructose sugar.

2. Breeding and genetics

Since endive is naturally self-pollinated, the primary breeding methods are selection and pedigree breeding. Goals include colour, type, leaf fineness and, rarely, disease resistance. Traditional cultivars of witloof chicory were developed by mass selection. Now most are **F_1** hybrids, developed to enable the production of uniform crops when forced in hydroponic systems without soil cover. Breeding goals include: uniform, tight, well-shaped chicons; seasonal production; tolerance to internal browning; reduced bitterness and resistance to premature **bolting**.

3. Development and production

Endive and chicory seed-like fruits are achenes, similar to lettuce but with a different shape (Colour Plate 6B). The achene is five-angled, truncated, with no beak, and a pappus of numerous scales. Endive achenes are usually very light brown or tan, with scales adhering to the crown of the seed, whereas chicory achenes range from light to dark brown and the scales usually do not adhere. Both are 2–3 mm long and 1 mm wide.

Both endive and chicory are treated as biennials for seed production. Seed can be sown in spring or early autumn, requiring the following winter for vernalization. Beehives are set in the field to facilitate pollination. Both flowering and seed maturing take place over 2 to 3 months. Plants may be desiccated and mowed, or mowed and dried on the ground, and combined for harvest.

4. Quality and treatment

Like lettuce, the germination of both endive and chicory may be subject to both **photodormancy** and **thermodormancy**. Seeds of both are often pelleted and space-planted on raised beds or flat ground, similar to lettuce, or, in some areas, on peat blocks in a nursery and transplanted to the field. The most common pathogen at the young seedling stage is *Alternaria cichorii*, which may be seedborne, and causes black leaf spot. Lettuce mosaic virus is not seedborne (as it is in lettuce), and is not a problem in young seedlings. (EJR)

Rick, C.M. (1953) Hybridization between chicory and endive. *Proceedings of the American Society for Horticultural Science* 62, 459–466.

Rubatzky, V.E. and Yamaguchi, M. (1997) *World Vegetables, Principles, Production, and Nutritive Values*, 2nd edn. Chapman and Hall, New York, USA.

Ryder, E.J. (1999) *Lettuce, Endive and Chicory*. CAB International, Wallingford, UK.

Schoofs, J. and De Langhe, E. (1988) Chicory (*Cichorium intybus* L.). In: Bajaj, Y.P.S. (ed.) *Biotechnology in Agriculture and Forestry*, Vol. 6, Springer-Verlag, Berlin, Heidelberg, New York, pp. 294–321.

Smartt, J. and Simmonds, N.W. (1995) *Evolution of Crop Plants*, 2nd edn. Longman Scientific and Technical, Harlow, UK, pp. 53–56.

CIMMYT

Centro Internacional de Mejoramiento de Maíz y Trigo

Situated in Mexico, CIMMYT is an international research centre for the breeding of wheat and maize. The wheat-breeding work of Nobel prizewinner, Norman Borlaug, at the Institute led to the **Green Revolution**. **(See: Dormancy – importance in agriculture)**

Cleaning

The purpose of seed cleaning is to remove impurities such as: trash, leaves, broken seeds, sand and grit, weed seeds and those of other plant species, and immature, shrivelled, unfilled and empty spikelets. Seed can be cleaned manually or by **winnowing**. Mechanical cleaning is commonly divided into the pre-cleaning stage (for example, using scalpers, brush machines, hammer mills, debearders, huller scarifier and/or delinter/deawners) and the cleaning stage (for example, on an air-screen machine, aspirators, gravity tables, destoners, disc-cylinder- or spiral-separators, and/or by surface texture using roll- draper- bumper-mills or vibratory or magnetic separators, or by colour sorting). **(See: Conditioning, I. Precleaning; II. Cleaning)**

Clone

A genetically uniform group of individuals derived originally from a single individual by vegetative propagation such as by cuttings, divisions, grafts, layers or apomixes.

In molecular biology, a recombinant DNA molecule containing a gene or other DNA sequence of interest. Also, the act of generating such a molecule. Used in the context of 'to isolate a clone' or 'to clone a gene' (**see: Transformation**). In **tissue culture**, refers to a population of identical cells arising from the culture of a single cell of a certain type.

Clovers

Cool-season legumes of the genus *Trifolium*. Most species are native to north temperate or subtropical regions, and all the American cultivated forms have been introduced from Europe. Red clover (*T. pratense*) was the leading leguminous hay crop of the northeastern USA regions until it was surpassed by **alfalfa**. It is frequently seeded with timothy grass. Swedish, or alsike, clover (*T. hybridum*) is similarly used. The common white, or Dutch, clover (*T. repens*) is also cultivated at times but is considered a weed in fields and pastures, where it spreads rapidly. Other plants are sometimes called clover, e.g. the related melilot, or sweet clover. Clover seed is very small (1.4 million seeds per kg) and is harvested for replanting. It is persistent in the soil. (**See: Legumes – forage**) (JDB)

CMS

See: Cytoplasmic male sterility

Coat - seed

See: Seedcoat

Coat-imposed dormancy

A form of **primary dormancy** imposed by the covering structures (e.g. testa, pericarp, endosperm, perisperm, megagametophyte, glumes) surrounding the embryo. (**See: Dormancy – coat-imposed**)

Coating

In seed technology, a term encompassing the family of **enhancements** or 'functional **treatments**' that use filler materials and binders to modify seed form and weight and change or obscure its shape to various degrees, and to apply other seed treatment materials such as nutrient and pesticide formulations (**see: Treatments – pesticides**). The term embraces 'pelleting' 'encrusting' and '**filmcoating**', and also is sometimes used as an alternative name for one or other of the last two individually.

Pelleting and encrusting are primarily used commercially to improve the accuracy of seed mechanical singulation of small or irregularly shaped seeds of field root and salad crops. In particular these techniques are applied to seeds that need to be precision sown in defined patterns or spacings to optimize yield and harvest quality, and in sowing under cover to raise seedling **transplants**: to help ensure reliable filling of the metering mechanisms and reduce the risk of double placement in **planting equipment**. These two coating techniques as well as filmcoating are used to apply other seed treatment materials on to seeds, such as formulations of **rhizobia** or chemical pesticides, in a more secure, uniform and accurate way than can be achieved using conventional application techniques. They are also used to aid the flow of seed that may otherwise be impeded by sticky or rough seed surfaces. Coatings are often distinctly coloured to identify and present seed varieties and treatments for sale in an attractive way, and to aid their visibility in the soil in order to monitor planting depth and spacing efficiency.

Pelleting characteristically produces the most rounded shapes, increasing seed weight by a factor of approx. 2–50-fold (or more in the case of tiny-seeded species like tobacco), compared with a factor of approx. 0.1–2 for coating and encrusting, and <0.1 for filmcoating (Colour Plate 11A). However, the weight build-up factor varies greatly between species, depending on seed geometry, seed size in relation to that of the desired coated product, and the packing density and porosity of the blend of coating materials.

Conventional commercial pelleting, coating, and film-coating types are designed to impose minimal mechanical or physiological barriers on germination. Some pellet types are designed to disintegrate rapidly or split after imbibition to expose the seed. Also, materials used can be tailored to modify seed water availability and gaseous exchange, and so control the timing of germination and emergence, such as to avoid poor seedling establishment. Water-attracting materials can be incorporated to aid imbibition and give more intimate seed–soil contact, or may retain moisture in the vicinity of the seed as soils dry, such as with non-ionic surfactants or hydrophilic gels. Alternatively, hydrophobic materials may be incorporated into coatings to delay imbibition, to allow seeds to germinate that would otherwise be susceptible to imbibitional **chilling injury** and avoid soaking injury in wet conditions (in vulnerable crops, such as certain cultivars of large-seeded grain legumes and super-sweet corn), or to artificially impose 'exogenous coat dormancy'. Oxygen-generating materials (such as Ca or Mg peroxide) can be used to supply more oxygen in waterlogged environments.

By way of distinction, the term '**encapsulation**' (though sometimes used as alternative names for both encrusting and filmcoating) usually refers to the technology of constructing so-called **synthetic seed** from embryos produced by **tissue culture** after **somatic embryogenesis**, or from meristems or buds, using a semi-solid matrix such as alginate.

For the other historical uses of coating techniques **see: Treatments – brief history**. (PH)

Halmer, P. (2000) Commercial seed treatment technology. In: Black, M. and Bewley, J.D. (eds) *Seed Technology and its Biological Basis*. Sheffield Academic Press, UK, pp. 257–286.

Scott, J.M. (1989) Seed coatings and treatments and their effects on plant establishment. *Advances in Agronomy*, 243–283.

Tsujimoto, T., Sato, H. and Matsushita, S. (1999) Hydration of seeds with partially hydrated super absorbent polymer particles. *US Patent No.* 5930949.

Ni, B.-R. (2001) Alleviation of seed imbibitional chilling injury using polymer film coating. In: Biddle, A. (ed.) *Seed Treatment: Challenges and Opportunities*. British Crop Protection Council, Farnham, UK, 76, 73–80.

Cob

In **maize**, this is most frequently used in the sense of 'corn on the cob': the structure that remains after the **kernels** have been removed. The **rachis** or main axis of the inflorescence to which the many **florets** are attached. Each of the ovules contained within them develops into a kernel following fertilization, and they remain attached to the cob at their base, where the embryo lies. During its development, nutrients pass through the vascular system of the cob to the kernel. In maize seed and grain production, seed is removed from the cob by mechanical or manual **shelling** of the dry dehusked ears.

Cobnut

An alternative name for **hazelnut**, filbert.

Cocoa

See: Cacao

Coconut

1. The crop

The monocotyledonous coconut palm (*Cocos nucifera*, 2n = 32, Palmae or Arecaceae) is grown in about 80 countries situated within 20° of the equator. (**See: Crop Atlas Appendix, Map 18**) The coconut ranks sixth in terms of world vegetable oil production (**see: Oilseeds – major**, Table O.1). The kernel or seed within the **nut** contains copra (the dried solid endosperm of the nut) that yields on extraction 65 to 72% **oil** (**see: Oilseeds – major**, Table O.2). About 84% of the copra is produced in Asia, primarily in the Philippines, Indonesia and India, with lesser amounts harvested in Mexico, Papua New Guinea, Vietnam and a host of other countries. In 2002, 53 million t of coconut were produced.

2. Origin

It has not been possible to pinpoint the centre of origin of the coconut palm due to its wide distribution and absence of wild populations. However, most likely it originated in Southeast Asia from where it was distributed to other regions by ocean currents and explorers. Due to the palm's heterozygous nature it has not been possible to segregate populations into identifiable races, but two distinct classes are known, the tall, cross-pollinated *typica* type and the primarily self-pollinating dwarf *nana* form.

3. The fruit and seed

The fruit is a 15–30 cm oval-shaped, fibrous **drupe** weighing about 1 kg and containing one of the largest seeds known in the plant kingdom. A smooth epidermis covers the 5–10 cm thick fibrous **mesocarp** composed of pigmented parenchyma cells in which fibrous strands containing vascular tissue are embedded. The **endocarp** or shell, about 3–5 mm thick, consists mostly of stone cells and some vascular tissue. The outer cells of the thin brown **testa** are mostly spindle-shaped with pitted walls while those of the inner testa layer are thin-walled and yellow-brown in colour. The 1–2 cm thick **endosperm** surrounds a cavity about 300 ml in volume (see Colour Plate 12C). Liquid endosperm (coconut milk) partially fills the cavity. The cells of the endosperm contain **oil bodies** and **protein bodies** with prominent **crystalloids**. The minute embryo is embedded in the endosperm opposite one of the three germ pores.

4. Uses

The coconut palm not only provides oil but also food, feed, **beverage** and shelter. The oil is used in margarine, shortening and soap manufacture. The desiccated pulp is used in confectionery and cooking. The fibres of the mesocarp are used to make mats, brushes and ropes. The shell and husk are used for fuel in copra processing. (**See: Coir**) (KD)

Canapi, E.C., Agustin, Y.T.V., Moro, E.A., Pedrosa, Jr, E. and Bendano, L.J. (1996) Coconut oil. In: Hui, Y.H. (ed.) *Bailey's Industrial Oil & Fat Products*, Vol. 2, 5th edn. John Wiley, New York, USA, pp. 97–124.

Vaughan, J.G. (1970) *The Structure and Utilization of Oil Seeds.* Chapman and Hall, London, UK.

Coffee

There are about 25 species of *Coffea* (Rubiaceae) but only three, from whose seeds the beverage coffee is made, are of commercial importance. Coffee is an extremely important crop, and follows oil as the most highly traded and valued commodity on the international markets. *C. arabica* (arabica coffee), yielding superior-flavoured coffee, accounts for about 90% of world production: many varieties exist, such as moka and typica. *C. canephora* (=*C. robusta*) (robusta coffee) makes up about 9% and *C. liberica* (liberica coffee), used mainly as a filler in other coffees, about 1% of production. *C. arabica* has a chromosome number of 44 as opposed to 22 in *C. canephora*: the former has been suggested to be a mutant type.

1. The plant

Plants of the three important *Coffea* spp., all tropical or subtropical, are evergreen shrubs or small trees, *C. arabica* reaching a height of about 5 m and *C. canephora* about 10 m. The former tolerates higher altitudes whereas the latter grows at lower altitudes in African equatorial forests. Fragrant, white flowers are borne in the leaf axils. After fertilization the fruits (strictly **drupes** but generally called berries) take 7–9 months to ripen, becoming red in the process (Colour Plate 7E). *C. arabica* is self-fertile, a feature that contributed to its transfer from the original habitat and spread throughout colonial possessions (see section 2).

2. Origin and history

C. arabica, the major coffee, originated in what is now the Yemen and in the southern highlands of Ethiopia (in the province of Caffa). Legend has it that the plant aroused human attention in about AD 850 when a goatherd noticed that his animals became especially lively after nibbling the berries, an effect now attributed to the ingestion of caffeine. By at least 1000 years ago the ground, roasted seeds were being used in Arabia to produce a beverage. Later, the trees came under cultivation and the seeds ('beans') were carried throughout the Middle East by Arab merchants based in the port of Mocha, the Yemen. Coffee first became firmly established in the Islamic world, reached Venice in the early 15th century and began its spread through Europe in the 17th century. The Dutch broke the Arab monopoly on coffee by obtaining viable seeds from which cultivation began in the East Indies and later in India, Sri Lanka and the Philippines. In the early 18th century coffee was introduced by the colonial powers into the Caribbean and then Central and South America.

3. Production

Most of the world's coffee is produced in Central and South America but with substantial amounts in Southeast Asia and Indonesia (Table C.18; **also see: Crop Atlas Appendix, Map 20**). Major exporters of green beans are, in descending order, Brazil, Vietnam, Colombia, Indonesia, Guatemala, Uganda and Germany (re-exports): the total value of exports in 2002 was over US$5 billion (FAOSTAT).

4. Seeds

The drupe ('berry') usually contains two seeds ('beans') in the horny **endocarp** ('parchment') surrounded by a fleshy, carbohydrate-rich **mesocarp**. Seeds are flattened on the inner surface where they are adpressed within the fruit. Each seed has a thin testa (seedcoat) which during seed development becomes fragmented as the inner tissues grow: these islands of testa are known as the silver skin. The bulk of the seed is **endosperm**: the small, curved embryo represents less than 1% of the seed mass. The green or blue-green *C. arabica* seeds are 9–16 mm long and 6–8 mm wide: *C. canephora* seeds are somewhat smaller and are yellow or yellow-brown. Arabica seeds are **intermediate** in storage behaviour between **recalcitrant** and **orthodox**. Some other *Coffea* species whose natural habitats have high humidity and rainfall are truly recalcitrant.

5. Seed composition

Green (i.e. in the 'berry') seeds of *C. arabica* contain about 6% water, 12% protein (similar in *C. robusta*), 11% **oils** (slightly lower in *C. robusta*), 70% total carbohydrate, 4% minerals (somewhat higher in *C. robusta*), all on a fresh weight basis. Sucrose makes up about 8% of the fresh weight but the major component of the endosperm is hard, cell-wall **mannans** comprising the main fraction of the seed carbohydrate. Early during seed development the mannan is highly galactosylated (galactose:mannose between 1:2 and 1:7) but as the seed matures the synthesized **mannans** are harder because of the presence of fewer galactose side chains (**see: Galactomannans**); the ratio can reach as low as 1:40. The mannans in the endosperm surrounding the radicle tip may be degraded by endo-β-mannanases prior to radicle emergence (**see: Germination – radicle emergence**). Lipids present in the seed, other than oils, include sterols, sterol esters, triterpenes, waxes and partly-hydrolysed oils. Palmitic, oleic and linoleic are the main **fatty acids**. Potassium is the main mineral element, representing about 20% of the total mineral content. Also present are the vitamins thiamine, niacin and riboflavin. Coffee seeds contain relatively high amounts of phenolics, flavonoids, anthocyanins and alkaloids, the most important of which is caffeine (0.9–1.2% fw in *C. arabica*, 1.6–2.4% fw in *C. robusta*), accounting for the stimulant properties of coffee (**see: Pharmaceuticals and pharmacologically active compounds**). Interestingly, genetic screening of large samples of *C. arabica* has revealed a few individual trees whose seeds have very low concentrations of caffeine (mean 0.076%). It is thought that these plants are **mutants** deficient in the enzyme caffeine synthase which converts theobromine to caffeine: theobromine in fact accumulates in these seeds to reach a concentration of about 0.6%. These mutants may be a source of low-caffeine beans as an alternative to solvent decaffeination. Other seed compounds are trigonelline, chlorogenic acid and the carboxylic acids, citric, malic, oxalic and tartaric acids: these contribute to the unique flavour; and many more are generated by the roasting process (see below).

Table C.18. Production of coffee beans, 2003.

Country	Tonnes
Brazil	1,970,010
Colombia	695,000
Costa Rica	731,126
Ethiopia	220,000
Guatemala	210,000
India	275,000
Indonesia	702,274
Mexico	310,861
Vietnam	771,499
WORLD	7,798,150

Data from FAOSTAT.

6. Processing

Seeds are removed from the berries either by wet or dry processing. In the former (which produces beans with a superior flavour) berries are pulped and allowed to undergo a non-alcoholic fermentation in water for 12–24 h, after which the collected seeds are dried and the endocarps (parchments) are stripped off mechanically. In the dry process, which is generally used for *C. robusta* beans, removal of the mesocarps and endocarps is entirely mechanical. After collection, the beans are dried in the sun for about 7 days, graded and shipped as 'green' beans. They are later roasted for 10–15 min at 200–230°C, the time adjusted to give light or highly-flavoured dark roasts. Roasting brings oily substances to the bean surface and induces many chemical changes including starch hydrolysis, sugar caramelization, the formation of brown pigments (melanodione), the generation of gases that cause the beans to swell, and also cell disruption. Many other chemical reactions occur to produce a vast range of aromatics that contribute to the flavour. More than 500 such substances have been detected including acetic acid, alcohols, aldehydes, esters, furans, guaiacols, hydrocarbons, ketones, mercaptans, pyrazine, pyridine, pyrroles, thiazoles and sulphides.

Decaffeinated coffee is produced usually by solvent extraction (e.g. with ethyl acetate) of steamed, green beans to remove the caffeine prior to roasting. The alkaloid is then partitioned into water and crystallized. Nearly 1 million kg of caffeine is extracted each year for use in the pharmaceutical and soft drinks industries. (MB)

Clarke, R.J. and Vitzthum, O.G. (2001) *Coffee: Recent Developments*. Blackwell Scientific, Oxford, UK.

Clifford, M.N. and Willson, K.C. (eds) (1985) *Coffee: Botany, Biochemistry and Production of Beans and Beverage*. Croom Helm, London, UK.

Dentan, E. (1985) The microscopic structure of the coffee bean. In: Clifford, M.N. and Willson, K.C. (eds) *Coffee: Botany, Biochemistry and Production of Beans and Beverage*. Croom Helm, London, UK, pp. 284–304.

Simpson, B.B. and Ogorzaly, M.C. (2001) *Economic Botany: Plants in Our World*, 3rd edn. McGraw-Hill, New York, USA, pp. 315–322.
Soccol, C.R., Pandey, A., Roussos, S. and Sera, T. (eds) (2000) *Coffee Biotechnology and Quality*. Kluwer Academic, Dordrecht, The Netherlands.

Coir

1. Botany

Coir is a coarse, multicellular, ligno-cellulosic **fibre**, 50–300 μm in diameter and up to 35 cm in length, derived from the thick fibrous **mesocarp** or husk of the **coconut** palm fruit, *Cocos nucifera* L., family Arecaceae. The fruit matures 12–15 months after fertilization and fruits are harvested every 2 months throughout the year. The husk is 35% of the fruit weight. Individual cells are 1 mm long and 5–8 μm in diameter. The surface of the yellowish-white to dark brown fibre may be covered with pores that are occluded by a silicon-rich substance (**see: Fibres**, Fig. F.5A).

2. Chemistry

The high lignin content of coir confers greater resilience and slower decomposition compared with most natural fibres (**see: Fibres**, Table F.1). Coir fibres will stretch 25% without breaking, making coir rope highly elastic. Coir has the greatest resistance to decay in sea water of all the natural fibres. **Phenolic** compounds are abundant, conferring the fibre's dark colour.

3. Origins

The use of coir is prehistoric in islands and coastal regions of the Indian Ocean and Polynesia. Aided by water dispersal, human intervention and domestication, coconut palms are now distributed throughout the tropics. The origin of coconut palms is uncertain and is an active subject of enquiry.

4. Location

Nearly all of the total world production of coconut, estimated to be 5×10^{10} nuts/year, comes from Asian and Pacific countries. The majority of coir is utilized in the countries of origin with India, Sri Lanka, the Philippines, Indonesia and Thailand being the largest producers. There is also some coir production in western Africa.

5. Production

There are three types of coir. The longest fibres, made from immature (green) coconut, are called 'mattress or white fibre' and can be spun into yarns for twine, rope and matting. Of the two types of fibre made from dried coconut ('brown fibre') one is very coarse and is suitable for brush bristles while the other is a short staple fibre used for stuffing mattresses and furniture. One thousand coconuts will produce 18 kg of coir bristle fibre and 72 kg of mattress fibre. Coir is also being used in non-traditional ways for erosion control mats and other geotextile applications.

6. Processing

To produce coir fibre, husks are first soaked in fresh or salt water for 3–9 months (retting) until **pectins** and polyphenolic compounds can be partially degraded by microbial action. This makes it easier to divide the husks into component fibres, which are then beaten with wooden paddles or rolled to strip the fibres away from the pith. On a smaller scale, mechanical fibre extraction or dry milling is also used. Decortication occurs when metal bars, beating at high speeds, separate fibre from pith and other components. Fibres are sun-dried and graded for staple length, colour, and amount of pith remaining. Mattress fibres are often twisted by hand or machine into long coiled ropes. Bristle fibres may be hackled or combed through steel spikes to reduce the short fibre content.

7. Economics

Only 10% of the coconuts produced in the world are used to extract fibre. In 2002, world production of coir was estimated to be approx. 637,000 t. Less than 150,000 t of coir is traded globally. (BAT)

Bhat, J.V. (1983) The fibre from coconut fruits. In: Nayar, N.M. (ed.) *Coconut Research and Development*. Wiley Eastern, New Delhi, India.
Maiti, R. (1997) *World Fiber Crops*. Science Publishers, Inc. Enfield, NH, USA.
Orpeza, C., Verdeil, J.L., Ashburner, G.R., Cardeña, R. and Santamariá, J.M. (1999) Current advances in coconut biotechnology. *Current Plant Science and Biotechnology in Agriculture*, Vol. 35. Kluwer Academic, Dordrecht, The Netherlands.
Varma, D.S., Varma, M. and Varma, I.K. (1984) Coir fibers. Part I. Effect of physical and chemical treatments on properties. *Textile Research Journal* 54, 827–832.
Woodruff, J.G. (1979) *Coconuts: Production, Processing, Products*, 2nd edn. Avi Publishing Company, Westport, CT, USA.

Cola (kola)

The *Cola* genus (Sterculiaceae) includes 90 species, 50 of which are endemic to West Africa, especially Nigeria. Only two species, however, are of commercial importance, *C. nitida* and *C. acuminata*, seeds of which, on account of their stimulant properties, are used as a masticatory or as the base of a **beverage**: the former species is the more important. Seeds are known as kola or cola nuts, *guru*, *bissy* nuts, and, in parts of West Africa, *banja* and *abata*. Kola nuts have been an important trade commodity in sub-Saharan Africa for thousands of years.

In the same family and closely related to **cacao**, trees of *Cola* are taller, reaching up to 18 m. The species are native to tropical West Africa but are now cultivated also in the Caribbean, Central and South America, and Southeast Asia. The two species are closely similar but can be distinguished by their seeds (see below).

1. Seeds

Cola fruits (referred to as capsules or pods) are leathery or woody, about 15 cm long, and contain approximately eight almost globose white or reddish seeds (kola 'nuts'), about 3 cm in diameter (Colour Plate 7F). The embryos of *C. nitida* bear two cotyledons while those of *C. acuminata* frequently reach six or more.

2. Seed composition

On a fresh weight basis *Cola* seeds contain about 9% protein, 1% **lipid**, 10% **starch** and some sucrose but their most important components, as far as seed use is concerned, are the psychoactive compounds caffeine (2–2.5% fw), traces of theobromine, theophylline, catechine and the glucoside kolanin (**see: Pharmaceuticals and pharmacologically active compounds**). Several phenolics are present such as kolatin which is enzymatically converted to the red phlobaphene (kola red) as the seeds are dried.

3. **Processing and uses**

The coats are removed and the kernels are allowed to ferment ('sweat') for several days during which the characteristic flavour develops: they are then dried when they redden. The dried kernels are chewed or powdered for preparation with water into a beverage.

Cola 'nuts' are used because of their pharmacological effects, due largely to the relatively high caffeine content. They have properties as stimulants, aphrodisiacs, to allay hunger and fatigue, and to aid digestion. When used as a masticatory the nuts also create an illusion of sweetness thus pleasantly adding to the taste of food and water. Cola is commonly used socially in Africa and other parts of the world among friends and guests. It has wide usage in Muslim countries perhaps because of the prohibition of alcohol. The nuts also have medicinal uses especially for skin ailments and infections. Extracts of cola nuts are added in certain soft drinks (e.g. colas). (MB)

Duke, J.A. (1989) *CRC Handbook of Nuts*. CRC Press, Boca Raton, Florida, USA.

Colchicine

An alkaloid ($C_{22}H_{25}NO_6$) extracted from the seeds or corms of the autumn crocus or meadow saffron (*Colchicum autumnale*), highly toxic to animals (**see: Pharmaceuticals and pharmacologically active compounds**), and used in plant breeding programmes to induce polyploidy. The resulting nucleus subsequently undergoes normal mitosis, giving rise to polyploid tissue. Colchicine acts by dissociating the mitotic spindle and preventing the migration of daughter chromosomes to opposite poles; where the resulting nucleus undergoes subsequent normal mitosis, a polyploid tissue is formed. (**See: Polyploids**)

Cold test

Physiological vigour test based on the application of stress during germination to induce **chilling injury** in sensitive species. The forms of test commonly used in AOSA procedures for maize and vegetable sweetcorn seed, for example, involve the assessment of **normal seedlings** after incubation at 10°C followed by a further period at 25°C in the dark (optionally covered with a thin layer of field-soil that may be mixed with sand, **vermiculite** or **peat**, to provide microbial activity) either on wetted rolled paper towels or on paper in trays, using seeds that have received their **treatment pesticides**. Cold germination protocols are also commonly used in the testing of soybean and cotton seed. (**See: Cool test; Vigour tests – physiological**)

Cold vigour tests should not be confused with the cold pre-chilling treatments that are necessarily applied in seed germination testing to break **dormancy** in certain species, such as trees and shrubs. (PH)

(**See: Dormancy breaking; Germination testing; Stratification**)

Coleoptile

The embryonic tissue that covers the shoot meristem in cereal grains. The coleoptile develops from the proembryo as a primordium adjacent to the shoot meristem and later encircles the shoot meristem (**see: Development of embryos – cereals**). Through congenital fusion of its margin, the coleoptile fully encloses the shoot. After germination, the coleoptile is the first etiolated, leaf-like structure to appear above the soil, and its base forms the first node of the vegetative plant. The green, fully differentiated leaf that it surrounds breaks through the coleoptile, after which it dies. The coleoptile is thought to function as a tissue protecting the shoot following germination, during its passage through the soil. Consistent with this function, growth of coleoptile after emergence is known to respond to various environmental signals such as light. Some authorities support the hypothesis that the coleptile corresponds evolutionarily to one of the two cotyledons of dicot plants, the other being the **scutellum**. (**See: Hypogeal**, Fig. H.9; **Structure of seeds**) (O-AO, HS)

Coleorhiza

A sheath-like structure that covers the radicle in Poaceae grains (e.g. cereals). Unlike the coleoptile, the coleorhiza develops as a structure surrounding the whole of the radicle from the start of development. (**See: Development of embryos – cereals**)

Collection

Scientific seed collection is the harvesting of seeds or fruits of adequate quality and quantity for research, breeding, propagation, **conservation** or restoration purposes (Fig. C.9). A large quantity (e.g. over 5000) seeds in a collection is an advantage because seed may be required for initial testing and periodic viability monitoring, for propagation or distribution. In **gene bank** storage, losses can also be anticipated due to **ageing**. Collectors of seed use visual cues (a 'search image') to maximize the harvesting of seed of high physical and physiological quality. Specific population sampling approaches can be used to obtain the desired genetic variation in the seed collection.

1. **Seed quality**

Quality of a collection includes visible elements such as the absence of empty, infested or malformed seeds (physical quality), as well as non-visible, physiological attributes such as **viability**, **desiccation tolerance** and potential **longevity**.

(a) *Physical quality*. This is determined by the frequency of empty, infested, damaged or malformed seeds in a seed lot, and can be estimated by performing a cut test or by making an X-ray image of a sub-sample. Cut tests are transverse and/or longitudinal sections of seeds typically performed prior to seed collection or harvesting, and in the course of **viability testing**, e.g. to assess non-germinants at the conclusion of a germination test (Fig. C.10).

Empty (i.e. unfilled or non-embryonic) seeds are produced when a fertilized ovule fails to develop into a filled, mature seed. The phenomenon is thought to be largely due to a deficiency in available resources (nutrients) from the parent plant during seed development, and is prevalent in non-domesticated

Fig. C.9. Collecting tree seed using pruners and tarpaulin. Images © Royal Botanic Gardens, Kew reproduced with permission.

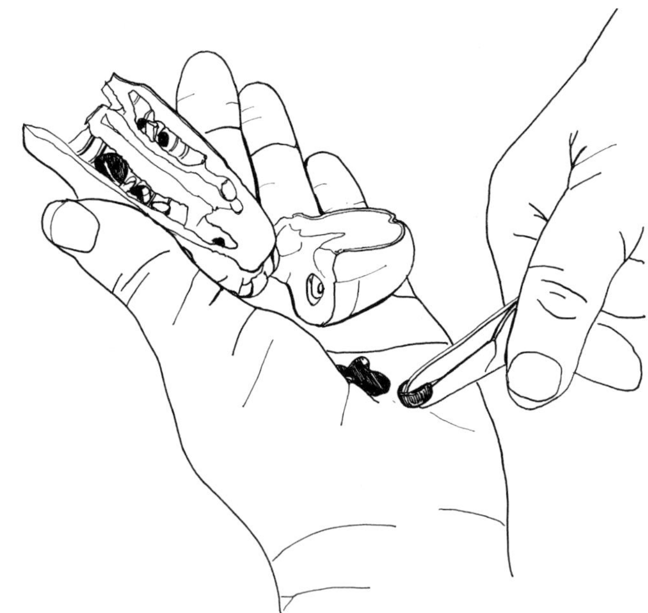

Fig. C.10. A cut test being performed on seed of *Yucca* sp. prior to collection. Images © Royal Botanic Gardens, Kew reproduced with permission.

species, especially of the Poaceae, Cyperaceae and Asteraceae. Empty seeds are characterized by near-normal seedcoat but an absent or under-developed embryo or storage tissue. All empty seeds, of course, are non-viable in germination tests.

Infested seeds are occupied, predated or damaged by invertebrate pests or microbial pathogens (**see: Pathogens – seedborne infection**). Such seeds will often be inviable. Infestation by pests can be initiated at any stage in the life cycle from flowering (e.g. the moth *Tegeticula maculata* Riley [Prodoxidae] which oviposits into flowers of *Yucca whipplei*) to post-dispersal infestation (e.g. by grain storage weevils [Coleoptera]) including after incorporation into the **soil seed bank**. Seed pests also include **weevils** and bruchid beetles, Lepidoptera (moths and butterflies) and Hemiptera (bugs). Microbes (bacteria, viruses and fungi) can also cause significant losses to seed harvests but, as in the case of insect predation, the proportion of seed affected can vary greatly according to the species, the year and the environment. In the most severe infections, discoloration and/or malformation of the seedcoat or embryo will be evident. In addition to the reduced viability and **vigour** expected for these seeds, they may act as agents for transmission of the pathogens into the remainder of the seed collection. It is known, for example, that endophytic fungi can remain viable under seed gene bank conditions.

(b) *Physiological quality*. For **orthodox** species, maximum physiological quality is normally reached around the time of natural **dispersal** and this is the optimum moment to collect seeds for long-term conservation. In certain circumstances it may be necessary to collect fruits prior to dispersal whose seeds are at the post-abscission stage. Such collections should then be held under ambient conditions in order to complete seed ripening processes and maximize potential longevity.

The physiological quality of a potential collection may be estimated by assessing the stage of seed development and ripening. This is done by observing the colour and hardness of the seedcoat, the size of the embryo and the texture of storage tissues. Fruit characteristics such as colour, dehiscence (in the case of dry fruits) and pericarp colour, texture and odour (fleshy fruits) may also be useful indicators.

Non-domesticated plant species often present non-synchronous seed dispersal, characterized by population heterogeneity, **indeterminate** inflorescences and **shattering** seed heads. Thus, pre-collection assessment of physiological seed quality needs to be carried out at the level of the seed, the individual plant, and the seed population.

2. Sampling (genetic)

This is the harvesting of seeds in a manner likely to result in a collection that is genetically representative of the source.

Initial sampling captures monomorphic **alleles** (i.e. those which are always present at a given locus) and other common alleles. As the expected number of alleles in a seed sample increases in proportion to the logarithm of sample size, there is a declining genetic benefit of continued sampling from the same source. If gene flow is significant between all populations of a plant species, random and even sampling of seed from a single population is likely to result in a collection representative of the genetic diversity of the species. Such seed collections are useful for breeding or ***ex situ* conservation**.

In practice, there are four main sampling approaches that may be used to ensure that the collection is truly representative of the variation sought:

- Simple random and even (each individual has an equal chance of selection).
- Biased (selecting seed from individuals that display desired attributes, e.g. for breeding or propagation).
- Stratified (combining sub-samples taken from each distinct sub-population identified by the collector).
- Systematic (consistent sampling from a series of transects across the sampling source).

(See: Tree seeds)

(MW)

Bonner, F.T., Vouo, J.A., Elam, W.W. and Land, S.B. (1994) *Tree Seed Technology Training Course, Instructors Manual.* USDA Forest Service General Technical Report SO-106.

Smith, R.D. (1995) Collecting and handling seeds in the field. In: Guarino, L., Ramanatha Rao, V. and Reid, R. (eds) *Collecting Plant Genetic Diversity*. CAB International, Wallingford, UK, pp. 419–456.

Way, M.J. (2003) Collecting seed from non-domesticated plants for long-term conservation. In: Smith, R.D., Dickie, J.B., Linington, S.H., Pritchard, H.W. and Probert, R.J. (eds) *Seed Conservation: Turning Science into Practice*. Royal Botanic Gardens, Kew, UK.

Colour sorting

This quality-upgrading **conditioning** technique removes defective individual seeds, one at a time, based on differences in reflected colour or brightness, under one or two selected visible or infrared light wavelengths. It also can remove seed-sized foreign material, such as small pieces of glass, husks, mud balls, rocks, sticks and stones. Both commodity grain traders and food industry processors use colour-sorting to remove gross or often subtle discolorations caused by insect or disease damage or by differences in seed maturity, for example, before, between or after the rice hulling, milling and polishing, nut shelling and blanching, or coffee roasting steps (Table C.19). The technique is also used to upgrade seed for sowing, especially high-value flower and vegetable seed species – for example, to remove off-colour individuals.

Seeds are fed from a vibratory feed in single file down a slide channel or between two counter-rotating smooth rollers, and drop through an optical inspection zone (for example, containing up to three photoelectric cells, with associated mirrors, lenses and light sources) positioned above an air ejector. Seeds that differ according to preset adjustable discriminator settings, based on reflectance and spectral composition, trigger the ejector to blow them one by one into the reject spout with a short blast of compressed air, while acceptable seeds drop into the collection spout. Key engineering parameters include the presentation of seeds one at a time to the detector. Colour sorting equipment is usually operated in parallel or series (for example, to reclaim high-value seed from previously colour-rejected material fractions), and individual machines may have from about five to 80 independent sorting channels. Such processing arrays can be capable of very high sorting throughputs for large seed, for example, at up to 10 t/h, depending on the model and seed type.

In conventional colour sorting, immature seeds are rejected on the basis of their pale green colour. A new variation on colour sorting uses laser-induced fluorescence to detect the residual level of chlorophyll in seedcoats, at levels that may not be visible to the naked eye. This new technique, also known as 'chlorophyll sorting', is of commercial value in upgrading

Table C.19. Examples of off-type seeds removed by colour sorting.

Seed (and variety/types)	Typical seed contaminants removed[a]
Bean, lentil, pea	Discoloured seeds, frost damaged, sprouted
Coffee (*arabica* and *robusta*)	Red, slightly mottled, yellow/grassy, insect-damaged, black or spotted, unroasted
Maize ([a] white and yellow, popcorn [b] seed corn)	[a] Field corn, damaged and mouldy kernels, also [b] other seeds and immature kernels
Peanut (groundnut) (redskins, blanched, roasted, splits, in-shell and seed for sowing)	Damaged discoloured nuts, in-shell (nubs), nut grass, maize, raisins
Rice (brown, white or parboiled)	Damaged, red, chalky, stained
Soybean (processing seed and edible)	Discoloured, immature, mosaic virus- or downy mildew-damaged, and foreign seeds
Sunflower (shelled meats and in-shell seeds)	Sclerotia
Tree nuts (for example, pecans, walnuts, hazelnuts, pistachios, pine nuts, cashew nuts, macadamia pieces and whole nuts, and raw, in-shell and blanched almonds)	Shell, discoloured, insect-damaged, chipped and scratched
Vegetable and flower seed (for example, beans and peas, carrot, cucurbits, lettuce, tomato)	Discoloured or off-coloured, immature and foreign seeds

[a]Other foreign material removed may include seed-size pieces of: glass, husks, mud balls, rocks, shells, soil, sticks and stones.

tomato, pepper, leek, cucumber and cabbage seed lots, as well as flower seeds, and specialist equipment has been designed to do this on a production scale.

Near infrared spectroscopy (NIRS) – a non-destructive quantitative analytical technique widely used for rapid chemical composition measurement of various agricultural products, such as in commodity **grain inspection** – also offers the prospect of characterizing and selecting out empty, non-viable, or insect-damaged seeds through their compositional differences, or their differential ability to take up moisture in seeds with hard and impermeable seedcoats. (PH)

(See: Conditioning, II. Cleaning, other sorting principles)

Jalink, H., van der Schoor, R., Frandas, A., van Pijlen, J.G. and Bino, R.J. (1998) Chlorophyll fluorescence of *Brassica oleracea* seeds as a non-destructive marker for seed maturity and seed performance. *Seed Science Research* 8, 437–443.

Tigabu, M. and Oden, P.C. (2003) Near infrared spectroscopy-based method for separation of sound and insect-damaged seeds of *Albizia schimperiana*, a multipurpose legume. *Seed Science and Technology* 31, 317-328

Coma

A seed appendage assisting wind dispersal (anemochory); one- or two-sided tufts of hairs in plumed seeds, usually formed by the testa in the region of the exostome (exostome aril) and/or chalaza; e.g. Apocynaceae-Asclepioideae (with exostomal tuft), Onagraceae (*Epilobium* with chalazal tuft). **(See: Seedcoats – dispersal aids)**

Compatibility

A self-compatible species, variety or line capable of producing progeny as a result of **self-fertilization**, leading to inbreeding when in isolation.

By contrast, self-incompatibility restricts self-fertilization and inbreeding and promotes cross-fertilization and outbreeding. Incompatibility may be caused by the failure of the pollen tube either to penetrate the stigma, or to grow along the entire length of the style or at a sufficient elongation rate. Incompatibility is widespread in nature, and is present in cultivated species of legumes (such as clover and alfalfa), brassicas, sugar beet, tobacco, composites (such as sunflower) and some grasses and cereals (such as fescue, ryegrass and pearl millet), amongst others. The so-called **self-incompatibility system (SI)**, has been exploited in the breeding of hybrids, particularly of some diploid horticultural brassicas. **(See: Brassica – horticultural)**

Compatible solutes

Low molecular weight organic compounds that are uncharged or zwitterionic (have properties simultaneously of an acid and a base) and do not interfere with the cellular structure and function. They accumulate in cells exposed to hypertonic conditions, and include non-reducing sugars, polyols, amino acids (proline) and amino acid derivatives. They play a role in stabilizing proteins against water stresses occurring while the protein's hydration shell is still present. **(See: Desiccation tolerance – protective mechanisms)**

Composite

In plant breeding: a mixture of genotypes from several sources maintained by normal **pollination**.

In seed production: a mix of samples representing an identifiable area of bulk grain or a bin, or seed lot being processed for sowing or processing. **(See: Sampling)**

In plant taxonomy: the traditional name for species belonging to the family Compositae (for which Asteraceae is now the recommended name), so-called after their characteristic flower head consisting of multiple florets composed of a ring of female ray flowers around the outside and disk flowers in the centre, in various proportions from one species to another. The Asteraceae contribute **sunflower**, **lettuce**, **chicory/endive** and **safflower** as major world crops, and marigolds and chrysanthemums to the cultivated ornamental flowers.

Compost

The product resulting from the managed aerobic decay and rotting of feedstock materials from vegetable and animal sources, used to germinate and propagate seeds and bedding or plants (as 'potting mix') or to **mulch** or condition soil. The term is strictly defined, in the words of the USA EPA for example, as the 'controlled biological decomposition of organic material that has been sanitized through the generation of heat and "processed to further reduce pathogens" (PFRP), and stabilized to the point that it is beneficial to plant growth'.

Less precisely, the term is generally and informally applied to the range of 'growing medium' substrates – made up variously of loam, **peat** or peat-substitute, mixed with **sand**, grit, **coir**, bark, **perlite**, **vermiculite**, with appropriate particle sizes, pH, salt and plant-available micro and macro nutrient contents and biological stability. Such proprietary mixtures are commercially available as germination media for **transplant** and **hydroponic production** (which also use **rockwool**), such as to fill **cell trays**.

Conditioning

The broad term used to describe the processing of seed or grain, embracing the procedures, operations and equipment used to remove contaminants, weeds, inert matter and poor quality seed (Fig. C.11): equivalent to the term 'dressing'. Conditioning takes advantage of differences in physical properties that are great enough to separate 'good' from 'inferior' seed and its impurities – such as size, weight (specific gravity), shape, length, surface texture and colour. As a general rule of thumb, each step in the conditioning process is operated at the expense of removing a few good seeds along with the discarded components.

The term 'conditioning' is also applied to some **priming** physiological enhancements, such as 'osmoconditioning' (**osmopriming**), and is also used in an entirely different context to describe the **hardening** process of seedling transplants grown under protected conditions before transfer to harsher field conditions, by reducing watering, shelter and fertilizers, and so on. **(See: Transplants)**

The ability to provide high quality seed that meets statutory and commercial requirements, in the quantities required, depends on the equipment, design, layout and management of seed conditioning plants. Facilities range from small on-farm units to large commercial installations, with a high degree of

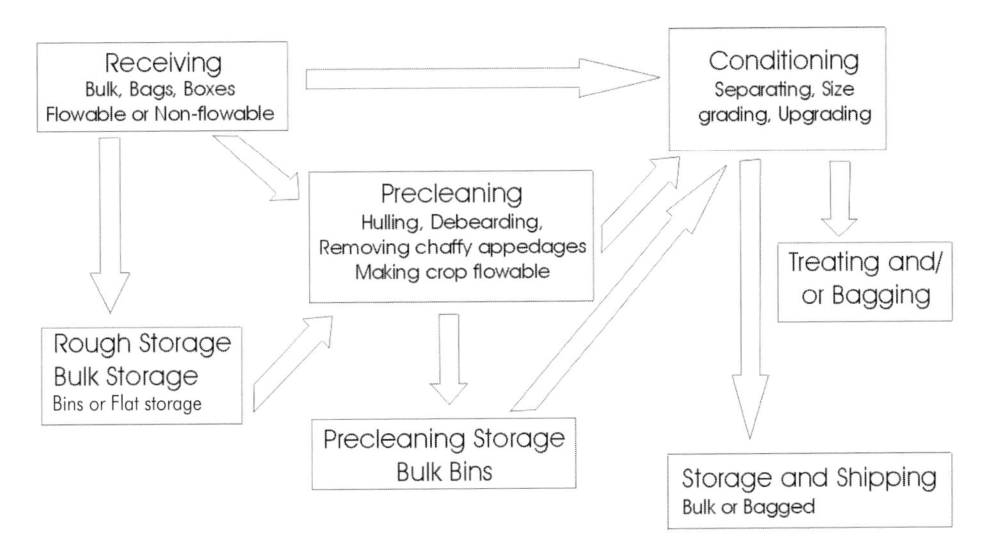

Fig. C.11. Flow diagram to show the steps involved between the receipt of seeds after harvesting and the bagging of seed for sale and re-sowing. The steps involved in Precleaning and Conditioning are detailed.

mechanization and many processing steps. The most commonly and widely used equipment comprises air-screen machines, indented-disc and -cylinder length separators, gravity tables, gravity spiral separators and roll mills. These are supplemented for processing of some particular species with dimensional sizers, surface texture separators and 'electronic eyes', amongst others. (**See: II. Cleaning**) During conditioning, seed may be graded into one or several size and shape fractions to facilitate subsequent **pelleting** and mechanized sowing. (**See: Planting equipment – agronomic requirements**) The conditioning sequence may require seeds to be precleaned to maximize efficiency (**see: I. Precleaning**) and, in the case of some contaminants, an initial separation must be made in the field.

Conditioning plants must be designed to ensure that seeds are handled in the proper sequence to remove all contaminants and maximize **pure live seed** percentage, without bottlenecks. The production line needs to be flexible so that it can be changed or organized to handle each kind or lot of seed in the most efficient manner. Facilities should be easily cleanable, with adequate receiving areas – such as a bulk bin, pit, or flat storage floor – in which seed can be efficiently unloaded and held in short-term storage. A complex combination of machines requires coupling with surge bins, **conveyors** and elevators to move the seed efficiently throughout the plant and to the bagging or bulk bin storage stage, and gently to avoid mechanical injury. Multi-storey facilities, where seed flows by gravity from one stage to the next, are desirable but not always possible. Fig. C.12 shows an example of the sequences of machines that are used to condition seeds, in this case grass seeds, but some of the steps are not necessary for all crops (e.g. for soybean no delinting or debearding is required, and separation requires a different type of separator, a spiral separator – see Table C.20 in **II. Cleaning. 5. Dimensional sizing**). Some of the same conditioning equipment is used in industrial feed and food grain processing plants, though these installations usually operate at appreciably higher throughput flow rates than seed conditioning plants, which require more and higher quality standards. (**See: Quality Testing**)

I. Precleaning

Rough cleaning (also known as preconditioning) of seed that is hard to handle because of its natural attachments or its content of a large amount of rubble, trash, straw, green leaves, weed seeds and insects, to increase the capacity and efficiency of the subsequent drying process or the cleaning line (**see: II. Cleaning**) and reduce the volume to be stored. The precleaning operation also may include removal of **awns**, hairs (**trichomes**), **lint**, spines and **chaff** from such as oat, barley, and some umbellifers and grasses, and hulls or pods from crambe, crownvetch, canola and birdsfoot trefoil, to complete the **threshing** operation started in the field.

'Special conditioning equipment' can also be conveniently classified within the precleaning stage. For instance, some seed may have to be subjected to **scarification**, such as by abrading or scratching the hard coat. Also dry indehiscent pods and fruits must be broken mechanically to extract the seeds, notably in some **tree** species, using a variety of mechanical techniques depending on their size and strength – such as beating, flailing, rolling with drums, or tumbling with blocks of wood or in grinding mills.

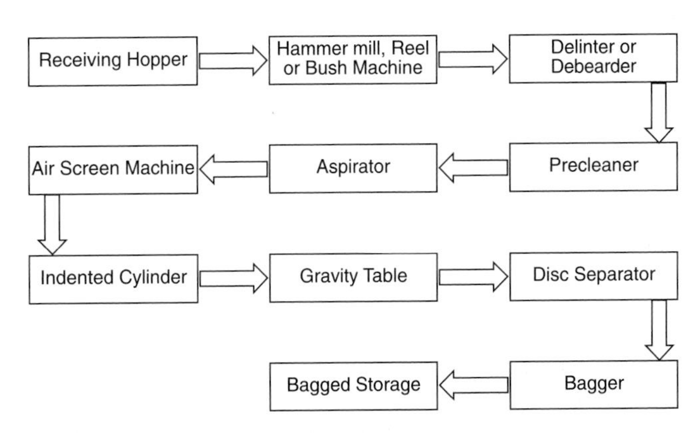

Fig. C.12. Steps involved in the conditioning of Kentucky bluegrass seed. As used in Habstritt's conditioning plant, Roseau, MN, USA.

(a) *Scalpers* remove large foreign material and weed seeds with screens that either have a shaking or rotating action. At its simplest, large material is screened off, 'scalped', while the desired seed material passes through the screen or sieve. Small weed seeds, along with large trash, can be almost completely removed in more complete precleaners such as simple air-screen machines (see: **II. Cleaning**), equipped with scalping and grading screens as well as aspiration. Other designs employ a rotating perforated metal screen, called a reel, through which seed is fed and tumbled. Scalping may make further precleaning unnecessary but some crops may require further de-awning, delinting, hulling or scarifying to make the product flowable so that it can move freely through the final cleaning equipment.

The 'Scalperator' is a more elaborate machine incorporating scalping and aspiration principles, and is most commonly used to preclean field-run material (such as wet paddy rice, maize, wheat and beans, at up to about 30 m^3 per hour) and help protect against flow problems or machine damage in subsequent drying, conveying and conditioning equipment. As well as removing large trash, the machine removes some insects and fine impurities to help minimize infestation with storage pests.

(b) *Brush machines* are equipped with rotating nylon brushes that sweep the inner surface of a stationary wire mesh basket screen into which seed is fed. As the product is brushed against the mesh, soft hulls are removed, soil lumps are broken up, and the surface of the seed is polished; the cleaned material is discharged at the end of the basket for further conditioning. The brush machine is used as a speciality machine to improve appearance and flowability, particularly with more costly seed, such as for de-awning flower seeds and native grasses, polishing vegetable seeds, and separating grass seed doubles. It is also used in some bluegrass conditioning plants to delint seed, which is collected through perforations in the wire mesh basket.

(c) *Hammer mills* are used to complete the threshing process of species such as grasses (e.g. wheatgrass, wild rye, bluegrass) and birdsfoot trefoil by removing awns, hairs, lint, chaffy appendages, and break seed loose from pods and seed clusters. The de-awned or trimmed seed is rubbed and rolled through screen openings, perforated with either slotted or round holes. The efficiency and damage limitation depends on the cylinder speed, size of the openings, and the rate of feed and the moisture condition of the seed.

(d) *Debearders.* Seed is passed between fixed and variably rotating beater arms, whose rubbing action breaks appendages, removes hulls, and threshes seed clusters or pods, while auguring seed to the discharge gate. Designed mainly for barley, debearders are now also used for oats, to hull or remove 'whitecaps' from wheat, to decorticate sugarbeet seed, to break up flax balls, legume pods and seed heads, and to de-awn or delint some grass seed to complete the threshing process.

(e) *Delinters (or Deawners)* have a horizontal drum into which bolts protrude by about 10 cm, mounted inside a housing equipped with a suction fan. The drum rotates at about 900 rpm to fluff-up the seed mass while the bolts beat and move it through in an auguring fashion, and the loosened lint or awns are drawn off in the air stream. This equipment is used in the bluegrass industry as a conditioning step between the hammer mill and the air-screen precleaner. **See also Delinting** for the specialist process used in cotton.

(f) *Huller-scarifier.* Some seeds of the grass and legume families need to be hulled (e.g. bermudagrass), scarified (e.g. alfalfa, white clover) or both (e.g. crown vetch, sweet clover). Hulling is the removal of the outer coat or husk on the seed while the seed itself remains unscathed. This makes the seed permeable to water, improves planting characteristics, or makes it easier to condition. Scarification is the scratching of the seed coat to make seeds permeable to water so they can germinate promptly and evenly. These processes may be accomplished separately or in a combination huller-scarifier. Seeds fall on a rotating distributing disc, which throws them against an abrasive carborundum or hard rubber stone ring. Scarification is a delicate operation, usually carried out after seed has been conditioned on an air-screen machine, and the severity of the abrasion or force of the impact must be controlled to prevent seed damage. (**See: Scarification**)

(g) *Dewingers.* Wings and, more broadly, the spines, hairs and some aril types attached to the dry fruits of wind-dispersed species (notably in tree species, such as conifers – **see: Seedcoats – dispersal aids**; Fig. G.35) are removed to reduce seed bulk and for convenience in handling. Methods vary considerably depending upon the strength or delicacy of the structures – which vary from very thin and membranous to hard and woody in most samaras – and may involve abrasion of slightly wetted seed tumbling in a drum with or without sand and gravel, abrasion by brushes or between brushes revolving against wire mesh, dehuskers or hammer mills or manual clipping.

II. Cleaning

The purpose of seed cleaning is to remove impurities such as trash, leaves, broken seeds, sand and grit, weed seeds and those of other plant species, and immature, shrivelled, unfilled and empty spikelets. Traditionally, and still today, seed is cleaned manually by winnowing, and in developed supply systems through one or several machines set in proper sequence to maximize the pure live seed content. The air-screen machine is usually the first cleaner (after any necessary precleaning, **see: I. Precleaning**), and removes the bulk of foreign material from the lot. Final separations are made on finishing machines in which separation is based mainly on one physical characteristic, such as length or width. Small capacity versions of equipment based on the same operating principles are used in many laboratory and small production environments.

1. Air-screen machine

Almost every kind of seed, ranging in size from grasses to large-seeded beans, must be cleaned over an air-screen machine before any other separations can be performed. Indeed some crop seeds can be cleaned completely using this equipment, and no further conditioning is necessary. Through a combination of airflow and oscillating screens, seed is separated on the basis of size, width, thickness, specific gravity and resistance to airflow. Air-screen machines, which were developed from simple hand-powered fanning mills (**see: Winnow**), are therefore the basic cleaner in most seed conditioning plants throughout the world.

The air-screen machine (Fig. C.13) uses three cleaning principles: (i) aspiration, whereby light material is removed

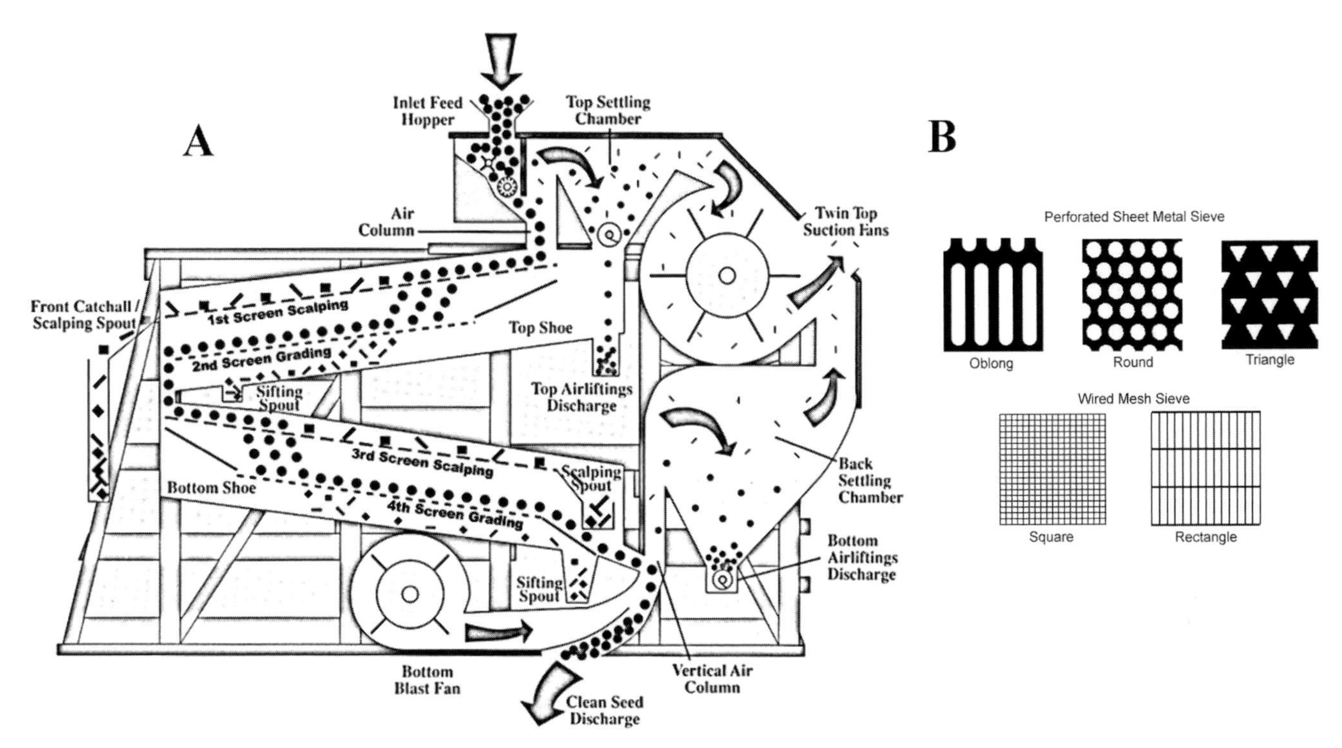

Fig. C.13. (A) A four-screen example of an air-screen machine, with (B) screen types. Seed passes through an upper air suction chamber, which removes light seed, chaff and dust before landing on the top rough scalping screen, which removes long material such as sticks, straw, and long contaminating seeds. The perforations in the second (grading) screen are just large enough to allow small seeds, weeds and dirt to drop through and be removed. The third close-scalper screen removes remaining foreign matter that is larger than the desired seed. Seed passing over the fourth close-grading, or sifting, screen drops through an upwardly-moving air current which lifts away remaining light seed and trash (along with a few good seeds). Screen perforations are kept open by tappers or knockers, or from underneath by brushes that travel back and forth or by rubber balls that continually bounce up against them. Courtesy of Cliffton, Bufton, IN, USA.

from the seed mass in a current of air; (ii) scalping, using a screen large enough to allow good seed to drop through it, but small enough to remove larger contaminants; and (iii) grading or sifting, with a screen small enough to allow the desired seed to pass over while small contaminants drop through. The scalping and grading screens, typically mounted in pairs at shallow angles (4–12°), are vibrated or shaken. Types of hoppers, screen arrangements and numbers, and air system (suction or blast) vary, and machines are made in many sizes ranging from small two-screen models, used to clean seed samples, to large industrial cleaners, with seven or eight screens and three or four air separation stages and handling capacities of up to about 2.75 t of seed per hour.

2. Aspirators (vertical air-column separators)

Aspiration is widely applied in the conditioning of wheat, barley, sunflowers, flax, bluegrass or timothy grass. The common principle is that, when the seed mixture is introduced into a confined upward-moving air stream, light particles with a low terminal velocity are carried along with the flow whereas heavy seed falls through; the light material is conveyed into an expansion chamber where the air speed drops considerably, so that the material drops down and is recovered (Fig. C.14).

In 'tabletop' cleaning equipment, 'Seed Blowers' utilize airflow through a column to trap lightweight seed and chaff, and 'Air Blast Seed Cleaners' are used for final cleaning of grain from large debris, chaff and light debris. Aspiration and a combination of riddles and sieves are used to remove readily separable foreign matter in **grain inspection** procedures, to assess grading and **dockage** content.

3. Gravity table (grading deck)

Specific-gravity separators separate seeds that are similar to the desired seed fraction in terms of size, shape and seedcoat characteristics but differ in unit weight or specific gravity. The separator not only removes undesirable seed (such as insect-damaged, deteriorated, rotten or mouldy seeds, and sterile florets present in chaffy grasses), weeds, 'other crop' seed and inert matter, but is also the best machine for upgrading quality in a seed lot. It is used only after air-screen and dimensional sizing equipment.

Several types and styles all use the same principle of separation, as seeds flow across an inclined porous reciprocating surface. Seed flowing on to one side of the deck is fractionated and discharged at different positions along the opposite open edge (Fig. C.15). A triangular deck is good for separating a small fraction of heavy seed from a large fraction of lighter seed, whereas a rectangular deck is good for reducing the amount of middlings and removing a small fraction of light seed from a large fraction of heavier seed.

A variation on this sorting principle is used in the oscillating table separator used in cleaning some tree seeds. A slightly inclined table with zig-zag partitions along its length is shaken sideways so that seed placed in the middle of the table is struck

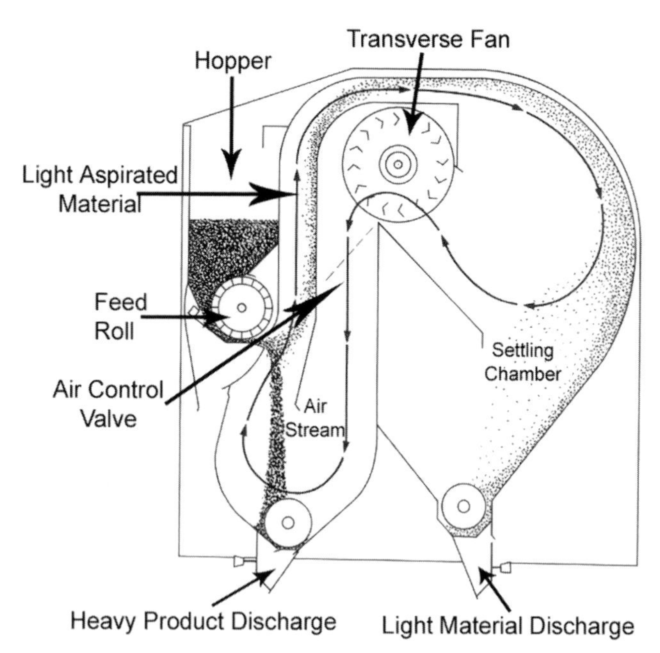

Fig. C.14. Aspirator machine. The volume and velocity of the air determines the cut-off point of the separation at which light material is lifted. 'Closed circuit' models, as this one, are completely self-contained and re-circulate the airflow within the expansion chamber. 'Open circuit' models, which separate products with comparatively large impurities (especially for extremely dirty or dusty material) or poor aerodynamic characteristics, require considerably more airflow because they are vented outside. Courtesy of Carter-Day, Minneapolis, MN, USA.

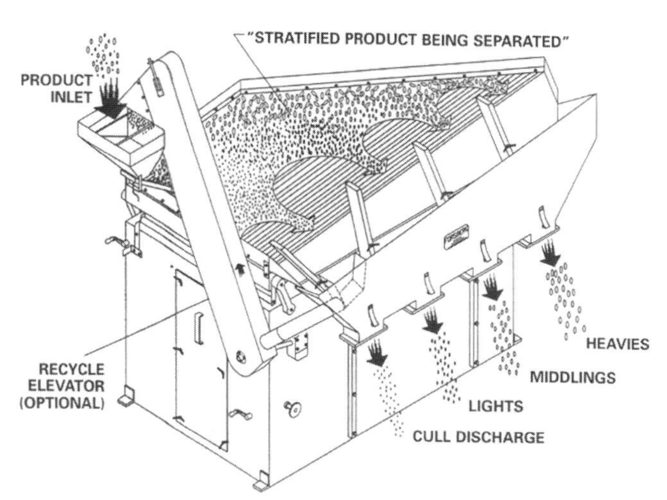

Fig. C.15. Pressure-style gravity separator. Fans force or draw air up through an oscillating seed separation deck (made for example, of canvas cloth, woven wire screen, or perforated sheet metal). The airflow, combined with the up-and-down motion of the deck, floats and layers the lighter seeds (which drift to the lower end of the deck) above the heavier seeds (which move uphill because of the back-and-forth motion of the deck). Medium-weight seed or 'middlings', discharged near the middle of the deck, are either retained or returned for another separation. Courtesy of Forsberg Inc., Thief River Falls, MN, USA.

by the partitions. Lighter seeds tend to move upwards, as heavier or rougher seed slides or rolls downwards.

4. Destoners

Destoners are intended to remove a small amount of heavy material, such as dirt, glass, rock, or stones, from a large volume of seed. Their design is modified from the gravity table, and they separate seeds on the same principles except that they produce only two fractions. A uniform layer of seed is fed on to the centre of the long dimension of the tilted porous deck, and desirable seeds flow to the lower end for collection, while the heavier material is discharged at the top end through an adjustable gate. There are two general types: the vacuum type, which is totally enclosed for a complete dust-free operation, and the open deck or pressurized type.

5. Dimensional sizing (separation)

Equipment that takes advantage of physical difference in the seeds' length, width, thickness, or ability to roll (Table C.20). Disc or cylinder separators are used to separate seeds that have practically the same width or thickness and would be impossible to separate were it not for differences in length. Seed that can be sized or separated on both their width and thickness may require the precision grader. Seed with little variation in either thickness or length, except that one is more round than the other, may be separated using a spiral separator.

(i) *Indented disc separators* exploit differences in seed length to effect separations (and are unaffected by seedcoat texture, density or moisture content). A series of vertically mounted cast-iron or urethane discs (about 40–65 cm diameter) revolve together on a shaft in and out of a trough through which the seed mass is being conveyed. Each disc contains hundreds of undercut pockets on each side in a wide range of size–shape relationships. As the discs revolve these recessed pockets lift out short seeds, which are dropped into a side trough, and reject the longer ones, which remain in the bottom of the machine and are conveyed to the discharge point. Trapdoors allow the 'liftings' from discs to be rejected or returned to the input end for recleaning (Fig. C.16). Disc separators are arranged in parallel or in series in conditioning lines for different separation strategies, and are well suited to handling large seed volumes.

(ii) *Cylinder separators (uniflow or indented)* accurately separate seeds by length, though the coat texture, weight and moisture of the seed also contribute. The product is fed into a long rotating cylinder or shell, lined with hundreds of 'hemispherical' teardrop-shaped or spherical pockets or indents. As the cylinder rotates between 50 and 60 rpm, short seeds that become embedded into the pockets are lifted out of the seed mass due to a combination of centrifugal force, specific gravity, weight, moisture and indent size (Fig. C.17). Cylinder separators come in many pocket diameters and depths, and cylinder shapes and sizes, and are used in conditioning lines either as single units or arranged in parallel for large-throughput operations, or in series to move the product through cylinders with a different size indents. They are more suited to handling heavier seed (small grains, maize and soybeans rather than grasses).

(iii) *Precision graders* separate seeds by width or thickness using slotted- or round-hole cylinder shells (for example, 30 or 45 cm diameter). Seed is fed into one end of the rotating cylinder (about 55 rpm) where it rolls and tumbles upon itself. A combination of

Table C.20. Examples of common uses of dimensional sizing equipment. For more details see information under separate headings.

Equipment	Used to (a) remove or (b) size-grade:
Disc separators (indented)	(a) Wild oats from wheat; sticks from sunflower; quackgrass from bluegrass; smartweed from flax; shrunken and broken seeds from whole kernels; dodder, dock, sorrel, or plantain from fescue and rye grass mixture.
Cylinder separators (Uniflow or Indented)	(a) Stones from peas; sticks from sunflower or sugarbeet seeds; (b) wheat, oat, barley, rice, grass seeds, lentils, legumes; length-grade hybrid maize.
Precision graders	(a) 'Splits' from soybeans, edible beans, peanuts; red rice from long grain rice; wild oats from barley; cheatgrass/rye brome from wheat; (b) barley, oats, wheat, maize, sunflowers.
Spiral separators	(a) Vetch from wheat; splits from soybeans; mustard from red clover; maize from soybeans; **sclerotia** from canola

gravity, centrifugal force and product pressure acts to force each particle into perforations while materials smaller than the perforations fall through the screen. The depth of the material and rolling action places a gentle pressure on the particles to give a mild press-fit for more accurate and uniform sizing. By screening the product through a cylinder rather than a flat screen, the centrifugal force and the weight of the seeds achieve 95–99% sizing efficiencies. The round openings can be recessed to help turn the seeds on end to either go through or pass over the openings. The oblong screens can be ribbed to help turn the seed on edge so that they can be graded according to thickness.

These precision sizers are as effective in removing contaminating weed seed and other crop seed as they are in sizing maize, sunflowers, or peanuts. The sizers can make extremely sensitive, or precise, separations of particles according to their width and thickness dimensions. Cylinder machines can be used singly, or two or three high, and arranged either in parallel or series as required.

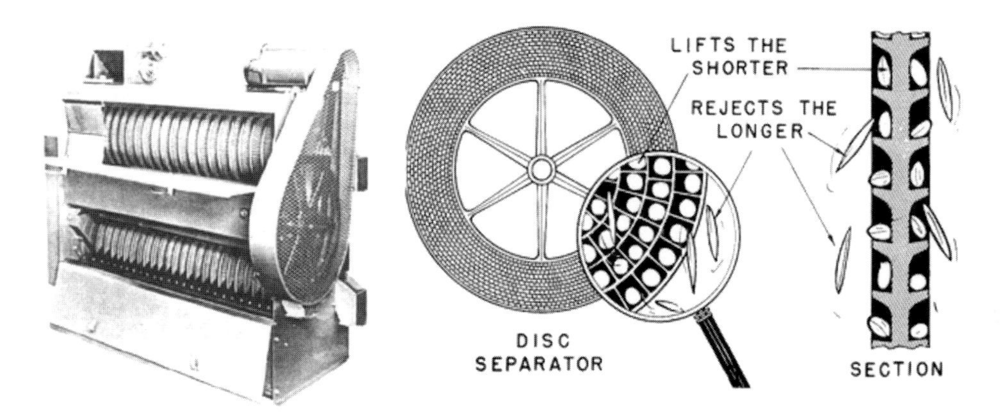

Fig. C.16. Carter indented disc separator, with diagram of disc face and a cross- section of a single disc. Disc pockets are made in three basic shapes and there are more than 75 different size–shape relationships. An R ('rice') shape has a flat horizontal lifting edge and round leading edge to lift tubular and short elongated seed. A V ('vetch') shape has a round, lifting edge and a horizontal leading edge, which picks up short, round-shaped seed. Square pocket shapes rapidly lift out short particles to reduce the mass in the machine, and produce fractions for resizing in separate operations. Several different pocket sizes and types can be used in one machine, grouped into sections with the smaller sizes closest to the feed inlet.

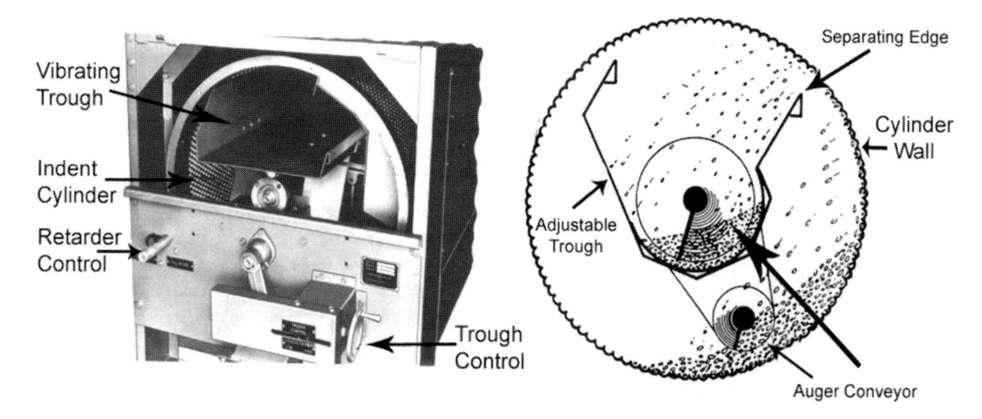

Fig. C.17. Cylinder separator. The selected seed is dropped into a trough mounted inside the cylinder and discharged by means of a vibratory or screw conveyor. Kernels longer than the indent diameter are rejected and remain inside the cylinder where they flow to the discharge end. Seed flow rate and bed depth are important operational variables. Rotation is anti-clockwise. Courtesy of Clifton, Bufton, IN, USA (left) and Carter-Day, Minneapolis, MN, USA (right).

(iv) *Spiral separators* are used to separate seeds that have little variation in thickness and length, according to shape, density, and the degree of roundness or the ability to roll. Spiral separators are made up of multiple flights that wrap around a central core (Fig. C.18). Machines may come as single or double spirals and as open or enclosed models. Seed is fed from a large hopper at the top and runs by gravity over a cone divider, which evenly spreads it to each of the spirals. The seed rolling down the inclined flights pick up speed based on their shape; round seeds travel faster, and centrifugal forces carry them over the edge of the inner spiral and drops them into the outer spiral where they are discharged. The non-round material remains on the small, inner flights and slides down to a separate spout at the bottom. Since there are no moving parts, no power is required to operate the machine. When the feed has been properly set, the separator continues to operate automatically as long as there is seed in the hopper.

Fig. C.18. Spiral separator showing different flightings used to separate seeds of different roundness.

6. Friction separation

A range of separators, designed to take advantage of differences in surface texture between seeds, are used as finishing machines on seed that has already been conditioned on an air-screen cleaner and/or other machines. They are used, for example, to clean smooth seed such as clovers, alfalfa, beans and sunflowers that are contaminated with weed seeds that have rough seedcoats (such as dodder), irregular shapes or sharp angles (such as dock) or are shrivelled or wrinkled, or where there are broken, chipped or damaged seeds of the harvested species, such as sunflowers that have been hulled or partly hulled (Table C.21).

Surface texture can be modified in mucilaginous seeds by spraying with water and mixing thoroughly with sawdust before air-screen or gravity separation, for example, to remove sticky buckhorn plantain contaminants from clover seed.

(i) *Roll mill (dodder mill, rice mill)*. A pair of cylindrical rollers covered with velvet-like material placed side by side, at a shallow inclination (of about 10°), close enough to touch lightly, and rotating in opposite directions outwardly from each other when viewed from the end or top (Fig. C.19). The

Table C.21. Examples of common uses of friction separation equipment. Details on each one in the following text.

Equipment	Used to remove
Roll mill (Dodder mill, Rice mill)	Dodder from alfalfa; dock and white cockle from clover; cocklebur from sunflowers; dehulled from hulled lespedeza; and many other rough weed seeds from smooth crop seeds
Inclined draper mill	Clover from grass seed; vetch from oats
Bumper mill	Alsike clover, Canada thistle, red sorrel, ryegrass, buckhorn, white cockle and other weeds, all from timothy grass
Vibratory separator	Curly dock from crimson clover; dog fennel and white cockle from timothy; ergot and plantain from bentgrass
Magnetic separators	Weed seeds like dodder, buckhorn, plantain and mustard from clovers and alfalfa

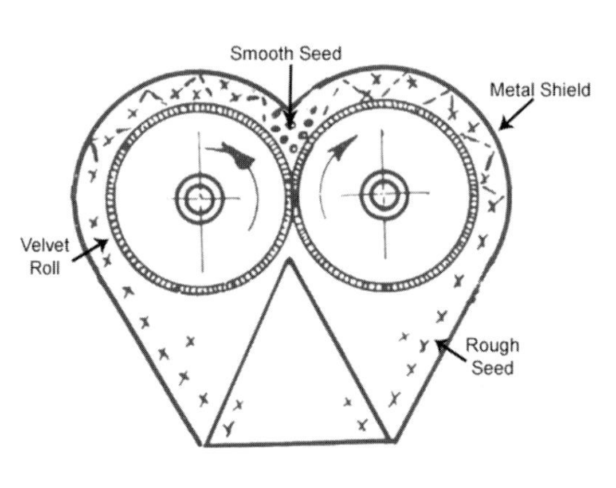

Fig. C.19. A roll mill, and a cross section of one pair of rollers. Courtesy of Rice Seed Company, Jerseyville, IL, USA.

roller assembly is mounted under a shield that provides a clearance equal to 1.5 times the diameter of the seed being cleaned (typically about 6 mm). As seed travels downhill in the groove between the rotating rollers, rough items are caught on the velvet surface, thrown up against the shield and deflected away, whereas smooth seeds continue bouncing downhill between the rolls and are collected at the bottom.

(ii) *Inclined draper mill* is designed to separate rough seed from smooth but it will also separate flat or elongated seeds from round seeds. The equipment exploits different rolling and sliding characteristics, using a flat adjustably-tilted surface with an endless rough- or smooth-surfaced belt that moves in an uphill direction (Fig. C.20). Seed is fed on to the centre of the inclined moving belt and spreads across its width in a thin layer. Rough seeds are caught by the belt and are carried uphill and discharged over the end, while smooth or round seeds roll or slide downhill under gravity and fall off the lower end. Extreme differences in surface texture, such as the roughness of dodder or the smoothness of alfalfa are not necessary to make a good separation; sharp edges, points, or projections may be sufficient.

(iii) *Bumper mill* (also sometimes called Timothy bumper mill) is a special machine designed to remove contaminant seeds from timothy grass (*Phleum pratense*), based on differences in shape, surface texture and weight. The machine consists of two identical, superimposed, interconnected and slightly inclined decks suspended in a rigid frame. A knocking action gives all the seeds an uphill motion. The plumper timothy seeds have a tendency to roll downhill between each bumping cycle, and so travel uphill a shorter distance than irregularly shaped seeds. By the time seeds have moved to the discharge point, timothy seeds are separated far enough from the contaminants to be discharged in a separate spout.

(iv) *Vibratory separators* are finishing machines that are designed to separate seed based on their shape and surface texture. The machines consist of one or more coated vibrating decks (for example, of metal, or coated with sandpaper) mounted in a rigid frame, which may be tilted at a wide range of inclinations. As seed is introduced near the centre of the deck, rough or flat individuals move up the slope, while more spherical ones slide or roll down.

(v) *Magnetic separators* take advantage of the surface texture of seed to make a separation. Finely-ground iron powder in the presence of moisture adheres selectively to rough, cracked and sticky seeds in the bulk. The mixture is then fed on top of a horizontal, revolving magnetic drum. Seeds that are smooth or slick will be free of powder and fall off the drum, while rough or sticky seeds coated with the powder are held on to it until they are removed by a rotary brush or a break in the magnetic field.

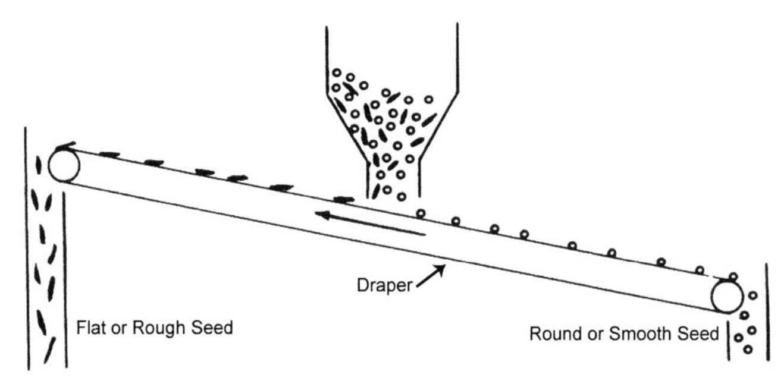

Fig. C.20. Inclined draper mill showing the separation of seeds of different shapes.

7. Other sorting principles

A range of techniques has been developed for particular separation tasks:

(i) *Density sorting*. Seeds are separated into fractions of different quality obtained by the differential buoyancy of imbibed seeds in water or aqueous solutions, with a density between those of good and poor quality seed, particularly to separate empty from full tree seeds, and to sub-fractionate primed seed lots (**see: Priming**). A pressure vacuum treatment (**PREVAC**), applied mostly to conifer tree seed lots, uses vacuum-infiltration of water to remove seeds with cracked or split coats. Flotation in water is a well-established technique in small-scale farming situations, and is probably of great antiquity (**see: Treatments – brief history**). For example flotation in brine may be used in India in wheat affected with tundu disease, to remove nematode galls, light kernels and debris from the sound kernels, which sink. Lightweight debris can be removed by placing seed lots in water, a technique used, for example, in onion seed cleaning.

Mixtures of non-aqueous solvents, such as chloroform and hexane, are used to upgrade high-value horticultural and ornamental seed lots, though care must be taken that there is no harmful effect of immersion on seed viability or storability. (**See: Density sorting**)

(ii) *Colour sorting*. Vision classification equipment used by food processors and in high-value seed conditioning to remove seed-sized foreign material and discoloured seeds, based on inspecting seeds individually for differences in reflected colour or brightness under visible or infrared light wavelengths. (**See: Colour sorting**)

(iii) *X-ray sorting*. X-ray analysis is a technique for inspecting seed internal structures and anatomy used, for example, as an adjunct in tree and sugarbeet seed processing to determine damaged internal structures or immature embryos. But in future X-radiography might become the basis of a rapid real-time automated sorting method by image analysis, analogous to colour sorting. (CHH)

(**See: Tree seeds; X-ray analysis,** Fig. T.10)

Copeland, L.O. (1976) *Principles of Seed Science and Technology*. Burgess Publishing, Minneapolis, MN, USA, pp. 239–256.

Copeland, L.O. and McDonald, M.B. (2001) *Seed Science and Technology*. Kluwer Academic, MA, USA, pp. 254–267.

Harmond, J.E., Brandenberg, N.R. and Klein, L.M. (1968) *Mechanical Seed Cleaning and Handling*. Agricultural Handbook No. 354. Agricultural Research Service, Washington, DC, US Department of Agriculture in cooperation with Oregon Agricultural Experiment Station.

Klein, L.M., Henderson, J. and Stoess, A.D. (1961) Equipment for cleaning seeds. In: Stefferud, A. (ed.) *Seeds, the Yearbook of Agriculture*. US Department of Agriculture, Washington, DC, pp. 307–321.

McDonald, M.B. and Copeland, L.O. (2001) *Seed Production – Principles and Practices*. Chapman and Hall, New York, USA, pp. 78–103.

Vaughan, C.E., Gregg, B.R. and Delouche, J.C. (eds) (1968) *Seed Processing and Handling, Handbook No. 1.* State College, Mississippi, Seed Technology Laboratory, Mississippi State University, USA.

Conductivity test

A seed vigour test, validated and commonly used for pea **vigour testing** but also used for other large-seeded legumes, based on measuring the degree of solute **leakage**, particularly potassium ions, into water from freshly imbibed seeds. Loss of solutes from seeds reflects the integrity of cell membranes and the presence of dead tissues within the seed, or both. (**See: Germination – physical factors, 1. Membranes and water; Vigour testing, biochemical, 1. Conductivity test**)

Cone

Also called strobilus. A unisexual reproductive structure of gymnosperms, e.g. conifers and cycads, typically consisting of a central axis around which there are scaly, overlapping, spirally-arranged sporophylls that develop pollen-bearing sacs or naked ovules, the latter after fertilization and embryogenesis becoming the seeds. (**See: Gymnosperm seeds**)

Conifers

See: Gymnosperm – seeds; Tree seeds

Conservation

1. Principles

Seed conservation is the use of seed storage as a means of ensuring the future availability of plant **germplasm**. Because the seeds are stored in seed **gene banks** (seed banks), away from where the original plant populations grow in nature or cultivation, it is a method of ***ex situ* conservation**. Long-term seed conservation is only practical for **orthodox**-seeded species (alternative gene bank methods need to be adopted for those species with seeds displaying **recalcitrant** or **intermediate** storage behaviour). It is used widely for the conservation of crop germplasm and, increasingly, for helping to ensure the future survival of non-domesticated species.

Most seed banks conserve crop germplasm. Indeed, the ability of their seeds to survive storage in a dry, cool state is thought to have played a major role in the early development of many crops by enhancing their survival between growing seasons and by assisting with the retention of selections. Modern crop seed banks keep seed for either a few years (short-term), about a decade (medium-term) or many decades to centuries (long-term). Short-term banks include community banks that help ensure continued availability of local varieties to farmers in poorer countries. Most banks, however, operate at either an international or at a national level under the auspices of international agencies or governments. Such banks have arisen over the past 50 years to conserve well-adapted, but relatively low-yielding landraces (or primitive cultivars) that have been displaced over much of the globe by relatively uniform and high-yielding modern varieties bred to meet the demands of intensive cultivation (**see: Heritage seeds**). Breeding programmes in the public and commercial sectors have drawn upon the banked landraces (and other material within crop gene pools) as a source of disease resistance and other useful characters in order to maintain the rapid turnover of these short-lived modern varieties.

There is a strong political dimension to seed banking. The recognition of national sovereignty over genetic resources is a central tenet of the **Convention on Biological Diversity** (CBD), nearly every banked collection has potential economic value, and there is interdependence of nations on plant species for food, forestry, drugs, etc. Consequently, collecting should be carried out under a material transfer agreement that sets out the terms of access to target germplasm and the requirements for any subsequent benefit-sharing relating to its exploitation. For many crop species and for parties to the **International Treaty for Plant Genetic Resources for Food and Agriculture** (ITPGRFA), such bilateral arrangements are replaced by a multilateral system by which access is facilitated on the basis of an internationally agreed benefit system.

2. History

Formal programmes of crop conservation were established in the late 19th and early 20th centuries in Russia (including the outstanding work of N.I. Vavilov), the USA, Canada and elsewhere. Such collections were maintained by an annual grow-out of seed. A forerunner of the modern generation of seed banks was that established at Fort Collins, Colorado, USA, in 1958. The establishment of a number of other such facilities followed both internationally (e.g. at the **International Rice Research Institute**, Philippines in 1961, a **Consultative Group on International Agricultural Research**, CGIAR, Institute) and nationally (e.g. Japan in 1966). Through the efforts of the **Food and Agriculture Organization of the United Nations** (FAO) there was increased global awareness of the accelerating loss of crop **germplasm** during the 1960s. In 1974, the International Board for Plant Genetic Resources (now the **International Plant Genetic Resources Institute**, IPGRI) was established to help promote and coordinate the conservation of plant **genetic resources** and the following two decades saw an intensive period of collecting and the establishment of new facilities. In 1996, the FAO review 'The State of the World's Plant Genetic Resources for Food and Agriculture' led to the development of a Global Plan of Action for the Conservation and Utilisation of Plant Genetic Resources for Food and Agriculture. This was adopted by 150 countries as the basis for a programme of expanding and sustaining crop conservation activities.

In 2001, the ITPGRFA was internationally adopted. It was born out of a desire to see greater facilitated access to collections held in crop seed banks and greater sharing of benefits resulting from the agricultural development of banked germplasm with the providers of that germplasm (frequently, financially poorer but genetically richer countries).

By late 2002, FAO's World Information and Early Warning System listed some 5.4 million accessions (samples) in some 482 medium- and long-term seed banks. However, there is significant duplication and perhaps less than 2 million of the accessions are unique. Many of these collections date back decades. But some expansion of activity has occurred since 1996 despite a background of decreased budgets for such conservation. Concern for the long-term financial security of many crop collections has led to the establishment of the Global Conservation Trust that aims to address this issue through an endowment fund. Apart from funding, the challenges for crop seed conservation include the improved implementation of storage protocols; an increasing number of accessions require urgent regeneration. The extra collecting target for crop germplasm set out in the **Global Strategy for Plant Conservation** (GSPC), will be a significant burden to an already stretched crop genetic network. Set against this is the possibility that the role of crop seed banks will start to be challenged by genetic engineering. Reliance by plant breeding on a given crop's stored germplasm could be reduced, as wide crossing with material outside that crop's gene pool becomes possible. However, for the foreseeable future, traditional plant breeding, whereby small step-by-step improvements are made using closely related material, is still likely to be practised in many parts of the world, and links between breeders and the main seed banks maintained. These links to banks may be fairly weak in the case of breeding companies which perhaps place greater importance on elite lines held in their own private collections than on material held in the main crop germplasm network. One consequence of the increased release of genetically modified varieties is that it may become increasingly difficult to protect related crop germplasm, maintained on farms as an ***in situ* conservation** measure, from dramatic genetic change. In this case, the role of crop seed banks as maintainers of landrace genetic integrity may become even more important.

Seed banks are used for the conservation of many non-crop species. There are many short-term forest-seed banks around the world and the same technology is used for large-scale habitat restoration programmes. Increasingly, seed banks are used as a means of conserving species that have local economic value (e.g. many medicinal plants) and rare and threatened plant populations. Wild plant populations are at risk as a result of increasing human intervention in natural ecosystems. This intervention takes the form of agricultural development, urbanization, logging and other unsustainable use of plants, pollution, the introduction of alien species and climate change. Perhaps half of all species will become so genetically eroded (loss of populations and reduction in population sizes) that they will be vulnerable to extinction, resulting not only from continued pressure but also from chance events.

Ex situ conservation of endangered wild plant populations has been encouraged by Article 9 of the CBD and given added impetus by the GSPC in 2002. This strategy aims to get 60% of threatened plant species in accessible *ex situ* collections by 2010. Some 150 botanic gardens were reported as having some form of seed bank in 1996 though probably less than 20,000 species (less than 8% of the minimum world total) are currently banked. The aim of such 'biodiversity' seed banks is to provide material for re-introduction, habitat restoration and research (including that aimed at improving *in situ* conservation). Two of the first such banks were the ones run by the Universidad Politecnica de Madrid, Spain (1966) and by the Royal Botanic Gardens, Kew, UK (late 1960s). With many wild crop relatives held in both crop and biodiversity banks, the distinction between the two types of banks is often blurred.

3. Procedures

The banking procedures for both crop and non-domesticated seeds are similar. Apart from species, the key factors determining the **longevity** of orthodox seeds are initial quality, moisture content and temperature; procedures try to optimize these factors. Following collecting, seeds arriving at a seed bank are accessioned (collecting or 'passport' data is entered on to a database). The collections are then cleaned (often with tests for aborted seeds, purity and health), dried, moisture status assessed, **viability** tested, packaged (**see: Hermetic storage**) and banked at low temperature. Following characterization (which includes confirmation of identification) and, in many cases, evaluation (of useful traits), samples are made available from seed banks for a range of uses. Such supply is often subject to material transfer agreements that specify the terms under which seeds may be used. The order of operations may vary between banks. In some, collections are grown out soon after arrival for study and multiplication. Cleaning sometimes occurs after drying and, in some cases, initial **viability testing** takes place after banking. Monitoring of the viability and moisture status of the seed lots is an essential part of collection maintenance. As seeds age, genetic damage accumulates and vigour (important in field establishment) is reduced. To limit these effects, collections undergo periodic **regeneration**, before viability has significantly dropped, in order to produce new high quality seed lots. The frequency with which this potentially selective process occurs is reduced by banking high-quality seed and by maximizing its longevity in store. Processing, maintenance and usage data are also recorded for each seed lot. Increasingly, data are recorded to a common standard on international database systems for improved access, e.g. the System-wide Information Network for Genetic Resources, SINGER, for collections held within the CGIAR centres (singer.cgiar.org/) or the US Department of Agriculture's Germplasm Resources Information Network, GRIN (www.ars-grin.gov/npgs/aboutgrin.html).

With so much valuable germplasm in their charge, it is important that seed banks operate at least to internationally-agreed baseline standards developed using the best available scientific data. FAO and IPGRI produced a set of standards in 1994 and, substantially, these are applicable to both crop and non-domesticated material. Currently, it is recommended that an **equilibrium relative humidity** (eRH) in the range 10–15% at 10–25°C is used for **drying of seed**. The lowest of the resultant **equilibrium moisture content** (emc) values lie within the range for **ultra-dry storage**. However, there is some suggestion that the recommended eRH drying conditions should be slightly revised upwards when seeds are to be subsequently stored at temperatures below 0°C. In sealed storage at sub-zero temperatures, although the emc of a seed lot will remain unchanged, its eRH and hence its **water potential** will be significantly reduced (**see: Sorption isotherms**). There is some evidence that storage is sub-optimal at these low eRH levels. It is recommended that for long-term storage, dried, packaged seeds be held at −18°C or cooler (**see: Cryostorage**). Under these latter conditions, seeds of many orthodox species have a potential longevity stretching to decades and, in some cases, centuries or even millennia. The more extreme storage lives are predicted by extrapolation from the **viability equations** that have been developed for a number of species using percentage viability data (transformed to **probits**) from controlled ageing tests. Furthermore, studies on **ancient seeds** suggest that seeds of many species are capable of surviving very long periods.

Storage may be organized either within or between seed gene banks such that material for day-to-day usage is held in active (medium-term) stores, where access is easy, and material stored as a security measure is held in base (long-term) stores. While one advantage of a seed bank is that it concentrates large amounts of genetic diversity in one facility, this is also a potential weakness with respect to risks from catastrophic damage. Consequently, most banks duplicate collections with a bank at a different location. Long-term banks vary in scale from local ones that dry seeds over silica gel and store in domestic refrigerators (few hundred collections, part-time staff) up to regional or international ones with large, insulated, walk-in drying rooms and freezers (many thousands of collections, ten or more staff) (see Fig. C.21).

A bank housing seeds of all the world's crop plants is planned to open in 2007. This will be built deep inside a sandstone mountain lined with permafrost on the Norwegian Arctic island of Spitsbergen. (RJP, SHL)

FAO (1997) *The State of the World's Plant Genetic Resources for Food and Agriculture*. Food and Agricultural Organization of the United Nations, Rome, Italy.

Linington, S.H. and Pritchard, H.W. (2000) Gene banks. In: Levin, S.A. (editor-in-chief) *Encyclopedia of Biodiversity*, Vol. 3. Academic Press, USA, pp. 165–181.

Smith, R.D., Dickie, J.B., Linington, S.H., Pritchard, H.W. and Probert, R.J. (eds) (2003) *Seed Conservation: Turning Science into Practice*. Royal Botanic Gardens, Kew, UK.

Consultative Group on International Agricultural Research

CGIAR is an association of public and private members, founded in 1971, that support 16 centres working in more than 100 countries carrying out and encouraging both strategic and applied agricultural research aimed at reducing hunger and poverty, improving nutrition and health, and protecting the environment. This organization, funded by industrial and developed countries, foundations, and international and regional organizations, employs several thousand scientific staff. The 16 centres, known as 'Future Harvest Centres', include International Agricultural Research Centres that are home to some of the world's largest seed **gene banks**. Examples are those at the **International Rice Research Institute (IRRI)** in the Philippines (90,000

Fig. C.21. The Millennium Seed Bank. (A) The Wellcome Trust Millennium Building, located at the Royal Botanic Gardens, Kew, Wakehurst Place, West Sussex, UK, houses the Millennium Seed Bank. (B–D) The Seed Bank's collections of wild plant species are held in trust for its international partners in underground vaults at −20°C.

accessions), and the International Maize and Wheat Improvement Center (**CIMMYT**) in Mexico (190,000 accessions). Another centre is the **International Plant Genetic Resources Institute (IPGRI)**, based in Italy. (RJP, SHL)

Controlled deterioration

A **vigour testing** procedure developed to detect differences in the vigour of small-seeded vegetable species. A set of seed lots is accurately raised to the same predetermined moisture content by imbibition on moist paper, and equilibrated; after being held at 45°C for 24 h, seeds are then subjected to a standard germination test. (**See: Vigour testing – ageing**)

Convention on Biological Diversity (CBD)

This United Nations convention has as its objectives: the conservation of biological diversity (ecosystem, species and within-species diversity); the use of this diversity in a sustainable manner; and the equitable sharing of benefits arising from the use of **genetic resources**. The Convention recognizes that among other things: conservation of diversity is a common concern of humankind; there is a close and traditional dependence of many indigenous and local communities on such diversity; and that biological diversity is reduced by certain human activities. Furthermore, it recognizes that governments have sovereign rights over the diversity within their national borders. With respect to these measures, the Convention recognizes that the fundamental requirement is for ***in situ* conservation**. However, ***ex situ* conservation** measures, preferably in the country of origin, are also seen to have an important role to play. The Convention is clear that no measures should be postponed by lack of full scientific certainty. It was opened for signature in 1992 and, by May 2003, only six countries were not party to it. The governing body of the Convention is the Conferences of the Parties that meets every 2 years to keep under review the implementation of the Convention and to steer its development. (RJP, SHL)

Conversion

In plant tissue culture, the successful germination of a **somatic embryo** and its development into a plant. A measure of the efficiency of **somatic embryogenesis** is the number of embryos that are converted to plants. Low efficiency conversion is due either to developmental arrest of embryos, or poor shoot development. Conversion efficiency is not only affected by conditions at the time of germination, but by longer-term effects of treatments during somatic embryogenesis that only manifest themselves later during early plant growth. Conversion is also strongly influenced by genotype. (**See: Tissue culture**) (PvA)

Conveyors

A collective term for the equipment designed to transport or carry seed by air, belts, augers, buckets and vibration, vertically, horizontally or up inclines as required. Conveyors are classified as pneumatic (air) conveyors, bucket elevators, vibratory conveyors, belt conveyors, screw conveyors, and chain conveyors.

Seed and grain **conditioning** plants must have conveyors to move seed in bulk efficiently from receipt, between conditioning machines and until it has been placed in storageor packed. Conveyors should minimize physical seed damage, which can occur during any handling operation from harvest to planting. Particularly vulnerable are heavy seeds such as bean, maize, pea and soybean, which may be damaged accumulatively by hard objects or by minor falls onto firm surfaces, more so if they contain too much or too little moisture. All mechanical handling equipment must be designed and maintained to minimize the risk of overheating or electrical sparks, which are the frequent ignition sources in grain **dust** explosions, should be relatively quiet, require little maintenance and be self-cleaning. Seed conveyor designs are available to meet all these requirements, and are available in many sizes and capacities, but no single conveyor performs all the functions.

(a) *Pneumatic conveyors* can be designed to move anything from chaffy non-flowable material to granular seed through a closed duct system by air. The system is flexible, as pipes can be placed in any direction, has few mechanical parts and can handle non-flowable seed that would plug other conveyors. However, it has a high power requirement, and may damage some seed, and pipe elbows may need to be reinforced to protect against abrasion by seed flows. The physical impact can have a beneficial side effect of killing some vulnerable species of storage pests.

(b) *Bucket elevators* (legs) consist of a vertically aligned endless belt or chain hung with evenly-spaced buckets made of metal, plastic or fibreglass. They are widely used in seed conditioning plants to move seed to overhead storage or surge bins, and in bulk grain storage facilities.

(c) *Vibratory conveyors* move seed for short horizontal distances through a shaking trough whose rapid pitching action moves the seed toward the discharge end.

(d) *Belt conveyors* are endless belts that can handle practically any type of material. Most belt conveyors operate on a horizontal plane but can operate up an inclination of up to 45°.

(e) *Screw (or auger) conveyor*. This conveyor is in the form of a helix around a shaft. Though excellent for moving bulk material these are not used on easily damaged seeds (for example, soybeans).

(f) *Chain conveyors* move material horizontally or up inclines of 45° by drags of varying shapes mounted on a chain drive. (CHH)

Cool test

A physiological **vigour test** based on the application of stress during germination developed specifically for cotton seed, involving the assessment of **normal seedlings** on wetted rolled paper towels after a period in the dark at 18°C (which is below the optimum temperature for germination). The test in effect measures the speed of germination. (**See: Cold test; Vigour tests – physiological**)

Cool-season legumes

The cool season pulses are characterized by **hypogeal** seedling growth, i.e. the **cotyledons** remain beneath the soil surface. This initially keeps much of the seedling below ground, thus possibly reducing the effects of freezing and other desiccating environmental conditions. This is followed by a period of rapid vegetative growth, followed by flowering when daylength becomes progressively longer. (**See: Legumes**)

Muehlbauer, F.J. and Kaiser, W.J. (1994) *Expanding the Production and Use of Cool Season Food Legumes.* Kluwer Academic, Dordrecht, The Netherlands.
Summerfield, R.J. (ed.) (1988) *World Crops: Cool Season Food Legumes.* Kluwer Academic, Dordrecht, The Netherlands.

Copenhagen tank

An apparatus used in **germination testing**, also known as the Jacobsen tank, in which seeds are placed on top or between paper blotter pads on metal shelves stretched across a thermoregulated water bath. Strips of paper in contact with the blotters dip down into the tank, so as to draw moisture up to the blotters by a wick action. Each individual blotter may be covered by an inverted glass funnel, and the whole tank assembly with a lid to maintain internal atmospheric humidity.

Coriander

Coriander (*Coriandrum sativum*, Apiaceae) is one of the earliest known spices (**see: Spices and flavours,** Table S.12). The plant is an annual herb, long prized for its spicy, aromatic 'seeds' (really fruits) (Fig. C.22) used either whole or in ground form as a main ingredient in curry and other food and flavour products. Freshly cut leaves are one of the world's most widely used culinary herbs.

The plant is an erect, herbaceous annual. When grown for the fruit, the crop duration varies from 100–200 days. The optimum time to harvest varies widely depending on the location, season and the cultivar. The aromatic fruit is an ovoid to globose **cremocarp** with two one-seeded **mericarps**, each 3–5 mm in diameter, yellow-brown when ripe with ten longitudinal ribs.

Ripe dried fruit contains 1–2% **essential oil** (obtained by steam-distillation) and 16–28% fixed oil. The flavour of coriander is contributed by the essential oil whose main constituents are shown in Table C.22 (see Fig. S.44). The oil is used in the food and perfume industries.

Indian households use huge quantities of coriander powder in culinary preparations. Ground coriander is a major component in spice mixes worldwide, curry powders and garam masalas. Coriander is used in flavouring meat preparations. It has antimicrobial properties and is also used in treatments for colic, neurologia and rheumatism. Coriander paste is applied to skin and mouth ulcers. Its fruits are used in gastrointestinal complaints such as dyspepsia and flatulence. It is useful as a poultice for ulcers and carbuncles. The essential oil is widely used in the food industry and for flavouring alcoholic beverages, as well as in perfumery. (KVP, NB)

Table C.22. Major components of coriander essential oil.

Components	Percentage (%)
α-Pinene	2.5
γ-Terpinene	3.4
p-Cymene	1.6
Trans-linalool oxide (furanoid)	0.4
Linalool	78.1
Octanol	0.9
Terpinen-4-ol	0.5
α-Terpineol	0.9
Geranylacetate	3.8
Cuminaldehyde	0.5
Geraniol	1.4
(E)-2-Dodecanal	0.5
Hexadecanoic acid	2.1

Atta-ur-Rahman, M.I., Choudhary, A., Farooq, A., Ahmed, A., Demirci, B., Demirci, F. and Baser, K.H.C. (2000) Antifungal activities and essential oil constituents of some spices from Pakistan. *Journal of the Chemical Society of Pakistan* 22, 60–65.

Fig. C.22. Dried coriander fruits (image by Mike Amphlett, CABI).

Peter, K.V. (ed.) (2001) *Handbook of Herbs and Spices.* Woodhead Publishing, Cambridge, UK.
Weiss, E.A. (2002) *Spice Crops.* CAB International, Wallingford, UK.

Corn

In North America the name is used synonymously with maize, whereas in the UK a general non-scientific name for cereals such as barley and wheat.

Historically, a more general term for seeds, rather than cereal seeds alone. The word is etymologically derived (as are the words 'grain' and kernel') from a root-word meaning a 'worn-down particle'.

Corn belts

Maize is cultivated in a wider range of environments than the other two major cereals – wheat and rice – because of its greater physiological adaptability. The crop is grown in several major so-called 'corn belt' regions, which together account for about half the world crop area, and are ranked in the following order by size.

- The Midwestern USA (centred in Iowa and Illinois) – also the principal world soybean production zone.
- The plains of northern China (mainly Jilin, Shandong, Heilongjiang, Hebei, Henan, Liaoning and Sichuan).
- Southeastern Brazil (mainly Parana, Rio Grande do Sul, Sao Paulo and Minas Gerais).
- The Danube basin, from southwest Germany through Hungary, Serbia and Romania to the Black Sea.

The Po valley of Italy and northeastern Argentina are also termed corn belts, although other larger crop-growing geographical areas, such as Mexico, India and Nigeria, are not usually referred to in this way. (**See: Crop Atlas Appendix, Map 7**) (PH)

Cotton

1. Botany

Cotton (*Gossypium hirsutum*, *G. barbadense*, *G. herbaceum*, *G. arboreum*, family Malvaceae) is economically the world's leading natural **fibre**. The long seed hairs or **trichomes** of seed cotton, known as cotton fibre or **lint**, develop from the **ovule** epidermis, as many as 13,000 to 21,000 fibres per ovule (for information on seed development **see: Cotton – cultivation, 2. Development**; see also Fig. C.23). Fibres are single, greatly elongated cells, ranging from 2.2 to 6.0 cm long, and about 20 μm wide, depending on the species and growing conditions. Expansion of fibre cells begins on or near flower opening. During elongation, a large central vacuole forms leaving a thin cytoplasmic area between the cell wall and the vacuole membrane. A waxy cuticle surrounds the elongating primary cell wall. Another class of cottonseed trichome, fuzz or **linter**, begins elongating from the ovule epidermis 5–7 days after the lint fibres have started growing. At 2–6 mm, fuzz fibres are considerably shorter than **lint** fibres, ultimately have thicker cell walls and may be pigmented. Approximately 2 weeks after lint fibre elongation starts, a secondary cell wall of highly crystalline cellulose is deposited until the wall is 2.5 to 6 μm thick. Four to six weeks later, at maturity, the **boll** dehisces, exposing the thin fibres to sun and air (Fig. C.24). The cytoplasmic components dry and adhere to the inner surface of the fibre wall forming a lumen in place of the central vacuole. Upon drying, the fibre wall collapses into a kidney-shaped ribbon structure with frequent twists called convolutions (**see: Seedcoat – structure**, Fig. S.23; **Fibres**, Fig. F.5C). During cotton processing, convolutions help adjacent fibres to interlock thereby forming a yarn.

There are over 40 recognized species of cotton including diploids (2n = 2x = 26 chromosomes) and allotetraploids (2n = 4x = 56 chromosomes). Most (>90%) of the world's cotton

Fig. C.23. Delinted cotton seed (image by Mike Amphlett, CABI).

is upland cotton (*G. hirsutum*) with smaller amounts of extra-long-staple cotton also called Egyptian or Pima cotton (*G. barbadense*) (8%), and Desi cotton (*G. arboreum*, *G. herbaceum*). All the commercially important species except *G. arboreum* continue to grow naturally in the wild. Plant breeders have developed hundreds of varieties of the domesticated species that are uniquely adapted for their respective growing areas. Additionally, cotton varieties improved by genetic engineering enjoy widespread adoption in the USA, China, Australia, South Africa, Mexico, India and Argentina. Over 70% of the cotton grown in the USA in 2003 contains one or more genes for insect resistance (the Bt toxin from *Bacillus thuringiensis*), or herbicide tolerance (to glyphosate or bromoxynil). Nearly 30% of worldwide cotton in 2002 was transgenic. (**See: Cotton – cultivation**; and inset in **Crop Atlas Appendix, Map 14**)

2. Chemistry

The chemical composition of cotton fibre is unique compared with other seed fibres since, at maturity, it is approx. 88–96% by

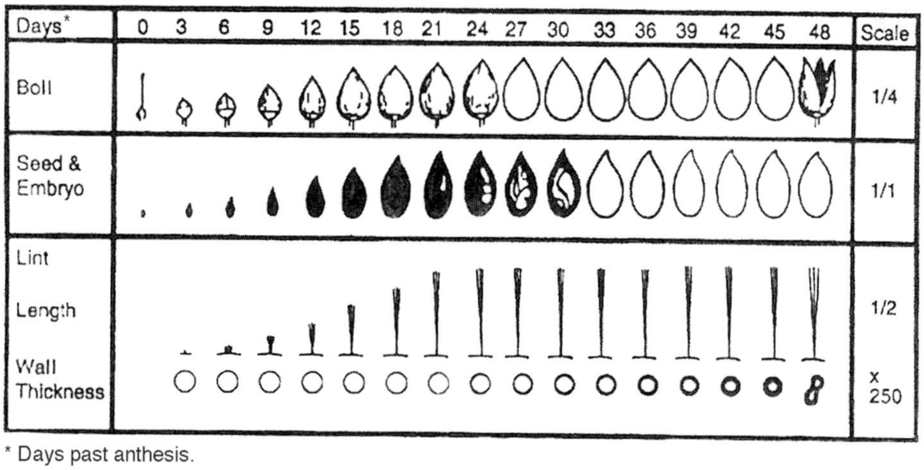

Days* 0 3 6 9 12 15 18 21 24 27 30 33 36 39 42 45 48 Scale

Boll 1/4

Seed & Embryo 1/1

Lint Length 1/2

Wall Thickness x 250

* Days past anthesis.

Fig. C.24. Diagrammatic representation of the developing boll in relation to corresponding changes in development of the seed and fibre, from flowering (0 days post **anthesis**, dpa) to the open boll stages (48 dpa). From Balls, W.L. (1915) *The Development and Properties of Raw Cotton.* A.C. Black, London, UK.

weight **cellulose** and is not lignified (**see: Fibres,** Table F.1). Cellulose microfibrils are highly organized and crystalline in mature cotton fibre, although less crystalline areas or amorphous regions also occur naturally. The primary cell wall and cuticle surrounding the mature fibre are composed of **pectins, hemicelluloses,** proteins, randomly oriented cellulose microfibrils, and waxes. Cultivated cottons are varying shades of white, but naturally coloured variants have existed for thousands of years. Brown cottons result from the accumulation of condensed tannins in the fibre lumen. The green shades arise from layers of suberin sandwiched between layers of cellulose microfibrils in the secondary wall.

3. Origins

The genus *Gossypium* was probably first cultivated in Africa and the Indian subcontinent. Archaeological excavations at Nubia in the Sudan and Mohenjo-Daro in Pakistan uncovered examples of cotton textiles dating from 2700 BC. Cotton bolls and textiles have been found also in sites in Peru and Mexico that are over 4500 years old. Recent phylogenetic studies suggest that the four cultivated cotton species arose independently from wild relatives. The fossil record combined with molecular phylogenetic analysis suggests that the progenitors of modern cultivated cottons diverged from a common ancestor 5–10 million years ago. Hybridization followed by polyploidization of two diverged diploid ancestors is estimated to have occurred 1–2 million years ago in the middle of the Pleistocene era. Most of the existing wild species are diploid and are distributed throughout Africa, Asia, South and Central America, Turkey, Australia, Hawaii and the Galapagos Islands.

4. Location

China is the largest producer of 80 countries producing cotton fibre, followed by 17 states of the southern USA. Other leading producers are Brazil, China, India, Pakistan, Turkey, Uzbekistan and West Africa (Table C.23). (**See: Crop Atlas Appendix, Map 14**)

Table C.23. International production and export value of cotton.

Country	Production (thousands of bales)[a] (2003)	Export value (US$ millions)[b] (2000)
China	25,500	307
USA	16,939	1,947
India	12,000	53
Pakistan	8,350	206
Uzbekistan	4,500	884
West Africa	4,315	N/A
Brazil	4,250	36
Turkey	4,200	72
Greece	1,700	305
Australia	1,300	888
Egypt	1,050	27

N/A, not available.
[a]National Cotton Council, EconCentral. One bale = approx. 0.2 tonnes.
[b]COMTRADE database, United Nations Statistics Division.

5. Production

Production methods vary widely depending on the extent of mechanization. Seeds are sown when the average soil temperature is above 18°C. Cotton is well adapted to semi-arid conditions, however irrigation is used in China, Egypt, the southwestern USA and California. The plant is naturally a perennial but is grown commercially as an annual. Fibre is harvested 120–150 days after sowing except for some of the *G. barbadense* varieties that may require up to 210 days. Annually, about 90 million bales or 19.6 million t of cotton **lint** are produced worldwide.

6. Processing

Fibre processing begins with harvesting cottonseed from the field when its moisture content is below 12% either by mechanical harvester in developed countries or by hand in many parts of the developing world. Harvested cottonseed is separated into lint fibre, linter fibre and seed at the cotton gin (**see: Ginning**). Every kilogram of cottonseed yields 300–400 g of lint fibre. In the USA, ginned cotton lint is packaged into large bales at a density of 360–450 kg/m^3. Most bales are sold to textile mills where the fibres will be woven or knitted into garments or domestic textiles. Scouring and bleaching facilitate the uptake of dyes and other chemical finishing treatments by the fibre. Cotton linters are processed into high quality paper, furniture batting, or industrial products.

A federal government agency measures six parameters of fibre quality of all USA-produced cotton by an automated high-volume instrument (HVI): fibre length, length uniformity (**see: Staple length**), micronaire (fineness), colour, and trash (non-fibre) content, which are highly correlated with production performance and market value. Outside the USA, Australia and South Africa, cotton is primarily graded manually. The fibre physical properties that are measured differ between different cotton-producing nations. (BAT)

For information on the particular contributions of the seed to seed science **see: Research seed species – contributions to seed science.**

Basra, A.S. (ed.) (1999) *Cotton Fibers: Developmental Biology, Quality Improvement, and Textile Processing*. Haworth Press, Binghamton, NY, USA.

Kim, H.J. and Triplett, B.A. (2001) Cotton fiber growth *in planta* and *in vitro*. Models for plant cell elongation and cell wall biogenesis. *Plant Physiology* 127, 1361–1366.

Ruan, Y.-L. (2005) Recent advances in understanding cotton fibre and seed development. *Seed Science Research* 15, 269-280.

Wakelyn, P.J. (2002) *Cotton. Kirk-Othmer Encyclopedia of Chemical Technology*. John Wiley, New York, USA. http://www.mrw.interscience.wiley.com/kirk/articles/cottbert.a01/frame.html

Wakelyn, P.J., Bertoniere, N.R., French, A.D., Zeronian, S.H., Nevell, T.P., Thibodeaux, D.P., Blanchard, E.J., Calamari, T.A., Triplett, B.A., Bragg, C.K., Welch, C.M., Timpa, J.D. and Goynes, W.R. (1998) Cotton fibers. In: Lewin, M. and Pearce, E.M. (eds) *Handbook of Fiber Chemistry*, 2nd edn. Marcel Dekker, New York, USA, pp. 577–724.

Cotton – an oilseed

The world production of cottonseed is second only to soybean and, like soy, the extracted **oil** is a by-product. The crop is primarily grown for the **fibres** produced on the seed surface (**see: Cotton**), but the low 16% oil content of the delinted seed (by definition, 'cottonseed'), and the feeding in some countries of unprocessed seed to cattle, makes cotton the sixth most important edible oil source (**see: Oilseeds – major**, Table O.1).

Cotton is grown around the world in tropical and subtropical latitudes and as far north as 45°N in China. Cotton plants are perennials but cultivated forms are grown as annuals. Major producing countries in order of importance are China, USA, India, Pakistan, the former USSR, Turkey, Brazil and Australia, but many other countries produce significant quantities. (**See: Cotton – cultivation; Crop Atlas Appendix, Map 14**)

Four cotton species are commercially grown (**see: Cotton**).

The use of cottonseed oil on a commercial scale is of relatively recent origin, although it was used medicinally and for lamp oil in ancient India and China. The invention of the cotton seed dehuller by William Fee in 1857, and the introduction in 1880 of bleaching clays to remove the characteristic red colour from cottonseed oil, revolutionized the cottonseed oil extraction process and economics.

For details of cottonseed structure **see: Cotton**; well-differentiated **cotyledons**, containing **oil bodies** and resin ducts, curl around each other and the **hypocotyl** and constitute the major portion of the **embryo**. The hypocotyl, shoot apex and **radicle** with root cap are well defined. After **delinting**, the hull is removed and the seed processed to yield about 16% oil, 45% high protein meal, 9% **linters** and 25% hull. The remainder is waste and processing loss.

Cottonseed oil (**see: Oilseeds – major**, Table O.2, for composition) is used as a cooking oil, in margarine and shortening blends and when winterized as a salad oil. The linters are used in the manufacture of mattresses, cushions and plastics. Cottonseed meal is used as a protein-rich (30%) feed for ruminants and the hulls are often added for roughage. However, the presence of gossypol (**see: Pharmaceuticals and pharmacologically active compounds**) in the meal can be toxic to swine and poultry. Heat treatment of the meal will bind up the free gossypol, but the treatment also binds up some of the available lysine. Plant breeders have developed gossypol-free varieties with improved oil and meal quality but the area sown to these is still limited. Cottonseed also contains **allergenic** proteins (**see: Storage proteins – intolerance and allergies**). (KD)

(**See: History of seed research**, Fig. H.7c)

Jones, L.J. and King, C.C. (1996) Cottonseed oil. In: Hui, Y.H. (ed.) *Bailey's Industrial Oil & Fat Products*, Vol. 2, 5th edn. John Wiley, New York, USA, pp. 159–240.

Kohel, R.J. (1989) Cotton. In: Robbelen, G., Downey, R.K. and Ashri, A. (eds) *Oil Crops of the World*. McGraw-Hill, New York, USA, pp. 518–532.

Vaughan, J.G. (1970) *The Structure and Utilization of Oil Seeds*. Chapman and Hall, London, UK.

Cotton – cultivation

Cotton bolls are made up of three to five sections called locules or carpels, each of which contains eight to ten seeds attached to a central column or placenta. For information on cultivated species of cotton (predominantly *Gossypium hirsutum*, upland cotton, as well as *G. barbadense*), its economic importance, origins, and the long and short fibres (**lint and linters**) with which mature seeds are covered, **see: Cotton; Fibres**.

1. Genetics, breeding

Most cultivated species are allotetraploids (2n = 4x = 56 chromosomes). Agronomic breeding objectives include higher yields, increased number of seeds per boll and number of fibres per surface area of seed, improved fibre quality, earliness, and disease and insect resistance, improved seed vigour, resistance to seed–seedling diseases and absence of unnecessary or unwanted seed components such as gossypol, which limits feed use to ruminant animals (**see: Pharmaceuticals and pharmacologically active compounds**). Although morphological traits, biotic and abiotic stress resistance, yield and its components, and fibre quality are determined genetically, they may be markedly affected by environmental factors such as plant population and moisture.

Hybrid vigour was exploited commercially for the first time in 1970 in India, and to a much lesser extent in the USA during the 1980s. However, substantial **heterosis** was found to be inconsistent in F_1 hybrids of upland cotton. The problems and practicalities of hybrid seed **production**, including costly and laborious hand-pollination and high labour costs in countries, have so far limited appreciable commercialization to India.

Genetically modified (GM) or transgenic cotton has caused enormous changes in cotton production, which used to require extensive insecticide and herbicide applications. Transgenic varieties enjoy widespread adoption in the USA, China, Australia, South Africa, Mexico, India and Argentina. (**See** inset in: **Crop Atlas Appendix, Map 14**) In 2004, 28% of the worldwide cotton crop was transgenic, including about 80% of the crop grown in the USA and 66% of that in China. At present, varieties expressing Bt toxins from ***Bacillus thuringiensis*** are very widely used for the control of some lepidopterous insects, and varieties resistant to glyphosate or bromoxynil herbicides are also available; these traits are available in combination ('stacked') in some varieties. In the future, GM cotton with enhanced fibre qualities, improved herbicide and insect resistance, and enhanced quality in respect of the protein and oil in the seed are anticipated.

2. Development

The self-pollinated cotton flower opens up at dawn and remains open for only 1 day. Pollination occurs soon after the flower opens and fertilization 12–24 h later, depending on temperature. Three weeks after fertilization seed attains full size but does not reach maturity until the **boll** opens 40–45 days after pollination. Fig. C.24 illustrates the time course and morphology of developing seed and boll, and fibre growth and maturity, in relation to each other.

Thousands of individual **seedcoat testa** cells on the surface of each seed begin to elongate about the time the flower opens, forming fibres on the immature ovules (or motes) that grow in length for about 21 days and then in thickness for approximately 20 days. At maturity, suture lines in the ovary wall between the locules of the boll dehisce and the boll wall, now called the bur, continues to dry and curve away from the central axis, exposing the so-called 'seed cotton' for harvest. Cultivars differ in seed size, number of seeds per locule, number of locules and number of fibres per seed. For each 50 kg of lint produced, the cotton plant yields approximately 80 kg of seed. About 660 bolls are required to produce 1 kg of lint; in other words, to produce a 227 kg bale of cotton lint (along with an approximately 363 kg of seed) requires about 680 kg (or 145,000 bolls) of picked seed cotton, or about 1067 kg of stripped seed cotton (due to the extra plant debris trash it contains).

During the boll maturation and seed development phase, as a result of sucrose depletion, new vegetative root and leaf tissue formation in the mother plant stops and photosynthetic efficiency declines during the so-called 'cutout' period, and finally boll fibre development is curtailed. The mature seed with all its fibres removed is ovoid and dark brown: **cotyledons** constitute about 60% of the seed weight, the seedcoat about 32%, and the **embryonic axis** about 8% (consisting of the epicotyl that will form the main stem, the **hypocotyl** that will expand to push the shoot out of the soil during emergence, and the **radicle** (**see: History of seed research,** Fig. H.6c; **Seedcoats – structure**, Fig. S.23). Mature seeds are typically composed of 20% protein, 20% oil and 3.5% starch (on a weight basis).

3. Production

Cotton seed separated from cotton lint in the **ginning** process is considered a separate crop because of its distinctive uses and economic importance, though secondary in value to the fibre. Less than 5% of the seed is used for planting the next year's crop. The remaining seed is sold by the ginner as raw material for the processing industry or as feed for cattle.

The increasing dominance of genetically modified varieties has tended to decrease the amount of farm-saved seed, in favour of production by the seed industry.

Moisture, temperature, mechanical injury during harvesting, handling and ginning, and chemical treatment are important in producing high quality planting seed. Seed deterioration begins when open bolls are exposed to prolonged wet conditions in the field or when damp seed are stored after harvesting. Seed should be stored at a uniform moisture level of 6 to 7%; high temperatures in the presence of high moisture exacerbate deterioration.

Cotton is harvested either laboriously by hand or mechanically by spindle picker or stripper-type equipment; pickers remove only the seed cotton, leaving the burs and the plant intact, whereas strippers remove the complete boll (seed cotton and bur) from the plants. Much of the cotton harvested in the USA today is stored in the field in a free-standing stack of seed cotton called a module, consisting of 15 to 20 bales or approximately 4000 kg. The modules, covered with plastic in areas where rain is likely, may remain in the field for a considerable length of time before being transported to the gin; this temporary storage system frees producers to devote time to harvesting. In the ginning process seed cotton is cleaned, dried, and the usable lint fibres and seed separated from each other, leaving clumps of '**fuzzy seed**'.

Seeds are held at the gin on a short-term basis and then transported to oil mills, or sold to cattle or dairy ranchers as feed. In either case, cotton seed have to be stored properly and processed immediately before they deteriorate due to the high moisture content. At the oil mill, seed are analysed for moisture and free **fatty acid** content, which indicate storage life and **oil** quality. Most seed in the USA is bought on the basis of grade. Cotton seed oil is extracted by two techniques: mechanically using a screw press or expellers or chemically using hexane. The oil receives a higher price than regular vegetable oils like **soybean** and **canola** because of its desirable flavour and higher stability due to very low content of trisaturated fatty acids. The farm value of cotton seed averages about 15% of the total farm value of the cotton crop. (**See: Cotton – an oilseed; Oilseeds – major**)

For sowing purposes, to ensure uniform flow without clumping in planting equipment, **delinting** processes remove the short fibres that remain on fuzzy seed after ginning, using dilute sulphuric acid or hydrochloric acid vapour followed by pH neutralization, which also kills disease pathogens borne on the seed surface. The resulting delinted ('black') seed is cleaned and conditioned by gravity separation and air-screen grading to remove immature seeds and hence enhance vigour, before seed treatments such as fungicides and insecticides are applied (see section 6; Fig. C.23).

4. Quality, vigour and dormancy

The quality of plantingseed is characterized by several factors: seed maturity, seedcoats free of damage, moisture content less than 10%, free fatty acids less than 1%, no internal or external infection by seed-borne pathogens, uniformity of seed size, and ability to germinate and emerge over a range of environments. After **physiological maturity**, **viability** and, more rapidly and earlier, **vigour** both start to decline. Deterioration in the field usually occurs between the time of boll opening and harvesting, due to exposure to moisture and heat, causing physical, physiological, and biochemical changes. In addition, factors such as early **desiccation** or defoliation, early frost, or diseases that cause premature leaf loss will cause seed to be immature at harvest.

Two tests for evaluating cotton seed quality predominate. The standard germination test evaluates at either alternating 20 to 30°C, or constant 30°C for 7 days; in the protocol adopted by the **Association of Official Seed Analysts** only seeds with radicles 3.75 cm long after 7 days are regarded as being 'germinated'. The cool germination test at a constant 18°C for 7 days is the most widely used measure of germination and vigour. Rapid evaluation tests for seed quality include inspection for seedcoat maturity and visible mechanical damage; cutting combined with a **tetrazolium test**; and determination of free fatty acid and seed moisture contents.

Seeds collected immediately after boll opening can display innate, primary post-harvest **dormancy**; the period of **afterripening** storage necessary to overcome this may be several months. Such dormancy can be often be overcome in the laboratory by treatment with hot water (50°C) for 24 h, which softens the **chalazal** plug, or by **scarification**. After sowing, secondary dormancy can also be induced by low soil temperature and/or moisture. Seeds for sowing can be conditioned before planting by exposing to limited amounts of moisture and heat.

Prior to the recent development of cotton cultivars, a proportion of the seeds exhibited **hardseededness**, especially in Pima types, in which impermeability to water prevents germination. This trait has been eliminated from modern commercial cultivars through breeding and selection, although it may occasionally occur.

5. Sowing

Seeding rates expressed on a weight can vary greatly due to seed size; some cultivars have less than 8000 seeds/kg, whereas others can have over 13,000 seeds/kg. Planting in soil with temperatures less than 13°C can result in poor vigour and seed or seedling disease problems. Indeed, seeds are susceptible to imbibitional **chilling injury** below about 20°C; the maximum sensitivity is experienced at the very beginning of imbibition and again about 1 day later. Seed emergence can occur within 5 to 7 days after sowing under the near optimal conditions. In practice, only 75–90% of planted seed may establish a stand due to combination of seed quality, soil conditions and seed–seedling pathogens. (**See: Cool Test; Vigour tests – physiological**)

6. Treatment

The most common diseases that attack the seed and young seedlings during or after germination are caused by *Rhizoctonia solani*, *Pythium* spp., *Thieviopisis basicola* and *Fusarium* spp. pathogens (**see: Pathogens**). Cool, wet soils are conducive to the development of diseases which, though they do not usually kill the entire population, result in uneven, slow-growing stands. Symptoms include pre-emergence rotting and decay of the seed or seedling. The 'soreshin' phase of *Rhizoctonia* seedling disease, for example, is characterized by reddish-brown, enlarging sunken lesions on the stem and root at or below ground level, which girdle the stem, thus destroying the plant's vascular system. Damaged seedlings that emerge are pale, stunted, slower growing, and sometimes die within a few days; often because the taproot is destroyed, only shallow-growing lateral roots remain to support the plant. Seed treatment fungicides are usually sufficient for controlling these diseases, unless seed quality or weather conditions are unfavourable for germination. Systemic insecticides (such as the neonicontinoids, imidacloprid and thiamethoxam, and thiodicarb) are currently used as seed treatments for protecting plants against early-season insect pests such as thrips, cutworms, leafminers and aphids. Commercial seed is sold with one or more of these seed treatments, either as standard within a growing area or customized to the request of the grower. (PT, CWS)

Association of Official Seed Analysts (1983) *Seed Vigor Testing Handbook*. Contribution No. 32 to the Handbook on Seed Testing, Las Cruces, NM, USA, pp. 63–64.

Hopper, N.W. and McDaniel, R.G. (1999) The cotton seed. In: Smith, C.W. and Cothren, J.T. (eds) *Cotton: Origin, History, Technology and Production*. John Wiley, New York, USA, pp. 289–317.

Oosterhuis, D.M. (2001) Development of a cotton plant. In: Seagull, R. and Alspaugh, P. (eds) *Cotton Fiber Development and Processing, An Illustrated Overview*. International Textile Center, Texas Tech University, Lubbock, Texas, USA, pp. 7–31.

Smith, C.W. and Cothren, J.T. (eds) (1999) *Cotton: Origin, History, Technology and Production*. John Wiley, New York, USA.

Cotton seed and its Products (2002) National Cotton seed Products Association. www.cottonseed.com/publications/cotton seedanditsproducts.asp

Cotyledon

The first leaf (monocotyledons) or pair of leaves (dicotyledons) of the embryo. Sometimes called the 'seed leaf'. Cotyledons are frequently the major storage tissue in the seed, and may or may not emerge following germination; in some cases they become **haustorial**. (**See: Development of embryos – dicots; Embryo; Epigeal; Hypogeal; Structure of seeds**)

Cowpea

1. World importance and distribution

Cowpea (black-eyed pea) (*Vigna unguiculata*) is a **warm-season legume**. The seeds are an important food throughout the tropics and subtropics in savanna regions; this includes Africa, Asia, and Central and South America. Southern Europe and the USA are also important producers. Cowpea is an essential component of cropping systems in drier regions of the tropics due to it being fast growing, its ability to fix nitrogen, its tolerance to drought, shade, soils with high or low pH, heat and nematodes plus its ability to curb erosion by covering the ground. Around 9 million ha are harvested in the world (FAOSTAT 2002), 75% of which are in Central and West Africa, mainly in the dry savanna and semi-arid agroecological zones. The principal producing countries are Nigeria (5 million ha – 57% of the total cowpea crop area in Africa), Niger, Burkina Fasso (more than 300,000 ha each), Senegal, Ghana and Mali. Nigeria is the largest producer (2,389,000 t) and consumer (2,617,400 t) of cowpea in the world. To cover the deficit, substantial amounts of cowpea come to Nigeria from other African countries. Niger is the largest cowpea exporter in the world (estimated at 215,000 t annually). South America, mainly Brazil (1–1.8 million ha), ranks second in world cowpea production (21%). (**See: Legumes; Crop Atlas Appendix, Map 3**)

2. Origin

Ethiopia is considered to be the sole centre of origin of cowpea which later spread to the African savanna, through Egypt and Arabia to Asia and the Mediterranean. Cowpea is a variable species composed of wild perennials, wild annuals and cultivated forms. The cultivated cowpea is closely related to annual wild cowpea which may include the likely progenitor of the cultivated form.

3. Plant types

Plants have either a **determinate** or an **indeterminate** growth habit, i.e. the flower either does or does not terminate growth of the main stem, respectively. The tendency of indeterminate cultivars to ripen fruits over a long time makes them more amenable to subsistence rather than to commercial farming. Determinate types are more suited to a monocultural production system.

4. Seed characteristics

The globular to reniform (kidney-shaped) seeds are 2–12 mm long, smooth to wrinkled, and red, white, brown, green, buff or black as dominant colours (Colour Plate 1E). They may be plain, spotted, marbled, specked, eyed or blotched; seed weight varies from 5 to 30 g/100 seeds.

5. Uses

Cowpeas are cultivated for the seeds, shelled green or dried, and the pods and leaves that are consumed as green vegetables. The plants are also used for pasturage, hay, ensilage and green manure.

6. Nutritive value

The composition of cowpea is in general similar to that of most edible legume seeds, with a range of contents of protein, starch, and sugars in the different types. Table C.24 shows the nutrient content range of cowpeas based on several reports in the literature.

Cowpea protein is rich in lysine, but it is deficient in sulphur amino acids. Its relative contents of methionine and tryptophan are higher than those of the other legumes.

7. Anti-nutritional factors

Cowpeas contain relatively high amounts of flatulogenic sugars particularly **raffinose,** the most active in this respect. The concentrations of **trypsin inhibitors** and **phytin** are half of those in soybean.

8. Market classes

To develop cowpea markets it is essential to know consumer preferences. Seed colour and size and usage are important factors. The following types of cultivars are recognized:

- 'Crowder peas': globose, crowded in pods, black, speckled brown, brown-eyed.
- 'Brown Crowder': brown.
- 'Black-eyed': white seeds with black-eye around the hilum; not crowded in pods.
- 'Cream': cream seeds, not crowded in pods.
- 'Purple Hill': deep purple, mature pods and buff or maroon-eyed seed.
- Snap: good for use as vegetable.
- Persistent-green: new market class of cowpea for the freezing industry.
- Dual-purpose: provide food (dry seeds) and forage.

The West African consumer in general prefers larger seeds. In some countries, Cameroon and Ghana for example, the eye colour is an important influence on the price: in Ghana consumers pay a premium for black-eyed cowpea while in Cameroon they expect a discount. In Ghana, there may be a price reduction or a premium on white grains according to the region.

9. Marketing and economics

Cowpea marketing in West Africa is based on the comparison in food production between the humid coastal zone, where it is relatively easy to produce carbohydrates but not vegetable protein, and the semi-arid interior where low rainfall creates good conditions for cowpea and groundnut.

In the Sudano-Sahelian zone there is a well developed network of village buyers who assemble small amounts from farmers into 100 kg quantities and the merchants who transport and store these large batches. Ghana imports cowpeas in this way from Burkina Faso and Niger. Cowpeas from the latter – large, rough-coated grains – sell for a premium, but they need to be marketed quickly as they do not store well in the humid coastal climate. There is an active trade among Benin, Niger, Togo, Nigeria and Gabon.

In 2001 the USA exported 4487 t of the black-eyed type of cowpea, mostly produced in California, for a value of $2,217,000. Countries in Latin America (e.g. Ecuador and Peru) also export small quantities of cowpea for canning purposes.

(OV-V)

Table C.24. Cowpea seed composition.

Component	Content (% fresh weight)
Protein	20.3–32.5
Crude fibre	2.7–6.9
Carbohydrates	59.7–71.6
Soluble sugar	5.9–8.3
Starch	39.1–54.9
Ash	2.9–3.9
Oil	1.1–3.0

Singh, B.B., Mohan Raj, D.R., Dashiell, K.E. and Jackai, L.E.N. (eds) (1997) *Advances in Cowpea Research*. International Institute of Tropical Agriculture (IITA) and Japan International Research Center for Agricultural Sciences (JIRCAS). IITA, Ibadan, Nigeria.

Singh, S.R. and Rachie, K.O. (eds) (1985) *Cowpea Research, Production and Utilization*. John Wiley, New York.

Som, M.G. (1993) Cowpea, *Vigna unguiculata* (L.) Walp. In: Kalloo, G. and Bergh, B.O. (eds) *Genetic Improvement of Vegetable Crops*. Pergamon Press, Oxford, UK, pp. 339–354.

Crambe

1. The crop

Crambe (*Crambe abyssinica*, 2n = 90, Cruciferae or Brassicaceae) is a much branched annual that is related to rapeseed and mustard. Over the past 45 years crambe has been introduced as a new speciality industrial oil crop in Poland, Canada, the EU and USA with little or no lasting commercial success. Subsidized production in North Dakota, USA, occupies about 5000–10,000 ha. Other names are Abyssinian mustard or kale, colewort or datran.

2. Origin

The crop is native to the Turko-Iranian region.

3. Fruit and seed

Each self-pollinating (90%) flower produces a single seed within a paper-thin capsule (**siliqua**) that is attached to a relatively long pedicel. Initially there are usually two **embryos** per flower but only one develops into a spherical greenish-brown to brown seed about 0.8–2.6 mm in diameter (Fig. C.25). The **pericarp** makes up about 40% by weight of the harvested 'seed'. The remaining weight is divided among the testa (4%), hypocotyl (6%) and the cotyledons (50%). For the histology of the seed **see: Brassica oilseeds**.

4. Uses

Crambe is grown for its high **oil** content (30–35% hulled, >50% dehulled) that contains about 55% erucic acid (**see: Oilseeds – major**, Table O.2). Oils high in erucic acid are used as biodegradable lubricants and as a source of erucimide, an important slip agent in plastics manufacture as well as in plasticizers and polymers. Worldwide the market for erucic

Fig. C.25. Crambe 'seeds' (one capsule near the bottom left corner is broken open revealing the true seed; scale = mm). (Photograph kindly produced by Ralph Underwood of AAFC, Saskatoon Research Centre, Canada.)

acid is estimated to be about 57,000 t. The presence of the capsule around every seed has limited Crambe's success since the added volume makes transport of seed expensive. Disease susceptibility (*Alternaria* sp.) and the high content of the **anti-nutritional** glucosinolates (**see: Pharmaceuticals and pharmacologically active compounds**) in the seed and meal is also a limiting factor. Crambe meal has a well-balanced **storage protein** (whole seed meal 22–28%, dehulled meal 48–50% protein) but because of the **glucosinolates** can only be fed to ruminant animals. (KD)
(See: Oilseeds – major)

Hirsinger, F. (1989) New annual oil crops. In: Robbelen, G., Downey, R.K. and Ashri, A. (eds) *Oil Crops of the World*. McGraw-Hill, New York, USA, pp. 518–532.

Vaughan, J.G. (1970) *The Structure and Utilization of Oil Seeds*. Chapman and Hall, London, UK.

Weiss, E.A. (2000) Crambe, Niger and Jojoba. In: *Oilseed Crops*. Blackwell Science, Oxford, UK, pp. 287–325.

Cremocarp

A fruit, typical of the Apiaceae, such as many spice 'seeds', consisting of two bilocular carpels. Each carpel forms a one-seeded, indehiscent capsule or **dispersal** unit pendant from a supporting axis (**see: Mericarp**).

Critical water content (Critical water potential)

The amount of water that marks a change in physiological activity, molecule conformation or mobility. In seeds, critical water content expresses the concentration of water, usually as grams of water per gram fresh or dry mass of the seed. Critical water potential describes the pressure (energy per volume) of water and is usually expressed as megaPascals (mPa) or bars. Water content and water potential can be inter-converted using **water sorption isotherms**. Critical water contents or potentials have been studied in seeds in relation to germination (threshold water potentials that allow germination), changes in patterns of protein synthesis and induction of stress proteins, **desiccation tolerance** (lowest water content before damage is measurable), **respiration** (water content above which oxygen consumption and electron transport are supported) and **ageing** (water content below which further drying does not slow deterioration or, in some cases, further drying increases deterioration rate) (Fig. C.26) (**see: Deterioration kinetics**). The occurrence of critical water contents or potentials supports the hypothesis that water controls the nature and kinetics of metabolism within hydration levels. Though critical water contents vary among species and temperature, critical water potentials are similar for diverse species at a wide range of temperatures. Changes in physiology at critical water contents are analogous to breaks in Arrhenius behaviour (**see: Arrhenius plot**), suggestive of a moisture-induced change in the **phase** behaviour of cellular constituents. Critical water contents for physiological activity correspond to changes in water properties and have led to the idea that critical water contents result from changes in moisture-dependent changes in the mobility of water in cytoplasm (**see: Glass**) or on macromolecular surfaces (**see: Water binding**). (CW)

Cross

In **angiosperms**, cross-pollination (a term synonymous with **allogamy**) is the transfer of pollen to the stigma from the anther of plants of different clones or lines, and cross-**fertilization** is the resulting union of an egg and sperm (gametes) to form a zygote. Extensive use is made of controlled crossing techniques in breeding programmes to produce and characterize breeding lines and their progeny. A crossing step is also the key basis of the production of $\mathbf{F_1}$ hybrid seed **varieties**. (Contrast with **Self-pollination**.)
(**See: Production for sowing, III. Hybrids**)

Cruciferin

A 12S globulin **storage protein** that is present, along with the albumin **napin**, in the embryos of *Brassica* oilseeds.

Crusting

Also known as capping. The formation in seedbeds of certain soil types and unstable structures of transient mechanically strong soil-surface layers (ranging from a few mm to a few cm thick) that are denser, structurally different, or more cemented than the material immediately beneath, and so can prevent the emergence of seedlings after sowing. (**See: Seedbed environment, 5. Soil structure**) Soil conditioning with various proprietary polysaccharide, polyacrylamide and polyelectrolyte materials is used to mitigate the tendency of susceptible soils to crust.

Crusting is also the phenomenon seen in grain **storage management**, whereby moisture condenses on the upper surface of the seed mass, due to the development of convection currents, favouring the proliferation of spoilage fungi and even germination.

Cryostorage

The storage of seed or tissue in or over liquid nitrogen (Fig. C.27). Cryostorage is often employed for **recalcitrant** seeds that are not suited to the conventional techniques used for **orthodox** seeds. In this case, isolated **embryos** are prepared for cryostorage by controlled drying to below the high

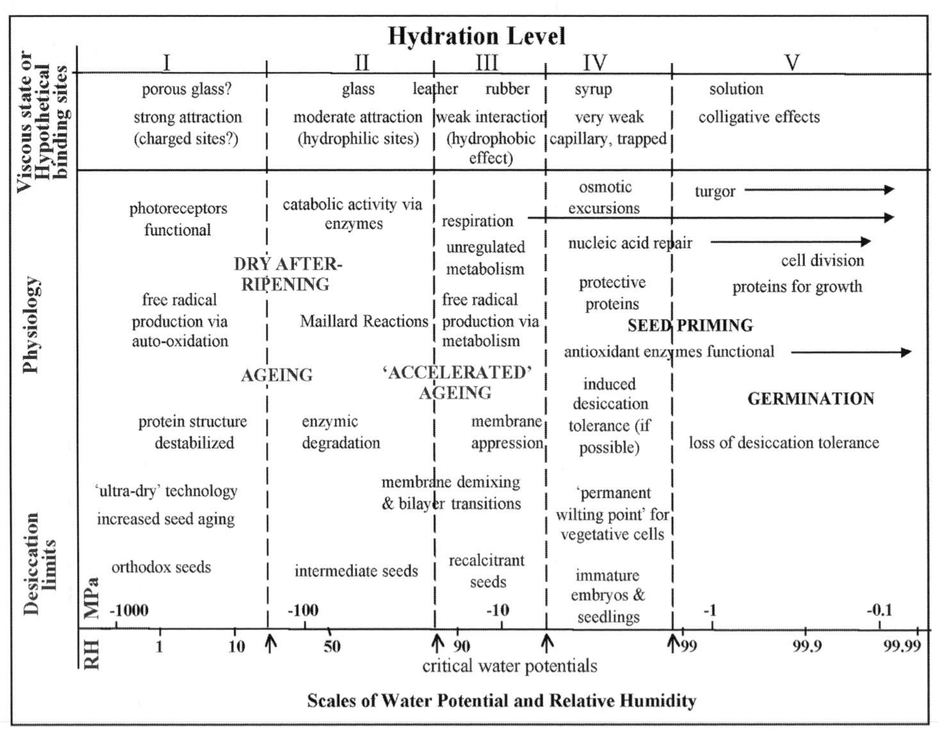

Fig. C.26. Critical water potentials of seeds with respect to their physical and biological behaviour. Five stages of metabolic activity can be distinguished according to the water availability, and water potentials at the transition of stages are considered critical. Water is not limiting at the highest stage, allowing for sufficiently integrated metabolism and developmental changes (e.g. germination). Reactions are progressively restricted and the potential for lesions to biomolecules change with drying. The figure provides a mechanistic understanding of the different classification of seeds (**orthodox**, **recalcitrant** and **intermediate**) based on the types of damage they are likely to experience when dried below a critical level. The critical water potentials that mark changes in biological behaviour correspond to changes in the physical properties of water within the seed suggesting that a change in viscosity or water binding activity drives the change in biological behaviour. Critical water contents are not provided in the figure since they vary with species and temperature. For soybean seeds at 25°C, critical water contents corresponding to transitions between hydration levels I and II, II and III, III and IV and IV and V are approximately 0.06, 0.22, 0.40 and 0.60 g H_2O/g dry mass, respectively.

moisture content freezing limit (HMFL) (corresponding to an **equilibrium relative humidity** [eRH] of 80–90%). After controlled drying, embryos are plunged into liquid nitrogen to maximize the rate of cooling, thus minimizing the risk of damage due to ice formation. On recovery of stored embryos, a rapid thawing rate is also employed, often through the use of a water bath at 35°C.

Short-lived orthodox seeds such as orchids are also suited to cryostorage because of their small size and predicted increase in longevity at −196°C compared to conventional storage temperatures. Cryostorage may also be employed for **intermediate** seeds as well as for other plant tissues such as pollen and meristems. However, the limited capacity of liquid nitrogen containers and the difficulties of ensuring a reliable

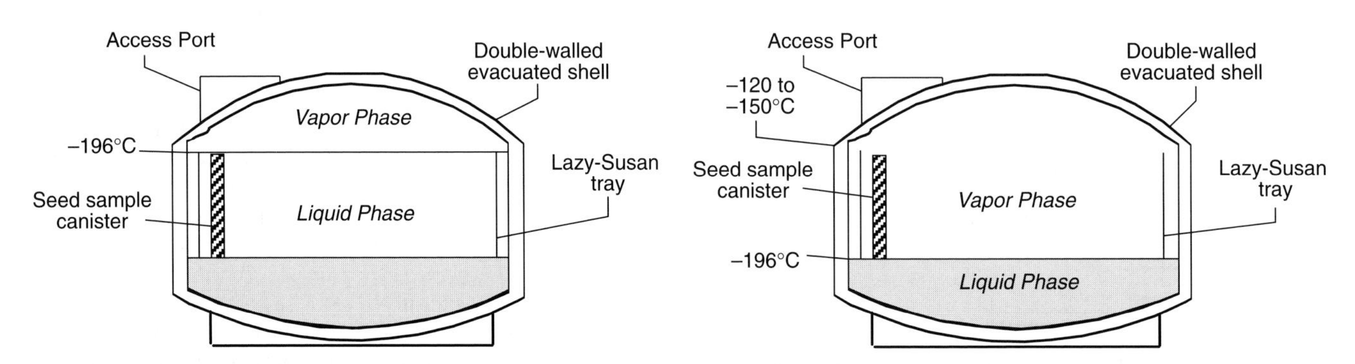

Fig. C.27. Schematic diagrams of cryovats used for cryogenic storage of seeds. The left-hand diagram illustrates storage in liquid nitrogen at −196°C, which may be necessary for the hydrated cells of recalcitrant seeds. The right-hand diagram illustrates storage in the vapour phase of liquid nitrogen, usually below −135°C, which is feasible for orthodox seeds. A Lazy-Susan tray is a circular platform that rotates horizontally on a central axis. (CW)

supply of liquid nitrogen in some countries mean that cryostorage techniques have not been routinely adopted in seed **gene banks** worldwide. (RJP, SHL)

Cryptochrome

A developmental photoreceptor absorbing predominantly in the blue (450–480 nm) and ultraviolet (UV) (*ca.* 375 nm) regions of the spectrum and to a much lesser extent in the green. Cryptochrome was so named because of its cryptic nature. Attempts to identify the blue light receptor in plants were not successful until 35 years after the isolation of the red light photoreceptor, **phytochrome**. Cryptochrome may be responsible for some blue-light effects in seeds.

Plant cryptochromes are thought to be derived from blue/UV-activated DNA repair enzymes known as type II photolyases that are present in both plants and animals. Activated photolyases can mediate repair of the thymine dimers often formed as a result of UV light damage. Cryptochrome, like the photolyases, binds two **chromophores**, a light-harvesting pterin and a catalytic **flavin** (Fig. C.28). Two cryptochromes exist in higher plants: cry1 and cry2. Cry1 is light-stable while cry2 is light-labile, being rapidly degraded in blue light (B). Evidence for the roles of the individual cryptochromes has come primarily from the study of loss-of-function **mutants** in the model plant *Arabidopsis thaliana*. Cry1 mediates responses to higher intensities (fluence rates) of B, whilst cry2 is very sensitive to low fluence rates of B but is rapidly lost at higher fluence rates of B.

Cryptochromes affect development throughout the life of the plant, e.g. seedling establishment, plant architecture and timing of flowering. They probably do not directly mediate an induction of germination but they may participate in its inhibition; a blue light-specific photoreceptor has been demonstrated in seeds. For example, germination of wild-type tomato (*Lycopersicon esculentum*) seed is inhibited by relatively low concentrations of the hormone **abscisic acid** (ABA) or by high osmoticum, but predominantly in B or white light (W). In darkness or red light there is less effect of ABA or osmoticum, implicating the involvement of a specific blue-light photoreceptor. Notably, inhibition of **hypocotyl** elongation in tomato by W or B occurs partially due to ABA accumulation. Therefore, a blue-light photoreceptor may amplify the responsiveness of tomato seeds and hypocotyls to ABA. In addition, since part of the inhibitory action of prolonged illumination on germination is due to the blue wavelengths, absorption by cryptochrome may be responsible.

Pterin (MTHF)

Flavin ($FADH^-$)

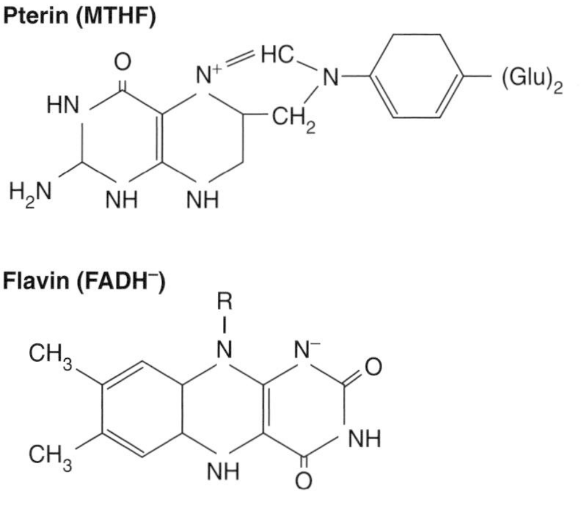

Fig. C.28. The two chromophores of cryptochrome. Cryptochrome binds both pterin and flavin chromophores. The structure of the pterin: 5,10-methenyltetrahydrofolylpolyglutamate (MTHF) and the flavin: flavin adenine dinucleotide ($FADH^-$) are shown.

The mechanism of cryptochrome action is poorly understood, although it is known to involve the induction of changes in gene expression. Close to 20% of the *Arabidopsis* genome shows response to blue light in de-etiolation with about 15% of the genome specifically regulated by cry1 and cry2.

Pertinent to this, cryptochromes are nuclear-localized. In fact, cry2 co-localizes and directly binds with the red light photoreceptor, phytochrome B, in nuclear speckles that possibly represent the sites of activation of gene expression. Furthermore, cry1 binds to the Constitutively Photomorphogenic 1 (COP1) protein, a repressor of light-regulated **gene expression** (**see: Light – dormancy and germination; Phytochrome**). (PD)

Briggs, W.R. and Huala, E. (1999) Blue-light photoreceptors in higher plants. *Annual Review of Cell and Developmental Biology* 15, 33–62.

Cashmore, A.R., Jarillo, J.A., Wu, Y.J. and Liu, D. (1999) Cryptochromes: blue light receptors for plants and animals. *Science* 284, 760–765.

Devlin, P.F. and Kay, S.A. (1999) Cryptochromes – bringing the blues to circadian rhythms. *Trends in Cell Biology* 9, 295–298.

Lin, C. (2002) Blue light receptors and signal transduction. *Plant Cell* 14 (Suppl.), S207–S225.

Crystalloid

An inclusion that is present in **protein bodies** and contains densely packed **storage proteins**. In **castor bean** the crystalloid is an aggregation of 11S **globulin** protein.

Cucumber

See: Cucurbits

Cucurbits

For over 10,000 years, cucurbits (Cucurbitaceae) have been cultivated to provide food and other products for humankind. The main present-day cultivated species include cucumber, gherkins, melons, muskmelon, watermelons, pumpkins and squashes. The cucurbits are a significant world food crop, mostly on account of their fruits, though the edible plant parts also include flowers, leaves, shoot tips, storage roots and seeds. Although tissues of many species contain bitter terpenoid compounds, cucurbitacins (**see: Poisonous seeds**), the plant parts low in cucurbitacin may safely be eaten fresh, processed (pickled, frozen, or canned) or dried. The immature fruit of *Cucurbita* spp. are consumed as a fresh vegetable ('summer squash') because they are perishable and have a limited storage life. Fully mature fruit develop a hard protective rind that allows for long-term storage and later use as 'winter squash'. Cucurbits also have medicinal uses. In parts of Southeast Asia, dried bitter melon fruit (*Momordica charantia*) is used as a medicine for treatment of haemorrhoids, gout, arthritis, skin disorders, burns and parasites. The fruit may also be used as animal feed. Estimated world production in 2003 of cucumbers, melons, watermelons, pumpkins and squash exceeded 8.3 million ha and 177 million t (FAO statistics). Cucurbits are extensively cultivated as annuals from seed or transplants in temperate regions, and have long growing seasons, whereas many cucurbits in tropical regions are perennials. Melon seed production in 2003 alone comprised over 608,000 ha with a total yield of over 584,000 t (FAO statistics).

In addition to their use as propagules for cultivating the crop, some cucurbit seeds are also used themselves as food. Watermelon and **pumpkin** seeds are popular snack foods in Southeast Asia and the Middle East, and cucurbit seeds have been used to make tea and edible oil. Watermelon (*Citrullus lunatus*) seed has been used to treat urinary tract infections, poor circulation and other disorders, and seeds of *Cucumis melo* have been used as a treatment for stomach cancer. Some cultivars of *Cucurbita* do not develop seedcoats and can be consumed without decoating.

1. History and origins

The cucurbits are largely tropical in origin with genera that are native to Africa, tropical America and Southeast Asia. Rapid **indeterminate** growth, developmental plasticity with regard to sex expression in response to environmental conditions, and genetic diversity helped wild cucurbits adapt to **domestication**, through which in turn, the size of both fruits and seeds increased, bitterness and seed dormancy was reduced, fruit dehiscence was decreased and flavour improved.

2. Genetics and breeding

The family Cucurbitaceae consists of about 120 genera and more than 800 species, between which sexual crosses are usually not possible. In *Cucurbita* however, plant breeders have made a number of successful interspecific crosses.

F_1 hybrids now dominate the market for most commercial cucurbit crops and represent the bulk of seed produced, particularly in developed countries because of their greater vigour, uniformity and higher yield. Also, seed cannot be propagated for a second generation from hybrid seed, protecting corporate investments in cultivar development. In hybrid cucumber production, genetic manipulation is used to grow ‘mother plants’ with only female flowers (gynoecious plants). Normal genotypes are planted in adjacent rows and provide pollen for bees to transfer from male to female flowers. In hybrid squash production, **ethylene**-releasing compounds are applied to suppress male flower production, so that treated plants produce only female flowers (see next section); the adjacent rows of the male parent lines are left unsprayed. In both the genetic and chemical methods, only the female rows are harvested for hybrid seed. However, hybrid seed production by these methods is not currently feasible for cantaloupes, mixed melons (honeydew, crenshaw, casaba), or watermelon because they have bisexual flowers and sex expression is not reliably changed by exogenously applied **hormones**. Hybridization of these crops requires extensive hand labour to emasculate flowers and make hand crosses.

Seedless watermelons are produced using **colchicine** to increase the chromosome number of the female parent from the normal **diploid** state to the tetraploid. Crossing a normal diploid male parent with the tetraploid produces F_1 triploid seed for growing the commercial crop. Because these triploid plants are highly self-sterile, planting 15% of the field with a normal diploid pollinator cultivar pollinates the seedless watermelon plants, in which the seeds normally abort shortly after pollination, leaving white empty seedcoats within the triploid fruit.

Agrobacterium-mediated **transformation** protocols have been developed for several major cucurbits. Several crops have been successfully genetically transformed, primarily to improve disease resistance. For example, cucumber plants over-expressing chitinase genes from rice were found to be more resistant to grey mould (*Botrytis cinerea*). Muskmelon cultivars have been genetically engineered for reduced ethylene production and thus delayed ripening. Genetically engineered cultivars of zucchini (courgette) and summer squash with enhanced virus disease resistance have been approved for commercial production in the USA but are not widely grown because of public concerns about **genetically modified** food.

3. Development

Most commonly, cucurbits are monoecious (separate male and female flowers are produced), but other forms can be induced, including hermaphrodite, andromonoecious, gynomonoecious, trimonoecious, gynecious and dioecious. For many cultivated cucurbits staminate flowers are produced first, followed by somewhat random production of pistillate and staminate flowers. This cycle tends to be repeated on lateral branches. Development of the first female flower is an important determinant of early yields. The ratio of male to female flowers is usually heavily skewed to favour maleness, and ratios of 25 male to 1 female are common. Although sex expression is an inherited trait, the environment also affects floral gender. For example, high temperatures and long days usually favour male flower production, whereas low temperatures and short days favour female flowers.

All cucurbit seeds have similar seed anatomy (e.g. Colour Plate 6.C), in which endocarp tissue forms a transparent layer that adheres to the seed coat. The outermost, true seed tissue is the testa with successive layers of epidermis, sub-epidermal layer, sclerenchyma, parenchyma and inner epidermis (**see: Seedcoats – structure**). The **endosperm** is of the nuclear type; shortly after fertilization the lower portion of the **embryo sac** grows downward and, after wall formation occurs in the upper embryo sac, the elongated structure is referred to as the endosperm proper. During the first 20 days following fertilization, the endosperm enlarges rapidly and has a gelatinous consistency. The rapidly growing **embryo** consumes the endosperm, which ultimately collapses. The **nucellus** gradually disintegrates before the developing endosperm proper leaving a thin layer of tissue surrounding the mature embryo. In muskmelon, a single layer of dense cytoplasm tissue completely encloses the embryo. A deposit of callose and suberin forms 30 to 35 days after **anthesis** (DAA) on the outer surface of the endosperm tissue and makes the endosperm semipermeable. In these seeds, it appears that the endosperm tissue has evolved to prevent electrolyte leakage, slow the rate of fungal attack, and protect seeds against fruit decomposition products.

Inside maturing and senescing fruits, physiological seed maturation continues after seed mass maturity because of the high moisture content and high temperature. This environment also favours rapid ageing if seeds are not harvested and dried once full physiological maturity is achieved. In the field, the transition from high quality to aged or even dead seeds occurs swiftly over a matter of a few days particularly at high temperatures (see: **Deterioration kinetics**). For example, an additional 10 days within the fruit beyond the point of optimum quality caused a fraction of 65 DAA muskmelon seeds to age and die. In over-mature fruit, seeds aged because of long-term hydration at low water potential, combined with high temperature, insufficient oxygen, and high CO_2 within the fruit.

Seed morphology has been used for identifying species, particularly within *Cucurbita* where seed colour, **testa** margin and texture, as well as the shape of the **chalaza** define each species. Watermelon seeds for example vary in size, number, and colour from white through tan, dark brown and black.

4. Production

Cucurbit seeds both for food use and propagation are produced commercially in warm, dry regions with long growing seasons, in the Middle East, North America, Asia and Africa to avoid **disease** problems.

Sex expression can be modified by exogenously applied **plant growth regulators**, and this is commonly done both for breeding and hybrid seed production purposes. Ethylene promotes female flower formation whereas **gibberellins** promote maleness, with GA4+7 being the most effective; silver nitrate and silver thiosulphate also induce male flowers. Soil fertilization and developing fruit inhibit the frequency and number of subsequent flowering, particularly of female flowers. Frequent harvesting extends flower production. **Parthenocarpic** fruit development occurs, particularly in cucumber, where cultivars are grown for processing and greenhouse fresh-crop production. Parthenocarpic fruit apparently have less influence on subsequent plant growth or flowering.

Honeybees are the most reliable and cost-effective way to achieve **pollination** of cucurbits. One to two hives per acre are introduced when 5% of the plants have open flowers, or more if in weedy areas or close to other flowering crops clustered around the periphery of fields, with additional hives placed inside larger fields. Multiple bee visits are required to transfer the 300 to 500 pollen grains needed for successful pollination to occur. For many species, flowers are only receptive to pollination for a morning and afternoon of 1 day, while other species remain open for longer periods. Inadequate pollination causes aborted flowers and misshapen or underdeveloped fruit.

Viruses transmitted by aphids and beetles are a serious pest problem, causing a variety of foliar symptoms, generally stunted plant growth and fruit have raised, blistered areas, variegated patches and are malformed. Plants infected early often fail to produce flowers or set fruit. To maintain healthy and vigorous plants, fields should be monitored often for unusual growth, disorders must be quickly and correctly identified, and affected plants must be removed or treated.

Most cucurbits are frost-sensitive, although some tolerate low temperatures better than others, and fruits are especially damaged by temperatures below 15°C. Cucurbit crops are harvested when the majority of fruit are mature, as generally indicated by yellowing. The times from plant emergence to harvest for irrigated desert regions of California, for example, are 78–90 days for muskmelon and mixed melons; 85–100 days for watermelons; 100–120 days for summer squashes; 120–130 days for cucumbers; and over 130 days for winter squashes. Optimal seed **quality** coincides with fruit maturity. Seed quality can decline if harvest is delayed, particularly if temperatures are high and the fruit are exposed to the sun, as indicated by sunburned fruit. Combining seeds from different stages of physiological maturity in this way may adversely affect quality and increase seed variability, but this is the most efficient way to harvest large-scale, commercial fields.

Cucurbit seed is mechanically harvested with specialized equipment. In a destructive harvest, **windrowed** fruit are carried into the harvester on a chain conveyer and crushed to release the seed. The seed and pulp travel through an inclined rotating drum screen that allows the seed to fall through and carries the fruit rind, pulp, dirt clods and trash out of the rear of the harvester. Screens of various sizes are used for different vineseed crops. The seed are then augered from a storage hopper into bins for transport to a washer-dryer. To prevent contamination or mixing of seed lots, the harvester must be thoroughly cleaned between fields of different cultivars or species. Hand-harvesting cucurbit fruits when fully mature improves seed quality and F_1 hybrid seed crops derived from hand-crossing are often hand-harvested and fed into a stationary wet-seed extractor.

Fermentation was once a standard, post-harvest practice, but mechanical washing of newly harvested seeds has replaced it. Harvested seed must be washed and dried promptly to avoid precocious germination, and rapid microbial fermentation in the slurry of seeds, juice and fruit tissue can expose seeds to excessive temperatures. Seeds are spray-washed as they pass through inclined rotating drum screens to remove fruit and other debris. Post-harvest washing also improves germination performance. Muskmelon seeds that were vigorously washed in tap water for 3 h had greater 4-day root lengths during standard germination testing at the time of harvest and displayed greater vigour after 6 years in storage when compared to unwashed controls.

The clean seed can be dried in a rotating drum with ambient or heated air. Final seed moisture should range from 5 to 8%. Seed with higher moisture stores poorly, while drier seed may be physiologically damaged. Seed moisture and temperature must be measured frequently to prevent under- or over-drying that can reduce vigour. Cucurbit seeds dry rapidly and can be sun-dried in warm dry areas if spread flat and mixed frequently. Seed temperature should not exceed 35°C because rapid drying using heated air may damage seeds. Once the seed has been cleaned and dried it is graded, and tested for purity and germination percentage. The presence of weed seed, variability in seed sizes and excess foreign material may require additional seed cleaning or conditioning at a seed-processing facility.

Cucurbit seeds are long-lived and can be successfully stored in airtight containers at 5 to 8% moisture content and 3 to

5°C for 20 years or more. The seeds may be successfully frozen at −12°C if their moisture content is sufficiently low (5 to 6%) to further extend storage life. Cryopreservation by seed storage (**cryostorage**) at −196°C is possible for ultra long-term storage of germplasm.

5. Quality, vigour and dormancy

As already mentioned, the stage of fruit development at harvest plays a major role in determining seed quality (**viability** and vigour), dependent largely on cultivar and environmental conditions. For example, optimum quality 'Top Mark' muskmelon seed produced in the Central Valley of California, based on **germination rate**, low temperature germination, and resistance to **controlled deterioration**, occurred when fruits were harvested 50 to 55 DAA, approximately 10 days past edible maturity but before severe fruit decomposition. This is also approximately 20 days past mass seed maturity.

Some cucurbit embryos can germinate if removed from the fruit and incubated on water before the intact undissected seeds are germinable. At 35 DAA, the time of mass maturity (maximum dry weight accumulation) in muskmelon, isolated embryos germinate in water and develop into normal seedlings but intact seeds are not fully viable until after 45 DAA suggesting that before then the surrounding endosperm and testa might maintain developmental metabolism.

Cucurbit seed development is likely to be under hormonal control. The highest concentrations of **abscisic acid** (ABA) in muskmelon seeds occur 25 DAA in embryonic axis and endosperm tissue, while cotyledons and testa have much lower ABA contents. As viability increases, ABA concentrations fall to very low levels in all tissues after 40 DAA when seeds are fully germinable and vigour increases (**see: Development of seeds – hormone content**).

Precocious germination occurs in some cultivars of cucurbit fruit, particularly during long-term storage of fully mature fruit, but multiple factors apparently serve to prevent this happening. Most cucurbit seeds display some degree of seed inhibition early in development, the degree varying widely among species. Also, the low osmotic potential in developing fruit, particularly in muskmelon and other types that accumulate sugars, limits seed hydration thus preventing precocious germination. This phenomena has been called 'osmotic inhibition' of germination. Precocious germination occurs more frequently in Armenian cucumber (*Cucumis melo*) for example, because these fruits accumulate less sugar and therefore have a less negative fruit osmotic potential.

When extensive solute leakage occurs during seed ageing, the testas of muskmelon seeds frequently tear apart upon full hydration, and these dead seeds are often termed 'fish-mouthed' or 'osmotically distended' (OD), because the damaged embryo cells leak solutes that are trapped inside the endosperm envelope and attract water by osmotic swelling resulting in splitting of the seedcoats. Cucurbit seed lots frequently contain a high percentage of OD seeds when harvest is delayed and seeds have advanced from physiological maturation into accelerated ageing. Not all dead seeds are fish-mouthed; however, all OD seeds are dead.

Embryos of muskmelon seeds do not fully hydrate because the **testa** and **endosperm** exert a pressure opposing water uptake. When these tissues are broken, the water content of the embryonic axis increases by almost 50%. (**See: Germination – physical factors, 2. Water uptake by the seed**) Weakening and rupture of the endosperm tissue by enzymatic degradation is required for the initiation of radicle growth and thus for germination to be completed. Glucanase, endo-β-mannanase and chitinase activity each increase in endosperm tissue adjacent to the radicle prior to protrusion and are correlated with endosperm weakening as measured by Instron analysis (**see: Germination – radicle emergence**).

In cucurbits, after completion of germination the **hypocotyl** elongates and the **cotyledons** are pulled free of the seedcoat, which often adheres to a peg-like appendage at the base of the hypocotyls. Seedling establishment is **epigeal**.

Many cucurbit seeds are dormant at harvest (the depth varies among species) and **afterripening** is required before they reach maximum germinability and vigour. These are routinely stored for 1 year prior to sale to allow these processes to occur. Seed vigour, as measured by germination rate and tolerance of low water potential, increases during afterripening even after muskmelon seeds become fully viable.

Embryo ABA concentrations correlate positively with the weak dormancy displayed by developing muskmelon seeds. Some cucurbit seeds exhibit dormancy when placed in pure water but germinate rapidly under slight water stress (-0.2 MPa) or low concentrations of ABA. In some cucurbits, seeds harvested and maintained in the dark fail to germinate, but do so after exposure to red light, whereas exposure to far red light inhibits it, implicating the involvement of **phytochrome** (**see: Light – dormancy and germination**).

6. Sowing

Cucurbits are warm-season plants that grow best from 21 to 32°C; freezing kills plants and cool weather below 15°C slows or stops growth. In long-season areas, cucurbit crops are direct-seeded. Plants are commonly grown on raised (1.5 m) beds, with a single row in the middle of the bed (vine types) or two rows for short-internode cultivars. Rows or hills can also be planted on flat surfaces or mounds. Seed are typically planted 2.5–5 cm deep into preirrigated beds. Seeds can germinate and emerge in as little as 4 days at a soil temperature of 25°C and from 6 to 12 days at 20°C. Most cucurbit seeds do not germinate well below 15°C. **Damping-off** (seedling mortality caused by *Phytophthora*, *Fusarium*, *Pythium* and *Rhizoctonia* fungi) may be a problem in cool, moist soils. Later in the season, *Fusarium* wilt in watermelon, *Phytophthora* root rot in winter squash and pumpkin, bacterial wilt in cucumber and muskmelon in humid regions, or fruit rots caused by fungi can occur.

An excess of seeds is sown and after establishment seedlings are thinned to final stands. In short-season crop areas, cucurbits are transplanted in the 3- to 5-leaf stage, often on to raised beds covered with plastic **mulch**. Plug-tray **transplants** are preferable to bare-root transplants since plants are sensitive to transplant shock and do not easily produce adventitious roots. **Seedless** watermelons are more frequently grown from transplants because their seeds are expensive, germination is slower, and higher temperatures are required for optimum germination.

Plants are sometimes grafted on to disease-resistant rootstalks to prevent disease. For example, greenhouse cucumber scions at the cotyledon stage are grafted to *Cucurbita ficifolia* rootstocks to improve resistance to soil fungal diseases and improve cold tolerance (**see: Grafted seedlings**). Depending on the species and cultivar, vining types may be untrained, trained, or trellis-supported.

7. Treatments

Pelleting is generally not required since most cucurbit seeds are rather large. Seed **coating**, increasingly **filmcoating**, is used to colour-code seeds, deliver pesticides, particularly fungicides, and to provide small amounts of fertilizer to stimulate seedling growth. Biological agents have been introduced on cucurbit seeds, using biopriming or as inoculants to improve seed performance and protect against disease. Some evidence suggests that systemic acquired resistance (SAR), the expression of natural resistance genes in plants, can be induced through pregermination treatment using salicylic acid derivatives. Other studies show that SAR is only a postgermination plant response that protects seedlings from certain pathogens. Chitinase, an enzyme produced as part of systemic acquired resistance response, has been found in both developing and germinating muskmelon seeds.

Priming (by matrix, osmotic and drum methods) is routinely used to improve cucurbit seed performance. Priming advances physiological maturity by substituting for after-ripening, thus making immature cucurbit seeds perform as if they were more mature. The **water potential** inside developing muskmelon fruits is approximately −1.3 MPa, similar to an **osmopriming** solution. Apparently, muskmelon seeds are primed *in situ* during the later stages of development after seed mass maturity. Therefore, seed lots comprised of immature seeds tend to respond more to priming than do seed lots comprised of older seeds. Priming also initiates the process of endosperm weakening prior to radicle emergence. Priming may also repair damage incurred during seed storage, although studies with muskmelon show that priming will not restore viability to seeds subjected to preharvest ageing.

Fungicide treatments are routinely applied to protect against fungal attack after planting. Foliar diseases may also cause localized problems. Angular leaf spot (*Pseudomonas syringae*) on cucumbers is a seedborne bacterium often spread through sprinkler irrigation. Bacterial fruit blotch (*Acidovorax avenae* ssp. *citrulli*) can be a serious problem on watermelon, causing water-soaked lesions on fruit. As the disease can be seedborne, close monitoring of seed production fields and testing for contamination are required for certified seed. Hot-water treatment is sometimes used to treat such bacterial seedborne diseases. (GEW)

Oluoch, M.O. and Welbaum, G.E. (1996). Effect of postharvest washing and post-storage priming on viability and vigour of 6-year-old muskmelon (*Cucumis melo* L.) seeds from 8 stages of development. *Seed Science and Technology* 24, 195–205.

Robinson, R.W. and Decker-Walters, D.S. (1997) *Cucurbits*. CAB International, Wallingford, UK.

Rubatzky, V.E. and Yamaguchi, M. (1996) *World Vegetables – Principals, Production, and Nutritive Values*, 2nd edn. Chapman & Hall, Westport, CT, USA.

Welbaum, G.E. (1993) Water relations of seed development and germination in muskmelon (*Cucumis melo* L.). VIII. Development of osmotically distended seeds. *Journal of Experimental Botany* 44, 1245–1252.

Welbaum, G.E. (1999) Cucurbit seed development and production. *HortTechnology* 9, 341–348.

Welbaum, G.E. and Bradford, K.J. (1991) Water relations of seed development and germination in muskmelon (*Cucumis melo* L.). VII. Influence of after-ripening and aging on germination responses to temperature and water potential. *Journal of Experimental Botany* 42, 1137–1145.

Whitaker, T.W. and Davis, G.N. (1962) *Cucurbits: Botany, Cultivation and Utilization*. Interscience Publishers, New York, USA.

Cultivar

An official international term for a commercial variety ('*culti*vated *var*iety', abbreviated cv.) of a plant species, usually one raised specifically for agricultural or horticultural cultivation; commonly used synonymously with '**variety**'.

Cumin

Cumin (*Cuminum cyminum*, Apiaceae) (**see: Spices and flavours**, Table S.12) is the dried fruit (commonly called a seed), either a single, one-seeded **mericarp** or the **cremocarp**. The ridged, pale brown mericarps measure approximately 5 × 2 mm (Fig. C.29).

Cumin seeds contain 2–5% **essential oil** whose major components are shown in Table C.25.

The characteristic flavour of cumin is due to cuminaldehyde (**see:** Fig. S.44 in **Spices and flavours**) and monoterpenes. The odour and taste of cumin is described as strong, warm, pungent and persistent.

Fig. C.29. Dried cumin 'seeds' (image by Mike Amphlett, CABI).

Table C.25. Composition of cumin volatile oils.

Component	Per cent
Cuminaldehyde	15–40
Menthadienal	30
Cineole, p-Cymene, Limonene	18–29
γ-Terpinene, α-Pinene	1.4
β-Pinene	2.0

Baser, K.H.C., Kurkcuoglu, M. and Ozek, T. (1992) Composition of Turkish cumin seed oil. *Journal of Essential Oil Research* 4, 133–138.

The most important products are the dried seeds, used as a spice and the essential oil used for flavouring. Cumin seed has a strong and persistent aromatic odour, with a hot spicy and somewhat bitter taste. It is a major ingredient in curry powders, speciality bakery products, processed meat products, pickles, soups, salad dressings, etc.

The seeds have long been considered as a stimulant and carminative and are used for urinary and gastric troubles. Cumin is widely used in traditional medicine to treat flatulence, digestive disorders, diarrhoea and wounds, possibly because of antimicrobial properties. The **essential oil** is used as a stimulant, antispasmodic, carminative, aphrodisiac and has a light anaesthetic action. It also has uses as a fungicide, insecticide and in veterinary medicine. (KVP, NB)

Guenther, E. (1978) *The Essential Oils*, Vol. 2. Robert E Krieger Publishing, New York, USA.

Peter, K.V. (ed.) (2001) *Handbook of Herbs and Spices*. Woodhead Publishing, CRC Press, UK.

Weiss, E.A. (2002) *Spice Crops*. CAB International, Wallingford, UK.

Cuticle

Layer of cutin covering the outer wall of epidermis cells. A cuticle serves as a protection against water loss or uptake and enhances the mechanical durability of the epidermis cells. The cuticle is often also covered with wax. In the ovule and, often later in the seed, cuticles are present on both the inside and the outside of the integument(s) as well as on the outside of the nucellus. (**See: Seedcoats – structure; Structure of seeds**)

Cutin

Fatty or waxy substances deposited on or within plant **cell walls**. Chemically they are esters of fatty acids, secreted in the liquid phase through the wall of the developing cell. (**See: Cuticle**)

Cyclitols

See: Galactinol series oligosaccharides

Cypsela

Commonly defined as a single-seeded, dry, indehiscent fruit that develops from a single-loculed, inferior ovary. Sometimes included with **achenes**, although these are derived from a superior ovary. Cypselas are typical fruits of the Dipsacaceae and Compositae, e.g. *Taraxacum officinale* (dandelion). They often bear longitudinally oriented **awns**, bristles, feathery staminodia (staminodium), or similar structures derived from accessory parts. (**See: Dispersal**)

Cytokinesis

Cell division. In plants, as **mitosis** enters its final stage (telophase) and replicated chromosomes separate, the cell is divided. A cross-wall assembles in the cell, in a plane that separates the two daughter nuclei. (**See: Cell cycle; Germination – metabolism**)

Cytokinin (CK)

A type of plant **hormone** that has diverse effects upon plants, most notably in promoting cell division. (**See: Germination, influences of hormones**)

Although several artificial compounds, including **kinetin**, have cytokinin activity, the name should strictly be reserved for those compounds that occur naturally in plant cells, e.g. zeatin (Fig. C.30). With the exception of the artificially produced phenylurea derivatives such as thidiazuron, all cytokinins are derivatives of the purine base adenine or its nucleoside and nucleotide derivatives, adenosine and adenosine 5′-monophosphate.

Cytokinins are synthesized primarily in root tips and are transported mainly through the xylem vessels to the aerial growing points of the plant. Although it was once thought that this was the only way in which cytokinin could accumulate in shoot tips, it is now known that shoot tips can also synthesize their own cytokinin. The third major site of occurrence is in developing **embryos**.

The major step in cytokinin biosynthesis is the addition of an isopentenyl side chain to the N^6 position of adenosine 5′-monophosphate to give the N^6-(Δ^2-isopentenyl)adenosine 5′-monophosphate (isopentenyl-AMP). Δ^2-Isopentenyl-pyrophosphate and adenosine 5′-monophosphate seem to be the only substrates for this reaction, which is carried out by cytokinin synthetase (also known as isopentenyl transferase). Thereafter a biosynthetic network, rather than a single pathway, seems to operate. Isopentenyl-AMP is readily converted into its pure base form, isopentenyl adenine via its nucleoside form, isopentenyl adenosine. These three forms, the base, nucleoside and nucleotide forms are so readily inter-converted in plant cells that it is difficult to ascertain whether they each have cytokinin activity or only the free base form. The isopentenyl chain may be hydroxylated to give rise to zeatin and the side chain of zeatin may have its double bond reduced to give rise to dihydrozeatin. Both zeatin and dihydrozeatin may be readily inter-converted into their nucleoside and nucleotide forms.

The main route of degradation of cytokinins *in vivo* is by removal of the isoprenoid side chain by cytokinin oxidases. Cytokinin oxidase activity is increased by treatment with artificial cytokinins and by other treatments that raise cytokinin contents. It is likely, therefore, that cytokinin oxidase is one of the main homeostatic mechanisms for regulating the amount of cytokinin. Diphenyl urea inhibits cytokinin oxidase and causes effects similar to external cytokinin treatment. Cytokinin oxidases do not seem to be active on any ribotides, O-glycosides or dihydrozeatins.

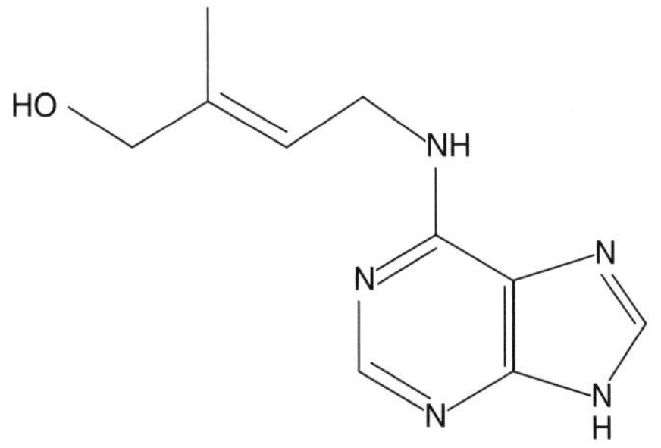

Fig. C.30. A common cytokinin, zeatin (systematic name, (*E*)-2-methyl-4-(1*H*-purin-6-ylamino)but-2-en-1-ol).

Conjugation of cytokinins occurs in several ways. Glycosylation of the hydroxyl group of zeatin, dihydrozeatin and their ribosides and ribotides (O-glycosylation) appears to be reversible. The function of these is not clear. They may be for transport or storage but they are readily converted back to the unconjugated form, so much so that in some tissues, the O-glycosylated form seems to be more active than the free form, possibly because it is resistant to cytokinin oxidase. All of the natural cytokinins in their free base, nucleoside and nucleotide forms may be irreversibly N-substituted to give inactive conjugates. This is achieved by the addition of glucosyl, ribosyl, xylosyl, alanyl and other groups to the nitrogens of the purine ring at the 6, 7 or 9 position. Conjugation of this type at the 3 position seems to be reversible.

Benzyladenine (BA) was thought to be a completely artificial cytokinin but BA and its hydroxylated derivatives, *meta*- and *ortho*-topolin, have now been found to occur naturally in plants. Nothing is known about the biosynthesis of these compounds but they are known to undergo N-conjugation.

Some tRNA molecules contain modified adenine nucleotides, particularly isopentenyl derivatives that have cytokinin activity. These are more common in bacterial cells than in plant cells and it is not clear whether or not this is an alternative pathway for cytokinin biosynthesis in plants.

Although cytokinins are known mainly from plants, certain plant pathogenic bacteria (*Pseudomonas syringae pv. savastanoi* and *P. amygdali*) produce methylzeatin. *Agrobacterium* spp. transfer an isopentenyl transferase (*ipt*) gene into plants to induce the plants to produce cytokinin. This gives rise to tumours and other effects. Interestingly, although not reported from mammals, cytokinins such as kinetin, isopentenyladenine and benzyladenine and their ribosides have anti-ageing effects on skin cells and cause redifferentiation and apoptosis in leukaemia cells. (GL, JR)

(See: Development of seeds – hormone content; Signal transduction – hormones)

Crozier, A., Kamiya, Y., Bishop, G. and Yokota, T. (2000) Biosynthesis of hormones and elicitor molecules. In: Buchanan, B.B., Gruissem, W. and Jones, R.L. (eds) *Biochemistry and Molecular Biology of Plants*. American Society of Plant Physiologists, Rockville, MD, USA, pp. 850–929.

Mok, D.W.S. and Mok, M.C. (2001) Cytokinin metabolism and action. *Annual Review of Plant Physiology and Plant Molecular Biology* 52, 89–118.

Moore, T.C. (1989) *Biochemistry and Physiology of Plant Hormones*, 2nd edn. Springer, New York, USA.

Cytoplasmic male sterility (CMS)

A natural mechanism, of quite widespread occurrence in nature, that prevents production of viable pollen. The discovery of cytoplasmic male sterility in onion about 60 years ago provided an artificial mechanism to force the crossing in self-pollinated angiosperm species, which has been exploited by breeders since to control pollination in the production of **hybrid** varieties. In essence, at the seed production stage, a male-sterile version of one strain (the 'seed parent', or 'female parent') is inter-planted in an isolated field with a male-fertile version of another strain (the 'pollen parent'), and only the seed produced on the male-sterile is harvested, which must therefore be an F_1 hybrid between the two strains. The CMS trait has been used extensively to produce female parental lines in hybrid canola,

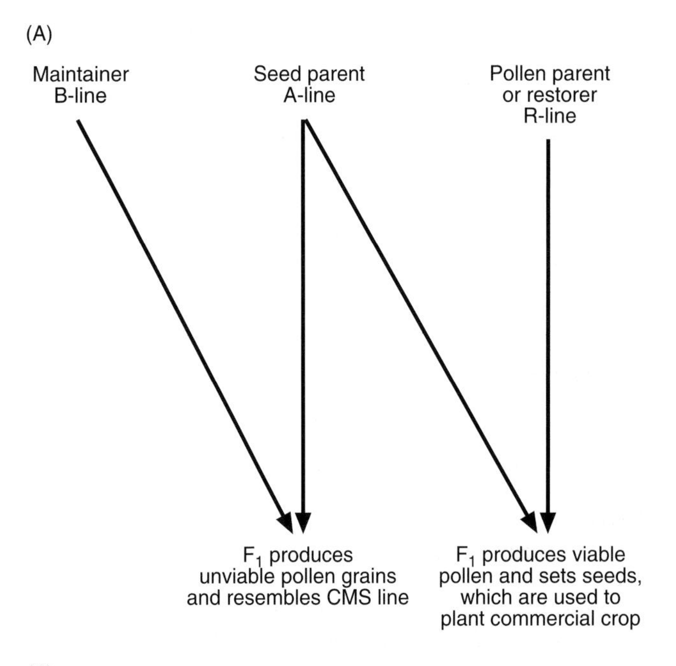

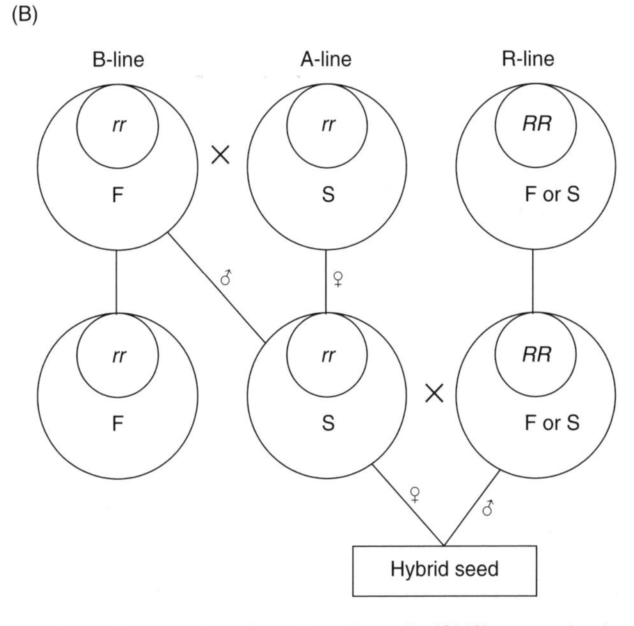

Fig. C.31. The cytoplasmic genic male sterile (CMS) system for the production of F_1 hybrid seed.
A. Breeding lines, showing the roles of the Seed Parent (A-line), Maintainer (B-line) and Restorer (Pollen Parent, R-line) in making crosses. The B-line and R-line are propagated by selfing.
B. The genetic status of elements in the system, in relation to the nucleus (inner circle) and the cytoplasm (outer circle). F (or N), the normal male-fertile factor; S, the male-sterile factor. R, dominant restorer allele; r, recessive restorer allele (respectively designated *Rf* and *rf* in maize and sunflower, and *MS* and *ms* in onion, sorghum and pearl millet). The R-line may have either F or S cytoplasm, as pollen only transmits a naked nucleus to the hybrid progeny; furthermore, R nuclear restorer genes are unnecessary in non-grain crops, because the hybrid itself need not be fertile (see text).

maize, pearl millet, sorghum, sunflower, sugarbeet and many vegetable varieties (such as carrot and onion), amongst other crops.

CMS **lines** are characterized by the inability to produce viable pollen: a trait under so-called cytoplasmic control, due to the presence of the sterility factor (*S*) expressed by a **mitochondrial** genome. The *S* factor causes male infertility only providing that the nuclear (chromosomal) fertility-restorer gene, *R*, is present in its homozygous double-recessive form (*rr*). (For this reason, CMS is strictly termed 'gene-cytoplasmic male sterility' or cytoplasmic/gen(et)ic male sterility.) The *S* character is transmitted to the zygote by the female parent in a cross, through the cytoplasm of the egg cell. The pollen parent cannot transmit the character because its cytoplasm does not survive the fertilization process. The effect of the *S* factor can be overcome, however, if a dominant *R* gene allele is present in the pollen nucleus.

The CMS system for hybrid seed production relies on three component breeding-lines (Fig. C.31A,B).

(a) *Male-sterile line (CMS-, A-line): ♀ female (or 'seed') parent line.* An inbred or single-cross hybrid line that cannot produce viable pollen, used as the female parent for pollination by the restorer line. High seed yields depend on the desirable floret and stigma characteristics of the CMS line (their number, receptivity to pollination, and so on). Male-sterile lines are themselves propagated by pollination from maintainer parent lines.

(b) *Maintainer line (B-line): ♂ male (or 'pollen') parent line.* A maintainer line is genetically similar to the male-sterile A-line (including the *rr* gene) except that it can produce viable pollen grains (because it contains the N cytoplasmic factor, instead of the S factor). In other words, it is the fertile counterpart (F) of the male-sterile line. The maintainer line is used to maintain the male-sterile female parental line, without restoring pollen fertility to the next generation of seed.

A variant on this procedure is to use an unrelated maintainer line, at the last maintenance stage. The resulting hybrid male-sterile may be more vigorous and consequently have greater potential for seed production. When crossed with another male-fertile line, a so-called 'three-way hybrid' cross is produced, though some of the generic uniformity of an F_1 between two inbreds is lost in the process.

Leeks differ because male sterility in the female line is derived from genomic DNA. Therefore the female line has to be propagated vegetatively and does not require a maintainer line.

(c) *Restorer line (R-line) – or alternatively, the C-line: ♂ male ('pollen') parent line.* An inbred or single-cross hybrid line that enables seed to be set, and permits restoration of fertility to the F_1 progeny, when it is crossed with male sterile lines. The restorer line is used as the pollinator for the CMS female parent line in the hybrid seed production stage. Fertility in the next generation is restored, if required, by including the homozygous dominant nuclear fertility gene (*RR*), which overcomes the effect of the cytoplasmic S factor inherited from the CMS line: so-called 'nuclear fertility restoration'. In the production of hybrid grass, onion and sugarbeet seed, by contrast, it is unnecessary to restore fertility because the product to be harvested is a vegetative organ (a leaf, bulb or root, rather than seed); in this case, the pollen parent line is usually called the 'C-line', rather than the R-line.

Desirable characteristics of both Maintainer and Restorer lines include large anthers with many pollen grains that are shed only after complete exertion from the floret, and reasonable synchronous flowering with the CMS line to facilitate **nicking**. (**See: Cross; Production for Sowing, III. Hybrids**)

CMS systems can be more complex: in maize for example, three major CMS cytoplasms are known, each requiring a different nuclear restorer *R* gene or genes. For an account of the complex CMS systems used in carrot, **see: Umbellifers, 3. Genetics and breeding**.

In practice CMS systems can be problematic. Fertility restoration can be sensitive to environmental conditions, and additional modifier genes are necessary. Notably in maize, the 'Texas' CMS variant is inseparably associated with susceptibility to Southern corn leaf blight (caused by the fungus *Bipolaris maydis*), which reached epidemic proportions in the USA in 1970-71 and devastated a substantial part of the crop. For this reason, CMS use was discontinued in much hybrid maize variety production, and the practice of **detasselling** resumed for a number of years until alternative male-sterility lines could be developed. (PH)

Edwardson, J.R. (1970) Cytoplasmic male sterility. *Botanical Reviews* 36, 341-420.

Havey, M.J. (2004) The use of cytoplasmic male sterility for hybrid seed production. In: Daniell, H. and Chase, C. (eds) *Molecular Biology and Biotechnology of Plant Organelles*, Springer, Dordrecht, The Netherlands, pp. 617-628.

Kaul, M.L.H. (1988) Male sterility in higher plants. In: Frankel, R., Grossman, M. and Maliga, P. (eds) *Monographs. Theoretical and Applied Genetics* Volume 10. Springer, Berlin Heidelberg, New York.

Shivanna, K.R. and Sawhney, V.K. (eds) (1997) *Pollen Biotechnology for Crop Production and Improvement.* Cambridge University Press, Cambridge, UK.

D

Damping-off

Diseases of young seedlings common to many species, mainly caused by the 'seedling disease complex'. Major causal pathogens are soil-living fungi such as *Rhizoctonia solani*, *Pythium* spp. and *Fusarium* spp., which thrive in wet flooded soil conditions, and infect plants by producing motile, swimming spores. The characteristic symptom is a rotting at the stem base near the soil surface (sometimes called wirestem) and in the root, leading to collapse and death by 'pinching-off', or even before seedlings have emerged. Damping-off is a growers' term, deriving from the moist conditions that can exist among seedlings crowded in a box, and that can favour the attack and spread of such pathogens. The disease can be partly controlled in many crops by fungicide seed treatments (**see: Disease; Treatments – pesticides**). (PH)

Day-degree

A common form in which **heat units** are expressed: the mathematical product of the difference between the ambient temperature and the **base temperature** of the physiological process under consideration and the chosen time unit (such as a single day), integrated over the period during which the process is completed.

Debearding

A general term for the seed conditioning processes used to break-off or remove appendages, seed heads, clusters, pods or hulls, to decorticate sugarbeet, or to de-**awn** or **delint** certain grass seed species. (**See: Conditioning, I. Precleaning**)

Decortication

The removal during seed precleaning of cortex tissue from dry fruits, such as sugar beet, to standardize seed size and shape, and reduce some residual **coat-imposed dormancy** mechanisms (such as compounds that slow or inhibit germination). (**See: Conditioning, I. Precleaning**)

Defoliants

Pesticides that cause leaves to be shed prematurely without killing the plants. As an alternative to **desiccants**, which by contrast can kill the whole plant, defoliants are applied by spraying to make harvesting a crop easier by reducing plant debris, or to advance the time of harvest by manipulating the rate and degree of seed drying. For example, defoliants are often used on **cotton**, **soybean** and **sunflower**, where green leaves can remain late in the season due to the indeterminate-growth nature of the plant (and, in the case of cotton, can stain the fibres). Many defoliants have either herbicidal or hormonal activity, and work by increasing **ethylene** synthesis in the plant, resulting in abscission-zone formation. Defoliation in some cases, however, is carried out without the use of chemical defoliants (**see: Grasses and legumes – forages**). (PH)

Dehiscence

See: Shattering

Dehydrins

A subset of the **LEA** (Late Embryogenesis Abundant) proteins. They include the LEA D-11 family and some denoted RAB (Responsive to ABA) in rice. Dehydrin genes exhibit a flexible expression repertoire; the protective role of dehydrins in the survival of water loss is purported to be dual: during **maturation** drying of the developing seed, and following germination/growth of the mature seed (i.e. in seedlings or plant vegetative tissues undergoing mild water stress). Dehydrins are hydrophilic as a result of a biased amino acid composition (e.g. a high percentage of glycine and glutamic acid). Most are characterized by a conserved 15-amino acid, lysine (symbol, K)-rich domain near the carboxyl terminus (EKKGIMDKIKEKLPG); this part of the polypeptide has been proposed to form an **amphiphilic α-helix** which may serve among other functions as an ion trap in dehydrating cells, sequestering ions as they become concentrated. (**See: Proteins and amino acids**) (ARK)

Delinting

A process applied to cotton to remove the linters (the short **fibres** left after **ginning** has removed the long ones). Delinting is also a conditioning step to remove **lint** or **awns** (deawning) from bluegrass (*Poa* spp.). (**See: Conditioning, I. Precleaning**)

Cottonseed is processed to remove the first- and second-cut linters (the first-cut short fibres and second-cut fuzz, which contain nearly 100% cellulose), usually by a mechanical method similar to those in cotton saw **gins**.

For the preparation of planting seed, acid delinting processes remove all linters, either using dilute sulphuric acid or hydrochloric acid vapours, before neutralization with calcium carbonate or ammonia gas, respectively, followed by seed conditioning, storage and fungicide and insecticide treatment as required. Acid-delinted cottonseed is also used for ruminant feeding, sometimes after rolling or cracking to improve its nutrient density and flowability. (**See: Cotton; Cotton – cultivation; Lint and linters**) (PH)

DELLA

See: Signal transduction – some terminology

Density sorting

Flotation techniques are used to separate seeds into fractions on the basis of differential buoyancy in aqueous or non-aqueous liquids.

(i) *Water.* At its simplest, water buoyancy is an inherent principle used during the extraction of 'wet seed' species, such as vine fruits from the Cucurbitaceae and Solanaceae, a process also known as 'fluming'.

- Physically-damaged tree seeds are removed from sound seed by a pressure-vacuum technique, called **PREVAC**, which applies a vacuum to submerged seed followed by atmospheric pressure, to remove those with cracked or split coats.
- Dead seeds of a few tree species can also be successfully removed by various separation means after a pretreatment technique called **IDS** ('Imbibition-Drying-Separation'), which exploits transient density differences due to differential absorption and drying rates of water-imbibed seed.
- Flotation in aqueous solutions of various densities (for example, using maltodextrins; **see: Germination – field emergence models**) can also sort imbibed seed, including directly after **priming**.

(**See: Production for sowing, V. Drying, 1. Pre-drying**)

(ii) *Organic solvents.* Mixtures of polar organic solvents that have different specific gravities (such as chloroform and hexane) are used to separate seed batches by flotation (Table D.1). Equipment has been engineered to handle these materials in a way that is safe for the seed, with very short exposure times, and the operator, with no escaping vapours. Although throughput is relatively slow, the technique is now used commercially for high value horticultural and ornamental seeds. Test fractions are made of a representative sample of the seedlot using a series of solvent density mixtures, followed by germination testing of each fraction to determine the most appropriate fractionating density for the production separation stage. (PH)

Hill, H.J., Taylor, A.G. and Min, T.-G. (1989) Density separation of imbibed and primed vegetable seeds. *Journal of the American Society of Horticultural Science* 114, 661–665.

Taylor, A.G., McCarthy, A.M. and Chirco, E.M. (1982) Density separation of seeds with hexane and chloroform. *Journal of Seed Technology* 7, 78–83.

Table D.1. Specific gravity of some liquids used for separation of seed by flotation.

Medium	Specific gravity
Pure water	1.00
Absolute alcohol (ethanol)	0.791
95% Aqueous ethanol	0.806
50% Aqueous ethanol	0.90
Diethyl ether	0.714
Petroleum ether	0.657
n-Pentane	0.626
95% Ethanol:n-penthane, 3:1 mixture	0.76
95% Ethanol:n-penthane, 12:13 mixture	0.71
Linseed oil	0.93

Deserts and arid lands

1. The desert biome

The extreme environmental conditions in deserts and arid lands present particular difficulties to seeds in respect of their survival and germination. For example, in the 6000-km-long belt of the northern region of the Sahara and Arabian deserts, of which the Negev Desert of Israel is part, the growing season is in winter, with 25 to 170 mm of rain (though unpredictable), mild temperatures and relatively short daylength. The summer is long, hot and dry, the evaporation rates are very high and the daylength is relatively long. In such deserts, there are years when there is very limited seed production but high pressure of seed consumption by seed-eaters. Many of the more common ephemeral plants, of the approximately 200 species that occur in such extreme deserts and belong to unrelated taxa, have developed complementary sets of adaptations and survival strategies (Figs D.1, D.2) as ecological equivalents. Important among these are: (i) seed **dispersal** and protection strategies, including the linking of dispersal to water availability; and (ii) strategies to link germination with sufficient water to support seedling establishment and survival. Most, if not all of the seed features on which these strategies depend are instituted during seed development due to maternal or environmental influences (**see: Germinability – maternal effects; – parental habitat effects**). The data on which this and the following accounts of desert seeds are based come largely from studies made in the Negev region though the principles are generally applicable.

2. Adaptations and survival strategies of ephemerals occurring in deserts

Many groups of annual species have developed different sets of complementary survival adaptations and strategies. As ecological equivalents with anatomical, morphological and physiological resemblance, these can be found in many species of ephemeral plants throughout the different stages of their life cycles. The adaptations and strategies, and in particular phenotypic plasticity of seed germination (PPSG) include regulation of seed dispersal and germination, and retention of a large number of seeds in the long-living **soil seed bank**. The PPSG may also ensure that only a portion of the aerial or soil seed bank of a particular species is ready to germinate after a particular rain event in the proper season. Strategies of dispersal and of PPSG are identifiable (Figs D.1, D.2).

The PPSG is influenced during seed development and maturation as well as by environmental factors post-maturation and during seed wetting and germination. These strategies have developed in two main directions: (i) the 'escape strategy' of production and dispersal of large numbers of tiny dust-like seeds into the soil seed bank; and (ii) seed bank of relatively small numbers of larger seeds which may remain protected in lignified structures of the dry mother plant below the soil surface until they germinate *in situ*. In other species, seeds may remain for years on the dry and lignified plant before their dispersal by rain (**serotiny**). According to the species, such seeds may germinate immediately, or they may first enter the soil seed bank. In others, seed groups are dispersed in lignified dispersal units

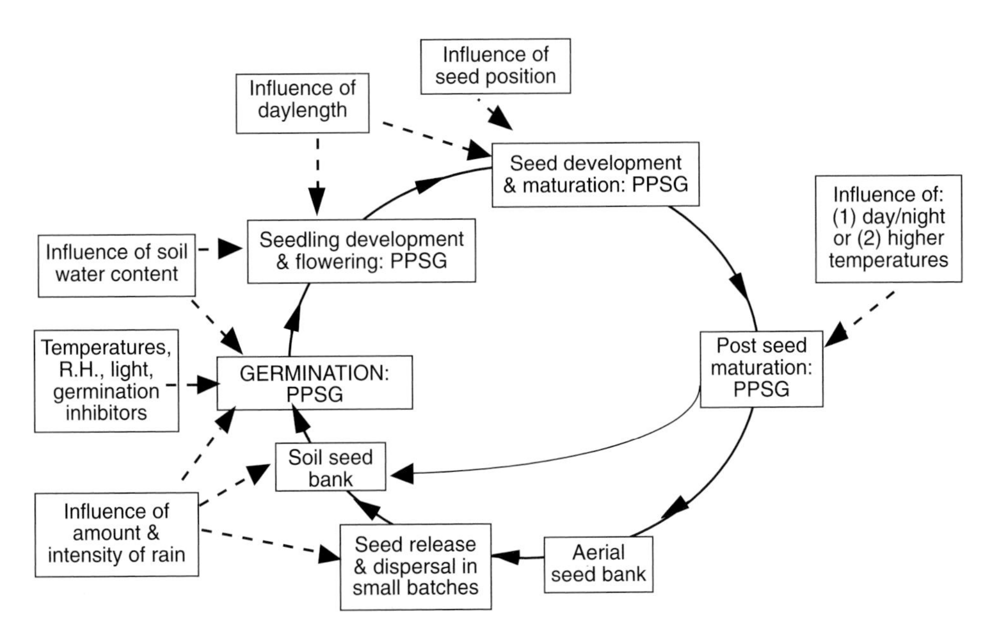

Fig. D.1. The influences of environmental and maternal factors on species' adaptation and survival, and strategies of phenotypic plasticity of seed germination (PPSG), during the different stages of the annual life cycle of desert ephemerals (Gutterman, Y. (2002) *Survival Strategies of Annual Desert Plants*. Adaptations of Desert Organisms. Springer, Berlin).

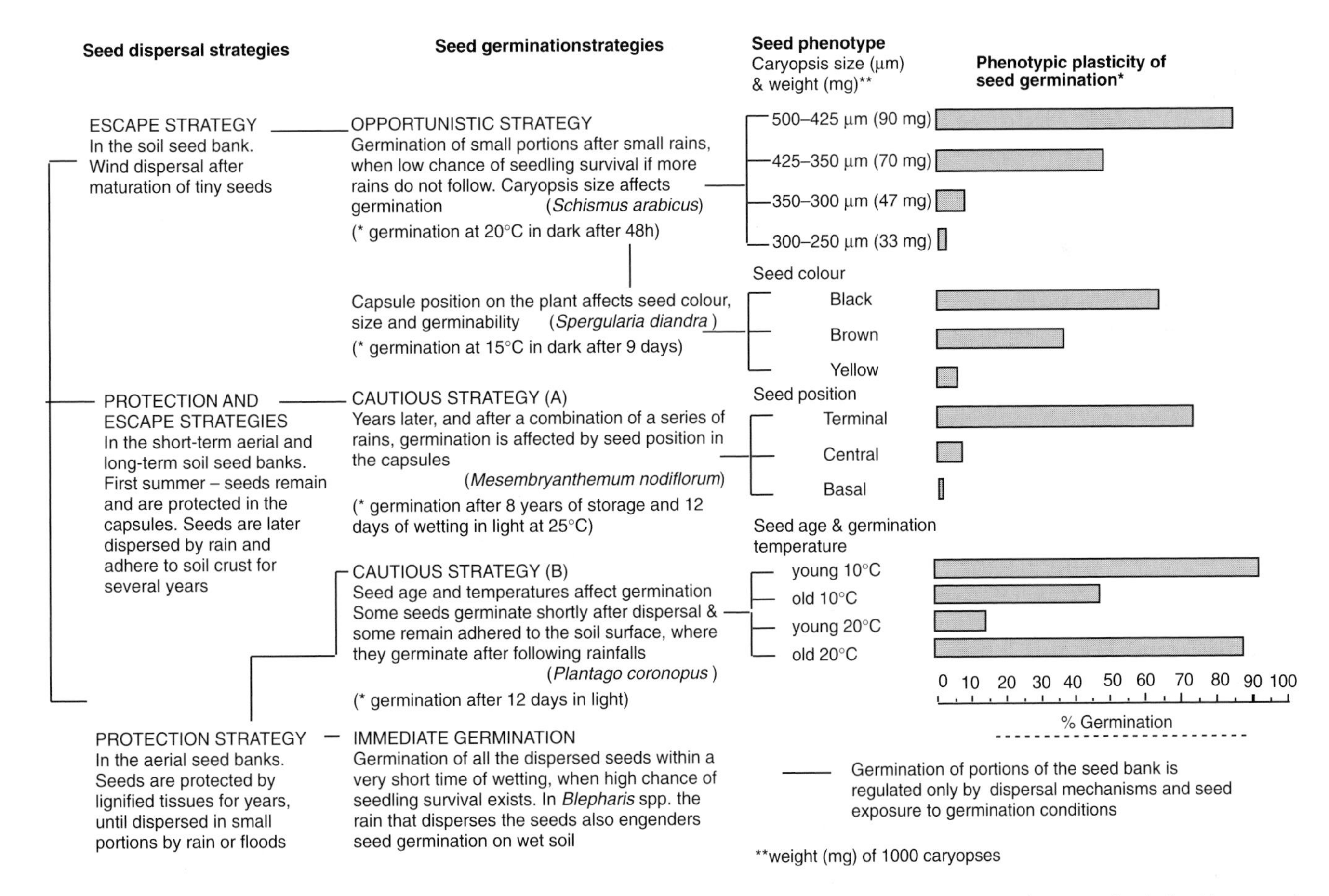

Fig. D.2. Strategies of seed dispersal and phenotypic plasticity of seed germination (PPSG) in some plant species occurring in the Negev, and in areas of the Saharo-Arabian phytogeographic region (Gutterman, Y. (2000) Environmental factors and survival strategies of annual plant species in the Negev Desert, Israel. *Plant Species Biology* 15, 113–125).

from which, usually, only one of the seeds may germinate in one particular year.

3. Seed dispersal strategies and mechanisms

Many desert creatures collect massive numbers of seeds as food. In many of the more common plant species, seed protection or seed escape strategies connected with **dispersal**, singly or together, combat such **predation.**

(a) *Protected seeds.* In some cases, seeds are protected by lignified structures of the lignified and dry mother plant and are located below the soil surface from the time of their maturation until their germination, e.g. seeds (propagules) in the subterranean inflorescences of the **amphicarpic** plant *Emex spinosa*, **achenes** in *Gymnarrhena micrantha*, and the subterranean **caryopses** (seeds) from the lower part of the lignified spike of *Ammachloa palaestina*.

Lignified dispersal units may be distributed by wind and the seeds germinate later, for example the spiny pod of *Medicago laciniata*, the inflorescence of *Pteranthus dichotomus*, and of *Aegilops kotschyi*, and *A. geniculata* (= *A. ovata*) in which the dispersal unit is a **spike** with two to four spikelets. Seed position in these dispersal units influences germinability. In 1 year, one or very occasionally two seeds may germinate. This increases the chance of species survival by preventing competition among seedlings emerging at the same place, and by spreading germination over time. In *A. geniculata*, the higher spikelets have greater amounts of water-soluble inhibitors. The multi-seeded pods of *Trigonella arabica* contain **germination inhibitors** (e.g. coumarin), and the fruits of *Zygophyllum dumosum* contain salts. These leachable inhibitors may regulate the timing of germination of the seeds in these dispersal units by acting as rain gauges.

The aerial protected seed banks in lignified **capsules, pods** or inflorescences of some species are gradually dispersed by rain. In *Trigonella stellata*, the distal side of the pod opens when wetted, through which some of the seeds are pushed out. In *Carrichtera annua* and *Anastatica hierochuntica*, after about 1.5 h of wetting the dehiscence zone between the two valves of the pod is weak enough for some additional drops of rain to open the pods. The dispersed mucilaginous seeds adhere to the soil surface near the dead mother plant, or may be carried by runoff and flood water into depressions, or further. In *Asteriscus hierochunticus* (= *A. pygmaeus*) the mature achenes in the capitulum of the tiny lignified plants are covered by two sets of lignified bracts. These open when wetted and a few achenes may be dispersed by drops of rain and are carried by the wind into depressions or splits in the soil crust. In perennial or annual species of *Blepharis*, found in many worldwide deserts, a very sophisticated set of mechanisms has developed in order to disperse, during an efficient rainfall, only a portion of the seeds from the aerial protected seed bank on the lignified dry plant. The capsule contains two seeds separated by a lignified septum along which is a dehiscence zone. After a certain degree of wetting, the basal tissues of the covering bracts and sepals swell, causing both to open, eventually wide enough for rain drops to reach the capsule. If the 'lock area' at the top of the capsule is wetted first and thus weakened, the capsule explodes, and the two seeds may be ejected to more than 1 m. In about 70% of the capsules this mechanism fails and only after drying and further rewetting will some of the capsules explode and the seeds be dispersed. If the dry seeds fall on to the wet soil, their mucilaginous hairs will adhere to the surface and germination may follow within as little as 50 min. This delayed seed dispersal by rain, and rapid germination of only a small portion of the seeds of the aerial seed bank of the plant after a long rain event, present a 'cautious' strategy of seed germination. Almost all the seeds that are dispersed develop into plants (Fig. D.2). If the seeds fall into excess water the mucilaginous layer surrounding the seeds reduces the oxygen reaching the embryo, preventing germination until the excess has diminished. Many of the *Blepharis* spp. occurring in the deserts of North and South Africa, as well as the Saharo-Arabian Desert and Iran, and Thar Desert of India, have similar mechanisms of seed dispersal and germination.

(b) *Intermediate strategies of seed protection followed by escape mechanisms.* One mechanism ensures the escape into the soil seed bank of seeds from different locations in the capsule at different times, even during different rain events. In *Mesembryanthemum nodiflorum*, distributed in the deserts of South Africa, for example, the tiny seeds remain enclosed in the dry and lignified capsules on the lignified plant as an aerial seed bank. During the winter(s) the capsules open after wetting. Wetting for 15 min is required for the upper about 20 seeds to be separated and dispersed by the drops of rain, for approximately 200 min for about 20 seeds that adhere at the centre part of the capsule, and for about 320 min for the approximately 20 seeds at the lower part of the capsule. The tiny seeds adhere to the wet soil surface immediately after they have settled. After some years of primary **dormancy**, the germinability of these seeds is affected by the position of the seeds in the capsule during seed **maturation** (Fig. D.2). This PPSG is still evident even 30 years later. The seeds of this species, found on salty soils, germinate only after some rainfalls have diluted the salts from the upper few cm of the soil. Another intermediate strategy is in *Neotorularia torulosa* (= *Torularia torulosa*) where the tiny seeds from the pods on the upper part of the plant are dispersed by wind following the separation of the two pod halves shortly after seed maturation. The seeds in the lower part of the lignified, dry plants are protected by lignified pods during the summer following seed maturation. These seeds are dispersed by rain only after the pods are wetted during the following season with rain.

(c) *Seed escape strategies.* In many common species, such as *Schismus arabicus*, *Spergularia diandra*, *Diplotaxis acris* and *Nasturtiopsis coronopifolia* ssp. *arabica*, the very small, dust-like seeds mature in great numbers and are dispersed by wind into small cracks in the soil crust to remain as a soil seed bank and possibly escape predation (Fig. D.2). Small portions of the seed bank of some species may germinate after a rainfall of as little as about 10–15 mm.

(d) *Mucilaginous seeds – advantages in a desert environment.* Mucilage on the seedcoat or achene causes adherence to the soil, when wetted by rain in winter or dew in summer. This adhesion may delay seed predation, which in the Negev is mainly by ants, and may also enhance seed water absorption during germination. The mucilage may penetrate and lubricate the soil to a certain depth, possibly aiding the passage of the roots of the seedling into the soil. Many common annual species in hot deserts produce mucilaginous seeds. Seeds dispersed by rain in winter include *Blepharis* spp., *Carrichtera annua*, *Anastatica hierochuntica* and *Plantago coronopus*. Those dispersed by wind and wetted by dew in summer include some *Plantago* spp. and *Diplotaxis acris*, and also the mucilaginous seeds of many perennial plants, such as *Artemisia* spp., *Helianthemum* spp. and *Diplotaxis harra* Forssk.

4. Maternal and environmental influences during seed development and maturation

An extremely important property of seeds of desert and arid lands that contributes to their success in the extreme environment is the range of germinabilities and dormancy exhibited by the offspring of individual plants of any one species – variability that contributes to seed survival and the production of seedlings. This phenotypic plasticity of seed germination is established on the mother plant through maternal or environmental influences.

(a) *Maternal influences.* Seeds in different inflorescences or capsules, or even in different locations in one capsule or inflorescence, may have different germinabilities (Figs D.1, D.2). The differently-germinable seeds of **amphicarpic** plants such as *Emex spinosa* and *Gymnarrhena micrantha* develop in aerial or subterranean inflorescences. Germinability may also be influenced by the position of the capsule: and here there are both genotypic and phenotypic effects. For example, in many *Spergularia diandra* populations in the central part of the Negev Desert, there are three genotypes that differ in the seed hairiness, dispersability and germinability. Within each genotype, the highly germinable, larger, black seeds are in the first capsules to mature that terminate the main stem. Later on, brown seeds, which are intermediates, develop on lateral stems. The smallest, yellow seeds, showing the highest percentage of dormancy and poorest germination, mature in capsules on senescing plants. Therefore, in one *S. diandra* population there are nine types of seeds with differing germinability. In addition, the day length during seed maturation also affects germinability of these seeds (Figs D.1, D.2).

The position of grains in the spike and spikelet (e.g. in *Aegilops geniculata* and *A. kotschyi*) affects germinability. In *Pteranthus dichotomus*, the one to seven 1-seeded fruits (pseudocarps) of the inflorescence are dispersal units: seeds germinate according to their position in the inflorescence, the terminals being first. The position of the **achenes** in the capitulum, as in *Asteriscus hierochunticus* (= *A. pygmaeus*), affect achene dispersal and germination.

In the Namaqualand desert of South Africa, several ephemeral species such as *Dimorphotheca polyptera* and *D. sinuata* produce, in one capitulum, **polymorphic** diaspores with differing dispersability and germinability.

Seeds in protected aerial seed banks may have PPSG according to their position in the capsule, as in *Mesembryanthemum nodiflorum* (Fig. D.2). In *Medicago laciniata*, the seeds in the spiny pods, the dispersal units, have different germinability according to their location in the pods.

Plant age at seed maturation may also influence seed **size** and germinability (Fig. D.2). For example, in *Schismus arabicus*, even at constant daylengths, most of the caryopses of the first harvest were larger than those harvested 2 months later. The temperature range and germination percentage of the two sizes differ (Figs D.2, D.3). The yellow seeds of *Spergularia diandra*, which mature the latest when the plants senesce, are also the smallest, with the highest percentage of dormancy. In these species, the daylength during seed maturation also affects seed germinability (Figs D.1, D.2). *Ononis sicula* Guss. and *Trigonella arabica* produce **hard seeds** under long days. Seeds maturing on senescing plants have less developed seedcoats, typical of seeds that mature under short days. The seedcoats of such seeds are more permeable to water thus affecting germinability. (**See: Germinability – maternal effects**)

(b) *Environmental factors.* In many species daylength is the main environmental factor affecting PPSG as well as flowering, influencing seed development and **maturation** and the subsequent seed germinability. The seeds of different species are affected in three ways: (i) the longer the daylength to which plants are exposed during seed development and maturation, the higher is the percentage of germinability reached and the lower the dormancy level, e.g. *Carrichtera annua*, *Plantago coronopus* subsp. *commutata*, *Polypogon monspeliensis*, *Schismus arabicus* and the black hairy seeds of *Spergularia diandra*, and the perennial plant *Cucumis prophetarum*; (ii) beneficial effects of maturation under short days, such as *Portulaca oleracea*; and (iii) both short- or long-day germinability responses to the daylength, i.e. when the daylength should be shorter or longer than certain values: examples are *Ononis sicula* and *Trigonella arabica* in which the

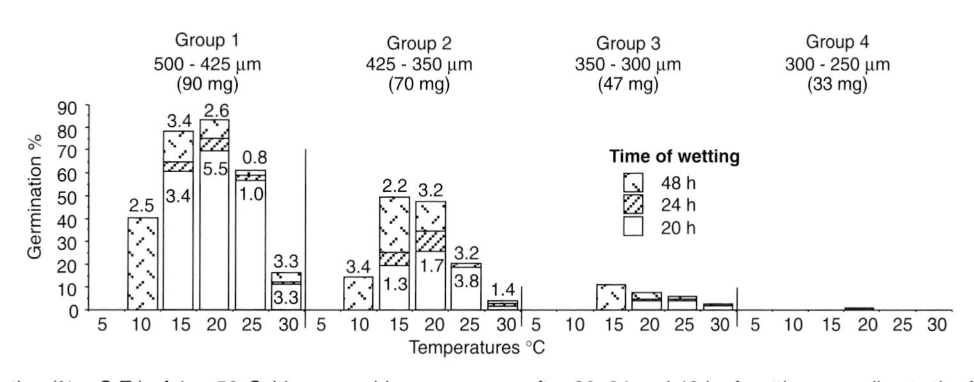

Fig. D.3. Germination (% ± S.E.) of 4 × 50 *Schismus arabicus* caryopses after 20, 24 and 48 h of wetting according to the four caryopsis size groups (μm) and the weight (mg) of 1000 caryopses of each group. The caryopses were harvested from natural populations in the Negev Desert on 28 April 1997; experiments started 11 May 1999. (The number in or on the columns denotes ± S.E.). (Adapted from Gutterman, Y. (2001) Phenotypic germination plasticity related to caryopsis size in *Schismus arabicus*. *Seed Science Research* 11, 173–178.)

seedcoat colour and permeability to water are affected by the daylength (Figs D.1, D.2). (**See: Germinability – parental habitat effects**)

5. Germination and seedlings

Various aspects of dormancy and germination physiology serve to adapt desert and arid land seeds to the particular challenges of the extreme environment. These include temperature- and light-dependent loss of dormancy and sensitivity to various amounts of rainfall. The newly-produced seedling can also display properties that support its survival.

(a) *Post-maturation environmental effects of high temperatures during the dry summer.* An important component of the strategies of desert seeds is primary dormancy, which prevents the freshly matured seeds from germinating after a late rainfall shortly before the summer, when the plants have no chance to develop and complete their life cycle. A long period of exposure to high temperatures during summer may decrease the primary dormancy by **afterripening** which thus allows germination to occur after a rain during the following winter with mild temperatures (Fig. D.1).

(b) *The minimum amount of rain for germination: opportunistic and cautious strategies of germination.* Species in the Negev desert have developed at least two germination strategies: (i) the opportunistic strategy, e.g. caryopses of *Schismus arabicus* may germinate in depressions after as little as 7.1 mm intensive rain: *Hordeum spontaneum* (wild barley) caryopses may germinate after as little as 10 mm of irrigation in winter, and its young seedlings might survive a few weeks of drought; (ii) the 'cautious' germination strategy, whereby seeds such as *Blepharis* spp. and *Anastatica hierochuntica* require a relatively longer period (at least a few hours) of rain, to be dispersed and to germinate at the same rain event, as described above.

Germination of seeds in the soil seed bank may be affected by air relative humidity (RH) as well as by rainfall. The higher the RH during the days between rainfalls and seedling emergence, in the mild winter temperatures, the greater the number of seedlings that emerge, and the lower the amount of rain required for germination. Rainfall and temperature are the main factors for the germination and development of winter or summer annuals in the Joshua Tree National Monument, California, a desert which receives rain in winter and thunderstorms in summer. Winter annuals germinate in winter at fluctuating day/night temperatures of 18/8°C. Summer annuals germinate at day/night temperatures of 27/26°C, and higher. In summer, *Pectis papposa* seeds germinate after a rainfall of 15 mm, but not after 10 mm. Many summer-germinating annuals may do so after at least 25 mm of rain. However, in gullies, where water accumulates, *Pectis papposa* and *Bouteloua* spp. may germinate after only 10 mm of rain. No germination of winter annuals occurs if the first rainfall of the season is less than 10 mm, but if a second rainfall of even as little as 6 mm follows, germination of many species may be abundant. Achenes of the small shrub, *Artemisia sieberi*, germinate after 16 days of wetting, when they adhere to the soil surface by their mucilage. Mass germination (stimulated by light) might happen only once in 10 to 20 years, when the soil surface is wetted for such a long period. There is evidence that water-soluble inhibitors in seeds, dispersal units, or in the soil (e.g. salts) feature in the mechanism enabling seeds to measure rainfall. Germination is permitted only after there has been sufficient rainfall to leach out the inhibitors. This is viewed as a type of rain gauge.

(c) *Light-regulated germination.*

(i) *Light inhibits germination.* Light-inhibited seeds or caryopses on or near the soil surface, or buried, may germinate to high percentages within hours after a night rain (e.g. *Schismus arabicus*) but not after the same amount of daytime rain. Seeds of the sand-dune perennial shrub *Calligonum comosum* are inhibited by light and high temperatures. Only seeds in the fruit located about 10 cm deep in the sand, or oriented with the seed micropyle down, even at a shallow depth, can germinate well. Seeds of the perennial geophyte, *Pancratium maritimum*, germinate only in darkness when they are buried in the sand, but after the high salinity near the surface has been diluted by rain.

(ii) *Light-requiring seeds.* Germination of such seeds is affected by their depth in the sand. For example, the deeper are the seeds of the common shrub in the Saharo-Arabian region, *Artemisia monosperma*, from 0–20 mm, the lower the germination percentage when water is not a limiting factor. Optimum conditions for germination are at 9 mm, and the deepest from which seedlings emerge is 18 mm. The upper layer of sand dries quickly after a rainfall, therefore from 0–9 mm, the deeper the seed is situated, the higher is the soil water content, and there is still enough light for germination. From 9–18 mm, the deeper the seed, the more limited is the light for germination. The depth from which these seeds germinate is regulated by two limiting factors – the lack of light at too great a depth, and lack of water when seeds are too shallow. Other *Artemisia* species from the sand deserts of China have a similar stategy. (**See: Cryptochrome; Light – dormancy and germination; Phytochrome**)

(d) *Seedling drought tolerance.* An important strategy that complements the opportunistic strategy of germination under the extreme desert conditions, where the small amounts and distribution of rain are irregular, is that of seedling survival. For example, seedlings of the annuals *Schismus arabicus* and *Hordeum spontaneum* in the Negev Desert, or the perennial grasses *Psammochloa villosa* and *Leymus racemosus* (wild rye) from the Ordos sand dunes of China, may survive even more than 6 weeks of drought and develop into normal plants. The later the stage of growth of the seedlings (the length of the roots and **coleoptiles**), the smaller is the amount of **endosperm** that remains, and the lower is the percentage of seedlings that survive. In *Schismus arabicus*, the month of germination also affects the percentage of seedling 'revival' after a week of drought. In different genotypes of *Hordeum spontaneum*, the higher the percentage of caryopses with primary dormancy, the higher is the percentage of seedlings that 'revive' after a period of drought.

(e) *Gene flow.* Because of the spread of germination over different years gene flow between populations and generations can occur. As a result of PPSG, in any one year, only some of the seeds in the species' seed bank produce plants, together with some of the seeds from several previous years. These different populations pollinate each other. Because of the PPSG, each year different genetic components may appear even in one population, habitat or area. Such an event may increase the genetic diversity of a species, possibly an advantage for its survival under the extreme desert conditions (Figs D.1, D.2). (YG)

Baskin, C.C. and Baskin, J.M. (1998) *Seeds – Ecology, Biogeography, and Evolution of Dormancy and Germination.* Academic Press, San Diego, USA.

Evenari, M., Shanan, L. and Tadmor, N. (1982) *The Negev: The Challenge of a Desert*, 2nd edn. Harvard University Press, Cambridge, MA, USA.

Gutterman, Y. (1993) *Seed Germination in Desert Plants.* Springer, Berlin.

Gutterman, Y. (2000) Environmental factors and survival strategies of annual plant species in the Negev Desert, Israel. *Plant Species Biology* 15, 113–125.

Gutterman, Y. (2001) *Regeneration of Plants in Arid Ecosystems Resulting from Patch Disturbance. Geobotany 27.* Kluwer Academic, Dordrecht, The Netherlands.

Gutterman, Y. (2002) *Survival Strategies of Annual Desert Plants. Adaptations of Desert Organisms.* Springer, Berlin.

Juhren, M., Went, F.W. and Phillips, E. (1956) Ecology of desert plants. IV. Combined field and laboratory work on germination of annuals in the Joshua Tree National Monument, California. *Ecology* 37, 318–330.

van Rheede van Oudtshoorn, K. and van Rooyen, M.W. (1999) *Dispersal Biology of Desert Plants. Adaptations of Desert Organisms.* Springer, Berlin.

Zhang, F. and Gutterman, Y. (2003) The trade-off between breaking of dormancy of caryopses and revival ability of young seedlings of wild barley (*Hordeum spontaneum*) C. Koch. *Canadian Journal of Botany* 81, 375–382.

Desiccants

In agriculture: pesticides that speed up the drying of leaves, stems or vines by inducing water loss, and so kill plants. As an alternative to **defoliants**, desiccants are applied by spraying a crop as seed approaches maturity to make mechanical harvesting easier by reducing plant debris, or to manipulate the rate and degree of seed drying. Desiccants are often used in North America, for example, before harvesting **cotton**, **sorghum** and **soybean** seed and main grain crops, and in the production of **alfalfa** or red clover seed. Examples of registered desiccants include diquat, dinitro, sodium chlorate and glyphosate. In seed production, care must be taken to avoid damaging the seed, resulting in poor or **abnormal seedling** production. (See: **Treatments – pesticides**)

In seed (and general) drying technology, in a different sense of the term: water-absorbing chemical desiccants such as silica gel are used, for example, to prepare small 'laboratory' samples for the long-term **conservation of seed.** (See: **Drying of seed (for storage)**)

Desiccation damage

Two non-mutually exclusive types of damage can be envisaged as occurring during seed desiccation: (i) desiccation damage *sensu stricto* resulting directly from the removal of water; and (ii) metabolically-derived damage. The former encompasses damage associated with mechanical strain and removal of the hydration shell of molecules, whereas the latter is caused by uncontrolled chemical reactions and/or perturbations of metabolism, resulting in the accumulation of by-products to toxic levels. For both types of damage, their cellular targets are identical and include **membranes**, **proteins** and **nucleic acids**. It is sometimes difficult to make a distinction between both types and to determine their relative importance when evaluating desiccation sensitivity. Furthermore, the occurrence of each type will depend on a range of factors including the ecophysiological origin of the species, anatomy and composition (e.g. the nature of storage reserves) of the seed, its cellular structure (e.g. lipid membrane composition), the degree of development of its tissues, and the rate at which water is lost (minutes vs days). Unfortunately, the literature does not always fully appreciate the time component of desiccation-induced damage. Furthermore, most of the studies assess damage as being irreversible. Damage incurred by dehydration but repaired during rehydration has not been extensively studied in seeds (**see Desiccation tolerance – repair mechanisms; Deterioration and longevity**).

1. Damage incurred by removal of water

As water dissipates, the cells progressively shrink and their plasma membranes are submitted to mechanical strains. If these strains are not reduced by changes in surface-to-area ratios, they can lead to membrane lesions; however, this has not been demonstrated in the context of desiccation damage. The decrease in cell volume causes molecular crowding of the cellular components. The **chromatin** condenses and the cytoplasm becomes increasingly viscous. As a result, solubility and diffusion of metabolites and oxygen within the cytoplasm and organelles decreases progressively.

2. Metabolically-derived damage

Seed **mitochondria** experience a decreased availability of oxygen during drying below about −6 MPa. Various studies on desiccation-sensitive seeds show that increasing amounts of ethanol and acetaldehyde evolve from sensitive tissues during enforced desiccation before the loss of membrane integrity. In contrast, fermentation processes are not switched on in desiccation-tolerant tissues. This indicates that metabolism *per se* is apparently not adapted to drying in desiccation-sensitive tissues, resulting in perturbation in metabolic fluxes, which in turn can induce an accumulation of by-products of metabolism to toxic levels.

Oxidative stress is another important symptom of deranged metabolism that occurs during drying of desiccation-sensitive seeds (**see: Desiccation tolerance – protection against oxidative damage**). Evidence for a link between oxidative stress and desiccation damage comes from the observation that *in vitro* **free radical**-generating systems promote similar membrane damage as those measured *in vivo*. Dried desiccation-sensitive tissues exhibit a variety of symptoms of free radical-induced injury such as **phospholipid** de-esterification and various forms of lipid **peroxidation** (**see: Deterioration reactions**). The chemistry of lipid peroxidation in plants and seeds varies from one species to another and depends on the nature and origin of the free-radical molecule and the lipid composition. By analogy with other stresses such as water deficit and freezing, the cause of this peroxidative damage is thought to originate from an increased production of **reactive oxygen species (ROS)** as a result of the impairment of electron transport chains within the mitochondria, chloroplasts and microsomal membranes during drying. Under normal conditions, the steady-state concentrations of ROS are low and mostly proportional to the intensity of electron transport (respiration and photosynthesis). In germinating seeds of soybean, mitochondria are the major source of superoxide molecules. Their steady-state levels increase sevenfold during the first 30 h of imbibition, concomitantly with the loss of desiccation tolerance. Generally, desiccation-sensitive tissues respire at comparatively greater rates than tolerant tissues at the same water content (**see: Desiccation sensitivity or intolerance**). These observations illustrate that desiccation-sensitive tissues are more exposed to oxidative stress during drying.

Metabolically derived damage is dependent on two intrinsically linked factors: the rate of water loss and the tissue water content. Metabolically derived damage is dependent on the time component during drying because the loss of water slows chemical reactions. Seeds, when dried rapidly, should sustain less metabolic perturbation and change in the chemistry of their cells. Thus, the extent of metabolism-induced desiccation damage can be viewed as a function of the rapidity at which the cascade of degradative reactions occur. It follows that the nature of desiccation damage and thus the cause for desiccation sensitivity will vary according to the drying rate. Also, drying **recalcitrant** seeds within an hour allows them to survive greater loss of water, but attempts to obtain live recalcitrant embryos in the dry state have failed. This suggests that metabolically derived damage and desiccation damage *sensu stricto*, resulting directly from the removal of water are equally important. The water content at which metabolic perturbations and oxidative stress occur is high. As water dissipates, metabolism slows down. Consequently, damage produced by a faulty metabolism occurs at a slower rate. Eventually, when the tissues are in the dry state, the presence of the cytoplasmic glass does not allow metabolism to occur (**see: Glass**). In these conditions, damage arises from non-enzymatic degradative reactions which can occur, albeit slowly, both in the lipid phase and in the cytoplasm. The chemistry of these reactions is complex and poorly characterized in seeds. It includes random fatty acid oxidation to various stable and unstable hydroperoxides, which can further degrade into harmful ketones, aldehydes and alcohols. These breakdown products as well as reducing sugars (fructose and glucose) can also react with certain amino acids, which lead to the degradation of proteins. So far, experiments aimed at determining whether these deteriorating reactions are cause or effect of loss of viability have produced conflicting results.

Desiccation damage *sensu stricto* occurs mostly at the molecular scale when the loss of water from membranes (Fig. D.4), proteins and nucleic acids modifies the hydrophobic (water repelling) and hydrophilic (water attracting) interactions determining the structure and function of these molecules. The effect of water removal on the physico-chemical properties of the phospholipid bilayer is well characterized *in vitro* using membrane model systems. With the loss of the hydration shell from the polar head groups, there is a reduction in the spacing between them, resulting in a cascade of structural changes. These include a thermotropic phase transition from liquid crystalline to gel phase and an increase in packing density of the acyl chains of the **phospholipids** and in membrane rigidity (**see: Desiccation tolerance – protection by stabilization of macromolecules**). These changes alter the functional properties of membranes, which can be assessed during rehydration because they temporarily or permanently lose their permeability to cytoplasmic solutes. The relative quantities of solutes and the rate at which they leak out of the rehydrating tissues are considered to give a good indication of the loss of membrane integrity. The structural defects at the boundary between the coexisting phases in the phospholipid bilayer that were brought about by dehydration are thought to induce the leakage of cytoplasmic solutes upon rehydration of dry tissues. Another cause that might explain changes in the conformational properties of membranes is the desiccation-induced **partitioning** of endogenous **amphiphilic** (water repelling and attracting) substances from the aqueous cytoplasm into the lipid phase during dehydration and vice-versa during rehydration. The nature of these compounds is unknown but might include **phenolic** acids and flavonoids. If amphiphilic substances present in the cytoplasm move into the lipid phase at high water content during drying and perturb membranes that harbour the electron transport chains, there is the likelihood of enhanced production of **free radicals**. However, there is no experimental evidence supporting this hypothesis in seed tissues.

Certain proteins such as phosphofructokinase and lactate dehydrogenase are labile to desiccation *in vitro*, whereas others such as phospholipase are not. Similar to membranes, the loss of water may alter the hydrogen-bonding requirements for native folding according to the polypeptide sequence and tertiary structure of the protein. Symptoms of protein denaturation and aggregation occur in dried desiccation-sensitive **somatic embryos** and in seeds of maturation-defective *Arabidopsis* **mutants**. In desiccation-intolerant (recalcitrant) seeds, the cytoskeleton, a network of fibrous elements made of cross-linked proteins, is disrupted at relatively high water contents during drying.

Damage to DNA and chromatin has been mostly studied *in vitro* and its impact on desiccation tolerance is poorly understood. Chromatin extracted from desiccation-sensitive seedlings of maize appears to lose its capacity to recondense during drying. Damage also involves the depletion of nucleoproteins, revealing portions of naked DNA. The conformation of DNA is determined by the combination of base sequence and the presence of water molecules. It is hypothesized that the removal of water leads to conformational changes that alter binding sites for regulatory proteins (e.g. **transcription** factors) and nucleases. In sensitive tissues, DNA might also be exposed to chemical modifications (oxidation, addition of reducing sugars to ribonucleoproteins) which would alter gene regulation and/or affect genome integrity. A correlative link between genome integrity and desiccation tolerance is described in **Desiccation tolerance – repair mechanisms**.

Even when seeds have safely reached the dry state, subsequent rehydration can be hazardous since water uptake (**imbibition**) can injure membranes. Damage occurring during the first minutes of rehydration of dry seed tissues is referred to as imbibitional injury (**see: Germination – influences of water**). It is related to desiccation damage in the

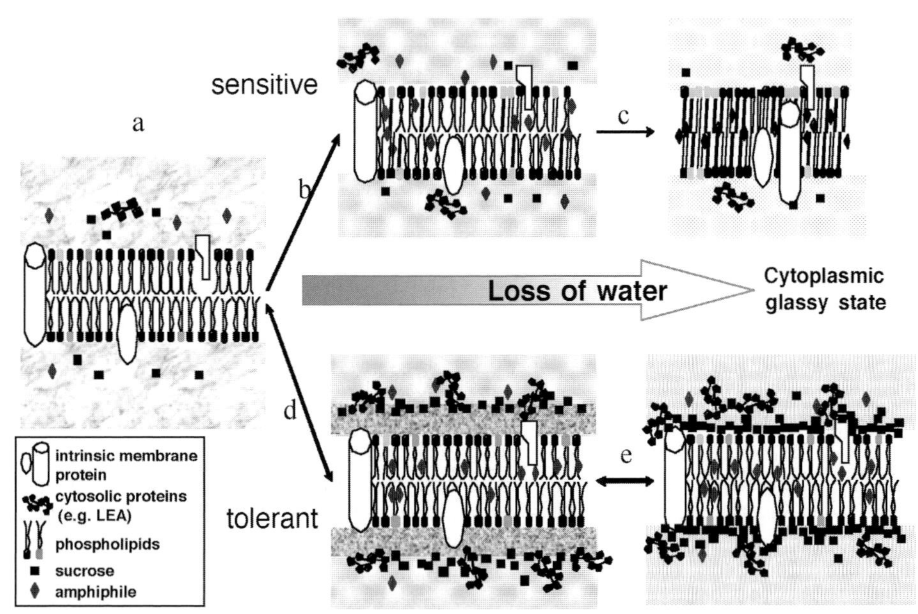

Fig. D.4. Cellular membrane behaviour during loss of water in seeds, as inferred from studies on *in vitro* model systems. In the fully hydrated state (a, grey area), phospholipids are in the liquid crystalline state and the membrane remains fluid. Upon loss of water (b–c, d–e), the amphiphilic compounds (diamonds) increase in concentration and preferentially migrate into the membrane. This causes membrane disturbances in both desiccation-sensitive (b–c) and -tolerant (d–e) cells. Preferentially excluded solutes (e.g. non-reducing sugars such as sucrose (squares) and perhaps **LEA** proteins) are present in high amounts in tolerant tissues. At the intermediate water contents (d), they maintain a certain hydration level (dark grey area) around the phospholipid head groups (grey and black circles), thereby preventing membrane damage in desiccation-tolerant cells. In the dry state (e), the hydration shell has disappeared. Sugars then replace the water molecules, thereby maintaining the spacing between phospholipid molecules and maintaining a fluid liquid-crystalline state. The double-headed arrows indicate the reversibility of the process. In contrast, sensitive tissues are not endowed with large amounts of sugars and the loss of water molecules from the hydration shell of the membrane results in the packing of phospholipid molecules and phase transition into a gel phase (c). In turn, membrane fluidity decreases and lateral phase separation and demixing may also occur. These changes in membrane physico-chemical properties are 'immobilized' by the glass formation and considered to induce irreversible damage to membrane. Upon rehydration, the functional and physical integrity is not restored, resulting in a permanent leakage of cytoplasmic solutes. The nature of amphiphilic molecules, and whether preferential exclusion has an influence on their partitioning into membranes, is not known. (Adapted from Hoekstra, F.A., Golovina, E.A. and Buitink, J. (2001) Mechanisms of plant desiccation tolerance. *Trends in Plant Science* 6, 431–438.)

sense that the hazards of rehydration must be prevented ahead of time (i.e. during dehydration) because the first minutes of rehydration are critical to the success of prior dehydration and to the subsequent orderly restoration of cellular organization and functions. The main symptom of imbibitional damage is an extensive leakage of the cytoplasmic solutes and ruptured appearance of membranes, indicating that plasma membrane is the primary target of damage. Mechanical stresses imposed by the influx of water and the subsequent increase in cell volume, as well as changes in the physical and conformational properties of membranes, are likely to be the causes of imbibitional injury. Several important crops, particularly those of tropical origin, suffer from imbibitional damage. The factors inducing imbibitional injury are the very low water content of the seed, chilling temperatures (**see: Chilling injury**) and rapid water uptake. (OL, JB)

Hoekstra, F.A., Golovina, E.A. and Buitink, J. (2001) Mechanisms of plant desiccation tolerance. *Trends in Plant Science* 6, 431–438.

Osborne, D.J., Boubriak, I. and Leprince, O. (2002) Rehydration of dried systems: membranes and the nuclear genome. In: Black, M. and Pritchard, H.W. (eds) *Desiccation and Survival in Plants: Drying without Dying*. CAB International, Wallingford, UK, pp. 343–364.

Pammenter, N.W. and Berjak, P. (1999) A review of recalcitrant seed physiology in relation to desiccation-tolerance mechanisms. *Seed Science Research* 9, 13–37.

Walters, C., Farrant, J.M., Pammenter, N.W. and Berjak, P. (2002) Desiccation stress and damage. In: Black, M. and Pritchard, H.W. (eds) *Desiccation and Survival in Plants: Drying without Dying*. CAB International, Wallingford, UK, pp. 263–291.

Desiccation sensitivity or intolerance

Desiccation-sensitive or -intolerant seeds are those which, at maturity, cannot withstand a partial or total loss of water. Furthermore, during **maturation**, they are not subjected to a water loss of the same magnitude as in desiccation-tolerant (**orthodox**) seeds and retain metabolic activity. In some tree species like *Quercus robur*, seed maturation (i.e. seed filling) is indeterminate and only stops at shedding. In other species like mangrove trees, the seed passes directly from **embryogenesis** to germination within the maturing fruit, a phenomenon called **vivipary** or viviparity.

Desiccation-intolerant seeds are usually named **recalcitrant**. Originally, the term was coined because of their behaviour during storage. Since recalcitrant seeds are desiccation-sensitive, they cannot be stored in dry and cold conditions like desiccation-tolerant (orthodox) seeds. Even when they are stored in moist conditions, their life span is short and only occasionally reaches several months. Thus, recalcitrant seeds are not an amenable germplasm for long-term **conservation** and require alternative methods of preservation such as cryopreservation (**cryostorage**). Examples of recalcitrant seeds of mostly economically important tropical species include **cacao**, tea and rubber. Intolerant seeds may also be chilling-sensitive if they are of similar origin.

The sensitivity to drying varies greatly among species. In some species such as mangroves, the sensitivity to an enforced desiccation is so great that a short water deficit can cause serious damage (**see: Desiccation damage**). The other extreme occurs in species like coffee and neem, the seeds of which can be dried to water contents close or similar to those of orthodox seeds when treated especially carefully. In between these extremes lie the vast majority of recalcitrant species, which can endure some water loss before dying (between 25 and 40% on a fresh weight basis). The variability in data on desiccation sensitivity of recalcitrant seeds has led to mixed interpretations. Some researchers suggest that the water content at which they can be dried without loss of viability forms a continuum among species. Others argue that recalcitrance can be classified in five, or more, discrete levels. It is clear that recalcitrance cannot be considered as an 'all or nothing feature'. It should be noted that it is the sensitivity to drying rather than the ability to survive drying (i.e. desiccation tolerance), which is a quantitative feature.

The sensitivity to drying among recalcitrant seeds depends on a number of experimental variables. These include the inherent characteristics of the seed (morphology, composition, whether one considers the intact seed or the isolated embryo), the seed developmental stage, the drying conditions (rate of water loss and temperature) and the rehydration conditions. Often, slow drying is more deleterious than fast drying. It is not that the rapid loss of water induces desiccation tolerance; rather the tissues when dried fast enough, reach low water contents and a glassy state (**see: Glass**) before sufficient time has elapsed for damage to propagate. Although some recalcitrant embryos can withstand a substantial loss of water after flash drying, there is no example of recalcitrant seeds that remain alive and storable at water contents similar to those of orthodox seeds. To quantify the extent of desiccation sensitivity, the **critical water content** is determined.

The nature and mechanisms of damage in recalcitrant seeds during drying are varied and range from an exacerbated water stress to structural and metabolically derived damage that are initiated at different water levels (**see: Desiccation damage**). The desiccation-sensitivity of recalcitrant seeds is often explained as a lack of one or more protective mechanisms that have been described in orthodox seeds (**see: Desiccation tolerance – protective mechanisms**). Mangrove seed species lack, for example, homologues of dehydrins (group 2 **LEA**) whereas these proteins have been immunologically detected in a range of other recalcitrant species including the temperate trees, *Quercus robur*, *Acer* spp., *Aesculus hippocastanum*, *Castanea sativa* and the tropical species *Castanospermum australe*. (OL)

Pammenter, N.W. and Berjak, P. (1999) A review of recalcitrant seed physiology in relation to desiccation-tolerance mechanisms. *Seed Science Research* 9, 13–37.

Desiccation tolerance

In the vast majority of seeds, a pre-programmed desiccation phase occurs as the final stage of development. Seeds that are subject to **maturation drying** and acquire desiccation tolerance beforehand are termed **orthodox**. Maturation drying is both an important developmental transition and a trait of major adaptive importance to the survival and **dispersal** role of the species. Maturation drying promotes germination capability, results in the cessation of storage reserve deposition and brings the seed embryo into a glassy state (**see: Glass**), which consequently leads to metabolic **quiescence**, storability and resistance against detrimental environmental conditions.

In contrast, there are many species that produce seeds that do not undergo maturation drying, remain indeterminate in their development and desiccation-sensitive at the shedding stage. These seeds are termed **recalcitrant** because they are not amenable to long-term storage. The water content at which recalcitrant seeds of different species can survive dehydration ranges from that equivalent to the wilting point of leaves to almost complete drying.

The ability to survive complete water loss (or a water potential at least below −150 MPa) requires several biochemical adaptations that are related to tolerance or hardiness rather than to avoidance. Desiccation tolerance should be discriminated from drought tolerance, which is based on the ability to avoid or strongly reduce the loss of water by altering growth and adjusting the osmotic potential during stress. In contrast, desiccation tolerance refers to the mechanisms that enable organisms to survive in spite of the loss of water. (OL)

(See: articles related to Desiccation tolerance; Deterioration; Maturation)

Black, M. and Pritchard, H.W. (eds) (2002) *Desiccation and Survival in Plants: Drying without Dying.* CAB International, Wallingford, UK.

Vertucci, C.W. and Farrant, J.M. (1995) Acquisition and loss of desiccation tolerance. In: Kigel, J. and Galili, G. (eds) *Seed Development and Germination.* Marcel Dekker, New York, USA, pp. 237–272.

Desiccation tolerance – acquisition and loss

At some point in their development seeds of the majority of species begin to lose water and may dry down to 5–10% water content (fresh weight) or less than 0.1 g water/g dry weight. This loss of water does not kill them and they are therefore regarded as being desiccation-tolerant. Young developing seeds are generally intolerant of water loss and the change from desiccation sensitivity to a desiccation-tolerant state generally occurs approximately midway through development (Fig. D.5A). Acquisition of desiccation tolerance depends on the rate of drying and the water content achieved after desiccation (**see: Maturation – effects of drying**). The faster the drying rate, the later during maturation does the developing seed become able to tolerate the water loss. For example, seeds of French or **common bean** isolated from pods at 26 days after pollination will not survive if the water is lost within hours. However, seeds harvested at the same age and slowly dried over a week are able to germinate afterwards.

In this case, the effect of a slow removal of water can be confounded with further development. Indeed, the premature enforced drying is slow enough to promote the germination ability and allow the synthesis of protective mechanisms (**see: Desiccation tolerance – protection**). When the maintenance of cellular integrity after drying is used as an indicator of desiccation tolerance, the acquisition of desiccation tolerance is a developmental programme that is not necessarily contingent on other programmes like **embryogenesis** and reserve deposition. In *Arabidopsis thaliana*, embryos with severe defects in embryogenesis might still become desiccation-tolerant. Similarly, in wheat, the competence to acquire desiccation tolerance is not dependent on the seed morphological development. In developing wheat embryos, a slow drying that is enforced a few days after anthesis leads to cellular changes that are associated with desiccation tolerance while embryogenesis is not yet completed.

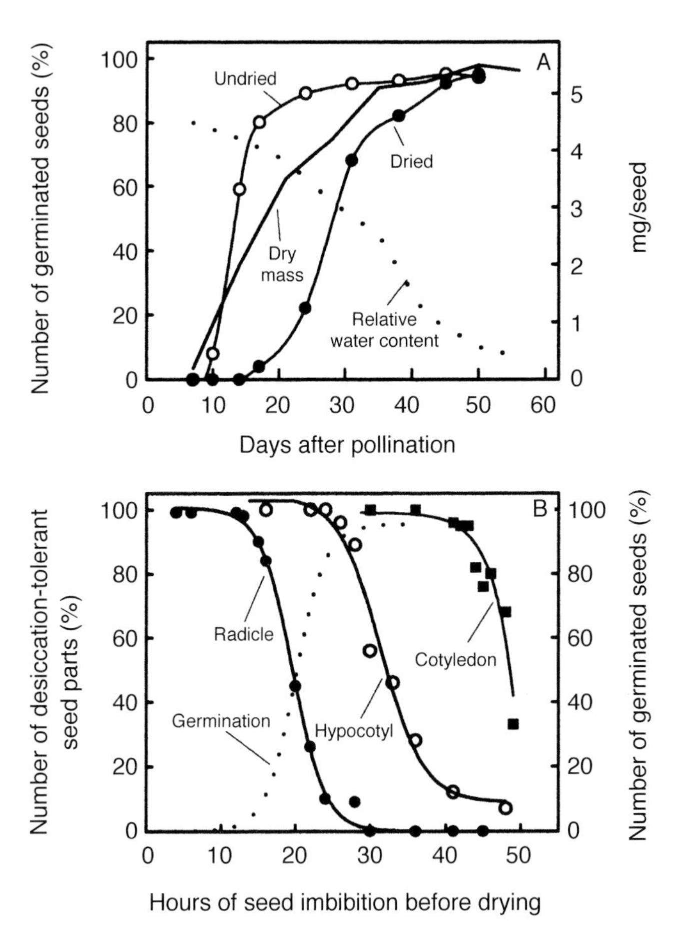

Fig. D.5 (A). Time course of seed development in the mustard *Sinapis alba*. The ability of isolated immature seeds to complete germination when sown directly in water (undried, open circles) precedes the acquisition of desiccation tolerance (which occurs later than 15–20 days after pollination), scored as the percentage of seeds capable of germination after fast drying (dried, closed circle). The solid line and dotted line represent the changes in seed dry mass and the relative water content, respectively. Adapted from Fisher, W., Bergfeld, R., Plachy, C., Schäfer, R. and Schopfer, P. (1988) Accumulation of storage materials, precocious germination and development of desiccation tolerance during seed maturation in mustard (*Sinapis alba* L.). *Botanica Acta* 101, 344–354. (B) Time course of the loss in desiccation tolerance of different seed parts of cauliflower (radicles (closed circles), hypocotyls (open circles) and cotyledons (closed squares)) during imbibition. The dotted line represents the percentage of germinated seeds that was scored as the number of seeds that exhibited a protruding radicle. Data from Hoekstra, F.A., Haigh, A.M., Tetteroo, F.A.A. and van Roekel, T. (1994) Changes in soluble sugars in relation to desiccation tolerance in cauliflower seeds. *Seed Science Research* 4, 143–147.

Mature seeds, when hydrated initially remain tolerant to re-imposed desiccation. Desiccation tolerance is progressively lost in germinated seeds sequentially in the **radicle**, in the **hypocotyl** and **epicotyl**, and finally in the **cotyledons** and **coleoptile**. As for maturation, these organs exhibit a differential sensitivity to the water content reached after drying (Fig. D.5B). The ability of the radicle to survive the loss of water is lost as it emerges from the seedcoat. In cereal seedlings that are dried at an advanced stage (i.e. several days) after the completion of germination, the coleoptile region resumes growth upon rehydration and initiates secondary root formation to compensate for the desiccation-induced death of the primary root. In general, imbibed seeds that are safely dried back to their original water content do not return to their original mature state. This is illustrated by the fact that the total imbibition time required to complete germination is longer in those seeds that have been subjected to drying during imbibition. Some events occurring during imbibition prior to dehydration do not have to be repeated upon rehydration, but not all are entirely stable to desiccation (**see: Desiccation damage**). For example, a pea seed that is desiccated and rehydrated after 16 h of imbibition (i.e. during germination, which takes 23 h before radicle emergence) essentially takes 8 h longer to complete germination in total time than a seed that is not desiccated and rehydrated during germination. Germinated, desiccation-sensitive radicles of several horticultural crops can be rendered desiccation-tolerant by exposing them to a controlled water loss treatment brought about by slow drying or by an osmoticum such as polyethylene glycol. The ability to impose imbibition and drying cycles has led to many horticultural applications (such as **priming**) aimed at improving the germinative quality and the commercial value of seedlots.

The developmental cues that trigger the **signal transduction** chain leading to the acquisition of desiccation tolerance remain to be elucidated. Two factors have been identified so far: **abscisic acid** (ABA), a plant growth hormone (**see: Hormones**) and osmotic stress. In developing seeds, ABA content generally peaks half-way through development and declines upon maturation drying (**see: Developing seeds – hormones**). Application of ABA confers desiccation tolerance upon young zygotic embryos (e.g. barley) and on sensitive somatic embryos (e.g. lucerne [alfalfa], oilseed rape). *Arabidopsis* mutants that are both insensitive to and incapable of synthesizing ABA (**see: Mutants**) produce desiccation-sensitive seeds. Ectopic expression of the **transcriptional** activator, ABI3, that controls seed maturation leads to the induction of desiccation tolerance in non-embryogenic cells of carrot in the presence of exogenous ABA. Furthermore, many genes involved in the protective mechanisms, such as those for LEA proteins (**see: Late embryogenesis-abundant proteins**), are controlled by ABA (see: **Desiccation tolerance – mechanisms**). The promotive effect of an osmotic stress on desiccation tolerance is inferred from experiments showing that when germinating seeds or developing seeds that are detached from the mother plant are submitted to a partial loss of water for several days, desiccation tolerance is acquired. This beneficial treatment occurs in developing seeds of the ABA-deficient mutant of *Arabidopsis*, suggesting that ABA and osmotic stress act independently. Furthermore, there is no evidence that loss of water provokes ABA synthesis, as it does in leaves submitted to a water deficit. (**See: Maturation – effects of drying**) (OL)

Bewley, J.D. (1995) Physiological aspects of desiccation tolerance – a retrospect. *International Journal of Plant Sciences* 156, 393–403.

Black, M. and Pritchard, H.W. (eds) (2002) *Desiccation and Survival in Plants: Drying without Dying*. CAB International, Wallingford, UK.

Vertucci, C.W. and Farrant, J.M. (1995) Acquisition and loss of desiccation tolerance. In: Kigel, J. and Galili, G. (eds) *Seed Development and Germination*. Marcel Dekker, New York, USA, pp. 237–272.

Desiccation tolerance – protection

There are at least two main strategies that confer desiccation tolerance.

The first strategy is the avoidance of the accumulation of damage that is induced by the loss of water and in the dry state (**see: Desiccation damage**). This avoidance relies on the presence of protecting factors, which include **non-reducing sugars**, stress proteins such as **late embryogenesis abundant proteins** (LEAs) and **heat shock proteins (HSPs)** and **free radical**-processing systems (**see: Desiccation tolerance – protection against oxidative damage; Desiccation tolerance – responses of metabolism; Desiccation tolerance – protection by stabilization of macromolecules; Deterioration prevention**). These protecting factors can be present prior to or synthesized during drying. They act in synergy and sequentially both during desiccation and in the dry state. For example, sucrose does not exert protective effects on dried membranes that have been damaged by free radicals prior to drying.

The second strategy is based on the activation of repair mechanisms upon rehydration (**see: Desiccation tolerance – repair mechanisms**). Desiccation tolerance must be considered as a multifaceted trait. Seeds are only able to survive drying if all the mechanisms and conditions necessary for desiccation tolerance are present. This implies that there is no obvious relationship between the level of desiccation sensitivity/tolerance and the amount or nature of any one tolerance factor. For example, in a maturation-defective double **mutant** of *Arabidopsis* that produces desiccation-sensitive seeds, induction of desiccation tolerance in embryos by an **abscisic acid** analogue does not lead to accumulation of sucrose. Finding correlative evidence between desiccation tolerance and amounts of protective substances is further complicated by the fact that researchers use various ways to enforce desiccation as well as different parameters to score desiccation tolerance. These may be based on physiological assessments (ability to germinate or to grow after drying, ability to produce normal seedlings, *in vitro* tissue culture), membrane integrity assays (leakage measurements; **see: Membrane damage**), and cellular activities using vital staining.

The different protective mechanisms are presented in sequence of hydration level at which they are thought to be active as illustrated in Fig. D.6. During desiccation, **anhydrobiotes** will first experience a water deficit resembling an osmotic stress. To protect macromolecules from moderate water loss, many plants and microorganisms accumulate **compatible solutes**, the concentration of which reflects the degree of tolerance to stress. *In vitro* studies have shown that the mode of action of compatible solutes is to counteract the destabilizing effects of molecules and ions (such as Na^{2+}; amino acids [arginine]) that concentrate in the cytoplasm as a result of loss of water. Compatible solutes are preferentially excluded from the surface of proteins, thus keeping these macromolecules hydrated. This protecting mechanism is thought to take place in **orthodox** seeds during drying at water content above 0.3 g water/g dry weight (approximately from −10 to −70 MPa). Putative compatible solutes in seeds are sucrose, **oligosaccharides** and **LEA** proteins (**see: Desiccation tolerance – protection by stabilization of macromolecules**). **Dehydrins** (group 2 LEA proteins) are typically induced in vegetative tissues by stresses that alter the cell osmotic balance such as salinity, cold and chilling. They are also correlated with several quantitative loci associated with traits related to tolerance of these stresses. These observations together with their apparent lack of structure and high hydrophilicity suggest that dehydrins retain water around other proteins at low water potentials. Upon further removal of water, preferential exclusion becomes impossible because there is not enough water left. Instead, hydrogen bonding and **glass** formation, both involving **non-reducing sugars** and specific proteins are likely to be the mechanisms by which macromolecules will be structurally and functionally preserved at these water contents. (OL)

Buitink, J., Hoekstra, F.A. and Leprince, O. (2002) Biochemistry and biophysics of tolerance systems. In: Black, M. and Pritchard, H.W. (eds) *Desiccation and Survival in Plants: Drying without Dying*. CAB International, Wallingford, UK, pp. 293–318.

Hoekstra, F.A., Golovina, E.A. and Buitink, J. (2001) Mechanisms of plant desiccation tolerance. *Trends in Plant Science* 6, 431–438.

Desiccation tolerance – protection against oxidative damage

Characteristically, **anhydrobiotes** suffer less from oxidative damage during drying than sensitive organisms (**see: Desiccation damage**). In seeds, two mechanisms that are not mutually exclusive exist to cope with the danger of **free radicals** and **reactive oxygen species** (ROS): (i) increased efficiency of antioxidant defences, and (ii) metabolic control of energy-producing and -consuming processes. The antioxidant defences cover a wide array of enzymes (**superoxide dismutase**, catalase, peroxidase, enzymes that are involved in regenerating molecular antioxidants), and water-soluble (homoglutathione, glutathione, ascorbate) and lipid-soluble (tocopherols, quinones, phenols) molecules. Their role is to destroy free radicals and ROS as soon as they are formed or to stop the propagation of free radical-induced damage.

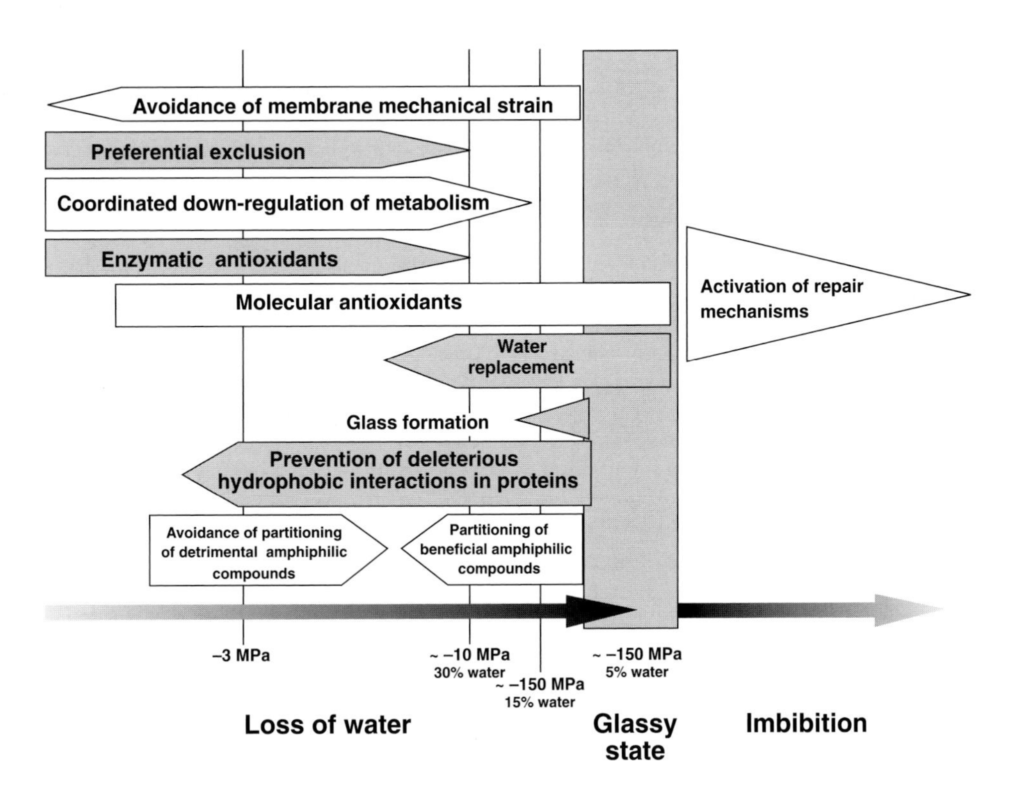

Fig. D.6. Schematic describing two strategies conferring desiccation tolerance: (i) protection during drying and (ii) repair of desiccation-induced damage during rehydration. The various protective mechanisms are represented in white and grey boxes together with the water content range at which they are likely to be effective or operate during drying and rehydration. Grey boxes indicate those mechanisms for which molecular components (non-reducing sugars, LEA proteins, HSPs) and regulatory mechanisms have been identified. White boxes indicate those mechanisms for which the various components remain to be identified. Water contents are given both as a mass ratio (% of dry weight) and water potential. The dry state and the glassy state are also indicated.

The regulation and mechanisms of action of antioxidant defences are complex. Although their involvement against environmental stresses is well characterized, their role in desiccation tolerance is not firmly established. Their action likely depends on the tissue water content. Free radical-processing systems relying on enzyme activities should mainly operate at high water content during drying and rehydration, whereas molecular antioxidants should play a dominant role in the reduced hydration conditions (**see: Desiccation tolerance – protection,** Fig. D.6). Considering that faulty electron transport within **mitochondria**, chloroplasts and microsomal membranes is responsible for ROS formation (**see: Desiccation damage**), a second mechanism to avoid oxidative stress is to decrease the chances of forming ROS. In mitochondria, this can be achieved by optimizing oxygen concentration near sites of oxidative phosphorylation and by uncoupling the electron transport from energy production. Stimulation of metabolic rate *per se* is favourable to oxygen reduction. Thus, it has been surmised that desiccation-tolerant tissues can reduce and adapt their metabolic activities early during drying to decrease the chances of ROS formation. By analogy with dormant animals that can resist extreme environmental conditions, a coordinated down-regulation of metabolism could be essential to confer desiccation tolerance in seeds. Evidence supporting this hypothesis is increasing but indirect. Desiccation-tolerant radicles generally exhibit much reduced **respiration** rates before drying compared with sensitive radicles. Similarly, **recalcitrant** seeds (**see: Desiccation sensitivity or intolerance**) also have high respiration rates that often remain unabated during desiccation, even after death of the tissues (**see: Desiccation tolerance – responses of metabolism; Deterioration prevention**). (OL)

Inzé, D. and Van Montagu, M. (1995) Oxidative stress in plants. *Current Opinion in Biotechnology* 6, 153–158.

Oliver, A.E., Leprince, O., Wolkers, W.F., Hincha, D.K. and Heyer, A.G. (2001) Non-disaccharide-based mechanisms of desiccation tolerance. *Cryobiology* 43, 151–167.

Desiccation tolerance – protection by stabilization of macromolecules

1. Membranes

At reduced water contents, the need for macromolecular stabilization becomes preponderant. Non-reducing sugars and **late embryogenesis abundant** (LEA) **proteins** and **heat-shock proteins (HSPs)** are among the molecules that protect the **phospholipid** bilayer of **membranes**, as well as **proteins** and **nucleic acids** from the deleterious removal of water. Acquisition of desiccation tolerance in developing seeds is characteristically accompanied by an accumulation of large quantities of sucrose and/or **oligosaccharides** (**raffinose family**, or derivatives of cyclitol). The reverse occurs during the loss of desiccation tolerance when the seeds germinate (**see: Desiccation tolerance – acquisition and loss**). The protecting mechanisms of these molecules have been elucidated using biochemical and biophysical studies on *in vitro* model systems. Using liposomes (artificial membranes formed from phospholipids), it has been established that during desiccation, **non-reducing sugars** interact with the polar head group of the phospholipids and replace the water molecules. The mechanism of action is referred to as **water replacement hypothesis**. The desiccation-induced membrane damage, such as phase transition and lateral phase separation (**see: Phase**) is thus circumvented by the hydrogen-bonding interactions between sugars and polar head groups of phospholipids (**see: Desiccation damage,** Fig. D.4).

The *in vivo* mechanisms by which membranes are protected are more complex. Drying experiments with liposomes in the presence of disaccharides show that a mass ratio of 5:1 sugar/phospholipid satisfies the head group hydrogen-bonding capacity. However, biochemical assays of sugar and phospholipid contents in **anhydrobiotes** suggest that the effective availability of disaccharides to phospholipids *in situ* is less than expected. Yet, *in situ* measurements in dry organisms show that the membrane phase is similar to that of hydrated samples.

Two schools of thought propose explanations that may not be mutually exclusive as to why membrane phase transition in dry anhydrobiotes is prevented in spite of a low disaccharide content. (i) *In vitro* studies indicate that besides sugars, **amphiphilic** molecules aid in preventing changes in the physico-chemical properties of membranes by a desiccation-induced **partitioning** effect. During drying, amphipathic molecules that are present in the aqueous cytoplasm in hydrated conditions move to the hydrophobic lipid phase. As amphipathic compounds are transferred from the aqueous cytoplasm to the phospholipids during drying, the membrane phase and fluidity are maintained. (ii) The alternative explanation considers that a mechanical stress imposed by the removal of water is conducive to changes in membrane phase transition. Then, instead of offering hydrogen bonding to phospholipids, the sugars and other solutes instead provide mechanical resistance against, or hinder the physical strain imposed on the bilayers by, the lack of water.

Research in food science has demonstrated that sugars can easily form a glassy state (e.g. hard candy). Cellular solutes of **orthodox** seeds also form a **glass** during drying. Glass formation is a prerequisite for desiccation tolerance. However, the water content at which glass formation occurs in seeds during drying at room temperature (in the order of 10% moisture [dry weight basis] or approximately -50 to -100 MPa) is much lower than the water content at which desiccation-sensitive seeds die. Therefore, it follows that glass formation is not a mechanism that initially confers desiccation tolerance during drying. It is, however, indispensable for surviving in the dry state since glasses contribute to the stability of macromolecular structural components during storage. Glass formation appears to participate in membrane preservation in the dry state. Anhydrobiotes contain large amounts of sucrose, a good glass former, as well as oligosaccharides. *In vitro* protection assays of liposomes show that di- and oligosaccharide glasses are capable of preventing the fusion of membranes upon dehydration. Thus, non-reducing sugars appear to play a dual role in protecting membranes in the dry state: timely glass formation during drying, and interaction with the polar head groups. In contrast to monosaccharides and sugar polymers, dissaccharides like sucrose and trehalose combine these two properties within the one compound.

2. Proteins

In addition to their protective action on membranes, non-reducing sugars also exert a stabilizing effect on dried proteins. The protecting effect relies on the dual properties of non-reducing sugars: glass formation and water replacement. An analogous mechanism to the water-replacement hypothesis is thought to protect proteins from desiccation-induced denaturation and aggregation. The protective role of sugars is demonstrated when a desiccation-labile enzyme is dried in the presence of a specific sugar without losing its activity upon rehydration. It has also been suggested that the formation of an intracellular glass ensures the fixation of the native secondary structure of proteins in the dry state. This is illustrated by the observation that protein secondary structure in dry desiccation-tolerant seeds is maintained over a long period of storage, even after the embryo is dead.

As well as non-reducing sugars, a number of proteins can also participate in the protection and stabilization of macromolecules. LEA proteins were originally identified in seeds as being predominantly synthesized during the late stages of maturation (**see: Maturation – controlling factors**). These proteins are also produced during the early hours of seed imbibition and germination, after which they decline. There is widespread distribution of LEA proteins among several plant species and five to seven groups have been described and arbitrarily categorized on the basis of their amino acid sequence homologies (**see: Late embryogenesis abundant proteins,** Table L.1). LEA proteins accumulate in all cell compartments, mainly in the cytoplasm and nucleus. At the tissue level, the localization pattern varies according to the cell type, species, stress and organs. For example, Em proteins and **dehydrins** (LEAs) are often detected in epidermal and vascular tissues of the embryos. Their accumulation pattern correlates well with the phase of desiccation tolerance during maturation and germination. They are also present in dried vegetative tissues of resurrection (desiccation-tolerant) plants. This correlative evidence indicates that LEA proteins are important for the survival in the dry state. However, LEA proteins are also detected in an increasing number of **recalcitrant** seeds (**see: Desiccation sensitivity or intolerance**) exhibiting a wide range of desiccation sensitivities. They can also be induced in immature embryos as a result of the detachment from the mother plant, a manipulation that does not necessarily induce desiccation tolerance. Moreover, certain LEA proteins accumulate in desiccation-sensitive leaves and roots as a response to an array of stresses such as water deficit, salinity, chilling and freezing. The expression of LEA genes can also be induced in immature seeds or vegetative tissues by exogenous **abscisic acid** (ABA), a plant hormone involved in the response to water stress (**see: Hormones**). Thus the role of LEA proteins is not limited to desiccation tolerance.

Genetic and molecular studies have established that LEA proteins are pivotal to the resistance against environmental stresses that perturb the cellular water balance. For example, the presence of a 35 kDa dehydrin protein is genetically linked to chilling tolerance during seedling establishment of cowpea. Overexpressing LEA proteins in transgenic plants and yeast confers an improved tolerance to salt stress and freezing stress (**see: Late embryogenesis abundant proteins,** Table L.1). In transgenic spring wheat, overexpression of a barley group 3 LEA results in an increased water use efficiency under drought stress. Thus, LEA proteins appear to be involved in the protection against or control of the removal of water, regardless of how the water loss is achieved and perceived by the cell (evaporative or osmotic dehydration, ice formation). But the contribution of LEA proteins is not sufficient to provide full protection against desiccation.

There are several possible modes of protection offered by LEA proteins. They are generally inferred from their biochemical and physico-chemical properties and structural features, which have been predicted from their peptide sequence or determined by *in vitro* protection assays using purified or recombinant LEA proteins. They remain generally soluble in solution after boiling, indicating that they are unable to denature in solution upon heating. This is due to the absence or to low amounts of hydrophobic amino acids, which results in a reduction in heat-induced deleterious hydrophobic interactions. The majority of LEA proteins are highly hydrophilic and contain predominantly alanine, glycine, glutamine and threonine, whereas tryptophan and cystine are absent. Various protective functions have been proposed for the different groups of LEA proteins. LEA proteins accumulate in cells to concentrations far above those of enzymes, and along with their high structural flexibility, this rules them out from having an enzymatic role. In general, LEA proteins are thought to provide an interface between the cytoplasm and membranes/proteins during drying and most likely stabilize macromolecules during removal of water (**see: Desiccation tolerance – protection,** Fig. D.6).

Some LEA proteins (particularly those of group 2 and 3) might prevent desiccation-induced damage of cellular membranes. This is suggested by their propensity to form amphipathic helices, their localization near the plasma membranes, and their ability to bind to lipid vesicles that contain acidic phospholipids. However, direct experimental evidence is lacking to ascertain the protective role of LEA proteins on membranes. Not all group 2 and 3 LEA proteins have been assigned such a stabilizing role. For example in soybean seeds, rGmD19, a group 1 LEA protein and rGmDHN, a group 2 LEA, do not interact with membrane lipids. In this case, since these LEA proteins exhibit large regions of unordered conformation, this is considered to promote an efficient interaction of the proteins with water, which prevents or slows down the cellular water loss. Some dehydrins ameliorate enzyme activity under reduced water conditions or protect enzyme structure from the deleterious effects of freezing and drying. Group 3 LEA protein secondary structure allows ion binding, and some dehydrins can bind metals. These observations suggest that LEA proteins aid in sequestering ions, whose intracellular concentrations might be otherwise unacceptably high in the dehydrating cell. This kind of protection corresponds to a preferential exclusion-type mechanism. Consequently, LEA proteins might also act as compatible solutes when the cells are exposed to water deficits. LEA proteins might play a protecting role not only during drying but also in the dry state. (**See: Desiccation tolerance – protection,** Fig. D.6) When LEA proteins are embedded into a sugar glass, they stabilize the glassy matrix, by changing the hydrogen-bonding properties within the model sugar system towards those of intracellular glasses.

Thus, the roles of LEA proteins in desiccation tolerance include preferential exclusion to minimize water loss, protein and membrane stabilization, ion sequestration and amelioration of the stability of sugar glasses (**see: Desiccation tolerance – protection,** Fig. D.6; **Glass**). It is likely that LEA proteins are present both during drying and in the dry state. However, the exact functions of the different groups of LEA proteins in desiccation tolerance and stress *in vivo* remain to be elucidated.

The ubiquitous heat shock proteins (HSPs) also participate in the protection of dehydrating cells of desiccation-tolerant tissues. In developing seeds, small HSPs are distinctly regulated by developmental cues and stress (e.g. heat, drought). They appear during reserve synthesis at mid-maturation and increase in abundance as the seed dehydrates, just like LEA proteins. During germination, the developmentally regulated HSPs are relatively abundant during the first few days after the start of imbibition and then decline quickly. There is a reduction in small HSP protein content in the desiccation-intolerant phenotype of several seed development **mutants** of *Arabidopsis*. For example, HSP17.4 (a small HSP) content of seeds of maturation-defective mutants (*abi* and *lec2-1*), and embryo-defective mutants (*emb266*) that are desiccation-tolerant, is similar to that in wild-type seeds (**see: Development of seeds – an overview**). In contrast, mature seeds of the desiccation-sensitive *abi3-6*, *fus3-3* and *lec1-2* mutants have less than 2% of wild-type HSP17.4 amounts. HSPs are ubiquitous proteins and are therefore present in recalcitrant seeds. This indicates that HSPs may be necessary but not sufficient to confer desiccation tolerance. The protective function of HSPs in desiccation tolerance is inferred from *in vitro* protection assays. HSPs act as chaperones: they interact with target proteins to minimize the probability that hydrophobic regions illegitimately interact within the protein or with neighbouring proteins, an interaction that causes protein denaturation or aggregation.

Other types of proteins (e.g. aquaporin, 1-*cys*-peroxiredoxin, psp54 [a member of the **vicilin** family]) may play a role in desiccation tolerance because of a possible correlation between changes in gene expression pattern and water contents during seed development. (OL, JB)

(See: Deterioration prevention)

Buitink, J., Hoekstra, F.A. and Leprince, O. (2002) Biochemistry and biophysics of tolerance systems. In: Black, M. and Pritchard, H.W. (eds) *Desiccation and Survival in Plants: Drying without Dying*. CAB International, Wallingford, UK, pp. 293–318.

Hoekstra, F.A., Golovina, E.A. and Buitink, J. (2001) Mechanisms of plant desiccation tolerance. *Trends in Plant Science* 6, 431–438.

Vertucci, C.W. and Farrant, J.M. (1995) Acquisition and loss of desiccation tolerance. In: Kigel, J. and Galili, G. (eds) *Seed Development and Germination*. Marcel Dekker, New York, USA, pp. 237–272.

Wise, M.J. and Tunnacliffe, A. (2004) POPP the question: what do LEA proteins do? *Trends in Plant Science* 9, 13–17.

Desiccation tolerance – repair mechanisms

Apart from protecting cellular structures during drying (**see: Desiccation tolerance – protective mechanisms**), **anhydrobiotes** have also developed another desiccation-tolerance mechanism, which consists of repairing the desiccation-induced damage during rehydration. The repair aspect of desiccation tolerance has been mostly demonstrated in vegetative anhydrobiotes such as mosses and resurrection plants. Molecular studies on these organisms have identified a series of genes (the so-called rehydrins) that are activated upon rehydration. One of them shows high similarity with polyubiquitin, a small protein involved in targeting and degrading damaged proteins. A similar gene product has also been detected during imbibition of grass seeds, indicating that a ubiquitin-based repair mechanism is activated during germination. L-isoaspartyl protein methyltransferase, a ubiquitous enzyme that facilitates the conversion of abnormal L-isoaspartyl residues to normal L-aspartyl residues could be another important enzyme repairing proteins damaged during seed drying and storage. This enzyme accumulates in wheat seeds during the late stages of development and is active in germinating embryos of sacred lotus, the oldest demonstrably viable seed species (up to 1200 years; **see: Ancient seeds**). These studies suggest that damaged proteins that accumulate during drying and/or storage must be effectively repaired upon imbibition to ensure survival. DNA repair processes are also present in seeds during rehydration to maintain genome integrity during desiccation/rehydration cycles. In rye embryos, single- and double-strand breaks induced in the dry state by irradiation are repaired during imbibition and do not compromise survival. When DNA repair is blocked by certain drugs, these embryos fail to survive upon rehydration. In seeds of vegetable crops, invigorating treatments such as **priming** reduce the frequency of chromosomal aberrations and decrease the frequency of **abnormal seedling** after storage in detrimental conditions. The physiological action of invigorating treatments is viewed as a means of inducing repair mechanisms or of allowing them to proceed early during **imbibition** before radicle emergence. (OL)

Osborne, D.J. and Boubriak I.I. (1994) DNA and desiccation tolerance. *Seed Science Research* 4, 175–185.

Osborne, D.J., Boubriak, I. and Leprince, O. (2002) Rehydration of dried systems: membranes and the nuclear genome. In: Black, M. and Pritchard, H.W. (eds) *Desiccation and Survival in Plants: Drying without Dying*. CAB International, Wallingford, UK, pp. 343–364.

Desiccation tolerance – responses of metabolism

Several observations support the hypothesis that a coordinated regulation of anabolic and catabolic pathways is an adaptive mechanism of tolerance. First, the utilization of carbon sources must be regulated to allow both the accumulation of protective sugars and production of energy to synthesize protective proteins without leading to cell starvation. Second, the tissues must cope with the decreased diffusion of metabolites as a result of the increase in cytoplasmic viscosity during drying (**see: Desiccation damage**). Third, activities of enzymes and metabolic pathways are not uniformly and simultaneously shut down during drying because they require different hydration levels to function. At the enzyme level, *in vitro* studies aimed at determining the water content that stops enzyme activity show that in reduced hydration conditions (less than 10% moisture), only lipid-hydrolysing enzymes (lipase) remain active. At an intermediate level, a much wider array of enzymes are active, including some hydrolytic enzymes (amylases) and enzymes involved in oxidation/reduction (dehydrogenases).

When bulk water is present, all enzymes are able to perform their catalytic activities. Enzymes for different metabolic pathways are not equally affected by the loss of water. For example, non-invasive measurements of oxidative phosphorylation (O_2 uptake) and oxidation of organic molecules (CO_2 release) show differential sensitivity to water content in various seed species. The respiratory quotient (RQ) is a measure of how much CO_2 is released per O_2 consumed and is around one (RQ = 1) in normal physiological conditions when sugars are the carbon source to supply cellular energy. In desiccation-tolerant seeds of soybean during drying RQ = ~1, until the tissues reach 0.25 g H_2O/g dry weight. Thereafter, RQ decreases to approximately 0.2 at 0.1 g H_2O/g dry weight. Therefore, metabolic fluxes must be repressed in a coordinated way during drying to take into account the limits of each enzyme and metabolic pathway in terms of hydration level. The signals and biochemical mechanisms leading to reduction of metabolism and their importance in desiccation tolerance are unknown. Metabolic activities during drying show a cut-off at around 0.15 to 0.25 g H_2O/g dry weight, below which the regulation of metabolism as a desiccation-tolerance mechanism becomes irrelevant (**see: Critical water content; Desiccation tolerance – protection,** Fig. D.6). (OL)

Leprince, O., Harren, F.J.M., Buitink, J., Alberda, M. and Hoekstra F.A. (2000) Metabolic dysfunction and unabated respiration precede the loss of membrane integrity during dehydration of germination radicles. *Plant Physiology* 122, 597–608.

Oliver, A.E., Leprince, O., Wolkers, W.F., Hincha, D.K. and Heyer, A.G. (2001) Non-disaccharide-based mechanisms of desiccation tolerance. *Cryobiology* 43, 151–167.

Destoner

Equipment, modified in design from the **gravity table**, used in the seed conditioning process to remove a small amount of heavy material, such as dirt, glass, rock, or stones, from a large volume of seed. (**See: Conditioning, II. Cleaning**)

Detasselling

The usually manual or sometimes mechanical removal of the immature male inflorescences ('tassels') from the monoecious female ('seed') parent before the start of anthesis to ensure complete cross-pollination in **hybrid** maize breeding and seed production, as a critical step impacting both seed yield and purity. Though since the late 1960s **cytoplasmic male sterile** lines have increasingly been used to make this very arduous practice unnecessary, maize detasselling remains widespread worldwide – employing about 100,000 people for a few weeks each summer in the USA alone, for instance.

The process of tassel removal is a monumental challenge that needs to be done completely, regardless of weather conditions. Although manual removal is practised in many locations, mechanical or combinations of mechanical and manual removal are widely used. In addition to timing, it is important that as little damage as possible is inflicted on the parental plants. The mechanical systems involve either cutting or pulling off the tassels prior to the shedding of pollen. A popular system utilizes a cutting of the female parent plants just prior to tassel exertion, and then when the tassel extends the female tassels are pulled with a mechanical puller. Final clean up involves hand removal of the remaining tassels. This assault on the female parent plant is not without consequence and the degree of leaf removal can impact the final seed **yield**. (JSB)

Deterioration and longevity

The process by which seeds lose **vigour** and eventually die. Deterioration is most often defined as a deleterious change with time (**ageing**) but it may also be considered as a change that occurs with lowered water content (**desiccation damage**) or high and low temperatures (denaturation or freezing damage, respectively). Although **deterioration reactions** occur in the ungerminated seed, the effects are most often detected when the seed is placed in germination conditions. Upon **imbibition**, deteriorated seeds may leak cellular contents, grow slowly, exhibit abnormalities (necrosis or deformed structures) and eventually fail to produce an emergent radicle. In the most severe cases, deteriorated seeds have only a few viable cells that are incapable of organized growth, or have little or no **viability** (Fig. D.7), because critical cellular constituents are seriously degraded.

Deterioration is caused by the combined effects of lesions to macromolecules, accumulated as a consequence of **desiccation** or over time following **maturation**, and the progressive inability to repair the lesions. Unrepaired damage prevents cells from replicating their genetic material which, in turn, delays or prohibits cellular changes needed for the completion of germination, and eventually leads to cell dysfunction and death.

Seed deterioration has important economic and ecological consequences. Lowered seed quality affects stand **establishment** and the ability of seeds to persist in soils affects the population dynamics of ecosystems (**soil seed banks**). The duration that seeds remain viable in a **gene bank** determines the cost of their conservation as well as the ability for curators to maintain the genetic identity of the sample (**genetic erosion**).

The conditions in which seeds are stored greatly affect their deterioration rate, and hence their ability to survive in storage (**longevity**) (**see: Deterioration kinetics**). **Relative humidity**, which determines the water content of seeds, is generally considered to be the most important environmental factor affecting the rate at which seeds lose viability, especially at higher temperatures. Seeds deteriorate faster as relative humidity (RH) increases above about 20%, and the water content corresponding to this is often referred to as the **critical water content** for seed storage because maximum longevity occurs at this water content for a given storage temperature. Reducing the storage temperature increases the maximum longevity, and cryogenic storage (**cryostorage**) of seeds is now routine in many gene banks (**see: Deterioration kinetics**). Gases surrounding seeds during storage are also proposed to affect deterioration rates, although information is sparse. High oxygen partial pressures (concentrations) appear to accelerate deteriorative reactions; ambient (atmospheric) oxygen pressures may not be damaging and, in hydrated seeds, are deemed necessary to allow for respiration to occur. Volatile organic molecules such as acetaldehyde produced by seeds in storage may accelerate their deterioration, particularly if the containers in which they are stored are not periodically flushed. Reports of remarkable longevity of seeds stored under negative pressures perhaps indicate that vacuum sealing of seeds enhances their longevity, although confirming studies are needed where other environmental factors are controlled.

1. Factors affecting deterioration in storage

In addition to the storage environment, the longevity of seeds is determined by several factors intrinsic to the seed that are collectively called seed quality. Different species of seeds have characteristic rates of deterioration (Table D.2) (**see: Ancient seeds**) but the range of deterioration rates within a species can also vary broadly depending on genetic factors and growth conditions.

The most important factor determining seed longevity is the degree to which the seed survives desiccation. Many seeds acquire the ability to survive extreme drying during maturation, the final stage of seed development, and lose this ability during germination. Drying itself slows many of the chemical reactions that lead to seed deterioration (**see: Deterioration kinetics; Glass**). In addition, the ability to survive drying is associated

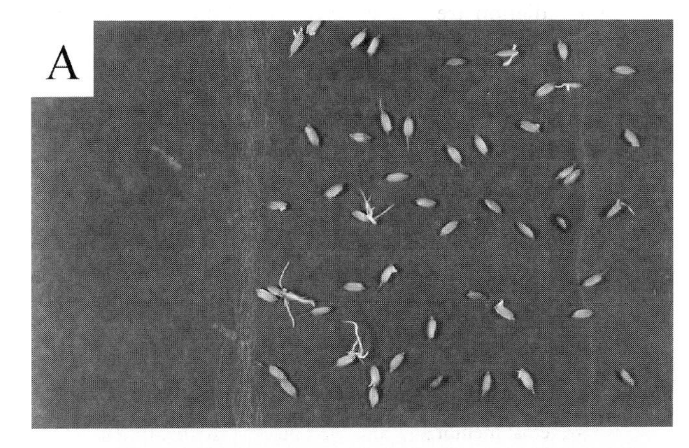

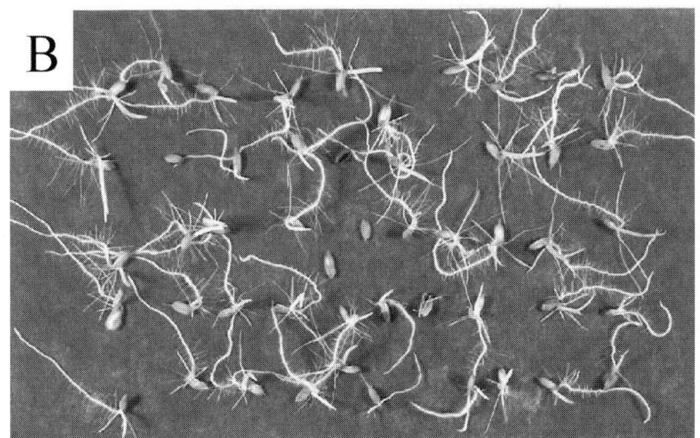

Fig. D.7. Germination test on rice seed following 20 years of storage. (A) Storage at 5°C; (B) storage at −18°C.

Table D.2. Some orthodox seeds classified by their average longevity in storage.

Species	P_{50} in controlled storage (5°C, 5 ± 2% water)	P_{50} in open storage (temperate climate)
Short-lived		
Abies procera	9	
Ulmus americana	10	
Lactuca sativa	22	6.4
Allium cepa	18	5.4
Petroselinum crispum	24	3.4
Glycine max	30	3.4
Medium-lived		
Secale cereale	36	4.5
Lotus corniculatus	52	6.7
Helianthus annuus	53	5.4
Ricinus communis	58	13.3
Linum usitatissimum	61	8.75
Spinacia oleracea	69	12.7
Long-lived		
Lycopersicon esculentum	>80	24.5
Pisum sativum	>80	15.9
Medicago sativa	>80	10.6
Vigna radiata	>80	19.5

Seeds are examples of species that maintain viability for limited to long times under dry (seeds contained between 3 and 7% water, depending on species), refrigerated conditions (unpublished data from the USDA National Seed Storage Laboratory, NSSL), and in uncontrolled conditions where the mean temperature is 10°C and average RH is about 60–75% (from Priestley, D.A. (1986) *Seed Aging, Implications for Seed Storage and Persistence in the Soil.* Cornell University Press, Ithaca, New York, USA). Longevity is expressed as the average number of years for their viability to fall to 50% (P_{50}).
Data from the NSSL are estimated by extrapolation of sigmoidal curves fitted to time versus survival data of 10 to 500 accessions of a species stored for 40–60 years.

with a level of quiescence acquired through cellular changes that limit metabolic activity, e.g. the synthesis and deposition of large molecular weight constituents (**storage reserves**) in cells and the decrease of membrane surface area. These cellular changes are complete at **physiological maturity**, which also marks the time at which the potential longevity of an **orthodox seed** is maximized. A seed's longevity is compromised when it is harvested either prematurely or its harvest is delayed. The composition of the reserves accumulated within seeds may also affect seed longevity, although consistent relationships between them are hard to demonstrate. Generally, crops that contain high amounts of **starch** (e.g. **cereals** and **pulse** crops) store well, perhaps a co-correlate of domestication. There is little support for the anecdote that high-oil seeds store poorly. For example, alfalfa (lucerne) and tomato seeds, which often contain more than 40% oil, may have exceptional longevities, yet onion seeds which contain less than 10% oil are notoriously difficult to store. Nonetheless, the chemistry and arrangement of oil within seeds may affect the susceptibility of seeds to deterioration. Based on our understanding of **deterioration reactions** involving **peroxidation,** it is possible that seeds with a higher polyunsaturated **fatty acid** content are more prone to deterioration; however, reducing the proportion of linolenic and linoleic acids in seed oils (through genetic manipulations) does not enhance longevity and sometimes results in seeds with lower overall quality. Because small molecular weight carbohydrates probably stabilize macromolecular structures (**see: Desiccation tolerance – protection by stabilization of macromolecules; Deterioration prevention**) the amount and composition of sugars has also been linked to storage behaviour. When a comparison is made between the sugar content of different species and their longevity, this link is not supported, but it is supported by observations of poor longevity in immature or germinating seeds, which have low di- and trisaccharide content but high monosaccharide (e.g. glucose and fructose) content, or seed lines that have been genetically manipulated to produce low amounts of trisaccharides, e.g. the **raffinose-series oligosaccharides**. High amounts of reducing sugars, usually found in immature or germinating seeds, are associated with rapid deterioration. It is contended that in some seeds, e.g. soybean, the ratio of sucrose to the raffinose-series oligosaccharides is important in maintaining cellular integrity. Morphological features of the seed, such as a hard seedcoat, appear to enhance longevity, perhaps because they behave as a partially impermeable barrier to water and harmful gases. **Dormancy** is also claimed to extend seed lifespans. However, evidence linking dormancy and low rates of seed deterioration is scant. The effects of seed quality factors on longevity have not been quantified, resulting in unreliable predictions of survival times for a seed population or an individual seed under specific storage conditions.

Because factors that contribute to seed longevity are under genetic control (e.g. **desiccation tolerance**, maturity, dormancy, morphology, composition, stress tolerance), deterioration rates are likely to vary among individuals in a heterogeneous population. Mortality associated with seed deterioration may cause shifts in the genetic composition of a seed population affecting its phenology, morphology, composition and dormancy traits. The evolutionary significance of seed longevity, especially as it pertains to **domestication** of crops, remains to be explored.

Research on seed deterioration often distinguishes between 'natural' and 'accelerated' ageing, terminology that reflects the conditions of storage more than the mechanisms of deterioration. Since natural ageing is relatively slow (many months to years) and asynchronous within a population of seeds, some laboratories, in an attempt to study ageing on a more convenient time scale, place seeds under conditions that lead to rapid loss of viability, i.e. **accelerated ageing**. Most often these conditions are high humidity (≥ 75% RH) and high temperature (≥ 35°C), although some laboratories have employed conditions of moderate dryness and extremely high temperatures (65°C) or extreme dryness (< 1% RH) and warm temperatures (35–45°C). High humidity–high temperature ageing has also been called 'controlled deterioration'. In contrast, conditions that cause so-called natural ageing vary widely depending on context, from fluctuating conditions of a **soil seed bank** or granary to extremely regulated conditions of a **gene bank**. As the driving forces for deterioration reactions vary with moisture availability and temperature, predictions of ageing kinetics or relative longevity among seedlots based on simulations in the laboratory are fraught with uncertainty. Because seed quality and seed longevity are inextricably linked, accelerated ageing or controlled deterioration tests have also been used to quantify other aspects of seed quality such as vigour. (CW)

Justice, O.L. and Bass, L.N. (1978) *Principles and Practices of Seed Storage*. Agriculture Handbook 506. USDA, Washington, DC, USA.

Deterioration kinetics

The pattern of **viability** loss with storage time is consistent among species and storage regimen. At first, there are no symptoms of deterioration, but signs of this become progressively apparent, leading to a more rapid decline in viability. Deterioration time courses, therefore, follow a sigmoidal curve with an initial lag period and then, though seeds germinate, there is a sharp change in seedling growth rate, followed by their tolerance to stress, and finally ability to germinate. The sigmoidal shape is usually reported for a population of seeds, where per cent seeds afflicted or average seedling size is measured, and has been developed into models that predict the per cent germination given storage time and conditions (**see: Viability equations**). Because ageing is usually measured as a property of a population of seeds, it is difficult to monitor the performance of an individual seed, or to link changes in chemical or physical attributes to seeds that have perished or are in the process of dying. Another difficult aspect of understanding the kinetics of ageing in seeds is understanding the processes that occur during the initial lag phase that is itself asymptomatic, but inevitably leads to the more rapid deterioration phase. This type of reaction kinetic is typical of autocatalytic or cooperative reactions – reactions that occur extremely slowly until accumulated products reach a threshold. Avrami kinetics, where the factor ln (total/germinated seeds) = (storage time/ϕ)^n was developed to describe cooperative reactions, and the coefficient ϕ and the exponent n describe the duration of the lag phase and the steepness of the cataclysmic phase. As the exponent decreases to 1, the Avrami equation assumes the familiar form of a first-order reaction that is mostly driven by the concentration of the substrates and products.

1. Storage temperature

Although all seeds deteriorate with time, their rate of deterioration can be controlled by the storage conditions. Reducing storage temperature reduces ageing rate almost exponentially, with a Q_{10} of approximately 2 for most species, a value that is typical when substrate diffusion is the limiting factor. According to early 'Thumb Rules,' (**Harrington's Thumb Rules**) a 6°C reduction in storage temperature doubles the life span of a seedlot. However, experimental and theoretical approaches demonstrate that the effect of temperature is not constant through all temperature ranges, and there may be a diminished effect of extremely low temperature on ageing rate.

In an Arrhenius plot (Fig. A.18 in **Arrhenius plot**), the changing effect of temperature is characterized by a curvilinear response of ln(ageing rate) with the reciprocal of temperature (in Kelvin). The slope of the Arrhenius plot gives the temperature coefficient, also known as the activation energy (Ea), which describes the amount of energy molecules need for the reaction to proceed. Reactions are faster at higher temperatures because more molecules have the critical amount of energy. Photochemical and **free radical**-mediated reactions have low activation energies because there is already sufficient energy in the system for them to proceed. Arrhenius plots of seed ageing are sometimes described as two intersecting lines with the lower temperature coefficient at the lower temperature range (Fig. A.18 in **Arrhenius plot**). The point of intersection has been attributed to a change in molecular mobility, at one time thought to result from a transition in the aqueous matrix from a fluid to a super-viscous **glass** at the glass transition temperature (Tg). More likely, curvilinear Arrhenius plots result when a series of reactions are involved, all with different temperature coefficients, or from a change in the temperature dependency of mobility within the glass, a property known as the glass fragility. There appear to be few differences among short- and long-lived seeds in their glass transition temperatures; however, there may be considerable differences among them in substrate levels or glass fragility that could affect the characteristics of temperature dependency. Consistent with models of cooperative reactions or glass fragility, the temperature dependency of ageing reactions changes as ageing progresses. The lag phase in ageing time courses, where visible signs of ageing are not apparent, is dependent on temperature; but, as the storage time increases, the reaction kinetics become increasingly independent of temperature. The finding that ageing reactions cannot be stopped by low temperatures once they have been initiated highlights the importance of initial quality and post-harvest treatments for subsequent seed performance.

Buitink, J., Leprince, O., Hemminga, M.A. and Hoekstra, F.A. (2000) Molecular mobility in the cytoplasm: an approach to describe and predict lifespan of dry germplasm. *Proceedings of the National Academy of Sciences USA* 97, 2385–2390.

Franks, F. (1985) *Biophysics and Biochemistry at Low Temperatures*. Cambridge University Press, Cambridge, UK.

Walters, C. (2004) Temperature-dependency of molecular mobility in preserved seeds. *Biophysical Journal* 86, 1–6.

2. Storage water content

The effects of water on seed deterioration kinetics are more complex than those of temperature and less well understood. Traditionally, the effects of water have been considered independently of the effects of temperature (**see: Viability equations**). According to the early 'Thumb Rules' a 1% drop in water content doubles the life span of a seed. However, the effect of changes in water content on deterioration rate is dependent on the water content range, the temperature and the seed species. Hydrolytic reactions are facilitated at high water contents and lead to increases in the concentration of sugars, free fatty acids and other substrates that promote deterioration cascades. Thus, reducing water contents to where hydrolytic reactions are not promoted has profound effects on ageing kinetics. Reducing water content also has the added benefit of reducing the mobility within aqueous matrices of cells, slowing down reactions that are largely driven by diffusion. Drying seeds beyond a certain water content tends to increase molecular mobility, perhaps because the removal of water makes the aqueous matrix less dense, or promotes motion in molecules that were constrained by hydrogen bonding. Extremely dry conditions can cause oxidation of metals or remove water that protects macromolecular surfaces from chemical or physical changes, promoting further chemical degradation of molecules. Water content affects seed ageing kinetics in a similar way as it affects molecular mobility; ageing slows with drying to a specific water content and further drying has no effect on the kinetics when seeds are stored at high temperatures and often leads to more rapid deterioration if seeds are stored at ambient or lower temperatures (see: **Ultradry storage**). Hydration regions delineate the different effects of water content on the nature and kinetics of reactions in seeds (and other dry biological systems). (See: Fig. C.26 in **Critical water contents**) The thermodynamic and motional properties of water differ within different hydration regions, leading to the concept that changes in physiological activity result from how water interacts with solutes and macromolecular surfaces. Water interactions with cellular constituents are also known as **water binding**, and reflect the degree to which water mobility is constrained by other molecules in the mixture. Water contents defining hydration regions are known as critical water contents.

Because both water content and temperature affect the kinetics of reactions through changes in molecular mobility and substrate activity, they have an interacting effect on deterioration kinetics. This is well known from studies of glass transition temperatures which increase in drier material. Similarly, critical water contents that define metabolic activity and water interactions change with temperature, decreasing as temperature increases.

While water contents that delineate hydration regions vary with temperature and species, the relative humidities corresponding to these critical water contents are fairly constant. For example, respiration becomes facilitated in seeds at a variety of water contents that correspond to 85–90% **relative humidity** (RH). For most cells and mixtures, hydrolytic reactions occur at RH of 75% or greater, which is the threshold for high humidity-accelerated ageing or controlled deterioration tests (see: **Deterioration and longevity**). The critical water content achieved by drying that results in maximum longevity, ranges from 3 to 8% water for most seed species, which consistently corresponds to about 18–25% RH among species for a given storage temperature. The relationship between water content, temperature and RH is described by a family of water **sorption isotherms**, collected at different temperatures, for a particular seed. Isotherms show the general trend of increasing water content with decreasing temperature and constant relative humidity. Because hydration regions correspond to a critical relative humidity, the value of the critical water content increases with decreasing temperature. Using isotherm relationships, one can determine critical water contents for maximum longevity for any species at a given temperature. These relationships suggest there are a variety of drying strategies possible to obtain the critical water content.

Walters, C. (ed.) (1998) Ultra-dry seed storage. *Seed Science Research* 8, 223–244.

3. Water content–temperature interactions

The interactions between water content, temperature and seed quality on seed deterioration kinetics can be broadly summarized by the generic equation describing spontaneous reactions: $J = -\Delta G \div R$, where the reaction rate (J) equals the driving force ΔG divided by the resistance R. The driving force is an energy term that is roughly governed by the concentration of substrates and products of the reaction (see: **Deterioration reactions**): it defines whether a reaction is possible. The term ΔG is useful to determine how changes in storage conditions or seed quality affect the overall likelihood of a reaction. For example, exothermic reactions (those that are driven through a loss of enthalpy) are typically favoured when the temperature is lowered. Hence, many of the physical reactions implicated in seed deterioration (i.e. crystallization events of polar lipids, sugars and water) are thermodynamically more feasible at reduced temperatures. Conversely, endothermic reactions (those that are driven by increased entropy) are more likely when the temperature is raised (i.e. protein denaturation, partitioning to hydrophobic regions). The chemical reactions implicated in seed ageing, that both release heat and increase entropy, are affected by temperature through changes in the equilibrium coefficients of product and reactant concentrations. Like temperature, water content affects the driving force of reactions by affecting the concentration of substrates and the likelihood of physical changes such as crystallization events (promoted at lower water contents) and disorder (promoted at higher water contents). Critical water contents or transition temperatures mark a change in the likelihood of reactions (see: **Arrhenius plot**). Critical water contents are more appropriately expressed as critical relative humidities because RH is a thermodynamic expression for the concentration of water. RHs that are critical for seed deterioration mechanisms are 20–25% and 75–85%. Ageing rates increase as water content decreases below the critical RH of 20–25%, suggesting that reactions involved in deterioration at this water level are exothermic and so will also be promoted by decreasing temperatures. The unavoidable conclusion is that excessive drying or cooling will eventually promote deterioration, and that seed life spans are therefore finite. The exact temperature and moisture content that provide maximum seed longevity remain to be defined, though it is likely less than −18°C and between 5 and 20% water, depending on species.

The resistance factor (R) is also important in defining the rate at which seed deterioration occurs. Although certain cellular reactions are thermodynamically favoured during deterioration, they may not occur because there are physical or thermodynamic barriers to prevent them. The temperature coefficient or activation energy (Ea) is a thermodynamic barrier that defines how much energy (or mobility) molecules must have to allow a spontaneous reaction to occur (see: **Arrhenius plot**). Critical seedcoat coverings that block moisture or oxygen diffusion, or membranes that sequester substrates, are examples of physical barriers. In addition to their effects on the driving force of reactions (ΔG), temperature and water content make a significant contribution to the resistance factor (R) because they directly control molecular mobility. Development of this idea of a resistance factor has contributed to our current understanding of the role of molecular glasses in slowing seed deterioration. Glasses form when complex mixtures are concentrated, cured or cooled and the intermolecular associations cause the material to become hard. The change in mechanical properties is relatively abrupt and occurs at the glass transition temperature (Tg). Further cooling of the glass leads to further constraints of molecular motion to the extent that chemical and physical reactions cannot be measured on a practical time scale. The temperature at which molecular motion is hypothetically zero is called the Kauzmann Temperature (T_K) and the ratio of Tg and T_K defines the temperature coefficient for molecular mobility (the glass fragility). (**See:** Fig. G.20 in **Glass**) In mixtures, molecular mobility is never zero, and so ageing is inevitable. Much of the research on glass behaviour in seeds has identified the relationship between water content and Tg (**see: Glass; Phase**), which exhibits little difference between seeds that die immediately upon drying, relatively rapidly with storage time, and very slowly. This suggests that the glassy state, per se, does not protect cells from deterioration, but rather that molecular mobility within the glassy state is the operative barrier to deteriorative reactions. Seeds that are stored wet under cryogenic conditions (cryostorage, cryopreservation) face the risk of lethal ice formation (since crystallization reactions are promoted at low temperatures). Cooling and warming rapidly, adjusting water content, or adding solutes protect cells from ice formation on the principle that molecular mobility can be sufficiently constrained before water nucleates and crystals grow. Achieving a vitrified state (i.e. glass) is a necessary first step in cryoprotection. Materials must be stored at temperatures well below Tg (preferably below T_K) to limit recrystallization and maintain viability. As with other chemical and physical reactions that lead to cell death, the kinetics of ice formation under cryopreservation have not been quantified, making it difficult to predict lifespans. The viability of cryopreserved cells from diverse life forms following up to 40 years of storage provides reasonable assurances that timespans for these reactions is indeed long. (**See:** Fig. C.27 in **Cryostorage**) (CW)

Buitink, J., Claessens, M.M.W.E., Hemminga, M.A. and Hoekstra, F.A. (1998) Influence of water content and temperature on molecular mobility and intracellular glasses in seeds and pollen. *Plant Physiology* 118, 531–541.

Walters, C. (2004) Principles for preserving germplasm in genebanks. In: Guerrant, E., Havens, K. and Maunder, M. (eds) *Ex Situ Plant Conservation: Supporting Species Survival in the Wild.* Island Press, Covelo, CA, USA, pp. 113–138.

Deterioration prevention

Seeds that acquire the ability to survive the initial stress of **desiccation** naturally possess the ability to survive in the desiccated state (**see: Desiccation damage**). Thus, most mechanisms for desiccation tolerance are also implicated for seed longevity (**see:** entries under **Desiccation tolerance**). Usually, putative protectants of life in a desiccated state are ubiquitous among life forms, providing an evolutionary context for the challenges of prolonging life and the forces that oppose it. Protective mechanisms are broadly divided into strategies, reducing the driving force of ageing reactions (ΔG) or obstructing the inevitable progress of these reactions (R). (**See: Deterioration kinetics**) The concentration of substrates for ageing reactions can be reduced directly or by competing reactants that are less harmful to the cell. For example, an effective means to slow the **Maillard reaction** is to reduce the amount of **reducing sugars** through polymerization of sugars into **starch** or **raffinose-series oligosaccharides**, a procedure that occurs during latter stages of **embryogenesis**. This stage of embryogenesis is also marked by dedifferentiation of organelles leading to lowered metabolic capacity. The resulting quiescence lowers the concentration of high energy intermediates from respiration, thereby reducing reactive oxygen species. Antioxidant enzymes such as catalase, peroxidase and **superoxide dismutase** catalyse the removal of **reactive oxygen species** from cells prior to and following storage. During storage, **antioxidants** such as tocopherol defuse reactive oxygen species and free radicals by quenching. Tocopherol is oxidized by the free radicals, but does not perpetuate the oxidation cascade because the unpaired electron is delocalized within the ring structure of the molecule. When oxidized tocopherol is reduced by ascorbate, it is available again to scavenge.

Protection of macromolecular structures by chaperone-like molecules also reduces the amount of substrates that are susceptible to deteriorative reactions. In organisms other than plants, the most widely researched protectants are a class of topoisomerases, consisting of small, acid-soluble proteins that bind to DNA, protecting the genome. These proteins have mostly been studied in bacteria, but the finding of mRNPs (messenger ribonucleoproteins) in slowly dried, desiccation-tolerant mosses present the further possibility that stabilization and sequestration of transcripts is a prerequisite for survival of plants (including seeds) in the dry state. Research on molecular-stabilizers in plants has mostly focused on solutes that accumulate when cells become tolerant of stress. Molecules such as trehalose (not in seeds), sucrose and perhaps LEA (**late embryogenesis abundant**) proteins stabilize some proteins and polar lipids in model solutions, but protection through specific interactions (**see: Water replacement hypothesis**) in the cells of seeds has not been demonstrated.

Protective barriers that slow reactions can be potent inhibitors of ageing reactions. For example, hard seedcoats limit the diffusion of water and air making seeds with these impermeable outer coverings more long-lived, especially under fluctuating storage conditions (**see: Ancient seeds**). Drying and cooling naturally slow down reactions by reducing molecular mobility (**see: Deterioration kinetics**), but sugars may have an additional protective effect by virtue of their strong glass-forming capabilities. (CW)

Deterioration reactions

All cytoplasmic constituents of the cell degrade during seed storage and the degree to which each affects seed **viability** depends on the importance of the particular constituent to cell function, the kinetics of its degradation, the potential for cascading reactions (cooperativity or autocatalytic), and the presence of protective and repair mechanisms. There are many potential physical and chemical reactions that contribute to seed ageing by adversely affecting molecular function. The diversity of the by-products produced, however, makes it difficult to directly measure the progress of deteriorative reactions.

(**See: Desiccation damage**; entries under **Desiccation tolerance**)

1. Physical reactions involved in deterioration

Physical reactions alter the structure of macromolecules or the distribution of molecules within cells. These changes are driven by temperature and moisture availability, and hence their occurrence is mostly described in the equilibrium state using **phase diagrams**. The occurrence must also be evaluated by the time to reach the equilibrium state, which can be prolonged for large molecules in complex mixtures. Physical reactions are often reversed when seeds are removed from storage, and so their importance to seed deterioration is frequently overlooked unless there are cooperative chemical changes that can be monitored.

Phase changes of the polar lipids of **membranes** are the most recognized physical change within seed cells during storage. Mechanisms are similar to those described for **desiccation damage** and can result in the loss of membrane function over time either through fusion among organelles, inactivation of membrane-associated proteins, or change in membrane selective-permeability.

Protein denaturation leads to loss of enzyme function and is driven by high temperature and ionic concentration, but slowed by dry conditions. During seed storage, many enzymes lose activity and proteins become less extractable. However, direct links between enzyme loss and protein denaturation are logistically difficult to demonstrate.

Sugar molecules purportedly protect polar lipids and proteins from structural changes, and crystallization of sugars can remove the protective effect. Thus, slow crystallization of sugars, especially sucrose, has been implicated in ageing, and molecules that retard crystallization, for example **raffinose-series** sugars, may promote seed **longevity**. This suggestion is based on behaviours of sugars in relatively pure solutions where crystallization is promoted. Crystallization of sugars in the complex mixtures of the cytoplasm is less likely and remains to be demonstrated during seed deterioration. **Starch**es form gels when exposed to high humidity and high temperature conditions reminiscent of **accelerated ageing**. The physical structure of starch grains affects the kinetics of starch depolymerization and the nature of starch–protein or starch–lipid complexes.

Water availability and temperature also affect hydrophilic and hydrophobic interactions which, in turn, affect the partitioning of **amphiphilic** molecules into aqueous and lipid-rich domains. Diffusion of toxic compounds to susceptible regions of the cell, or protective molecules to regions where they are no longer effective, can have profound effects on the nature and kinetics of deteriorative reactions.

The slow crystallization of water molecules during cryostorage of hydrated seeds will cause progressive deterioration via freezing damage. This form of ageing is expected in **recalcitrant** seeds that remain hydrated during **cryostorage**, but remains to be shown since cryopreservation of recalcitrant seeds is a recent endeavour.

2. Chemical reactions involved in deterioration

Chemical reactions that occur during seed storage can be classified as hydrolytic, oxidative and peroxidative. These reactions are promoted at high, intermediate and low relative humidities, respectively, but the balance of different types of reactions changes as substrate concentrations change within the cell. Most of the deteriorative reactions lead to fragmentation of large molecules into smaller ones.

(a) *Hydrolytic reactions.* These are known to occur during seed storage, especially at higher humidity, and can result in rancidity (i.e. accumulation of free **fatty acids**) or a cascade of oxidizing reactions that lead to further scission of polymers, or to cross-links within polymers. Examples of hydrolytic reactions include the de-esterification of lipids into free fatty acids and glycerol, the depolymerization of starches into simpler sugars, and the initial step of the **Maillard reaction** (the later steps are oxidizing), all of which require the presence of water molecules.

(b) *Oxidative reactions.* Degradation occurs through oxidation–reduction reactions that remove electrons from carbon or through the direct insertion of oxygen into the molecule (**peroxidation**). Oxidative degradation can occur through non-enzymatic mechanisms, catalysed by oxygen, heat, light and metals. However, in cells, oxidation–reduction reactions are usually catalysed by enzymes, making use of oxidizing agents such as **NAD**, metal ions or sulphydryl (-SH) groups, to retrieve energy from stored reserves (i.e. β-oxidation of fatty acids or glycolysis) or control the redox potential. The extent to which these catabolic enzymes participate in ageing reactions in seeds remains unclear, though consumption of reserves is a long-standing hypothesis used to explain the loss of viability during storage. Incomplete or imbalanced metabolism, arising from changes in enzyme activity, shifts in substrate concentrations, or uncoupled phosphorylation steps of respiration (electron transport without **ATP** production) is promoted when cells are water-stressed (water potentials becomes less than -2 MPa). This results in the accumulation of partially reduced oxygen or other high-energy intermediates. These by-products of oxidation reactions, also known as **reactive oxygen species** (ROS), are **free radicals** or readily lead to the production of free radicals. Seeking an electron for the unpaired one, free radicals are extremely electrophilic and attack electron-dense regions present in all organic polymers, creating more free radicals and further degradation in an autocatalytic reaction cascade called auto-oxidation. DNA is particularly sensitive to free radical oxidation of its bases and sugar phosphate backbone which, in turn, leads to single-stranded nicks; these, in turn, often lead to increased oxidative damage of bases or to double-stranded breaks.

(c) *Peroxidation reactions.* The oxygen molecule itself is extremely electrophilic (it sometimes exists as a free radical), and can form hydroperoxides (R-O-O-H) by directly combining in a peroxidative reaction with electron-dense moieties, such as unsaturated substrates, e.g. fatty acids. Because the peroxide bond is weak (it is a single bond rather than the double bond of molecular oxygen), it is easily broken by reduced agents to form the strongest known oxidant, the hydroxyl radical, initiating the auto-oxidation cascade (**see:** Fig. P.7 in **Peroxidation**). Peroxidative reactions are catalysed by metals, heat and light. The numerous mechanisms by which macromolecules can be degraded through oxidation reactions result in a myriad of by-products. By-products of lipid peroxidation depend on the fatty acid and catalyst, but may include pentane, ethane, ethylene, aldehydes and ketones. **Malondialdehyde** and hydroxy-trans-nonenal are by-products of lipid oxidation that are known to damage DNA and cause cross-links in proteins. Oxygenation of fatty acids and loss of double bonds cause increased membrane leakiness and decreased fluidity. (CW)

Buitink, J., Hoekstra, F.A. and Leprince, O. (2002) Biochemistry and biophysics of tolerance systems. In: Black, M. and Pritchard, H.W. (eds) *Desiccation and Survival in Plants: Drying Without Dying*. CAB International, Wallingford, UK, pp. 293–318.
Halliwell, B. and Gutteridge, J.M. (1999) *Free Radicals in Biology and Medicine*. Oxford University Press, Oxford, UK.
Smith, M.T. and Berjak, P. (1995) Deteriorative changes associated with the loss of viability of stored desiccation-tolerant and desiccation-sensitive seeds. In: Kigel, J. and Galili, G. (eds) *Seed Development and Germination*. Marcel Dekker, New York, USA, pp. 701–746.
Walters, C. (1998) Understanding the mechanism and kinetics of seed ageing. *Seed Science Research* 8, 223–244.
Walters, C., Farrant, J.M., Pammenter, N.W. and Berjak, P. (2002) Desiccation stress and damage. In: Black, M. and Pritchard, H.W. (eds) *Desiccation and Survival in Plants: Drying Without Dying*. CAB International, Wallingford, UK, pp. 263–292.

Determinate flowering

The type of development in which a plant flowers and ripens all of its seeds at approximately the same time, characterized by an inflorescence in which the terminal flower opens first, thereby arresting the elongation of the floral axis (Fig. D.8). Some forages have definite flowering periods, whilst others flower continuously through the year. Most species with a determinate growth habit have a distinct flowering period and produce only one seed crop a year. (**See: Grasses and legumes – forages**)

Indeterminate growth, by contrast, is characterized by flowering over a long period, where developing and ripe seeds, and some blossoms and vegetative shoots may all be present on the plant at the same time. (**See: Production for Sowing, IV. Harvesting**) Species that have an indeterminate growth habit may also have a distinct flowering period but commonly continue to produce flowers over a longer period and may produce several seed crops annually. In indeterminate flowering, the terminal flowers on racemes tend to be the last to open, so that the floral axis may be prolonged indefinitely by a terminal bud.

In **canola**, for example, the inflorescence is formed from multiple indeterminate racemes, and pollination, seed filling and maturation continues on lateral racemes as long as adequate photosynthate is available or until frost interrupts development. Cultivated **tomato** varieties are divided into two types, based on growth habit: indeterminate types are normally trained to a single stem (side shoots or suckers removed), and suspended from high wires in the greenhouse; determinate types are bush types, typically used in field cultivation. There are determinate and indeterminate types of stem growth habit and floral initiation in soybean, which are adapted to long and short growing seasons, respectively.

(PH)

Fig. D.8. A determinate and an indeterminate flower on an avocado plant, a less common phenomenon where a single plant produces both flower types. (From: http://ucavo.ucr.edu/AvocadoWebSite)

Development of embryos – cereals

1. Structure–function overview

The embryo is that part of the seed which germinates and grows into the vegetative plant. In cereal grains, embryos (germs) often accumulate nutritious metabolites such as oil and vitamins (**see: Cereals – composition and nutritional quality**). Structurally distinct tissues are formed in the mature embryo of cereals: these include embryo **axis**, **scutellum** and **suspensor**. Shoot and root structures are initiated from the embryo axis and further develop after germination, as the seedling becomes established. The development of cereal embryos differs from that of typical dicot embryos, which form a symmetrical structure with two cotyledons.

A general description of embryo formation is found under the heading **Embryogenesis**, but details pertaining to cereals are presented here. (**See also: Development of embryos – dicots; Embryo**)

Based on morphological criteria, development of the embryo in cereals can be divided into several stages, as illustrated for rice:

(a) *Proembryo stage*: the initial stage of embryo development. Proliferation of the terminal cell leads to the formation of a globular-shaped embryo which then elongates to a club-like shape showing radial symmetry (Fig. D.9.a, A–C).

(b) *Coleoptile stage*: the shape of the developing embryo becomes bilaterally-symmetrical through differentiation of the scutellum and **coleoptile**. This stage also marks the establishment of the embryo axis (Fig. D.9.a, D, E).

(c) *Leaf stage*: shoot and root meristems become visible. Several leaf primordia differentiate from the shoot meristem. Seed storage products, such as **oil bodies**, are deposited (Fig. D.9.a, F, G).

(d) *Maturation stage*: embryonic cells cease dividing, the embryo is complete and desiccates to become **quiescent** (Fig. D.9.a, H. Fig. D.9.b). Some cereal grains also develop **dormancy**.

These events are illustrated for embryogenesis in wheat as a series of micrographs in Fig. D.10.

The duration of embryogenesis varies among different species. For instance, under normal growing conditions a typical **rice** embryo completes development to the final three-leaf stage in 12 days after fertilization, whereas a **maize** or **wheat** embryo requires more than 20 days to reach an equivalent stage. Embryo development of monocot plants, specifically those of the grass family represented by cereal crops, significantly differs from that of dicot plants after the proembryo stage. A typical dicot embryo, e.g. of a legume, forms a symmetric structure with two cotyledons and the

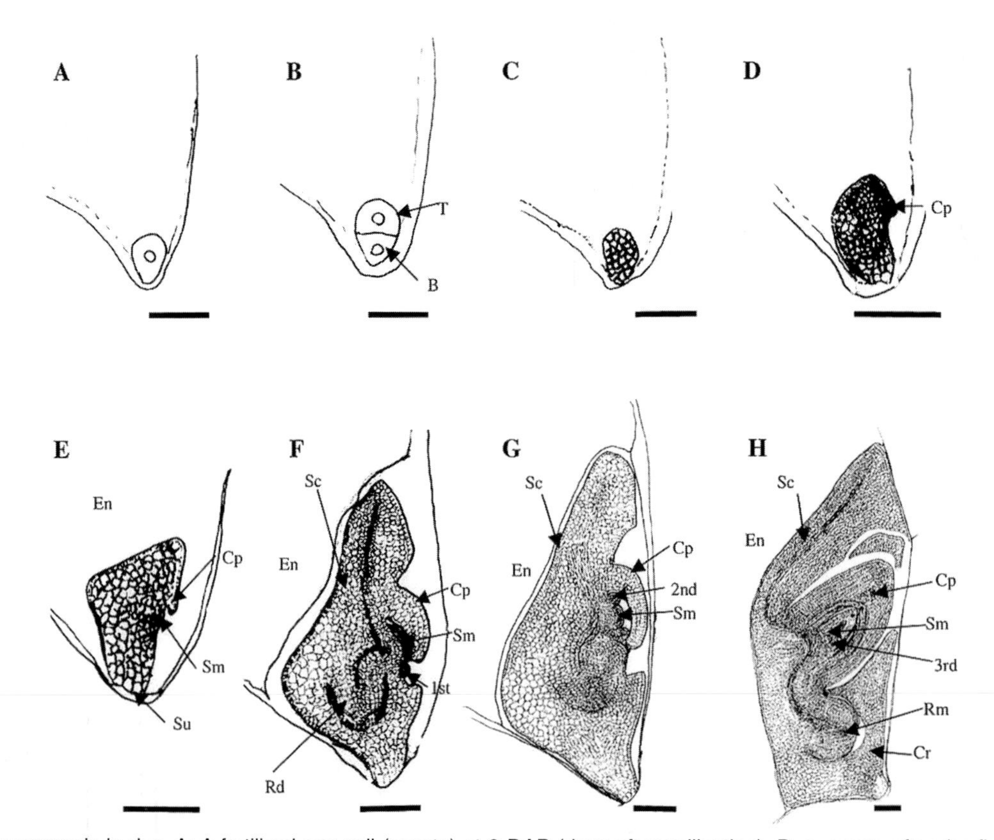

Fig. D.9.a. Embryogenesis in rice. A: A fertilized egg cell (zygote) at 0 DAP (days after pollination); B: a zygote after the first cell division (0 DAP); C: a multicellular zygote at the proembryo stage (2 DAP); D: at the early coleoptile stage (3–4 DAP); E: at the late coleoptile stage; F: at the 1st leaf stage (5–6 DAP); G: at the 2nd leaf stage (7 DAP); H: a mature embryo.
T: terminal cell; B: basal cell; Cp: coleoptile; Sm: shoot meristem; Su: suspensor; Sc: scutellum; 1st , 2nd, 3rd: 1st, 2nd and 3rd leaf; Rd: radicle; Rm: root meristem; Cr: coleorhiza; En: endosperm. Bar = 30 μm (A, B) and 100 μm (C–H).

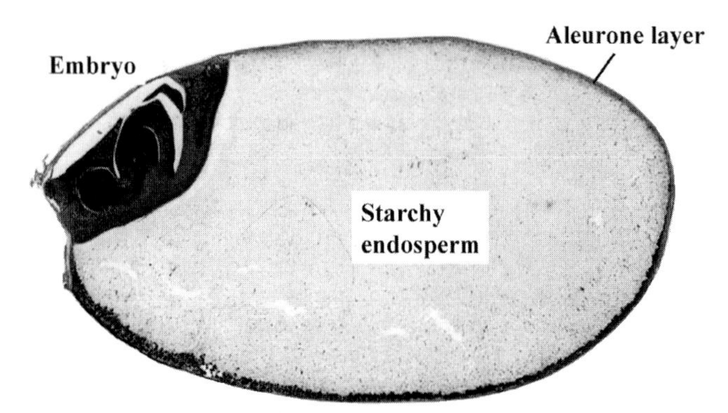

Fig. D.9.b. A mature rice grain (longitudinal section) showing the position of the embryo and starchy endosperm, surrounded by the aleurone layer, visible as darker stained cells.
Drawings and photograph courtesy of Dr Nobuhiro Nagasawa.

primordia with true leaves (**see: Development of embryos – dicots**).

Various **hormones** play a role in embryogenesis. Auxin is involved in embryo axis and scutellum formation. Applications of **auxin** polar transport inhibitors result in multiplied embryonic axes and supernumerary scutella. **Abscisic acid** (ABA) and **gibberellin** are involved in dormancy and germination. The lack of ABA or disruption of ABA **signal transduction** pathways results in a **viviparous** (premature germination) phenotype.

2. Embryonic mutants and genes

In order to dissect the pathways that control embryogenesis, genetic studies have been performed by investigating **mutants** that show aberrant developmental patterns. Among these, mutants that show arrested embryo development are of particular interest because they reveal genes that are required for the development of specific embryonic stages. In maize, each embryonic stage can be blocked by mutations in more than one gene, indicating the complexity of the genetic pathways underlying each stage of embryo development. Several mutants lacking the shoot or radicle have been isolated from maize and rice, suggesting that the specification of shoot identity is genetically distinct from that of radicle identity. Interestingly, radicle-less mutants are not lethal, and still produce adventitious roots after germination. This shows that development of adventitious roots does not fully require the genetic pathways that create the embryonic radicle, i.e. the primary root.

In monocots, a few regulatory genes controlling embryogenesis have been identified and their functions are currently under investigation. In dicot plants, especially in the model species *Arabidopsis thaliana*, functions of several

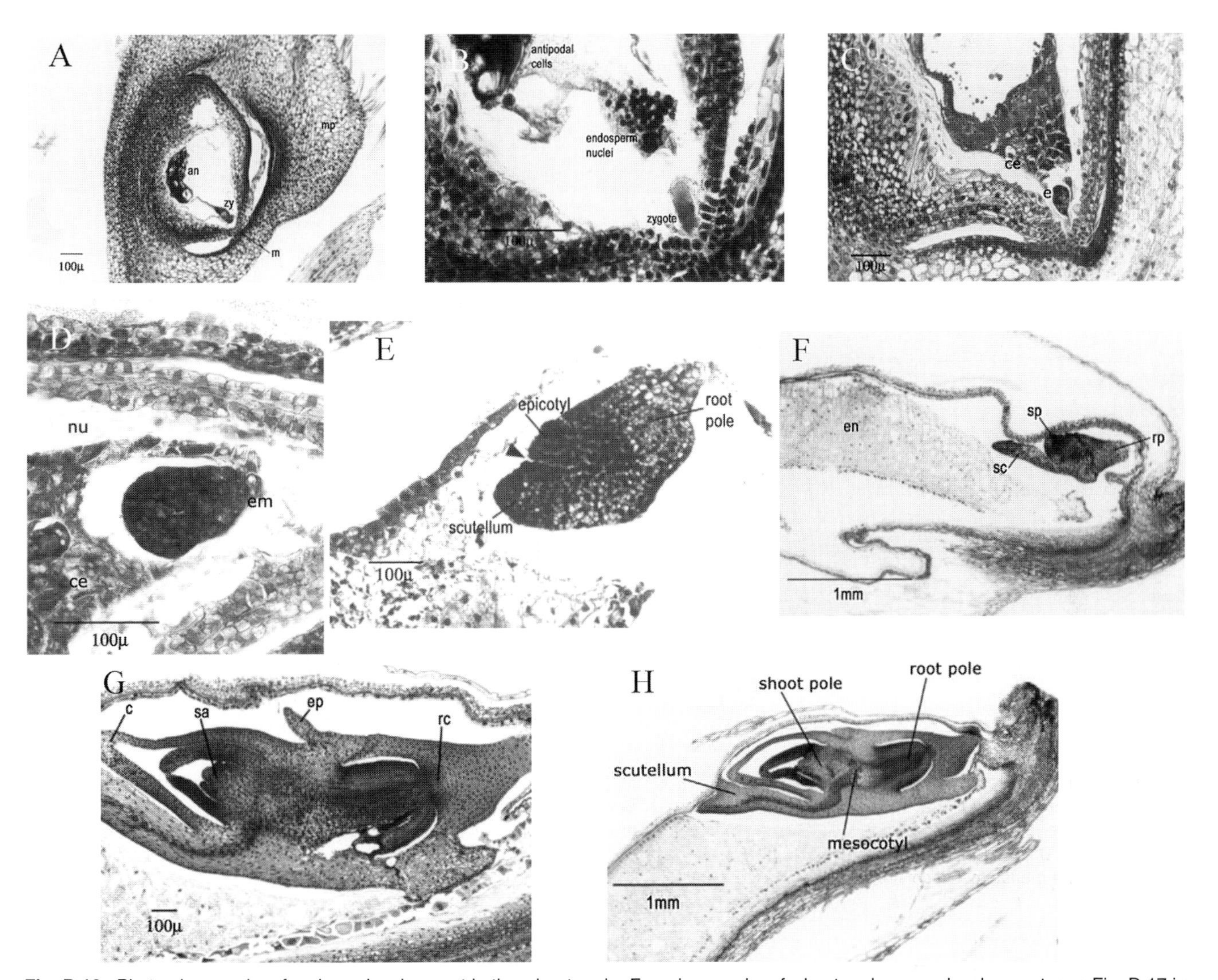

Fig. D.10. Photomicrographs of embryo development in the wheat grain. For micrographs of wheat endosperm development, **see** Fig. D.17 in **Development of endosperms – cereals**. All micrographs are longitudinal sections of developing grains.

(A) Embryo sac within hours of fertilization. The diploid zygote (zy) and antipodal cells (an) are visible, but the triploid endosperm nucleus lying close to the zygote is not discernible. m, Micropyle through which the pollen tube entered into the embryo sac to effect fertilization; mp, maternal pericarp.

(B) 1 DAF (day after flowering, fertilization). The undivided zygote remains at the micropylar end of the embryo sac. The endosperm nucleus has migrated towards the centre of the embryo sac and has undergone several nuclear divisions without cell wall formation (free nuclear stage). The antipodal cells remain undivided.

(C) 3 DAF. Cellular endosperm (ce) has formed. The embryo (e) has undergone one cell division.

(D) 7 DAF. Embryo (em) at the globular stage. The root pole has begun to differentiate. nu, Nucellus; ce, cellular endosperm.

(E) 11 DAF. Patterning of cells at the root pole is established, and other regions of the embryo have become distinct. The cleft (arrowed) on the dorsal side of the embryo defines the point of separation of the epicotyl (primordial shoot which will form the shoot apex) from the scutellum.

(F) 15 DAF. The scutellum (sc), embryonic root (rp) and shoot (sp) primordia have differentiated. The region of the endosperm (en) closest to the scutellum (sc) has become depleted as the embryo grows.

(G) 20 DAF. The shoot apex (sa) is associated with leaf primordia and is enclosed by the coleoptile (c). The root is composed of the root cap (rc) and files of differentiated cells which form the vascular tissue and the cortex. It is enclosed by a coleorhiza. A small flap of tissue, the epiblast (ep), is visible; this tissue has no known function.

(H) ~22 DAF. A near-mature embryo prior to maturation drying.

All pictures taken from: www.wheatbp.net. *Wheat: The Big Picture*. Grain growth 1–4, 4–10, 11–16, 17–21, 21–30 days.

such regulators have been characterized. For instance, *LEAFY COTYLEDON 1* (*LEC1*), which encodes a HAP3-like transcription factor, is active in early embryogenesis, specifying cotyledon identity. Ectopic expression of *LEC1* in vegetative leaves leads to the formation of somatic embryos, indicating the fundamental role of *LEC1* in inducing embryonic identity even in differentiated cells. On the other hand, embryonic identity is repressed in vegetative tissues through the activity of genes including *PICKLE* (*PKL*), which encodes a putative chromatin-remodelling factor. As for genes that control the embryonic body pattern, mutations in *MONOPTEROS* result in the lack of the basal structure including the root meristem. *MONOPTEROS* codes for an auxin response transcription factor, providing a link between auxin and embryogensis. On the contrary, the *TOPLESS* (*TPL*) gene is required for specifying **shoot meristem** identity. The lack of *TPL* activity results in homeotic transformation of shoot to root tissues. Since homologues of these *Arabidopsis* genes are found in the near-completely sequenced rice genome, it is possible that similar genetic pathways are also present in monocots.

Among the regulatory genes that function as **transcription** factors, the homeobox gene type are known to control various aspects of development in eukaryotes. In plants, the *Knotted 1* homeobox gene in maize is required for activity and maintenance of meristematic cells in shoots. In rice, several other homeobox genes such as *OSH1* are expressed in distinct tissues during embryogenesis. Reverse genetics of these genes identifying loss-of-function mutants should prove to be illuminating. (O-AO, HS)

Clark, J.K. and Sheridan, W.F. (1991) Isolation and characterization of 51 embryo-specific mutations of maize. *Plant Cell* 3, 935–951.

Hong, S.-H., Aoki, T., Kitano, H., Satoh, H. and Nagato, Y. (1995) Phenotypic diversity of 188 rice embryonic mutants. *Developmental Genetics* 16, 298–310.

Development of embryos – dicots

During **embryogenesis** the basic future body plan, such as shoot and root meristems, is defined, in addition to cell types such as the ephemeral suspensor and provascular tissue components. Later during embryogenesis, maturation occurs and the embryo is generally made ready for desiccation. (**See: Development of seeds – an overview**) In recent years both embryo and endosperm development have been illuminated by genetic analyses, mainly in *Arabidopsis* (**see: Development of endosperms – dicots**).

The *Arabidopsis* embryo, which is viewed as a model of dicot embryo development, passes through a series of morphological stages, as indicated in Figs D.11–D.13. Asymmetric division of the **zygote** creates two cells, an apical and a basal cell. The two-celled embryo undergoes a number of divisions that lead to the formation of the **suspensor** and root precursor cells, as well as a proembryo, consisting of two tiers each of four cells. The suspensor consisting of six to eight cells, is thought to act as a conduit for the passage of nutrients from the endosperm to the growing embryo. The uppermost cell of the suspensor eventually produces the root meristem (Fig. D.11). The embryo passes through several distinct stages, from the globular to the heart and torpedo stage, when the tissue regions become defined, and finally to the mature (cotyledonary stage) embryo (Fig. D.12). As the embryo matures, the suspensor disintegrates due to **programmed cell death**. Failure of normal signals from the suspensor results in abnormal embryo development, and the *sus* mutation in *Arabidopsis* results in the formation of **callus** on the suspensor, its swelling and its failure to degenerate.

Development of the globular, torpedo and cotyledonary stages within the seed, accompanied by diminution of the endosperm, are shown as a series of micrographs in Fig. D.13.

A large number of **mutants** has been described which help in our understanding of embryo development. During embryogenesis, through a series of regular cell divisions, the embryo establishes an outer protoderm layer, and establishes two 'tiers' of inner cells with distinct developmental fates. The apical tier produces the cotyledons, and shoot meristem, while the lower tier produces the hypocotyl and root meristem (Figs D.11 and D.12). Several mutants have been isolated that affect the pattern of embryo development, and among these only the *GNOM/EMB30* affect the apical/basal polarity of the embryo. In the *gnom* zygote the first cell division appears to be symmetrical. Mutation in the *FACKEL* gene specifically reduces the hypocotyl resulting in a seedling where the cotyledon appears to be linked to the roots. Seedlings homozygous for the *monopteros* mutation lack hypocotyls and roots.

As the embryo grows to the heart stage, the cotyledons, hypocotyl, embryonic root and root meristems are delineated. *MONOPTEROS* encodes a protein similar to *AUXIN RESPONSE FACTOR1*, a **transcription** factor, and is implicated in polar **auxin** flow. Thus, polar auxin flow might be important for delineating the body plan. Other genes important for regulating embryogenesis include *KEULE, KNOLLE, LEC1* and *PICKLE*. Some genes that have been identified as being important in the regulation of *Arabidopsis* embryo, by determining the effects of mutations to them, are outlined in Table D.3. For a more detailed list of genes see www.seedgenes.org. This website contains information on a large number of genes that regulate the various stages of *Arabidopsis* seed development. Information is available on the chromosome locus, gene class and resultant phenotype, and is linked to the TAIR and TIGR databases. From these can be obtained information on regulatory genes and their products, the tissue(s) affected by mutations, and the resultant phenotypes. An example is given below of some of the

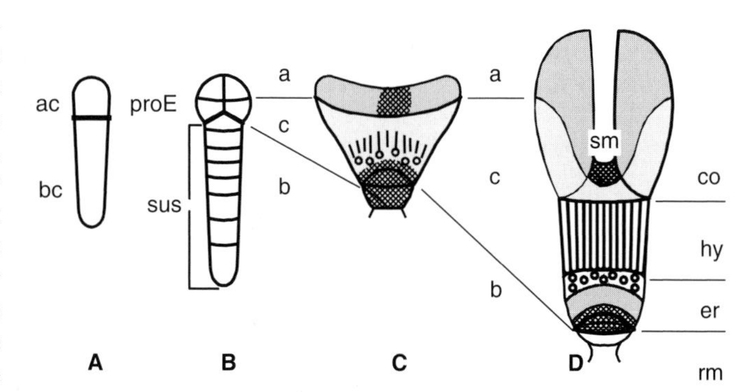

Fig. D.11. Development of the *Arabidopsis* embryo. (A) Asymmetric division of the zygote. ac, Apical cell; bc, basal cell. (B) Eight-cell stage. The proembryo (proE) consists of two tiers each of four cells (a,c). sus, Suspensor; a, apical; c, central; b, basal region. (C) Heart stage embryo. (D) Torpedo stage embryo. Primordia of seedling structures : co, cotyledons; hy, hypocotyl; er, embryonic root; rm, root meristem; sm, shoot meristem.

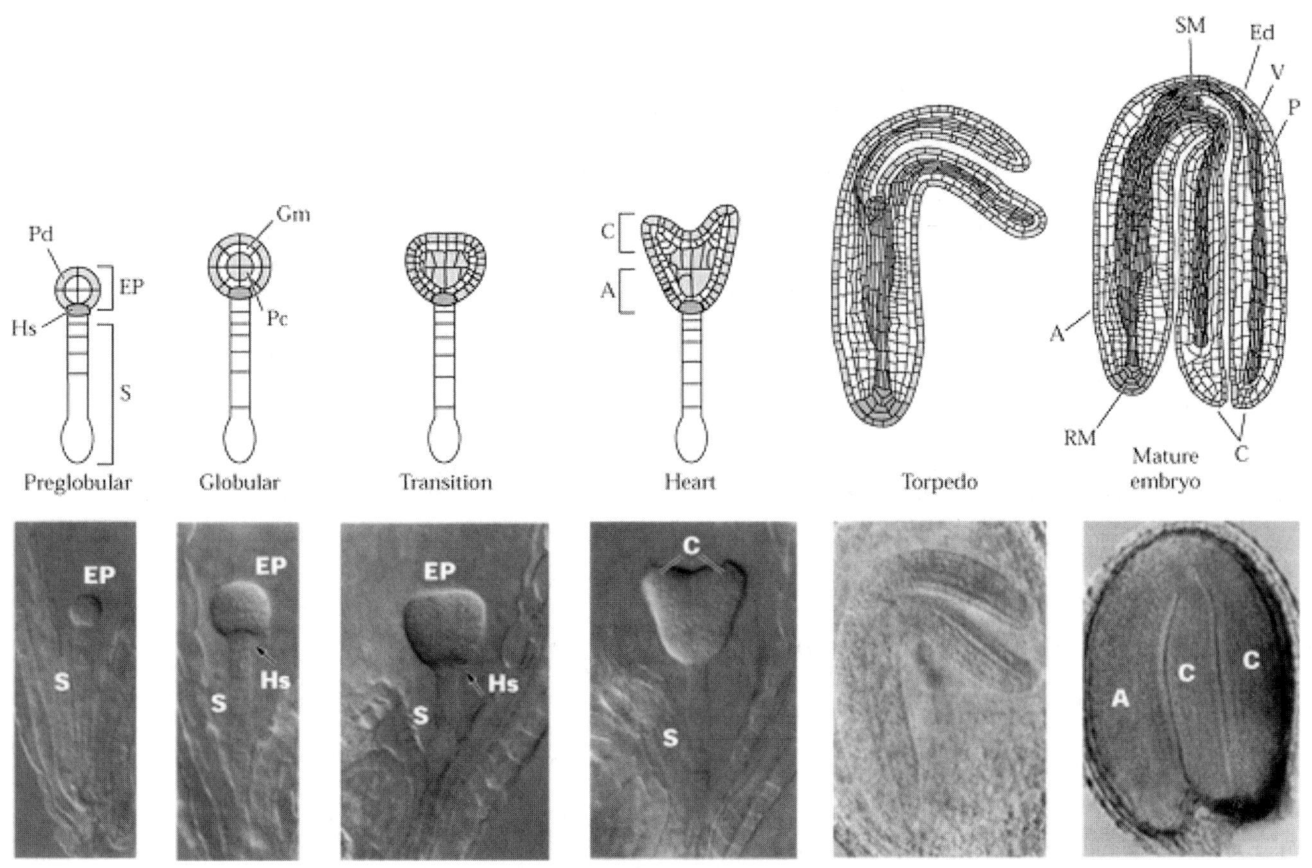

Fig. D.12. Stages of development of the embryo of *Arabidopsis*, from stage B onwards of Fig. D.11. The apical cell (stage A in Fig. D.11) divides to become the proembryo (embryo proper, EP) at the preglobular stage, attached to the suspensor (S). The protoderm (Pd) develops into the epidermal layer (Ed) in the mature embryo. The ground meristem (Gm) in the globular stage develops into the storage parenchyma cells (P) in which the reserves are deposited and the procambium (Pc) differentiates into the vascular conducting tissue (V). The hypophesis (Hs) develops into the root and shoot meristems (RM, SM). A, axis; C, cotyledons. The testa and endosperm surrounding the embryo are not shown.
From: Bewley, J.D., Hempel, F.D., McCormick, S. and Zambryski, P. (2000) Reproductive development. In: Buchanan, B.B., Gruissem, W. and Jones, R.L. (eds) *Biochemistry and Molecular Biology of Plants*. American Society of Plant Physiologists, Rockville, MD, USA, pp. 988–1043. Reproduced with permission.

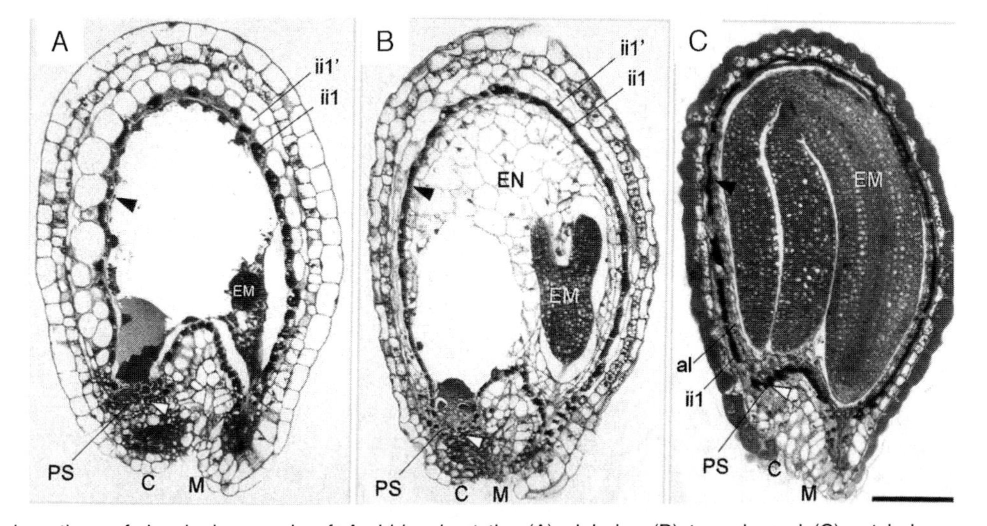

Fig. D.13. Stained sections of developing seeds of *Arabidopsis* at the (A) globular, (B) torpedo and (C) cotyledonary, mature stages of development. PS, Pigmented strand and C, chalazal region indicate where the developing seed is attached via the funiculus to the pod. M, Micropylar region. EN, Endosperm, which is resorbed as the embryo (EM) grows into it. ii1 and ii1' are regions of the inner integument of the seedcoat. al, Aleurone layer, a one cell layer which is all that remains of the endosperm at maturity. Bar: 80 μm. From Debeaujon, I., Nesi, N., Perez, P., Devec, M., Grandjean, O., Caboche, M. and Lepiniec, L. (2003) Proanthocyanidin-accumulating cells in *Arabidopsis* testa: Regulation of differentiation and role in seed development. *The Plant Cell* 15, 2514–2531. Reproduced with permission.

Table D.3. Some mutants that affect embryo development in *Arabidopsis*. For more details, type in the name of the mutant in www.seedgenes.org

Mutant	Phenotype of mutant	Molecular function of wild-type gene
Apical-basal patterning		
gurke	lacks shoot	unknown
fackel	lacks hypocotyl	sterol C-14 reductase
monopteros	lacks hypocotyl and root	transcription factor (IAA24) similar to auxin-responsive factor 1
gnom	lacks cotyledons and root	unknown
Radial patterning		
knolle	cell division inhibited, multinucleate cells, poor cell wall formation	cell division-specific syntaxin, which localizes cell plate
keule	as *knolle*	as *knolle*
Shoot meristem		
stm	no shoot meristem	class 1 knotted-like homeodomain protein
Embryo development		
pickle	embryo development traits expressed after germination	chromatin remodelling factor; responsive to gibberellin
lec1	desiccation intolerant; precocious germination; trichomes (hairs) on cotyledons	transcription factor activator[a]
lec2	similar to *lec 1*	B3 domain (DNA motif)[a] transcription factor
fus3	abnormal cotyledons; abnormal late embryo development	transcriptional activator[a]
aba1	premature germination of mature seed; fails to synthesize abscisic acid (ABA)	zeaxanthin epoxidase, an enzyme involved in ABA synthesis[b]
abi3	unresponsive to ABA; precocious germination	transcription factor[a]

[a]**See: Dormancy genes**; [b]**see: Abscisic acid.**
Some examples of the phenotypes of the mutants are shown in Fig. D.14.

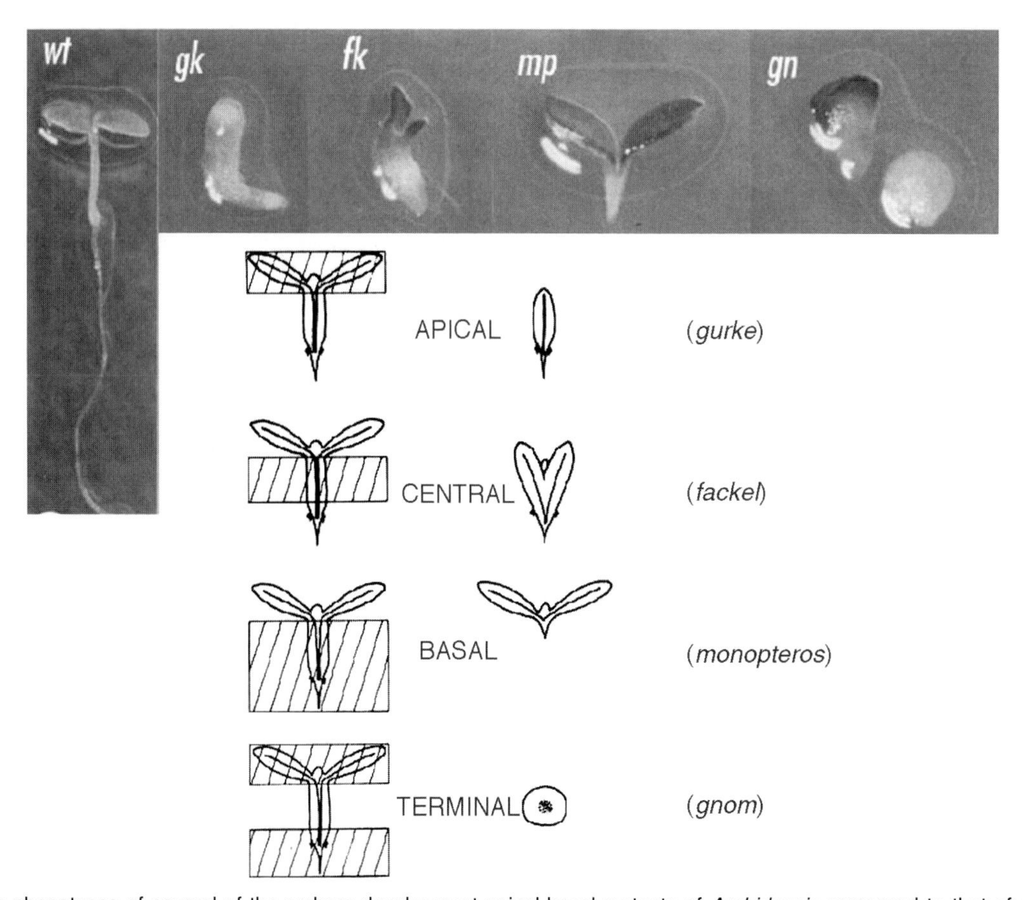

Fig. D.14. The phenotypes of several of the embryo development apical-basal mutants of *Arabidopsis* compared to that of a wild-type (wt) embryo. The sites of the mutations are shaded in the diagrams of the embryo. From: Mayer, U., Ruiz, R.T.A., Berleth, T., Miséra, S. and Jürgens, G. (1991) Mutations affecting body organization in the *Arabidopsis* embryo. *Nature* 353, 402–407. Reproduced with permission.

information obtained for one developmental gene from the TAIR database:

Locus: AT1G21970

Date last modified	2003–05–02
TAIR Accession	Locus:2201163
Representative Gene Model	AT1G21970.1
Other names	EMB 212, EMB212, LEAFY COTYLEDON 1, LEC1
Other Gene Models	
name	LEC1
description	Transcriptional activator of genes required for both embryo maturation and cellular differentiation. Required for the specification of cotyledon identity and the completion of embryo maturation. Sequence is similar to CCAAT-box binding factor HAP3. Mutants are desiccation intolerant, have trichomes on cotyledons and exhibit precocious meristem activation
source	GenBank
date	1998–07–02

J Harada Laboratory
2001–02–25

Seed development is under the control of **hormones**, particularly those entering from the parent plant. Mutations that affect the synthesis or responsiveness of the seeds to these hormones, particularly **abscisic acid** (ABA), influence development of both the embryo and endosperm, the ability of the embryo to tolerate desiccation and to resist germinating precociously while developing (**see: Desiccation tolerance; Vivipary**). (JDB, AC)
(See: Development of endosperms – dicots)

Chaudhury, A.M., Koltunow, A., Payne, T., Luo, M., Tucker, M.R., Dennis, E.S. and Peacock, W.J. (2001) Control of early seed development. *Annual Review of Cell and Developmental Biology* 17, 677–699.

Schwartz, B.W., Yeung, E.C. and Meinke, D.W. (1994) Disruption of morphogenesis and transformation in abnormal *suspensor* mutants of *Arabidopsis*. *Development* 120, 3235–3245.

Vielle-Calzada, J.P., Basker, R. and Grossniklaus, U. (2000) Delayed activation of the paternal genome during seed development. *Nature* 404, 91–94.

Weijers, D., Geldner, N., Offringa, R. and Jurgens, G. (2001) Seed development: early paternal gene activity in *Arabidopsis*. *Nature* 414, 709–710.

Development of endosperms - cereals

1. Structure–function overview

In grass species, including the cereals, the endosperm is a major source for food, feed and industrial raw materials. In **cereals** the endosperm typically constitutes the majority of the grain, such as in **maize**, where the endosperm is 90% of the dry weight of the grain. The fully developed endosperm consists of four major cell types, the starchy endosperm, the **aleurone layer** (see: section 2c), **transfer cells** (see: 2a) and the cells of the embryo-surrounding region (ESR) (Fig. D.15). Functionally, these cell types play distinctive and different roles during seed development, germination and early seedling establishment. During the period following fertilization, the endosperm is assumed to play a support role for the developing embryo, endosperm development (of the nuclear type, **see: Development of endosperms – dicots**) initially proceeding at a faster pace than embryo development (**see: Development of embryos – cereals**). Although their exact function remains to be identified, the cells of the ESR are assumed to mediate this supportive role of the endosperm. The transfer cells play an important role during grain filling in transferring sugars from the source tissues to the endosperm and the embryo.

During the grain-filling period, which normally takes from 40 to 60 days, the starchy endosperm expands rapidly and is filled with **starch granules** and prolamin **storage proteins**. In some cases, for instance in **oats**, the starchy endosperm also accumulates globulins. Towards the end of the grain maturation period, the cells of the starchy endosperm dehydrate and die. In contrast, the cells of the **aleurone layer**, which varies in thickness from one cell layer in maize and **wheat**, to three cell layers in barley, are **desiccation tolerant**, and are living, **quiescent** cells in the mature, dry grain. Upon imbibition, the embryo produces the **hormone gibberellin**, which diffuses into the endosperm where it induces the production of glucanases, amylases and proteases that break down cell walls, starch and proteins, respectively, in the starchy endosperm. The resulting sugars and amino acids support seedling growth until this is sustained by photosynthesis. (**See: Mobilization of reserves – cereals**)

For convenience and clarity, development of the cereal endosperm is considered below in the different regions of this tissue, although the events noted occur concurrently.

2. Endosperm fertilization, cellularization and differentiation

The endosperm develops from the central cell in the **ovule** after **fertilization** of its diploid nucleus by one of the **gametes** released from the pollen tube (Fig. D.16). The central cell is part of the **embryo sac**, or **megagametophyte**, which is the product of the female meiosis in angiosperms. This is initiated in the meiocyte derived from a

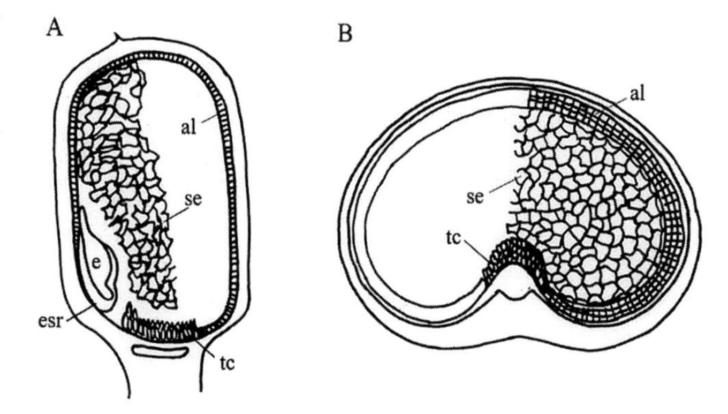

Fig. D.15. Tissue composition of grains of maize (A) and barley (B). al, Aleurone layer; esr, embryo-surrounding region; e, embryo; se, starchy endosperm; tc, transfer cells.

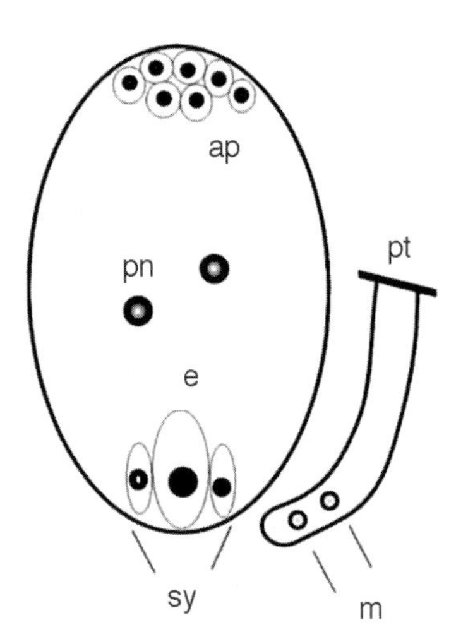

Fig. D.16. The *Polygonum*-type embryo sac region of the ovule. This contains the egg cell (e), antipodal cells (ap), synergids (sy) and polar nuclei (pn) in the centre of the embryo sac (central cell). During the process of double fertilization, the two haploid male nuclei (m) released from the pollen tube (pt) fuse with the haploid egg cell nucleus to form the diploid zygote, which gives rise to the embryo, and with the diploid fused polar nuclei, the triploid product of which gives rise to the endosperm.

nucellus parenchyma cell, resulting in one surviving haploid meiospore. In the so-called *Polygonum* type of embryo sac, which occurs in most monocots, three rounds of nuclear divisions result in eight nuclei that migrate to opposite ends of the cell of the embryo sac. Of these, one nucleus from each pole, called the polar nuclei, migrates to the centre of the embryo sac. Subsequent cellularization and differentiation result in the egg cell (n), the synergids (n) and the antipodal cells (n) (formed by additional mitotic divisions). The central cell is formed by the remainder of the embryo sac and includes the two polar nuclei (Fig. D.16). Importantly, the embryo sac is a differentiated, highly polarized structure that develops under the influence of signals from the maternal tissue that surrounds it. The details of this influence are poorly understood. (See: **Reproductive structures, 1. Female**)

A double fertilization process occurs in angiosperms, in which one of the two sperm nuclei (n) carried by the growing pollen tube fertilizes the egg cell (n) to initiate embryo formation; the second nucleus (n) fertilizes the central cell containing the polar nuclei (2n) initiating endosperm development (Fig. D.16). The resulting primary endosperm nucleus (3n) is situated in the proximal bulk of central cell cytoplasm (Fig. D.17, A). The unusual nature of the early phase of endosperm development is seen already in the first mitotic divisions of the primary endosperm nucleus, which lack the cell wall that usually forms between the separating daughter nuclei in most plant tissues (Fig. D.17, B and C). The molecular basis for the suppression of cell wall formation during this nuclear endosperm development is unknown. In maize, the nuclear divisions are arrested when the number of nuclei is between 256 and 512 (Fig. D.17, D). As shown in Fig. D.15, the fully developed cereal endosperm is a cellular structure, and initiation of the cellularization process is marked by the

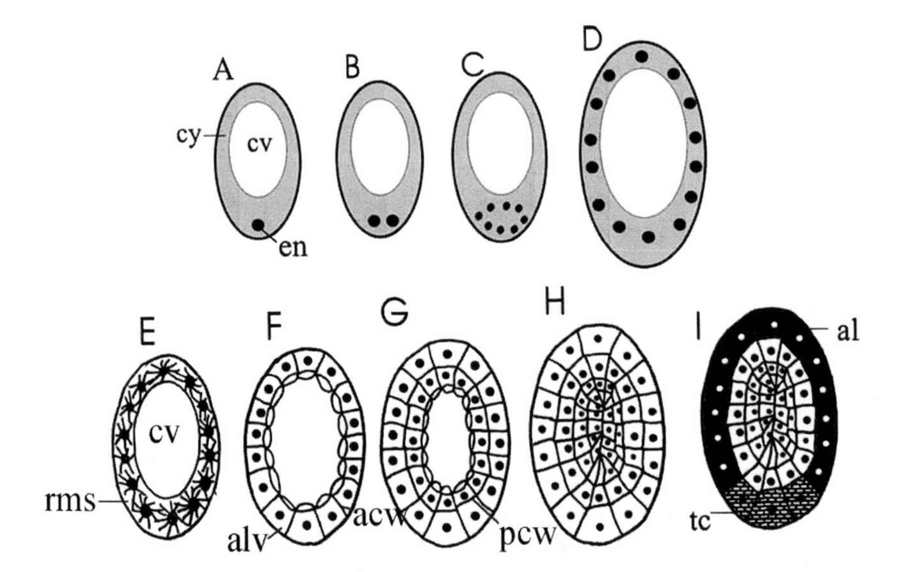

Fig. D.17. The cereal endosperm coenocyte and the cellularization process. (A) The central cell contains the diploid endosperm (fused polar) nucleus (en) in the basal cytoplasm. The cytoplasm of the central cell (cy) surrounds a large central vacuole (cv). (B and C) The primary endosperm nucleus divides without cell wall formation. (D) The endosperm coenocyte (many nuclei within a single cytoplasm) contains evenly-spaced nuclei throughout the entire cytoplasm. (E) The process of endosperm cellularization is initiated by the formation of radial microtubular systems (rms) on all nuclear surfaces. (F) Anticlinal cell walls (acw) are formed around each nucleus and form a tube-like structure (the alveolus, alv) with its open end towards the central vacuole. (G) Each of the nuclei divides, and a pericinal cell wall (pcw) is formed between the separating daughter nuclei. (H) A repetition of the process of cell wall formation in G leads to a cellular endosperm. (I) Developmental domains of the endosperm include the transfer cell layer domain (tc), the aleurone cell layer domain (al) and the central starchy endosperm (white). Following many further mitotic cell divisions, the structure of the mature grain, as illustrated in Fig. D.15, is achieved. See also Fig. D.18 for micrographs of the various stages of development in wheat.

formation of nuclear radial microtubules on the surface of all endosperm nuclei (Fig. D.17, E). Microtubular arrays from neighbouring nuclei form inter-zones in which cell walls are deposited such that a tube of cell wall material is formed around each nucleus, with its open end pointing inwards towards the central vacuole (Fig. D.17, F). Subsequent to the formation of these walls, the mitotic arrest is released, all nuclei then divide synchronously, and a cell wall is formed between the separating nuclei, parallel to the outer wall of the endosperm (periclinal orientation). The result of this round of mitosis is one peripheral layer of cells and a new layer of interior cells similar to the original layer of nuclei, surrounded by a tube-like wall structure with its opening pointing in towards the central vacuole (Fig. D.17, G) (see: section 2c). The same process is repeated until the whole cavity of the endosperm is filled with columns of cells, which in maize takes three or four rounds of divisions and occurs approximately 3 DAP (days after pollination).

Similar to the situation in many other plant tissues, specification of the developmental fate of the four endosperm cell types is thought to occur by so-called positional signalling, i.e. a cell type becomes destined for a certain fate based on its location, not on its developmental history, or lineage. In the endosperm, the different locations are those corresponding to the transfer cell layer, the aleurone layer, and the starchy endosperm (Fig. D.17, I).

(a) *Transfer cell layer.* Transfer cells develop in the basal endosperm over the main vascular tissue of the maternal plant (Fig. D.15, A and B), where they facilitate solute transfer, mainly of amino acids, sucrose and monosaccharides across the plasmalemma between cells of the maternal plant and those in the endosperm compartments. In maize, the *miniature1* (*mn1*) **mutant** has reduced grain size and their transfer cells lack normal levels of type 2 cell wall invertase, strongly suggesting this enzyme contributes to the establishment of a sucrose concentration gradient between the maternal vascular tissue and endosperm by hydrolysing sucrose to glucose and fructose. The transfer cell region is a separate developmental domain from the main vascular tissue of the maternal plant, even by the time the endosperm has reached the syncytial stage, and this is fully established at the completion of cellularization (Fig. D.17, I). The molecular mechanism underlying transfer cell differentiation is unknown. The endosperm transfer cells are typified by heavy secondary wall ingrowths that provide a larger surface area, presumably to facilitate nutrient uptake. Typically, cereals have two or three cell layers of transfer cells. Molecular markers for transfer cells include *END1* from barley with unknown function, and the maize genes *Bet 1–4* that may play a role in defence against pathogens, and *Incw2*, a soluble acid invertase that may function to hydrolyse sucrose to fructose and glucose.

(b) *The embryo-surrounding region (ESR).* The embryo develops in a cavity within the endosperm, and is also completely surrounded by the endosperm in mature grains, although the aleurone layer that surrounds the embryo in mature grains is typically reduced (Fig. D.15, A). At an early stage during endosperm development, the cells of the ESR line the cavity where the embryo develops. Maize ESR cells have dense cytoplasmic contents and express *Esr1–3* transcripts that are detectable already at 4 DAP. Although the function of the ESR region is undetermined, it is generally assumed that these cells play a role in the transfer of nutrients to the embryo and/or in establishing a physical barrier between the embryo and the endosperm during seed development. (**See: Development of seeds – nutrients and water import**) The molecular mechanism involved in transfer cell fate specification remains unknown. No mutant that specifically affects the ESR region has been reported. (**See: Development of embryos – cereals**)

(c) *Aleurone layer cells.* In fully developed grains, the aleurone layer covers the perimeter of the endosperm except in the basal transfer cell region (Fig. D.15). Based on morphogical differences in cytoskeletal arrays, aleurone layer cell fate is specified already after the first periclinal mitotic cell division that results in two layers of endosperm cells (Fig. D.17, G). The number of cell layers in maize and wheat is one, one to several layers in rice, and three layers in barley. In barley, the aleurone layer consists of approximately 1,000,000 cells, compared to an estimated 250,000 cells in maize. The aleurone layer cell cytoplasm is dense and granular due to the presence of many aleurone grains, small vacuoles with inclusion bodies. These aleurone grains may be lytic vacuoles or **protein bodies**, both of which may contain two major types of inclusion bodies, the globoid bodies which contain a crystalline matrix of **phytin**, protein and lipid, and protein–carbohydrate bodies. Lipid droplets surround the aleurone grains. The endoplasmic reticulum of these cells is well developed, and a high number of **mitochondria** are also present. Mature aleurone layer cells appear cuboidal in section, and contain anthocyanins, responsible for the colourful grains of so-called Indian corn (**see: Maize**). Barley aleurone layer cells are highly **polyploid**, but the ploidy of these cells in other cereals is unknown. Molecular markers for aleurone layer cells include the barley transcripts *Ltp2*, *Ltp1*, *B22E*, *pZE40*, *ole-1*, *per-1* and *chi33*, and *C1* in maize.

The first sign of aleurone layer cell differentiation occurs after the first division of endosperm nuclei encased by cell walls (Fig. D.17, F) resulting in a cell wall that is parallel to the outer wall of the endosperm (Fig. D.17, G). The formation of the aleurone layer is a gradual process, and a morphologically well-defined cell layer is observable by light microscopy in maize grains around 12 DAP. At early developmental stages, the outermost aleurone layer cells divide both anticlinally (cell walls form perpendicular to the maternal cell wall surrounding the endosperm, leading to the expansion of the surface of the aleurone layer; Fig. D.17, F) as well as periclinally (cell walls form parallel to the maternal cell wall, contributing to the inner cell mass of the starchy endosperm, Fig. D.17, G). A clear demonstration of the role of positional signalling in aleurone layer cell fate specification comes from studies on wheat, in which the innermost daughter cells from periclinally dividing aleurone layer cells lose their aleurone cell identity and become starchy endosperm cells.

The mechanisms underlying aleurone layer cell fate specification and development are being investigated using mutants that are perturbed in aleurone layer cell formation and development. In the maize *crinkly4* (*cr4*) mutant, white patches are formed on the surface of purple grains due to the lack of aleurone layer cells producing the dark anthocyanin pigment.

The *cr4* gene encodes a Tumor Necrosis Factor Receptor (TNFR)-like receptor **kinase** that is predicted to be targeted to the plasma membrane with its external receptor domain on the extracellular side of the membrane. A model has been proposed in which the CR4 receptor is activated by an activator (ligand) that represents the positional signal for aleurone layer cell differentiation as described above. According to this model, activation of the CR4 receptor kinase leads to transcription of genes necessary for aleurone layer cell formation. The proposed ligand remains unidentified. In the maize *defective kernel mutant1* (*Dek1*), aleurone layer cells are almost entirely absent from mature grains. The *Dek1* gene encodes a 240 kDa protein of the calpain super gene family, *DEK1* being the only plant member of this family. *DEK1* has a complicated domain structure, containing the calpain cysteine proteinase domain at the C-terminus, and 21 membrane-spanning segments, plus a domain predicted to be an extracellular loop region in the N-terminal half of the protein. Analysis of mutant sectors in wild-type endosperm has demonstrated that the *Dek1* gene is necessary for aleurone layer cell maintenance; cells losing *Dek1* function convert to starchy endosperm cells, even if they are positioned in the periphery of the endosperm. The third maize mutant gene that affects aleurone layer cell formation is *supernumerary aleurone layers 1* (*sal1*). Homozygous mutant grains possess an aleurone layer that is up to seven cells thick, in shrunken kernels. The *sal1* gene encodes a homologue of the human *Chmp1* gene, a member of the conserved family of the class E vacuolar protein-sorting genes implicated in membrane vesicle trafficking. In mammals, CHMP1 functions in the pathway targeting plasma membrane receptors and ligands to lysosomes for proteolytic degradation, thereby regulating the sensitivity of cells carrying such receptors to hormone signalling, but in the aleurone layer its function is unknown.

In addition to the three mutants described here, many other maize mutants that affect aleurone cell development have been isolated, including *extra cell layers 1* (*xcl1*), *naked* (*nkd*), *collapsed2-o12* (*cp2-o12*), *paleface* (*pfc*), *Defective aleurone pigmentation* (*Dap*) and *etched*. Undoubtedly, these mutants will contribute valuable insight into the molecular mechanisms of aleurone layer cell formation as the genes underlying the mutant phenotypes are cloned.

(d) *Starchy endosperm cells*. These represent the largest cell mass in the endosperm, and consists of an estimated 80,000–90,000 cells in barley, and 60,000 in wheat. In the mature grain, the cells of the peripheral layer of the starchy endosperm, juxtaposed to the aleurone layer, are commonly referred to as the sub-aleurone layer. This consists of two or three layers of small cells with a higher density of protein, and a lower number of starch granules than the central starchy endosperm cells. Most likely, a high proportion of the sub-aleurone layer cells is derived from aleurone layer cells, following their periclinal division and redifferentiation. As the name signifies, the starchy endosperm cells contain **starch** in the form of granules synthesized within **amyloplasts** (**see: Starch–synthesis**). In addition, starchy endosperm cells accumulate prolamin storage proteins, which are specific for starchy endosperm cells (**see: Storage protein**). These storage protein genes were among the first nuclear-encoded genes to be cloned, and the basis for their endosperm-specific pattern of expression has been extensively studied.

After completion of cellularization in the young endosperm, the starchy endosperm cells develop from the inner part of the cell files located at the endosperm closure, which are all derived from the interior daughter cells of the first periclinal division in the endosperm alveoli (Fig. D.17, G). In contrast to aleurone layer cells, those of the starchy endosperm divide in randomly oriented planes. Because of this, the cell files extending from the periphery and inwards are soon lost. Cell divisions in the interior of the starchy endosperm cease relatively early, being mainly restricted to the peripheral layers. One characteristic of starchy endosperm cells in maize is that they undergo **endoreduplication**, a process in which the nuclear DNA is replicated without nuclear divisions; the significance of this is unclear. Towards the end of the seed-filling period, the starchy endosperm nuclei undergo a process that bears resemblance to **programmed cell death**.

Little is known about the mechanisms that specify which particular cells will be of the starchy endosperm type. However, based on the phenotype of the *cr4* and *dek1* mutants, in which lack of aleurone layer cells leads to the formation of starchy endosperm cells in the peripheral cell layer of the endosperm, starchy endosperm cell formation appears to be the default fate of endosperm cells. Mutants that specifically affect the formation of starchy endosperm cells have not been reported. However, *defective kernel* (*dek*) mutants have been reported in maize, in which they are more or less shrunken. In most cases, the developmental lesion in these mutants has not been identified, however. Genes that have been cloned from such mutants include *discolored1* (*dsc1*), a gene of unknown function, and *empty pericarp2* (*emp2*), encoding a Heat Shock Binding Protein (ZmHSBP1), a negative regulator of the heat shock response. In addition to these, there are barley shrunken endosperm mutants similar to the maize *dek* mutants, and mutants with defective or reduced endosperm that are maternally inherited, but the function of these mutant genes in barley endosperm development is unknown.

Fig. D.18 is a series of photomicrographs showing the development of the endosperm in wheat. (O-AO, HS)

Becraft, P.W. (2001) Cell fate specification in the cereal endosperm. *Seminars in Cell and Developmental Biology* 12, 387–394.

Berger, F. (2003) Endosperm: the crossroad of seed development. *Current Opinion in Plant Biology* 6, 42–50.

Brown, R.C., Lemmon, B.E. and Nguyen, H. (2002) Endosperm development. In: O'Neill, S.D. and Roberts, J.A. (eds) *Annual Plant Reviews; Plant Reproduction*. Sheffield University Press, Sheffield, UK, pp. 193–220.

Olsen, O.-A. (2003) Nuclear endosperm development in cereals and *Arabidopsis thaliana*. *The Plant Cell* 16, S214–S227.

Thompson, R.D., Hueros, G., Becker, H.A. and Maitz, M. (2001) Development and functions of seed transfer cells. *Plant Science* 160, 775–783.

Development of endosperms – dicots

The genetic and physiological balance involving the endosperm, the embryo, and the maternal tissues is important for proper seed development. A crucial role is played by dicot endosperm, even though, in contrast to the cereal endosperm, it is ephemeral in many species, and therefore is not a major component of the mature seed (**see: Endosperm; Non-endospermic seed**). Two views have been put forward to explain the evolution of

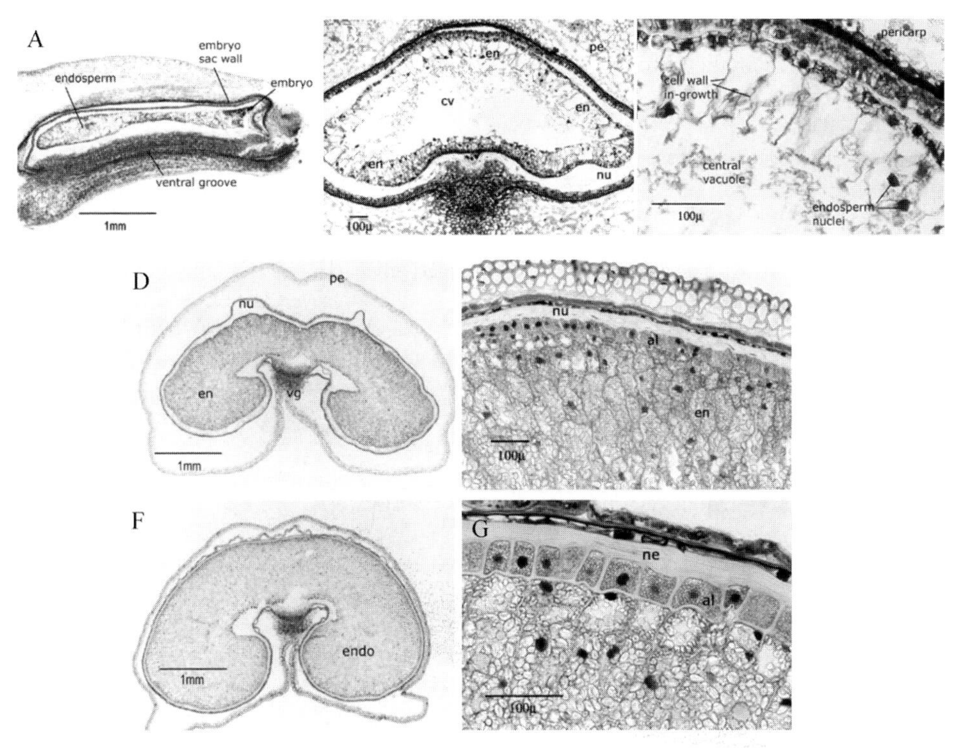

Fig. D.18. Photomicrographs of endosperm development in the wheat grain.
For early stages of endosperm development within the embryo sac, **see** Fig. D.10 in **Development of embryos – cereals**. A is a longitudinal section of a developing grain; B–G are transverse sections.
(A) 6 DAF (days after flowering, or fertilization). Free dividing coenocytic nuclei of the endosperm start to populate the embryo sac.
(B) 6 DAF. The large central vacuole (cv) of the embryo sac is surrounded by the developing endosperm (en). Cell walls are starting to grow inwards from the nucellus (nu). pe, Pericarp.
(C) 6 DAF. Detail of the area surrounding the central vacuole. Some endosperm nuclei have migrated to the periphery and are contained within newly formed cells. Other nuclei are still in the freely dividing state. Cell formation commences near the nucellar projection of the ventral groove and differentiation spreads laterally to the dorsal side of the grain.
(D, E) 15–16 DAF. The endosperm (en) is expanding but does not completely fill the nucellus (nu), which is becoming compressed. There is differentiation of the endosperm into regions where starch and protein are being synthesized (starchy endosperm, en) and the aleurone layer (al) is becoming differentiated as the outermost layer of the endosperm. pe, Pericarp (fruit coat); vg, ventral groove.
(F, G) 21 DAF. Storage reserves are being deposited within the endosperm which is expanding around the ventral crease, through which nutrients are entering the developing grain via the vascular tissues. The single-cell-thick aleurone layer (al) is differentiated, to the inside of which is a sub-aleurone layer and the starchy endosperm, the cells of which now contain prominent starch grains. Nuclei (darkly stained areas) are still present in some starchy endosperm cells, but they fragment later in development, which takes approx. 30 DAF to be completed. The nucellus (nucellar epidermis, ne) is reduced to being the walls of crushed cells.
All pictures taken from www.wheatbp.net. *Wheat: The Big Picture.* Grain growth 1–4, 4–10, 11–16, 17–21, 21–30 days.

endosperm. In one view, the endosperm is derived from a primitive embryo, whose original role was to supply nutrients to the embryo proper. In the second view, the endosperm is thought to be comparable with the **megagametophyte**, the haploid nutritive tissue present in the gymnosperm seeds.

Endosperm development in monocots (**see: Development of endosperms – cereals**) and in dicots have important similarities. In both plants endosperm development proceeds through four stages: syncytial, cellularization, differentiation and death.

There are two basic patterns of endosperm development in dicots and monocots: another type (helobial) is present in only a very few monocot families (**see: Endosperm**). (i) The cellular type of endosperm development is typical of many dicot families, and also some monocots. Here all nuclear divisions are accompanied by cell division. (ii) The nuclear pattern of endosperm development is also common in both monocots and dicots, in which the nucleus undergoes several divisions before cell wall formation occurs. Usually cell division commences towards the periphery of the endosperm and continues inwards, but it may not become completely cellular (e.g. in *Arabidopsis*, which retains free nuclei in the region close to the chalazal region, Fig. D.19). There is nuclear-type endosperm development in maize and other cereals, but here cellularization of the nuclei is complete.

In *Arabidopsis*, the fertilized endosperm cell nucleus undergoes mitosis to form a single cell with several nuclei, the syncytial stage (Fig. D.19). In the multinucleate syncytium, the cytoplasm is compartmentalized into nuclear-cytoplasmic domains (NCDs) defined by a microtubule system. Cellularization begins in the embryo-surrounding regions and then travels towards the chalazal region. **Endoreduplication** may occur in nuclei in the endosperm cells in this latter region. Eventually the embryo grows into the endosperm, which degrades due to programmed cell death, leaving only a single surrounding layer, often called the aleurone layer.

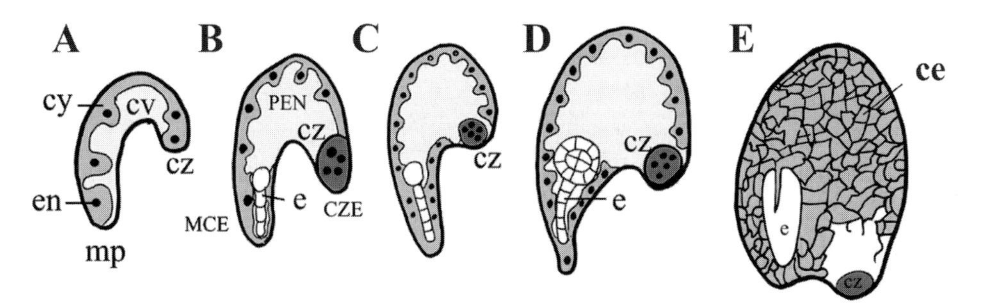

Fig. D.19. The *Arabidopsis* triploid central cell undergoes cell division, and (A) nuclei (en) migrate from the micropylar region (mp) of the cell towards the chalazal end (cz), eventually lying in the cytoplasm (cy) throughout the coenocyte (multinucleate cell), which has a large central vacuole (cv). (B and C) As endosperm development progresses, the endosperm coenocyte develops three distinct regions: that surrounding the embryo (MCE), the central or peripheral endosperm (PEN) and the chalazal endosperm (CZ, CZE). (D) At the end of globular endosperm development, the embryo (e) is completely surrounded by cytoplasm. (E) There is almost complete cellularization of the endosperm at the torpedo stage of embryo development, except near the chalazal region. Later the embryo grows to occlude the endosperm (see: Fig. D.13 in Development of embryos – dicots). From: Olsen, O.-A. (2004) Nuclear endosperm development in cereals and *Arabidopsis thaliana*. *The Plant Cell* 16, S214–S227. Reproduced with permission.

A number of important **mutants** have been described that illuminate dicot endosperm development. In *Antirrhinum*, the deletion of a gene for a GTPase (a multifunctional family of GTP hydrolases that can act as molecular and metabolic switches) results in an anomalous syncytial endosperm with large nuclei. Mutants grouped under the TITANI family also show impaired endosperm development in *Arabidopsis*; they have giant polyploid endosperm nuclei and cellularization is blocked. The *FIS* class of genes has revealed control of endosperm development without fertilization. Mainly through the analyses of *fis* class mutants it is now thought that early endosperm development is under the control of polycomb (*PcG*) class genes (whose products are known to repress [silence] other genes, and thus regulate transcription). Important members of this class of genes include *MEA/FIS1*, *FIS2* and *FIS3/FIE*, since they are regulatory genes controlling the initiation and proliferation of endosperm cells. When either of the *FIS* class genes is mutated, a diploid endosperm is formed without fertilization (**see: Apomixis**). After fertilization in the *fis* class mutants, the endosperm undergoes excessive proliferation at the free nuclear stage of growth and is not cellularized properly.

Genetic imprinting, i.e. the unequal influence of one parental set of genes, also has an effect on development of the endosperm. While at the time of double fertilization the egg cell and the central cell express genes derived only from the female genome, after fertilization gene expression in both the embryo and endosperm should have a contribution from the paternal genome as well. However, certain genes, or even the whole paternal genome, are silenced during early stages of seed development. Thus the maternal genome directs early seed development and particularly endosperm development.

The process by which the paternal genome appears to be silenced in the developing embryo or endosperm is known as imprinting, which occurs in mammals. Imprinting effects on the seed embryo are generally the result of changes in the development of the endosperm. This is a phenomenon where a DNA sequence can have a conditional behaviour depending whether it is maternally or paternally inherited; thus the same gene (allele) may have a different effect on the offspring, depending upon its parental source. During imprinting, transcription is impaired in one genome (the imprinted locus), often due to an epigenetic (non-inherited) change in its DNA, such as by methylation. (**See: Imprinting**)

There is little known about endosperm development in dicot species in which it persists, although up to the stage of complete cellularization it is likely to be the same as in species from which it is lost. Presumably the programme for cell death is repressed at least until after germination, since in most dicot species with persistent endosperms the cells are living at seed maturity. In species that have endosperms with thickened cell walls, due to the deposition of **hemicelluloses**, this occurs following cellularization, and limited hydrolysis of the cells may occur to provide a central cavity into which the embryo can grow. Deposition of reserves within persistent endosperms is frequently uneven, and in castor bean, for example, storage protein deposition is completed in the micropylar and peripheral regions of the endosperm several days before it is completed in the regions close to the cotyledons.

(AC, JDB)

Berger, F. (1999) Endosperm development. *Current Opinion in Plant Biology* 2, 28–32.

Berger, F. (2003) Endosperm: the crossroad of seed development. *Current Opinion in Plant Biology* 6, 42–50.

Boisnard-Lorig, C., Colon-Carmona, A., Bauch, M., Hodge, S., Doerner, P., Bancharel, E., Dumas, C., Haseloff, J. and Berger, F. (2001) Dynamic analyses of the expression of the Histone: YFP fusion protein in *Arabidopsis* show that syncytial endosperm is divided in mitotic domains. *Plant Cell* 13, 495–509.

Costa, L.M., Gutierrez-Marcos, J.F. and Dickinson, H.G. (2004) More than a yolk: the short life and complex times of the plant endosperm. *Trends in Plant Science* 9, 507–514.

Gehring, M., Choi, Y. and Fischer, R.L. (2004) Imprinting and seed development. *The Plant Cell* 16, S203–S213.

Greenwood, J.S. and Bewley, J.D. (1985) Seed development in *Ricinus communis* cv. Hale (castor bean). III. Patterns of storage protein and phytin accumulation in the endosperm. *Canadian Journal of Botany* 63, 2121–2128.

Development of seeds – an overview

Upon double fertilization of the egg and the endosperm (polar) nuclei (**see: Reproductive structures**) two different structures are formed within the angiosperm seed, the **embryo** from the **zygote**, and an **endosperm** which may or may not persist until seed maturity. While in most **monocots**, endosperm cells define the bulk of the grain or seed, in dicots, the bulk of the seed is often the embryo, and endosperm cells are restricted to one or a few cell layers in mature seed (**see: Non-endospermic seed**). There are notable exceptions however, e.g. in the endospermic legumes, **coffee**, **coconut**, **castor bean**, etc. (**See: Endosperm**)

Embryo and endosperm result from a complex interaction of genes, derived from the haploid pollen grain, the **megagametophyte**, the haploid female gamete and the diploid **sporophytic** body of the mother. Thus genes specific to each of these three genomes play pivotal roles in the generation of the seed.

Seed development, although a continuum of events, can be conveniently divided into three stages (Fig. D.20). Initially (Phase I) there is the formation of the different tissue types within the embryo, endosperm and surrounding seed as a result of extensive cell division; this occurs during the stage of histodifferentiation. A large increase in fresh weight (fw) is achieved by the time this stage is completed. After this, little cell division occurs, so the number of cells into which reserves can be deposited is fixed. Then follows the stage of cell enlargement or expansion (Phase II), when the major reserves within the embryo and storage tissues are laid down (**see: Hemicellulose – synthesis; Oils and fats – synthesis; Starch – synthesis; Storage protein – synthesis**). The rate of synthesis of individual storage reserves is not always constant, so the concentration of oil, starch and protein can change during the linear growth phase. As reserve deposition occurs, water is displaced from the seed and is replaced by the insoluble polymeric reserves. Hence the dry weight of the seed increases, and its proportional water content declines. During the third phase dry matter accumulation slows and stops at **physiological maturity**. Seeds of most species follow this pattern of development, making it possible to derive a general description of seed growth that is not species-specific. (**See: Development of seeds – growth phases**; note that two growth phases are recognized, which are the same as the first two phases of development noted here. The final developmental phase, loss of water and cessation of metabolism, is not regarded as a growth phase.)

Synthesis of the major storage reserves in the seed takes place from a few basic raw materials (primarily sucrose, amino acids and mineral nutrients) supplied by the parent plant. Thus the supply of carbon and nitrogen plays a major role in determining the rate and duration of seed growth and, thus, the final weight per seed (seed size). (**See: Charles-Edwards model; Development of seeds – nutrient and water import**)

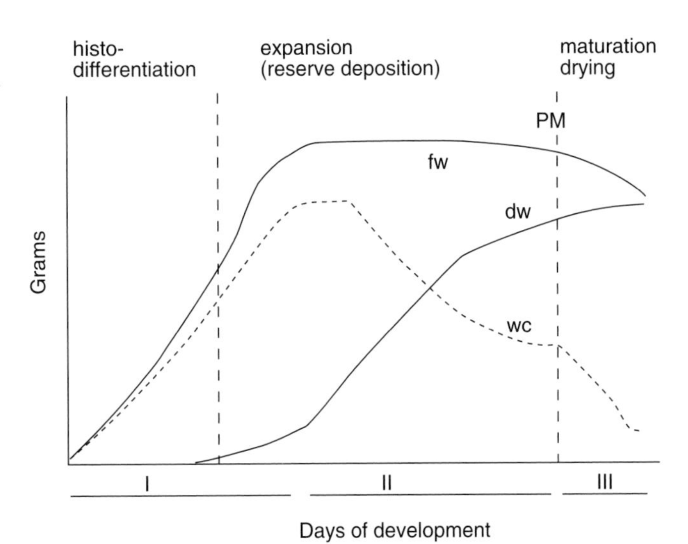

Fig. D.20. Pattern of seed development to show the changes in whole-seed fresh weight (fw), dry weight (dw) and water content (wc). Three phases of development are noted: Phase I, gain in fresh weight because of cell division and expansion; Phase II, gain in dry weight due to enlargement of storage cells and the deposition of insoluble stored reserves therein; Phase III, loss of fresh weight as the seed undergoes maturation drying. The time taken to complete development varies between species, cultivars and with the ambient environmental conditions. Physiological maturity (PM) is the time at which reserve deposition virtually ceases. After: Bewley, J.D. and Black, M. (1994) *Seeds. Physiology of Development and Germination.* Plenum Press, NY, USA.

The water concentration at maximum seed dry weight (**see: Physiological maturity**) varies among species (e.g. soybean ~ 550 g/kg; wheat ~ 430 g/ kg; maize ~ 350 g/kg) and is always well above the concentration at which grain is usually harvested. Water loss continues after maturation until the seed reaches a harvestable moisture level (**see: Maturation – changes in water status; Maturation – effects of drying**) and the rate of drying varies among species and environments. **Respiration** continues during the early stages of this drying phase but there is probably no significant loss of dry matter.

The seed water content (mg H_2O/seed) increases to a maximum and then declines as the seed approaches maximum seed dry weight (Fig. D.20). Water movement into the seed drives the increase in cell volume that accounts for most of the increase in seed size during Phase II. There is some evidence that the ability of the seed to continue increasing in volume plays a role in maintaining seed growth, e.g. the final maturation events in soybean may be triggered by continued accumulation of dry matter after cell expansion stops. In cereals, the vast majority of the reserves are stored within the endosperm, which is non-living at maturity (**see: Development of endosperms – cereals**). In many dicot seeds, the major site of storage is the cotyledons, and during seed development the endosperm is resorbed, and its degraded components used as a source of carbon and nitrogen to support the synthesis of reserves in the embryo. In species with persistent endosperms, these become a major site of stored reserves, e.g. as oil and protein in castor bean, or as **hemicelluloses** in the endospermic legumes. In the final phase of seed development in most species, that of maturation drying, there follows the disconnection of the seed from the parent plant (shrivelling of the funiculus), the seed becomes desiccation tolerant, dries down to about 7–15% moisture content, and is shed. At this stage the seeds are **quiescent**, i.e. with imperceptible metabolic activity, and some may also have acquired **dormancy** during their development. In some species, such as mangroves, growth of the embryo continues into germination while still on the parent plant, there being no maturation drying phase. The duration of each phase of development may last from several days to many months, depending upon the species and prevailing environmental conditions. (JDB, DBE, AC)

(See: Development of seeds – dry matter accumulation; Development of seeds – influence of external factors)

Bewley, J.D., Hempel, F.D., McCormick, S. and Zambryski, P. (2000) Reproductive development. In: Buchanan, B.B., Gruissem, W. and Jones, R.L. (eds) *Biochemistry and Molecular Biology of Plants*. American Society of Plant Physiologists, Rockville, MD, USA, pp. 988–1043.

Srivastava, L. (2002) *Plant Growth and Development. Hormones and Environment*. Academic Press, New York, USA, pp. 75–92.

Development of seeds – dry matter accumulation

Seed growth is frequently characterized by the rate of dry matter (seed storage reserve) accumulation during Phase II of development, the linear phase of growth, and the duration of growth. (**See:** Fig. D.20 in **Development of seeds – an overview**). Seed growth rate is usually estimated by linear regression analysis of seed dry weight vs time. More complex non-linear models have been used, but by predicting a constantly changing growth rate during Phase II, they misrepresent the seed growth process and are not appropriate.

Estimating the duration of seed growth is more difficult, for it is not easy to determine when dry matter accumulation starts or stops. The effective filling period (EFP, Eq. 1) is

$$EFP = \text{Mature seed weight/seed growth rate} \qquad \text{Eq. 1}$$

a concept which is widely used because it provides an estimate of duration that avoids determining when the seed starts and stops accumulating dry matter, and it is relatively easy to determine. The EFP can be defined on an area (yield/total seed growth rate in $g/m^2/day$) or an individual seed basis (mature weight per seed/individual seed growth rate in mg/seed/day).

A regression model of the complete growth curve (Fig. D.20 in **Development of seeds – an overview**) can be used to estimate the time from a beginning point (e.g. 5% of maximum seed weight) to an end point (e.g. 95% of maximum seed weight).

Various whole plant growth stages can also be used to estimate, non-destructively, seed-fill duration on a whole plant or plant community basis. Anthesis to maximum seed dry weight is a convenient measure in many crops, while the beginning of seed formation to physiological maturity is frequently used in soybean.

All estimates of seed-fill duration provide only a relative indication of the length of seed filling and comparisons among species may be misleading, especially for estimates based on plant growth stages. However, all measures can be used to characterize genotype and environmental effects.

Seed growth rate and seed-fill duration are under genetic control in most crop species and they are also influenced by environmental conditions. Evidence for genetic control includes demonstrations that seed growth rate and seed-fill duration are heritable characteristics in several crop species and many examples of genotypic differences that are consistent across environments. Genetic differences in seed growth rate seem to be controlled by the number of cells in the **endosperm** or **cotyledons** in many crop species. The environment also affects cell number and the subsequent seed growth rate. Not as much is known about the regulation of genetic differences in seed-fill duration, probably because potential regulation by both the seed (ability to continue cell expansion and increase in volume) and the parent (through leaf senescence during seed filling) creates a more complex system.

Most genetic differences in mature seed **size** (weight per seed) within a species are related to differences in seed growth rate, i.e. large seeds have high seed growth rates and small seeds have low seed growth rates. It is possible, however, to find variation in mature size that is a function of seed-fill duration, i.e. seed growth rate is constant. Differences in seed size among species are also usually associated with seed growth rate: crops with large seeds (e.g. maize and soybean) have much higher seed growth rates (typically 4–15 mg/seed/day) than small-seeded crops such as wheat or rice (*Oryza sativa*) (typically 1–2 mg/seed/day). (DBE)

Development of seeds – growth phases

After setting, seed development is comprised of two growth phases. A pre-storage phase, dominated by cell division, precedes a storage phase when most of the seed's dry matter (nutrients) is accumulated. Initially cell division, cell expansion and storage overlap sequentially. Storage phase onset is marked by cessation of cell division and induction of the metabolic machinery and associated organelles linked with polymer storage. During transition from pre-storage to storage functions, seed absolute growth rate (net nutrient import) increases to a plateau reached coincidentally with attainment of final cell number/size and hence seed volume. Linked to cell expansion is a net import of phloem water. Thereafter, absolute growth rate remains constant for the remainder of seed fill, before declining precipitously as seed maturity is approached. Coincidentally, phloem-imported water is re-exported through the xylem (**see: Development of seeds – an overview; Development of seeds – nutrient supply**) (MT-H, CEO, JWP)

Development of seeds – hormone content

Developing seeds were the first discovered higher-plant sources of most of the known plant hormones. While the seed is accumulating its major storage reserves, changes are also occurring in its content of the **hormones – auxins, abscisic acid, cytokinins, gibberellins, ethylene, brassinosteroids** and **jasmonic acid**. These regulators are involved in the control of certain aspects of seed development, as well as in fruit growth, though during the course of development seeds change in their sensitivity to these substances. There are numerous **mutants** affecting content and responsiveness to hormones which are especially important in studies analysing the roles of hormones in seed development.

The types, kinetics and locations are considered below.

1. Abscisic acid (ABA)

ABA has been isolated from immature seeds of many species both free and conjugated as the glucosyl ester and the glucoside. The free form can reach high concentrations particularly in legume seeds (e.g. between 2 and 5 mg/kg fresh weight in soybean), but generally its maximum concentrations are 0.1 to 1 mg/kg fresh weight in others. These contents, if converted into approximate micromolar concentrations, fall well within the range known for physiological actions of applied ABA. ABA contents in seeds increase during development. In some cases accumulation exhibits one or two maxima towards the end of dry matter deposition and generally parallels seed dry weight increase (Fig. D.21). There is commonly a substantial and rapid loss of ABA at seed maturity and dehydration with little carry-over into the dry seed. Metabolites of ABA, phaseic and dihydrophaseic acid, also can be present at high concentrations. Both free and bound ABA occur in various parts of the seed but their concentrations are much higher in the **embryo** than elsewhere (e.g. in sorghum, embryo ABA concentration is three- to sixfold higher than in the **endosperm**). This might account for the higher concentrations in legume seeds (which are mostly embryo) compared to cereal seeds (which are mostly endosperm). The ABA may result both from synthesis in the embryo and translocation from other parts of the plant (e.g. leaves in plants subjected to stress). ABA contents can be artificially lowered either genetically (**see: Hormone mutants**) or by chemicals that interfere with any of the steps of its biosynthesis, such as fluridone which inhibits the formation of the carotenoids that are precursors of ABA (**see: Abscisic acid**). ABA is involved in the control of several processes that take place in the developing seed: (i) prevention of precocious germination; (ii) reserve deposition; (iii) acquisition of desiccation tolerance; (iv) prevention of reserve mobilization; (v) imposition of dormancy (**see: Desiccation tolerance; Dormancy – onset; Late-embryogenesis abundant proteins; Maturation; Mobilization of reserves – cereals; Storage reserves synthesis – regulation**).

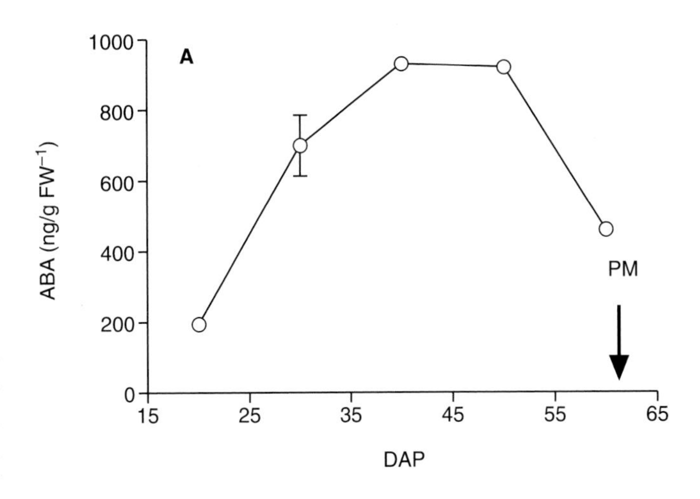

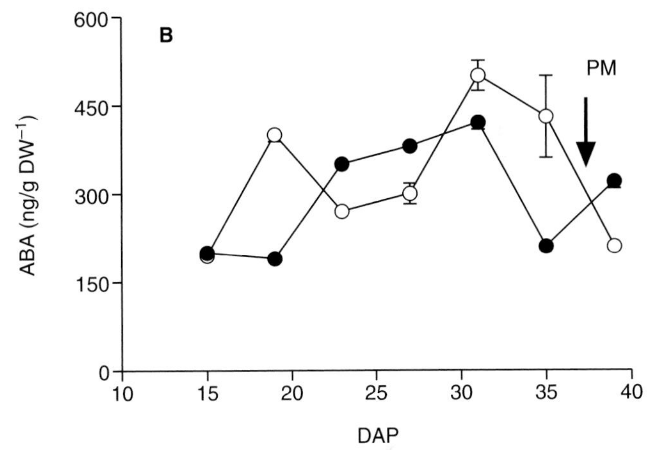

Fig. D.21. ABA content in developing embryos of (A) wheat cultivar Kitakei-1354 (adapted from Kawakami, N., Miyake, Y. and Noda, K. (1997) ABA insensitivity and low ABA levels during seed development of non-dormant wheat mutants. *Journal of Experimental Botany* 48, 1415–1421) and (B) two lines of *Sorghum bicolor* (adapted from Steinbach, H.S., Benech-Arnold, R.L., Kristof, G., Sánchez, R.A. and Marcucci-Poltri, S. (1995) Physiological basis of preharvest sprouting resistance in *Sorghum bicolor* (L.) Moench. ABA levels and sensitivity in developing embryos of sprouting-resistant and -susceptible varieties. *Journal of Experimental Botany* 46, 701–709). PM = Physiological maturity.

2. Gibberellins (GAs)

Sensitive physicochemical methods of analysis, principally combined gas chromatography and mass spectrometry (GC-MS) has revealed that gibberellins are a large group of natural products, with 126 different compounds currently known in higher plants, fungi and bacteria. Most of these GAs, moreover, are present in the majority of developing seeds and vegetative and floral tissues at low concentrations (0.1–100 ng/g fresh weight). The abundance of their biosynthetic enzymes is similarly low, precluding their purification from these sources. However, the developing seeds of several species (e.g. *Pisum sativum*, *Vicia faba*, *Phaseolus vulgaris*) have exceptionally high rates of GA biosynthesis and these have been the materials of choice both for the characterization of the enzymes and as a traditional biochemical route to isolate the genes.

GAs biosynthetic pathways are usually dissected into three sequential stages according to the nature of the enzymes involved, also reflecting their subcellular compartments (**see: Gibberellins**). The first step involves the production of *ent*-kaurene from the two-step cyclization of geranylgeranyl diphosphate (GGDP) and is carried out by soluble enzymes (terpene cyclases) localized in plastids. In the second stage of the pathway, *ent*-kaurene is oxidized by Cyt P450 mono-oxygenases associated with the endoplasmic reticulum to form GA_{12}-aldehyde. Finally, GA_{12}-aldehyde is converted to bioactive GAs (GA_1, GA_4, GA_3 and GA_7) by 2-oxoglutarate-dependent cytosolic dioxygenases. The latter include GA 7-oxidase, GA 20-oxidase; GA 3-oxidase, and GA 2-oxidase (which is responsible for the formation of inactive C-19 GAs – GA_{51}, GA_{29}, GA_{34} and GA_8, that are subsequently transformed by the same enzyme into catabolites). Most of the precursors of biologically active GAs have been found in developing seeds (e.g. geranylgeranyl pyrophosphate, *ent*-kaurene, GA_{12}-aldehyde, GA_5, GA_6, GA_8, GA_{17}, GA_{19}, GA_{20}, GA_{28}, GA_{34}, GA_{37}, GA_{38} and GA_{44} in *P. sativum*, *P. vulgaris* and *Phaseolus coccineus*) as well as the active ones (GA_1, GA_4 and GA_3) and their catabolites (GA_{29}-catabolite, GA_{51}-catabolite, GA_{34}-catabolite and GA_8-catabolite). Several gibberellin glycosides are formed by covalent linkage between GA and a monosaccharide. These GA conjugates are particularly prevalent in seeds of some species (e.g. *P. vulgaris*, *P. coccineus*). Glycosylation may represent a form of inactivation and/or a source of biologically active, free GAs after de-conjugation. Table D.4 shows some examples of GAs that have been detected in developing seeds.

Gibberellins are involved in: (i) seed development (and later are present in mature seeds and seedlings); (ii) fruit growth and development; (iii) seed germination; and (iv) mobilization of stored reserves, particularly in cereals (**see: Mobilization of reserves – cereals**). There is a good correlation between the content of the late precursors GA_9 and GA_{20} (which can be converted into bioactive GAs by the action of the enzyme 3-β-hydroxylase or oxidase) and seed growth: concentration peaks for these two GAs coincide with the active growth phase of the seeds (Fig. D.22). Similarly, GA_1 is a major component at early stages of embryo development and its role in early embryo growth has been suggested. GAs produced by the developing seed are implicated in fruit growth and development. For example, fruit development in *P. sativum* is impaired when the immature seeds are killed; fruit growth is restored by application of GA_3. Conversion of inactive GA compounds into active GAs by the enzyme GA 3-oxidase (hydroxylase) (the last step in the biosynthesis of bioactive GAs) occurs in lettuce seeds stimulated to germinate by red light (**see: Light-dormancy and germination; Phytochrome**), but it has not been demonstrated in developing seeds.

GAs may be distributed unequally in the seed. In pea, for example, the various GAs occur at very different concentrations in the testa, cotyledons and embryonic axis. In maize, the GA_1 content of the embryo is about 40 times higher than in the endosperm. It might be that GAs are transported around the seed and may be subjected to differential metabolism at various sites. (**See: Gibberellins**)

Table D.4. Some examples of GAs detected in developing seeds.

Gibberellin	Species
GA_1	*Pisum sativum, Phaseolus coccineus, Zea mays*
GA_3	*P. sativum, P. coccineus, Z. mays*
GA_4	*P. sativum*
GA_5	*P. sativum, P. coccineus*
GA_8	*P. sativum, P. coccineus, Z. mays*
GA_{17}	*P. sativum, P. coccineus*
GA_{20}	*P. sativum, P. coccineus, Z. mays*
GA_{29}	*P. sativum, Z. mays*
GA_{37}	*P. sativum*
GA_{44}	*P. sativum, P. coccineus, Z. mays*
GA_6	*P. coccineus*
GA_{19}	*P. coccineus, Z. mays*
GA_{28}	*P. coccineus*
GA_{34}	*P. coccineus*
GA_{38}	*P. coccineus*

3. Auxins

The **auxin**, indole-3-acetic acid (IAA) accumulates differently among the **embryo**, **suspensor** and **integument** during *Phaseolus* seed development: highest total (free and bound) concentrations (13 and 33 μg/g fw, respectively) are in the embryo during early-heart stage, and decrease substantially by the later heart stage of **embryogenesis**. In *Arabidopsis* also IAA is implicated in various stages of embryogenesis. In maize kernels, IAA synthesis and accumulation occurs mainly in the endosperm; total IAA contents increase between 10 and 12 DAP (days after pollination) and stabilize at 20 DAP (1.5 and 120 μg/g kernel fw for free acid and conjugated IAA). IAA stimulates nuclear DNA **endoreduplication, amyloplast** differentiation and storage material accumulation in the maize endosperm.

Physiological levels of IAA are regulated by conjugation (amides and esters) with sugars, amino acids and peptides. In developing maize endosperm after 12 DAP as much as 2% of total IAA is in the free acid form while 98% is as IAA ester

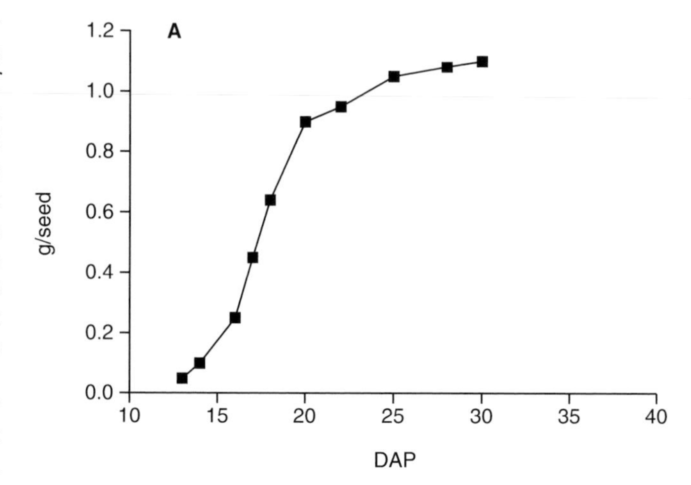

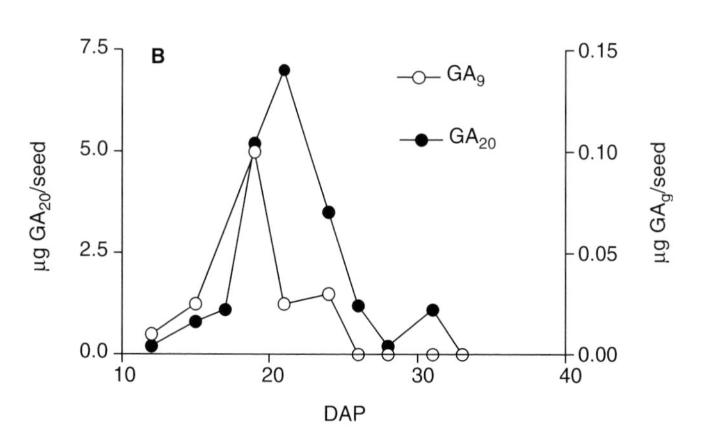

Fig. D.22. Seed weight (A) and content (B) of gibberellin GA_9 and GA_{20} (both precursors of bioactive GAs) during development of *Pisum sativum* seeds. (Adapted from Frydman *et al.*, 1974, in Khan, A.A. (1982) Gibberellins and seed development. In: Khan, A.A (ed.) *The Physiology and Biochemistry of Seed Development, Dormancy and Germination.* Elsevier Biomedical Press, Amsterdam, The Netherlands, pp. 111–136.)

conjugates. Ratios of free to conjugated IAA vary widely among seed parts throughout the development of bean (*Phaseolus vulgaris*) seeds. Total IAA content in mature maize seeds is about 1000 times higher than in bean seeds (160,000 vs 170 ng/seed), and this can be related to the fact that maize seedlings are totally reliant on IAA from conjugates stored in the seed, while *de novo* synthesis provides the IAA for developing bean seedlings.

Tryptophan (Trp) is the main precursor of IAA biosynthesis in maize endosperm and is converted to IAA via deamination and decarboxylation. Other possible routes also occur in plants, which may or may not involve Trp as a precursor. (**See: Auxins**)

Synthesis of IAA in seeds has been demonstrated to occur in developing maize endosperm. In developing seeds of dicots, *in situ* synthesis of IAA has not been proved and transport of IAA from the mother plant into the developing seeds cannot be ruled out.

4. Cytokinins (CKs)

Cytokinin biosynthesis in developing seeds was first shown to occur in *Lupinus albus*: but it is possible that in some species there is CK transport from the mother plant into the seed.

Cytokinin content in developing maize kernels and other cereal grains shows a transient increase between 6 and 12 DAP (2 nmol/g fw) in the endosperm. This peak coincides with the highest nuclear division and amyloplast differentiation rates in the endosperm. In both *L. albus* and *Cicer arietinum* seeds, highest CK concentrations occur in the liquid endosperm during onset of **cotyledon** expansion and endosperm depletion (40–46 DAP) (Fig. D.23). In these and other species CKs appear to regulate seed sink capacity by maintenance of cell division in the storage tissue.

Physiological levels of active CKs are likely to be regulated mainly by cytokinin oxidase (CKO)-mediated cleavage of the side chain and release of the inactive adenine or adenosine. Reversible CK conjugates may play a role in hormone homeostasis, but highly variable rates of CKO activity observed among different tissues and throughout development suggest a key role of this enzyme in regulation of CK levels. Also, the biological activity of the multiple forms of CK remains to be characterized for us to understand their role in seed development. The main CK types in developing *L. albus* seeds are listed in Table D.5.

Most CKs are N^6-substituted adenine derivatives, and include **zeatin** (Z), dihydrozeatin (DHZ), and their nucleotide, nucleoside and riboside forms. All these can be conjugated to sugars or amino acids; some conjugates are reversible (like *O*-glucosyl) while others are not (*N*-glucosyl). The isopentenyl-transferase (ipt) enzyme adds the isopentenyl side chain to the purine ring, and several genes encoding ipt have been cloned in *Arabidopsis* and other plant species. Several enzymes involved in the interconversion of CK forms have also been identified. Knowledge about the *in vivo* role of different CK types is very limited. (**See: Cytokinins**)

5. Ethylene

A variety of patterns of **ethylene** (ethene) production in developing seeds has been reported for several species.

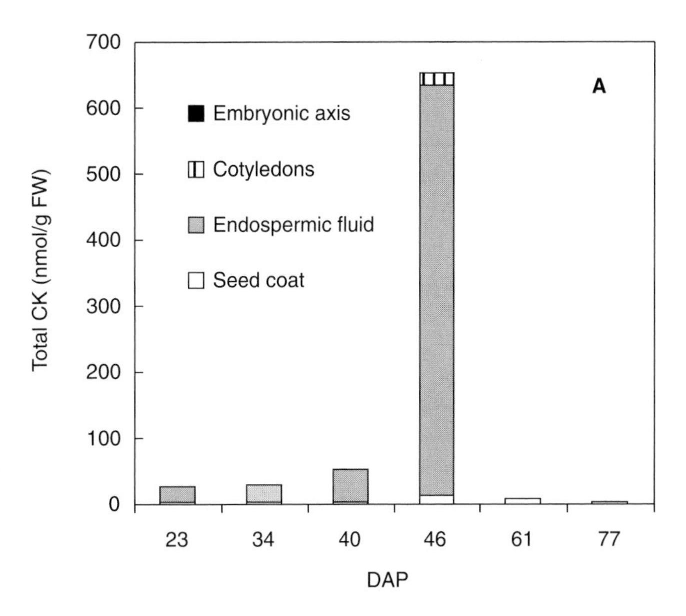

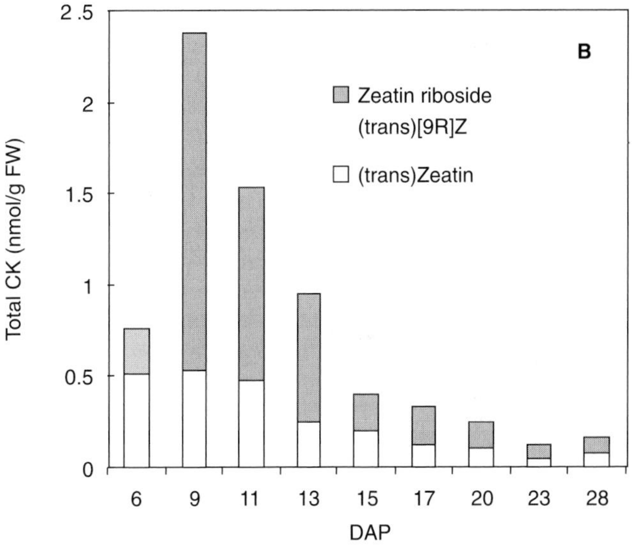

Fig. D.23. (A) Total cytokinin concentration measured in seed tissues during *Lupinus albus* seed development. (From: Emery, R.J.N., Ma, Q. and Atkins, C.A. (2000) The forms and sources of cytokinins in developing white lupine seeds and fruits. *Plant Physiology* 123, 1593–1604.) (B) Cytokinin concentration in developing maize endosperms. (Adapted from Lur, H-S. and Setter, T.L. (1993) Role of auxin in maize endosperm development. Timing of nuclear DNA endoreduplication, zein expression, and cytokinin. *Plant Physiology* 103, 273–280.) DAP, days after pollination.

Ethylene production decreases during seed maturation in sunflower, while in *Brassica* seeds it decreases from 600 to 100 nl/g fw/h between 15 and 35 DAP, and then peaks transitorily during the desiccation phase (around 40 DAP).

The ethylene biosynthesis pathway is well understood: methionine is sequentially converted to SAM, ACC and finally ethylene (for an explanation of these and subsequent abbreviations see: **Ethylene**). Key enzymes ACC-synthase and ACC-oxidase are rate limiting in ethylene biosynthesis, and their activities are specifically regulated in plant tissues. The ACC pool can be controlled also through reversible

Table D.5. Cytokinin content measured in seedcoat, endospermic fluid and cotyledons of 46 DAP *Lupinus albus* seeds, and in 61 DAP embryonic axis. Physiological maturity at 77 DAP.

	Cytokinin concentration (pmol/g fresh weight)			
	Seedcoat	Endospermic fluid	Cotyledons	Embryonic axis
(*trans*) Zeatin riboside	4,737	23,9414	7,323	9,054
Dihydro-zeatin riboside	3,932	10,9083	5,004	4,686
(*trans*) Zeatin riboside *O*-glucoside	2,642	16,7522	3,849	Nd
(*trans*) Zeatin	248	23,692	1,558	972
(*trans*) Zeatin *O*-glucoside	181	15,729	Nd	Nd
(*cis*) Zeatin-riboside	351	14,364	189	443
Dihydro-zeatin	166	6,669	1,060	718
(*cis*) Zeatin riboside *O*-glucoside	274	5,835	175	167
(*trans*) Zeatin nucleotide	428	4,731	1,591	966
Dihydro-zeatin nucleotide	377	2,873	576	645
(*cis*) Zeatin-nucleotide	16	598	860	119
Dihydro-zeatin *O*-glucoside	24	948	110	Nd
Dihydro-zeatin riboside *O*-glucoside	Nd	Nd	Nd	Nd

Adapted from: Emery, R.J.N., Ma, Q. and Atkins, C.A. (2000) The forms and sources of cytokinins in developing white lupin seeds and fruits. *Plant Physiology* 123, 1593–1604.
Nd, not detected.

conjugation to MACC or GACC (glutamylACC) by specific enzymes. ACC-synthase and ACC-oxidase mRNA contents and enzyme activity in *Cicer arietinum* embryos reach their maxima during mid-stages of development and decline thereafter. Ethylene and ACC levels peak at 16 and about 36 DAP during maize kernel development.

The function of ethylene during early embryogenesis is not clear but, among other effects, it may be involved in chlorophyll loss. Ethylene climateric production, in combination with low auxin content, induces cell separation in the abscission zone of *Brassica* siliques (pod shatter). Exogenous ethylene induces **vivipary** when introduced into developing *Phaseolus* pods, and hastens grain development in wheat; ACC in *Lupinus* seeds inhibits assimilate accumulation in the embryo. In maize kernels, endogenous and applied ethylene induce **programmed cell death** in the endosperm. In these cases ethylene seems to counteract ABA-mediated maintenance of the developmental mode.

6. Brassinosteroids (BRs)

Brassinosteroids are present in developing seeds of several species (e.g. *Brassica campestris*, *Arabidopsis thaliana* and *Pisum sativum*). The biosynthesis of brassinolide originates in campesterol and involves several intermediates: teasterone, typhasterol castasterone in the early C_6 oxidation pathway, or 6-deoxoteasterone, 6-deoxotyphasterone and 6-deoxocastasterone in the late C_6 oxidation pathway. Brassinolide and castasterone are among the BRs with higher biological activity.

Brassinosteroids are required for normal seed development, and probably in dormancy and germination, since BR can restore the germination phenotype in GA-deficient **mutants**. Decreasing BR signalling (**see: Signal transduction**) increases germination sensitivity to inhibition by ABA.
(See: Brassinosteroids)

7. Jasmonates

Jasmonic acid (JA) and its methyl ester (methyl jasmonate [MeJA]) derive from linolenic acid (LA). Hydroperoxidation of LA by lipoxygenase is the initial synthetic step followed by reactions catalysed by allene oxide synthase and allene oxide cyclase, forming 12-oxo-phytodienoic acid. Then by a reduction and three steps of β-oxidation (+)-7-iso-JA is produced. MeJA has been found in developing seeds of, for example, *Quercus robur* and *Fraxinus excelsior*: in soybeans the content of JA is low in young developing seeds (*ca.* 0.1 ng/g FW) and increases in older seeds (0.5 ng/g FW). When the recalcitrant seeds of *Q. robur* are dried, JA and MeJA content increase before loss in viability. However, the production of jasmonates might simply be a consequence of lipid peroxidation and may not necessarily be involved in regulation of germination of these seeds. JA can either promote germination (in dormant seeds) or inhibit it (in non-dormant seeds). For example, during development, JA might decrease the effect of ABA on dormancy inception and in mature seeds the addition of JA during imbibition might enhance the synthesis of or sensitivity to ABA.
(See: Jasmonates)

8. Sensitivity to hormones

Tissue sensitivity to hormone action can be expected to change throughout seed development (**see: Hormones**). However, only in the case of two hormones have these changes been assessed.

Embryo sensitivity to abscisic acid (ABA) is very high at early stages of development: concentrations of between 0.5 and 1 μM ABA are enough, not only to suppress embryo germination, but also to allow the expression of genes that are clearly associated with a seed development programme. Sensitivity to ABA gradually decreases throughout seed development and considerably higher concentrations (5 to 50 μM ABA) are usually required to maintain the embryo in a developmental mode (including suppression of germination) (Fig. D.24, and **see** Fig. M.9 in **Maturation – controlling factors**). Desiccation at the end of the maturation period in orthodox seeds greatly reduces embryo responsiveness to ABA. Sensitivity to ABA can be modulated by the action of other hormones such as auxins, brassinosteroids and ethylene (i.e. the ABA-hypersensitive *era1* mutant with defective ethylene response). Embryo sensitivity to ABA during development can also be modified by the environment experienced by the mother plant (i.e. water or nutrient availability). Low sensitivity to ABA may be involved in susceptibility to preharvest sprouting **(see: Pre-harvest sprouting – mechanisms)**.

The **aleurone layer** in wheat is insensitive to gibberellins during seed development. Only after the grain has undergone desiccation are these cells able to respond to applied GA_3 and hence to produce α-amylase. The mechanism(s) by which sensitivity is modulated is unknown but it could involve changes in receptor availability, in transduction chain components or in changes at the gene level or in hormone turnover **(see: Signal transduction – hormones)**.

(RLB-A, RAS, MVR)

Creelman, R.A. and Mullet, J.E. (1997) Oligosaccharins, brassinolides, and jasmonates: nontraditional regulators of plant growth, development and gene expression. *The Plant Cell* 9, 1211–1223.

Finkelstein, R.R., Gampala, S.S.L. and Rock, C.D. (2002) Abscisic acid signalling in seeds and seedlings. *The Plant Cell* S15–S45 (Supplement 2002).

Haberer, G. and Kieber, J. (2002) Cytokinins. New insights into a classic phytohormone. *Plant Physiology* 128, 354–362.

Hedden, P. and Phillips, A.L. (2000) Gibberellin metabolism: new insights revealed by the genes. *Trends in Plant Science* 5, 523–530.

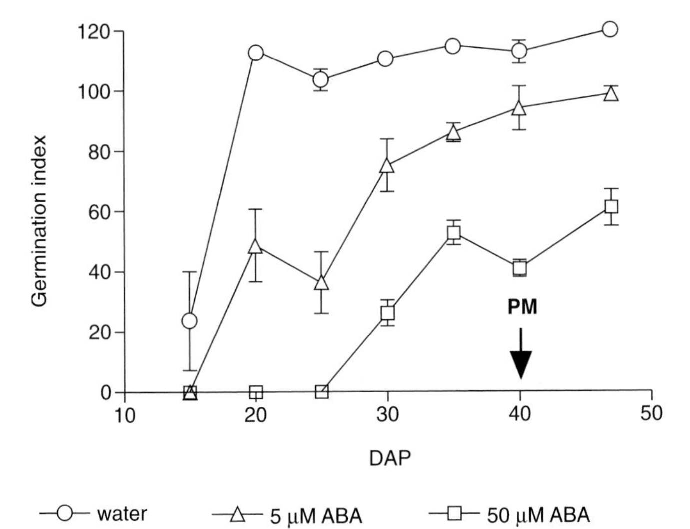

Fig. D.24. Loss of sensitivity to the germination-inhibiting effect of ABA in developing *Sorghum bicolor* embryos. Germination index is a measure of the rate and amount of germination. From: Steinbach, H.S., Benech-Arnold, R.L., Kristof, G., Sánchez, R.A. and Marcucci-Poltri, S. (1995) Physiological basis of preharvest sprouting resistance in *Sorghum bicolor* (L.) Moench. ABA levels and sensitivity in developing embryos of sprouting-resistant and -susceptible varieties. *Journal of Experimental Botany* 46, 701–709.
PM, Physiological maturity. (**See** also: Fig. M.9 in **Maturation – controlling factors**)

Lur, H-S. and Setter, T.L. (1993) Role of auxin in maize endosperm development. Timing of nuclear DNA endoreduplication, zein expression, and cytokinin. *Plant Physiology* 103, 273–280.

Matilla, A.J. (2000) Ethylene in seed formation and germination. A Review. *Seed Science Research* 10, 111–126.

Nambara, E. and Marion-Poll, A. (2003) ABA action and interactions in seeds. *Trends in Plant Science* 8, 213–217.

Picciarelli, P., Ceccarelli, N., Paolicchi, F. and Calistri, G. (2001) Endogenous auxins and embryogenesis in *Phaseolus coccineus*. *Australian Journal of Plant Physiology* 28, 73–78.

Schrick, K., Mayer, U., Martin, G., Bellini, C., Kuhnt, C., Schmidt, J. and Jürgens, G. (2002) Interactions between sterol biosynthesis genes in embryonic development of *Arabidopsis*. *Plant Journal* 31, 61–73.

Young, T.E., Gallie, D.R. and DeMason, D.A. (1997) Ethylene-mediated programmed cell death during maize endosperm development of wild-type and *shrunken2* genotypes. *Plant Physiology* 115, 737–751.

Development of seeds – influence of external factors

The environment can affect seed growth directly by regulating the ability of the seed to accumulate dry matter, or indirectly by affecting the ability of the parent plant to supply raw materials to the seed. The effect of the environment on the mature seed can be best understood by considering its effect on the individual processes that combine to produce the final product. Considering the mature seed as a product of a rate of growth expressed over a finite duration provides an excellent framework to evaluate genetic and environmental effects on seed growth and development. This model can be applied to all plant species because the characteristics of seed growth and development are remarkably consistent across species, in spite of the variation in seed structure, seed composition and other plant characteristics. Although it is not always easy to separate direct and indirect effects, it seems that indirect effects may be more common than direct effects in many plant species. **(See: Yield determination)**

Temperature affects growth rate of a seed directly by controlling its metabolic activity. Seed growth rates of most crop species show a relatively broad temperature optimum, with rate decreasing at high (> 30°C) or low temperatures (< 20°C), although these critical temperatures vary among species. Seed-fill duration generally increases as temperature drops below 30°C with the most rapid increase at temperatures below approximately 20°C. Again, this response varies among species, with some showing little change between 20 and 30°C. The effect of temperature on seed-fill duration could involve both direct and indirect (via leaf senescence) effects, as it seems that seed-fill duration is regulated at the seed and leaf levels.

Surprisingly, water stress seems to have little direct effect on the seed, but there are major indirect effects. Seed growth rate is relatively insensitive, but stress during seed filling accelerates leaf senescence and, indirectly, shortens the seed filling period. This response is not obvious without a well-watered plant for comparison, as the senescence and seed maturation processes under stress appear to be completely normal, they just occur sooner.

Other environmental factors that affect plant growth (e.g.

nutrient supply, plant population, solar radiation, CO_2 concentration) probably affect seed growth indirectly by modifying the supply of raw materials to the seed. These indirect effects are frequently limited by the tendency of crop plants to adjust seed number to changes in assimilate supply (**see: Charles-Edwards model**), thereby maintaining a relatively constant supply of assimilate and N to the individual seed. Variation in environmental conditions that cause large differences in plant growth and yield may have minimal effects on the growth of the individual seed since the plant adjusts sink size to maintain a constant source–sink ratio, thereby minimizing changes in the assimilate supply to the individual seed. This adjustment process insulates the individual seed from environmental effects that occur during vegetative growth, flowering and seed set. The environment is much more likely to affect the supply of assimilate to the seed during seed filling when seed number can no longer adjust. For example, reducing canopy photosynthesis by shading the crop during the entire reproductive period will reduce yield and seed number, but will have little effect on the growth of an individual seed. Shade only during seed filling after seed number is fixed, however, will probably reduce seed growth rate, mature seed size and yield. A sudden improvement in the external conditions during seed filling that increase photosynthesis after seed number is fixed will increase seed growth rate, unless the seed is already growing at its maximum rate. Seed-fill duration may increase if the rate does not respond, producing larger seeds and higher yield.

The location of the seed on the plant or in the ear, head or panicle may affect seed growth rate and seed-fill duration. It is not clear if location effects are due to the physical location of the seed or to the time of flowering, which is frequently confounded with position. For example, seeds from late developing flowers may be smaller because of shorter seed-filling periods – shorter because the seeds start growth later, but mature at nearly the same time as the other seeds on the plant.
(See: Development of seeds – dry matter accumulation; Development of seeds – growth phases) (DBE)

Egli, D.B. (1998) *Seed Biology and the Yield of Grain Crops*. CAB International, Wallingford, UK.

Development of seeds – nutrients supply

See: Development of seeds – principles of nutrient and water import; Development of seeds – sources of nutrients and water

Development of seeds – nutrients and water import

The key structural features of the pathways of nutrient and water import into developing seeds and the types of plasma membrane transporters are shown in Figs D.26 and D.27, and schematics of a **grain legume** and **wheat** seed are shown in Fig. D.25. In the grain legume (Fig. D.25A), the maternal tissue is comprised of the seedcoat in which the transport tissues (phloem and xylem) are located in one or more veins embedded in ground tissues. The enclosed filial (embryonic) tissue consists of two large storage cotyledons and the root and shoot axes (not shown). Nutrient efflux to the embryo from the seedcoat occurs across the entire inner surface of the latter, from ground tissues, and influx into the cotyledons is principally by their epidermal cells abutting the seedcoat. In wheat (Fig. D.25B), the maternal tissue is comprised of the **pericarp/testa** and **nucellus**, with import of nutrients occurring through a single crease vein running the length of the seed. Beneath the crease vein the nucellus extends (nucellar projection) into a cavity (endosperm cavity) that forms between the maternal and enclosed filial (embryonic and endosperm) tissues prior to seed fill. A cuticle separates the maternal pericarp/testa and nucellus from the filial **aleurone layer** and starchy endosperm tissues and extends between the crease vein and nucellar projection. This region is referred to as the pigment strand. This structural organization isolates the apoplasms of the maternal and filial tissues and directs exchange of nutrients from maternal to filial tissues through the endosperm cavity. The maternal nucellar projection cells are those from which there is the efflux of nutrients and the aleurone layer cells bordering the cavity are responsible for nutrient influx.

Detailed descriptions of the components of this nutrient and water transport system into developing seeds are presented below.

1. Cellular connections and compartments

(a) *Apoplasm and apoplasmic barrier*. The apoplasm (or apoplast) is the extracellular compartment located outside the plasma membrane of plant cells (Fig. D.26). It is principally

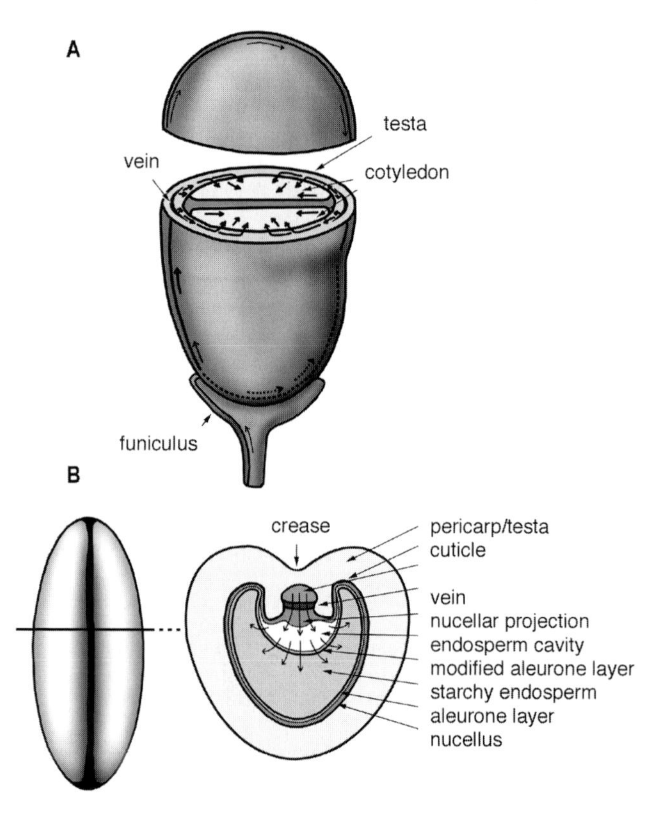

Fig. D.25. Key structural features of developing seeds. Schematics of a grain legume (A) and wheat (B) seed showing routes of nutrient transport (arrows). Redrawn from Patrick, J.W. and Offler, C.E. (2001) Compartmentation of transport and transfer events in developing seeds. *Journal of Experimental Botany* 52, 551–564.

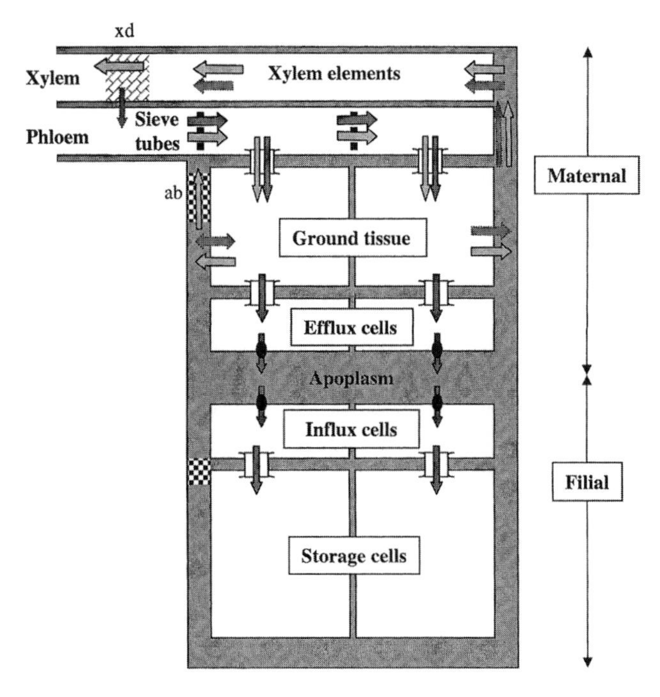

Fig. D.26. Scheme illustrating nutrient and water transport pathways into developing seeds. Nutrients (⇓) and water (⇓) enter seed maternal tissues through the phloem. Nutrients are unloaded from the phloem symplasmically (via plasmodesmata, ⫫). Thereafter the path for water transport is uncertain but nutrients move symplasmically to transport cells located proximal to the maternal/filial interface. The maternal transport cells support nutrient release to the seed apoplasm. Nutrients are retrieved from the seed apoplasm by juxtaposing filial transport cells. Both nutrient release and retrieval are mediated by transporters (⬇) located in the plasma membranes of the transport cells. Excess water is returned to the maternal plant via the xylem, while loss of nutrients from the seed is prevented by xylem discontinuities (xd, ▒) and apoplasmic barriers (ab, ▩). Redrawn from Patrick, J.W. and Offler, C.E. (2001) Compartmentation of transport and transfer events in developing seeds. *Journal of Experimental Botany* 52, 551–564.

comprised of cell walls but also includes the fluid occupying spaces in cell walls, surface films around intercellular spaces and lumens of dead xylem elements. The fluid-filled component of the apoplasm is referred to as the free space. The seed apoplasm acts as a component of the nutrient transport route from maternal to filial tissues of developing seeds. Nutrient concentrations of free-space fluid collected from developing seeds are high and generate osmotic concentrations (**see: 3. Osmotic pressure and concentration** in **Development of seeds – principles of nutrients and water import**) in the range of 300 to 400 mOsmol/kg.

Barriers to nutrient transport through the apoplasm are formed by impregnation of the cell wall by suberin, sometimes accompanied by lignification. The formation of such barriers in strategic locations regulates accumulation and distribution of nutrients by directing or restricting flow to symplasmic compartments.

(b) *Symplasm and symplasmic transport*. The symplasm (or symplast) is the cytoplasmic continuity of adjoining cells interconnected by plasmodesmata (Fig. D.26). Since the majority of cells are symplasmically connected, the symplasm offers a route for intercellular transport of nutrients. Exceptions occur at symplasmic disjunctions, for example, between filial and maternal tissues in seeds, mandating transmembrane exchange of nutrients between cells via the seed apoplasm.

Symplasmic transport of nutrients occurs through plasmodesmata and cytosols of adjoining cells. Nutrient import into developing seeds is characterized by transport through extensive symplasmic pathways in both maternal and filial tissues (Fig. D.26).

Other than in exceptionally large cells, symplasmic transport appears to be rate-limited by transport through interconnecting plasmodesmata. This movement of nutrients occurs by diffusion but, in some cases, is supplemented by bulk flow (**see: Development of seeds – principles of nutrient and water import**). The relative contribution of diffusion and bulk flow is determined by differences in nutrient concentrations or hydrostatic pressures, respectively, between importing sieve elements (SEs) and recipient sink cells (**see: 5. Post-sieve element transport pathways**). For a set concentration or pressure difference, rates of nutrient transport through the symplasm are regulated by changes in plasmodesmal conductances. Plasmodesmal conductances of symplasmic pathways are determined by plasmodesmal numbers summed over their hydrodynamic radii (**see:** equations describing ***bulk flow*** and ***diffusion*** in **Development of seeds – principles of nutrient and water import**). In this context, plasmodesmata located at the sieve element/vascular parenchyma interfaces are considered to be key sites of control of seed loading, as suggested by their relatively low frequency and large declines in sucrose concentrations (200 to 400 mM) across the cell interfaces they interconnect.

In developing seeds, nutrients are probably moved from the importing sieve elements to the surrounding vascular parenchyma cells by bulk flow of phloem sap, and then by diffusion through plasmodesmata interconnecting the cells of the ground tissues to the transport cells involved in membrane transport of nutrients to the seed apoplasm (Fig. D.26). In addition to delivering nutrients, bulk flow of sap from the phloem provides a mechanism to meet water requirements for growth in volume of expanding seeds (once seeds reach full expansion all phloem-imported water is recycled to the parent plant – **see: 3b. Xylem export** and Fig. D.26) and to dissipate a pressure build-up in the sieve tubes that would otherwise compromise phloem import (**see: 3a. Phloem import**).

2. Vascular transport

(a) *Phloem*. Along with the xylem, this tissue is part of the plant's vascular system – the conduit for water and nutrient transport (Fig. D.26). Phloem tissue is responsible for translocation of nutrients over long distances (up to 100 m in the tallest trees) requiring specialized transport cells. The transport cells, sieve elements (SEs), are organized into longitudinal files called sieve tubes and in most instances are accompanied by companion cells forming sieve element–companion cell (SE–CC) complexes, which are surrounded by vascular (sometimes called phloem) parenchyma cells. In developing seeds, phloem is arranged in discrete vascular bundles that may or may not branch.

Sieve elements (SE) and their accompanying companion cells (CC) are ontogenetic and functional complexes (SE–CC complexes). Key features of sieve element specialization for transport function are a sparse and peripheral cytoplasm interconnected into a continuous symplasm via large pores (sieve pores) formed in their end walls (sieve plates). Sieve elements are symplasmically connected by complex branched plasmodesmata to their accompanying companion cells. These cells are characterized by a dense organelle-rich cytoplasm that contributes to maintaining functional integrity of their accompanying sieve element and to aid in loading of nutrients for long-distance transport. In contrast to leaves, companion cells in sink regions (**see: Yield determination, 2. Sink**) are reduced in size and in some instances, including developing seeds, do not accompany each sieve element. These reductions in companion cells have been linked with a decreased requirement for nutrient loading of sieve elements in sinks.

(b) *Xylem and xylem discontinuity.* This component of the plant's vascular tissue system (Fig. D.26) is composed of water-conducting xylem elements (tracheids and/or vessel elements), vascular parenchyma and xylem fibres. Tracheids and vessel elements lack protoplasts at maturity and are arranged in longitudinal files to support water transport through their hollow lumens in the transpiration stream. Transport of water between adjoining xylem elements is facilitated by narrowed regions of the shared cell walls (pits) in tracheids and by partial or complete dissolution of end walls in vessel elements. The principal vascular bundles of developing seeds contain only a relatively small amount of xylem tissue consisting of two to six vessel elements. Indeed, in some cases, xylem is absent from vascular bundles, such as those in lateral veins located proximal to the embryonic axis in both monocot and dicot seeds.

In a wide range of developing seeds, the vasculature located in organs connecting the maternal seed tissues to the parent plant is characterized by xylem tissues containing greatly diminished numbers of vessel elements (e.g. funicles of grain legumes). In some cases, xylem elements are absent (e.g. pedicels of wheat grains) resulting in a complete xylem discontinuity. These structural modifications probably contribute to the independence of seed water relations from the parent vegetative plant body: a feature central to regulating phloem import.

3. Import and export from vascular tissues

(a) *Phloem import.* Nutrient transport through the phloem is by bulk flow (**see: Development of seeds – principles of nutrient and water import**). The bulk flow is driven by turgor pressure differences (**see: Turgor**) generated by water movement into and out of the phloem. Water movement into the phloem occurs down osmotic gradients (**see: 3. Osmotic pressure and concentration in Development of seeds – principles of nutrient and water import**) caused by nutrients loaded into the phloem at the source (**see: 1. Source in Yield determination**) end of the phloem path. Dissipation of turgor pressure at the sink (**see: 2. Sink in Yield determination**) ends of phloem pathways results from phloem unloading of water driven by hydrostatic or osmotic gradients. Flow rate of a nutrient species through the phloem is governed by the bulk flow rate of water and the concentration of the nutrient species in the flowing sap. In contrast, nutrient transport rates are not limited by conductances (**see: 1. Bulk flow in Development of seeds – principles of nutrient and water transport**) of mature phloem pathways to support bulk flow.

The pressure-flow hypothesis of phloem transport describes the import rate (R_f) of a nutrient species as:

$$R_f = L_p\,(P_{source} - P_{sink})\,A\,C$$

Differences in hydrostatic pressure (P) between source (leaf minor veins and **see: 1. Source in Yield determination**) and sink (**see: 2. Sink in Yield determination**) ends of the phloem path, modulated by its hydraulic conductivity (Lp) and cross-sectional area (A), determines volume flow of water carrying nutrients at certain concentrations (C) (**see: 1. Bulk flow in Development of seeds – principles of nutrient and water import**). At quasi water equilibrium, water potential (ψ) of sieve element (se) sap approximates that of the surrounding apoplasm (a) such that:

$P_{se} - \pi_{se} = P_a - \pi_a$ and, as a consequence,
$P_{se} = (\pi_{se} - \pi_a) + P_a$

where π is sap osmotic pressure (**see: 3. Osmotic in Development of seeds – principles of nutrient and water import**) and P hydrostatic pressure (**see: Turgor**).

Hydrostatic pressures in sieve elements of source leaves (P_{source}) are largely set by π_{se} resulting from phloem loading of major osmotic (**see: 2. Osmotic in Development of seeds – principles of nutrient and water import**) species into leaf minor veins. Variations in evaporative loss that alter leaf P_a do not appear to impact on P_{source} since phloem loading is turgor-regulated.

Phloem sap concentration (C) of each transported nutrient is determined by its phloem loading in source leaves and along the axial path following remobilization of reserves (**see: Development of seeds – sources of nutrients and water**). Hydraulic conductivities (L_p) and cross-sectional areas (A) of the phloem path are determined by sieve element numbers and sieve pore dimensions (limiting radius – **see: 1. Bulk flow in Development of seeds – principles of nutrient and water import**), respectively.

P_a has not been measured in developing seeds but is likely to exert minimal influence on P_{se} relative to π_a. This has been shown by reproducing seed import rates into attached seedcoat halves with solution osmotic pressures (π_a) that match those of seed apoplasmic saps and hence seedcoat turgors. Seed water potentials and turgor pressures are controlled by seeds independent of the water relations of vegetative organs as a result of their hydraulic isolation by structural barriers.

(b) *Xylem export.* The sites at which phloem-imported water is recycled and the export routes it follows have not been resolved in developing seeds. However, it is clear that unloaded nutrients are not returned to the parent plant in recycled water or by any other means. This outcome is considered to be achieved by strategically positioned selectively permeable apoplasmic barriers in maternal seed tissues and/or efficient retrieval mechanisms located along xylem export routes. Examples of the latter include diminished xylem paths (funicles of grain legumes), or complete discontinuities (pedicels of wheat grains; **see: 2.(b) Xylem and xylem discontinuity**), with closely juxtaposed xylem and phloem elements with or without intervening transfer cells (see: below).

4. Transport and transfer cells

Transport cells are specialized for the plasma membrane exchange of nutrients. Their specialization is manifest as a plasma membrane enriched in transporter proteins and a dense cytoplasm rich in **mitochondria**, **endoplasmic reticulum** and vesicles. In some transport cells, the plasma membrane surface is amplified by deposition of elaborate invaginated wall ingrowths usually polarized towards the direction of nutrient flow. These cells, termed **transfer cells**, often differentiate at 'bottlenecks' within nutrient transport pathways. Their differentiation is developmentally regulated and they form immediately prior to the onset of seed fill in response to signals that include sugars. In developing seeds of cereals, grain legumes and grasses transfer cells may be completely absent or differentiate in tissues at one or both sides of the maternal/filial interface. Commonly, at least two rows of influxing cells become transfer cells, but the extent of transfer cell differentiation can favour either the filial or maternal tissues. In species where transfer cells are absent, the cells of the maternal/filial interface are nevertheless specialized for nutrient transport (e.g. French [common, kidney] bean and rice). Why the wall ingrowth/plasma membrane complex of transfer cells occurs in only some transport cells is unknown.

5. Post-sieve element transport pathways

These pathways for accumulation of nutrients in sinks (**see: Yield determination, 2. Sink**) can be apoplasmic, symplasmic, or symplasmic interrupted by an apoplasmic step (seeds follow only one of these possible pathways). In developing seeds, a symplasmic discontinuity between maternal and filial tissues necessitates membrane efflux of nutrients from maternal tissues and subsequent influx by filial tissues (Fig. D.26). Transport pathways for cereal and grain legume seeds share common features, namely extensive symplasmic routes located on either side of the maternal/filial interface where exchange of nutrients occurs via the seed apoplasm. In cereals (wheat, maize) this exchange is restricted to a relatively short length of the maternal/filial interface (Fig. D.25B) while in grain legumes (pea, bean) the entire interface is involved (Fig. D.25A). During the storage phase of seed development, plasmodesmatal frequencies and their spatial organization canalize nutrient flow from the importing sieve elements to the effluxing cells. For wheat and barley, suberization and lignification of the pigment strand prevents apoplasmic transport, thus dictating that the nucellar projection transfer cells (Fig. D.25B and **see: 4. Transport and transfer cells**) are the sole cells responsible for membrane efflux. For pea and bean, the cellular sites of proton-coupled efflux (**see: Plasma membrane transporters**) from the seedcoats are located at (pea, broad or faba bean), or adjacent to (French or common bean), the maternal/filial interface. At this interface, plasma membrane proton-ATPases co-localize with sucrose plasma membrane transporters in transport cells of maternal and filial tissues. Accumulated nutrients are moved symplasmically through high frequencies of plasmodesmata for storage in endosperm cells of cereals or cotyledon storage parenchyma cells of grain legumes (Fig. D.26).

Post-sieve element transport through a symplasmic compartment allows for imported nutrients to be sequestered into short-term storage pools that function to buffer variation in phloem import rates. The small extra-phloem sucrose pool in maternal tissues of developing wheat grains offers minimal buffering capacity (sucrose turnover time of 1.3 h) but this is compensated by sucrose supplies accumulated in the endosperm that are capable of supporting starch biosynthesis for approx. 24 h. In maternal tissues of grain legume seeds, some 70 to 90% of amino acids and sucrose are located in readily-exchanged vacuolar pools that are estimated to meet embryo demand for 4 to 12 h. Starch turnover also buffers sucrose levels, but relative to vacuolar sucrose, contributes little to the overall sucrose flux.

6. Transport into and between cells

(a) *Plasma membrane transporters*. These are proteins present in the membrane that surrounds the protoplasm of cells, i.e. the plasma membrane. This membrane, and all other biological membranes, is impermeable for most solutes, including many that are important nutrients for seed development, e.g. sucrose, amino acids and potassium. However, these solutes are allowed membrane passage via membrane transporters (Fig. D.27). Most membrane transporters are specific for a certain substrate, i.e. a sucrose transporter mainly transports sucrose, but may, to some extent, transport molecules that are similar to sucrose, e.g. other disaccharides. Membrane transporters fall into three different classes, namely, channels, primary active transporters and secondary active transporters. Channels, or porins, catalyse energy-independent facilitated diffusion, whereas primary and secondary active transporters are energy dependent. Primary active transporters are driven by e.g. electrical or chemical energy. Secondary active transporters use energy derived from e.g. the proton motive force or concentration gradients of substrates, and function as uniporters, antiporters or symporters. A uniporter transports only one substrate, down its concentration or electrochemical gradient across the membrane. An antiporter transports two substrates in opposite directions, whereas symporters transport them in the same direction. This normally involves transport of one of the substrates down its concentration or electrochemical gradient, and the energy derived from that transport is used to drive the transport of the other substrate against its concentration or electrochemical gradient.

Transporters that play important roles in seeds are aquaporins, ion channels, sugar and amino acid transporters (Fig. D.27). Aquaporins and ion channels are porins, and catalyse the facilitated diffusion of water and ions, respectively. The main function of aquaporins is to regulate the water balance of cells, but they may also be involved in transporting molecules other than water, e.g glycerol. Several poorly selective ion channels, transporting univalent cations such as potassium, sodium and ammonium ions, are present in seeds. Sugar and amino acid transporters are important in nutrient transport, leading to seed filling. All sugar and amino acid transporters responsible for uptake of nutrients into the filial seed tissue are symporters driven by the proton motive force

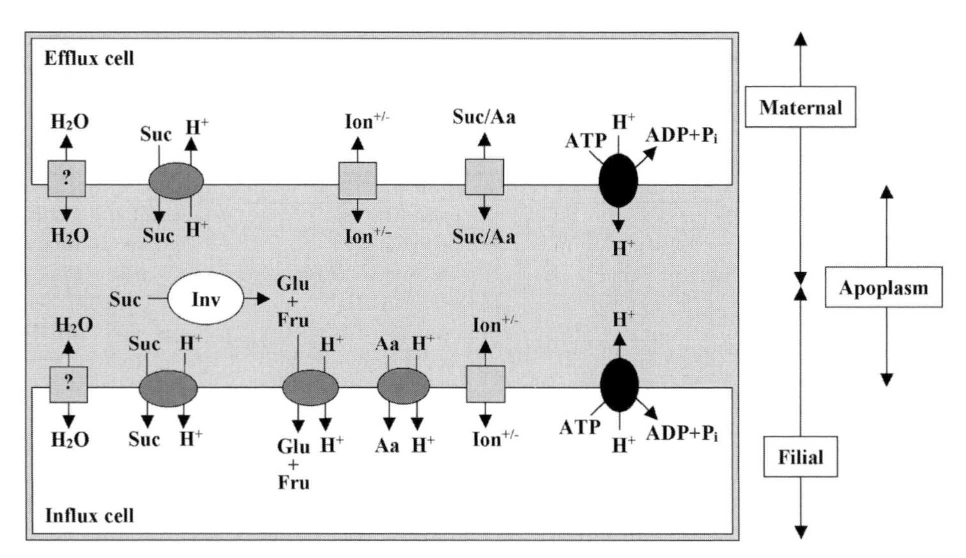

Fig. D.27. A schematic drawing of some important plasma membrane transporters found in efflux and influx cells of developing seeds is shown in Fig. D.26. Sucrose (Suc) is released from the maternal tissue to the apoplasm, possibly via a sucrose/proton (H^+) antiporter and/or a poorly selective channel, which may also transport amino acids (Aa). In the apoplasm sucrose may be transported directly into the filial tissue via a sucrose/proton symporter or be degraded to glucose (Glu) and fructose (Fru) by an invertase (Inv) (see: **Invertase (extracellular)**). The hexoses (glucose and fructose) can then be taken up by the influx cells via a hexose/proton symporter. Apoplasmic amino acids are transported into the filial tissues via amino acid/proton symporters. Also shown are poorly selective ion channels, water transporting (aquaporins), and proton-pumping ATPases, responsible for building up the proton gradient across the plasma membranes.

(i.e. the proton gradient that exists across many types of biological membranes). The proton motive force is built up by proton-pumping ATPases, which are primary active transporters. Sucrose release from seedcoats is driven by either sucrose/proton antiport and/or facilitated diffusion via a porin, which may also transport amino acids. High densities of plasma membrane transporters have been localized in maternal and filial cells proximal to the seed apoplasmic space.

(b) *Plasmodesmata.* Plasmodesmata (singular plasmodesma) are plasma membrane-lined micropores (usually 40 to 50 nm in diameter) that interconnect cytosols of adjacent cells via shared cell walls (Fig. D.26). A strand of compressed endoplasmic reticulum (ER, desmotubule), continuous between ER of adjacent cells, partially occludes the plasmodesmal pore to leave a ring of cytoplasm (cytoplasmic annulus). Nutrients move through the cytoplasmic annulus by diffusion and possibly in some cases by bulk flow (**see: Development of seeds – principles of nutrient and water import**).

7. Nutrients and the regulation of nutrient transport

(a) *Nutrients transported to the developing seed.* Most nutrients are imported into seeds through the phloem (see Fig. D.26). Sap composition varies according to plant species and genotype, plant nutrition (± nitrogen fixation), seed development and site of sampling along the axial phloem path. However, a consistent feature of phloem sap composition is that the relative contributions of solute species to phloem sap osmotic concentration (**see: Development of seeds – principles of nutrient and water import, 3. Osmotic pressure and concentration**), in descending order of significance, are sugars (approx. 50% and mostly sucrose) and potassium (interchangeable with sucrose and amino compounds), amino compounds (approx. 15% and rising to approx. 40% during seed filling), keto acids and inorganic compounds and mineral elements. Iron, manganese and zinc are translocated in chelated forms.

Amides (asparagine and glutamine) are the principal forms in which amino nitrogen is translocated in most species. Their amino acid derivatives (aspartate and glutamate) and, at much lower sap concentrations, various species-specific combinations of alanine, arginine, homoserine, serine and threonine are the main amino acids present in phloem saps. In some legume species (e.g soybean), **ureides** are present but these are rapidly metabolized to amino acids and amides before reaching developing seeds.

Bulk flow through the phloem means that developing seeds exert little control over rates at which minor osmotic species are imported. Thus, metabolic interconversion is critical to meet nutritional requirements of the filial tissues. For example, imported amides undergo considerable metabolism to a spectrum of amino acids in maternal seed tissues and concomitantly increase the molar ratio of amino nitrogen released to filial tissues. (**see: Development of seeds – sources of nutrients and water**).

(b) *Regulation of nutrient transport.* Phloem import rates of minor osmotic (**see: Development of seeds – principles of nutrient and water import, 3. Osmotic pressure and concentration**) species, such as mineral elements and individual amino nitrogen compounds, rely solely on sieve element concentrations and hence sieve element loading. Genotypic variation in seed iron and storage protein amounts is positively related with phloem sap composition of the precursors.

Hydraulic conductances ($L_p \times A$, **see: Development of seeds – principles of nutrient and water import, 1. Bulk flow**) of differentiating phloem may impose limitations on nutrient supply during seed set and cell division phases of seed development. Here, nutrients are imported symplasmically through differentiating vascular strands of relatively low hydraulic conductances. That path conductance could be a key factor in regulating nutrient supply to young developing seeds comes from the observed increases in seed set and size under atmospheres enriched in carbon dioxide (see: **Yield determination, 6. Source/sink limitation**). Here, photoassimilate production exceeds the small sink demand suggesting some form of regulation by the differentiating phloem path. In contrast, sink-limitation for photoassimilates during seed fill implies that phloem conductance does not control photoassimilate supply. Thus, seed control of import rates must be exercised through regulating hydrostatic pressures in sieve tube elements unloading assimilates into developing seeds (i.e. P_{sink}; **3.(a) Phloem import**).

Integration between phloem import, maternal export and filial import of nutrients could be achieved by feedforward and feedback signals operating along the phloem unloading pathway to regulate plasmodesmal and membrane transport with metabolic accumulation in the filial tissues. In this context, activities of amino acid and sucrose transporters (see: **6.(a) Plasma membrane transporters**) in developing cotyledons of grain legume seeds are regulated by substrate feedback de-repression providing a positive link with nutrient metabolism. In maternal seed tissues of developing wheat grains, plasmodesmal conductances could exert significant control over phloem unloading at the sieve element/vascular parenchyma interface. For seeds of grain legumes, transport control appears to be shared between symplasmic and membrane transport. Release by facilitated diffusion from maternal seed tissues provides a mechanism specific for each nutrient species to respond to filial demand. However, for major osmotic (see: **Development of seeds – principles of nutrient and water import, 3. Osmotic pressure and concentration**) species, a decline in their apoplasmic pool sizes translates into significant shifts in osmotic concentration that would result in elevated **turgor** pressures of maternal cells thus slowing phloem import. Grains of temperate cereals avert this outcome by the presence of apoplasmic barriers separating their endosperm cavities from maternal phloem. These barriers dampen any influence of turgor alterations in the nucellar projection transfer cells on pressures in the crease phloem as shown by phloem import being unaffected when endosperm cavities are perfused with solutions of differing osmolalities. An alternative strategy is based around turgor-dependent release of nutrients from maternal tissues of tropical cereals and grain. Small differences in nutrient concentrations between maternal symplasm and apoplasm ensure that small changes in amounts of apoplasmic nutrients cause large changes in their transmembrane concentration (and osmotic (see: **Development of seeds – principles of nutrient and water import, 3. Osmotic pressure and concentration**)) differences and hence cell turgors. If cell turgor rises above a set point, this produces a hypothetical error signal that causes an enhanced rate of nutrient release to the seed apoplasm release restoring cell turgor and meeting increased demand by the filial tissues. (MT-H, CEO, JWP)

Patrick, J.W. and Offler, C.E. (1995) Post-sieve element transport in developing seeds. *Australian Journal of Plant Physiology* 22, 681–702.

Patrick, J.W. and Offler, C.E. (2001) Compartmentation of transport and transfer events in developing seeds. *Journal of Experimental Botany* 52, 551–564.

Weber, H., Borisjuk, L. and Wobus, U. (1997) Sugar import and metabolism during seed development. *Trends in Plant Science* 2, 169–174.

Weber, H., Heim, U., Golombek, S., Borisjuk, L. and Wobus, U. (1998) Assimilate uptake and the regulation of seed development. *Seed Science Research* 8, 331–345.

Development of seeds – principles of nutrient and water import

1. Bulk flow

Bulk flow occurs when a solvent (e.g. water) and dissolved solutes (e.g. nutrients) move together in the same direction at the same velocity, most often in response to a pressure gradient. In plants, water and nutrients are transported by bulk flow over long distances and, in some circumstances, over short distances. The rate of solute transport (R_f) by bulk flow is the product of flow velocity (V), cross-sectional area of the transport pathway (A, such as lumens of xylem elements or sieve pores of phloem sieve elements) and concentration of the transported solute (C). That is:

$$R_f = VAC$$

Velocity or volume flux (J_v) of pressure-driven bulk flow through a plant system is normally modelled according to Poiseuille's Law that describes volume flux through a tube in terms of tube radius (r), viscosity (η) of the flowing solution and the pressure gradient ($\Delta P/\Delta x$) as:

$$J_v = (\pi r^4/8\eta)(\Delta P/\Delta x)$$

(See: **Development of seeds – nutrient and water import**)

2. Diffusion

Diffusion describes the free random movement of molecules (e.g. water; nutrients) from a region of higher to lower concentration. This net movement results from an increased probability of more particles moving toward the region of lower concentration than vice versa. In a complex solution, the diffusion rate (R_d) of each molecular species is independent of all other solution constituents and is described by Fick's First Law of Diffusion:

$$R_d = DA(\Delta C/\Delta x)$$

where D is the diffusion coefficient (a combined property of the medium through which diffusion occurs and the diffusing substance), A is the cross-sectional area of the diffusion pathway and $\Delta C/\Delta x$ the concentration gradient of the diffusing substance.

3. Osmotic pressure and concentration

Osmotic pressure (potential) is the effect of solutes in aqueous solutions on decreasing the activity of water molecules and hence their ability to diffuse (see: **2. Diffusion**). In dilute solutions, osmotic pressure (measured using an osmometer) is directly proportional to the number of solute particles present in the solution (i.e. osmotic concentration) and independent of their chemical properties. Hence water molecules undergo diffusion from regions of low to high osmotic concentrations with the gradient in osmotic concentration maintained either side of a selectively-permeable membrane. (MT-H, CEO, JWP)

Nobel, P.S. (1991) *Physicochemical and Environmental Plant Physiology*. Academic Press, San Diego, USA.

Development of seeds – sources of nutrients and water

Seed set is supported largely by current photoassimilates produced by subtending leaves (present at the same node) with secondary contributions from vegetative parts of inflorescences (cereals) and fleshy pod walls (grain legumes). Nitrogen for seed filling is derived largely from source (**see: Yield determination, 1. Source**) leaves undergoing senescence such that maintenance of a continued photoassimilate supply relies increasingly on non-structural carbohydrates remobilized from stems and inflorescences. For example, as the photosynthetic rate of the flag leaf (that closest to the flowering head) in wheat drops, grain growth relies increasingly on remobilization of stem reserves that may contribute upward of 60% of final seed biomass. (**See: Development of seeds – nutrient and water import**) (MT-H, CEO, JWP)

Dewinging

The removal during seed precleaning of **wings** (and, more broadly, the spines, hairs and some **aril** types found in some wind-dispersed species) from the dry fruits to reduce seed bulk and for convenience in handling. (**See: Conditioning, I. Precleaning**)

Diaspore

Also called a disseminule, it is a part of a plant that becomes separated and dispersed for reproduction, e.g. seeds, fruits and spores. (**See: Deserts and arid lands; Dispersal**)

Dibbling

Dibbling is the practice of making small holes in the ground for sowing individual seeds or plants. Modern dibbling devices are usually boards with pyramid or round-type protrusions to match the cell alignments in trays used to raise seedling transplants that provide consistent depressions for seed placement in growing media, such as for sowing tobacco pellets. (**See: Transplants**)

Dibbling is an old practice – used in England, for example, until the late 19th century – where farmers set seeds in evenly spaced holes made with the aid of a peg or spike roller, for example by walking backwards along the furrow holding a double-pronged 'dibbling iron'. The dibbling principle is used in certain modern machine designs, such as punch-type openers used for sowing mulch beds. (**See: Planting equipment – metering and delivery; Planting Equipment – placement in soil; Tillage, 5. Raised-bed tillage**)

In horticulture dibbling provides for placement of the seed in the centre of the cells (**see: Cell trays**) for uniform radial root growth. The tray also provides a consistent depression in material (potting soil, peat moss, etc.) allowing for uniform planting depth across the entire tray. This promotes uniform emergence. (PH)

Dicotyledons (dicots)

(Greek: *di* = two; *cotyledon*: sucker) A group of the angiosperms characterized by embryos with two **cotyledons** instead of only one (**monocotyledons**, moncots) (**see: Historical aspects of seed research**). Dicotyledons show numerous other differences from the monocotyledons, e.g. such as the long-lived primary root, a ring-like rather than scattered arrangement of the vascular bundles in the stem, the presence of secondary thickening by a cambium, flower parts (e.g. petals) usually in fours or fives, and leaves with usually reticulate venation.

Traditionally the dicots have been treated as a class, originally called the Dicotyledoneae, but more recently called Class Magnoliopsida after the type genus *Magnolia*. Some botanists, however, divide the dicots into two or more classes in an effort to create monophyletic (from a common ancestor) groups. The Eudicots (Eudicotyledons) make up least 65% of angiosperm species and include major families such as the Asteraceae, Fabaceae, Rubiaceae and Euphorbiaceae. They are set apart most notably by the production of tricolpate pollen which has three or more pores set in furrows called colpi. In contrast, most other flowering plants and gymnosperms produce monosulcate pollen, with a single pore set in a differently oriented groove called the sulcus.

The dicots contain many orders, of which the following contain several commercially important species (a few examples are given): Rosales (apple, pear, cherry, plum, peach, strawberry, raspberry), Leguminales (legumes), Fagales (beech, birch, oak), Cactales (cacti), Caryophyllales (sugar-beet, carnations), Cruciales (rape, *Arabidopsis*, vegetable crops such as cabbage, turnip, cauliflower), Umbellales (carrot, celery, anise, caraway), Asterales (lettuce, endive, sunflower), Solonales (tomato, potato, tobacco, chilli pepper, belladonna), Laminales (lavender, mint and other aromatic herbs). (JD, WS, JDB)

Dihaploid

A haploid produced either from the **sporophyte** developmental stage of a **tetraploid** plant, or by spontaneous or chemically induced chromosome doubling of a haploid line produced from a **diploid** plant. Haploids are used as a rapid route in breeding strategies to produce **homozygous** dihaploid lines, such as in the breeding of potato. (**See: Haploid; Ploidy**)

Dill

Dill (*Anethum graveolens*, Apiaceae) was one of the herbs and spices used as a flavouring in dynastic Egypt and the Mesopotamian, Greek and Roman civilizations (**see: Spices and flavours,** Table S.12). Indian dill is currently described as *Anethum graveolens* ssp. *sowa*. It is an erect annual herb with long fine leaves, compound umbels with small yellow flowers and small pungent fruits. All plant parts have the characteristic strong smell when crushed. Commercial dill 'seed' are the separated one-seeded mericarps (fruits), measuring approximately 3.5 × 2.5 mm. They are brown with yellowish, membranous wings (Fig. D.28).

The **essential oil** distilled from the seeds ranges from 2.5 to 5% and sometimes up to 8% fw. The main oil components accounting for the odour (flavour) are shown in Table D.6. (**See** Fig. S.44 in **Spices and flavours**)

Dill is an important spice used for flavouring in the

European kitchen especially with fish and egg dishes, in salads and soups. Both the seed and the oil are well known for their medicinal properties in the treatment of many conditions such as halitosis, dyspepsia, intestinal worms, skin diseases, hepatopathy and haemorrhoids. The oil is strongly antiseptic and has anticarcinogenic activities in mice. Dill, combined with fennel, is the active ingredient in 'gripe water', the infant colic remedy taken around the world in the former British Empire. (KVP, NB)

Guenther, E. (1978) *The Essential Oils*, Vol. 2. Robert E. Krieger Publishing, New York, USA.
Peter, K.V. (ed.) (2001) *Handbook of Herbs and Spices*. Woodhead Publishing, CRC Press, UK.
Weiss, E.A. (2002) *Spice Crops*. CAB International, Wallingford, UK.

Diploid (2n)

A plant having two sets of chromosomes in its somatic (vegetative) cells. (**See also: Amphidiploid; Haploid; Ploidy**)

Direct drilling (seeding, or sowing)

A planting system in which seeds (either **pregerminated** or dry) are sown directly in field soil without any tillage or mechanical seedbed preparation since the harvesting of the previous crop: also known as conservation ('no-till') **tillage**

Fig. D.28. Dill 'seeds' (image by Mike Amphlett, CABI).

Table D.6. Main constituents of Indian dill seed oil.

Component	Per cent (fw)
Limonene	45.0
Carvone	23.1
Dill apiole	20.7
cis-Dihydrocarvone	5.2
trans-Dihydrocarvone	4.2
α-Phellandrene	0.1
α-Pinene	0.1
β-Pinene	0.1
Linalool	0.1
(E)-Anethole	0.1
Myristicin	0.1

systems. The term also covers sowing directly into an established pasture, also known as **sod seeding**.

Disease

Seed- and seedling-associated diseases – a term broadly defined as 'factors that interfere with normal functions or structure' – may result from infection by fungal, bacterial or viral **pathogens** acquired during seed development, maturation and harvest, or from the soil environment soon after germination and emergence. Some diseases can have non-pathogenic origins, including adverse environmental factors or nutrient deficiencies during seed production causing blemishes (such as marsh spot in pea, due to manganese deficiency), or congenital disorders (such as **hollow heart** or **silk-cut** in maize), or mechanical injury during threshing, cleaning and other seed processing steps. Parasitic pests may cause **blind seed** or **embryoless seed** diseases. (Traditionally, plant diseases are called by various permutations of the plant name, the plant part affected and the nature of the symptom; in some cases the name of the causal pathogen is incorporated into, or becomes, the name of the disease.)

The symptoms of the diseases caused by seedborne or soilborne pathogens include the prevention of germination, the inhibition or just the weakening of seedling and plant development. Conspicuous diseases affecting seedling growth notably include **'damping-off'**, seedling-**blight** or **-wilts**, root **rot** (caused by the **'seedling disease complex'** of fungi), rots and wirestem. Symptoms characteristically include the development of water-soaked lesions on the seedling roots or shoots, causing pale, stunted or slow growth or collapse. Depending on the mode of action of the pathogen and severity of infection, seeds or seedlings may die before or soon after they emerge above the soil surface, or seedlings may grow with damage or reduced vigour leading to poor subsequent plant development. In some cases, symptoms may not be expressed at the seedling stage at all, and pathogens may infect the plant systemically, such as those that cause **smuts** and **bunts** in cereals. Some seedborne diseases may be extremely virulent, causing destructive epidemics in the crop, such as some caused by bacterial and viral pathogens.

Seed transmission of bacterial pathogens is of particular importance in cultivated annuals, such as *Phaseolus* bean (*Corynebacterium flaccumfaciens*, *Xanthomonas phaseoli* and *Pseudomonas phaseolicola*), soybean, pea, cucumber (*Pseudomonas lachrymans*), horticultural brassicas (*Xanthomonas campestris*), cotton, tomato and pepper (*Xanthomonas vesicatoria*) and cereals. Most species cause parenchymatous and vascular infections, or soft rots, including seed abortion, rotting or abortion and 'slime disease'. Bacterial and halo blight caused by *X. phaseoli* and *P. phaseolicola*, for example, is characterized by water-soaked lesions on bean pods from which developing seeds can be invaded. Seeds of cabbage and cauliflower infected by seedborne *X. campestris* form black rot lesions which appear in the **cotyledons** after germination. Surface-borne bacteria may remain alive for perhaps a year, but those harboured within seed tissues may last far longer.

Some diseases directly affect seed development, causing flower sterility or seed abortion, or seeds that are more wrinkled, shrivelled, or substantially smaller than normal, or have reduced viability, all affecting seed **yield** and physiological quality. Fungal pathogens may have considerable direct impact on seed in this way. Some directly affect seed primordial (the regions of growth in the root and shoot apices) development and seed maturation, causing abortion (such as smuts and **ergots**), and shrunken or rotted seed. A variety of seed discolorations may also result: many seedborne parasitic, saprophytic or pathogenic fungi can produce black, brown to grey necroses or infections (such as kernel **blight** or 'black point', and heavy incidences of smut or bunt disease), and some fungi may produce pigments such as **pink seed** in pea, and **purple seed stain** in soybean. Viruses too can affect seed development in many crops in similar ways, such as in wheat and barley (barley stripe mosaic virus [MV]), sorghum (sugarcane MV), squash (squash MV), onion (onion dwarf virus) and tomato (tomato aspermy virus). Others affecting legumes such as soybean (Colour Plate 10B), pea, broad bean, cowpea and mungbean, make them unsuitable for processing.

Crop disease may also be transmitted by superficial seed **infestation** (as opposed to **infection**). Examples are *Sclerotinia* and *Claviceps* (ergots) whose **sclerotia** become deposited in the soil subsequently at the time of sowing, and may infect the next plant generation from there. Superficial infestations with pathogenic or saprophytic fungi – *Alternaria*, *Cladosporium*, *Fusarium*, *Penicillium* and *Aspergillus* spp. – that invade or contaminate seed surfaces during seed maturation or **swathing**, can grow on inadequately stored grain, leading to a decrease in germination, seed discoloration (such as **blue eye** in maize), heating, weight loss, **mycotoxin** production and other biochemical changes (**see: Storage fungi**). To some extent such seed infestations with fungi, or with animal parasites such as nematodes, can be reduced at the **conditioning** stage by screening and aspiration.

Species of five genera of plant-parasitic nematodes are important seed-transmitted pathogens, carried either on the seed itself or in soil that may be mixed with it. *Anguina tritici*, for example, causes ear-cockle of wheat, a worldwide problem of economic importance. The larvae remain as ectoparasites around the growing point until the floral initials develop; they then invade the floral tissues, eventually forming galls in place of seeds. Other examples include: *Ditylenchus*, stem and bulb nematodes, which infect young seedlings in certain tropical areas, for example, rice and lucerne, and may infest seed of **broad beans**; *Heterodera*, cyst nematodes, in soybean and pea; a species of *Aphelenchoides*, the bud and leaf nematodes, which infect under the hull of rice grains; and a species of *Rhadinaphelenchus*, an endoparasite of coconut. Three genera are vectors of seedborne and other viruses: *Longidorus*, needle nematodes; *Xiphinema*, dagger nematodes; and *Trichodorus*, stubby root nematodes. Other species may enter a synergistic relationship with pathogenic seedborne fungi and bacteria, aggravating the disease-complex produced. Species of root knot nematodes (*Heterodera*) may aggravate *Fusarium* wilts in cotton, for example. Tundu disease, caused jointly by *Rathayibacter tritici* (*Corynebacterium tritici*) and the nematode *Anguina tritici* as the transmitting vector, is an important widespread disease of wheat in India, whereby a yellow slimy bacterial ooze is formed on the surface of the leaves and grains of affected plants. Another pathogenic nematodal-bacterial association produces galls within the developing seedheads of infected annual ryegrass, and a toxic slime that can be fatal to grazing livestock (**see: Grasses – forage and turf, 4. Quality and dormancy**).

Lastly, crop disease in general can have major practical and economic consequences in the production of seed and grain. Seed and seedling diseases that are non-lethal at the young seedling stage, together with the wide range of soil-, air- or insect-transmitted diseases that afflict young or mature plants (such as cankers, rots and wilts), affect crop biomass and hence may indirectly impair seed yield and quality.

Major symptomatically named seed- or soilborne diseases that may affect seed development or post-germination seedling growth are briefly described in separate articles – **see: Anthracnose; Blackleg; Blights; Bunts; Damping-off; Mildew; Rots; Rust; Seedling disease complex; Smuts; Soil sickness; Wilts**; and **Wirestem**. Major genera containing causal pathogens of such diseases are also briefly described in the article on **pathogens**: fungal genera – *Alternaria*, *Aphanomyces*, *Ascochyta*, *Aspergillus*, *Botrytis*, *Claviceps*, *Cochliobolus*, *Colletotrichum*, *Diaporthe/Phomopsis*, *Fusarium*, *Helminthosporium*, *Microdochium*, *Peronospora*, *Phoma*, *Phomopsis*, *Phytophthora*, *Plasmopara*, *Puccinia*, *Pyrenophora*, *Pythium*, *Rhizoctonia*, *Sclerotinia*, *Septoria*, *Tilletia*, *Ustilago* and *Verticillium*; and bacterial genera – *Acidovorax*, *Clavibacter*, *Erwinia*, *Pseudomonas* and *Xanthomonas*. Most of these pathogens can be at least partially controlled by forms of seed treatment. (**See: Treatments – pesticides**) For more details refer to the CABI Crop Protection Compendium (a relational database, that contains plant pest and disease information): www.cabicompendium.org/cpc (PH)

Disinfection

The control or elimination of seedborne **pathogens**, carried as spores or other resting structures deep inside the seed. (**See: Fungicides; Treatments – pesticides**)

Disinfestation

The control or elimination of seedborne **pests** or **pathogens**, carried as spores or other resting structures on the surface of seed; also referred to as seed-surface **disinfection**. (**See: Fungicides; Treatments – pesticides**)

Dispersal

Dispersal is the re-distribution of seeds following release from the parent plant. Virtually all seeds travel some distance from the parent, and the distribution of dispersal distances is often presented as the **seed shadow** or dispersal curve (Fig. D.29). This shows the density of seeds with distance from the plant(s) and can generally be described by a curve which increases to a mode and then decays to approach zero seed density. The mode is often close to the plant and simple mathematical functions have been used to represent the decay in seed density, e.g. negative exponential, Gaussian and inverse power. Ecologists are often interested in understanding the longest dispersal distances, seen in the tail of the dispersal curve, and the simple functions can underestimate these and describe the curve poorly (see Fig. D.29). Recently, more complex functions have been used to capture the tail, e.g. log-normal, mixtures of functions (e.g. exponential + power), and Clark's 2Dt.

It is rare for a seed to simply drop to the ground. Some plants actively throw seeds by the fruit exploding through tension in dead or living tissues (explosive or ballistic dispersal), e.g. *Impatiens* spp., *Acanthus* spp., squirting cucumber (*Ecballium elaterium*). More often, seeds are carried (passive dispersal) by one of a wide range of dispersal vectors, and many seeds and fruits have structures (developed from a wide range of flower, fruit or seed structures) which are thought to facilitate dispersal by particular vectors. The **dispersule (diaspore)** is the part of the plant which disperses with the seed(s). (**See: Seedcoats – dispersal aids**)

Dispersal by wind (**anemochory**) is common, and is aided by structures which slow the fall of the dispersule. These are generally of two types: hairy plumes (e.g. thistles *Cirsium* and *Carduus* spp., willows *Salix* spp., willow herbs *Epilobium* spp.) and wing-like extensions, e.g. maples (*Acer* spp.), pines (*Pinus* spp.) and docks (*Rumex* spp.). In a few plants the fruit forms an inflated balloon-like structure, e.g. *Anthyllis* vetches, and bladder sennas (*Colutea* spp.). These structures slow falling by increasing the ratio of surface area to mass, but in some cases are more efficient and act like parachutes or airplane wings which can even allow seeds to be lifted in the wind or to take off again after landing. Tumbleweeds are species, frequent in steppe regions, in which the fruit or part or whole of the plant forms a rounded shape which is torn off by the wind and blown along the ground, e.g. Russian thistle or tumbleweed (*Salsola kali*) and rose of Jericho (*Anastatica heirochuntica*).

Dispersal by floating in fresh- or seawater (hydrochory) is aided by reduction of dispersule specific weight by airspaces or corky tissues, e.g. sea kale (*Crambe maritima*), water lilies (*Nymphaea* spp.) and marsh marigold (*Caltha palustris*). **Hairs**, which increase surface area, may aid floating as well as dispersal by wind, e.g. bulrushes (*Typha* spp.). Many seeds die after even a short time in seawater and some species have an impermeable layer protecting the embryo. The coconut (*Cocos nucifera*) is well-adapted for sea-dispersal: its **endocarp** is impermeable, its fibrous **mesocarp** aids flotation and the 'coconut milk' allows seedling establishment on beaches with little fresh water.

Dispersal by animals (**zoochory**) involves two fundamental adaptations. Dispersules can attach to animals (epizoochory) by spines and hooks, usually on fur or feathers, e.g. cleavers (*Galium aparine*), burdocks (*Arctium* spp.) and threeawn (*Aristida* spp.) or by a sticky coating, e.g. plantains (*Plantago* spp.), *Pisonia* spp. and *Paspalum* spp. (grasses). The African genus *Dicerocaryum* has fruits flattened on one side and with two large projecting spines on the other, which stick into the feet of large mammals.

Dispersal through ingestion and subsequent defaecation by animals (endozoochory) is common. Many plants have edible fruits, which tempt certain animals to eat them (frugivory), combined with tough **seedcoats** for protection from mouthparts, gizzards and gut acids. In fact, passage through animal guts increases both the percentage and rate of seed germination of most plants with fleshy fruits, probably through abrasion of the seedcoat.

There are many examples of plants with edible fruit, including species cultivated by humans (e.g. orange, tomato, melon, peach, apple, blackcurrant, blackberry, quince, banana). The pulp of such fruits tend to have higher oil and sugar (non-structural carbohydrates) concentrations than leaves, but lower amounts of protein. Fruits may be eaten whole or parts torn off, with the seed being eaten in both cases. Or animals discard seeds while eating the fruit. Many animals eat fruits as part of a wider diet, but certain species rely on this food source. Some 16% of bird families are strictly frugivorous, while 40% consume a large quantity of fruit. Mammals are the second major group of frugivores: fruit bats and many primates are almost totally frugivorous, while many others (e.g. bears, some ungulates, elephants and lemurs) eat a lot of fruit. Some reptiles are also important frugivores, e.g. some tortoises, lizards and iguanids. While many plants are dispersed by a range of frugivores, others have a more specific interaction. Mistletoe (*Viscum album*) fruit is eaten regularly by thrushes (*Turdus* spp.), which regurgitate or defaecate the sticky seed,

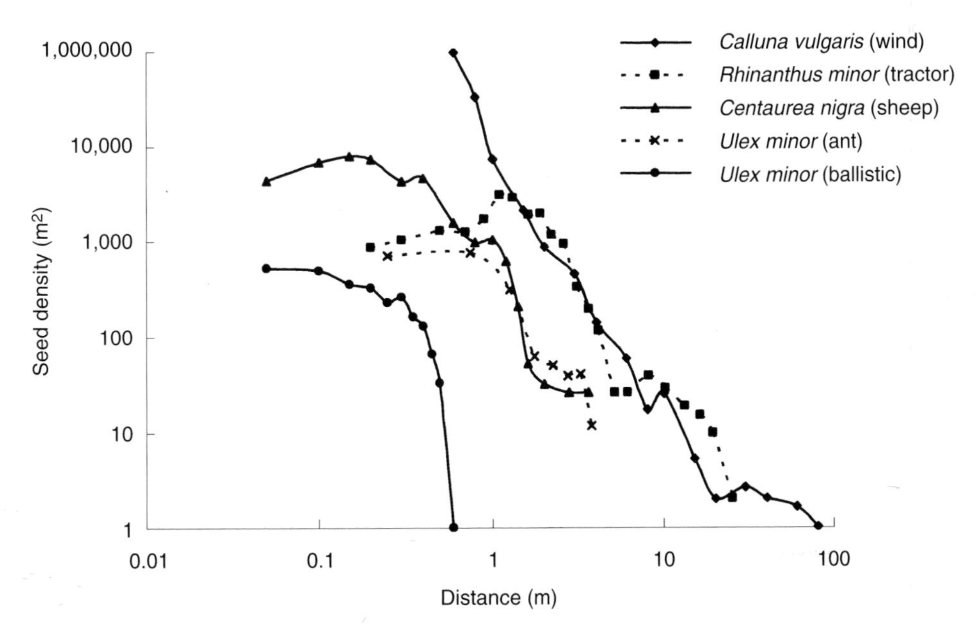

Fig. D.29. Examples of dispersal curves of herb and shrub species in the UK, with different dispersal mechanisms. Distance in metres from the parent plant is recorded. (Data from J.M. Bullock)

which adheres to branches. The **Brazil nut** (*Bertholletia excelsa*) is dependent on the agouti (*Dasyprocta aguti*) which is the only animal with teeth strong enough to break into the pod, after which it eats the pulp and buries the seeds.

Insects, and particularly ants (**myrmecochory**) do something similar to frugivory. Certain plants, e.g. violets (*Viola* spp.), brooms (*Cytisus* spp.), gorses (*Ulex* spp.) and some sedges (*Carex* spp.) have seed appendages called **elaiosomes** which are rich in nutrients. These attract ants that carry the dispersule to the nest, remove the elaiosome and discard the seed.

Dispersal by animals is linked to **predation**. The agouti stores Brazil nuts for consumption, but some are not eaten and so can form new trees. Seeds of plants including the oak (*Quercus* spp.), beech (*Fagus* spp.), pines (*Pinus* spp.) and bitterbrush (*Purshia tridentate*) are gathered and stored (caching or hoarding) by mammals such as squirrels, chipmunks, mice and other rodents, and birds such as nutcrackers (*Nucigfraga* spp.) and many woodpeckers (*Picidae* spp.). The hoarding animal eats many of these seeds, but a small proportion (generally much less than 10%) survive to germinate. Predators may drop and lose seeds, e.g. the seed-eating ant (*Tetramorium caespitum*) drops 13% of gorse seeds during transport to the nest. Seeds that are eaten whole, either accidentally ('the foliage is the fruit' hypothesis) with other food or intentionally, may survive passage through the gut. For example, in the UK, seeds of species without edible fruit, such as pearlwort (*Sagina procumbens*), meadow grass (*Poa annua*) and bittercress (*Cardamine hirsute*) have been found in sheep dung, and nettle (*Urtica dioica*), rushes (*Juncus* spp.) and willowherb or fireweed (*Chamerion angustifolium*) in rabbit dung.

Some species show adaptations for multiple forms of dispersal. Primary dispersal from the plant is by one mechanism and secondary dispersal is by another. For example, many species with elaiosomes (myrmecochory) leave the plant initially by ballistic dispersal (e.g. gorse, broom, violets).

The mechanism by which seeds are dispersed tends to affect the distance seeds travel, i.e. the mode and the length of the tail of the dispersal curve. In general, seeds with adaptations for dispersal by wind or water travel further than seeds dispersed ballistically or by invertebrates or which have no obvious adaptations. Vertebrate-dispersed seeds travel somewhat intermediate distances, but some can go very far.

However, it can be misleading to concentrate solely on apparent adaptations for dispersal. Many seeds have no obvious adaptations for dispersal, but are dispersed, seemingly accidentally, by the vectors described above. Wind disperses many seeds without wings or plumes (especially in storms). Epizoochory can occur in mud or tangled in fur or feathers on animals. Accidental endozoochory is described above. Simple seeds or fruit often float in rivers or sea, especially during floods, either by themselves or still attached to branches and stems or on any variety of flotation aids, such as driftwood, ice, pumice or leaves. In fact, many species are dispersed by multiple vectors and even a single seed may be moved in sequence by a range of vectors. For example, rattle (*Rhinanthus minor*) seems to have no dispersal adaptations but has been found to be dispersed well by wind, by sticking to livestock and farm machinery, by floods and even in earthworm guts. Oilseed rape (*Brassica napus*) seeds are relatively large and spherical, but are carried short distances by the wind, and longer distances in rivers and by spillage during transport of harvested seeds.

Humans have a major role in dispersing seeds. This again is generally unrelated to apparent dispersal adaptations, but humans may disperse seed far further than can any natural process. Humans transport seed intentionally of crop, forestry and ornamental plants, some of which have become problem species when introduced to other countries (called exotics), e.g. rhododendron (*Rhododendron ponticum*) in the UK, *Eucalyptus* spp. in several countries including South Africa, and broom (*Cytisus scoparius*) in California, but there are many mechanisms of accidental dispersal. Long distance dispersal between countries and continents can occur in ship ballast, as contaminants in cargo such as grain, timber or wool, on livestock, or in the soil of imported garden plants. Vehicles often disperse seed over shorter distances, e.g. in mud on tyres, on farm machinery, or spillage from crop seed transport. Humans themselves often act just like other large mammals and transport seed stuck to clothing or ingested with food (fertilizer derived from human sewage often produces large crops of tomatoes).

Seed **size** is often related to dispersal. Very small seeds rarely have structures relating to dispersal (e.g. plumes or hooks), but those < 0.1 mg have a large surface area to weight ratio and so disperse well by wind and water, and are also more likely to avoid damage by mouthparts if ingested by animals. Furthermore, because plants with smaller seeds generally produce a greater **number**, more seeds will disperse long distances for a given dispersal curve. Seeds > 100 mg tend to have adaptations for vertebrate dispersal, possibly because other dispersal mechanisms are not effective for such large seeds.

There are a number of measured ecological consequences of seeds dispersing some distance from the parent. Some of these suggest benefits of dispersal and therefore reasons for the evolution of observed adaptations for increased dispersal distances. An important idea (the Janzen-Connell hypothesis) is that survival of a seedling is increased with increasing distance from the parent plant. This may be because it either: (i) escapes competition (for resources such as light, nutrients and water) with the parent or with sibling seedlings which are in higher numbers near to the parent; or (ii) avoids seed **predators** or **pathogens** which are often concentrated near existing plants. Another possible benefit of dispersal happens because the environment often varies in space and time, and so the location of sites allowing **regeneration** may be scattered and unpredictable. So a plant dispersing its seeds over a larger area may have a greater chance of its seeds falling into favourable sites. This is taken a step further in the phenomenon of directed dispersal, whereby the dispersal process delivers seeds into favourable conditions: e.g. ant-dispersed seeds end up at ant nests which may have increased soil nutrient and lower vegetation cover; or frugivores deposit seeds in a pile of dung which forms a nutrient-rich seedbed; and birds deposit sticky mistletoe (*Viscum* spp., *Loranthus* spp.) seeds on branches.

A second class of consequences of increased dispersal distances relates to the distribution and abundance of a species at large spatial scales. The metapopulation theory is based on the observation that appropriate habitat of plants may occur in patches (e.g. chalk grassland, ancient woodland, or ponds) with uninhabitable areas in between. Populations in each patch are liable to go extinct and dispersal between patches has been shown both to rescue declining populations and to lead to re-colonization in patches in which the species has gone extinct. Another scenario is where a species is spreading across an area, e.g. an invasive plant is colonizing a country, or re-colonization is happening following habitat destruction by fire or volcanic eruption, or a species is recovering following conservation measures. Models and empirical studies have shown that higher dispersal distances tend to enhance persistence of metapopulations and increase the rate of population spread.

Dispersal, and the role of humans in transporting seed, is of great environmental importance. Invasive plants introduced and spread by humans have major ecological and economic impacts in many parts of the world (e.g. California, Australia, the Mediterranean). Dispersal is the major process determining the ability of species to shift their distributions in response to climate change, and to maintain metapopulations as habitat destruction leads to fragmentation of populations. (See: **Deserts and arid lands, 3. Seed dispersal strategies and mechanisms**) (JMB)

Bullock, J.M., Kenward, R.E. and Hails, R. (2002) *Dispersal Ecology*. Blackwell Science, Oxford, UK.

Forget, P.M., Lambert, J.E., Hulme, P.E. and Vander Wall, S.B. (2004) *Seed Fate: Predation, Dispersal and Seedling Establishment*. CAB International, Wallingford, UK.

Levey, D.J., Silva, W.R. and Galetti, M. (2002) *Seed Dispersal and Frugivory*. CAB International, Wallingford, UK.

Levin, S.A., Muller-Landau, H.C., Nathan, R. and Chave, J. (2003) The ecology and evolution of seed dispersal: a theoretical perspective. *Annual Review of Ecology and Systematics* 34, 757–604.

Pakeman, R.J., Digneffe, G. and Small, J.L. (2002) Ecological correlates of endozoochory by herbivores. *Functional Ecology* 16, 296–304.

Ridley, H.N. (1930) *The Dispersal of Plants Throughout the World*. L. Reeve & Co., Ashford, Kent, UK.

van der Pijl, L. (1982) *Principles of Dispersal in Higher Plants*, 3rd edn. Springer, Berlin.

Wenny, D.G. (2001) Advantages of seed dispersal: A re-evaluation of directed dispersal. *Evolutionary Ecology Research* 3, 51–74.

Willson, M.F. (1993) Dispersal mode, seed shadows, and colonization patterns. *Vegetatio* 108, 261–280.

Dispersule (disseminule)

Alternative names for **diaspore**. (See: **Deserts and arid lands; Dispersal**)

Distinctness, Uniformity and Stability (DUS)

The set of criteria by which a new plant **variety** is defined and assessed.

(a) *Distinctness*. A variety must show some phenotypic difference from similar already known varieties, either by one characteristic that is important, precise and subject to little fluctuation, or by several characteristics the combination of which is such as to give it the status of a new variety.

(b) *Uniformity/homogeneity*. A variety must be sufficiently genetically uniform and homogeneous in its relevant characteristics.

(c) *Stability* is linked to uniformity and means continued uniformity over different generations; that is, remaining uniform and true over time to a description of its essential characteristics after successive cycles of propagation or multiplication.

DUS trials are based on field experiments lasting for 2 or 3 years, and conducted according to methods specified by the International Union for Protection of New Varieties of Plants (**UPOV**). These trials involve the detailed comparative study of individual plants sown in small plots alongside a wide range of recognized varieties, and are mainly based on morphological traits: multi-state characters recorded using numerical keys, and quantitative characters as dimensions or counts.

DUS information can be used for two different purposes.

- Including the variety on a National List (and the Common Catalogue as well in the EU), which is a register of what can be marketed in a country. The listing is usually compulsory for a new plant variety to be registered but is restricted to crops of economic importance (e.g. food crops) and usually includes a performance requirement in **Value for Cultivation and Use** (VCU) trials.
- Granting of ownership rights to the breeder of the variety (Plant Variety Rights), which is a voluntary option. (See: **Laws and Regulations**)

(RD, RT)

Dividers

A range of equipment used to reduce the size of the composite sample of seedlots for subsequent laboratory **quality testing** and **health testing**, without introducing more variation than would be expected by simple random sampling. Most sample reduction methods (such as the Conical, Soil/riffle, Centrifugal and Rotary types) are based on the principles that the sample is thoroughly mixed and then divided. (See: **Sampling, 3. Sample reduction**)

DNA

See: **Nucleic acids**

Dockage

Dockage is defined as non-grain foreign material and all matter lighter, larger, or smaller than grain that can be identified in a representative sample of a traded grain consignment by size grading analysis during **grain inspection**; in the words of the United States Grain Standards Act, dockage includes 'plant fragments and any underdeveloped, shrivelled, unthreshed or small kernel pieces that cannot be reclaimed by rescreening'. Official procedures for traded grain in the USA and Canada, for example, regulate the testing of wheat, maize, barley, rye, flaxseed, rice and grain sorghum consignments in this way: grading either by hand or by mechanical sieving, using a combination of aspiration and passage over a sequence of screens, such as in the approved 'Carter dockage tester'. (PH)

Dodder mill

Seed conditioning equipment consisting of a pair of cylindrical rollers covered with velvet-like material, used to separate rough seed from smooth (also known as a rice mill or roll mill). (See: **Conditioning, II. Cleaning**)

Domestication

Domestication of wild species as crops began at least 10,000 years ago in six major regions (Table D.7). Many of the species were brought into cultivation for their seeds which furnished foods; and most of those valued for their vegetative parts were grown from seeds: seeds therefore had a pivotal role in domestication.

During domestication plants underwent many genetic, physiological and morphological changes, among the most important of which were those occurring in the seeds. These are listed in Table D.8 and several examples are considered in more detail below.

In some cases, the seed characteristics that became dominant during domestication were deliberately selected by humans. Many of these arose as mutations (**see: Mutants**), such as non-**shattering** in **cereals**, loss of seed dispersal in **legumes**, and **endosperm** characters in **maize**. On the other hand, it is likely that changes would also have occurred without deliberate human intervention, resulting from the selection pressures imposed by cultivation. Seeds would be favoured that have the competitive advantage over neighbours, such as those better endowed with storage reserves or a higher germination rate, both of which would enhance seedling establishment and competitiveness. Thus, these two genetically determined characteristics would slowly become fixed in the population even without selection by humans. An indication of such changes in gene frequency is given by an experiment in which mixed seeds of 28 barley varieties were sown together and the resulting plants were allowed to reproduce naturally. The seeds produced were not selected artificially and random samples were sown in successive years. Over several decades grain yield, spike size and seed number steadily increased.

Table D.7. Domestication of some major seed crops.

Crop[a]	Place	Time (1000 years ago)
Cereals		
Barley	Near East[b]	8–12
Finger millet	Africa	4–8
Maize	Mesoamerica	4–8
Oats	Europe	up to 4
Pearl millet	Africa	4–8
Rice	China	8–12
Rye	Europe	4–8
Sorghum	Africa	4–8
Wheat	Near East[b]	8–12
Legumes		
Chickpea	Near East[b]	8–12
Common bean	South America	up to 4
Cowpea	Africa	4–8
Faba bean	Near East[b]	4–8
Groundnut	South America	up to 4
Pea	Near East[b]	8–12
Soybean	China	4–8
Others		
Cacao	Mesoamerica	up to 4
Coconut	Southeast Asia	4–8
Coffee	Africa	up to 4
Cotton	Mesoamerica, Africa	4–8
Oil palm	Africa	up to 4

[a]There are separate articles about all of the listed crops. [b]'Fertile crescent'.
Adapted from Table 7.1 in Hancock, J.F. (2004) *Plant Evolution and the Origin of Crop Species.* CAB International, Wallingford, UK.

Table D.8. Seed traits associated with the domestication syndrome.

Increase in reproductive allocation (e.g. inflorescence size, number)
Increase in seed size
Loss of shattering and dehiscence
Increase in germination rate
Increase in germination uniformity
Reduction or loss of dormancy
Increase in palatability (e.g. increased protein, starch; reduction of toxin content)
Changes in seed colour
Reduction or loss of glumes, etc.
Increase in seed survival properties

1. Loss of dispersal mechanisms

This is arguably the most marked feature that distinguishes domesticated seeds from wild relatives. Grains of cultivated cereals do not shatter (**see: Shattering**) as do the related, wild grasses; and pods of crop legumes and **siliques** of Brassicas are indehiscent, unlike fruits of the wild relatives. The normal dispersal mechanism loses its function in domesticated crops, where seeds are collected, and also early cultivators would have selected those seeds that were easy to harvest directly from the mother plant.

Shattering in grasses occurs because the **rachis** of the **spikelet** abscises and breaks off, the grains falling from the stalk when shaken by wind, animals or contact with neighbouring plants. In domesticated cereals, the rachis is tough and resistant to abscission so grains do not 'shatter' (Fig. D.30). The non-shattering character occurred as a consequence of mutations. Evidence for **einkorn** wheat (from which the non-shattering trait in **wheat** is derived) is that only two genetic loci are implicated in the development of this characteristic and it seems in several species that two orthologous genes (i.e. descended from one ancestral gene) are involved: in some species three or four loci participate. It has been estimated that at the rate of one mutation in a million plants, the tough, resistant rachis could become universally established after about 200 years of selection by early agriculturalists.

Another feature concerning the ease of release of grains from the inflorescence is their threshing quality. Where grains separate from the inflorescence stalk below the bracts (lemmas, paleas, glumes) of the **floret** or **spikelet** the caryopsis remains invested by the hull, presenting difficulties for ready use of the grains as food. Grains which are more easily removed from the enveloping bracts by threshing – 'free-threshing' types – would have been selected by humans, and now modern cereals generally have this property (Fig. D.31). What happened was that during domestication, selection took place for mutants in which the bract bases easily detach. This simple Mendelian trait arose in **emmer** wheat, for example, and is now present in durum and bread wheats.

2. Germination and dormancy

As mentioned above, cultivation from seed introduces pressures whereby rapid germinators are self-selected so that

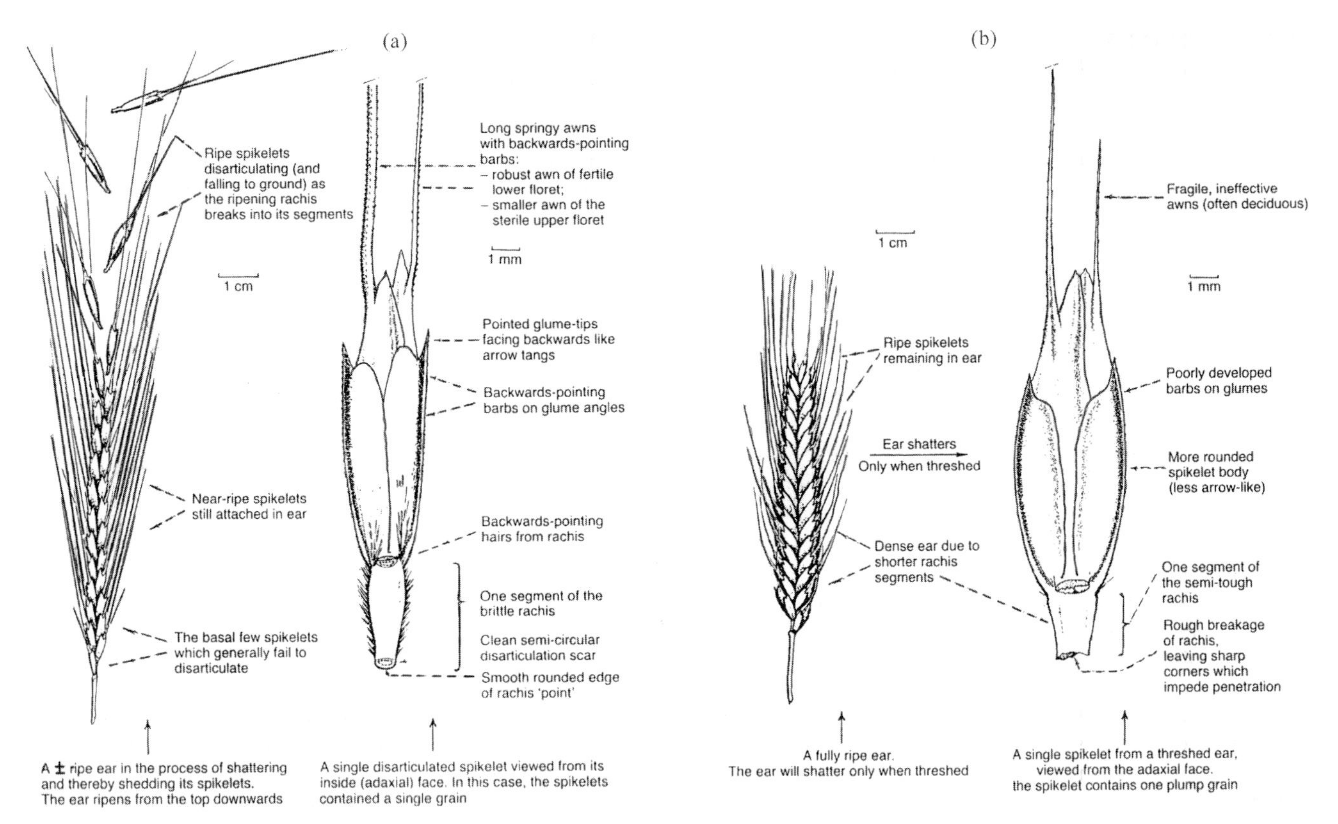

Fig. D.30. Shattering and other dispersal characteristics in wild (a) and domesticated (b) einkorn wheat. In the wild einkorn the end of the rachis is smooth having been detached by the abscission layer. The rachis of the domesticated form, having neen forcibly broken off, is rough-edged. The shape of the wild dispersal units (grain with 'hull') favours penetration into surface litter and cracks in the earth: this is aided by the barbs on the glumes. These dispersal aids are lacking in the fatter, smoother units of the domesticated form. From Davies M.S. and Hillman, G.C. (1992) Domestication of cereals. In: Chapman, G.P. (ed.) *Grass Evolution and Domestication.* Cambridge University Press, Cambridge, UK, pp. 199–224. Thanks to CUP.

over time such seeds become prominent in the crop. Present-day cereals generally show rapid, uniform germination. The thick seedcoats (**hardseededness**) that delay germination in many wild legumes are a rarer occurrence in cultivated types because they have been selected against.

Uniformity and capacity for germination are also determined by **dormancy**, an ecophysiologically important property in very many wild species, including grasses. This also has been reduced by self-selection, and probably also by human intervention, so that modern cereals lack appreciable levels of dormancy. One disadvantageous consequence in modern agriculture of this reduction in dormancy, however, is the propensity to **preharvest sprouting**, which can present problems in wheat, rice and sorghum, for example.

3. Seed productivity

Enhanced seed productivity during domestication is brought about by increases in seed size (Colour Plate 13E) and in seed number. Larger seeds would have been 'self-selected', as already discussed, and also selected by humans, especially if the seeds were to serve as food.

Increase in seed quantity or availability results from the operation of several factors such as: (i) loss of photoperiodic control, liberating the plant from one seasonal constraint on productivity; (ii) increase in number of inflorescences and flowers, either by more branches or tillering (e.g. in wheat); (iii) increase in inflorescence size (e.g. in **foxtail millet**); and (iv) increase in flower/floret number (e.g. maize).

(a) *Reduction in photoperiodic control.* This partly releases the plant from one limitation to its productivity and extends the opportunity in time for flowering and seed formation. An example is the common bean, where flowering in the wild species is markedly delayed by photoperiods of 16 hours or longer while the domesticated form is unaffected.

(b) *Increase in available inflorescence number.* For example, in cereals such as wheat, **barley** and **rye**, this is brought about by higher degrees of simultaneous tillering, leading to synchrony of seed production.

(c) *Increase in inflorescence size.* This commonly occurs in cereals and many cultivated crops – a good example is foxtail millet (Fig. D.32).

(d) *Increase in flower/floret number.* This occurs in a wide range of species but just two examples will be discussed – barley and maize.

Two rows of grains per stalk are found on all the wild and primitive barleys but the later domesticated types have six rows. Each side of the node of the inflorescence stalk in all barleys bears a group of three, one-floret spikelets, i.e. six florets per node. In the wild and primitive types only one spikelet in each group is fertile, hence after fertilization each

Fig. D.31. Free-threshing in wheat. The ear and the loose grains on the left are non-free-threshing; those on the right are free-threshing. Note how in the non-free-threshing type the glumes still closely adhere to the caryopses while in the free-threshing type the glumes have come away in several of the grains. Image kindly donated by M. Nesbitt, Royal Botanical Gardens, Kew, UK.

Fig. D.32. Inflorescences (seed heads) of wild ancestor (left) and domesticated (right) foxtail millet. (From: http://agronomy.ucdavis.edu/gepts/pb143/lec08/pb143l08.htm)

node bears two grains; all spikelets are fertile in the modern, domesticated barleys, hence the six rows of grain. Fertility is restored by two recessive mutations at one locus.

The wild ancestor of domesticated maize is widely held to be a teosinte, possibly *Zea mays* ssp. *mexicana* (syn. *Euchlaena mexicana*) (according to some authorities the common name teosinte actually applies to four botanical *Zea mays* subspecies and three other *Zea* species) (**see: Maize**). Here, spikelets are borne in deep segments on the floral axis (cupules) of which there are two rows. Two spikelets occur in each cupule but only one is fertile, producing one grain. The floral axis thus bears two rows of grain, each one covered by a hard glume. In domesticated maize, however, there are between four and ten rows of cupules, each with two fertile spikelets, so maize can

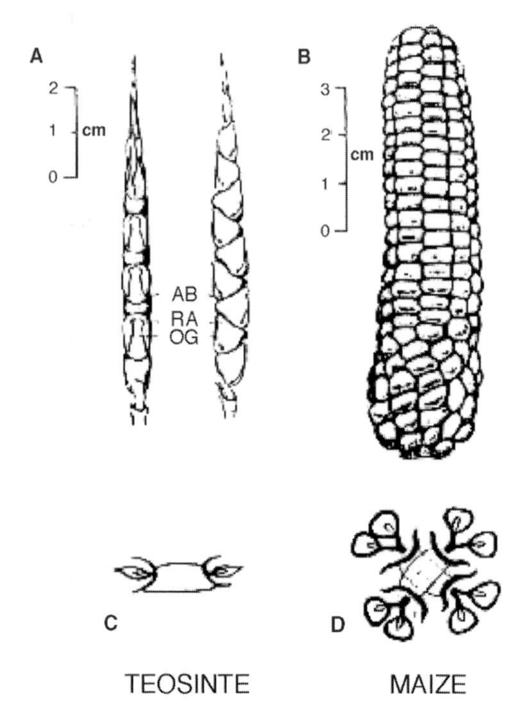

Fig. D.33. Female inflorescences (seed heads) of teosinte (A and photo on right) and domesticated maize (B, D). In (A): AB, abscission layer; RA, rachis internode; OG, outer glume. (C) Diagrammatic cross section of teosinte showing two cupules, each with one fertile spikelet. (D) Diagrammatic cross section of a maize ear showing four cupules, each with two fertile spikelets. (From: Doebley, J., Stec, A., Wendel, J. and Edwards, M. (1990) Genetic and morphological analysis of a maize-teosinte F_2 population: implications for the origin of maize. *Proceedings of the National Academy of Sciences USA* 87, 9888–9892. Reproduced with permission.) The photo shows intact and dissected immature teosinte (*Zea diploperennis*) ears: note the hard, woody 'fruitcases' (glumes around grains) on the far left, two at the bottom cracked open to reveal the grains (photo by Hugh Iltis, from www.wisc.edu/teosinte/images.htm).

have 8–20 rows of grains, each lying outside of a relatively soft glume (Fig. D.33). Genetic studies indicate that these changes result from mutations in four primary genes each separately located on five of maize's ten chromosomes. Other genes, on different chromosomes, also participate through modifying, pleiotropic effects. The morphology of domesticated maize also differs profoundly from that of teosinte. It is thought that in the evolution of the crop some lateral branches and ears of teosinte became suppressed and the remaining terminal male inflorescences on the much reduced primary branches were feminized, to produce the pattern of female inflorescences (becoming 'cobs') on a single, main stem. It has been suggested that included in domestication was the selection of phenotypes that displayed strong suppression of ear branching, an effect brought about by the *ramosa-1* gene. Superimposed on these changes were an increase in grain size, various endosperm mutations and development of pigmentation, to give an array of cob and grain types (Colour Plate 13F).

4. Genetics of domestication

Genes controlling the different seed traits of domestication, some of which have been mentioned above, appear to lie in clusters or to be linked. In the common bean, for example, genes for seed **dispersal** and dormancy are in one genomic location while in a second location are genes for seed size. Clustering has also been found in maize, rice, pearl millet and sorghum (Table D.9). Interestingly, other domestication traits such as spike morphology, flowering time and growth habit occur in the same clusters as the seed traits. Domestication therefore seems to be controlled by a small number of chromosomal loci, presenting a genetic condition that would favour rapid change. The occurrence of seed **size**, dispersal and photoperiodic sensitivity in a small number of loci in, for example, maize, sorghum and rice would enable domestication to occur in a relatively short time, possibly in a timespan of only 100 years.

Table D.9. Domestication traits and loci.

Locus	Species	Trait
pl1/sd1/sp1/pl7	Rice	Inflorescence size
nn1, nn2, nn3, nn5, nn2/10,	Rice	Seed size
nn1, nn2, nn3, nn9	Rice	Shattering
LoS7/Dens7	Pearl millet	Inflorescence size
nl6	Pearl millet	Shattering
nn1, nn3, nn4, nn5, nn7	Maize	Seed size
nn1, nn4	Maize	Shattering
nnC	Sorghum	Shattering
nnA, nnB, nnC, nnE	Sorghum	Seed size
Ph5	Common bean	Seed size
nn2, nn4	Mung bean	Seed size
nn2, nn7	Cowpea	Seed size

Letters preceding numbers or upper-case letters are named loci except *nn* (not named); numbers and upper-case letters signify chromosome number or linkage group. Slashes (/) indicate putative orthologues.
Adapted from Table 2 in Frary, A and Doğanlar, S. (2003) Comparative genetics of crop plant domestication and evolution. *Turkish Journal of Agriculture and Forestry* 27, 59–69.

5. Other changes occurring in domestication

The emphasis in the above discussion has been on seeds. It should be noted, however, that other important changes to plants occurred during their domestication. These include, for example, changes in morphology to a more compact growth habit, synchronization of ripening and a shift from a perennial to an annual habit (e.g. in rice). One important consequence of domestication, profoundly visible in agricultural practice in the developed world, is the trend towards monocultures with the accompanying reduction in genetic diversity.

(MB)

See: Archaeobotany

Gepts, P. (2004) Crop domestication as a long-term selection experiment. *Plant Breeding Reviews* 24 (Part 2), 1–44.

Hancock, J.F. (2004) *Plant Evolution and the Origin of Crop Species.* CAB International, Wallingford, UK.

Hillman G.C. and Davies, M.S. (1990) Domestication rates in wild-type wheats and barley under primitive cultivation. *Biological Journal of the Linnean Society* 39, 39–78.

Koinange, E.M.K., Singh, S.P. and Gepts, P. (1996) Genetic control of the domestication syndrome in common-bean. *Crop Science* 36, 1037–1045.

Poncet, V., Lamy, F., Enjalbert, J., Joly, H., Sarr, A. and Robert, T. (1998) Genetic analysis of the domestication syndrome in pearl millet (*Pennisetum glaucum* L., Poaceae): inheritance of the major characters. *Heredity* 81, 648–658.

Simmonds, N.W. and Smartt, J. (1995) *Evolution of Crop Plants*, 2nd edn. Longman, London, UK.

Wang, R.L., Stec, A., Hey, J., Lukens, L. and Doebley, J. (1999) The limits of selection during maize domestication. *Nature* 398, 236–239.

Zohary, D. and Hopf, M. (1988). *Domestication of Plants in the Old World.* Clarendon Press, Oxford, UK.

www.wisc.edu/teosinte/

Domestication syndrome

The tendency of selecting specific traits when wild populations of diverse species become domesticated. Though some morphological traits are common among crops (e.g. branching habit, erectness), most domestication traits pertain to seed characteristics such as **size**, colour, **hardseededness**, **dormancy**, time to maturity, and **dispersal**. **(See: Domestication)** (CW)

Dormancy

1. General

Dormancy is a property of a seed that prevents its germination even under external conditions that are adequate to support the germination process itself. Because of the various experiences that a dormant seed must undergo to lose its dormancy, the phenomenon serves to operate as a means for preventing germination in environments generally unfavourable for subsequent vegetative plant growth and reproduction of the next generation. **(See: Dormancy breaking; Dormancy – ecology; Dormancy – ecophysiology)** For example, when chilling is required for the removal of dormancy, the seed cannot germinate until after the cold winter season has passed. While the presence of dormancy is a selective advantage that ensures survival of many wild species, seed dormancy is a chronic source of most weed control problems in agriculture. Dormancy in crop species at the time of planting is generally considered undesirable, since a population of seeds with some dormant individuals will prevent rapid and uniform stand establishment: dormancy also interferes in some industrial uses of seeds (**see: Dormancy – importance in agriculture; Malting** entries). However, there is some evidence that dormant seeds retain greater subsequent vigour and viability for longer periods of time in dry storage.

Dormancy is controlled by the intrinsic structural or physiological properties of a seed. A viable seed is only termed dormant if it fails to germinate under adequate conditions of water, oxygen, etc., after dispersal from the mother plant, i.e. dormancy is predominantly a postharvest or post-dispersal phenomenon. However, since dormancy sets in during seed **maturation** (**see: Dormancy – acquisition**), it follows that seeds still on the mother plant are in the dormant state for a relatively short period before dispersal (**see: Pre-harvest sprouting**). Growth arrest or the absence of **vivipary** during seed development are generally not considered to be dormancy. Seeds such as **maize, pea** and other crop species that only require **imbibition** for germination to occur over a broad range of environmental conditions are non-dormant; in the dry, unimbibed state, such seeds are termed **quiescent**. It is also important to recognize that the failure of imbibed, viable seeds to germinate may not necessarily be due to dormancy but might occur because certain conditions, e.g. temperature, are unsatisfactory for germination: again, such seeds are best described as quiescent. Slightly delayed (of several hours to days duration) germination of a population of non-dormant seeds may reflect the presence of residual components of a prior dormant state or a slight loss of vigour, and these seeds should not be classified as dormant.

Dormancy is a highly specific block to germination and not a general lowering of vital activity of the seed. Imbibed, dormant seeds are fully capable of a wide range of metabolic functions including **respiration**, **nucleic acid** and **protein** synthesis, membrane generation, etc. while nevertheless remaining unable to germinate. Seeds that have entered secondary dormancy (see below) undergo a gradual decline in intensity of metabolism, but this does not occur during primary dormancy.

2. Primary and secondary dormancy

A seed can exhibit primary or secondary dormancy. Primary dormancy is induced as part of the genetic programme of seed development and maturation, and can be conserved in mature, dry seeds by appropriate storage conditions or when buried in the soil. In contrast, already mature, imbibed seeds are induced into dormancy, termed secondary dormancy, by unfavourable environmental conditions after seed **dispersal**. Secondary dormancy can be induced in non-dormant seeds or in seeds that initially possess primary embryo dormancy. As noted, dormancy should not be confused with the failure of a non-dormant seed to germinate due solely to external imposition of unfavourable growth conditions. For example, a non-dormant, imbibed seed may not germinate because soil temperatures exceed the optimum for germination; such **thermoinhibition** is relieved as soon as temperatures return to a favourable range (**see: Germination – influences of temperature**). In contrast, prolonged excessive temperatures may induce a non-dormant seed into a state of secondary dormancy (**thermodormancy**). A seed induced into thermodormancy will not readily germinate when temperatures simply return to a favourable range. Seeds whose germination is inhibited by light may enter a secondary dormancy called **photodormancy** which persists in darkness and can be removed only by special treatments. It is not clear whether the physiological, biochemical and molecular bases for primary embryo dormancy are the same as those for secondary dormancy.

Primary dormancy can be caused by the maternally derived seed covering structures (**coat-imposed dormancy**) or by embryonic factors (embryo dormancy), acting individually or in combination (**see: Dormancy – coat imposed; Dormancy – embryo**). Seedcoat-imposed dormancy has been attributed to at least four phenomena. (i) The seedcoat may be completely impermeable to water and prevent imbibition. Primary dormancy of this type is commonly called '**hardseededness**', and can be readily demonstrated by the failure of a seed to increase substantially in fresh weight when incubated in water. (ii) The seedcoat may also maintain dormancy by preventing or limiting gas exchange with the environment. Restricted gas exchange, as a cause of dormancy, has been inferred from experiments in which dormant seeds can be induced to germinate when incubated in an oxygen atmosphere greater than that of air, or by 'surgical' treatments (e.g. puncturing, scratching) of the coat. However, there are few direct measurements of seedcoat impermeability to gases (**see: Germination – influences of gases**). (iii) The seed-covering structures (**testa, pericarp** or **glumes**) may prevent or restrict leaching of **germination inhibitors** from the embryo. (iv) The seed-covering structures also may mechanically restrain elongation and emergence of the embryonic axis. If the coat is the only source of dormancy, scarification (abrasion) or removal of seed-covering structures will elicit germination. Dormancy due to (iii) and/or (iv) is usually inferred when germination is observed after removal of the surrounding structures from a water-permeable seed. However, care must be taken to ensure that the embryo or **caryopsis** has not been physically damaged by such surgery, since injured living tissue will exhibit a wound response that evolves **ethylene** and carbon dioxide (from a respiratory burst), both of which can act as dormancy-breaking agents. Dormancy resulting from (iii) and (iv) would seem to comprise interactions between coat-imposed and embryonic factors. Although the ultimate causes of coat-imposed dormancy are the various constraints imposed by the enclosing tissues, properties of the embryo must also be involved, since the embryo evidently lacks the ability to overcome these constraints. (**See: Dormancy – embryo**)

In many cases the seedcoat appears to play no role, and primary dormancy of the seed can be attributed entirely to embryo properties. One or more of the following phenomena may be responsible. The seed may be shed with the embryo in an underdeveloped state and, therefore, some time must elapse before it reaches a germinable stage. Most commonly, however, the dormant seed has fully matured anatomically on the mother plant, and after shedding, it fails to germinate. There are several possible causes of this failure. (i) There is a hormonal imbalance within the embryo: concentrations of germination inhibitors (such as **abscisic acid**) are too high or growth promoters (predominantly **gibberellins, ethylene**, or **cytokinins**, depending upon the species) are too low. The effectiveness of a given hormonal balance also can be modulated by embryonic sensitivity to these substances. The molecular basis of physiologically modified hormone sensitivity has not been fully defined. (**See: Development of seeds – hormone content; Dormancy breaking – hormones**) (ii) Differences in the status of hormones would affect expression of genes: those whose expression is required for germination may remain inactive or may be repressed, and expression of others might actively inhibit germination. (**See: Dormancy breaking – cellular and molecular aspects**) (iii) Some aspects of the molecular regulation of germination and/or dormancy might occur without the overall control of **hormones**. (iv) There may be deficiencies in a metabolic process that is especially required for germination. (v) Cell **membranes** may be in a state that does not support certain signal transduction events. Seeds with embryo dormancy can be stimulated to germinate by a variety of species-dependent treatments including cold or warm stratification, light or darkness, daily alternating temperature regimes, plant growth-promoting hormones, dry storage at ambient temperatures, wood smoke, as well as by a multitude of low molecular weight organic and inorganic chemicals under laboratory conditions (**see: Dormancy breaking** sections).

In nature, the loss of primary dormancy due solely to coat factors is an irreversible process resulting in germination. However, if embryo dormancy is present alone or in combination with coat dormancy, the situation is much more fluid. Environmental requirements necessary for the breaking of embryo dormancy may be partially or fully met, but germination does not occur, e.g. due to suboptimal temperature, lack of water, or depth of seed burial in the soil. In such situations a germinable seed may be re-induced into secondary dormancy. Once a seed has entered into this state, it can dynamically interact with varying degrees of sensitivity to environmental factors that can regulate germination (seasonal dormancy-cycling). This dynamic sensitivity to environmental factors has been best documented by changing responsiveness to light and/or nitrate. (**See: Dormancy – ecophysiology**) While there is controversy among specialists as to whether these latter stimuli represent true dormancy-breaking factors or, rather, triggers of germination, it is nonetheless clear that many species of seeds with embryo dormancy possess changing sensitivities (or depths/intensities) of dormancy to environmental factors that finely tune the probability of successful germination in a favourable environment for continued growth.

It can be argued that primary, secondary, embryo and coat-imposed dormancy are the fundamental categories of dormancy. But in an effort to organize our knowledge about the different forms of dormancy in relation to seed ecology and physiology, numerous detailed dormancy classification systems (analogous to the anatomical taxonomy of plants) have been formulated. While these schemes (e.g. Table D.10) are based primarily upon the descriptive phenomenological behaviour of seeds, they could act as an organizing guide to the molecular dissection of processes that control the induction, maintenance, or breaking of seed dormancy.

3. Evolution of dormancy

Breeders of agriculturally important plants have selected against the dormant phenotype, which can pose a barrier to rapid crop establishment. However, seed dormancy endows undomesticated plants with a selective advantage in some environments, since it enhances the probability that a species will survive to produce subsequent generations, despite the vagaries of the environment. While an individual seed either germinates or does not, a population of seeds contains individuals with differing depths (or intensities) of dormancy, such that germination can be spread over both space and time.

Table D.10. Types of dormancy identified in some classifications.

Dormancy Type	Characteristic
1. PHYSIOLOGICAL	Includes embryo and coat-imposed dormancy.
a)Deep	Embryo dormancy: isolated embryos do not germinate or if they do, produce abnormal (e.g. dwarf) seedlings: usually require long periods of chilling for dormancy breakage: application of **GA** may not be effective.
b) Intermediate	Coat-imposed: isolated embryos germinate to produce normal seedlings: usually require chilling for dormancy breakage: applied growth regulators frequently effective.
c) Non-deep	Coat-imposed: sometimes expressed only at certain temperatures: relatively short period of chilling breaks dormancy, as do growth regulators, certain chemicals.
2. MORPHOLOGICAL	Embryos not fully developed or differentiated at the time of seed dispersal: further development occurs in the dispersed seed.
3. MORPHOPHYSIOLOGICAL	Combines features of morphological and physiological dormancy; embryo growth and dormancy breakage needed for germination to occur.
4. PHYSICAL	Impermeable seedcoats prevent water entry.
5. PHYSICAL/PHYSIOLOGICAL	Combines 1 and 4.
6. CHEMICAL	Dormancy imposed by inhibitors, especially in coats.
7. MECHANICAL	Hard, woody covering prevents germination.

Based on the classification in Baskin, C.C. and Baskin, J.M. (1998) *Seeds. Ecology, Biogeography, and Evolution of Dormancy and Germination.* Academic Press, San Diego, USA.

The physiological basis for such 'bet hedging' against environmental variation is not known. Presumably, the different types and intensities of seed dormancy (even a complete lack of dormancy, reported in representatives of the 14 vegetation types presently extant) have evolved to exploit unique features of ecological niches where plants grow.

Some understanding as to the environmental selection pressures likely to have been involved in the evolution of dormancy can be gained by examining the present species distribution of seed dormancy in different environments. When worldwide precipitation and temperatures are taken into account, the pattern that emerges is of an increasing incidence of dormancy in tropical and subtropical regions as precipitation and temperature decrease: in temperate and arctic zones the average number of species with dormancy is about four times that without dormancy. These findings are consistent with the view that dormancy evolved as a device to maximize plant establishment and survival under demanding environmental conditions (Fig. D.34). There is, however, insufficient evidence to place the evolutionary origin of seed dormancy in a particular geological period. It is possible that the several types of dormancy originated at different times according to prevailing selection pressures.

Our understanding of the evolution of seed dormancy in plants is relatively modest and is based upon descriptive seed behaviour in the context of what is known about general plant phylogeny, the seed fossil record and seed ecology of extant species. Four general trends have been identified by Baskin and Baskin (1998) regarding the evolution of seed dormancy: (i) dormancy due to underdeveloped embryos is considered as the most primitive form; (ii) coat-imposed dormancy due to water-impermeable seed covering structures, physiological dormancy, and their combinations have evolved from this base; (iii) dormancy due to water-impermeable seed coverings and combinations of this type of coat-imposed dormancy with physiological dormancy are restricted to angiosperms and are not present in gymnosperms; and (iv) physiological dormancy is the most widely evolved form. Selection pressure for the evolution of dormancy is thought to have been generated by whether or not seeds form a **soil seed bank**, the extent of seedling death due to excessive competition, seed **predation** and modes of seed **dispersal**, among other factors. The application of **genomics** may help in the understanding of the molecular evolution of genes that regulate both coat-imposed and embryo dormancy. (**See: Evolution of seeds**) (MAC)

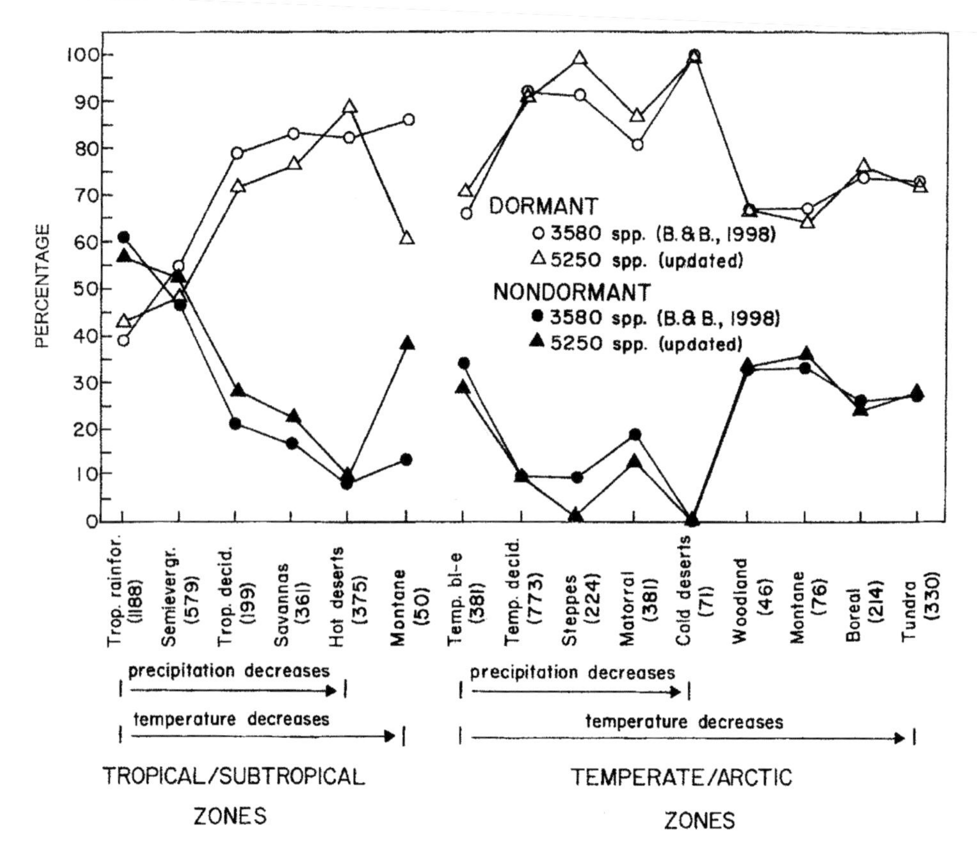

Fig. D.34. World distribution of dormancy in relation to environment. Number of species is shown in brackets. The percentage of dormant and non-dormant species in a particular geographical region is shown. Abbreviations: rainfor., rainforest; evergr., evergreen; trop., tropical; decid., deciduous; temp., temperate; bl-e., broad-leaved evergreen. Circles (B&B, 1998) from Baskin, C.C. and Baskin, J.M. (1998) *Seeds. Ecology, Biogeography, and Evolution of Dormancy and Germination.* Academic Press, San Diego, USA. Triangles; updated, 2003. (Reproduced with permission from Baskin, J.M. and Baskin, C.C. (2003) Classification, Biogeography, and Phylogenetic Relationships of Seed Dormancy. In: Smith, R.D., Dickie, J.B., Linington, S.L., Pritchard, H.W. and Probert, R.J. (eds) *Seed Conservation: Turning Science into Practice.* Royal Botanic Gardens, Kew, UK, pp. 519–544.)

Baskin, C.C. and Baskin, J.M. (1998) *Seeds. Ecology, Biogeography, and Evolution of Dormancy and Germination*. Academic Press, San Diego, USA.

Baskin, J.M., Baskin, C.C. and Li, X. (2000) Taxonomy, ecology and evolution of physical dormancy in seeds. *Plant Species Biology* 15, 139–152.

Baskin, J.M. and Baskin, C.C. (2004) A classification system for seed dormancy. *Seed Science Research* 14, 1–16.

Bewley, J.D. and Black, M. (1994) *Seeds. Physiology of Development and Germination*, 2nd edn. Plenum Press, London, UK.

Nikolaeva, M.G. (1999) Patterns of seed dormancy and germination as related to plant phylogeny and ecological and geographical conditions of their habitat. *Russian Journal of Plant Physiology* 46, 369–373.

Dormancy - acquisition

Dormancy has a hereditary component (**see: Dormancy – genetics**), as has been demonstrated for a large number of species, including many crops. Also the degree, or depth of dormancy is hereditary. Many different genes are associated with dormancy (**see: Dormancy genes**) and obviously their expression is involved in the onset of dormancy in the immature seed. Most, but not all of these participating genes are connected with **hormones**. In many species the onset of dormancy is coupled to morphological or physiological characteristics, e.g. seed colour and seedcoat thickness and to positional effects on the mother plant. Hybridization studies with dormant and non-dormant pure lines have revealed that seed dormancy may be controlled by both the maternal and paternal parent genotype. The environment in which seeds develop has a very strong influence on the acquisition of dormancy. (**See: Germinability – parental habitat effects**)

1. Hormones

There is good evidence that the onset of dormancy requires the action of **abscisic acid** (ABA), perhaps only transiently, during seed development. ABA accumulates at certain stages in the developing seed (**see: Development of seeds – hormone content**) and dormancy sets in shortly after peak concentration is reached. Interference with the content of ABA or the sensitivity of the seed tissues to it disrupts the development of dormancy. Hence, seeds of ABA-deficient and ABA-insensitive mutants of *Arabidopsis* (*aba* and *abi*, respectively), tomato (*sitiens*) and *Nicotiana plumbaginifolia* do not become dormant or have a reduced degree of dormancy. Some seeds, e.g. **sorghum**, have less dormancy when treated during their development with fluridone, an inhibitor of ABA biosynthesis. Enhancing the biosynthesis of ABA by over-expression of an ABA biosynthetic gene in tobacco on the other hand leads to the production of more deeply dormant seeds. Furthermore, there is evidence, for example in wheat, sunflower and *N. plumbaginifolia*, that once the seed has become dormant the subsequent production of ABA in the hydrated seed is needed in order to maintain that dormancy. The capacity to achieve this is disrupted by the inhibitor of ABA biosynthesis, fluridone, and is lost naturally during **afterripening**.

Interference with ABA signalling pathways also modifies the onset of dormancy. Mutations affecting several transcription and other factors in signalling impair the development of dormancy in *Arabidopsis*. Onset of dormancy is also modified by other mutations not immediately related to ABA but which alter **gibberellin** content or perception and embryo or testa characteristics.

But although the evidence is very persuasive that ABA participates in the imposition of dormancy in a developing seed, how it does so is not known. The ABA acts upon one or more genes but the gene product(s) that actually causes dormancy is obscure. Since the failure to germinate, i.e. dormancy, represents an inability to initiate cell extension, usually in the radicle, it is possible that the ultimate action of ABA may be found there. The other side of the coin is this: since developing seeds of virtually all species appear to contain ABA why do they not all develop dormancy? **See: Dormancy genes** for an account of mutations affecting the acquisition of dormancy; **Preharvest sprouting-mechanisms**.

2. Environmental factors

Environmental factors that may affect dormancy onset include temperature, light quality, daylength, soil moisture and mineral nutrition. However, effects of these factors on the extent of dormancy are not readily predictable and depend on the species.

(a) *Temperature*. The temperature experienced during seed development can very strongly influence the degree of dormancy that sets in. A good example of this is *Avena fatua*, wild oats (Fig. D.35). Temperature conditions also affect **preharvest sprouting** in cereals such as wheat and barley, which depends on the intensity of dormancy in the immature grain. Relatively low temperatures early in grain development lead to a more intense dormancy while grains experiencing higher temperatures at maturation time have little or no dormancy and tend to sprout. The effect of the later high temperature is probably to accelerate afterripening in grains still on the mother plant: this illustrates that pre-mature dormancy is a fluid, variable phenomenon.

(b) *Light*. The wavelength composition received by developing seeds can influence the onset of dormancy. Experimentally, exposure to light relatively rich in far red leads to the formation of more dormant seeds, e.g. in *Arabidopsis* and cucumber. This can be attributed to the lowering of **phytochrome** Pfr content by the far red. Seeds of some species mature while enclosed in green (chlorophyllous) covering tissues which attenuate the red wavelengths while not preventing the transmission of far red. These seeds, too, develop as dormant, light-requiring seeds (**see: Light – dormancy and germination**).

The acquisition of dormancy in some species is apparently determined by the daylength. Seeds of *Chenopodium album*, for example, developing under long daily photoperiods mature with a deep dormancy. The seedcoat structure in this species and others (**see: Deserts and arid lands, 4. Maternal and environmental influences during seed development and maturation**) is also affected by the daylength, the coats being thicker, which could account for the deeper dormancy. The mechanism by which daylength exerts these effects is not understood.

Seeds of the same species can have different dormancy characteristics according to their provenance. It is likely that environmental factors acting upon the developing seeds at least to some extent account for these differences. (**See: Germinabilty – parental habitat effects**)

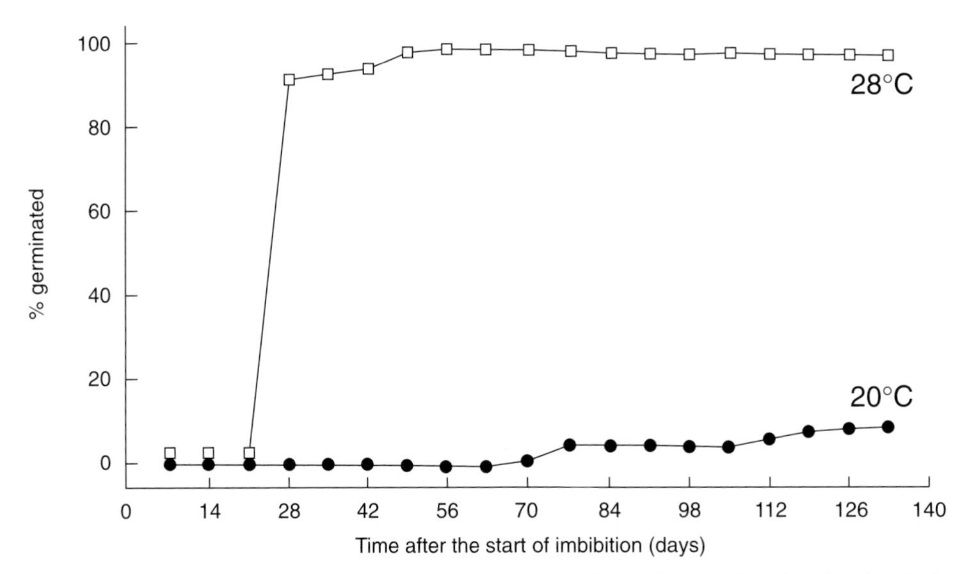

Fig. D.35. Effect of temperature on the acquisition of dormancy. Plants of an inbred line of *Avena fatua* bearing developing grains were held at 20 or 28°C until just before grain maturation/drying. They were then tested for germination at 4°C. Grains developing at the lower temperature have a long-lasting dormancy. (From Bewley, J.D. and Black, M. (1982) *Physiology and Biochemistry of Seeds in Relation to Germination.* Vol. 2. *Viability, Dormancy and Environmental Control.* Springer, Berlin.)

3. Maternal effects

Seeds from the same parent plant do not always show equal degrees of dormancy so it can be assumed that in these cases, called **heteroblasty** or **polymorphism**, some kind of maternal influence(s) determines the onset of dormancy. For example, in many members of the cereal and grass family (Poaceae) **caryopses** from different **spikelets** or from the **florets** within can have different levels of dormancy: two cases are *Avena ludoviciana* and *Aegilops kotschyi*. Interestingly, removal of one developing caryopsis can affect the dormancy of those remaining. Seeds formed in the disc (i.e. central) and ray (i.e. petalloid) florets in inflorescences of some **Compositae** species develop with varying intensities of dormancy. Of the two seeds in the dispersal unit of *Xanthium* spp. the upper one is deeply dormant while the lower one has a very shallow dormancy. It seems clear that correlative effects are operating here, that is plant parts are influencing each other. The distribution of nutritional, hormonal or other factors among the developing seeds might account for the differences in the acquisition of dormancy in neighbouring seeds. (**See: Germinability – maternal effects**)

Interactions between the environment and the genome in relation to dormancy are very complex. Environmental factors may modify seed size, weight and coat thickness, which may have an impact on seed dormancy.

4. Secondary dormancy

Hydrated, non-dormant, mature seeds under certain conditions may be forced into a state of dormancy termed **secondary dormancy**. In some cases, seeds which possess some primary dormancy might also undergo the same change (**see: Dormancy**). Seeds whose germination is inhibited by relatively high temperatures or by high photon fluxes might enter secondary dormancies called, respectively, **thermodormancy** and **photodormancy**. Cultivars of lettuce seed are examples that show both of these phenomena. The mechanism by which secondary dormancy is induced is obscure though there is evidence in some cases to suggest that ABA is implicated.

(MB, HWMH)

Baskin, C.C. and Baskin, J.M. (1998) *Seeds. Ecology, Biogeography, and Evolution of Dormancy and Germination.* Academic Press, San Diego, USA, pp. 181–237.

Bewley, J.D. and Black, M. (1982) *Physiology and Biochemistry of Seeds in Relation to Germination.* Vol. 2. *Viability, Dormancy and Environmental Control.* Springer, Berlin, Germany.

Bewley, J.D. and Black, M. (1994) *Seeds: Physiology of Development and Germination.* Plenum Press, New York, USA.

Nambara, E. and Marion-Poll, A. (2003) ABA action and interactions in seeds. *Trends in Plant Science* 8, 213–217.

Dormancy – coat-imposed

While the term 'coat'-imposed is in common use, this is applied loosely in a morphological sense, since it has been employed to refer to any structure that surrounds the embryo, whether it be the seedcoat (**testa**) or fruit coat (**pericarp**), or non-coat structures such as the **endosperm** and **perisperm** in angiosperms and the **megagametophyte** in gymnosperms.

Embryo coverings include seed and fruit tissues that may be composed of living or non-living cells (**see: Seedcoats – structure**). Their contribution to dormancy may be of a physical, chemical or mechanical nature, or combinations of these. Many mature seeds contain an endosperm. This triploid tissue, which is of maternal and paternal (2:1) genetic origin, functions as a storage tissue. In many species, particularly the cereals, it is mostly a dead tissue in which the majority of its cells are packed with reserves (starch and protein). In other species it is a living tissue that may significantly contribute to the dormancy and germination behaviour of the seed. The relative contribution of the endosperm to seed size varies greatly, from only a few cell layers surrounding the embryo as in many Cruciferae and Compositae, to an almost complete occupation of the seed as in the Poaceae (**see: Endosperm; Non-endospermic seeds; Structure of seeds**). The latter is often observed in seeds with a very small immature embryo, e.g. within the Ranunculaceae family. Very often, endosperm tissue contains thickened cell walls. This provides a considerable mechanical protection to the embryo but it is also a source of reserve carbohydrates, in the form of polysaccharides (**see: Hemicelluloses**). In a number of species a perisperm is present, a diploid endosperm-like tissue of maternal origin that is derived from the **nucellus**.

Before discussing the properties of the seedcoat that might account for coat-imposed dormancy it is important to note that since factors that break the dormancy of many types of such seeds act upon the embryo (e.g. chilling, alternating temperatures, chemicals), it is likely that even in coat-imposed dormancy deficiencies of the embryo are also involved so that it is unable to overcome the constraints imposed by the coat.

1. Physical inhibition; impermeability to water

In this type of dormancy the seedcoat (testa, derived from the integuments) and/or fruit tissues (pericarp) delay germination. At least 15 angiosperm families contain species with physical dormancy. The delay or lack of germination (dormancy) is usually the result of impermeability to water and/or oxygen. Before water and oxygen can penetrate to the embryo, the seedcoat requires a modification according to the nature of the block.

Seedcoat impermeability is usually caused by the presence of one or more layers of palisade cells. These palisade cell layers are composed of sclereid cells with thick lignified secondary cell walls. Fruit tissues may also contain sclereid cells. (**See: Seedcoats – structure**) Apart from lignin in secondary cell walls, a number of other components in these types of cells are water repellent, including cutin, quinones, suberin, waxes, callose, **phenolics** and hydrophobic (lipid-like) substances. Some seeds contain mucilage layers that function as a water 'gauge'. Too much water makes the layer impermeable to oxygen, e.g. in *Blepharis* (**see: Deserts and arid lands, 3. Seed dispersal strategies and mechanisms**) and too little water only hydrates the mucilage layer to a certain extent and sufficient water will not reach the embryo.

In addition to the overall impermeability of seed and fruit coats, many seeds contain specialized structures that regulate the uptake of water. These structures are generally derived from tissues that close the natural openings in the seed or fruit coat, such as the **micropyle**, **hilum** and **chalazal** area (**see: Seedcoats – structure; Structure of seeds**). For example, a chalazal cap is present in members of the Malvaceae and Cistaceae families, a strophiole (between hilum and chalaza) in the Papilionoideae and Mimosoideae, and an operculum (derived from the micropyle) in the Musaceae. The development of seedcoat impermeability is controlled by the **relative humidity** of the air during the **maturation** phase of seed development. Evidently, there is also a genetic component determining the degree of impermeability.

Physical dormancy is broken when the seed or fruit coat becomes permeable to water. In most cases this seems to occur by unplugging of the natural opening of the seed or fruit. However, depending on the moisture content of the seed and the relative humidity of the air, imbibition of water may also occur across the whole seedcoat. In a large number of species with physical dormancy, seedcoat impermeability increases as seed moisture content decreases. Moisture contents at which seedcoats become fully impermeable range from about 2 to 20% (fresh weight basis). Under natural conditions physical dormancy is usually broken by exposure of the seeds to extremes in temperature and temperature fluctuations. Detailed studies of a considerable number of species, including *Stylosanthes humilis*, *S. hamata* and *Heliocarpus donnell-smithii* seeds suggest that (extreme) temperature fluctuations rather than absolute temperature are decisive for the unplugging of seed openings. The required day/night temperature amplitudes may range from 15 to 40°C. It is generally assumed that the perception of high fluctuating temperature represents a mechanism by which gaps in the overlying vegetation are detected, since plant material dampens the amplitude of the fluctuations. The best-known condition of exposure of seeds to extreme heat is fire, which is a natural component of many ecosytems. The regeneration of many plant species after a fire is due to the breaking of physical dormancy of seeds in the soil. Seeds cannot withstand fire temperatures of over 600°C but a steep temperature gradient in the soil top layer ensures more moderate temperatures at a depth of a few centimetres. Many seeds lose physical dormancy when exposed to temperatures between 50 and 100°C for a limited time. In addition, burnt vegetation leaves gaps and open spaces which create greater day/night temperature fluctuations. Apart from unplugging of the seedcoat openings, exposure to dry heat also causes cracks in the coat, often starting from the opening. High heat, or fire softens or cracks the seedcoat in some species of legumes; there is cracking of the strophiole in *Lupinus varius*, or in the micropylar region of *Rhus ovata*, or disruption of the chalazal plug in cotton. In *Albizia lophantha*, the strophiolar plug is ejected by heat and probably by fire.

Storage of hard-coated seeds in liquid nitrogen at −196°C can also cause small cracks to appear in their coats, allowing better water penetration when the seeds are returned to germination temperatures. It is likely that this type of damage not only increases the permeability of the seedcoat but also decreases its mechanical restraint. (**See: Germination – influences of water**)

Although frequently mentioned, there is no conclusive evidence concerning microbial action in the loss of physical inhibition. Similarly, there are hardly any data available on the possibility of dormancy breaking by passage through the digestive tract of animals. Although a number of studies have shown that passage through an animal enhances germination, it is not known how this is achieved. Within the animal, dormancy may be broken by mechanical scarification or by acid. However, passage through a digestive system may indirectly break the dormancy, because the animal's waste material is usually dropped on the surface, thereby exposing seeds to higher (fluctuating) temperatures, as well as elevated temperatures as a result of fermentation of the faecal material.

2. Chemical inhibition

Numerous chemical compounds derived from seeds or dispersal units have been reported to inhibit germination (**see: Germination – influences of chemicals**). The chemicals have been extracted from all seed and fruit parts, including seedcoats. **Abscisic acid** (ABA) seems to be omnipresent in seed and fruit tissues. It is plausible to assume that the ABA present in mature seeds is a remnant of the ABA pools that are involved in the regulation of seed development. If the ABA content in the seed is high enough to inhibit germination, this compound has to be leached out and/or catabolized before germination can proceed. This likely represents a mechanism within the seed to spread germination in time. The coat may act, at least partially, as a barrier to leaching. Other inhibitory compounds that have been identified include phenolic acids, tannins and coumarins (**see: Phenolic compounds**). However, the specific role of these compounds in the inhibition of germination is not certain (but see Section 4). In most cases a direct relationship is lacking between their presence within a seed and their physiological action (**see: Germination – influences of chemicals**). On the other hand, germination of seeds containing inhibitory chemicals may be accelerated by extensive rinsing with water of the seeds prior to germination ('leaching'), for example in *Beta vulgaris* (**see: Deserts and arid lands, 5. Germination and seedlings**). It is possible that many of these inhibitory substances are primarily present to avoid **predation** or microbial infections.

3. Mechanical inhibition

Tissues surrounding the embryo will impose a certain mechanical restraint to its expansion. Their removal or partial removal will often lead to normal germination and embryo growth, indicating that the block to germination is located in the embryo coverings. Very often, the mechanical restraint will add up to possible blocks to uptake of water and/or oxygen. The embryo may overcome the mechanical restraint by the generation of sufficient thrust but, alternatively, the restraint may be weakened by an active process, controlled by the embryo, but with no apparent changes in the embryo growth potential. This has been demonstrated for a number of species in which the endosperm is digested by hydrolytic enzymes, e.g. tomato, *Datura ferox*, tobacco and white spruce (**see: Germination – radicle emergence**).

Although mechanical inhibition of germination is usually associated with seeds surrounded by thick, hard seedcoats or hard, woody fruit walls, germination of thin-coated seeds may (partially) depend on the properties of the relatively thin seedcoat (testa). In many species, including *Arabidopsis thaliana* and tomato (partial) removal of the seedcoat at least increases the rate of germination. In tomato, the testa imposes a mechanical force of approximately 0.15 Newton to the embryonic radicle tip, which is approximately 15% of the total resistance. This resistance is overcome by the growing radicle after the micropylar end of the endosperm (embryo cap) has been digested. Removal of the micropylar part of the testa increases the rate of germination and enables the seeds to germinate at a more negative osmotic potential of the germination medium than can the untreated seeds.

4. Resistance to oxygen entry

Seedcoats may also restrict the movement of oxygen to the embryo and, thus, slow down or inhibit germination (**see: Germination – influences of gases**). This may be caused by a low permeability of the seedcoat caused by its anatomy, but also by chemical compounds within the seedcoat that bind oxygen. Phenols and phenol oxidases are the most notable examples. Most of the evidence for these features of the seedcoat comes from studies in which seedcoats are punctured or split, which often enhances germination. Indirect evidence comes from the beneficial effects on dormancy breakage of subjecting seeds of certain species to high oxygen concentrations. (HWMH)

(See: Seedcoats – functional aspects)

Baskin, C.C. and Baskin, J.M. (1998) *Seeds. Ecology, Biogeography, and Evolution of Dormancy and Germination*. Academic Press, San Diego, USA.

Bewley, J.D. and Black, M. (1994) *Seeds: Physiology of Development and Germination*, 2nd edn. Plenum Press, New York, USA.

Werker, E. (1997) *Seed Anatomy. Encyclopedia of Plant Anatomy*, Spezieller Teil, Band X, Teil 3. Gebrüder Borntraeger, Berlin, Stuttgart.

Dormancy – cycles

In embryos exhibiting physiological dormancy, a dynamic system operates that potentially over many years: (i) enables dormancy status to fluctuate without irreversibly initiating germination, and (ii) maintains cell viability until germination can proceed. In these circumstances dormancy is alternately relieved and reimposed (Fig. D.36, and **see: Dormancy – importance in agriculture**, Fig. D.41). This system might involve hormonal and genetic interactions in the embryo, including changes in sensitivity to hormones with a possible contribution by parts of the covering tissues in some species.

Seeds of many species held within the soil seed bank only germinate when brought to the soil surface and exposed to increasing ranges of light and temperature. However, such exposure only results in germination at certain times of the year showing dormancy status to have a seasonal periodicity. This periodicity synchronizes the dormancy status of the population with the environment regardless of the position of a seed within the soil. This increases the germination potential of the population coincident with environmental conditions favourable for germination and seedling growth. Seeds within that soil seed bank may undergo several dormancy cycles before experiencing favorable environmental conditions. However, for this periodicity to be initiated, in most cases primary dormancy must first be broken. Freshly matured seeds that fail to germinate under a wide range of suitable conditions have primary dormancy. At this point only environmental conditions capable of delivering an intense dormancy-breaking stimulus will be effective in the bulk of the population. As primary dormancy decays, less extreme dormancy-breaking environmental conditions will be increasingly effective and the range of conditions in which germination will proceed increases until the seeds are non-dormant. Non-dormant seeds that do not germinate due to a lack of the appropriate environmental signals enter a state of **relative** (or conditional) **dormancy**. This is a transitional period where the conditions under which seeds germinate progressively narrow until they will not germinate under any conditions, at which point they have attained the state of **secondary dormancy**. It should be noted that some species only cycle between conditional dormancy and non-dormancy.

Dormancy cycles have been well described in the winter annual, *Lamium purpureum* and the summer annual, *Polygonum*

persicaria (**see:** Fig. D.37 in **Dormancy – ecology**). Secondary dormancy of the winter annual is caused by the chilling temperatures of winter and is lost in summer. Thus, germination occurs in autumn. In the summer annual, secondary dormancy develops at the relatively high summer temperatures and is broken in winter. Consequently, seeds germinate in springtime. *Orobanche* seeds lose their responsiveness to their germination stimulants (**see: Parasitic plants**) when preconditioned for too long at low or high temperatures, and this is considered as a secondary dormancy. Seeds with such a secondary dormancy regain their sensitivity to the germination stimulants during a further adequate preconditioning, and then also show annual dormancy cycles. (**See: Dormancy – ecophysiology**) (SF, MH, DC, FC)

Baskin, C.C. and Baskin, J.M. (1998) *Seeds: Ecology, Biogeography, and Evolution of Dormancy and Germination.* Academic Press, London, UK.

Probert, R.J. (2000) The role of temperature in germination ecophysiology. In: Fenner, M. (ed.) *Seeds: The Ecology of Regeneration in Plant Communities*, 2nd edn. CAB International, Wallingford, UK, pp. 261–292.

Dormancy – ecology

The three fundamental types of seed dormancy (morphological, physical and physiological) and their combinations (e.g. morphophysiological dormancy or MPD) can all play important ecological roles. Physiological dormancy is repeatedly reversible, while other types are generally not, so physiological dormancy permits a more flexible response to the environment. (**See: Dormancy**)

Breaking MPD requires growth or differentiation of the embryo, while breaking physical dormancy requires actual rupture of the seedcoat. In both cases, both dormancy and dormancy breaking are readily observable phenomena. Physiological dormancy is not so visibly evident, can vary continuously in depth and is also usually reversible. The elusive nature of physiological dormancy has led some ecologists to define any non-germinating seed (in the **soil seed bank**, for example) as dormant, which is incorrect (**see: Dormancy; Dormancy breaking**).

In an attempt to remove possible confusion, Vleeshouwers *et al.* (1995) have suggested the following definition of dormancy: 'a seed characteristic, the degree of which defines what conditions should be met to make the seed germinate'. Thus, dormancy actually represents a seed's fastidiousness about the conditions it requires for successful germination to be initiated, while germination is what happens when the environment meets these and other basic requirements such as temperature and oxygen. (**See: Germination – influences of gases; Germination – influences of temperature**) This definition makes a clear distinction between changes in dormancy and the process of germination itself. If seeds are kept under natural conditions (e.g. buried in soil in the field or an unheated greenhouse) and germination is tested at intervals in a range of conditions, it is commonly found that the conditions which permit germination change over time. In seeds with physiological dormancy, this widening and narrowing of germination requirements is caused only by temperature. Other factors that do not change a seed's dormancy, but are often essential for germination itself, are germination triggers or cues.

Germination occurs when dormancy-breaking is followed by exposure to a suitable germination cue or cues. To some extent, the distinction between the ending of dormancy and the triggering of germination depends on where one chooses to draw

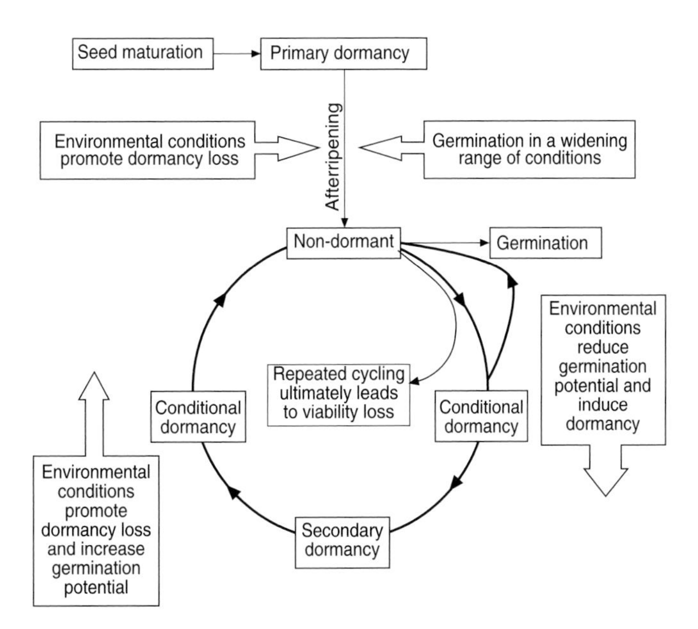

Fig. D.36. The change in dormancy status of a mature seed as it loses primary dormancy and enters a dormancy cycle. Once primary dormancy is lost the non-dormant seed will germinate in a wide range of conditions. The absence of such conditions leads to conditional dormancy, then secondary dormancy. As secondary dormancy declines the seed returns to a non-dormant state. Repeated cycles ultimately lead to reduced viability. (Adapted from Baskin, C.C. and Baskin, J.M. (1998) *Seeds: Ecology, Biogeography, and Evolution of Dormancy and Germination.* Academic Press, London, UK.)

the line between these two processes. One could legitimately argue that exposure to light (or nitrate ions, or fluctuating temperatures) changes the internal conditions of the seed, so that it can now germinate in darkness (**see: Dormancy – ecophysiology**). The final resolution of the debate awaits a better understanding of dormancy at the molecular level. One would merely observe that the immediate response of a seed to germination cues is germination itself (by definition), while the immediate response of a seed to a change in dormancy status is not necessarily germination at all, unless the change in dormancy brings the seed's germination requirements within the range of current environmental conditions. Indeed a seed may cycle in and out of dormancy for years or even decades without germinating (**see: Dormancy – cycles**). Until we know more, it seems only reasonable to separate such metabolic changes from those that result in immediate and obligate germination.

The common ecological division of dormancy into innate, induced and enforced is both wrong and misleading. Seeds in 'enforced' dormancy are currently being prevented from germinating by some environmental constraint (too cold, too dry, too anoxic, etc.). Enforced dormancy, therefore, is not only an attribute of the environment rather than the seed, but also too vague to be useful, including as it does all seeds not truly dormant and not actually in the act of germinating. A seed that remains ungerminated because the minimum requirements for germination are lacking is better described as **quiescent**. The terms 'innate' and 'induced' dormancy have now largely been replaced by **primary** and **secondary dormancy**.

1. Dormancy breaking in field and laboratory
In the laboratory, seeds can sometimes be persuaded to germinate by treatments that clearly do not operate in the field. For example, physiological dormancy can often be broken by treatment with **gibberellins** (**see: Dormancy breaking – hormones**). Similarly, in the laboratory it is possible to break physical dormancy by physically or chemically abrading the seedcoat. Perhaps for this reason, many textbooks and reviews suggest that similar processes (physical abrasion or microbial attack) break dormancy in the field, although there is no evidence for this. In the field, high temperatures or very large temperature fluctuations seem to be crucial. There is good evidence that physical dormancy is a specialized signal-detecting system, involving highly specialized seedcoat anatomy (**see: Seedcoats – functional aspects**), and is not normally broken simply by generalized damage to the seedcoat. This distinction is crucial to the function of seed dormancy. Seed dormancy is only adaptive if it improves plant fitness, by increasing the chance of germination in circumstances that result in the highest probability of subsequent survival and reproduction. This requires that dormancy is broken by specific changes in the environment, and not by random and unpredictable processes such as microbial activity or movement of soil particles.

But microbial attack can help to break some forms of dormancy, though not physical dormancy. Dormancy breaking in some temperate species requires moist chilling, preceded by warm moist storage (**stratification**). This behaviour is most often linked to MPD, with different temperature optima for

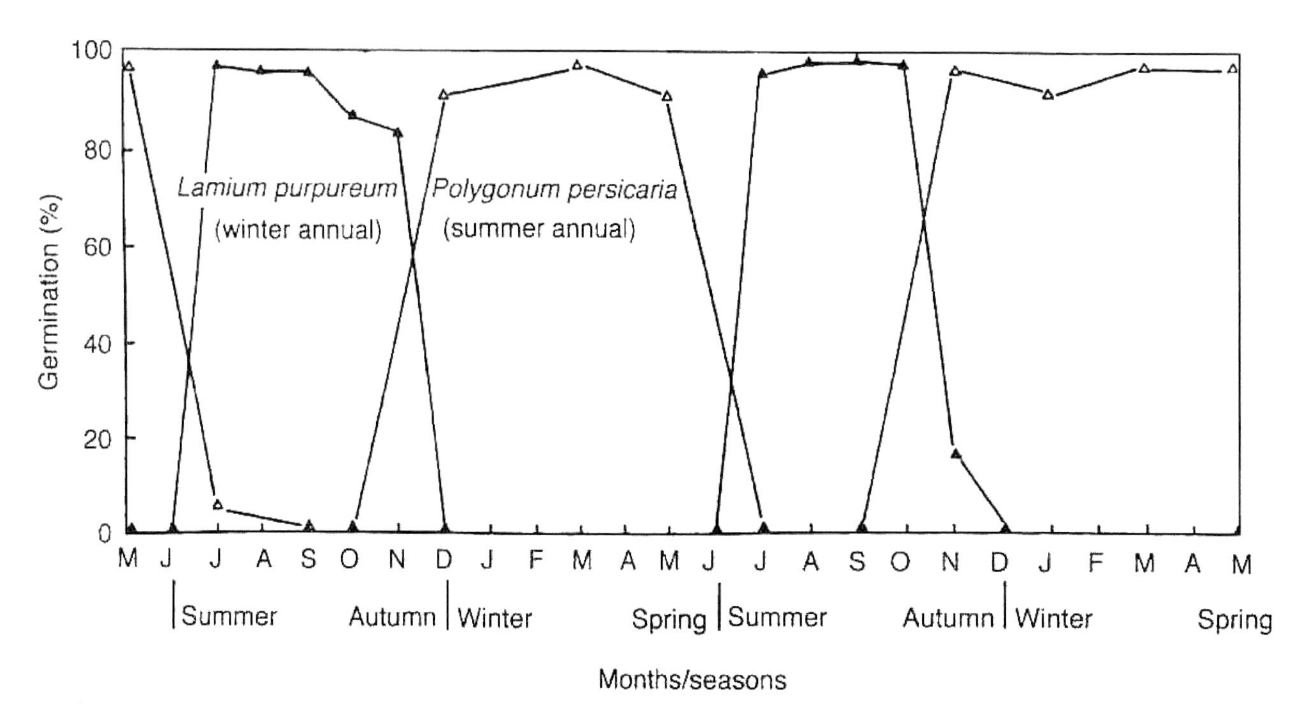

Fig. D.37. Seasonal changes in the germination behaviour of buried seeds of the winter annual *Lamium purpureum* (▲) and summer annual *Polygonum persicaria* (△). Seeds of *L. purpureum* were buried in May in Lexington, Kentucky, USA, and germination of exhumed seeds was tested at 25/15°C. Seeds of *P. persicaria* were buried in December in Wageningen, The Netherlands, and tested at 20°C. All germination tests were conducted in the light. Figure reproduced from Probert, R.J. (2000) The role of temperature in the regulation of seed dormancy and germination. In: Fenner, M. (ed.) *Seeds: the Ecology of Regeneration in Plant Communities*, 2nd edn. CAB International, Wallingford, UK, pp. 261–292.

embryo growth (usually warm) and breaking of physiological dormancy (usually cold). In some species, however, including several *Rosa* spp., these requirements are not associated with MPD. Physiological dormancy in *Rosa corymbifera* is broken only after the thick fruit coat is breached by microbial attack during the warm period. If sterile seeds are incubated, dormancy is not broken, and the highest germination percentage follows incubation with nutrients that promote bacterial growth. Seeds incubated with nutrients but without microbes do not germinate, and re-inoculation of sterile seeds with microbes restores germination to its previous level. It is not known how widespread is this dormancy-breaking mechanism, but it may occur in a number of woody Rosaceae.

2. Seasonal cycles of dormancy

Temperature regulates dormancy but can also act as a germination cue. For example, dormancy of summer annuals is broken by low temperatures, but germination itself usually requires much higher temperatures. In *Polygonum persicaria*, seeds lose primary dormancy during the winter and may germinate in spring if exposed to light (Fig. D.37). If they remain in the dark, the same temperatures that permit germination will also begin to re-impose secondary dormancy, although the two processes occur at very different rates. Conversely, in winter annuals (e.g. *Lamium purpureum*), dormancy is broken by high temperatures, with dormancy breaking usually accompanied by a progressive reduction in the minimum temperature required for germination. Together with re-imposition of dormancy by low temperatures, this ensures that germination of winter annuals, typically plants of habitats with mild winters and hot, dry summers, occurs only in autumn. Seed of many species with persistent seed banks, including both winter and summer annuals, frequently undergo seasonal cycles of dormancy (**See: Dormancy – cycles**), which can continue for many years if germination is not triggered. In most winter- or summer-annuals, dormancy is broken and induced by quite different temperature ranges, but there is no reason why the two ranges should not overlap. Therefore sometimes, and perhaps quite frequently, apparently rather complex behaviour may arise from the simultaneous induction and breaking of dormancy. Finally, even more complex non-linear responses to temperature may arise from the simultaneous induction and breaking of dormancy and loss of viability, each with a unique response to temperature. Note that secondary dormancy is liable to be induced whenever imbibed or partially-imbibed seeds are prevented from germinating, although the inhibition of germination may arise from a wide variety of causes, e.g. anoxia, prolonged white light, or a temperature unsuitable for germination. (**See: Germination – influences of temperature; Light – dormancy and germination**)

3. Dormancy and habitat

To varying extents, the different dormancy types constrain the habitats and climates that species possessing them can occupy. Breaking morphological dormancy and its more common variant, MPD, requires embryo growth and/or differentiation, and the seed must be imbibed for this to happen, although the physiological part of the dormancy can sometimes be broken in the dry seed (**afterripening**). MPD is therefore frequent in parts of the world with moist seasonal climates, and is particularly common in plants of woodlands or damp grasslands. *Trillium*, *Erythronium* and *Heracleum* are typical MPD genera. Breaking of physical dormancy, on the other hand, needs high temperatures, large temperature fluctuations or fire, and of course does not require the seed to be imbibed. It is therefore common in many habitats with a distinct dry season, including tropical deciduous forests, savannas, hot deserts, steppes and matorral. Embryos of seeds with physical dormancy seem more resistant to high temperatures than are those of other species. Physiological dormancy is common everywhere, although species with no seed dormancy are in the majority in tropical evergreen forest. (**See: Deserts and arid lands**) Habitats with the highest frequencies of non-dormancy (evergreen rainforest, semi-evergreen rainforest, and moist warm temperate woodland) (**see**: Fig. D.34 in **Dormancy**) also have the largest proportions of species with recalcitrant (desiccation-sensitive) seeds. (**See: Desiccation tolerance; Recalcitrant seed**) However, the association does not appear to be causal, since both recalcitrant and **orthodox** species in these habitats have high frequencies of non-dormancy, and the majority of non-dormant species also have orthodox seeds. It seems likely that both recalcitrance and non-dormancy confer a fitness advantage (particularly on large seeds) in aseasonal environments by reducing the length of time that seeds are exposed to attack by predators or pathogens.

4. The function of dormancy

The primary function of seed dormancy is sometimes regarded to be to prevent germination during periods that are unsuitable for germination and establishment. In fact, though, dormancy is not required for such a purpose; seeds will not germinate during a dry season, and a requirement for moderately high temperatures for germination is sufficient to prevent germination during winter. The crucial function of dormancy is to prevent germination when conditions are suitable for germination, but the probability of survival and growth of the seedling is low. In temperate regions, this often means delaying germination from autumn until spring, by requiring a period of low temperatures (chilling or stratification) before germination can occur. (**See: Dormancy breaking – temperature**) Surprisingly, the most exacting requirements for dormancy breaking need not be found among species that experience the longest and most severe unfavourable periods. For example, seeds of *Sorbus aucuparia* from high altitudes in Scotland needed only 6 weeks' cold to break dormancy, while seeds from low altitudes needed up to 18 weeks. This apparently paradoxical behaviour is because seeds at low altitudes, if non-dormant, might be persuaded to germinate during mild spells in mid-winter, with potentially lethal consequences for the seedlings. At high altitudes, where winters are consistently severe, there is no danger of seeds germinating before spring, because the low temperatures do not support the process of germination itself.

In cool temperate climates the summer is always the favourable growing season, but over longer climatic gradients, the identity of the favourable season itself may change, with striking consequences for seed dormancy. In the eastern Mediterranean, *Pinus brutia* occurs over a wide latitudinal range, from Crete to northern Greece. In the southern parts of this range, the climate is typically Mediterranean, with mild, damp winters and hot, dry summers. Thus *P. brutia* seeds from Crete have no dormancy and germinate in autumn. Thrace, in northern Greece, has a much more continental climate, with frost on most days in winter. Seeds from there have deep dormancy, fully broken only by 3 months of chilling, and therefore germinate almost exclusively in spring. Between these two extremes, for example in the northern Aegean, no single germination season is ideal, and seeds from such climates have intermediate germination behaviour.

(KT)

Baskin, C.C. and Baskin, J.M. (1998) *Seeds: Ecology, Biogeography and Evolution of Dormancy and Germination*. Academic Press, San Diego, USA.

Skordilis, A. and Thanos, C.A. (1995) Seed stratification and germination strategy in the Mediterranean pines *Pinus brutia* and *Pinus halepensis*. *Seed Science Research* 5, 151–160.

Vleeshouwers, L.M., Bouwmeester, H.J. and Karssen, C.M. (1995) Redefining seed dormancy: an attempt to integrate physiology and ecology. *Journal of Ecology* 83, 1031–1037.

Dormancy – ecophysiology

Depth of dormancy and the occurrence, or absence, of germination are seed responses to environmental factors, such as light, temperature, and soil and atmospheric components. Seeds are able to sense the continuous streams of information about the suitability of the environment for successful seedling emergence and subsequent plant growth. The seed's responses to this information may be of an immediate nature, e.g. to commence germination, or of a long-term nature, such as the seasonal regulation of dormancy of seeds in the **soil seed bank**, so-called **dormancy cycling**. The ecophysiology of dormancy and germination deals with the perception of environmental factors by the seeds and how these are translated into signalling inside the seed tissues (**see: Dormancy – ecology; Dormancy breaking**).

By far most progress in elucidating the principles of processing by seeds of external information has been made with the large group of arable weeds of the temperate zones (**see: Dormancy – importance in agriculture, 2. Weeds**). In these regions emergence is restricted to certain periods of the year, and in most cases (**summer annuals**) within a limited period in the spring, sometimes followed by additional flushes in summer. Examples of this group of summer annuals are *Ambrosia artemisiifolia*, *Polygonum persicaria*, *Chenopodium album*, *Spergula arvensis* and *Sisymbrium officinale*. Species originating from climates with a hot dry summer and a cool humid winter, such as *Arabidopsis thaliana*, mainly germinate in autumn, surviving the winter as rosette plants (**winter annuals**).

Dicotyledonous annual weed species form large persistent soil seed banks of which the size outnumbers the annual input of seeds. Survival of seeds in these seed banks may be as long as decades to centuries (**see: Ancient seeds**). Emergence from seed banks is strongly stimulated by disturbance or wetting of the soil, e.g. by cultivation, but the seasonal timing of emergence is not significantly affected. Disturbance or wetting only appear to influence the number of emerging seedlings. The timing of emergence can be explained by changes in seed dormancy. Seasonal fluctuations in temperature are the main regulatory factor of annual dormancy cycling. Evidently, in the temperate zones, temperature fluctuations over the year remain within narrow ranges, unlike, for example, rainfall. The range of conditions over which germination and emergence can occur widens during the alleviation of dormancy whereas it narrows during dormancy induction, indicating that dormancy is a relative phenomenon.

Apart from seasonal temperature fluctuations, seed emergence is influenced by the actual field temperature, light and nitrate. Together with water and oxygen, light (through **phytochrome**) and nitrate may be considered the two most important naturally occurring factors required for the germination of many species (**see: Dormancy breaking – chemicals; Germination – influences of chemicals; Light – dormancy and germination**). However, it is unlikely that light and nitrate play a role in the regulation of annual dormancy cycles. Light can only penetrate to a very limited depth in the soil, whereas soil nitrate amounts show a large variability between successive years and thus are not associated with the strict seasonal timing of seedling emergence. This implies that alleviation of dormancy, and subsequent germination, must be regarded as separate phenomena. However, alternative concepts of dormancy and germination are widely in use. For example, extensive studies with *Avena fatua* seeds have resulted in a concept of dormancy being comprised of several blocks to germination. These blocks may vary in number and intensity and each one may require a different external stimulus to be eliminated. There may also be a specific order of removal of these blocks.

The breaking of dormancy is characterized by a widening of the germination temperature optimum (temperature 'window') and a reduction in the requirement for other germination promoters (e.g. light and nitrate) (**see: Dormancy breaking – temperature; Germination – influences of temperature**). In the field, germination occurs when the germination temperature window overlaps with the actual field temperature (Fig. D.38). In summer annuals the window is very narrow or even closed in dormant seeds at the end of the summer season and, hence, germination will not occur. When dormancy is terminated, as a result of the cool winter temperatures, the germination temperature window widens and will eventually overlap with the (increasing) field temperature in spring. However, germination will only occur when the requirements for germination, such as oxygen, water, soil components and possibly light are met. Conversely, induction of dormancy will be associated with a narrowing of the germination temperature window and an increase in requirement for environmental factors. Thus, dormancy cycling is a process of alternating perceptiveness of seeds to environmental factors that promote germination. The ecological relevance of dormancy cycling is to prevent seeds from germinating during short spells of favourable conditions in an otherwise non-favourable season.

Nitrate is the most important inorganic soil component that stimulates seed germination (**see: Dormancy breaking – chemicals; Germination – influences of chemicals**). This property of the nitrate ion has been known since the early 20th century. Since then, numerous wild and crop species, both monocots and dicots, have been found to be sensitive to the stimulatory action of nitrate. In addition, the

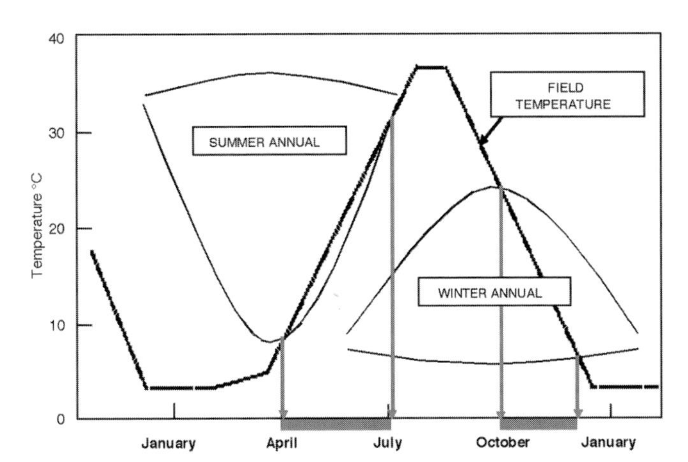

Fig. D.38. Opening and closing of the germination temperature windows of a summer and winter annual and the occurrence of overlap with the actual field temperature, allowing seedlings to emerge. Beginning and ending of seedling emergence is indicated by vertical arrows. Dotted line represents field temperature and uninterrupted lines the borders of the germination windows. For examples of species, **see** Fig. D.37 in **Dormancy – ecology**.

germination of fern spores may also be promoted by nitrate. Since nitrate is central to the nitrogen cycle, most soil types contain nitrate, often within the range of concentrations that are effective for germination.

Light and nitrate display a strong interaction in the dormancy and germination of many weed species. In seeds of *Avena fatua* and *Sisymbrium officinale* the effectiveness of nitrate depends on the level of active **phytochrome** (Pfr). In *S. officinale* the dependency on light is absolute. However, seeds may contain Pfr at the moment of shedding. This pre-existing phytochrome may persist for long periods in the dry seeds. If the amount of pre-existing Pfr is above the threshold level for germination and if sufficient endogenous nitrate is present, seeds may germinate in the dark.

A flush of germination of weeds in a crop field after disturbance of the soil indicates that light may be the limiting factor for germination in nitrogen-rich soils. During soil disturbance, a very short exposure to light (even less than a second) may induce germination. Germination may even proceed when the seeds are later returned beneath the soil surface to depths of up to 10 cm where light cannot penetrate. This phenomenon may have far reaching implications for weed control and management in agriculture. Soil cultivation in the dark may greatly reduce the emergence of weeds in the field. In addition, soil disturbance may also result in considerable release of nitrate ions from the soil. This is probably a significant factor in the promotion of germination of weeds, such as *Avena fatua*, in North America where summer fallowing is commonly practised. In *Sisymbrium officinale* and *Arabidopsis thaliana*, the effect of light and nitrate on seed germination may be reciprocal. Thus, the light requirement of seeds in a soil with low nitrate levels may be higher than that of seeds in nitrate-rich soils. Whether this has any ecological meaning remains to be shown.

Attempts have been made to summarize the respective actions of temperature, light and nitrate in the regulation of dormancy and germination in arable weeds in descriptive models (**see** Fig. D.39 **in Dormancy – embryo (hypotheses)**). Predictions of weed emergence in the field based upon these models are encouraging but are still far from possible application in weed management and control.

Although the ecological significance of sensing mechanisms for light and nitrate seems obvious, further studies are required to provide convincing evidence. Obviously, a plant that will eventually grow from a seed requires light and nitrogen for optimal development and growth. However, no relationship is known between the amounts of light and nitrate that promote seed germination and the amounts required by the growing plant. A plausible explanation for the presence of such sensing mechanisms is the ability of seeds to detect local disturbances of light and nitrate amounts in the immediate environment. Seeds that are shaded by the foliage of neighbouring plants receive light with a lower ratio of red to far-red light than direct sunlight. This is because the leaf chlorophyll absorbs red light more effectively than it does far-red light. Consequently, the amounts of Pfr that are established in the shaded seeds will be lower. In this way, germination beneath established plants may be prevented and competition avoided. Similarly, established plants may lower the nitrate content of the soil around their root systems. Nitrate is consumed and nitrification may be inhibited. As a result, the seeds in the immediate environment are depleted of nitrate and germination will be reduced.

The soil temperature and diurnal temperature fluctuations are influenced by the thickness of neighbouring vegetation. Diurnal soil temperature fluctuations are much less pronounced than in soils with no vegetation. Many weed species require diurnal temperature fluctuations for successful germination. Thus, this is also an effective mechanism to detect the proximity of competitors. (HWMH)

Hilhorst, H.W.M., Derkx, M.P.M. and Karssen, C.M. (1996) An integrating model for seed dormancy cycling; characterization of reversible sensitivity. In: Lang, G. (ed.) *Plant Dormancy: Physiology, Biochemistry and Molecular Biology*. CAB International, Wallingford, UK, pp. 341–360.

Probert, R.J. (2000) The role of temperature in germination ecophysiology. In: Fenner, M. (ed.) *Seeds: The Ecology of Regeneration in Plant Communities*, 2nd edn. CAB International, Wallingford, UK, pp. 285–325.

Simpson, G.M. (1990) *Seed Dormancy in Grasses*. Cambridge University Press, Cambridge, UK.

Dormancy – embryo

Embryo dormancy is recognizable because the viable embryo fails to germinate when it is freed of its covering structures. In some cases the failure is because the embryo is structurally immature: in others the embryo is fully formed but an internal block arrests germination. Some examples of the two types are given in Table D.11.

1. Immature embryos

Several plant families have genera with species whose dispersed seeds have undifferentiated embryos. The best known are the Orchidaceae and Orobanchaceae families (**see: Orchids; Structure of seeds – identification characters**). Their seeds are usually small and their embryos may consist of

Table D.11. Some species showing embryo dormancy.[a]

Immature embryos	Mature embryos
Apium graveolens	*Acer platanoides*
Anemone coronaria	*Acer tartaricum*
Annona squamosa	*Avena fatua* 'Montana'
Conium maculatum	*Corylus avellana*
Fraxinus excelsior	*Crataegus mollis*
Heracleum spondylium	*Euonymus europaea*
Ilex opaca	*Impatiens parviflora*
Panax ginseng	*Malus arnoldiana*
Pastinaca sativa	*Malus sylvestris*
Pulsatilla slavica	*Prunus persica*
Ranunculus biternatus	*Rosa* spp.
Rhododendron ferrugineum	*Rhodotypos kerrioides*

[a]Information from Baskin, C.C. and Baskin, J.M. (1998) *Seeds. Ecology, Biogeography, and Evolution of Dormancy and Germination.* Academic Press, San Diego, USA and Bewley, J.D. and Black, M. (1982) *Physiology and Biochemistry of Seeds in Relation to Germination*, Vol. 2. *Viability, Dormancy and Environmental Control.* Springer, Berlin.

only 2 to 100 cells. Since these seeds do not contain sufficient food reserves, external sources of nutrition are required for growth, e.g. from higher plants (on which they may be parasitic), or as mycoheterotrophs, where they form an association with a **mycorrhizal** fungus, which acts as a provider of carbon. In these seeds, developmental arrest appears to occur at an early stage of morphogenesis (histodifferentiation). Other species do complete their morphogenetic phase and have a differentiated embryo, but do not appear to enter the embryo growth/maturation phase, i.e. they do not expand, and are underdeveloped. Seeds of this type usually contain relatively large amounts of **endosperm** tissue in which the small embryo is embedded. These embryos have to grow inside the dispersed seed prior to the initiation of germination. This has been studied particularly well in celery (*Apium graveolens*) in which the surrounding endosperm is digested during embryo growth. Although seeds with undifferentiated or underdeveloped embryos are not dormant in the strict sense they are usually ranked as morphologically dormant (**see: Dormancy**).

2. Fully mature embryos

In the majority of species with dormancy located in the fully developed embryo, dormancy mechanisms appear to be related to reversible metabolic processes. This is often referred to as physiological dormancy (**see: Dormancy**). In contrast, **coat-imposed dormancy** can be released but not induced again, which makes it an irreversible process. The reversible nature of dormancy of developed embryos allows the occurrence of repeated dormancy cycles (**see: Dormancy – cycles; Dormancy – ecology; Dormancy – ecophysiology**). Seeds whose primary dormancy is broken may re-enter the state of dormancy under certain conditions. This feature is the basis of 'dormancy cycling', which is highly relevant for seed emergence under natural conditions. Although several genes have been identified that are associated with embryo-located dormancy control, little is known about the mechanisms involved.

Most of our views on the mechanism of dormancy have been generated by studies on the breaking of dormancy (**see: Dormancy breaking** entries). The expectation is that if we understand the changes that occur when dormancy is removed by various factors we can infer that the feature or process that is altered is likely to be implicated in the dormancy mechanism itself.

There are two common conditions by which **primary dormancy** may be broken: (i) dry-**afterripening**, either on the plant or in dry storage, particularly at elevated temperatures; and (ii) stratification by chilling of the fully imbibed seeds. (*Editors' note: Since in many cases dormant seeds can be made to germinate by exposure to light or alternating temperatures these two factors are often considered as dormancy-breaking agents.*) It is not generally known which seed compartments are affected by these conditions. Some seeds with embryo dormancy are sensitive to afterripening and/or chilling, which means that the embryo must be affected. However, species with coat-imposed dormancy, such as tomato, also respond to these treatments; thus, even in **coat-imposed dormancy**, deficiencies of the embryo are also likely to be involved. Observed changes accompanying the termination of dormancy by these conditions are an increased sensitivity to light, nitrate and **gibberellins** (GAs), and leakage of inhibitory compounds, including **abscisic acid** (ABA) in the case of chilling. (**See: Germination – influences of chemicals; Light – dormancy and germination**)

When seeds are released from dormancy but not stimulated to germinate because of unfavourable conditions, such as sub-optimal temperature, darkness or absence of chemical stimulants, they will become dormant again (**secondary dormancy**). The mechanism of this reversible nature of (embryo-located) dormancy is largely unknown. However, inhibitors of ABA biosynthesis, such as norfluorazon, may promote germination of dormant *Arabidopsis* seeds or embryo-dormant sunflower. Furthermore, ABA content does not decrease upon imbibition of dormant barley seeds whereas this occurs rapidly in non-dormant seeds, suggesting that ABA synthesis and degradation are involved in the transition between the dormant and non-dormant state.

There is very strong evidence implicating abscisic acid (ABA) and gibberellin in the mechanism of embryo dormancy (**see: Dormancy – genes**). But despite the abundance of information on ABA signalling in plants, the mechanism of ABA-induced dormancy remains unknown. Extensive analysis of the five *abi* loci known to control the seed's responsiveness to ABA strongly suggests this hormone activates the relevant genes through different secondary messengers and different ABA-response elements within the promoter regions of these genes (**see: Signal transduction – hormones**). However, the gene products that are directly responsible for the onset of dormancy remain to be identified. Analysis of the *COMATOSE* (*CTS*) locus as a major control point in the transition between dormancy and germination strongly suggests that *CTS* is an important target for loci repressing germination during embryo development, including *ABI1-1*, *RDO2*, *ATS* and *TTG1-1*. *CTS* has been identified as a peroxisomal **ATP**-binding-cassette transporter that plays an important role in lipid mobilization, but how this is involved in dormancy is unclear.

A wide array of compounds is known that break dormancy of dehulled red rice (*Oryza sativa*) seeds (Table D.12). Many of

Table D.12. Selection of organic compounds that break dormancy of red rice (*Oryza sativa* L.) caryopses.

Organic acids (pH-dependent)	
formic acid	glycolic acid
acetic acid	lactic acid
propionic acid	succinic acid
butyric acid	trimethylacetic acid
valeric acid	benzoic acid
caproic acid	salicylic acid
heptanoic acid	dimethadione
octanoic acid	anthracene-9-carboxylic acid
isobutyric acid	cyclohexanemonocarboxylic acid
isovaleric acid	
Esters	Aldehydes
methyl formate	acetaldehyde
ethyl acetate	propionaldehyde
methyl propionate	cyclohexanecarboxaldehyde
Alcohols	
methanol	hexanol
ethanol	heptanol
propanol	octanol
isopropanol	cyclohexyl-methanol
butanol	2-cyclohexylethanol
pentanol	2-butanol
Enols	Ketones
1,5 pentanediol	2-butanone
2-propen-1-ol	acetone
2-buten-1-ol	2-pentanone
cis-2-penten-1-ol	3-pentanone
cis-2-hexen-1-ol	
trans-2-hexen-1-ol	
Miscellaneous	
fluridone (pH-dependent)	
liquid smoke	
phenyl arsine oxide	

From Cohn, M.A. (1997) QSAR modelling of dormancy-breaking chemicals. In: Ellis, R.H., Black, M., Murdoch, A.J. and Hong, T.D. (eds) *Basic and Applied Aspects of Seed Biology*. Kluwer Academic, Dordrecht, The Netherlands, pp. 289–295; and M.A. Cohn, personal communication.

these compounds are non-metabolic and there is a high degree of structural dissimilarity among them. These compounds all have a certain lipophilicity in common which can be expressed as a partitioning coefficient. These coefficients correlate well with the effectiveness of dormancy relief. The organic acids and alcohols (after metabolism to the corresponding weak acid) give rise to acidification of the embryonic tissues, although further metabolism of the compounds cannot be excluded. This implies that the deficiency in the embryo that normally prevents the seed from germinating may be linked with the internal pH conditions of the embryo cells. (HWMH)

(See: Germination – influences of chemicals)

Bewley, J.D. and Black, M. (1994) *Seeds: Physiology of Development and Germination*. 2nd edn. Plenum Press, New York.

Footitt, S., Slocombe, S.P., Larner, V., Kurup, S., Wu, Y., Larson, T., Graham, I., Baker, A. and Holdsworth, M. (2001) Control of germination and lipid mobilization by *COMATOSE*, the *Arabidopsis* homologue of human ALDP. *EMBO Journal* 21, 2912–2922.

Kornneef, M., Bentsink, L. and Hilhorst, H. (2002) Seed dormancy and germination. *Current Opinion in Plant Biology* 5, 33–36.

Kucera, B., Cohn, M.A. and Leubner-Metzger, G. (2005) Plant hormone interactions during seed dormancy release and germination. *Seed Science Research* 15, 281-308.

Nambara, E. and Marion-Poll, A. (2003) ABA action and interactions in seeds. *Trends in Plant Science* 8, 213-217.

Dormancy – embryo (hypotheses)

There have been several hypotheses put forward to explain dormancy in mature embryos within the seed. These generally fall into two categories, those involving the state of membranes, and those invoking some aspect of respiration. Some of the hypotheses have indirect and suggestive support, but all lack definitive evidence. The concepts for the most part stem from observations on the breaking of dormancy, reasoning that if changes occurring in the release from dormancy can be identified, an indication could thereby be obtained as to the 'status quo' responsible for dormancy. (**See: Dormancy acquisition; Dormancy breaking** entries)

1. Membranes

Cellular membranes may be involved in transitions between the dormant and non-dormant states. This is based on the observations that: (i) temperature plays a decisive role in the regulation of dormancy and germination, and that membranes are highly sensitive to changes in temperature; and (ii) certain groups of chemicals, including anaesthetics, are capable of breaking dormancy and that these may directly act on cellular membranes and/or membrane components. The 'membrane hypothesis' of dormancy is that a cell membrane(s) is not in a favourable state to allow germination to be initiated.

Membranes are frequently referred to as the primary target for perception of temperature at the cellular level, and they adapt their order and composition to changes in the environmental temperature. In plants, this acclimation is attained by changes in protein and lipid components of cellular membranes. One aspect of these changes is homeoviscous adaptation. Generally, membranes adjust to maintain their fluidity. Membrane fluidity depends on the temperature as well as on the degree of saturation of the fatty acid tails of the membrane phospholipids. Alterations in the temperature will change the membrane's fluidity and, as a response, the degree of saturation may be adjusted to redirect the fluidity to its original extent. Membrane-associated proteins, including receptors (e.g. of **hormones**), may also be altered directly or indirectly by changes in temperature, which may change their tertiary structure or location within the membrane. In this way, receptors may become accessible to substances that bind to them, i.e. to their ligands. The inference is that in the dormant state the receptors in the membrane are not accessible.

The observation that anaesthetics can break dormancy of grasses such as *Panicum capillare* and *Echinochloa crus-galli* has also added evidence that membranes are involved, and has led to a membrane hypothesis of dormancy. Anaesthetics are known to affect cell membranes and their components by decreasing their packing density. By doing so, anaesthetics exert effects on membranes quite similar to temperature. However, germination of *Avena sativa* seeds can only be stimulated by those alcohols that are metabolized by the enzyme alcohol dehydrogenase. Thus, besides exerting possible anaesthetic effects, alcohols may affect germination and dormancy through (unknown) metabolic modifications. (**See: Dormancy breaking – chemicals, 3.(c) Membranes**)

Based on these characteristics a number of generic models involving cellular membranes have been described for temperature responses of seeds, mediated by changes in fluidity and order of membranes (Fig. D.39). These changes would sensitize a receptor, for example one for **phytochrome,** that is located in or on the membrane. However, evidence for such a mechanism is largely circumstantial.

2. Respiration

Besides several alcohols, a large number of other chemical compounds can influence dormancy and germination. These include organic acids, esters, aldehydes and inorganic weak acids (**see: Dormancy – embryo,** Table D.12; **Dormancy breaking – chemicals, 3.(b) Respiration; Germination – influences of chemicals**). Organic acids might lower intracellular pH, thus activating H^+-ATPase, resulting in cell wall acidification, enabling cell expansion and hence the completion of germination to occur.

Inhibitors of glycolysis, citric acid cycle and terminal oxidation reactions of the mitochondrial electron transport chain can promote germination. It is hypothesized that dormant seeds are deficient in an alternative oxygen-requiring process essential for germination, which is depleted of oxygen because of its lower affinity for this gas than the cytochrome pathway of respiration. The pentose phosphate pathway is suggested as the alternative oxygen-requiring process essential for germination.

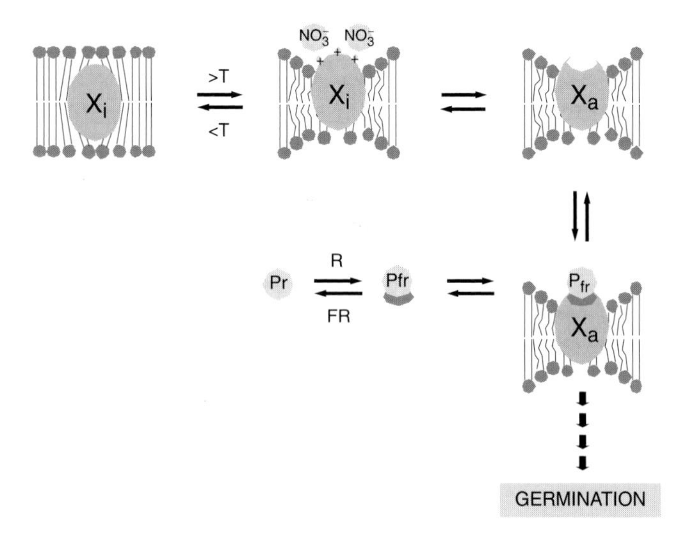

Fig. D.39. Generic model for the regulation of dormancy and germination in an annual temperate weed that requires light and nitrate to germinate. X_i and X_a are inactive and active phytochrome receptors; Pr and Pfr are inactive and active forms of phytochrome, respectively; R, red light; FR, far-red irradiation; T, threshold temperature for transition in membrane order/fluidity. Multiple arrows indicate multiple steps towards germination. Redrawn from Hilhorst, H.W.M. (1998) The regulation of secondary dormancy. The membrane hypothesis revisited. *Seed Science Research* 8, 77–90.

Another hypothesis also emphasizes the importance of the pentose phosphate pathway but assumes that re-oxidation of $NADPH_2$ is attained through peroxidase action. Both hypotheses have received extensive criticism because of a lack of conclusive evidence that activity of the pentose phosphate pathway is correlated with the breaking of dormancy (**see: Dormancy breaking – chemicals** for more details). Nitroxide is an important signalling molecule in animals and plants, and is a product of the reduction of nitrite by nitrate reductase. It has been postulated that nitrate-stimulated germination is mediated by nitroxide. However, nitrate may promote germination of *Sisymbrium officinale* seeds without the involvement of nitrate reductase. (**See: Dormancy – genes**)

(HWMH)

Bewley, J.D. and Black, M. (1994) *Seeds: Physiology of Development and Germination*, 2nd edn. Plenum Press, New York, USA.

Hallett, B.P. and Bewley, J.D. (2002) Membranes and seed dormancy: beyond the anaesthetic hypothesis. *Seed Science Research* 12, 69–82.

Roberts, E.H. and Smith, R.D. (1977) Dormancy and the pentose phosphate pathway. In: Khan, A.A. (ed.) *The Physiology and Biochemistry of Seed Dormancy and Germination*. North Holland, Amsterdam, The Netherlands, pp. 385–411.

Dormancy - genes

1. Overview

The use of model plants for molecular genetics, particularly maize and *Arabidopsis*, has provided a wealth of information about the types of biochemical pathways and interactions that control dormancy breaking. Analysis of the behaviour of mutant alleles of loci controlling the transition from dormancy to germination has led to the hypothesis that this event is controlled by two non-overlapping sets of loci with opposing effects. One set acts to enhance germination potential and activate the transition from embryo state to seedling (loci promoting germination, Table D.13), and includes those related to growth regulators known to enhance dormancy breaking (for example **gibberellin** and **ethylene**), metabolism (the *COMATOSE* locus, *CTS*) and others related to **signal transduction**. Seeds containing mutant alleles of these loci show reduced germination potential, and imbibed seeds mimic the dormant seed state (**see: section 3. Gibberellin-related genes**). In some cases this can be overcome, for example by the application of GA to *ga* mutants, or by the removal of surrounding **testa** and **endosperm** layers. A second set of loci acts to repress germination potential and enhance dormancy, inhibiting the transition from embryo to seedling (loci promoting dormancy/repressing germination, Table D.13). For this loci set, the mutant phenotype includes increased germination potential (seeds display no dormancy, even when freshly harvested) and for some alleles seedling characteristics are presented by mutant embryos. Mutated loci associated with **abscisic acid** (ABA) biosynthesis and signal transduction exert this effect of decreased dormancy capacity, confirming the role of this hormone in repressing germination. Mutations at several loci, and combinations of mutant alleles from this gene set have extreme effects on embryo phenotypes. (SF, MH)

(**See: section 2. Abscisic acid-related genes**)

2. Abscisic acid-related genes

The onset of **primary dormancy** is associated with the presence of the plant hormone abscisic acid (**see: Dormancy – acquisition**). Generally, ABA content rises during the first half of development and declines during late maturation (**see: Development of seeds – hormone content**). ABA is present in both seed and fruit tissues and has also been associated with other developmental processes, including storage protein synthesis (**see: Maturation – controlling factors; Storage reserves synthesis – regulation**) and acquisition of **desiccation tolerance.** Typically, ABA-deficient mutants (**see: *ABA* genes; Abscisic acid; Hormone mutants**) (e.g. in *Arabidopsis thaliana* and tomato) produce non-dormant seeds that may also be desiccation-intolerant, depending on the severity of the mutation (Table D.13). Dormancy is absent from ABA-deficient tomato and *Arabidopsis* seeds when the embryo lacks the dominant *ABA* allele (i.e. *aba* mutants). Maternal ABA (i.e. located in testa and fruit tissues) has no influence on dormancy. In addition, over-expression in tobacco of the gene encoding for zeaxanthin oxidase, one of the enzymes of the ABA-synthetic pathway, results in more dormant phenotypes, whereas suppression of the gene yields less dormant phenotypes. Hence, a (transient) rise in embryonic ABA content during seed development is required to induce dormancy. (**See: Dormancy – acquisition**)

Sensitivity to ABA also contributes to the expression of dormancy. The ABA-insensitive mutants *abi1* to *abi3* of *Arabidopsis thaliana* all show considerable reduction in seed dormancy (**see: *ABI* genes; ABI proteins**). Conversely, seeds from the ABA-hypersensitive mutant '*enhanced response to ABA*' (*era1*) display enhanced seed dormancy. Several cultivars of wheat and maize exhibit **vivipary** or **precocious germination**, and also **preharvest sprouting** under humid conditions. These sprouting-susceptible cultivars have a reduced sensitivity to ABA. Analysis of viviparous mutants in *Arabidopsis* and maize has identified *ABI3* and *VP1*, respectively, as the ABA-responsive genes that are responsible for these phenotypic characteristics. These **orthologous genes** encode for transcription factors of the B3 domain family (a group of ABA-responsive transcription factors characterized by one of three conserved -COOH terminal basic DNA-binding domains, B1, B2, B3) that activate the transcription of the ABA-inducible genes regulating gene expression during seed development. *ABI3* is involved in maintaining the developmental state in seeds and suppressing transition to the vegetative or growth stage. Thus, seeds of mutants lacking this gene will show characteristics of (germination) growth.

In addition to the ABA-hypersensitive *era* mutants, the ethylene-insensitive *ein2* and *etr* mutants in *Arabidopsis* are also hypersensitive to ABA. The *ctr1* mutant, which is characterized by a constitutive ethylene response, enhances the ABA-insensitive mutant *abi1-1*, and *ctr1* monogenic mutants are also slightly ABA insensitive. Thus, ethylene may suppress seed dormancy by inhibiting ABA action. Furthermore, *ctr1* has a sugar-insensitive phenotype, and *etr* a sugar-hypersensitive phenotype. This suggests the presence of cross-talk between ABA, sugar and ethylene signalling which interact at the level of dormancy, germination and early seedling growth. Coincidentally, many sugar-signalling mutants are ABA-biosynthesis mutants or alleles of *abi4* and *abi5*, which represent a subclass of the ABA-insensitive mutants, with seedling phenotypes (**see: Sugar sensing**).

Other mutants are known with altered dormancy characteristics, but with normal ABA-content and sensitivity throughout seed development. Examples are the *leafy cotyledon* (*lec1*, *lec2*) and *fusca* (*fus3*) mutants of *Arabidopsis* (Table D.14). The phenotypes of these mutants display characteristics of the vegetative state, including reduced tolerance to desiccation, active meristems, expression of germination-related genes and absence of dormancy. The *ABI3*, *LEC1*, *LEC2* and *FUS3* loci have partially overlapping functions in the overall control of seed maturation. Deletion of these gene functions yields genotypes that are defective in many aspects of seed maturation (including dormancy) (**see: Maturation**). The *LEC1*, *LEC2* and *FUS3* loci probably regulate developmental arrest, because mutations in these genes (i.e. *fus*, *lec*) cause a continuation of growth of immature embryos. *ABI3* is also active during vegetative quiescence processes in other parts of the plant in which it suppresses meristematic activity.

Since the maturation phase is largely abnormal in the *abi3*, *lec1*, *lec2* and *fus3* mutants, no dormancy is initiated and the seeds may germinate precociously, particularly when combined with ABA-deficiency. The *fus3*, *lec1* and *abi3* mutants differ in their sensitivity to ABA, yet this does not seem to correlate with the extent of vivipary of each of these genotypes. However, the occurrence of vivipary strongly depends on the relative humidity (RH) of the air, which compromises clear effects. Furthermore, double mutants of these genotypes with **gibberellin** (GA)-deficient mutants behave differently. The dependency of germination on GA is maintained in the *fus3* mutant but not in *lec1*, which suggests that these mutants affect the germination potential of seeds in different ways. Thus, *ABA*- and *ABI*-controlled dormancy may represent a different mechanism, which occurs later and is additive to the developmental arrest controlled by *LEC1* and *FUS3*.

Table D.13. Genetic loci identified in *Arabidopsis* as affecting embryo dormancy status and germination.

Functional classification	Loci promoting dormancy and/or repressing germination	Loci promoting germination
Hormone biosynthesis/signalling		
ABA	*ABA1, ABA2, ABA3, ABI1, ABI2, ABI3, ABI4, ABI5*	*ERA1, AFP*
GA	*GIN5 RGL1, RGL2*	*GA1, GA2, GA3, GAI, SLY1, SPY1, PKL*
BR		*DET2, BRI1*
Ethylene	*CTR1*	*ETR, EIN2*
Leafy cotyledon class	*LEC1, LEC2, FUS3*	
Increased dormancy class		*CTS*
Reduced dormancy class	*RDO1, RDO2, RDO3, RDO4*	

See Table D.14 for details on the gene products. (SF, MH)

The *rdo1–rdo4* mutants of *Arabidopsis* all have normal ABA contents and sensitivity but they display a reduced dormancy. These mutants show mild **pleiotropic** adult plant effects, indicating that the genes are not specific for dormancy but may affect other events as well. *RDO2*, *RDO3* and probably *RDO1* may affect the GA requirement for germination, albeit in a less profound way than ABA, perhaps due to redundancy of the function in the dormancy control of the various genes.

The ABA-deficient *sitiens* (*sit*^{w}) mutant of tomato produces non-dormant seeds, yet it initially displays a developmental arrest as in the wild-type seeds. In tomato, developmental arrest is usually terminated by the induction of dormancy, concurrent with a transient rise in ABA content. In the *sit*^{w} mutant there is no induction of dormancy and, hence, developmental arrest is not maintained. This results in viviparous germination within the ripe fruits. However, at this stage, wild-type seeds also possess undetectable amounts of ABA but do not display vivipary, and the osmotic potentials of the locular tissues of the fruits of both genotypes are similar. Apparently, the mutant seeds possess a greater growth potential. The mature mutant seeds also display a stronger resistance to osmotic inhibition *in vitro*. The testa of the mutant seeds contains only one cell layer, compared to four or five in the wild type, thereby reducing the coat-imposed restraint to germination. Since the testa is a maternal tissue it is likely that maternal ABA indirectly affects the dormancy status of the seed by controlling development of the **integuments**.

More maternally determined mutants are known with altered dormancy characteristics (but with normal ABA content and sensitivity). Examples are the *aberrant testa shape* (*ats*), *transparent testa* (*tt*) and *transparent testa glabra* (*ttg*) mutants of *Arabidopsis thaliana*. The *ats* mutant has a reduced testa thickness and germinates more, and faster, than wild-type seeds. This mutant produces ovules in which the integuments do not develop properly. This resembles the situation in the ABA-deficient *sit*^{w} tomato mutant. Seeds of the *tt* mutants (*tt1* to *tt17*, *ttg1 and ttg2* and *banyuls* (*ban*), Table D.14) show defects in the flavonoid pigmentation of the testa, and their seed colour ranges from yellow to pale brown. The *ttg1* mutant also lacks **mucilage** and **trichomes** and is affected in the morphology of the outer layer of the seedcoat (**see: Seedcoats – development**). The *ban* mutant accumulates pink flavonoid pigments in the endothelium of immature seeds and produces greyish-green, spotted, mature seeds. Both physical and chemical modifications of the testa may affect the uptake of water and oxygen, and the leaching of inhibitory substances from the seed. In addition, changes in testa dimensions may reduce the mechanical restraint to embryo growth. (**See: Dormancy – coat-imposed**)

Another mutant with a maternally determined genotype is *dag1*. The *DAG1* gene, which encodes a DOF transcription factor (a member of the GA-responsive '*D*NA-binding with *O*ne *F*inger' family of transcription factors), is expressed in the vascular tissues. It is genetically derived from the mother plant and is passed on to the developing seeds. The *dag1* mutant has a reduced dormancy and is affected both in the light requirement for germination and the structure of the testa. A related gene, named *DAG2*, has a similar (vascular) expression pattern as *DAG1*, but the *dag2* mutant shows increased dormancy.

Unravelling the mechanism of action of ABA with respect to the regulation of dormancy is extremely complex. Genome-wide, gene-expression profiling in *Arabidopsis* has demonstrated that more than 1000 genes may be up- or down-regulated by ABA. This is because ABA-signalling is involved in many plant functions, including development, stress responses, energy and metabolism, protein synthesis, transcription and transport. (**See: Dormancy breaking – hormones**)

3. Gibberellin-related genes

Gibberellins (GAs), a large group of plant hormones, play a crucial role in promoting seed germination. GA-deficient mutants (Tables D.13 and D.14) are unable to germinate without exogenous GAs. Also inhibitors of GA biosynthesis, such as paclobutrazol and tetcyclacis prevent germination of wild-type seeds. Thus, *de novo* biosynthesis of GAs is required during imbibition to promote germination. GA has a role in the induction of flowering, and hence the flower parts, but the hormone does not seem to be involved in any of the subsequent seed developmental processes since seeds develop normally in GA-deficient mutants. By inference, it is therefore unlikely that GA content or type of GA is involved in the control of dormancy during seed development. However, GA-biosynthesis is not required for germination when ABA is absent, e.g. in ABA-deficient mutants. This suggests that ABA imposes upon the seed a requirement for GA, i.e. it is required to overcome the ABA-induced dormant state. This concept has been extended to primary dormancy in the so-called **hormone balance theory** of dormancy and germination, which explains the loss of dormancy or promotion of germination by the net effect of the action of the antagonists ABA and GA.

Applied GAs replace the dormant seed's requirement for afterripening and cold, which are the common conditions to break primary dormancy (**see: Dormancy breaking**). This suggests that such environmental factors induce GA biosynthesis during the early phases of germination. However, a cold treatment does not stimulate GA biosynthesis directly but rather increases the sensitivity of a seed to GAs; there is some evidence, however, that the capacity for later GA synthesis may be enhanced by chilling. On the other hand, light (through **phytochrome**) induces one of the two 3-β-hydroxylases (oxidase) involved in the biosynthesis of GAs, encoded by the *GA4H* gene, in germinating lettuce and *Arabidopsis* seeds. In a number of species, particularly those belonging to the summer annuals in the temperate regions, seeds require both a cold treatment and, subsequently, light to promote their germination. Afterripening or a cold treatment increases their sensitivity to light. This has led to the hypothesis that the afterripening or cold treatment breaks dormancy, and that it is light that stimulates germination (through promotion of GA-biosynthesis). In some cases the light requirement is lost during the dormancy-breaking treatment. This can be explained by the assumption that very low (light-independent), pre-existing amounts of active phytochrome are sufficient to saturate the light response in the sensitized seeds. These studies imply that the breaking of dormancy and the induction of germination may be distinct

Table D.14. *Arabidopsis* mutants with altered seed dormancy/germination properties.

Mutant	Gene	Germination phenotype	Encoded protein
Seed development mutants			
abi	*ABI3*	+	B3 domain protein with B1 and B2 domain
fus3	*FUS3*	+	B3 domain protein with B2 domain
lec1	*LEC1*	+	HAP3 subunit of CCAAT box binding protein
lec2	*LEC2*	+	B3 domain transcription factor
Abscisic acid-biosynthesis and -signalling mutants			
abi1	*ABI1*	+	serine/threonine phosphatase 2C
abi2	*ABI2*	+	serine/threonine phosphatase 2C
abi4	*ABI4*	+	APETALA2 domain protein
abi5	*ABI5*	+	basic leucine zipper transcription factor
aba1	*ABA1*	+	zeaxanthin epoxidase
aba2	*ABA2*	+	xanthoxin oxidase
aba3	*ABA3*	+	molybdenum cofactor sulfurase
era1	*ERA1*	–	famesyl transferase
Gibberellin-biosynthesis and -signalling mutants			
ga1	*GA1*	–	copalyl diphosphate synthase
ga2	*GA2*	–	ent-kaurene synthase
ga3	*GA3*	–	ent-kaurene oxidase
spy	*SPY*	+	threonine-O-linked N-acetylglucosamine transferas
sly1		–	unknown
Seedcoat mutants			
ats		+	unknown
tt1		+	unknown
tt2	*TT2*	+	R2R3 MYB domain protein
tt3	*DFR*	+	dihydroflavonol-4-reductase
tt4	*CHS*	+	chalcone synthase
tt5	*CHI*	+	chalcone isomerase
tt6	*F3H*	+	flavonol 3-hydroxylase
tt7	*F3'H*	+	flavonol 3′-hydroxylase
tt8	*TT8*	+	basic helix-loop-helix domain protein
tt9		+	unknown
tt10		+	unknown
tt11		+	unknown
tt12	*TT12*	+	MATE family protein
tt13		+	unknown
tt14		+	unknown
tt15		+	unknown
t16	*TT16*	+	ARABIDOPSIS BSISTER (ABS) MADS domain protein
ttg1		+	WD40-repeat protein
ban	*LAR*	+	leucoanthocyanidin reductase

Data taken from: Bentsink, L. and Koornneef, M. (2003) Seed dormancy and germination. In: Somerville, C.R. and Meyerowitz, E.M. (eds) *The Arabidopsis Book.* American Society of Plant Biologists, Rockville, MD, USA, doi/10.1199/tab.0050, www.aspb.org/publications/arabidopsis/
Germination phenotype: + indicates faster and – indicates slower germination than the wild type. See Table D.13 for a summary of the genes that promote dormancy or germination. (HWMH)

events (see: **Dormancy – ecophysiology; Dormancy breaking – cellular and molecular aspects; hormones; Light – dormancy and germination; Phytochrome; Signal transduction**). (HWMH)

Bewley, J.D. (1997) Seed germination and dormancy. *The Plant Cell* 9, 1055–1066.

Finkelstein, R.R., Campala, S.S.L. and Rock, C.D. (2002) Abscisic acid signaling in seeds and seedlings. *The Plant Cell* (Suppl.), S15–S45.

Hilhorst, H.W.M. (1995) A critical update on seed dormancy. I. Primary dormancy. *Seed Science Research* 5, 61–73.

Holdsworth, M., Kurup, S. and McKibbin, R. (1999) Molecular and genetic mechanisms regulating the transition from embryo development to germination. *Trends in Plant Science* 4, 275–280.

Hoth, S., Morgante, M., Sanchez, J.-P., Hanafey, M.K., Tingey, S.V. and Chua, N.-H. (2003) Genome-wide gene expression profiling in *Arabidopsis thaliana* reveals new targets of abscisic acid and largely impaired gene regulation in the abi1-1 mutant. *Journal of Cell Science* 115, 4891–4900.

Koornneef, M., Bentsink, L. and Hilhorst, H.W.M. (2002) Seed dormancy and germination. *Current Opinion in Plant Biology* 5, 33–36.

Kucera, B., Cohn, M.A. and Leubner-Metzger, G. (2005) Plant hormone interactions during seed dormancy release and germination. *Seed Science Research* 15, 281-308.

Nambara, E. and Marion-Poll, A. (2003) ABA action and interactions in seeds. *Trends in Plant Science* 8, 213–217.

Dormancy - genetics

Genetic variation for seed dormancy and germination within a species exists among accessions of both non-domesticated plants and cultivars of domesticated plants. Dormancy is generally considered a genetically complex or quantitative trait. This complexity and associated variation arises from segregation of alleles at many interacting loci, called **quantitative trait loci** (QTLs), and from sensitivity of these alleles to environmental factors and genotype × environmental interactions. Seed components are either of maternal or zygotic origin and vary in their genetic constitution and ploidy level. Components most likely to affect dormancy are the **embryo**, **endosperm**, **perisperm**, **testa**, **pericarp** and **hull**. Individual species may lack one or more of these components or tissues and their prominence and proximity to the embryonic axis vary tremendously. For example, the hull, which is derived from lemma and palea of a grass **floret**, is obviously present only in grass seeds. The embryo is almost always diploid as a result of fertilization between a male and female gamete and has the genetic constitution of a hybrid, F_1, under usual conditions where the gametes are genetically different. The endosperm is triploid as a result of fusion of two polar nuclei of the female gametophyte with one male gamete. The testa, pericarp, perisperm and hull are diploid and of maternal origin and genetic constitution. Thus, the embryo and endosperm of the seed or **caryopsis** in a cereal are of the same generation, but the testa, pericarp and hull are of the previous (maternal) generation. The generation of the seed is generally designated as the generation of the embryo and this designation is important, but not always clear in dormancy research. The component structures or tissues of a seed may act independently and together to determine dormancy and germination. For example, a seed might be dormant due to pericarp/testa **(coat)-imposed dormancy**, or **embryo dormancy**, or both. To prevent confounding of genetic effects due to the different types of dormancy, it is important to investigate the role of individual components as a prelude to investigating inheritance. The role of individual components can be determined by germinating intact seeds, dehulled seeds, dehulled seeds with the pericarp/testa removed, and excised embryos. Basically, the dormancy phenotype is measured by differences in germination of dormant or partially afterripened seeds, caryopses, or embryos compared to the same cultivar or accession that is non-dormant or fully afterripened and was grown under the same environmental conditions.

Genetic investigations of seed dormancy are not exhaustive in any species, but model plant species have been used, such as Arabidopsis (*Arabidopsis thaliana*) and **rice** (*Oryza* spp.); selected cultivars of cereal grain crops, e.g. **barley** (*Hordeum vulgare*), **sorghum** (*Sorghum bicolor*), rice (*Oryza sativa*) and **wheat** (*Triticum aestivum*); other agronomic, horticultural, or ornamental crops such as **groundnut** (*Arachis hypogaea* L.), cucumber (*Cucumis sativus*), **faba (fava) bean** (*Vicia faba*), snapdragon (*Antirrhinum* spp.), lisianthus (*Eustoma grandiforum*), poppy (*Papaver* spp.) and *Cuphea* spp.; and some weeds and non-domesticated plants, e.g. wild oat (*Avena fatua*), wild mustard (*Sinapsis arvensis*), downy brome (*Bromus tectorum*) and bitterbrush (*Purshia tridentata*). The genetics of dormancy related to preharvest sprouting of cereal grain crop species is covered in **Pre-harvesting sprouting – genetics**. The focus here is on *Arabidopsis* and wild oat, a dicotyledonous and a monocotyledonous weed, respectively.

Arabidopsis is a small cruciferous plant with a short, annual growth cycle. It has many advantages for plant biology and genetic research because it is a diploid with a small, fully sequenced genome (130 Mbp) distributed over five chromosomes (Fig. D.40). Both natural and mutagenesis-induced variation for dormancy and germination exist. *Arabidopsis* mutants have been used extensively to identify genes involved in the regulation and signalling of dormancy and germination (**see: Dormancy – genes**), but the focus here is on the investigation of natural variation for dormancy using a molecular genetic approach (**see:** Fig. P.22 in **Pre-harvesting sprouting – genetics**). The accessions or ecotypes 'Landsberg *erecta*' (L*er*) and 'Columbia' (Col), which are most frequently used in *Arabidopsis* research, show low levels of seed dormancy. Other ecotypes such as 'Cape Verde Island' (Cvi) show stronger dormancy. A recombinant inbred line (RIL) population for genetic mapping, derived from the L*er*/Col cross, has identified 14 QTL with small effects that account for differences in germination behaviour under several maternal environment and germination conditions. A second QTL analysis, using a RIL population derived from the L*er*/Cvi cross, has identified additional loci and/or novel or stronger alleles derived from the strongly dormant Cvi parent. The dormancy phenotype was measured as the **afterripening** requirement, with L*er* and Cvi seeds generally requiring 6 and 15 weeks of afterripening, respectively, under warm, dry conditions, for 100% germination. Similar to other species, the level of dormancy in *Arabidopsis* is influenced by environmental conditions during seed development. Maternal genetic effects on dormancy variation were not observed from phenotypic comparisons of reciprocal F_1 and F_2 generation seeds. Phenotypic values for the afterripening requirement indicate overall co-dominance of the L*er* and Cvi alleles. However, germination curves indicate a partial dominance of the strong Cvi dormancy during the first 3 weeks of afterripening, whereas the L*er* germination phenotype appears to be partially dominant after periods beyond 10 weeks. Genotype by afterripening time interactions are significant, showing that lines respond differently to afterripening.

Seven QTLs, named *delay of germination* (*DOG*) 1 to 7, have been identified that control 61% of the natural L*er*/Cvi phenotypic variation (Fig. D.40). Three QTLs, *DOG1* to *DOG3*, each explain more than 10% of the phenotypic variation, with remaining QTLs accounting for less variation. The Cvi alleles increase dormancy, except that the Cvi allele at *DOG2* decreases dormancy, similarly to the L*er* alleles. **Epistatic** effects also occur. For example, *DOG3*-Cvi alleles in a *DOG1*/*DOG7*-L*er* background increase dormancy, but reduce dormancy in a *DOG1*/*DOG7*-Cvi background. Clearly, these observations concerning genetic and environmental

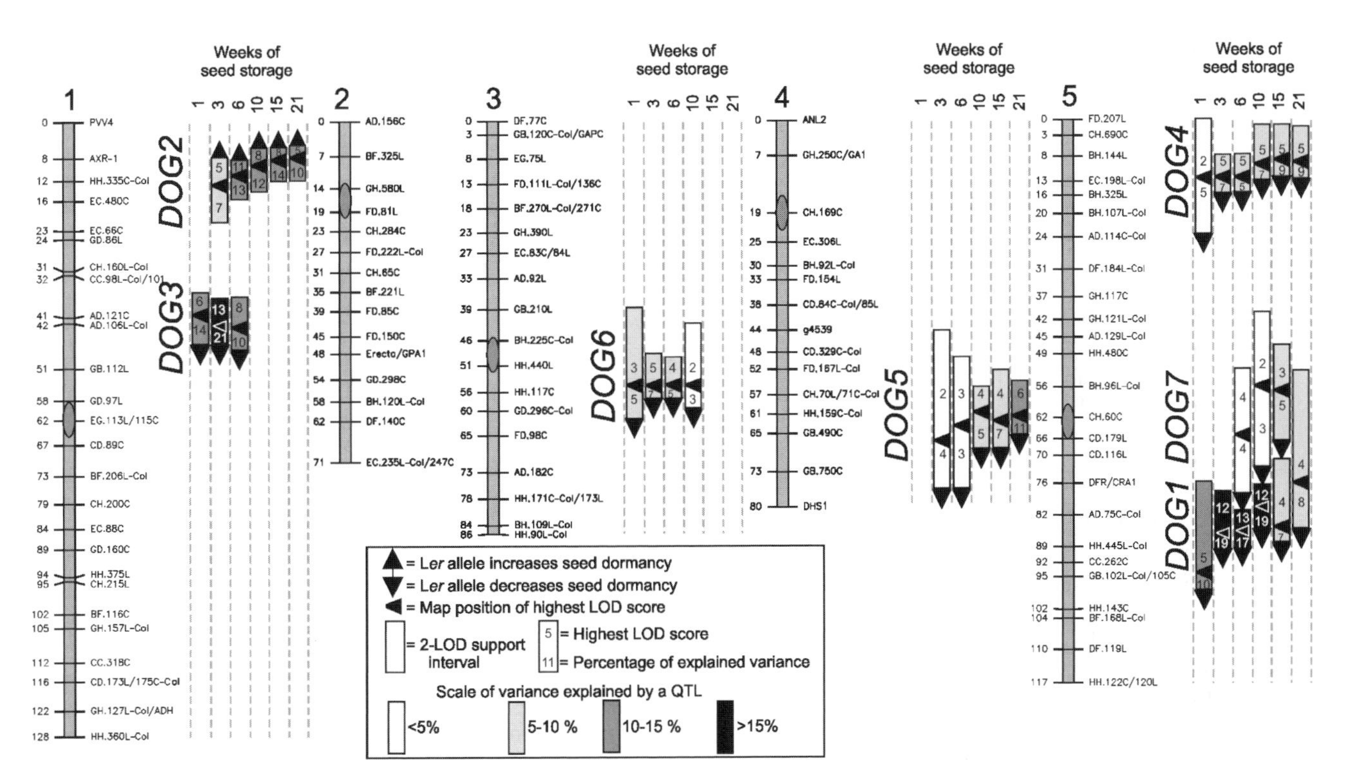

Fig. D.40. Quantitative trait loci (QTL) mapping for germination of the *Arabidopsis* 'Ler'/'Cvi' recombinant inbred lines at different times of seed storage, e.g. afterripening. The significant QTLs named delay of germination 1 to 7 (*DOG1–7*) are depicted close to the genetic map of the five linkage groups used to mark the location of QTLs on genetic linkage maps. The box arrows denote the chromosomal interval containing a QTL and if the *Ler* allele increases or decreases dormancy. Numbers in the arrows indicate the statistical level of support for the interval, i.e. LOD score (total relative probability, on a log scale, that a linkage relationship exists between selected loci) and the per cent of the phenotypic variation explained by a QTL. From Alonso-Blanco, C., Bentsink, L., Hanhart, C.J., Blankestijn-deVries, H. and Koornneef, M. (2003) Analysis of natural allelic variation at seed dormancy loci of *Arabidopsis thaliana*. *Genetics* 164, 711–729. Reproduced with permission of the journal.

factors and their interaction indicate why seed dormancy and germination are biologically very complex physiological traits.

As a step towards characterization and cloning of *Arabidopsis* dormancy QTLs, introgression lines have been developed. These near-isogenic lines (NILs), which carry one or two Cvi genomic fragments around the *DOG* QTLs in an otherwise *Ler* genetic background, were developed using backcrossing procedures combined with phenotypic and genotypic, i.e. marker-assisted, selection. The NILs were used to confirm the *DOG* QTLs and to discover that cold treatment is more effective than nitrate, the hormone **gibberellin** $GA_{(4+7)}$, or inhibitors of the hormone **abscisic acid** (ABA) in the overcoming of *Arabidopsis* seed dormancy. In addition, NILs were used to demonstrate that *Arabidopsis* displays both testa- and embryo-imposed dormancy, and to facilitate the genetic fine mapping of the strongest QTL, *DOG1*. This QTL maps to an approximately 75 kb region on chromosome 5 which is predicted by genome sequencing to contain 22 open reading frames or putative genes. None of the genes is known to be involved in seed dormancy or germination, so no obvious candidate gene is apparent. In time, map-based cloning procedure will reveal the gene or genes in regions containing QTLs that directly affect seed dormancy and germination.Wild oat is a serious weed in many semi-arid regions where wheat and barley are produced. It is a difficult weed to control because of its dormant seeds in the **soil seed bank.** Thus, wild oat seed (caryopsis) has been used extensively as a model to investigate many aspects of dormancy, afterripening, and germination (**see** Fig. D.43 in **Dormancy – importance in agriculture**). As with most species, wild oat contains natural variation for seed dormancy. Inbred lines with differing levels of dormancy have been selected in the field and used extensively to investigate environmental and genetic factors that regulate dormancy. In terms of the environment, cool, moist conditions during seed development enhance the level of dormancy compared with warm, dry conditions. Such conditions also impact on dormancy in barley and wheat in a similar way. Temperature also influences germination (**see: Germination – influences of temperature**). Seeds from non-dormant lines germinate rapidly over a range of temperatures, while seeds from dormant lines germinate slowly in a narrow range from about 8 to 12°C. The temperature window for germination widens as seeds receive increasing amounts of afterripening under warm, dry conditions (**see** Fig. D.42 in **Dormancy – importance in agriculture**). Studies on this genotype by environmental interaction are aimed at marking dormancy QTLs of wild oat that respond to different germination temperatures. Several genetic models for dormancy in *Avena* spp. have been developed based on inter- and intra-specific crosses. In general, wild oat displays embryo- and pericarp/test-imposed dormancy, non-dormancy is dominant over dormancy, at least three genes control the dormancy trait, a dominant dormancy-conferring allele exists at one or more loci, and epistatic interactions occur between loci. Based on germination data for several generations derived from the 'M73' (dormant)/'SH430' (non-dormant) cross, loci *G1* and *G2* promote early germination, and the *D* locus promotes late germination or dormancy.

Quantitative trait loci analysis for seed dormancy, such as that in *Arabidopsis* and cereal grain species (**see: Preharvesting sprouting – genetics**), requires certain resources that are not often developed for weeds, for example, a genetic linkage map and a permanent mapping population, such as RILs to replicate seed germination assays and repeat experiments (**see** Fig. P.22 in **Preharvest sprouting – genetics**). The absence of any tools precludes a typical QTL analysis to mark chromosomal regions corresponding to dormancy, but a technique called bulked segregant analysis has been used to identify two random amplified polymorphic DNA markers associated with early germination of wild oat. The two independent markers explain approximately 13 and 7%, respectively, of the M73/SH430-F_2 phenotypic variance for germination at 15°C. Although unusual, a RIL population has been derived from the M73/SH430 cross. This population is being used to develop a genetic linkage map to identify markers more closely linked with the QTLs, to identify additional QTLs, and to investigate the genotype by germination temperature interaction. Marking the dormancy QTL in species like wild oat will not lead to map-based cloning as in *Arabidopsis*, because wild oat, a hexaploid with a large complement of nuclear DNA (11,315 Mbp), is not a tractable system for such an approach. However, progress in cloning dormancy QTL in more tractable systems like rice should facilitate cloning in other grass species if **orthologous loci** regulate dormancy and germination in cereal grain species. Then, cloned rice genes or DNA fragments could be used to probe genomes of wild oat, barley, wheat, etc. segregating for dormancy to test the hypothesis that orthologous loci regulate dormancy in these species. Ultimately, cloning and characterization of dormancy QTLs from *Arabidopsis*, rice, wild oat and other species will be an important step in understanding signals, pathways and mechanisms that regulate dormancy, afterripening and germination of seeds. (MEF)

Fennimore, S.A., Nyquist, W.E., Shaner, G.E., Doerge, R.W. and Foley, M.E. (1999) A genetic model and molecular markers for wild oat (*Avena fatua* L.) seed dormancy. *Theoretical and Applied Genetics* 99, 711–718.

Flintham, J.E. (2000) Different genetic components control coat-imposed and embryo-imposed dormancy in wheat. *Seed Science Research* 10, 43–50.

Mauricio, R. (2001) Mapping quantitative trait loci in plants: uses and caveats for evolutionary biology. *Nature Reviews Genetics* 2, 370–381.

Dormancy – gymnosperms

In general, dormancy is imposed during seed and embryo development so that varying degrees and types of dormancy may be exhibited by the seed during the later stages of development. Seeds of several conifer species exhibit deep and prolonged dormancy at maturity (e.g. western white pine (*Pinus monticola*), yellow-cedar (*Chamaecyparis nootkatensis*) and true fir (*Abies*) species). Seeds of *Picea* species are generally much less dormant and some conifer seeds are non-dormant at maturity (e.g. western red cedar, *Thuja plicata*). When excised from their surrounding tissues (the seedcoat, nucellus and megagametophyte), the embryos of dormant conifer species (e.g. white spruce (*Picea glauca*), yellow-cedar, *Pinus* spp. and others) can germinate, suggesting that the surrounding seed tissues contain factors which block germination of the embryo (**see: Dormancy; Dormancy – coat-imposed**). Prevention of germination by the tissues surrounding the embryo may be related to the action of one or more factors including: (i) interference with water uptake; (ii) mechanical restraint; (iii) interference with gas exchange; (iv) prevention of exit of inhibitors (e.g. **abscisic acid**, ABA) from the embryo; (v) supply of inhibitors to the embryo, or encouraging ABA biosynthesis within the embryo; and (vi) enhanced turnover of inhibitors in the embryo and/or surrounding tissues. Some of the important factors for dormancy maintenance of conifer seeds have been elucidated, primarily by examining changes during and following the termination of dormancy.

1. Changes in the mechanical restraint of the megagametophyte

For some conifer seeds (e.g. yellow-cedar, white spruce and western white pine) the megagametophyte appears to act as a mechanical barrier to prevent radicle protrusion. In a dormancy mechanism involving mechanical restraint, weakening of the cell walls of the megagametophyte, involving the induction of cell wall hydrolases, is proposed to be a prerequisite for germination (**see: Germination – radicle emergence**). In western white pine seeds, dormancy termination is accompanied by a decreased mechanical restraint of the megagametophyte in the micropylar region; structural changes revealed by scanning electron microscopy (SEM) indicate that the nucellar cap becomes more porous. In yellow-cedar and white spruce seeds, the micropylar megagametophyte decreases in mechanical strength following a **dormancy-breaking** treatment and, during germination, the cells of the megagametophyte in the area immediately surrounding the radicle of yellow-cedar seeds exhibit a loss of their internal structure, that would represent significant weakening to allow radicle emergence. Concurrently, the embryo exhibits increased turgor and a reduced sensitivity to low osmotic potentials. There is a predominance of pectins in the cell walls of tissues surrounding the embryo of yellow-cedar seeds. Induction of activity of the cell wall hydrolase, pectin methyl esterase, is correlated with dormancy breakage of yellow-cedar seed; moreover, the amount of enzyme activity produced is strongly correlated with the ability of seeds to germinate. In germinating white spruce seeds there is an increase of endo-β-mannanase activity, another cell wall hydrolase, in the micropylar region of the megagametophyte, which declines following splitting of the testa.

2. ABA and dormancy

Abscisic acid (ABA) in gymnosperms was first identified in *Taxus baccata* (English yew) seeds and was correlated with dormancy termination of *Taxus* embryos cultivated in a liquid medium. *Pinus pinea* (Stone pine) seeds washed for 24 hours in water before being placed in vermiculite exhibit a higher germination capacity and rate than their non-soaked counterparts, which has been attributed to ABA being leached from the seedcoat. In *Pinus sylvestris* (Scots pine) seed, dormancy-breaking treatments that include either white or red light, decrease ABA prior to radicle protrusion; seeds subjected to a far-red light pulse after red light do not exhibit as great a decline in ABA nor is dormancy relieved (**see: Phytochrome**). During dormancy breakage of yellow-cedar seeds, there is about a twofold reduction of ABA in the embryo. In the megagametophyte, ABA does not change; however, the embryos exhibit a marked decrease in their sensitivity to ABA. Dormancy termination is further associated with changes in the embryo's metabolism of ABA, in which 8′ hydroxylation becomes rate limiting (**see: Abscisic acid**). In Douglas-fir (*Pseudotsuga menziesii*) seed, the structures enclosing the embryo (i.e. the seedcoat and megagametophyte) appear to maintain the seed in a dormant state by limiting ABA metabolism and by preventing the leaching of ABA; they may also facilitate accumulation of *de novo* synthesized ABA. ABA decreases in both the megagametophyte and embryo during a 7-week moist chilling period (**stratification**); further, the longer the duration of moist chilling, the faster the rate of ABA decline during subsequent germination. In western white pine seeds dormancy termination, requiring prolonged moist chilling, results in a marked decrease in embryo and megagametophyte ABA. The decline of ABA during dormancy breakage is coincident with an increase in the germination capacity of seeds. In addition, changes in ABA flux, i.e. shifts in the ratio between biosynthesis and catabolism, occur at three distinct stages during the transition from dormant seed to seedling; i.e. during moist chilling, germination, and post-germinative growth. Catabolism occurs via several routes, depending on the stage and the seed tissue (embryo or megagametophyte) and includes 8′ hydroxylation of ABA (accumulation of phaseic acid and dihydrophaseic acid), 7′ hydroxylation of ABA (generation of 7′-hydroxy ABA) and ABA conjugation (accumulation of ABA-glucose ester). These studies provide strong support that ABA turnover in part controls the dormancy-to-germination transition of conifer seeds.

In seeds of certain angiosperms (e.g. **oat, sorghum**) expression of the *ABI3/VP1* gene is correlated with the degree of embryo dormancy and may be important for maintaining ABA-controlled metabolism in the dormant imbibed seed (**see: Dormancy – genes; Preharvest sprouting – mechanisms**). A similar mechanism may exist for the maintenance of dormancy of yellow-cedar seeds. For example, down-regulation of *CnABI3* gene expression and synthesis of CnABI3 protein (a transcription factor that regulates ABA-responsive genes) is positively correlated with dormancy breakage. Characterization of the functional attributes of proteins that interact with the conifer ABI3 protein may contribute to our understanding of the mechanism by which ABI3 regulates the dormancy to germination transition. The pattern of accumulation of an ABI3-related protein of western white pine seeds (degradation during dormancy termination and re-synthesis during early post-germinative growth) also suggests a potential role for ABI3 in seedling establishment. (**See: Dormancy** entries; **Signal transduction – hormones; Tree seeds**)

3. Role of water uptake

Effective dormancy-termination of several conifer seeds appears to require the maintenance of a relatively high moisture content, particularly during moist chilling of seeds. Many industry protocols for dormancy termination include a prolonged water soak, prior to the imposition of moist chilling. The implementation of **nuclear magnetic resonance spectroscopy** (NMR) and related technologies will likely contribute to our understanding of the importance of water distribution within the seed as well as the route of water uptake during imbibition. (ARK)

Corbineau, F., Bianco, J., Garello, G. and Côme, D. (2002) Breakage of *Pseudotsuga menziesii* seed dormancy by cold treatment as related to changes in seed ABA sensitivity and ABA levels. *Physiologia Plantarum* 114, 313–319.

Feurtado, J.A., Ambrose, S.J., Cutler, A., Ross, A., Abrams, S. and Kermode, A.R. (2004) Dormancy termination of western white pine (*Pinus monticola* Dougl. Ex D. Don) seeds is associated with changes in ABA metabolism. *Planta* 218, 630–639.

Kermode, A.R. (2004) Seed germination. In: Christou, P. and Klee, H. (eds) *Handbook of Plant Biotechnology Online*. John Wiley, Chichester, UK, Chapter 33.

Ren, C. and Kermode, A.R. (2000) An increase in pectin methyl esterase activity accompanies dormancy breakage and germination of yellow cedar seeds. *Plant Physiology* 124, 231–242.

Schmitz, N., Abrams, S.R. and Kermode, A.R. (2002) Changes in ABA turnover and sensitivity that accompany dormancy termination of yellow-cedar (*Chamaecyparis nootkatensis*) seeds. *Journal of Experimental Botany* 53, 89–101.

Zeng, Y., Raimondi, N. and Kermode, A.R. (2003) Role of an ABI3 homologue in dormancy maintenance of yellow-cedar seeds and in the activation of storage protein and Em gene promoters. *Plant Molecular Biology* 51, 39–49.

Dormancy – importance in agriculture

Seed **dormancy** contributes to the survival and persistence of domesticated and non-domesticated plants in managed and natural ecosystems. In agriculture and agroecosystems, crop and weed seed dormancy impacts on the production of plants for food, fibre and shelter.

1. Crop species

Uniform and rapid germination is generally desirable for crop seeds. Therefore, seed dormancy has been removed from most crops by selection during the **domestication** process. In some cases, however, there are advantages to maintaining a degree of dormancy in crop seeds. Plant breeders of varieties of **barley** (*Hordeum vulgare*), **wheat** (*Triticum aestivum*), **rice** (*Oryza sativa*) and **sorghum** intended for production where environmental conditions at harvest are conducive to **preharvest sprouting**, often select for some, albeit low, dormancy. Four of the seven mega-environments for which plant breeders at the International Maize and Wheat Improvement Center (**CIMMYT**) develop wheat germplasm have environmental conditions conducive to **preharvest sprouting**.

While selection for low levels of dormancy guards against economic losses due to such sprouting, the residual dormancy that reduces the speed and uniformity of postharvest germination can be problematic for farmers, plant breeders and maltsters. For example, unless environmental conditions during storage are conducive for **afterripening** (warm and dry for barley and wheat), dormancy sometimes persists until planting time. Even low levels of postharvest dormancy in crops can interfere with stand establishment and, ultimately, yield. If cultivars are too dormant to use for immediate planting, because the period between harvest and planting is too short to break dormancy, older seeds may be used, but this could lead to problems with stand establishment if **vigour** has been reduced by prolonged storage. Low degrees of post-harvest dormancy that persist in cultivars of malting barley may necessitate periods of controlled storage to remove the dormancy and thereby to maximize extract yield and other malting quality traits (**see:** entries in **Malting**). Finally, dormant crop seeds in the soil can also create volunteers, i.e. a weed problem in subsequent crops of other types.

The seed dormancy trait has not been substantially altered in all agronomic, horticultural and forestry crops because practical or economic constraints may deter intensive selection for the whole range of desirable traits. Seeds of cool- and warm-season forage grasses (Poaceae) and **legumes** (Fabaceae), flowers and trees display various types and levels of dormancy. **Alfalfa** (*Medicago sativa*) is a forage legume that, depending on the cultivar and growth conditions, can have a special type of coat-imposed dormancy called **hardseededness**. Hard seeds are impermeable to water under normal germination conditions but impermeability can be overcome by physical or chemical scarification treatments. However, the duration and intensity of a scarification treatment is critical because if too little the treatment is ineffective; if too much the embryo is damaged, leading to reduced vigour and stand establishment. Hardseededness is also typical in some families of weeds, e.g. Malvaceae and Fabaceae, which is overcome naturally and gradually in the soil.

Dormant seeds of cool season crop and **tree seed** species often respond to **stratification** or chilling temperatures between 2 and 10°C under moist conditions. The duration of treatment depends on the species and depth of dormancy, but often extends from 2 to 12 weeks, to trigger maximum germination once seeds are later exposed to higher temperatures for germination. The mechanisms by which chilling overcomes dormancy are not known (**see: Dormancy breaking**). There are clear indications, however, that seeds and embryos of deciduous fruit trees forced to germinate without adequate chilling develop into abnormal seedlings called physiological dwarfs, as though some dormancy is retained. While chemical treatments, for example, the application of gibberellic acid (**gibberellin**), or physical treatments such as seedcoat removal or wounding induce germination of dormant seeds, they too may not end the dormancy programme despite triggering germination (**see: Dormancy – coat-imposed**).

Factors that affect seed dormancy, afterripening and germination are special considerations in enhancing the characteristics and distribution of some crops. Potato (*Solanum tuberosum*) is a cool season crop propagated vegetatively by tubers. There is interest in using sexually produced potato seed in subtropical and tropical regions where production is feasible, but where the production and maintenance of pathogen-free tubers is problematic. Potato seed is very dormant. It may take 18 months of afterripening under warm, dry conditions to 'open' the germination temperature window to achieve rapid and uniform germination at the supra-optimal temperatures generally found in tropical regions. However, potato seeds afterripen quickly under very high temperatures (45°C) and low seed moisture conditions (3%) and techniques that accelerate afterripening may help in developing methods for growing potato in subtropical regions. Although potato may germinate uniformly at high temperatures, seeds of some crop species, e.g. lettuce, show **thermodormancy** at elevated temperatures, which limits production to certain seasons or areas with favourable temperatures for germination. Ways of avoiding this (e.g. **priming** at lower temperatures) have been devised.

Oat (*Avena sativa*) is a spring-seeded grain crop in northern latitudes. The earlier that oat is sown in the spring the higher it generally yields. However, wet weather, poor soil drainage and other considerations often delay sowing. If germination could be delayed until spring, sowing oats in late autumn would have advantages such as higher yields because of disease avoidance and the longer growing season as a result of very early spring emergence. However, seedlings of autumn-sown winter oats lack winter hardiness to survive in those areas where spring oats are cultivated.

Attempts, therefore, have been made to develop a new crop called **dormoat** by crossing non-dormant cultivated oat with dormant wild oat. Seeds of dormoat would overwinter in a dormant state and emerge in the spring as much as 3 weeks before spring oat. But these attempts to develop dormoat and other dormant-seeded crops have not been widely successful because the genetic and environmental factors that control seed dormancy and afterripening, and thus spring emergence, are not sufficiently understood to rationally develop these novel crops (**see: Dormancy – genetics**).

Determining the germination potential of crop or weed seeds in a population has important practical implications. For example, the **cereal** industries would benefit from inexpensive and rapid biochemical or molecular tests that could be conducted at the grain elevator to detect non-visible signs of preharvest sprouting or dormancy before purchase of a batch of wheat or barley. Currently, methods to determine germination potential or seed dormancy require germination or other tests, e.g. **tetrazolium test** for **viability**, that are time consuming and sometimes destructive. As insight is gained into the physiological basis or biochemical and molecular processes for dormancy and germination, there is potential to develop novel single-seed and non-invasive tests to determine the physiological state of a seed or seeds in a population.

2. Weeds

(See: Dormancy – ecophysiology)

Weeds are undesirable because they interfere with the growth of crop and forage species through competition for space, light, water and nutrients. Chemical, cultural and mechanical weed control practices are aimed at managing the incidence of weeds. In the absence of weed seed germination in the soil in a particular year, control measures are undermined since they are normally aimed at emerging and established seedlings or pre-flowering stage plants. Weed seeds in the soil form a repository called a **soil seed bank** for future infestations. Seed banks can contain a mixture of transient and persistent species. The seeds from transient species may give rise to weed infestations, but their seeds do not remain viable in the soil for more than 1 year after shedding. Persistent seed banks are most characteristic of plant communities that are subjected to frequent and unpredictable disturbance, such as in agroecosystems. The persistent seed bank contains **quiescent** and dormant seeds, which protect the species against germination in periods unfavourable for seedling emergence and growth (**see: Dormancy – ecology**). The **primary** and **secondary dormant** seeds in the soil contribute to weed infestations over relatively long periods of time. Thus, seed dormancy is a key component governing germination, seedling emergence, and the demographics of weeds.

Seeds of many plants display germination periodicity in that they are not always ready to germinate even if environmental conditions are favourable. This periodicity gives rise to seasonal patterns of seedling emergence. For example, emergence of summer annuals occurs in spring to early summer, and of winter annuals in late autumn to early spring. Several field studies using artificially buried and natural seed populations have shown that seasonal patterns of seedling emergence are due to cyclical changes in secondary seed dormancy (**dormancy cycles**).

Thus, secondary seed dormancy also helps to ensure that germination is restricted to seasons with the most favourable conditions for seedling and plant survival. These cycles are influenced by external temperature fluctuations (Fig. D.41) and to some extent soil moisture. A genetic component to secondary seed dormancy has been proposed, but not verified.

Temperature is a major factor governing loss of primary dormancy, cycling of seeds in and out of secondary dormancy, and induction of germination in partially afterripened and non-dormant seeds (**see: Dormancy breaking – temperature**). Seeds in a population can have multiple and overlapping responses to their thermal environment, which makes prediction of seed behaviour difficult. At a specific time, diurnally fluctuating temperatures may affect seeds in a population in a variety of ways. For example, relief of dormancy may proceed in a fraction of the population at the lowest temperature, whereas at low to moderate temperatures some seeds are germinating, and as the temperature increases above some threshold for dormancy relief or germination other seeds in the population are being forced into secondary dormancy. The range of temperatures over which germination can proceed is narrow in primary and secondary dormant seeds: characteristically the range widens as primary and secondary dormancy are relieved (Fig. D.42); (**see Dormancy – Genetics**). This attribute ensures that field germination of non-dormant or partially afterripened seeds takes place when the required and actual temperatures overlap. There is no direct measure for relief or induction of dormancy at any particular temperature or set of conditions. The defining measure for the level of dormancy or the amount of afterripening that seeds in a population have incurred is the length of time from the start of imbibition to the onset of germination, and the rate of germination. Thus, non-dormant or fully afterripened seeds display a relatively more rapid onset and rate of germination compared with dormant and partially afterripened seeds.

Development of bioeconomic and comprehensive models of population dynamics to improve weed management practices depends, in part, on a fundamental understanding of environmental and genetic factors that regulate seed dormancy, longevity in the soil, afterripening and germination. Some research is aimed at understanding how seeds integrate and process environmental cues such as temperature and hydric conditions to promote the relief of dormancy and the induction of germination. Predicting the time course of dormancy relief and germination using thermal and hydric conditions as key determinants has been approached empirically by measuring seed bank densities and proportional emergence, and mathematically using the **hydrothermal time** model. This model proposes that seed germination rates are proportional to the amount by which temperature and water potential exceed base or threshold values. It accounts for many characteristic features of primary

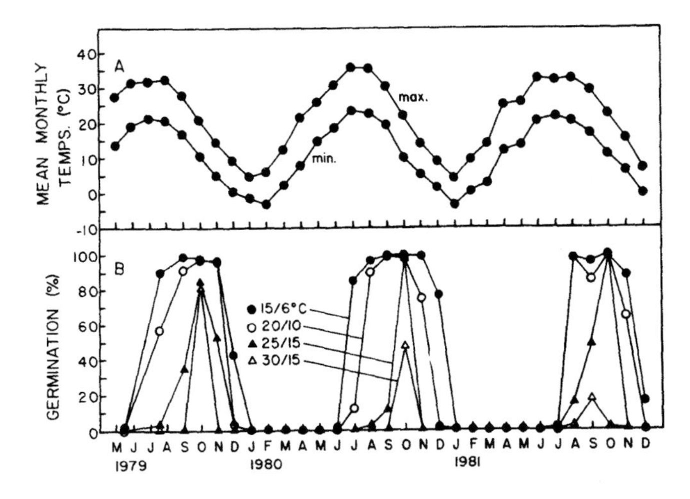

Fig. D.41. Effect of seasonal soil temperature changes on germination of *Arabidopsis thaliana* seeds. Seeds were germinated in light with a 14-h photoperiod under optimal and sub-optimal conditions of alternating temperature. Adapted from Baskin, J.M. and Baskin, C.C. (1985) The annual dormancy cycle in buried weed seeds: a continuum. *BioScience* 35, 492–498.

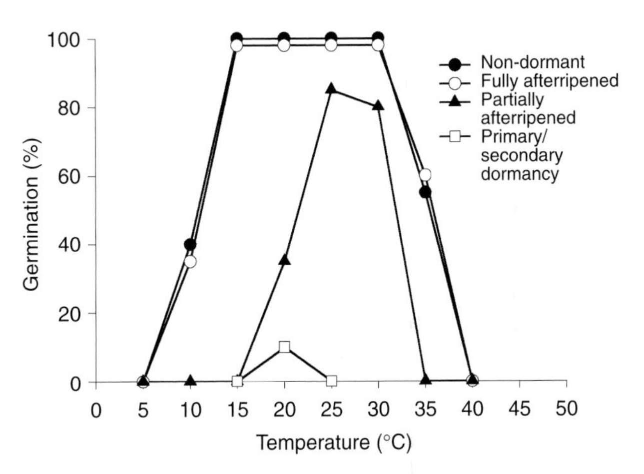

Fig. D.42. A generalized model showing comparative germination of dormant, afterripened, and non-dormant seeds grown and germinated under nearly identical conditions. Increasing amounts of afterripening broaden the germination temperature window to mimic the germination of seeds from a non-dormant line. Adapted from Foley, M.E. (2001) Seed dormancy: an update on terminology, physiological genetics, and quantitative trait loci regulating germinability. *Weed Science* 49, 305–317.

and secondary seed dormancy and provides a quantitative way to characterize germination potential of seeds in a population, given certain environmental conditions. Eventually, models may be improved by including other variables such as light, **tillage**, seed position in the profile, maternal plant environment and genetic determinants.

The agricultural practices of crop rotation, tillage, herbicide usage and genetic engineering of crops may impact on the incidence of weed seed dormancy. Dormant seeds in the seed bank serve as a genetic repository that protects the population and enhances colonizing success by hosting favourable gene combinations that have been selected over long periods of time. These seeds allow weed populations to evolve and adjust under changing agricultural conditions. For example, as compared with continuous cropping, summer fallow of crop land enhances the relative abundance of dormant wild oat (*Avena fatua*) at the expense of non-dormant types. This pattern fits the expectation that dormant types will be more prevalent in the less predictable environment provided by a crop rotation with a fallow period (Fig. D.43).

Tillage affects the distribution of weed seeds in the soil (Fig. D.44). In no-till systems, the majority of seeds remain at or near the soil surface. Tillage moves the seeds from the surface into the soil profile and alters the physical conditions that effect seed survival, seedling emergence and cyclical changes in secondary dormancy. In situations where weed management is primary to economic consideration, knowledge about periodicity of weed seed germination can be used to postpone tillage and planting past peak germination. This practice should deplete seed storage reserves and effectively control emerged weed seedlings. Some seeds require light for the promotion of their germination (**see: Light – dormancy and germination; Phytochrome**). Tillage methods that exclude light or limit tillage to dark periods (i.e. moonless nights) to prevent germination have met with limited success as a weed management strategy. Chemicals or **hormones** that break dormancy or promote germination, such as nitrate fertilizer, substituted phthalimides, and **gibberellic acid**, lack the consistency to promote nearly complete germination of a mixed population of dormant and quiescent weed seeds in the field (**see: Germination – influences of chemicals**). However, the plant hormone **ethylene** has been effective in stimulating germination of the parasitic witchweed (*Striga asiatica*) in efforts aimed at its eradication in the USA (**see: Parasitic plants**).

Recurrent treatment of susceptible weed populations with some herbicides leads to instances of herbicide resistance, and the persistent seed bank may regulate the timing and incidence of resistance. If the herbicide's selection pressure on the weed is low, and many dormant seeds of herbicide-susceptible plants exist in the soil, there may be a delay in the occurrence of resistance. From the mid-1970s, the incidence of herbicide resistance has significantly increased. Research into a variety of herbicides and weeds has shown there are some instances where low or delayed seed germination, implying seed dormancy, is associated with the resistant biotype, and others in which rapid seed germination is associated with the resistant biotype. The presence of seed dormancy could affect the evolution of traits not directly associated with dormancy and germination. For example, as **transgenes** that impart herbicide resistance in crop species spread into related weeds through cross-pollination, such as from canola to weedy mustards, dormancy could influence the rate of spread through the weed population and give rise to adaptative combinations of germination and post-germination traits, e.g. herbicide resistance. After all, dormant seeds that germinate in an environment to which their post-germination traits are adapted will have a selective advantage. (MEF)

Baskin, J.M. and Baskin, C.C. (1985) The annual dormancy cycle in buried weed seeds: a continuum. *BioScience* 35, 492–498.

Batlla, D., Kruk, B.C. and Benech-Arnold, R.L. (2004) Modelling changes in dormancy in weed soil seed banks: implications for the prediction of weed seed emergence. In: Benech-Arnold, R.L. and Sánchez, R.A. (eds) *Handbook of Seed Physiology: Applications to Agriculture*. Food Products Press, New York, USA, pp. 245–270.

Bradford, K.J. (2002) Applications of hydrothermal time to quantifying and modeling seed germination and dormancy. *Weed Science* 50, 248–260.

Dyer, W.E. (1995) Exploiting weed seed dormancy and germination requirements through agronomic practices. *Weed Science* 43, 498–503.

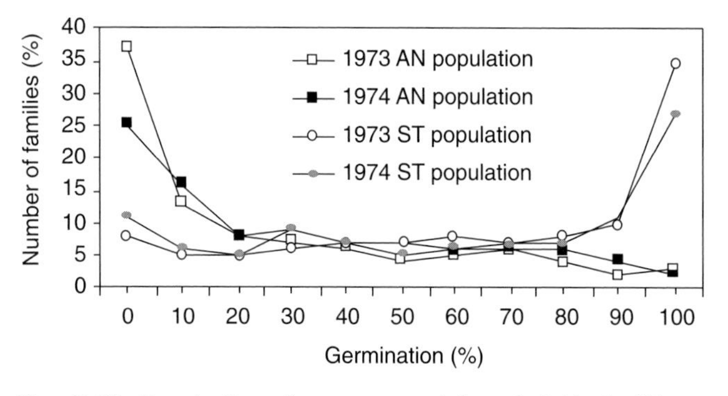

Fig. D.43. Germination of progeny seed from individual wild oat plants collected in fields with different cropping systems. The AN population had a 20-year history of cultivation involving a 3-year rotation with 2 years in crop followed by 1 year in summer fallow. The SN population had a 12-year history of continuous cropping. More dormant individuals are present in the former, less predictable cropping system. Adapted from Naylor, J.M. (1983) Studies on the genetic control of some physiological processes in seeds. *Canadian Journal of Botany* 61, 3561–3567.

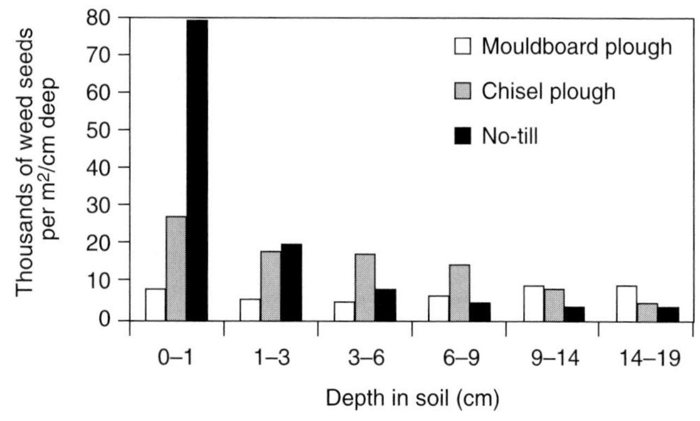

Fig. D.44. Effect of tillage method on the distribution of weed seeds in the soil. Adapted from Yenish, J., Doll, J.D. and Buhler, D.D. (1992) Effects of tillage on vertical distribution and viability of weed seed in soil. *Weed Science* 40, 429–433.

Fenner, M. (1995) Ecology of Seed Banks. In: Kigel, J. and Galili, G. (eds) *Seed Development and Germination*. Marcel Dekker, New York, USA, pp. 507–528.

Dormancy breaking

Primary dormant seed will not germinate even under optimal germination conditions without first receiving a dormancy-breaking stimulus (**see: Dormancy**). The agent providing this stimulus is not required for germination itself indicating that the two processes are separate events. Furthermore, the dormancy-breaking agent must stimulate the seed above a threshold value in order to initiate the dormancy-breaking process. This threshold value may be achieved incrementally by repeated stimulation, gradually by continuous low-level stimulation, or rapidly by a saturating stimulatory event. The magnitude of this threshold value is determined by two interacting components: (i) the genetic contributions of the parents (**see: Dormancy – genetics**); and (ii) the environmental factors impacting on the seed during development (**see: Dormancy – acquisition**). This results in a distribution of a range of seed dormancy levels within a local population, and between ecotypes of a species (distinct within-species populations). The resultant effect is to distribute dormancy breaking and hence germination through time. This is illustrated by the old gardening adage 'One year's seeds, seven years weeds', reflecting the propensity of weed seeds to lose dormancy intermittently over a long time period. The evolutionary success of this strategy is evident in the human effort expended on weed control.

The efficiency of this evolutionary adaptation in dispersing germination through time presents particular problems when trying to unravel the processes that control the maintenance and loss of dormancy. From a practical point of view, interest has centred more on factors relieving dormancy (Table D.15) than on dormancy-breaking mechanisms. These factors can broadly be classified as environmental effects (such as light, temperature and certain chemicals), maternal effects (such as seedcoat dormancy) and embryo effects, all of which contribute to the phenomena of **dormancy cycling**, which occurs in **soil seed banks.** (**See** also: **Tree seeds, 8. Seed dormancy treatment**, Table T.8.)

Dormancy in a particular seed type can generally be relieved by any one of a number of factors and it is highly unusual that only one agent is effective. Also, most of the dormancy-breaking factors act only on the hydrated seed. Only in the case of afterripening, and possibly in hard seeds with physical dormancy, can dormancy be removed in a 'dry' seed.

Many environmental factors effect dormancy-breakage to some extent, depending on the type of dormancy. The majority of species have seeds with physiological dormancy, while others have dormancy imposed by inhibitors, an impermeable seedcoat, physical restraint of the embryo, or by embryo immaturity (**see: Dormancy**).

The loss of both **primary** and **secondary dormancy** has evolved to increase the coincidence of the non-dormant state with periods optimal for germination and seedling establishment. The mature dormant seed is exposed to an array of environmental conditions that provide the key inputs into the process of dormancy loss. Also, dormancy can be terminated by agents that do not normally occur in the environment but are primarily experimental, laboratory treatments (e.g. certain chemicals). Many have considerable

Table D.15. The environmental factors that overcome (break) dormancy, the seed factors that maintain dormancy, and the common variants of each factor.

Source	Factor	Variant
Environmental	Dormancy-breaking	*Temperature*
		Dry afterrippening
		Alternating temperature (Warm/Cold cycles)
		Chilling (Cold stratification)
		Warming (Warm stratification)
		Light
		Alternating Light (Light/Dark cycles) **(see: Light – germination and dormancy)**
		Single doses of light **(see: Light, germination and dormancy)**
		Chemicals
		Smoke **(see: Germination – influences of smoke)**
		Inorganic **(see: Germination – influences of chemicals)**
		Organic **(see: Germination – influences of chemicals)**
Maternal	Dormancy maintaining	*Seedcoat*
		Seedcoat strength
		Seedcoat inhibitors
Embryo		Seedcoat permeability
		Endosperm
		Restraint of embryo growth **(see: Germination – radicle emergence)**
		Hormones
		Abscisic acid/Gibberellin antagonism
		Genetics
		Interaction of dormancy-promoting and germination-repressing loci with germination-promoting loci (see Table D.13 in **Dormancy – genes**).

practical importance, such as the use of **gibberellin** or seed abrasion in **malting**, and certain non-hormonal chemicals are employed experimentally to probe into dormancy-breaking mechanisms.

The receptiveness of a seed to a particular factor is again affected by genetic and environmental components, which give rise to noticeable differences in response even between ecotypes of a single species. These interacting components have an impact on the dormant state from the moment the seed reaches maturity on the mother plant. Although in a laboratory situation these components can be studied in isolation, in nature they are continually interacting.

When it comes to understanding dormancy-breaking mechanisms, the distribution (variation) of the dormancy-breaking response within a seed population presents a problem. Increasing doses of a dormancy-breaking factor provide information on the population distribution of a response. But under such conditions changes in seed physiology at the molecular and biochemical level are too subtle to observe. In order to investigate the mechanistic basis of a response, a saturating dose is required, so that the whole population responds to the same degree. This allows for investigation of the perception (signal perception) of a dormancy-breaking factor and the potential involvement of cellular signals, metabolism and gene expression (molecular effects) in the dormancy-breaking process. (SF, MH)

(See: articles on Dormancy breaking; Dormancy; and the influences of factors on Germination)

Baskin, C.C. and Baskin, J.M. (1998) *Seeds. Ecology, Biogeography, and Evolution of Dormancy and Germination.* Academic Press, San Diego, USA.

Bewley, J.D. and Black, M. (1982) *The Physiology and Biochemistry of Seeds in Relation to Germination*, Vol. 2. *Viability, Dormancy and Environmental Control.* Springer, Berlin.

Bewley, J.D. and Black, M. (1994) *Seeds: Physiology of Development and Germination*, 2nd edn. Plenum Press, New York, USA.

Dormancy breaking – cellular and molecular aspects

Although the environmental factors effecting dormancy breaking have been largely identified (**see: Dormancy breaking – light; – temperature**) the questions of their mechanism of action and how dormancy is broken are more elusive. A dormancy-breaking factor is in most cases an environmental signal: a receptor is required in the physiologically dormant embryo for perception of this external signal. This in turn is transduced to an effector protein then onwards through secondary messengers to the target proteins inducing the response (Fig. D.45) (**see: Signal transduction**). There are several possible components of the transduction pathway and the final response in the removal of dormancy and the promotion of germination: (i) Effects on membranes (**see: Dormancy – embryo (hypotheses)**), (ii) changes in transduction messengers, (iii) changes in hormonal status (**see: Dormancy breaking – hormones**) and (iv) changes in gene expression. In addition, dormancy breaking may require certain alterations in metabolism.

In general terms, however, it is important to recognize that dormancy breaking in the embryo is controlled by the interaction of genetically controlled biochemical processes in the imbibed seed with the external environment. The genetics of dormancy and the transition from the dormant to the germinative state are considered in **Dormancy – genes.**

1. Effects on membranes

Dormancy-breaking signals make first contact with embryo cells at their plasma membrane. **Membranes** are fluid mosaics consisting of a fluid **phospholipid** bilayer, within which are membrane proteins. Both lipids and proteins are able to move through this matrix. Through their action on molecular motion of phospholipids, changes in temperature lead to decreased (chilling; cold **stratification**) and increased (warm stratification) membrane fluidity and cross-sectional area; spacing of the phospholipid head groups is also altered. At those temperatures effective in breaking dormancy the resulting head group spacing may allow specific peripheral membrane proteins (PMPs) to bind with the membrane thus enabling completion to occur of a **signal transduction** pathway. Once a critical population of a specific protein has bound to the membrane a potential signal threshold may be crossed initiating germination. A possibility is that the important changes in fluidity might occur after the dormancy-breaking chilling. This could happen because the low temperature provokes an adaptive change in membrane phospholipid composition (homeoviscous adaptation): this leads to a relatively large change in fluidity when the temperature increases to those at which germination is promoted.

It should be noted that **hormones** or **plant growth regulators** (some of which can break dormancy) are known to interact with specific receptor proteins within the plasma membrane to which they bind, or inside the cell, initiating a cascade of reactions involving a signal transduction pathway. The **phytochrome** light receptors are soluble cytosolic proteins that redistribute to the nucleus in response to red light. Phy B contains a nuclear localization signal (NLS), with Phy A possessing multiple weak NLSs. The NLS enables protein docking with a nuclear pore complex that facilitates their import to the nucleus where gene expression is effected. The requirement for interaction with a membrane environment may explain the temperature sensitivity of red light-induced dormancy breaking. **(See: Dormancy breaking – chemicals, 3.(c) Membranes; Dormancy – embryo (hypotheses))**

2. Secondary messengers, signal transduction

It is likely that once external dormancy-breaking factors have been perceived, secondary messengers then carry this information through the signal transduction pathways – complex transfer systems from receptors to effectors. These pathways are complex, being non-linear with a high degree of interaction. The secondary messengers carrying this information must be simple, easily transmitted around the cell, and easily dissipated.

The main secondary messenger in many transduction systems is calcium (Ca^{2+}). Calcium's normally low cytoplasmic concentration increases transiently during signalling. The length and source of these transients, resulting from Ca^{2+} import through ion channels or via release from intracellular stores (e.g. the **endoplasmic reticulum**, **mitochondria** and vacuole), affect the range of targets and the type of response. Transients localized to the plasma membrane provoke different outcomes from those impacting the nucleus. The primary Ca^{2+} receptor is the protein, **calmodulin**. On binding Ca^{2+}, it then binds with and activates its target proteins, many of which are Ca^{2+}-dependent protein kinases: together, these sense, amplify and transduce the Ca^{2+} signal further downstream. Many calmodulin-independent **kinases** also respond to Ca^{2+} transients via their calmodulin-like Ca^{2+} binding domains.

Protein kinases catalyse the **phosphorylation** of target proteins regulating metabolic activity, gene expression and ion fluxes. These kinases are counterbalanced by specific protein phosphatases that dephosphorylate the target proteins thus regulating activity. Cellular or organellar pH changes may also occur (**see: Dormancy breaking – chemicals**).

Though on conceptual grounds the participation of such a cascade is likely, there is no widespread evidence that it occurs during release of seeds from dormancy. There is evidence, however, that an ABI1, ABI2-type phosphatase (**see: Signal transduction – hormones**) is up-regulated in beech seeds (*Fagus sylvatica*) by cold stratification: the consequence of this would be to reduce sensitivity to internal

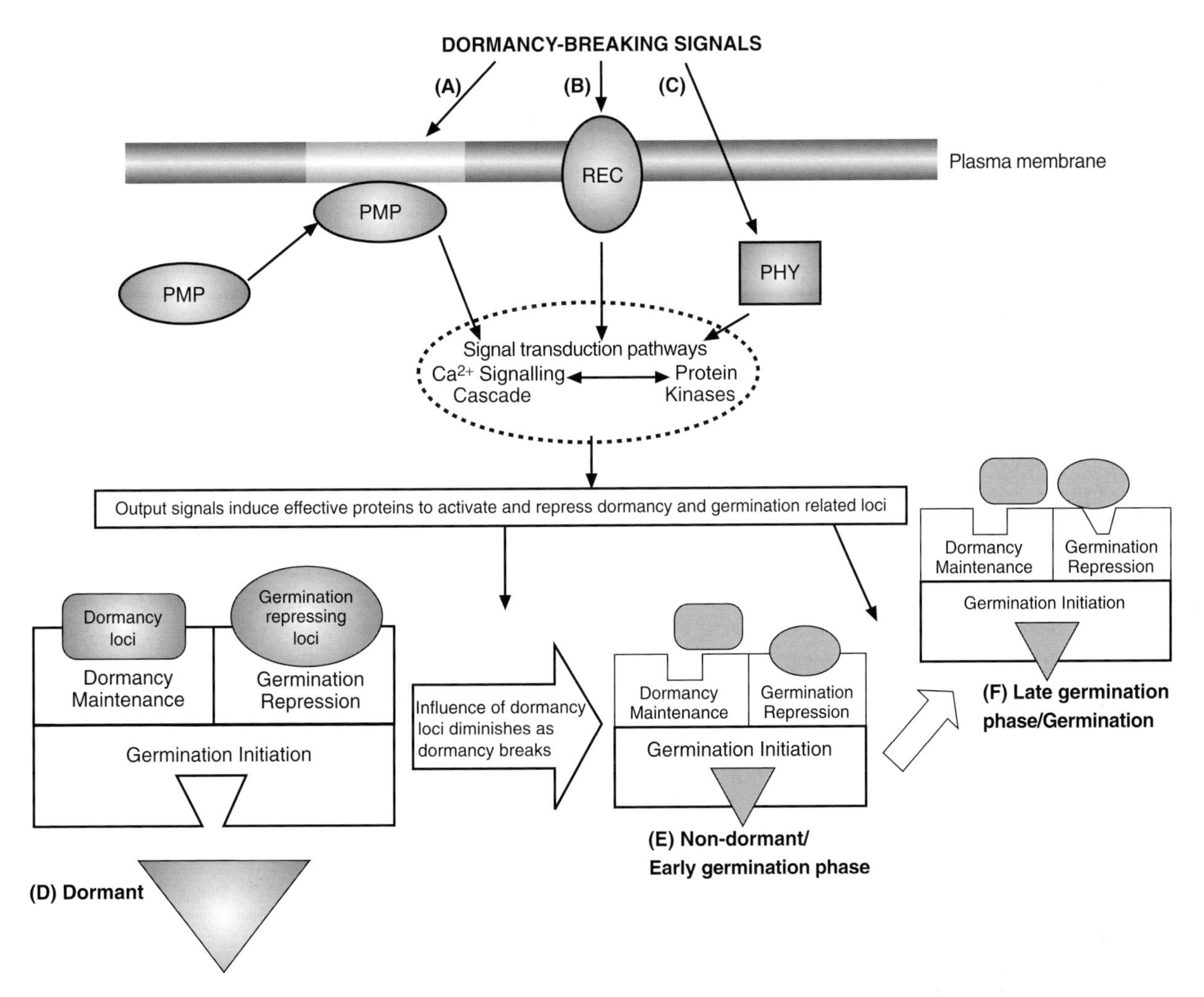

Fig. D.45. Potential routes of dormancy-breaking signals. The initiation of signal cascades leading to the termination of dormancy may be initiated by: (A) Temperature changes and lipophilic chemicals permeating the plasma membrane (dark grey membrane) induce changes in membrane fluidity (light grey membrane). This allows binding of a peripheral membrane protein (PMP), a component of a **signal transduction** pathway, to the membrane enabling activation of a signal cascade. (B) Organic and inorganic chemicals may bind to membrane-spanning receptor proteins (REC) to initiate a signal. (C) Red light initiates signalling through **phytochrome** (PHY). Once initiated, signals are amplified through signal transduction pathways to effector proteins that activate and repress dormancy and germination related loci. (D) In the hydrated dormant seed the processes that maintain dormancy and repress germination are opposed by those activating germination. The expression of dormancy-promoting (rectangle) and germination-repressing (oval) loci maintain the operation of the dormancy- and germination-repressing processes by blocking the expression of loci that initiate germination (triangle). (E) As dormancy is lost the dormancy loci are unable to maintain dormancy due to the increasing effectiveness of those loci that activate germination and the now non-dormant seed enters the early germination phase. At this point the influence of germination-repressing loci also decreases as germination-activating loci fully activate the germination process. (F) In the absence of **secondary dormancy**, the action of germination-activating loci culminates in the attainment of germination, the terminal point of the germination phase.

ABA and thereby to disrupt the maintenance of dormancy. Though not in dormancy, there is a well-established case of Ca^{2+} signalling in seeds that operates in the **aleurone layer** cells of cereals following germination. Here **gibberellin** induces a steady state increase in Ca^{2+} via the inositol triphosphate (IP3) signal pathway. This ultimately results in α-amylase secretion. **Abscisic acid** blocks IP3 formation in aleurone layer cells, preventing the increase in Ca^{2+}.

(SF, MH)

(See: Mobilization of reserves – cereals)

3. Gene expression

Changes in gene expression that are involved in the transition from dormancy and quiescence to germination following seed hydration are points at which germination may be regulated. **(See: Germination)** In primary dormant seeds it is likely that the vast majority of the metabolic events associated with cell metabolism proceed, such as respiration, nucleic acid and protein synthesis, but genes associated specifically with the completion of germination are blocked. Very little is known about the nature of these genes. It is well established that plant hormones, particularly abscisic acid (ABA) and gibberellin (GA), are key participants in the regulation of dormancy and germination so hormone-sensitive genes are involved in dormancy breaking.

Synthesis of and sensitivity to ABA are essential for the establishment of dormancy during seed development, as mutations in genes coding for ABA biosynthetic enzymes (e.g. *ABA1*, *ABA2* in *Arabidopsis*) or in the ABA signalling pathway (e.g. *ABI3*) result in **viviparous** germination or lack of dormancy (Fig. D.46) (see section 1, above). As mentioned above, there is some evidence that genes involved in sensitivity to ABA (ABI1,2-like genes) may be involved in the effects of cold in dormancy breakage (**see: *ABI* genes**). The *ABI3* gene encodes a transcription factor, a protein that regulates the activity of genetic promoters determining when and in which tissues particular genes are expressed. In addition, *ABI4* and *ABI5* genes have been identified by their **mutants** as affecting the sensitivity of germination to inhibition by ABA (Fig. D.46). These genes also encode transcription factors that appear to act later in the germination process, possibly even after radicle emergence, to determine whether embryo growth will continue. Application of ABA or reduced **water potential** block embryo growth, and ABI4 and ABI5 are involved in this inhibition. These proteins, and their homologues in other species, bind to and regulate the activity

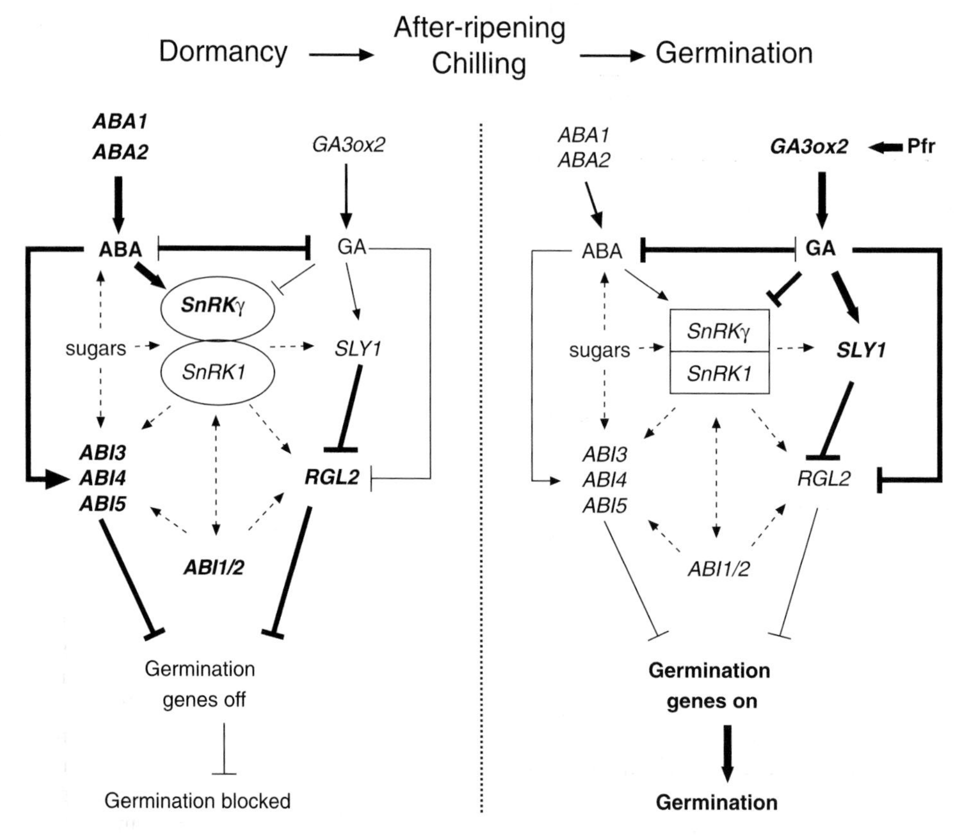

Fig. D.46. Model of changes in ABA and GA regulatory pathways in dormant versus germinating seeds. The transition from dormancy to germination often requires dry afterripening or moist chilling. Since the balance between ABA and GA is critical, all pathways are shown for both conditions, but bold text and bold lines and arrows indicate genes and pathways that are predominant under either condition. Arrows indicate promotive action, while bars indicate inhibition. Dashed arrows indicate interactions that are still hypothetical and require additional experimental support. The change in shape of *SnRKγ* and *SnRK1* from ovals to rectangles indicates that this complex may still be active under both conditions, but with altered substrate specificity or cellular localization depending upon the specific ligands associated with the SnRK1 kinase. Pfr = active phytochrome; other symbols defined in the text. The regulatory events associated with the *RGL2* gene occur only following the completion of germination in *Arabidopsis*, and do not occur during induction of germination in gibberellin-deficient tomato seeds by GA.

of other genes that are associated with either seed maturation or germination. In addition, sugars such as sucrose, glucose and mannose can inhibit the germination of some seeds, and the **sugar-sensing** pathway by which this occurs intersects with the ABA signalling pathway (Fig. D.46). Overall, when ABA contents and sensitivity are high, molecular and biochemical processes associated with seed maturation, reserve accumulation or conservation, dormancy, and inhibition of embryo growth predominate. It seems likely that much of this is coordinated by a network of key transcription factors that transduce internal (hormonal) and external (e.g. water or salt stress) signals into specific gene expression patterns.

Gibberellins are essential for germination, evident from the failure of seeds carrying defects in GA synthesis or perception to complete germination. (**See: Dormancy – genes**) GA and ABA act largely in opposition to each other in regulating germination, and the balance between them determines the developmental fate of an individual seed (**see: Dormancy breaking – hormones**). Experiments with application of GA or ABA and with mutants deficient in synthesis or perception of these hormones indicate that there is a critical balance between the effects of these regulators that can be tipped either way depending upon the seeds' developmental history and current environment. Red light, for example, acts through the phytochrome system to stimulate the expression of a specific GA 3-β-hydroxylase (oxidase) gene (*GA3ox2*) encoding an enzyme involved in the synthesis of active GAs (Fig. D.46) (**see: Light – dormancy and germination**). In *Arabidopsis* and lettuce, this gene (again, a specific member of a larger gene family) is expressed in the embryo in response to light, resulting in the synthesis of active GAs and stimulation of germination. Expression of the gene in *Arabidopsis* is also induced by cold. Several transcription factors have also been identified in the GA signalling pathway. Their genes (the *GAI/RGA/SLN1* or *GRAS* family) are involved in the regulation of stem elongation, flowering and other developmental events in addition to germination. (**See: Dormancy – genes**) GA also promotes expression of the *SLEEPY1* or *SLY1* gene that is involved in the degradation of proteins through the ubiquitination pathway (**see: Signal transduction – hormones**).

A good example of the antagonism at gene expression level between GA and ABA is seen in cereal grains. Here, GA stimulates gene expression for reserve mobilization processes in the scutellum and aleurone layer and these responses are countered by ABA (**see: Mobilization of reserves – cereals;** entries in **Malting**).

The activity of many enzymes and transcription factors is regulated by **protein kinases** and phosphatases that either phosphorylate or de-phosphorylate specific sites in the proteins (**see:** section 3, above; **Signal transduction**). Two genes known to be involved in ABA signalling encode protein phosphatases (*ABI1* and *ABI2*). While the specific substrates for these phosphatases are unclear, they could be involved in regulating the phosphorylation status of transcription factors (Fig. D.46). Another potential regulator of both enzymes and transcription factors is the SNF1-related kinase (SnRK1) kinase complex. This kinase is conserved in yeasts, plants and mammals, and plays various roles in different organisms. It is involved in regulating stress responses in yeast, energy balance in animals, and carbon balance in plants. The SnRK1 complex consists of three subunits, the kinase (α-subunit), a localization (β) subunit and an activating (γ) subunit. The activating subunit (*LeSNF4*) is expressed during late tomato seed development and in dormant seeds or seeds exposed to ABA, but is suppressed in non-dormant seeds or seeds exposed to GA (Fig. D.46). The presence or absence of this SnRKγ subunit could alter the activity of the SnRK1 kinase to shift seed metabolism from a maturation/dormancy mode to a germination/reserve mobilization mode. Another related kinase in wheat, PKABA1, has a similar expression pattern and phosphorylates a transcription factor having similarity to ABI5. The activity of the SnRK1 kinase complex is also regulated by sugars, suggesting a mechanism for the interaction of ABA and sugars in regulating germination (Fig. D.46). Since supplying sugars and amino acids can at least partially overcome the inhibition of germination by ABA in some seeds, ABA may act by restricting early germinative reserve mobilization (**see: Germination**) or energy generation in the embryo.

SnRK1 kinases can also directly modify histones associated with DNA to alter gene transcription; such chromatin remodelling has been implicated in dormancy. Although data are sparse on the biochemistry of SnRK1 action in seed development, dormancy and germination, it could have a significant regulatory role.

Many additional genes have been shown to affect seed dormancy or germination, but whether they directly or indirectly affect these events remains uncertain. In addition to ABA and GA, **ethylene (ethene)**, **brassinosteroids** and other hormones also are involved in regulating germination in some species. A complex network of enzymes and transcription factors, regulated by hormones, by kinases and phosphorylases, and by sugars appears to be involved in the transition from seed quiescence and dormancy to germination.

For a review of gene expression associated with radicle expansion and weakening of structures surrounding this, **see Germination – molecular aspects**. (KJB)

4. Metabolism

Comparisons of the metabolism of dormant and non-dormant seeds have revealed some differences which might be causally related to the relief of dormancy. Some of these differences reflect the activation of metabolic pathways as dormancy is lost.

At the biochemical level, pathway activation can take the form of allosteric control, covalent modification and enzyme interactions. During metabolic derepression and entry into the germination phase the levels of fructose-2,6-bisphosphate increase, for example in red rice. This compound is an allosteric activator of the glycolytic enzyme pyrophosphate:fructose 6-phosphate 1-phosphotransferase which is favoured under the low oxygen conditions which may persist in the embryo. Covalent modification of enzyme activity during metabolic activation is likely to involve phosphorylation as seen in the activation of photosynthetic enzymes and during metabolic derepression of estivating ('dormant') animals. Control by enzyme interactions is through binding of enzymes to the cytoskeleton or particulate fraction of the cell. In dormant tubers that are subsequently activated, the distribution and activity of glycolytic enzymes changes from the soluble to the particulate cell fraction, a phenomenon also seen in other organisms.

Activation of the pentose phosphate pathway has long been considered possibly to have a role in the loss of seed dormancy but the evidence is ambivalent. For discussions on the possible role of this pathway **see: Dormancy breaking – chemicals, 3.(b) respiration; Germination – influences of chemicals.**

(KJB, SF, MH)

Bradford, K.J., Downie, A.B., Gee, O.H., Alvarado, V.Y., Yang, H. and Dahal, P. (2003) Abscisic acid and gibberellin differentially regulate expression of genes of the SNF1-related kinase complex in tomato seeds. *Plant Physiology* 132, 1560–1576.

Cheng, S.H., Willmann, M.R., Huei-Chi Chen, H.C. and Sheen, J. (2002) Calcium signaling through protein kinases. The Arabidopsis calcium-dependent protein kinase gene family. *Plant Physiology* 129, 469–485.

Finkelstein, R.R., Gampala, S.S.L. and Rock, C.D. (2002) Abscisic acid signaling in seeds and seedlings. *The Plant Cell* 14, S15–S45.

Gazzarrini, S., and McCourt, P. (2001) Genetic interactions between ABA, ethylene and sugar signaling pathways. *Current Opinion in Plant Biology* 4, 387–391.

Kornneef, M., Bentsink, L. and Hilhorst, H. (2002) Seed dormancy and germination. *Current Opinion in Plant Biology* 5, 33–36.

Merlot, S., Gosti, F., Guerrier, D., Vavasseur, A. and Giraudat, J. (2001) The ABI1 and ABI2 protein phosphatases 2C act in a negative feedback regulatory loop of the abscisic acid signalling pathway. *The Plant Journal* 25, 295–303.

Ogawa, M., Hanada, A., Yamauchi, Y., Kuwahara, A., Kamiya, Y. and Yamaguchi, S. (2003) Gibberellin biosynthesis and response during Arabidopsis seed germination. *The Plant Cell* 15, 1591–1604.

Peng, J. and Harberd, N.P. (2002) The role of GA-mediated signaling in the control of seed germination. *Current Opinion in Plant Biology* 5, 376–381.

Rolland, F., Moore, B. and Sheen, J. (2002) Sugar sensing and signaling in plants. *The Plant Cell* 14, S185–S205.

Toyomasu, T., Kawaide, H., Mitsuhashi, W., Inoue, Y. and Kamiya, Y. (1998) Phytochrome regulates gibberellin biosynthesis during germination of photoblastic lettuce seeds. *Plant Physiology* 118, 1517–1523.

Yamaguchi, Y., Ogawa, M., Kuwahara, A., Hanada, A., Kamiya, Y. and Yamaguchi, S. (2004) Activation of gibberellin biosynthesis and response pathways by low temperature during imbibition of *Arabidopsis thaliana* seeds. *The Plant Cell* 16, 367–368.

Dormancy breaking – chemicals

The soil environment exposes seeds to a multitude of inorganic and organic chemical compounds, some of which promote germination, possibly by reducing dormancy. Many of these chemicals are biological in origin being breakdown/waste products of the associated flora and fauna. The effectiveness of these chemicals in breaking dormancy depends on factors such as dormancy level and moisture content. Also the concentration and duration of exposure to chemicals can have inhibitory as well as stimulatory effects. In most cases the chemicals act upon the embryo enabling it to overcome the resident embryo dormancy or the dormancy imposed by the seedcoat. Light and temperature regimes also interact with chemical influences.

But of the many different chemicals that provoke germination, seeds are highly unlikely to meet most of them in the natural environment (**see**: Table G.4 in **Germination – influences of chemicals**). As pointed out in that entry, because of the material and conditions used in the reported experiments it is not always possible to distinguish between effects on germination, on dormancy, or on both: the reader should be aware of this uncertainty. And this uncertainty explains why the same chemicals are considered here in relation to dormancy and elsewhere in the context of germination. The chemicals include various respiratory inhibitors (e.g. malonate, cyanide, azide), electron acceptors (e.g. methylene blue), oxidants (e.g. peroxide) and several others, such as nitrate, alcohols, chloramphenicol and thiourea. All of these chemicals have been used experimentally as a means of probing into different biochemical processes that might be implicated in the dormancy-breaking mechanism. (**See: Dormancy – embryo**)

1. Inorganic chemicals

Dormancy-breaking inorganic chemicals are generally but not exclusively weak acids in solution (e.g. carbon dioxide, nitrite and nitrate) and are ubiquitous in the soil. Their use in artificial dormancy breaking has been common since ancient times with animal dung and other concentrated sources of nitrogenous compounds being widely used. Their activity in removing dormancy is pH-dependent and related to molecular size, in contrast to the lipophilicity of organic chemicals (see below). Some compounds may be active because of chemicals that they generate: one effect of nitrate, for example, may be attributed to the production of nitrogen oxides.

2. Organic chemicals

Dormancy-breaking organic chemicals appear far more varied and show a range of similar activity not exclusively restricted to seeds. The activity of organic compounds is closely related to their lipophilicity (i.e. the ability to partition into a lipid membrane). This is further modified by a functional group effect with the activity of a chemical series increasing as an alcohol (e.g. propanol) is successively oxidized to its parent acid (e.g. propionic acid).

Smoke breaks dormancy of many species adapted to seasonally arid regions where germination occurs only following fire. Together with high temperatures, smoke is a dormancy-breaking component of fire, presumably effective because of organic chemicals that are present such as **butenolides** (**see: Germination – influences of smoke**).

It is interesting that many of these organic and inorganic compounds may also break dormancy or terminate developmental arrest in animals, fungi and lower organisms.

3. Mechanisms of action

(a) *pH*. One response of cells is a change in internal pH, which decreases when dormancy is broken in Jerusalem artichoke buds and red rice embryos. Changes in internal pH in red rice may strongly affect gene expression. **Abscisic acid** (ABA), essential for dormancy induction (**see: Dormancy – acquisition**), induces an increase in internal pH and the expression of dormancy-related genes such as *ABI3* (**see: Dormancy – genes**). Dormancy-breaking weak acids depress internal pH and down-regulate ABA-induced gene expression. In contrast, **gibberellin** (GA)-induced gene expression does not appear to be pH sensitive. Internal pH may act as an internal modulator of some regulatory signals.

(b) *Respiration*. A group of chemicals which appears to break dormancy in many species includes respiratory inhibitors and other similar metabolic inhibitors. Inhibitors of the mitochondrial electron transport chain (cyanide, azide, hydrogen sulphide, sodium sulphide) and inhibitors of the Kreb's tricarboxylic acid cycle are effective. Several hypotheses have been put forward to explain these effects which are discussed in **Germination – influences of chemicals**.

An interesting organic compound that apparently removes dormancy in numerous seed types is ethanol, also considered in **Germination – influences of chemicals**. Several hypotheses can be put forward to explain this action of ethanol. Suffice it to note in the present context that ethanol may have metabolic effects connected with respiration or it might interact with membranes. (**See: Anaesthetics**)

Nitrate promotes seed germination in many species. Two hypotheses for the mechanism of action of nitrate have been proposed. The first hypothesis is based on observations that inhibitors of glycolysis, citric acid cycle and terminal oxidation reactions of the mitochondrial electron transport chain can promote germination. An assumption of the hypothesis is that dormant seeds are deficient in an alternative oxygen-requiring process essential for germination. This oxygen-requiring process is depleted of oxygen because of its lower affinity for oxygen than the cytochrome pathway of respiration. Inhibition of the latter by such inhibitors as azide and cyanide directs the flow of oxygen to the alternative pathway, thus breaking dormancy. The pentose phosphate pathway is suggested as the alternative oxygen-requiring process essential for germination. Germination stimulants such as nitrate and nitrite are assumed to re-oxidize $NADPH_2$ to NADP, thus stimulating the operation of the pentose phosphate pathway.

The second hypothesis also acknowledges the importance of the pentose phosphate pathway but assumes that re-oxidation of $NADPH_2$ is attained through peroxidase action. This assumption is based on the observation that germination in lettuce and pigweed (*Amaranthus albus*) seeds is correlated with inhibition of catalase activity. This inhibition by nitrite, thiourea and hydroxylamine provides peroxidase with substrate for the $NADPH_2$, requiring reduction of H_2O_2 to H_2O. Both hypotheses have received extensive criticism because of a lack of conclusive evidence that activity of the pentose phosphate pathway is correlated with the breaking of dormancy.

(SF, MH, DC, FC, HWMH)

(c) *Membranes*. Environmental signals that break dormancy, such as temperature and chemicals, make first contact with the embryo at the plasma membrane. Membranes are fluid mosaics consisting of a fluid phospholipid bilayer containing membrane proteins. Both lipids and proteins are able to move through this matrix.

Some dormancy-breaking chemicals (Table G.4 in **Germination – influences of chemicals**) are hypothesized to act by interacting with membranes. The dormancy-breaking activity of **amphipathic** compounds (those with polar and non-polar constituents such as the hydroxyl group of an alcohol and its alkyl chain) increases with the length of the alkyl chain: this determines the ability of the chemical to partition into a lipid layer (lipophilicity). While the alkyl chain interacts with the acyl chain region of the membrane lipids, the polar hydroxyl group of an alcohol may interact with a carbonyl or phosphate group by hydrogen bonding with oxygen. The resulting change in head-group spacing may be equivalent to that of a change in temperature (**see: Dormancy breaking – cellular and molecular aspects**, section 2). This change in charge distribution on the membrane surface will influence the binding of peripheral proteins which might then initiate a **signal transduction** pathway culminating in the stimulation of germination. The increased dormancy-breaking activity of carboxylic acids compared with alcohols may be due to their greater hydrogen-bonding ability, thus increasing the perturbation of head-group spacing (**see: Anaesthetics; Dormancy – embryo; Dormancy – embryo (hypotheses); Germination – influences of chemicals**). (SF, MH)

Cohn, M.A. (1997) QSAR modelling of dormancy-breaking chemicals. In: Ellis, R.H., Black, M., Murdoch, A.J. and Hong, T.D. (eds) *Basic and Applied Aspects of Seed Biology*. Kluwer Academic, Dordrecht, The Netherlands, pp. 289–295.

Footitt, S. and Cohn, M.A. (2001) Developmental arrest: from sea urchins to seeds. *Seed Science Research* 11, 3–17.

Hallet, B.P. and Bewley, J.D. (2002) Membranes and seed dormancy: Beyond the anaesthetic hypothesis. *Seed Science Research* 12, 69–82.

Hilhorst, H.W.M. (1998) The regulation of secondary dormancy. The membrane hypothesis revisited. *Seed Science Research* 8, 77–90.

Dormancy breaking – hormones

An important concept relating to the mechanism of dormancy is the **hormone balance theory**. According to this, a determinative factor in the regulation of dormancy is the presence and action of hormones or growth regulators that promote germination (**gibberellins**, GAs) and those that inhibit it, or maintain dormancy (**abscisic acid**, ABA). Dormancy breaking, therefore, might involve changes in the status of these hormones, for example a decrease in effective ABA and/or an increase in the efficacy of gibberellin. Such changes may be effected by alterations in content of the hormones, by changes in sensitivity to the hormones, or by both.

1. Hormone content

There is evidence linking hormonal changes with the breaking of dormancy. In dormant seeds such as tomato, ABA content decreases during afterripening. Light treatment of lettuce seeds and application of dormancy-breaking chemicals to barley embryos have been reported to cause reductions in their ABA content. The changes are relatively small, however, but combined with other changes induced by light, or those occurring during afterripening, additive or synergistic effects could lead to dormancy-breakage. In lettuce seeds, light, operating through **phytochrome**, can elevate their active gibberellin content. Phytochrome Pfr acts to increase the expression of the gibberellin 3-oxidase gene, the resultant enzyme then provoking the synthesis of the active GA_1 that drives germination. Chilling may also enhance the capacity of a seed for GA biosynthesis, again by the same mechanism. Synergy between light and GA in causing germination is also known (**see: Germination – influences of hormones**).

Since GA can promote processes involved in seed germination it is not difficult to understand how enhanced production of the hormone might break dormancy. One scenario in which GA is implicated is in the production of enzymes such as endo-β-mannanase or β-1,3-glucanase which weaken the tissues enclosing the embryo of seeds of some species (e.g. *Datura ferox*, tomato, tobacco, white spruce), allowing the radicle to emerge. GA might also enable the embryo to develop the growth potential required for the completion of germination. And in both of these cases ABA antagonizes the action of GA thus maintaining dormancy (**see: Dormancy; Germination – molecular aspects; Germination – radicle emergence**).

2. Sensitivity to growth regulators and hormones

The ability of ABA and GA to counteract one another, leading to loss of dormancy, may also involve changes in sensitivity to hormones. The reduction in the intensity of dormancy that precipitates **preharvest sprouting**, for example in wheat and sorghum, is associated with a lowering of the sensitivity of the embryo to its own ABA.

As dormancy gives way to germination the sensitivity of the embryo to both ABA and GA changes. The embryos of cereal varieties with different levels of preharvest sprouting have similar ABA contents during development but exhibit altered sensitivity to ABA on maturity. This indicates that sensitivity is not dependent on hormone concentration but is related to perception and interaction between signal transducing pathways. Once an ABA-derived signal has been generated in the cells of the seed, signal attenuation is required to control the response. Attenuation of ABA signals involves a protein farnesylation gene (*ERA1*), an inositol signalling gene (*FRY1*) and RNA processing genes (*ABH1*, *SAD1* and *HYL1*). Furthermore, the transcription factor genes *ABI3*, *ABI4* and *ABI5* appear to interact in a complex fashion to determine the overall levels of sensitivity to ABA. Expression of *ABI3* and *ABI5* are closely linked in the seed, with some evidence for protein interaction between them and regulation of *ABI5* by *ABI3* (**see: Dormancy – genes; Signal transduction – hormones**). The phosphatase genes *ABI1* and *ABI2* act as negative regulators of ABA signalling, reducing sensitivity to ABA. This implicates the phosphorylation states of key proteins in the control of ABA signalling. The multiple dormancy-breaking signals impinging on a seed potentially affect the transcription or translation of all of these genes, or the activity/stability of their proteins, thus reducing sensitivity to ABA. For example, in wild oats, **afterripening** reduces, and **secondary dormancy** increases, expression of *AfVP1*, an **orthologue** of ABI3. The ABI3 protein, possibly by interaction with the ABI4 and ABI5 proteins, may control the activity of ABI1 and ABI2. Hence, when dormancy is broken the loss of ABI3 expression and activity (and potentially ABI5) may result in reduced ABA sensitivity through the increased action of ABI1 and ABI2.

Decreases in ABA brought about by dormancy-breaking treatments such as red light, chemicals and GA are thought to be through ABA catabolism or conjugation. Dormancy-breaking through chilling increases the subsequent content of active GA in the seed, for example in hazel and *Arabidopsis*, but overall GA_1 concentrations do not appear to correlate with germination potential. If dormancy-breaking treatments reduce both ABA contents and sensitivity in the embryo, the consequent reduction in repression of ABA-sensitive genes and a relative increase in sensitivity to GA may allow the initiation of germination. This is borne out by the observation that when sensitivity is reduced through mutation of the GA signalling gene *SLY* the residual sensitivity to GA is still sufficient to allow germination in the *Arabidopsis* ABA-insensitive mutant, *abi1-1*.

Growth regulators other than GAs and ABA, such as **ethylene**, are involved in germination of seeds, and might also play some role in their release from dormancy. (SF, MH)

Gazzarrini, S. and McCourt, P. (2003) Cross-talk in plant hormone signalling: what Arabidopsis mutants are telling us. *Annals of Botany* 91, 605–612.

Kucera, B., Cohn, M.A. and Leubner-Metzger, G. (2005) Plant hormone interactions during seed dormancy release and germination. *Seed Science Research* 15, 281-308.

Merlot, S., Gosti, F., Guerrier, D., Vavasseur, A. and Giraudat, J. (2001) The ABI1 and ABI2 protein phosphatases 2C act in a negative feedback regulatory loop of the abscisic acid signalling pathway. *Plant Journal* 25, 295–303.

Ogawa, M., Hanada, A., Yamauchi, Y., Kuwahara, A., Kamiya, Y. and Yamaguchi, S. (2003) Gibberellin biosynthesis and response during Arabidopsis seed germination. *The Plant Cell* 15, 1591–1604.

Sanchez, R.A. and Mella, R.A. (2004) The exit from dormancy and the induction of germination: physiological and molecular aspects. In: Benech-Arnold, R.L. and Sánchez, R.A. (eds) *Handbook of Seed Physiology: Applications to Agriculture*. Food Products Press, New York, USA, pp. 221–243.

Dormancy breaking – light

Germination of many kinds of dormant seed is promoted by light which may be acting by its impact on dormancy status (though direct effects upon germination itself may also be involved).

Imbibed seeds perceive light through the red light photoreceptor **phytochrome**. Biosynthesis of an active **gibberellin** (GA) is promoted by phytochrome through the enhancement of expression of the gene for the enzyme GA 3-oxidase. The elevation of GA_1 content is responsible for the breaking of dormancy/promotion of germination.

Light only penetrates the top few millimetres of soil, so the majority of seeds in the **soil seed bank** are in darkness. Seeds held in the dark for long periods at temperatures non-permissive for germination will enter **secondary dormancy**. When light-sensitive seeds are exposed to light as a result of soil disturbance or by the development of a vegetation gap, dormancy will be progressively lost. In some species this can be accelerated by diurnal cycles of light and dark. Daily cycles of light and temperature are found to interact in the dormancy-breaking process. (SF, MH)

(**See: Cryptochrome; Light – dormancy and germination; Phytochrome; Signal transduction – light and temperature**)

Dormancy breaking – temperature

The effect of temperature on dormancy breaking differs between dry and imbibed seeds. Perhaps the most extreme cases of dormancy breaking by temperature in dry seeds are seen in the actions of fire on the coats of hard, water-impermeable seeds (**see: Dormancy – coat-imposed**, section 1). But dry seeds (typically 5–15% moisture content) more commonly may experience gradual dormancy loss through dry **afterripening**, which commences on the mother plant and continues following shedding to the soil. This process, which is enhanced as temperatures increase, is most obvious in species with short life cycles whose seeds often exhibit shallow dormancy, an attribute that allows rapid afterripening in warm summers enabling the production of more than one generation a season. For example, dormancy in cereal grains is maintained for a relatively long time (at least 5–10 years) when stored at −18 to −30°C, whereas they become non-dormant in about 1 month at 30°C.

As dormancy reduces the range of external conditions under which germination can occur, removal of dormancy by afterripening increases the temperature range over which seeds can germinate. Release from dormancy results in a widening of these conditions. For example, freshly harvested seeds of cereals from temperate climates (wheat, barley, oat) are regarded as being dormant because they germinate poorly at 20°C or higher temperatures; but they germinate well at lower temperatures, the thermal optimum being around 10°C. A few months of afterripening in dry storage breaks their dormancy, and they then become able to germinate at temperatures ranging from 0–5 to 30–35°C (Fig. D.47A),

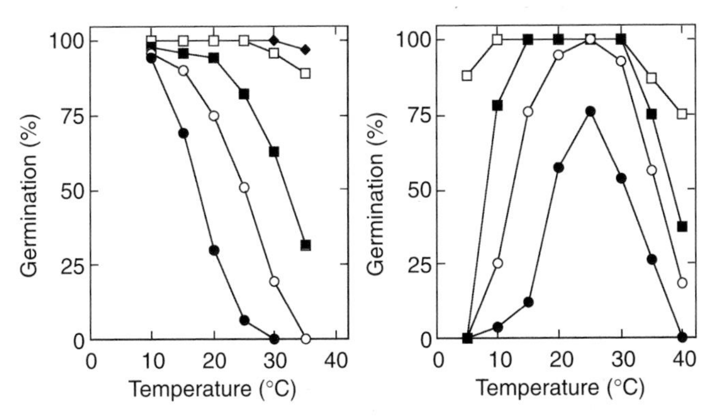

Fig. D.47. Relative dormancy in oat (A) and sunflower (B) seeds. (A) Influence of temperature on the germination percentages obtained after 7 days with freshly harvested (dormant) *Avena sativa* (cv Moyencourt) seeds (●) and with seeds stored dry at 20°C for 3 weeks (○), 2 months (■), 6 months (□) and 12 months (♦). All seeds germinated at temperatures below 10°C, but the germination rate was reduced and 100% germination was not obtained within 7 days (not shown in the figure). From Corbineau, F. and Côme, D. (1993) The concept of dormancy in cereal seeds. In: Côme, D. and Corbineau, F. (eds) *Fourth International Workshop on Seeds. Basic and Applied Aspects of Seed Biology.* ASFIS, Paris, Vol. 2, pp. 581–589. (B) Influence of temperature on the germination percentages obtained after 7 days of freshly harvested (dormant) *Helianthus annuus* (cv Mirasol) seeds (●) and with seeds stored dry at 20°C for 1 month (○), 2 months (■) and 6 months (□). From Corbineau, F. and Côme, D. (1990) Sunflower (*Helianthus annuus* L.) seed dormancy and its regulation by ethylene. *Israel Journal of Botany* 39, 313–325.

their germination rate is enhanced at all temperatures, and the thermal optimum approaches 25°C.

The mechanism of afterripening is obscure, but interesting information is emerging from research on tobacco (*Nicotiana tabacum*) seeds. It appears that here the water content at the radicle tip region of the seed is high enough, in air dry seeds, to allow the gradual production of the enzyme β-1,3-glucanase. This enzyme participates in the normal germination process by weakening the seedcoat, permitting the radicle to emerge (**see: Germination – molecular aspects; Germination – radicle emergence**). It is possible, therefore, that the gradual loss of seed dormancy in afterripening in this species occurs because of the slow production of a coat-weakening enzyme. There is also some evidence from another *Nicotiana* species, *N. plumbaginifolia*, that ABA content declines during afterripening.

In cereals, as in many other species, the inability of dormant seeds to germinate at relatively high temperatures may result from the presence in the seedcoats of high amounts of **phenolic compounds**, the **polyphenol oxidase**-mediated oxidation of which limits the flux of oxygen reaching the embryo, and the higher the temperature the greater this limitation (**see: Germination – influences of gases**). Breaking of dormancy during dry storage in some cases may decrease the efficiency of the barrier to oxygen diffusion through the seedcoats, thus allowing a better oxygenation of the embryo. Sunflower produces dormant seeds that germinate much better at 25–30°C than at lower and higher temperatures. The inability to germinate at the higher temperatures results principally from a seed**coat-imposed dormancy**, whereas poor germination at temperatures below 25°C is mainly due to **embryo dormancy**. Dry storage improves germination at temperatures ranging from 5 to 40°C by eliminating both the seedcoat-imposed and the embryo dormancy (Fig. D.47B). Embryo dormancy is also alleviated by ethylene (**see: Germination – influences of gases**). During dry storage the sensitivity of the embryo to this gas gradually increases.

Once the seed is on or in the soil (the **soilseed bank**), seed moisture content equilibrates to that of the surroundings. As moisture content increases, the afterripening effect is reduced. At the low moisture contents (15–30%) that are unable to support germination, seed viability may decrease. At higher moisture contents, in conditions unfavourable for germination, secondary dormancy may develop. Hence, afterripening to produce germinable seeds occurs only over a narrow, low range of seed water contents.

Seeds may also experience an alternating temperature cycle, i.e. warm day/cold night as part of the diurnal cycle. As seed moisture content increases this cycle can become increasingly effective in breaking dormancy in the imbibed seed. Those of several species, such as *Cynodon dactylon*, *Typha latifolia*, *Lycopus europaeus* and *Rumex obtusifolius* (in the dark), seem to be able to germinate only at alternating temperature. The effectiveness of this depends on both the duration and temperature of the respective high and low phases, the difference between them, i.e. the amplitude (e.g. 1–10°C), the rate of cooling and heating, the number of cycles, and the time between imbibition and the start of the cycle.

Low and high constant temperatures also impact on the dormancy status of imbibed seeds. Chilling (1–15°C, optimum approx. 4°C), generally for several days or weeks (cold **stratification**) is effective in breaking dormancy in seeds of species adapted to germinate in the higher temperatures of late spring and summer. It is commonly effective in seeds of temperate woody species (**see: Tree seeds**). In species that germinate in the cooler temperatures of late summer and overwinter in the vegetative phase, warming or warm stratification (15–35°C) is effective in breaking dormancy.

Embryo dormancy is primarily localized in the embryonic axis (though cotyledons can also show *quasi* dormancy effects). In most cases, it is displayed by the absence of growth or abnormal growth of the **radicle**, and the **epicotyl** does not seem to be much affected. Cold treatment breaks dormancy of the radicle and allows its normal growth. In a few species however, such as peach, it is the epicotyl and not the radicle that is unable to develop normally without cold treatment. This is known as epicotyl dormancy. In this case, the embryo germinates without chilling and the root grows normally, but the stem remains dwarf and forms a rosette of leaves. When the embryo is given cold treatment at the beginning of imbibition, the epicotyl dormancy is removed and the young plant grows normally. The same result is obtained by applying cold treatment to the dwarf young plant after root development. Seeds of *Paeonia suffruticosa* and some *Lilium* and *Viburnum* species also show epicotyl dormancy without radicle dormancy, but cold has no effect on epicotyl dormancy if applied to the embryo before germination; the presence of roots is necessary for the epicotyl to respond to chilling treatment. In natural conditions, the complete development of the young plants produced by such seeds requires 2 years. The roots grow in the first year but the epicotyl remains dormant. The shoots develop only after winter. A more complicated case is that in lily-of-the-valley (*Convallaria majalis*), *Trillium erectum* and *Caulophyllum thalictroides* seeds. These show a double dormancy, i.e. a dormancy localized in both the radicle and the epicotyl. They require chilling treatment to enable the roots to develop, but the epicotyl does not grow. A second period of exposure to cold following root growth is necessary to obtain shoot development. Two successive winters are required in natural conditions for obtaining the complete emergence of the plant.

In seeds of several species there are various substitutes for cold temperatures. For example, apple embryo dormancy is broken by chilling, but it is also fully removed by placing the embryos for 2–3 weeks at 20°C in an atmosphere completely free of oxygen (pure nitrogen), or in the presence of respiratory inhibitors (KCN or NaN_3) at relatively high concentrations. Apple seeds containing dormant embryos do not germinate at 30 or 35°C, but dormancy is gradually broken when the embryo is subjected to anoxia inside the seedcoats at such high temperatures (**see: Germination – influences of gases**). Removal of cotyledons of dormant embryos also leads to normal germination of isolated embryo axes, possibly by removing a source of **abscisic acid**; and **gibberellins** provoke germination of dormant embryos in the absence of a cold treatment. In some woody species, such as *Crataegus* spp., *Fraxinus* spp. and *Euonymus* spp., a 1- to 3-month incubation period of the hydrated seeds at 20–25°C prior to chilling is necessary for the latter to be effective. In *Fraxinus* species, the warm period (about 20°C) allows growth of the immature embryo and renders it responsive to chilling. In seeds of a few species (*Narcissus bulbicoidium*, *Hyacinthoides non-scripta*), which germinate at relatively low temperatures (10–15°C), dormancy is released by several weeks of warm treatment (25–30°C) in a moist medium, but chilling is not required.

In contrast to cold, temperatures that are too high for germination often induce a **secondary dormancy** or **thermodormancy**. For example, apple embryos whose primary dormancy has been removed by appropriate cold treatment fail to germinate at 30°C and gradually lose their ability to germinate when transferred to 15–20°C. A further chilling treatment is required to enable them to germinate at these temperatures. Primary dormant, dehulled oat seeds germinate at 20°C but not at 30°C. If they are maintained for 1–2 days at the latter temperature, they lose their ability to germinate later at 20°C, because they enter a thermodormancy that appears to be a reinforcement of the primary dormancy. Curiously, this secondary dormancy diminishes during a prolonged period at 30°C (Fig. D.48). In some cases (much less common than thermodormancy) temperatures below the minimum for germination also induce a secondary dormancy, as for example in seeds of *Rumex* spp., *Phacelia dubia*, *Polygonum pennsylvanicum*, *Torilis japonica* and *Veronica hederofolia*. Yet more rare is the case in which secondary dormancy is induced by both too low and too high temperatures, as in *Taraxacum megalorhyzon* seeds. Similar behaviour arises for seeds of the root holoparasites *Orobanche* spp. These seeds germinate only in the vicinity of a suitable host plant in response to germination stimulants present in the exudates from host roots. However, they are able to respond to the stimulants only after a period of incubation in a wet medium, called conditioning or preconditioning, which is temperature-dependent. The thermal optimum for preconditioning of *Orobanche* seeds varies from 20 to 30°C

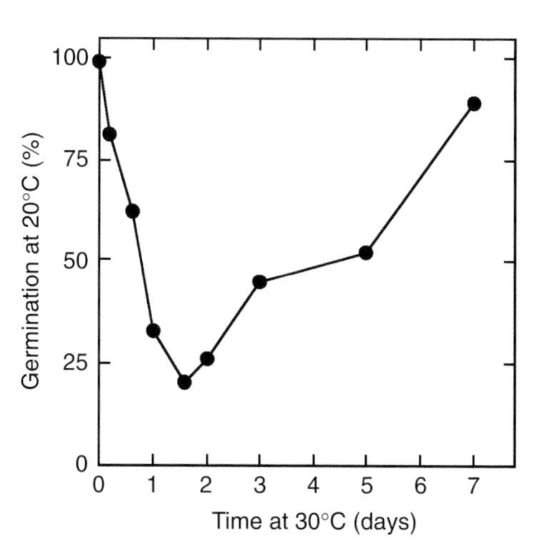

Fig. D.48. Induction of thermodormancy by high temperature in *Avena sativa* (cv Moyencourt) seeds. After various durations (0 to 7 days) of incubation at 30°C, dehulled, hydrated dormant seeds were transferred to 20°C and germination was recorded after 3 days. From Corbineau, F., Black, M. and Côme, D. (1993) Induction of thermodormancy in *Avena sativa* seeds. *Seed Science Research* 3, 111–117.

depending on the species. When seeds are preconditioned at sub- or supra-optimal temperatures, their responsiveness to the germination stimulants first increases and then decreases. These phenomena have been considered as loss of primary dormancy and induction of secondary dormancy, respectively (see: **Parasitic plants**).

All these effects of temperature are due to action upon the embryo. Seeds with an embryo or a coat-imposed dormancy, or with a degree of both, are affected. In all these cases the dormancy-breaking factor enhances the ability of the embryo to proceed through germination.

(DC, FC, SF, MH)

For a discussion of the possible mechanism of action of temperature in dormancy breaking **see: Dormancy breaking – cellular and molecular aspects**.

Baskin, C.C. and Baskin, J.M. (1998) *Seeds. Ecology, Biogeography, and Evolution of Dormancy and Germination*. Academic Press, London, UK.

Bewley, J.D. and Black, M. (1982) *Physiology and Biochemistry of Seeds in Relation to Germination*, Vol. 2, *Viability, Dormancy and Environmental Control*. Springer, Berlin, Germany.

Côme, D. and Corbineau, F. (1989) Some aspects of metabolic regulation of seed germination and dormancy. In: Taylorson, R.B. (ed.) *Recent Advances in the Development and Germination of Seeds*. Plenum Press, New York, USA, pp. 165–179.

Probert, R.J. (2000) The role of temperature in germination ecophysiology. In: Fenner, M. (ed.) *Seeds: The Ecology of Regeneration in Plant Communities*, 2nd edn. CAB International, Wallingford, UK, pp. 261–292.

Dormoat

This is a cross between wild oat (*Avena fatua*) and cultivated oat (*A. sativa*) designed to introduce the seed dormancy of the former into the latter. (**See: Dormancy – importance in agriculture**)

Double fertilization

The **fertilization** of the egg nucleus (n) to form the zygote (2n), and of the polar nuclei (2n) to form the endosperm nucleus (3n), each by a separate haploid nucleus (n) from the pollen grain. An event unique to the **angiosperms**.

Double haploid

A method of producing plants with fully homogenic genomes from genetically heterogeneous parental plants. Production of inbred cultivars can be significantly shortened by taking the double haploid approach, which takes advantage of a natural phenomenon of haploid seed development from diploid plants. Occasionally, a failure in fertilization of the haploid egg cell by the pollen sperm nucleus leads to divisions resulting in a haploid embryo. Spontaneous occurrence of maternal haploids in **maize** is in the order of 1 in 10^3 ovules. The occurrence is significantly increased to 1–12% in certain genetic backgrounds such as Stock 6 and RWS strains in maize. In order to produce seed, haploid plants need to be diploidized so that gametes are created through meiosis. Diploid cells can be induced by **colchicine** treatment, which inhibits chromosome segregation during cell divisions.

(O-AO, HS)

Draper mill

Seed conditioning equipment consisting of a moving tilted rough- or smooth-surfaced belt, used to separate rough seed from smooth, or flat or elongated seed from round. (**See: Conditioning, II. Cleaning**)

Dressing

A seed technology term for a **pesticide** or fertilizer formulation applied as a seed treatment before planting, and for the application process itself. (**See: Production for Sowing; Treatment**). Also occasionally used to describe the process of cleaning seed (using sieves, indented drums, gravity tables, blowers) and to describe the treatment of seeds with pesticides. (**See: Conditioning, II. Cleaning**)

Drill (drilling)

The general term for the machines and the practice of sowing or planting seed in soil at a controlled depth and in specified amounts, usually in furrows. (**See: Direct drilling; Planting equipment – agronomical requirements; Planting equipment – placement in soil**)

(a) *Drill seeding*, strictly defined, is the planting pattern resulting from the random dropping (and subsequent covering) of seeds in furrows to give definite rows of randomly-spaced plants. (**See: Planting equipment – agronomic requirements**, Fig. P.15.)

(b) *Drilling window* is the period of time during which crop sowing should ideally take place. If drilling is delayed later than this date, the yield of the crop may be affected. The start of the window is determined by consideration of likely stresses affecting germination, such as soil condition and the likelihood of extreme cold or wet weather following sowing.

Drupe

A **fruit**, generally one-seeded, in which the pericarp is differentiated into three regions – the outer, relatively thin exocarp, the more-or-less fleshy mesocarp and the hard, woody or stony endocarp enclosing the seed. (**See: Fruit – types; Pyrene**)

Dry bean

Edible beans that are harvested dry for subsequent consumption. (**See: Beans; Pulses**)

Dry weight

The mass of the seed after artificial drying: the fresh weight of the seed minus the seed water content. Dry weight is generally determined by weighing the seed after a period of oven drying (e.g. 24 h at 100–150°C) though care must be taken in practice that seed components, such as **oils and fats**, **essential oils** or other volatiles, are neither lost by evaporation nor gain in weight through oxidation. The **International Seed Testing Association** (ISTA) Rules prescribe procedures to determine seed moisture content in **quality testing**, as do **grain inspection** rules: such as stipulating whether a 'low' or 'high' constant temperature method, or whether a preliminary grinding or predrying step, is required. Compare with **fresh weight**. (**See: Maturation – effects of drying**)

Drying of seed (for storage)

Most seeds are hygroscopic; hence the water content of a seed depends upon the relative humidity of the surrounding air at a given temperature, unless the seed-covering structures are impermeable. Thus water will be lost from moist seeds by diffusion if they are surrounded by drier air until the **water potential** of the seeds and air are at equilibrium. The rate of drying slows down exponentially as the seeds approach equilibrium. Depending on temperature, seeds will attain an **equilibrium moisture content** at any given **relative humidity** of the surrounding air and the relationship between these parameters is described by characteristic reverse sigmoid **sorption isotherms**.

Drying is the most important step in the processing of **orthodox** seeds for storage. Drying seeds from the moisture content levels in equilibrium with ambient air at the time of harvest to the levels in equilibrium with the low relative humidity conditions (≅15% RH) used in seed **gene bank** dry rooms can result in a 1000-fold increase in subsequent **longevity**.

Seed drying rate increases with increasing temperature mainly because the water-holding capacity and hence relative humidity of air is temperature dependent; the relative humidity of air at a given water content falls as the temperature rises. Raising the temperature, therefore, will increase the water potential gradient between a wet seed and the surrounding air thus speeding up the rate of drying. However, care must be taken to minimize the attendant risk of loss in seed **viability** when raising the temperature for drying. Air movement also speeds up drying and small seeds dry more quickly than large seeds.

Although warm-air drying is an essential component of commercial crop seed drying, especially if the seeds are not intended for replanting (e.g. for industrial use), (see: **Production for sowing, V. Drying; Storage management**), cool-air systems employing artificial desiccants such as lithium chloride are the preferred choice for seed **gene banks** involved in long-term **conservation of seed**.

Seeds can be dried without significant loss of quality under natural conditions using the combined effects of the sun and wind. However, care must be taken to avoid seeds overheating in direct sunlight. Well-ventilated partial shade conditions, raised above the ground, are preferred. Seeds should be taken indoors at night to avoid predation and moisture gain as the air cools and relative humidity rises.

For drying small samples of seeds, artificial desiccants such as silica gel, 'Drierite' (calcium sulphate), calcium chloride, charcoal or even toasted seeds such as rice may be used. In all cases, the desiccant must itself be dried to practically zero water content before it can be used to dry seeds. The ratio of desiccant to seeds required to achieve successful drying, without the need for regeneration of the desiccant, will vary depending on the water-holding properties of the desiccant and how wet are the seeds to be dried. As a rough guide for silica gel, three parts dry silica gel to one part seeds will be adequate in most cases. (RJP, SHL)

DUS

See: Distinctness, Uniformity and Stability

Dust

Grain dust is produced when seed- and grain-production crops, notably cereals and grasses, are harvested, dried, moved, stored and processed by conditioning, cleaning, treating and milling. Stored conditioned grain typically contains 0.1–0.5% dust, by weight. The dust, which is comprised of bacteria, fungi, insects and dry plant particles such as dislodged **chaff** and undigested seed fragments produced by the action of some grain **storage pest** species, is associated with two main hazards: irritation or diseases caused by human inhalation, and explosions.

Inhaling grain dust can cause ill health: for example, asthma, bronchitis and 'grain fever'. Some people can become sensitized, and subsequent low-level exposure can trigger irritation and attacks. The main cause of human aspergillosis is *Aspergillus fumigatus*, a common **storage fungus** in grain. This disease mostly affects agricultural workers exposed frequently to massive doses of the spores, and can take an infectious form in which masses of living hyphae cause lesions in tissues of the lung, skin, ear, eye or nose. Grain dust may contain mould spores of aerobic actinomycetes, often less than one micron in diameter, that if inhaled can cause the potentially fatal allergic disease, 'Farmer's Lung'.

Explosion hazards usually exist during grain transfer and handling, where a critical concentration of suspended dust can become entrained and suspended in the air (the critical concentration for wheat dust, for example, is more than 50 mg/m^3). Particularly vulnerable points are within the grinding equipment in grain mills, and **conveyors** generally (such as **bucket elevators** or enclosed belts) where small particles become dislodged due to the agitation, impact and redirection of fast-flowing grain. Dust-removal and -suppression techniques implemented in grain elevators include pneumatic controls and fine-mist water-sprays to capture dust, and the application of food-grade-quality oil to stick the fine particles on to seed surfaces. (PH)

Dust seeds

In ecology and taxonomy, the term 'dust seeds' is used to describe minute seeds (generally those smaller than 0.3 mm) that are adapted to wind dispersal. According to the fossil record, dust seeds dominated the first stage of the **evolution of seed** in angiosperms, through most of the Cretaceous period. **(See: Structure of seeds; Structure of seeds – identification characters)**

Dust-off test

A measurement of the degree to which either natural dust or an applied seed treatment material can flake off during seed handling. A commonly used test protocol involves drawing an air current under slight vacuum through a seed mass that is agitated for a short period, and weighing the dust retained on a filter. Dust-off tests are important during the development of new seed treatment fungicides, insecticides, colorants and coating formulations, so that regulatory authorities can set dust-off standards as part of the registration process, and are also used in routine quality control of treated seed bulks. (PH)

(See: Dust; Treatments – Pesticides)

E

Ear rows

A type of field plot that is useful when testing cereal varieties, for instance in **Distinctness, Uniformity and Stability** trials. Seed is threshed out of at least 100 ears, and the seeds from each ear are planted in separate rows.

Ecology of seeds

Plants are sedentary organisms, and at the time of germination every plant has the opportunity to fix where it will spend the rest of its life. Seeds are thus a key stage in the ecology of plants. Much of seed ecology is concerned with **dispersal**, with the various mechanisms that determine if and when **germination** takes place, and with avoiding hazards such as **predation**. Seed output depends partly on reproductive allocation, i.e. the proportion of available resources devoted to seeds, and partly on **size of seeds**. There is a trade-off between the **number of seeds** produced and the size of individual seeds.

Freshly shed seeds may be dormant, i.e. incapable of germination until some physiological change has taken place, until an immature embryo has fully developed, or until an impermeable seedcoat has been breached. Dormancy prevents germination when basic conditions are suitable, but the probability of successful establishment and survival is low, e.g. immediately before a severe winter. Seeds of many species exhibit **dormancy cycles**, in which primary dormancy is broken (e.g. by high temperatures in winter annuals) and **secondary dormancy** is subsequently re-imposed (by low temperatures in **winter annuals**) (**see: Dormancy; Dormancy – ecology**). Many seeds possess structures to promote dispersal by wind, water (abiotic dispersal), animals (biotic dispersal) or other vectors including humans, and elaborate typologies exist to classify dispersal mechanisms on this basis (**see: Seedcoats – dispersal aids**). Although there is a mutualistic relationship between plants and their animal seed dispersers (particularly fruit eaters, or frugivores), close coevolution is rare: most animals disperse the seeds of several different species, and most plants are dispersed by a variety of animals. Despite the widespread occurrence of obvious dispersal adaptations, most seeds are dispersed only short distances, and the precise function of dispersal is not always clear: for example, is it to escape from the neighbourhood of the parent, or to locate rare 'safe sites' for germination, or both? The distribution of seeds around the parent is described as the **seed shadow**, and the seeds (of all species) arriving at a given location are described as the **seed rain**. Both before and after dispersal, many seeds are eaten by animals. Seeds often possess physical or chemical defences to deter predators, but some predators are also important dispersers.

Once dormancy is broken (**see: Dormancy breaking**), seeds germinate in response to specific cues, including light, specific absolute temperatures, temperature alternations and chemicals such as nitrate ions (**see: Germination – influences of chemicals; Germination – influences of temperature; Light – dormancy and germination**). Responses to such cues often detect seasonal changes or gaps in vegetation, i.e. sites free from competition from established plants. **Parasitic plants** may germinate in response to specific chemical signals from their hosts. Germination may be inhibited by allelopathic chemicals from other plants or seeds, although the ecological significance (and even the existence) of **allelopathy** is controversial. The specificity of germination requirements varies widely between species, with small seeds often having both the greatest need for gaps and correspondingly stringent germination requirements.

If seeds do not germinate immediately and avoid being eaten, they may enter a **soil seed bank**. The densities of seeds in the seed bank and in the seed rain are usually expressed on a per square metre basis. A short-lived (< 1 year) seed bank is described as transient, a long-lived one as persistent. Persistence is unrelated to dormancy. Burial appears to be a first crucial step in the formation of a persistent seed bank. In most floras, small seeds are more persistent than large ones. In some habitats, particularly those prone to fire, viable seeds are retained on the plant for several years and released after a fire. (**See: Serotiny**) (KT)

(**See: Deserts and arid lands; Evolutionary ecology; Pathogens; Regeneration – ecology**)

Baskin, C.C. and Baskin, J.M. (1998) *Seeds. Ecology, Biogeography and Evolution of Dormancy and Germination.* Academic Press, San Diego, USA.

Fenner, M. (ed.) (2000) *Seeds: The Ecology of Regeneration in Plant Communities*, 2nd edn. CAB International, Wallingford, UK.

Fenner, M. and Thompson, K. (2005) *The Ecology of Seeds.* Cambridge University Press, Cambridge, UK.

Forget, P.M., Lambert, J.E., Hulme, P.E. and Vander Wall, S.B. (eds) (2004) *Seed Fate: Predation, Dispersal and Seedling Establishment.* CAB International, Wallingford, UK.

Egg apparatus

The egg cell plus the two **synergid** cells at the **micropylar** end of the **embryo sac** (the **megagametophyte**) of the angiosperms. (**See: Fertilization; Reproductive structures, 1. Female**)

Eggplant

See: Pepper and Eggplant (aubergine)

Einkorn

Einkorn (Fig. E.1) is the domesticated version of the wild diploid wheat *Triticum monococcum* (AA, **see: Wheat**, Table W.1). There are 'wild' einkorns which are different species of *Triticum*. The non-**shattering** einkorn has a tough **rachis** which differentiates it from its wild ancestors, but the grains are covered with a **hull** that is not removed during the threshing operation. Einkorn has a single fertile **floret** per **spikelet** at each node in the spike. It was one of the first grains domesticated in the Fertile Crescent region of the Middle East around 10,000–12,000 years ago, but has been practically abandoned as a cultivated crop in favour of the free-threshing tetraploid and hexaploid wheats as they came into cultivation. (**See: Cereals; Domestication**)

Einkorn is not grown commercially to any extent and there is no significant breeding effort in this crop. It is grown occasionally by organic farmers and for historical reference.

(DF)

Leonard, W.H. and Martin, J.H (1963) *Cereal Crops*. Macmillan, New York, USA.
Simmonds, N.W. (ed.) (1976) *Evolution of Crop Plants*. Longman, New York, USA.
Stallknecht, G.F., Gilbertson, K.M. and Ranney, J.E. (1996) Alternative wheat cereals as food grains: einkorn, emmer, spelt, kamut, and triticale. In: Janick, J. (ed.) *Progress in New Crops*. ASHS Press, Alexandria, VA, USA, pp. 156–170.

Fig. E.1. Ears of einkorn.

Elaiosome

An appendage of a seed, rich in oil, not essential for the viability of the seed but attractive to animals (especially ants) which carry the seeds to their nest, thereby facilitating their dispersal (**myrmecochory**). It is a synonym of **aril** but mostly referring to those of myrmecochorous seeds. (**See: Seedcoats – dispersal aids**)

Electromagnetic enhancement

Advantageous germination and seedling growth responses can result after exposing dry seeds for up to a few minutes to either continuous, intermittent or rapidly-pulsed static or alternating magnetic fields (typically up to about 0.25–1.0 Tesla) or electric fields (typically of up to 100 kV/m). These phenomena have received sparse research attention in the seed scientific literature, and consistent results about their usefulness and reproducibility as practicable seed **enhancement** treatments are lacking. Possibly, the responses only occur under specific environmental or seed physiological states. Underlying mechanisms are possibly associated with **free radical** chemistry and the enhancement of endogenous **antioxidant** systems.

Bombardment of seeds with electrons has been developed as a phytosanitary measure, using the biocidal effect of low energy-accelerated electrons (up to about 10 kGy) to eliminate seedborne **pathogens** from cereals and other species, e.g. the eradication of *Tilletia caries*, *Septoria nodorum* and *Fusarium* spp. associated with winter wheat. In an unrelated physical sense of the term, 'magnetic separators' are used to selectively remove rough, cracked and sticky seeds (**see: Conditioning, II. Cleaning, friction separation**). (PH)

Phirke, P.S., Patil, M.N., Umbarkar, S.P. and Dudhe, Y.H. (1996) The application of magnetic treatment to seeds: methods and responses. *Seed Science and Technology* 24, 365–373.
Schröder, T., Röder, O. and Lindner, K. (1998) E-dressing – a unique technology for seed. *ISTA News Bulletin* 118, 13–15.
Stephenson, M.M.P., Kuschalappa, A.C. and Raghavan, G.S.V. (1996) Effect of selected combinations of microwave treatment factors on inactivation of *Ustilago nuda* from barley seed. *Seed Science and Technology* 24, 557–570.

Elevator

A term used to describe certain mechanisms designed to lift grain to a higher level during conditioning (such as **bucket elevators, see: Conveyors**). Also in some countries the large storage structures and associated unloading, loading and handling systems, or the entire facility, that receives and temporarily stores grain in bulk: either directly from producers for storage or forwarding ('primary' elevators), or for manufacture or processing into other products ('process' elevators); or for official inspection, weighing, cleaning to remove **dust** and **chaff**, storing, treating and shipping ('terminal' or 'transfer' elevators). Large elevator storehouse facilities are designed to handle a range of different commodities and grades. (**See: Storage management, 1. Structures, containers and packages**) (PH)

ELISA (enzyme-linked immunosorbent assay)

ELISA is an immunological test based on the principle of antibody–antigen interaction. The antibody is enzyme-tagged so that its presence can be detected because it can be induced to catalyse a colorimetric reaction. The test is usually carried out in plastic microtitre plates, containing a numbers of wells. In seed **health testing**, seeds or seedlings presumed to contain the **pathogen** are extracted in a sterile liquid medium (often water or saline buffer). The extract is bound to the walls of the well, and the enzyme-tagged antibody is added; the wells are washed to remove surplus soluble antibody and the substrate for the enzyme added, producing a colour change if the insoluble pathogen–antibody complex is present. The intensity of the colour change gives a quantitative measure of the pathogen. There are three main types of ELISA: the direct method (double antibody sandwich, DAS ELISA), the indirect method and the competitive assay. The direct method is most often used in seed health assays. (**See: Immunoassay**)

(VC)

Elite seed

A term used informally in some countries as equivalent to **Basic Seed** (and, similarly, 'super-elite' for Pre-Basic Seed), although it has no official meaning in the context of botanical true seed. It is used, however, as a name for certain production 'nuclear-stock' classes of disease-free 'seed potato' tubers and, more generally, for the selection of superior lines in breeding programmes, such as in forestry.

In some European countries certain grades of seed are marketed at higher standards than the minimums required by EU Directives. The official term for this higher standard seed is 'Higher Voluntary Standard'. Though seed companies may market such seed using terms such as 'elite', 'vigour-tested', 'premier' or 'eco-friendly', these terms have no legal status.

(RD)

Emasculation

Removal of anthers from a bud or flower before pollen is shed, as a preliminary step in crossing to prevent **self-pollination** in hand-pollinated hybrid crops such as tomato and pepper, or removal of male flowers in monoecious plants, such as the **detasselling** of maize. In a crop like sorghum, however, with many tiny, bisexual flowers grouped in one big inflorescence, emasculation by hand with tweezers is extremely laborious. By and large, chemical methods of emasculation are not yet suitable for large-scale application. Alternatively, genetic systems are used to regulate the ability to produce or release functional pollen, such as **cytoplasmic male sterility**, or to prevent the formation of male sex organs.

Embryo

(Greek: *embryon* = unborn fetus, germ) Young **sporophyte** of the Embryophyta (embryo-bearing plants) developing from the egg cell in the **embryo sac** after **fertilization**, generally consisting of a **hypocotyl–root–embryonic axis**, the **radicle**, one (**monocotyledons**) or two **cotyledons** (**dicotyledons**) and the **plumule** (shoot apex).

The morphology and underlying anatomy of the embryo in the mature seed varies considerably among species. Embryo development begins in the embryo sac, following fertilization and division of the zygote (but see below, and **Apomixis**). Since its pattern varies widely among species and often provides valuable taxonomic characters, e.g. at family level, this variation is the basis of the study of comparative embryology, largely beyond the scope of this account, which is mainly concerned with the mature seed.

The very first cell division of the zygote results in the differentiation of the pro-embryo (apical cell) from the **suspensor** (basal cell), an organ that anchors the developing embryo in maternal tissue and is involved in nutrient transport and synthesis of **hormones** (growth substances, growth regulators). The suspensor usually ceases development early in **embryogenesis**, and is mostly absent from, or barely recognizable in the mature seed; but in some species (e.g. *Pinus*) it is retained and visible at maturity as a thread-like tissue. In many dicotyledonous species, the pro-embryo typically undergoes rapid divisions to pass through several, usually well-defined stages, referred to as globular, heart (appearance of two cotyledonary lobes) and torpedo, related to the general shape of the growing embryo (**see: Development of embryos – cereals; Development of embryos – dicots**). When organogenesis is completed, histogenesis (histodifferentiation) usually takes place, to give most, but not all embryos protoderm, ground meristem and procambium.

In general, the mature **angiosperm** embryo consists of an embryonic axis, with either a single cotyledon (monocotyledons), or a pair of cotyledons (dicotyledons). There are apparent exceptions, e.g. in some species of Apiaceae and Ranunculaceae (dicots) the cotyledons are fused at maturity (pseudomonocotyledonous). In angiosperms, the number of cotyledons rarely exceeds two, though individuals of some species may develop three or more cotyledons (e.g. *Magnolia grandiflora*), and in other species three or more cotyledons may be the norm (e.g. *Degeneria vitiensis*, *Cola acuminata* – **see: Cola**). **Gymnosperm** embryos frequently have more than two cotyledons, sometimes as many as 12 (they are polycotyledonous), e.g. *Pinus pinea*, depending on species. The embryonic axis bearing the cotyledons usually shows a polarity from the earliest stages of embryogenesis, with the (proximal) end towards the **micropyle**, being the **radicle** containing the root meristem, and that at the distal end the shoot meristem (plumule), sometimes with recognizable leaves. The radicle is usually adjacent to the micropyle, through which it emerges on completion of germination. The first shoot segment above the cotyledons is the **epicotyl**, and the region of the axis between the radicle and the point of insertion of the cotyledons is the hypocotyl. While the cotyledons are often well developed and serve as food storage organs in **non-endospermic** seeds (see below), and may contain similar or different reserves from the **endosperm** in endospermic seeds, there are examples where other organs, usually the hypocotyl, are adapted to provide embryonic food storage, e.g. *Bertholletia excelsa* (**Brazil nut**).

In all embryos that depend on tissues external to them (e.g. endosperm) for nutrition, particularly after germination, the cotyledons become absorptive organs. In most dicotyledons, there is no particular morphological specialization for this function, but in many Loranthaceae the cotyledons fuse to form a conical absorptive organ, which exhausts the endosperm following germination and dies with it. Adaptation of the cotyledon to form some kind of absorptive organ or haustorium is much more apparent in the monocotyledons, such as in asparagus and date (*Phoenix*), or the **scutellum** of grasses, e.g. that of **wild oat**, which expands and almost fills the starchy endosperm as it is hydrolysed. (**See: Cereals; Mobilization of reserves – a monocot overview; – cereals**)

In the grasses (Poaceae), the embryo of the mature 'seed' (**caryopsis**) has been well described, at least in the economically important cereals (Fig. E.2). In all of this advanced family the embryo appears to be highly differentiated and provides a number of taxonomically useful characters, including scutellum shape. The scutellum is a highly specialized secretory and absorptive organ, generally held to be derived from the cotyledon; this is disputable, however, and some have interpreted it as homologous with the first foliage leaf, or even the true embryonic axis. Involved in secretion and absorption, it has extensive vascularization and epithelial cells adjacent to the endosperm. These have dense cytoplasm and are tightly packed and radially elongated, often with heavily folded cell walls to increase surface area ('transfer cells'). The shoot apex is enclosed within and protected by a sheath-like leaf – the **coleoptile**, possibly homologous to a foliar leaf, but that interpretation is challenged by some authors. Similarly, the radicle or primary rootlet is enclosed by a sheath, again of disputed homology – the **coleorhiza**, which is presumed to protect the radicle from mechanical damage as it emerges into the soil, and may also protect it from transient dehydration. The epiblast is a scale-like appendage found opposite the scutellum in some members of the Poaceae, interpreted by some as the vestigial second cotyledon.

The variation in embryogenesis among species results in a wide variation in disposition, size, shape and degree of differentiation of the embryo at seed maturity. At one extreme, embryos are very well developed, occupying all or most of the seed, and having easily recognizable root (radicle) and shoot (plumule) with large fleshy (Fig. E.3, and **see Legumes,** Fig. L.1) or folded cotyledons (Fig. E.4) (modified first leaves packed with food storage compounds) and little or no endosperm (see below). In dicot seeds, cotyledons can show early differentiation into spongy and palisade mesophyll regions or layers. A root cap is usually present on the radicle. Examples of this type are found in the Fabaceae, *Impatiens* (lateral root primordia present), *Ceratophyllum* (elongated plumule with several leaf primordia) and *Nelumbo* (young leaves differentiated into petiole and lamina) (Fig. E.5). At the other extreme, the mature seeds of a number of groups have embryos that are relatively under-developed, being small, or undifferentiated, or both, together with a relatively large amount of endosperm (e.g. Annonaceae, Aquifoliaceae,

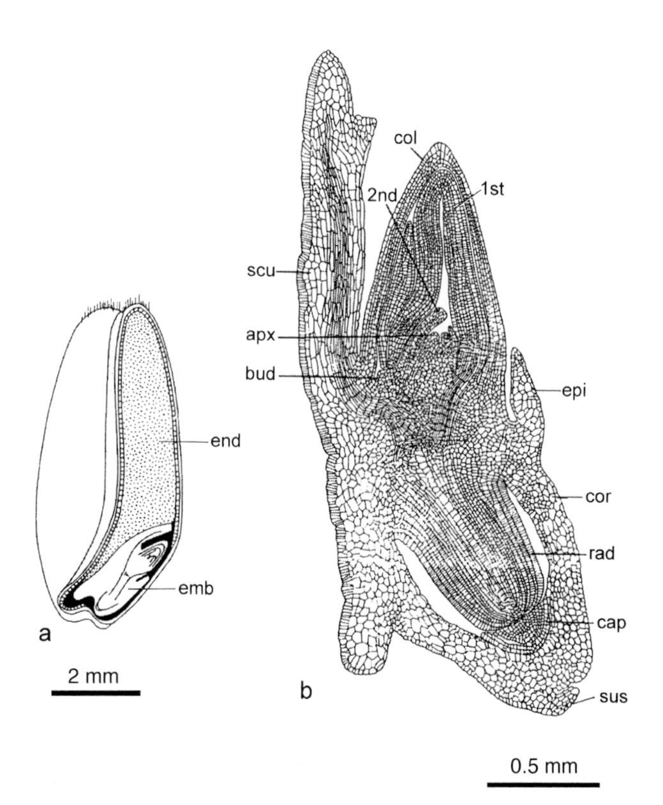

Fig. E.2. Fruit (caryopsis) and embryo of *Triticum* (wheat, Poaceae). (a) Longitudinal section of caryopsis in the region of the crease. Adapted from Esau, K. (1965) *Plant Anatomy*, 2nd edn. John Wiley, New York. (b) Longitudinal section of embryo. Adapted from Hayward, H.E. (1938) *The Structure of Economic Plants.* Macmillan, New York. apx = Shoot apex; bud = bud in axil of coleoptile; cap = root cap; col = coleoptile; cor = coleorhiza; emb = embryo; end = endosperm; epi = epiblast; rad = radicle; scu = scutellum; sus = remains of suspensor; 1st = first foliage leaf; 2nd = second foliage leaf.

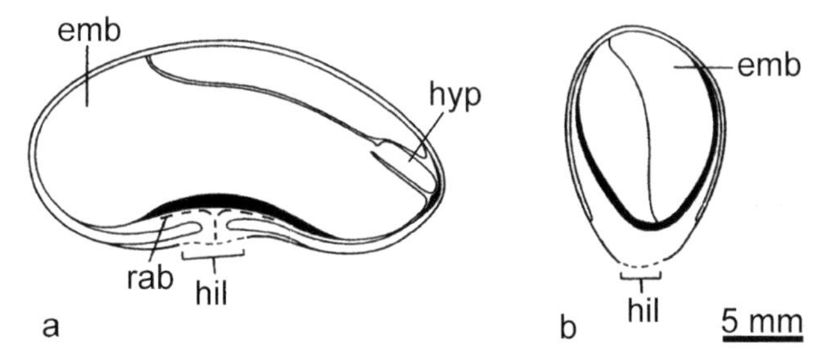

Fig. E.3. *Amanoa bracteosa* (Phyllanthaceae). Transmedian longitudinal section (a) and cross section of seed (b) showing the large storage embryo with two cotyledons (from Stuppy, W. (1996) *Systematic Morphology and Anatomy of the Seeds of the Biovulate Euphorbiaceae.* PhD Dissertation, Dept of Biology, University of Kaiserslautern, Germany [in German].). emb = Embryo; hil = hilum; hyp = hypocotyl; rab = raphe bundle.

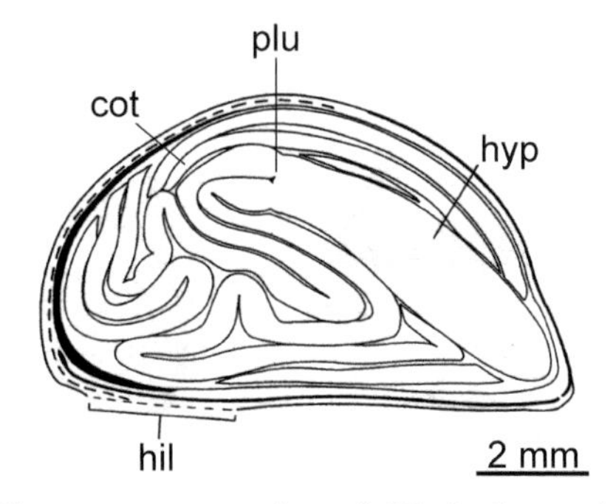

Fig. E.4. *Discocarpus essequeboensis* (Phyllanthaceae). Transmedian longitudinal section of seed showing the large embryo with folded cotyledons (from Stuppy, W. (1996) *Systematic Morphology and Anatomy of the Seeds of the Biovulate Euphorbiaceae.* PhD Dissertation, Dept of Biology, University of Kaiserslautern, Germany [in German].). cot = Cotyledons; hil = hilum; hyp = hypocotyl; plu = plumule. Dotted line is the vascular tissue in the testa. (**See** also Fig. S.23 of **Seedcoats – structure** for a diagram of the cotton seed, which has similarly much-folded cotyledons.)

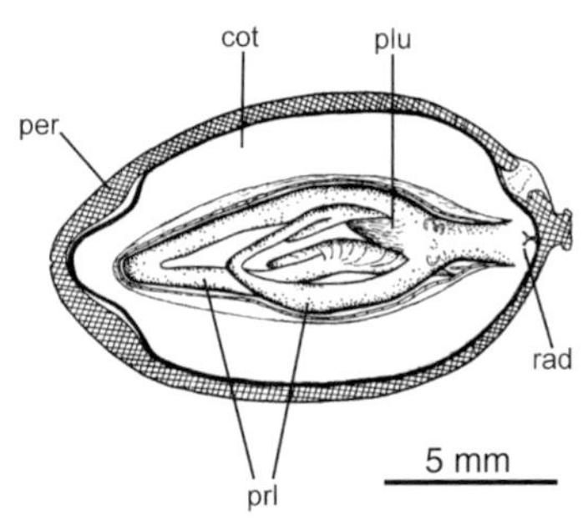

Fig. E.5. *Nelumbo nucifera* (Nelumbonaceae). Sacred lotus. Transmedian longitudinal section through nutlet showing the peculiar-shaped embryo: two basally-connate (fused) cotyledons arise as separate lobes from an annular primordium forming a sheath around the green plumule in the mature seed. The primary leaves are very well developed with their long petioles bent in the middle (from Takhtadzhyan (= Takhtajan), A. (ed.) (1988) *Anatomia Seminum Comparativa*, Vol. 2. Nauka, Leningrad [in Russian].). cot = Cotyledon; per = pericarp; plu = plumule; prl = primary leaves; rad = radicle.

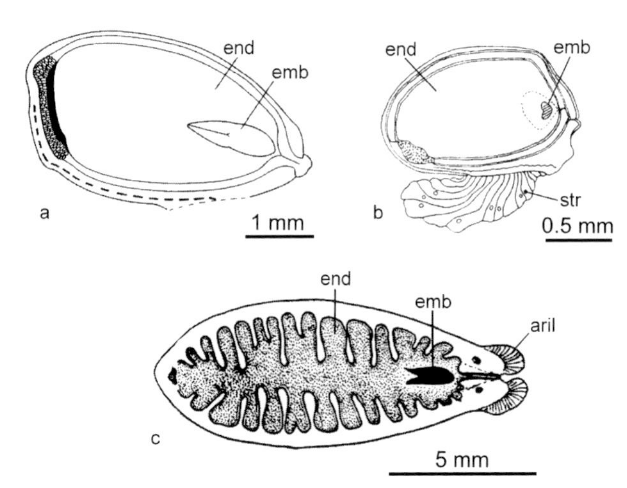

Fig. E.6. Seeds with small embryos embedded in copious endosperm. Transmedian longitudinal sections of: (a) *Centroplacus glaucinus* (Pandaceae) (from Stuppy, W. (1996) *Systematic Morphology and Anatomy of the Seeds of the Biovulate Euphorbiaceae.* PhD Dissertation, Dept of Biology, University of Kaiserslautern, Germany (in German).); (b) *Chelidonium majus* (Papaveraceae) (from Takhtadzhyan (= Takhtajan), A. (ed.) (1988) *Anatomia Seminum Comparativa*, Vol. 2. Nauka, Leningrad. [in Russian]; original from Crété, P. (1937) Etude sur la strophiole du *Chelidonium majus* L. *Bulletin de la Societé botanique de France* 84, 196–199.); (c) *Annona squamosa* (Annonaceae), also showing rumination (irregularity) of the endosperm (from Periasamy, K. (1962) The ruminate endosperm: development and types of rumination. In: *Plant Embryology: a Symposium.* CSIR, New Delhi, pp. 62–74). aril = Exostome aril; emb = embryo; end = endosperm; str = strophiole.

Magnoliaceae, Papaveraceae and Ranunculaceae) (Fig. E.6, also **see: Seedcoats – dispersal aids,** Fig. S.9b). Cotyledons are frequently underdeveloped (e.g. *Aquilegia*), or even lacking completely (e.g. **orchids** and Orobanchaceae), development being arrested at the heart and globular stage, respectively. Rather than being primitive, little-developed embryos are often the result of secondary reduction, with the least developed embryos of all being found in seeds of saprophytic, mycotrophic or **parasitic plants**. In some species the state of development at maturity may vary among the organs of the embryo; e.g. in *Arabidopsis thaliana* the radicle, hypocotyl and cotyledons are relatively well developed and differentiated, but the plumule consists of as few as 2–4 cells with no histological differentiation. Seeds with under-developed embryos are likely to exhibit morphological dormancy, with the embryos requiring weeks or months of slow post-dispersal development before germination can proceed normally. By contrast, seeds with well-developed embryos usually complete germination relatively more quickly under appropriate environmental conditions, if they are non-dormant or have only relatively weak physiological dormancy (**see: Dormancy**). When the embryo is extremely well developed, with advanced differentiation of leaves, buds, roots and vascular system it may be able to by-pass the usual resting or quiescent period altogether, so that the seedling rapidly becomes well established. This trend culminates in **vivipary** seen, for example, in some genera of the tribe Bambusoideae and mangroves such as *Rhizophora*.

The most important criteria for description of the internal morphology of mature seeds are relative size, shape and position of the embryo. Martin (1946) proposed a classification system (Fig. E.7), based on the gross internal morphology of the seeds of over 1000 genera, that is still widely used. His system is a quantitative one, with respect to the relative size of embryo (and reciprocally the endosperm). It consists of five size-designations, representing approximately quarter-unit volumetric proportions of embryo to endosperm: small (less than 25%), quarter (25%, or more, but less than half), half (50% or more, but less than 75%), dominant (75% or more), and total (100%). Martin's quantitative classification thus avoids applying vague terms for embryo size, nor does it automatically accept that a small embryo implies copious endosperm and copious endosperm a small embryo. According to differences in shape, size or position of the embryo, Martin's classification of seeds according to their internal morphology distinguishes 12 seed types, falling into two divisions, as outlined in Table E.1.

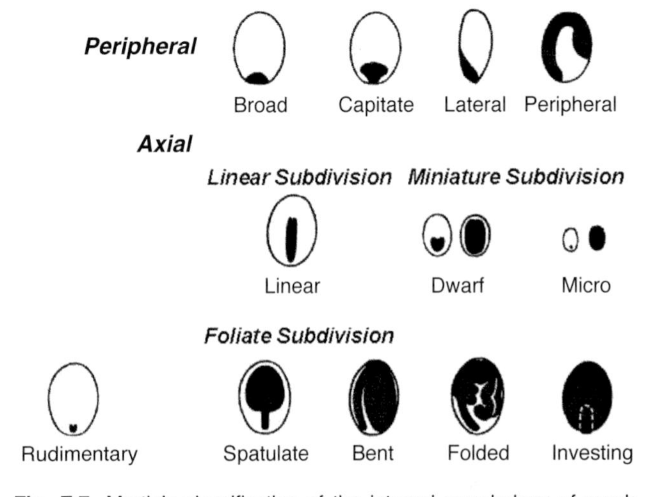

Fig. E.7. Martin's classification of the internal morphology of seeds. Based primarily on the position, size and shape of the embryo, Martin (1946) distinguished 12 seed types which he assigned (apart from one, the Rudimentary Type) to two divisions, Peripheral and Axial, the latter of which is further split into three subdivisions, Linear, Foliate and Miniature. For details of the definition of the 12 types, see Table E.1.

Table E.1. Seed internal morphology – embryo types according to Martin (1946).

Division	Subdivision	Type
A. PERIPHERAL: Embryo small and restricted to the lower half of the seed or elongate and large, quarter to dominant; endosperm (or perisperm in the Peripheral type) abundant and generally starchy; cotyledons minute or larger and then either narrow or expanded.	None recognized	**1. Broad:** Embryo as wide as or wider than high, peripheral or nearly so; Monocotyledons and Dicotyledons. **2. Capitate:** Embryo expanded above into a head-like form; Monocotyledons only. **3. Lateral:** Embryo basal-lateral or lateral, inclined to expand in the plane of the periphery; small or half or rarely larger. This type includes only the Poaceae, but it represents much diversity in embryo size. **4. Peripheral:** Embryo ordinarily elongate and large, quarter to dominant, contiguous in part at least to the testa and often curved; perisperm conspicuously starchy, central or in a few instances lateral; in several cases one of the cotyledons is abortive; Dicotyledons only.
B. AXILE (or Axial): Embryo small to total, central (axial) straight, curved, coiled, bent or folded; endosperm not starchy except in five Monocotyledon families of the Linear type. Well represented in Gymnosperms, Monocotyledons and Dicotyledons.	**Linear:** see characters for Linear type, the only one in the subdivision.	**5. Linear:** Embryo generally several times longer than broad, straight, curved or coiled; cotyledons not expanded; seeds normally not minute.
	Miniature: Seeds small to minute with embryos that are stocky or minute; seed covering generally delicate and often cellular-reticulate; endosperm not starchy.	**6. Dwarf:** Embryo variable in relative size, small to total, generally stocky, usually oval to elliptic or oblong, cotyledons inclined to be poorly developed; seeds small, generally 0.3–2 mm long exclusive of testa, often nearly as broad as long. **7. Micro:** Seeds minute, usually less than 0.2 mm long exclusive of testa, generally globular and consisting of relatively few cells, approximately 50–150 within the testa; embryo minute to total.
	Foliate: Embryo large, generally quarter to total, central rather than peripheral; cotyledons expanded; seeds generally medium to large; endosperm not starchy.	**8. Spatulate:** Embryo erect; cotyledons variable, thin to thick and slightly expanded to broad. **9. Bent:** Embryo spatulate but bent in jackknife form; cotyledons generally thick. **10. Folded:** Embryo with cotyledons usually thin, extensively expanded and folded in various ways. **11. Investing:** Embryo erect and with thick cotyledons overlapping and encasing the somewhat dwarfed stalk for at least half its length; endosperm wanting or limited.
Rudimentary (not formally assigned the status of a division by Martin)	None recognized	**12. Rudimentary**: Embryo small, globular to oval-oblong; seeds generally of medium size or larger; cotyledons are usually rudimentary and obscure but sometimes they are evident, making the embryos appear like miniatures of the Linear or Spatulate types. The group is not entirely clear-cut since most of the families concerned have some genera that merge into the Linear type and a few incline towards the Broad. Both in Monocotyledons and Dicotyledons.

Peripheral, bent and sometimes also partly folded embryos are usually the result of a curvature of the longitudinal axis of the **ovule** and seed. Such ovules and seeds with a curved longitudinal axis, where the micropyle and **chalaza** are not at opposite ends of the seed, are known as **campylotropous** (see: **Reproductive structures, 1. Female**). Embryos other than straight can become much longer than the actual seed and give rise to a larger seedling, presumably with an improved chance of survival.

Embryos of many species are green (chlorophyllous) at least at some stage in their development, e.g. some members of the Fabaceae, Sapindaceae, Brassicaceae, Celastraceae, Ceratophyllaceae. Some mature embryos are green, e.g. *Acer*, *Viscum*, while in others there is a colour change late in maturation, e.g. those of *Arabidopsis thaliana* turn green at the torpedo stage, before mostly turning pale or brownish yellow at the final stage of **maturation**. Following emergence, re-greening of cotyledons may occur, particularly in those that emerge above the ground (**epigeal** seedlings). Perhaps surprisingly, those species having green embryos at maturity include some with hard, thick seedcoats or **endocarps**, e.g. *Pistacia vera* (Anacardiaceae).

While the **diploid** embryos of the seeds of most species result from the sexual union of male and female gametes (sperm and egg cells), about 400 species in over 40 families are able to reproduce asexually through seeds, by the process of **apomixis**. Apomixis is thought to have evolved independently multiple times from sexual ancestors, and the mechanisms leading to apomictic reproduction are diverse. The common feature is that **meiosis** (meiotic division) is by-passed, so that an unreduced egg cell develops into an embryo without fertilization (**parthenogenesis**), or the cells that surround the embryo sac may form a diploid embryo. In some apomictic species, the central cell of the embryo sac (containing the two polar nuclei) also initiates endosperm development autonomously, but in most apomicts fertilization of the central cell is required for normal seed development. The outcome of the various apomictic mechanisms is that the seeds formed contain embryos with a chromosome and genetic constitution identical to that of the parent plant.

Seeds of some species may have more than one embryo (**polyembryony**). This phenomenon is broadly classified into 'simple' and 'multiple' depending upon whether it results from the presence of one (simple) or more (multiple) embryo sacs in the same ovule. Within each category the polyembryony may have more than one cause, e.g. cleavage of the proembryo (simple), or several embryo sacs in the nucellus of the same ovule (multiple). (JD, WS)

(See: Development of embryos – cereals; Development of embryos – dicots; Development of endosperms – cereals; Development of endosperms – dicots)

Boesewinkel, F.D. and Bouman, F. (1984) The seed: structure. In: Johri, B.M. (ed.) *Embryology of Angiosperms*. Springer, Berlin, pp. 567–610.

Bouman, F. (1984) The ovule. In: Johri, B.M. (ed.) *Embryology of Angiosperms*. Springer, Berlin, pp. 123–157.

Johri, B.M., Ambegaokar, K.B. and Srivastava, P.S. (1992) *Comparative Embryology of Angiosperms*. Springer, Berlin.

Martin, A.C. (1946) The comparative internal morphology of seeds. *American Midland Naturalist* 36, 513–660.

Natesh, S. and Rau, M.A. (1984) The embryo. In: Johri, B.M. (ed.) *Embryology of Angiosperms*. Springer, Berlin, pp. 377–443.

Werker, E. (1997) *Seed Anatomy*. Gebrüder Borntraeger, Berlin, Stuttgart (*Handbuch der Pflanzenanatomie = Encyclopedia of plant anatomy*, Spezieller Teil, Bd. 10, Teil 3).

Embryo dormancy

A form of primary or secondary dormancy induced and maintained by physiological conditions (such as elevated **abscisic acid** content, enhanced sensitivity to abscisic acid, or low **gibberellin** content) within the embryo; sometimes referred to as 'physiological' dormancy. **(See: Dormancy – embryo)**

Embryo rescue

A procedure of cultivating a **zygote**, which is removed from the **ovule**, on synthetic media to produce a plantlet *in vitro*. Often, this procedure is employed for breeding new **hybrids** between distant cultivars, which would suffer from aborted embryogenesis in the seed, or **seedless fruit** varieties of certain species. Embryo rescue provides a way to circumvent the influence of maternal tissues or endosperm incompatibility that could detrimentally affect embryo development. Experimentally, the embryo rescue procedure is used for studying embryo development and for identifying external factors that influence **embryogenesis**.

Embryo sac

A structure within the **ovule** that contains the female gamete (egg) and several other cells, including those which after **fertilization** give rise to the **endosperm**. The seed plants' homologue of the **megagametophyte** of their prehistoric progenitors. **(See: Reproductive structures, 1. Female)**

Embryogenesis

The process of seed development to produce the final form of the **embryo** that has the capacity to germinate. Many embryos become **desiccation tolerant** during the late stages of development, and are **quiescent** in the mature dry state (**orthodox** seeds). In addition, they may have **dormancy**. Despite the same term, embryogenesis in plants is distinct from that in animals because the embryo formed in the seed does not possess the entire pattern of the mature plant body. The formation of the embryo begins with fertilization of the egg cell by the pollen nucleus creating one zygotic cell in the **ovule**. Beside the **zygote** cell, another fertilization takes place in the ovule, creating an **endosperm** founder cell through the fusion of one male gamete with two central (polar) nuclei. This process, called double **fertilization**, is conserved among flowering plants (**angiosperms**), and produces a **diploid** zygote and a **triploid** endosperm. Genetically speaking, the zygotic and endosperm nuclei are constituted with the identical genetic information, the only difference being the dosage of male and female **genomes**. The zygote has an equal contribution from the both parent genomes, but the endosperm has a double dosage of the female genome and a single dosage of the male genome. This genomic dosage, along with positional cues in the ovule, appears to create two distinct structures in the seed: the embryo and endosperm.

After fertilization, the zygote cell first divides through the transverse plane. This results in the formation of a basal (proximal) cell and a terminal (distal) cell. The terminal cell gives rise to the embryo proper, and the basal cell further divides in the same plane to form a column of cells called the **suspensor**. This connects the embryo proper to maternal ovule tissues, and is thought to function as an anchor as well as in nutrient transfer to the developing embryo. The embryo proper undergoes many cell divisions to form an embryo structure distinct to each species. (O-AO, HS)

(See: Chromosome number; Development of embryos – cereals; Development of embryos – dicots; Development of seeds – an overview; Embryo; Somatic embryogenesis)

Embryoless seed

Various **umbellifer** crops (Apiaceace) are prone to producing normal-looking seeds in which the embryo is replaced with a cavity while the endosperm and seedcoat remain undamaged. This injury is caused by various species of *Lygus* insect bugs, which feed on the developing seeds. **(See: Disease)** Another example of the diseased formation of a seed without an embryo arises from the parasitosis of the infertile Douglas fir ovule by the developing egg of the chalcid insect, *Megastigmus spermotrophus*, which induces the megagametophyte to develop into storage tissue (**see: Gymnosperm seeds, I. Anatomy and morphology**)

Embryonic axis

Part of the embryo that bears the root and shoot meristems and the structures in between excluding the **cotyledons** or **coleoptile**: such as, variously, the **hypocotyl**, **epicotyl** and **radicle**.

Emergence

Radicle (or embryo) emergence through the seedcoat is in practice the first visible sign that germination is completed, and subsequent events are associated with seedling growth. By contrast, seed analysts define germination as the radicle emergence event itself (and do not include the events leading up to this), although germination cannot be confirmed until a normal seedling has been produced – **see: Germination testing**. However, in seeds of some species, the hypocotyl, coleoptile or cotyledons are the first part(s) of the embryo to emerge. (**See: Germination**)

Seedling, crop or field emergence refer to the appearance of aerial parts of the seedling above the surface of the soil (in practical terms, what many farmers and growers regard as the end point of 'germination'). (**See: Establishment**)

Emmer

Emmer is the domesticated version of the wild tetraploid wheat *Triticum turgidum* (AABB, **see: Wheat,** Table W.1 and **Einkorn,** Fig. E.1). Emmer wheat was domesticated in roughly the same region and same time frame as **einkorn** wheat in the Fertile Crescent. The progenitor species often occur together in wild stands in the Middle East and are not easily distinguished from each other. Emmer has a tough **rachis** and so does not **shatter** at maturity, but has a **hull** that is not easily removed during threshing. Durum wheat (**see: Wheat**) is likely derived from cultivated emmer wheat and is free-threshing. Durum replaced much of the emmer in the drier areas of the Middle East about 2000 years ago, while the hexaploid common wheat became dominant in the moister areas and cooler regions in ancient times. (**See: Domestication**)

Emmer is not grown commercially to a significant extent and there is no significant breeding effort in this crop. It is grown by organic farmers and for historical reference. The durum wheat that has been derived from emmer is widely grown for the production of semolina which is used in pasta and noodle production. (DF)

Leonard, W.H. and Martin, J.H. (1963) *Cereal Crops*. Macmillan, New York, USA.

Simmonds, N.W. (ed.) (1976) *Evolution of Crop Plants*. Longman, New York, USA.

Stallknecht, G.F., Gilbertson, K.M. and Ranney, J.E. (1996) Alternative wheat cereals as food grains: einkorn, emmer, spelt, kamut, and triticale. In: Janick, J. (ed.) *Progress in New Crops*. ASHS Press, Alexandria, VA, USA, pp. 156–170.

Encapsulation

Somatic embryos and shoot buds are encapsulated by **coating** them in alginate, which forms a semi-solid matrix around these propagules. To improve rates of germination or rooting of such propagules, a variety of substances may be added to the alginate, including macronutrients, micronutrients, carbohydrates (sucrose), as well as fungicides and other antimicrobial agents. The final product is an artificial or **synthetic seed.** (**See also: Encrusting**)

Encrusting

A type of seedcoating **enhancement**, also variously called 'mini-pelleting' or '**encapsulation**' or 'coating', in which materials are applied in sufficient quantities to make irregularly shaped seed more round, to upgrade size grades or to add weight (such as for **aerial** or **broadcast seeding**), or to apply nutrient, **rhizobia** or pesticide seed treatments. Encrusted seed are similar products to pelleted seed, except that in the former the original seed shape is still more or less visible whereas in the latter proportionally more material is added to produce still rounder shapes that disguise the underlying seed. Their manufacturing technologies are also closely related. (**See: Coating; Pelleting and encrusting**) (PH)

Endive

See: Cichorium (Endive and Chicory)

Endocarp

The inner of three layers of the **pericarp** that may be differentiated into thick, hard tissue as in **drupes**. (**See: Epicarp; Mesocarp; Nuts; Seedcoats – structure**)

Endoplasmic reticulum (ER)

A network of membranes, the ER or endomembrane system, throughout the cell cytoplasm. It is present as rough or smooth ER depending on whether or not protein-synthesizing ribosomes are arranged on the membrane surface. (**See: Cells and cell components**)

Endoreduplication

The replication of **DNA** without subsequent mitosis and cell division. Particularly studied during maize **endosperm** development, but also occurs during and after germination in **embryos** and living endosperms (**see: Cell cycle; Development of endosperms – cereals; Germination**).

Endosperm

The nutritive tissue in the developing and mature seeds of the **Spermatophytes**. In **gymnosperms** the endosperm is represented by the haploid **megagametophyte** (also called 'primary endosperm'), while in **angiosperms**, it is a triploid (3n) tissue (also called 'secondary endosperm') as the result of the **double fertilization** (**embryo sac**).

Endosperm generally develops from the fusion of the two polar nuclei (sometimes more) with one of the male nuclei from the pollen tube, and is usually triploid (3n), although n = 5 in *Lilium*, where one of the two polar nuclei is triploid and the other **haploid** (**see: Reproductive structures**). The fusion nucleus is often called the primary endosperm nucleus. Frequently, though not always, divisions of the fusion nucleus begin before those of the zygote. Three main types of endosperm formation have been recognized. In the nuclear, or non-cellular type, free nuclear division gives rise to many nuclei, and walls may or may not form afterwards. In the cellular type, cell walls form immediately after the first nuclear division and continue to do so as the endosperm grows. The third type, sometimes regarded as intermediate, is restricted to some **monocotyledons** and known as 'helobial' (from Helobiae). Here, the embryo sac is divided into two unequal chambers after the first mitosis, with the larger one (**chalazal**) developing into non-cellular endosperm, and the smaller (**micropylar**) having a rather variable fate. Deviations from these main patterns include *Cocos* (**coconut**), where the free nuclei initially suspended in a clear fluid later become associated with cytoplasm as free spherical cells. Migration of these cells (and any remaining free nuclei) to the periphery is followed eventually by cellular endosperm formation, with the central cavity remaining filled with sap (coconut milk). Liquid endosperms at maturity also occur in some Poaceae (e.g. *Limnodea arkansana*, Ozark grass), and several members of the grass tribe Aveneae.

The structure of fully developed endosperm in the mature seed is variable and has not been as widely investigated as has its early development. In the **dicotyledons** and some monocotyledons the endosperm of the mature seed usually consists of living cells in which nuclei can be distinguished. In the cereals of the Poaceae only the outermost cell layer of the endosperm (the **aleurone layer**) is alive, and the majority of the cells which contain storage **starch** and **storage protein** are dead.

In the early developing seed, the endosperm consists of thin-walled cells with large vacuoles and no reserve materials. The endosperm in many species is partly or completely degraded, and the products of this absorbed by the developing embryo, leaving none at maturity, or a restricted (sometimes referred to as scanty, or residual) endosperm of only one or a few cell layers, e.g. lettuce, tomato, soybean and some cacti (**see: Germination – radicle emergence; Non-endospermic seed**). In many other cases, e.g. *Ricinus* (**castor bean**), (endosperm surrounds the embryo) and the Poaceae (endosperm adjacent to the embryo in the typical cereal arrangement), the endosperm functions as storage tissue. In the endosperm of the former, the stored substances are mainly **oils** and storage proteins, in the latter starch grains and proteins. The endosperm often consists of thin-walled cells, with the reserve material located within the cell, but in some species there are thick-walled cells, with the walls themselves constituting the reserve substance as described below (**see: Hemicellulose**). For information on the relative sizes of endosperm and embryo **see: Embryo**. Also **see: Perisperm; Rumination**.

Proteins in the endosperm are sequestered in **protein bodies**, but in non-living endosperms such as of those of many of the cereals, these bodies lose their outer membrane as the seed matures and dries, and the protein forms an amorphous matrix (e.g. **glutens** in wheat). The protein bodies remain intact in the outermost (living) layer of endosperm cells, the aleurone layer, although the type of storage protein in this tissue is different from that in the non-living starchy endosperm. In *Ricinus*, the protein bodies occur throughout the whole endosperm.

The outermost cell layers of endospermic legumes, e.g. **fenugreek**, have also been called the aleurone layer. Apart from in this outer layer the cell walls of the endosperm are so thick with hemicellulose that the cell contents become occluded and the mature cells are dead. The origin of the aleurone layer in the Brassicaceae has been the subject of controversy, but it now appears to originate from the endosperm, and not the inner integument, as was once contended. Where endosperm contains little or no starch there may be abundant oils (triacylglycerols), in **oil bodies** (or oleosomes), often together with protein bodies. While the storage of oil and starch are often held to be mutually exclusive, there are examples where it is not the case, e.g. in Myristicaceae the endosperm stores both starch and oil as reserve materials, as do some storage cotyledons (**see: Embryo**) e.g. those of **peanut** and **soybean**. In soybean, starch is initially synthesized, and as development proceeds much of it is mobilized and the resultant sugars used to synthesize oils. When cell walls are the reserve material, they usually consist of hemicelluloses, frequently **mannans** and related **carbohydrate** polymers, e.g. in some Arecaceae such as ivory nut (*Phytelephas macrocarpa*) (**see: Jewellery, arts and crafts**), the endospermic legumes, Zygophyllaceae, *Diospyros*, *Strychnos* and many liliaceous families, as well as in the cotyledons of *Impatiens*, *Lupinus* (and many other Fabaceae) and *Tropaeolum*.

In *Phoenix* (date) endosperm the walls may have in addition up to 6% cellulose, and these cells have a very thick secondary wall. In seeds with hard endosperms the cell walls are thickened in the region surrounding the **cotyledons** (**lateral endosperm** region), but in the region surrounding the radicle (**micropylar endosperm**) the walls are usually thin (e.g. in fenugreek, Chinese senna, **carob**), allowing for easier penetration of the radicle. In the date, the radicle penetrates by pushing out a flap (**operculum**) in the endosperm (**see: Structure of seeds – identification characters**, Fig. S.70), due to the presence of a ring of thin-walled cells in the micropylar region in the otherwise copious thick-walled cells.

Mucilaginous endosperm occurs in the seeds of some species, e.g. *Ceratonia* (carob), whose endospermic cell walls have stratified thickening. The endosperm cells are hard when dry, becoming more or less mucilaginous on imbibition, providing both a swelling and a nutritional tissue following germination. The endosperm of fenugreek, which has a high affinity for water because of the **galactomannan**-rich cell walls (**see: Hemicellulose**), is considered to play an important role in regulating the water balance of the embryo. Interestingly, many members of the Trifolieae, which have this type of mucilaginous, water-retaining endosperm are thought to originate in the dry regions of the Mediterranean (**see: Deserts and arid lands; Rumination**). (JD, WS)

(See: Development of embryos – dicots; Development of endosperms – cereals; Development of endosperms – dicots)

Dore, W.G. (1956) Some grass genera with liquid endosperm. *Bulletin of the Torrey Botanical Club* 83, 335–337.

Esau, K. (1965) *Plant Anatomy*, 2nd edn. John Wiley, New York, USA.

Werker, E. (1997) *Seed Anatomy*. Gebrüder Borntraeger, Berlin, Stuttgart (*Handbuch der Pflanzenanatomie = Encyclopedia of plant anatomy*, Spezieller Teil, Bd. 10, Teil 3).

Endosperm cap

Also called the micropylar region of the endosperm, or micropylar endosperm. It is the region of endosperm tissue enclosing the radicle tip such that in seeds of some species it acts as a restraint to radicle emergence (**see: Dormancy – coat imposed; Germination – radicle emergence**). Weakening of this region is required to permit germination (and to break dormancy) in seeds such as those of lettuce, tomato, tobacco and *Datura ferox*. Some seeds with thick-walled cells in the **lateral endosperm** have only thin, non-restricting walls in the endosperm cap, e.g. Chinese senna, carob and fenugreek (**see: Endosperm**). In date seeds, the endosperm cap is an **operculum**, which is pushed off as the radicle emerges. This definition does not apply to the endosperms of cereals since these do not surround the embryo.

Gong, X., Bassel, G.W., Wang, A., Greenwood, J.S. and Bewley, J.D. (2005) The emergence of embryos from hard seeds is related to the structure of the cell walls of the micropylar endosperm, and not to endo-β-mannanase activity. *Annals of Botany* 96, 1165-1173.

Enhancement

A general term in seed technology, used both in industry and in the scientific literature more or less synonymously with 'functional seed treatment', to describe the range of practical beneficial **treatment** techniques performed on seeds after harvesting and **conditioning** to improve their physical or physiological performance after sowing; *but excluding*, by consensus, those treatments that control seedborne **pathogens** or that are conventional crop protection chemicals, as considered in their own right. (For more on this distinction, **see: Treatment**) The term also describes the physical and biological responses to the application of these treatment techniques.

Broadly speaking, seed enhancements are aimed at: (i) improving germination or seedling growth, such as manipulating the **vigour** or physiological status of seeds by hydration treatments such as **priming, steeping, hardening** and **pregermination** or chemicals to trigger systemic acquired resistance (**see: Systemicity**) or environmental stress tolerance, or **antioxidants** (some authors use the term 'enhancement' narrowly in this sense of 'physiological enhancement'); (ii) facilitating the planting of seeds (see: **Planting equipment – agronomic requirements**), such as by **pelleting and encrusting**; (iii) delivering materials required at the time of sowing (other than pesticides), such as major, minor or micronutrients and inoculants, using techniques such as pelleting, encrusting or **filmcoating**; (iv) removing weak or dead seeds by non-traditional **conditioning** 'upgrading' techniques; (v) identifying or otherwise 'tagging' seeds with visible pigments or other detectable materials or marker devices, to help in traceability and **identity preservation**; and (vi) potentially in future, the pre-sale application of the chemical inducer agents that will be integral to Genetic Use Restriction Technologies (**see: GMO – Patent protection technologies**).

Enhancement techniques of most of these types are now extensively used in the cultivation of high-value, low-volume horticultural and ornamental crops. Amongst higher-volume crops, filmcoating is also widely applied, particularly in higher-value agronomic crop seed species such as cotton, maize, sunflower and oilseed rape, partly due to the advent of high-value crop protection treatments (see: **Treatments – pesticides**) and varieties (such as those transformed through **genetic modification**). Coating is well established for small-seeded legumes and some turf grasses, and priming techniques are beginning to be more widely used as seed value increases. In commercial practice, seed companies usually perform such treatments, frequently on a just-in-time basis, and quite often using proprietary methodologies; some are patented whereas others are not, but in either case detailed operational procedures are often kept secret. (PH)

Halmer, P. (2000) Commercial seed treatment technology. In: Black, M. and Bewley, J.D. (eds) *Seed Technology and its Biological Basis*. Sheffield Academic Press, UK, pp. 257–286.

Taylor, A.G., Allen, P.S., Bennett, M.A., Bradford, K.J., Burris, J.S. and Misra, M.K. (1998) Seed enhancements. *Seed Science Research* 8, 245–256.

Enzyme-linked immunosorbent assay

See: ELISA

Epicarp

The outer of three layers of the **pericarp** that may be differentiated into a skin-like tissue, as in **drupes**. (**See: Endocarp; Mesocarp; Nuts; Seedcoats – structure**)

Epicotyl

In seedlings that have an **epigeal** mode of **emergence**, the first shoot segment (internode) immediately above the **cotyledons**, at the apex of which is the **plumule** (shoot meristem).

Epigeal

Seedlings in which the **cotyledon** (**monocot**) or cotyledons (**dicot**) emerge from above the soil surface following germination, due to extension of the **hypocotyl** (Fig. E.8, Table E.2). In one form of epigeal seedling **establishment**, initially both the **radicle** and hypocotyl grow downwards to maintain contact with moisture in the surface layers of the soil, and then a hook forms in the hypocotyl to facilitate its upward growth. When above ground, the cotyledons often become photosynthetic as the stored reserves are being depleted. Eventually they shrivel and are shed from the seedling as the first true leaves are formed and the seedling grows.

Variations on this epigeal form of seedling establishment occur (Fig. E.8). In castor bean and coffee (dicots), the endosperm is also carried above the ground by the cotyledons as they consume its food reserves, and it may be days to weeks before the cotyledons expand and become photosynthetic. A similar pattern occurs in onion (monocot), although the absorptive tip of the single cotyledon remains buried in the endosperm while the rest becomes green and photosynthetic. In other species, seedling emergence is semi-epigeal (e.g. *Peperomia peruviana*, dicot), for one cotyledon remains below the ground as an absorptive structure, the **haustorium**, and the other emerges above the soil and becomes photosynthetic (by contrast **see: Hypogeal**).

Sometimes the term epigeal germination is used, particularly with respect to field emergence; but strictly this is an incorrect term since emergence from the soil is an indication of seedling establishment, and not of germination *per se* (**see: Germination**). (JDB)

Epistase

A cap-like structure (usually derived from the nucellus epidermis) occluding the **micropyle** in the mature seed, often stained deep reddish-brown by tanniniferous materials. The epistase has similar probable functions to the **hypostase**, i.e. to seal the micropylar opening to reduce dehydration of the seed and to prevent penetration by harmful bacteria and fungi. (**See: Structure of seeds**)

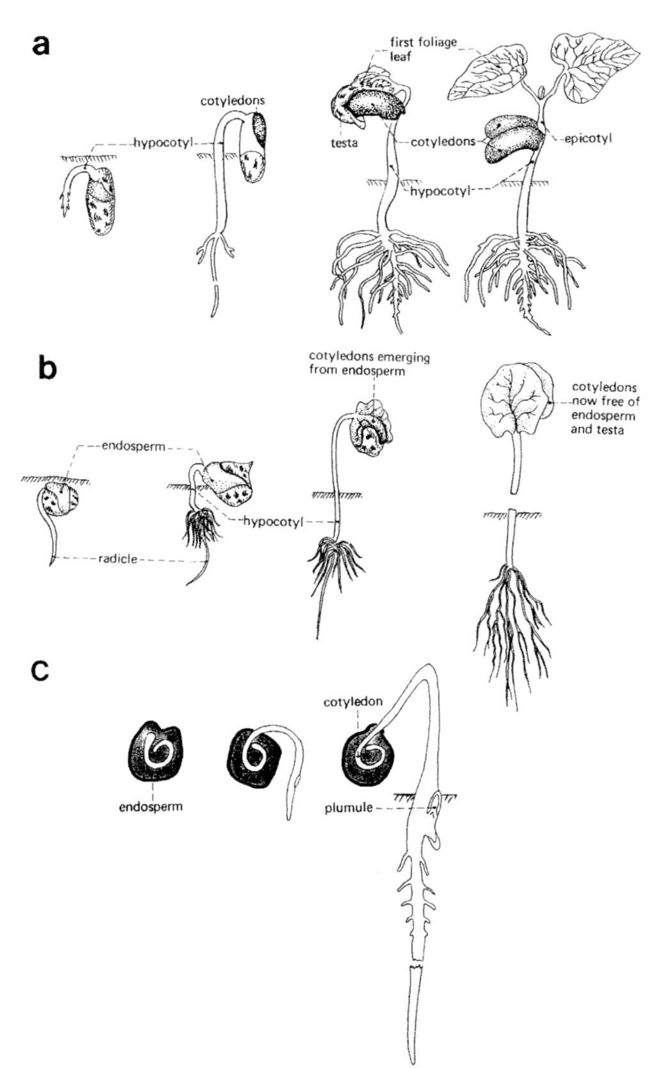

Fig. E.8. Examples of the epigeal form of seedling establishment. (a) *Phaseolus vulgaris* (common, French). The cotyledons swell only a little and turn green. They are shed when the storage reserves are depleted. (b) *Ricinus communis* (castor bean). The cotyledons become a leaf-like structure (seed leaves) and conduct photosynthesis until the first true leaves are produced by the shoot apex, when they die and are shed. (c) *Allium cepa* (onion). The cotyledon emerges above the soil but remains embedded in the endosperm, acting as an haustorium through which the hydrolysed storage reserves are imported into the growing seedling. It disintegrates after the reserves are depleted. Growth of the seedling is from the basal plumule.

Epistatic genes

Epistasis is when the effects of one gene (epistatic gene) override or mask the phenotype of a second gene. It is not the same as dominance since epistasis occurs when one gene masks the expression of a different gene for a different trait, whereas dominance is when one **allele** masks the expression of another allele of the same gene. **(See: Dormancy – genetics; Preharvest sprouting – genetics)**

Equilibrium moisture content

Seeds (with permeable **seedcoats**) are hygroscopic. When moved to conditions with different **relative humidity** and temperature (e.g. during **seed drying**), a point is reached

Table E.2. Examples of species whose seeds exhibit the epigeal form of seedling establishment.

Endospermic	Non-endospermic
Allium cepa (onion)	*Arachis hypogaea* (peanut)
Arabidopsis thaliana[a]	*Crambe abyssinica*
Coffea arabica (coffee)	*Cucumis sativus* (cucumber)
Fagopyrum escultentum (buckwheat)	*Cucurbita pepo* (pumpkin)
Glycine max[a] (soybean)	*Phaseolus vulgaris* (French bean)
Helianthus annuus[a] (sunflower)	*Sinapis alba* (mustard)
Lactuca sativa[a] (lettuce)	
Lycopersicon esculentum (tomato)	
Ricinus communis (castor bean)	
Rumex spp.	
Trigonella foenum-graecum (fenugreek)	

[a]Species with a non-storage endosperm, which is only one to a few cells thick.

where there is no net movement of moisture into or out of the seeds. At this point, the seeds' **water potential** matches that of the air and the corresponding percentage of moisture weight per seed weight is termed the equilibrium **moisture content** (% emc) for those atmospheric conditions. The % emc can be expressed as the percentage of moisture weight either per seed fresh (or wet) weight (% emc, fresh weight basis) or per seed dry weight (% emc, dry weight basis). At low % emc values, the two expressions are effectively the same for many practical purposes. The % emc value at a given **equilibrium relative humidity** and temperature will depend on the seeds' chemical composition and particularly on their oil content; the higher the seed oil content, the lower the % emc. (RJP, SHL)
(See: Sorption isotherms)

Equilibrium relative humidity

The percentage **relative humidity** at which a given **equilibrium moisture content** is expressed (at a specified temperature). Because seeds placed within a chamber with limited air will control the humidity that surrounds them, measurement of the equilibrium relative humidity and temperature within the chamber gives a non-destructive determination of the moisture status of those seeds. This moisture status is expressed either as seed % eRH, **water activity (see: Relative humidity)** or, by computation, **water potential**. If seed oil content is known, the equilibrium moisture content can be roughly estimated using the equation presented by Cromarty *et al.*

$$\text{Moisture content (\% dry weight)} = ((1 - X)\sqrt{(-440 \times \ln(1 - R))}) / (1.1 + (T / 90))$$

where X = seed oil content (weight of oil / dry weight of seed), R = relative humidity (decimal value) and T = temperature (°C). (RJP, SHL)
(See: Sorption isotherms)

Cromarty, A.S., Ellis, R.H. and Roberts, E.H. (1990) *The Design of Seed Storage Facilities for Genetic Conservation*. International Board for Plant Genetic Resources, Rome, Italy.

Eradicants

Chemical, **treatment** or phytosanitary measures applied to eliminate a pest or disease **pathogen** from the affected organism or environment, such as are used in seed treatment technology. Assessment of the degree of elimination of a seedborne pathogen is determined in practice by the sensitivity of the seed **health testing** procedure used. (**See: Treatments – pesticides, 1. Modes of action**)

Ergot

An infectious disease affecting grasses and the grains of cereal crops, especially rye, where it is caused by the ascomycete fungal **pathogen** *Claviceps purpurea*, which directly affects seed primordial development and seed **maturation**, leading to seed abortion and the replacement of the grain by large purple-black sclerotial 'ergot' bodies (**see:** illustration in **Sclerotia**). The sclerotia become deposited in the soil at the time of sowing, from where they may infect the next plant generation. Samples of seed may be visually examined for the presence of ergots during seed **health testing** and **purity testing**, which can sometimes be removed by **conditioning** techniques.

Ergot fungi produce indole alkaloids, based on lysergic acid, which cause assorted poisoning and noxious hallucinatory, convulsive and gangrenous effects, and death in cattle and humans that eat infested grain. 'Ergotism', whose true source was not deciphered until the mid-19th century, occasionally reached epidemic proportions in Europe in the Middle Ages, where it became known as St Anthony's Fire and caused much ignorant havoc and economic devastation in affected populations. Ergot extracts are used in prescribed medicines: ergotamine, for migraines and other causes of headaches, and ergonovine to induce contraction of the uterus and control haemorrhage after childbirth. (**See: Mycotoxins; Pharmaceuticals and pharmacologically active compounds**) (PH)

ESA

See: European Seed Association

Essential amino acids

Seeds are major sources of dietary protein for humans and livestock. (**See: Crop Atlas Appendix, Map 25**) Although **cereals** are relatively low in protein (on average about 12%) their vast production means that they account for about 70% of the total protein harvested, amounting to about 250 million t per year. In addition, over 100 million t of protein come from more protein-rich **legumes** and **oilseeds**. The most important non-cereal source of protein is **soybean**, which contains about 40% protein, with a total annual yield of about 60 million t. However, the total content of proteins is not the sole dietary consideration: protein quality is also important.

Whereas animals can derive energy from several types of compounds, a source of dietary protein is essential to enable them to grow and replace proteins lost by breakdown. This is because they are only able to synthesize about half of the 20 commonly occurring amino acids. The remaining ones must be provided in the diet and are therefore termed 'essential'. The only exception to this requirement is by ruminants (e.g. cattle), in which the microflora present in the rumen can synthesize all 20 amino acids. This can provide for the dietary requirements of the animal except under high production conditions. Providing a balanced diet of amino acids to non-ruminants is important, since if only one essential amino acid is limiting the remaining ones will be broken down. This leads to poor growth and release of nitrogen into the environment.

The ten essential amino acids are listed in Table E.3. Although cysteine and tyrosine are listed these are not truly essential since they can be synthesized from methionine and phenylalanine, respectively. Hence, combined values for cysteine + methionine and tyrosine + phenylalanine are often given. Similarly, histidine is often omitted from lists, as it is essential for human children but not adults.

Table E.3 also compares with the WHO (World Health Organization) recommended levels the proportions of essential amino acids in typical legume (soybean), cereal (wheat) and oilseed (oilseed rape) crops. The sulphur-containing amino acids cysteine and methionine are particularly low in legumes, and lysine and threonine in cereals. These deficiencies result from a low amount of these amino acids in the major storage protein fractions in seeds. Combining legume seeds (which are rich in lysine) and cereal grains (which have higher proportions of sulphur-containing amino acids) provides an adequate balance of amino acids, a strategy often adopted in the production of livestock feeds. (PRS)

(**See: Proteins and amino acids; Storage protein**)

Table E.3. Comparison of the essential amino acid contents of a typical legume (soybean), cereal (wheat) and oil-storing (rape) seed compared with the WHO dietary recommended levels. Values used vary but all are approximately equivalent to % by weight.

Amino acid	Soybean	Wheat	Oilseed rape	WHO recommended level
Cysteine	1.3	2.6	2.9	3.5
Methionine	1.3	1.3	2.1	
Lysine	6.4	2.0	5.9	5.5
Isoleucine	4.5	3.6	4.7	4.0
Leucine	7.8	6.7	7.3	7.0
Phenylalanine	4.9	5.1	4.1	6.0
Tryrosine	3.1	2.6	3.0	
Threonine	3.9	2.7	5.0	4.0
Tryptophan	1.3	1.1	1.2	1.0
Valine	4.8	3.7	6.0	5.0
Histidine	2.5	2.2	3.6	–

Essential fatty acids

The 18-carbon polyunsaturated linoleic and α-linolenic acids present in most edible seed **oils and fats**. These **fatty acids**, also termed omega-6 and omega-3 (ω3, ω6) fatty acids, are regarded as being essential for the human diet because they cannot be synthesized by humans and some other animals and yet are precursors of long-chain polyunsaturated fatty acids such as arachidonic acid, eicospentaenoic acid and docosohexaenoic acid. These very long chain fatty acids in turn play a vital role in human immune function, inflammatory responses and brain function. In addition, linolenic acid is thought to be directly involved in skin function in humans.

(AJK)

(**See: Oilseeds – major; Oilseeds – minor**)

Essential oil

Mixture of volatile constituents of plants generally obtained by steam distillation. Components are highly aromatic, consisting largely of benzene and terpene derivatives. They are present in seeds of many species. Contrast with **fixed oils**. (See: Spices)

Establishment

The stage at which seedling **emergence** from the soil or growing medium is complete (allowing for any subsequent post-emergence losses), usually assessed at the point of development of an agreed number of true leaves.

Ethene

An alkene gas with hormone activity that is produced by plant cells. Although often still described within the plant science community as **ethylene**, ethene is now the systematic name preferred by the International Union of Pure and Applied Chemistry (IUPAC).

Ethnobotany

1. Introduction

Ethnobotany is the study of traditional uses of plants by humans. It is a broad field, encompassing both the use of wild plants and of indigenous varieties of cultivated plants. As a field of study, ethnobotany may be said to have begun when the expanding European empires first colonized Africa, Asia and the Americas. Early colonial histories and travel books dating from the 16th to the 18th centuries are still a rich source of information on uses of plants, in cultures now disappeared or greatly altered. The aim of most ethnobotanical work in the last 200 years has been to draw on traditional knowledge for wider economic development, particularly through the development of new crops and medicines.

The Convention on Biological Diversity (The Rio Earth Summit of 1992) was a milestone in its recognition of national and local rights to plants, animals and indigenous knowledge. In the last decade ethnobotanists have learned to work in close partnerships with local communities, aiming to further sustainable harvesting of wild plants with benefits for local economies and for plant conservation. The methodologies of ethnobotany have also changed, with a strong emphasis on use of quantitative techniques drawn from **ecology**. While ethnobotany still straddles multiple disciplines, including botany, anthropology, nutrition and medicine, it is now emerging as a discipline in its own right. This article focuses on ethnobotanical aspects of wild foods as this represents the main use of seeds (here broadly defined to include fruits and **nuts**). Traditional management of seed crops is covered in the appropriate entries.

2. Nutritional value of seeds

The quality and quantity of plant foods varies considerably, but must be assessed in the context of overall consumption. For example, 'edible greens' (the leaves of wild plants) are low in energy, but are an important resource for vitamins, minerals and dietary diversity in the spring, when other foods are scarce. Energy values are also affected by the energy required for processing, such as shelling, dehusking or detoxifying nuts. Studies on hunter-gatherers rank seeds, nuts and roots and tubers roughly equal in terms of net energy yield, well below meat, but above foliage. In practice, choice of foodstuffs in foraging societies must balance the energy return of the food against the ease and predictability with which it can be obtained: unlike animals, plants are not mobile and, after harvest, can be easily stored.

There are also differences between wild and domesticated plants. The latter have often been selected for higher concentrations of carbohydrates, and reduced elements such as toxins that affect palatability. In contrast, wild foods are richer in fibre, vitamins, minerals and a range of phytochemicals, many with nutritional and medicinal significance (**see: Pharmaceuticals, medicines and biologically active compounds – an overview**).

The likely health benefits of wild foods, including wild animals which typically have fewer, healthier fats, have led some scholars to propose reversion to a 'Palaeolithic diet', replicating the characteristics of prehistoric hunter-gatherer diets. Ironically, such a diet would be expensive and difficult for most people to achieve today, given the high productivity of the carbohydrate-rich crops that are so central to our diet, and its likely impact on health is controversial.

3. Diet in foraging societies

Living hunter-gatherers have been much studied as case studies of pre-agricultural foraging societies. The most extensive studies are of the Kung San bushmen of the Kalahari desert, who collect fruits, roots and other parts from some 200 plant species (**see: Legumes, 1. Origins and domestication**). Overall, about 70% of Kung San dietary energy derives from plants (about half of this from the mongongo nut, *Schinziophyton rautanenii*) and the remainder from animals. It is likely that plants dominated hunter-gatherer diets, except in high latitudes where peoples such as the Inuit until recently subsisted almost entirely on hunting meat.

One major obstacle to the use of ethnographic data is that hunter-gatherers are now mainly restricted to areas too arid or too cold for farming. In areas like this, in which plant resources are scarce, they tend to follow plant resources as they ripen: people move, rather than resources. Such societies are highly mobile, and are often characterized by relatively little investment in management of wild resources. Observations of the ample leisure time available to Kung San bushmen led anthropologists to coin the phrase 'The Original Affluent Society' in the 1960s. However, older ethnographic records show that hunter-gatherer groups in more favourable environments (now occupied by farmers) were sedentary and territorial, with permanent villages forming the base from which resources were collected, and at which they were stored. Well-known examples are the salmon societies of British Columbia and the acorn-based economies of Californian native Americans in the 19th century. Like mobile hunter-gatherers, they consumed a wide range of species, though with some, such as wild grass seeds or nuts, often dominant. Archaeological evidence shows that sedentary hunter-gatherer societies were widespread prior to the beginning of farming.

4. Evolution of human diet

The changing dietary importance of seeds as food is thought to have played an important role in human evolution, although the exact nature and timing of dietary shifts and their importance for human behaviour are much debated and remain controversial. Three types of evidence are used: non-human primate diet, recent hunter-gatherer societies, and hominid fossils.

Current-day apes and monkeys obtain up to 95% of their dietary energy from plant foods, with the remainder derived from invertebrates, rather than meat. Lower quality, easy-to-acquire plant foods such as leaves are a major food resource; higher quality plant foods derive from ripe fruits, rather than the more difficult to acquire seeds, nuts and tubers. Overall, compared to apes and monkeys, the hunter-gatherer diet (and human diet in general) is characterized by higher quality, harder-to-acquire resources, including meat, roots, hard seeds and nuts, with a much smaller role for forage, ripe fruit and insects. Fossil evidence from teeth suggests that this dietary shift began with the first hominids, *Australopithecus* of 4–2 million years ago. The 'Expensive-tissue hypothesis' suggests that the shift to higher quality foods during human evolution is linked to increasing relative brain size. Since both the brain and the gut have high metabolic requirements, the relatively large brains of human primates could only evolve if the gut shrank – only possible through the consumption of higher quality foods. In addition to more sophisticated foraging techniques, hominids also used food processing tools and cooking to enhance nutrient availability, although there is no evidence for human control of fire prior to 400,000 years ago, and grinding stones first appeared in the time of anatomically modern *Homo sapiens*, from 20,000 to 5000 years ago in different parts of the world.

5. Wild seeds in human diet

The role of wild seeds in agricultural societies follows three patterns: integral staple, famine (or emergency) food, or luxury food. Prior to the widespread adoption of modern agricultural techniques in the 1940s, many rural communities depended on wild harvests as a major supplement to farm produce, for example grains of lyme grass in Iceland, or acorns in eastern Turkey. Wild grasses still maintain this role in large parts of Africa. Famine foods are those that are less preferred, perhaps because of taste and/or toxicity, but which become important at times of food shortage (**see**, for example: **Chickling pea**). Their role in famine relief has only recently been recognized by emergency nutritionists, which has led to a substantial increase in research into their nutritional quality and availability. Luxury foods are wild plants gathered because they are valued primarily for their taste properties or cultural associations; examples include sloe fruits (*Prunus spinosa*) for flavouring alcoholic drinks in Europe, and Australian plant foods, once staples for hunter-gathering aborigines, now high-value foods for urban consumers, e.g. the **macadamia** nut.

6. Grasses

The **caryopses** (grains) of grasses are easily collected, by beating stands of plants, and form a stable, storable food resource, mainly composed of **starch**, with **storage protein** contents of 5–10%. Although harvesting of wild grass seeds was often abandoned after the introduction of domesticated cereals, they are or were recently important staple foods in some farming communities, mainly in arid zones. On the fringes of the Sahara, *Panicum turgidum* (*merkba*, *afezu*), *Stipagrostis pungens* and *Cenchrus biflorus* (*kram-kram*) are used by the Tuareg and other peoples. In sub-Saharan Africa wild grains are harvested from seasonally flooded lakes and swamps, including species of *Panicum*, *Eragrostis*, *Paspalum* and *Echinochloa*, as well as **wild rice** species indigenous to Africa (e.g. *Oryza barthii* and *O. longistaminata*).

In 19th-century North America, wild grains were harvested from *Panicum sonorum* and *P. hirticaule* in the Sonoran desert, and *Panicum*, *Eragrostis* and *Oryzopsis* from the arid Great Basin of California. Here the Owens Valley Paiute sometimes irrigated wild stands of grasses. Today, only the wild rice indigenous to North America (*Zizania palustris*) is widely harvested, in the Great Lakes region (**see: Rice**). Both the wild form and the non-**shattering** form developed in the 1960s command high prices as a luxury food. In 19th-century Australia wild grasses such as *Panicum* and *Eragrostis* were harvested on a large scale in the arid interior. Wild grasses were important in some parts of Europe until the mid-20th century. Manna grass (*Glyceria fluitans*), named on account of its sweet taste, was extensively harvested from marshes in central and northern Europe. Crabgrass (*Digitaria sanguinalis*) was both collected and cultivated in eastern Europe, while lyme grass (*Leymus arenarius*) was an important food staple in Iceland.

Grass caryopses are usually well protected by **hulls** (the **lemma**, **palea** and **glumes**), which in part account for their keeping qualities. The husks were often made brittle by parching (contact with dry heat on a metal plate or in an oven), and then threshed with a flail or in a mortar and pestle. Wild grains were often consumed as a gruel or porridge, some ground into flour for bread.

7. Nuts

Botanists use the term nut to refer to one-seeded dry fruits, with a woody pericarp, such as the acorn, **hazelnut** (filbert) and beechnut. In everyday use, nut refers to any edible kernel within a hard shell, such as **walnut**, **Brazil nut**, **almonds** and **pine nuts**. Most nuts in trade are now derived from domesticated trees grown in plantations, but the relative late domestication of many nut trees, and easy availability of wild nut trees in their native habitat, means that wild nuts have not been entirely replaced by cultivated sources (**see: Nut**).

Both the acorn and **chestnut** are rich in starch, and have been staple foodstuffs. Wild chestnuts (*Castanea sativa*) were widely consumed in southern Europe, often mixed with wheat flour for bread. Acorns have a long history as a food staple, with archaeological finds dating to 21,000 years ago at Ohalo II in Israel, and were staple foods in North America, particularly in California, the Mediterranean and Near East, and Japan and Korea. Some species have sweet acorns, but most acorns contain bitter tannin and must be roasted or leached before use.

The other nuts have high **oil** contents (50–70% by weight). These nuts are more often highly appreciated supplementary foods than staples on the scale of acorns, although the hazelnut (*Corylus avellana*) is found in huge quantities in Mesolithic, pre-agrarian settlements in northern Europe. The **pecan** nut (*Carya illinoinensis*) was an important wild resource for native Americans in the southern USA. Many pine species bear edible nuts: the best known is the Mediterranean stone pine, *Pinus pinea*. Other nuts that were locally used, and are now traded, include *Pinus edulis*, in the southwest USA, and *P. koraiensis* in China. The similar nuts of another conifer, the monkey puzzle tree, *Araucaria araucana*, are collected in Chile. The best known nut of Australia is the macadamia nut (*Macadamia ternifolia* and *M. tetraphylla*), now cultivated, but with a long history of aboriginal use. Tropical nuts include the Brazil nut, *Bertholletia excelsa*, still mainly harvested from wild trees in Amazonia.

8. Fruit

As with nuts, domesticated fruits, whether juicy, starchy or oily, have largely displaced wild fruits as important elements of diet. The continuing role of wild fruits owes much to the ease with which they can be used as flavouring for food or beverages, or be dried or fermented as a means of storage. Nutritional properties such as high vitamin C are now less important. Not all fruits are sweet, and many wild fruits in the Rosaceae family contain cyanogens, which can be removed by cooking.

Collection of wild berries is still popular in Scandinavia and eastern Europe, where lingonberry (*Vaccinium vitis-idaea*), bilberry (*V. myrtillus*), cranberry (*V. oxycoccus*) and bearberry (*Arctostaphylos uva-ursi*), all in the Ericaceae family, are popular. Jam-making is a common use in Europe for sour fruits such as these, or brambles (*Rubus* sp.), or cornelian cherry (*Cornus mas*) in the Balkans. *Vaccinium* species were also used by native Americans, as cakes of fruit, or pounded with meat, but commercial cranberry (*V. macrocarpon*) and blueberry (*V. corymbosum*) are both domesticated today.

A very wide variety of fruits is used in other regions. Some well known examples include the noni fruit (*Morinda citrifolia*) of Polynesia, now marketed as a health food in North America, the latex-rich cow tree of Amazonia (*Brosimum utile*), and the baobab tree of western Africa (*Adansonia digitata*).

(MN)

(See: Archaeobotany; Poisonous seeds; Psychoactive seeds)

Cotton, C.M. (1996) *Ethnobotany: Principles and Applications*. John Wiley, New York, USA.

Etkin, N.L. (ed.) (1994) *Eating on the Wild Side: the Pharmacologic, Ecologic, and Social Implications of using Non-cultigens*. University of Arizona Press, Tucson, USA.

Harris, M. and Ross, E.B. (eds) (1987) *Food and Evolution: Toward a Theory of Human Food Habits*. Temple University Press, Philadelphia, USA.

Pieroni, A. (2005) Gathering food from the wild. In: Prance, G. and Nesbitt, M. (eds) *Cultural History of Plants*. Routledge, New York, USA, pp. 29–43.

Tanaka, T. (1976) *Tanaka's Cyclopedia of Edible Plants of the World*. Yugaku-sha, Tokyo, Japan.

Ethylene

Ethylene is a hormone made by all seed plants and also by some bacteria and fungi. It was discovered by Dimitri Neljubow in 1886 after he discovered that the gas used for lighting affected the growth of plants. It is the simplest of the alkenes with the chemical formula C_2H_4. Although **ethene** is now the systematic name preferred by chemists, plant scientists continue at present to refer to it as ethylene.

Ethylene is made in most tissues of plants. It may be induced developmentally, for instance in germinating seeds, seedlings (particularly in the plumular hook), in senescent leaves and ripening fruit. It is also induced by environmental stresses and other factors such as diurnal cycles, drought, flooding, high and low temperature, physical stresses (touch and pressure), wounding and attack by pests and pathogens. Finally, ethylene may be induced by auxin, which may be further enhanced by cytokinin and or ABA. Ethylene may also regulate its own synthesis either positively or negatively, depending upon the tissue. **(See: Germination – influences of gases)**

Ethylene is biosynthesized from 1-aminocyclopropane-1-carboxylic acid (ACC), which is in turn synthesized from S-adenosyl-L-methionine (SAM). In addition to being a methyl donor in numerous other enzyme reactions, SAM is the common precursor of both ethylene and **polyamines**. The first committed step in ethylene biosynthesis, therefore, is the conversion of SAM to ACC and 5′-methylthioadenosine. The valuable methylthio- group of methylthioadenosine is recycled back to methionine and SAM by the Yang cycle, named after Shang Fa Yang, who did much of the early work on it (Fig. E.9).

The formation of ACC is carried out by the enzyme ACC synthase (ACS), which requires pyridoxal phosphate as a cofactor (and hence is sensitive to inhibitors of pyridoxal phosphate such as aminoethoxyvinyl glycine and aminooxyacetic acid) and probably functions as a dimer. It is encoded by a small family of genes each member of which is regulated by different signals. It has been generally accepted that the production of ACC is the rate-limiting step in ethylene production. However, this may not be universally true because several treatments that promote ethylene production also induce ACC oxidase genes.

ACC oxidase (ACO), which was formerly known as ethylene-forming enzyme (EFE), carries out the final step in ethylene biosynthesis: the conversion of ACC to ethylene. It is a dioxygenase, requiring Fe^{2+} ions, ascorbate and molecular oxygen. There is also a requirement for carbon dioxide and the enzyme is inhibited by Co^{2+}. The reaction produces cyanide, which is detoxified to β-cyanoalanine. The *Aco* gene family is small and there is again some evidence that the genes are differentially induced by different stimuli.

When subjected to high concentrations of externally applied ethylene, plant cells will convert it to ethylene glycol via the intermediate of ethylene oxide. However, it is not clear that this happens at normal physiological concentrations of the gas. The important control is to remove the immediate precursor of ethylene by converting it to its malonyl conjugate (MACC). This seems to be irreversible *in planta*. Germinating peanut seeds that contain considerable amounts of MACC seem to make ethylene from newly synthesized ACC. However, inhibition of the enzyme that makes MACC, ACC N-malonyl transferase, does increase ethylene synthesis so malonylation is a method of regulating ethylene production. Exogenously applied ethylene-generating compounds, e.g. ethephon, can have practical applications in seed enhancement treatments (**See: Priming – technology, 2. Techniques and terminologies**).

(GL, JR)

(See: Signal transduction – hormones)

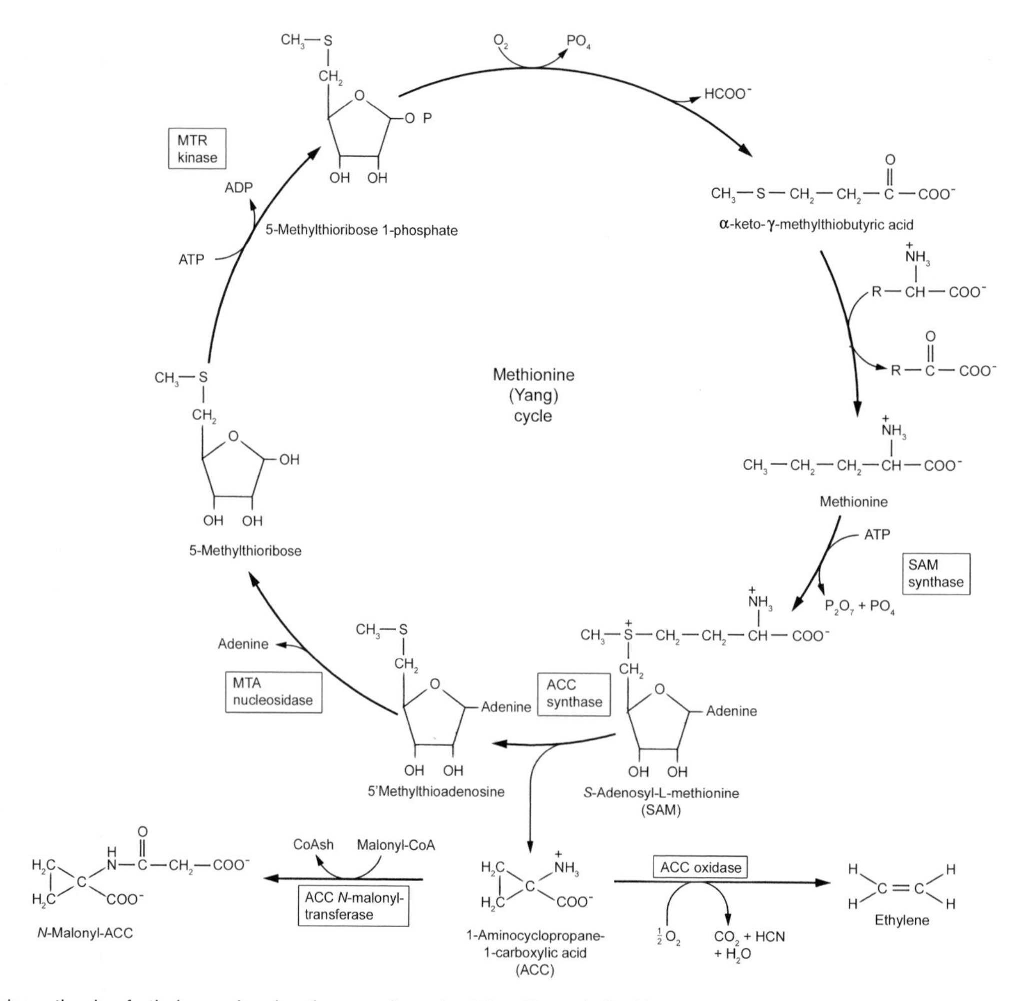

Fig. E.9. The biosynthesis of ethylene, showing the recycling of sulphur through the Yang cycle. Modified from Crozier, A., Kamiya, Y., Bishop, G. and Yokota, T. (2000) Biosynthesis of hormones and elicitor molecules. In: Buchanan, B.B., Gruissem, W. and Jones, R.L. (eds) *Biochemistry and Molecular Biology of Plants*. American Society of Plant Physiologists, Rockville, MD, USA.

Bleeker, A.B. and Kende, H. (2000) Ethylene: A gaseous signal molecule in plants. *Annual Review of Cell and Developmental Biology* 16, 1–18.

Imaseki, H. (1999) Control of ethylene synthesis and metabolism. In: Hooykaas, P.J.J., Hall, M.A. and Libbenga, K.R. (eds) *Biochemistry and Molecular Biology of Plant Hormones*. Elsevier, Amsterdam, The Netherlands, pp. 189–207.

McKeon, J.C., Fernández-Maculet, J.C. and Yang, S.F. (1995) Biosynthesis and metabolism of ethylene. In: Davies, P.J. (ed.) *Plant Hormones: Physiology, Biochemistry and Molecular Biology*. Kluwer Academic, Dordrecht, The Netherlands, pp. 118–139.

Evening primrose

Evening primrose (*Oenothera biennis*, 2n = 14, Onagraceae) is native to North America but was introduced into Europe in the 17th century, where it became a weed. Many seeds are borne within a 3 cm long pod or **siliqua**. The seeds are tiny, 1 mm long × 0.5 mm wide, flattened on three sides and rounded on the fourth (Fig. E.10). There is a thin, soft **seedcoat** surrounding the white **embryo** and **endosperm**. **Oil** content is about 24% of which some 9% is gamma linoleic acid (GLA), an important **essential fatty acid** (**see: Oilseeds – major**, Table O.2). The oil's main use is in the cosmetic industry and as a nutritional supplement. (KD)

Brandle, J.E., Court, W.A. and Roy, R.C. (1993) Heritability of seed yield, oil concentration and oil quality among wild biotypes of Ontario evening primrose. *Canadian Journal of Plant Science* 73, 1067–1070.

Fig. E.10. Seeds of evening primrose (scale = mm) (photograph kindly produced by Ralph Underwood of AAFC, Saskatoon Research Centre, Canada).

Evolution of seeds

The life cycle of land plants follows the basic pattern known as the alternation of generations. One generation, the **sporophyte**, is diploid and produces **haploid** spores by meiosis. The spores germinate to give a haploid **gametophyte** which bears the **gametes**. After fertilization of the female gamete by the male, a new sporophyte – the diploid number of chromosomes now restored – is formed (Fig. E.11). Two generations therefore alternate but one of them represents the dominant form of the species. During evolution, various modifications have occurred which can be traced as giving rise to the seed habit. These include: (i) the sporophyte assumes dominance; (ii) separate male and female spores (**micro-** and **megaspores**) are produced (**heterospory**), and hence separate male (micro-) and female (**mega-**) **gametophytes**; (iii) reduction in size and complexity of the gametophytes; (iv) retention of the megaspore (therefore the megagametophyte) on the sporophyte; and (v) loss of the necessity for water for motility of the male gamete and for fertilization. It is possible, from the fossil record, to allocate these events to approximate times in the geological past (Fig. E.12). These modifications, culminating in the evolution of the seed habit, had profound effects on the plant life cycle and accounted for the wide success of plants in the biosphere. Reproduction was freed from the requirement for external water which helped plants to spread from the limitation of a wet environment into drier habitats. Embryos (the new sporophytes) were protected during their development within the seed on the sporophyte mother plant, and were endowed with a supply of food to aid their establishment as a free-living individual. These features thus enabled plants to survive the environmental rigours which might threaten an especially critical stage in their life cycle.

1. The alternation of generations and the development of the seed habit

In the liverworts and mosses the identifiable plant is the haploid gametophyte bearing male and female sex organs in which gamete formation occurs. After fertilization of the female egg cell within the female organ by the motile, swimming male gamete the **diploid** sporophyte develops, remains embedded in the gametophyte, never becomes free living and is relatively short-lived: hence the gametophyte is the dominant generation. Morphologically identical, haploid spores produced by the sporophyte germinate into new gametophyte plants. In the ferns, however, the sporophyte, which initially is borne by the relatively insignificant, short-lived gametophyte, persists as the dominant generation – the plant recognized as a fern. All species higher than the ferns on the evolutionary scale have a dominant sporophyte.

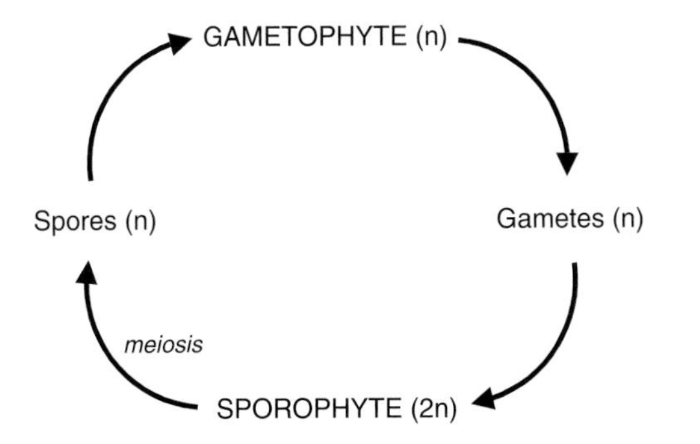

Fig. E.11. The alternation of generations. n, Haploid number of chromosomes; 2n, diploid number of chromosomes.

An important evolutionary development was the production of two types of spores that are morphologically and sexually differentiated (heterospory) in different spore-producing organs (sporangia). Several extant species such as the club mosses (e.g. *Selaginella*) illustrate this. Here, the terminal parts of the diploid sporophyte plant are given over to producing two types of sporangia, the **micro-** and **megasporangia** bearing the small, haploid male microspores and the larger, haploid female megaspores, respectively. When released, the microspore becomes the much-reduced (just a few cells) male gametophyte, and the megaspore, the larger (though only barely visible to the naked eye) female or megagametophyte. This structure contains some storage reserves, a feature that later is developed to an advanced degree in seeds. Both gametophytes are free-living and water is needed for the male gamete to reach the egg cell.

Four megaspores are formed from each megaspore mother cell (megasporocyte) within the megasporangium. The fossil record indicates that by the Devonian period (354–417 million years ago) heterosporous forms had appeared in which three megaspores aborted leaving one functional representative (Fig. E.13). The next evolutionary step was the retention of the single megaspore within the megasporangium of the sporophyte, thereby enabling the female gametophyte to remain within the sporophyte. This led to the evolution of the ovule of the seed plants by the sheathing of the megaspore by protective covers developed from appendages (telomes) that became the integuments or **seedcoat**. Progressive closure of the covers has resulted in there remaining only a small pore between them, the **micropyle** (Fig. E.14). The sex organs of the female gametophyte gradually became reduced during evolution, exemplified in extant forms by the cycads, conifers and finally the **angiosperms** in which only a few cells are present within the embryo sac (**see: Reproductive structures**). The microspore remains as a morphologically simple structure, becoming the pollen grain which is released from the sporophyte of the seed plants.

Double fertilization is thought to have evolved in a common ancestor of the flowering plants (angiosperms) and the Gnetales (considered by some authorities to be a gymnosperm with close affinities to the angiosperms) possibly about 200 million years ago. In the present-day Gnetales (e.g. *Gnetum*) double fertilization produces two diploid embryos, only one of which survives to maturity. Double fertilization in the angiosperms gives rise to the diploid embryo and the **triploid** nourishing tissue, the **endosperm**.

2. Seeds in the fossil record

The first seed plants in the fossil record were two groups of early gymnosperms present in the Carboniferous forests (360–290 million years ago). These were the pteridosperms or seed ferns (plants with fern-like leaves) and the Cordaitales, both of which were predominantly trees. In the

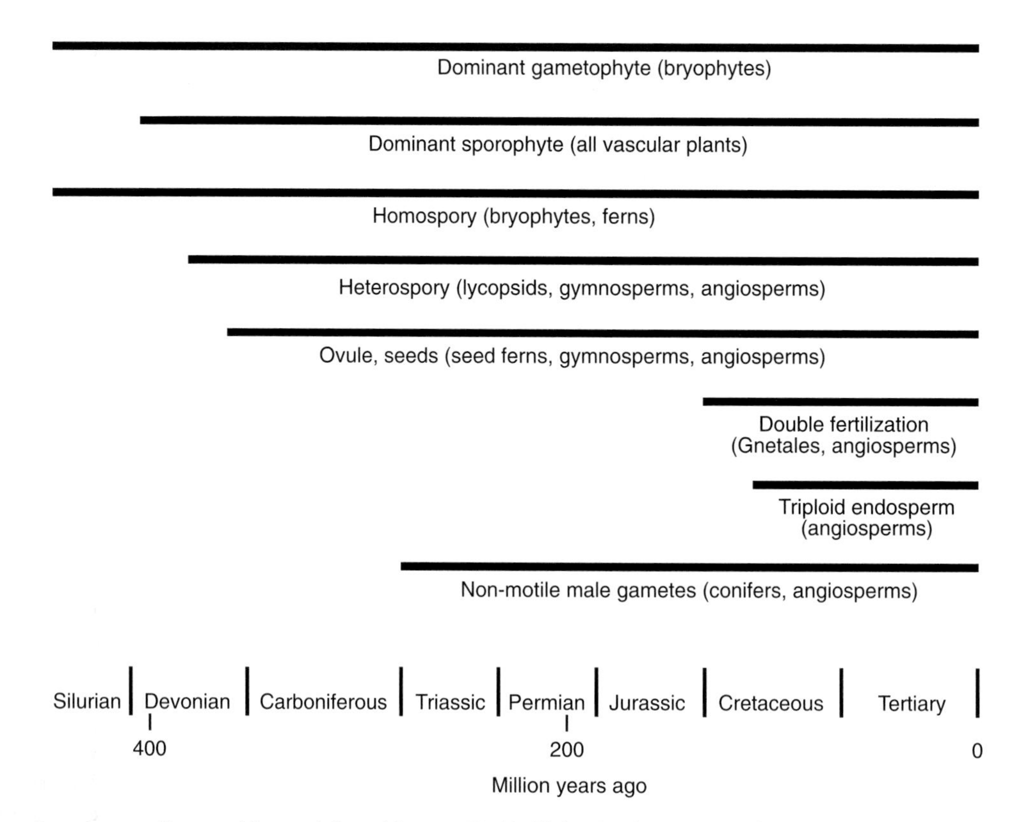

Fig. E.12. Alternation of generations and the evolution of the seed habit. Major developments as they occurred in geological time.

Permian (50–80 million years later) seed plants dominated the world flora. Angiosperms are thought to have evolved about 125 million years ago and authentic angiosperm seeds are present in the fossil record from about 90 million years ago (the Cretaceous).

3. Evolution of angiosperm seed characters

The many features of modern angiosperm seeds such as development of **embryo** and endosperm, **dispersal** strategies, **germination** and establishment strategies, **dormancy** and resistance to **desiccation** are, of course, the products of evolution. Fossil evidence is available which enables contrasts to be made between early Cretaceous and later Cretaceous and Tertiary (Palaeogene, 35–65 million years ago) representatives, suggesting the evolutionary trends that have occurred.

(a) *Structural features*. Throughout most of the Cretaceous the angiosperm fossil seed flora is dominated by very small seeds – **dust seeds** (or micro seeds) (**see: Structure of seeds**). This implies that the storage reserve content was much more limited as compared with that in the larger seeds of the later Palaeogene. Many of these larger seeds had extensive ruminate endosperms (**see: Rumination**) suggesting that the embryos of many seed types were relatively small. Indeed, some preserved embryos are poorly differentiated, possibly presenting early examples of morphological dormancy (**see: Dormancy**). Some fossil samples from the middle Eocene (approx. 50 million years ago) indicate the presence of **perisperm**. There are several examples from different families of seeds with relatively large, differentiated embryos and little endosperm, dating from the middle Eocene. Cases with fully differentiated embryos and virtually no endosperm (e.g. members of the Sapindaceae and *Juglans*, the walnut genus) also occur in the Eocene. Inferences have been made from these structural features relating to the possible germination behaviour (e.g. slow *vs* fast) of the types but these, of course, are somewhat speculative.

(b) *Coat structure*. Seedcoats of fossil remains from the Cretaceous and Palaeogene have been examined microscopically, revealing evidence of tissue structure that in modern seeds is associated with coat hardness (**see: Hardseededness**). Hence, many seeds of these periods may have possessed 'dormancy' caused by impeded water uptake. These structures are rather more common in Cretaceous than in Palaeogene seeds. Widespread in the latter are various structural devices for permitting water entry such as plugs or valves. This is suggestive of germination regulatory mechanisms that are sensitive to fire. Chemical analyses of fossil samples indicates the presence in the cuticle and sclerotic tissues of chemicals found in present-day relatives, such as **hemicellulose** and **lignin** derivatives, **phenolics**, alkanes, alkanones and others.

(c) *Dispersal devices*. The plumes that are common in extant seeds (**see: Seedcoats – dispersal aids**) were lacking in Cretaceous seeds and of limited occurrence in the Palaeogene flora. Winged forms were rare in the Cretaceous but were widespread and abundant in the Palaeogene. The diversity of winged structures found in fossil seeds of this period bears comparison with the range evident in extant seeds. This information suggests that wind dispersal was an important

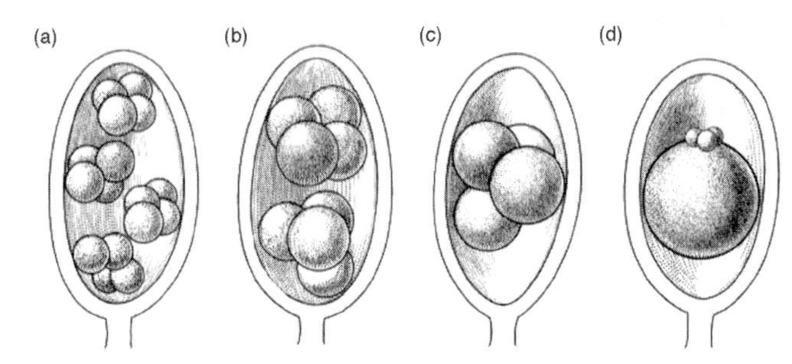

Fig. E.13. The probable evolution of a single functional megaspore. (a,b,c,d) Sequence depicting reduction of megaspore number to a single spore within the megasporangium. Tetrads of haploid megaspores are shown. From Willis, K.J. and McElwain, J.C. (2002) *The Evolution of Plants*. Oxford University Press, Oxford, UK.

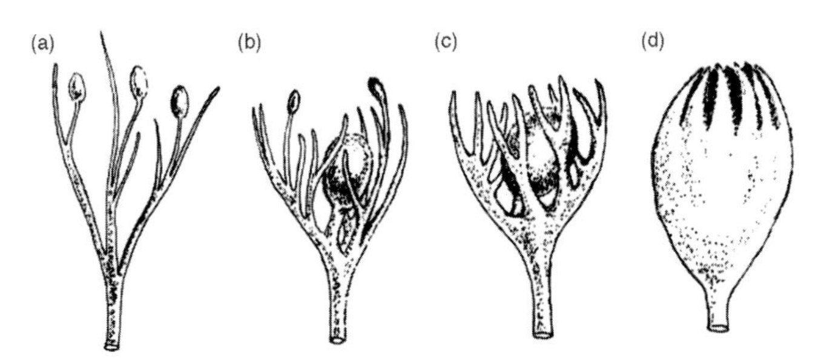

Fig. E.14. Possible evolution of integument and ovule. (a) Shows megasporangia on branches; (b) (c), sheathing of a megasporangium by sterile appendages (telomes) forming the integument, until (d), the megasporangium is enclosed by the integument to become the ovule with the incipient micropyle at its apex. There is fossil evidence (*Archaeosperma arnoldii*) for such an early ovular structure. From Willis, K.J. and McElwain, J.C. (2002) *The Evolution of Plants*. Oxford University Press, Oxford, UK.

strategy. On the other hand, devices such as spines and hooks were rare in both the Cretaceous and Palaeogene fossil seeds suggesting that dispersal by animals (especially mammals) was uncommon.

(d) *Desiccation tolerance, recalcitrance and vivipary*. There is no fossil or phylogenetic evidence pointing to the evolution of desiccation tolerance but it seems likely that it arose early in seed evolution and probably developed independently a number of times in different plant groups. Based on its occurrence in different plant families of varying ancestry, it has been suggested that recalcitrance is an ancestral condition in the angiosperm seed. Fossil evidence shows that **vivipary** was exhibited by members of the mangrove family by the early Eocene. (MB)

(See: Structure of seeds – identification characters)

Baskin, C.C. and Baskin, J.M. (1998) *Seeds: Ecology, Biogeography and Evolution of Dormancy and Germination*. Academic Press, San Diego, USA.

Collinson, M.E. and van Bergen, P.F. (2004) Evolution of angiosperm fruit and seed dispersal biology and ecophysiology: morphological, anatomical and chemical evidence from fossils. In: Hemsley, A.R. and Poole, I. (eds) *The Evolution of Plant Physiology. From Whole Plants to Ecosystems*. Elsevier Academic Press, Amsterdam, The Netherlands, pp. 344–375.

Moles, A.T., Ackerly, D.D., Webb, C.O., Tweddle, J.C., Dickie, J.B., Westoby, M. (2005) A brief history of seed size. *Science* 307, 576-580.

Pammenter, N. and Berjak, P. (2000) Evolutionary and ecological aspects of recalcitrant seed biology. *Seed Science Research* 10, 301–306.

Willis, K.J. and McElwain, J.C. (2002) *The Evolution of Plants*. Oxford University Press, Oxford, UK.

Evolutionary ecology

The ecology and characteristics of seeds of different species can only be understood by combining studies of the fate of seeds in nature with understanding how this translates into evolutionary pressures which lead to selection for particular combinations of seed traits (a strategy). Seed traits of ecological importance (i.e. which may affect the number and performance of plants produced from the seed output of an individual) include **number of seeds, size of seeds**, adaptations for **dispersal** or for defence against **predators** or pathogens, requirements for regeneration (including **germination** triggers), **dormancy** attributes, and the type of **soil seed bank**. **(See: Dormancy – ecology; Pathogens – ecology; Regeneration – ecology)**

The combination of seed traits shown by a species can be linked to traits exhibited by a plant through other parts of the life cycle, resulting in an overall life history strategy. Such strategies are studied in two ways: by comparative studies of correlated variation in traits among species (and in relation to the phylogeny – the evolutionary relationships among species); and by theoretical analyses of evolutionary stable strategies (ESS), where an ESS is the most successful strategy for a particular set of environmental conditions.

Fundamental to both approaches is the concept of trade-offs. Limitations on resources and on the biochemical, physiological and morphological pathways in a plant (e.g. a consequence of the development of a meristem into a flower is that it cannot form a new branch) mean that a plant cannot express all the most advantageous traits, and there has to be a process of natural selection by which one advantageous trait is expressed instead of another. This also means that a plant well adapted to one environment may be poorly adapted for another. Trade-offs are described both in terms of overall life history (e.g. colonization versus competitive ability), or simple traits (e.g. growth of roots versus shoots).

Ecologists have suggested many seed-related trade-offs, but most have not been demonstrated or have been disproved subsequently. The only convincing trade-off is that of size versus number: whereby, for a fixed allocation of resources to seed production, the more seeds a plant produces, the smaller those seeds must be. This trade-off has many consequences, such that smaller seeds show better dispersal and increased longevity in the soil seed bank, but poorer establishment and ability to compete against other plants during regeneration. A variation on this trade-off is that adaptations for dispersal (**samaras**, hooks, fruit, etc.), or defence against predation (e.g. thick **seedcoat**) reduce the seed number.

The seed size/number trade-off is linked to simple classifications of plant life history strategies: competitor versus ruderal, K versus r, or competitor versus colonizer. In each pair the first strategy describes plants that compete well in dense vegetation and have long lives, tall stature, and a relatively low annual investment in reproduction. Plants in the second category can colonize bare ground rapidly, but decline as the vegetation becomes dense. They have a short life span and good seed production (especially annuals or monocarpic perennials, i.e. those that flower only once, which put all their resources into seed production and then die). Besides a greater reproductive allocation, colonizer plants tend to have smaller seeds. This means competitor species are better able to regenerate in dense vegetation, but colonizers disperse seeds over greater distances and at greater densities, increasing the chances of some seeds landing on bare ground.

Traits vary within species. Natural selection leads to differences among populations which are growing in different environmental conditions. For example, for the ash (*Fraxinus ornus*) spreading through a landscape in France, new populations have seeds with better dispersal ability than older, dense populations. Within populations, or even within individuals, an advantageous strategy may be to show some variation in a trait, to allow a range of responses to environmental conditions. Many individuals produce a wide range of seed sizes, which may mean some can disperse long distances while others are more competitive during regeneration. In fact, some species take this one stage further by producing more than one type of seed; fruits of ragwort (*Senecio jacobaea*) produced in the centre of the flower head are lighter and have both plumes and hairs which aid dispersal by wind and animals, while those at the edge of the flower head are heavy and have no obvious dispersal adaptations. (JMB)
(See: Evolution of seeds)

Leishman, M., Wright, I.J., Moles, A.T. and Westoby, M. (2000) The evolutionary ecology of seed size. In: Fenner, M. (ed.) *Seeds: the Ecology of Regeneration in Plant Communities*. CAB International, Wallingford, UK, pp. 31–57.

Moles, A.T., Ackerly, D.D., Webb, C.O., Tweddle, J.C., Dickie, J.B., Westoby, M. (2005) A brief history of seed size. *Science* 307, 576-580.

Thompson, K., Rickard, L.C., Hodkinson, D.J. and Rees, M. (2002) Seed dispersal – the search for trade-offs. In: Bullock, J.M., Kenward, R.E. and Hails, R. (eds) *Dispersal Ecology*. Blackwells, Oxford, UK, pp. 152–172.

Exalbuminous

Seeds without **endosperm** or **perisperm** at maturity. **Non-endospermic** (exendospermic) is the term used more commonly with respect to the former.

Ex situ conservation

The conservation of plants, animals and microbes at a site away from where they occur in a wild state or, in the case of domesticated biota, at a site away from where husbandry usually takes place. The aim of this activity is to conserve **germplasm** samples that comprise a genetic representation of the diversity threatened *in situ*. Conservation *ex situ* acts as a complement to that carried out *in situ* by acting as an insurance against loss and by providing an accessible source of material for use. In the case of wild populations, it provides material for conservation research and education, and for species and habitat restoration. For plants, this 'off-site' conservation includes the use of **gene banks** and botanic gardens. (RJP, SHL)
(See: Conservation; *In situ* conservation)

Excised embryo test

A test to determine the **viability** of certain tree seeds, such as *Fraxinus* spp. and *Prunus* spp., that germinate slowly or show deep levels of **dormancy**. **Embryos** are excised and incubated on moist filter paper for 5 to 14 days at 25°C. Viable embryos either remain firm and fresh or show evidence of growth (for example, expansion, elongation or greening) or growth and differentiation (for example, **radicle** and lateral root formation; and **epicotyl** and first leaf formation), whereas non-viable embryos show extreme brown or black discoloration, have off-grey or white colour with watery appearance or show signs of decay. **(See: Tree seeds)**

Exostome

The **micropylar** opening of the outer integument. **(See: Structure of seeds,** Fig. S.61)

An exostome aril is a special case of a localized **aril** represented by a proliferation of the exostome. A small, disc-like exostome aril is present in many seeds of the Euphorbiaceae, and is traditionally called a **caruncle**.

Expansins

Cell wall-associated proteins that can induce wall expansion during cell growth and may be involved in **germination**. They also play a role in other processes where cell wall modification or disassembly is involved, such as abscission, fruit ripening, and tissue weakening or softening. Their role is probably to catalyse the breaking of hydrogen bonds between cellulose microfibrils and cross-linking glycans, such as **xyloglucan**, thus reducing rigidity and allowing the **cellulose** microfibrils to slide over each other. **(See: Germination – molecular aspects; Germination – radicle emergence)**

Extensin

A structural **cell wall** protein which covalently binds to itself (and probably other proteins) to form an interlocking network conferring strength to the wall after its expansion. They are not present in the cell walls of a large sub-group of the monocots, including the Poaceae (grasses, cereals).

F

F_1, F_2, F_3

Symbols that designate the first, second, or third generation after a **cross**. An F_1 hybrid is the first filial generation resulting from a cross of two selected parental lines that are genetically different. Many commercial hybrid crops are F_1. The F_2 generation is obtained by self-fertilization of one or more F_1 hybrid plants, the F_3 generation self-fertilization of the F_2 generation and so on. Both may be used in some breeding strategies. (**See: Production for sowing, III. Hybrids**)

Fabaceae

See: **Legumes**

Fanning mills

Hand-powered machines for seed cleaning, from which modern air-screen machines have been developed. (**See: Conditioning, II. Cleaning**)

FAO

See: **Food and Agriculture Organization of the United Nations**

Farm(er)-saved seed

That portion of harvested grain kept back for local planting; and also by extension, seed of non-grain crops that are cultivated specifically for such a purpose. Farmer-saved seed (FSS) is also variously called 'common', 'home-grown', 'bin-run', 'elevator-run' or 'brown bag' seed, depending on circumstances.

The practice of keeping FSS, which dates from the prehistoric origins of agriculture and crop **domestication**, is still the starting point for a substantial part of world crop production. By some estimates, the 'informal' seed supply system accounts for about 90% of agriculture in sub-Saharan Africa and 70% in India. About 20% of world maize is grown from FSS, including more than half the crop in south Asia, Central America and west-central Africa; and nearly all rice in Asia is grown from FSS. Substantial quantities of FSS are used in developed countries too, including much wheat and barley and some soybean, for at least a few seasons until the farmer buys commercial **variety** seed again. (**See: Industry – supply systems**)

Seed can be successfully saved for re-use from open-pollinated species (including most small-grain cereal, legume and cotton, and some maize varieties), which inherit the identity of the original variety, but not from hybrid or synthetic varieties whose traits genetically segregate in succeeding generations. (**See: Production for sowing**) Because of its variable quality, there is a greater risk of poorer crop establishment and yield with FSS than with commercial certified seed. In some countries FSS is **quality tested** for purity, germination and health, **certification** and **treatment**, but not by obligation or as a general rule.

The principle of a farmer's right to save, clean, condition and store seeds for sowing without economic or legal consequence is one factor at the centre of the ongoing worldwide debate over variety ownership. In many countries, this 'farmers' privilege' (or 'exemption') is constrained by the intellectual property rights granted to the owners who have bred, discovered or developed the varieties. Plant Variety Protection or Plant Breeders' Rights laws typically allow the production of FSS of eligible varieties for use within farmers' own businesses, with certain limitations. (**See: Laws and regulations, 3. Intellectual property protection**)

Increasingly, farmers in developed countries are legally obliged to provide accounts of the seed they save. In the EU, permitted FSS is subject to royalty payment, depending on farm size. In the USA, newly released varieties may be accorded different levels of legal protection. For some varieties, the amount of FSS must be only enough to sow on the same farm holding; for others, FSS may be shared or exchanged with another farmer; for others, FSS may be sold by variety name, though permission may be needed from the owner and seed may have to be officially certified. FSS is illegal in the USA if a variety is patented or contains a patented gene, such as those engineered through **genetic modification** (GM); but there is no such intellectual property protection for GM soybean in Argentina, which is mostly grown from FSS. In practice, illegal 'brown bagging' (where common seed is sold by distributors without naming the variety) can continue where there are no effective means to enforce intellectual property rights. The commercial implementation of some Genetic Use Restriction Technologies would prevent FSS from germinating or expressing its traits (**see: GMO – patent protection technologies**). (**See also: Heritage seeds**)

(PH)

Fats

See: **Oils and Fats**

Fatty acids

Fatty acids are chains of 2 to 30 or more carbon atoms terminated at one end with a carboxy group (-COOH). Most fatty acids in seed **oils** range from 10 to 22 carbons in length. The carbons may be joined by a carbon–carbon single or double bond. If a fatty acid contains one double bond it is referred to as unsaturated or monounsaturated; if it contains two or more double bonds then it is polyunsaturated. Fatty acids with no double bonds are saturated. A common convention used to describe fatty acids uses the number of carbons in the chain followed by the number of double bonds (e.g. 16:0, 18:1, 18:3, etc.) (Fig. F.1).

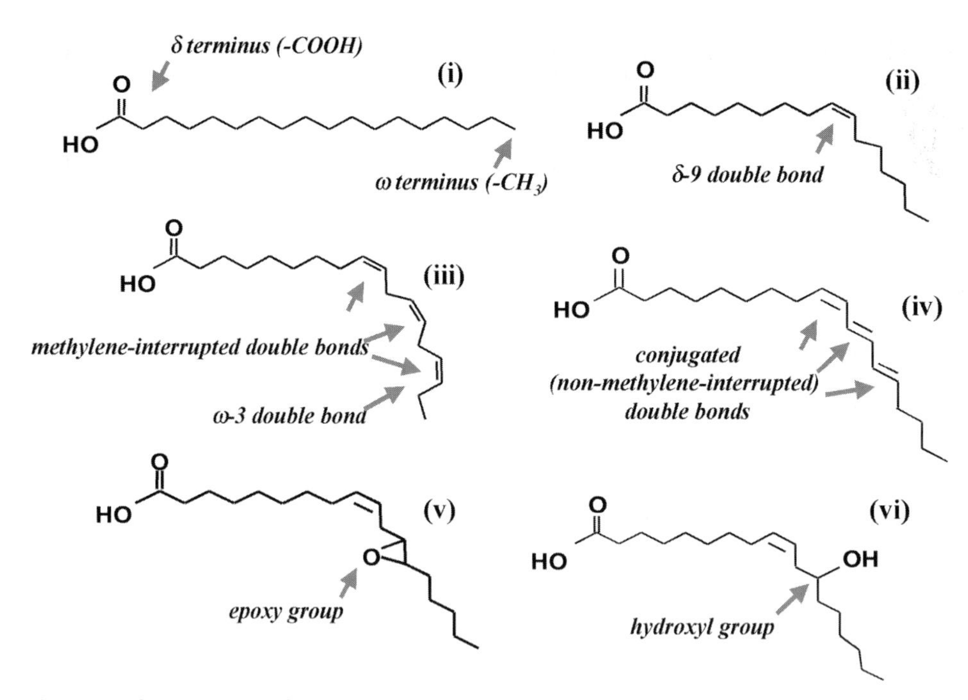

Fig. F.1. Structures of some 18 C fatty acids. (i) Stearic acid, a saturated fatty acid. With the exception of the δ terminal –COOH group and the ω-terminal –CH_3 group, the chain is composed of –CH_2 groups, joined by a single C–C bond. (ii) Oleic acid, a monounsaturated fatty acid; (iii) α-linolenic acid, a polyunsaturated fatty acid; (iv) α-eleostearic acid, a conjugated fatty acid; (v) vernolic acid, an epoxy fatty acid; and (vi) ricinoleic acid, an hydroxyl fatty acid.

Double bonds in polyunsaturated fatty acids are normally separated by a methylene (-CH_2-) group; polyunsaturated fatty acids that are not separated by a methylene group and thus have double bonds between each of two adjacent pairs of carbons are called conjugated fatty acids. The position of the double bond is usually specified by counting from the carboxyl (-COOH), or delta (Δ or δ) end of the molecule (e.g. 18:2Δ9,12). It is the delta end that is condensed with a glycerol moiety as an ester to form a fat or **oil** (triacylglycerol).

Sometimes fatty acids are named based on the position of a double bond from the terminal methyl (omega, ω) group (the other end from the delta end). The most common examples of this are the so-called omega-6 and omega-3 fatty acids. These are unsaturated fatty acids of various carbon chain lengths and double-bond numbers that always have a double bond either six (n-6) or three (n-3) carbons from the methyl (ω) end of the chain, and no other double bond between it and the methyl group. Thus linoleic (18:2Δ9,12) and γ-linolenic (18:3Δ6,9,12) are omega-6 (n-6) fatty acids whereas α-linolenic (18:3Δ9,12,15) is an omega-3 (n-3) fatty acid. Linoleic and linolenic acids are also known as **essential fatty acids** (EFAs). Fortunately, most plant oils used for food purposes are rich in linoleic acid and also contain some linolenic acid. The other major fatty acids in common edible oils are palmitic (16:0), stearic (18:0) and oleic (18:1) acids. Common edible seed oils include those of **soybean, brassica oilseeds** such as canola (rapeseed), **cotton, maize, safflower, sesame** and **palm**. (**See: Oils and Fats; Oilseeds**) Double bonds in common seed oil fatty acids, such as oleic, linoleic and linolenic acids, are in the *cis* configuration. That is, the two hydrogen molecules on adjacent carbons are both on the same side of the carbon–carbon double bond. There are, however, many examples of plant oils that contain fatty acids with *trans* double bonds, having adjacent hydrogens on opposite sides of the carbon–carbon bond. These oils include that of **tung** (*Aleurites fordii*, *A. montana*) seed, which contains about 80% wt/wt of the conjugated *trans* fatty acid α-eleostearic (18:3Δ9*cis*11*cis*,13*trans*), and is used in wood refinishing and protection. Another example is *Impatiens* seed oil, which contains up to 30% α-parinaric acid (18:4Δ9*cis*11*trans*,13*trans*,15*cis*) but is not commercially produced. *Trans* fatty acids, however, are not present in edible seed oils in their natural state. When the term '*trans* fatty acids' is used in the context of dietary oils it normally refers to the *trans* isomer of oleic acid (18:1Δ9*trans*) formed during the partial chemical hydrogenation of *cis*-polyunsaturated fatty acids such as linoleic and α-linolenic acids. Oils rich in linoleic and linolenic acid, such as from soybean and canola seed, are often chemically hydrogenated after refining to increase their oxidative stability and impart the functionality of a solid fat. Partially hydrogenated seed oils may also contain other *trans* isomers, including 18:2Δ9*trans*12*trans* and 18:2Δ9*cis*12*trans*.

Seed oil fatty acids may also contain triple (acetylenic) bonds and other functional groups such as hydroxyl, epoxy and keto groups, in addition to double bonds. Indeed, there are many examples of seed oils rich in a single fatty acid with an interesting functional group or structure. Only a few of these non-standard oils are currently produced, mostly for non-food applications (**see: Oils and fats; Oilseeds – minor**). Castor (*Ricinus communis*) oil, rich in the hydroxylated fatty acid ricinoleic acid (18:1Δ9-OH) was once used mainly as a human purgative or laxative but is now used in paints, lubricants, plastics, textiles, vegetable inks and industrial adhesives. The seed oil of several *Limnanthes* species, especially meadowfoam (*Limnanthes alba*), is rich in long chain (C_{-20} and C_{-22}) fatty acids with a Δ^5 double bond. The most abundant fatty acid in these seeds is Δ^5-eicosenoic acid (20:1Δ^5), which accounts for 60% of the total fatty acids and is used in the manufacture of cosmetics, surfactants and lubricants. Oiticica oil, which comes from the tree *Licania rigida* and contains the keto fatty acid α-licanic (18:4Δ4-keto, 9*cis*,11*trans*,13*trans*), is used in the manufacture of printing inks, linoleum, paints and enamels. Epoxy fatty acids, such as vernolic acid (C18:2Δ9*cis*,12-epoxy), are useful in the manufacture of resin transfer moulded composites and as PVC plastisizers/stabilizers. For this reason seed oils rich in these fatty acids, such as the oil obtained from *Vernonia galamensis* and *Euphorbia lagascae* seeds, are considered as potential targets for commercialization. (AJK)

(See: Fatty acids – synthesis; Oils and fats – synthesis; Oilseeds – major; Oilseeds – minor; Pharmaceuticals and pharmacologically active compounds)

Badami, R.C. and Patil, K.B. (1981) Structure and occurrence of unusual fatty acids in minor seed oils. *Progress in Lipid Research* 19, 119–153.

Gunstone, F.D., Harwood, J.L. and Padley, F.B. (1994) *The Lipid Handbook*, 2nd edn. Chapman & Hall, London, UK.

Fatty acids – synthesis

Some key components involved in the synthetic processes are described below and detailed in Fig. F.2. For incorporation of fatty acids into oils and fats (triacylglycerols), **see: Oils and fats – synthesis**. For the terminology of fatty acid chemistry, **see: Fatty acids**.

Seed triacylglycerols (fats, oils) are derived from glycerol-3-phosphate and fatty acyl-CoA (coenzyme A; fatty acids with a terminal CoA replacing the carboxyl, -COOH, group) molecules in association with the **endoplasmic reticulum** (ER) membrane system of **oilseeds**. However, the initial site of fatty acyl chain synthesis is the **plastid**. (The term fatty acyl is normally used instead of fatty acid during its synthesis, when it is attached to other groups, e.g. CoA, ACP, prior to its combination with glycerol-3-P to form triacylglycerols.)

Saturated and monounsaturated fatty acid chains (up to 18 C long) are synthesized from plastidal *acetyl-CoA*, which may be derived from a number of cytosolic metabolites (ultimately synthesized from sucrose imported into the seed from the parent plant) including glucose 6-phosphate, malate, phosphoenol pyruvate (PEP) and pyruvate. These metabolites are imported into the plastids of seed cells by specific transporters where they are converted to acetyl-CoA by various enzymes including glucose 6-phosphate dehydrogenase, PEP carboxylase and pyruvate dehydrogenase (PDH). Since pyruvate is the common precursor for acetyl-CoA synthesis from all these metabolites it has often been suggested that PDH controls the entry of carbon into oil biosynthesis and hence regulates its flux relative to the synthesis of other storage components such as carbohydrates.

Plastidal acetyl-CoA is converted to malonyl-CoA by the action of an enzyme complex known as *acetyl-CoA carboxylase*. This ACCase is thought to play an important role in the regulation of the rate of oil biosynthesis in seeds since it is the first committed step for fatty acyl-chain synthesis.

Since the growing fatty acyl chain is solubilized by

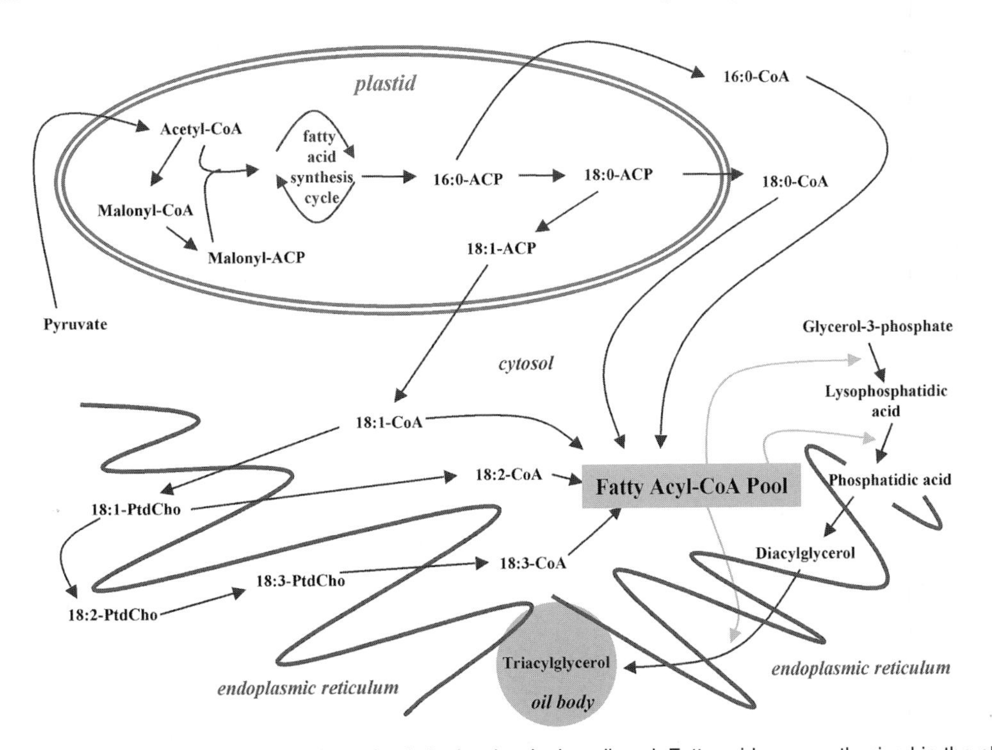

Fig. F.2. Predominant pathways contributing to oil biosynthesis in the developing oilseed. Fatty acids are synthesized in the plastid and further modified in the endoplasmic reticulum. Assembly of fatty acids into triacylglycerol begins in the cytosol and is completed in the endoplasmic reticulum (depicted as wavy lines). The oil body containing the triaclyglycerol is formed from budding of the endoplasmic reticulum (**see: Oil bodies** for micrograph). ACP: acyl carrier protein; PtdCho: phosphatidyl choline. For details of the enzymes involved, see text.

attachment to an *acyl carrier protein* (ACP), the malonyl-CoA derived from the ACCase complex is first converted to malonyl-ACP by the action of a malonyl transacylase enzyme. The chain elongation then proceeds by the stepwise condensation of C2 units from acetyl-CoA with a growing fatty acyl-ACP chain, starting with malonyl-ACP (4:0-ACP) and extending all the way to stearoyl-ACP (18:0-ACP). This is followed by a desaturation of 18:0-ACP to oleoyl-ACP (18:1Δ9*cis*-ACP). This elongation of a fatty acyl-ACP by two carbons is also a cycle of four reactions: an initial condensation reaction followed by reduction, dehydration and then a second reduction, catalysed by a series of enzymes collectively called fatty acyl *elongases* (FAE). The condensation reaction is catalysed by a β-ketoacyl synthase (KAS) enzyme. The first condensation of the growing chain, that of acetyl-CoA with malonyl-ACP to form 3-ketobutyl-ACP and CO_2 is catalysed by an enzyme called KAS III. Subsequent condensations to C16 are catalysed by KAS I and the final C16 to C18 condensation by KAS II. Seeds that have medium-chain (8:0–12:0) fatty acids in their seed oil also have a condensing enzyme (KAS IV) with high specific activity toward medium-chain fatty acyl-ACP substrates.

While all of these KAS-type condensing enzymes are specific to particular chain lengths of fatty acyl-ACP, the other elongase enzymes of the elongation cycle, 3-ketoacyl-ACP reductase, 3-hydroxyacyl-ACP dehydratase and 2,3-*trans*-enoyl-ACP reductase, are common to all the elongation steps from two carbons up to the elongation of palmitoyl-ACP (16:0-ACP).

In most oilseeds, 18:0-ACP is desaturated in the plastid to 18:1Δ9*cis*-ACP, a reaction catalysed by a fatty acyl-ACP *desaturase*. This desaturase is encoded by a gene known as AAD, which is part of a gene family encoding soluble fatty acyl-ACP desaturases. Almost all other desaturase gene families encode membrane-bound desaturases. This soluble desaturase gene family includes a number of other members which encode desaturases that are different from the standard 18:0-ACP Δ9 desaturase in their substrate specificity, regiospecificity (i.e. which specific bond they desaturate), or both. Examples of divergent desaturases encoded by *AAD* genes include a delta-4 desaturase from carrot and coriander, a Δ6 16:0-ACP desaturase from *Thunbergia alata* and a Δ9 16:0-ACP desaturase from *Doxantha* spp. and from *Asclepias syriaca*. As a result, these species contain oils with monounsaturated fatty acids that have double bonds in positions other than the ninth carbon from the carboxyl end group. For example, seeds of carrot and coriander contain oils rich in petroselenic acid (18:1Δ6*cis*).

A fatty acyl-chain is released from ACP by the action of fatty acyl-ACP thioesterases. Once released, the fatty acyl chain can no longer be modified in any way in the plastid and is released into the cytoplasm wherein it attaches to co-enzyme A (CoA) to form fatty acyl-CoA. There are two distinct gene-families which encode these fatty acid thioesterase enzymes: *FAT A* and *FAT B*. *FAT A*-encoded enzymes hydrolyse predominantly 18:1-ACP with minor activities towards 18:0-ACP and 16:0-ACP. *FAT B*-encoded enzymes usually hydrolyse saturated C14–C18 ACPs, preferentially 16:0-ACP, but they will also hydrolyse 18:1-ACP. The production of medium-chain (C8–C12) fatty acids in seeds such as those of *Cuphea* spp. is the result of FAT B enzymes that have chain length specificities for medium chain fatty acyl-ACP rather than 16:0-ACP and 18-carbon fatty acyl-ACPs. These medium-chain FAT B thioesterases are present in many species with medium-chain fatty acids in their oil, including California bay laurel, coconut and elm. The high content of stearic acid (over 50%) in the seeds of mangosteen (*Garcinia mangostana*) is the result of a FAT A-type thioesterase with high specificity towards 18:0-ACP rather than 18:1Δ9*cis*-ACP.

When fatty acids leave the plastid and become esterified to CoA they can be substrates for triacylglycerol synthesis. However, they may be further modified before becoming part of a triacylglycerol. These modifications occur within the endoplasmic reticulum (ER). Some are modified directly as fatty acyl-CoA. For example, some plant oils, such as oilseed rape (*Brassica* spp.), contain very long chain, unsaturated fatty acids (VLCFAs) of 20–26 carbons. These are the result of the elongation of oleoyl-CoA by an ER-membrane-bound elongase complex analogous to the fatty acyl-ACP elongase complex described above. As is the case for fatty acyl-ACP elongation, the condensing enzyme (FAE 1) confers the specificity for elongation, in this case for 18:1Δ9-CoA. The long chain fatty acids in the liquid waxes in the seeds of **jojoba** (*Simmondsia chinensis*) are synthesized by an analogous 18:1Δ9-CoA elongation complex. The condensing enzyme of this complex, KCS, is closely related to FAE 1. The fatty alcohol component of liquid waxes is the result of a microsomal fatty acyl reductase. **(See: Wax – storage)**

The unique oil from seeds of *Limnanthes* spp. (e.g. meadowfoam), which contains 20:1Δ5*cis* fatty acids, is the result of the cytosolic elongation of 16:0-CoA which contains a 16:0-CoA specific thioesterase also similar to FAE 1. The Δ5 double bond is inserted into 20:0-CoA by a membrane-associated desaturase belonging to the same family as mammalian and yeast 18:0-CoA desaturases.

Most other modifications of fatty acyl chains after they leave the plastid occur while the acyl chain is esterified with the sn-2 position of phosphatidylcholine (PtdCho), a phospholipid present within the ER membrane (**see: Membrane lipids**). The synthesis of linoleic and linolenic acids, catalysed by two specialized microsomal membrane-associated desaturases, is a good example of this type of modification. But, even in this well-studied example, the mechanism by which fatty acyl chains are attached to membrane lipids for modification is not fully understood. It is thought that 18:1Δ9*cis*-CoA from the cytosolic pool is transferred to lysoPtdCho by lysophosphatidylcholine acyltransferase, an enzyme activity detected in cell-free extracts of oilseeds. The lysoPtdCho substrate is presumably generated by the action of a phospholipase A2, which releases a previously polyunsaturated fatty acyl chain from the sn-2 position of PtdCho making the space available for a new 18:1Δ9*cis* chain. This phospholipase activity has to be coupled with a fatty acyl-CoA synthetase to activate the free fatty acid for further utilization in oil biosynthesis.

The newly created 18:1Δ9*cis*-PtdCho is the substrate for two membrane-associated fatty acid *desaturases*, a Δ12 (FAD 2) and ω3 (FAD 3) desaturase, which introduce the second and third double bonds respectively into this fatty acyl chain. The ω3 desaturase is sometimes referred to as a Δ15 desaturase, especially in older publications, although it is now known that FAD2 is a true Δ12 desaturase while FAD 3 is of the ω-x type and is thus designated an ω3 desaturase.

Polyunsaturated fatty acyl chains with two and three double bonds are returned from PtdCho to the fatty acyl-CoA pool, a reaction catalysed by an unknown fatty acyltransferase or fatty acyl hydrolase. Polyunsaturated fatty acyl-CoA molecules are then available for incorporation into triacylglycerol. Polyunsaturated fatty acyl chains esterified with PtdCho may also be directly incorporated into triacylglycerol, a reaction catalysed by the enzyme phosphatidylcholine-diacylglycerol acyltransferase (PDAT). The result of either pathway is that the seed triacylglycerol becomes enriched with linoleic and linolenic acids.

Many seeds contain fatty acids with non-methylene-interrupted bonds and other types of functional groups instead of double bonds (**see: Oils and fats**). Many of these novel fatty acyl groups are the result of the activity of evolutionarily diverged membrane desaturases and are members of the *FAD2*-gene family. These diverged desaturases include fatty acid Δ12-hydroxylases from *Ricinius communis* and *Lesquerella fendleri*, fatty acid Δ12-epoxygenases from *Vernonia galemensis* and *Crepis palaestina* and a fatty acid Δ12-acetylenase from *Crepis alpina*. (**See: Fatty acids**)

Some key components of the fatty acid synthesis pathway:

(i) Acetyl-CoA

The carbon source for fatty acid synthesis in the plastids of developing oilseeds. Acetyl CoenzymeA is converted to malonyl-CoA by an acetyl-CoA carboxylase enzyme complex. Malonyl-CoA is an energetically favourable carboxyl donor for fatty acid elongation (after transacylation to malonyl-ACP). Acetyl-CoA is also a substrate for, or product of, many reactions involved in cellular metabolism, e.g. the citric acid cycle, amino acid catabolism and **gluconeogenesis**.

(ii) Acetyl-CoA Carboxylase (ACCase)

A biotin-containing enzyme present in the plastids of seeds. It is a key enzyme in fatty acid synthesis since it provides the primary source of carbon molecules used for fatty acyl-chain elongation. Its role is to carboxylate acetyl-CoA to malonyl-CoA and thus provide an energetically favourable substrate for the condensation of 2-carbon units with the growing fatty acyl chain. The reaction is a two-step process. Biotin is first carboxylated and the resulting carboxybiotin then donates a carboxyl group to the 2C acetyl-CoA to form the 3C malonyl-CoA.

In most seed plastids these reactions are catalysed by a 230 kDa heteromeric protein complex that can easily dissociate into four distinct protein components, each with enzyme activity. These four activites are: α-carboxyltransferase, β-carboxyltransferase, biotin carboxylase and biotin carboxyl carrier protein. The β–carboxyltransferase protein is encoded by the plastid genome and the other three subunits are nuclear encoded. In grass (Poaceae) seeds the plastidal ACCase is a multifunctional, 250 kDa homomeric enzyme that contains a single carboxyltransferase activity in addition to biotin carboxylase and biotin carboxyl carrier protein domains. A homomeric ACCase is also located in the cytosol of **dicot** seed cells and in those of some **monocots**, including grasses. This cytosolic ACCase is not known to be involved with seed triaclyglycerol biosynthesis except to provide malonyl-CoA for extra-plastidal fatty acid elongation in species that produce very long chain fatty acids (e.g. high erucic acid rapeseed) (**see: Brassica – oilseeds**).

(iii) Acyl Carrier Protein (ACP)

A small (80 amino acid-) central protein co-factor for fatty acid biosynthesis in the plastids of developing oilseed cells, which becomes activated by the attachment of a phosphopantetheine group near the centre of its peptide chain. During their synthesis, fatty acids are linked to the terminal sulphydryl of this phosphopantetheine group.

(iv) Desaturases (FAD)

Enzymes that remove hydrogens from a fatty acyl chain to yield a double carbon-carbon bond. Most fatty acyl desaturases are membrane bound; however, the synthesis of oleic acid in the plastids of developing oilseeds is catalysed by a desaturase that belongs to a class of soluble dehydrogenation enzymes. All desaturases have a specific regioselectivity, a method of counting to determine double bond placement. Delta (Δx) desaturases introduce a double bond a set number of carbons from the carboxyl end of a fatty acyl chain, while omega (ω-x) desaturases introduce a double bond a set number of carbons from the methyl terminus. A third type, (ν+x) desaturases, dehydrogenate a set number of carbons from a preexisting double bond. The nomenclature of *desaturases* is often based upon its regioselectivity. This convention is not always rigorously followed in the literature, however, especially since the regioselectivity of a desaturase is not usually understood when the enzyme or the reaction it catalyses is first discovered.

(v) Elongases (FAE)

A general term for a complex of four enzymes that catalyse the addition of two carbons to a growing fatty acyl chain during fatty acid synthesis. The term is sometimes used to refer specifically to the condensing enzyme (β-ketoacyl synthase) that catalyses the initial step of fatty acid elongation. However, this initial condensation reaction is followed by reduction, dehydration and then a second reduction reaction all of which are part of the elongation process. The other enzymes that catalyse these elongation steps are a 3-ketoacyl reductase, 3-hydroxyacyl dehydratase and a 2,3-*trans*-enoyl reductase. During the elongation process the fatty chains are attached to either fatty *acyl carrier protein* (in the plastid) or to CoA (in the cytosol). The fatty acyl-ACP elongases in the plastid are all soluble enzymes, whereas the cytosolic fatty acyl-CoA elongases are multi-enzyme complexes associated with cellular membranes.

(vi) Thioesterases (FAT)

A fatty acyl hydrolase whose substrate can be fatty acyl-ACP or fatty acyl-CoA. In the plastids of developing oilseeds fatty acid thioesterases (FAT A, B) terminate chain elongation by releasing fatty acyl groups from ACP. The free energy of this hydrolysis is similar to that generated by the conversion of ATP to ADP. The relative specificity of different fatty acyl-ACP thioesterases determines the ultimate fatty acid composition of oil triacylglycerol. (AJK)

Harwood, J.L. (2005) Fatty acid biosynthesis. In: Murphy, D.J. (ed.) *Plant Lipids: Biology Utilisation and Manipulation.* Blackwell, Oxford, UK. pp. 27-66.
Ohlrogge, J.B. and Jaworski, J.G. (1997) Regulation of fatty acid synthesis. *Annual Review of Plant Physiology and Plant Molecular Biology* 48, 109–136.
Rawsthorne, S. (2002) Carbon flux and fatty acid biosynthesis in plants. *Progress in Lipid Research* 41, 182–196.
Shanklin, J. and Cahoon, E.B. (1998) Desaturation and related modifications of fatty acids. *Annual Review of Plant Physiology and Plant Molecular Biology* 49, 611–641.
Voelker, T. and Kinney, A.J. (2001) Variations in the biosynthesis of seed storage lipids. *Annual Review of Plant Physiology and Plant Molecular Biology* 52, 335–361.

Favism

A haemolytic anaemia caused by the glycosides **vicine** and covicine in the seed (or occasionally the pollen) of the **broad bean** (fava bean). It affects people who have an inherited absence of the enzyme glucose-6-phosphate dehydrogenase in their red blood cells. (**See: Pharmaceuticals and pharmacologically active compounds**)

FELAS

FELAS, the Latin American Federation of Seed Associations (Federación Latinoamérica de Asociaciones de Semillas, based in Bogotá, Colombia) gathers the National Seed Associations of 11 Latin American countries, to promote their functioning and the development of intra-regional trade. Every 2 years, FELAS organizes the Pan-American Seed Seminar, the most important meeting of the regional seed industries and institutions. www.felas.org/ (BlB)
(**See: Industry – structure**)

Fennel

Fennel (*Foeniculum vulgare*, Apiaceae – an **umbellifer**) is an aromatic herb best known for its seeds, which are sold commercially as a spice (**see: Spices and flavours**, Table S.12). The 'seeds' are actually one-seeded, dry, indehiscent, green-brown fruits (**mericarps**), measuring approximately 6 × 2mm – virtually the dimensions of the seeds (Fig. F.3). The plant is a tall erect, glabrous, biennial or perennial herb with yellow flowers. It is cross-pollinated, coming to flower in 80–110 days and requires 100–115 days to mature before harvest.

Fig. F.3. Fennel 'seeds' (image by Mike Amphlett, CABI).

Fennel seeds have a fragrant odour and pleasant licorice-like aromatic taste due largely to anethole (**see:** Fig. S.44 in **Spices and flavours**) in the **essential oil**. The oils are pleasantly fresh, warm, spicy and slightly camphoraceous. Two types of fennel are recognized. Seeds of common fennel *F. vulgare* var. *amara* usually contain 2.5–6.5% volatile oil. The oil is colourless to pale yellow and contains α-phellandrene, pinenes, anethole and methyl chavicol. Bitter fennel oil is obtained from *F. vulgare* var. *vulgare*, cultivated in Europe. Sweet fennel *F. vulgare* var. *dulce*, is mainly cultivated in France and Italy. The essential oil of this is a yellowish-green liquid with characteristic anise odour. The main constituents are anethole (50–70%) and fenchone (10–20%).

Fennel oils and **oleoresins** are used to flavour food and confectionery, alcoholic and non-alcoholic beverages, and for seasoning of processed meat. The oil is also used to scent soaps, perfumes and flavour carminative medicines. Ground fennel is an important constituent in meat and curry mixes. The fruits have medicinal properties, used in gastro-intestinal disorders and as an expectorant. Anethole has stimulant, carminative and oestrogenic properties. (KVP, NB)

Guenther, E. (1978) *The Essential Oils*, Vol. 2. Robert E. Krieger Publishing, New York, USA.
Peter, K.V. (ed.) (2001) *Handbook of Herbs and Spices.* Woodhead Publishing, CRC Press, UK.
Weiss, E.A. (2002) *Spice Crops.* CAB International, Wallingford, UK.

Fenugreek

Fenugreek is the dried seed of the annual **legume** herb *Trigonella foenum-graecum, T. arabica* (Fabaceae). It is cultivated all over the world as a leafy vegetable, condiment and as an important medicinal plant (**see: Spices and flavours**, Table S.12). The plant is erect, up to 60 cm tall, stiff, strongly scented, annual, with trifoliate leaves and propagated by means of seeds. The crop comes to harvest in about 70–120 days. Fruits are straight to sickle-shaped glabrous pods containing 10–20, 4–6 mm-long, deeply furrowed, hard, yellowish-brown seeds (Fig. F.4). After harvesting, the crop is dried for 4–6 days and seeds are separated by beating, followed by **winnowing** and sun-drying to reach 9% moisture level.

Fenugreek possesses a substantial amount of **endosperm**, most cells of which are composed solely of **hemicellulose** cell walls rich in **galactomannans**.

The bitter, mucilaginous seeds contain a very small quantity of highly odorous **essential oil** (< 0.02%) and fixed oil at 7%. Alkaloids are present (e.g. trigonelline) which are responsible for the bitter taste. Other important chemical constituents are anethole and methyl chavicol (**see: Spices and flavours**, Fig. S.44).

Seeds are used whole or ground as condiments or flavouring agent in curries, pickles, mayonnaise. It is basically bitter but in appropriate quantities enhances food taste. It is also used medicinally as a digestive and to promote lactation in both women and cows. Seed or extracts are commonly used in syrups, baked foods, condiment, chewing gums and meat

Fig. F.4. Fenugreek seed (image by Mike Amphlett, CABI).

seasoning and as an emollient and flavouring agent in pharmacy. It has high nutritive value having a high protein content. The seeds have many medicinal uses such as for bronchitis and colonitis or in topical applications as poultices. Traditionally, fenugreek is used to help control blood sugar levels in the case of diabetes, protein deficiency and anaemia. (KVP, NB)

For information on the particular contributions of the seed to seed science **see: Research seed species – contributions to seed science**.

Peter, K.V. (ed.) (2001) *Handbook of Herbs and Spices*. Woodhead Publishing Co., Cambridge, UK.

Purseglove, J.W., Brown E.G., Green C.L. and Robbins, S.R.J. (1981) *Spices*, Vol. 1. Tropical Agricuture Series, Longman, New York, USA.

Weiss, E.A. (2002) *Spice Crops*. CAB International, Wallingford, UK.

Fertility-restoring genes

Nuclear genes that act to restore fertility in plants of **cytoplasmic male-sterile** lines, and hence enable the production of fertile F_1 hybrids.

Fertilization

The fusion of the haploid male **gamete** (sperm cell, from the pollen) with the haploid female gamete (egg) to form a **diploid zygote**. In the angiosperms, however, this definition is broadened to include what is termed '**double fertilization**' in which a second male gamete from the pollen fuses with two female nuclei to form a **triploid endosperm** nucleus (**See: Development of embryos – cereals; Development of embryos – dicots; Development of endosperms – cereals; Development of endosperms – dicots**). In the **gymnosperms** only a single fertilization event occurs when the pollen nucleus fuses with an egg nucleus.

In vitro fertilization of haploid egg cells to produce diploid embryos, and of diploid endosperm (central) cells to produce triploid endosperms has been achieved, using microdissection and micromanipulation techniques combined with tissue culture methods adapted for the culture of single cells. Zygotes which are isolated after *in vivo* pollination develop into embryos and fertile plants in culture. This has been particularly successful using maize gametes. Attempts to genetically cross related species using this method, thus overcoming incompatibility associated with **pollination**, have generally been unsuccessful.

In vitro-produced zygotes and endosperms are able to self-organize in culture independently from maternal tissue. Many steps of early development *in vitro* of both the embryo and endosperm are comparable to the situation *in planta*, making this technique a useful model for studying embryogenesis and endosperm development, which otherwise is very difficult because these tissues are buried in the ovule.

In plant nutrition: fertilization is the application of major, minor and trace nutrients to soil to promote growth and development of plants. In seed technology, certain minor or trace elements are delivered to crops as seed treatments (**see: Treatments – Micronutrients**), and major nutrients may be applied as a **starter fertilizer** near to seed at sowing time, or in **band drilling**. (JDB)

Kranz, E. and Dresselhaus, T. (1996) *In vitro* fertilization with isolated higher plant gametes. *Trends in Plant Science* 1, 82–89.

Fibres

Seed fibres are among the most useful of the plant-derived fibres. Whether originating from the seed epidermis (**cotton**, **kapok** and **milkweed**) as **trichomes**, or the inner layer of the **pod** (kapok), or the thick seed-covering (**coconut**), plant fibres are flexible materials characterized by being substantially longer than wide. Cotton, kapok and milkweed are unicellular and represent some of the longest plant cells known. **Coir** is multicellular with individual cells held together by **pectins** and **hemicelluloses**. The fibres differ in dimensions and in appearance (Fig. F.5). The principal components of plant fibres are **cellulose** and varying amounts of **lignin** (Table F.1). In the plant, these two biopolymers simultaneously provide strength and flexibility to the plant **cell wall**. As processed fibres, the cellulose and lignin contents are very important in determining how the fibres can be utilized. (**See: Seedcoats – structure**)

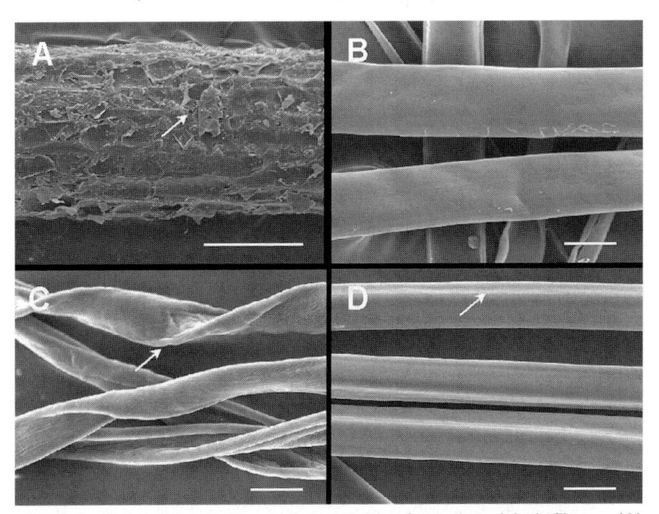

Fig. F.5. Scanning electron micrographs of seed and fruit fibres. (A) Coir, arrow indicates pore, (B) kapok, (C) cotton, arrow indicates convolution, (D) milkweed, arrow indicates ridge. The scale bar equals 100 μm in (A) and 20 μm in (B), (C) and (D).

Table F.1. Chemical composition of seed and fruit fibres.

Property	Coir[b]	Cotton[c]	Kapok[b]	Milkweed[d]
Cellulose (wt %)	36–43	88–96	43–64	34–60
Cellulose DP		4700	3300	5700
Cellulose crystallinity[a] (%)	29	90	67	73
Hemicellulose (wt %)	0.15–0.25	5.7	23–32	N/A
Pectin (wt %)	3–4	0.7–1.2	6–23	N/A
Lignin (wt %)	41–45		13–15	12–23
Oils/waxes (wt %)		0.4–1.0		0.3–3.0
Water soluble (wt %)	some	1.0		N/A

N/A, not available.
[a]Proportion of cellulosic material that is crystalline by X-ray diffraction as opposed to amorphous. The ratio of crystalline to amorphous regions in a fibre will dictate the physical properties.
[b]Batra, S.K. (1998) Other long vegetable fibers. In: Lewin, M. and Pearce, E.M. (eds) *Handbook of Fiber Chemistry*, 2nd edn, International Fiber Science and Technology Series, no. 15. Marcel Dekker, New York, USA, pp. 505–575.
[c]Wakelyn, P.J. (2002) *Cotton. Kirk-Othmer Encyclopedia of Chemical Technology*. John Wiley, New York, USA, www.mrw.interscience.wiley.com/kirk/articles/cottbert.a01/frame.html
[d]Privileged communication, Natural Fibers Corporation, Ogallala, Nebraska.

Seed and fruit fibres are used in the production of clothing and domestic products, paper, composite materials, non-woven products, brushes, and batting or stuffing. Natural fibres from plant sources have mechanical properties that are similar to manufactured fibres such as Nylon (Table F.2). In general, seed and fruit fibres are lighter, less expensive, biodegradable, and more easily renewable than synthetic fibres. Among the disadvantages of plant fibres is that they usually have a large variation in physical properties including length, are not fire-resistant, and are sensitive to moisture absorption.

Seed fibres probably evolved to aid in the wind dispersal of seeds (**see: Seedcoats – dispersal aids**). After **domestication** and **selection** by humans over thousands of years, plants such as cotton produce many more seed fibres than do wild species. Cotton has the largest production volume among the seed fibres. In addition to cotton, kapok, milkweed, and coir, there are several other seed fibres that are used in localized regions of the world (Table F.3).

Table F.2. Physical properties of seed, fruit and synthetic fibres.

Property	Coir[b]	Cotton[c]	Kapok[b]	Milkweed[d]	Nylon-6,6[e]
Tensile strength (N/tex)[a]	0.15	0.18–0.37	0.25	0.20	0.26–0.64
Young's-modulus (GPa)	6	12	31		2–5
Extension at break (%)	15–25	3–10	1.2	1.2	4.8–300
Moisture absorption (wt %)	10	8–25	10	7–13	0.97–8.5

[a]Newtons/tex. In textile manufacturing tex is used to describe g fibre per 1000 m yarn.
[b]Batra, S.K. (1998) Other long vegetable fibers. In: Lewin, M. and Pearce, E.M. (eds) *Handbook of Fiber Chemistry*, 2nd edn, International Fiber Science and Technology Series, no. 15. Marcel Dekker, New York, USA, pp. 505–575.
[c]Wakelyn, P.J. (2002) Cotton. *Kirk-Othmer Encyclopedia of Chemical Technology*. John Wiley, New York, USA, www.mrw.interscience.wiley.com/kirk/articles/cottbert.a01/frame.html
[d]Privileged communication, Natural Fibers Corporation, Ogallala, Nebraska.
[e]Cook, J.G. (1984) *Handbook of Textile Fibres*, Vol. 1 *Natural Fibres*. Woodhead Publishing, Cambridge, UK.

Table F.3. Additional seed fibres.

Common name	Species	Family
Tree cotton, simal, red silk cotton	*Bombax ceiba*	Bombacaceae
Indian kapok	*Bombax malabaricum*	Bombacaceae
Akund	*Calotropis procera*	Bombacaceae
Balsa, Corkwood tree, Downtree	*Ochroma pyramidale*	Bombacaceae
Silk floss tree	*Chorisia speciosa*	Bombacaceae
Pochote fibre	*Ceiba aesculifolia*	Bombacaceae
Silk-cotton tree, Yellow silk cotton	*Cochlospermum gossypium*	Bixaceae
Herald's trumpet	*Beaumontia grandiflora*	Apocynaceae
Kombé	*Strophanthus kombé*	Apocynaceae
Cattail	*Typha latifolia, T. angustifolia*	Typhaceae

The word fibre is also used in reference to diet, as dietary fibre, also called roughage. This is the undigested remains of plant material, particularly indigestible plant carbohydrates, such as the **hemicelluloses**. Bread, flour, breakfast cereals, rice and pasta – cereal grain products – are major sources of insoluble dietary fibre. Among the seeds, **pulses**, **oats**, **barley** and **rye** are rich sources of soluble fibre. (BAT)

Cook, J.G. (1984) *Handbook of Textile Fibres*, Vol. 1 Natural Fibres. Woodhead Publishing, Cambridge, UK.

FAOSTAT: http://apps.fao.org/page/collections?subset=agriculture

Witt, M.D. and Knudsen, H.D. (1993) Milkweed cultivation for floss production. In: Janick, J. and Simon, J.E. (eds) *New Crops*. John Wiley, New York, USA, pp. 428–431.

Field capacity

The amount of water, expressed as a percentage of the oven-dry soil weight, that is retained in the soil after the excess of water (termed gravitational water) has drained away under gravity (typically for 2 days, without rain). (**See: Seedbed environment**)

Field inspection

In the context of seed production, the inspection of a potential seed crop by a crop inspector or seed company representative, such as to meet the requirements of Seed **Certification** or **Quality-Declared Seed** Systems. (**See: Production for sowing, I. Principles**)

Filbert

An alternative name for **hazelnut**, cobnut.

Filmcoating

In seed technology, thin-filmcoating – sometimes called 'polymer coating' – is the process of applying materials on to seeds, such as colourants, formulations of **rhizobia** or chemical pesticide treatments. The technology is used to apply materials that are more firmly adhered on to seed surfaces, and in a more uniform and accurate way than can be achieved using conventional application techniques, and to present high-value seed for sale in an attractive cosmetic fashion. Filmcoating also facilitates smoother seed flow through planting equipment, and minimizes dust-off losses during seed handling, particularly in seed treated with pesticide formulations (**see: Planting equipment – agronomic requirements; Treatments – pesticides application**).

Characteristically, a uniform, dry, dust-free, water-permeable, thin coating membrane surrounds the seed, more or less evenly covering the seed surfaces. The quality of coverage over individual seeds depends on the amounts of materials used, on seed shape and contours (whether wrinkled or concave) and the nature of the seedcoat (whether mucilaginous, fibrous, chaffy, or with hairs or hooks – **see: Seedcoats – structure; Structure of seeds – identification characters**). The seed weight-increase is typically small (in the range 1–10%) and the underlying seed shape and size remains easily distinguishable, though the surface texture may or may not be greatly altered.

Since the early 1980s filmcoating technology has become progressively well established for high-value horticultural and ornamental seed species, and more recently is being adopted for treating some higher-volume crops, such as **maize**, **cotton**, **sunflower**, **canola**, **alfalfa**, **clover**, and some **grasses**. Filmcoating is also widely used to apply insecticides, fungicides and colourants on to dry pelleted seed; in some cases this is a preferred method of application to minimize phytotoxic effects, especially where treatments have to be applied at very high loading rates. Materials that can form relatively water-impermeable layers (including some that have temperature-dependent properties) are starting to be used to control the rate of seed imbibition: for example in maize to delay germination until more favourable moisture and temperature conditions have developed in the soil, and to manipulate **nicking** in hybrid seed production.

Aqueous suspensions of a binder, treatment formulations, pigments, opacifiers and adjuvants (such as surfactants and rheology agents) – commercially available in a variety of ready-mixed formulations – are sprayed onto or mixed in with the seed mass. Types of binders used include derivatized soluble **starches** and **celluloses** (e.g. with hydroxypropyl or methyl-substitution), polyvinyl acetate, polyvinyl alcohol or polyvinylpyrrolidone. Filler materials, such as talc or mica, may be included in the sprayed formulation or added separately as powders during the process, to aid seed flow or produce cosmetic effects.

A range of equipment is used according to the crop, seed throughput, treatment application rates and the required quality standard. Conventional seed treatment equipment (such as mist, auger or rotary coaters – **see: Treatments – pesticides application**) or **pelleting and encrusting** equipment are appropriate where the amount of applied liquid is relatively low and can be satisfactorily absorbed by the seed itself – a situation known as 'self drying' – or by added absorbent powders. If necessary, a separate drying stage can be included.

Where the amount of materials to be applied is relatively large, or where a higher analytical standard of seed-to-seed distribution or cosmetic finish is required, large amounts of liquid are usually needed, which involves concurrent spraying and drying. Batches of seed are presented to the spray system many times for treatment in an enclosed chamber to build up an even film layer, or apply a coat of several layers so as to separate one component from another within the filmcoat or to protect those who handle seed, or to control the release of an **active ingredient** after planting. Depending on conditions, a certain amount of cross-transfer of material from seed to seed occurs before drying. Two design principles (both derived from technology used for coating pharmaceutical tablets with delayed-absorption or colourant coatings) are in wide commercial use.

(a) *Spouted or fluidized bed systems*. Seed is held in vertical, cylindrical or inverted conical vessels; solutions are sprayed from below or above into a vigorous upward-moving stream of air, which stirs and dries the mass more or less immediately after the liquid is applied (Fig. F.6).

(b) *'Side-vented pan coaters'*. Seed is held within a horizontally inclined, perforated rotating drum, equipped with baffles or riser-blades; solutions are sprayed on to the surface of the seed mass, which is stirred and mixed, as drying air is drawn across the drum and through the seed (Fig. F.6).

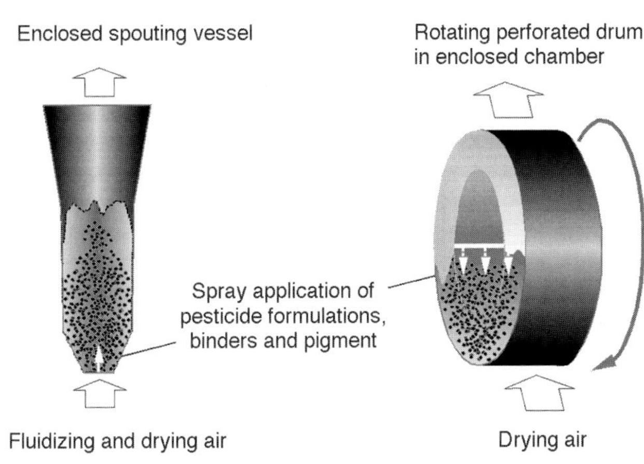

Fig. F.6. The principles of batch seed filmcoating using spouted bed (left) and side-vented pan coater systems (right). The geometrical proportions of commercially available equipment varies, with typical capacities from less than 1 kg up to about 25 kg and 250 kg seed, respectively. From Halmer, P. (2000) Commercial seed treatment technology. In: Black, M. and Bewley, J.D. (eds) *Seed Technology and its Biological Basis*. Sheffield Academic Press, Sheffield, UK, pp. 257–286.

There is increasing interest in using filmcoating polymer techniques as barrier layers to delay **imbibition**, and hence ensuing **germination** and emergence, until climatic conditions become suitable for continued crop growth, or for use in species that are vulnerable to rapid imbibition or in individual seedlots in which seedcoats are abnormally thin or damaged. Polymers with *in vitro* temperature-dependent water-permeability properties are being marketed to coat seeds for early planting so that the seeds can imbibe only when favourable moisture and temperature conditions have developed. Amongst the objectives of this technology is the regulation of emergence, so that subsequent vegetative growth results in the coordination of the flowering of parental lines (nicking) planted at the same time, such as for hybrid maize seed production (**see: Production for sowing, III. Hybrids**). Somewhat similarly, a water-resistant polymer coating has been evaluated as a tool to give a wider window of opportunity for sowing canola in the autumn in northern American latitudes, just before soils freeze over the winter; this results in earlier emergence in the spring, crop maturation and greater yields compared to the conventional spring sowing time (**See: Brassica oilseeds – cultivation, 4. Sowing and treatment**). (PH)

Halmer, P. (1988) Technical and commercial aspects of seed pelleting and film-coating. In: Martin, T.J. (ed.) *Application to Seed and Soil*. British Crop Protection Council Monograph, No.39, Farnham, Surrey, UK, pp. 191–204.

Stewart, R.F. (1992) Temperature sensitive seed germination control. *US Patent No*. 5129180.

Finger millet

Finger millet (*Eleusine coracana* subsp. *africana* is also known as African-, *kurrakan-*, *koracan-*, red millet, *akuma* (Uganda), bird's foot, *bulo* (Uganda), *coracana*, *dagusa* (Ethiopia and Eritrea), *kal* (Uganda), *nachni* (India), *oburo* (Uganda), *osgras* (South Africa), *ragi* (India), *rapoko* (Zimbabwe), *telebun* (Sudan) and *wimbi* (Swahili). The inflorescence has the appearance of a number of fingers at the apex of the stem, hence the name.

1. World distribution, economic importance

Finger millet grows in warm and temperate regions, in the cool, high-altitude regions of eastern, central and southern Africa (900–2000 m) and Asia (e.g. India, Indonesia, Japan, Australia). In parts of East and Central Africa it is a very important staple of the diet. This fourth-ranking millet (after **pearl**, **foxtail** and **proso** millets) comprises some 8% of the cultivated area under millet and 11% of the world millet production. Annual national grain yield is 500–1500 kg/ha. (**See: Cereals; Millets**)

2. Origin

African archaeological agricultural discoveries date the cultivation of finger millet back to 5000 years ago. The species existed in India nearly 3500 years ago, and reached Europe around 2000 years ago. *E. coracana* evolved to a polyploid (36 chromosomes) from *E. indica* subsp. *indica* (18 chromosomes) either in Africa (the supposed centre of origin is Uganda or Ethiopia) or India, with larger seeds than *E. indica* and **spikelets** that do not **shatter** at maturity.

3. Seed structure and composition

A spikelet has up to seven seeds. The seed of 1–2 mm in diameter is globose, smooth or rugose, with a depressed black **hilum**, one side slightly flattened. Seed colours are white, orange red, reddish brown, dark brown, purple to nearly black. The **pericarp** remains distinct during development and becomes a papery structure surrounding the seed. African highland types have long spikelets, long **glumes**, short **lemmas**, and grains enclosed within the **florets**. Afro-asiatic types have short spikelets, short glumes, short lemmas, and mature grains exposed outside the florets.

Seeds contain typically (fresh weight basis) about 8% **storage protein**, 73% carbohydrate (mostly **starch**), 1.5% **oil** (triacylglycerol), and calcium and iron at 350 and 4 mg/100g respectively. B vitamins are present. (**See: Cereals, 4. Basic grain anatomy**)

4. Uses

Finger millet is a food grain, mainly produced for household consumption and local trade often providing extra income for the women. It is an important famine food because the seed stores well for many years. Fresh or stored grain is used for malting and brewing of traditional beer. Ground germinated grain is boiled in water to make a thin gruel. Its flour, obtained after grinding the grain, is used for beverages, weaning food and stiff porridge. Before grinding, cassava chips or flour may be added to cleaned grain. In India, grain soaked and germinated for 7 days is dried, roasted and ground. The plant is used as a vegetable (Indonesia), for domestic use (e.g. thatching, granary food containers, bracelets), traditional medicine, and chemicals (hydrocyanic acid). The straw can be used as fodder and the field can be grazed after harvest. (WAJdeM)

Dendy, D.A.V. (ed.) (1994) *Sorghum and Millets: Chemistry and Technology*. American Association of Cereal Chemists. St. Paul, MN, USA.

Finger millet – cultivation

Finger millet (*Eleusine coracana* sp. *africana*) and its wild relatives (in South Africa, e.g. *E. indica*, *E. multiflora* and *E. tristachya*) are members of the Poaceae, subfamily Chloridoideae, which also includes teff. It is an allotetraploid with 36 chromosomes, and a chromosome number, x = 9. The crop is grown in warm and temperate regions, as well as in the cooler, high-altitude regions of parts of Africa, India, Indonesia, Japan and Australia.

Finger millet is a **C4 photosynthetic** plant, adapted to many soil types, and requiring an annual rainfall between 0.5 and 1.25 m and minimum temperatures ≥18°C , though it may tolerate low rainfall (130 mm if well distributed) and dry spells during early crop development. The plant grows up to about 1.3 m, and forms a dense, shallow and fibrous root system. It can be a sole crop or used in a mixed cropping with **legumes** (e.g. **pigeon peas**), vegetables or trees (fruits). Yield is highest after early, thorough weeding on well-drained, fertile soils; irrigated high-input crops yield between 1500 and 5000 kg/ha. Two crops per year may be produced in areas with a bimodal rainfall, or by means of irrigation. However, it is an important cereal on infertile soils in slash-and-burn (chitemene) cultures, e.g. in Zambia.

1. Genetics and breeding

Globally, breeding inputs have been low compared to other cereals, though selection breeding programmes exist in India, Ethiopia, Uganda and Kenya, and improved cultivars have been released from them. Several sizeable germplasm collections exist in India, Kenya, Ethiopia and in the USA. Named cultivars are not necessarily uniform, however; in Uganda, for example, they may have plant mixtures of three to six distinct head types. The incidence of cross-pollination is low (1%). No hybrids have been released, possibly because of lack of **cytoplasmic male sterility**. A white-seeded, sterile mutant has been developed by induced mutation, as an alternative to the natural coloured seed forms. The first genetic maps allow comparisons of the molecular genetics across the grasses.

2. Development and production

Finger millet is a **short-day plant**, with an optimum photoperiod of 12 h. **Florets** are hermaphrodite, sterile, or male only. Heads consist of four to nine spikes (also called fingers), bearing spikelets with up to seven seeds each. Flowering of the **spike** is from the top downwards, and in the **spikelet** from the bottom to the top, and takes 8 to 10 days. The time from sowing to harvest is from 110 days to 6 months, depending on cultivar, location, and time of planting (temperature and daylength). When a mixture of cultivars is grown, ripening may be uneven.

In cultivars with low susceptibility to **shattering**, harvesting is delayed until full ripeness; but timing is critical if susceptibility is high, and three or four pickings may be required. Often this is by labour-intensive hand-harvesting of individual heads with a knife. Piling of the heads in heaps or stacks for a few days, or several months, is used for fermentation and to ease removal of the grain. Threshing of heads is done by beating with sticks, or by bullocks or stonerollers. After thorough drying, seed is stored in sacks, bins or grain bins, though unthreshed heads are also stored. Grain can be stored for 10 years or more without deterioration or damage by stored grain insects; consequently finger millet is an important famine food. Dormancy is normally not an important issue.

3. Sowing

Finger millet is a labour-intensive crop – in terms of seedbed preparation, thinning, weeding, bird scaring, harvesting and threshing. A fine, firm seedbed is required and may be compacted after sowing by animals. Seed can be sown (< 20 kg/ha), broadcast (up to 35 kg/ha), dry planted or transplanted (at 3–4 weeks). Drilling and planting in lines facilitates inter-row weeding.

4. Treatment

Finger millet is susceptible to a wide range of diseases and pests, but only a few are important. The most important disease is blast, in particular head blast, caused by the fungus *Pyricularia grisea* (teleomorph *Magnaporthe grisea*), causing infection from seedling to grain formation in most varieties and causing yield losses from 10–90%. Resistant cultivars appear to have a higher total phenol and tannin content than susceptible ones, compact heads, and/or dark seeds. Leaf and seedling blight is caused by the fungus *Helminthosporium nodulosum* (synonym *Dreschlera nodulosum*).

Seed treatments are not commonly used, even though low seed rates make it convenient for such treatments to be applied. They may have modest value for control of blast and may give additional control of seedborne fungi, pre-emergence **damping off**, and seedling **blight**, especially in young plant productions. (WAJdeM)

Belton, P.S. and Taylor, J.R.N. (eds) (2002) *Pseudocereals and Less Common Cereals. Grain Properties and Utilization Potential.* Springer, Berlin and New York.

Dendy, D.A.V. (ed.) (1994) *Sorghum and Millets: Chemistry and Technology*. American Association of Cereal Chemists, St Paul, MN, USA.

FIS

The French acronym for the International Seed Trade Federation, which in 2002 merged with **ASSINSEL** to become the International Seed Federation. (**See: ISF**)

Fixed oil

An oil that does not evaporate on warming, e.g. those from maize, safflower and peanut. Such oils, consisting of a mixture of fatty acids and their esters, are classified as *solid* (chiefly stearic acid-based), *semisolid* (chiefly palmitic acid-based) and *liquid* (chiefly oleic acid-based). They are also classified as *drying*, *semidrying* and *nondrying*, depending on their tendency to solidify when exposed, in a thin film, to air. (**See: Oils and fats**)

Contrast with **essential oils**, which are aromatic liquid oils that usually smell pleasant, or have an essence: hence 'essential'.

Flavin

From the Latin *flavus* meaning yellow: a yellow water-soluble nitrogenous pigment derived from isoalloxazine (molecular formula $C_{10}H_6N_4O_2$). Examples include riboflavin (vitamin B2). The coloured nature of flavins allows them to act as **chromophores** for **chromoproteins** such as **cryptochrome**, which may be involved in light-inhibited germination (**see: Light, dormancy and germination**).

Flax
See: Linseed

Floret
A flower in a grass spikelet (see Fig. F.7). (**See also: Domestication**, Fig. D.30; **and Wheat – cultivation**, Fig. W.6)

Floss
The fine, long seed hairs of the silk-cotton tree, milkweed, kapok and swallowwort plants. (**See: Fibres; Kapok; Milkweed**)

Flotation
Techniques used in seed **conditioning** to separate lighter from heavier seeds. (**See: Density sorting; Fluming**) The principle is also used in several **pre-germination** procedures.

Flower structure
See: Determinate flowering; Indeterminate flowering; Inflorescence; Reproductive structures

Flowering plants
See: Angiosperms

Fluid drilling
A seed enhancement, sometimes referred to as fluid sowing or gel seeding, where just-germinated seeds are suspended in an aqueous gel and transferred to a seedbed. This crop establishment technique is devised for use primarily in a field environment to overcome problems associated with conventional dry-seed sowing. For fluid drilling, seeds are pre-germinated under ideal conditions, avoiding negative seedbed environmental effects such as insufficient moisture or high temperature, so seedling emergence is more rapid and synchronous. Faster seedling emergence should lessen the likelihood of soil crusting or attack by soilborne pathogens.

Fluid drilling is an integrated system that involves: (i) germination of seeds before sowing; (ii) separation of germinated and nongerminated seeds; (iii) storage of germinated seeds; (iv) preparation of the gel for suspending seeds; and (v) planting of germinated seeds suspended in gel using specialized equipment. Several types of gel materials are used: synthetic mineral clays, starch-polyacrylonitrile polymers, derivatized **cellulose** polymers, copolymers of potassium acrylate and acrylamide, and natural **starch**. A good gel should suspend seedlings of various sizes for at least 24 h and yet be easily pumped through delivery tubing. Gels should be nonphytotoxic and mix easily with water of different pH and mineral content. Gel material should be relatively inexpensive, not dry to form a skin, and rapidly breakdown in the soil. Fungicides, herbicides, safening agents (to reduce any

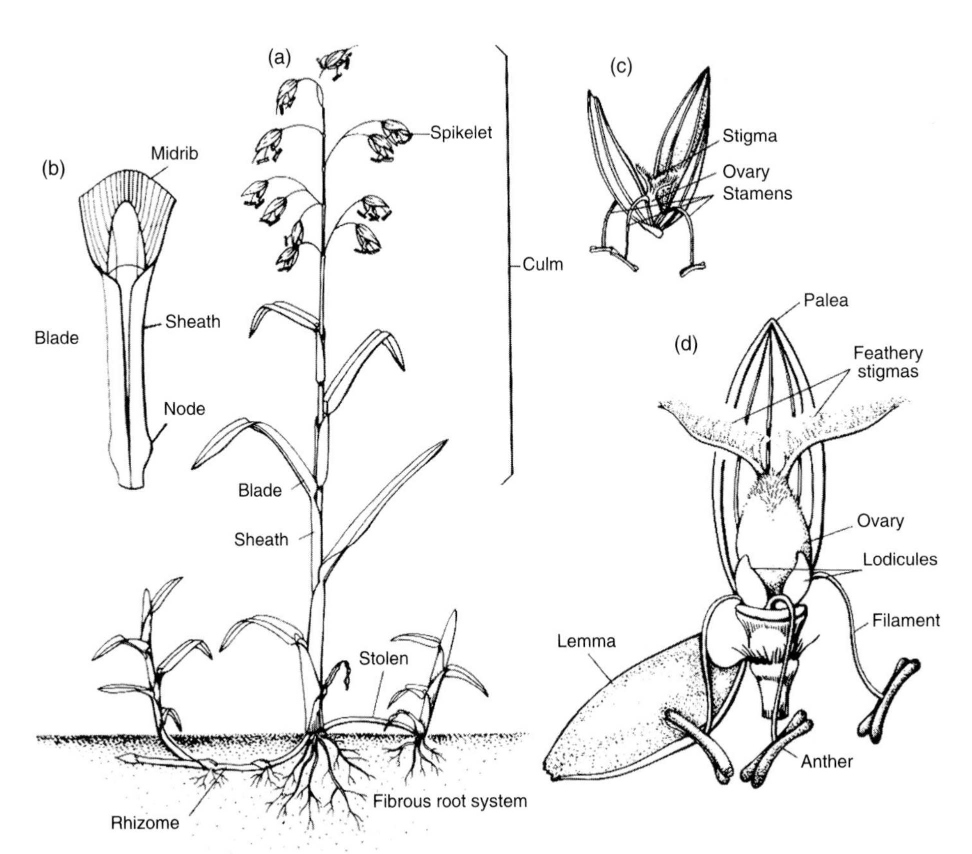

Fig. F.7. A generalized grass plant (a) with leaf base (b), closed (c) and open (d) floret with detached lemma. The ovary after fertilization becomes the **caryopsis** (grain) of the grass (cereal) often with the closely-adhering **lemma** and **palea** as the hull. (Note the palea lacks an **awn**.) Adapted from Fig. 5.2 in Simpson, B.B. and Ogorzaly, M.C. (2001) *Economic Botany*. McGraw Hill (International edition). With thanks to Dr M. Ogorzaly and the publishers.

undesirable effects of either of these two types of chemicals, **see: Safeners**), beneficial **rhizosphere microorganisms**, and minerals are some of the common gel additives that have been reported to improve establishment.

Despite the benefits of fluid drilling, there are challenges as well. Specialized planting equipment is required, extra time is needed to prepare for planting, and seeds must be pregerminated. The use of **priming** and plug **transplants** may deliver comparable performance to fluid drilling with less effort and expense. Once germinated, seeds have limited shelf life. If adverse weather prevents field establishment, germinated seeds must be stored with minimal radicle growth and loss of vigour. They have been successfully preserved in cold humid air, cold aerated water, aerated osmotic solutions, in plastic bags under partial vacuum, or nitrogen gas at 7°C, or in cold hydroxyethyl cellulose fluid drilling gel. Although fluid drilling helps mitigate unfavourable soil conditions, planting pregerminated seeds into dry soils may result in poorer stands than if dry seeds were used. As the **radicle** of pregerminated seeds elongates, seeds progressively lose desiccation tolerance. Thus, the importance of adequate soil moisture at planting is essential for successful establishment.

Seedlots that have variable germination rates are problematic and result in germinated and nongerminated seeds mixed in the same batch of seeds, which has an adverse effect on the uniformity of field establishment particularly if the nongerminated seeds are not removed. Various seed treatments are applied to increase the speed and uniformity of germination in preparation for fluid drilling. These enhancements include sorting to obtain uniform seed size or remove nongerminated seeds, abscisic acid treatment, cold treatment, priming, and controlled desiccation of germinated seedlings. (GEW)

Finch-Savage, W.E. (1981) Effects of cold-storage of germinated vegetable seeds prior to fluid drilling on emergence and yield of field crops. *Annals of Applied Biology* 97, 345–352.

Finch-Savage W.E. and McQuistan, C.L. (1989) The use of abscisic acid to synchronize carrot seed germination prior to fluid drilling. *Annals of Botany* 63, 195-199.

Pill, W.G. (1991) Advances in fluid drilling. *HortTechnology* 1, 59–65.

Pill, W.G. (1995) Low water potential and presowing germination treatments to improve seed quality: preplant germination and fluid drilling. In: Basra, A.S. (ed.) *Seed Quality: Basic Mechanisms and Agricultural Implications*. Food Products Press, New York, USA, p. 342.

Fluming

In seed **conditioning**, the removal of lightweight and immature seed by flotation in water, as part of the seed extraction activity. A technique used in 'wet seed' species such as vine fruits of the Cucurbitaceae and Solanaceae. (**See: Production for Sowing, V. Drying**)

Follicle

A fruit derived from a single **carpel** dehiscing along one (usually the ventral) suture, e.g. *Cercidiphyllum* (Cercidiphyllaceae), *Grevillea* (Proteaceae). (**See: Fruit – types**)

Food and Agriculture Organization of the United Nations (FAO)

Founded in 1945, this organization, which is one of the largest United Nations agencies, has the role of raising nutrition and standards of living, and improving agricultural productivity and the condition of rural human populations. Its Seed and Plant Genetic Resources Service manages programmes concerned with seed policies, seed improvement and security, and germplasm exchange. It also supports the FAO Global System on Conservation and Utilization of Plant Genetic Resources.

The FAO publishes FAOSTAT (faostat.fao.org) – international time-series databases that include statistics on seed and grain, for instance on Production, Trade, Food Balance Sheets, and Exports by Destination. Selected information is cited in many articles in this Encyclopedia, and is the basis for the **Crop Atlas Appendix** maps.

Food seeds

Seeds are the staples of the human diet. Important for hunter-gatherers (**see: Ethnobotany**), they also became the foundation crops where agriculture developed (**see: Cereals; Domestication; Legumes; Pulses**). The major food seeds are the cereals and legumes. In 2001, for example, their combined world production totalled nearly 2.5 billion t. It is estimated that about 70% of human food is consumed directly as seeds, and a high percentage of important dietary components are manufactured from seeds, such as breads, pastas, noodles and tofu. A substantial annual tonnage of **oilseeds** provides the culinary oils. (**See: Crop Atlas Appendix, Maps 5–19, 21, 22, 25**)

Of lesser quantitative importance worldwide are the **pseudocereals** such as **amaranth**, **buckwheat** and **quinoa**. Other relatively minor food seeds used, for example, as snack foods and in confectionery are **sunflower**, **pine nuts**, **poppy**, **pumpkin**, **squash**, **sesame**, **watermelon** and **linseed**. **Nuts** are important for direct consumption or as a minor source of edible oils. Numerous other types of seeds are of local use. The many **spice** seeds are used as food flavours or as medicinals.

Seeds of many species are an important food source for numerous wild and domesticated animals. Sprouted seeds also serve as food (**see: Sprouted seeds – food**). (MB)
(**See: Beverages; Malting**)

Kiple, K.F. and Ornelas, K.C. (eds) (2000) *The Cambridge World History of Food*, Vols 1 and 2. Cambridge University Press, Cambridge, UK.

Vaughan, J.G. and Geissler, C.A. (1997) *The New Oxford Book of Food Plants*. Oxford University Press, Oxford, UK.

Forages

Crops grown or maintained in pastures or natural grasslands, consisting predominantly but not entirely of grass and legume species, to feed livestock animals, by grazing or by harvesting to produce hay or silage. (**See: Grasses and legumes – forages**)

Forensic uses of seeds

The identification of seeds or seed parts is sometimes an important contribution to evidence presented in criminal investigations and legal cases. (**See: Archaeobotany; Structure of seeds – identification characters**)

Foundation (stock) seed

An official term used in North America for the first major seed multiplication of an improved strain or variety, handled so as to maintain its superior genetic qualities: a term equivalent to 'Basic Seed' in the **OECD Seed Scheme**. Seed stocks are multiplied from **Breeder Seed**, for subsequent use in producing **Certified Seed**, either directly or through **Registered Seed**. Pre-foundation seed usually refers to small supplies of seed planted to produce Foundation seed. Seed Associations or Institutes, whose work is regulated by government agencies, usually carry out the production of Foundation seed. (**See: Production for sowing**)

Foxtail millet

Foxtail (German, Italian, or Siberian) millet (*Setaria italica* (L.) Beauv.) is an annual warm-season grass which grows 0.5–1.5 m tall under cultivation. Foxtail millet is divided into two subspecies: *S. italica* subsp. *maxima* and *S. italica* subsp. *moharia*. The former is higher yielding therefore more suitable for grain production. In terms of quantity produced it is the second most important millet (**see: Cereals; Millets**).

1. Distribution and importance

It is grown in some Asian countries (mostly China, India, Indonesia, the Korean peninsula and Japan) for food and/or feed. China ranks first in worldwide production where it is grown for both food and feed: it is the most important millet in Japan. Russia has decreased the total area sown with this crop. Foxtail millet is widespread in the USA through the Great Plains where it is used for high quality hay, pasture and green fodder. In some parts of southern Europe foxtail millet can be found mainly for feed purposes.

2. Origin

Foxtail millet was cultivated in China at least 5000 years ago, possibly originating from *S. viridis* in East Asia. Seeds have been found in remains in Austria and the Swiss lake dwellings of the late stone age and early bronze age. (**See: Domestication**, Fig. D.32).

3. Seed characteristics and composition

Seeds, borne on a terminal panicle, are small, 2–3 mm long, and somewhat narrower. Colour may be white, yellow, brown, red or black. Panicles do not readily **shatter**.

Typical **storage protein, starch** and **oil** (triacylglycerol) contents (fresh weight basis) are respectively 11, 63 and 4%. There is a higher proportion of essential amino acids and B vitamins than in pearl millet. (**See: Cereals, 4. Basic grain anatomy**)

4. Uses

Foxtail millet is used for both human and animal food. Grains are often utilized for the preparation of different types of porridges, and can be used to make beer (in Russia) or distilled liquor ('*Awamori*' in Japan). (MB)

Baltensperger, D.D. (1996) Foxtail and Proso Millet. In: Janick, J. (ed.) *Progress in New Crops*. ASHS Press, Alexandria, VA, USA, pp. 182–190.

Dendy, D.A.V. (ed.) (1994) *Sorghum and Millets: Chemistry and Technology*. American Association of Cereal Chemists, St Paul, MN, USA.

Free radical

An atom or molecule with one or more unpaired electrons. Seeking to pair the unpaired electron, free radicals either abstract (oxidize) or donate (reduce) electrons from and to other molecules, frequently leading to the production of new radicals in a reaction cascade that degrades lipids, proteins and nucleic acids (**see: Peroxidation**). Electron-dense portions of biomolecules are particularly susceptible to free radical-mediated oxidation. For example, the hydrogen atoms on acyl chains of lipids (**see: Fatty acids**) are a ready source of electrons, which are even more accessible in unsaturated acyl chains since the carbon–carbon double bonds weaken the carbon–hydrogen bond. Free radicals are usually highly reactive even at low temperatures and moisture levels and thus have been implicated in seed **ageing** and **desiccation damage** (**see: Deterioration reactions**). An important source of free radicals is molecular oxygen which can be converted to **reactive oxygen species** (ROS) with ionizing radiation, UV light, and transition metals, or as a consequence of metabolism (e.g. incomplete oxidation of water during photosynthesis or reduction of oxygen during respiration). Free radicals can also form from reactions of metals with hydroperoxides to form alkoxyl radicals, peroxides to form peroxyl radicals or aldehydes to form ROS (and a dicarbonyl compound). **Antioxidants** inhibit free radical-mediated reactions by a variety of mechanisms. (CW)

Fresh weight

The total mass of the seed (or other plant structure) in the 'fresh' state. It includes all the 'dry' material in the seed and any water that is present. (**See: Dry weight; Water content**)

Frost seeding

The surface broadcast placement of seed into existing vegetation in late winter or very early spring while the soil is still frozen but the top soil is dry. Used in North America, where it is also known as overseeding. Due to lack of uniform germination and emergence, frost seeding is more suited to permanent pastures than hay fields. Success of this seeding method is dependent on the timing of soil freeze–thaw cycles, late snowfall, spring rain, and the management given to the existing vegetation prior to and after seeding. (**See: Grasses and legumes – forages**)

Also used to describe the practice of sowing into wet soil where the top layer is temporarily frozen, so that the ground can support heavy machinery.

Fructans

Linear and branched polymers of fructose, with a terminal sucrose (i.e. a single glucose). They are present as a storage form of **carbohydrate** in about 15% of all flowering plant species, but only in vegetative tissues. However, transgenic seeds (e.g. **maize**) have been generated which synthesize and store fructans in the developing endosperm vacuole, though the amount produced in dent corn was only about 1% of mature kernel dry weight.

Fruit

Commonly defined as the product of a flower or a gynoecium (female **reproductive structures**), i.e. the fully matured **ovary**, containing seeds. However, this simplistic definition causes difficulties in cases where, for example, entire inflorescences form the dispersal unit (**diaspore**), e.g. figs, pineapple. Spjut, R.W. (1994) (A systematic treatment of fruit types. *Memoirs of the New York Botanical Garden* 70, 1–182) offers a much more complex but more logical definition of the fruit: a fruit is a propagative unit consisting of one or more mature **ovules** and their **megasporophylls**, or megasporophyll-scale complexes (conifers), or parthenocarpic ovaries, in a strobilus, cone, gynoecium, concrescent gynoecia, or gynoecia that disseminate together, at the time it or its seed disperses from the plant, or just prior to germination on the plant, including any attached scales, bracts, modified branches, perianth, or inflorescence parts. (See: **Seedless fruits**)

Fruit – types

The seed-bearing structures in **angiosperms** formed from the **ovary** after flowering.

A typical fruit is surrounded by the fruit wall, the **pericarp**. The outer wall of this is the **epicarp** (exocarp, ectocarp), to the inside of which are the **mesocarp** and **endocarp**.

Fruits may be ***simple***, if derived from one ovary, ***aggregate***, if from a single flower with many ovaries, or ***multiple*** if from a fusion of independent flowers borne on a single structure. They are classified primarily on origin, texture and dehiscence, some of which are noted below, and illustrated in figures after each section.

1. Dry indehiscent fruits (fruits that do not split open at maturity)

Achene. A one-seeded, dry, indehiscent fruit with seed attached to fruit wall at one point only, derived from a one-loculed superior ovary, e.g. **buckwheat** (Polygonaceae) and Ranunculaceae.

Calybium. A hard one-loculed dry fruit derived from an inferior ovary, e.g. *Quercus* (oak).

Capsule, indehiscent. Dry fruit derived from a two- or more loculed ovary, e.g. boab or lemonade tree.

Caryopsis or grain. A one-seeded dry, indehiscent fruit with the seedcoat adnate (joined by having grown together) to the fruit wall, derived from a one-loculed superior ovary. As in grasses and **cereals**.

Cypsela. An achene derived from a one-loculed, inferior ovary typical of composites (Asteraceae), e.g. **sunflower**.

Nut. A one-seeded, dry, indehiscent fruit with a hard pericarp, usually derived from a one-loculed ovary, e.g. **hazelnut**, **walnut**.

Nutlet. A small nut; most are dry, but some such as lychee are surrounded by a fleshy **aril** (the term is also applied to the stone of a cherry or peach).

Samara. A winged, dry fruit, as in maple, elm and ash.

Utricle. A small, bladdery or inflated, one-seeded, dry fruit; present in some *Amaranthus* spp.

2. Dry dehiscent fruits (fruits that split open at maturity)

Capsule, dehiscent. Dry, dehiscent fruit derived from a compound ovary of two or more carpels. The **poppy** dehisces through pores.

Follicle. A dry, dehiscent fruit derived from one carpel that splits along one suture, e.g. delphinium, **milkweed**.

Legume. A usually dry, dehiscent fruit derived from one carpel that splits along two sutures. Many **legumes**, e.g. **peas** and **beans** (Fabaceae), form this type of fruit, also known as a pod, although those of **carob** and honey locust are indehiscent.

Loment. A legume that separates transversely between seed sections, e.g. crown **vetch**.

Silicle. A dry, dehiscent fruit derived from two or more

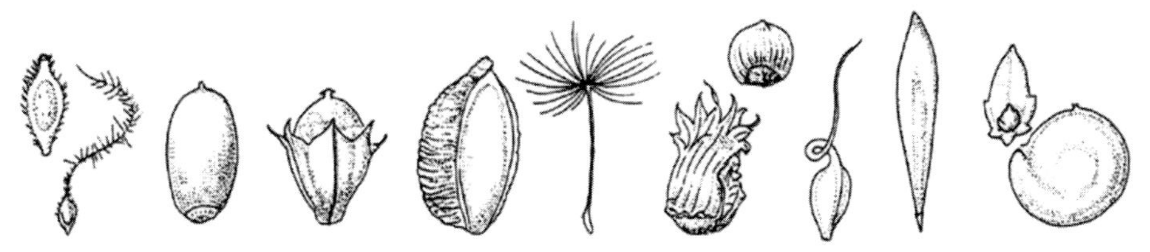

Fig. F.8. Dry indehiscent fruits.

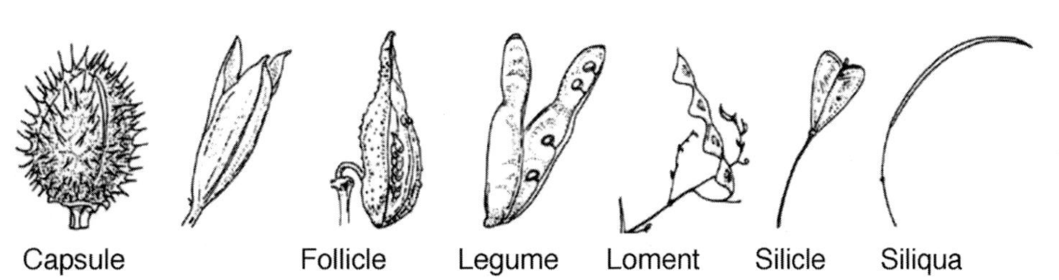

Fig. F.9. Dry dehiscent fruits.

carpels that dehisce along two sutures and which has a persistent partition after dehiscence and is as broad as, or broader, than long, such as in *Capsella*.

Siliqua. A silicle-type fruit that is longer than broad; typical of the **Brassica** family.

3. Fleshy fruits

Berry. Fleshy fruit, with succulent pericarp, as in **tomato**, **eggplant**, blueberries, grape.

Drupe (stone). A fleshy fruit with the seed surrounded by a stony endocarp, as in **almond**, plum, peach, cherry, **olive**. As *pyrene*.

Drupelet. A small drupe, as in *Rubus*.

Hesperidium. A thick-skinned septate berry with the bulk of the fruit derived from glandular hairs, as in *Citrus* spp.

Pepo. A berry with a leathery non-septate rind derived from an inferior ovary, as in **cucurbits** (cucumbers, melons, squash).

Pyrene. Fleshy fruit with the seed surrounded by a stony endocarp, as *drupe*, but defined by the fact that in fruits bearing more than one seed, each is surrounded by a stony endocarp, as in holly (*Ilex*).

4. Accessory fruits

Bur (Involucre). Cypsela enclosed in dry involucre, as in cocklebur (*Xanthium*).

Coenocarpium. Multiple fruit derived from ovaries, floral parts, and receptacles of many coalesced flowers, as in pineapple.

Diclesium (Calyx). Achene or nut surrounded by a persistent calyx, as in four o'clock (*Mirabalis*).

Glans (Involucre). Nut subtended by a cupulate, dry involucre, as in *Quercus*.

Hip or Cynarrhodion. An aggregate of achenes surrounded by an urceolate receptacle and hypanthium, as in roses.

Pome (Receptacle and Hypanthium). A berry-like fruit, adnate to a fleshy receptacle, with cartilaginous **endocarp**, as in apple, pear, quince.

Pseudocarp (Receptacle). An aggregation of achenes embedded in a fleshy receptacle, as in strawberry.

Pseudodrupe (Involucre). Two–four-loculed nut surrounded by a fleshy involucre, as in black walnut.

Syconium (Receptacle, possibly peduncle). Multiple fruit surrounded by a hollow, compound, fleshy receptacle, as in figs (*Ficus*).

Tryma (Involucre). Two–four-loculed nut surrounded by a dehiscent involucre at maturity, as in most species of *Carya*, e.g. mockernut.

Winged Nut (Bract). Nut enclosed in a winglike bract, as in *Carpinus* spp. (hornbeam).

Figures based on:

Radford, A.E., Dickison, W.C., Massey, J.R. and Bell, C.R. (1974) *Vascular Plant Systematics*. Harper and Row, New York, USA.

Fruit coat

See: Fruit – types; Pericarp

Fumigants

Volatile solid, liquid or pressurized-gaseous pesticides are used to control **storage pests** in food and other commodities, and to sterilize soil during seedbed preparation. Grain and grain products are directly treated while in bulk store or in transit; and grain bins, handling equipment or entire buildings are pretreated before they are filled. (**See: Cereals – storage; Storage management; Treatments – pesticides**) Fumigants are used to sterilize soils by injection, sometimes in conjunction with **mulching**, to eradicate weed seeds and fungal, insect and other pests before sowing seeds or to ensure more rapid transplant growth in **greenhouse transplant production** systems.

The active ingredients in fumigants are either liquids that are volatile at atmospheric pressure (such as chloropicrin) or when packaged under high pressure (such as phosphine and methyl bromide, MeBr), or are solids formulated as pellets or tablets (such as aluminium or magnesium phosphide, which generate phosphine) or liquids (such as metam-sodium, which generates methyl isothiocyanate) that release gases under high humidity conditions. Other fumigant active ingredients include sulfuryl fluoride, carbonyl sulphide, ethyl formate, and 1,3-dichloropropene or propylene oxide (used under low pressure or in CO_2-enriched atmospheres). Such chemicals are used singly, or sometimes in combinations. The effective, safe and economic fumigation of seed or seedbeds depends on such factors as temperature control, gas-tight sealing and efficient air circulation by fans, preliminary evacuation, post-treatment ventilation, and the avoidance of explosion and operator exposure hazards. For fumigating malting barley, chemicals that have no adverse effects on seed germination must be used, such as phosphine.

For many years, MeBr has been the soil fumigant and sterilant of choice, due to its efficacy in many different growing situations, soil types, temperatures and moistures, and its season-long activity. MeBr can virtually eliminate soilborne pathogens, nematodes and weeds, such as *Fusarium*, *Phytophthora*, *Pythium* and *Verticillium* species (**see: Pathogens**). It has been widely used, for example, in tomato and pepper production systems under polyethylene mulches, in raising tobacco seedlings in nursery seedbeds, and to

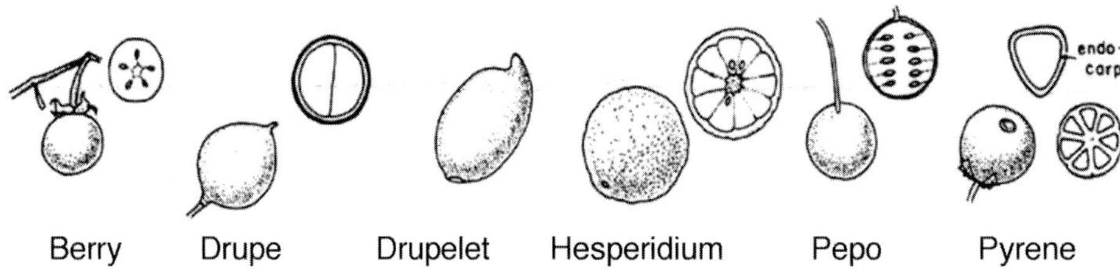

Fig. F.10. Fleshy fruits.

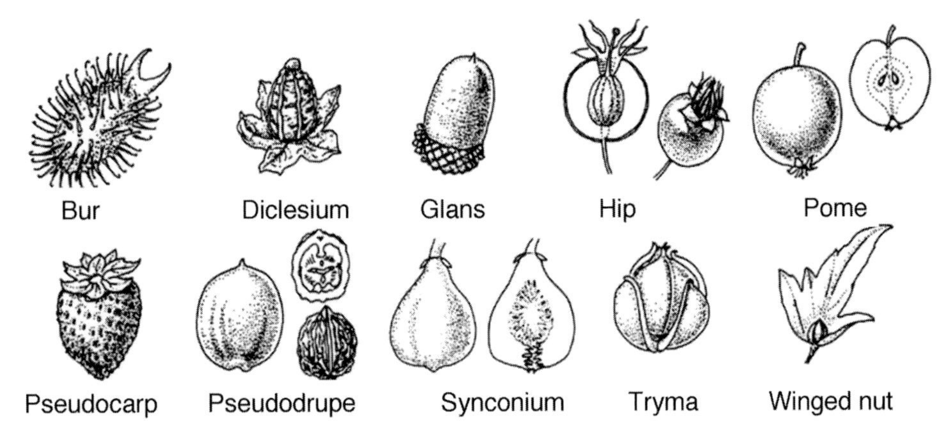

Fig. F.11. Accessory fruit types.

fumigate stored commodity grain, storage structures and transport vehicles. However, MeBr has been implicated as a major atmospheric ozone-depleting substance and its use is banned in developed countries from 2005 and in developing countries from 2015. Alternative fumigants are now being evaluated or re-evaluated to match its efficacy. (PH)

Functional genomics

See: **Genomics**

Fungi

See: **Arbuscular mycorrhiza; Mycorrhiza; Pathogens; Rhizosphere microorganisms; Storage fungi**

Fungicides

Chemical, physical or biological agents intended for preventing, destroying, repelling or mitigating the action of **fungi**, such as by killing or inhibiting the development (fungistatic) of spores or mycelia of **pathogenic** species. Microorganisms used in this way are termed biofungicides. Treatments aimed at bacterial and viral pathogens are sometimes, though inaccurately, called 'fungicides' in a broad sense. (**See: Rhizosphere microorganisms; Treatments – pesticides, 3. Fungicides**)

Funiculus (funicule)

The stalk of the **ovule** by which it is attached to the **placenta**. (**See: Reproductive structures, 1. Female**)

Furrow

A shallow groove made in the soil surface into which seeds are sown, and which is formed manually or mechanically by a variety of furrow opener implements mounted in front of the seed drill. Alternatively, a shallow channel cut in the soil surface, usually between planted rows, for controlling surface water and soil loss, or for conveying irrigation water. (**See: Planting equipment – placement in soil; Tillage**)

(a) *Furrow (or band) application* is a term usually applied to the placement of pesticides or fertilizers near or alongside seed at time of sowing in furrows. (**See: Starter fertilizer**)

(b) *Furrow irrigation*, or 'surface irrigation' (in contrast to sprinkler irrigation or sub-irrigation), applies water between crop rows in furrows, such as between ridge-tillage or raised-bed systems, using a system of ditches, pipes, siphons and pumps. Crops sown in dry regions are often furrow-irrigated shortly after sowing.

Fusion protein

See: **Biopharmaceuticals; Recombinant protein**

Fuzzy seed

Linted cottonseed, often referred to as 'whole cottonseed'. After ginning the long fibres from Upland varieties of **cotton** (*hirsutum* type) an appreciable amount of short fibres (linters) remains attached causing clumping, and needs to be removed before seed can either be sown or processed. (**See: Cotton; Delinting; Lint and Linters**)

G

G proteins

See: **Phytochrome; Signal transduction – hormones; Signal transduction – some terminology**

GA

Symbol for **gibberellin**. Each gibberellin (of which there are over 135) has a number assigned to it, e.g. GA_3 (gibberellic acid), GA_9, GA_{27}, which to a large part is related to the order in which they were identified. Most have no known biological activity.

GADA test

A proposed test for storage potential of some seeds, mainly used in seed research, based on the observation that glutamic acid decarboxylase (GADA) enzyme activity decreases in stored seedlots before overall **viability** begins to fall. (**See: Vigour testing – biochemical**)

Galactinol series oligosaccharides

Galactosyl oligomers of the cyclitols *myo*-inositol, D- *chiro*-inositol and D-pinitol (3-O-methyl-D-*chiro*-inositol) (Fig. G.1). They are soluble sugars and are generally regarded as being a source of stored respirable substrate, present in the cytoplasm for use during germination. In many seeds, especially those of legumes, they are present in much lower quantities than the **raffinose series oligosaccharides** (Table G.1).

Galactinol ((mono-) galactosyl *myo*-inositol, Fig. G.1), is synthesized by the addition of galactose to *myo*-inositol, the enzyme involved being galactinol synthase.

UDP-galactose + *myo*-inositol → UDP + galactinol

Galactinol itself is a galactose donor in the production of the raffinose series oligosaccharides; it is therefore prevalent during their synthesis in developing seeds. Di- to tetra-galactosyl *myo*-inositol may be present in some seeds.

Galactopinitols (Fig. G.1) are common in legume seeds; the mono-, di- and tri-galactosyl oligomers of pinitol are called galactopinitol, ciceritol or digalactopinitol, and trigalactopinitol. There are two series of galactopinitols (A and B) that differ in the position of the methyl group in the pinitol ring. Galactose is removed from the galactopinitol oligomers during germination by α-galactosidase. Similar removal in the colon of monogastric animals appears to occur more slowly than from the raffinose series oligosaccharides, and thus galactopinitols are less prone to cause flatulence.

Table G.1. Content (in mg/g dry mass) of the major galactinol- and, for comparison, total raffinose-series oligosaccharides in the axes and cotyledons of some seeds.

	Galactinol		Galactopinitols		Ciceritol		RSOs	
Species	A	C	A	C	A	C	A	C
Legumes								
Soybean	1.1	0.04	7.9	3.9	0.75	tr	248	55
Faba bean	1.9	0.4			1.3	1.3	170	67
Lentil	6.9	0.4	5.8	2.3	39.2	19.0	132	46
Chickpea	4.9	1.9	16.4	4.0	36.4	24.4	60	32
Others								
Castor bean	16.4	12.7		0.1			13	1
Lettuce	tr						12	
Sesame		0.4					0.5	
Buckwheat	1.6	0.9					2	tr

Based on data in Horbowicz, M. and Obendorf, R.L. (1994) Seed desiccation tolerance and storability: dependence on flatulence-producing oligosaccharides and cyclitols – review and survey. *Seed Science Research* 4, 385–405.
Galactopinitols are the combined values of A and B and RSOs (raffinose series oligosaccharides, the combined values of raffinose, stachyose and verbascose, rounded to the nearest whole number).
A = axes. C = cotyledons.
Lettuce, whole seed amounts are given. Castor bean, cotyledons and endosperm were analysed together.
Blank space denotes none detected.
Small amounts of galacto-*chiro*-inositol are present also in some seeds, but in buckwheat it is very abundant: A, 60.43 mg/g, and C, 26.4 mg/g.

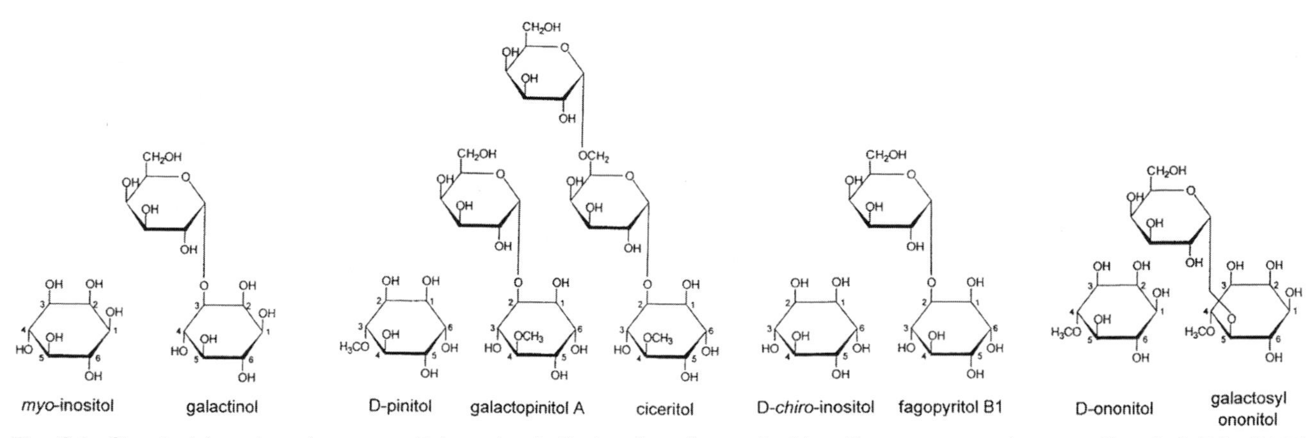

Fig. G.1. Chemical formulae of some cyclitols and galactinol series oligosaccharides. For more examples, see Obendorf, R.L. (1997) Oligosaccharides and galactosyl cyclitols in seed desiccation tolerance. *Seed Science Research* 7, 63–74.

Fagopyritols occur in legume seeds, sugarbeet, castor bean, and particularly in buckwheat (**see: Pseudocereals**) (up to 40% of soluble sugar content). They are the mono-, di- and tri-galactosyl oligomers of D-*chiro*-inositol, and are called fagopyritol B1, B2 and B3. They are synthesized during seed development and have been associated with the development of **desiccation tolerance**. Their hydrolysis by α-galactosidase occurs during germination.

Galactosyl ononitols are oligomers of galactose (mono- to tri-) and ononitol (an intermediate in the synthesis of pinitol) but are generally present in trace amounts.

The importance of members of the galactinol series oligosaccharides in seeds is thought to include: as a readily respirable carbohydrate substrate during germination; protection against **free radical** damage, as a scavenger; increasing water stress and desiccation tolerance. (JDB)

Peterbauer, T. and Richter, A. (2001) Biochemistry and physiology of raffinose family oligosaccharides and galactosyl cyclitols in seeds. *Seed Science Research* 11, 185–197.

Galactomannans

Polymeric carbohydrates (**hemicellulose** polysaccharide) with a backbone of β-1,4-linked mannose residues and single α-1,6-linked galactose side chains (Fig. G.2). Galactomannan increases in solubility in water with increasing substitution (addition) of single galactose side chains. It occurs in seeds as a cell wall carbohydrate storage reserve (Fig. G.3 and Table M.4 in **Mannans**), although in some seeds (e.g. **fenugreek**) it may play a role in water storage for the embryo following **imbibition**, by protecting it from desiccation if the soil dries out. Galactoglucomannan also occurs in some seeds, and this has galactose side chains α-1,6-linked to mannose residues in the β-1,4-linked mannose- and glucose-containing backbone.

Carob (22–24% galactose) and **guar** (38–40% galactose) seed galactomannan are of particular importance commercially; of these, the latter is more widely and reliably available. Some 90–100 t of galactomannan is used each year in processed foods as a thickener and stabilizing agent, and chemically modified galactomannan is used in the pharmaceutical, cosmetic, textile, paper, explosives and mining industries.

Carob gum (also called locust bean gum) is produced by extracting galactomannan from the separated **endosperms** of the carob seed in warm to hot water; gel formation is achieved by the addition of cross-linkers such as borax. Synergistic gel formation in the presence of a second polysaccharide (e.g. carageenan, an algal polysaccharide) is preferable for the food industry, where it is used as a stabilizer in ice cream, cheese spread and other dairy products with a high water content (**see:** Table G.16 in **Gums**). It is also used as a toothpaste thickener. In combination with xanthan (a bacterial polysaccharide) it is becoming important in the pharmaceutical industry as a hydrophilic matrix system in tablets from which drugs can be released in a controlled manner.

Guar gum is extracted from separated endosperms of the guar seed using warm/hot water extraction. Its higher galactose content makes it less useful than carob gum in the production of elastic gels when mixed with other polysaccharides. Hence

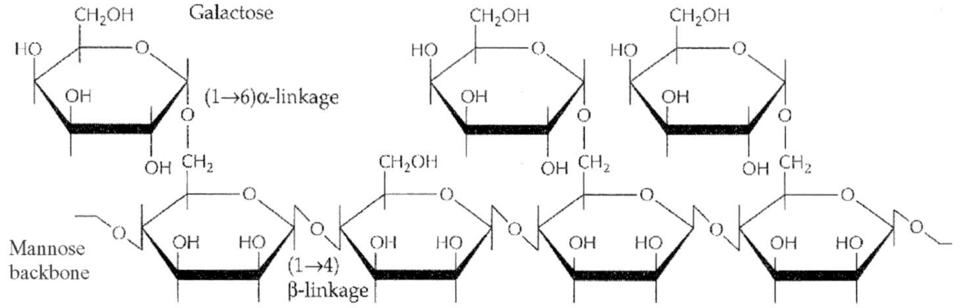

Fig. G.2. Structure of galactomannan, showing linkages in the mannose backbone and galactose side-chains.

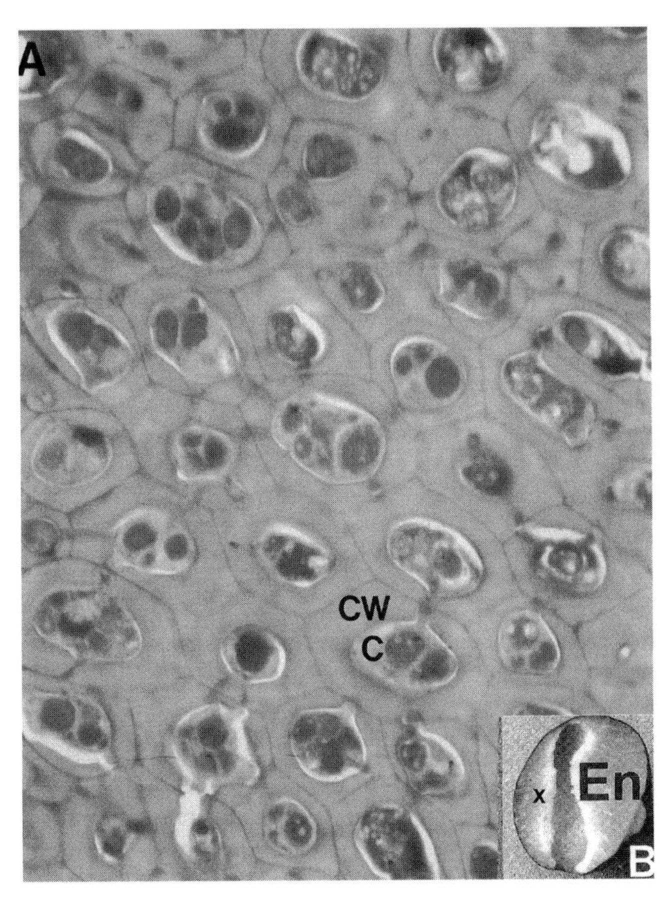

Fig. G.3. (A) Light micrograph of thickened galactomannan-containing cell walls in the cells of the lateral endosperm of asparagus seed. In one cell the cell wall (cw) and living cytoplasm (c) are marked. Inset (B) shows the endosperm (En) of an asparagus seed, and the location of the cells pictured in A is marked with an x. The embryo has been removed from the central cavity. From Williams, H.A., Bewley, J.D., Greenwood, J.S., Bourgault, R. and Mo, B. (2001) The storage cell walls in the endosperm of *Asparagus officinalis* L. seeds during development and following germination. *Seed Science Research* 11, 305–315.

chemical modifications of guar gum, such as controlled removal of galactose side chains by α-galactosidase, is being attempted. The thickening and water-binding properties of guar gum make it useful as a: drilling lubricant; suppressant of turbulent flow in piping operations; food hydrator and stabilizer; thickener in shampoos and conditioners; constituent of freezer-packs for food preservation and shipment; dye thickener in the carpet and fabric manufacturing industry; waterproof layer for stick dynamite; and in the formation of biodegradable mulches for hydroseeding. In the paper industry it is used to increase bonding strength, drainage and retention properties of cellulosic paper fibres in the making of paper and paperboard. The gum is considered to be a good dietary fibre to add to food (e.g. bread) which can lower blood cholesterol and aid glycaemic control in patients with diabetes. (JDB)
(See: Glucomannans)

Tombs, M.P. and Harding, S.E. (1998) *An Introduction to Polysaccharide Biotechnology*. Taylor and Francis, London, UK.

Gamete

A reproductive cell or germ cell (in seed-bearing plants the ovum and sperm cells of the germinated pollen grain), usually haploid in chromosome number, and capable of uniting with another gamete of the opposite sex to form the next generation of plant. (**See: Reproductive structures**) When the female's germ cells are generally larger than the male ones, the terms megagametes and microgametes apply, respectively. In contrast to spores, gametes generally can only give rise to a new individual or generation after they fuse with a gamete of the opposite sex. (**See: Apomixis; Parthenogenesis**)

Gametophyte

(Greek: *gametes* = spouse; *phyton* = plant) The **haploid** generation (e.g. the prothallium of the ferns or the embryo sac and pollen grain of the angiosperms) of the plant which produces the gametes. Evolutionarily, the gametophyte generation in the advanced **angiosperms** is very much reduced compared with in the lower plants, where is it the predominant generation (e.g. algae, mosses and liverworts). (**See: Evolution of seeds; Reproductive structures**)

GAMYB

See: Mobilization of reserves – cereals; Signal transduction – hormones; Signal transduction – some terminology

GARE

See: Mobilization of reserves – cereals; Signal transduction – hormones; Signal transduction – some terminology

Gene

The physical and functional unit of heredity, which carries information from one generation to the next. In molecular terms, a gene is the entire DNA sequence necessary for the synthesis of a functional polypeptide or RNA molecule. In addition to coding regions most genes also contain non-coding intervening sequences (introns) and transcriptional-control regions (promoter sequences). (**See: Alleles; Genomics; Nucleic acids**)

Gene banks

Collections of plants, animals and microbes developed and maintained for the purposes of *ex situ* **conservation** and/or for increased ease of access. Gene banks can also be collections of DNA assembled for research and monitoring. For plants, gene banks comprise collections of either seeds (seed gene banks, usually referred to as '**seed banks**' though note the use of this term for **soil seed banks**), growing plants (field gene banks), pollen, spores, *in vitro* vegetative samples, or DNA. Collections of plants growing in botanic gardens might also be included in this list though they tend to contain rather poor genetic representation of the wide range of species grown. Most crop **germplasm** collections kept in gene banks are held as seed. This is because seed gene banks offer a good compromise between ability to store intra-specific variation, applicability to a wide range of species, potential sample longevity, recovery of gene products, and technical input requirement. (**See: Conservation of seed**) (RJP, SHL)

Gene expression

The active production of mRNA transcripts (transcription) and protein (translation) corresponding to a particular gene. (**See: Nucleic acids**)

Gene-switch technologies

A name for a **genetically modified** system in which transcription of a gene is regulated in response to a specific external chemical stimulus, such as by activating a second gene that encodes a protein transcription factor promoter for the first gene. (**See: Promoters – chemically induced**)

Genetic erosion

The attrition of genetic diversity that occurs in small or fragmented populations as a consequence of selection (directed change), genetic drift (random change) or introgression from hybridizing populations. Genetic erosion during *ex situ* conservation can lead to loss of genetic integrity when small captive populations of seeds deteriorate during storage, or are regenerated. Inadvertent selection of wild seed populations for longevity, easier harvest and uniform germination is a form of genetic erosion called the **domestication syndrome**. (CW)

Genetic modification

Though open to considerable discussion and interpretation, 'genetic modification' can be broadly defined as the application of several molecular biological techniques to introduce one or a few specific **genes** (commonly called 'transgenes') into a target host from an unrelated species, by the transfer and integration of DNA. Often abbreviated to GM (which according to context can also stand for 'Genetically Modified', as in GMO for '**genetically modified organism**'), the term is also used more or less interchangeably with 'genetic engineering' or 'genetic enhancement' (GE), and with 'genetic transformation', 'transgenic-' or 'gene-technology'. GE or transgenic are arguably more appropriate terms, because GM wrongly implies there were no genetic modifications before this technology was developed.

GM is one of the portfolio of techniques categorized under the general heading of biotechnology, and depends centrally on recombinant-DNA or gene-splicing platform technologies – such as gene isolation, the purification and engineering of promoters, and **transformation** by means of a bacterial plasmid, virus or other vectors. GM also relies on conventional 'classical' breeding strategies, such as gene mapping, regeneration, marker-assisted selection and hybridization. Traditional breeding technology is also necessary to ensure that a new GM variety retains all its other agronomically desirable traits – such as extensive progeny testing to identify which 'integration events' are consistent with a high level of stable expression, and backcrossing to introduce the transgenes into the desired genetic background. **GMO patent protection technologies**, themselves based on GM technology, have been devised both to protect intellectual property rights and the environment from the unintentional release of engineered genes, but are not yet commercialized.

There are currently two main target categories of GM in agricultural crop breeding:

- *Agronomic (input) traits* aimed at improving the profitability of primary agricultural production by making the crop easier and cheaper to grow, such as herbicide tolerance and resistance to the action of insect pests, to reduce inputs, increase yield or reduce risk. Major GM-crop varieties of this type became available to farmers in the mid-1990s, initially in the USA: insect-resistant ('Bt') cotton, and herbicide-tolerant cotton, soybean and maize (commonly marketed as 'Roundup Ready®', RR). More recently, insect-resistant (Bt) maize and herbicide-tolerant canola were introduced.

  Bt insect resistance uses genes from ***Bacillus thuringiensis*** that encode proteins that are toxic to plant-feeding insects. RR technology uses plants transformed with a gene encoding an enzyme (mEPSPS) that makes the plant tolerant to glyphosate, the **active ingredient** in the herbicide. Herbicide tolerance to another herbicide, glufosinate, has also been introduced, as 'Liberty-Link®'. (Some **lectin** genes, to resist insect predation, are amongst the potential future candidates for transgenic introduction.)
- *Quality (output) traits* that alter the nature of the harvested product, through compositional changes or enhanced nutritional value, for the benefit of food manufacturers, processors or consumers. Early examples commercialized in the mid-1990s were the 'FlavrSavr®' tomato, engineered with an anti-sense gene to polygalacturonase, the major pectin-softening enzyme, which made the fruits soften more slowly than conventional varieties; and a different processing tomato variety used for making tomato paste, which was engineered with a non-functional poly-galacturonase gene. Many such traits are under development, such as are high oleic or low linoleic **oilseeds** for enhanced stability cooking oils or low saturates, to reduce the need for hydrogenation and create 'healthier' food products lower in trans-lipids. Another example is high-carotene **'Golden' rice**, engineered with two genes introduced from daffodil and the bacterium *Erwinia uredovora*, which produces provitamin A. (**See: Oils and fats – genetic modification**)

The first year of significant commercialization of GM varieties was 1966, with 1.7 Mha sown worldwide. In 2003, five principal countries were growing 98% of the estimated 68 Mha global transgenic crop area: the USA 63% (soybean, maize, cotton, canola), Argentina 21% (soybean, maize, cotton), Canada 6% (canola, maize, soybean), Brazil 4% (soybean) and China 4% (cotton). The principal transgenic crops in that year were: soybean, occupying 61% of the global GM area; maize, 23%; cotton, 11%; and canola, 5%. Or, expressed as proportions of their respective total global crop areas, transgenic varieties then represented: 55% for soybean, 21% for cotton, 16% for canola and 11% for maize. In 2002, the estimated global market value of transgenic seed (the sale price plus any technology fees) represented about 14% of the global crop-protection on the one hand and the global commercial-seed markets on the other, which were each estimated to be worth about US$30 billion. (Data from the International Service for the Acquisition of Agri-biotech Applications: www.isaaa.org.) (**See** inserts in: **Crop Atlas Appendix, Maps 7, 13, 14, 16**)

The commercialization of GM technology is a very politically sensitive matter in many countries. There are widespread and complex public concerns about a combination of issues – including concerns about fundamental ethics (what is regarded as the 'unnatural nature' of the technology, especially where the crosses are not possible under natural selection or through traditional plant breeding methods), food health and safety, environmental contamination (gene flow from crop to crop, and from the crop to its wild relatives, through possible outbreeding) and threats to biodiversity, and the socio-economic impacts. These are being reflected in regulatory systems in some countries that govern maximum acceptable proportions of GM in the food chain.

Genetic modification – identity testing

Consumer resistance to GM products has led to the development in some countries of identity preservation systems which track and segregate GM products from the conventional food supply chain. This involves testing food and feed for trace levels of GM contamination, notably in products derived from commodity maize and soybean grain. Seed companies are also concerned to protect their intellectual property, particularly in open-pollinated GM soybean and cotton varieties in the USA for instance, which technically can be easily multiplied by farmers from one year's crop to the next.

There are several commonly used methods to detect GM in seeds of field crops such as maize, soybean, canola and cotton. Identity preservation has generated a need for quick tests, conducted by kits (immunoassays and DNA-based tests to detect foreign DNA or protein). These include protein-based enzyme linked immunosorbant assay (**ELISA**) tests and biological tests (bioassays, such as germinating herbicide tolerant varieties in the presence of the herbicide). There is also a broad range of DNA-based **polymerase chain reaction (PCR)** tests routinely used for the qualitative detection of the gene constructs inserted in GM varieties. The accuracy and reproducibility of these test results depends greatly on the methodology and on the equipment used, which is making standardization on an international scale very difficult. The establishment of relevant, reliable and economical methodology for detection, identification and quantification of GMO content in conventional seed lots continues to challenge. Many research institutes, inspection agencies and companies are developing strategies and methods for GMO testing, and the topic is the subject of a GMO task force by the **International Seed Testing Association**. Technical issues include statistical sampling theory, false-positive and false-negative rates, and the limits of detection and quantification. (PH)

Union of the German Academies of Sciences and Humanities (2006) *Literature database on genetically modified crop plants*. www.akademienunion.de/publikationen/literaturs ammlung_gentechnik/ (A downloadable selection of papers and reports from many sources.)

Genetic resources

Germplasm that is of value to humankind either now or in the future. Because future use cannot be predicted, nearly all species might be considered to be genetic resources. Consequently, germplasm and genetic resources are terms that are sometimes used interchangeably. (**See: Gene banks**)

Genetically modified organism (GMO)

Generally defined as an organism in which the genetic material has been altered in a way that does not occur naturally by mating and/or natural recombination, i.e. employs transformation technology. However, some argue that any organism that is deliberately bred by humans for new traits, even using conventional breeding techniques, is genetically modified. (**See: Biopharmaceuticals; Genetic modification; Transformation**)

Genome

The entire genetic complement of a prokaryote (e.g. bacterium), virus, DNA-bearing organelle such as a mitochondrion or chloroplast, or the haploid nuclear genetic complement of a eukaryotic (plant or animal) species. In a looser sense, it is used to convey the entire genetic information present in the nucleus of a typical cell of a eukaryote. This, strictly, is the nuclear genome, since there is nuclear material within a mitochondrion, for example the mitochondrial genome. The entire nuclear DNA sequence of several plants is being, or has been, determined, with *Arabidopsis* being first to be completed, followed by rice. Information on the *Arabidopsis* genome sequence can be found on the TAIR database at www.arabidopsis.org, and on all genome databases in www.tigr.org. (**See: Genomics**)

Genomics

Studies of the genome. During the early years of the genomics era, researchers concentrated on accumulating DNA sequence information (both genomic and EST or expressed sequence tag data, i.e. sequences obtained relatively close to inserted DNA of known base composition) from a range of economically important and model plants. While most of the advanced work has been done on *Arabidopsis thaliana*, public domain genomic sequences exist for many major crops, including rice, maize, soybean, cotton and sorghum. The accuracy and completeness of the genome sequence record is of fundamental importance for all subsequent analyses. Nearly all genes in multicellular eukaryotes contain regions coding for mRNA (exons) (**see: Nucleic acids**) separated by non-coding regions (introns), and in the absence of experimental data such as full-length **cDNA** sequences, protein sequences are derived from such genes by probabilistic modelling. These analyses have revealed many novel protein sequences in plants, e.g about 30% of the genes in *Arabidopsis thaliana* have no significant similarity at the protein level to genes present in any other organism, and a significant proportion of these proteins can be expected to contribute to plant-specific processes.

As genomics information has become available on a broad range of organisms, a new post-genomics era has arisen in which data can be used as a resource base to characterize gene expression under different conditions. Because of the emphasis on gene function, this research is often referred to as 'functional genomics'. The purpose of functional genomics is to use the information made available upon sequencing a genome to quantitatively determine the spatial and temporal accumulation patterns of specific mRNAs, proteins and important metabolites using high-throughput technologies. Functional genomics relies heavily on three levels of high-throughput analyses: **transcriptomics** (or RNA profiling) for measuring amounts of mRNA, **proteomics** for determining concentrations of individual proteins, and **metabolomics** (metabolite profiling) for determining amounts of important metabolites.

Tradeoffs exist among the analytical power of these systems in the amounts of data generated and the usefulness of the results obtained. These tradeoffs can readily be visualized in the context of the central dogma of molecular biology, in which DNA is transcribed into mRNAs, mRNAs translated into proteins, and proteins act to catalytically interconvert metabolites. While DNA sequences indicate what genes are present in a plant, mRNA measurements show which of these genes are expressed. Similarly, protein measurements identify those specific mRNAs that are being translated and the amount of the specific enzymes that are present. This may or may not be reflected at the mRNA level. Finally, the amount of metabolite present (particularly if flux information can be deduced) may be more important than determining the potential for product information as estimated by measuring enzyme activities. For metabolic engineers who are largely interested in using living organisms to produce proteins and metabolites for commercial purposes, functional genomics provides new tools and approaches for understanding, modelling, and ultimately manipulating seeds. (AJR)

Germinability

The potential to complete germination of a single seed or of a seed population, in the latter case expressed as the maximum attainable percentage, usually under defined conditions. It is determined by **dormancy**, **viability** and factors such as light, temperature, oxygen and chemicals as well as the developmental history of the seed. It is sometimes referred to as germination capacity. (**See: Germinability – maternal effects; Germinability – parental habitat effects**; articles in **Germination – influences** etc.)

Germinability – maternal effects

Differences in seed germinability (e.g. **dormancy** and **germination rates**) often arise as a result of influences of the mother plant. Their origins frequently can be ascribed to positional effects, that is where on the plant the seeds are formed. In some species seeds produced by different inflorescences, different parts of the same inflorescence, different florets in a **spikelet**, or different parts of a fruit, might vary in germinability (Table G.2). Physiological age of the mother plant might also determine subsequent germinability of the seeds. The possession of different germination properties by seeds of the same plant or of different plants of the same species is described as **heteroblasty**. Frequently, such seeds also differ morphologically, for example in size, colour or coat thickness, to which the term **polymorphism** is applied. The two terms are sometimes used synonymously. The causes of such variability are unknown but could include unequal partitioning to the developing seeds of assimilates, hormones and other substances participating in development, growth, onset of dormancy, and maturation. (MB)

(See: Deserts and arid lands; Dormancy – acquisition; Germinability – parental habitat effects; Germination)

Baskin, C.C. and Baskin, J.M. (1998) *Seeds. Ecology, Biogeography, and Evolution of Dormancy and Germination.* Academic Press, New York, pp. 181–237.

Germinability – parental habitat effects

In very many species differences have been observed in the **germinability** of seeds produced by a plant species in different habitats. Seed differences are manifest in: (i) germination responses to external conditions such as temperature, **water potential**, light/dark, (ii) germination rate, (iii) depth of **dormancy** and the conditions for **dormancy breaking**, such as chilling, light (Table G.3). In some cases the factor(s) at the parental location that apparently relates to subsequent germinability can be named but in many instances it is unknown. It seems likely, however, that those environmental factors that have been identified to affect the onset of dormancy would account for some of the differences. These include spectral light quality, water stress conditions, temperature and possibly daily photoperiod (**see: Dormancy – acquisition**). Some of the differences in seed germinability between populations are ascribable to genetically determined ecotypic variation. (MB)

(See: Deserts and arid lands; Germinability – maternal effects; Germination; Light – dormancy and germination)

Baskin, C.C. and Baskin, J.M. (1998) *Seeds. Ecology, Biogeography, and Evolution of Dormancy and Germination.* Academic Press, New York, USA, pp. 181–237.

Germination

Strictly defined as those events occurring between the start of uptake of water by a seed (**imbibition**) and the emergence of the embryonic axis through its surrounding structures (usually the **radicle** penetrating the **testa** or **pericarp**).

Table G.2. Some examples of maternal effects on germinability.

Species	Affected seed property	Factor
Aegilops ovata	Dormancy	Spikelet position
Apium graveolens	Dormancy	Different inflorescence
Cryptomeria japonica	Germination rate	Maternal age
Daucus carota	Germination rate	Different inflorescence
Emilia sonchifolia	Dormancy	Position on inflorescence (capitulum)
Oldenlandia corymbosa	Dormancy	Maternal age
Petroselenium crispum	Germination capacity	Position on plant
Xanthium pennsylvanicum	Dormancy	Position in dispersal unit

Data from Baskin, C.C. and Baskin, J.M. (1998) *Seeds. Ecology, Biogeography, and Evolution of Dormancy and Germination.* Academic Press, New York, USA.

Table G.3. Some effects of parental habitat on subsequent seed germinability.

Species	Affected seed property	Possible factor responsible
Abutilon cephalonica	Dormancy	Rainfall
Amaranthus retroflexus	Dormancy	Moisture
Artemisia tridentata	Chilling requirement	Elevation
Avena fatua	Dormancy	Latitude
Betula alleghaniensis	Germination rate	Latitude
Campanula punctata	Germination rate	Latitude
Chenopodium bonus-henricus	Dormancy	Elevation
Dactylis glomerata	Germination temperature, light/dark requirement	Latitude, temperature
Liquidambar styraciflua	Germination rate	Elevation, latitude
Medicago sativa	Dormancy	Elevation
Nemophila insignis	Germination temperature	Soil moisture, temperature
Nicotiana tabacum	Germination capacity	Mineral nutrition
Picea mariana	Dormancy	Latitude
Pinus strobus	Chilling requirement	Latitude
Poa trivialis	Light quality requirement	Light quality

Based on data in Chapter 8 and Tables 8.2, 8.4 in Baskin, C.C. and Baskin, J.M. (1998) *Seeds. Ecology, Biogeography, and Evolution of Dormancy and Germination.* Academic Press, New York, USA.

A seed from which the radicle has emerged is regarded as having germinated, and subsequent events are associated with **seedling** growth and establishment. Some looser definitions of germination appear in published works, e.g. the emergence of the radicle from a seed may be termed germination or visible germination (the strict pragmatic definition used by seed analysts in official **germination testing**, subject to there also being the subsequent development of a 'normal seedling' – although this event should be termed: completion of germination), and in field/greenhouse situations germination may be regarded as being completed when a seedling emerges from the soil (although in this case the term incorporates both germination and early seedling establishment).

In the seeds of some species it is not the radicle which emerges first to complete germination, but may be the **cotyledons** or the **hypocotyl**. This occurs particularly in some members of the families Bromeliaceae, Chenopodiaceae, Onagraceae, Palmae, Saxifragaceae and Typhaceae. In some Gramineae, the **coleoptile** emerges before the radicle (e.g. *Oropetium tomaeum*).

A time course of the major events associated with germination and early seedling growth is outlined in Fig. G.4. The time for these events to be completed varies with species and ambient conditions, and may take from several hours to many weeks or months.

Availability of water to the seed is obviously critical. Several components are involved: (i) Is water present? (ii) What is its water potential? Factors involved here are the matric potential, determined in the field by the soil substratum, and the osmotic potential, determined by the solutes therein. (iii) What contact does the seed make with the soil particles, which are surrounded by water? Particle size and shape and relative seed size and shape are important parameters to be considered. (**see: Germination – influences of osmotica and salinity; Germination – influences of water**).

Uptake of water by a seed is triphasic, commencing with imbibition (phase I), followed by a plateau phase (phase II) when the seed has swollen to maximum size. This latter phase has been called by some, **germination *sensu stricto***. The third phase (phase III) occurs only after germination is completed, as the cells of the **embryonic axes** elongate and undergo cell division to form the young seedling (**see: Germination – molecular aspects; Germination – physical factors; Germination – radicle emergence**). Because dormant and non-**viable** seeds do not complete germination they cannot enter phase III (**see: Dormancy**). Seeds which are shed in a hydrated state, and do not undergo **maturation** drying (e.g. **recalcitrant seeds**) may exhibit little or no water uptake and hence phase I is not prominent.

During the initial rapid uptake of water into a seed there is leakage into the surrounding medium of solutes, such as sugars, amino acids, organic acids and ions from the cells of the embryo and surrounding structures. **Leakage** from cells is

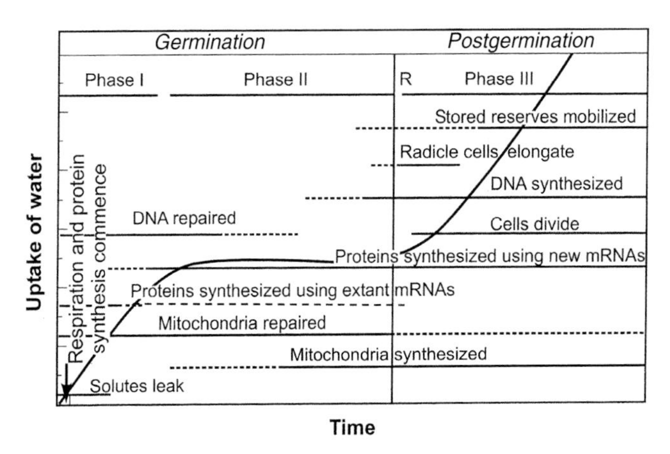

Fig. G.4. Time course of major events occurring during germination and subsequent post-germinative growth. The time for events to be completed varies from several hours to many weeks, depending on plant species and germination conditions. The curve shows the course of water uptake. Vertical line R denotes the time of radicle protrusion from the seed. Modified from Bewley, J.D. (1997) Seed germination and dormancy. *The Plant Cell* 9, 1055–1066. With permission of the American Society of Plant Biologists.

due to a rearrangement of the structure of their **membranes** (plasmalemma or plasma membrane), and may be diagnostic of the viability or vigour of a seed. (**See: Chilling injury; Germination – influences of water; Germination – physical factors; Imbibition**) An early event in germination is the restitution of the integrity of the membranes, but details on how this occurs remain vague. Germination is sensitive to several factors such as temperature, hormones and several chemicals (**see: Germination – influences; Germination – influences of hormones; Germination – influences of temperature**). (JDB)

Bewley, J.D. (1997) Seed dormancy and germination. *The Plant Cell* 9, 1055–1066.

Bewley, J.D. and Black, M. (1994) *Seeds. Physiology of Development and Germination*. Plenum, New York, USA.

Obroucheva, N.V. (1999) *Seed Germination. A Guide to the Early Stages*. Backhuys Publishers, Leiden, The Netherlands.

Germination – field emergence models

This article describes some of the mathematical models that have been developed to describe the time course of germination and subsequent seedling **emergence**. These models serve as an aid to understanding the physiological processes involved, and also aim to monitor and predict seedlot performance after sowing. In most situations the principal environmental factors considered are temperature, **water potential** and the variable strength of soil through which seedlings grow (**see: Seedbed environment**). Other factors are also important for particular species and situations, such as the availability of oxygen, light and nitrate. (**See: Germination – external factors**) For modelling purposes, the time taken for seedling emergence can be conveniently divided into two phases: during germination and after the completion of germination (protrusion of the embryo, usually the **radicle**, from the surrounding seed structures). Although these two phases are confounded in most studies, they are each uniquely affected by environmental conditions and must be included separately when modelling the impact of the environment on seedling emergence. The large numbers of both empirical and mechanistic modelling approaches that have been adopted to describe these environmental effects are too diverse to cover in detail here. However, emphasis has been placed on the use of population-based threshold models, because it is likely that they both have physiological significance and also provide a framework for developing a generic understanding of seed and seedling responses to the environment.

1. Germination – population-based threshold models

Germination is completed at different times for each individual seed within the population, leading to a distribution of germination times and hence a characteristic cumulative germination curve for any seedlot.

Population-based threshold models provide a useful framework to understand and describe the interaction of these distributions with temperature, water potential and other environmental factors. Within these models, the rate of germination is held to increase above a base (threshold) value for a given factor (e.g. temperature, water potential, hormone concentration). In many cases, **germination rate** over the range of interest is approximately linearly related to the magnitude of the factor above its base value. Below this base value, germination will not progress to completion. The base values, however, may differ between individuals in the population, which are therefore important in describing differences in population responses to the factor concerned.

Within these models it is generally assumed that seeds complete germination in a set order and that this order is not affected by germination conditions. Each seed can therefore be assigned a value of G, which determines the time order at which it completes germination relative to other seeds in the population – in other words the germinated fraction of the population to which it belongs: thus, for example, $G = 50$ denotes the 50th percentile.

(a) *Temperature and thermal time*. Non-dormant seeds can usually germinate over a wide range of temperatures that is characteristic for a given species or cultivar, but the maximum percentage germination attained is typically reduced at the extremes of this range (**see: Germination – influences of temperature**). Members of the seed population therefore have different levels of tolerance (thresholds) to both high and low temperatures.

For any individual seed under constant temperature with unrestricted water supply, germination rate (the reciprocal of germination time) increases in relation to temperature from a base threshold (when it is infinitely slow) to an optimum, above which it decreases to a threshold ceiling temperature that is the upper temperature limit of its tolerance. This response to temperature can often be described by two linear relationships where the base and ceiling temperatures for a given percentile are defined by mathematical extrapolation to calculate the intercepts on the temperature axis where rate tends to zero (Fig. G.5). Experimentally on this basis, in many cases there appears to be little variation in these base temperatures, ($T_b(G)$), amongst individuals within a seedlot whereas there is significant variation amongst ceiling temperatures ($T_c(G)$).

- Thus at sub-optimal temperatures, germination rate $1/t(G)$ is linearly related to temperature, but with a different slope for each percentile:

  $$1/t(G) = (T - T_b)/\ \theta_{T1}(G) \quad (1)$$

  where, $\theta_{T1}(G)$ is the **thermal time** to germination of percentile G, T is the temperature, T_b is the base temperature for all seeds and $t(G)$ is the time taken for germination of percentile G.

- In contrast, at supra-optimal temperatures the rate of germination declines in a series of parallel lines. The intercepts therefore differ in the population and thus:

  $$1/t(G) = (T_c(G) - T)/\ \theta_{T2} \quad (2)$$

  where θ_{T2} (thermal time) is assumed constant for all G, because variability is modelled using differing ceiling temperatures, $T_c(G)$.

Thus by incorporating an appropriate distribution function for θ_{T1} and T_c into equations (1) and (2), time to germination under any constant temperature regime can be calculated for any percentile of the population.

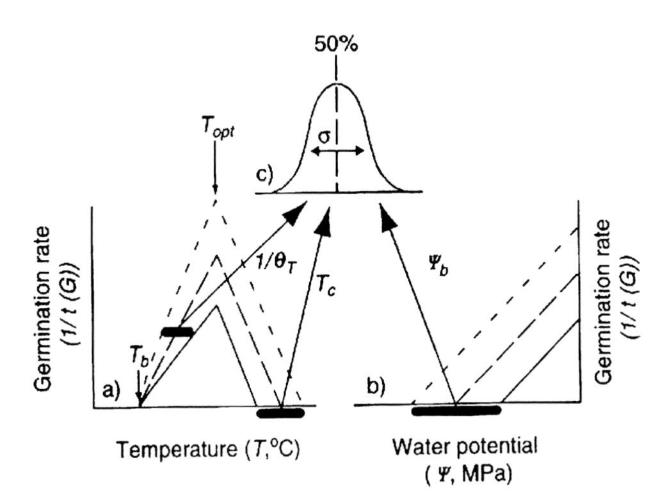

Fig. G.5. Schematic illustration of the effects of temperature (a) and water potential at sub-optimal temperature (b) on the rate of germination. *G10* (dotted lines), *G50* (dashed lines) and *G90* (solid lines) represent individual seeds in the population at percentiles 10, 50 and 90, respectively. Germination rate increases linearly with temperature above a base (T_b). The slopes of these lines are the reciprocal of the thermal times to germination ($1/\theta_T$). As temperature increases above an optimum (T_{opt}), rate of germination decreases to a ceiling temperature (T_c). Rate of germination also decreases linearly with water potential (ψ in MegaPascals, MPa) to a base (ψ_b). T_b is common to all seeds in the population, but $1/\theta_T$, T_c and ψ_b vary between seeds in a normal distribution (c). For further explanation of these parameters see the text.

This commonly used approach describes data where there are linear relationships between germination rate and temperature above and below an optimum temperature. However, in many other data sets, a plateau or curved relationship has been reported near to the optimum and, in some cases, near to the base and ceiling temperatures also. This plateau response has been accommodated in a further development of the thermal time model, based on Gaussian curves, which describes the germination response across both sub- and supra-optimal temperature ranges.

(b) *Hydro(thermal) time* is a concept which models water potential and temperature in combination. It extends the thermal time approach to consider germination time $t(G)$ as a function of the extent to which the constant water potential (ψ) and constant sub-optimal temperature (T) of each seed (G) exceed thresholds (bases; ψ_b, T_b) below which germination will not occur.

$$\theta_{HT} = (\psi - \psi_b(G))\,(T - T_b)\,t(G) \qquad (3)$$

Here hydrothermal time (θ_{HT}) and the base temperature (T_b) are regarded as constant and it is assumed that the distribution of the germination times of individual seeds within the population is effectively determined by the base water potentials (ψ_b), which vary with G (Fig. G.5). Water potential (ψ) is the sum of the matric potential and the osmotic potential in the environment surrounding the seed, and is usually expressed in units of pressure, MegaPascals (MPa) (**see: Seedbed environment**). The response of seeds to water potential at a given constant temperature can be illustrated by an hydraulic analogy (Fig. G.6). Accepting the assumptions above, it is possible to describe seed

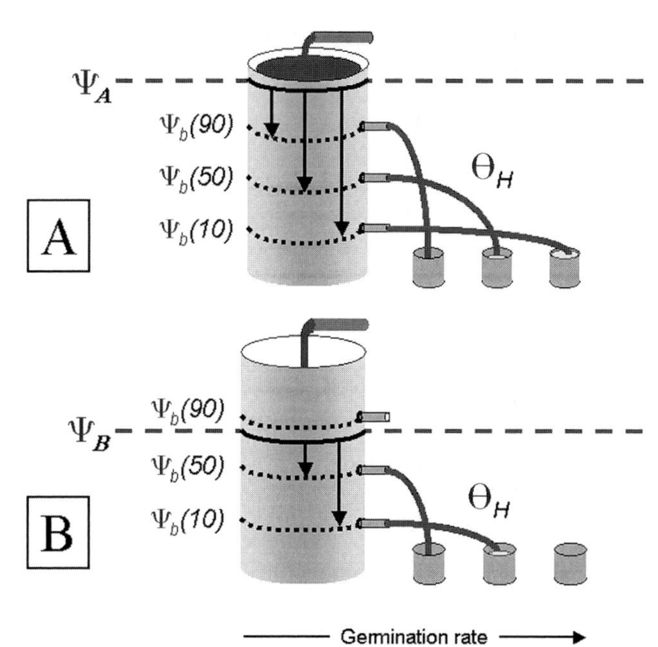

Fig. G.6. Hydraulic analogy of the hydrotime model of seed germination of a seedlot below its supra-optimal temperature. (Redrawn after Bradford, 2002.) The Panels A and B illustrate the responses of the seedlot to two different total external water potentials (ψ_A and ψ_B), represented by the height of water in the tanks. The outlets at different heights on the tank indicate values for germination base threshold potentials ($\psi_b(G)$) for the 10th, 50th and 90th percentiles of the germination order in this seed population. The water level in the tank is maintained by the flow from the inlet pipe, which is adjusted by a mechanism (not shown) to exactly match the flow from the outlets (in other words, ψ is held constant). The hydrotime required for complete germination of each fraction is represented by the filling time of cups of the same size by the flow of water from the different outlets (the actual time taken will depend on the temperature conditions). The greater the height of water above an outlet, the faster the flow and the quicker the cup is filled. The hydrotime constant (θ) can be visualized as the viscosity of the water: the higher θ is, the more viscous the water becomes, and the slower it fills the cups. When the water in the tank is lowered to a new constant level (ψ is reduced, as in panel B), some outlets stop flowing because the water level falls below them: in other words, germination does not occur for that fraction because the water stress has become too great. The rate of flow from other outlets is reduced in proportion to the lower water height, taking a longer time to fill the cups. Note that the same effect on outflows shown in the panel B would be attained if the outlets in panel A were raised by the same amount that the water level is lowered in panel B (representing the performance of a second seedlot, which had higher base potentials).

response of the whole population in a single equation by incorporating a suitable distribution (usually a normal distribution, Fig. G.5 and Equation 4) of base water potentials within the population:

$$\text{Probit}\,(G) = ((\psi - \theta_{HT} / (T - T_b)\,t(G)) - \psi_b(50)) / \sigma_{\psi b} \qquad (4)$$

where $\psi_b(50)$ is the base water potential of the 50th percentile and $\sigma_{\psi b}$ is the standard deviation of ψ_b within the population. For the purposes of calculation, the probit statistical transformation is used to linearize the normal distribution of data.

At lower (sub-optimal) temperatures, $\psi_b(50)$ and $\sigma_{\psi b}$ are constant; however, as temperature increases and enters the supra-optimal range, $\psi_b(50)$ increases linearly with temperature, while $\sigma_{\psi b}$ remains constant, until at the ceiling temperature ψ_b = 0 MPa (that is, the seeds cannot complete germination in pure water). By incorporating this linear shift in ψ_b at higher temperatures, hydrothermal models can be extended to include supra-optimal temperatures. However, these models are not flexible enough to describe all germination behaviour. Examples have been shown where $\psi_b(G)$ varied with T, both above and below the optimum; T_b can also have a normal distribution and vary with ψ. At present these data have been described by modelling thresholds and rates separately.

(c) *Hydrothermal priming time* extends the hydrothermaltime concept to describe the germination advancement that occurs in dryer soils and during **priming**. Although germination does not occur below ψ_b, seed priming studies have shown that advancement can occur below that base water potential. The rate of progress towards germination under these conditions can be modelled in a similar fashion to that described above by accumulating time in proportion to the amount by which the water potential exceeds the minimum for metabolic advancement ψ_{min} (which is lower than ψ_b). In this way, hydrothermal priming time accumulates to take account of advancement below ψ_b. This model is separate from the hydrothermal time model but may be coupled to it to predict when germination will be completed at higher water potentials.

An alternative population-based threshold approach has been developed to aid simulation of the effects of a changing environment in a single model, known as the Virtual Osmotic Potential (VOP) model, which utilizes the concepts of ψ_b and T_b, but also incorporates progress towards germination at water potentials below ψ_b and above ψ_{min}.

(d) *Dormancy* is a complex phenomenon that is affected by a wide range of environmental factors; consequently many different approaches to modelling dormancy have been used (**see: Dormancy; Dormancy breaking**). Population-based threshold models are now being developed to quantify the effects of those environmental factors that modify dormancy status, and also those that terminate dormancy where they are different. Changes in appropriate temperature and water potential thresholds have been shown to relate to the widening and closing of the environmental range that is permissive for germination within seed lots as dormancy status changes. Thermal **afterripening** time and also hydrothermal afterripening time models can describe the effect of environmental factors on the rate of change in these thresholds, and therefore dormancy status, in a manner similar to that described above. Thus germination is predicted to take place when ambient environmental conditions coincide with the range permissive for germination, which itself is continually changed by the ambient conditions experienced. These models become increasingly complex as other factors such as the amplitude of temperature change and light are considered.

(e) *Imbibition* is also not considered separately within current models, which creates a significant limitation to predictions under conditions that vary: seeds are assumed to instantly attain the same water potential as the surrounding substrate. But imbibition can have an important influence on the prediction of germination and emergence times under agricultural conditions when non-dormant seeds are sown into dry soils or when the contact between seed and soil is poor. (When dormancy delays germination, however, the imbibition factor is less important.) In future it may be possible to account for imbibition directly, so that estimates of seed water potential, rather than measurements of external water potential, drive the models.

Seed imbibition under variable conditions of moisture availability is likely to be a complex function to model. Nevertheless, a number of models have been published in the research literature that make a range of different assumptions. In general, mechanistic models propose that the movement of water into the seed is driven by gradients of water potential (hydraulic conductivity theory) between the seed and the surrounding soil. A reduction in water potential of the surrounding soil will therefore reduce the rate of water uptake by the seed because the gradient between them is less. However, the effect on imbibition rate is not directly proportional to changes in the water potential gradient because hydraulic conductivity also varies as a function of the permeability of the seed and surrounding soil, of the extent of contact between them and of the temperature. Water concentration (diffusivity theory) has also been used to model imbibition, but should only be applied in homogeneous environments.

2. Models of pre-emergence seedling growth and emergence

To predict seedling emergence, germination models must be coupled to models that accurately describe environmental effects on pre-emergence seedling growth and the effect of seed burial depth. A large number of studies have been made of the response of seedling growth to temperature, and in many cases a thermal time approach has been adopted to take account of fluctuations in soil temperature. Fewer attempts have been made to model the interaction between temperature and water potential.

(a) *Simple models.* In practice, it is likely that germination will only be completed in conditions of moisture that are favourable to seedling growth, so that as the root extends downwards the seedlings may not experience significant water stress. In this case, simple models that use only a distribution of thermal times for seedling growth to emergence in conjunction with a hydrothermal prediction of germination time could be successful. A further simplification has proved successful in some circumstances, whereby the germination and the seedling growth stages are each regarded as progressing in unconstrained thermal time but are separated by a water potential 'switch'. In this model, the completion of germination and the start of pre-emergence seedling growth will only take place if water potential at that time exceeds an appropriate base level.

(b) *Models allowing for soil strength and structure.* Even when seedlings do not experience water stress, the soil surface can become very dry. Soil resistance increases as it dries and can become the limiting factor to upward shoot growth. (**See: Capping**) Therefore reasonable predictions are possible from a simulation that accounts for soil moisture, temperature, soil resistance to growth, and time. Few attempts have been made to include the effects of seedbed structure in emergence models.

However, one detailed model has been developed that includes the effect of aggregate size and organization in the seedbed and crust development on hypocotyl growth. Non-emergence can be accounted for in these models by incorporating seed weight and a function describing the exhaustion of reserves with time, in particular when emergence is delayed due to increasing soil resistance.

A range of models therefore exist to predict germination and seedling emergence in the field. However, progress towards putting these models into practice is limited at the present time by the accuracy with which appropriate soil properties can either be predicted or measured. (**See: Germination – influences of osmotica and salinity; Germination – influences of temperature; Germination – influences of water; Seedbed environment**)

(WFS, WRW)

Bradford, K.J. (1995) Water relations in seed germination. In: Kigel, J. and Galili, G. (eds) *Seed Development and Germination*. Marcel Dekker, New York, USA, pp. 351–396.

Bradford, K.J. (2002) Applications of hydrothermal time to quantifying and modeling seed germination and dormancy. *Weed Science* 50, 248–260.

Finch-Savage, W.E. (2004) The use of population-based threshold models to describe and predict the effects of seedbed environment on germination and seedling emergence of crops. In: Benech-Arnold, R.L. and Sánchez, R.L. (eds) *Seed Physiology: Applications to Agriculture*. Haworth Press, New York, USA, pp. 51–96.

Forcella, F., Benech Arnold, R.L., Sanchez, R. and Ghersa, C.M. (2000) Modeling seedling emergence. *Field Crops Research* 67, 123–139.

Rowse, H.R. and Finch-Savage, W.E. (2002) Hydrothermal threshold models can describe the germination response of carrot (*Daucus carota* L.) and onion (*Allium cepa* L.) seed populations across both sub and supra-optimal temperatures. *New Phytologist* 158, 101–108.

www2.warwick.ac.uk/fac/sci/hri2/research/seedscience/simulation/ (downloadable application)

Germination – influences

Seed germination is affected by several external factors: by 'external' is meant any factor that acts from outside of the seed. Broadly, these are of two kinds: (i) factors that are found in the field situation; and (ii) factors imposed experimentally, generally in the laboratory. Some factors, however, fall within both groups.

The external factor on which germination of most seed types entirely depends is water, necessary for the assumption of the physiology, metabolism and molecular processes that drive germination (**see: Germination – influences of water**). The availability of water in the soil is critically important and the readiness with which it is taken up by the seed can be influenced by the osmotic factor (**see: Germination – influences of osmotica and salinity; Hydrotime; Seedbed environment for germination**).

The most important physical factor is the temperature, active in both the field and in experimental, laboratory situations. Temperature can determine whether or not seeds germinate and the rate at which they do so (**see: Chilling injury; Germination – influences of temperature; Hydrothermal time; Thermotime**).

Another important physical factor is light which has both promotive and inhibitory effects. Promotion of germination is generally held to be through the breaking of dormancy though direct effects on germination itself might also operate. Light prevents or delays germination in seeds of many species, acting at different stages of the germination process and on incipient radicle emergence. In these effects **phytochrome** acts in its various modes and **cryptochrome** is also involved. (**See: Light – dormancy and germination**)

Numerous chemicals promote or inhibit germination or influence germination rate (**see: Germination – influences of chemicals**). Some of these chemicals occur in the soil (e.g. nitrate) but many others (e.g. various organic acids, nitrogenous compounds and electron acceptors) feature only in the laboratory situation, where they are usually used as an experimental tool to probe into germination biochemistry and germination mechanisms.

Germination is sensitive to a variety of growth regulators (**see: Germination – influences of hormones**). Some of these occur as a natural external factor, in the soil environment (e.g. **ethylene**), while others are applied to seeds in the laboratory or the field. However, the activity of the latter exogenous regulators reflects their importance as naturally occurring **hormones** (e.g. **abscisic acid, gibberellins**) that control several aspects of the germination process. In some cases, for example **parasitic plants**, germination relies upon **stimulants** in the soil derived from the host plants. In other cases, effects on germination might come from plant secretions or from chemicals that leach out of plant material (**allelopathy**): however, there is no agreement as to the significance of allelopathy in the natural situation.

Gases in the soil or used experimentally (e.g. oxygen, carbon dioxide, ethylene (see above)) act upon germination, as does smoke derived from burning vegetation (**see: Germination – influences of gases; Germination – influences of smoke**). (MB)

Germination – influences of chemicals

A wide variety of chemicals stimulate or inhibit germination of non-dormant seeds, or cause dormant seeds to germinate, but it is not always clear from the observations reported in the literature whether dormancy or germination proper has been affected. In this account, however, we will consider all chemicals that promote or inhibit germination, but the reader should keep in mind the possibility that the chemicals may be acting on dormancy or germination, or both (**see: Dormancy – embryo; Dormancy – embryo (hypotheses); Dormancy breaking – chemicals**). Most of the active chemicals are not encountered by seeds in their usual environment or have an effect only when applied at concentrations that are not likely to be present in natural germination media. The exception is nitrate, which may play an important ecological role in different soils and vegetation types, mechanistically possibly through the generation of nitrogen oxides. **Hormones** (gibberellins, cytokinins, abscisic acid, ethylene, brassinosteroids, jasmonic acid) are well known to strongly affect germination of dormant or non-dormant seeds, (**see: Germination – influences of gases; Germination – influences of hormones**).

Germination stimulators are very numerous (Table G.4). Some are of great interest because their use may contribute to an understanding of the biochemical mechanisms involved in the germination process or dormancy breakage. An example is nitrate, which interacts with light: indeed, models have been proposed to explain light effects in germination in which nitrate:phytochrome interaction has been invoked (**see: Dormancy – embryo; Dormancy – embryo (hypotheses)**).

The first important group of chemicals which strongly stimulate germination of dormant embryos (e.g. apple) or seeds with a **coat-imposed dormancy** (e.g. wild and cultivated oat, barley, rice and lettuce) includes respiratory inhibitors and other similar metabolic inhibitors. The most effective are inhibitors of the mitochondrial electron transport chain (cyanide, azide, hydrogen sulphide, sodium sulphide). Some studies have suggested that the promotive action on germination of a chemical that inhibits **respiration** implies the participation in the seed of another respiratory pathway – the cyanide-insensitive pathway of respiration. (**See: Germination – metabolism, 1. Respiration**) This is not so, however, at least in barley and oat seeds, or in apple embryos, since an inhibitor of this pathway (salicylhydroxamic acid, SHAM) has no effect when applied alone, and it does not alter the stimulation of germination induced by cyanide or azide. A long-standing hypothesis is that respiratory inhibitors stimulate germination by diverting glucose oxidation to the pentose phosphate pathway. Although this hypothesis depends only on indirect evidence and is very contentious, it is given some weight by the fact that inhibitors of the tricarboxylic acid cycle (malonate) or glycolysis (sodium fluoride), and stimulators of the pentose phosphate pathway (nitrites, nitrates, methylene blue) have the same stimulatory effect as cyanide and azide. In the apple embryo, for example, all these chemicals at high concentration stimulate germination and have an inhibitory effect on **radicle** growth. In red rice (*Oryza* spp.) the dormancy-breaking response to nitrite, azide, cyanide and hydroxylamine is pH-dependent. Optimal efficiency of these chemicals occurs at pH values which favour the uncharged forms, so it has been suggested that many of the dormancy-breaking chemicals act by lowering the pH of seed tissues, though how this change affects dormancy is not known. (**See: Dormancy – embryo (hypotheses)**)

Of the many organic compounds that stimulate germination of numerous seed types an interesting example is ethanol, which acts on dormant seeds of lettuce, barley, wild and cultivated oat and various others. Several hypotheses can be put forward to explain this action of ethanol, although none meets with particular favour. Ethanol might modify membrane characteristics, as do other organic compounds that have **anaesthetic** properties, such as chloroform, ethyl ether and acetone. It might also be involved metabolically as a respiratory substrate, since it increases oxygen uptake by *Avena fatua* and *A. sativa* seeds. Lastly, it might activate glycolysis through an increase in fructose 2,6-bisphosphate, as shown in *Avena sativa* seeds. In fact, in this species, the stimulatory effect of ethanol requires its metabolism through alcohol dehydrogenase (ADH). Indeed, other alcohols have the same action as ethanol, but only if they are good substrates of ADH such as butanol-1, propanol-1 and 2-propen-1-ol, but alcohols which cannot be oxidized by ADH (propanol-2, methanol) have practically no effect on germination. In red rice the efficacy of dormancy-breaking compounds is an inverse function of their lipophilicity, as measured by octanol/water partition coefficients. However, sufficient lipophilicity alone is not enough to promote germination of dormant seeds since alkanes are ineffective. Furthermore, the stimulatory effect of chemicals with small enough molecular dimensions, such as cyanide, azide, nitrite and formic acid, are related to their size rather than to their lipophilicity.

Potent oxidants (hydrogen peroxide, hypochlorite) encourage germination of various seeds with coat-imposed dormancy. These compounds might modify the structure of the **seedcoats**, but their action results more probably from the oxidation of phenolic compounds which act as inhibitory chemicals because they limit oxygen diffusion to the embryo (**see: Phenolics**).

Nitroprusside and nitroglycerine, both of which stimulate germination, e.g. in *Paulownia* and *Lupinus*, might act by generating nitric oxide (NO), which is emerging as a signalling molecule in plants (**see: Signal transduction – hormones**). As mentioned above, the promotive effects of nitrate might also be due to the nitric oxide that can be produced from this compound.

Fusicoccin, a diterpene glucoside produced by the fungus

Table G.4. Chemicals (excluding hormones) which stimulate germination or break dormancy of seeds.

Classes	Compounds
Respiratory inhibitors and other metabolic inhibitors	cyanide, azide, hydrogen sulphide, sodium sulphide, carbon monoxide, dinitrophenol, monofluoracetate, malonate, iodoacetate, sodium fluoride
Nitrogenous compounds	hydroxylamine, thiourea, mercaptoethanol, dithiothreitol, nitric oxide, nitroprusside, nitroglycerine
Oxidants	hydrogen peroxide, hypochlorite
Alcohols	ethanol, butanol-1, propanol-1, 2-propen-1-ol
Anaesthetics	chloroform, ethyl ether, acetone, ethanol
H acceptors	nitrites, nitrates, methylene blue
Fungal products	fusicoccin, cotylenol, cotylenin
Root exudates	strigol and other sesquiterpene lactones, hydrobenzoquinones, growth regulators

See also: Table D.12 in **Dormancy – embryo**.

Fusicoccum amygdali, and the similar compounds cotylenol and cotylenin E isolated from *Cladosporium* spp., stimulate germination of dormant lettuce seeds and hasten the germination of the non-dormant maize and radish seeds, but their mechanism of action is not known. (**See: Stimulants**)

Seeds of the parasitic angiosperms *Striga* spp. and *Orobanche* spp. germinate only in the vicinity of a suitable host in response to germination stimulants present in exudates from host roots. Various stimulants have been isolated from these exudates. Predominantly, they are strigol, sorgolactone, orobanchol and alectrol which were identified as sesquiterpene lactones, but others, such as dihydrosorgoleones and the oxidized forms sorgoleones (hydrobenzoquinones), also promote the germination of *Striga* seeds. Synthetic strigol analogues of the GR series (named after Gerry Roseberry, who first synthesized them) also induce the germination of *Striga* and *Orobanche* seeds, GR 7 and GR 24 being as efficient as strigol. However, these seeds are able to respond to the germination stimulants only after a suitable period of incubation in a wet medium, called conditioning. Ethylene may substitute for the root exudates, and hence the latter may be acting by causing the production of ethylene by the seeds which then induces germination. (**See: Parasitic plants**)

Germination inhibitors have been described in fleshy fruits, dehiscent and indehiscent fruits, all parts of seeds (coats, **endosperm, embryo**) and various vegetative organs (leaves, roots, bulbs, tubers). These chemicals have been studied largely in the years 1930–1940 and reviewed by Evenari (1949) and Roberts and Smith (1977). A great variety of compounds has been mentioned as germination inhibitors. These include gases, heterosides, aldehydes, organic acids, aromatic acids, unsaturated lactones among which coumarin is the best known, essential oils, alkaloids, amino acids such as proline, and many others (Table G.5). But the physiological role and the mechanism of action of most of them have never been determined. Those which have been shown to be present in succulent fruits and supposed to inhibit the germination of the seeds inside these fruits were termed blastokolins by Koeckeman in 1934 without any idea of their chemical nature.

The presence of germination inhibitors in plant organs has been shown in various ways, but the most common one consists of germinating seeds on organ extracts or juices. Various seeds that are particularly sensitive to inhibitors can be used for this purpose since the inhibitors are not specific, the most frequently selected ones being wheat, oat, lettuce and cress. But extracts from a plant species which inhibit germination of wheat or lettuce seeds can be without effect when tested on seeds of the species under investigation. For example, beet seeds germinate well in the presence of a methanol extract of the same seeds, and the germination of apple seeds is not altered by apple juice, whereas beet seed extract and apple juice both strongly inhibit germination of lettuce seeds. Moreover, the presence of an inhibitor in a seed or fruit extract does not constitute evidence that it is functional in germination or dormancy. A true germination inhibitor is a compound which inhibits the germination of the embryo. Excepting **abscisic acid** (ABA) (**see: Germination – influences of hormones**), none of the inhibitors mentioned in the literature is able to do this. Apple seeds possess embryo dormancy and a seedcoat-imposed dormancy and their coats contain high amounts of phenolic compounds, which are considered to be germination inhibitors at high concentrations. The inhibition of germination imposed by the seedcoats might then be attributed to these chemicals. This is not the case, however, since isolated embryos, the dormancy of which has previously been broken by cold, germinate easily in the presence of saturated solutions of various phenolic compounds. These chemicals do not inhibit directly the germination of the embryo, but they constitute a barrier to oxygen diffusion through the seedcoats. Diverse germination inhibitors therefore certainly exist, but their biological action is far from clear. They are mostly invoked when there is not a more rational explanation for the failure to germinate. For example, seeds usually do not germinate inside succulent fruits. Seeds do not germinate in apples because the embryos are dormant and the seedcoats strongly inhibit germination by limiting oxygen supply, and not because the fleshy fruit contains blastokolins. When embryo dormancy in apple seeds within the fruit has been broken by placing the fruit at a chilling temperature (0–4°C) for 2–3 months, seeds germinate readily if oxygen is introduced into the core of the fruits. (DC, FC)

Bewley, J.D. and Black, M. (1982) *Physiology and Biochemistry of Seeds in Relation to Germination*. Vol. 2, *Viability, Dormancy and Environmental Control*. Springer, Berlin.

Cohn, M.A. (1996) Chemical mechanisms of breaking seed dormancy. *Seed Science Research* 6, 95–99.

Cohn, M.A. and Hilhorst, H.W.M. (2000) Alcohols that break seed dormancy: the anaesthetic hypothesis, dead or alive? In: Viémont, J.-D. and Crabbé, J. (eds) *Dormancy in Plants. From Whole Plant*

Table G.5. Chemicals which inhibit germination.

Classes	Compounds
Gases	H_2S, NH_3, Cl_2, SO_2
Heterosides	allyl-isothiocyanate, β-phenethyl-isothiocyanate
Aldehydes	acetaldehyde, benzaldehyde, salicylic aldehyde, cinnamaldehyde, etc.
Organic acids	malic acid, citric acid, acetic acid, etc.
Aromatic acids	cinnamic acid and its derivatives, phenolic acids (caffeic acid, ferulic acid, vanillic acid, *p*-coumaric acid, *p*-hydroxybenzoic acid)
Unsaturated lactones	parasorbic acid, anemonin, coumarin, osthenol, phthalides
Alkaloids	cocaine, physostigmine, caffeine, quinine, cinchonine, strychnine, codeine
Amino acids	proline
Various	essential oils, naringenin, saponin, juglone, tryptophan, oxalate, fatty acids (short-chain), tannins, inorganic ions

Behaviour to Cellular Control. CAB International, Wallingford, UK, pp. 259–274.
Côme, D. and Corbineau, F. (1989) Some aspects of metabolic regulation of seed germination and dormancy. In: Taylorson, R.B. (ed.) *Recent Advances in the Development and Germination of Seeds*. Plenum Press, New York, USA, pp. 165–179.
Evenari, M. (1949) Germination inhibitors. *Botanical Review* 15, 153–194.
Hallett, B.P. and Bewley, J.D. (2002) Membranes and seed dormancy: beyond the anaesthetic hypothesis. *Seed Science Research* 12, 69-82.
Roberts, E.H. and Smith, R.D. (1977) Pentose phosphate pathway and germination. In: Khan, A.A. (ed.) *The Physiology and Biochemistry of Seed Dormancy and Germination*. North Holland, Amsterdam, The Netherlands, pp. 385–411.

Germination – influences of gases

The main gas affecting seed germination is oxygen, but ethylene and, to a much lesser extent, carbon dioxide can also do so. Volatile metabolites, such as nitrogen oxides, acetaldehyde and ethanol, and smoke may also regulate the germination of some seeds (**see: Germination – influences of chemicals; Germination – influences of smoke**). The composition of the gaseous phase of the soil at the level of the seedbed depends mainly on the soil structure, the diffusion and solubility coefficients of the gases, as well as on oxygen uptake and the release of CO_2, ethylene and other gases by seeds and various other organisms (fungi, bacteria, earthworms, etc.). The concentration of oxygen in the soil atmosphere does not usually fall below 19%, but it can decrease to 1% or even less in flooded soils. It can also be below 10% when a crust is formed at the soil surface. The concentration of CO_2 does not usually exceed 0.5–1%, but it can increase to 5–8% in clay or in muddy soils with decomposing organic matter. It fluctuates during the seasons and is always higher in warm soils than in cool ones. **Ethylene** in the soil results from the activity of living organisms. Its concentration is usually between 0.05 μl/l (0.05 ppm) in winter and 1.5 μl/l in summer, but can reach 10 μl/l in compacted soils. Since it acts as a **hormone**, it may affect seed germination at such low concentrations. Nitrogen oxides are among the soil trace gases: nitric oxide (NO) and nitrogen dioxide (NO_2) are generated in association with soil nitrification and denitrification processes carried out by microbes. It is now becoming clear that nitric oxide can play an important role in **signal transduction** in plants, possibly in seeds.

Seeds of only seven plant species (rice, four species of *Echinochloa*, *Erythrina caffra*, *Chorisia speciosa*) have been mentioned as being able to germinate in complete anoxia. It is most probable that all seeds require oxygen for their germination, however. Indeed, **imbibition** of seeds reactivates their metabolism, particularly their respiration, and thus induces a requirement for oxygen. But various seed species germinate better under low oxygen concentrations, as under water, than in the air. Examples are *Alisma plantago*, *Cynodon dactylon*, *Echinochloa turnerana*, *Leersia oryzoides*, *Poa compressa* and *Zizania aquatica*. There are also seeds (*Oldenlandia corymbosa*) or embryos (apple) that can be induced to germinate in reduced oxygen concentrations and not at all in air when they are dormant. Oil-storing seeds (lettuce, sunflower, radish, turnip, cabbage, flax, soybean) usually require more oxygen for completion of germination than starchy seeds (rice, wheat, maize, sorghum, pea). Table G.6 presents data on the sensitivity of seeds to oxygen deprivation.

The oxygen requirement for seed germination greatly depends on other environmental factors, such as temperature, light or water potential and on treatments applied to the seeds (e.g. **coating, pelleting, priming**). For many species, the higher is the temperature the richer in oxygen the atmosphere must be for seed germination. Indeed, when the temperature rises the embryo has less oxygen at its disposal to support its increased respiration as the solubility of oxygen (and therefore its availability) decreases. Tomato and sunflower seeds require more oxygen for germination when sown in a medium with low water potential. In the case of the photoinhibitable seeds of *Amaranthus caudatus* and *Bromus rubens*, the presence of continuous white light reinforces their sensitivity to hypoxia (**see: Light – dormancy and germination**). Coated seeds and particularly pelleted seeds are more sensitive to oxygen deprivation than naked seeds. In contrast, priming reduces the oxygen requirement for seed germination.

But a critically important phenomenon involved in the requirement for oxygen for germination is the part played by the seedcoats in the limitation of oxygen supply to the embryo. This limitation, which is largely under the control of temperature, is reflected by the fact that naked embryos usually require much

Table G.6. Sensitivity to oxygen of non-dormant seeds of some species at the approximate optimal germination temperature.

Species	Temperature (°C)	$O_{2(0)}$ (%)	$O_{2(50)}$ (%)
Allium porrum (leek)	20	1	4–5
Amaranthus caudatus (love-lies-bleeding)	25	1	7–8
Avena sativa (oat)	20	0.5	0.8–1
Beta vulgaris (sugarbeet)	20	1	4.5–5
Brassica napus (rape)	20	2–3	5
Brassica oleracea (cabbage)	20	3	7
Brassica rapa (turnip)	20	1–2	3–8
Cichorium intybus (chicory)	20	1	3
Cucumis melo (melon)	25	1	3–5
Cyclamen persicum (cyclamen)	15	5	8–10
Daucus carota (carrot)	20	3	9–10
Glycine max (soybean)	25	2	6
Helianthus annuus (sunflower)	25	3	7–8
Hordeum vulgare (barley)	20	0.5	1–3
Lactuca sativa (lettuce)	20	2	5.5
Linum usitatissimum (flax)	20	1–3	3–8
Lycopersicon esculentum (tomato)	20	1	3–4
Pisum sativum (garden pea)	25	0.02	0.9
Primula acaulis (primrose)	20	5	7–10
Raphanus sativus (radish)	20	3	7.5
Sorghum bicolor (sorghum)	25	0.015	0.5
Triticum aestivum (wheat)	20	0.5	1–3
Valerianella locusta (lamb's lettuce)	20	3	7–8
Vigna radiata (mung bean)	25	0.25	0.5–1
Zea mays (maize)	25	0.25	0.5–1

$O_{2(0)}$, Oxygen concentration at which no seed germinates; $O_{2(50)}$, oxygen concentration at which 50% of the maximum germination is obtained. From Corbineau, F. and Côme, D. (1995) Control of seed germination and dormancy by the gaseous environment. In: Kigel, J. and Galili, G. (eds) *Seed Development and Germination*. Marcel Dekker, New York, USA, pp. 397–424.

less oxygen for germination than do whole seeds. The seed-covering tissues (the **testa** and other structures such as the **pericarp** of **achenes**, and **caryopses** or **glumes** of some cereal grains such as **oats** and **barley**), are generally composed of dead cells (**see, for example: Cereals, 4. Basic grain anatomy**). During imbibition, water penetrates by capillarity within the intercellular spaces and the empty cells, and completely invades the covering structures. These imbibed structures constitute a continuous wet layer around the embryo, and as the solubility of the gas in water is low, relatively little oxygen reaches the embryo. If no other phenomenon were involved in oxygen diffusion through the coats, the flux of oxygen reaching the embryo of a given seed would depend only on temperature. The higher the temperature, the lower is the solubility of oxygen and hence the oxygen availability to the embryo. Taking into account the diffusion coefficient of oxygen in water, the thickness of the seedcoat and the respiratory activity (i.e. oxygen consumption) of the embryo, it has been calculated for *Sinapis arvensis* seeds on water in air that the concentrations of oxygen at the surface (under the coat) and at the centre of the embryo are 2 and 0.4%, respectively, after 4 h of imbibition at 25°C. Such results clearly show that the embryo is in severe hypoxia under the seed-covering structures. However, most naked embryos require no more than 1 to 3% oxygen, or even less, in the atmosphere for germination. The flux of oxygen reaching the embryo decreases as the thickness of the seedcoats, or of the layer of water surrounding the seeds, increases. It is the reason why the germination of many seeds is impaired in the presence of an excess of water. For the same reason seeds with mucilaginous coats and pelleted seeds are usually very sensitive to oxygen deprivation. But in addition to all these effects a limitation of oxygen supply to the embryo can also result from the presence, within the coats of numerous seeds, of large amounts of phenolic compounds (**see: Phenolics**) and of the enzymes, mainly **polyphenol oxidases**, responsible for their oxidation. This polyphenol oxidase-mediated oxidation of phenolic compounds fixes part of the oxygen which dissolves in the seedcoats and reduces the flux of oxygen reaching the embryo. This barrier to oxygen diffusion, and hence the inhibitory effect of the coats, increases with temperature since oxygen solubility decreases and oxidation of phenolic compounds is more intense. When dormancy results from such a limitation of oxygen diffusion through the seedcoats, high oxygen can sometimes break it. But elevation of the external oxygen concentration may have no effect because the extra oxygen is simply consumed by the phenolic compounds and, therefore, there is no marked increase in available oxygen to the embryo.

When hypoxia inhibits germination a **secondary dormancy** in non-dormant seeds of some species may be induced (e.g. in *Xanthium pennsylvanicum*, *Viola* spp., *Veronica hederifolia*, *V. persica*). In contrast, complete lack of oxygen may actually provoke subsequent germination by breaking dormancy, such as in apple embryos and seeds of *Oldenlandia corymbosa*, subterranean clover and sunflower. Complete deprivation of oxygen also prevents the development of secondary dormancy in apple embryo and seeds of lettuce, *Xanthium pennsylvanicum*, *Rumex crispus* and *Sisymbrium officinale*. The effect of oxygen deprivation on breaking of dormancy or in preventing induction of secondary dormancy might result from ethanol production since this compound is known to stimulate germination of various dormant seeds (**see: Germination – influences of chemicals**). Anoxia or pronounced hypoxia might also intervene in seed germination or breaking of dormancy via ethylene biosynthesis. Oxygen deprivation strongly inhibits ethylene production, since the conversion of 1-aminocyclopropane 1-carboxylic acid (ACC) to ethylene by ACC oxidase requires oxygen (**see: Ethylene**). As a consequence, ACC accumulates which then gives rise to a burst of ethylene when seeds are re-exposed to air.

Exogenous ethylene stimulates the germination of various dormant and non-dormant seeds. This is the case for **parasitic plants** (*Orobanche* spp., *Striga* spp.), weeds (*Amaranthus caudatus*, *A. retroflexus*, *Chenopodium album*, *Rumex crispus*, *Spergula arvensis*), cultivated plants (lettuce, peanut, sunflower) and **trees** (apple, beechnut) (Table G.7). This gas can break **coat-imposed dormancy** (*Xanthium pennsylvanicum*, *Rumex crispus*) or **embryo dormancy** (apple, sunflower) and overcome **thermodormancy** (lettuce) or secondary dormancy (*Xanthium pennsylvanicum*). But it can also inhibit the germination of some seeds (*Chenopodium rubrum*, *Plantago major*, *P. maritima*), or have no significant effect in either direction. Seeds whose germination is stimulated or inhibited by ethylene are often photosensitive seeds. Ethylene does not overcome the light requirement of seeds (**see: Light – dormancy and germination**), but it enhances its action. However, it is not clear if its effect is mediated by phytochrome (**see: Phytochrome**). Ethylene also interacts with various other factors (e.g. temperature, oxygen tension, osmotic agents and **abscisic acid**).

The concentrations of exogenous ethylene which stimulate germination are in the range of 0.1 to 200 μl/l. However, the sensitivity (or the response) of seeds to ethylene depends on numerous factors, particularly on their depth of dormancy. Breaking of dormancy increases the effectiveness of ethylene in stimulating germination. Ethylene does not necessarily have to be present continuously for its maximal effect and application for a relatively short duration (one to a few hours) is very often enough, provided it occurs at the right time during imbibition, depending on the species. For dormant sunflower seeds, for example, a 1- or 3-h treatment with ethylene is optimal 2 days after the beginning of imbibition.

Table G.7. Some plant species whose seed germination is stimulated by ethylene. Compilation from various publications.

Amaranthus caudatus (love-lies-bleeding)
Amaranthus retroflexus (green amaranth)
Arachis hypogea (peanut)
Chenopodium album (fat hen)
Fagus sylvatica (beechnut)
Helianthus annuus (sunflower)
Hypochaeris radicata (hairy catsear)
Lactuca sativa (lettuce)
Malus domestica (apple)
Orobanche ramosa (broomrape)
Portulaca oleracea (common purslane)
Rumex crispus (curled dock)
Spergula arvensis (spurry)
Striga lutea, S. asiatica (witch weed)
Striga hermonthica
Trifolium repens (white clover)
Trifolium subterraneum (subterranean clover)
Xanthium pennsylvanicum (cocklebur)

Various experiments using inhibitors or stimulators of ethylene biosynthesis or action have shown that ethylene synthesized by the seeds themselves is involved in germination in various species, but its mechanism of action in seeds remains unknown. For more information on ethylene perception within plant cells **see: Signal transduction – hormones**.

Of the nitrogen oxides, nitrogen dioxide and nitric oxide stimulate germination, the former in *Oryza sativa* and the latter in *Emmenanthe pendulifolia* and *Arabidopsis thaliana*. When present in soil, therefore, these gases may have important effects on seeds. Nitric oxide participates in signal transduction processes so this could be the basis of its action on seeds.

Carbon dioxide at the concentrations present in soils also promotes seed germination in numerous species, but it probably acts in association with ethylene or more probably by enhancing the synthesis of the latter. Indeed, the stimulatory effect of CO_2 on the germination of dormant sunflower embryos is counteracted by inhibitors of ethylene biosynthesis or ethylene action. (DC, FC)

Beligni, M.V. and Lamattina, L. (2000) Nitric oxide stimulates seed germination and de-etiolation, and inhibits hypocotyl elongation, three light-inducible responses in plants. *Planta* 210, 215–221.

Corbineau, F. and Côme, D. (1995) Control of seed germination and dormancy by the gaseous environment. In: Kigel, J. and Galili, G. (eds) *Seed Development and Germination*. Marcel Dekker, New York, USA, pp. 397–424.

Matilla, A.J. (2000) Ethylene in seed formation and germination. *Seed Science Research* 10, 111–126.

Germination – influences of hormones

Since the involvement of endogenous hormones in seed germination and dormancy is treated in other sections (**see: Development of seeds – hormones; Dormancy – embryo; Dormancy – acquisition; Dormancy breaking hormones; Germination – cellular and molecular processes**), only the effects of their application (exogenous supply) will be examined. Also, **ethylene** is considered under gases (**see: Germination – influences of gases**).

Applied **gibberellins** (GA) and **cytokinins** can promote seed germination in numerous species, whereas **abscisic acid** (ABA) is a strong **germination inhibitor**. **Gibberellins**, usually GA_3 (gibberellic acid), GA_4 and GA_7, have the widest spectrum of stimulatory action, cytokinins, and particularly **auxins**, being less effective. However, their action depends on other external factors, including temperature, oxygen and light, and they interact among themselves. In particular, GA and cytokinins can often reverse the inhibitory effect of ABA. Since these growth regulators are naturally present in seeds, the effects of their application on germination have given rise to the **hormone balance theory** of the regulation of dormancy (**see: Dormancy**).

The earliest reports on effects of cytokinin on seed germination concern the synergistic stimulatory action of **kinetin** and light on lettuce seeds. When applied in complete darkness to dormant lettuce seeds, kinetin has only a weak promoting effect, but even though it induces germination in the presence of light, abnormal seedlings arise, the **cotyledons** emerging from the seeds before the **radicles**. The promoting effect of cytokinins is clearest in their capacity to reverse the inhibition of germination of isolated embryos by ABA. Similar effects of kinetin have been observed in intact lettuce seeds in darkness, but only in association with GA_3. Synergistic or additional effects of cytokinins and ethylene (or ethephon) have also been noted in connection with the release of **thermoinhibition** of lettuce seeds and breaking of dormancy of cocklebur and Indian rice grass seeds.

Applications of GA_3 strongly stimulate germination of seeds with **coat-imposed dormancy**, such as cereals (barley, oat, wheat), or with **embryo dormancy** (e.g. apple, beech, hazel). GA_3 or GA_4+GA_7 also improve germination in the dark of light-requiring seeds of numerous species (Table G.8). Research on seeds of *Lactuca sativa* and *L. scariola* was the first to show that GA substitutes for light in promoting germination. GA can also reverse the inhibition of germination caused by light in negative-photosensitive (negatively **photoblastic**) seeds of *Cyclamen persicum*, *Phacelia tanacetifolia* and others. In *Kalanchoë blossfeldiana*, GA is only effective in breaking dormancy when seeds have previously received some light, suggesting that here the growth regulator is only efficient in the presence of Pfr (**see: Light – dormancy and germination; Phytochrome**). Synergy between light and GA has also been observed in lettuce, *Oldenlandia corymbosa* and *Arabidopsis thaliana* seeds. In the latter, seeds of the GA-deficient **mutant** *ga1* cannot germinate unless supplied with GA, and they do not respond to light in the absence of GA unlike the wild-type seeds which do. In fact, light increases their sensitivity to exogenous GA.

Exogenous GA_3 markedly reduces the sensitivity of dormant seeds to temperature and **water potential** of the germination medium. It also strongly improves germination in hypoxia of seeds of various species such as oat, barley, *Capsicum annuum*, *Primula malacoides* and *Oldenlandia corymbosa*.

Effective concentrations of GA generally lie within the range of 10^{-5}–10^{-3} M. The stimulatory action of high concentrations of GA_3 has been attributed in part to the low pH of the solutions (pH = 3.8 for an aqueous solution of 10^{-3}

Table G.8. Responsiveness in darkness of light-requiring seeds of some species to exogenous GA.

Species with light-requiring seeds whose germination is:	
Improved by GA in darkness	Not improved by GA in darkness
Apium graveolens (celery)	*Begonia evansiana*
Arabidopsis thaliana (mouseear cress)	*Juncus maritimus* (sea rush)
Barbarea vulgaris (bitter wintercress)	*Kalanchoë blossfeldiana*
Chenopodium album (fat hen)	*Rumex crispus* (curled dock)
Lactuca sativa (lettuce)	*Spergula arvensis* (spurry)
Lactuca scariola (prickly lettuce)	
Lamium amplexicaule (henbit deadnettle)	
Nicotiana tabaccum (tobacco)	
Oldenlandia corymbosa	
Primula malacoides (fairy primrose)	
Primula obconica (German primrose)	

Compilation from various publications.
See: Light – dormancy and germination.

M GA_3). A reinforcement of the action of GA_3 by low pH has been noted for lettuce, celery and *Oldenlandia corymbosa* seeds, but low pH does not have such an effect on oat and cyclamen seeds. (**See: Dormancy breaking – chemicals**)

A short treatment (sometimes 1 h or even less) with GA_3 at the beginning of imbibition is sufficient to stimulate later germination on water. This has been taken by some to mean that the growth regulator acts at the early stage of germination, or that its influence is maintained after removal of the seed from it (although enough of the hormone may enter during the short period in which the seed is exposed to it, to exert later promotory effects). Interestingly, this stimulatory effect remains after drying back the seeds. These results can lead to interesting treatments for enhancing seed performance in various species (**see: Enhancement**).

An important industrial role for applied GA is found in malting where its physiological action is to accelerate endosperm modification (**see: Malting – process**).

Exogenous ABA strongly inhibits the germination of many seeds at concentrations ranging from around 10^{-6} to 10^{-3} M depending on the species. This inhibitory effect largely depends on whether the seed has any dormancy: seeds or embryos with some dormancy are much more sensitive to ABA than non-dormant ones. In cereals, isolated embryos (which can germinate even though the intact grains are dormant) progressively lose their sensitivity to ABA when seeds from which they are excised are stored in dry conditions which break their dormancy (**see: Afterripening; Dormancy breaking – temperature**). In oat for example, 10^{-6} M ABA is sufficient to inhibit the germination of 50% of the embryos isolated from dormant, freshly harvested seeds, whereas embryos isolated from seeds stored dry for 4–5 weeks at 30°C germinate almost completely in the presence of 10^{-4} M ABA. The responsiveness of certain cereal embryos to ABA is highly correlated with resistance of the seeds to sprouting (**see: Pre-harvest sprouting – mechanisms**); it is much weaker in grain varieties susceptible to sprouting (less dormant ones) than in those which are resistant (more dormant ones). The inhibitory effect of ABA is often temperature-dependent. Thus, the embryos isolated from dormant cereal seeds are about 1000 times less sensitive to ABA at 10°C, a temperature at which caryopses break dormancy and germinate, than at 30°C, a temperature at which dormancy is strongly expressed.

The inhibitory action of exogenous ABA is generally reversed when seeds are transferred on to water. Pretreatment of non-dormant seeds of carrot and tomato with ABA for 7 to 12 days even stimulates further germination, similarly to a priming treatment (**see: Priming**), suggesting that ABA does not inhibit the realization of **germination *sensu stricto*** but more likely radicle elongation. In apple, however, a prolonged application of ABA at a concentration which inhibits the germination of non-dormant embryos induces a secondary dormancy. Exposure for at least 15 h to ABA (0.5 μM) of embryos isolated from dormant oat (*Avena sativa*) seeds results in synthesis of new proteins and causes subsequent inhibition of germination. Preincubation of dormant *Helianthus annuus* embryos with ABA (1 mM) results also in a decrease in their sensitivity to ethylene. In all cases, the inhibitory action of ABA on seed germination can be reversed, at least partly, by GA and sometimes also by cytokinins.

Other growth regulators have been recorded as stimulators or inhibitors of seed germination. These are **jasmonates, salicylic acid** and **brassinosteroids**. Jasmonates are represented by jasmonic acid (JA) and its volatile derivative methyl jasmonate. Like ABA, they inhibit the germination of seeds of various species such as lettuce and sunflower. Similarities in ABA and JA effects suggest that they have a synergistic action. Salicylic acid has also been reported to inhibit seed germination. Since it is also known to inhibit ethylene biosynthesis it could have inhibitory effects on seeds whose germination requires ethylene. Brassinosteroids (BRs) represent a group of over 60 kinds of compounds isolated first from pollen of rape (*Brassica napus*), which promote seed germination. In *Arabidopsis*, they rescue the germination phenotypes of severe GA-biosynthesis **mutants** and of the GA-insensitive mutant *sleepy1*. In tobacco, they promote germination in the light and in the dark of non-photodormant seeds but, contrary to GAs, they cannot break dormancy of **photodormant** seeds. In both species, they counteract the inhibitory effect of ABA on seed germination in a GA-independent manner. BRs and GA would therefore promote germination by distinct signalling pathways (**see: Signal transduction**).

In all these effects, the applied growth regulators can be presumed to exert their actions in a similar manner to the endogenous hormones. A consideration of the mechanism of action of these hormones can be found in the entries listed at the beginning of this article and in **Signal transduction – hormones**. (DC, FC)

Bewley, J.D. and Black, M. (1994) *Seeds: Physiology of Development and Germination*, 2nd edn. Plenum Press, New York.

Davies, P.J. (1995) *Plant Hormones: Physiology, Biochemistry and Molecular Biology*, 2nd edn. Kluwer Academic, Dordrecht, The Netherlands.

Kucera, B., Cohn, M.A. and Leubner-Metzger, G. (2005) Plant hormone interactions during seed dormancy release and germination. *Seed Science Research* 15, 281-308.

Steber, C.M. and McCourt, P. (2001) A role for brassinosteroids in germination in *Arabidopsis*. *Plant Physiology* 125, 763–769.

Germination – influences of light

See : Cryptochrome; Light – dormancy and germination; Phytochrome

Germination – influences of osmotica and salinity

Water uptake by seeds, the essential initial step towards germination, is controlled by the difference in water potential between the seed and the external medium, e.g. the soil water (**see: Germination – influences of water; Germination – physical factors; Seedbed environment; Water potential**). Water potential results from the sum of three main components, osmotic potential (ψ_π), matric potential (ψ_m) and pressure potential (ψ_p). Under laboratory conditions, where sufficient water is made available to the seed, ψ_π of the external water is the critical component and ψ_m, ψ_p can be disregarded. When set on osmotica, seeds of most species can complete germination normally (i.e. to **radicle** emergence) in solutions of water potentials down to about -1 MPa but not below about

−2 MPa. In the soil, ψ_m plays the major role (because of effects of the soil particles on water potential), except under saline conditions where ψ_π may also be important. Saline soils contain predominantly NaCl, but they can also accumulate sulphates and carbonates of Na, Mg and Ca, particularly in steppe and desert areas. Salinity is usually given in terms of electrical conductivity (EC_e) of the aqueous saturation extract of the soil, in S/m, which is linearly proportional to osmotic potential (1 mS/cm = – 0.036 MPa).

Soil salinity may affect germination by decreasing the external osmotic potential, thus reducing water uptake by the seeds, and/or through a direct specific effect of ions which may be toxic. Colonization of saline habitats depends on salt tolerance of the germinating seeds and the seedlings. Seed germination may be a limiting step, because it is generally 10 to 100 times more sensitive to NaCl than is the growth of seedlings. Seeds of the majority of halophytes germinate best in fresh water. NaCl inhibits germination, but inhibitory concentrations vary among the species. Seeds of *Atriplex hastata*, *A. littorale*, *A. patula*, *A. halimus*, *A. canvescens*, *Helianthemum ventosum*, *Puccinellia distans* and *Spergularia media*, do not germinate in salinities above 1–1.5% NaCl (approx 0.16–0.25 M), whereas those of *Atriplex prostata*, *A. triangularis*, *Cochlearia danica*, *Eurotia lanata*, *Limonium vulgare*, *Salicornia pacifica* and *Spartina patens* can germinate in NaCl concentrations up to 3–3.5% (approx 0.5 M). A few species (*Salicornia brachiata*, *S. europea*, *S. bigelovii*, *Suaeda linearis*, *Tamaris pentandra*) are even able to germinate in up to 5% NaCl (approx 0.8 M).

However, seed sensitivity to NaCl depends on temperature. For example, seeds of *A. halimus* and *A. canvescens* are more sensitive to NaCl at 20°C than at 15°C (Fig. G.7). In *Salicornia bigelovii*, at 26.6°C seeds germinate best in the absence of NaCl whereas at lower temperatures (4.4 and 15.5°C) germination is optimal in 4% NaCl. The sensitivity to NaCl depends also on the seed origin, as exemplified by *Lepidium perfoliatum* in which seeds originating from plants growing in a saline area germinate to higher percentages in a NaCl solution at −0.25 MPa than those from non-saline habitats.

Salt tolerance at the germination stage is based on the ability of seeds to germinate at relatively high salinity, but also on the seed capacity to remain viable during imbibition under saline conditions. In fact, in numerous species such as *Salicornia europea*, *Spergularis marina*, *Suaeda depressa* and *S. linearis*, inhibition of germination by high concentrations of NaCl is overcome after transfer of the seeds on to water, indicating that there is no permanent ion toxicity. On the other hand, NaCl pretreatment of seeds of *Suaeda depressa*, *Juncus maritimus*, *Limonium bellidifolium*, *L. humile* and *L. vulgare* stimulates further germination in fresh water. (DC, FC)

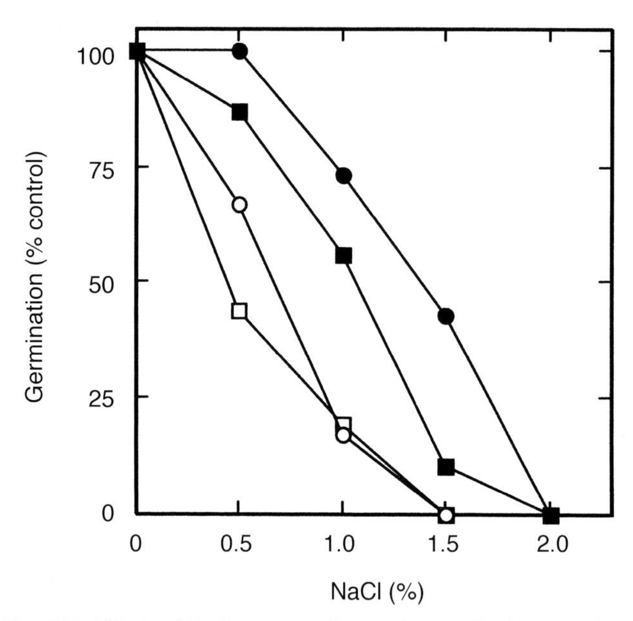

Fig. G.7. Effects of NaCl concentration on the germination capacity of seeds of *Atriplex canvescens* (●, ○) and *A. halimus* (■, □) at 15°C (●, ■) and 20°C (○, □). Germination is expressed as percentage of the germination capacity of control seeds incubated on water.

Baskin, C.C and Baskin, J.M. (1998) *Seeds. Ecology, Biogeography, and Evolution of Dormancy and Germination*. Academic Press, London, UK.

Bennett, M.A. (2004) Seed and agronomic factors associated with germination under temperature and water stress. In: Benech-Arnold, R.L. and Sánchez, R.A. (eds) *Handbook of Seed Physiology: Applications to Agriculture*. Food Products Press, New York, USA, pp. 97–123.

Ungar, I.A. (1995) Seed germination and seed-bank ecology in halophytes. In: Kigel, J. and Galili, G. (eds) *Seed Development and Germination*. Marcel Dekker, New York, USA, pp. 599–628.

Germination – influences of smoke

Smoke is an emanation composed of gases and small solid particles, the diameter of which is generally less than 1 μm. It can result from condensation of vapours or from exothermic chemical reactions. Ecological studies by de Lange and Boucher in 1990 were the first to demonstrate that smoke produced by burning plants stimulates the germination of dormant seeds of *Audouinia capitata*. This stimulatory effect of smoke on seed germination has been confirmed by further investigations in about 100 other fynbos (heathlands in South Africa) species from various families (Asteraceae, Ericaceae, Proteaceae and Restionaceae) and numerous native Western Australian species of the genera *Geleznowia* (Rutaceae), *Actinostrobus* (Cupressaceae), *Stirlingia* (Proteaceae) and *Verticordia* (Myrtaceae). Although most studies concern species native to fire-prone areas, improvement of seed germination by smoke has been reported in several non-fire-dependent species including cultivated plants such as lettuce and celery, and in different weeds such as wild oats. While all of the active compounds of smoke have not been identified, they are known to be water soluble. Smoke-derived extracts obtained by bubbling smoke through water indeed stimulate germination and commercial preparations of 'smoke water' for the promotion of germination have been marketed in Australia and South Africa. **Ethylene** contained in smoke is not involved in its stimulatory effect, at least in numerous fynbos species, but it is possible that nitrogen oxides are among the major active constituents. Analysis of smoke components has also revealed the presence of chemical compounds, **butenolides**, that have extremely potent germination-promoting action on several seed species (including lettuce) at concentrations at the level of parts per billion. Interestingly enough, the improvement of germinability by seed fumigation or imbibition in aqueous smoke extracts is maintained during dry storage, suggesting that smoke during fire may be an important factor of germination in the fire-prone environments.

(DC, FC)

Adkins, S.W., Davidson, P.J., Mathew, L., Navie, S.C., Wills, D.A., Taylor, I.N. and Bellairs, S.M. (2000) Smoke and germination of arable and rangeland weeds. In: Black, M., Bradford, K.J. and Vázquez-Ramos, J. (eds) *Seed Biology: Advances and Applications*. CAB International, Wallingford, UK, pp. 347–360.

Brown, N.A.C. and van Staden, J. (1997) Smoke as a germination cue: a review. *Plant Growth Regulation* 22, 115–124.

de Lange, J.H. and Boucher, C. (1990) Autecological studies on *Audouinia capitata* (Bruniaceae). Plant-derived smoke as a seed germination cue. *South African Journal of Botany* 56, 700–703.

Flematti, G.R., Ghisalberti, E.L., Dixon, K.W. and Trengove, R.D. (2004) A compound from smoke that promotes seed germination. *Science Express* www.scienceexpress.org/8 June 2004 / Page 1/10.1126/science.1099944

Minorsky, P.V. (2002) Smoke-induced germination. *Plant Physiology* 128, 1167–1168.

Germination – influences of temperature

1. Temperature, overview

Temperature affects the germination of seeds either directly, through action on germination itself, or indirectly, by affecting **dormancy** and **viability**. Characteristic of each species, there is a minimum and a maximum temperature below and above which seeds cannot germinate. That is, each species, is characterized by a temperature range over which germination is possible. When time is factored into germination, there is another effect of temperature – upon the rate at which seeds proceed through the germination process. Temperature controls the germination rate, in a pattern characteristic of each species: and in each case a temperature can be identified at which the rate is at its highest (**see: 4. Germination rate**).

Expression of dormancy is generally temperature-dependent, i.e. in many seed types there is a temperature range (which might change with time) over which primary dormancy is exhibited (**see: Dormancy**). In mature seeds, **secondary dormancy** is induced by relatively low or high temperatures, according to species (**see: Dormancy; Dormancy – ecophysiology; Dormancy cycles**). Temperature is also important in the loss of dormancy. Afterripening, the gradual loss of dormancy in dry seeds with time, is temperature-determined. And in imbibed mature seeds, dormancy is broken frequently by relatively low temperatures (chilling) or in some species by warm temperatures (**see: Dormancy breaking – temperature**).

The retention of viability by seeds is partly determined by temperature, higher temperatures favouring the loss of viability (**see: Viability**).

2. Temperature maxima, minima and optima

For each seed population, there exist three cardinal temperatures for germination (expression introduced by J. Sachs in 1860): the maximum, minimum and optimum. These temperatures are typical of each species, but they can vary according to the variety, the cultivar, the ecotype, the geographical origin of the seeds, the year of harvest, the duration of storage, the extent of any dormancy or the treatments (e.g. **priming**, **hormones** such as **gibberellins**) applied to the seeds. To determine these three temperatures it is important to allow the maximum possible time for germination. With increasing germination time the minimum and maximum temperatures shift to lower and higher values, respectively.

Maximum temperature: the temperature above which no seed from a population can complete germination.

Minimum temperature: the temperature below which no seed from a population can complete germination.

Optimum temperature: the temperature at which the germination rate of a population of seeds is highest.

3. Temperature range

Temperature is a highly important factor affecting seed germination because it has a strong influence on all biochemical reactions but also, and often mainly, because it regulates the flux of oxygen reaching the **embryo** through the **seedcoat** (**see: Germination – influences of gases**). The temperature range within which germination occurs varies substantially among the species (Table G.9, Fig. G.8) and even the varieties or the cultivars of a single species and depends on the geographical origin of the species. Moreover, the sensitivity of seeds to temperature can be strongly modified by many other factors intervening during seed development or maturation, or after harvest. It is also greatly affected by dormancy. In this section only the temperature ranges allowing the germination of non-dormant seeds are considered, and also only the intact seeds, with their coats present, because isolated embryos are often less sensitive to temperature than the whole seeds.

The temperature range allowing the germination of seeds of a given species is related to the climatic and ecological conditions to which the species is adapted. In the field, seeds germinate at temperatures and other external conditions (mainly rainfall) that are favourable to subsequent seedling growth and establishment. It is possible roughly to classify non-dormant seeds in three categories according to the temperature range within which they can germinate (Table G.9). Some seeds germinate only at relatively cool temperatures, the thermal optimum being between about 10 and 20°C and sometimes less (tulip for example). These are seeds from temperate climates. In contrast, other seeds germinate only at relatively high temperatures, the thermal optimum being sometimes close to lethal temperatures (35–40°C). Almost all these seeds are from tropical or subtropical plants. Other seeds are able to germinate in a very large temperature range. Their thermal optimum depends on the species, but in tropical seeds it is always relatively high. Species can therefore have widely different maximum and minimum temperatures for germination. These temperatures are usually determined using controlled-temperature incubators, but it is difficult to secure sufficient facilities to cover a wide temperature range at small intervals, and the thermal regulation of incubators is not always very

Table G.9. Approximate temperature range and thermal optimum of germination of seeds from various species. Dormancy in some cases has been broken.

Type of seeds	Species	Temperature range (°C)	Thermal Type of optimum (°C)
Seeds germinating well only at cool temperatures	*Allium porrum* (leek)	5–25	15–20 relatively
	Apium graveolens (celery)	5–25	10–15
	Chrysanthemum coronarium (chrysanthemum)	5–30	15–20
	Consolida regalis (field larkspur)	5–20	10–15
	Cyclamen persicum (cyclamen)	10–20	15–18
	Dahlia variabilis (dahlia)	5–25	15–20
	Malus domestica (apple)	0–25	15–20
	Nemesia strumosa (pouch nemesia)	5–20	10–15
	Nigella damascena (love-in-a-mist)	5–25	15–20
	Primula acaulis (primrose)	5–25	20
	Schefflera paraensis (arbre de St. Jean)	10–25	15–20
	Tropaeolum majus (nasturtium)	5–30	15–20
	Tulipa gesneriana (tulip)	0–12	5
Seeds germinating well only at relatively high temperatures	*Amaranthus caudatus* (love-lies-bleeding)	10–40	30–35
	Castanospermum australe (black bean)	17–40	30–35
	Cedrela odorata (West Indian cedar)	15–35	25–30
	Ceiba pentandra (kapok)	15–40	30–35
	Chrysanthemum parthenium (feverfew)	15–35	25–30
	Cucumis sativus (cucumber)	15–40	30–35
	Dianthus barbatus (bearded-pink)	10–30	20–25
	Ficus capensis (Cape ficus)	10–35	25
	Gomphrena globosa (globe amaranth)	15–35	25–30
	Gyrocarpus americanus (propeller tree)	17–40	30–35
	Lycopersicon esculentum (tomato)	10–35	25
	Oldenlandia corymbosa	25–45	35–40
	Pennisetum americanum (pearl millet)	15–45	35–40
	Portulaca oleracea (common purslane)	15–40	35–40
	Santalum austrocaledonicum (sandalwood white)	17–35	27–30
	Sorghum vulgare (sorghum)	10–40	35
	Symphonia globulifera (chewstick)	15–35	25–30
	Terminalia superba (myrobolan)	20–40	30
	Zea mays (maize)	12–40	25–30
Seeds germinating in a large temperature range	*Avena sativa* (oat)	0–30	25–30
	Beta vulgaris (sugarbeet)	5–35	20–25
	Brassica oleracea (cabbage)	5–40	25–30
	Camelina sativa (falseflax)	5–35	25–30
	Cichorium intybus (chicory)	5–35	25–30
	Daucus carota (carrot)	5–40	25–30
	Gerbera spp. (gerbera)	10–35	20–25
	Glycine max (soybean)	5–40	25–30
	Helianthus annuus (sunflower)	5–40	25–30
	Hopea odorata (merawan)	5–40	30–35
	Hordeum vulgare (barley)	0–35	25–30
	Mangifera indica (mango)	5–45	30–35
	Pisum sativum (garden pea)	2–35	25–30
	Shorea roxburghii	5–40	30–35

Compilation from various publications.

precise (often ± 1 or 2°C). To solve this problem, **thermogradient bars** have been developed. It is also possible to determine these temperatures with graphs representing the variations of the reciprocal of time t_g ($1/t_g$) to reach the germination percentage g (usually 50%) as a function of temperature (**see: 4. Germination rate**).

The temperature range allowing germination can be changed by various seed treatments. It is usually extended when the treatments increase the **vigour** of the seeds (e.g. priming, gibberellin) or reduced with treatments or in conditions that can sometimes lower seed vigour (ageing, **coating, pelleting**, chemical treatments). (**See: Thermal time** and entries referred to therein for further discussion of the effects of temperature and germination.)

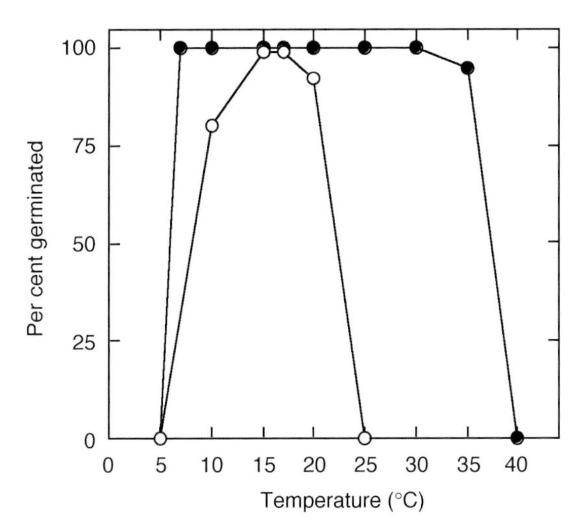

Fig. G.8. Example of two different temperature ranges within which germination occurs. ●, Cauliflower seeds; ○, cyclamen seeds.

Bennett, M.A. (2004) Seed and agronomic factors associated with germination under temperature and water stress. In: Benech-Arnold, R.L. and Sánchez, R.A. (eds) *Handbook of Seed Physiology: Applications to Agriculture*. Food Products Press, New York, USA, pp. 97–123.

Bewley, J.D. and Black, M. (1982) *Physiology and Biochemistry of Seeds in Relation to Germination*. Vol. 2, *Viability, Dormancy and Environmental Control*. Springer, Berlin.

Bewley, J.D. and Black, M. (1994) *Seeds: Physiology of Development and Germination*. Plenum Press, New York, USA.

Probert, R.J. (2000) The role of temperature in germination ecophysiology. In: Fenner, M. (ed.) *Seeds: The Ecology of Regeneration in Plant Communities*, 2nd edn. CAB International, Wallingford, UK, pp. 261–292.

Roberts, E.H. (1988) Temperature and seed germination. In: Long, S.P. and Woodward, F.I. (eds) *Plants and Temperature*. Symposia of the Society for Experimental Biology 42, 109–132.

Thompson, P.A. (1973) Geographical adaptation of seeds. In: Heydecker, W. (ed.) *Seed Ecology*. Butterworths, London, UK, pp. 31–58.

4. Germination rate

A population of seeds is always more or less heterogeneous. This heterogeneity within a single population has many origins. The germinative properties of seeds depend on their maturity stage at harvest or shedding, their size, their position on the mother plants or within the inflorescences, their degree of **dormancy**, the quality of pollination, the treatments applied to the seeds after harvest, the conditions and the duration of storage, etc. As a consequence, all the seeds from one population do not complete germination at the same time. The rate of germination corresponds to the time needed for a given germination percentage or the maximum percentage (i.e. the germination capacity – **see: Germinability**) to be reached. This time depends on the species and is highly temperature dependent. It is also markedly dependent on dormancy. The thermal optimum can be about the middle of the temperature range allowing germination, or close to one extremity of this range. In the first case, germination progresses at temperatures that fall off from both sides of the optimum. In the second case, the advancement of germination arises mainly towards the minimum or the maximum temperature. At their thermal optima, seeds exhibit a wide range of germination rates (Table G.10).

A **thermal time** approach can be used to normalize the times for germination at different sub-optimal temperatures. The thermal time (θ_T *(g)*) to completion of germination at temperature *T* of percentage *g* of the seed population is calculated according to the equation $\theta_T(g) = (T - T_b)\, t_g$, where T_b is the minimum or base temperature that permits radicle protrusion, and t_g is the time to completion of germination of percentage *g* of the seed population. In general, T_b is considered to be equal for all the seeds; the thermal time is then constant for a given percentage of germination. Fig. G.9 shows that there exists a linear relationship between germination rate (i.e. $1/t_g$) for a specific germination percentage *g* and the temperature, but the thermal time increases with increasing percentage of germination. At supra-optimal temperatures, germination rate decreases linearly with temperature, but unlike T_b, the ceiling or maximum temperature allowing germination T_c varies in a normal distribution among individual seeds, resulting in a series of parallel lines with different intercepts on the temperature axis ($T_c(g)$). (**See: Germination – field emergence models**)

Theoretical cardinal temperatures for germination (**see: section 2**) are often determined by germinating the seeds over a large range of temperatures, for example from 5 to 40°C at intervals of 5°C or less, sometimes using a thermogradient bar and measuring the time to reach 50% germination (t_{50}). There exists a positive linear relationship between the rate of germination, expressed as the reciprocal of t_{50} ($1/t_{50}$) and temperature up to the thermal optimum, and a linear negative relationship at higher temperatures. The thermal optimum corresponds to the intersection of the two regression lines. Extrapolation of the same lines to the temperature axis allows the calculation of the minimal and maximal temperatures for germination. Fig. G.10 illustrates this mode of calculation with cauliflower seeds. If the seeds do not all germinate because some of them are dead or empty (without embryo), or dormant, the same calculations are possible by replacing t_{50} by the time to reach 50% of the maximum germination percentage of the particular population under study.

The time to reach the maximum germination percentage and t_{50} (or its reciprocal) are good expressions of the germination rate. But other expressions are often used, such as the mean time to germination (MTG) which is calculated as follows:

$$MTG = (N_1 \times 1 + (N_2 - N_1) \times 2 + (N_3 - N_2) \times 3 + \ldots + (N_n - N_{n-1}) \times n) / n$$

where N_1, N_2, $N_3 \ldots N_{n-1}$ and N_n are the percentages of germinated seeds at times (usually days) 1, 2, 3... *n*-1 and *n*.

The time course of germination is usually represented by germination curves which correspond to the evolution of cumulated germination percentages over time. The shape of these curves, which is usually sigmoidal, shows how many seeds in a population can complete germination and the rate at which germination progresses. Another possibility for expressing the germination rate is to represent the percentages of seeds which complete germination every day or every other period of time. This is shown in Fig. G.10 for cauliflower seeds.

Table G.10. Time in days necessary to obtain the maximum germination percentage at the thermal optimum (in °C) of some non-dormant seeds (data from several sources).

Species	Family	Thermal optimum (°C)	Time (days)
Allium cepa (onion)	Liliaceae	25–30	4–5
Allium porrum (leek)	Liliaceae	15–20	8–10
Amaranthus caudatus (love-lies-bleeding)	Amaranthaceae	30–35	1–2
Atriplex halimus (saltbush)	Chenopodiaceae	20–25	2–4
Avena sativa (oat)	Poaceae	25–30	2–3
Beta vulgaris (sugarbeet)	Chenopodiaceae	20–25	4–6
Brassica napus (rape)	Cruciferae	25	3–4
Brassica oleracea (cabbage)	Cruciferae	25–30	2
Cedrela odorata (West Indian cedar)	Meliaceae	30–35	5–7
Cheiranthus cheiri (wallflower)	Cruciferae	20	5–7
Cichorium intybus (chicory)	Compositae	25–30	3–4
Cyclamen persicum (cyclamen)	Primulaceae	15–18	15–20
Daucus carota (carrot)	Umbelliferae	25–30	4–6
Dicorynia guianensis (basralocus)	Caesalpinioideae	25–30	4–5
Glycine max (soybean)	Papilionoideae	25–30	3
Helianthus annuus (sunflower)	Compositae	25–30	2–3
Heritiera utilis (niangon)	Sterculiaceae	30–35	7–10
Hopea odorata (merawan)	Dipterocarpaceae	30–35	4–5
Hordeum vulgare (barley)	Poaceae	25–30	1–2
Lycopersicon esculentum (tomato)	Solanaceae	25	3–5
Malus domestica (apple)	Rosaceae	15–20	15–20
Mangifera indica (mango)	Anacardiaceae	30–35	3
Nicotiana tabacum (tobacco)	Solanaceae	30	2–3
Oldenlandia corymbosa	Rubiaceae	35–40	1
Pennisetum americanum (pearl millet)	Poaceae	35–40	5–8
Pisum sativum (garden pea)	Papilionoideae	25–30	2–3
Primula acaulis (primrose)	Primulaceae	20	7–10
Pseudotsuga menziesii (Douglas fir)	Pinaceae	25	50–70
Pyrethrum spp.	Compositae	10–15	18–20
Santalum austrocaledonicum (sandalwood white)	Santalaceae	27–30	60–70
Shorea roxburghii	Dipterocarpaceae	30–35	12–15
Sorghum vulgare (sorghum)	Poaceae	35	2–3
Symphonia globulifera (chewstick)	Guttiferae	25–30	40–55
Tagetes patula (French marigold)	Compositae	20	4–5
Terminalia superba (myrobolan)	Combretaceae	30	12–15
Triticum aestivum (wheat)	Poaceae	25–30	1–2
Tulipa gesneriana (tulip)	Liliaceae	5	60–70
Valerianella locusta (lamb's lettuce)	Valerianaceae	15–20	7–8
Vigna mungo (mung bean)	Papilionoideae	30–35	2–3
Zea mays (maize)	Poaceae	25–30	2–3

Here, only constant temperatures have been considered. But seeds of some species which germinate poorly at constant temperatures germinate much better (faster and to higher percentages) if the temperature fluctuates (alternating temperatures).

The mechanism of action of temperature at the embryo level is not known, and it is not clear whether cell membranes can be considered as the thermal receptors, or if conformational effects on specific proteins are involved, for example. The latter would require there to be beneficial effects as the temperature rises to the optimum and then deleterious effects as it passes the optimum. Temperature obviously can affect germination through its action on metabolic activity of the embryo; in particular, the respiratory activity of the latter increases as temperature rises: but the relationships between temperature and metabolism and germination rate do not suggest a causal link. When the embryo is naked and non-dormant, its consumption of oxygen can be a good indication of its germination rate, but such a relationship very often does not exist with intact seeds, i.e. when the seedcoats cover the embryo (**see: Germination – influences of gases**). Moreover, temperature is not the only external factor regulating the germination rate. Its action can be largely modulated by all the other factors involved in germination, e.g. oxygen, ethylene, light and water potential of the germination medium. (DC, FC)

Covell, S., Hellis, R.H., Roberts, E.H. and Summerfield, R.J. (1986) The influence of temperature on seed germination rate in grain legumes. I. A comparison of chickpea, lentil, soybean, and cowpea at constant temperatures. *Journal of Experimental Botany* 37, 705–715.

Probert, R.J. (2000) The role of temperature in germination ecophysiology. In: Fenner, M. (ed.) *Seeds: The Ecology of Regeneration in Plant Communities*, 2nd edn. CAB International, Wallingford, UK, pp. 261–292.

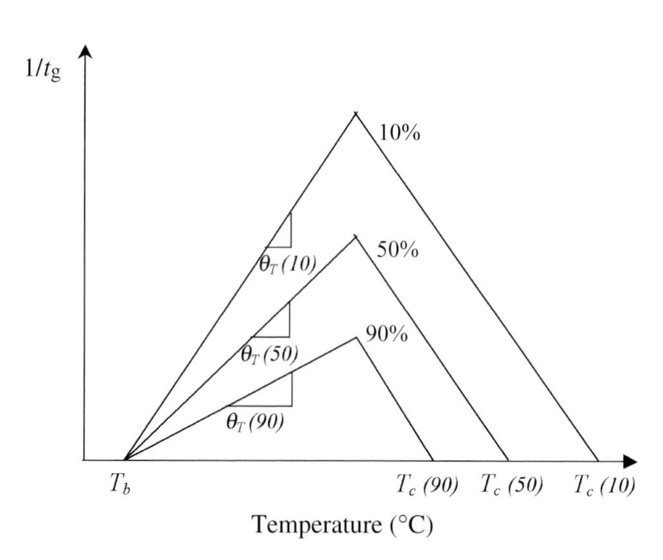

Fig. G.9. Relationships of germination rates for specific percentages ($1/t_g$) to germination temperature (theoretical curves). T_b, base (minimum) temperature allowing germination; T_c *(g)*, ceiling (maximum) temperature allowing germination of percentage *g* (10, 50 or 90%) in the seed population; θ_T *(g)*, thermal time to completion of germination of percentage *g* (10, 50 or 90%) in the seed population. From Bradford, K.J. (1995) Water relations in seed germination. In: Kigel, J. and Galili, G. (eds) *Seed Development and Germination.* Marcel Dekker, New York, USA, pp. 351–396.

Roberts, E.H. (1988) Temperature and seed germination. In: Long, S.P. and Woodward, F.I. (eds) *Plants and Temperature*. Symposia of the Society for Experimental Biology 42, 109–132.

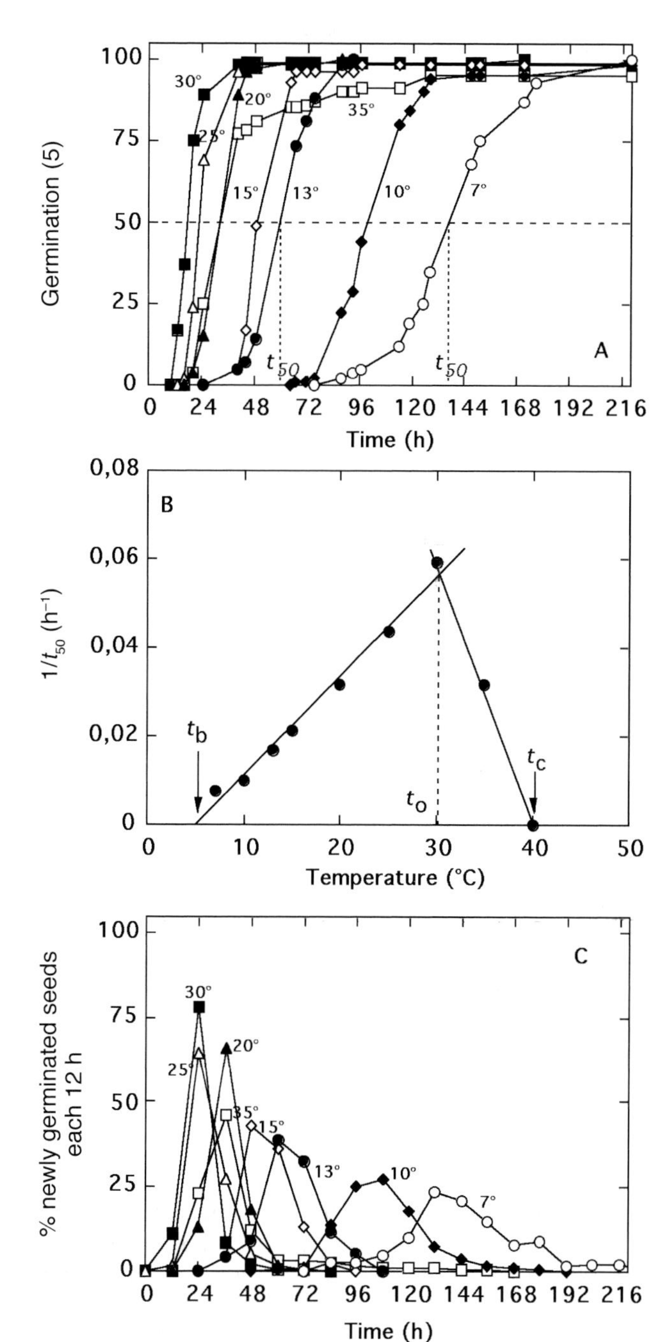

Fig. G.10. Effects of temperature on the germination rate of cauliflower seeds. (A) Time course of germination at 7, 10, 13, 15, 20, 25, 30 and 35°C; the curves obtained allow the calculation of the time to reach 50% germination (t_{50}), as indicated on the figure for 7 and 13°C. (B) Effects of temperature on the germination rate expressed as the reciprocal of t_{50} ($1/t_{50}$); the straight regression lines allow the calculation of the theoretical optimum temperature (t_o), minimum temperature (t_b) and maximum temperature (t_c). (C) Distribution of germination with time (derived from A).

Germination – influences of water

The water status of seeds is central to their survival and their germination. **Orthodox seeds** are desiccation tolerant and are usually highly dehydrated at maturity. Their moisture content is about 10–15% dry weight basis. They survive well in this dehydrated state and can even withstand more intense desiccation. Owing to their low moisture content, they also tolerate very low temperatures (even that of liquid nitrogen: −196°C. **See: Cryostorage**). Therefore, the lower their moisture content and the temperature the longer they survive in storage. Most of them tolerate ultra-drying and even freeze-drying. Their germination of course requires water, but they lose their desiccation tolerance as soon as they have germinated, or a little later (**see: Desiccation tolerance – acquisition and loss**).

Recalcitrant seeds remain rich in water at shedding and are intolerant of dehydration. They rapidly die if they lose too much water. The moisture content under which they do not survive varies from about 20 to 50% dry weight basis depending on the species. Most of them are from trees of tropical rainforests, tropical plantation crop or timber species and mangrove swamps, but some large-seeded hardwoods from temperate climates also produce such seeds (**see: Recalcitrant seeds,** Table R.2). They must germinate as soon as they fall to the ground or remain on moist soil until they germinate, otherwise they do not survive. They usually do not require external water for germination since their natural water content is sufficient for them to complete germination, but later water supply is of course necessary for subsequent growth of the seedlings. Some are even capable of germinating on the parent trees before shedding (species from mangrove swamps such as the Rhizophoraceae *Rhizophora* spp., *Bruguiera gymnorrhiza* and *Ceriops tagal*, and *Avicennia marina*) or in the fleshy fruits within which they are enclosed (mango). Their storage is very difficult because of their desiccation intolerance (**see: Desiccation sensitivity or intolerance**). They must be maintained in a wet medium, but they germinate since they usually do not possess any **dormancy**.

In the case of non-dormant, orthodox seeds, i.e. seeds that germinate in both light and dark over a wide temperature range, the determinant of germination is the presence of sufficient available water, whose uptake by the seeds is the initial step toward germination (**imbibition**). However, an excess of water is generally harmful which is why most seeds do not germinate when immersed in water. Even a thin layer of water over or around the seeds is sometimes enough to strongly inhibit their germination. This is the case, for example, for sugarbeet seeds. In the **malting** process, germination of barley is affected by excess water, so-called water sensitivity. The excess of water intervenes indirectly by depriving the embryo of oxygen (**see: Germination – influences of gases**). But there are some seeds which germinate well when covered with water, such as seeds from aquatic plants (*Typha latifolia* and **rice** for example). Some of the seeds of the tropical weed *Oldenlandia corymbosa* are able to germinate only when completely immersed.

A very rapid consequence of seed imbibition is the non-respiratory release of gases. This phenomenon (called the wetting burst) is immediate and lasts only a few minutes. It occurs by the release of adsorbed atmospheric gases (oxygen, nitrogen, carbon dioxide) retained in the dry porous structures of the seedcoats. Water penetrates these structures by capillarity driving off the adsorbed gases.

Water uptake by orthodox seeds is very fast as soon as they are placed in a moist medium. This initial period of imbibition induces an immediate and rapid **leakage** of solutes, such as sugars, organic acids, amino acids, proteins and ions, from the seed tissues but it rapidly decreases and becomes negligible within about 30 min to 1 h (**see: Germination – physical factors; Imbibition; Vigour testing – biochemical** for information on leakage and soaking injury). It results from reorganization of cell membranes during imbibition. In hydrated cells, **membranes** are composed of proteins set in **phospholipids** arranged in a fluid bilayer (**see: Desiccation damage**), which is selectively permeable and retains the cell solutes. This liquid-crystalline membrane configuration requires a cell water content of at least about 25%. During **maturation** drying of the seeds on the mother plant, part of the phospholipid molecules become rearranged into a hexagonal phase and proteins are displaced. As seeds take up water, the membranes spontaneously recover the bilayer configuration, but as long as they are not completely reconstituted they are not selectively permeable and leakage occurs. In high-vigour seeds, imbibition rapidly restores the membrane structures and leakage is reduced, but in low-vigour seeds the repair mechanisms are weakened and leakage is longer and more abundant (**see: Vigour testing**). Ultra-dried seeds are often injured during imbibition probably because membranes are more damaged or water uptake is too fast. It is advisable to rehydrate them slowly for one or a few days in an atmosphere at high relative humidity before sowing.

When there is no limitation to water availability, the rate of imbibition is determined by the **water potential** of the seed and its permeability to water. This permeability depends mainly on the structure and the composition of the coats. Various species produce seeds with impermeable testa. These seeds, called hard seeds (**hardseededness**), are common in the Leguminosae (Fabaceae), but they are found also in other plant families such as Cannaceae, Chenopodiaceae, Convolvulaceae and Malvaceae. In hard-coated seeds, hindrance to water entry is then the limiting factor of germination. As such seeds cannot imbibe they usually survive long periods of time, and their germination can be delayed for many years in natural conditions. In a population of hard-coated seeds there exist various degrees of waterproofing from one seed to another, and in several species, mainly in cultivated plants, only a fraction of the seeds is impermeable to water. As a consequence, germination can be spread over months or years, which is an effective mechanism contributing to the survival of the species. Germination of hard seeds requires that their coats be softened, and experimentally this can be achieved by biological (microbial), chemical or mechanical actions. It is not clear, however, how coat-softening occurs in natural conditions. Hard seeds which are ingested by birds or mammals are not digested, but their coats are more or less degraded in the alimentary canal and when ejected with the excrement they have become permeable to water. High temperatures and temperature fluctuations can also result in coat softening. Forest or savanna fires play an important role because they destroy the seed surface but they do not have time to burn the embryo. High temperatures, and possibly fire, are also known to crack the micropylar region of *Rhus ovata* seedcoat and to eject the strophiolar plug of *Albizia lophantha* seeds, thus allowing water to enter later (**see: Seedcoats – functions; Seedcoats – structures**). Artificial treatments have been developed for softening hard seeds, all of which have as the objective to damage the seedcoats. These include treatment with concentrated sulphuric acid, rubbing with sandpaper, dipping in boiling water, mechanical scarification by percussion, or immersion in liquid nitrogen. (**See: Scarification**)

As orthodox seeds are usually highly dehydrated, their water potential (Ψ) (**see: Water potential**) is very low (between about −350 and −50 MPa depending on the species and the degree of dehydration). The consequent difference in water potential between dry seeds and water (Ψ = 0) or moist soil being very large, seeds with permeable coats absorb water from their surrounding medium. **Imbibition** is at first very fast but, as seeds absorb water, the gradient for water uptake decreases and the seed moisture content reaches a maximum, which depends on seed composition and water potential of the germination medium. In fact, water uptake by a germinating seed develops in three phases consisting of imbibition (Phase I), plateau (Phase II) and growth (Phase III) (**see: Germination**). Maximum seed water uptake can be easily modulated by the water potential of the imbibition medium. This is the basis of the priming treatments (**see: Priming**).

In soil, in addition to the difference in water potential between seed and soil solution, the degree of contact of the seed with soil moisture plays an important role in the rate of imbibition. Water uptake is determined by water availability in the immediate vicinity of the seed, but is also markedly influenced by seed **size**, nature, form and **mucilage** content of seed surface, and size and nature of the solid particles. Moreover, seeds can absorb water only from a short distance (about 10 mm) through soil, irrespective of soil water content. (**See: Seedbed environment**)

Similarly to the thermal time model used for temperature responses of seeds (**see: Germination – influences of temperature; Thermal time**), a hydrotime concept has been developed to characterize the effects of reduced water potential on germination (**see: Hydrotime**). A hydrothermal time model has also been proposed to combine the responses of seeds to both temperature and water potential (**see: Hydrothermal time**). (**See: Germination – field emergence models**) (DC, FC)

Bewley, J.D. and Black, M. (1978) *Physiology and Biochemistry of Seeds in Relation to Germination*. Vol. 1. *Development, Germination and Growth*. Springer, Berlin.

Bradford, K.J. (1995) Water relations in seed germination. In: Kigel, J. and Galili, G. (eds) *Seed Development and Germination*. Marcel Dekker, New York, USA, pp. 351–396.

Germination - metabolism

Metabolism commences in the seeds as soon as their cells are hydrated. **Respiration** and protein synthesis have been recorded within minutes of **imbibition**, using components conserved in the dry seed. This is followed by synthesis of RNA, and DNA repair and synthesis. Numerous enzymes are synthesized during germination, including some that appear to be particularly associated with this process. Many genes expressed during germination have been identified, some of which are regulated by **hormones** or light (**see: Light – dormancy and germination**). The final event in germination is the expansion of the cells of the radicle, which precedes cell division (**mitosis**). Although seed metabolism is an integrated and controlled process, aspects of this will be considered under different headings, for convenience.

1. Respiration

Three respiratory pathways operate in a seed during and following germination, glycolysis, the pentose phosphate pathway, and the citric acid cycle. They produce key intermediates in metabolism, energy in the form of adenosine triphosphate (**ATP**), and reducing power in the form of reduced pyridine nucleotides, the nicotine adenine dinucleotides (**NADH** and **NADPH**). Some seeds can complete germination under water, in conditions where oxygen concentrations are low, or under controlled conditions of no oxygen, and **anaerobic respiration** occurs. These include seeds adapted to an aquatic environment, e.g. rice, wild rice, cat-tail (bulrush), barnyard grass, and some which are not. Most terrestrial seeds will perish if maintained under water for an extended period, and although germination may be completed, the seedling will not become established (**see: Germination – influences of gases; Germination – influences of water**). Even seeds germinating in the soil often undergo temporary anaerobiosis because oxygen uptake, which may be limited by the coat, is insufficient to meet the demands of respiration. Temporary production of ethanol or lactate may result, which are removed by enzymic (alcohol or lactate dehydrogenase) degradation before cellular damage by these products can occur. Seeds of submerged rice may develop abnormal **mitochondria** under conditions of limiting O_2, whereas those of barnyard grass develop intact mitochondria, although in neither do they operate efficiently.

Efficient synthesis of ATP by oxidative phosphorylation during germination requires the presence of active mitochondria. The site of initial ATP synthesis in imbibing seeds is not well understood; in some species it appears that mitochondria are active soon after imbibition, whereas in others, the early source of ATP synthesis may rely more upon the less efficient cytoplasmic substrate level phosphorylation in the respiratory pathway, glycolysis. Mitochondria generally exhibit damage following **maturation** drying and imbibition, and two distinct modes of their regeneration have been identified. In some species, particularly those which are oil-storing, there are few distinct mitochondria in the dry seed and there is predominantly biogenesis of new ones, incorporating proteins encoded by both the nuclear and the organellar genome. In many **starch**-storing seeds, however, mitochondria appear to be present following maturation drying, but have a poor internal structure; these are extensively repaired during germination. To what extent these different patterns of restitution of these organelles are unique to **oil**- and starch-storing seeds is unknown, for mitochondria in maize embryos, which store lipids, undergo both re-differentiation and biogenesis during germination.

An alternative oxidase (cyanide-insensitive) pathway of respiration may operate in some seeds during germination when the mitochondria are operating inefficiently. The flow of electrons from NADH to O_2 is diverted from the electron transport chain through an alternative oxidase, thus preventing the synthesis of ATP.

Haemoglobin is present in grass seeds during and following germination, and in barley it increases in the **radicle** and **aleurone layer**. While its function is not fully understood, haemoglobin could enhance O_2 availability to seed tissues, thus resulting in more efficient synthesis of ATP at a time when metabolism is increasing.

The **raffinose series oligosaccharides**, along with sucrose, are early sources of respiratory substrate, and the enzymes associated with their hydrolysis and conversion to sugars that enter the glycolysis pathway (glucose, fructose) are present in the dry seed and throughout germination.

2. Protein and RNA synthesis

Protein synthesis resumes within minutes following hydration of the cells of the seed. The components of the protein synthesizing complex (ribosomal RNA (rRNA), transfer RNA (tRNA), some messenger RNAs (mRNA) (**see: Cells and cell components; Nucleic acids**) and specific cytoplasmic proteins, e.g initiation and elongation factors) are stored in a stable form in the dry seed. The stability of these components has allowed for the development from dry wheat germ of the *in vitro* protein synthesizing system (now available commercially from molecular biology supply companies in kits) to which mRNA fractions can be added and translated (for commercial purposes, endogenous mRNAs are removed).

Initial protein synthesis involves **translation** of messenger RNAs (mRNAs, messages) stored within the dry seed (long-lived or stored messages), as polyribosomes (**polysomes**) are formed. The mRNAs might be stabilized in the dry seed by being in association with protective proteins, forming a messenger-ribonucleoprotein (mRNP) complex. Over several hours, as these stored messages are released from the complex and used in protein synthesis all, or the vast majority, of them are also degraded and are either replaced by newly synthesized mRNAs which encode the same proteins (this process is termed mRNA turnover) or by those for new proteins. Many of the proteins which are synthesized during germination are so-called house-keeping or growth maintenance proteins which are required to maintain the metabolic integrity of the cells of the germinating seed. There may be some residual messages, usually in small amounts, which were associated with seed development (e.g. for storage proteins), and which were not destroyed during maturation drying (**see: Maturation – effects of drying**), but their translation is brief and minimal. While there are several proteins synthesized that are unique to germination, including those associated with the hydrolysis of cell walls of the **endosperm** surrounding the radicle in some species (**see: Germination – molecular aspects; Germination – radicle emergence**), none has been shown to be essential for the completion of germination. No universal germination proteins or their mRNAs, fundamental to the completion of this process, have been identified. One protein, termed germin, was once thought to be integral to germination, but is now known to be the enzyme oxalate oxidase, and is synthesized post-germinatively.

Analysis of the pattern of protein synthesis during germination has been achieved using two-dimensional polyacrylamide gel electrophoresis (2-D PAGE), by which proteins extracted from germinating seeds are first separated on the basis of their size by SDS-PAGE (sodium dodecyl sulphate polyacrylamide gel electrophoresis) in the first dimension, and on the basis of their charge in the second dimension (IEF, isoelectric focusing). Newly synthesized proteins can be identified if the germinating seeds are incubated in a radioactive amino acid precursor (e.g. ^{3}H-leucine or ^{35}S-methionine) and the extracted and separated proteins are subjected to fluorography which allows their position on the gel to be located because they cause blackening of a juxtaposed X-ray film. An example of the changing pattern of protein synthesis during germination of pea axes is shown in Fig. G.11.

Both ribosomal RNA synthesis and synthesis of ribosomal proteins occurs during germination, and also the synthesis of tRNAs. Enzymes essential for RNA synthesis, the RNA polymerases, have been identified in seeds during germination.

(JDB)

3. Proteomic analysis of germination

Functional **genomic** approaches (e.g. **proteomics**), when combined with the functional characterization of proteins, will increasingly contribute to our understanding of germination. A proteomic analysis of germination in *Arabidopsis* (ecotype Landsberg erecta) involving two-dimensional gel electrophoretic separation of proteins derived from seeds at different stages (mature-dry, and mature seeds imbibed for 1, 2 and 3 days) shows that germination (1-day imbibed seed) is characterized by changes in the abundance of about 40 proteins. Correlating with the resumption of cell elongation and cell cycle activity is the accumulation of an actin 7 (potentially required for germination and/or hypocotyl elongation), tubulin subunits and a WD-40-repeat protein (containing a repetitive segment of 40 amino acids ending in Trp-Asp). This latter protein resembles receptors of activated **protein kinase** C (having approx. 80% homology) and it may play a role in signal transduction and hormone-controlled cell division. Many of the proteins accumulated during very early hydration are derived from components of the mature dry seed (e.g. conserved mRNAs or proteins). The post-germinative phase is characterized by changes in the abundance of 35 proteins. Many are linked to the establishment of photosynthesis, the mobilization of reserves or protective mechanisms (i.e. pathogen and herbivore defence-related). Many of the changes in abundance of proteins are probably undetectable using these methods, in part due to the higher abundance of seed storage proteins and limitations associated with 2-D gel

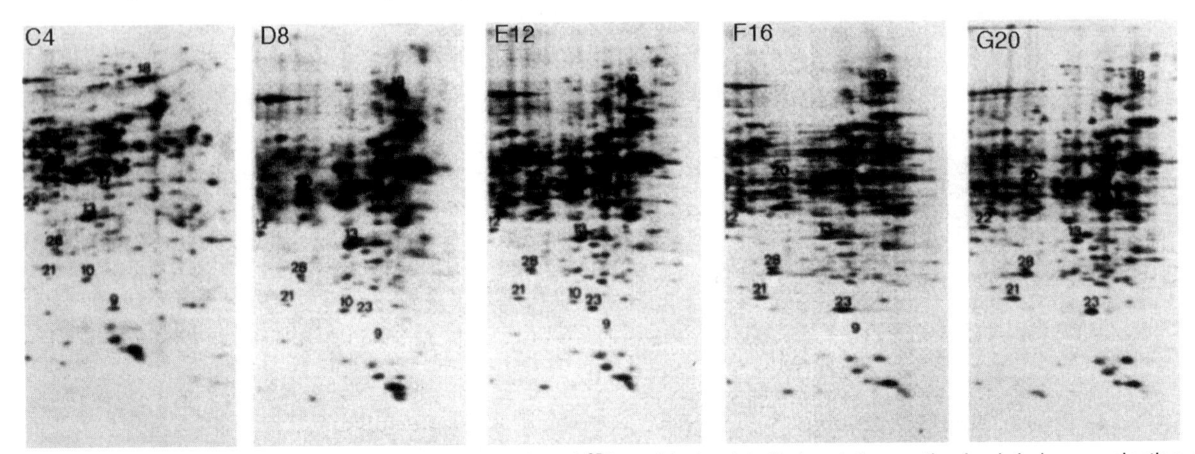

Fig. G.11. Fluorographs showing two-dimensional gel separation of ^{35}S methionine-labelled proteins synthesized during germination of pea axes. Proteins were extracted and separated at 4 h intervals up to the start of radicle emergence at 16 h (gels left to right) and following its elongation at 20 h. Some proteins are numbered and show a decline in their synthesis during germination (e.g. 9, 10) or an increase (e.g. 21, 23, 28), or that their synthesis remains unchanged (e.g. 22). From Lalonde, L. and Bewley, J.D. (1986) Patterns of protein synthesis during germination of pea axes, and the effects of an interrupting desiccation period. *Planta* 167, 504–510.

separation techniques (e.g. hydrophobic proteins do not separate well).

'Phenomic' approaches have been initiated and will extend proteomic and other functional genomics approaches to include **mutants** that are disrupted in hormone biosynthesis/response. For example, a proteomics analysis to compare wild-type *Arabidopsis* seeds with seeds of the **gibberellin** (GA)-deficient mutant (in which radicle emergence is completely dependent on exogenous GA) has been undertaken. This study revealed that of the 40 or so proteins whose abundance changes during germination, only one, the cytoskeleton component α-2,4 tubulin, appears to require GA. The abundance of several proteins associated with later stages, including metabolic control of seedling establishment, was dependent on GA, and this included an increase in S-adenosyl-methionine synthetase, a 'housekeeping' enzyme that catalyses the formation of S-adenosyl-methionine from ATP and methionine. Notably, GAs also appeared to control the abundance of the cell wall hydrolase, α-glucosidase, which possibly mediates embryo cell wall loosening involved in cell elongation and radicle extension.

(ARK)

4. DNA synthesis and the cell cycle

Two separate phases of DNA synthesis occur in the cells of the radicle after imbibition. The first, which occurs early during germination, probably involves the repair of DNA damaged during maturation drying and rehydration, as well as the synthesis of mitochondrial DNA. Part of the repair processes that occur following imbibition may involve the new synthesis of telomeric DNA, using the enzyme telomerase. Telomeres are **nucleotide** sequences located at each end of a DNA molecule, which also become shorter with each replication of the molecule, and must be periodically extended by telomerase. Failure to do so reduces the ability of DNA to replicate, and to produce RNA. Delay, or failure to repair the DNA ends may affect the speed and efficiency of germination. Drying of the mature seed leads to more frequent single-strand breaks in the DNA molecule, either by endonuclease cleavage, **free-radical** damage or spontaneous base loss. These breaks are usually effectively repaired in viable seed during imbibition, due to the activities of DNA polymerases and ligases (**see: Ageing**).

Replication of DNA commences usually within the few hours prior to the completion of germination, and is followed by the continuing synthesis which is required for nuclear and cell division that is associated with radicle and **plumule** growth during subsequent seedling development. DNA polymerases, necessary for DNA repair and replication, are present in the dry seed, but in maize an additional polymerase activity appears at the time of DNA replication; this enzyme seems to be activated (by **phosphorylation**) rather than *de novo* synthesized. Consideration of replication of DNA during and following germination is best conducted in relation to the **cell cycle**. Cells in a dry seed embryo contain two sets of double-stranded DNA (2C state), and the majority of cells remain with this state throughout germination. Following DNA repair, during germination, DNA synthesis occurs (S phase), without cell division, to produce double (4C) the normal complement of DNA (replication). Then mitosis occurs, when the chromosomes become visible as their chromatin condenses and they and their constituent DNA divide (M phase, mitotic metaphase to anaphase). The 2C state of the DNA is restored as cells of the embryo divide and then elongate, contributing to growth of the seedling. In a limited number of species, cells in the dry embryo may also be in the 4C state (in *Phaseolus vulgaris* and *Spinacea oleracea* embryo root tips have 55% and 29% 4C cells, respectively (**see:** Table C.4 in **Cell cycle**)), and as germination is completed, and cell division begins, these complete mitosis and return to the 2C state. Generally, dry seeds with cells in the 4C state are regarded as being more susceptible to stresses such as **desiccation** and free radical damage during storage, and hence the 2C state is most commonly found (**see: Ageing**).

In sugarbeet embryos, as an example, an increase in DNA content from the 2C to the 4C state occurs during germination, up to 60 h, which is indicative of preparation of the cells for mitosis (Table G.11). The G_2/G_1 ratio shows about half of the cells are in a pre-mitotic state (4C) at the time of radicle protrusion. Thereafter, the number of cells in the 4C state increases as mitotic cell divisions occur during seedling growth. Some cells reach 8C and above, which is likely due to **endoreduplication** of DNA, when the DNA continues to duplicate, without concomitant cell division. The value of this to the growing seedling is unclear (**see: Cell cycle**). In maize

Table G.11. The proportion of cells with different DNA contents, and the G_2/G_1 ratio of dry, germinating and germinated embryos of diploid sugarbeet.

Time from the start of imbibition (h)	Nuclear DNA content (%)					G_2/G_1 ratio
	2C	4C	8C	16C	32C	
0	92	8	0	0	0	0.09
24	88	10	2	0	0	0.15
48	64	28	7	1	0	0.53
60	61	30	7	1	0	0.69
72	42	39	14	4	1	1.14
96	24	47	18	9	2	1.94

Radicle emergence commences at about 60 h from the start of imbibition. The G_2/G_1 ratio is an indication of the number of cells in which there is a pre-mitotic doubled (4C) DNA (G_2) content compared to those cells which contain DNA in the post-mitotic 2C state (G_1). Data from Śliwińska, E. (2000) Analysis of the cell cycle in sugarbeet seed during development, maturation and germination. In: Black, M., Bradford, K.J. and Vázquez-Ramos, J. (eds) *Seed Biology. Advances and Applications*. CABI Publishing, Wallingford, UK, pp. 133–139.

embryos, the majority of cells also remain in the G1 phase during early germination, and the DNA replication (S) phase later in germination is accompanied by the initiation of histone synthesis, proteins that are associated with DNA in the structure of chromosomes. Thymidine kinase, an enzyme associated with DNA synthesis, is present in low amounts early in germination, when it might be involved in DNA repair, and then increases greatly in activity later during DNA synthesis.

DNA replication in the living **endosperm** of some seeds may also occur, but without cell division. A triploid endosperm cell in which the DNA is 3C (G_1) or 6C (G_2) at the beginning of its development (fusion of the diploid maternal polar nuclei and the haploid nucleus of the pollen of the paternal parent) can become 6C at maturity, or even larger if endoreduplication occurs.

In embryos of seeds that are polyploid, the cell cycle pattern during germination is similar to that of diploid cells, e.g. triploid sugarbeet embryos contain mainly cells with DNA in the 3C state (resulting from a cross between the gametes from a diploid maternal parent and a tetraploid paternal parent) before it replicates to the 6C state during germination, prior to post-germinative mitosis.

The DNA content of cells can be determined by flow cytometry. For seeds, nuclei are extracted from cells of the tissue being analysed in an appropriate buffer and the DNA in the nuclei is stained with a fluorescent dye. In a flow cytometer, as the nuclei pass through a laser beam they emit fluorescence, which is directly proportional to the amount of DNA present in each nucleus. This is recorded and analysed by computer.

During mitosis, separation of the chromosomes involves a complex cellular machinery, including the involvement of the mitotic spindle. This is built mainly from microtubules and is the structure with which chromosomes are moved to separate. Protein components of these microtubules include tubulins, and in germinating tomato seeds β-tubulin is not present in the dry seed, but increases in the radicle tip at about the time there is the transition of the nuclei from a 2C to a 4C state. A further increase in β-tubulin occurs after germination as the cell cycle progresses through G_2, mitosis and cell division.

5. Radicle extension

Elongation of the cells of the radicle, such that they penetrate their surrounding structure and complete germination, is a turgor-driven process which requires the walls of the cells of the radicle to yield and expand. The causes of the increase in turgor, and the yielding of the cell walls are unknown. There is no evidence for a change in osmotic potential (Ψ_π) in the cells of the radicle at the time of germination that would account for increased water uptake, nor is there evidence for radicle cell-wall weakening at this time. The enzyme **xyloglucan** endotransglycosylase (XET) and the protein **expansin** have been implicated in cell wall extension in elongating vegetative tissues. There is the possibility that in some seeds, particularly those which exhibit coat-enhanced dormancy, the structures surrounding the radicle tip weaken, thus allowing the tip to elongate (**see: Germination – molecular aspects; Germination – radicle emergence**).

Following the completion of germination, during seedling establishment, the cotyledon(s) may be raised above the soil, or remain below. This has been termed **epigeal** or **hypogeal** germination, respectively. However, these terms are misnomers since they refer to the mode of post-germinative seedling establishment, rather than the completion of germination (**see: Epigeal seedling establishment; Hypogeal seedling establishment**). (JDB)

Bewley, J.D. (1997) Seed dormancy and germination. *The Plant Cell* 9, 1055–1066.

Bewley, J.D. and Black, M. (1994) *Seeds. Physiology of Development and Germination.* Plenum, New York, USA.

Black, M., Bradford, K.J. and Vázquez-Ramos, J. (eds) (2000) *Seed Biology. Advances and Applications.* CAB International, Wallingford, UK.

Gallardo, K., Job, C., Groot, S.P.C., Puype, M., Demol, H., Vandekerckhove, J. and Job, D. (2001) Proteomic analysis of Arabidopsis seed germination and priming. *Plant Physiology* 126, 835–848.

Rajjou, L., Gallardo, K., Debeaujon, I., Vandekerckhove, J., Job, C. and Job, D. (2004) The effect of α-amanitin on the Arabidopsis seed proteome highlights the distinct roles of stored and neosynthesized mRNAs during germination. *Plant Physiology* 134, 1598–1613.

Van der Geest, A.H.M. (2002) Seed genomics: germinating opportunities. *Seed Science Research* 12, 145–153.

Vázquez-Ramos, J.M. and Páz Sanchez, M. (2003) The cell cycle and seed germination. *Seed Science Research* 13, 113–130.

Germination – molecular aspects

Early studies on mRNA profiles and protein expression established that following desiccation and rehydration, seed metabolism and biochemistry shift quickly from a residual developmental mode to a germinative mode (**see: Maturation**). Stored mRNAs expressed initially may represent some developmental processes occurring during seed maturation, as well as coding for proteins that may be needed immediately, such as for repair of DNA or other cellular constituents. However, gene expression and protein synthesis patterns quickly shift to ones characteristic of germination (**see: Germination – metabolism**). Much of the 'housekeeping' biochemistry (e.g. respiration, ion transport, ribosomes, etc.) is not unique to germination, but other genes are expressed specifically in association with germination. Even if the same enzyme or protein is produced during both germination and seed development or seedling growth, specific members of a gene family may be expressed uniquely during each event. Thus, studies using conventional biochemistry tend to have difficulty in identifying processes specific to germination. However, molecular studies are revealing that even though similar biochemical processes may be involved in both germination and other stages of plant development, the actual genes expressed may differ. An example of this is the **expansins**, and to some extent various cell wall hydrolases. These genes and enzymes, among others, are involved in germination.

1. Expansins

These proteins were first identified as being involved in cell wall relaxation associated with vegetative cell growth. Although they have little enzymatic activity themselves, expansins are able to cause cell walls to stretch, possibly by loosening the hydrogen bonds that link the various components of the cell wall together (particularly between cellulose microfibrils and **xyloglucans**). They may also act in concert with cell wall enzymes to modify wall composition or function for specific purposes, such as softening during fruit ripening or separation during abscission of plant organs. Genome sequencing and cloning have identified large families of expansin genes (>30 per species) in a variety of plants. It appears that specific members of these expansin families are expressed in association with particular developmental processes. Thus, one way that plants may regulate their development is to express different complements of expansins (and associated enzymes) that modify the cell walls in unique ways to result in the desired characteristics.

With respect to seed germination, there are at least two processes in which expansins are involved. Completion of germination requires embryo growth (initiated by cell expansion), so it is likely that expansins are involved in this process. In addition, in many seeds the embryo is enclosed within endosperm and/or testa tissues that must be weakened before the radicle can penetrate them. In imbibed tomato (*Lycopersicon esculentum*) seeds, specific expansin genes are expressed in association with both processes. One such gene (*LeEXPA4*) is expressed within 6–12 h of imbibition exclusively in the endosperm cap (**micropylar endosperm**) tissue enclosing the radicle tip of the **embryo** (Fig. G.12). (The same gene is also expressed during early tomato fruit expansion; expression in multiple, but tissue-specific, locations seems to be characteristic of expansin genes.) This expansin is likely to be involved in the cell-wall weakening and degradation that occurs in this tissue in association with its physical weakening to allow radicle emergence. Shortly after *LeEXPA4* is expressed, a second expansin gene (*LeEXPA8*) is turned on specifically in the cortical tissue of the radicle (Fig. G.12). This gene continues to be expressed in the elongation zone of the radicle as it emerges and grows. Another related gene (*LeEXPA10*) is also expressed throughout the embryo, but is most abundant during early embryogenesis, suggesting that it is associated with general growth processes (Fig. G.12). Additional expansins may be involved in the **mobilization of the reserves** stored in the cell walls of the lateral endosperm of tomato seed following radicle emergence.

2. Cell wall hydrolases

The case of expansins is likely to be the norm rather than the exception. For example, endo-β-mannanase exhibits a similar tissue-specific expression pattern during tomato seed germi-

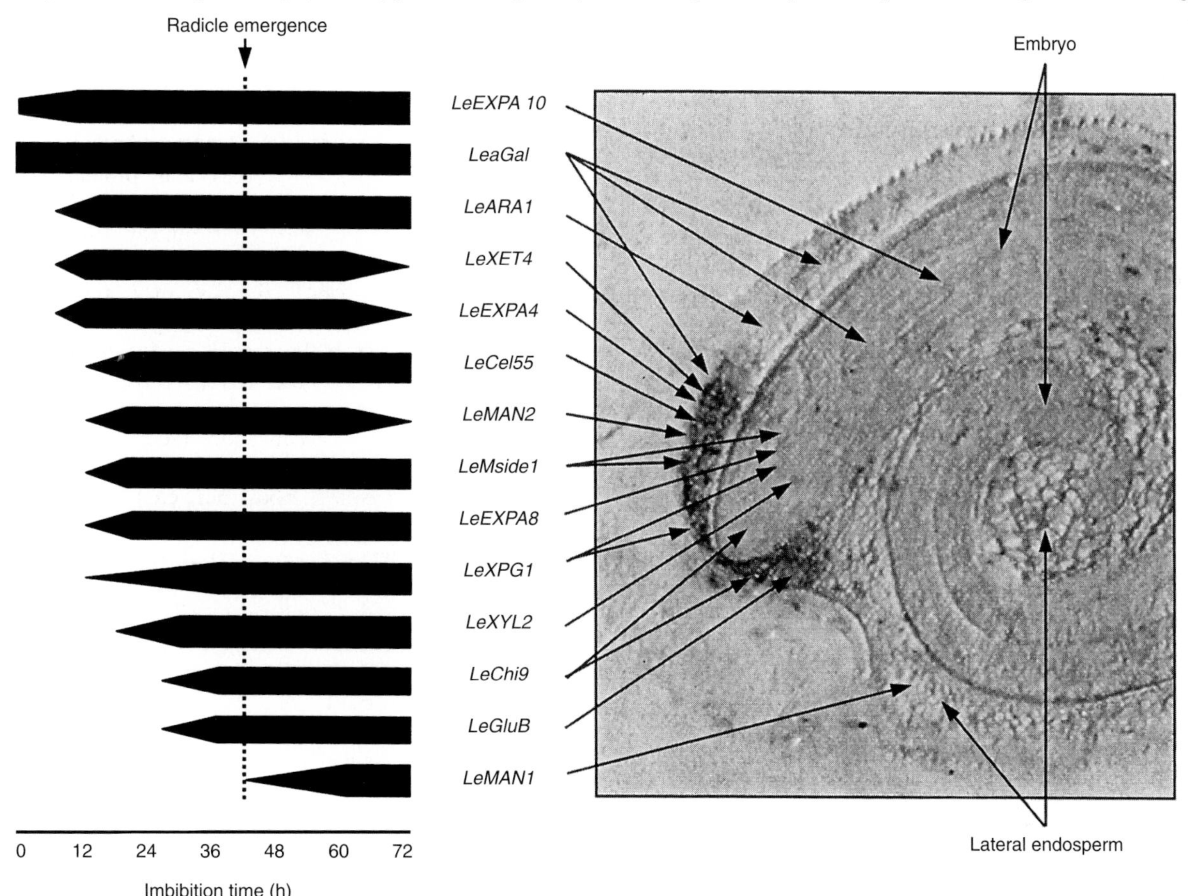

Fig. G.12. Time course and tissue localization of expression of germination and growth-associated genes in tomato seeds. The bars at left indicate when expression of the indicated gene occurs following imbibition. The tissue print at right shows the localization of genes expressed specifically in the endosperm cap tissue (shaded area) (**micropylar endosperm**), and the outline of the embryo and **lateral endosperm** are visible. Arrows indicate the tissues (endosperm cap, lateral endosperm or embryo) in which expression occurs. *LeARA1*, arabinofuranosidase; *LeCel55* β-1,4-glucanase (cellulase); *LeEXPA4*, *LeEXPA8*, *LeEXPA10*, expansins; *LeMANI*, *LeMAN2*, endo-β-mannanase; *LeMsideI*, β-mannosidase; *LeGluB*, β-1,3-glucancase; *LeChi9*, chitinase; *LeaGal*, α-galactosidase; *LeXET4*, xyloglucan endotransglycosylase; *LeXPG1*, exo-polygalacturonase; *LeXYL2*, xylosidase. This figure is modified from one in Nonogaki, H., Chen, F. and Bradford, K.J. (2006) Mechanisms and genes involved in germination *sensu stricto*. In: Bradford, K.J. and Nonogaki, H. (eds) *Seed Development, Dormancy and Germination*. Blackwell Publishing, Oxford, UK. Also see Colour Plate 6E and Fig. G.17 in **Germination – radicle emergence**.

nation. The cell walls of the endosperm are rich in **galactomannans**, polysaccharides that contribute to the rigidity of this tissue. During endosperm weakening, the galactomannans of the cell walls are disassembled by the combined action of endo-β-mannanase, which cleaves the long mannan chains, α-galactosidase, which removes the galactose side groups, and β-mannosidase, which breaks down the smaller mannan chains resulting from endo-β-mannanase action (**see: Hemicellulose**). Endo-β-mannanase activity arises first in the endosperm cap, and then in the lateral endosperm following radicle emergence. This is due to the tissue-specific expression of two distinct genes encoding similar forms of the enzyme: *LeMAN2* is expressed specifically in the endosperm cap prior to radicle emergence, and *LeMAN1* is expressed in the lateral endosperm after radicle emergence (Fig. G.12). Thus, while from a biochemical point of view, endo-β-mannanase enzyme activity simply spreads from one end of the endosperm to the other during and following germination, at the molecular level this is due to the sequential tissue-specific expression of two different genes. The lettuce (*Lactuca sativa*) seed also expresses an endo-β-mannanase gene in a tissue-specific manner, in the endosperm, following germination. β-Mannosidase (*LeMside1*) is expressed in the tomato endosperm cap and in the lateral endosperm in a pattern similar to that for endo-β-mannanase (although this appears to be a single gene expressed in both tissues), while α-galactosidase (*LeaGal*) is constitutively present in tomato seeds (Fig. G.12).

A number of additional cell wall hydrolases have been identified in tomato and other seeds during and after germination, including β-1,4-glucanase (cellulase), polygalacturonase, xyloglucan endotransglycosylase (XET; now termed xyloglucan endotransglucosylase-hydrolase (XTH)), arabinosidase, β-1,3-glucanase (callase) and chitinase (Fig. G.12). For those for which information is available, the pattern of seed-specific expression of certain gene family members is continued. A single XET gene (*LeXET4*) of a larger gene family is expressed co-incident with *LeEXPA4* in the tomato endosperm cap prior to radicle emergence. Evidence from non-seed tissues suggests that XET and expansins cooperate to relax or disassemble cell walls. Also, since endo-β-mannanase activity alone is not sufficient to allow for the radicle to penetrate the endosperm of tomato, so it is likely that a suite of cell wall hydrolases is involved in the endosperm weakening process.

Another role may be played by specific class I β-1,3-glucanase and chitinase genes that are expressed in the endosperm cap tissue just prior to radicle emergence. In tomato, both of these genes (*LeGluB*, *LeChi9*) are expressed coordinately during germination (Fig. G.12), although they are regulated differently. β-1,3-Glucanase is induced primarily by **gibberellins** (GAs) and is inhibited by **abscisic acid** (ABA). In tobacco, expression of the gene for this enzyme appears to be associated with germination. Chitinase, on the other hand, is not expressed in germinating tobacco seeds. The relative expression of β-1,3-glucanase and chitinase appears to vary among different species of the Solanaceae and also in other species (e.g. muskmelon (*Cucumis melo*)). In tomato, chitinase expression may be regulated primarily by **jasmonic acid** and is induced by wounding of the seed. The late expression of both β-1,3-glucanase and chitinase just prior to radicle emergence suggests they are involved in a prophylactic defence against fungi that could potentially invade the endosperm following its rupture. Such a role has often been suggested, since these enzymes are known to be associated with pathogenesis and wounding responses and are often found in germinating seeds. (KJB)

(See: Germination – radicle emergence)

Bradford, K.J., Chen, F., Cooley, M.B., Dahal, P., Downie, B., Fukunaga, K.K., Gee, O.H., Gurusinghe, S., Mella, R.A., Nonogaki, H., Wu, C.-T. and Yim, K.-O. (2000) Gene expression prior to radicle emergence in imbibed tomato seeds. In: Black, M., Bradford, K.J. and Vázquez-Ramos, J. (eds) *Seed Biology: Advances and Applications*. CAB International, Wallingford, UK, pp. 231–251.

Chen, F., Dahal, P. and Bradford, K.J. (2001) Two tomato expansin genes show divergent expression and localization in embryos during seed development and germination. *Plant Physiology* 127, 928–936.

Leubner-Metzger, G. (2003) Functions of β-1,3-glucanases during seed germination, dormancy release and after-ripening. *Seed Science Research* 13, 17–34.

Mo, B. and Bewley, J.D. (2003) Mobilization of galactomannan-containing cell walls of tomato seeds: where does β-mannosidase fit into the picture? In: Nicolás, G., Bradford, K.J., Côme, D. and Pritchard, H.W. (eds) *The Biology of Seeds. Recent Research Advances*. CAB International, Wallingford, UK, pp. 121–129.

Nonogaki, H., Chen, F. and Bradford, K.J. (in press) Mechanisms and genes involved in germination sensu stricto. In: Bradford, K.J. and Nonogaki, H. (eds) *Seed Development, Dormancy and Germination*. Blackwell Publishing, Oxford, UK.

Sanchez, R.A. and Mella, R.A. (2004) The exit from dormancy and the induction of germination: physiological and molecular aspects. In: Benech-Arnold, R.L. and Sánchez, R.A. (eds) *Handbook of Seed Physiology: Applications to Agriculture*. Food Products Press, New York, USA. pp. 221-243.

Germination – natural stimulants

Chemical compounds secreted by plants or possibly other organisms, generally in the soil, that stimulate germination. **(See: Parasitic plants; Stimulants of germination)**

Germination – physical factors

The physical factors affecting germination considered here relate to the ability of the seed to absorb water and of the embryo subsequently to expand. During seed development, constituents of the seed such as sucrose, **oligosaccharides**, **late embryogenesis abundant** (LEA) proteins, and **antioxidants** are deposited to allow survival of desiccation and enable the cells of the seed to persist in the dry state for extended periods of time. The tissues covering the embryo, including the **endosperm** or **perisperm**, **testa** and **pericarp**, are also involved in protecting the seed from physical damage and may serve other purposes (e.g. **dispersal**). **(See: Seedcoats – dispersal aids; Seedcoats – functions)** In most cases, the testa or seedcoat is dead at maturity. In many seeds, the seedcoat is relatively permeable to water and readily cracks upon uptake of water, presenting little or no hindrance to embryo swelling due to water uptake. In others, such as many **legume** seeds, once it has dried to low moisture contents, the seedcoat presents an impermeable barrier to water penetration, effectively preventing germination by impeding hydration of the embryo (**hardseededness**). (**See: Germination – influences of water**)

1. Membranes and water

The permeability of the testa and the rate of water uptake have a significant effect on subsequent germination. Many large seeds (e.g. maize, beans) exhibit imbibitional damage if they absorb water too rapidly, particularly if the temperature is also low (Fig. G.13) (**see: Chilling injury**). Part of the damage is purely physical due to uneven stresses caused by swelling of seed constituents. Proteins in particular swell considerably upon hydration, whereas carbohydrates and lipids swell relatively little. In a large seed, as water enters from the seed periphery, the hydrated outer tissues begin to swell while internal tissues remain dry, resulting in physical tensions between these layers which can result in cracking of dry tissues and extrusion of cellular contents from partially hydrated cells. Rapid water uptake exacerbates these tensions, resulting in imbibitional damage. Slow penetration of water through a relatively impermeable seedcoat allows internal water redistribution to keep pace with water uptake, ameliorating the internal tensions. This principle has been applied in practice by **coating** seeds with polymeric materials that slow the entry of water. Such coatings can improve germination and seedling emergence under adverse field conditions or when seedcoats have been damaged during production or harvesting.

Imbibitional damage is exacerbated by low initial seed water contents and low temperatures. This is now understood to be due to physical changes that occur in the **phospholipids** that comprise the internal and peripheral cellular bilayer **membranes**. These phospholipids have hydrophilic head groups and hydrophobic **fatty acid** chains. The polar head groups of the two layers of the membrane orient towards the cellular contents and the external cell wall, while the fatty acid chains associate in the middle of the membrane bilayer. This structure is maintained primarily by the presence of water and the hydrophilic/hydrophobic interactions of the membrane components. In the dry seed, most of the water has been removed, and has been replaced by sugars such as sucrose, the hydroxyl groups of which act as surrogates for water to maintain the bilayer membrane structure (**water replacement**). However, the membranes also undergo further changes in internal structure related to their water content and temperature. When dry and cold, the membrane lipids are in the gel phase in which the head groups are closer together and the movements of the lipid chains are restricted (Fig. G.14A). Membranes in the gel state are not good barriers to the passage of cellular constituents, and rapid water uptake can allow leakage to occur as the membranes change into the liquid-crystalline state characteristic of hydrated cells. The transition from gel phase to liquid-crystalline phase is dependent upon both the temperature and the water content (Fig. G.14). Low temperatures and low water contents promote the gel phase, while higher temperatures and water contents result in liquid-crystalline phase membranes. If dry seeds are first warmed before imbibition, then the membranes melt into the liquid-crystalline phase before liquid water enters, and damage is reduced (Fig. G.14B). Similarly, if seeds are hydrated slowly via the vapour phase, the membrane phase transition can occur even at lower temperatures, and again damage is ameliorated when imbibed in cold water (Fig. G.14C). These physical properties of membranes appear to largely explain why imbibition damage is most severe when dry seeds are imbibed at cold temperatures (Fig. G.13). (**See: Desiccation damage**)

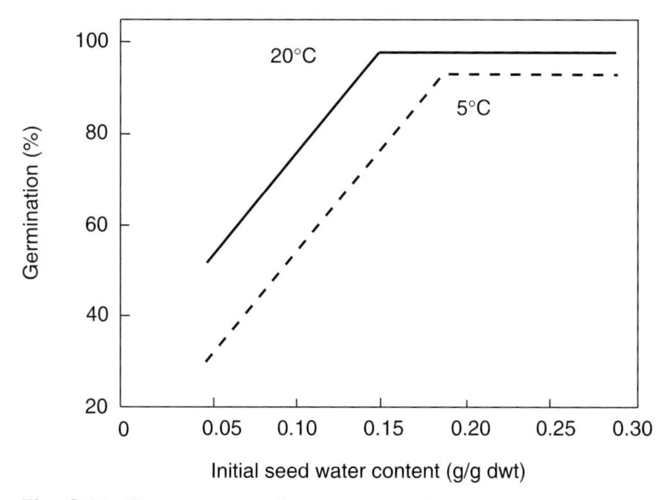

Fig. G.13. Temperature and water content interact to determine the occurrence of imbibitional damage in seeds of *Phaseolus vulgaris*. Seeds at higher initial water contents are relatively unaffected by imbibition at 5°C, i.e. are able to germinate. However, as the initial water content is reduced, seeds imbibed at 5°C exhibit greater damage than do seeds imbibed at 20°C. At the lowest seed water contents, even seeds imbibed at 20°C suffer imbibitional damage. Adapted from Wolk, W.D., Dillon, P.F., Copeland, L.F. and Dilley, D.R. (1989) Dynamics of imbibition in *Phaseolus vulgaris* L. in relation to initial seed moisture content. *Plant Physiology* 89, 805–810.

2. Water uptake by the seed

Assuming that there are no permeability barriers, uptake of water by seeds generally follows a three-phase pattern (**see: Germination**, Fig. G.4). The initial rate of water uptake is rapid due to the large **water potential** (Ψ) gradient between the inside of the seed and the external solution and the physical absorption of water by the seed constituents (i.e. matric absorption). This Phase I occurs in both living and dead seeds, and dead seeds often absorb more water than viable ones (**see: Imbibition**). This is because when water is absorbed by living cells, the final stages of water uptake are due to the selectively permeable membranes that retain solutes and therefore attract water osmotically into the cell. This osmotic absorption of water continues until the membranes are constrained by the relatively rigid cell walls, which exert an opposing force that results in generation of pressure, or **turgor** (or pressure potential), which acts in opposition to the osmotic gradient and reduces the further entry of water into the cell. In dead seeds, the membranes are not intact, and therefore no turgor pressure develops to offset the osmotic gradient for water uptake. However, in some dead seeds, e.g. of lettuce and muskmelon, in which the embryo is enclosed within an endosperm or perisperm, this outer region retains solutes released from the embryonic cells within the outer envelope, which then swells, or 'osmotically distends'. The total water uptake due to this extracellular volume makes the water content of the dead seeds higher than that of the living seeds. Viable seeds generally enter a second phase in which the water content is relatively stable or only slowly increases (**see: Germination**, Fig. G.4). During this Phase II, repair mechanisms are active and metabolism associated with the initiation of growth begins. The length of this phase can vary widely depending upon the water potential of the surrounding medium, the temperature, whether dormancy is present and the age of the seeds.

In seeds in which the tissues enclosing the embryo are weak or break upon initial swelling, it is the cell walls of the embryo (and endosperm or perisperm, if living) that contain the turgor pressure generated by osmotic water uptake. In many seeds, however, either an inelastic seedcoat or a rigid endosperm surrounds the embryo, restricting its water uptake and expansion. In this case, it is the tissue external to the embryo that is restricting further water uptake and embryo expansion. Consequently, after water uptake by the whole seed has reached a plateau, removal of the embryo and placing it on water allows an immediate increase in its water content (and Ψ). When removed from the enclosing tissues, the restraining force is no longer present and the embryos can absorb water until further uptake is limited only by their own cell walls. The situation is analogous to that of a rubber tube inside a tyre. Pressure inside the tube presses it against the walls of the more rigid tyre enclosing it, which prevents the tube from expanding further. If the outer tyre is cut, then the tube can bulge through the opening and the internal pressure decreases due to the increase in volume of the tube. Similarly, some seeds (e.g. Solanaceae) are physically restrained from completing germination by rigid enclosing tissues that must first be softened before the radicle can emerge. Once the radicle is no longer restrained by the enclosing tissues, it rapidly absorbs water until the turgor pressure in its cells balances the osmotic gradient for water uptake.

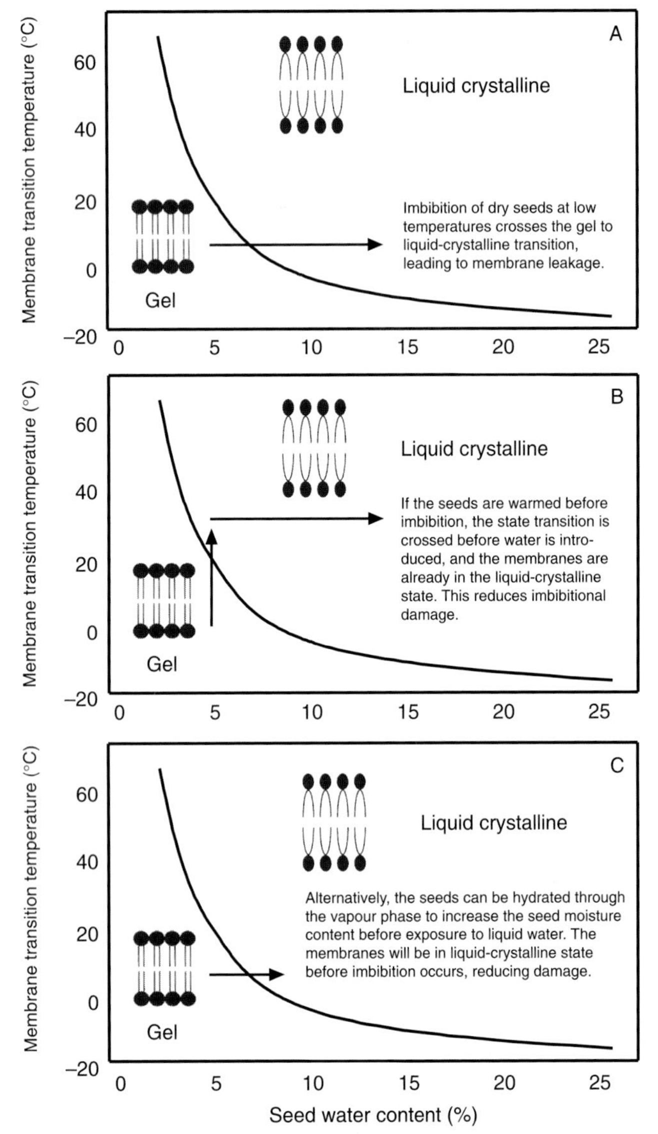

Fig. G.14. Membrane state transitions during seed imbibition. (A) Rapid imbibition at low temperature causes membrane transition from gel to liquid-crystalline state, resulting in leakage and damage. (B) Rapid imbibition at warmer temperatures is not damaging since membranes are already in the liquid-crystalline state due to the higher temperature. (C) Even at lower temperature, seeds can be hydrated through the vapour phase, causing the membrane state change before liquid water is introduced and reducing damage.

Tissues enclosing the embryo can also restrict the movement of gases such as oxygen that are required for germination (**see: Germination – influences of gases; Seedcoats – function**). In general, respiration rates follow a pattern similar to that shown for water uptake. Tissues inside large seeds having relatively impermeable seedcoats can be limited for oxygen following imbibition. Excess water can have the same effect by reducing the rate of diffusion of oxygen to internal tissues. Some **Cucurbit** seeds, such as muskmelon or cucumber, are very sensitive to excess water or restricted oxygen during germination (**see: Germination – influences of water**). Genotypes of these species having more permeable seedcoats are much less susceptible to excess water or to cool temperatures during germination. For most seeds, however, oxygen availability must be reduced considerably (to less than 10% compared to 21% in air) before seed respiration rates are limited. Seeds of most terrestrial species (rice is an exception) do not germinate well when completely submerged in water due to oxygen limitation. Nonetheless, it is likely that a combination of **aerobic respiration** and fermentation occurs during the early stages of imbibition in many seeds. (**See: Germination – metabolism**)

Phase III of water uptake at the completion of germination is associated with the growth of the embryo (**see: Germination**, Fig. G.4). If the enclosing tissues or dormancy prevent completion of germination, the water content reaches a constant level and remains there indefinitely. For germinable seeds, however, the completion of germination is marked by the expansion and growth of the embryo through enclosing tissues. Since plant cells grow by taking up water to expand their volume, growth associated with completing germination is accompanied by increasing water content. At this point, other physical factors that can affect seedling growth come into play, such as soil density or compaction, but these are post-germination phenomena after radicle emergence has occurred. (**See: Desiccation damage; Germination – influences of temperature; Germination – influences of water; Humidification; Imbibition**)

3. Embryo cell expansion

A consequence of water uptake by the seed is that the cells of the composing tissues, the embryo and surrounding structures expand. Expansion of the radicle (**see: Germination – radicle emergence**) is essential for germination to be completed, followed thereafter by cell division as the seedling tissues grow.

The physiological consequence of the action of the genes regulating embryo growth potential is to initiate expansion of the embryo (**see: Germination – molecular aspects**). This probably involves two components affecting cell water potential: increasing the solute content of the cells to attract water, and relaxing the cell walls to allow expansion in response to water influx. Both components are required if the growth potential of the embryo is to increase (Fig. G.15). Solute generation or accumulation in the cell lowers its osmotic potential, creating a gradient for water uptake from its surroundings. Mobilization of stored reserves can generate additional solutes for this purpose, including sugars, amino acids and mineral ions. There is mixed evidence, however, regarding the extent to which such changes in osmotic potential occur prior to radicle emergence. In some species (e.g. lettuce), solute accumulation is claimed to be associated with increased growth potential, but there is no evidence for changes in osmotic potential in others (e.g. tomato). For seeds not enclosed in a restraining tissue, cell-wall relaxation may be the key regulated step, since water uptake in response to an osmotic gradient will increase growth, rather than just turgor, only if the cell walls are able to expand in response to this

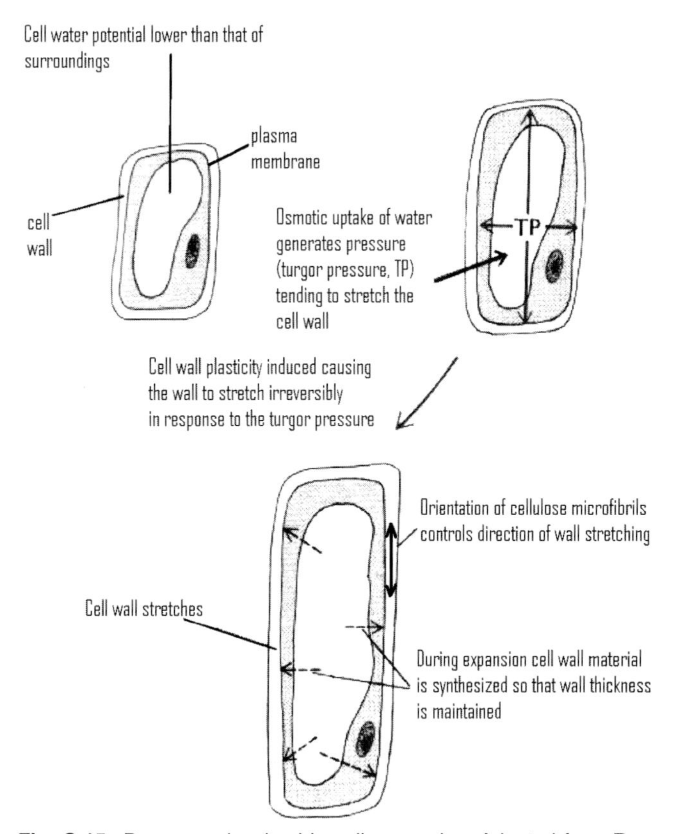

Fig. G.15. Processes involved in cell expansion. Adapted from Ray, P.M., Steeves, T.A. and Fultz, S.A. (1983) *Botany*. Saunders College Publishing, Philadelphia, USA.

pressure. Once the cell has expanded due to wall relaxation, the solutes will be diluted and additional solutes will be needed, by uptake or generation, to maintain water influx and growth.

Wall relaxation probably involves the coordinated action of **expansins** and hydrolases that can modify the cell wall structure to allow it to stretch. (**See: Germination – molecular aspects**) Expansins are proteins that have the ability to cause cell walls to stretch under tension, possibly by breaking the hydrogen bonds between wall polymers. They may act in concert with hydrolases such as **xyloglucan** endotransglycosylase to rearrange cell wall **hemicellulose** polymers and allow the cellulose microfibrils to move apart while retaining strength to contain the cell turgor. At the same time, new wall material is added to maintain wall structure and direct the orientation of cell expansion.

For seeds enclosed in a restraining tissue, the situation is somewhat different. If the embryo is truly restrained from expanding by the enclosing tissue, then the pressure generated by water uptake into the embryo will be exerted against that tissue. That is, the cell walls of the embryo must be relaxed such that the pressure is transferred to the external tissue. In this case, when the external tissue weakens, the pressure is reduced, a gradient for water uptake is created, and water will flow into the embryo to expand it. This will continue until the embryo penetrates the enclosing tissue, at which point the external pressure will go to zero, and water will rapidly enter the cells until their walls exert a counteracting pressure. In tomato seeds, for example, embryos excised from seeds fully imbibed on water have a relatively low water potential (due to removal of the external restraint) and will rapidly take up water. Similarly, if the endosperm cap (**micropylar endosperm**) is excised, the radicle will generally protrude rapidly from the cut due to water uptake, and then subsequently begin sustained growth. For both endospermic and **non-endospermic seeds**, both solute accumulation and cell wall relaxation must be involved in the completion of germination. With respect to **abscisic acid** (ABA) for example, this **hormone** reduces the ability of the embryo cell walls to expand.

Another factor directly affecting embryo growth potential is the water potential (Ψ) of the seeds' surroundings, i.e. the source water. Any reduction in the Ψ of this water will have a direct effect on the Ψ of the embryo, reducing the pressure it can exert against any restraining barriers. Thus, high source-water Ψ will tend to promote germination, while decreasing Ψ will first delay and then prevent germination (Fig. G.15). In fact, the effect of Ψ on both the timing and extent of germination is so consistent that mathematical models have been developed that can accurately predict germination behaviour over time based upon the external Ψ and knowledge of the water potential thresholds for germination within the seed population (**see: Germination – field emergence models; Hydrotime**). Low external Ψ can also inhibit the expression of some of the genes for the cell wall hydrolases associated with endosperm weakening. Thus, seeds are very sensitive to the water availability of their environment and often exhibit what could be seen as a conservative strategy. That is, seeds will not germinate at water potentials at which their seedlings will not grow (**see: Deserts and arid lands**). Most seeds are prevented from completing germination at Ψ of approximately -1.0 MPa, while roots in particular can continue to grow at this stress. This probably reflects the fact that once a seed has germinated, it is committed to grow, so it makes sense to be conservative with respect to future water availability. For a seedling root, however, the necessity for survival is to continue to grow in search of additional water deeper in the soil. Interestingly, in germinated seedlings, ABA actually helps maintain root growth at lower Ψ rather than inhibiting growth.

For most seeds, the initial expansion required to complete germination can occur without cell division. However, the root and shoot meristems must quickly begin cell division for organ growth to continue. The timing of when the cell cycle begins relative to radicle emergence can vary among seeds. Some seeds that are shed with very immature embryos will require extensive cell division associated with embryogenesis prior to the completion of germination. Other seeds with more mature embryos do not initiate cell division until after radicle emergence. In some cases, such as imbibition at reduced Ψ that occurs during **priming**, for example, an intermediate situation has been observed. In this case, the **cell cycle** can progress through the DNA synthesis phase, then arrest in G2 prior to cell division. This may contribute to the rapid initiation of growth when primed seeds are rehydrated. (KJB)
(See: Germination – influences of water)

Bradford, K.J. (1995) Water relations in seed germination. In: Kigel, J. and Galili, G. (eds) *Seed Development and Germination*. Marcel Dekker, New York, USA, pp. 351–396.

Chen, F., Dahal, P. and Bradford, K.J. (2001) Two tomato expansin genes show divergent expression and localization in embryos during seed development and germination. *Plant Physiology* 127, 928–936.

Crowe, J.H., Crowe, L.M., Hoekstra, F.A. and Wistrom, C.A. (1989) Effects of water on the stability of phospholipid bilayers: the problem of imbibition damage in dry organisms. In: Stanwood, P.C. and McDonald, M.B. (eds) *Seed Moisture*. CSSA Special Publication No. 14, Crop Science Society of America, Madison, WI, USA, pp. 1–14.

De Castro, R.D., van Lammeren, A.A.M., Groot, S.P.C., Bino, R.J. and Hilhorst, H.W.M. (2000) Cell division and subsequent radicle protrusion in tomato seeds are inhibited by osmotic stress but DNA synthesis and formation of microtubular cytoskeleton are not. *Plant Physiology* 122, 327–335.

Ni, B.R. and Bradford, K.J. (1992) Quantitative models characterizing seed germination responses to abscisic acid and osmoticum. *Plant Physiology* 98, 1057–1068.

Welbaum, G.E., Bradford, K.J., Yim, K.-O., Booth, D.T. and Oluoch, M.O. (1998) Biophysical, physiological and biochemical processes regulating seed germination. *Seed Science Research* 8, 161–172.

Germination - radicle emergence

In most situations this is an irreversible commitment by a seed to complete germination, since there is no going back once the seedling is emerged.

Once the hydrated seed has repaired accumulated damage and sensed a favourable environment, physiological processes directed towards germination are initiated. Since completion of germination requires growth of the embryo, processes related to cell growth are expected to be involved (**see: Germination; Germination – physical factors**). In some seeds in which the testa or other covering tissues are weak or split upon imbibition, initiation of embryo growth is essentially the only process required for completion of germination. However, in many seeds the embryo is enclosed within the **endosperm**, **testa** or **pericarp** that can present a barrier to its expansion (Fig. G.16). In these cases, softening or weakening of the enclosing tissues is required in order to allow radicle emergence to occur. The most detail is known about this for seeds of the Solanaceae, particularly tomato (*Lycopersicon esculentum*), tobacco (*Nicotiana tabaccum*) and Datura (*Datura ferox*), and seeds of other families such as lettuce (*Lactuca sativa*) and muskmelon (*Cucumis melo*). The lettuce endosperm envelope enclosing the embryo was essential for the expression of **thermoinhibition** of germination. S.P.C. Groot and C.M. Karssen, using a tomato **mutant** (*gib-1*) deficient in the ability to synthesize **gibberellins** (GAs), showed that in the absence of added GA, its seeds failed to germinate, and that the endosperm cap (or **micropylar endosperm**, the thin-walled endosperm tissue directly enclosing the radicle tip) did not weaken. When GA was supplied, however, the physical strength of the endosperm caps (as measured by the force required to puncture them) declined prior to emergence of the radicle (Fig. G.16). This and related work demonstrated that in the case of seeds enclosed within a rigid endosperm or a strong membrane, weakening of

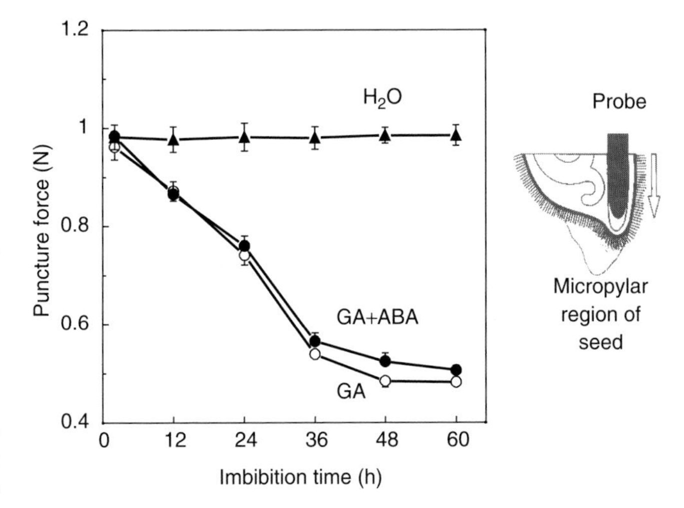

Fig. G.16. Puncture force analysis of *gib-1* tomato seeds. Puncture force was measured at different times after imbibition in water (▲), 100 μM GA (○) or 100 μM GA plus 100 μM ABA (●). The endosperm cap (micropylar endosperm) region was excised, the enclosed radicle tip was removed, and the force required to puncture the cap (+testa) with a probe was measured (diagram at right). In the absence of GA, the puncture force does not decline and the seeds do not germinate. GA induces both endosperm cap weakening and germination. ABA, at the concentration used here, does not block endosperm weakening, but does counteract the effect of GA on germination. Error bars indicate ± SE (n = 24) when they exceed the size of symbols. Reprinted from Chen, F. and Bradford, K.J. (2000) Expression of an expansin is associated with endosperm weakening during tomato seed germination. *Plant Physiology* 124, 1265–1274, with permission of the American Society of Plant Biologists, Rockville, MD, USA. Diagram from Groot, S.P.C. and Karssen, C.M. (1987) Gibberellins regulate seed germination in tomato by endosperm weakening: a study with gibberellin-deficient mutants. *Planta* 171, 525–531.

the enclosing tissue, presumably by enzyme action, must occur before the embryo can penetrate it.

In tomato, the primary component of the barrier to radicle emergence is in the endosperm cap tissue, whose cell walls are rich in **galactomannans**. These polymers are the major site of enzyme attack in endosperm cap weakening but other hemicelluloses are also involved. For details of the other polymers and enzymes active in wall softening **see: Germination – molecular aspects; Hemicelluloses**. The testa of tomato also contributes to the restraint on the embryo, and in *Arabidopsis thaliana* also the properties of the testa are important in imposing and maintaining **dormancy**. However, since the testa is a dead tissue at maturity, its effects must be primarily physical, although enzymes secreted from the embryo may be involved in weakening this structure also. **Endosperm** cells in tomato, on the other hand, are living after hydration and actively participate in the softening of this tissue by producing the necessary enzymes under the influence of GA. This hormone induces a number of cell wall hydrolases specifically in the endosperm cap of tomato that are involved in degradation of the galactomannan-rich cell walls of this tissue (**see: Germination – molecular aspects**). In addition, the storage reserves in the endosperm cap cells are mobilized rapidly and the cells vacuolate prior to radicle emergence, whereas these processes do not occur in the lateral endosperm until after radicle emergence. Interestingly, while **abscisic acid** (ABA) blocks germination, it does not prevent the induction of hydrolases in the endosperm cap or the majority of weakening of this tissue (Fig. G.16). It may block a late stage of endosperm weakening, or act in the embryo to reduce its growth potential. Because of its promotion of GA biosynthesis Pfr, the active form of **phytochrome**, also promotes endosperm weakening in tomato and other species.

Of course, the embryo must also be capable of exerting a growth force or growth potential against the weakened restraining tissues. Thus, germination is controlled by the balance between the thrust generated by the embryo resulting from osmotic uptake of water and the resistance offered by any enclosing tissues (Fig. G.17). The growth of the radicle to generate the thrust is promoted experimentally by GA, and endogenous GA may also play a role. Radicle growth can also be inhibited by several chemical inhibitors and also by relatively high light quantum fluxes (**see: Light – dormancy and germination**). (KJB)

Bewley, J.D. (1997) Breaking down the walls – a role for endo-β-mannanase in release from seed dormancy? *Trends in Plant Science* 2, 464–469.

Debeaujon, I., Léon-Kloosterziel, K.M. and Koornneef, M. (2000) Influence of the testa on seed dormancy, germination, and longevity in Arabidopsis. *Plant Physiology* 122, 403–413.

Kucera, B., Cohn, M.A. and Leubner-Metzger, G. (2005) Plant hormone interactions during seed dormancy release and germination. *Seed Science Research* 15, 281-308.

Nonogaki, H., Gee, O.H. and Bradford, K.J. (2000) A germination-specific endo-β-mannanase gene is expressed in the micropylar endosperm cap of tomato seeds. *Plant Physiology* 123, 1235–1245.

Germination – seedbed

See: Seedbed environment

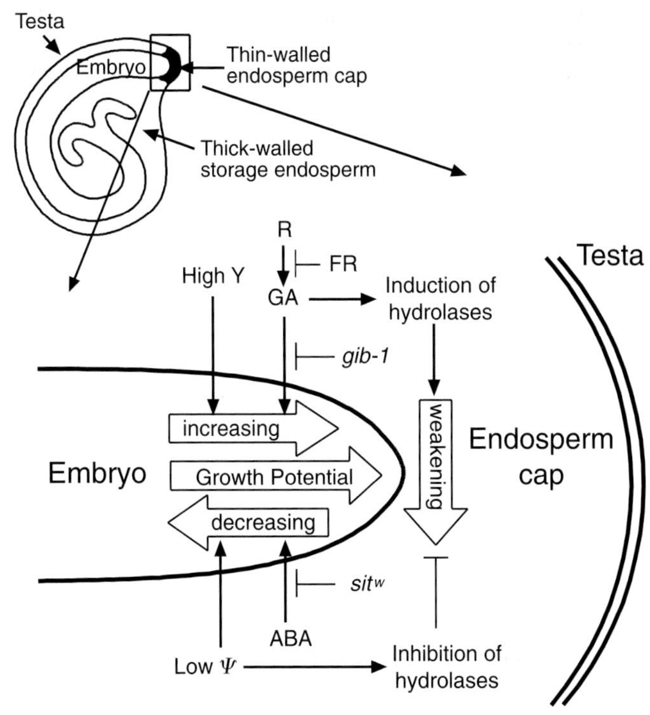

Fig. G.17. Diagram of a tomato seed and regulatory interactions involved in germination. Tomato seeds contain an embryo completely enclosed by a thick-walled lateral endosperm and a thin-walled endosperm cap (upper left). The testa surrounds the entire endosperm. Weakening of the endosperm cap (micropylar endosperm) following imbibition, combined with the expansive force (growth potential) of the embryo, results in radicle protrusion through the endosperm cap to complete germination. Some of the factors known to be involved in regulating endosperm cap weakening and embryo growth potential are shown (see text for details). Arrows indicate promotion of an action, while bars indicate inhibition of the action. ABA = abscisic acid; FR = far-red light; GA = gibberellin; *gib-1* = gibberellin-deficient mutant; R = red light; sit^w = ABA-deficient tomato mutant; Ψ = water potential. Diagram based on: Groot, S.P.C. and Karssen, C.M. (1987) Gibberellins regulate seed germination in tomato by endosperm weakening: a study with gibberellin-deficient mutants. *Planta* 171, 525–531.

Germination – species characteristics

Seeds of every species have characteristic conditions or requirements for germination. Even within species, e.g. cultivars of crop plants or ecotypes in wild species, there may also be characteristic requirements. Provided that water and oxygen are easily available, the prime factors that determine the extent and rate of germination in non-dormant seeds are temperature and possibly light (**see: Germination – influences of temperature**): there is a characteristic temperature range over which germination occurs, with distinct minimum and maximum temperatures and there is a temperature optimum at which germination rate is greatest. In some species, seeds must germinate in darkness because they can be inhibited by light. Seeds of many species have primary **dormancy** and certain factors or experiences are required to release seeds from this state (**see: Dormancy breaking**). These include: (i) light (single or repeated doses, at certain fluence rates, perhaps of certain spectral quality) (**see: Light –

dormancy and germination); (ii) alternating temperatures (with minimum temperature differentials and durations at each temperature); (iii) chilling (characteristic temperature and duration) (**stratification**); (iv) 'warm' temperatures (of certain values and durations); and (v) **afterripening**. Each species has a characteristic factor or several alternative factors which are efficacious. And there are special cases, for example where seeds have very hard coats (**hardseededness**), when under natural conditions the coat must slowly become more permeable or must experience heat or experimentally be abraded, or where seeds need to be stimulated by other plants (**see: Parasitic plants**). In some instances, seeds have been forced into **secondary dormancy**: relief of dormancy in such cases involves the same factors as with primary dormancy.

If the conditions needed to obtain germination of viable seeds are unknown, the effects of different temperatures must be tested, both constant and alternating, most conveniently carried out on a **thermogradient bar**. Should germination be unsuccessful, an assessment should be made of the dormancy status of the seed by testing the effects of seedcoat removal, of light treatments, of chilling, of warm stratification, of the application of hormones (e.g. **gibberellins** or **ethylene**) or other compounds (**see: Germination – influences of chemicals**) and of afterripening. In some seeds, **scarification**, either physical abrasion or with strong acids, might be applied. Special cases such as seeds of parasitic plants should be treated accordingly.

The germination requirements of many species, especially crops, have been described and are recorded in the literature. Data regarding wild species are available in many seed banks and as part of conservation projects. It is not feasible in this encyclopedia to give details covering individual species. Information concerning many cases may be found in the following references. (MB)

(See: Germination testing)

Baskin, C.C. and Baskin, J.M. (1998) *Seeds. Ecology, Biogeography, and Evolution of Dormancy and Germination*. Academic Press, San Diego, USA.

Ellis, R.H., Hong, T.D. and Roberts, E.H. (1985) *Handbook of Seed Technology for Genebanks I. Principles and Methodology. II. Compendium of Specific Germination Information and Test Recommendations*. International Board for Plant Genetic Resources, Rome: www.ipgri.cgiar.org/publications/HTMLPublications/52/index.htm

Royal Botanic Gardens, Kew, database: www.rbgkew.org.uk/data/sid/

Simpson, G.M. (1990) *Seed Dormancy in Grasses*. Cambridge University Press, Cambridge, UK.

Germination curve

The characteristic kinetic cumulative germination pattern of a population of individual seeds following **imbibition**, usually measured by repeated counting of the proportion or number of seeds showing **radicle** emergence. This best is depicted as a graph showing the number of seeds germinated at a sequence of time points. For an example of germination curves **see:** Fig. G.10(A) in **Germination – influences of temperature, 4. Germination rate**. (**See also: Germination rate**)

Germination energy

A measure of the rapidity of germination expressed as the percentage of seeds germinating within a given time under defined conditions or as the time (days) required for a given percentage to germinate. Accordingly, it reflects germination rate, uniformity, vigour and viability. The measure is commonly used in the malting and forest tree industries where rapid, uniform, high germination is an aim. **See: Malting – an overview**, section 2; **Vigour testing**

The term has also been used to describe the energy (heat) given off by germinating seeds as measured by the temperature rise in an enclosed, heat-retaining system.

Germination index

A quantitative expression of germination that relates the daily germination rate to the maximum germination value. Parameters of germination rate are used mathematically in several ways to produce several types of these indexes. (**See: Germination curve; Germination rate; Vigour testing – physiological**)

Germination inhibitors

Chemicals, generally naturally occurring in the seed, the soil or plant material, that inhibit germination. (**See: Germination – influences of chemicals; Germination – influences of hormones**)

Germination rate

Measures used in seed research and **vigour testing** that express the concept of the inverse of the time taken for seeds just to complete **germination** (generally as marked by radicle-**emergence**) in a defined fraction of the population under defined conditions: so that the higher the value, the faster the germination rate.

It is potentially confusing that the terms 'germination rate', '**germination speed**' and 'germination velocity' are in effect used interchangeably in the scientific literature, and at times imprecisely too. Also, units of germination time, rather than the inverse, should be avoided because by definition the terms rate and speed imply the passage of a process in time, not the time taken for the process to occur. Mathematically, this is the ratio of a number to a time period.

The term 'rate' has two other different meanings in the scientific literature, however. One is to signify the 'cumulative germination percentage' (the proportion of the number of seeds sown that finally germinate), or formula-combinations of two or more germination values obtained under different conditions: mathematically, a ratio of one number-fraction to another. The other is to describe the slope of the cumulative **germination curve** – the number of seeds that complete germination in a unit of time: mathematically, a number divided by time.

One frequently used rate index is the reciprocal of the time ($t(G)$) from the start of **imbibition** for an arbitrarily chosen percentage (G) of seeds to just complete germination under defined conditions. Effectively, this measures the time taken by the 'notional seed' that occupies the Gth percentile rank in the population. G may be expressed either as a proportion of the total number of seeds sown, or of the total number of seeds that show **viability** (discarding the non-germinable seeds for the computation, thereby making the index independent of viability). Commonly the '50% value' (50th percentile) is used. The $t(G_{50})$ (often abbreviated to t_{50}) is determined after making a series of successive counts by simple interpolation or by modelling the germination curve; and the germination rate is calculated as $1/t_{50}$. (**See**, for examples: Fig. G.5 in **Germination – field emergence models**; and Fig. G.10 in **Germination – influences of temperature**). Where the base temperature of germination is known, the rate can be expressed instead as the reciprocal of the **day-degrees** taken.

Amongst the equations used, with varying degrees of success, to model germination curve data, are the following.

- The logistic function, which has the form:
  $P = a + c/(1 + \exp(-b(t-m)))$,
  where P is the cumulative percentage of seeds germinated at time t.
- The cumulative Weibull function, in its three-factor version:
  $P = F(1 - \exp(-((t-L)^d/k)))$,
  where F is the final germination percentage, L is the lag before the first seed has completed germination, k is a spacing function (proportional to the time between the start of germination and when cumulative germination reaches a certain point, and d determines the shape of the curve (exponential, or simulated normal or skewed-normal distributions).

Once the values of the constants are derived by statistical fitting of the equation to the data, a germination rate can be calculated (to attain half the maximum cumulative germination, $t(G_{F/2})$ for example).

An alternative index, the Coefficient of the Rate of Germination (proposed by Kotowski in 1926, who used the word 'velocity') expresses the mean rate of germination as the reciprocal of the 'mean germination time' (MGT) in days (**see: Vigour Testing – physiological. 1. Germination rate and related parameters**), calculated as:

$$\mathrm{CRG} = \frac{\Sigma n}{\Sigma(D.n)}.100$$

where n equals the number of seeds completing germination on day D, reckoning the day of sowing as 0.

The Timson Index uses a very simple expression, of the form:

$$\frac{\Sigma G}{\Sigma t}$$

where G is the percentage seed germination measured at regular intervals, and t is the total germination interval period studied. Different maximum values are possible depending on the number of counts made during the period. (PH)

(See: Germination – field emergence models; Germination testing; Vigour testing)

Bridges, D.C., Wu, H., Sharpe, P.J.H. and Chandler, J.M. (1989) Modeling distributions of crop and weed seed germination time. *Weed Science* 37, 724–729.

Brown, R.F. and Mayer, D.G. (1988) Representing cumulative germination. 2. The use of the Weibull function and other empirically derived curves. *Annals of Botany* 61, 127-138.

Heydecker, W. (1973) Glossary of terms. In: Heydecker, W. (ed.) *Seed ecology*. Butterworths, London, UK.

Timson, J. (1965) New method of recording germination data. *Nature* 207, 216–217.

Germination *sensu stricto*

A term introduced by M. Evenari in 1957 and developed by D. Côme. They regarded this as the phase of germination that starts at the end of water uptake by an **orthodox seed** (imbibition phase, Phase I), and terminates with the initiation of growth of the embryo axis (usually radicle), i.e. Phase II of germination (**see: Germination**, Fig. G.4). However, germination might best be considered as a continuum from the start of imbibition to emergence of the radicle. Indeed, many researchers use the term *sensu stricto* to denote this complete process of germination.

Germination speed (or velocity)

Alternative, though arguably less appropriate, names for germination rate: a term properly defined as the reciprocal of the time from sowing to the completion of germination for a defined fraction of the population. (**See: Germination rate**)

Germination testing

Official germination testing is concerned with establishing the percentage by number of pure seeds that have the potential to produce established seedlings in the field. Testing procedures and evaluation rules are produced by the **International Seed Testing Association (ISTA)** or the **Association of Official Seed Analysts (AOSA)**, and are the technical basis for seed-testing laboratories throughout the world. New or modified official procedures are adopted into the formal Rules after extensive testing and validation, under the supervision of expert committees.

More generally, a variety of germination testing procedures (some derived from official procedures) are used informally in seed research laboratories throughout the world, though usually these tests are not carried out to the same exacting standards – such as, for example, for the rigorous categorization of 'abnormal seedlings' (see below). Though determination of the rate of germination is not an explicit objective of official germination testing, often a measure of kinetics is incorporated, for vigour testing purposes (**see: Germination – field emergence models; Germination rate; Vigour testing – physiological**).

Biologically speaking, seed germination is a complex series of physiological processes, beginning with the seed taking up water by **imbibition** and ending with the emergence of the embryonic axis through its surrounding structures (usually the **radicle** penetrating the **testa** or **pericarp**). However, in the official seed quality testing laboratory, germination is taken to begin with the embryonic axis piercing the seedcoat, and to end when the seedling has developed to the stage at which it can be evaluated as normal or abnormal. This is because, from the practical quality testing point of view, no assessment about whether the seed has germinated can be made until the embryonic axis protrudes. The ISTA 'seed quality testing' definition of germination, accordingly, is: 'The percentage by number of pure seeds which in the course of a germination test produce seedlings which have developed those structures that indicate the ability to develop further into a mature plant under favourable conditions in the field'.

Official seed germination testing rules (ISTA Rules and AOSA Rules) stipulate the near-optimal germination conditions that need to be provided for individual species: suitable substrate(s), moisture, aeration, temperature regime and, in some cases, the nature of light irradiance, and any required **dormancy-breaking** techniques. The Rules also prescribe the timing of assessments, and the criteria for judging 'normal seedlings' (see below). In practice, the range of equipment and methods used contributes to variation in results obtained within and between different laboratories. By prescribing equipment, substrates and procedures, the Rules aim to minimize between-laboratory variation to a statistically acceptable level. Tolerance tables are used to decide whether differences between test and replicate results (both within and between laboratories) are significant.

1. Substrates

Germination tests use primarily either paper or sand as the germination media, though under special circumstances soil or artificial compost may be used. Paper and sand quality are critical to results, and detailed specifications of their composition, texture, water holding capacity, pH range and freedom for fungi, bacteria and toxic substances are given in both ISTA and AOSA Rules.

(a) *Paper* substrates are used in several variants:

- TP (on *t*op of *p*aper), also known as the 'blotter test'. Seeds are sown on top of one or more layers of moist paper that are placed on the **Copenhagen** (or Jacobsen tank) apparatus, into transparent boxes or Petri dishes, or placed in open trays in a humidified germinator cabinet.
- BP (*b*etween *p*aper). Seeds are placed between two layers of moist paper – for example, under a cover of a flat filter paper, in folded paper envelopes or in rolled towels (also known as the 'ragdoll' test). These are incubated as in the TP systems, except that rolled towels are usually wrapped first in plastic bags.
- PP (*p*leated *p*aper). Seeds are placed in an accordion-like paper strip (typically with two seeds in each of the 50 pleated folds); after planting, a flat strip of paper is wrapped around the pleats, which are placed in boxes and after moistening incubated within a germination cabinet.

(b) *Sand* substrate is used in two ways:

- S (in *s*and). The seeds are placed on a layer of moist sand and covered with another layer.
- TS (*t*op of *s*and). The seeds are merely pressed into the surface of the sand.

2. Moisture

ISTA Rules prescribe that: 'The substrate must at all times contain sufficient moisture to meet the requirements of germination'. Sufficient moisture depends on the species being tested. *Trifolium pratense* and *Pinus sylvestris* are sensitive to excessive moisture of the substrate; the same is true of species with very small seeds, such as *Begonia*, *Kalanchoë* and *Nicotiana*. If the germination substrate is too wet in these cases, such species produce weak, glassy-looking seedlings and the root tips of seedlings may become brown and decayed. By contrast, other species, such as *Pinus palustris*, require markedly wet conditions for normal germination and seedling growth; if insufficient water is available, the primary root and **hypocotyl** curl, and growth is hampered.

Water quality can have a critical effect on germination, and AOSA and ISTA Rules give detailed specifications for the water that is used in germination tests in terms of cleanliness, quality, pH range, and the quality control measures that should be adopted. A permitted range of water amounts is allowed in TP and pleated paper systems, leaving discretion to the testing laboratory. Rolled towels are usually saturated by submersion in water, and then squeezed to remove enough of the excess. Blotter water levels in the Copenhagen tank are continuously regulated by the wicking action of the paper strip that connects the blotters to the water reservoir in the bath beneath.

3. Light

Seeds of most cultivated species will germinate either in light or in darkness (**see: Light, dormancy and germination**). However, except for a few species where light may be inhibitory to germination (for example, *Phacelia tanacetifolia*), illumination of the substrate from an artificial source (typically fluorescent rather than incandescent light, to minimize the emission heat and infra-red irradiation wavelengths) or by indirect daylight is generally recommended. This leads to better developed seedlings that are more easily evaluated, because light prevents excessive elongation, promotes the formation of chlorophyll and gives seedlings a natural look. Grown in continuous darkness, seedlings become etiolated; they become susceptible to saprophytic fungal attack, and albino seedling off-types cannot be distinguished.

4. Temperature

The temperature(s) to which the seed is exposed are prescribed for each species (**see: Germination – influences of temperature**). Temperature control should be as uniform as possible throughout the germination apparatus (cabinet or room), and the variation from the prescribed should not be more than ±2°C. Where alternating temperatures are prescribed, the lower temperature should be maintained for 16 h and the higher for 8 h: a gradual changeover between the alternating phases (up to 3 h) is permitted, but a more abrupt change (in no more than 1 h) is recommended for seed that is likely to be dormant.

5. Dormancy-breaking techniques

The 'seed testing' objective of the germination test is: 'To determine the maximum germination potential of a seed lot, which in turn can be used to compare the quality of different lots and also estimate the field planting value'. To meet this objective it is important that as many seeds as possible complete germination and that the seedlings can be assessed within the test period. The most important reasons why seeds will not germinate, even though they are exposed to favourable germination conditions, are **dormancy** in the embryo and due to **hardseededness**. These obstacles to **germination** may be removed using a selection of **dormancy-breaking** treatments, permitted by the seed testing rules, of which the most frequently used treatments are as follows:

(a) *Pre-heating*. Before planting, dry seeds are exposed to temperatures of 30–35°C, for up to 7 days. For certain tropical and subtropical species, temperatures of 40–50°C are recommended (for example, peanut at 40°C and rice at 50°C).

(b) *Pre-chilling*. Before exposing the seeds to the prescribed germination temperature they are kept on the moist substrate at a low temperature (usually between 5 and 10°C) for initial period of up to 7 days. Imbibed tree and shrub seeds are often pre-chilled at temperatures of between 1 and 5°C, for 2 weeks to a year, depending on species. (**See: Stratification; Tree seeds**)

(c) *Potassium nitrate* (KNO_3). The germination substrate is moistened with a 0.2% solution of KNO_3 instead of water.

(d) *Gibberellic acid* (GA_3) is mainly used to break the dormancy of temperate cereals and a 0.02 to 0.1% solution of GA_3, instead of water, is used to moisten the substrate.

(e) *Sealed polythene (polyethylene) envelopes*. Seed of *Trifolium* spp. are germinated in sealed polythene envelopes. Carbon dioxide from the respiring seed and ethylene from the natural breakdown of the polythene bag stimulate the germination of dormant seed.

(f) *Alternating temperatures*. The seed is germinated using an alternating temperature regime, with the higher temperature having a duration of 8 h and the lower temperature, 16 h.

(g) *Light*. When light is used as a dormancy-breaking stimulus, it should have an intensity of 750 to 1250 lux. Tests should be illuminated for a continuous period of at least 8 hours in every 24-hour cycle, and this should coincide with the high temperature phase if an alternating temperature regime is used. Where an isothermal temperature regime is used, continuous illumination of at least 12 h in each 24 h period is recommended. (**See: Light, dormancy and germination**)

(h) *Pre-washing*. Naturally occurring inhibitors (germination inhibitors) in such seeds as those of beets (*Beta vulgaris*), may be removed by washing them in running water at a temperature of 25°C. After washing, the seed is dried back at no more than that temperature, before planting on or in the germination substrate.

(i) *Removal of enveloping seed structures*. For certain species in the Poaceae, removal of outer structures such as an involucre of bristles, or the lemma and palea, promotes germination. (**See: Seedcoats – structure**)

(j) *Removing hardseededness*. Hardseededness (that is, structural impermeability of the seedcoat to water and gases) is a widespread phenomenon in species of the Fabaceae and other plant families, including tree seeds. For many such species no attempt is made to induce germination of any seeds that remain ungerminated at the end of the test period, and they are reported as 'hard seeds'.

Special measures must be taken where a fuller assessment is required – before the start of the germination test or (if it is suspected that the treatment may adversely affect non-hard seeds) on any hard seeds remaining after the prescribed test period, which are then returned after treatment for a further germination test period.

Three different methods can be used, as follows:

- Soaking in water for up to 24 or 48 h, and starting the germination test immediately afterwards.
- Mechanical **scarification**. Careful piercing, chipping, filing or sandpapering of the seedcoat may make the seed permeable to water. Care must be taken to avoid damage to the **embryo** and resultant seedling; the most suitable site for attack is immediately above the tips of the **cotyledons** and as far away from the tip of the **radicle** as possible.
- Acid scarification. Digestion of the seedcoat in concentrated sulphuric acid is effective in some species such as *Brachiaria* sp. and *Macroptilium* sp. The seeds are soaked until the seedcoat becomes pitted. Before the start of the germination test the seed must be thoroughly washed in running water.

6. Seedling evaluation

Seed quality testing deals with eight morphologically distinct types of seedlings of **gymnosperms** and **angiosperms** (**monocotyledons** and **dicotyledons**). These types are distinguished on the basis of which seedling parts emerge above the surface, including whether the cotyledons do or do not emerge (**epigeal** or **hypogeal** types), and how long the cotyledon remains within the seed.

When the seedlings have reached a defined stage of development, they are assessed with regard to the 'normality' of their essential parts. Sometimes two or more successive counts are required before all seeds have germinated and reached the stage of development required, before assessment is possible. By the end of the germination test, seedlings are assessed as 'normal' or 'abnormal' and ungerminated seeds as 'fresh' (dormant), 'hard' or 'dead'. As a general rule, seedlings must not be removed from the test before all their essential structures have developed to such an extent that they can be assessed reliably and without doubt. This implies that for the majority of seedlings in a test, depending on the morphological type being tested:

- the cotyledons have freed themselves from the seedcoat (for example, lettuce and *Brassica* species); or
- the primary leaves have expanded in sand tests, or have protruded from between the cotyledons in rolled-towel tests (for example, *Phaseolus*); or
- the first leaf has emerged from the coleoptile in Poaceae.

(a) *Normal seedlings* are those assessed to have the potential for continuing their development into satisfactory plants, when grown in good quality soil and under favourable conditions of moisture, temperature and light. This potential is regarded as being dependent on the soundness and correct functioning of the essential seedling parts during germination and early seedling stages. To be classified as normal, therefore, a seedling must conform to one of the following categories.

- *Intact seedlings*, with all their essential structures well developed, complete, in proportion to each other and healthy.
- *Seedlings with slight defects or deficiencies* to their essential structures, provided they show an otherwise satisfactory and balanced development comparable to that of intact seedlings in the same test.
- *Seedlings with secondary infection*, which it is evident would have conformed to either of the two preceding conditions,

but which have been infected by fungi or bacteria from sources other than the actual seed itself (usually from a neighbouring seed in the same test).

(b) *Abnormal seedlings*, by contrast, are those that do not show the potential for continued development into satisfactory plants when grown in the same favourable conditions, because one or more of their essential structures are either missing or irreparably defective and thus are judged incapable of normal functioning. The most frequent causes of damage and deficiencies on seeds, which may result in abnormal seedlings, are shown in Table G.12.

(c) *Ungerminated seed*. At the end of the germination test any ungerminated seed are assessed as dead, fresh or hard according to the criteria in Table G.13. (RD)

ISTA (2006) *International Rules for Seed Testing, Edition 2006*. International Seed Testing Association, Switzerland.

McDonald, M.B. (1998) Seed quality assessment. *Seed Science Research* 8, 265–275.

Miles, S.R. (1963) Handbook of Tolerances and Measures of Precision in Seed Testing. *Proceedings of the International Seed Testing Association* 28, 520–686.

Table G.12. Causes and symptoms of seedling abnormalities.

Cause	Examples of symptoms
Mechanical injury of the embryo	Split, stunted or missing primary roots; cracks and splits in the seedling axis or in the cotyledons; broken or separated cotyledons or parts of the axis; broken or otherwise damaged coleoptile
Deficiencies in the physiological make-up of the seed or embryo	Retarded growth of the seedling as a whole, or of individual parts of it; short, stubby or spindly primary or seminal roots; short and thick or otherwise deformed seedling axis or coleoptile; curled, discoloured or necrotic cotyledons (or primary leaves); negative geotropism (for example, roots growing upward); yellow or white albino seedlings with chlorophyll deficiency; and spindly or glassy seedlings
Primary infection and disease of the seedling	Decay of individual seedling parts or of the whole seedling (which eventually leads to the death of the seedling)

Table G.13. Assessment of non-germinated seeds.

Criterion	Dead	Fresh (dormant)	Hard
Seed has not germinated	√	√	√
Seed has imbibed water	√	√	x
Embryo is not discoloured	x	√	√
Seed has not decayed	x	√	√
Seed is firm to the touch	x	√	√
Embryo is alive	x	√	√

Germplasm

At its broadest, any living material and, at its narrowest, the 'genetic material' constituting the heritable characteristics of that material. However, in most plant genetic conservation contexts, it refers to propagating material including that of a vegetative nature, seeds (**see: Conservation of seed**) and pollen.

A germplasm bank is an organized collection of accessions of seed. The name derives from an early theory of inheritance which advanced the notion that hereditary characters were contained in an immutable 'plasm' (literally, a thing cast or moulded), transmitted unchanged from parent to offspring.

Gibberellic acid

GA_3: one type of **gibberellin**. This was one of the first gibberellins to be discovered, as a product of the pathogen of rice, *Gibberella fujikuroi* (*Fusarium moniliforme*). Since, commercially, it the most easily produced and affordable of the active gibberellins, it is often used in studies of the actions of this hormone on seeds, as a dormancy-breaking technique for certain species within official **germination testing** rules procedures, in industrial grain malting (**see: Malting – process**), to manipulate floral development (**see** for example: **Gymnosperms – seed production; Production for sowing, III. Hybrids, 4. Pollination management; Lettuce, 4. Production**), and in pre-sowing seed **enhancements**, such as some **priming** procedures and the treatment of semi-dwarf rice varieties (**see: Rice – cultivation**) and of some **ornamental flowers**.

Gibberellins

A large family of compounds that can regulate seed **germination, mobilization of reserves** in cereal grains and stem growth. The first gibberellin identified (**gibberellic acid** – GA_3) was isolated from an extract of the fungus (an ascomycete) *Gibberella fujikuroi* in the 1930s. Rice plants infected with this grow so tall that they lodge and become senescent ('foolish seedling' disease). The first plant gibberellin was isolated in 1957 and termed GA_1. Since then over 125 naturally occurring gibberellin-like compounds have been isolated from an array of higher and lower plants including mosses and algae.

Gibberellins (GAs) are cyclic diterpenes and are derived from an *ent*-gibberellane ring structure comprising 20 carbon atoms. The biologically active gibberellins are termed C19-GAs (Fig. G.18) as the C20 methyl group has been replaced by a lactone. Many of the gibberellins that have been isolated are thought to be biosynthetic intermediates and there may be only a few (e.g. GA_1, GA_3, GA_4 and GA_7) naturally occurring compounds with biological activity.

In cereals, gibberellins play an important role in regulating the degradation of storage reserves by promoting the expression of genes encoding key hydrolytic enzymes specifically within the aleurone layer cells of the grain (**see: Mobilization of reserves – cereals**). Gibberellins also play a role in the endosperm weakening that occurs in germination of several seed types. (**See: Germination – radicle emergence**)

Evidence from mutants of *Arabidopsis* and tomato supports the hypothesis that the timing of seed germination may be coordinated by a balance between the stimulatory effects of gibberellins and the inhibitory properties of another plant hormone, abscisic acid (ABA) (**see: Hormone balance theory**). Seeds of GA_1-deficient mutants such as *ga1* (*Arabidopsis*) and *gib-1* (tomato) remain dormant unless treated with gibberellins, while GA-insensitive mutants such as *gai*

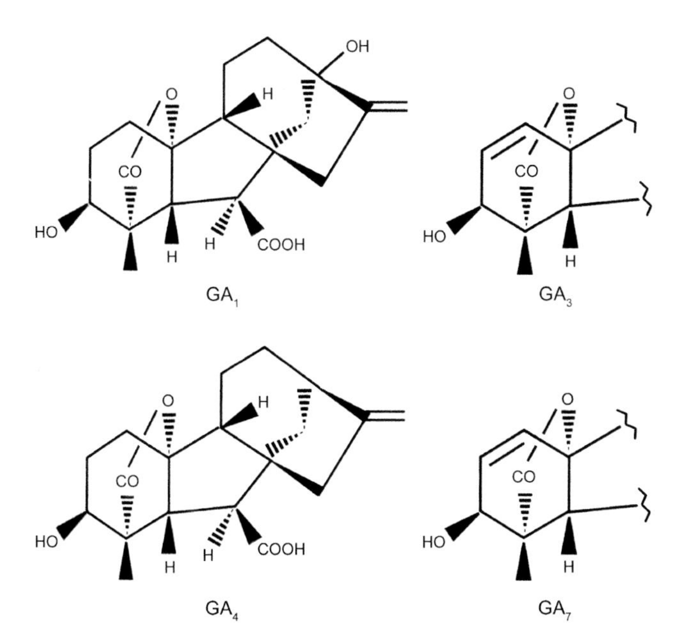

Fig. G.18. Structures of four key C19-gibberellins. Modified from Srivastava, L.M. (2002) Gibberellins. In: *Plant Growth and Development. Hormones and Environment.* Academic Press, New York, USA, pp. 171–190.

(*Arabidopsis*) require a lower dosage of ABA for germination to be inhibited. Some types of seed dormancy may also be the consequence of changes in gibberellin contents or sensitivity. The germination of photoblastic seeds such as lettuce and *Arabidopsis* is promoted by exposure to red light and this treatment has been shown to elevate the transcription of certain isoforms of 3β-hydroxylase (oxidase), an enzyme that is critical for the synthesis of the biologically active gibberellin GA_1. (See: **Germination – radicle emergence; Light – dormancy and germination; Phytochrome; Signal transduction – light and temperature**)

Gibberellins play a central role in coordinating the elongation growth of tissues such as stems and inflorescences. Evidence to support this assertion comes from the study of dwarf mutants that are commonly deficient in their ability to synthesize GA_1 and from application of compounds such as AMO-1618 that block gibberellin biosynthesis and result in a substantial retardation of plant height.

The biosynthetic pathway is complex (Fig. G.19): many of the original *in vivo* labelling studies to discover the pathway were carried out on developing pea or pumpkin seeds as these had been shown to contain substantial quantities of gibberellins (see: **Development of seeds – hormone content**). The process comprises three stages with conversion from geranylgeranyl diphosphate (GGPP) to GA_{12}-aldehyde (Stages 1 and 2) occurring in all plants examined. Stage 1 is thought to occur in plastids with *ent*-kaurene being transported into the cytoplasm where conversion to GA_{12}-aldehyde takes place via the action of a series of P450 monooxygenase enzymes associated with the endoplasmic reticulum. The final stage exhibits species-specific variations giving rise to characteristic gibberellin profiles in different groups of plants. The most frequently observed alternatives are shown and these comprise: oxidation of the C-7 aldehyde group to give GA_{12}; sequential oxidations of the C-20 group to form a lactone and give rise to C19 GAs; and hydroxylation of GA_{12} at C-13 to generate GA_{53} or a non-hydroxylation route via GA_{15}.

Gibberellin-deficient mutants of *Arabidopsis*, maize, pea and tomato, identified by their dwarf habit or their inability to germinate without applied GA, have proved particularly valuable in determining the biosynthetic pathway and clarifying which compounds have biological activity. In addition, they have enabled some of the genes encoding key regulatory enzymes to be cloned and then used as probes to isolate **orthologous genes** in species that may have particular agricultural or horticultural significance. Database searches of the *Arabidopsis* and rice genomes have revealed that biosynthetic enzymes such as GA_{20}-oxidase and 3β-hydroxylase (oxidase) are represented by sizeable gene families and there is evidence that different members are expressed in different organs. In general, expression of these genes is highest in young tissues such as developing fruits and seeds.

For biological activity, gibberellin molecules must have a lactone group at C-19 and a carboxylic acid group at C-6. Furthermore, the ability of a gibberellin to promote α-amylase activity and stem growth is dependent on the presence of a β-hydroxyl group at the C-3 position. By conjugation with sugar residues such as glucose at C6 it is possible to temporally inactivate a gibberellin. Such glucosyl esters have been identified in a range of tissues including maturing seeds where they may act as a gibberellin store to support growth during seedling development. Conjugates may also provide an opportunity to locate gibberellins in different sub-cellular compartments by changing their polar nature and as a way of transporting the plant hormone from sites of synthesis to sites of response.

Gibberellin contents are highly coordinated in plant tissues and 2β-hydroxylation of biologically active compounds provides a mechanism for irreversibly converting biologically active to inactive compounds. This may be one of the principle processes for regulating gibberellin metabolism and mutants such as *slender* (pea) that lack the ability to carry out this biochemical event exhibit excessive elongation growth as a consequence of elevated amounts of GA_1. (GL, JR)

(See: Dormancy breaking – hormones; Germination – influences of hormones; Signal transduction – hormones)

Heden, P. and Kamiya, Y. (1997) Gibberellin biosynthesis: enzymes genes and their regulation. *Annual Review of Plant Physiology and Plant Molecular Biology* 48, 431–460.

Ritchie, S. and Gilroy, S. (1998) Gibberellins: regulating genes and germination. *New Phytologist* 140, 363–383.

Srivastava, L.M. (2002) *Plant growth and development. Hormones and environment.* Academic Press, New York, USA. (Gibberellins pp. 171–190.)

Ginning

The cotton gin separates useable lint fibres from the seed, after drying and cleaning, by one of two methods. The process leaves **fuzzy seed** with residual linter fibres, which is usually then delinted before oil extraction or sowing.

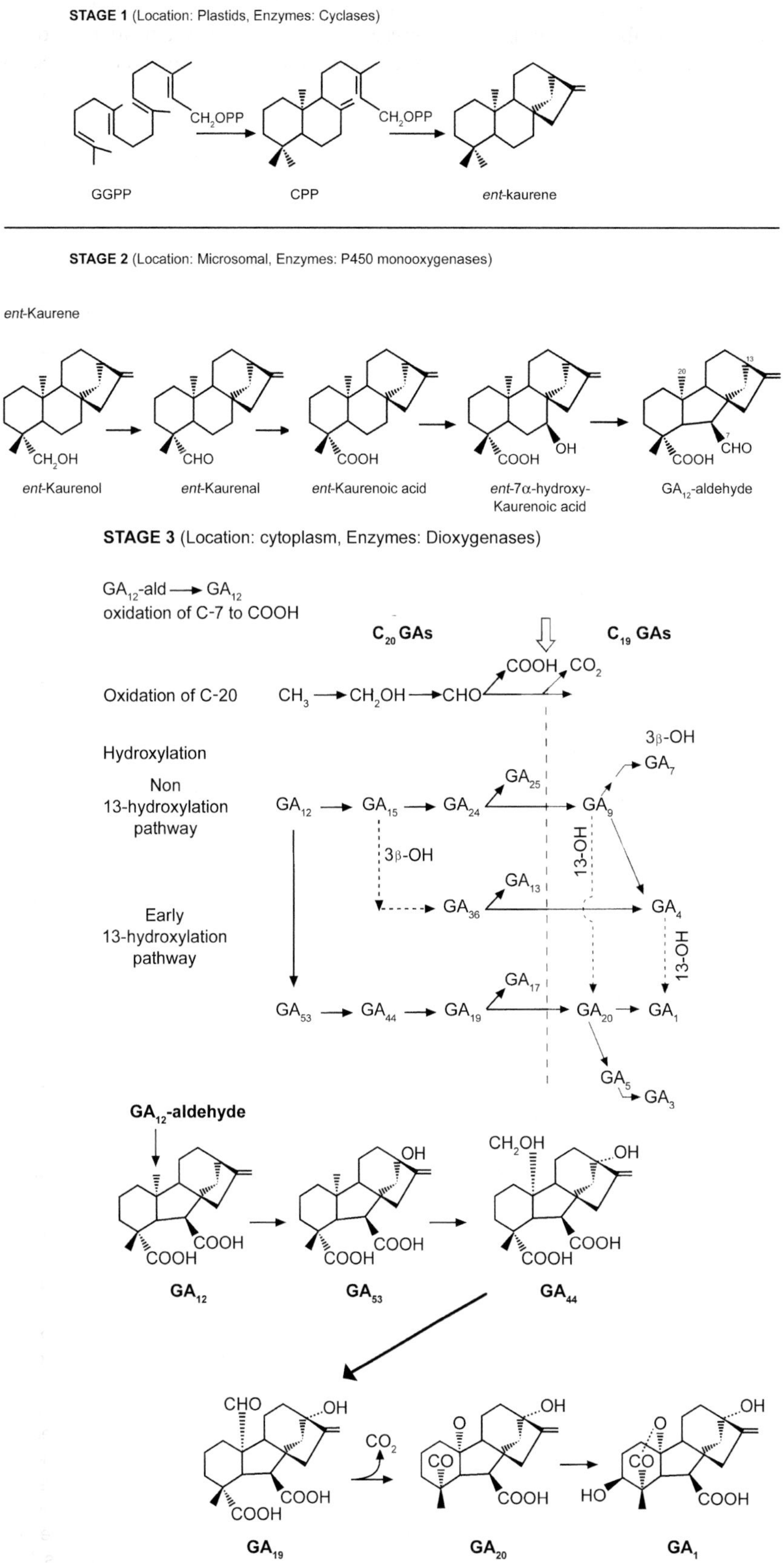

STAGE 1 (Location: Plastids, Enzymes: Cyclases)

CH_2OPP → CH_2OPP →

GGPP CPP *ent*-kaurene

STAGE 2 (Location: Microsomal, Enzymes: P450 monooxygenases)

ent-Kaurene

CH_2OH → CHO → COOH → COOH OH → COOH CHO

ent-Kaurenol *ent*-Kaurenal *ent*-Kaurenoic acid *ent*-7α-hydroxy-Kaurenoic acid GA_{12}-aldehyde

STAGE 3 (Location: cytoplasm, Enzymes: Dioxygenases)

GA_{12}-ald → GA_{12}
oxidation of C-7 to COOH

C_{20} GAs C_{19} GAs

Oxidation of C-20 CH_3 → CH_2OH → CHO → COOH, CO_2

Hydroxylation

Non 13-hydroxylation pathway: GA_{12} → GA_{15} → GA_{24} → (GA_{25}) → GA_9 → 3β-OH → GA_7

3β-OH

Early 13-hydroxylation pathway: GA_{36} → (GA_{13}) → GA_4 → 13-OH

GA_{53} → GA_{44} → GA_{19} → (GA_{17}) → GA_{20} → GA_1

GA_5 → GA_3

GA_{12}-aldehyde

GA_{12} **GA_{53}** **GA_{44}**

CO_2

GA_{19} **GA_{20}** **GA_1**

Fig. G.19. Biosynthetic pathway of gibberellins showing the putative cellular locations where the events take place. For convenience the pathway has been broken into three discrete stages. GGPP, geranylgeranyl pyrophosphate; CPP, copalyl pyrophosphate. Modified from Srivastava, L.M. (2002) Gibberellins. In: *Plant Growth and Development. Hormones and Environment.* Academic Press, New York, USA, pp. 171–190.

- Cotton varieties with shorter staple or fibre lengths are ginned with circular saws that grip the fibres and pull them through narrow slot openings that are too large for the seeds to pass through, so that the fibres are pulled free.
- Long-fibre cottons must be ginned in a roller gin to avoid damage to their delicate fibres. Rough rollers grab the fibre and pull it under a rotating bar with gaps too small for the seed to pass. Long-staple cottons, like Pima varieties, separate from the seed more easily than Upland varieties.

(See: Cotton; Delinting; Fibres; Lint and linters)

Glans

(Latin for acorn). An indehiscent type of **fruit** composed wholly or partially enclosed by a fruiting-cupulate (**aril**-like) involucre (cup) that is derived from a swelling of certain flower parts, e.g. the bracts, receptacle, or perianth, as present in *Quercus robur* (oak) and *Anacardium occidentale* (**cashew**).

Glass

A glass is a fluid with such extreme viscosity that it has mechanical properties similar to a solid (it is hard and strong), but is not considered a true solid, which is crystalline. Because of the solid-like behaviour in a fluid-like structure (sometimes called amorphous structure because molecular arrangements are random compared to a crystal), a glass is distinguished from the other **phases** (or states) of matter, namely solid, liquid and vapour. Glasses are ubiquitous, being formed from most compounds and mixtures and the study of their properties is fundamental to many disciplines in chemistry, physics, engineering and biology. In seeds, aqueous glass formation is an important natural consequence of drying that has led to hypotheses on the nature of **desiccation tolerance** (**see: Hydration force explanation**) and seed **longevity** (**see: Deterioration kinetics**), although differences in glass properties among seeds remains to be demonstrated. Aqueous glass formation is also critical to successful cryopreservation (**cryostorage**), where lethal ice is prevented by dehydration, rapid cooling, or exogenous protectants that limit nucleation or crystal growth (also known as **vitrification**).

During a glass transition, the structure and energy of a supercooled liquid become partially fixed because the extreme viscosity slows molecular rearrangements. Further cooling or concentrating results in only small changes in the structure and energy of the mixture and the new relationship between temperature and structure or energy causes a discontinuous change in the heat capacity of the substance at the glass transition temperature (Tg) (Fig. G.20A). While the high viscosity slows molecular rearrangements, it does not prevent them, and with time the structure and energy of the glass revert to those of the supercooled liquid (also called relaxation). The cytoplasm of dry seeds exists in a glassy state. In contrast to the situation with most glass-forming substances, the intracellular glasses in seeds can be formed simply by drying, due to their complex composition that prevents crystallization. In general, drying at room temperature leads to intracellular glass formation in seeds at about 10% moisture (Fig. G.20B).

Glass formation is deemed fundamental to the preservation of biological material because it provides a mechanism to restrict motion while maintaining structure. However, because changes in glasses are inevitable, it is important that preservation is considered relative to the timescale in which the glass relaxes. Numerous models predict a double exponential relationship between relaxation time or viscosity and temperature near the glass transition temperature (Tg), with viscosity rising to infinite levels at a temperature below

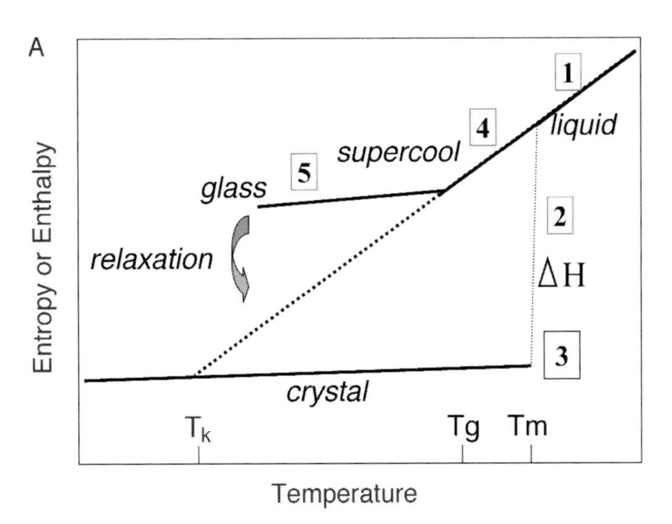

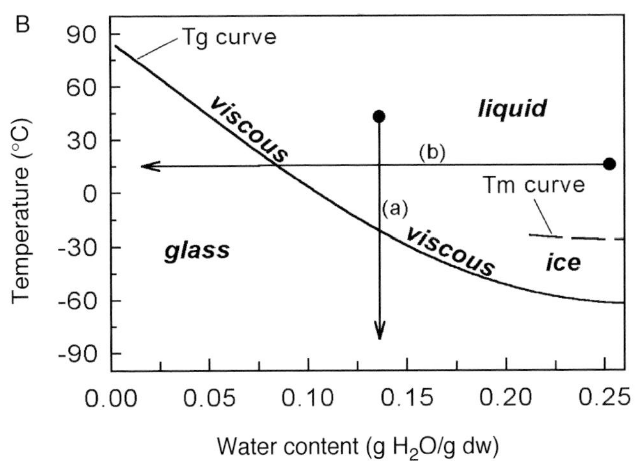

Fig. G.20. Glass formation. (A) Schematic of energy changes during phase transitions. As a liquid cools (1) it may undergo a large decrease in enthlapy (2) and crystallize (3) at Tm in a first-order transition. If crystallization is prevented (by rapid cooling, lack of nucleation, low growth of nucleators, restricted mobility, etc.), the liquid will supercool (4), and eventually the structure of the liquid will be 'frozen' into a glass (5) at Tg in a second-order transition. Once in a glass, restructuring of the liquid to maintain equilibrium (dashed line) occurs very slowly in a process called relaxation. At the hypothetical Kauzmann temperature, T_k, the glass reaches maximum viscosity. (B) State-phase diagram describing the interaction between temperature and water content on the glass transition temperature (Tg) and ice melting temperature (Tm) in a starchy seed species. Also indicated are the physical states of water (i.e. glass, liquid and viscous) and the liquid phase. Tm corresponds to the melting temperature of ice present in the tissues after a cooling/heating cycle. The limit of ice formation (broken line) corresponds to the unfrozen water content. Below this value, the cytoplasm becomes too viscous to allow ice crystal formation in a reasonable time scale. A decrease in temperature (path a) or water content (path b) results in the formation of the glassy state inside the tissues.

Tg known as the Kauzmann temperature (T_k) (Fig. G.20A). In reality, mobility within glasses deviates from the double exponential relationship to **Arrhenius** behaviour at a temperature between Tg and T_k. A measure of how steeply temperature affects the viscosity is a fundamental property of the glass described as 'fragility' and can be approximated by the ratio of T_k to Tg. In the most fragile glasses, T_k/Tg approaches 1 such that temperature has a profound effect on mobility. Glasses within seeds appear to be relatively 'strong', implying that intermolecular viscosity never approaches the theoretical infinite limit, but that structures remain relatively unperturbed when temperature (and probably moisture) fluctuates. There may be evolutionary significance to fragility in seed glasses: a fragile glass may provide quiescent seeds with a mechanism to monitor the extent of environmental changes, while a strong glass would guard against abrupt changes in physiology and maintain seed quality over a broad range of conditions. Seed constituents that affect glass fragility are yet to be discovered.

The glass transition temperature is a function of the relative concentrations of constituents in the glass, and the relationship between concentration and Tg is expressed in a **phase diagram**. Anhydrous sugars form glasses at relatively high temperatures, with larger molecules having higher Tg, e.g. trimaltose (76°C), sucrose (52°C), glucose (31°C). Raffinose, which has received attention in the seed literature as a glass-former or stabilizer, is unstable at high temperatures making it difficult to measure Tg. Water, on the other hand, has a very low Tg at about −135°C. When a little water is added to sugars Tg decreases; thus, water is a plasticizer. Conversely, addition of solutes to water raises the Tg of the dilute solution. Proteins form glasses only at relatively low temperatures and so are often disregarded as 'good glass formers'. Without regard to fragility, high Tgs are considered better for seeds than low Tgs, since this implies stabilization without a requirement for lowered temperature. However, the prediction that phase diagrams vary for long- and short-lived seeds does not hold: Tgs for orthodox and recalcitrant seeds, alike, are between 20 and 30°C for seeds dried to about 6% water (35–50% RH at 25°C) (T_k −40°C).

A role of sugars in seed glasses has been hypothesized to explain their protective effect on membrane structure during desiccation (**see:** entries on **Desiccation tolerance**) and their accumulation in response to stress. Though sucrose stabilizes macromolecules, it tends to crystallize when simple solutions are dried. Growth of sucrose crystals in model systems is inhibited by raffinose-family oligosaccharides allowing the solution to become supersaturated and highly viscous. This model of protection by **raffinose oligosaccharides** is problematic since sucrose crystallization is already limited in the heterogeneous cytoplasm because of low concentrations, obviating the need for additional solutes such as raffinose. Raffinose alone shows no protective effects and is easily crystallized. Comparisons of model sugar glasses with intracellular glasses show that several cellular components are implicated in the glassy state. Despite the reputation of sucrose and oligosaccharides as being good glass formers, they do not appear to be the major factors determining the intracellular glass properties. A change in the oligosaccharide content in seeds does not affect the Tg and the cellular viscosity. Artificial glasses that resemble most the intracellular glasses in seeds are those made of a mixture containing sugars and **late embryogenesis abundant** (LEA) **proteins**. Thus, other cytoplasmic components, such as proteins (the types have not been discovered) interact in a complex way to form and stabilize the glassy matrix within the cells. The temperature at which glasses are formed is strongly dependent on the presence of a diluent, with water being the most ubiquitous one. Increasing the water content results in a decrease in Tg as indicated above.

The main role of glasses therefore is to maintain the structure and functional integrity of macromolecules. Due to their high viscosity, glasses permit the retention of the activity of enzymes and the conformation of proteins, and are thought to inhibit membrane fusion (**see: Desiccation tolerance – protection; Desiccation tolerance – protection by stabilization of macromolecules; Deterioration reactions**). Intracellular glasses are indispensable for seed survival in the dry state and play an important role in storage stability by severely slowing down seed ageing. In general, conditions that promote the formation of the glassy state, such as low temperatures and water contents, increase the longevity of seeds. Measurements of the molecular motion within intracellular glasses of various seed species show that the mobility of molecules is highly restricted because of the extreme high viscosity of the glassy matrix. The rates of seed ageing were found to be correlated with the molecular mobility within the intracellular glass over a wide range of water contents and temperatures. Therefore, the rates of deteriorative reactions leading to seed ageing and thus the life span of seeds are influenced by the molecular stability of the cytoplasm. (CW, JB)

Buitink, J., Hemminga, M.A. and Hoekstra, F.A. (2000) Is there a role for oligosacharides in seed longevity? An assessment of the intracellular glass stability. *Plant Physiology* 122, 1217–1224.

Buitink, J., Leprince, O., Hemminga, M.A. and Hoekstra, F.A. (2000) Molecular mobility in the cytoplasm: an approach to describe and predict lifespan of dry germplasm. *Proceedings of the National Academy of Sciences USA* 97, 2385–2390.

Leopold, A.C., Sun, W.Q. and Bernal-Lugo, I. (1994) The glassy state in seeds: analysis and function. *Seed Science Research* 4, 267–274.

Walters, C. (2004) Temperature-dependency of molecular mobility in preserved seeds. *Biophysical Journal* 86, 1–6.

Williams, R.J., Hirsh, A.G., Takahashi, T.A. and Meryman, H.T. (1993) What is vitrification and how can it extend life? *Japanese Journal of Freezing and Drying* 39, 3–12.

Xiao, L.M. and Koster, K.L. (2001) Desiccation tolerance of protoplasts isolated from pea embryos. *Journal of Experimental Botany* 52, 2105–2114.

Gliadin

Along with glutenin this is a prolamin storage protein present within the endosperm of wheat. (**See: Storage proteins**)

Global Strategy for Plant Conservation (GSPC)

This strategy was agreed to in 2002 by the Sixth **Convention of Biological Diversity** Conference of the Parties. It set out

a set of global strategy targets relating to improvements in understanding, documentation, conservation, sustainable use, awareness, and capacity building with respect to plant diversity. The targets are to be met by 2010 with countries contributing according to their capacities. With respect to seeds, targets 8 and 9 are pertinent in that, respectively, they relate to the ***ex situ* conservation** of threatened plant species, and to the conservation of crops and socioeconomic plants.
(RJP, SHL)

Globoid

An inclusion present in **protein bodies** that is electron-dense due to the presence of **phytin**.

Globulins

See: Osborne fractions; Storage protein

(β-1,3),(β-1,4)-Glucan (mixed-linkage glucan, MLG)

Unbranched (corkscrew-shaped) polymers of glucose; trimers and tetramers of β-1,4-linked glucose units spaced by β-1,3-links. They are present in the cell walls of **endosperms** of cereals, particularly in **oats** and **barley** (up to 75% of wall content, but only 3–10% of grain weight, and less than 1% in most other cereals), although environmental factors during grain growth can affect the amount present, e.g. moist conditions can result in a lowered content of mixed-linkage glucans. They are soluble in warm/hot water, forming solutions of high viscosity.

Both the mixed-linkage glucans and **arabinoxylans** in barley endosperm cell walls are problematical because of their viscosity: in the baking industry causing dough gumminess, and in **malting** by interfering with filtration and causing beer hazing. Sticky faeces results from feeding barley to monogastric animals, e.g. pigs and fowl, and for the latter this increases the cost of egg production because of the additional washing steps required. Beneficial effects of the mixed-linkage glucans, especially in oat bran consumed by humans as soluble dietary fibre, include reduction in serum cholesterol content.
(JDB)

Glucomannan

A polymeric carbohydrate (**hemicellulose** polysaccharide) with a backbone of β-1,4-linked mannose residues and single α-1,6-linked glucose side chains (see the structure of **galactomannans**, which instead of glucose have a single galactose side chain); it is prevalent in the cell walls of some monocotyledonous seeds. Uses include: as a slimming agent in food since it expands about 50-fold when consumed with water, as a dietary fibre (bulk-form laxative) and in the production of soft sponges. The common source of glucomannans is the vegetative plant (e.g. Konjac roots), not seeds. (JDB)
(See: Hemicellulose; Mannans)

Gluconeogenesis

Utilization of the storage energy in seed **oils and fats** (triacylglycerols) following germination involves the generation of free **fatty acids** and glycerol by the action of triacylglycerol lipases, conversion of the free fatty acids back to acetyl-CoA by **β-oxidation** and the conversion of acetyl-CoA to fructose 1,6-bisphosphate by the **glyoxylate cycle** and a reversal of glycolysis. Gluconeogenesis then converts the fructose 1,6-bisphosphate to sucrose by a number of reactions, the control step of which is sucrose phosphate synthase. The sucrose formed by gluconeogenesis is transported to and used by the cells of the growing seedling. (AJK)
(See: Glyoxysome; Oils and fats – mobilization)

Glucosinolate

Glucosinolates are secondary metabolites present in the Brassicacae and some other families. They are derived from the amino acids methionine, phenylalanine and tyrosine, or tryptophan, with glucose linked to the nitrogen-containing part of the molecule. They are grouped into aliphatic, aromatic and indolic glucosinolates, depending upon which amino acid they are synthesized from. *Arabidopsis* accumulates at least 23 different glucosinolates in both seeds and leaves.

Glucosinolates are stored in the vacuole and come into contact with the enzyme myrosinase upon breakage of the cells. Myrosinase de-glucosylates the glucosinolates and the breakdown products (e.g. isothiocyanates, nitriles and thiocyanates) give the typical sharp taste and odour that are identified with species such as radish, cauliflower, Brussels sprouts, broccoli and mustard. The glucosinolate/myrosinase system protects plants against herbivore attacks, and is implicated in host-plant recognition by specialized predators.

Plant breeding strategies have concentrated on reducing the glucosinolate content of rape seeds, to improve the feed quality and acceptability of rapeseed meal, and to meet increasingly stringent requirements from the processing industry. (JDB)
(See: Brassica – horticultural; Brassica oilseeds; Brassica oilseeds – cultivation; Mustard)

Grubb, C.D. and Abel, S. (2006) Glucosinolate metabolism and its control. *Trends in Plant Science* 11, 89–100.

Glumes

The two bracts at the base of a grass spikelet. **(See: Floret)**

Glutelins

Storage proteins that are present in cereal grains, which were once thought to be a distinct class of proteins on the basis of their solubility. This is no longer held to be the case; they are a type of prolamin. **(See: Osborne fractions)**

Gluten

Gluten is a mixture of prolamin **storage proteins** present in the **wheat** grain, and also in those of rye, triticale, barley and oats. **(See: Prolamins)** In wheat, the prolamin **gliadin** has been identified as a cause of coeliac disease, a serious disorder of the intestine; gliadin antibodies, for example, are commonly found in human immune complexes associated with this disease. A variety of other diseases and symptoms have been identified in patients who are gluten-intolerant.

Gluten is an important component of wheat flour, involved in the rising dough; it helps bread and other baked goods bind, and prevents crumbling, a feature that has made gluten widely used in the production of many processed and packaged foods.

A strict gluten-free diet is necessary for those who have allergies to it, and protein sources from rice, sorghum and millet, and non-cereals such as **pseudocereals** and **legumes**, which contain mostly **globulin** storage proteins in their seeds, are considered to be suitable alternatives. (JDB)

(See: Storage proteins – processing for food)

Glutenin

Along with gliadin this is a prolamin storage protein present within the endosperm of wheat. (**See: Storage proteins, 3. Prolamins**)

Glyoxylate cycle

This is involved in the conversion of acetyl-CoA (from fatty acid β-oxidation) to sucrose during seed **oil and fat** mobilization. The pathway is located in the **glyoxysomes** and **mitochondria** of germinated seeds. The first step is a condensation of acetyl-CoA with oxaloacetic acid to form citric acid, which is then converted to isocitrate and cleaved to form succinate and glyoxylate. Malate synthase then catalyses the formation of malate from the glyoxylate and a second acetyl-CoA molecule. The malate then leaves the glyoxysome for **gluconeogenesis** in the cytosol. The succinate produced by isocitrate lyase is regenerated to form more oxaloacetic acid in the glyoxysome after first being transported to the mitochondria for conversion to α-ketoglutarate and aspartate by the citric acid cycle. The mitochondrial α-ketoglutarate and aspartate are moved back into the glyoxysome where they are converted back to oxaloacetic acid. This oxaloacetic acid is then condensed with another acetyl-CoA as the cycle continues. (AJK)

(See: Glyoxysome; Oils and fats – mobilization)

Glyoxysomes

Subcellular organelles that represent a specialized class of **peroxisomes**. Glyoxysomes are commonly found in storage tissues or organs of germinated **oilseeds** where they play a central role in the mobilization of reserve oils and wax esters (Fig. G.21). They are present also in dark-grown suspension cultured cells, pollen grains and senescent organs, in which they are responsible for the mobilization of membrane lipids.

Glyoxysomes play an essential role in fatty acid catabolism, and are the site in the cell where the enzymes of the glyoxylate cycle and β-oxidation are located. (**See: Oils and fats – mobilization**) They possess (with the exception of aconitase) all of the enzymes of the **glyoxylate cycle**, a metabolic pathway responsible for the conversion of two molecules of acetyl-CoA, derived by β-oxidation of the fatty acids of stored **triacylglycerols**, into succinate. This end product of the glyoxylate cycle is eventually converted to hexose and used as an energy source by the growing seedling. A number of other notable glyoxysomal enzymes, such as ascorbate peroxidase and monodehydroascorbate reductase, are membrane-bound and are implicated in the regeneration of NAD^+ and protection of the membrane from toxic reactive oxygen species during seedling growth. Glyoxysomes also possess other enzymes characteristic of all peroxisomes, including catalase and hydrogen peroxide-producing flavin oxidases.

Glyoxysomes contain a coarsely granular matrix (usually with large crystalline inclusions) delineated by a single membrane. They range in size depending upon the organ or tissue type. For instance, in the cotyledons of oilseeds, glyoxysomes enlarge greatly in diameter from ~0.2 μm in the mature seed to 2–4 μm following germination. Glyoxysomes also display remarkable structural variability, ranging from

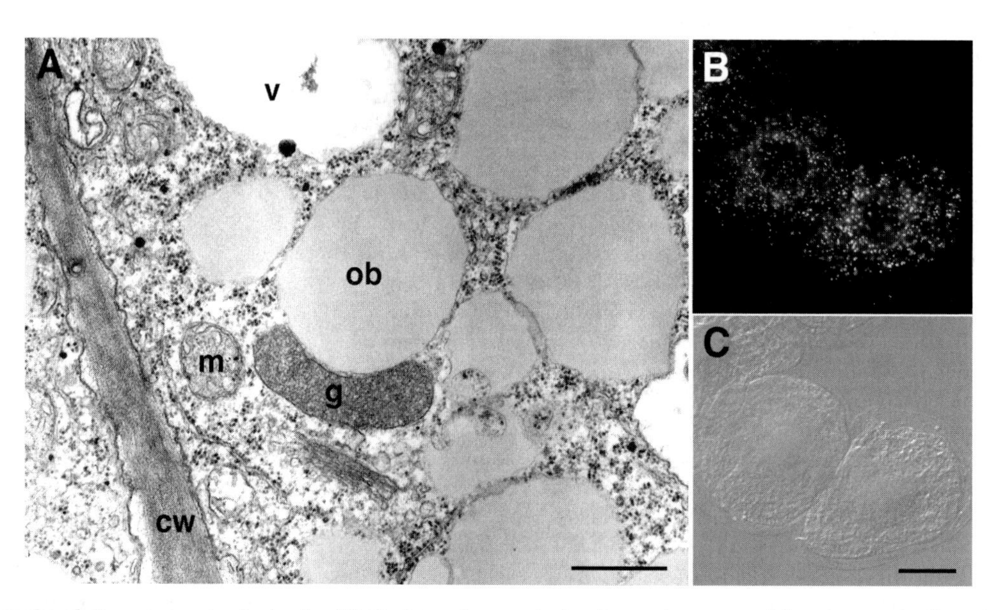

Fig. G.21. Micrographs of glyoxysomes in plant cells. (A) Electron micrograph showing a glyoxysome (g) and several adjacent oil bodies (ob) in cotyledon cells of dark-grown cucumber seedlings. Abbreviations: cw, cell wall; m, mitochondrion; v, vacuole. (B) Immunofluorescence micrograph of several tobacco BY-2 suspension-cultured cells illustrating the punctate fluorescence pattern attributable to the binding of antibodies raised against the glyoxylate cycle enzyme isocitrate lyase in the glyoxysomes. (C) Differential interference contrast image of the same group of BY-2 cells shown in B. Bar in A = 0.5 μm; in C = 10 μm. Courtesy of (A) Prof. R.N. Trelease, Arizona State Univ. and (B, C) Yeen-Ting Huang, Univ. Guelph.

nearly spherical to ovoid and they are frequently pleomorphic (of many shapes within a cell) with large invaginations, dependent upon their placement among adjacent organelles, particularly oil bodies.

1. Biogenesis of glyoxysomes

How glyoxysomes are formed and maintained within the cell has long been a matter of debate. While there is considerable evidence for the enlargement of pre-existing glyoxysomes during seedling growth, there is no information on whether they initially proliferate into nascent organelles during embryogenesis or following seed imbibition. One model for glyoxysome biogenesis in oilseed cotyledons (e.g. those of **cotton** and cucumber) is that the organelles in the mature seed begin to enlarge, but not proliferate, during post-germinative growth. During this enlargement phase of glyoxysome biogenesis, membrane lipids, including phospholipids and certain non-polar lipids, are acquired from **oil bodies** and newly synthesized matrix and membrane proteins are acquired from the cytosol. The acquisition of **membrane** proteins by the glyoxysome may also occur in an indirect manner, i.e. some newly synthesized membrane proteins can first be inserted into specialized regions of the **endoplasmic reticulum** (ER) and then be sorted to glyoxysomes via ER-derived vesicles. This proposed role for the ER as the membrane template for developing glyoxysomes, however, remains contentious.

In oilseeds where the **cotyledons** emerge from the soil and become green and photosynthetic (**epigeal**), the glyoxysomes are eventually transformed directly to leaf-type peroxisomes, and peroxisomal enzymes responsible for photorespiration replace those involved in the glyoxylate cycle. This glyoxysome-to-peroxisome transition does not occur, however, in oil-storing cotyledons that remain below the soil (**hypogeal**), nor in endosperm tissues that senesce following germination (e.g. castor bean seed endosperm and pine **megagametophytes**). Rather, glyoxysomes in these tissues are eventually degraded (by an unknown mechanism) following the depletion of **storage reserves**.

2. Import of enzymes into the glyoxysome

Glyoxysomes (like all peroxisomes) do not possess their own genome and lack ribosomes. As a consequence, matrix and membrane-destined glyoxysomal proteins are nuclear encoded, synthesized on free cytosolic ribosomes, and targeted post-translationally to the organelle in a regulated manner. At least two types of evolutionarily conserved peroxisomal targeting signals (PTSs) are present in the newly synthesized proteins which are capable of directing them to the glyoxysomal matrix. The type 1 PTS (PTS1) is an uncleaved C-terminal tripeptide motif, i.e. small-basic-hydrophobic residues, or variants thereof, that is found in most glyoxysomal matrix-destined proteins. The type 2 PTS (PTS2) is a nonapeptide motif (-Arg/Leu-X6-His/Gln/Ala-Leu/Phe-; where X indicates any amino acid) located in the N-terminus of another set of matrix proteins that are proteolytically processed after import into glyoxysomes.

Studies of the matrix protein import pathways including the identification and characterization of proteinaceous receptors (those recognizing PTS1- and PTS2-containing proteins) and several components of the translocation apparatus have revealed at least two distinct pathways that converge at the peroxisomal boundary membrane. The proteins involved in these import processes, as well as other proteins involved in different aspects of glyoxysomal (peroxisomal) biogenesis, are termed peroxins. Interestingly, glyoxysomes and other peroxisome types from various organisms and/or cell/tissues are competent to import all peroxisomal proteins, suggesting that the import pathways are highly conserved.

The mechanisms involved in the targeting and insertion of proteins into the glyoxysomal membrane appear to be distinct from those involved in matrix protein import. Glyoxysomal membrane proteins do not possess a PTS1 or PTS2 but rather utilize one or more membrane peroxisomal targeting signal (mPTS) that typically consists of a cluster of basic amino acid residues immediately adjacent to a membrane-spanning domain. Peroxins that participate in the insertion of glyoxysomal membrane proteins appear to be different from those involved in matrix protein import. (**See: Microbody**)

(RTM)

Baker, A. and Graham, I. (eds) (2002) *Plant Peroxisomes: Biochemistry, Cell Biology and Biotechnological Applications.* Kluwer Academic, Dordrecht, The Netherlands.

Huang, A.H.C., Trelease, R.N. and Moore, T.S., Jr (1983) *Plant Peroxisomes.* American Society of Plant Physiology. Monograph Series, Academic Press, New York, USA.

Mullen, R.T., Flynn, C.R. and Trelease, R.N. (2001) How are peroxisomes formed? The role of the endoplasmic reticulum and peroxins. *Trends in Plant Science* 6, 256–261.

GMO

See: Genetically modified organism

GMO – patent protection technologies

Particular forms of **genetic modification** technologies for use in transgenic (genetically engineered) crops that are designed to accomplish two goals: (i) the protection of intellectual property rights for the inserted **transgene** (inserted engineered genes); and (ii) the protection of the environment from the unintentional incorporation of transgenes into closely related non-crop species. Such genetic technologies have been generally classified as Genetic Use Restriction Technologies, or GURTs and are sometimes described as 'gene switching'. The overall strategy for any GURT is to make a value-added transgenic technology, e.g. genes for herbicide tolerance and pesticidal proteins, available to a farmer/producer for only a single generation at a time and to make that technology incapable of transfer to another plant or generation (see Fig. G.22).

Two types of GURTs have been described: variety-level GURTs (V-GURTs) and trait-specific GURTs (T-GURTs). V-GURTs protect intellectual property rights and prevent transgene escape by making the transgenic plant incapable of producing viable offspring (i.e. viable seeds), thus achieving protection at the varietal level. T-GURTs accomplish the same goals either by the specific removal of the inserted transgenic technology prior to the production of a second generation, or by its inactivation, thus achieving protection at the trait

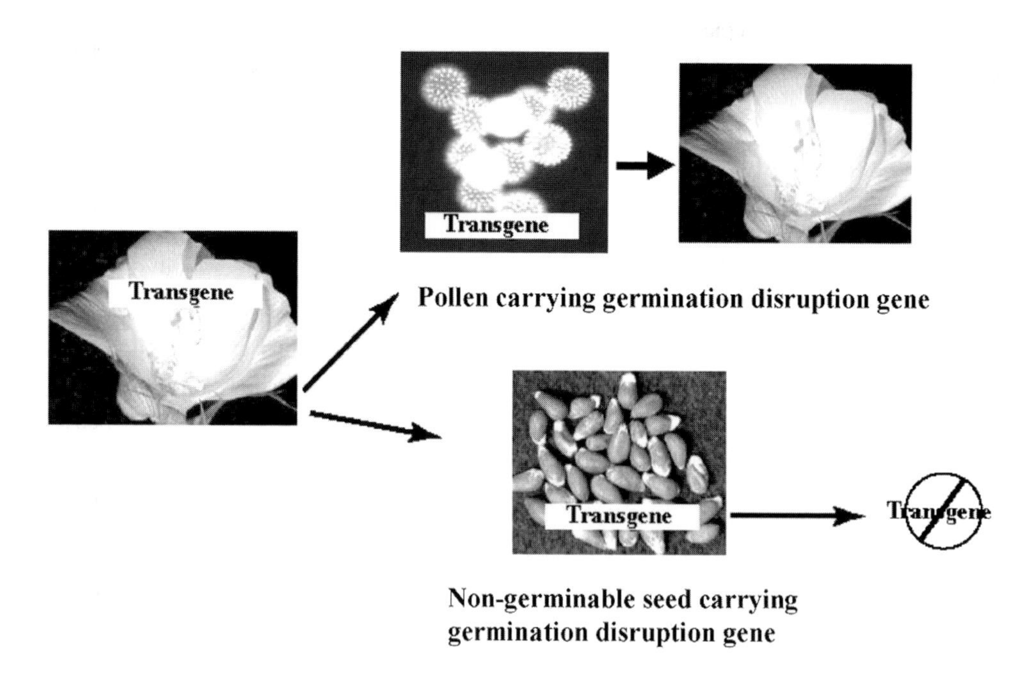

Fig. G.22. A general V-GURT strategy. Targets for this are pollen or seeds, which are only viable for one generation. Illustrated here with cotton.

(transgene) level. Patent protection technologies do not, in themselves, offer any advantageous traits to a farmer, and are only useful in crops that carry value-added transgenic technology.

GURTs in general have to be capable of induction either prior to planting, or while in the farmer's field. In the case of V-GURTs, induction activates the transgene escape prevention technology and in T-GURTs it activates the transgene and later removes or deactivates it. The seed company that sells the farmer seed has to be able to develop elite transgenic lines and generate seed stocks for sale without the GURT genetic components becoming active, since they are designed to remove or inactivate the inserted technology.

To turn GURTs on 'at will' these genetic systems take advantage of **promoters** or chemically reversible repressor protein systems to effect external control (Figs G.23–G.25). Application of a specific chemical to either the imbibing seed or growing plant results in the activation or derepression of a gene associated with the GURT resulting in the activation of a delayed germination-disruption gene, the removal of engineered genes, the activation of a value-added transgene, or any combination of these events. The physical removal of genetic elements associated with GURTs is achieved by the use of genes encoding **site-specific recombinases**, i.e. enzymes that can remove (or invert) DNA that is located between two precisely defined recognition sites. The use of a site-specific recombinase allows the chemical signal used to activate the GURT to be translated into a permanent genetic alteration – that is, activation. This has the utility of activating the GURT in a sustained fashion without the need for the constant presence of the chemical inducer.

In V-GURTs, the seed producer activates the genetic system to ensure that the seed sold will only produce one generation of plants by preventing them from producing viable seed. In the case of T-GURTs, the farmer can only receive the benefits of the inserted transgenic technology if the GURT system is activated. Ideally, the activated T-GURT will also remove the technology after the benefit has been accessed in order to prevent the possible transfer of genes to other species.

1. V-GURT strategies

In the example of a V-GURT shown in Fig. G.23 there are (a) several genes linked together (a (trans) gene cassette), including *tet-R*, which encodes a tet-repressor protein. A separate gene *CRE* (b), encodes CRE recombinase, but it is inactivated (repressed) by the tet-repressor protein that binds to the promoter region of the gene. When the chemical inducer, tetracycline (tet), is added to an imbibing seed the tet-repressor protein becomes released from the repressed *CRE* gene, and an active *CRE* gene results (c). Once the *CRE* gene is activated (derepressed) it directs the synthesis of the site-specific CRE recombinase enzyme (d). This enzyme finds its specific recognition sites in the gene cassette (a), called LOX sites (Left and Right), and physically removes the repressive *tet-R* gene DNA situated between them (e). With the removal of the repressive *tet-R* gene, the '*RIP*' gene in the modified gene cassette is now capable of encoding an enzyme that disrupts protein synthesis and thus cellular integrity. This cellular disruption results in germination of the seed being permanently inhibited (f). Examples of products of an '*RIP*' gene are cellular disruptive enzymes such as the plant ribosomal inhibitory protein, saporin, which affects the structural integrity of ribosomes, and the bacterial ribonuclease, Barnase, that degrades RNA, preventing its use in protein synthesis. An additional feature of this technology is that it must only be effective to prevent germination;

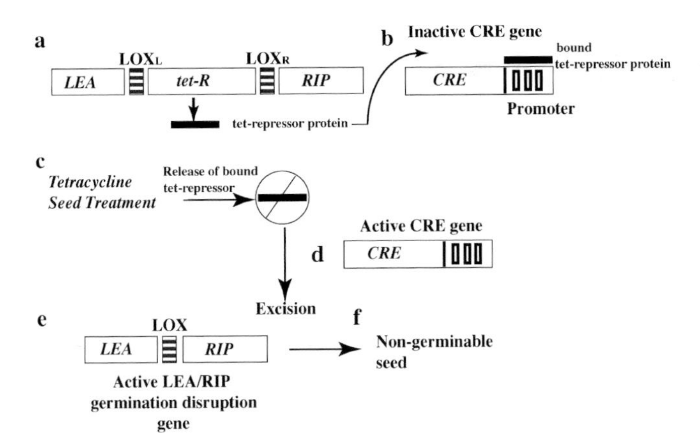

Fig. G.23. A repressible V-GURT strategy (see text for explanation).

activation of the genes at any other time by the spraying of tetracycline, for example during seedling establishment, would kill the plant. Hence the activity of the '*RIP*' gene is also under the control of a promoter that has been isolated from a **late embryogenesis abundant protein** (LEA) gene. This promoter is developmentally controlled and is active only at the very last stages of seed maturation on the parent plant, resulting in newly matured seeds that are incapable ever of germinating. Since the V-GURT constructs are carried in all the cells of the plant, the pollen it produces will also carry the germination-disruption gene. Flowers fertilized by the pollen from an activated V-GURT plant will therefore also produce seed that cannot germinate. This, in effect, makes an activated V-GURT plant an evolutionary dead end (both seed and pollen are effectively non-viable) and incapable of spreading transgenes into the environment.

Fig. G.24 presents a simpler V-GURT strategy that involves a gene promoter which can be activated directly by the application of a chemical inducer. Here the chemical inducer directly and transiently activates the promoter of the CRE recombinase gene that, in turn, by the removal of the blocking sequence, creates a permanent germination-disruption gene. The blocking sequence in this example can be any DNA fragment since its only purpose is to physically separate the promoter from the coding sequence of the gene that prevents activity. In a practical example the blocking sequence would be a selectable marker gene, useful in plant transformation protocols, that could be removed from the seed during the induction of the system prior to planting.

2. T-GURT strategies

Because they do not target the viability of seeds, T-GURT strategies are faced with the problem of either deactivating an inserted transgene, or removing it after the farmer has received the benefit of the technology. However, T-GURT strategies have the advantage of offering the farmer the possibility of 'technology on demand', i.e. the ability to activate the transgene only when required. For instance, where an inserted transgene confers insect resistance to a crop, the farmer may not be faced with insect infestations within a particular growing season. By not activating the transgene the farmer does not have to pay for a trait that he does not need. If, however, an insect infestation were to occur the farmer

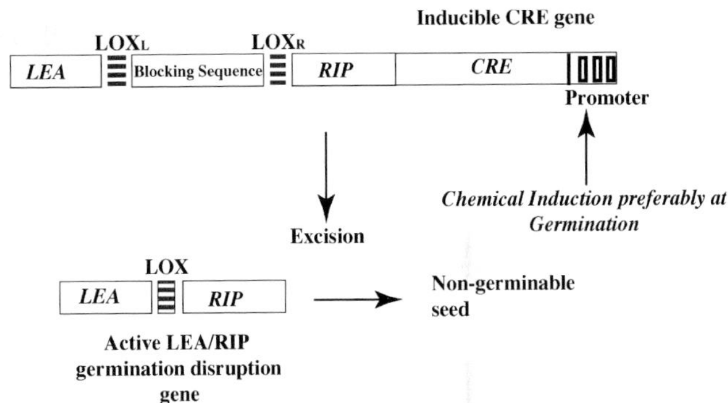

Fig. G.24. An inducible V-GURT strategy (see text for details).

would pay for the particular proprietary chemical capable of activating the transgene and generate insect resistance in the crop.

To prevent transgene escape to other species, a T-GURT should ideally also cause the permanent removal of the inserted technology and a scheme by which this might be achieved is presented in Fig. G.25. In this scheme it is conceived that there is an undefined value-added (trans)gene (VAG) attached to a constitutive plant promoter (CPP). If this gene is spliced out by a CRE recombinase, then the added value for the farmer is lost and there is no chance of the transfer of the gene to another plant. The strategy is to supply the farmer with seed already treated with a chemical (chemical inducer) that drives the production of a repressor protein during seed germination, whose gene expression is controlled by a promoter that is induced by the chemical (Fig. G.25A). This repressor protein, in turn, binds to a germination-specific promoter (GSP), which has been engineered specifically to contain recognition sites for it and thereby turns off expression of the CRE recombinase gene: the value-added gene remains active, and the crop is produced with that trait expressed.

Once the seed has completed germination and the seedling has become established, the chemical inducer is no longer present at an effective concentration, either due to its inherent instability, or dilution in the growing plant. At this point, the repressor protein is no longer induced, the germination-specific promoter is no longer repressed, and the CRE recombinase gene is capable of being produced (Fig. G.25B). However, because the CRE recombinase gene is under the control of a germination-specific promoter it is still inactive during the normal growth and production of the crop. Seed and pollen produced by the crop now contain the complete T-GURT but, the chemical inducer that controls the production of the repressor protein preventing CRE gene activation is absent, unless added by the farmer. Thus in these seed, or in seed produced by fertilization from the pollen from the T-GURT crop, the CRE recombinase is produced during germination and the transgene, located between the two LOX sites, is spliced out permanently, thus removing it from all cells in the growing plant. The producer can only avoid this outcome by buying the proprietary inducer chemical to treat the seed prior to planting. The T-GURT system, unlike the V-GURT system, requires that the transgene be closely linked to

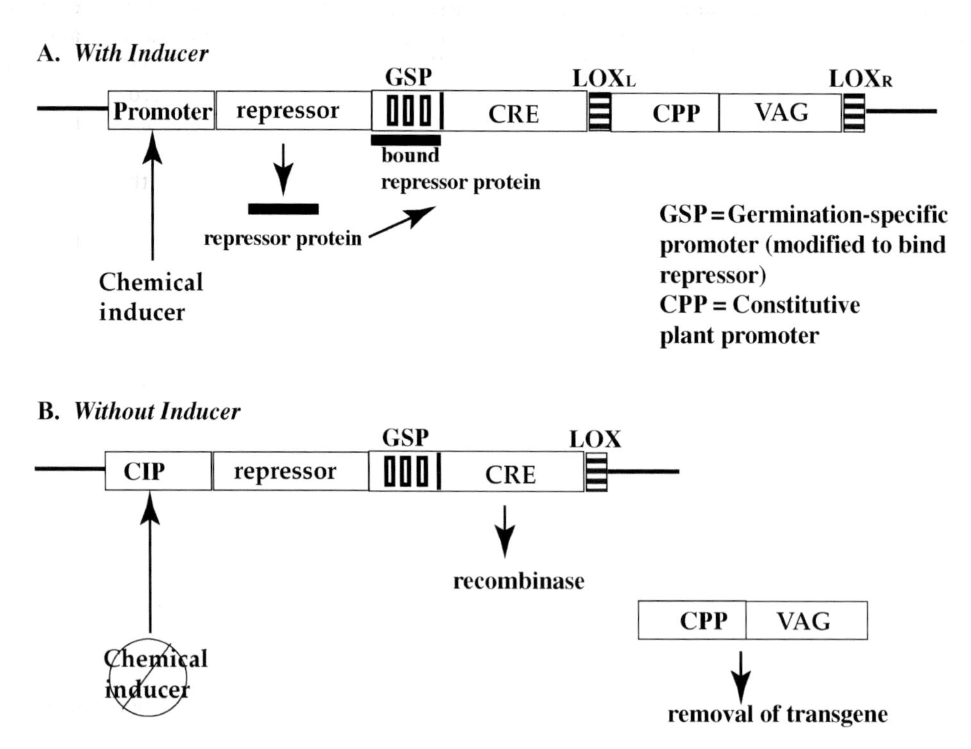

Fig. G.25. A T-GURT strategy (see text for explanation).

the T-GURT in order to prevent the transgene from segregating away from the recombinase during pollen and seed production.

3. **Controversy over GURTs**

The introduction of GURTs into the 'tool box' of agricultural biotechnology has not been without controversy. The agricultural biotechnology community has promoted GURTs as a means to prevent the spread of transgenes into the environment (biosafety issues), protect the intellectual property rights of biotechnology developers, stimulate research into improving crops that have traditionally not been profitable for breeding programmes and, by virtue of the ability to protect intellectual property, to stimulate interest in the biotechnology industry. Advocates of GURTs argue that the ability to protect technology in an agricultural setting, where such property is easily transferable as seed, will enable seed companies to introduce valuable technologies to impoverished farmers in countries where such technology is needed. The anticipation is that such technologies will generate a more productive agriculture that could replace the subsistence farming that is characteristic of many poor nations.

Opponents of GURTs have a different view of what these technologies will lead to. Environment groups have labelled these technologies as Terminator (V-GURTs) and Traitor (T-GURTs) genes. They contend that GURTs will have a negative impact on the world's food supply and on agriculture in under-developed countries. They argue that GURTs would put the control of the world's food supply into the hands of a few multinational companies, will reduce agricultural biodiversity by limiting access of farmer-based and public breeding effort to elite germplasm or desirable traits, and cause negative effects on yield in neighbouring crops as a result of outcrossing of the GURT (primarily for V-GURTs). The controversy has reached the level of the **FAO** who have commissioned several studies on the potential impact of GURTs on world agriculture, biodiversity and food security.

Many of the objections to GURTs can be overcome by regulatory means, e.g. in the USA existing regulations and policies make it difficult and economically unfeasible to use GURTs to simply protect an elite germplasm unless a value-added transgenic trait is present. This protects the access of public breeding programmes to elite germplasm for the improvement of local varieties. Some of the biological concerns will be addressed as new GURT designs are generated, although opposition to them will no doubt continue to be targeted by those opposed to genetically modified organisms (GMOs) in general. (MJO)

(See: Genetic modification; Transformation)

Jepson, I., Greenland, A.J. and Thomas, D.R.P. (1997) Cysteine protease promoter from oil seed rape and a method for the containment of plant germplasm. Patent Cooperation Treaty (PCT) patent application WO97/35983, World Intellectual Property Organization. www.wipo.int/ipdl/en.

Oliver, M.J., Trolinder, N.L.G., Keim, D.L. and Quisenberry, J.E. (1998) Control of plant gene expression. US Patent #5,723,765.

Golden rice

Half of the world's population eats **rice** (*Oryza sativa*) daily and depends on it as their staple food. Rice, however, is a poor source of many essential micronutrients and vitamins. In Southeast Asia, 70% of children under the age of five suffer from vitamin A deficiency, leading to vision impairment and increased susceptibility to disease. In tropical areas, rice is typically milled to remove the oil-rich **aleurone layer** to reduce rancidity during storage. The remaining edible portion, the **endosperm**, like the germ (embryo), lacks provitamin A (the plant carotenoids that are vitamin A precursors). Genetic engineering (**genetic modification**) techniques have thus been undertaken to produce rice grains containing β-carotene, the major precursor of vitamin A.

Immature rice endosperm can synthesize the intermediate compound in carotenoid biosynthesis, geranylgeranyl diphosphate (GGPP) (Fig. G.26). GGPP can be used to produce phytoene, an uncoloured carotene, by the enzyme phytoene synthase. The synthesis of β-carotene from phytoene requires complementation with three additional plant enzymes: phytoene desaturase and β-carotene desaturase, each catalysing the introduction of two double bonds, and lycopene β-cyclase, encoded by the *lcy* gene. The β-carotene biosynthetic pathway was introduced into the rice endosperm via *Agrobacterium*-mediated introduction of the genes for these enzymes: the genes for phytoene synthase, phytoene desaturase and lycopene β-cyclase were obtained from daffodil and the gene for carotene desaturase was of bacterial origin.

These engineered seeds could not provide enough provitamin A to satisfy the recommended dietary allowance with a normal daily ration of rice. But β-carotene production has been increased 23-fold over the first engineered values by incorporation of phytoene synthase from maize rather than from daffodil. The production of various carotenoids other than β-carotene could provide additional health benefits since carotenoids have been implicated in reducing the risk of certain types of cancers, cardiovascular disease and age-related macular degeneration. Fortunately, excess dietary β-carotene, in contrast to excess vitamin A, has no harmful effects, so plants with enhanced β-carotene content should be a safe and effective means of vitamin delivery. (AJR)

(**See: Transformation technology**)

Al Babili, S. and Beyer, P. (2005) Golden rice – five years on the road – five years to go? *Trends in Plant Science* 10, 565–573.

Fray, R.G., Wallace, A., Fraser, P.D., Valero, D. and Hedden, P. (1995) Constitutive expression of a fruit phytoene synthase gene in

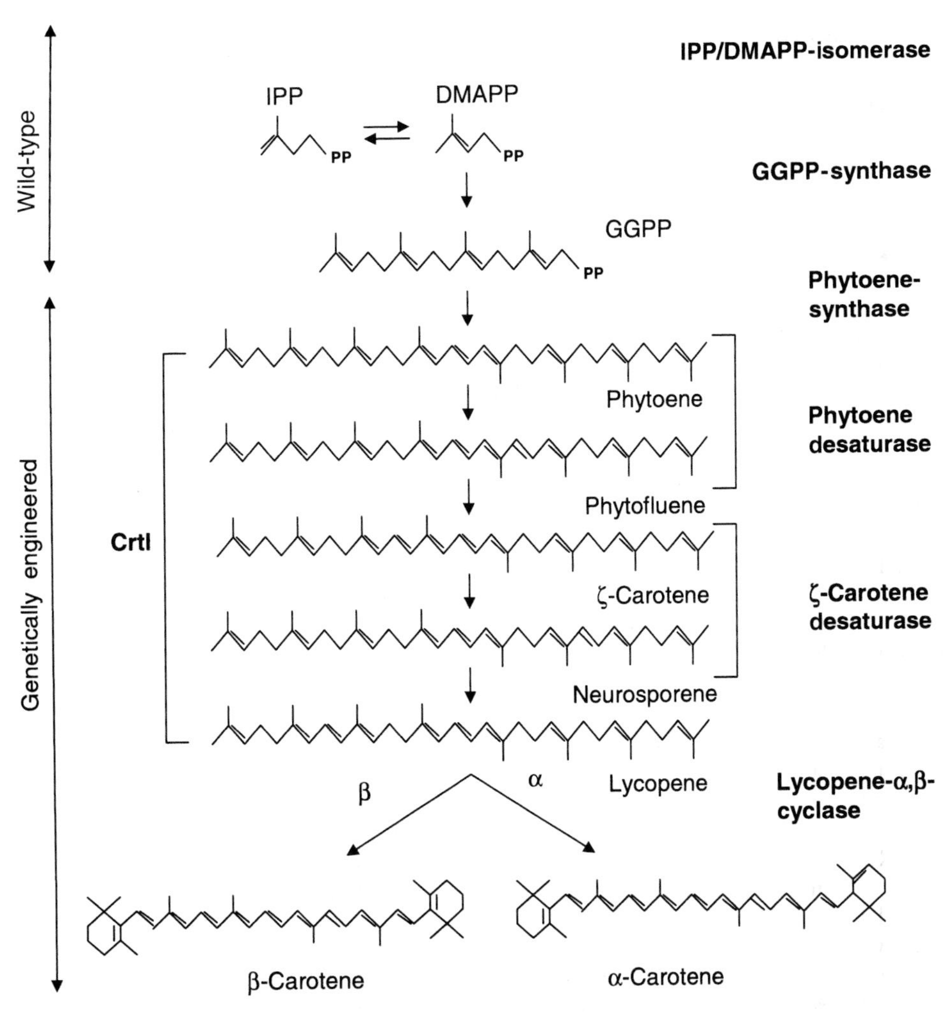

Fig. G.26. The metabolic pathway for the synthesis of β-carotene in Golden Rice, showing the constitutive enzymes (wild type) present in the rice grain, and those introduced by genetic engineering. Based on Ye, X., Al-Babili, S., Kloti, A., Zhang, J., Lucca, P., Beyer, P. and Potroykus, I. (2000) Engineering the provitamin A (β-carotene) biosynthetic pathway into (carotenoid-free) rice endosperm. *Science* 287, 303–305.

transgenic tomatoes causes dwarfism by redirecting metabolites from the gibberellin pathway. *The Plant Journal* 8, 693–701.
Paine, J.A., Shipton, C.A., Chaggar, S., Howells, R.N., Kennedy, M.J., Vernon, G., Wright, S.Y., Hinchcliffe, E.J., Adams, J.L., Silverstone, A.L. and Drake, R. (2005) Improving the nutritional value of Golden Rice through increased pro-vitamin A content. *Nature Biotechnology* 23, 482–487.
www.goldenrice.org

Grading

In seed and grain processing technology, the separation of produce into different categories according to size, weight, colour, quality. The screening and other processes for separating seeds of uniform size from oversize and undersize fractions in seed conditioning, for example, using perforated screens in an air-screen machine. The Precision Grader is a particular design of equipment that separates seeds by width or thickness using slotted- or round-hole cylinder shells. (**See: Conditioning, II. Cleaning**)

Grafted seedlings

In seed technology, a technique for the production of seedlings by attaching together two varieties to take advantage of their respective traits and properties as scion and rootstock in the growth, development, survival and yield of the joined plant.

Grafting techniques have long been used in the propagation of long-lived rubber, ornamental, vine and top-fruit shrubs or trees, using either **clonal** or seedling rootstocks. Techniques devised for young seedlings became prominent in the cultivation of annual horticultural greenhouse-grown fruit-bearing vegetables in Japan about two decades ago, and in very recent years also have become popular in Europe and elsewhere. Their main objective is to avoid soilborne diseases such as *Fusarium* wilt in **cucurbits** and bacterial wilt, nematodes and other soilborne diseases in Solanaceae, especially in intensive growing situations where cropping is repeated frequently or where there is a risk of flooding in outdoor production systems. A number of varieties have been specifically bred for use as rootstocks. Inter-specific grafting is generally applied, such as with eggplant and tomato. But inter-generic grafting is also used, such as for cucumber grafted on to pumpkin rootstock, watermelon grafted on to bottle gourd, melon grafted on to white or wax gourd, and tomato grafted on to eggplant.

Grafting is typically done when young plants have developed two to three true leaves – that is, typically at about 2 weeks old for cucumber and melon, 3 to 4 weeks old for tomato, or somewhat older still for eggplant. Tomato plants are, for example, produced by 'cleft grafting': the stem of the scion is cut into a tapered wedge, which is fitted into a cleft cut in the end of the rootstock. Alternatively 'tube grafting' is used for younger vegetable seedlings grown in plug trays: the scion and rootstock are each cut at a slant and spliced together. 'Tongue grafting' is used for cucumber and melon plants: the hypocotyl of the scion remains attached to its root until the graft union has formed, after which the connection is cut. Seeds of the scion and rootstock varieties may need to be sown up to about a week apart to make sure that their stems are of similar diameter, according to their relative germination and seedling growth rates. In some cases it is possible to obtain more than one scion from a single seedling, such as in the propagation of hybrid tomato and pepper varieties. Success depends upon maximizing the area of the cut surfaces being spliced, upon pressing and holding the cut surfaces securely together, usually with a plastic clip, and upon a high degree of cleanliness to avoid infections. Grafted plants are generally 'healed and acclimated' under cover in shade and high relative humidity, which are gradually reduced over a period of about a week. Efforts have been made to mechanize parts of these laborious and time-consuming manual processes. The field management of grafted plants is generally similar to the management of non-grafted plants, providing that for example the graft union is kept above the soil-line. (**See: Transplants**)

(PH)

Black, L.L., Wu, D.L., Wang, J.F., Kalb, T., Abbass, D. and Chen, J.H. (2003) Grafting tomatoes for production in the hot-wet season. *Asian Vegetable Research and Development Center*, Publication 03–551.
Lee, J.M. (1994) Cultivation of grafted vegetables. 1. Current status, grafting methods, and benefits. *HortScience* 29, 235–239.

Grain amaranth

See: Pseudocereals

Grain and seed

It is not always clear in what sense 'grain' is used in the literature, in relation to 'seed'. According to the commodity-trading definition used by the FAO, for instance, 'grain' is mature seed, commonly of the **cereals** and **legumes**, that is used for consumption by humans or animals, directly or for malting, milling or other further processing, and not for planting. In this sense, grain may or may not be capable of germination under favourable environmental conditions and of growing into a normal plant. Some authors and legislatures preferentially reserve the term 'seeds' to that which is marketed and used for sowing. But in seed research circles, the two words are often used interchangeably though grain may be preferred if used to describe a dispersal unit such as a **caryopsis** (i.e. strictly a fruit).

Within the cereals, 'coarse grains' exclude rice and wheat, which are called 'fine grains'. In the Indian subcontinent, whole unsplit pulses (that is 'whole grains' of **grain legumes**) are collectively named 'gram' (a word derived from the Portuguese for grain, *grão*): Bengal gram (chickpea), green or golden gram (mung bean), red gram (pigeon pea) and white gram (soybean). Cereal kernels were the original basis of the 'grain' standards in **weights and measures**, and the unit is retained still for precious metals.

For the food and health industries and its consumers, 'whole grains' have been defined (by the American Association of Cereal Chemists, 1999) to consist of 'the intact, ground, cracked or flaked caryopsis, whose principal anatomical components – the starchy endosperm, germ and bran – are present in the same relative proportions as they exist in the intact caryopsis'. (PH)

Grain inspection

As a condition of national and international trade, dry bulk cargoes of grain are routinely inspected by authorized personnel, who examine and test official samples and lot records, and certify results. For example, the Official USA Standards for Grain stipulate the **sampling** equipment (probe samplers and triers), patterns, ratios and sizes, and procedures for measuring **dockage**, debris, infestation with injurious **storage pests** (such as **weevils**, grain borers, insect larvae, bran bugs and moths). Heating in the grain is indicated by sour or musty odours, resulting from excessive respiration, mould growth or fermentation, leading to deterioration in grain quality and value. (**See: Storage management**) Quality Grade standards are specified in terms of size (minimum weights per volume), degrees of damage, and foreign or broken material (Table G.14). **Wheat**, for example, is divided into eight classes based on colour and length of the kernel and other characteristics, such as the shapes of the kernel, germ, crease and brush (Table G.15). (PH)

Wrigley, C.W. (ed.) (1995) *Identification of Food Grain Varieties*. AACC Press, St Paul, MN, USA.

Grain legumes

Those species of legumes whose seeds are used for human food and in some cases as animal feed. Examples are the **beans, peas, soybeans, chickpeas, lentils (see: Legumes)**.

Summerfield, R.J. and Roberts, E.H. (eds) (1985) *Grain Legume Crops*. William Collins, London, UK.

Grapeseed

1. The crop

The grape plant (*Vitis vinifera*, 2n = 38, Vitaceae) is a perennial vine primarily grown (90%) for wine making. Other species are used for table grapes, sources for jam and juice or dried to produce raisins. Grapeseed is a by-product of the wine industry and contains 6–20% oil, depending on the cultivar and growing environment (**see: Oilseeds – major**, Table O.2). Oil is produced locally in vine-producing regions but no statistics on world production are available. The largest production is from countries bordering the Mediterranean, including Italy, France, Spain and Turkey. In addition, Australia, USA, China, Iran, Argentina and Chile are among the world's top producers.

2. Origin

Domestication of the cultivated grape vine is thought to have occurred about 4000 BC from a single wild species inhabiting an area from northeastern Afghanistan to the southern shores of the Black and Caspian seas. From there the vine was exported throughout the Mediterranean, Europe, China, India and the New World. However, European vines did not establish well in early America and native species were selected and cultivated. These North American species also provided insect-resistant rootstocks for European vines and parents for hybrid varieties.

3. Fruit and seed

The fruit is a juicy, round, smooth berry, black, red or yellow-green in colour normally containing less than four seeds. The fleshy **pericarp** consists of a thin epicarp and hypoderm of tannin containing polygonal cells, a thick **mesocarp** of thin walled parenchyma, fibro-vascular bundles and blast cells with a thin **endocarp** of parenchyma cells.

Grape seed is brown, pear-shaped and about 5–6 mm long, with two distinct grooves on one side running between the **raphe** and the **chalaza** (Fig. G.27). The **micropyle** is on the opposite side and a fatty cuticle covers the epidermis. The soft outer region of the testa is made up of parenchyma, some with raphides, others with brown pigment while vascular tissue occurs in the raphe region. The hard region of the testa contains radially elongated, pitted brown stone cells. The **perisperm** consists of a single layer of flattened thin walled cells. The **endosperm**, with its **oil body**- and **protein body**-containing parenchyma, fills most of the ovule. The small **embryo** is embedded in the endosperm at the **hilum** end. The cells of the embryo are smaller but similar to the endosperm in type and content.

4. Uses

Grapeseed oil, after refining, is mainly used for edible purposes, as a salad oil or sprayed on raisins to improve their appearance and pliability. (KD)
(**See: Oils and fats; Oilseeds – major** Table O.2, for composition)

Olmo, H.P. (1976) Grapes. In: Simmonds, N.W. (ed.) *Evolution of Crop Plants*. Longman, London, UK. pp. 294–298.
Vaughan, J.G. (1970) *The Structure and Utilization of Oil Seeds*. Chapman and Hall, London, UK.

Table G.14. Quality grade standards for wheat as determined by GIPSA (Grain Inspection, Packers and Stockyards Administration), USA.

	Basis of determination		
Lot as a whole	Factors determined before the removal of dockage	Factors determined after the removal of dockage	Factors determined after the removal of dockage and shrunken and broken kernels
Distinctly low quality	Distinctly low quality	Ergot	Class
Heating	Heating	Kind of grain	Damaged kernels
Infestation	Infestation	Odour	Foreign material
Odour	Kind of grain	Protein	Heat-damaged kernels
	Moisture	Shrunken and broken kernels	Subclass
	Odour (including smut)	Smut	Wheat of other classes
	'US Sample Grade' factors	Stones	
	Other unusual conditions	Test weight	
		Treated	

Table G.15. Official USA Inspection Standards for Grain.

Classes	Subclasses	Number of numerical grades into which each Class and Subclass is divided[a]
Wheat		
Hard Red Spring	based on proportion of dark, hard and vitreous kernels	5
Hard Red Winter Soft Red Winter	(no subclasses)	
Durum	based on proportion of hard and vitreous kernels of amber colour	
Hard White Soft White Unclassed Mixed	(no subclasses)	
Barley		
Malting Barley	Six-rowed, Six-rowed Blue, Two-rowed	5 and special grades[b]
Barley	Six-rowed, Two-rowed, and 'Barley'	
Soybean		
Yellow Mixed	(no subclasses)	4
Sorghum		
Sorghum Tannin sorghum White sorghum Mixed sorghum	(no subclasses)	4
Oats **Rye** **Triticale**	(no classes or subclasses)	4 and special grades[b]
Canola		3
Sunflower		2

[a]And also 'Sample Grade'.
[b]To emphasize special qualities or conditions affecting value added to and made a part of the grade designation.

Grasses – forage and turf

The family Poaceae can be divided roughly into two groups that contain the species that are recognized today as 'forage and turf' grasses: the temperate, cool-season tribes and the warm-season, tropical and subtropical species. Fossil evidence from grazing animals suggests that grasslands were evolving from the Eocene through to the Pleiocene age. These ancestral grasses evolved initially under tropical conditions, and developed into plains, steppe and desert forms as climates changed. For a brief account of the uses of forage or herbage grasses, which are often grown in combination with forage legumes, **see: Grasses and legumes – forages.**

(a) *Temperate grasses* are used throughout the world, and the main genera include *Agropyron*, *Agrostis*, *Arrhenatherum*, *Bromus*, *Cynosurus*, *Dactylis*, *Festuca*, *Holcus*, *Lolium*, *Phalaris*, *Phleum* and *Poa*. Species include the prostrate and rhizomatous (e.g. browntop) types, those that are compact with prolific tiller production (e.g. perennial ryegrass), those which have both many small prostrate tillers, some plants with a few large, erect tillers (e.g. cocksfoot), and erect large-tillered species (e.g. prairie grass). All are used for pasture production; perennial ryegrass, Kentucky bluegrass, chewings fescue, red fescue, hard fescue, tall fescue, browntop and colonial bent are

Fig. G.27. Grapeseed (image by Mike Amphlett, CABI).

also used for turf and amenity purposes. A forage cultivar may have good spring and/or summer performance, resistance to crown rust and winter hardiness, while a turf cultivar of the same species may tolerate close mowing, have good wear resistance and a dark green colour.

The international forage seed trade is dominated by ryegrasses (*Lolium* spp., 50–70% of total annual production), fescues (*Festuca* spp.) and bluegrasses (*Poa* spp.). Over 90% of the temperate grass seeds marketed internationally are produced in North America (USA and Canada), Europe (mainly Denmark, Germany and The Netherlands), New Zealand and Australia. Temperate grass seeds are also produced in South America (Argentina and Brazil), the Middle East (Turkey and Iran), Asia (India, Japan, China) and Russia.

Most temperate genera (e.g. *Lolium*, *Dactylis*, *Festuca*, *Poa*, *Holcus*, *Phleum*) are native to Europe, but some (e.g. *Bromus*, *Cynosurus*, *Agropyron*) are sourced to Eurasia, or are North American in origin (some species of *Agropyron* and *Poa*), or are native to South America (some *Bromus* spp.). These grasses range from species with 100 or more different cultivars (e.g. perennial ryegrass) to those with less than ten (e.g. prairie grass).

(b) *Tropical forage grasses* are mainly used in Australia, tropical Latin America (Brazil, Colombia, Venezuela, Cuba), the southern USA, Asia (India, Thailand, Japan) and Africa (Kenya, Zimbabwe, South Africa, Côte d'Ivoire). These countries also produce tropical forage grass seed crops, with Australia, Brazil, Thailand and Kenya predominating internationally. There are many genera, including *Andropogon*, *Axonopus*, *Bothriochloa*, *Brachiaria*, *Cenchrus*, *Chloris*, *Cynodon*, *Dichanthium*, *Digitaria*, *Echinochloa*, *Eragrostis*, *Hemarthria*, *Hymenachne*, *Melinis*, *Panicum*, *Paspalum*, *Pennisetum*, *Setaria*, *Sorghum*, *Stenotaphrum*, *Tripsacum* and *Urochloa*. But for many of these genera only a single species and/or cultivar has been commercially released. In the international seed trade, the most important species are *Brachiaria brizantha*, *B. decumbens*, *B. ruziziensis*, *Cenchrus ciliarus*, *Chloris guyana*, *Cynodon dactylon*, *Eragrostis tef*, *Panicum maximum*, *Paspalum notatum*, *Pennisetum clandestinum*, *P.* hybrids, and *Sorghum halepense* (**see: Sorghum**). Some, though (e.g. *Pennisetum clandestinum*), are propagated vegetatively rather than by seed.

The tropical grasses range from 3–4 m tall (e.g. *Andropogon gayanus*) down to 50 cm or less (e.g. *Digitaria didactyla*) with intermediates 1.5–2 m tall (e.g. *Chloris gayana*, *Setaria sphacelata*). Most are tussock-forming, but some spread by stolons (e.g. *Bothriochloa insculpta*) or underground rhizomes (*Pennisetum clandestinum*). The majority are strongly perennial. All are used for pasture production, but *Paspalum notatum* (bahiagrass) and *Pennisetum clandestinum* (kikuyu) are also used as turf grasses. Nearly all the tropical grasses are native to Africa, particularly East Africa. The exceptions are the *Paspalum* spp. which are South/Central American in origin, and *Cenchrus ciliaris* which has been sourced to Southeast Asia as well as Africa.

1. Genetics, breeding

In the latter part of the 20th century, extensive breeding of temperate forage grasses has occurred, mainly by private companies, and large numbers of cultivars are available. Many of these cultivars are protected by Plant Variety Rights (PVR) legislation, and their genetic purity is guaranteed through seed certification (such as the Herbage Scheme of the **OECD Seed Scheme**).

Most temperate forage grasses are cross-fertilizing, natural **polyploids**, and are **auto(allo)polyploids**. *Lolium perenne* and *L. multiflorum* (perennial and annual/Italian ryegrass) are natural **diploids**, but induced autotetraploids have been developed and bred into cultivars. From the seed production point of view, these grasses are still considered to have several 'wild' characters: small seed size, seed shedding and uneven ripening. While there is ample genetic variation for seed yield, selection cannot be at the expense of vegetative yield and quality.

Tropical forage grasses are mostly either apomictic or cross-pollinating polyploids. **Apomixis** is widespread and may be obligate or facultative; high levels tend to be associated with high levels of polyploidy. Over 65 tropical grass taxa have been commercialized, but one-third of these are represented by a single cultivar. Relatively little plant breeding has occurred. Many 'cultivars' are local genotypes or unselected/informally released material, and in some cases are genetically identical. The formal release of new cultivars is relatively recent. Few are included in seed certification schemes (as there is little risk of genetic drift in apomictic species), but at least in Australia, new proprietary cultivars are protected under PVR.

2. Development

(a) *Floral induction.* The transition from vegetative to reproductive growth in temperate grasses is controlled primarily by photoperiod and temperature. Most perennial species require short days and/or **vernalization** before they can respond to increasing daylength and temperature in the spring. Some species (e.g. cocksfoot) must pass through a seedling juvenile stage (a certain number of leaves or shoot size) before they can respond, whereas others (e.g. perennial ryegrass) can be induced during the germination phase. The time at which the processes of floral differentiation occur depends firstly on species, then cultivar, and finally region of production, being delayed as latitude decreases.

In tropical grasses, daylength responses depend on the environment from which the species originated. Qualitative **short-day plants** require a specific daylength (such as 11 h), whereas a quantitative daylength response allows plants to initiate flowering at daylengths that approach the specific daylength. In addition, flowering responses can be modified by temperature. Requirements for floral induction vary among species, and often also among cultivars of the same species. The time from floral initiation to ear emergence is often more rapid than in temperate species.

(b) *Pollination and seed-set.* Cross-pollination occurs in most temperate forage grasses, although self-pollination occurs in *Bromus tectorum* and *Lolium temulentum* (and at low frequency in other species), and some *Poa* spp. are apomictic. The anemophilous (wind-pollinated) flowers are often small and inconspicuous, producing large quantities of dry, low-density pollen. Pollen release and transfer are reduced by lodging of the crop and environmental conditions, such as low temperature. Pollination and fertilization may be only 30%

successful. This loss of potential yield is further compounded by ovule degeneration (seed abortion), which occurs within 7 days following fertilization, probably because of genetic or cytological factors, but also possibly because of resource limitations: losses may be up to 30% of seeds initially set.

Tropical forage grasses are mostly either cross-pollinating or apomictic (partial or facultative; obligate or near-obligate) though some annuals are self-pollinating. In *Pennisetum* spp. cross-pollination is favoured by non-synchrony of pollen release and stigma receptivity. Wind pollination predominates, but there is insect pollination in some species, including by bees. Pollen is released in large quantities, and the large, branched grass stigmas maximize the chance of receiving pollen. Seed set is highly variable, and often is very low.

Seed development proceeds in three stages: growth (I), a food reserve accumulation (II) and maturation or ripening (III) (**see: Development of seeds – an overview**). The duration of each stage will depend on species, cultivar and the environment, but for perennial ryegrass the duration of stage I is around 10 days, stage II 10–14 days, and stage III 3–10 days, and **physiological maturity** is therefore reached at around 25 days. The time taken for seed development in tropical grasses varies: for example, it is around 7 days for *Panicum maximum*.

3. Production

For species that require vernalization, autumn sowings must be early enough for plants to become established and tiller strongly before cold temperatures inhibit growth or cause frost damage. Slow establishing species such as tall fescue, cocksfoot and Kentucky bluegrass struggle to produce sufficient tillers if autumn sown, and are normally sown in spring or late summer for harvest 12–15 months later.

Crops can be drilled into a conventionally prepared seedbed, or may be direct-drilled (no-tillage). Sowing rates vary from 5 to 25 kg/ha (similar yields are produced from a range of sowing rates), and row spacings vary from 15 to 60 cm depending on species and equipment design and availability. In countries with severe winters, a companion crop (e.g. cereal) is often used to provide protection for the grass seedlings during establishment.

For tropical species, sowing time usually relates to the reliability of rainfall during the growing season, so that spring/summer sowings are common. Crops can be drilled, but stoloniferous grasses such as *Chloris gayana* can also be broadcast. Sowing rates vary with species, but a general recommendation is for double the normal pasture-sowing rate. Row spacings are also species-dependent, ranging from 20 to 100 cm. (Note also that some tropical grass seed crops are harvested as opportunistic stands from established pasture.)

Most crops are grown as certified seed, and the field must therefore meet all certification requirements (previous crops and isolation distances). Seedbed preparation involves removal of existing vegetation (cultivation or herbicides), and cultivation to provide a firm, well-worked soil, free from large clods.

(a) *Crop management.* If needed, fertilizer may be applied during seedbed preparation or immediately prior to sowing. Nitrogen (N) may or may not be required at this stage, depending on the previous crop and species to be sown, but may be required during vegetative growth, particularly for N-depleted fields or in countries with a short growing season depending on the species and field history and, for example, upon plant-N status at around the time of stem elongation. In slow-establishing tropical crops or those where establishment has been poor, tiller or stolon production can be encouraged by N application.

In temperate crops, if vegetative growth is abundant prior to winter, the crop may be either grazed or cut. Reproductive growth in temperate grasses begins in early spring with floral initiation and the development of spikelet primordia at the shoot apex, which is still at the base of the plant. All **defoliation** (grazing or cutting) should stop before the stem elongates, to avoid damaging the reproductive meristem. If lodging is likely to occur in the seed crop, the use of a registered **plant growth regulator** (PGR) can prevent or delay it by reducing stem internode length.

The management required for tropical grasses often depends on species, or sometimes cultivar, because of the different growth habits. However, the overall aim is to produce a synchronized and short flowering period, which in most cases involves a combination of defoliation (grazing or cutting) and N fertilizer (which most commonly is all applied at closing, when defoliation is stopped). Defoliation requirements range from severe to lax depending on the species. With day-neutral grasses, defoliation should be timed to match expected growing conditions (e.g. minimize risk of moisture stress); short-day grasses only flower at specific times of the year, and defoliation must be timed so that the maximum number of reproductive tillers are available at this time.

Control of weeds in the establishing crop is important for both eventual seed yield and quality. There is a near total reliance on either pre- or post-emergence herbicides in temperate grass seed production, which are also commonly used for tropical species, although in some countries weeds are controlled by hand weeding and hoeing. Foliar and stem diseases which occur between ear emergence and seed physiological maturity may severely reduce seed yield, and fungicide either at or just before anthesis may be necessary.

(b) *Yield management.* Seed production crops should aim to establish and utilize the seed **yield potential**, defined as the number of florets per m^2 at anthesis. Usually the greatest potential yields (the total seed weight per unit area) are obtained by maximizing seed number, which in turn depends on the number of fertile tillers produced per unit area, the number of **spikelets** per tiller, the number of **florets** per spikelet, and the FSU. Utilizing the potential yield depends on the success of floret site utilization (FSU – the proportion of florets present at anthesis that produce a seed) and final seed size.

In common with other gramineous crops, seed yield in forage grasses increases as a function of increasing fertile tiller numbers up to a maximal point above which there is no further appreciable response. The numbers required depend on species, and range from 600–900/m^2 in tall fescue and cocksfoot to 1800–3000/m^2 for agricultural or 2000–3500/m^2 for turf ryegrasses. Crop management should be aimed initially at ensuring that these numbers are not limiting factors: fertile tiller production is increased by N fertilization, but can be decreased by late sowing, very high sowing rates, severe and/or late defoliation, lodging, weed competition and moisture stress. Cultivars within a species may also differ markedly in their tillering ability. Seed yield is not usually greatly influenced by the number of spikelets or florets per tiller, although spring-formed tillers will have fewer spikelets and florets than those formed in the autumn, because the plant has had a reduced time to accumulate spikelet and floret primordia. Late sowing and defoliation can therefore reduce spikelet and floret number. Cultivars also vary in the numbers of spikelets and florets produced and the number of florets shed from the spikelet between anthesis and harvest.

Low FSU is the most important limiting factor for high seed yield: it commonly ranges from 10 to 30%, and is often <20%. Losses occur during pollination, fertilization and because of poor seed set. Management practices that avoid stress (e.g. nitrogen shortages, moisture deficits, pathogen attack or lodging) on the plant during seed set and development can lead to increased FSU. Soil water deficits of more than 100 mm after stem elongation can reduce floret site utilization and prevent responses to nitrogen; if irrigation is available, scheduling should be such that moisture stress is prevented, particularly between ear emergence and the end of anthesis. Excess water, however, promotes the production of vegetative tillers, which compete with reproductive growth and seed development.

(c) *Harvesting, drying and storage.* For temperate forage grasses harvesting should begin when the yield of viable seed is at a maximum, before any seed **shattering** occurs. Delay of only a few days can result in substantial seed losses, as many grass species have poor seed retention. Seed moisture content (mc) of around 40–45% is the most reliable indicator.

Grass seed crops are cut and left in the **swath** to dry, either for 2–3 days during which time mc will fall to around 25%, or for 4–6 days during which mc will fall to around 14%, before threshing. For the former, seeds will need to be dried post-threshing. Counter-rotating tined belt pickups in a range of designs are attached to combines for grass seed crops. The crop may also be direct-combined, usually at around 43% mc, after which seed must be dried to a storage moisture content of <14%.

Drying can be achieved by blowing either ambient or heated air through the seed lot; the maximum safe drying temperature is around 32°C when mc is >20% and it should not exceed 40°C at 11–13% mc. For temperate grasses, a mc of 8–12% is considered 'safe' for storage for many years providing temperature does not exceed 25°C, and insect and rodent pests are excluded. If seed is harvested at >15% MC, it may be dry on its outside, but have a wetter interior which can encourage the activity of storage fungi (*Aspergillus* and *Penicillium* spp.), whose metabolic activities cause rapid heat production, raising the seed lot to around 55°C, when germination losses occur. A seed lot can be killed in 12–15 h under these conditions. Seed harvested at 13–14% mc on a hot, sunny day may be 10–12°C hotter than the ambient air temperature (radiant heat); if such seed is stored in bulk, heat is retained in the mass, with subsequent seed damage through the activities of storage fungi. (**See: Storage management**)

In tropical grasses, flower heads may be produced over a 2–6 week period, individual heads may flower progressively over several days to weeks, and mature seeds are shed progressively as they ripen. Actual yields are therefore often only half of the potential seed harvest. Harvest should occur when large numbers of seeds on the standing crop are close to shedding. A single, destructive direct harvest of the standing crop is most commonly used, either with a combine or by hand cutting with sickles. Alternatively, the crop can be cut and left in a swath or stook for a few days to dry before threshing. In Australia and Brazil mechanized harvesting is usual, but in countries such as Kenya and Thailand hand harvesting is used. Alternatives to a single destructive harvest include multiple hand harvests from the standing crop (leaving immature seeds to ripen), and manual sweeping to recover shed seed from the soil surface.

Most freshly harvested seeds require drying, and direct combined seed lots in particular may have a mc of 50–60%, which must be reduced quickly to avoid rapid deterioration. In smallholder systems, seed is sun-dried by spreading it out in shallow layers followed by regular turning to avoid heating. Artificial drying using heated air is used for large seed lots; drying air temperature should not exceed 35–40°C. Tropical forage grass seed can be stored for several years providing MC does not increase and temperature is kept below 25°C. Unlike a temperate environment where ambient conditions can provide a natural storage environment, tropical species usually require storage in a controlled environment if viability is to be maintained. Seeds of some tropical forage grasses are naturally relatively short-lived (2–3 years).

Seed weight is traditionally recorded as **thousand seed weight** (TSW), and as it is recorded after seed processing has removed small and light seeds it is relatively constant and not significantly related to either seed number or yield.

(d) *Postharvest management.* In perennial grasses, reproductive tillers for the next season's harvest are produced in the autumn and early winter, and accumulated straw, debris or stubble from the previous harvest can seriously impair new tiller development. Removal of this material, by hard grazing, cutting and removing stubble, or burning, reduces shading, encourages new vegetative growth, and may help to control pests and diseases. In large clumpy grasses such as tall fescue and cocksfoot, gapping (removal of plants by cultivation or herbicide) can sometimes benefit seed yield in the second and subsequent harvest years by reducing stand density and hence competition among tillers. In dry areas, irrigation may be required to stimulate new vegetative growth, and nitrogen fertilizer may also be applied. Likewise, many tropical grass seed crops become less productive with age, because of reduced tiller fertility and 'sod-bound' stands. Such fields may be rejuvenated by burning (e.g. *Brachiaria humidicola*) or using glyphosate herbicide to remove 20–40% of the plants (e.g. *Setaria sphacelata*). For other species, N application soon after harvest encourages new tiller production.

4. Quality and dormancy

For many temperate grass species, seed quality assurance includes seed certification (as under the **OECD Seed Scheme**) which offers primarily an assurance of cultivar purity, but may also include standards which must be met for analytical purity, weed seed contamination, seed size, seed health and germination. Requirements differ from country to country, but there is widespread use of internationally standardized methods developed by the **International Seed Testing Association** (ISTA). Quality problems occur most commonly through failure to meet analytical purity standards because of the presence of undesirable (e.g. noxious) weed seeds, excess other weed seeds or excess inert matter (soil, sclerotes, straw, etc.): many weed seeds are similar in size to the grass seeds and therefore difficult to remove from seed lots. Germination problems may occur as a result of disease (e.g. **blind seed** disease of ryegrass and fescue caused by the seed-borne fungus *Gloeotinia temulenta*), poor harvest conditions and machinery settings, or failure to dry seed properly. What is yet to be explained is why seed vigour differences occur in high (>90%) germinating seed lots: poor vigour can reduce field emergence, storage ability and seed survival during international transit.

A number of grasses, including tall fescue and perennial ryegrass, contain a fungal endophyte which has a beneficial relationship with the grass host (e.g. *Neotyphodium coenophialum* of tall fescue). Though very beneficial to tall fescue plants, this endophyte produces chemicals which are toxic to a variety of animals such as horses and cattle. The endophytes spread through infected seed produced by infected plants. Where seed is used for lawn and turf purposes toxicity does not present a problem, but some authorities (e.g. The Oregon Department of Agriculture, USA) have an endophyte-labelling programme to describe quality in respect of the infection. (**See: Health Testing, 2. Fungal seedborne pathogens**)

Annual ryegrass toxicity (ARGT), or 'staggers', is an often fatal animal disease that occurs, mainly in western and southern Australia and South Africa, in livestock that are grazing pasture or eating hay infected by a nematode (*Anguina* sp.) and its symbiotic bacterium (*Corynebacterium* sp./ *Rathayibacter toxicus*). The nematodes invade the developing florets of the ryegrass (or other susceptible grasses), producing dark enlarged galls in place of the seeds, in which the adults mate and lay eggs and which are then shed and lie dormant in the soil to reinfect young seedlings. Within the galls, the bacteria multiply and produce a hardening orange-yellow toxic slime (containing corynetoxin, a stable tunicamycin-like nerve poison) that may spread out onto the seed-heads. Preventative practices include the **health testing** of seed produced in suspect areas, the careful inspection of seedheads as they develop on the grass, or mowing to prevent their formation.

Seed certification is little used for tropical grass species, partly because there is minor risk of genetic drift in apomictic or self-pollinating species, but also because, in Australia at least, companies have relied on their own internal quality assurance procedures for maintaining genetic purity. Because of their growth habit and/or the long period over which seed heads are produced, both analytical purity and germination are often low in tropical grasses. *Chloris gayana* for example typically has a germination of between 20 and 50%, while seed lots may contain from 40 to 60% pure seed. Because of this, the concept of pure live seed (PLS) is often applied for tropical grass seed lots, where PLS = pure seed (%) × germination (%)/100.

Most temperate grass seeds are dormant immediately after harvest, and require time (up to 3 months) for this **dormancy** to dissipate (**afterripening**). Germination testing of freshly harvested grass seeds therefore requires the breaking of dormancy (e.g. by chilling) (**see: Dormancy breaking**). Postharvest dormancy is also a feature of many tropical grasses.

5. Treatment

Harvested seed may require a pre-treatment to facilitate subsequent cleaning and handling, for example the removal of awns, e.g. by clipping or flaming (*Bromus*). For tropical grasses that have long awns, entwined appendages (e.g. *Stylosanthes hamata*) or excessive amounts of **chaff** and straw after harvest, pre-cleaning may be necessary using aspirators, coarse screens, scarificers or modified hammer mills. (**See: Conditioning, I. Precleaning**)

Seeds may be fungicide- or insecticide-treated, though this is not a common practice across the range of forage grass species, but may be used where specific problems arise: for example with the neonicotinoid systemic insecticide, imidacloprid, to control soil-borne insect larvae. (**See: Treatments – Pesticides**) Insecticide seed treatment may be used for protection in storage against the tropical warehouse moth.

Seed encrusting or **pelleting** (perhaps including additives) is sometimes used for forage grass seeds, for example to improve the ballistic properties of very small seeds such as browntop (*Agrostis capillaris*) for precision or aerial sowing. Grass seeds are also sown by **hydroseeding**. Pelleting has not been widely used with tropical grass seeds. Pre-sowing hydration treatment (seed **priming**) to improve the uniformity and rate of establishment has been used successfully in turf grasses. (JGH)

Fairey, D.T. and Hampton, J.G. (eds) (1997) *Temperate Forage Seed Production*, Vol. I. *Temperate Species*. CAB International, Wallingford, UK.

Loch, D.S. and Ferguson, J.E. (eds) (1999) *Temperate Forage Seed Production*, Vol. II. *Tropical and Subtropical Species*. CAB International, Wallingford, UK.

Grasses and legumes – forages

Forage or herbage grasses and legumes have a multiplicity of uses: as feed for livestock in pasture grazing, hay and silage; for turf/amenity purposes in lawns, sports fields, parks, road-sides; as ornamentals in landscaping; for ecological repair in soil conservation, cover and reclamation; and for assorted medicinal, pharmaceutical and industrial fuel and fibre purposes. According to FAO statistics, in 2001 the total world land area used to grow mixed grasses and legumes was about 102 million ha (compared to 671 million ha for all the cultivated cereals, 222 million ha for all the primary oilcrops, 65 million ha for the total pulses and 46 million ha for all vegetables and melons).

Some countries (e.g. Denmark and New Zealand) export around 90% of their total seed production, while other countries (e.g. UK) import over 50% of their requirements. Data for individual countries are not always available, but in the USA alone, forage seed is a multi-billion dollar industry. The global annual international trade in forage seeds is estimated at US$450 million (that is, approximately 10% of the total export seed trade); over 80% of this trade is in temperate species. Forage seed is also an important component of the world's forage-based commodity production of meat, milk and wool; its value to such systems often goes unrecognized.

Temperate forage seed production of both grass and legume species began in 19th-century Europe following demand for pasture development, but it was not until the 1920s that cultivar improvement programmes began in many countries. This led to the establishment of forage plant breeding and seed production companies, the establishment of the first seed quality standards (via **seed certification** and seed legislation), and the rapid expansion of international trading in forage and (more recently) turf seeds.

Tropical forage seed production began in the early 1900s in Australia, in the 1940s in Anglophone Africa and in the 1990s in tropical Latin America and Southeast Asia. Much of this production was for local consumption, and it is only in the last 25 years that a small international seed trade has developed. In contrast to the temperate species, the number of commercialized tropical species is large, but the number of available cultivars is low; seed certification is not a universal factor, partly because of the difficulties of producing quality seed. (JGH)
(See: Grasses – forage and turf; Legumes – forage)

Fairey, D.T. and Hampton, J.G. (eds) (1997) Temperate Forage Seed Production, Vol. I. Temperate Species. CAB International, Wallingford, UK.

Loch, D.S. and Ferguson, J.E. (eds) (1999) *Temperate Forage Seed Production*, Vol. II. *Tropical and Subtropical Species*. CAB International, Wallingford, UK.

Gravity table

Separators used in seed conditioning are to separate seeds that are similar to the desired seed fraction in terms of **size**, shape, and **seedcoat** characteristics but that differ in unit weight or specific gravity, using the principle of separation as seeds flow across an inclined reciprocating surface. **(See: Conditioning, II. Cleaning)**

Green revolution

The term used to describe the great increase in crop **yields** (initially wheat, and then rice) resulting from the development of new cultivars and also agricultural technology under the joint sponsorship by certain charitable foundations and government agencies. Much of the basic research on wheat was carried out under the direction of Norman Borlaug at **CIMMYT**. The increase in productivity has allowed several developing countries mainly in Asia, but also in Latin America and Africa to become self-sufficient, and in some cases to become exporters.

One feature of the improved cultivars of wheat is a shorter stem ('dwarf' wheat), now known to be due to the introduction, by breeding, of mutated *Rht* (reduced height) genes; this results in the progeny having a reduced response to the growth-promoting hormone **gibberellin**. Normally, a **DELLA** protein, which represses GA-regulated events, is encoded and synthesized by the wild-type tall plant; production and stability of this protein is reduced by the GA present in the plant, and as a result the repression is overcome, and there is GA-induced growth. In the dwarf plant, the DELLA is defective, due to a **mutation** in the *Rht* gene; hence the protein is largely insensitive to GA, it continues to suppress GA-induced growth, and consequently the plant fails to grow to full height even in the presence of the hormone (**see: Signal transduction – hormones**). A consequence of this mutation is less damage to the crops by wind, and a greater allocation of carbon to grain filling and less to stem growth. (MB, JDB)

Groats and Grits

Old and closely related terms in food grain preparation. Groats are whole cereal kernels, sometimes crushed or toasted, from which the **hull** has previously been mechanically removed, by scouring or flinging the seed against a hard surface; grits are a coarsely ground version. The word usually refers to oat grains when used alone, but also to barley or the **pseudocereal** buckwheat. (Oat varieties are now available in which the seed threshes free from the husk, to serve this market.) Similarly, grits – especially in the southern states of North America — are dehulled, crushed and more or less finely ground maize grain (sometimes prefaced by 'hominy', a word derived from the native-American Algonquian word for 'parched corn'). **(See: Cereals; Oat)**

Groundnut

See: Bambara groundnut; Peanut

Grow-out test

The general procedures of testing genetic (phenotypic), purity or health quality of a seed lot, by raising and inspecting young plants, either under cover or in the field. Also known as 'growing-on' tests. Evidence of disease symptoms can then be confirmed by molecular diagnostic tests, such as **ELISA** and **PCR**. **(See: Certification schemes; Health testing)**

GSPC

See: Global Strategy for Plant Conservation

Guar

1. World importance and distribution

Guar (*Cyamopsis tetragonolobus*), also called cluster bean because of the manner in which its pods are clustered together, is an extremely drought-tolerant legume growing in semi-arid regions. Its seeds are an important **industrial legume**. Most of the world's guar is grown in India (especially Rajasthan) and Pakistan (see section 6). **(See: Legumes)**

2. Origin

Guar is generally understood to be a native plant of India though a wild relative, *C. senegalensis*, is found in Africa. It has been postulated that this species was taken by Arab traders as horse fodder to southwest Asia.

3. Seed characteristics

The guar bean seed is spherical. Unlike the majority of legume seeds it has a large endosperm: the weight of the seed (20–30 mg/seed) is made up of 14–17% testa, 35–42% endosperm and 43–47% embryo. Mature, dry seeds vary from dull-white to pink to light grey or black and measure about 8 mm in diameter (Fig. G.28). There are 5–12 seeds in the pods which range from 3.5–10 cm long.

Fig. G.28. Guar seeds (image by Elly Vaes and Wolfgang Stuppy, Millennium Seed Bank, RBG, Kew).

4. Nutritional value and composition

Seeds contain approximately 30% fresh weight (fw) **storage protein** (low in methionine), 45% fw carbohydrate, most of which is **galactomannan gum**, and 2% fw **oil**. After extraction of the gum, guar meal contains approximately 35% protein, which is about 95% digestible.

5. Uses

In Asia guar is grown for green fodder and the bean pods are used a vegetable for human consumption. The primary importance of guar is the commercial value of its seed **gum**, made from the processed **endosperm**. After extraction and processing the gum is a white to yellow, water-soluble powder, with a slight odour. It is a hydrocolloidal galactomannan with mannose and galactose in a 2:1 ratio. Guar gum has a wide variety of food and non-food uses – as a thickener and emulsifier in commercial food processing, emulsifier of salts in bakery and baking mixes, a cereal, jams, jelly and dairy products stabilizer, a firming agent in fat and oil, a thickener in gravies, sauces, ice cream and yogurts, a stabilizer/thickener in processed vegetables and juices, soups and many other food categories. Guar gum has almost eight times the thickening power of maize starch. It is also used in paper manufacturing, textiles, printing, cosmetics and pharmaceuticals, oil-well drilling muds, explosives, a host of other industrial applications, and in **hydroseeding** in agriculture. The gum is considered to be a good dietary fibre which can lower blood cholesterol and aids glycaemic control in patients with diabetes.

6. Marketing and economics

The annual world market is estimated to be around 150,000 t, 70% of which is produced in India and Pakistan, who export much of the crop as partially processed endosperm material. Following harvest, the seed is graded for size and cleaned to remove shrunken seed and crop residues. Blackening and small shrunken seed are the principal factors that decrease quality. White seed is preferred for many food applications, and black seed is often discounted. Small seed contains less endosperm and therefore is less desirable for milling: the preferred seed size is 4 mm diameter.

7. Plant type

Guar is a coarse, upright, drought-resistant, bushy annual, growing to 1–2.7 m high: there are dwarf and tall cultivars. Plants have single stems, fine branching or basal branching. Racemes are distributed on the main stem and lateral branches. (OV-V)

Reid, J.S.G. and Edwards, M.E. (1995) Galactomannans and other cell wall storage polysaccharides in seeds. In: Stephen, A.M. (ed.) *Food Polysaccharides and their Applications*. Marcel Dekker, New York, USA, pp. 155–186.

Whistler, R.L. and Hymowitz, T. (1979) *Guar: Agronomy, Production, Industrial Use and Nutrition*. Purdue University Press, West Lafayette, USA.

Guarana

Seeds of guarana (*Paullinia cupana*, Sapindaceae) are used for their stimulant and related effects. The name is taken from the Guarani people of the Brazilian Amazon with whom the seeds are long associated.

In the Amazon forests, where the species originated, the plant grows as a climbing, woody vine often reaching 8–10 m in height, but it has a shrub habit when cultivated. The ovoid or pear-shaped fruits, about the size of a grape and red when ripe, contain 1–3 seeds. Each seed is sub-spherical, about 1 cm in diameter with a shiny, red-brown testa, covered by a pale brown-white, cupular **aril** at the base (Fig. G.29).

The most important component of the seeds as far as their use is concerned is caffeine, normally *ca*. 5% on a fresh weight basis, much higher than in **coffee** or **cola**, plus traces of theophylline and theobromine (**see: Pharmaceuticals and pharmacologically active compounds**).

When ripe, the seeds are removed from the fruits, allowed to undergo slight fermentation for 2–3 days, dried, the arils

Fig. G.29. Guarana seeds. Note the pale-coloured aril on the seed (image by Elly Vaes and Wolfgang Stuppy, Millennium Seed Bank, RBG, Kew).

removed, and then roasted for 5–6 h, after which the kernels are freed from their testas. Guarana is available as whole roasted kernels or a powder. A common processing treatment is to mix the ground kernels with water (sometimes with added cassava flour) into a dough which is moulded in a sausage shape on to a stick, then dried (and sometimes smoked) over a moderate fire until hard. This type of processing was practised among Amazonians and the sticks are still a common form of guarana available in the markets. Powder is scraped from the 'sausage' and mixed with water to make a beverage. A syrup, used for soft drinks, is made from powdered kernels.

Guarana is consumed widely in South America and in various forms it is second to coffee in Brazil. It is effective as a stimulant and is reputed to counter fatigue, aid mental alertness and promote stamina and physical endurance. It also has uses as a health tonic, to combat premature ageing, dispel flatulence and dyspepsia and to deter arteriosclerosis. (MB)

Bempong, D., Houghton, P.J. and Steadman, K. (1991) The caffeine content of guarana. *Journal of Pharmacy and Pharmacology* (suppl.) 43, 125.

Erickson, H.T., Pinheiro, M., Correa, F. and Escobar, J.R. (1984) Guarana (*Paullinia cupana*) as a commercial crop in Brazilian Amazonia. *Economic Botany* 38, 273–286.

Gums

Polymeric carbohydrates (other than of glucose) of high molecular weight, soluble in water but precipitable from aqueous solutions by acidified alcohol; this general term can include **hemicelluloses** (e.g. guar gum, a galactomannan) and **pectins**. Is also used loosely, but incorrectly, with reference to latexes (e.g. chicle, rubber), which are composed of many isoprene units and are insoluble in water.

Commercial gums are important in the food industry (Table G.16) as emulsifiers and binding agents, as lubricants, and in the cosmetics, paper and textile industries. Gums are only partially digested by humans, and as inert substances within the digestive system can be used in diets as low-calorie food, as dietary fibre, as toothpaste gels, and as inert fillers in the mixing and encapsulation of pharmaceuticals. They may be obtained from plant sources (e.g. gum arabic or gum acacia from the bark of the *Acacia* tree species, gum tragacanth from locoweed, *Astralagus gummifer* bushes), or from seeds. Common seed gums are the legume seed gums (**carob** and **guar, see: Galactomannans**). Also, Psyllium (a plantain, *Plantago ovata*) seedcoat gum (mucilage) containing arabinose, xylose, galacturonic acid and rhamnose, is used as a laxative, purgative and in cosmetics and hair-setting lotions; quince (*Cydonia vulgaris, C. oblongs*) seedcoat gum (**mucilage**), with polysaccharides composed of xylose, arabinose and uronic acids, is used in cosmetics and hair-setting lotions and as an emulsifier and stabilizer in pharmaceutical preparations; and tamarind seed gum is a cotyledon cell wall **xyloglucan**, used as textile sizing agent to make threads somewhat stiff during weaving before being washed out of the finished product.

Linseed (flax) gums, usually called mucilages, are a complex mixture of oligo- and polysaccharides, rich in **arabinoxylans**, and are present in the seedcoat from which they are exuded as they swell upon hydration. They account for 5–12% of the seed by weight. Their presence may help in water acquisition and retention by the seed, as has been suggested for galactomannans in other seeds, e.g. fenugreek. Flax seeds are mainly harvested for their oil content, although the mucilage may be of limited value as a laxative or bulking agent. (**See: Pharmaceuticals and pharmaceutically active compounds** for further information on medicinal uses) (JDB, PH)

Table G.16. Some major uses of polysaccharides from 'gum'-producing seeds in the food industry.

Function	Application	Guar	Locust bean	Psyllium (husks)	Quince
Bulking agent	Dietetic foods			X	
Emulsifier	Salad dressing	X			
Gelling agent	Desserts, aspics		X		
Stabilizer	Mayonnaise, ice cream, spreads	X	X		
Suspending agent	Chocolate milk		X		
Swelling agent	Processed meats	X	X	X	X
Thickening agent	Jams, sauces, gravies	X	X		

Based on: Sanford, P.A. and Baird, J. (1983) Industrial utilization of polysaccharides. In: Aspinall, G.O. (ed.) *The Polysaccharides*, Vol. 2. Academic Press, New York, USA, pp. 411–490.
Whistler, R.L. and BeMiller, J.N. (1983) *Industrial Gums*, 2nd edn. Academic Press, New York, USA.

Gymnosperm

(Greek: *gymnos* = naked; *sperma* = seed) An inhomogenous, polyphyletic (having more than one ancestral origin) group of the Spermatophytes (seed plants) comprising the three recent monophyletic (single common ancestor) entities: Coniferatae (conifers), Cycadatae (cycads) and Gnetatae. In contrast to **angiosperms**, gymnosperms do not bear their ovules within closed **megasporophylls**, instead, they lie 'naked' on the megasporophylls and are openly exposed to the environment. (**See: Gymnosperm seeds; Gymnosperms – seed production**)

Gymnosperm seeds

1. Anatomy and morphology

Gymnosperms are vascular plants in which the **ovules** and seeds are exposed on scales or similar structures, rather than enclosed within an ovary, as is the case in the **angiosperms**. Among the gymnosperms (cycads, ginkgo, conifers and gnetophytes), the conifers are the dominant group. They have ovules and pollen grains clustered in strobili or cones of different shapes and sizes; they are predominantly monoecious (i.e. the male and female reproductive structures are in different cones). Fertilization is generally not a dual event (that is, unlike the double fertilization in angiosperms). Exceptions are *Ephedra* and *Gnetum* (both belonging to the order Gnetales); however, in these genera double fertilization results when two sperm nuclei are released from a pollen tube and each fuses with a nearby undifferentiated female nucleus to form a **zygote**. Thus, there is the formation of two diploid zygotes per pollen tube and each is viable and initiates embryogenesis. The female gametophyte serves as an embryo-nourishing tissue. (**See: Evolution of seeds**)

At some stage during its development the gymnosperm seed (Fig. G.30a, b) is composed of the following.

(a) *The embryo* – the result of the fertilization of the egg nucleus in the female **megagametophyte** by a gamete released from the pollen tube. The formation of several embryos is also a widespread phenomenon in the gymnosperms, but usually there is only one well-developed embryo in the seed, the others degenerating during development. The embryo consists of the **radicle, hypocotyl** and **cotyledons**. The mature embryo may contain one cotyledon (*Ceratozamia*), two (most Cycads, *Ginkgo*, *Cupressus*, *Podocarpus*, *Auraucaria*, Taxaceae, *Ephedra*, *Welwitschia* and *Gnetum*), three (*Encephalartos*, occasionally *Ginkgo*, Taxodiaceae, Taxaceae and *Gnetum*), four (*Sequoia*) or multiple (Pinaceae, e.g. about 12 in *Pinus pinea*). In *Welwitschia* and *Gnetum* the embryo develops a special haustorium-like structure, the 'feeder' that is a lateral outgrowth of the hypocotyl that is more prominent than the embryo itself (Fig. G.30c). Between the various cotyledons lies the shoot apex, which is either an undifferentiated meristem or a plumule differentiated into leaf initials. Unlike in some angiosperm seeds, gymnosperm seeds are not shed with undifferentiated embryos.

(b) The megagametophyte – this surrounds the embryo and is functionally equivalent to the true endosperm of angiosperm seeds; it provides protection and serves as the major storage organ of the seed. In gymnosperms there is no fusion of the male and polar nuclei; the megagametophyte is haploid and originates from the maternal ovule (i.e. it is a modified megagametophyte, the tissue in which the eggs or **archegonia** were embedded).

(c) The **seedcoat**, which is derived from maternal tissues, and therefore is diploid. The seedcoats of many gymnosperm seeds are composed of three distinct layers: an outer (sarcotesta), a middle (sclerotesta) and an inner layer (endotesta), and like those of angiosperms, they provide a protective barrier between the embryo and the external environment. But there can be one (cycads, conifers), two (*Ephedra*, *Welwitschia*) or three (*Gnetum*) integuments which have an opening (**micropyle**) at the apex, through which pollination is effected (see below). In conifer seeds, there is great variation in seedcoat colour, shape, texture and the presence of resin vesicles. Some of these features can vary even within a single seed lot of a species. True firs (*Abies* spp.), hemlocks, and western red cedar have resin vesicles in their seedcoats. Seedcoat colour has been implicated in germinability (degree of seed **dormancy**) and in susceptibility to **damping-off** fungi, although definitive studies are lacking. The seedcoat

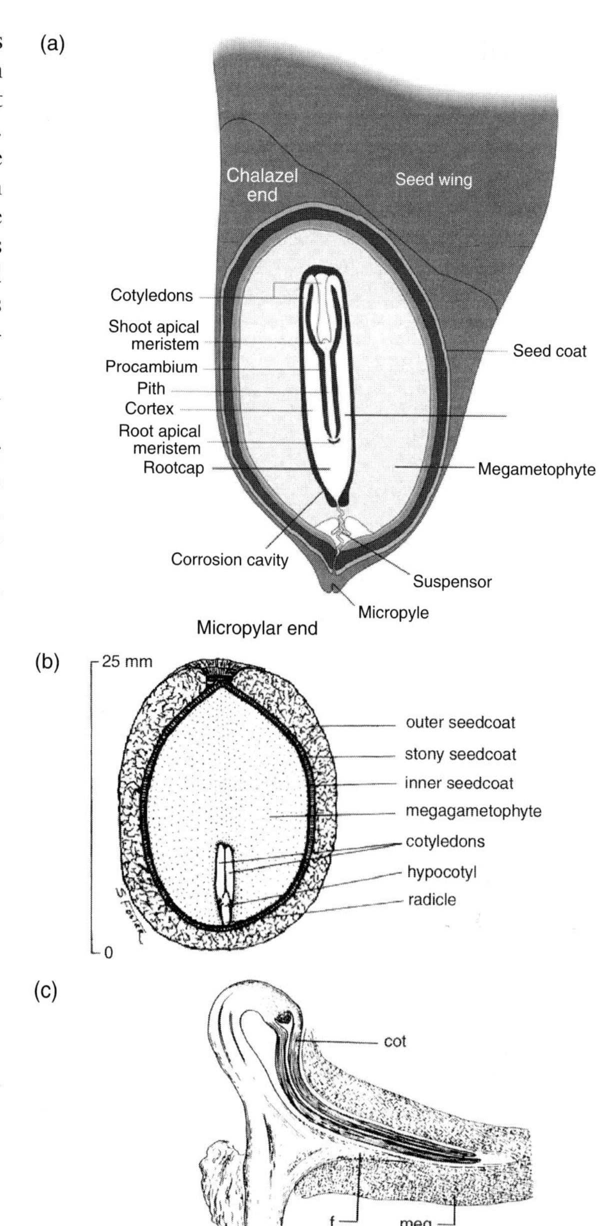

Fig. G.30. Structure of conifer seeds. (a) Pine species, in which the seed develops within a cone, and when shed has a 'wing' structure attached that aids in its dispersal by wind. In the conifer seed, it is the haploid megagametophyte that stores the majority of the seed's reserves; this tissue is living at seed maturity. (From: Kolotelo, D. (1997) *Anatomy and Morphology of Conifer Tree Seed.* Forest Nursery Technical Series 1.1. B.C. Ministry of Forests, Surrey, B.C. Canada. With permission from B.C. Ministry of Forests, Canada.) (b) Section through the seed of *Ginkgo biloba* which is borne in a fleshy fruit, not a cone, to show the composition of the seedcoat consisting of a stony layer which is embedded between the outer and inner fleshy layer. (From *Seeds of the Woody Plants of the United States* (1974) Forest Service, USDA, Washington, DC, USA, pp. 429–430.) (c) *Welwitschia mirabilis.* Section through a germinated seed to show the hypocotyl acting as a haustorium within the megagametophyte (cot: cotyledon; meg: megagametophyte; f: haustorium or 'feeder'). (From: Martens, P. and Waterkeyn, L. (1964) Etudes sur les Gnetales. VII. Recherches sur Welwitschia mirabilis. IV. Germination et plantules. Structures, fonctionnement et productions du meristeme caulinaire apicale. *Cellule* 65, 7–68.)

of cycads contains mucilage canals. (**See: Seedcoats – structure**)

Two other structures derived from maternal tissues are evident in many gymnosperm seeds: the **nucellus** and **megaspore** membrane. The nucellus is the multi-celled inner tissue within which the megagametophyte develops. In some conifer seeds, this tissue becomes compressed during seed development to form a papery covering (the nucellus membrane) outside the megaspore membrane; in the micropylar end of the seed, the hardened nucellus forms the nucellar cap, which may play a role in controlling germination. The megaspore membrane surrounds the megagametophyte and is a lipid-rich multi-layered tissue; in Scots pine and Norway spruce, this tissue may restrict water uptake.

It is not clear what regulates the growth and development of the megagametophyte into a storage tissue. This process normally only occurs if an embryo is present, so it can be assumed that some influence emanates from the zygotic cells during embryogenesis. However, there are circumstances when the megametophyte develops into a storage tissue even in the absence of an embryo. This happens in infertile ovules of Douglas fir parasitized by the insect (a chalcid) *Megastigmus spermotrophus*. Eggs oviposited into unfertilized ovules induce the megagametophyte to grow and lay down reserves, whereas unparasitized, infertile ovules degenerate. Clearly, the developing insect larva mimics the effect of the embryo, through a mechanism which is not understood.

One special characteristic of the nucellus is that it forms a pollen chamber in the Cycadaceae, *Ginkgo* and *Ephedra*, which is essentially a space at the apex of the nucellus. Apart from a few members of the Pinaceae, which show a special stigmatic micropyle ('pollen cushion'), and a few more (*Araucaria*, *Agathis*, *Tsuga dumosa*) in which the pollen grains do not land on the micropyle, all other gymnosperms produce a sugary exudate at the time of pollination. The fluid serves as a receptor of wind-borne pollen as well as a vehicle for transporting it to the egg cell. In families lacking a pollen chamber, at the time of pollination the nucellus produces a 'pollen cushion' whose function can be compared with that of the stigma in the angiosperms (**see: Reproductive structures: 1. Female**). Pollen grains are deposited and embedded here and subsequently form pollen tubes. Exceptions are certain Pinaceae (*Tsuga*, *Cedrus*, *Pseudotsuga* and *Larix*), and *Araucaria* where the pollen is not deposited on the apex of the nucellus. (**see:** section **3. Embryogenesis**)

(ARK, WS, JD)

Biswas, C. and Johri, B.M. (2001) Reproductive biology of gymnosperms. In: Johri, B.S. and Srivastava, P.S. (eds) *Reproductive Biology of Plants*. Springer, Berlin and Narosa Publishing House, New Delhi, India, pp. 215–236.

Schnarf, K. (1937) Anatomie der Gymnospermen-Samen. *Handbuch der Pflanzenanatomie = Encyclopedia of Plant Anatomy*. 10, 1. Borntraeger, Berlin.

Von Aderkas, P., Rouault, G., Wagner, R., Rohr, R. and Roques, A. (2005) Seed parasitism redirects ovule development in Douglas fir. *Proceedings of the Royal Society*, *B* 272, 1491–1496.

2. Storage reserves

Oils (storage triacylglycerols) are the most abundant storage reserve in the megagametophyte of coniferous seeds. During seed development they accumulate in **oil bodies**, their breakdown following germination providing a source of energy and carbon to support early post-germinative seedling growth. **Storage proteins** and carbohydrates account for the remaining reserve compounds, with the latter comprising only 2–6% (Fig. G.31). In seeds of *Pinus* species, oil constitutes up to 40–60% of the dry weight of the seed; the three most abundant **fatty acids** in western white pine seed are oleic, linoleic and pinolenic (an isomer of linolenic) acids. Similarly in Douglas fir seeds, the dry weight composition of the megagametophyte is 60% oils.

3. Embryogenesis and seed development

In conifers, the egg nucleus lies within the female megagametophyte and is fertilized by a gamete released from the pollen tube. The period of post-pollination/pre-fertilization may be weeks to years and includes pollen germination, pollen tube growth, and penetration of the nucellus, as well as archegonium and gamete development. Many conifers such as *Pinus* species, have pollen adhering to the integument; later, a pollination droplet forms which is exuded by the ovule. During evaporation of the droplet, the pollen is mobilized to the base of the micropylar canal where it finds the egg cell. In other species (*Larix occidentalis* and *Pseudotsuga menziesii*), the pollen is captured and detained by a stigmatic extension with many trichomes. Male gametes of conifers lack any specialized means of locomotion but a unique feature of the cycads and *Ginkgo*, is multiflagellate sperm. The sperm cell of *Zamia pumila* (a cycad), for example, has many thousands of motile flagellae on its surface, and after

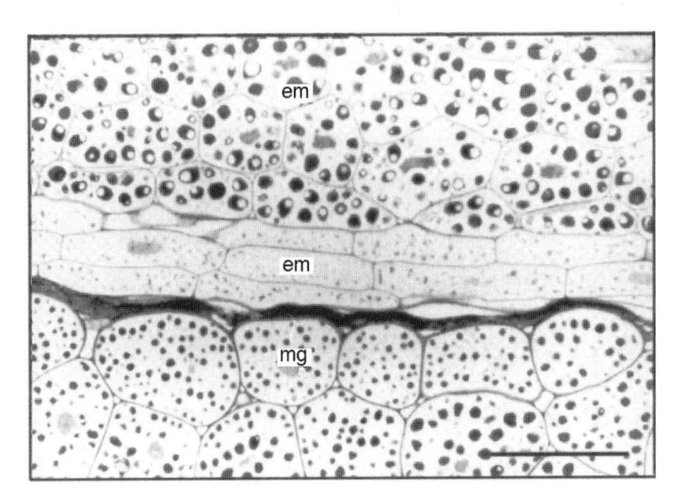

Fig. G.31. Parenchyma cells of the megagametophyte (mg) and embryo (em) of the yellow-cedar seed. The cells are stained darkly with methylene blue to show numerous protein bodies with storage proteins; oil bodies that store triacylglycerols are packed in between the protein bodies. The outermost cell layers of the embryo do not contain reserves. Bar = 0.05 mm. From Ren, C. and Kermode, A.R. (1999) Analyses to determine the role of the megagametophyte and other seed tissues in dormancy maintenance of yellow cedar (*Chamaecyparis nootkatensis*) seeds: morphological, cellular and physiological changes following moist chilling and during germination. *Journal of Experimental Botany* 50, 1403–1419, with permission from Oxford University Press.

being brought close to the ovule by the pollen tube it is released and moves towards it.

Following fertilization, the resultant zygote undergoes a free nuclear stage in which several free nuclei are produced. Later, cell walls are laid down to form the proembryo (Fig. G.32). Subsequent cell divisions give rise to the embryonal and suspensor cells. These may develop to form a single embryo and an elongated suspensor: however, often in the conifers there is cell separation to form four embryos (polyembryony). Only one embryo continues its development, while the others degenerate. Within the seed, cells in the central portion of the megagametophyte (the haploid female gametophyte), containing primarily **starch**, break down to form the corrosion cavity into which the embryo later expands. Following the very early stages of seed development associated with pattern formation and differentation of the basic body plan of the embryo (histodifferentiation), is the expansion (or maturation) stage. It is during this stage that the megagametophyte cells (primarily storage parenchyma cells – large, thin-walled and spherical in shape) accumulate storage reserves: oil, protein, and to a lesser extent, starch. This stage involves expression of genes for cell expansion, and genes that encode enzymes and other proteins required for the synthesis of reserves and products important for survival in the dry state. During the final stage of seed development, **maturation drying**, water is lost from seed tissues; there is a gradual reduction in metabolism and the embryo passes into a metabolically inactive, or **quiescent** state. Not all gymnosperm seeds can withstand drying, i.e. some are **recalcitrant**, e.g. Ginkgo (*Ginkgo biloba*), a temperate species.

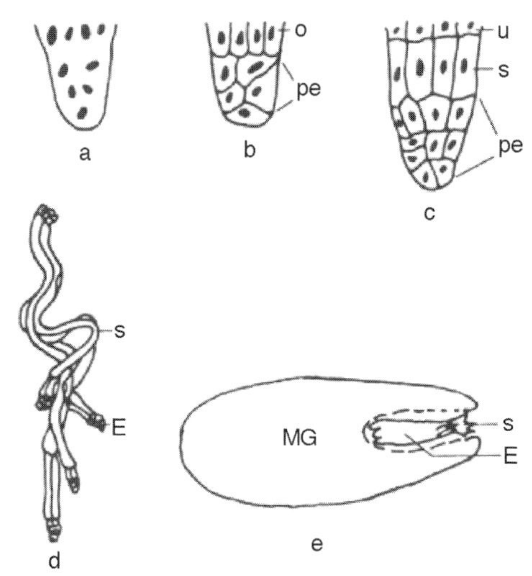

Fig. G.32. Embryo development in conifers. (a) Free nuclear stage. (b) Proembryo (pe) showing internal division of cells, with open tier (o) cells above. (c) Three-tiered mature proembryo (pe) beneath the suspensor tier (s), which elongates to form the primary suspensor, and the upper tier (u), which degenerates. (d) Polyembryonic stage. Elongated suspensor (s) bearing the embryonal masses of cells (E). (e) The single maturing embryo (E) elongates into a cavity in the megagametophyte (MG); the suspensors (s) degenerate. From Bewley, J.D. and Black, M. (1994) *Seeds. Physiology of Development and Germination*, 2nd edn. Plenum Press, New York, USA, with permission from Kluwer.

Most recognized among the seed-bearing structures of the gymnosperms are the cones, or strobili, in which the seeds are attached to the inside of their scales until shed, often up to several years after fertilization, and sometimes following intense heat generated by forest fires. After fertilization the female cones become enlarged and woody. There is considerable variation in size and shape of the seed-bearing cones, from the commonly recognized structures of conifers such as the pines (*Pinus* spp., Fig. G.33), cedars (*Cedrus* spp.), firs (*Abies* spp., Fig. G.33) and spruces (*Picea* spp.) which, depending upon the species, are usually in the range of 4–15 cm in length, to the huge cycad cones, which in *Cycas revoluta* (sago palm) bear large orange seeds, and in *Zamia amplifolia* can reach 50 cm in length and 12 cm in diameter. Seeds of *Gnetum* and *Ginkgo* (Fig. G.30b) are fleshy, rather than cone-like, and the rotting smell of this latter seed after it has been shed can be obnoxious. Hence the planting of female *Ginkgo* trees in gardens is not encouraged in some urban areas. Seeds of yew (*Taxus* spp.) produce seeds that are partly surrounded by a red flesh **aril**, which attracts birds that eat it and disperse the seed; it is toxic to humans and although rarely fatal it is a cause of poisoning in children.

Conifer seeds are often dormant when dispersed from the parent tree at maturity and some require several months to terminate **dormancy**. This effectively distributes their germination in time and space. Dormancy inception occurs during the seed expansion/maturation stage.

4. Control of seed maturation

It is likely that the mechanisms controlling maturation processes in seeds of angiosperms are also common to seeds of

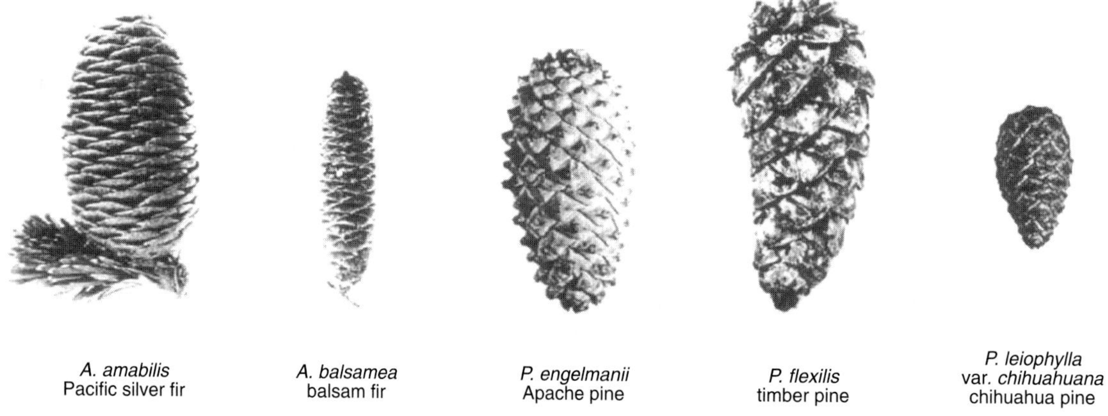

Fig. G.33. Female cones of some *Pinus* (pine) and *Abies* (fir) species (× approx. 0.25).

gymnosperms. Support for this comes from the isolation of gymnosperm genes that represent homologues/orthologues of several angiosperm genes known to participate in regulating seed development. Examples are homeobox genes that control embryo radial pattern formation and *APETALA-2*-like genes implicated in meristem/floral identity and in ovule and seed development, genes encoding aquaporin proteins with transporting and nutritive functions during seed development, genes implicated in the acquisition of **desiccation tolerance** (**late embryogensis abundant**, *LEA* genes) and genes encoding proteins involved in cell expansion. Although few in number, functional studies (e.g. expression of gymnosperm genes in angiosperm seeds) generally support the evolutionary conservation of genes that regulate seed maturation. Analyses of cycad expressed sequence tags (ESTs) have uncovered both conserved and potentially novel genes.

As in the seeds of angiosperms, **abscisic acid** (ABA) plays an important role in several processes central to seed development (**see: Desiccation tolerance – acquisition and loss; Dormancy – acquisition; Dormancy – genes; Late embryogenesis abundant proteins; Maturation – controlling factors; Maturation – effects of drying**). ABI3/VP1 proteins are members of a large group of **transcription** factors that act as intermediaries in regulating ABA-responsive genes during seed development. Processes associated with the later stages of seed development appear to be controlled by ABI3/VP1 proteins; notably these include reserve deposition, prevention of **vivipary**, dormancy imposition and the acquisition of a tolerance of seed tissues to desiccation. An *ABI3* gene **orthologue** (*CnABI3*) has been cloned from the gymnosperm *Chamaecyparis nootkatensis* (yellow-cedar). This gene encodes a 794-amino acid protein containing all four regions that are typically conserved and shares a similar role to its angiosperm counterparts in relation to the control of seed developmental gene expression (e.g. activation of storage-protein and LEA gene expression, **see: Storage reserves synthesis – regulation**). Although it shares many of the characteristics of other *ABI3/VP1* genes of angiosperms, the conifer gene/protein also has unique characteristics. When expressed in *Arabidopsis*, the gymnosperm gene is able to functionally complement an *abi3* null mutant following stable transformation. The seeds of this severe *abi3* mutant accumulate reduced amounts of storage proteins, fail to degrade chlorophyll, remain non-dormant and display an impaired capacity to express various mRNAs regulated by different temporal programmes during seed development. Several of the visible mutant phenotypes of the severe mutant (e.g. production of green seeds due to a lack of chlorophyll breakdown) are fully restored to those of the wild-type, and *Arabidopsis* seeds transformed with the gymnosperm *ABI3* gene also acquire desiccation tolerance and exhibit restored accumulation of several seed proteins (including seed storage proteins, α-tonoplast intrinsic protein and **oleosin**). However, some phenotypes of the mutant seeds are only partially restored by expression of the gymnosperm *ABI3* gene, notably the degree of ABA sensitivity as far as the inhibition of germination is concerned. These observations reveal the degree of conservation of control processes between gymnosperms and angiosperms. (**See: Signal transduction – hormones**)

5. Dormancy

See: Dormancy – gymnosperms

6. Germination and seedling establishment

The visible signs of the completion of germination (the emergence of the radicle from the seed) are generally preceded by a splitting of the seedcoat at the micropylar end. The radicle emerges through this opening causing further splitting of the seedcoat. Conifer seedlings are **epigeal** in their growth, since their cotyledons emerge from the soil.

In most conifer seeds, mobilization of the major protein and lipid storage reserves commences within the embryo during germination and is likely to be important for early seedling establishment. In the megagametophyte it commences after radicle elongation, i.e. it is a post-germinative event. The breakdown products are exported by the megagametophyte and are taken up by the developing seedling where they are used as a nutritive source. Free amino acids, especially arginine, derived from storage protein mobilization, accumulate in both the megagametophyte and seedling and this is accompanied by a marked increase in arginase activity within the cotyledons and **epicotyl** of the seedling. Thus, as in some angiosperm seeds, it is likely that arginase supports post- germinative growth of conifer seedlings by assimilating arginine into metabolic/biosynthetic pathways. Lipids in the megagametophyte are metabolized by lipases to yield free fatty acids and glycerol. Most of the glycerol is converted to sucrose for export to the developing seedling. The fatty acids are metabolized by **β-oxidation** and **gluconeogenesis** (via the **glyoxylate cycle** and reverse glycolysis), which ultimately generates sucrose to be similarly exported (**see: Oils and fats – mobilization**).

Following post-germinative mobilization of reserves, the megagametophyte undergoes **programmed cell death**; this process involves several proteases and nucleases (as in angiosperms) and may be triggered in part by the accumulation of reactive oxygen species.

7. Genome projects

Genomic projects on conifer species (including *Pinus*, *Picea* and *Cycas*) are likely to yield a wealth of information; in particular functional **genomics** approaches hold promise for elucidating genes and proteins that control complex physiological processes in gymnosperm seeds including development, dormancy and germination. Cycads are ancient seed plants with origins in the Palaeozoic. They exhibit characteristics intermediate between vascular non-seed plants and the more derived seed plants. This is borne out by genomic analyses in which expressed sequence tags created from *Cycas* (*Cycas rumphii*) uncovered both conserved and novel genes – 1718 cycad 'hits' are to angiosperm genes (primarily those genes involved in development and signal transduction in present-day flowering plants), 1310 match genes of other gymnosperms and 734 are similar to genes of lower (non-seed) plants. (ARK)

(See also: **Gymnosperms – seed production; Tree seeds**)

Brenner, E.D., Stevenson, D.W., McCombie, R.W., Katari, M.S., Rudd, S.A., Mayer, K.F.X., Palenchar, P.M., Runko, S.J., Twigg, R.W., Dai, G., Martienssen, R.A., Benfey, P.N. and Coruzzi,

G.M. (2003) Expressed sequence tag analysis in Cycas, the most primitive living seed plant. *Genome Biology* 4, R78.

Carmichael, J.S. and Friedman, W.E. (1995) Double fertilization in Gnetum gnemon: The relationship between the cell cycle and sexual reproduction. *The Plant Cell* 7, 1975–1988.

Flores, E.M. (2002) Seed Biology. In: Vozzo, J.A. (ed.) *Tropical Tree Seed Manual*. US Department of Agriculture Forest Service, Washington DC, pp. 1–118.

He, X. and Kermode, A.R. (2003) Proteases associated with programmed cell death of megagametophyte cells following germination of white spruce (*Picea glauca*) seeds. *Plant Molecular Biology* 52, 729–744.

King, J.E. and Gifford, D.J. (1997) Amino acid utilization in seeds of loblolly pine during germination and early seedling growth (I. Arginine and arginase activity). *Plant Physiology* 113, 1125–1135.

Plomion, C., Cooke, J., Richardson, T., Mackay, J. and Tuskas, G. (2003) Report on the forest trees workshop at the plant and animal genome conference. *Comparative and Functional Genomics* 4, 229–238.

Von Aderkas, P., Rouault, G., Wagner, R., Rohr, R. and Roques, A. (2005) Seed parasitism redirects ovule development in Douglas fir. *Proceedings of the Royal Society*, B 272, 1491–1496.

Zeng, Y., Raimondi, N. and Kermode, A.R. (2003) Role of an ABI3 homologue in dormancy maintenance of yellow-cedar seeds and in the activation of storage protein and Em gene promoters. *Plant Molecular Biology* 51, 39–49.

Gymnosperm seeds – dormancy

See: **Dormancy – gymnosperms**

Gymnosperms – seed production

1. Factors affecting cone production and seed yields

The tree seed orchard industry accounts for a significant and growing amount of stock for reforestation. Problems can occur when environmental conditions do not mimic those in natural stands, particularly factors such as elevation and temperature, which can have a negative impact on seed development. **Tree seed** orchards involved in the production of seeds of lodgepole pine (*Pinus contorta*), yellow cedar and other species often yield seeds that are shrivelled (indicative of poor seed fill) and exhibit poor germination rates and impaired seedling growth. Spraying with **gibberellin** (GA) to induce male and female flowers is a common practice in tree seed orchards and may contibute to resource limitation (inadequate allocation of carbon and nitrogen resources from the parent tree to developing seeds and **ovules**) and thereby reduce seed fill or viability.

Insects may attack the reproductive structures of conifers throughout their development – from cone bud initiation to seed maturity, and even during cone collection and seed extraction. Forest insects that feed directly on conifer cones and/or seeds are termed conophytes; only about 100 of 50,000 species of insects known in Canada are conophytic and far fewer have an economic impact. In conifer seed orchards, which are normally managed to produce cone crops on a more regular basis than what would occur in natural stands, conophytic insects tend to be more prevalent and cause more extensive damage.

It should be noted, however, that insect infestation does not always cause damage to seeds. For example, parasitism of infertile Douglas fir ovules by *Megastigmus spermotrophus* induces megagametophyte development into storage tissue (**see: Gymnosperm seeds, 1. Anatomy and morphology**).

The frequency of seed-borne *Fusarium* in British Columbia, Canada, is equivalent regardless of the source of seeds (i.e. from tree seed orchards or natural stands). Seeds and cones harbouring this fungus can contaminate processing equipment, and spread to previously uncontaminated seeds. For contaminated seed lots, persistence of the fungus during imbibition and moist chilling of seeds is common; thus, the implementation of a hydrogen peroxide treatment (e.g. 30% v/v for less than 1 h or 3% for 4–8 h) is usually warranted.

2. Seed dormancy and seedling vigour

In plant species whose seeds exhibit dormancy (**see: Dormancy – gymnosperms; Dormancy breaking** entries), there is a block to germination such that the seed does not germinate, even when the appropriate germination conditions are provided. For the forest industry, problems can arise because the seeds of certain forest species develop pronounced dormancy, exhibiting a low capacity for germination upon dispersal. Seeds of yellow-cedar are a prime example. Only a small percentage of seeds will germinate the first year after dispersal with the majority requiring another year to break dormancy, in which seeds are maintained in a moist state usually at low or fluctuating temperatures. During this time, the seed numbers can potentially decline greatly as a result of consumption by birds and rodents or deterioration caused by fungal attack. Seeds of conifers that require little or no period to break dormancy, e.g. Sitka spruce (*Picea sitchensis*), western hemlock (*Tsuga heterophylla*) and western red cedar (*Thuja plicata*), obviously have a distinct advantage in quickly occupying sites following logging. Other species exhibiting deep dormancy at dispersal are the true firs (e.g. Pacific silver fir, subalpine fir, noble fir; *Abies amabilis*, *A. lasiocarpa* and *A. procera*, respectively) and western white pine (*Pinus monticola*). Reforestation efforts are becoming increasingly important, yet the forest industry is greatly affected by problems associated with seed dormancy and seed quality. Forest nursery operations must accommodate the lengthy treatments (generally 3–4 months in duration, and sometimes not optimized) to break dormancy. Poor germination rates and impaired seedling growth occur frequently. This causes a significant loss of valuable seed; increased operational costs to nurseries can also occur because of irregular seedling performance. In British Columbia, up to 40% of true fir (*Abies* species) seeds planted in nurseries never germinate or are not commercially viable when they do. For some species, depending on the specific seed lot, germination percentages are even lower than this. Multiple seeding to compensate for the loss leads to increased costs and decreased efficiency of operations. Techniques for breaking dormancy need to be rigorously optimized and tested on different seed lots or clones. Forest tree seeds generally have an inherent high genetic variability which results in great heterogeneity in their behaviour and, in particular, in their germinability following dormancy-breaking procedures. Seed dormancy can vary considerably between different clones, from seed lot to seed lot, and among seeds within one seed lot.

Some progress has been made towards modifying traditional dormancy-breaking protocols used by the forest industry (generally 3–4 months in duration), for yellow-cedar, true fir and western white pine seeds, such that dormancy is effectively terminated in a shorter period. These seed treatments lead to synchronous and high rates of germination, even for some of the poorer performing seed lots, and also yield vigorous growth under nursery greenhouse conditions and after out-planting at natural stands.

As noted, certain seed lots of conifer species may also exhibit poor performance following germination such that seedlings are stunted in their early growth or cease growing altogether, again posing problems for the forest nurseries. Dormancy termination appears to be a multi-step process; some of the important events following receipt of the appropriate signals by the seed are the down-modulation of **abscisic acid** (ABA)-regulated metabolism and the up-modulation of **ethylene/gibberellin** (GA)-regulated metabolism (Fig. G.34). Part of this process is likely to involve controlled proteolysis of certain signal transduction components to mediate the attenuation of ABA signalling and to promote GA/ethylene signalling (**see: Signal transduction – hormones**); moreover, several events may need to proceed in a sequential fashion. Later on in the process, the induction of cell wall hydrolases may weaken surrounding seed structures (e.g. the megagametophyte) to allow radicle protrusion, a factor which may also involve regulation by ABA and other hormones such as gibberellins (through regulation of cell wall rigidity) (**see: Germination – radicle emergence**). Post-germinative growth may also depend on cell wall hydrolases that play a role in reserve mobilization or participate in cellular growth/extension. It is possible that effective dormancy breakage is a prerequisite, not only for high germination rates, but also for optimal post-germinative growth. Thus, when germination/emergence conditions are not optimal, or in cases where conifer seeds germinate, but fail to undergo normal post-germinative growth, what events prevent vigorous growth of seedlings? These questions can only be addressed by further studies to elucidate factors that control the dormancy-to-germination transition, as well as the germination-to-growth transition.

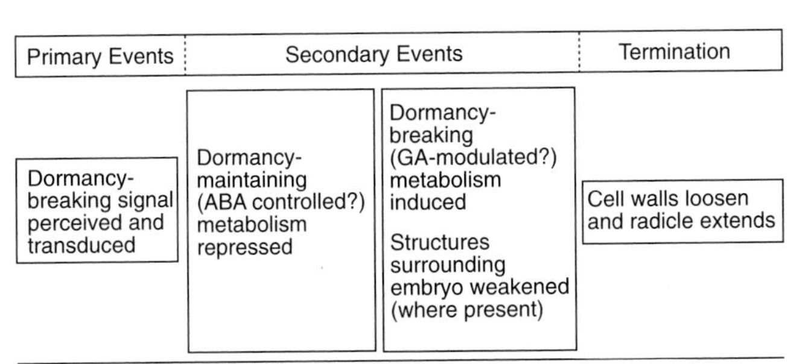

Fig. G.34. Time course of major events associated with germination and subsequent post-germinative growth. The time for events to be completed varies from several hours to many weeks depending on the plant species and the germination conditions. From: Bewley, J.D. (1997) Seed germination and dormancy. *The Plant Cell* 9, 1055–1066, with permission from American Society of Plant Biologists.

3. Cone and seed processing, seed deterioration during long-term storage

Cone processing includes cone conditioning, cone cleaning, kilning and extraction of seeds (Fig. G.35). Part of seed processing involves seed cleaning in which a multi-screened vibrational seed cleaner uses metal screens of various mesh sizes, shapes and arrangements to separate seeds from debris (**see: Cleaning**). Typically the moisture content of seeds is then reduced and non-viable seeds are removed; two types of equipment are used for this final step: aspirators or the gravity table. The seed lot is then blended to ensure a homogeneous product. For the B.C. Ministry of Forests Tree Seed Centre, registration requirements for purity (97%+) and moisture content (4.9–9.9%) must be met for a seed lot before it is placed into long-term storage at −18°C.

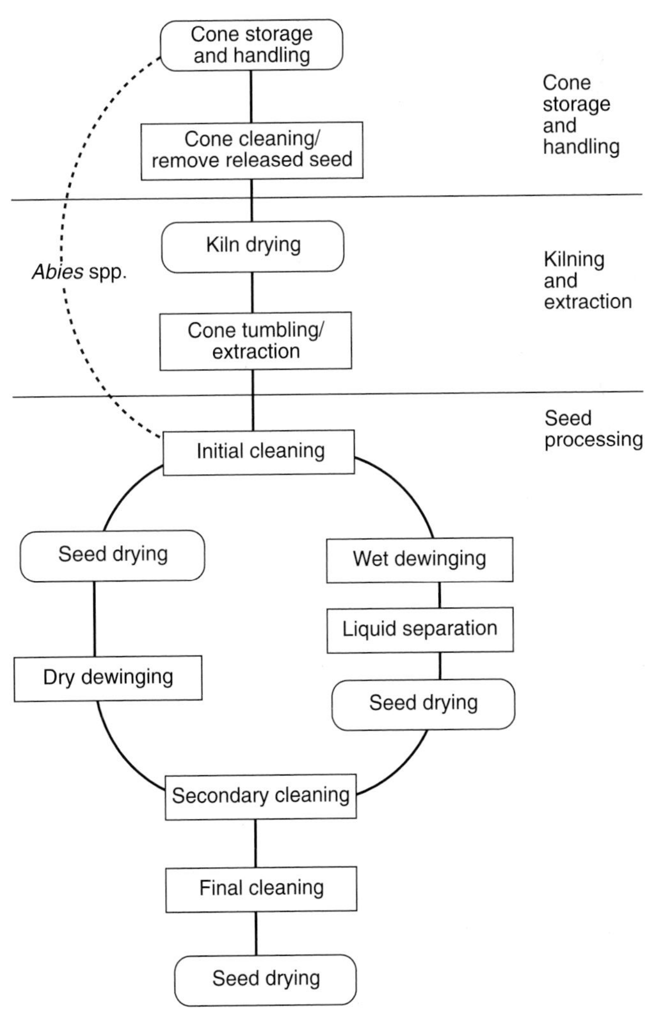

Fig. G.35. Typical cone and seed processing steps and sequence of activities. Sequence may vary and all steps may not be performed for each seed lot. From: Kolotelo, D., Van Steenis, E., Peterson, M., Bennett, R., Trotter, D. and Dennis, J. (2001) *Seed Handling Guidebook*. British Columbia Ministry of Forests, Tree Improvement Branch, Surrey, B.C., Canada, with permission from British Columbia Ministry of Forests.

The irregularity of good tree seed production makes long-term storage necessary to ensure that a consistent supply of high quality seeds is available. Additionally, long-term storage of the plant gene pool is required for conservation, breeding and improvement purposes, and this is best achieved by maintaining dry seeds under suitable conditions (**see: Conservation of seed**). Seed stocks of some conifer species (e.g. western red cedar and western hemlock) exhibit a significant reduction in their capacity for germination following a few years of cold storage at −18°C.

Some of the characteristics of imbibed deteriorated seeds (e.g. electrolyte leakage) indicate that membrane damage, potentially caused by the generation of reactive oxygen species, is one cause of reduced viability and germination. The rate of water penetration into the seed is critical to the success of germination. If water uptake is too slow, then germination is reduced because seeds may deteriorate; if water uptake is too rapid, seeds may suffer excess imbibitional damage (**see: Vigour testing**). NMR microimaging (**nuclear magnetic resonance**) may serve as a useful tool for predicting conifer seed viability.

4. Somatic embryogenesis and germplasm preservation

Maximizing the efficiency of seed storage to secure high levels of seed longevity, quality and genetic diversity is a key endeavour. Several coniferous species can be propagated via **somatic embryogenesis**, allowing the commercial propagation and long-term storage of superior genotypes (e.g. high genetic gain clones exhibiting rapid growth and rust/pathogen-resistant lines). The success of generating viable somatic embryos for a given species relies upon knowledge of how the seed environment (particularly the hormonal environment) promotes embryo maturation. In essence, researchers who develop specific culture procedures for maturation of somatic embryos are attempting to mimic the *in situ* environment generated by tissues surrounding zygotic embryos such that exogenously supplied compounds or treatments are added to promote reserve accumulation, accumulation of stress-protectants, and acquisition of germinability. There is also research into techniques for inducing desiccation tolerance so that embryos may be dried for storage and distribution.

Cryopreserved tissues (storage of 'induced' somatic tissue) can be considered as a clone bank or gene conservation method allowing preservation of unique genotypes and provides tree breeders and foresters the time required to conduct clonal testing (**see: Cryostorage**). (ARK)

Bates, S., Lait, C., Borden, J.H. and Kermode, A.R. (2001) Effect of feeding by the western conifer seed bug, *Leptoglossus occidentalis* (Heteroptera: Coreidae) on the major storage reserves of developing seeds and seedling vigor in Douglas-fir (Pinaceae). *Tree Physiology* 21, 481–487.

Cyr, D.R. (2000) Seed substitutes from the laboratory. In: Black, M. and Bewley, J.D. (eds) *Seed Technology and its Biological Basis*. Sheffield Academic Press, Sheffield, UK, pp. 326–374.

El-Kassaby, Y.A. (2001) Genetic improvement and somatic embryogenesis. *Canadian Tree Improvement Association, Tree Seed Working Group. News Bulletin* 33, 9–11.

Kermode, A.R. (2004) Seed germination. In: Christou, P. and Klee, H. (eds) *Handbook of Plant Biotechnology Online*. John Wiley, Chichester, UK, Chapter 33.

Gynophore

The stalk on which the developing **pod** is carried underground as in various types of groundnut (**see: Bambara groundnut, Peanut**).

Hair

See: Trichome

Half-seed technique

A non-destructive technique used in oilseed breeding programmes, involving the excision of one of the cotyledons from a single seed for chemical analysis of **fatty acid** composition, which does not affect the viability of the remaining parts of the embryo. Initially developed for oilseed rape (**canola**), variations of this technique have been developed for most oilseeds, including **sunflower** where a small portion of one cotyledon distal to the embryonic axis is removed and analysed.

Haemoglobin

Composed of a haem group (an iron-containing porphyrin) and a globin protein. Forms of haemoglobin (Hb) occur in plants including in seeds, for example barley **embryos** and **aleurone layers**; an increase in Hb transcripts occurs soon after imbibition, which is enhanced in conditions of limited oxygen. The function of Hb in seeds is not entirely clear but they are implicated in the maintenance of cell metabolic rates and **ATP** production under conditions of low oxygen availability, in limiting fermentation under such hypoxic conditions, and in nitrate and nitric oxide metabolism, including scavenging of the latter to maintain a favourable redox potential and energy balance. (MB, JDB)

Igamberdiev, A.U. and Hill, R.D. (2004) Nitrate, NO and haemoglobin in plant adaptation to hypoxia: An alternative to classic fermentation pathways. *Journal of Experimental Botany* 55, 2473–2482.

home.cc.umanitoba.ca/~rhill/

Hammer mill

Seed conditioning equipment used to complete the **threshing** process of some grasses, etc., by removing **awns**, **hairs**, **lint**, **chaffy** appendages, and loosening seed from pods and seed clusters. (**See: Conditioning, I. Precleaning**)

Haploid

A term indicating one-half the normal somatic diploid complement of chromosomes. The germ cells (gametes) in diploid plants normally have one set of chromosomes, and are therefore haploid.

Haploid plants contain the gametic chromosome number. If produced from a diploid species they are called monohaploid, or if produced from a **polyploid** plant they are called polyhaploid. Haploid plants occur spontaneously at low frequencies in many species, or can be artificially produced. They are used in breeding programmes, for several purposes in basic research including the identification of useful recessive mutants through **quantitative trait locus** (QTL) detection techniques, and have been used in creating cultivars of barley, canola, maize and tobacco amongst other crops. Haploid plants are also used to produce 'doubled haploid' with twice the chromosome content, induced by chemical treatment with **colchicine**. In this way highly homozygous plants are produced in one generation, as an alternative to conventional inbreeding taking several generations. For example, embryos produced *in vitro* from the young pollen grains (microspores) of a set of F_1 plants can be used to create an array of double haploids, each carrying a unique combination of traits of both parent plants. (**See also: Cross; Diploid; Heterosis; Ploidy**)

Haploid embryogenesis

When haploid tissues of plants are cultured on inductive media, haploid embryos can be produced (Colour Plate 11D). The developmental process that produces pollen-derived embryos is called androgenesis. When megagametophytes or embryo sacs are induced, it is called gynogenesis. Androgenesis is the preferred route because pollen is both abundant and easily isolated. In **angiosperms**, androgenesis is more easily induced than gynogenesis. The opposite is the case for **gymnosperms**. Androgenesis can be achieved by directly culturing anthers or pollen, but some species require pretreatment. This is usually an applied stress, such as cold treatment or heat shock. Subsequently, male parts are placed on induction-enhancing medium. Haploid embryos produce plants that may spontaneously diploidize, resulting in homozygous plants that represent uniform genetic backgrounds of great genetic and experimental value. Haploid cells may also be induced to diploidize by treatments, of which the most common is application of **colchicine**. Haploid cultures also supply plant material that can be used in protoplast fusion experiments to create interspecific and intergeneric hybrids of value (there are a number of such products in the Brassicaceae). (**See: Double haploid**) (PvA)

Hard seeds

A hard **seedcoat** (structural impermeability of the seedcoat to water and gases) is a widespread phenomenon in species of the Fabaceae and other plant families, including **tree seeds** (**see: Seedcoats – structure**). In testing of such species no attempt is made to germinate any seeds remaining at the end of the test period, which are reported as 'hard seeds'. Hard seeds is also a term used in relation to the resistance to cooking of some legume seeds such as chickpea and common bean. (See: **Bean – common; Dormancy; Germination testing; Hardseededness**)

Hardening

In seed technology: a term (coined by the Russian scientist P.A. Henckel in the first half of the 20th century) used for physiological seed **enhancement** treatments based on successive (typically 2 to 3) cycles of **steeping** with water and drying, whose benefits can include improved tolerance of adverse environmental conditions, such as drought, as well as germination advancement. Hardening techniques of this type are not in widespread commercial use, probably because these manipulations are time-consuming. In a different sense of the term, seedlings are drought-hardened prior to **transplanting** by exposure to full sunlight and transient water stress to ensure their greater survival.

Heydecker, W. and Coolbear, P. (1977) Seed treatments for improved performance – survey and attempted prognosis. *Seed Science and Technology* 3, 353–425.

Hardseededness

Seeds of many species possess hard **seedcoats** that are poorly permeable to water, a phenomenon sometimes referred to as hardseededness. It is especially common in the Fabaceae (Leguminosae), and also found in the Cannaceae, Chenopodiaceae, Convolvulaceae and Malvaceae. Hard seedcoats are relatively thick, with layers of lignified sclereids, often impregnated with tannins and pigments and covered with cutin (Fig. H.1). Because water entry into the seeds is impeded their germination may be delayed for substantial periods of time, sometimes for years. This might be considered as a type of **dormancy**. (MB)
(See: **Dormancy – coat-imposed; Germination testing; Seedcoats – structure**)

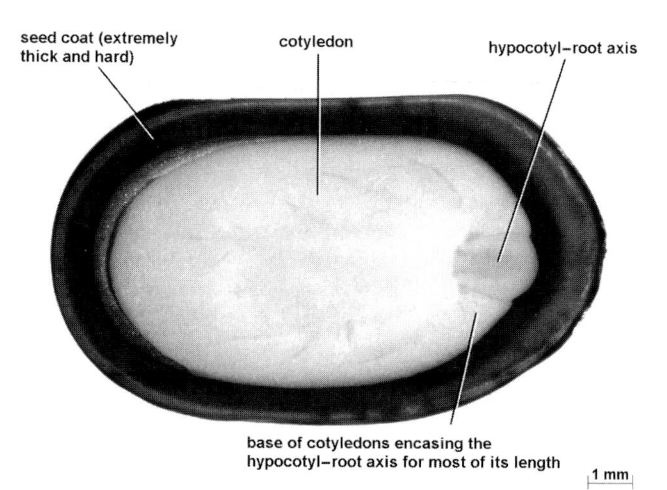

Fig. H.1. Longitudinal section of seed of *Delonix floribunda* (Fabaceae) showing the hard, thick seedcoat (photo by Wolfgang Stuppy, Millennium Seed Bank, RBG, Kew).

Harrington's Thumb Rules

Also known as Harrington's Rules of Thumb, these are guidelines developed by seed technologist James Harrington in the 1950s, to describe seed storage environments that were safe for commercial seed producers or use. The Rules prescribe a 1% drop in water content or a 10°F decrease in temperature (6°C) to double seed lifespans. Harrington's 'Hundreds Rule' says that the sum of storage RH and temperature in °F should be less than 100 to maintain seed **viability** for about 5 years. (CW)
(See: **Viability equations**)

Harvest index

The proportion of the biological **yield** that is grain or seed. (See: **Yield components; Yield determination; Yield penalty; Yield potential**)

Harvest maturity

The stage of **development** at which a seed, or the majority of a seed population, is best suited to harvesting in high **quality** and yield, considering its safe storage, its handling characteristics to minimize mechanical injury, and potential field losses due to its inefficient collection by harvesting equipment. In most cases harvest maturity is reached after **physiological maturity**, when seed has first attained its maximum **viability** and **vigour**, and which is generally achieved at seed **moisture contents** that are too high for the purposes of safe **storage management** and to minimize potential **deterioration and longevity** and spoilage due to **storage pests** and **storage fungi**. **Swathing** may be appropriate to allow curing in the field, or to avoid undue losses due to **shattering**. However, the longer it remains in the field after physiological maturity, the more seed is subject to environmental factors that can result in harvest losses. In practice, harvest maturity dates and storage requirements vary among crops depending on weather conditions, and even within a crop if it is being used for different purposes, such as for sowing or for food or industrial processing. (PH)
(See: **Production for Sowing, IV Harvesting**)

Haustorium

Generally refers to an organ (e.g. an outgrowth of stem or root) that functions as a sucker to provide the plant or a specific organ with water and/or nutrients. The term is usually applied in connection with parasitic or hemiparasitic plants, but haustoria can also be formed in developing seeds, e.g. by the **endosperm**. Endosperm haustoria can be micropylar, chalazal or both and facilitate the absorption of nutrients from surrounding tissues. They can be persistent in the mature seed and then serve in blocking the micropyle or chalaza, e.g. *Marcgravia* spp. In some dicots, usually one or both of the **cotyledons** may remain embedded in the endosperm and act as a conduit for the products of reserve mobilization to the growing seedling, e.g. in asparagus, onion, yucca (Fig. H.2) and *Peperomia peruviana*. In the germinated seed of the monocotyledonous date and other palms the cotyledon acts as a haustorium, absorbing nutrients from the endosperm. In *Welwitschia* the hypocotyl acts as a haustorium (**see**: Fig. G.30 in **Gymnosperm seeds**) (See: **Mobilization of reserves – a monocot overview**) (JD, WS)

Hawaiian baby wood rose

Seeds of this species (*Argyrea nervosa*, Convolvulaceae) are used because of the hallucinogenic effects caused by the indole-type

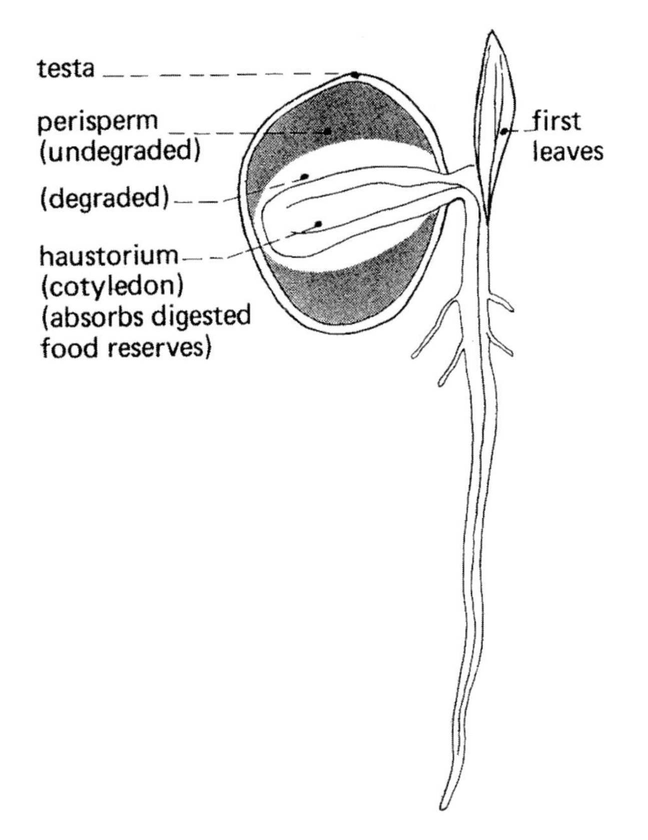

Fig. H.2. An emerged seedling of yucca, with one cotyledon remaining within the nutritive endosperm, as a haustorium.

alkaloids, structurally related to lysergic acid diethylamide (LSD), that they contain (**see: Pharmaceuticals and pharmacologically active compounds; Psychoactive seeds**).

Hazelnuts

The most important species are *Corylus avellana* and *C. maxima* L. (Betulaceae), native to temperate climates of Europe and Asia Minor, though there are several other *Corylus* spp. in different parts of the world. Other names are cobnuts and filberts.

The hazelnut is a true nut (**see: Nut**) with a woody pericarp that loosely surrounds the seed, consisting of a thin testa and embryo (kernel). The globoid nuts which measure approx 1.5 × 2.0 cm long are borne within a green, leafy husk or involucre. The seeds are approx. 1.2 cm × 1.4 cm wide: filbert seeds are somewhat larger (Fig. H.3).

According to FAOSTAT, approximately 697,000 t of hazelnuts were produced in 2003, principally in Turkey (70% of world production), Italy (12.5%) and the USA (4.5%). (**See: Crop Atlas Appendix, Map 4**)

The kernels are used in bakery and confectionery, chopped, ground and whole. Hazelnut oil (**see: Nut**, Table N.2) is used for culinary purposes. (VP, MB)

Vaughan, J.G. and Geissler, C.A. (1997) *The New Oxford Book of Food Plants*. Oxford University Press, Oxford, UK, pp. 30–31.

Fig. H.3. Hazelnuts. (A) Whole nuts (image by Mariko Sakamoto) and (B) filbert kernels (seeds) (image by Mike Amphlett, CABI).

Health testing

Seed health is one of the main aspects of the seed **quality testing** system that underpins commerce and trade in seed for sowing. Seed health refers primarily to the presence or absence of disease-causing organisms – **pathogens** such as fungi, bacteria, viruses, nematodes and of disease-insects, though physiological conditions such as trace element deficiency may be causal elements as well. The objective of testing is to determine the pathogen infection status of a seed sample, and by inference that of the seedlot from which it was taken, which is important for two main reasons:

- for quality assurance – in evaluation of planting value, in **certification schemes**, and to support decisions about the need for seed **treatment**; and
- for the detection of plant quarantine organisms, such as to meet phytosanitary regulations for seed that is exported for trade.

Additionally for grain, tests determine the presence of **storage fungi**, which can affect the storage quality or its value for feed and human consumption.

To be worthwhile, seed health testing methods must deliver:

- specificity – the ability to recognize the target pathogen from all other organisms present on the seeds;
- sensitivity that is fit for the purpose of the test;
- information relating to the field performance of the crop (except in cases of quarantine inspections);
- test results that are repeatable within, and between, samples of the same seedlot, regardless of who performs the test, within statistical limits and sample variability;
- and they must be cost effective.

It is often difficult to meet all of these requirements, because seed health testing is not only dependent on the pathogens themselves but also on the presence of other microorganisms on the seed, which may be antagonistic to, or synergistic with, each other in some test methods such as incubation tests (see below).

1. Qualitative and quantitative tests

There is a wide range of seed health testing methodologies, including relatively simple direct inspection methods for macroscopic fungal bodies, incubation tests, growing-on tests, bacterial extraction and bacterial identification, and the more complex immunoassays and molecular assays, such as **ELISA** and polymerase chain reaction (**PCR**). The test used depends on the organism that is being tested for, and the purpose of the test.

Qualitative assays establish whether seed is infected with a plant pathogen, with a very low probability of 'false negatives'. Qualitative tests are necessary in the case of plant quarantine, where the aim is to prevent non-indigenous diseases establishing in the importing country. However, a test can also be used for quality control purposes, at pre-defined tolerance and accuracy levels.

Quantitative tests are designed to estimate the true level of infection in a seedlot: that is, how many individual seeds are infected, and the level of infection or contamination. Such information can inform decisions about whether or not seed treatment is justified. Most fungal pathogens are currently determined using quantitative methods. Bacterial and viral pathogens, where severe epidemics can result from relatively low numbers of infecting microorganisms and there are few control measures in the field, are commonly determined using qualitative methods.

2. Fungal seedborne pathogens

(a) *Direct inspection methods*. Samples of seed may be visually examined for the presence of **ergots** (*Claviceps purpurea* in cereals and grasses), other **sclerotia** (*Sclerotinia sclerotorium* in brassicas) and **smut** balls (bunts caused by *Tilletia tritici* in wheat). Although this type of test provides information quickly and may be readily applied in seed testing stations in conjunction with purity testing, only a few diseases are adequately detected in this way.

Seed samples may be immersed in water or other liquid to make fungal bodies (for example, pycnidia) or symptoms (for example, **anthracnose** on seedcoats) more visible, or to encourage the liberation of fungal spores, hyphae and so on that are attached to or carried within the seed. Examination of soaked seeds is by means of a stereoscopic microscope or, where seeds are immersed in liquid and shaken, the liquid is examined at a higher magnification using a compound microscope.

Only a few fungi are adequately detected using these methods, such as *Septoria* on celery seed. Moreover, the viability of any detected fungi is uncertain, and additional tests may be required. Examination of seed washings is only valuable if the spores are known to lead to field infection, for example, in the case of common or stinking **bunt** (*Tilletia tritici*).

Where seedborne infection is more deep-seated within the seed tissues, such as the embryo, fungal staining methods are used. Tissues are normally extracted or macerated in sodium hydroxide, stained using a combination of lactic-acid:lactophenol and aniline:trypan-blue dyes, and examined for fungal mycelium under a stereoscopic microscope. Examples are tests for loose smut in barley, *Neotyphodium* fungal endophytes in tall fescue and ryegrass, and downy mildew in maize.

(b) *Incubation tests – agar plate and blotter substrates*. The agar plate test gives an indication of the viable fungal or bacterial **inoculum** present in an infected seed sample, and is best used for high-incidence pathogens that occur in seed samples with greater than 1% infection. Some examples of pathogens tested for using the agar plate method are *Microdochium nivale* on wheat, *Ascochyta pisi* on pea and *Phomopsis* spp. on soybean. Sample sizes are normally from 200 to 400 seeds, which are evenly placed on to solidified sterile agar medium in lidded plates (normally Petri dishes). Potato dextrose agar and malt agar are commonly used to encourage the growth of seedborne fungal pathogens, but there are many variations: for example, acidic agars may be used to reduce bacterial contaminants, and media may be made semi-selective by the addition of specific chemicals, antibiotics or fungicides. The planted seeds are incubated at a fixed temperature in the dark for a specified number of days, usually 7. Near-infrared light may be used to encourage the development of fruiting bodies. After incubation the characteristic mycelial growth of the fungi is used to identify the seedborne pathogen(s) under test. Where growth of fungal saprophytes from the seed is excessive, to the extent that the pathogen is concealed, or where it is desirable to identify internal infection only, seeds may be surface sterilized by soaking in 1% available chlorine bleach.

The blotter (paper) test gives an indication of seedborne fungi on the seed, as shown by the presence of mycelium and fruiting bodies and, in some tests, by symptoms on young plants. Tests differ depending on the fungus being tested for, but most involve sowing seeds in suitable containers on moist absorbent paper blotters (usually 7 days in Petri dishes or boxes at a specified temperature, and normally under alternating cycles of light and dark) and detection of fruiting bodies by microscopic examination. In some cases seed germination is suppressed or seeds are killed using chemicals or by deep-freezing, to increase the ability of pathogens to grow. *Phoma lingam* in brassicas, *Alternaria dauci* in carrots, *Colletotrichum lindemuthianum* on *Phaseolus* beans and *Bipolaris oryzae* in rice are all examples of seedborne pathogens tested for using different variations of the blotter test. The blotter test is limited by the fact that it provides excellent conditions for the development of a large number of fungi not just the required pathogen and it is not always possible to determine pathogenicity. As in the agar method, surface disinfection of seed can be used to reduce surface contaminants.

3. Bacteria and viruses

Because of the virulence of some of the many bacterial and viral pathogens, infection thresholds are very low (**see: Disease**), and for statistical reasons their testing requires large numbers of seeds. Samples of 10,000 seeds for testing *Xanthomonas* on horticultural brassicas and 30,000 seeds for seedborne lettuce mosaic virus, are common for example.

(a) *Grow-out/Growing-on tests*. These tests are performed on plants grown from seed samples beyond the seedling stage in a greenhouse, controlled-environment chamber or the field and the seedlings/plants are observed for symptoms of the pathogen. They have been used for testing for the effects of many seedborne bacteria and viruses, for example lettuce mosaic virus (LMV) and *Pseudomonas syringae* pv. *phaseolicola*, and for fungi, for example, anthracnose in lupin. Grow-out tests give an indication of potential transmission from external and internal seed inoculum under the environmental conditions used. However, both fluctuations in those conditions and viability of the seed stock may influence the test result. The tests are also time consuming and require a considerable amount of space where large numbers of seeds require testing for the presence of low-incidence pathogens.

(b) *Laboratory tests*. Laboratory methods for the detection of seedborne bacteria involve three stages: (i) their extraction from seeds; (ii) isolation into culture; and (iii) identification.

Bacteria are extracted from seeds or from 'seed flour' (obtained by grinding or milling untreated seed) in a liquid medium – usually sterile pH-buffered saline, though buffered sterile water with various other enrichments can be used. The volume and agitation of the liquid medium together with the duration and temperature of soaking are all critical to the optimum recovery of the target pathogen, which may also be affected by the saprophytic microflora of seeds and inhibitor compounds in seeds. Some seedborne bacteria can be identified directly following isolation on general plating agar media; for example, *Pseudomonas syringae* pv. *phaseolicola* is cultured on 'Kings B' medium. These non-selective media methods are most effective when high levels of pathogen and low levels of saprophytes are present in extracts. Where high levels of saprophytes and low levels of pathogen are present, semi-selective media can be used, in which chemical agents are used to reduce the growth of saprophytes. Further identification may be achieved by morphological and biochemical tests, by immunoassays or other molecular biological methods (see below). All of these tests give presumptive diagnoses and most require confirmation by a host pathogenicity test.

(c) *Host pathogenicity test*. Either the pure culture of a bacterium or the crude seed extract medium is inoculated into the host plant or seeds to test the pathogenicity of the bacteria. Plants are inoculated by a number of different methods, including injection, spraying following leaf abrasion and vacuum infiltration.

4. Molecular test methods

The relative expense of traditional seed health testing methods, and the capacity requirements for trained personnel has led to interest in using new techniques to improve the speed, accuracy and sensitivity of diagnosis, concentrating on the use of immunodetection using polyclonal or monoclonal antibodies in **immunoassays** (e.g. ELISA), and PCR mainly for viruses and bacteria. Once developed and validated, molecular assays can provide relatively simple, quick, high-throughput diagnosis. However, these methods tend to be expensive to set up and apply, and advantages have to be considered against the cost of cheaper methodologies.

Although a number of commercial tests now exist, including PCR tests for *Pyrenophora* spp. in barley, anthracnose in lupin, *Tilletia tritici* and *Microdochium nivale* in wheat and *Pseudomas syringae* pv. *pisi* in peas, the application of DNA molecular techniques in routine seed health testing has not been fully exploited. The majority of viruses have RNA genomes that at first made them unsuitable for PCR, but this has been overcome by the development of reverse-transcription PCR assay (using reverse transcriptase to synthesize the corresponding DNA from the viral RNA, immediately before PCR), which has been the basis of a successful method, for example, in the detection of pea seedborne mosaic virus (PSbMV).

5. Relationship between seedborne infection and disease transmission

The relationship between the level of pathogen infection or contamination in a seedlot and the incidence of disease in the resulting crop is complex, and varies according to the pathogen, its location in or on the seed and the environmental conditions in which the crop is grown – and seed treatment practices, such as with fungicides. Temperature and moisture are probably the key environmental factors, but soil type, the composition of soil microflora and the effects of host genotype may also be important. If the pathogen is carried within the seedcoat tissues, infection of seedlings will usually be greatest when germination or emergence is slowed down by adverse conditions, for example at low or very high seedbed temperatures. If the pathogen is carried within the embryo, however, the relationship between seedborne infection and the incidence of disease in the field will usually be close.

Permissible infection standards depend on knowledge of the epidemiology of the disease, and may be set at national or regional regulatory levels as appropriate. If the inoculum is in the form of fungal spores or bacterial cells on the seed surface, the percentage of seeds infected may be very high but disease incidence in the crop may be low because, for example, spore viability has declined during storage or conditions during plant growth may be unfavourable for infection. On the other hand, seedborne infection may be critical at very low infection levels if the pathogen has the potential to multiply rapidly within the crop (such as subspecies of *Xanthomonas campestris*, black rot in horticultural brassicas, and of *Acidovorax avenae*, bacterial fruit blotch in cucurbits) if there is a risk of introducing a pathogen into a new area or country or if the seed is to be used for multiplication over several generations. For instance, seedborne infection of *Phaseolus vulgaris* bean by *Pseudomonas phaseolicola* at the rate of one infected seed in

1000 is sufficient to start an epidemic within the growing crop. Depending on the environmental conditions in which the seed is grown, standards may differ across regions. For example in *X. campestris* and lettuce mosaic virus, laboratories use thresholds ranging from 'nil in 10,000' or 'nil in 30,000' seeds: more infected seedlots are rejected.

Other diseases, such as *Microdochium nivale* (seedling blight), in which there is no multiplication within the growing crop or from year to year, may have higher standards or thresholds: in the UK, for instance, an advisory standard of 10% infection is permitted, above which seed treatment is advisable to control the disease. Different standards may also be applied to seed categories depending on the risk, such as to earlier generations in a seed multiplication programme compared to the seed sown to produce the crop for consumption. Appropriate cost–benefit analysis is required for each case. Practical economic disease contamination thresholds are also a pressing issue in **organic seed** production where for many seedborne diseases there are no technically or economically feasible organically acceptable control methods available. (VC, PH)

(See: **Thermotherapy; Treatment – pesticides, 3. Fungicides**)

Hutchins, J.D. and Reeves, J.C. (eds) (1997) *Seed Health Testing. Progress Towards the 21st Century*. CAB International, Wallingford, UK.

Mathur, S.B. and Kongsdal, O. (2003) *Common Laboratory Seed Health Testing Methods for Detecting Fungi*. International Seed Testing Association, Basserdorf, Switzerland.

Heat shock proteins (HSPs)

A set of proteins that are synthesized as a response to heat stress and/or during seed development in the absence of significant changes in temperature. Known as chaperones, they comprise five conserved nuclear-encoded gene classes that have related functions: to overcome the problems associated with protein misfolding and aggregation through class-specific mechanism of action (degradation, refolding, prevention from aggregating). (See: **Desiccation tolerance – protection by stabilization of macromolecules; Storage protein – synthesis**)

Heat units

A concept commonly used in describing the **thermal time** requirement for plant physiological processes to be completed, including the germination behaviour of seed populations, **bolting**, **vernalization**, and seed development and maturation (such as days to flowering in **soybean**). The developmental event occurs after a characteristic required number of units have been accumulated, in relation to a characteristic base temperature for the species or cultivar. Heat units are often expressed as **'day-degrees'**. (See: **Germination – field emergence models**)

Hemianatropous

A type of **ovule**. **Orthotropous** and **anatropous** ovules are connected by a complete range of transitional stages depending on the insertion of the funiculus. If this meets the ovule or seed at an angle of about 90 degrees, the ovule or seed is called hemi(ana)tropous. (See: Fig. R.3 in **Reproductive structures, 3. Female**)

Hemicellulase

A general term used for enzymes that hydrolyse hemicelluloses. (See: **Hemicellulose – mobilization**)

Hemicellulose

Hemicelluloses are large, mixed (hetero)polymers of several sugars which are extracellular, often as components of cell walls. Although hemicelluloses is a widely used term it is inaccurate since they are not related chemically or biosynthetically to **cellulose**, as was assumed when they were so named over a century ago.

They are principally composed of combinations of five sugars, the hexoses glucose, galactose and mannose, and the pentoses xylose and arabinose, comprised of one or two sugars in a long backbone chain with short side chains (branches, substitutions) varying in number and spacing, usually of one or two sugars different from the backbone (Table H.1). Correct naming of the individual hemicelluloses often refers first to the side chain sugar(s) and then the backbone sugar, e.g. galactomannan is composed of a mannose backbone with single (unit) galactose residues as side chains, while arabinoxyloglucans have a glucose

Table H.1. The most common cell wall storage hemicelluloses: structure, synthesis and mobilization.

Hemicellulose	Main chain sugar	Side chain sugar (rare)	NDP-sugar used in synthesis	Mobilizing enzymes	Examples where present in seeds
Mannan	mannose	none (galactose)	GDP-Man	endo-β-mannanase	coffee, ivory nut, caraway
Galactoglucomannan	mannose, glucose	galactose	GDP-Man, UDP-Gal, UDP-Glc	endo-β-mannanase, β-glucanase, α-galactosidase	uncommon, sometimes mixed with other mannan polymers
Galactomannan	mannose	galactose	GDP-Man, UDP-Gal	endo-β-mannanase, α-galactosidase	lettuce, tomato, legumes
Xyloglucan	glucose	xylose, (galactose), (arabinose)	UDP-Glc, UDP-Xyl, UDP-Gal	α-xylosidase, β-glucanase	legumes
Arabinogalactan	galactose	arabinose	UDP-Gal, UDP-Ara	α-arabinosidase, exo-β-galactanase	legumes, esp. lupins

Adapted from: Buckeridge, M.S., dos Santos, H.P. and Tiné, M.A.S. (2000) Mobilisation of storage cell wall polysaccharides in seeds. *Plant Physiology and Biochemistry* 38, 141–156.

backbone with side chains of single xylose residues or of arabinose and xylose. The nomenclature is less clear with respect to hemicelluloses with mixed backbones however, e.g. galactoglucomannans, which have both a glucose and mannose in the backbone, and occasional unit side chains of galactose on the mannose residues.

The major storage hemicelluloses in seeds are polymers of the sugar mannose, particularly as mannans, glucomannans or galactomannans, which are reserves of carbohydrate present in the **endosperm** and **perisperm** cell walls of some species. More rarely, seeds store xyloglucans (**amyloids**) in the **cotyledon** cell wall, e.g. tamarind, nasturtium and mustard, or arabinogalactans, e.g. lupins. Hemicellulose-rich seeds do not usually store **starch**.

Non-storage hemicelluloses occur as the structural elements in the endosperm of cereals (e.g. arabinoxylans and mixed-linkage β-glucans). Hemicellulose cell walls are considerably less complex than the cellulose- and pectin-rich primary cell walls present in non-storage tissues, and require fewer enzymes for their dissolution; this is advantageous for their rapid degradation to permit the penetration of other enzymes involved in the mobilization of stored cellular reserves. (**See: Mobilization of reserves – cereals**) In the seedcoat and **pericarp** hemicelluloses are often present as **mucilages** or **gums**, although the latter term is applied more widely in industry to any viscous polysaccharide hemicellulose, including galactomannans and xyloglucans (Table G.16, in **Gums**).

Galactomannans are particularly prevalent in endospermic legumes (such as **guar**, **carob** and **fenugreek**), polymers rich in mannans in hard-seeded date palm, **coffee** and ivory nut, and glucomannans in seeds of the lily (e.g. asparagus) and iris families. The relatively thin endosperms of some species, e.g. tomato, lettuce and soybean, also contain galactomannans and in the former two are debatably involved in **coat-enhanced dormancy**. Rarely, the mannan-rich cell walls completely fill the cell, thus obliterating the cytoplasm, e.g. in the fenugreek and guar endosperm, and only a surrounding one-cell-thick living **aleurone layer** remains. In most species the thickened cell wall surrounds a living cytoplasm and no aleurone layer is present.

In addition to their storage role, galactomannans may play a protective role for the germinating embryo. They swell and become mucilaginous as the seed imbibes water and thus maintain a high water environment for the embryo in times of temporary drought. (JDB, PH)

(See: Amyloid; Arabinogalactan; Arabinoxylan; Cellulose; Galactomannan; (β-1,3),(β-1,4)-Glucan; Glucomannan; Gums; Hemicellulose – mobilization; Hemicellulose – synthesis; Mannans; Mucilages; Pectin; Xyloglucan)

Bewley, J.D. and Reid, J.S.G. (1985) Mannnans and glucomannans. In: Dey, P.M. and Dixon, R.A. (eds) *Biochemistry of Storage Carbohydrates in Green Plants*. Academic Press, London, UK, pp. 289–304.

Carpita, N. and McCann, M. (2000) The cell wall. In: Buchanan, B.B., Gruissem, W. and Jones, R.L. (eds) *Biochemistry and Molecular Biology of Plants*. American Society of Plant Physiologists, Rockville, MD, USA, pp. 52–108.

Reid, J.S.G. (1985) Galactomannans. In: Dey, P.M. and Dixon, R.A. (eds) *Biochemistry of Storage Carbohydrates in Green Plants*. Academic Press, London, UK, pp. 165–288.

Hemicellulose – mobilization

A process which requires the activity of hydrolytic enzymes that are specific to the sugar(s) and nature of the bond(s) in a hemicellulose polymer. These may include endo-enzymes, which cleave the backbone chain randomly into smaller oligomers, and exo-enzymes which remove sugars one at a time from the chain terminus, or the short side-chains. Enzymes which specifically cleave the released oligomers may also be involved. An example is the enzymes required for the mobilization of galactomannans, shown in Fig. H.4.

The enzymes involved in hemicellulose degradation are given the general term of **hemicellulases**, and trivial general names such as xylanases, arabinases, glucanases, to indicate their substrates are polymers of xylose, arabinose or glucose, for example. Their specific (systematic) names indicate the substrate they hydrolyse, as shown above.

In many seeds, cells of the **cotyledons** or **endosperms** which store hemicelluloses in their walls remain metabolically active, and following germination synthesize and secrete hemicellulases into the **cell walls**. The products of hydrolysis are absorbed into the cells and converted to sucrose. Conversion may occur in the endosperm and is followed by their absorption into the embryo, and/or conversion of the products of hydrolysis may occur within the embryo itself. In some endospermic legumes where the contents of the cells are occluded by the cell wall hemicelluloses, e.g. by galactomannans in fenugreek, a living aleurone layer surrounds the endosperm and synthesizes and secretes the hemicellulases to effect wall hydrolysis. Galactose and mannose are absorbed by the cotyledons and converted to sucrose, although the metabolic steps involved in this process have not been studied in detail. Conversion must occur quickly upon absorption into the cells of the cotyledons because accumulation of mannose and galactose can be toxic to

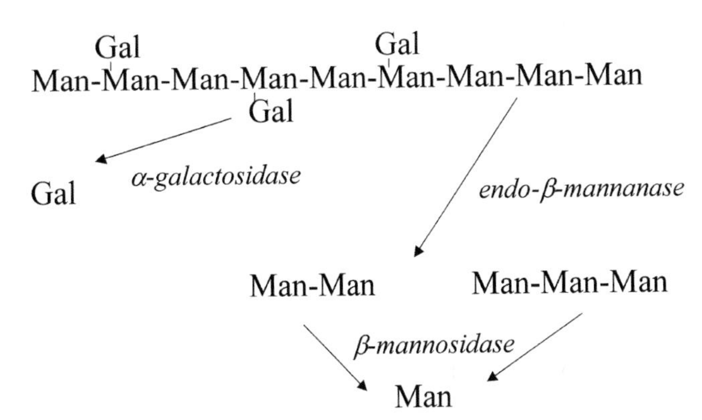

Fig. H.4. Enzymes required for the mobilization of galactomannan: endo-β-mannanase [1→4-β-D-mannan endohydrolase], β-mannosidase [β-D-mannoside mannohydrolase] and α-galactosidase [α-D-galactosidase galactohydrolase]. Endo-β-mannanase and α-galactosidase concurrently hydrolyse the galactomannan polymer to yield galactose and mannobiose and mannotriose, the latter two being hydrolysed to mannose by β-mannosidase.

metabolism. If excess sucrose is produced in the cotyledons it is temporarily converted to starch, and remobilized after endosperm hydrolysis is completed.

The site of synthesis of hemicellulases within the cell is generally accepted to be the endoplasmic reticulum, which are then secreted from the cell via vesicles, but there is little definitive evidence.

Major mobilization of hemicellulose-rich cell walls of the endosperm or cotyledons (**see: Xyloglucan**) occurs after germination is completed, although in some seeds weakening of the hemicellulose-rich cell walls in the region of the radicle may be required to allow germination to be completed (**see: Dormancy; Germination – radicle emergence**). The control of hemicellulase synthesis is not clearly understood. In some seeds, e.g. tomato and lettuce endosperms, there is suppression of hemicellulase production (endo-β-mannanase) by **abscisic acid**, and possibly in fenugreek, although seedcoat inhibitors (e.g. saponins) have been proposed to play a role in the latter. **Gibberellins** promote an increase in endo-β-mannanase activity in the endosperms of tomato and lettuce seeds. An increase in enzymes to degrade the arabinoxylan- and mixed-glucan-rich cell walls of the endosperm of barley grains (arabinases, xylanases, β-1,3-, β-1,4-glucanases) which are produced and secreted by the **aleurone layer**, occurs in response to gibberellic acid. In many seeds no response in hemicellulase activity to plant growth regulators has been reported. (**See: Mobilization of reserves – cereals**) (JDB)

Bewley, J.D. and Black, M. (1994) *Seeds. Physiology of Development and Germination*, 2nd edn. Plenum Press, New York, USA.

Buckeridge, M.S., dos Santos, H.P. and Tiné, M.A.S. (2000) Mobilisation of storage cell wall polysaccharides in seeds. *Plant Physiology and Biochemistry* 38, 141–156.

Hemicellulose – synthesis

A complex process which occurs during seed development and involves many enzymes and intermediates. Storage hemicelluloses are synthesized within the cell and deposited within the surrounding cell wall. In **cotyledons** and many **endosperms** the cells retain their cytoplasmic and metabolic integrity during and following hemicellulose synthesis, but in the endosperms of some species the ingrowing cell wall may occlude the cytoplasm and the cells are non-functional at seed maturity. (**See: Hemicellulose**)

In general, hemicellulose synthesis in seeds requires the import of sucrose from the parent plant, and its conversion to glucose-1-phosphate (Glc-1-P). The Glc moiety is attached to a nucleotide triphosphate (NTP), either uridine triphosphate (UTP) or guanidine triphosphate (GTP) to form a nucleotide diphosphate sugar (UDP-Glc or GDP-Glc) and inorganic phosphate (PPi). The NDP-Glc may be used in hemicellulose synthesis, or converted to other NDP-sugars by enzymic modification of the Glc. GDP-Glc can be converted to GDP-mannose, and then to GDP-fucose. UDP-Glc can be converted to UDP-Gal, as can their respective uronic acids. UDP-arabinose and UDP-xylose can be formed from UDP-glucuronic acid. If the individual sugars are released from hemicellulose polymers, they can be converted by a salvage pathway to their sugar-1-phosphate form and combined with NTP to produce the NDP-sugar, which can then be subjected to further interconversions, as above (Fig. H.5).

The synthesis of polymeric hemicelluloses requires donation of the sugar moiety from the NDP-sugar to a short primer to form the backbone and the short (often single) side chains (Table H.1.). Enzymes specific for each NDP-sugar are involved in the synthesis of the polymer. Mannan synthase (GDP-Man dependent mannosyl transferase) makes the β-1,4-linked mannan backbone from GDP-Man and α-galactosyltransferase adds galactosyl residues to this from UDP-Gal. The site of synthesis probably is the Golgi apparatus (β-mannan synthase is located therein in guar seeds, for example) from where the polymers are transported from the cell cytoplasm via vesicles and released to the inner side of the surrounding cell wall. (JDB)

Carpita, N. and McCann, M. (2000) The cell wall. In: Buchanan, B.B., Gruissem, W. and Jones, R.L. (eds) *Biochemistry and Molecular Biology of Plants*. American Society of Plant Physiologists, Rockville, MD, USA, pp. 52–108.

Dhaugga, K.S., Barreiro, R., Whitten, B., Stecca, K., Hazebroek, J., Randhawa, G.S., Dolan, M., Kinney, A.J., Tomes, D., Nichols, S. and Anderson, P. (2004) Guar seeds β-mannan synthesis is a member of the cellulose synthase super gene family. *Science* 303, 363–366.

Herbaria – seed

Published descriptions of seeds are not as common or comprehensive as those of plants and a seed herbarium is a useful tool when trying to identify an unknown species of seed found during analytical purity and other seed determination tests. The size of the herbarium required by a seed testing station depends on the range of seed material encountered, and collections vary from 100 to 10,000 specimens, usually arranged phylogenically by plant families and then according to species within the family. (**See: Structure of seeds – identification**)

The US National Seed Herbarium is the world's largest taxonomic seed collection. http://www.ars.usda.gov/is/np/systematics/seeds.htm, http://nt.ars-grin.gov/sbmlweb/collections/SeedHerbarium/Index.cfm. (The Germplasm Resources Information Network (GRIN) web server, operated within the US Department of Agriculture's Agricultural Research Service, also provides germplasm information about plants (and equivalent programmes for animals, microbes and invertebrates)) www.ars-grin.gov/npgs/aboutgrin.html)

Online 'Virtual Herbaria' are maintained by the **International Seed Testing Association** (www.seedtest.org/PUR/VirtualHerbarium.cfm) and the Ohio State University (www.oardc.ohio-state.edu/seedid/). (RD)

Herbicide tolerance

See: Genetic modification

Heritage seeds

Some plant varieties (and especially those of vegetables and culinary herbs) have been passed from generation to generation of gardeners and small-holders. Although some of these heritage, or 'heirloom', varieties originated in trade, under some circumstances it may now be illegal to sell them.

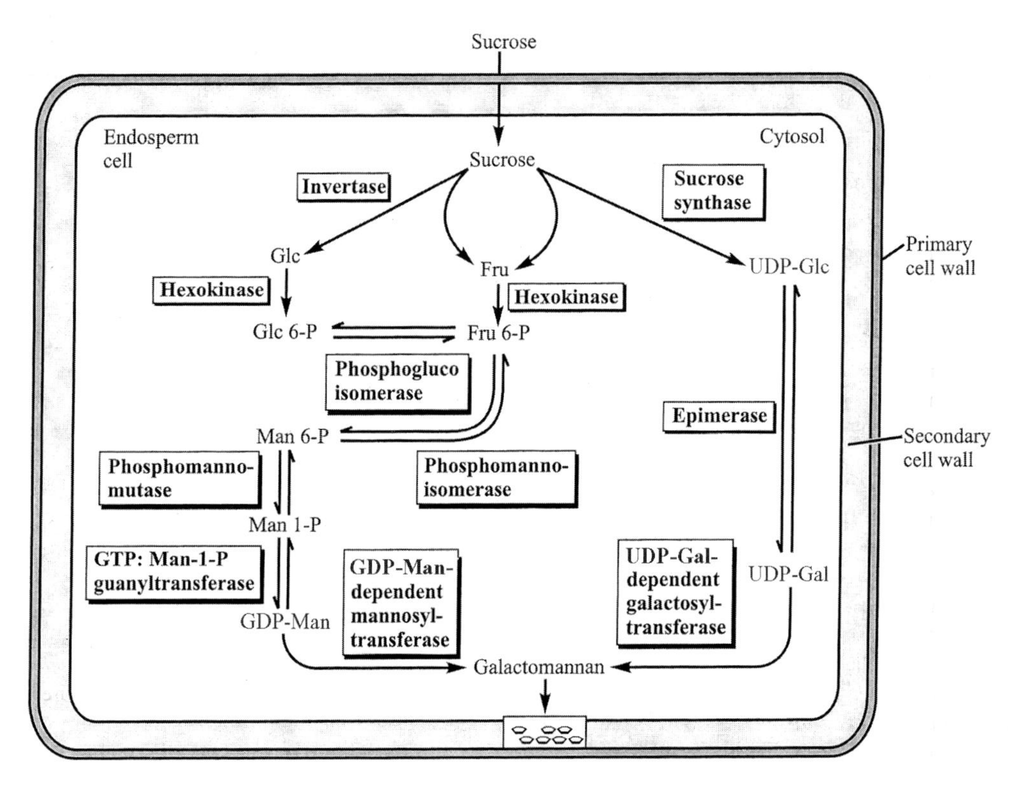

Fig. H.5. Pathway for the biosynthesis of galactomannan in the endosperm of developing endospermic legumes. The galactomannan is synthesized in the Golgi apparatus and transported to the inner side of the primary cell wall, to form the secondary cell wall. The enzyme GDP-man-dependent mannosyl transferase is also known as β-mannan synthase.

Consequently, most are exchanged and this is helped by the establishment of seed guardian networks such as Seed Savers Exchange in the USA, www.seedsavers.org. The Heritage Seed Library of the Henry Doubleday Research Association, UK, www.gardenorganic.org.uk/hsl provides the opportunity for members to try out old varieties. In return, members are requested to send back seed to replenish the library's seed stocks. The heritage vegetable varieties are usually grown because, for example, they have an exceptional taste or have historical interest. Further, their maintenance through continual cultivation means that they have adaptations to local conditions. The number of heritage vegetable varieties has declined for reasons akin to those for the decline of agricultural landraces (**see: Conservation of seed**). Heritage varieties have been displaced by newer commercial varieties with features that have been attractive to many gardeners, such as more consistent appearance due to their high genetic uniformity, and higher yield (though the latter may be most apparent under higher chemical input). A price of the uniformity is that the newer varieties may lack the diversity to provide at least some yield when attacked by a new disease strain. Therefore, they have shorter useful lives than the heritage varieties. Because many of the newer commercial varieties are F_1 hybrids, they do not breed true, thereby negating seed saving and affording the plant breeder control over supply. It would seem that this greater control, the marketing of varieties regionally rather than locally, and plant variety legislation have all contributed to an overall reduced diversity within vegetable species. Thus, the conservation of the surviving heritage seed varieties has some significance. The best long-term prospects for this conservation would appear to be through their wide use (promoted by the seed saver networks), underwritten by storage in **gene banks**.

(RJP, SHL)

Cherfas, J., Fanton, M. and Fanton, J. (1996) *The Seed Savers' Handbook*. Grover Books, UK.

Hermetic storage

Storage inside air-tight containers. Such storage is essential for maintaining the low moisture content required for successful long-term storage of dried **orthodox** seeds in seed **gene banks**. In reality, few containers are completely air-tight particularly over prolonged periods. A key route for leakage is through the sealed opening. The most air-tight, commercially available containers include lever-lidded glass jars with natural rubber gaskets, high quality heat-sealed laminated aluminium foil bags, and heat-sealed glass tubes. Glass containers, although breakable, allow the viewing of included silica gel beads that indicate by colour change if inward moisture leakage has occurred. The type of container determines the ease of access to the stored seed collection. (RJP, SHL)

Heteroblasty

A term describing the possession of different germination properties by seeds from the same plant or from different plants of the same species. The term **polymorphism** is sometimes used synonymously. (**See: Germinabilty – parental habitat effects; Germinability – maternal effects; Deserts and arid lands; Dormancy – acquisition**)

Heterogamy

The phenomenon of sexual dimorphism, such as the production of a distinct male gamete (e.g. in pollen) and female gamete (egg) (e.g. in embryo sac). Also known as anisogamy. (See: **Reproductive structures**)

Heteromorphism

The production of two different forms of a plant part, e.g. seeds (**heterospermy**); fruits, (heterocarpy). (See: **Polymorphism**)

Heteropyle

Also called the 'false micropyle'; the chalazal (**chalaza**) opening in the mechanical tissue of the seedcoat to allow access of the water- and nutrient-conducting tissues of the vascular bundle(s). In the ripe seed the heteropyle is usually closed by the **hypostase**. (See: **Development of seeds – nutrient and water import**)

Heterosis

A term used synonymously with hybrid **vigour** describing the increased vigour, growth, size, yield or function and resistance to diseases and pests of a hybrid progeny over its genetically unlike parents, by controlled cross(es) between two or more lines. Heterosis is a phenotypic property that depends on the method of measurement and on its environmental and genetic background. Though plant breeders and agronomists have utilized heterosis as a means of improving crop productivity, the biological and genetic basis of the superiority of the hybrid over its parents is largely unknown, partly due to the complexity of characters determining traits such as yield, and the number of genes responsible for generating quantitative variation. (See: **Hybrid**) (PH)

Heterospermy

The phenomenon when the same plant produces two or more morphologically different types of seeds, e.g. as occurs in some Brassicaceae and Fabaceae. (See: **Polymorphism**)

Heterospory

Formation of spores that differ with respect to their size and sexual differentiation. In angiosperms the larger, female spore is called the **megaspore** (which produces the **megagametophyte**, constituting the **embryo sac**), the smaller, male one is the **microspore** (which produces the **microgametophyte**, the pollen grain). (See: **Evolution of seeds; Reproductive structures**)

Heterozygous

Having unlike **alleles** located on corresponding loci of homologous chromosomes. An organism may be heterozygous for one, several or many genes. In plant breeding and seed production, depending on the degree of relatedness of the parental **lines**, an **F_1 hybrid** is typically highly heterozygous. Heterozygosity in populations may be reduced by successive backcrossing to one of the parental lines. (See: **Homozygous**)

High irradiance response

A response to prolonged irradiation with red or far-red light. The HIR can trigger germination in some cases and also inhibit germination and axis elongation as part of de-etiolation. It is mediated by light-labile phytochrome A and so is most apparent in dark-grown seedlings. It has an action maximum at 712 nm, at which wavelengths a large pool of phytochrome A is retained, and is not red/far red-reversible. The HIR shows dependence on light intensity (fluence rate) rather than on the total amount of light given, and this may be a response to repeated photoconversion between the two forms of phytochrome. (See: **Light, dormancy and germination**)

Hill dropping

The pattern resulting from the accurate placement (and subsequent covering) of groups (or hills) of seed, at about equal intervals in **furrows** to give definite rows of almost-equally-spaced groups of plants. (See Fig. P.15)

Hilum

Point of attachment of the seed to the **funiculus** or (if a funiculus is absent) to the placenta. Hilar seeds are characterized by a large, expanded hilum constituting part of the **seedcoat**. They are usually flattened and campylotropous with a short **raphe** and **antiraphe** (found in some Meliaceae, Fabaceae like *Mucuna*, *Erythrina*). (See: **Reproductive structures, 1. Female; Seedcoats – structure; Structure of seeds**)

HIR

See: High irradiance response

History

See: History of seed research: History of seeds in Mexico and Central America; Industry supply systems, I. Formal and **II. Informal, Historical development; Sowing methods – brief history; Treatments – brief history**

History of seed research

Seeds have a long and important history in culture. They have featured prominently at least in ancient Chinese, Egyptian, Hebrew, Hindu, Greek, Aztec and Mayan religions, myths, customs and practices. Long used in medicine, they were held to cure diseases, stimulate the libido and increase athletic prowess. Many ancient myths centred on seeds. In Greek myth, for example, pomegranates were associated with the underworld, and when Persephone ate the seeds proffered by Hades the gods condemned her to remain for half of each year with him in the underworld. While she was there, her mother Demeter, the goddess of fertility and giver of wheat (later to be Ceres in the Roman pantheon) refused to feed the world and instead brought on the winter. The abundance of seeds in the pomegranate was used in ancient times as a fertility symbol: and there is a rabbinic reference to these seeds, the number of which in the fruit (held to be 613) corresponds to the number of commandments in the Bible. Seeds as food and symbols thus featured in myth. And myth also extended to seed physiology. For example, to the Greeks, parsley was associated with death: the notorious slow germination rate of the seeds was because they had to journey to the underworld and back several times before they could begin to germinate.

Theophrastus (371–287 BC), the most important botanist of

antiquity, can arguably be regarded as the founder of the scientific as opposed to the mythical or medicinal approach to seeds. His writings cover germination, dormancy, longevity, viability, seed treatments and other aspects of seeds (Table H.2) and many of his observations and commentaries make good sense in the context of modern studies. He was a native of Eresos, a small town on Lesbos, became a student of Plato in the Athens Academy and later joined with Aristotle to establish and teach in the Lyceum. Theophrastus must have been aware of Aristotle's view that 'plants have no other task to perform than the generation of the seed' (Aristotle, *On the Generation of Animals*, 1, 23). In the botanical writings of Theophrastus, *Historia Plantarum* (HP) and *De Causis Plantarum* (CP), seeds are treated extensively. CP's Book I was headed by the translators 'The modes of generation', Book IV deals with agricultural aspects of seeds, and Book V.18 with the death of seeds in storage (**see: Viability**). A detailed study on the history of germination research, including Theophrastus's contribution, was written by Evenari (1980/81).

Concerning the nature of the seed and its contents, Theophrastus crystallized the subject by citing the pre-Socratic philosopher Empedocles:

> The seeds of all contain within themselves a certain amount of food, which is brought together with the starting-point, as in eggs. Thus Empedocles has put it badly when he says: 'the tall trees lay their egg', since (he says) the nature of seeds is close to that of eggs. He should however have spoken ... of all plants, since every seed contains in itself a certain amount of food. This is why they are able to survive for some time, and do not, like the seed (semen) of animals, perish directly on separation from the parent ... (CP I. 7.1). (**See: Storage reserves**)

Though he seemed to have no doubts about the biological importance of seeds, Theophrastus did not ascribe a similar generative significance to the 'pericarpion' (seed 'vessels'):

> Of the two ripenings this of the seed is the more important for reproduction, that of the pericarpion the more important for human requirements. To which of the two ripenings we are to assign the greater achievement by the tree of its goal is another question. Indeed if we assign it to the ripening of the pericarpion we should have to say that in plants whose leaves (or ... roots) we use alone, as vegetables, the concoction of these parts is the more important; and yet the goal lies here, in their seeds, which we do not use for food at all. (CP I. 16.6).

Theophrastus is somewhat ambivalent about the possibility of spontaneous generation but he concedes that it might occur 'either after spells of rain or when some other special condition has arisen in the air or the ground'. But he later states:

> And if the air too provides seeds which it carries down with the rain as (another pre-Socratic philosopher) Anaxagoras says, the rainy spells will be all the same prolific, since they would then produce an additional set of starting-points possessing supplies of food. Rivers again and collections of water and streams bursting forth from the ground would do so too, importing from many sources seeds both of trees and of woody plants (which is why rivers that shift their course make many regions wooded that were unwooded before). These last forms of generation, however, would not appear to be spontaneous, but a kind of propagation by sowing seeds ... or setting pieces, in the ground.

The modern mind would perhaps invoke the **soil seed bank** and **dispersal** mechanisms to interpret these speculations.

The writings of Theophrastus cover many diverse aspects of seed physiology (Table H.2) and one can only be impressed with his awareness and understanding of seed properties and behaviour. His Roman 'successor' was Pliny The Elder (AD 32–79), who drew from Greek, Roman and Carthaginian sources, while adding to the original observations of Theophrastus (Table H.2). But Pliny took much of his material from Theophrastus (Morton, 1986). An example of interest to the seed biologist concerns germination of mistletoe. Theophrastus (CP 2.17. 1–7) had already commented on the fastidious behaviour of this seed and Pliny made similar

Table H.2. Seed biology known to Theophrastus (372–287 BC) and Pliny (AD 23–79).

Process or feature of seeds	Theophrastus	Pliny
Storage reserves from mother plant	+	
Beneficial effect of drying		+
Environmental effects during seed development (e.g. 'quality')	+	
Seed and fruit coats	+	
Seedcoat hardness of leguminous seeds: soil and environmental effects	+	
Aspects of dispersal	+	
Roles of water, seasons, temperature, seed age, habitat in germination	+	
Species differences in germination behaviour	+	
Differences in seed longevity	+	+
Germination rates	+	+
Seed presoaking (priming?)	+	
Afterripening	+	
Seed size and crop yield	+	
Passage through gut promotes germination (mistletoe) (zoochory)	+	+
Seed pelleting (in dung)		+
Seed pretreatment against pests		+

Adapted from Evenari, M. (1984) Seed physiology: Its history from antiquity to the beginning of the 20th century. *The Botanical Review* 50, 119–142.

observations when he later wrote (Book XVI of 'Natural History'):

> But universally when mistletoe seed is sown it never sprouts at all, and only when passed in the excrement of birds, particularly the pigeon and the thrush: its nature is such that it will not shoot unless it has been ripened in the stomach of birds.

Both scholars also presented information on the germination rate (mainly of cultivated plants), for example, about which the Greek had written: 'Not all herbs germinate within the same time ... some are quicker, others slower, namely those which germinate with difficulty'.

The two authors give values for the germination rates of many cultivated species. They agree that in celery, for example, the time to emergence is 39–50 days: interestingly, in its 1976 rules **The International Seed Testing Association** recommends germination tests on this seed at 10 and 28 days after planting, so there has been some increase in rate since classical times!

After the relatively few contributions of Roman authors, little of note for the seed scientist was produced for many centuries. In the 13th century Albertus Magnus commented on the inhibitory effect on germination of fruit flesh (**see: Germination inhibitors**) and in 1664 John Evelyn recorded the requirement for low-temperature **stratification** to promote germination of some **tree seeds**. But we have to await the 17th century, when scientific learning in Europe began to flourish, for the next significant explorations into the nature of seeds. This period saw the development of magnifying glasses and the simple microscope that enlarged objects up to 100 times. Notable among scientists who used the emerging optical technology were the Italian, Marcello Malpighi (1628–1694), the Dutchman, Antoni van Leeuwenhoek (1632–1723) and the Englishmen, Robert Hooke (1635-1703) and Nehemiah Grew (1641–1712); to this illustrious quartet must be added another Englishman, John Ray (1627–1705), who while not particularly renowned for his microscopy is much admired for his original observations and investigations in botany. All five carried out extensive studies on seed and seedling structure (see Fig. H.6). In 1664, Hooke published his microscopic investigations in *Micrographia*: Grew and Malpighi both presented their enquiries to the Royal Society in 1671, published later by Grew in his *Anatomy of Plants* (1682) and by Malpighi in *Anatome Plantarum* (1675) (Fig. H.6b). It is interesting that the nature of sexual reproduction in plants and the origin of seeds was being established at about the same time. Rudolf Jakob Camerarius (1665–1721) showed, in 1694, that in order for plants to bear fruits and seeds, the pistils in female and hermaphrodite flowers had to be provided with pollen (Morton, 1981: 214ff).

Hooke reported detailed observations on seeds of several species, often remarking on their similarity under the microscope to other objects:

> There are divers Kinds of Seeds which imitate the Shape of much larger Bodies: The Seed of *Scurvy-Grass* nearly resembles the Form of a *Concha Venerea*, or Sort of *Porce- lain Shell*: Those of *Sweet-Marjoram* and *Pot-Marjoram* represent Olives. *Carrot-Seeds* are like the Cleft of a Cocoa Nut Husk: The Seeds of *Succory* like a Quiver full of Arrows (from *Micrographia Restaurata*).

Using bean seeds, for convenience on account of their size (Fig. H.7a), Grew observed:

> If then we take a bean and dissect it, we shall find it cloathed with a double vest or coat. These coats, while the bean is yet green are separable and easily distinguished. Or in an old one, after it hath lay'n two or three days in a mallow soil; or been soaked ... in water... When 'tis dry, they cleave so closely together, that the eye not before instructed, will judge them but one; the inner coat (which is of the most rare contexture) so far shrinking up, as to seem only the roughness of the outer, somewhat resembling wafers under Maquaroons.

Then later, 'The inner coat in its natural state, is every where twice, and in some places thrice as the outer. Next to the radicle ... it is six or seven times thicker, and encompasses the radicle about, as in the same figure appears ... ' and,

> At the thicker end of the bean in the outer coat, a very small Foramen presents it self, even to the bare eye ... In dissection 'tis found to terminate against the point of that part which I call the radicle. It is of that capacity as to admit a small virginal wyer; and is most of all conspicuous in a green bean. Especially, if a little magnified with a good spectacle-glass. This foramen is not a hole casually made ... but designedly formed, for the uses hereafter mentioned. It may be observed not only in the great garden-bean, but likewise in the other kind ... [here Grew listed a number of pulses] and in many seeds not reckoned of this kindred, as in that of fonugreek [*sic*], Medica Tornata [*Medicago tornata*], Goats-Rue [*Galega officinalis*] and others. ... ... .That this foramen is truly permeable, even in old Setting-Beans [presumably meaning bean seeds used for the production of 'sets' – plants for growing further], and the other seeds above named, appears upon their being soaked for some time in water. For then, taking them out, and crushing them a little, many small bubbles will alternately arise and brake upon it. [The 'foramen' was later (1821) renamed 'micropyle' (Gk. 'little gate') by De Candolle and Sprengel.]

It is worthwhile recording Grew's description of the bean embryo:

> The main body is not one entire piece, but always divided, lengthwise, into two halves or lobes, which are both joyn'd together at the basis of the bean. These lobes in dry beans, are but difficultly separated or observ'd; but in young ones, especially boil'd, they easily slip asunder ...

Grew used the terms radicle and plume (plumule) but not cotyledons. This was introduced as a botanical term in 1751 by Linnaeus who imported it from zoology and physiology where it is applied to the absorptive villi of the fetal chorion of ruminants: hence, the absorptive, nutritive role was the analogy, not the shape (Gk. *kotyle*, a hemispheric drinking vessel). We might note, in this context, that Linnaeus, Malpighi and Grew were all physicians and frequently resorted to analogy between animal and plant functional anatomy.

Leeuwenhoek (1685, in *Collected Letters*) also described the structure of several seeds, for example cassia, coffee, date and cotton (Fig. H.6c). In the latter, he examined the internal

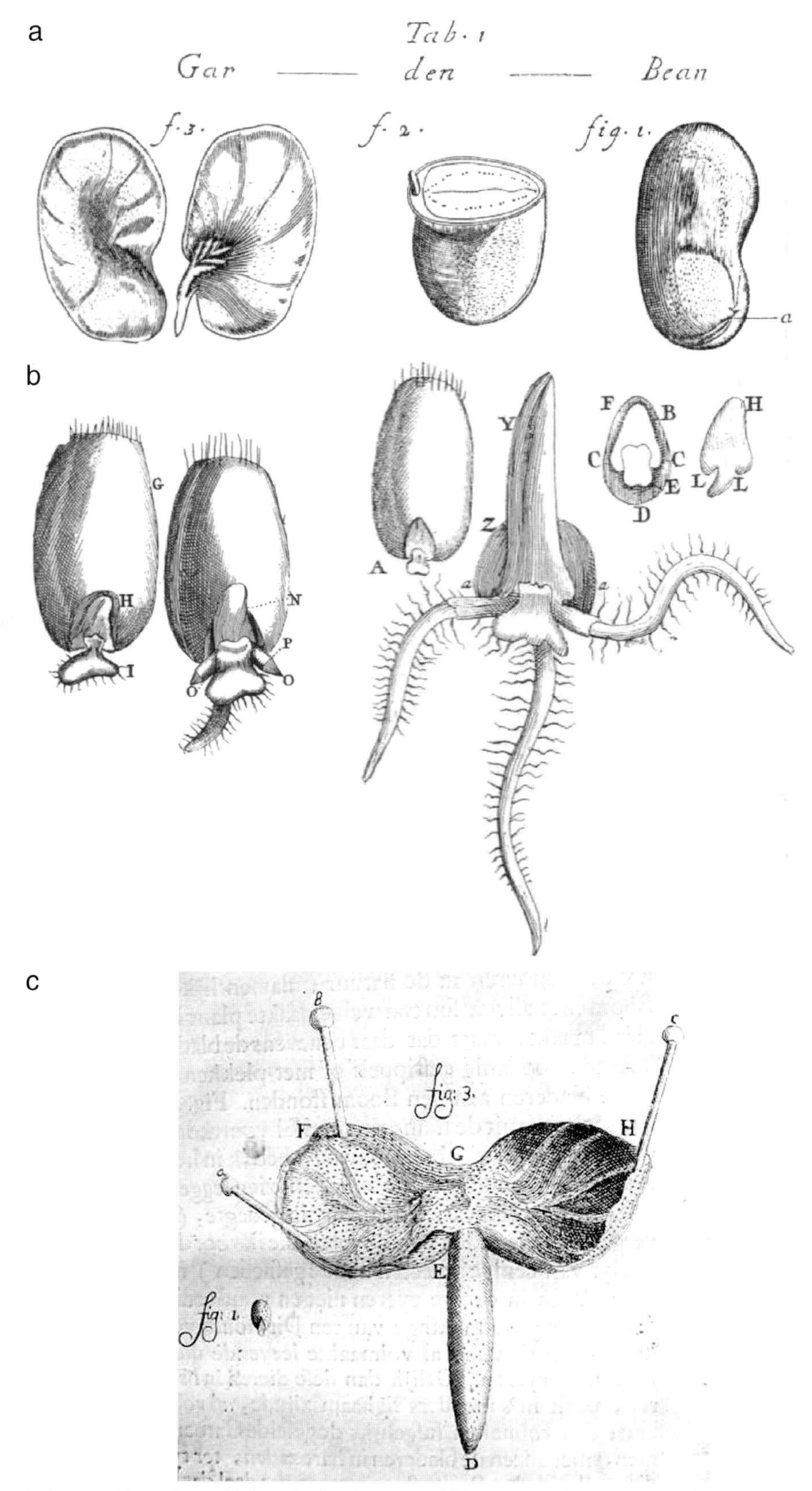

Fig. H.6. (a) Bean seeds (adapted from Grew, N. (1682) *The Anatomy of Plants,* 2nd edn. London; reprint edition edited by C. Zirkle (1965) Johnson Reprint Corporation, New York). (b) Wheat seed germination and seedling (taken from Malpighi, M. (1675) *Anatome Plantarum* (On the anatomy of plants) London). (a and b by permission of The Linnean Society of London.) (c) Cotton seed and seedling (Leeuwenhoek). The cotyledons of the newly germinated seed are displayed pinned to a surface.

structure by cutting 'one of them into twenty-five to twenty-six round slices and the other into twenty-eight to twenty-nine round slices', a technique that we would now recognize as serial sectioning. Leeuwenhoek comments that the 'leaves' of the cassia seed

> have been made so exceedingly large in order to provide nourishment for that part of the root and for the beginning of the young plant, which root, by comparison with the two leaves, is very short. In order to satisfy myself on this point, I have laid the cassia seed to sprout in sand moistened with common rain-water, until the root had grown as long as the width of my thumb, when the two aforesaid leaves had been pushed outside the earth, having between them the beginning of the young plant, which before that could not be discerned.

John Ray, too, was highly interested in the seminal leaves (cotyledons) noting (*Novus Methodus Plantarum*, 1682) that seeds have either one or two of these organs. His view was that:

> A general distinction of plants is able to be deduced from this division of seeds and this distinction, in my judgement, is the first and best by far: between those that have double-leaved seminal plants and those which have a seminal plant analogous to the adult plant

so foreseeing the basic taxonomic division of the angio-sperms into the Monocotyledons and Dicotyledons. To discover the function of these seminal leaves Ray proposed that

> it would be a simple experiment to find out by cutting the leaves or lobes as soon as the little plant comes forth from the earth and to compare it to another plant of the same age whose leaves or lobes remain. For if the mutilated plant is equal to the intact one in strength and size it is certainly clear that leaves of the plant are not necessary to supply nutrition.

Ray learned that Malpighi, 'a most sagacious and incomparable philosopher', had actually performed this experiment which he (Malpighi) describes thus in his *Anatome Plantarum* (1675):

> Also, in the month of May, I put other seminal plants of Faba and Phaseoli for incubation with two seminal leaves removed in the same way, of which only one little plant of Faba sprouted, the bud lifted a little above the earth and underneath a small leaf grew, so that the whole plant did not exceed the length of my little finger. After twenty one days the plant wasted away. I planted afresh many other seminal plants of Faba with every seminal leaf pulled off, of which none grew.

Malpighi experimented with seeds of several species finding that 'when they were deprived of seminal leaves were not achieving any increments of growth, but either were swiftly extinguished or they survived briefly without significant growth'. He concludes, 'Therefore, from these things it is possible to conjecture that the two-fold leaves, as fat as possible, which cling to the seminal plant perform the service of uterine placenta.'

Endospermic and non-endospermic seeds were clearly described by Ray when he drew attention to seeds in which 'no pulp or pith is contained' and 'those in which besides the seminal plant other things are also contained'. He realized that in grains and grasses it is the endosperm (though not so named by him) that initially provides nourishment for the young plant: 'The seminal plant in these (*Hordeum*) equals in magnitude hardly a tenth part of the pulp or pith of the seed. The remaining pulp or grain of the plant, while somewhat tender, serves to nourish, even after the roots have grown.' Then later,

> For if you uproot the plant when it has first sprouted, you discover the pulp in the grain nearly intact; but at the same time, if you continue to pluck sewn plants from day to day, you will observe the pulp or flour to be perceptibly diminished daily until nothing is left except the empty little follicle clinging to the bottom of the plant.

This describes what we now recognize as endosperm mobilization (**see: Mobilization of reserves – a monocot overview**) and Ray hints at some of the biochemical changes that are taking place when he continues, 'Moreover the pulp or grain of the seed after germination combining with the moisture of the earth percolated through the pores of the covering turns into plant juice not unlike the sap extracted from a plant.' It has to be added, however, that Ray had a mistaken belief that the 'pulp' was a 'globular leaf', originally so-called by (according to Ray) 'a man never praised enough, Marcellus Malpighi'.

The wonder and excitement generated by these scientists' investigations is transmitted by Ray when he writes 'Moreover, if anyone desires to see the seminal plant he should dissect a seed. Anyone who has done so must marvel at the artifice of nature in making seminal plants so carefully complicated and must be carried away with delight at such a pleasing spectacle.'

Though progress in botanical science including physiology began to accelerate rapidly in the 18th century it was not until the 19th century that the foundations of modern seed science, especially seed physiology and biochemistry, were truly laid. An excellent account of the history of discovery is given by Evenari (1984). Table H.3 lists the some of the major contributions made at this time. (MoN, MB)

Evenari, M. (1980/81) The history of germination research and the lesson it contains for today. Israel Journal of Botany 29, 4–21.

Evenari, M. (1984) Seed physiology: its history from antiquity to the beginning of the 20th century. *The Botanical Review* 50, 119–142.

Leeuwenhoek, A. van (1685) *Collected Letters* (cited in Ford, B. *From Dilettante to Diligent Experimenter: A Reappraisal of Leeuwenhoek as Microscopist and Investigator* (www.sciences.demon.co. uk/a-avl01.htm).

Morton, A.G. (1981) *History of Botanical Sciences*. Academic Press, London.

Morton, A.G. (1986) *Pliny on Plants. His Place in the History of Botany*. In: French, R. and Greenway, F. (eds) *Science in the Early Roman Empire: Pliny the Elder, his Sources and Influence*. Croom Helm, London, UK, pp. 86–97.

Pliny (1950) *Natural History*, with an English translation by H. Rakham, W.H.S. Jones and D.E. Eichholz. Harvard University Press, Cambridge, MA, USA; Heinemann, London, UK.

Ray, J. (1682) *Novus Methodus Plantarum*, translation by S. Nimmis, K. Schantz and M. Vincent (www.vroma.org/~snimis/ methoduspreliminary.html).

Table H.3. Some major discoveries in seed biology in the 19th and early 20th centuries.

Finding	Author
Ovule	Treviranus 1815
Embryogenesis	Amici 1847; Hofmeister 1849 *et seq*
Double fertilization	Nawashin 1898
Light and seed germination	Ingenhousz 1779; Caspary 1860
Action spectrum; red, far-red effects	Cieslar 1883; Flint and McAlister 1935
Coat impermeability	Nobbe 1876
Seed size and vigour	Lehmann 1871
Temperature and germination	Sachs 1860
Temperature and after-ripening	Atterberg 1899; Gassner 1910
Aleurone and scutellar physiology	Brown 1831; Sachs 1862; Ebeling 1885; Haberlandt 1890
Endosperm mobilization	Brown and Morris 1890
Biochemistry of protein mobilization	Sachs 1887
Biochemistry of oil, starch mobilization	De Vries 1877
Germination inhibitors	Wiesner 1894
Longevity	Haberlandt 1874
Seed proteins	Osborne 1924

Adapted from Evenari, M. (1984) Seed physiology: its history from antiquity to the beginning of the 20th century. *The Botanical Review* 50, 119–142, in which further details may be found.

Theophrastus (1916, 1926) *Enquiry into Plants*, with an English translation by Sir Arthur Hort. Harvard University Press, Cambridge, MA, USA; Heinemann, London, UK.

Theophrastus (1976, 1990) *De Causis Plantarum*, with an English translation by B. Einarson and G.K.K. Link. Harvard University Press, Cambridge, MA, USA; Heinemann, London, UK.

History of seeds in Mexico and Central America

Introduction

Mexico and its surrounding Central America regions were recognized as one of the seven world centres of origin of cultivated plants by the Russian biologist Vavilov. North and Mesoamerica are well-documented centres of origin of important crops including **maize** (*Zea mays*), squash (*Cucurbita pepo*), **sunflower** (*Helianthus annuus*), **common bean** (*Phaseolus vulgaris*), **cacao** (*Theobroma cacao*) and **amaranth** (*Amaranthus* spp.). There is evidence from fossil pollen that maize predecessors were present in central Mexico 60,000–80,000 years ago, before the arrival of humans in America; but the earliest evidence of maize **domestication** was found in Central (Puebla and Oaxaca) and North (Tamaulipas and Chihuahua) Mexico, dating from only 7000–3000 years ago. Even though maize is currently the most important cereal in the region, there is convincing evidence that squash, still an important crop, was domesticated almost 10,000 years ago, and it has been proposed also that cultivation of *Setaria geniculata*, a relative of today's grain **millets**, and perhaps common bean, also preceded maize domestication. Compared to **wheat**, **rice** and millet, maize has a highly productive metabolism (**C4 photosynthesis**), which supports an efficient carbon assimilation at reduced water loss, allowing a high biomass yield in a short life cycle. Therefore, it seems no accident that the evidence of the settlement of cultural groups relying mostly on maize agriculture coincides with the expansion and the later splendour of Toltec, Aztec and Maya civilizations.

Not only did seeds of various kinds form a substantial part of the basic diet in pre-Columbian times but they were also central to many traditions and religious practices. In this regard, four of the most important seeds are considered.

1. Maize

By 800 BC, maize had become the most important source of carbohydrates to almost every Central and North American human settlement. Its high productivity demands nitrogen and phosphate, rapidly impoverishing the soil; paradoxically, this condition may have contributed to the eventual decline of the Mayan empire, which thrived on the remarkably shallow layer of soil found in the tropical rainforest. In this respect, it is likely that wise agricultural practices, like mixing or alternating planting of maize and beans in the same location, helped the Aztec splendour to last until the conquest of Mexico by Spain. The importance of maize seeds as food in Mexican cultures reflects on the outstanding variety of recipes using maize in the Mexican prehispanic and modern cuisine, and in the over 100 words that refer either to the plant, its seed, its cultivation, or its use. It is also strongly represented in the Aztec religion, where there is a maize goddess named *Centeotl* who represented the mature and dry corn cob, and was also referred to as Iztacenteotl, Tlatlauhquincenteotl, Tzinteotl and Tonacayohua, according to the stage of development and colour of the grain. The developmental stages of the maize plant were linked to celebrations in the calendar, and the goddess of the maize cultivars was offered sacrifices as Chicomecoatl in the dry season (Fig. H.7), symbolizing the infertility and famine in the drought and as Chalchiuhcihuatl in the spring, for abundance and rejoicing.

Maize featured very prominently in Mayan mythology and religion. To the Maya, maize was the grain upon which the civilization was built. Their creation story is that the gods first tried making humans from animals, clay and wood but failed, so they tried maize and succeeded in creating mankind from maize dough and divine blood.

The modern maize is a domesticated relative of Teozintle or teosinte (includes *Zea mays* ssp. *mexicana* and other *Zea* spp.

Fig. H.7. A representation of an Aztec maize goddess, Chicomecoatl.

– see: **Domestication**) and its evolution is no doubt linked to the development of human civilization. The respected Mexican anthropologist Arturo Warman has described maize as an artefact, a species that does not exist in the wild and survives thanks to human intervention. This last view has been further supported by the analysis of the drift of genetic traits of several cultivated plants from their ancestors. It is evident that human social development was also highly dependent on maize seeds. Maize seeds have a major nutritional drawback, however – their low lysine content. During the Spanish colonization, domination of the indigenous peoples limited their access to fish and animal protein and the resulting malnutrition caused by over-reliance on maize is considered amongst the important reasons for the significant mortality of the Mayans observed in those times. (**See: Cereals; Cereals – composition and nutritional quality; Maize**)

2. Amaranth

Amaranths are dicotyledonous plants of the family Amaranthaceae, widely used by American indigenous cultures for more than 5000 years. *Amaranthus hypocondriacus*, *A. cruentus* and *A. caudatus* were known as *Huautli* by the Aztecs and valued for their highly nutritious seeds. The seeds contain a high proportion of **starch** (50–60%), accumulated in the **perisperm**, and are rich in **storage protein** (up to 17%), with a high lysine content; protein accumulates mainly in the **embryo** and **endosperm**. For these reasons it has been considered as a **pseudocereal**. *A. dubius*, *A. hybridus* and *A. spinosus* were appreciated for their abundant edible foliage (*quelites*) and their tasty inflorescence (*huauzontle*). Amaranths have probably been cultivated in America for more than 5000 years by Aztecs, Chichimecas, Toltecas, Mayas, Incas, and other Central and South American cultures, although most data concerning its use come from Aztec traditions. It is repeatedly mentioned in the *Mexicayotl* chronicle as one staple food cultivated even before the foundation of Mexico-Tenochtitlan (today Mexico City). *Huautli* was so precious to the Aztec culture that it was demanded as a tribute from the conquered towns and formed part of ritual and religious food. With flour from the grains, mixed with maguey (*Agave* sp.) syrup (and human blood in some cases), Aztecs prepared a dough, called *tzoalli*, and elaborated figures of their Gods which was eaten at the end of the religious ceremonies, sharing them among participants, probably officials and priests. Amaranth *tortillas* (baked flat bread), *tamales* (steamed bread) and *pinole* (sweetened amaranth powder) were also food products used by the Aztec population. The religious use of amaranth was seen by the Spanish conquerors as a dangerous imitation of the rituals of the Catholic Mass and its cultivation was therefore prohibited and the cult was banned. Amaranth cultivars would have disappeared during the Spanish rule, were it not for the strong ties that some native farmers felt for the seed. The contempt the Spaniards felt for this seed still remains in the common and somehow offensive Spanish idiom 'me importa un bledo', literally 'I don't give an amaranth seed' meaning 'I could not care less'.

The practice of mixing amaranth grains with syrup did not die, and the product, called *alegría* ('joy') is consumed in high amounts by Mexicans, constituting a very tasteful and nutritious snack. The difference from the Aztec *tzoalli* is that grains are 'popped' by roasting, not ground.

Amaranth plants conduct C4 photosynthesis and can grow under adverse conditions of temperature, salinity or drought and also they show a high resistance to pests. These cultivated varieties tend to be tall with large inflorescences. The leaf-producers are generally wild, grow as weeds, and are shorter in stature. The seeds are of a creamy or wine colour, lenticular in shape and approximately 1 mm in diameter.

Nowadays, amaranth is cultivated not only in Mexico, but also in China, India and the USA, where it is very appreciated both as a good source of protein for human nutrition and as a source of an excellent skin oil. (**See: Grain amaranth; Pseudocereals**)

3. Cacao

Cacao is a plant native to the Americas, and was used in Mesoamerica and South America; however, most knowledge about its uses and traditions has been gathered from Mesoamerica. The word *cacao* probably originated with the Olmeca people (Gulf of Mexico) and spread to all Mesoamerica; Aztecs (or Mexicas) and Mayans also called it, or the beverage, *cacahuatl*. The centre of domestication is in doubt; some authorities claim that it is South America, whereas others are of the opinion it is Mexico or the Caribbean area. The wild ancestors of cacao found in Mexico are genetically distinct from those of other regions. Mayan people cultivated cacao trees more than 2000 years ago.

According to the Mayan and Aztec religions, cacao had divine origins. In both cases, the Plumed Serpent God (*Quetzalcoatl*) discovered cacao, among other delectable foods, and gave it to humans. Mayans and Mexicas celebrated an annual festival to honour their cacao God (or Gods), in which there were cacao offerings and other gifts, and animals or sometimes humans, were sacrificed. Aztecs also offered a cacao beverage to the prisoners to be sacrificed to the gods, on the eve of the ceremony. The Mexica people perceived cacao as an intoxicating food and thus it was not suitable for women and children; however, it was an invigorating and prestigious food for adult males such as priests, high government and military officials and distinguished warriors. Cacao beans or 'almonds' were also used as money in Aztec markets to purchase luxury goods.

Although apparently Cristobal Colón (Christopher Columbus) and Cortés knew about the existence of cacao seeds, it was probably in Moctezuma's court where Cortés and his men observed the preparation and consumption of *chocolatl*, the cacao beverage. *Chocolatl*, or chocolate, was more a beverage than food, since it also had medicinal properties. The medical use of cacao/chocolate is well described in several Aztec documents, such as the Badianus Manuscript, the Florentine Codex and the Princeton Codex. Diseases such as stomach and intestinal complaints, infections, diarrhoea, fever and weakness, cough and many other ailments were relieved or 'cured' by using cacao mixed with several other distinct flavourings such as chilli peppers, rhubarb and **vanilla**.

Cacao was introduced to the Spanish court in 1544 when Dominican friars took Maya nobles to be introduced to Prince Philip. Within a century, chocolate was consumed in all Europe, mixed first with honey and then with sugar, giving the beverage the taste that it has nowadays. But it was in France, in the court of King Louis XIV where the Spanish Princess Maria Teresa introduced the habit of drinking a lighter chocolate beverage, not thick as in Spain, and this use was extended to all Europe. Also in France, the chocolate was first blended with milk and sugar, to prepare a beverage, or pieces of solid chocolate called 'bon bons'.

The scientific name for cacao, *Theobroma cacao*, (food of the Gods) combines Greek and Mayan etyma. Cacao is a fatty seed, now recognized as a mild stimulant with anti-depressive effects, due to its elevated content of long chain N-acylethanolamines (the natural chemical messengers for the cannabinoid neuro-receptors). The cacao fat is the most expensive vegetable fat in the market, for it is delicious and literally melts in the mouth (at 37°C). **(See: Cacao; Oilseeds – major)**

4. Beans

The seeds of **common bean** (*Phaseolus vulgaris*) have a long history as an extremely important food in Mexico and Central and South America. The plant is an annual legume whose seeds are rich in carbohydrates, with a protein content of up to 20%; however, as in many other legumes, beans possess antinutritional factors such as lectins, tannins and protease inhibitors (**see: Protease inhibitors**). It is low in methionine and tryptophan, and tends to rapidly lose vigour and nutritional value during storage. Early evidence of the domestication of common bean comes from sites in Tamaulipas, north-east Mexico, and has been dated as early as 5000–2200 BC. Later on, its cultivation extended far north to the USA and southwards reaching Central America, but there is evidence of the independent domestication of **lima bean** (*P. lunatus*) in South America. But surprisingly, only rarely are beans mentioned in the chronicles, or the mythology of important cultures like the Aztecs, Mayans or Incas. However, they appear in songs and myths from many tribes and minor cultures established in the north of Mexico and the USA, who range from gatherers and hunters to completely settled farmers, but who did not develop writing. The common beans gained additional relevance in the diet of the American natives when their consumption of animal protein became limited during the Spanish domination and thereafter. Beans were, and continue to be, a very valuable agricultural product in Mexico and Central America due to their symbiosis with nitrogen-fixing bacteria (*Rhizobium etli*). Beans are commonly planted together with, or alternated with, maize and/or hot peppers to enrich the soil. Although it can be found in some dishes of ancient tradition, like the delicate 'bean soup' from Oaxaca, beans are more represented in the intercultural gastronomy of modern American countries frequently mixed with rice or meat as in the Cuban dish 'Arabs and Christians', in the Texan 'chilli con carne', or in the Brazilian 'feijoada'. **(See: Legumes; Pseudocereals)** (RRS, JVR)

Dillinger, T.L., Barriga, P., Escárcega, S., Jiménez, M., Salazar Lowe, D. and Grivetti, L.E. (2000) Food of the Gods: cure for humanity? A cultural history of the medicinal and ritual use of chocolate. *The Journal of Nutrition* (Suppl.) (Erdman, J.W., Jr, Wills, J. and D'Ann Finley, D., eds) 130, 2057s–2072s.

Larsen, C.S. (2000) Alimentación y salud de los indígenas en la colonias americanas. *Investigación y Ciencia* (Barcelona, Spain) 287, 42–47.

Millar, M. and Taube, K. (1997) *An Illustrated Dictionary of the Gods and Symbols of Ancient Mexico and the Maya*. Thames & Hudson. London, UK.

Prem, H.J. (1997) *The Ancient Americas. A Brief History and Guide to Research*. Translated by Kornelia Kurbjuhn. University of Utah Press, Salt Lake City, USA.

Robelo, C.A. (1980) *Diccionario de Mitología Nahuatl*. Editorial Inovación S.A., México.

Vavilov, N.I. (1992) *Origin and Geography of Cultivated Plants*. Cambridge University Press, Cambridge, UK.

Hollow heart

A disorder affecting pea seed development, also called **cavitation**, caused by high ambient temperature during seed maturation or by drying them while immature. Affected seeds have a sunken or cracked area in the centre of the adaxial cotyledon faces and as a consequence produce weak plants. **(See: Disease; Pea – cultivation, 3. Quality, vigour and dormancy)**

Holoprotein

A conjugated protein or protein complex. A holoprotein can be formed by binding of multiple proteins (polypeptides, subunits) or by conjugation of a non-protein moiety to a previously-unmodified apoprotein such as occurs when **phytochrome** binds to its **chromophore**. **(See: Storage protein)**

Homologous genes

Alternative name homologues. Genes that have recognizably evolved from a common ancestor. Includes **orthologues** and **paralogues**.

Homozygous

Having like **alleles** located on corresponding loci of homologous chromosomes. An organism may be homozygous for one, several or many genes. In plant breeding and seed production, homozygosity in populations is the characteristic outcome of inbreeding in **self-pollinated** species. (**See: Heterozygous**)

Hopper

Hoppers control the rate of feed of seed (or other material) into **conveyors**. Hopper types used in conjunction with seed conditioning equipment include: 'roll-feed', 'roll-feed brush', 'metering', 'corn', 'vibrating' or 'cottonseed'.

Hordein, hordenin

The **prolamin** storage proteins present within the **protein bodies** of the **endosperm** of barley grains. (**See: Storage proteins**)

Hormone

Plant hormones are chemical compounds in plants that are involved in the regulation of growth, development and certain biochemical processes. The word 'hormone' was originally devised in the context of animal function and, coming from the Greek, *hormein*, 'to stir up', defined a substance that is secreted by one tissue or organ and acts upon another at some distance away. Such a long-distance effect was indeed the basis of the discovery of the first known hormone in plants, **auxin**, found to be the signal by which the apical region of a plant regulates the growth of cells lower down in the phenomenon of phototropism. One property of a hormone, then, is that it is a means of communication or signalling between cells, tissues and organs. This feature is shown particularly well by one hormone-regulated system in seeds – the regulation by the embryo of certain cereal grains of hydrolytic enzyme production in the peripheral aleurone layer tissue (**see: Mobilization of reserves – a monocot overview; Mobilization of reserves – cereals**). But action from a distance is now not critical to the definition of a plant hormone since many cases are known where these substances act within the cell, tissue or organ in which they are produced.

Hormones are also sometimes referred to as plant growth regulators especially when used exogenously. Synthetic equivalents of hormones are also growth regulators.

Plant hormones and **plant growth regulators** are active at very low concentrations. When applied to a seed they are effective in the micromolar range or in some cases less, and it can be assumed that the concentration reached within the seed may be at least an order of magnitude lower. Active concentrations of endogenous (i.e. seed-derived) hormones are of a similar low order.

Seven classes of hormones are known to be active in seeds. They belong to chemically diverse groups, including indoles (**auxin**), diterpenes (**gibberellins**), a sesquiterpene (**abscisic acid**), substituted purines (**cytokinins**), a gaseous alkene (**ethylene** or **ethene**), steroids (**brassinosteroids**) and a cyclopentenone (**jasmonate**). Auxins are implicated in the regulation of certain stages of **embryogenesis**. The gibberellins (GAs) regulate processes in **germination**, **dormancy** and **mobilization of storage reserves**. Control of dormancy, **maturation**, **desiccation tolerance**, deposition of certain **storage protein** and **oil** reserves and germination is effected by abscisic acid (ABA). Early events in seed development where extensive cell division is taking place involve the action of cytokinins. Ethylene, jasmonate and the brassinosteroids all participate in germination. Another signalling molecule present in adult plants, **salicylic acid**, also has activity when applied to seeds but it is not clear if it is an endogenous regulator. (**See: Germination – influences of hormones**)

A fundamental action of the hormones is on gene expression, by which effect they regulate growth, developmental and biochemical processes, though some have additional effects such as regulation of ion fluxes. The primary site of interaction of a hormone with a cell (e.g. in a seed) is the hormone receptor located either at the surface membrane or internally. Binding of the hormone signal to the receptor then sets in train a series of molecular changes generally involving switches in structural conformation of proteins, known as the **signal transduction** chain, eventually culminating at an effect on the gene(s) itself. The consequent promotion or inhibition of gene expression elicits the hormone-controlled response, such as promotion or inhibition of germination.

Whether or not a particular hormone-regulated process takes place in a seed depends not only on the presence or absence of the hormone but also on the sensitivity to it. Sensitivity may be determined by components of the signal transduction pathway: elements in the transduction chain may be missing, perhaps by a failure in biosynthesis or because of mutations. The latter are particularly important in seeds in relation to experimental approaches towards unravelling control processes in maturation, dormancy and germination. Mutations can also affect hormone biosynthesis and such mutants have also contributed greatly to our understanding of the role of hormones in seed biology. Inactivation or turnover also influences sensitivity. (**See: Hormone mutants; Mutants**) (MB)

Davies, P.J. (ed.) (1995) *Plant Hormones: Physiology, Biochemistry and Molecular Biology*. Kluwer Academic, Dordrecht, The Netherlands.

Napier, R. (2004) Plant hormone binding sites. *Annals of Botany* 93, 227–233.

Srivastava, L.M. (2002) *Plant Growth and Development. Hormones and Environment*. Academic Press, New York, USA.

Trewavas, A. (2000) Signal perception and transduction. In: Buchanan, B.B., Gruissem, W. and Jones, R.L. (eds) *Biochemistry and Molecular Biology of Plants*. American Society of Plant Biology, Rockville, MD, USA, pp. 930–987.

Hormone-balance theory

The concept that dormancy and germination are regulated by the internal actions of hormones, predominantly **gibberellin** (promotive) and **abscisic acid** (inhibitory). (**See: Dormancy; Dormancy – acquisition; Dormancy – embryo (hypotheses); Dormancy – genes; Dormancy breaking; Dormancy breaking – cellular and molecular aspects; Dormancy breaking – hormones; Dormancy breaking – temperature; Germination – influences of hormones; Germination – molecular aspects; Germination – radicle emergence**)

Hormone mutants

1. Abscisic acid

ABA-deficient (*aba*) and ABA-insensitive (*abi*) mutants were first isolated in *Arabidopsis* and tomato (e.g. the deficient *sit* mutant) in the 1980s, and since then dozens of **mutants** with defects in **abscisic acid** (ABA) response have been described. The genetic screens and selections used for isolation of mutants with altered ABA response have produced seeds exhibiting **vivipary** or defective germination, loss of **desiccation tolerance**, altered embryonic development, altered reserve deposition, loss or gain of sensitivity to ABA at germination, seedling growth or root growth, incorrect expression of **reporter genes** and screens for suppressors or enhancers of GA-deficient, non-germinating lines or ABA-INSENSITIVE lines. Additional mutants with defects in responses to multiple signals, including ABA, have been isolated via non-ABA-based screens such as salt-resistant germination, sugar-resistant seedling growth or gene expression, or defects in responses to **auxin, brassinosteroids,** or **ethylene (ethene)**.

Mutations in the *ABI3*, *ABI4* and *ABI5* loci have similar qualitative effects on seed development and ABA sensitivity but null mutations in *ABI3* are more severe than those in *ABI4* and *ABI5*. These loci display similar genetic interactions: digenic mutants combining the leaky *abi3-1* **alleles** with severe mutations in either *ABI4* or *ABI5* produce seeds that are only slightly more unresponsive to ABA than their monogenic parents, whereas mutations at any of these three loci greatly enhance the lack of ABA response of *abi1-1* mutants. Null alleles of *ABI3* (*abi3-3* and *abi3-4*) or double mutants combining the weak *abi3-1* allele with ABA deficiency (the *aba1-1* mutant) have more severe defects in seed maturation than any of the other *abi* mutants. These plants produce green seed that fail to lose chlorophyll, to accumulate **storage proteins**, or to attain desiccation tolerance, and they contain a high amount of denatured protein. Although this and other mutants combining ABA-insensitivity with ABA-deficiency (i.e. *aba/abi3*) or suppression of early and late ABA peaks (i.e. *fus3/aba3*) display vivipary, no ABA-insensitive (*abi*) or even digenic ABA-deficient (*aba3* and *aao3*) *Arabidopsis* lines are viviparous. In contrast, maize mutants with single defects in either ABA response (*vp1*, an **orthologue** of *abi3*) or synthesis (other *vp* mutants) are viviparous (**see: Abscisic acid**, Table A.1). Most *vp* mutants are also deficient in carotenoids (a precursor of ABA), resulting in a white pigmentation. Few of these mutants can form viable seedlings since carotenoid deficiency generally leads to photobleaching of chlorophyll in the newly formed leaves, and hence failure of photosynthesis. In contrast, the *vp8* mutation results in a reduction in ABA content without reducing the amount of carotenoids present in the kernels, which remain yellow. The effects on seed development of the mutation to the *Vp1* gene are pleiotropic. This gene encodes a protein which may affect the expression of several genes related to development, as a transcriptional factor.

(**See:** ***ABA*** **genes;** ***ABI*** **genes; ABI proteins; Dormancy – genes; Dormancy breaking – hormones; Maturation – control factors; Mutants; Storage reserves synthesis – regulation**)

2. Gibberellins

Several gibberellin-deficient mutants of *Arabidopsis* have been isolated. Reduction in gibberellin (GA) content may result from mutation at any of the different genes encoding enzymes of the biosynthesis pathway: *ga1*, *ga2*, *ga3*, *ga4* and *ga5* have mutations of the genes encoding, respectively, for *ent*-copalyl disphosphate synthase, *ent*-kaurene synthase, *ent*-kaurene 19-oxidase, GA_{12} aldehyde 7-oxidase, GA 3β-hydroxylase and GA 20-oxidase (**see: Gibberellins**). Some mutants are defective for more than one enzyme (i.e. *ga1-2*). Severely GA-deficient mutants (e.g. *ga1-3* and *ga2*) fail to germinate and are dwarfed and male sterile. Similarly, tomato GA-mutant *gib-1* seeds require exogenous GAs for fruit development and seed germination.

The *Arabidopsis GIBBERELLIN INSENSITIVE (GAI)* family contains five members, *GAI, RGA, RGL1, RGL2* and *RGL3*. *GAI, RGA* and *RGL1* are negative regulators of stem elongation, including in early seedling growth. Alleles of *rgl2* confer strong resistance to the inhibitory effects of paclobutrazol (a GA-biosynthesis inhibitor), because the seeds are already insensitive to GA, whereas *rgl1-1*, *gai-t6* and *rga-t2* do not. Loss-of-function mutations in *RGL2* also completely restore germination (in the absence of exogenous GA) in *ga1-3* mutants.

3. Auxin

Few seed development mutants have been identified with defective **auxin** content or transport. Defective kernel (*dek*) maize mutants with substantially lower IAA contents than wild-type counterparts undergo less endoreduplication, which can be partially restored in this mutant by applied auxin. *Arabidopsis pin1-1* mutants, defective in auxin polar transport, resemble the altered phenotype obtained in *Brassica juncea* zygotic embryos treated with auxin transport inhibitors. Experiments based on the use of auxin polar transport inhibitors in several dicot and monocot species support a major role of auxins in the establishment of embryonic polarity in zygotic **embryogenesis**. (**See: Development of embryos – dicots**)

4. Cytokinins

No mutants with reduced **cytokinin** content have been isolated. The *Arabidopsis* mutant *cre1* (cytokinin receptor) has reduced cell division in provascular tissue, and embryos do not develop phloem tissue.

5. Ethylene (ethene)

Ethylene overproduction in maize *shrunken* (*sh2*) mutants induces early **programmed cell death** in the **endosperm**. Ethylene response mutants in *Arabidopsis*, e.g. *ein2* (ethylene insensitive, high seed dormancy) and *ctr1* (constitutive triple response, low seed dormancy) also exhibit altered sensitivity to ABA. Ethylene production in seeds may also be implicated in stress responses, and interactions between ethylene and other hormones during seed development probably occur.

6. Brassinosteroids

Mutations in at least eight loci of *Arabidopsis* produce plants with characteristics of **brassinosteroid** deficiency. Mutations occur both in the general sterol biosynthesis pathway (cycloartenol to campesterol) and the BR-specific pathway (campesterol to brassinolide). Most BR-deficient mutants have reduced fertility. However *dwf5-1*, a BR mutant with wild-type fertility, produces seeds that do not develop normally and that require exogenous BR for full germination. BR in the incubation medium can rescue the germination phenotype of GA synthesis mutants, and *det2-1* (BR biosynthetic mutant) and *bri*1-1 (BR insensitive mutant) increase the sensitivity of germination to inhibition by ABA.

7. Jasmonic acid

The *Arabidopsis* mutants: *fad3-2*; *fad7-2*; *fad8* and *AtLox* JA and tomato *def1* have reduced capacity to synthesize **jasmonic acid**. *Arabidopsis* mutants defective in the perception of jasmonates (*jin4*, *jar1*, *coi1*) produce seeds with increased sensitivity to ABA: JA may modulate dormancy and germination, interacting with ABA. It is somewhat paradoxical that while germination of seeds from plants with reduced sensitivity to jasmonates are more inhibited by ABA than the wild type, MeJA and ABA synergistically inhibit germination when included in the incubation medium, an effect dependent on *COI1* (a gene involved in jasmonate perception).

(RLB-A, RAS, MVR)

(**See: Mutants; Signal transduction – hormones;** the separate entries on the above hormones) For other reading see under: **Dormancy – genes; Dormancy breaking – cellular and molecular aspects; Germination – radicle emergence.**

Choe, S., Tanaka, A., Noguchi, T., Fujioka, S., Takatsuto, S., Ross, A.S., Tax, F.E., Yoshida, S. and Feldmann, K.A. (2000) Lesions in the sterol delta reductase gene of Arabidopsis cause dwarfism due to a block in brassinosteroid biosynthesis. *Plant Journal* 21, 431–443.

Srivastava, L.M. (2002) *Plant Growth and Development. Hormones and Environment*. Academic Press, New York, USA.

Hull

Dried bracts (**lemma** and **palea**) surrounding the mature caryopsis (grain) in certain cereals. Sometimes described as husk. In maize and some other species the **pericarp** is referred to as the hull. (**See: Floret**, Fig. F.7)

Huller

Also sometimes called a dehuller, this equipment is used to remove the outer coat or hull (husk) on the seed while the seed itself remains unscathed, as part of the seed conditioning process. (**See: Conditioning, I. Precleaning; Shelling**)

Humidification

In seed technology: the adjustment of atmospheric **relative humidity** during storage to achieve a desired seed **moisture content**. Such manipulations are important in the storage of large-seeded legumes, to decrease susceptibility to mechanical stresses and injury during harvesting, handling and hydration damage due to rapid water uptake during **imbibition**, both of which result in decreased normal germination. (**See: Germination – physical factors; Storage management**)

Husk

See: Hull

Hyacinth bean

This bean, also known as the dolichos bean, the Egyptian bean or the bonavista bean (*Dolichos lablab*, syns. *Lablab niger*, *L. purpureus*), is a tropical legume, thought to originate in tropical Asia, with a long history of cultivation in India. It was introduced into tropical Africa about 1000 years ago, from which it was carried, with the slave trade, to the Caribbean and South America. The plant is a twining vine which can reach 6 m in height (length) and can become woody. The seeds, used as a food, are purple to almost black, brown, white or mottled with a hard **testa** and a prominent, white **hilum** (Fig. H.8). Seed weight is 250–500 mg. (**See: Legumes**)

The seeds contain approximately 25% fw **storage protein**, 1% **oil** (triacylglycerol) and about 60% **carbohydrate**. A cyanogenic glucoside is present (**see: Pharmaceuticals and pharmacologically active compounds**).

The bean is an important food in Asia and Africa. The immature pods are consumed as a vegetable, especially in India. The green or dried beans may be cooked (boiled) as a component of various dishes (*dhal* in India) or used as an adjunct to salads, but boiling must include changes of the water so as to remove the dangerous cyanogens. In Asia, mature seeds are used to make *tofu* or fermented to produce *tempeh*.

(MB)

Smartt, J. (1990) *Grain Legumes: Evolution and Genetic Resources*. (The hyacinth bean, pp. 294–298.) Cambridge University Press, Cambridge, UK.

Fig. H.8. Hyacinth bean (image by Elly Vaes and Wolfgang Stuppy, Millennium Seed Bank, RBG, Kew).

Hybrid

Hybrid seed is produced as a first filial generation **cross** (F_1) between two compatible parent **lines** or species, either of the same genus or different genera. In plant breeding, hybrid **varieties** are produced by controlled crosses to combine traits from selected parental lines that are genetically different from each other, and also to take advantage of **heterosis** ('hybrid vigour'). Modern hybrid varieties are based on single, double, or three-way crosses. Much of the hybrid vigour exhibited by F_1 hybrid varieties is lost in the next generation and the traits segregate out; consequently, the farmer purchases new seed each year from seed companies. (**See: Production for Sowing, III. Hybrids**)

Hybridization

The process of crossing related individuals of dissimilar genetic constitution. Interspecific hybridization occurs naturally during plant evolution, such as between closely related native and introduced wild or **domesticated** species. Hybridization is thought to have played an important role in the early development of most of the major crop species, and also in some circumstances to lead to the evolution of more effective agronomic weeds. Hybridization techniques are used in breeding programmes to obtain genetic recombination for a variety of purposes, including the production of pure-breeding populations from F_1 generation plants by backcrossing. (**See: Cross**)

In its molecular-biological sense, the term for the process of binding complementary nucleotide chains in **nucleic acids**: used as the basis for a wide range of analytical and experimental research techniques, such as the hybridization of cDNA or RNA probes or fragments to specific DNA sequences in **Southern blot** or **northern blot analysis**, in the **PCR** and, generally, in **transcriptomics**.

Hydration force explanation

A proposed explanation for the forces that cause macromolecular perturbations during drying and how small molecules protect structures. The premise for this hypothesis is that solutes in the interstitial space between macromolecules (such as membranes) can limit their close approach during drying (e.g. in seed desiccation) thereby limiting the tendency for **phase** transitions or denaturation. With sufficient drying, the interstitial solution forms a **glass** that provides mechanical resistance to macromolecule compression. This model accounts for the greater protection of some solutes in model mixtures as a balance between molecular size (smaller molecules are more osmotically active and are less easily excluded from narrow spaces) and mechanical resistance (larger carbohydrates are generally more viscous and have higher glass transition temperatures). Resistance to drying in the interstitial glass may account for the high sorption enthalpy that is traditionally explained by strong binding sites (**see: Water binding**) and provides a rationale for the greater energy needed to dehydrate desiccation-tolerant versus intolerant seeds. The mechanical failure of the interstitial glass, either by time or loss of water molecules with further drying may account for faster ageing rates of some seeds when dried below a **critical water content** (**see: Deterioration kinetics; Ultradry storage**). An alternative, though not mutually exclusive, explanation of macromolecular stabilization by hydrophilic solutes is given by the **water replacement hypothesis**. (CW)
(**See: Desiccation tolerance – protection by stabilization of macromolecules**)

Bryant, G., Koster, K.L. and Wolfe, J. (2001) Membrane behaviour in seeds and other systems at low water content: the various effects of solutes. *Seed Science Research* 11, 17–25.

Hydrochory

Dispersal of seeds and fruits by water.

Hydroponic production

'Hydroponics' is a nutriculture technology for growing plants without soil (derived from the Greek for 'water' and 'work').

Seeds are often germinated hydroponically in cubes of inert support media, made for example from rockwool or plastic foam, in a propagation area or nursery that is periodically supplied with nutrient solution or water. Most preformed germination blocks or cubes are available in several sizes, designed usually with small holes so that seeds can be sown directly (with, for example, **vermiculite** sprinkled over the seed to maintain moist conditions), and subsequently can be easily transplanted into larger cubes with holes made for that purpose, and grown to maturity without a check in growth. Seeds may also be germinated in trays or pots of well-drained potting mix. Plants can be transplanted when they have four or five true leaves. The final crop growing media should be properly leached, moistened and warmed, and irrigated with nutrient solution immediately after transplanting. Alternatively, seeds can be direct seeded into the ultimate crop production media.

Commercial systems have been developed, over the last 50 years or so, using: (i) water, in which complete necessary mineral nutrients are dissolved (salts of the macronutient elements N, P, K, Ca, Mg, S, and the micronutrients Fe, Mn, Zn, Mo, B and Cu, at pH 6–6.5), that is constantly circulated across the root zone in such a way as to maintain aeration; (ii) with or without a solid aggregate growth medium (such as sand, gravel, **perlite**, vermiculite, or **rockwool**) to anchor and provide mechanical support for the roots. Hydroponic systems produce plants in high densities, and are usually operated in covered glasshouses, and are most popularly used for growing cucumber, lettuce, tomato and pepper and flower crops, and houseplants. Various water and nutrient delivery techniques have been developed, including the nutrient film technique (NFT, growing plants using a very shallow film of nutrient solution constantly pumped over the roots), ebb and flow, and drip systems. NASA has experimented with the combination of an 'aeroponic' mist-spray system (where roots hang down into an enclosed box) and NFT, in order to conserve nutrients for the use of hydroponics in space (**see: Seeds in space**).

Separate transplant production greenhouses are often used to reduce the carry-over of pests and diseases and to optimize environmental conditions such as temperature. It is important to thoroughly moisten the growing medium to ensure uniform germination. Flood and drain (ebb and flow) systems are often used for germination because they provide better control of watering. In these systems, nutrient solution or water floods a shallow tray containing the sown cubes, pots, or trays, providing moisture from the bottom, which spreads through the root zone by capillary action. The tray is then drained, to promote root zone aeration. (GEW)

Morgan, L. (1999) *Hydroponic Lettuce Production and Growing*. Casper Publications, Sydney, Australia.

Staff of the Ontario Ministry of Agriculture and Food (2003) *Greenhouse Vegetables*. Publication 371, Ontario Ministry of Agriculture and Food. Toronto, Ontario, Canada.
Staff of the University of Arizona. *Growing Tomatoes Hydroponically*. http://ag.arizona.edu/hydroponictomatoes

Hydropriming

Seed **enhancement** treatment techniques based either on the 'continuous or staged addition of a limited amount of water' or 'imbibition in water for a short period' (as in **steeping**) with or without subsequent incubation in humid air, and drying. (**See: Priming – technology**)

Hydroseeding

A sowing technique of spraying a slurry of seeds, **mulch** (hydromulching) and fertilizers in water on to bare ground, which is used to establish broad ground-cover cultivations, such as lawns, in landscaping and land reclamation, especially on terrain that is steeply sloping or otherwise difficult to access. The components are mixed with a so-called 'tackifier', which acts as a glue to hold them all in place on the soil until seedlings have become established; **gum** from **guar** is commonly used for this purpose. Hydroseeding can be distinguished from the **fluid-drilling** technique, in which sprouted seeds are suspended in a viscous gel and sown by extrusion into the soil (**see: Pregermination**). (The term is distinct from '**water seeding**', which refers to the practice of aerial sowing of pre-imbibed seed.) (PH)

Hydrothermal time

The hydrothermal time constant (θ_{HT}) models the response of seeds to both temperature (T) and **water potential** (Ψ). It is defined as $\theta_{HT} = (T - T_b)\,(\Psi - \Psi_b(g))\,t_g$, where T_b is the base temperature permitting radicle emergence, $\Psi_b(g)$ is the base water potential that just prevents germination of percentage g in the seed population, and t_g is the time to germination of percentage g. This concept can be adapted to explain the response of germination to external factors, and be extended to describe the basis of **priming** effects (**see: Priming – physiological mechanisms, 1. Mathematical models**), and to predict seedling emergence in various agro-ecological conditions. (DC, FC)
(**See: Germination – field emergence models; Hydrotime; Thermotime**)

Bradford, K.J. (1995) Water relations in seed germination. In: Kigel, J. and Galili, G. (eds) *Seed Development and Germination*. Marcel Dekker, New York, USA, pp. 351–396.

Hydrotime

The hydrotime model was introduced by Gummerson (1986) to characterize the effects of reduced **water potential** (Ψ) on germination. The hydrotime constant (θ_H) is the total hydrotime, in MPa.h, needed for each seed to complete its germination and is defined according to the equation $\theta_H = (\Psi - \Psi_b(g))t_g$ where Ψ is the seed water potential, $\Psi_b(g)$ is the base water potential that just prevents germination of percentage g in the seed population, and t_g is the time to complete the germination of the same percentage g. The germination rate ($1/t_g$) increases linearly with Ψ above $\Psi_b(g)$. The concepts of hydrotime and **thermal time** have been combined in the development of **hydrothermal time** models of germination. (DC, FC)
(**See: Germination – field emergence models; Germination – influences of osmotica and salinity**)

Bradford, K.J. (1995) Water relations in seed germination. In: Kigel, J. and Galili, G. (eds) *Seed Development and Germination*. Marcel Dekker, New York, USA, pp. 351–396.
Gummerson, R.J. (1986) The effect of constant temperatures and osmotic potential on the germination of sugar beet. *Journal of Experimental Botany* 37, 729–741.

Hypocotyl

Stem-like region of the (generally dicot) embryo delimited by the radicle at one end and by point of insertion of the **cotyledons** at the other. (**See: Epigeal; Structure of seeds**)

Hypogeal

Also sometimes hypogeous seedling establishment or hypogeal 'germination'. Seedlings in which the **cotyledon** (monocot) or cotyledons (dicot) remain below the soil after germination and the first true leaves are borne on the epicotyl (Fig. H.9, Table H.4). When the term hypogeal 'germination' is used, particularly with respect to field emergence, strictly from the seed physiological point of view this is an incorrect term since emergence from the soil is an indication of seedling establishment, not of germination *per se* (**see: Germination**). The cotyledons of dicots become depleted of reserves and ultimately are totally degraded in the soil. The single cotyledon of cereals (monocot), the **scutellum**, serves to transfer nutrients from the degradation of reserves within the endosperm to the growing seedling, after its own oil reserves have been mobilized. In some cereals, e.g. wild oat, the scutellum is **haustorial** and grows into the endosperm to increase the surface area for absorption of the sugars and amino acids released from the degraded reserves. Cotyledons also act as haustoria in other species, such as in the date palm and asparagus, where they remain embedded in the hard **hemicellulose-** (galactomannan-) containing endosperm and absorb the products of its mobilization to support growth of the seedling. (JDB)
(**See: Epigeal**)

Table H.4. Examples of species whose seeds exhibit the hypogeal form of seedling establishment.

Endospermic	Non-endospermic
Asparagus officinalis (asparagus)	*Aponogeton* spp.
Avena fatua (wild oat)	*Phaseolus multiflorus* (runner bean)
Hevea spp.	*Pisum sativum* (pea)
Hordeum vulgare (barley)	*Tropaeolum* spp.
Phoenix dactylifera (date palm)	*Vicia faba* (broad bean)
Tradescantia spp.	
Triticum aestivum (wheat)	
Zea mays (maize, corn)	

Hypostase

A cup-, disc- or plate-like, sometimes even globular or pyriform tissue of the nucellus present in the region of the

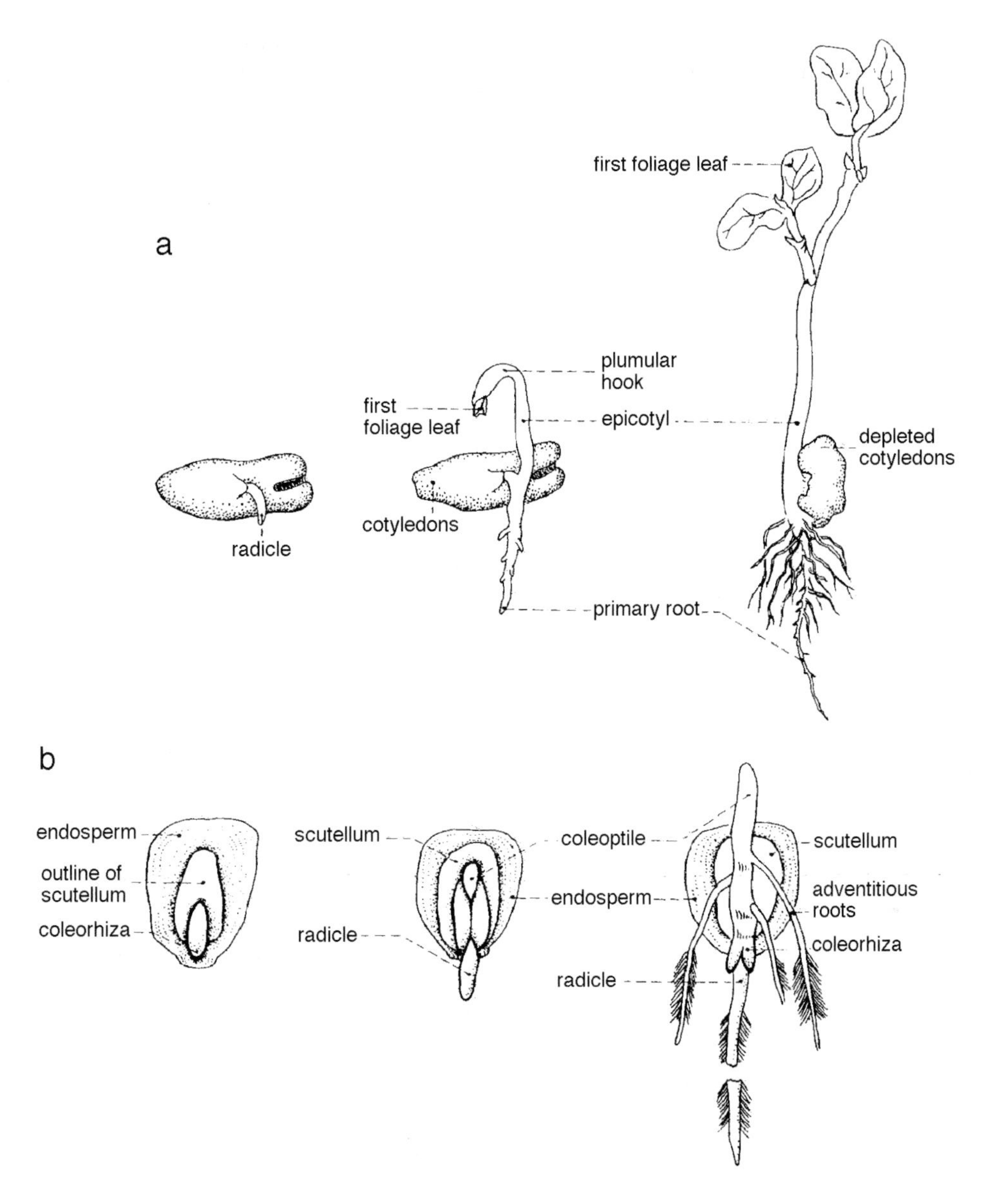

Fig. H.9. Examples of the hypogeal form of seedling establishment. In both the dicot (a) (*Vicia faba*, horse bean, broad bean) and the monocot (b) (*Zea mays*, maize, corn) the seed containing the two cotyledons (*Vicia*) or the single cotyledon, the scutellum (*Zea*) remains below the ground. The cotyledon(s) are degraded when the stored reserves have been depleted.

chalaza at the level of origin of the **integuments**. The cells forming the hypostase differ from the remaining nucellar tissue in their size, shape, contents and cell walls. In the mature seed, the cells are often filled with tanniniferous materials letting the hypostase appear as a dark reddish-brown 'chalazal plug'. One of the potential functions of the hypostase is to seal the chalazal opening (**heteropyle**) against dehydration or microbiological and mechanical damage, especially when the cells are thick-walled. (**See: Structure of seeds**)

Hysteresis

The lagging of an effect behind its cause: manifested in chemistry, for example, by the phenomenon of different apparent equilibrium product concentrations arising as a result of reactions starting with high reactant concentration and proceeding to product and starting with high product concentration and proceeding to reactants. Likewise, in the physics of seed hydration, the different water contents often observed when a seed is dried or hydrated at a certain **relative humidity** is an example of hysteresis and is symptomatic that the reaction in at least one of the directions did not reach equilibrium. This is important when considering **water sorption isotherms**, which are only valid representations of the water content–relative humidity relationship at equilibrium. Achieving equilibrium may be slowed during seed drying by impermeable seed coverings (seedcoats) or high viscosities in concentrated solutions. Hysteresis may be observed upon hydration because the added water molecules allow surfaces to relax and expand to expose more surface. Loss of hysteresic effect is a good indication that the equilibrium concentration (water content in the drying example used here) is achieved. (CW)

I

ICARDA

The International Center for Agricultural Research in the Dry Areas conducts research and training projects in dry areas of the developing world, in partnership with governments and non-governmental organizations, institutions, universities and research institutes internationally. Projects include improvement of barley, lentil and fava bean in subtropical and temperate dry areas of the world, and of bread wheats, chickpea and pasture and forage legumes in CWANA countries (Central and West Asia and North Africa). Established in 1977 and based in Aleppo, Syria, ICARDA is one of the 15 centres supported by the **Consultative Group on International Agricultural Research** (CGIAR). See: www.icarda.org.

ICRISAT

The International Crop Research Institute for the Semiarid Tropics. From its headquarters in Patancheru near Hyderabad, India, ICRISAT works in 48 developing countries on biotechnology and related-science projects, focussing on pearl millet, sorghum, chickpea, pigeon pea and groundnut crops. Resources include a **gene bank** with more than 113,000 accessions. See: www.icrisat.org

Identity preservation

Identity preservation (IP) is the system of crop and raw material management that preserves the integrity, distinctiveness and traceability of the source or nature of materials, including seed and grain, throughout the supply chain, and channels them to their specific markets.

At the simplest level, the majority of large-volume commodity-traded agricultural grain is governed by systems that classify according to both variety or type and purity, in official **grain inspection** procedures. Thus within each of the different classes of **wheat** (hard and soft), **maize** (flint, dent, white and yellow), **barley** (feed and malting) and **rice** (round, medium or long grain japonica, and basmati), there can be quality grades determined by prescribed degrees of absence of contamination. The purchasers of many agricultural grain products, such as in the breakfast-cereal, bread, beer-brewing and starch manufacturing industries, often impose additional requirements concerning content or composition that are incorporated into articles of trade. These kinds of class and quality specifications represent basic forms of IP.

Superimposed on this level of IP is the safeguarding of grain varieties that provide higher or more consistent food quality or unique product quality traits, which are of added value to end-users and hence command premium prices. Examples include high oleic **sunflower**, low-linolenic or high-laureate or high erucic acid canola/oilseed rape, **soybean** varieties with high sucrose or protein or for tofu or soya-milk production, and waxy, white, hard-endosperm or high-oil maize. Rigorous steps must be taken to segregate such premium commodities from 'mainstream' grain products. Conversely, high erucic acid canola for industrial use and certain **genetically modified** maize varieties approved for animal feed must be systematically kept out of the human-food market supply chain. Likewise, on the basis of fundamental principle, certified organic production systems exclude both 'conventional' and 'genetically modified' (GM) products or ingredients, including **organic seed**. Such specialized market segments require more sophisticated IP systems: keeping grain production separate at all stages including preserving minimal isolation distances from potential sources of contaminating pollen, and segregation throughout harvest, cleaning, transport, storage, processing (in some cases through dedicated facilities) and distribution.

For several countries, currently, transgenic grain varieties must be kept apart to satisfy consumers' rights to the information to exercise choice. The IP of GM-related products is tending towards the labelling throughout the whole food chain, from farm level through to the shippers and processors, to avoid commingling with the commodity-traded crop. Main commodity soybean, maize and oilseed rape grain, and food or feed products made from them, that are traded into markets where this is an issue of concern are tested for the adventitious presence of specific DNA or protein derived from GM genes, to guarantee there is less than the regulatory maximum allowed for them to carry the 'non-GM' label.

At the crop-growing level, farmers rely for the IP of their seed inputs on the well-established industry systems of variety testing and verification that are intrinsic to modern seed breeding, production, multiplication and distribution (**see: Certification schemes**). The main sources of additional risk of loss of IP on the farm are pollen drift (particularly in open-pollinated crops such as canola, maize and sorghum, but appreciably too in self-pollinated crops such as soybean), the appearance of 'volunteer' or feral GM plants in subsequent crops, as well as adventitious commingling at harvest or in store. For this reason, grain production contracts typically stipulate practices such as the cleaning of equipment, non-adjacent crop plantings to avoid cross-pollination or harvesting errors, and the keeping of accurate records: procedures on much the same lines as are routine in seed production (**see: Production for sowing, I. Principles**).

(PH)

Huffman, W.E. (2004) Production, identity preservation and labeling in a market place with genetically modified and non-genetically modified foods. *Plant Physiology* 134, 3–10.

IDS

An abbreviation of Imbibition (or Incubation)/Drying (or Dehydration)/Separation: a three-stage technique used to remove filled-dead seeds from a mixture of filled-live and filled-dead seeds, mostly applied to **tree seeds**.

In addition to filled-live seeds, most tree species habitually produce relatively large proportions of filled-dead and empty seed. **Flotation**, air-column and specific gravity separation **conditioning** techniques make it possible to eliminate, or at least greatly reduce, the proportion of empty seeds, but until the early 1980s it was not possible to remove filled-dead seeds. The breakthrough came in Sweden, when M. Simak sequentially applied imbibition followed by dehydration and finally separation. It is important to appreciate that IDS is a convenient label for a family of processes, and each step can be conducted in any of a number of ways; it is not a unique combination of precisely defined steps linked into one unalterable process.

Living and dead seeds at normal storage moisture contents (6–8% fresh weight basis) have similar densities and other physical properties; also when fully imbibed (to 30–65% moisture contents, depending upon species) the physical properties are indistinguishable. The technique of IDS is to fully imbibe a seedlot containing full-live plus full-dead seeds and then partially re-dry the mixture. The two categories of seeds dry at different rates, which in turn brings about transient density differences between them. With careful timing these differential drying-rates can be exploited to effect separation.

In practice, all empty seed must be removed from the seedlot , and the percentage germination and percentage of filled-dead seed is determined before IDS is applied. The seedlot is then imbibed and samples are taken regularly during drying, and separated by whatever means is available, and when two fractions obtained from the separation technique are in the same ratio as the germination percentage to filled-dead seeds, it is assumed that the separation has been achieved. In theory, the IDS treatment of a 10 kg seedlot with 60% germinated and 40% dead seed should yield 6 kg with 100% germinated and 4 kg of completely dead seed. But in practice this level of efficiency is rarely achieved. It is wise, therefore, to try the technique on a relatively small quantity of a seedlot first and, having established the likely losses and an acceptable recovery rate, a decision on whether to proceed to separation of the main seed bulk. (PG)

Simak, M. (1984) A method for the removal of filled-dead seeds from a sample of *Pinus contorta*. *Seed Science and Technology* 12, 767–775.

Imbibition

The initial uptake of water by a dry seed from the surrounding medium. This uptake is determined by the water potential gradient between the medium and the seed (**see: Germination – physical factors**). Initially the matric potential (Ψ_m) (**see: Water potential**) of the seed is very low (very negative, >−100 MPa) and water is absorbed on to the dry surfaces of the structures that surround the embryo (e.g. **pericarp, testa, endosperm**), and into the cell walls and polymeric storage compounds (e.g. protein, starch). This also causes the release of air that was absorbed on to the dry surfaces, resulting in a 'wetting burst' which may last for several minutes.

When the dry surfaces become hydrated, the matric potential of the seed is zero, and further water uptake depends on the osmotic potential (Ψ_π) of the seed being more negative than the Ψ_m and the Ψ_p and Ψ_π of the surrounding medium, e.g. soil (**see: Germination – influences of osmotica and salinity; Water potential**).

Leakage of low molecular weight solutes (e.g. sugars, amino acids, organic acids, ions) occurs from seeds during imbibition due to disruption of the outer cell membrane as water enters and permeates the dry cells. This leakage is transient, and ceases as the membranes return to their stable liquid crystalline state from the dry gel state (**see: Membrane damage**). The degree of imbibitional leakage increases as seeds deteriorate, reflecting cellular and metabolic degeneration; this effect is the basis of the seed vigour conductivity test in large-seeded legumes (**see: Vigour testing – biochemical, 1. Conductivity test**). The leaked solutes from seeds in the soil might stimulate the growth of bacteria and fungi, which in turn could invade the seed or young growing seedling and bring about its deterioration. Larger molecular weight compounds, including large polymers such as proteins and starch, may leak from cells disrupted by imbibition, e.g. the outer cells of legume cotyledons, and some polymers may also may leak from cell walls, particularly those of the seedcoat. In soybean, the leakage of **proteinase inhibitors** and **lectins** could function to protect seeds against microbial or insect attack. Damaged seedcoats, as might occur during harvesting or planting, can result in more rapid uptake of water into the seed than into seeds with intact coats, leading to damage to the embryos and poor germination and growth. In some seeds, there are structural modifications to the seedcoat (**see: Seedcoats – structure**) that aid in water uptake and initially direct the water towards certain areas within the seed, e.g. the **embryo**, or **embryonic axis**.

The early stages of imbibition/germination may be critical in determining the subsequent course of events. For example, in lettuce (*Lactuca sativa*) seeds the temperature experienced during the first 6 h of imbibition influences whether germination will proceed to completion. Seeds imbibed at higher temperatures (above 25–30°C for many cultivars) will exhibit **thermoinhibition** and will be essentially dormant until cooler temperatures prevail. However, if lettuce seeds are imbibed at lower temperatures for only the first 6 h, then transferred to higher temperatures, germination proceeds rapidly to completion. Similarly, only partial hydration is required for seeds to be sensitive to light through the phytochrome system (**see: Light – dormancy and germination; Phytochrome**). Thus seeds are highly sensitive to their physical environment during the initial period of hydration (**see: Germination – influences of external factors**). They are 'testing the waters' to determine if it is an opportune time and place in which to initiate seedling growth. (JDB, KJB)

Imbibition lid

Synonymous with **operculum**, germination lid, seed lid, micropylar collar. (**See** Fig. S.70 in **Structure of seeds – identification characters**)

Immunoassays

In seed science and technology, immunoassays are used to detect mainly viruses and bacteria in seeds and seedlings. The methods are based on the principle that foreign molecules (immunizing agents or antigens, including those carried on foreign microorganisms and viruses), injected into the blood stream of mammals, stimulate their immune system to produce specific antibodies that will recognize and bind to the antigens, and can be purified. Two types of antibodies can be produced for use in seed health assays, most commonly raised against pathogen-specific proteins: polyclonal antibodies which recognize many chemical sites, known as epitopes, on target antigens; and monoclonal antibodies, which recognize one epitope on the target antigen, and can therefore be more specific. The two most widely used **immunoassays** in seed **health testing** are enzyme linked immunosorbent assay (**ELISA**) and immunofluorescence; other assays include agglutination and immunodiffusion. False-positive and -negative results are a continual problem with immunoassay tests, due to antisera either being too specific to react with all strains of the pathogen or not specific enough to prevent cross-reaction with other organisms. Dead cells can also result in false negatives. (VC)

Imprinting

Genomic imprinting, the parent-of-origin-specific non-expression or silencing of genes, plays an important role in the development of the **endosperm**. Because different sets of genes are imprinted and hence silenced in paternal and maternal **gametophyte** (pollen and egg) genomes, the contributions of the parental genomes to the offspring are not equal. Deviation from the normal 2 maternal:1 paternal ratio in the endosperm genome has a severe effect on endosperm development, and often leads to seed abortion. Imprinted genes on either the maternal or paternal side are marked and silenced in a process involving DNA methylation and chromatin condensation. In addition, on the maternal side, imprinted genes are most probably under control of the *FIS* genes (**see: Apomixis**).

The best-known example of genetic imprinting is the *R* gene control of anthocyanin production in the **aleurone layer** of **maize** endosperm. When the RR female (red endosperm) is crossed with the rr (colourless endosperm) male all of the kernels are red, but when the reciprocal cross is made, the kernels are mottled due to irregular anthocyanin distribution. This mottling is not due to a dosage effect in the triploid endosperm (e.g. RR/r versus rr/R, female/male) but is due to the mode of inheritance of the R allele. Kernels are mottled regardless of the number of R alleles inherited paternally and always solid coloured if an R allele is inherited maternally. Alleles of other maize genes, e.g. those for the synthesis of one of the prolamin **storage proteins** (α-zein) are also imprinted, when inherited maternally. In *Arabidopsis* the *MEA* gene plays a role in genetic imprinting. For example, a *MEA/mea* female when crossed with a wild-type male results in about 50% seed abortion, but in the reciprocal cross all of the seeds are normal and viable. The paternal allele cannot rescue a seed that has inherited the female *mea* gene. The *MEA* gene is imprinted exclusively in the endosperm, and only the maternal allele is expressed therein; *MEA* is expressed from both alleles in the vegetative plant.

FWA genes, as well as *MEA* genes are silenced when derived paternally in the developing endosperm. *FWA* is silenced by DNA methylation in its promoter region, and only *MEA* (*MEDEA*) genes of maternal origin are active during *Arabidopsis* **embryogenesis**. Inheritance of a maternal loss-of-function *MEA* allele results in embryo abortion and prolonged endosperm production, irrespective of the genotype of the paternal allele. *FWA*, which encodes a homeodomain transcription factor, is imprinted in the endosperm of *Arabidopsis* and is expressed only from the maternal allele. Expression of *FWA* in the endosperm coincides with an overall reduction in the amount of methylated cytosine residues in the direct repeat sequences of the 5′ region of this gene. The amount of *FWA* methylation remains high in other tissues of the plant, preventing transcription of this gene in these tissues. (JDB, AC)

(See: Endosperm development – dicots; Xenia)

Alleman, M. and Doctor, J. (2000) Genomic imprinting in plants: observations and evolutionary implications. *Plant Molecular Biology* 43, 147–161.

Gehring, M., Choi, Y. and Fischer, R.L. (2004) Imprinting and seed development. *The Plant Cell* 16, S203–S213.

In situ conservation

At the species level, the active conservation of individual species where they naturally occur or, in the case of domesticated biota, where husbandry usually takes place (on-farm conservation). However, most *in situ* conservation is directed towards habitats and ecosystems. Such conservation regimes will also passively conserve many species and much of their diversity. Included within such diversity is the soil seed bank. (**See: *Ex situ* conservation**) (RJP, SHL)

Inclusions

An aggregation of molecules that are present within organelles in a cell. Examples are **phytin**, **crystalloids** and crystalline inclusions of calcium oxalate, termed druses, which can be present in **protein bodies** within seeds.

Incompatibility

See: Compatibility

Indeterminate flowering

The continuation of vegetative growth and further flowering after flowering is initiated in the plant. Considerable differences occur between the initial and final flowering, causing non-uniformity in seed maturity (such as in silique pods in Brassicaceae, and the umbels in Apiaceae). However, the indeterminacy trait is desirable in some crop situations, such as in varieties for the short-season production of **soybean** and the season-long production of fresh **tomato** fruits, and many forage species (**see: Legumes – forage**). (**See:** Fig. D.8 in **Determinate flowering**)

Induced dormancy

While all forms of dormancy can be broadly considered as 'induced', the term is generally used as a synonym for **secondary dormancy**, i.e. dormancy induced in a mature seed.

Industrial legumes

Legumes whose importance lies largely in the use of their seeds in a wide range of industrial, food, pharmaceutical and agricultural products. Examples are **soybean**, **groundnut** (peanut) (which two are sometimes included in the **grain legumes**), **carob** and **guar** (**see: Legumes**).

Industry – current structure

The modern global seed industry is comprised of a wide range of small-, medium- and large-size private-sector commercial operations, which operate alongside public-sector enterprises, see: **Industry Supply Systems, I. Formal**. The annual value of worldwide commercial seed trade generated by the industry – i.e. excluding seed produced and saved on the farm (**farm-saved**) – is estimated at US$30 billion, of which about US$4 billion is created through international, as distinct from internal country, trade. To this may be added the estimated US$0.9 billion value of the global agrochemical seed treatment market.

1. Companies

Most seed companies started as very modest operations and until recently and with few exceptions most have been small and medium enterprises, specialized in a limited range of crops, niche markets and knowledge of local conditions. Seed production does not exhibit large economies of scale, and it has always been possible for the smaller companies to prosper alongside larger competitors.

But during the past three decades or so the trend towards consolidation that had been underway throughout the 20th century increased markedly, with the growth of multinational seed companies. Two main waves of concentration are worth mentioning. In the 1970s, chemical and oil companies such as ICI, Shell, Elf-Aquitaine, Ciba-Geigy and Upjohn acquired seed companies, but most disinvested a few years later as the synergies between the agrochemical market and the seed market were not as great as expected. In the 1990s, strategic diversification of pharmaceutical and chemical companies led to the development of so-called 'life science' company conglomerates, which gathered together human, animal and plant science activities – such as Monsanto, DuPont, Novartis, Dow Chemical, Rhone-Poulenc, Bayer CropScience, BASF. Of late, this concept is also in decline, because most of the pharmaceutical companies have spun off their crop-science activities.

Several of the largest companies have been the subject of mergers and take-overs, so that in 2001 the main ten companies controlled about a quarter of the global seed trade, nearly double the equivalent proportion 10 years earlier (Table I.1). Most major international seed companies have headquarters in the USA, the European Union or Japan. Despite these changes in the complexion of the industry, there is still a wide diversity of smaller firms, and there is much less concentration compared to the seed crop protection industry, for instance, where a substantial number of mergers and acquisitions have also taken place in recent years, and which is currently dominated by four companies (Bayer CropSciences, Syngenta, Crompton-Uniroyal and BASF – of which the first two are also amongst the largest global seed companies). However, simplified aggregate global analysis does not reflect the segmentation of discrete seed markets into crop or species groups, in which typically only a few companies dominate. For example, Syngenta and Limagrain are involved in a wide range of ornamental, vegetable and field crops; Seminis and Rijk Zwaan are specialized in vegetables, Sakata and Takii in vegetables and ornamentals, Pioneer and Monsanto in field crops, or Barenbrug and DLF-Trifolium in grass and turf.

2. Trade associations

Today seed industry representation is structured in four tiers: companies and national, regional and international trade associations, which promote the use of good quality seed and defend trade interests (Table I.2).

ASTA, the American Seed Trade Association, was one of the first of the national seed trade associations, established in 1883, and many now exist. At the beginning of the 20th century, as the seed trade started to become truly international, mainly at first for forage and vegetable crop seed, the need to standardize trade conditions led to the first international seed congress in London in 1924, which gave birth to the International Seed Trade Federation (FIS, Fédération Internationale du Commerce des Semences) (**see: ISF**).

Later, with the development of regional trading blocks, Regional Seed Trade Associations were established, which now promote the interests and inter-communication of regional seed industries in various ways, such as working with governmental and non-governmental organizations (NGOs) and national seed associations, and promoting education and training activities. The first one was COSEMCO, the Seed Trade Association of European Union Members, created in 1961, followed in 1977 by COMASSO, the Plant Breeders' Association of the European Community; these two merged to become the **ESA** in 2000. In 1986, **FELAS**, the Latin America Federation of Seed was created, followed by **APSA**, the Asia Pacific Seed Association in 1994 and **AFSTA**, the African Seed Trade Association in 2000.

In addition, at the initiative of the public or the private sectors, less structured organizations known as Seed Networks came into being: WANA, the West Asia and North Africa Seed Network in 1992, WASNET, the West Africa Seed Network in 1998 and EESNET, the Eastern European Seed Network in 2001. (BlB)

ISF (2002) *Seeds for Mankind, Plant Breeding, Seed and Sustainable Agriculture*. ISF, Nyon, Switzerland.

Le Buanec, B. (1996) Globalization of the seed industry: current situation and evolution. *Seed Science and Technology* 24, 409–417.

Le Buanec, B. (1999) Public and private sector investments in the seed industry. *Proceedings of the 1999 World Seed Conference*. ISTA, Zurich, pp. 225–230.

Le Buanec, B. and Heffer, P. (2002) The Role of International Seed Associations in International Policy Development. In: *Seed Policy Legislation and Law*. *Journal of New Seeds* Vol. 4, The Haworth Press.

www.seedquest.com

Industry Supply Systems

I. Formal

An underlying feature of seed enterprises is that seed provision involves a number of distinct operations that may be carried out by separate enterprises: breeding, field production, processing and treatment, wholesaling and retailing. These

Table I.1. The evolution since 1985 of the largest private seed companies supplying seed for sowing, ranked by market share.

	1985	1991	1996	2001
Ranking by sales volume (in approximate descending order)				
1st–5th	Pioneer	Pioneer	Pioneer	Pioneer
	Sandoz	Sandoz	Novartis	Monsanto
	Dekalb	Limagrain	Limagrain	Syngenta
	Upjohn-Asgrow	Upjohn-Asgrow	Advanta	Limagrain
	Limagrain	ICI	Seminis	Advanta
6th–10th	Nickerson-Shell	Cargill	Takii	Seminis
	Takii	Takii	Sakata	Takii
	Ciba	Dekalb	KWS	KWS
	VanderHave	VanderHave	Dekalb	Sakata
	CACBA	Orsan	Cargill	Bayer CropScience[a]
11th–15th	Sakata	Cebeco	Pau Euralis	DLF-Trifolium
	Orsan	KWS	Monsanto	Dow AgroSciences
	Cargill	Sakata	Sigma	Barenbrug
	Lubrizol	Ciba	Saaten-Union	In Vivo
	Volvo	Sanofi	RAGT	Pau Euralis
16th–20th	ICI	Sigma	Svalöf-Weibull	BayWa
	Royal Sluis	Lubrizol	Cebeco	RAGT
	Cebeco	Volvo	DLF-Trifolium	Svalöf-Weibull
	KWS	Royal Sluis	Barenbrug	Saaten Union
	Clause	RAGT	Mycogen	Rijk Zwaan
Percentage of the total market, by aggregate sales turnover				
Largest 5	8%	11%	13%	20%
Largest 10	12%	14%	18%	25%
Largest 15	15%	18%	20%	28%
Largest 20	17%	19%	23%	31%
Approximate total global market size (US$ billion)				
Total	20	26	30	30
International trade	1.3	–	3.3	3.9

This table gives a global view of seed companies, ranked by their approximate sales volume (turnover). However, some companies may be missing or incorrectly placed, due to difficulties in obtaining reliable and comparable data, and to changes in currency exchange values; the ranking is also complicated by the history of mergers, partial divestitures and cross-ownerships within the industry, and because some companies have been parts of larger groups involved in other industry sectors. Not shown are independent companies involved in genetic engineering technology. Source: International Seed Federation, ISF. (Subsequent developments in the period up to the end of 2005 have resulted in the acquisition of Seminis by Monsanto, and the acquisition of parts of Advanta by Syngenta, Limagrain and Deprez.)

[a] Finalized in 2002.

activities may be undertaken by a single enterprise or shared between more than one, depending on the crop and the market.

1. The modern private and public sectors

Private sector breeders – popularly known as 'seed companies' – develop and promote new proprietary varieties or selections. Growth of the private seed industry occurs in instances where farmers acknowledge the superiority of commercially produced seed over **farm-saved** seed. This may occur when it is difficult to save seed, where processing and conditioning guarantees higher quality seed, or where new varieties – in particular, **hybrids** – become available. Hence there is a strong link between advances in plant breeding and the growth of the seed industry. Most horticultural crops are not grown to provide seed as the end product for consumption, and root, bulb, leaf and head plants are not grown even to the flowering stage. Consequently little farm-saved seed results from the crop (**legumes** and some herbs are the main exceptions), and separate plots must be cultivated specifically for the production of seed. Similar considerations apply to forage grass and legume crops (**see: Grasses and legumes – forage**). Where seed saving is convenient and varieties are changed infrequently, the seed industry has less of a presence, even in industrialized countries. (**See: II. Informal**)

Some seed companies specialize in certain crop seed species while others have a diversified range. Almost all seed companies subcontract farmers to multiply seed rather than owning the land for production themselves (**see: Production for sowing**). Seed processors may be subcontracted for the physical seed **conditioning**, **treatment**, storage and distribution operations. Retailers (seed merchants, dealers and distributors) or the seed companies' own internal staff, handle sales and delivery to the farmer. Companies also vary in the degree of their direct involvement in marketing provision

Table I.2. The ISF Network of Seed Trade and Plant Breeders Associations.

World level	ISF			
	Africa and Middle East	*Asia/Pacific*	*Europe*	*Americas*
Regional level	AFSTA WANA WASNET	APSA	ESA EESNET	FELAS
National level (abbreviations for country in parenthesis)	AMSP (MA) ASPLANTE (LB) ESAS (EG) ISA (IL) SANSOR (ZA) STAK (KE) USTA (UG) ZSTA (ZW)	ASI (IN) CNSTA (CN) HKSTA (CN) JASTA (JP) KSA (KR) NZGSTA (NZ) NZPBRA (NZ) SAI (IN) ASF (AU) TURK-TED (TR)	AIC (UK) AIS (IT) AMSOL (FR) ANSEME (PT) APROSE (ES) ASSINSEL (BE) ASSINSEL (FR) ASSOSEME (IT) BDP (DE) BSPB (GB) CFS (FR) CMSSA (CZ) DAPB (DK) DF (DK) EEPES (GR) FDH (DK) FFSF (FR) FCSGBS (FR) FNPSP (FR) FSTA (FI) GESLIVE (ES) GSPA (GR) HSTA (HU) ISTA (IE) PIN (PL) Plantum (NL) RNSA (RU) SASTAB (SK) SEMZABEL (BE) SEPROMA (FR) SISP (CH) SS (SI) SSTA (SE) VPSO (AT) VSSJ (CH) YUSEA (YU)	ABRASEM (BR) AMSAC (MX) ANPROS (CL) ASA (AR) ASTA (US) BRASPOV (BR) CSBC (AR) CSTA (CA) CUS (UY)

For explanation of abbreviations, see: www.worldseed.org/associations.html.

chains, and many intermediate arrangements exist. Some may begin by concentrating on a few operations and evolve to a more integrated operation, or some may remain specialized in plant breeding and provide **foundation seed** to smaller companies who concentrate on production and distribution, or license varieties to a wholesaler or a cooperative. Generally, for high-volume low-value seed such as the grain **cereals** and legumes, such task separations still more or less exist, whereas for low-volume high-value seed, such as hybrid vegetables or flowers, private companies often take care of all the links in the supply chain (except often the retailing stage).

In the USA in 1995, for instance, there were over 300 companies selling hybrid maize seed; the seven largest controlled 70% of the market, but there were many other (usually local) firms, some of which did not have their own plant breeding capacity but depended on foundation seed companies to provide them with inbred lines. In the main wheat-producing areas of the USA, seed growers associations and small commercial operations predominate, but for groundnuts (a relatively unprofitable seed crop), some of the seed supply comes from commercial grain merchants who contract certified seed growers and dedicate part of their processing and storage facilities to the seed business. Private seed companies (and seed producer cooperatives, such as in France) have long been important in many European countries, even for crops with high rates of farmer seed saving, such as wheat.

In both the developed and developing world, the private sector functions alongside state-run 'public' organizations (research institutes, corporations and the like) who supply varieties, particularly of the high-volume and low margin non-hybrid seeds, such as cereals or seeds of 'public varieties'.

There is often significant public–private sector interaction and the mix can again be quite complex. Commercial seed provision may still depend upon access to the plant breeding capacity and new genetic material that public institutions have traditionally supplied in the last century, especially for field crops, with seed production, sale and distribution almost entirely in the hands of private firms and cooperatives.

2. Historical development

Farmers have been exchanging seed since the origins of agriculture, and it is difficult to identify a precise date for the emergence of the formal seed trade. Early specialization concentrated on crops whose seed farmers have difficulty saving, such as forages and vegetables as already mentioned, where the private seed industry took an early and leading role, including responsibility for plant breeding.

(a) *Europe and North America*. In Europe, the 16th and 17th centuries saw a growing demand for a range of vegetable crops and the expansion of market gardening. In England, for example, seed produced by specialist growers was sold by a variety of petty merchants and itinerant pedlars. But purchasers often had few guarantees or recourse in cases of bad quality seed. This was remedied by the appearance of specialist seed merchants in larger towns and cities who established their reputations through attention to quality and invested in promotional activities such as extensive catalogues. The first seed companies of the modern period were established in the 18th century, mainly in vegetable crops; one of the first was Vilmorin-Andrieu, in Paris in 1743. Companies developed in 19th-century Europe to produce seed for their colonies, as for example in Einkhuizen in The Netherlands, which still houses several major horticultural seed companies. Farmers often had little time to dedicate to seed production and storage and preferred to rely on commercial sources; by the early 20th century, for example, most farmers in the USA purchased seed for forage crops such as alfalfa and sorghum. But some in Europe and North America were selecting and developing named varieties from crops in their fields, and several entered the informal and formal trade, specializing as plant breeders and small and medium-sized family-owned companies, alongside public research institutions. The applications of scientific plant breeding to commercial agriculture led to a rapid expansion of the industry from the late 19th century. Both public and private organizations played a growing role in variety development and seed provision. In the case of wheat, for instance, private breeders did much of the early work in France whilst in The Netherlands public plant breeding played the leading role; in North America, most wheat varieties were produced by public plant-breeding institutions.

A major impetus to seed trade was provided by the development of hybrid maize varieties, which became a focus of increasing public and private investment in the USA beginning in the 1920s. Most of the early plant breeding for hybrid maize was done by public universities, but private seed companies (which were initiated by individual farmers, farmer associations or businessmen) came to play an increasingly dominant role in hybrid seed production and sale. Many of these companies soon developed their own plant breeding capacity. In the 1930s, for example, nearly 200 maize hybrid seed companies were formed in the USA. The extension of hybridization techniques to other crops (such as rice, cotton, sorghum, sunflower, oilseed rape/canola, pearl millet, sugar-beet and a series of vegetables including table beet, carrot, tomato, melon and onion) led to a significant growth in private seed trade. The establishment of plant breeders' rights regimes beginning in the late 1960s (**see: Laws and regulations, 3. Intellectual property protection**) provided a further impetus for more investment in private plant breeding. But even for a highly commercial crop like hybrid maize, half of the inbreds in commercial production in the USA as late as the 1980s were the products of public plant breeding.

(b) *Africa, Asia and Latin America*. In developing countries, the origins of the formal seed trade have usually featured government (parastatal) seed companies, but recent years have seen a significant shift from state enterprises towards the private sector. A major exception is the vegetable seed trade where foreign private firms (and domestic companies in some countries) have long played the leading role.

Many countries in North and sub-Saharan Africa and the Middle East established government-owned seed production operations for the major food crops in the 1970s and 1980s, usually based on hybrid maize, and were given exclusive rights to the products of public plant breeding programmes. In the 1990s liberalization policies began the sale of these assets and/or the opening of markets to private domestic and foreign competition, although in many economies farmers often have limited incentives to use the commercial market to replace farm-saved seed. In South Africa, by way of exception, private seed companies have played an important role since the 1960s.

In Asia, the state has also played a leading role in seed trade development. A major factor was the Green Revolution in **rice** and **wheat**, beginning in the 1960s, which led to the development of seed production and distribution infrastructures. The fact that many Asian countries' agriculture is based on rice, a crop with relatively little private breeding investment and relative ease of seed saving, has meant that some state involvement in that crop continues to be prevalent there. More recently, domestic and foreign private investment are becoming important factors in other crops, though. By the late 1990s, for example, more than half of India's seed trade was in the hands of private companies of all sizes, particularly in hybrid **sorghum**, **pearl millet**, **maize** and **cotton**. In China, an extensive network of local and provincial seed companies and a national seed corporation is responsible for field crop seed production, but significant privatization is anticipated.

In Latin America, the evolution of seed trade has followed several paths, depending on national agricultural economies. Argentina, and to a lesser extent Brazil, have a tradition of private seed trade, including for wheat. Mexico and the Andean nations depended on parastatal seed enterprises, often with monopoly status. But policies were modified in the 1980s and today there are few government-run seed enterprises.

For a description of the modern seed industry, **see: Industry**. (RT)

Kloppenburg, J. (1988) *First the Seed.* Cambridge University Press, Cambridge, UK.

Louwaars, N.P. and van Marrewijk, G.A.M. (1996) *Seed Supply Systems in Developing Countries.* CTA, Wageningen, The Netherlands.

Louwaars, N.P. (ed.) (2002) *Seed Policy, Legislation and Law: Widening a Narrow Focus.* The Haworth Press, New York, USA (also published as *Journal of New Seeds* Vol. 4, No.1/2).

Morris, M. (ed.) (1998) *Maize Seed Industries in Developing Countries.* Lynne Rienner, Boulder, CO, USA.

Tripp, R. (ed.) (1997) *New Seed and Old Laws: Regulatory Reform and the Diversification of National Seed Systems.* Intermediate Technology Publications (on behalf of the Overseas Development Institute), London, UK.

Mumby, G. (1994) Seed Marketing. FAO Technical Paper No. 114. www.fao.org/docrep/V4450E/V4450E00.htm

II. Informal

An analysis of farm-level seed management is important for understanding the emergence of the modern seed industry (**see: I. Formal; Industry – structure**) as well as for appreciating the crucial role that farmers in many parts of the world still play in ensuring an adequate seed supply. Farm-level seed management encompasses both the maintenance and improvement of plant genetic resources and the storage and distribution of seed – described generally as the 'informal seed sector'. (The same term is also used more narrowly by some to describe situations where professional farmers specialize to produce seed according to formal standards and norms, in well-organized though unofficial networks.)

1. Historical development

Farmers have been responsible for seed management since the beginnings of agriculture more than 10,000 years ago. Until the advent of modern plant breeding, farmer seed selection was the major factor determining the development and diversification of crop varieties. Initially, seed selection was probably largely the product of serendipitous discoveries by the earliest hunter-gatherers, related to the adaptation of plant species to human exploitation (**see: Ethnobotany**); later, they saved seeds and clonal material of superior types for replanting. Soon seed began to be transported, traded and tested in new environments, leading to the diffusion of crop cultivation between settlements. The early **domestication** of plants involved selection for qualities that aided harvesting, such as reduced **shattering** and uniform maturity, and food preparation. As cultivation techniques and their cultural accompaniments developed, variety selection and diversification began to involve more sophisticated criteria, including compatibility with crop management techniques (such as rotations and intercropping), resistance to the various pests and crop diseases that were an increasing challenge for cultivated fields, the requirements of cuisine (that surely co-evolved within the parameters provided by crop genetic variation), and various ritual and ceremonial uses. Crop domestication took place in a relatively few centres, but the results spread through trade, migration and conquest. The process of crop diffusion was accelerated after the European discovery of the New World and the colonization of large parts of Africa and Asia, which provided farmers with the opportunity of adapting many new species to their farming environments and cultures.

2. Seed supply management

(a) *On-farm.* The management of the physical supply of seed at farm level begins with the selection criteria and techniques that determine varietal diversity. In only a minority of cases do farmers plant a separate plot specifically for seed (the main exception being those cases where the crop is normally harvested before seed maturity, such as for most vegetables). More commonly, seed is selected from the general grain crop. Selection techniques vary by crop and farming system. In some cases farmers begin scouting the standing crop for appropriate plants for seed; in other cases seed is selected after harvest and before storage; and in still other cases no selection is done until just before the next season's planting, and seed may be whatever is left from the household store. If seed selection begins in the field, agronomic characteristics may play an important role, and farmers may look for particularly robust and disease-free plants or those that are early maturing (where this is a relevant criterion). Even if farmer seed selection does not give conscious attention to the maintenance of characters such as disease or pest resistance, the healthiest plants will most likely be selected and will hence contribute more to grain production. Seed selection characteristics after harvest are more likely to feature qualities such as grain type and colour, or ear size. These qualities may be related to food preparation, aesthetics or other attributes (e.g. colour has in some cases come to be used as a marker for maturity).

If seed is selected before crop storage, it is usually accorded special care. A wide range of traditional structures and methods has been described. Ears of maize or heads of sorghum or millet destined for seed may be hung above the cooking area for drying and allowing the smoke to discourage insects. Legume seed may be placed in sealed pots or jars. A variety of substances (such as **neem** leaves or wood ash) may be mixed with stored seed, or if insecticide is available it may be purchased for use on the household seed supply. (**See: Storage management**)

(b) *Off-farm.* Although most farmers endeavour to store seed for the following season, a number of circumstances account for the fact that, even in the most traditional farming system, a proportion of farmers each year go outside of their farm for their seed supply. There simply may be insufficient seed – the household may have harvested little grain the previous season and/or consumed all of its stored supplies; inadequacies in seed threshing, storage or growing conditions may make the farmer concerned about seed physical quality, or varietal purity or vigour; or the farmer may seek a new variety.

For these reasons there is a significant amount of seed giving, exchange and sale within farming communities. The seed provided may simply be grain from the household store or it may be the product of some of the seed selection procedures described above. The other major informal source of seed is to visit the local market at planting time and buy grain suitable for sowing – selecting what looks like adequate seed. This is an important source if farmers or market traders are able to distinguish variety and seed quality (such as freedom from pest damage) by visual inspection, as in the case of beans. For instance, grain markets are an important source of **cowpea** seed in many areas of densely populated West Africa, and much soybean seed in Indonesia is acquired from traders who move selected grain from regions where the harvest is recently completed to others where planting is about to begin. On the other hand, it is much less common to acquire seed for rice sowing in grain markets.

3. Local varieties and landraces

The evolution of farming practices and the movement of crops by exchange and sale has provided all farming communities over the years with a significant repertoire of crop genetic resources. Farmers usually, but not always, distinguish among these with individual names. They may be regarded as the equivalents of **varieties** (or **landraces**, to indicate their heterogeneity relative to many of the products of modern plant breeding).

The great diversity of farmers' varieties is widely recognized, but that diversity varies by crop and environment. There are records of farming populations that recognize an exceptional number of varieties; in the 1950s the Hanunoo peoples of the Philippines were reported to be able to distinguish over 90 varieties of rice, for example. On the other hand, there are traditional farming systems where only a few varieties of even the important staple crops are recognized, and there is relatively little local interest in nomenclature. In very approximate terms, the degree of varietal diversity for a particular crop and farming system depends not only on the amount of genetic diversity available to farmers but also on the diversity of the environments in which the crop is grown and the diversity of uses for the crop. For instance, the farmers in an area that contains several different types of field will likely manage more varieties of a crop. Or, if a crop is the focus of a diverse culinary tradition, correspondingly more varieties are likely to be found. Even where significant varietal diversity exists, knowledge may not be evenly spread among farmers, and in any case it is relatively unusual to find an individual farmer growing more than a few named varieties in one season.

There is a common misunderstanding that equates traditional farmer varieties with an unchanging genetic makeup. It is now generally agreed that, whilst farmer varietal management may preserve certain characteristics, the situation is best characterized as a dynamic one, in which farmers manage varietal selection in response to changing biotic and abiotic circumstances as well as changes in the socioeconomic environment.

There is justified concern about the degree of genetic erosion in local farming systems. Blame is most often cast at the replacement by modern varieties (MVs), but the situation is more complex. Farmers simplify their varietal repertoire not only because of the availability of new high-yielding varieties, but also because of increasing market dependence and the simplification or loss of environments (e.g. the expansion of irrigation in dryland areas) and customs (e.g. the abandonment of time-consuming food preparations) that once elicited greater diversity. Adoption studies have shown that the acceptance or rejection of MVs is based on farmers' consideration of a complex set of criteria and that MVs do not gain acceptance unless they are compatible with farmers' circumstances.

4. Summary

Farmer-level seed management systems are complex and varied. Local seed supply systems may be imperfect, but in general they are much more robust than is commonly supposed. Recent investigations regarding the conduct and impact of emergency seed programmes (operating in times of drought, flood or civil disorder) have found that seed donations have often been delivered to areas where farmers or local grain traders have adequate supplies of seed (though those hardest hit may not have the financial resources to acquire it). On the other hand, it is important to resist the temptation to idealize local seed systems. They have been responsible for crop varietal development and seed provisioning over the centuries, but they are not always equitable, or perfectly efficient at preserving seed stocks, or at providing information about what seeds or varieties are available. (RT)

Richards, P. (1986) *Coping with Hunger: Hazard and Experiment in an African Rice Farming System*. Allen and Unwin, London, UK.

Sperling, L. and Longley, C. (eds) (2002) Special issue: beyond seeds and tools. *Disasters* 26.

Tripp, R. (2001) *Seed Provision and Agricultural Development*. James Currey, Oxford, UK.

van der Heide, W., Tripp, R. and de Boef, W. (1996) *Local Crop Development: An Annotated Bibliography*. IPGRI/CPRO-DLO/ODI, Rome/Wageningen/London.

Zimmerer, K.S. (1996) *Changing Fortunes: Biodiversity and Peasant Livelihood in the Peruvian Andes*. University of California Press, Berkeley, CA, USA.

Infection

The introduction or entry of a parasite or microorganism **pathogen** into a susceptible host seed (or more generally, any other plant part or animal), whether or not this causes detectable pathological symptoms or effects of overt disease within the body of the host. The first infection following a resting or dormant stage of a pathogen is termed a primary infection. Strictly, the term is distinct from both 'contamination' and '**inoculation**', neither of which may lead to infection. An infected host is usually said to be diseased only when symptoms become evident. (**See: Disease; Health testing; Pathogens**) (VC)

Infestation

The introduction or entry of a pathogen or pest into the environment of a host (or, in reference to soil, contamination by fungi, worms, insects or the like, in large numbers). Seed **pathogen** infestation does not imply disease on the seed itself, though insect infestation may cause damage to seed, as with several major seed **storage pests** and fungi. Infestation is not to be confused with seed **infection**, which means the foreign organism is present *within* the body of the host. (**See: Conditioning – precleaning; Disease; Grain inspection; Health testing**) (VC)

Inflorescence

A cluster of flowers, all of which arise from the main stem axis or peduncle. Inflorescences are classified into:

Determinate inflorescence – mature flower formation results in the cessation of growth of the central axis of the plant.

Indeterminate inflorescence – lateral flower formation occurs without arrest of growth of the central axis of the plant and further flower development. (**See: Determinate flowering; Indeterminate flowering**)

There are a variety of classifications of these inflorescence types (Fig. I.1), including:

Corymb: often a flat-topped indeterminate cluster of flowers due to the older ones having a longer pedicel than the younger. Typical of the **Brassica** family, for example. A variation on this is the *cyme*, which is a determinate flower in which the

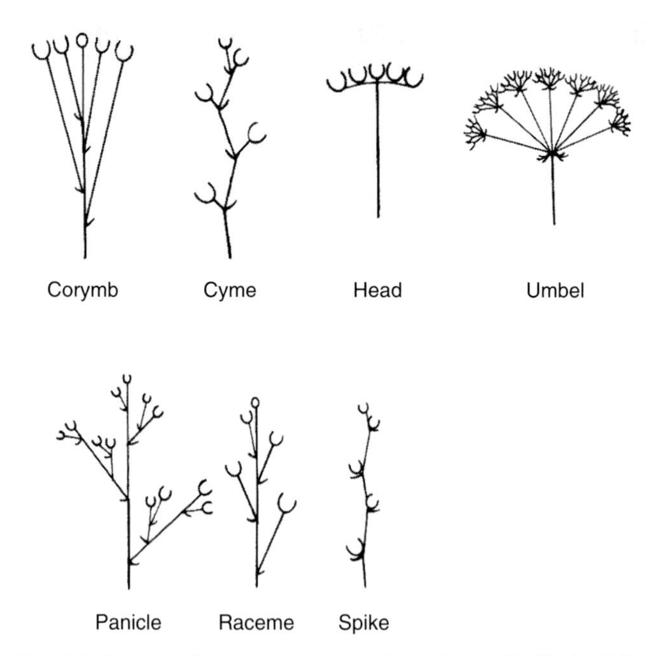

Fig. I.1. Some inflorescence types. Based on: Radford, A.E., Dickison, W.C., Massey, J.R. and Bell, C.R. (1974) *Vascular Plant Systematics*. Harper and Row, New York.

central or terminal flower opens earliest, as in cabbage, grain amaranth and the **buckwheat** family.

Head (capitulum): a shortened compact cluster of flowers so arranged that the whole gives the effect of a single flower as in **clover** or members of the family Compositae (Asteraceae), e.g. **sunflower**. May be determinate or indeterminate.

Panicle: an indeterminate compound raceme or branched cluster of flowers (as present in **millets, rice, oats**).

Raceme: an indeterminate inflorescence, usually an elongate cluster of flowers along the main stem in which the flowers at the base open first. Examples are the grape flower and lily-of-the-valley.

Spike: an indeterminate inflorescence bearing sessile flowers on an unbranched axis. The tightly clustered spike (spikelet) is typical of the grasses of the Poaceae (including **wheat** and **barley**) and sedges.

Umbel: a flat-topped or rounded inflorescence (can be determinate or indeterminate) characteristic of the **Umbellifers** (Apiaceae) in which the individual flower stalks arise from about the same point; youngest flowers are at the centre.

Some terminology associated with inflorescences:

Bract: modified, usually reduced, leaf in the inflorescence.

Involucre: a group or cluster of bracts subtending an inflorescence.

Pedicel: individual flower stalk.

Peduncle: main stalk for entire inflorescence.

Rachis: major axis within an inflorescence.

Spathe: a sheathing leaf subtending or enclosing an inflorescence. (JDB)

Inoculum (Inoculation)

In seed technology: living bacterial or fungal material that is artificially applied (inoculated) on, with or alongside seed, either as a pretreatment or at the time of sowing, such as **rhizobial inoculation** of legume seeds, or generally placement of beneficial rhizosphere microorganisms on to seeds before planting them, or into soils or planting media. (**See: Rhizobia**)

In seed pathology: the infective material which is transferred into the 'infection court' – usually microorganisms (fungal mycelial fragments, propagules, spores, etc., bacterial cells and spores, virus particles) that naturally contaminate seeds during their development, harvest or storage, or are present in the soil. Strictly speaking, many of the standard seed **health testing** methods detect seedborne pathogens, rather than determine their effective inoculum loadings – their ability to transmit infection to plants. But most methods have been validated by relating the one to the other.

'Inoculum potential' describes the interaction of the infective inoculum present on infected seeds and the capacity of the environment to produce disease, for a stated host and pathogen. Too much or too little inoculum can result in non-transmission of disease: high quantities may impair seed germination so that the disease is not manifested in the next plant generation, whereas merely superficial contamination may be eliminated by soil microflora when seeds are sown. (PH)

Maude, R.B. (1996) *Seedborne Diseases and their Control: Principles and Practice*. CAB International, Wallingford, UK.

Insect pests (in storage)

See: Storage pests

Insecticides

Chemical, physical or biological agents intended for preventing, destroying, repelling or mitigating any member of the class Insecta or allied classes in the phylum Arthropoda. Microorganisms used in this way are termed bioinsecticides. (**See: Fumigants; Storage management; Treatments – pesticides, 4. Insecticides**)

Integument

On its outside, the **nucellus** is usually covered by one or two layers, the integuments. The **ovules** of the choripetalous (polypetalous, having distinct petals) **dicotyledons** and the **monocotyledons** usually have two integuments (hence bitegmic ovules) while the sympetalous (gamopetalous, having petals fused at their edges) dicotyledons usually have only a single (hence unitegmic) ovule. The integument(s) usually give rise to the **seedcoat** unless the seed is **pachychalazal**. (**See: Seedcoats – structure**)

Intermediate seeds

The least well represented of the three categories used to describe seed storage behaviour at the species level. Intermediate seeds are best regarded as a sub-set of **orthodox** seeds. They can tolerate partial desiccation to about 8% moisture content (fresh weight basis), equivalent to a water potential of around −100 MPa. **Longevity** is often greater at temperatures above freezing with **viability** quickly declining at conventional seed **gene bank** temperatures (−20°C). Examples include tropical, **endospermic** oil seeds such as **coffee** and **oil palm** (**Table I.3**). The incomplete desiccation

Table I.3. Some intermediate seeds.

Family	Species	Common name
Passifloraceae	*Passiflora edulis*	Passion fruit
Zingiberaceae	*Elettaria cardamomum*	Cardamom
Musaceae	*Musa acuminata*	Banana
Strelitziaceae	*Ravenala madagascariensis*	Traveller's tree
Oxalidaceae	*Averrhoa carambola*	Carambola
Rutaceae	*Citrus sinensis*	Orange
Salicaceae	*Populus tremula*	European aspen
Lecythidaceae	*Bertholletia excelsa*	Brazil nut
Rubiaceae	*Coffea arabica*	Coffee
Piperaceae	*Piper nigrum*	Black pepper
Lauraceae	*Cinnamomum subavenium*	Cinnamon
Orchidaceae	*Dactylorhiza maculata*	Common spotted orchid
Arecaceae	*Elaeis guineensis*	Oil palm
Poaceae	*Zizania palustris*	Wild rice
Araucariaceae	*Araucaria columnaris*	Cook pine

tolerance displayed by immature orthodox seeds can resemble intermediate (or even **recalcitrant**) storage behaviour. Thus to avoid potential classification errors, comparative studies aimed at defining seed storage behaviour must be conducted on fully mature seeds at or close to the point of natural seed dispersal.

International Center for Agricultural Research In The Dry Areas

See: ICARDA

International Plant Genetic Resources Institute (IPGRI)

A **Consultative Group on International Agricultural Research** centre with a mandate to advance the conservation and use of plant genetic diversity. Its activities are directed mainly at plant genetic resources (PGR) for food and agriculture. It was founded in 1974, is mainly funded by developed-country donor and development agencies, and has a staff of over 170, based at 15 sites. Its main objectives are to help developing countries meet their PGR needs, to strengthen international collaboration in the conservation and use of PGR, and to develop and disseminate knowledge and technologies relevant to the improved conservation and use of PGR. (RJP, SHL)

International Plant Protection Convention (IPPC)

This Convention, comprising 113 participating countries, serves as the foundation for all basic plant quarantine laws. These laws are applied mainly to prevent the spread and introduction of quarantine pests involved in the international trade in seeds, plants and plant products. The IPPC requires each country to issue and accept proper phytosanitary certificates, and establishes a National Plant Protection Organization (NPPO) to ensure that only safe material enters and leaves the country. It also cooperates to establish Regional Plant Protection Organizations, of which there are currently nine globally, which coordinate more effective plant protection activities and promote information on the occurrence of pests in their region. www.ippc.int

International Rice Research Institute (IRRI)

A non-profit research and training institute in the Philippines dedicated to the improvement of **rice** and its cultivation.

International Seed Federation

See: ISF

International Seed Health Initiative (ISHI)

The International Seed Health Initiative, www.worldseed.org/ishis.htm, assesses, develops and publishes information on suitable seed health testing methods for economically important seedborne pathogens of vegetable, herbage and field crops. The first ISHI group concerns Vegetable Crops (ISHI-Veg, founded in 1994), and its membership includes companies from The Netherlands, France, USA, Israel and Japan representing the production of over 75% of the world's vegetable seed supply. Two more initiatives for herbage (ISHI-H) and field (ISHI-F) crops were established in 1997 and 1998, respectively. ISHI is a private sector programme administered by the ISF (International Seed Federation). In terms of its aims and scientific responsibilities ISHI is the counterpart of the **ISTA Seed Health Committee**. (VC)

Meijerink, G. (1997) The International Seed Health Initiative. In: Hutchins, J.D. and Reeves, J.C. (eds) *Seed Health Testing: Progress Towards the 21st Century*. CAB International, Wallingford, UK, pp. 87–94.

International Society for Seed Science (ISSS)

A society dedicated to education and communication among seed scientists. Its official journal is *Seed Science Research*. www.seedscisoc.org

International Seed Testing Association (ISTA)

The International Seed Testing Association, founded in 1924, is an international non-profit organization based in Switzerland whose primary purpose is to develop, adopt and publish standardized procedures for seed **sampling** and testing, and to promote the uniform and accurate application of these procedures worldwide for evaluation of seeds for sowing, as they move in trade. This is accomplished through the publication, at intervals since 1931, of the International Rules for Seed Testing ('ISTA Rules' see below), based on comprehensive method validation through technical committees, a laboratory accreditation system, and the running of training workshops.

The secondary purposes of the Association are:

- to actively promote research in all areas of seed science and technology, including production, sampling, testing, storing, processing, and distributing seeds;
- to encourage variety (cultivar) certification;
- to participate in conferences and training courses and liaise with other organizations having common or related interests in seed.

Seed Science and Technology is the scientific journal of ISTA, launched in 1973 (as a successor to the *Proceedings of the International Seed Testing Association*, which was first published in 1925).

ISTA holds a triennial Congress, comprising Technical Committee and Task Force Meetings sessions, which discuss scientific and technical work performed, and a Seed Symposium, with presentations in the fields of seed science and technology. New proposals and amendments to the Rules are discussed and decided at the Ordinary Meeting. The ISTA Technical Committees organize workshops, seminars and symposia at other times.

ISTA works in close cooperation with national and international organizations to prevent the duplication of work in seed quality control and facilitate a uniform approach to seed quality evaluation with regard to the international trade of seedlots. (**See: Industry**)

(a) *International Seed Analysis Certificates* can be issued by accredited ISTA member laboratories that are authorized by their respective governments for seedlots that they test, to serve as a 'seedlot passport' for international trade.

The three types of International Seedlot Certificates issued under the auspices of ISTA are:

Orange	Sampling and testing are carried out by the same accredited member laboratory
Green	Sampling and testing are carried out by two different accredited member laboratories in different countries
Blue	Only the issuing accredited member laboratory tests the sample as submitted

(b) *ISTA Rules*. The most recent edition, published by the International Seed Testing Association – *International Rules for Seed Testing* (2006) – lays down detailed standard techniques and procedures and describes principles and definitions in detail, approved and amended at ISTA Ordinary and Extraordinary Meetings on the basis of advice tendered by the ISTA Technical Committees. The Rules are designed for the principal crop species of the world, but apply in general, if not in detail, to any species of crop plant. This latest edition is to be supplemented with updates published each year, including additions or replacements of existing pages. (**See: Germination testing**)

(c) *ISTA Technical Committees*. Sixteen Technical Committees, comprising over 500 seed scientists and experts from a variety of institutions and countries, conduct comparative studies and surveys through working groups. The 16 Standing Committees of ISTA (as at 2006) are:

Bulking and Sampling	Purity
Flower Seed	Rule
Forest Tree and Shrub Seed	Seed Health
Germination	Statistics
GMD Task Force	Storage
Moisture	Tetrazolium
Nomenclature	Variety
Proficiency Test	Vigour

These committees are responsible for the further development of the ISTA Rules and publications through the ISTA Method Validation Programme, and organize frequent regional training workshops and proficiency testing around the world to ensure consistency in application of the Rules.

The Seed Health Committee (SHC) for example, aims to develop, adopt and publish, in the International Rules for Seed Testing, standard methods for seed health testing, using training courses and quality assurance protocols. The SHC also actively promotes research in all areas of seed health through a triennial Seed Health Symposium. The **International Seed Health Initiative (ISHI)** is a private-industry-led organization with similar aims.

(d) *ISTA Publications*. A wide range of handbooks is written by the Technical Committees, which give background information on seed testing procedures. Recent titles include: *Flower Seed*, *Germination*, *Nomenclature*, *Purity*, *Sampling*, *Seed Health*, *Statistics*, *Tetrazolium*, *Tree and Shrub Seed*, *Variety* and *Vigour*. There are also information documents and position-papers providing background or reference or information on a range of topics of current interest, and guidance documents focussed on Accreditation and Quality Assurance, as well as the journal *Seed Science and Technology* and the *International Rules for Seed Testing* and a twice-yearly news bulletin, *Seed Testing International*. See www.seedtest.org for current publications. (RD)

International Treaty on Plant Genetic Resources for Food and Agriculture (ITPGRFA)

This was adopted in 2001 at the 31st Food and Agricultural Organization of the United Nations Conference. Its objectives are to ensure the conservation and sustainable use of plant genetic resources for food and agriculture (PGRFA) and the fair and equitable sharing of the benefits arising out of their use. The treaty covers the Global Plan of Action on Plant Genetic Resources, the *ex situ* collections of PGRFA held by the International Agricultural Research Centres (**see: Consultative Group on International Agricultural Research**) and other international institutions, the international plant genetic resources networks, and the Global Information System on PGRFA. The treaty has been harmonized with the **Convention on Biological Diversity**. It contains a list of crops to be covered under the multilateral system of access and use. (RJP, SHL)

Introgression

The introduction of a few selected **genes** from one species or breeding line into another species or line, either through natural evolution or by such processes as backcrossing in breeding programmes. (**See: Cross**)

Invertase (extracellular)

Invertases catalyse the irreversible exothermic hydrolysis of sucrose into its hexose moieties, glucose and fructose:

$$\text{Sucrose} + H_2O \rightarrow \text{glucose} + \text{fructose}$$

Extracellular invertases are acid invertases (EC 3.2.1.26) located in, and ionically bound to, the cell wall matrix. During early seed development of grain legumes, an extracellular invertase is specifically expressed in their seedcoats. Hydrolytic activity is lost as the expanding cotyledons crush the innermost layer of seedcoat cells expressing these enzymes. In tropical grasses, an extracellular invertase is expressed in juxtaposing maternal (coat) and filial (embryo and endosperm) cells throughout seed development.

Acid invertases are characterized by pH optima of 4 to 5.5 commensurate with pH values found in apoplasmic spaces of developing seeds. Extracellular invertases display K_m values of 2–6 mM sucrose, accounting for the low apoplasmic sucrose concentrations in developing seeds that express this enzyme. These low amounts of apoplasmic sucrose provide transmembrane sucrose concentration differences that drive sucrose unloading from maternal seed tissues mediated by facilitated diffusion (**see: Development of seeds – nutrient and water import; Storage reserves synthesis – regulation**). (MT-H, CEO, JWP)

IPGRI

See: **International Plant Genetic Resources Institute**

IPPC

See: **International Plant Protection Convention**

IRRI

See: **International Rice Research Institute**

ISF

The International Seed Federation is a non-profit organization based in Nyon, Switzerland, with membership composed of national seed trade associations, individual seed company and industry service providers of breeding, multiplication and distribution from 70 countries, whose interests it promotes and contributions it publicizes. Its main objectives are to:

- develop and facilitate the free movement of seed through fair regulations;
- promote the establishment and protection of intellectual property rights for seeds, plant varieties and associated technologies;
- facilitate marketing by publishing rules for the trading of seed in international markets and for the licensing of technology.

Work is organized through Crop Sections (on 'Forage and Turf', 'Vegetables and Ornamentals', 'Industrial Crops', 'Cereals and Pulses', 'Maize and Sorghum') and Committees (on 'Breeders', including 'Intellectual Property and Sustainable Agriculture'; 'Trade and Arbitration Rules'; and 'Phytosanitary' matters). The ISF has also since 1994 created a series of **International Seed Health Initiatives** (ISHIs) to develop seed health testing methods through technical groups established on a crop-by-crop basis, working in close cooperation with the ISTA-PDC. The Federation organizes an annual world seed congress, which serves as a discussion and decision forum for the seed industry, alongside representatives of governmental and other non-governmental organizations.

The ISF resulted from the merger in 2002 of two previous organizations: FIS, the International Seed Trade Federation (Fédération Internationale du Commerce des Semences); and ASSINSEL, the International Association of Plant Breeders for the Protection of Intellectual Property (Association Internationale des Sélectionneurs pour la Protection de la Propriété Intellectuelle), who had shared a joint Secretariat since 1977. FIS was established in 1924, at the same time as ISTA, when a need for harmonization for seed trade and seed testing had become obvious, which led to the adoption of International Rules for Seed Testing in 1931, and the International Seed Testing Certificate, known as the 'Orange Certificate'. (**See: International Seed Testing Association**) ASSINSEL was established in 1938, at a time of intensive international discussions on the protection of the results of breeding work, and was instrumental in the establishment of the Union for the Protection of New Varieties of Plants (**UPOV**) in 1961 and the adoption of the Convention for the Protection of Plant Varieties.

ISF publications include: 'Rules and Usages for the Trade in Seeds for Sowing Purposes'; 'Procedure Rules for Dispute Settlement for the Trade in Seeds for Sowing Purposes' and 'Management of Intellectual Property'; and leaflets on the roles of plant breeding and the seed industry, and on food security and sustainability. www.worldseed.org (**See: Industry**) (BlB)

Le Buanec, B. (1998) Fédération Internationale du Commerce des Semences (FIS), International Seed Trade Federation. In: Kelly, A.F. and George, R.A.T. (eds) *Encyclopaedia of Seed Production of World Crops*. John Wiley, UK.

Leenders, H. (1967) The function of the International Seed Trade Federation (FIS) in the international seed trade. *Proceedings of the International Seed Testing Association*, Vol. 32.

ISHI

See: **International Seed Health Initiative (ISHI)**

Isolation distance

The spatial separation required between a seed field and other sources of mechanical and genetic contamination, especially between cross-pollinated varieties, to ensure purity (**see: Purity testing, 2. Varietal purity**) and **identity preservation** in the progeny seed and grain. (**See: Production for Sowing, II. Agronomy, III. Hybrids**)

ISSS

See: **International Society for Seed Science**

ISTA

See: **International Seed Testing Association**

ISTA Seed Health Committee

The Seed Health Committee (SHC) is a technical working group of the **International Seed Testing Association** (ISTA), which develops, validates and publishes standard methods for seed health testing, and promotes their uniform application using proficiency testing, training courses and quality assurance protocols. The SHC also promotes research through a triennial Seed Health Symposium. **International Seed Health Initiative** (ISHI) is a private-industry-led organization with similar aims.

ITPGRFA

See: **International Treaty on Plant Genetic Resources for Food and Agriculture**

J

Jack bean

1. World importance and distribution

Jack bean (*Canavalia ensiformis*) is a leguminous plant grown in the tropics and sub-tropics (central and south America, India, North and East Africa, Far East) for the seeds, which serve as food for human consumption, or for fodder or green manure. The latter two are its main uses in the USA, for example. (**See: Legumes**)

2. Origins

The species originates and was domesticated in Central America and the Caribbean: remains dated about 5000 years old have been found in archaeological sites in Mexico.

3. Seed and pod characteristics

The white, ellipsoid seeds are 1.5–2 cm long and almost as wide, with a large, brown **hilum**. Between three and 18 seeds are present in each pod, which can reach 25–35 cm in length when fully mature but when to be eaten as food are generally harvested at about half this size.

4. Composition and nutritional quality

Seeds contain 23–34% fresh weight **storage protein**, up to 55% carbohydrate (mostly **starch** but also about 2% **oligosaccharides**) and 1–2.4% **oil** (triacylglycerol) and are a good source of calcium, copper, magnesium, nickel, phosphorus and zinc. The major storage protein is the vicilin, canavalin. A relatively high concentration of the haemagglutinin, concanavalin A, is one of the anti-nutritional factors that are present: others are trypsin inhibitors and cyanogenic glucosides (**see: Lectins; Pharmaceuticals and pharmacologically active compounds; Protease inhibitors**). The seeds are rich in the enzyme urease, which the biochemist Sumner used for the first crystallization of an enzyme in 1926. Other interesting compounds in the seed are the non-protein amino acid canavanine, and a protein showing strong homology with bovine insulin in the **seedcoat**.

5. Uses

Immature pods are used as a vegetable. Mature seeds are dried for later culinary use but caution has to be exercised in food preparation because of the toxic elements in the seed. The bean yields several industrial products such as protein concentrates and isolates, starch, flakes, grits and flours. The plants, pods and seed meals are used as livestock feed but because of the harmful factors in the seeds a maximum of 30% should comprise feed for cattle. Heat-treatment renders harmless the seeds and pods. Because of the relatively high urease content of the seeds they should not be added to feed that contains urea.

6. The plant

This is a bushy, partially twining annual, fast-growing, usually erect, up to 1 m in height, with runners that can reach 10 m long. It is deep rooted and drought resistant. (MB)

Smartt, J. (1990) *Grain Legumes*. Cambridge University Press, Cambridge, UK, pp. 301–309.

Jasmonates

Also known as **oxylipins**, jasmonates are a group of compounds that are thought to play an important role in protection against stress due to wounding or pathogen attack. They have been isolated from angiosperms and gymnosperms as well as some algae and fungi. Jasmonates have been proposed to play a number of other roles including the regulation of **germination**, vegetative storage reserve synthesis, anther dehiscence and thigmomorphogenesis of tendrils. (**See: Germination – influences of hormones**)

The two principal members of this class of compounds are jasmonic acid (JA) (Fig. J.1) and its volatile methyl ester (MeJA), both of which promote their own synthesis. JA is produced via the octadecanoid pathway from the polyunsaturated fatty acid linolenic acid (LA) that may be freely available in the cell or derived from membrane lipids. Conversion of LA to an intermediate in the pathway, 12-oxophytodienoic acid, occurs in a wide range of plant tissues including the **cotyledons** of germinating seeds and there is evidence that the initial events take place in **plastids**. A number of JA synthesis **mutants** have been isolated from tomato and *Arabidopsis* and these exhibit a greatly reduced ability to tolerate attack by insects unless supplied exogenously with JA or MeJA.

A spectrum of JA-metabolites has been isolated from plants with many of these being derived from substitutions at C-1,

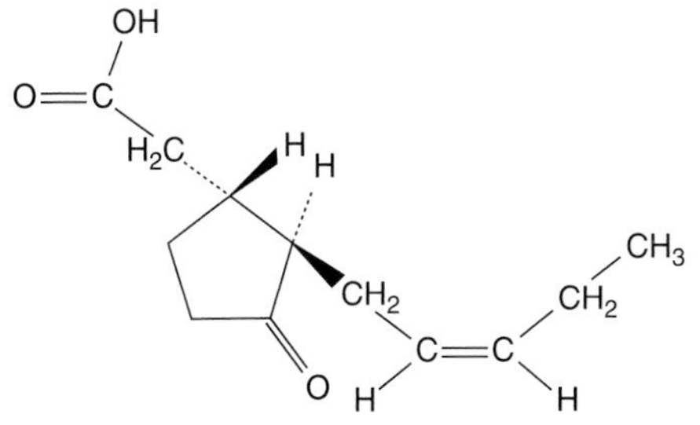

Fig. J.1. Jasmonic acid. Modified from Srivastava, L.M. (2002) Jasmonates and other defense-related compounds. In: *Plant Growth and Development. Hormones and Environment*, Academic Press, New York, USA, pp. 251–268.

C-6, C-11 and C-12 sites within the molecule. Reduction of the C-6 ketone group generates cucurbic acid, a compound, as its name suggests, that can be isolated from cucumber seeds. Cucurbic acid can inhibit growth and, like tuberonic acid (12 OH-(+)-7-iso-JA), can also promote tuberization in potato.

(GL, JR)

(See: Hormones; Signal transduction – hormones)

Creelman, R.A. and Mullet, J.E. (1997) Biosynthesis and action of jasmonates in plants. *Annual Review of Plant Physiology and Plant Molecular Biology* 48, 355–381

Srivastava, L.M. (2002) Jasmonates and other defense-related compounds. In: *Plant Growth and Development. Hormones and Environment*, Academic Press, New York, USA, pp. 251–268.

Jewellery, arts and crafts

Because of their attractive colours, patterning, shapes and sizes, seeds of some species have long been used for jewellery and in the decorative arts. They feature prominently in the craftwork of many ethnic groups in the Americas, Africa and Asia and aboriginal jewellery products are highly valued in many parts of the world.

Necklaces, bracelets and pendants are made from seeds. Many types of seeds that are used serve exclusively in jewellery but crop seeds such as maize and beans are also employed for this purpose. Jewellery seeds are of many size, shape and colour combinations. A relatively small one is the wild tamarind, *Leucaena leucocephala*, a legume originating in the tropical and sub-tropical Americas. The shiny, brown, flattish, ovoid seeds weigh about 40 mg – close to that of a wheat grain. The seeds are strung in necklaces together with larger types. Towards the other end of the scale in size is the so-called sea heart, *Entada gigas* (Fabaceae), a blue-black, heart-shaped seed measuring several centimetres across. These seeds are produced in extremely long pods (over 100 cm) on vines in tropical Africa and the Americas. Growing by streams and rivers, the dispersed seeds are carried down to the sea and can travel across from one continent to another, hence their name – the sea heart. This is just one of many 'sea beans' which travel the oceans and are collected for use in jewellery. Inhabitants of the Hebrides (islands off the coast of Scotland) collect white 'Indian nuts' for use in amulets. These seeds have been carried in the Gulf Stream from the Caribbean. More commonly known as nickernuts (*Caesalpinia* spp., Fabaceae) seeds of different colours – grey, yellow and chocolate brown – are used for necklaces and earrings in the Caribbean, Mexico and Central America.

Colours of the jewellery seeds are generally striking, and some have highly attractive patterns. Seeds of the coral trees (*Erythrina* spp.), leguminous shrubs from Central and South America and Africa, are shiny, very bright red, almost spherical, and ideal for stringing into necklaces. The rosary bean (*Abrus precatorius*, Fabaceae) is also bright red but is distinguished by having a jet-black, cap-like marking. They are extremely poisonous (**see: Poisonous seeds**) and several cases of serious illness or accidental death caused by the seeds have been recorded. Some cases follow the inhalation or ingestion of the powder produced when the seeds are drilled for necklace manufacture; others occur among children who are attracted by the bright red, shiny appearance. The 'hamburger beans' are so called because they are about the shape of a hamburger looking like the meat (actually the dark brown almost circular hilum) sandwiched between two lighter brown outerlayers. These seeds (*Mucuna* and *Dioclea* spp., both Fabaceae) are also 'sea beans', originating in Mexico, Central America and the Caribbean. Necklaces are made from them. Some seed jewellery of Amazonian origin is shown in Colour Plate 13A–D, together with items from the extensive collection of botanical jewellery in the Economic Botany Collection in the Royal Botanical Gardens, Kew, UK.

Many of the jewellery seeds also feature in various decorative objects and craftwork. Other seeds often used in craftwork are the ivory nut, *Phytelephas macrocarpa*, and the closely related Tagua nut, *Phytelephas aequatorialis*, both originating in tropical South America. The trees, in the palm family (Arecaceae), produce nuts of variable size (the average about that of a **walnut**) which have an extremely hard, ivory-like **endosperm** composed mainly of **mannans**, which lend themselves well to carving. They are shaped into faces, animals, *netsukes*, chess pieces, dice, dominoes, *mah-jongg* tiles, religious figurines and toys, among other objects. Unscrupulous merchants attempt to pass off carvings in ivory nut and Tagua as real ivory. Unlike genuine ivory though, these carvings soften in hot water, and harden again on cooling. Before the advent of plastics a significant percentage of buttons was manufactured from these materials. (MB)

See: waynesword.palomar.edu/ww0901.htm for an account of botanical jewellery including excellent photographs. **See**: www.kew.org/collections/ecbot/jewel.htm for botanical jewellery in the Economic Botany Collection of the Royal Botanic Gardens, Kew, UK.

Job's tears

Also called adlay. This minor cereal (*Coix lachryma-jobi*), grown in India, China and Southeast Asia, is eaten in parts of these regions. It is sometimes grouped with **millets**.

The seed, larger than those of most other cereals, is surrounded by a hard shell which before use is broken by pounding or rough grinding.

Typical composition (fresh weight basis) is **storage protein** 18.8%, **oil** (triacylglycerol) 6.2%, carbohydrate (mostly **starch**) 59.5%, fibre 1.28% and ash 3.4%.

In parts of India it is either mixed with shama millet or maize and made into bread or it is prepared like rice. It is said to taste like wheat.

Job's tears is perhaps better known for a non-edible form used in botanical **jewellery**. (MB)

Jojoba

1. The crop

Jojoba (*Simmondsia chinesis*, 2n = 52, Buxaceae) is a perennial evergreen shrub, native to the Sonora Desert of California, Arizona and Mexico. It was unknown to commercial agriculture until the 1975–1980 period. It is now grown in vegetatively propagated plantations in Argentina (4800 ha), USA (1900 ha), and a few hundred hectares each in Israel, Mexico, Peru and Australia. Total oil production is about 4 million t. About one quarter of the USA production comes from harvesting wild plants.

2. Fruit and seed

Jojoba fruit is a dehiscent capsule containing 1 to 3 **ovules**. Jojoba seeds are about 14 mm long and have a smooth red-brown surface with three ridges, one larger than the others (Fig. J.2). The outer epidermis consists of large, brown, thick-walled palisade cells. The remaining testa is composed of loose, mostly pigmented parenchyma. The **embryo** completely fills the inner seed cavity; there is little or no **endosperm**. The **cotyledon** parenchyma cells contain liquid wax droplets and **protein bodies**.

Fig. J.2. Seeds of jojoba (scale = mm) (photograph kindly produced by Ralph Underwood of AAFC, Saskatoon Research Centre, Canada).

3. Uses

Jojoba seeds yield about 50% clear golden oil that is unique in the plant kingdom. Unlike other vegetable oils, it contains no glycerides (e.g. triacylglycerols) and is technically a wax composed almost entirely of esters of straight, long chain (C20, C22, C24) unsaturated **fatty acids** and alcohols. Its composition and properties are similar to sperm whale oil. The high-end cosmetic market is presently the major user of the natural oil. Application of different chemical reactions yields a broad spectrum of useful compounds for pharmaceuticals, lubricants and surfactants. The oil-extracted meal contains up to 30% relatively well balanced protein, except for its low methionine content. However, the presence of toxic simmondsin compounds (cyanogenic glucosides) limits its use as an animal feed. **(See: Wax – storage)** (KD)

Benzioni, A. and Forti, M. (1989) Jojoba. In: Robbelen, G., Downey, R.K. and Ashri, A. (eds) *Oil Crops of the World*. McGraw-Hill, New York, USA, pp. 448–461.

Hassall & Associates Pty. Ltd (1999) *Australian Jojoba Products*. Rural Industry Research & Development Publication No. 99/85.

Vaughan, J.G. (1970) *The Structure and Utilization of Oil Seeds*. Chapman and Hall, London, UK.

Kamut

Kamut became known to cereal scientists and growers, particularly in the USA, in the 1970s. It has been grown in Egypt for many years and indeed its origins are thought to go back several thousand years. When its cultivation began in the USA it was given the registered trade name kamut from an ancient Egyptian word for wheat. There has been some confusion about its taxonomy but it is now generally agreed to be *Triticum turgidum*, ssp. *durum*, genomic constitution AABB (**see: Wheat,** Table W.1 and **Einkorn,** Fig. E.1) similar to an Egyptian cultivar 'Egiptianka'.

Kamut grains on average are approximately double the size of the modern bread wheats, with a distinct humped appearance. Protein content is higher than in the bread wheats.

It is grown as a speciality grain in the USA where its products include whole grain flour, breads, hot and cold cereals, and pastas which are said to have a mild, nutty flavour. Individuals who show certain allergic reactions to products from bread wheat grains (**see: Storage protein – intolerance and allergies**) do not react adversely to kamut. (MB)

Quinn, R.M. (1999) Kamut®: ancient grain, new cereal. In: Janick, J. (ed.) *Perspectives on New Crops and New Uses*. ASHS Press, Alexandria, VA, USA, pp. 182–183.

Stallknecht, G.F., Gilbertson, K.M. and Ramey, J.E. (1996) Alternate wheat cereals as food grains: Einkorn, emmer, spelt, kamut, and triticale. In: Janick, J. (ed.) *Progress in New Crops*. ASHS Press, Alexandria, VA, USA, pp. 156–170.

Kapok

1. Botany

The kapok (**see: Fibres**) or silk-cotton tree (*Ceiba pentandra*, family Bombacaceae also known as *Eriodendron anfractuosum*) is a large, deciduous, tropical tree that may grow to 40 m in height. The fruit is a smooth, elliptical capsule, 7–15 cm long. Seeds are entangled by a mass of long silky **floss** (fibres of 30–36 μm diameter) arising from the epidermal layer of the inner pod wall. Fibres are 19 mm long, hollow, highly lustrous, from white and light grey to yellowish-brown (**see: Fibres,** Fig. F.5B). The cell wall is very thin, resulting in a very large empty cell.

2. Chemistry

Kapok is a ligno-cellulosic fibre with a higher **cellulose** than **lignin** content (**see: Fibres,** Table F.1). The linear density (weight per unit length) of kapok fibre is 83% less than that of **cotton**. Though brittle and unsuitable for spinning by itself, the fibre is highly buoyant and moisture-resistant because of a highly water-resistant waxy cuticle and air trapped inside the fibre lumen. The outer layer contains a high mineral content that also influences fibre physical properties.

3. Origins

Kapok probably originated in tropical America but has spread to West Africa, Polynesia, Indonesia and Southeast Asia. The importance of the kapok tree to ancient Mayans is evident in their origin-of-man and after-life myths where kapok trees played a central role.

4. Location

Kapok trees are the largest in Africa and often dominate the tropical rainforests. Due to their size, kapok trees are planted as shade trees in villages throughout the tropics. Kapok is especially abundant in secondary forests in tropical climates, spread by fruit bats that are also the primary pollinators.

5. Production

In Madagascar, selective breeding has led to hybrids capable of producing 1.5 kg fibre by the third year after planting. By the tenth year, trees can produce 6–8 kg fibre per annum and mature trees will yield 15–20 kg. Trees are productive for at least 30 years. For the newer hybrid varieties, 1 kg of kapok fibre is produced for every 1.5 kg seeds.

6. Processing

Kapok pods are manually collected, broken open, and air-dried. Fibre and seeds are picked out and separated by hand. Due to its compressibility and resilience, kapok was once favoured for stuffing mattresses, life preservers, sleeping bags, and furniture. Competition from less expensive synthetic materials has reduced the demand for kapok.

7. Economics

Kapok is graded according to growing location, colour, and degree of contamination with non-fibrous material. World production of kapok fibre was 124,400 t in 2002. Thailand supplies 66% of the world's production with Indonesia responsible for 16%. Kapok markets in Japan, China, Europe, and the USA are responsible for US$11 million in trade. (BAT)

Berger, J. (1969) *The World's Major Fibre Crops and Their Cultivation and Manuring*. Centre d'Etude de l'Azote, Conzett and Huber, Zurich, Switzerland.

Hori, K., Flavier, M.E., Kuga, S., Lam, T.B.T. and Iiyama, K. (2000) Excellent oil absorbent kapok [*Ceiba pentandra* (L.) Gaertn.] fiber: fiber structure, chemical characteristics, and application. *Journal of Wood Science* 46, 401–404.

Sunmonu, O.K. and Abdullahi, D. (1992) Characterization of fibres from the plant *Ceiba pentandra*. *Journal of the Textile Institute* 83, 273–274.

Tapia, M.J. (1986) *Kapok. Plant Industry Production Guide*, No. 61. Ministry of Agriculture and Food, Bureau of Plant Industry, Manila, The Phillipines.

Kernel

This can refer to the whole seed of a **cereal** or, more generally, the mature **ovule** or the inner softer edible part of a seed, fruit or nut within the shell or stone (equivalent to the term seed **meat** in food technology). The word is derived from the Old English for a 'little seed', and is related to the word 'corn'.

Kersting's groundnut

A member of the Fabaceae (legumes), *Kerstingiella geocarpa* (= *Macrotyloma geocarpum*) is a sub-Saharan species grown in a restricted range of West Africa: no wild members of this particular species are known. The flattened reniform seeds in a typical legume pod are variously whitish, brown, black, spotted and speckled, somewhat smaller than the **peanut**. An important food, they are eaten locally often after being cooked in various ways.

Kinase

See: **Signal transduction – some terminology**

Kinetin

The first compound to be discovered with **cytokinin** activity. Isolated from old, partially degraded herring-sperm DNA by Folke Skoog and co-workers, it is an artificially produced adenine derivative (N^6-furfuryladenine) that, in conjunction with **auxin**, will induce plant cells to divide. It can promote germination in some species although in lettuce it tends to promote 'abnormal germination', resulting in the **cotyledons** emerging first from the seed. (**See: Germination – influences of hormones; Hormones**)

Kohl rabi

See: **Brassica – horticultural**

Kola

See: **Cola**

Labelling

See: Certification schemes, 3. Labelling; Laws and regulations; OECD Seed Scheme; Quality-declared seed; Treatments – pesticides

Labyrinth seed

Ruminate seeds that show an intricate network of lobes usually on the **endosperm** when the seed is cut in any plane. (**See: Rumination**, Figs R.7 and R.8)

Landrace

See: Race

Late embryogenesis abundant proteins

The late-embryogenesis-abundant (LEA) proteins are a highly abundant set of hydrophilic proteins that exhibit temporal regulation during seed development. They generally accumulate to their highest content during the mid-expansion stage of embryo development and just prior to **desiccation**. Since their first description in cotton embryos, messages homologous to the LEA cDNAs of cotton (representing at least five conserved families of corresponding proteins) have been found in abundance in mature dry embryos and storage organs of many diverse plant species and during drying of xerophytic species. (**See: Desiccation tolerance – protection by stabilization of macromolecules**)

LEA proteins are classed into five groups by virtue of sequence similarities (Table L.1). The least studied of the LEA proteins are those in Groups 4 and 5, which are somewhat atypical. Group 5 LEA proteins are more hydrophobic than other LEA proteins and are not resistant to high temperature.

A subset of the LEA proteins has been termed **dehydrins**; they exhibit some common features in their structure that may be important for their putative protective function. Dehydrin genes exhibit a flexible expression repertoire; the protective role of dehydrins in the survival of water loss is purported to be dual: during **maturation drying** of the developing seed and following germination/growth of the mature seed (i.e. in seedlings or plant vegetative tissues undergoing mild water stress). **Abscisic acid** (ABA) likely plays a central regulatory role in the expression of *LEA* genes within the developing seed; severe ABA-insensitive (*abi*) **mutants** are generally disrupted in late maturation programmes, including the accumulation of LEAs. The bZIP **transcription** factor ABI5 plays a role in regulating the expression of certain *LEA* genes in seedlings exposed to water stress. In vegetative tissues under water-deficit stress, *LEA* gene expression involves both ABA-dependent and ABA-independent signal transduction events and this may also be the case in seeds.

At the subcellular level, LEA proteins are localized primarily, but not exclusively, within the cytosol. The maize RAB 17 protein is distributed between the cytoplasm and nucleus of maize embryos. This dehydrin has the ability to bind peptides with nuclear localization signals, a property which is dependent upon **phosphorylation** of some of the dehydrin's serine residues. Thus RAB17 may mediate the transport of specific nuclear-targeted proteins during stress.

A general feature of LEA and dehydrin proteins is their hydrophilicity as a result of a biased amino acid composition. Most LEA proteins are lysine- and glycine-rich. These features may be central to their putative role in protecting cells against potential damage as a result of water loss.

The strict conservation of amino acids in the lysine-rich motif of dehydrins (EKKGIMDKIKEKLPG) (see Table L.1 for key to letters) implies conservation of some higher order structure, and this part of the polypeptide has been proposed to form an **amphiphilic** α-helix which may serve among other functions as an ion trap in dehydrating cells, sequestering ions as they become concentrated. Certain LEA proteins have the capacity to bind large numbers of phosphate ions and their counterions. They also have a tendency to electrostatically coat membrane surfaces in the desiccated state, thus indicating a role in providing a 'surrogate' water film to stabilize membranes of seeds in a near dry state.

Dehydrins also have detergent and chaperone-like properties and may interact with **compatible solutes** (e.g. sucrose, proline and glycine-betaine) in plant vegetative tissues to serve as structural stabilizers of macromolecules under conditions of water deficit. The current model is that hydration of the hydrophilic regions of dehydrins results in the formation of an envelope of ordered water, which operates through a preferential exclusion mechanism (especially in the presence of compatible solutes) to inhibit protein denaturation.

Dehydrin proteins accumulate during seed development, and in response to seed drying, in a number of **recalcitrant** species. However, dehydrin proteins are not detected in mature undried axes of species at the extreme end of the spectrum of desiccation sensitivity (e.g. in several tropical wetland species). (ARK)

(See: Desiccation tolerance; Desiccation tolerance – protection; GMO – patent protection technologies; Hormones)

Alpert, P. and Oliver, M.J. (2002) Drying without dying. In: Black, M. and Pritchard, H.W. (eds) *Desiccation and Survival in Plants: Drying without Dying*. CAB International, Wallingford, UK, pp. 3–43.

Close, T.J. (1996) Dehydrins: Emergence of a biochemical role of a family of plant dehydration proteins. *Physiologia Plantarum* 97, 795–803.

Table L.1. General characteristics of LEA protein groups.

Group	Predicted structural features, protein signatures and consensus pattern	Representative examples	Protective effect
1	Most of the protein is random-coiled. 83 to 153 amino acid residues. G-(EQ)-TVVPGGT is the sequence with the highest degree of conservation (20 amino acids).	Em (early methionine labelled protein) which is seed specific: *Arabidopsis* (GEA6, maize Emb564). LEA D-19 (cotton). ATEM 6 (*Arabidopsis*)	Overexpression of a wheat Em protein in yeast confers improved growth in saline and osmotic conditions. Extensive regions of random coil promote water binding; maintains solvation state of cellular constituents during drying. May buffer water loss during maturation.
2	Largely unstructured with 10–15% with α-helices. Dehydrins are predicted to form amphipathic α helices. Diverse combinations of consensus patterns (segments): S-segment: S(5)-(DE)-x-(DE)-G-x(1,2)-G-x(0,1)-(KR)(4) K-segment: (KR)-(LIM)-K-(DE)-K-(LIM)-P-G Y-segment: (V/T)DEYGNP. The number of repeats of specific segments is variable.	100 dehydrins (DHN in many species). LEA-D11 (cotton). RAB28, RAB21 (responsive to abscisic acid).	In barley, three *DHN* genes map in a QTL for winter hardiness. Their expressions in different cultivars correlate with freezing tolerance. Overexpression of a tomato group 2 protein gene (*le4*) in yeast partially ameliorates detrimental effects of salt and freezing stress. Amphipathic helical repeats bundle together to provide surface for sequestering ions (protect proteins and other cellular components from increasing ionic strength in the cytoplasm during drying).
3	Predicted to contain α helices. Exhibit a 11-mer repeating peptide sequence. Consensus pattern TA(K/Q)AAE(Q/D)K(T/A)xE.	D-7 (cotton). MLG3 (maize). HVA1 (barley).	Ectopic expression of the barley HVA1 in transgenic rice leads to increased biomass under conditions of water drought and salinity. Inter- and intra-molecular repeat interactions help the protein to function as an anchor in the structural network of the cytoplasm during drying and in the dry state. Stabilizes sucrose glass formation; may confer long-term stability to cytoplasm.
4	N-terminal α helices and random coil at the C-terminus. Conserved N-termini.	LEA D-113 (cotton). Le25 (tomato). ds11 (sunflower).	Overexpression of a tomato group 4 protein gene (*le25*) in yeast increases growth rate in the presence of KCl.
5	Adopt a globular conformation.	LEA D-34 (cotton).	

Groups 1, 2 and 3 are the best characterized. Several authors consider one or two additional groups of LEA (e.g. PvLEA-18 in *Phaseolus vulgaris* seedlings) that are not included here. Protein signatures and consensus patterns[a] were determined from PROSITE, a database of protein families and domain that can be found at www.expasy.org. Consensus patterns of Groups 4 and 5 have not been identified. (OL, JB, ARK)

[a]Letters signify different amino acids: A, alanine; D, aspartic acid; E, glutamic acid; G, glycine; I, isoleucine; K, lysine; L, leucine; M, methionine; N, asparagine; P, proline; Q, glutamine; R, arginine: S, serine; T, threonine; V, valine; x, any amino acid; Y, tyrosine.

Dure, L.S. III (1993) A repeating 11-mer amino acid motif and plant desiccation. *The Plant Journal* 3, 363–369.

Kermode, A.R. and Finch-Savage, W. (2002) Desiccation sensitivity in orthodox and recalcitrant seeds in relation to development. In: Black, M. and Pritchard, H.W. (eds) *Desiccation and Survival in Plants: Drying without Dying*. CAB International, Wallingford, UK, pp. 149–184.

Wise, M.J. and Tunnacliffe, A. (2004) POPP the question: what *do* LEA proteins do? *Trends in Plant Science* 9, 13–17.

Lateral endosperm

That region of the endosperm which surrounds the **cotyledons**; the **micropylar endosperm** is to the outside of the **radicle**. In some seeds, e.g. Chinese senna, **carob** and **fenugreek** the thickness of the cell walls in the two regions is different, being much thicker in the lateral endosperm than in the micropylar. The two regions may also differ in respect of the production of some cell wall-degrading enzymes, e.g. endo-β-mannanase and in the sensitivities of this process to hormones, such as in the tomato endosperm. (**See: Endosperm cap; Germination – radicle emergence**)

Lathyrism

A neurotoxic disease, sometimes culminating in complete paralysis, resulting from the consumption of **chickling pea** (*Lathyrus sativus*) seeds. The causative agent is a non-protein amino acid, 3-N-oxalyl-L-2,3-diaminopropionic acid or β-ODAP. (**See: Pharmaceuticals and pharmacologically active compounds; Poisonous seeds**)

Latin American Federation of Seed Associations

See: FELAS

Laws and regulations

Most countries control by law the production and marketing of seed for sowing, to protect the buyer and seller from uncertain quality and from fraudulent practices, and thereby improve agricultural productivity. These controls are based upon three major elements, on which there is broad international agreement:

- **OECD** (Organisation for Economic Co-operation and Development)-**Seed Scheme**, which sets standards for the **certification** of seed moving in international seed trade (mainly concerned with varietal purity);
- International Rules for Seed Testing (**ISTA** Rules), which standardize **quality testing** methods and are also used to facilitate seed trade; and
- **UPOV** (International Union for the Protection of New Varieties of Plants) system, which harmonizes variety protection (intellectual property) rights.

Additional bodies of law directly affect the seed trade, including: the control of certain seedborne pathogens by **phytosanitary certification**, which affects seed export and import of some species; the use of seed treatment **pesticides**; and biosafety regulations related to the genetic modification of organisms. Distinct from these again is the legal regulation of the trade in harvested grain.

1. Genetic and physical quality control

One of the salient features of seed is that its sowing quality is not obvious from superficial examination, apart from such conspicuous features as size and colour, and farmers often cannot recognize inadequate seed until well after planting or even until harvest. Both genetic quality (that seed of one variety might be sold as something else) and physical quality (germination capacity, cleanliness, etc.) are potential problems.

The study of seeds and the evaluation of its basic properties for sowing purposes based on botanical and scientific principles did not start until the second half of the 19th century. Up until then knowledge of seed morphology and physiology was limited and, as a result, it was relatively easy for unscrupulous merchants to sell seed bulked up with sand and other impurities. Such problems of product quality and identification became so common and widespread (and remain still in some parts of the world) that laws were passed to protect farmers from such fraudulent practices and seed testing procedures were established, which have become the basis of modern legal systems. Specific penal laws were enacted in various countries throughout the 19th century, and governments in several European countries began to establish seed testing and certification systems towards its end, while in the USA many State universities established their own certification programmes, though seed companies resisted the establishment of mandatory inspection. The first seed-testing laboratory was established in Saxony (1869) and soon followed in The Netherlands (1877); within 50 years, laboratories had been set up all over Europe and North America. In 1924, the **International Seed Testing Association** (ISTA) was formed, and has grown into a worldwide intergovernmental organization with the primary purpose of developing, adopting and publishing standardized procedures for the quality testing of seeds.

Present-day laws, rules and regulations govern the seed trade, both at domestic as well as for import and export, by focusing on:

- registration of the varieties that are offered for sale as seed, and
- assurance of the physical quality and genetic purity of commercial seed.

The laws establish government responsibilities, appropriate regulatory authorities and enforcement mechanisms, while giving the broad principles of how the legislation should work and rapid ways of amending regulations. The details of variety registration and seed quality are described in regulations rather than in seed law itself. Regulations may govern, for example: how official seed samples are to be obtained; what information must be included on seed **labels**; what standards, if any, are to be applied to the parameters of germination, analytical and varietal purity and levels of disease; and what tolerances and limits of variation should be applied to the acceptability of results. In countries such as the USA and India there are both federal and state-level seed laws: the lack of harmonization between the two may cause some barriers to seed trade even within a single country. In New Zealand, however, there is no seed law, but rather a voluntary industry code of conduct.

There are two basic means by which seed quality control is achieved in law: by enforcing a set of minimum standards, or by declaration of quality – the so-called 'truth-in-labelling'. The respective advantages and disadvantages of such schemes are set out in Table L.2.

Table L.2. Advantages and disadvantages of the two basic quality control systems for seeds.

Minimum-standards systems	
Advantages	Easy to check and enforce. Easy for farmers to understand.
Disadvantages	No choice given to farmers; all have to buy seed of above the minimum standard. Not very flexible: when a poor harvest results in seed of low quality, it may be difficult to find enough seed meeting the standards to satisfy market demands.
Truth-in-labelling, or Declaration of quality	
Advantages	The farmer has a wide choice of quality of seed. The system is flexible and ensures that most of the time there will be enough seed to satisfy needs. High-quality seed is expected to drive low-quality seed off the market.
Disadvantages	Farmers find it more difficult to understand the system. Farmers have to be aware of the relative importance of quality and cost. It is more difficult to check and enforce than minimum standards.

(a) *Minimum standards* are set for each aspect of quality, such as: field crop-inspection standards for varietal purity, isolation distances, species purity, freedom from noxious weeds, diseases, and seed standards for analytical and varietal purity and for germination capacity. To be eligible for sale, seed has to be certified officially to show it has met these standards. Certification agencies may be public or private entities, and seed testing laboratories may be separate from them. Usually each seed bag or container will carry a label or tag that in itself guarantees the quality of the seed. This is the system used in the European Union (EU), where Directives set minimum requirements to be implemented by member countries, which may impose stricter disease thresholds on seed produced in their own territories as well as other stricter quality requirements.

(b) *Truth-in-labelling*, or *Declaration of quality*, allows seed of any quality to be sold as long as full details of the quality are given to the purchaser. A Quality-Declared Seed (QDS) system has been developed by the **FAO**, which sets out the principles involved.

The results of tests are made known on the labels on each bag, on test certificates, on invoices or perhaps in seed catalogues. Labels may be colour-coded to show the category of seed but do not in themselves guarantee the quality in the way that 'minimum standards' do. It is the responsibility of the purchaser to ensure the appropriate quality of seed is obtained, by using all the information available. This is the system that is operated in the USA, for example, whose seed law merely directs that seed be sold with a truthful label, without setting minimum standards (although individual state laws may mandate additional requirements).

(See: Certification schemes; Germination testing; Health testing; Purity testing; Quality testing)

The management of variety and seed quality control regulation varies significantly between countries. For example there is a sharp contrast between European countries and the USA. In the former, all field crop varieties must be registered and tested for performance. A variety that has been approved in one EU country is eligible to be listed in a Common Catalogue that allows its sale in other member states. All seed sold in EU countries must also be certified, either by public agencies or private agencies licensed by the government. In the USA, on the other hand, there are no mandatory variety registration or performance requirements, and seed certification is managed voluntarily by independent agencies. It is thus possible to have a certification system without having a seed law.

Commercial seed regulation also can differ significantly between developing countries and is undergoing significant reform in many of them. Most established systems have been based on public plant breeding and para-statal seed production, supported by mandatory variety registration, seed certification and seed performance testing regulations. Some countries maintain less strict public control and with more flexibility, such as by making regulations voluntary for private varieties and/or by specifying a minimum set of standards. The wide range of growing environments in many such countries makes it particularly difficult to adequately test variety performance. The management of seed quality control is similarly problematic; the FAO's QDS system offers a more accommodating approach, which specifies reasonable certification standards and suggests a system of spot checks rather than universal inspections.

2. Variety regulation

One of the early problems that plagued the seed trade was the uncoordinated nomenclature of the varieties in the market. A single variety might be sold under several different names, or one name might be used for similar varieties. The first attempts at regulation were based on a register of varieties, which included morphological characterization and performance testing, which became codified in seed laws. In the USA, for example, some states created seed laws in the early 20th century and the Federal Seeds Act was enacted in 1939; in the UK, a National Seeds Act was passed in 1964, establishing a voluntary index of approved varieties. Describing a variety automatically gives it some protection, whether or not it is given ownership rights, because once it is in 'common knowledge' no one else can demonstrate that they have selected or bred it. Ownership rights imply that someone has bred a new variety, and allow some financial recompense to be obtained for its use. In some cases when a crop is important for export, variety regulation may be instituted to protect or promote a country's export markets.

(a) *Registration and national listing*. It is an important requirement of any plant variety protection system that 'novelty' is shown. This means that a variety must be distinct from all other known varieties in common knowledge. Registration involves two issues: distinctiveness and performance. The first is represented by assessing its **Distinctness, Uniformity and Stability (DUS)**: to be registered, a new variety must exhibit some phenotypic distinctness from other varieties, be genetically uniform, and maintain its characteristics over successive cycles of propagation. The second issue is often expressed as **Value for Cultivation and Use (VCU)**, and involves some type of performance testing to ensure that the new variety is superior, or at least acceptable, under local growing conditions.

Where variety registration and testing is in force, a new variety receives authorization through a variety release committee or similar body. DUS information is used for inclusion on National Variety Lists, which may be either compulsory (that is, only the varieties on the list may be sold as seed) or recommended (merely giving guidance), see Table L.3.

3. Intellectual property protection

A separate aspect of seed legislation relates to the intellectual property protection of plant varieties. The International Union for the Protection of New Varieties of Plants (UPOV) has established and progressively amended a convention that

Table L.3. Examples of seed variety registration where National List (NL) and/or Plant Variety Protection (PVP) apply.

NL + PVP	PVP only
Cereal crops	Ornamentals
Pulses	Fruit
Oilseeds	Trees
Parental lines	

provides a system for plant variety protection, of which most OECD countries are members. Its basic rules allow plant breeders, public or private, to register their varieties and protect them from unauthorized use in seed production. DUS information is used in the granting of these ownership rights, that is 'Plant Variety Rights'.

The degree to which other breeders are able to use a protected variety and the so-called 'Farmers' Privilege' to **farm save seed** (of open-pollinated varieties) are restricted in law. Indeed, the agreement on Trade-Related Intellectual Property Rights (TRIPS) requires all members of the World Trade Organisation (WTO) to provide some type of intellectual property protection for new plant varieties. This may be through utility patents for plant varieties (used in some instances in the USA), by accession to UPOV, or by an acceptable system of its own. There are also additional ways of protecting plant varieties. The ability to protect the identity of the inbreds used in **hybrid** formation allows them to function as trade secrets. In the USA, some seed is now sold subject to an agreement in which the purchasing farmer is prohibited from saving or selling the harvest as seed.

The notion of 'Farmers' Rights' was introduced by the FAO International Undertaking on Genetic Resources for Agriculture, both as a possible stimulus to conserve biodiversity and to offer a possible means of compensation (to a national fund or to individual communities) for the use of local germplasm in breeding commercial varieties. This voluntary undertaking is now a formal treaty, though as of 2005 it remains to be ratified by the requisite number of countries.

4. Impact of biotechnology

The advent of agricultural biotechnology places additional requirements on national legal and seed regulatory systems. The breadth and scope of patenting systems for the genes, processes and products of biotechnology is a subject of continuing debate. In addition, the introduction of transgenic crops (**genetically modified organisms**), bearing locally developed or imported modifications, is requiring the creation of a biosafety capacity to assess and contain related risks, where they are judged necessary, including understanding their potential interactions with the environment and food safety concerns (**see: Identity preservation; Genetically modified organisms – identity testing**). EU legislation, for example, considers biological safety, food and feed, traceability and labelling, deliberate release and marketing, coexistence, seed, and environmental liability. (RT, RD)

Louwaars, N.P. (ed.) (2002) *Seed Policy, Legislation and Law*. Haworth Press, New York, USA.

Tripp, R. (ed.) (1997) *New Seed and Old Laws*. Intermediate Technology Publications, London, UK.

LEA

See: Late embryogenesis abundant proteins

Leakage

Dry seeds, when placed in water, undergo **imbibition**, a process that is accompanied by leakage of ions and low molecular weight metabolites from their cells. The extent of imbibitional leakage is used in some species as a measure of deterioration of a seed in storage. (**See: Desiccation damage; Germination; Germination – physical factors; Vigour testing – biochemical**)

Lectins

Plant lectins are defined as plant proteins, often of seed origin, possessing at least one non-catalytic domain that binds reversibly to a specific mono- or **oligosaccharide**, generally in animal cell membranes. Plant lectins strongly differ from each other with respect to their molecular structure, sugar-binding specificity and biological activities. However, based on sequence and structural data, the apparently very heterogeneous group of plant lectins can be subdivided into seven families of structurally and evolutionarily related proteins. These families are, in alphabetical order: the amaranthins, the chitin-binding lectins comprising hevein domain(s), the (cucurbit) phloem lectins, the jacalins, the legume lectins, the monocot mannose-binding lectins and the ribosome-inactivating proteins.

Most plant lectins are targeted against O- and N-glycans of animal glycoconjugates, and accordingly are thought to play a role in plant defence against plant-eating (phytophagous) invertebrates and/or herbivorous higher animals. Some lectins are valuable candidates for genetic engineering into plants as resistance factors against **predation** by insects (**see: Genetic modification**). Due to their high binding affinity for animal glycans, many plant lectins bind on glycoconjugates exposed on the surface of animal and human cells. Some lectins are noxious or even toxic upon oral uptake, e.g. **ricin** in castor bean seeds is extremely poisonous, and **phytohaemagglutinin** is responsible for the high toxicity of raw kidney beans. Other lectins are considered antinutrients (e.g. **soybean** lectin, wheat germ agglutinin). Only some lectins can be inactivated by heat treatment, e.g. **soybean lectin** (and all other legume lectins), ricin. Others are fairly heat stable (e.g. wheat germ agglutinin). (EVD, WJP)

(**See: Poisonous seeds; Storage proteins**)

Peumans, W.J. and Van Damme, E.J.M. (1996) Prevalence, biological activity and genetic manipulation of lectins in foods. *Trends in Food Science Technology* 7, 132–138.

Van Damme, E.J.M., Peumans, W.J., Barre, A. and Rougé, P. (1998) Plant lectins: a composite of several distinct families of structurally and evolutionary related proteins with diverse biological roles. *Critical Reviews in Plant Sciences* 17, 575–692.

Leeks

See: Alliums

Legume

A typical dehiscent fruit (pod) of the Fabaceae (Leguminosae) developed from a monomerous ovary (a single carpel) opening along two sutures (dorsally and ventrally) with the seeds attached to the ventral suture, e.g. *Acacia* spp., *Medicago sativa* (alfalfa), *Glycyrrhiza glabra* (liquorice), *Delonix regia* (flamboyant), *Pisum sativum* (**pea**), *Phaseolus coccineus* (**runner bean**), *Wisteria sinensis*. The non-botanical use of the term legume refers to plants, pods or seeds of the family, often in the context of them being edible to humans and animals. (**See: Fruit – types; Legumes**)

Legumes

The legume family (Leguminosae or Fabaceae) includes some 12,000 species distributed throughout the world and adapted to a great variety of habitats. Members of the family are easily recognizable particularly from three characteristics: compound leaves, an irregular butterfly-like flower and a fruit that is a pod. For the purpose of our discussion we have chosen a conventional classification of legume species into two groups (setting aside the third major group – forage species that are not primarily cultivated for grain, **see: Legumes – forage**):

1. Species mostly used as food: **grain legumes** – leguminous plants, the seeds of which are used primarily for human or livestock consumption or which are known to be edible. Although this definition includes a number of trees and shrubs, the latter are treated separately for convenience. The annual leguminous food crops that are harvested for dry seeds are called **pulses**.
2. Species whose seeds yield important industrial (including food) products: **industrial legumes** (two of these, **peanut** (groundnut) and **soybean**, are sometimes included in 'grain legumes', and are two of the world's most important crops). (**See: Oilseeds – major**)

Where agriculture developed, the food support systems generally consisted of cereal and legume grains. The two are complementary, the legumes providing the proteins, and in some cases the **oils**, that occur in much smaller quantities, weight for weight, in the cereals. But it is not only in the quantity that the importance of legume seed protein lies but also in its quality, for it has some of the **essential amino acids** that are low in cereals: and cereals contain those in which most legumes are deficient. An important feature of almost all legume species is that the plants have root nodules housing nitrogen-fixing *Rhizobium* bacteria (**see: Rhizobial inoculation**). The plants therefore have available a relative abundance of organic nitrogen, a property that relates to the laying down of relatively massive amounts of protein as a seed storage reserve (**see: Storage protein**).

Grain legumes, then, are an extremely important source of human food second to the cereals (in many countries they are the major, sometimes the only, source of dietary protein), and many of them also serve as animal feed. The dried seeds are

Table L.4. The most important species of edible grain and industrial legumes.

Subfamily	Species	Name[a]	Main producer[b]
		Warm-season grain legumes	
More important species:			
Papilionoideae	*Phaseolus vulgaris*	**Common bean** (numerous types)	Brazil, Mexico
	Vigna unguiculata	**Cowpea**	Nigeria
	Vigna angularis	**Adzuki bean**	China
	Cajanus cajan	**Pigeon pea**	India
Minor species:			
	Phaseolus lunatus	**Lima bean**	NA
	Vigna radiata	**Mung bean**	NA
	Phaseolus acutifolius	**Tepary bean**	NA
	Phaseolus polyanthus	Year bean	NA
	Dolichos lablab	**Lablab, Hyacinth bean**	NA
	Canavallia ensiformis	**Jack bean**	NA
		Cool-season grain legumes	
Important species:			
Papilionoideae	*Cicer arietinum*	**Chickpea**	India
	Pisum sativum	**Pea**	India, China
	Lens culinaris	**Lentil**	India
	Vicia faba	**Broad or fava bean**	China
Minor species:			
Papilionoideae	*Lupinus mutabilis*	**Andean lupin**	NA
	Phaseolus coccineus	**Runner bean**	USA
	Lathyrus sativus	**Chickling pea**	India
		Industrial legumes	
Papilionoideae	*Glycine max*	**Soybean**	USA Brazil Argentina
	Arachis hypogaea	**Peanut, Groundnut**	China, India
	Ceratonia siliqua	**Carob**	Spain
	Cyamopsis tetragonolobus	**Guar**	India, Pakistan

NA, Not available.
[a]See separate entries for those in **bold** lettering.
[b]FAOSTAT (2002).

consumed throughout the year but a significant proportion of the seeds are eaten as green vegetables. Although all food legumes are essentially similar in protein quality and quantity, each is unique in physical characteristics, chemical composition and anti-nutritional components; the methods of primary as well as secondary processing are thus different for individual grain legumes. Industrial legumes can be used as grain legumes for human or animal feed but they also have major importance in furnishing a variety of industrial, agricultural, food and pharmaceutical products.

The most important legume seeds are shown in Table L.4. They fall into: (i) **Warm-season legumes**, which show **epigeal seedling emergence**, flowering when daylengths become progressively shorter, preceded by a period of rapid growth; (ii) **Cool-season legumes**, which have **hypogeal seedling emergence**, rapid growth and flowering when daylengths become longer; (iii) **Industrial legumes**. Six genera of the family comprise the most important contributors of edible species of grain legumes. The common bean (*Phaseolus vulgaris* L.) is the most important pulse in the world, chickpea (*Cicer arietinum* L.) being the second. As already noted, the 'industrial' soybean and groundnut are extremely important in the world economy.

Lesser known or less common edible legume seeds are African yam bean (*Sphenostylis stenocarpa*), asparagus pea (*Tetragonolobus purpureus*), **bambara groundnut, Kersting's groundnut**, marama bean (*Tylosema esculentum*), **sword bean, tamarind**, velvet bean (*Mucuna deeringiana*), **winged bean** and ye-eb (*Cordeauxia edulis*).

Worldwide production of the pulses is more than 60 million t, and if soybean and groundnut are added, grain legumes exceed 270 million t, with export values greater than US$16 billion (Table L.5). (**Crop Atlas Appendix, Maps 3, 19**)

1. Origins and domestication

Legumes have evidently long been an important component of the human diet and before the start of agriculture seeds of the wild ancestors of modern cultivated species must have been consumed. Non-cultivated legumes feature prominently in the diets of present-day hunter-gatherers of southern Africa. Seeds of *Tylosema esculentum* and *Acacia albida* are eaten by indigenous peoples, and those of the shrub *Bauhinia petersiana* and *Gioubourtia coleosperma* and *Schotia afra* are dietary components for bushmen of the Kalahari.

Based on studies of the occurrence of wild relatives, six 'centres of origin' of grain legumes have been described, four in the Old World and two in the New World (Table L.6).

Archaeological evidence points to the association of several present-day, cultivated legumes with early human settlements

Table L.5. Production and export values of legume seeds in 2002.

Seed	Production (1000 t)	Export value (1000 US$)
Beans (dry)	19,244	1,329,497
Beans (green)	5,559	221,686
Broad (faba) beans (dry)	4,257	116,790
Broad (faba) beans (green)	1,036	47,521
Carob	214	55,149
Chickpeas	7,858	328,336
Cowpea (dry)	3,777	2,164
Groundnuts	33,735	123,039 (in shell)
		655,220 (out of shell)
Lentils	2,857	359,231
Lupins	707	192,987
Peas (dry)	9,592	552,897
Pigeon peas	3,049	151
Soybean	180,552	10,545,058
String beans	1,705	34,077
Vetches	893	NA
Pulses (total)[a]	60,153	2,859,012

[a]Excluding soybean, groundnut. From FAOSTAT.

Table L.6. Major centres of origin of grain legumes.

Centre of origin	Type
Africa (equatorial)	Cowpea, lablab (hyacinth bean), Bambara groundnut,[a] Kersting's groundnut[a]
China	Adzuki bean, soybean
Near East ('Fertile Crescent', Mediterranean basin)	Broad (faba) bean, chickpea, lentil, lupin, pea
Southeast Asia (India, Indochina, Pacific Islands)	Jack bean, pigeon pea, moth bean,[a] winged bean[a]
Central America (Mexico, Guatemala)	*Phaseolus* beans
South America (Andes)	*Phaseolus* beans, groundnut, Jack bean, lupin

[a]Not grown commercially.
Adapted from Chrispeels, M.J. and Sadava, D.E. (2003) *Plants, Genes and Crop Biotechnology.* Jones and Bartlett, Sudbury, MA, USA, and ASPB.

(**see: Archaeobotany**). In the Old World, peas, lentils, broad bean, chickpea and vetch appear to have a simultaneous or near simultaneous domestication with early 'primitive' cereals, **emmer, einkorn** and **barley**. Lentils have been found in sites in Greece and the Near East dated to be 9500–13,000 and 9500–10,000 years old, respectively: and there is evidence that the species spread over Europe about 8000 years ago. Interestingly, lentils found at about 9000 years old, Near East sites measured 2.5–3 mm long, whereas those at younger sites (7000–7500 years ago) were 4–4.5 mm in length (see below: effects of **domestication**). Peas have been found in large amounts with **wheat** and barley seeds in early Neolithic villages in the Near East (8000–9500 years old) and there is evidence that this species had reached central Germany by about 6400 years ago. A plant described in Sumerian records of approx. 2350 BC is thought to be cowpea. There is evidence for the domestication of broad (faba) bean in the Mediterranean basin approximately 6500 years ago and for introduction of the species into China by about 2200 years ago. Small chickpeas were found in 9500-year-old sites in the Near East and larger seeds in Jericho (8500 years old).

In general, the records of domestication in the New World are later than for the Old. There is evidence for the occurrence of scarlet runner beans north of Durango, Mexico, about 1300 years ago and more than 2000 years ago in the central highlands of Mexico and Guatemala. Evidence for domestication of the common bean comes from ancient sites (5000–2200 BC) in Tamaulipas, Mexico (the north-east) (**see: History of seeds in Mexico and Central America**). Tepary beans have been found in settlements dated at 2300 years old in central Mexico, and 1200 years old in Arizona. Bean pods, at least 2300 years old, were recorded in the Tehuacan valley of Mexico, and possibly 8000 years old in the Peruvian Andes, though dating of the latter is controversial.

Legume seeds were carried by migrating peoples from their centres of domestication to outlying regions, and in more recent times by explorers, adventurers and traders. The spread of *Phaseolus* beans is an interesting example that has been traced using knowledge of the structure of the **storage protein**, phaseolin, and its DNA. Analysis of the frequency of different types of the protein and the DNA suggests that the cultivars now grown in western Europe, parts of Africa and northeastern USA originated in the Andes. The Central American centre (Table L.6) does not appear to have been the source of European *Phaseolus*.

Domestication led to several important changes in fruit and seed characteristics as compared with wild progenitors, namely an increase in seed size (see the lentil and chickpea examples above), a reduction or loss of hard-coat-imposed **dormancy**, and the loss of pod dehiscence (**shattering**) (**see: Domestication**).

2. Structure and composition

Seeds of the majority of species possess no or very little **endosperm** (non-endospermic legumes) whose structure can be illustrated by the common bean (*Phaseolus vulgaris*) (Fig. L.1). Almost all the mass of the seed is **embryo**, consisting of two bulky **cotyledons** containing the **storage reserves** which can account for over 80% of the seed weight. Attached to the cotyledons is the axis comprising **radicle** and **hypocotyl**: the latter carries the **plumule** in which the true leaves are carried on the **epicotyl**. The whole is surrounded by the testa on which the **hilum** and **micropyle** are generally discernible. The various types of legume seeds differ as to the actual and relative sizes of these main parts and in the extent of development, for example, of the plumule and the thickness of the testa. The seeds of some species of legumes are endospermic and in most cases the cotyledons are relatively smaller than in the non-endospermic types. Important commercial examples are **carob** (**see: Carob**, Colour Plates 2E, F), **guar** and **fenugreek**.

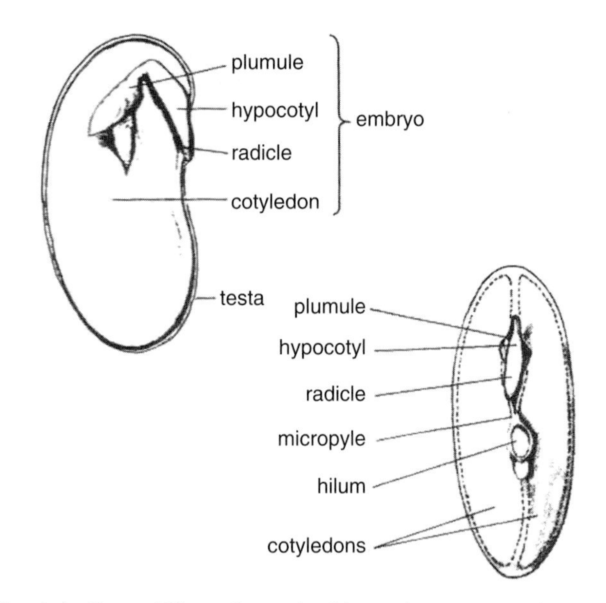

Fig. L.1. Bean (*Phaseolus vulgaris*) seed, an non-endospermic legume. Left: The seed cut open to reveal the inner face of one cotyledon and the embryonic axis. Right: External view of one side of the seed with the internal cotyledons delineated (seed length approx. 15 mm) (**see: Carob**, Colour Plate 2F, for an endospermic legume).

The major storage reserves in the cotyledons are **starch**, protein (mostly globulins, and in some species also albumins; **see: Osborne fractions**) and **oil** (triacylglycerols) in different proportions according to species (Table L.7). Although the seed storage protein is relatively abundant it is deficient in certain amino acids that are essential to the human and livestock diet, especially the sulphur-containing methionine and cysteine. However, there has been some success in elevating the

Table L.7. Reserves of some legume seeds.[a]

Seed	Protein	Oils	Starch
Chickpea	21.3	5.4	43.8
Common bean (haricot)	21.4	1.6	42.7
Ground(pea)nut	26.0	47.5	18.6[b]
Lentil	24.3	1.9	44.5
Mung bean	23.9	1.1	40.9
Pea	24.1	1.3	60.3
Soybean	34.1	17.7	33.5[b]

[a]Per 100 g edible portion.
[b]Total carbohydrate.
Taken from Vaughan, J.G. and Geissler, C. (1997) *The New Oxford Book of Food Plants*. OUP, Oxford, and Simpson, B.B. and Ogorzaly, M.C. (2001) *Economic Botany*. McGraw Hill, New York.

methionine content by introducing the gene for a methionine-rich 2S albumin from sunflower into lupin, whose seeds are used as animal feed. This points the way to important possibilities for genetically engineered improvement in legume seed nutritive value.

The seeds also contain relatively high amounts of potassium (e.g. approx. 1.2 g/100 g fresh weight in broad beans – *Vicia faba*) as well as calcium, iron and sodium. Vitamins A, B1, B2, C, E and folate are also present. Legume seeds are relatively rich in galactose-containing **oligosaccharides**, such as **raffinose**, stachyose and verbascose (e.g. approx. 3% dry weight (dw) in common bean, 5–6% dw in soybean) located primarily in the axis. These oligosaccharides are the cause of the flatulence often associated with the consumption of legume seeds as they act as a readily available substrate for the gas-producing gut bacteria. An aim of food and crop scientists is to reduce or eliminate these carbohydrates from the seeds, but as these compounds might contribute to seed **desiccation tolerance** or be required as the first reserve to be mobilized during germination, their reduction could have undesirable effects on seed biology.

The endosperm of most endospermic legumes is rich in reserve **mannans**, in the cell walls, which have different degrees of galactosylation according to the species.

Several so-called 'anti-nutritionals' occur in legume seeds. **(See: Amylase inhibitors; Lectin; Phytohaemagglutinin (PHA); Protease inhibitors; Soybean lectin; Storage proteins – intolerance and allergies)** These include cyanogenic glycosides, amylase and protease inhibitors, haemagglutinins, saponins, tannins, alkaloids and **mycotoxins**. Some of these can be removed by the cooking process or by denaturation in the gut. Flavanoids – the isoflavones or phytoestrogens – may also be present, sometimes in relatively high abundance (e.g. in soybean). These can have deleterious effects in animal feed but they have also received much attention in relation to their possible heart disease- and cancer-preventative properties in humans (**see: Pharmaceuticals, medicines and biologically active compounds**).

3. Seed storage

Stored legume seeds may become infested with numerous pests. They are described in detail in **Storage pests**.

(OV-V, MB)

For information on the particular contributions of seed to seed science **see: Research seed species – contributions to seed science**.

Broughton, W.J., Hernandez, G., Blair, M., Beebe, S., Gepts, P. and Vanderleyden, J. (2003) Beans (*Phaseolus* spp.) – model food legumes. *Plant and Soil* 252, 55–128.

Gepts, P. (1998) Origin and evolution of common bean: past events and recent trends. *HortScience* 33, 1125–1130.

Pearman, G. (2005) Nuts, seeds and pulses. In: Prance, G. and Nesbitt, M. (eds) *Cultural History of Plants*. Routledge, New York, USA, pp. 133–152.

Summerfield, R.J. and Bunting, A.H. (eds) (1980) *Advances in Legume Science*. Her Majesty's Stationery Office, London, UK.

Summerfield, R.J. and Roberts, E.H. (eds) (1985) *Grain Legume Crops*. William Collins, London, UK.

Vaughan, J.G. and Geissler, C.A. (1997) *The New Oxford Book of Food Plants*. Oxford University Press, Oxford, UK.

Zohary, D. and Hopf, M. (1993) *Domestication of Plants in the Old World*. Clarendon Press, Oxford, UK.

Legumes - forage

Temperate forage legumes are grown either as pure stands (such as lucerne, alfalfa) or as components of a pasture mix (such as clovers) with forage or herbage grasses for livestock production (**see: Grasses and legumes – forages**). Seed production locations are dominated by Europe (mainly Italy, France and Spain), North America and New Zealand, though production also occurs in Argentina, Brazil, Turkey, Iran, India, Japan, China and Russia. In tropical areas, forage legume seed production and use are on a much smaller scale than for grass seeds (**see: Grasses – forage and turf**). Seed is mainly produced in Australia, Brazil, Thailand, India and Ethiopia, mostly for domestic consumption.

Temperate species range through the prostrate and stoloniferous (e.g. white clover), the erect and rhizomatous (e.g. Caucasian clover), the prostrate to erect and bushy (e.g. red clover), and those that have erect stems produced from a crown (e.g. alfalfa or lucerne). They are grown as components of pastures (e.g. **clovers**) or as a pure stand (e.g. **alfalfa**) for both grazing and hay. There are more than 100 different white clover and alfalfa cultivars worldwide, each with attributes specific to a particular environment or use.

The tropical legumes include several annuals (e.g. *Stylosanthes humilis*) but most are perennials, some short-lived (e.g. *S. hamata*). They range from tall shrubs >3 m (*Leucaena leucocephala*) down to short stoloniferous clover-like plants that are >20 cm (e.g. *Lotononis bainesii*). They may be erect, sprawling, twining or creeping. Species which are erect are usually **determinate** (e.g. *Stylosanthes guianensis*), while those that are twining are usually **indeterminate** (e.g. *Macroptilium atropurpureum*). They are used as the legume component of pastures or grown as pure stands for forage and hay production.

Of the temperate forage legumes, *Medicago sativa* originated in Iran/northern Turkey, while *Trifolium pratense* is thought to be native to Russia. The Mediterranean region is the home of most other temperate types. The tropical forage legumes are native to Central America and northern South America, mainly Mexico and Brazil. Major genera are shown in Table L.8.

Though some Fabaceae were present by the lower to mid-Oligocene age, it was not until the Miocene age that forage legume forms evolved. Their development then closely followed the evolution of the forage grasses, with a similar

Table L.8. Major forage legume genera.

Temperate	*Coronilla, Hedysarum, Lotus, Medicago (sativa), Onobrychis, Ornithopus, Trifolium (pratense, repens), Vicia.*
Tropical	*Aeschynomene (americana), Alysicarpus (vaginalis), Arachis, Calopogonium, Centrosema, Desmanthus, Desmodium, Lablab (purpureus), Leucaena, Lotononis, Macroptilium (atropurpureum), Pueraria, Stylosanthes (guianensis, hamata, scabra), Vigna.*

The most important species are shown in parentheses.

separation into temperate and tropical/subtropical tribes. The Papilionoideae subfamily provides almost all the temperate and tropical forage legumes (but note that the tropical genera *Leucaena* and *Desmanthus* come from the tribe Mimoseae, subfamily Mimosoideae).

2. Genetics, breeding

Most temperate forage legumes are cross-fertilizing, natural **polyploids**, although *Trifolium pratense* and *T. hybridum* are natural **diploids**. Realized seed yields are low, but potential seed yield is high, providing the potential for increasing the efficiency of the reproductive system. Extensive breeding effort in the last 30 years has produced many cultivars, many of which, like the temperate forage grasses, are protected by Plant Variety Rights (**see: Laws and regulations**) and have an assurance of genetic purity through seed **certification**.

Although most tropical forage legumes are self-pollinating, many contain genotypes that possess a degree of cross-pollination. Polyploidy is not widespread. Around 50 tropical legume taxa have been commercialized, but half are represented by only one cultivar each. Indeed, there has been relatively little cultivar development within tropical legume species, the exceptions being *Stylosanthes guianensis*, *Leucaena leucocephala* and *Vigna unguiculata*. (**See: Grasses – forage and turf**)

3. Development

In temperate types, the principal factors controlling the transition from vegetative to reproductive growth are photoperiod and temperature. Flower initiation usually occurs in late spring, with flower heads emerging around 6 weeks later. Most temperate forage legumes flower in the sowing year, if photoperiod is adequate, after plants have passed the juvenile stage, that is obtaining a certain number of leaves or shoot size. The transition can be induced in white clover whilst seeds are germinating, and in red clover and lucerne in seedlings 5 weeks after sowing. The critical photoperiod requirement is around 14 h, although this may vary and may be modified by temperature.

Cross-pollination occurs in most temperate perennial forage legumes (though *T. subterraneum* is self-fertilizing). Pollination for many species is by honeybees, but these are not very effective for lucerne so colonies of other species (such as alfalfa leaf-cutting bee, long-tongue bumblebee) may need to be introduced to production fields. Insect-attracting (entomophilous) flowers are large, often brightly coloured and scented, with nectaries. Pollen grains are relatively few in number, and seed set is often low and unpredictable due to both environmental (low temperature and shading of flowers under the canopy) and resource limitations (inadequate nutrient partitioning). Altogether 10–50% of ovules may develop into harvestable mature seed (around 5% in lucerne and 25% in white clover).

Most tropical species also have a juvenile stage requirement. While some are day-neutral, most are **short-day** plants (quantitative or qualitative). The responses may be modified by temperature, and flowering tends to increase at cooler day/night temperatures (e.g. 25°C day/21°C night). Tropical legumes are intolerant of shade, which reduces the rate of flower appearance and decreases the duration of flowering. The majority are **self-pollinating** although some, such as *Desmodium* spp. and *Lablab*, can also be cross-pollinated. In the former, flowers are cleistogamous (sexually mature before the flower opens). For most species, insect pollinators are not a prerequisite for successful pollination, although flowers are often large and brightly coloured. Seed set is variable, and maybe as low as 2%, and generally is higher for self- than for cross-pollinating species. **Ovules** closer to the stylar end of the ovary have a greater chance of developing a seed than more proximal ovaries.

Seed yield depends on the number of fertile shoots per m^2, inflorescences per shoot, flowers per inflorescence, ovules per flower, and ovule site utilization (OSU). What is important is that fertile shoot numbers are adequate to allow maximum floral expression, because it is the number of inflorescences per unit area that is the major determinant of seed yield. Management should therefore have the objective of producing an intense, short flowering with as many flowers as possible above the plant canopy and therefore available to pollinators. Flowers per inflorescence and ovules per flower depend on the species, and can be negatively affected by decreasing temperature, moisture stress and shading. Sometimes there is compensation between the number of fertile shoots and yield per shoot depending on plant density. OSU depends first on the success of pollination and fertilization, and the extent of ovule abortion in the first week after fertilization. A shortage of assimilates following competition from vegetative growth has been proposed as a determinant of eventual seed number. Initial pod production may be up to five times greater than the number eventually retained, but the reasons for pod abortion are not clear.

Seed development proceeds in three stages – growth (I), a food reserve accumulation (II) and maturation or ripening (III) (**see: Development of seeds – an overview**). The duration of each stage will depend on species, cultivar and the environment, but for white clover the duration of stage I is around 10 days, stage II 10–14 days, and stage III 3–10 days. **Physiological maturity** is therefore reached between around 25 days (white clover) to 40 days (lucerne) after pollination.

4. Production

The field must meet all seed certification requirements (which for white clover, for example, may require a 5-year break between white clover crops plus meeting a standard for the occurrence of buried (hard) seed).

Seedbeds should be a fine tilth and firm surface and, in fields where disturbance of buried seed through cultivation could pose a contamination problem, **direct drilling** may be a better option. Species requiring **vernalization** must be well established before winter, and this requires either a spring or late summer/early autumn sowing. A companion crop (grass or cereal) may provide protection in severe winter climates, but usually the highest yields come from pure stands. Sowing rates vary from 2–10 kg/ha depending on species and establishment method (higher if sown with a companion grass). Similar yields can be obtained from different row spacings, but typically 30–60 cm spacings are used.

For tropical species sowing time usually relates to the reliability of rainfall during the growing season, so that spring/summer sowings are common. Annual legumes such as *Stylosanthes humilis* require a relatively high sowing rate in comparison with perennial species. Sprawling legumes such as most *Stylosanthes* tend to yield better following drilling in rows, but **broadcast seeding** is an option for legumes with a twining (e.g. *Macroptilium atropurpureum*) or creeping (e.g. *Lotononis bainesii*) growth habit. Legumes tend to establish more readily from burial 10–20 mm in the soil than from surface sowings.

Temperate forage legumes, because they are indeterminate plants, continue to produce vegetative growth at the time reproductive initiation is occurring in the spring. If such growth is excessive, flower production may be reduced or flowers become shaded under the canopy, so that both pollination and seed set are poor. For many species, therefore, spring defoliation is a requirement, either by grazing or cutting. The objective is to remove leaf material and not damage stolons or branches, so that defoliation height is species dependent. Generally, **defoliation** should occur around 10 days after the emergence of the first reproductive buds. Tropical forage legumes include both determinate and indeterminate species, with growth habits ranging from erect to sprawling to twining to creeping. Management is therefore very species-specific. As with temperate legumes, the prime objective is to maximize the number of reproductive units. For some, this involves defoliation to remove excessive vegetative growth and to expose axillary buds to the light to promote flowering. Moisture stress during flowering should be avoided.

For temperate species, deciding on the optimum harvest time is often difficult; because the indeterminate growth habit can spread flowering over several weeks, the seed crop will not ripen uniformly and a single harvest will include seed at all stages of development. For a species such as white clover, which can be managed to produce a single peak of flowering, the optimum harvest date is around 5–6 weeks after peak anthesis, but in big trefoil (*Lotus uliginosus*), for example, where umbels mature over several weeks, the optimum time is when the majority of pods are light brown. Seed crops are cut, preferably with a double reciprocating knife mower to reduce seed **shattering**, and left for 5–14 days in the **swath** to dry before being combined. Lifters or drapers are necessary to feed the crop into the combine auger platform. The use of **desiccants** (diquat) to reduce vegetative bulk prior to cutting is an option; in upright, desiccated red clover and lucerne crops, direct combining can be used.

In temperate types, stand density may increase in perennial species, because of volunteer plants growing from shed seed and regrowth of old plants, to a level where too much interplant/interstem competition can reduce seed yields. Where legume crop residues are allowed to be extensively grazed, inter-row cultivation may then be an option for removing volunteer plants (as well as weed seedlings). In very dense stands, gapping by cultivation or using a herbicide can reduce plant density, often leading to increased flower production in the subsequent crop.

Tropical forage legumes, in addition to the indeterminate growth-habit problems described for temperate forage legumes, have a great diversity of morphological types, so that pods and seeds may be formed inside a mat of vegetative growth, below the soil surface, or on aerial branches. The decision on harvest time is based primarily on pod development stage – such as pods that are black or brown, with some beginning to open. Harvesting methods include single destructive harvest, multiple harvesting of the standing crop, and the recovery of mature seeds shed by the standing crop, either mechanically or by hand. For information on drying and storage of seeds **see: Grasses – forage and turf**. The productive life of tropical legume seed crops may be from 1 to 4 years; yields tend to decline as soil-N accumulates over time.

5. Quality, dormancy

The seed quality components described for forage grasses also apply to temperate forage legumes. In tropical types the relatively free flowing legume seeds can be cleaned and graded to relatively high levels of analytical purity, unlike many of the tropical grasses. (**See: Grasses – forage and turf**)

Temperate forage legume seeds have a **dormancy** imposed by a hard seedcoat (**see: Hardseededness**), a character which is heritable but also strongly environment-dependent. In nature, the effects of alternating winter temperatures and soil microorganisms eventually reduce hard seed; in commerce mechanical scarification can be used, but it is often difficult to avoid physically damaging the seeds. Hardseededness is a feature of lotus, lucerne and red clover, but is less frequent in white clover, possibly because some scarification occurs during machine harvest. Hardseededness has also been recorded in all commercial tropical forage legume species except *Arachis hypogaea*. While small quantities of seed have been successfully hot-water treated or acid scarified, mechanical scarification is not yet widely used. For both temperate and tropical forage legumes, the percentage of hard seed recorded during a germination test is reported separately on the seed analysis certificate.

6. Treatments

The most common treatment to the seed of temperate forage legumes is the addition of *Rhizobium* bacteria to promote nitrogen fixation (**see: Rhizobia; Rhizobial inoculation**). The inoculant may be applied directly to the seedcoat, as a slurry of bacteria in a peat carrier, or via seed **pelleting**. The *Rhizobium* species used must be compatible with the legume species to be effective. Seed inoculated by the former method should be used immediately, while for the latter, the bacteria will remain viable for around 3 months.

Seed treatment is little used in tropical legumes, and inoculation may not be necessary in many soils because (unlike the temperate legumes) many nodulate freely with a wide range of nitrogen-fixing root nodule bacteria in the soil, particularly *Bradyrhizobium* spp. In acid soils however, responses to inoculation have been reported in species such as *Centrosema*, *Desmodium* and *Stylosanthes*. (JGH)

Fairey, D.T. and Hampton, J.G. (eds) (1997) *Temperate Forage Seed Production*, Vol. I. *Temperate Species*. CAB International, Wallingford, UK.

Loch, D.S. and Ferguson, J.E. (eds) (1999) *Temperate Forage Seed Production*, Vol. II. *Tropical and Subtropical Species*. CAB International, Wallingford, UK.

Legumes – grain and vegetable

Grain legumes contain a wide range of crop species in the Fabaceae or Leguminosae including the **pea** (*Pisum sativum*), **bean (common)** (*Phaseolus vulgaris*), field or **broad bean**

(*Vicia faba*), **chickpea** (*Cicer arvensis*), **lupin** (*Lupinus* spp.), **cowpea** (*Vigna unguiculata*), **lentil** (*Lens culinaris*), **soybean** (*Glycine max*) and **pigeon pea** (*Cajanus cajan*). Peas, green and broad beans are also harvested as a fresh vegetable, or for canning or freezing. They are grown in many areas of the world and represent a significant source of food both for human consumption and animal feed. Legume seeds that are eaten directly (often after dry storage) are the **pulses**. For information on legume seeds, **see: Legumes**.

Legumin

A storage protein (11S globulin) that is present in the seeds of many species. (**See: Osborne fractions; Storage proteins**)

Lemma

One of the two bracts in the grass floret. (**See: Floret**, Fig. F.7)

Lens

A raised zone in the area of the **chalaza** of the seeds of the Fabaceae which externally often appears as a lens-shaped structure near the **hilum**, also called *marca rapheale* or *macula raphae*. This structure has also been included by some authors under the term **strophiole** which, however, refers to a localized aril assisting the dispersal of the seed by animals (**see: Seedcoats – dispersal aids**). *Lens* is also the generic name for **lentils**.

Lentil

1. World importance and distribution

Lentil (*Lens culinaris*) (a **cool-season legume**) is an ancient crop that has been grown for more than 8500 years. Production of lentils spread during the Bronze Age from the Near East to the Mediterranean area, Asia, Europe and finally in relatively modern times to the western hemisphere. Around 60% of current world production is from India and Turkey, the former being by far the largest producer. Canada is the world's third largest producer. Turkey, Canada and the USA are the major exporters of lentils. (**See: Crop Atlas Appendix Map 3**)

2. Origin

Lentil is probably the earliest of the grain legumes to be domesticated, in the Fertile Crescent (**see: Legumes**). Archeological evidence, together with morphological and cytogenetic comparisons, suggest that *Lens culinaris* was derived from *L. orientalis*, one of the four wild subspecies recognized in the genus *Lens*. Earliest carbonized remains of lentils are about 10,000 years old. The pottage (lentils) for which Esau sold his birthright (Genesis Ch. 25) is an early reference to the grain legume.

3. Seed types

Seeds (Fig. L.2) are lens-shaped, weigh 2–8 g/100, round, biconvex, with dimensions in the range of 4–9 mm × 2.2–3 mm. They are produced in small pods usually containing 1–2 seeds per pod. **Testa** colours range from pale green to tan to brown and black, with purple and black mottles in some cultivars. Lentil seeds are often divided into two types, *macrosperma* and *microsperma*. The large *macrosperma* (diameter 6–9 mm) is found mainly in the Mediterranean region and the Americas. **Cotyledons** are normally yellow or green, and flowers or vegetative structures have very little or no pigmentation. *Microsperma* (seeds 2–6 mm) is found mainly in the Indian subcontinent and the Near East. Cotyledons are red, orange, or yellow. Plants are shorter than the *macrosperma*, more pigmented with smaller pods, leaves and leaflets.

4. Nutritional quality

Lentil is a highly nutritious food legume. Seed protein content (mainly globulins and some albumins; **see: Storage protein**) is in the range 22–35% fresh weight (fw). It contains significant concentrations of lysine, arginine, leucine, and sulphur amino acids. Though lentils have the richest content of important amino acids among the cool-season legumes they are nevertheless deficient in methionine and cystine to meet dietary needs. Seed **starch** content ranges from 35 to 53% fw with **amylose** varying from 20.7 to 38.5% of the starch. Lentils are a good source of B vitamins, particularly thiamine, riboflavin, nicotinic acid, folic acid, pantothenic acid, biotin and pyridoxine as well as choline and inositol. **Oil** (triacylglycerol) content is approx. 1.5% fw: oleic, palmitic and linoleic are the major **fatty acids**. Total **oligosaccharide** content (**raffinose**, stachyose and verbascose) is approx 4.5%.

5. Uses

Seeds are used as food in soups, stews, casseroles and salad dishes. In India, lentil is mostly eaten as *dhal* (seed that is decorticated and split). Lentil flour can be mixed with cereals to make bread and used as a food for infants. The whole plants can serve as animal feed.

6. Marketing

Lentils for the export market are graded mostly according to sieve size, used to determine the price. In the USA the grades are: for red lentils, 50% of the product should remain on a 4.35 mm round hole-perforated screen and 100% remain on a 3 mm screen, while for yellow lentils, 50% should remain on 7 mm and 100% on the 5 mm screens.

World area planted with lentils in 2002 amounted to 3,653,000 ha with a production of 2,857,000 t. The largest

Fig. L.2. Lentil seeds. *Macrosperma* (left) and *microsperma* (right) types are shown (image by Mike Amphlett, CABI).

importers in 2001 according to FAO figures were Egypt, Turkey, India, Pakistan and Colombia: main exporters according to the same source were Canada, Australia, Turkey, India and the USA (Table L.9). (**See: Lentil – cultivation**) (OV-V)

Muehlbauer, F.J. and Kaiser, W.J. (eds) (1994) *Expanding the Production and Use of Cool-season Food Legumes*. Kluwer Academic, Dordrecht, The Netherlands.

Summerfield, R.J. (ed.) (1988) *World Crops: Cool Season Food Legumes. A Global Perspective of the Problems and Prospects for Crop Improvement in Pea, Lentil, Faba Bean and Chickpea*. Kluwer Academic, Dordrecht, The Netherlands.

Summerfield, R.J. and Roberts, E.H. (eds) (1985) *Grain Legume Crops*. William Collins, London, UK.

Webb, C. and Hawtin, G.C. (eds) (1981) *Lentils*. Commonwealth Agricultural Bureau, Farnham Royal, UK.

Lentil – cultivation

See: Lentil for information on economic importance, origin, seeds and uses. One of man's earliest food crops, lentils originated in the Near East and are now produced in Asia, which grows around 60% of the world production, eastern Europe, Africa and North America. They are best adapted to temperate climates, but the maturity is extended by northern latitudes. The plants are generally bushy and pods are oblong and bulging over the seeds. Up to four pods can be produced from each flower peduncle, but on average one pod per peduncle is common.

1. Production

Lentil flowers from the bottom of the plant, and progresses upward. Lentils will not produce acceptable seed yields in hot, wet areas, where vegetative growth is promoted and seed production reduced. Combining is carried out at between 18 and 22% seed moisture to prevent harvest losses and cracked **seedcoats**. There is a risk of excessive **shattering** losses when harvesting is done at higher temperatures. Most varieties of lentils should reach maturity within 85 days of emergence.

In most of the developing countries where lentils are produced, a significant amount of harvesting is carried out by hand. The crop is cut and left in small heaps in the field until dry. The crop is then threshed by hand or by a machine. Where mechanized harvesting is carried out, the usual method is by direct combining of the crop when it is desiccated. Seed is more susceptible to physical damage than peas or faba beans and therefore the drum speed must be reduced.

2. Sowing and treatment

Lentils may be sown as an autumn or spring crop, depending on area and winter conditions and rain availability in the summer season. Freshly harvested seeds show hard seed dormancy (**hardseededness**), which may last up to 3–4 weeks. Non-dormant seeds imbibe quickly and germination is rapid (8–9 days) in warm moist soil. Seedling establishment is **hypogeal.**

Lentils grow well on soils ranging from heavy clay to loamy sands. The crop performs well on land of moderate fertility but not on waterlogged soils. Germination requires a moist soil as uptake of water by the seed begins immediately after planting and seed can absorb to the extent of 100% of their dry weight in 24 h. The optimum temperature for germination is between 15 and 20°C and emergence takes place in 5–6 days. Seed yield is dependent on the plant density and vegetative growth but normally 100 plants per m^2 appears to be the optimum.

In the absence of *Rhizobium* populations in the soil, seed can be inoculated with *Rhizobium leguminosarum*. It is not usual to use fungicidal seed treatments although control of seedborne fungi such as *Ascochyta lentis* may require such treatment. (AB)

Summerfield, R.J. and Roberts, E.H. (eds) (1985) *Grain Legume Crops*. William Collins, London, UK.

Webb, C. and Hawtin, G.C. (eds) (1981) *Lentils*. Commonwealth Agricultural Bureau, Farnham Royal, UK.

Lettuce

Lettuce (*Lactuca* spp.) is worldwide an important commercial crop and garden vegetable, used primarily as a salad vegetable and incidentally in other uses, such as in cigarettes, as a source of cooking oil from its seeds, and for medicinal purposes.

Table L.9. Lentil exports and imports in 2001.

Imports			Exports		
World/ Country	(thousand t)	Value (million US$)	World/ Country	(thousand t)	Value (million US$)
World	1098	436	World	1169	462
Egypt	113	50	Canada	491	148
Turkey	99	38	Australia	218	68
India	87	31	Turkey	159	85
Pakistan	68	25	India	106	51
Colombia	50	14	USA	99	37
Spain	47	17			
Algeria	47	16			
France	32	12			
Mexico	31	9			
Morocco	29	10			

Source: FAOSTAT 2001.

Cultivated lettuce is in the species *Lactuca sativa* L. in the family Asteraceae. The crop is grown in diverse environments over large geographical areas and is adapted to most climates, although because high temperatures may induce premature **bolting** it is most commonly grown in cooler areas or seasons. The USA is the largest producer of the world's lettuce crop, followed by Spain, Italy, Japan and France.

1. Origins

The cultivated species probably originated in western Asia, which has the greatest number of related species. It was first recorded, however, in Egyptian tomb paintings about 2500 BC. From there, it was taken around the Mediterranean area, where it was recorded as being popular in Greece and Rome, and then to China, to elsewhere in Europe and to the Americas.

2. Breeding

Lettuce exists in several 'head' and 'leaf' types. Crisphead lettuce forms a large dense head; Romaine (cos) lettuce forms an upright head of elongated coarse textured leaves; and butterhead lettuce is small, with soft oily leaves. Leaf lettuces form a rosette of leaves in a variety of leaf shapes and colours. In each group, there are a number of cultivars – varying in size, disease resistances, earliness and other traits: many suited to specific and fairly narrow environmental niches – by season, geographical and regional location, soil type, salinity, temperature, daylength, moisture availability and presence of specific diseases and insects.

Lettuce inbreeds naturally due to its floret structure composed of fused anthers that dehisce from the inside as the stigmas and style elongate within the tube, which ensures self-fertilization. To make a cross for breeding purposes, the easiest and most efficient technique is to wash the pollen off the male parent as the stigmas emerge. Pedigree breeding, backcross breeding and selection are the most frequently used breeding methods.

Of the 100–150 species of *Lactuca*, only a few can be crossed to cultivated lettuce. *Lactuca serriola* (prickly lettuce) is most closely related; it is distributed worldwide, probably having been introduced as a weed accompanying lettuce.

3. Development

Lettuce is an annual crop. When vegetative growth reaches a mature stage, the stem begins to elongate, producing an inflorescence composed of multiple flower heads. Each head includes 12 to 20 perfect and self-fertile florets, each consisting of a single petal, a double **carpel**, an elongated style and a divided stigma, with the ovary located beneath the corolla. The flowers open only once for about 1 h in the morning.

Lettuce seed is strictly an **achene** – a dry indehiscent single-seeded fruit (Figs L.3 and L.4) – consisting of an embryo, surrounded by successive layers: endosperm, integument and

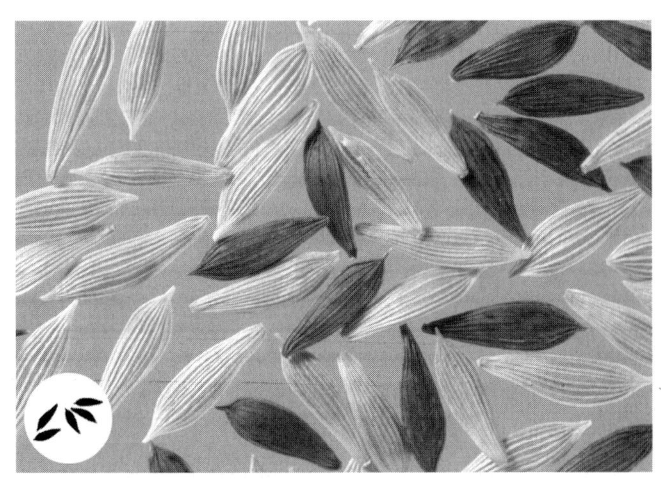

Fig. L.3. Lettuce seeds (achenes) (image by NIAB, Cambridge). Inset circle represents 1 cm diam.

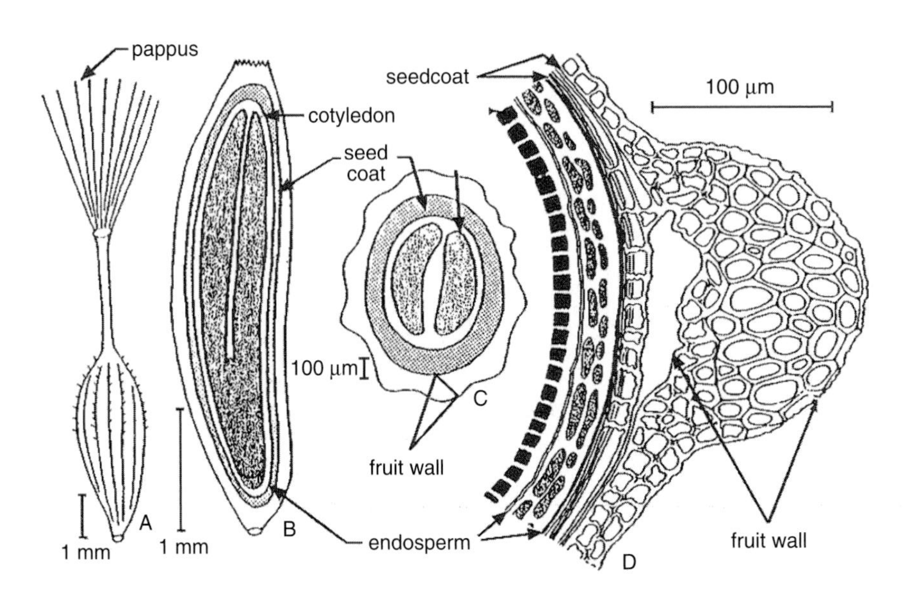

Fig. L.4. Lettuce achene. (A) Entire achene with beak and pappus. (B) Longitudinal section. (C) Cross section. (D) Portion of fruit wall and endosperm.

pericarp. The achene, weighing 0.8–1.2 mg, is wider than thick, ribbed, obovate, up to 6–8 mm long, 1 mm wide, with a beak topped by a pappus. The coat colour may be black or dark grey, brown, yellow or whitish depending upon variety.

Prior to anthesis, in the young ovary, the nucellar tissue has almost completely disintegrated, while the integument consists of 12–18 layers of parenchyma cells with thick cell walls. After anthesis, disintegration of layers of both the pericarp and integument takes place and the endosperm enlarges and then becomes partially resorbed, as the embryo enlarges. The **pericarp** becomes ribbed and lignified, while the integument becomes primarily a single, translucent membranous layer, and the **endosperm** forms two layers of thick-walled cells, except around the radicle tip, where it may be three or four cells thick.

4. Production

The two principal areas for lettuce seed production are the San Joaquin Valley of California, sown in early May and harvested in September, and New South Wales, Australia, sown in late November and harvested in May.

Reproductive growth begins with elongation of the growing point into a stalk, which expands at the top into the flower-bearing seed head. Upon reaching vegetative maturity, non-crisphead cultivars are then simply allowed to produce a seed stalk, but to prevent damage to the emerging seedstalk in crisphead lettuces, the top of the dense head is cut. Some lettuce fields are treated with **gibberellic acid** to encourage seedstalk formation before heading takes place.

Seed fields are frequently inspected, and off-type or diseased plants are removed. In particular, the presence of lettuce mosaic virus in plants at the flowering and seed maturity stages not only may reduce seed yield but the surviving seeds may carry the virus and serve as a source of primary infection when the seedlings emerge in a field planting. Lettuce seedlots must also be tested for seedborne mosaic virus, using samples of 30,000 seeds. (**See: Health Testing, 3. Bacteria and viruses**). If no virus is found, the seedlot is labelled as 'mosaic indexed'. If mosaic virus is found, the seed cannot be sold in areas where local ordinance demands seed that is essentially virus-free. Other possible seedborne pathogens are *Michrodochium panattoniana*, the cause of anthracnose, *Septoria lactucae* and *Pseudomonas cichorii*.

The seed is sized on an air-screen device and the portion with the lightest seeds is discarded. Germination percentage typically approaches 100%.

5. Germination

The optimum germination temperature is 18–21°C. There are three types of **dormancy** in lettuce; postharvest dormancy occurs in some types, which is relieved by **afterripening** in storage; and there are two forms of environmentally induced dormancy – **photodormancy** and **thermodormancy**. Red or white light promotes germination, and far-red light inhibits it – a response mediated by **phytochrome**, whose physiological attributes were first characterized in studies of lettuce seed germination. Prolonged, high-fluence rate light can inhibit germination, and cause photodormancy. (**See: Light – dormancy and germination**)

At 26°C or above, depending on variety, germination is subject to **thermoinhibition** and does not occur in darkness and eventually not even in light. Generally, crisphead cultivars can germinate at higher temperatures than butterhead, cos or leaf types. For this reason, wherever possible, growers avoid sowing at the hottest time of the day, and **transplant** raisers use cooler parts of the greenhouse.

Seed populations exposed to supraoptimal temperatures in the first few hours after imbibition exhibit thermodormancy when returned to previously favourable temperatures, causing a marked reduction in the speed and uniformity of germination. The upper temperature limit can be raised by a range of applied chemicals, such as **gibberellin**, **cytokinins** and thiourea, which interact with light. (**See: Germination – influences of hormones**) In order to complete germination, the **radicle** must penetrate the tough endosperm that encloses the embryo, and which is responsible for the expression of thermoinhibition. (**See: Germination – radicle emergence**)

6. Sowing and treatments

Lettuce seed is commonly **pelleted** to modify the narrow angular seed to a more rounded unit, which can be handled more easily by planting equipment. Direct planting in the field, in rows either on raised beds or on flat ground, is usually on a belt meter with holes punched to give the desired spacing, usually 5–6 cm (**see: Planting equipment – metering and delivery**). Surplus emerging seedlings can then be hoed out easily, leaving the remaining plants at the desired final spacing, which is usually about 25–30 cm. Alternatively, lettuce in many European latitudes is sown in peat blocks or plugs in nurseries, and seedlings are transplanted into the field at the required final spacing. However, in recent years, rather than forming heads, some lettuce varieties are sown in very dense plantings, sometimes with different types (and sometimes different species, including endive, radicchio and spinach) in alternating groups of rows or even intermixed, and harvested by direct mowing for ready-to-use loose leaf salad mixtures ('baby leaf' production).

Lettuce seed is often treated by **priming** both for faster emergence, to circumvent the requirement for light and to reduce the risk of induced thermodormancy.

Seeds and seedlings may be affected by several diseases and insect pests. Diseases include lettuce mosaic; and seedling **damping-off**, due to *Pythium* spp., and Botrytis, caused by *B. cinerea*, which both cause wilting at an early growth stage. Insect pests include flea beetles, crickets and cutworms, each of which can destroy young seedlings. In some countries seeds are treated with systemic insecticide (imidacloprid) directed against lettuce root aphid and foliar aphids. (EJR)

(See also: Cichorium)

For information on the particular contributions of the seed to seed science **see: Research seed species – contributions to seed science**.

Borthwick, H.A. and Robbins, W.W. (1928) Lettuce seed and its germination. *Hilgardia* 3, 275–305.

Davis, R.M., Subbarao, K.V., Raid, R.N. and Kurtz, E.A. (eds) (1997) *Compendium of Lettuce Diseases*. APS Press, St Paul, MN, USA.

Rubatzky, V.E. and Yamaguchi, M. (1997) *World Vegetables, Principles, Production, and Nutritive Values, 2nd edn. Chapman and Hall,* New York, USA.

Ryder, E.J. (1999) *Lettuce, Endive and Chicory*. CAB International, Wallingford, UK.
Smartt, J. and Simmonds, N.W. (1995) *Evolution of Crop Plants*, 2nd edn. Longman Scientific and Technical, Harlow, UK, pp. 53–56.

Leucine zipper protein

A dimeric protein in which two α-helical protein chains are linked together ('zipped') by hydrophobic interactions between their leucine residues. It is a characteristic of many **transcription** factors active in seeds, e.g. in abscisic acid signal transduction and in the regulation of expression of some **late embryogenesis abundant** (LEA) genes. (**See: ABI proteins; Signal transduction – hormones**)

LFR

See: Low Fluence Response

Light line

A region in the outer layers of the **seedcoat**. (**See:** ***Linea lucida***)

Light – dormancy and germination

Light influences seed germination in two ways – it can promote or inhibit. Seeds of many species are stimulated to germinate by light but there is some debate as to whether this represents a direct action on germination itself or rather the breaking of **dormancy**. This is a fine point which need not be pursued further here: in the present context it is sufficient to accept that some species of seeds are incapable of germinating in darkness, or do so only to a low percentage, while after receiving illumination their germination proceeds. Sensitivity to light – both promotion and inhibition – is extremely important ecologically as it provides a mechanism through which seedling establishment can be achieved in the most favourable circumstances.

1. Promotion of germination

(a) *The role of phytochrome*. Some seeds whose germination is promoted by light are shown in Table L.10. Whether or not seeds have an absolute requirement for light depends on several factors such as the temperature, the depth of dormancy, the presence of certain chemicals (e.g. nitrate), the integrity of the seedcoat, the provenance and the experiences during development and maturation. The 'classical' light-requiring seed, the Grand Rapids cultivar of lettuce ('classical' because it was the subject of much basic research on seed photosensitivity) illustrates some of these points. The light-requirement is evident only at higher temperatures (e.g. 18–25°C), in fully dormant seeds (i.e. not **afterripened**) and when the **seedcoat** and **endosperm** are intact.

The photoreceptor responsible for most types of light-sensitivity is phytochrome. The photoreceptor exists in two photoreversible forms, the red-absorbing Pr (inactive form) and the far-red-absorbing Pfr (active form) (**see: Phytochrome**). It is synthesized in darkness as Pr while Pfr is the form that promotes germination. The action spectrum for germination shows peak promotive activity (generating Pfr) at 660 nm (orange-red light) and peak inhibitory action (reversion of Pfr) at 730 nm (far red – barely visible to the human eye) but promotive photoconversions can occur over a wide spectral range (Figs L.5, 6). A photoequilibrium mixture of Pr and Pfr is established at different wavelengths of light depending on the energy absorption properties of phytochrome at the particular wavelength (Fig. L.6) and also in mixed wavelengths such as sunlight. In the latter about 55% of phytochrome is in the form Pfr. A certain proportion of Pfr is required for the promotion of germination: the actual value depends upon the species but in lettuce it is about 0.5, in tomato 0.2–0.4 and in cucumber about 0.15.

Pfr slowly reverts to Pr over several hours of darkness after the light is removed. Hence, after a short 'pulse' of light there is a fall in the amount of Pfr available to promote germination. This explains why in some species repeated exposure to light is necessary. The Pfr form of one type of phytochrome, phyA (see below), is extremely labile and is fairly rapidly degraded. This property clearly influences the mode of operation of phyA in germination.

Photoconversions of phytochrome from Pr to Pfr can take place only when the seed is hydrated, in lettuce to above about 18% water content: and in the laboratory or the field, seeds drier than this are less photostimulated. On the other hand, seeds that are dried to below 8% water content immediately after Pfr formation retain the Pfr which can promote germination even in darkness when the seed is later hydrated. The conversion between Pr and Pfr takes place in several steps, only the initial ones of which are driven by light. The subsequent stages, the phytochrome intermediates, do not require light to go to completion, but they do require that they are hydrated. Thus, as a developing seed undergoes **maturation drying** in the light, these intermediates increasingly become trapped until, when a certain low water content is reached, complete conversion, say from Pr to Pfr,

Table L.10. Some species that produce photosensitive seeds.

Light promoted	White-light inhibited	Non-light-requiring when mature but containing Pfr
Amaranthus retroflexus	*Amaranthus caudatus*	*Cucumis sativus*
Begonia evansiana	*Brassica juncea* (under water stress)	*Lactuca sativa* cv. Great Lakes
Betula pubescens	*Calligonum comosum*	*Lycopersicon esculentum*
Chenopodium album	*Cucumis sativus* (under water stress)	*Raphanus sativus*
Kalanchoë blossfeldiana	*Lamium amplexicaule*	
Lactuca sativa cv. Grand Rapids	*Nemophila insignis*	
Nicotiana tabacum	*Phacelia tanacetifolia*	
Pinus sylvestris	*Raphanus sativus* (under water stress)	

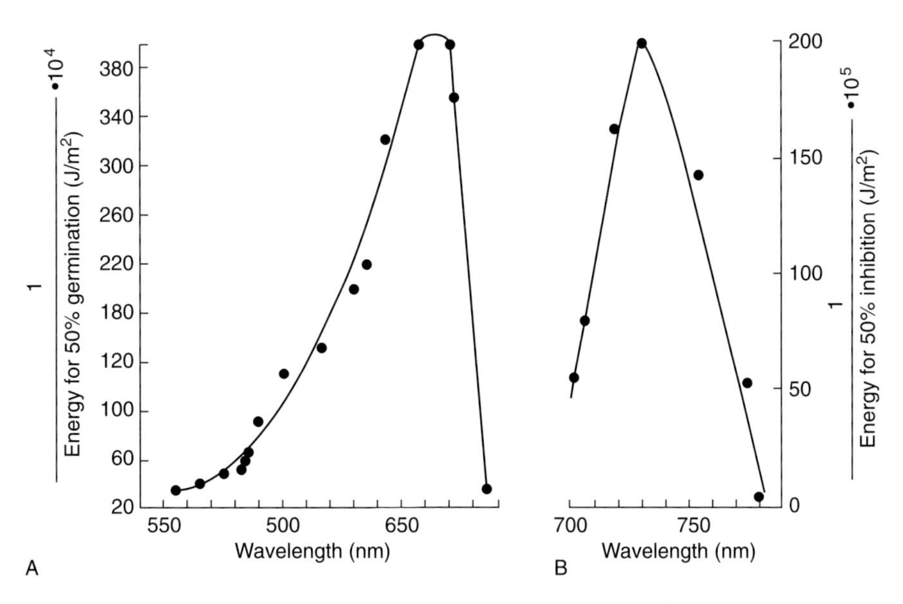

Fig. L.5. Spectral effectiveness (action spectrum) (A) for the promotion of germination, and (B) for the reversal of light promotion. These were determined by measuring the energy required at different wavelengths to give 50% of the maximum effect. The lower the required energy the more effective is the wavelength. To show the curves in a more accessible shape the reciprocals of the energy requirements were used for the plot. From Bewley, J.D. and Black, M. (1982) *Physiology and Biochemistry of Seeds in Relation to Germination. 2. Viability, Dormancy and Environmental Control.* Springer, Berlin, Germany.

can no longer occur. The completion of the conversion can take place, however, when the matured seeds are later rehydrated (see below).

As mentioned above, to achieve the promotion of germination it is necessary that the proportion of phytochrome existing as Pfr is above a certain value. This always occurs in white light (e.g. sunlight) but not necessarily if the light has first been filtered through green leaves, because the chlorophyll absorbs the red wavelengths: such light is rich in far red. Leaf-transmitted light can establish a Pfr:total phytochrome ratio as low as 0.05, inadequate to satisfy the requirements for germination of many species. Hence, many types of light-requiring seeds will not germinate in leaf shade, under conditions where the seedling would be deprived of light suitable for photosynthesis, i.e. rich in the red wavelengths. This is a device that encourages germination and seedling establishment in less competitive situations, e.g. in a forest light gap.

Small seeds do not have sufficient food storage reserves to support much growth (seedling stem elongation) in darkness before photosynthetic self-sufficiency is accomplished. Hence, such seeds are at a disadvantage if they germinate at too great a soil depth. The vast majority of light-requiring seeds are small: the light requirement ensures that germination occurs near the soil surface at a depth that does not demand excessive stem elongation before emergence into the light.

Seeds of many species that can germinate in the dark are, nevertheless, in the strict sense Pfr-requiring (Table L.10). But in these cases the generation of the Pfr for germination has taken place during seed development and the mature seed relies on this legacy from maturation. Germination of such seeds can be prevented experimentally by a dose of far-red light which forces the promotive Pfr back to the inactive Pr. And in some

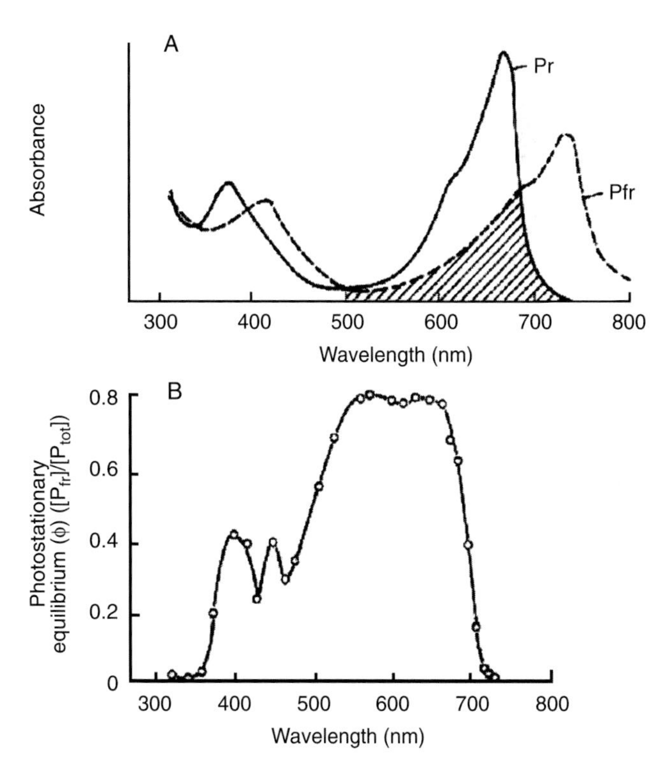

Fig. L.6. Photoconversion of phytochrome. In (A) the wavelength absorbances (absorption spectra) of Pr and Pfr are shown. The hatched zone is where there is major overlap in absorbance of the two forms in the green–yellow–red region of the spectrum. In (B) the photoequilibria (i.e. Pfr/total phytochrome) established at different wavelengths are shown. From Bewley, J.D. and Black, M. (1982) *Physiology and Biochemistry of Seeds in Relation to Germination. 2. Viability, Dormancy and Environmental Control.* Springer, Berlin, Germany.

cases repeated 'pulses' of far red are needed when phytochrome intermediates, trapped by the drying of the seed, gradually succeed in completing their transformation to Pfr as the mature seed hydrates. It is likely that the far red transmitted through a leaf canopy is also effective in photoreversing the Pfr. The generation of sufficient Pfr or of intermediates en route to Pfr can only occur if active light (usually red) reaches the developing seed. This is not the case when the seed matures while still covered by green tissues (the chlorophyll absorbs the red light) and indeed many species of such seeds mature with inadequate Pfr and their light requirement must be satisfied later, i.e. after dispersal from the mother plant.

(b) *Phytochrome types*. Phytochrome is a family of chromoproteins where the protein part (phytochromes A–E, and possibly F in tomato) of the molecules are encoded by different genes. PhyA is a form that appears after several hours of seed hydration (e.g. in *Arabidopsis*) and is responsible for sensitivity to very low fluence rates which generate a relatively low proportion of phyA Pfr. Indeed, the low amount of phyA Pfr that can stimulate germination is also produced by far-red light (the absorption spectrum in Fig. L.6 shows that some Pr $\rightarrow$ Pfr conversion can occur even in far-red light). Germination of some species is therefore permitted even in canopy light under the influence of phyA operating in the very-low-fluence-rate mode (see section 4). PhyB is produced during seed development: any phyB Pfr formed as a consequence would be responsible for the later germination of the mature seed in darkness. But if this amount of Pfr is insufficient to support germination a further light-driven input is needed. PhyB photoconversions to generate sufficient active Pfr are brought about by higher fluence rates than are active on phyA but are nevertheless in the **low fluence response** range. There is also evidence in *Arabidopsis* that phyE functions similarly.

2. How does light-generated Pfr promote germination?
Pfr enhances embryo growth potential and the weakening of the tissues enclosing the radicle tip (**see: Germination – radicle emergence**). One mechanism by which these are achieved is through the promotion of active **gibberellin** (GA_1) formation in the seed: and there is also evidence that sensitivity to GA is provoked by Pfr. It is therefore understandable why application of GA promotes germination of light-requiring seeds in darkness. Lettuce and *Arabidopsis* seeds show a substantial light-promoted increase in the active GA_1 through an elevation in the mRNA transcripts for the enzyme GA 3-oxidase (formerly hydroxylase) that converts GA_{20} to GA_1. Pfr therefore acts on the expression of the GA 3-oxidase gene (**see: Signal transduction – light and temperature**). There is also evidence that GA_1 degradation by GA 2-oxidase is prevented by Pfr. PhyB is active in these effects but it is not known if phyA exercises a similar role.

Studies to locate the GA 3-oxidase mRNA transcripts have shown them to be in the cortical cells of the **embryonic axis** of *Arabidopsis*, so this is presumably where the production of the active GA_1 occurs. In lettuce seeds, microbeam irradiation indicates that the perception of light lies in the **hypocotyl**. It would seem, therefore, that the action of light to promote active GA production resides in the hypocotyl cortical cells of the embryo.

The GA_1 production provoked by Pfr has two actions – on embryo growth potential and on weakening of the tissues enclosing the radicle tip. Endo-β-mannanases in tomato and *Datura ferox*, β-mannosidase in the latter and β 1,3-glucanase in tobacco, enzymes that participate in endosperm weakening, all increase in response to Pfr: genes for these enzymes are all regulated by GA.

It is not clear, however, that phytochrome acts in all cases through stimulation of GA biosynthesis. The double mutants of *Arabidopsis*, for testa character and the ability to produce GA, nevertheless require light (and chilling) for dormancy breaking, which may indicate that here light might have an effect other than to promote GA biosynthesis.

Finally, it should be mentioned that there is some evidence, for example in lettuce seeds, that **abscisic acid** (ABA) content is reduced by Pfr. It is to be expected that this would increase the embryo growth potential and possibly also the sensitivity to GA of the weakening-enzyme genes.

3. Inhibition by light
(a) *Far-red light*. Seeds of many species that normally germinate in darkness are inhibited by exposure to prolonged (i.e. several hours) irradiation with far-red light given either continuously or as a series of pulses: examples are tomato, *Amaranthus caudatus* and some cultivars of lettuce. What is happening here is that Pfr is slowly appearing in darkness (i.e. by non-photochemical conversions) from those phytochrome intermediates that were trapped en route to Pfr when the seed underwent maturation drying on the parent plant. The far red drives the Pfr back to the non-promotive Pr.

Prolonged far red also inhibits via the **high-irradiance reaction (HIR)**. Here, continuous far red over a period of several hours arrests germination of both dark-germinating seeds and seeds which have been potentiated to germinate by Pfr. The effect is not due to simple reversion of Pfr but possibly to Pr $\leftrightarrow$ Pfr cycling: the most effective wavelength to drive this is about 710 nm. It is not clear what type of phytochrome is responsible for this HIR in seeds but in seedlings phyA is implicated.

(b) *White light*. Prolonged exposure to relatively high fluence-rate white light (sunlight or lamp light) also inhibits germination of many species (Table L.10). Illumination for many hours is generally required, extending over the period when germination would be completed, i.e. when the radicle is about to emerge. The far-red HIR is unlikely to be responsible but instead the effect can be attributed to the blue region of the spectrum operating through **cryptochrome**. This photoinhibition is magnified when seeds are under water stress, for example at very negative water potentials, and indeed in some cases (e.g. cucumber, radish, mustard) it is manifest only under such conditions. There is evidence from studies of tomato seeds that ABA biosynthesis is promoted at low water potentials and that blue light increases the sensitivity to the ABA.

(c) *Photodormancy*. In some cases (e.g. *Nemophila insignis*, *Phacelia tanacetifolia*) seeds whose germination has been inhibited by light are forced into a **secondary dormancy** called **photodormancy** (unfortunately, this term is sometimes used incorrectly to describe the condition in seeds whose primary dormancy is removed by light). Photodormant seeds can be made to germinate by several **dormancy-**

breaking treatments such as low temperature or seedcoat abrasion or removal.

4. Light and seed ecology

Several points relating to the significance of seed photosensitivity in their ecology have been made above and they are summarized here.

(a) *Burial.* Germination at too great a depth (a danger for small seeds) is prevented if seeds have a light-requirement for germination. The phytochromes involved are phyB and in some cases phyA. This effect is responsible for the flush of weed seedlings that appears after tillage (**see: Dormancy – importance in agriculture**).

(b) *Avoiding competition.* Germination under vegetation canopies is disadvantageous as the resulting seedlings face extreme competition for light for photosynthesis. Germination in such a situation is discouraged because the far-red-rich light transmitted through green leaves does not establish adequate Pfr ratios in the seeds and also because of the far-red HIR. Gaps in the canopy resulting, for example, from the death of overlying plants, permit the entry of light richer in the red wavelengths so provoking a higher seed Pfr ratio. It has been reported, however, that phyA, activated by very low fluence rates, permits germination of *Arabidopsis* even in canopy light.

(c) *Avoiding water stress.* Successful seedling establishment cannot occur unless an adequate supply of water for growth is available. This is not assured in arid habitats, for example. Photoinhibition of germination by intense sunlight (the blue wavelengths) which is magnified by threateningly low water potentials would prevent seedling emergence under potentially drying conditions. Seeds of several desert species use this mechanism to secure germination in the protective shade or in cracks in the soil substrate (**see: Deserts and arid lands**).

(MB)

Bewley, J.D. and Black, M. (1994) *Seeds. Physiology of Development and Germination*, 2nd edn. Plenum Press, New York, USA.

Casal, J.J. and Sanchez, R.A. (1998) Phytochromes and seed germination. *Seed Science Research* 8, 317–330.

Fellner, M. and Sawhney, V.K. (2002) The *7B-1* mutant in tomato shows blue-light-specific resistance to osmotic stress and abscisic acid. *Planta* 214, 675–682.

Frankland, B. and Taylorson, R. (1983) Light control of seed germination. In: Shropshire, W. and Mohr, H. (eds) *Photomorphogenesis*. Vol. 16A. *Encyclopedia of Plant Physiology*. Springer, Berlin, pp. 428–456.

Franklin, K.A. and Whitelam, G.C. (2004) Light signals, phytochromes and cross-talk with other environmental cues. *Journal of Experimental Botany* 55, 271–276.

Toyomasu, T., Kawaide, H., Mitsuhashi, W., Inoue, Y. and Kamiya, Y. (1998) Phytochrome regulates gibberellin biosynthesis during germination of photoblastic lettuce seeds. *Plant Physiology* 118, 1517–1523.

Yamaguchi, S. and Kamiya, Y. (2002) Gibberellins and light-stimulated seed germination. *Journal of Plant Growth Regulation* 20, 369–376.

Lignin

A rigid component of secondary (lignified) cell walls, characteristic of vascular plants, composed of a very complex network of aromatic compounds (mostly hydroxycinnamoyl, *p*-coumaryl, coniferyl and sinapyl alcohols) called phenylpropanoids, cross-linked to cell wall polysaccharides. Present in some seedcoats. (**See: Seedcoats – structure**)

Lima bean

1. World importance and distribution

Lima bean (= butter bean) (*Phaseolus lunatus*), a **warm-season legume**, is cultivated in warm temperate, subtropical and tropical environments throughout the world. The beans are an important food particularly in Central and South America. (**See: Legumes**)

2. Origin

Two centres of diversity have been postulated: Mesoamerica (southern Mexico and Central America) for the smaller, 'sieva' type and South America (Peru, Ecuador, Bolivia) for the larger bean.

3. Seed types and provenance

Seeds are easily distinguished from other legumes by their half-moon shape, striated from the hilum. The exception is a group of cultivars from the Caribbean with spherical seeds. There are two main genetic stocks domesticated from two separate wild forms:

- Large-seed limas: flat, half-moon shape, 54 to 280 g per 100 seeds;
- Small-seed limas ('sievas'): flat, round, 24 to 70 g per 100 seeds.

Seeds of commercial cultivars are almost white (Fig. L.7).

The large-seeded cultivars are grown almost exclusively on the Peruvian coastal region. This material appeared 5000 years ago on the coast of Peru where they were of great nutritional and cultural value, particularly for the people of the *Mochica* and *Nazca* cultures. The small-seeded type, sometimes identified as baby limas or sieva, are the most common, dating back in Mesoamerica (Mexico) only 1200 years.

4. Nutritional quality

Storage protein, **starch** and soluble carbohydrate, and oil content of lima beans are approximately (fresh weight basis) 25%, 60% and 1.5%, respectively. The seeds also contain vitamins A and C and dietary fibre.

Wild *P. lunatus* seeds have a relatively high concentration of cyanogenic glucosides which yield toxic hydrocyanic acid when the seeds are bruised or chewed (**see: Pharmaceuticals and pharmacologically active compounds**). Modern domesticated varieties, particularly those with white seeds, have minimal quantities and are not dangerous. Cooking in boiling water also destroys the cyanogens.

5. Uses

Immature seed are used like peas in soups and stews. The dried mature seed must be thoroughly cooked before being eaten. The sprouted seeds are cooked and used in Chinese dishes. The dried seed can be ground into powder then used as thickener in soups. Young pods are steamed and used as a side dish with rice, or added to soups and stews.

A

B

Fig. L.7. Lima beans. (A) Smaller, sieva type; (B) larger type (from http://waltonfeed.com/self/ beans.html).

6. Types of plants

Two growth habits are recognized, the **indeterminate** (vine) type, with a broader, ovate leaf pattern and the **determinate** (bush) type, with lanceolate leaflets. Lima beans are perennial, but are usually grown as annuals, even in the tropics. (OV-V)

Baudoin, J.P. (1993) Lima bean, *Phaseolus lunatus* L. In: Kalloo, G. and Bergh, B.O. (eds) *Genetic Improvement of Vegetable Crops*. Pergamon Press, Oxford, UK, pp. 391–403.

Office of International Affairs (1989) *Lost Crops of the Incas: Little-known Plants of the Andes with Promise for Worldwide Cultivation. Part III Legumes*, Washington DC, USA, pp. 162–163.

Limit dextrin

A glucan polymer that results from the hydrolysis of **starch** or **amylopectin** by β-amylase or α-amylase, producing a β-limit dextrin or an α-limit dextrin, respectively. β-Amylase acts by cleaving sequential **maltose** units from the **non-reducing** ends of linear α-D-glucan chains, and α-amylase randomly cleaves internal α-(1→4) linkages. Neither enzyme hydrolyses α-(1→6) branch linkages in the polymer, so digestion stops in regions adjacent to branches. The molecules that remain are termed 'limit dextrins', because the activity of the enzyme has reached its limit. Starch often is converted to limit dextrins to achieve different paste viscosities for industrial use. (**See: Starch – mobilization**) (MGJ)

Line

A group of plants from a common ancestry: more narrowly defined than a strain or **cultivar** or land**race**, and integral to modern plant breeding systems.

(a) *Pure-lines (inbred lines)* are plants that 'breed true' or produce sexual offspring that closely resemble their parents. A true-breeding line consists of a single genotype in which all loci are homozygous and whose variability is thereby effectively eliminated. Pure lines and cultivars are traditionally developed from the progeny of a single plant, usually originating by **self-pollination** or other inbreeding techniques such as **sib**-pollination or back**crossing**, and subsequent selection of desired traits such as visual characteristics, yield and quality. (**See: Multiline**)

(b) *Female and male parental lines*. Hybrid seed production is based on **hybridization** or **crossing** of pure **parent lines** that have each been produced through inbreeding (**see: Production for sowing III. Hybrids**). Some systems rely on incorporating genetic infertility mechanisms into the parental lines to ensure crossing; in these cases special temporary fertility-restoration techniques are required to propagate the lines themselves.

- The widely used system of **cytoplasmic male sterility (CMS)** is based on preventing the production of viable pollen by the female parent; the CMS system involves three lines: a Male sterile (A) line, Maintainer (B) line and Restorer (R) line.
- In horticultural brassicas, especially cabbage and cauliflower, almost all commercial F_1 hybrids are produced by utilizing the **self-incompatibility system** (SI), which is based on interference in the pollination process on the stigma; in this system, the female parent lines are usually propagated by **bud-pollination** (or -selfing). (PH)

Linea fissura (linea sutura)

A specialized valve-like structure on seeds. (**See: Pleurogram**)

Linea lucida

Light line. Malpighian cells in the outer seedcoat usually possess a zone that differs in its light refraction and therefore appears brighter than the rest. (**See: Seedcoats – structure**)

Linseed

1. The crop

Linseed or flax (*Linum usitatissimum*, 2n = 20, Linaceae) is an annual plant but in some regions can also be grown as a winter annual. It is cultivated for the fibre in the stem or the oil in the seed or both. The hectarage sown to the tall, low seed and oil-yielding fibre flax has been declining in the major producing countries of Russia, Ukraine, Belarus and Lithuania where the linseed area is almost equally divided between fibre and oilseed varieties. Some stable production of linseed fibre remains in Belgium, France and the UK. Major producers of the shorter, high seed and oil yielding varieties are Canada, India, China and the USA (**see: Oilseeds – major**, Table O.1). In all, about 16 countries produce the crop.

2. Origin

Linseed is one of the oldest of cultivated plants with production in the Fertile Crescent dating back to 7000 BC. Remains left by Stone Age people included flax fibre and linen. The origin and centre of diversity is thought to be the Indian subcontinent from where the crop spread north to Asia Minor, the Mediterranean and east to China. Of the more than 20 *Linum* species, linseed is the only species with non- or semi-dehiscent capsules suitable for commercial farming.

3. Fruit and seeds

The **ovary** is **syncarpous** with five locules that form an almost round 6–9 mm capsule-like fruit commonly called a boll. Each carpel forms two septa creating ten lodicules. Thus, ten is the maximum number of seeds per boll but seven seeds per boll is more common. Seeds are flattened, oval, rounded at the base, pointed at the tip and 3 to 6 mm long, weighing about 5.5 g/1000 seeds. The surface is smooth and generally shiny. Seed colour varies from pure yellow to dark brown and olive (Fig. L.8); variegated seeds may also occur. The **testa** or **seedcoat** consists of five distinct layers all originating from the **integuments**. The outer epidermal cells are filled with **mucilage**. Upon hydration the mucilage expands very rapidly lifting the cuticle and unfolding the walls of the mucilage cells until their outer surfaces break. Below the epidermis are several layers of parenchymatous tissue, which, because of their circular appearance in surface view, are called ring or round cells. The round cells may protrude into the several rows of underlying fibre cells that run parallel to the length of the seed. Below the fibres are several rows of thin-walled elongated cells called 'cross cells' that are oriented at right angles to the length of the fibre cells.

The fifth structural layer is made up of pigment-filled cells. These square- or polygonal-shaped cells originate from the innermost layer of the inner integument. The tannin pigments they contain determine the seed colour. Yellow seeds contain very little or no pigment. Attached to the pigment cells are the remnants of the parenchymatous endosperm. The **endosperm** may vary in thickness from three to many layers along the seed axis. These cells contain **oil bodies, protein bodies** with **crystalloids** and **globoids**. The endosperm surrounds the two large cotyledons of the **embryo** that occupy more than two-thirds of the inner seed volume, the remainder being endosperm tissue. The cotyledons are white or yellowish in colour and contain many oil and protein bodies. Together the embryo and endosperm contribute to the 35–44% oil content and 23–28% protein of the seed.

4. Uses

The extracted oil, because of its high content of linolenic acid (**see: Oilseeds – major,** Table O.2), is primarily used as an industrial drying oil in paints, floor coverings, printing inks, as a concrete preservative, in brake linings and environmentally friendly adhesives.

Although linseed oil was widely used by ancient civilizations as a food and medicine, its tendency to rapidly oxidize and polymerize, causing off flavours, has limited its edible use in the modern world. However, the discovery of the desirable nutritional properties of α-linolenic acid (ALA) has stimulated interest in cold pressed, specially processed, linseed oil as a functional food.

Plant breeders have also bred linseed varieties that produce oil with less than 3% linolenic acid (called 'solin' in Canada) that are now grown and processed for the edible oil market.

Linseed has been used as a laxative because it contains both insoluble and soluble (mostly mucilage) fibre. The seed- and oil-free meal are also fed to pets and livestock both as a protein source and to produce a shiny coat for show purposes. Linseed meal is considered a good source of thiamin, riboflavin, nicotinamide, pantothenic acid and choline in ruminant diets. However, due to its high mucilage content and low levels of the **essential amino acids** lysine and tryptophan it is little used in monogastric rations. (KD)

(See: Fatty acids; Oils and fats)

Lay, C.L. and Dybing, C.D. (1989) Linseed. In: Robbelen, G., Downey, R.K. and Ashri, A. (eds) *Oil Crops of the World.* McGraw-Hill, New York, USA, pp. 416–430.

Muir, A.D. and Westcott, N.D. (2003) *Flax, the Genus* Linum. Taylor and Francis, London, UK.

Vaughan, J.G. (1970) *The Structure and Utilization of Oil Seeds.* Chapman and Hall, London, UK.

Fig. L.8. Linseed (scale = mm) (photograph kindly produced by Ralph Underwood of AAFC, Saskatoon Research Centre, Canada).

Lint and linters

Lint is the long **cotton fibres** separated from the seed during **ginning**, that are baled and shipped to textile mills to produce yarn or cloth. The fuzzy shorter cotton fibres, known as 'linters', left on the seed after ginning are removed by the **delinting** process. The higher grade, more resilient first-cut linters are used, for example, in manufacturing twine, and the second-cut short fibres, or fuzz, are incorporated in products used in various foods, toiletries and paper. Cottonseed that is delinted in one step instead of two produces 'mill run' linters. Cottonseed from Pima varieties of extra-long staple cotton is naturally devoid of linters. (PH)

(See: Cotton – cultivation; Fuzzy seed)

Lipids

A structurally diverse group of molecules that have in common a preferential solubility in non-aqueous solvents such as chloroform. They include many fatty-acid-containing compounds, such as **oils and fats,** as well as many compounds that are metabolically unrelated to fatty acid metabolism, e.g. many pigments.

Lipids that are present in cell membranes are polar lipids, e.g. **phospholipids** and glycolipids, and with the exception of the **oil body** they are arranged into bilayers. Storage lipids, as oils and fats, are **triacylglycerols**, and are classified as neutral lipids. Polar lipids are soluble in polar solvents such as methanol and ethanol, whereas neutral lipids are only soluble in non-polar (non-aqueous) solvents such as benzene, ether and chloroform. (JDB)

Locule

A cavity within the **ovary** or **fruit** in which the **ovules** are borne, and in which the seed develops, e.g. in the tomato fruit.

Locus

In genetics, the position occupied by a region of DNA on a chromosome that has a specific influence on the phenotype of an organism. Different **alleles** of the locus can have different effects on phenotype.

Lodging

The usually undesirable bending, breaking or beating to the ground of a plant before harvest so that it cannot stand upright again, especially in cereals damaged by wind and rain, affecting the efficiency of seed recovery and, possibly, seed quality. (**See: Production for sowing, IV. Harvesting**)

Long-day plant

Plants that initiate flowers best under long-day (short-night) regimes. Even though their flowering is initiated by long days, they can produce flowers under subsequent short days.

Longevity

Refers to the life span of a seed, or population of seeds, usually in air-dry storage but sometimes also in the context of seed survival under natural conditions in, for example, **soil seed banks**. Maximum potential seed longevity is acquired during the post-abscission phase of seed development usually reaching a maximum at the time of natural seed **dispersal**. After dispersal, or seed **collection**, seed longevity is mainly affected by moisture and temperature (**see: Drying of seed**). The longevity of desiccation-intolerant (**recalcitrant**) seeds is limited to a few years at best unless **cryostorage** techniques are used. Partially dried **intermediate** seeds could probably survive for up to a decade or so at cool temperatures (e.g. 15°C), but the longevity of desiccation-tolerant **orthodox** seeds might extend to hundreds of years in some cases under seed **gene bank** conditions. However, even under optimum conditions for long-term storage, potential longevity varies from species to species indicating the ultimate genetic control of this trait.

The longevity of orthodox seeds can be related to seed moisture status and the process of seed drying by reference to the three main water-binding regions delineated by **sorption isotherms**. Moist seeds that contain bulk water (region III) are reasonably long lived provided that they are aerated so that metabolic activity and repair mechanisms can function. At the boundary between region III and II (85–90% **equilibrium relative humidity** (eRH)), seeds are shortest lived because respiration is greatly reduced and damage accumulates rapidly. However, as seeds are dried through region II, seed longevity approximately doubles for every 10% reduction in eRH reaching a maximum at the boundary of region I. Removal of **bound water** (region I) does not lead to a further increase in longevity and may actually reduce it.

The longevity of orthodox seeds in storage can be predicted using the so-called **viability equations** after the determination of viability constants by controlled ageing experiments. Predictions suggest that seed longevity varies from a few decades at best for the shortest lived species to over 1000 years for the longest lived. Soluble sugars and **antioxidants** have been implicated in protecting seeds from the effects of ageing (**see: Deterioration and longevity; Deterioration in storage**). (RJP, SHL)

Lot (seedlot)

A uniform batch of certified seed: a homogeneous quantity of seed of one cultivar derived from known material and treated uniformly throughout its production harvesting and processing. A 'blended lot' is comprised of two or more seedlots of the same cultivar. (**See: Production for sowing, IV. Harvesting**)

For the purposes of **sampling**, a seedlot or grain bulk is a specified quantity that can be physically identified. For agricultural seed for sowing, maximum seedlot sizes depend on seed size, according to **ISTA Rules** and the **OECD** Certification Programme (see Table L.11). Equivalently, national grain inspection rules set equivalent maximum bulk sizes of harvested grain that can be represented by a single submitted sample.

Table L.11. Maximum seedlot sizes permitted under the OECD Certification Programme.

Crops	Maximum lot size (kg)
Seed sizes smaller than wheat	10,000
Seed sizes equal to wheat or greater	20,000
Exemptions:	
Soybeans, Lupins, Peas, Beans, *Vicia*	25,000
Maize	40,000

Low fluence response

A response to red light mediated by light-stable **phytochrome** (predominantly phytochrome B). LFRs include promotion of germination, de-etiolation and shade avoidance. The LFR exhibits reciprocity and can be fully reversed if red light is immediately followed by subsequent irradiation with far-red light. (**See: Light – dormancy and germination**)

Lucerne

See: Alfalfa

Lupin

1. World importance and distribution

Seeds of several species of *Lupinus* have long been used as food, such as *L. albus* (white lupin), *L. luteus* (yellow lupin) and *L. angustifolius* (blue or narrow-leaved lupin) in the Mediterranean region. The recent development of these as crops originated from alkaloid-free or sweet types in Germany in the 1920s. Cultivation over substantial areas only started within the last 50 years. Eastern European production has been extensive, but latterly the crop has been developed in Western Australia where it is a major source of animal feed. The crop can be grown either for the grain or the foliage, which is ensiled for winter feeding. In recent years, eastern European selected varieties have been developed for western European production and some crop is also now grown in the western USA. However, yields of *luteus* are usually lower than those of the other two species. *L. mutabilis* (Andean lupin) is an edible species widely grown in temperate cold areas (2000 to 3650 m above sea level) in the Andean zone of South America which includes Venezuela, Colombia, Ecuador, Peru, Bolivia, Chile and northern Argentina, where it is known as *chocho*, *lupino* or *tarvi*. Though more significant as human food than other lupins it is nevertheless a species of relatively minor importance among the grain legumes.

Lupins are important as forage plants: for this purpose, world production in 2002 was about 507,000 t (FAOSTAT). **(See: Crop Atlas Appendix, Map 3; Legumes; Lupin – cultivation)**

2. Origin

There is evidence of the domestication of *L. albus*, *L. luteus* and *L. angustifolius* in the Mediterranean region about 3000 years ago. The Andean lupin was domesticated by the ancient inhabitants of the central Andean region from pre-Incan times, as indicated by seeds found in tombs of the *Nazca* culture and the plant's representation on *Tihuanaco* pottery. Wild annual or perennial plants from the same genus are often called lupines.

3. Seed characteristics

The seeds of *Lupinus* spp. vary in colour, size and shape. The *albus* types are usually larger, with a grey to cream coloured **seedcoat** and a seed size of around 400 g per 1000 seeds. *Angustifolius* is usually smaller, light green to brown, some varieties having dark brown flecked coloration of the testa. *Luteus* is usually a similar seed size with light brown to cream coloured seed. The seeds are usually slightly flattened, oval in shape and comprise two cotyledons containing the embryo axis and covered in a testa. Seeds of the Andean lupin are lenticular, 8–10 mm long and 6–8 mm wide. The weight of 1000 seeds is 200–250 g. The colour varies – black and white through bay, dark or light grey and greenish yellow.

4. Nutritive value

L. angustifolius and *L. albus* typically contain 30–35% fresh weight (fw) storage protein, *L. luteus* up to 40%, while seeds of the Andean lupin are somewhat higher at 42% fw. **Oil** (triacylglycerols) reaches approximately 16% fw. **Starch** is also present but the content of flatulogenic **oligosaccharide** carbohydrates is low. Several antinutritionals occur in the seeds, the most important and dangerous of which are alkaloids (**see: Pharmaceuticals and pharmacologically active compounds**) which impart a bitter taste. Seeds are soaked in running water, often for several days, to remove these factors.

5. Uses

Seeds of the Andean lupin are used for human consumption in stews, soups and various desserts, after the bitter taste has been removed. They are an industrial source of a flour which is used as a 15% addition in breadmaking. This mixture has the advantage of considerably improving the protein and energy value of the product and also allows the bread to be kept longer. The flour is also added to papaya juice to make a soft drink. The seeds' alkaloids (sparteine, lupinine, lupanidine, etc.) find use to control ectoparasites and intestinal parasites of animals; and the cooking water may be used by farmers as a laxative and to control pests and diseases. The plant is used as green manure.

In the Mediterranean region lupin seeds no longer have a culinary use but they may still be found, soaked in brine and eaten as snack food.

The main use for lupin worldwide is in livestock rations as a rich source of protein. (OV-V, AB)

Caligari, P.D.S., Romer, P., Rahim, M.A., Huyghe, C., Neves-Martins, J. and Sawicka-Sienkiewcz, E. (2000) The potential of *Lupinus mutabilis* as a crop. In: Knight, R. (ed.) *Linking Research and Marketing Opportunities for Pulses in the 21st Century*. Kluwer Academic, Dordrecht, The Netherlands, pp. 569–574.

Gladstones, J.S., Atkins, C. and Hamblin, J. (eds) (1998) *Lupins as Crop Plant: Biology, Production and Utilization*. CAB International, Wallingford, UK.

Lupin – cultivation

See: Lupin for other information on economic importance, origins, seeds and uses. Commercially grown lupins are confined to three species: *Lupinus albus*, *L. luteus* and *L. angustifolius*. A long-established species in Andean South America is the Andean lupin, *Lupinus mutabilis*, mostly of local importance.

1. Breeding

A large breeding programme is being carried out in Australia but at present most of the varieties grown in Europe are from eastern European breeding. Almost all varieties in use are of the sweet type, i.e. low alkaloid, whose presence reduces the nutritive value to livestock. Evaluations regarding suitability of the different types for grain or forage crops in Europe are currently being assessed.

Genetic modification has been carried out by the introduction of sunflower seed albumin, to improve the sulphur content of the seed protein for animal feed, but the product is not in agricultural use at present.

2. Production

L. albus tends to be later maturing, although recent selections have brought the harvest date for European production to within the normal harvest schedule for arable cropping. At present, spring-sown varieties are produced in northern Europe with some autumn-sown types in France. Lupins may be left in the field once mature to allow natural drying to take place before harvest without the risk of seed decay or pod **shatter**. Yield variability is experienced between species and variety, but often this can be related to environmental effects.

3. Sowing and treatments

Seed is sown in medium to light textured soils at a depth of around 4 cm. Germination takes place rapidly when soil temperatures are above 10°C. Seedling establishment is **epigeal**, the first trifoliate leaves emerging from between two fleshy cotyledons shortly after emergence. Lupins prefer soils which are acid to neutral although some *L. albus* varieties will tolerate more alkaline soils.

Seed treatments containing thiabendazole or iprodione, to control seedborne fungi such as *Colletotrichum acutatum*, can be used; and in some seasons insecticidal protection against seed fly larvae (*Delia platura*) may be necessary, especially in autumn-planted varieties in Europe. A large proportion of seed is treated by **rhizobial inoculation** with strains of the specific symbiont, *Bradyrhizobium* sp. (*Lupinus*), also known as *Rhizobium lupini*. (AB)

Gladstones, J.S., Atkins, C. and Hamblin, J. (eds) (1998) *Lupins as Crop Plant: Biology, Production and Utilization.* CAB International, Wallingford, UK.

Lygus

An insect **pest** affecting seed development by larval feeding, leading to the development of **embryoless seed** in umbellifers, such as carrot, and the formation of misshapen or shrivelled or low-vigour seeds in lucerne (**alfalfa**) or **sugar-beet**.

M

Macadamia

Macadamia has the distinction of being the only native Australian plant from which an important crop has been developed. There are two species of macadamia used for their edible nuts (also known as the Queensland nut), the smooth-shelled macadamia (*Macadamia integrifolia*) and the rough-shelled macadamia (*M. tetraphylla*) (Proteaceae). Hybridization occurs between them. *M. integrifolia* and *M. tetraphylla* are native to the rain-forests of southeastern Queensland, and southeastern Queensland and northeastern New South Wales, respectively, both growing in or close to water. Trees reach 20 m in height with a spreading habit. Macadamia has been used since ancient times by Australian aboriginal peoples and was brought under cultivation in the middle of the 19th century. Shortly afterwards, it was introduced into Hawaii and has become an important crop there. Macadamias are also commercially important in Australia, South Africa and Central America.

The flowers occur in groups of three or four, each giving rise to a fruit with a fleshy green husk enclosing a single seed – the nut: hence macadamia is not a true nut in the botanical sense (**see: Nut**). Each nut is almost spherical (diameter 1.3–2.5 cm) with a tough, fibrous shell: the embryo (kernel) is somewhat smaller (Fig. M.1).

The kernel contains about 3% fw water, 8% protein, approx. 75% **oil** (triacylglycerol, high in monounsaturated **fatty acids**), 15.5% total carbohydrate, Ca, P, Fe, K, thiamine, riboflavin and niacin (**see: Nut**, Table N.2).

Fig. M.1. Macadamia nuts (seeds) and kernels (embryos, white) (from http://farrer.riv.csu.edu.au/ASGAP/m-int.html, with thanks to Brian Walters).

Macadamia nuts are highly valued because of their fine flavour: they are eaten raw or roasted. The extracted oil is used as a salad oil. (MB)

Rosengarten, F., Jr. (1984) *Book of Edible Nuts*. Walker and Co., New York, USA.

Samson, J.A. (1986) *Tropical Fruits*, 2nd edn. Longman Scientific and Technical, London, UK, pp. 282–284.

Vaughan, J.G. and Geissler, C.A. (1997) *The New Oxford Book of Food Plants*. Oxford University Press, Oxford, UK, pp. 34–35.

Mace

See: Nutmeg and mace

Macrosclereid

A relatively large, columnar non-living cell with lignified walls, often in the palisade layer, constituting the mechanical layer in many seedcoats (Malpighian cell). (**See: Seedcoats – structure**)

Macula raphae

A seedcoat structure. (**See: Lens**)

MADS

MADS is an acronym taken from the code names of four transcription factors which possess the same highly-conserved sequence motif (domain). These transcription factors are MCM1, AGAMOUS, DEFICIENS and SRF (serum response factor), encoded by the *MADS* box genes. (**See: Maturation – changes in water status; Shattering**)

Maillard reaction

A complex series of reactions that are ubiquitous in nature and ultimately lead to browning and degradation of biomolecules. It has been suggested that these reactions play a role in seed **deterioration**, though whether they are a primary effect (leading to loss of **viability**) or a secondary effect (biochemical degradation of dead cells) is debatable. Maillard reactions are also called glycation or glyoxidation reactions because the initial steps involve the bonding of sugars to macromolecules containing amine groups. Hence, Maillard reactions are promoted in high concentrations of **reducing sugars** (e.g. glucose and fructose). Seeds that contain high amounts of glucose or fructose are often susceptible to rapid deterioration during drying or storage; however, most seeds have low concentrations of reducing sugars unless they are immature, highly **recalcitrant**, germinating, or have received prior exposure to warm, wet conditions. In addition to susceptibility to Maillard reactions, the poor storage characteristics of seeds with high amounts of reducing sugars has been attributed to metabolic imbalances, insufficient protectants, or **glass**-forming ability. The reaction cascade can be inhibited by some **antioxidants**. Maillard reactions are initiated slowly at room temperatures, and so are significant in biological systems when there is excess reducing sugar or low protein turnover (such as during seed storage). The kinetics of Maillard reactions are strongly affected by water availability since water is an initial substrate and then serves as a medium that controls the concentration and diffusion rates of other substrates. In simple solutions and foods, Maillard reactions occur at water activities between 0.4 and 0.7 (RH between 40 and 70%), corresponding to **water contents** common in seed storage warehouses. Maillard reactions have been widely studied in food chemistry, since they influence flavour and stability of most foods, especially during cooking.

The initial step of the Maillard reaction is the formation of a **Schiff's base**, the product of the reaction between an aldehyde (usually a reducing sugar) and an amine (usually the N-terminal end of a protein). In its straight chain form, glucose behaves as an aldehyde, and so the initial kinetics of the reaction are affected by the slow and reversible isomerization of glucose to the cyclized form. Amine groups are most prevalent on proteins, but also form the structure of phosphatidylethanolamine, a prevalent membrane **lipid**, and certain DNA bases. Loss of a water molecule in the Schiff's base leads to the equilibrium formation of isomers, a glycosylamine and a non-cyclized form known as an **Amadori product**. Amadori products may also form directly from reactive carbonyl (C=O) by-products of **peroxidation**. The intermediate stage of the Maillard reaction involves non-reversible oxidation (sometimes via reactive oxygen species, ROS) that cleaves the Amadori product into a variety of glycation products (depending on time, temperature, water availability and pH) that influence colour and odour. Continued degradation of the molecule leads to an amine (that perpetuates the Maillard reaction) and reactive dicarbonyl compounds that oxidize other molecules to form monosaccharides, aldehydes, CO_2, and AGEs (advanced glycation end-products that are usually fluorescent and cross-linked). By-products of Maillard reactions and lipid peroxidation may be similar (e.g. **malondialdehyde**) and many are considered mutagenic. Glycosylated DNA is not amplified in **polymerase chain reactions** (PCR), accounting for difficulty in detecting and using ancient DNA found in seeds at archaeological sites. (CW)

Wettlaufer, S.H. and Leopold, A.C. (1991) Relevance of Amadori and Maillard products to seed deterioration. *Plant Physiology* 97, 165–169.

Maintainer line

See: Cytoplasmic male sterility

Maize

1. Importance and economics

As far as production quantity is concerned maize (corn), *Zea mays*, is the world's foremost cereal: it is third in the area cultivated (**see: Cereals,** Tables C.8, C.9). Major growers are China, France and the Americas (Argentina, Brazil, Mexico), with the USA by far outstripping all other producers. With the exception of Mexico, these countries are the main exporters and in 2002 world export trade was valued at nearly US$10 billion (FAOSTAT). Dent corn is the predominant type for cultivation and trade. (**See: Crop Atlas Appendix, Maps 1, 7, 21, 23, 25**)

2. Origin and domestication

The wild ancestor of maize, teosinte (possibly *Zea mays* subsp. *mexicana*), was taken into **domestication** in the highlands of southern Mexico by 8000 years ago. (Note: The taxonomy of teosinte is complex. It probably consists of four *Zea* species, though originally it was named *Euchlena*. See: www.wisc.edu/teosinte/taxonomy/htm) The earliest, generally accepted archaeological evidence comes from caves near Puebla, in which were found cobs up to 25 mm long. Three thousand years ago maize spread from Mexico to the Andes and to eastern North America, and after 1492 rapidly through Africa and Asia, and much more recently into Europe. Genetic work has begun to elucidate the evolution of the distinctive multi-rowed cobs (made up of female flowers) from the two-rowed cobs of teosinte.

Maize is a particularly interesting example of how several features of the seed became changed during **domestication**, such as loss of **dormancy** and **shattering**, huge increase in grain size, and lengthening of the grain stalk. (**See: Domestication; History of seeds in Mexico and Central America**)

Maize has **C4 photosynthetic** metabolism, which is typical of some tropical grasses such as sugarcane, as well as drier area crops like **sorghum** and the **millets**. This is a more efficient form of photosynthesis, due to modifications in biochemical pathways and leaf anatomy, and results in more plant biomass being produced per unit of water consumed.

3. Grain characteristics

There are six main types of grain grown commercially, differing in size, shape and in features determined principally by the nature and behaviour of the **endosperm** during development and maturation (they are, in fact, different endosperm **mutants**) (Colour Plate 8B-C): (i) *Pop*. This is probably the original domesticated type. The endosperm core is floury and surrounded by a hard flinty shell. The moisture trapped in the floury starch expands upon heating causing the hard shell to burst, creating the popular confection. This type represents less than 1% of commercial production. (ii) *Dent*, the most abundant type (73%), in which the greater contraction during drying of the soft, floury endosperm in the core and crown of the grain results in a small indentation appearing on the end. (iii) *Flint*, which is smaller than *dent* but larger than *popcorn*, with a hard endosperm (a property dependent on the type of **starch**) (Fig. Colour Plate 8C); it represents 14% of commercial production. (iv) *Floury*, with a soft and mealy endosperm. This accounts for 12% of commercial production. (v) *Sweet*, which have mutations in their biosynthetic pathway for starch, resulting in an endosperm that contains a high proportion of sucrose and little starch; in some mutants an intermediate sugar polymer called phytoglycogen is present which adds to the sweetness (1% of commercial production). (vi) *Waxy*, in which the starch is composed entirely of **amylopectin**.

Grains are of different colours, white, yellow, brown, blue, purple or speckled, selectively preferred for various purposes (Colour Plate 8C). The colours are conferred by the endosperm, the aleurone layer or the pericarp. (**See: Imprinting; Xenia**)

Since dent corn is the most abundantly grown this type has been selected for a more detailed description. The maize grain is the largest of cereal grains: it is 10–15 mm long and 8–10 mm at its widest. The basal part (embryo end) is narrow, the apex broad. The **embryonic axis** and **scutellum** are relatively large. The **pericarp** (fruit coat) is thicker and more robust than that of the smaller grains. It is known as the hull, and the part of the hull overlying the embryo is known as the tip-cap. A hairless epidermis is present as the outermost layer. No cellular testa (seedcoat) layer is present but a cuticular skin persists to maturity.

In spite of the great size of maize endosperm, individual aleurone layer cells are comparable in size with those of oats and rice. One layer of these cells is present. In blue varieties, it is the aleurone layer cells that provide the coloration, due largely to the presence of anthocyanins. In the starchy endosperm many small **starch granules** (average 10 μm diameter) occur. **Storage protein** (zein) also occurs in tiny granular form within protein bodies. Horny endosperm occurs as a deep cap surrounding a central core of floury endosperm. The dent is not found in other types of maize such as flint maize, popcorn, and sweetcorn. (**See: Cereals, 4. Basic grain anatomy**)

4. Composition and nutritional quality

There are differences in grain composition among different types, cultivars, provenances, etc. but typical values (% fw) are: **storage protein** 11%, **starch** 65%, **oil** (triacylglycerol) 5%. These are not distributed equally among the grain parts. The endosperm contains 80% of the protein (the remainder is in the embryo and the aleurone layer) and almost all the starch. Most of the oil is in the embryo (scutellum), which on a fresh weight basis is about 33% triacylglycerol. The pericarp contributes most of the fibre in the grain and is itself approx. 86% crude fibre.

Almost 60% of the endosperm protein is the prolamin, zein and about 34% is glutelin, a mixed fraction of prolamins, globulins and structural proteins (**see: Osborne fractions**). Since prolamins are seriously deficient in the essential amino acid lysine, maize protein is nutritionally poor. Several spontaneous mutants of maize have been isolated in which the zein content is reduced in favour of other protein fractions, the consequence being (e.g. in the *opaque-2*, *floury-2* mutants) a 32% increase in lysine content. This trait has been introduced into elite lines of maize that are now popular in several parts of the world.

Maize grains are low in a B vitamin, niacin, and because of this, compounded by the low tryptophan in maize protein, an unbalanced diet depending on maize can lead to the deficiency disease pellagra (but see section 5).

5. Uses and processing

Although most (80–90%) of the annual maize grain crop is used for animal feed (as dent corn) it is nevertheless an important food for humans in many regions of the world (e.g. tortillas, polenta, porridges, stews) and an important source of various food and industrial products. Sweetcorn is used as a vegetable and popcorn is a 'recreational' snack food.

Processing of maize grains prior to use was developed in Mesoamerica. Here, the practice is to cook the grains for several hours in a solution of lime. Grains are then soaked and washed, producing *nixtamal* which is then stone ground to make a dough (*masa*). This processing treatment improves the quality in several ways. **Pericarp** removal is facilitated, the rheological properties of the dough are improved (alkaline hydrolysis releases **gums** from the pericarp and saponified lipids from the embryo into the dough), calcium content is elevated, protein quality is improved, **mycotoxin** contamination is reduced and niacin is released from bound forms. The latter effect explains why pellagra was very rare in pre-Columbian civilizations even though the readily available niacin in untreated grains is low.

Most of the maize used as food or for industrial products is first either dry or wet milled.

Abrasion features in dry milling, where it is achieved most commonly in the Beall degerminator, as a first treatment in a process leading to separation of the embryo which is valuable because of its high oil content, and 'grits', which are coarse particles of endosperm, used mainly in the production of corn flakes, although they also feature as an ingredient in many commercial and domestic recipes. Flour is also produced by this process. Flows that incorporate the Beall are said to use the TD (tempering-degerming) system. Grains are damped to 20% moisture content before entering the Beall in which passage between coaxial rotating screens with protrusions removes the embryo, pericarp and seedcoats, and breaks the endosperm into grits and finer particles. Emerging stocks are dried, cooled and sieved before being subjected to further treatments (Fig. M.2). Germ extraction may be achieved by passing through a screw press, but more frequently solvent extraction is used.

Maize is the major source of 'industrial' starch, produced by wet milling. The first stage is steeping for 1 or 2 days in warm water containing sulphur dioxide. The wet grains are then subjected to pressure and abrasion to release the embryo, which is separated from the denser fragmented components by flotation or, more efficiently, by hydrocyclone. The remaining stock contains starch fibre and protein (denatured by the sulphur dioxide in the steep water). It is ground to facilitate separation of the fibrous elements, which remain large, from the starch and protein by screening. Starch and protein are then separated by centrifugal means in which the dense starch (approx 1.5 specific gravity) is readily separated from the less dense and soluble protein elements. Starch and protein (known as gluten although it has none of the cohesive and extensible properties of wheat gluten) are then dried as separate end-products. Corn gluten is used as a component of stock feeds. Starch has many uses, such as a thickener in various food and confectionery products, in adhesives, for brewing of maize beer but the main one is production of glucose syrups.

Bourbon is a whisky made from milled grain, of which at least 51% must be maize (although commonly it is 65–75%). It was first distilled by early settlers in some southern states of the USA, but a common legend is that it originated in 1789 from maize ground at a mill in Georgetown, Bourbon County,

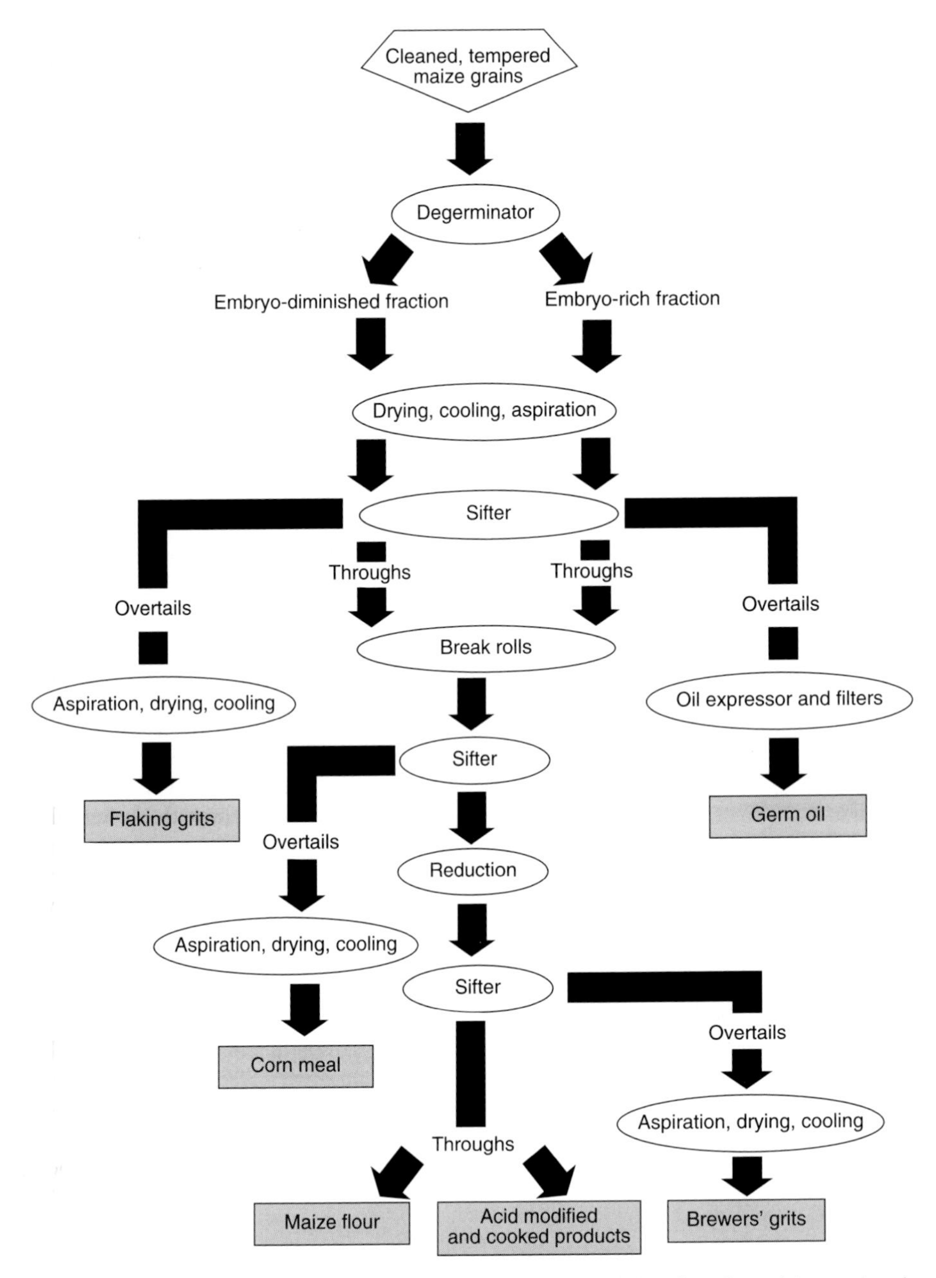

Fig. M.2. Simplified schematic diagram to show the passage of maize during processing in a flow-through tempering-degerming plant.

Kentucky; hence Bourbon whisky. But this has been called into question since there are some historical inconsistencies in this account. (AE, MN)

6. Oil

Maize embryos yield the important culinary maize or corn oil. Over 45 countries produce maize oil with the leading producers, in order of importance, being the USA, the EU, Japan, China, Brazil, South Africa and Canada. (**See: Oilseeds – major,** Table O.1)

The maize kernel contains only 3.1 to 5.7% oil but nearly all of it is concentrated in the **scutellum** region of the small embryo (83%) and endosperm (15%) located at the base of the maize kernel. Maize oil is a by-product of the starch wet-and-dry milling process in which the maize germ (embryo) is separated from the kernel. The wet milled germ yields some 44 to 50% oil while the dry milled germ, due to endosperm milling losses, contains only 33% oil. The oil is extracted using the expeller and/or solvent process. About 50% of the USA production is used for salad oil and frying, 30 to 35% in margarine manufacture and the remainder in processed foods. (**See: Oilseeds – major,** Table O.2, for oil composition; **Oils and fats; Oils and fats – genetic modification)** (KD)

For information on the particular contributions of the seed to seed science **see: Research seed species – contributions to seed science.**

Maiti, R. and Wesche-Ebeling, P. (1998) *Maize Science*. Science Publishers, Enfield, UK.
Strecker, L.R., Bieber, M.A., Maza, A., Grossberger, T. and Doskocznski, W.J. (1998) Corn oil. In: Hui, Y.H. (ed.) *Bailey's Industrial Oil & Fat Products*, Vol. 2, 5th edn. John Wiley, New York, USA, pp. 125–158.
Vaughan, J.G. (1970) *The Structure and Utilization of Oil Seeds*. Chapman and Hall, London, UK.
White, P.J. and Johnson, L.A. (eds) (2003) *Corn Chemistry and Technology, 2nd edn*. AACC Press, St. Paul MN, USA.

Maize - cultivation

Maize (*Zea mays*) is a major cereal that can be cultivated in temperate and tropical parts of the world. Utilizing the C4 photosynthetic pathway, the crop is an efficient converter of carbon dioxide and water to carbohydrate.

1. Genetics and breeding

As a cross-pollinated species, maize has broad morphological variability and geographic adaptability. Varieties may range from 0.5 to 5 m in height at flowering, mature in 60 to 330 days from planting, produce one to four ears per plant, 10 to 1800 kernels per ear and yield from about 0.5 to 22 t of grain per hectare. Kernels may be colourless (white) or yellow, red, blue or variegated which may occur in mottled or striated patterns (**see: Imprinting; Xenia**). The crop, which is produced from 50° latitude N to 40° S, is adapted to a wide range of environments, ranging from desert to high-rainfall areas, and at elevations ranging from sea level to 4000 m above, and is now grown in six major world **corn belts**. (**See: Crop Atlas Appendix, Maps 1, 7, 21, 24**) For the geographic origins, evolution, history, world distribution and economic importance **see: Maize**.

Many maize hybrids are produced – both conventional (based only on inbred lines) and non-conventional (where at least one parent is not an inbred line). Single-crosses are popular in the developed world because of their high yield performance and uniformity, whereas three-way crosses are the most common hybrid types grown in much of the developing world (**see: Production for sowing, III. Hybrids, crosses**). Recent advances in biotechnology have allowed the insertion of novel new traits into maize hybrids by **genetic modification** (GM) techniques; currently, varieties with herbicide resistances (against either glufosinate or imidazolinone) and insect resistance using the Bt protein (**see: *Bacillus thuringiensis***) (such as against the European corn borer in North America, which leads to ear dropping), or combinations of these as 'stacked traits', have been approved for widespread production. In addition to the range of conventionally bred varieties with special kernel quality traits (such as 'white' and hard endosperm varieties favoured for human consumption in various countries), many other GM 'output traits' are under development including pharmaceutical, industrial and nutritional modifications. The challenges of **identity preservation**, containment and isolation because of the potential for pollen dispersal or contamination are a serious concern in maize.

2. Development

Maize is a tall, **determinate**, **short-day**, open-pollinated annual plant, which unlike other grasses, produces separate **reproductive structures** on the same (monoecious) plant: male staminate **inflorescences** (tassels) at the stem apex and female pistillate inflorescences (ears) at the apex of one or more lateral branches about midway on the stalk, producing silks which function as both stigma and style. Normally about 95% of ovaries are cross-pollinated and 5% self-pollinated. After fertilization by wind-blown pollen, the female spikelet matures into a maize **kernel**, botanically a **caryopsis**, containing a single seed fused to the inner tissues of the fruit case (cob).

In the blister stage, the content of the kernel is clear, and jellylike. During the milk stage, the kernel content is soft and sweet: the end of this stage is the ideal time for picking green maize. During the soft and hard dough stages the moisture content of the grain drops rapidly and the kernel content changes to a paste; the formation of starch increases and the weight of the grain is about half the actual yield. The dent stage of development is complete when all the kernels have formed a dent on the crown; the moisture content of the grain decreases quickly, and should be around 50%. Physiological maturity, indicated by the formation of a black layer at the tip of the kernel, is the last stage of the grain-filling period and takes 15 to 20 days while the moisture content of the grain decreases from 50% to below 40%. At this stage, there are no soft portions of grain at the tip of the kernel (where they are joined to the cob), and the hard starch line, also known as the milk line, has moved through the kernel from the crown of the kernel to the attachment point. The kernels at the base of the cob develop first and the kernels at the tip last. In the grain-drying phase this process is reversed, with the kernels at the tip of the cob losing moisture first.

3. Production

The agronomic practices associated with the production of open-pollinated genotypes are relatively straightforward and similar to grain production. However, hybrid production, based on **cytoplasmic male sterility** (CMS), requires the maintenance of inbred seed male and female parental lines. Population and row spacing needs to be adjusted to maximize the performance of both parents and, because flowering time of the seed parents may be different, various strategies are employed to ensure adequate pollination. Practices to improve the '**nick**' include: variable depth of planting; application of **starter fertilizers** to modify vegetative development; application of foliar or supplemental fertilizer often in the irrigation water; seedcoatings, including temperature and time release agents that adjust the germination process (see below); vegetative modification, either by clipping or flaming; utilization of soil **mulches** or plastic coverings; and **priming** of one of the parent seed stocks. The effectiveness of these practices is quite variable, and depends on when in the growing season they are applied and the environmental conditions that both precede and follow their application (**see: Production for sowing, III. Hybrids**).

In addition, where CMS systems are not being used, **detasselling** is a critical step in hybrid seed production that impacts on both seed yield and purity, and needs to be done completely and so as to inflict as little damage as possible on the female parental plant. Mechanical and manual removal, or combinations of these approaches, are used in different locations.

Until the late 1980s, maize seed harvest technology paralleled that used for the grain harvest, evolving from hand harvesting to powered ear-corn pickers designed to harvest and dehusk in the field. However, following the introduction of sweetcorn harvest equipment to that industry in the early 1980s, the 'husk-on' harvest system rapidly came to be used – removing the ears from the plant with a minimum amount of damage and leaving the husk intact to further protect the ears in transport. Combine harvesting remains an option for double-, three- and four-way crosses, especially for flint-type hybrids.

Harvest moisture content varies with harvest system. Ideal harvest moisture varies with inbred lines, production scheme and harvest system, and production research is required to identify what is appropriate for specific genotypes. Combine harvesting is typically carried out when the moisture content is between 18 and 20% (indicated by the formation of the 'black layer' mentioned above), while ear harvesting begins at approximately 28 to 35% moisture content to avoid undue shell corn losses, and is normally completed before the moisture content drops below 25%.

Weak **pericarps** in certain cultivars can lead to two types of disorder, which can break the seed at its weakest point, and open it to fungal infection: popped kernels, and **'silk cut'** where the pericarp is cut or split over the sides of the kernel.

The husking step (removal of the **husks**) is critical to the success of the drying operation, and the sorting step allows for a final improvement in genetic purity by the removal of selfed or off-type ears and the reduction in seedborne disease by the removal of damaged or infected ears. Although some automation of these steps has been introduced, the final sorting remains a labour-intensive, visual operation.

Drying systems range from simple air-drying of a few ears to thin-layer drying of the shelled seed following harvest with a combine harvester, or industrial scale batch-drying of the husked ears, varying in concept from simple single-pass systems to single-pass-reversing to computer-controlled multiple-pass systems. Systems that utilize a nearly closed system and a heat-pump dehumidification process perform more predictably and thus may result in consistently higher quality seed (**see: Production for sowing, V. Drying, 3. Drying methods (h)**, for a more detailed description). When seed is harvested at higher moisture contents (>25%), then both the rate and drying temperature may have an impact on seed quality. Significant differences exist between various genotypes too (an example is shown in Table M.1).

Following drying, ears are moved to a stationary **shelling** system that removes the seed from the cob, essentially by rubbing, and effectively pre-cleans the seed prior to storage. During **storage management**, the seed is aerated to reduce the seed temperature and to minimize moisture changes prior to processing and bagging. After shelling, seed is pre-cleaned using an **air-screen cleaner** system, to remove cob pieces, husks, silks and other debris, and then transferred to bulk storage where it is aerated to remove the latent dryer heat prior to **conditioning** by which, depending on the ultimate use, seed is sized by passage through a series of air-screen separators or a series of round-hole cylinder separators. These operations result in cross-sectional sizes, which can then be length-sized prior to final conditioning. Typically the final step includes gravity separation in which the seed is separated based on density and, if needed, **colour sorting** may be used to improve the final physical quality. Most maize seed receives a seed treatment prior to bagging (**see: Units**) and storage.

Storage of most maize seed is of short duration – that is, for the few months that precede planting – but in order to provide insurance against a seed crop failure, a significant portion of the annual sales is carried over by the seed industry. In temperate regions the seed is held in ambient warehouse conditions. However, to reduce the impact of deterioration, a significant quantity of seed is held in controlled environment (temperature and humidity) storage.

4. Quality, vigour and dormancy

In addition to the important traits such as composition, yield and disease and pest resistance, physiological seed quality has become an important characteristic of modern maize hybrids. In general, ancestral lines, open-pollinated populations and most double-cross hybrids produce high quality seed, which expresses good seedling **vigour**. Modern single-cross hybrids,

Table M.1. The effect of drying temperature at different moisture contents on germination percentage of maize seed grown from different inbred parents (predicted values).

	Inbred Seed Parental line[a]							
	A632				B 73			
	Drying Temperature (°C)				Drying Temperature (°C)			
Seed moisture at harvest (%)	50	45	40	35	50	45	40	35
	Germination (%)							
45	87	99	99	99	41	92	99	100
40	88	99	99	99	57	95	99	100
35	89	99	99	99	72	98	99	100
30	91	99	99	99	88	100	99	99
25	92	99	100	99	100	100	99	99

[a]Least significant difference ($p < 0.05$) = 9.
Source: Burris, J.S. and Navratil, R.J. (1980) Drying high-moisture seed corn. *Proceedings of the Annual Corn Sorghum Research Conference* 35, 116–132.

however, because they are produced on relatively weak inbred parents and have been heavily selected for hybrid performance, may produce seed that can perform poorly in the field.

Primitive ancestral lines often exhibit **dormancy**, but modern maize hybrids generally do not. However, seed from some genotypes harvested at high moisture levels and tested immediately after drying can express a reduced performance in a cold **vigour test**, although they perform well in standard 'warm' germination tests; such short-term postharvest effects typically improve after about a month in storage.

5. Sowing

In general, maize will germinate at temperatures above 10°C when provided with adequate moisture and oxygen. Seedling emergence is hypogeal: the **coleoptile** enclosing the first true leaf provides the physical strength to penetrate the seedbed. (**See: Hypogeal**, Fig. H.9b) Seed is usually planted at a 2 to 5 cm depth and seedlings may emerge in 3 to 14 days, depending upon depth and seedbed conditions. In general, growers are advised not to plant hybrid seed before the soil temperature at 5 cm depth has reached 10°C. Maize is susceptible to imbibitional **chilling injury** and seedlots are often tested using 'cold test' protocols (using, for instance, 7 days at 10°C, followed by 4 to 7 days at 25°C in contact with soil, **see: Vigour tests – physiological, *cold test***). In general, approximately 100 heat units are required for emergence (the day-sum of the mean daily minimum and maximum temperatures, expressed on a base temperature, T_b, of 10°C) (**see: Germination – field emergence models**).

The most important determinant of seed and seedling disease expression and severity is the **seedbed environment**. Thus, soil and climatic conditions unfavourable to seed germination and growth are often favourable to seedling **disease** development as well. Fortunately, in most maize production areas these conditions occur only rarely in all fields but may frequently occur in fields with adverse soil conditions. Field-to-field variation can result in large differences in the amount of disease present and seed quality may also influence the impact of seed and seedling diseases.

In the production of seed for sowing, because of the generally lower vigour of seed **parental lines**, planting is usually delayed until after soils have warmed beyond the 10°C threshold. However, to avoid the early frosts of autumn and to maximize the utilization of harvest and drying facilities, the seed producer cannot wait too much beyond the normal grain planting period. The seed producer must also use delays in planting between the male and female parent to ensure optimum **nicking**. Usually these delays are based on the best estimate possible of **heat unit** difference between the flowering period for the two parents. However, because field conditions after emergence can dramatically alter the rate of growth of a given genotype, the 'best' emergence delays may not guarantee a perfect nick, and guidelines must be refined based on field experience to ensure efficient production.

In the developing world, maize is still planted by hand or, in rare instances, is transplanted from greenhouses or cold frames to allow for maturation of full-season genotypes. In general elsewhere, however, maize is machine planted to ensure good seed-to-soil contact and to provide accurate singulation, typically also including fertilizer and insecticide applicators. Finger pick-up and vacuum or pressure plate metering systems are increasingly replacing rotating plates; but all metering systems operate more accurately with carefully sized seed (**see: Planting equipment – metering and delivery**). Sizing is thus an important seed processing step; the USA **Corn Belt** demands just five to six sizes, and some production regions demand as many as 18. Seed producers must often plant parent seed lines that vary dramatically in size, shape and vigour. Because male and female rows are often planted on different days, the planter configuration must be flexible and easily modified. (**See: Planting equipment – metering and delivery**)

6. Treatment

Seed and seedling diseases caused by fungi, bacteria and nematodes may occur in most types of maize and sweetcorn, and result in stand reductions, initial seed rot and seedling infections. Most pathogens are common soil organisms; fungi cause the majority of early-season seed and seedling diseases, with soilborne *Pythium*, *Fusarium* and *Rhizoctonia* the more common genera.

Although several pathogens are commonly seedborne, in general this is not considered to the main source of disease transmission. *Diplodia* and *Colletotrichum* are common leaf- and ear-rot diseases that can be seedborne and result in seedling decay; but significant levels of either are rare and, if present in the seed production field, infected ears are removed at sorting, prior to drying and shelling, or the infected kernels can often be removed during seed processing. *Penicillium* species, often associated with storage deterioration losses, is found commonly on sweetcorn and may be present on field maize too, due to improper handling following harvest or inappropriate storage conditions. However, seed infected with *Erwinia*, a bacterial pathogen which causes Stewart's wilt, may be seedborne and is considered a transmission source for the disease and is thus regulated by most importing countries.

Due to the widespread distribution of seed and seedling fungal pathogens of maize, most, if not all, maize seed is treated with fungicides prior to sale which, under most planting conditions, perform well in protecting plants for the first 10 to 14 days after planting from diseases caused by *Pythium*, *Fusarium*, *Rhizoctonia* and *Phomopsis* for example (**see: Treatments – pesticides**). But under extreme conditions, such as excessive or sustained high soil moisture and/or prolonged low soil temperatures, protection may be reduced and serious short-term (seed decay and **damping-off**) and long-term (impaired root development) disease effects may result. Maize does not have a great ability to compensate for reduced field stand (unlike soybean, for example), and a reduced number of plants will usually result in a reduced yield. Consequently, the risk of impact from seed and seedling fungal diseases is high; relative to total seed cost, the cost of seed-applied fungicides is low and fungicide treatment is generally recommended.

Insect and pests, including seedcorn maggot (*Delia platura*), wireworms, *Aeolus* sp., cutworms, *Agrotis* sp., and the seedcorn beetle (*Stenolophus lecontei*) may attack the seed as it germinates and the seedling as it emerges and well into early seedling growth. Control is most often associated with seed- or planter-applied insecticides or, in the case of cutworms, with insecticide baits applied post-emergence. Pathogenic soilborne nematodes may also attack the germinating seed and seedling roots and may be severe enough to reduce the final stand or the health of the developing seedlings. Seed-application of non-systemic insecticides (such as chlorpyrifos) provides pre- and post-emergence seed/seedling insect control, and in recent years systemic insecticides (such as the neonicotinoids, imidacloprid and thiamethoxam) have become widely used to provide additional early season insect protection. The requirement to apply these treatments in relatively high doses involves the use of **filmcoating** techniques.

Functional seed **enhancements** are not widely used in maize at present, perhaps due to the logistical constraints of the processes for such a large-volume crop. Filmcoatings that have some activity in altering water uptake are used by some seed producers to provide delays in emergence in male or female parental lines to improve the nick. Priming has not found wide acceptance with maize seed producers, and its application to vegetable sweetcorn, though technically successful, remains little used. **Coating** to achieve better plantability has not been used to any significant extent, though the technique is now employed as an inventory-management technique to upgrade one seed size grade to another. (JSB)

Kirkpatrick, T.L. and Rothrock, C.S. (eds) (2001) *Compendium of Cotton Diseases, Second Edition*. APS Press, St Paul, MN, USA.

McDonald, M.B. and Copeland, L.O. (1997) Corn. In: *Seed Production, Principles and Practices*. Chapman and Hall, New York, USA, pp. 193–205.

Poehlman, J.M. and Sleper, D.A. (1995) Breeding Corn (Maize). In: Poehlman, J.M. and Sleper, D.A. (eds) *Breeding Field Crops*, 4th edn. Iowa State University Press, USA, pp. 321–344.

White, P.J. and Johnson, L.A. (eds) (2003) *Corn Chemistry and Technology, 2nd Edition*. AACC Press, St Paul, MN, USA.

Male-sterile line

See: **Cytoplasmic male sterility**

Malondialdehyde

MDA ($H_2C(HC{=}O)_2$) is a dicarbonyl by-product formed during **peroxidation** of polyunsaturated fatty acids or DNA, or the advanced stages of the Maillard reaction. Assessment of MDA concentration using the thiobarbituric acid (TBA) test is one of the most common assays for measuring lipid peroxidation because the fluorescent product is easily quantified. The test itself produces numerous artefacts, but the largest drawbacks for quantifying lipid peroxidation using MDA are its low production (except in liver cells), its unspecific origin (from oxidized sugars or lipids) and its high reactivity with other molecules (**see: Maillard reaction**). Carbonyl groups (C=O) react with amines to form **Amadori products** that enter into the Maillard reaction cycle of non-enzymatic oxidation. MDA (and other dicarbonyl compounds) react with amine groups of proteins and nucleic acid bases to from intra- and intermolecular crosslinks. (CW)
(See: Deterioration reactions)

Malpighian cell

A highly specialized, palisade-like macrosclereid cell with unevenly thickened (lignified) walls, with the narrow remaining lumen, devoid of cytoplasm, usually at the distal end of the cell; there are usually hexagonal facets (in surface view), often with a *linea lucida*. A Malpighian palisade layer is a common feature of the exotesta of Cannaceae, Fabaceae and Rhamnaceae, or of the exotegmen of Bombacaceae, Celastraceae, Cistaceae, Euphorbiaceae and Malvaceae. (**See: Seedcoats – structure**)

Malting

1. Overview

All cereal grains, **barley** being the preferred one, can be transformed into malt by the malting process. Traditional African, opaque beer is made from malted sorghum and traditional European-type lager, ale or stout beers from malted barley. There is evidence that barley was malted in ancient Egypt and was likely to have been used to make some kind of ancient beer. A substantial amount of malted wheat is used to make Weiss beer. (**See: Malting – non-barley**)

The malting process is usually divided into three stages: steeping, germination and kilning. Steeping hydrates the grain, while germination permits the steeped grain to grow and develop hydrolytic enzymes which hydrolyse the substrates of the starchy endosperm to produce what is termed 'green malt'. Kilning converts the 'green malt' into a friable dried malt. In the production of traditional African sorghum malt, the 'green malt' from the germination process is sometimes dried in the sun before being processed into beer.

The production of malted barley is controlled by a complex of physiological processes (**see: Malting – barley**, Fig. M.3) which includes production of **gibberellins** (primarily GA_1, GA_3) by the germinated embryo. These hormones are transported to the **aleurone layer** where endosperm degrading enzymes such as endo-β-1,3:1,4-glucanases, endo-β-1,3-glucanases, pentosanases, endo-proteases, α-amylase and limit dextrinases are produced (**see: Malting – barley**, Table M.2; **Mobilization of reserves – cereals**). During this process, β-amylase and carboxypeptidases are activated in the starchy endosperm. The control of this process of embryo growth and endosperm modification are primary areas of research and development which support the production of barley by farmers and the production of malt by maltsters for the brewing, distilling and food industries.

The malting process is a simple germination/growth process but the underlying physiology and biochemistry are complex. The malting, brewing and distilling industries have been, for centuries, very supportive of research that has developed malting technology. Scientists such as Watt, Pasteur, Sorensen, Kjeldahl, Beaven, Horace Brown, O'Sullivan, Sandegren, Lovibond, Gosset, Greiss, Preece and MacLeod have been involved in research work that has not only developed malting and brewing technology, but has extended scientific knowledge of the structure and function of yeast and cereal grains. The pro-active approach of the industry to research and development is exemplified by the observation that the malting industry was using **gibberellic acid** as a malting aid in the commercial production of malt in 1960, before the physiological and biochemical concepts of gibberellic acid function in enzyme synthesis in the aleurone layer were developed. A close working relationship between the industry and scientific institutions is essential, if significant developments in technology are to be achieved.

2. Malting terms and definitions

Several terms are used specifically in relation to malting practice. Not all of these are included in the articles dealing with malt and malting, but since they will be encountered in connection with this important industrial use of seeds they are collected and presented as follows:

Acrospire (shoot) growth

The growth of the shoot along the dorsal surface of the malting grain is used as an indication of extent of enzymic modification of the underlying endosperm. Growth to the distal end of the grain indicates that the malt is well modified.

Additives

Substances added to malting barley to improve malting performance: gibberellic acid (to accelerate enzymic modification of the endosperm during malting), potassium bromate (to reduce malting loss by reducing respiration rate and root growth, and proteolysis by reducing protease production), formaldehyde (to reduce microbial activity), dilute sulphuric acid (to reduce root growth and increase proteolysis), dilute sodium hydroxide (to reduce microbial activity and malt phenols) and dilute calcium hydroxide also to reduce malt phenols.

Air resting

Prescribed periods when the steeping water is removed from the grain. Air resting accelerates the germination (root-emergence) process and overcomes water sensitivity. Duration of steeping and steeping temperatures are usually below 50 h and 20°C, respectively.

Balanced specification

A proposal by G.H. Palmer that modification analyses should relate to each other, reflecting the normal progression of enzymic hydrolysis of the endosperm during malting. Balanced specification encourages uniform processing of malt in the brewhouse. In contrast, unbalanced specification relates to specifications where modification analyses can run counter to each other, such as when soluble nitrogen ratio is high and β-glucan content is high. Such imbalance often reflects uneven modification or injudicious blending of barleys or malts.

Curing temperature

Specialized temperature treatments applied at end of standard kilning periods. For example, 82°C for 2 h for traditional lager malts, 100°C for 2 h for traditional ale malts.

Dextrinizing Unit (DU)

Measure of α-amylase activity, the major enzyme involved in the hydrolysis of grain starch.

Diastatic Power (DP)

Enzyme activity in malt assessed by the Fehling's (reducing sugar test) method. This method also determines small amounts of α-amylase activity, but it primarily determines β-amylase activity. Diastatic Power is therefore an assessment mainly of the β-amylase potential of the malt.

Dormancy assessment

The per cent germination results of the **Germination Capacity** test minus corresponding results of the **Germination Energy** Test (see: Germination tests below).

Excess gravity

Excess weight caused by soluble materials in the wort. Used to calculate hot water extract.

Fermentable extract

That sugar portion of the hot water extract that can be fermented by yeast into alcohol. It is calculated as: Hot water extract % × % **Fermentability of wort/100**.

Germination (process)

The extended growth phase (*ca*. 4–6 days) of the malting process at a relative humidity of around 100% and temperatures below 20°C, usually 15–18°C. During the germination process, moisture levels in the grain fall from about 46% to about 44%.

Germination tests

Germination Energy Test is a test for the rate of germination of 100 grains in 4 ml of water, at 18°C, over a period of 3 days.

Germination Capacity Test is a test for optimal percentage of germination of 200 grains in 200 ml of 0.75% hydrogen peroxide, at 18°C, over a period of 3 days.

Water Sensitivity Test is a test for the potential of 100 grains to germinate in 8 ml of water, at 18°C, over a period of 3 days.

Tetrazolium (Chloride) Test for grain viability: 100 grains are cut longitudinally, then 100 half grains are incubated in a solution of tetrazolium chloride. Red-pink embryos indicate that the grains are likely to germinate. **See: Tetrazolium test**

β-Glucanase activity

Enzyme activity in malt assessed directly or indirectly in terms of β-glucan breakdown.

Green malt (unkilned malt)

Cereal grains, e.g. of barley, that have gone through the steeping and germination processes.

Hot spots

Localized deterioration of grain when in bulk. Hot spots spread as a result of the products of microbial activity, i.e. heat, moisture, and further the development of fungi and bacteria.

Hot water extract

Hot water extract is a measurement of the volume of soluble material in the wort. Units of hot water extracts are: litre degree of extract per kilogram of malt (l°/kg) or %

extract. Litre degrees of extract multiplied by 0.263 give % extract.

Kilning
Final stage of the malting process. The temperature is increased to 60°C or higher. The moisture falls over 24 h from 44 to 5%. Enzyme activites decline as malt colour and malt flavour develop

Malt
Barley grains or other cereal grains that have gone through the malting process.

Malt specification
Specific group of analyses set by brewing, distilling and food industries to ensure that production requirements are met. However, these analyses do not guarantee production performance.

Malt types
Malt types listed in terms of increasing colour-producing potential – Lager, Distilling, Ale, Vienna, Munich, Light crystal, Crystal, Dark crystal, Caramel, Chocolate, Black, Roasted barley. Distillers also use peated malts to give phenolic and smoky flavours to Scotch whisky. Peated (distilling) malt is produced by burning peat and passing the smoke through the malt during kilning

Malted barley
Barley that has gone through the malting process.

Malting barley
Barley that can be processed into malt.

Malting loss
Percentages of materials lost from barley grains during the malting process: steeping loss about 1%, respiratory loss about 3%, wort weight loss about 3%.

Malting plant process
Events that involve: steeping of barley (or other cereal) grains; germination (growth) of grains; kilning of grains; removal of roots from kilned grains. **(See: Malting – process)**

Modification
See: Malting – process, 7. Malt modification

NIR
Near infrared reflectance spectroscopy, used to measure grain moisture, proteins, hot water extract and malt modification.

Proteolytic activity
The activity of proteases assessed indirectly as soluble proteins in worts.

Recommended analyses
Official British (Institute of Brewing), European (European Brewing Congress, EBC) and American (American Society of Brewing Chemists) methods of barley, malt, wort and beer analyses.

Specific gravity
The weight of a given volume of material in the wort, compared with an equal volume of water.

Spirit yield
The amount of pure alcohol which can be produced from a tonne of malt. Spirit yield is related to level of fermentable extract. Spirit yield = % fermentable extract × 6.06.

Steeping
Soaking (hydration) of cereal grains in water under controlled conditions to increase the moisture of grains to an optimal level of about 45–46%. Submersion (steeping) of the grain in water is usually broken by air-rest periods.

Thousand corn weights
Dry weight of 1000 cereal grains. A high 1000 corn weight and low nitrogen suggest that the hot water extract potential of the malted grain will be high.

Water sensitivity
Failure of barley grain to germinate in an excess of water.

Wort (wash)
The brown-coloured filtrate collected from a mash of malt flour and hot water. This filtrate is called wort by brewers and wash by distillers. Wort contains sugars and dextrins produced from hydrolysed gelatinized starch. It also contains soluble proteins and amino acids, minerals, fatty acids and vitamins such as vitamin B. The solid matter contents of worts (hot water extract) is calculated from specific gravity values and °Plato (1.040 specific gravity is equivalent to 10° Plato). Specific gravity of water is 1.000.

(GHP)

(See: Malting – barley; Malting – process)

(The author thanks Miss C. Brown BA(Hons) and Dr Annie Hill for help with preparation of all of the articles on Malting)

Brissart, R., Brauminger, U., Hydon, S., Morand, R., Palmer, G.H., Savage, R. and Steward, B. (2000) *Malting Technology*. European Brewing Manual of Good Practice. Hans-Carl-Fach, Nurnberg, Germany.

Clark, C. (1998) *The British Malting Industry since 1830*. Hambledon Press, UK.

Institute of Brewing, London (1986) *Centenary*. CWP, Dorset, UK.

Palmer, G.H. (1989) Cereals in malting. In: Palmer, G.H. (ed.) *Cereal Science and Technology*. Aberdeen University Press, Aberdeen, UK.

Palmer, G.H. (1995) Structure of ancient cereal grains. *Journal of the Institute of Brewing* 101, 103–112.

Malting – barley

1. The grain
Barley is a grass which, like other cereals, belongs to the Poaceae (Gramineae) family. It has a wide geographical range

but grows best in temperate climates. There are two types of barleys, spring and winter. In various parts of the UK, for example, spring barleys are usually planted in March and harvested in June whereas winter barleys are usually planted in September and harvested in July. Both types may be either 2-rowed or 6-rowed. This morphological difference is reflected in the inflorescence/ear or 'head' of the plant. In 2-rowed barleys (*Hordeum distichon*) two grains out of six, in close proximity, develop into mature grains, giving the ear a flat appearance of two rows of grains. In 6-rowed barleys (*H. vulgare*) all six grains develop and mature, giving the ear an open (lax) appearance.

The grains of 2-rowed barleys tend to be more uniform in size than 6-rowed barleys, whose grains vary greatly. Of the grains of 6-rowed barleys, often only one of six is of a size suitable for malt production. Therefore, although 6-rowed barleys produce a greater number of grains than 2-rowed barleys, the latter give a higher yield of starch extract per hectare. Optimal starch contents are important in barley and malt because the sugars produced from starch are the primary raw materials that are converted during fermentation, by yeast, into alcohol. (**See: Malting, 2. Malting terms and definitions; Malting – process**)

Barleys are usually recommended for malting and are graded with regard to malting and agronomic quality. The UK's National Institute of Agricultural Botany's variety grading system is 1 (the poorest) to 9 (the best). In conventional plant breeding, it takes about 10 years to breed a new malting barley. Yield trials and malting trials are conducted after about 5 years into the breeding programme. About 8 years into the breeding programme, potential varieties are tested for **distinctiveness, uniformity** and malting quality. At present (2004), no **genetically modified** malting barley has been produced for commercial use.

World production of barley is about 140 million t, of which only about 21 million t is considered suitable for producing the 17 million t of malt required by the industry worldwide. Of the total malt produced, about 94% is required to produce 1.6 billion hectolitres of beer; about 4% is required for distilling (e.g. Scotch whisky, Irish whiskey and Bourbon). Scotch malt whisky, like German Reinheitsgebot (pure food law) beers, is made using 100% malted barley which should not have been treated with gibberellic acid. Other whiskies and beers are made using malted barley and different proportions of unmalted cereal grains, such as maize, rice and wheat.

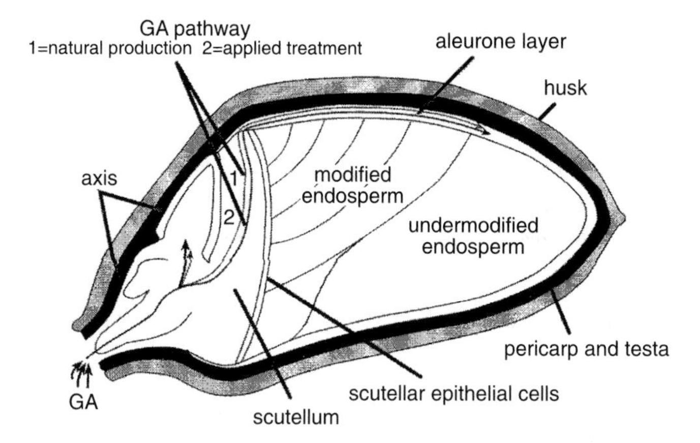

Fig. M.3. Relationship between transport of gibberellic acid (GA), aleurone layer activity, and enzymic modification of the endosperm.

2. Malting

This describes the events that take place within the malting barley grain as it is being malted. For an overview of the industrial process of malting, **see: Malting – process**.

In the barley grain (Fig. M.3) about 10% of the dry weight is husk, 2–3% is **embryo**, 3–4% is **pericarp/testa**, 5–6% is the **aleurone layer**, and the starchy **endosperm** is 77–80%. The **husk** comprises two overlapping leaf-like structures called the **palea** and **lemma**. The shape of the point of attachment of the husk to the stalk of the inflorescence (ear) can be used to identify barley varieties, as can husk colour and texture. The husk contains a high content of silica and is very abrasive. It absorbs a considerable amount of water during steeping and is important as a filter aid during the mashing stage of the brewing process. Microbial contamination is confined mainly to this region. Physical damage can cause the husk to separate from the grain. Husk damage is usually taken to indicate embryo damage so samples which contain significant husk-damaged grains are usually regarded as unsuitable for malt production. Fungal infection can damage the germination potential of the grain, can cause diseases such as Farmer's Lung (e.g. *Aspergillus*) and the production of dangerous mycotoxins which can damage human health and kill farm animals (**see: Mycotoxins**). Grain infections are as field fungi (e.g. *Alternaria* and *Fusarium*), or as **storage fungi** (e.g. *Aspergillus*). Field fungi are unavoidable, storage fungi are generally unacceptable. Some mycotoxins from *Fusarium* can induce gushing of beer. The husk also contains polyphenols which can associate with proteins in the beer to form hazes.

Various recommended tests are applied at purchasing operations to ensure that the **germinative energy** (rate) and the **germinative capacity** (overall vitality) of grain samples are optimal. Embryo vitality is also assessed using the tetrazolium (chloride) test. The embryo contains about 15% sucrose, 8% raffinose and about 20% lipids, about one-third of the total lipids in the grain. In terms of malting, the main function of the germinated embryo is to produce **gibberellic acid** in the growing axis. The hormone is then transported through the vascular strands of the **scutellum** to the **aleurone layer** during the germination process (**see: Mobilization of reserves – cereals**). The scutellum (Fig. M.3) is attached to the axis which has a shoot (acrospire) and a **coleorhiza** (chit) that encloses a root system of four or five seminal roots. (**See: Cereals**)

Germination and root growth are important in the malting process because they are associated with the production of a **gibberellin**, gibberellic acid. But maltsters try to limit root growth during malting because it is a part of the dry weight loss that occurs (malting loss). Expert control of the malting process can still result in a 7% malting loss, of which about 3% is respiratory loss and 3% is root growth loss. Malting loss reflects an economic loss of starch extract.

The coat of the grain is composed of **pericarp** and **testa**. The pericarp (but not the testa) prevents gibberellic acid from entering the grain directly. During malting, applied gibberellic acid enters the embryo through the **micropyle** but it can

Table M.2. Enzyme development in barley endosperm.

Gibberellic acid-dependent	*Not established as gibberellic acid-dependent*	
Aleurone layer	Aleurone layer	Starchy endosperm
α-Amylase	Endo-β-1,3-glucanase	β-Amylase
Endo-β,1,3: 1,4-glucanase	Phytase (acid phosphatase)	Carboxypeptidase
Endo-protease	Phospholipase	β-Glucan solubilase
Pentosanase	Lipases	
Limit dextrinase[a]	Carboxypeptidase	
Cellobiase (β-glucosidases)	β-Glucan solubilase	
Laminaribiase		

[a]Also produced in developing grains and is secreted into the starchy endosperm where it is inactivated. Production in the germinated grain is GA-dependent.

enter the underlying aleurone cell directly if the pericarp is damaged or abraded (**see: Malting – process**).

The triploid (21 chomosomes) aleurone layer of the barley grain is the living, outer layer of the starchy endosperm. It is about two to three cells deep and constitutes about 5% of the total dry weight of the grain, with cell walls about 3 μm thick containing about 30% β-glucans and about 60% pentosans. The cell walls of the starchy endosperm are about 2 μm thick and contain around 70% β-glucans and 25% pentosans.

Like the embryo, the aleurone layer contains significant amounts of lipids (about 30%) and proteins (about 20%). B-group vitamins, phytic acid, minerals and sucrose are also present. In the malting process, the vitality of the aleurone is as crucial as the vitality of the embryo, both of which may be assessed by the **tetrazolium test**. The aleurone layer (**see: Mobilization of reserves – a monocot overview**, Fig. M.13A) is induced by gibberellic acid to synthesize and release endosperm-degrading enzymes such as endo-β-1,3:1,4-glucanase, endoprotease and α-amylase. α-Amylase, endoprotease and limit dextrinase are synthesized *de novo* in the aleurone layer after treatment with gibberellic acid. Table M.2 shows the relationship between the action of gibberellic acid and the development of endosperm-degrading enzymes in the endosperm of barley during the malting process.

The stimulation of enzyme production along the aleurone layer, from the embryo to the distal end of the grain, is promoted by the transport of gibberellic acid activity, through the plasmodesmata system of this tissue. Optimal concentrations of gibberellic acid and their transport along the aleurone layer are important features of the malting process. Sub-optimal moisture content can impair this transport and thereby restrict enzyme production, distribution and action in the starchy endosperm, resulting in the production of poor quality, under-modified malt. Because sub-optimal amounts of gibberellic acid can also cause under-modification, where permitted, maltsters can add commercially-produced hormone (derived from the imperfect form of the fungus *Gibberella fujikuroi*) to optimize the concentration required to produce desired amounts of endosperm-degrading enzymes. Modification is the term used to describe the degree to which the starchy endosperm of the malted grain is hydrolysed by endosperm-degrading enzymes during malting. In this regard, the endosperm may be under-modified, modified or well-modified. In general, brewers' and distillers' malts tend to be well-modified. (**See: Malting – process**)

The starchy endosperm is about 75% of the total weight of the grain. The centrally-placed cells are elongated, whereas the outer cells are larger and isodiametric in shape. Barley grains contain 60–65% starch and 10–13% protein, depending on prevailing conditions while developing on the parent plant (**see: Cereals – composition and nutritional quality**). All the starch and most of the protein are in the cells of the starchy endosperm, the walls of which contain about 5% protein, 70% β-glucans (70% β-1,4 and 30% β-1,3 linkages) and about 25% pentosans. The pentosans are arabinoxylans, containing β-1,4-linked xylans and arabinose side chains linked to the xylan backbone through α-1,2 and α-1,3 links (**see: Hemicellulose**). It has been suggested that there are no consecutive β-1,3 links in β-glucans, although the endo-β-1,3-glucanase enzyme will hydrolyse high molecular weight β-glucans. The β-glucans of poor quality barleys may have more β-1,3 links than the corresponding β-glucans of high quality barleys. The total β-glucan content of malting barleys is about 3–4%, and the pentosan content is about 9%. Both are very viscous in water and can cause brewhouse problems related to slow filtration and haze. Hence, one important requirement of the malting process is that β-glucans of the starchy cell walls of the endosperm are broken down from about 3–4% to less than 0.4%. Pentosans pose fewer problems than β-glucans and undergo limited hydrolysis during malting.

The protein reserves of the grain constitute a mixture of three components: albumin, globulin, the prolamin, hordein, plus what was called the glutelin, hordenin, fraction, which is now regarded as a type of hordein also (**see: Osborne fractions; Storage proteins**). The storage proteins of the starchy endosperm exist in the grain as a matrix and are mainly hordeins. Albumins and globulins are extracted using 6% sodium chloride, and prolamins by hot 70% ethanol or hot 60% isopropanol. All the proteins are extractable in 4% sodium hydroxide.

There is a relationship between the protein and starch content of the starchy endosperm, which maltsters, brewers and distillers use to assess the starch extract potential of barleys. The concept is that the higher the protein content, the lower the starch content of the grains and vice versa. Therefore, barley is purchased on the nitrogen (protein) content of the grain. Within specified percentages of proteins, low protein grains tend to be the most expensive. Factors that increase the protein content (i.e. nitrogen content × 6.25) of the grain are: high nitrogen in the soil, heavy soil conditions, barley variety, and time of rainfall during grain development. Hordein proteins, which are rich in proline and glutamine, increase significantly as the total nitrogen content of the grain increases.

Structurally, endosperms of barley grains can be either mealy or steely. Steely endosperms have higher protein levels and lower starch contents than corresponding mealy grains. Steely grains tend to malt (modify) slower than mealy grains but may have higher activities of enzymes such as β-amylase. A test has been developed to identify mealy (high turbidity) and steely samples of grain by selective milling and turbidity analyses of released flours. During malting, grain protein is solubilized by proteases to produce polypeptides, peptides and α-amino nitrogen (e.g. amino acids) (**see: Storage protein – mobilization**). The soluble protein fraction of malt extract is referred to as total soluble nitrogen (TSN). The ratio of TSN/TN (total nitrogen) is used as an index of protein modification. A modified malt has a value of about 38–40 (**see: Malting – process**, Table M.3). Protein solubilization during malting is effected by endoproteases and carboxypepidases, the latter producing most of the amino acids present in the malted grain. Solubilized polypeptides can form hazes, and specific polypeptides of molecular weights 40,000 and 9,000 are important in beer foam development. These polypeptides tend to be hydrophobic. Amino acids and small peptides are utilized by yeast during its conversion of sugars to alcohol during the fermentation stages of beer and whisky production.

Large, A type (20–25 μm diameter) and small, B type (1–5 μm diameter) **starch granules** of the starchy endosperm are embedded in the protein matrix (Fig. M.4), and both types are associated with lipoprotein materials on their surfaces. About 90% of the starch granules are small, and about 10% are large. However, because of the size of the large starch granules, they account for about 90% of the weight of the starch in the endosperm. Large starch granules are composed of 25% **amylose** and 75% **amylopectin**. Small starch granules are reported to have similar contents of amylose and amylopectin, but the very small starch granules (<4 μm) can contain as much as 40% amylose. An important processing difference between the large and small starch granules is that the gelatinization temperature of large starch granules is about 62°C, compared to about 80°C for the small granules. During malting, about 10% of the starch is broken down, mainly the small starch granules, about 50% of which are degraded. Starch degradation during malting is effected by α-amylase, although significant quantities of β-amylase and limit dextrinase are present. α-Amylase can attack raw (ungelatinized) starch but β-amylase cannot. About 70% of the soluble proteins of the wort is solubilized during the malting process. Starch is hydrolysed mainly during mashing.

In general, the conversion of the hard, hydrated food reserves of the starchy endosperm of barley (Fig. M.4), into the friable endosperm of malt (Fig. M.5), is effected by hydrolytic enzymes such as β-glucan solubilase (i.e. proteases and β-glucanases), endo-β-1,3:1,4-glucanase, endo-β-1,3-glucanase, proteases and α-amylase. The induction and control (Fig. M.6) of this process of enzymic modification of the food reserves of the starchy endosperm is the raison d'être of the malting process. (GHP)

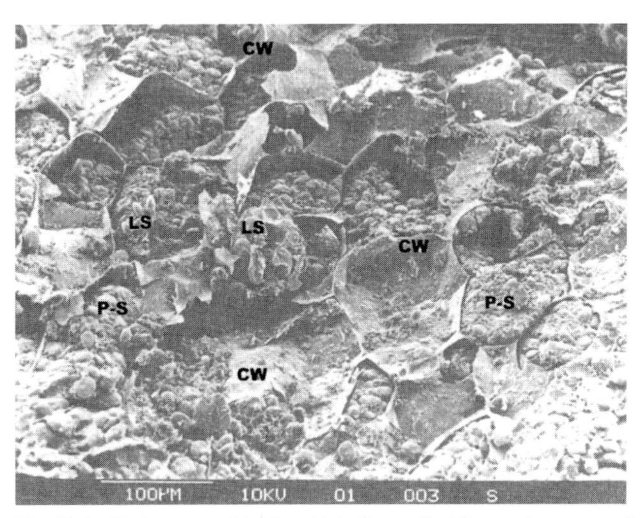

Fig. M.4. Endosperm of barley. Note the wall of the endosperm cells and the association of large and small starch granules and the protein matrix. P-S = protein matrix and small starch granules; LS = large starch granules; CW = cell wall.

Briggs, D.E. (1998) *Malts and Malting*. Blackie Academic and Professional, London, UK.

Broadbent, R.E. and Palmer, G.H. (2001) Relationship between β-amylase activity, steeliness, mealiness, nitrogen content and nitrogen fractions of barley grains. *Journal of the Institute of Brewing* 107, 349–354.

Koliatson, M. and Palmer, G.H. (2003) A new method to assess mealiness and steeliness of barley varieties and relationships of mealiness with malting parameters. *Journal of the American Society of Brewing Chemists* 61, 114–118.

Palmer, G.H. (ed.) (1989) *Cereal Science and Technology*. Aberdeen University Press, Aberdeen, UK.

Savin, R., Passarella, V.S. and Molina-Cano, J.L. (2004) The malting quality of barley. In: Benech-Arnold, R.L. and Sánchez, R.A. (eds) *Handbook of Seed Physiology: Applications to Agriculture*. Food Products Press, New York, USA, pp. 429–457.

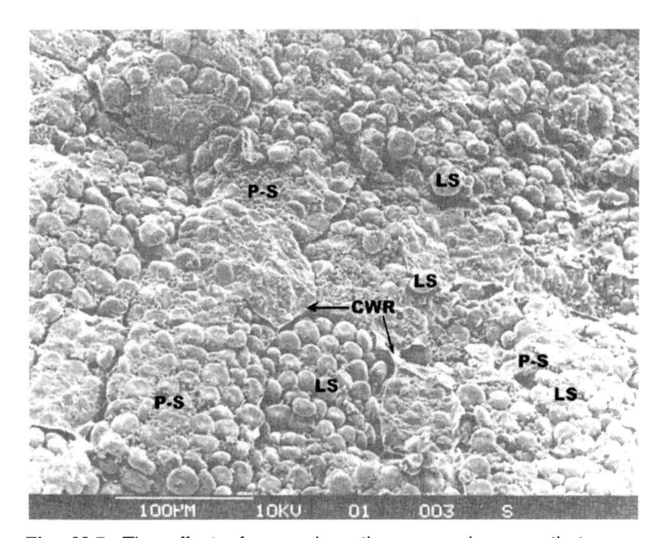

Fig. M.5. The effect of enzymic action on endosperm that occurs during malting of barley. Enzymic action has degraded the cell wall and protein matrix and has reduced the number of small starch granules (endosperm well modified). CWR = cell wall residue; LS = large starch granules; P-S = protein matrix-small starch granule residue.

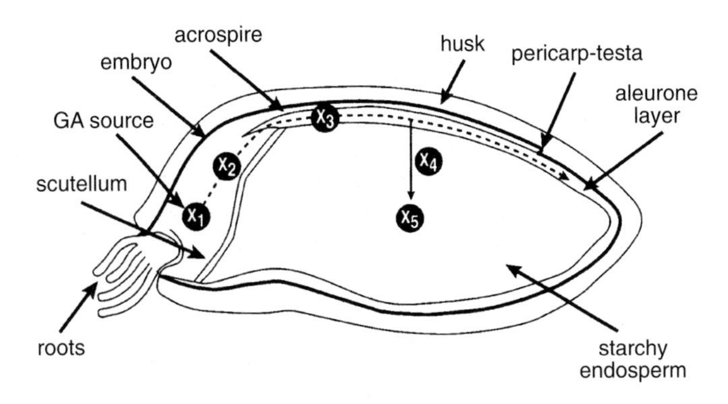

Fig. M.6. Points of grain modification failure during malting (X_1, X_2, X_3, X_4, X_5). X_1 = Gibberellin (GA) production; X_2 = GA transport out of embyro; X_3 = GA transport through aleurone layer; X_4 = secretion of enzymes from aleurone layer; X_5 = enzyme distribution in starchy endosperm.

Malting – non-barley

Speciality malts are produced primarily from **wheat** and **sorghum** grains. Smaller quantities of such malts are also produced from **millet, rye, rice** and **oats**. Unlike in the case of barley malt, the production procedures for malt from these cereals are not clearly defined.

1. Wheat malts

These are used mainly to make 'Weissbier' and are also used to produce standard beers to improve foam stability. Speciality breads also contain small quantities of wheat malts. In contrast to the production of barley malts, where the grain is steeped to 45–46% moisture (**see: Malting – process**), wheat grains are steeped for approximately 26 h at 13–16°C to achieve a water content of about 40%. Extract yields of wheat malt tend to be higher than those of barley malts but the activities of α-amylase and β-amylase in both kinds are similar. The α-amylase potential of wheat malt can be increased by **gibberellin**. Wheat malts are more difficult to process because the grains have no husk, and endosperm storage proteins and pentosans are more viscous than those in barley malts.

2. Sorghum malts

These are produced from red or white sorghum grains. In contrast to the production of barley and wheat malts, sorghum grains are steeped and germinated at 25–30°C. Embryo growth during malting is very rapid and the malting loss (**see: Malting, 2. Malting terms and definitions**) is high at 20% compared with the 7% expected for barley malt. α-Amylase production occurs in the **scutellum** and is not increased by application of gibberellic acid. The water uptake levels and the malting time of sorghum malt are equivalent to those of wheat malts. The extract potential of sorghum malt is similar to those of barley and wheat malts. However, the α-amylase activities in sorghum malts are significantly higher than those of the other two but β-amylase is significantly lower. The reasons for these differences are not known. Sorghum malts are used in Africa to make traditional opaque beers. Proteolysis in sorghum malts is comparable to that in wheat and barley malts. Sorghum malts, like wheat malts, can cause difficulties during the production of standard European-type beers because the grains have no husk, and gelatinization temperature of their starch (>70°C) is higher than that of wheat starch (*ca.* 58°C) or barley starch (*ca.* 62°C). The β-glucans of sorghum malts can also contribute to processing problems. In contrast to the difficulties caused by sorghum malt, unmalted sorghum is used successfully as a brewing adjunct in different parts of the world. (GHP)

Briggs, D.E. (1998) *Malts and Malting*. Blackie Academic and Professional, London, UK.

Brissart, R., Brauminger, U., Hydon, S., Morand, R., Palmer, G.H., Savage, R. and Steward, B. (2000) *Malting Technology*. European Brewing Manual of Good Practice. Hans-Carl-Fach, Nurnberg, Germany.

Palmer, G.H. (1989) Cereals in malting. In: Palmer, G.H. (ed.) *Cereal Science and Technology*. Aberdeen University Press, Aberdeen, UK.

Malting – process

This article considers the processes that convert malting grains into malt. Since barley is by far the most used grain for malting the present account deals with this species. Several component parts of the process are involved, as described below. Weight for weight, there is an increase in value in this conversion. Depending on the season, a tonne of barley can cost about £100 and a tonne of malt can cost about £200. (**See also: Malting; Malting – barley; Malting – non-barley**)

1. Drying and storage of barley grains

An important pre-requisite of the malting process is that the moisture content of harvested grains be at about 10–12% so that the grain can be stored for long periods: if moisture reduction is required a grain drier is used. Freshly harvested barleys should be dried as quickly as possible to avoid deterioration of the grain. It is essential that the heat during drying does not damage the embryo and **aleurone layer** tissue. Damage is avoided by using air-on temperatures of between 50 and 60°C and the in-grain temperature should not exceed 40°C. Airflow should ensure that grain temperatures in the drier do not exceed temperatures recommended by maltsters. The dried barley (10–12% moisture) may be stored at 30–40°C for short periods, with regular monitoring of **germinative energy**. Such periods of warm storage tend to break **dormancy**. The expert management of stored barley is important. For example, high moisture areas of the grain bed can develop 'hot spots' of microbial activity, and here the viability of the grain declines rapidly. Aeration of the grain is crucial during long term storage at ambient temperatures.

2. Steeping

This is the soaking (hydration) of cereal grains in water under controlled conditions to increase the moisture of grains to an optimum of about 45–46%. Submersion (steeping) of the grain in water is usually broken by air-rest periods.

During steeping, usually at about 16°C, barley grains should germinate to over 96%. Steeping vessels vary in shapes and sizes. They usually have aeration facilities and the grain can be aerated during submersion or during air-rest periods when the water has been drained from the germinating grain. The moisture content of the grain increases during steeping from its starting value (about 11%) to about 40–46%. Grain at the lower values of moisture content are usually sprayed with water during subsequent germination. In the UK, the capacities of steeping vessels vary from 25 to 500 t per batch. About 800–1000 gallons (approx. 3600–4500 l) of water are required to steep 1 t of barley. The treatment of contaminated steep water (effluent) is an important aspect of the malting process and is an expensive additional cost to malt production. Steeping washes the grain and removes about 1% of extraneous material. Carbon dioxide is also released during steeping and germination. Its removal from the germinating grain helps to promote germination.

3. Germination

By the end of steeping the grain has usually germinated. It is then transferred to a germination vessel. The germination/growth process can last for as long as 5 days, whereas the steeping process does not usually extend beyond 48 h. Some vessels are designed to carry out both steeping and germination. During germination/growth, humidified air (RH *ca.* 100%) is passed through the growing grains to control grain temperature and to restrict water loss. Germination temperature is usually 16–19°C, or 12–16°C depending on the malting plant location.

During the germination/growth, the grain is turned daily. In some malting plants, **gibberellic acid**, at about 0.2 ppm with respect to grain weight, is sprayed on the germinated (chitted) grain within 24 h of the grain entering the germination vessel. Gibberellic acid accelerates the rate at which endosperm-degrading enzymes are produced in, and secreted from, the **aleurone layer** into the starchy endosperm during malting. As a result, malting time is reduced. Gibberellic acid treatment can cause excessive proteolysis during malting. Potassium bromate can be added at about 100 ppm, with gibberellic acid, to reduce proteolysis.

During germination/growth, endosperm-degrading enzymes hydrolyse cell wall β-glucans (about 90%), solubilize the protein matrix (about 38–40%) and degrade about 10% of the starch. This enzymic action converts the hard endosperm of barley into the soft and friable endosperm of malt. At the end-malting, the malted grain is called 'green malt' (about 40% moisture) because it has not yet been kilned.

4. Kilning

The treatment of heating germinated grain is called kilning. As the grain dries, enzyme activity declines and colour and flavours develop.

Kilning can be carried out in a separate vessel or in the germination vessel. Single-unit plants in which the grain is steeped, germinated and kilned are also in operation in different parts of the world. During kilning the temperature is raised from 16°C to about 60°C for about 24 h. An additional 2–3 h kilning at 82°C for lager malt and 100°C for traditional ale malts may also be applied. Water is driven from the grain and the moisture of the kilned malt is usually about 5%. Kilning increases the colour of the malt but decreases enzyme activities in the grain such as those of the heat-labile enzymes β-amylase, limit dextrinase, endo-β-1,3:1,4-glucanase and endoproteases.

Lager malts and distillers' malts are usually kilned at lower temperatures than ale malts and have lower colours and higher enzyme levels (Table M.3). Malt colour development is effected through a complex series of reactions, the **Maillard reactions**, between amino acids and sugars. The primary colour compounds are melanoidins, pyrazines, pyrroles, furans and thiophenes. Malts or barleys intended for stouts or dark beer production are kilned longer than lager or ale malt at higher temperatures ranging from 75°C (colour 35° EBC) to 230°C (colour 1550° EBC) (**see: Malting, 2. Malting terms and definitions**, Recommended analysis). Black or dark malts have no enzyme activity but add different flavours and colours to beers.

Lager malts tend to be less modified than ale malts. This is indicated by the extract and β-glucan levels, the soluble nitrogen ratio and the friability score (Table M.3). However, lager malts have more S-methylmethionine (S-MM) which is the precursor of the lager flavour compound dimethyl sulphoxide (DMS). During kilning, if the fumes of low sulphur natural gas are passed through the malt to dry it, carcinogenic nitrosamines (e.g. nitrosodimethylamine, NDMA) can develop from the reaction between nitrogen oxides and hordenin of the embryo of the malt. This problem is now controlled by indirect heating or by passing combustion fumes of fuels which contain sulphur through the malt. Sulphur inhibits the production of nitrosamines. An essential difference, in Scotland, between distilling malt and other malts is that peat is burned in the kiln and phenolic materials are deposited on the malt which give peaty and smoky flavours to many Scotch whiskies.

Table M.3. Some properties of unmalted barley and lager and ale malts.

Analyses	Barley grain	Lager malt	Ale malt
Moisture %	11.0	5.0	4.5
Starch %	65.0	60.0	58.0
Nitrogen %	1.6	1.6	1.6
Total soluble nitrogen %	0.3	0.62	0.65
Soluble nitrogen ratio	19.0	39.0	41.0
α-Amino nitrogen (mg/l)	0.05	0.15	0.17
Hot water extract %[a]	39.0	80.0	81.0
Sucrose %	1.0	2.0	2.0
β-Glucans %	3.5	0.3	0.2
Minerals %	1.5	3.0	3.0
Colour °EBC	1.0	2.0	5.0
Diastatic power (β-amylase °L)[b]	20.0	70.0	60.0
Dextrinizing unit DU (α-amylase)	5.0	40.0	30.0
Endo-β-glucanase (IRV units)[c]	100.0	700.0	500.0
Friability %	Very low	88.0	92.0

See: Malting, 2. Malting terms and definitions, for explanation of some analyses.
[a]% extract as by European Brewing Convention (EBC) analytical methods;
[b]L=Lintner; [c]Initial reciprocal viscosity.

5. Pattern of enzymic modification of the starchy endosperm

There are two main views regarding the progression of enzymic modification of the endosperm during malting. One is that the scutellum of the embryo secretes endosperm-degrading enzymes which break down the endosperm symmetrically from embryo to distal end of the starchy endosperm. The alternative view states that the pattern of modification is asymmetric, emanating initially from the dorsal (upper side) of the aleurone layer (**see: Malting – barley**, Fig. M.3). The asymmetric production of endosperm-degrading enzyme is induced by an asymmetric transport of gibberellic acid in the scutellum, along an asymmetric arrangement of vascular strands, to the dorsal part of the aleurone tissue envelope in which the starchy endosperm is contained.

It appears, however, that enzyme production in the scutellum is in fact very low and is not increased by gibberellic acid. Also, methods (for example, those of Heineken and Carlsberg breweries), which have been developed to assess endosperm modification by sectioning, have had to adapt their sectioning, procedures to suit an asymmetric pattern of enzymic breakdown of the starchy endosperm.

6. The abrasion process

This process was developed to reduce the malting rate and ensure that even the slowest of malting barley can malt at commercially accepted rates. The basic science of this relies upon damaging the permeability properties of the pericarp, which surrounds the grain, allowing gibberellic acid to enter the aleurone layer directly rather than from the embryo alone, as occurs in normal grains (**see: Malting – barley**, Fig. M.3). An abrading machine was developed to effect limited damage to the pericarp, especially at the usually slow modifying distal (non-embryo) end of the malting grain. Abraded malt was used extensively in the 1970s and the 1980s to produce malt for the brewing industry.

7. Malt modification

Describes the degrees to which the storage reserves of the starchy endosperm are solubilized (hydrolysed) during malting by endosperm-degrading enzymes such as endo-β-glucanases which degrade cell walls, proteases which degrade storage proteins and α-amylases which degrade starch granules (**see: Malting – barley**, Figs M.4 and M.5). The stages of modification are as follows: Under-modified, Modified, Well-modified and Over-modified. Under-modified malt grains give low hot-water extract and processing problems. Well-modified grains process efficiently and give high hot-water extracts. Some common tests used to measure the degree of modification follow.

Fine–coarse extract difference. This is the value for hot water extract of finely milled malt minus the value for hot water extract of the same malt, coarsely milled. A high percentage difference, for example, >2%, reflects under-modification. A low percentage difference, for example 1%, indicates a malt of good modification.

Friability and homogeneity. Both results are gained using the friabilimeter. Friability is determined by assessing the amount of flour released from the malt as it is turned in a rolling milling drum for a specified period of time. Friability is assessed as a percentage. Homogeneity of malt modification is determined as the total percentage of particles less than 2.2 mm released from the milled malt. A well-modified malt has a friability of over 90% and a homogeneity of over 98%.

Heineken's sanded grain test. Malt grains are sanded longitudinally and stained with methylene blue solution. The starchy endosperm of modified grains will absorb the blue stain; under-modified endosperms do not absorb the blue stain.

Carlsberg's calcofluor sanded grain test. Malt grains are sanded as in the Heineken test and then stained using calcofluor solution. Under-modified β-glucan walls of the starchy endosperm fluoresce in under-modified malts.

Soluble nitrogen ratio. The total soluble nitrogen (TSN) of the hot water extract divided by the total nitrogen (TN) of the malt, gives the soluble nitrogen ratio, sometimes called the Kolbach Index or the index of malt modification. This ratio is really an index of protein modification of the malted grain.

8. The concept of homogeneity of malt modification

It has been suggested that many malt-related brewhouse problems are caused by inhomogeneity of endosperm modification in a population of grains. Many of the standard analyses used today cannot detect this kind of inhomogeneity of malt modification. For example, a malt sample with overall 0.3% β-glucan was found to have 0.7% β-glucan in 20% of the grains and 0.2% β-glucan in the remaining 80% of the grains. Such differences in modification rate can be caused by differences in the physicochemical structure of the starchy endosperm. One of the most effective methods of identifying this kind of inhomogeneity of malt modification is hand-sectioning of single grains. New analytical approaches are being developed to assess inhomogeneity of modification by analysing β-glucan and protein breakdown in single grains of malt. (GHP)

(See: Malting; Malting – barley; Malting – non-barley)

American Society of Brewing Chemists (1992) *Methods of Analysis*. St Paul, MN, USA.

Briggs, D.E. (1998) *Malts and Malting*. Blackie Academic and Professional, London, UK.

Brissart, R., Brauminger, U., Hydon, S., Morand, R., Palmer, G.H., Savage, R. and Steward, B. (2000) *Malting Technology*. European Brewing Manual of Good Practice. Hans-Carl-Fach, Nurnberg, Germany.

European Brewery Convention (1997) *Analytica-EBC*. Zurich, Switzerland.

Institute of Brewing (1997) *Recommended Methods of Analysis*. London, UK.

Marins de Śa, R. and Palmer, G.H. (2003) Assessing malt modification using single grain analyses. *Proceedings of the 29th Congress of the European Brewing Convention*, Fachverlag, Hans Carl, Nurnberg, Germany, pp. 213–215.

Palmer, G.H. (ed.) (1989) *Cereal Science and Technology*. Aberdeen University Press, Aberdeen, UK.

Palmer, G.H. (1999) Achieving homogeneity in malting. *European Brewing Congress Proceedings*. Cannes, France, pp. 323–363.

Palmer, G.H. (2000) Malt performance is more related to inhomogeneity of protein and β-glucan breakdown than to standard analysis. *Journal of the Institute of Brewing* 106: 189–192.

Palmer, G.H. (2003) Maintaining progress in malting technology. *Proceedings of the 29th Congress of the European Brewing Convention*, Fachverlag, Hans Carl, Nurnberg, Germany, pp. 133–148.
Palmer, G.H. and Sattler, R. (1996) Different rates of development of α-amylase in the distal endosperm ends of germinated/malted Chariot and Tipper barley varieties. *Journal of the Institute of Brewing* 102, 11–17.
Paul's Malt Brew Room Book (1998–2000) Morton Hall Press, Suffolk, UK.

Maltose

A disaccharide comprised of two D-glucosyl units joined by an α-(1→4) linkage. In plants, maltose is a product of transient starch (**transitory starch**) degradation during the dark cycle in leaves, and is thought to be a primary molecule that is transported from the **amyloplast** to the cytosol during **starch mobilization**. Industrially, maltose has utility in the brewing, speciality foods and pharmaceutical industries, and is used as feedstock for the production of mannitol and other chemical compounds. (**See: Carbohydrates**) (MGJ)

Manilla Tamarind

See: Tamarind

Mannans

Polymeric carbohydrates (**hemicellulose** polysaccharides) with a backbone of β-1,4-linked mannose residues, which are prevalent as a carbohydrate reserve in the cell walls of the **endosperm** of some species, including coffee (Table M.4). Pure mannan is rare, is insoluble and the extreme hardness of ivory nut seeds is due to the extensive deposition of mannan to form the thick cell walls of the endosperm. The presence of an endosperm which is highly resistant to digestion may be of value to seeds which are dispersed by ingestion and elimination by mammals and birds. Commercially important mannan-rich polymers possess galactose side chains, and are categorized as galactomannans. (JDB)
(**See: Galactomannans; Hemicellulose**)

Mass maturity

In seed development and production, the point of maximum seed **dry weight**. (**See: Harvest maturity; Physiological maturity**)

Mast crop (or) year

In seed production: a mast year is one in which a large amount of forest tree seed happens to be formed and collected, such as when there is mass synchronized flowering in a population of a species with pronounced periodicity. 'Mast' is a collective name for the fruit of forest trees, such as once was gathered and fed to pigs: the word shares its derivation with **'meat'**, in its related sense denoting the edible parts of fruits or seeds. (**See: Tree seeds**)

Matrix priming

A seed **enhancement** treatment technique based on slow imbibition to reach an equilibrium hydration level, determined by the water potential of the moisture adsorbed on the particles with which the seed is mixed, followed by incubation, sieving and drying. Also known as 'solid matrix priming', 'matripriming' and 'matriconditioning'. (**See: Priming – technology**)

Mats

Formats for sowing seed impregnated on or between sheets. (**See: Tapes**)

Maturation – changes in water status

The majority of seeds are referred to as **orthodox** in which desiccation (**maturation drying**) occurs as a pre-programmed and final stage in their development (Fig. M.7). Seeds of the orthodox type are unique in the extent of water loss tolerated; as much as 90–95% of the original water is removed during the final stages of their development. In this dehydrated state, the seed can survive the vagaries of the environment and, unless dormant, will resume full metabolic activity, growth and development when conditions conducive to germination are provided.

Table M.4. Some examples of species that store cell-wall-mannan polymers in the endosperms of their seeds.

Species	Hemicellulose	Comments
Endospermic legumes		
Fenugreek	galactomannan: all approx. 55%	nl endosperm, al present
Crimson clover	man	nl endosperm, al present
Alfalfa (lucerne)		nl endosperm, al present
Carob	galactomannan. man:gal ~ 4:1	
Guar	galactomannan. man:gal ~ 2:1	nl endosperm? al present
Others		
Asparagus	glucomannan. man:glc ~ 3:2	
Ivory nut	97% man, 2% gal, 1% glc	nl endosperm?
Date palm	92% man, some gal present	
Coffee	98% man, 2% gal	
Caraway	mannan rich	

Storage cells contain a living cytoplasm unless indicated as nl: non-living cells, cytoplasm occluded by the storage mannans deposited in the cell wall. al: living aleurone layer (usually one cell thick) surrounds non-living storage cells. N.B. The aleurone layer is part of the endosperm, but does not store mannan polymers. man: Mannose, gal: galactose, glc: glucose, ?: unknown.

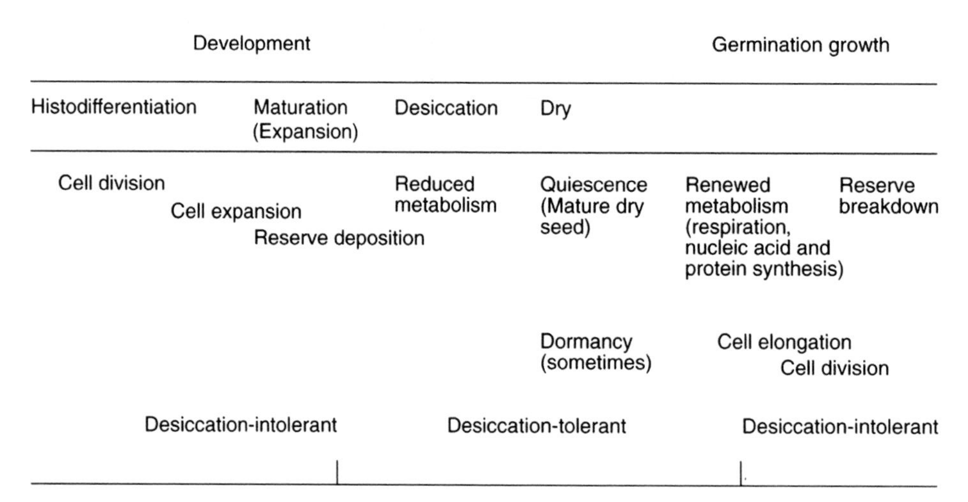

Fig. M.7. Some events associated with seed development, germination and growth. From Kermode, A.R. (1990) Regulatory mechanisms involved in the transition from seed development to germination. *Critical Reviews in Plant Sciences* 9, 155–195, with permission of CRC Press Inc.

The three major phases of seed development, characteristic of orthodox seeds (viz. histodifferentiation, expansion, and maturation drying; Fig. M.7) are marked by distinctive changes in **fresh weight**, **dry weight** and **water content** (**see**: Fig. D.20, **Development of seeds – an overview**). During histodifferentiation and early cell expansion, there is a rapid increase in whole-seed fresh weight and water content (**see: Development of embryos – cereals; Development of embryos – dicots; Embryogenesis**). Generally, a period of rapid dry-weight gain follows (when whole-seed fresh weight is relatively stable); this takes place during the later part of the seed expansion phase of development. Most seeds lose water during this phase as reserves are deposited primarily within storage tissues, displacing water from the cells. This decline in water content slows as the seed approaches its maximum dry weight. Then, as the seed undergoes maturation drying and approaches **quiescence**, there is a period of fresh weight loss accompanied by a rapid decline in whole-seed water content (**see**: Fig. D.20, **Development of seeds – an overview**).

Little is known about the mechanism and route of water loss from orthodox seeds. For many seeds (e.g. **castor bean** and **soybean**), the desiccation period is most likely initiated by the severing of the vascular supply to the seed (**funiculus** detachment) and senescence of the pod or capsule. Similarly, pectic substances in the lumina of xylem elements of the **rachis** of **wheat** and **barley** (laid down during the final stages of grain maturation) may lead to the progressive dehydration of the ear by cutting off its water supply (**see: Endosperm development – cereals**). Another possibility is that there is a relocation of water from the seed to the parent plant by a metabolically active process, i.e. the plant actually 'pumps' the water from the seed. The use of techniques such as NMR microimaging should contribute to our knowledge of the mechanisms and route of water loss from seeds of different species. When overexpressed in transgenic *Arabidopsis*, the gene *AGL15* (*AGAMOUS-like 15*), for a **MADS domain protein**, delays fruit senescence and slows the process of chlorophyll loss and seed desiccation. The effect is specific to the genotype and physiological state of the maternal tissues and there appears to be a strong connection between senescence/maturation of maternal tissues and seed desiccation.

Recalcitrant seeds do not undergo maturation drying, nor are they capable of withstanding water loss of the magnitude of that experienced by orthodox seeds. The seeds are shed at relatively high moisture contents and are highly susceptible to desiccation injury; in order to remain viable, they must not undergo any substantial change in moisture. Generally the development of recalcitrant seeds is similar to that of orthodox seeds up to the stage of maximum dry weight (mass maturity). For example, when adjacent trees of the sympatric species *Acer pseudoplatanus* (recalcitrant) and *A. platanoides* (orthodox) are compared, there is a strong temporal correlation in developmental events such as the accumulation of storage reserves and the acquisition of germinability. Although a few temperate recalcitrant species are dormant at shedding (e.g. *Aesculus hippocastanum*), **viviparous** germination (germination of the maturing seed on the parent plant) is a common event in many tropical recalcitrant species such as *Telfairia occidentalis*. Indeed, some of the unique characteristics of recalcitrant seed development are likely to be important for ensuring rapid germination and seedling establishment – features of significant adaptive value in certain habitats. In general, recalcitrant seeds at shedding have had almost no net loss of water; they can still be metabolically active and increasing in dry weight, remain desiccation sensitive and have no requirement for drying to stimulate subsequent germination. (ARK)

(**See: Desiccation tolerance** entries; **Desiccation sensitivity or intolerance; Recalcitrant seeds**)

Kermode, A.R. and Finch-Savage, W. (2002) Desiccation sensitivity in orthodox and recalcitrant seeds in relation to development. In: Black, M. and Pritchard, H.W. (eds) *Desiccation and Survival in Plants: Drying without Dying*. CAB International, Wallingford, UK, pp. 149–184.

Maturation – controlling factors

1. Hormonal control

Whether a seed is **dormant** or **quiescent** at maturity, its

quality and **vigour** rely heavily on processes that occurred during seed development: reserve deposition (accumulation of **storage proteins** and storage **oils** or **starch**), suppression of precocious germination, and development of stress tolerance. Control of seed maturation, in turn, is mediated by key interactions between different hormone signalling pathways and other regulatory cues provided by the seed environment.

During seed maturation there are changes in the concentration of the plant **hormones**, **auxins**, **gibberellins** (GAs), **cytokinins** and **abscisic acid** (ABA) (**see: Development of seeds – hormone content**). Several generalizations can be made regarding the complexity of this regulation by these hormones: (i) the amounts of a given hormone can change considerably during seed development; (ii) knowledge concerning hormone thresholds for the induction of certain processes and temporal changes in sensitivity or responsiveness to the hormone provides a more complete picture of hormonal regulation *in situ* than does monitoring changes in concentration; (iii) an apparent decreased sensitivity of seed tissues to exogenous hormone may not reflect a change in hormone perception, but a greater capacity for hormone turnover/catabolism, or less uptake; (iv) changes in hormone content give us little information about the physiological status of the embryo or seed; rather hormone flux and the balance between biosynthesis and turnover is of greater significance; (v) molecular analyses of **mutants** in which seeds are deficient in hormone biosynthesis or response have uncovered key components of hormonally induced **signal transduction** pathways; (vi) pathways of hormone catabolism can differ in different tissues of a seed, and hormone conjugates (e.g. glucosides or glucose esters) may be sequestered in separate compartments such as the vacuole. In some cases, the conjugates can later be metabolized to release active hormone; in other cases, the conjugates represent irreversibly inactivated hormone; (vii) in some cases, hormone metabolites themselves are able to induce physiological changes or changes in gene expression; and (viii) there is mounting evidence for hormonal 'cross-talk' in seeds and interaction between sugar and hormone signal transduction networks.

The regulation of seed maturation processes by ABA serves as an excellent example of some of these generalizations (Fig. M.8). An important role of the seed environment is to maintain embryos in a developmental mode until they are fully formed and have accumulated sufficient reserves to permit successful germination and subsequent seedling establishment. ABA was first implicated as a regulatory factor by studies on the effects of exogenous ABA on immature embryos excised from the seed and on the patterns of accumulation of endogenous ABA during seed development. Isolated embryos treated with ABA are inhibited from germinating precociously and they continue maturation processes, including the synthesis of storage proteins and storage lipids. Typically the content of ABA in seeds is low during early development (i.e. during the histodifferentiation and early pattern formation stage). It increases thereafter and usually peaks around mid-maturation and then declines precipitously during late development, particularly during the maturation drying phase (**see: Development of seeds – hormone content**). However, several factors add to the complexity of this pattern, including developmental changes in the sensitivities of embryo and seed tissues to ABA and the differential thresholds for the initiation and maintenance of developmental events (e.g. the expression of certain developmental genes and the induction and maintenance of dormancy). In alfalfa (*Medicago sativa*), sensitivity to ABA decreases linearly during development, and mature embryos (i.e. those that have undergone maturation drying *in planta*) require a high concentration of exogenous ABA to prevent their germination. In experimental transgenic tobacco seeds, **chimeric genes** containing the **vicilin** or **napin** storage protein gene promoter are responsive to exogenous ABA during seed development; however, after premature drying, ABA induction of gene expression is

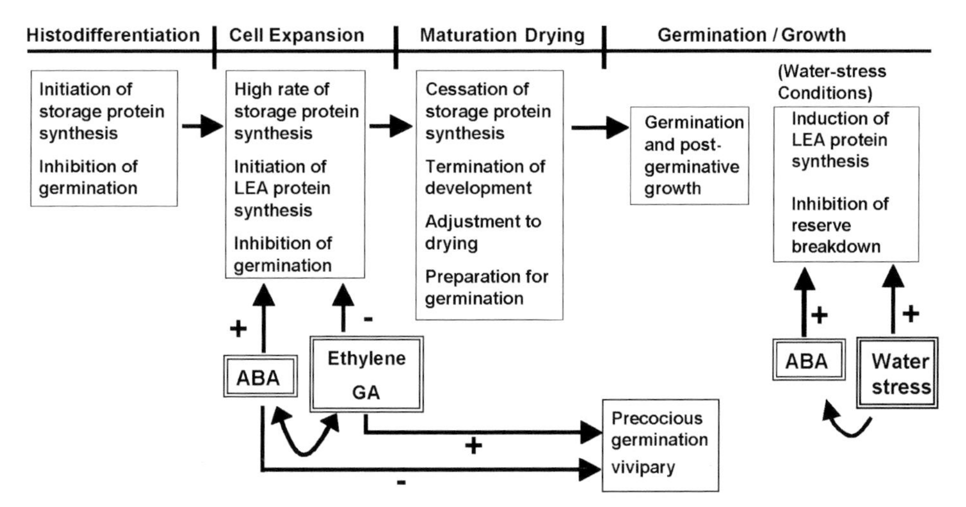

Fig. M.8. Events during the development and germination/growth of seeds that are affected by ABA and other hormones. Desiccation tolerance is generally acquired by orthodox seeds around mid-maturation, when LEA proteins and other protective substances are synthesized. Induction of a subset of LEA proteins also occurs following the transition to germination and growth, i.e. in seedlings and plant vegetative tissues when they are subjected to water-deficit-related stresses. Based on Kermode, A.R. (1995) Regulatory mechanisms in the transition from seed development to germination: interactions between the embryo and the seed environment. In: Galili, G. and Kigel, J. (eds) *Seed Development and Germination*. Marcel Dekker, New York, pp. 273–332. With permission of Marcel Dekker Inc.

virtually abolished. Embryos of alfalfa that are removed from the seed during development will germinate without being dried. These embryos can be re-induced to synthesize storage proteins, usually an exclusively developmental event, if treated with ABA, even in the seedling stage; this never occurs in the germinating and germinated embryos of dried seeds. In general, sensitivity of seeds to ABA declines during development, especially during desiccation.

Seed quality undoubtedly has a genetic basis, and is also strongly affected by environmental conditions (e.g. water stress) during seed formation, seed filling and harvest. The environment during seed development (e.g. light, temperature and water availability) can strongly influence the ABA content and sensitivity of the mature seed. In wheat, low temperatures can effect changes in ABA content or sensitivity, which in turn influence the degree of dormancy during development and in the mature grain. Likewise, water stress imposed during the development of *Sorghum bicolor* seeds decreases both their ABA content and sensitivity, and the seeds have an increased capacity for germination during development (**see: Pre-harvest sprouting – mechanisms**).

Mutants of maize and *Arabidopsis* that are either deficient in ABA, or exhibit a relative insensitivity to ABA (so-called 'response mutants') have been invaluable for elucidating the role of ABA and for defining the components of the ABA signalling pathways that control maturation events and interact with other signalling pathways (**see: Hormones; Hormone mutants; Mutants; Signal transduction – hormones**). Screens to identify suppressors of ABA-signalling mutants have also uncovered interesting evidence for cross-talk between hormone signalling pathways. These and other approaches have led to the following conclusions regarding the control of maturation:

(1) ABA, as well as promoting storage reserve synthesis and deposition (**see: Storage reserves synthesis – regulation**) also plays a role in the acquisition of desiccation tolerance during seed development. For example, ABA has a promotive role on the accumulation of desiccation-protectants (including soluble sugars, HSPs, small **Heat Shock Proteins** and LEAs, **Late Embryogenesis Abundant proteins**) (**see: Desiccation tolerance – protection**).

(2) ABA plays a key role in dormancy inception during seed development (**see: Dormancy – acquisition**). Mutants with defects in ABA signalling pathways have reduced dormancy, which is generally accompanied by disruption of seed maturation and precocious expression of germinative/post-germinative genes.

(3) Characterization of the ABA-insensitive mutants of maize and *Arabidopsis* led to the cloning of novel transcription factors. Lesions in the genes encoding these factors underlie some of the phenotypic characteristics of these mutants that are disrupted in seed maturation. For example, the *VP1* gene of cereals and the *ABI3* gene of *Arabidopsis* encode transcription factors that play a role in the expression of ABA-responsive genes during seed development (e.g. storage-protein-, *HSP*- and *LEA* genes).

(4) Redundancy in developmental and physiological response pathways is the rule rather than the exception in seeds and is exemplified by the emerging and mounting evidence for cross-talk between developmental and hormone response pathways. These studies have determined that there are key interactions between the ABA, **ethylene** and GA signalling pathways (**see: Signal transduction**) in controlling seed maturation and the inception of dormancy. Both ethylene and GA appear to act as antagonists to ABA function during seed development and germination (Fig. M.8). Interestingly, when a viviparous, ABA-deficient mutant of maize (*vp5*) is manipulated either genetically or via biosynthesis inhibitors to induce GA-deficiency during early seed development, **vivipary** is suppressed in developing kernels and the seeds acquire desiccation tolerance and storage longevity. Major accumulation of GA_1 and GA_3 occurs in wild-type maize kernels, just prior to a peak in ABA content during development. These GAs may induce a developmental programme that leads to vivipary in the absence of normal amounts of ABA; a reduction of GAs re-establishes an ABA/GA ratio appropriate for suppression of germination and induction of maturation.

(5) The later stages of maturation, including the acquisition of desiccation tolerance, inception of dormancy and loss of chlorophyll, are disrupted in seeds of certain *Arabidopsis* mutants (e.g. *leafy cotyledon1*, *leafy cotyledon2* and *fusca3* mutants and null *abi3* mutants). Hence these seeds are less viable and storable at maturity. Synergistic interactions between several proteins (including ABI3, FUS3, LEC1 and LEC2) control various key events, including accumulation of chlorophyll and anthocyanins, sensitivity to ABA and expression of individual members of the 12S storage protein gene family. The proteins encoded by two genes – ABI3 and GRS (the latter defective in the mutant *grs*, *green seed*) – have been pinpointed as major determinants affecting the long-term storability of seeds and loss of chlorophyll.

(6) ABI5, a member of the family of basic **leucine zipper (bZIP)** transcription factors regulates a subset of *LEA* genes during seed development and in vegetative tissues in the presence of ABA. This factor plays a key role during post-germinative stages of seedling development; more specifically it re-induces seedling '**quiescence**' under adverse environmental conditions and here ABA signalling interacts with **sugar-sensing**/signalling pathways. During germination and early post-germinative seedling development/growth, sugars can repress nutrient mobilization, **hypocotyl** elongation, **cotyledon** greening and expansion, and shoot development. High sugar accumulation during early seedling development may elicit a protective mechanism (i.e. developmental arrest), as it likely reflects poor growth conditions at a stage when the young seedling is highly vulnerable. As regards seedling viability, so far the influential genes uncovered encode several proteins associated with chloroplast function.

The maturation drying period that terminates the development of most seeds is important for overcoming the constraints imposed by the seed environment, including those mediated by ABA. For many seeds, desiccation promotes the transition from development to germination and is also important for vigorous post-germinative growth (**see: Maturation – effects of drying**).

2. Osmotic environment

Mechanisms whereby the seed environment retains the embryo in a developmental mode and suppresses germination have implicated not only ABA but also restrictions on water uptake by the embryo. One component of the chemical environment surrounding the developing embryo is the osmotic environment. This specialized environment maintains a highly negative osmotic potential during development, which is thought to suppress precocious germination of embryos by restricting water uptake. This osmotic restriction diminishes during maturation. As already noted, embryos of alfalfa seeds exhibit a progressive decline in their ABA responsiveness during development and maturation, with early-stage embryos having the greatest sensitivity. In contrast, sensitivity of the embryos to osmoticum is greatest at a relatively late stage of development (Fig. M.9). The combined inhibitory effect of ABA and osmoticum on germination during development is greater than their individual effects; again, this combined effect is stage dependent. There may also be species differences in respect of the relative importance of these two developmental cues; for example, the osmotic environment within the tissues of the tomato fruit appears to play a greater role than endogenous ABA in preventing precocious germination of developing tomato seeds.

Vivipary is characterized by germination of the embryo within the fruit on the parent plant. There is an uninterrupted progression from embryogenesis to germination with little or no cessation of growth (**quiescence**) and in most cases, little or no dehydration (maturation desiccation). Vivipary is a normal occurrence in *Rhizophora mangle* (a mangrove), in which seeds and seedlings exhibit a relative insensitivity to ABA. In other cases, sensitivity of embryos to osmotic potential is a controlling factor. The ABA-deficient *sitiens* mutant of tomato has approximately 10% the ABA content of its wild-type counterpart. This mutant exhibits reduced dormancy and seeds germinate viviparously in overripe fruits. However, low endogenous ABA may not directly contribute to vivipary in this case, since the embryos also exhibit a reduced sensitivity to the highly negative osmotic potential of surrounding seed tissues. Germinated *viviparous-1* (*vp1*) maize embryos are not only less sensitive to growth inhibition by ABA than are wild-type embryos, but are also less sensitive to inhibition by high osmoticum. Moreover, these *viviparous-1* embryos still germinate viviparously when surrounded by wild-type endosperm. The different ABA levels (or sensitivity) at various stages during seed development may induce a different sensitivity to osmotic stress, which permits vivipary under appropriately hydrated conditions. Thus the suppression of germination may involve a complex interplay between ABA and the osmotic environment. The conclusion provoked by these findings for viviparous embryos is that normal maturation processes involve the participation of ABA and osmotic controls. (ARK)

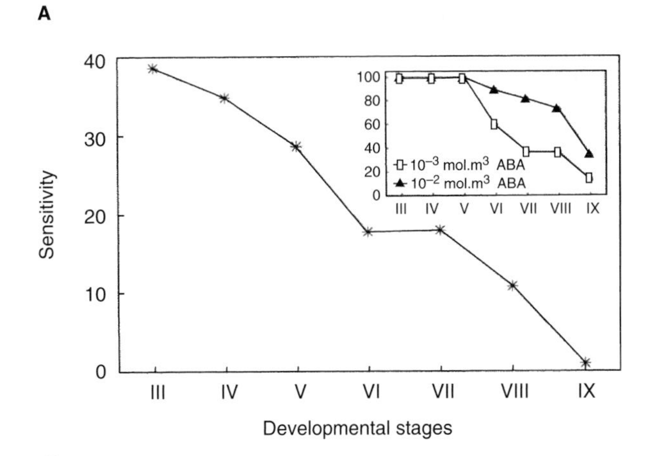

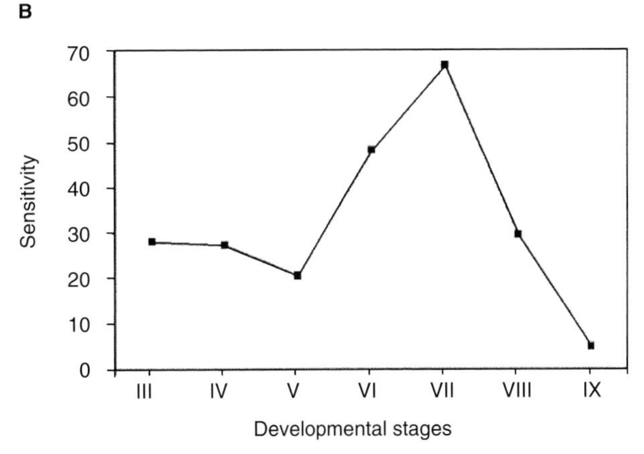

Fig. M.9. Sensitivity of developing alfalfa embryos to ABA and to high osmoticum. (A) Changes in embryo sensitivity to ABA (10^{-4} mol/m^3 (0.1 μM) in Murashige and Skoog medium) expressed in terms of a germination index. Sensitivity decreases almost linearly with the advancement of development, with the mature dry embryo (stage IX) being completely insensitive. Note that alfalfa seeds undergo maturation drying during stage VIII (31–36 DAP) to produce the mature dry stage IX seed. The inset shows sensitivity of developing embryos to 10^{-3} mol/m^3 and 10^{-2} mol/m^3 (1.0, 10 μM) ABA in Murashige and Skoog medium. **B.** Changes in embryo sensitivity to osmoticum (−1.01 MPa in Murashige and Skoog medium) expressed in terms of a germination index. The sensitivity of embryos to osmoticum is low at early stages of development, increases to a peak at stage VII, but then falls again at maturity (stage IX). Osmotic potentials which are more negative than −2.34 MPa are needed to prevent germination of alfalfa embryos excised from the mature seed. The combined inhibitory effect of ABA and osmoticum on germination during development is greater than their individual effects; furthermore, this combined effect is stage-dependent. From Xu, N. and Bewley, J.D. (1991) Sensitivity to abscisic acid and osmoticum changes during embryogenesis of alfalfa (*Medicago sativa*). *Journal of Experimental Botany* 42, 821–826, with permission of Oxford University Press.

Berry, T. and Bewley, J.D. (1992) A role for the surrounding fruit tissues in preventing the germination of tomato (*Lycopersicon esculentum*) seeds. A consideration of the osmotic environment and abscisic acid. *Plant Physiology* 100, 951–957.

Bewley, J.D. and Black, M. (1994) *Seeds. Physiology of Development and Germination*, 2nd edn. Plenum Press, New York, USA.

Finkelstein, R.R., Gampala, S.S.L. and Rock, C.D. (2002) Abscisic acid signaling in seeds and seedlings. *The Plant Cell (Suppl)*, S15–S15.

Gazzarina, S. and McCourt, P. (2001) Genetic interactions between ABA, ethylene and sugar signaling pathways. *Current Opinion in Plant Biology* 4, 387–391.

Kermode, A.R. (2004) Developmental Traits. Germination. In: Klee, H. and Christou, P. (eds) *Handbook of Plant Biotechnology*. John Wiley, UK.

White, C.N., Proebsting, W.M., Hedden, P. and Rivin, C.J. (2000) Gibberellins and seed development in maize I. Evidence that gibberellin/abscisic acid balance governs germination versus maturation pathways. *Plant Physiology* 122, 1081–1088.

Maturation – effects of drying (desiccation)

The experience of desiccation that normally occurs in **orthodox** seeds during maturation drying can effect changes that are important in determining the seed's post-maturation behaviour. These changes include the acquisition of **germinability** and the induction of expression of several genes. Experimentally, this aspect of seed maturation is explored by drying seeds of different developmental ages at various rates and to different extents followed by a determination of germinability and/or changes in gene expression (when the seed is later rehydrated). The ability of developing seeds to respond to premature desiccation and to undergo these changes upon subsequent rehydration depends strongly on the conditions of dehydration. The experimental approach of subjecting seeds to premature desiccation is presumed to mimic normal maturation drying and is a direct approach to establishing the nature of changes specifically induced by drying. It allows the researcher to distinguish between changes in gene expression that are pre-programmed during late maturation (and hence would take place even in the absence of seed desiccation) versus those that are directly induced by drying.

1. Desiccation acts as a 'switch' to terminate development and promote the transition to a germination and growth programme

The mature, quiescent, orthodox seed is able to commence germination immediately when hydrated under the appropriate conditions. The changes in mRNA subsets that underlie the metabolic events associated with seed development and germination/growth, imply the involvement of a switch, which effects the transition from development to germination. The switch is normally triggered in a precise temporal fashion to allow completion of developmental events that are important for successful germination and survival of the young seedling (e.g. the accumulation of sufficient reserves and desiccation protectants). There is substantial evidence that the final stage of seed development, maturation drying, enables seeds to switch off or down-regulate the transcription of genes associated with seed development (e.g. **storage protein** genes). Moreover, maturation desiccation promotes a normal transition to a germination and post-germinative programme, i.e. a switch in cellular activities from an exclusively developmental programme to an exclusively germination/growth-oriented programme. The role of maturation drying was first shown experimentally by imposing desiccation prematurely at certain stages prior to the completion of seed development. When imposed during a desiccation-tolerant stage of development (**see: Desiccation tolerance – acquisition and loss**), drying elicits a switch in synthetic events upon rehydration; events unique to development, such as synthesis of storage proteins, are terminated, while those associated with germination and growth are initiated.

In effecting this switch in gene expression, desiccation appears to act primarily at the transcriptional and post-transcriptional levels. The ability of the seed to synthesize storage protein mRNAs normally declines during desiccation; the subsequent hydration phase leads to the degradation of residual developmental messages that survive desiccation and at this time that genes associated with germination and growth are transcribed.

Elements associated with transcriptional control of genes are largely associated with the 5′ upstream region (sequences that are 'upstream' of the **transcription** start site and coding region of the gene). Down-regulation of storage protein gene expression, evident after rehydration, is by direct action upon the 5′ upstream regions of these genes; drying may also induce the synthesis of a transcriptional repressor of storage protein gene transcription. In transgenic tobacco seeds, transcription of chimeric genes driven by the **napin** and **vicilin** 5′ upstream regions is down-regulated by desiccation/rehydration while a chimeric gene driven by the constitutive (viral 35S) promoter is not (Fig. M.10).

Maturation drying may be critical for overcoming constraints by the maternal environment that maintain seeds in a developmental mode (**see: Maturation – controlling factors**). A decline in ABA sensitivity or attenuation of ABA signalling may be an important factor in the cessation of storage protein gene expression during late seed development or upon subsequent rehydration following natural or imposed drying. In transgenic tobacco seeds, chimeric genes containing the vicilin or napin storage protein gene promoter are responsive to exogenous abscisic acid (ABA) during development; however, after premature drying, ABA induction of gene expression is virtually abolished (**see: Storage reserves synthesis – regulation**).

Prior desiccation may also be important for the transition from germination to growth. Vigorous growth of seedlings may well require maturation drying; desiccation is required for the induction of enzymes involved in reserve mobilization during post-germinative seedling development. For example, isocitrate lyase, an enzyme involved in **gluconeogenesis** from **oils**, is absent from developing **castor bean** seeds, but after drying, the rehydrated seeds produce the enzyme, i.e. even though they are much too young normally to contain the enzyme. In developing aleurone layers of **wheat** and **barley** (cv. Golden Promise) only premature desiccation is capable of inducing the GA-responsivity typical of the **aleurone layer** of mature grain (**see: Mobilization of reserves – a monocot overview; Mobilization of reserves – cereals**). **Northern blot analyses** and transient expression of chimeric gene constructs in immature aleurone layer cells of half-grains suggest that drying may act, in part, by up-regulating the α-amylase gene promoter in response to GA. In immature barley aleurone layer cells, premature desiccation induces GUS reporter gene expression from a chimeric gene containing the minimal α-amylase promoter elements required for normal hormonal regulation of this gene.

The transition from development to germination (and from germination to growth) appears to involve extensive cross-talk between hormone signalling pathways (e.g. those induced by ABA, **gibberellin** (GA) and **ethylene** and other hormones) (**see: Signal transduction**). Post-transcriptional and post-translational controls (such as those mediated by RNA-binding proteins and by the proteasome-mediated degradation of signal transduction components) can promote (or attenuate)

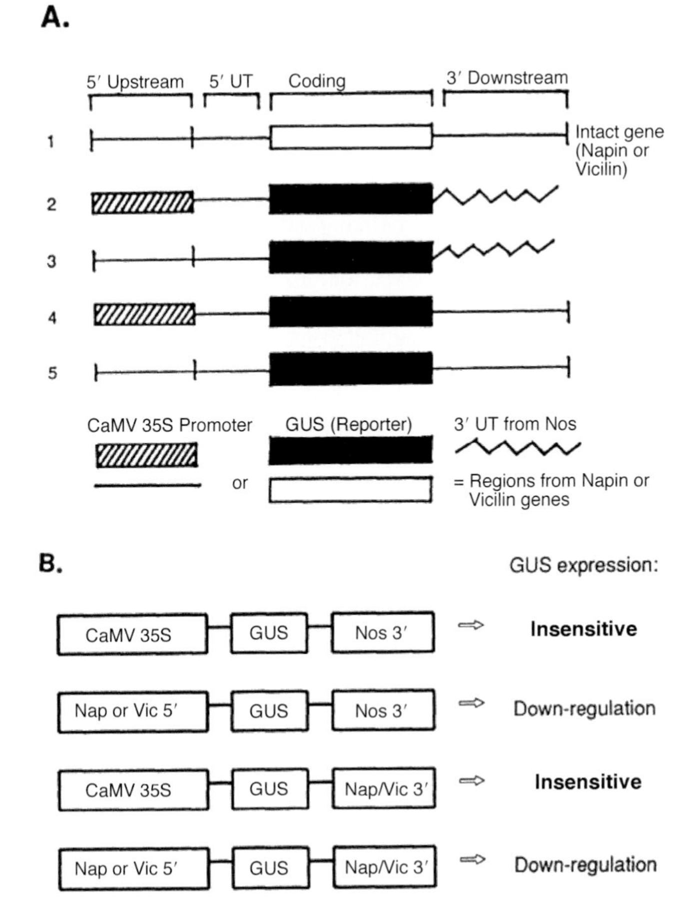

Fig. M.10. (A) Chimeric gene constructs were introduced into tobacco to determine the desiccation-responsive regions in vicilin and napin seed storage protein genes. All chimeric genes contained the coding region of the reporter gene for the bacterial enzyme β-glucuronidase (GUS). 5′ Upstream = promoter region; 5′ UT = 5′-Untranslated region. (B) Mature (i.e. dried) seeds were germinated and tested for expression of the GUS reporter gene. No GUS was expressed when this coding region was attached to the napin or vicilin promoters (Down-regulation) but was expressed when attached to the 35S viral constitutive promoter (Insensitive). Drying inactivated the promoters of napin and vicilin, which are seed-developmentally expressed genes, but not the viral promoter, whose activity is unrelated to development or germination. From Jiang, L., Downing, W.L., Baszczynski, C.L. and Kermode, A.R. (1995). The 5′ flanking regions of vicilin and napin storage protein genes are down-regulated by desiccation in transgenic tobacco. *Plant Physiology* 107, 1439–1449, with permission of the American Society of Plant Biologists.

hormone action. It is quite possible that desiccation (or the subsequent rehydration phase) acts through some of these mechanisms to effect a switch in cellular activities from an exclusively developmental programme to an exclusively germination/growth-oriented programme.

2. Effects of the rate and extent of desiccation

Seeds of orthodox species acquire desiccation tolerance (and can therefore undergo the maturation-induced changes described above) at a certain stage during their development. If drying is imposed during the tolerant stages of development, germination is elicited upon subsequent rehydration. However, the rate at which drying is imposed during early development is critical for the subsequent expression of germinability, and thus, when it is stated that a seed acquires a tolerance of desiccation at a particular stage during its development, it is necessary to define the rate of water loss to which it is subjected (**see: Desiccation tolerance – acquisition and loss**). Whole seeds of several **legumes** and *Ricinus communis* (**castor bean**) are unable to withstand rapidly imposed drying during most of development, prior to maturation drying, and do not germinate upon subsequent rehydration. This contrasts with seeds at the same stage of development dried slowly over saturated salt solutions or air-dried while enclosed in the pod, where full germinability is evident. Tolerance of rapid drying generally occurs only at or near the completion of reserve deposition (i.e. the attainment of maximum **dry weight**) just after the onset of natural drying, although there are exceptions (see subsequent discussion). Hence, it seems likely that the maturation processes important for tolerating water loss can only proceed normally if drying is slow and/or is imposed late in seed development.

Gradual water loss may allow protective changes to occur and hence increase the seed's resistance to disruption by dehydration (**see: Desiccation damage**). Rapid drying presumably would not allow such protective changes to take place and may cause considerable disruption to cellular membranes and internal structures. As a result, the seed may require time for metabolic readjustment (i.e. repair) following rapid drying. Repair cannot take place during drying itself because the seed reaches a critical dry (and quiescent) state before the repair processes can be initiated (**see: Desiccation tolerance – repair mechanisms**). Such repair is also impeded upon imbibition because of a too-rapid influx of water that cannot be accommodated by the weakened or damaged structural components of the cell. In fact, rapid rates of drying may predispose seeds to imbibitional injury, as indicated by increased rates of solute leakage, a symptom of cellular membrane disruption. However, slowing the rate of hydration of rapidly-dried seeds may allow time for repair to occur. Since seeds become capable of surviving rapid water loss during later development, they must acquire a greater cellular and metabolic resistance to this event or have an enhanced ability to effect repair during the early stages of imbibition.

The capacity to withstand rapid or slow desiccation during the tolerant phase of orthodox seed development varies between species. An intolerance of rapid desiccation occurs during the first 55 days of development of *R. communis* seeds, but they can tolerate slow drying as early as 25 DAP (days after pollination). Seeds (and isolated embryos) of the Gramineae (Poaceae), on the other hand, can survive and germinate following a severe drying treatment (to 5% water) at relatively early stages of development. The reasons for such differences among species are not known.

3. Premature desiccation can elicit changes normally invoked by maturation drying when imposed at stages earlier than physiological maturity

The time at which artificial drying can be tolerated and is effective is clearly strictly limited, but not necessarily to the time of physiological maturity. During very early development, seeds are generally intolerant of drying, but they later undergo a transition to a desiccation-tolerant state at a particular time (**see: Desiccation tolerance**). In many cases, drying at a desiccation-tolerant stage of seed development promotes germination upon subsequent rehydration, i.e. the switch is effected. Air-dried grains of wheat not only become germinable at an earlier stage of development than do non-dried grains, but when they later germinate they may do so at a faster rate than their non-dried counterparts. Seeds of *Phaseolus vulgaris* (common, French bean) undergo a transition to a desiccation-tolerant state around 26 DAP (approximately half-way through development) but not younger. The younger seeds eventually deteriorate because of a failure to recover cellular and metabolic integrity following the premature drying treatment.

A similar situation exists for the castor bean seed (*Ricinus communis*). Here germinability is not achieved *in planta* until 50 to 55 DAP (around the time of physiological maturity and the achievement of maximum dry weight). But imposed, premature drying will promote the germinability of seeds at least as young as 25 DAP. For both *P. vulgaris* and *R. communis*, the seeds acquire a tolerance of desiccation and the induction of germinability around 25 days after development has begun. There is a similar pattern in other species such as soybean, maize, barley and *Agrostemma githago*. Tolerance of desiccation is gained over only a few days of development well before the completion of major developmental events such as reserve deposition and the commencement of normal maturation drying.

During development of many **recalcitrant** species, some tolerance to desiccation increases throughout reserve accumulation, a stage associated with a progressive decrease in percentage moisture content as occurs in seeds of orthodox species. However, in recalcitrant seeds there appears to be no clear end point to development. For example, from year to year, seeds of *Quercus robur* may be shed from the same tree at different moisture contents; those shed at the lowest moisture content are most tolerant to desiccation. The percentage of seeds remaining viable after drying is highly correlated with their moisture content at harvest (premature and at shedding). This contrasts with orthodox species, where desiccation tolerance continues to increase after the acquisition of maximum seed dry weight and physiological maturity (i.e. during maturation drying). *Avicennia marina*, on the other hand, will tolerate little drying and can be considered at the extreme sensitive end of the tolerance continuum across species: developmental age has little influence on the desiccation sensitivity of seeds. In some other species, maximum desiccation tolerance is reached at a point before shedding (e.g. *Acer pseudoplatanus*) and tolerance may then subsequently decline (e.g. *Litchi chinensis*, *Clausena lansium* and *Coffea arabica*). This decrease in tolerance may be due to the initiation of germination, and in recalcitrant seeds there is a gradual decrease in tolerance as germination proceeds, similar to that occurring in orthodox seeds.

4. The effects of premature desiccation during the tolerant and intolerant stages of orthodox seed development

The processes that are induced by normal maturation drying fail to occur if seeds are subjected to dehydration at relatively early stages in their development. An analysis of the deleterious effects of early water loss can give some clues as to the qualitative changes that must take place in the seed to render it resistant to drying during normal maturation. Premature desiccation during the early developmental stages of *P. vulgaris* (up to 22 DAP, i.e. during the desiccation-intolerant stage) drastically reduces the metabolic and cellular integrity of the axis upon subsequent rehydration. Particularly evident is a loss in the capacity to re-form polyribosomes and to resume protein synthesis. Damage is suffered by cellular organelles (including protein bodies and mitochondria) and the nuclear membrane (**see: Desiccation damage**). In contrast, such severe perturbations do not occur following desiccation at a tolerant stage closer to normal maturation drying, e.g. at 32 DAP. Moreover, the limited damage that is sustained during drying at this stage is rapidly reversed following rehydration; cells regain their normal appearance within a very short time.

Studies on the effects of desiccation during the intolerant, pre-maturation stages of seed development suffer the defect of not distinguishing between the causes of desiccation intolerance and the changes occurring during the death of cells resulting from desiccation. Nevertheless, several studies (particularly those comparing the effects of drying at intolerant and tolerant stages) have provided some useful information on the cellular sites and/or metabolic processes that are most susceptible to damage during desiccation/ rehydration, and hence requiring protection for retention of viability. As implied earlier, the integrity of membranes is of crucial importance to the maintenance of viability; any undue disruption during drying is likely to be of immediate consequence once the seed imbibes (**see: Desiccation damage**). It is probable that some changes in membrane structure are provoked by desiccation (even during the tolerant stages of development), but the physical nature of such changes is unclear. While drastic changes to membranes have been observed upon their drying *in vitro*, the evidence suggests that membranes (in the desiccation-tolerant state) are protected against certain major alterations (e.g. transformation of the bilayer arrangement to hexagonal-type arrangements) (**see: Desiccation tolerance – protection**). The importance of a preservation of basic membrane integrity during drying is obvious, for cells would surely perish without the prompt re-establishment of functioning membranes upon rehydration when they are challenged by a swiftly changing hydration environment.

Fourier transform infrared microspectroscopy has been useful for elucidating some of the changes to membranes and other cellular constituents (e.g. proteins) following desiccation at the tolerant and intolerant stages of seed development. Isolated immature maize embryos tolerate slow drying from 18 DAP onwards and rapid drying from between 22 and 25 DAP. Lower membrane permeability upon rehydration is associated with rapid drying at the tolerant stages in contrast to almost complete loss of membrane integrity in embryos rapidly dried at an intolerant stage. In addition, there is a greater proportion of α-helical protein structures in embryos rapidly dried at a tolerant compared to an intolerant stage. The proportion of α-helical protein structures increases in the axes of embryos during slow drying of 20 and 25 DAP seeds (as compared to that within fresh developing seeds at these stages) and this factor coincides with the acquisition of additional tolerance of desiccation. (ARK)

Bewley, J.D. and Black, M. (1994) *Seeds. Physiology of Development and Germination*, 2nd edn. Plenum Press, New York, USA.

Kermode, A.R. (1995) Regulatory mechanisms in the transition from seed development to germination: interactions between the embryo and the seed environment. In: Galili, G. and Kigel, J. (eds) *Seed Development and Germination*. Marcel Dekker, New York, USA, pp. 273–332.

Kermode, A.R. and Finch-Savage, W. (2002) Desiccation sensitivity in orthodox and recalcitrant seeds in relation to development. In: Black, M. and Pritchard, H.W. (eds) *Desiccation and Survival in Plants: Drying without Dying*. CAB International, Wallingford, UK, pp. 149–184.

Maturation drying

A process that occurs in **orthodox** seeds during the final stages of their development, when they lose water (desiccate) and become quiescent. Water content declines from 40–60% to about 10–12% in many seeds as a consequence, rendering the seed storable. (**See: Desiccation tolerance; Development of seeds – an overview**; entries under **Maturation**)

Maturity

In an **orthodox** seed, maturity is attained when the seed has completed all stages of its development; it has achieved maximum dry weight and has undergone **maturation drying**. Maturity in **recalcitrant** seeds is less clear, and at shedding they may have lost very little water, be metabolically active and still increasing in dry weight. (**See: Harvest maturity; Mass maturity; Physiological maturity**)

Maturity group

The classification of varieties on the basis of suitable harvest maturity date for the cultivation purposes in different geographical latitudes or at different sowing dates, including species in which flowering or maturation is daylength-sensitive, such as soybean (**see: Soybean – cultivation**), oat and onion.

Meadowfoam

Meadowfoam (*Limnanthes alba*, 2n = 18, Limnanthaceae), a small herbaceous winter annual, is native to the Pacific Coast region of North America. Contract commercial production on limited hectares occurs in Oregon, USA. The seeds are held within sepals, normally in groups of three nutlets, each capable of producing five seeds. The seeds are pear shaped, about 4 mm long, 6 mm in circumference at the widest point, and weigh 8–9 g/1000 seeds (Fig. M.11). The mid-brown coloured seeds have a deeply ridged **pericarp** over a thin brown **testa** that closely adheres to a crushed layer of **endosperm** and the underlying **cotyledons**.

The seed yields about 29% oil with a unique fatty acid composition (**see: Oilseeds – major**, Table O.2). About 90% of the **oil** is made up of **fatty acids** with carbon chain lengths of C20 to C24, nearly all of which are monounsaturated and have the double bond in the Δ5 position. The oil is of interest to the cosmetic and oleochemical industry. The tendency to **shatter** seed is a constraint to commercial expansion as is the presence of glucosinolates in the residual, high protein meal. (**See: Oils and fats – genetic modification**) (KD)

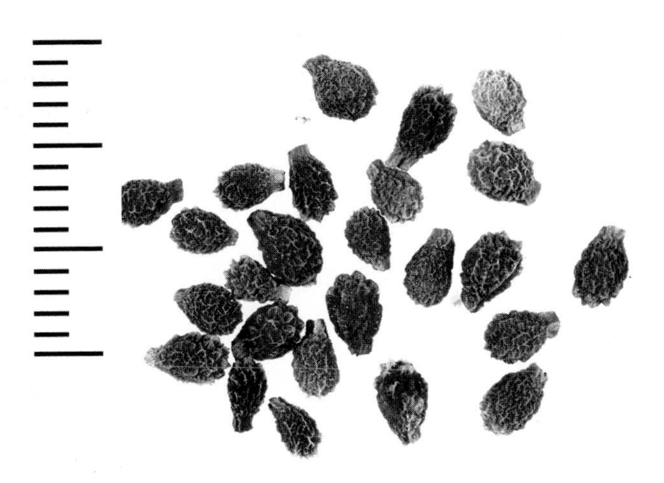

Fig. M.11. Seeds of meadowfoam (scale = mm). Photograph kindly produced by Ralph Underwood of AAFC, Saskatoon Research Centre, Canada.

Hirsinger, F. (1989) New annual oil crops. In: Robbelen, G., Downey, R.K. and Ashri, A. (eds) *Oil Crops of the World*. McGraw-Hill, New York, USA, pp. 518–532.

Kleiman, R. (1990) Chemistry of New Industrial Oilseed Crops. In: Janick, J. and Simon, J.E. (eds) *Advances in New Crops*. Timber Press, Portland, Oregon, USA, pp. 196–203.

Mean germination time

A measure used in calculating one expression of **germination rate**, often used as a measure of seed vigour. (**See: Germination rate; Vigour tests – physiological**)

Mean germination value

An assessment of the relative rate of germination, used as a measure of seed vigour of some tree species. (**See: Germination rate; Vigour tests – physiological**)

Meat

A broad term for the edible part of anything – which, in the context of seeds, is used in some countries for the embryo of dehulled **nuts**, **sunflower** and **cotton** kernels, for example.

Medicines

See: Pharmaceuticals, medicines and biologically active compounds – an overview

Megagamete

The larger, female gamete formed by organisms that produce unequal-sized gametes-anisogamous (anisogamy), e.g. the ovum, egg cell. (**See: Reproductive structures, 1. Female**)

Megagametophyte

Also macrogametophyte, macroprothallium. A female structure that develops from the **megaspore**. In the evolutionarily less advanced gymnosperms it gives rise to the **archegonia** and primary endosperm. The homologue of the megagametophyte in the angiosperms is the **embryo sac**. In seeds of **gymnosperms** the megagametophyte (primary 'endosperm') is also used generally as the term for the haploid (n) maternal storage tissue that remains after fertilization of the egg, and which remains to surround the embryo. It is a persistent tissue formed at the time of the female gamete. (See: **Evolution of seeds; Gymnosperm seeds**)

Megasporangium

Also called macrosporangium. The organ (sporangium) in which the **megaspores** are formed, the spermatophytes' (seed plants) homologue of which is the nucellus. (See: **Evolution of seeds; Reproductive structures, 1. Female**)

Megaspore

Also called the macrospore; the larger spore formed by **heterosporous** plants that gives rise to a female (mega) gametophyte. (See: **Evolution of seeds; Reproductive structures, 1. Female**)

Megasporophyll

Also called the macrosporophyll; the fertile leaf bearing the **megasporangium**, called the carpel in seed-bearing plants. (See: **Evolution of seeds; Reproductive structures, 1. Female**)

Meiosis

The type of cell division in eukaryotes by which germ cells or **gametes** (egg and sperm) are produced. Meiosis (reduction division) involves two successive nuclear divisions with only one round of DNA replication. As a consequence, the chromosome number within the resultant cells is half that of the parent cells, i.e. the **diploid** (2n) number of chromosomes becomes **haploid** (n). During meiosis there can also occur an exchange of genetic material between chromosomes, due to a process known as crossing-over. This, and the eventual fusion of two haploid gametes from different parents accounts for the variations in the chromosomes between the parent and offspring generations. Contrast with **mitosis**, a key event of the **cell cycle**. (See: **Reproductive structures**)

Melons

See: **Cucurbits**

Membrane

The structures that define the boundary of each cell and delimit cellular compartments. Consisting of **phospholipids** (usually arranged in a bilayer) and proteins, membranes serve as a barrier to the diffusion of molecules and as signal receptors/transmitters, maintain electrochemically distinct environments, and provide mechanical links between the cytosol and the cell wall. Specialized membranes, or membrane regions, are also involved in protein synthesis, secretion and electron transport. Chloroplast membranes are rich in glycolipids. (See: **Cells and cell components; Desiccation damage**)

Membrane damage

Membranes are particularly sensitive to drying and subjected to two types of damage: structural and chemical. Both lead to changes in the physico-chemical properties of membranes, which consequently affect their functional integrity and result in loss of compartmentation within the cell. Excessive leakage upon rehydration (e.g. during **imbibition**) is the most commonly observed symptom of membrane damage, such as normally occurs as one of the **deterioration reactions**; this effect is used as the basis for assessing vigour is some species (**see: Vigour testing – biochemical, 1. Conductivity tests**). Water is the driving force that governs the intermolecular associations between polar lipids, thereby conferring the bilayer structure and functional integrity to membranes. The removal of water may lead to increased packing and rigidity of **phospholipids, phase** transition from a liquid crystalline to a gel phase, lateral phase separation (demixing) and exclusion of membrane proteins, membrane fusion and, sometimes, formation of non-bilayer structure such as a hexagonal phase. Chemical damage arises from a **free radical** attack to which lipids are particularly sensitive. Symptoms of free radical injury include lipid peroxidation and increased free fatty acids. (OL)

Mericarp

One of the two dispersal units of a **cremocarp**. Typical of the Apiaceae in which family are many spice seeds. (See: **Spices and flavours**)

Meristem

(Greek: *merizein* = divide) A higher-plant tissue in which cells (meristematic cells) are capable of division mitotically and to give rise to new cells, tissues and organs. Meristems are generally located at shoot and root apices (apical meristems, e.g. in the **plumule** and **radicle** of seeds), in leaf axils, in developing fruits as well as in the stems of plants with secondary thickening (cambium and cork cambium). (See: **Embryonic axis**)

Mescal bean

Plants of this species (*Sophora secundifolia*, Fabaceae) bear **psychoactive seeds** that contain the hallucinogenic alkaloid, cytisine (**see: Pharmaceuticals and pharmacologically active compounds**).

Mesocarp

The middle of three layers of **pericarp** that in **drupes** becomes relatively large and fleshy. (See: **Endocarp; Mesocarp; Nut**)

Mesotegmen

In the mature seedcoat it is the middle layer between the inner and outer epidermis of the inner integument. (See: **Seedcoats – structure**)

Mesotesta

In the mature seedcoat it is the middle layer between the inner and outer epidermis of the outer integument. (See: **Seedcoats – structure**)

Metabolomics

In parallel to the terms transcriptome and proteome, the set of metabolites synthesized by an organism constitute its metabolome. For metabolic engineers who seek to optimize the production of valuable biochemicals, this is often the target for manipulation. One of the elements of metabolomics is the science of determining the concentrations of metabolites in the tissue of interest at a given time (i.e. metabolic profiling). Metabolic profiling provides a snapshot of the chemical composition of that tissue. By comparing metabolite profiles between two tissue states, separated either in time or in space, or by genetic variation, differences in genome functionality can be assessed.

However, some theoretical considerations limit direct interpretation of metabolic networks generated from metabolic profiling. First, any subcellular compartmentalization is lost in the process of sample preparation. Although mRNA or protein expression can sometimes be ascribed to plant compartments on the basis of their target sequences, there is a high degree of uncertainty about the actual location of metabolites, many of which occur simultaneously (and for potentially different purposes) in different locations and in varying amounts. Therefore, at best, metabolomic information can be interpreted on the multicellular, tissue or organ level. Another element of metabolomics is identification of the biochemical and/or genetic mechanisms that regulate the flux through metabolic pathways. Knowing how the flux through a metabolic pathway is regulated provides metabolic engineers with insights as to how to modulate that flux in order to optimize the concentration of the metabolites that are derived from that pathway.

Developments in functional **genomics** have propelled the need for global profiling of gene expression at the level of metabolites. The analytical technologies that are being utilized combine chromatographic procedures for separating metabolites, based upon their physical and chemical properties, coupled with mass spectral-based identification of each metabolite. A variety of chromatographic methods can be used for separation purposes, including gas–liquid chromatography (GLC), liquid chromatography (LC), or capillary electrophoresis (CE). The mass spectrometric-based identification of metabolites has expanded with the development of different ionization capabilities. In particular, the advent of electrospray ionization and matrix-assisted laser desorption ionization are the two biggest success stories in this regard, making possible the coupling of LC and CE separation methods to mass spectrometry (MS). Another analytical tool also being utilized in the analysis of plant metabolism is nuclear magnetic resonance (NMR). Although this technology is not as sensitive as the chromatographic-based methods for profiling metabolites, NMR holds the promise of elucidating the regulation of metabolism and possibly imaging of metabolites in intact tissues. (AJR)

Ratcliffe, R.G. and Shachar-Hill, Y. (2001) Probing plant metabolism with NMR. *Annual Review of Plant Physiology and Plant Molecular Biology* 52, 499–526.

Microarrays

Miniature grids ('DNA chips') containing specific DNA sequences in up to thousands of spots that can be hybridized with cDNA made from expressed mRNAs. This technique allows for the simultaneous identification and quantification of differences in the expression of 10,000 or more genes between two samples. It is an important technique for the study of gene expression in seeds. (**See: Transcriptomics**)

Microbody

A class of ultrastructurally defined subcellular organelles that ranges in size from 0.2 to 1.7 μm in diameter, and in shape from spherical to tubular. Microbodies possess a single boundary membrane and a matrix that is usually coarsely granular or fibrillar, although amorphous or crystalline inclusions are occasionally present. They are devoid of DNA and a protein synthetic machinery. Originally identified morphologically in rat liver cells in the early 1950s, microbodies have been since been found in virtually all other types of eukaryotic cells. All microbodies contain the enzyme catalase and several hydrogen peroxide-generating oxidases. They also possess remarkable functional versatility, housing a wide array of enzymes depending on the organism, cell/tissue type and/or environmental conditions. The term microbody is a general name not implying any particular function and has given way to a more universal and functional nomenclature of peroxisome including the several subclasses of plant peroxisomes: **glyoxysomes** in oilseed storage tissues, leaf-type and root peroxisomes, and unspecialized **peroxisomes**. (RTM)
(See: Cells and cell components)

Beevers, H. (1979) Microbodies in higher plants. *Annual Review of Plant Physiology* 30, 159–193.

Huang, A.H.C., Trelease, R.N. and Moore, T.S., Jr (1983) *Plant Peroxisomes*. American Society of Plant Physiology. Monograph Series, Academic Press, New York, USA.

Microgamete

Smaller, male gametes formed by hetero- or anisogamy, a phenomenon of producing distinct gametes of different sizes. (**See: Evolution of seeds; Reproductive structures, 2. Male**)

Microgametophyte

Develops from the **microspore**; in some lower plants (cryptogams) the prothallus giving rise to male gametes: in spermatophytes, the pollen grain producing the male gametes. (**See: Evolution of seeds; Reproductive structures, male**)

Micropylar endosperm

See: Endosperm cap

Micropyle

Opening of the integument(s) at the apex of the ovule and later the seed, usually acting as a passage for the pollen tube. (**See: Seedcoats – structure**)

Microsporangium

Organ (sporangium) in which the microspores (early pollen grains) are produced, called the pollen sac in flowering plants. (**See: Evolution of seeds; Reproductive structures, 2. Male**)

Microspores

Male pre-gametophytes produced in **microsporangia** of **heterosporous** plants (those with male and female reproductive structures). Microspores that have evolved a tube to deliver the male gamete to an egg are known as pollen. Using temperature treatments timed to coincide with specific

stages of pollen or microspore development, the pollen can be induced to produce **haploid embryo** cultures when plated on to specifically designed media. This process is called **androgenesis**. (PvA)
(See: Fertilization; Pollination; Reproductive structures, 2. Male)

Microsporophyll

Fertile leaf bearing the microsporangia, called stamens in the flowering plants. **(See: Evolution of seeds; Reproductive structures, 2. Male)**

Microsymbionts

Soil-living organisms that form symbiosis with plant roots. The three types that are important for cultivated plants are mycorrhiza (including **arbuscular mycorrhiza**), **rhizobia** and *Frankia*. **(See: Rhizosphere microorganisms)**

Mildews

Diseases in which parasitic fungi of the family Peronosporaceae cover infected plant surfaces with a white, grey or brightly coloured dense growth of mycelia and spores – including seedlings of some species that are growing in moist soil, thereby reducing or delaying field emergence. Soybean downy mildew (*Peronospora manshurica*), for instance, is transmitted in the seedcoat, which can appear dull white and cracked because it is so heavily encrusted with oospores; infected seeds also may be lighter and smaller than healthy seeds. Sunflower downy mildew, an endemic worldwide disease of high economic importance, is primarily caused by a soilborne pathogen (*Plasmopara halstedii*) that infects underground tissues of young seedlings and, depending on severity, causes pre- or post-emergence seedling **damping-off**, seedling **blight**, and stunted plants that produce heads with few, under-developed, colourless or otherwise poor quality seed; a latent (symptomless) seedborne form of the disease quite often occurs also, in which one or two generations may be grown before infection becomes evident. Seedborne **pathogens** can usually be eradicated by a fungicide seed treatment. (PH)
(See: Treatments – pesticides, fungicides)

Milkweed

1. Botany

Milkweed floss (mass of fibres) is composed of **trichomes** arising from the micropylar end of seeds of *Asclepias syriaca*, *A. speciosa* or *Calotropis gigantea* in the family Asclepediaceae. The elliptical follicles (seed pods) are usually about 10 cm long and 3–4 cm thick. Packed inside each are hundreds of seeds each bearing a small tuft of about 900–1500 fibres, perfectly designed to aid air dispersal (**see: Seedcoats – dispersal aids**). The hollow, highly lustrous, yellow-white, unicellular fibres may be 2.5–4 cm long and 30 μm in diameter. Thick ridges, giving the appearance of longitudinal striations, are a unique feature of this seed fibre (**see: Fibres** Fig. F.5D).

2. Chemistry

Milkweed floss is a ligno-cellulosic fibre (**see: Fibres**, Table F.1). The lower cellulose and higher lignin content than cotton makes the fibres soft but brittle. Cell wall thickness is 1.2–1.5 μm. Milkweed fibres are exceptionally buoyant and water repellent due to cutin and wax deposition in the cell wall.

3. Origins

A. syriaca or common milkweed is an indigenous plant of North America where there are more than 100 identified species. *C. gigantean* is a native milkweed of southern Asia and Africa.

4. Location

Until recently, milkweed pods were only collected from native stands in the USA and Asia. As a perennial, deep-rooted plant, milkweed grows in old fields and meadows, roadsides, and in habitats not grazed by livestock. There have been some recent efforts to develop milkweed as a cultivated crop in the Midwestern USA. A drawback is that only 1–3% of flowers produce mature pods.

5. Production

During World War II, American children collected milkweed pods for a government-sponsored programme to produce life vests and insulated aviator jackets. Over 11 million kg of pods were collected, yielding enough fibre to fill 1.2 million emergency flotation devices. Since 1987 there has been a single commercial operation for processing milkweed floss in North America producing a floss for use as pillow and comforter stuffing. Hybrids between *Asclepias speciosa* and *Asclepias syriaca* show improved fibre yields of 350 kg/ha with fibre representing 24% of the total follicle weight, and with more pods per stem (2–3 pods) than the parent species (1–2 pods).

6. Processing

For commercial production, unopened milkweed follicles are harvested with a modified corn picker. They are transferred to a rolling mill to open them and are then dried. When the moisture content is less than 10%, floss is separated from other pod components. The yield of fibre from the inter-specific hybrids is 1 kg from 227 follicles. Milkweed floss is too brittle to spin into yarn by itself. Cotton-milkweed blends have been produced that show promise as novel textiles.

7. Economics

Milkweed floss (US$20 per kg) is an inexpensive and hypoallergenic alternative to white goose down (US$66 per kg). There is a projected annual use of 1.27 million kg of milkweed in the pillow and comforter filling market. Research programmes in the UK and Germany are investigating alternative uses of milkweed floss as a novel, natural fibre. Smaller potential markets exist in non-woven materials, in blended yarns, and in pulp and paper manufacturing. New uses for other milkweed seed components such as oil and meal are being developed and should stimulate the market for floss. (BAT)

Drean, J.Y.F., Patry, J.J., Lombard, G.F. and Weltrowski, M. (1993) Mechanical characterization and behavior in spinning processing of milkweed fibers. *Textile Research Journal* 63, 443–450.

Knudsen, H.D. and Zeller, R.D. (1993) The milkweed business. In: Janick, J. and Simon, J.E. (eds) *New Crops*. John Wiley, New York, USA, pp. 422–428.

Silva, B. (1999) Milkweed-crop of the 20th century. *AgVentures* 3, 45–49.
Von Bargen, K., Jones, D., Zeller, R. and Knudsen, P. (1994) Equipment for milkweed floss-fiber recovery. *Industrial Crops and Products* 2, 201–210.
Witt, M.D. and Knudsen, H.D. (1993) Milkweed cultivation for floss production. In: Janick, J. and Simon, J.E. (eds) *New Crops*. John Wiley, New York, USA, pp. 428–431.

Millets

'Millet' is a broad category of any number of small-seeded grasses. Millets include: pearl millet (*Pennisetum glaucum*); proso, browntop, or broomcorn millet (*Panicum miliaceum*); little millet (*P. sumatrense*); foxtail millet (*Setaria italica*); finger millet or ragi (*Eleusine coracana*); teff (*Eragrostis tef*); fonio (*Digitaria* spp.); guinea millet (*Brachiaria deflexa*); barnyard or japanese millet (*Echinochloa crus-galli*); jungle rice millet (*E. colonum*); kodo millet (*Paspalum scrobiculatum*); and Job's tears (*Coix lacryma-jobi*) (Table M.5) (Colour Plate 8D).

Millets are mainly cultivated rain-fed for their grain, often by small-scale farmers or pastoralists, as regional or national products, primarily on marginal lands in semi-arid areas in temperate, subtropical and tropical regions, throughout Africa, Asia (together producing 94% of global output) and Europe. In the USA, Central and Latin America, some species are also grown irrigated by large-scale farmers.

The species differ in seed traits, soil and climatic requirements, and growth duration. Grain yields, often below 1 t/ha, are produced in environments unsuitable for the dominating food suppliers, **wheat**, **rice** and **maize** (each ± 600 million t grain globally), and thus contribute to food security, but are currently not utilized to their potential. The grain is without **gluten** but may have specific surmountable nutritional shortcomings. Processing into commonly acceptable or attractive products is laborious and complicated by the small seed size, shape and structure. Most millets have a **C4 photosynthetic** pathway (like maize), have different base chromosome numbers even within the genus, are **diploid**

Table M.5. Millet species.

Common names and synonyms[a]	Botanical name	Botanical synonyms	Approximate worldwide grain production (t)	Areas grown	Additional notes
Pearl millet (bulrush millet)	*Pennisetum glaucum*	*P. americanum* *P. typhoides*	15 million	Africa (mainly west), India	The major millet. Probably originated in West Africa about 5000 ya.
Foxtail millet (Italian millet, German millet, Hungarian millet, Siberian millet)	*Setaria italica*	*Panicum italicum* or *Chaetochloa italica*	±5 million	Asia (mainly China) and Europe	> 1000 BC the grain size was used in China to define the measure of the foot and the inch, and the length (tone) of the 27 strings of a musical instrument
Proso millet, Common millet (hog millet, broomcorn millet, Russian millet)	*Panicum miliaceum*		±4 million	Eastern Asia (Mongolia, Manchuria, Japan), India, eastern and central Russian Federation (Kazhakstan and the Ukraine), the Middle East and USA	Bird seed, also grown by nomads because of short growing cycle. Together with pearl millet , more heat tolerant than sorghum and maize
Finger millet	*Eleusine coracana*		< 4 million		
Others below 3 million t					
Fonio, black fonio, acha, fundi, hungry rice	*Digitaria exilis*			Isolated pockets in Senegal to Cameroon	
White fonio, jungle rice	*Digitaria iburua*	*Echinochloa colona*		Throughout sub-Sahelian West Africa. Important in Mali	Crosses with *E. crus-galli* and *E. haploclada*
Barnyard millet	*Echinochloa crusgalli*	*E. subverticillata*		Thoughout the tropics in Asia (e.g. Java), Egypt, California	Grown in swampy places
Japanese barnyard millet, savanna millet, sanwa millet	*Echinochloa frumentacea*	*Panicum frumentaceum*		Asia (e.g. China, India and Japan) and the USA	Salt tolerant
Teff	*Eragrostis tef*	*E. abyssinica*		Main producers: Ethiopia and Eritrea	
Little millet	*Panicum miliare*	*P. sumatrense*		India	
Kodo millet, creeping paspalum, dronkgras (South Africa)	*Paspalum scrobiculatum*			India	

[a]See separate entries for types in **bold** lettering. ya: years ago.

or **polyploid** (two or more genomes (sets of chromosomes) per nucleus, often resulting from hybridization followed by chromosome doubling).

Sorghum (*Sorghum bicolor*) is confusingly called great millet. Its global production is ≈60 million t grain. The major millet species is **pearl millet,** or bulrush millet (*P. glaucum*), which accounts globally for ≈15 million t grain of the ≈28 million t millet grain. (**See: Crop Atlas Appendix, Maps 1, 10, 21, 25**) (WAJdM)

Dendy, D.A.V. (ed.) (1994) *Sorghum and Millets: Chemistry and Technology*. American Association of Cereal Chemists, St Paul, MN, USA.
FAO Yearbook (2002) *Production 2000*. Vol. 54, pp. 71, 89, 91.

Mitochondria

Plural of mitochondrion. A cellular organelle composed of an outer and inner **phospholipid** membrane, the latter being invaginated to form cristae, and an aqueous matrix. The citric acid (Kreb's) cycle takes place within the matrix and **ATP** synthesis by oxidative phosphorylation occurs on the electron transport chain present within the inner membrane (**see: Respiration**). It is a semi-autonomous organelle which contains its own genome and protein synthetic capacity to make a limited number of mitochondrial proteins (including, for example, the 'cytoplasmic' factor in **cytoplasmic male sterility** systems). (**See: Cells and cell components**)

Mitosis

The division of the nucleus of eukaryotic cells (M phase of the **cell cycle**) during which replicated DNA is condensed into visible chromosomes, each of which divides into two. The confluent stages of mitosis are termed prophase, metaphase, anaphase and telophase, which lead up to division of the cell (**cytokinesis**), where the resultant daughter cells contain identical sets of chromosomes. **Haploid, diploid,** or **polyploid** cells can undergo mitosis. (**See: Cell cycle**)

Mobilization of reserves

See: Hemicellulose – mobilization; Mobilization of reserves – a monocot overview; Mobilization of reserves – cereals; Mobilization of reserves – dicots; Oils and fats – mobilization; Starch – mobilization; Storage proteins – mobilization

Mobilization of reserves – a monocot overview

1. Overview

Fruits and seeds of monocots are an important source of food and dietary fibre for humans and livestock. **Cereal** grains make up almost 90% of all seed crops grown worldwide, and **wheat, rice** and **barley** make up three-quarters of cereal grain production. Other monocot fruits/seeds that are important agricultural commodities include those of oil **palm, coconut** and date: relatively few studies have been carried out on the reserves of these seeds and the physiology, biochemistry and regulation of their mobilization. Nutrient reserves in monocot fruits and seeds are mobilized to support the growth of the embryo and to ensure seedling establishment and the success of the next generation. Reserves of carbon, nitrogen and minerals are stored as large complexes and require enzymatic hydrolysis so that sugars, amino acids or small peptides, and ions can be taken up by the growing **embryo**. Nutrient uptake by the **scutellum** is especially important in cereals: a specialized organ, the **haustorium,** fulfils this function in many other monocots, e.g. the date palm, asparagus and coconut. Reserve mobilization is absolutely dependent on the living cells of the seed. The dead starchy **endosperm** of cereals stores the bulk of the nutrient reserves in the grain, but plays only a small role in the mobilization of those reserves. Instead, nutrient reserves are hydrolysed by secreted enzymes that in cereals are produced in the **aleurone layer** and scutellum epithelium and secreted into the starchy endosperm. These enzymes include α-amylases, proteases, nucleases, xylanases, glucanases and phosphatases. The transcription of genes for secreted hydrolases and the synthesis and transport of secretory proteins is preceded by the synthesis of another set of hydrolases. These are required for the mobilization of stored reserves in the cereal aleurone layer and scutellum. The products of these hydrolases are used within the cells of the aleurone layer and scutellum to synthesize the secreted enzymes. Nutrient reserve mobilization in cereals concludes with the **programmed cell death** of the aleurone layer and scutellum. (**See: Mobilization of reserves – cereals**)

2. Fruits and seeds of monocots, morphology

Cereal grains are fruits (**caryopses**) where the **pericarp** surrounds the outer layers of the seed (**see: Seedcoats – structure**). The pericarp consists of crushed cell layers derived from the ovary wall, and these are fused to the outside of the seedcoat (**testa**). Crushed remnants of the **nucellus** are fused to the inside of the testa (**see: Cereals**, section 4). Together, these dead, crushed cell layers are a barrier that permits the uptake of water but prevents the uptake of most other compounds. The seedcoats enclose a **diploid** embryo and an initially **triploid** endosperm. The cereal endosperm is subdivided into the outer aleurone layer (Fig. M.12) and the inner, much larger starchy endosperm. In barley the aleurone layer is three or four cells thick, in rice it is from one to four layers, but in the other small grain cereals (wheat, **oats, rye**) and in **maize** (*Zea mays*) it is mostly a single layer of cells. The starchy endosperm of the mature cereal grain is dead, but the aleurone layer remains alive until reserve mobilization is complete. For small grain cereals such as wheat and rice the endosperm makes up 83–92% of the total dry weight, the embryo 2–3%, aleurone layer 5–7% and pericarp 2–8%.

The embryo in cereal grains consists of the axis with the **coleorhiza** enveloping the **radicle** and the **coleoptile** enclosing the shoot meristem and primary leaves (Fig. M.15). The scutellum, a shield-shaped tissue thought to be a modified **cotyledon**, is adjacent to the starchy endosperm and is specialized for the synthesis and secretion of enzymes, and for the uptake and transport of nutrients. The scutellum of maize and **sorghum** is a relatively large structure appressed to the large flattened surface of the grain, but the scutellum of the small grain cereals is much less extensive. That part of the scutellum adjacent to the starchy endosperm is made up of a single layer of epithelial cells, whereas the tissue nearest to the embryo axis contains highly vascularized parenchyma. The epithelial cells of the scutellum in maize, sorghum, barley, rice, rye and wheat elongate about two- to fourfold during imbibition. Oat and several other grass genera have a more specialized scutellum. In the dry oat grain the scutellum is shield shaped, but during germination it elongates considerably pushing beneath the aleurone layer within the starchy endosperm.

In **palms**, the seeds are enclosed within **berries**, **drupes** or **nuts**. The pericarp of the palm fruit consists of an outer **exocarp**, a fleshy **mesocarp** and a hard inner **endocarp**. The endocarp becomes the hard 'shell' of the palm fruit. The oil palm seed has a thin testa within which is found the fleshy endosperm from which kernel oil is extracted. **Oil** is also stored in the mesocarp of the oil palm fruit and the mesocarp forms the bulk of the edible part of the date fruit. The 'stone' of the date is mostly seed having a very hard endosperm.

3. Nutrient reserves in fruits and seeds of monocots

(a) *Cereals*. Polysaccharides, proteins, lipids and minerals are the predominant nutrients stored in monocot seeds. In rice, wheat, barley and the other small-grain cereals, **starch** makes up about 80–90% of endosperm carbon reserves with cell wall polysaccharides making up about 5–10%. Endosperm starch is present as unbranched **amylose** and branched **amylopectin** and the ratio of amylose to amylopectin is highly variable and genotype dependent. In barley, endosperm starch is about 20–30% amylose and 70–80% amylopectin. Starch is also present in the living cells of the embryo axis and the scutellum (especially in the developing grain) where it is localized in **amyloplasts**. Starch represents only a minor component of embryo reserves, making up to about 8% of embryo dry weight.

Cell wall polysaccharides contribute a significant amount of stored carbon in cereals. The cell walls of the starchy endosperm are composed largely of **arabinoxylan** and mixed-linkage (1→3, 1→4)-β-**glucan**. Aleurone layer tissue is also an important reservoir of polysaccharides which are stored almost exclusively in the thick aleurone layer cell wall. This cell wall is made up of two distinct regions, the wall matrix composed largely of arabinoxylan and a thin inner wall layer adjacent to the plasma membrane that is mostly mixed-linkage β-glucan polymer. A small amount (~2% each) of **cellulose** and **glucomannan** are also present. **Phenolic** compounds, especially ferulic acid, are abundant in the cell walls of cereal grains, particularly in the aleurone layer cells.

Cereal grains contain an average of 10–12% dry weight **storage protein**. Although this is relatively low when compared to the seeds of **legumes**, where seed protein contents range from 25 to 40%, cereals are by far the largest source of dietary protein. Data from FAO for 2000 show that for all of the major seed crops grown worldwide, 70% of dietary protein is estimated to come from cereals. The protein content of cereals is species and genotype dependent. Proteins are especially abundant in the aleurone layer tissue and embryo and seeds having large embryos such as maize or multilayered aleurone layers such as barley generally contain relatively high amounts of protein. Maize endosperm protein can vary from a low of 5% dry weight to 10% and in barley total grain protein varies from 12 to 18% dry weight. The proportional protein content of the embryo (18% dry weight) and aleurone layer (22–28%) is much higher than that of the starchy endosperm.

The bulk of cereal grain proteins are members of the **prolamin**, 7S **globulin** and 11S globulin classes of storage proteins. Prolamins are alcohol-soluble, water-insoluble

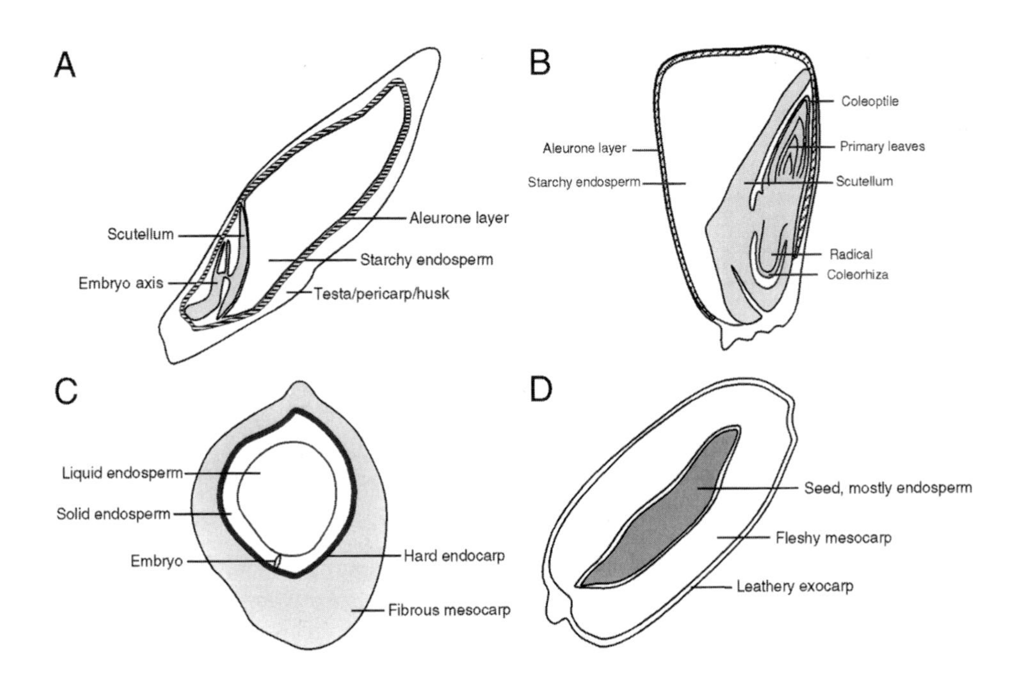

Fig. M.12. Seeds and fruits of monocot crops are shown with important tissues labelled. (A) Barley grain, (B) maize kernel, (C) coconut and (D) date.

proteins that are relatively rich in proline and glutamic acid. The zeins in maize, the hordeins in barley and the gliadins and glutenins in wheat are prolamins that make up most of the **storage proteins** in these species (**see: Osborne fractions**). In rice and oats, most of the storage proteins in the starchy endosperm are members of the 11S globulin class of storage proteins. Storage proteins are all deposited in **protein bodies** or protein storage vacuoles (PSV) during seed development. (**See: Storage protein – synthesis**)

Compartmentation of storage proteins is lost in the starchy endosperm of mature grain because the cells undergo a programmed cell death and membrane integrity is lost. Proteins of the 7S globulin class are stored in PSV where they are retained in the mature grain until they are broken down by **proteases** secreted from the scutellum and aleurone layer during germination and early seedling growth (**see: Thioredoxins**). Metabolic and structural proteins are also present in dry grain. Among the metabolic proteins present in the endosperm are chymotrypsin inhibitors, **trypsin/α-amylase inhibitors**, thionins and β-amylase.

Oils (triacylglycerols, TAG) are an important reserve in cereals which are stored in **oil bodies** (oleosomes) (Fig. M.13). Oil bodies are unique organelles because TAGs are deposited between the inner and outer leaflet of the endoplasmic reticulum (ER) membrane bilayer. The mature oil body is therefore surrounded by a half-unit membrane (**see: Oils and fats; Oils and fats – synthesis**). TAGs make up about 80% of cereal grain lipid, the remaining lipids being present as phospholipids and glycolipids. TAGs are largely absent from the starchy endosperm, but can make up as much as one-third of the dry weight of the scutellum (e.g. in maize) and 20% of the weight of the aleurone layer cells. Corn oil is derived largely from the embryo which may be 10–15% of the mass of the maize kernel. About 2 million t of corn oil are produced annually. Corn oil is comparable to that of olive oil with 83% of **fatty acids** being unsaturated, of which linoleic (~60%) and oleic acid (~20%) make up the majority of the fatty acyl chains. Oil is also extracted from the embryos/aleurone layer tissue of other cereals, most notably rice, from which about 1.1 million t of rice bran oil are produced annually. (**See: Cereals; Oilseeds – major; Oilseeds – minor**)

Mineral reserves are also stored in seeds often as insoluble complexes with oxalate or **phytin** (a salt of inositol hexaphosphoric acid). In mature barley grains more than 70% of K, Mg, Ca and P is in complexes of phytate located within the PSV of aleurone layer cells. Most of the other seed minerals such as K and Mg are present in the embryo. The starchy endosperm of the mature cereal grain contains only a small fraction of total grain minerals. Phytate is hydrolysed by the enzyme phytase. Genes encoding this enzyme have been cloned from germinated maize embryos and soybean cotyledons. Surprisingly, these two enzymes show marked differences. The maize gene encodes an enzyme of the histidine acid phosphatase class, whereas the soybean enzyme is a member of the purple acid phosphatase class of enzymes. Nothing is known about the subcellular distribution of these phytases. (**See: Phytin**)

(b) *Oil palm, coconut and date*. The fruits of palms are fleshy drupes or nuts. In the oil **palm**, TAGs are stored both in the red fleshy mesocarp of the fruit (about 45–55% oil by weight) and the seed (about 55% oil). Both mesocarp and seed are sources of palm oil, the mesocarp yielding red palm oil that is rich in carotenoids, and the seeds producing palm kernel oil which is less pigmented. Palm oil is as important as **soybean** oil as a source of vegetable oil, both being produced at about 25 million t per year. Coconut oil is also an important commodity with about 3.5 million t produced annually. Coconut palm oil is present in the fleshy part of the endosperm, and for commercial oil production the dried endosperm, known as copra, is pressed or rolled. Palm and coconut oil are widely used in confectionery and are an important part of the diet in the tropics. These oils are high in saturated fats; both palm kernel and coconut oil have more than 80% saturated fatty acids and less than 10% of unsaturated fatty acids. (**See: Oilseeds – major**)

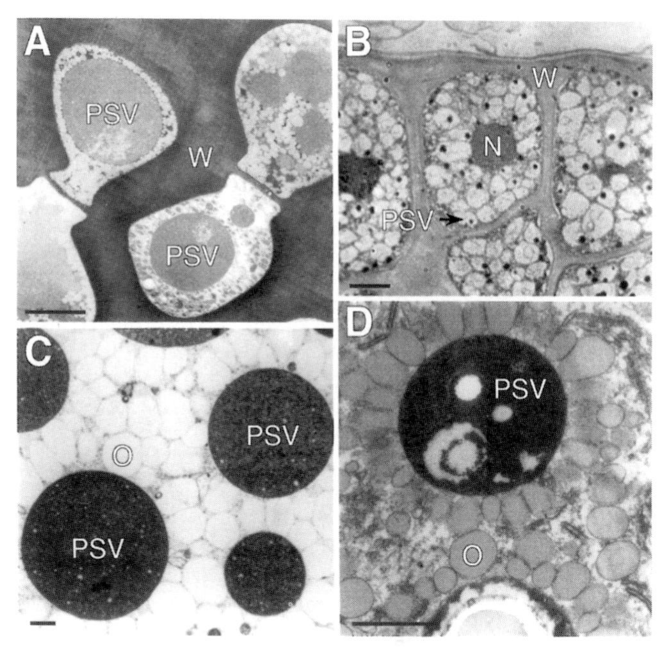

Fig. M.13. Endosperm storage tissues in date (A and C) and barley (B and D). (A) Cross section through the hard endosperm of date showing the thick cell walls (W) and individual endosperm cells. Each cell contains a few large protein storage vacuoles (PSV). (B) Cross section through the aleurone layer of barley. Each cell contains numerous PSV and a prominent nucleus (N). (C) Higher magnification of the PSV and oil bodies (O) in date endosperm (C) and barley aleurone layer (D). Scale bars in A and B are 10 μm and in C and D are 1 μm. Panels A and C are adapted from Demason, D., Sexton, R., Gorman, M. and Reid, J. (1985) Structure and biochemistry of endosperm breakdown in date palm, *Phoenix dactylifera*, seeds. *Protoplasma* 126, 159–167.

Date fruits have a fleshy mescocarp that stores mostly starch, which is converted to sugar as the fruit ripens. These sugar reserves, like the oil reserves in the mesocarp of oil palm fruit, are not readily available for use by the embryo which relies on seed reserves to initiate growth. The date seed is unusual in that seed storage reserves are present largely as **mannans** in the hard, endosperm cell wall. As in other palms, the date palm endosperm is composed of living cells in the mature seed, but these cells are unique in that the cell wall makes up as much as 50% of the volume of the endosperm. (PB, RJ)

Black, M. and Bewley, J.D. (eds) (1999) *Seed Technology and its Biological Basis.* Sheffield Academic Press, Sheffield, UK.

Fincher, G.B. (1989) Molecular and cellular biology associated with endosperm mobilization in germinating cereal grains. *Annual Review of Plant Physiology and Plant Molecular Biology* 40, 305–346.

Negbi, M. (1984) The structure and function of the scutellum of the Gramineae. *Botanical Journal of the Linnean Society* 88, 205–222.

Shewry, P. and Halford, N. (2002) Cereal seed storage proteins: structures, properties and role in grain utilization. *Journal of Experimental Botany* 53, 947–958.

Mobilization of reserves – cereals

The most detailed studies on seed reserve mobilization have been carried out on cereal grains. **Barley** features prominently, a status prompted by the importance of mobilization in the industrial process of malting (**see:** entries under **Malting**).

1. Mobilization of the endosperm reserves

Aleurone layer and **scutellum** epithelium cells synthesize and secrete a wide range of acid hydrolases to break down storage reserves (**see: Malting – barley; Starch–mobilization**). These reserves can be the starch and proteins in the starchy **endosperm**, **proteins** and **oils** (triacylglycerols, TAGs) in the **aleurone layer** cells, and polysaccharides in the walls of the aleurone layer and **scutellum**. α-Amylases, limit dextrinases, α- and β-glucosidases, exo- and endopeptidases, nucleases, phosphatases, β-glucanases, xylanases and other enzymes are all released into the starchy endosperm. α-Amylases work in conjuction with β-amylases, limit dextrinases, and β-glucosidases to hydrolyse starch to glucose. Endo- and exopeptidases hydrolyse storage proteins to peptides and amino acids. Nucleases digest DNA and RNA to oligonucleotides and nucleotides. Phosphatases remove phosphate from macromolecules. β-Glucanases and xylanases degrade cell wall polysaccharides. β-Amylases, enzymes that are crucial for starch breakdown, are synthesized during grain maturation in barley and **wheat** and exist in the starchy endosperm of mature grain in a bound form. In rice, however, this enzyme is produced only following germination. Other starch-degrading enzymes, including limit dextrinase (pullulanase) and α-glucosidase, are also synthesized during grain development and accumulate in the starchy endosperm. These preformed starch-degrading enzymes are present in the starchy endosperm as 'bound', inactive and 'free' forms. Some of the bound forms of these enzymes can be released by treatment with α-amylases or proteases. In the case of wheat, thioredoxin increases the activity of endosperm pullulanase.

Enzyme inhibitors are also released into the endosperm and these include the **amylase**/subtilisin **inhibitor** (ASI), a serine **protease inhibitor** of the subtilisin class. ASI inhibits α-amylases of the high pI family of enzymes but is ineffective in inhibiting α-amylases in the low pI family. ASI forms a complex with α-amylase that is dependent on the presence of Ca^{2+}. At low pH the ASI-amylase complex dissociates. In addition to ASI, inhibitors of limit dextrinases and proteases have been identified in cereal grains.

The pH optimum of most of the hydrolases secreted into the endosperm is low, generally around 4.5, as exemplified by the pH profile of the barley aleurone-layer-secreted protease EP-B (Fig. M.14). Hence a low pH in the endosperm as it is being degraded favours hydrolytic enzyme activity and might also favour the dissociation of enzyme inhibitors from α-amylases and other enzymes. The aleurone layer and scutellum of barley release organic and phosphoric acids into the starchy endosperm following germination and this helps to maintain the pH at 4.5 or below.

In **maize** the scutellum is the principal source of secreted hydrolases. In the small-grain cereals, the aleurone layer is the most important source of these enzymes, but the scutellum is a significant source as well, especially in **rice**. Regardless of their relative contributions to the pool of starchy endosperm-degrading enzymes, hydrolase production begins with the scutellum and with time progresses along the aleurone layer from the area proximal to the embryo to the distal end of the grain (**see: Malting – barley**). In barley and wheat, α-amylase and other hydrolase activities accumulate in the endosperm beginning about 2–3 days following imbibition. In the Morex cultivar of malting barley the accumulation of α-amylase activity in germinated grain is preceded by an increase in the accumulation of α-amylase mRNA (Fig. M.15).

Hydrolase synthesis and secretion have been studied extensively in the aleurone layer of small-grain cereals where they are controlled by **abscisic acid** (ABA) and **gibberellins** (GA). Both ABA and GA are synthesized by the cereal **embryo** and these **hormones** diffuse to the aleurone layer where they regulate the activity of this tissue. GAs stimulate the aleurone layer to produce the hydrolytic enzymes that break down the reserves stored in the starchy endosperm and ABA prevents their synthesis. The role of these plant hormones in regulating hydrolase synthesis by the scutellum is much less clear. Hydrolase synthesis is not increased when isolated embryos/scutella of wheat or maize are treated with GA, and inhibitors of GA biosynthesis do not prevent the

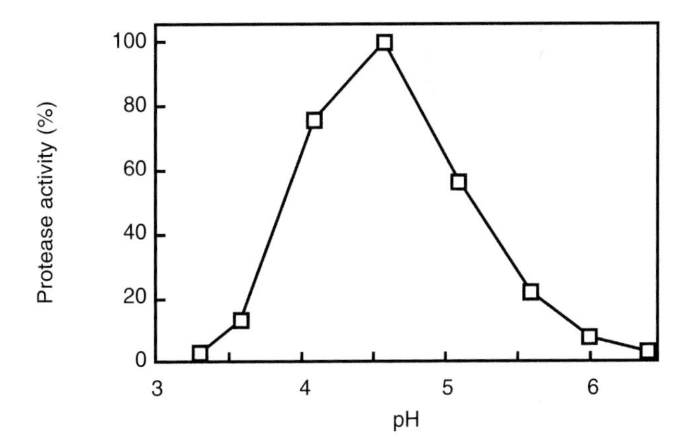

Fig. M.14. Effect of pH on protease activity. Relative protease activity of the secreted barley aleurone layer protease EP-B is highly dependent on pH. Activity is highest at pH 4.5 and this is the approximate pH of the starchy endosperm in imbibed barley grain. Redrawn from Koehler, S.M. and Ho, D.T.-H. (1988) Purification and characterization of gibberellic acid-induced cysteine endoproteases in barley aleurone layer. *Plant Physiology* 87, 95–103.

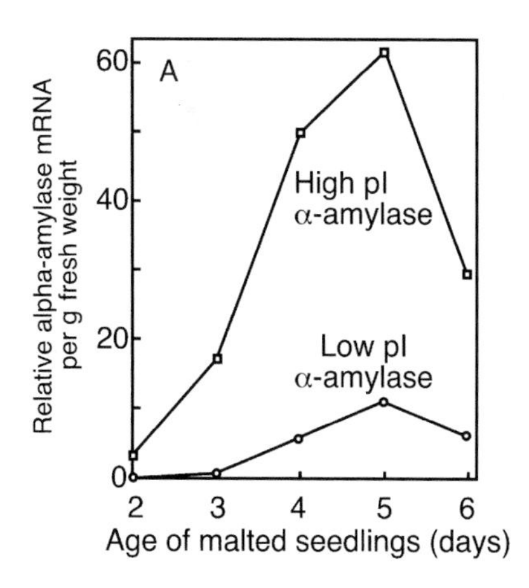

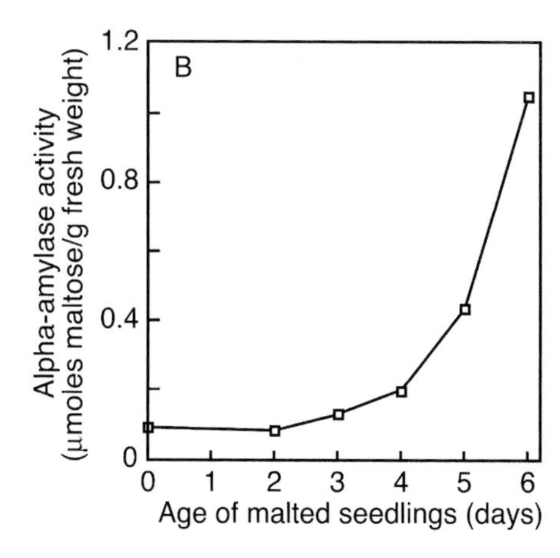

Fig. M.15. Production of α-amylase mRNA. The synthesis of mRNAs for α-amylase (A) in malted seedlings (i.e. germinated grains) of the barley variety Morex precedes the synthesis of α-amylase enzyme (B) as measured by an amylase activity assay that produces maltose from starch. Redrawn from Skadsen, R.W. and Tibbot, B.K. (1994) Temporal expression pattern of α-amylase isozymal genes in polysomal and total RNAs of germinating barley. *Journal of Cereal Science* 19, 199–208.

accumulation of α-amylase mRNAs in mature **wheat** embryos.

2. Mobilization of stored reserves in the aleurone layer and scutellum

The degradative enzymes produced by the aleurone layer and scutellum are synthesized *de novo* at the expense of proteins stored in the protein storage vacuoles (PSV) within these tissues. For *de novo* protein synthesis to occur, storage proteins are degraded to provide amino acid building blocks, and genes encoding secretory proteins are transcribed. Prior to this, however, the rate of cellular metabolism is increased and carbon substrates for energy production and biosynthetic reactions are produced. TAG (triacylglycerol) is the principal source of stored carbon in the aleurone layer, whereas the scutellum stores TAG, sucrose, the **oligosaccharide**, raffinose, and in some cases starch. TAG catabolism begins when a lipase converts TAG to glycerol and free fatty acids. Lipase activity is generally absent from dry seeds and increases following imbibition. In the scutellum of maize lipase is synthesized *de novo* following imbibition on non-membrane-bound ribosomes. The enzyme subsequently becomes associated with the surface of **oil bodies** where it initiates TAG hydrolysis.

Fatty acyl chains produced from TAG can be used for the synthesis of **membrane** lipids, such as those of the endoplasmic reticulum (ER) and Golgi apparatus. These membranes proliferate in aleurone layer and scutellum cells following imbibition and exposure to GA. Alternatively, fatty acids can be metabolized to acetyl Co-A by β-oxidation, a process that occurs in the **glyoxysome**. Once formed, acetyl Co-A can donate carbon to other anabolic reactions, be oxidized to provide energy, or enter the glyoxylate cycle and be used for gluconeogenesis. In castor bean seeds, as much as 70% of stored lipid in the endosperm is initially converted to sugars, but no information is available on the extent of gluconeogenesis from oils in monocot seeds. (**See: Oils and fats – mobilization**)

There is little information on the synthesis of enzymes of β-oxidation and **gluconeogenesis** in monocot seeds. In dicot seeds most of the enzymes of gluconeogenesis appear to be present in the dry seed, in low amounts, but greatly increase following germination due to *de novo* synthesis. In small-grain cereals, the synthesis of some **glyoxylate cycle** enzymes is hormonally regulated, at least at the transcriptional level. In barley and **rice** aleurone layers, **transcription** of genes for isocitrate lyase is stimulated by GA. mRNAs encoding some lipases are induced by GA in aleurone layer cells, but the location and role of these lipases is not known.

Catabolism of stored TAG, starch and other carbohydrates is used to supply most of the energy and some of the building blocks that are used for *de novo* enzyme synthesis in the aleurone layer and scutellum. The amino acids needed for the synthesis of secreted acid hydrolases are produced by breaking down **storage proteins**. Labelling experiments carried out in the 1960s showed conclusively that secreted enzymes such as α-amylase and ribonuclease are not present in the dry barley grain, but are synthesized in their entirety from amino acids following imbibition. Storage proteins are degraded by proteases already present in the PSV, but it is not known whether these proteases are also made *de novo*. Protein storage vacuoles from barley aleurone layer cells contain a spectrum of proteases, and their activity increases following treatment of these cells with GA. This may result from the activation of inactive precursors, or *de novo* synthesis. (**See: Storage proteins – mobilization**)

The proteases isolated from vacuoles of a wide range of seeds have been shown to have acidic pH optima for catalytic activity. The proteases of the barley aleurone layer PSV have pH optima for activity of pH 5 or less, and little or no activity at pH 7. Although the plant vacuole is usually an acidic compartment, the PSV in barley aleurone layer cells appear to have a pH of around 7 prior to receiving a GA signal. Following GA perception, however, the PSV lumen is acidified rapidly. ABA prevents rapid acidification of the PSV (Fig. M.16). When GA-stimulated acidification of the PSV lumen is blocked, storage protein breakdown is also inhibited. These observations suggest that acidification of the PSV lumen exerts an important control over the mobilization of vacuolar reserves.

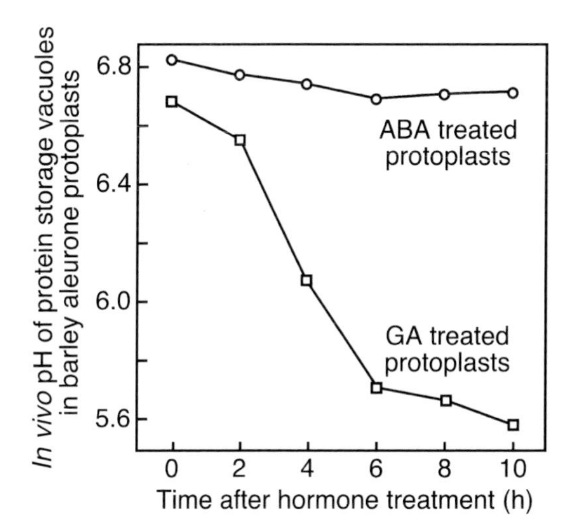

Fig. M.16. GA-induced acidification. The protein storage vacuoles in barley aleurone layer protoplasts are acidified rapidly when protoplasts are treated with gibberellin (GA) but not by abscisic acid (ABA). Redrawn from Swanson, S.J. and Jones, R.L. (1996) Gibberellic acid induces vacuolar acidification in barley aleurone. *Plant Cell* 8, 2211–2221.

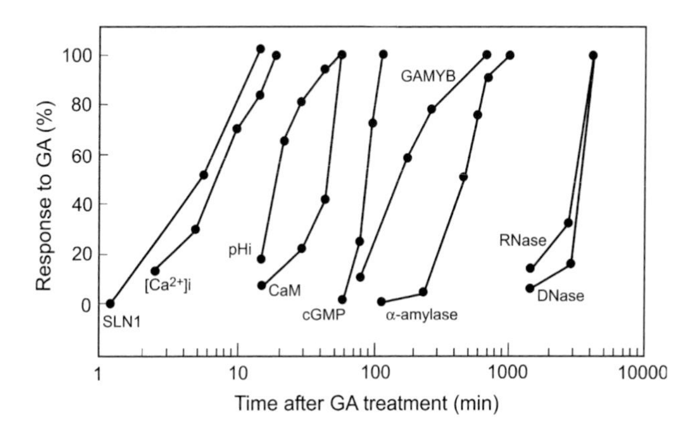

Fig. M.17. Changes in barley aleurone layer cells induced by gibberellin. Gibberellin (GA) perception by the aleurone layer of small cereals initiates a series of time-dependent events. Early events include degradation of the Slender1 protein (SLN1), an increase in the concentration of cytosolic free calcium (Ca^{2+}), and an increase in cytosolic pH (pHi). Intermediate events include an increase in transcript abundance for calmodulin (CaM), cyclic-GMP (cGMP) and GAMYB (a GA-promoted transcription activator) and α-amylases. Late events include the transcription of genes for ribonuclease (RNase) and deoxynuclease (DNase). From Sun, T.-p. and Gubler, F. (2004) Molecular mechanisms of gibberellin signaling in plants. *Annual Review of Plant Biology* 55, 197–223.

3. Transcription in the aleurone layer of genes for secreted hydrolases

The cereal aleurone layer tissue can be viewed as a secretory gland that is hormonally regulated by signals emanating from the embryo. The effect of GA on transcription of hydrolase genes in the aleurone layer of barley, wheat and rice is dramatic. In the barley aleurone layer it has been estimated that α-amylase transcripts make up as much as 25% of total cellular mRNA after GA treatment. Transcription of α-amylase and other secretory protein genes begins within 2–3 h of GA treatment of isolated aleurone layers (Fig. M.17) and is strongly suppressed by ABA. This has given rise to the notion that GA is simply an up-regulator of aleurone layer cell function and that ABA is an inhibitor of these processes, but RNA profiling experiments with barley and rice aleurone layer present a different picture of the effects of GA and ABA at the level of transcription. On a global scale, far more genes are up-regulated by ABA in the cereal aleurone layer than are down-regulated by it, whereas almost equal numbers of genes show up- and down-regulation in response to GA (Table M.6).

Aleurone layer cells perceive and respond to ABA and GA, but information about the receptors for these hormones and their signalling pathways is limited. (**See: Signal transduction – hormones**) Circumstantial evidence suggests that ABA and GA receptors are located on the plasma membrane (PM), but there are also indications that there are receptors for ABA in the cytosol. At least two signalling pathways downstream of the GA receptor regulate the production of hydrolases in aleurone layer cells. One of the pathways leads to the transcription of genes for secreted hydrolases. The other pathway involves increases in steady-state cytosolic Ca^{2+} concentrations and is required for synthesis of secreted hydrolases (Fig. M.17).

Table M.6. Number of genes up- or down-regulated by gibberellin (GA) or abscisic acid (ABA) treatment of barley aleurone layers as a function of time after hormone treatment. Transcriptional changes were measured using an oligonucleotide microarray that contained probes for approximately half of all barley genes.

Time (h)	GA up	GA down	ABA up	ABA down
3	35	0	63	0
6	74	22	81	18
12	189	116	133	32
24	139	182	130	33

The signalling pathway from the putative GA receptor to the transcription of secretory protein genes is known to involve regulatory proteins. Some of the most important are members of the GRAS family of proteins that contain the conserved amino acid sequence DELLA. **DELLA** domain proteins have been identified in barley, rice and wheat and they act as negative regulators of the GA response. In the barley aleurone layer, the nuclear-localized DELLA domain protein Slender1 (SLN1) undergoes **proteasome**-mediated degradation within minutes of treatment with GA. Once SLN1 is degraded, the transcriptional activator **GAMYB** can be transcribed as illustrated in Fig. M.18. GAMYB mRNA migrates to the cytosol, where it is translated, and the protein product migrates back into the nucleus. There it promotes the transcription of genes for secreted hydrolases by binding to the GA response element (**GARE**) present in the α-amylase and EPB-1 protease promoters. (**See: Signal transduction – hormones**)

The transcription of hydrolase genes is also regulated by ABA and a consensus ABA response element (**ABRE**) is also present in the promoters of the α-amylase and EPB-1 protease genes. ABA perception results in signals that prevent transcription from these promoters. ABA also induces the activity of a **protein kinase**, pkABA, that acts to down-

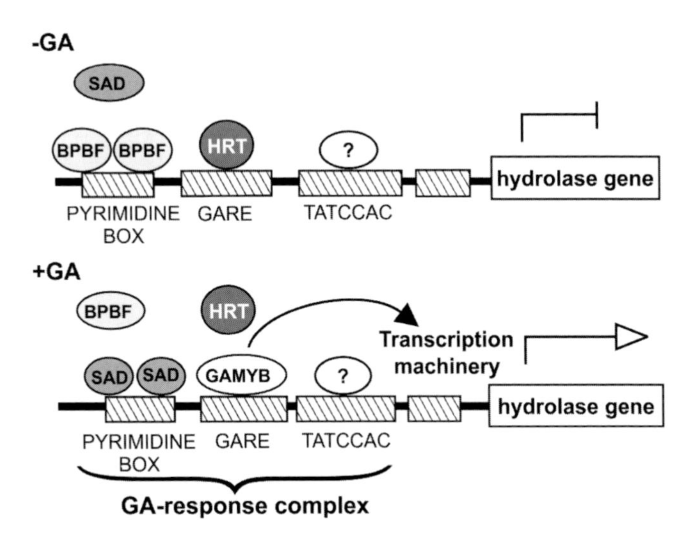

Fig. M.18. Hypothetical model for the regulation of some GA-responsive genes in the aleurone layer cells of small grain cereals. The pyrimidine box, GA responsive element (GARE) and TATCCAC are three regulatory domains in the promoters of some genes for secreted hydrolases. Transcription is inhibited by the binding of BPBF (barley prolamin-box binding factor) to the pyrimidine box and HRT (*Hordeum* repressor of transcription) to the GARE. After GA perception by the aleurone layer cell transcription is promoted when the SAD protein (scutellum and aleurone layer expressed DNA-binding with one finger) displaces BPBF and GAMYB displaces HRT. From Sun, T.-p. and Gubler, F. (2004) Molecular mechanisms of gibberellin signaling in plants. *Annual Review of Plant Biology* 55, 197–223.

regulate the expression of GA-regulated genes such as the α-amylases and EPB-1.

4. Synthesis and transport of secretory proteins from the aleurone layer

GA causes a rapid resetting of cytosolic Ca^{2+} in aleurone layer cells from a resting concentration of ~100 nM to ~200–300 nM. Several lines of evidence indicate that changes in Ca^{2+} are required to elicit a GA response in the aleurone layer cell, including the demonstration that chelation of intracellular Ca^{2+} prevents GA-induced responses. Withdrawal of GA or treatment with ABA results in a resetting of Ca^{2+} to resting concentrations. Calcium takes on an additional role in the aleurone layer cell because α-amylase is a Ca^{2+}-containing metalloprotein that must bind Ca^{2+} in the lumen of the ER to be active when secreted.

Ca^{2+} interacts with cellular Ca^{2+} sensors such as **calmodulin** (CaM) or **calcineurin** B-like proteins (CBLs), or with Ca^{2+}-activated enzymes such as calmodulin-domain protein kinases (CDPKs). CaM activates protein kinases of the CaM kinase class whereas CBLs activate kinases of the SIPK class. CDPKs and CBLs are involved in the GA response in the barley aleurone layer. GA stimulates the transcription of at least one CBL gene in the rice aleurone layer. This gene is one of the first to be expressed in response to GA and it plays a role in a **signal transduction** chain leading to vacuolation of the aleurone layer cell but not to the expression of α-amylase genes. Conversely, CaM and a CDPK are required for the synthesis of α-amylase in barley aleurone layers.

Secreted hydrolases are synthesized on membranes of the rough ER and are transported to the cell exterior along the secretory pathway via the Golgi apparatus (**see: Cells and cell components**) and secretory vesicles. Targeting of secretory proteins to the ER requires the presence of a signal peptide, which occurs at the N-terminus of all secreted hydrolases examined in cereals. Transport through the endomembrane system to the cell exterior appears to be along the classical default pathway described in animal cells. Movement of secretory proteins along the default pathway occurs because additional signals are not present to retain secretory proteins in the ER, Golgi or vacuoles.

Hydrolases may undergo post-translational modifications prior to their secretion from the aleurone layer or scutellum. The barley and wheat aleurone layer α-amylases, for example, are neither N- nor O-glycosylated. The corresponding enzymes secreted by the rice scutellum are N-glycosylated in the Golgi apparatus. The low pI α-amylases of the barley aleurone layer undergo a post-translational modification that results in the lowering of their isoelectric points. It is not known if these post-translational modifications alter the rate of movement along the secretory pathway, change enzyme activity or confer increased protein stability.

In mature grains, the thick aleurone layer cell wall may function as a barrier to penetration by large macromolecules and microbes. This cell wall barrier is breached within hours of exposing aleurone layers to GA as secreted endoxylanases break down the wall matrix. Histochemical staining shows that cell wall breakdown initially results in the formation of narrow channels in the wall, but eventually the entire cell wall matrix is digested. Enzymes such as α-amylase are released from the aleurone layers via cell wall channels, indicating that the wall matrix is also a barrier to release of secretory proteins.

A diagrammatic summary of the events taking place from GA perception to α-amylase secretion is shown in Fig. M.19.

5. Nutrient uptake by the scutellum

During the first few hours after the start of **imbibition**, cells of the **scutellum** epithelium elongate and separate from each other to form finger-like extensions that project into the starchy endosperm. Transporters in the plasma membrane of these epithelial cells participate in sugar, peptide and amino acid uptake. Some of the peptide and amino acid transporters have been especially well characterized, and at least one **peptide transporter** (HvPTR1) has been cloned from barley scutellum. Amino acid and peptide transporters are synthesized within 6 h of imbibition of barley grain, and transport activity reaches a maximum after 24 h. Surprisingly, mature barley grains contain significant amounts of small peptides and amino acids and it is thought that the rapid synthesis of peptide transporters is required for the uptake of these preformed pools. Peptide and amino acid transport is pH dependent and is likely to occur by a H^+-co-transport mechanism. The pH optimum of the barley scutellum peptide transporter is pH 3.8, pointing again to the importance of an acidic endosperm.

Sugars are also taken up by scutellum epithelium cells, and monosaccharide (glucose) and disaccharide (maltose and sucrose) transporters have been identified. Maltose is one of the hydrolysis products of starch, and sucrose can arise from

1. GA_1 from the embryo first binds to a cell surface receptor.

2. The cell surface GA receptor complex interacts with a heterotrimeric G-protein, initiating two separate signal transduction chains.

3. A calcium-independent pathway, involving cGMP, results in the activation of a signalling intermediate.

4. The activated signalling intermediate binds to DELLA repressor proteins in the nucleus.

5. The DELLA repressors are degraded when bound to the GA signal.

6. The inactivation of the DELLA repressors allows the expression of the *MYB* gene, as well as other genes, to proceed through transcription, processing and translation.

7. The newly synthesized MYB protein then enters the nucleus and binds to the promoter genes for α-amylase and other hydrolytic enzymes.

8. Transcription of α-amylase and other hydrolytic genes is activated.

9. α-Amylase and other hydrolases are synthesized on the rough ER.

10. Proteins are secreted via the Golgi.

11. The secretory pathway requires GA stimulation via a calcium–calmodulin-dependent signal transduction pathway.

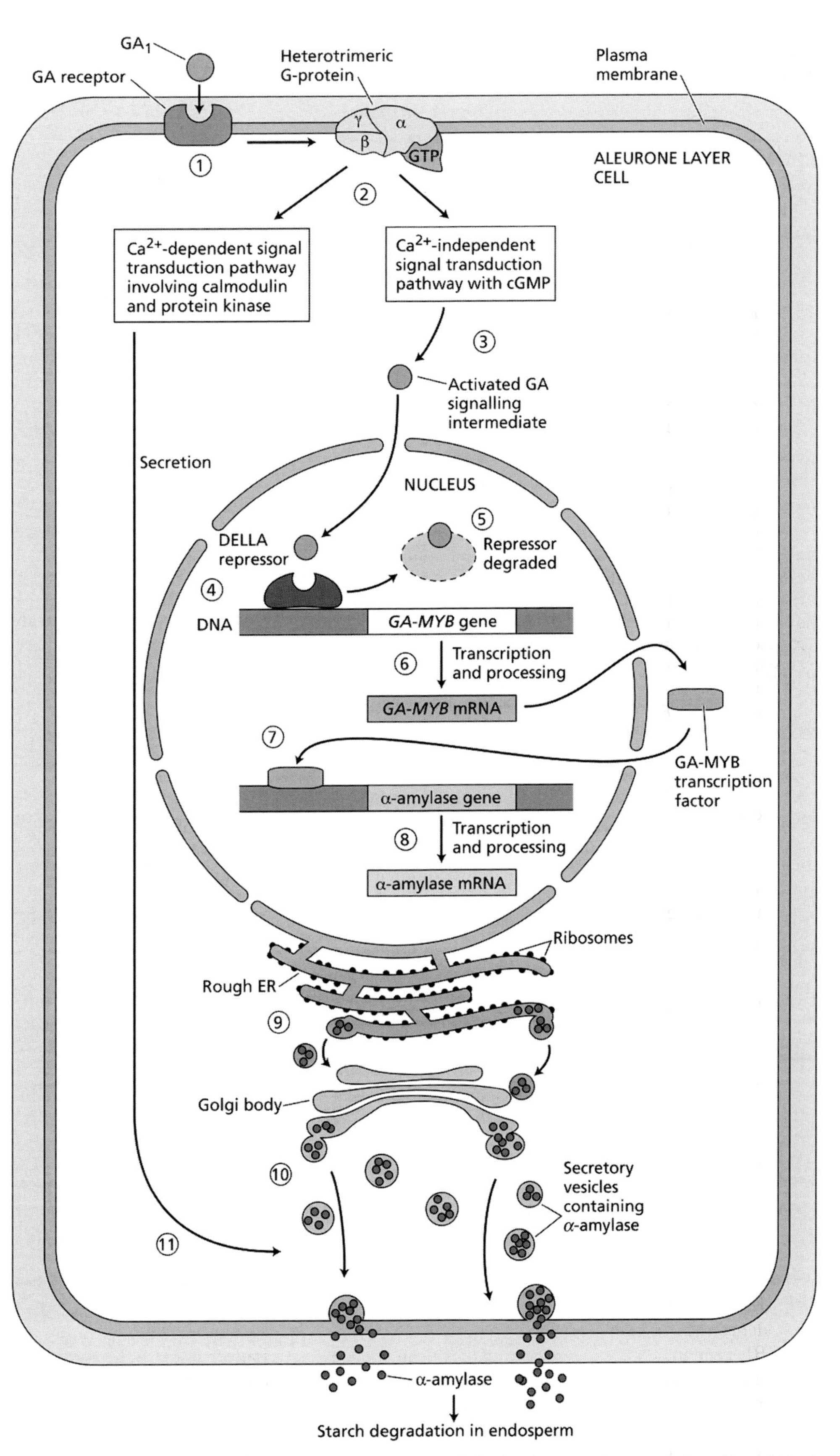

Fig. M.19. A diagrammatic summary of gibberellin action on an aleurone layer cell culminating in α-amylase secretion. From Taiz, L. and Zeiger, E. (2002) *Plant Physiology*, 3rd edn. Sinauer Associates, Sunderland, MA, USA. By kind permission of the authors and publishers.

gluconeogenesis in the aleurone layer. Sucrose synthesized in the aleurone layer is released into the endosperm. Sucrose synthesis occurs in the scutellum, possibly utilizing the maltose and glucose arising from endosperm starch hydrolysis and from **gluconeogenesis** after TAG hydrolysis. The sucrose is transported from the scutellum to, and throughout, the embryonic axis.

6. Death of the aleurone layer and scutellum

After the storage reserves of the endosperm have been mobilized, the aleurone layer and scutellum of the grain undergo **programmed cell death** (PCD), a mechanism that ensures that all remaining nutrients within the grain are available to the developing seedling. PCD of the aleurone layer cell is under strong hormonal control. GA stimulates PCD, and ABA prevents cell death. PCD in barley aleurone layers results from oxidative stress and can be delayed by antioxidants. Among the antioxidants that are effective in arresting PCD is nitric oxide (NO). NO donors could delay GA-stimulated PCD, whereas scavengers of NO accelerated death. These observations, taken together with experiments showing that aleurone layer tissue can synthesize NO, suggest that NO could play a role in regulating cell death *in vivo* in cereal grains.

There are at least two ways in which GA promotes the oxidative stress that leads to aleurone layer PCD. First, GA initiates TAG hydrolysis and stimulates β-oxidation of fatty acids. Enzymatic reactions in β-oxidation produce hydrogen peroxide, a **reactive oxygen species** (ROS). Additional ROS are produced by the electron transport chain in the mitochondria. GA also results in the down-regulation of enzymes that defend aleurone layer cells against ROS. These include catalase, ascorbate peroxidase and superoxide dismutase. These effects of GA bring about an increase in intracellular reactive oxygen and ultimately PCD. ABA, on the other hand, does not stimulate storage oil metabolism nor does it repress catalase, ascorbate peroxidase or superoxide dismutase expression. Rather, ABA increases catalase mRNA accumulation as well as bringing about an increase in catalase activity. These effects of ABA, among others, may allow for the maintenance of viability in **dormant** and **quiescent** seeds. (PB, RJ)

(See: Starch – mobilization)

Bethke, P.C., Swanson, S.J., Hillmer, S. and Jones, R.L. (1998) From storage compartment to lytic organelle: the metamorphosis of the aleurone protein storage vacuole. *Annals of Botany* 82, 399–412.

Fath, A., Belligni, V., Bethke, P., Spiegel, Y. and Jones, R. (2001) Signaling in the cereal aleurone: hormones, reactive oxygen species and cell death. *New Phytologist* 44, 255–266.

Fincher, G.B. (1989) Molecular and cellular biology associated with endosperm mobilization in germinating cereal grains. *Annual Review of Plant Physiology and Plant Molecular Biology* 40, 305–346.

Jacobsen, J.V., Pearce, D.W., Poole, A.T., Pharis, R.P. and Mander, L.N. (2002) Abscisic acid, phaseic acid and gibberellin contents associated with dormancy and germination in barley. *Physiologia Plantarum* 115, 428–441.

Olszewski, N., Sun, T.-p. and Gubler, F. (2002) Gibberellin signaling: Biosynthesis, catabolism, and response pathways. *Plant Cell* 14, S61–S80.

Ritchie, S., Swanson, S.J. and Gilroy, S. (2000) The cell and molecular biology of endosperm development and function. *Seed Science Research* 10, 193–212.

Sun, T.-p. and Gubler, F. (2004) Molecular mechanisms of gibberellin signaling in plants. *Annual Review of Plant Biology* 55, 197–223.

Waterworth, W.M., West, C.E. and Bray, C.M. (2001) The physiology and molecular biology of peptide transport in seeds. *Seed Science Research* 11, 275–284.

Mobilization of reserves – dicots

The biochemistry of the mobilization of the major storage reserves in seeds is covered in other entries (**see: Hemicellulose – mobilization; Oils and fats – mobilization; Starch – mobilization; Storage proteins – mobilization**). Here, attention will be given to the physiology of the process and to our understanding of how mobilization in dicot seeds is regulated.

The bulk reserves of **oils** (triacylglycerols, TAGS), **proteins**, **carbohydrates** and **phytate** (**see: Storage reserves**) in the storage organs (generally cotyledons or, less commonly, endosperm) are mobilized mostly after germination has been completed, i.e. after radicle emergence from the seed, when seedling growth ensues. Minor mobilization (in quantity but not in physiological importance) occurs within the embryonic axis itself to support the onset of growth. Reserves utilized here are usually the biochemically more readily accessible (i.e. less polymerized) compounds – oligosaccharides such as sucrose, raffinose and others – though in some seeds (e.g. vetch, *Vicia sativa*) the small amount of storage protein in the radicle is used in early elongation of that organ. The regulation of this initial mobilization is poorly understood. It is clear, however, that utilization of all reserves must be linked to the completion of germination and should be limited, for example, in a hydrated, metabolically active but still dormant seed.

1. Mobilization of the bulk reserves

Examples of the increase in mobilizing-enzyme activity and the decrease in storage reserves following germination are seen in Figs M.20, M.21 and M.22. The initial mobilization of proteins, carbohydrates (generally starch) and TAG are brought about, respectively, by proteolytic, amylolytic and lipolytic enzymes. Some of the products are respired away but most are converted into amino acids and peptides and sucrose for transport into the growing axis. There is a great number of enzymes and metabolic pathways involved in the conversion of the reserves into transport products and though it is not understood how all the components are regulated, their expression and operation must be highly coordinated.

In general, the major enzymes that initiate mobilization, such as proteases, amylases and lipases, are present in the dry seeds at barely detectable or extremely low levels of activity though there are a few cases where a limited range of stored enzymes is utilized: one such is the proteolysis of the small amount of storage protein in the radicle of vetch by stored cysteine proteinase. Almost all of the mobilizing-enzyme activity appears after germination has occurred (Fig. M.20). This is also true for the enzymes that operate after the reserves themselves have been attacked, such as those in the glyoxylate

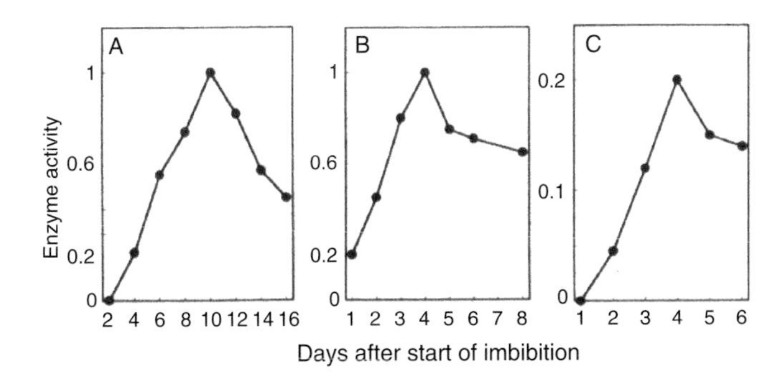

Fig. M.20. Changes in activity of mobilizing enzymes in dicot seeds. (A) α-Amylase, bean cotyledons (*Phaseolus vulgaris*); (B) lipase, cucumber cotyledons (*Cucumis sativus*); (C) carboxypeptidase, castor bean endosperm (*Ricinus communis*). From Bewley, J.D. and Black, M. (1994) *Seeds: Physiology of Development and Germination.* Plenum Press, New York, USA.

cycle participating in **gluconeogenesis** following TAG hydrolysis and fatty acid oxidation. There is good evidence, based largely on isotopic labelling experiments or with inhibitors of protein synthesis, that the increase in enzyme activity following germination is brought about by *de novo* synthesis.

2. Regulation of mobilization

Unlike in cereals, in dicots there is generally no separation of enzyme-producing tissue from the storage reserve tissue (cf. **aleurone layer** vs. starchy endosperm in cereal grains): one of the few exceptions is the aleurone layer tissue of fenugreek which produces endo-β-mannanase, β-mannosidase and α-galactosidase that attack the endosperm cell walls. Moreover, there is no definitive evidence for the operation of a controlling hormonal signal, as occurs in the cereals, though hormones may exercise some influence.

It is nevertheless apparent from many cases that the **embryonic axis** exerts a regulatory role over mobilization in the cotyledons or endosperm. When cotyledons are excised from the axis the activity of mobilizing enzymes is greatly reduced and mobilization is consequently severely affected. This is shown in Fig. M.21 for proteolysis and proteases in mung bean cotyledons: a similar situation exists in many other species, as well as for α-amylase, lipase, certain endo-β-mannanases and gluconeogenesis enzymes. During TAG mobilization in *Arabidopsis* endosperm, the **gluconeogenic** enzyme phosphoenolpyruvate carboxykinase (**see: Sucrose synthesis**) is produced in a wave-like pattern starting in cells adjacent to the embryo, suggestive of regulation by the latter.

Two types of influence of the embryonic axis have been considered – hormonal and relief of feedback inhibition.

(a) *Hormonal regulation.* Application of hormones to isolated cotyledons or endosperm in several cases causes an increase in mobilization activity (Table M.7). The supply of hormone generally only partially restores activity, however, and the role of the axis is incompletely replaced. Whether or not these are direct effects or indirect ones through, say, the promotion of cell growth which would act as a sink (see: (b) **Relief of feedback inhibition**), is not clear. In some cases, support for a possible hormonal involvement comes from the

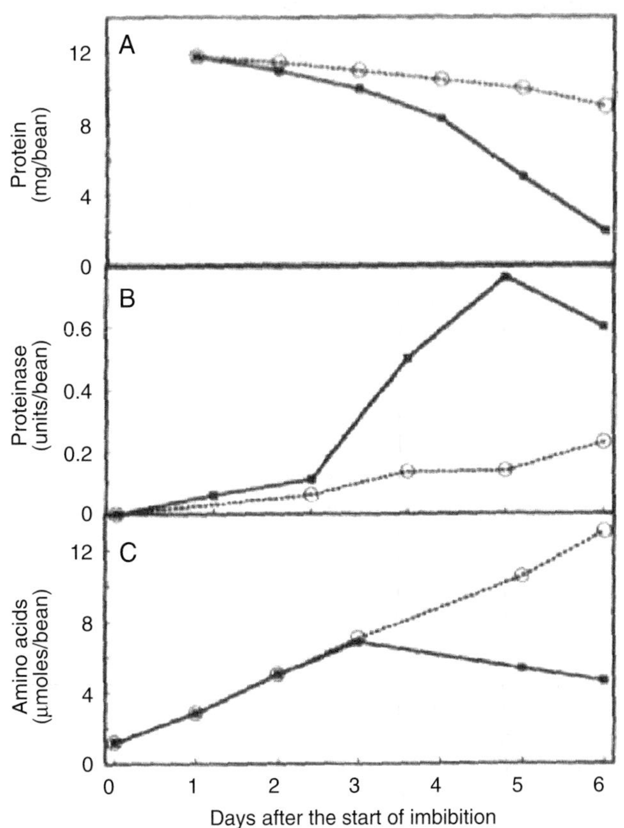

Fig. M.21. Effect of the axis on protein mobilization in mung bean cotyledons (*Vigna radiata*). Solid line, intact germinated seed; dashed line, isolated cotyledons. The proteinase is vicilin peptidohydrolase. From Bewley, J.D. and Black, M. (1994) *Seeds: Physiology of Development and Germination.* Plenum Press, New York.

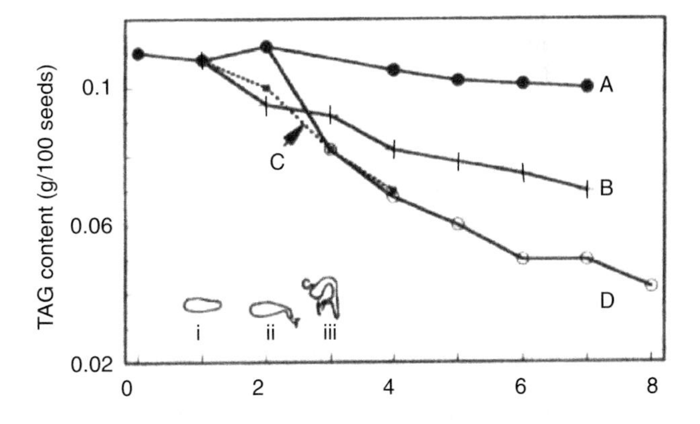

Fig. M.22. Mobilization of triacylglycerol in cucumber cotyledons. (A) Embryonic axis removed at day 0, testa present; (B) as A but testa and membrane removed; (C) testa removed from newly imbibed seed, axis intact; (D) fully intact seed. The appearance of the intact seed/seedling over the first 2.5 days is indicated on the bottom left: stage i, dry seed; stage ii, radicle emergence; stage iii, testa displacement (c, cotyledon; r, radicle; t, testa). From Bewley, J.D. and Black, M. (1994) *Seeds: Physiology of Development and Germination.* Plenum Press, New York.

Table M.7. Hormonal stimulation of reserve mobilization: some examples.[a]

Species and tissue	Reserve and/or enzymes	Hormone
Cicer arietinum (chickpea) cotyledons	Starch Protein	Cytokinin
Corylus avellana (hazel) cotyledons	Glyoxylate cycle	Gibberellin
Cucurbita pepo (marrow) cotyledons	Protein	Cytokinin
Hymenaea courbaril (Jatoba or locust tree)	Xyloglucan	Auxin
Lactuca sativa (lettuce) endosperm[b]	Endo-β-mannanase	Gibberellin Cytokinin
Pisum sativum (pea) cotyledons	Starch; α-amylase	Gibberellin
Vigna mungo (black gram) cotyledons	Papain-type proteinase	Gibberellin

[a]Isolated cotyledons or endosperms were exposed to the hormone.
[b]In ABA-treated endosperms.

effects of inhibitors of hormone biosynthesis (e.g. **gibberellin** and proteinase production by *Vigna mungo* cotyledons) or hormone transport inhibitors (e.g. **auxin** and xyloglucan mobilization in seedling cotyledons of *Hymenaea courbaril*, the Jatoba or locust tree) or from correlation with hormone content (e.g. **cytokinin** and amylolysis in chickpea). In no case, however, has it been convincingly shown that a hormone from the axis or embryo exerts overriding control over mobilization in the cotyledons or endosperm in a manner comparable with that in cereals (**see: Mobilization of reserves – cereals**).

There is some evidence that **abscisic acid** (ABA) can inhibit production of certain reserve-mobilizing enzymes, for example mannanases in lettuce seed endosperm, and can decrease the formation of transcripts for gluconeogenesis enzymes in *Arabidopsis*, but on the whole knowledge about the roles of this hormone in reserve mobilization is limited.

(b) *Relief of feedback inhibition.* Many enzymes are inhibited by the products of the reaction they catalyse, either in respect of their activity or their synthesis. This effect could account for the lowered reserve mobilization when the axis or embryo is removed – the accumulation of products from the initial mobilization inhibits further activity. For mobilization to proceed unchecked the axis must be present to draw off the enzyme products, i.e. to act as a 'sink'. There is evidence for this in several cases, such as TAG and protein breakdown in cucumber cotyledons where lipolytic and proteolytic enzymes are present in the isolated organs but reserve hydrolysis ceases. In *Vigna mungo*, part of the process for starch degradation – insertion of starch granules into lytic vacuoles as a prelude to autophagy (**see: Autophagic vacuoles**) – can occur in isolated cotyledons but starch breakdown itself is blocked, even though α-amylase is present. This might possibly be accounted for by feedback inhibition.

Breakdown of TAG in *Arabidopsis* is reduced by sugars such as glucose, an effect which is not osmotic or substrate-level inhibition but may be due to signalling perturbation. In the same species, relatively high carbon:nitrogen ratios in the medium surrounding germinated seeds have an inhibitory effect: this may reflect the situation *in vivo*.

(c) *Other controls.* The seedcoat can influence reserve mobilization. In cucumber cotyledons, TAG hydrolysis proceeds in the intact, germinated seed only when the testa and the inner membrane are removed. In normal germination this happens naturally when a peg of tissue on the radicle forces off the testa; if the radicle is cut off, mobilization is permitted by manual removal of the seedcoat (Fig. M.22). The testa appears to exercise a similar role in α-amylase production and amylolysis in germinated pea and mung bean.

At the molecular level, several genes have been implicated in the regulatory processes. For example, the cysteine proteinase gene in *Phaseolus vulgaris* (common bean) appears to be regulated by a *Vp1*-like factor (**see: Dormancy – genes**). In *Arabidopsis*, the *COMATOSE* gene – which encodes a transport protein that may participate in gluconeogenesis – is implicated in TAG mobilization. Seeds carrying the mutant form of this gene are able to germinate but TAG breakdown is inhibited.

3. Summary

A multiplicity of possible controls is clearly involved in the regulation of mobilization in dicot seeds and it is not possible to present an integrated picture of how they act. There is no firm evidence analogous to the embryo gibberellin/aleurone layer tissue complex in cereal grains. The presence of the axis, in the case of cotyledonary reserves, or the whole embryo, in respect of endosperm reserves, appears to be required: and in these cases hormonal and feedback controls might operate.

(MB)

(See: Mobilization of reserves – a monocot overview)

Davies, H.V. and Slack, P.T. (1981) The control of food mobilization in seeds of dicotyledonous plants. *New Phytologist* 88, 41–51.

Müntz, K., Belozersky, M.A., Dunaevsky, Y.E., Schlereth, A. and Tiedemann, J. (2001) Stored proteinases and the initiation of storage protein mobilization in seeds during germination and seedling growth. *Journal of Experimental Botany* 52, 1741–1752.

Rylott, E.L., Hooks, M.A. and Graham, I.A. (2001) Co-ordinate regulation of genes involved in storage lipid mobilization in *Arabidopsis thaliana*. *Biochemical Society Transactions* 29, 283–287.

Toyooka, K., Okamoto, T. and Minamikawa, T. (2001) Cotyledon cells of *Vigna mungo* seedlings use at least two distinct autophagic machineries for degradation of starch granules and cellular components. *The Journal of Cell Biology* 154, 973–982.

Model seed systems

Much of the knowledge in seed science has been gained from research on a relatively small number of 'model' species. These are seeds, mostly of crop species, which, for various reasons, are favoured for the study of various aspects of seed biology. (See: **Research seed species – contributions to seed science**)

Moisture content

See: **Water content**

Moisture migration

In bulk seed storage technology, the movement of moisture from one part of the seed mass to another by convection due to temperature and relative humidity gradients, leading to local development of spoilage fungi, bacteria and pests, and resultant heating in some cases. In large storage structures, moisture movement is controlled by a combination of design, loading, aeration and ventilation techniques. (See: **Storage management**)

Monocotyledonous plants (monocots)

(Greek: *monos* = single; *cotyledon*: sucker) A group of the **angiosperms** characterized by embryos with only one cotyledon instead of two as in the **dicotyledons** (see: **History of seed research**). In addition, monocotyledons show numerous differences from dicotyledons, e.g. a short-lived primary root, scattered rather than a ring-like arrangement of the vascular bundles in the stem, general lack of secondary thickening by a cambium, parts of the flowers usually in threes, and leaves usually with parallel venation.

Includes several orders in which there are commercially important species, the Graminales (cereals, grasses), Liliales (lily, asparagus, tulip), Amaryllidales (onion, daffodil) and Orchidales (orchids).
(See: **Cereals; Grasses, forage and turf**)

Monogerm

A term used to describe hybrid varieties of **sugarbeet** and other beet crops that are produced from seed parental lines that only produce a single flower at each node. Genetic monogerm varieties are now used for most of the sugarbeet crop to ensure even space planting in the field, without the need to single out superfluous plants. In the past, mechanical processing methods were used to produce 'mechanical monogerm seed', by reducing multigerm clusters to single-seeded pieces.

Monosomics

Meaning 'one chromosome', monosomics are a type of **aneuploid** in which the somatic cells lack a single chromosome, and thus in **diploids** have a **chromosome number** $2n - 1$. Plants that are tetraploids or hexaploids can often survive when one chromosome is missing in this way, but there is usually an associated phenotypic abnormality. Geneticists use viable plant monosomics (as well as **nullisomics**) to identify the chromosomes that carry the loci of recessive **mutant alleles** in a new **line**, by crossing it in turn with different monosomic lines, each of which lacks a different chromosome, and examining the resulting phenotype. They are also used to substitute a specific gene or genes contained in an entire chromosome that is introduced from another closely related cultivar or species. Complete monosomic sets have been established for wheat (that is, all the 21 that are possible), and partial sets for cotton, oat, tobacco and a few other **polyploid** crop species. (PH)

Morning glory

Seeds of this species (*Ipomea violacea*, Convolulaceae) contain the hallucinogenic indole-type alkaloid, lysergic acid amide (see: **Pharmaceuticals and pharmacologically active compounds**). The seeds have been used because of their pharmacological effects since ancient times, for example by the Zapotecs of Mexico. (See: **Psychoactive seeds**)

Mosaic indexing

Official labelling description applied to seed, for example lettuce, that has been tested and found free of seedborne lettuce mosaic virus.

Mucilages

A general term for polymeric **hemicelluloses**, other than of glucose, often used synonymously with **gums**. Mucilages in seedcoats and some **endosperms** swell upon hydration and may be important in water retention for the seed. (See: **Deserts and arid lands; Seedcoat – structure**)

Mulch

Mulches are materials spread or formed to cover the soil surface. Mulching improves control over soil moisture content by reducing evaporation losses, and greater uniformity when used with drip irrigation. Mulches have various specific uses in aiding seedling establishment, in either raised beds or flat ground: to increase soil and ambient-air temperatures; to reduce soil drying, compaction or crusting; or to prevent light from reaching the soil, thus preventing weed growth. (Mulches have a range of other uses later in crop development to protect the soil and/or plant roots: for example, polythene or straw layers to limit the action of heavy pounding rainfall on the soil, prevent soil freezing, to reflect light back on to the crop plant, and generally to help control weeds, insects and diseases.)

1. Plastic sheet ('Plasticulture')

(a) *Direct seeding*. Transparent plastic film mulches (commonly made of polyethylene or polypropylene) are laid in sheets snugly fitted on the soil, just after direct seeding. They are mainly used for high-value horticultural crops in cold soils, such as for the early sowing of carrot, or for pepper varieties which need relatively high soil temperatures (25–31°C) for optimal germination and emergence. Perforations in the film allow rainfall to penetrate through to the soil; in dry environments, drip irrigation systems are put in place before the film is laid. Once the seedlings have emerged, the film is removed from the field. Weed growth can often be a problem with transparent film mulches.

(b) *Transplants*. Transparent plastic film sheets are laid or stretched on frames over crops, at the time of **transplanting** of glasshouse-raised seedlings, to raise mean ambient temperatures by about 1 to 2°C. As the crop grows, the film is raised by foliage and is later removed, after about 1 or 2 months. Alternatively, tinted plastic sheets (in black or in other colours) are laid on the soil surface before transplanting, and then seedlings are inserted at intervals through small incisions.

(c) *Presowing*. Plastic sheets are also used for **solarization**, a presowing heating technique to sterilize soils.

2. Straw and paper mulching

Loose chopped straw or agglomerated pellets made from absorbent recycled paper, are spread or blown over all or part of the soil surface of a newly seeded bed to aid seedling establishment of turf grasses (for example, in parts of North America), for erosion control and to help retain soil moisture. **Coir** pith, the water-absorbent **lignin**-rich residue after fibre extraction from the coconut husk, is also used as a soil surface mulch. Alternatively, blankets or mats made of straw with a quick-degrading net sewn in, are pegged to the ground in situations where erosion is likely, such as for seeding steep slopes and ditches. The mats typically degrade within 2 months, but there are longer-lasting versions designed for use in low-maintenance areas. (**See: Mats**, in which seed is impregnated on or in paper sheets.)

3. Mulch tillage and strip-till planting

Mulch **tillage** prepares the soil in such a way that plant residues ('stubble mulch') or other materials are left to cover >30% of the surface with crop residue.

In strip-till planting, an area 30 to 50 cm wide is tilled sufficiently through a 'living mulch' or standing residue to form a seedbed for each row. At planting or at first cultivation, the remaining mulch in the row middle is cut loose, or chemically killed or retarded.

4. Hydromulching

An alternative name for the **hydroseeding** process used to establish grass lawns, or in landscaping and land reclamation.

(PH)

Multiline

A mixed population ('composite **variety**') of several almost genetically identical breeding lines of a **cultivar** (i.e. isolines, that have all but a few genes in common), used to increase the intra-specific diversity of a crop. For example, a mixture of isolines selected to differ in the quality and/or quantity of major gene resistance traits, although they are otherwise equal in agronomic characters, thereby confronting pests with a mixture of host gene-types.

Multiplication

In seed production, the process of producing large quantities of seed from stock seed: for example, the multiplication of **Breeders Seed** to produce **Foundation Seed**, which is in turn multiplied to produce **Certified Seed**. (**See: Production for sowing**)

'Multiplication rate' is the factor used to determine the area of production needed to meet the total volume for the seed production plan, based on the reproduction coefficient of the species.

Mung bean

1. World importance and distribution

Mung bean, or green gram (*Vigna radiata*, syn. *Phaseolus aureus*) is a warm-season legume cultivated most extensively in the India–Burma–Thailand region of southeastern Asia and also in Iran, Pakistan, Vietnam, China, the Philippines, Malaysia, Indonesia, and adjacent countries and the islands of eastern Asia and the South Pacific. It is also grown in the Middle East, Africa, South America and Australia, although not as a major crop. There are no official statistics on the world production of mung beans. FAO statistics include lima bean, mung bean, adzuki bean and other minor legumes under 'dry beans'. Estimates of production from various sources show around 3 million ha planted worldwide with mung bean. India is the largest producer with almost 2 million ha followed by Thailand with around 500,000 ha. Estimates suggest that the area planted with mung bean could be as large as that planted with **cowpea** or **lentil**. (**See: Legumes**)

2. Origin

The mung bean is native to the northeastern India–Burma region of Asia but its progenitor species is unknown. The closest wild relative is believed to be *Vigna radiata* var. *sublobata*, which may be found growing in wasteland in eastern India. Mung bean was carried from Asia by emigrants, or by traders, to the Middle East, Africa, Latin America and Australia.

3. Seed types

Seeds are nearly spherical to oblong, about 5 mm long, glossy or dull-coated with green, yellow, tawny brown, black, or mottled testae. Dull seeds are coated with a layer of the pod inner membrane which may be translucid or pigmented. If translucid, the seed is reticulated with numerous fine wavy ridges and cross walls. Seeds average about 22,000–30,000 per kg. The hilum is round, flat and white (Fig. M.23).

4. Nutritive value

The protein content of mung bean averages around 24% fw. It is relatively rich in lysine but deficient in amino acids such as methionine. Mung beans are a source of vitamin A, thiamin, riboflavin, niacin, and some other soluble vitamins and are rich in phosphorus, potassium and iron but are relatively low in calcium.

The nutritive value of legume proteins is generally adversely affected by the presence of toxic substances such as trypsin

Fig. M.23. Mung beans (see text for dimensions) (from www.foodsubs.com/Beans.html, with permission).

inhibitor, haemagglutinins (**see: Phytohaemagglutinin (PHA); Protease inhibitors**), or other inhibitors that exert a deleterious effect on growth by lowering the digestibility of protein or utilization of particular amino acids, including methionine. But mung bean is reported to be low in such anti-nutritionals compared with several other pulse grains. The concentrations of the **oligosaccharides, raffinose** and stachyose, possible flatus producers, are, respectively, 0.44 to 0.51 g/100 g seeds and 1.01 to 1.95 g/100 g seeds: these are much less than in soybeans.

5. Uses

Mung bean seeds may be eaten green with pods, cooked and used in soups, made into porridge, boiled and eaten with rice, or sprouted. **Starch** from mung bean is used for making noodles. Mung bean flour is used to fortify wheat flour, or to produce high protein supplements for feeding children. In India, the largest producer of mung bean, the grain is dehusked and split to produce a product locally known as *dhal*, which may be used as food in many ways.

6. Plant types

The mung bean is an annual, semi-erect to erect or sometimes twining deep-rooted herb, 25 to 100 cm tall. Emergence of the seedling is epigeal. From 10–25 flowers are borne in ancillary clusters or racemes. The seed pods are curved and pointed. When mature they are 5–14 cm long and 4–6 mm wide; each pod may contain 8–20 seeds. (OV-V)

Poehlman, J.M. (1992) *The Mungbean*. Westview Press, Boulder, CO, USA.

Mustard

The spice and condiment, mustard, is the dried seeds of three important species of *Brassica* (Cruciferae) viz., *Brassica alba* (*B. herta*, *Sinapis alba*), *B. juncea* and *B. nigra*, which give white, brown and black mustard, respectively (see: Spices and flavours, Table S.12). All are annual herbs. The fruit is a four-sided **siliqua** up to 3–5 cm long with a slender beak. Pods contain four to ten small, spherical seeds, of approximate diameter, 1 mm, 1.5 mm and 2 mm, respectively, in black (Colour Plate 3D), brown and white mustard. It has been suggested that the word 'mustard' arises from the ancient practice (in some European countries) of adding the seeds to unfermented grape juice, or 'must'. The suffix 'ard' is derived from the Latin *ardere* – to burn.

1. Constituents

Mustard seeds have a fixed **oil** (triacylglycerol) content of up to 30% (**see: Oilseeds – major**). The **fatty acids** of oil consist mainly of erucic (4.2%), oleic (32.5%), linoleic (17.8%), linolenic (2.9%) and behenic (3.9%) acids: quick and efficient drying is essential to prevent heating and rancidity. Seeds normally contain mucilage, the glucosinolate, sinigrin (110–140 µmol/g) (**see: Spices and flavours,** Fig. S.44), sinapic acid and sinapine. Sinigrin releases the aggressive, volatile allyl isothiocyanate which is responsible for the pungent taste and irritant properties of black and brown mustard. Enzymatic release is by the action of myrosinase which occurs when water is added to the seed powder. White mustard releases parahydroxy benzyl isothiocyanate which accounts for the milder taste of the seed. The essential oil (0.5–2.5%) is a yellowish mobile liquid with an acrid taste and high pungency. Consisting of over 93% allyl isothiocyanate, it is extremely powerful and toxic and was the raw material for mustard gas. It blisters the skin and is lachrymatory. The volatile oil also has antimicrobial and fungicidal properties.

2. Uses

Mustard seed is a highly important commercial product, which is the basic material for a wide range of flavourings and the mustard condiment. Mustard, either the whole, powdered or ground seeds, is an indispensable ingredient in cooking. The white seeds are used in pickling, while the brown seeds are used throughout India in curry powders. The condiment, English mustard, is traditionally powdered, decorticated black mustard seeds mixed with a little powdered yellow seed and wheat flour: increasingly, however, the black mustard seed is being replaced by the cheaper and less pungent white mustard. The dark French mustard from Bordeaux is made from non-decorticated seeds with various additives (e.g. tarragon and other herbs). The more pungent, lighter, Dijon mustard is made from decorticated black and yellow seeds, traditionally with no herb additives. Various flavourings are sometimes added in French mustards, such as wine, red pepper or fruit juices. The relatively mild mustard used in the USA and elsewhere to accompany 'hot dogs' is based on yellow seeds.

The edible oils extracted from mustard are classified as nutraceutical oils, since the omega-3-fatty acids are considered to reduce heart attacks and inflammatory diseases. The expressed oil is widely used as a flavour ingredient in processed meat products, pickles, gravies, sauces, confectionery and non-alcoholic beverages. The seed meal left after removal of oil contains 40–45% protein, with a well-balanced composition of amino acids. Hence it forms a valuable source of food. The presence of isocyanate has, however, a toxic effect but it can be inactivated by heat and selective extraction.

In traditional medicine, mustard meal mixed with water was used extensively as a plaster preparation and in mustard baths to treat skin ailments, arthritis, rheumatism and bruises. Hot water poured on crushed seeds makes a household remedy for headaches and colds and a stimulating footbath. Mustard oil stimulates hair growth. It is well documented as a virtual cure-all by the Romans, Early Greeks and Egyptians. In 530 BC Pythagoras recommended it as an antidote for snake bites and scorpion stings. A paste of washed mustard seed and honey was applied to bruises or neck pains. (KVP, NB)

(See: Brassica – horticultural; History of seed research; Weights and measures)

Peter, K.V. (ed.) (2001) *Handbook of Herbs and Spices*. Woodhead Publishing, CRC Press, Boca Raton FL, USA.

Purseglove, J.W., Brown, E.G., Green, C.L. and Robbins, S.R.J. (1981) *Spices*. Vol. 1. Tropical Agriculture Series, Longman, New York, USA.

Weiss, E.A. (2002) *Spice Crops*. CAB International, Wallingford, UK.

Windholz, M. (1983) *The Merck Index. An Encyclopedia of Chemicals, Drugs and Biologicals*, 10th edn. Merck and Co., Rathway, NJ, USA.

Mutants

The term mutant describes the altered gene and also any cell or organism in which a mutation or mutations have occurred. A mutation is strictly an alteration in the genetic material, more specifically in the amount or structure of the deoxyribonucleic acid (DNA) that comprises the genes: this very frequently leads to a change or changes in some property of the cell or organism. A gene, cell or organism without mutation is known as the wild type. Mutations existing in the germ cells (gametes) are inheritable by the offspring: those only in somatic (body) cells are not, except sometimes in cases where vegetative (non-sexual) reproduction occurs.

1. What are mutations?

There are several types of mutations:

(a) *Point mutation* – the most common type – is a change in a single base pair of DNA or in a very small sequence of base pairs. Such changes may result in a faulty codon so that an incorrect amino acid (or a small sequence of amino acids) is subsequently translated in the protein. Base pair changes may also cause frameshift mutation when the **transcription** of the entire DNA sequence malfunctions.

(b) *Transposition* – the movement of a sequence of DNA into a gene, by excision from one position and insertion into another, or by replication of a sequence and insertion of the copy. Such moveable sequences are called transposable elements or **transposons**. They were discovered by the Nobel-prize-winning geneticist, Barbara McClintock, in maize and are now known in all organisms.

(c) *Inversion* – a segment of chromosome is inverted.

(d) *Deletion* – a segment of chromosome is deleted.

(e) *Translocation* – a segment of one chromosome attaches to another.

As indicated above, one consequence of a gene mutation is the production of abnormal mRNA and protein in which the primary sequence of amino acids is altered and possibly also the secondary and higher levels of protein structure. Thus, the abnormal protein does not function correctly, e.g. as an enzyme. Transcription of an entire gene might be arrested also or there may be interference with normal expression and interaction of many genes. Among the effects are:

(a) *Loss-of-function mutation*. This is the most common type, when the product of the mutated gene has lost the function of the wild type. It is usually recessive and therefore evident phenotypically only when two copies of the gene are present.

(b) *Gain-of-function mutation*. This rarer type increases the activity of the gene product and is generally dominant. Hence, only one copy of the mutated gene is required for phenotypic expression.

(c) *Other mutations* which are generally less significant in the present consideration of seed characteristics are: (i) conditional – the phenotype is seen only under certain conditions, e.g. of temperature; (ii) null – the activity of the gene is completely abolished; (iii) dominant negative – the activity of another gene is blocked and a loss-of-function phenotype is induced; (iv) lethal; and (v) silent or neutral – usually a point mutation in which the codon change does not alter the protein product.

2. The production of mutants

Mutations occur spontaneously or can be induced (Table M.8). Spontaneous mutation rates vary greatly among organisms but many estimates lie in the range 10^{-4} to 10^{-6} mutations per gene per generation. Spontaneous mutation, by generating genetic diversity, is considered to be the driving force for evolutionary change. In seeds, many of the properties that were important for the process of **domestication** and more recently in agriculture arose as spontaneous mutations (Table M.9).

Mutants are produced experimentally first by treating plant parts, such as seeds or flowers, with the mutagenizing agent

Table M.8. Mutations: induction and characteristics.

Mutagen	Mechanism	Mutation type
Spontaneous mutations		
Deamination, oxidations (free radicals), alkylation, tautomerization, mutagens in environment	Errors in DNA replication, repair, base pairs	Base transitions, frameshift transitions, translocations
Induced		
Acridine orange	Intercalates in DNA	Frameshifts
Aminopurine	Base analogue of adenine	Base pair transitions
Bromouracil	Base analogue of thymine	Base pair transitions
Ethylmethane sulphonate (EMS)	Base alkylation (generates methyl guanine)	Base pair transitions
Ethidium bromide	Intercalates in DNA	Frameshifts
Ionizing radiation: X-rays fast neutrons gamma rays	Generation of reactive oxygen species, free radicals, DNA breaks, direct effects on bases	Base alterations, chromosome loss and damage
Nitroso methylurea (NMU)	Base alkylation	Base pair transitions
Nitrous acid	Oxidative deamination	Base pair transitions, deletions
T-DNA, transposons	Insertion of foreign sequences	Deletions, frameshifts
UV irradiation	Pyrimidine dimerization	Base pair transitions, deletions, frameshifts

Table M.9. Some seed mutants.

Gene type	I or S[a]	Effect of mutation	Encyclopedia entry
Seed utilization			
Abscission	S	Retention of cereal grain on stem	**Domestication; Shattering**
Endosperm	S	Endosperm character in maize	**Domestication; Maize**
Number of spikelets	S	Rows of barley grains	**Barley; Domestication**
Pod dehiscence	S	Retention of legume seeds on plant	**Domestication; Shattering**
Seedcoat thickness	S	Thin coat, ready germination	**Domestication**
Seed composition			
brittle2, shrunken2, sugary1,2, waxy	S	Low or no starch, high sucrose, maize	**Starch; Sweetcorn**
fama,b	I	Low linolenic acid, soybean	**Oilseeds – major; Oils and fats – genetic modification**
floury2, opaque2,7	S	Low zein protein, maize	**Maize**
hiproly	S	More glutelin (higher lysine), barley	**Barley**
lpa1,2	I	Low phytate, barley, maize, soybean	**Barley; Cereals – composition and nutritional quality**
rugosus	S	Low amylopectin, wrinkled pea	**Wrinkled pea**
rug3,4,5	I	Starch content, pea	**Starch**
Seed development			
Endosperm genes	I	Altered endosperm development, e.g. barley	**Development of endosperm – cereals**
Embryogenesis genes	I	Defective embryos, e.g. *Arabidopsis*	**Development of embryos – dicots; Embryogenesis**
Seed structure, appearance			
ats, ban, ttg, etc.	I	Altered testa structure, pigmentation, dormancy, *Arabidopsis*	**Dormancy – genes; Seedcoats – development**
bs, bks1–1,2	I	Brown, black testa, tomato	–
Dap1	S	Dappled aleurone layer, maize	–
Hormones			
aba, ga, gib, sit, vp2,5,7	I, S	ABA, GA deficiencies in *Arabidopsis*, tomato, maize. Effects on dormancy, germination, desiccation tolerance, maturation, reserve deposition, vivipary	**Abscisic acid; Desiccation tolerance – acquisition and loss; Dormancy – genes; Germination – radicle emergence; Gibberellins; Hormone mutants; Maturation**
abi1,2,3 etc, etr, spy, vp1	I, S	Insensitivity to ABA, ethylene, GA in *Arabidopsis*, maize. Effects on dormancy, germination, desiccation tolerance, maturation, reserve deposition, vivipary	**Abscisic acid; Desiccation tolerance – acquisition and loss; Dormancy – genes; Germination – radicle emergence; Gibberellins; Hormone mutants; Maturation**
Light perception			
phy a,b		Phytochrome mutants; affect seed responses to light	**Phytochrome**

[a]I, induced; S, spontaneous.
Note: The gene symbols for mutants use lower case lettering (e.g. *abi*); the wild type dominant is indicated by upper case lettering (e.g. *ABI*). Also **see: Seminal roots**.

(Table M.8). Commonly used mutagens are EMS, gamma rays, X-rays, fast neutrons, T-DNA and transposons (in the latter, natural transposition occurs). Plants are grown from the treated material, then self- or cross-pollinated according to the species and the resultant offspring screened for mutant phenotypes. Very many thousands of plants usually are involved in the screening process. After initial selection, chosen phenotypes are used for breeding further generations until 'pure' mutant lines have emerged.

3. Seed mutants

There are numerous mutants with altered seed compositional or functional properties that have arisen spontaneously (Table M.9). Mutation has been an extremely important force in crop evolution. In cereals, for example, the free-threshing and non-**shattering** characters probably arose as mutants, identified and multiplied by our agricultural ancestors. In legume seeds, thinner seedcoats (allowing readier germination) and non-dehiscent pods are thought to be mutants of the thicker-coated, dehiscent wild forms. The **wrinkled pea**, used by Mendel in his explorations into inheritance, has a spontaneous mutation of the gene encoding the starch-branching enzyme, greatly reducing the **amylopectin** content of the starch and making a seed with a higher water content that wrinkles upon drying. The different types of **maize** (e.g. dent, sweetcorn, etc.) are mutations affecting the endosperm composition. The generation of mutants to alter and improve seed composition (e.g. for nutritional or processing purposes) is an important goal in plant breeding. Generated mutants are an extremely important tool for investigation of all aspects of seed **development**, metabolism and function such as **hormone** action, **dormancy**, **desiccation tolerance**, **embryogenesis**, **vivipary**, **storage reserve** deposition and **germination**. A collection of examples is given in Table M.9 and in section 4.

4. Uses of mutants

Mutation ultimately affects the protein encoded by the gene. The normal encoded protein may have an enzymic or other function or it may be structural. Hence, affecting proteins' mutations may cause a vast range of different types of changes in seed properties. These changes are potentially valuable in the following ways:

(a) *Metabolism.* By mutating the appropriate genes the nature and extent of enzyme-regulated metabolic processes may be modified. For example: (i) mutations affecting enzymes involved in **starch** synthesis alter the amount and type of starch in a seed; (ii) mutation of a gene encoding an enzyme for the synthesis of a particular fatty acid leads to altered fatty acid composition of a seed oil (**see: Oils and fats – genetic modification**); (iii) mutation of a gene encoding a type of **storage protein** changes the storage protein composition of a seed; (iv) mutation of an enzyme involved in hormone biosynthesis may block the production of hormone in the seed; and (v) metabolic pathways may be redirected down a branch point by mutating the enzyme responsible for one particular step.

(b) *Characterization of seed function.* Mutations that lead to changes in seed functions give clues as to the factors or processes normally involved. Examples illustrating this are: (i) seeds of a mutant fail to produce **abscisic acid** (ABA) and to develop dormancy or desiccation tolerance thus implicating ABA in the onset of **dormancy** and **desiccation tolerance**; (ii) mutant seeds that are unable to produce certain **gibberellins** (GA) fail to germinate, suggesting that GA is required for germination; (iii) seeds of a mutant that lacks a particular type of **phytochrome** cannot respond to light indicating the type of phytochrome normally responsible for light perception; and (iv) mutant seeds that are insensitive to ABA lack a certain **signal transduction** factor, indicating the participation of that factor in the normal action of ABA on the seed. (**See: Dormancy – genes; Hormone mutants; Maturation – controlling factors, 1. Hormonal control**)

(c) *Identification of wild-type genes and proteins.* By means of the techniques of molecular genetics mutant genes can be isolated and then used to identify the DNA sequence of the wild-type counterparts and, by reference to a database if necessary, to identify the encoded protein. In this way, for example, the enzymes for various steps in normal starch synthesis in a seed can be characterized and the various signal transduction factors participating in hormone action can be identified. Similarly, by using the many different embryogenesis mutants in *Arabidopsis*, over 250 genes participating in normal embryogenesis have been isolated and many of their protein products (enzymes and other proteins) identified. The processes occurring at different stages of embryogenesis can thus be unravelled. (MB)

Mycorrhiza

The manifestation of a widespread symbiotic association between plant roots and fungi, and the fungi capable of this relationship: literally, 'fungus root'. (**See: Arbuscular mycorrhiza**)

Mycotoxins

1. Introduction

The term 'mycotoxin' combines the Greek word for mushroom (fungus) *mykes* and the Latin word *toxicum* meaning poison. It is usually reserved for the toxic chemical products formed by a few fungal species that readily colonize crops in the field, or after harvest, and pose a potential threat to human and animal health through the ingestion of food products prepared from these commodities.

Many of the food commodities susceptible to mycotoxin contamination are seeds (Table M.10). However, any seed is susceptible to attack both in the field and in store if conditions favour the growth of toxin-producing fungi. Thus, the possibility of mycotoxins should not be overlooked wherever seeds are used for food or for medicinal and pharmaceutical purposes.

Each mycotoxin is produced by one or more very specific fungal species (Table M.10), and so their occurrence is determined by the conditions that favour the growth of each of these. Thus the aflatoxins are most readily formed by *Aspergillus flavus* and *A. parasiticus* in tropical regions, while ochratoxin A is the principal product of *A. ochraceus* in tropical regions but of *Penicillium verrucosum* in temperate areas. The presence of a recognized toxin-producing fungus does not necessarily imply the presence of the associated toxin as many factors are involved in its formation, including a ready source of inoculum. Conversely, the absence of visible mould does not guarantee freedom from toxins as it may have died out leaving the toxin intact.

Mycotoxins are important because they have a diverse range of toxic effects in humans and animals (Table M.11), in part because their chemical structures are very different from each other. High amounts must be ingested to induce acute effects

Table M.10. Species of seeds used for food in which mycotoxins may be found, and the fungi responsible for their production.

Mycotoxin or mycotoxin group	Seeds affected	Principal producing fungi
Aflatoxins B_1, B_2, G_1, G_2	Peanuts, many other nuts, cereals, spices	*Aspergillus flavus, A. parasiticus*
Ochratoxin A	Cereals, legumes, coffee, cocoa, field beans	*Penicillium verrucosum, A. ochraceus*
Citrinin	Cereals	*P. verrucosum*
Deoxynivalenol, nivalenol	Cereals	*Fusarium culmorum, F. graminearum*
T2-toxin, HT2-toxin	Cereals	*F. poae, F. sporotrichioides*
Zearalenone	Cereals	*F. culmorum, F. graminearum*
Fumonisins	Maize, other cereals	*F. moniliforme, F. proliferatum*
Cyclopiazonic acid	Nuts, maize	*A. flavus, P. commune*
Alternaria mycotoxins	Sunflower, rape, cereals	*Alternaria alternata, Alternaria tenuis*
Ergot alkaloids	Wild grasses, all cereals, but particularly rye	*Claviceps purpurea* and other *Claviceps* species

Plate 1. A. Adzuki beans (image by Elly Vaes and Wolfgang Stuppy, Millennium Seed Bank, RBG, Kew). **B.** A collection of dry common beans (from http://agronomy.ucdavis.edu/gepts/pb143/pb143_hot_news.htm with thanks to Dr P. Gepts). **C, D.** Broad or field beans: small, round (C) and large, oval (D). Inset circles represent 1 cm diameter (images by NIAB, Cambridge). **E.** Cowpeas (scalebar = 1 cm; from http://aggie-horticulture.tamu.edu/seeds/cowpea.html, by kind permission of R.D. Lineberger). **F.** Tepary beans (blue speckled) (image by Scott and Zizi Vlaun/Seeds of Change) (for dimensions see text).

Plate 2. A. Carob pods and seeds (www.advsuroviny.cz/ krmiva.htm). **B.** Carob seeds, internal views. The embryo rests in the endosperm (dark). In the section (right) the cotyledons are each approximately 1 mm thick and the endosperm approximately 2 mm thick on each side (photo by Xuemei Gong, University of Guelph). **C.** Chickpeas, kabuli type (left) and desi type (right). Scale = mm (image by Mike Amphlett, CABI). **D.** Peas (vegetable, field and maple, round, wrinkled of various sizes and colours). Inset circle represents 1 cm diameter (image by NIAB, Cambridge). **E.** Drawing of a peanut plant showing underground fruits on gynophores (from http://nprl.usda.gov/AllAboutPeanut/GeneralInfo.htm with thanks). **F.** Scarlet runner beans. Inset circle represents 1 cm diameter (image by NIAB, Cambridge).

Plate 3. A. Black pepper: upper row, right to left – green, black and white dried peppercorns; lower row – green and red pickled peppercorns (from www.ang.kfunigraz.ac.at/~katzer/engl/Pipe_nig.html by kind permission of Dr G. Katzer). **B.** Black cardamom: whole fruit (left) and one cut open to display the seeds (image by Mike Amphlett, CABI). **C.** Green cardamom: whole fruits (left) and one cut open to display the seeds (image by Mike Amphlett, CABI). **D.** Black mustard seeds (image by Mike Amphlett, CABI). **E, F.** The spices nutmeg (right in F) and mace (left in F) (image by Mike Amphlett, CABI). Scales = mm.

Plate 4. A. Almond: the drupe is shown intact (left) and cut away to reveal the kernel (seed) (approx 2.5cm long) inside the hard endocarp, together constituting the nut. The endo-, meso- and exo-carp are regions of the fruit coat. **B.** Brazil nut capsules (fruits) (15–20 cm diameter) (image by Dr Carlos E. Velazco, kindly supplied through Resources Brazil). **C.** Cashew 'apples' (8–9 cm long) with the attached drupes (image by Dr Mark Rieger, University of Georgia). **D.** Cashew kernels (embryos) (2–2.5 cm long) (image by Mike Amphlett, CABI). **E.** Sweet chestnut: involucral bracts have opened to reveal the cluster of four nuts (image by Wolfgang Stuppy, Millennium Seed Bank, RBG, Kew). **F.** Pine nut that has been cracked open to show the thick seed coat and the internal kernel (embryo within megagametophytic tissue). Scale = mm (image by Mike Amphlett, CABI). **G.** Whole pine nut kernels (image by Mike Amphlett, CABI).

Plate 5. A. A collection of nuts. Included are (clockwise from the top) walnut, pecan (true nuts), two almonds (culinary nuts), hazelnut (true nut) and a Brazil nut (culinary nut). In the centre are two pistachios and several pine nuts (culinary nuts) (from http://dolly.cside.com/marine/sproutmix.htm with thanks for permission). **B.** Pecan, whole nuts (image by Mariko Sakamoto) (for dimensions see text). **C.** Pecan, two separated cotyledons of an embryo. Scale = mm (image by Mike Amphlett, CABI). **D.** Pistachio fruits (image by V. Polito) (for dimensions see text). **E.** Pistachio nuts showing the exposed kernels (green cotyledons of the embryos and the brown seed coat) within the endocarp (image by Mariko Sakamoto) (for dimensions see text). **F.** Walnut: nut within the dehiscing, green, fibrous husk covering (image by V. Polito).

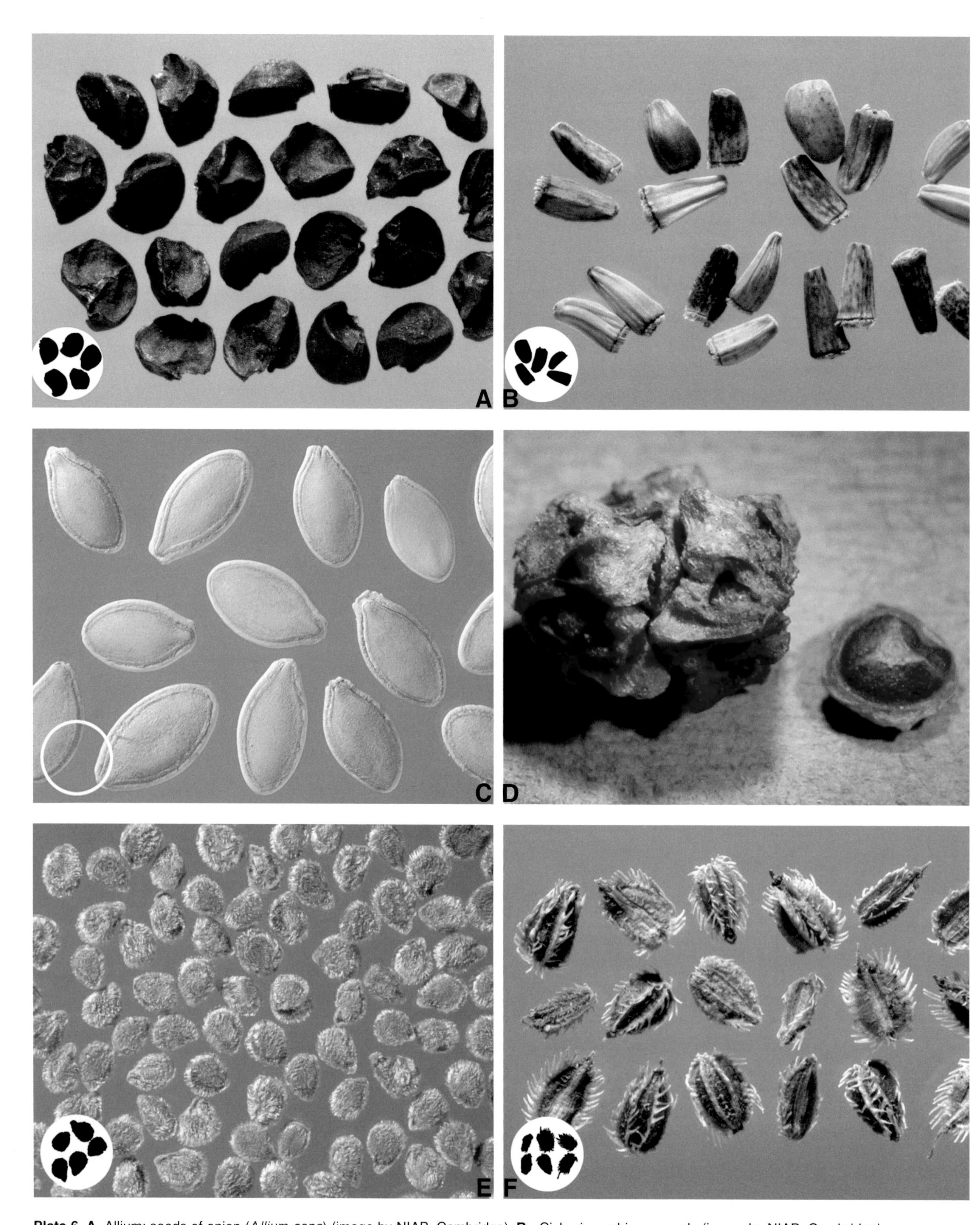

Plate 6. A. Allium: seeds of onion (*Allium cepa*) (image by NIAB, Cambridge). **B.** Cichorium: chicory seeds (image by NIAB, Cambridge). **C.** Cucurbits: seeds of *Cucurbita pepo* (courgette, marrow, pumpkin, some squashes) (image by NIAB, Cambridge). **D.** Sugarbeet: an unprocessed multigerm sugarbeet achene (left) and a rubbed and graded monogerm seed (approx. 3mm across) (right) showing the lid-shaped operculum through which the germinated embryo emerges. **E.** Tomato seeds (image by NIAB, Cambridge). **F.** Umbelliferae: carrot seeds (image by NIAB, Cambridge). Inset circles represent 1 cm diameter.

Plate 7. A. Annatto capsules and seeds. **B.** Cacao pods on the plant (image by Dr Carlos Velazco). **C.** Cacao: drawing of a pod with seeds and a trio of partly processed seeds. For approximate dimensions see text (from www.swsbm.com/Images/New10-2003.html with thanks to Michael Moore). **D.** Cacao beans. **E.** Coffee: drawing of shoot of *C. arabica*, with green, unripe and red, ripe 'berries' and seeds (from www.swsbm.com/Images/New10-2003.html with thanks to Michael Moore). **F.** Cola pods and seeds (from www.botgard.ucla.edu/html/MEMBGNewsletter/Volume5number4/pTX.html) (for dimensions see text).

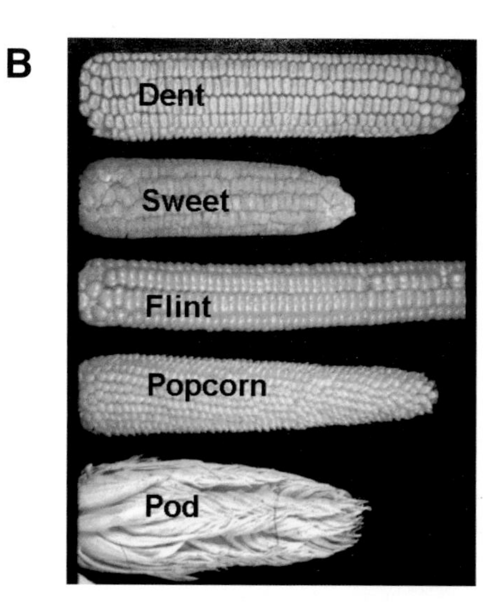

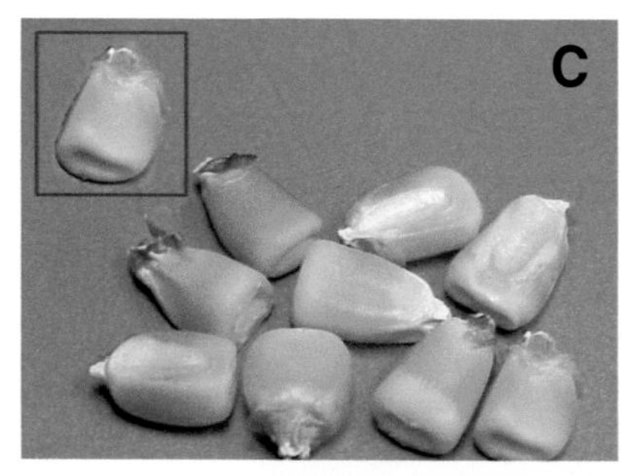

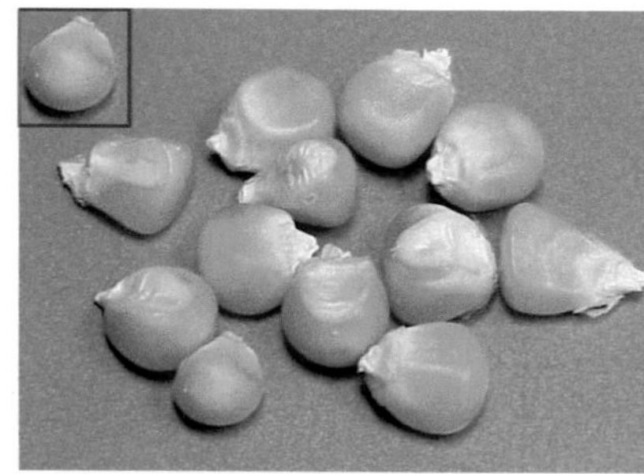

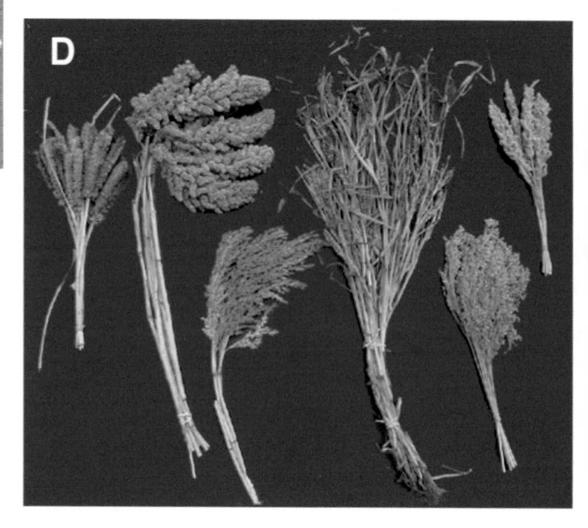

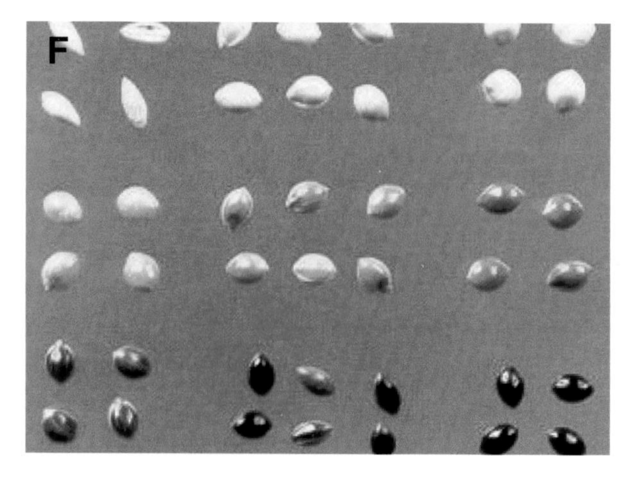

Plate 8. A. Hulled (right) and naked (left) barley grains. The ventral crease is visible on four hulless grains. The embryo is evident on the dorsal side at the more pointed end of the grains. Scale = mm (image by Mike Amphlett, CABI). **B.** Cobs of different maize grain types (from www.psu.missouri.edu/plsci274/images/images.html). For dimensions see text. **C.** Maize grains: dent (top left); flint (top right); coloured flint (bottom). The 'sharper' proximal end of the grain is where the radicle end of the embryo is present, and is the end that is inserted into the cob. Remnants of the funiculus connection to the cob can be seen as white and shrivelled. In many grains the embryo can be seen as an indentation along much of the dorsal side of the grain (www.oznet.ksu.edu/kansascrops/corn_class.htm). For dimensions see text. **D.** Millets, some with potential to become important crops in semi-arid production systems, are (left to right): finger millet, foxtail millet, little millet, proso millet, barnyard millet and kodo millet (from www.icrisat.org/text/research/grep/homepage/ archives_cd/archives/minormillets.jpg). **E.** Oat grains: the grains in the group on the left have intact hulls while those on the right are hulless. In these, note the ventral crease and, on the dorsal side, the outline of the embryo at the pointed end of the grain, particularly the elongated scutellum. Scale = mm (image by Mike Amphlett, CABI). **F.** Proso millet: common millet grains (www.gov.on.ca/OMAFRA/english/crops/facts/87-025.htm) (for dimensions see text).

Plate 9. A. Rye grains: a ventral crease is visible, as is the somewhat elongated, pointed embryo to one end of the dorsal surface. Inset circle represents 1 cm diameter (image by NIAB, Cambridge). **B.** Sorghum grains. Scale = mm (image by Mike Amphlett, CABI). **C.** Triticale: a ventral crease is visible, as is the embryo to one end of the wrinkled, dorsal surface. Inset circle represents 1 cm diameter (image by NIAB, Cambridge). **D.** Grains of bread wheat (*T. aestivum*): red-grained wheat on the right, white-grained on the left. Ventral creases are evident and an embryo terminates the dorsal surface of each grain. Scale = mm (image by Mike Amphlett, CABI). **E.** American wild rice grains. For dimensions see text.

Plate 10. A. Infected maize kernels with blue eye symptom (copyright Denis C. McGee, Iowa State University, from the CABI Compendium website: www.cabicompendium.org/cpc). **B.** Soybean seeds infected by soybean mosaic virus have a brown or blackish mottled coat and are substantially smaller than healthy seeds (photograph courtesy J.H. Johnson, from the *Compendium of Soybean Diseases*, www.apsnet.org/online/archive/1-20.htm). **C.** 'Starburst' symptom in maize, caused by *Fusarium* ear rot (photograph kindly provided by Albert Tenatu, Ontario Ministry of Agriculture and Food). **D.** Symptoms of *Phomopsis* seed decay in soybean. Infected seeds are shrivelled and cracked (photograph kindly provided by Albert Tenatu, Ontario Ministry of Agriculture and Food). **E.** Sclerotium (dark areas) and apothecia ('mushroom'-shaped structures) of *Sclerotinia sclerotiorum* (photograph courtesy R. Hall, from the files of G.J. Boland, www.apsnet.org/online/Archive/2000/Bean082.asp). **F.** Common smut on an ear of maize. Note dark masses of teliospores in galls (from www.apsnet.org/online/Archive/2000/IW000014.asp; photograph courtesy Mike Boehm).

Plate 11. A. The family of coating treatments illustrated on onion seed (centre): filmcoating; encrusting; and two degrees of pelleting (photographs courtesy of the Germain's Technology Group). **B.** Preharvest sprouting in wheat. The ear on the far left has no sprouted grains. The second and fourth ears show sprouting as it often occurs under field conditions, i.e. coleoptiles of the sprouted upper grains. The third ear from the left shows sprouting experimentally induced by repeated wetting of the mature ear. **C.** *Amaranthus* grains. Scale = mm (image by Mike Amphlett, CABI). **D.** Mature haploid embryo of European larch (*Larix decidua*) shown growing from an embryogenic culture. Numerous green cotyledons arise from a green hypocotyl, below which is the brown root cap.

Plate 12. A, B. Seeds of *Brassica* sp. (scale = mm) (photographs kindly produced by Ralph Underwood of AAFC, Saskatoon Research Centre, Canada). **C.** Coconut 'seed', whole and cut open to show the milky white solid endosperm (from www.caribbeanfoodemporium.com with permission). **D.** Cross section, oil palm fruit (right) and internal nut (photo by Elly Vaes and Wolfgang Stuppy, Millennium Seed Bank, RBG, Kew). **E, F.** Sunflower achenes: confectionery type (E) and oilseed type (F). Scale = mm (photograph kindly produced by Ralph Underwood of AAFC, Saskatoon Research Centre, Canada).

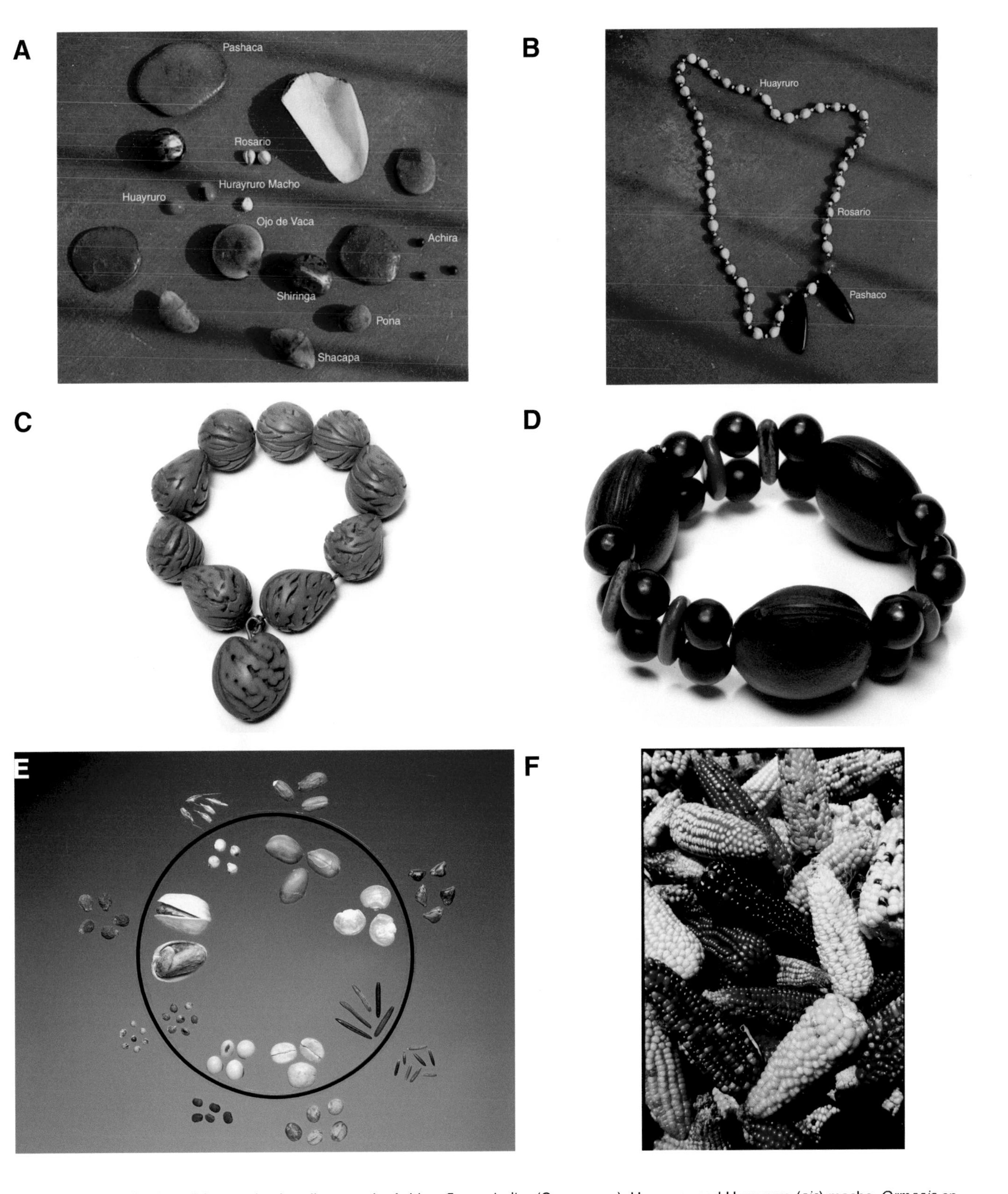

Plate 13. A. A collection of Amazonian jewellery seeds: Achira, *Canna indica* (Cannaceae); Huayruro and Hurayruro (*sic*) macho, *Ormosia* sp. (Fabaceae); Ojo de vaca, *Mucuna rostrata* (Fabaceae); Pashaca, *Microlobium acaciifolium* (Fabaceae); Pashaco, *Parkia igneiifolia* (Fabaceae); Pona, *Socratea exorrhiza* (Arecaceae); Rosario, *Coix lacryma-jobi* (Poaceae); Shacapa, *Parina* sp. (Poaceae); Shiringa, *Hevea brasiliensis* (Euphorbiaceae) (from www.biobio.com/Articles/craftplants.html). **B.** A necklace. **C.** Nineteenth century bracelet made of polished peach stones. Economic Botany Collection, Royal Botanic Gardens, Kew, Catalogue No. 57340. © The Board of Trustees of the Royal Botanic Gardens, Kew. **D.** Seed bracelet from Costa Rica. Large seeds: *Mucuna holtonii* (sea bean, Fabaceae); small spherical seeds: *Sapindus saponaria* (soapberry, Sapindaceae); brown seeds: *Delonix regia* (flame tree, Fabaceae); red seeds: *Cassia grandis*. Economic Botany Collection, Royal Botanic Gardens, Kew. © The Board of Trustees of the Royal Botanic Gardens, Kew. **E.** Domestication and seed size. Seeds from domestic crops (inner circle) are usually larger, lighter in colour and more uniform than their wild relatives. Clockwise from top: peanuts, corn (maize), rice, coffee, soybean, hops, pistachio and sorghum (image by Stephen Ausmus, USDA, ARS Photo Library). **F.** A selection of maize cob types (from www.mc.maricopa.edu/dept/d10/asb/anthro2003/lifeways/hg_ag/maize.html).

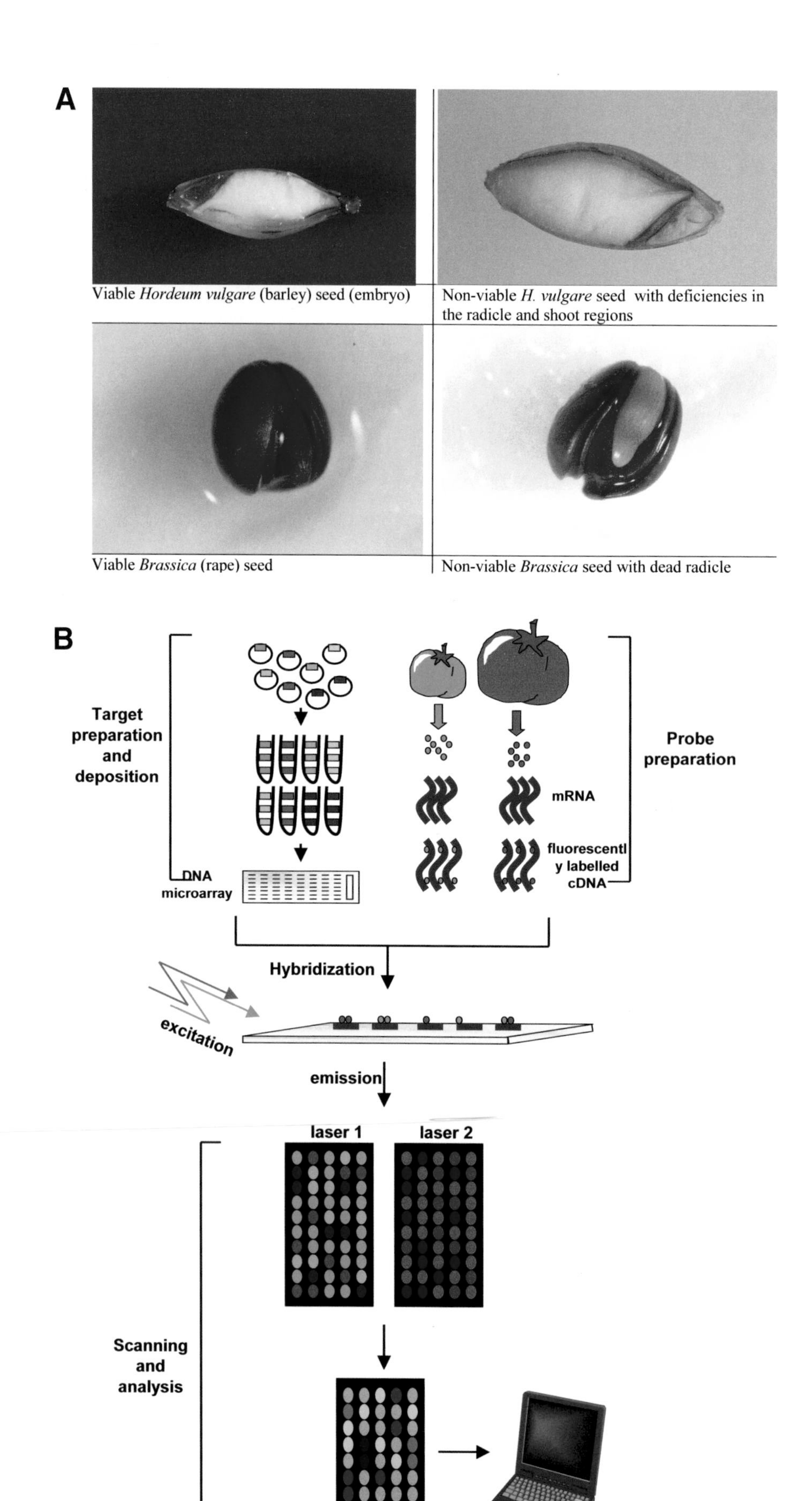

Plate 14. A. Examples of tetrazolium evaluation. **B.** Transcriptomics: scheme of a typical cDNA microarray for gene expression analysis.

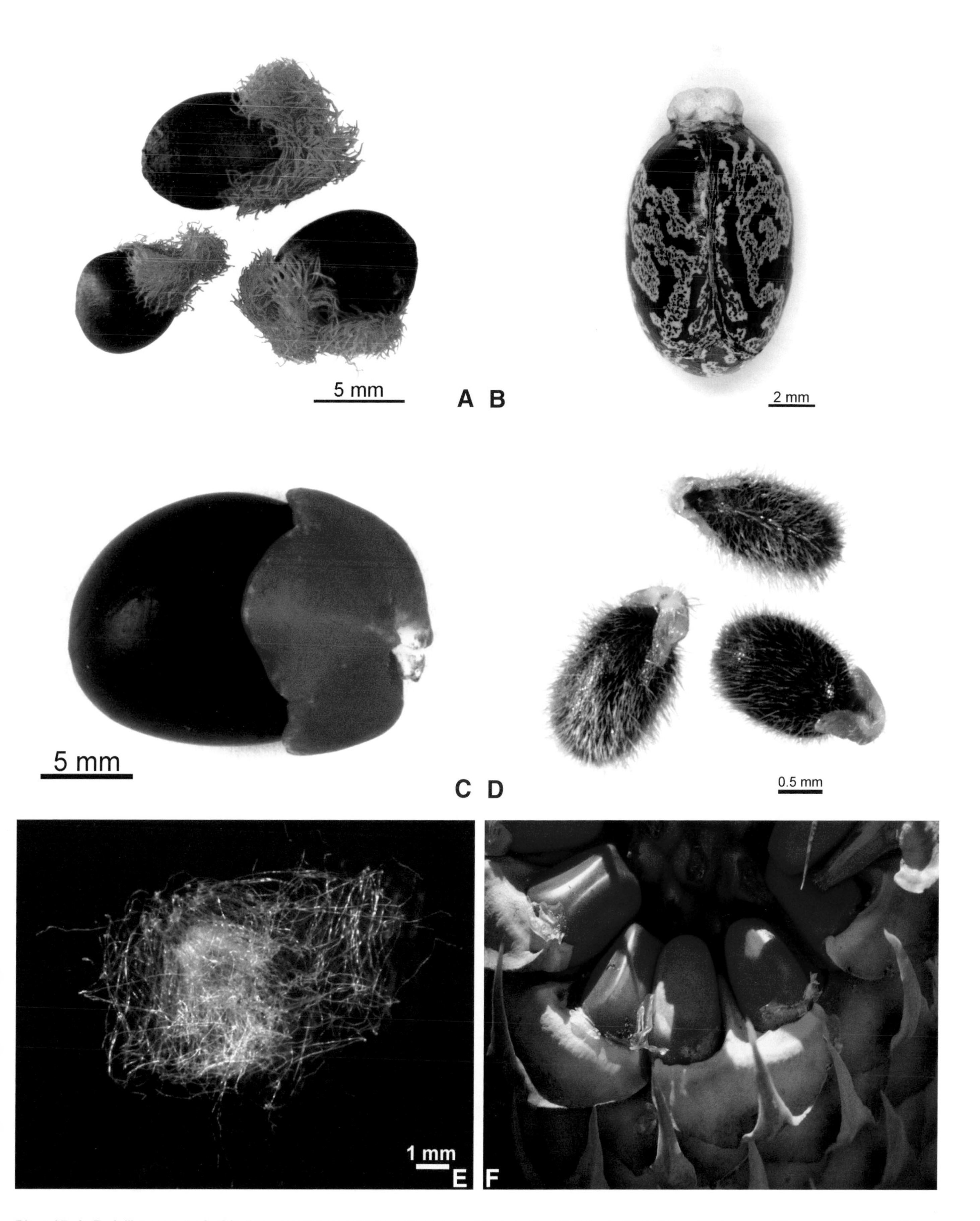

Plate 15. A–D. Arillate seeds: **A**. *Strelitzia reginae* (Strelitziaceae); seeds with complex aril (formed by both the funiculus and the exostome). **B.** *Ricinus communis* (Euphorbiaceae); seed with caruncle (exostomal aril). **C.** *Afzelia africana* (Fabaceae); seed with orange-red funicular aril. **D.** *Polygala vulgaris* (Polygalaceae); seeds with three-lobed exostomal aril. Hairs: **E.** *Populus angustifolia* (Salicaceae); seeds plumose at the micropylar end from funicular hairs. **F.** *Macrozamia fraseri* (Zamiaceae), mature female cone with seeds showing the bright red sarcotesta. Images by Wolfgang Stuppy, Millennium Seed Bank, RBG, Kew.

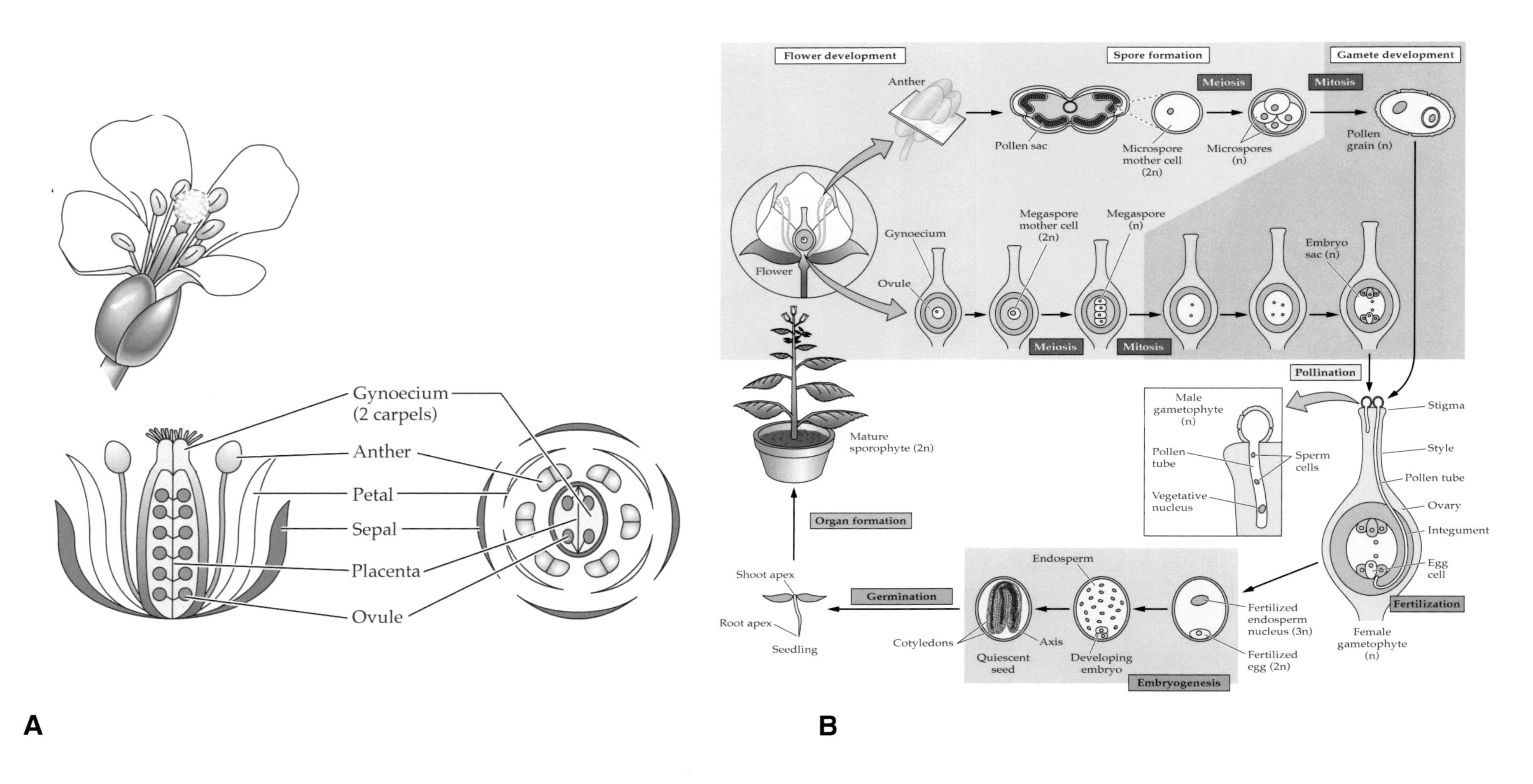

Plate 16. A. Diagram of an *Arabidopsis* flower to show the positioning of the whorls of sepals, petals, anthers and gynoecium, which is composed of two fused carpels that bear the ovules attached to the central dividing structure, the placenta. These whorls are positioned on a receptacle (not shown). To the right is a floral diagram which is used to illustrate the positioning and number of each of the parts of the flower: sepals, 4; petals, 4; anthers, 6; carpels, 2, each with two rows of ovules. From Bewley, J.D., Hempel, F.D., McCormick, S. and Zambryski, P. (2000) Reproductive development. In: Buchanan, B., Grissem, W. and Jones R.L. (eds) *Biochemistry and Molecular Biology of Plants*. American Society of Plant Physiologists, pp. 988–1043. With permission. **B.** The life cycle of a flowering plant, with the predominant diploid (2n, sporophyte) generation, the vegetative plant and the brief haploid (n, gametophyte) generation associated with the production of the male and female gametes, within the pollen and the egg respectively. From Bewley, J.D., Hempel, F.D., McCormick, S. and Zambryski, P. (2000) Reproductive development. In: Buchanan, B., Grissem, W. and Jones R.L. (eds) *Biochemistry and Molecular Biology of Plants*. American Society of Plant Physiologists, pp. 988–1043. With permission.

Table M.11. Some toxic effects of mycotoxins.

Mycotoxin	Effects
Aflatoxins B_1	Hepatotoxic, carcinogenic
Ochratoxin A	Nephrotoxic, genotoxic, immunosuppressive
Citrinin	Nephrotoxic
Deoxynivalenol	Emetic, feed refusal in animals
T2-toxin	Gastrointestinal problems, immunosuppressive
Zearalenone	Oestrogenic
Fumonisins	Equine leucoencephalomalacia, porcine oedema, human throat cancer
Cyclopiazonic acid	Neurotoxic
Patulin	Gastrointestinal problems, haemorrhagic, possibly carcinogenic

so that such incidents are usually restricted to the less developed regions of the world. The consumer in the more affluent regions is demanding of food free from, or low in, contaminants such as mycotoxins, and this has led to improved quality control or management of fungal infection including to livestock. Chronic effects that may result from ingesting much lower amounts are of concern for the long-term health of the human population. Most mycotoxins are quite stable under normal conditions, so they tend to survive storage and processing even when cooked to temperatures such as those reached during baking bread or producing breakfast cereals. Other products such as nuts are eaten without significant processing.

National and international organizations evaluate the risk that mycotoxins pose to humans, and there are various statutory or guideline maximum permissible limits for some mycotoxins. Many countries now have legal limits for the aflatoxins, the most widespread and toxic mycotoxins.

Fungi typically develop in a heterogeneous manner in crops. This results in a very uneven distribution of the mould and of any mycotoxin within a consignment. Hence great care is required to ensure that a sample taken for analysis is truly representative. Most mycotoxins are toxic at very low concentrations, so sensitive and reliable methods are required for their detection. Sampling and analysis taken together represent an extremely demanding challenge for the analyst. Failure to measure accurately the concentrations present can lead to unacceptable consignments being accepted or satisfactory ones being unnecessarily rejected. Rigorous quantitative analysis is usually carried out with high pressure liquid chromatography (HPLC), gas liquid chromatography (GLC) or mass spectrometry (MS), although thin layer chromatography (TLC) and immunological methods can sometimes be used. Cheap, rapid and reliable methods for monitoring are required and are constantly being developed and improved.

2. Major mycotoxins

(a) Aflatoxins. These are a group of 20 or more related mycotoxins of which only four, aflatoxins B_1, B_2, G_1 and G_2, occur naturally in seeds. They occur in a wide range of crops, particularly in warm regions, but a consequence of international trade is that aflatoxins are a worldwide problem. Aflatoxins M_1 and M_2 are the hydroxylated metabolites of aflatoxins B_1 (Fig. M.24), and B_2 is excreted in milk when ruminants ingest contaminated feed, and may then contaminate dairy products.

Producers, importers and food manufacturers routinely screen commodities for aflatoxins. These have been found at quantities of 1000 μg/kg or more in **peanuts**, and peanut butter may be contaminated. Other nuts particularly prone to contamination are **pistachios** and **Brazils**. Climate controls the conditions that encourage aflatoxin, so the problem varies in severity from year to year. Drought leading to crop stress followed by rain is particularly unwelcome for cereal producers since these conditions encourage fungal growth.

Aflatoxins are crystalline substances that are extremely stable in the absence of light and UV radiation, even above 100°C. They are freely soluble in moderately polar solvents such as chloroform, methanol and dimethyl sulphoxide and dissolve in water to the extent of 10–20 mg/l. They fluoresce under UV radiation. Some properties of the aflatoxins are shown in Table M.12.

Aflatoxins are both acutely and chronically toxic, aflatoxin B_1 being one of the most potent liver carcinogens known. The outbreak of so called 'Turkey-X disease' which caused the deaths of 100,000 turkeys and other poultry in the UK in 1960 was caused by extremely high concentrations of aflatoxins in imported groundnut (peanut) meal.

Acute aflatoxin toxicity has been demonstrated in a wide range of mammals, fish and birds. Age, sex and nutritional status all affect the degree of toxicity. For most species the LD_{50} (the amount of material, given all at once, that causes death of 50% of a group of test animals) is between 0.5 and 10 mg/kg body weight (bw). The liver is the principal target organ. Teratogenic effects have been reported in some species. Acute poisoning of humans by aflatoxins occurs occasionally and aflatoxins have

Table M.12. Some physical properties of the aflatoxins.

Aflatoxin	Molecular formula	Molecular wt	Melting point (°C)	UV absorption max (ε), nm, methanol: 265	360–362
B_1	$C_{17}H_{12}O_6$	312	268–269	12,400	21,800
B_2	$C_{17}H_{14}O_6$	314	286–289	12,100	24,000
G_1	$C_{17}H_{12}O_7$	328	244–246	9,600	17,700
G_2	$C_{17}H_{14}O_7$	330	237–240	8,200	17,100
M_1	$C_{17}H_{12}O_7$	328	299	14,150	21,250 (357)
M_2	$C_{17}H_{14}O_7$	330	293	12,100 (264)	22,900 (357)

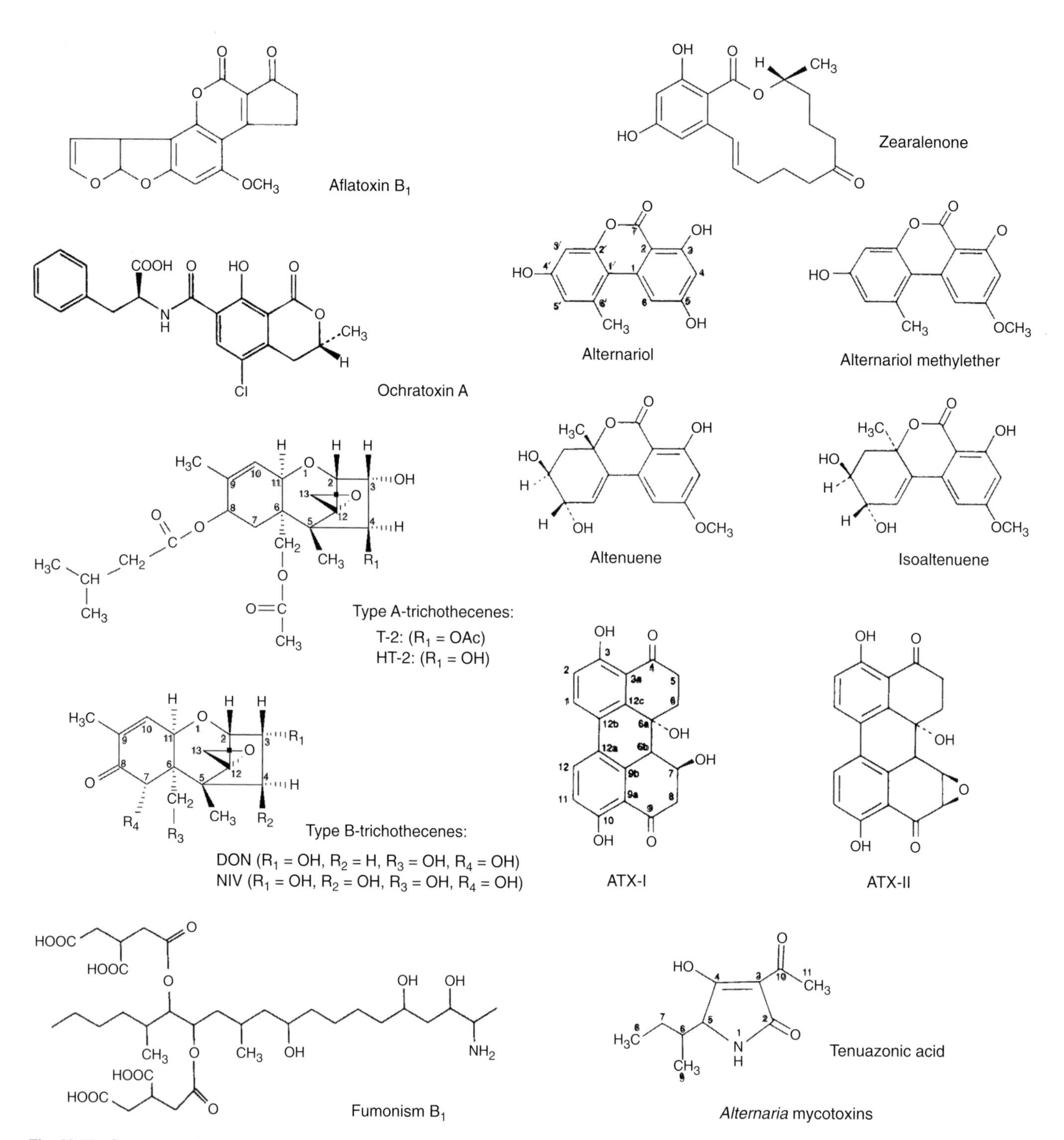

Fig. M.24. Structures of some mycotoxins.

been implicated in sub-acute and chronic effects in humans. These include primary liver cancer, chronic hepatitis, jaundice, hepatomegaly, cirrhosis and immunosuppressence.

Measurement is usually by HPLC, TLC or ELISA (enzyme-linked immunosorbent assay), using methods sensitive to below 1 μg/kg. Aflatoxins are susceptible to alkaline hydrolysis but if treatment is mild, acidification will reverse the reaction to reform the original aflatoxin. In acid, aflatoxin B_1 and G_1 are converted to aflatoxins B_{2a} and G_{2a}.

Aflatoxins are quite stable in many foods and are fairly resistant to degradation.

As carcinogens, amounts in food should be reduced to the lowest values that are technologically possible. Regulations have been set for human food and animal feed in many countries. The EC (European Commission) has established maximum permissible limits for aflatoxins in a range of commodities including **nuts**, dried fruit and **cereals** and for aflatoxin M_1 in milk and dairy products.

(b) Ochratoxin A. This mycotoxin, and the relatively non-toxic ochratoxin B, are formed by some *Penicillium* and *Aspergillus* species during storage of a wide range of seed crops.

Ochratoxin A is the most important of a structurally related group of compounds (Fig. M.24). It usually develops when products are poorly stored and a significant proportion of cereal samples may be contaminated in temperate regions. A colourless crystalline compound, it exhibits blue fluorescence under UV light. It crystallizes from benzene to give a product melting at 90°C containing one molecule of benzene removed under vacuum at 120°C. The sodium salt is soluble in water. As the acid, it is moderately soluble in polar organic solvents such as chloroform, methanol and acetonitrile and dissolves in dilute aqueous sodium bicarbonate. On acid hydrolysis it yields phenylalanine and ochratoxin α. Reaction in methanol plus hydrochloric acid yields the methyl ester, while methylation with diazomethane gives the O-methyl methyl ester.

Ochratoxin A is a potent toxin affecting mainly the kidneys in all investigated mammalian species. In acute toxicity studies, LD_{50} values vary greatly in different species, the dog being especially susceptible. Ochratoxin A is a potent teratogen in mice, rats, hamsters and chickens, is immunosuppressent and genotoxic, both *in vitro* and *in vivo*. Human exposure to ochratoxin A has been demonstrated by its detection in blood and breast milk. In biological systems, it will bind to serum albumin.

Measurement is by HPLC, TLC or ELISA using methods sensitive to below 1 μg/kg. Ochratoxin A is a moderately stable molecule so will survive most food processing and occur in consumer products. Estimates of daily intakes range from 0.7 to 4.7 ng/kg bw derived from analyses of food and from 0.2 to 2.4 ng/kg bw derived from blood samples. In Europe about 50% of this intake can be attributed to cereal and cereal products. The EU have recently introduced statutory maximum limits for ochratoxin A of 5 μg/kg in raw cereal grains and 3 μg/kg for cereal products or grains for direct human consumption. Limits for other products are being considered.

(c) Deoxynivalenol and trichothecenes. Deoxynivalenol (DON or vomitoxin) is one of about 150 compounds known as the trichothecenes formed by species of *Fusarium* and some other fungi that infect cereal grains and other seeds. It is toxic and is sometimes formed in high concentrations before harvest, often associated with cereal diseases. Trichothecenes are classified as Group A or B compounds depending on whether they have a side chain on the C7 atom (Fig. M.24). Group A includes T-2 toxin, HT-2 toxin, neosolaniol, monoacetoxyscirpenol and diacetoxyscirpenol and Group B, deoxynivalenol, nivalenol, 3- and 15-acetoxynivalenol and fusarenon X. The macrocyclic trichothecenes are more toxic and produced by mould species such as *Stachybotrys atra* and include satratoxins, verrucarins and roridins.

Deoxynivalenol is a polar trichothecene soluble in water and polar solvents while other Group B compounds are soluble in methanol, ethanol and acetonitrile. All trichothecenes containing an ester group are hydrolysed to their respective parent alcohols when treated with alkali and many of the alcohols are unaffected, even by hot dilute alkali. Trichothecenes are thus chemically stable and can persist for long periods once formed. Group A trichothecenes are soluble in ethyl acetate, acetone, chloroform, methylene chloride and diethyl ether.

When given orally or by intraperitoneal injection, the trichothecenes are acutely toxic at low concentrations (Table M.13). T-2 toxin and the macrocyclic mycotoxins (e.g. verrucarin A) are far more toxic than deoxynivalenol, but occur less commonly in agricultural products. Acute toxicity is characterized by gastrointestinal disturbances, such as vomiting, diarrhoea and inflammation, haemorrhaging, dermal irritation, feed refusal, abortion, anaemia and leukopenia. They are acutely cytotoxic and strongly immunosuppressive. Because metabolites occur together, the toxicology is complex with both synergistic and antagonistic effects observed. Alimentary toxic aleukia is a human trichothecene mycotoxicosis in which T2-toxin may be implicated.

Extraction solvents depend on the properties of the compound. GC, GC-MS or LC-MS (liquid chromatography with mass spectrometry) are the methods of choice for measurement of trace amounts because complex mixtures are often present, although immunoassay tests are available for DON, T-2 and HT-2 toxins.

The trichothecene structure is stable so compounds can contaminate processed cereal products, including beer in which deoxynivalenol causes the problem of gushing (**see: Malting**). DON analysis is performed to assess the gushing (fobbing) potential of malts. During milling, higher amounts of trichothecenes concentrate in bran while concentrations in white flour are reduced.

Few countries have imposed limits for trichothecenes. However, the EC has proposed levels at which action should be taken, for deoxynivalenol of 500 μg/kg for cereal products and 750 μg/kg for raw cereals.

(d) Fumonisins. These are a group of at least 15 closely related mycotoxins, the most important being fumonisin B_1, that frequently occur in **maize**, and which are produced by specific *Fusarium* species.

They were identified during the mid-1980s although their effects on horses have been recognized for at least 150 years. They are polar metabolites with structures based on a long hydroxylated hydrocarbon chain containing methyl and amino groups (Fig. M.24). Fumonisins often occur at concentrations of mg/kg together with other mycotoxins that can include, for example, aflatoxins, deoxynivalenol and zearalenol in maize. They can be found in other cereal grains, and seeds of other species, but in much lower concentrations.

Table M.13. LD_{50} (the amount of material, given all at once, that causes death of 50% of a group of test animals) values for mice (intraperitoneal route) of some trichothecenes.

Trichothecene	LD_{50} (mg/kg bw)
Deoxynivalenol	70
Diacetoxyscirpenol	23
Neosolaniol	14.5
HT-2 toxin	9.0
T-2 toxin	5.2
Nivalenol	4.1
Verrucarin A	0.5

bw: Body weight.

Pure fumonisin B_1 is a white hygroscopic powder that is soluble in water, acetonitrile-water or methanol. Fumonisins are soluble in polar solvents and fumonisins B_1 and B_2 are stable in methanol if stored at −18°C but degrade at 25°C and above.

The effect of fumonisins in horses is a sporadic fatal disease called equine leucoencephalomalacia (ELEM), characterized by oedema in the brain and liquefaction of areas within the cerebral hemispheres. They are considered to be toxic principally due to their effects on sphingolipid synthesis where effects are species related. In pigs, fumonisins induce pulmonary oedema, while rats fed culture material from *F. moniliforme* develop primary hepatocellular carcinomas. Fumonisin B_1 has also been shown to affect the foetus in pregnant rats. In humans, there may be a link with the occurrence of oesophageal cancer.

Measurement is usually by HPLC or mass spectrometry after extraction with aqueous methanol or acetonitrile. Dry milling of **maize** seed results in distribution of fumonisins into different milled fractions with high concentrations in bran, and lower amounts in other fractions. In experimental wet milling, fumonisin has been detected in steep water, gluten, fibre and germ, but not in the starch. Fumonisins are quite stable and are not destroyed by moderate heat, although an 80% reduction by heating at higher temperatures has been reported. However, fumonisins are degraded by alkali so are destroyed in foods such as tortillas. Fumonisins may be found in polenta, maize-based breakfast cereals, snack products and beer.

There are limited controls to protect susceptible animals such as horses (5 mg/kg) and pigs (50 mg/kg), or temporary guidelines to limit the exposure for humans.

(e) Zearalenone. Zearalenone is a phenolic resorcyclic acid lactone (Fig. M.24) produced by several species of *Fusarium* that colonize cereal grains: it develops particularly during cool, wet seasons. It may occur in **wheat, barley, rice,** maize, and other cereals, and in other crops, and can survive into consumer products. Closely related compounds or conjugated products occur in cereals and animal feeds and can be more potent oestrogens than the parent compound. Because zearalenone can be produced by the fungi that form deoxynivalenol and nivalenol, these mycotoxins can occur together.

Zearalenone is a white crystalline compound, which exhibits blue-green fluorescence when excited by long wavelength UV light (360 nm) and a more intense green fluorescence when excited with short wavelength UV light (260 nm). It is slightly soluble in hexane and progressively more so in benzene, acetonitrile, methylene chloride, methanol, ethanol and acetone. It is also soluble in aqueous alkali.

Zearalenone is oestrogenic so it affects the reproductive system. It causes infertility in sheep and hyperoestrogenism, particularly in swine. Symptoms include abortion, stillbirths, reduced litter size and reduced weight of piglets. Effects on dairy cows include vaginitis, prolonged oestrus, and infertility. Closely related metabolites of zearalenone possess similar properties. The oestrogenic potency of zearalenone has been compared with other plant-derived oestrogens and is one of the most potent natural xenoestrogens. There is limited evidence for carcinogenicity or genotoxicity.

Commonly used extraction solvents are aqueous mixtures of methanol, acetonitrile or ethyl acetate. Measurement is usually by HPLC with UV or fluorescence detection, GC-ECD (gas chromatography with electron capture detection), GC-MS or HPLC-MS. Concentrations down to 5 μg/kg can be detected. Zearalenone is only partly decomposed by heat so a significant amount remains in bread, noodles and extruded products. In dry milling of maize, zearalenone is concentrated in the bran and germ while wet milling concentrates it in the gluten fraction. Zearalenone is metabolized by yeasts during brewing, mainly to β-zearalenol, which can occur in beer.

Zearalenone has been evaluated on the basis of the dose that has no hormonal effect in pigs, the most sensitive animals. In Europe, dietary intakes suggest that a significant margin of safety exists. A few countries have guidelines ranging from a requirement for non-detectable amounts, to 1000 μg/kg.

(f) Alternaria *mycotoxins*. Species of the fungus *Alternaria* produce many toxic metabolites, most of which are phytotoxins, but about six mycotoxins have been found in seeds.

The most important mycotoxin-forming species of *Alternaria* is *A. alternata*, which is known to produce alternariol, alternariol monomethyl ether, altenuene, iso-altenuene, altertoxins I, II and II and tenuazonic acid (Fig. M.24). *A. alternata* infects many seeds, ripe fruits and vegetables causing major spoilage. Strains isolated from many of these commodities readily form some or all of these mycotoxins in culture but their natural occurrence in commercial seed and other foodstuffs is much less studied. The main foods affected are **sunflower** seeds, ***Brassica* oilseed** and cereals.

Isolates of many *Alternaria* species are toxic to animals but the pure toxin is probably somewhat less noxious than mycotoxins produced by other species. Tenuazonic acid often occurs in high concentrations and is probably one of the more acutely toxic compounds, with poultry being particularly sensitive to its presence. However, it also has anti-cancer producing qualities. Study of the toxicology of these compounds is made difficult because mixtures are usually present. There is limited evidence that some metabolites are teratogenic, mutagenic or carcinogenic.

Tenuazonic acid is a colourless, viscous oil, soluble in methanol and chloroform that can form crystalline salts. Alternariol and alternariol monomethyl ether form colourless crystals melting at 350°C and 267°C, respectively, soluble in most organic solvents; altenuene forms crystalline prisms melting at 190°C and altertoxin I is an amorphous solid melting at 180°C. Analytical methods for detecting *Alternaria* mycotoxins are not well developed. TLC methods are now superseded by HPLC. Most mycotoxins can be extracted using aqueous methanol with further clean-up, although tenuazonic acid needs special treatment. Concentrations of up to 15 mg/kg of tenuazonic acid have been found in sunflower seeds although alternariol and alternariol monomethyl ether usually occur at less than 1 mg/kg. When oilseeds such as sunflower or rapeseed are pressed, most of the mycotoxins seem to remain in the meal used for animal feed, while the oil used in human food has very little contamination.

No regulatory controls exist and the nature of the risk posed by *Alternaria* remains to be assessed.

3. Other mycotoxins infecting seeds
Several hundred toxic fungal metabolites have been identified in fungal culture but few, such as those listed here, have been shown to occur naturally in seeds used for food or feed.

(a) Citrinin. A nephrotoxic compound that can be formed by some *Penicillium* and *Aspergillus* species, often co-occurring with ochratoxin A.

(b) Cyclopiazonic acid. This mycotoxin is formed by some *Penicillium* and *Aspergillus* species in peanuts, maize and other seeds and can co-occur with aflatoxins.

(c) Ergot alkaloids. These are products related to lysergic acid formed in the fruiting bodies of *Claviceps purpurea* and other *Claviceps* species, most notably in wild grass seeds and **rye**, but also in grains of other cereals. Originally the cause of Holy fire, or St Anthony's fire in the middle ages, the main ergot alkaloids are ergotamine, ergocornine, ergocristine, ergocryptine, ergometrine and agroclavine (**see: Pharmaceuticals and pharmacologically active compounds**).

(d) Moniliformin. A low molecular weight metabolite formed in cereal grains by some *Fusarium* species; it can co-occur with trichothecenes and other *Fusarium* mycotoxins.

(e) Sterigmatocystin. This carcinogenic mycotoxin is structurally related to the aflatoxins that can occur occasionally in cereal grains, green **coffee** beans and other products.

(KAS)

Chelkowski, J. (ed.) (1991) *Cereal Grain Mycotoxins, Fungi and Quality in Drying and Storage*. Elsevier Science, Amsterdam, The Netherlands.

EMAN (European Mycotoxin Awareness Network) (2000) www.mycotoxins.org.

Scudamore, K.A. (1998) Mycotoxins. In: Watson, D.H. (ed.) *Natural Toxicants in Food*. Sheffield Academic Press, Sheffield, UK, pp. 147–181.

Sinha, K.K. and Bhatnagar, D. (eds) (1998) *Mycotoxins in Agriculture and Food Safety*. Marcel Dekker, New York, USA.

Smith, M.E., Lewis, C.W., Anderson, J.G. and Solomons, G.L. (1994) A literature review carried out on behalf of the Agro-industrial division, E2, of the European Commission Directorate-General XII for scientific research and development. *Mycotoxins in Human Nutrition and Health*. EUR 16048 EN.

(Some of the material in this article is adapted from the European Mycotoxin Awareness Network (EMAN) fact sheets prepared by the author (see www.lfra.co.uk/eman2/index.asp). Permission from Leatherhead Food International, the coordinators of this EU project, to use the material is gratefully acknowledged.)

Myrmecochory

(Greek: *mirmekos* = ant; *chorizein* = disperse) Dispersal of the diaspores of a plant (usually fruits and seeds) by ants. (**See: Dispersal**)

N

NAD, NADP

NAD (nicotinamide adenine dinucleotide). NADP (nicotinamide adenine dinucleotide phosphate). The oxidized and reduced forms are denoted as NAD, NADP and NADH, NADPH, respectively. NAD and NADP are coenzymes that participate in oxidation reactions by accepting a hydride ion (H^-) from a donor molecule to form NADH or NADPH. NADH is an important carrier of electrons for oxidative phosphorylation in **respiration**. NADPH is used extensively in biosynthetic pathways.

Napin

Storage proteins that are 2S **albumins** present, along with the globulin **cruciferin**, in *Brassica* oilseeds.

Necrotroph

An organism that rapidly kills host cells as it grows through them and then lives saprophytically on the dead tissues. These organisms frequently have a wide host range and can be grown in pure culture. (See: **Pathogens**)

Neem oil

Oil extracts from neem (*Azadirachta indica*) tree seeds (leaves, flowers and fruits are less potent sources) contain the active ingredient azadirachtin, a nortriterpenoid belonging to the lemonoids, which has a range of fungicidal, bactericidal and insecticidal properties. The latter affect insect growth regulation by disrupting the metabolism of ecdysone, the juvenile-moulting hormone. Neem is widely used by farmers in the Indo-Pakistan subcontinent and south Asia (where the plant grows abundantly) where food grain is stored in containers, by mixing seed with whole leaves or pre-treating it with extracts (**see: Industry, informal supply systems**). Neem extracts are also marketed as a stomach and contact insecticide for greenhouse and ornamental plants. (PH)

Nicking

Planting of male and female parent lines at appropriate times, or manipulating their subsequent development, to ensure sufficiently synchronous flowering and hence successful pollination and adequate seed-set in hybrid seed production, e.g. maize breeding. (See: **Production for sowing, III. Hybrids**)

Alternatively, an artificial physical **scarification** technique: the cutting of a thick or hard fruit or seedcoat to stimulate germination.

Nicotinamide adenine dinucleotide (phosphate)

See: NAD, NADP

Nigella

The plant of the seed spice nigella (*Nigella sativa*, Ranunculaceae), widely known as black cumin (**see: Spices and flavours**, Table S.12), is an approximately 60 cm-tall, erect herbaceous annual with blue-white flowers. (It should be noted that in Iran and parts of Asia and the Middle East, black cumin is the name given to another species, *Bunium persicum* – Apiaceae.) The fruit is a capsule with numerous, angular, black seeds, about 1.5 mm long (Fig. N.1). The intact seeds have little aroma but when rubbed they give a peppery smell with hints of oregano. Nigella is widely used in Indian cooking and in some confectioneries. It is extensively used in indigenous medicine.

Seeds contain about 0.5% **essential oil**, a yellow liquid with an unpleasant smell and that tastes like juniper berries. The main components are thymoquinone (25–50%), p-cymene (*ca.* 31%) and α-pinene (*ca.* 9%), traces of limonene, carvacrol, anethole, α-terpineol and others (**see:** Fig. S.44 in **Spices and flavours**). Seeds also contain nigellimine (an isoquinoline) and several related alkaloids and sterols. Also present is 1.5% of the glucoside melanthin, which on hydrolysis yields the toxic melanthogenin.

The seeds are important in Indian cuisine. They are also used medicinally because of various properties – carminative, diuretic, anodyne, antibacterial, anti-inflammatory, deodorant, digestive, anthelmintic, sudorific, febrifuge, expectorant, etc. They are used to treat (among others) skin diseases, haemorrhoids, rheumatic pain, asthma, jaundice, inflammation, fever, paralysis, ophthalmia, halitosis, diarrhoea, dysentery, amenorrhoea, dysmenorrhoea, helminthiasis, strangury, and vitiated conditions of *vāta* and *kapha*. Crushed seeds in vinegar

Fig. N.1. Nigella seeds (image by Mike Amphlett, CABI).

are applied in skin disorders such as ringworm, eczema and baldness. Pounded seeds are used against nausea and in parts of north India to induce abortion. (KVP, NB)

Peter, K.V. (ed.) (2001) *Handbook of Herbs and Spices*. Woodhead Publishing Co., CRC Press, Cambridge, UK.
Purseglove, J.W., Brown, E.G., Green, C.L. and Robbins, S.R.J. (1981) *Spices*, Vol. 1. Tropical Agricuture Series, Longman, New York, USA.
Weiss, E.A. (2002) *Spice Crops*. CAB International, Wallingford, UK.

Niger

1. The crop

Niger or noug (*Guizotia abyssinica*, 2n = 30, Asteraceae) is an important oil crop of the peasant farmer in Ethiopia and India. The plant is an annual herb that provides 50–60% of Ethiopia's edible oil needs and 2% of the total vegetable oil production in India. The crop is also grown in Sudan, Uganda, Tanzania and Malawi. The seed is processed and consumed locally.

2. Origin

Niger probably originated in the highlands of Ethiopia and migrated to surrounding countries and across the ocean to India.

3. Fruit and seed

Following fertilization, shiny black or brown seeds (**achenes**) develop. At maturity seeds are 3–5 mm long and 1.5–2 mm wide, weighing 3–4.8 g/1000 seeds (Fig. N.2). On processing the seed yields 36–42% pale yellow oil. The seed structure and histology is similar to that of sunflower achenes.

4. Oil and uses

The **fatty acid** composition is typically 75–80% linoleic acid, 7–8% palmitic and stearic acids, and 5–8% oleic acid, although Indian types may have less linoleic and more oleic acids.

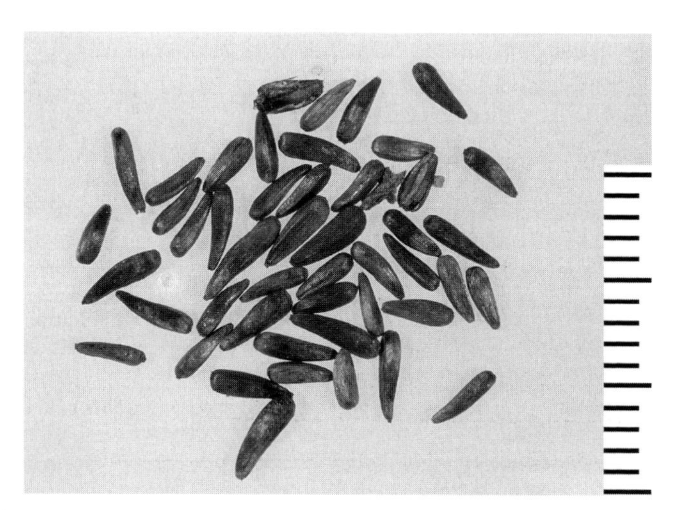

Fig. N.2. Achenes of niger (scale = mm). Photograph kindly produced by Ralph Underwood of AAFC, Saskatoon Research Centre, Canada.

The **oil** is used in cooking and parched seed is sometimes mixed with pulses as a snack food. The oil meal, with a protein content of 24–30% and a fibre content of 14–24%, may be used as livestock feed, fuel or as fertilizer. Whole seeds may be used as birdfeed. (KD)
(See: Oils and fats; Oilseeds – major)

Getinet, A. and Sharma, S.M. (1996) Niger *Guizotia abyssinica (L.f.) Cass*. International Plant Genetic Resources Institute, Rome, Italy.
Riley, K.W. and Belayneh, H. (1989) Niger. In: Robbelen, G., Downey, R.K. and Ashri, A. (eds) *Oil Crops of the World*. McGraw-Hill, New York, USA, pp. 394–403.
Vaughan, J.G. (1970) *The Structure and Utilization of Oil Seeds*. Chapman and Hall, London, UK.

Non-endospermic seed

This term, together with **exalbuminous**, is applied to mature seeds that lack an obvious, well-developed **endosperm**. There are rare cases in which an endosperm never develops because of various dysfunctions of the endosperm apparatus (**see: Reproductive structures, 1. Female**). In such cases, the developing embryo is supported by the nucellus and suspensor. In other truly non-endospermic seeds, the endosperm is completely resorbed during seed development leaving no trace at maturity. But many seeds that are described as non-endospermic do nevertheless possess the tissue, generally reduced to one or a few layers of cells. Examples of the three types are given in Table N.1. (WS)

Boesewinkel, F.D. and Bouman, F. (1984) The seed: structure. In: Johri, B.M. (ed.) *Embryology of Angiosperms*. Springer, Heidelberg, Germany, pp. 567–610.
Johri, B.M., Ambegoakar, K.B. and Srivastava, P.S. (1992) *Comparative Embryology of Angiosperms*, Vols 1 and 2. Springer, Berlin, Germany.

Non-reducing sugar

A di- or polysaccharide, of which the aldehyde group is unable to reduce copper (or other reagents) under alkaline conditions, most likely because it is utilized in forming the glycosidic bond of the dimer or polymer. Sucrose and the **raffinose series oligosaccharides** are all non-reducing sugars. (**See: Carbohydrates; Desiccation tolerance – protection; Reducing sugar**)

Normal seedlings

Seedlings that in an official germination test are considered to show the potential for continued development into satisfactory plants when grown in good quality soil and under favourable conditions of moisture, temperature and light. (**See: Germination testing**)

Northern blot analysis

A method by which the amount and size of a particular RNA can be assessed. Bulk RNA is isolated from tissue, denatured, fractionated by agarose gel electrophoresis under denaturing conditions and transferred to nitrocellulose paper (or nylon membrane). The membrane is incubated in the presence of a radiolabelled DNA and any RNA corresponding to the DNA will hybridize to it and be detected by autoradiography. The

Table N.1. 'Non-endospermic' seeds.

Endosperm never formed	Endosperm completely resorbed	Endosperm one to few layers of cells
Many Orchidaceae (in *Spiranthes australis* the two polar nuclei degenerate; in *Habenaria* spp. the primary endosperm nucleus degenerates)	Some Asteraceae (e.g. *Helianthus annuus*)	Some Asteraceae (e.g. *Lactuca* spp.)
Podostemaceae (the single polar nucleus degenerates)	Ceratophyllaceae	Begoniaceae
Trapa natans (Lythraceae) (endosperm nuclei degenerate after two to three divisions)	Clusiaceae	Cactaceae
	Most Fabaceae (e.g. *Phaseolus vulgaris, Pisum sativum*)	Caryophyllaceae
	Fagaceae (but not *Fagus*)	*Fagus*
	Lythraceae (e.g. *Punica granatum*)	Juglandaceae (e.g. *Carya, Juglans*)
	Ochnaceae	Myricaceae
	Onagraceae	Malvaceae (e.g. five or six at both ends of the seed in *Gossypium herbaceum*)
	Sapindaceae (e.g. *Aesculus hippocastanum*)	Lecythidaceae (e.g. two or three in *Bertholletia excelsa*)
	Salicaceae (sometimes one cell layer left)	

intensity of the 'signal' is a measure of the steady-state level of a given messenger RNA (a function of the rate of synthesis versus the rate of degradation); thus this method does not directly measure the rate of gene transcription. The method is named whimsically 'northern' in contrast to Southern (developed by Edward Southern) in which the presence of a sequence of DNA can be determined within a large mixture of DNA sequences (typically genomic DNA).

Nucellus

Oval central mass of tissue in the **ovule** containing the **embryo sac**. It is morphologically the seed plants' homologue of the **megasporangium**, present in their progenitors. (**See: Evolution of seeds; Reproductive structures; Structure of seeds**)

Nuclear fertility restoration

A genetic component of the **cytoplasmic male sterility (CMS)** system, used widely in the breeding and seed production of **hybrid** crops. A homozygous dominant nuclear fertility gene is included in the male parental line ('Restorer line' or 'R-line') used for crossing with the CMS female parental line, in order to overcome the male-sterile effect of the cytoplasmic *S* factor present in the latter, and hence enable the F_1 progeny to set seed.

Nuclear magnetic resonance (NMR) spectroscopy

This is a technique that can be used for revealing the distribution of water in seeds. The technique is also applied in the determination of the following in seeds: metabolic fluxes, distribution of compounds such as oils and sucrose, mineral ions, the structure and dynamics of cell membranes especially in relation to desiccation. **See: Desiccation damage; Water content**

Leprince, O. and Golovina, E.A. (2002) Biochemical and biophysical methods for quantifying desiccation phenomena in seeds and vegetative tissues. In: Black, M. and Pritchard, H.W. (eds) *Desiccation and survival in plants: drying without dying*. CAB International, Wallingford, UK, pp. 111–146.

Nucleic acids

Complex, high-molecular-weight macromolecules composed of **nucleotide** chains that convey genetic information. The most common nucleic acids are deoxyribonucleic acid (DNA) and ribonucleic acid (RNA). Nucleic acids are present in all living cells and viruses.

DNA is present within **chromosomes**, along with proteins, in the nucleus and is the basic hereditary material, as well as the template for RNA synthesis. It is composed of two antiparallel chains (hence DNA is described as being double stranded or a double helix), each of which has a backbone of sugar (deoxyribose) and phosphate molecules, to which the purine bases adenine (A) and guanine (G), and pyrimidine bases cytosine (C) and thymine (T), are attached. Each phosphate residue forms covalent links with the C3 position of one of its adjacent ribose residues and the C5 position of the other; hence each nucleic acid has a 5′ terminus and a 3′ terminus, both non-phosphorylated. The chains are linked by hydrogen bonds through complementation of the bases A-T and G-C. DNA contains many genes, which encode specific proteins; each gene contains a unique sequence of nucleotides, which are both synthesised (by DNA polymerase) and 'decoded' (by RNA polymerases and **polysomes** during **translation**) from the 5′-end towards the 3′-end direction.

Three major classes of RNA are present in the cytoplasm of cells. All are composed of a single backbone of sugar (ribose) and phosphate to which the purine and pyrimidine (uracil, U instead of thymine, T) bases are attached. They are transcribed from DNA. Messenger RNA (mRNA) contains the information from a gene encoding specific proteins, and this is translated in a protein-synthesizing complex in the cytoplasm which involves ribosomes (of which a major component are various size-classes of ribosomal RNA, rRNA) and low molecular weight transfer RNA (tRNA), which carries amino acids to the protein-synthesizing complex. (**See: Transcription; Transformation technology**)

Nucleotide

A molecule composed of a nitrogenous base (e.g. adenine, cytosine, guanine, thymine, uracil) linked to a sugar molecule, normally ribose or deoxyribose, which has 1–3 phosphate residues attached (e.g. ADP, ATP). (**See: Nucleic acids**)

Nullisomics

A type of **aneuploid** in which the somatic cells lack one type of chromosomes, that is with a **chromosome number** such as $n-1$ or $2n-2$ (where the 2 represents a pair of chromosome homologues). Although nullisomy is a lethal condition in

diploids, an organism such as wheat, which behaves during **meiosis** like a diploid although it is a hexaploid, can tolerate the loss of a specific pair of chromosomes, as can for example the allotetraploid tobacco. Plant geneticists use viable nullisomics (as they do **monosomics**, though nullisomics are less vigorous and fertile than these) to identify the chromosomes that carry the loci of recessive **mutant alleles**, to characterize a new **line** such as by **crossing** it in turn with different nullisomic lines, each of which lacks a different chromosome, and examining the resulting phenotype. (PH)

Number of seeds – ecological perspective

There is huge variation in the number of seeds produced by different plant species. For example, a bush of heather (*Calluna vulgaris*) can produce 3 million seeds in a year, while a buttercup (*Ranunculus repens*) plant may produce only one seed. In general, larger-sized species produce more seeds, but this relationship is complicated by the trade-off between seed **size** and number. Also important in determining seed number is the amount of resources a plant puts into reproduction (reproductive allocation), which is greater in annual plants than in perennials.

Species also differ in the pattern of seed production through the life cycle. Monocarpic or semelparous plants are those that flower and produce seed only once and then die. Annuals flower and die in 1 year. Other monocarpic species grow for 1 year or more before they flower, an extreme example being the giant bamboo (*Phyllostachys bambusoides*) which can wait 120 years. Polycarpic or iteroparous plants are those that survive following seed production and may flower for many seasons.

Within a species, younger or smaller plants produce fewer seeds. The seed production for *Calluna vulgaris* given above is from a bush 2 m in diameter and 13 years old. A 1-year-old plant, 4 cm in diameter, may produce fewer than 100 seeds. Seed production may also vary among or within populations because of differing degrees of attack by seed **predators** or **pathogens**, or by factors affecting plant growth, such as grazing, disease or low nutrient or water levels. Pollination limitation can also reduce seed production, whereby not all ovules are fertilized because insufficient pollen reaches the flowers, for example, due to a shortage of insect pollinators.

The same factors may cause yearly variation in seed production. However, a form of extreme yearly variation is seen in masting species (**see: Mast crop**), such as beech (*Fagus*) and oak (*Quercus*) trees in the temperate zone and trees of the family Dipterocarpaceae in the tropics. Masting involves production of huge numbers of seed in some years and virtually none in intervening years, and these patterns are synchronized among individuals over large areas. The most general explanation for this phenomenon is that it allows plants to avoid seed predation by producing more seeds than can be eaten, or to keep seed predator numbers low in poor production years. (JMB)

Fenner, M. (ed.) (2000) *Seeds: the Ecology of Regeneration in Plant Communities*, 2nd edn. CAB International, Wallingford, UK.

Shipley, B. and Dion, J. (1992) The allometry of seed production in herbaceous angiosperms. *American Naturalist* 139, 467–483.

Silvertown, J. and Charlesworth, D. (2001) *Plant Population Biology*, 4th edn. Blackwell Science, Oxford, UK.

Nut

In the strict botanical sense, a nut is a single-seeded, indehiscent fruit with a hard, dry **pericarp** (fruit coat). Authorities differ in their definition of a true nut. Some restrict the term to fruits in which the hard pericarp is fully exposed at maturity. Such fruits include hazelnuts (filberts) and acorns. Others include fruits where the stony pericarp is enclosed in accessory tissue derived from floral appendages, and/or the involucre (bracts and bracteoles). The accessory tissues in such cases comprise the hull or husk which is typically green and has a leathery or fibrous texture. By this interpretation, walnuts, pecans and chestnuts are considered as true nuts (Colour Plate 5A).

In common and culinary usage, however, the term 'nut' describes a diverse group of large, edible seeds with a 'nutty' texture and flavour that form within a hard, stony, fibrous or woody shell, many of which are not true nuts by botanical classification. It may refer to the edible seed, or **kernel**, directly or to the kernel and the shell together. In almonds, pistachios and coconuts, the hull or husk derives from the outer layers of the pericarp rather than accessory tissues. In almond, the inner tissue of the pericarp (the endocarp) forms the hard shell of the 'nut'. Such fruits are classified by botanists as **drupes** (**see: Structure of seeds,** Fig. S.63). Brazil 'nuts' form in a woody capsule and are actually seeds. Macadamia fruits are follicles. Peanuts are seeds in fruits called legumes. Pine nuts are the naked seeds of gymnosperms; their hard shells derive from the seedcoat. (**See** also: **Shelling**)

Nuts are sometimes grouped under those of temperate climates, e.g. **almond**, butternut, **chestnut**, **hazelnut** (cobnut, filbert), hickory, **pecan**, **pistachio** and **walnuts**, and those of warm climates, e.g. Australian chestnut, **Brazil nut**, **cashew**, **coconut**, **macadamia**, **peanut** (**groundnut**), **pine nuts**. There are, in addition, about 15 less significant edible nuts such as acorn (*Quercus* spp.), beech (*Fagus*) and paradise nuts (*Lecythis* spp.) that are mainly of historical or local importance.

In almost all cases the edible component of the nut is the **embryo**, principally the two swollen cotyledons. Exceptions are the Brazil nut whose embryo has a greatly distended **hypocotyl** and rudimentary cotyledons, and the pine nut, which is mostly megagametophytic tissue. **Endosperm** is absent from mature nuts since it is degraded and its products used by the embryo during development. All botanical and culinary nuts are oily apart from the chestnut which is richer in carbohydrate, mainly **starch**. The **oils** (triacylglycerols) in most nuts (an exception is coconut) have a relatively low content of saturated **fatty acids** and are relatively high in monounsaturated and polyunsaturated **fatty acids**. **Storage protein** contents lie in the approximate range 2–21% fresh weight and many nuts are therefore a good dietary alternative to meat. Mineral and vitamin contents are considered to be nutritionally favourable: all nuts are relatively rich in α-tocopherol vitamin E, an antioxidant vitamin (Table N.2). **Allergens** are present in several types of nuts, especially peanuts, and to a lesser extent in almonds, Brazil nuts, hazelnuts and pine nuts. The allergic response can be acute, even life threatening in children, and it may persist into adulthood (**see:** e.g. **Peanut; Storage proteins – intolerance and allergies**).

Table N.2. Constituents of the most common nuts.[a]

Nut[b]	Water	Protein	Oil	Starch	Sugars	Principal minerals	Principal vitamins
Almond (*Prunus dulcis*)	4.2	21.1	55.8	2.7	4.2	K, Ca, Na, Fe	E, B_2, folate
Brazil (*Bertholletia excelsa*)	2.8	14.1	68.2	0.7	2.4	K, Ca, Se	E, B_1, folate
Cashew (*Anacardium occidentale*)	4.4	17.7	48.2	13.5	4.6	K, Ca, Na, Fe	E, B_1, folate
Chestnut (*Castanea sativa*)	51.7	2.0	2.7	29.6	7.0	K, Ca	E
Hazel (*Corylus avellana*)	14.8	7.6	36.0	2.1	4.7	K, Ca	E, folate
Macadamia (*Macadamia* spp.)	1.3	7.9	77.6	0.8	4.0	K, Ca	E
Pecan (*Carya pecan*)	3.7	9.2	70.1	1.5	4.3	K, Ca	E, B_1, folate
Pine (*Pinus* spp.)	2.7	14.0	68.6	0.1	3.9	K, Ca, Mg	E, B_1, folate
Pistachio (*Pistacia vera*)	2.1	17.9	55.4	2.5	5.7	K, Ca	E, B_1, folate
Walnut (*Juglans regia*)	2.8	14.7	68.5	0.7	2.6	K, Ca	E, B_1, folate

[a]Values are percentage kernel fresh weight. Adapted from Vaughan, J. and Geissler, C. (1997) *The New Oxford Book of Food Plants*. Oxford University Press, Oxford, UK.
[b]See individual entries.

Nut production is small relative to cereals, oilseeds and legumes. For example, in 2003, world production of **coconuts** was 52.9×10^6 t, peanuts 35.6×10^6 t, and tree nuts 8.2×10^6 t as compared with approx. 560×10^6 t for wheat (FAOSTAT). (**See: Crop Atlas Appendix, Map 4; Ethnobotany, 7. Nuts**)

(VP, MB)

Duke, J.A. (1989) *CRC Handbook of Nuts*. CRC Press, Boca Raton, Florida, USA.

Pearman, G. (2005) Nuts, seeds and pulses. In: Prance, G. and Nesbitt, M. (eds) *Cultural History of Plants*. Routledge, New York, USA, pp. 133–152.

Simpson, B.B. and Ogorzaly, M.C. (2001) *Economic Botany: Plants in our World*, 3rd edn. McGraw-Hill, New York, pp. 70–74, 104–106.

Vaughan, J.G. and Geissler, C.A. (1997) *The New Oxford Book of Food Plants*. Oxford University Press, Oxford, UK, pp. 32–35.

Nutlet

An individual **carpel** (monocarp) of a fruit displaying the characteristics of a **nut**.

Nutmeg and mace

The nutmeg tree, *Myristica fragrans* Houtt (Myristicaceae) is unique among the spice plants as it produces two separate and distinct products – the nutmeg which is the kernel of the seed and the mace which is the dried **aril** that surrounds the single seed within the fruit (Colour Plate 3E). It is a dioecious, tall evergreen tree native to Moluccas Islands, from where it has spread to other tropical regions (**see: Spices and flavours,** Table S.12). Nutmeg is usually propagated by seeds and sometimes by grafting. The tree starts yielding when 5–8 years old for up to 40 years or more. Six–nine months after flowering the fruit ripens as a pendulous fleshy drupe. The succulent aromatic yellow **pericarp** splits into halves to expose the purplish, brown, shiny **testa** surrounded by a net-like red aril (Colour Plate 3E). The testa encloses an ovoid greyish brown kernel, 2–3 cm long and 1.4–2.0 cm broad. The exterior of the kernel is longitudinally wrinkled and consists of convoluted dark brown **perisperm**, a lighter coloured **endosperm** and a small **embryo**. Nutmeg and mace are traded whole or in ground form.

Nutmeg was especially prized in the 17th century as it was thought to cure the plague. Many of the squabbles and wars, especially between the Dutch and the English, were provoked by the search for dominance over the areas of production of the seeds, the East Indies. Because of nutmeg, in the early 1660s the English seized (and kept) Manhattan, in what later became New York, in revenge for the Dutch capture of the island of Run, then the major source of the spice. To end hostilities, the Dutch ceded Manhattan to the English, in exchange for Run: the former eventually formed one of the world's most prominent cities, while the latter failed to become re-established as a centre for spice production.

1. Constituents

The **oleoresins** are obtained by solvent extraction for use in flavouring processed food. Nutmeg oleoresins contain volatile (essential) oil ranging from 10 to 90%. Mace oleoresins contain 10–55% volatile oils. Nutmeg contains 25–40% **fixed oil**, extractable with solvents in the presence of steam, as a concentrate called nutmeg butter, which is highly aromatic orange-coloured fat. It consists of dry myristicin and high volatile oil content.

The organoleptic properties of both nutmeg and mace are similar and are principally determined by the constituents of aromatic **essential oils**. The oils have at least 40 constituents of which the major ones are shown in Table N.3. Monoterpene hydrocarbons predominate (e.g. sabinene, pinenes) followed by oxygenated monoterpenes and aromatic ethers (myristicin, safrole, elemicin) (**see: Spices and flavours,** Fig. S.44).

2. Uses

Dried nutmeg and mace are used directly as spices and also for their oils and oleoresins. Nutmeg butter is also used in a few pharmaceutical preparations. The pericarp of the fruit is also used in pickles and jellies. Nutmeg oil, a fresh, warm, spicy aromatic with a rich, sweet, spicy body note is used for flavouring processed food and soft drinks as well as for pharmaceutical formulations especially to combat bronchial disorders and to cure various gastro-intestinal complaints, psychological disorders and urinary diseases.

Nutmeg and mace are bitter, acrid, astringent, sweet, thermogenic, and have anti-inflammatory, anthelminthic, deodorant, carminative, expectorant, diuretic, antispasmodic, narcotic, anticonvulsant, antiseptic, aphrodisiac and other properties. They are used to treat vitiated conditions of *kapha*

Table N.3. Main constituents of essential oils in nutmeg and mace.

Components	Nutmeg (% wt)	Mace (% wt)
α-Pinene	4.9	5.3
β-Pinene	4.6	4.9
Sabinene	1.9	2.5
α-Terpinene	3.5	3.2
Limonene	3.2	2.7
β-Phellandrene	2.7	2.8
γ-Terpinene	7.8	5.6
p-Cymene	6.5	2.4
Terpinolene	2.4	2.0
Terpinen-4-ol	31.3	20.0
α-Terpineol	5.2	3.5
Methyleugenol	0.8	13.3
Elemicin	4.8	4.7
Safrole	2.0	3.4
Myristicin	7.1	14.4

Atta-ur-Rahman, M.I., Choudhary, A., Farooq, A., Ahmed, A., Demirci, B., Demirci, F. and Baser, K.H.C. (2000) Antifungal activities and essential oil constituents of some spices from Pakistan. *Journal of the Chemical Society of Pakistan* 22, 60–65.

and *vāta*, inflammations, many ailments of the digestive system and other conditions. The burnt seed kernel powdered and mixed with buttermilk forms a very specific remedy for diarrhoea and vomiting in children. Nutmeg can be used as a hallucinogen and in excess quantities it is a narcotic.

(KVP, NB)

Devon, T.K. and Scott, A.I. (1975) *A Hand Book of Naturally Occurring Compounds*, Vol. 1. Academic Press, New York, USA.

Guenther, E. (1978) *The Essential Oils*, Vol. 2. Robert E. Krieger, New York, USA.

Peter, K.V. (ed.) (2001) *Handbook of Herbs and Spices*. Woodhead Publishing Co., Cambridge, UK.

Purseglove, J.W., Brown, E.G., Green, C.L. and Robbins, S.R.J. (1981) *Spices*, Vol. 1. Tropical Agriculture Series, Longman, New York, USA.

Weiss, E.A. (2002) *Spice Crops*. CAB International, Wallingford, UK.

Windholz, M. (1983) *The Merck Index An Encyclopedia of Chemicals, Drugs and Biologicals*, 10th edn. Merck and Co., Rathway, NJ, USA.

Nutraceuticals

The term 'nutraceutical' was coined in 1989 by the Foundation for Innovation in Medicine (New York, USA), to provide a name for this rapidly growing area of biomedical research. A nutraceutical is defined as any substance that may be considered a food or part of a food and provides medical or health benefits including the prevention and treatment of disease. Nutraceuticals range from isolated nutrients, dietary supplements and diets to genetically engineered 'designer' foods, herbal products and processed products such as cereals, soups and beverages. This research (termed nutritional genomics or nutrigenomics) uses functional **genomics**, **proteomics** and **metabolomics**, to provide new molecular insights into nutrient metabolism and human health.

Many intrinsic plant compounds have been linked to positive health effects and would therefore be an attractive target for plant engineering. There is substantial evidence mainly from epidemiological studies that diets rich in vegetables, fruits and seeds (including so-called 'whole **grains**'), and their products, can prevent human cardiovascular diseases and cancer. The role of traditional soy-food in disease prevention and treatment has gained worldwide recognition because of its anti-diarrhoeal, hypolipidemic, anti-cancerogenic and anti-osteoporotic effects. Isoflavone phytoestrogens in seed-derived soy flour, such as daidzein and genistein, are known to be responsible for the beneficial biological effects (**see: Soybean**).

Joint research in epidemiology, nutrition and food toxicology is needed to select relevant compounds in seeds and other plant parts, and to demonstrate their beneficial action, although it might not be possible to identify single 'miracle' compounds given the complexity of foods and the interactions with living organisms. (AJR)

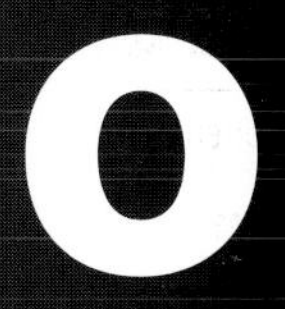

Oat

1. Distribution and importance

The common oat (*Avena sativa*) is widely grown in the temperate regions of the world as livestock feed and forage, and as a human food crop. It is generally better adapted to cooler, moister climates and higher latitudes or elevations than other cereals. The red oat (*A. byzantina*), a less-common oat, is adapted to warmer and more arid climates. There are five other species of *Avena* cultivated to varying extents, the most important being *A. nuda*, the naked or hulless oat. The oat crop is internationally the sixth or seventh most significant cereal: the main producer is Russia (**see: Cereals,** Table C.8). In 2002, worldwide exports were valued at US$324 million (FAOSTAT). **(See: Oat – cultivation; Crop Atlas Appendix, Maps 1, 11, 21, 25)**

2. Origin and domestication

Oats are thought to have developed initially as a weedy species in barley and wheat crops from which they were later selected and 'domesticated' to be grown as crops in their own right. Non-domesticated species of *Avena* have been found in archaeological deposits in the Near East ('Fertile Crescent') and eastern Mediterranean dating to 7,000–12,000 years ago. There is evidence of the occurrence of domesticated oat in central Europe 3000–4000 years ago. Interestingly, these oats had lost the **shattering** characteristic of the weedy types. Oats (especially *A. nuda*) have been cultivated in China for at least 2000 years.

3. Grain characteristics

See Cereals, section 4.

The grains are about 10 mm long and 3–4 mm at their widest, with a buff-pale brown colour (Colour Plate 8E). The whole grain of hulled oats is characterized by a rigid **lemma** and **palea** (hull) that are not removed during threshing. As they do not adhere to the groat (the name describing the actual caryopsis of the oat) within, they can be removed mechanically. Hulled oats are traded with the hull (husk) in place. In this condition they have an extremely elongated appearance and even with the husk removed groats are long and narrow (Colour Plate 8E). The groat's contribution to the entire grain mass can vary from 65 to 81% (average 75%). The uncommon naked oat (*Avena nuda*) readily loses its husk during threshing.

In the **endosperm** there is a one-cell deep **aleurone layer** with thinner cell walls than those in wheat and rye. Conversely, the starchy endosperm cells have thicker walls than those in wheat. **Starch granules** are compound consisting of many tiny granuli which fit together to form a spherical structure. Possibly 80 or more granuli up to 10 μm diameter constitute a single compound granule. Individual free granuli are also present in the spaces among the aggregates. Endosperm cells of the oat have a relatively high oil content.

3. Composition and nutritional quality

The dehulled grain (groat) is high in **oil** (triacylglycerol) (typically 7–8% fresh weight, fw, of which a high proportion occurs in the endosperm) and protein (typically 16–18% fw), with a good amino acid profile for most monogastric animals. Unlike most other cereals, apart from rice, the **storage protein** in the starchy endosperm is rich in **globulin** thus enhancing the nutritional quality. Carbohydrate, mostly starch, amounts to about 65% fw. The oat groat is also high in soluble fibre (β-glucans) and is gaining popularity as a component in many 'healthy foods' (reputed to lower blood cholesterol levels) for humans. The mineral content (calcium, copper, iron, manganese, magnesium and zinc) is similar to that of other cereals: selenium is also present. Oats contain substantial amounts of vitamins B and E, but not C.

4. Processing and uses

A very high proportion of cultivated oats is used for animal feed but this has declined in recent decades with the reduction of animal power, particularly horses, in agriculture. Because of the fibrous hull which adheres to the mature grain, it has a lower energy value, weight for weight, compared to most other feed grains. As food for humans, uses include porridge, oatcakes, grits, breakfast and baby foods.

The English scholar Samuel Johnson (1709–1784) described oats as a grain which in England is generally given to horses, but in Scotland supports the people. His Scottish colleague and chronicler James Boswell is purported to have suggested this is why England breeds such fine horses, and Scotland such fine men.

The forms in which oats are used include **groats** (whole oat caryopses, sometimes crushed), rolled oats, also called oatmeal (made by rolling whole groats that have been heated), bran and flour. The purpose of oat milling is to produce flakes of oatmeal. Because oats have a high oil content they are unstable in storage, especially when they have been ground, allowing exposure to air and to enzymes present in grain tissues. For this reason it is necessary to inactivate enzymes, notably lipase, by kilning before milling. Lipase resides in the outer tissues of the grain and it is thus those tissues that need to be heated. Traditional kilning is performed on grains with husks in place; however, less energy is required to heat only the groats, after husk removal, and this practice is now widely adopted.

Following kilning, groats are passed through a device that cuts them, at right angles to the long axis, into about four pieces. These 'pinhead groats' are then steamed and flaked (Fig. O.1). For consumption as porridge the flaked oats are boiled in water. (AE, MN)

For information on the particular contributions of the seed to seed science **see: Research seed species – contributions to seed science**.

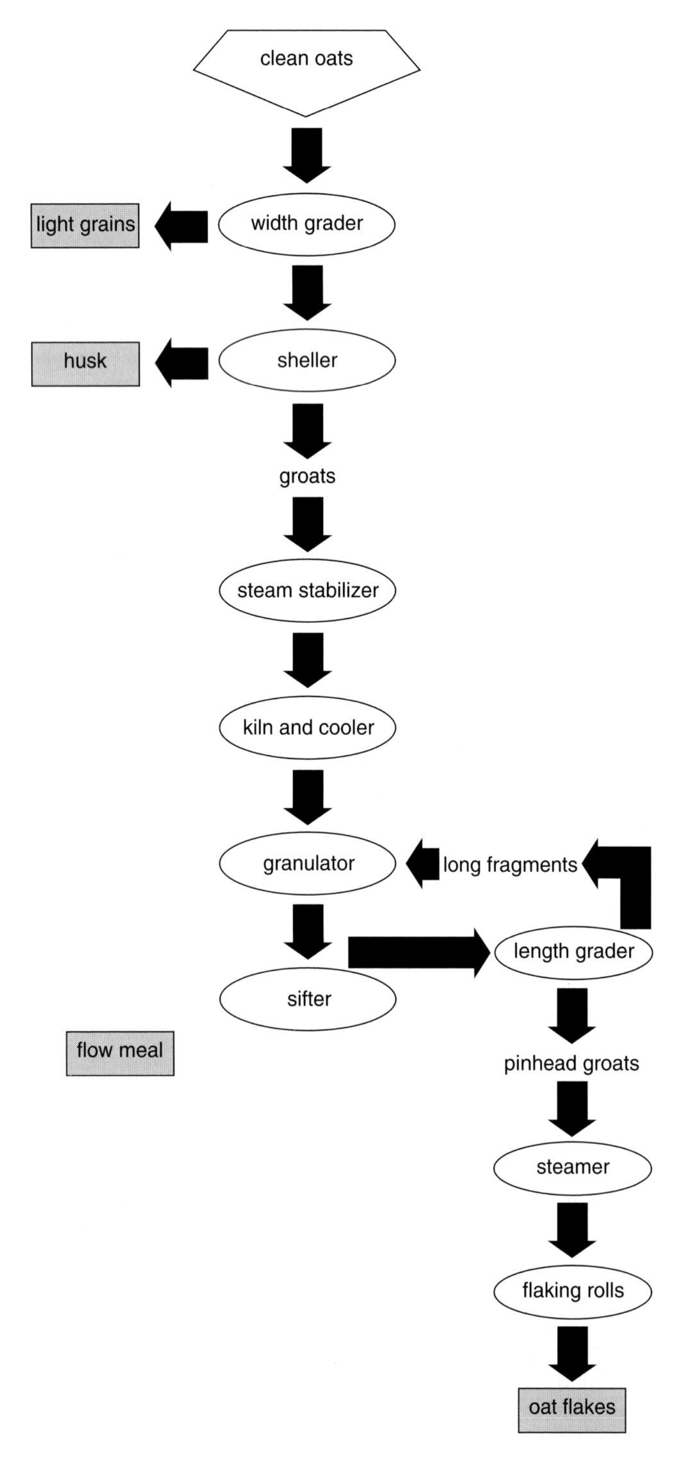

Fig. O.1. Processing of oats by milling for the production of flakes.

Peterson, M.P. (1992) Composition and nutritional characteristics; oat grain and products. In: Marshall, H.G. and Sollells, M.S. (eds) *Oat Science and Technology*. American Society of Agronomy, Madison, WI, USA, pp. 266–287.

Welch, R. (ed.) (1995) *The Oat Crop*. Chapman & Hall, London.

Youngs, V.L., Peterson, D.M. and Brown, C.M. (1982) Oats. In: *Advances in Cereal Science and Technology*, V. American Association of Cereal Chemists, St Paul, MN, USA.

Oat – cultivation

The oat is a cool-season grass belonging to the family Poaceae and the tribe Aveneae. *Avena sativa*, the major species cultivated in temperate parts of the world for animal and human food and forage, and is better adapted to cool, moist climates and higher latitudes or elevations than are other cereals. Long grown in the Near East and Europe both cultivated and as weeds, oat spread to the New World and the southern hemisphere with other cultivated grains as part of the migrations of European settlers in the 16th and 17th centuries, and is now grown in much of Canada and the northern USA and throughout the grain belts of Australia and South America. (**See: Crop Atlas Appendix, Map 11**) In Canada and the USA, oats are used primarily as mature grain in animal feed, or dehulled for human food, while the straw is used for animal bedding. In Australia, a significant part of the crop is harvested during early- or mid-grain fill as hay (dried forage) for domestic livestock feed and export. It is generally used as a feed and food grain in South America; however, it is widely grown as a 'roll crop' in Brazil where the immature crop is rolled down (which kills it) and is used as a weed-suppressing and moisture-conserving **mulch** for seeding a following crop (generally soybeans). For the geographic origins, history, and economic importance, composition and uses, **see: Oat**.

Oat is distinguished from the other small cereal grains (such as wheat, barley and rye) in having a **panicle** inflorescence rather than a **spike**. The panicle has a central **rachis** and rachis branches from several nodes, which may themselves be further branched. Every rachis branch is terminated by a **spikelet**, each of which is indeterminate with two to many florets. Each floret has a lemma and palea containing the single ovary and three anthers.

There are two cultivated hexaploid species of oats, *A. sativa* (white oat) and *A. byzantina* (red oat), along with the cultivated hexaploid hull-less oat *A. nuda*. The tetraploid *A. abyssinica* is also a cultivated form, but is only grown in and around Ethiopia. There are also two major weedy hexaploid oats *A. fatua* and *A. sterilis*. The taxonomy of oats is difficult to describe clearly, as there are numerous chromosomal rearrangements within and among the genomes in the many polyploid species of *Avena*. The current, most widely held view is that the weedy *A. sterilis* is the common ancestor of the other hexaploids, while *A. fatua* may have developed directly from *A. sterilis*, or from *A. sativa* or *A. byzantina*. *A. nuda* appears to be a free-threshing derivative of *A. sativa* or *A. byzantina* and occurs in a limited geographic area east of the Fertile Crescent. *A. sterilis* is widespread in the Middle East and Mediterranean and is often used for grazing livestock. It often forms dense stands in undisturbed habitat but can also be an aggressive weed in cultivated crops.

A. fatua is generally found in close association with cultivation and disturbed sites around the world, but seldom in the Middle East. The species is generally better adapted to cooler environments than *A. sterilis*. *A. fatua* and *A. sterilis* have seed **dispersal** mechanisms and **dormancy** factors that make them quite effective weeds in other cereal crops. The cultivated oats *A. sativa*, *A. byzantina* and *A. nuda* do not **shatter** at maturity and do not have significant dormancy problems. *A. sativa* and *A. byzantina* are differentiated from each other by the point at which the rachilla fractures during the threshing operation; this difference is due to one or two genes. (**See: Oat** for information on origins, economics and grains).

1. Genetics, breeding

Oats are highly self-pollinated and are generally handled in much the same way as wheat and barley in a breeding programme. Because cultivated oat is an allohexaploid, the genetics are often complicated. Interchanges within and among the genomes lead to further problems in determining the genetic nature of many traits. There are few traits that show simple, single gene inheritance and the genetic nature of many traits is unclear due to complementation among the genomes.

Most breeding programmes have used a relatively traditional approach of biparental crosses followed by several generations of selfing and selecting in a pedigree-type programme or a bulk system. Oats can also be handled readily using a **single seed descent** (SSD) system. There has been some use of interspecific crosses in oat breeding programmes, particularly to improve protein content and disease resistance. All the hexaploid species are fully cross compatible, so it has been relatively easy to introgress wild germplasm into the cultivated types. **Haploids** can be produced in oats via anther culture or with maize pollination followed by **embryo rescue**, but the efficiency is much lower than in most other cereal crops. Molecular marker maps are being developed for application in marker assisted selection (MAS) to enable pyramiding of disease resistance genes.

Progress in increasing yield in oat has lagged somewhat behind the other cereal crops. Much of this may be due to the much lower level of resources being applied compared to the more widely grown crops of wheat, maize and barley. Another factor may be the complex genetic nature of the crop and possibly the limited germplasm being utilized in the breeding programmes. Oats are generally quite daylength-sensitive and somewhat more narrowly adapted than barley and wheat.

2. Development

The oat seed is capable of germinating in quite cool soil and is usually planted in the spring as soon as the soil can be worked. Oats have been successfully grown from frost seeding (planting into soil when the surface is frozen enough to support the planting equipment) in many parts of the world. The germination of the oat seed is relatively rapid with emergence from 4–5 cm often occurring in 5–7 days, depending on soil temperature. Upon emergence, a crown is established about 2–3 cm below the soil surface from which the adventitious root system forms. Oats generally produce numerous tillers (shoots from axillary buds at the base of the main stem) if growing conditions are good. The vigorous oat seedling can generally compete well with most weeds and early plantings at reasonable populations seldom have weed problems. There is considerable anecdotal evidence that oats may have an **allelopathic** weed suppression effect, but it is very difficult to determine scientifically and there is little published information on this phenomenon.

Because most cultivated oats are fairly daylength-sensitive and require long days to flower, most varieties cannot be grown across a wide north–south geographic area. Some varieties have been developed from **landraces** originating in the Mediterranean region or from day-neutral wild oats that are insensitive and can be grown across a wide range of latitudes.

Winter oats having a **vernalization** requirement are grown in some regions. The winter hardiness level of oats is considerably less than that of winter wheat and rye, and somewhat less than winter barley. Winter oats are often used for grazing in the autumn and generally outyield spring oats in regions that they are adapted to. Some spring types can be grown as a winter crop in mild regions.

Once the flower has been pollinated, the grain develops rapidly with the grain-fill period varying from 35 to 50 days, depending on temperature and environmental conditions. The oat grain develops following fertilization as a typical grass seed. (**See: Development of embryos – cereals; Development of endosperms – cereals**) The **embryo** develops and matures well before the **endosperm** is filled. The embryo is nearly completely functional within 2 weeks of fertilization and can be dissected from the seed and cultured on media where it may germinate and grow into a 'normal' seedling. The endosperm takes from 35 to 50 days to fill completely, depending on climatic conditions and genotype.

3. Production

Oats can be harvested by swathing (cutting and **windrowing**) and threshing in separate operations in moist climates, or they can be direct combined with cutting and threshing in a single operation in drier conditions. Generally the crop can be dried to threshing and storage moisture (12–14%) standing in the field or in the swath. Oats are handled very much like wheat and barley using conventional augers, **elevators** and so on; low moisture and cool temperatures are the most critical factors in grain **storage management** to avoid fungal growth, the stimulation of insect development and the rapid enzymatic breakdown of lipids to cause rancidity, particularly in hull-less oats. Aeration with cool, dry air can reduce or eliminate most storage-related problems in grain.

4. Quality, vigour and dormancy

Seed quality in oat is usually closely related to the growing and harvesting conditions of the crop. Stress during grain filling can lead to small seeds and reduced vigour of the seed crop. Plumpness is a major consideration in milling oats and this is one of the first quality parameters to be affected by late-season stress of the maturing crop. Generally oat seeds have good seedling vigour and establish well under reasonable sowing conditions.

Some oat seeds have considerable **dormancy** and may not germinate readily immediately after harvesting. Cold or heat treatments have both been shown to reduce dormancy and promote germination shortly after harvesting. Gibberellic acid (a **gibberellin**) can also be used to break dormancy in some cases for seed testing purposes.

Oats are also susceptible to several viruses, some of which can be seed-transmitted.

5. Sowing
Oats are generally sown 2–6 cm deep, depending on soil type, soil moisture and soil temperature. Under dry surface conditions, it is better to plant deeper, if necessary, to reach moist soil and obtain rapid germination. Seed can be placed deeper in sandier soil and should be sown relatively shallow in wet clay soils.

Oats, because of the hull surrounding the **caryopsis**, need good seed-to-soil contact to absorb water rapidly to initiate germination. A firm seedbed (**see: Seedbed environment**) can keep the seed from being placed too deeply for proper germination. Generally there are no problems with direct drilling into crop residues, and they are equally adapted to being broadcast and lightly worked in with a harrow.

6. Treatment
Cultivated oats are not particularly prone to seed and seedling diseases. **Smuts** that can infect seeds are often found at low to moderate levels in crops that are not regularly treated for them, and seed is commonly treated with a systemic fungicide to prevent losses due to the seedborne pathogens. (DEF)

Baum, B.R. (1977) *Oats: Wild and Cultivated.* Thorn Press, Canada.

Duffus, C.M. and Slaughter, J.C. (1980) *Seeds and Their Uses.* John Wiley, New York, USA.

Evans, L.T. (1993) *Crop Evolution, Adaptation, and Yield.* Cambridge University Press, Cambridge, UK.

Leonard, W.H. and Martin, J.H. (1963) *Cereal Crops.* Macmillan Company, New York, USA.

Marshall, H.G. and Sorrells, M.E. (eds) (1992) *Oat Science and Technology*. American Society of Agronomy, Madison, WI, USA.

Sauer, D.B. (ed.) (1994) *Storage of Cereal Grains and Their Products*, 4th edn. American Association of Cereal Chemists, St Paul, Minnesota, USA.

Simmonds, N.W. (ed.) (1976) *Evolution of Crop Plants.* Longman Group, New York, USA.

OECD Seed Scheme

Since 1960, the Organisation for Economic Co-operation and Development has coordinated the development of agreed principles to authorize and label **certification** of seed varieties produced and processed moving in international trade. The OECD Schemes are the nearest equivalent to a completely international seed certification organization and have become the model basis for most modern national schemes, and have been applied in more than 50 countries.

The overall Scheme defines rules and directions for seven groups of species of cultivated plants: grasses and legumes; crucifers and other oil or fibre species; cereals; fodder and sugarbeet; subterranean clover and similar species; maize and sorghum; and vegetables. (**See: Cereals; Grasses and legumes – forages; Legumes; Oilseeds – major**) Each individual scheme designates or defines such matters as:

- types of variety and hybrids, and their parental **line** constituents;
- seed categories (such as Pre-basic, **Basic** and **Certified seed** and parental lines involved in hybrid seed production), and the conditions for their production and maintenance (controlling such matters as previous cropping history, isolation distances, field inspections, the number of generations permitted for cross-pollinating varieties, sampling and post-control tests – **see: Production for sowing**);
- the acceptance conditions (including minimum analytical varietal purity, and germination standards) and the laboratories that are to carry out the testing;
- the maximum size of seed lots or small packages;
- the means for the tamper-proof fastening and identification-labelling (including prescribed shape and colours) of seed containers.

The official OECD 'List of Varieties Eligible for Certification' is an annually revised country-based list of varieties and their **maintainer**(s), that have been accepted by the national designated authorities as eligible for certification in accordance with the Rules of the OECD Seed Schemes. These are varieties that have been officially recognized as having **Distinctness, Uniformity and Stability (DUS)** and having an acceptable **Value for Cultivation and Use (VCU)** in at least one participating country. (**See also: Laws and regulations**) (PH)

Bowring, J.D.C. (1998) Organisation for Economic Co-operation and Development seed schemes. In: Kelly, A.F. and George, R.A.T. (eds) *Encyclopedia of Seed Production of World Crops.* John Wiley, Chichester, UK, pp. 23–26.

OECD (2000) OECD Schemes for the Varietal Certification or the Control of Seed Moving in International Trade Council Decision C(2000)146/FINAL. www.oecd.org/agr/code/seeds/seeds1.htm

Off-type

An individual plant occurring in a seed production crop that is not of the cultivar or line being multiplied. A genetic variant, usually within a variety, which has phenotypic differences from the variety being produced and is undesirable. It is important to remove off-types in hybrid seed production, by **roguing**. (**See: Production for sowing, III. Hybrids**)

Oil bodies

Also sometimes called lipid bodies, oleosomes or spherosomes. Seed storage **oils and fats** (triacylglycerols) are stored in oil bodies, small (0.5–2.0 μm) spherical organelles bounded by a **phospholipid membrane** monolayer. Unlike other cellular membranes, the oil body monolayer is rich in negatively charged phospholipids, specifically phosphatidyl inositol and phosphatidyl serine. Oil bodies contain unique proteins of low molecular mass (15 to 25 kDa) called **oleosins**.

Oil bodies are most likely formed by a deposition of triacylglycerols within the bilayer of the outer **endoplasmic reticulum** (ER) membrane followed by a budding from the ER and then a separation of the oil body into the cytosol. However, the separation process itself has not been observed, and there is some dispute as to whether the oil is deposited within the membrane bilayer or into the ER lumen. But oil bodies can be seen in close association with the ER (Fig. O.2),

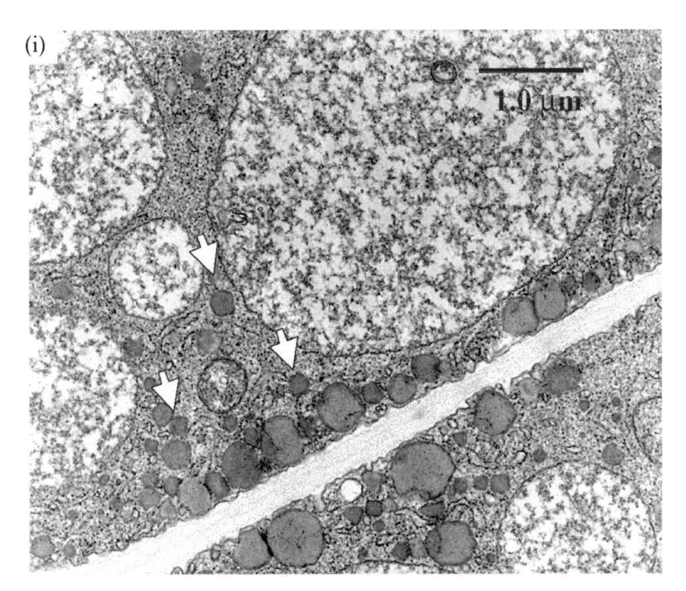

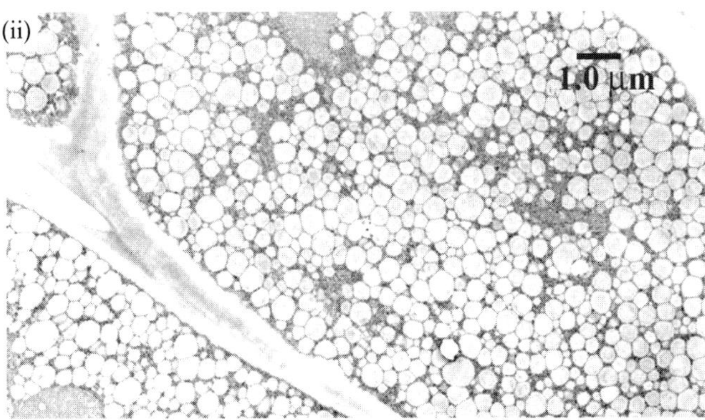

Fig. O.2. Electron micrographs of developing soybean seed cells in early–mid (i) and mid–late (ii) maturation. In early to mid maturation (i) oil bodies can be seen budding from the endoplasmic reticulum (some examples are marked with white arrows). By mid–late maturation (ii) the cells are densely packed with oil bodies. (Photos: T. Bourett, DuPont.)

as well as free within the cytosol, so it is reasonable to assume that they are derived from some kind of ER-budding process. It is not understood why the protein and **phospholipid** composition of the oil body monolayer is so different from that of the rest of the ER membrane, but it is known that there are distinct phospholipid sub-domains within the ER and these may be the sites of oil body formation. (AJK)

(See: Biopharmaceuticals; Oils and fats; Oils and fats – synthesis)

Murphy, D.J. (2001) The biogenesis and functions of lipid bodies in animals, plants and microorganisms. *Progress in Lipid Research* 40, 325–438.

Oil palm

See: **Palm**

Oils and fats

The oils and fats in seeds are usually **triacylglycerols**, three **fatty acids** esterified with a glycerol backbone. One exception is the liquid **wax** present in the desert shrub jojoba (*Simmondsia chinensis*). **Jojoba** liquid wax consists of esters of very long chain fatty acids and alcohols.

An oil is liquid at room temperature, a fat solid. An oilseed may contain from just a few per cent to over 60% wt/wt triacylglycerol. Most commercial sources of seed oils contain from about 5% (maize) to around 40–50% (canola (***Brassica***), **flax**, **peanut**, **palm kernel**, **sesame** and **sunflower**) oil. **Soybean** and **cotton** seeds have an oil content about midway between these extremes (15–20%) (**see:** Table O.1 in **Oilseeds – major**).

Seed oils and fats are used in a wide range of food applications including baking and frying, salad dressings and confectionary coatings. Some seed oils, such as castor oil and tung oil, are used in non-food applications. These applications include paints, coatings, finishes and ink, all of which benefit from the unique cross-linking abilities of the fatty acid components of the oil.

The five major fatty acids present in common edible oils, such as soybean and canola oil are palmitic (16:0), stearic (18:0), oleic (18:1), linoleic (18:2) and linolenic acids (18:3). Seed oil fatty acids may also contain triple (acetylenic) bonds and other functional groups such as hydroxyl, epoxy and keto groups, in place of or in addition to double bonds. For example, castor oil is rich in the hydroxy-fatty acid ricinoleic acid. These uncommon functional groups provide the bases for other industrial applications. For example, because of its ability to remove free chlorine radicals, an epoxy group is a good stabilizer for PVC. Since seed triacylglycerols are also good lubricants, oils rich in epoxy groups (such as *Vernonia* oil or chemically epoxidized soybean oil) are used for combination plasticizers/stabilizers in the manufacture of some PVC resins. (AJK)

For details on the synthesis, deposition, composition, mobilization and modification of oils and fats **see: Fatty acids; Fatty acids – synthesis; Glyoxysomes; Oils and fats – genetic modification; Oils and fats – mobilization; Oils and fats – synthesis; Oilseeds – major; Oilseeds – minor.**

Gunstone, F.D., Harwood, J.L. and Padley, F.B. (1994) *The Lipid Handbook*, 2nd edn. Chapman & Hall, London, UK.
USDA New Crops Database (www.ncaur.usda.gov/nc/ncdb/search.html-ssi)

Oils and fats – genetic modification

For background information on the biosynthetic pathways for fatty acids and oils and fats, **see: Fatty acids – synthesis; Oils and fats – synthesis.**

Great potential exists for extending the range of domesticated crops, such as **soybean** and canola (**See: Brassica oilseeds**), for novel food, biomedical and industrial purposes by the use of biotechnology. Genetic engineering techniques have been used to change the existing ratios of seed oil **fatty acids** and to produce new types of fatty acids in various seed oils. There are few examples of transgenic oilseed crops with modified fatty acid content that have been fully commercialized. This is partly due to the length of time it takes to develop a new transgenic crop and to the cost of meeting the regulatory requirements in the countries where plant seed oils are made and sold. In many cases there are still technical hurdles related to maintaining oil yield or seed viability when new fatty acid compositions are made. On the other hand, the number and types of modifications to plant seed oils that have been described are impressive.

Developing seed oils with increased monunsaturated fatty acids (specifically 18:1Δ9, oleic) and a concomitant decrease in polyunsaturates (18:2Δ9,12, linoleic 18:3Δ9,12,15, linolenic) has been a goal of oilseed breeders for many years. Reducing the polyunsaturated fatty acid content of plant oil increases its oxidative stability and eliminates the need for partial hydrogenation of a refined vegetable oil. This in turn eliminates certain *trans* fatty acids, which have been associated with cardiovascular disease in humans (**see: Oils and fats**). **Mutants** of maize, peanut, canola and sunflower with an oleic acid content ranging from 60 to 90% have been developed since the 1980s. With the exception of sunflower, however, these are usually recessive mutations involving several structural and non-structural genes. Many high oleic acid mutant lines have a variable phenotype, depending on the environment in which they are grown. Some of these exhibit poor germination or growth under some environmental conditions. An understanding of oleate desaturation in oilseeds has provided an explanation for these limitations and has shown they can be overcome by the use of biotechnology.

In the developing oilseed, cytosolic oleoyl-CoA is incorporated into membrane phosphatidylcholine (PtdCho) where it is desaturated to linoleoyl-PtdCho by a membrane bound Δ12 desaturase encoded by at least two different *FAD 2* genes (**see: Fatty acids – synthesis**). In most oilseeds one of these *FAD 2* genes is expressed mainly in the seed, the other throughout the plant, i.e. constitutively. Both contribute to seed oil polyunsaturation. The constitutive gene is also expressed later in seed development when compared with the seed-predominant gene. Thus, when only the seed-predominant *FAD 2* gene is mutated, the seeds show a variable oleic acid content depending on the length of seed fill time, which in turn is environmentally determined. If both *FAD 2* genes are knocked out by mutations, polyunsaturated fatty acid synthesis in roots is impaired, compromising plant vitality. Gene silencing techniques have been used to silence both genes in a seed-specific manner, thus achieving maximum, environmentally independent seed oleic acid content. The root and leaf fatty acid content in these transgenic plants remains unchanged and the plants germinate and grow in a similar manner to wild-type lines in all environments. High oleic acid soybeans produced in this way are one of few transgenic crops with modified oil profile that have been commercialized.

Mutants of some **oilseeds (major)**, such as soybean, have also been produced with a reduced saturated fatty acid content. Most of these mutants are recessive and are the result of mutations in several genes. In most low saturate mutants at least one of these altered genes encodes a palmitoyl-ACP thioesterase (FAT B). It has been possible to produce transgenic lines with a 50% reduction in total 16:0 by silencing *FAT B* genes in soybean. It has also been possible to combine high oleic and low saturate phenotypes by silencing genes encoding the desaturase FAD 2 and thiosesterase FAT B proteins in a single event.

Increasing the saturate content of plant oils may also be important for solid fat applications, such as margarine or cocoa butter substitutes. An oil is usually called a fat when it is solid at room temperature. Over-expression of *FAT B* genes can lead to increased palmitate and stearate in the seed oil. Saturates, especially stearic acid, may also be increased by blocking the desaturation of stearoyl-ACP to oleoyl-ACP. The gene encoding the desaturase that catalyses this reaction (*AAD 1*) can be silenced in a similar manner to the *FAD 2*. When this is done, the stearate content in canola and soybean can be increased from a few per cent to over 20% of the oil total fatty acids. There is also a specialized thioesterase involved in the production of stearate in the seeds of mangosteen (*Garcinia mangostana*). This thioesterase is a member of the oleoyl-ACP thioesterase (*FAT A*) gene family although the mangosteeen FAT A has a strong preference for stearoyl-ACP. When the mangosteen thioesterase gene (*FAT A*) is expressed in canola seeds, an oil stearic acid content of over 20% is also achieved.

Besides modifications to the relative ratios of the five major fatty acids in oilseeds, there are also numerous examples of the production of new fatty acids in the seeds of domesticated oilseed crops. One commercially produced example of this type is High Laurate Canola. In this transgenic canola line, over 40% of the seed oil fatty acids is lauric acid (C12:0). Lauric acid has numerous food and industrial uses and is currently obtained from **coconut** oil. The new canola line was made by expressing a *FAT B* gene from California bay laurel in canola. This, and other species such as *Cuphea* spp., have medium-chain length fatty acids (C8:0 to C14:0) in their seed oil, and there are specialized versions of FAT B-type fatty acyl-ACP thioesterases which redirect the common fatty acid synthase complex to produce medium-chain fatty acids (**see: Oilseeds – minor**). These species also contain specialized KAS IV-type condensing enzymes with high specific activity for medium-chain fatty acyl-ACP substrates. There is the potential to produce other types of medium-chain fatty acids in domesticated oilseeds by expressing in them *FAT B* and *KAS IV* genes from *Cuphea* and other medium-chain oilseed species.

Oils with novel unsaturated fatty acids have also been produced although none has been commercialized. For example, the seed oil of **meadowfoam** (*Limnanthes alba*), is rich in a long chain (C20) fatty acid with a Δ5 double bond, Δ5-eicosenoic acid (20:1Δ5). This fatty acid is used in the manufacture of cosmetics, surfactants and lubricants. Soybean oils containing 15–20% Δ5-eicosenoic acid have been produced by co-expressing *Limnanthes* genes encoding a fatty acyl-CoA condensing (elongase) enzyme (FAE 1) and a Δ5-desaturase. The soybean seeds do not have difficulty in tolerating this novel fatty acid and it is thought that commercial quantities may be achieved in soybean lines that also over-express FAT B, the palmitoyl-ACP thioesterase (16:0-CoA is the substrate for FAE 1).

Often, however, commercial production of novel unsaturated fatty acids in domesticated oilseeds is far more complex. For example, seeds of some **Umbellifer** species, such as carrot and **coriander**, contain an AAD-type desaturase that converts palmitoyl-ACP to Δ4 hexadecanoyl-ACP. The hexadecanoyl-ACP then elongates to petroselenoyl-ACP and as a result the oils of these species are rich in petroselenic acid (18:1Δ6). When expressed in oilseeds such as canola, this desaturase leads to the accumulation of petroselenic acid in the seed oil. To make commercially viable quantities of this novel fatty acid it is necessary to also express many other modified genes including a KAS I-type condensing enzyme with specificity for Δ4 hexadecanoyl-ACP, a FAT B-type thioesterase with activity towards petroselenyl-ACP and numerous other variant enzymes including fatty acyltransferases, ferredoxins and fatty acyl carrier proteins. This is a good illustration of some of the technical challenges that limit the commercial production of novel fatty acids in domesticated oilseed crops.

A similar challenge is faced in the production of industrial oils with novel functionality in domestic oilseed crops. There is a wide variety of plants that have oils containing non-methylene-interrupted double bonds or functional groups other than double bonds. These fatty acids have multiple non-food uses ranging from paints to polymers. They are produced in exotic oilseed plants, the result of the activity of desaturase-related enzymes that are evolutionarily diverged members of the *FAD-2* gene family. These desaturase-related enzymes include hydroxylases from *Ricinus communis* (**castor bean**) and *Lesquerella fendleri*, conjugases from *Momordica charantia*, *Aleurites* (**tung**) and *Calendula officinalis*, an acetylenase from *Crepis palaestrina* and an epoxygenase from *Vernonia galamensis*. A gene encoding a cytochrome P450-type epoxygenase from *Euphorbia lagascae* seeds has also been described. When any of these genes are expressed in a domesticated oilseed, such as soybean or canola, up to 20% of the novel fatty acid is produced in the transgenic seed oil. There is also always a concomitant increase in oleic acid in the seed oil and an accumulation of the novel fatty acid in seed membrane phosphatidylcholine in addition to in the triacylglycerol. As a result the transgenic seeds usually have reduced oil content and often do not germinate well. The usual interpretation of these observations is that the novel fatty acyl chain is not being properly channelled from the membrane to triacylglycerol after the new functional group has been inserted. Thus the whole desaturation cycle is slowed down and there is an accumulation of oleoyl-CoA in the cytosol.

In a situation analogous to the petroselenic acid example, the production of these oils in commercial quantities is likely to involve additional evolutionarily diverged enzymes from the species producing these unusual fatty acids. Genes encoding these additional enzymes will need to be co-expressed with the novel desaturase-related gene. Since these additional enzymes have not been identified, the timelines for the commercialization of these types of industrial oils remains unknown.

Perhaps one the most exciting possibilities made possible by genetic engineering is the production of very long chain polyunsaturated fatty acids in domestic oilseed crops. These fatty acids include arachidonic (C20:4), eicosopentaenoic (C20:5) and docosahexaenoic (C22:6) acids. The current source of these fatty acids in the human diet is certain fish and there are many well-documented beneficial effects for infants and adults associated with their intake in the diet. A less expensive, higher quality supply of these fatty acids would much extend their use in human foods, and seed oils are thus a very attractive source for producing them. Most of the genes that encode the enzymes involved in the pathways of very long chain polyunsaturated fatty acid biosynthesis have been cloned from marine organisms. Individually these genes are expressed well by most oilseeds. For example, expression of Δ6 and Δ12 desaturases from the marine protist *Mortierella alpina* in canola seeds results in a γ-linolenic acid content of over 40% of the total seed oil fatty acids. The challenge for plant biotechnology is to successfully express the multiple genes necessary to recreate these entire pathways in plant seeds. (AJK)

(See: Oilseeds – major; Oilseeds – minor)

Abbadi, A., Domergue, F., Bauer, J., Napier, J.A., Welti, R., Zahringer, U., Cirpus, P. and Heinz, E. (2004) Biosynthesis of very-long-chain polyunsaturated fatty acids in transgenic oilseeds: constraints on their accumulation. *The Plant Cell* 16, 2734–2748.

Domergue, F., Abbadi, A. and Heinz, E. (2005) Relief for fish stocks: oceanic fatty acids in transgenic oil seeds. *Trends in Plant Science* 10, 112–116.

Dyer, J.M. and Mullen, R.T. (2005) Development and potential of genetically engineered oilseeds. *Seed Science Research* 15, 255–267.

Hildebrand, D., Suryadevara, R. and Hatanaka, T. (2002) Redirecting lipid metabolism in plants. In: Kuo, T.M. and Gardner, H.W. (eds) *Lipid Biotechnology*. Marcel Dekker, New York, pp. 57–84.

Kinney, A.J. (1996) Development of genetically engineered soybean oils for food applications. *Journal of Food Lipids* 3, 273–292.

Kinney, A.J. (1997) Genetic engineering of oilseeds for desired traits. *Genetic Engineering* 19, 149–166.

Qi, B., Fraser, T., Mugford, S., Dobson, G., Sayanova, O., Butler, J., Napier, J.A., Stobart, A.K. and Lazarus, C.M. (2004) Production of very long chain polyunsaturated omega-3 and omega-6 fatty acids in plants. *Nature Biotechnology* 22, 739–745.

Thelen, J.J. and Ohlrogge, J.B. (2002) Metabolic engineering of fatty acid biosynthesis in plants. *Metabolic Engineering* 4, 12–21.

Oils and fats – mobilization

Utilization of the storage energy in triacylglycerols following seed germination involves a number of processes, including the generation of free **fatty acids** by the action of triacylglycerol lipases, conversion of the free fatty acids back to acetyl-CoA by **β-oxidation** and the conversion of acetyl-CoA to sucrose by the **glyoxylate cycle** followed by **gluconeogenesis** (Fig. O.3). The other product of the triacylglycerol lipase action is glycerol, which is also converted to sucrose via triose phosphate. Triacylglycerol lipases have been isolated from the germinated seeds of numerous oilseed species and almost all of them possess a very broad specificity for many types of fatty acids. A few lipases with higher substrate selectivity for medium-chain fatty acids have been isolated from species which contain these fatty acids in their seed oil, for example *Cuphea* spp.

In germinated seeds, β-oxidation of straight chain saturated and unsaturated fatty acids occurs predominantly or exclusively in **glyoxysomes**, which are a highly specialized form of peroxisome similar in size to **oil bodies**. The roles of the mitochondria in the β-oxidation of straight chain saturated and unsaturated fatty acids and of the glyoxysome in the β-oxidation of branched-chain 2-oxo acids derived from amino acid catabolism is unresolved. Sequencing of the *Arabidopsis* genome has led to the discovery of genes that are homologous to mammalian and yeast genes encoding α- and ω-oxidation enzymes. The function of these pathways in plants is unclear but it is unlikely that they are involved in seed oil reserve mobilization.

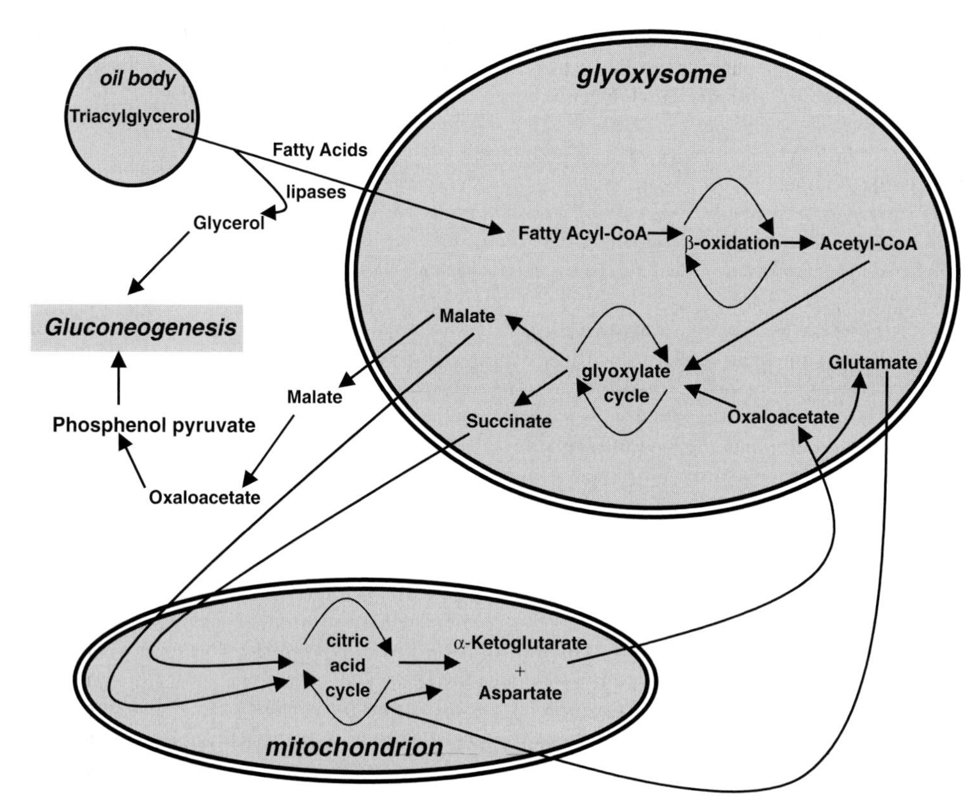

Fig. O.3. Fatty acid degradation and sucrose synthesis in germinated oilseeds. Triacylglycerol is converted to glycerol and fatty acyl chains by the action of lipases. After conversion to acyl-CoA, the chains are transported into the glyoxysomes for ultimate conversion to malate, the cytosolic source of carbon for sucrose synthesis. ATP generated during these cycles (not shown) is the energy source for sucrose synthesis. Additional carbon from the glyoxysome is recycled back into the glyoxylate cycle via the mitochondria.

Saturated and unsaturated free fatty acids released by lipases are first activated for β-oxidation by conversion to fatty acyl-CoA, a reaction catalysed by fatty acyl-CoA synthetase. This renders the fatty acid susceptible to an oxidative reaction at the β-carbon (i.e. the Δ3 carbon). These fatty acyl-CoA molecules are then imported into the glyoxysomes by specific transporter proteins that are related to peroxisomal ATP-binding cassette transporter proteins in mammals (ABC transporters). Once in the glyoxysome, the fatty acyl-CoA molecules are stripped of two carbon units in repeated cycles of what is, in effect, a reversal of **fatty acid synthesis**. The C-2 units are repeatedly removed from the thiol (CoA) end of the fatty acid until the chain is completely degraded. Each cycle involves reactions catalysed sequentially by fatty acyl-CoA oxidase, followed by a multifunctional protein and then 3-ketoacyl-CoA thiolase. Each of these cycles yields one molecule of acetyl-CoA. The intermediate enzyme of this cycle, the multifunctional protein, catalyses the hydration of 2-*trans*-enoyl-CoA to 3-hydroxyacyl-CoA and is composed of up to four separate catalytic activities, 2-*trans*-enoyl-CoA hydratase, L-3-hydroxyacyl-CoA dehydrogenase, D-3-hydroxyacyl-CoA epimerase and 3,2-enoyl-CoA isomerase.

It is not known if variant enzymes exist for the oxidation of the many uncommon fatty acids present in the seed oils of exotic plants. The presence of an unusual functional group, such as an epoxy or hydroxyl function, in place of a double bond in a fatty acyl-CoA chain could interrupt β-oxidation degradation unless there were enzymes for dealing with the unusual catabolic product. Oxidation of ricinoleic acid in germinated **castor beans** (*Ricinus communis*) seems to require such variant β-oxidation enzymes, but none has been characterized.

The acetyl-CoA molecules generated by β-oxidation enter the glyoxylate cycle, which is also located in predominantly the glyoxysome.

Malate leaving the glyoxysome is converted back to oxaloacetic acid in the cytosol which is in turn converted to sucrose via phosphoenol pyruvate and fructose 1,6-bisphosphate. The conversion of oxaloacetic acid to phosphoenol pyruvate is catalysed by phosphoenol pyruvate carboxylase. Fructose 1,6-bisphosphate is synthesized by a reversal of glycolysis and this metabolite is then converted to sugar. This conversion involves a series of reactions, the control step of which is sucrose phosphate synthase.

The sucrose and ATP formed from triacylglycerol oxidation can be transported and used by the growing seedling. (AJK)

Gerhardt, B. (1992) Fatty acid degradation in plants. *Progress in Lipid Research* 31, 417–436.

Graham, I.A. and Eastmond, P.J. (2002) Pathways of straight and branched chain fatty acid catabolism in higher plants. *Progress in Lipid Research* 41, 156–181.

Oils and fats – synthesis

See: Fatty acids – synthesis for details on this process, which occurs as a prelude to the synthesis of oils and fats (**triacylglycerols**).

In the endoplasmic reticulum (ER) membranes of developing seeds fatty acyl-CoAs are esterified to a phosphorylated glycerol backbone to form triacylglycerol by membrane-bound fatty acyltransferases.

Synthesis of the glycerol backbone occurs in the cytosol. Hexose phosphate (glucose-6-P) is converted to triose phosphate (dihydroxyacetone-P) and then to glycerol-3-P.

The first fatty acyltransferase reaction involves transfer of a fatty acyl chain from CoA to the *sn*-1 position of glycerol-3-phosphate to form lysophosphatidic acid (LPA). (**See:** Fig. F.2 in **Fatty acids – synthesis**) This reaction is catalysed by glycerol-3-phosphate acyltransferase (GPAT). The second reaction is the transfer of fatty acyl-chain from another CoA ester to the *sn*-2 position of LPA to form phosphatidic acid. This is catalysed by lysophosphatidic acid acyltransferase (LPAAT), which is encoded by two separate gene families, A and B. Both types of LPAAT enzyme are found in seeds that accumulate unusual fatty acids. Oilseeds that only produce the five major fatty acids have only type A LPAAT activity. It is thought that class B LPAATs are involved in the incorporation of unusual fatty acids into triacylglycerol.

The phosphate group is removed from phosphatidic acid by phosphatidic acid phosphatase to form diacylglycerol (DAG). A third fatty acyl group is then transferred to DAG in a reaction catalysed by diacylglycerol acyltransferase (DAGAT). There are also two different types of DAGAT, encoded by different gene families. One type of DAGAT is part of the larger fatty acyl CoA:cholesterol acyltransferase gene family but the other DAGAT family is not related to any other known genes. Both types of DAGAT contribute to triacylglycerol formation in oilseeds. (AJK)

Voelker, T. and Kinney, A.J. (2001) Variations in the biosynthesis of seed storage lipids. *Annual Review of Plant Physiology and Plant Molecular Biology* 52, 335–361.

Oilseed rape

See: Brassica oilseeds; Oilseeds – major

Oilseeds – major

1. Introduction

Oilseed crops produce seed or fruits that contain a substantial quantity of plant storage lipids. The extracted vegetable oil is used for edible and/or industrial purposes. The oilseed meal or cake remaining after oil extraction contains moderate to high levels of nutritionally well-balanced protein and, with a few exceptions, is an important livestock and poultry feed supplement (**see: Storage proteins**). In some countries the meal may be used as an organic fertilizer while processed meal, primarily from **soybean**, is used in foodstuffs. There is mounting commercial interest in using oilseeds such as soybean as oil feedstocks from which to manufacture 'biodiesel' fuels.

Oilseeds and their products are second only to petroleum as the most valuable commodities moving in world trade. The expanding world population and rising living standards in the developing world have fuelled a sustained growing demand for edible vegetable oils (Table O.1). (**See: Crop Atlas Appendix, Maps 2, 13, 15, 16, 17, 18**)

2. Oil composition

Fatty acids are the building blocks of seed storage oils (triacylglycerols). They constitute about 98% of the oil's weight

Table O.1. Estimated world production of vegetable oils from major oilseed and tree crops for selected years.

	Year				
	1963	1973	1983	1993	2003
Oilseed	million t				
Soybean	3.5	6.8	13.9	17.3	27.9
Palm	1.5	2.4	5.8	14.2	27.8
Palm kernel	0.4	0.5	0.9	1.9	3.2
Rape, Canola, Mustard	1.2	2.6	5.0	9.0	12.1
Sunflower	2.4	3.6	6.1	7.8	8.0
Groundnut (Peanut)	2.8	2.9	3.2	4.3	5.8
Cotton seed	2.5	3.0	3.1	3.7	3.7
Coconut	2.0	2.2	2.8	3.0	3.3
Olive	1.8	1.5	1.7	1.9	2.7
Linseed	1.0	0.8	0.8	0.6	0.5
Sesame	0.5	0.5	0.6	0.6	0.7
Castor	0.3	0.3	0.4	0.4	0.5
Safflower	0.2	0.1	0.2	0.2	0.2
Maize (Corn)	0.4	0.6	1.0	1.6	0.9
Tung	<0.1	<0.1	<0.1	<0.1	<0.1
Subtotal	20.6	27.9	45.6	66.6	97.7
Other seed oils	0.6	1.1	1.1	1.8	1.9
Total world	21.2	29.0	46.7	68.4	99.6

Source FAOSTAT.

Species in **bold** lettering have individual entries. **See also: Cacao; Crambe; Evening primrose; Grapeseed; Jojoba; Meadowfoam; Niger.**

Table O.2. Typical oil content and per cent fatty acid composition of major world vegetable oil crops. (First number of a fatty acid, i.e. 16:0, is the number of carbon atoms, the second number is the number of double bonds in the carbon chain.)

Oil Source[a]	Typical % oil	Fatty acid composition							
Tropical oils		**6:0+8:0**	**10:0**	**12:0**	**14:0**	**16:0**	**18:0**	**18:1**	**18:2**
Palm oil	56	–	–	0.2	1.9	44.0	39.2	0.4	10.1
Palm kernel	50	2.8	2.5	48.3	19.8	–	2.4	16.5	0.7
Coconut	65	7.8	6.7	47.5	18.1	8.8	2.6	6.2	1.6
Cacao 'butter'	40–50				0.1	25.4	33.2	32.6	2.8
Temperate oils		**16:0**	**18:0**	**18:1**	**18:2**	**18:3**	**Other**		
Soybean	19	10.7	3.9	22.8	50.8	6.8			
Canola (LEAR)[b]	46	??	2.0	61.3	20.1	9.3	2.0% C20+C22		
Low 18:3	46	5.0	2.0	61.3	27.1	2.1	2.0% C20+C22		
High 18:1	46	3.4	2.1	76.8	13.6	2.5	2.0% C20+C22		
Sunflower	40	6.0	4.0	19.5	68.5	0.5	1.0% C20+C22		
High 18:1	40	4.0	5.0	81.0	8.0	tr[d]	1.0% C20+C22		
Mid 18:1	40	4.0	5.0	65.0	26.0	tr	1.0% C20+C22		
Groundnut	45	9.9	2.5	46.9	34.5	tr	7.0% C20–C24		
Cotton	18	24.0	2.5	1.0	52.5	tr	1.0% C16:1		
Olive	20	9.0	2.5	78.0	8.3	0.8	tr of C20:0+C20:1		
Linseed	44	6.0	3.0	16.5	14.0	60.0			
Low 18:3	44	6.4	4.1	16.5	69.5	1.8			
Sesame	50	8.5	4.7	42.0	42.0	<1	<1% C20:0		
Castor bean	45	1.1	1.0	3.1	4.7	0.5	85% ricinoleic, 0.5% eicosenoic+0.7% dihydroxystearic acids		
Safflower	33	5.3	1.5	15.0	77.0	tr			
Rape (HEAR)[c]	46	3.0	1.0	10.0	13.5	9.8	1.25% other fatty acids C20:1, 6.8%; C22:0, 0.7%; C22:1, 53.6%; C24:1, 1%		
Tung	33	4.0	1.5	9.0	14.0	tr	75% eleosteric acid		
Crambe	50	1.7	1.3	16.7	7.8	6.9	C22:0, 1.3%; C20:1, 2.9%; C22:0, 2.7%; C22:1, 55.7%; C24:1, 2.9%		
Grape seed	12	7.0	3.7	22.8	57.0	ND[e]			
Meadowfoam	29	0.3	ND	1.8	ND	ND	C20:1, 66%; C22:1, 13.1%, C22:2, 17.0%		
Evening primrose	17	7.0	2.0	9.0	81.0	ND	9% of the 81% C18:3 is GLA[f]		
Maize germ	33	11.4	1.9	25.4	60.7	tr			

[a] See entries in Encyclopedia for each of the listed species.
[b]LEAR = Low erucic acid rape; [c]HEAR = High erucic acid rape; [d]tr = trace; [e]ND = not detected (< 0.1%); [f]Gamma linolenic acid.

and are composed of carbon chains terminated by a methyl group at one end and a carboxyl group at the other linked to a glycerol molecule (three fatty acids per glycerol) (**see: Oils and fats; Triacylglycerols**). Fatty acids are differentiated by the number of carbons and/or the number of double bonds in the chain. The presence and amount of individual fatty acids largely determines the value and use of a vegetable oil while the quantity of oil present affects the economics of extraction. Typical oil contents and fatty acid compositions of the various commercially grown oil crops are given in Table O.2. The oil content and fatty acid composition of an oilseed will vary with the production environment and the variety grown. Higher temperatures during the growing season are likely to produce oils that have a higher ratio of saturated to unsaturated fatty acids. Oils from crops that evolved in the tropics tend to have a very high content of saturated fatty acids (Table O.2). There is much plant breeding effort to modify the fatty acid content to suit nutritional and health requirements. Two examples are **rape (Brassica, canola)** where erucic acid content has been almost eliminated, and **soybean** where mutation breeding has lead to the production of low-linolenic-acid lines (the oil needs less hydrogenation and therefore generation of *trans* acids is reduced).

3. Tissue and cellular location

In most seeds, oils are located in the **embryo**, especially the **cotyledons**, as **oil bodies** (oleosomes or lipid bodies).

(KD)

(See: Oils and fats – genetic modification; Oils and fats – synthesis; Oilseeds – minor)

Hui, Y.H. (ed.) (1996) *Bailey's Industrial Oil & Fat Products*, 5th edn, Vol. 2. John Wiley, New York, USA.

Robbelen, G., Downey, R.K. and Ashri, A. (eds) (1998) *Oil Crops of the World*. McGraw-Hill, New York, USA.

Vaughan, J.G. (1970) *The Structure and Utilization of Oil Seeds*. Chapman and Hall, London, UK.

Weiss, E.A. (2000) *Oil Crops*, 2nd edn. Blackwell Science, Oxford, UK.

Oilseeds – minor

The major oilseeds (**see: Oilseeds – major**) represent a relatively small fraction of the species whose seeds yield useful **oils** (triacylglycerols). Seeds of many other species, which can be classed as minor oilseeds, are very important locally for culinary or other purposes but have little commercial impact on the international oilseed economy. Examples are shown in Tables O.3 and O.4. Some of these, however, are increasingly attracting attention for potential commercial exploitation. For example, false flax or gold-of-pleasure seeds, used in Europe in the Bronze age and later in Roman times, yield an oil with a fatty acid composition particularly suitable for some uses, and is cultivated in parts of Europe (e.g. Austria, Finland and on set-aside land (i.e. non-food use) in Germany, the UK), albeit in relatively minor quantities. It is under investigation in Europe and North America for possible development as an oilseed crop. Another example is the **physic nut**, whose oil can be used as biodiesel fuel and has potential importance as a commercial crop in several regions, such as West Africa (e.g. Mali) and the Middle East (e.g. Saudi Arabia). Commercial production of Cuphea oil is also being explored. (MB)

(See: Oils and fats; Oils and fats – genetic modification; Oils and fats – synthesis; Oilseeds – major)

Budin, J.T., Breene, W.M. and Putnam, D.H. (1995) Some compositional properties of camelina (*Camelina sativa* L. Crantz) seeds and oils. *Journal of the American Oil Chemists Society* 72, 309–315.

Coronel, R.E. (1996) *Pili nut*: Canarium ovatum Engl. International Plant Genetic Resources Institute, Rome, Italy.

Graham, S.A. (1989) *Cuphea*: a new plant source of medium chain fatty acids. *CRC Critical Reviews in Food Science and Nutrition* 28, 139–173.

Table O.3. Minor edible seed oils.

Oil source	Provenance	Fatty acids (average %)[a]	Uses
Argan (*Argania* spp.)	Morocco	Ln 4.6; Lo 31.5; M 4.3; O 46; P 13; S 5	Cooking oil; pastes
Babassu palm (*Orbignya* spp.)	Brazil, Mexico, Guyana	C 7; Cy 5; L 45; M 17; P 7; S 6; O 15	General; margarine
Balanites (*Balanites* spp.)	Sudan	Lo 44; O 31; Sat 24	Cooking oil
Borneo tallow nut, Illipe (*Shorea stenoptera*)	Sarawak, Borneo, SE Asia	A 1; O 37; P 18; S 43	Cocoa butter substitute; non-edible uses
Brazil nut (*Bertholettia excelsa*)	Brazil, Bolivia, Venezuela, Guyana, Peru	Lo 30; O 48; P 15; S 6	Cooking , salad oil
Buffalo gourd (*Cucurbita* sp.)	Mexico, USA	Lo 61; O 27; P 8; S 3	General; non-edible uses
Caryocar (*Caryocar* spp.)	Brazil, Guyana	Lo 3; O 46; P 48	Cooking oil
Chinese vegetable tallow (*Stillingia sebifera*)	China, India	L 2; Lo 1; M 3; O 25; P 65; S 5	General; non-edible uses
Cohune palm (*Orbignya cobune*)	Central, S America	C 6; Cy 7; L 46; M 16; O 10; P 9.5; S 3	Baking; margarine
Hazelnut (*Corylus avellana*)	Europe, New Zealand	Lo 19; O 70; S 7	Cooking; olive oil adulterant
Kokum (*Garcinia indica*)	Southern India	Lo 3; O 36; P 4; S 56	Cocoa butter substitute; cosmetics
Macadamia (*Macadamia* spp.)	Hawaii, Australia	A 2; E 2; Lo 2; O 60; P 9; Po 22; S 2	General
Mango (*Mangifera indica*)	India	O 40; P 7; S 45	Cocoa butter substitute
Noog (*Guizotia abyssinica*)	Niger, Ethiopia, India, Asia, Caribbean	A 2; Lo 60; Ln 2; M 2; O 20; P 8; S 7	General; non-edible uses
Perilla (*Perilla frutescens*)	China, India, Japan, Korea	Lo 14; Ln 60; O 19	General; non-edible uses
Pili nut (*Canarium ovatum*)	Philippines	O 50; P 35; others N/A	Cooking oil; non-edible uses
Rice bran (*Oryza sativa*)	Asia, Africa, Americas	Lo 35; O 25; P 15; S 2	Cooking oil; foodstuffs
Sacha inche (*Plukenetia volubilis*)	W Africa, Central and S America	Lo 37; Ln 45; O 10; P 4; S 3	Foodstuffs
Sal (*Shorea robusta*)	India	High in S	Cocoa butter substitute
Seje (*Jessenia* spp.)	S America, Caribbean	Lo 3; O 76; P 9; S 6	Cooking oil; foodstuffs; non-edible uses
Shea nut (*Butryospermum parkii* = *Vitellaria paradoxa*)	W Africa	Lo 5; O 50; P 7; S 35	Cooking; general; non-edible uses (medicinal, candles, soap)
Smooth loofa (*Luffa cylindrica*)	China, India, SE Asia, S America	Lo 64; O 7; P 10; S 19	General
Teaseed (*Camellia sasanqua*)	China, Vietnam, Assam	Lo 8; O 75; Sat 9	General
Walnut (*Juglans regia*)	France, Japan	Lo 55, Ln 14; P 11; S 5	'Gourmet' cooking

[a]Fatty acids: A, arachidic (eicosanoic), 20:0; C, capric, 10:0; Cy, caprylic, 8:0; E, eicosenoic; L, lauric, 12:0; Ln, linolenic; Lo, linoleic; M, myristic, 14:0; O, oleic; P, palmitic; Po, palmitoleic, 16:1; S, stearic; Sat, saturated. For other structural details **see: Fatty acids**.
N/A, not available.

Table 0.4. Minor seed oils, mainly non-edible uses.

Oil source	Provenance	Fatty acids (average %)[a]	Uses
Almond (*Prunus* spp.)	Mediterranean	Lo 23; O 66; P 8; S 2	Cosmetics
Chaulmoogra (mainly *Taraktogenos kurzli*)	Indian subcontinent, Myanmar, Nigeria, Uganda	Cyclopentene fatty acids, e.g. chaulmoogric and hydnocarpic acids	Medicinal
Cuphea (*Cuphea* spp.)	Central and S America	Cy 3–73; C 18–87; L 57	Source of short-chain fatty acids (for detergents)
Dimorphotheca daisy (*Dimorphotheca pluvialis*)	SW Africa	65% dimorphecolic acid (C 18 OH acid)	Cosmetics; lubricants; paints
False flax or gold-of-pleasure (*Camelina sativa*)	E Europe, SW Asia	E 10; Er 2; Ln 36; Lo 20; O 20; P 5; S 3	Cosmetics; lighting; detergents; paints; edible
Hemp (*Cannabis sativa*)	Wide	Lo 57; Ln 20; O 12; P 6; S 3	Cosmetics
Honesty (*Lunaria annua*)	N America, Europe	Er 41; Lo 9; Ln 3; N 22; O 18	Medicinal
Jatropha (physic nut) (*Jatropha curcas*)	S America, worldwide sub- and tropics	Lo 30; O 50; P 15; S 6	Biodiesel; soaps; lubricant; lighting
Karanja (*Pongamia glabra*)	East Indies, Philippines	Lo 11; O 71	Pharmaceutical; lighting; pesticide; soap
Lesquerella (*Lesquerella fendleri*)	N America	Lo 7; Ln 14; Lq 54; O 15; S 2	Cosmetics
Marigold (*Calendula officinalis*)	Europe	Cl 67; Lo 31	Paints; cosmetics
Neem (*Azadirachta indica*)	Indian subcontinent, tropical Australia, Africa	A 1; Lo 15; O 42; P 19; S 21	Medicinal; pesticide; soap
Sea buckthorn (*Hippophae rhamnoides*)	Europe, Asia	Rich in Ln, Lo, O, sterols, tocopherols	Cosmetic; medicinal; edible
Tonka bean (*Dipteryx odorata*)	Guyana, Venezuela	Lo 51; O 60; P 6; S 6	Perfumery
Ucuuba (*Virola* spp.)	Brazil, Guyana	L 16; M 73; O 6; P 5	Candles; cosmetic; medicinal
Wheat germ (*Triticum aestivum*)	N America, Europe, etc.	Lo 60; Ln 5; O 17; P 17; S 3	Cosmetic

[a]Fatty acids: A, arachidic (eicosanoic), 20:0; C, capric, 10:0; Cl, calendic, 18:3; Cy, caprylic, 8:0; E, eicosenoic; Er, erucic; L, lauric, 12:0; Ln, linolenic; Lo, linoleic; Lq, lesquerolic, 20:1 OH; M, myristic, 14:0; N, nervonic, 24:1; O, oleic; P, palmitic; Po, palmitoleic, 16:1; S, stearic; Sat, saturated. For other structural details **see: Fatty acids**.

Gunstone, F.D. (2004) *The Chemistry of Oils and Fats.* CRC Press, Blackwell Publishing, Oxford, UK, pp. 12–18.

Heller, J. (1996) *Physic nut:* Jatropha curcas *L.* International Plant Genetic Resources Institute, Rome, Italy.

Janick, J. and Simon, J.E. (eds) (1993) *New Crops: Exploration, Research and Commercialisation. Proceedings of the Second National Symposium.* John Wiley, New York, USA.

Koul, O., Isman, M.B. and Ketkar, C.M. (1990) Properties and uses of neem, *Azadirachta indica*. *Canadian Journal of Botany* 68, 1–11.

Salunkhe, D.K., Chavan, J.K., Adsule, R.N. and Kadam, S.S. (1992) *World Oilseeds. Chemistry, Technology and Utilization.* Van Nostrand Reinhold, New York, USA.

Vaughan, J.G. (1976) *The Structure and Utilisation of Oilseeds.* Chapman and Hall, London, UK. www.fao.org/docrep/X5043E/x5043E00.htm#Contents

Oleoresin

A mixture of resins and volatile **(essential) oils**. They are present in certain spice seeds. **(See: Spices and flavours)**

Oleosins

Oleosins are **amphiphilic** proteins associated with **oil bodies**. They are embedded in the **phospholipid** monolayer and quite possibly protrude into the hydrophobic interior of the organelle. Although the hydrophobic domain of the oleosin has a conserved amino acid structure among oilseed species there is not a great degree of overall amino acid conservation. The secondary structure of all oleosins, however, is highly conserved. This has led to the suggestion that they play a role in stabilizing the structure of the organelle, particularly during seed desiccation. Oil body proteins similar in structure to oleosins but which bind calcium (caleosins) to the surface of the organelle have also been described. **(See: Biopharmaceuticals; Oil bodies)** (AJK)

Oligopeptides

Short chains of amino acids that are degradation products of **proteins** (polypeptides) as a consequence of their incomplete hydrolysis by proteolytic enzymes, or are produced as polypeptides are synthesised on the ribosomal complex. **(See: Storage proteins – mobilization)**

Oligosaccharides

Compounds comprising two to ten monosaccharides. The best known group of these compounds, in the context of seed physiology, is the soluble α-galactosides, which are characterized by the presence of α(1–6) links between the galactose moieties and include: (i) the galactosyl derivatives of sucrose, referred to as the **raffinose family oligosaccharides**

(RFO) comprising raffinose, stachyose and verbascose; and (ii) the galactosyl cyclitols or **galactinol series oligosaccharides** (e.g. pinitol, ciceritol). Partial breakdown products of polysaccharides, such as **starch**, are also referred to as oligosaccharides.

Olive

1. The crop

The olive tree (*Olea europaea*, 2n = 46, Oleaceae), a perennial evergreen, is the ninth most important world source of edible oil. Ninety-eight per cent of the production is centred in the Mediterranean area with the EEC countries of Spain, Italy, Greece, Portugal and France contributing about 84% of the total, with North Africa and the Near East countries contributing 8.5 and 5.5%, respectively. The remaining 2% is produced in the Americas. (**See: Oilseeds – major,** Table O.1)

The oil is pressed from both the fleshy pulp and seed (woody pit) by various mechanical techniques and the remaining oil (8%) in the meal (pomace) is solvent extracted. Oil yields can vary from 15 to 35% depending on the cultivar, degree of ripeness and the extraction method used. There are various classes of olive oil. 'Virgin oil' is obtained by the exclusive use of a mechanical extraction process while 'pure olive oil' is a mixture of virgin and refined virgin oil. Olive-pomace oil, of which there are various categories, is solvent-extracted oil from the pomace. (**See: Oils and fats; Oilseeds**)

2. Origin

Its ancestors are not known but the cultivated olive is thought to have originated in the mountainous region of the eastern Mediterranean where a wide variety of **kernel** types that date back to the 4th millennium BC have been found. There is evidence that the olive oil industry was well established from Palestine and Syria to Greece in the mid- and late Bronze age. The trees are very long-lived and can regenerate by suckering from the root. Some Tunisian trees are thought to have persisted from Roman times.

3. Fruit and seed

The fruit is a fleshy, oval-shaped **drupe** containing a brownish-black pit or seed. About 50% of the weight of the fruit is water, contained mostly in the **mesocarp** or pulp. The epicarp cuticle adheres tightly to the mesocarp. Oil-containing parenchyma tissue forms the mesocarp with radially elongated cells adjacent to the endocarp. Also present are branched stone cells. Vascular tissue is present among the mostly flattened, brown pigmented parenchyma of the **testa**. The abundant **endosperm** consists of parenchyma containing **oil bodies** and **protein bodies** with the innermost cell layers flattened. The thin **cotyledons** of the **embryo** also contain many oil and protein bodies. Although the pits or seeds (Fig. O.4) may contain 12 to 28% oil, the pulp contributes about 97% of the extracted oil.

4. Uses

About 90% of the olives produced are utilized for oil production (**see: Oilseeds – major**, Table O.2, for composition). The remainder are treated, pickled and eaten as fruit. The oil is used primarily as a salad and cooking oil, but it is also utilized by the cosmetic and pharmaceutical industries. The pomace can be fed as animal feed but has a relatively low feeding value due to its low **protein** content and the presence of pits. (KD)

Brousse, G. (1989) Olive. In: Robbelen, G., Downey, R.K. and Ashri, A. (eds) *Oil Crops of the World*. McGraw-Hill, New York, USA, pp. 462–474.

Firestone, D., Fedeli, E. and Emmons, E.W. (1996) Olive oil. In: Hui, Y.H. (ed.) *Bailey's Industrial Oil & Fat Products*, Vol. 2, 5th edn. John Wiley, New York, USA, pp. 241–270.

Vaughan, J.G. (1970) *The Structure and Utilization of Oil Seeds*. Chapman and Hall, London, UK.

Ololiuqui

Known also as the Mexican morning glory (*Rivea corymbosa*, Convolvulaceae) seeds of this central American species contain lysergic acid amide and therefore have hallucinogenic properties (**see: Pharmaceuticals and pharmacologically active compounds**).

The seeds were used by the Aztec priests to provoke psychedelic effects for communication with the gods. Because the dosage is critical, the ground seeds were ingested only by experienced persons and less than eight seeds were utilized for psychedelic experiences, which last for 6 to 8 h. The traditional practice is to soak the finely ground or chewed seeds in a small amount of water for some hours after which both the water and seeds are consumed.

Onion

See: Alliums

Open pollination

See: Pollination

Operculum

Plug-like structure in the micropylar region of the seed which detaches during germination by circumscissile dehiscence; also called germination lid, imbibition lid, seed lid, micropylar (**micropyle**) collar, etc. Opercula are best known from **monocotyledonous** families (e.g. Araceae, Marantaceae, Commelinaceae, Arecaceae). (**See: Imbibition lid; Structure of seeds – identification characters**)

Fig. O.4. Seeds of olive (scale = mm). Photograph kindly produced by Ralph Underwood of AAFC, Saskatoon Research Centre, Canada.

Orchids

Orchids are considered to be the largest and one of the most advanced plant families because of their complex interactions with pollinators and **mycorrhizal** symbiosis. Orchids are found on every continent except Antarctica and in diverse habitats ranging from arctic tundra, to deserts and rainforests. Some of the best-known orchids are epiphytes but many species are terrestrial. There are several centres of diversity for sub-families within the Orchidaceae, but the most rapid and largest species radiation has occurred in tropical America, Africa and Asia. The evolution of these species can be largely attributed to the specific interaction of orchids with their pollinators. Another contributing factor to orchid evolution is their mycorrhizal association, which in its most advanced form has led to the evolution of achlorophyllous orchids that are fully parasitic on their fungal partner.

The popularity of orchids as ornamental plants is growing. The showiness of tropical orchid flowers has led to the development of a large cut flower and potted plant industry. **Vanilla**, produced from the long, thin capsules of the vanilla orchid is also a large industry in several South American countries.

The discovery of *in vitro* techniques for rapid propagation of tropical orchids has caused an explosion in the production of new hybrids. Orchids cross readily and inter-specific and inter-generic crosses are possible within closely related genera of subfamilies. Though crosses are possible within subfamilies, crosses between sub-families are often unsuccessful. **Hybrids** are not common in nature because many species are isolated due to pollinator specificity. Despite the highly evolved interaction between orchids and their pollinators, apomictic (**see: Apomixis**) and self-pollinating species are also known.

Orchid seeds are small (some extremely so) and notoriously difficult to germinate. They range in size from 50 μm to several mm in length. Weights are in the microgram range with the heaviest seeds averaging 40 seeds/g, the lightest about 1 million seeds/g. Their small size is an adaptation for long distance dispersal and a survival mechanism. (**See: Seedcoats – structure**, Fig. S.16; **Structure of seeds – identification characters**, Fig. S.69e)

At pollination, ovules are underdeveloped and only three placental ridges exist in the ovary; some species produce ovules in various stages of development. After pollination, **ovule** development proceeds so that fertilization can take place. Double fertilization occurs as in other angiosperms but the endosperm fails to develop (**see: Fertilization**). Even after fertilization, the **zygote** may rest for some time before division and embryo development: the time to maturation of orchid seeds in some cases can require over 1 year. At maturity, the seeds consist of a **testa** surrounding a tiny embryo in the globular stage with no defined cellular organization into embryonic tissues. The number of seeds produced in a **capsule** ranges from 50 or less in *Rhizanthella* to several million in the genera of *Stanhopianeae*. During germination, which can take from several days to many weeks to be completed, the embryo swells to form an ovoid protocorm and eventually rhizoids emerge from epidermal cells. At this time, all orchids are mycorrhizal, requiring infection through the rhizoids for continued development of the protocorm. The chalazal end of the embryo gives rise to the shoot apex, forming a leaf primordium, and adventitious roots that are contiguous with the protocorm, and eventually a rhizome. Orchid seed germination is regarded as the swelling and greening of the embryo to form the protocorm and not the formation of a rhizome.

In vitro germination requires that seeds be sterile, but sterilization is difficult with such small and delicate seeds. Those harvested before dehiscence of their enclosing capsule will germinate, and since seeds within the capsule are free of undesirable microorganisms surface sterilization of the capsule before seed removal is usually sufficient for acquiring sterile samples.

Long-term storage of orchid seeds is a valuable preservation technique. Some tropical orchids are intolerant of dry storage though many hybrids can withstand desiccation and storage with little loss in viability. For certain species 5% relative humidity and warm storage will reduce viability. Freezing (5% moisture content and temperatures below −40°C) can prolong storage for a number of species. In preparation for ultra-cold storage, seeds are dried over $CaCl_2$ or silica gel before submergence in liquid nitrogen (−196°C).

Seeds of some tropical orchids are **recalcitrant** and cannot withstand desiccation. Tropical orchid species which germinate readily on media *in vitro* have no dormancy. Terrestrial orchids often are dormant, which may be because of their lipid-rich, hydrophobic testa. Orchid seeds may be dormant because they have an immature embryo at the time of shedding from the parent plant. **Abscisic acid** in orchid seeds may play a role in physiological dormancy. Some orchids require **afterripening** to become fully germinable.

Several treatments reportedly improve the germination of temperate terrestrial orchids. Seeds of *Spiranthes cernua* respond favourably to cold storage while *Goodyera pubescens* do not. Warm incubation followed by cold storage promotes germination of *Epipactis palustris*. **Prechilling**, sodium hypochlorite, and **cytokinin** treatment may enhance the germination of *Cypripedium macranthos*. The treatment of seeds with an oxidizing agent such as sodium hypochlorite degrades the testa, allowing imbibition to occur. Cold and warm **stratification** may also soften the testa.

Since the discovery in 1922 by Knudson that it is possible to germinate orchids *in vitro* without the requirement for a fungal symbiont, this has become the preferred means of propagation for many species. Several different media have been developed, which in general are low in mineral salts and high in organics. Vitamins also seem to be important and many media include yeast extract as a source of B vitamins. Complex organics are also included in the form of coconut water (liquid endosperm) and ripe banana extracts.

Temperate terrestrial orchids do not germinate well on asymbiotic media and germination with a symbiotic fungus has been reported as a viable alternative. Orchid mycorrhizae are all Basidiomycetes with a large percentage in the teleomorphic genera of the form genus *Rhizoctonia*. The germination of a tropical epiphyte with a mycorrhizal fungus has also been reported. Unfortunately, difficulties producing orchids from seed have lead to the collection of adult plants, causing some species to become nearly extinct. (JIW, GEW)

Arditti, J. and Ghani, A.K.A. (2000) Numerical and physical properties of orchid seeds and their biological implications. *New Phytologist* 145, 367–421.
Knudson, L. (1922) Nonsymbiotic germination of orchid seed. *Botanical Gazette* 73, 1–25.
Pritchard, H.W., Poynter, A.L.C. and Seaton, P.T. (1999) Interspecific variation in orchid seed longevity in relation to ultra-dry storage and cryopreservation. *Lindleyana* 14, 92–101.
Rasmussen, H.N. (1995) *Terrestrial Orchids: From Seed to Mycotrophic Plant*. Cambridge University Press, New York, USA.

Organic seed

There are considerable technical challenges in the production of healthy organic seeds with high germination quality. Current regulations in many countries are moving towards the point where all planting material used for organic farming should be produced under approved organic methods, and a similar consideration also applies to materials used for seed **treatment, filmcoating** and **pelleting**.

Though definitions and interpretations vary in detail from country to country, 'organic' farming (most languages employ the alternative terms 'biological' or 'ecological') involves using cultural, biological and mechanical methods in preference to synthetic materials – avoiding artificial chemical fertilizers and dealing with pests and diseases through prevention and monitoring rather than treatment with **pesticides**. (Thus, by and large, 'organic' farming excludes using the products of synthetic 'organic' chemistry as direct crop inputs, and is also distinct from general 'organic' biochemistry: conflations of meaning that can sometimes confuse – with the reader left to judge ambiguity by context, as in this book.) Certification of organic production systems is a key requirement. **Genetic modification** (GM) cannot be used in certified processes, and discussion continues about whether organic seeds need lower thresholds for contamination with GM seeds than conventional seeds.

1. Seed production

Organic seed production has a greater risk of contamination with weed seeds, and seedborne diseases may accumulate to become a problem after several seed multiplication cycles. After the seed is sown, there is likely to be stronger competition from weeds and fertilization with organic manures, which mineralize slowly in the cold soils; this puts increased value on high seed **vigour** and faster developing root systems. For a large number of crops it is currently very difficult to reach the desired seed quality standards. Biennial vegetable crops, such as cabbages, carrot and onion, present particular difficulties because of the extended time needed for seed production and the need for **vernalization** to induce flowering, which both greatly increase the risk of latent **pathogen** infections. In Denmark, for example, the emergence of organically produced wheat seeds is often affected by seedborne *Fusarium*, which is treated with fungicides in conventionally produced seeds. In some situations, early harvest is a possible measure to improve seed health. At present in the EU, due to lack of sufficient organically produced material of the desired varieties, particularly of vegetable crops, a temporary derogation system continues to allow the sowing of organic seeds of some crops produced using conventional chemicals providing there is no postharvest chemical treatment.

2. Seed treatment

Several alternative organically acceptable seed sanitation methods are under investigation and development, for use singly or in combinations (see Table O.5). An important practical concern in some of these methods includes direct damage to seed or possible phytotoxicity affecting germination quality, which can vary between different seed lots. Another concern is the risk of eliminating naturally beneficial microorganisms as well as pathogens, leaving a more or less 'sterilized' seed, which weak soil-borne pathogens might exploit. **(See: Pesticides)**

3. Disease contamination thresholds

However, for many seedborne diseases there are no technically or economically feasible organically acceptable control methods. In practice, the current procedure is to analyse for seedborne pathogens and discard seed lots if disease infections exceed accepted threshold tolerance levels. In some recent years in Denmark up to 90% of the seed lots of certain crops

Table O.5. Acceptable seed sanitation techniques in the production of organic seed.

Type of treatment	
Hot water	Typically immersion in water at around 50°C for about 30 min. Precise control of exposure important. Used for many years, on a very limited scale in certain crops with small and high-value seeds, because of the cost and complication of drying afterwards and the need to avoid the risk of seed damage.
Hot humid air	The treatment of cereal seeds using a moist heat treatment (~2 min at 50–60°C, precalibrated for each seedlot).
Compounds of natural origin	Examples include: skimmed milk powder, wheat flour and powdered seaweed, mustard extracts, organic acids (in the chemical sense, such as acetic acid), essential oils (such as from thyme), to control certain seedborne bacteria and fungi. In organic agriculture systems, unrefined plant extracts tend to be preferred to single chemical compounds isolated from plants.
Salts	As an example, copper-salts (**Bordeaux mixture**, a mixture of copper sulphate ($CuSO_4$) and hydrated lime ($Ca(OH)_2$), which has been used as fungicides for two centuries) are acceptable as seed dressings in some countries.
Antagonistic microorganisms	Such as *Pseudomonas chlororaphis* approved in some countries. (**See: Rhizosphere microorganisms**)
Other approaches	Electron beams and microwaves.

have been discarded on this basis. (**See: Health testing**) In many cases, however, the same threshold values and analytical methods are used as those used for conventional seed production and, as a consequence, a large proportion of organically produced seed lots are rejected as unsuitable for sale. This is leading to recognition of the need to re-evaluate practical economic threshold tolerance values for seedborne pathogens for organic crop production: relating measured seed contamination levels to potential disease risks in practice.

(PH)

Borgen, A. and Kristensen, L. (2001) Use of mustard flour and milk powder to control common bunt (*Tilletia tritici*) in wheat and stem smut (*Urocystis occulta*) in rye in organic agriculture. In: Biddle, A.J. (ed.) *Seed treatment – challenges and opportunities*. Proceedings from BCPC Symposium No. 76. BCPC, Farnham, UK, pp. 141–150.

Codex Alimentarius Commission of the FAO (1999) *Guidelines for the Production, Processing, Labelling and Marketing of Organically Produced Foods*, CAC/GL 32. (Downloadable from www.fao.org/DOCREP/005/Y2772E/Y2772E00.HTM)

Lammerts van Bueren, E.T. (ed.) (2004) *Proceedings of the First World Conference of Organic Seed Production*. IFOAM, Bonn, Germany.

Olvång, H. (2004) Early harvest – a possible method for production of healthy seed for organic farming. *Seed Testing International, ISTA News Bulletin* 127, 22–25.

Ornamental flowers

1. History and origin

Humans probably have been attracted by the beauty and splendour of flowers as long as human civilization has existed. Flowers still enrich daily lives in many contemporary cultures. Domestication of flowering plants started with human civilization as early as Mesopotamia and the Roman Empire in the western world. In late medieval periods in Persia and Turkey, gardeners started to develop native bulb species. In that time in Japan and China large numbers of beautiful gardens had already arisen. It is because of these early activities of mankind that many present-day ornamental plant species have a long and distinguished past. Also from Europe, as early as in the 16th century, modern ornamental plant species may have arisen.

Pansy or *Viola* has a history which began in the early 19th century in Europe with **hybridization** of *V. tricolor* and *V. lutea*, both native to Britain. *Impatiens* arrived from Asia, Africa and North America around the late 19th century. *Begonia* was discovered by a Franciscan monk in 1690 in the West Indies. *Cyclamen persicum* reached us in the early 17th century from Persia. Early *Pelargonium* species came from Africa at the beginning of the 17th century to Holland and later to England. Petunia was discovered in Argentina during the early 19th century. *Primula*, as a native species, was already common as a garden plant in Britain, as early as the 15th century. Ongoing breeding of related species, including hybridization, has led to a countless number of contemporary varieties with many different flower shapes and colours. These varieties are the commercial pillars of the estimated US$ 250–350 million ornamental flower market, which comprises bedding plants (annuals and biennials), potted plants, cut flowers and perennials. Flower seeds are sold and used in almost every country of the world; however, the main markets are Europe, the USA and Japan. (**See also: Orchids**)

2. Seed structure and development

Ornamental flower seeds are primarily **dicotyledonous** seeds (**see: Structure of seeds**). The thickness of the **cotyledons** may vary among species and is often related to the amount of **endosperm** present. Seeds with endosperm only a few cell layers thick have thicker cotyledons. The seedcoat can consist of different layers (**see: Seedcoats – structure**). The seedcoat may also operate as a barrier against water and oxygen uptake (e.g. *Pelargonium*) during germination. The size of the seeds varies greatly, ranging from the larger *Cyclamen* seeds (2–3 mm) to the very fine *Petunia* (0.5–1 mm) and *Begonia* seeds (smaller than 0.5 mm).

3. Quality

Seed quality can be defined in relation to the usability of young plants obtained by growers, which is related to seed **vigour**. Seed quality is of high economic importance to the flower seed companies and growers. The usability of seed lots may vary greatly between different crops but also between batches within a crop.

Seed quality is influenced by a large number of factors, and is the net result of breeding, seed production and treatment, though in flower seeds it can be only partly controlled through these means. The chance of producing a vigorous seed batch can only be maximized but not predicted. The end results need to be established through testing before a seed batch can be sold to a grower. For this, several tests can be applied.

(a) *Soil*. Seed can be germinated on soil in trays and the seedlings grown for a designated number of days in greenhouses or germination cells, after which the usability and germination percentage are counted. This can be carried out manually or by so-called vision systems in which a camera and computer software counts the number of satisfactory plants. These tests give a usability number through germination percentage and uniformity or a combination of both in a vigour index.

(b) *Paper*. Seeds can also be germinated on paper to test the viability of a seed lot, and less of vigour.

(c) *Accelerated ageing*. The potential shelf life, longevity or storage life of seeds can be tested through an accelerated ageing test, like a **controlled deterioration** test. Seeds are kept at elevated temperature and relative humidity for several hours or days and germinated afterwards. This can be predictive for the stress resilience of a seed lot or the shelf life. However, for a precise determination of shelf life, a storage experiment, whereby seeds are stored for prolonged periods under low temperature and relative humidity, is more accurate.

4. Breeding

Flower seed companies aim at providing professional growers with new products fulfilling the requirements of the consumers and landscapers. The originator companies are developing varieties to meet the product profiles demanded by the markets.

The diversity in markets, crop genetics and available breeding tools determine the efforts and results of variety development. Crops with large potential markets, a high genetic variability and ease of hybridization have generated hundreds of commercial **F_1** hybrid varieties, and some F_2. Other crops with limited use-potential, a narrow colour range, or which are difficult to hybridize, have a lower breeding input and are available only as open-pollinated varieties.

The start of all breeding is the exploration of genetic diversity. In the case of seed-propagated flowers, the genetic variation might still be obtained by the use of species collections. Intra- and inter-specific crosses are being applied to generate new forms and colours. Recombination breeding of existing varieties is often used to combine wanted features in one phenotype.

The next step in the process is narrowing down the variation. In the case of open-pollinated varieties, different sibbing methods are used. In cases of line development for F_1 hybrid varieties, both selfing and sibbing are used. The choice of the line development method depends on the degree of inbreeding depression. Some crops suffer strongly from inbreeding after selfing and may lose vigour or fertility with each generation. The only option then is to use very extensive ways of sibbing and avoid genetic fixation. For some crops, the alternative is vegetative maintenance of clones to be used for hybridizing. Once lines are sufficiently stable and uniform, test-crosses are made to determine combining ability. In many cases, male sterility is used to develop female parent lines in order to prevent self-pollination during the hybrid production process (**see: Cytoplasmic male sterility**). Once the experimental varieties are tested positively for performance under a range of conditions and can be reproduced consistently, the decision can be taken to start commercial production.

The maintenance and reproduction of varieties requires continual attention. The fixation of inbred lines makes it easy to control the genetic uniformity. However, the open-pollinated varieties require a lot of selection effort to maintain variety uniformity and stability.

The use of advanced technologies in ornamental seed breeding is fairly limited compared to field crops. Inter-specific crosses using **embryo rescue** techniques are commonly used. Mutation breeding is not very common, due to the available genetic variation. The application of molecular marker techniques in the breeding process is gaining popularity. The development of transgenic, seed-propagated ornamentals is very rare. Many hurdles still have to be overcome before a genetically modified variety comes to market, if at all.

5. Production

Production of ornamental flower seeds is carried out at locations worldwide, depending on the crop and production costs, often in countries such as Indonesia, Chile, Guatemala and Kenya, which may be managed by flower seed companies or by vendors. Production in different parts of the world enables flower seed companies to choose the best climate for each crop and also to produce year round.

Production of seeds can be carried out both in greenhouses or the open field. Two different types of production can thereby be distinguished:

- F_1 hybrids: two parent lines (pure inbred lines) are grown. Parent plants may be emasculated by hand after which the parent lines can be hand crossed. Emasculation and crossing has to be carried out by skilled operators to prevent self-pollination of the mother lines, thus polluting the seed batches. For some crops like *Pelargonium* or *Impatiens*, male sterility on the mother plant can be used to eliminate the need for emasculation.
- Open-pollinated crops: plants are pollinated by bees or other insects and are grown in the field at appropriate **isolation distances** from other production sites of the same crop to avoid unwanted cross-pollination.

The parent lines of any production have to be well maintained as seed quality can depend on the health of the mother plant, and stress caused in the greenhouse by, for example, improper watering, soil quality, extreme temperatures or insufficient nutrient supply, or in the open field, rain and cold temperatures.

Harvesting of seeds can be performed by hand, which is done for pansy, *Impatiens* and petunia. Seeds may be extracted mechanically after drying the crop in the field or shaken off by hand. If seeds are harvested before full maturation this can lead to a decreased seed quality, for seeds are immature and undersized. Seed quality can also be compromised by late harvest, which can lead to seed loss (*Impatiens*) but also to deterioration due to weather circumstances, or infection with fungi. Some seeds are dried after harvesting, if not in the field. The combination of relative humidity (drying rate) and temperature, i.e. slow or fast drying, can have a profound effect on the quality of those seeds.

6. Treatments

Seeds may undergo several enhancements to improve seed quality and performance. These treatments may include seed cleaning and upgrading, seedcoat abrasion (**scarification**), **priming**, **pre-germination**, treatments against diseases (see: section 8), **dormancy breaking**, **pelleting**, **coating**, or combinations of different techniques.

Seed batches can be upgraded through various ways of cleaning the seeds, or sorted in different fractions based on seed calibre/size, density or shape. *Impatiens* seeds, for instance, can be calibrated in different size fractions, which may be sold separately for batches with greater uniformity of cotyledon sizes. Sorting seeds on a density basis is often used to upgrade batches to higher levels of germination. However, seed upgrading is often not sufficient to provide the growers with the seed performance they demand.

Additionally, seeds can be physiologically enhanced by hydration techniques including priming and drying back after treatment to the original water content. Priming can be achieved in several different ways, either through osmotic priming, solid matrix priming or through drum priming. Flower seeds that are primed show a faster and more uniform germination and may show reduced dormancy. Furthermore, they may germinate better under adverse conditions. Priming does have one major disadvantage, however, in that seeds might have shorter longevities compared to untreated seeds.

A different physiological enhancement is pre-germination, in which seeds are allowed just to germinate (the radicle tip protruding through the seedcoat) and then are separated from the non-germinators. The pre-germinated seeds are sold to the grower, either in a wet state or with induced **desiccation tolerance** as dried seeds. The wet seeds have a limited storage lifespan whereas dried seeds can be stored for periods over 6 months at the appropriate temperature and relative humidity. This pre-germination technique is offered commercially for *Impatiens* and pansy. Pre-germinated seeds have additional benefits compared to primed seeds – very high germination counts (up to 100%), fast plant stand, and high uniformity within and between batches.

Physiological treatments may also include dormancy breaking, which can be performed by chilling for periods of days or weeks, by (red) light (**see: Phytochrome**), or through addition of **hormones** such as **gibberellins**. The applied method depends on the type of dormancy and the crop. Certain degrees of dormancy can still be present in some hybrid varieties, but do not present a major problem. However, with the continuing effort in developing new varieties, a breeder may use species closer to wild species; in these circumstances it is likely that dormancy can become an issue. Dormancy is still present in many perennial species. Here, different types of dormancy can be observed, from deep physiological dormancy, in for instance *Lavandula*, or morphological dormancy combined with physiological dormancy in, for instance, *Delphinium*. Combinations of different types of dormancy frequently can be found. *Pelargonium* seeds have a dormancy imposed by a thick seedcoat; scarification with acid is used to remove this.

Physical seed treatment includes pelleting and coating, though the use is less common than in vegetable and field crops. Many flowers seeds are small, with *Begonia* and *Petunia* being among the smallest. To enable mechanical sowing at the grower level, seeds are encrusted or pelleted to increase their size, and also, in the case of pelleting, to change their shape. Seed **filmcoating** or **encrusting** can be applied to protect seeds against seedborne fungal pathogens. The number of active ingredients for crop protection that can be applied to flower seed is very limited due to legislation in different countries.

7. Young plant production – plug growing

Growers, in many different countries, vary from small family businesses to large-scale plant factories (20 ha of greenhouse space is no exception). During the past century, and especially from the 1950s onward, many growers who originally started on a small scale expanded their capacity and expanded into large-scale professional establishments.

Many professional growers have shifted from growing seeds in flats or prepared-ground beds to sowing seeds in plug trays, where seeds are sown in individual cells. For crops with a lower seed quality, the **cell tray** may be sown with up to eight seeds per plug. Many professional growers sow mechanically with an in-line automatic filling of the trays with soil and automatic watering.

For several reasons, many seed crops benefit from being covered with soil, cork, **vermiculite** or plastic media. Large seeds such as pansy, *Impatiens* and *Cyclamen* benefit from the moisture contained by the covering media, provided that there is a good oxygen supply to the seeds. Covering media may also limit light penetration to the seeds (**see: Light – dormancy and germination**). This is beneficial for seeds that require darkness for germination (e.g. *Cyclamen*), but it also regularizes root growth into media (roots may be sensitive to light and the direction of growth is upset) as in crops such as pansy and *Geranium*.

Different crops require different germination conditions: this may require the placing of trays directly in the greenhouse or stacking them, as the germination phase is carried out in germination cells, during which it is critical to keep the relative humidity and temperature in the most optimal range. This is easier to achieve in a germination cell than in a greenhouse environment. Young plants have to be treated with great care to maximize uniformity between plugs and control diseases or pests. During early growth or during germination, trays may still be covered with plastic (black or white) or porous types to keep a high moisture level. To secure compact, firm growth of the young seedlings, **plant growth regulators** may be applied. As young seedlings, plants may be transplanted to trays with larger cells to establish continuing growth. Gaps in trays can be filled: empty plugs are blown out and replaced with young plants from a 'donor' tray, resulting in trays with essentially 100% usable plugs. Often, machines with image analysis and robotics perform this process, thereby saving the high costs of human labour. The full-grown plugs are sold to growers who take the young plants to a more mature stage for sale to the end-consumer through dedicated channels such as garden centres and supermarkets.

8. Seed health

Research on seed health of flower seeds is limited. In general, numerous fungi can be present on seeds without being pathogenic. Infection with fungi may take place during seed growth and production. Three pathways of infection can be roughly distinguished.

1. Infections of *Verticillium* spp. or *Fusarium oxysporum* pathovars often go unnoticed if the seed-producing plants are well nourished and well watered. During ripening of the seeds it will be very difficult to spot the wilting due to infection or the normal decay of plants. Seeds harvested from such plants are likely to be infested by these fungi. An example is Cyclamen infested by *F. oxysporum* fsp *cyclaminis*. Despite the fact that the pathogenicity of this fungus has never been established, it may cause decreased quality of Cyclamen seeds.
2. Infections of true **pathogens** or weak pathogens can also take place during pollination and seed set. Bees or other insects can carry spores of fungi to the open flower. Pollination and fungal infection can go hand-in-hand. Examples of these late infections are *Colletotrichum* spp. for *Cyclamen*, *Lobelia* or *Gaultheria*, but also *Alternaria* spp. for *Lobelia*, *Viola* or *Impatiens*. Infections are usually located on the surface of seeds and thus fairly easy to control. But here also, pathogenicity is uncertain.
3. Infection of weak pathogens or saprophytes during post harvest is the third possibility. If seeds are not stored properly, infections of *Alternaria* spp., *Phoma* spp. or even *Botrytis* spp. (humid conditions) can occur readily. Also, ineffective cleaning of fruits bearing the seeds can lead to high infection rates by fungi such as non-pathogenic Mucorales, *Penicillium* spp. and other species. It is not likely that these fungi lead to **damping-off** in the seedling stage.

Based on these three possibilities, a cleaning strategy can be determined: spores of fungi of pathway 2 and 3 can be washed from seeds in running tapwater, treated in hot-water or rinsed with disinfectants. For seeds infected by *Verticillium* spp. or *Fusarium oxysporum*, more extensive efforts need to be taken since seeds are internally infected. (PS, WvK, PDR)

McDonald, M.B. and Kwong, F.Y. (2004) *Flower Seeds: Biology and Technology*. CAB International, Wallingford, UK.

Ornithochory

Dispersal of **diaspores** (seeds and fruits) by birds.

Orthodox seeds

Seeds that remain viable (**see: Viability**) when desiccated to 5% moisture content (fresh weight basis) or less, corresponding to a **water potential** of around −250 MPa. The **longevity** of orthodox seeds increases in a predictable way with decreasing moisture content and reduction of temperature. Orthodox seeds acquire the ability to withstand drying at around mass maturity (maximum dry weight). Prior to this, immature orthodox seeds are desiccation intolerant and therefore behave like **recalcitrant** seeds. Even when **desiccation tolerance** has been acquired, some individuals may be sensitive to the extent of desiccation and therefore behave like **intermediate** seeds. Maximum tolerance to drying, and maximum longevity, are usually acquired in orthodox seeds close to the point of natural seed dispersal. (RJP, SHL)

Orthologous genes

Also called orthologues. **Homologous genes** that are conserved in different species, and that have evolved from a single common ancestral gene. (**See: Paralogous genes; Preharvest sprouting – genetics**)

Orthotropous

Also called atropous. An **ovule** or seed is called orthotropous when the **funiculus, chalaza** and **micropyle** lie in a straight line, the hilum, as a result, lying adjacent to the chalaza. Such ovules have no **raphe**. They are typical for **gymnosperms** but they are also found in 20 angiosperm families (e.g. in Juglandaceae, Piperaceae, Polygonaceae, Urticaceae) and they are mostly in one-seeded fruits. (**See:** Fig. R.3 in **Reproductive structures, 1. Female**)

Osborne fractions

The concept of classifying proteins based on their extraction and solubility in a series of solvents developed during the 18th and 19th centuries but was formalized by T.B. Osborne who worked on the Connecticut Agricultural Experiment Station from 1886 to 1928. He published detailed studies of seed proteins from 32 plant species leading to the definition of four protein fractions that are extracted sequentially. The first fraction, termed albumins, is extracted in water; the second, globulins, in dilute saline solution (usually 0.5 M NaCl). However, it is also possible to extract these two groups of proteins together with dilute saline and then separate them by dialysis against distilled water, the globulins being precipitated and recovered by centrifugation while the albumins remain in solution and can be recovered by lyophilization. This is now regarded as the best method of preparation of these fractions since albumin fractions extracted directly with distilled water contain small amounts of contaminating globulins resulting from the presence of salts in the seed tissue. The third fraction, prolamins, is extracted with alcohol/water mixtures. Traditionally 60–70% (v/v) ethanol was used but it is now known that other alcohols (notably 50% (v/v) propan-1-ol) give more complete extraction, with a reducing agent (e.g. 2% (v/v) 2-mercaptoethanol) being required to extract prolamin subunits which are present in insoluble polymers stabilized by disulphide bonds. If a reducing agent is not included, these proteins are present in the fourth fraction, the glutelins. This fraction essentially comprises insoluble proteins that are only extracted after reduction of covalent disulphide bonds and denaturation. This is traditionally achieved using alkaline or acidic solvents but more recently with chaotropic agents (which disrupt the non-covalent interactions that stabilize the structures and interactions of proteins, notably urea) and/or detergents (notably sodium dodecylsulphate, SDS). Although the albumin and globulin fractions are complex mixtures they may also contain well-defined groups of storage proteins. Similarly, the prolamins comprise storage proteins in cereals and other grasses, but do not form a valid fraction in other species. The glutelins may contain prolamin or globulin storage proteins which are not efficiently extracted by the previous solvents, together with a mixture of structural and metabolic proteins.

Although Osborne fractionation is still widely used it must be borne in mind that the efficiency of extraction of components is not only affected by their intrinsic properties (e.g. the major storage protein of oats is present in the glutelin fraction but is essentially a poorly soluble globulin) but also the conditions used. For example, fineness of grinding, volume to weight ratio, stirring conditions, temperature and whether the tissue has been defatted prior to extraction. It is usual to use at least 10 ml solvent/g tissue and to repeat each extraction three times, the extracts being combined for quantitation and analysis. (PRS)

(**See: Proteins – non-storage; Storage proteins**)

Osborne, T.B. (1924) *The Vegetable Proteins*, 2nd edn. Longmans Green and Co., London, UK.

Shewry, P.R. and Casey, R. (1999) Seed proteins. In: Shewry, P.R. and Casey, R. (eds) *Seed Proteins*. Kluwer Academic, Dordrecht, The Netherlands, pp. 1–10.

Osmopriming

A seed priming **enhancement** technique widely used in both research and the seed technology industry, which involves imbibing seeds for a controlled time at a specific temperature in aerated aqueous solutions of low water potential (dissolved salts or other osmotic agents), followed by rinsing and drying. Also known as 'osmotic priming' or 'osmoconditioning' (not to be confused with the set of **conditioning** technologies that are used to mechanically clean, purify and fractionate on the basis of seed physical properties). (**See: Priming, 2. Techniques and terminologies**)

Ovary

The enlarged, usually lower portion of the pistil (carpel, or carpels) containing the ovules. (**See: Reproductive structures, 1. Female**)

Over-seeding

An alternative term for **broadcasting** seeds on the soil surface; also used for **frost-seeding**.

Ovule

The coated (by one or two **integuments**) **megasporangium** (**nucellus**) of the Spermatophytes (seed plants) containing the **megagametophyte** (**embryo sac**). The ovule generally consists of one or two integument(s), the nucellus, the **raphe**, the **chalaza** and the **funicle**. After **fertilization** within the embryo sac, the ovule develops into a seed. Several types of ovules are present within the carpels of angiosperms, depending upon their orientation with respect to the funiculus, e.g. amphitropous, anatropous, campylotropous, hemianatropous and orthotropous. (**See:** Fig. R.3 in **Reproductive structures, 1. Female**)

The term ovuliferous means bearing or containing ovules. (**See: Evolution of seeds; Structure of seeds**)

Ovum

The female gamete (egg, egg cell). (**See: Reproductive structures, 1. Female**)

Oxalic acid

Oxalic acid in the form of calcium oxalate occurs in many plant tissues, and in such high quantities in the vegetative tissues of some plants that it is toxic to humans, e.g. in rhubarb leaves. It does not appear to reach toxic concentrations in seeds. It is present as raphide or druse crystals, and in seeds it is sequestered in **protein bodies**. The functions of calcium oxalate are thought to be as a store for calcium and protection against herbivory.

β-Oxidation

During **oil and fat mobilization** in the **glyoxysomes** of germinated seeds, 2-carbon units are sequentially removed from fatty acyl-CoA chains to generate acetyl-CoA molecules, the carbon source for further metabolism. The term β-oxidation is derived from the observation that the diminishing fatty acyl-CoA chains are attacked by the initial fatty acyl-CoA oxidase enzyme at the β (or Δ3) carbon.

Branched-chain 2-oxo acids, which are actually products of branched-chain amino acid (rather than fatty acid) catabolism, likely undergo β-oxidation in the mitochondria. (AJK)

Oxylipins

A family of compounds that protect plants from **pests** and **pathogens**. They are synthesized from the breakdown products of polyunsaturated fatty acids.

P

Pachychalazaly

A form of seed development that occurs when through intercalary growth the **chalaza** replaces the **seedcoat** partly or entirely. Pachychalazaly is considered a derived condition and has developed independently in many families (e.g. Sapindaceae, Lauraceae, Meliaceae, Tropaeolaceae). (**See: Seedcoats – structure**)

Palea

One of the two bracts in the grass floret. (**See: Floret,** Fig. F.7)

Palisade cells

Radially elongated and usually thick-walled cells often found in **seedcoats**. Palisade layers are usually present either in the exotesta (e.g. Fabaceae) or the exotegmen (e.g. Malvaceae, Euphorbiaceae). (**See: Seedcoats – structure**)

Palm

1. The crop

The oil palm tree (*Elaeis guineensis*, 2n = 32, Palmae or Arecaceae) is a **monocot** that produces more oil per unit land area than any other oil-producing plant. Commercial palm oil production is a relatively recent development. Few plantations existed and little use was made of palm oil as late as the 1920s. However, after the Second World War, plantations and palm oil production rapidly expanded to become a major commodity in world trade. Today palm oil is the second most important source of vegetable oil after **soybean**, and its production is growing faster than any other vegetable oil source.

The oil palm is adapted to tropical areas within 12 to 15° of the equator. The tree occurs naturally in areas between the rain forest and savanna, or following forest clearings, although production is primarily from plantations. Malaysia, Indonesia, Zaire, Nigeria and Ivory Coast are the main producing countries, with Malaysia by far the largest producer, accounting for about half the world's exports. (**See: Crop Atlas Appendix, Maps 2, 18, 22; Oilseeds – major,** Table O.1)

The oil palm is thought to have originated in West Africa and was spread throughout the tropics in the 18th and beginning of the 19th centuries through world trade and exploration.

2. Fruit and seed

Two oils with different **fatty acid** compositions are extracted from two separate parts of the palm fruit: the fleshy mesocarp and the kernel or seed (**see: Oilseeds – major,** Table O.2 for compositions). About 6 months after fertilization each female inflorescence produces 1100 to 1500 fruits that together may weigh 15–20 kg. The individual fruit is a red to reddish-orange, oval sessile **drupe** varying in size (2.5–5 cm long, 2–4 cm diameter) and weight (3–30 g). An oil-rich mesocarp, with a thin leathery skin, surrounds the kernel or seed (Colour Plate 12D) and is made up of parenchyma cells containing **oil bodies**. Fibre rings run longitudinally through the oil-bearing tissue and make up 11 to 12% of the mesocarp or pulp. Oil content of the pulp is about 56%.

The kernel or **nut** consists of a shell with three germ pores corresponding to the tricarpellate **ovary**. However, only one seed is normally produced. Nut size varies (≤2 to 3 cm long) depending on the shell thickness, of which there are three types. One type lacks a shell (*pisifera*) but is not grown commercially. The *dura* type has a thick **hull** making up 20–40% of the kernel weight. The *tenera* type, resulting from a cross between the shell-less and *dura* types, produces kernels with shells that make up 5–20% of kernel weight. The *dura* nuts are 2–3 cm long and weigh 4 g or more. The length of the *tenera* nuts is normally 2 cm or less and weigh about 2 g.

The shell consists primarily of black sclerenchyma cells and longitudinal fibres that pass through and adhere to it. A plug of fibre occurs in each of the three germ pores with a plate-like structure on the inner shell surface. The mesocarp of the *dura* palms constitutes 35–55% of the drupe while in the *tenera* types it makes up 60–96% of the drupe weight. A dark brown **testa** surrounds the greyish-white, hard and oily **endosperm** that fills most of the nut. The small straight (3 mm long) **embryo** lies with its distal end opposite one of the germ pores, embedded in the heavily pitted cells of the endosperm. The embryo apex has two differentiated leaves and a third rudimentary one. The **radicle** is poorly differentiated. Upon germination the **cotyledon** remains within the shell and a petiole exits through the germ pore to establish the seedling. The cotyledon becomes enlarged and, as a **haustorium**, absorbs the food reserves of the endosperm. Upon processing the dried kernel or nut yields about 50% oil and an oil meal.

3. Processing and uses

A simple oil extraction method involves fermenting the fragmented fruit bunches in water for 2–3 days, followed by boiling, pounding, addition of water, then collection of the supernatant oil. Advanced processing involves pressing and milling. Newly-extracted oil contains the yellow β-carotene, which as the precursor of vitamin A is nutritionally important, but industrial processing, etc. destroys the carotene. Kernel oil is extracted from the nuts.

About 90% of palm oil production is used for edible purposes. However, it is not suitable as a salad oil since its high content of saturated fatty acids makes it a solid at ambient temperatures in temperate climates. The oil is widely used in margarine blends, shortenings, as a frying fat, in baking, biscuits, coffee creamers and in ice cream making. The remaining 10% is used industrially mainly in the production of soaps and the manufacture of oleochemicals. (KD)

Basiron, Y. (1996) Palm oil. In: Hui, Y.H. (ed.) *Bailey's Industrial Oil & Fat Products*, Vol. 2, 5th edn. John Wiley, New York, USA, pp. 241–270.

Gascon, J.P., Noiret, J.M. and Meunier, J. (1989) Oil palm. In: Robbelen, G., Downey, R.K. and Ashri, A. (eds) *Oil Crops of the World*. McGraw-Hill, New York, USA, pp. 475–493.

Hartley, C.W.S. (1988) *The Oil Palm*, 3rd edn. Longman Group, Burnt Mill, UK.

Vaughan, J.G. (1970) *The Structure and Utilization of Oil Seeds*. Chapman and Hall, London, UK.

Panicle

See: **Inflorescence**

Paper

In seed technology, absorbent paper materials are used in official germination testing in the form of flat blotters, pleated papers and creped paper (for the **rolled towel test**), all manufactured to high specifications (**see: Germination Testing, 1. Substrates**). Paper is also used in seed mat and **tape** formats, as 'paper pots' in some seedling **transplant** production formats, and in a form of **mulching**.

Fibrous materials from seeds and fruits, such as cotton **linters** and **peanut** skins, are used to make paper, as is **guar** seed gum to give a denser surface for better printing, writing, erasing and folding properties. Edible 'rice paper' used to wrap or bake food is made from rice flour and tapioca, or alternatively from potato starch and vegetable oil (but, the mistakenly named drawing paper is made from the pith of the stem of shrub-trees, *Tetrapanax papyrifera*, *Broussonetia papyrifera* or other plant fibres).

Paralogous genes

Genes within the same species that have arisen from a common ancestor by duplication and subsequent divergence. (**See: Orthologous genes**)

Parasitic plants

About 3000 plant species worldwide within 16 families are parasitic, meaning they form connections into the vascular system of host plants from which they extract nutrients and water. This can be from the roots or the stem, depending on the parasite species. Most species (80%) have no chlorophyll and are totally dependent on the host plant (holoparasitic), while the remainder can photosynthesize to some extent (hemiparasitic). Some species, e.g. witchweeds (*Striga* spp.), broomrapes (*Orobanche* spp.) and dodders (*Cuscuta* spp.) are severe weeds of crops.

For most species the seedling will die if it does not attach to a host plant (although some hemiparasites can survive for some time and even occasionally flower and set seeds without a host), and this is especially true of small-seeded species which have poor food reserves. This dependence seems to have led to the evolution of host-recognition systems in many parasitic plants. In many cases, this involves chemical exudates from host plants which are necessary to trigger **germination**. Thus, the parasite is ensured of germinating only when a potential host is nearby.

There is not a great deal known about the identity of exudates that trigger germination. Strigol, a tetracyclic sesquiterpene, and various analogues, was the first chemical identified, and it stimulates germination of a wide range of parasitic species. However, this general activity is in conflict with the host specificity of many parasites and suggests other, more specific, chemicals are important. These host-derived chemicals, or xenognosins, are generally sampled from aqueous exudates of host roots. Few have been described and all in relation to the weed genera *Striga* and *Orobanche*: sorgolactone and dihydrosorgolene, both from sorghum, and which trigger germination in witchweeds; alectrol from cowpea and clover, which acts on witchweeds and broomrapes; orobanchol, from clover, which acts on broomrapes (Fig. P.1). (**See: Germination – influences of chemicals; Stimulants of germination**)

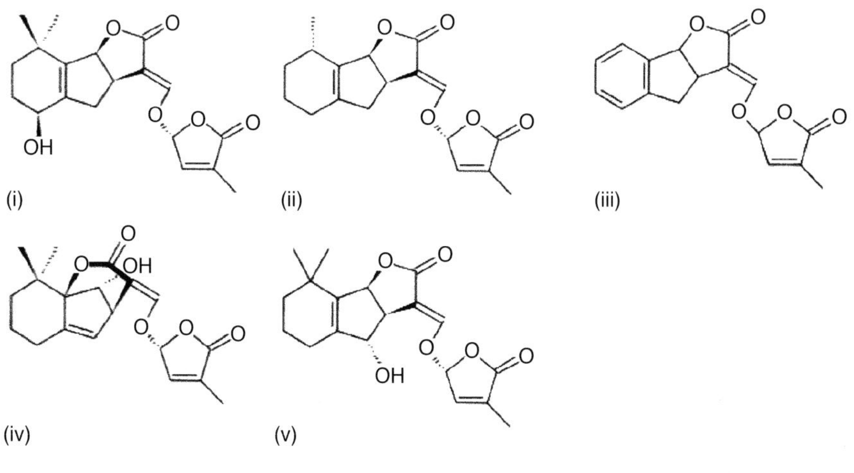

Fig. P.1. Stimulants of parasitic weed seeds. (i) (+)-Strigol; (ii) sorgolactone; (iii) a synthetic stimulant GR24; (iv) alectrol; (v) orobanchol. (Adapted from Matusova, R., van Moutik, T. and Bouwmeester, H.J. (2004) *Seed Science Research* 14, 335–344.)

Consistent with a process ensuring germination very near a host, these chemicals break down rapidly (e.g. by oxidation) and so are active over only short distances (e.g. 3–5 mm). It is not well understood how xenognosins trigger germination, but a process involving a redox reaction has been proposed, which perhaps induces biosynthesis of **ethylene** that in turn stimulates germination.

Seed **dispersal** strategies are linked to the parasitic life style. Holoparasites which specialize on particular host species tend to have much higher seed production (within the Rhinanthoideae 10,000–1 million) than hemiparasites with a range of host plants (20–10,000), with a resulting decrease in seed **size** (**see: Number of seeds – ecological perspective; Size of seeds**). This suggests a strategy within the specialist holoparasites which increases the chances of finding possibly rare and scattered host plants. Root parasites tend to be dispersed by wind, while many stem parasites, e.g. dodders (*Cuscuta* spp.) and mistletoes (*Loranthus* spp., *Viscum* spp.) are dispersed by fruit-eating animals, probably because seeds are more likely to be deposited on stems and branches. (JMB)

Baskin, C.C. and Baskin, J.M. (2001) *Seeds: Ecology, Biogeography of Dormancy and Germination.* Academic Press, San Diego, USA.

Bouwmeester, H.J., Matusova, R., Sun, Z.K. and Beale, M.H. (2003) Secondary metabolite signalling in host-parasitic plant interactions. *Current Opinion in Plant Biology* 6, 358–364.

Press, M.C. and Graves, J.D. (1995) *Parasitic Plants.* Chapman and Hall, London, UK.

Parental line

An inbred line or a single-cross **hybrid** used as either the male or female parent in a hybrid cross – also described, respectively, as the 'pollen parent' and the 'seed parent'. (**See: Line; Production for sowing, III. Hybrids**)

Parthenocarpy

A fruit that develops without **fertilization**, either producing seed that is derived entirely from the maternal tissues by **parthenogenesis**, or no seed at all. (**See: Seedless fruit**)

Parthenogenesis

A type of reproduction, also known as agamospermy, in which the egg or any cell of the female gametophyte is stimulated to undergo **embryogenesis** without **fertilization**, producing a seed whose genetic material is solely derived from the mother plant. Parthenogenesis can be variously induced by stimulation by pollen or by the exogenous application of **plant growth regulators**. In many *Citrus* and *Mangifera* (mango) species, for example, the **sporophytic** tissue surrounding the **embryo sac** can also be induced to form embryos, but these only mature if they are pushed into the sac at an early stage of their development (**see: Reproductive structures**). These **clonal** seedlings, which are identical with the mother plant, can be great nuisances in breeding strategies that are based on sexual progeny. (**See also: Somatic embryogenesis**) (PH)

Partitioning of molecules

Transfer of small molecules that possess both hydrophobic and hydrophilic characteristics from the polar aqueous phase (e.g. the cytoplasm) into the apolar lipid phase (e.g. **membranes** and **oil bodies**) and vice versa. Partitioning can play both a beneficial and detrimental role in **desiccation tolerance** according to the cell water status. (**See: Desiccation damage; Desiccation tolerance – protection; Desiccation tolerance – protection by stabilization of macromolecules**)

Patent protection technology

See: GMO – patent protection technology

Pathogens

Any family or species of a bacterial, fungal or nematodal microorganism or viral strain that lives and feeds parasitically on a seed or plant (or other larger host organism) and is capable of producing or inciting injury or disease or premature death under normal conditions of host resistance (**see: Pathogens – ecology**). More widely, a pathogen is *any* substance or factor causing disease although, by common acceptance, the term does not include insects; however, some seedborne nematodes and insects may be considered as pathogens.

Pathogens that infect plants may affect seed production indirectly by reducing biomass or photosynthetic capacity, or may directly affect seed development, causing abortion, rotting, or shrunken, discoloured or physiologically weakened seed (**see: Diseases**). Pathogens may thereby become externally or internally seedborne, and be spread to the next plant generation. Fungi (the majority of which are **necrotrophs**) and bacteria can inhabit the superficial or deeper tissues of the seed (as can non-pathogenic endophytes in grasses). Many viruses are located within the embryo, and hence are particularly difficult to eradicate without harming the seed. However, a pathogen may be seedborne without necessarily being the primary infection route for transmission of the associated disease. (**See: Pathogens – seedborne infection**)

Though it is common practice to identify plant pathogens by the disease they cause, strictly an infected seed, seedling or plant is said to be *diseased* only when symptoms of that disease become evident. High-, intermediate-, low- or non-virulent races, strains or isolates of the pathogenic species may exist, which differ in their ability to cause disease under defined conditions in the same host.

The agronomic importance of pathogens changes with time for various reasons, as climatic conditions alter, as seed is sown or crops are grown untreated on economic or ecological grounds, with the introduction of different levels of genetic tolerance or resistance in place of susceptible varieties, and as races mutate and adapt to that host resistance or to fungicides.

Following are brief individual descriptions of selected major genera containing fungal and bacterial pathogens that may affect seed development, germination or early seedling growth: *Acidovorax*; *Alternaria*; *Aphanomyces*; *Ascochyta*; *Aspergillus*; *Botrytis*; *Cercospora*; *Clavibacter*; *Claviceps*; *Cochliobolus*; *Colletotrichum*; *Diaporthe*; *Erwinia*; *Fusarium*; *Gaeumannomyces*; *Helminthosporium*; *Microdochium*; *Penicillium*; *Peronospora*; *Phoma*; *Phomopsis*; *Phytophthora*; *Plasmopara*; *Pseudomonas*; *Puccinia*; *Pyrenophora*; *Pythium*; *Rhizoctonia*; *Sclerotinia*; *Septoria*; *Tilletia*; *Ustilago*; *Verticillium*; and *Xanthomonas*. For additional comments on the effects of fungal, bacterial, viral and nematodal pathogens, see also the article on **disease**.

Fungal and bacterial taxonomy is complicated by the existence of alternative generic and specific epithet names as a consequence of revisions in binomial nomenclature, which can sometimes involve major changes. Also quite dissimilar names can be given to the sexual (teleomorph, or 'perfect state') and asexual (anamorph or 'imperfect state') stages of the fungal life cycle, which can both be used. For example *Pyrenophora graminea*, the seedborne cause of barley leaf stripe, is the sexual stage of *Drechslera graminea* – which was formerly called *Helminthosporium gramineum*. Here, an attempt has been made to use the most common nomenclature.

1. *Acidovorax*
Seedborne *A. avenae* subsp. *citrulli* is the major source of transmission of Bacterial Fruit Blotch in cucurbits, causing ruinous water-soaked lesions in mature fruit, with watermelon and melons being the most susceptible. Secondary spread, such as in the **transplant** house due to splash dispersal from overhead irrigation, can result in high numbers of infected seedlings in the field. For this reason, there is zero tolerance for contamination of less than one seed in 10,000 to 50,000 seeds, as determined in seedling grow-out tests. Another subspecies causes bacterial brown stripe in rice, and can cause occasional losses related to the inhibition of seed germination and to seedling damage (abnormal mesocotyl elongation) in nursery beds and boxes.

2. *Alternaria*
A genus containing pathogenic fungi causing various blights and rots in plants. *A. brassicae* and *A. brassicicola* are high-incidence seed-transmitted pathogens in many oil- and vegetable-brassicas as well as non-cruciferous plants, and a wider host range. Similarly *A. radicina* (causing black rot) is substantially seed-transmitted in carrot and other Apiaceae, where it can threaten seed production yields due to umbel or seed rotting. Heavy infection can result in reduced germination and seedling vigour and seedling death by pre- and post-emergence seedling **damping-off**. Most control methods concentrate on seed hygiene.

3. *Aphanomyces*
A. cochlioides: the soilborne causal fungal pathogen of blackleg in sugarbeet, causing seedling stunting or death, especially in warm and wet soils. Moderate infections can be controlled by fungicide seed treatment.

4. *Ascochyta*
A. pisi is an extensively seedborne fungal pathogen, causing various leaf, stem and pod spot of pea and other legumes, and also post- and pre-emergence seedling damping-off. It occurs together with *Phoma medicaginis* var. *pinodella* and *Mycosphaerella pinodes* as part of a complex of seedborne fungi; the complex can be controlled by fungicide seed treatments. Infected seeds may have brown areas on the seedcoat.

5. *Aspergillus*
A fungal genus notably containing the predominant species of seed **storage fungi**.

6. *Botrytis*
B. aclada is a fungal pathogen causing (bulb) neck rot, a systemic seedborne disease of alliums, mainly affecting onion, where it grows asymptomatically in leaves until they senesce. The fungus infects the cotyledon from infected seedcoats, which remain attached to the tip after emergence; seed treatments with fungicides provide very effective control of seedborne infection. *B. fabae* causes 'chocolate spot' disease on leaves in broad bean, which sometimes affects pods and seeds.

7. *Cercospora*
C. kikuchii, the fungal cause of the disfiguring **purple seed stain** disease in soybean. The genus also contains pathogens causing vegetative diseases in many crops, causing economic losses of grain.

8. *Clavibacter*
C. michiganensis causes bacterial wilts due to the production of a phytotoxic glycopeptide that causes plugging of the xylem vessels in the plant vascular system, leading to stunting of growth, wilting and withering. Contaminated seed is the primary source of infection of tomato bacterial canker (ssp. *michiganensis*), which causes serious losses worldwide in both greenhouse and field crops, though there are no visible symptoms either in germination or the ensuing seedling stand established infected seeds. Eradicative seed treatments include fermentation, hot water, hydrochloric acid (pectinase-HCl), and calcium or sodium hypochlorite. Externally and internally seedborne *C. michiganensis* (ssp. *nebraskensis*) is the causal pathogen of Goss's wilt of maize and sorghum, which causes severe local grain losses; early infection may cause seedlings to wilt, wither and die. Subspecies *insidiosus* causes lucerne and clover to produce seeds in reduced number and of low quality, usually shrunken and shrivelled.

9. *Claviceps*
An ascomycete fungal genus including *C. purpurea*, which leads to the formation of large purple-black **ergot** bodies (**sclerotia**) in **rye**, which fall onto the soil to infect plants in succeeding years. Infected florets secrete a sweet 'honeydew', which attracts insects that transmit the infection to florets on other plants. Though most prevalent on rye, with its open floral structure, *Claviceps* species infect other cultivated cereals (including barley, oats, sorghum, pearl millet and wheat) as well as many wild and cultivated grasses.

10. *Cochliobolus*
The fungal genus containing *Cochliobolus sativus* (the teleomorphic state of 16. *Helminthosporium sativum*), an economically important cause of root and foot rots, leaf spots, kernel infections (black point, kernel **blight**) and seedling blights in small-grain cereal crops (wheat, barley and rice). The pathogen, often associated with other fungi, causes dark brown to black discoloration of seeds. It may also cause root and foot rots in cereals and grasses, and '*Helminthosporium* leaf blight' and head blight in wheat: prematurely bleached spikelets that are usually sterile or contain shrivelled, discoloured seeds. The primary source of inoculum is from conidia that survive on crop debris or in the soil, though there are incidences of seedborne infections: *C. heterostrophus* (southern corn leaf

blight) kills, wilts or stunts developing seedlings, with brownish roots and coleoptiles; and *C. lunatus* is part of the complex of fungi that may produce grain mould in sorghum – a shiny, velvety black, fluffy surface growth – and turn rice seed aleurone and starch layers brown or black. The seedborne pathogens can be controlled by fungicidal seed treatment.

11. *Colletotrichum*
See: Anthracnose

12. *Diaporthe*
The teleomorphic state of the 21. *Phomopsis* fungal pathogen complex. *Diaporthe phaseolorum* may be seedborne causes of stem canker, or stem and pod blight in soybean.

13. *Erwinia*
A bacterial genus comprised of plant saprophytes and pathogens causing bacterial wilt diseases, related to 8. *Clavibacter*. Notably, maize seed may transmit *Erwinia stewartii* (*Pantoea stewartii*) within the endosperm, causing seedling blight shortly after emergence. Seed contamination is commercially regulated by most importing countries.

14. *Fusarium*
A genus of soilborne and sometimes seedborne fungal pathogens, such as *Fusarium oxysporum*, that can infect and parasitize the root cortex of a wide range of crop species (such as tomato, cucumber and watermelon) causing various wilt, blight and 'soil-sickness' diseases in pre- and post-emergent seedlings. The fungus produces conidia, which move passively upward in the xylem, where the plant vascular system responds by producing tyloses (plugs) that block water movement, resulting in tissue wilting. Also, 17. *Microdochium nivale* (formerly called *F. nivale*) is mainly seedborne in wheat (where high levels of infection can cause very poor crop establishment and significant yield losses) and *F. culmorum* (culm rot in cereals) is frequently seedborne in temperate cereals and grasses, and *F. moniliforme* causes rice bakanae disease (or 'foolish seedling' disease, due to the production of **gibberellins** by the fungus), the symptoms of which are too much stem growth, which makes it weak and liable to break. *Fusarium* can be controlled to some extent by seed treatment fungicides. Non-pathogenic strains of this species can be used to control the pathogenic – **see: Rhizosphere microorganisms**. The genus also includes ear and kernel rots which can cause direct damage in the field and losses and contribute to spoilage or **mycotoxin** production during storage, such as *Gibberella* ear rot: (starburst) (*F. graminearum*) (Colour Plate 10C), the same fungus that causes stalk rot of maize and scab of wheat.

15. *Gaeumannomyces*
G. graminis is the soilborne pathogenic cause of 'take-all' disease in cereals, of notable economic importance in wheat, virtually throughout the temperate world: blackened roots and stems appear especially in the second and third rotation where the crop is grown successively and when warm and wet seasons are followed by moisture stress. The disease can be partially controlled by seed treatment fungicides (fluquinconazole and silthiofam; **see: Treatment – pesticides**).

16. *Helminthosporium*
A fungal genus, the anamorphic state of 10. *Cochliobolus*, containing pathogens that cause shrivelled kernels and embryo staining in small grain cereals and grasses, amongst other diseases.

17. *Microdochium*
M. nivale is an important seedborne disease in wheat, see: 14. *Fusarium*.

18. *Penicillium*
A genus notably containing species of storage fungi that cause spoilage to high moisture grain, producing mycotoxins that can have detrimental effects on animal or human health, and in maize can produce the 'blue eye condition' when fungus fruits below the pericarp in the embryo. *P. oxalicum* can be transmitted on seed in maize, causing reduced germination and blight in which seedlings turn yellow and growth is retarded. In the field, it also causes ear rot and discoloured (bleached and streaked) seed.

19. *Peronospora*
A genus containing the primarily soilborne causal pathogens of **downy mildew**, that also may be seed-transmitted in some crops, such as *P. farinosa f.* sp. *spinaciae* on spinach seed and *P. manshurica* in soybean, where it proliferates in the inner pod wall and seedcoat during seed development, causing white coloration usually near the **hilum**, cracking and the formation of abnormally light and small seeds.

20. *Phoma*
Phoma medicaginis var. *pinodella*, occurs extensively as one part of a complex of seedborne fungi causing various leaf, stem and pod spot of pea and other legumes (see: 4. Ascochyta). *P. betae* is the seedborne cause of blackleg in sugarbeet. Both can be effectively controlled by fungicide seed treatments. *P. exigua* var. *linicola* is a soilborne and occasionally seedborne pathogen in flax causing foot rot disease, which may affect seedlings within a few weeks of emergence.

21. *Phomopsis*
The *Diaporthe*/*Phomopsis* complex (respectively the teleomorphic and anamorphic states of the fungus) includes seedborne pathogens that are of major economic importance in the soybean crop. Infected seeds are a minor source of inoculum compared to soybean residue in the field where the pathogen may over-winter; spores infect pods by water splashing, and the pathogen remains latent until the pods turn yellow (physiological maturity), when colonization and seed infection occurs. In the most severe infections, stimulated by warm, moist environmental conditions during seed development and maturation, '*Phomopsis* seed decay' (*Diaporthe phaseolorum* var. *sojae* and *Phomopsis longicolla*) results in shrunken and light seeds, with a white, chalky appearance due to mycelia, increased susceptibility to breakage (Colour Plate 10D), direct losses of both germination and vigour and a reduction in flour and oil quality. Because seeds also may be infected without showing symptoms, detecting the pathogen is a major target in soybean seed **health testing**. *P. longicolla*, depending on severity of infection, causes seedling

losses, pod and stem blight. *P. phaseoli*, soybean stem canker, causes wilting and death by disturbing nutrient translocation.

22. *Phytophthora*
A persistent soilborne pathogen causing pre- and post-emergence seedling damping-off, as well as various wilts and rots in older plants, especially in warm wet conditions, and controllable by seed fungicide treatment as well as by tolerant or resistant cultivars. Examples include *P. sojae* in soybean; *P. medicaginis* in alfalfa/lucerne; *P. palmivora* in palm plants (such as black pod rot in cocoa); and *P. nicotianae* in a large number of tropical and temperate agricultural and ornamental crops, with tobacco and tomato as primary hosts.

23. *Plasmopara*
P. halstedii is the causal pathogen of downy mildew in sunflower, an endemic disease worldwide, that may also cause pre- or post-emergence seedling damping-off and seedling blight.

24. *Pseudomonas*
A bacterial genus containing seed- (as well as soil- and water-) borne pathogens that cause a range of blights in cultivated crops, including at the seedling stage, and can cause virulent disease epidemics because of the ease with which they can be transmitted in the crop. In soybean, for example, seedborne *P. savastanoi* pv. *glycinea* leads to the development of black water-soaked spots on pods and discoloured, shrivelled seeds, amongst other symptoms – a disease common in cool, temperate growing regions worldwide; related pathogens affect *Phaseolus* (*P. phaseolicola*) and other beans and peas with halo blight, where they can be extensively seed-transmitted. In rice, seedborne *P. fuscovaginae* causes sheath brown rot (discoloured, deformed or empty grains), seedling rot and, as a consequence, substantial yield losses. *P. syringae* pv. *lachrymans* is a well-recognized seedborne pathogen of cucumber and other vegetable cucurbits, infecting cotyledons soon after germination leading to angular or bacterial leaf spot disease. The *Pseudomonas* genus also contains several non-plant-pathogenic species and strains (such as *P. fluorescens*) that have beneficial **plant growth promoting rhizobacteria** (PGPR) and fungal-antagonist biological-control properties, and are of interest for application as seed treatments in cultivated crops. (**See: Rhizosphere microorganisms**)

25. *Puccinia*
See: Rusts

26. *Pyrenophora*
P. graminea, primarily a seedborne fungal pathogen, causes leaf stripe in barley and can cause severe crop losses. The developing seed is infected and this infection is systemically passed on to the seedling. Infected plants are stunted, produce fewer ears and fewer seeds per ear, and seed produced may show severely reduced seed mass, discoloration, browning and lack of development, and may fail to germinate. The fungus survives as mycelia in the hull, pericarp and seedcoat, and can easily be controlled by fungicidal seed treatment. *P. teres*, the cause of barley net and spot blotch, is seedborne and also transmitted from field debris and stubble.

27. *Pythium*
A soilborne fungal genus, belonging to the water moulds (Oomyctes), which produces motile zoospores that swim towards a host plant in the soil. *P. debaryanum* and *P. aphanidermatum*, notably, cause seedling damping-off symptoms, and are components of the **seedling disease complex**, which affects a broad host range of native and cultivated plants. Symptoms may include water-soaked lesions on the roots or hypocotyls, causing pale, stunted or slow growth or collapse and death, or the rotting of seeds and seedlings before they emerge. Control measures include a combination of timely planting as the disease tends to be more severe during cool, damp weather, high-quality seed and good seedbed preparation and planting depth to reduce soil moisture in the root zone, **solarization**, and seed treatment and in-furrow fungicides. *P. ultimum* causes cavity spot in carrot – black lesions on the harvested root.

Non plant-pathogenic *Pythium* species can compete with or attack different soilborne fungi, and may therefore protect the seedling by biological control. *P. oligandrum* has been investigated as a rhizosphere biofungicide, and is considered to act both by outcompeting pathogenic soil fungi and also stimulating the growth of crops rendering them less susceptible to disease attack. The product is commercialized in several countries, targeted at a wide range of soilborne fungal pathogens that affect glasshouse- and field-grown vegetables and cereals, for application to the soil. (**See: Rhizosphere microorganisms**)

28. *Rhizoctonia*
R. solani is a soilborne fungal pathogen of high economic importance and worldwide distribution: the causal agent of many diseases with a very wide host range, including pre- and post-emergence seedling damping-off as a constituent of the seedling disease complex, seed rots (for example in cotton, lupin and peanut) and seedling blight (for example in alfalfa, clover and many vegetables), as well as many head-, root- and stem-**rots**. Seedling diseases can be controlled through fungicide seed treatment, or by sterilization of the planting medium in nurseries. *R. solani* (the anamorph counterpart of *Thanatephorus cucumeris*) is an 'aggregate' species with many other scientific names depending on the host, and includes many physiological strains that are differentiated by their hyphal fusion properties into, at present, 12 'anastomosis groups'. Some non-pathogenic strains have been shown to improve plant growth (**see: Rhizosphere microorganisms**).

29. *Sclerotinia*
S. sclerotiorum is a primarily soilborne fungal pathogen, causing diseases of high economic importance with an extremely wide host range, where it may cause a variety of blights, rots and wilts, including pre- or post-emergence seedling damping-off, and premature ripening that affects both seed yield and harvest quality. Typically, water-soaked lesions form, expand and become brown or bleached. The disease can, for example, cause considerable yield losses in canola/rapeseed, where contaminating sclerotia can result in rejection of infested grain lots. After heavy plant infections, *Sclerotinia* mycelia can also contaminate seeds in sunflower, soybean and French bean crops.

30. *Septoria*
S. apiicola, a primarily seedborne fungal pathogen of celery and celeriac, surviving as pycnidia and or mycelia in the pericarp and testa, causing destructive black blight and other diseases worldwide. Infection can spread from plant to plant by water splash droplets, leading to extensive infection in seedbeds. Control measures include fungicide or hot-water seed treatments.

31. *Tilletia*
A pathogenic seedborne fungal genus causing common **bunt** in cereal and grass inflorescences, a disease in which bunt balls full of spores severely taint grain quality with the odour of rotting fish. Following infection, mycelium passes into the crown and keeps pace with the growth of the apex until the ear is formed. Two closely related smut fungi, *T. laevis* and *T. tritici* (both with several synonyms), for example, cause wheat common bunt, a potential hazard for most of the world crop; both are readily controlled with fungicide seed treatments. *T. indica* (Karnal bunt of wheat) reduces the length of ears and the number of spikelets; in the case of mild infection, grain may be only partially destroyed, leaving the endosperm intact or, if severe, yield, seed quality and germination are adversely affected, as glumes open and grain drops to the ground.

32. *Ustilago*
A fungal genus containing pathogenic species that cause seedborne **smuts** in cereals, such as *U. hordei* (covered smut of barley or oats). The mycelium develops systemically, invading the flower primordia, and ultimately transforming individual mature grains into spore balls. Control measures include systemic seed treatment fungicides.

33. *Verticillium*
A fungal genus containing pathogenic species that infect the plant vascular system systemically, following attack from the seedling stage onwards, causing **wilt** diseases and consequent seed losses. Soilborne or internally seedborne *V. dahliae* has a very wide host range among economically important crops (including cotton and tomato), ornamental plants and native species, and is distributed very extensively worldwide. Seeds are also an important vehicle for spread of *V. albo-atrum* in alfalfa/lucerne.

34. *Xanthomonas*
A bacterial genus containing seed- (as well as soil- and water-) borne pathogens that cause a range of blights in all major crops, including at the seedling stage. Species are differentiated into a number of different pathovars, but have high phenotypic similarity. *X. campestris* pv. *campestris (Xcc)* (black rot), for example, is a pathogen of high economic importance in most cultivated cruciferous plants – cauliflower and cabbage are the most readily affected: infected seedlings show dark discoloured cotyledons, which later shrivel and drop off; in mature plants, vascular vessels become plugged, restricting the water flow and resulting in characteristic 'V'-shaped leaf lesions, wilt and necrosis; infections appearing just before seed maturation result in either seed abortion or gradations of infection on the surface or internally. *X. campestris* (or *X. hortorum*) pv. *carotae* is a seedborne pathogen of carrot, causing similar symptoms on leaves and umbels. The pathogens often cause virulent disease epidemics because of the rapidity with which they can be spread from plant to plant from a single contaminated seed, especially in wet conditions and in module trays in the nursery; seed health testing is therefore of crucial importance, using thresholds ranging from zero infected seeds in 10,000 to 60,000. Other diseases caused by seed-transmitted *Xanthomonas* include black chaff or leaf streak in wheat and barley, leaf blight in rice, and common blight in bean, all associated with symptoms including discoloured, shrivelled seed, and blighted seedlings. (PH)
(See: Pathogens – ecology; Pathogens – seedborne infections)

CABI Compendium of Plant Diseases www.cabicompendium.org/cpc
Neergard, P. (1977) *Seed Pathology*, Vols I and II. John Wiley, New York, USA.
Richardson, M.J. (1990) *An Annotated List of Seedborne Diseases*, 4th edn. International Seed Testing Association, Zürich, Switzerland.

Pathogens – ecology

As with **predation**, pathogens can severely decrease seed numbers, both pre- and post-**dispersal**. Seed diseases caused by fungi, e.g. ergot on cereals (*Claviceps purpurea*) and the general causes of seed rot (*Pythium*, *Fusarium*, *Diplodia*, *Rhizoctonia*, *Penicillium* spp.), viruses (e.g. bean pod mottle virus) and bacteria, e.g. pea bacterial blight (*Pseudomonas syringae*) seed diseases are common in agriculture, and attack seeds which are on the crop plant, sown into the soil, or stored (**see: Cereals – storage; Mycotoxins; Pathogens – seedborne infection**).

In natural systems, many diseases can reduce seed production by attacking flowers or developing fruits, e.g. anther smut (*Microbotryum violaceum*) on white campion (*Silene alba*). Pathogens may also attack seeds on the plant directly, e.g. ergot on a range of grasses, including cordgrass (*Spartina* spp.). However, seed diseases are most common in dispersed seeds lying in or on the ground where they kill the seed, inhibit germination or cause weakened or abnormal seedlings. In fact, disease may be the major cause of death of seeds in the **soil seed bank**. Death can be caused by generalist decomposing (saprophytic) fungi or bacteria that break down the seed, or by more specific diseases caused by pathogens with a specialist association with the plant species.

Post-dispersal seed losses can also be large. The African forest tree, *Strychnos mitis*, can lose >80% of the seed crop to fungal attack; and a range of fungi, including *Fusarium*, *Penicillium* and *Mucor* spp., caused 94% loss of sagebrush (*Artemisia tridentata*) and 50% loss of bluegrass (*Poa canbyi*) seeds through one winter in a USA shrub-steppe.

The potential for large seed losses to disease might be expected to lead to evolution of strategies to escape pathogens. The Janzen-Connell hypothesis suggests that dispersal away from the parent evolved as a strategy to avoid pathogens which tend to be concentrated near established plants. More direct strategies involve some sort of resistance to disease. These can be active or passive mechanisms. In the former, there is a direct physiological or biochemical reaction to the presence of the pathogen, while in the latter the plant is (largely) unaffected by the pathogen. Little is known about disease resistance in seeds, but passive resistance through a thick **seedcoat** or anti-pathogenic chemicals on the seedcoat (e.g. tannin in seedcoats has anti-fungal properties) exists. (**See: Phenolic compounds; Polyphenol oxidases; Seedcoats – structure**) (JMB)

Gilbert, G.S. (2002) Evolutionary ecology of plant diseases in natural ecosystems. *Annual Review of Phytopathology* 40, 13–43.

Pathogens – seedborne infection

Pathogens may be carried with, on or within seeds, with the inoculum located either superficially on **seedcoats**, or within the seedcoat or **embryo** tissues (Table P.1); detection is the primary objective of seed **health testing**. Though they have the potential to cause disease in the seedling or the mature plant (as 'seed-transmitted pathogens'), most do not actively attack or show pathological signs directly on the seed itself. Indeed, pathogens may be seedborne, without seed infection being considered to be the primary route of transmission of the disease that they can cause. (The suffix '-borne' relates to *carrying* pathogens, rather than expressing disease symptoms.) Some seedborne pathogens infest, as distinct from infect, seeds – that is, they are mixed and carried with them, rather than being closely attached to them. Seeds also are colonized by many other saprophytic microorganisms, which have little or no influence on seed or plant quality, such as fungal endophytes of grasses, which complete their life cycle in symptomless symbiosis with the host. Seedborne saprophytes may however contribute to the population of **storage fungi** that can cause spoilage losses in grain. For details on specific species, **see: Pathogens**.

1. Modes of infection and spread

(a) *Infection*. Seedborne pathogens usually infect developing seeds prior to harvest. Colonization depends on the presence of an **inoculum** source, climatic conditions and the susceptibility of the crop.

Direct infection via the xylem of the mother plant is common for many viruses, but fewer fungi and bacteria. For example, the fungus causing covered **smuts** of cereals (*Tilletia tritici*, stinking smut or **bunt** of wheat, for example) invades the **coleoptile** of young seedlings and produces a systemic mycelium, which moves with the developing tissues of the shoot and eventually grows into the flowers, replacing the ovary tissues with teliospores which are enclosed by the seed pericarp (Fig. P.2); at harvest when the plants are threshed, the teliospores are released and contaminate healthy grain. Seedborne onion neckrot, caused by *Botrytis allii*, systemically infects cotyledons from inoculum on the seedcoat and lives within the green leaves of the growing plant causing no disease until it enters senescing tissues, so that bulbs eventually rot in store.

Indirect infection can be systemic via the stigma to the seed embryo as in the case of virus-infected pollen or wind blown teliospores of the smut fungi *Ustilago segetum* var. *tritici* of wheat and barley. A large number of microorganisms indirectly infect seed by penetrating the ovary wall tissues (Fig. P.2). The pathogens attack the flower, fruit or pod tissues with the funicle and the micropyle being the most common entry points for fungi and the **hilum** for bacteria. *Alternaria brassicicola* of brassicas, *Phomopsis* spp. of soybean and *Pseudomonas syringae* pv. *pisi* of peas have been shown to infect seed in this way.

(b) *Air- and insect-borne transmission*. The majority of bacterial and many of the fungal pathogens whose spores are contained in pycnidia, acervuli or sporodochia are spread by wind-driven rain and splash on to seedlings and eventually on to the aerial parts of the plant if the microclimate is favourable. Overhead irrigation can also contribute to their proliferation. Most seedborne pathogens that are splash-dispersed cause infection over relatively short distances from the inoculum source, whereas those with windborne spores may travel large distances. Some fungi require rain or high humidity to allow the release of ascospores, and a few seedborne pathogens are dispersed by 'dry' conidiospores whose liberation is often stimulated by falling humidity and physical displacement, particularly at harvest time. Conidia are spread by wind or may hang in the air above or within the crop canopy.

Although most plant viral diseases are not seedborne, seed transmission is rather frequent in some plant genera, notably in legumes and composites. Lettuce mosaic virus, for example, is an extremely harmful pathogen (leading to defective hearting, leaf disease and necrosis in main crops, as well as reduced seed yield and viability in seed production crops) for which seeds can serve as a source of primary infection; the pathogen is mainly controlled by a rigorous combination of seedlot screening and field inspections. Most viruses are transmitted by insects and for seedborne viruses, aphids are by far the most important vector, though mites, nematodes, fungi and bacteria act as vectors for a few, as can simple mechanical contact. Viruses may survive and remain infective in seeds from a few months to many years, particularly if they are embryo-borne.

(c) *Multiplication and spread*. Some seedborne pathogens survive only in growing plants or in harvested seed. Because they remain within their host for most of their existence there is no spread of disease to other plants within the growing crop (so-called monocyclic disease): for example, loose smut of barley, *Ustilago nuda*. (**See: Smuts**) However, multiplication

Table P.1. The location and status of pathogens on or in seeds.

Organism	Status	Location on/in seeds
Viruses	Biotrophs	**Embryonic axes** (most species); **testa** and **endosperm** occasionally
Fungi	Biotrophs	Embryonic axes (most **smut** species); superficial (most **bunts**, a few rusts and **downy mildews**)
	Necrotrophs	Superficial; in testa/pericarp tissue (most species); in embryonic axes (very few species)
Bacteria	Necrotrophs	As for necrotrophic fungi but almost none in embryonic axes

Reproduced from: Maude, R.B. (1996) *Seedborne Diseases and their Control: Principles and Practice.* CAB International, Wallingford, UK.

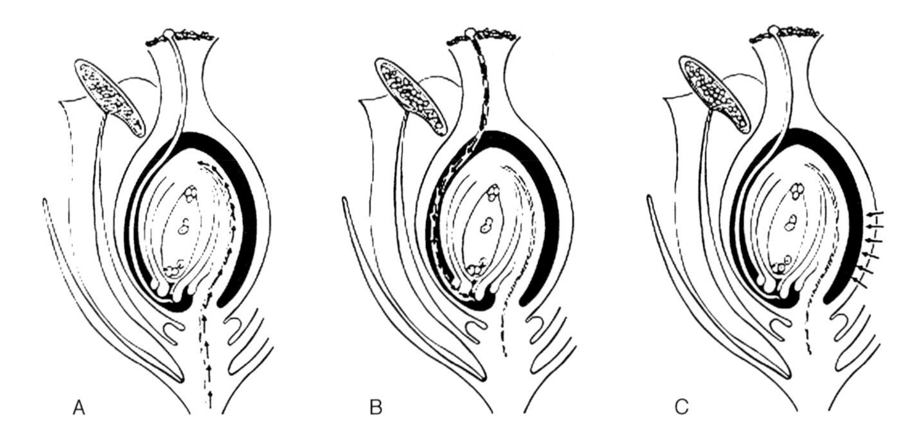

Fig. P.2. Direct and indirect routes of infection of pathogens. (A) Invasion of the mother plant tissues via the xylem – some bacteria and fungi. (B) Indirect invasion of the embryo via the stigma – viruses in pollen tubes; fungi as their hyphae grow down through the style to invade the embryo. (C) Indirect, by penetration of the ovary wall tissues – many organisms. Reproduced from: Maude, R.B. (1996) *Seedborne Diseases and their Control: Principles and Practice.* CAB International, Wallingford, UK.

of pathogens from infected plants within a crop can be significant. In Scotland, for example, two crops of barley which had fewer than 1% tillers infected with *Pyrenophora graminea* (leaf stripe) yielded a crop with more than 50% infection of harvested seed. Other seed-transmitted pathogens once established can multiply and spread during the crop growth cycle whenever conditions are favourable (polycyclic disease), such as leaf spot of celery, *Septoria apiicola*. Polycyclic seedborne pathogens may also survive for long periods on infected crop debris, volunteers or other host plants increasing the complexity of the disease cycle. In this way pathogens may be carried over from one growing season to another and spread from one area to another.

Some seed-transmitted viruses are non-persistent in the crop. Lettuce mosaic virus, for example, although a very serious pathogen, is only retained and spread for a short time and over small distances.

2. Control of seedborne pathogens

Controlling the incidence of seed infection is effective where seedborne inoculum is the only or principal cause of plant disease, and its adverse effects on the subsequent crop can be limited or controlled. On the other hand, if the disease is also soilborne or readily carried over from one season to another in crop debris, seedborne infection may be of limited importance.

Infected or infested seedlots of cultivated plants are identified by seed health testing and either discarded or remedially treated, such as with fungicides or other eradication treatments (**see: Treatment – pesticides**). Seeds of some crops are also treated prophylactically as a precaution against soilborne disease, such as seedling **damping-off**.

Control measures may include: exclusion (quarantine) to guard against the introduction of non-indigenous pathogens and pests; selecting disease-free seed for sowing, based on rigorous seed **health testing** and setting standards; by treating seedlots with an appropriate chemical seed treatment; by using resistant cultivars and by altering cultural practices. Changes in climatic conditions, cultural practices or the use of more susceptible varieties can very quickly increase the inoculum levels of some seedborne pathogens (e.g. *Claviceps purpurea*, *Fusarium* spp. in cereals), causing economic loss. Disease control in these cases may require changes in rotational and/or cultivation practices as well as different varieties, though for many crops the choice may be limited, as breeding for resistance to seedborne pathogens may not be considered a priority. (VC)

Agarwal, V.K. and Sinclair, J.B. (1997) *Principles of Seed Pathology*. CRC Press, Boca Raton, Florida, USA.

Cockerell, V. (1997) New and priority seedborne diseases in Western Europe. In: Hutchins, J.D. and Reeves, J.C. (eds) *Seed Health Testing: Progress Towards the 21st Century*. CAB International, Wallingford, UK, pp. 1–10.

Sheppard, J.W. (ed.) (1997) *Seedborne Fungi.* 2nd reprint in 1997 with revised nomenclature. ISTA, International Seed Testing Association, Zürich, Switzerland. (A description of 77 species.)

Singh, D. and Mathur, S.B. (2004) *Histopathology of Seedborne Infections*. CRC Press, Boca Raton, Florida, USA.

PCR (Polymerase chain reaction)

A biochemical method of **DNA** amplification of target sequences by up to a million-fold, into detectable quantities: a Nobel prize-winning invention made by Mullis in 1987 that has become the basis for the development of powerful molecular biological technologies including DNA cloning, sequencing and forensic analysis. Through its capacity to multiply specific nucleic acids and its high sensitivity and accuracy, the PCR technique is widely and increasingly used to detect pathogens, including seedborne fungi and bacteria in **health testing**, to detect the DNA constructs used to transform germplasm in **genetic modification**, such as in testing for adventitious contamination in **identity preservation** systems for seed, grain and foodstuffs produced from them, and in molecular genetic investigations of seed function and processes.

A thermostable DNA (Taq) polymerase enzyme and specific DNA probes (oligonucleotide primers, which anneal to specific sequences in the genome) are used to amplify the base sequences between the primers by repeated thermal cycling. Each cycle of the PCR doubles the number of copies of the amplified DNA due to the action of the polymerase enzyme, and the result is very large numbers of identical short DNA fragments (usually 200–2000 base-pairs long) between the primers, which subsequently can be separated and analysed by sequencing on an electrophoretic gel, after specific cleavage of the products using selected **restriction enzymes** (restriction fragment length polymorphism, RFLP).

In the case of **RNA** genomes, such as in the majority of viruses, a reverse-transcriptase enzyme is used immediately before the PCR is started, to transcribe the RNA first into DNA: a technique known as reverse-transcription PCR.

Random amplification of polymorphic DNA (RAPD) uses arbitrarily chosen oligonucleotide primers to produce DNA fragments by the PCR. RAPD marker techniques, in several variations that improve specificity, are used in genetic **purity testing** and to identify the parental pedigree and relatedness of **hybrid** seed cultivars. RFLP analyses of PCR-amplified DNA are used to identify and distinguish pathogen strains.

(PH)

Smith, O.P., Peterson, G.L., Beck, R.J., Schaad, N.W. and Bonde, M.R. (1996) Development of a PCR-based method for identification of *Tilletia indica*, causal agent of Karnal bunt of wheat. *Phytopathology* 86, 115–122.

Zhang, J., McDonald, M.B. and Sweeney, P.M. (1996) Random amplified polymorphic DNAs (RAPDs) from seeds of differing soybean and maize genoytpes. *Seed Science and Technology* 24, 513–522.

PDC (Plant Disease Committee, ISTA)

The former name of the **ISTA Seed Health Committee**.

Pea

1. World importance and distribution

Pea (*Pisum sativum*) is an annual, **cool-season legume**. The type includes field peas (*P. sativum* ssp. *arvense*, dry peas), processing peas (fresh peas) and Austrian winter peas (used as cover crop and green manure). The production of field pea on a world basis is on the increase especially in Europe, Canada and Oceania, for expanded use in animal feeding. France (1.6 million t), Canada (approx 1.4 million t) China and Russia (approx 1.2 million t each) are the major producers of the world dry pea crop: India (approx. 3.8 million t) and China (approx. 1.6 million t) are the main producers of green peas (FAOSTAT data for 2002). (**See: Crop Atlas Appendix, Map 3; Legumes**)

2. Origin

Peas were domesticated in southwest Asia and cultivated with the cereals as early as the seventh millennium BC. From the Near East, peas spread to central and northern Europe, and post Columbus they were introduced into the western hemisphere.

3. Seed characteristics

Seeds are spherical, round or slightly wrinkled, green or yellow. The green pea type has a light green **seedcoat** and dark green **cotyledons**; the yellow pea has a light yellow seedcoat and deep yellow cotyledons. Both colour types are considered primary classes suitable for human food though green pea varieties can be more difficult to market because of their susceptibility to bleaching. Green pea diameters average 7.5–8.3 mm and weigh 16–19 g per 100 seeds. Yellow peas are slightly larger, averaging 8.0–9.0 mm in diameter and weighing 20–21 g per 100 seeds, and larger still are marrowfat peas (Colour Plate 2D). Brown peas exist but they are rarely used for human consumption. Blue, white and dun peas are being developed for the speciality market.

4. Nutritive value

'Dry', mature pea seeds contain approximately 21–25% fresh weight (fw) protein (mainly **globulin, see: Storage protein**), high concentrations of carbohydrates (e.g. 45–50% fw starch), and are low in fibre. In round peas the starch consists of 45–49% **amylose** while in the wrinkled type it comprises about 86% (**see: Starch**). **Wrinkled peas** have a mutation in an enzyme which converts amylose to **amylopectin** (the branching enzyme, **see: Mutants; Starch synthesis**), and this results in less starch, more sucrose and a higher water content in the seeds. Consequently, when the seeds dry out, they shrivel and wrinkle. Their higher sugar content makes them more desirable to eat, and they are used for canning. Relatively high concentrations (> 5% fw) of soluble sugars are found in some varieties: the flatulogenic **raffinose-series oligosaccharides**, predominantly stachyose, can reach 5% fw. **Oil** (triacylglycerol) content is about 1.4% fw. Peas have relatively high amounts of the **essential amino acids**, lysine and tryptophan.

Antinutritionals such as trypsin inhibitor (**see: Protease inhibitors**) and phytin also occur, but at acceptable dietary levels. **Lectins** are also present.

5. Uses

Dry peas are marketed as the shelled product for either human or livestock food, whereas fresh peas are as a freshly picked, or frozen or canned vegetable. Both green and yellow peas are used for canning, soup, and as ingredients in food processing. The round, green- and yellow-seeded varieties are used for human consumption as dry split field pea. Because peas are an excellent source of dietary protein, the flour is an important item in the food industry. Marrowfat peas are used for canning and for the traditional British 'mushy peas'.

A growing use of field peas is for livestock feed. The relatively high contents of tryptophan and lysine, essential amino acids normally deficient in small cereal grains, make peas valuable as feed and an excellent protein supplement for all classes of livestock, including poultry. Peas also can be a partial or complete replacement for soybean meal in cattle, lamb and swine rations. Peas contain 5–20% less of the trypsin inhibitors than soybean so they may be fed directly to livestock without having to go through the extrusion heating process. Field peas are often cracked or ground and added to cereal rations. Peas grown together with barley, oat, triticale, or wheat are highly productive and nutritious.

All field pea varieties may be considered for feed peas, but only selected varieties are acceptable for either the green or yellow human edible market. Types of dried peas rarely used as food are the coloured peas (round, brown seeds) and the wrinkled peas (though they may be used for canning when harvested before maturity). Peas are also used as a forage crop.

6. Marketing

The following are the current market classes:

- Food peas: yellow or green, large, round, and smooth seeds with a uniformly coloured seedcoat.
- Feed peas: low-grade food pea types are acceptable in the feed market: seed shape, cotyledon colour, and seedcoat pattern and colour are variable. These are unacceptable as food types.
- Speciality peas: pea types such as Alaskan, maple, blue peas and Austrian winter peas are grown for speciality confection, birdseed and forage markets.

Premium prices are associated with the human food and seed markets. To be considered when marketing grade peas are the market class, seed size and shape, splitting potential, harvest moisture, and seed damage factor (bleached, cracked seedcoats, split, shrivelled seeds, etc.). Among the nations which consistently import field pea for human consumption are Colombia, Venezuela, Brazil, UK, Taiwan and Japan. The European Union is the main importer of feed peas, used as livestock protein supplement (Table P.2).

7. Plant types

Field pea shows **hypogea**l emergence of the seedling from the soil. There are two main types: one has an indeterminate growth habit, aggressive climbing and continual blooming throughout the summer until the temperatures and moisture become limiting; the other has a determinate growth habit, bushy or dwarf, with a shorter flowering period and earlier maturation. Stem length is 60–150 cm; vines are prostrate at maturity. Pods contain 4–9 seeds. Some pea varieties have a semi-leafless growth habit. The tendrils of adjacent plants intertwine to provide better resistance to lodging for the entire canopy.

Green varieties tend to yield approximately 20% less than yellow varieties under similar growing conditions. (OV-V)

For information on the particular contributions of the seed to seed science **see: Research seed species – contributions to seed science**.

Davies, D.R., Berry, G.J., Heath, M.C. and Dawkins, T.C.K. (1985) Pea (*Pisum sativum* L.) In: Summerfield, R.J. and Roberts, E.H. (eds) *Grain Legume Crops*. William Collins, London, UK.

Deshpande, S.S. and Adsule, R.N. (1998) Garden pea. In: Salunkhe, D.K. and Kadam, S.S. (eds) *Handbook of Vegetable Science and Technology: Production, Composition, Storage, and Processing*. Marcel Dekker, New York, USA, pp. 433–456.

Kalloo, G. (1993) Pea, *Pisum sativum* L. In: Kalloo, G. and Bergh, B.O. (eds) *Genetic Improvement of Vegetable Crops*. Pergamon Press, Oxford, UK, pp. 409–425.

Muehlbauer, F.J. and Kaiser, W.J. (eds) (1994) *Expanding the Production and Use of Cool-season Food Legumes*. Kluwer Academic, Dordrecht, The Netherlands.

Summerfield, R.J. (ed.) (1988) *World Crops: Cool Season Food Legumes. A Global Perspective of the Problems and Prospects for Crop Improvement in Pea, Lentil, Faba bean and Chickpea*. Kluwer Academic, Dordrecht, The Netherlands.

Pea – cultivation

The pea is an annual with seeds that can be harvested in the immature state for fresh vegetable production (vining) or in the dry mature state (combining), when they may be stored for several months before using for both human and animal nutrition. For information on origins, economic importance, seed structure, composition and uses, **see: Pea**. Varieties include coloured flowered types, but more common are the white flowered types, which produce white or green coloured seeds.

1. Breeding

Characteristics that the breeder is striving for include quality attributes for flavour, texture and maturity determinacy for the

Table P.2. World data and major countries exporting and importing peas (2001).

World/Country	Imports (thousand t)	Value (million US$)	World/Country	Exports (thousand t)	Value (million US$)
Green peas					
World	235	156	World	121	7
Belgium	66	15	France	25	5
India	36	8	UK	18	5
Japan	22	29	USA	7	10
USA	17	19	Belgium	6	2
Dry peas					
World	3364	703	World	3495	620
India	849	297	Canada	1970	319
Spain	523	72	France	565	87
Belgium	415	61	Australia	337	58
Bangladesh	260	48	Belgium	112	19
Netherlands	165	24	Ukraine	108	13
Pakistan	110	20	USA	106	33

FAO STAT 2001.

vegetable peas, whilst yield, straw strength and earliness are features of the dry-harvesting types. In both types, disease resistance for both foliar- and root-infecting pathogens is an objective. Breeders are incorporating the afila characteristic into most new varieties, which produces plants with a normal, or reduced sized stipule at the leaf axis but the compound leaf is reduced to a compound tendril. This assists in stem strength and standing ability, whilst reducing the foliage area and risk of foliar diseases, which can cause problems in wet seasons. However, because of the incipient **shattering** problem of currently available varieties, afila peas have not yet been accepted by the industry.

2. Production

Seed is produced in most parts of the world where peas are normally grown, although certain countries and specific areas are favoured for seed production because their weather conditions are generally drier as the peas mature. In North America most of the vining pea seed production is centred in the western USA, particularly Washington State and Idaho, and also in south-west Canada. In Europe, production is mainly in Hungary. New Zealand is also an important centre. Often breeders can make use of two generations per year by growing in the northern and then in the southern hemisphere. Seed for combining peas is grown in those countries with the greatest pea production, that is, France, UK, USA and Canada. Because the multiplication rate of peas is only about tenfold, there is a relatively large area of seed production in the world in proportion to the total area of peas grown.

Pea seed is harvested when fully mature and the moisture content is around 15% or less; however, below 12% the risk of mechanical injury to the seeds increases. Peas can be harvested using a conventional combine harvester with only minor adjustments to the drum speed and concave settings to allow for the large seeds. Vining peas are susceptible to mechanical injury and, to prevent seedcoat cracking, seed is best handled carefully using rubber belt **conveyors** and avoiding blown air or augers. Indeed, leguminous seeds in general are particularly susceptible to thresher injury, especially cultivars with thinner seedcoats, and machinery must be adjusted according to the moisture content of the seed. Typically, pea seed weight ranges from 200 to 350 g per 1000 seeds. After harvest, the seed may be dried with ambient or slightly warmed air ensuring that the drying air temperature does not exceed 43°C. To prevent mould growth during storage it is essential to reduce the seed moisture content to 15%.

3. Quality, vigour and dormancy

Most types of combining pea seed, i.e. round seeded or marrowfat types, do not suffer from problems with seed **vigour** but wrinkle-seeded vining peas are very susceptible. Low vigour results in high pre-emergence losses in cold, wet seedbeds and all seed of early sown varieties should be of high vigour. Seed vigour is determined using an electrical conductivity test; the greater the amount of electrolytes that have been leached from imbibing **cotyledon** cells during 24 h of soaking in water, the lower is the seed vigour (**see: Vigour tests – biochemical**).

A major cause of low vigour is seedcoat cracking, which allows a rapid inrush of water to cotyledon cells causing them to rupture. When in the soil, such areas of damaged tissue are invaded by soilborne pathogenic fungi such as *Pythium ultimum*, causing pre-emergence decay. Peas may also suffer from a physiological disorder known as hollow heart or cavitation, where the cells in the centres of the adaxial surfaces of the cotyledons are dead, and when imbibition has been completed, the centre of the seed contains a cavity. The dead cells produce a growth toxin, which inhibits germination and reduces seedling survival. It is thought that high temperatures occurring during seed maturation predispose pea seed to hollow heart.

Pea seed rarely is seriously affected by dormancy, although occasionally newly harvested seed is slow to germinate.

4. Sowing and treatments

Peas are sown in a well-worked seedbed and no deeper than 5 cm. Most cereal seed drills are suitable for peas. Spring sown peas are usually sown when soil temperatures have exceeded a mean of 5°C.

Fungicide seed treatments, such as thiram, are used to protect seed during the early stages of germination from soilborne damping-off diseases, such as seedling blight caused by *Fusarium*, *Rhizoctonia* and *Pythium* (see: **Pathogens**) in dry (including field) peas. Seed-borne fungi such as those in the Ascochyta complex, *Ascochyta pisi* and *Mycosphaerella pinodes*, are often found in seed harvested in areas where rainfall has encouraged foliar infection in the summer. These pathogens can also be controlled by systemic fungicides, such as thiabendazole or fludioxinyl. In addition, seed treatments such as metalaxyl or cymoxanil can be applied to protect the seedlings from soilborne infection of downy mildew (*Peronospora viciae*). (AB)

Davies, D.R., Berry, G.J., Heath, M.C. and Dawkins, T.C.K. (1985) Pea (*Pisum sativum* L.). In: Summerfield, R.J. and Roberts, E.H. (eds) *Grain Legume Crops*. William Collins, London, UK.

Deshpande, S.S. and Adsule, R.N. (1998) Garden pea. In: Salunkhe, D.K. and Kadam, S.S. (eds) *Handbook of Vegetable Science and Technology: Production, Composition, Storage, and Processing*. Marcel Dekker, New York, USA, pp. 433–456.

Kalloo, G. (1993) Pea, *Pisum sativum* L. In: Kalloo, G. and Bergh, B.O. (eds) *Genetic Improvement of Vegetable Crops*. Pergamon Press, Oxford, UK, pp. 409–425.

Peak germination value

An assessment of the rate of germination, often used as a measure of seed vigour of tree species. (**See: Germination rate; Vigour tests – physiological**)

Peanut

1. World importance and distribution

The peanut, or groundnut (*Arachis hypogaea*) is an extremely important legume seed in world economy, being used directly as food for humans and livestock and as a source of vegetable oil. The major growing countries are China, India, Senegal, Nigeria, Myanmar, Zaire, Sudan, USA, Argentina and Indonesia. The largest area of peanut cultivation is in India with close to 8 million ha followed by China (see section 6). (**See: Crop Atlas Appendix, Maps 2, 3, 15, 22; Legumes**)

In terms of seed oil production peanut is the fourth most important crop (**see: Oilseeds – major,** Table O.1). About two-thirds of the world's production is processed for the oil, the remainder consumed directly as food. The higher priced edible market requires large seed with a low oil and high protein content. In contrast, the oilseed industry favours high oil content, high shelling percentage, long postharvest seed dormancy and good storability.

The seed oil content varies from 45 to 52%. (**See: Oilseeds – major,** Table O.2 for the oil composition). The oil is primarily used as cooking oil, salad oil and in margarine manufacture while the meal is used as a high protein feed supplement and sometimes as a fertilizer. (**See: Oils and fats**) Consumer desire for enhanced oil quality in peanut products has resulted in breeding programmes worldwide striving to alter peanut fatty acid composition (**see: Oils and fats – genetic modification**).

2. Origin

Peanut is native to South America – Bolivia, Brazil and Peru having been postulated as centres. Of the 60 related wild species of *Arachis*, only 20 have been described, and all are only found in the area east of the Andes, between the Amazon River in the north and the Rio de La Plata in the south. The oldest archaeological evidence comes from Peru where peanut remains dating from about 3000 years ago have been found. The Spanish took the peanut to Spain (and later to the Philippines, from where it spread over Asia), and the Portuguese brought the plant to the Molucca islands, to India and to Africa. The multiple introductions to Africa from widely separated regions of South America spawned a great deal of variation, establishing Africa as a secondary centre of diversity. It is thought that the peanut was carried over to North America with the slave trade and did not travel northwards from Peru.

3. Seed characteristics

The seeds are ovoid with a reddish brown **testa** and can reach 2 cm long and 1 cm wide, depending on the type (Fig. P.3; Colour Plate 2E). Each seed consist of two large, fleshy **cotyledons** attached to a smaller **embryonic axis,** surrounded by a thin **testa**. The colour of the peanut testa ranges from dark purple to white and may be a single colour or mottled and there are strong preferences for a specific testa colour within market classes (see below). The embryonic axis consists of a **radicle, hypocotyl** and **epicotyl**. The epicotyl is well developed, and the mature seed embryo may contain nine, or more, young leaves on its main and lateral axes. The radicle tip protrudes well below the protective cotyledons, making it especially vulnerable to mechanical injury during handling. The thin testa is also vulnerable to damage during drying.

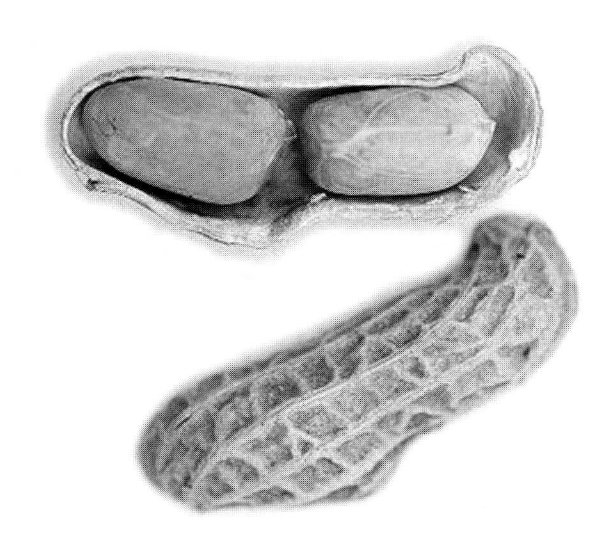

Fig. P.3. Peanut seeds within the pod (see section 3 for dimensions).

There are four basic market types each distinctive in size, flavour and nutritional composition. (i) *Runner* (*Virginia* and *Peruvian*) is the most common type. It is a medium-sized 'nut' of uniform kernel size. Fifty four per cent is used for peanut butter. (ii) *Virginia*: often called cocktail nuts, they have the largest kernels and account for most of the peanuts roasted and processed in the shell. Isolated kernels are sold as snack peanuts. (iii) *Spanish*. These have smaller kernels, are used predominantly in peanut candies, with significant quantities used for snack nuts and peanut butter. They have a higher oil content than the other types making them particularly suitable as a source of oil. (iv) *Valencia*. These are small kernels with a bright-red testa, usually three or more in a pod. They are sweet and are usually roasted and sold in the shell but are also excellent when boiled. Spanish and Valencia are considered as a subspecies, *fastigiata*.

4. Composition and nutritive value

Peanuts are rich in **oil** (triacylglycerols) (35–55% fresh weight, fw) and **storage protein** (21–36% fw): the major storage proteins are arachin and conarachin. Peanut varieties have been developed with elevated oleic acid and reduced linoleic and/or linolenic acid. High oleic peanuts have been developed with oleic/linoleic ratios in excess of 40, compared to the ratio of 2 to 3 in traditional varieties (**see: Oils and fats**). Carbohydrate is about 15% fw, of which **starch** comprises approx. 7%. Of the vitamins, thiamin and niacin are relatively high (approx. 0.85 mg and 18 mg/100 g fw, respectively). Riboflavin and ascorbic acid are also present. Analyses of calcium and iron content report these elements at 50–90 mg and 2.5–3.8 mg/100 g fw, respectively.

Antinutritionals. The most important factors are the **allergens (see: Storage proteins – intolerance and allergies)**. Peanut allergy is variously manifest by acute hives, facial swelling, bronchospasm and anaphylactic shock. Two major allergenic proteins have been identified, araH1 and araH2 and also the minor araH3 and araH4. AraH1 is a 65 kDa glycoprotein, 30–45% similar in amino acid identity to other **vicilins** – from soybean, peas and beans. AraH2 is also a trypsin inhibitor (**see: Protease inhibitors**) whose allergenicity is enhanced by roasting. It has been suggested that the much lower incidence of peanut allergy in China may be attributed to the absence of roasting in the Chinese use of peanuts. Interestingly, an 18 kDa **oleosin** is also strongly allergenic. Some individuals, especially babies and young children, show allergenic responses to the vanishingly small quantities of protein carried over into peanut oil. Other adverse effects arising from consumption of peanuts may be due to teratogenic and carcinogenic aflatoxins produced by fungal contamination of damaged seeds (*Apergillus* spp.) (**see: Mycotoxins**).

5. Uses

The most common use of peanuts is for oil production (**see: Oilseeds – major**). In India, peanut is the most important oilseed crop. Other uses of peanuts are as a component of cooked dishes, as a snack food, in candy and to make peanut butter. About 10% of the peanut crop in the USA is sold as in-shell peanuts, usually the Virginia and Valencia types. Most often, snack peanuts are **shelled**, blanched, roasted and salted (although Spanish peanuts are usually roasted with their testas on). Peanuts may be roasted in oil or by a dry process. About one-half of all edible peanuts produced in the USA are used to make peanut butter and peanut spreads. Peanuts are used in candy-making in a large number of ways. Many kinds of candy bars (six of the top ten sold in the USA) contain peanuts (whole, chopped or as butter) combined with chocolate, nougat, marshmallow, caramel, other nuts and dried fruits. The high protein content of peanuts make them ideal for high energy snacks. There are also some non-food uses for peanuts. The shells, skins (testas) and kernels may be used to make several products, for example the shells in wallboard, fireplace logs, fibre roughage for livestock feed and cat litter; and the skins in paper making. Peanuts are often used as an ingredient in products such as detergents, salves, metal polish, bleach, ink, axle grease, shaving cream, face creams, soap, linoleum, rubber, cosmetics, paint, explosives, shampoo and medicines.

6. Marketing and economics

Major producing countries and exports are shown in Tables P.3 and P.4. In 2002 world production was nearly 34 million t. Less than 6% of the crop was traded internationally but world export values reached nearly US$800 million, the main exporters being Argentina, China, USA, Vietnam and the Netherlands. The latter was by far the main importer and a high proportion of the material was re-exported. (**See: Crop Atlas Appendix, Map 15 Peanut cultivation**)

(OV-V, MB, KD, JS, DLJ, JFS)

Ahmed, E.M. and Young, C.T. (1982) Composition, quality, and flavor of peanuts. In: Pattee, H.E. and Young, C.T. (eds) *Peanut Science and Technology*. American Peanut Research and Education Society, Yokum, Texas, USA, pp. 655–688.

Coffelt, T.A. (1989) Peanut. In: Robbelen, G., Downey, R.K. and Ashri, A. (eds) *Oil Crops of the World*. McGraw-Hill, New York, USA, pp. 319–338.

Holbrook, C.C. and Stalker, H.T. (2003) Peanut breeding and genetic resources. *Plant Breeding Reviews* 32, 297–356.

Maiti, R. (2002) *The Peanut Crop*. Science Publishers, Enfield, NH, USA.

Vaughan, J.G. (1970) *The Structure and Utilization of Oil Seeds*. Chapman and Hall, London, UK.

Woodruff, J.G. (1981) Peanuts, *Arachis hypogaea*. In: McClure, T.A. and Lipinsky, E.S. (eds) *CRC Handbook of Biosolar Resources*, Vol. II. *Resource Materials*. CRC Press, Boca Raton, Florida, USA.

Table P.3. Major producers of peanuts, 2002.

Country	Production (thousand t)
Argentina	517
China	14,895
Ghana	520
India	5,200
Indonesia	722
Myanmar	700
Nigeria	2,699
Senegal	501
Sudan	1,267
USA	1,506
World	*33,735*

FAOSTAT

Table P.4. Major world peanut exports, 2002.[a]

Country	Exports (thousand t)	Exports (US$ million)
Argentina	119	67
China	521	264
India	68	37
Nicaragua	48	25
Netherlands	57	50
South Africa	50	26
Vietnam	107	50
World	*1341*	*778*

[a]In- and out-of-shell data combined.
FAOSTAT

Peanut – cultivation

The peanut, or groundnut (*Arachis hypogaea*, Fabaceae) is an extremely important food seed for humans and livestock and an important oilseed. For information on economic importance, origins, composition and uses, etc. **see: Peanut**.

1. Genetics and breeding

Peanut is an indeterminate, essentially self-pollinating crop, with natural out-crossing occurring at less than about 2.5%. Breeders throughout the world rely on the use of plant introductions, artificial hybridization (by manual cross-pollination) and pure-line selection to improve yield, quality, food value (such as increased oleic and reduced linoleic and/or linolenic acid contents) and pest resistance. Molecular technologies, plant regeneration systems and transformation techniques have been identified, but to date no commercial varieties have been developed through the use of these technologies. There is great interest, however, in exploring the use of genetic engineering to enhance peanut fungal and viral disease resistance (**see: Genetic modification**). Researchers have successfully transformed peanut lines to be resistant to tomato spotted wilt tospovirus and peanut stripe virus; however, no plans seem to be currently in place to pursue registration of these varieties.

2. Plant types

Plants are relatively short statured with an upright main axis, 20–40 cm long. The primary branches bear flowers in the leaf axils on the fruiting branches; flowers open for only half a day from sunrise to the afternoon. During this period pollination occurs after which the fertilized ovary begins to enlarge, and is carried away from the plant on a small stem (the **gynophore**) which extends downward, pushing the developing fruit into the soil. Prior to penetration into the soil, the fruit is referred to as 'the peg'. Once the peg enters the soil it turns horizontally to the surface and begins to mature. Pods become evident at about 60 to 70 days after pollination (DAP), with maximum numbers occurring at about 100 to 120 DAP, depending on cultivar and environment. Pod weight increase is nearly linear during pod fill after a short lag period. The plant eventually produces some 40 or more underground mature fruits (Colour Plate 2E).

There are basically two types of growth habit: (i) *alternate type*: alternate branching, long growth cycle, and inflorescence absent from the main axis; semi-spreading or spreading growth habit; (ii) *sequential type*: sequential branching, strictly annuals, short growth cycle, and inflorescence usually present in the main axis; upright, erect growth habit with lateral branches that do not extend higher than the main axis. The Virginia (ssp. *hypogaea*) and Peruvian runner (ssp. *hirsuta*) have no floral axes on the main axis and alternate vegetative and floral axis along the lateral branches. The Virginia type has short branches with few hairs while the Peruvian runner has long, hairy branches. The Valencia (ssp. *fastigiata*) and Spanish (ssp. *vulgaris*) types have floral axes on the main axis and continuous runs of multi-floral axes along the lateral branches. The Valencia and Spanish types differ in the degree of branching, with the Spanish type being more branched.

3. Development

Flowers typically begin to form 25 to 30 days after planting (DAP), but this can vary among cultivars. The number of flowers produced daily increases to a maximum between 2 and 4 weeks after flower initiation and then declines to near zero during pod fill. Fertilization ultimately results in the development of the fruit underground – hence the name 'groundnut', and the epithet '*hypogea*'. The fertilized egg undergoes three to four divisions then stops development until after soil penetration.

The flower stalks curve downwards, and meristematic tissue adjacent to the basal ovule becomes active and forms a pedicel structure (botanically, a **gynophore**, commonly referred to as a 'peg'), which elongates by positive gravitropism and grows into the soil, thereby pushing the ovary to a depth of a few cm. For further information on inflorescence production and plant types **see: Peanut, 7. Plant types**.

Peanut yield is influenced by the number of plants per unit area, the number of seeds per pod, the number of pods per plant (ranging from one to five) and the weight per seed; the last two factors appear to be the most influential yield components. Partitioning of photosynthate to the fruit during pod fill has been shown to be a critical factor contributing to seed size and weight. In addition, the duration of pod fill, which differs among cultivars, has a profound influence on yield. All yield components can be heavily influenced by climate, cultural practices, weed infestation, and disease and insect pressure.

4. Sowing

The peanut crop grows best on light-textured, well-drained soils. Heavy soils that are prone to crusting can reduce peg penetration during pod development and may cause pegs to break during harvesting and vine inversion, which can both reduce seed yield and quality. Soils high in organic matter may also adversely affect yield and quality by encouraging pod rot and other seed diseases. The spacing between row-to-row and plant-to-plant varies with the type.

Planting dates vary by production area and cultivar planted, the most critical factors being soil temperature and soil moisture content. The optimum temperature for the most rapid germination and seedling development is about 30°C. Generally, peanuts should not be planted until soil temperature at 10 cm depth is 18°C or higher for 3 consecutive days, and a favourable air temperature is expected. Planting in cool and wet soils can reduce germination and increase seedling death. Time from planting until radicle emergence varies with seed size, occurring within 24 h for vigorous small-seeded Spanish-seed types, but requiring 36 to 48 h in large-seeded Virginia types. Emergence is **epigeal**; however, unlike most dicots exhibiting this mode of emergence, peanuts lack a hypocotyl arch to pull the cotyledons to the soil surface; instead, cotyledons are pushed up through the soil by the growing hypocotyl. The cotyledons open when they reach the soil surface and the epicotyl elongates, pushing the primary leaves upward. Seedlings are sensitive to salts in the soil, and fertilizers are therefore applied several months before planting.

Calcium is a critical element in peanut seed production, especially for large-seeded cultivars. Lack of Ca uptake results in aborted or shrivelled fruit, and seedlings grown from Ca-deficient seed often exhibit weak and watery hypocotyls in which vascular water conducting tissues have collapsed ('collar rot'), and deformed plumules; at least 450 ppm Ca is needed in the seed for rapid, normal seedling growth and development. Ca taken up by the roots does not move physically downward in the plant, and therefore does not supply the nutrient to the subterranean developing fruit. Rather, the peg and the fruit pod themselves must absorb calcium directly from the fruiting or pegging zone (the upper 7 to 10 cm of soil). Ca should be applied at flowering time, such as gypsum powder or granules; application before planting is ineffective.

5. Production

Determining the optimum time to harvest ('dig') peanut seed should be based on pod maturity, soil characteristics, soil moisture content and frost probability. The **indeterminate** flowering pattern requires an assessment of when the maximum number of pods in the field is at the correct maturity stage, typically indicated by the development of a dark tan colour inside the shell. Harvesting prior to optimum maturity can result in small, immature, low quality seeds, whereas later harvesting may lead to pod loss and increased seedborne diseases. Maximum seed quality is most often achieved when the seed crop is harvested approximately 1 week before commercial food and feed stock peanuts are dug. Seed moisture content at this time is approximately 40%.

In developed countries, commercial peanut seed crops are mechanically dug, and the plants inverted and placed in **windrows** in the field, with pods on top of the plants, where they are allowed to remain until seed moisture is between 20 and 25% before mechanical harvesting. In parts of the world with high-intensity direct sunlight or that lack the means for artificial drying, inverted plants are stacked in various ways to prevent the damage caused by too rapid drying. After that stage, plants are lifted from the ground into the combine harvester, where the peanuts are then stripped from the vines, the foreign matter separated from them, and deposited in bulk tanks or wagons. Care must be taken to achieve optimal separation from the vines, without causing loose-shelled seed or hull damage. In developing countries, harvesting and threshing the crop is typically an extremely labour-intensive process, whether by age-old manual or animal-power assisted processes or mechanically powered systems.

At this point, peanuts must be dried to a moisture level below 10% on a fresh-weight basis (frequently referred to as 'curing'). This must be done slowly, within a 2 to 3 day period, to prevent seed quality losses that can arise due either to biochemical processes that cause rapid viability loss during storage, or to physical factors – notably the formation of hard kernels, or **testa** slippage that can lead to excessive skinning and splitting during **shelling**. Similarly, curing is a very critical factor in establishing the basic flavour qualities after harvest of the main crop. Seeds for planting are stored from harvest to planting the next season (6 to 9 months) at approximately 9% moisture. Often seeds are kept in the pod in a cool, dry environment for about 4 months, before shelling and seed sizing is carried out.

Sheller machines use beaters that revolve inside a drum and crush the pod against the ridges of grates, spaced to allow seeds to pass through with minimal injury. However, this aggressive process typically damages a proportion of seeds, in which **cotyledons** are bruised or broken away from the embryonic axis, or the delicate papery testa is split. Subsequent **conditioning** uses air-screen cleaners or gravity tables, and indent cylinder sizers. In less developed regions of the world, simpler manual or powered machines are used, and on smallholder farms pods are basically shelled by hand – or mouth.

6. Quality, vigour and dormancy

Peanut seed quality is most frequently associated with germination, vigour and health. Seed purity, other than genetic purity, is not often a concern in commercial seed production. **Testa** colour can be an important varietal identification characteristic. Germination and vigour are greatly influenced by maturity at harvest and by seed production environment. Maximum seed vigour of several Virginia market-type cultivars has been shown to occur at **physiological maturity**. Germination of immature seed (such as when they have achieved only 50% of their final dry weight) can be high if conditions during seed development were relatively stress-free; however, the vigour of such immature seeds is frequently low, regardless of production environment. Germination and vigour are also closely related to curing practices; curing at high temperatures can cause physiological damage and physical skin (testa) slippage.

Freshly harvested peanut seed of some varieties can be dormant. This is especially true for the Virginia market types, though other varieties can exhibit dormancy traits too, depending upon their genetic background, production environment and curing conditions. Postharvest dormancy is broken naturally during a short afterripening period in ambient conditions, or by the addition of **ethylene** gas or ethephon during germination testing in the laboratory. (**See: Germination – influences of gases**)

7. Treatments

Several seedborne diseases adversely affect peanut germination and seedling survival, including *Rhizopus* spp., *Aspergillus flavus*, *A. niger* and *Fusarium* spp. (**see: Pathogens**). Seed infection by *Rhizopus* is minimized by proper harvesting and curing to reduce seed injury. *Aspergillus* spp. infection is favoured by dry weather during pod and seed development and by mechanical damage during handling. Seed treatments have been successful in managing most seed or seedling diseases. Chemical combinations of both contact and systemic fungicides that target seed and soilborne fungi are successful in preventing stand losses in most production areas. **Rhizobial inoculation** of the seed or the soil is usually not necessary in fields previously cropped with peanuts. (JS, DLJ)

Holbrook, C.C. and Stalker, H.T. (2003) Peanut breeding and genetic resources. In: Janick, J. (ed.) *Plant Breeding Reviews*, Vol. 22. John Wiley, Hoboken, NJ, USA, pp. 298–356. media.wiley.com/product_data/excerpt/14/04712154/0471215414-1.pdf

McDonald, M.B. and Copeland, L.O. (1997) Peanut. In: *Seed Production – Principles and Practices*. Chapman and Hall, New York, USA, pp. 273–281.

Nautiyal, P.C. Groundnut Post-harvest Operations. National Research Centre for Groundnut (ICAR) www.fao.org/inpho/compend/text/Ch21sec1.htm

Pattee, H.E. and Stalker, H.T. (eds) (1995) *Advances in Peanut Science*. American Peanut Research and Education Society, Stillwater, OK, USA.

Pattee, H.E. and Young, C.T. (eds) (1982) *Peanut Science and Technology*. American Peanut Research and Education Society, Yokum, Texas, USA.

Simpson, C.E., Krapovickas, A. and Valls, J.F.M. (2001) History of *Arachis* including evidence of *A. hypogaea* L. Progenitors. *Peanut Science* 28, 78–80.

Spears, J.F. (2000) Germination and vigour response to seed maturity, weight, and size within the Virginia-type peanut cultivar, VA-C 92R. *Seed Technology* 22, 23–33.

Spears, J.F. and Sullivan, G.A. (1995) Relationship of hull mesocarp colour to seed maturity and quality in large-seeded Virginia-type peanut. *Peanut Science* 22, 22–26.

Pearl millet

Pearl or bulrush **millet** (*Pennisetum glaucum* (syn. *P. americanum* and *P. typhoides*)) is also known as bajra (India), babala, cumbu, dukhn, gero, sajje, sanio, souna or cat-tail-, spiked millet. Pearl millet is drought resistant and is an important crop on land too dry for **maize** or **sorghum**. (**See: Pearl millet – cultivation**)

1. World distribution, economic importance

Pearl millet is a staple crop of very many millions of people in hot, dry parts of the world where agriculture is possible. It is also grown as a forage and feed grain crop, for example in the south-east USA, though major producers of pearl millet are Africa and India. Africa grows 49% of the global production on 55% of the global area. West Africa (mainly Nigeria, Niger and Burkina Faso) produces ± 80% of the African grain, and east and north-east Africa ±12%. National annual yields are ≤1 t/ha, even in India. In India, ±50% of the area is grown with hybrids and improved varieties; yield and production are increasing, but the total cultivated area is decreasing. There is very limited international trade in pearl millet. Senegal exports some seeds to Senegalese living in the USA.

2. Origins

Pearl millet is thought to have originated about 5000 years ago in tropical West Africa. Domestication of the crop is likely to have started by disruptive selection after hybridization and introgression and selection against **shattering**. The migrating Bantus took it to eastern (around 2000 years ago) and southern Africa (approx. 1500 years ago). It reached India some 2000 years ago, Europe in 1566, and the USA in 1850. In the 20th century, the **gene banks** in India, Italy and the USA facilitated its global access.

3. Seed structure, composition

The seeds of pearl millet are small (length 3–5 mm, width 2.1–3 mm), exposed, intermediate or enclosed (**see: Cereals, 4. Basic grain anatomy**). The **hilum** is short, marked by a black dot. The embryo is large (half the length of the grain) and elliptical. The grain shape is obovate, lanceolate, elliptical, hexagonal, or globular. The colour is ivory, cream, yellow, grey, deep grey, grey brown, brown, purple, purplish black or orange (Fig. P.4). No cellular seedcoat is present but a membranous cuticle is. A single layer of aleurone layer cells surrounds a starchy endosperm with horny and floury regions. The aleurone layer cells have conspicuous knobbly thickening. Similar cells have been noted in foxtail millet and assigned a transfer function.

At 12% moisture there is 55–75% carbohydrate (mostly **starch**), 8–16% **storage protein**, 3–9% fibre and 3–5% **oil**. The grain is rich in certain vitamins (thiamine, riboflavin, niacin, panthothenic acid and folate) and minerals (copper and manganese). The high amylase produced by the grains is favourable for **malting**. It is low in lysine and vitamins A and C. The protein and lysine content can likely be improved by breeding.

4. Uses

Pearl millet is a food grain mainly produced for household consumption and local trade. In Africa it is consumed as a porridge or gruel, and in India as flat unleavened bread.

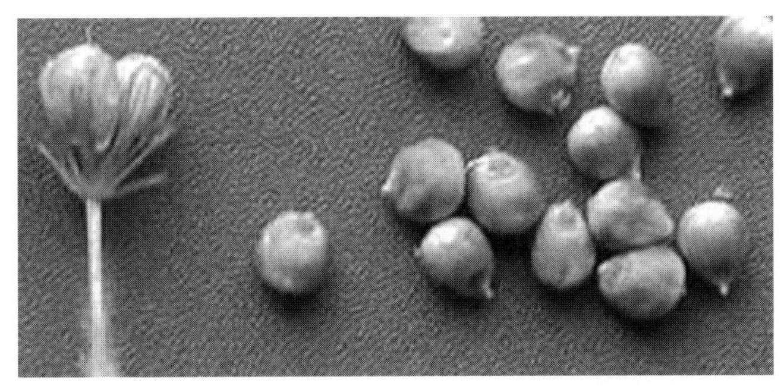

Fig. P.4. Pearl millet floret and grains (for dimensions, see text) (www.d1.dion.ne.jp/~tmhk/yosida/hana.htm) (by kind permission of T. Yoshida).

Commercial production is risky for economic and environmental reasons. Pearl millet farmers often also have livestock and the cattle may obtain additional feed from the millet. It is used as food, drink (alcoholic and non-alcoholic), fuel for domestic use, bedding, roofing, fencing, feed and cultivated pasture. Grain compounds may contribute to goitrogenicity. (WAJdeM)

(See: Cereals; Millets)

Andrews, D.J., Rajewski, J.F. and Kumar, K.A. (1993) Pearl millet: new feed grain crop. In: Janick, J. and Simon, J.E. (eds) *New Crops*. Wiley, New York, USA, pp. 198–208.

Andrews, D.J., Hanna, W.W., Rajewski, J.F. and Collins, V.P. (1996) Advances in grain pearl millet: utilization and production research. In: Janick, J. (ed.) *Progress in New Crops*. ASHS Press, Alexandria, VA, USA, pp. 170–177.

Dendy, D.A.V. (ed.) (1994) *Sorghum and Millets: Chemistry and Technology*. American Association of Cereal Chemists, St Paul, MN, USA.

Maiti, R.K. (1997) *Pearl Millet Science*. Science Publishers, Enfield, NH, USA.

Pearl millet - cultivation

Pearl or bulrush millet is *Pennisetum glaucum* – for botanical synonyms and other vernacular names **see: Pearl millet**. The species utilizes the C4 photosynthetic pathway and being characteristically drought-resistant is therefore an important cereal crop where maize or sorghum cannot be cultivated, because of dryness. (**See: Millets**)

1. Genetics, breeding

Grain hybrids of pearl millet (*Pennisetum glaucum* (mainly single cross)) have been made in India (since the 1960s) and the USA; they tend to be short, early maturing, responsive to fertilizers, irrigation and yield >5 t grain/ha. Forage **hybrids** tend to be double-cross hybrids and tall. Seed **shattering** can occur after outcrossing.

The production of **dihaploids** is possible, and is used in India and Japan, but not easy and is not used as a tool for breeding. Molecular markers have been developed, but this technology tends to be too expensive for common use. Despite pearl millet outcrossing with wild relatives, genetic modifications are being discussed, such as in South Africa, but sustainable economic incentives are still relatively small.

2. Development

Plants grow to 1–4 m with profuse tillers, but only some (e.g. two out of ten) produce a harvestable head. Short cycle, photoperiod-insensitive varieties may become mature from 65–90 days. The medium-cycle types require up to 130 days. Long-cycle types and photoperiod-sensitive types in inclement daylengths may take 180 days to harvest. Pearl millet is cross-pollinated. The panicle is hermaphrodite, 10–150 cm long, and female flowering starts at the top of the panicle. Within a few days, the anthers start to appear in the middle of the panicle. **Thousand seed weights** vary between accessions from 3 to 19 g.

3. Production

To maintain seed purity, production under isolation is essential. For the production of hybrid seed, use is made of transplanting to secure **nicking**. High quantity of pollen throughout the female flowering period is not only important for good seed setting but also reduces the risk of **ergot** infection. Germinating ergot reaching the female flower before the pollen tube prevents fertilization. Outcrossing during seed production is a serious risk, and can make seed quality unacceptable. (Because of outcrossing with wild relatives, transfer of any **genetically modified** traits would also be a hazard.) When farmers continue with the seed from hybrids based on **cytoplasmic male sterility**, uniformity is lost and often sterility problems occur, leading to lower yields. Hybrid production therefore requires recurrent commercial purchases. Harvesting, handling and storage are as for finger millet. (**See: Finger millet – cultivation**)

4. Quality, vigour and dormancy

Wet harvesting of pearl millet may reduce germination by ≥90%. The seed emerges relatively quickly compared to larger seeds (70% at 4 days after sowing, versus 40% for sorghum and 25% for maize). Four years after storage, South African **landraces** had 79% emergence, varying per entry from 8–97% (due to insect damage deterioration, and inherent problems). In the fifth year, several entries still had over 90% emergence. So, with control of stored grain insects, seed can be used for over 5 years. **Dormancy** is not normally an issue of importance.

5. Sowing

Pearl millet can be sown (≤15 kg/ha), broadcast (up to 20 kg/ha), dry planted, or **transplanted** (at 3–4 weeks). Good seedbed preparation is not necessarily crucial to obtain good yields. It is also grown in rows (up to 5 cm deep, 25–100 cm between rows, 10–100 cm within rows) on the flat or on ridges. Drilling and planting in lines facilitates inter-row weeding. Birds sometimes cause a complete crop loss in a few days. In those areas, the crop should not be grown without bird scaring but labour to do so may not always be available.

Transplanting pearl millet has several advantages. Planting of the farmer's favourite seed can be done in relation to rainfall and replanting with growing plants requires no delaying germination period. Transplanting may also help to obviate or reduce soilborne pests and diseases. Stand establishment of pearl millet tends to be good, and time to harvest tends to be reduced by 2 weeks, and thus it becomes a premier product in a time of food scarcity when maize is hardly available. Transplants are considered to be of the vegetable type and therefore the income goes to the woman in traditional societies. Transplanting has been proved to double the yields over several years in tests with poor volunteers in southern and West Africa (Zimbabwe, Ghana). It facilitates the introduction of treatments and new technologies.

Pearl millet can be the sole crop, sometimes with varieties of different maturity, or in a mixed crop with **legumes** (e.g. **pigeon peas**), vegetables or trees (fruits). Manure and fertilizer increase yields. Pearl millet is a labour-intensive crop requiring weeding, bird scaring, harvesting and threshing. Mechanization is limited to land preparation and sowing.

6. Treatment

Over 100 diseases reportedly affect pearl millet. The five most economically important fungal diseases are: downy **mildew** (caused by *Sclerospora graminicola*), **rust** (*Puccinia substriata* var. *indica*), **smut** (*Moesziomyces penicillariae*, syn. *Tolyposporium penicillariae*), ergot (*Claviceps fusiformis*), and pyricularia leaf spot (*Pyricularia grisea*). Landraces appear to have a more durable resistance than hybrids or inbred cultivars. Long-term experiments over more than a decade indicate that under both traditional and high-input farming practices seed treatments against downy mildew and/or soil and seedling insects increased yields by some 30%. Thousands of subsistence millet farmers in the Sahel of West Africa use a seed treatment product, a first example of a partnership between the crop protection industry and subsistence farmers.

On-farm priming – using **steeping** techniques – to improve germination rates and uniformity has been sporadically practised in Africa and India since the mid-1990s.

(WAJdeM)

Acland, J.D. (1971) *East African Crops*. Food and Agriculture Organization of the United Nations, FAO, pp. 27–28, 114–116, 247.

Belton, P.S. and Taylor, J.R.N. (eds) (2002) *Pseudocereals and Less Common Cereals. Grain Properties and Utilization Potential.* Springer, Berlin and New York.

Levy, A.A. and Feldman, M. (2002) The impact of polyploidy on grass genome evolution. *Plant Physiology* 130, 1587–1593.

Mukiibi, J.K. (2001) *Agriculture in Uganda, Crops*. Vol. II. Fountain Publishers, CTA, and NARO, Kampala, Uganda, pp. 1–54.

Purseglove, J.W. (1972) *Monocotyledons. Tropical Crops*. Longman, London, UK, pp. 118–159, 198–214, 256–259.

Peat

Soil materials in which the original plant parts are recognizable. A common and often predominant constituent of horticultural growing media, with a wide range of uses in seed germination and potting mixes. Peat is naturally formed in wetlands and occurs abundantly in certain parts of the world, especially at cooler latitudes, but there are major research efforts to find alternative materials, particularly in countries with minor indigenous resources and in response to socioeconomic and environmental concerns. International Peat Society www.peatsociety.fi (**See: Transplants – greenhouse production; Compost**)

Pecan

The most important native American nut tree (*Carya pecan* syn. *C. illinoinensis*, Juglandaceae), from Mexico and the southwest USA (Colour Plate 5B,C). The pecan tree can reach great heights and one is reported to be the largest tree in the state of Georgia, USA. The pecan is a true nut (**see: Nut**): like the walnut the nut itself is surrounded by green accessory tissues which at maturity split to reveal the nut. The nuts are 3–4 cm long and about 2.5 cm wide and the internal seed is slightly smaller. The **cotyledons** of the embryo are infolded though not as markedly as is the walnut.

The kernel can contain up to 17% fresh weight **storage protein** and almost 80% **oil** (triacylglycerol) which has a high proportion of oleic acid (**see: Nut**, Table N.2). The kernels have wide use in confectionery, bakery, as snack food and in ice cream.

Hickory nuts and butternut are closely related members of the same genus.

(MB)

Vaughan, J.G. and Geissler, C.A. (1997) *The New Oxford Book of Food Plants*. Oxford University Press, Oxford, UK, pp. 32–33.

Pectin

A structural **cell wall** component which is a mixture of heterogeneous, branched and highly hydrated polysaccharides rich in galacturonic acids linked by α-1,4-bonds (homogalacturonans, xylogalacturonans and rhamnogalacturonans). Extractable with hot water and calcium chelators. Pectins perform many functions in the cell wall: determining its porosity, providing charged surfaces that modulate pH, regulating cell–cell adhesion, as recognition molecules for pathogens and binding cell wall enzymes. Fruit, but not seed, pectins are used commercially as gelling agents, e.g. in the making of jams and jellies. (JDB)

Pedigreed seed

A term used in the Canadian seed certification system for agricultural seed that has been granted **Breeder**, Select, **Foundation**, **Registered** or **Certified** crop status, under the auspices of the Canadian Seed Growers Association for most crops.

Pelican sampler

A device for taking primary samples from a free-flowing vertical stream of seed, for example at the outlet of a **conveyor** belt. (**See: Sampling, 2. Seed stream samplers**)

Pelleting and encrusting

In seed **enhancement** technology: coating techniques that are used to improve drill performance by making seeds rounder, to change size grading, to carry nutrients and growth-stimulating materials and pesticide seed treatments. (**See: Treatments – pesticides**)

Pelleting is usually carried out to make irregularly shaped seed ovoid and smooth, or to make small seeds larger. Pellets are made to narrow-size tolerances, which can differ for a species between markets. In comparison, encrusting (also sometimes called 'minipelleting' or simply 'coating') applies less material, so that the original seed shape is still more or less visible; where the amount of material added is very small, the resulting product may be hard to distinguish from filmcoated seed. It is not always possible just by inspecting or analysing a particular coated seed product to tell how it has been made.

Species pelleted in substantial commercial amounts include monogerm sugarbeet (globally, by far the largest use), carrot, celery, chicory and endive, leek, lettuce, onion, pepper, tobacco, tomato, and to a lesser extent some *Brassica* species and 'super-sweet' maize varieties, and certain flower species, particularly those with tiny seeds. Encrusting is alternatively used in these crops as well, and also for the preinoculation with **rhizobia** of small-seeded legumes such as **alfalfa (lucerne)**, to reduce size grade variation in crops such as **maize** and **sunflower**, and to add weight to seeds to avoid drift during sowing, such as in turf **grasses**.

Seed pelleting was developed commercially in the second half of the 20th century, using manufacturing techniques derived from 'dragée' confectionery panning, which is still used for coating or 'enrobing' **nuts** and other sweetmeat centres with sugar, syrup and gum mixtures. Seed contained in revolving pans or drums of various designs (Fig. P.5) is wetted and blends of powdered materials (various mixtures of chalk, clays, diatomaceous earths, gypsum, lime, mica, peat, perlite, pumice, quartz, talc, **vermiculite**, wood fibres as well as water-attracting or hydrophobic materials, **see: Coatings**), binders (such as those used in **Filmcoating**), are progressively added, along with more water, until the desired weight or size increase is reached. Thereafter, the wet-coated seed is dried with heated air, usually in separate equipment. The tumbling action distributes and moulds the blend to give good size distribution, and prevents the formation of 'empty' seedless pellets or seeds sticking together (though special techniques can produce multi-seeded pellets if required). Though automated for some species, pelleting of others is a skilled manual process, usually performed on a batch or batch-continuous basis – typically from about 250 g to 100 kg seed per batch – though some systems operate by continuous throughput. Seed can be encrusted in the same equipment as used for pelleting, or in machines of simpler design, similar to those used for conventional seed treatment, where seed and a slurry of water, binder and powdered solid coating material are stirred together in a trough or a vortex mixer in a single stage addition. A more recent innovation is the use of rotary coaters (**See: Treatments – pesticides application, 2. Equipment**) as an alternative to the dragée method for pelleting, as well as encrusting: powders and liquids are added directly into the spinning seed mass, resulting in a much more rapid build-up of the coated seed. Pesticide formulations and other materials (major and minor nutrients) may be applied throughout this manufacturing process, or at stages as the layers are built-up, or by filmcoating after drying. Coated seed is commonly coloured for product identification and easier visibility in the soil. (**See: Colour Plate 11A**)

For the historical uses of pelleting and coating techniques **see: Treatments – brief history**. In a rare use of the term, seeds are embedded into or stuck on to extruded *cylindrical* 'pellets', using technology similar to that used in the production of animal feed. (PH)

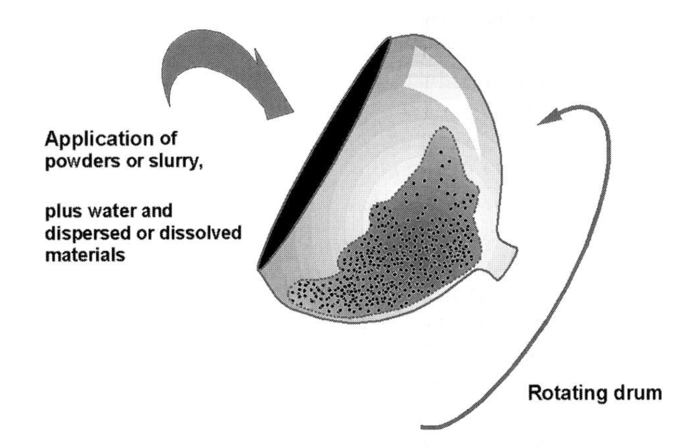

Fig. P.5. Drum coating pan (or mill) for the batch pelleting of seed. A range of geometrical drum shapes is used in commercial practice, including flattened and extended spheroids and cylinders. From Halmer, P. (2000) Commercial seed treatment technology. In: Black, M. and Bewley, J.D. (eds) *Seed Technology and its Biological Basis.* Sheffield Academic Press, Sheffield, UK, pp. 257–286.

Pepper and eggplant (aubergine)

Both pepper and eggplant are grown as vegetables around the world. Pepper is known by a plethora of common names around the world including chile, chilli, chili, *aji*, pimento, cayenne, *jalapeno* and paprika. (Peppers, incidentally, are not related to *Piper nigrum* the source of the spice **black pepper**.) Eggplant, also known as *brinjal* and aubergine, is one of the most popular vegetables in many tropical and subtropical countries. According to 2002 FAO statistics, the crops were grown on approximately 1.6 million ha worldwide; global pepper production was 22 t, while eggplant was 29 t.

The pods of pepper and the fruit of eggplant are technically **berries**, which in both cases are the plant parts utilized by humans as a food source. While peppers have a great many uses, eggplants are predominantly consumed as a vegetable. Peppers are harvested in both the 'mature green' and 'fresh mature' stages. They are canned, dehydrated, frozen, or consumed fresh. They are used as a vegetable (notably the non-pungent bell pepper) or as a spice and flavour-enhancer and, historically, have had medicinal uses. They have approximately 15 pod types and forms, various colours, and are even used as ornamentals. The chemical compounds responsible for colour and pungency in pepper (the latter due to the capsaicinoid alkaloids that are unique to the *Capsicum* genus) may be extracted to produce **oleoresin**. If pungent peppers are used, the extract is called 'oleoresin capsicum', and if non-pungent, or paprika, peppers are used, the extract is called 'oleoresin paprika'. Oleoresins are used as colouring agents in food and cosmetics, as pharmaceuticals, and in self-defence sprays. **See also: Tomato**, in which the seed has close similarities.

1. History and origins

Pepper, like most solanaceous crops, originated and was first domesticated in the Americas. Five species, *Capsicum annuum*, *C. baccatum*, *C. chinense*, *C. frutescens* and *C. pubescens*, were domesticated by indigenous cultures independently and in geographically different parts of tropical and subtropical South America. Today, *Capsicum annuum* is the species with the greatest economic importance, and there are over 25 wild species within this genus. The worldwide spread, adoption, and cultivation of pepper occurred relatively quickly after Christopher Columbus introduced them to Europe following his maiden voyage to the Americas and were later disseminated around the globe via Spanish and Portuguese shipping and trading routes. Within a century, this crop was being grown on virtually every continent.

Eggplant has been grown in India and China for countless generations. Eggplant and closely related species, also in the Solanaceae, probably originated in Asia and Africa, but exactly where is unknown. The most widely grown species of eggplant, *Solanum melongena*, was first domesticated in a region from Burma to Indo-China. Other species of eggplant such as *Solanum aethiopicum* and *S. macrocarpon* are cultivated mainly in Africa. There are approximately 200 wild species related to domesticated eggplant. In the 7th century, it was introduced to the Mediterranean Basin from the east where it was quickly adopted in such countries as Syria, Turkey and Persia; and in the 8th century, it was introduced to Japan from China. Eggplant however was not adopted in Europe as quickly as pepper, where it was referred to as 'mad apple' and said to cause insanity.

2. Seed structure and composition

Pepper seeds are flat and disc-like in shape, with a deep **chalazal** depression. (Seed structure closely resembles that of **tomato**, another solanaceous crop.) The average *C. annuum* seed is about 1 mm thick, 5.3 mm long, and 4.3 mm wide, and the **thousand seed weight** is about 6–7 g. They develop from a **campylotropous** ovule. Embryos are surrounded by a well-defined **endosperm**, part of which lies directly around the **radicle** and is seven to nine cells thick, having cell walls with **mannan**-containing polysaccharides. The cells are bordered by the internal epidermis, are angular in shape, have slightly thickened walls, and include oil and aleurone granules of crystalloid content. **Storage reserves** within seeds of *C. annuum* are mainly **storage protein** and **oil** (triacylglycerol). A cuticle covers cells of the embryo protoderm cells and the cells of the outermost layer of the endosperm next to the **seedcoat**. The growing point or plumule is between the two tightly appressed **cotyledons**. The seedcoat is derived from a single **integument** (**see: Seedcoats – structure**), and is straw-coloured, yellow, tan, or black depending on species. Seeds are attached to the placental wall predominantly towards the calyx end. Eggplant seeds have a kidney shape, are light brown in colour, and are relatively small; the thousand seed weight is about 4 g.

3. Genetics and breeding

Most species of pepper and eggplant are diploid, with 24 chromosomes ($2n = 2x = 24$). Pepper has one or two pairs of acrocentric chromosomes with ten or 11 pairs of metacentric or submetacentric chromosomes. Pepper flowers are complete and most cultivated species are self-pollinated, or **autogamous**, although cross-pollination, or **allogamy** can range from 2 to 90%. Therefore, measures must be taken in order to prevent unwanted cross-pollination. Peppers lend themselves to a variety of breeding methods including mass selection, pedigree, backcross, and recurrent selection, and do not exhibit inbreeding depression.

Eggplant is also autogamous with variable rates of allogamy. Eggplant is susceptible to many pests and diseases while wild relatives possess a great deal of resistance: however, it has proved difficult to hybridize these types with domestic eggplant, though efforts are continuing in this direction.

F_1 hybrids are rapidly replacing traditional varieties; as with pepper, F_1 seed is produced via hand emasculation and hybridization.

In both crops **genetic modification** work has been done to a limited extent, but not on a commercial level.

4. Development and production

Seed development in both pepper and eggplant parallels both fruit development and maturity. Therefore, seed should be taken from ripe, mature fruit to ensure good quality, mature seed. Pepper fruit average 50 to 60 days to ripen following anthesis. Eggplant take 30 to 40 days to ripen following anthesis.

Growing in isolation, or in enclosed structures or insect proof cages to prevent cross-pollination allows pure open-pollinated seed of both pepper and eggplant to be obtained. Seed should be clean, and free from pathogens: therefore, cultural practices concerning seed harvest, preparation and storage should be followed which allow for optimum quality in these respects.

At present, most F_1 hybrid seed is produced by hand. Although **cytoplasmic male sterility** and genetic male sterility systems that could be used to produce F_1 hybrid seed have been described, these systems have not been stable under various environmental conditions and no description of an efficient male sterility system has been made public.

Seed is extracted from either dry fruit (peppers) or from fresh ripe fruit by means of wet extraction in water (eggplant); seed of both is of highest quality when extracted from mature fruit. Generally, heavier, thicker and non-discoloured seed are of the best quality.

Seed can be dried at ambient temperatures or in driers below 35°C. Seed of eggplant and pepper can be stored well under ambient conditions for short periods of time, but as with many seeds longevity is increased if conditions are cool and humidity low. Pepper seeds may remain viable for 2 to 3 years at ambient temperatures; eggplant seed rather less so.

5. Quality, vigour and dormancy

In both pepper and eggplant, seedling **emergence** is **epigeal** and takes about 14 days to complete. Light is not a factor in the germination of pepper seed, whereas eggplant seeds germinate better in darkness. Freshly harvested seed of both can exhibit **dormancy**, which is relieved by being kept dry at room temperature for about a 6-week **afterripening** period. The best daily temperature regime to promote germination of dormant pepper seed is 30/15°C (16 h/8 h) for 14 days, whereas a regime of 30/23°C (8 h/16 h) has been reported as being sufficient to promote germination of fresh and dormant eggplant seeds (**see: Germination – influences of temperature**). Dormancy can be broken in both by treatments such as potassium nitrate at 2 g/l (seeds soaked for 4 h before planting), or **gibberellin** (100 ppm and 1000 ppm gibberellic acid used daily when watering). (**See: Germination – influences of chemicals; Germination – influences of hormones**)

In pepper, the appearance of the endosperm changes 1 day before radicle emergence, when the endosperm at the radicle enlarges and protrudes outward. This change is accompanied by weakening and degradation of the endosperm directly in front of the radicle (**see: Germination – radicle emergence**).

6. Sowing

Peppers and eggplants are warm-season crops that require similar growing conditions. In the field, both may be direct-seeded or **transplanted** with greenhouse-grown seedlings. They also may be grown in greenhouses in soil or **hydroponically** in soil-less media. **Grafted seedlings** are sometimes used. Long, frost-free growing seasons allow for best germination and yields. Soil temperatures of 24–30°C allow for accelerated pepper seed germination, and temperatures between 20 and 30°C have been reported as being optimal for eggplant. Under good conditions, seedlings of both eggplant and pepper take about 14 days to emerge after sowing. Factors that can adversely affect complete, uniform and rapid plant establishment of direct-seeded fields include temperature extremes, high concentrations of soluble salts in the soil, aridity, crusting, or over-watering. Pepper and eggplant germinate, grow, flower and fruit best in deep, well drained, medium-textured sandy loam or loam soil that holds moisture and has some organic matter. Most peppers grow best in soils with a pH 7.0–8.5. Transplanting seedlings presents many benefits. Seeds can be started in a greenhouse under controlled environmental conditions thus allowing for better germination percentage as well as more even germination. Therefore, fewer seeds need to be planted compared to direct seeding of an entire field. Additionally, transplanting helps guarantee a well-distributed stand. Many aspects of direct seeding and transplanting can be mechanized.

7. Treatment

Seed of pepper may be treated with sodium hypochlorite in order to surface sterilize seed of pathogen contamination; the treatment may also promote germination. **Priming** has been shown to improve per cent germination and rate, emergence, seedling growth, uniformity and yield in pepper, even at low temperatures. **Osmoconditioning** solutions have included potassium salts and polyethylene glycol. Plug-mix, **fluid drilling** and gel-mix delivery systems have been used to deliver non-treated, primed and pre-germinated seeds to the field. **Pelleting** seed facilitates precision planting, but may have a tendency to delay germination.

Treating seeds of both pepper and eggplant with fungicides before planting is a proven method to protect from pre-emergence **damping-off**. After planting, cultural practices that reduce seedling death and disease should be followed, including cultivation of weeds, good seedbed drainage, monitoring of insect and other pests, and monitoring of plant health. (EJV)

Bosland, P.W. and Votava, E.J. (2000) *Peppers: Vegetable and Spice Capsicums*. CAB International, Wallingford, UK.

Daunay, M.C., Lester, R.N., Gebhardt, Ch., Hennart, J.W., Jahn, M., Frary, A. and Doganlar, S. (2001) Genetic resources of eggplant (*Solanum melongena*) and allied species: a new challenge for molecular geneticists and eggplant breeders. In: van den Berg, R.G., Barendse, G.W.W., van der Weerden, G.M. and Mariani, C. (eds) *Solanaceae V. Advantages in Taxonomy and Utilization*. Nijmegen University Press, The Netherlands, pp. 251–274.

Doijode, S.D. (2001) *Seed Storage of Horticultural Crops*. Chapter 34. Eggplant: *Solanum melogena*, Tomato: *Lycopersicon esculentum*, and Peppers: *Capsicum annuum* L. The Hawthorne Press, Binghampton, NY, USA, pp. 157–180.

Peptidase

See: Protease; Storage protein – mobilization

Peptide transport

The transport of small peptides across the plasma membrane via peptide transporters plays a key role in the nitrogen nutrition during the early phases of seedling growth. Peptide transport has been studied mostly in monocot seeds, and in particular in barley (*Hordeum vulgare*), in which, following germination, storage proteins in the starchy **endosperm** are mobilized by a combination of endopeptidases and carboxypeptidases (**see: Storage protein – mobilization**). This storage protein mobilization results in a mixture of amino acids and short **oligopeptides** which are taken up by the absorptive **scutellum**, the region of the embryo adjoining the endosperm. Peptide transport into the barley scutellum is most important for the nitrogen nutrition of the seedling during its early growth since it precedes amino acid transport by at least 1 day. (**See: Mobilization of reserves – cereals**)

Peptide transport is also important in dicot seedlings. Mobilization of the storage proteins in the **protein bodies** also results in a mixture of amino acids and short oligopeptides that are transported across the protein body membrane into the cytosol of the cotyledonary parenchyma cells. Once in the cytosol the peptides are then hydrolysed to free amino acids by aminopeptidases and other peptidases.

Peptide transporters are the membrane proteins that move short oligopeptides across membranes. Mobilization of storage proteins in the endosperms of cereal grains results in a pool of short oligopeptides (and free amino acids). The plasma membrane of the epithelial cells of the scutellum in barley contains peptide transporters, which are synthesized within hours of the start of imbibition, and conduct the uptake of these oligopeptides into the embryo.

Experiments with barley scutella dissected free from the starchy endosperm have allowed the definition of peptide transporters, which are distinct from the amino acid transporters in the same plasma membrane. Dipeptides and tripeptides are most efficiently transported, with uptake of some tetrapeptides occurring at a much reduced rate. There can be a single peptide transporter, or multiple species with broadly overlapping specificities, e.g. the transporter is most active with peptides composed of L-amino acids, whereas those with D-amino acids at the N-terminus are transported at a much lower rate. In barley there may be two transporter peptides with molecular masses of 66 and 42 kDa. Transport is pH dependent, and inhibited by conditions and agents interfering with cellular respiration (anoxia, azide, cyanide), by thiol-reactive agents such as p-chloromercuribenzenesulphonate (pCMBS), and by uncouplers of the proton (H^+ ion) motive gradient across the plasma membrane, such as dinitrophenol. This suggests that the peptide transporter is coupled to the proton concentration gradient across the epithelial cell plasma membrane, and may be a peptide/proton symport. (KAW)

Waterworth, W.M., West, C.E. and Bray, C.M. (2001) The physiology and molecular biology of peptide transport in seeds. *Seed Science Research* 11, 275–284.

Pericarp

The fruit coat. The wall of the **ovary** at the fruiting stage, consisting of **epi-** (exo-)**carp**, **mesocarp** and **endocarp** (these three layers are typically differentiated in **drupes**, but not in fleshy and dry fruits where the pericarp is more or less homogeneous). In some fruits the pericarp develops relatively little and remains more or less closely adhered to the single seed; for example in the **achenes** and **caryopses** of cereals. (**See: Fruit – types; Nut; Seedcoats – structure**)

Perisperm

In some taxa, development of the **nucellus** gives rise to a (diploid) storage tissue, the perisperm. Among the monocotyledons, perisperm is present in many Zingiberales (e.g. Zingiberaceae, Fig. P.6c), but some dicotyledons also have this type of storage tissue (e.g. Piperaceae, Nymphaeaceae, most Caryophyllales, e.g. Cactaceae, Fig. P.6a, b; Amaranthaceae, Fig. P.6d; and Polygonaceae, Fig. P.6e). The presence of perisperm as the only nutritive tissue in the seed is rare. It is usually present together with **endosperm** in various proportions, locations and shapes. In the majority of families that contain perisperm, its main reserve material is **starch**. In Chenopodiaceae, the small amount of endosperm present in the mature seed is living, while the perisperm making up the predominant food store is non-living. (JD, WS)
(**See: Endosperm; Rumination**)

Boesewinkel, F.D. and Bouman, F. (1995) The seed: structure and function. In: Kigel, J. and Galili, G. (eds) *Seed Development and Germination*. Marcel Dekker, New York, USA, pp. 1–24.
Esau, K. (1965) *Plant Anatomy*, 2nd edn. John Wiley, New York, USA.
Werker, E. (1997) *Seed Anatomy*. Gebrüder Borntraeger, Berlin, Stuttgart (*Handbuch der Pflanzenanatomie = Encyclopedia of Plant Anatomy*, Spezieller Teil, Bd. 10, Teil 3) .

Perlite

Expanded perlites are used as components of sterile soil-less horticultural and as growing media in **hydroponic production** systems (such as in mixtures with **peat**), to provide aeration and substantial water retention combined with the ready drainage of any excess. They are also used: as soil conditioners for heavy clay soils, to reduce surface crusting; as covering layers for seed sown for raising transplants; as a top-dressing on growth media to reflect solar radiation back on to seedlings; as formulation carriers for **fertilizers** and **pesticides**; and for **pelleting** seed.

Technically, perlite is a generic term for a type of naturally occurring siliceous rock. Its distinguishing feature from other volcanic glasses is that, when quickly heated to above about 870°C, it expands from four to 20 times its original volume to form a white, very lightweight powder (pH 7), due to the presence of 2–6% combined water in the crude rock creating tiny bubbles. Expanded perlite grades have many industrial applications, due in part to their very low density and ability to hold substantial amounts of water. (**See also: Vermiculite**)

Peroxidation

Reactions resulting from oxidative stress or hyperoxia that cause the degradation of macromolecules. Peroxidation of lipids, proteins and DNA is one of the most commonly cited mechanisms of ageing in all biological materials, including seeds (**see: Desiccation damage; Deterioration reactions**). During peroxidation, molecular oxygen (O_2) is inserted into a macromolecule to form an O–O single bond (peroxide) (**See: Reactive oxygen species**). The peroxide bond is easily broken, creating organic **free radicals** capable of propagating damage in a reaction cascade known as autoxidation.

Peroxidation, also known as oxygenation, occurs directly when singlet oxygen reacts with the ene group (carbon double bond) of unsaturated **fatty acids** to form hydroperoxides (ROOH). It also occurs as a secondary reaction when free radicals abstract a hydrogen atom from a molecule and the resulting organic radical reacts with triplet oxygen to form a peroxyl radical, a strong oxidant that perpetuates the chain reaction (Fig. P.7). Because they are electron-dense, fatty acids

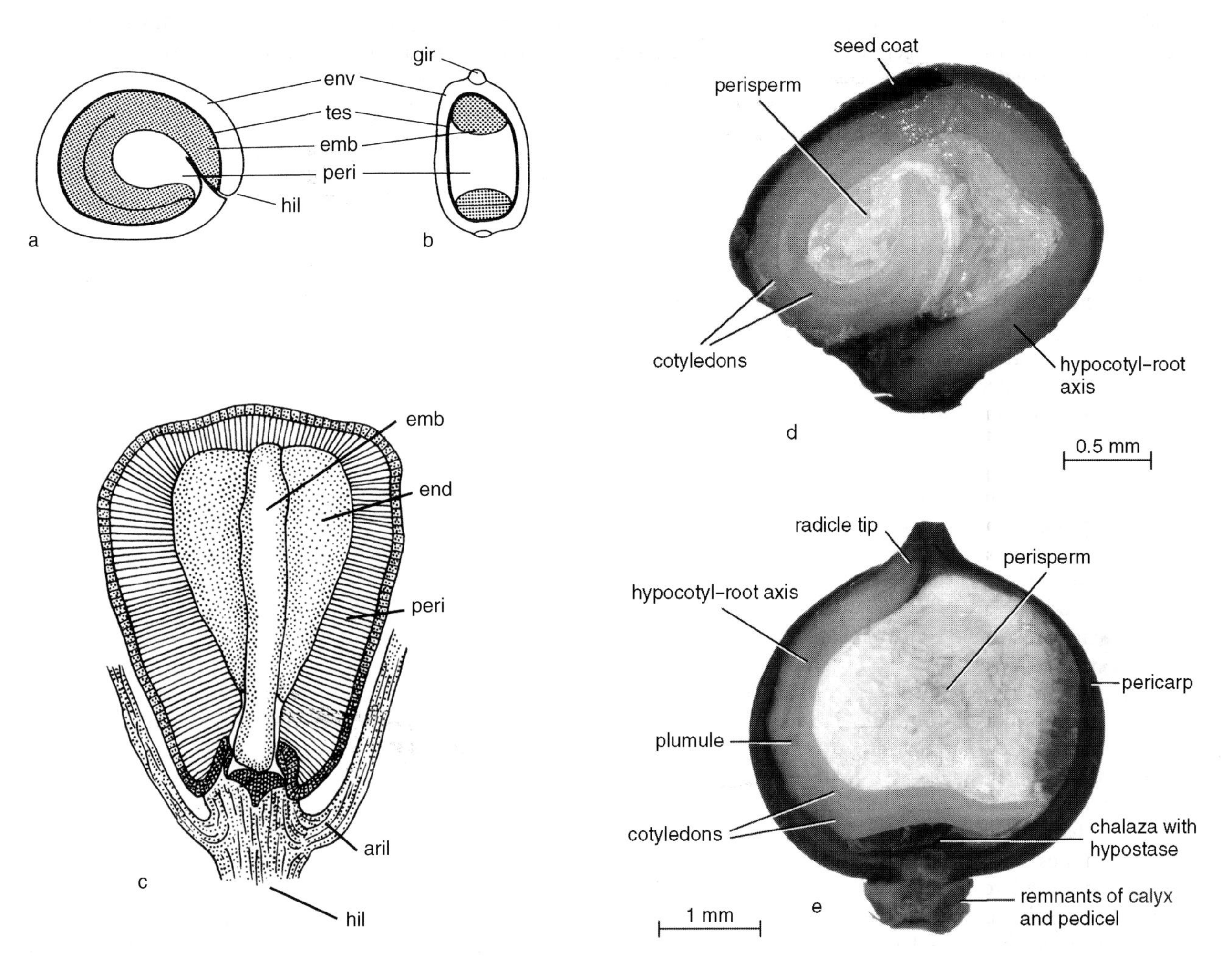

Fig. P.6. (a), (b) *Cylindropuntia* sp. (Cactaceae); transmedian longitudinal and cross section of seed showing the peripheral embryo bent around the central perisperm (the endosperm consists of only one single cell layer surrounding the embryo); on the outside the testa is covered by a woody aril derived from the funiculus (called 'funicular envelope'), a unique feature only found in the subfamily Opuntioideae and nowhere else in the angiosperms. From Stuppy, W. (2002) Seed characters and generic classification of Opuntioideae. *Succulent Plant Research* 6, 25–58. (c) *Elettaria speciosa* (Zingiberaceae); transmedian longitudinal section of seeds showing the linear embryo which is surrounded by both endosperm and perisperm. After Tschirch, A. (1891) Physiologische Studien über die Samen, insbesondere über die Saugorgane derselben. *Annales du Jardin Botanique de Buitenzorg* 9, 143–183. (d) *Pupalia lappacea* (Amaranthaceae); transmedian longitudinal section of seed (photo: Wolfgang Stuppy, copyright RBG Kew). (e) *Persicaria senegalensis* (Polygonaceae), transmedian longitudinal section of seed (photo: Wolfgang Stuppy, copyright RBG Kew). tes = Testa, emb = embryo, end = endosperm, peri = perisperm, gir = funicular girdle, env = funicular envelope, hil = hilum.

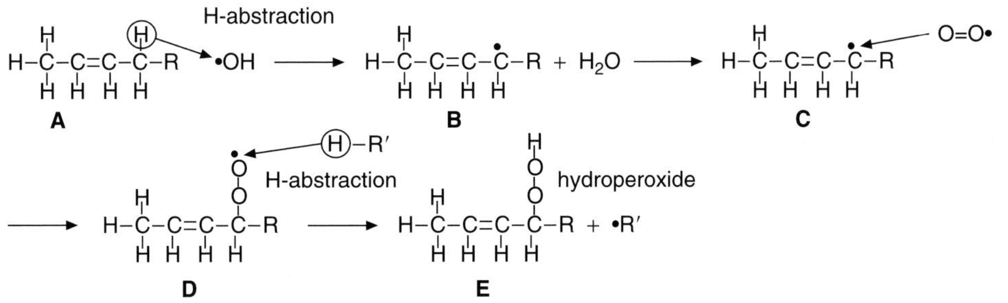

H-abstraction

A: H–C(H)(H)–C(H)=C(H)–C(H)(H)–R + •OH ⟶ B: H–C(H)(H)–C(H)=C(H)–Ċ(H)–R + H_2O ⟶ C: H–C(H)(H)–C(H)=C(H)–Ċ(H)–R ← O=O•

⟶ D: H–C(H)(H)–C(H)=C(H)–C(H)(O–O•)–R ← (H)–R′ (H-abstraction) ⟶ E: H–C(H)(H)–C(H)=C(H)–C(H)(O–O–H)–R (hydroperoxide) + •R′

Fig. P.7. Free-radical mediated peroxidation of fatty acids. The hydroxyl radical (R-OH) initiates the reaction by removing a hydrogen atom (A), leaving an unpaired electron on the fatty acid (B) and making the fatty acid more susceptible to attack by molecular oxygen, itself a free radical (C). Molecular oxygen is inserted into the molecule where the hydrogen atom was originally, forming a peroxyl radical (D). Seeking to pair its unpaired electron, the peroxyl radical removes a hydrogen atom from another fatty acid to become a hydroperoxide (E) and create a new free radical that can be degraded by molecular oxygen (C) to perpetuate the degradation. Unsaturated fatty acids (A) are more susceptible to hydrogen abstraction because the double bonds of the acyl chain weaken the adjacent methyl groups.

in lipids are susceptible to peroxidation, and unsaturated fatty acids are particularly susceptible because the carbon double bond weakens the hydrogen attachment on the neighbouring methyl group. In polyunsaturated fatty acids, the initial hydrogen abstraction causes a molecular rearrangement to form conjugated double bonds (double bonds separated by a single bond), which can be detected photometrically. The overall effects of lipid peroxidation are oxygenation of fatty acids, loss of double bonds and cleavage of the acyl chain. Lipid peroxides and hydroperoxides decompose further in a variety of reactions depending on the substrates, radiation, temperature and the presence of metal ions and this can lead to multiple by-products, including hydrocarbons (e.g. pentane, ethane), epoxides, saturated and unsaturated aldehydes (e.g. hexanal and hexenal), ketones (e.g. butanones). Highly reactive carbonyl (C=O) compounds (e.g. **malondialdehyde** and hydroxy-*trans*-nonenal) are cytotoxic. The enzyme lipoxygenase catalyses peroxidation of unsaturated fatty acids without producing free radical intermediates, and there is no evidence that the enzyme promotes the vicious cycle of damage observed with free-radical-mediated peroxidation. Substrates for lipoxygenase are almost exclusively free fatty acids and the products are thought to form important plant **hormones**. DNA and proteins may be directly damaged by peroxidation from free radical attack in the presence of oxygen or secondarily damaged through **Maillard reactions** with carbonyl by-products of lipid peroxidation. The sugar-phosphate backbones of **DNA** and **RNA** are fragmented by the hydroxyl radical to form sugar peroxyl radicals that can crosslink with proteins or degrade further into carbonyl and dicarbonyl compounds. (CW)

Peroxisomes

A subcellular organelle typically ranging in diameter from 0.3 to 1.5 μm and part of the class of organelles called microbodies. Peroxisomes are devoid of **DNA** and ribosomes, are bound by a single membrane, and are present in virtually all types of eukaryotic cells. These organelles typically contain catalase and hydrogen peroxide-producing oxidases as well as enzymes necessary for the **β-oxidation** of **fatty acids**. They also house a wide range of other enzymes involved in specific metabolic processes during different stages of a higher plant life cycle. For instance, peroxisomal enzymes catalyse reactions involved in the **glyoxylate cycle** in germinated **oilseeds** (**see: Oils and fats – mobilization**) and senescent tissues, one- and two-carbon metabolism in photosynthetically active tissues, and nitrogen transport in the nodules on legume roots. Transcriptional regulation of these enzymes appears to determine which of them are sequestered in the organelle at a given time and in a specific cell/tissue type. These differences in the contents define the several different classes of plant peroxisomes including glyoxysomes, leaf-type and root peroxisomes, and unspecialized peroxisomes. (RTM)

(See: Glyoxysomes; Microbody)

Baker, A. and Graham, I. (eds) (2002) *Plant Peroxisomes: Biochemistry, Cell Biology and Biotechnological Applications.* Kluwer Academic, Dordrecht, The Netherlands.

Huang, A.H.C., Trelease, R.N. and Moore, T.S., Jr (1983) *Plant Peroxisomes.* American Society of Plant Physiology, Monograph Series. Academic Press, New York, USA.

Pesticides

See: Treatments – pesticides

Pests – storage

See: Storage pests

Pharmaceuticals and pharmacologically active compounds

A variety of chemical types forms the active pharmaceutical and pharmacological constituents of seeds (**see: Pharmaceuticals, medicines and biologically active compounds – an overview**). These are considered in the following sections covering alkaloids, amino acids, amines and proteins, glycosides, phenolics, volatile oils, and polysaccharides used in pharmacy. (**See: Ethnobotany; Poisonous seeds; Psychoactive seeds**)

1. Alkaloids

Alkaloids are nitrogenous substances which are basic in chemical nature. Of limited distribution in plants and some other organisms, most are tertiary amines although some secondary and quaternary amines are also classed as alkaloids. Many alkaloids have a very complex molecular structure, usually comprising one or more heterocyclic rings. The rigidity conferred by this structure, together with the uneven electronic distribution, may explain their wide range of potent biological activities since these enable such molecules to mimic very closely receptors or enzyme-binding sites to produce agonist or antagonist effects.

Many seeds contain alkaloids, possibly as a protective mechanism against pests. Alkaloids can be classified according to the heterocyclic skeleton of the molecule, which is the system used below. (**See: Pharmaceuticals, medicines and biologically active compounds – an overview**, Table P.5)

(a) *Purine alkaloids.* The best known of these is caffeine, which is present in seeds of several species, many of which are used to make beverages because of the stimulatory effect of the caffeine on the central nervous system (CNS). **Coffee** 'beans' *Coffea arabica* and *C. robusta* are well-known, with *C. robusta* generally containing a higher content of caffeine (**see: Beverages**). Also well-known are drinks based on extracts of the kernels of the **cola** nut *Cola nitida and C. acuminata* and related species. These originate from West Africa but are now cultivated in other tropical regions. Another soft drink is made from the kernels of *Paullinia cupana*, a Brazilian bush, whose crushed kernels are made by local peoples into a pulp, which is then dried to produce the substance known as **guarana**, which may contain up to 5% caffeine. Caffeine is extensively used as a mild CNS stimulant, particularly in patent cold and 'flu remedies, and in preparations designed to maintain a state of wakefulness. It also raises blood pressure because of its stimulant effect on the heart and has a diuretic effect. The seeds of the **cocoa** tree *Theobroma cacao* contain caffeine and appreciable amounts of another purine alkaloid, theobromine, which has a stronger diuretic effect but has little effect on the CNS. The seeds are mostly encountered as the fermented product which forms the basis of cocoa and chocolate. (**See: Cacao**)

The shredded seeds of the areca nut *Areca catechu*, mixed with spices and lime and wrapped in the leaves of *Piper betel*, are widely used as a masticatory in south Asia and some areas of the Pacific (the so-called '**betel nut**'). Over 400 million people are estimated to use this preparation for its mild stimulant effect. The active ingredient is an alkaloid, arecoline, which also causes an increase in salivation and kills intestinal worms. It has been used for the latter property in veterinary medicine.

(b) *Tropane alkaloids*. Examples are hyoscine and hyoscyamine, widely used as pharmaceuticals especially to reduce crampings and movement of the gastrointestinal tract and to suppress CNS activity to bring about relaxation and reduce nausea. Although they are more usually derived from leaves and roots, a source of hyoscyamine (the racemic mixture of which is known as atropine) is the seeds of *Datura stramonium* (**thornapple**).

(c) *Indole alkaloids*. Several of these are used in pharmacy but only one of importance is derived from seeds. Physostigmine, from the seeds of the Calabar or ordeal bean (*Physostigma venenosum*), used in Nigeria for trials by ordeal involving the drinking of poison, was isolated in the 19th century and found to inhibit cholinesterase. It has been used in ophthalmology and in the treatment of the muscle weakness disease, myasthenia gravis. (**See: Poisonous seeds** and **Structure of seeds – identification characters** for further information on the calabar bean) More recently, rivastigmine, a synthetic compound based on the structure of physostigmine, has been introduced as a drug to treat early symptoms of Alzheimer's disease because of its cholinesterase inhibitory properties which increase the levels of the neurotransmitter acetylcholine in the CNS. Strychnine, obtained from the Nux Vomica seed *Strychnos nux-vomica*, is now of historical interest only as a pharmaceutical appetite stimulant, but it is still employed as a toxin for killing some pests, such as moles. Another type of indole alkaloid is based on lysergic acid and is found both in the group of fungi known as **ergots**, which are common infections on the grains of cereal crops, and in the seeds of the Convolvulaceae, especially **morning glory** (*Ipomoea tricolor*) and related genera. The ergot alkaloids cause poisoning in cattle and noxious effects on humans through eating ergot-infested grain, especially rye (**see: Mycotoxins; Pathogens**). One of the symptoms of poisoning is hallucinations, and lysergic acid is related to the synthetic hallucinogen LSD025. The seeds of the Convolvulaceae containing similar alkaloids were used in pre-Hispanic Mexico as a hallucinogen and gained some publicity during the 1960s as one of the many hallucinogenic substances used at that time in the USA (**see: Morning glory; Ololiuqui**).

(d) *Other alkaloids*. **Colchicine** is an alkaloid derived from the seeds of the Autumn crocus, *Colchicum sativum*, and has antimitotic and anti-inflammatory properties. It is an anomalous alkaloid since the nitrogen is not contained in the rings of the molecule and it also contains a 7-carbon ring. It is used in plant breeding to produce **polyploidy** and was one of the first agents to be used to reduce the growth of tumours, although it is not now employed for this purpose. It is still prescribed to alleviate the pain of gout.

Cevadilla seeds (*Schoenocaulon officinale*) contain steroidal alkaloids which have a minor use as insecticides and have been applied to rid the hair of lice. There has been much recent interest in the sugar analogue alkaloids such as castanospermine from the seeds of *Castanospermum australe*. These alkaloids inhibit β-glucosidase and there has been clinical interest in their application in the treatment of HIV infection since this enzyme participates in the correct glucosylation of proteins that have a key role in the assembly of new virions.

Alkaloidal pyrimidine glycosides (e.g. vicine and convicine) are found in the seeds of **faba (fava) beans** (*Vicia faba* – Leguminosae (Fabaceae)), which are often eaten as part of the diet. In the majority of the population, these have little effect, but in some populations, especially those in the eastern Mediterranean and Middle East, where a relatively high proportion of individuals is deficient in the enzyme glucose-6-phosphate dehydrogenase (GPDH), there is a severe response to the beans. The two alkaloids are metabolized to give divicine and isouramil which are damaging to GPDH-compromised individuals causing a type of haemolytic anaemia, **favism**. Formation of reduced glutathione is affected leading to the accumulation of hydrogen peroxide and **free radicals**. Malaria and faba beans have mutually exacerbating effects through action on GPDH deficiency.

An alkaloid, coniine, is also responsible for the toxic nature of the fruits of hemlock *Conium maculatum*, whose most famous application was in the death of Socrates.

Some members of the Leguminosae produce toxic quinolizidine alkaloids in their seeds which are of importance since the species involved, e.g. laburnum (*Laburnum anagyroides*) and broom (*Cytisus*) and related species are common wild and horticultural plants in temperate Europe and hence the seeds are available for careless ingestion.

2. Amino acids, amines and proteins

Many leguminous seeds contain appreciable amounts of non-essential amino acids and amines. 3,4-hydroxy-L-phenylalanine (L-DOPA) is found in several species and some, e.g. *Vicia faba*, *Mucuna* spp., have been used as commercial sources. L-DOPA is metabolized in the brain to the catecholamines, such as dopamine, and it is used for the treatment of Parkinson's disease, which is characterized by dopamine deficiency in some areas of the brain. (**See: Pharmaceuticals, medicines and biologically active compounds – an overview**, Table P.5).

Toxic amino acids (and proteins, see below) are present in many seeds and presumably have a protective role against insect, mammalian and other predators. Unusual amino acids and amines are found in the seeds of many leguminous species and toxic ones occur especially in Jequirity beans, *Abrus precatorius*. Another aspect of toxicity is shown by the unusual amino acids, known collectively as lathyrogens, such as L-γ-diaminobutyric acid found in *Lathyrus* species (e.g. *L. sativus*, *L. cicera*), also a member of the Leguminosae (Fabaceae). The seeds of this genus are consumed as food, especially *L. sativus* (**chickling pea** or vetch, grass pea), but non-cultivated species contain much higher amounts of the amino acids and are eaten in times of famine. The amino acids cause neurotoxic and skeletal abnormalities which together present as a characteristic condition known as lathyrism. An unusual cyclic amino acid hypoglycin A, which has hypoglycaemic properties, occurs in the seed arils of akee (*Blighia sapida*) particularly in unripe fruit. Akee is a plant from the West Indies and the fleshy parts of the fruit are eaten as a food. Hypoglycin, however, is also a teratogenic toxin so it is important that the seeds are removed completely from the fruits when food is prepared.

A protein of particular note is **ricin** from the **castor bean** oil plant (*Ricinus communis*), which is extremely toxic and has become suspected as a potential means of terrorist attacks (**see: Poisonous seeds**).

3. Glycosides

These consist of one or more sugars attached to a non-sugar part of the molecule, which is called the aglycone. The chemical nature of the aglycone varies immensely and can range from simple molecules such as benzaldehyde to various types of phenols, steroids and alkaloids. There is much less variation among the sugars, although some of them are rarely found free, such as the 2-deoxy sugars occurring in cardiac glycosides. Glycosides can be classified according to their pharmacological effects, e.g. cardiac glycosides, or according to the chemical nature of their aglycone, e.g. anthraquinone glycosides. (**See: Pharmaceuticals, medicines and biologically active compounds – an overview**, Table P.5)

(a) *Cyanogenic and thiocyanate glycosides.* The former release hydrogen cyanide upon hydrolysis and are characteristic of the cotyledons of seeds of some species in the Rosaceae. They are in high concentrations in the seeds of the bitter almond *Prunus amygdala*, but poisoning due to ingestion of excessive amounts occurs only rarely. They occur in lower concentrations in apple, peach, apricot, wild cherries and in some non-Rosaceous seeds such as linseed meal. Thiocyanate glycosides, e.g. sinigrin, are also known as glucosinolates and are characteristic of the Brassicaceae (Cruciferae) and, on hydrolysis, they produce unpleasant odorous, sulphydryl compounds. **Glucosinolate** compounds are hot to the taste and have been used as condiments, e.g. powdered **mustard** seed, *Brassica nigra* (**see: Spices and flavours**), and traditionally as powdered poultice applied externally to produce inflammation and relieve pain. In recent years there has been interest in the health-promoting role of these glycosides in the diet as antioxidants and as preventative agents against the development of cancer. **Antioxidants** are compounds which protect enzymes, cell membranes, DNA and other vital biochemical structures from damage by oxygen free-radicals and other active oxygen species. In the normal, healthy state, a balance of oxygen free-radicals is maintained since they are also produced as part of the body's defence mechanisms. Excessive free-radicals, however, are implicated in the etiology of many diseases and the hypothesis is that such damage can be reduced by dietary or medical intake of antioxidants.

(b) *Cardiac glycosides.* These contain an unusual steroid with a lactone ring as the aglycone, with 2-deoxysugars comprising at least one of the attached sugars. These sugars, e.g. cymarose, occur only as part of this type of molecule. Cardiac glycosides have a strong inotropic effect on the heart and are still extensively used for regularizing and strengthening the heartbeat. Although digoxin from the leaves of *Digitalis* is the most common cardiac glycoside in clinical use, these compounds are also found in the seeds of the same genus. Ouabain, which is also used clinically, is derived from the seeds of *Strophanthus* species and these have a long history of use as the active ingredients of arrow poisons in parts of Africa.

(c) **Saponins**. These are naturally occurring detergents which consist of several sugars attached to an aglycone which is either a steroid or triterpenoid. The steroidal saponins from plants are important commercially as the major source of pharmaceutical steroids used extensively as oral contraceptives, and to treat inflammatory diseases and several other conditions. Although steroidal saponins are obtained from different parts of plants, seeds are an important renewable source. **Soybean** seeds (*Glycine max*), as well as being used for food, are also a source of steroids. There has been interest in the past in using seeds of **fenugreek**, *Trigonella foenum-graecum*, for the same purpose since they contain appreciable amounts of diosgenin glycosides, and diosgenin is a useful starting point for the synthesis of other steroids. Triterpenoid glycosides have a variety of pharmacological activities and often occur in a plant as a mixture. Extracts containing these compounds, rather than the isolated compounds, are used in pharmacy. The seeds of horse chestnut, *Aesculus hippocastaneum*, produce an extract whose chief component is aescin, which is used extensively in continental Europe to reduce the inflammation associated with varicose veins. A crude water extract containing the triterpenoids of *Phytolacca dodecandra*, native to Ethiopia, where it is known as 'endod', has had some use as a molluscicide to kill the water snails which are the vectors of the parasite responsible for bilharzia. The seeds from *Swartzia madagascarensis*, which also contain triterpenoid saponins, have been used for similar purposes in Malawi and surrounding regions.

(d) *Flavonoids.* These are phenolic compounds, with a characteristic carbon skeleton (**see: Pharmaceuticals, medicines and biologically active compounds – an overview**, Table P.5), which most commonly occur as glycosides. No flavonoids are used as isolated chemical compounds as orthodox drugs but there is increasing interest in their role as active compounds in many plants used in traditional medicine. Members of the Leguminosae (Fabaceae, e.g. **soybean**) produce flavonoids known as isoflavones. These compounds act in the body in a similar way to the female sex hormones, the oestrogens (and are often termed phytoestrogens) and were first noticed because of the toxic abortifacient effects in livestock eating Leguminous plants. There has been interest, however, in their potential benefits to health as cancer preventative agents, especially because of epidemiological evidence that populations with diets containing a high intake of soybean products have a reduced incidence of breast cancer. Soybean contains two major isoflavones known as genistein and daidzein, which have an oestrogenic effect, but the results from studies carried out to test their cancer preventative properties have been equivocal. Flavonolignans are a different type of flavonoid and are the major active constituents of the seeds of the milk thistle *Silybum marianum*. An extract of these seeds is widely used in Europe for the treatment of liver ailments and to protect the liver against damaging oxygen free radicals produced by agents such as alcohol and environmental pollutants.

4. Phenolics, volatile oils, and polysaccharides

(**See: Pharmaceuticals, medicines and biologically active compounds – an overview**, Table P.5)

(a) *Phenolics.* **Coumarins** are a group of phenolic substances widely distributed throughout the plant kingdom. Coumarin itself is obtained from Tonco beans (*Dipteryx odorata*) and is used as a perfume base. The furanocoumarins, especially those from various Apiaceae (Umbelliferae) species such as *Ammi majus* and *Psoralea* spp., form the basis of the use of the **cremocarps** in traditional remedies from China and India in the treatment of skin diseases such as vitiligo and psoriasis. In the presence of sunlight, furanocoumarins become cytotoxic and prevent proliferation of skin cells associated with these diseases. 8-Methoxypsoralen is a coumarin isolated from *Ammi majus* and is used in Western medicine in combination with UV light to treat psoriasis. Gossypol is a polyphenolic dimeric naphthalene present in **cotton** (*Gossypium* spp.) seeds. It is derived from sesquiterpenes and has the empirical formula $C_{30}H_{30}O_8$. Occurring in the glands of the seed it is considered, by virtue of its toxicity, to afford some protection against insects. The seed, a by-product of cotton processing, is often used as a cattle feed but can have cardiac toxicity and so precautions have to be taken to minimize this. Gossypol has also found use as a male contraceptive, especially in China. The compound exists as two isomers. The (+) form has greater activity as a cardiac toxin while the (−) form is more effective as a contraceptive. One advantage is that the contraceptive action is reversible; nevertheless this use of the chemical is declining because of undesirable side effects. Other pharmacological properties are as an anti-viral, anti-herpes, anti-psoriasis, anti-keratitus and anti-neoplastic agent.

(b) *Oils*. Those in plant seeds fall into two major categories, the **fixed oils** (also including storage fats and waxes, **see: Oilseeds – major; Oilseeds – minor**) and the **volatile oils** (these latter also being known as **essential oils**) (**see: Oils and fats**). Although both types share the common property of immiscibility with water, they differ quite radically in their chemical and pharmacological properties. Fixed oils have use as formulating agents, but some have definite biological effects that are exploited in their pharmaceutical uses. Volatile oils are complex mixtures of quite small molecules consisting of mono- or sesquiterpenes or phenylpropanoids.

Pharmacological uses of fixed oils may be based either on substances dissolved in the oil or on the fatty acids in the triacylglycerols. Oil-soluble vitamins, especially vitamin E, an example of a tocopherol, are found in the germ oil of the seeds of various **cereal** crops, e.g. **wheat** germ oil from *Triticum* spp. Vitamin E is important for proper development of the spinal cord in the foetus and for a variety of body functions including wound healing. **Sesame** oil, from seeds of *Sesamum indicum*, is used as a food but it contains a lignan, sesamin, which is a synergist for pyrethrum insecticides, so it is often mixed with pyrethrum extracts when it is used for this purpose. Croton oil from *Croton tiglium*, has fallen from its use in Victorian times as a drastic laxative because of the toxicity of the contained phorbol diterpenoids, which were also responsible for the purgative effect. The seeds of *Nigella sativa*, sometimes known as **black cumin**, are extensively used in the Middle East and in other societies with a predominantly Islamic culture (**see: Spices and flavours**). One of the uses of the oil is as a treatment for arthritic pain and this is partly due to thymoquinone, the major component of the volatile oil, which is dissolved in the fixed oil.

Some fixed oils from seeds contain high amounts of relatively uncommon **fatty acids** that form the basis of pharmaceutical usage of the oils. Castor oil from *Ricinus communis* contains very large amounts of ricinoleic acid, a hydroxylated analogue of oleic acid that has laxative properties because of its irritant effect on the intestine. The use of castor oil as a laxative has now largely ceased, although the oil is still used extensively in the formulation of soothing creams. A more recent introduction as a product widely sold and used, is **evening primrose** oil derived from the seeds of *Oenothera biennis*. This oil contains large amounts of γ-linolenic acid which is metabolized in the body to prostaglandins and leukotrienes. These have a variety of protective and beneficial effects, especially in some inflammatory and pain-associated conditions. In the late 20th century, it was licensed in the UK as a treatment for breast pain and for eczema, but the licence was later withdrawn. It is also commonly reputed to help alleviate premenstrual symptoms in many women, but evidence based on clinical studies is not very convincing. Seeds of blackcurrant *Ribes nigrum* and borage *Borago officinalis*, also contain high amounts of γ-linolenic acid and have also been used in dietary supplements promoted for similar beneficial effects. Fatty acids are known to have an antimicrobial activity. The oil from the seeds of *Hydnocarpus* species, known in India as Chaulmoogra oil, contain unusual fatty acids with a cyclic terminal group which are strongly active against *Mycobacterium* species, particularly those which are the cause of leprosy. This oil has been used topically to treat this disease in India.

Volatile oils are widely used in **spices** and as perfumes and are the active components of many medicines. Only a few volatile oils are obtained from seeds alone, although many important oils and pharmaceuticals are obtained from the fused seeds and fruits, known as cremocarps, which are characteristic of the Apiaceae (**umbellifers**). These include the oils of **caraway** (*Carum carvi*), **fennel** (*Foeniculum vulgare*), **dill** (*Anethum graveolens*) and **anise** (*Pimpinella anisum*). All of these oils are used as carminatives, a term which describes the aid to digestion caused by a relaxant effect on the smooth muscles of the stomach sphincter and intestinal wall. Oil of **nutmeg**, from the endosperms of *Myristica fragrans*, has a similar effect but has also been used as a hallucinogen. The phenylpropanoid myristicin is thought to be the compound responsible but its mode of action has not been clarified (**see: Spices and flavours**).

(c) *Polysaccharides*. These consist of sugar units joined to form long strands or branched polymer molecules. Many types of polysaccharides are present in seeds, although **starch** is the major one in many instances. In some seeds the polysaccharides are classified as **gums** since they swell and dissolve in water to form viscous solutions. Some of these are exploited in pharmacy as formulation agents for dispersion of liquids and fine solids, e.g. **carob** gum from *Ceratonia siliqua*, while others have a quasi-clinical role. A prime example is **guar** gum from *Cyamopsis tetragonolobus*, which reduces the uptake of glucose from the intestinal tract and is added to food to control diabetes. The **seedcoats** of some species contain **mucilage**, and the ability of this substance to swell considerably in the presence of water has been utilized in various ways. Some soothe irritation in membranes, e.g. **linseed** (*Linum usitatissum*) which has been used to treat bronchitis, whilst the seed husks from *Plantago* species, such as psyllium (*P. afra* syn., *P. psyllium*) and ispaghula (*P. ovata*) are frequently used as mild laxatives, since swelling in the colon produces reflex peristalsis and contraction. (PJH)

(See: Psychoactive seeds)

Table P.5. Biologically active compounds found in seeds: chemical types, examples and their uses.

Chemical type	Carbon skeleton of molecule	Example	Source	Activity
Alkaloids – purine	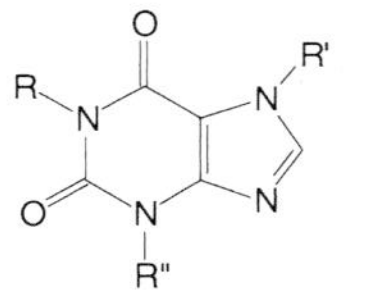	Caffeine 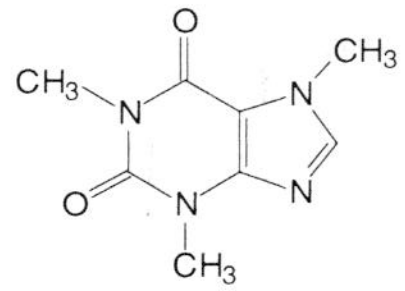	*Coffea arabica*	CNS stimulant
Alkaloids – tropane	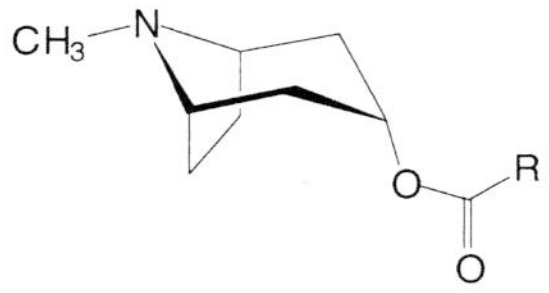	Hyoscyamine 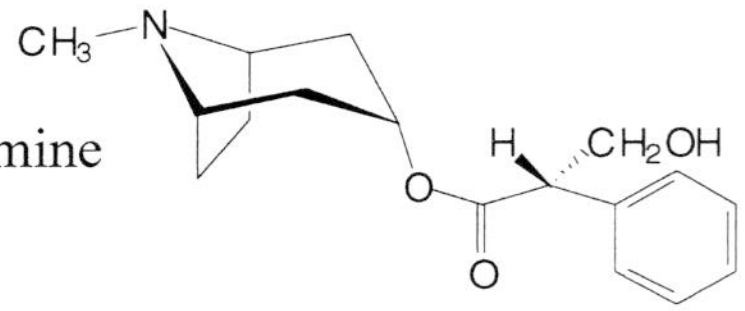	*Datura stramonium*	Anticholinergic – reduces smooth muscle spasm
Alkaloids – indole	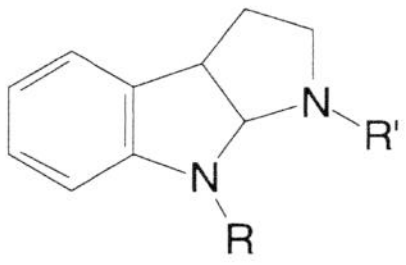	Physostigmine 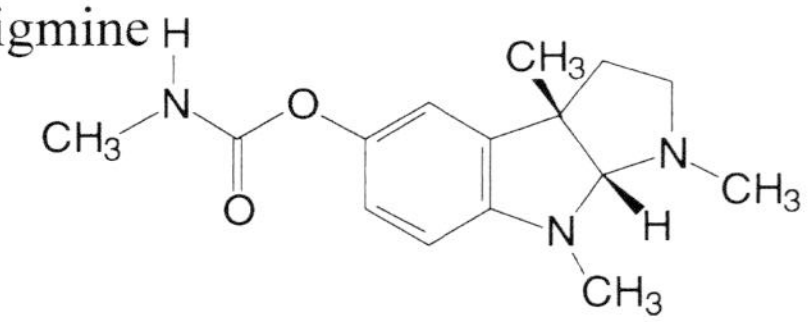	*Physostigma venenosum*	Cholinesterase inhibitor – used for myasthenia gravis.
Amino acids	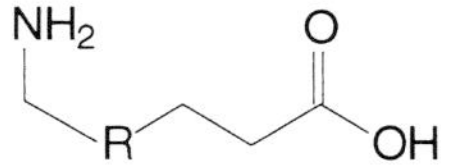	L-γ-Diaminobutyric acid 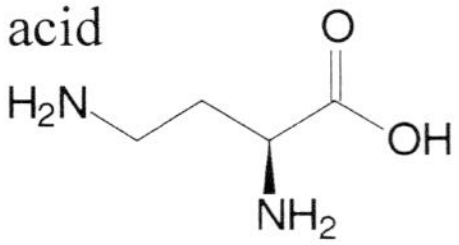	*Lathyrus sativus*	Toxic – causing paralysis
Glycosides – glucosinolates	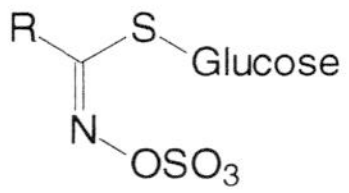	Sinigrin 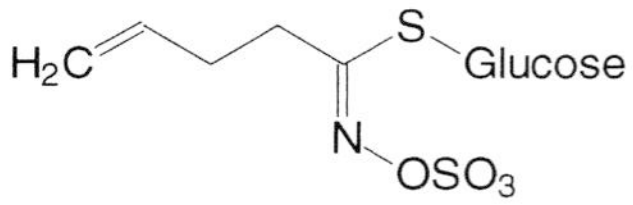	*Brassica nigra*	Pungent taste on hydrolysis, causes increased blood flow through skin.

Chemical type	Carbon skeleton of molecule	Example	Source	Activity
Glycosides – cardiac	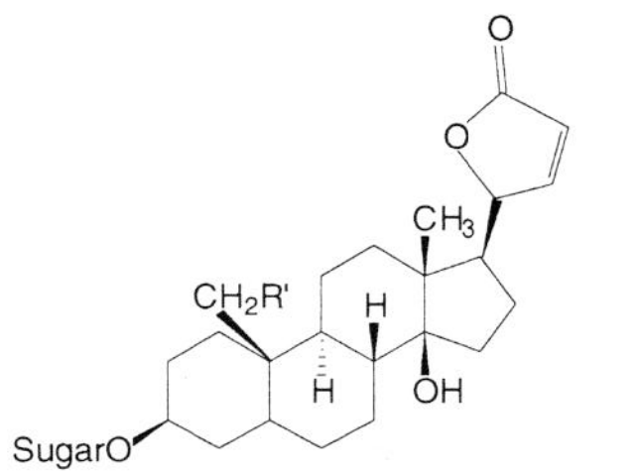	Ouabain 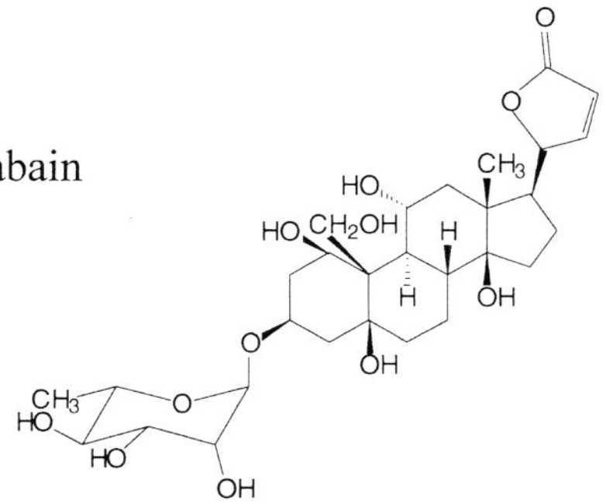	*Strophanthus gratus*	Stimulation of cardiac muscle
Glycosides – saponins (steroidal)	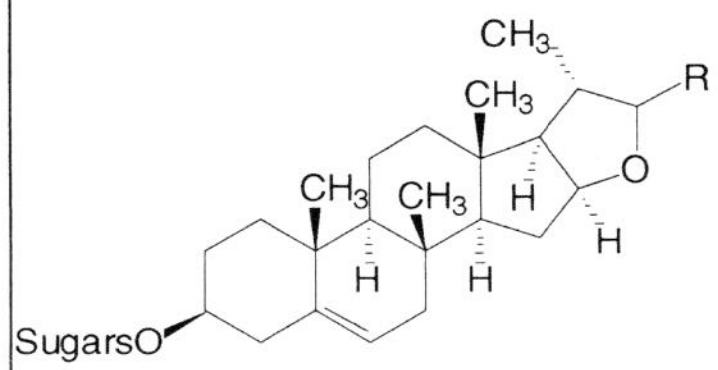	Dioscin 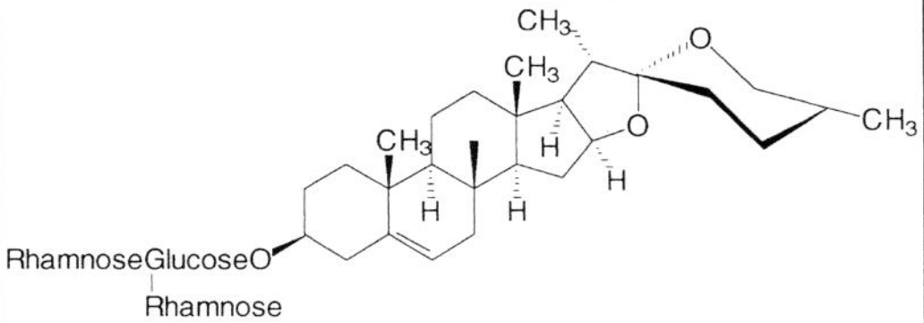	*Trigonella foenum-graecum*	Source of steroids used as antiinflammatories and female contraceptives
Glycosides – saponins (triterpenoid)	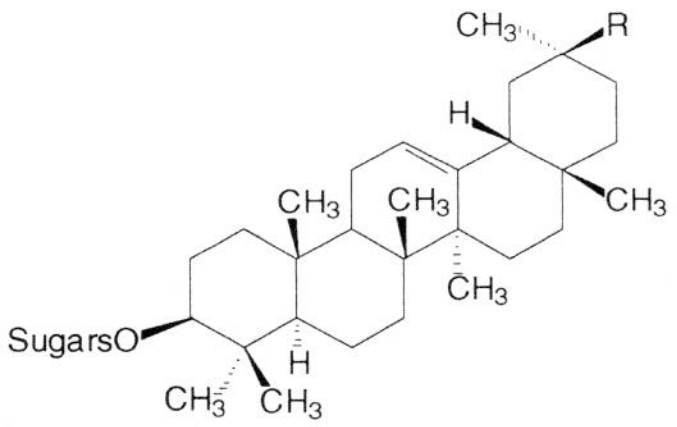	Aescin 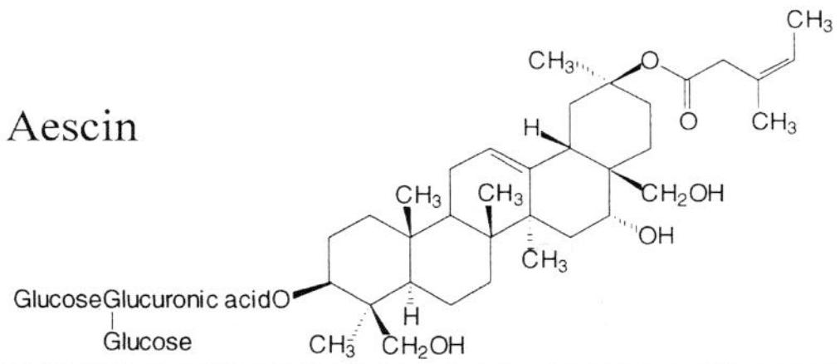	*Aesculus hippocastaneum*	Antiinflammatory for varicose veins

Table P.5. *Continued*

Chemical type	Carbon skeleton of molecule	Example	Source	Activity
Isoflavones	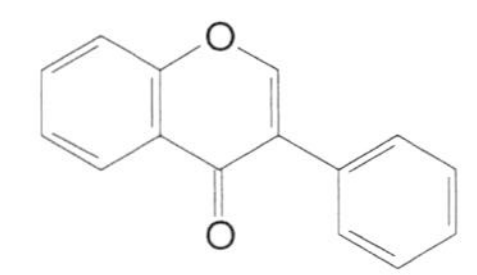	Daidzein 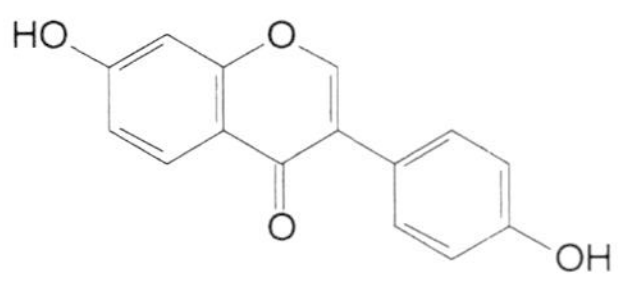	*Glycine max*	Weak oestrogenic effects, possible cancer preventative effect
Furanocoumarins	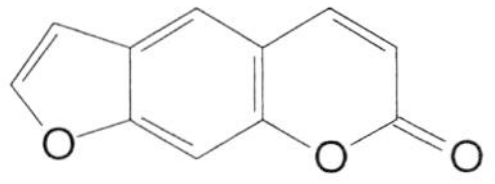	8-Methoxypsoralen 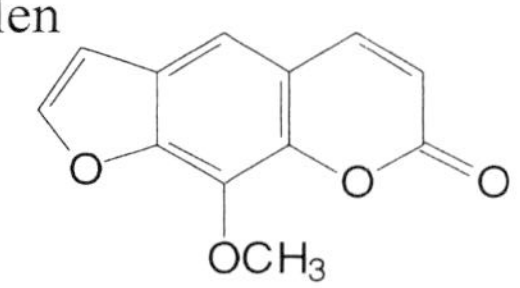	*Psoralea corylifolia*	With UV light used to treat psoriasis
Phenolic naphthalene	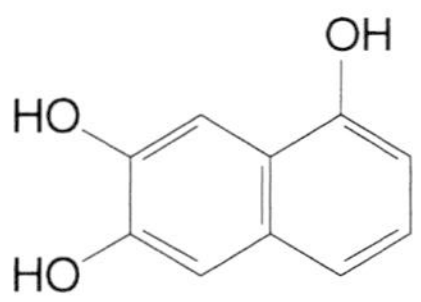	Gossypol 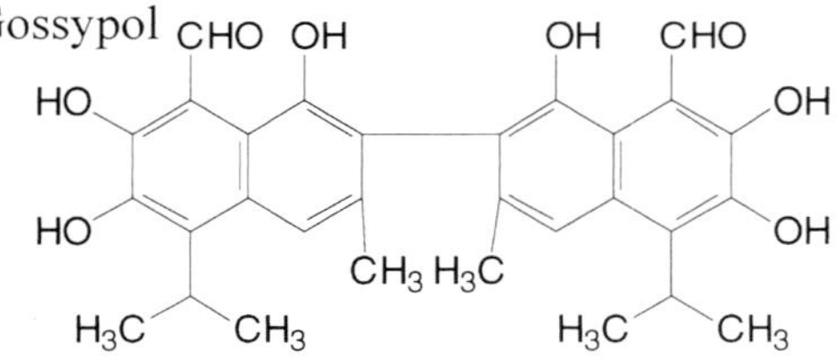	*Gossypium* spp.	Male contraceptive effect
Fixed oil acyltriglyceride	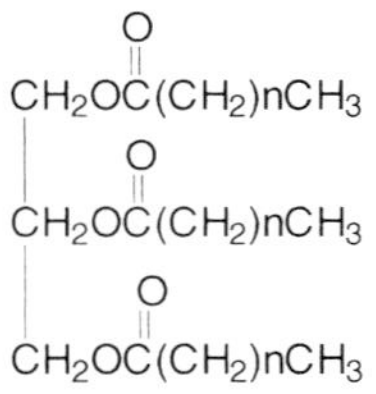 $CH_2OC(O)(CH_2)nCH_3$ $CH_2OC(O)(CH_2)nCH_3$ $CH_2OC(O)(CH_2)nCH_3$	Ricinoleic acid triglyceride 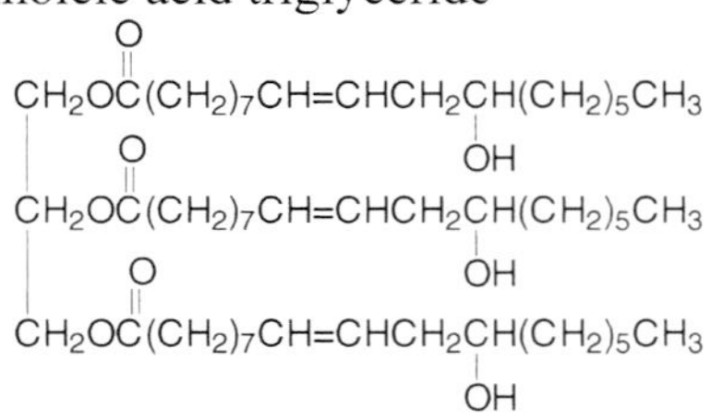 $CH_2OC(O)(CH_2)_7CH{=}CHCH_2CH(OH)(CH_2)_5CH_3$ $CH_2OC(O)(CH_2)_7CH{=}CHCH_2CH(OH)(CH_2)_5CH_3$ $CH_2OC(O)(CH_2)_7CH{=}CHCH_2CH(OH)(CH_2)_5CH_3$	*Ricinus communis*	Laxative

Chemical type	Carbon skeleton of molecule	Example	Source	Activity
Volatile oil monoterpenoids	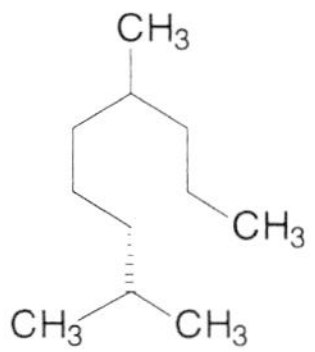 CH_3 CH_3 CH_3 CH_3	(-) Carvone CH_3 O CH_3 CH_2 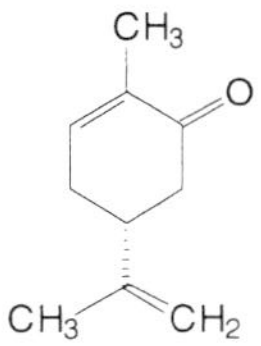	*Carum carvi*	Antispasmodic
Volatile oil phenylpropanoids	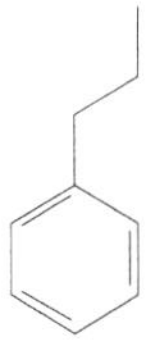	Myristicin 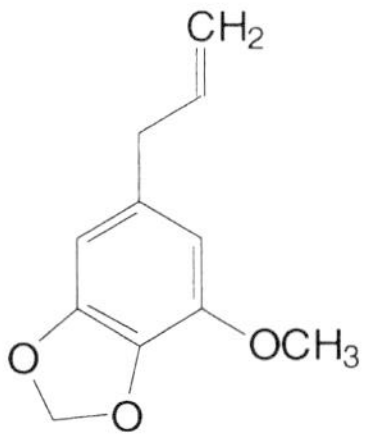 CH_2 OCH_3 O O	*Myristica fragrans*	Carminative, possible hallucinogen
Polysaccharides	-Sugar-Sugar-Sugar-	Guar gum component -Galactose-Mannose-Galactose-Mannose-	*Cyamopsis tetragonolobus*	Decreases uptake of sugar from intestine – used orally to control blood sugar levels in diabetes

Dewick, P. (2002) *Medicinal Natural Products*, 2nd edn. Wiley, Chichester, UK.
Evans, W.C. (2002) *Trease & Evan's Pharmacognosy*, 15th edn. Saunders, London, UK.
Samuelsson, G. (1999) *Drugs of Natural Origin*, 4th edn. Swedish Pharmaceutical Press, Stockholm, Sweden.

Pharmaceuticals, medicines and biologically active compounds – an overview

Seeds, in common with many other plant parts, are extensively used in most societies and cultures because of the biological activity of the chemical substances which they contain. A widespread use of seeds and their constituents is as pharmaceuticals. The seed itself, an extract or a chemical compound extracted from it, has been used in the preparation of medicines, either as the active ingredient or in formulation. Medicines can be defined as substances that alleviate or cure disease, and pharmaceuticals are the substances that form the components of a medicine. The pharmacological activity of pharmaceuticals from seeds relies on the chemical substances present. In most instances a mixture of chemically related compounds is found in the seed, but that present in the greatest amounts, or having the greatest pharmacological effect, is considered to be the 'active' ingredient. The variety of chemical types forming the active constituents (Table P.5) are considered in detail in sections (1) alkaloids, (2) amino acids, amines and proteins, (3) glycosides, (4) phenolics, volatile oils, and polysaccharides used in pharmacy in: **Pharmaceuticals and pharmacologically active compounds**.

1. Active pharmaceutical ingredients

It is widely recognized that plants produce biologically active compounds as a survival mechanism, either to attract beneficial organisms, e.g. pollinators and fruit **dispersal** agents or, more usually, to deter, destroy or limit organisms which compete with the plant. Many such compounds affect mammalian physiology in quite small doses and they are applied as pharmaceuticals to treat disease or physical deficiencies. Many of these compounds, however, can have adverse effects, and are poisonous substances or toxins. The boundary between pharmaceuticals and toxins is indistinct, however, and is determined largely by the dose involved. A special group of biologically active compounds called hallucinogens affect normal perceptions and sensations of the environment. These effects often mimic religious experiences, and plants that contain hallucinogens are often incorporated into religious rituals, in a wide range of societies throughout the world. As distinct from poisons, few hallucinogens have therapeutic applications. (**See: Poisonous seeds; Psychoactive seeds**)

2. Formulations

Pharmaceuticals are either used in formulation to improve the appearance, delivery and stability of the active constituent, or as drugs that have a pharmacological effect on the body systems. Many fixed oils, usually obtained by pressing or grinding the **endosperm** or **cotyledons** of seeds, are used in formulations, especially in topical preparations such as massage oils, ointments and creams. The drug is dissolved in the oil which assists penetration through the lipophilic layers of the skin. Common seed oils used in this way are **coconut** oil, **castor** oil, arachis (**peanut**) oil and **almond** oil (**see: Oilseeds**). A more specialized use of an oil in formulation is found with theobroma oil, from **cocoa** butter, which melts at approximately body temperature and was formerly extensively used in the preparation of suppositories. The fractionated kernel oil of seeds of the oil **palm** (*Elaeis guineensis*), which melts at the same temperature, is more commonly now used as a suppository base. Many of these fixed oils are used as emollients, i.e. they make the skin smoother and more flexible. Most substances having this effect consist mainly of **oils**, i.e. long chain acids esterified with glycerol, but an exception is **jojoba** from *Simmondsia chinensis*, which is a liquid wax, i.e. it consists of esters of long chain fatty acids and long chain alcohols.

Other constituents found in seeds which are useful as formulating agents include the volatile (or essential) oils which are used for flavouring and perfuming medicinal (and other) products. Examples of seeds which provide such agents are the **cremocarp** fruits found in the Apiaceae (umbellifers) such as **dill** (*Anethum graveolens*), **aniseed** (*Pimpinella anisum*) and **fennel** (*Foeniculum vulgare*) (**see: Spices and flavours**).

Seeds may contain **mucilage** in their **testa**, possibly to retain or absorb moisture. This has been utilized pharmaceutically to produce gels, especially in the intestinal tract, where the mucilage-containing seedcoats of linseed (*Linum usitatissimum*) and *Plantago* spp., e.g. psyllium, attract water and increase the gut contents, thereby producing a reflex evacuation (**see: Gums**). (PJH)

Dewick, P. (2002) *Medicinal Natural Products*, 2nd edn. Wiley, Chichester, UK.
Evans, W.C. (2002) *Trease & Evan's Pharmacognosy*, 15th edn. Saunders, London, UK.
Samuelsson, G. (1999) *Drugs of Natural Origin*, 4th edn. Swedish Pharmaceutical Press, Stockholm, Sweden.

Phase diagram/transition/separation

The structural (volume, conformation, miscibility), mechanical (strength, elasticity, permittivity) and thermodynamic (enthalpy, entropy, heat capacity) properties determined by a combination of inter- and intramolecular interactions. Phase diagrams (Fig. P.8) delineate phase transition temperatures given pressure, composition or other factors. The equilibrium transition temperature is characteristic for a substance and provides an analytical tool for identification as well as assessment of purity. For example, the transition temperature for pure water is 0°C (liquid water ↔ ice) and solutes reduce this temperature through known colligative relationships. There are two types of phase transitions that are distinguished by how the energy of the substance changes with temperature.

In a first order transition, there is a discrete change in the free energy function at the transition temperature, causing an abrupt change in entropy, enthalpy and volume (first derivatives of free energy) and a characteristic change in form to a solid, liquid or gas. The change in energy causes a release or consumption of heat (exothermic versus endothermic, respectively). The substance is said to supercool if it fails to crystallize below the equilibrium freezing temperature. Supercooling is promoted by rapid cooling, impurities, and no nucleation sites and can lead to the formation of **glasses**,

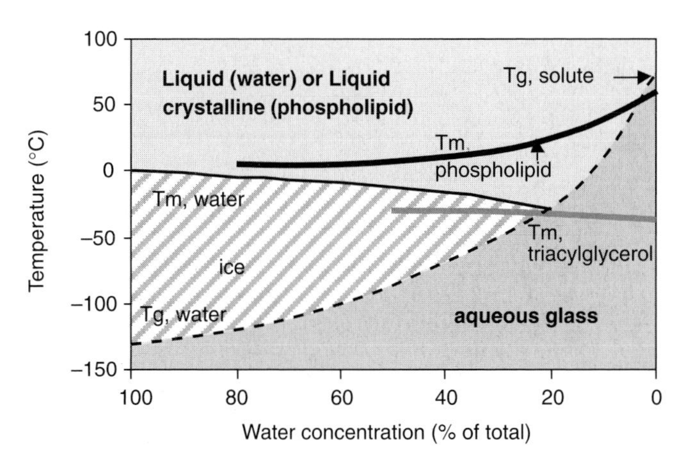

Fig. P.8. Phase diagram of water and lipid showing the interrelationships between water content (a variable of composition) and the transition temperature for water or lipids. The diagram shows that the first order transition for water (liquid ω ice, also known as freezing and melting) (Tm, water) is 0°C for pure water (100% water) and decreases to about −25°C when water content decreases (by adding solutes) to about 20%. At water concentrations less than 20%, freezing and melting transition are rarely observed suggesting that water remains unfrozen (less accurately called unfreezable water). Above Tm, water is fluid (grey shading) and below Tm, water is crystalline (cross-shaded area). The temperature of the second order transition for pure water (Tg, water) is about −134°C, but approaches the Tg of the dry solute as the solution becomes increasingly concentrated. Above Tg, the structure of water will be in its equilibrium state (fluid or crystalline); while below Tg, water will be 'frozen' into a structure reminiscent of a higher temperature (**see: Glass**). The phase behaviour of polar lipids (Tm, phospholipids) is greatly affected by water concentration, while phase behaviour of nonpolar lipids (Tm, triacylglycerol) is less affected.

which are representative of the second type of phase transition.

In a second order transition, the free energy function does not change discretely at the transition temperature, but the volume, enthalpy or entropy functions do, leading to an abrupt change in the heat capacity (second derivative of free energy), marked changes in mechanical properties but no changes in form.

Phases have also been used to describe miscibility of substances, most often on the basis of polarity but also on the basis of molecular geometry. For example, water-soluble materials partition to the aqueous (also known as hydrophilic) phase of the cytoplasm and non-polar materials partition to the non-aqueous (also known as lipophilic) phases of the cell. The hydrophilic and hydrophobic interactions of amphipathic molecules (molecules that have both hydrophilic and hydrophobic regions) are fundamental to the establishment of cell organelles and emulsions. A first-order phase transition from a liquid to a solid can reduce the miscibility of substances, causing them to partition out in a process known as demixing or phase separation.

Phase behaviour has been implicated in many temperature-dependent processes of seeds because of the expected profound change in molecular structure at physiological water contents and temperatures. First order phase transitions of **membrane**, **lipids** and triacylglycerols may cause **desiccation damage**, imbibitional damage, low temperature sensitivity during storage, and demixing of lipid-soluble **antioxidants** (i.e. Fig. P.8 shows that membranes of drying or rehydrating seeds would undergo a phase transition at ambient temperatures). Second order phase transitions of the aqueous milieu in seeds to **glasses** is attributed to the storage stability of dried seeds (Fig. P.8 shows that seeds containing about 5–10% water and stored at ambient temperatures would be near Tg, but those stored at < 0°C would be protected from damaging ice formation since it is restricted by the glass). Changes in water properties are regarded as phase changes and may contribute to water content effects on the nature and kinetics of metabolic activity and stress tolerance in seeds (**see: Glass; Water binding**). Thermotropic seed coatings, developed by the seed technology industry to allow early planting, are impermeable to water at temperatures below their melting point (**see: Filmcoating**). (CW)

Phaseolin

Also called glycoprotein II and GI globulin, it is a storage protein that is a 7S **globulin** present in the embryo of the common bean *Phaseolus vulgaris*. It is a much-studied protein; it was among the first to be extensively characterized using electrophoretic separation techniques, for its messenger RNA to be isolated and translated in a cell-free system, and for its cDNA and genomic sequences to be determined. Its gene was also the first to be introduced into seeds of other species, as a **transgene**, and shown to respond therein in a developmentally regulated manner.

Some landmark publications are:

Sengupta-Gopalan, C., Reichert, N.A., Barker, R.F., Hall, T.C. and Kemp, J.D. (1985) Developmentally regulated expression of bean β-phaseolin in tobacco seed. *Proceedings of the National Academy of Science USA* 80, 1897–1901.

Sun, S.M., Buchbinder, B.U. and Hall, T.C. (1975) Cell-free synthesis of the major storage protein of the bean, *Phaseolus vulgaris* L. *Plant Physiology* 56, 780–785.

Sun, S.M., Slightom, J.L. and Hall, T.C. (1981) Intervening sequence in the plant gene – comparison of the partial sequence of cDNA and genomic DNA of French bean phaseolin. *Nature* 289, 37–41.

Phaseolus beans – cultivation

This group includes dwarf French or common bean, navy bean and haricot bean (*Phaseolus vulgaris*), lima bean (*Phaseolus lunatus*), runner bean (*Phaseolus coccineus*) and other closely related types. *P. vulgaris* is the most commonly grown species and many of the characteristics are common to all *Phaseolus* species. (**See: Bean – common; Lima bean; Scarlet runner bean** for information on economic importance, origins, seeds, uses, etc.)

Beans can be grown in a wide range of climates, as a warm-season annual in subtropical areas to temperate regions, although they require enough heat to develop rapidly within the short growing period. The plants are generally bushy and grow up to 1 m high, although *P. coccineus* is mainly single-stemmed and has a climbing habit, twisting the stem around its neighbours. Leaves are trifoliate although the first pair of leaves are single and entire. The flowers are borne on axillary racemes originating from thickened nodes. The flowers range in colour from white to purple or red. The pods are linear and have two valves as with other legume species.

With all species, breeding is being carried out to develop a wide range of characteristics. For common beans grown as a fresh vegetable (or harvested mechanically and frozen as green pods) disease resistance, pod characteristics and flavour are prime objectives for improvement. Dry bean breeding is aimed at producing **determinate flowering** types, yield improvement and uniformity of seed size and colour, whilst disease resistance is a high priority especially for rust (*Uromyces appendiculatus*) and white mould (*Sclerotinia sclerotiorum*).

Seed is very fragile and is susceptible to mechanical injury during the handling and seed-processing operations, especially the fracturing of the plumule. Such damaged seed produces a seedling shoot without a growing point, known as 'bald head'. In the USA, *Phaseolus* beans may become too dry for combining; to prevent this, bean **windrows** may be sprayed with water to prolong or induce seedcoat toughening.

Soil temperatures need to be above 10°C before germination begins and the optimum soil temperature should be nearer to 20°C to allow rapid early growth. Seedling establishment is **epigeal**. (**See: Epigeal**, Fig. E.8a) The crop is usually grown in rows having been drilled to a depth of 4 cm. Beans are favoured by light to medium soil textures and an adequate supply of soil moisture either by irrigation or natural rain enhances the emergence. They are very susceptible to frost.

During the season, fungicides such as vinclozolin, or iprodione can be used to protect against *Botrytis cinerea*, which causes a pod rot and against white mould (*Sclerotinia sclerotiorum*). Seed production of the vegetable types is mainly in irrigated, arid areas of western USA or Canada where the disease risk is lower. Several bacterial pathogens are considered serious, including halo blight (*Pseudomonas syringae* pv. *phaseolicola*) and common blight (*Xanthomonas phaseoli*) and strict hygiene measures are enforced in the seed production areas. In most intensive agricultural systems, insecticidal seed treatments such as chlorpyrifos is essential to control damage by bean seed fly larvae (*Delia platura*).

When mature, the seed crops are mechanically harvested but the combine harvesters contain rubber belts which rub the beans out of the pods to reduce the risk of mechanical injury. Beans bred for their pod quality characteristics also tend to produce seeds with a thin and therefore easily damaged testa. Special care is required when handling and harvesting these types.

Phaseolus beans are also susceptible to pre-emergence seed rot following infection by *Pythium ultimum* in cold wet soils which particularly affect low-vigour seeds. Seed **vigour** is therefore a factor particularly in regard to white-seeded bean types and various methods of seed **vigour testing** are in use. The electrical conductivity test has been adopted in the same way as for peas and biochemical staining of the cotyledon tissue using tetrazolium chloride also provides a useful method of assessing seed vigour (**see: Vigour testing – biochemical**). (AB)

For reading, see the lists given for the species in this article, in **Legumes** and in **Grain legumes**.

Phenolics

A wide range of aromatic metabolites that possess (or formerly possessed) at least one hydroxyl group attached to the phenyl ring. Various phenolic compounds, such as phenolic acids (*p*-hydroxybenzoic, vanillic, gallic, *p*-coumaric, caffeic, ferulic, sinapic, chlorogenic acids, etc.), coumarins, flavonoids and tannins, are present in many seeds, particularly in their coats. Their oxidation products give the brown colour which frequently appears in the coats of mature seeds (and their oxidation and condensation are responsible for various browning reactions affecting plant tissues and organs). They may act as **germination inhibitors** or regulate germination through their chemical or enzymatic oxidation which results in a decrease in oxygen supply to the **embryo**.

(DC, FC)

(See: Chilling injury; Germination – influences of chemicals; Germination – influences of gases; Polyphenol oxidases)

Côme, D. and Corbineau, F. (1998) Semences et germination. In: Mazliak, P. (ed.) *Physiologie Végétale II. Croissance et Développement.* Hermann, Paris, France, pp. 185–313.

Croteau, R., Kutchan, T.M. and Lewis, N.G. (2000) Natural products (secondary metabolites). In: Buchanan, B., Gruissem, W. and Jones, R. (eds) *Biochemistry and Molecular Biology of Plants.* American Society of Plant Physiologists, Rockville, MD, USA, pp. 1250–1318.

Phospholipids

The predominant lipids present in most **membranes**, usually composed of two **fatty acid** chains esterified to two of the carbons of glycerol phosphate, with the phosphate esterified to one of various polar groups (e.g. serine, inositol, ethanolamine, choline). Phospholipids are **amphiphilic** molecules characterized both by hydrophobic fatty acids and a polar, charged head group, which confers affinity for water. This property allows phospholipids to form a bilayer. (**See: Cells and cell components; Desiccation damage**)

Phosphorylation

The addition of a phosphate residue, and hence a negative charge, to a compound, for example to specific amino acid residues in a protein; the introduction of negative charges can change the conformation and substrate binding capacity of a protein, a means by which a signal is passed along a **signal transduction** pathway. (**See: Dormancy breaking – cellular and molecular aspects; Late Embryogenesis Abundant Proteins**)

Photoblastic

Applied to seeds whose germination is sensitive to light. Seeds whose germination is promoted by light are said to be positively photoblastic, and those whose germination is inhibited by light are negatively photoblastic. (**See: Cryptochrome; Light – dormancy and germination; Phytochrome**)

Photochromicity

The ability of a light-absorbing molecule to switch back and forth between two isomeric forms as a result of absorbance of a photon of light. Each isomeric form has a different optimum wavelength for light absorbance. An example is **phytochrome**. **See: Light – dormancy and germination**

Photodormancy

A form of **secondary dormancy** induced by light conditions that are unfavourable for germination (e.g. high photon fluxes). (**See: Light – dormancy and germination**)

Physic nut

Also known as purging nut, these are not true nuts but are seeds of *Jatropha curcas* (Euphorbiaceae). The species is thought to have originated in Central America but now grows widely in the tropics. The ovoid, orange-brown seeds (3–4 per capsule) are about 8 mm long and 4 mm wide. They contain about 35% non-edible oil which attracts attention for development as a possible bio-fuel (**see: Oilseeds – minor**). The residue left after extracting the oil is rich in protein but is toxic and therefore cannot be utilized for animal feed. Because of the toxicity (**see: Poisonous seeds**) the raw seeds are dangerous for human consumption, but after roasting, which removes or greatly reduces the toxic components, they are eaten in Veracruz and some other parts of Mexico. Non-toxic provenances of seed occur in Mexico that are consumed locally. (MB)

Physical dormancy

A form of dormancy maintained by seed covering structures (**seedcoats**) due to a lack of permeability to water or gases, or to mechanical resistance of the covering structures that prevents emergence of the embryo from the seed. (**See: Dormancy – coat imposed; Hardseededness**)

Physiological maturity

The stage of **development** at which a seed, or the majority of a seed population, has reached its maximum **viability** and **vigour**. This, however, is not usually the stage of maturity at which seed should be harvested from the plant, whether it is being produced for sowing the next crop or for grain processing, since seeds generally achieve physiological maturity at moisture contents that are too high for safe handling and storage. However, the seed may still be suitable for **swathing** and curing in the field. (**See: Harvest Maturity; Production for sowing, IV. Harvesting**)

Phytin

A storage form of phosphate and mineral ions deposited in seeds during development and mobilized following germination. A minor reserve, usually accounting for 0.5–2% of seed dry weight.

Phytin (phytate) is the insoluble mixed potassium (K^+), magnesium (Mg^{++}) and calcium (Ca^{++}) salt of *myo*-inositol hexaphosphoric acid (phytic acid) (Fig. P.9). The negatively charged sites on the phosphate groups of the *myo*-inositol hexaphosphoric acid, of which there are 12, bind the mineral ions (cations).

Phytin is located in **protein bodies** of the **aleurone layer** in cereal grains, in those of the **endosperm** and **cotyledons** of dicot seeds, and of the **megagametophyte** in gymnosperms. Phytin forms a distinct structure, the **globoid**, within

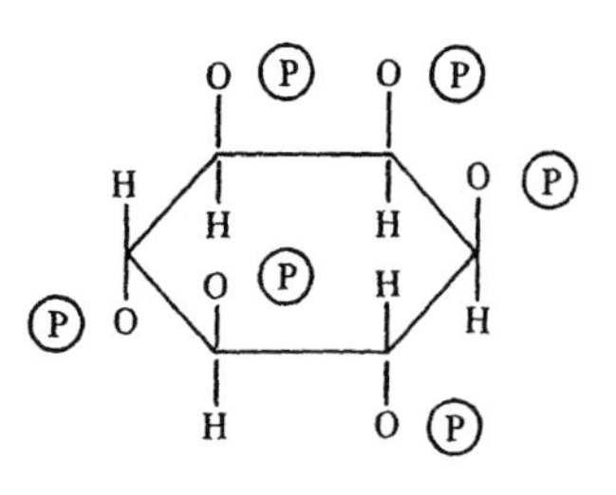

Fig. P.9. Phytic acid molecule.

the **protein body**, which is electron-dense due to the presence of the cations (Fig. P.10). These can be measured quantitatively in fixed sections of tissue using an electron microscope customized with energy-dispersive X-ray spectrometry (EDAX). While the major ions present in phytin are Mg^{++}, Ca^{++}, and K^+ (Table P.6), others also may be present in very small amounts, e.g. barium (Ba^{++}), iron (Fe^{+++}), manganese (Mn^{++}) and zinc (Zn^{++}).

The phytin content of seeds and its mineral ion composition is influenced by the nutrient status of soils. Seeds of plants growing in soils containing high phosphate synthesize an increased amount of phytin; as much as a four-fold increase occurs in soybean under conditions of high phosphate fertilization. However, there also can be a substantial variation in phytin content (as great as 80%) between cultivars of the same species (e.g. soybean) growing under identical conditions. Differences in the mineral content of soils appear to have a relatively smaller influence on the cation composition of phytin in developing seeds.

Phytin is synthesized on the endoplasmic reticulum of developing seed storage cells and is transported therefrom in

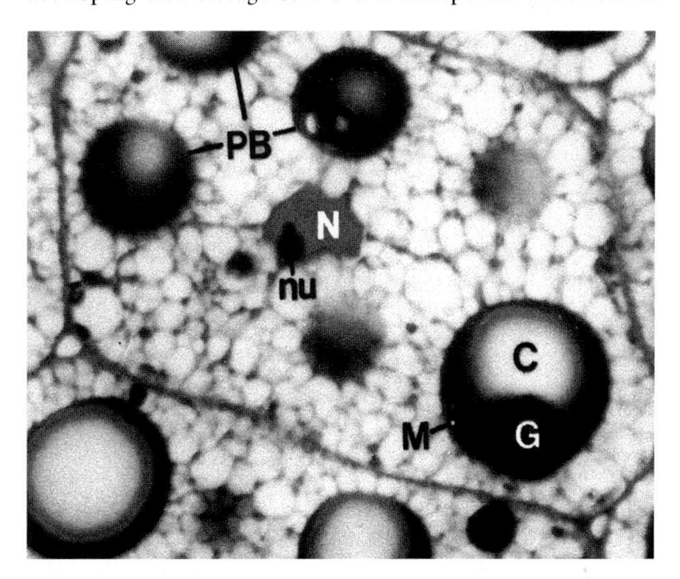

Fig. P.10. Light micrograph of a cell in the endosperm of a castor bean seed showing the presence of protein bodies (PB) surrounded by many smaller lipid bodies. One protein body is labelled to show the presence of insoluble crystalloid storage protein (C) and a phytin-containing globoid (G) embedded in more soluble storage protein matrix (M). N, nu: nucleus and nucleolus. Courtesy J.S. Greenwood, Univ. Guelph.

Table P.6. Mineral content in the phytin of selected seeds.

Species	Region	Mg^{++}	Ca^{++}	K^{+}
Castor bean	E+Em	0.41	0.03	0.48
Brazil nut	Em	0.17	0.64	0.83
Sunflower	Em (C)	0.21	0.11	0.56
Pea	C	0.1	0.8	1.04
Soybean	C	0.22	0.13	2.18
Barley[a]	Al	48.8	8.4	164.2
	Em	17.5	4.9	59.5

Some variability in mineral ion content occurs between cultivars and lines of a species, and also is influenced by the mineral content of the soil. Amounts expressed are as a % of seed dry weight, except in barley[a], for which the values are expressed in μg/seed part.
Major region in which phytin is deposited: C, cotyledon; E, endosperm; Al, aleurone layer; Em, embryo.
Data from: Lott, J.N.A., Randall, P.J., Goodchild, P.J. and Craig, S. (1985) Occurrence of globoid crystals in cotyledonary protein bodies of *Pisum sativum* as influenced by experimentally induced changes in Mg, Ca and K contents of seeds. *Australian Journal of Plant Physiology* 12, 341–353; and Stewart, A., Nield, H. and Lott, J.N.A. (1988) An investigation of the mineral content of barley grains and seedlings. *Plant Physiology* 86, 93–97.

vesicles to the protein bodies. The biosynthetic pathway from *myo*-inositol-1-phosphate to the hexaphosphoric acid (phytic acid) is unknown. Following germination the phytin is hydrolysed by phytase to release the cations, phosphate and *myo*-inositol which are used to support early growth of the seedling.

Phytin is an anti-nutritional agent because it is not metabolized by animals and can bind essential dietary minerals in the gut (e.g. Ca^{++}, Fe^{+++}, Zn^{++}) thus reducing their absorption and mineral deficiency, particularly in children, is an identified problem in the developing world. (**See: Cereals – composition and nutritional quality**) The extensive processing of foods for human consumption in developed countries diminishes their phytin content to a non-problematical level. In domestic animals fed on seeds of cereals and legumes, phytin reduces mineral availability, and is eliminated in their waste. This results in increased phosphate in run-off water (eutrophication), especially where there is intensive farming. Phytin in grains, when fed in excess, can cause thinning of bones, especially in fowl, because of its ability to bind Ca^{++}. Molecular technology is being used to remove phytin from seeds. One approach has been to introduce genes for phosphatase (phytase) enzymes that are expressed during seed development. So far this has not replaced the practice of introducing preparations of bacterial phosphatases into animal feed so that the phosphate is released from the phytin during digestion. Phytase enzymes have been genetically engineered into pigs, which are released in their saliva, in an attempt to remove utilizable phosphate from phytin. (JDB)

Lott, J.N.A., Ockenden, I., Raboy, V. and Batten, G.D. (2000) Phytic acid and phosphorus in crop seeds and fruits: a global estimate. *Seed Science Research* 10, 11–33.

Phytochrome

A photoreceptor with peak absorption in the red (*ca.* 660 nm) and far red (near infrared) (*ca.* 730 nm) regions of the spectrum. The name originates from a combination of the Greek words for plant and colour. Phytochrome controls many aspects of plant development and metabolism, and is important in the regulation of seed germination and dormancy.

Plants, as sessile organisms, need to be able to display considerable developmental plasticity in order to accommodate to and take advantage of the environment in which they grow. Light is a very important factor in the life of a plant and hence plants have evolved an extensive range of photoreceptors to perceive it, including phytochrome and the blue light receptor, **cryptochrome**. These photoreceptors allow plants to detect the quantity, quality, duration and direction of incident light. Phytochrome regulates a range of developmental responses throughout the life of a plant from **germination** and de-etiolation through regulation of plant architecture to controlling time of flowering. Regulation of germination is possibly the most crucial role of phytochrome. The best developmental strategy for a seed falling in an unfavourable environment may be to delay germination until the environmental situation is more favourable. (**See: Light – dormancy and germination**)

Phytochrome is a reversible photochromic pigment. It is synthesized in a biologically inactive, predominantly red-absorbing, Pr form and on illumination with red light (maximal activity at 660 nm) is converted to an active, predominantly far red-absorbing, Pfr form. Subsequent illumination with far-red light (maximal activity at 730 nm) results in conversion back to the Pr form (Fig. P.11). Alternatively, in long periods of darkness a process known as dark reversion can also lead to the relatively slow conversion of Pfr back to Pr. Both forms of phytochrome absorb to some extent over the whole spectrum from near ultra-violet to near infra-red but peak absorptions (and therefore major activities) are of red and far-red wavelengths by Pr and Pfr, respectively. Furthermore, the absorption peaks of these two forms overlap to some extent meaning that at a given wavelength of light, rather than securing a complete conversion to one form or another, a particular ratio (photoequilibrium) of Pr to Pfr is set up. This photoequilibrium is a dynamic one in that while the ratio of Pr to Pfr remains constant at a given wavelength, transitions will still be occurring between the two forms. A dynamic photoequilibrium is also established in light of mixed wavelengths (e.g. sunlight) (**see: Light – dormancy and germination**, Fig. L.6).

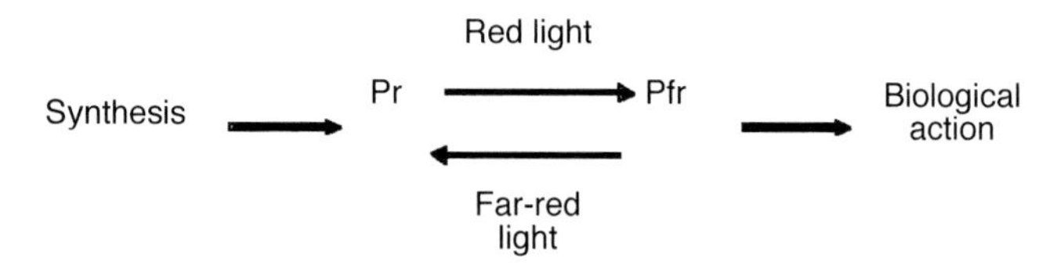

Fig. P.11. Phytochrome photoconversion. The phytochrome chromoprotein displays a red/far-red-reversible photochromicity. Phytochrome is synthesized in the inactive, red-absorbing Pr form (peak absorption 660 nm). In red light Pr is converted to the active, far-red absorbing Pfr form (peak absorption 730 nm), initiating a range of developmental responses. In far-red light this Pfr form is converted back to the Pr form.

In small seeds with a low mass of storage reserves, which are insufficient to support growth of the seedling when buried too deeply, light activation of phytochrome is a key requirement to ensure that germination occurs close to the surface.

Studies of the effect of light on germination gave us the first clues as to the nature of phytochrome. Like many seeds, imbibed seeds of some cultivars of lettuce (*Lactuca sativa*) show a strong promotion of germination in response to light. In 1935, Lewis Flint and Edward McAlister demonstrated that illumination with red light (R) promoted germination but far red inhibited it. Later, Harry Borthwick, Sterling Hendricks and colleagues discovered that red and far red could reverse each other's effects and in 1952 they proposed a radical model in which a single pigment with two photo-interconvertible forms could be responsible for the induction or suppression of germination. By 1959, this prediction had allowed Warren Butler to complete the first isolation of the pigment, phytochrome.

Phytochrome is a **biliprotein**. The protein moiety (the **holoprotein**) consists of a 124 kDa protein with a covalently attached linear **tetrapyrrole chromophore** derived from biliverdin (Fig. P.12). Two such holoproteins then combine to form a dimer, the state in which phytochrome occurs in the cell. Five types of phytochrome exist in higher plants, (phyA–E). They have the same chromophore but different proteins which are the products of a family of five genes, *PHYA–PHYE*. PhyA, the product of the *PHYA* gene, is light labile in its Pfr form: it is sometimes referred to as type I phytochrome. It is the predominant phytochrome in dark-grown seedlings but amounts of phyA rapidly fall upon illumination with red (or white) light. PhyB–phyE are light stable (sometimes referred to as type II phytochrome) with phyB forming the most predominant phytochrome in light-grown seedlings.

Phytochrome responses can be categorized according to the amount of light required to initiate a response: the **Very Low Fluence Responses (VLFR)**, to a very dim pulse of light; the classical R/FR reversible **Low Fluence Responses (LFR)**; and the **High Irradiance Responses (HIR)**, to prolonged irradiation. The majority of our understanding of the roles of the various phytochromes in these responses has come from the study of mutants of the model plant *Arabidopsis thaliana* deficient in one or more pigment types. PhyA has been demonstrated to be responsible for the VLFR and also the FR HIR. PhyB–phyE participate in the LFR.

In examining the role of the phytochromes in germination (**see: Light – dormancy and germination**), experimenters typically measure the percentage of germination of a population of seeds as opposed to observing the germination responses of individual seeds. Hence a measurement of the degree of promotion of germination can be obtained. VLFR, LFR and HIR responses can promote germination of *Arabidopsis* seeds. Wild-type *Arabidopsis* seeds show the classical R/FR reversible promotion of germination observed in lettuce (LFR). This response is eradicated in the *phyB* mutant, implicating phyB as the major player in this response. *phyB* mutant seeds, however, still show some promotion of germination in response to a very dim pulse of light (VLFR) and a pronounced promotion of germination in response to prolonged FR irradiation (the FR HIR). The absence of these responses by *phyA* mutants implicates phyA as the photoreceptor involved in these two responses. Interestingly, the promotion of germination by the FR HIR in *Arabidopsis* is more pronounced in a *phyB* mutant, indicating some moderation of this response by phyB. In the absence of both phyA and phyB, action of a further phytochrome can still be observed. Double mutant seeds, *phyA phyB*, show a promotion of germination in response to prolonged R irradiation (>24 h), which is mediated by phyE. The above-mentioned promotion of germination by the FR HIR in *Arabidopsis* is, however, atypical. Many species show an inhibition of germination in response to prolonged FR. In these cases, prolonged FR light is thought to act to inhibit incipient axis elongation rather than to affect the breaking of dormancy.

Germination can be promoted by phytochrome stored in mature dry and dormant seeds. Germination of imbibed seeds in darkness is high if the seeds are collected from parent plants

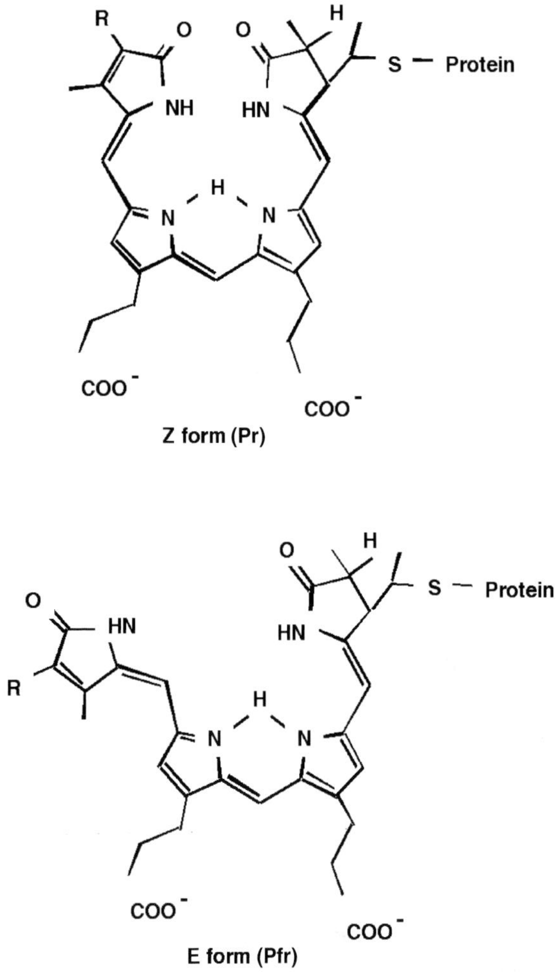

Fig. P.12. The phytochrome chromophore, phytochromobilin, a linear tetrapyrrole. As a result of absorbing a photon of light, there is a change in conformation, due to a *cis–trans* isomerization about the double bond between rings C and D. The red and far-red-absorbing Pr and Pfr forms (Z and E forms) of the chromophore are shown. Isomerization between these two forms results in an associated conformational change in the phytochrome protein moiety.

grown in a high R:FR ratio. In such seeds phytochrome, therefore, must be maintained in the active Pfr form in the dry seed ready to promote germination following imbibition. Seeds collected from plants grown in a low R:FR ratio, such as occurs under canopy shade, show a much reduced germination because of their low Pfr value. Such seeds will remain dormant until no longer shaded, when the light environment will be more favourable for growth. Observations in phytochrome mutants have demonstrated that phyB is the main seed-stored phytochrome.

Phytochromes are localized throughout the adult plant. At the cellular level, for all phytochromes of the phyA–E type, the Pr form is localized within the cell cytoplasm. However, upon conversion to the Pfr form, the phytochromes move into the nucleus where they localize in discrete 'speckles', possibly sites of activation of gene expression. This phenomenon results in a diurnal regulation of phytochrome subcellular localization. The intracellular localization of phytochromes in seeds, however, is not understood.

The mode of action of phytochrome in photomorphogenesis is being elucidated, and an important component of this is its binding to one of several proteins. Phytochrome **kinase** Substrate 1 (PKS1) binds to phyA and phyB in the cytoplasm and is **phosphorylated** upon phytochrome photoconversion. In the nucleus, phyA and phyB bind to Phytochrome Interacting Factor 3 (PIF3), a transcription factor that is responsible for the light regulation activation of a number of genes. Interaction between phytochrome and Nucleotide Diphosphate Kinase 2 (NDPK2) has also been demonstrated suggesting that phytochrome action also occurs via a kinase cascade. Finally, a physical interaction between phytochromes and blue light photoreceptors has been shown for both phyA and phyB. PhyA directly interacts with the blue light photoreceptor cryptochrome 1, whilst phyB directly interacts with cryptochrome 2 (cry2) resulting in a co-localization of phyB and cry2 to the same nuclear speckles. The involvement of any of the phytochrome-interacting factors is unknown.

The action of phytochromes also involves classical intracellular signalling mechanisms such as **G proteins**. Further involvement of **cyclic GMP** and calcium/calmodulin occurs in divergent phytochrome-induced responses. (**See: Signal transduction – an overview; Signal transduction – light and temperature**) No studies have been made specifically on seeds, although signalling mechanisms are likely to be the same as in adult plants.

Many **mutants** of *Arabidopsis* have been discovered that display a loss of all or a selection of phytochrome-mediated responses. A number of these mutants are the result of disruption in genes encoding transcription factors, and it has been proposed that a major mechanism by which phytochrome triggers changes in plant development is via changes in gene expression. In support of this, microarray analysis has demonstrated that about one-third of the genes in the *Arabidopsis* genome are regulated by light during seedling establishment with close to 10% specifically affected by phyA or phyB signalling.

A particularly interesting class of photomorphogenic mutants are the Constitutively Photomorphogenic (*cop*) mutants. These show a light-grown phenotype in almost all respects when grown in complete darkness, likely because the COP1 protein is a repressor of photomorphogenesis in wild-type plants. COP1 represses activation of light-regulated genes. Thus for many genes, light activation actually involves the removal of a repressor of gene expression. In darkness, COP1 is nuclear-localized, where it binds to the HY5 transcription factor, normally an activator of gene expression. Binding of COP1 to HY5 causes HY5 to be degraded, preventing it from activating gene expression. In light, COP1 is exported from the nucleus releasing HY5 to activate light-regulated genes. However, COP1 does not seem to play a major role in seeds, since *cop1* mutant seed of *Arabidopsis* still show a wild-type, phytochrome-mediated light responsiveness for germination.

Despite our knowledge of the changes in gene expression mediated by phytochrome, the ultimate mechanisms by which phytochromes specifically affect germination and dormancy are unclear. Some clues can be gained by looking at the genes that are specifically-regulated by phytochrome in de-etiolating seedlings or in established plants. Many such phytochrome-regulated genes are transcription factors, particularly those showing rapid induction in response to light. However, several other categories of genes are regulated by phytochrome, including those involved in **hormone** biosynthesis, hormone transport or hormone response pathways, as well as genes involved in cell expansion and cell division, phenomena which are important in the promotion of germination. **Gibberellin** is the key hormone involved in promotion of germination and there is strong evidence for phytochrome regulation of gibberellin 3-oxidase in germinating seeds of *Arabidopsis*. This enzyme catalyses the last step in the formation of active gibberellins (GA_1 in most species, GA_4 in *Arabidopsis*), and this is consistent with the observation in lettuce seeds that the amounts of such gibberellins are regulated in a R/FR-reversible manner. It also explains why application of GA promotes germination of light-requiring seeds in darkness, i.e. the light requirement is bypassed. (**See: Dormancy breaking – hormones; Germination – influences of hormones; Light – dormancy and germination**)

A number of ion-channel genes, including one encoding a potassium channel, are phytochrome-regulated. Changes in membrane permeability due to such regulation of ion channels could be important in generating changes in turgor pressure required for germination as the embryo breaks through the surrounding and constraining endosperm. This finding also correlates well with observations that phytochrome can alter membrane permeability. Phytochrome has a rapid effect on leaf angle in *Samanea saman*, a nyctinastic leguminous plant of the Mimosa family. *Samanea* opens its leaves and leaflets in response to light and closes them in darkness. In this case light, acting through phytochrome, triggers movement of potassium ions across the membranes of cells in the pulvinus, a specialized motor organ located at the junction of leaf and stem. Changes in the osmotic potential in the cells in this region result in changes in turgor pressure (potential) which generate the mechanical forces that determine leaf orientation. (PD)

Casal, J.J. and Sanchez, R.A. (1998) Phytochromes and seed germination. *Seed Science Research* 8, 317–329.

Ma, L., Li, J., Qu, L., Hager, J., Chen, Z., Zhao, H. and Deng, X.W. (2001) Light control of Arabidopsis development entails

coordinated regulation of genome expression and cellular pathways. *The Plant Cell* 13, 2589–2607.
Nagy, F. and Schafer, E. (2000) Control of nuclear import and phytochromes. *Current Opinion in Plant Biology* 3, 450–454.
Neff, M.M., Fankhauser, C. and Chory, J. (2000) Light: an indicator of time and place. *Genes and Development* 14, 257–271.
Quail, P.H. (2002) Phytochrome photosensory signalling networks. *Nature Reviews in Molecular Cell Biology* 3, 85–93.
Smith, H. (2000) Phytochromes and light signal perception by plants – an emerging synthesis. *Nature* 407, 585–591.
Toyomasu, T., Kawaide, H., Mitsuhashi, W., Inoue, Y. and Kamiya, Y. (1998) Phytochrome refulates gibberellin biosynthesis during germination of photoblastic lettuce seeds. *Plant Physiology* 118, 1517–1523.
Whitelam, G.C., Patel, S. and Devlin, P.F. (1998) Phytochromes and photomorphogenesis in *Arabidopsis*. *Philosophical Transactions of the Royal Society London (Biology)* 353, 1445–1453.

Phytoglycogen

A highly branched, water-soluble α-D-glucan polymer, present in some plants, that is similar in structure to animal and bacterial glycogen. Phytoglycogen molecules have approximately 10% branch linkages, and a fairly uniform average chain length of 14–15 glucosyl units. Phytoglycogen accumulation occurs in seeds of maize, rice and barley that are homozygous for mutations in genes encoding an isoamylase-type starch debranching enzyme, suggesting that deficiency of the isoamylase results in synthesis of a more highly branched glucan polymer. In *sugary1* **sweetcorn** varieties, the presence of phytoglycogen in the kernel contributes a desirable creamy texture. (MGJ)
(See: Starch)

Phytohaemagglutinin (PHA)

A trivial name for the **lectin** present in kidney (French or common) bean (*Phaseolus vulgaris*) seeds (**see: Bean – common**). It is an abundant protein accounting for up to 10% of the total seed protein in some cultivars. PHA occurs as a natural mixture of isoforms of approximately 120 kDa, which arise by the random association of a leucocyte agglutinating L-subunit (252 amino acids) and an erythrocyte agglutinating E-subunit (254 amino acids) into tetramers. Neither the L- nor the E-subunit binds simple sugars but react exclusively with complex glycans. By virtue of this particular specificity, PHA is highly reactive with most differentiated mammalian cells that present glycoconjugates containing complex N-glycans on their plasma membrane. PHA and especially PHA-L is widely used as a mitogen (stimulator of mitosis) in lymphocyte cultures. A high dietary intake of PHA is nutritionally toxic for most higher animals including birds, rats, ruminants and humans. The lectin survives passage through the gastro-intestinal tract where it binds avidly to the epithelial cells and induces hyperplastic (increased tissue) growth. For this reason PHA is the major antinutrient of kidney bean. At high doses, PHA can be lethal, as is illustrated by accidental poisoning by raw or insufficiently cooked beans. In principle, PHA is inactivated by heat treatment, but it is fairly heat-stable, especially in dry beans. PHA is a reasonably abundant protein (representing 2–10% of the total seed protein) in most commercial kidney bean cultivars. Lectins very similar to PHA have been identified in seeds of a few other *Phaseolus* species (e.g. *P. acutifolius*, *P. lunatus*). However, these lectins are not called PHA but *P. acutifolius* and *P. lunatus* lectin, respectively. (EVD, WJP)

Mirkov, T.E., Wahlstrom, J.M., Hagiwara, K., Finardi-Filho, F., Kjemtrup, S. and Chrispeels, M.J. (1994) Evolutionary relationships among proteins in the phytohemagglutinin-arcelin-α-amylase inhibitor family of the common bean and its relatives. *Plant Molecular Biology* 26, 1103–1113.

Phytomelan(in)

A black, polyphenolic deposit laid down in the **testa** or **seedcoat** of some seeds. (**See: Seedcoats – structure; Sunflower**)

Phytosanitary certification

A system of official rules (legislation and regulation) to prevent the introduction and/or spread of quarantine pests and **pathogens** across international boundaries, or to limit the economic impact of regulated non-quarantine pests, including establishment of phytosanitary procedures (a term derived from the Greek for 'plant cleaning') and their certification.

These regulations cover seedborne pathogens – as well as those carried on whole plants or plant material (**see: Pathogens – seedborne infection**). Many importing countries specify such phytosanitary import requirements and require a combination of import permits and phytosanitary certificates for the international movement of a seedlot, even if it is destined for re-export (after storage, and possible splitting up, combination with other consignments, or repackaging); otherwise the lot may be destroyed. The exporting country issues phytosanitary certificates as a result of seed health testing to ascertain that seeds are free from general or specifically regulated diseases or pest organisms. The certificate might only be issued after quarantine or disinfection. Possible additional requirements are documenting that growing-season inspections have been performed, or declarations that seed has been produced in an area free from the pests and pathogens.

To help achieve international harmonization of phytosanitary measures, and avoid the use of unjustifiable measures as barriers to trade, the **International Plant Protection Convention** has introduced International Standards for Phytosanitary measures (ISPMs) as well as guidelines and recommendations on a number of issues including 'Guidelines for Phytosanitary Certificates' and a standard on the 'Principles of Plant Quarantine as related to International Trade'.

Problems are created in the seed industry where importing and exporting countries use different protocols to conduct a test, and get different results. Countries have established standardized protocols and annual proficiency tests, such as in the USA. (**See: Certification; Health Testing; International Seed Health Initiative (ISHI); ISTA Seed Health Committee**) (VC)

Anon. (2001) *Guidelines For Phytosanitary Certificates*. FAO, Rome.

Phytotoxicity

Liable to damage or kill (especially, higher) plants: a term used in a restricted sense to describe adverse side-effects caused by fungicides, insecticides or formulating agents to the crop. Some seed-treatment **active ingredients** slightly inhibit the speed of germination, and may lead to impaired **vigour** and seedling **emergence** in certain circumstances. (**See: Treatments – pesticides**)

Pigeon pea

1. World importance and distribution

Pigeon pea (red gram) (*Cajanus cajan*), a **warm-season legume** widely grown in the tropics for its edible seeds and pods, is one of the oldest food crops and ranks fifth in importance among edible legumes of the world. A little more than 4 million ha are planted in Asia, Africa and Latin America, almost 80% of which is in India. In the latter, pigeon pea represents around 10% of the total pulse crops. Other important producers are Malawi, Mozambique, Kenya, Uganda, Tanzania, Ethiopia, Myanmar, Thailand, Indonesia, the Philippines and Australia. In Latin America, pigeon peas are important in the Caribbean islands, Venezuela and Panama. (**See: Crop Atlas Appendix, Map 3; Legumes**)

2. Origin

The most accepted theory is that India is the home of the pigeon pea from where it was taken to Malaysia, then to East Africa and across to West Africa. Seeds are believed to have travelled the slave route from Zaire or Angola to Bermuda, the West Indies, the Guianas and Brazil. An origin in tropical Africa from where it was carried by traders to India or Ceylon has also been suggested.

3. Seed characteristics

The smooth seeds are nearly spherical to ovoid, green when immature but variable when mature – white, grey, yellow, sometimes mottled with purple-red, light or dark brown (Fig. P.13), entirely red or black or black speckled with white, 6–9 mm in diameter, weighing 4–25 g/100 seed. The pod contains three to ten seeds and is depressed between them.

Fig. P.13. Pigeon pea seeds (image by Elly Vaes and Wolfgang Stuppy, Millennium Seed Bank, RBG, Kew).

4. Nutritive value

Pigeon pea seeds are rich in high quality **storage protein** (averaging 7% fresh weight, fw, in unripe seeds; from 18 to 32% in mature seeds). The main proteins are two **globulins**: cajanin (58% of the total N) and concajanin (8%). Cajanin contains most of the methionine, cystine and tyrosine but is low in tryptophan, threonine and lysine. The mature seeds contain 57.3 to 58.7% (fw) carbohydrates, 1.2 to 8.1% crude fibre and 0.6 to 3.8% **oil** (mostly triacylglycerol), the **fatty acids** being linolenic (5.7% of the total), linoleic (51.4%), oleic (6.3%) and saturated fatty acids (37.7%). Pigeon peas are low in antimetabolites (though inhibitors of trypsin and chymotrypsin are present; **see: Protease inhibitors**) and flatus-inducing sugars. Phytin-phosphorus (75–89% **phytin**) is approx. 200 mg per 100 g of seed. Pigeon pea is a good source of vitamins A, B and C.

The green seed is more nutritious than the dry seed because it has higher proportions of protein, sugar and fat and its protein is more digestible. There are also lower quantities of the flatulogenic sugars in the green seeds; the dried seeds contain somewhat more minerals (Table P.7).

5. Uses

In India, the pigeon pea is the second most important food legume and in the Caribbean it is almost an indispensable food. Pigeon peas are used as human food, fodder, browse plants, green manure, and a grain cash crop. The most common use for food is as split seeds but also whole, as dry beans, dehulled or as a flour. In the Caribbean region as well as in western and eastern Africa immature seeds are eaten in soups. The vegetable types, generally large-podded with large, sweet-tasting green seeds, are the most preferred. Canned pigeon peas are marketed in certain parts of the world. In India, the largest consumer, the split, decoated, dry seed is processed into *dhal*. If left intact, the leathery seedcoat reduces cooking time. The plant's woody stems are valuable as firewood, thatch and fencing. The leaves are an important source of organic matter and nitrogen.

Table P.7. Comparison of some nutritional constituents of green and mature pigeon peas on a dry weight basis.

Constituent	Green seed	Mature seed
Protein (%)	21.0	18.8
Protein digestibility (%)	6.8	58.5
Trypsin inhibitor (units/mg)	2.8	9.9
Starch (%)	44.8	53.0
Starch digestibility (%)	53.0	36.2
Amylase inhibitor (units/mg)	17.3	26.9
Soluble sugars (%)	5.1	3.1
Flatulence factors (g/100g sol. sug)	10.3	53.5
Crude fibre (%)	2.3	1.9
Oil (%)	2.3	1.9
Minerals and trace elements (mg/100 g)		
Ca	113.7	122.0
Mg	1.4	1.3
Cu	4.6	3.9
Fe	2.5	2.3

From data in www.tropical-seeds.com/tech_forum/veg_herbs/pigeon_pea.html Dr M. Price.

6. Plant types

The plant is an erect annual or short-lived perennial shrub reaching a height of 4 m. The leaves have three leaflets that are green and hairy above and a silvery greyish-green with longer hairs on the underside. The flowers are yellow with reddish brown lines on the outside, present in axillary or terminal pedunculate racemes. Most of the flowers are thought to be self-pollinating. The crop can come to maturity over a wide range (80–250 days), greatly affected by temperature and photoperiod. The first harvest can be expected 4–6 months after sowing. **ICRISAT** has, however, developed short duration (100–105 day to maturity) and short-statured types that have broadened the cultivation of this legume into new environments. It is a deep-rooted plant with a reputation for drought-resistance. Pods are compressed, similar to English **peas**, green and pointed with slight reddish mottling; they contain two to nine seeds.

Cultivars of pigeon peas fall roughly in two main divisions: (i) *C. cajan* var. *bicolor* (called *arhar* in India) that includes primary large, long-lived, late-maturing plants bearing red- or purple-stained flowers and hairy, purplish pods with four or five seeds, usually purple-mottled or dark-coloured; and (ii) *C. cajan* var. *flavus* (called *tur* in India) that includes mostly smaller, early-maturing plants, with yellow flowers and light-green pods having only two or three seeds. There are intermediate forms and crosses which display characters of both.

7. Cultivation

Breeding programmes have produced new varieties with a shortened growing season, and an extended range in which the crop will grow. Resistance to disease and pod-boring pests is currently being improved, as well as drought tolerance.

Short duration varieties take between 100 and 140 days to mature and the introduction of these types has improved the harvestability of the crop and allows it to be grown as part of a wheat rotation. The perennial nature of pigeon pea also enables the production of multiple harvests in tropical areas. In most areas, a **rhizobial inoculation** is used. (OV-V, AB)

Nene, Y.L., Hall, S.D. and Sheila, V.K. (eds) (1990) *The Pigeon Pea*. CAB International, Wallingford, UK.

Fig. P.14. (A) Whole pine nuts (image by Mariko Sakamoto) and (B) pine kernels (image by Mike Amphlett, CABI). Bars = 1 cm

Pine nuts

Several species of pines (*Pinus* spp.) produce edible 'nuts' but the principal pine nut is that of *P. pinea*, the stone pine. This is a Mediterranean tree found from Portugal to the Middle East. The pine nuts called *piñons* are from *P. monophylla* and *P. cembroides*, in the southwestern USA and Mexico and *P. edulis* in southwest USA. Pine nuts are not nuts in the botanical sense but are seeds in which the **testa** ('shell') surrounds the kernel consisting of the **megagametophyte** tissue enclosing the **embryo** (Colour Plate 4F, G; **see: Nut**). The kernels, called nuts, are the items of commerce (Fig. P.14).

There is considerable variation in the composition of the different species of pine nuts, but as an example, reported values for *P. pinea* are: water 5.6% fresh weight, storage protein 31.1%, **oil** (triacylglycerol) 47.4%, carbohydrate 11.6% (**see: Nut**, Table N.2). Some types are rich in Mg, K, P and the vitamins niacin, thiamin and folate.

Pine nuts are frequently used raw or roasted as snack food, in pastas, soups, breads, pastries and salads, and in meat, poultry and fish dishes. The Italian *pesto* is a blend of basil, garlic, pine nuts and olive oil. (MB)

Vaughan, J.G. and Geissler, C.A. (1997) *The New Oxford Book of Food Plants*. Oxford University Press, Oxford, UK, pp. 34–35.

Pink seed

Pink seed of dry **pea** (where it also greatly reduces seed size and grading quality, as well as seedling emergence and vigour, and can be mistaken for fungicide-treated seed) and of durum wheat and common wheat are all caused by the bacterial **pathogen** *Erwinia rhapontici*. A pink seed colour is also a symptom of various other diseases. *Fusarium monilifome* can cause a pink discoloration in **maize** seed, as can *Trichoconis padwickii* in rice. (**See: Disease**)

Pistachio

This temperate-climate 'nut', *Pistacia vera* (Anacardiaceae), is not a true nut in the botanical sense (**see: Nut**) but a **drupe** (Colour Plate 5D, E). The nut of commerce is the seed within the **endocarp** (inner region of the fruit coat: shell): it measures about 2 cm long and approx. 1.2 cm wide, while the seed is slightly smaller. The surrounding **mesocarp** pulp is removed after harvesting. The species (a dioecious tree) is thought to have originated in the desert areas of central Asia. According to FAOSTAT approx. 540,000 t were produced worldwide in 2003, the leading producers being Iran (57% of world production), Turkey (16%), USA (9%) and Syria (9%). (**See: Crop Atlas Appendix, Map 4**) The **shell** splits from the apex at maturity exposing a seed (kernel) that is readily eaten out of hand as a snack food. The kernels have a distinctive flavour and are valued for use in confectionery, bakery and ice cream. For information on composition **see: Nut**, Table N.2. (VP, MB)

Vaughan, J.G. and Geissler, C.A. (1997) *The New Oxford Book of Food Plants*. Oxford University Press, Oxford, UK, pp. 32–33.

Pistil

The complete female reproductive structure of the **angiosperms**, composed of one or more **carpels** and consisting of the **ovary**, style and stigma. (**See: Reproductive structures, 1. Female**)

Placenta

A region within the **ovary** where the **ovules** are formed and stay attached (usually via a funiculus) until seed maturation. (**See: Reproductive structures, 1. Female**) Placentation is the arrangement of the placentas, and hence of ovules, within the ovary.

Plant growth promoting rhizobacteria (PGPR)

Species and strains of soil-inhabiting bacteria (including many *Pseudomonas* and *Bacillus* species, and *Burkholderia cepacia*), which promote plant growth by forming associations with roots, particularly at the seedling stage, by a variety of mechanisms. 'Plant growth promoting microorganism' is a wider term, encompassing fungal as well as bacterial rhizosphere organisms. (**See: Rhizosphere microorganisms**)

Plant growth regulators

In seed (and other crop) technology, any substance or mixture of substances applied to a seed or a plant with an intended physiological action of accelerating or retarding the rate of growth or rate of maturation, or for otherwise altering the behaviour of plants or to alter normal plant processes in some way. (**See: Priming – technology, 2. Techniques and terminologies**) Plant growth regulators are classified as pesticides in regulatory laws. (**See: Treatments – pesticides**)

In seed physiology, the term plant growth regulator is also used as an alternative name for plant **hormones**, or for endogenous compounds when there is uncertainty about whether they qualify as hormones. Growth regulators may be chemically identical to a naturally occurring hormone (such as **gibberellins**), or may be a chemically modified form of them, or may be chemically different (such as the **defoliants** and some **desiccants** used to manipulate seed harvests).

Plant variety protection (PVP)

See: Laws and regulations

Planting equipment – agronomic requirements

The planting, seeding or sowing operation is probably the most important cultural practice associated with crop production. Increases in crop yield, cropping reliability, cropping frequency and crop returns are all largely dependent upon the uniform and timely establishment of optimum plant populations. While the planting operation can influence all stages of crop production, its major impact is on the germination, emergence and establishment phases, that is, from time of sowing seeds to the time when healthy seedlings exist.

There are two broad areas of responsibility for optimization of plant establishment. First, plant breeders, seed growers and seed merchants have a responsibility to provide quality seed, with high genetic and physical purity, high viability, etc. (**see: Production for sowing; Quality testing**). Secondly, farm managers must be aware of the agronomic requirements for the successful germination, emergence and establishment of the crops to be grown and be able to use this information in soil tillage (**see: Seedbed environment**) and to assist in the selection, setting and management of all farm machinery, planters in particular.

In this and the following two articles, the overall performance of mechanical planters is generally considered on the basis of the economic and timely establishment of uniform seedlings at the required population and spacing. In this article, the agronomic requirements for successful crop establishment and cropping systems are briefly discussed with a view to establishing the functional requirements of 'complete' planting machines. The next two articles (**Planting equipment – metering and delivery; Planting equipment – placement in soil**) discuss the components, their principles of operation and particular requirements, and the range of typical planting devices in greater detail. Both manufacturers and users of planting machinery must be aware of these requirements and principles to successfully design and/or select, set and manage their equipment. (**See also: Sowing methods – brief history**)

1. Seedlot requirements

(a) *Size and shape*. The size and shape of seed, and variations in them within a particular seedlot, influence the performance of seed metering devices to varying degrees. Some, such as plate meters, require uniformity in both respects (such as 'graded' or pelleted seed) for optimum performance. Others, such as vacuum disc meters, tolerate a range of shapes and sizes without a significant reduction in metering performance. Uniformity of planter depth control is more critical when the seeds being sown are small (such as cabbage) rather than large (such as maize). Similarly, small seeds generally need a finer seedbed tilth (that part of the cultivated surface of the seedbed turned over for planting) for optimum establishment.

(b) *Fragility*. The germination percentage of fragile seeds, such as peanut, can be significantly reduced if some are split or otherwise damaged when subjected to aggressive seed metering systems, such as those in horizontal plate meters.
(c) *Surface characteristics*. Seeds that tend to cling together (**chaffy seed**, such as buffelgrass, *Cenchrus ciliaris*), pack together, or otherwise reduce the 'flowability' of the seedlot, may require the use of specialized seed metering or agitation devices. Pre-sowing treatments, such as de-awning, **delinting**, or **pelleting** are available to modify seed surface characteristics and thereby reduce metering difficulties with some species (**see: Conditioning, I. Precleaning**).
(d) *Physical purity*. Foreign matter in the seedlot (soil, organic matter) and some seed **dressings** may adversely influence the performance of some metering systems. For example, the holes in vacuum or pressurized disc seed metering systems may become blocked.
(e) *Seed treatments* can affect the size and shape of seeds, their flow characteristics and the amount of foreign material in the seedlot. All affect the performance of seed metering systems to some degree.

2. Germination, emergence and establishment
The seeding rate required to establish a given plant population is influenced by the germination percentage and **vigour** of the seedlot. If the germination percentage is low, the planting rate has to be increased to obtain the desired plant population. The performance of most precision-type seed metering systems declines as the rate of metering increases and in some cases dictates the type of seed meter that can be used.

Crop performance is highly dependent on the complex interaction of seed (type, quality, etc.), soil (physical, chemical, biotic, etc.) and climatic (temperature, rainfall, evaporation, etc.) factors. Most of these interactions are affected by, and have implications for, planter design, selection, setting and management.

For example, soil physical factors (temperature, aeration, moisture, light, strength, etc.) are affected by the method of opening the **furrow**, depth of planting, method of covering the seed and the method of firming the seedbed. (**See: Seedbed environment**) Soil chemical (soil pH, nutrients, toxic substances, etc.) and biotic (weeds, insects, disease, etc.) factors can be substantially changed by the application of fertilizer and pesticides and by the nature and extent of the soil disturbance, etc., at time of planting. Adapting planters to operate on raised beds, to improve drainage in wet conditions or to plant to depth to access moisture under dry conditions, can modify the impacts of adverse climatic conditions. The planter's primary job is to place seed where it will germinate and grow. Proper seed spacing minimizes competition for the light, nutrients and soil moisture essential for crop growth.

Several factors influence planter performance, including adjustments and correct operation. Traditionally, producers use equipment that created a well-tilled, residue-free seedbed for planting. (**See: Tillage**) Many producers now are adopting conservation tillage methods that have fewer tillage operations and leave a protective residue cover on the soil surface. Although effective in reducing erosion, these practices have increased concern about seed placement and general planter performance. Tillage systems currently available range from the traditional moldboard plough to conservation tillage methods such as no-till. No-till systems disturb only a narrow strip of soil, leaving most of the residue on the soil surface, which offers the best erosion control.

3. Plant population and spacing
Many factors have to be considered when determining the optimum population (that is, the number of established plants per area) and spacing (that is, distance between rows of established plants and the spacing of the plants within the row) requirements for a particular crop under given environmental conditions. With respect to **yield potential**, population and spacing requirements depend upon the interaction of crop species, cultural practice and harvest machinery as briefly discussed below.

Under given environmental conditions the optimum population and spacing is largely dependent upon the species' inter- and intra-plant response to competition for growth resources such as moisture, nutrients and sunlight. With some crop species (usually uni-culm types, i.e. with a single stem, such as sunflower and maize), there is a comparatively narrow range of plant populations from which optimum yields can be expected. For others (particularly those that have the ability to tiller or otherwise compensate, such as wheat and canola), there appears to be a wider range of populations over which potential yield does not vary appreciably, above a certain minimum value for the particular conditions. The spacing of plants, both within and between rows, also may or may not be an important factor determining yield. Many crop species (such as wheat, oats and barley) can tolerate reasonable variation in the uniformity of spacing without loss of yield potential, provided the overall population is within the required range. For others, however, yield potential can be improved by uniformity of spacing within the optimum population range, such as sunflower, maize, grain sorghum and sugarbeet.

Plant populations and row spacing may also influence the nature and extent of weed growth, the degree of crop **lodging** (breaking of the stalk so that the crop lies flat), the size of seed heads, all of which may have implications for crop yield and ease of harvest. In many horticultural crops the quality of the harvested plant part is very dependent on uniform spacing in the field – such as in the cole and head salad crops.

Cultural practices have a significant influence on populations and spacings. Where soil moisture and plant nutrients are limiting factors, irrigation and fertilizer application can be used to substantially modify optimum plant populations and increase yield potential. However, under these intensive crop production systems, row spacing may need to accommodate furrow irrigation, inter-row cultivation, the application of crop chemicals and so on. In fact, in crops such as cotton and sugarcane, the row space is often dictated by the design of the harvest machinery.

With respect to planting machinery, therefore, the combination of population and spacing requirements for a cropping system dictates the types of seed metering systems that can be used, and the flexibility in row spacing. These combinations give rise to distinct planting patterns (Fig. P.15), which form a broad basis for classifying planting equipment.

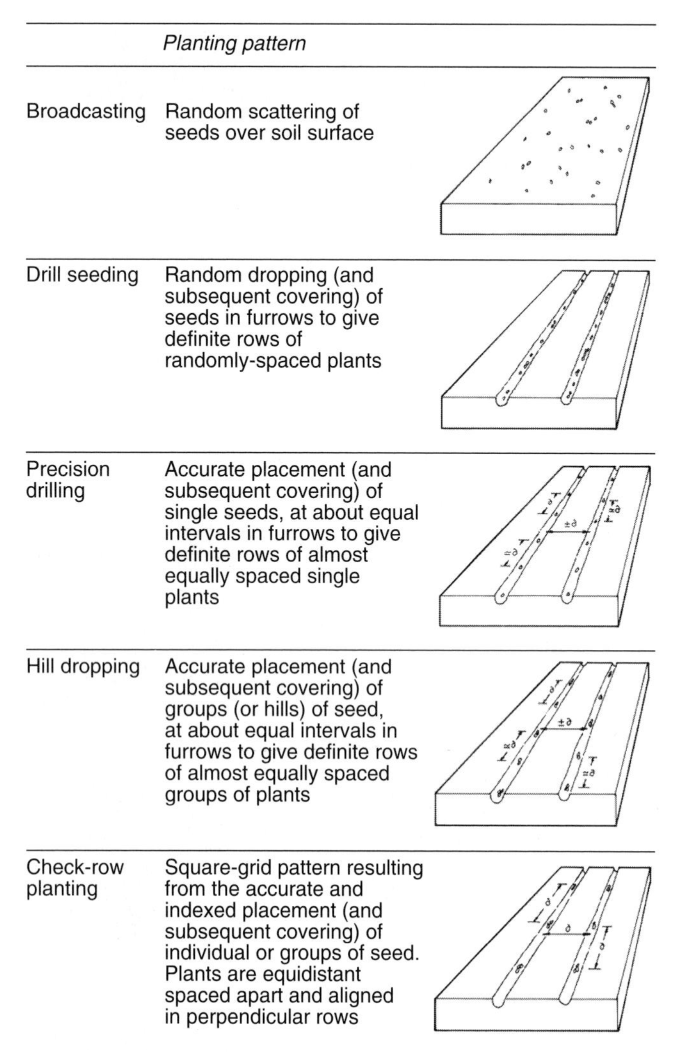

Fig. P.15. Classification of the planting patterns resulting from different population and spacing requirements.

4. Planter components

On the basis of these agronomic requirements for crop establishment, the functional requirements for a 'complete' planting machine may be to:

- open a furrow (or hole) in the soil;
- meter the seed;
- deliver the seed to, and place the seed appropriately in, the furrow (or hole);
- cover the seed;
- firm the seedbed;
- perform other functions as necessary – for instance, cut or otherwise manipulate surface residue prior to opening the furrow, and apply crop chemicals (both fertilizer and/or pesticide), or granular **rhizobial inoculation** formulations.

A considerable variety of machines is available, based on combinations of the wide range of mechanical devices that meet these needs. Not all planting machines are capable of performing all functions. (For example, broadcast planters do not open a furrow, cover the seed or firm the seedbed.) Overall performance depends on the functioning of each component and their interactions. The assembly must operate as the planter moves at appropriate forward speed and over the range of soil conditions likely to exist at planting time. (For example, the ability to plant through surface residues is a prime requirement for planters used in conservation tillage systems.) It is an appropriate generalization to say that, the more specialized and effective a planter is, the greater the emphasis placed upon the design and flexibility of the devices that perform its key functions. (JRM)

(See: Planting equipment – metering and delivery)

Planting equipment – metering and delivery

This article outlines the internal mechanical designs used to deliver functional requirements for a 'complete' planting machine metering the seed; and delivering the seed to, and placing the seed appropriately in, the furrow or hole.

1. Classification

The five planting patterns discussed in **Planting equipment – agronomic requirements** (Fig. P.15) can be used broadly to classify equipment as follows:

(a) *Broadcast planters* that randomly distribute seed on the soil surface. In mechanized systems, the aerial sowing of pasture species or the use of tractor-mounted spinner-type fertilizer spreaders (Fig. P.16A) to distribute seed are examples of broadcast seeders. In non-mechanized systems seed may be spread by hand or with the assistance of a small, hand-held and powered, spinner-type distributor.

Where it is necessary to subsequently cover the seed with soil this is usually done in a separate operation. In mechanized systems this is usually accomplished by a single pass of a tine or disc harrow, set at a shallow depth to stir and thereby incorporate the seed in the surface layer of the soil. In non-mechanized systems this may be done by, for example, hand raking or the use of animals to trample the seed into the soil.

While broadcast seeding is extensively used in developing countries, its use in developed agricultural systems is generally limited to sowing cereal or pasture crops under very favourable (wet) conditions or where topography or field obstructions (such as trees, stumps and stones, etc.) restrict the use of more sophisticated planter operations.

(b) *Drill planters* that randomly drop, and subsequently cover, seeds in furrows to form definite rows of 'randomly' spaced plants. In mechanized systems this type is typified by the so-called 'drill planters' but includes the 'air seeders' (based on scarifiers and cultivators) and the Australian 'combine' planters. These machine types are extensively used for the planting of both winter and summer crops where there is no need for accurate placement of single seeds equidistantly down the rows. While the seeds are metered in a random fashion by 'mass flow' metering systems, reasonably accurate control over the planting rate per hectare is obtained. Drill planters are often called 'swath crop' planters because of the narrow row spacing (75–300 mm) typically used.

In non-mechanized planting systems, drill planting can be achieved by dribbling seeds into furrows by hand and subsequently covering the seeds in a separate hand- or machine-powered operation.

A typical tractor-drawn three-point linkage mounted drill

planter (with one depth wheel removed) used for planting rice and wheat in Pakistan is shown in Fig. P.16B.

(c) *Precision planters* accurately place, and subsequently cover, single seeds or groups of seed, in furrows (or holes) to form definite rows of 'equally' spaced plants. They are commonly called 'precision'-, 'unit'- or 'row crop'-type planters and are generally used to sow crops that have a yield response to spacing uniformity. Crops that fall into this category include maize, sunflower, sorghum, soybean, cotton and most horticultural crops. A typical multi-row 'precision planter' is shown in Fig. P.16C.

Because of the requirement for accurate depth control, each row usually has its own self-contained and independently mounted planting 'unit'. Most crops planted by precision planters have wide row spaces to allow for furrow irrigation and/or inter-row cultivation for weed control or fertilizer application during their growth period. Hence, the planting pattern, unit type construction and wide row spacing give rise to the various names used to identify this planter type.

Precision planter seeding patterns include the **precision drilling, hill drop** and **check-row** types. Recent advances in plant breeding, seed selection and seed treatment and precision-type seed metering systems have almost removed the need for hill dropping as a technique to ensure optimum plant establishment. Hence, machines capable of hill dropping are only used for specialized applications and check-row planters are almost non-existent.

In non-mechanized systems, precision planting can be achieved by hand-placing seeds equidistant along furrows or by the use of hand-operated punch or 'dibbler'-type planters.

(d) *Specialized planters* are those that do not plant seeds. They may, for example, plant seed potatoes, sugar cane sets or whole plants (e.g. transplants).

2. Seed metering

The major functional requirements of a seed metering system are to:

- meter the seed at a pre-determined rate (e.g. kg/ha or seeds/m of row length);
- maintain an acceptable spacing between seeds along the row;
- impart minimal damage to the seed during the process.

The operational requirements therefore include:

- the ability to meter the range of seed types to be planted by the machine;
- the ability to meter these seeds over the range of seeding rates necessary to meet individual crop and/or particular environmental conditions;
- the ability to maintain the predetermined rate (output) and spacing (accuracy) over the range of field surface conditions and field speeds used, etc. at the time of planting;
- operational reliability.

In addition, the size and ease of filling and emptying the seed-box, and the ease of calibrating, cleaning and adjusting the metering system, affect the overall performance efficiency of the sowing operation.

Seed meters can mostly be classified as either precision or mass-flow types, depending primarily upon their principle of operation, and hence the planting pattern that results from their use.

Fig. P.16. Different types of planting equipment. (A) Broadcast planter. (B) Drill planter. (C) Precision planter. For details refer to appropriate sub-headings in the text.

(a) *Mass flow meters* do not attempt to meter individual seeds, but rather attempt to meter a consistent volume of seed per unit time, such that average seed spacing is equal to the desired seed spacing, i.e. broadcast or drill planting patterns. In general, mass flow meters are used for sowing crops that:

- are planted at relatively high seeding rates;
- are planted in relatively narrow rows;
- can tolerate considerable variations in both the seeding rate and uniformity of seed spacing without a significant loss in yield.

'Swath' crops such as barley, wheat and oats fall into this category.

While an extensive range of types exist, the three types of mass flow metering systems in common use are:

- the external force feed or fluted roller;
- the internal force feed or double run;
- the peg roller.

While all three meter types are still commonly used on drill planters, the peg roller type predominates because of its relatively simple design and ability to meter both seed and granular fertilizer. The general form of each type and their method of adjustment is given in Table P.8.

(b) *Precision-type seed meters* are those that attempt to select single seeds and deliver them at a pre-set time interval. The resultant planting pattern is equidistant placement along the furrow: that is, precision drilling, hill dropping or check-row planting patterns. In general, precision meters are used for sowing crops that:

- are planted at relatively low seeding rates;
- are planted in relatively wide rows;
- have a relatively narrow range of populations from which optimum yields can be expected;
- have a yield response to evenness of seed spacing along the row.

Examples include row crops such as sorghum, sunflower, sugarbeet and beans and many root and head vegetables.

The precision metering systems in common use include: plate-type meters, belt meters and vacuum disc-type meters. Over the past 20 years the vacuum disc-type has become the standard unit for use in horticultural production, and is becoming increasingly popular for use on field crop precision planter types.

Plate-type meters. Common to all these metering systems is the seed plate that rotates at a speed relative to the planter's forward speed. The plate has a number of holes, cells or cups around its periphery (edge) and a portion of its circumference is exposed to the seed held in the seed box. If it is of the appropriate size, a single seed falls into the cavity as the plate passes through, and is then moved to a point where it is either positively ejected, or falls into the seed tube. A metal pawl, brush or soft roller is used as a cut-off to ensure that no seeds other than those in the hole, cell or cup pass from the seed box. Plate meters can be further sub-divided as horizontal, inclined or vertical plate types (Fig. P.17A–C).

Accuracy of metering with all plate-type meters requires the following:

- The seed is graded so that all seeds are of a more or less similar size and shape; otherwise, two small seeds may be picked up.
- The size of the holes, cells or cups suits the particular seedlot; different plates are needed for different seed types or for different size grades.
- The seeds have sufficient opportunity to enter the holes, cells or cups. The two major factors of influence are the rotation speed of the plate and the exposure distance; high speeds and short distances reduce metering performance.

Belt-type meters have similar operational principles, characteristics, requirements and operational performance to plate-type meters, except that the plate is replaced by a belt

Table P.8. Summary of common mass flow seed meter types and their method of adjustment.

	Meter Type		
Characteristics	External Force Feed (Fluted Roller)	Internal Force Feed (Double Roller)	Peg Roller (Studded Roller)
General form			
Seeding pattern:	Drill planting	Drill planting	Drill planting
Method of adjusting metering rate:	(1) Increase or decrease roller speed. (2) Expose more or less of the fluted roller to seed. (3) Adjust the position of the feed gate. (Note: a feed gate is not provided on all meters of this type.)	(1) Select the fine or coarse side of the disc. (2) Increase or decrease disc speed. (3) Adjust the position of the feed gate. (Note: A feed gate is not provided on all meters of this type.)	(1) Increase or decrease roller speed. (2) Various brand-specific options, e.g. (a) change roller type; (b) adjust feed gate; or (c) adjust metering width by the use of blank roller, etc.

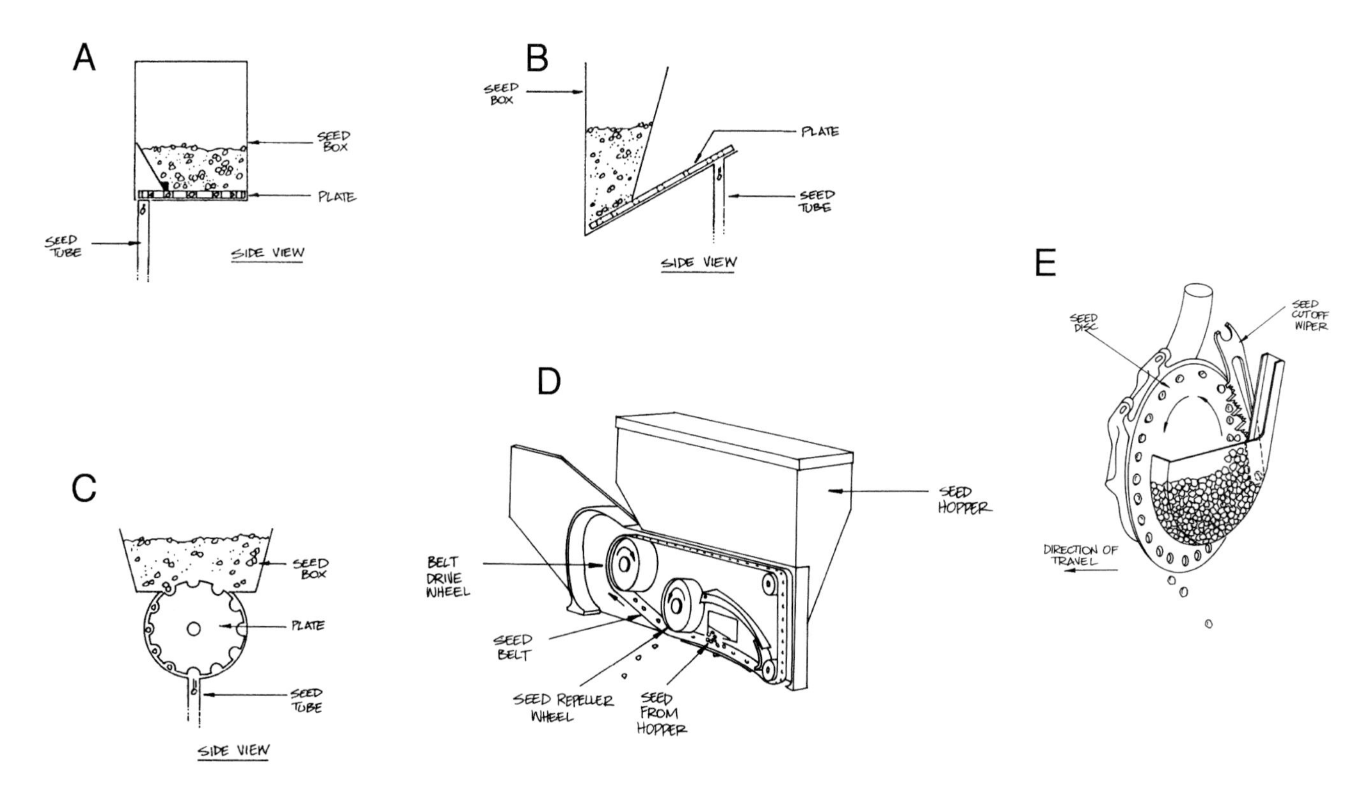

Fig. P.17. General form of (A) horizontal, (B) inclined, (C) vertical, (D) belt and (E) vacuum disc-type plate seed meters.

with a row of centrally punched holes to act as the cells for seed singulation. In operation, seed falls from the seed box into the chamber where it is exposed to a portion of the belt (Fig. P.17D). The seed falls into the cells formed by the combination of belt hole diameter and the width and depth of a groove in the base plate. Both have to be selected for a given seedlot. As the belt moves it carries the seed to the end of the base plate where it is dropped into the furrow. The repeller wheel ensures that seed, other than that in the belt holes, remains in the metering chamber. Metering accuracy is dependent upon the uniformity of seed size, the matching of cell size to seed size and the speed of the belt.

Vacuum disc-type meters. Essentially, this type of metering system consists of a seed box, a split housing, a vertical rotating disc which has a row of holes located around its circumference and a fan to create the vacuum (Fig. P.17E).

In operation the disc rotates between the two halves of the housing. A reduced pressure ('vacuum') is applied on one side of the housing and the seed is held in the lower portion of the other. As the disc rotates, each hole passes through the seedlot and picks up a seed or seeds as a result of the pressure difference across the disc. The wiper is set so that one seed remains on each hole; seeds dislodged by the wiper fall by gravity back into the seedlot. At a point near the base of the meter, the pressure difference is removed and the seeds fall into the seed tube.

The performance of vacuum disc-type meters is largely dependent upon:

- selecting the correct hole size for a given seedlot;
- selecting the correct level of vacuum, to generate the correct pressure difference across the disc;
- the setting of the wiper cut-off;
- the speed of the disc.

Because seeds do not have to be precisely matched to a seed cell, grading of seed is not essential, although still preferable for maximum accuracy of metering.

In addition to being a sowing system for producing seedling transplants for field crops, the vacuum metering principle is also used for the production of bedding plants. Seeds or pellets (in the case of irregularly shaped seeds) are held by suction against countersunk holes in a rectangular flat plate, or an array of nozzles, until the pressure is released. A wide selection of seeder plate sizes and configurations are readily available to fit chosen seed trays. Industry suppliers offer, for example, arrays of between about 25 and 500 holes, in a choice of at least seven different hole diameters, to grip the seed and provide the highest singulation accuracy for a wide range of small flower and vegetable seed species. Similar devices are commonly used in seed testing laboratories, to set seeds on flat paper blotters or towels at the start of **germination testing**.

The method of adjusting seeding rate and general performance characteristics of plate, belt and vacuum disc meters in the field are given in Table P.9.

(c) *Meter selection.* From the previous discussion, it is obvious that no one seed metering system is capable of meeting the requirements of all crops. Compromises have to be made and, in some cases, at least two planting machines may be necessary – one for drill seeding and the other for precision seeding.

Irrespective of the cropping programme, an informed decision can only be made if the following information is known for each of the crops to be planted:

Table P.9. Summary of common precision seed meter types and their method of adjustment.

Meter type	Method of changing meter rate	Comments
Plate	(1) Changing the plate speed relative to ground speed, vertical types. (2) Selecting plates with more or less holes, cells or cups around their circumference.	Different plates required for each seed size/shape. Potential for damage to fragile seeds is high, particularly with horizontal and vertical types. The use of graded seed is required for accuracy of metering.
Belt	(1) Changing the belt speed relative to ground speed. (2) Selecting belts with more or less holes per unit length.	Cell size is a function of belt hole size and the surface profile of the base plate. High potential to damage fragile seeds. The use of graded seed required for accuracy of metering.
Vacuum disc	(1) Changing the speed of the disc relative to ground speed. (2) Selecting discs with more or less holes around the circumference.	Metering performance depends upon matching the hole size and pressure/vacuum for each seed type and the setting of the wiper cut-off. The use of graded seed is recommended for high accuracy of metering. These meters have a lower potential for seed damage.

- the established population desired and the expected germination and field emergence percentage;
- the range of agronomically acceptable row spacings;
- the sensitivity of crop yield to the evenness of seed spacing along the row;
- the physical properties of the seed (e.g. seed size, seed shape, seed fragility, etc.).

3. Seed delivery systems

The functional requirements of a seed delivery system are to:

- convey the seed from the seed meter discharge to the furrow opener or seed boot;
- maintain the metering accuracy during conveyance;
- deposit the seed in the furrow in an appropriate manner (i.e. appropriate with respect to both seed placement within the furrow and seed spacing along the row).

Ideally the delivery system should deposit the seed on the firm, moist base of the furrow (unless it needs to be placed elsewhere for a specific reason). (**See: Seedbed environment**) Furthermore, the spacing of seeds along the furrow should be proportional to that at time of metering.

Most delivery systems can be classified as either gravity drop or pneumatic.

(a) *Gravity drop systems.* Seeds simply fall by gravity, directly or through a seed tube, to the furrow. To maintain metering accuracy (seed spacing), particularly if precision metering systems are used, the seed tube should be as short, straight and rigid as possible, have the smallest adequate cross-section, and have a smooth interior surface so as to reduce the potential for seed bounce within the tube. If one seed bounces down the tube while the next falls straight through, the final spacing between seeds will be adversely affected. If the tube has a rearward deflection at its base so that the seeds leave the tube with approximately zero forward velocity, bouncing and rolling when seeds hit the furrow base are both reduced.

(b) *Pneumatic delivery systems.* The use of these systems is essentially confined to planting machines where the seed meter or meters are centrally located on the machine. Seeds are delivered by air velocity through tubes of varying lengths to the remotely located openers.

- In the 'delivery only' systems, seed is metered directly into an individual delivery tube for each opener.
- The 'delivery and distribution-type' systems form the basis of the so-called 'air seeders' which are available as discrete machines or as seeding attachments for chisel ploughs, scarifiers and cultivators, and have considerable variations in design. The seed for a number of openers (usually six to eight) is metered (usually by mass flow-type meter) into a single tube that conveys the airstream to a 'dividing head', which splits the flow equally to a number of outlets (or to other dividing heads) located symmetrically around it along smaller diameter tubes. In general, these systems are only used for drill planting and, if appropriately designed, set and maintained, their performance can compare favourably to traditional drill planters that use a mass-flow meter and gravity drop delivery. (JRM)

Planting equipment – placement in soil

This article outlines the external mechanical designs used to deliver functional requirements for a 'complete' planting machine (**see: Planting equipment – agronomic requirements**): opening a furrow or hole in the soil; covering the seed; and firming the seedbed. These operations result in the seed physically placed at the desired sowing depth in a seedbed of the desired aggregate size. (**See: Seedbed environment; Tillage**)

1. Furrow opener devices

The function is to open the furrow or hole into which the seed is placed. The requirements are to:

- open the furrow (or hole) to the required depth;
- maintain uniformity of depth along the length of each furrow across the width of the planter;
- cause the minimum necessary disturbance to the seedbed;
- avoid smearing or over-compaction of the walls and base of the furrow;
- prevent soil flow back into the furrow before seed placement; and
- promote the flow of displaced soil back over the seed after its placement.

To meet these needs, furrow openers should:

- be as narrow as possible to reduce soil disturbance, restraining force and interference with the operation of adjacent openers;
- be held as rigid as possible to maintain depth and allow for accurate seed placement; and
- have provision for vertical and horizontal adjustment to enable alteration of planting depth and row spacing.

The general features of the major types of openers are briefly described in Table P.10, and illustrated in Fig. P.18A–D. Most furrow openers can be broadly classified as a runner, shoe, concave disc, disc-coulter (single, double and triple types, in various diameters, alignments and conformations), tine or punch types. Some opener designs result from a combination of types: for example, the 'Bioblade'™ type opener, which combines attributes of vertical, aligned disc coulter and the 'inverted "T"' narrow tine type. The capability of each opener is largely dictated by the soil engaging action that is used to create the furrow, and the method of depth control. The resulting shape or profile of a furrow depends largely on the soil conditions and type of opener.

2. Pre-opener devices

A diverse range of devices mounted in front of the furrow opener to cut and/or otherwise manipulate soil and/or remove soil residues and clods is available. These devices can be used to substantially improve opener performance under less than favourable conditions. For example, a blade can be used to displace dry soil into the inter-row space and allow the opener to access the more favourable moisture conditions deeper in the seedbed. A single vertical disc-coulter mounted in front of a tine-type opener can be used to cut residue to reduce 'blockage' caused by residue accumulation around the tine standard. Finger wheels can be mounted in front of openers to remove residue from the row area to facilitate furrow opening or improve soil temperature.

Table P.10. Types of furrow openers and their typical applications.

Classification type	Mode of action	Suitability	Disadvantages
Runner and Shoe (Fig. P.18A)	Presses soil downwards and outwards, increasing its strength and density to form a neat furrow of uniform depth, with firm walls and base, and slight overall disturbance to seedbed. (i) Runner types have angled blade that widens and splits towards the rear, through which cavity seeds are dropped. (ii) Shoe types are more compact.	(i) Deep, well-prepared seedbeds (good tilth below planting depth, free from weeds and residue) in more 'frictional' soil types (sands to loams). Widely used for intensive horticultural crop production. Or for use in conservation cropping systems where surface residues are retained and/or reduced/no-till techniques are used. (ii) Well-prepared seedbeds devoid of surface residues. Commonly used on drill type planters in Europe.	(i) Performance generally reduced or unsatisfactory in moist cohesive/adhesive soil types (heavy clays) or in shallow or unprepared seedbeds. (ii) Of limited application in reduced-tillage systems.
Concave disc (Fig. P.18B)	Cuts and digs, displacing soil upwards and outwards to one side. Plain-edged, slightly concave disc drawn with a rolling action at an angle to the direction of travel.	Ideally suited for crops, particularly pastures, with soil obstructions (stones etc.). Penetrates better than runner, shoe and most disc-coulter types.	Generally unsuited to conservation cropping systems: relatively high soil disturbance and small disc diameter.
Disc coulter (Fig. P.18C)	Cuts, then digs or presses. One or more flat rolling disc-coulters, in three broad types: (i) single – drawn at angle to direction of travel, (ii) double – aligned to direction of travel, offset, and/or inclined in many designs, touching on the lower leading edge, displacing soil downwards and outwards, producing firm furrow walls and base or (iii) triple – vertical aligned disc, in front of double-disc assembly, cuts soil and residue.	Wide range of soil types and seedbed conditions, with scrapers fitted in moist clay soils. Double discs most popular; used for row crops (irrigated and dryland, conventional and conservation cropping). Single and triple disc types tend to cut residue better.	
Tine (Fig. P.18D)	Digs, displacing soil upwards and outwards, depositing on both sides of the furrow. Wide range of interchangeable types, broadly classified as narrow ('point', 'knife' or 'inverted "T"') or wide (also provide full-width cultivation for weed control on narrow row spacing). Small rake angles.	Operate over the widest range of soil types and conditions. Favoured for drill-type planters particularly in the USA, Canada and Australia, for pasture and sward crop. Wide-types used where need to cultivate at planting time (e.g. Australian dryland cereals). Narrow types very popular on precision planters.	Prone to blockage under heavy surface residue conditions and narrow row spacings, without preceding device to remove (e.g. a row cleaner) or cut (e.g. a vertical disc coulter) residue.
Punch	Creates a row of holes (not a continuous furrow), by either a pressing or digging action.	Popular for transplanting or planting seeds into moist, deep and well-prepared horticultural seedbeds covered by plastic **mulch**.	Use in extensive field crop production is largely limited by cost, technical complexity and/or reliability

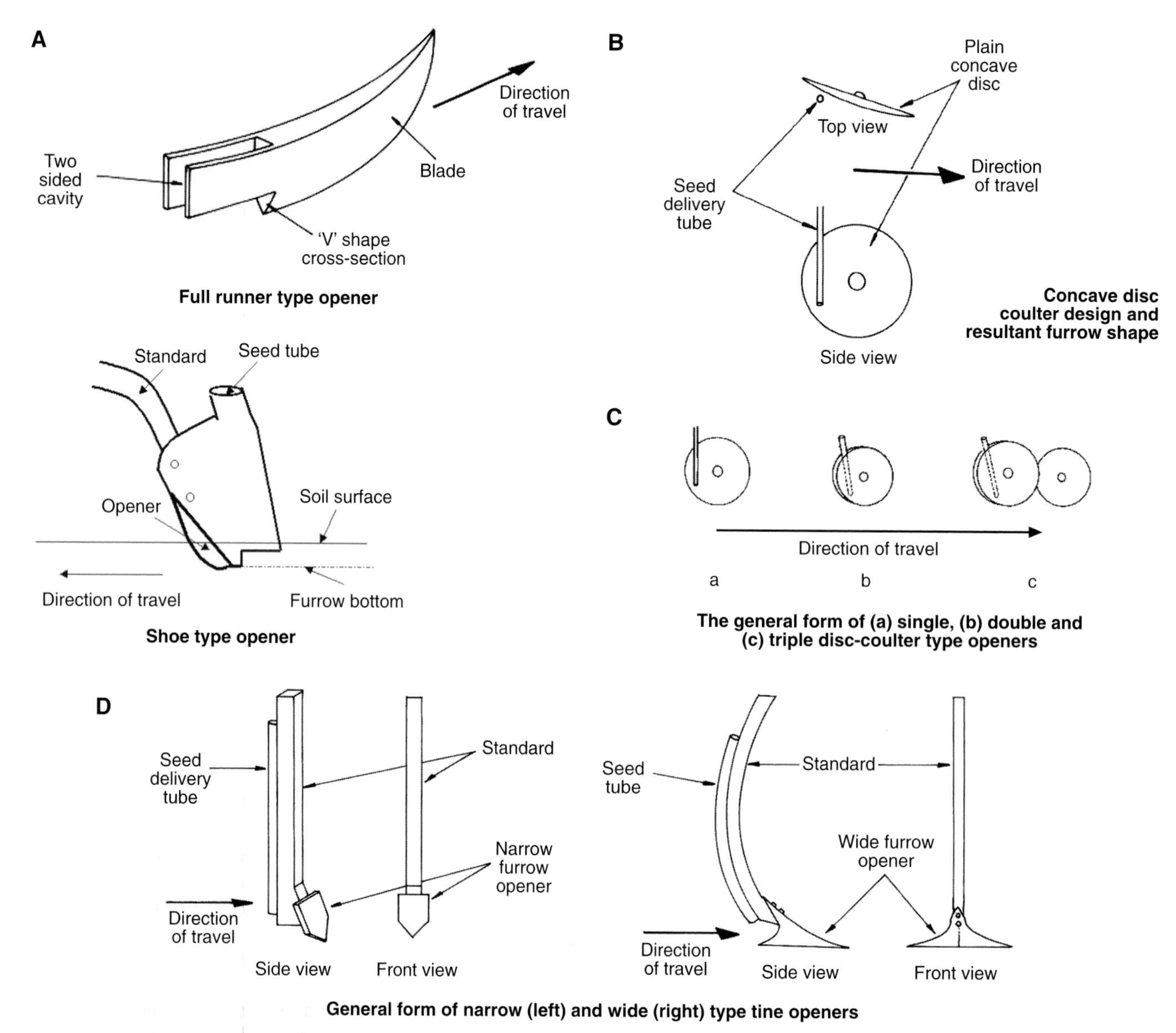

Fig. P.18A–D. Diagrammatic representations of the types of furrow openers referred to in Table P.10.

3. Seed-covering devices

Once placed in the furrow, the seed has to be covered with soil to:

- assist in the provision and stabilization of an appropriate seed environment;
- assist in the protection of seed from predators such as birds, mice and insects.

The functional requirement of seed-covering devices is therefore to transfer displaced surface soil back into the furrow for these purposes. Operational requirements include:

- that the covering devices can be selected and/or adjusted to enable them to operate effectively over the range of field conditions likely to exist at time of seeding;
- that the depth of cover be uniform and appropriate for the species sown;
- that seeds in the furrow are not displaced during the covering process;
- that the soil covering the seed is left in a condition that does not impede shoot emergence.

The need for, and the design of, the covering device on a planter depends mainly upon: soil type and condition, the design of the furrow opener, the type and amount of surface residue, and the speed of operation. A diverse range of seed-covering devices exists to cater for different conditions. On drill planters, harrows are commonly used, usually drawn directly behind, and extending across the full width of the machine to assist in levelling the seedbed. Some drill planters, and most precision planters, have individual row-type covering devices, varying in complexity from narrow individual harrow sections to tines, knives, discs or a short length of chain drawn behind each opener (Fig. P.19).

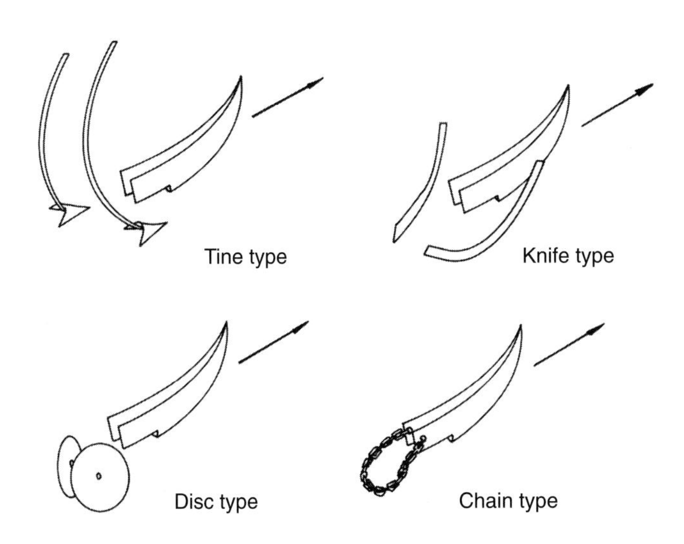

Fig. P.19. Common types of covering devices used on precision planters.

4. Seedbed firming devices

Under almost all field conditions, firming the soil in the seed zone has been shown to improve both seedling **emergence** and growth. The significant improvement in seedling emergence (commonly 15–20%) as a result of seedbed firming is generally attributed to one or more of the following:

- the general effect soil firming/compaction has on the stabilization of the seed environment;
- improved moisture availability/transfer to the seed as a result of improved seed-to-soil contact;
- a reduction in the depth of cover and hence the amount of growth needed to achieve seedling emergence depth;
- a reduction in the amount of light penetrating the seedbed – which has been shown to reduce the potential for sub-surface leaf growth, particularly in the case of heavy clay soils;
- a reduction in insect damage – since firm soil has been shown to restrict the subsurface movement of some insects;
- beneficial effects as a result of alteration to the micro-relief of the seedbed.

The last point arises because most seedbed firming devices leave a depression or furrow above the seeded row. This furrow tends to concentrate the runoff from rainfall and has been shown to improve the moisture status in the seed zone. While this is obviously beneficial in 'dry' seasons, excessive accumulation of moisture in 'wet' seasons can be detrimental to seed germination and/or seedling emergence, establishment and growth. Further, during heavy rainfall events, soil may be washed into the furrow. This may reduce emergence (due to excessive soil cover/depth above the seed) and promote the incidence of disease if it covers the stems of establishing plants.

While seedbed firming has been shown to improve emergence and establishment over a wide range of conditions, the effect is most apparent as soil conditions, particularly moisture, become limiting. The use of seedbed firming devices, therefore, enables:

- crops to be established under conditions that would otherwise be called marginal or unsuitable for sowing when using machines that do not have seedbed firming devices;
- extension of the available planting time after an effective rainfall event.

Most of the benefits attributed to seedbed firming devices accrue from optimization of soil compaction in the seed zone. However, over-compaction of the seedbed can have disastrous effects on seedling development, and emergence in particular.

The range of soil firming devices can be broadly classified as either rollers or press wheels.

(a) *Rollers.* Rollers are generally wide and typically of rigid, one-piece construction (Fig. P.20A). When drawn across the seedbed, their action tends to close and firm the soil surface, tending to reduce light penetration (particularly on heavier clay soils) and level the surface.

(b) *Presswheels.* Presswheels are the preferred seedbed-firming device, compared to rollers, because the latter result in:

- reduced water infiltration, and promoted runoff, increasing the potential for soil erosion;
- benefits the prospects for germination of weed seeds in the inter-row spaces;
- insufficient ability to adjust pressure on rows to derive maximum benefits in the seed zone, unless seedbeds are well-prepared and level.

The diverse range of types can be classified as over-centre types, zero pressure types and twin inclined types. The general form and relative pressure zones created by each type is shown diagrammatically in Fig. P.20B, and described in Table P.11.

The ability of seedlings to emerge through compacted layers is somewhat species-dependent, the particular form of germination being a major determinant. Monocotyledonous cereals and grasses, which exhibit **hypogeal** seedling establishment (i.e. the slender coleoptile emerges above the soil surface), can emerge through compacted layers more easily than dicotyledonous plants that exhibit either **epigeal** seedling establishment (where the **hypocotyl** but not the bulky **cotyledons** emerges above the surface, such as in bean) or have a hypogeal mode of establishment (where the **epicotyl** and the cotyledons both emerge, as in pea). For crop species known to be more sensitive to compaction, presswheel pressure must be reduced accordingly and, where possible, over-centre type presswheels should be avoided.

Soil type and condition can influence both the presswheel pressure and type to be used. Less pressure is necessary in a well-prepared friable seedbed to give the same degree of seed/soil contact than in a less-well-prepared, cloddy seedbed. Where the level of soil moisture or structure is such that firming causes or induces a hard setting layer above the seed, presswheel pressure should be reduced and zero pressure or twin-wheel systems used.

Presswheels are also often used as the depth control mechanization for furrow openers. Further, they are often selected and positioned so as to confer other benefits such as: limiting the degree of opener disturbance, maximizing cover from a particular opener action, and reducing interference

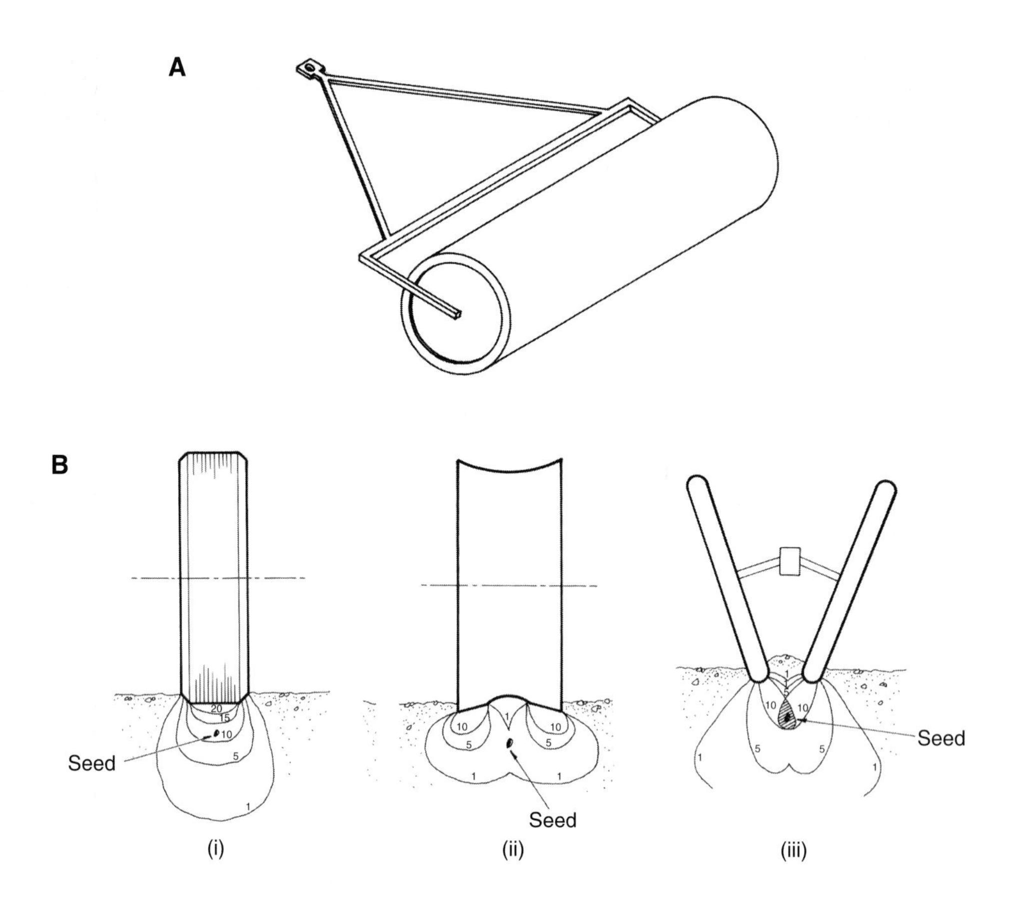

Fig. P.20. (A) Roller type of seedbed firmer. (B) Presswheel type of firmer, showing the pressure zones created in the soil by three types: (i) over-centre; (ii) zero pressure; and (iii) twin-inclined.

Table P.11. Relative characteristics of different presswheel types of seedbed firming devices.

Presswheel type	Description	Location of maximum pressure for a given loading (force/unit width)
Over-centre	Convex profile.	At the soil surface directly above the seed. Favoured for reduced/no-till drilling.
Zero pressure	Concave profile, mostly formed by non-pneumatic rubber tyre, with plain or ribbed outer surface.	Above but to the side of the seed zone. Serves intermediate role between the other two types.
Twin inclined ('Closing wheels')	Usually two narrow wheels with solid rubber tyres and inclined axles. Tend to close the seed furrow as well as firm the seedbed.	Around the seed, but leaving the surface soil uncompacted. Favoured for for well-prepared, friable seedbeds.

with adjacent openers. The size, shape, location, orientation and type of wheel used is therefore often a compromise between the firming and depth control functions. (JRM)

Breece, H.E. (1975) *Planting.* John Deere, Moline, Illinois, USA.

Jacobs, D.O. and Harrell, W.R. (1983) *Agricultural Power and Machinery.* McGraw Hill, New York, USA.

Shippen, J.M., Ellin, C.R. and Clover, C.H. (1980) *Basic Farm Machinery*, 3rd edn. Pergamon Press, New York, USA.

Plasmid

Small, circular extrachromosomal **DNA** molecule capable of autonomous replication in a cell. Commonly used as a cloning vector. **(See: Transformation)**

Plastids

Plastids are plant organelles bound by double **membranes** (envelopes) that control the import and export of molecules. Three general plastid types are described: (i) colourless leucoplasts, which include **amyloplasts** that accumulate **starch** and **oil bodies** (elaioplasts) that accumulate **oils**; (ii) chromoplasts, which accumulate red, yellow and orange pigments that confer colour to fruits, flowers and leaves; and (iii) chloroplasts, which are sites for photosynthesis and contain chlorophyll and often starch. All plastids are derived from proplastids and have the same genome, but develop specialized structures and functions as they mature in accordance with both intrinsic programmes of cell differentiation and environmental signals.

Chloroplasts are in all green tissues of the plant, especially leaves, but also during the development of some seeds and many **pods**, and are specialized for photosynthesis. They contain a unique, highly folded internal membrane system, the thylakoid membrane, which produces the **ATP** and **NADPH** needed for photosynthetic carbon fixation. In the light, chloroplasts fix carbon in the form of transitory starch. Other plastids such as amyloplasts do not have thylakoid membranes. Amyloplasts are specialized for reserve starch accumulation, and occur most frequently in storage tissues such as in starchy seeds (cereal grains) and potato tubers. Amyloplasts also are present in root cap cells, where they may function in gravitropism.

Plastids of photosynthetic plant cells are thought to have arisen from symbiotic associations between photosynthetic bacteria and non-photosynthetic eukaryotic organisms. Subsequently, much of the plastid genome was lost, so plastid functions require the coordinated expression of plastid and nuclear genes. (MGJ)
(See: Cells and cell components; Starch synthesis)

Pleiotropic

Applied to a **gene** that has an effect on more than one character in an organism. An example is the *rugosus* (wrinkled) **mutation** in pea, which normally produces a seed that is round at maturity. The mutation results in a failure of the developing pea seed to synthesize one enzyme in the **starch synthesis** pathway, starch branching enzyme I (SBEI). This not only causes the expected reduction in **amylopectin** synthesis, but an increase in oil, a reduction in storage protein, an increase in sucrose, accumulation of water during development, and the wrinkling of the pea seed upon maturation drying.

Pleurogram

Also called *linea fissura* and *linea sutura*. A specialized structure on the seeds of the Mimosoideae and Caesalpinioideae and some Cucurbitaceae, appearing externally on both sides of the seed as a horseshoe-shaped depression or furrow which is open towards the **hilum**, the enclosed area being called the areole. It has been suggested that the pleurogram functions as a hygroscopic valve. **(See: Structure of seeds)**

Plicate

Meaning folded; in seeds referring to folded **cotyledons** (e.g. in the Convolvulaceae, Malvaceae).

Ploidy

Refers to the number of **allelic** sets of chromosomes present in the somatic cells of plants, derived from the Greek for 'fold' (as in a multiple of something). In the ordinary alternation between the **haploid gametophyte** and **diploid sporophyte** plants the chromosome numbers are referred to as n and 2n (**see: Evolution of seeds**). In the genomic formula, x represents the basic (or monoploid) chromosome number, n is the gametic chromosome number (the number in the ordinary pollen or egg cells) and 2n the somatic chromosome number. Thus, a haploid has one set of chromosomes ($n = x$), a **diploid**, two sets ($2n = 2x$), a tetraploid, four sets ($2n = 4x$). (In these cases '2n' and 'n' become just the conventional symbols for the zygotic and gametic chromosome numbers.) Plants with more than two chromosomes sets (x) are termed **polyploid**. In **triploids** ($2n = 3x$) meiosis is not possible (a property exploited in the production of some **seedless fruit crop** varieties). **(See: Chromosome number)**

Plugs

Small units of growing media, formed into different sizes and shapes, into/onto which seeds are sown to produce seedlings. **(See: Transplants)**

Plumule

In cereal embryos, the apical part of the embryonic shoot inside of the coleoptile from which the primordia for young leaves develop. In dicot embryos, the meristematic region often located between the cotyledons at the apex of the hypocotyl, which gives rise to the shoot after germination. **(See: Shoot meristem; Structure of seeds)**

Pod

A multiseeded dehiscent **fruit**, sometimes meaning only the fruit wall (**pericarp**). The pod is derived from maternal **diploid ovary** tissue. Many genera of plants produce seeds enclosed in pods, including the legumes, such as peas and beans; hence used generally for the fruits of leguminous plants (**see: Legume; Legumes**).

Pod shatter

See: Shattering

Poisonous seeds

Seeds of very many species are poisonous to humans and other animals, i.e. they can have noxious, sometimes fatal effects. The severity of their action varies according to the species. One seed of the rosary or Jequirity bean (*Abrus precatorius*), for example, if ingested can kill an adult, and there are even cases of ill effects on people who have accidentally taken in some of the chippings generated when the seeds have been drilled to make necklaces (**see: Jewellery, arts and crafts**). And two *Wisteria floribunda* seeds, if eaten, can cause child death. **Ricin**, a protein in the castor bean seed (*Ricinus communis*) is exceedingly toxic. If ingested, it may cause serious illness, possibly not death, but if the protein directly enters the blood stream it will be fatal. Microgram or milligram quantities are enough to kill. The protein was used for assassination purposes in London in the deepest days of the Cold War; and there is great concern that it may become a weapon for terrorist action. Close to the other end of the scale are the cyanogenic apple seeds: it has been calculated that about 150 g would have to be consumed for ill effects to result, and even then fatality may not occur.

The latter example illustrates two important points. First, because there is a very wide range in the degree of toxicity in seeds, the quantities that bring about deleterious effects vary enormously with the species. When ingested in sufficient amounts, probably most seeds will provoke undesirable consequences. Secondly, many seeds which are foods or are associated with foods have potentially damaging components especially if taken in suitable quantity. Examples are

chickling pea (the cause of lathyrism), the **faba bean** (favism) and even soybean. A substantial number of food seeds contain so-called anti-nutritionals and **allergens** (e.g. many legumes and cereals) which might qualify them to be classed as poisonous, at least in certain situations where they impair health when eaten.

Organisms affected by seed poisons include domestic animals (mammals and birds) and therefore care must be taken to protect against ingestion of noxious seeds by preventing contamination of pasture and food. Two examples of offending weed seeds are pokeweed (*Phytolacca dioica*) and darnel ryegrass (*Lolium temulentum*), whose seeds contain a toxic protein and alkaloids, respectively. Sensitivity to some toxins can vary among different species. For example, ruminants are much more affected by ingestion of cyanogenic seeds such as apricot or almond than are monogastric animals including birds. This is because the pH of their rumen (close to neutral) favours activity of the β-glucosidase that releases cyanide from the glucoside while the highly acidic pH of the stomach of monogastric animals does not.

Wild animals can also be affected, but in many cases it appears that they reject poisonous seeds. For example, experiments with the South American rodent, the red-rumped agouti, showed that they collect and hoard seeds of *Ormosia arborea*, which contain quinolizidine alkaloids, but preyed upon them significantly less than on non-poisonous species: palatable seeds treated with extracts containing the alkaloid also survived predation. The seed toxin therefore protected against predation of the seeds but did not prevent their dispersal.

Predation of seeds by insects might also be discouraged by seed components. Extracts of **neem** tree (*Azadirachta indica*) seeds are toxic to several insect species, deter feeding and ovipositing, and inhibit growth. Larvae of several species are affected by amylase and trypsin inhibitors: the toxicity of a galactorhamnan polysaccharide from the seedcoat of *Canavalia ensiformis* (**jack bean**) to the **cowpea** weevil (*Callosobruchus maculatus*) is an example of deterrence of an adult insect.

It is indeed considered that many of the toxic constituents of seeds evolved to protect against predation. The ecological role of seed toxins is not always clear, however. For example, the rotenoids, deguelin and tephrosin in seeds of *Millettia dura* (Fabaceae) showed high toxic activity against second-instar larvae of the mosquito, *Aedes aegypti*, which do not predate the plant: possibly, predatory larvae of other species are similarly affected. For similar reasons it is also difficult to understand the ecological significance of the potency of *Millettia thonningii* seed extracts (the active principles of which are isoflavonoids) as a molluscicide (against water snails) and cercaricide (i.e. against the cercariae stage of the parasite on humans, *Schistosoma*).

Toxic substances from the environment might accumulate in seeds. For example, wheat grown in parts of west USA where soil selenium content is relatively high accumulate the element in the grains so that cattle fed substantially on grain from this source are adversely affected.

Toxins from seeds are or have been widely exploited by various peoples, for example for self-harm or murder, in religious and 'judicial' ceremonies or in hunting (arrow poisons – curares). It is said that Cleopatra, when contemplating suicide, considered the seeds of *Strychnos nux-vomica* as a possible agent and tested them on her servants. But on witnessing the seizures and facial contortions induced by the seeds she decided against them, choosing a reptilian means of death as a preferable alternative. Preparations made from seeds of *Strychnos* spp. and *Paulinnia pinnata*, for example, are used as arrow poisons in South America and Africa. Poisonous seeds featuring in judicial or ordeal ceremonies include *Physostigma venenosum* (Calabar bean) in West Africa, and *Tanghinia venenifera* in Madagascar. There is evidence that the former was responsible for the ritual death of a boy in London in 2001. And regarding the latter, the French in their occupation of Madagascar were so appalled by the trials using the seeds that they ordered the destruction of the trees bearing them.

Poisons can arise through the action of pathogens during seed development, such as the formation of **sclerotia** in rye **ergot**, which physically resemble the seed shape and size so closely that they and their fragments are hard to remove completely from infected batches of grain, which are therefore condemned from sale.

The constituents of seeds that render them poisonous are of a very varied chemical nature and physiological action (Table P.12). There are numerous different kinds of alkaloids and other compounds having pharmacological effects. Cyanogenic and other glycosides are present in many seeds, for example in members of the Rosaceae such as the bitter almond (*Prunus amygdalus* var. *armara*) where 20 seeds are a lethal dose for adults. Noxious proteins occur, acting as allergens, **lectins**, **phytohaemagglutinins**, and **protease inhibitors** and **amylase inhibitors**: the most toxic of all are the ribosome-inactivating proteins. Various non-protein amino acids are present in many species. Other types of compound are listed in Table P.12. (MB)

(See: Ethnobotany; Pharmaceuticals and pharmacologically active compounds; Pharmaceuticals, medicines and biologically active compounds; Psychoactive seeds; Soybean lectin; Storage proteins – intolerance and allergies)

Cooper, M.R. and Johnson, A.W. (1994) *Poisonous Plants and Fungi: An Illustrated Guide*. CAB International Bureau of Animal Health, Weybridge, UK.

Harborne, J.B. and Baxter, B. (eds) (1996) *Dictionary of Plant Toxins*. Wiley, Chichester, UK.

Lampe, K.F. and McCann, M.A. (1985) *AMA Handbook of Poisonous and Injurious Plants*. American Medical Association, Chicago, IL, USA.

Schwarting, A.E. (1963) Poisonous seeds and fruits. *Progress in Chemical Toxicology* 18, 385–401.

Polar nuclei

Two **haploid** (n) nuclei that fuse with one of the pollen nuclei (n) on seed **fertilization** within the **embryo sac**; the resultant **triploid** (3n) nucleus develops into the **endosperm**.

(See: Reproductive structures, 1. Female)

Pollination

The transfer of pollen from the anther to the stigma either from the same flower, plant or clone (**self**-pollination) or from plants of different clones (**cross**-pollination), leading to

Table P.12. Some seeds containing poisonous or noxious substances.

Species	Active Constituent(s)
Abrus precatorius (rosary bean)	Ribosome-inactivating protein (abrin)
Achras sapota (sapodilla tree)	Saponin
Aconitum napellus (monk's hood, poison aconite)	Alkaloids (diterpenes, e.g. aconitine, mesoaconitine)
Argemone mexicana (Mexican poppy)	Alkaloids (isoquinolines)(sanguinarine)
Brassica spp. (wild rapeseed, etc.)	Fatty acid (erucic acid)
Castanospermum australe (Morton Bay chestnut)	Indolizidine (castanospermine); saponins
Caulophyllum thalictroides (blue cohosh)	Alkaloids (e.g.N-methylcytisine, baptifoline, anagyrine)
Chenopodium quinoa (quinoa)	Saponins
Cinnamomum porrectum (camphor tree)	Ribosome-inactivating protein (porrectin)
Cucurbita spp. (e.g. pumpkin, squash)	Polyphenolic triterpene (e.g. cucurbitacin)
Cycas spp. (cycads)	Azoxyglycosides (cycasin)
Datura stramonium, D. ferox (Jimson weed, thornapple)	Tropane alkaloids (hyoscyamine, scopolamine)
Delphinium virescens (prairie larkspur)	Alkaloids (delphinine, ajacine)
Ginkgo biloba	Methylpyridoxine
Glycine max (soybean)	Lectins, protease inhibitors, protein toxins (e.g. soyatoxin)
Gossypium spp. (cotton)	Polyphenolic naphthalene (gossypol)
Heliotropium europaeum (common heliotrope)	Alkaloids (pyrrolizidines)
Jatropha curcas (physic nut)	Ribosome-inactivating protein (curcin)
Laburnum spp. (golden chain tree)	Alkaloid (cytisine)
Lathyrus sativus (chickling pea)	Non-protein amino acid (β-N-oxalylamino-L-alanine)
Millettia spp.	Flavonoids (e.g. rotenoids)
Physostigma venenosa (Calabar bean)	Alkaloid (physostigmine)
Prunus spp. (e.g. almond, apricot, cherry)	Cyanogenic glucosides (e.g. amygdalin)
Ricinus communis (castor bean)	Ribosome-inactivating protein (ricin)
Senecio vulgaris (groundsel)	Alkaloid (pyrrolizidines)
Sesbania cannabina (sesbania pea)	Non-protein amino acid (canavanine)
Solanum dulcamara (bittersweet nightshade)	Alkaloid (solanine)
Strychnos nux-vomica (crow fig)	Alkaloids (strychnine, brucine)
Thevia peruviana (yellow oleander)	Cardiac glycosides (thevetin A,B, thevetoxin, neriifolin, peruvoside
Vicia faba (fava, broad bean)	Alkaloids (glycosides, vicine and convicine)

See: Table P.5 in **Pharmaceuticals, medicines and biologically active compounds** for molecular structures of some substances in the above table.

fertilization (the union of the egg and pollen gametes). (**See: Reproductive structures**)

In the **angiosperms**, this is defined as the transfer of pollen from the anther (the site of pollen production) to the stigma of the female reproductive structure, the carpel (**see: Reproductive structures, 1. Female** (and) **2. Male**). The mode of this transfer varies between species, and can be by wind, water, insects (e.g. bees, moths) and even by animals (e.g. bats).

Pollen is released from the anthers in a partially dry state, and when it lands on a stigma of the same (or in some cases very closely related) species it hydrates, and a pollen tube grows down the style towards the ovule, bearing the male gametes, to effect double fertilization (**see: Development of embryos – cereals; Development of embryos – dicots; Development of endosperms – cereals; Development of endosperms – dicots; Fertilization**). While the structure of most flowers favours the pollen landing on the **stigma** of the same flower, there are often mechanisms in place to encourage or ensure out-crossing. These include the production of anthers and carpels on different flowers, or the sexual organs becoming mature at different times when on the same flower. Another mechanism is self-incompatibility, which is when pollen is incapable of fertilizing carpels in flowers of the same plant, but pollen from other plants of the same species are capable. Self-incompatibility is genetically controlled by a single locus (S), and in the most common form, gametophytic self-incompatibility, it is determined by the haploid pollen genotype of this locus. (**See: Self-incompatibility system**)

In the **gymnosperms**, pollination is achieved by the transfer of the pollen, usually by wind, from the male cone directly on to the ovule of the female cone (**see: Gymnosperms**). In conifers a pollen tube conveys the male gamete to the egg, but in more primitive gymnosperms such as *Ginkgo* and the cycads there is a short pollen tube which is **haustorial**. This grows into the **nucellus** close to the egg, and after several months it bursts and releases multiflagellate, swimming sperm cells, one of which will fuse with the egg cell.

(a) *Open pollination* is a breeding and seed production term describing natural cross-pollination, where plants are allowed to inter-pollinate freely within the field producing heterogeneous non-hybrid populations – in contrast to hybrid seed production, where controlled crosses are used. (**See: Compatibility; Production for sowing, III. Hybrids**)

(b) *Bud pollination* is the process used to achieve self-fertilization to propagate the self-incompatible inbred lines used for the production of hybrid vegetable brassicas, in breeding systems that rely on sporophytic self-incompatibility system. (**See: Bud pollination**) (JDB)

Polyamines

Polyamines have **hormone**-like activities in plants, though quite high concentrations are required for these activities. Some effects on seeds have been described but their possible natural roles are unclear. The polyamine putrescine is synthesized in plants from both arginine and ornithine. Spermidine and spermine are made from putrescine, a reaction that requires S-adenosyl-L-methionine (SAM). Since SAM is also required for synthesis of the hormone **ethylene**, there is competition for this precursor. For the main part these two plant hormones have opposite effects.

Polyamine biosynthesis is induced by a range of environmental factors. White light will induce arginine decarboxylase, whereas far-red light induces it in pea buds but represses it in pea hypocotyls. The same enzyme is induced by potassium deficiency and osmotic stress. Putrescine production is also induced by chilling, whereas heat and drought stress cause production of the 'thermopolyamines', norspermidine and norspermine. The other major factors inducing polyamine production are hormonal. **Auxin** induces polyamine production in sunflower explants, germinating barley grains and rice embryos. **Gibberellic acid** also induces ornithine decarboxylase in barley aleurone layers and **gibberellin**-induced elongation of pea internodes is accompanied by a rise in arginine decarboxylase activity and increased polyamine levels. **Cytokinin** increases polyamine biosynthesis in lettuce and cucumber **cotyledons**. Ethylene inhibits polyamine production over and above the effect of competition for SAM by inhibiting arginine decarboxylase. Polyamines also inhibit the enzyme that produces ethylene, ACC oxidase.

Polyamines also occur as the amide conjugates of hydroxycinnamic acids such as *p*-coumaric acid, ferulic acid and caffeic acid (**see: Phenolics**). These have significant activity in flower, seed and fruit development and in the hypersensitive response to infection. The breakdown of putrescine is initiated by diamine oxidase and that of the other polyamines by polyamine oxidase. This process is not simply a means of destroying these compounds. Both of these enzymes produce peroxide and polyamine oxidase that is very active in the cell walls of cereals, where the peroxide is thought to be involved in lignin formation. Putrescine has another non-hormonal role as the precursor for alkaloids such as nicotine. Auxin inhibits this reaction by promoting the conversion of putrescine to its conjugate. (GL, JR)

Crozier, A., Kamiya, Y., Bishop, G. and Yokota, T. (2000) Biosynthesis of hormones and elicitor molecules. In: Buchanan, B.B., Gruissem, W. and Jones, R.L. *Biochemistry and Molecular Biology of Plants*. American Society of Plant Physiologists, Rockville, MD, USA, pp. 850–929.

Galston, A.W. and Kaur-Sawhney, R. (1995) Polyamines as endogenous growth regulators. In: Davies, P.J. (ed.) *Plant Hormones: Physiology, Biochemistry and Molecular Biology*. Kluwer Academic, Dordrecht, The Netherlands, pp. 118–139.

Matilla, A. (1996) Polyamines and seed germination. *Seed Science Research* 6, 81–94.

Polyembryony

The formation of multiple embryos in one **ovule**. One form of polyembryony is that which originates from one **zygote**. The inhibition of **auxin** polar transport by inhibitors such as N-1-naphthylphthalamic acid (NPA) and 3,3′,4′,5,7-pentahydroxyflavone (quercetin) during early **embryogenesis** *in vitro* often produces more than one **embryo axis**, causing polyembryony. The other form of polyembryony is produced by adventitious embryogenesis, where embryos develop from somatic maternal tissues such as **nucellus**. The latter form, which is also referred to as **apomixis**, may occur at the same time as a normal embryo is developing from the egg cell. (**See also: Gymnosperm seeds**) (O-AO, HS)

Polymerase chain reaction

See: PCR (Polymerase chain reaction)

Polymorphism

A term describing the production of morphologically different seeds (e.g. colour, size, **seedcoat** thickness) by the same plant or different plants of the same species. Polymorphism may exist in seeds of different or the same inflorescence or **fruit**. Polymorphism may have an environmental or maternal cause. Polymorphic seeds often vary in their germination properties, i.e. they show **heteroblasty**. The two terms are often used interchangeably. Synonymous with **heteromorphism, heterospermy**. (**See: Deserts and arid lands; Dormancy – acquisition; Germinability – maternal effects; Germinability – parental habitat effects**)

Polyphenol oxidases

Enzymes that mediate the oxidation of **phenolic** compounds in the presence of oxygen. Various enzymes are involved in this oxidation. Monophenol oxidases convert monophenols into *o*-diphenols, which are subsequently oxidized to *o*-quinones by catechol oxidases. Laccases may accept both *o*-diphenols and *p*-diphenols. In both cases, quinones are later polymerized into brown pigments by a non-enzymatic mechanism. Phenolic compounds are abundant in **seedcoats** and their oxidation, either enzymic or non-enzymic, reduces oxygen supply to the embryo and may result in the inhibition of germination in many seeds (**see: Germination – influences of chemicals; Germination – influences of gases**). The polyphenol oxidase:phenolics system is responsible for production of many of the dark pigments of the **testa** and other parts of the seedcoat. (DC, FC)

Côme, D. and Corbineau, F. (1998) Semences et germination. In: Mazliak, P. (ed.) *Physiologie Végétale II. Croissance et Développement*. Hermann, Paris, France, pp. 185–313.

Macheix, J.J., Fleuriet, A. and Billot, J. (1990) *Fruits Phenolics*. CRC Press, Boca Raton, FL, USA.

Polyploid

Plants in which the somatic cells have multiples of the complete basic allelic chromosome set (x) more than two (the diploid number) – such as triploid (3x), tetraploid (4x) and so on. (**See: Ploidy**)

Amphipolyploids possess the total chromosomal complement of the parents; **aneuploids** vary from amphiploids by the addition or deletion of specific chromosomes. Furthermore, polyploid plants may arise as: **autoploids**, or autopolyploids, from the duplication of genomes of a single species, or as **alloploids**, or allopolyploids, from the combination of genomes from two or more unrelated species.

Many commonly cultivated crop species have evolved in nature as polyploids, mostly as allopolyploids, including cotton, oat, tobacco and wheat. Various horticultural brassica species, for example, are allotetraploids (also known as amphidiploids), derived from combining two different diploid species.

Polyploidy is of special significance in plant breeding programmes because it permits greater expression of existing genetic diversity. However, its effects on phenotype are varied and difficult to predict, and **vigour** is not always improved, leading to the concept of 'optimal ploidy' levels for each species. Autopolyploidy can be induced by environmental shock or artificially by chemicals that disrupt normal chromosome division, of which **colchicine** (an alkaloid derived from the seeds or corms of the autumn crocus, or a synthetic equivalent) is the most widely used. Autopolyploids can have desirable characteristics, and have been successfully introduced for sugar and fodder beets, where pure triploid cultivars are produced by pollinating a cytoplasmic male-sterile diploid with a tetraploid. **Triticale** is the most prominent modern example of an artificially bred, induced allopolyploid – a new grain crop derived by extensive breeding efforts to combine wheat and rye chromosomes. A unique feature of seed formation in the **angiosperms**, due to the **double fertilization** process, is that the endosperm is a triploid tissue, consisting of one paternal and two maternal genomes. (PH)

Polysomes

Shortened version of polyribosomes. The site of protein synthesis (translation) within cells, it is the structure formed by the binding of many ribosomes and other essential translational components to messenger RNA (mRNA). Polysomes may be associated with the **endoplasmic reticulum** (bound polysomes), e.g. in the synthesis of proteins secreted into the vacuole (**storage proteins**), or secreted from the cell (α-amylase), to form the rough endoplasmic reticulum (RER), or they may occur within the cytoplasm as free polysomes.

Pome

Usually simply defined as an indehiscent, fleshy **fruit**, whose flesh is derived primarily from the receptacle (e.g. apple). (**See: Fruit – types**)

Pomegranate

The fruit of pomegranate (*Punica granatum*, Punicaceae – native to semi-arid regions of the Near East and Mediterranean) is divided into several chambers, each of which contains many transparent vesicles of red-pink, juicy pulp surrounding an angular, elongated seed, 5–9 mm long. The seeds can be removed and dried when they become dark-red to black in colour and slightly sticky. The seeds are astringent with a sweet-sour taste. In Indian cooking, pomegranate seeds are used as a souring agent rather like **tamarind**. In Middle Eastern cuisines crushed seeds are sprinkled in some cooked dishes and they can be used as an additive in various salads. The seeds have medicinal uses, for example in Indian practice, in gargles, to ease fevers and to counteract diarrhoea.

Pomegranate seeds featured prominently in ancient classical myth and in religion and folklore (**see: History of seed research**). (MB)

Popcorn

Purported to date back at least 6000 years, this variety of **maize**, *Zea mays everta*, has hard kernels that burst to form white, irregularly shaped puffs when heated. Commonly, the term is used with respect to the edible popped kernels of this variety of maize. Structurally the popcorn maize kernel has a ring of hard, protein-rich cells of the endosperm (horny endosperm) and a tough outer **hull** surrounding starch-rich cells (starchy endosperm) that contain more moisture. Heating the kernel causes the water to turn to steam, creating immense internal pressure, which is released by the explosive rupture of outer endosperm and hull.

Poppy

Edible seeds of poppy (*Papaver somniferum*) are produced in abundance in the capsules of the opium poppy. When allowed to ripen, the capsules and seeds contain relatively little of the opium alkaloids. Ripe seeds are dark blue–dark grey in colour, more or less kidney shaped, measuring 0.8–1 mm in length and weighing on average about 0.5 mg. The species is thought to be of Mediterranean origin but has long been cultivated in Asia Minor, India and China. There are records of its cultivation in Europe since Neolithic times. The main producers of poppy seed for culinary purposes are now Turkey, the Netherlands, Czech Republic, Hungary, Romania, Austria, France and Germany (FAOSTAT).

Poppy seeds have a somewhat nutty flavour. They are used to flavour and decorate various European and Middle Eastern baking products and also are ground into a paste for use in desserts and cakes. In Eastern Europe they are incorporated with fish and vegetables while in India, especially in the north, the ground seeds are utilized in sauces. Poppy seeds are popular in Japanese cuisine.

The seeds contain an **oil** (triacylglycerol), at 40–50% fresh weight, which is extracted by cold pressing. It is composed mostly of unsaturated **fatty acids** – 60% linoleic acid, 30% oleic acid, 3% linolenic acid. The oil is used as an edible cooking oil and also to make paints, varnishes and soaps. Several allergenic compounds are present in the seeds. Since the alkaloids of the unripe poppy capsules occur at such low concentrations in the ripe seeds these have little or no pharmacological effect: but several cases have been reported of detectable amounts of morphine in the urine of subjects who have consumed a small number of poppy seed-dressed bakery products.

Seeds of several *Papaver* species are stimulated to germinate by light (**see: Light – dormancy and germination**). It is held that the disturbance of soil by shells and shrapnel exposed seeds of the corn poppy (*P. rhoeas*) to light causing the profusion of red-flowered poppy plants in the battlefields of Flanders in World War I. In many countries the red poppy is a symbol of remembrance of this and other wars. (MB)

(**See: Pharmaceuticals and pharmacologically active compounds**)

Potato

Solanum tuberosum, the most important cultivated edible potato species worldwide, is a member of the Solanaceae, which also contains tobacco, tomato, pepper and eggplant. The crop is considered to have been domesticated about

13,000 years ago in the central Andes, in the high plateau region of present-day Peru and Bolivia, adapted to the cool short tropical days and selected to produce tubers with low alkaloid contents. A few landraces were taken to and spread throughout Europe in the 16th century, and to North America in the 17th. Based on later introductions, modern varieties have been produced that are adapted to a wide range of climates in tropical and temperate environments. The cultivated species is an autotetraploid (2n = 4x = 48), but diploid, triploid and pentaploid landraces are still grown in South America (**see: Polyploid**). Almost all related wild species can be crossed with some ease; indeed, in the wild there is a high degree of natural **hybridization**, which has contributed to the complicated ancestry of the cultivated tetraploid species.

Commercial production of the ware potato crop throughout the world is almost completely based on asexual vegetative propagation of so-called 'seed' tubers, which ensures a high degree of genetic uniformity. Such clonal propagation has some disadvantages however: vulnerability to disease transmission, particularly of tuber-borne viral pathogens, which requires 'seed' production in areas isolated from the main crop; liability to premature sprouting before planting where cold storage is unavailable; and the expense of producing, storing and transporting disease-free tubers in the quantities necessary to sow the annual crop – costs that can be prohibitive for small-scale subsistence farmers in developing countries. For these reasons amongst others, there has been continuing interest since the technique was developed in the late 1950s in growing the crop from 'true potato seed' (TPS), free from the major potato diseases. The technology has been commercially implemented in many countries (including marginal farming areas in Peru, Nicaragua, India, Sri Lanka, Bangladesh, Nepal, China, Vietnam and Indonesia) and is under consideration in others. TPS is used either to produce seedling **transplants**, or for high-density nursery production of 'seed tubers' to be planted into the field in the second year. Only about 100–250 g TPS are required per hectare, compared to the 2–3 t of tubers needed to sow the same area. The third option of direct sowing in the field is not currently practical due to the high variability in germination and poor seedling survivability, and its unsuitability in the labour-intensive small-field potato production systems used in developing countries. Some disadvantages of TPS technology are the relatively small and more variably sized tubers and yields produced in the maincrop in some areas compared to tuber-propagated crops, and the tendency towards later harvest maturity. Type, size and quality are also affected because of the degree of genetic variability found in open-pollinated seed, due in part to its tetraploidy.

True seed production is also necessary in breeding. Potatoes are mainly **self-pollinated**, with some cross-pollination by bumblebees. Hybrids are accordingly produced by hand emasculation, though male sterility systems are under development. Completely homogeneous progeny is best constructed by the use of 4x families from 4x × 2x crosses (rather than 4x × 4x crosses) in which the 2x male or female parent produces 2n gametes due to mutants in **meiosis** that give rise to 'unreduced **gametes**'.

Flowering strongly depends upon environment, being favoured by long daylengths (around 16 h) and cool temperatures (around 15°C). Though newer cultivars flower abundantly, older ones seldom do, or produce pollen poorly (if at all) or have reduced seed set, due to male sterility or incompatibility. Seeds develop about 6–8 weeks after pollination within 1–3 cm-diameter round berries (the 'seed ball' or 'apple'). Depending on cultivar, berries contain 50–400 round seeds that vary in diameter from about 1.3 to 1.8 mm and weight from about 0.5 to 0.8 mg. Harvested berries are stored until they soften, and seeds extracted with a grinder and dried after debris has been removed, such as by flotation in water. TPS is viable for up to about 2 years or more but is liable to prolonged postharvest dormancy (**see: Dormancy – embryo**), which typically lasts for about 4–6 months until the maximum germination level is attained. Chemical treatments that can break dormancy include soaking in solutions of KNO_3, K_3PO_4, or **gibberellic acid** (GA_3). Germination generally is completed 8–10 days after sowing, but varies within and between genotypes. The temperature optimum is typically between 12 and 15°C, and shading is therefore a good practice in higher temperature sowing conditions. (PH)

Almekinders, C., Chilver, A. and Renia, H. (1996) Current status of the TPS technology in the world. *Potato Research* 39, 289–303.

Simmonds, N. (1997) A review of potato propagation by means of seed, as distinct from clonal propagation by tubers. *Potato Research* 40, 191–214.

The International Potato Center, Peru (Spanish acronym, CIP) www.cipotato.org/projects/PF03_truepotatoseed.htm

Potential seed number

A theoretical concept describing the number of seeds that would be produced if all reproductive structures, flowers or flower primordia present at a specific time, continued development and produced normal mature seeds. Potential seed number is an estimate of the reproductive potential of a plant or plant community and recognizes that this potential may change during reproductive growth. Its value is zero before reproductive development begins and increases to a maximum in most plant species as the number of reproductive structures per plant or per unit area increases (Fig. P.27). Potential seed number may then decline as some structures abort and it is eventually equal to the actual seed number. The critical period for seed number determination, the period at the beginning of reproductive growth when the number of seeds is determined, is also the period when potential seed number changes. By definition, potential seed number and the number of seeds the plant will produce at maturity are equal at the end of the critical period (Fig. P.27).

The maximum potential seed number is larger than final seed number in many plant species (excess capacity > 0, Fig. P.27), so seed number is determined by reducing the potential until it equals the final number. This adjustment is at least partially related to the availability of assimilate from photosynthesis (**see: Charles-Edwards model**). The excess reproductive capacity (maximum potential seed number minus the actual seed number at maturity) can be quite large as more than 50% of the flowers do not produce mature reproductive structures in many crop species. Excess capacity may be larger in species that increase potential seed number in favourable

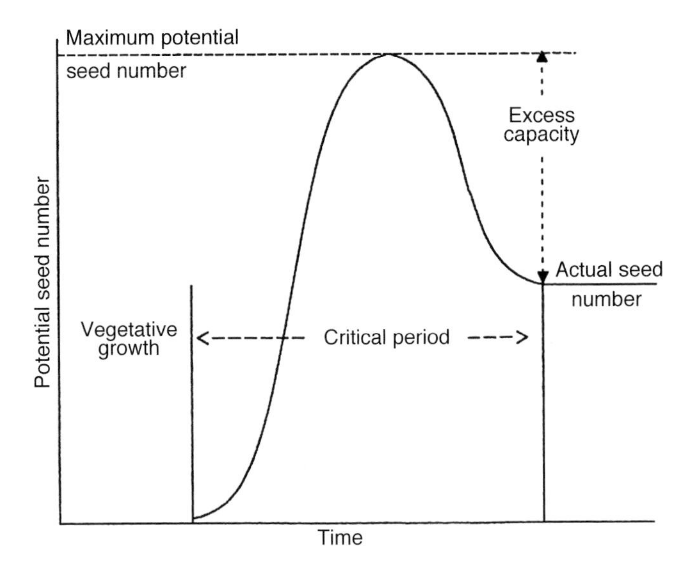

Fig. P.21. Theoretical changes in potential seed number during the critical period. Potential seed number is zero at the beginning of the critical period and is equal to actual seed number at the end of the critical period when the latter can no longer change. The excess reproductive capacity represents the difference between maximum potential seed number and actual seed number. From Egli, D.B. (1998) *Seed Biology and the Yield of Grain Crops*. CAB International, Wallingford, UK.

environments by branching or tillering, e.g. wheat (*Triticum* species), soybean (*Glycine max*), rice (*Oryza sativa*). Species that lost their ability to branch or tiller during domestication and improvement, such as maize (*Zea mays*), may have less reproductive plasticity and their excess reproductive capacity may approach zero in favourable environments. Seed number and yield of these species will be more sensitive to plant density than those that tiller or branch at low densities.

Seed **number** and **yield** will not be limited by the number of flowers when the plant has excess reproductive capacity, so that the plant or crop can adjust its yield to the maximum supported by the environment. If flower number is limiting (maximum potential seed number and actual seed number are equal), yield may be less than potential for that environment. Maize grown at a low plant density is an example of this scenario, where every flower produces a seed and the plant cannot produce more flowers. (DBE)

Egli, D.B. (1998) *Seed Biology and the Yield of Grain Crops*. CAB International, Wallingford, UK.

Pre-basic seed

An official term usually referring to the small supplies of seed planted to produce **Basic seed**, such as used in North America. (**See: Certification schemes**)

Prechilling

A seed processing term (still also commonly known by the old term, **stratification**) for a treatment applied to break the **dormancy** of several tree and garden species. The incubation of moist seeds at a cold temperature (usually 0–5°C) for a sufficient period of time, usually several weeks or months, as an artificial means of overcoming dormancy and hence promoting germination. (**See: Dormancy breaking – temperature**)

Prechilling, for up to 7 days, may be carried out according to **ISTA** Rules to break dormancy before germination testing, of both tree and certain other species. (**See: Germination testing**)

Precision drilling

The pattern of row planting resulting from the accurate and almost equally spaced placement (and subsequent covering) of single seeds in **furrows**. (See Fig. P.15.)

Precleaning

In seed conditioning, operations concerned with the removal of plant debris (such as **awns**, appendages, beards, **hulls**), **scarification** of the seed and removal of the coarse non-seed foreign material. Depending on the scale and the location of the operation, precleaning methods include hand sieving, **winnowing**, and precleaning scalping and screening machines before the drying, cleaning and finishing machines. (**See: Conditioning, I. Precleaning**)

Precocious germination

The germination of the embryo in the developing seed before it reaches maturity: synonymous with **vivipary**. It differs from **preharvest sprouting** as the latter includes seeds which have matured on the parent plant but which germinate in humid conditions because of their very low or total lack of **dormancy**. (**See: Development of seeds – an overview; Embryogenesis; Maturation – controlling factors; Preharvest sprouting**)

Pre-curing

Seed processing stage used before the seed extraction step for some forest tree species for afterripening of fruits (which were deliberately or incidentally collected immature), or to overcome **serotiny** for **case-hardening**. Techniques may consist of either freezing, moistening, slowing the drying process or accelerating it with higher temperatures, or any combination of these in sequence.
(**See: Tree seeds**)

Predation

Predation is the eating of seeds by animals and is also known as granivory. It is a major cause of seed loss and seems to be important in the **evolution** of seed characteristics. Pre-dispersal predation takes place while the seeds are developing on the plant. Post-dispersal predation happens after seeds are shed from the parent and continues to be a threat (even through years in the **soil seed bank**) until the seeds germinate.

Many terrestrial and freshwater animal groups contain seed eaters, but the most important are birds, mammals (especially rodents), insects, worms (annelids), and slugs and snails (molluscs). Seeds have good food value, being higher in proteins than most plant tissues. However, they can be a problematic food source, often having tough coats and other defences (see below) and are generally highly seasonal in abundance. Predators deal with these problems in different ways. Specialists, which feed on the seeds of one or a few related species, are generally invertebrate, pre-dispersal predators. They may spend part or the whole of the life cycle feeding on seeds, e.g. only the early larval stages of the large blue butterfly (*Maculinea arion*) feeds on seeds of thyme (*Thymus praecox*), while all larval stages of the weevil (*Apion ulicis*) feed on gorse (*Ulex europaeus*) seeds. The life cycle of such specialists is tuned to that of the host plant, so that seed-feeding stages are present at the time of seed production. Furthermore, the fruit can often protect these small animals against predation or exposure, and many specialists exploit this by remaining on a single plant (e.g. *Maculinea arion*) or even within a single fruit (e.g. *Apion ulicis*) for the whole seed-feeding period. Many agricultural pests are specialist seed predators, such as the cotton bollworm (*Pectinophora gossypiella*) and the coffee-bean weevil (*Araecerus coffeae*).

Generalist predators either eat only seeds of a wide range of species, e.g. many finches, some rodents such as the harvest mouse (*Micromys minutus*), and granivorous ants (such as *Tetramorium* spp.), or they eat seeds along with other types of food (e.g. members of the crow family, rodents such as squirrels, scavenging ants (such as *Formica* spp.), and humans) . Some of these are pests of crops or stored seed, e.g. wireworms (*Limonius* spp.) and the Indian mealmoth (*Plodia interpunctella*). Post-dispersal predation is generally by generalists, but these may also eat seeds and fruit on the plant. The various predators can combine to remove a huge proportion of seeds from a plant. For example, a gorse plant can lose >90% of seeds to pre-dispersal predation from specialists such as *Apion ulicis* and a great number of generalists such as the moth *Mirificarma mulinella*, and a large portion of the remainder can be lost post-dispersal to generalists such as the ant *Tetramorium caespitum* and the wood mouse (*Apodemus sylvaticus*).

A predator may break into a seed and eat its contents, thus killing it, and many seed predators have strong mouthparts to do this. For example, the agouti (*Dasyprocta aguti*) is reportedly the only animal with teeth strong enough to break into the incredibly tough pod of the **Brazil nut** (*Bertholletia excelsa*). Seeds can also be eaten whole, especially when the seed is small relative to the predator and is ingested within a fruit or with other plant parts. Many seeds do not pass unharmed through the animal's digestive tract, but some are defecated whole and so are **dispersed** rather than predated.

Plants have many strategies that help prevent seed mortality through predation. Access to seeds on the plant can be hindered by strategies preventing herbivory in general, such as leaves and stems with spines or toxins. Spiny (e.g. sweet **chestnut**, *Castanea sativa*) or mucilaginous (e.g. **flax**, *Linum usitatissimum*) fruits or seeds may deter some predators, although these are usually associated with dispersal. Toxic or distasteful seeds, e.g. yew (*Taxus baccata*), laburnum (*Laburnum anagyroides*) and **castor bean** (*Ricinus communis*) or (less commonly) fruits, e.g. deadly nightshade (*Atropa belladonna*) and henbane (*Hyoscyamus niger*) are more obvious deterrents, although some toxins are only effective on particular animal groups (**see: Pharmaceuticals and pharmacologically active compounds; Poisonous seeds**).

Seeds may be protected by a woody fruit, e.g. pods on gorse, the husk of the **coconut** (*Cocos nucifera*) and oak acorns (*Quercus* spp.), by a thick **seedcoat** (many **legumes**), or both (e.g. the **Brazil nut**). Seeds with thick coats are also protected during passage through an animal's gut. The timing of seed set can be important: gorse plants which set seed in late spring suffer much less loss of seeds to weevils (6%) than those which set seed during the peak of weevil abundance in summer (60%). Seeds may become hidden from predators post-dispersal, e.g. by incorporation into the **soil seed bank** (and smaller seeds are incorporated more rapidly) or by removal by **dispersal** agents into caches or nests. **Masting (see: Number of seeds – ecological perspective)** may allow a large proportion of seeds to escape because seed predator numbers are kept low by poor seed production in non-mast years and they cannot fully exploit the huge number of seeds in mast years. (JMB)

(See: Deserts and arid lands)

Crawley, M.J. (2000) Seed predators and plant population dynamics. In: Fenner, M. (ed.) *Seeds: the Ecology of Regeneration in Plant Communities*, 2nd edn. CAB International, Wallingford, UK, pp. 167–182.

Forget, P.M., Lambert, J.E., Hulme, P.E. and Vander Wall, S.B. (2004) *Seed Fate: Predation, Dispersal and Seedling Establishment*. CAB International, Wallingford, UK.

Hulme, P.E. (2002) Seed-eaters: seed dispersal, destruction and demography. In: Levey, D.J., Silva, W.R. and Galetti, M. (eds) *Seed Dispersal and Frugivory*. CAB International, Wallingford, UK, pp. 257–274.

Pre-germination

Seed **enhancement** technologies based on the concept of allowing seeds to just germinate (**chit**) for subsequent sowing, in order to produce 'high viability' seedlots with a very high germination percentage and speed and uniformity of seedling emergence. In some applications, only those individuals that are at a specific stage of **radicle emergence** are selected for sowing.

Fluid drilling is a technique of this sort, in which seed is allowed to complete germination in an aerated medium of relatively high water potential and, after removal of ungerminated individuals, such as by density separation, the sprouted seeds are suspended in a viscous gel and sown by extrusion into the soil. Low water potentials or plant growth inhibitors, such as **abscisic acid**, can be used to synchronize the pre-germination process, or keep the seedlings in a suspended growth stage for a period of time. Fluid drilling is not widely used commercially, however, in part because of the difficulty in achieving precision spacing and seedbed conditions to ensure the even development of the young seedlings. The technique is best suited to situations where crop production follows a fixed plan and is not likely to be interrupted by bad weather.

In another modern type of pre-germination treatment, development is suspended just after radicle emergence for conventional distribution and sowing. Fully imbibed seeds are germinated to the point where radicles are just visible, sorted by machine vision, **flotation**, or other means to remove ungerminated seeds, and gradually dried to induce desiccation tolerance. This can produce either damp pre-germinated seed (30–55% moisture content) with a storage life of a few weeks at ambient temperature, or dry seed that is viable for a few months. Although technically applicable to various species, this treatment is commercially available at present only for high-value flower seeds, because of its expense.

At its very simplest, pre-germination derives from on-farm **steeping** and sowing of wet seed – practices probably of considerable antiquity in some crops – and is related to **malting** techniques. Texts from 6th-century China, for example, advocate the pre-germination of rice (**see: Treatments – brief history**). Versions of this technique are still practised today to avoid impaired germination if seed is sown under water that is too deep or too muddy, and over the years has probably been applied to a number of other crops including vegetable seeds, and in the **water-seeding** of rice.

(PH)

Bruggink, G.T. and van der Toorn, P. (1995) Induction of desiccation tolerance in germinated seeds. *Seed Science Research* 5, 1–4.

Far, J.J., Upadhyaya, S.K. and Shafii, S. (1994) Development and field evaluation of a hydropneumatic planter for primed vegetable seeds. *Transactions of the American Society of Agricultural Engineers* 37, 1069–1075.

Finch-Savage, W.E. and McQuistan, C.I. (1989) The use of abscisic acid to synchronize carrot seed germination prior to fluid drilling. *Annals of Botany* 65, 195–199.

Pill, W.G. (1991) Advances in fluid drilling. *HortTechnology* 1, 59–65.

Preharvest sprouting

The adaptive significance of **dormancy** is evident for plants living in the wild. However, its presence has mostly been seen as a complication in seeds from plants that are grown as crops. Indeed, a persistent dormancy prevents the utilization of a seedlot either for the generation of a new crop or for industrial purposes (e.g. **malting**). For this reason, crops that originally must have had dormancy have been selected heavily against throughout their **domestication**. In some cases, this selection pressure has gone too far and the seeds are germinable even prior to crop harvest. This situation, combined with rainy or damp conditions prevailing during the last stages of maturation, may lead to germination on the mother plant, a phenomenon that is known as preharvest sprouting. Note that preharvest sprouting is distinct from **vivipary**, which is the germination of developing seed while still on the mother plant, prior to seed maturation.

Though its occurrence has been reported for several cultivated plants, preharvest sprouting is predominantly a feature of cereal crops. White wheat (*Triticum aestivum*) and barley (*Hordeum vulgare*) are two of the major crops affected in different areas of the world (Colour Plate 11B). Most modern barley genotypes have low resistance to sprouting because low dormancy at harvest is a characteristic required by the malting industry and, therefore, a major goal of breeding. Rye (*Secale cereale*) is particularly susceptible to preharvest sprouting because the flowers are cross-pollinated, and the open structures of the glumes allow water to reach the grain. This is also the case in triticale (x *Triticosecale*). Sprouting of oat (*Avena sativa*) occurs episodically in some areas, but has little effect on the quality of the grain for animal feed. *Japonica* rice (*Oryza sativa*), which is usually grown in more temperate regions, sprouts more easily than the *indica* rice of the tropics, which is highly resistant to preharvest sprouting. Sprouting is not a problem in maize (*Zea mays*) because the grain is protected by the husk from the moist conditions that promote germination in other cereals. Sorghum (*Sorghum bicolor*) and pearl millet (*Pennesitum glaucum*) are crops usually grown in semi-arid regions and for that reason sprouting episodes have been rarely reported; however, when they are cultivated in more humid areas (i.e. sorghum grown in the humid Pampa of Argentina), grains are particularly prone to sprouting if conditions are appropriate.

The inception of dormancy in cereal grains takes place very early during development. **Embryos** are usually fully germinable from the early stages (i.e. 15–20 DAP, days after pollination) if isolated from the entire grain and incubated in water: the entire grain, however, reaches the capacity to germinate only well after it has been acquired by the naked embryo. This coat (**testa** plus **pericarp** plus glumellae and/or **glumes**)-imposed dormancy is the barrier preventing untimely germination and its duration depends on the genotype and on the environment experienced during **maturation** and beyond. Therefore, though cases of embryo dormancy occur in barley and other cereal crops, sprouting-susceptible cultivars are usually those whose coat-imposed dormancy is terminated well before harvest maturity (**see: Dormancy – coat-imposed**).

Nonetheless, despite a high dormancy level prior to harvest, sprouting can still occur if damp conditions prevail. With the exception of seeds that show deep dormancy and consequently do not germinate at any temperature, it is a common feature that seed dormancy, including in cereal grains, is expressed at certain temperatures and not at others. In summer cereals like sorghum, dormancy is not expressed at high temperatures (i.e. 30°C), while in winter cereals like wheat and barley it is not expressed at low temperatures (i.e. 10°C or lower) (**see: Relative dormancy**). This lack of expression of dormancy at, for example, high temperatures in grains from summer cereals, implies that in years when damp conditions are combined with high air temperatures around harvest time, both resistant (high dormancy) and susceptible (low dormancy) cultivars might be expected to sprout.

The environment experienced by the mother plant during seed development may sometimes modulate the rapidity with which seeds are released from dormancy and thus the sprouting behaviour of a crop. Among the different factors acting while the seeds are developing, temperature appears to be mainly responsible for year-to-year variation in grain dormancy of a particular genotype. Positive relationships between temperature experienced during development and extent of release from dormancy have been established for some cereals (i.e. the higher the temperature during grain filling, the lower the dormancy level).

Preharvest sprouting has a wide range of consequences, all of them adverse. These range from the immediate loss of seed **viability** upon subsequent desiccation (**see: Desiccation tolerance**), to a marked reduction in seed **longevity**. But, overall, the completion of germination triggers the synthesis of enzymes that promote reserve mobilization. Following germination, the growing embryo synthesizes and secretes **gibberellins** into the starchy endosperm (**see: Mobilization of reserves – a monocot overview; Mobilization of**

reserves – cereals). The gibberellins diffuse into the **aleurone layer** and induce the synthesis of α-amylase. This endoamylase starts the degradation of the stored starch in the endosperm since it is the only enzyme that can hydrolyse the raw **starch granules**. α-Amylase in cereals is commonly divided into two types, an endogenous late maturity, green, or low pI group and a high pI group that is associated with sprouting (pI refers to the charge on the protein molecules). Sprout damage may be assessed by the method of the Hagberg Falling Number (FN), which serves as a gauge for α-amylase activity and the deleterious starch degradation in a flour preparation. The FN test measures primarily the change in viscosity due to enzymatic breakdown of starch and is defined as the time in seconds required to stir and to allow a viscometer stirrer to fall a measured distance through a hot aqueous meal, flour or starch gel undergoing liquefaction due to activity of the enzyme. In addition to α-amylase, many types of proteolytic enzymes – endopeptidases, carboxypeptidases, aminopeptidases, etc. – are associated with sprouting. The consequences of this reserve hydrolysis directly depend on the types of products for which the cereal is intended and on the processing methods used (Table P.13). Even a small percentage of sprouted grains in the harvested material is a serious contaminant, producing enough enzymes to cause eventual spoilage of the flour. (RLB-A)

See: other entries on **Preharvest sprouting**

Paulsen, G.M. and Auld, A.S. (2004) Preharvest sprouting of cereals. In: Benech-Arnold, R.L. and Sánchez, R.A. (eds) *Handbook of Seed Physiology: Applications to Agriculture*. Food Products Press, New York, USA, pp. 199–219.

Preharvest sprouting – economic importance

Preharvest sprouting in cereals is a problem in many parts of the world and, in some **wheat** production areas, occurs in 3 to 4 years out of 10. Grain sprouting has been reported in the USA, Canada, northern and western Europe, New Zealand and Australia, as well as in portions of central South America and the southern parts of Africa. From 1978 to 1988, average annual losses in 37 countries totalled over US$450 million, mostly to wheat. However, the major cereal-producing countries of China, India, USSR, and Argentina were not included in the survey, and estimates were not available from the USA and several other countries. It is likely that total worldwide direct annual losses approach US$1 billion.

Direct economic losses to producers from preharvest sprouting occur in several ways. The yield may be reduced due to carbohydrate **respiration** which, at the same time, generates a favourable environment for **saprophytic** attack (i.e. by fungi and bacteria); the volume density (test weight) may decrease from loss of dry matter and irreversible swelling of the grain, and its suitability (as flour) for many food products may be diminished (**see: Cereals, 6. Grain quality**). All of these effects result in a reduction of the farmers' income since sprouted grains are often down-graded or even completely rejected by flour millers, for example. Indeed, regardless of the grading system prevailing in each country, all of them tend to ensure that top grades of wheat contain minimal amounts of α-amylase (**see: Preharvest sprouting**). Due to the relatively warm and dry conditions around harvest time, preharvest sprouting in wheat is not a frequent problem in Argentina; however, a severe sprouting episode took place in 1996 which affected 10% of the total production (1.5 million t) and compelled the government to create an, until then, unexisting grade, the 'forage wheat grain' to be used for livestock feed only. This grade, in turn, was divided in three sub-grades, depending on sprouting severity. The **malting** industry is particularly strict with sprouted lots and often they reject grain even if FN values show moderate incidence.

Indirect losses arise from the fact that some high value crops cannot be grown in areas where conditions around harvest time are suitable for sprouting to occur. For example, Chinese farmers are forced to cultivate red wheat instead of the more valuable white type because of the hazard of sprouting damage. (RLB-A)

See: other entries on **Preharvest sprouting**

Wahl, T.I. and O'Rourke, A.D. (1993) The economics of sprout damage in wheat. In: Walker-Simmons, M.K. and Ried, J.L. (eds) *Preharvest Sprouting in Cereals*. American Association of Cereal Chemists Inc., St Paul, MN, USA, pp. 10–17.

Preharvest sprouting – genetics

The preponderance of genetic investigations of **dormancy** have been conducted on cereal grain crops, such as **barley**, **rice**, **sorghum** and **wheat**, to impart resistance to preharvest sprouting. This resistance is highly correlated with seed dormancy and most genetic research on preharvest sprouting is directly applicable to this phenomenon (**see: Dormancy – genetics**). For example, when resistance to preharvest

Table P.13. Consequences of using sprouted grains on the quality of different end products.

End product	Consequences of sprouting
Breads	Increases stickiness of the dough Weakens the dough strength Decreases the amylographs' peak viscosities Causes poor handling and machining properties Alters rheological properties to proteolytic enzymes
Pasta	α-Amylase weakens the dough in dry noodles and they break during drying Enzymes produced during sprouting affect colour, texture and brightness in wet and Cantonese noodles
Cakes	Cake volume decreases at high levels of sprouting Poor baking
Alcoholic beverages	Sprouting in barley lowers conversion of the malt and extractability of fermentable material

sprouting is determined by laboratory assays of seed germination in a Petri dish, the test directly determines the level of postharvest seed dormancy. Conversely, if developing seeds on the flowering head are placed in a mist chamber and sprouting in the head is evaluated, or studies are done in the field, factors for resistance to preharvest sprouting other than dormancy, e.g. spike morphology, drying rate, may be included.

1. Rice

Rice (*Oryza sativa*) is the world's most important cereal crop for human consumption, and it is a model experimental system for Poaceae (grass) species (**see: Dormancy – genetics**). Rice is a diploid with a relatively small genome (430 Mbp) distributed over 12 chromosomes. Its genome is the base for comparative mapping among cereal grains and other grass species because there is a high degree of co-linearity in gene content and order among cereal genomes. The genome has been sequenced in both the *japonica* and *indica* rice subspecies. With a few exceptions, genetic investigations of dormancy in rice have been limited to domesticated cultivars because of problems with preharvest spouting. In general, cultivars belonging to the *indica* subspecies have greater dormancy than those belonging to the *japonica* subspecies, but dormancy is strongest in weedy rice. Weedy rice must contain additional genes or stronger allelic variants that are not present in domesticated cultivars.

Dormancy in rice cultivars is a quantitative trait with dormancy being dominant or incompletely dominant. Nuclear factors control rice seed dormancy. Cytoplasmic factors do not seem to contribute to the genetic regulation of rice seed dormancy, unlike in seeds of some other plant species. Seed dormancy in rice is imposed by maternal tissues, the **hull**, **pericarp/testa**, or both, and nearly independent genetic factors appear to regulate each type. It is unclear whether **embryo dormancy** occurs in rice. Heritability is estimated at 12 to 42% for germination of rice seed grown under field conditions. For two weedy strain-derived F_2 populations grown under controlled conditions, heritability for germination or dormancy of dispersal units (intact grains) (hull-imposed) or **caryopses** (pericarp/testa-imposed) was 76% and 82%, respectively. As with other species, dormancy in rice is influenced by both genetic and environmental factors, for example, temperatures during seed development and afterripening: temperature also affects germination.

Plant geneticists are seeking molecular makers tightly linked to dormancy **quantitative trait loci** (QTLs) in cereal grain species to conduct marker-assisted breeding for resistance to preharvest sprouting. Also, this molecular genetic approach is important for understanding mechanisms underlying dormancy since it could lead to map-based cloning and characterization of genes that directly regulate seed dormancy and germination (Fig. P.22). Although not always done, it is advantageous to cross a non-dormant cultivar and a strongly dormant accession in order to detect a QTL that explains a major portion of the phenotypic variance for dormancy, and to set the stage for developing additional germplasm for map-based cloning. The **domestication** process might have eliminated strong dormancy **alleles** that exist in weedy accessions or wild progenitors. Strong alleles may enhance the dormancy phenotype in the segregating population and make it easier to measure variance due to genetic factors against the background variance due to environmental and other factors.

A QTL analysis for seed dormancy in rice (*Oryza* spp.) has been conducted using a recombinant inbred line population derived from the cross *Oryza sativa* ssp. *indica* cv. 'Pei-kuh'/*O. rufipogon* strain 'W1944'. *O. rufipogon* is a wild progenitor to Asian rice. Factors for hull- and pericarp/testa-imposed dormancy were analysed independently. Pei-kuh seeds grown under field conditions displayed germination of > 90%. Dormancy in W1944 seed varied depending on the test conditions. The population segregated for both hull-imposed and pericarp/testa-imposed dormancy, and at 30 days after heading (i.e appearance of the flowering head) 12 QTLs for dormancy with complete grains (hull-imposed dormancy) and two QTLs for dormancy with caryopses (pericarp/testa-imposed dormancy) were detected. The contribution of individual loci to total variance ranged from 7 to 21%. (**See: Dormancy – coat-imposed**)

2. Wheat

Wheat (*Triticum aestivum*) is one of the world's three most important cereal crops (**see: Cereals**). White-grain wheat cultivars are particularly susceptible to preharvest sprouting, as are some red-grain cultivars. The inheritance of seed dormancy in several white- and red-coloured cultivars has been examined in detail. Because research is conducted on many different genotypes under various environmental conditions, it is difficult to generalize about the genetic aspects of dormancy for any species, including wheat; and interested readers should seek out specifics on individual cultivars and segregating populations derived from various non-dormant and dormant parents. Nevertheless, pericarp/testa-imposed seed dormancy in wheat is generally considered a quantitative trait controlled by one or two major genes and an unknown number of minor genes. Dormancy in wheat mostly appears to be dominant or partially dominant, but it is recessive in some white-grain cultivars. Heritability for resistance to preharvest sprouting under field conditions ranges from low values to as high as 92%. As in other species, genotype by environmental interactions are common among loci within a population. Seven segregating wheat populations have been used to mark QTLs for dormancy or resistance to preharvest sprouting.

Loci have been detected on 20 of the 21 hexaploid wheat linkage groups with individual loci generally accounting for 11% or less of the observed phenotypic variation. One major gene, *Phs*, on the long arm of chromosome 4A displays Mendelian inheritance. While wheat is a less tractable system for map-based cloning because its genome size is 40 times larger than rice, this red-grain derived gene might be a candidate for map-based cloning using aneuploid lines (those having a chromosome number that is not an exact multiple of the **haploid** number) and chromosome substitution techniques.

Comparative genetics and mapping have been used in an attempt to identify candidate loci that play a direct role in dormancy, particularly preharvest spouting in wheat. For example, several studies have suggested a relationship between the well-characterized maize *Viviparous1* (*Vp1*) gene (**vivipary**) and seed dormancy. **Orthologues** of *Vp1* have

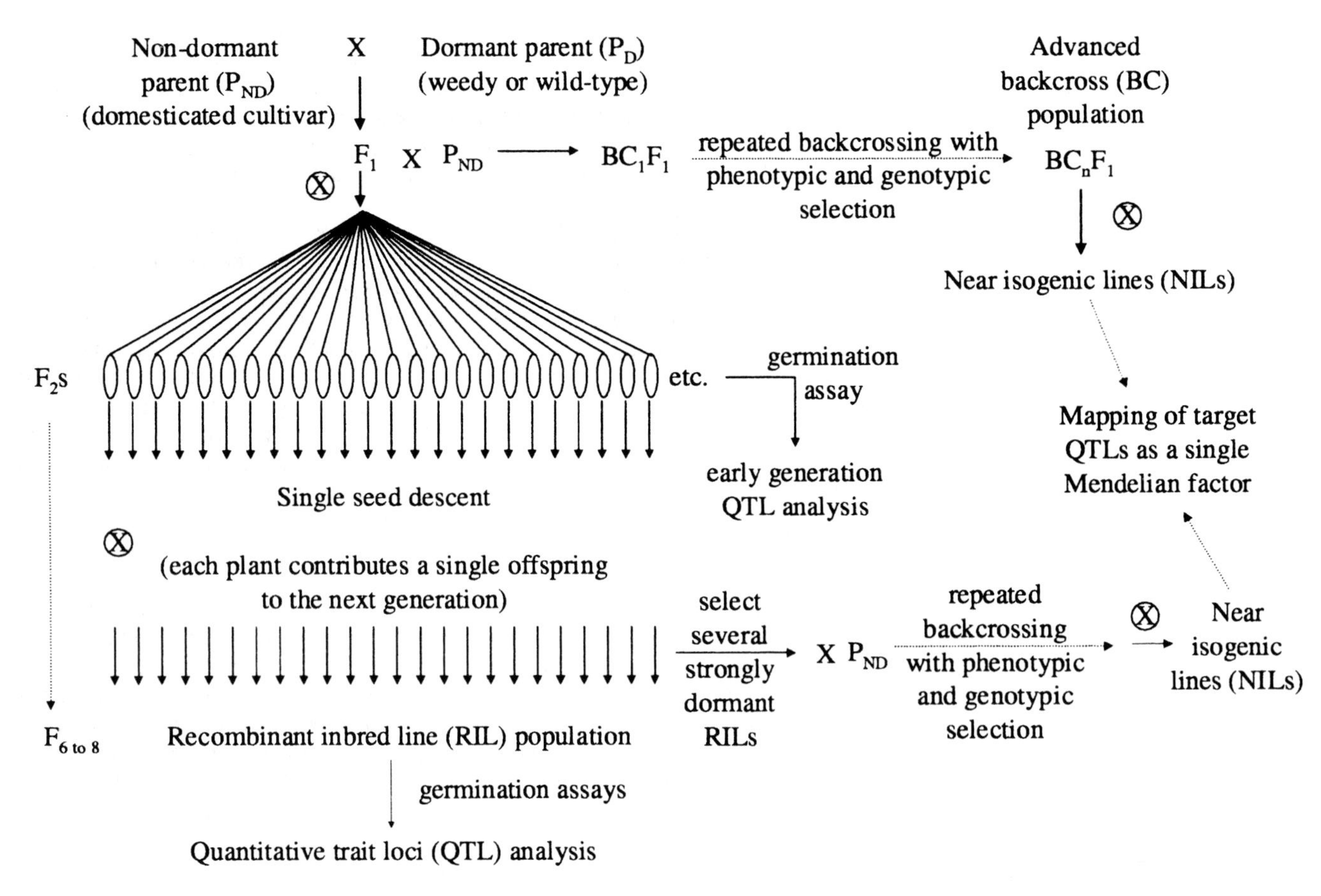

Fig. P.22. Identification of quantitative trait loci (QTLs). General scheme for development of populations to identify quantitative trait loci (QTLs) and for fine-scale mapping of QTLs towards the goal of molecular identification of genes that directly regulate dormancy and germination. Recombinant inbred lines (RILs) derived from a population segregating for dormancy represent a permanent mapping population that allows replication of germination assays on a large number of genetically identical seeds for each line. Near isogenic lines (NILs) are developed to verify and characterize individual QTLs. Using backcrossing procedures, individual dormancy-conferring alleles are introgressed into the genetic background of the non-dormant parent to separate a QTL from the rest of the segregating loci. Ideally, a NIL differs from the non-dormant parent only for alleles in a narrow genomic region around the QTL. Once an individual QTL can be mapped as a single Mendelian factor and fine-scale mapping has been done, map-based cloning may proceed. Phenotypic and genotypic selection denotes germination assays and marker assisted selection, respectively. An x represents cross-pollination; ⊗ self-pollination; → steps omitted.

been identified in many other species. Genetic mapping experiments have determined that the *Vp1* gene in wheat does not map to the region of a dormancy QTL. In any event, comparative mapping can be used to screen known genes as candidate QTLs and discard those that fall outside the QTL. However, those that lie within a QTL interval need not be the target gene because, depending on the interval, it could contain several to many hundreds of genes. A great deal of additional research goes into proving that a candidate gene is the target gene. In another use of comparative mapping, in this case between wheat and barley, it was suggested that the *Phs* gene in wheat is orthologous with the barley seed dormancy QTL, *Seed Dormancy4* (*SD4*). This hypothesis is based on similar positions on a comparative genetic map, and verification would be possible when the QTL is cloned from one species or another.

3. Barley

Dormancy in barley (*Hordeum vulgare*) is also a quantitative trait with nearly complete dominance for non-dormancy. Estimates for heritability range from 69 to 80%. Four QTLs affecting germinability of barley have been detected using a mapping population derived from the cross of a non-dormant and moderately dormant cultivar. The QTL *SD1* accounts for 55% of the phenotypic variation for germinability and is relatively insensitive to the environment. Although barley is a **diploid** with a genome size 12 times that of rice, map-based cloning has proved feasible for some Mendelian genes for disease resistance. However, map-based cloning of a gene regulating a quantitative trait such as dormancy from cultivated barley is likely to be very difficult, because the dormancy phenotype is weak and *SD1* may contain a cluster of genes that regulate dormancy.

4. Sorghum

A molecular genetic approach has also been used to investigate resistance to preharvest sprouting in sorghum (*Sorghum bicolor*). Preharvest sprouting-resistant and -susceptible inbred lines were crossed and germination of F_2-generation seed was determined. Two QTLs that each accounted for about 53% of the phenotypic variation were discovered. Moreover, based on this cross, it appears that resistance to preharvest sprouting is partially dominant, that **epistatic** interactions occur between loci, and that there are genotype by environmental interactions relating to annual differences in environmental conditions during seed development.

Using genetic approaches to discover novel dormancy QTLs and markers that are tightly linked to these QTLs will be useful in marker assisted selection when developing cultivars in locations where preharvest sprouting is a problem. Moreover, map-based cloning and characterization of dormancy QTLs from rice or other tractable grass species will be an important step in understanding signals, pathways, and mechanisms that regulate resistance to preharvest sprouting in cereal grain species. (MEF)

See: other entries on **Preharvest sprouting**

Cai, H.W. and Morishima, H. (2000) Genomic regions affecting seed shattering and seed dormancy in rice. *Theoretical and Applied Genetics* 100, 840–846.

Flintham, J., Adlam, R., Bassoi, M., Holdsworth, M. and Gale, M. (2002) Mapping genes for resistance to sprouting damage in wheat. *Euphytica* 126, 39–45.

Gu, X.-Y., Chen, Z.-X. and Foley, M.E. (2003) Inheritance of seed dormancy in weedy rice (*Oryza sativa* L.). *Crop Science* 43, 835–843.

Han, F., Ullrich, S.E., Clancy, J.A. and Romagosa, I. (1999) Inheritance and fine mapping of a major barley seed dormancy QTL. *Plant Science* 143, 113–118.

Li, C., Ni, P., Francki, M., Hunter, A., Zhang, Y., Schibeci, D., Li, H., Tarr, A., Wang, J., Cakir, M. *et al.* (2004) Genes controlling seed dormancy and pre-harvest sprouting in a rice–wheat–barley comparison. *Functional and Integrative Genomics* 4, 84–93.

Preharvest sprouting – mechanisms

1. Water uptake

One primary aspect to be considered when analysing mechanisms conferring a particular sprouting behaviour on a crop is the rate of absorption of moisture by **kernels**. Indeed, water for germination is usually available for limited periods during most sprouting episodes; thus, a high rate of water absorption or, conversely, the possibility of remaining moistened for a long time after damp conditions have ended, might add to sprouting susceptibility. Absorption of moisture by kernels is influenced by morphology of the inflorescence and characteristics of the **seedcoat**. **Imbibition** is increased by features associated with **awns** in wheat and is affected by waxiness, pubescence, and angle of the inflorescence in barley. Also implicated in controlling the rate of water absorption are grain hardness, colour, restriction by the seedcoat, thickness of the **testa** and other layers, size, and surface-to-volume ratio.

2. Dormancy

Overall, the primary cause for sprouting susceptibility is a low **dormancy** prior to crop harvest, and status of the phytohormone and germination inhibitor **abscisic acid** (ABA) (**see: Development of seeds – hormone content; Hormones**). ABA-deficient or -insensitive **mutants** of *Arabidopsis* and maize germinate precociously (**see: Dormancy – genes**). Also, the application of the ABA-synthesis inhibitor fluridone reduces dormancy in developing seeds of several species. In cereals, the imposition of dormancy on the embryo by the structures that surround it is likely to be mediated by the high endogenous ABA existing in the embryos during grain development. The ABA content of embryos is usually low until about 15 DAP. From that time onwards, ABA content goes up coinciding with the acquisition of the capacity of the embryo to germinate if isolated from the rest of the grain; hence, one possibility is that precocious germination would be prevented by the surrounding structures by impeding the leaching of ABA from the embryo. The ABA content has been reported to peak at around **physiological maturity** and to decline thereafter. However, no correlations have been found between embryonic content of ABA during seed development and susceptibility or resistance to sprouting. In other words, although inhibiting ABA synthesis (either genetically or through chemicals) accelerates the termination of dormancy, sprouting-susceptible genotypes do not have a lower ABA content during grain development than resistant ones. Different sprouting behaviour between genotypes is better explained by differences in embryo responsiveness to ABA. Embryo sensitivity to ABA is measured as the embryo capacity to overcome the inhibitory action of a certain concentration of the hormone. Higher concentrations of ABA are required to block germination of embryos isolated from grains of sprouting-susceptible cultivars than are needed to inhibit embryo germination of resistant ones. In most cases differences in embryo sensitivity to ABA between sprouting-resistant and -susceptible cultivars are maintained throughout most of development. Fig. P.23 illustrates this with an example in which the embryonic sensitivity to ABA of two barley cultivars – one sprouting-suceptible ('B 1215') and the other sprouting-resistant ('Quilmes Palomar') – was followed throughout grain development: embryos from cv. B 1215 started to lose sensitivity to ABA well before those from cv. Quilemes Palomar. The nature of the low sensitivity to ABA in embryos from sprouting-susceptible genotypes remains unclear but it might be due either to a high rate of degradation of the hormone in the outside walls of the embryo or to alterations in the ABA **signal transduction** pathway. The gene *Vp1*, for example, encodes a transcription factor whose involvement in the control of embryo sensitivity to ABA has been evident since the identification of maize *vp1* mutants which are insensitive to ABA and exhibit **vivipary**. Preharvest sprouting in cereals is very similar phenotypically to the *vp1* mutation in maize, raising the possibility that preharvest sprouting in cereals is caused, in part, by the physiological disruption of the *Vp1* function. Genes homologous to *Vp1* from barley and other Gramineae (Poaceae) such as rice, sorghum and *Avena fatua*, have been cloned and sequenced but only in some cases were there close correlations between expression of the gene and dormancy. Manipulation of *Vp1* function in those cases where a relationship with sprouting behaviour has been found, could offer an opportunity for adjusting the timing of dormancy release.

The exit from dormancy in developing cereal grains can be altered by inhibiting **gibberellin** (GA) synthesis, thus suggesting that this event (and consequently the sprouting behaviour of the crop) depends on the extent to which ABA action as a dormancy imposer is counterbalanced by the effect of GAs. Indeed, applications of the inhibitor of GA synthesis

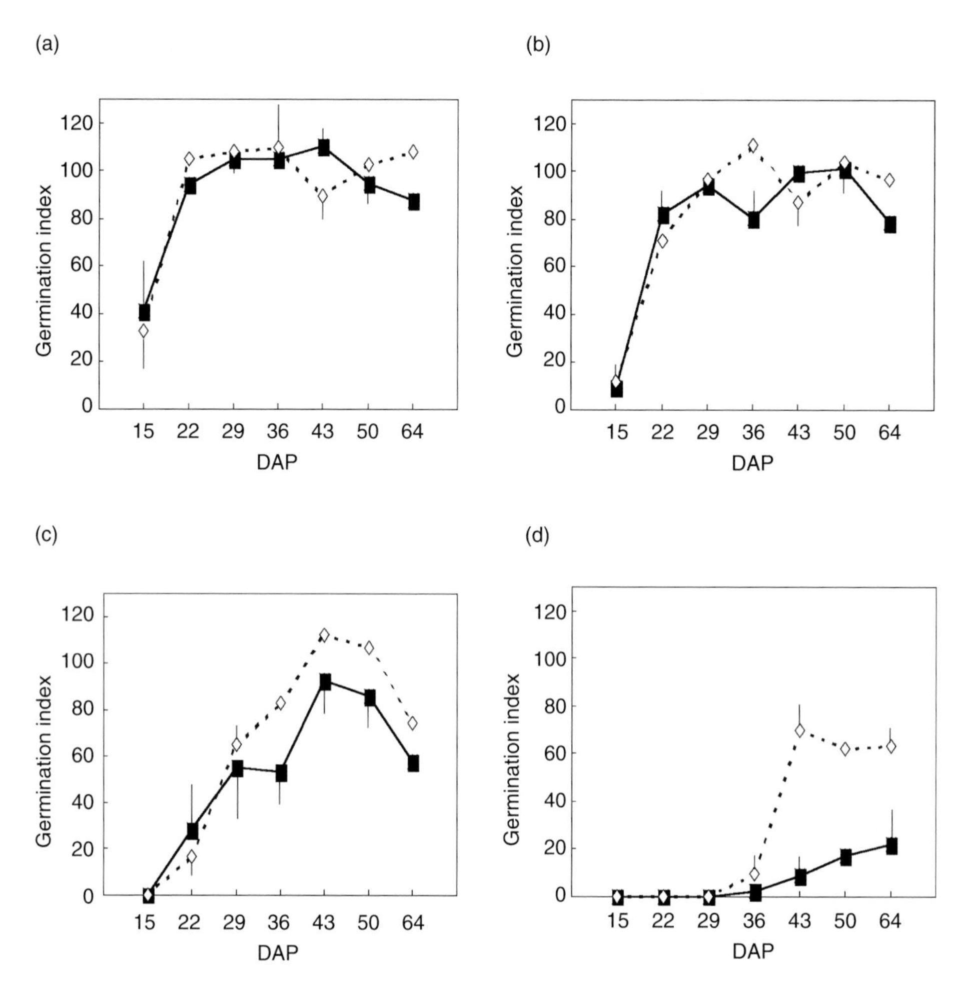

Fig. P.23. Sensitivity to ABA in B 1215 (◇) and Quilmes Palomar (■) barley embryos. Embryos were isolated from grains at different times (days after pollination, DAP). Germination index was measured for ten embryos in three replicates treated in distilled water (a), in 0.5 μM (b), in 5 μM (c) and 50 μM ABA (d). Vertical bars are ± SE. PM and HM: physiological maturity and harvest maturity, respectively. Adapted from Benech-Arnold, R.L., Giallorenzi, M.C., Frank, J. and Rodriguez, V. (1999) Termination of hull-imposed dormancy in barley is correlated with changes in embryonic ABA content and sensitivity. *Seed Science Research* 9, 39–47.

paclobutrazol almost immediately after anthesis of a sprouting-susceptible sorghum variety prevents exit from dormancy of the grain. Nonetheless, similar to the relationship between ABA content and sprouting susceptibility, sprouting-resistant genotypes do not appear to have a lower GA content during development. But perhaps lowering GA content through genetic means will result in genotypes with an extended dormancy and, consequently, with resistance to sprouting.

In addition to ABA and GAs, other substances have been implicated in the physiological control of sprouting. The **glumes** (bracts) of wheat, for instance, contain an unknown inhibitor that delays sprouting and is simply inherited. Similarly, the well-known resistance to preharvest sprouting of red wheats, relative to white wheats, has been attributed to precursors of the pigment phlobaphene in the **testa** layer of the former. These compounds, catechin- and tannin-like materials, inhibit germination: and they occur in lower amounts in white wheats than in red wheats, in which they decline during **afterripening** to permit germination. Tannin-containing **sorghums**, though less palatable than white sorghums, are very resistant to sprouting. Pigments in the seedcoat may be part of a two-factor system that inhibits germination directly or by interfering with gaseous exchange (**see: Germination – effects of gases**). (RLB-A)

See: other entries on **Preharvest sprouting**

Benech-Arnold, R.L. (2002) Bases of preharvest sprouting resistance in barley: physiology, molecular biology and environmental control of dormancy in the barley grain. In: Slafer, G.A., Molina-Cano, J.L., Araus, J.L., Savin, R. and Romagosa, I. (eds) *Barley Science. Recent Advances from Molecular Biology to Agronomy of Yield and Quality*. Food Product Press, New York, USA, pp. 481–502.

Black, M. (1991) Involvement of ABA in the physiology of developing and mature seeds. In: Davies, W.J. (ed.) *Abscisic Acid Physiology and Biochemistry*. Bios Scientific Publishers, Oxford, UK, pp. 99–124.

Derera, N.F. (1989) The effects of preharvest rain. In: Derera, N.F. (ed.) *Preharvest Field Sprouting in Cereals*. CRC Press, Boca Raton, FL, USA, pp. 1–25.

Gubler, F., Millar, A.A. and Jacobsen, J. (2005) Dormancy release, ABA and pre-harvest sprouting. *Current Opinion in Plant Biology* 8, 183–187.

McCarty, D.R., Hattori, T., Carson, C.B., Vasil, V., Lazar, M. and Vasil, I.K. (1991) The *viviparous-1* developmental gene of maize encodes a novel transcriptional activator. *Cell* 66, 895–905.

McKibbin, R.S., Wilkinson, M.D., Bailey, P.C., Flintham, J.E., Andrew, L.M., Lazzeri, P.A., Gale, M.D., Lenton, J.R. and Holdsworth, M.J. (2002) Transcripts of *Vp-1* homeologues are misspliced in modern wheat and ancestral species. *Proceedings of the National Academy of Sciences USA* 99, 10,203–10,208.

Paulsen, G.M. and Auld, A.S. (2004) Preharvest sprouting of cereals. In: Benech-Arnold, R.L. and Sánchez, R.A. (eds) *Handbook of Seed Physiology: Applications to Agriculture*. Food Products Press, New York, USA, pp. 199–219.

Steinbach, H.S., Benech-Arnold, R.L. and Sánchez, R.A. (1997) Hormonal regulation of dormancy in developing sorghum seeds. *Plant Physiology* 113, 149–154.

Pretreatment

In seed testing: a physical or chemical treatment of mature dry seed prior to sowing, given solely to facilitate seed health testing, e.g. **prechilling**, preheating, or pre-washing to remove water-leachable germination inhibitors, or dormancy-breaking treatments. (**See: Germination testing**) In seed technology: a term synonymous with **treatment**.

PREVAC (Pressure-Vacuum)

A **flotation** treatment (in the presence of water) applied mostly in the processing of conifer seeds, which applies a vacuum followed by return to atmospheric pressure to remove seeds with cracked or split coats from seedlots containing a mixture of intact and mechanically damaged seeds.

Some stages during the extraction and processing of conifer seeds can cause visible and even microscopic physical damage to the **seedcoat**, which adversely affects germination. The separation of mechanically damaged from sound seed is therefore desirable. When placed in water, mechanically damaged seeds sink quickly, but they do not all sink before some undamaged seeds also begin to precipitate: no effective separation can be obtained between the two seed types on this basis. Subsequently PREVAC developed as a quicker and better separation method in the presence of water, obtained through partial evacuation followed by re-pressurization.

A seedlot containing a mixture of dry-intact and mechanically damaged seeds is placed in water in a container capable of withstanding partial vacuum. The seeds are held beneath the surface of the water (but not on the bottom), for example by means of a retaining screen such as a weighted gauze. The vessel is evacuated, and any air inside damaged seeds escapes more quickly than from intact seeds. When the vacuum is released, water enters those seeds with cracked or split coats more quickly than sound seeds, making the damaged seeds denser. After the retaining screen is removed, the less dense intact seeds float to the surface and the heavier seeds sink and are discarded. In some instances the evacuation and re-pressurization cycle may be repeated.

In practice, PREVAC separation is applied to seedlots that are at normal storage moisture contents (6–8% fresh weight basis), at which intact, dry-stored seeds have about the right density to float on water. However, in some situations, glycerol solutions of different densities have been used to effect separation. It is generally held that temperature and treatment time are relatively unimportant, but that the minimum pressure and the number of cycles are the most important factors. (PG)

(See: Tree seeds)

Bergsten, U. and Wiklund, K. (1987) Some physical conditions for removal of mechanically damaged *Pinus sylvestris* seeds using the PREVAC method. *Scandinavian Journal of Forestry Research* 2, 315–323.

Primary dormancy

Dormancy induced during seed development and maturation and which is observed in seeds just before and after their dispersal from the mother plant. (**See: Dormancy – an overview**)

Priming

In seed technology: a broad term describing methods of physiological **enhancement** of seed performance, that use a range of seed hydration techniques followed usually by drying. The term also describes the biological processes that occur during these treatments, and the characteristic improvements in germination speed and/or uniformity that result after seed is sown. Priming is of considerable interest in seed physiological research, in part as a tool for understanding **germination** processes, and is of considerable importance in industry as a tool to 'invigorate' seedlots and hence improve their commercial value and help crop establishment. (**See: Pre-germination; Priming – physiological mechanisms; Priming – technology**)

1. Terminology

Originally, the increased speed and synchrony responses were distinguished in research circles by the terms 'advancement' and 'priming', respectively. Priming thus had the specific connotation of 'increased germination synchrony', resulting from bringing as many individual seeds as possible to the brink of radicle emergence. But in recent years the meaning of the term priming in both academic and industry circles has changed from its original sense, and instead is now commonly used in an operational sense to describe seed pre-sowing controlled-hydration methodologies in general, without discriminating between the means of imbibition and hydration during treatment, or the specific germination kinetics that result after subsequent sowing. For example, in recent years seed soaking followed directly by sowing has been called 'on-farm priming' (**see: Steeping**).

2. Agronomic uses

Priming responses, taken together, contribute in field-planted crops to better crop stands and hence to improved yield and harvest quality, especially under sub-optimal conditions at the time of sowing, such as cold, wet or high-temperature; in plants cultivated under cover, priming response can improve the efficiency of producing useable seedling **transplants** (plugs). In both situations, the final number and uniformity of spacing of plants and/or also the speed and uniformity of seedling emergence (whether in the field or in the greenhouse prior to transplanting) are all important quality factors that can have a major impact on the efficiency of subsequent management of the crop, harvest quality (and the time to harvest and the number of harvest operations, in the case of crops such as head lettuce), and hence the ultimate profitability of the crop.

Primed seeds are currently used commercially in the production of many high-value crops where reliably uniform germination is important. It is most extensively commercialized in the field seeding or plug production of leek, tomato, pepper, onion and carrot, and the production of several potted or bedding ornamental herbaceous plants, like cyclamen, begonia, pansy (*Viola* spp.), *Polyanthus* and primrose (*Primula* spp.), and several culinary herbs; it is also used for some large volume field crops, such as sugarbeet and some turf **grass** species. Priming is also a valuable tool in circumventing induced or secondary **thermodormancy**, such as in susceptible cultivars of lettuce, celery and pansy, due to its ability to raise the upper temperature limit for germination and hence reduce the risk of **thermoinhibition**. (PH)

Heydecker, W. and Coolbear, P. (1977) Seed treatments for improved performance – survey and attempted prognosis. *Seed Science and Technology* 3, 353–425.

Priming – physiological mechanisms

Our understanding of the physiological, biochemical and cytological processes underlying priming has primarily come through investigations of: (i) mathematical modelling of germination kinetics, using **hydrotime** concepts (**see: Germination – field emergence models**); (ii) **DNA** replication and the preparation for cell division (**see: Cell cycle**); and (iii) **endosperm** weakening by hydrolases in species where that tissue encloses the embryo and mechanically restrains its expansion (**see: Germination – radicle emergence**). Explanations for the reduced longevity of primed seed may come through an understanding of the mechanisms of **desiccation tolerance**, such as changes in the degree of protection of cytoplasmic and membrane structures, of DNA damage and repair, and of free radical-associated oxidative damage during drying, air-dry storage, reimbibition and germination (**see: Desiccation damage; Desiccation tolerance**, all entries; **Germination – metabolism**).

There is considerable research interest in identifying marker signals that correlate well with the degree of advancement and/or the loss of desiccation tolerance. Studies involving priming have been used to gain an understanding of the biochemical and cytological mechanisms involved in germination (**see: Germination – metabolism; Germination – molecular aspects**). From the seed technology point of view, these marker signals might provide the seed industry with new means to assess the potential effectiveness of priming of a particular seedlot and to determine priming parameters quickly and reliably before the treatment is started; they are also useful as experimental tools in evaluating or developing new priming processes. In practical terms, such tests should give precise and reliable information across all varieties and seedlots, and also be fast and convenient to perform.

A note of caution is appropriate about taking care in drawing conclusions from the scientific literature about the general mechanisms of priming. Priming methods differ considerably in terms of their mode of application of water and the water potential used, such as osmopriming, matrix-priming and hydropriming, and their variants, and because conditions are varied by choice for a given seedlot. (**See: Priming – technology**) Furthermore, seeds display great variety of internal structure and morphology that dictate how embryos enlarge and emerge from the seed; this is likely to be reflected in different physiological and biochemical germination and priming mechanisms from one species to another.

1. Mathematical models

(a) *Germination responses to priming*. In terms of the mathematical hydrothermal time model (**see: Germination – field emergence models**), seed enhancement has one or more of the following objectives:

- to raise the optimum temperature and the upper limit for germination;
- to lower the distribution of base water potentials in the seed population (which determines when and whether a given seed will germinate under specific environmental conditions);
- to minimize the *thermal time*, **hydrothermal time** and hydrotime constants and the distribution of the base water potential and the thermal time constant which each can determine the rate and uniformity of germination.

Responses to priming differ, in practice. For instance, osmopriming of tomato at lower water potentials (below −0.5 MPa) shifts the base water potential to lower values allowing subsequent germination to occur at a water potential that initially would have blocked radicle emergence; but when priming occurs at higher water potentials (above −0.5 MPa) there is no shift in base water potential.

(b) *Priming process*. The fact that germination can be substantially advanced by priming seeds with water potentials below the base water potential for germination led to the concept of 'hydrothermal priming time' – the idea that seeds can also accrue hydrotime and/or heat units in relation to a different base water potential and temperature (lower than the base water potential and temperature for germination).

However, although this concept has proved capable of explaining a large part of the variance in the germination of tomato seed primed over a range of water potentials, detailed studies have concluded that it does not predict the responses very precisely. Refinements may improve the value of the concept, or alternative ones may be applicable, such as the Virtual Osmotic Potential (VOP) model which uses a different variable to integrate the effect of constant or varying water potentials, without having to distinguish between priming and germination potentials (**see: Germination – field emergence models**).

Despite their imperfections, the apparent general validity of the hydrotime and hydrothermal priming time models has focused research on relating temperature and, in particular, water potential thresholds to the forces that drive and/or hold back radicle emergence from seeds, and to their biological determinants.

2. Embryo cell division

Seeds such as carrot and celery have rudimentary immature embryos, which grow before the radicle emerges by a combination of cell division and cell expansion, and both processes also occur, to a lesser degree, during priming at lower water potentials. By contrast, in seeds with proportionally large embryos the general case seems to be that cells do not complete mitosis, or even expand appreciably, before the embryo structures emerge from the seed. Priming seems to have no appreciable effect on either cell size or number in leek and onion, for instance.

Studies of cell-cycle events in tomato, pepper and sugarbeet using flow cytometry and other techniques have revealed that the beneficial effects of priming are associated with the onset of replicative DNA synthetic processes in radicle meristem nuclei during priming, leading to cells stably arrested in the G2 (4C) phase after drying, preceded by the accumulation of β-tubulin. **(See: Cell cycle; Germination – metabolism)** However, the relationship of cell cycle activity to the degree of subsequent priming response is not at all straightforward, as the following examples illustrate.

- In tomato and pepper, the degree of change in DNA replication varies considerably among similarly osmoprimed seedlots, even of the same cultivar, though all display more rapid radicle emergence. In some seedlots, the frequency of 4C nuclei increases in proportion to the accumulated hydrothermal priming time, while in other seedlots there is no increase.
- When a single tomato seedlot is primed at up to 25°C and above −1.5 MPa, there is a positive linear relationship between the frequency of 4C nuclei (which can be as high as about 30%) and the improvement in median subsequent germination time. But there is no correlation between these two parameters after priming at higher temperatures or at −2.0 MPa: germination rates are improved without generating any increase in 4C signals.
- In pepper, a range of osmotic treatments induces different frequencies of radicle tip nuclei to enter the synthetic phase despite producing very similar effects on germination rate. Similarly, there is no consistent relationship between the frequencies and rates in cauliflower seeds after aerated hydration or osmopriming.

In these three species at least, therefore, entry into G2 is not essential for germination advancement, especially in 'suboptimal' priming conditions, and 4C:2C ratios are not a general measure of the efficiency of priming. Indeed, considerable increases in germination rates can be observed in the absence of any increases in 4C nuclei.

3. Endosperm cell-wall degradation

There is considerable evidence that hemicellulases (e.g. endo-β-mannanase) or glucanases (β-1,3-glucanase) are involved in – though are not necessarily the only determinants of – the weakening of the endosperm cell walls in species where the endosperm envelops the embryo and restricts radicle emergence, such as lettuce, tomato and tobacco **(see: Dormancy – coat-imposed; Germination – molecular aspects, 2. Cell wall hydrolases; Germination – radicle emergence; Hemicelluloses; Hemicellulose mobilization)**. The relationship between priming responses with respect to weakening of the endosperm region to the outside of the radicle and the degree of endo-β-mannanase activity has been studied in tomato, but neither the presence of the enzyme nor weakening of the endosperm appear to be a consequence of priming that results in faster or more uniform germination. While priming alleviates thermoinhibition of lettuce seed germination it is not known if this is because it lowers the embryo yield threshold sufficiently to compensate for the increased endosperm resistance, or lowers the resistance required to puncture the enveloping tissues.

4. Other metabolism

Very few studies have been conducted about the specific effects of priming on metabolism, as distinct from studies of germination. Osmopriming effects in tomato may be retained after drying with more intense respiratory metabolism during the first hours of subsequent imbibition in water. During germination and priming, the content and composition of intracellular soluble carbohydrates changes, e.g. the oligosaccharide:sucrose ratio is reduced in primed pea and cauliflower. These, and other changes in metabolism, e.g. to general protein synthesis or **storage protein mobilization**, have not been essentially linked to enhanced germination, however. **Proteomic** analysis is a potentially useful approach to identify unique proteins that accumulate during priming. **(See: Germination – metabolism, 3. Proteomic analysis of germination)** **(PH)**

Bradford, K.J. and Somasco, O.A. (1994) Water relations of lettuce seed thermoinhibition. I. Priming and endosperm effects on base water potential. *Seed Science Research* 4, 1–10.

Cheng, Z.Y. and Bradford, K.J. (1999) Hydrothermal time analysis of tomato seed germination responses to priming treatments. *Journal of Experimental Botany* 50, 89–99.

Gallardo, K., Job, C., Groot, S.P.C., Puype, M., Demol, H.,Vandekerckhove, J. and Job, D. (2001) Proteomic analysis of Arabidopsis seed germination and priming. *Plant Physiology* 126, 835–848.

Gurusinghe, S.H., Cheng, Z.Y. and Bradford, K.J. (1999) Cell cycle activity during seed priming is not essential for germination advancement in tomato. *Journal of Experimental Botany* 50, 101–106.

Halmer, P. (2004) Methods to improve seed performance in the field. In: Benech-Arnold, R.L. and Sánchez, R.A. (eds) *Handbook Of Seed Physiology: Applications To Agriculture.* The Haworth Press, New York, USA, pp. 125–166.

Priming - technology

Priming techniques manipulate seed water-relations, temperature and duration so as to allow some of the metabolic and cellular changes associated with **germination** to occur. They also exploit the ability of mature **orthodox seeds** to survive one or more cycles of imbibition and drying without severe deterioration of quality **(see: Desiccation damage)**. Industrial seed priming is carried out in such a way that after becoming fully imbibed most individual seeds in the seedlot remain tolerant to re-imposed desiccation. The seedlot is dried (before or after **coating** and pesticide application **treatment**, as required), packed, distributed, handled and planted in essentially the same way as unprimed seeds. (In the related **pre-germination** technique, the hydration stage is extended to the point soon after the completion of germination, and seeds may or may not be dried before they are sown.)

1. Responses

When subsequently reimbibed, primed seeds retain sufficient of the metabolic and cellular changes that have occurred prior to dehydration that the result is faster germination and emergence; these are completed over closer time-spreads (i.e. with greater **uniformity**, or synchrony) and over wider temperature ranges (**see: Germination – influences of temperature**). Typically, times to reach 50% of maximum radicle emergence (t_{50}) can be decreased by as much as a third, and sometimes more, depending on the degree of priming applied to the seedlot and the environmental conditions after sowing. Seedlots differ in the magnitude of their response to a standard priming treatment: in general, but not always, slower germinating lots exhibit greater proportional responses.

One practical drawback in many species is that primed seeds can often, though not always, suffer from faster **deterioration kinetics** during storage under normal (air-dry) conditions, compared to untreated seeds. Symptoms include the onset of a reduced rate, uniformity and final percentage of germination, and an increased proportion of **abnormal seedlings** and susceptibility to **accelerated ageing** – although the degree of disadvantage in vulnerable species varies between seedlots and storage conditions. A related problem is the increased injury and decreased storage life that occurs if priming proceeds too far. This reflects the fact that seeds' ability to survive drying and the dry state for extended periods of time is progressively lost as germination proceeds, at first in the **radicle**, then in the **hypocotyl** or **epicotyl**, and finally in the **cotyledons** in dicot seeds or the **coleoptile** in cereals. (**See: Desiccation tolerance – acquisition and loss**).

2. Techniques and terminologies

Seeds are imbibed or partially imbibed to a water content and/or for a period of time less than that required for them to complete germination (radicle/embryo emergence from the seed), and then usually dried (Fig. P.24). During priming, therefore, seeds are held essentially in Phase II of imbibition (**see: Germination – physical factors, 2. Water uptake by the seed**), and are prevented from proceeding into emergence and seedling growth (Phase III), either by restrictions in water potential or because of insufficient time.

Water is either made freely available to the seed, as in **steeping** or soaking – which some call hydropriming (i.e. at water potentials close to 0 MPa) – or is restricted to a predetermined moisture content, or sequence of moisture contents, usually employing external water potentials (ψ) equivalent to that in the seed between −0.5 MPa and −2.0 MPa. When priming is carried out without an excess of water, some seeds, e.g. sugarbeet and umbelliferous species, benefit from a prewashing (steeping) and drying procedure to remove endogenous germination inhibitors. Some positively **photoblastic** species (such as lettuce) benefit from treatment in the presence of light of appropriate wavelengths. It is possible to include other materials such as nutrients and growth regulators with the water. In general, the period of treatment typically ranges from less than 1 day to about 2 weeks, using temperatures in the range 15–25°C.

If a seedlot becomes 'overprimed', due to too long an exposure at the chosen water potential and temperature, more and more individual seeds typically display damaged primary radicle **meristems**. Although they may complete germination after subsequent sowing, in the strict sense that the radicles emerge, seedlings may not develop properly (more abnormal types being detected in the statutory germination test) and so may develop plants with a weakened root system or a damaged shoot tip, or they may die before they emerge from the soil. Even though only a small percentage of seeds in the seedlot may be affected in this way, such losses are unacceptable for high-value seed in commercial practice. It is therefore valuable to determine safe limits for priming to minimize or preferably avoid these handicaps.

Because of the variability in response from one batch of seeds to another, the optimum priming conditions used in commercial production practice – choosing the balance between the most rapid germination and the longest storage life – often need to be determined on a case-by-case basis for individual cultivars or seedlots. This can be done, for example, by conducting pilot priming runs on small samples, and testing germination responses.

Essentially three basic systems are employed to deliver and restrict the amount of water and to supply air to seed during priming: submersion in water or in solutions of osmotica, mixing with moist solid particulate materials, and controlled-imbibition – respectively known as osmopriming, matrix-priming and hydropriming. All tend to be conducted as batch processes, and commercial processing systems can handle quantities ranging from tens of grams to several tonnes at a time. These systems are engineered to ensure adequate aeration and to prevent the formation of temperature or moisture gradients within the seed mass. An array of technical

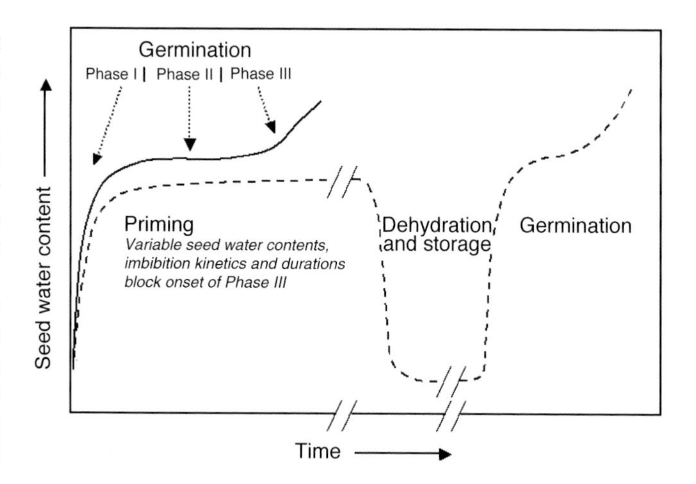

Fig. P.24. A schematic representation of the hydration status of seeds during and after priming and germination. Phase I is **imbibition**, when metabolic activity commences upon uptake of water into the seed; Phase II is marked by little increase in water content, but a continuation of metabolic activity; Phase III only occurs in seeds whose germination is complete. In primed seeds Phase II is maintained, and interrupted by dehydration and storage, and Phase III is achieved upon subsequent rehydration. Modified from Bradford (personal communication). (For details on the events occurring within the seed during these phases, **see: Germination**.)

variations on these basic priming approaches has been developed in recent years, and new descriptive names and variations have proliferated in the scientific and technical literature.

(a) *Osmopriming* or 'osmotic priming' of seeds (also known as 'osmoconditioning') describes contacting seeds with aerated solutions of low water potential, which are rinsed off after completion of priming. Mannitol or inorganic salts (such as KH_2PO_4, $KH(PO_4)_2$, K_3PO_4, KCl, KNO_3, $Ca(NO_3)_2$, and various mixtures of these) have been used extensively as osmotica but, because of their low molecular size, these are capable of being absorbed by the seeds and are associated with toxic side-effects in some cases. Instead, polyethylene glycol (PEG) is now a preferred osmoticum by many in research and the seed industry, most commonly used in 6000–8000 molecular weight fractions, whose large size prevents it from entering the living cells of the seed. Where only small quantities are involved, seeds are either placed on the surface of paper moistened with these solutions, or immersed in columns of them. Continuous aeration is usually important to ensure adequate gas exchange with submerged seeds and, for larger seed batches, stirred **bioreactors** may be used to achieve this.

Alternatively, a rotating tube with an outer jacket has been devised for osmopriming small amounts of small-seeded flower species, in which the seed is separated from the osmoticum by a selectively permeable membrane. A related engineering approach (membrane tube invigoration, still in the experimental stage) has been developed for conifer seeds, which are generally not amenable to PEG-priming without detrimental effects on seed germination, both in time and percentage of emergence. Seeds are placed in tubes (the ends of which are covered by Gore-Tex membranes), and hydrated in a high relative humidity to 30% moisture content.

(b) *Matrix-priming*. 'Solid, or dry, matrix priming' ('matripriming' and the closely-allied 'matriconditioning' technique) layers or mixes seeds in a mixture of water and solid insoluble matrix particles (such as exfoliated vermiculite, diatomaceous earth or highly water-absorbent polymers) in predetermined proportions. Seeds slowly imbibe to reach an equilibrium hydration level, determined by the reduced matrix potential of the water adsorbed on the particle surfaces. After the incubation period, the moist solid material is removed by sieving or may be partially incorporated into a coating process.

(c) *Hydropriming* is currently used both in the sense of 'imbibition in water for a short period' (**see: Steeping**) and also the sense of the 'continuous or staged addition of a limited amount of water', with or without subsequent incubation in humid air. These approaches have the practical economic advantages that the production of waste materials associated with osmopriming or matripriming is avoided, and that the relatively modest amounts of water involved are removed by drying.

Slow imbibition is the basis of the patented 'Drum Priming' and related experimental techniques, which evenly and slowly hydrate seeds up to a predetermined moisture content – typically about 25–30% on a fresh-weight basis – by misting, condensation or dribbling. Tumbling in a rotating cylindrical drum ensures that seedlots are evenly hydrated, aerated and temperature-controlled during the damp incubation stage. Another patented system uses direct hydration from a humid atmosphere (RH > 98%) to control the final stage of imbibition and maintain the moisture content in a static seed mass.

(d) *Biopriming* techniques involve including beneficial rhizosphere microorganisms in the priming processes, either as a mechanism to deliver them to the crop or to control pathogen proliferation during priming itself. (**See: Rhizosphere microorganisms**) The use of techniques such as matrix priming and hydropriming, for example, is associated with increase in indigenous microbial population on seeds. Such techniques are also suitable delivery mechanisms for introduced beneficial microorganisms, being akin to solid-state fermentation. Biopriming is not in evident commercial use, however.

(e) *Promotive and retardant substances*. There are many studies reporting the benefits of combining priming with the **plant growth regulators** or hormones (**gibberellins**, **ethylene** and/or **cytokinins** like benzyl adenine) that may strongly affect germination of seeds (**see: Germination – influences of gases; Germination – influences of growth regulators**). Also, treatment with growth retardants has been advocated to dwarf the growth habit of transplants, such as bedding plants, which tend to develop an etiolated growth habit, especially if grown in low light environments. For instance, priming tomato or marigold seeds with a triazole (50 ppm paclobutrazol) produces seedlings that are shorter, greener, more uniform, with stronger thicker stems and higher root:shoot weight ratios than non-primed controls. Indeed, there is a wider assortment of promoting agents and extracts from plant species that can be considered for inclusion in priming methodologies (**see: Germination – influences of chemicals**). **Abscisic acid**, for example, can be employed to synchronize germination prior to fluid drilling (**see: Pre-germination**).

(f) *Drying*. The technique and rate of drying after priming is important to subsequent seed performance. Slow drying at moderate temperatures is generally but not always preferable. Various manipulations can extend the storage life of primed seeds. For example, controlled moisture-loss treatments by slow drying or by an osmoticum are effective in rendering desiccation-sensitive seeds tolerant, and hence extending their longevity, e.g. by 10% or more in hydroprimed tomato seeds. Heat shock is another tactic, and greater longevity can be obtained by keeping primed seeds under a mild water and/or temperature stress for several hours (tomato) or days (*Impatiens*) before drying. These methods are very similar to those used successfully to induce desiccation tolerance in just-germinated seeds. (**See: Desiccation tolerance – acquisition and loss**)

(PH)

Bruggink, G.T. and van der Toorn, P. (1995) Induction of desiccation tolerance in germinated seeds. *Seed Science Research* 5, 1–4.

Halmer, P. (2004) Methods to improve seed performance in the field. In: Benech-Arnold, R.L. and Sánchez, R.A. (eds) *Handbook Of Seed Physiology: Applications To Agriculture*. The Haworth Press, New York, USA, pp. 125–166.

McDonald, M. (2000) Seed priming. In: Black, M. and Bewley, J.D. (eds) *Seed Technology and its Biological Basis*. Sheffield Academic Press, Sheffield, UK, pp. 287–325.

Rowse, H.R. (1996) Drum-priming – a non-osmotic method of priming seeds. *Seed Science and Technology* 24, 281–294.

Taylor, A.G., Klein, D.E. and Whitlow, T.H. (1988) SMP: solid matrix priming of seeds. *Scientia Horticulturae* 37, 1–11.

Wright B., Rowse, H. and Whipps, J.M. (2003) Microbial population dynamics on seeds during drum and steeping priming. *Plant and Soil* 255, 631–640.

Probes

In seed technology, one of several designs of devices that are used in **sampling** of seedlots, to obtain a representative sample for **quality testing** after insertion into the seed container. Perforated probes (**'triers'**) have apertures that can be opened and closed after insertion into the seed mass. (**See: Sampling, 1. Spears and Probe samplers**)

In molecular biology, a labelled sequence of **DNA** that is used to detect the presence of messenger RNAs in **northern blot analysis**.

Probits

A statistical procedure involving the mathematical transformation of data that is expressed as a ratio, such as percentages, in order to impose a 'normal distribution' upon the data, and thereby convert it into an appropriate form for further statistical analyses.

A normal equivalent deviate (ned) with the numerical value of five added to it. Seed biologists transform germination percentages to probit values when seed population traits are normally distributed such as individual seed life-spans (in the context of seed **viability**) or sensitivity of individuals to a dormancy factor. The probit transformation linearizes otherwise sigmoidal response curves enabling a form of linear regression analysis to be made, called probit analysis. (RJP, SHL)
(**See: Viability equations**)

Processing

In seed and grain technology, a collective term for a wide range of seed handling methods and operations, after harvest or collection and through to storage, which enhance the physical properties of a seedlot whether it is intended for sowing crops or for feed, food or industrial uses. Processing steps include one or more of the following, depending on the nature of the seed species: **shelling**, preconditioning, drying, cleaning, size grading, upgrading, soaking, crushing, grinding or milling. Industrialized seed production uses conditioning or processing plants to prepare seed for market, where the sequence of machines and operations is usually referred to a 'processing line'. In the case of production of seed for sowing, the term 'processing' is used almost interchangeably with 'conditioning', but it is often kept distinct from the term **'treatment'**, which broadly refers to dressing of seed with agrochemicals (**see: Treatments – pesticides**), **rhizobial inoculation**, and physiological enhancement (such as **priming**), **coating** and **pelleting**. (**See: Production for sowing**) (CHH)

Production for sowing

I. Principles

The availability of high quality seed is fundamental to the successful production of all crops, whether for agronomic, vegetable, horticultural, forage, turf, agroforestry or ornamental uses. Seed buyers must be assured of the seed they purchase and plant, and the preservation of its identity is fundamental to all successful seed production. The concept of seed **certification**, first developed in the early 1900s, established the key principles of segregation, labelling, record keeping and testing to assure varietal integrity (principles now also adopted in the **identity preservation** of added-value grain varieties).

Great variability exists throughout the world in systems of seed **variety** development, production and distribution. (**See: Industry – supply systems**) Most private-sector proprietary varieties are produced under contract with experienced seed producers who have the appropriate combination of land, equipment, facilities, resources and experience necessary to produce high quality seed. For hybrid crops, the contracting company often provides the necessary technical expertise and equipment to perform critical activities, such as **roguing** and **detasselling** in hybrid seed maize production. For low-volume high-value hybrid varieties and because of the complex systems used, it is common for seed companies to closely supervise production, or produce the seed in-house. Close technical help and equipment inputs by the contracting company are not as likely to be critical for production of non-hybrid seed crops, but may be supplied when needed, especially for seed production of certain vegetables to help avoid contamination by seedborne diseases. In public breeding programmes, certified seed growers produce seed of the varieties for sale, either to nearby wholesale outlets for conditioning or retail directly to local farmers, such as in the USA system. For the special set of challenges presented in agroforestry, **see: Trees – seed production and cultivation**.

For complementary detail on seed production methods for particular crops, **see: Alliums; Brassica – horticultural; Cotton – cultivation; Cucurbits; Grasses – forage and turf; Gymnosperms; Legumes – forage; Maize – cultivation; Ornamental flowers; Pepper and eggplant; Rice – cultivation; Soybean – cultivation; Sugarbeet; Sunflower – cultivation; Tobacco; Tomato; Tree Seeds; Umbellifers**.

1. Starting material

(a) *Seed-to-seed production*. Whenever possible, seed should be produced by planting the seed rather than rootstocks, bulbs or other propagules. This is usually the most efficient arrangement because it saves labour, is easily mechanized, and is most economical because it results in the maximum seed with the least amount of inputs and/or costs. Seed-to-seed production is the method used for the production of most agronomic crop seed, both open-pollinated species such as cereal grains, soybean, grasses and hybrid species such as maize, sorghum and sunflower, and of most vegetable and ornamental flower seed crops (Table P.14). In biennial plants, such as onion, the seed is planted in summer, and carried through the winter as small bulbs in the field. Although most production occurs on a relatively large scale, it may also be conducted in smaller plots or under cover, depending on the crop, the size of the market, and the amount of seed stock available.

(b) *Transplants*, **Roots or Bulbs**. Seed produced from vegetative propagules (Table P.14) such as whole plants, roots

Table P.14. Most common methods of seed production for selected crops.

Crop	Primary production method	Type of pollination	Pollination requirement (if cross-pollinating)	Type of varieties (hybrid/open pollinated)	Relative isolation distances
Alfalfa	Seed to seed	Cross	Insect (honey, leaf cutter or alkali bees)	Both	Moderate
Barley	Seed to seed	Self		Open-pollinated	Minimal
Beans, garden or field	Seed to seed	Self		Open-pollinated	Minimal
Beet, sugar	Seed to seed; root to seed	Cross	Wind	Hybrid	Long
Beet, table	Root to seed	Cross	Wind	Both	Long
Bluegrass, Kentucky	Seed to seed	Self		Open-pollinated	Minimal
Cabbage	Plant to seed	Cross	Insect (honeybees)	Both	Long
Canola	Seed to seed	Cross	Insect (honeybees)	Both	Long
Cantaloupe	Seed to seed	Cross	Insect (honeybees)	Both	Long
Carrot	Seed to seed; root to seed	Cross	Insect (honeybees)	Both	Long
Corn, field or sweet	Seed to seed	Cross	Wind	Hybrid	Long
Cotton	Seed to seed	Self (primary)		Open-pollinated	Minimal
Cucumber	Seed to seed	Cross	Insect (honeybees)	Both	Long
Fescue, tall	Seed to seed	Cross	Wind	Open-pollinated	Minimal
Lettuce	Seed to seed	Self		Open-pollinated	Minimal
Onion	Bulb to seed	Cross		Both	Long
Peanut	Seed to seed	Self		Open-pollinated	Minimal
Peas, English	Seed to seed	Self		Open-pollinated	Minimal
Rice	Seed to seed	Self		Open-pollinated (some hybrid)	Minimal
Rye	Seed to seed	Cross	Wind	Open-pollinated	Moderate
Ryegrass, perennial	Seed to seed	Cross	Wind	Open-pollinated	Minimal
Sorghum	Seed to seed	Cross	Wind	Hybrid	Long
Soybean	Seed to seed	Self		Open-pollinated	Minimal
Spinach	Seed to seed	Cross	Wind	Hybrid	Long
Sunflower	Seed to seed	Cross	Insect (honeybees)	Hybrid	Long
Tomato	Seed to seed	Self	Some indeterminate hybrids greenhouse produced using manual emasculation	Both	Minimal
Wheat	Seed to seed	Self		Open-pollinated	Minimal

or bulbs (known as *'root-to-seed'* or *'bulb-to-seed'* production) offers distinct advantages compared to seed-to-seed production, although being more labour-intensive the cost is usually much higher. Advantages include efficient utilization of a quantity of stock seed, the opportunity for plants to be selected for genetic uniformity of characters before replanting, or at least to discard any obvious off-types by roguing, and the more uniform growth of vegetative material after transplanting. It also permits flexibility, by deferring final decisions between which varieties to produce. The production of the plant, root or bulb propagules themselves usually requires planting of stock seed in separate plant beds, nurseries or in greenhouses, often some distance away from the primary seed production area, for disease control purposes. Outdoor plant beds are utilized for biennial crops, such as table- (red) and sugarbeet transplants (**stecklings**), carrot roots and onion bulbs. Bulbs are harvested and stored through the winter, when they can receive any cold **vernalization** requirement, and planted in the spring. Alternatively, greenhouse production of transplants provides more optimum environment for rapid and uniform growth of seedlings. For a further example, **see: Onion**.

(c) *Foundation stock seed.* Stock seed production in seed certification programmes usually involves the successive steps of **Breeder**, **Foundation** and **Registered** seed production (for public varieties in USA) or, equivalently, **Pre-basic**, **Basic**, and First-generation Certified seed (in countries that have adopted the **OECD Seed Scheme** terminology). (**See: Certification; Certified seed**) Private varieties in the USA may follow a comparable but self-regulated seed production scheme under the quality assurance programme of the seed company concerned, to maintain genetic identity of the original released material. Production involves propagating parent seed that will be used to produce commercial seed for the market.

Field sizes for stock seed production are typically smaller than those for commercial seed (grain) production crops. To ensure that superior genetic seed quality is produced and to maintain genetic identity of the original released material, close supervision and management are required and usually includes involvement of the plant breeder or organization responsible for the release of the variety or parent lines. (**See: Purity testing, 2. Varietal purity**) Often stock seed production fields are placed in remote areas, making them more difficult to monitor and manage. One reason for this remote placement is because increased isolation distances are required away from potential sources of outcrossing, such as other seed production fields and commercial production, as well as gardens, nurseries, 'volunteer' escape plants or related wild species. Relatively remote isolation sometimes results in stock seed production that is under higher insect and disease pressure, and generally may cause these hazards to be more difficult to monitor and manage. Occasionally, this results in lower physiological seed quality in the interest of ensuring superior genetic purity.

(d) *Hybrid seed parental lines.* The inbred parental lines required in hybrid seed production systems need to be maintained individually. (**See: III. Hybrids**) Systems based on **cytoplasmic male sterility** (CMS), for example, require the reproduction of three separate lines: the male-sterile female parent A-line; the counterpart 'isogenic' fertile maintainer B-line, to propagate the male-sterile female line; and the restorer male parental line used to produce the hybrid seed. In certification schemes, each of these parental lines is produced as Basic Seed from Pre-basic Seed. Where relatively small quantities of seed are required, they are often produced in isolation under bags in the breeding nursery, in greenhouses or in screened cages under fine mesh tent covers, to prevent entry of insects bearing pollen from outside and to control pollination by artificially introducing insect pollinators such as honey bees or blowflies. Alternatively the maintenance of some lines requires hand pollination.

Leeks differ; the female line has to be propagated vegetatively and therefore does not require a maintainer line.

For parental lines which are difficult to propagate by seed (such as some male-sterile or self-incompatible lines), plants may be multiplied using tissue culture techniques, using embryo, hypocotyl, stem, leaf or shoot explants or undifferentiated callus tissue. The efficiency of regeneration depends on genotype, source and age of the explant, as well as the culture media and techniques used.

2. Choice of location

The ability to consistently produce high quality seed at an economical cost and high yields are important requirements of any seed production area. Choices of a location depend on weighing together factors such as climatic requirements, direction of prevailing winds, and site histories of seed yield and quality (which together constitute the marketable yield per unit area) and any crop failures. Often a production area may provide excellent seed yields but the quality may not be acceptable, due to low germination or vigour, the presence of noxious weeds or other contaminants, the degree of genetic outcrossing from commercial production fields or related wild plants, or infestation by seedborne diseases. Also of importance are production costs, the local availability of Foundation or Stock seed, conditioning facilities, and distribution and delivery schedule to customers. For seed production of high-value crops of certain flowers and vegetables, especially those that are manually pollinated and harvested, high costs in traditional areas have driven seed production to countries such as Chile, China, Costa Rica, Kenya and Thailand, where less costly land and manual labour are available. The demand for high quality precision seed for horticultural crops has led to the development of seed production entirely undercover, for example in northern Europe using 'polythene tunnels'.

(a) *Logistics.* Conditioning facilities are normally located in or near production and distribution areas to improve efficiencies and lower expenses such as freight and other related costs. This is particularly important for large-seeded crops such as hybrid seed maize and soybeans. Small seeded crops such as alfalfa, vegetable seeds or flower seeds, however, tend to have central seed conditioning facilities, often far removed from where the seed will be distributed and planted.

If it can be successfully achieved as well, it is usually preferable to produce seed relatively near to where the commercial crop is to be grown, to minimize transportation and other costs. Local seed production is generally followed for a wide range of crops such as small grains and soybeans (also **see: Industry – supply systems**). However, for a variety of reasons, it may be better to produce seed in special areas far removed from the main crop.

On a practical level, it is common for seed producers in the same geographical area to cooperate with each other in their overall seed production programmes. There are many instances of agreements to avoid planting certain kinds of seed crops in recognized seed production areas that would represent a threat for genetic contamination in order to prevent or minimize potential outcrossing, such as table (red) beet seed production in established sugarbeet seed production areas.

(b) *Organic seed.* Regulations in this rapidly developing market sector are evolving towards the goal that the seed planted to produce organic crops is itself produced entirely under certified organic production standards covering production field, harvesting, handling and conditioning procedures. Practical requirements include a high degree of sanitation, the use of disease-free basic seed, attention to roguing throughout plant development and stringent isolation of production fields, including from the risk of contamination with pollen from genetically modified varieties. The control of seedborne diseases is of vital significance, and there is an ongoing need for developing new methods for prevention, monitoring and control and for re-evaluating disease thresholds. (**See: Organic seed; Health testing**)

(c) *Climate.* Climatic advantages are often related to **vernalization** or photoperiod requirements and drier conditions for reduced incidence of seedborne disease, more favourable harvesting weather and higher seed quality. Examples include vegetable seed production in the western USA and Mediterranean-type climates of southern Europe, Chile and South Africa. These areas typically have moderate winters and dry summers, ample supplies of irrigation water and favourable pollination conditions. For cross-pollinated crops, climatic conditions during flowering should favour outcrossing and seed set, that is, be generally warm and reasonably humid but not rainy. Climatic conditions for hybrid parental-line production should be conducive to favourable seed yields, even if the hybrid seed may be produced for another geographical area.

(d) *Disease incidence.* A key consideration, especially to reduce the potential and risk of seedborne disease infection. For example, in the USA, soybean seed of each **maturity group** is typically produced in the northernmost part of its adapted range, where seed matures in a cooler environment and hence is under lower infection potential from pod and stem blight, resulting in higher germination quality. Another example is the production of garden and dry bean seed in drier climates of the western USA, Chile, New Zealand and Africa.

(e) *'Genetically modified' varieties.* In principle, seed production of genetically modified crops is similar to that for conventional crops with the exception that additional precautions are required related to isolation distances, to address genetic purity requirements and prevent any adventitious mixtures with conventional varieties, and to accord with intellectual property rights. Thus, seed production contracts may contain additional precautions and agreements.

3. Supply and risk management

One of the most challenging aspects of seed production involves the decisions about how much should be produced, considering sales projections. Since production is usually very expensive and field isolation may be difficult to achieve, often a given parental line or variety is produced in quantities large enough to supply enough seed for more than 1 year. Typically also, an inbred line may be used in more than one hybrid. On the other hand, seed inventory must be closely managed to avoid oversupply of parent material or varieties that may become obsolete.

When a large volume of supply is critical, production is typically split between several growers or growing areas. Large seed companies may divide up seed production across several states, regions, countries or hemispheres to ensure that an adequate supply of seed is available to meet market demands and timing needs – for instance, for crops such as hybrid maize or vegetable seeds. On a smaller scale, an individual grower with a large hectarage of a given variety may utilize several seed fields and planting dates to minimize risks.

Copeland, L.O. and McDonald, M.B. (2001) *Principles of Seed Science and Technology*, 4th edn. Kluwer Academic, Norwell, MA, USA.

Desai, B.B. (2004) *Seeds Handbook: Biology, Production, Processing and Storage*, 2nd edn. Marcel Dekker Ltd., New York, USA.

George, R.A.T. (1999) *Vegetable Seed Production*, 2nd edn. CAB International, Wallingford, UK.

Kelly, A.F. (1988) *Seed Production of Agricultural Crops.* John Wiley, New York, USA.

McDonald, M.B. and Copeland, L.O. (1997) *Seed Production, Principles and Practices.* Chapman and Hall, New York, USA.

II. Agronomy

Although preparation of seed production fields and planting the seed crop may be similar to commercial production, additional considerations are often required. Since stock seed is costly and often limiting, a primary objective is to perform the necessary field preparation activities to promote adequate stands that will produce good yields of high quality seed. This includes soil testing, pre-plant fertilization, sampling the soil or plant for nutrient availability during the growing season, weed control and insect control practices to allow the crop to grow without hindrance. Thorough cleaning of all planting equipment is a necessity to avoid contamination from other varieties or crop species.

1. Sowing depth and plant population

Proper placement of the seed is critical, especially when soil moisture may be limiting or excessive. Often seed fields are planted at lower seeding rates to maximize seed yield per plant and to allow better expression of the plant phenotype for easier removal of off-type plants. For example in California, alfalfa (lucerne) seed producers achieve best yields at seeding rates as low as 1 kg/ha, compared with seeding rates as high as 15.5 kg/ha for forage production. In hybrid seed production, the parent lines may be planted at different plant population densities or may even be subsequently thinned out to achieve desired populations. For example, the male parent in sorghum seed production may be planted at higher densities than the female, to promote faster flowering and pollen shedding. Timing of planting is also important, especially with hybrid seed production where one of the parent lines may require earlier planting, followed by the other parent at a later date to ensure adequate nicking during flowering and pollination. **(See: III. Hybrids, 4. Pollination management in this article)**

2. Water and fertilizer management

Many seed crops require timely irrigation to achieve adequate seed yield, seed size and overall seed quality, especially those produced in the more arid areas such as western USA, Chile or South Africa. However, excessive irrigation can delay crop maturity and result in reduced seed quality, especially in some biennials. In some cases, extended irrigation or untimely rain may cause the seed crop to 're-bloom' causing a wide range in seed maturity as in indeterminate flowering small-seeded vegetable seed crops such as cabbage and spinach.

Nitrogen fertilizer is often added in irrigation systems during seed development to enhance seed yield and quality in sorghum or maize. However, additional nitrogen in other crops may cause re-bloom and a delay in seed maturation, especially again in species with indeterminate flowering. Some crops may require additional micronutrients, such as boron or zinc to improve seed germination and vigour.

3. Pollination management

Non-hybrid open-pollinated crops, which are allowed to interpollinate freely within the field, are divided between those that are normally self-pollinated and those that are normally cross-pollinated (Table P.14), with a few intermediate crops whose pollination behaviour lies in between (such as broad bean, cotton and pigeon pea). The nature of seed production differs greatly between the two categories. For the specific pollination issues involved in producing hybrid seed, see: III. Hybrids, 4. Pollination management in this article.

(a) *Self-pollinated crops.* Seed production of self-pollinated crops is the simpler and less problematical, because they generally represent minimum risks of outcrossing – typically below 1%. Thus seed fields of crops such as soybean, wheat, or rice can be planted as close as 2 m away from other varieties, which is the minimum distance adequate to avoid adventitious mechanical mixing. However, some self-pollinated crops, such as field and garden beans, cotton and pigeon peas, can experience significant outcrossing when there are high numbers of insect pollinators present. In the bean seed production areas of Idaho, for example, barriers of maize are sometimes planted between seed fields of beans to help minimize this potential.

(b) *Cross-pollinated crops.* Seed producers of cross-pollinated crops must be concerned about pollination management, regardless of whether the pollen is dispersed by wind or by insects, and whether the varieties are open-pollinated or hybrids. The principle in all cases is to prevent outcrossing from uncontrolled pollen sources representing other known or unknown varieties or related wild plants. Seed production crops therefore need to be isolated by distance or time from such potential sources (see next section).

Many kinds of insects are used by seed producers for crops that require this means of pollination (Table P.14). Honeybees, leaf cutting bees, and alkali bees are especially useful, as well as bumble bees and multiple lesser-known bees. Various kinds of flies are also effective, especially for certain vegetable crops. Wild populations of insect pollinators seldom exist in adequate numbers to completely satisfy seed production needs, so it is usually necessary to bring in domesticated colonies of pollinating insects during flowering time to obtain adequate pollination and seed set, such as honeybees from professional apiarists. In addition, some seed producers (such as alfalfa) may supplement wild (leaf-cutting-, alkali-, and bumble-) bee activity by 'domesticating' their own colonies. (See Fig P.32.)

Wind pollination is usually easier to manage for seed producers than insect pollination, requiring only specified spatial isolation distances from sources of foreign pollen, including consideration of the direction of prevailing winds, to minimize outcrossing. An example is the control of wild-type plants such as Johnsongrass (*Sorghum halepense*) in or around sorghum seed production fields.

4. Isolation

(a) *Open fields*. A major factor during the course of seed production is to ensure that the possibility of cross-pollination between plots or fields containing different cross-compatible plants is minimized. This is achieved in open field situations by ensuring that other crops that are likely to cross-pollinate are not flowering at the same time, or that a suitable distance isolates them. Isolation also helps minimize the transmission of pest and pathogens from alternative host crops.

Effective isolation distances to produce seed with acceptable levels of genetic purity have been established for many crops from research over many years and are well-documented by seed companies, producers and certification agencies (Table P.15). However, distances for species are sometimes determined rather more from production experience and historical records regarding acceptable genetic purity for market requirements and the class of seed. Distances for fields producing certified seed or private-brand seed that will be sold for producing commercial crops are generally not as great as those required for fields producing seed stocks (**Breeder**, **Foundation** or **Registered**) grown for commercial seed production.

Table P.16 shows isolation standards specified for different classes of maize seed produced in North America and reveals at least two additional factors that influence isolation requirements. One of these factors is field size. Statistical probabilities alone show that flowers of plants in large fields have a higher likelihood of being fertilized by pollen from plants in the same field than do equivalent plants in smaller fields. Thus, isolation requirements are greater for smaller seed fields than for larger ones. Secondly, the greatest likelihood of out-crossing within any field occurs within the outside, or border, rows facing the source of the foreign pollen. If border rows representing the same variety are established in that direction, the isolation distance may be reduced. (**See also: III. Hybrids, 2. Isolation in this article**)

The isolation distance between seed production fields also varies depending on the degree of difference in genetic make-up from potential outcross contaminants. Examples include the need to separate white from yellow maize varieties, or a grain sorghum hybrid from others with dissimilar grain colours, maturation, or plant heights or from forage sorghum or sorghum × Sudangrass hybrid production fields in the high plains of west Texas. Production of open-pollinated onion, for instance, may require an isolation distance of about

Table P.15. Minimum genetic standards in seed production (from AOSCA Certification Handbook).

	Foundation				Registered				Certified			
Crop	Land[1]	Isolation[2]	Field[3]	Seed[4]	Land[1]	Isolation[2]	Field[3]	Seed[4]	Land[1]	Isolation[2]	Field[3]	Seed[4]
Alfalfa	4	600	1000	0.1	3	300	400	0.25	1	165	100	1.0
Barley	1	0	3000	0.05	1	0	2000	0.1	1	0	1000	0.2
Hybrid	1	660	3000	0.05	1	660	2000	0.1	1	330	1000	0.2
Birdsfoot trefoil	5	600	1000	0.1	3	300	400	0.25	2	165	100	1.0
Clover (all kinds)	5	600	1000	0.1	3	300	400	0.25	2	165	100	1.0
Maize												
Inbred lines	0	660	1000	0.1								
Foundation single-cross	0	660	1000	0.1								
Hybrid									0	660		0.5
Open-pollinated									0	660	200	0.5
Sweet									0	660		0.5
Cotton	0	0	0	0	0	0	35000	0.01	0	0	7000	0.1
Cowpea	1	10	1000	0.1	1	10	500	0.2	1	10	200	0.5
Crambe	1	660	2000	0.05	1	660	1000	0.1	1	660	500	0.25
Crown vetch	5	600	1000	0.1	3	300	400	0.25	2	165	100	1.0
Beans, field and garden	1	0	2000	0.05	1	0	1000	0.01	1	0	500	0.2

[1] Number of years between establishment of variety of same crop.
[2] Distance (feet) from any contaminating sources.
[3] Minimum number of plants/heads in which a single plant/head of another variety or off-type is permitted.
[4] Maximum percentage of other varieties or off-types permitted in seed.

Table P.16. Minimum isolation standards for hybrid maize seed production specified by the Association of Official Seed Certification Agencies (AOSCA).

Minimum distance from other maize crops		Field Size	
		1–20 acres (0.4–8 ha)	20 acres (8 ha) or more
		Minimum number of border rows	
feet	m		
410	125	0	0
370	112.5	2	1
330	100	4	2
290	87.5	6	3
245	75	8	4
205	62.5	10	5
165	50	12	6
125	37.5	14	7
85	25	16	8
0		–	10

5 km from a field of hybrid onion production or of an onion variety of another colour, but can be closer to one with a similar colour (Table P.17).

(b) *Enclosed cages and tunnels.* Isolation of plants in cages, plastic-clad tunnels or greenhouses is also used to exclude foreign pollinating insects for the production, for example, of parental lines or basic seed, with pollination by hand or appropriate introduced insects, unless the crop is largely self-fertilizing. Caged isolation may also help to eliminate infection with insect-transmitted viruses, such as the production of lettuce seed in aphid-proof structures to minimize the transmission of lettuce mosaic virus disease. This isolation technique is also used in breeding programmes.

5. Pre-flowering management

It is important to monitor fields closely for weeds, diseases and insects before the seed crop begins to flower and to implement control measures as necessary, and remove off-type or volunteer plants of the same or related species that could otherwise contaminate the crop. Wind- or insect-pollinated crops require the monitoring of all areas within the required isolation distance for potential sources of outcrossing. Sources of contamination include other fields of the same species, wild related plants and home gardens. In areas of hybrid seed maize production, home gardens may need to be monitored for sweet maize; sorghum seed production areas require the elimination of Johnsongrass and shattercane in and around seed production fields.

(a) *Weeds and contaminant crops.* Since plant populations in seed fields are often lower than in fields producing commercial crops, seed production fields can become very weedy, especially those of biennial crops with long production cycles. Field inspection during seed maturation and prior to harvest is important to identify problem weeds or other crop contaminants that are difficult to remove during conditioning, and fields may need to be rogued prior to harvest to remove off-type individuals. (**See: Conditioning; Roguing**; and III. Hybrids in this article).

(b) *Diseases.* Another critical factor for production of stock seed or commercial seed is freedom from seedborne pathogens (**see: Pathogens – seedborne infection**), since the occurrence of **infestation** can render the seed crop useless and too risky for use in commercial production. Practices to control or minimize this include the use of disease-free seed stocks, crop rotation and sanitation of field equipment, and field inspection prior to harvest. In certified seed production of field beans, for example, if either bacterial bean **blight** or halo blight is found, the field most likely will be rejected and may need to be destroyed in order to protect the area for future seed production. Likewise seedlots of cabbage, cauliflower, broccoli and other brassica species contaminated with the very infectious black rot (*Xanthomonas campestris*) are a high commercial liability (**see: Health testing, Bacteria and viruses**). Although there are seed treatment procedures

Table P.17. Recommended isolation distances for onion seed production (O.P.: open-pollinated).

Production fields	Isolation distance (km)
Hybrid Onions	
(according to male parent)	
From hybrids of different colour	4.8
From O.P. of different colour	4.8
From hybrid or O.P. of same colour but different shape (e.g. globe shape vs. flat)	3.2
From O.P. of same colour and shape	3.2
From hybrid of same colour but different type (e.g. globe vs. Spanish)	3.2
From hybrid of same colour, shape, and type (e.g. yellow Spanish)	1.6
From bunching onions (*Allium fistulosum*), chives, or leek	None
Open-pollinated Onions	
From hybrid of different colour or shape	4.8
From O.P. of different colour	4.8
From O.P. of same colour but different shape (e.g. yellow globe vs. yellow high globe)	3.2
From hybrid of same colour and shape	3.2
From O.P. of same colour but different type (e.g. yellow Spanish vs. yellow globe)	2.4
From O.P. of same colour, type, and shape (e.g. yellow Spanish)	1.6
From bunching onions (*Allium fistulosum*), chives, or leek	None

Adapted from Thornton, M.K., Mohan, S.K., Wilson, D.O., Beaver, R.G. and Colt, W.M. (1993) *Onion and Leek Seed Production.* Pacific Northwest Extension Publication 0433. Washington State University, Pullman, Washington, USA.

to reduce this pathogen, the emphasis is to focus on production of 'disease free' stock seed and commercial seed. **(See: Diseases; Pathogens)**

Crop rotation is another critical factor in minimizing the incidence of plant or seedborne diseases. It is common for seed companies to have minimum rotation cycles to reduce disease pressure, typically 3 to 5 years; however, some species require longer. For example, certain parental lines of hybrid spinach may require a rotation of 15 or more years due to susceptibility to certain races of *Fusarium*.

Disease control practices during seed maturation are also critical for many crops, especially for biennials. Cabbage, for example, is planted in the late summer or early fall and seed is harvested the following summer (see I. Principles in this article), and disease pressure from *Sclerotinia* and *Alternaria* can build up to the point where yield or seed quality can be severely reduced. Such crops may even be sprayed with a fungicide after the crop is cut and while it is curing in the **windrow** to control the spread of *Alternaria*.

(c) *Insect pests*. Insects can have a devastating effect on the yield and quality of seed during development and maturation and, therefore, must be controlled for high quality seed production. The source of harmful insects can be from other crops or undesirable host plants near or surrounding the seed crop. Knowledge of insect biology and observation of all host crops in the seed production area is important to disease-free seed production.

To give four examples:

- It is important in hybrid seed production of both field- and sweetcorn to control maize earworm, which may not only damage the maturing seed but also permit infection by *Fusarium*, *Penicillum* and other fungal and bacterial pathogens that are detrimental to seed quality.
- Head worm infection in sorghum during seed development can be devastating to seed yield and quality, and continual monitoring of the crop is important during seed maturation.
- Producers of carrot seed near to fields where an alfalfa hay crop has been cut monitor production fields for ***Lygus*** insects prior to and during pollination as well as during seed development, and timely control measures are critical to maximize seed germination and yield and avoid the formation of **embryoless seed**.
- Aphids can be vectors in the transmission of virus diseases such as lettuce mosaic virus.

Copeland, L.O. and McDonald, M.B. (2001) *Principles of Seed Science and Technology*, 4th edn. Kluwer Academic, Norwell, MA, USA.

George, R.A.T. (1999) *Vegetable Seed Production*, 2nd edn. CAB International, Wallingford, UK.

Kelly, A.F. (1988) *Seed Production of Agricultural Crops*. John Wiley, New York, USA.

McDonald, M.B. and Copeland, L.O. (1997) *Seed Production, Principles and Practices*. Chapman and Hall, New York, USA.

III. Hybrids

Multiplication of **hybrid** seed is biologically limited to one generation. The hybrid-seed industry is therefore based on the production of new seed each year, by the controlled **crossing** of selected parents to produce the desired combination of characters in the progeny. Production requires strict mechanisms of pollination control, and well-synchronized flowering of the parental lines, which must be produced separately as pure lines or, in some cases, as hybrids themselves. Because of the complex systems used, it is common for breeders and seed companies to closely supervise the production of hybrid seed. (See: I. Principles, Starting material in this article)

Various means of genetic male sterility have been developed for both wind- and insect-pollinated crops to ensure that pollen that could otherwise result in inbreeding of the female seed parent is not available. The most common type used, **cytoplasmic male sterility** (CMS), is based on the interaction between cytoplasmic and nuclear genes. The sporophytic **self-incompatibility system** (SI), used for the production of some horticultural brassicas, is based on inhibition of pollen tube growth on or in the stigma. Some hybrids are produced using highly labour-intensive hand emasculation and pollination (see below).

The development of hybrid seed maize – a fundamentally new pattern of plant breeding introduced commercially in the USA in the late 1930s – was one of agriculture's most important success stories during the 20th century, exploiting the advantages of **heterosis** and the discovery of the CMS system. Early maize hybrids represented almost entirely double-cross rather than single-cross hybrids (see following subsection), because the generally poor vigour of inbred maize lines reduced their effectiveness as female parents and made single-cross seed production costs prohibitive. However, present-day plant breeders in all crops generally have a good supply of higher yielding inbred lines to support the production of all three types of crosses if desired. Even with the availability of superior yielding inbred lines, double-cross hybrids still tend to produce better seed quality than single-cross hybrids. Amongst the other advantages of double and three-way cross hybrids are a generally longer pollination period, which can help in the production of higher seed yields and hence lower seed costs.

Though the economic scale of success of hybrid maize has not been equalled elsewhere, hybrid seed varieties have been commercialized in scores of other agronomic, vegetable and flower crops and are an important segment of the modern seed industry. Major hybrid crop examples include canola (oilseed rape), pearl millet, sugarbeet, sunflower and sorghum.

Hybrid seed production in naturally self-pollinating crops such as soybeans, cotton or cereals (other than maize) is difficult and hybrids are not widely available, though an exception is rice, where Chinese scientists have developed successful, but very labour-intensive, hybridization methods, allowing hybrid rice to account for over 50% of total production in China. There is also hybrid cotton production in India and China. In high-value crops such as tomatoes and peppers, removal of anthers and pollination by hand are used to produce hybrid seeds (see section 6, following).

1. Crosses

Conventional hybrid types are constructed as single, double or three-way crosses, which use different permutations of parental lines, as follows.

(a) *Single-cross* is the first generation (F_1) hybrid progeny of a cross between two inbred lines (A × B). Single-cross hybrids tend to have the advantages of better phenotypic characteristics in terms of plant vigour and height, as well as greater uniformity in maturity and higher levels of insect and disease resistance, since only two inbred parents are involved. They also tend to respond better to high-yield environments than do double-cross hybrids. In practice most modern hybrid field- and sweetcorn in the USA is produced as single-cross hybrids. A 'Modified Single-cross' is obtained when a single-cross of two related inbred lines is in turn crossed with an unrelated inbred line ((A1 × A2) × B)

(b) *Double-cross* is the hybrid progeny of a cross between two single-cross (F_1) hybrids, such as in maize. Four inbred lines are first crossed separately in pairs (A × B, and C × D), and then the two F_1 hybrids are crossed again ((A × B) × (C × D)) . Double-cross hybrids may have several advantages over single-cross hybrids. Commercial seed is produced on the highly productive single-cross (A × B) female 'seed' parent, rather than on a possibly poor-yielding inbred as in single-cross hybrids, and they generally have a longer pollination period, thus increasing yields and reducing production costs, although the maintenance of the appropriate lines and crosses increases the isolation requirements. Even with the availability of superior yielding inbred lines, double-cross hybrids still tend to produce better seed quality. Since they are not all genetically identical, double-crosses tend to be buffered somewhat against environmental stresses, and provide the opportunity to combine variable, diverse characteristics into one variety. The double-cross method is used, for example, to produce hybrid oilseed rape/canola varieties, by exploiting four different self-incompatibility (SI) alleles.

(c) *Three-way cross* is a hybrid progeny of a cross between an inbred or pure-line cultivar as the male parent and a single-cross (F_1) hybrid as the female parent ((A × B) × C) . When used for example in maize, seed yield and quality tend to resemble double-cross hybrids, and the advantages and disadvantages are midway between those of single-cross and double-cross hybrids. Three-way cross hybrids are the most common maize hybrid types grown in much of the developing world.

(d) *Non-conventional hybrids*, by contrast, have at least one parent that is not an inbred line, and can be classified into four major categories: varietal hybrids, family hybrids, topcross hybrids, and double topcross hybrids. Because non-conventional hybrids often do not include an inbred line, they do not have the difficulties associated with producing conventional hybrids; however, they are often less uniform and lower yielding than their counterparts.

2. Isolation

The consequences of outcrossing are usually greater for hybrid seed production than for most non-hybrid seed crops. Thus, in CMS systems, it is important to prevent outcrossing so that the male-sterile female 'seed' parent (or A-line) is only pollinated by the male parent chosen as the Maintainer (B-) line or the Restorer (R-) line, as the case may be. (**See: Hybrid seed; Line**). In open field situations, this may require as little as about 50 m isolation for wind-pollinated crops with heavy pollen such as maize to as much as several kilometres for grain sorghum, which has lighter pollen. Distances established for insect-pollinated crops are related to insect activity, such as in onion (see: II. Agronomy in this article). For new hybrids, additional distance may be required due to the smaller field size and the commercial importance placed on their successful introduction.

Male pollinator rows are often planted on the ends of seed production fields, such as for sorghum, to provide some buffer from wind-blown foreign pollen, to improve genetic purity, and to allow better pollination and seed set of the seed crop. Likewise maize seed companies sometimes plant additional male pollinator rows alongside a seed field, within the required isolation distance from another maize field.

3. Planting row ratios

The desirable number of rows of the male or pollen parent (B- or R-line) in relation to that of the female or seed parent (A-line) in a CMS seed production plot varies greatly even within a species, depending on the compatibility of the female parent and the amount of viable pollen available from the male parent. The row ratio and interplanting layout also vary from region to region, depending on weather, management and parental line characteristics (such as height, vigour and pollen production capacity). Preferably, rows in seed production plots should be perpendicular to the prevailing wind direction expected at flowering time of the parents. Ultimately, the male parent rows are destroyed or harvested first and kept entirely separate from the female parent rows, which are harvested, threshed, and dried to produce the hybrid seed for conditioning and sale.

Alternatively, in some situations hybrid seed is produced by a so-called 'mixed method'. An example is the production of 'synthetic' canola rapeseed hybrid varieties in Canada, in which up to about 20% of male-paternal seeds are mixed into the maternal seeds during sowing, and both are bulk harvested together to produce the commercial hybrid for sale. Although the **heterosis** is inferior to that obtained in pure hybrids, yields are not greatly reduced in crops grown from seed produced in this way. The method relies on the parental line plants having comparable growth stages, height and vegetative vigour, and very high purity.

4. Pollination management (nicking)

Pollination management refers to practices by the seed producer to ensure that adequate pollen is available to fertilize the female flowers during the time when they are most receptive, in order to maximize seed set and hence the yield of the hybrid seed (or the multiplied CMS parental line). This is accomplished by providing maximum overlap in flowering of male and female lines, a phenomenon known as the 'nick', so that adequate pollen from the R- or B-line is available to the A-line during the desired length of the flowering period. Success depends on synchronizing the development of the seed parent and the pollen parent in the field, even though they may have different growth durations. Nicking management is particularly important in wind-pollinated crops because of the short viability of windblown pollen. In insect-pollinated crops, nicking management also depends on combining an adequate ratio of male to female flowers in addition to assuring an adequate population of insect pollinators. Typically the male parent is manipulated, by various means.

(a) *Differential seeding times*. These are used for optimizing the nick in many hybrid agronomic and vegetable seed crops, such as maize, sorghum, sunflower, brassicas, spinach and cucurbits. The most common method is split planting – that is, adjusting the seeding dates of the parents. In addition, male parents may be planted at more than one date, separated from each other by a few days. Both measures help ensure adequate pollen availability over a longer period of time during the flowering period of the female parent. The optimal relative planting times are determined by conducting testing programmes called 'nicking trials', through which seed companies catalogue the number of days (or **day-degree** units) after planting at which flowering occurs in the seed production environment for all their inbred male and female parental lines. The degree of staggered planting differs from hybrid to hybrid, season to season, and place to place, however.

(b) *Transplantation*. For some crops (such as rice, broccoli, cabbage and carrot) lines may be transplanted as seedlings to give proper plant spacing, to gain advantages of uniform growth, optimum crop stand for maximum yield, and to make field operations such as weeding, spraying, fertilizing and roguing easier. The age of seedlings at time of transplanting may advance or delay flowering.

(c) *Mechanical manipulations* are used in some cases to ensure wide dispersal of the pollen grains. For example, supplementary pollination in rice may be encouraged by artificially shaking, or dragging across, the canopy of the male parent at flowering to cause the anthers to shed all their pollen. Removal of the flag leaf (that closest to the flowering head) by cutting also can enhance uniform pollen dispersal.

(d) *Growth promotion and retardation*. Even with highly coordinated and controlled planting times, nicking can still be a problem. Various ways may be employed to slow down or stimulate the development and flowering of one of the parental lines when it is deemed necessary.

- In sorghum the male parental line may be planted at higher densities to promote faster flowering and pollen shedding.
- Some hybrid maize seed producers use 'flaming' of the male parent during early plant development, usually in the seedling stage, to delay development and the onset of flowering. Others use mechanical clipping techniques during the seedling stage to delay plant development to delay the onset of flowering.
- In hybrid *Brassica* seed production, one of the parents may be 'topped' or 'pinched' to adjust the nick (Fig. P.25).
- Plant development of one of the parental lines can either be promoted or delayed by either supplying or withholding nitrogen or other plant nutrients, or irrigation, or by using growth regulators such as gibberellic acid to manipulate plant height or floral development, such as in rice.

(See: **Conditioning**; IV. Harvesting and V. Drying in this article)

5. Other flowering manipulation techniques

Aside from manipulating planting date and controlling plant development by various techniques (flaming, cutting, etc.) and the use of cytoplasmic or genetic male sterility, cited above, seed producers have other means of pollination control to ensure that the seed produced represents the desired cross between the selected parental lines.

(a) *Emasculation* may be used to prevent inbreeding – such as in hybrid maize seed production, to remove the entire staminate flower, although hand labour is still required to remove any missed or partial tassels before they can shed any pollen

(b) *Growth regulators*. In certain hybrids of the cucurbits (such as squash and pumpkin), growth regulators are used to eliminate or minimize the production of male flowers on the female parental line. This method allows the production of hybrid seed at a reasonable cost compared to removing the male flowers with hand labour.

6. Manual pollination

Time-consuming hand techniques are only cost-effective for high value crops, and production often takes place in areas where labour costs are low. The approach is widely used in tomato and pepper hybrid seed production, using forceps to remove staminate flower parts (anthers) prior to pollination (**see: Reproductive structures**). Hand pollination is also used for producing hybrid cotton in India, commonly using a thumbnail technique to strip off the stamen column. The techniques and timing of the steps (emasculation, pollen collection and transfer, and isolation bagging of the fertilized buds, if necessary), as well as the careful labelling of plants and individual flowers, are all critical to assuring adequate seed set and desired hybridity.

Pollen preserved in cryostorage is used for manual pollination in a few situations, both for **hybridization** in breeding programmes and for hybrid seed production. In practice this approach is mostly used at present for the solanaceous crops (chiefly tomato, pepper and tobacco), where the viability of the stored pollen after unfreezing is suitably robust. Supplying pollen from a central location cuts field and greenhouse resource costs at expensive seed production sites, avoiding the need for staggered plantings of parental lines to synchronize flowering, and reducing the need for the frequent maintenance planting of male parental lines. This approach

Fig. P.25. Hybrid cabbage in peak bloom. One parent (the female) in this field was 'hand-pinched' prior to bloom to improve nicking. The male and female rows are alternated, as is evident from the difference in density of their floral parts. The male pollinator line is present in the right-hand rows, and the nicked female line in the rows to their left. Beehives on the edge of the field are present to promote cross-pollination. Courtesy of Alf Christianson Seed Company, Mt Vernon, WA, USA.

also helps protect seed companies' intellectual property, and provides greater flexibility in decision-making and operations timing. The methodology is advantageously used, for example, to hasten the commercial introduction of new varieties, sometimes by a season, because the pollen of a new male parental line can be made available appreciably sooner than the seed.

Basra, A.S. (1999) *Heterosis and Hybrid Seed Production in Agronomic Crops*. Food Products Press, New York, USA.
George, R.A.T. (1999) *Vegetable Seed Production*, 2nd edn. CAB International, Wallingford, UK.
Lovic, B.R. and Hopkins, D.L. (2003) Production steps to reduce seed contamination by pathogens of Cucurbits. *Hortechnology* 13, 50–54.
Schreiber, A and Ritchie, L. (1993) *Washington Minor Crops*. Food and Environmental Quality Lab, Richland, WA, USA.

IV. Harvesting

1. Preharvest activities

The periods of seed **development** and **maturation** are the next especially critical stages in the production of high quality seed. It is important that routine procedures be put into place to monitor seed fields during this period for any disease, insect and weed pressures that can otherwise influence the quality of the harvested seed. Other preharvest activities include the removal of male rows in hybrid seed production and **roguing** of off-type plants as well as other crop contaminants and troublesome weeds. Finally, it is important to make a proper determination of seed maturity to ensure the optimum harvest time for the highest physiological quality of seed possible.

(a) *Genetic roguing*. Good seed production practices usually include the removal of any genetic off-types or other varieties from seed fields prior to harvest. Some off-type plants only become distinguishable during the seed maturation phase of crop development by observation of such factors as pubescence colour in soybean, grain colour in sorghum or chaff colour in wheat. In sorghum, taller mutant plants become visible during seed development and can be removed prior to harvest. Roguing of certified seed fields is essential if certification standards are to be achieved at field inspection. It is a common requirement of programmes to rogue Breeder, Registered or Foundation seed fields of cereals, soybeans, dry edible beans amongst others. An important factor in determining the need for roguing a seed crop is the percentage of genetic off-type plants or other varieties present. Genetic standards are usually established by the seed certification agencies or by private companies. Many crops have several categories of genetic off-types or other varieties, of varying strictness, with lower tolerances allowed in seed production fields. But any level of highly objectionable off-types is disallowed, and removal prior to harvest is required. Table P.15 in II. Agronomy, Field preparation in this article shows an example of the genetic standards for Foundation, Registered, and Certified seed in North America.

(b) *Hybrid seed – removal of male rows*. For many hybrid seed crops, the removal of the male (pollen) parental rows is important prior to harvesting the female (seed) parent rows, to prevent physical contamination. Removal is also important where **lodging** may occur as in sugarbeet, table beet, spinach, sorghum-sudangrass, sunflower and carrot. In other crops, removal of the male rows allows easier access through the field for the final roguing activities and reduces disease and insect pressures. In agronomic crops such as sorghum or maize, the male parental plants are normally harvested and sold as commercial grain or are used as feed for livestock.

(c) *Final field inspection*. Prior to **swathing** or direct harvest of most seed crops, this field inspection will normally include observation and documentation for trueness to type, presence of other varieties or genetic off-types, other crops, noxious and common weeds and diseases present. Typically a yield estimate is determined and is used to project the quantity of potential marketable seed.

Seed crops designated for export may require a **phytosanitary** field inspection, depending on the country to which the seed will be shipped. Many are inspected for seedborne and other important diseases in the field. Suspect plants or fruits may be submitted for further analysis for key disease **pathogens**. For example, **cucurbits** may be field-inspected for Bacterial Fruit Blotch (BFB), a drastic seedborne disease that can impair seed quality. A seed production field may be rejected by the certification agency or private company unless such problems are resolved by the seed grower.

(d) *Cleaning of equipment*. The cleanliness of harvesting, transportation and handling equipment, as well as receiving, drying and storage facilities, are crucial in maintaining the integrity and genetic purity of the seed and avoiding contamination and are often required as a part of quality assurance protocols.

Whether harvesting equipment is a modified version of that designed for threshing commercial grains or is specially engineered for the particular crop, a certain amount of dismantling is often required for easier access for air blowing or washing. In some cases, the harvesting, drying or handling equipment may need to be disinfected between seed fields to prevent the spread of seedborne disease. For example, thorough washing and disinfection of harvesting equipment is necessary for controlling BFB in cucurbits.

2. Considerations for harvest

The methods of harvest depend on a myriad of factors, including the biology of the crop (e.g. the nature and position of the seed-bearing organs (**see: Inflorescences; Reproductive structures; Structure of seeds**), the growing area or environment and availability of equipment and labour. Susceptibility to mechanical damage and moisture content of the seed are also important considerations.

Seed growers with more than one field of the same variety often try to complete the harvest of all fields of one variety before moving on to another. However, this is not always practical due to differences in crop maturity or other reasons. Whenever possible, it is best to minimize down-time through proper planning and scheduling during periods of unfavourable harvesting weather.

(a) *Seedlot separation*. Differences in seed quality often are encountered within seed production fields due to many causes, including non-uniformity of seed maturation, delays in harvest due to weather conditions, the degree of weed contamination and insect or disease pressure in certain areas, or even excessive crop residue. Differences in genetic quality can occur because of inadequate pollination or outcrossing from within or outside the seed field. These can all cause problems of heterogeneity within harvested seedlots, especially from large fields.

Quality assurance programmes should endeavour to achieve uniform quality within all seedlots. Thus, it can be helpful to subdivide seed from large production fields into more than one seedlot. The seed producer usually requires that seed with any apparent differences in quality be kept separate in the field until quality testing is complete, in order to prevent the entire seedlot from becoming unmarketable. Seed crops of most high-value species are often 'split lot' at harvest until testing indicates that the seed may be safely blended at a later point in the handling or **conditioning** processes.

Conversely, when heterogeneity is not a problem, seed from several seed fields may be consolidated into one large seedlot. This is especially common for cereal grains such as wheat, barley, rice and oilseed crops such as soybeans or rapeseed with large hectarages and similarities in seed quality. A dilemma of seed production can occur, when multiple small fields of the same hybrid or variety are produced, about how to consolidate harvests of similar quality. For example, the production of hybrid melon seed in China involves many growers with small hectarages, different pollination procedures, planting and harvest dates (whereas similar hectarages in California might be produced in only one or two very large fields). This requires a high level of tracking and considerable expense in keeping seedlots separate during cleaning and testing for physiological, genetic and pathological quality before decisions are made about blending.

(b) *Seed moisture.* Seed moisture content is one of the most important considerations in determining the time of harvest. Most cereal grain or oilseed crops such as wheat, barley, oats, soybean and canola are harvested when the seed moisture drops to a critical moisture level safe for storage. This is preferred since it avoids the need for drying large volumes of seed. However, conditions and circumstances may require the seed to be harvested at higher moisture contents and dried prior to storage.

Rice, for example, is often harvested at moisture levels unsafe for storage, and thus local drying facilities must be available for this seed crop. Peanuts are harvested in the 20 to 25% moisture content range and are dried in portable trailers. 'Wet seed' crops such as cucumber, melon, pumpkin or squash also require the availability of drying equipment immediately after the seed has been extracted and washed.

Hybrid dent and flint field and vegetable sweet maize are nearly all machine-picked at higher moisture levels than commercial grain, ranging from 15 to 50%, and dried to minimize mechanical damage to the seed during subsequent conditioning. This helps avoid the risk of lower seed quality due to early-autumn frosts, crop lodging and disease infections. Large production volumes, dryer availability and turn-around time are also factors that influence the decision to harvest at higher seed moisture levels. Dent or flint maize harvest, for instance, usually begins at seed moisture contents of around 35%, depending on the genetic background of the female parent.

Some crops, e.g. peanuts, garden beans and many field beans, may be pulled out of the soil after physiological seed maturity while the crop still retains some foliage, and then allowed to dry in the **windrow** until the seed has reached safe storage levels and can be adequately harvested (see next section). These beans are very susceptible to mechanical damage during harvest if the seed moisture level is 10% or less; higher seed germination and vigour is usually obtained when the seed is harvested at higher moisture contents of 11–13%.

(c) *Seed physiological and harvest maturity.* The timing of swathing or direct harvesting a seed crop can directly influence the quality of the seed. For many seed crops, it is important to know when physiological maturity has been reached across the crop as a whole prior to starting the harvest. Visual indicators of physiological maturity are very helpful since quick scheduling decisions often must be made. Seed maize producers, for instance, often look for the development of the 'black layer', the formation of the milk line within the kernel, or seed moisture content to determine when to harvest. However, often seed companies start harvesting maize at a seed moisture content of around 35%, or even higher for certain female parental lines used for hybrid seed production. **Umbellifers** are typically harvested when seed on the primary or secondary umbels are beginning to turn brown, or stems become dry and brittle. Colour indicators of maturity are used in judging the best time to collect forest tree seed species. **Brassica oilseed** and horticultural brassica (canola) seed harvesting must be finely judged to limit **shattering** losses.

The scheduling and managing of a large number of varieties or hybrids and fields is a challenge, and sometimes taking risks may be necessary to minimize financial losses. Weather conditions, available harvest equipment, labour and drying facilities all must be considered in successfully managing a large crop plan. Crops with **indeterminate** flowering patterns (e.g. certain vegetable, herb and flower species) can present a difficult challenge in deciding when to windrow or harvest because of the wide range of maturities present within the seed crop, and shattering starts at different times. In some cases it may be possible to harvest individual seed heads by hand as they ripen. Therefore, decisions about when to harvest to achieve the best yield of high quality seed become a combination of art, science, economics and experience.

(d) *Mechanical harvesting.* Both conventional and highly specialized equipment are used to harvest seed crops. Conventional harvesting equipment designed for threshing commercial grains, rather than seed, is often modified in order to minimize mechanical damage when harvesting sensitive seed crops. Specialized equipment is used for field seed maize, which is adapted from equipment used to harvest sweetcorn (maize) for food processing. For harvesting cucurbits such as cucumber, melon, pumpkin and squash, highly specialized equipment crushes and separates the rind and most of the pulp from the seed.

Operation of conventional equipment at slower ground speeds, slower cylinder speeds and wider concave openings are examples of some of the operational adjustments that can be made to reduce mechanical damage to seed. This is especially important in harvesting seed that is mostly composed of embryo tissues, such as **legume** seeds; soybean, garden and field beans and peanut, for example, are particularly susceptible to thresher injury, especially those cultivars with thinner **seedcoats**. It is important to monitor seed moisture content as air temperature and relative humidity changes throughout the harvest day, and to adjust procedures and settings as necessary to help avoid the detrimental effects of aggressive cylinder threshing action.

Seed handling equipment is also adapted for reducing mechanical damage in vulnerable crops. Thus, belt **conveyors** are used for handling soybean, garden and dry edible bean seed produced for sowing, rather than the unloading augers used for commercial soybean oilseed production. Likewise, field beans have no appreciable seedcoat and any impact they receive is more likely to reduce seed quality. 'Bean ladders' are often utilized for receiving harvested seed of these crops, to avoid undue mechanical damage that can occur when low-moisture seed is dropped into empty storage bins.

(See: **Conditioning**; **Conveyors**)

Fig. P.27. Direct harvesting of Group V soybean seed production in southeast Missouri, USA.

3. Harvest methods (Table P.18)

(a) *Direct harvesting*. Direct harvesting (Fig. P.27) is the most common method used for small-grain seed crops such as wheat, barley, rice and sorghum and oilseed crops such as soybean, canola and sunflower. It is usually the most cost-effective method of harvest as it minimizes the number of trips across the field. Use of commercial grain combine harvesters may require additional attention to obtain the highest seed quality, as outlined in the previous section. The success of direct harvesting also depends on the production environment and crop condition.

Small-grain crops that are severely lodged usually cannot be direct-harvested successfully, for example. Peas and lentils may be direct-harvested in some production areas, but may need to be windrowed in other areas to enable the crop to cure uniformly (see next section).

Table P.18. Harvest methods of selected crops.

Crop	Harvest method	Required curing period	Typical postharvest handling and safe storage
Alfalfa	Direct or swathed	No	Scalping
Barley	Direct or swathed	No	None
Beans, garden or field	Swathed	Yes	None
Beet, table	Swathed	Yes	Scalping/drying
Beet, sugar	Swathed	Yes	Scalping
Bluegrass, Kentucky	Swathed	Yes	Scalping
Cabbage (open-pollinated)	Swathed	Yes	Scalping/drying
Cabbage (hybrid)	Hand cut	Yes	Scalping/drying
Canola	Direct or swathed	No	Scalping
Cantaloupe (hybrid)	Hand picked	No	Washing/drying
Cantaloupe (open-pollinated)	Direct	No	Washing/drying
Carrot	Swathed	Yes	None
Corn, field or sweet	Machine picked	No	Husking/sorting/drying
Cotton	Direct	No	None
Cucumber	Direct	No	Fermentation/washing/drying
Fescue, tall	Swathed	Yes	Scalping
Lettuce	Hand cut	Yes	None
Onion	Hand cut or swathed	Yes	Scalping/drying
Peanuts	Digging and inverting	Yes	Drying
Peas, English	Direct or swathed*	Yes*	None
Rye	Direct	No	None
Ryegrass, perennial	Swathed	Yes	Scalping
Sorghum	Direct	No	None
Soybean	Direct	No	None
Spinach	Swathed	Yes	Scalping/drying
Sunflower	Direct	No	Scalping
Tomato (hybrid)	Hand picked	No	Fermentation/washing/drying
Tomato (open-pollinated)	Direct	No	Fermentation/washing/drying
Wheat	Direct	No	None

High quality standards for germination and seed vigour are crucial in the seed maize industry, and germination levels lower than 92% often are considered unmarketable or substandard in the US and European markets. Seed companies have focused on the importance of production and harvest strategies to obtain the highest possible seed quality. Harvesting methods have changed from the primary removal of husk on the ear in the field using traditional 'maize pickers' to the removal of the entire ears with equipment adapted from commercial sweetcorn harvesters. This switch in practice has resulted in less mechanical damage and higher physiological quality seed for both field and sweet maize.

(b) *Windrowing or swathing*. Many seed crops, such as grasses and most vegetables especially those with indeterminate flowering patterns, are windrowed, or swathed, and allowed to 'cure' in the field until the crop is ready for threshing. Windrowing also helps prevent lodging and loss from shattering, and allows the crop to be consolidated for more efficient harvesting.

The curing process allows green plant material to dry down so the crop is more easily threshed with commercial harvesting equipment; otherwise, the wet crop material is less flowable through the combine and would require more aggressive cylinder action, resulting in excessive mechanical damage. Windrowing results in more uniformly dry uncleaned seed which allows safe storage until it can be conditioned; even minimal amounts of green crop material in harvested seed can otherwise cause storage problems and lower seed quality. Thus, windrowing can minimize or eliminate the need for artificial drying.

Windrowing depends both on the condition of the crop and growing environment. To give some illustrative examples:

- Dry, garden and lima beans are windrowed, e.g. in the western USA where a slow curing time is preferred for best germination quality.
- Grasses are usually windrowed to minimize shattering loss and to allow existing green foliage to dry down.
- Sugar and table (red) beets are windrowed when the plants are still somewhat green but after the seed has reached physiological maturity (Fig. P.28).
- Some cereal grain crops such as barley and oats may be windrowed where there is a tendency to lodge during seed maturation.
- Peanuts are dug and inverted into windrows for curing, and after the seed reaches 20 to 25% moisture, the crop is harvested with specialized equipment.
- 'Wet seed' crops such as watermelon and other cucurbits are windrowed and then harvested through a wet-seed threshing machine.

In practical terms, it is important to accumulate the cut crop in volumes large enough to fill the combine cylinder for more efficient threshing and to minimize seed damage. In some seed production areas, strips of paper or similar material are placed beneath the windrows of certain crops (e.g. onion) to retain seed that shatters during the curing process. Upon harvest, the seed is separated from the paper and crop residue, which is returned to the field.

(c) *Handcutting or picking*. Some seed crops produced in small production fields or plots are hand cut and placed in a windrow or on a drying sheet nearby, or transported to a separate location for curing and threshing. Hand cutting can also reduce or eliminate contamination from troublesome or noxious weeds.

Fig. P.28. Swathing open-pollinated seed production of table beet in the Skagit Valley of Washington, USA. Courtesy of Alf Christianson Seed Company, Mt Vernon, WA, USA.

- Hybrid cabbage, broccoli, cauliflower and other vegetable brassica seed crops are usually treated in this way due to their high economic value, compared to their counterpart open-pollinated varieties which are usually machine-swathed. These crops are typically cut when the plant structure and seedpods are still somewhat green, but after the seed has reached physiological maturity (**see: Brassicas – horticultural**).
- In some carrot seed production, the 'king' primary umbels are selectively hand cut, and remaining umbels allowed to continue development for later harvest (**see: Umbellifers**).
- In some instances, the crop is placed on a tarpaulin sheet or similar smooth drying surface to permit recovery of shattered seed and additional curing and drying until threshing – such as umbels of onion seed or umbellifers.
- Small hectarages of **stock seed** of a wide range of species are typically harvested by hand to minimize losses that can occur during machine harvesting.
- Hybrid wet seed crops such as **tomato, pepper and eggplant** are typically hand picked. The fruits are collected when they reach the proper stage of seed maturity, followed by the crushing, seed separation and washing operations.

V. Drying

The drying and storage of seeds are often essential steps in the production and maintenance of high quality seed. Excess water is removed, either in the field by natural means (sun and wind, by **solarization** or windrowing) or in a bin or dryer with the aid of a fan and heater. Seeds rapidly lose viability when dried with air at excessively high temperatures, or when stored under excessive temperatures and high moisture contents (**see: Desiccation damage; Harrington's Thumb Rules; Viability equations**), and such conditions must be avoided in the production of high-quality seed. Sensitivity to high temperatures and high moisture varies among different species and varieties or hybrids. Most of the diverse group of **recalcitrant** and intermediate seed species that are important in agroforestry, in addition to being intolerant of desiccation, undergo **maturation drying** and need to be brought to individual minimum safe moisture contents; the rate of drying may also be important for subsequent storability (**see: Trees – seeds**). Thus, the design and operation of a drying and/or storage system depends on the kind of seed and location of production.

1. Pre-drying

Many seed crops require prompt pre-drying after harvest to stabilize the seed crop and maximize seed quality. Any delay at higher than acceptable seed moisture can cause declines in physiological quality to the point of the seed being unmarketable. Drying trailers are used for peanuts and onion, for example. A seed crop may also require other activities prior to drying, bulk storage and/or seed conditioning, such as husking or scalping.

(a) *Aspiration or scalping*. An important principle of seed drying is to have a uniform seed mass. Harvested seed is often roughly cleaned by being aspirated or 'scalped' to remove crop debris, soil and weeds to permit a uniform seed mass for airflow, which if restricted will cause inadequate or uneven drying and seed quality problems. Many vegetable seed crops such as table beet, spinach and cabbage may have a high percentage of crop residue present in the harvested material received from the seed grower, and the uncleaned seed is usually scalped for this reason. (**See: Conditioning, I. Precleaning**)

(b) *Husking and sorting of maize*. Husking and sorting is necessary before seed maize can be properly dried and shelled prior to conditioning. Hybrid seed maize is harvested 'on the ear', usually with the husks attached to minimize mechanical damage of the seed kernels. (See: IV. Harvesting, this article) The harvested crop is transported to a central location where it is directed over a series of husking beds to remove the husk from the ear. Proper adjustment of the dehusking equipment is critical to remove the husks and to minimize mechanical damage to the seed. Any husks remaining on the ears will prevent adequate airflow, resulting in poor drying. The seed maize ears are also sorted by hand, usually directly after husking, before being placed into drying bins to remove any of the genetically off-type ears, diseased ears, crop residue, or other elements that may interfere with drying.

(c) *Fermentation, extraction and washing of 'wet seeds'*. Vegetable species from the Cucurbitaceae (cucumber and melons) and Solanaceae (tomato and eggplant) families are considered as wet seeds due to their method of harvest and handling requirements prior to seed drying. Wet seeds are extracted from the fruit either mechanically, by hand or a combination of both methods. Cucurbit seed from large production fields, such as in California or Colorado for instance, is harvested mechanically with highly specialized equipment, whereas production in Chile, China and Thailand is usually hand-harvested due to the small size of the production fields and the availability of low cost labour. Multiple hand harvests are common for high-value seed crops such as hybrid tomato, pepper and cucurbit species or stock seed grown in many field areas or in the greenhouse.

Hand-harvested fruit is sometimes manually cut up to remove the seed or may be placed in a stationary or portable crusher for an initial seed extraction process carried out during harvesting. In this step, fruit is crushed and the seed separated from the rind and most of the pulp. The collected seed slurry is typically transported in bins to a separate location for fermentation, washing, further seed extraction and drying.

Some species, such as cucumber, watermelon and tomato, require a fermentation period after harvest to allow the gelatinous or mucilaginous **pectin** material surrounding the seed to breakdown before it can be removed during the washing and further extraction process. The length of the fermentation period depends on the species, temperature of the seed slurry and the length of time after harvest. Squash, pumpkin, cantaloupe, pepper and eggplant, however, are sensitive to fermentation and require washing and seed extraction as soon as possible, typically within a day after the harvest.

During the washing process, the seed passes through a revolving cylinder to remove the materials from the seed surface. As part of the seed extraction activity, the seed may be 'flumed' in water to remove light and immature seed by **flotation**. After the seed has been washed and inspected, it is ready for drying.

2. Drying parameters

These key terminologies and concepts of drying are used to describe the moisture properties of air and seed, and airflow characteristics.

(a) *Air humidity*. The medium in which seed is dried and stored is a mixture of dry air and water vapour. The temperature of the moist air may refer to the dry-bulb temperature (i.e. measured with an ordinary thermometer) or to the wet-bulb temperature (i.e. measured with a thermometer covered with a wet wick). The amount of water vapour contained in the drying air can be expressed in terms of vapour pressure (the partial pressure exerted by the water vapour molecules in moist air) or the relative humidity (the ratio of the vapour pressure in the air to the saturated vapour pressure at the relevant dry-bulb temperature).

(b) *Equilibrium moisture content (EMC)*. The moisture content of the seeds after they have been exposed for a long period of time to ambient conditions and is a function of the relative humidity and temperature of the storage environment, and the species and variety of seed.

(c) *Airflow*. When air is forced through seed in drying and storage systems, its flow rate depends on the characteristics of the fan(s) used, the species and the depth of the seed layer. Airflow is a function of the resistance by the seed (the so-called static pressure) causing friction and turbulence. The characteristics of a fan are conventionally expressed in terms of the airflow rate at various static pressures, e.g. 500 m^3/min at 5 cm of water (column).

(d) *Maximum temperature*. In drying seed, it is important to know the maximum temperature at which the viability of the seed is not impaired. The maximum temperature is a function of the moisture content and seed type. For maize at 24% moisture content, the maximum temperature for safe drying is about 61°C; at 18%, it is about 67°C. It should be noted that the maximum seed temperature and not the maximum drying-air temperature is specified since, in seed dryers, the two temperatures are not necessarily the same because of the latent heat of evaporation. Fig. P.29 shows the influence of the drying temperature on the viability of seed maize at 32% moisture content in a shallow-bed (<5 cm) bin dryer.

3. Drying methods

Seed can be dried in various ways, including sun drying (solarization), bin drying, portable batch drying, wagon bed drying, continuous-flow/crossflow drying, and rotary drying. In addition, seed maize is dried in specially designed ear-maize dryers. Each of these dryer types is explained in detail in texts on grain drying, and will only be outlined here.

(a) *Sun drying*. Sun drying is the practice of spreading moist seeds in a 3–6 cm layer on the ground or on a concrete floor and exposing them to ambient conditions. The practice is still commonly used in the tropics and subtropics, especially in developing countries. The process requires intermittent stirring or raking of the seed layer to prevent non-uniform drying and overheating of the seed at the surface. During night hours (and when it rains) the seed is raked together and covered. The time required for sun drying depends on the ambient weather conditions, the thickness of the seed layer, the initial moisture content, type of seed, and the frequency of raking or stirring. For example, on Java (Indonesia) about 3 days are required during the non-rainy season to dry rice seed from 20–22% down to 14%, assuming the 4 cm layer is raked twice daily.

(b) *Bin batch-dryer*. In bin batch-dryers the seed is placed in a (usually round) bin, and a fan blows ambient or slightly heated air through it (Fig. P.30). The maximum thickness of the seed layer in the bin depends on the initial moisture content, the type of seed, the air temperature and relative humidity, and the power of the fan. To obtain a uniform airflow through the seeds, a fully perforated floor is required.

After the seed in a bin has reached the acceptable average moisture content, a moisture gradient will remain from the top to the bottom of the seed. The surface layer will have a moisture content above the average and the bottom layer of the bin will be lower than average. Thus, proper mixing of the seeds is essential before further storage or packaging. This can be addressed by installing one or more grain stirrers to mix the entire content of a bin for 3–12 h.

(c) *Wagon batch-dryer*. A grain-transport wagon can be transformed into a batch-dryer by equipping it with a plenum chamber (a space in which the pressure is slightly increased by the airflow through it) below a perforated floor, with a fan and optional heater unit attached to it by a canvas transition. The drying principles are similar to those of a bin batch-dryer. Wagon batch-dryers are most frequently used for drying fragile seeds such as large-seeded legumes (e.g. field or garden beans and peanuts).

(d) *Bin layer-dryer*. In the process of bin batch-drying the entire seed batch is dried together. By contrast, in bin layer-drying, seed is dried a part at a time; successive layers of moist seed are added periodically to the drying bin, after the previous layers have been partially dried. The final seed depth in the layer-drying bin is much greater than in a bin batch-dryer.

Bin layer-drying requires a thorough understanding of the in-bin drying process, since the depth of the successive layers of seed to be added depends on the type of seed and its moisture content, the drying-air conditions and the fan rating. (Protocols can be complex. For example, one recommendation for layer-drying seed maize from 20% down to 14% moisture content, using air at 25–30°C and 55% RH over 13 days, is that 1.3 m, 1.1 m and 0.6 m of damp seed be added successively after 3.9, 3.8 and 3.5 days at which times airflow should be adjusted to 7.1, 7.9 and 12.8 m^3/m^2 screen area, respectively.)

(e) *Column batch-dryer*. In a column batch-dryer, drying air is forced upwards from an air chamber through the wet seed held in a relatively narrow perpendicular column (0.25–0.45 m diameter), see Fig. P.31. After the seeds have been dried to near the desired average moisture content, the seeds are cooled by ambient air after turning off the heating unit. (Representative airflows for maize range from 15 to 30 m^3/min of slightly heated air (35–40°C) per m^2 of screen area. The total time required for drying and cooling from 20 to 14% moisture content is about 3.5 h, depending on the seed

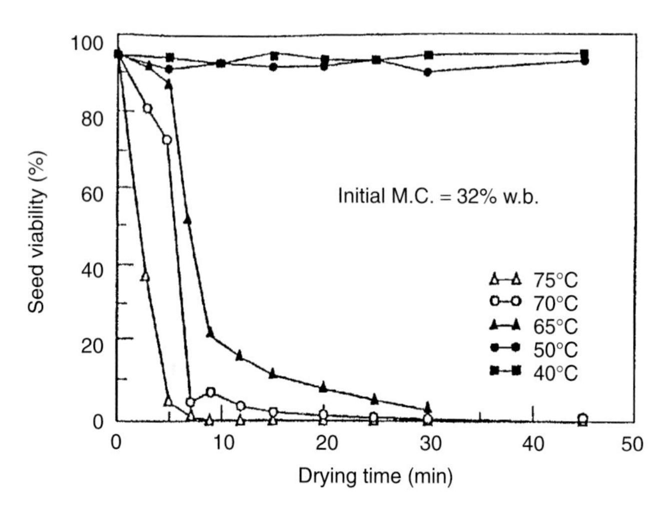

Fig. P.29. The influence of drying temperature on the viability of maize seed at 32% moisture content. From: Brooker, D.B., Bakker-Arkema, F.W. and Hall, C.W. (1992) *Drying and Storage of Seeds and Oilseeds*. Van Nostrand Reinhold, New York, USA.

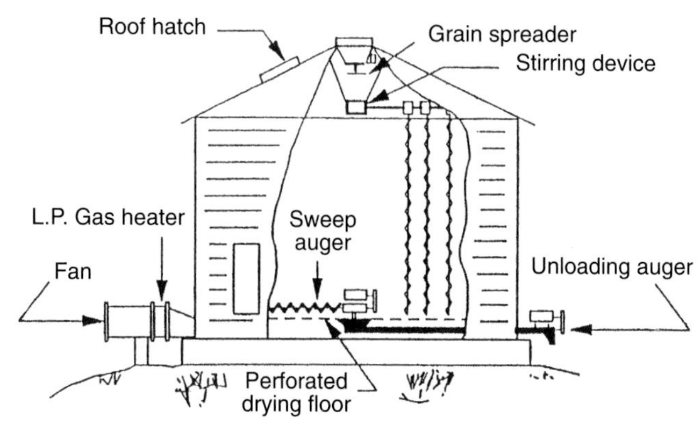

Fig. P.30. Bin batch-dryer.

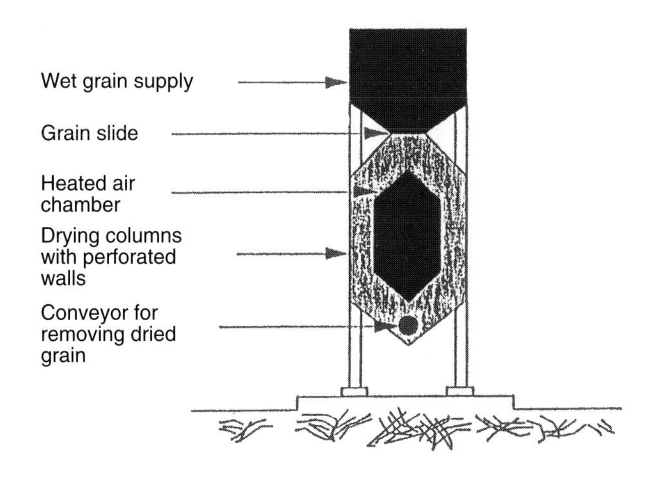

Fig. P.31. A column batch-dryer.

type, air temperature, and airflow rate.) As in other batch-type dryers, the seeds in a column batch-dryer are not uniformly dried and have to be well mixed prior to further storage or packaging.

(f) *Continuous-flow cross-flow dryer*. The basic designs of the column batch-dryer and the continuous-flow cross-flow dryer are similar. In both dryers, heated air is forced from a heated-air plenum through the moist seeds. In the column batch-dryer, the seed is stationary, whereas in the continuous-flow cross-flow dryer the seed moves continuously, first through the drying section and next through the cooling section. The diameter of the seed columns in the two dryers is similar, as are airflow rates, drying-air temperatures and moisture gradients across the seed column. The residence time in the continuous-flow cross-flow dryer of seed dried from about 20% down to 14% at 40–45°C is about 2–4 h. Fig. P.32 shows a typical continuous-flow cross-flow dryer. Such dryers are usually designed to operate at 80–100°C for the drying of seed maize. Since such temperatures can be deleterious to seed viability, continuous-flow cross-flow dryers are seldom used for drying seeds in general.

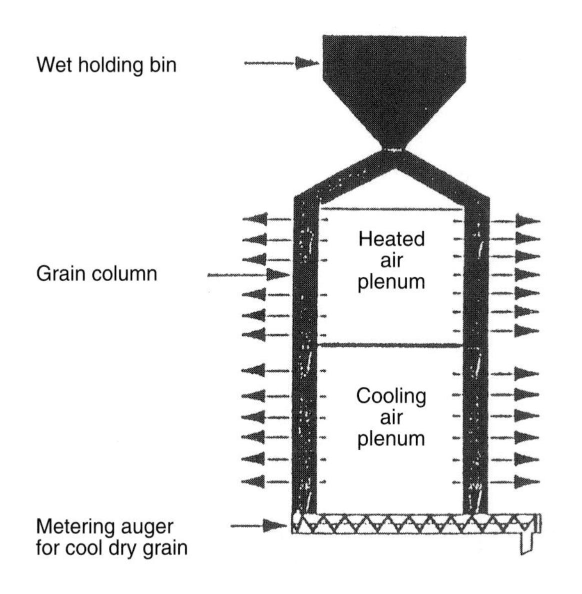

Fig. P.32. A continuous-flow cross-flow dryer. From: Loewer, O.J., Bridges, T.C. and Bucklin, R.A. (1994) *On-Farm Drying and Storage Systems*. American Society of Agricultural Engineers, St Joseph, MI, USA.

(g) *Rotary dryers*. Most rotary dryers such as those used for drying many vegetable seeds consist of a slightly inclined long drum (shell), which slowly rotates (4–8 revolutions/min). The moist seed and the drying air are continuously introduced at one end of the shell, and the dried seeds and exhaust air exit at the other end. The inside of the shell often has a set of flights, which cascade the seeds. Fig. P.33 shows a concurrent (continuous)-flow cascading rotary dryer for seeds.

Rotary seed dryers are also frequently used as batch type units for drying small quantities of seed, particularly small lots of high moisture-content vegetable and/or flower seeds. Such dryers typically consist of a perforated rotating drum in which the seeds are tumbled while being heated with drying air, which is injected perpendicularly to the drum axis. The great advantage of this dryer type is that seeds are dried uniformly, unlike in sun/bin/wagon/portable batch dryers. However, it has relatively high fixed and maintenance costs.

(h) *Ear maize drying*. Ear maize requires a totally different dryer design. Not only must the maize seed be dried, but the cob as well – typically from moisture contents of 20–40% to 11–12%.

An ear maize dryer can be considered as a modified bin batch-dryer with a perforated 45° angled floor. Fig. P.34 depicts a typical ear maize dryer used in the hybrid seed maize industry; it consists of two drying bins, which operate in tandem. The drying air (at about 35°C and 65–75% relative

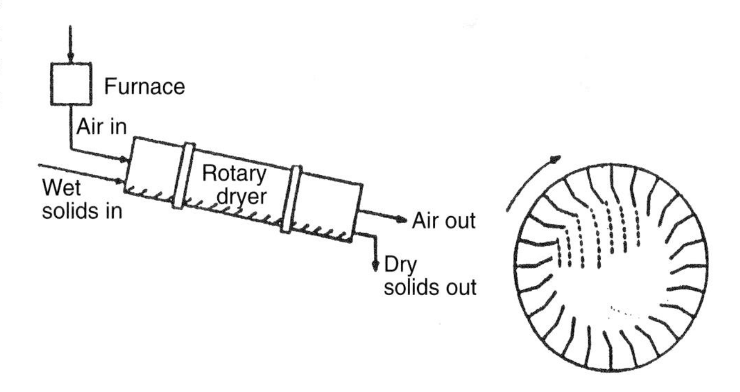

Fig. P.33. A concurrent-flow cascading rotary seed dryer. The diagram on the right is a cross-section.

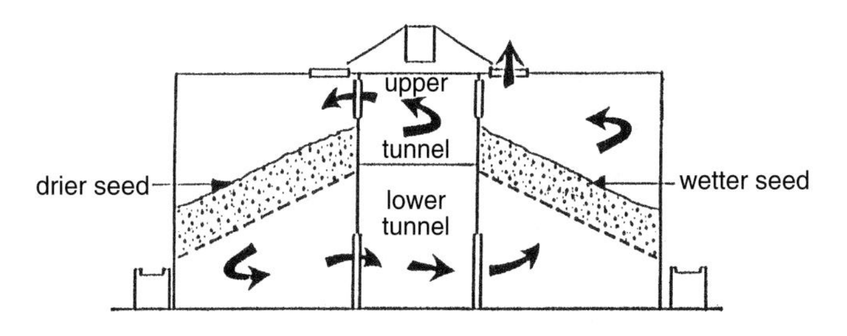

Fig. P.34. A modified bin batch-dryer for maize grain on the ear. By Cabrera, E. (1999) Personal communication. Pioneer Hi-Bred International, Inc. Johnston, IA, USA.

humidity, which has already been passed through the partially dried bed of ear maize in the adjoining bin) moves upwards through a bed depth of 60–90 cm. The movement of the air (now at 35–45°C and a relative humidity of 15–30%, depending on the ambient conditions) is reversed to downwards when about 65–70% of the moisture to be removed has been extracted. The total drying time is 50–100 h. At large ear maize installations, the number of tandem units may be between 15 and 30. Management of such systems is a challenge, i.e. the logistics of filling/emptying the bins, of the reversal time of the air, of the choice of the air temperature, etc.

(See: Cereals – storage; Storage management)

Copeland, L.O. and McDonald, M.B. (2001) *Seed Science and Technology*. Kluwer Academic, MA, USA, pp. 268–276.

Justice, O.L. and Bass, L.N. (1978) *Principles and Practices of Seed Storage*. USDA Agriculture Handbook, No. 506. United States Department of Agriculture, Washington, DC, USA.

McDonald, M.B. and Copeland, L.O. (2001) *Seed Production – Principle and Practices*. Chapman and Hall, New York, USA, pp. 104–121.

4. Storage principles

The objective of seed storage is to preserve seed quality (viability) throughout the storage period. Although favourable storage conditions cannot improve seed viability, incorrect storage can reduce its germination potential, usually due to physiological deterioration or mould development. **(See: Deterioration and longevity; Storage fungi)**

Moulds can be prevented from developing in stored seeds, thus maintaining their viability, as long as the temperature and seed moisture content are controlled within narrow limits. In general, most seed stores best at 12–13% moisture content and 10–13°C temperature. In tropical regions, seed chillers must be used to attain these conditions. If a chiller is not available, seed in the tropics should be stored at 10% (or below) moisture content.

When bulk seed is stored in bins, intermittent aeration (cooling of the seed with ambient air at a low airflow rate) may be necessary in order to offset the non-uniform heating/wetting of the seed due to natural convection currents occurring in bins due to changes in the ambient conditions. Manual control of aeration fans is possible, but automatic control is recommended. If properly done, this will prevent moisture migration and the development of wet or 'hot spots', which can otherwise lead to seed deterioration (**see: Storage management**). (LOC, TL)

Sauer, D.B. (1992) *Storage of Cereal Grains and their Products*. American Association of Cereal Chemists, St. Paul, MN, USA.

Sullivan, G.A. and Reusche, G.A. *Six Steps to High Quality Peanut Seed*. The North Carolina Agricultural Extension Service. Publication no. AG-320.

www.ilri.cgiar.org/html/trainingMat/Tropical/sld_show/slide01.htm

Programmed cell death (PCD)

A process in higher plants that has been widely observed in predictable patterns throughout development and in response to pathogenic infection. It is a means by which cells self-destruct, resembling either the common form of PCD in animals called apoptosis (in which there is a characteristic pattern of degradation of **DNA**), or a morphologically distinct form of cell death, in which DNA degradation ('laddering') may not occur. In both there is a characteristic production of cysteine **proteases** (enzymes with this amino acid present in the active site). Events in which PCD plays a role include the formation of xylem (xylogenesis), senescence, **endosperm** and **megametophytic** tissue degeneration during seed **development** or following **mobilization of reserves**, **aleurone layer** degeneration associated with synthesis and secretion of hydrolases, and seedcoat development. PCD is frequently regarded as a means whereby the materials which compose cell components can be recycled for utilization by other tissues. (JDB)

(See: Germination and seedling establishment; Gymnosperms; Mobilization of reserves – a monocot overview; Mobilization of reserves – cereals)

Lam, E., Fukuda, H., Greenberg, J., Lam, E. and Fukuda, H. (2001) *Programmed Cell Death in Higher Plants*. Kluwer Academic, Dordrecht, The Netherlands.

Prolamins

See: Osborne fractions; Storage proteins

Promoter

A **DNA** sequence, normally found upstream (towards the 5′ deoxyribose end of the nucleic acid backbone chain, from which direction **transcription** starts) of a gene, which determines where, when and to what extent a gene is expressed. The promoter contains specific DNA sequences that **transcription factors** bind to and direct transcription. **(See: Promoters – chemically inducible; Signal transduction – hormones; Storage reserves synthesis – regulation)**

Promoters – chemically inducible

These are gene **promoters** that regulate transcription in response to a specific external chemical stimulus. The term chemically inducible promoters is not strictly accurate as chemical induction of gene expression in general requires two genes: one that encodes a protein transcription factor with which the chemical interacts, and a second one whose activity is controlled by the binding of that transcription factor in the presence or absence of the chemical inducer. A more precise term would be a chemically inducible promoter system.

Such promoter systems offer the ability to activate a particular trait on demand and only when needed. This attribute makes such promoter systems particularly useful in plant biotechnology. Chemically inducible promoter systems, to be of use, must have no residual activity in the absence of the inducing chemical and must respond to a chemical that is not normally generated within the plant or encountered within its environment. In general, most chemically inducible promoter systems that have been developed for use in plant biotechnology have come from non-plant sources, including bacteria, fungi and animals. One example is an ethanol-inducible system that utilizes a fungal (*Aspergillus*) alcohol-regulated transcription

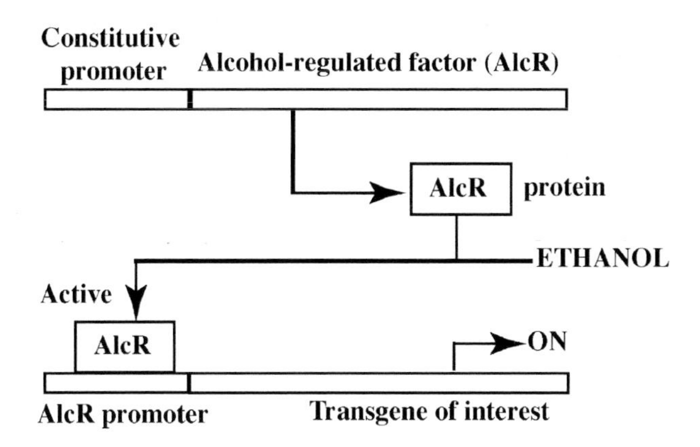

Fig. P.35. The ethanol-inducible promoter system: a model of chemically inducible promoter systems. One gene is introduced into a plant that produces a protein with which the inducing chemical interacts (ethanol with the AlcR protein). The activated AlcR then binds to the promoter region of the second introduced gene causing it to be expressed.

factor AlcR (see Fig. P.35). Here the gene for an alcohol-regulated (alcohol-activated) protein (produced under the control of a constitutive promoter) is introduced into a plant, and in the presence of ethanol this protein can bind to the specific promoter to induce the second introduced gene of interest. There are two chemically inducible promoter systems that have been developed from plant sources: the first is based on a benzothiadazole-inducible pathogen-related promoter (from genes that respond to pathogen attack), and the second is a **safener**-inducible promoter (proprietary chemicals that are used in herbicide mixes to reduce – or safen – chemical damage to crop plants). Although currently undergoing much experimental development, no such system has yet been implemented into a commercial variety of any crop. (MJO)
(See: GMO – patent protection technologies)

Padidam, M. (2003) Chemically regulated gene expression in plants. *Current Opinion in Plant Biology* 6, 169–177.

Proso or common millet

1. Distribution and importance

Common (proso, broomcorn, hog, Russian, Indian, hershey, white) millet (*Panicum miliaceum*) is an annual grass, extensively cultivated in India, China, Russia, in the Middle East including Iran, Iraq, Syria and Turkey, and also in Afghanistan and Romania. It is the only millet grown as a grain crop in the USA. The plant is well suited to dry climates such as central Russia, the Middle East, northern India, Africa, Manchuria, and the Great Plains area of North America. This is the only millet of quantity involved in world trade. **(See: Cereal; Millets)**

2. Origin

There is evidence for the cultivation of proso millet at least 4000 years ago and it is thought to have been **domesticated** in central and eastern Asia, spreading to India, Russia, the Middle East and Europe. Remains have been found in Swiss lake dwellings (late Stone, early Bronze Age). Some authorities include it among the earliest of cultivated grasses.

3. Seed characteristics

The 15–30 cm long panicle is much branched (at the top of the stem) and loose or compact according to the population density. Each **spikelet**, borne singly at the ends of the branches, is about 5 mm long with one fertile and one sterile **floret** enclosed in the **glumes**. The ovoid grain is surrounded by the hard, shiny **hull** (**lemma** and **palea**). These units are about 3 mm long and 2 mm wide, coloured from almost white, through red-orange to brown-black (Colour Plate 8F). **(See: Cereals**, section 4)

4. Composition

The content of protein, **oil** and carbohydrate (mostly **starch**) is, respectively (fresh weight basis), 10, 4 and 70%. Calcium, phosphorus and iron occur at (mg/100 g) 14, 206 and 3.5, respectively.

5. Uses

The grain is used as both human and animal food. As livestock feed, common millet is similar to oats and barley in nutritional value. A common use of the millet in North America and parts of Europe is as a component of pet bird food. (MB)

Dendy, D.A.V. (ed.) (1994) *Sorghum and Millets: Chemistry and Technology*. American Association of Cereal Chemists, St Paul, MN, USA.

Protandry

This is the development or maturity of the male **reproductive structures** (anthers) before the female (carpels) to avoid **self-pollination** and fertilization.

Protease

Hydrolytic enzymes that catalyse the hydrolysis (hydrolytic cleavage) of peptide bonds between amino acids in a protein or (poly)peptide chain (**see: Protein and amino acids**). The term protease is thus synonymous with the terms proteolytic enzyme and peptidase. The latter is the general term recommended by the Enzyme Commission, under subclass E.C. 3.4. Proteases are further subdivided into the endopeptidases (i.e. enzymes cleaving internal peptide bonds in a polypeptide, synonymous with proteinases) and the exopeptidases (enzymes cleaving off one to three amino acid residues at a time from either the N- or C-terminus of a protein or peptide).

1. Endopeptidases

Proteolytic enzymes that catalyse the cleavage (hydrolysis) of one or more internal peptide bonds in a protein or polypeptide (in contrast to an exopeptidase). Endopeptidases (also known as proteinases) are classified into four major groups based upon their mechanism of catalysis: (i) serine endopeptidases, which have an active site that contains a 'catalytic triad' consisting of a seryl, histidyl, and aspartyl residues, where the seryl hydroxyl group carries out the initial nucleophilic attack on the peptide bond; (ii) cysteine endopeptidases, however, utilize a cysteinyl thiol group for a similar attack; whereas

(iii) aspartic endopeptidases have two active site aspartyl residues; (iv) metalloendopeptidases are a diverse group of enzymes that are united by the presence of a catalytically active metal ion (most often Zn^{2+}) at their active site.

2. Exopeptidases

Enzymes which catalyse the cleavage (hydrolysis) of one to three amino acids at a time from either the N-terminus (aminopeptidases) or the C-terminus (carboxypeptidases) of proteins and peptides. Seeds and seedlings contain multiple exopeptidases. Proteases involved in the mobilization of storage proteins are usually present as free monomeric molecules, or complexed with their protein substrates. However, a number of proteases in plants occur as large multi-subunit complexes, e.g. the proteosome, which functions in protein turnover, and the Clp protease complex of the chloroplast. The increasing availability of amino acid sequence data has allowed Barrett and coworkers to propose a classification of the proteolytic enzymes based upon sequence homology groups.

The two main types of exopeptidases are:

(a) *Aminopeptidases* are a common constituent of tissues involved in storage protein mobilization, and often there are multiple forms with differing specificity for amino acid cleavage. In most seeds at least two different aminopeptidases are present. They generally are active at neutral or somewhat alkaline pHs, and exhibit a broad specificity with respect to the N-terminal amino acid they remove, although cleavage of substrates with hydrophobic N-terminal residues is generally favoured.

(b) *Carboxypeptidases* are likewise common participants in storage protein mobilization. All plant carboxypeptidases are serine carboxypeptidases (as opposed to the Zn^{2+} metallo-enzymes found in the vertebrate digestive tract) and exhibit relatively little preference for the C-terminal amino acid(s) they remove. As with the aminopeptidases, multiple forms of carboxypeptidases are generally present in seeds. (KAW)

(See: Storage proteins – mobilization)

Barrett, A.J., Rawlings, N.D. and Woessner, J.F. (eds) (1998) *Handbook of Proteolytic Enzymes*. Academic Press, New York, USA.

Protease inhibitors

A large number of seed proteins have been identified that inhibit the catalytic activity of one or more animal **proteases**. Based on their amino acid sequences these inhibitors can be classified into a number of structurally and evolutionarily related families of proteins (Table P.19). Multiple serine protease inhibitors in the same homology family may be present in a seed and in some species this is due to multiple allelic forms of the inhibitor gene, or multiple related gene loci. Multiplicity of inhibitor forms may be due also to differing post-translational processing of the initial inhibitor gene product during seed maturation, or during germination and seedling growth.

Bowman-Birk inhibitors are small proteins (8–9 kDa) with seven intra-chain disulphide bridges, and they specifically inhibit serine proteases (e.g. trypsin, chymotrypsin and elastase). Most of them are double-headed due to the occurrence of two reactive sites that bind to the same or different serine proteases. Bowman-Birk inhibitors are common proteins in seeds of many legumes (e.g. soybean, pea, chickpea, cowpea, beans, other *Phaseolus* spp., *Dolichos biflorus*, *Macrotyloma axillare*). Depending on the species and variety (or accession) the inhibitors account for up to 5% of the total seed protein and accordingly are major antinutrients (e.g. in soybean).

Table P.19. Families of protease inhibitors present in seeds, classification based upon amino acid sequence homology.

Family	Present in:
Kunitz Trypsin inhibitor family	Gramineae (Poaceae) Leguminosae (Fabaceae)
Bowman-Birk inhibitor family	Cucurbitaceae Gramineae (Poaceae) Leguminosae (Fabaceae)
Squash Trypsin inhibitor family	Cucurbitaceae
Mustard Trypsin inhibitor family	Cruciferae (Brassicaceae)
Potato Proteinase inhibitor I family	Amaranthaceae Cucurbitaceae Gramineae (Poaceae) Leguminosae (Fabaceae) Polygonaceae
CM protein/napin protease inhibitor family	Cruciferae (Brassicaceae) Gramineae (Poaceae)
Protein Z/serpin family	Gramineae (Poaceae)
Maize bifunctional inhibitor/thaumatin family	Gramineae (Poaceae)
Phytocystatin family	Cruciferae (Brassicaceae) Compositae (Asteraceae) Fagaceae Gramineae (Poaceae) Leguminosae (Fabaceae) Papaveraceae Rosaceae Umbelliferae (Apiaceae)

Table modified from Wilson, K.A. (1997) The protease inhibitors of seeds. In: Larkins, B.A. and Vasil, I.K. (eds) *Cellular and Molecular Biology of Plant Seed Development*. Kluwer Academic, Dordrecht, The Netherlands, pp. 331–374.

Inhibitors of approximately 14 kDa that arose by a duplication of the typical 7 kDa Bowman-Birk inhibitor have been identified in cereals (e.g. rice **aleurone layer**, rice and wheat germ and *Coix lachryma-jobi* (**Job's tears**) seeds).

Kunitz protease inhibitors consist of polypeptides of approximately 21 kDa with two intra-chain disulphide bonds. They are widely distributed in legumes (e.g. winged bean (*Psophocarpus tetragonolobus*), soybean, *Erythrina variegata*, *Acacia confusa*, *Canavalia lineata*) and are often very abundant proteins representing up to 10% of the total seed proteins (usually about 3% in some soybean cultivars). Since they inhibit serine proteases (trypsin, chymotrypsin and subtilisin) Kunitz inhibitors are considered to be the major seed antinutrients of many legumes (e.g. soybean). Close relatives of the legume Kunitz-type inhibitors are also present in the embryos and aleurone layers of cereals such as rice, wheat and barley. These inhibitors are bifunctional since they act as α-amylase/subtilisin I inhibitors.

A subgroup of the family of the cereal trypsin/α-amylase inhibitors (**see: Amylase inhibitors**) possesses only trypsin inhibition activity. This group is typified by the 120-amino acid residue barley inhibitor BTI-Cme. Similar inhibitors occur also in rye, maize and finger millet. Only one of these proteins, namely the 122-amino acid residue homologue from finger millet (RBI; RATI) is also a trypsin/α-amylase inhibitor.

Some seed protease inhibitors belong to the potato I protease inhibitor family. This typical chymotrypsin/subtilisin inhibitor was originally discovered in potato tubers. However, orthologues have also been identified in seeds of barley and *Momordica charantia*. In barley two such inhibitors CI-1 (83 amino acid residues) and CI-2 (84 amino acid residues) occur. Both are mixtures of two isoforms. In contrast to the previous inhibitors, the structure of the barley inhibitors CI-1 and CI-2 is not stabilized by intra-chain disulphide bond(s).

Phytocystatins, which are inhibitors of cysteine endoproteases (e.g. papain), are a small subgroup of a large family of cysteine protease inhibitors. They have been isolated from seeds of cereals (e.g. rice and maize) and legumes (e.g. cowpea and *Wisteria floribunda*), and are perhaps the most widespread from a taxonomic standpoint. The best studied example, oryzacystatin from rice, consists of the isoforms oryzacystatin I (102 amino acid residues) and oryzacystatin II (107 amino acid residues). Seed cystatins are not considered antinutrients, and they most probably play a role in plant defence against invertebrate pests (e.g. cyst nematodes).

The squash trypsin inhibitors are small (27–33 amino acids) proteins present in seeds of squash and numerous other cucurbit species. They are among the most powerful inhibitors of serine proteases including trypsin, chymotrypsin, subtilisin, elastase, thrombin and kallikrein. Squash-type inhibitors are of little or no concern as antinutrients because they do not occur in seeds used for food or feed production.

In addition to these common inhibitors, there are minor types of protease inhibitors in seeds. Barley seeds contain a protein (called protein Z) that belongs to the serpin superfamily of serine protease inhibitors. It consists of a 399-amino acid residue polypeptide chain with no intra-chain disulphide bond. It inhibits several serine proteases (including trypsin, chymotrypsin, thrombin and kallikrein) but is hydrolysed by subtilisin as a substrate. An orthologue of protein Z occurs in wheat. Seeds of white mustard (*Sinapis alba*) contain two different types of inhibitors, MTI-1 and MTI-2, respectively. MTI-1 is an 18 kDa serine protease inhibitor that inhibits trypsin but it has a low activity towards chymotrypsin. MTI-2 is a small (63 amino acid residue) protein stabilized by four intra-chain disulphide bridges that inhibits both trypsin and chymotrypsin. None of these inhibitors is considered anti-nutritional.

Only one metalloendopeptidase inhibitor, from buckwheat seeds (*Fagopyrum esculentum*), has been described. This inhibitor (BWI-2b) and a processing product lacking the C-terminal tripeptide (BWI-2a) share a significant sequence similarity with the N-terminal region of the vicilin-type storage proteins, but it has not been assigned to an inhibitor homology family. No aspartic protease inhibitors from seeds have been described.

A number of functions has been suggested for the protease inhibitors in seeds: (i) as storage proteins (especially for sulphur-containing amino acids); (ii) as defensive molecules against the digestive proteases of microbial and invertebrate pests; and (iii) as regulators of endogenous seed proteases. In some cases an inhibitor may fulfil two or more of these functions at the same time or at different stages of seed development, germination and post-germination. For example, the Bowman-Birk type inhibitors, with their very high content of half-cystine, probably function as storage proteins, and in protecting the seed from the extracellular proteases of microbial pathogens and the gut proteases of insects. (EVD, WJP, KAW)
(See: Storage protein – mobilization)

Shewry, P.R. and Casey, R. (eds) (1999) *Seed Proteins*. Kluwer Academic, Dordrecht, The Netherlands (several chapters on seed inhibitor proteins).

Proteasome

A large multiprotein complex responsible for the degradation of proteins that have been targeted for this fate by the small protein, ubiquitin. Recruitment into this complex is by an F-box protein. (**See: Mobilization of reserves – cereals; Signal transduction – hormones**)

Protection

The embracing term for technologies and practices concerned with protecting cultivated plants from the action or effects of pests, diseases and weeds. For seed-applied crop protection technology, **see: Diseases; Pathogens; Rhizosphere microorganisms; Treatments – pesticides**. For grain protection technology, **see: Storage fungi; Storage management; Storage pests**.

Protein and amino acids

Proteins (polypeptides) are polymers of amino acids. They are linked by covalent peptide bonds between the carboxyl- and amino- groups of adjacent amino acids. Proteins may be composed of a single polypeptide (also called peptide) chain, or of a number of associated chains of similar or different sizes and composition, which are linked by non-covalent hydrogen bonds between amino acids of adjacent chains. Chains may also be covalently linked by disulphide (-S-S-) bonds between cysteine residues (interchain links) or such bonds can form within a single chain (intrachain links). Polypeptide chains composing a protein may also be termed subunits.

Individual proteins can have up to four levels of structure. The primary structure is the sequence of amino acids from the end where the amino group is exposed (amino- or N-terminus) to that where the carboxyl is exposed (carboxyl- or C-terminus). The three-dimensional structure of protein is determined by the other levels of structure. Secondary structure occurs over certain regions of the protein due to hydrogen–oxygen bonds between amino acids in the same polypeptide chain, resulting in helical-shaped (α-helices) or sheet-shaped (β-sheets or β-strands) configurations. The tertiary structure describes the completed folded polypeptide chain. Quaternary structure refers to proteins in which two or more polypeptides are arranged into a multi-subunit or oligomeric protein.

Protein structure can be disrupted experimentally using chemicals to break the hydrogen and oxygen bonds between amino acids in the same and adjacent chains (e.g. by high-salt buffers), but disulphide bonds require reducing agents to break them (e.g. mercaptoethanol or dithiothreitol).

Proteins may be positively or negatively charged, depending on the make-up of their component amino acids, and have a distinct isoelectric point (pI) determined by their overall charge. Their size is determined largely by their amino acid composition, and can be expressed with respect to the weight of a hydrogen atom, as molecular weight (e.g. MW or Mr 50,000), or as molecular mass, 50 kDa (kiloDaltons). Some proteins are encoded by more than one gene (multigene families, which are not identical) and the resultant isozymes exhibit small differences in amino acid composition, and hence differences in MW and pI. Others may be synthesized from the same gene, but undergo modifications during or following synthesis (e.g. methylation, glycosylation), resulting in proteins with small changes in size and charge also; these are isoforms of the same protein.

Depending upon their amino acid composition, proteins, or regions of proteins, may be hydrophobic (exclude water), hydrophilic (preferentially soluble in water) or amphiphilic, amphipathic (have both properties).

Approximately 20 different amino acids make up the composition of proteins, and are so called because they are amino derivatives of carboxylic acids. The amino group and carboxyl group are bonded to the same α-carbon atom, to which a side chain (R group) is attached. The nature of the side chain determines the structure and properties of the amino acid. Amino acids with aliphatic R groups include glycine (Gly, G) where the R group is simply a hydrogen atom, alanine (Ala, A), valine (Val, V), leucine (Leu, L) and isoleucine (Ile, I). Amino acids with aromatic R group are phenylalanine (Phe, F), tyrosine (Tyr, Y) and tryptophan, (Trp, W); Phe is strongly hydrophobic, and the others less so. Sulphur-containing R groups occur in methionine (Met, M) and cysteine (Cys, C), and disulphide bonds between residues of the latter stabilize the three-dimensional structure of some proteins (e.g. storage and extracellular proteins). R groups that are alcohols are present in serine (Ser, S) and threonine (Thr, T). Basic R groups are found in histidine (His, H), lysine (Lys, K) and arginine (Arg, R), which are hydrophilic and positively charged. Aspartate (Asp, D) and glutamate (Glu, E) are dicarboxylic amino acids with acidic R groups, and are negatively charged. Their amides, asparagine (Asn, N) and glutamine (Gln, Q) are uncharged, but are highly polar and interact with water molecules. Of these, the most hydrophobic amino acids are I, F, V, L and M, and the most hydrophilic are H, E, N, Q, D, K and R. (JDB)

(See: Proteins – non-storage; Proteomics; Storage proteins)

Protein body

Also called a protein storage vacuole. An organelle present in the storage tissues of seeds that contains **storage proteins**, and frequently **phytin**. **(See: Storage proteins – synthesis)**

Protein kinase

An enzyme that will transfer the terminal phosphate from **ATP** or a similar nucleoside triphosphate to a protein, frequently modulating the activity of the phosphorylated protein. There are three main classes of protein **kinase**: those that will phosphorylate the amino acids serine or threonine, those that will phosphorylate tyrosine, and also **histidine kinases**. Different protein kinases recognize consensus sequences within their target proteins. **(See: Germination – metabolism; Signal transduction – light and temperature)**

Protein superfamilies

Comparisons of amino acid sequences show that many of the major seed storage proteins fall into two large groups or superfamilies.

The first is the cupins, which includes the 7S and 11S storage globulins, as well as a range of other non-storage plant proteins (notably oxalate oxidase or germin and auxin-binding protein) and microbial proteins.

The second major group is the prolamin superfamily. Proteins in this family have a characteristic pattern of cysteine residues, which was initially identified in the non-repetitive domains of the S-rich and HMW (high molecular weight) prolamins of the Triticeae (barley, wheat and rye). Wider comparisons showed a similar pattern of cystine residues in the β-, γ- and δ-zeins of maize, and in the prolamins of oats (avenins). It is also present in a number of other small sulphur-rich proteins, most but not all of which are characteristic of seeds. They include the 2S albumin storage proteins from dicot seeds, non-specific lipid transfer proteins (LTPs) from various tissues (including seeds) and several other groups of proteins present in cereal **endosperms**: puroindolines, grain softness protein (GSP) and **trypsin inhibitors** and **α-amylase inhibitors**. The three-dimensional structures determined for a number of these proteins are also similar, all comprising α-helices arranged in a right-handed superhelix **(see: Protein and amino acids)**. It is considered that the prolamin superfamily evolved from a single small globular protein with the addition of repetitive sequences being the major event that separated the prolamins from the other family members. A number of the important groups contain major allergens, notably the 2S albumins, LTPs and cereal inhibitors. (PRS)

(See: Osborne fractions; Storage proteins)

Dunwell, J.M., Culham, A., Carter, C.E., Sosa-Aguirre, C.R. and Goodenough, P.W. (2001) Evolution of functional diversity in the cupin superfamily. *Trends in Biochemical Sciences* 26, 740–746.

Shewry, P.R. and Tatham, A.S. (2000) The characteristics, structures and evolutionary relationships of prolamins. In: Shewry, P.R. and Casey, R. (eds) *Seed Proteins*. Kluwer Academic, Dordrecht, The Netherlands, pp. 11–33.

Proteins – non-storage

Although **storage proteins** may account for half or more of the total amount of protein present in seeds, they only account for a small proportion of the different types of proteins that are present. For example, 2-D electrophoretic separations of proteins from mature wheat grains show the presence of about 2000 proteins, while molecular analyses indicate that about two to four times as many different mRNAs (corresponding to expressed genes) are transcribed during grain development.

These include proteins which contribute to cellular structures (e.g. cell membranes and walls, **see: Cells and cell components**) or metabolism (e.g. enzymes, transporters), and many are present only transiently during development, or in low amounts, and consequently have little or no impact on the composition and properties of the mature seed. One enzyme which is present in the mature grain of barley and has a major impact on starch breakdown during **malting** is β-amylase. This enzyme also contains approx 5% lysine and hence contributes to the elevated lysine content of the high lysine barley line 'Hiproly'. Seeds also often contain substantial amounts of proteins whose role is to provide protection against pests and pathogens This reflects the fact that seed storage reserves are an attractive source of nutrients for a range of animals, including mammals, birds and insects, and for fungal **pathogens**. A number of such protective proteins may be present within the seeds of a single species and in most cases they appear to contribute to a broad spectrum but relatively low level of resistance rather than targeted resistance to a specific organism. In many cases they appear to be stored in protein bodies together with two major storage proteins and hence they may play a secondary role as storage proteins. In some cases the amounts present are sufficiently great to also affect the nutritional quality and processing properties of the grain. (PRS)

(See: Amylase inhibitors; Lectins; Phytohaemagglutinin; Poisonous seeds; Protease inhibitors; Soybean lectin; Thionins)

Proteomics

The systematic analysis and documentation of all protein species of an organism or a specific type of tissue is termed proteomics, analogous to **genomics**. Proteomics addresses analytical questions about the abundance and distribution of proteins in the organism, the expression profiles of different tissues and the identification and localization of individual proteins of interest. These questions are closely connected with more functional ones, which aim to elucidate interactions between different proteins, or between proteins and other molecules, and may reveal the functional role of proteins.

Accurate measurement of peptide masses and tandem mass spectrometry (MS-MS) experiments that produce peptide sequence data allow correlation with genomic data using software that translates genes and calculates peptide mass and/or fragment mass data. Measurement of intact protein masses is insufficient to allow proper identification of all proteins, and thus enzymatic (trypsin) or chemical (CNBr) cleavage is used to break, in a sequence-dependent fashion, whole gene products into manageable pieces, some of which completely match a portion of a translated gene. Many proteomics studies rely on separation of proteins by two-dimensional gel electrophoresis (2DE, separation by charge, and then by mass, Fig. P.36), followed by identification of individual protein spots after excision from the gel, cleavage reactions, extraction of proteins, and mass spectrometry (MS) with database searches.

A 'shot-gun' approach can also be used whereby whole-cell protein extracts are immediately cleaved and the peptide mixtures separated before MS to generate peptide sequence data. Multidimensional chromatography is used to enhance fractionation of the complex peptide mixtures from whole-cell digests, giving rise to the acronym MudPIT (Multidimensional Protein Identification Technology). Because 2D-SDS-PAGE and MudPIT use complementary and independent separation methods for the resolution of proteomic components, the integration of datasets obtained using both technologies should provide improved proteomic coverage.

As an example of the increasing power of proteomic technology, in 2002, Koller *et al.* performed an analysis of protein extracts from rice leaf, root and seed separated by 2D-SDS-PAGE and analysed by tandem mass spectrometry. This resulted in the detection and identification of 556 unique proteins (1509 peptides): 348 different proteins from leaf, 199 different proteins from root, and 152 different proteins from seed. A repeat of the analysis of rice proteins using MudPIT yielded 2363 unique proteins (5189 peptides), whereby 867 different leaf proteins, 1292 different root proteins and 822 different seed proteins were detected. An integration of both datasets identified 2528 unique proteins (6296 peptides): 1022 different proteins from leaf, 1350 different proteins from root and 877 different proteins from seed. A total of 165 proteins were detected only via 2D-SDS-PAGE (29.7% of all proteins identified via 2D-SDS-PAGE), whereas 1972 proteins were uniquely detected via MudPIT.

The seed **storage proteins** detected by the Koller *et al.* proteomic survey included seven different globulins, ten different prolamins and 13 different glutelins. Whereas the globulins and prolamins were present only in the seed, five of the glutelins were also detected in the leaf. Interestingly, the seed sample contained four known allergenic proteins, which belong to the **α-amylase inhibitor/trypsin inhibitor** gene family. The ability to detect known allergens illustrates the utility of proteomic approaches to proteotype food sources for the presence of allergenic proteins. It will be crucial to study how changes in metabolic profiles (**metabolomics**) are related to changes in individual proteins (including isoforms), so the mechanisms by which gene products control metabolic patterns in seeds and vegetative tissues can be determined. (AJR)

(See: Germination – metabolism, 3. Proteomic analysis of germination)

Cahill, D.J., Nordhoff, E., O'Brien, J., Klose, J., Eickhoff, H. and Lebrach, H. (2000) Bridging genomics and proteomics. In: Pennington, S. and Dunn, M. (eds) *Proteomics*. BIOS Scientific Publishers, Abingdon, UK, pp. 1–17.

Koller, A., Washburn, M.P., Lange, B.M., Andon, N.L., Deciu, C., Haynes, P.A., Hays, L., Schieltz, D., Ulaszek, R., Wei, J., Wolters, D. and Yates, III, J.R. (2002) Proteomic survey of metabolic pathways in rice. *Proceedings of the National Academy of Sciences USA* 99, 11969–11974.

Smith, R.D., Anderson, G.A., Lipton, M.S., Pasa-Tolic, L., Shen, Y., Conrads, T.P., Veenstra, T.D. and Udseth, H.R. (2002) An accurate mass tag strategy for quantitative and high throughput proteome measurements. *Proteomics* 2, 513–523.

www.expasy.org/sprot

www.ncbi.nlm.nih.gov

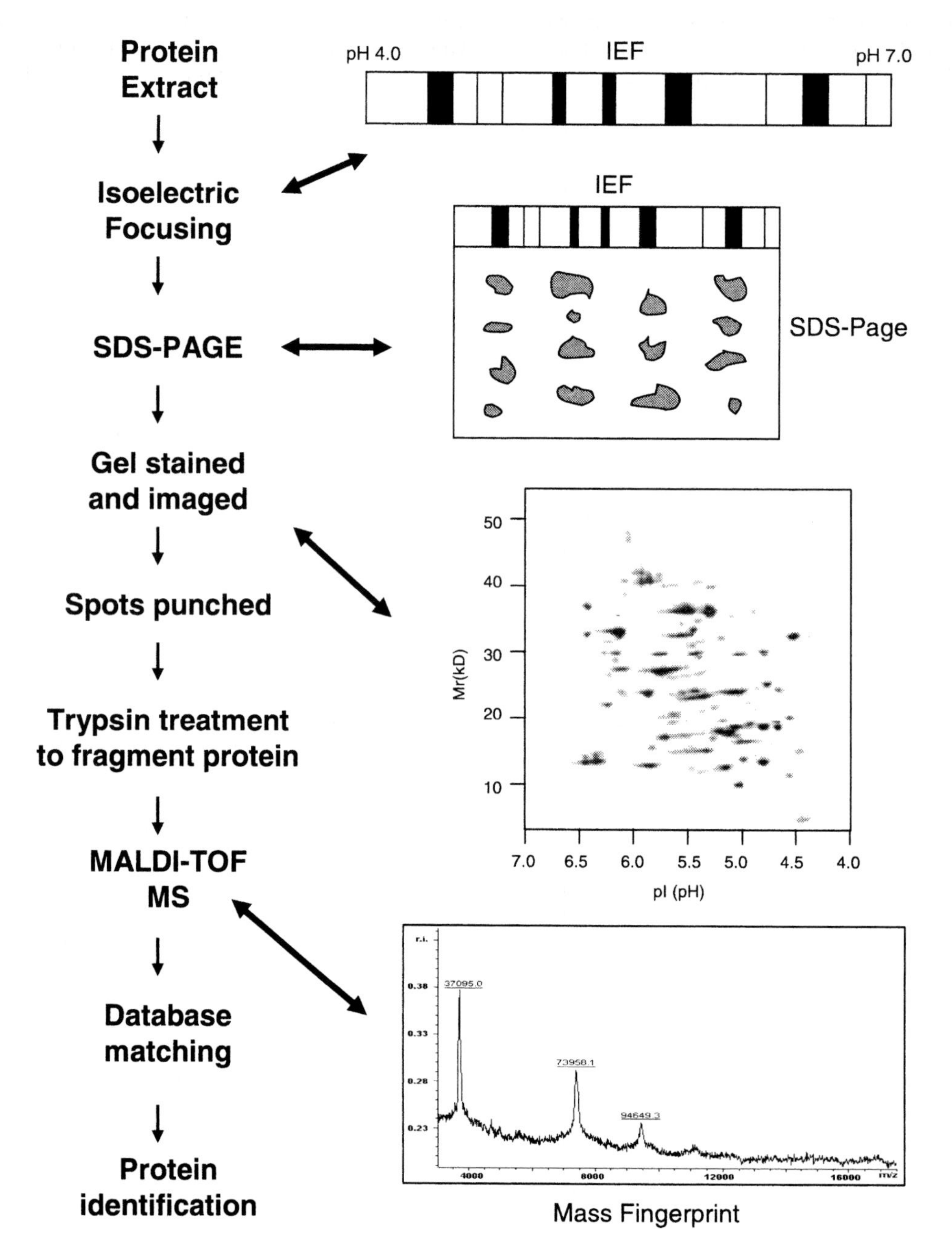

Fig. P.36. Separation of proteins from seeds and other tissues for analysis and identification. Proteins are first extracted and separated by two-dimensional gel electrophoresis, based on charge (IEF, isoelectric focusing) and molecular mass (SDS-PAGE). Individual proteins are identified by staining, removed from the gel, fragmented by proteases and subjected to mass spectrometry (MALDI-TOF-MS). Sequences are matched to those of known proteins in databases for identification.

Protoplast

A cell bounded by the plasma membrane and which contains all cell parts, except the extracellular matrix (**cell wall**). It is possible to isolate protoplasts, the 'living' part of the cell, by mechanical treatment, but the method of choice is to use combinations of cell wall-degrading enzymes such as cellulases, **hemicellulases** and pectinases. Protoplasts are able to not only regenerate their cell walls, but to controllably and directly form either specific cell types, **callus**, tissues or **somatic embryos**. Protoplasts are used to study the properties of the plasma membrane, cell wall regeneration and the regulation of cell division (**see: Cell cycle**). Protoplasts are particularly important in certain areas of seed biology especially in studies of **gibberellin** action on **aleurone layer** cells. (**See: Mobilization of reserves – cereals; Signal transduction – hormones**)

They are also used in fusion experiments to create novel genotypes that would otherwise be impossible due to natural reproductive barriers. Protoplasts are also the preferred cell biological system for **transformation** when microinjection is performed. Since it is possible to create clones from isolated individual protoplasts, they are also important tools in developmental biological studies. A protoplast may be processed so that it loses its nucleus while retaining its cytoplasmic contents, including plastids and/or **mitochondria**. Such cytoplasts may be used in cytoplasmic fusion experiments in which the nucleus of one species is introduced into the cytoplasmic background of another. (**See: Cells and cell components**) (PvA)

Provenance

A term used to define a tree seed source in forestry: the place from which the seedlot is produced, or a natural seed obtained. (**See: Tree seeds**)

Pseudocarp

Meaning false **fruit**, a term commonly used where the simplistic definition of a fruit is adopted to describe fruits in which parts other than the **ovary** are included such as, for example, the receptacle or parts of the perianth (e.g. **pome**). (**See: Fruit – types**)

Pseudocereals

Annual crops which superficially resemble **cereals** in general appearance and growth habit and in their production of abundant dry seeds containing **starch** and **storage protein** as the main reserves are referred to as pseudocereals. Their cultivation, harvesting, postharvest treatment and economic utilization are also similar to those of cereals. Botanically they are fundamentally different, being **dicotyledons** which show many structural differences from the cereals which are **monocotyledons** belonging to the family Poaceae (Gramineae). **Job's tears** (*Coix lacryma*) and **wild rice** (*Zizania aquatica*) are both members of the Poaceae and will not be treated here as pseudocereals though they are sometimes included in that category.

The pseudocereals possess inflorescences comprising many small flowers each with a single **ovary** of two or three **carpels** each enclosing one seed. The **embryos** are curved and embedded in reserve tissue variously described as **perisperm** and/or **endosperm**, a sufficiently diagnostic feature for the pioneering taxonomists Bentham and Hooker to include the three families containing pseudocereals in the 'Curvembryae', a group comprising seven flowering plant families. The three main types of pseudocereal are buckwheat (Polygonaceae), quinoa (Chenopodiaceae) and grain amaranth (Amaranthaceae), of which buckwheat is of Asian origin and is thought to have been in cultivation for about four millennia while the other two are more ancient having been in cultivation in Central and South America for at least seven millennia. All have been categorized in the present day as underutilized crops in that their potential advantages have not been fully exploited and that improvements via plant breeding seem feasible. Their advantages include ability to grow in locations and climates unfavourable to cereals, good protein contents and amino acid composition for human nutrition, and suitability for special foods as the grains lack certain cereal constituents such as **gluten** (**see: Storage protein – intolerance and allergies**). They are useful as break crops in cereal cultivation as they are unaffected by cereal pests and diseases. Their underutilization is also evident from **FAO** (United Nations Food and Agriculture Organization) estimates of world production in 2002 of a little over 2.2×10^6 t of buckwheat and 5.5×10^4 t of quinoa in comparison with total world cereal production of 2×10^9 t, of which wheat production amounted to 5.73×10^8 t. The FAO statistics did not include any figures for grain amaranth production. Comparison of the major nutrients of cereals and pseudocereals shows similarities but quinoa and grain amaranth are marginally richer in protein, substantially richer in **oils** (triacylglycerol) and marginally poorer in carbohydrates (Table P.20).

Table P.20. The major nutrients of pseudocereals and cereals expressed as a percentage of dry matter.

	Carbohydrate	Protein	Oil
Maize	81.0	12.8	3.7
Quinoa	72.1	18.3	7.8
Rice	88.0	8.5	2.2
Buckwheat	81.9	13.1	2.7
Grain amaranth	69.2	16.9	7.8
Wheat	77.6	15.7	2.5

1. Buckwheat

Common buckwheat (*Fagopyrum esculentum*) has been cultivated in China since the first or second century BC, though some sources date cultivation earlier and state that it may have been introduced to Japan 3000 years ago via the Korean peninsula. It is cultivated in temperate climates but will tolerate tropical climates at higher altitudes. Approximately 60% of world production was in China in 2002 with another 27% in the countries of the former Soviet Union and significant amounts in France, Poland, North America, Brazil and Japan, a total of 22 countries reporting production to the FAO.

A second species, bitter buckwheat (*F. tartaricum*), is also cultivated in India and China under cooler and harsher conditions but has the disadvantage of a bitter taste which necessitates extraction with boiling water to achieve palatability. Buckwheat is also important as a honey-plant and there are a number of minor cultivated species. Continued selection is needed in both major species to reduce seed **shattering** and seed **dormancy** and to encourage an annual growth, more progress in these respects having been made with *F. esculentum*.

F. esculentum grows to 60–80 cm in height, has a growing season of 70–90 days and produces dry three-sided **achenes**, some varieties with wings, and a seed weight of 25–30 mg (*F. tartaricum* 15–25 mg). As ripening is not uniform the crop is cut when 75% of the grains are mature, careful handling being required to minimize grain shattering which can lead to losses of up to 22%. Seeds are pointed, broad at the base, and triangular to nearly round in cross section (Fig. P.37). They vary in size from about 4 mm at maximum width and 6 mm long to 2 mm wide and 4 mm long. The seed consists of an outer layer or hull, an inner layer, the seedcoat proper, and within this a starchy endosperm and the embryo. As with the cereals, the buckwheat seed endosperm is **starchy**, with

Fig. P.37. Buckwheat seeds (achenes) (image by Mike Amphlett, CABI).

protein, which is surrounded by a single-cell-thick **aleurone layer**. However, the embryo has two distinct **cotyledons**.

Protein content is 13–15% fw, about 50% of which is **globulin**, although there is (allergenic to some) gluten (**glutelin**)-type protein also present. Overall, the amino acid composition of buckwheat protein makes it reasonably well-balanced from a dietary standpoint, e.g. unlike cereals it is not deficient in lysine, and unlike legumes it does not lack methionine or cystine. Starch (67–75% fw) is the major carbohydrate. Oil (triacylglycerol) content is 1.5–4% fw, most highly concentrated in the embryo but also present in the endosperm. Vitamins B1, B2 and niacin occur at relatively high concentrations compared with most of the cereals. The grains are rich in fagopyritols (galactosyl derivatives of inositol – **see: Galactinol series oligosaccharides**) which are thought to have beneficial effects on insulin activity and blood glucose levels.

Most of the crop is milled and the flour is used alone or mixed with wheat, barley, maize or rice flours. Japan is a major consumer in the form of noodles, involving, according to a 1995 report, over 6,500 factories and 41,500 restaurants. There is also human consumption of buckwheat tea, confectioneries, liquor, and buckwheat in the grain form, which is served as a vegetable cooked like rice but tastes somewhat bitter. It is used as an animal food in the USA and Russia.

2. Quinoa

Chenopodium quinoa is thought to have been domesticated in the Altiplano region of the Andes surrounding Lake Titicaca more than 9000 years ago. It was grown by the Incas throughout their Empire, by the Araucanian Indians of Argentina and Chile and by the Chilcha Indians of northern Colombia. In Inca and pre-Inca times its cultivation likely ranged from Bogota in Colombia (5°N) southwards through Peru and Bolivia to the island of Chiloe in Chile (42°S) and southeastwards to Cordoba in Argentina. After their observation of the use of quinoa in religious rituals by the Incas, the Spanish discouraged the cultivation of quinoa, replacing it with cereals. A resurgence of interest has led to its growth as a sole crop on about 20,000 ha in Peru and 40,000 ha in Bolivia as well as small scale cultivation elsewhere in South America. The FAO recorded production of 31,000 t in Peru in 2002, 23,500 t in Bolivia and 320 t in Ecuador.

Quinoa tolerates poor conditions, typically being grown as a subsistence crop at an altitude of 3000–4000 m, where the mean temperature is 7–10°C and frosts and droughts are common, so that yields of about 500 kg/ha are obtained. Under mechanized agriculture with combine harvesters, yields of up to 6.5 t/ha have been recorded. As a 'break-crop' in cereal rotations, control of gramineous weeds and of cereal pests and diseases can be achieved. Quinoa is described as an annual gynomonoecious plant which achieves a height of 0.4–3.0 m in a 150–220 day growth cycle. The small flowers are contained in a dense inflorescence in which one-seeded indehiscent fruits, 1–3 mm in diameter develop (Fig. P.38). The embryo is curved and it surrounds a floury perisperm in which most of the starch (up to 70% fw) is located. A two cell-layered endosperm surrounds the embryo. While the perisperm is rich in starch, the endosperm and embryo are rich in protein, with **phytin** being present in the protein bodies; **oil bodies** are also abundant. Protein accounts for 10–22% of the seed with a nutritionally good amino acid balance. The two-layered pericarp is adpressed to the two cell-thick seedcoat, both outer layers being rich in saponins, of which 15 types of molecule have been identified, derivatives of three main triterpene aglycones, namely phytoaccagenic acid, hederagenin and oleanolic acid (**see: Pharmaceuticals and pharmacologically active compounds**). The saponins are described as anti-nutritional and are removed or reduced in content prior to human consumption either by washing or by toasting the fruits and then grinding off the outer layers. The seeds may be ground into flour which can be used for a coarse bread or an additive, or boiled and added to soups.

3. Grain amaranth

There are three species, *Amaranthus hypochondriacus* and *A. cruentus* native to Mexico and Guatemala and *A. caudatus* native to the Andes of Peru. All have edible leaves and *A. cruentus* is an important vegetable crop throughout the tropics

Fig. P.38. Quinoa grains. Black or brown grains also occur (image by Mike Amphlett, CABI).

and also in temperate regions. All 60 species of the genus possess the 'Kranz' anatomy diagnostic for the more efficient C4 photosynthetic pathway and the NAD-malic enzyme type from that pathway has been identified in the genus. The grain amaranths are now grown widely in the hill regions of south Asia and *A. hypochondriacus* and *A. caudatus* are reported to have become major crops of north-west India.

They are substantial annual plants, *A. hypochondriacus* reaching up to 3 m in height while the other two species reach 2 m. Under traditional cultivation wide variation in yield has been noted (200–2000 kg/ha) with up to 3000 kg/ha resulting from the best procedures and seed. In Latin America and Asia manual harvesting, sun drying (**solarization**) and manual **threshing** and **winnowing** are usual. For mechanical harvesting a killing frost followed by the drying down of the whole plant in about 10 days is needed and yields would be improved if varieties resistant to shattering could be produced. The fruit is a capsule yielding a single seed of about 1 mg with a curved embryo surrounding a central mass of mealy endosperm containing 62–69% of starch and 14–19% of protein, relatively rich in lysine and leucine. The seed has approx. 8% oil. The grain amaranth seed is quite small (0.9–1.7 mm diameter) and seed weights vary from 1000 to 3000 seeds/g. Seed colours can vary from cream to gold and pink to black.

Grain amaranth had featured as a standard food for long before the Spaniards colonized the Americas. But its use in various religious rituals, for example by the Aztecs who ground the seed with honey or human blood and shaped it into idols, threw the grain into disrepute and it was banned by the Spaniards (**see: History of seeds in Mexico and Central America**). The commonest use of grain amaranth in Latin America today is a snack of popped seeds with molasses, with its use as a flour coming second in importance. Used on their own in flour, the amaranth starches lead to poor quality breads and cakes but the flour is a good supplement to cereal flours. Grain amaranth is also used for special foods for people who are allergic to wheat products. The unused parts of the grain serve as animal feedstuffs. Grain amaranth is exported by China, Kenya and Mexico. (JWB)

Belton, P.S. and Taylor, R.N. (eds) (2002) *Pseudocereals and Less Common Cereals*. Springer Verlag Berlin-Heidelberg-New York.

Flemming, J.E. and Galwey, N.W. (1995) Quinoa (*Chenopodium quinoa*). In: Williams, J.T. (ed.) *Cereals and Pseudocereals*. Chapman and Hall, London, UK, pp. 3–83.

Joshi, B.D. and Rana, R.S. (1995) Buckwheat (*Fagopyrum esculentum*). In: Williams, J.T. (ed.) *Cereals and Pseudocereals*. Chapman and Hall, London, UK, pp. 85–127.

Prego, I., Maldonado, S. and Otegui, M.S. (1998) Studies on quinoa seed: seed structure and localization of seed reserves. *Annals of Botany* 82, 481–488.

Williams, J.T. and Brenner, D. (1995) Grain amaranth (*Amaranthus* species). In: Williams, J.T. (ed.) *Cereals and Pseudocereals*. Chapman and Hall, London, UK, pp. 129–186.

Psychoactive seeds

Seeds of many species contain chemicals that have pharmacological effects as stimulants and hallucinogens. Most of these seeds have been employed since ancient times as aids to divination, spiritual experiences, stress relief, enhanced perception, enhanced energy, hunger and fatigue control, etc.

Some, such as **coffee**, **cola**, **chocolate** and **guarana**, are in very or fairly common use in present-day society while others are more limited in their appeal. Moreover, many seeds that are generally not considered to be psychoactive contain substances that at the right concentration are effective. An example is anethole which occurs in the spices **anise** and **fennel**; and nutmeg which can have hallucinogenic or narcotic effects when consumed in suitably excessive quantities.

The active chemicals are generally alkaloids (**see: Pharmaceuticals and pharmacologically active compounds**). **(See also: Betel nut; Hawaian baby woodrose; Mescal bean; Morning glory; Ololiuqui; Thornapple; *Voacanga africana*; Yopo)**

Psychrometric chart

Also known as a Mollier diagram, it is a graphical representation (Fig. P.39) of the properties of air and water mixtures that is an essential tool for temperature and humidity control technologies. The interrelationships between temperature, relative humidity, concentration of water (also called specific humidity), vapour pressure, dew point temperature and enthalpy are described and if two parameters are known, the others can be determined. Psychrometric charts represent for air what **water sorption isotherms** represent for seeds. Relationships between **water content**, **relative humidity**, temperature and enthalpy are opposite in psychrometric charts and sorption isotherms because they represent opposite reactions. For example, at a constant water content, increases of temperature will result in decreases in activity (relative humidity ÷ 100) in the air, but increases in water activity on the sorbent. (CW)

Pterin

From the Greek *pteron* meaning wing, since it is a component of the pigments of butterfly wings. A yellow compound that contains the bicyclic ring system characteristic of pteridine

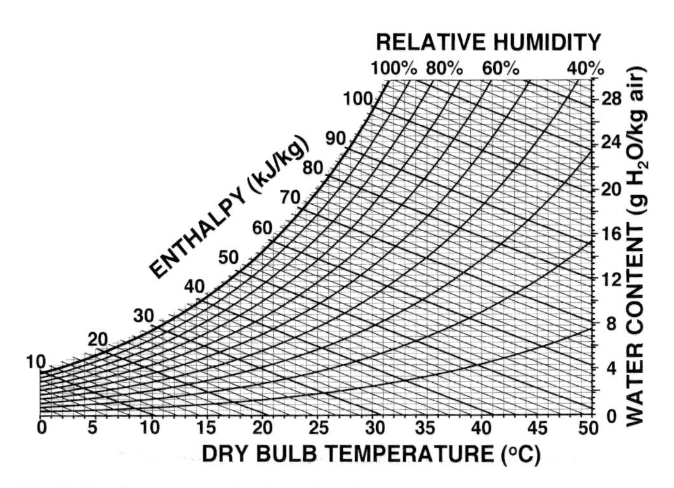

Fig. P.39. Psychrometric chart showing the interrelationships between the various factors that determine the properties of air and water mixtures. If two of the determining parameters are known, the others can be calculated from the chart.

(molecular formula $C_6H_4N_4$). The coloured nature of pterins allows them to act as chromophores for chromoproteins such as **cryptochrome**. (**See: Light – dormancy and germination**)

Pulses

Seeds of grain legumes that are stored, generally in the dry state, and directly used for food (**see: Legumes**).

In the Indian subcontinent, a distinction is often but not always made between whole pulse grains, called gram (a word derived from grão, the Portuguese for grain), and split pulse grains (after the seedcoat **testas** have been removed), called *dal/dahl* (a name also given to culinary dishes prepared from these ingredients).

Pumpkin

Seeds of pumpkin (*Cucurbita pepo*) are eaten largely as a snack food and generally after roasting. The seeds are flat, oval, 1.5–2 cm long and 1 cm at the widest. They contain 40–45% **oil** (triacylglycerol) (predominantly linolenic, linoleic and oleic **fatty acids**), about 35% **storage protein** and approx. 13% carbohydrate, all on a fresh weight basis. They are relatively rich in several mineral elements. They also contain cucurbitacin, a triterpene (**see: Poisonous seeds**). Apart from their culinary uses they are also employed medicinally, to treat benign prostatic hyperplasia and prostatic cancer and as a tapeworm purgative. (MB)

(**See: Cucurbits**)

Pure live seed (PLS)

In seed conditioning, the percentage of pure seed in a lot that has the ability to germinate. PLS = (% pure seed × % germination) / 100. (**See: Quality testing**)

Purity

In seed quality testing, pure seed is the component of a seed**lot** that consists of seeds of the designated species. According to **ISTA rules**, this includes not only mature, undamaged seeds but also undersized, shrivelled, immature and germinated seeds provided they can be positively identified as the designated species, and also pieces of seed resulting from breakage which are more than half their original size. (**See: Purity testing**)

Purity testing

An aspect of seed quality testing that deals with two distinct aspects: the proportion of seed of the correct type by visual inspection, and the 'genetic' purity of the variety.

1. Analytical purity

The objective of the analytical purity test is to determine the proportion of pure seed in relation to unwanted contaminant seeds and material that is not seed at all. It is perhaps the most complex and exacting of all the seed quality tests to conduct, because it requires analysts with a comprehensive knowledge of seed structure and function as well as the ability to identify a wide array of different seed species using visible morphological characteristics (**see: Structure of seeds – identification characters**). However, this test does not discriminate between seed of different sizes and **vigour**, or the live and the dead, or necessarily between different varieties (though skilled analysts can often distinguish varieties of major cereal varieties, for instance, by subtle differences in seed shape and structure).

(a) *Basic Method.* The physical composition of a sample of approximately 2500 seeds is analysed by separating into three fractions, as follows, each of which is weighed; results are expressed as percentages of the combined weights. Separations are usually carried out by hand, using sieves, forceps, a hand-lens or binocular microscope, and a balance. For grasses a seed blower can be used to separate empty florets and other inert material, and a light-box to determine whether a caryopsis is present inside the floret, and its size. Diagrammatically, the stages in the separation of each individual item of material in a sample are as in Fig. P.40.

(i) *Seed of the species being analysed, including all its botanical varieties and cultivars.* In cases where species are very similar, no attempt is made to separate them. For example, of the various forms of cultivated *Brassica* species, only brown (*B. juncea*) and black (*B. nigra*) can be easily identified using visual characteristics; no attempt is made to distinguish seeds of other cultivated *Brassica* species.

(ii) *Seed of 'other species'.* This fraction contains seeds of all species other than the pure seed species under consideration, though a further separation into 'crop' and 'weed' seed categories can be made if required. Each species found must therefore be identified, and a seed herbarium is an essential tool in identifying unknown species. Where the other species are so similar that precise identification is impossible it is sufficient to identify the genus alone, reported with the suffix 'species', normally abbreviated to sp. (singular) or spp. (plural).

(iii) *Inert matter.* This fraction contains everything in the sample that is neither 'pure' nor seed of 'other species', and it can be separated into four source categories:

1. material derived from seeds, e.g. broken seed and empty **florets**;
2. material derived from other parts of the plant, for example, leaves, flowers and stalks;
3. non-plant organic material, for example, **ergot** (*Claviceps purpurea*) **sclerotia**, **bunt** balls, insects, mites and nematodes; and
4. material not derived from living organisms, for example, sand, soil and stones.

The classification of material as seed (pure seed or other seed species) or inert matter is prescribed by the *Handbook of Pure Seed Definitions* and in the *ISTA International Rules for Seed Testing*. In summary, the ISTA Rules state that 'pure seed' includes:

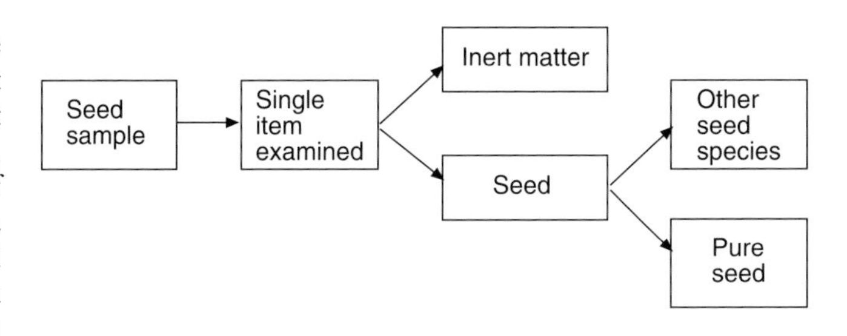

Fig. P.40. Stages in the separation of components of a seed sample during an analytical purity test.

- immature, undersized, shrivelled, diseased and sprouted seed provided they can be identified as of that species and have not been transformed into fungal sclerotia, **smut** balls or nematode galls;
- free caryopses of Poaceae and florets with an obvious caryopsis containing endosperm;
- broken pieces of seed larger than one-half the original size irrespective of whether an **embryo** is present.

Particular rules apply to:

- the Fabaceae, where seed with coats that are entirely removed and separated cotyledons are all regarded as inert matter; and
- certain genera of Poaceae, where a minimum size of **caryopsis** may be required; for *Elytrigia*, *Festuca*, *Lolium* and ×*Festulolium* the caryopsis must be at least one-third the length of the **palea**, measured from the base of the rachilla; for all other Poaceae there is no size requirement provided the caryopsis is present and contains endosperm.

(b) *'Other seed' determination.* The objective of this quality test is to determine the number of seeds of other species, either generally (for example, all other species), by reference to a particular category of seeds (for example, species scheduled as noxious in a country's legislation), or specifically (for example, a particular species) in a sample of seed. The result is expressed as the number and weight of 'other seeds' belonging to each stated species or category in the weight of seed examined. The test requires a much larger sample of seed – typically 25,000 – than in the basic method of the analytical purity test (described above), but otherwise the techniques used in the analysis and the evaluation of contaminant seeds and disease organisms are the same.

The ten-times larger sample is necessary to avoid missing significant impurities altogether. For the sake of illustration: 750 seeds/ha of a contaminant species, such as wild oats (*Avena fatua*) or couch (*Elytrigia repens*), could become the source of a serious weed problem in temperate cereals, or 750 ergot sclerotium bodies or 'other seed' species could seriously reduce the value of the crop; but, at this level of contamination, sampling theory dictates that a random 2500 seed sample (which is equivalent to about 1:1500 of the quantity of seed sown on a hectare) has a less than 50% probability of containing any seed contaminants at all, whereas with a 25,000 seed sample the probability is over 99%.

2. Varietal purity

The production of seed for sale to the farmer involves taking a small quantity of seed from the breeder, which is true-to-variety and 100% pure, and multiplying it over one or several generations, or in the case of hybrid seed multiplying and then crossing parental lines. During multiplication, deterioration in varietal purity can occur due to genetic factors, such as cross-pollination with other varieties. Also during multiplication, harvesting and processing, physical contamination with seeds of other closely related varieties can also occur which, generally, is not easily removable by conditioning. **(See: Conditioning; Production for Sowing)**

Varietal – or 'genetic' – purity differs fundamentally from analytical purity because, as a general rule, varieties cannot be identified by visual examination of their seeds. It may be possible to say that the seed belongs to a group of varieties according to grain morphological characteristics, but it is rarely possible to identify the exact variety. A variety can be identified with greater certainty through the examination of growing plants, and **certification schemes** have been established in which mother plants are examined at the seed-production farm and in plots sown from the same seed. Varietal purity determined in such a manner cannot give the same assurance as laboratory tests for seed quality carried out on seed harvested from the mother plants.

(a) *Laboratory tests for varietal identification/purity.* Until the advent of biochemical and image-analysis tests, laboratory tests for varietal identification were somewhat limited.

- Chemical tests can be used to distinguish and identify some varieties. For example, the phenol test is used for wheat, barley, oat, ryegrass and bluegrass, based on colour reaction with the pericarp. Similarly: a sodium hydroxide test is used to distinguish between red and white wheat; a hot alkaline **bleach test** is used in official **grain inspection** protocols, to determine pigmented wheat and sorghum varieties that contaminate 'white' variety seedlots, or the degree of heat damage; and an iodine/potassium iodide staining procedure (I_2/KI) used for detecting 'waxy' maize varieties, in which the **starch** is composed entirely of **amylopectin**.
- Fluorescence tests of seeds or seedlings are used to distinguish ryegrass varieties, or to estimate contamination of them with annual ryegrass.
- **Chromosome number** counts are used in **ploidy** determinations in grasses, clovers and beets, counting squashed cells of seedling root tip cells under a microscope after staining, or using flow-cell cytometry.
- Electrophoretic tests have been developed for some species, as an accurate (though relatively costly) method of identification.
- In the case of herbicide-tolerant varieties produced by **genetic modification**, such as maize, the introduced trait can be tested in germination or seedling grow-out tests by wetting or spraying with an appropriate concentration of the herbicide in question.
- In the future and subject to cost, it might be possible to identify a variety at the **genomic** level. (RD)

Purple seed stain

A fungal disease of **soybean**, caused by *Cercospora kikuchii*, that can affect most parts of the plant including discoloration of the seedcoat, varying from pink to dark purple, often accompanied by cracking. Though there is little effect on ultimate crop yield, severe (>5%) **infection** can reduce the commercial grade of harvested grain. First symptoms of inconspicuous seedborne infection are usually dark purple **cotyledons** with premature drop, followed by the stunting of seedlings with purple to red angular lesions along the outer margins of upper leaves, though **germination** and **vigour** are not usually reduced. (PH)

(See: Disease; Pathogens)

Pyrene

Greek, referring to the stone of a **fruit**, i.e. a seed surrounded by a hard, bony **endocarp**, the term often employed when a **drupe** contains more than one stone; sometimes also used to describe the fruit (a multi-seeded drupe) itself. (**See: Fruit – types**)

Pyxidium

A capsular **fruit** dehiscing by loss of a lid, or by a transverse suture across the cells, or through apical or basal pores on each **carpel**, that enlarge and unite at maturity to a single pore, e.g. in many Lecythidaceae, *Amaranthus hybridus*; *Ecballium* spp., *Eucalyptus* spp., *Callistemon* spp., *Anagallis arvensis*, *Hyoscyamus niger*, *Reseda odorata*. (**See: Dispersal**)

Q locus

In **wheats**. A dominant **allele** at this locus is thought to be responsible for the square-headed **spike**, the free-threshing characteristic and the non-fragile **rachis** as well as various **pleiotropic** effects. (**See: Domestication; Einkhorn; Emmer; Spelt**)

Quality

In the context of the production and testing of seed for sowing, 'quality' is commonly regarded as a multiple-component property that includes: germination ability and **vigour**, varietal (genetic) purity, analytical purity (the absence of contaminant foreign species and inert matter), **'pure live seed'**, seed health (the absence of seedborne **pathogens**), **moisture content**, and the uniformity of mixing and blending. The quality of treatment with fungicides and insecticides also may be assessed, in terms of loading analysis. Such quality parameters reflect the manner in which a seedlot has been produced, conditioned, conveyed, treated and stored. Seed physiological quality also reflects some aspects of genetic composition, such as the phenomenon of **heterosis**.

Quality grade parameters and standards of grain consignments are expressed variously in terms of seed size, length, colour, odour, degrees of contamination and damage due to the presence of foreign or broken material, or infestation with injurious **storage pests** or **storage fungi** or the damage or taints they have caused. Grain intended for specific food and beverage processing or industrial production purposes may be subject to additional quality standards, such as concerning content and composition and the adventitious presence of varieties that have been subject to **genetic modification**.

For articles covering the diverse aspects of seed quality management and assessment, **see: Cereals – storage; Conditioning; Conveyors; Grain inspection; Identity preservation; Malting; Pathogens – seedborne infection; Preharvest sprouting; Production for sowing; Quality-Declared Seed; Quality testing; Storage management; Treatment; Treatment – pesticides**. (PH)

Quality-Declared Seed

The Quality-Declared Seed (QDS) scheme has been developed by the UN Food and Agriculture Organization (FAO) to provide quality control during seed production, and is designed for countries where compulsory certification is beyond the resources of the seed production and quality control system. The QDS scheme is more flexible and less bureaucratic than seed **certification schemes**, as the emphasis is placed on actual seed quality rather than administrative procedures designed to ensure that standards are met. Quality standards need not be relaxed from the certified seed standards, however, and the QDS scheme is no less rigorous than a certification scheme.

A quality-declared seed scheme is based on four principal requirements:

- a list of varieties eligible to be produced as QDS;
- seed producers to register with an appropriate national authority;
- a national authority to check 10% of the seed crops;
- a national authority to check 10% of seed offered for sale under the QDS designation.

The scheme relaxes the requirement for limited generations of multiplication from the **Breeder's seed**. It allows QDS to be produced either from certified seed or from other QDS, providing the origin can be traced to the original breeder and the applicable quality standards are met. Responsibility for quality control is placed with producers (private companies), who are encouraged to develop a quality control capacity. The government plays a monitoring role and remains independent of the production process, so the cost of the QDS scheme is much lower than certification. Like voluntary certification, its success depends on farmers recognizing the label as a sign of quality. (**See: Laws and Regulations**) (RD)

FAO (1993) Plant Production and Protection. Paper No. 117.

Quality testing

Several aspects of seed quality are tested routinely to underpin commerce and trade to minimize the risk of sowing seedlots that do not have the capacity to produce the desired crop. Tests have been developed to measure the following ten parameters that combine to determine the planting value of a seedlot.

(a) *Analytical purity* measures the amount of material in a seed sample that is of the desired species, expressed as the percentage by weight that is 'pure seed' of the species being analysed. The percentages of 'other seeds' species and inert matter are also determined. (**See: Purity testing, 1. Analytical purity**)

(b) *'Other Seed' determination* measures the numbers of each kind of 'other seed' in a prescribed weight of seed. (**See: Purity testing, 1. Analytical purity**)

(c) *Germination*, in seed testing terms, is defined as the percentage by number of pure seeds that have the potential to produce established seedlings in the field. (**See: Germination testing**)

(d) *'Pure Live Seed'* is defined as the percentage of pure seed in a lot that has the ability to germinate, and is calculated by multiplying the 'Analytical Purity' by the 'Germination' values.

(e) *Moisture content* measures the amount of water in the seed expressed as a percentage by weight of the original seed, on a 'fresh weight' basis.

(f) *Varietal purity* measures the percentage of pure seed in a sample that is of the desired variety. (**See: Purity testing, 2. Varietal purity**)

(g) *Seed health* is concerned with the presence and identity of organisms that may lead to disease of the crop. (**See: Health testing**)

(h) *'Seed size'* is an expression of the weight of a given number or volume of seed: usually measured as the mass of seeds (**Thousand seed (or grain) weight**) or the mass contained in a specified capacity (for example, kilograms per hectolitre, or pounds per bushel). A minimum 'test-weight' limit, expressed on a similar volume basis, is one criterion commonly used in establishing official trading quality grades during **grain inspection**. (Note that this expression of 'seed size' does not directly measure geometrical size grading, though it tends to reflect it.)

(i) *Vigour* is a measure of the sum total of those properties of the seed that determine the level of activity and performance of the seed during germination and seedling emergence. Seeds that perform well are termed high vigour seed. (**See: Vigour testing**)

(j) *Uniformity* is a measure of the effectiveness of mixing and blending to ensure that the different components are evenly distributed throughout the seedlot. Every lot of seed is to some extent a mixture of different components, such as 'pure seed', 'other seed' and 'inert matter', and different geometric sizes and shapes. There may be variation within the harvested bulk as a consequence, for example, of parts of the seed production field having different levels of disease or weed contamination. There will be differences in the distribution of geometric seed size within seed lots that have been size graded, even though they happen to share the same thousand-seed weight.

See also **X-ray analysis** – test procedures that are used to detect internal quality of certain seeds (abnormalities, absence of embryos and other parts, mechanical and internal damages and insect infestation). (RD)

For the quality test procedures applied during grain malting processes **see: Malting – process, 7. Malt modification; also see: Grain inspection; Mycotoxins**.

Quantitative trait loci (QTLs)

A chromosomal segment, potentially encompassing many hundreds of individual loci (genes), one of which is involved in the expression of the quantitative trait, e.g. seed dormancy. Such traits are typically affected by more than one gene, and also by the environment. DNA-based molecular markers are frequently used to mark the location of QTLs on genetic linkage maps. Mapping QTLs is not as simple as mapping a single gene that affects a qualitative trait (e.g. flower colour). (**See: Dormancy – genetics; Preharvest sprouting – genetics**)

Quarantine

Holding a seed lot from sale or movement usually for a period of time until concerns about its **phytosanitary certification** requirements have been addressed.

Quiescence

Sometimes used synonymously with 'rest'.

Of a seed, one that is not germinating, e.g. a dry seed or (sometimes) a hydrated one under conditions that do not support germination, such as unsuitable temperature, oxygen, or moisture. Quiescence is generally characterized by reduced metabolic activity. Quiescence is distinct from **dormancy**. A quiescent seed may not be dormant but a seed possessing dormancy may also be quiescent.

Quinoa

See: Pseudocereals

R

Race

A subdivision of a species or variety, or a physiologically specific form of a microorganism. In plants: a **cultivar** or special form of a cultivar. Also applied, for example, to **pathogens** that cause plant diseases.

A **landrace**, in plants, is an early-cultivated or locally adapted form of a crop species, evolved from a wild population and containing a range of genotypes, and derived as a result of the selection pressures in specific environments. Pure **lines** were isolated as components of landraces, and largely displaced them by the end of the 19th century in more advanced agricultural markets. Many present-day local **sorghums**, **pulses** and tropical **rices** are still landraces, notably in Asia and Africa, though their use in agriculture is declining fast as they are displaced by modern cultivars. Landraces still serve as important sources of useful genes in breeding programmes, and in practice are now mainly made available to plant breeders from the pool of accessions made in **gene banks**. (PH)

(See: **Conservation**)

Rachis

The stalk of a grass (cereal) **spikelet**. (See: **Floret**; **Inflorescence**)

Radicle

Part of the **embryo axis** that gives rise to the embryonic root, i.e. primary root system, after germination is completed. (See: **Germination – radicle emergence; Structure of seeds**)

Radish

See: **Brassica – horticultural**

Raffinose series oligosaccharides (RSOs)

Also called: Raffinose family oligosaccharides (RFOs). Galactosyl sucroses.

A family of oligomeric sugars in which an increasing number of galactose units are attached to a terminal sucrose (Fig. R.1). The major members of the family in seeds are raffinose (galactosyl sucrose), stachyose (digalactosyl sucrose), and verbascose (trigalactosyl sucrose).

RSOs are widely present in mature dry seeds, mostly in the **embryo**, and usually account for 2–6% of the seed dry weight (Table R.1); they are generally assumed to be located in the cytoplasm. They decline during germination since they are used as an early source of respirable substrate. Their hydrolysis requires the activities of: (i) α-galactosidase (which in the developing and dry seed is located in **protein bodies**), an enzyme which breaks the α-1,6 bonds between galactose and sucrose, and between adjacent galactose sugars; and (ii) invertase, which cleaves the sucrose molecule to produce the hexoses glucose and fructose.

Synthesis of RSOs in the developing seed initially involves the synthesis of galactinol from UDP-galactose and *myo*-inositol, catalysed by the enzyme galactinol synthase (GolS). The galactose from the galactinol is then transferred to sucrose by raffinose synthase to form raffinose.

Galactinol + sucrose → *myo*-inositol + raffinose

A further galactosyl residue may be transferred by stachyose synthase to form stachyose, and a third by verbascose synthase to produce verbascose.

RSOs are usually regarded as being anti-nutritional, although their beneficial role as pre-biotics, stimulating the growth of remedial bacteria (e.g. lactic acid-degrading bacteria) in the colon, thus maintaining a desirable gut microflora, is being investigated. Humans and monogastric animals do not produce enzymes to degrade the α-galactosidic linkages. Thus RSOs escape digestion in the small intestine; bacteria within the colon degrade them, however, to produce hydrogen and carbon dioxide which build up and cause

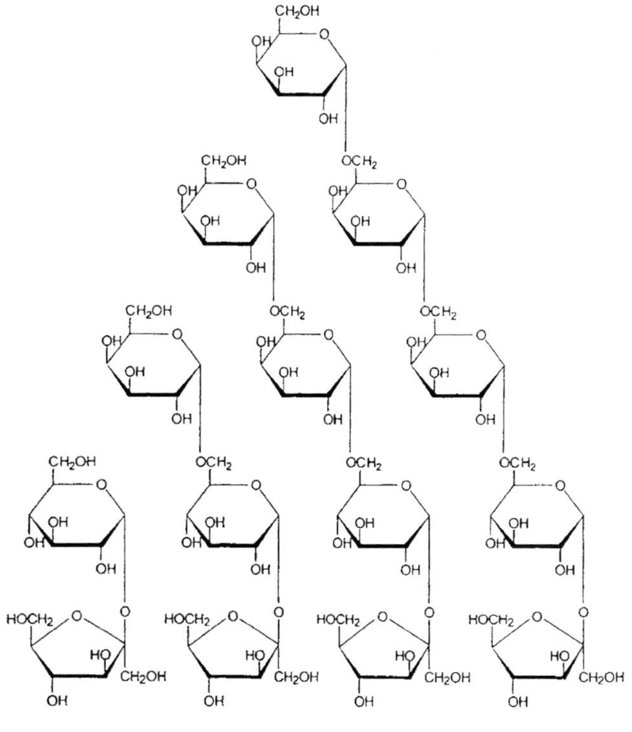

sucrose raffinose stachyose verbascose

Fig. R.1. Raffinose series oligosaccharides.

Table R.1. Raffinose oligosaccharide content of select species of seeds. Sucrose content is shown for comparison. Species are grouped according to family. Amounts are expressed as mg/g of defatted whole mature seed meal.

Species	Raffinose	Stachyose	Verbascose	Sucrose
Cotton	69.1	23.6	tr	16.4
Garden pea	11.6	32.3	19.1	62.3
Soybean	12.6	43.4	tr	64.2
Mung bean	3.9	16.7	26.6	22.7
Lima bean	6.9	30.3	tr	13.9
Green bean	2.5	34.3	tr	19.4
Broad bean	4.3	10.7	11.4	20.7
Peanut	3.3	9.9	tr	81.0
Sunflower	30.9	1.4	–	65.0
Pumpkin	6.5	16.3	–	28.8
Barley, wheat	7.2–7.9	tr	–	13.8–14.2
Maize	3.1	–	–	14.2
Rice	tr	–	–	8.4

For more species see: Kuo, T.M., VanMiddlesworth, J.F. and Wolf, W.J. (1988) Content of raffinose oligosaccharides and sucrose in various plant seeds. *Journal of Agricultural and Food Chemistry* 36, 32–36. tr = Trace, – = undetectable. Note that there are differences in RSO and sucrose content between cultivars of any species, and thus the values should be regarded as approximate.

flatulence, and sometimes diarrhoea. Attempts are underway to reduce seed RSO production, particularly in **legumes**, using conventional breeding and molecular technologies. The addition of α-galactosidase to legume seed meal fed to monogastric animals, e.g. pigs, may also be useful in removing RSOs, and enhancing their growth. (JDB)

Downie, B., Gurusinghe, S., Dahal, P., Thacker, R.R., Snyder, J.C., Nonogaki, H., Yim, K., Fukanaga, K., Alvarado, V. and Bradford, K.J. (2003) Expression of a *GALACTINOL SYNTHASE* gene in tomato seeds is up-regulated before maturation desiccation and again after imbibition whenever radicle protrusion is prevented. *Plant Physiology* 131, 1347–1359.

Obendorf, R.L. (1997) Oligosaccharides and galactosyl cyclitols in seed desiccation tolerance. *Seed Science Research* 7, 63–74.

Peterbauer, T. and Richter, A. (2001) Biochemistry and physiology of raffinose family oligosaccharides and galactosyl cyclitols in seeds. *Seed Science Research* 11, 185–197.

(See also: articles on **Deterioration; Galactinol series oligosaccharides; Germination – metabolism)**

Rapeseed

The brassica crop, also known as oilseed rape and, in Canada as Canola™. **(See: Brassica oilseeds; Brassica oilseeds – cultivation; Oilseeds – major)**

Raphe

The continuation of the **funiculus** of the ovule that runs parallel to the nucellus and ends in the **chalaza**. In fact, the funiculus, raphe, chalaza and nucellus form a continuous tissue and no sharp demarcation lines can be drawn between them. Can be seen as a ridge in the seedcoat. **(See: Seedcoats – structure; Structure of seeds – an overview)**

Reactive oxygen species (ROS)

Molecules derived from molecular oxygen (O_2) that are highly reactive and usually electrophilic, attacking other molecules to abstract electrons. ROS are produced from reactions of molecular oxygen, which exists naturally as a moderately reactive **free radical**, with ionizing radiation, UV light and transition metals or as a consequence of metabolism (e.g. incomplete oxidation of water during photosynthesis or reduction of oxygen during respiration). In its least reactive form, O_2 has two unpaired electrons (so it is a free radical), both with similar spins, and is called triplet oxygen because of its spectral signature. Photoexcitation can convert triplet oxygen to singlet oxygen by causing a spin reversal in one of the electrons, creating an even more reactive molecule because of the vacant orbital. Other ROS are formed by partially reducing (adding electrons) triplet oxygen to a superoxide radical (one electron added), hydrogen peroxide (two electrons added), hydroxyl radical (three electrons added) (fully reduced O_2 is H_2O, water). Partially reduced oxygen is highly reactive because of the free radical status (superoxide and hydroxyl radicals) and because the O=O double bond of molecular oxygen is progressively weakened as each electron is added. Cytotoxicity varies among species based on the oxidizing capacity, cellular substrates and membrane permeability. Hydroxyl radicals are considered the strongest oxidant known and will attack all organic molecules to form peroxyl radicals in the presence of molecular oxygen (**see: Peroxidation**). The electron transport chain and reactions of triplet oxygen with a transition metal create partially reduced oxygen, and much of cellular metabolism is designed to allow these potentially dangerous reactions to occur under controlled conditions. For example, cytochrome oxidase binds partially reduced oxygen until reduction is complete. ROS can also be produced when the carbonyl group (HC=O) of **reducing sugars** abstracts an electron from metal ions.

Evidence is accumulating for the involvement of ROS in seed development, maturation and germination, including their participation in signaling and growth processes. (CW)

(See: Desiccation tolerance – protection against oxidative damage; Deterioration reactions)

Bailly, C. (2004) Active oxygen species and antioxidants in seed biology. *Seed Science Research* 14, 93–108.

Schopfer, P., Plachy, C. and Frahry, G. (2001) Release of reactive oxygen intermediates (superoxide radicals, hydrogen peroxide, and hydroxyl radicals) and peroxidase in germinating radish seeds controlled by light, gibberellin, and abscisic acid. *Plant Physiology* 125, 1591–1602.

Recalcitrant seeds

Recalcitrant seeds are intolerant of desiccation. Unlike **intermediate** and **orthodox** seeds, recalcitrant seeds rapidly lose **viability** when bulk, or freezable, water is removed by drying (**see: Drying of seed for storage**). This corresponds to water-binding region III on water **sorption isotherms**.

Plant species that inhabit moist tropical biomes and which produce large, spherical, fleshy seeds are more likely to be recalcitrant than smaller-seeded species from the drylands. The relatively small seeds of some submersed aquatic plants such as *Zostera* are also recalcitrant. However, a survey of 38 UK aquatic plants found only three that did not have desiccation tolerant, orthodox seeds. Clearly, there is no certain, quick diagnostic and in some genera, such as *Acer*, orthodox and recalcitrant behaviour occurs in related species. The term 'recalcitrant' was first used by E.H. Roberts in recognition of the fact that these seeds were difficult to store compared to the overwhelming majority of seeds which he called 'orthodox'. Based on (admittedly biased) data for over 9000 species (www.rbgkew.org.uk/data/sid/index.html), it is now estimated that around 7% of the world's flora produce recalcitrant seeds. See Table R.2. for examples.

Recalcitrant storage behaviour is not strongly related to phylogeny and has evolved at least six times during angiosperm evolution (**see: Evolution of seeds**). An ecological trade-off seems likely in those species that have lost the ability to withstand desiccation during their evolution. Such species are unlikely to possess dormancy and are more likely to be adapted for germination soon after dispersal. To maintain the viability of recalcitrant seeds they must not be allowed to dry and they must be kept aerated. Tropical recalcitrant seeds are often susceptible to **chilling injury** and therefore must be kept above about 15°C to avoid damage. Temperate recalcitrant seeds on the other hand can be cooled to just above 0°C. Long-term conservation options for recalcitrant seeds involve **cryostorage** of isolated embryos. (RJP, SHL)

Table R.2. Some examples of species which produce recalcitrant seeds.

Family	Species	Common name
Nymphaeaceae	*Nuphar lutea*	Water lily
Fabaceae	*Castanospermum australe*	Australian chestnut
Clusiaceae	*Garcinia mangostana*	Mangosteen
Malvaceae	*Durio zibethinus*	Durian
Malvaceae	*Theobroma cacao*	Cocoa
Sapindaceae	*Nephelium lappaceum*	Rambutan
Sapindaceae	*Aesculus hippocastanum*	Horsechestnut
Sapindaceae	*Acer pseudoplatanus*	Sycamore
Sapindaceae	*Acer saccharinum*	Silver maple
Anacardiaceae	*Mangifera indica*	Mango
Fagaceae	*Quercus rubra*	Red oak
Cucurbitaceae	*Telfairia occidentalis*	Fluted pumpkin
Sapotaceae	*Pouteria campechiana*	Egg-fruit
Euphorbiaceae	*Hevea brasiliensis*	Para rubber
Moraceae	*Artocarpus altilis*	Breadfruit
Fagaceae	*Castanea sativa*	European chestnut
Arecaceae	*Cocos nucifera*	Coconut
Zosteraceae	*Zostera marina*	Eel grass
Araucariaceae	*Araucaria hunsteinii*	Klinkii pine
Araucariaceae	*Araucaria angustifolia*	Brazilian pine

Pammenter, N.W. and Berjak, P. (1999) A review of recalcitrant seed physiology in relation to desiccation-tolerance mechanisms. *Seed Science Research* 9, 13–37.

Pammenter, N.W. and Berjak, P. (2000) Evolutionary and ecological aspects of recalcitrant seed biology. *Seed Science Research* 10, 301–306.

Recombinant DNA

Any DNA molecule formed by joining DNA fragments from different sources. They are commonly produced by cutting DNA molecules with **restriction enzymes** and then joining the resulting fragments from different sources with DNA ligase. (**See: Genetic modification; GMO – patent protection technologies; Transformation**)

Recombinant protein

Any protein that can be expressed in a non-native or heterologous environment. Recombinant proteins are most commonly expressed in experimental and production biotechnology in *Escherichia coli* bacteria, but plants can also be used for their expression. Some recombinant proteins are fusion proteins, where the genes for more than one protein are joined together and expressed simultaneously, resulting in proteins that are likewise joined. (**See: Biopharmaceuticals**)

Redox reactions

Oxidation–reduction reactions in which there is a transfer of electrons from an electron donor (reducing agent or reductant) to an electron acceptor (oxidizing agent or oxidant). (**See: Reactive oxygen species; Respiration; Thioredoxin**)

Reducing sugar

One that acts as a mild reducing agent, and can reduce various inorganic ions. Fehling's and Benedict's reagents, which contain acidified inorganic cupric ions, turn from blue to orange, by forming the cuprous ion in the presence of reducing sugars. Monosaccharides (e.g. glucose and fructose) and most disaccharides (not sucrose) are reducing sugars because of the presence of an aldehyde group in their structure, although fructose has a ketose group. Reducing and non-reducing sugars are important in relation to desiccation tolerance of seeds. An oligosaccharide or polysaccharide molecule may have either a single 'reducing' terminus, or none. (**See: Carbohydrates; Non-reducing sugar**)

Regeneration – ecology

The production of the next generation in plants requires reproduction either by seeds or by vegetative means, such as rhizomes, stolons, or bulbils. Virtually all seed plants do produce seeds, and so this is an important process in the ecology of a species. In short-lived plants, especially annuals, yearly production of seeds can be vital to maintain a population, although the **soil seed bank** can ensure a population does not go extinct in years of poor seed production. For longer-lived species, frequent seed production is less important, because survival of the plants themselves maintains the population between years.

Seedling establishment involves germination of the seed and initial growth and survival of the seedling. This is a very risky stage in the life cycle and often a large proportion of seeds die before they can establish seedlings, e.g. 0.06% of the 3 million seeds from a heather (*Calluna vulgaris*) bush and 9–35% of the 60 seeds from a yellow rattle plant (*Rhinanthus minor*) produce seedlings. Where conditions are particularly favourable, such as weeds in an arable field, the resulting high proportion of seedling establishment can result in huge plant densities. More generally, many seeds succumb to **predators** and **pathogens** either on the plant or in the soil and few of the remainder end up in sites suitable for germination or seedling survival, and so die or enter the soil seed bank (where they will die over time if they cannot germinate). This is probably the main evolutionary cause of the large seed production of many plants (**see: Evolutionary ecology; Number of seeds – ecological perspective**).

Sites or microsites in which seedling establishment can take place are often defined by a set of values of particular environmental variables (the niche). The first process is that a seed encounters particular environmental conditions that trigger germination. Studies have determined a number of these triggers which are found in many species: water, sufficient light, a high red:far-red ratio in the light spectrum (**see: Light – dormancy and germination; Phytochrome**), high temperature, increased temperature fluctuations over time and high soil nitrate concentrations. The precise values of these variables needed to trigger germination vary among species. (**See: Dormancy breaking;** entries under **Germination**) These responses to triggers are thought to have been determined by evolution such that a seedling emerges into conditions that enhance its survival and growth. In general for plants, favourable conditions are sufficient resources – water, nutrients and light. These determine the suitability of a habitat for a species (e.g. very low soil water in an arid area, or low nutrients in sandy soil prevent establishment by certain species). They also determine which microsites are suitable for seedling establishment within a habitat occupied by a species. Within-habitat suitable microsites can be summarized by the concept of the gap. This is an area free of competing plants which otherwise reduce resources below that needed by the seedling. The germination triggers described above are all characteristic of a gap compared to closed vegetation. A leaf canopy reduces the light intensity and its red:far-red ratio, decreases the temperature and insulates a microsite against temperature fluctuations through the day. Plant roots in a microsite will also extract water and nitrate from the soil.

Occasionally the gap concept does not work; in some dry habitats, plant cover maintains a high humidity that favours seedling establishment compared to the aridity of the gap. In most habitats, however, all species show increased seedling establishment in gaps, and larger gaps increase establishment. There is some evidence that species show differences in their responses to gaps. It is a popular textbook contention that species with small seed **size** show stronger responses to increased gap size, suggested by the idea that they have fewer seed resources which might support growth in unfavourable conditions. (**See: Mobilization of reserves** entries) In fact, there is a theory which links the maintenance of high diversity in communities such as tropical forests to this suggested seed size–gap size relationship. However, there is only slight evidence to support this simple relationship and the theory is in dispute.

Some species have more specialized regeneration requirements and these are again linked to conditions which trigger germination. **Parasitic plants** germinate in response to chemicals derived from host plants (**see: Stimulants of germination**). Seeds of some species with fleshy fruits which are dispersed by animals will only germinate after passing through an animal's digestive system (**see: Dispersal**). Seeds with **elaiosomes** (appendages rich in nutrients which induce dispersal by ants) show better germination after ants have removed them. (**See: Seedcoats – dispersal aids**)

Many ecosystems are burnt regularly in wild fires, e.g. Californian chaparral and sage-scrub, pine forests worldwide, Mediterranean scrub, Australian *Eucalyptus* woodlands and South African fynbos. These create large disturbances, analogous to huge gaps, and many species respond to the standard gap-associated germination triggers. However, many species only regenerate after fires, presumably exploiting the open ground, and respond to particular triggers associated with fires. These include: fruits or cones (held on the plant or on the ground) which open and release seeds only after severe heating in a fire (**serotiny**, e.g. in many *Pinus* spp. and *Protea* spp.); germination stimulated by extreme heat-shock (>100°C, e.g. hoary ceanothus (*Ceanothus crassifolius*), rock rose (*Cistus laurifolius*); or germination triggered by charred wood or smoke, presumably through chemicals in these substances (e.g. *Emmenanthe penduliflora*, *Syncarpha vestita*). (**See: Germination – influences of smoke; Germination – influences of water**) (JMB)

Baskin, C.C. and Baskin, J.M. (1998) *Seeds: Ecology, Biogeography of Dormancy and Germination.* Academic Press, San Diego, USA.

Fenner, M. (2000) *Seeds – the Ecology of Regeneration in Plant Communities*, 2nd edn. CAB International, Wallingford, UK.

Regeneration – seed collections

The growing out of a seed gene bank collection to generate fresh seed when its viability has reduced to a level (often 75 or 85%), the regeneration standard, below which there is a serious acceleration of viability loss with associated genetic damage and reduction in seed vigour. One aim of regeneration is to ensure that the alleles present in the initial seed collection are carried forward to the new one, particularly by minimizing genetic selection and genetic drift (loss of alleles through random chance). Another aim is to isolate the flowering individuals during regeneration from contaminant pollen. Some definitions of regeneration include the growing out of a seed collection to generate a larger seed collection (multiplication). (RJP, SHL)

Registered seed

An official term used in North America for the progeny of **Breeder** or **Foundation** seed, that is to be used as the source of **Certified** seed, and is approved and certified by an official seed certification agency to comply with specified standards of purity and quality. A term equivalent to '1st generation Certified Seed' in the **OECD Seed Scheme**.

Relative dormancy

A form of **primary** or **secondary dormancy** that is expressed in a given species under a specified set of environmental conditions, such as elevated temperatures. (**See: Dormancy – acquisition**; Fig. D.47 in **Dormancy breaking – temperature**)

Relative humidity (RH)

The ratio of water vapour pressure in the air to the saturation water vapour pressure of that air, at a given temperature and atmospheric pressure, and expressed as a percentage. Where not expressed as a percentage, the value is termed 'water activity'. Relative humidity is closely similar to percentage saturation, which is the percentage ratio of the moisture content of the air to the saturation moisture content of that air, at a given temperature. Control of relative humidity is an important factor in **drying of seed for storage**.

(RJP, SHL)

(**See: Equilibrium relative humidity**)

Reporter genes

These are genes typically derived from microbes, insects or jellyfish that have been engineered to function in a transgenic plant host. They are frequently used in experimental research studies of gene expression in seeds. The reporter gene encodes a protein (often an enzyme) that can be readily assayed; in a transgenic plant, its activity is readily distinguishable from endogenous activities. The β-glucuronidase (*GUS*) gene is a common reporter gene derived from *E. coli*. In order to study transcription, a common strategy is to create a **chimeric gene** that fuses the coding region of a reporter gene (e.g. *GUS*) to transcriptional elements (the 5′ region upstream of the genetic code for a particular protein/enzyme) of a gene of interest. The construct is introduced by transgenic technology into the plant host. Sensitive assays allow for detection of *GUS* activity in transgenic cell extracts (the fluorometric assay) and in cells, tissues and organs of a transgenic plant host (the histochemical assay, in which a blue precipitate forms in the presence of *GUS* activity). Thus, it is possible to assess both quantitative aspects of gene transcription (e.g. the amount of gene transcription driven by the **promoter**) as well as qualitative aspects of gene transcription (e.g. the temporal-, cell-, tissue- and organ-specificity of gene transcription).

(ARK)

(**See: Transformation; Transgene**)

Reproductive structures

In the **angiosperms** (the flowering plants), the reproductive structures are borne within the flowers. These are highly variable in their composition, and this variability is important in the taxonomy of families and species of flowering plants. A complete flower (i.e. one with all of its major components represented) is composed of four major structures set in whorls on a receptacle, a structure present at the end of the flower stalk (pedicel) (Colour Plate 16A). The outer two whorls are made up of sterile (accessory) structures that do not form sexual gametes.

These are the outermost sepals, structures that cover the petals until their expansion; the petals are in the next innermost whorl. In some species separate petals and sepals cannot be distinguished and there is just one whorl, the tepals.

The next innermost whorl is the stamens, which are the male reproductive structures, producing the male gamete-containing bodies, the pollen, and the innermost whorl is the gynoecium, in which the egg(s) is produced.

In some species the flowers do not include one or more of the sterile structures. Also, there are flowers that are either male or female; if both entirely male and female floral structures occur separately within the same plant the species is monoecious, or if they occur in different plants, it is dioecious.

1. Female

The terminology associated with the female reproductive structures is subject to some inconsistencies, and hence a somewhat simplistic overview is presented here. The female reproductive apparatus is a modified leaf (sporophyll), and is termed the carpel. Although there is debate among evolutionary plant biologists about the appropriateness of this term, it is still widely used. Hence the carpel is that part of the flower that bears the ovules, in which the egg resides, and includes the stigma and style which are important in pollination. The carpel itself is composed of several parts (Fig. R.2):

- funiculus (funicule): cavity in which the ovules reside;
- ovary: ovule-bearing part of the carpel;
- ovule: consists of the nucellus, embryo sac, in which the egg cell resides, and integuments; this gives rise to the seed.
- placenta: ovule-bearing region of the ovary wall;
- stigma: pollen receptive surface;
- stipe: basal stalk to which it is attached to the receptacle (absent from some species); and
- style: region between the stigma and the ovary (absent from some species).

There are numerous variations of carpel, stigma and style structures and types, each of which has its own specialized terminology, which can be found in the reference in Fig. R.2.

Frequently there is more than one carpel in a flower, and they may be separate, or more commonly fused, sometimes with a common stigma and style, or with separate ones. The total aggregation of carpels within the flower is called the pistil, or the gynoecium. A simple pistil contains one carpel in the central whorl of the flower, and a compound pistil more than one. The ovary may be borne on top of the receptacle, in which case it is called a superior ovary, or if within the receptacle, below the level of insertion of the other whorls, an inferior ovary. The ovule is composed of several parts (Fig. R.2):

- chalaza: distal end from the micropyle;
- embryo sac: female gametophyte in which the egg (gamete) is borne;
- integuments: outer covering of the ovule, with an inner and outer integument, which form the seedcoat;

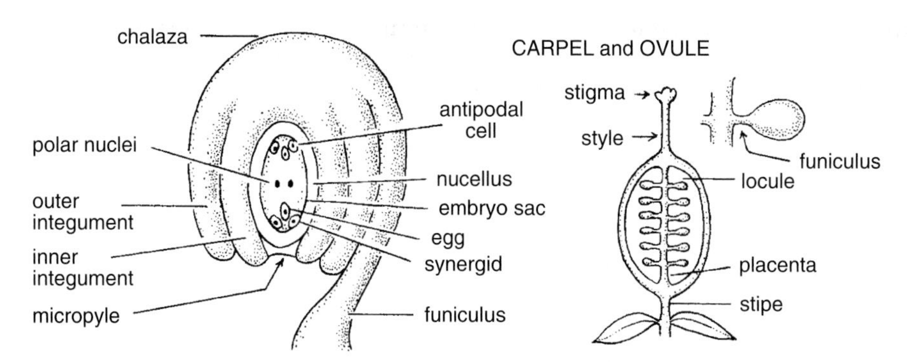

Fig. R.2. A carpel (right) with insert showing attachment of an ovule to the placenta, and an ovule (left) with its component parts. From Radford, A.E., Dickison, W.C., Massey, J.R. and Bell, C.R. (1974) *Vascular Plant Systematics.* Harper and Row, New York, USA.

- micropyle: gap in the integuments through which the pollen tube grows and effects fertilization; and
- nucellus: central mass of tissue in which the embryo sac is embedded (evolutionarily, the female (mega)sporangium).

Several types of ovules are identified, depending upon the orientation of the ovule body in relation to the funiculus and micropyle (Fig. R.3).

Formation of the embryo sac within the nucellus starts with a single diploid (2n) cell, the megasporocyte (now less frequently called the **megaspore** mother cell), undergoing reduction division (meiosis) to produce four haploid (n) megaspores (Colour Plate 16B). Three of these abort, and the remaining cell undergoes mitosis to form two nuclei; mitosis then continues to produce four and then eight haploid nuclei. One of these nuclei forms the female gamete, and lies within the egg cell, which is located at the micropylar end of the embryo sac, flanked by two other haploid cells, the synergids. Three nuclei migrate to the opposite end of the embryo sac, and form the antipodal cells, which play no further role in reproduction. These cells represent the evolutionary remnants of the **gametophyte** generation, which is predominant in the less advanced or lower plants such as the mosses (**see: Evolution of seeds**). The two central nuclei within the embryo sac are the polar nuclei. This structure of embryo sac is called the *Polygonum* type and is typical of more than 70% of the angiosperms. But there are a wide number of variations on this mature embryo sac structure. The common element is that they all contain a single egg cell, but just to detail two of the ten variations, in some species the antipodal cells degenerate and the two central nuclei fuse to form a diploid maternal nucleus (*Oenothera* type, as in *Arabidopsis*), and in others the antipodals divide to form up to 11 cells (*Drusa* type, as in maize).

The ovule develops into the seed after fertilization. Fertilization is achieved when a sperm nucleus (male gamete) from the pollen fuses with the egg cell to restore diploidy, and a double fertilization is unique to the angiosperms when a second sperm nucleus fuses with the polar nuclei to form a triploid (3n) nucleus, with one paternal and two maternal complements of chromosomes (Colour Plate 16B). The egg develops into the embryo and the triploid nucleus into the endosperm (**see: Development of embryos – cereals; Development of embryos – dicots; Development of endosperm – cereals; Development of endosperm – dicots**). The integuments develop into the seedcoat (**see: Seedcoats – structure**), and the ovary develops into a fruit coat (**see: Fruits**).

Development of the female gametophyte in the gymnosperms is considerably different, with the eggs being borne in archegonia within the megagametophyte in the ovule, which develops into the seed after fertilization (**see: Gymnosperm seeds**).

2. Male

The male gametes are formed in the pollen grains which are produced within the anthers, specialized structures borne at the end of a stalk-like structure, the filament; both structures together compose the stamen. Each anther is made up of four chambers, the pollen sacs, in which are present many microsporocytes (now less frequently called the **microspore** mother cells). These diploid (2n) cells undergo meiosis to produce four haploid (n) microspores (a tetrad), each of which becomes a pollen grain (i.e. the microgametophyte). Late in the development of each grain the nucleus undergoes a further mitotic division, to produce two (haploid) cells, one of which

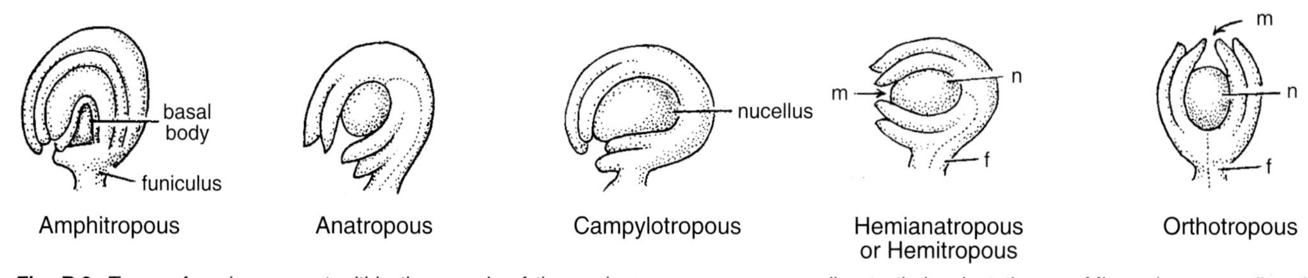

Fig. R.3. Types of ovules present within the carpels of the angiosperms, names according to their orientation. m: Micropyle; n: nucellus; f: funiculus. From Radford, A.E., Dickison, W.C., Massey, J.R. and Bell, C.R. (1974) *Vascular Plant Systematics.* Harper and Row, New York, USA.

will subsequently divide to produce the two male gametes or sperm cells.

Thus from each of many diploid microsporocytes within the pollen sac, four haploid pollen grains are produced, and each of these contains two haploid nuclei, one in the gamete-forming generative cell, and one in the tube cell (Colour Plate 16B). The pollen grains now mature, forming a surrounding two-layered coat (the intine and exine layers), and become packed with starch or oils. Pollen grains are variable in size (20–250 μm in diameter), shape, and surface features, characteristics that can be used to identify species. Pollen grains are released by dehiscence of the anthers, and may be dispersed by wind, insects, water or even animals. Upon landing on the receptive surface (stigma) of a compatible female gynoecium (**pollination**), the pollen grain puts out a pollen tube through pores in the pollen wall which grows towards the egg cell. The two cells migrate down the pollen tube towards the egg, which is located within an embryo sac, and in the process, the generative cell undergoes a further mitotic division to form two sperm cells. In the literature, the terms nuclei and cells are used synonymously with respect to the components of the pollen and embryo sac. It is the nuclei that fuse, and hence the more common use of terms like generative and sperm nuclei; but it is generally held that they are contained within cells with very thin walls. **Fertilization** is effected when one of the sperm nuclei fuses with the nucleus of the egg cell to produce the diploid zygote, from which the embryo develops (Colour Plate 16B) (**see: Development of embryos – cereals; Development of embryos – dicots**). Double fertilization in the angiosperms results in the fusion of the nucleus from the other sperm cell with the polar nuclei to form the triploid nucleus, from which the endosperm develops (**see: Development of endosperm – cereals; Development of endosperm – dicots**). Gymnosperms also produce pollen, but they do not form a second sperm nucleus, since there is no double fertilization, and because there is no equivalent flower structure to the angiosperms, the mode of pollination is such that much less pollen tube growth is required (**see: Gymnosperm seeds**).

(JDB)

Research seed species - contributions to seed science

Seeds of many of the crop and other species are particularly significant in seed science or plant biology because of the findings that have emanated from their experimental use. A selection of these species is listed in this article, in relation to topics covered elsewhere in this encyclopedia. It should not be concluded, however, that the particular process or phenomenon was researched exclusively with the species mentioned, or that key discoveries were not made using other species.

1. Alfalfa

Alfalfa has been used for the study of somatic embryogenesis, storage and deterioration, and the participation of **abscisic acid** and the osmotic environment in late seed development and in maturation. (**See: Deterioration and longevity; Maturation – controlling factors; Somatic embryogenesis**)

2. *Arabidopsis*

Since the mid-1970s *Arabidopsis* has emerged as the major model, experimental plant in molecular genetics and molecular plant physiology research (**see: Genomics; *Arabidopsis***), including discoveries that have contributed greatly to our understanding of seeds. In virtually all cases, the particular value of this species has rested on the vast array of **mutants** that has been generated, and has led to the identification of wild-type genes and the processes that they regulate.

(a) *Embryogenesis, seed development, vivipary.* **See: Development of embryos – dicots; Development of endosperm – dicots; Development of seeds – an overview; Embryogenesis**

(b) *Seedcoat development and function.* **See: Dormancy – genes; Seedcoats – development**

(c) *Acquisition and maintenance of dormancy.* **See: Dormancy – acquisition; Dormancy – genes**

(d) *Maturation, desiccation tolerance.* **See: Desiccation damage; Desiccation tolerance – acquisition and loss; Desiccation tolerance – macromolecular stabilization; Maturation – changes in water status; Maturation – controlling factors**

(e) *Hormone biosynthesis.* **See: Abscisic acid; Brassinosteroids; Gibberellins; Hormone mutants; Jasmonates**

(f) *Transduction of light, temperature and hormonal signals.* **See: Signal transduction – hormones; Signal transduction – light and temperature**

(g) *Phytochrome and its action in seeds.* **See: Light-dormancy and germination; Phytochrome**

(h) **Source of gene sequences, expressed sequence tags (EST), genomics, proteomics. See: Germination – metabolism**

(i) *Fruit (pod) dehiscence.* **See: Shattering**

3. Barley

Because it is the major malting grain, barley has long been the subject of investigations on endosperm mobilization and modification. Also, as an important animal feed grain, its protein storage reserves have been much studied. Major research contributions arising from the use of barley grains are:

(a) *Synthesis and mobilization of aleurone layer cell and starchy endosperm reserves and endosperm modification.* **See: Malting – barley; Malting – process; Mobilization of reserves – a monocot overview; Mobilization of reserves – cereals; Peptide transport; Starch – mobilization; Starch – synthesis**

(b) *Gibberellin signal transduction.* **See: Mobilization of reserves – cereals; Signal transduction – hormones**

(c) *Storage proteins, mutants.* **See: Mutants; Protein superfamilies; Storage protein; Storage protein – intolerance and allergies**

Barley is one of the first species in which embryogenesis from microspores (pre-pollen) were obtained (**androgenesis**) (**see: Haploid embryogenesis; Microspores**).

4. Brassica

The brassicas include important oilseeds (**see: Brassica oilseeds**), so much research has been carried out on them, most of which is concerned with breeding. But the genetic work on qualitative modification of the fatty acid composition connects directly with seed science (**see: Oils and fats – genetic modification**). Studies of storage protein have also employed brassica seeds (**see: Essential amino acids; Storage protein**). Molecular studies on the regulation by maturation drying of the expression of a storage protein gene from *Brassica* (napin) contribute to knowledge about the role of dehydration in seed biology (**see: Maturation – effects of drying**).

The Brassicaceae family also contains *Arabidopsis* (see section 1).

5. Castor bean

Castor bean has attracted interest because of its seed storage oil composition, which contains an unusual fatty acid, ricinoleic acid. Pioneering work on oil mobilization was carried out with these seeds, which have also featured in studies of maturation drying and the regulatory effects of drying on seed physiological processes.

(a) *Oils and fatty acids.* **See: Fatty acids; Fatty acids – synthesis; Oils and fats – mobilization**

(b) *Maturation drying and related processes.* **See: Maturation – changes in water status; Maturation – effects of drying**

(c) *Phytin synthesis.* **See: Phytin**

6. Cotton

Important information about seed development and maturation comes from research with cotton seeds. Particularly significant is the discovery of the late embryogenesis abundant proteins (LEAs) (**see: Late embryogenesis abundant proteins**). Also, contributions to knowledge about chilling injury were made using these seeds (**see: Chilling injury**).

7. Fenugreek

The endosperm cell walls of this endospermic legume are rich in galactomannans. The seed has therefore featured prominently in investigations of their biosynthesis and mobilization. (**See: Galactomannans; Hemicellulose – mobilization; Hemicellulose – synthesis**)

8. Legumes (e.g. common or French bean, broad bean, soybean, peanut)

Seeds of different legume species, because of their relatively high content of protein and oil reserves, have featured conspicuously in studies of these compounds, reaching from the early classification of types of storage protein to the molecular biological analysis of the relevant genes and their regulation. Indeed, legume seeds were arguably responsible for the birth of plant molecular biology. Because of the intensive synthesis of reserve protein in developing legume seeds it was expected that their mRNAs would also be abundant and that the seeds would therefore be a good source of these compounds. Legume-seed (particularly *Phaseolus vulgaris* and soybean) reserve protein mRNAs were therefore among the first to be isolated from plants and the corresponding cDNAs the first to be cloned, preceding the isolation and characterization of the genes themselves and the identification of promoter sequences involved in the regulation of their expression. Early work in transgenic technology was carried out with storage protein genes.

(a) *Storage protein types, evolution and deposition.* **See: Osborne fractions; Storage protein; Storage protein – synthesis**

(b) *Regulation of storage protein genes.* **See: Storage reserve synthesis – regulation**

(c) *Nutritional and other uses.* **See: Legumes; Soybean; Storage protein – processing for food**

(d) *Allergens, antinutritional proteins.* **See: Amylase inhibitors; Lectins; Phytohaemagglutinin; Protease inhibitors; Soybean lectin; Storage protein – intolerance and allergies**

(e) Much of our knowledge about the role of ethylene (ethene) in germination and dormancy comes from research on peanut seeds. (**See: Germination – influence of gases**)

9. Lettuce

Lettuce has been a key species for investigations of seed germination and dormancy since at least the 1920s, and was for many years almost the 'standard' experimental material for such studies. Effects of the seedcoat, of various chemicals and physical factors and of applied hormones were all followed principally using lettuce, but its great distinction lies in its pivotal role in the discovery of phytochrome. With this seed was elucidated the stimulatory action of red light on germination, the inhibition by far-red, red–far-red photoreversibility, the phytochrome action spectrum, the escape time, dark reversion and the high-irradiance response (HIR) – all of which gave birth to modern views of photomorphogenesis in plants. The possible mechanisms of phytochrome action, especially the involvement of gibberellins, have also been investigated using lettuce seeds. Current perceptions of the action of the endosperm in regulating radicle emergence and of endosperm weakening through hydrolysis of cell wall mannans are founded on information coming from research initiated on these seeds.

(a) *Light effects, phytochrome.* **See: Light – dormancy and germination; Phytochrome**

(b) *Effects of chemicals, applied hormones.* **See: Germination – influences of chemicals; Germination – influences of hormones**

(c) *Coat effects.* **See: Dormancy – coat imposed**

(d) *Endosperm and radicle emergence.* **See: Germination – radicle emergence; Hemicelluloses**

10. Maize

One of the most important contributions to biology in which maize grains featured is in the discovery of **transposons** – not strictly a topic of seed science but rather of genetics. Sometimes referred to as 'jumping genes', transposons are sections of DNA that can move to different positions in the genome of a cell, thus altering the amount of DNA in a part of the genome and possibly also causing mutations. Their existence was deduced in the 1940s by Barbara McClintock, at first through cytogenetic observations on the patterns, frequency and timing of chromosome breakage in maize, and then through genetic studies of the pattern of expression of unstable mutants affecting seed coloration. It was already known that when a certain genetic line of maize that normally produces pale whitish grains was crossed with another line with similar grains, the developing grains of the offspring gradually begin to show coloured patches (e.g. pink, red, brown, purple, dark blue, due to the production of different types of anthocyanins) – groups of cells that have arisen from a single cell. She explained the basis of this variegation in terms of transposable insertions that disrupted genes, but were excised frequently during development, thus restoring gene function: a factor from one of the parents has modified (reactivated) a gene in the other, inducing pigment production. Evidence later emerged for the widespread occurrence of transposons in plants, animals and microorganisms (see, for example, section 12 below). Though the importance of McClintock's work was not recognized for over 30 years, she received the Nobel prize for these discoveries in 1983.

Maize, and its mutants, have also been prominent in studies of grain storage protein and starch. The *vp* mutants were among the first **viviparous** mutants to be found. Their discovery helped to initiate our perceptions of the importance of **abscisic acid** biosynthesis and signal transduction in seed biology. Maize mutants also contribute to studies of embryogenesis.

(a) *Transposons.* See: **Mutants**

(b) *Protein, starch.* See: **Amylose; Phytoglycogen; Starch; Starch – synthesis; Storage protein; Storage protein – synthesis; Sweetcorn**

(c) *Viviparous mutants.* See: **Maturation – controlling factors**

(d) *Embryogenesis.* See: **Development of embryos – cereals**

11. Oat

The cultivated oat (*Avena sativa*) has been employed in several studies of dormancy but its close relative, the wild oat (*A. fatua*), has received special attention because of several different genetic lines, developed by J.M. Naylor and his colleagues, that show a range of dormancy types, ranging from a relatively shallow coat-imposed dormancy to a deep embryo dormancy. Studies on the inheritance in the dormancy types have contributed to knowledge about the genetics of the phenomenon. (See: **Dormancy – genetics**)

12. Pea

The most important contribution of the pea seed in biology undoubtedly arises from its use by Mendel to investigate inheritance, from which the basic laws of genetics were derived. Interestingly, the famous 'wrinkled' pea of Mendel links to modern seed science through research on starch synthesis. The 'wrinkled' allele was characterized relatively recently as a mutation (by a transposon – see section 10 above) in the starch-branching enzyme responsible for amylopectin synthesis. In present-day seed science the pea seed, including several mutants, is employed in studies of several aspects of starch biochemistry. (See: **Amylopectin; Wrinkled pea**)

13. Rice

Rice accompanies barley as one of the pair of cereal grains most intensively studied in respect of reserve mobilization. The grain has also been prominent in investigations on the types, synthesis and processing of storage proteins. (**See: Mobilization of reserves – cereals; Osborne fractions; Storage protein; Storage protein – synthesis**) Rice is the first crop to have its genome sequenced (**see: Rice – cultivation**) and because of this it is likely to gain in popularity as a model seed system.

14. Sorghum

Sorghum grains have been used for investigations of storage protein types and synthesis. (**See: Storage protein; Storage protein – synthesis**) Studies with this species are also contributing to the elucidation of preharvest dormancy and sprouting. (**See: Preharvest sprouting – mechanisms**)

15. Soybean

As a very important oil- and protein-storing seed, soybean has been used for investigations of several aspects of these storage reserves, such as protein characterization (**see: Storage protein**), protein processing (**see: Storage protein – processing for food**), protein allergenicity and anti-nutritional properties (**see: Soybean lectin; Storage protein – intolerance and allergies**) and modification of fatty acid composition (**see: Oils and fats – genetic modification**). In seed physiology, studies using soybean have contributed to our understanding of maturation, storage, deterioration and chilling injury (**see: Chilling injury; Deterioration and longevity; Maturation – changes in water status**).

16. Sugarbeet

Sugarbeet is a well-known example of a seed whose germination is severely delayed by naturally occurring inhibitors. (**See: Germination – influences of chemicals; Germination inhibitor**)

17. Tobacco

Knowledge about endosperm weakening involving the action of β-1,3-glucanase, including the action of light, gibberellin and afterripening thereon has been advanced through research in which tobacco seeds have been utilized (**see: Dormancy breaking – temperature; Germination – radicle emergence**). Studies with the closely related *Nicotiana plumbaginifolia* have contributed to perceptions of the role of abscisic acid in dormancy (**see: Dormancy – acquisition**). Tobacco plants are frequently used in transgenic technology. Aspects of seed science that have been elucidated by means of this technology include the effect of maturation drying on reserve protein gene expression (**see: Maturation – effects of drying**).

18. Tomato

Much of the research that is clarifying the roles of gibberellin, abscisic acid and hydrolytic enzymes (e.g. mannanases) in the weakening of the endosperm cell walls that allows radicle emergence to proceed has been done using tomato seeds. Mutant seed – abscisic acid or gibberellin deficient – are particularly valuable for these studies, as is the ease with which seeds can be dissected. Molecular studies on the endosperm and embryo in relation to germination have been carried out with tomato seeds. (**See: Germination – molecular aspects; Germination – radicle emergence**)

19. Wheat

As a starchy seed, wheat has featured in studies of the synthesis and mobilization of this reserve (**see: Mobilization of reserves – cereals; Starch; Starch – synthesis**). Research on many aspects of storage protein has been done with wheat (**see: Osborne fractions; Storage protein; Storage protein – intolerance and allergies; Storage protein – processing for food; Storage protein – synthesis**). (MB)

Respiration

Respiration in the presence of oxygen (**aerobic respiration**) involves the controlled oxidation of organic compounds to CO_2 and H_2O. Numerous compounds are used as substrates for respiration, including carbohydrates, lipids, proteins, amino and organic acids. A large amount of free energy is released during respiration, which is conserved as adenosine triphosphate (**ATP**). This chemical bond energy can be used to drive metabolic reactions required for seed germination and seedling establishment.

The primary pathways of respiration are glycolysis (Embden-Meyerhof pathway), the pentose phosphate pathway (hexose monophosphate shunt) and the citric acid cycle (tricarboxylic acid (TCA) or Krebs' cycle) to which is linked the electron transport chain by which oxidative phosphorylation occurs. Glycolysis and the pentose phosphate pathway are present in the cytoplasm of the cell (and the latter in **plastids**), whereas the citric acid cycle and electron transport chain occur within the **mitochondria** (**see: Cells and cell components**). The pentose phosphate pathway provides sugar intermediates for the synthesis of other compounds (e.g. ribose for RNA), reducing power in the form of **NADPH**, and 6C (six carbon) and 3C sugars for conversion by glycolysis to 3C pyruvate.

The hexose sugars glucose and fructose (**see: Carbohydrates**), derived by hydrolysis of sucrose, are common initial respiratory substrates, and these are converted to 3C sugars by glycolysis and pyruvate (pyruvic acid) by enzymes of the glycolysis pathway (and to a lesser extent the pentose phosphate pathway). ATP is produced by 'substrate-level' phosphorylation and, like NADH, is formed in the absence (**anaerobic respiration**) and presence of oxygen (aerobic respiration). In anaerobic conditions, pyruvate is converted by alcoholic fermentation (fermentation metabolism) to ethanol and CO_2. In aerobic conditions pyruvate is decarboxylated to acetyl CoA (2C), which enters the citric acid cycle in the matrix of the mitochondrion, after being condensed with the 4C oxaloacetate to form citrate. This 6C citrate is decarboxylated to form the 4C succinate which is converted to oxaloacetate to complete the cycle. Reducing power in the form of NADH is released and this enters the electron transport chain in the inner membrane of the mitochondrion; this chain catalyses the flow of electrons from NADH to oxygen. NAD is reproduced by oxidation and the electrons are transferred to oxygen in a reaction (**redox reaction**) coupled with the synthesis of ATP (oxidative phosphorylation). Aerobic respiration yields 32–36 molecules of ATP per molecule of glucose or fructose respired, but anaerobic respiration yields only two.

The efficiency of ATP synthesis (from adenosine diphosphate, ADP) with respect to O_2 consumption is called the ADP:O ratio and varies with the initial substrate that is respired (ratios may be in the range 1.6–2.7). Lower ratios can be indicative of deficiencies in the phosphorylating ability of mitochondria. Another measurement of the efficiency of respiration (as with the ADP:O ratio it involves using isolated mitochondria) is the respiratory control ratio (RCR), an indication of how tightly respiration is controlled by ADP availablity.

Respiratory quotient (RQ) indicates the molar ratio of CO_2 released by respiration with respect to the amount of O_2 consumed. The complete oxidation of hexose to CO_2 and H_2O results in an RQ of 1, and of fatty acids an RQ of 0.8. Most tissues which do not use fatty acids or organic acids as their major respiratory substrate have an RQ of about 1.0.

The Energy Charge (EC) or adenylate energy charge (AEC) is thought to be an indication of the metabolically available energy which is momentarily present in the adenylate system (ATP, ADP and AMP concentrations) of a living cell, tissue or organ. More recently, however, the concentrations of ADP and Pi available in the cytosol have been regarded as the most important factors in the regulation of mitochondrial respiratory rates. (JDB)

Rest

See: **Quiescence**

Restorer line

See: **Cytoplasmic male sterility**

Restriction enzymes

A class of enzymes that recognizes specific DNA sequences (usually palindromic sequences four, six, eight or 16 base pairs in length) and produces cuts on both strands of DNA containing those sequences only. Used widely in molecular biological research and analysis (for example **see: PCR**).

Rhizobia

An encompassing term for soil bacteria of the genera *Rhizobium*, *Bradyrhizobium*, *Sinorhizobium*, *Allorhizobium*, *Mesorhizobium* and *Azorhizobium*, commonly known as rhizobia. These are of great environmental and agricultural importance because their symbioses with legumes are responsible for most of the atmospheric nitrogen fixed on earth. These microorganisms are soil bacteria able to induce the formation of nodules on most species of the family Fabaceae, and on the non-leguminous *Parasponia*, in which they reduce atmospheric nitrogen to ammonia to the benefit of the host plant. In the absence of leguminous plants, populations of free-living rhizobia are commonly found in soils where they can survive **saprophytically**. Host specificity is summarized in Table R.3. (PMW)

Table R.3. Host (macro-symbiont) specificity with rhizobial micro-symbionts.

Micro-symbiont	Macro-symbiont
Rhizobium leguminosarum biovar. *trifolii*	Clover
Rhizobium leguminosarum biovar. *phaseoli*	*Phaseolus* beans
Rhizobium leguminosarum biovar. *viceae*	Pea, vetch, faba bean
Bradyrhizobium japonicum	Soybean
Bradyrhizobium sp. (*Arachis*)	Peanut
Bradyrhizobium sp. (*Vigna*)	Cowpea, mung bean
Bradyrhizobium sp. (*Lupinus*)	Lupin
Sinorhizobium meliloti	Alfalfa
Mesorhizobium loti	Birdsfoot trefoil

Rhizobial inoculation

Effective nodulation of legume roots occurs in soils that contain a native, or naturalized population of specific bacteria for the legume being planted (**see: Rhizobia**). Many soils, particularly those from which leguminous plants have been absent, are devoid or contain only low numbers of these bacteria, surviving **saprophytically**; and if present, they may be only moderately effective in their nitrogen-fixing capacity. In cases where nodules are absent or appear to be ineffective, or where the native rhizobial strains are sub-optimal in their nitrogen-fixing capacity, maximum fixation levels, and thus optimum crop yields, can be ensured by the introduction of highly effective strains via the use of inoculants.

Rhizobium inoculants are normally applied to seed but granular formulations and some liquid formulations are applied in-**furrow** at the time of sowing. Commercially available inoculant formulations are summarized in Table R.4.

(a) *Peat-based inoculants* have constituted the majority of cultures marketed over the last decades, and their development was primarily due to their capacity to maintain high rhizobial concentrations over extended storage periods. Peat-based inoculants are produced by growing single or mixed strains of selected rhizobia in fermenters and either aseptically injecting the cultures into previously sterilized peat sachets, or by mixing the cultures with non-sterilized peat, and subsequently packaging the mix. Most inoculant manufacturers now produce using the 'sterile' route as the rhizobium populations are generally higher ($>10^9$/g) and product shelf life may be extended to more than one agricultural season. Inoculants produced with non-sterilized peat generally have lower rhizobial numbers and product shelf life is compromised due to growth of non-rhizobium contaminants in the product. Peat-based inoculants are generally sold in sachets ranging from 200 to 1600 g, and application rates for large seed-legumes such as soybean range from 200 to 400 g per 100 kg of seed, and rates for small seeded legumes (alfalfa) range from 200 to 400 g per 50 kg of seed. Such application rates normally deliver 10^5 viable cells on each soybean seed, and 10^3 on each seed of forage legumes – these populations providing effective nodulation and maximum yields in the crop if soil conditions are ideal when the seed is sown.

In most countries where inoculation is a common practice, the regulatory authorities set minimum standards in terms of rhizobial concentration and acceptable levels of contamination. In Canada, for example, soybean inoculants when used at their recommended rate of application must deliver 10^5 viable rhizobia per seed, whereas the French authorities stipulate 10^6 per seed. Such standards have been set on the basis of dose/yield responses in field trials, and concern over the possible presence of pathogens in inoculants has prompted the French authorities to insist that all inoculant products must contain pure rhizobial cultures.

(b) *Liquid inoculants* have been developed primarily for their perceived convenience of application. Early formulations were produced using standard culture media and were generally unstable during storage. More recently, however, liquid formulations have been developed which maintain high-count rhizobial populations over a 2-year period. Such inoculants are gaining popularity in many countries throughout the world, and are now used extensively in North America. They are generally packaged in a plastic film that allows some diffusion of oxygen, and the on-seed stability of such cells suggests a change of physiological state as compared to those cultured in standard growth media. Pack sizes range from 200 ml to 12 l, and application rates vary between 200 and 400 ml/100 kg of soybean seed. In most countries, liquid inoculants are mixed with seed on-farm and the general recommendation is that such seed is planted within a few hours' post inoculation. In North America, some farmers have the equipment for in-furrow application.

(c) *Granular inoculants* have been popular with farmers for their application convenience. Granular formulations are often produced from natural or extruded peat or clay granules by mixing with liquid cultures. The rhizobial concentration attained in such formulations is approximately 5×10^8/g, but this lower specification is compensated by application rates which average 10 kg/h. Granule pack sizes vary from 10 to 20 kg and

Table R.4. Rhizobial inoculant formulations used in agriculture.

Inoculant formulation	Application method/timing
Carrier-based inoculants (peat, clay, coir, etc.)	On-farm application to seed just prior to planting
Liquids	On-farm application to seed just prior to planting or in-furrow application at planting
Granular (peat, clay)	In-furrow application at planting
Pre-inoculants based on carrier-based formulations ± film-forming polymers	Applied to seed by seed distributors, up to several months prior to planting
Carrier-based or liquid co-inoculants containing *Rhizobium* and *Bacillus subtilis*	On-farm application to seed just prior to planting

application is facilitated by means of a granule applicator which feeds the product into the seed furrow of each row through the **planting equipment**.

Commercial products based on freeze-dried rhizobial cells have met with little success. Although freeze-dried cells can maintain viability for extended periods, the drying process has been found to greatly reduce field performance.

Seed inoculation

Seed inoculation is carried out by the farmer by simply mixing the inoculant and the seed together, by hand, using a batch mixer or by metering inoculant on to bulk seed as it is screw-augered into planting equipment. Adhesion of the inoculant to the seed is often enhanced by mixing the inoculant with a sticker such as carboxymethylcellulose or other adhesive, but more recent formulations contain polymers which not only enhance peat adhesion, but also greatly extend the on-seed survival of the rhizobial cells, thereby extending the planting window for farmers from a few hours' post inoculation to 2–3 days. The protective nature of such polymers is equally important in maintaining cell viability when inoculated seed is treated with potentially toxic agrochemicals or in situations where inoculated seed is planted into hot, dry soils where nodulation failures can occur if poor quality products are used.

Pre-inoculant formulations are designed to be applied to seed well ahead of planting, and the process is often performed by seed distributors or seed coaters rather than on-farm. The formulations must therefore be easy to apply and provide on-seed stability of the rhizobial cells for several months prior to being planted. The pre-inoculation of soybean seed has been commercialized in Canada and Austria using a slurry of a high titre peat-based inoculant and film-forming polymers which slow the desiccation process of the rhizobial cells on the seed surface. (**See: Filmcoating**) Such pre-inoculants maintain 10^5 rhizobial cells per seed for approximately 3 months. In Europe, North America and elsewhere, alfalfa seed is commonly coated with mixtures of peat-based or clay-based inoculants, and manufacturers claim that 10^3 viable rhizobial cells can be maintained for 12–18 months. (PMW)

McQuilken, M.P., Halmer, P. and Rhodes, D.J. (1998) Application of microorganisms to seeds. In: Burges, H.D. (ed.) *Formulation of Microbial Pesticides: Beneficial Microorganisms, Nematodes and Seed Treatments*. Kluwer Academic, Dordrecht, The Netherlands, pp. 255–285.

Rhizosphere

The region immediately around plant roots in which the microbial population is enhanced by root exudates. The zone inside the root is termed the 'endorhizosphere'; that on the root surface, the 'rhizoplane'; and that around it, the 'ectorhizosphere'. The microorganisms that develop may be derived from those in the soil, or already present within the seed or on the seed surface. Soil-derived organisms include pathogens (such as those constituting the **seedling disease complex**) or species beneficial to plants (such as **rhizobia** and plant growth-promoting rhizobacteria). (**See: Diseases; Rhizosphere microorganisms**)

Rhizosphere microorganisms

When seeds are planted a population of microorganisms establishes on the seed surface. This population may originate from microorganisms in the soil, those already present within the seed, or on the seed surface (**spermosphere**) itself, and as the seeds complete germination the population develops further, associated with root growth (**rhizosphere**).

The bacterial population that develops in the rhizosphere is typically 10–20 times greater in size than that present in the bulk soil but may be greater than 100 times depending on the plant concerned. The increase in fungi is generally smaller than for bacteria but still occurs consistently.

The driving force for the development of the spermosphere and rhizosphere microbial populations is the loss of carbon compounds from the plant. During imbibition, carbon losses from the seed can be quite large and as germination and emergence occurs, carbon flow from the shoots to the roots and into the soil continues to drive microbial development in the rhizosphere. Levels of carbon loss from the roots of mature plants have been estimated to lie between 10 and 30% of the total carbon fixed by photosynthesis for annual plants but may be even more for some perennial species, illustrating the consistent, significant effect that this carbon loss has on rhizosphere development. The compounds lost from the root include: water soluble exudates such as sugars, amino acids and organic acids; secretions such as polymeric carbohydrates; root debris and dead cells; and gases such as **ethylene** and CO_2.

Of particular importance in horticulture and agriculture are those microorganisms that are beneficial to plant growth. The value of mycorrhizas (**see: Arbuscular mycorrhiza**) and symbiotic nitrogen fixers such as species of *Rhizobium*, *Bradyrhizobium* (**see: Rhizobia**) and *Frankia* are well known, but in the last 20 years it has become clear that some groups, of what have been traditionally considered merely free-living, saprotrophic rhizosphere microorganisms, can have a beneficial effect on plant growth, particularly when applied to the seed or to the rhizosphere. These microorganisms form the focus of this article.

1. Indigenous beneficial microorganisms

The number of beneficial free-living bacteria and fungi reported to improve plant growth continues to increase. Originally, such bacteria, sometimes called plant growth-promoting rhizobacteria (PGPR) were thought to be largely *Pseudomonas* species but now include a huge number of species with *Bacillus* species and *Burkholderia* species featuring strongly (Table R.5). Nevertheless, the most frequently reported genus containing PGPR remains *Pseudomonas*. Similarly, *Trichoderma* remains the fungal taxon most frequently shown to promote plant growth, but numbers of reports of other such beneficial fungi have also increased. Overall, bacteria are still the most important free-living microorganisms reported to promote plant growth.

Depending on the plant species considered, growth promotion has been expressed in various ways but most commonly as increases in germination, emergence, fresh or dry weight of the roots or shoots, root length, yield or flowering.

Table R.5. Examples of free-living bacteria and fungi that improve plant growth.

Microorganism group	Genus or species
Bacteria	*Acetobacter, Aeromonas hydrophila, Azotobacter, Arthrobacter citreus, Bacillus amyloliquefaciens, Bacillus cereus, Bacillus licheniformis, Bacillus polymyxa, Bacillus subtilis, Burkholderia cepacia, Clostridium, Enterobacter cloacae, Flavobacterium, Hydrogenophaga, Kluyvera ascobata, Pseudomonas fluorescens, Pseudomonas putida, Serratia liquefaciens, Serratia proteamaculans, Streptomyces griseoviridis.*
Fungi	*Phoma, Piriformspora indica, Pythium oligandrum, Rhizoctonia solani* (binucleate/nonpathogenic), Sterile red fungus, *Talaromyces flavus, Trichoderma koningii, T. harzianum, T. viride.*

2. Microbial inoculants

Beneficial microorganisms represent a considerable potential resource for improving crop growth and a huge amount of experimental work has been done to demonstrate that application of such microorganisms to seeds, roots, cuttings, tubers, corms, bulbs, soil and growing media can lead to disease control or plant growth promotion. Formulations of these microorganisms are termed microbial inoculants and those that have a disease biocontrol activity are often called biopesticides or bioprotectants. Production, formulation and application of these beneficial microorganisms has been explored widely with the view of producing commercially successful, cost-effective microbial inoculants. Both bacterial and fungal production can be done in liquid fermentation and, more recently, the use of solid-state fermentation has been employed for production of fungal biomass. Having obtained the cells or biomass required there is generally a need to dry and formulate the material unless it is being used immediately. If large-scale field applications are envisaged there are often practical problems in producing the amount required to reach the rhizosphere and achieve efficacy, as well as concerns over costs. Consequently, there has been much interest in the application of beneficial microorganisms to seeds provided they are competent to grow, survive and spread in the spermosphere or rhizosphere. Nevertheless, the relatively harsh environmental conditions present on the seed during commerical seed pelleting and coating processes often kill vegetative cells and this is reflected by the frequent use of bacterial and fungal spores. This in itself can lead to problems as, although the propagules survive, they may not germinate well and allow establishment to occur fast enough to provide control of some pathogens.

One way of improving the colonization of seeds by microbial inoculants may be to incorporate the microorganisms during the **priming** process (which some have termed 'biopriming'). This has been done successfully in the USA for *Trichoderma* species applied through **solid matrix priming** and recently in the UK through drum priming (**hydropriming**) for several bacterial species, including *Pseudomonas fluorescens*. Importantly, some of the bacterial species tested survived the standard drying process that follows drum priming whereas they did not survive direct application through coating procedures. This is a significant breakthrough for addition of microbial inoculants on to seed. The efficacy of microbial seed treatments applied through drum priming for improving plant growth is now being explored.

The number of commercially available microbial inoculants that are applied to seed or soil to improve plant growth or control disease is increasing (Table R.6). Most seed-applied organisms are either drenched on at planting, mixed with the growing medium or added as some form of slurry or dust to the planting box. Only two products, Cedomon™ (*Pseudomonas chlororaphis* MA342 applied to cereal seed) and Rhizovit™ (*Streptomyces* sp. DSM212424 applied to seed of lambs lettuce and sugarbeet) appear to be applied as a coating, but details of the actual procedures used to apply these microorganisms are not clear, presumably because of commercial sensitivities.

The number of microbial inoculants registered as **biopesticides** is considerably less than the number of microbial inoculants sold as plant growth promoters, even though in some cases they work through biocontrol. This reflects the fact that all products claiming disease control capability must undergo full registration in the same way as a chemical pesticide. This costly and time-consuming regulatory requirement is frequently not commercially acceptable for products that will be used in niche markets. Not surprisingly, therefore, some products sold as plant growth promoters or plant strengtheners in some countries actually work by biocontrol but no formal claims of activity via this mechanism are made. Nevertheless, this does then question the safety and reliability of such products, as no independent toxicity and efficacy data are generated.

Environmental pressure to reduce chemical inputs and provide sustainable horticultural and agricultural practices will continue to drive research for microbial inoculants that promote plant growth regardless of the mode of action involved. Numerous potentially useful microorganisms have already been identified and techniques to apply them successfully to seed and soil have now been demonstrated. Important areas for future consideration must be compatibility with existing insecticide and fungicide treatments, especially on seeds, and the potential to apply combinations of more than one microbial inoculant to provide multiple modes of action and broaden activity. This approach may also be applicable in organic farming (**see: Organic seed**). A key aspect in all these areas concerns the regulations controlling the use of microbial inoculants. Currently, it would seem that eventually all microbial inoculants that promote plant growth might soon come under the same sets of regulation as those for chemical pesticides (**see: Treatments – pesticides, 8. Registration and labelling**). Whilst providing a safe approach, the costs of registration may well hinder commercial uptake and use of such beneficial microorganisms in the developed world.

Brief details on some of the beneficial rhizosphere microorganisms (Table R.6) are given below.

(a) *Bacillus*: soil-inhabiting bacterial genus of several species (*B. amyloliquefaciens*, *B. cereus*, *B. licheniformis*, *B. polymyxa* and *B. subtilis*) have individually been shown to be antagonistic

Table R.6. Some plant growth promoting microorganisms (PGPs) or biocontrol agents on or near the market for soil or seed application.

Microorganism	Target pathogen/activity	Host	Method of application	Product name	Distributor/ country available
Bacteria					
Bacillus laterosporus	PGP: phosphate solubilizing organism	Various plants	Seed soak or and drench into sterile potting mix	Bio M8	Bio Organics Ltd, UK
Bacillus pumilus GB34	Soilborne pathogens causing root diseases	Soybean	Dry powder/slurry mix for hopper box	YieldShield	Gustafson, USA
Bacillus subtilis FZB24	*Alternaria, Fusarium, Rhizoctonia, Sclerotinia, Streptomyces scabies*	Potatoes, maize, vegetables, ornamentals	Water dispersable granule added to seed or as soil drench or dip	Rhizo-Plus	KFZB Biotechnik, Germany
Bacillus subtilis GB03	*Alternaria, Aspergillus, Fusarium, Rhizoctonia*	Cotton, legumes	Dry powder, slurry mix with seed	Kodiak	Gustafson, USA
Burkholderia cepacia	*Fusarium, Pythium, Rhizoctonia*	Alfalfa, barley, beans, clover, cotton, peas, vegetables, wheat	Peat-based biomass/ liquid applied to seed in planter box	Deny	Stine Microbial Products/Helena Chemicals, USA
Pseudomonas chlororaphis MA342	*Bipolaris sorokiniana, Drechslera avenae, D. graminea, D. teres, Fusarium nivale, Tilletia caries, Ustilago hordei*	Barley, oats and other cereals	Seed coating	Cedomon	BioAgri, Sweden
Streptomyces sp. DSM212424	*Phoma, Pythium, Rhizoctonia*	Lambs lettuce, sugar-beet	Seed coating	Rhizovit	Prophyta, Germany
Streptomyces griseoviridis K61	*Alternaria brassicicola, Botrytis, Fusarium, Phomopsis, Phytophthora, Pythium*	Seed, root and stem rots; wilt diseases; field crops, ornamentals and vegetables	Drench, spray or through irrigation	Mycostop	Verdera Oy [Kemira], Finland
Fungi					
Fusarium oxysporum (non-pathogenic)	*Fusarium oxysporum*	Asparagus, basil, carnation, cyclamen, gerbera, tomato	Spores, microgranules	Fusaclean	Natural Plant Protection, France
Gliocladium catenulatum j1446	Seed, stem and root diseases	Ornamentals, vegetables and trees	Powder; drench or spray through irrigation system	Primastop (previously Gliomix and Prestop)	Verdera Oy [Kemira], Finland
Gliocladium virens GL-21 (*Trichoderma virens*)	Damping-off and root rots; *Rhizoctonia, Pythium*	Ornamentals, protected vegetables	Granules into soil/ potting mix prior to seeding	SoilGard (formerly GlioGard)	Certis Inc., USA
Penicillium bilaii	PGP; phosphate solubilizing organism	Various plants	Wettable granule		Prophyta, Germany
Pythium oligandrum	*Alternaria, Aphanomyces, Botrytis, Fusarium, Gaeumannomyces, Phytophthora, Pseudocercosporella, Pythium, Rhizoctonia, Sclerotium cepivorum, Tilletia*	Vegetables, fruits, legumes, cereals, oilseed rape, trees and ornamentals	Wettable powder, root and stem drench, spray	Polyversum (formerly Polygandrum)	Biopreparaty, Czech Republic
Trichoderma harzianum	*Fusarium, Pythium, Rhizoctonia, Sclerotium rolfsii*	Flowers, vegetables	Spores + peat + other organic material mixed into potting material at planting or transplanting	Root Pro, RootPotato	Efal Agri, Israel
Trichoderma viride	*Fusarium, Pythium, Rhizoctonia* and many others	Various plants	Powder applied as a dry or wet form to seed or drench; spread/ broadcast over field	Trieco	Ecosense Labs, India

to one or more plant-pathogenic fungi (such as *Alternaria*, *Aspergillus*, *Fusarium*, *Phytophthora*, *Pythium*, *Rhizoctonia* and *Sclerotinia*) or to be capable of forming symbiotic associations with some higher plants. Some strains are commercially available in some countries as biofungicides for application on or with seeds at sowing time.

(b) *Burkholderia cepacia*: soil-inhabiting bacterial species of certain strains have been shown to be antagonistic to one or more plant-pathogenic fungi (such as *Fusarium*, *Pythium* and *Rhizoctonia*). Some of these strains are commercially available in some countries as biofungicides for application on or with seeds at sowing time.

(c) *Gliocladium*: soil-inhabiting fungal genus of several species (*T. koningii*, *T. harzianum* and *T. viride* and related *Gliocladium catenulatum* and *G. virens*) and strains have been shown to be antagonistic to one or more plant-pathogenic fungi (such as *Fusarium*, *Pythium* and *Rhizoctonia*). Some of these strains are commercially available in some countries as biofungicides for application on or with seeds at sowing time.

(d) *Streptomyces*: soil-inhabiting bacteria of several species and strains have been shown to be antagonistic to one or more plant-pathogenic fungi. Some of these strains are commercially available in some countries as biofungicides for application on or with seeds at sowing time.

(e) *Trichoderma*: soil-inhabiting fungi of species (such as *G. catenulatum* and *G. virens*, and closely related *Trichoderma koningii*, *T. harzianum* and *T. viride*) and strains have been shown to be antagonistic to one or more plant-pathogenic fungi (such as *Fusarium*, *Pythium* and *Rhizoctonia*). Some of these strains are commercially available in some countries as biofungicides for application on or with seeds at sowing time.

(JMW)

Benizri, E., Baudoin, E. and Guckert, A. (2001) Root colonization by inoculated plant growth-promoting rhizobacteria. *Biocontrol Science and Technology* 11, 557–574.

Harman, G.E. and Björkman, T. (1998) Potential and existing uses for *Trichoderma* and *Gliocladium* for plant disease control and plant growth enhancement. In: Harman, G.E. and Kubicek, C.P. (eds) Trichoderma *and* Gliocladium *Vol. 2. Enzymes, Biological Control and Commercial Applications.* Taylor & Francis, London, UK, pp. 229–265.

Lynch, J.M. (ed.) (1990) *The Rhizosphere.* John Wiley, Chichester, UK.

McQuilken, M.P., Halmer, P. and Rhodes, D.J. (1998) Application of microorganisms to seeds. In: Burges, H.D. (ed.) *Formulation of Microbial Pesticides: Beneficial Microorganisms, Nematodes and Seed Treatments.* Kluwer Academic, Dordrecht, The Netherlands, pp. 255–285.

Whipps, J.M. (1997) Developments in the biological control of soilborne plant pathogens. *Advances in Botanical Research* 26, 1–134.

www.oardc.ohio-state.edu/apsbcc

Rice

Rice belongs to the genus *Oryza*. The genus includes 20 wild species, widely distributed in the humid tropics and subtropics, and two cultivated ones, *O. sativa* and *O. glaberrima*. The former is now grown commercially in 112 countries on all continents but Antarctica, and provides food to more people than any other single species, while the latter is confined to West Africa. (**See: Rice – cultivation**)

O. sativa rice is characterized by **races** or subspecies, the primary ones being *indica* and *japonica*. These races share some characteristics although it is apparent that the rices of the USA are neither pure *japonica* nor *indica* but result from crosses of these two types. In addition, *japonica* rice is divided into temperate *japonica* and tropical *japonica*. Often, the term *indica* is used to refer to the long-grain, non-sticky rices while the term *japonica* is used to refer to round, sticky rices. Other characteristics of *indica* and *japonica* exist but the grain characteristics are the ones most stressed. There are very many cultivars of the two types.

O. glaberrima, sometimes called African red rice, is cultivated in West and central tropical Africa, but is being replaced by *O. sativa*.

Note that the **wild rice** of the Great Lakes region of Canada and the USA is in a different genus, *Zizania palustris*, though in the same tribe as rice.

(**See: Cereals**)

1. Distribution and economic importance

As far as production quantity is concerned, rice is now second in the cereal 'big three', maize, rice and wheat (**see: Cereals**, Table C.8). Most of rice production is in Asia though substantial amounts are grown in Brazil, Egypt and the USA. It is the main staple food of about 40% of the world's population. In Southeast Asia it provides about 60% of the food intake: in east and south Asia the figure is about 35%. Only about 5% of the total world production enters the international market. In 2001, world exports totalled 1.5 million t at a value of US$241 million (FAOSTAT). (**See: Crop Atlas Appendix, Maps 1, 6, 21, 25**)

2. Origin and domestication

The wild ancestor of rice, *Oryza rufipogon*, grows throughout south and Southeast Asia, and finds of rice at early sites in the region may derive from harvesting of wild grain. The earliest securely dated domesticated rice (*O. sativa*) is from the Yangtze river valley of southern China, dating to about 6000 BC although some scholars are of the opinion that it was cultivated in both China and northern India about 10,000 years ago; there are claims for it being grown about 14,000 years ago, in Korea. Genetic evidence suggests that the two main groups of rice varieties, *japonica* and *indica*, may have been independently domesticated. Rice reached other parts of India 4500 years ago, and Japan 2400 years ago.

3. Seed characteristics and types

See: Cereals, section 4.

The grain is harvested as rough rice or paddy rice indicating the grain encased within the hull. The **hull** is composed of the **lemma** and **palea** (with a high silica concentration by grain maturity) which are removed from the grain only with difficulty, since they are locked together by a 'rib and groove' mechanism. The proportion of hull (by weight) in the rice grain averages about 20%. The **glumes** on rice still remain on a mature grain but they are relatively small. Unlike the other small-grained cereals, the rice grain does not have a ventral crease. Its surface is longitudinally indented where broader ribbed regions of the palea restricted expansion during development. The outer covering of the brown rice grain (section 5) consists of the **pericarp** and **seedcoat** proper. Cells of the epidermis have wavy walls and the testa has one cellular layer, with an external cuticle. The **aleurone layer** in rice is one to seven cells thick, depending on the part of the seed and the type of rice. Together, the aleurone layer and the subaleurone layer contain cellulose, protein, starch and oil. **Starch granules** in the starchy **endosperm** cells are compound, similar to those of **oat**. The **embryo** of the mature rice grain is not firmly attached to the endosperm.

The major varietal types classified (e.g. in the USA) by grain sizes and shapes are long-grain, medium-grain and short-grain rice. Grains of the *japonica* type are short, broad and thick, with a round cross-section; whereas *indica* grains are long, narrow and slightly flattened in shape (Fig. R.4). There is also a range of grain colour types.

The definitions of these different-sized grain types are set by law in the USA and have to do with grain dimensions and proportionalities. Long-grain rice from the USA has long, slender kernels (7–7.5 mm) about 3.5 times their width and weigh 16–20 mg. Medium-grain kernels are 5.9–6.1 mm long, 2.5 to 3 times their width (weight, 18–22 mg) and short-grain kernels are 5.4–5.5 mm long, 1.1 to 2 times their width (22–24 mg). But grain dimensions in all available types cover a wide range: basmati, for example, can be up to 18 mm long. In addition to the main varietal types (long-, medium- and short-grain) there are aromatic rices (e.g. basmati), glutinous rices (which either lack **amylose** or have very low amylose (1%) in the endosperm starch) and various other speciality rices, e.g. arborio, a risotto rice. Glutinous rice is also called sweet rice since it is used to prepare sweetened dishes because of its gelatinization properties which allow it to be moulded as it cools after cooking.

4. Composition and nutritional quality

See: Cereals – composition and nutritional quality

There is such a wide range of different grain types that only generalizations can be made here. Typical contents (fw basis) of **storage protein**, **oil** (triacylglycerol) and carbohydrate (mostly **starch**) are 7, 2.5–3 and 75%, respectively. As in all cereal grains most of the protein is in the starchy endosperm but unlike other cereals, except oat, the endosperm protein is 70–80% **globulin**, which in some respects (e.g. lysine content) has a more favourable amino acid composition. The oil resides in the embryo (**scutellum**) and the endosperm, principally in the aleurone layer cells. Important differences in the nature of the starch – the **amylopectin/amylose** ratio – occur among the different types. At higher percentages of amylopectin, the greater is the water absorption capacity of the starch and the lower is the temperature at which the starch gelatinizes. The proportion of amylose is highest in long-grain rices, typically 23–26%. Because of this, long-grain rice is fluffier after cooking, with separate grains that tend not to adhere. The amylose proportion is lower in medium, short and waxy rice (15–20%, 18–20% and 0% respectively): this makes these varieties more tender with a greater stickiness after cooking. Glutinous rice has an even lower proportion of amylose.

Several vitamins and mineral elements are present in the outer, bran, layers of the grain but during processing these cells are often removed with consequent effects on nutritional quality.

5. Processing and uses

In contrast to maize and wheat, virtually all the rice grains that are produced are used as human (i.e. not livestock) food, principally as whole boiled grains or in processed form such as noodles.

Rice is milled before use. Rice milling (Fig. R.5) requires the removal of outer coverings to produce whole, pure endosperms known as ‘white rice’. To achieve this, whole

Fig. R.4. Rice. Left, intact grains with hulls; centre, polished long-grain type; right, polished short-grain type (image by Mike Amphlett, CABI).

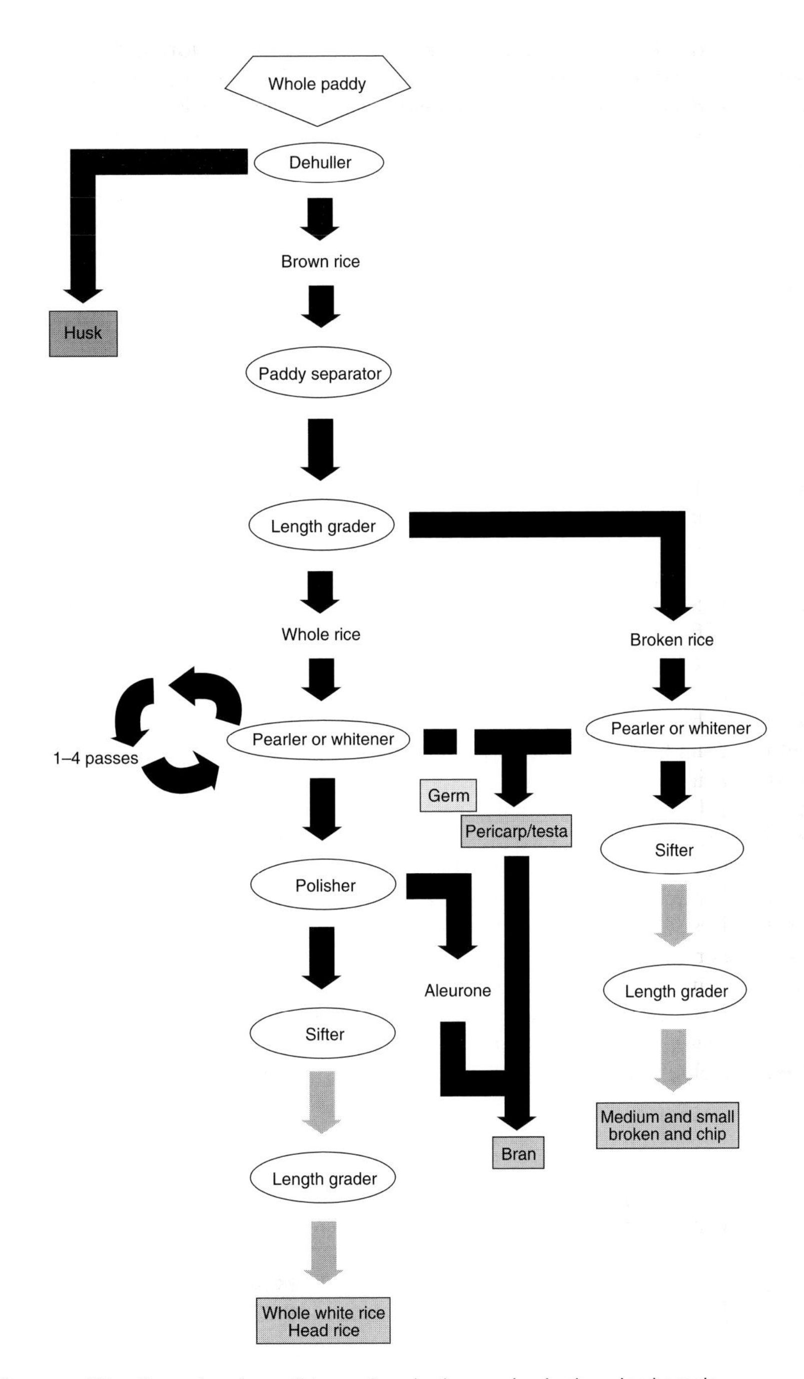

Fig. R.5. Simplified schematic diagram of flow through a rice mill to produce broken and unbroken rice kernels.

grains of 'paddy' (grains as harvested, with lemma and palea, or hull, attached) are passed through rubber rollers to remove the hull. The resulting dehulled or 'brown' rice grains then pass through a series of abrasive mills to remove the bran and embryo. The abrasive agent employed may be in the form of perforated screens or hard sharp particles such as carborundum stuck to a backing. The product, 'white rice', then passes through a polisher, in which the abrasion is much more gentle, to remove remains of the aleurone layer and to give the rice a shiny appearance. Removal of the aleurone layer tissue, with its high fat content, improves storage potential. The milled rice kernel, the primary form of human food, consists primarily of the starchy endosperm, predominantly cellulose, starch and protein. The unbroken milled kernel (or a kernel $\frac{3}{4}$ or more of its original length) is referred to as a head or as head rice.

A process known as 'parboiling' is sometimes applied to paddy before milling. This not only loosens the hull but also causes water-soluble nutrients present in the outer tissues to migrate into the endosperm, improving the nutritional value of the polished product. A parboiled grain retains more of its aleurone layer but the treatment inactivates enzymes and hence improves storage potential.

Removal of the bran layer during polishing results in the depletion of protein from the aleurone layer tissue that contains certain essential amino acids. The bran also contains digestible fibre as well as minerals and vitamins not present in white rice. The incidence of beriberi was dramatically reduced when diets where changed from white rice to brown rice, or even parboiled rice that contains more thiamin. Because of the loss of minerals and vitamins when the bran is removed the practice is often adopted of fortifying white rice with vitamins and minerals. But despite its nutritional value, brown rice is less popular because of its longer cooking time and its texture and taste: the latter deteriorates because the bran layer becomes rancid when the hull is removed. Cracks may develop in the rice kernel either in the field, in drying or from milling. Such kernels may break during milling – in the USA 12 to 24% are affected – but most of the broken rice is removed during milling. That which remains becomes mushy during cooking thus lowering the quality. The collected broken rice is diverted towards use in brewing or animal food.

Sake is a wine fermented from rice grain that is traditionally associated with Japan, and one which there goes back at least until the 3rd century BC. But there is evidence that it was also produced in China at least 2000 years ago, from analysis of containers unearthed from the tombs in Xi'an containing the terracotta warriors. In both countries it was used also in ceremonies and ritual sacrifices. A yellow rice wine (Shaoxing wine) is currently produced in China. There are several classes of Japanese *sake*, for which variables include whether alcohol is added (for cheaper brands), the amount of highly polished rice used as a source for fermentation, and the use of rice-koji, a mould that surrounds the grain and converts the starch to sugars.

6. Market quality of rice

The properties of rice that determine its market value are: anatomy, grain length and width, grain weight, hull and bran colour, kernel whiteness, cracking, percentage of whole grains present, and cooking and eating characteristics – amylose proportion, gelatinization temperature, viscosity, texture and flavour and aroma. (AE, MN, PC, MB)

For information on the particular contributions of the seed to seed science **see: Research seed species – contributions to seed science**.

See: Golden rice

Juliano, B.O. (1985) *Rice Chemistry and Technology*, 2nd edn. American Association of Cereal Chemists, St Paul, MN, USA.

Juliano, B.O. (1998) Varietal impact on rice quality. *Cereal Foods World* 43, 207.

Kohlwey, D.E. (1994) *Rice Science and Technology*. Dekker, New York, USA and Basel, Switzerland.

Luh, B.S. (ed.) (1991) *Rice*, 2nd edn. AVI, New York, USA.

Rice – cultivation

The major cultivated rice species are *Oryza sativa* (Asian rice, subspecies *indica* and *japonica*) and *O. glaberrima* (African rice). Cultivated rice is a warm season crop, grown predominantly in tropical and subtropical climates as a semiaquatic annual grass, although in the tropics it can survive as a perennial, producing new tillers from nodes after harvest (ratooning). More than 90% of the world crop is produced and consumed in Asia. Paddy rice is grown enclosed by bunds or levees, using either irrigated (or flooded, usually <15 cm deep) or rain-fed (or lowland, where water depth can progressively increase to 50–100 cm) production methods, which account for about a half and a quarter of the global rice area, respectively. Deep-water (or 'floating') rice is grown in lowland areas that become uncontrollably submerged to 1–5 m during the rainy season, such as the river basins of Southeast Asia; and upland (or dryland) rice is grown with natural rainfall alone: production methods that represent about 10% and 15% of the world crop area. Distinct from *Oryza sativa* is the so-called '(Indian) **wild rice**', a grass species (*Zizania aquatica* or *Zizania palustris*) native to North America that grows in shallow lakes; it is unrelated to the 'wild' species of *Oryza barthii* and *O. rufipogon*, which can interbreed with cultivated *Oryza sativa* varieties, or contaminate seed production crops. Another unrelated species is 'Indian ricegrass', *Achnatherum hymenoides*. For the geographic origins, evolution, history, world distribution and economic importance **see: Rice**.

1. Genetics and breeding

Rice breeding is extensively conducted in many countries, by both public and private organizations. As it is a highly self-pollinating crop (typically >99%), most breeding strategies have utilized crosses of different parents followed by selection throughout the process of segregation into homozygous lines, using the pedigree method, the bulk method and backcrossing. The breeding of high productivity traits into Asian rice was one mainstay of the **Green Revolution**, and currently these high yielding traits are being crossed into the rugged African landraces/varieties, using **embryo rescue** techniques to produce true-breeding 'NERICA' lines ('New Rice for Africa'). **Hybrids** have also been developed, and are widely used in China, as well as in India and elsewhere. In the USA hybrids are being used but only on a relatively small land area. The primary hybrid production methods exploit **cytoplasmic male sterility**, though a more limited hectarage has been produced using chemical male gameticides applied during microsporogenesis to disrupt the production of male gametes in the female parent plants. Thermosensitive genic male sterility (TGMS), which is currently in the research stage, has the potential to revolutionize hybrid rice production.

In many tropical situations, particularly in upland rice, weed growth can be prolific, and considerable manual weeding is required to manage this major constraint on yield. Breeding efforts have produced cultivars that are resistant to non-selective herbicides. Imidazolinone-resistant rice was produced by EMS-induced mutation and is commercially available. Rice with other herbicide-resistances (which are likely to be commercially available in the near future) and other valuable traits (including reduced **allergens** or enhanced nutritive value) may be produced by mechanisms including **genetic modification**. (**See: Genetically modified organism; Golden rice**)

Due to both its small genome size (about 430 Mb) and its major importance as a world staple crop, rice has become a model plant for genetic mapping and analysis, and restriction fragment length polymorphism (RFLP)- and polymerase chain reaction (**PCR**)-based markers are now available for a wide range of important genes. The International Rice Genome Sequencing Project (IRGSP) in 2005 published the sequence of the *japonica* subspecies (the temperate type). Draft sequences of *indica* subspecies (the tropical type) have also been published independently by both the Beijing Genomics Institute and Syngenta.

2. Development and production

At maturity the plant has a main stem and a number of productive tillers, each of which carries a terminal flowering head or panicle, bearing many single-flowered spikelets. Plant height varies with variety and environmental conditions, from about 0.4 m to >5 m in some floating rice varieties. Most rice cultivars produce 10 to 22 leaves on the main stem. In each structural plant unit (phytomer) that develops during the vegetative phase of crop development, the leaf emerges first with elongation of the blade followed by the elongation of the sheath, elongation of the internode and production of tillers and tiller buds. To some extent, rice can compensate for thin but uniform stands by increased tillering.

The reproductive phase – the initiation, development and exsertion (or 'heading') of the panicle, in growth stages R0 to R3 – is coordinated with the development of the final five leaves and their associated internodes on the culm. Fertilization usually occurs in growth stage R3 and anthesis in stage R4. Caryopsis elongation begins during R5, first attaining maximum grain length, then increasing to its maximum width and, finally, thickness during R6. During R7, the hull becomes yellow, as chlorophyll in the lemma and palea is degraded. At R8 grain drying begins on the plant, reducing grain moisture content from >80 to <20%. Growth stage R9 is reached when all full-sized grains on the panicle have become brown (Fig. R.6). A developing grain normally reaches its mature size about 25 days after it begins to grow, though this varies somewhat with temperature and variety.

Critical to effective seed rice production is the avoidance and removal of off-types and noxious weeds (such as red rice in the USA, a feral type of *Oryza sativa* with extensive **dormancy** mechanisms, and wild rice *O. rufipogon*, in many locations). Considerable efforts must be made to eliminate these by a combination of cultivation techniques and by **roguing**, including hand weeding after the paddy field has been drained before harvest. Irrigation water from a field of one cultivar is not usually allowed to enter a field of another. **Isolation distances** vary from about 5 m in drill-plantings to 15–30 m for broadcast sowing, mainly to avoid the hazard of mixing cultivars at harvest. (The presence of mixtures of grain types – long-grain mixed with medium-grain rice, for instance – can lead to rice being unmarketable.) For hybrid rice, isolation of 40 m from other pollen sources is the usual practice. Current parental lines, however, suffer from low seed-set, due to limitation in pollen production and receptivity. Labour-intensive procedures are practised in China to stimulate pollen shedding, such as pulling and tightening a cord across the field.

3. Harvesting

Rice grain is generally harvested at 18 to 22% moisture content to maximize both 'rough rice yield' and quality, which is determined as 'head rice yield' (the proportion remaining after milling as kernels that are unbroken or at least $\frac{3}{4}$ of their original length). Prompt harvest allows the maximum head rice yield, but harvesting too early (with immature grains in the harvested crop) or too late (thus having grains which have rewetted after reaching their **equilibrium moisture content**) can greatly reduce milling quality. Harvest practices vary greatly. In many less-developed agricultural systems, rice is cut by hand, placed in bundles or shocks, allowed to dry somewhat, and carried to a stationary thresher; further drying is frequently done on hard surfaces, such as on the edge of asphalt highways. In countries with highly mechanized agricultural systems, rice is cut and threshed in combine harvesters, then either dried on the farm or transported to a mill or storage facility where it is dried further.

Physical stresses during drying of rice can be extremely large, leading to damaged grains, and care must be taken to prevent cracking, or checking, of the grain. Rice grain can be safely stored below 13% moisture.

Normally rice seed for sowing is harvested, dried in air at ambient temperatures <40°C, conditioned, stored, bagged, assessed for germination at standard temperatures and planted in the following spring.

4. Quality, vigour and dormancy

Seeds of rice are routinely checked for germination and purity (such as red rice contamination, by examining hulled grain for a red **pericarp**). **Vigour** tests are not usually performed, however. Domesticated white rice generally lacks dormancy but postharvest dormancy can occur, which is usually removed by normal storage during the autumn and early winter. For a discussion of the genetics of rice dormancy, **see: Preharvest sprouting – genetics, 1. Rice**.

5. Sowing

Unlike many cereal seeds, radicle emergence from the seed may be suppressed in rice, depending on the conditions around the root zone. In relatively dry, aerated conditions the radicle emerges first through the **coleorhiza**, followed by the **coleoptile** and the emergence of the prophyll leaf. Under anaerobic conditions, such as when the seed is submerged in water, the coleoptile emerges first, and roots only develop when it reaches aerated regions.

Rice is planted in two ways: directly into the grain production field (direct-seeded), either into a dry seedbed (dry-seeded) or into a flooded field (water-seeded); or into an outdoor or covered nursery seedbed followed by transplanting into the field. A transplanted crop establishes a crop canopy that will inhibit weeds in less time than a direct-seeded crop. The labour demands of transplanting and manual weeding have encouraged the move to direct seeding in irrigated and rainfed lowland rice in many locations.

(a) *Dry-seeded rice.* Seed is sown by broadcasting, in many parts of Asia by hand, and incorporated into the soil. Watering can either be rainfed or, if irrigated, fields must often be temporarily soaked, or flushed, by progressive flooding for several hours followed by draining, to allow the coleoptile to

Growth stage	R0	R1	R2	R3	R4
Morphological marker	Panicle development has initiated	Panicle branches have formed	Flag leaf collar formation	Panicle exertion from boot, tip of panicle is above collar of flag leaf	One or more florets on the main stem panicle has reached anthesis
Illustration	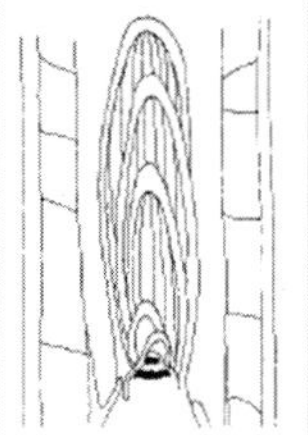	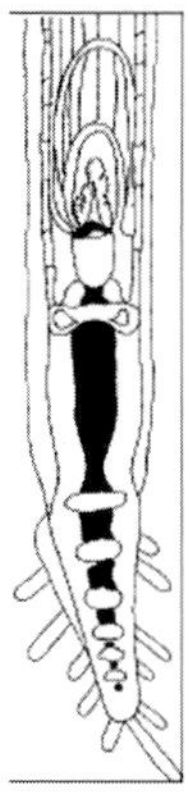		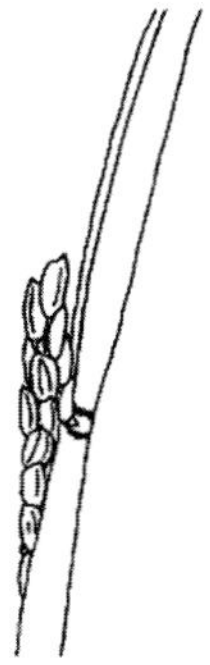	

Growth stage	R5	R6	R7	R8	R9
Morphological marker	At least one caryopsis on the main stem panicle is elongating to the end of the hull	At least one caryopsis on the main stem panicle has elongated to the end of the hull	At least one grain on the main stem panicle has a yellow hull†	At least one grain on the main stem panicle has a brown hull‡	All grains which reached R6 have brown hulls
Illustration					

Fig. R.6. Rice reproductive growth stages with morphological markers. Reprinted from Counce, P.A., Keisling, T.C. and Mitchell, A.J. (2000) A uniform and adaptive system for expressing rice development. *Crop Science* 40, 436–443.

† Determinations of physiological maturity or cessation of dry matter accumulation are difficult or impossible to make in the absence of any known morphological marker.

‡ The brown hull indicates the grain has begun to dry.

emerge. Failure to do this at the proper time can lead to uneven or inadequate plant stands that are insufficient to produce optimum crops. The emergence in large-scale dry-seeded rice fields can sometimes be less than 20% of the seeds sown.

(b) *Water-seeded rice, or 'Wet Direct Seeding'.* This is sown 'into the flood', under which conditions the seed will germinate and emerge fairly well. Prior to this, seed is usually soaked in water for 1 day, allowed to drain for another day, and is then sown by hand, using drum seeders or, in large-field systems where affordable, by aeroplane. One problem with water seeding is that radicle elongation is suppressed so that the plant is not properly anchored to the soil and can be moved by the wind, producing an uneven stand. In some places, therefore, floods are removed to create the relatively dry conditions that stimulate radicle growth.

(c) *Transplanted rice.* Seedlings are grown in outdoor or indoor nurseries until 3–7 leaves have developed on the main stem (the V3–V7 stages), sowing pre-soaked seed into dry soil that is then watered. In the 'Mat Nursery' technique the soil medium is placed on plastic sheets or banana leaves, to minimise the effort in later transport of seedlings to the field and the damage caused to the root system: about 5–10% of the crop area to be transplanted is needed for such a nursery. A 'Reduced Area' nursery requires only 1% of the crop area: seeds are sown at high densities into a flooded and drained seedbed, and seedlings are pulled, bunched and transported to the field. Transplanting is generally done into the flood, and the period between flooding and planting is kept short to give less opportunity for weeds (other than aquatic or semi-aquatic plants) to emerge. There is a fairly severe shock to the plant in transplanting, but roots begin to regrow within a day. In many parts of the world, transplanting is an arduous and labour-intensive manual operation. In Japan, however, mechanized transplanters are now used on almost all the land devoted to rice cultivation, using seedlings raised undercover in box trays, after germination in temperature-controlled incubators.

6. Treatments

In some countries, though more rarely in Asia, both fungicide and insecticide seed treatments are frequently used to enhance stand establishment. Fungicides (various combinations of metalaxyl, thiram, carboxin, fludioxonil and trifloxystrobin) provide protection against **damping-off**, seed **rot** (*Pythium*) and seedling **blight** diseases (*Fusarium*, *Rhizoctonia*, *Helminthosporium*), especially where seed is planted early or environmental conditions are unfavourable to germination and seedling growth (**see: Pathogens; Treatments – pesticides**). Seeds in the USA are treated with a systemic insecticide (fipronil) to control the serious rice water weevil pest, which feeds upon the developing roots in its larval stage along with larvae of the stalk borer, chinch bugs, billbug and the lespedeza worm. In some places, zinc or other micronutrients are included as a seed treatment, and in some places seed is inoculated, after soaking, with *Azospirillum* to stimulate nitrogen fixation.

Gibberellin (gibberellic acid) soaking is applied to much of the rice seed sown in the southern USA, to increase uniformity of emergence, particularly where there is deep seed placement during direct-drilling or early seeding dates. The widely grown semi-dwarf varieties suffer from slow and low final seedling emergence because of their short coleoptile and mesocotyl. (Gibberellic acid, one of the first gibberellins to be discovered, was originally found as a metabolite of the fungus *Gibberella fujikuroi* (*Fusarium moniliforme*), a seedborne pathogen of rice that causes Bakanae or so-called 'foolish seedling' disease, in which leaves and stems elongate so rapidly that the plant collapses.)

As mentioned already, **imbibition** for about a day before seeding – a form of on-farm seed **priming** – is widely carried out, especially for water-seeded rice. On-farm liquid **density sorting**, is a technique of some antiquity that is still performed today – including, for example in Japan, using dilute brine to float off weak seeds before sowing to raise nursery transplants.

(PC)

International Rice Research Institute, Philippines. www.irri.org

Khush, G.S. and Toenissen, G.H. (eds) (1991) *Rice Biotechnology*. CAB International, Wallingford, UK.

McKenzie, K.S., Bollich, C.N., Rutger, J.N. and Moldenhauer, K.A.K. (1987) Rice. In: Fehr, W.R. (ed.) *Principles of Cultivar Development*, Vol. 2. Macmillan, New York, USA.

Smith, C.W. and Dilday, R.H. (eds) (2003) *Rice Origin, History, Technology and Production*. John Wiley, New York, USA.

Rice mill

Seed conditioning equipment consisting of a pair of cylindrical rollers covered with velvet-like material, used to separate rough seed from smooth (also known as a dodder mill or roll mill). (**See: Conditioning, II. Cleaning**)

Ricin

A toxic protein present only in **castor bean** seeds. It belongs to the family of type-2 ribosome-inactivating proteins and consists of an N-terminal A chain (32 kDa) with RNA N-glycosidase activity linked by a disulphide bond to a C-terminal domain (34 kDa) with carbohydrate-binding activity. Ricin is extremely toxic to human and animal cells. The **lectin** moiety of ricin binds to galactose/N-acetylgalactosamine-containing glycoconjugates exposed on the surface of the plasma membrane. After binding, the protein is endocytosed and enters the cytoplasm where the A- and B-chains dissociate, and the free A chains catalytically inactivate the ribosomes. This inactivation relies on the removal of a single adenine residue, which causes a conformational change in the 28S rRNA of the large (60S) ribosomal subunit and prevents binding of elongation factor EF2 to the ribosome. As a result, protein synthesis is arrested and the cell dies. Ricin is lethal for cells at concentrations below 1 ng/ml. Less than 1 mg is enough to kill an adult human by injection but much larger amounts are needed when the toxin is ingested or inhaled.

(EVD, WJP)

(See: Poisonous seeds)

Lord, J.M., Roberts, L.M. and Robertus, J.D. (1994) Ricin: structure, mode of action, and some current applications. *FASEB Journal* 8, 201–208.

R-line

See: Cytoplasmic male sterility

RNA

See: Nucleic acids

Rocket

See: Brassica – horticultural

Rockwool

A mineral material used in horticulture as a hydroponic solid aggregate growth medium, on crops such as tomato, pepper and cucumber. (**See: Hydroponic production**) Rockwool, or 'stone wool', is made from coke and limestone that has been melted at 1600°C, spun into fibres and then in a wool-like fleece woven into mats or slabs, or moulded into preformed blocks or plugs for raising seedling transplants. (**See: Transplants**) (The material is also used for a variety of thermal insulation, fire protection, acoustic control and environmental purposes.)

Roguing

A system of negative selection during plant breeding and seed production. The identification and removal of off-types or other unwanted plant types (e.g. deformed, sick and insect-damaged individuals, or volunteers from an earlier crop) from a population of plants or a cultivar to prevent undesirable cross-pollination during seed production. Roguing is for instance an important step in hybrid seed production plots using cytoplasmic male sterility schemes to ensure crossing between only true-to-type A-line and R-line parents to maintain the purity of the progeny seed. Roguing is also carried out in open-pollinated crops at the final stages of multiplication to maintain cultivar purity. Off-type rogues can be removed whenever they appear during crop development, which can be most crucial before flowering and/or just before harvest. (**See: Production for Sowing, II. Agronomy; Production for Sowing, IV. Harvesting**)

Roll mill

Seed conditioning equipment consisting of a pair of cylindrical rollers covered with velvet-like material, used to separate rough seed from smooth (also known as a dodder mill or a rice mill). (**See: Conditioning, II. Cleaning**)

Rolled towel test

A type of germination test, conducted between sheets of moist creped paper, also known as the 'ragdoll' test. (**See: Germination testing, 1. Substrates**)

Root meristem

A group of cells that are located at the distal end of the **radicle** and adventitious roots. The root meristem serves as the site of root proliferation, producing new cells that differentiate into specific root tissues, i.e. epidermis, cortex, endodermis, pericycle and procambium, and the root cap which protects and lubricates the root as it grows in the soil. Root hairs are produced from the epidermis after germination through interactions with the cortex. (**See: Seminal roots**)

Rots

The disintegration of plant tissue as a result of the action of invading fungi or bacteria; and diseases (head-, root- and stem-rots) characterized by brown or sunken lesions or cankers, often covered with a fungal mycelium or sclerotia. Bacterial soft rots, for example, are associated with seed abortion and 'slime diseases'. Seed rots are characterized by rotting of seed before germination. Fungal foot rots, arising from infected seed or seeds growing in infested soil, affect **hypogeal** seedlings leading to rotting of the lower part of the hypocotyl immediately above the seed in the pre-emergence stage or shortly after emergence. Examples include seedborne foot rot of flax (*Phoma exigua* var. *linicola*) and pea (*Phoma pinodella*, which is part of a complex of fungi associated with *Mycosphaerella pinodes* and *Ascochyta pisi* – the '*Ascochyta* blight leaf and foot rot complex'). Infected seedlings may recover, but with damaged root structures throughout the life of the plant. (PH)

(**See: Disease**)

Rucola

See: **Brassica – horticultural**

Rumination

A mature seed in which the surface of any part is irregular or uneven is called 'ruminate', which means 'chewed'. The original concept was that the endosperm is eaten away by ingrowths of the periphery of the seed; this is not correct, but the term has remained in general use. It usually relates to the **endosperm**, but similar irregularities may occur in the **perisperm**, **covering structures (seedcoat)** or **embryo**; and ruminations can occur on both the inner and the outer surface of the seed. Therefore, each part of the seed (endosperm, **nucellus** or seedcoat) can become ruminate due to uneven growth of its own cells, or by adapting to an adjacent ruminate tissue, or by both.

Ruminate seeds are found in about 30 families, mainly **dicotyledonous**. Seven types of rumination have been distinguished:

(1) **Passiflora type** (Passifloraceae, Oxalidaceae): in bitegmic ovules (having two integuments), the mature seed shows unequal radial elongation by any one of the layers of the seedcoat. In members of the Passifloraceae the inner layer of the outer integument, or the outer layer of the inner integument, or a combination of both, is responsible for this (Fig. R.7b).

(2) **Verbascum type** (Scrophulariaceae): this type resembles the Passiflora type, but occurs only in unitegmic (single integument) ovules. In members of the Scrophulariaceae the endothelial cells show unequal radial elongation.

(3) **Myristica type** (Myristicaceae, Palmae) the pluri-layered seedcoat shows, by localized meristematic activity, definite ingrowths or infoldings which are vascularized (Fig. R.7c, d).

(4) **Annona type** (Annonaceae, Aristolochiaceae, Degeneriaceae, Dipterocarpaceae, Ebenaceae, Menispermaceae, Vitaceae and Palmae): this type resembles the Myristica-type, but the ingrowths, in contrast to the integument, are not vascularized (Fig. R.7a).

(5) **Coccoloba type** (Polygonaceae): this type occurs in one-layered seedcoats of bitegmic ovules (Fig. R.7e).

(6) **Spigelia type** (Apocynaceae, Loganiaceae, Araliaceae, Caprifoliaceae, and Rubiaceae): the rumination is the result of meristematic activity in the tissues of the seedcoat of unitegmic ovules.

(7) **Elytraria type** (Acanthaceae): the rumination is the result of unequal peripheral growth activity of the endosperm during later stages of development.

Ruminate seeds that show a particularly intricate network of lobes when the seed is cut in any plane are called 'labyrinth seeds' (e.g. *Kingiodendron*, Fig. R.7 f, g; *Picrodendron baccatum*, Fig. R.8 a, b). Such seeds, with a complicated network of interconnected lobes formed by the seedcoat are found, for example, in Annonaceae, Myristicaceae and Rubiaceae. Whether rumination brings any advantage to the developing or germinating seed is not clear, although it has been suggested that rumination of the endosperm enlarges the surface between the seedcoat and endosperm which might assist in the uptake of water and/or nutrients from the coat. Ruminate seeds have been found in 32 families of both dicotyledonous and monocotyledonous plants. (JD, WS)

(**See: Seedcoats – structure**)

Corner, E.J.H. (1976) *The Seeds of Dicotyledons*. Cambridge University Press, Cambridge, UK.

Heel, W.A., van (1971) Notes on some more tropical labyrinth seeds. *Blumea* 19, 109–111.

Takhtajan (Takhtadzhyan), A. (ed.) (1985, 1988, 1991, 1992, 1996, 2000) *Anatomia Seminum Comparativa*. Vol. 1–6. Nauka, Leningrad (in Russian).

Werker, E. (1997) *Seed anatomy*. Gebrüder Borntraeger, Berlin, Stuttgart (*Handbuch der Pflanzenanatomie = Encyclopedia of Plant Anatomy*, Bd. 10, Teil 3: Spezieller Teil).

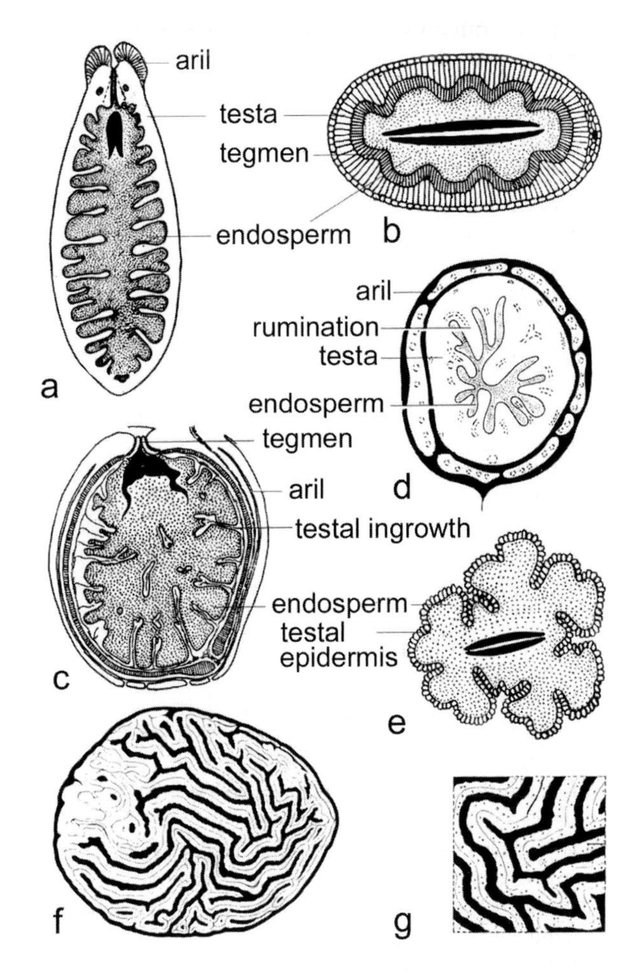

Fig. R.7. Different types of rumination. (a) Transmedian longitudinal section of seed of *Annona squamosa* (Annona type). (b) Transverse section of seed of *Passiflora calcarata* (Passiflora type). (c) and (d) Transmedian longitudinal and transverse section of seed of *Myristica fragrans* (Myristica type). (e) Transverse section of seed of *Coccoloba uvifera* (Cocoloba type). (f) *Kingiodendron pinnatum*, longitudinal section of labyrinth seed and (g) detail of the testa intruding the cotyledons. Plate reproduced from Boesewinkel, F.D. and Bouman, F. (1984) The seed: structure. In: Johri, B.M. (ed.) *Embryology of Angiosperms.* Springer, Berlin, pp. 567–610, with illustrations orginally from Periasamy, K. (1962) The ruminate endosperm: development and types of rumination. In: *Plant Embryology: A Symposium.* CSIR, New Delhi, pp. 62–74 (a, b and e); Voigt, A. (1888) Untersuchungen über Bau und Entwicklung der Samen mit ruminiertem Endosperm aus den Familien der Palmen, Myristicaceen und Annonaceen. *Annales de Jardin Botanique de Buitenzorg* 7, 150–190 (c); and Heel, W.A., van (1970) Some unusual tropical labyrinth seeds. *Proceedings Koninklijke Nederlandse Akademie van Wetenschappen*, Ser. C 73, 298–301 (f and g); d was adapted from Mohana Rao, P.R. (1974) Nutmeg seed: its morphology and developmental anatomy. *Phytomorphology* 24, 262–273.

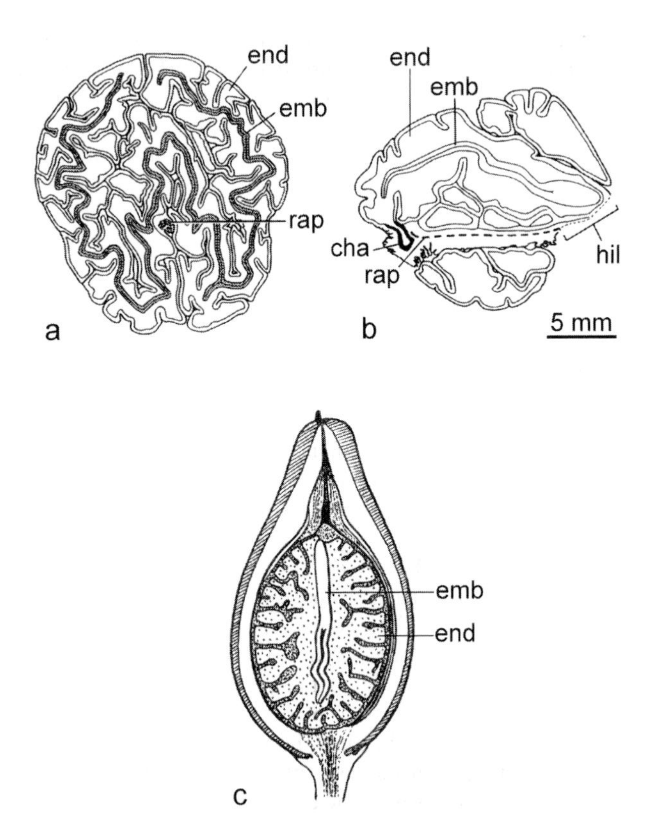

Fig. R.8. Ruminate seeds. (a) and (b) *Picrodendron baccatum* (Picrodendraceae); cross and transmedian longitudinal section through labyrinth seed. (c) Transmedian longitudinal section of the single-seeded fruit (drupe) of *Scytopetalum kleineanum* (Lecythidaceae) (emb = embryo; end = endosperm; cha = chalaza; rap = raphe bundle; hil = hilum). a and b from Stuppy, W. (1996) *Systematic Morphology and Anatomy of the Seeds of the Biovulate Euphorbiaceae.* PhD Dissertation, Dept of Biology, University of Kaiserslautern, Germany [in German]; c from Engler, A. (1897) Scytopetalaceae. In: Engler, A. and Prantl, K. (eds) *Die natürlichen Pflanzenfamilien, Nachträge zum II.–IV. Teil.* Wilhelm Engelmann, Leipzig, Germany, pp. 224–245.

Rusts

Diseases affecting the stem and leaves of cereals and turf grasses, and the pathogenic fungal species of the Uredinales that cause them, notably *Puccinia* spp. such as: *P. arachidis* (groundnut leaf rust), *P. graminis* (stem rust of cereals), *P. hordei* (brown barley rust), *P. purpurea* (rust of grasses and sorghum), *P. sorghi* (common corn rust), *P. striiformis* (yellow rust) and *P. triticina* (brown rust). Rusts generally can be controlled by seed treatment with systemic **fungicides**, or with resistant cultivars. However, achieving and maintaining adequate resistance is difficult due to the ongoing appearance of new **races**. (PH)

Rutabaga

See: Brassica – horticultural

Rye

Rye (*Secale cereale*) is closely related to *Triticum* (**wheat**) (**see: Cereals**). It is cultivated typically in northern regions having a dry summer and a very cold, dry winter. Rye is free-threshing and the only cross-pollinated small-grain cereal crop. There are many legends surrounding rye and its association with St Anthony's fire (a very painful burning sensation of the limbs and extremeties), abortion, witchcraft and poisoning due to contamination of the rye grain with **ergot** sclerotia produced by the fungus *Claviceps purpurea*. (**See: Pathogens**)

1. Distribution and importance

As far as production quantity is concerned, rye is a relatively minor cereal (**see: Cereals**, Table C.8) but it is important in Scandinavia, central and eastern Europe. It is grown almost exclusively in Europe: nearly 95% of worldwide production is in the northern part of the area between the Ural mountains and the North Sea. Main producers are the Russian Federation, Belarus, Poland and Germany. Of the total production in 2002, approximately 21 million t, only about 2 million t were exported at a value of US$162 million (FAOSTAT). **(See: Crop Atlas Appendix, Map 1, 12, 21, 25)**

Because it is a high biomass crop, there are also rye varieties for forage, and there is increasing interest is using the crop for compost production.

2. Origin and domestication

The origins of rye are in the Near East. There is evidence of its cultivation in Neolithic times, but it is not clear if it was a crop in its own right or simply a weedy contaminant of wheat and barley. By about 6000 years ago it had reached Europe, probably as a weed.

3. Seed characteristics

The **hull** is removed during threshing. Rye grains are more slender and pointed than **wheat** grains but they also have a crease and indeed share many of the features described for wheat. The bead-like appearance of the cell walls of the pericarp is less distinct than in wheat. The **caryopsis** measures about 9 mm long and about 3–4 mm wide. The colour is light brown with a bluish tinge (Colour Plate 9A). The starchy **endosperm** constitutes about 80–85% of the weight of the caryopsis (two populations of **starch granules** are present as in wheat), the embryo 2–3% and the outer layers about 10–15%. The aleurone layer is one cell thick. **(See: Cereals, 4. Basic grain anatomy)**

4. Composition and nutritional quality

On a dry weight basis the composition of the grain is protein 10–15%, **oil** (triacylglycerol) 2–3%, **starch** 55–65%, total fibre 15–17% and soluble fibre (β-glucan, arabinoxylan) 3–4%.

The **aleurone layer** is rich in minerals (e.g. manganese, iron, copper, zinc, selenium, magnesium) and vitamins, especially B-vitamins.

5. Processing and uses

More than half of the rye is consigned to animal feed; but the remainder is used as food for humans – pumpernickel, the rye breads and crackers. Rye flour produces a much denser, coarser loaf than wheat and it makes a rather sticky dough due to the pentosans in the grain. Because of the properties of its gluten, rye dough does not rise in the same way that wheat dough does.

Rye flour-milling techniques are similar to the process described for bread wheats (**see: Wheat**). Rye stocks are stickier and fluted reduction rolls are better suited to these. For the same reason purifiers are less used. The granularity of rye flours is greater than that of wheat flour.

After hulling, which generally occurs during threshing, the grains are used whole, cracked or flaked, or they are ground to make flakes or flour. Three grades of flour are produced consisting of decreasing proportions of the outer coat (and therefore 'ash' – mineral elements) which imparts the colour. The dark pumpernickel, rye breads and crisp bread are made of whole grain rye flour (ash content about 2%), where all outer caryopsis layers are included.

Rye whisky must contain at least 51% rye in the fermented mash. It is likely to have been a predecessor of bourbon amongst the early settlers in North America, and George Washington was a notable rye whisky producer. Nowadays its production is very low. Canadian whisky is sometimes mistaken to be a rye whisky, and while it does contain some alcohol derived from this cereal, this is no more than 6 or 7% of the grain used.

The presence of ergot sclerotia, which are very difficult to remove completely, is sufficient to condemn a grain lot as unacceptable for food use. However, in seed produced for sowing, ergot residues can sometimes be sufficiently removed and controlled by a combination of **conditioning** and fungicide seed treatment.

High-yielding hybrid rye varieties dominate the crop in some countries, and they can be sown with a small proportion of conventional varieties to improve pollination and hence grain quality and yield stability.

(See: Triticale – a synthetic rye/wheat cross.)

(AE, MN, DF, JDB)

Bushuk, W. (ed.) (2001) *Rye: Production, Chemistry and Technology*. American Association of Cereal Chemists, St Paul, MN, USA.

Bushuk, W. (2001) Rye production and uses worldwide. *Cereal Chemistry* 46, 70–73.

S

Safeners

In seed **treatment** technology, herbicide safeners are chemicals applied with the intention of reducing or eliminating the phytotoxic effects of specific herbicides that are to be soil-applied close to the time of sowing, without substantively affecting control achieved on the targeted weeds. (In broadest definition terms, safeners are the diverse class of chemicals that reduce the undesired harmfulness of another chemical, on either plants or livestock.)

A major current seed treatment example is of grain- and forage-sorghum varieties that are slow to emerge and hence very susceptible to competition by weeds during the seedling growth stage. Seed treatment with fluxofenim or oxabetrinil enables soil-incorporated metolachlor herbicide to be used to control grass weeds without harming the establishment of that crop. Alternatively, a safener may be co-formulated with its associated pesticide, such as benoxacor in combination with metolachlor to increase the tolerance of emerging maize seedlings to the side effects of the latter, particularly under cool wet conditions.

Many current herbicide safeners act by enhancing the plant's ability to metabolize the target herbicide, such as by enhancing the activity of cytochrome P450 mono-oxygenases and glutathione-*S*-transferases. This type of safening technology provides an alternative to the breeding of herbicide-tolerance traits into varieties, such as by **genetic modification**. (Parenthetically – one of the currently proposed **chemically inducible gene promoter** systems is based on using a herbicide safener as the external trigger.)

Seeds can also be safened against certain phytotoxic effects of seed-applied fungicide and insecticide active ingredients on germination by interleaving barrier-coating layers, using **pelleting** or **filmcoating** techniques to separate the chemicals from direct contact with the seed. (PH)

(**See: Treatments**)

Davies, J. and Caseley, J.C. (1999) Herbicide safeners: a review. *Pesticide Science* 55, 1043–1058.

Safflower

1. The crop

Safflower (*Carthamus tinctorius*, 2n = 24, Compositae or Asteraceae) is a much branched annual plant, related to **sunflower**, but with many heads (5–50) that vary in diameter from 1.2 to 4.0 cm containing 20–180 **florets**. The crop's production is limited to those regions that have rains or irrigation for the early part of the growing season followed by dry atmospheric conditions during flowering and maturation. Such conditions are needed to avoid serious losses from foliar diseases. Thus, production is limited to such areas in the major producing countries of India, USA, Mexico, Ethiopia, Argentina and Australia. (**See: Oilseeds – major** Table O.1)

2. Origin

Safflower is thought to have originated in the desert areas of the Middle East and from there spread down to the horn of Africa and across northern India to southern China. The Spaniards introduced safflower to Central and South America while its introduction to Australia is of recent origin. The crop has a long history of cultivation, primarily for the orange-red dye extracted from its florets. Traditionally the crop was grown on field borders, in small plots or around homesteads. Although India, the world's largest producer, has grown safflower for oil for over 100 years the crop was not seriously investigated as an oil source until the USA became interested in the 1930s and 1940s, first as an industrial and then as an edible oil source.

3. Fruit and seed

Safflower seed is a small (up to 10 mm long) irregularly pear-shaped **achene** with a smooth, grey to white, slightly shiny surface (Fig. S.1). Hairs or bristles may occur on the seed borne at the flower's centre. Seed weight and oil content vary widely from 11 to 105 g/1000 and 11 to 47%, respectively, with commercial varieties averaging about 33% oil. The **epicarp** of the **hull** consists of longitudinally elongated, thick-walled cells, free of hairs and pigment. Below the epicarp are nine or more rows of irregularly shaped stone cells. Similar cells occur within the irregular network of the **phytomelan** layer. The **endocarp** is made up of an overlapping layer of elongated, thick-walled stone cells. The **testa**, which constitutes about 15% of the hull weight, consists of two to three layers of loosely

Fig. S.1. Safflower achenes (scale = mm) (photograph kindly produced by Ralph Underwood of AAFC, Saskatoon Research Centre, Canada).

packed, pitted parenchyma with squashed cells below. The **endosperm** and **embryo** have the same histological features as sunflower seeds.

4. Uses

Safflower oil gained prominence in the 1960s on reports that its high linoleic acid content (77%), the highest of any vegetable oil, was nutritionally desirable (**see: Fatty acids; Oilseeds – major,** Table O.2 for composition). Selection of varieties with a thin hull and resulting higher oil content improved the crop's economics. The development of high oleic acid (75%) varieties also created an additional niche edible oil market. The seed is normally processed without dehulling and the oil is marketed primarily as a speciality salad and cooking oil or for use in pharmaceutical products. The high fibre meal, with about 25% protein, is fed to cattle. (KD)
(**See: Oils and fats**)

Knowles, P.F. (1989) Safflower. In: Robbelen, G., Downey, R.K. and Ashri, A. (eds) *Oil Crops of the World*. McGraw-Hill, New York, USA, pp. 363–374.

Smith, J. (1996) Safflower oil. In: Hui, Y.H. (ed.) *Bailey's Industrial Oil & Fat Products*, Vol. 2, 5th edn. John Wiley, New York, USA, pp. 411–456.

Vaughan, J.G. (1970) *The Structure and Utilization of Oil Seeds*. Chapman and Hall, London, UK.

Salicylic acid

Now recognized as a potential **growth regulator**. Its presence in the bark of willow tree (*Salix*) is at the origin of its name. It belongs to the large group of **phenolic compounds** and is likely synthesized from *trans*-cinnamic acid. Exogenously supplied salicylic acid or its close analogue aspirin inhibit seed germination in several species, e.g. barley grains that have some dormancy, and **soybean** and **rice** seeds (**see: Germination – influences of hormones**), though it is reported to promote germination as a seed treatment in carrot, and to induce systemic acquired resistance in **cucurbits**. (DC, FC)

Samara

A winged **achene** or more accurately a dry, indehiscent **fruit**, whose **pericarp** (fruit coat) extends around the fruit in the form of a wing, e.g. *Ulmus* spp., *Fraxinus* spp. (**See: Fruit – types; Seedcoats – dispersal aids**)

Sampling

Seeds and grain produced for sale to farmers, growers, food and industrial manufacturers and the general public. Seeds are subject to **quality testing** at several stages in their production – at harvest, processing, transfer, storage, loading and delivery – such as for moisture content to establish the required degree of drying after harvest and changes throughout storage, for quality control and to ensure that they meet regulatory and contractual standards and customer expectations. Similarly, **grain inspection** samples are used in determining product density, moisture content, foreign material, commodity quality and other essential factors needed for classification and pricing. As it is not possible to examine all the seed from a seed**lot** or a grain bulk (a 25 t wheat lot, for example, contains about 500 million seeds), it is necessary to remove a small sample for analysis. No matter how accurately any analysis is carried out, it can only show the quality of the sample submitted for testing, so every effort must be made to ensure that the sample submitted represents the seedlot from which it has been drawn. The objective is therefore to obtain a sample of a size suitable for tests, in which only its level of occurrence in the seedlot or grain bulk determines the probability of a constituent being present. A representative sample provides results that can be reliably used to predict the likely upper and lower limits of the quality of a seedlot or grain bulk. Field sampling procedures are also integral to seed production under **certification schemes**.

There are several types of seed samples that are used for quality testing, and these are detailed in Table S.1.

ISTA Rules and AOSA Rules (**See: International Seed Testing Association; Association of Official Seed Analysts**) give detailed procedures for sampling seed in both the warehouse and the laboratory covering:

- sealing the seedlot;
- marking/labelling the seedlot;
- sampling intensities/frequency;
- instruments to be used to draw primary samples;
- sample size to be submitted;
- procedures to be used to reduce the size of the composite sample;
- marking, sealing and packing the sample to be submitted;
- laboratory procedures to reduce the submitted sample to working samples required for the different seed quality tests that are to be carried out. (**See: Quality testing**)

Seed Marketing Regulations make reference to 'International Sampling Methods' or quote from the ISTA Rules when specifying procedures. Likewise, Official Grain Inspection Regulations prescribe sample procedures and sample method and size, sample ratio, sampling frequency and pattern, identification, sealing, storage procedures, proportionality of blending, sample bags, etc.

The sampling instrument used (Table S.2) depends on whether the seed is in containers (and if so, what type) or sampled on-stream. The instrument used must be capable of obtaining seed from all parts of the container or bulk.

1. Spears and probe samplers

For free-flowing seed, perforated probes or 'triers' of various designs, long enough to sample all portions of the container, are used to take samples by spearing the seed mass.

Table S.1. Types of seed samples used for official quality testing.

Name	Description
Primary	A small portion drawn from a single position in a seedlot or grain bulk.
Composite	Formed by combining and mixing all the primary samples taken from a seedlot or grain bulk.
Submitted	The sample submitted to a testing laboratory, which may comprise either the whole or a representative sub-sample of the composite sample.
Working	A representative sub-sample taken from the submitted sample in the laboratory, on which tests are carried out.
Sub-	A representative portion of a sample, obtained by a laboratory sample-reduction method.

Table S.2. ISTA-approved instruments for sampling seedlots stored in containers. For diagrams of these instruments see Figures below.

Container type	Approved instruments
Closed bags or sacks	Dynamic spear (Nobbe trier) Walking stick (pointed end)
Small open bags or sacks	Deep Bin Probe (small Neate sampler) Walking stick (pointed end or blunt end)
Bulk or large containers (over 100 kg)	Deep Bin Probe (large Neate sampler) (+ extension) Cargo sampler (+ extension)
Seed stream	Automatic sampler

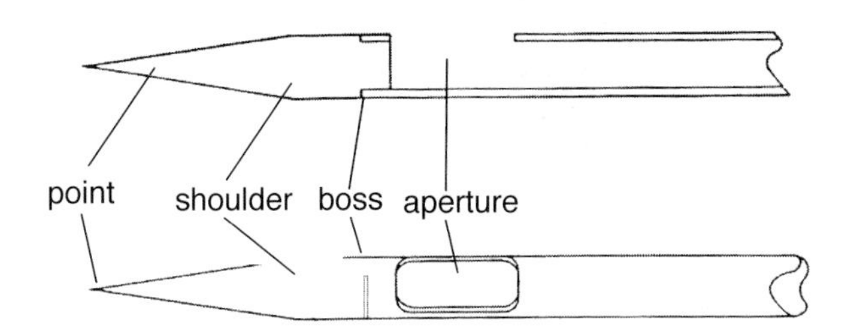

Fig. S.2. The Dynamic Spear is inserted in a near horizontal position into a sack of seed with the aperture facing down, rotated to point the aperture upwards, and then withdrawn at specified rates, to allow seed to pass down the tube and exit at the handle end. It is preferable that the point of the spear passes to the opposite side of the bag from insertion, and is withdrawn at a uniform rate. Different dimensions (of the aperture, point, outside diameter, and so on) are specified for different kinds of seed.

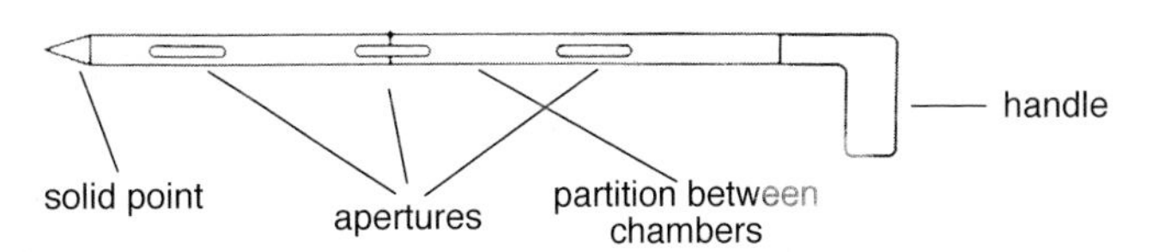

Fig. S.3. The Walking Stick Sampler probe is inserted diagonally into sacks or containers (<1 m deep) – usually in a vertical plane to sample seed in open sacks, but also in a horizontal plane position on closed sacks. Another version of this instrument has a rounded tip for sampling from open bags or open containers.

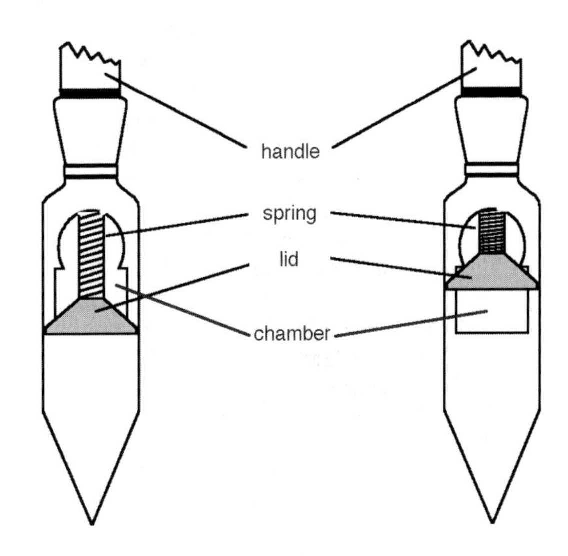

Fig. S.4. The Cargo Sampler consists of a long metal or wooden pole at the end of which is a large metal collecting chamber, with an aperture that opens and closes by means of a spring-loaded lid. When the instrument is pushed vertically into a bulk of seed (>30 cm deep) the lid is kept closed by the pressure of the spring and the aperture remains closed; on withdrawal the lid opens and seed enters the chamber.

(a) *The dynamic spear (Nobbe trier)*, suited for sampling bags and sacks, is essentially a tapered hollow metal tube with a solid point at one end and a short slot opening or aperture along one side; turning the tube, so that the hole is uppermost, allows seed and any contaminants to enter the tube and flow out of the end.

(b) *Stick samplers* consist of two tubes, one inside the other, with matching slot openings arranged in a straight line or spirally, which permit seed to flow into the inner tube as the holes are aligned. Further turning of the inner tube closes off the holes again, the probes are withdrawn and the collected seed emptied for examination. In principle, the instrument should be capable of penetrating every part of the container. The Single Chamber Type, also called the Deep Bin Probe or 'Neate Sampler', can be used for sampling seed in open sacks, large containers or in bulk. In the Multi Chamber type or Walking Stick Sampler (Fig. S.3), seed enters through a number of apertures into separate compartments; the content of each separate chamber is a Primary Sample. Large versions of these triers, about 1.5–3.5 m long, with up to about 20 partitioned inner-tube compartments, are used for sampling grain from trucks and rail-cars for grain inspection.

(c) *Cargo samplers – Bullet or Torpedo probes*, sometimes called deep bin cups (Fig. S.4), are used in grain inspection and consist of a short hollow tube or container that collects a small amount of grain (about 400 ml is typical). A quick upward pull opens the sampling chamber so it can fill with grain from one narrow layer.

(d) *Vacuum probes* are used in grain inspection in commercial stores and at grain shipping terminals and, although expensive and massive, are often the only way to effectively sample the bottom of bins, silos, tanks, trucks, rail-cars and other bulk containers. Equipment may be hydraulically operated, and in some cases samples are automatically delivered by pneumatic tube systems.

2. Seed stream samplers

Seed stream samplers, whether automatic or manual, must uniformly sample the entire cross section of the seed stream without any subsequent loss of the seeds that have entered the sampler. The seed stream should flow uniformly and constantly and there must be enough space around to move the instrument through it. Timing devices on automatic devices should be set so that the frequency and duration of sampling meets requirements of any rules or regulations governing the sampling operation. In systems with an open flow of the seed at the end of processing sometimes manual sampling is the most convenient solution, for example at the outlet of a **conveyor**.

(a) *Pelican samplers* are scoops used for taking primary samples from a free-flowing vertical stream, for example at the outlet of a conveyor belt. One or two handles are attached to a frame, which holds open the mouth of a narrow canvas bag, and the pelican is swung completely through the entire seed stream in one continuous motion. Larger-scale automated pelican samplers are mounted in grain spouts, at the end of belts, or at the head of **elevator** legs, to draw samples periodically under the control of a timing device, which are then passed to a secondary sampler to reduce the sample size.

(b) *Hoe samplers* are simple devices to sample from a horizontal, constant stream, such as a rubber belt conveyor, based on moving a straight-edged hoe perpendicular to the movement of the seed stream from one side to the other in one movement, to obtain a primary sample. This sampling technique can only be used when the conveyor belt has a smooth surface and is not running too fast.

(c) *Cup or jug samplers* are manual sampling devices to obtain samples from a falling stream of grain, such as hand-held scoops constructed of lightweight aluminium, such as the so-called Ellis cup.

3. Sample reduction

The composite sample, obtained by combining the primary samples taken from different parts of the seedlot, is often too large to be sent directly to the laboratory and has to be reduced in size to give the submitted sample. In the laboratory the submitted sample has to be further subdivided to give the working samples used for various laboratory tests. The sample reduction methods used must not introduce more variation than would be expected by simple random sampling. Most sample reduction methods are based on two principles: the sample is thoroughly mixed and then divided.

(a) *Conical divider* or *Boerner divider* consists of an inverted pointed cone over which the seed falls under gravity and on to a series of baffles formed in a circle of alternate channels and spaces of equal width, which ultimately directs the seed into two spouts.

(b) *Soil/riffle divider* consists of a series of channels or chutes arranged side by side, that direct alternately into two equal portions in collecting pans on opposite sides. This divider is available in different sizes and designs.

(c) *Centrifugal divider* uses centrifugal force to distribute the seed across a stationary baffle, which divides the sample into two parts.

(d) *Rotary dividers* subdivide seed into 6–10 equal sub-samples, either by feeding it on to a distributor within a rotating crown (100 rpm) from which it is dispensed to collecting bottles simultaneously, or by feeding directly into the inlets of the collecting bottles rotating below the inlet cylinder. Similar designs are used to subdivide grain samples.

(e) *Manual methods*. If at all possible, mechanical sampling reduction methods are preferred to hand methods because they are more independent of the person and therefore more robust. However, in the case of species that are very chaffy and uncleaned seed, mechanical reduction methods may not be possible. In addition, in the case of seed health testing for pathogens that might cross-contaminate other samples, hand reduction methods are essential. Examples of hand methods include modified halving, the spoon method and as an example, the 'hand halving method', in which seed is poured evenly on to a smooth clean surface, thoroughly mixed into a mound with a flat-edged spatula, divided with a knife or ruler and recombined in such a way that the sample size is repeatedly halved. (RD)

Anon. (1997) *Grain Inspection Handbook*. US Department of Agriculture (Grain Inspection, Packers And Stockyards Administration, Federal Grain Inspection Service).

Sand

Graded and washed sands are components of some seedling growing media/'**compost**' and commercial **pelleting** blends, and are used by themselves as planting media in some **germination testing** protocols.

Geologically, sand is defined as a sedimentary material, finer than a granule and coarser than silt, comprised of rounded or angular particles between 0.06 and 2.0 mm in diameter, which is formed by the weathering and decomposition of rocks. As a major component of soils, sand is a major determinant of textural classification, and a major influence on the mode of seedling growth in soils under environment stresses. (**See: Soil types; Seedbed environment**). The most abundant mineral constituent is silica (SiO_2), usually in the form of quartz, though many other minerals are often present in small quantities.

Saprophyte

An organism that feeds on dead or decaying organic matter (derived from the Greek meaning 'rotten plant'). (**See: Diseases; Health Testing; Storage fungi**)

Sarcotesta

If the outer integument or testa partly (e.g. only the mesotesta, as in Magnoliaceae) or entirely (e.g. *Baccaurea*) differentiates into a fleshy tissue it is called a sarcotesta. (**See: Seedcoats – structure**)

Saurochory

Dispersal of the diaspores (seeds and fruits) of a plant by reptiles.

Scalping

A process of precleaning, in equipment that shakes or rotates screens with openings large enough to allow the desired seed to pass through and larger foreign material (crop debris, soil) and weed seeds to be removed. Harvested seed is often aspirated or scalped (using a 'scalperator') to permit uniform airflow through the seed mass, which if restricted will cause inadequate or uneven drying and seed quality problems. (**See: Conditioning, I. Precleaning; Production for Sowing, V. Drying, Pre-drying** activities)

Scarification

In seed processing and seed testing: any treatment that abrades or otherwise alters a fruit or seedcoat, and so stimulates the germination of viable, dormant seeds. (**See: Dormancy – coat imposed; Dormancy breaking; Germination**

testing; Tree seeds) The term is also used in the different context of loosening compacted soil during seedbed preparation followed, for example, by broadcast seeding – **see: Planting equipment – metering and delivery**.

Many dry fruits or seeds possess a coat whose structure or composition prevents, or at the very least significantly retards, the germination of otherwise live seeds. Such coats can inhibit germination for several reasons:

- act as an impermeable barrier to water uptake;
- act as an impermeable barrier to respiratory gas exchange;
- provide mechanical resistance against embryo growth and emergence;
- contain chemical inhibitors;
- act as an impermeable barrier to the escape of inhibitors from the embryo;
- modify the quality of light reaching the embryo;

or any combination of the above. (**See: Seedcoats – functions**)

It is generally held that the first three causes of seed (or fruit, as appropriate for the species) coat inhibition are the most common, and that the majority of scarification techniques act by: (i) increasing the permeability of seedcoats to water; (ii) increasing the permeability of seedcoats to respiratory gases; or (iii) lowering the mechanical resistance of the seedcoat to embryo emergence. Less commonly, treatments may reduce or remove the influence of chemical inhibition: for example, removing a seedcoat which contains germination inhibitors or, in other instances, rendering a seedcoat permeable to the escape of chemical inhibitors from the embryo within. Occasionally, scarification techniques overcome the light filtering properties of some seedcoats which impede germination.

(a) *Natural scarification*. In nature, seedcoat scarification can take place in several ways. For example, when seeds are eaten, teeth can fracture the seedcoat. As seeds pass through a bird or animal digestive tract, their coats can be subjected to mechanical abrasion (in the bird crop), chemical digestion (by stomach acids) and microbial breakdown (in the animal hind gut). Other natural scarification processes include fire, fungal and microbial breakdown (in soil, though there is no compelling evidence for this), and repeated freezing/thawing and wetting/drying cycles. (**See: Dormancy – ecology**)

(b) *Artificial scarification*. Numerous artificial scarification techniques are used to improve seed performance. One method of describing, classifying and presenting these treatments is shown in Table S.3, which divides them into 'physical' methods (either 'mechanical' or 'thermal') that remove, thin, split, disrupt or puncture a seedcoat; and 'chemical', 'biochemical' and 'biological' methods that dissolve or degrade the seedcoat or open up chemical or cellular plugs in seedcoat apertures such as the hilum, micropyle, lens and strophiole. Some, such as nicking, are more suited for germination tests, or small valuable research seedlots. (PG)

(**See: Conditioning, I. Precleaning**)

Scarlet runner bean

1. World importance and distribution

The scarlet runner bean (*Phaseolus coccineus*) is grown in North and South America, Europe, Asia and Africa. It is found predominantly in inhospitable situations where the more demanding string bean (*Vigna unguiculata*) does not thrive. Among the grain legumes it is a species of relatively minor importance. (**See: Legumes**)

Table S.3. A method of 'classifying' scarification methods and their likely action.

Scarification method	Action on seedcoat or sealed aperture
Physical treatment	
Mechanical	
Cut, nick, chip, clip, pierce, puncture	Localized removal/disruption of seedcoat
File, grind, sandpaper, abrade, scarify	More general removal/disruption
Shake, tumble (+/- abrasive)	General disruption/fracture of seedcoat
Heat	
Diurnal fluctuation(s), warm oven, microwave	Split seedcoat, disrupt cellular or chemical plug
Fire, hot-ashes, dying embers, hot oven	General burning/fracture of seedcoat
Soldering iron, hot wire	Localized burning of seedcoat
Chemical treatment	
Acids	
Sulphuric acid (concentrated or diluted, usually <1 h), and washing	General chemical degradation of seedcoat, or a specific sealed aperture
Solvents	
Water, ethanol, methanol, acetone (and many others)	General dissolution of waterproof properties of seedcoat, or sealed aperture
Biochemical treatment	
Purified enzyme	Degradation of entire seedcoat or specific sealed aperture
Biological treatment	
Microbial/fungal	Degradation of entire seedcoat or specific sealed aperture
Combined treatments	
Heat + solvent	
Boiling/hot water, sometimes followed by steeping	Degradation/softening of entire seedcoat or specific sealed aperture

2. Origin

Runner beans are native to the cool, misty, mountain regions of tropical Central America and Mexico. There is evidence that they were gathered, probably for food, about 9000 years ago and cultivated at least 2000 years ago in Mexico.

3. Seed characteristics

Seeds are relatively large (>2 cm long) with a large, white, flattened and somewhat incurved **hilum**. Seed colours include white, shining black to violet-black with deep red (oxblood to carmine) mottling (Colour Plate 2F).

4. Nutritional quality

Seeds contain approx. 20% fw protein, 62% total carbohydrate (mostly **starch**), 1.8% **oil** (triacylglycerol), the B vitamins and several mineral elements (e.g. calcium, iron, phosphorus). Antinutritionals are also present, such as **protease inhibitors**, but their action is minimized by boiling the seeds.

5. Uses

Both the pods and the green shelled bean are edible in the fresh stage. Dry seeds, especially the white ones, are used for human consumption.

Because of their very decorative red, pink, white, red and white flowers, runner beans are sometimes grown as ornamentals.

6. Plant types

Runner beans have **hypogeal** emergence. They are perennials with a vigorous climbing, twining habit and can reach 3.5–4.5 m. The flowers can self-pollinate but they need to be 'tripped' by insects for this to happen. The flowers produce a fair amount of pollen and nectar in order to encourage insect visits but this also leads to cross-pollination. (OV-V)

Kalloo, G. (1993) Runner bean, *Phaseolus coccineus* L. In: Kalloo, G. and Bergh, B.O. (eds) *Genetic Improvement of Vegetable Crops*. Pergamon Press, Oxford, UK, pp. 405–407.

Smartt, J. (1990) *Grain Legumes*. Cambridge University Press, Cambridge, UK, pp. 85–139.

Schiff's base

The product of the reaction between an aldehyde (or other carbonyl group) and an amine group (usually from amino acids). The reaction is ubiquitous in biochemistry and is fundamental to macromolecular synthesis as well as degradation in reaction cascades such as the **Maillard reaction**, which involves an amino acid (or protein) and a **reducing sugar**. The reversible reaction, catalysed by aldolases and transaminases, forms a 2R-C=N-R′ bond, and further reaction in the presence of water results in an amino group and an α-keto acid. (CW)

Schizocarp

Fruits derived from a simple, two- or more locular compound ovary (schizocarpous gynoecium) in which the locules separate at fruit maturity.

Sclereid

A relatively short non-living cell with thick and often lignified cell walls. Present in seed and fruit coats. (**See: Seedcoats – structure**)

Sclerenchyma

Mechanical tissue in plants composed of sclereids and/or fibres. (**See: Seedcoats – structure**)

Sclerotium

A mycological term describing a hardened body formed by certain fungi at their mature stage (plural: sclerotia). This is a resting structure of hyphal tissue bearing no spores, from which apothecia arise, the spore-bearing structures (Colour Plate 10E). A notable example is the firm purple-black **ergot** body produced in rye by the ascomycete *Claviceps purpurea* (**see: Pathogens**), which may be larger than a wheat kernel and also may contain some host tissue. The elimination of sclerotia contaminating seed production may require special seed **conditioning** techniques, and are specifically assessed in **purity testing**.

Screens – seed cleaning

Screens used in seed conditioning equipment and laboratory size analysis are usually made of perforated sheet metal such as stainless steel or of woven wire mesh on a wooden frame. The openings in perforated screens are either round, oblong (slotted, with the longer dimension in line across or diagonal to the direction of seed flow) or triangular, whereas those in wire mesh screens are square or rectangular. Each type of screen is available in a wide range of standard hole-sizes: dimensions in perforated screens are conventionally measured either in 64ths of an inch or in millimetres, whereas wire mesh screens are denoted by the number of openings per inch or centimetre in each direction. (**See: Conditioning, II. Cleaning**, Fig. C.13B; **Grain inspection**)

Scutellum

A specific region of the cereal embryo that serves as a reservoir for storage products such as **oil**. The scutellum is thus functionally similar to the **cotyledons** of dicot **embryos**, but it does not emerge from the seed and develop into a leaf after germination but rather degrades within the seed when mobilization of the stored reserves is completed. In some cereals, e.g. wild oats, the scutellum grows into the starchy endosperm as the reserves therein are hydrolysed to aid in the absorption of the products of starch and protein mobilization. (**See: Mobilization of reserves – a monocot overview; Mobilization of reserves – cereals**)

Development of the scutellum occurs early in embryogenesis concomitant with, or prior to, coleoptile development. As the embryonic region that contacts the endosperm, the scutellum plays an important role in establishing interactions between the embryo and endosperm, and in transferring food materials from the endosperm to the embryo. Also, the scutellum contributes to the embryo size. Often, enlarged embryos, such as those seen in high oil **maize** kernels, are mainly due to the increased scutellum size without altering the size of the embryo axis. (**See: Development of embryos – cereals**) (O-AO, HS)

Seakale

See: **Brassica – horticultural**

Second messenger

See: **Signal transduction – some terminology**

Secondary dormancy

A form of dormancy that is induced by environmental conditions after a seed has been released from the plant. (**See: Dormancy; Induced dormancy**)

Seed

A ripened ovule of **gymnosperms** and **angiosperms**, which develops following **fertilization** and contains an embryo surrounded by a protective cover. Other food reserve-storing tissues (e.g. **endosperm**) may be present in the mature seed. A propagule that contains the next generation of a plant. The words 'seed' and 'sow' share the same linguistic root. (**See: Grain and seed; Structure of seeds**)

Seed Analysis Certificates (ISTA)

See: **International Seed Testing Association**

Seed banks

See: **Gene banks; Serotiny; Soil seed banks**

Seed crop

Either the whole of a seed-bearing plant, or the seed **yield** from a crop grown for its seed.

Seed rain

The seeds added to the **soil seed bank** from overlying plants.

Seed shadow

The areas to which seed-bearing plants disperse their seeds. (**See: Dispersal; Ecology; Seed rain**)

Seed storage

See: **Cereals – storage; Conservation; Storage fungi; Storage management**

Seedbed environment

In an ideal situation the seedbed will support the seed at the desired sowing depth, while the soil structure remains mechanically weak during seedling growth, drains well and helps to minimize water loss by evaporation. But the seedbed is a complex environment in which seeds and seedlings are exposed to multiple stresses. In essence, seeds and seedlings are not at all sensitive to **soil types**, condition or water content (pests, diseases, salinity and chemicals apart), but they are extremely sensitive to the physical stresses that a soil, whatever its type or condition, imposes on them as the seedlings grow. The enormous scientific literature on the effects of soil physical conditions on seedling **emergence** contains conflicting accounts of what leads to either good or poor results. The key to understanding the effects is to decipher how soil type and condition determine the soil physical stresses, how the physical stresses interact with each other, and how damage due to compaction or heavy irrigation changes them.

Soil is often described in terms of its water content, and this can be an informative description of how we may expect the seedbed to perform as an environment for seedling growth. However, it is paradoxical that neither the seed nor seedling is sensitive to water content in itself, but both are sensitive to almost all other soil physical characteristics that change with water content. This is perhaps why water content is a useful guide; it is, however, no more than a guide. This article explores how the soil physical stresses to which seeds and seedlings are sensitive vary with water content: water stress, mechanical impedance, oxygen and temperature. It also discusses how soil structure can mediate the levels of physical stress when soils are either drying or wetting and, because they usually occur in combination, how the seedbed physical stresses interact with each other.

1. Soil water characteristics

(a) *Water, osmotic and matric potential.* Models for seed germination are often expressed in terms of temperature and water potential. The water potential term most relevant to germination, ψ, is given by:

$$\psi = \psi_\pi + \psi_m$$

where ψ_π is the osmotic potential and ψ_m is the matric potential. The matric potential is due to the capillary sorption of soil water into the soil matrix. In saturated soils ψ_m has a value close to zero and becomes increasingly negative as the soil dries. Though strictly defined in terms of energy per unit volume, ψ_m is more commonly expressed in units of pressure, which are dimensionally equivalent.

This is the term (ψ) that is referred to in **hydrothermal** time models for seed germination, and that also determines the growth rate of seedlings at a given temperature and oxygen availability and under given mechanical impedance.

With respect to the rate of initial water uptake, the matric potential of the seed relative to that of the soil is a determining factor, the former being more negative. After the matrices of the seed are saturated, then further uptake during **imbibition** is determined by the ψ_m of the soil, although in saline soils the osmotic potential can be of sufficient magnitude to affect water uptake. (**See: Germination – field emergence models,** *e. Imbibition;* **Germination – influences of osmotica and salinity**)

(b) *Soil water retention characteristics* describe the relationship between soil matric potential and water content, as a very powerful tool for comparing the physical characteristics of different soil types (Fig. S.5). For example, at a matric potential of −1 MPa, which is close to the base water potential for the germination of many seeds, there can be quite a wide variation in **water content** between different soil types.

2. Soil strength

Though soil strength is very unlikely to affect the germination of seeds (in the sense of the events leading to radicle emergence), it has a considerable effect on the rate of elongation of seedling roots and pre-emergent shoots.

(a) *Penetrometer pressure.* Soil strength is often estimated from the pressure that is needed to push a steel needle of defined dimensions through the soil, usually measured in a downward direction using instruments called penetrometers. Designs vary according to the shape and size of the conical tips that the needles have to deform the soil. Soil strength estimated in this manner as 'penetrometer pressure' is proportional to, but is generally expected to be higher than, the pressure that roots have to exert to deform the same soil, because of differences between both the degree of soil–root and soil–metal friction and the way that roots and penetrometers

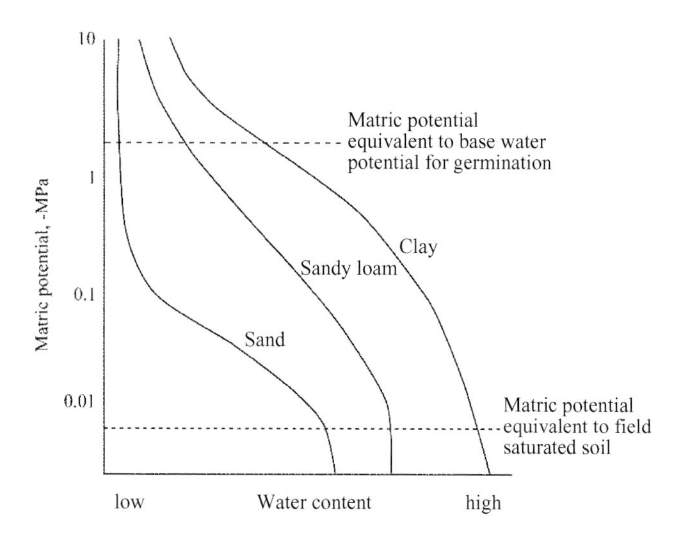

Fig. S.5. Water retention characteristics for three soil types. Note that different soil types can have very different absolute water contents at the base potential for germination. The amount of water available for uptake by seedlings is given by the difference between water content at the saturation point (field capacity, lower dotted line) and the **wilting point** (very close to the base potential for germination, shown in the upper dotted line). Thus, although clay soils tend to be wetter than sandy soils, they often have much less water available for plant uptake.

deform soil. Experiments have shown that root elongation decreases with increasing penetrometer pressure in a relationship that is non-linear and differs between crop and soil type.

As an indicative guide, a resistance to penetration of greater than 3 MPa is likely to restrict root growth. There are fewer studies on the penetrometer strength of seedbeds that hinder shoot emergence, but the available evidence suggests that pre-emergent shoots are far more sensitive than roots, that is, they are severely restricted at much lower penetrometer pressures. Indeed, in some cases high soil strength can physically damage emerging seedlings, such as those of sorghum. However, some caution is needed in using penetrometer pressure as a guide to shoot emergence, because upward elongation near the soil surface is likely to result in a different type of soil failure to that which occurs when a penetrometer is pushed downwards. Although penetrometers that move upwards and break though the surface of the seedbed have been developed, practical difficulties in their use have resulted in a much smaller dataset than that relating root elongation to downward-acting penetrometers.

At a matric potential of −100 kPa the penetrometer strength of the soil can be of the order of 1 MPa or more, which is sufficient to significantly reduce root elongation (although such a matric potential might have very little effect on germination – see: **Germination – influences of osmotica and salinity**). Therefore as the seedbed dries it is more likely that mechanical impedance will dominate the response of the seedling to its environment.

(b) *'Effective Stress' model – soil capillary pressure*. In the seedbed the strength of soil tends to be due to the capillary pressure of water in the pores holding the soil particles together. Even in non-compacted seedbeds, the strength derived from capillary pressure can result in a soil that is difficult for the roots and shoots to deform. This concept has been formalized in the Effective Stress model of soil strength, as applied to unsaturated soils. Here soil tensile strength, Y is given by

$$Y = c - \chi\psi_m$$

where χ is the factor that represents the proportion of water films at a matric potential ψ_m, and c is the cohesion factor due to chemical inter-particle bonds. In wet soils this can be written as

$$Y = c - \left(\frac{S\psi_m}{f_s}\right)$$

where S is the degree of saturation and f_s is a factor that takes into account pore shape and tends to have a value close to 2.

The concept of Effective Stress is useful in understanding the interaction between soil strength and water stress. For instance, clay soils tend to have a higher degree of saturation at a given matric potential than sandy soils: because their Effective Stress is higher, clay soils tend to be stronger than sandy soils. The Effective Stress model is also useful for understanding interactions in 'poorly structured' soils that have connected capillary networks (see below). In 'well-structured' soils, by contrast, where the capillary networks are less well connected, soil strength increases more slowly with drying, and seedlings can elongate between the soil aggregates and push them apart.

3. Soil oxygen stress

In general, the germination of seeds of monocot species is less sensitive to reduced oxygen availability than those of dicot species, and seeds with high oil content are more sensitive than those with high starch content. For example, *Brassica* species, which are lipid-storing dicots, all appear to have similar oxygen requirements with, for example, *B. oleracea*, showing reduction in germination to 50% at an oxygen concentration of 7% (vol/vol). Carrot seeds are sensitive to oxygen stress. Differential sensitivity is also expressed in terms of germination rate, which has a linear relationship with the logarithm of pO_2 in a range of species; extrapolation is used to determine the lower pO_2 limit at which germination can be observed (pO_2 threshold), which also differs between species.

Oxygen stress depends on the balance between supply and consumption in a soil. As the rate of diffusion of oxygen is approximately 10,000 times greater in air than in water, the supply of oxygen is determined by the degree of soil saturation. The demand for oxygen by seeds is likely to be smaller than the demand from soil microbes, whose activity tends to increase with both water content and temperature. Thus in warm wet soils, where demand is high, oxygen availability can be low, but in cold seedbeds low oxygen availability is far less likely because, although the supply can be low, so too is the demand from soil microbial activity.

As a rule of thumb, therefore, oxygen stress should only be considered as a factor that will give poor seedling emergence in hot wet conditions. Otherwise, in practice, seedbeds do not generally have sufficiently low oxygen availability to cause problems, except where there is surface crusting following heavy rain, which tends to lower the rate of atmospheric oxygen diffusion into the soil mass. Oxygen stress is more likely to be an issue for plants that are already established and which have roots growing deep into the soil. (**See: Germination – influences; Germination – influences of gases**)

4. Soil temperature

Temperature is one of the key variables influencing germination, and plays a prominent part in models of germination and field emergence germination and seedling growth. Solar radiation dominates all other factors that determine the seedbed temperature, though this can be influenced as well by **tillage**, irrigation and **mulching** practices. For a given irradiance, in most circumstances, soil water content influences soil temperature. In dry soil the loss of heat by evaporation uses only a small proportion of the incoming radiation, and heat is conducted away slowly: thus dry seedbeds warm up quickly in early spring. However, dry surfaces also cool down quicker and so are prone to frost at night. (**See: Germination – field emergence models; Germination – influences of temperature**)

5. Soil structure

Soil structure describes the arrangement of soil particles and the pores between them, and is probably one of the most widely used terms in describing a seedbed. However, unlike the water stress, soil strength, oxygen stress and temperature factors discussed above, structure is not easily measured nor is it in itself a stress.

As noted, different soil types tend to be associated with different types of structure and water retention characteristics. When they are wet, clay soils are plastic and difficult to modify and, when dry, the aggregates are strong and harder to reduce in size; seedbeds in clay soils, however, tend to have stable structures. Very sandy soils, by contrast, are more easily worked into a seedbed, but the resulting seedbed can be very unstable.

A 'well-structured' soil will provide a seedbed that is relatively weak when dry but relatively strong when wet. This is possible because one effect of the presence of aggregates is to disrupt the continuity of capillary-sized pores in soil. Where this happens, soils can dry without developing large Effective Stresses at the aggregate scale, and thus they remain weak.

Generally the development of a good and stable structure depends on the presence of organic matter, which stabilizes the soil so that it does not disintegrate when wetted rapidly. Unfortunately many intensively cultivated soils have degraded structures that are low in organic matter, and are therefore unstable. They tend to disintegrate into soils with a well-connected capillary network. Such seedbeds dry quickly and become mechanically strong, causing problems that can prevent the emergence of seedlings.

(a) *Soil crusting*. Crusts, or caps, are usually formed when soil structure at the surface of the seedbed disintegrates during heavy rainfall or irrigation, and the soil particles sediment to form thin but dense layers. When dry, the soil crusts give rise to considerable mechanical impedance, which can hinder penetration of the seedlings and so delay crop emergence and reduce final emergence. There is plentiful observational evidence of buckled or bent but unemerged germinated shoots lying trapped beneath crusts formed on seedbed composed of large aggregates. Such buckling is less likely in seedbeds with small aggregate size, where the shoot is supported rather than impeded by the soil.

(b) *Seed placement depth*. An important function of the seedbed is to hold the seed physically at the desired sowing depth (**see: Planting equipment – placement in soil**), and to achieve this it is preferable that **tillage** should produce an aggregate size similar to or smaller than the seed size. In some cases very poor seedbed preparation can lead to seeds falling through the pore matrix to a much deeper seed depth than is ideal. Table S.4 summarizes how aggregate size affects some of the principal properties of a seedbed, and it suggests that very coarse and very fine seedbeds should be avoided. The presence of stable aggregates is very beneficial because in this condition the soil will tend to remain weaker when it dries.

(c) *Seed-to-soil contact and imbibition*. Frequently soil structure is discussed in the context of seed-to-soil contact. It is often said that good contact is important in facilitating water uptake by seeds. However, recent evidence shows that seeds can uptake water effectively in the vapour phase and that seed–soil contact has little effect on water imbibition.

6. Effect of multiple soil stresses

In the field environment seeds and seedlings are exposed to a combination of physical stresses that vary with soil water content and have relationships that differ depending on soil

Table S.4. Desirability scores for different aggregate size ranges for some principal physical properties of seedbeds. 1 = undesirable, 2 = acceptable, 3 = desirable. The higher the sum of the properties, the more efficacious the aggregate size.

Aggregate size range (mm)	Lowest evaporation	Greatest intra-aggregate aeration	Greatest inter-aggregate aeration	Least wind and water erosion	Lowest compactability	Sum of all properties
0.1–0.2	1	3	1	1	1	7
0.2–0.5	2	3	1	1	1	8
0.5–1.0	3	3	2	2	2	12
1.0–2.0	3	3	3	2	2	13
2.0–4.0	2	2	3	3	2	12
4.0–8.0	2	2	3	3	2	12
8.0–16.0	1	1	3	3	2	10

From Braunack, M.V. and Dexter, A.R. (1989) Soil aggregation in the seedbed: a review. II. Effect of aggregate sizes on plant growth. *Soil and Tillage Research* 14, 281–298.

structure – particularly in respect of soil strength. Water stress and increased mechanical impedance – the stresses most likely to limit germination and emergence in the seedbed – almost always occur simultaneously. But these act differently on seedlings: root and, probably, shoot elongation both tend to decrease as a linear function of water stress but as a non-linear function of soil strength. Thus as the soil dries, not only does soil strength itself increase rapidly but the seedling's expansive growth also becomes far more sensitive to that increase; in contrast, seedlings have a lesser response to the accompanying changes in water stress. (WRW, WFS)

Gliński, J. and Lipiec, J. (1990) *Soil Physical Conditions and Plant Roots*. CRC Press, Boca Raton, Florida, USA.

Smit, A.L., Bengough, A.G., Englels, C., Van Noordwijk, M., Pellerin, S. and Van De Geijn, S.C. (2000) *Root Methods: A Handbook*. Springer, Berlin.

Smith, K.A. and Mullins, C.E. (2000) *Soil and Environmental Analysis: Physical Methods*. Marcel Dekker, New York, USA.

Sumner (2000) *Handbook of Soil Science*. CRC Press, Boca Raton, Florida, USA.

Whalley, W.R., Finch-Savage, W.E., Cope, R.E., Rowse, H.R. and Bird, N.R.A. (1999) The response of carrot (*Daucus carota* L.) and onion (*Allium cepa* L.) seedlings to mechanical impedance and water stress at sub-optimal temperatures. *Plant, Cell and Environment* 22, 229–242.

Wuest, S.B. (2002) Water transfer from soil to seed: the role of vapour transport. *Soil Science Society of America Journal* 66, 1760–1763.

Seedcoats – development

In angiosperm seeds, the **embryo** and **endosperm** or **perisperm**, where present, are surrounded by a seedcoat (testa) (**see: Seedcoats – structure; Structure of seeds**). This coat has important functions (**see: Seedcoats – functions**) including protecting the embryo from biotic stresses, providing a conduit for nutrients for the embryo and endosperm during their development (**see: Development of seeds – nutrient supply**), providing a means for seed dispersal (**see: Seedcoats – dispersal aids**) and it can delay germination by restricting oxygen and water uptake (**see: Dormancy – coat imposed**). It is also a primary host structure for complex seedborne microfloral populations, including **saprophytic** and **pathogenic** species. Seedcoat development commences after fertilization and the two integuments (inner and outer) of the ovule develop into the mature seedcoat layers (**see: Reproductive structures**). Therefore, the seedcoat is of maternal origin. Its component tissues are initially undifferentiated but rapidly undergo programmed developmental events that end in the seedcoat features that are characteristic of a species (**see: Structures of seeds – identification characters**). While the same general pattern of development occurs throughout the seedcoat of a particular species, there may be regional differences in the number of cell layers or thickness of cell walls, e.g. near the site of attachment (**funiculus/hilum**) to the parent plant.

The initiation of seedcoat development from the integuments seems to be triggered by endosperm development. For example, in embryo-lethal or *Fertilization Independent Seed (FIS)* mutants, seedcoat development is initiated in the absence of pollination and the endosperm develops autonomously, but there is no embryo development. The seedcoat is developmentally transitory, and there are a number of tissues present during early and mid-development which do not persist into the mature seedcoat. This is illustrated for the development of soybean and *Arabidopsis* seedcoats.

In the soybean (Fig. S.6), around the time of fertilization (post-anthesis) the outer integument consists of an epidermis of cuboidal-shaped cells surrounding several layers of thin-walled parenchyma cells. To their inside is the inner integument (endothelium) which is one-cell thick (Fig. S.6A). The inner integument divides to form three to five layers of cells, which by 9 days have become thick-walled and surrounded by an increased number of layers of parenchyma cells of the outer integument (Fig. S.6B). Also the outer integument has differentiated into an outer layer of thin-walled parenchyma cells and an inner layer of thicker-walled cells. Two epidermal layers, the epidermis and hypodermis, are evident at this time. By 15 days, the cells of the inner integument are diminished as they become stretched and compressed (Fig. S.6C). At this time the vascular conducting tissues (xylem and phloem) in the coat are starting to differentiate. From 12–15 days the epidermis develops into a thick-walled palisade layer and the thin-walled cells of the hypophesis expand and take on an hour-glass appearance; the walls of this layer subsequently thicken. The vascular conducting tissue becomes fully differentiated and the parenchyma cells in the inner regions of the outer integument develop large air spaces (aerenchyma), and those to the inside of them are thick-walled and small. The inner integument is crushed and barely discernible (Fig. S.6C). As the cotyledons continue to develop and expand from the inside, the endosperm and most of the layers of the seedcoat are crushed, and in the mature seed (Fig. S.6D) only a few flattened layers of aerenchyma cells remain beneath the prominent palisade and hour-glass cells (called osteosclereids). Immediately beneath the seedcoat and surrounding the embryo is a layer of thick-walled living cells, the remnant of the endosperm, sometimes called the aleurone layer.

A similar pattern of development of the seedcoat occurs in pea (*Pisum sativum*), although there is no aleurone layer evident in the mature seed. There appear to be no studies on the development of the seedcoat in cereals, although this is a diminutive structure since the seed is surrounded by a fruit coat (the development of which is also largely unstudied).

Development of the seedcoat of *Arabidopsis* also involves the specialization and reduction of seedcoat layers, but in this the outermost cell layer becomes laden with mucilage. The outer integument of the developing seed is derived from the incipient ovule epidermis and remains two cells thick throughout its existence (Fig. S.7a–e). During seedcoat development, these layer cells undergo marked differentiation, including cell morphogenesis and the synthesis, secretion of mucilage and starch deposition. During the late globular and heart-shaped stage of embryogenesis (**see: Development of embryos – dicots**), starch granules appear in the outer epidermis (Fig. S.7b, c). They are degraded as the cells produce a reinforced wall surrounding the columnar protoplast (Fig. S.7c, d). Now, from the early torpedo stage, the outer layer, the only cells producing mucilage, deposit it between the primary cell wall and plasma membrane (Fig. S.7c, d). This secretion forces the cytoplasm into a columnar shape (columellae) in the centre of the cell (this also occurs in the seedcoats of other mucilage-producing species). As the seed matures with embryo development the starch grains decline, mucilage accumulates in the outermost layer, although the layers become less distinct, and there is the formation of thick cell walls (Fig. S.7e). The innermost layer eventually contributes to the brown pigment layer (see below).

In a related crucifer, *Brassica rapa* ssp. *pekinensis* (Chinese cabbage), by the torpedo stage of embryo development, a palisade layer is beginning to form in the inner region of the seedcoat, possibly from the inner layer of the outer epidermis. As the seed matures the cells of the palisade layer thicken in the basal and radial walls, and the epidermal and sub-epidermal layers external to this degenerate. The thin-walled layers of the inner integument likely undergo **programmed cell death**, as in the coats of *B. napus*, and are crushed and become pigmented. A precociously germinating line of this species forms a palisade layer with regions of much thinner cell walls; cracks appear in these regions, and the less rigid coat may contribute to the ability of the seeds to germinate during their development.

The inner integument of *Arabidopsis* (not shown in Fig. S.7) starts as three layers of cells. Initially the innermost layer is

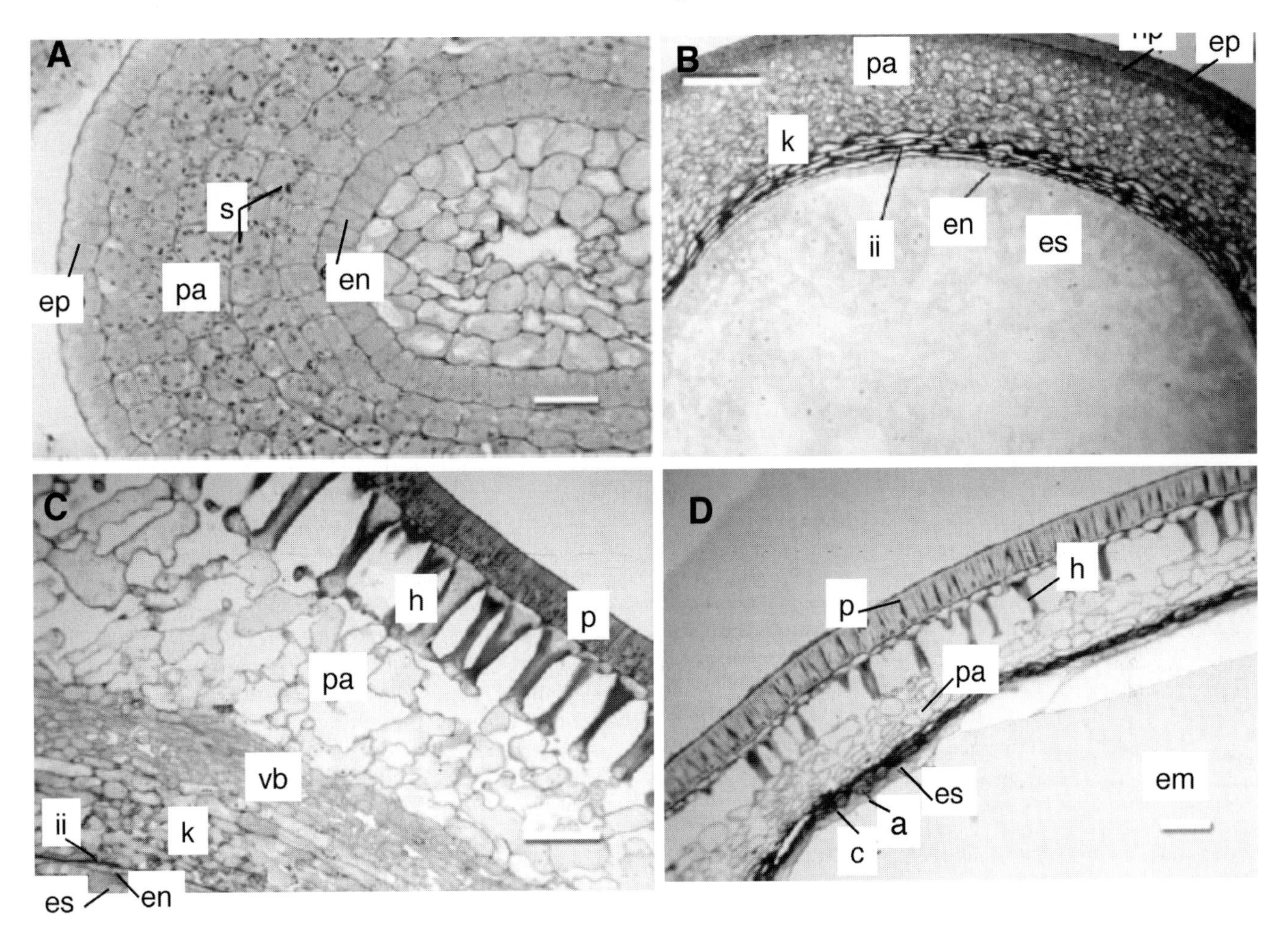

Fig. S.6. Development of the seedcoat of soybean (*Glycine max*). (A) 3 days post-anthesis (dpa). The seedcoat consists of an inner (en, endothelium) and outer (ep, epidermis) cell layer with several layers of parenchyma cells (pa) between them, containing starch granules (s). (B) 9 dpa. Immediately beneath the epidermis (ep) the hypodermis (hp) is starting to differentiate. The thin-walled parenchyma cells (pa) are still present, although the inside layers are beginning to develop thicker cell walls (k). There are several elongated cell layers of the inner integument, and the endothelium (en) lies against the developing endosperm (es). ii, Inner integument. (C) 15 dpa. The outer epidermal layer is differentiating into a thick-walled palisade layer (p) containing starch granules. The thicker-walled hour-glass shaped cells (osteosclereids) of the hypophesis (h) are fully formed to the outside of the parenchyma layer (pa) which is developing air spaces and becoming aerenchyma. Vascular conducting tissue is well formed (vb). Thick-walled parenchyma of the outer integument is present (k) but the inner integument (ii) and endothelial cells (en) are crushed and lie against the endosperm (es). (D) Mature seedcoat (45 dpa) showing the thick-walled outer palisade layer of cells (p) with the hour-glass cells of the hypophesis (h) to the inside. The aerenchyma (pa) is compressed and there are only crushed remnants of the endothelium and thick-walled parenchyma (c). The aleurone layer (a) derived from the endosperm, which is now only a remnant (es), is to the inside of the seedcoat, surrounding the embryo (em). Bar in A: 20 μm; B and C: 50 μm; D: 100 μm. From Miller, S.S., Bowman, L.-U.A., Gijzen, M. and Miki, B.L. (1999) Early development of the seedcoat of soybean (*Glycine max*). *Annals of Botany* 8, 297–304. With permission, Oxford University Press.

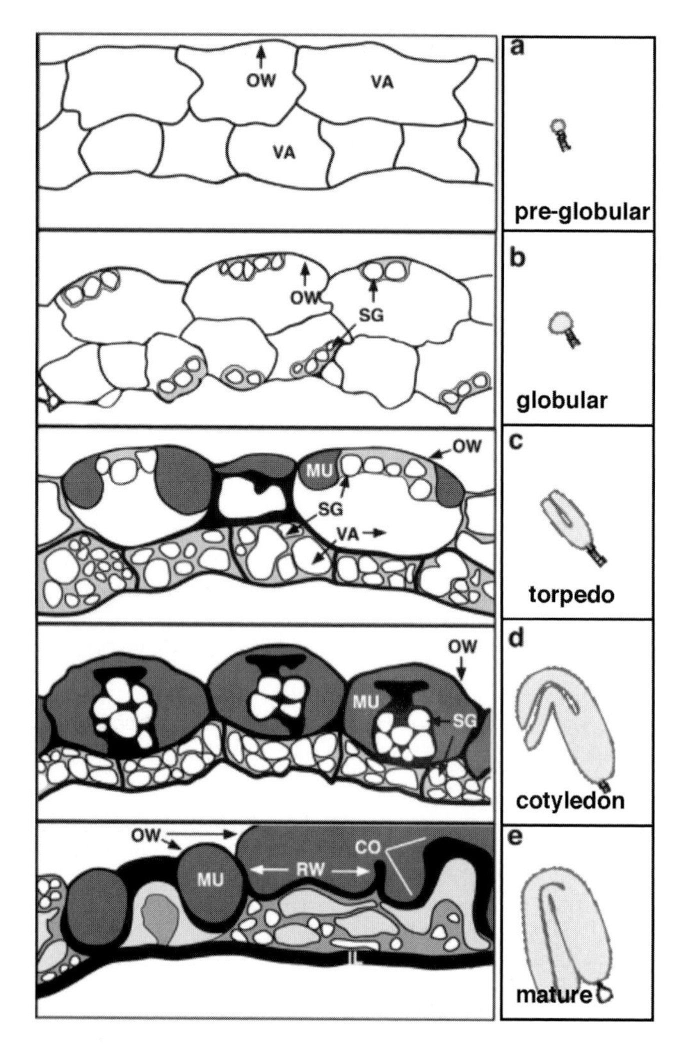

Fig. S.7. Differentiation of the outer integument of developing *Arabidopsis thaliana* seedcoats to form the mucilage layer. The stages of embryo development are shown to the right of the diagrams. (a) Large vacuolated cells with cytoplasm compressed against the cell walls. (b) Both cell layers produce prominent starch granules. (c) The outer layer of cells only produces and secretes mucilage into the region between the cell wall and the cytoplasm. Inner cell layer starts to compress against the outer. (d) Mucilage production in the outer cell layer is complete, forcing the cytoplasm into a columnar shape in the centre of the cell. The vacuole is no longer visible and the starch granules start to degrade; likewise in the inner cell layer. (e) Secondary thickening of the cell walls of the inner and outer layers occurs, particularly in the basal region of the cells. Hydration of the mucilage can cause the distal cell walls to break down as it swells (depicted to the right of the diagram). Inner layer of cells becomes compressed. OW: thin outer cell wall of the outer layer of cells; SG: starch granule; MU: mucilage; VA: vacuole; CO: columella; RW: radial cell walls; IL: inner layer. From: Windsor, J.B., Symonds, V.V., Mendenhall, J. and Lloyd, A.M. (2000) *Arabidopsis* seedcoat development: morphological differentiation of the outer integument. *Plant Journal* 22, 483–493. With permission.

heavily pigmented, but pigmentation is lost as the cell contents are degraded and the walls become thickened. This layer is called the endothelium. The parenchyma cells in the two outermost layers of the inner integument initially swell but eventually are crushed as the embryo expands. Along with the inner layer of cells of the outer integument they make up the thick amorphous brown pigment layer that gives the brown colour to the coat of the mature seed as new pigments are deposited therein.

Several genes have been identified with respect to seedcoat development in *Arabidopsis*. The *RHM2* gene and *MUM* genes, which are involved in the synthesis of pectinaceous rhamnogalacturonan I (RG I) have a specific function in the synthesis of seed mucilage and in the formation of normal seedcoat structure. *MYB61* appears to function upstream of *RHM2* to promote the accumulation of linear RG I during seedcoat development. The production and deposition of mucilage affects seedcoat structure; seeds of non-mucilage-producing **mutants** of *Arabidopsis*, e.g. of *rhm2*, *mum1-5* and *myb61-1* mutants have a distorted seedcoat morphology and a strong reduction in the amount of mucilage present. The *ttg1* and *glabra2* (*gl2*) mutants have similar defects in mucilage production. The stages of mucilage cell development (with respect to the embryo development stages in Fig. S.7), and mutations to the genes which affect mucilage synthesis at these stages, can be summarized as:

Ovule → Stage a: Coat expansion (*ant*, *ats*, *bell1*, *ino*) → Stage b: Amyloplast accumulation (*ap2*) → Stages c, d: Mucilage synthesis and secretion, cytoplasmic column production (*ttg1*, *gl2*, *mum4* and *mum3*, *mum5*) → Stage e: Secondary cell wall production (*mum1*, *mum2*).

Another group of mutants such as *inner no outer* (*ino*), *aintegumenta* (*ant*), *aberrant testa shape* (*ats*) and *bell1* (*bel1*) also show a phenotype with an aberrant seedcoat (fewer layers) and seed shape. These mutations affect the initiation, development or differentiation of the integument layers early in development and result in abnormally shaped seeds. This suggests that the seedcoat rather than the embryo determines the shape of the seed. These mutants also lack the ability to synthesize mucilage. Another mutant *apetela 2* (*ap2*) which lacks a functional *APETALA2* (*AP2*) gene, is deficient in the development of columellae (Fig. S.7d, e) and the biosynthesis of mucilage. Strong *AP2* mutant alleles lead to a weakened and misshapen seedcoat. However, they apparently do not lack any seedcoat layer. Seeds of *ttg*, *ats* and *aba* mutants lack dormancy and have a reduced ABA content (**see: Dormancy – genes**). Perhaps the absence of mucilage from the seedcoats of these mutants also allows more oxygen to diffuse into the seed, and this higher oxygen level might be able to release it from dormancy. (**See: Dormancy – coat-imposed**)

There are other seedcoat mutants in *Arabidopsis* that affect flavonoid pigmentation and are represented by *transparent testa* (*tt*), *transparent testa glabra* (*ttg*) and *banyule* (*ban*) (Table S.5). The seedcoat colour of *tt* and *ttg* mutants ranges from yellow (*tt1-tt5*; *tt8*; and *ttg1*) to pale brown (*tt6*, *tt7*, and *tt10*) or greyish brown (*tt9*). Some *tt* mutants affect different genes involved in pigment synthesis (*tt3* affects the synthesis of the enzyme DFR; *tt4*, CHS; *tt5*, CFI; *tt6*, F3H; *tt7*, F3′H; see Fig. S.8 for location of these enzymes in the biosynthetic pathway). Others influence the expression of genes involved in this pathway, e.g. *TT2* (MYB) and *TT8* (bHLH) synergistically specify the correct expression of the *BANYULS* gene (*BAN*, which encodes a DRF-like protein) and *TT16* is necessary for the expression of the *BAN* gene. At least two mutants that affect other epidermal cell-fate and cell differentiation

Table S.5. Mutants that affect the colour of the seedcoat during development in *Arabidopsis*.

Mutation	Wild type gene product, or function	Seedcoat colour of mutant
tt1	WIP protein	Yellow
tt2	R2R3-MYB transcription factor Regulator of *BAN, DFR*	Yellow
tt3	Dihydroflavonol reductase (DFR)	Yellow
tt4	Chalcone synthase (CS)	Yellow
tt5	Chalcone isomerase (CI)	Yellow
tt6	Flavanone 3-hydroxylase (F3H)	Pale brown
tt7	Flavonoid 3′-hydroxylase (F3′H)	Pale brown
tt8	bHLH transcription factor Regulator of *BAN, DFR*	Yellow
tt9	nd	Greyish
tt10	nd	Pale brown
tt12	Vacuole transporter	Pale brown
tt15	nd	Pale greenish-brown
tt16	MADS domain protein	Yellow
tt19	Glutathione S-transferase (GST)	Pale brown
ttg1	WD40-Repeat	Yellow
ttg2	WKRY transcription factor	Pale fawn, less mucilage
ban	Leucoanthocyanidin reductase (LAR)	Greyish green

For the location of enzymes in the pigment biosynthesis pathway, see Fig. S.8.

processes also affect the differentiation of the outer seedcoat. The mutants *transparent testa glabra1* (*ttg1*) and *glabrous 2* (*gl2*) affect anthocyanin production and mucilage production during seedcoat development. All known *tt* mutations are recessive and show maternal inheritance of the seed phenotype.

While in *Arabidopsis* it is the presence of recessive mutations that leads to the lack of pigmentation in the seedcoat, in soybean (and some other legumes) it is the presence of dominant genes that result in lack of colour. Thus, although the same genes are involved in flavonoid pigment synthesis (Fig. S.8) during seedcoat development in both species, the mechanism controlling their expression is different. In commercially grown soybean varieties there is no pigment deposition in the innermost layer cells of the integument during seedcoat development; hence these seeds have a yellow seedcoat. However, seeds of some soybean varieties accumulate anthocyanins within the epidermal layer of the seedcoat. This pigment deposition and seed colour are controlled by at least three different gene loci, *I*, *T* and *R*. The dominant *I* allele prevents seedcoat pigment accumulation, resulting in the yellow seed colour, because it reduces the amount of mRNA for chalcone synthase (CHS). This likely leads to a deficit of narigenin chalcone (Fig. S.8), a required substrate for subsequent flavonoid, proanthocyanin and anthocyanin pigment synthesis. The mutant *i* genotype has higher CHS activity, and pigment accumulation occurs in the coat. The *T* locus, which likely encodes a flavonoid 3′-hydroxylase, affects coat colour, the recessive *t* producing a grey seed. The *R* locus determines the colour of the coat pigmentation, with *R*/- giving rise to a black coat and *r*/*r* resulting in a brown one. There is also a correlation between the *I* genotype and abundance of a cell wall protein, PRP1. *I*/*I* genotypes which produce a yellow seedcoat have an abundant soluble PRP1 in immature seedcoats. In contrast, the homozygous recessive *i* genotypes with unpigmented, black- or brown-coated lines have a decreased amount of soluble PRP1. This implies a novel influence of a gene for an enzyme in the flavonoid pathway on an unrelated protein associated with seedcoat wall structure. (XG, JDB)

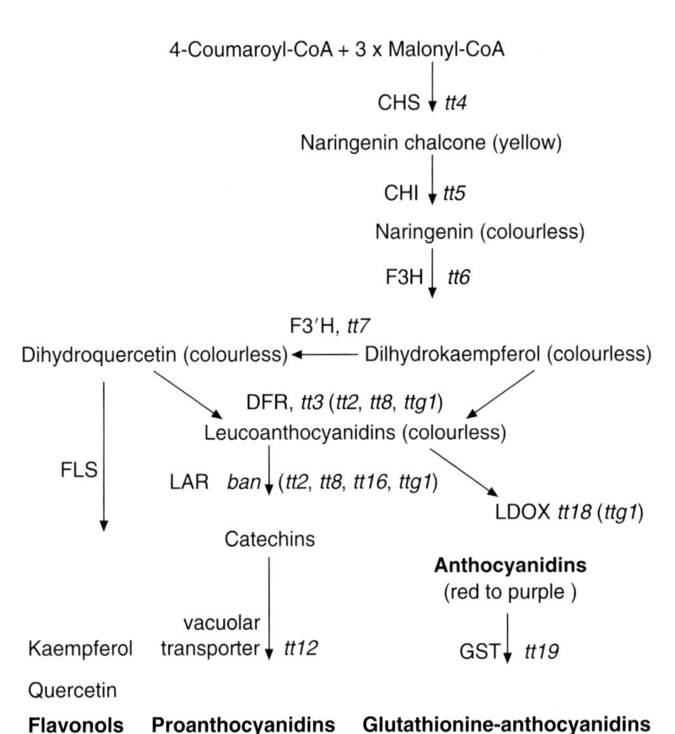

Fig. S.8. The major pathways of flavonoid biosynthesis and identified mutations of genes (*italics*) that affect seedcoat pigmentation in *Arabidopsis*. CHI: chalcone isomerase; CHS: chalcone synthase; DFR: dihydroxyflavonol-4-reductase; F3H: flavonone-3-hydroxylase; F3′H: flavonoid-3′-hydroxylase; FLS: flavonol synthase; GST: glutathione-S-transferase; LAR: leucoanthocyanidin reductase; LDOX: leucoanthocyanidin dioxygenase. Modified from: Nesi, N., Debeaujon, I., Jond, C., Stewart, A.J., Jenkins, G.I., Caboche, M. and Lepiniec, L. (2002) The *TRANSPARENT TESTA 16* locus encodes the ARABIDOPSIS BSISTER MADS domain protein and is required for proper development and pigmentation of the seedcoat. *The Plant Cell* 14, 2463–2479.

Beeckman, T., De Rycke, R., Viane, R. and Inzé, D. (2000) Histological study of seed coat development in *Arabidopsis thaliana*. *Journal of Plant Research* 113, 139–148.

Debeaujon, I., Peeters, A.J.M., Léon-Kloosterziel, K.M. and Koornneef, M. (2001) The *TRANSPARENT TESTA12* gene of *Arabidopsis* encodes a multidrug secondary transporter-like protein required for flavonoid sequestration in vacuoles of the seed coat endothelium. *The Plant Cell* 13, 853–871.

Haughn, G. and Chaudhury, A. (2005) Genetic analysis of seed coat development in Arabidopsis. *Trends in Plant Science* 10, 472–477.

Moise, J.A., Han, S., Gudynaite-Savitch, L., Johnson, D.A. and Miki, B.L. (2005) Seed coats: structure, development, composition, and biotechnology. *In Vitro Cellular and Development Biology – Plant* 41, 620–644.

Percy, J.D., Philip, R. and Vodkin, L.O. (1999) A defective seed coat pattern (Net) is correlated with the post-transcriptional abundance of soluble protein-rich cell wall proteins. *Plant Molecular Biology* 40, 603–613.

Ren, C. and Bewley, J.D. (1998) Seed development, testa structure and precocious germination of Chinese cabbage (*Brassica napus* subsp. *pekinensis*). *Seed Science Research* 8, 385–397.

Seedcoats - dispersal aids

Besides its mechanical (**see: Seedcoats – structures**) and other functional aspects (**see: Seedcoats – functions**) the seedcoat in angiosperms and gymnosperms can also develop characteristics or specialized structures that assist **dispersal**.

1. Angiosperms

As an attraction mainly for birds, the outer integument (testa) can become partly or entirely fleshy and is then called a sarcotesta (Fig. S.9). Sarcotestal seeds are usually brightly coloured and basically fulfil the same function as a berry or a drupe with respect to endozoochory (dispersal by animals through ingestion). With the fleshy part on the outside, the inner layers of the seedcoat (e.g. endotesta or exotegmen) will still provide mechanical protection for the **endosperm** and the **embryo**. Sarcotestal seeds are found in 17 different angiosperm families across both **dicotyledons** and **monocotyledons**, e.g. in Annonaceae, Cucurbitaceae, Euphorbiaceae, Flacourtiaceae, Liliaceae, Magnoliaceae, Meliaceae and Palmae. Another adaptation to endozoochory is a mucilaginous testa epidermis (e.g. *Ribes sanguineum*, *Linum usitatissimum*, Fig. S.10), which facilitates the passage through dispersing animals' intestines.

Another structure assisting seed dispersal is the **aril** (Fig. S.11). There has been confusion about the definition the term aril, and a plethora of specialized terms for arils of different origins and locations have been coined since the 18th century. Corner (1976) suggested to define aril as a 'pulpy structure which grows from some part of the **ovule** or **funiculus** after fertilization and invests part or the whole seed'. Other authors sometimes distinguish so-called localized arils or 'arillodes' which develop from some part of the ovule (e.g. exostome, **raphe**, or **chalaza**) from 'true' funicular arils. An aril of double origin, i.e. arising from both the funiculus and the testa, is called a 'complex' aril (e.g. *Euonymus europaeus*, *Myristica fragrans* – **nutmeg, mace**).

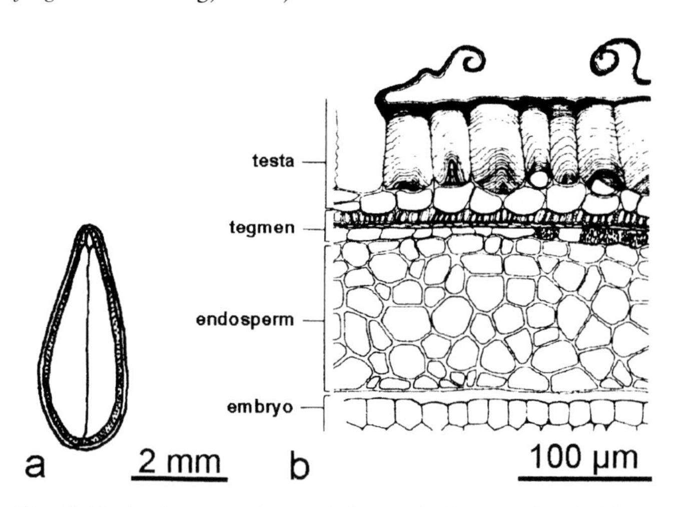

Fig. S.10. Anatomy and morphology of the seed of *Linum usitatissimum*. (a) Transmedian longitudinal section through seed showing the large spatulate embryo surrounded by a small amount (2–6 cell layers only) of endosperm (from Vaughan, J.G. (1970) *The Structure and Utilisation of Oil Seeds*. Chapman & Hall, London, UK). (b) Cross section through seedcoat (from Boesewinkel, F.D. (1980) Development of ovule and testa of *Linum usitatissimum* L. *Acta Botanica Neerlandica* 29, 17–32). The testa is mucilaginous.

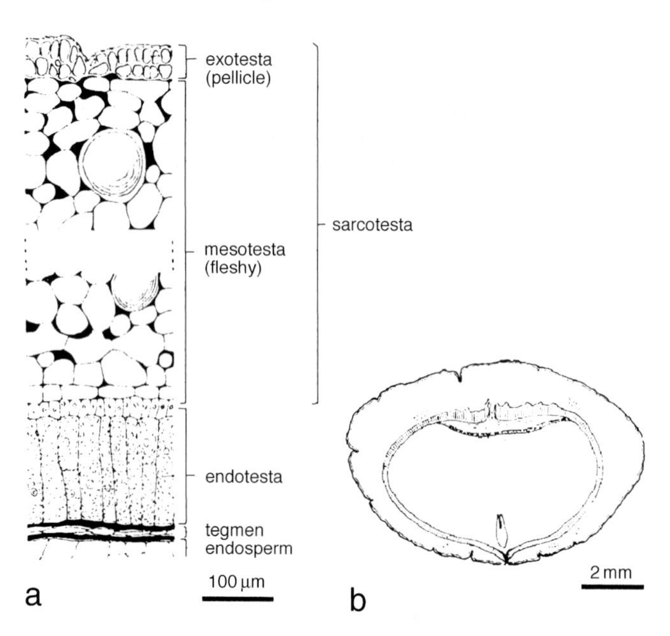

Fig. S.9. Anatomy and morphology of the seed of *Magnolia soulangeana*. (a) Cross section through seedcoat. (b) Transmedian longitudinal section through seed showing the small embryo embedded in copious endosperm (from Corner, E.J.H. (1976) *The Seeds of Dicotyledons*. Cambridge University Press, Cambridge, UK).

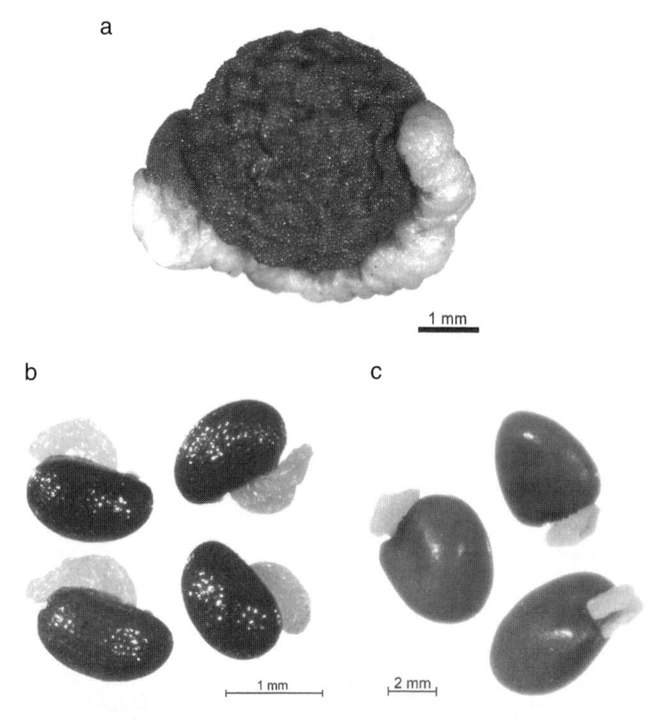

Fig. S.11. Arillate seeds. (a) *Iris popovii* (Iridaceae); seed with stophiole (swollen raphe). (b) *Chelidonium majus* (Papaveraceae); seeds with strophiole (swollen raphe). (c) *Ulex europaea* (Fabaceae); seeds with funicular aril. (Photos: Wolfgang Stuppy, copyright RBG Kew.)

One well-known example of an aril is the exostome aril of many members of the Euphorbiaceae, which is still commonly addressed by the specific term 'caruncle' (e.g. **castor bean**, Colour Plate 15B). The caruncle usually consists of an oil-containing elaiosome to attract ants that carry the seeds away to feed later on the aril. Exostome arils are also found in other families such as Violaceae and Polygalaceae (Colour Plate 15D). Both exostome and funiculus contribute to the aril of *Strelitzia reginae* (Colour Plate 15A). Similar elaiosomes ('food bodies') can be formed by the chalaza (e.g. *Luzula campestris*), the raphe (then sometimes often also called '**strophiole**', e.g. in, *Iris popovii* (Fig. S.11a) *Chelidonium majus* (Fig. S.11b), *Corydalis cava*, *Helleborus niger*) or the funiculus (as e.g. *Acacia* spp., *Afzelia africana* (Colour Plate 15C), *Cardiospermum halicacabum*, *Moehringia trinerva*, *Passiflora edulis*, *Ulex europaea* (Fig. S.11c)).

Some plants acquire great economic importance due to their pleasantly flavoured, edible arils. One well-known example is mace. Another is the so-called Durian or civet fruit (*Durio zibethinus* Murray), a bombacaceous fruit that is very popular in Malaysia and other parts of Southeast Asia (Fig. S.12). Designed to attract mammals (especially orang-utans) for their dispersal, the large (up to 30 cm) spiny fruits emit one of the most pungent odours of any fruit in the world, resembling a mixture of scatole and onions. Inside the fruit, which easily splits between its septae when ripe, are few large seeds entirely wrapped in a thick, creamy funicular aril. Despite the foul smell of the fruit, many people regale the deliciously tasting aril with its unique flavour sometimes described as a mixture of nuts, spices, banana, custard cream, vanilla and onions. Other fruits which gained economic importance because the seeds bear edible arils are the passion fruit, *Passiflora* spp. (e.g. *P. edulis*, Passifloraceae) and a number of members of the Sapindaceae, such as the litchi, *Litchi* (*Nephelium*) *chinensis* ssp. *chinensis*, the rambutan, *Nephelium lappaceum*, the longan or longyen, *Dimocarpus longan*, also called 'dragon's eyes', and akee, *Blighia sapida*, also called 'vegetable brain'.

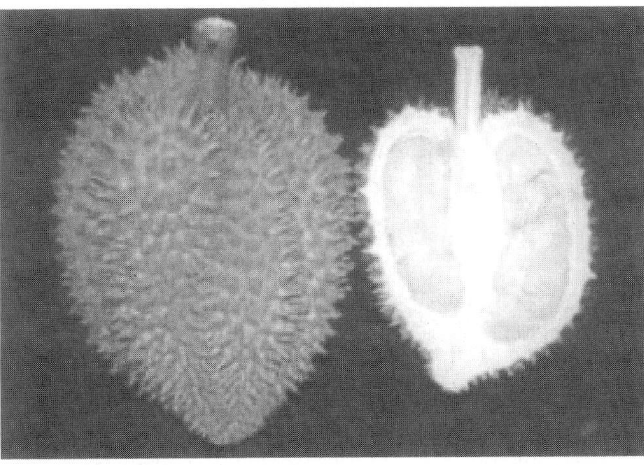

Fig. S.12. (a) Roadside stall in Malaysia offering Durian fruits (*Durio zibethinus*, Malvaceae) for sale (photo: Wolfgang Stuppy, copyright RBG Kew). (b) Section of durian fruit showing seeds surrounded by arils (from www.fftc.agnet.org/library/image/pt2001010f1.html).

While fleshy arils are indicators of **zoochory**, wind-dispersed (**anemochorous**) seeds often possess wings (e.g. many Bignoniaceae, Casuarinaceae, Cucurbitaceae, Hippocrateaceae, Ranunculaceae). Wings can also be formed by the fruit (i.e. the ovary and/or other parts of the flower), e.g. by the ovary in *Acer*, or by accrescent (enlarging after flowering) sepals in Dipterocarpaceae (Fig. S.13a–d); however, in contrast to the fruit wing, the seed wing (Fig. S.14) is rarely provided with vascular bundles. Corner (1976) describes seed wings as 'a local outgrowth of the testa or, in the unitegmic seed, of the seedcoat'. Such wings may be completely peripheral as in Bignoniaceae (Fig. S.14a), Caryophyllaceae (Fig. S.14b) or *Newtonia* (Fabaceae; Fig. S.14c), or restricted to the raphe, chalaza, antiraphe, **hilum**, funiculus, and even along the three angles of a plump seed (e.g. *Moringa*).

Fig. S.13. Fruits of Dipterocarpaceae. (a) *Shorea macrophylla*. (b) *Shorea ovalis*. (c) *Dryobalanops aromatica*. (d) *Dipterocarpus cornutus*. The fruits of the Dipterocarpaceae are characterized by a calyx of five sepals of which two to five are strongly accrescent (i.e. enlarging) and persist in the mature fruit to serve as wings. The fruits of *Shorea* possess three very large sepals and two smaller ones. Those of the genus *Dryobalanops* are characterized by five equally strongly accrescent sepals; only two sepals enlarge and develop into wings in the fruits (photos: Wolfgang Stuppy, copyright RBG Kew).

Other specialized structures assisting wind dispersal are the trichomes (hairs) of various kinds produced by the seedcoat, the funiculus or even the placenta (e.g. *Populus*, Colour Plate 15E). In some cases, specialized hairs can support dispersal through water (hydrochory), e.g. through an enlargement of the seed surface, or epizoochory (adhesion to animals by hooked hairs). Seeds can be entirely covered in hairs (e.g. *Gossypium* (**see: Cotton**), *Ipomoea* (Fig. S.14f)) or show one or two-sided tufts (so-called 'coma') and crowns of hairs. For example, within the Apocynaceae (*sensu lato*), the members of subfamily Asclepiadoideae usually only possess a micropylar tuft of hairs (coma) while members of subfamilies Apocynoideae and Rauvolfioideae can have a coma at both the exostomal and chalazal end; a chalazal tuft of hairs is also found in *Epilobium*.

An interesting adaptation to water dispersal is displayed by the seeds of *Victoria amazonica* (Nymphaeaceae), the giant water lily. They possess an aril developed from the funiculus that invests the whole seed and serves as an air float (Fig. S.14g).

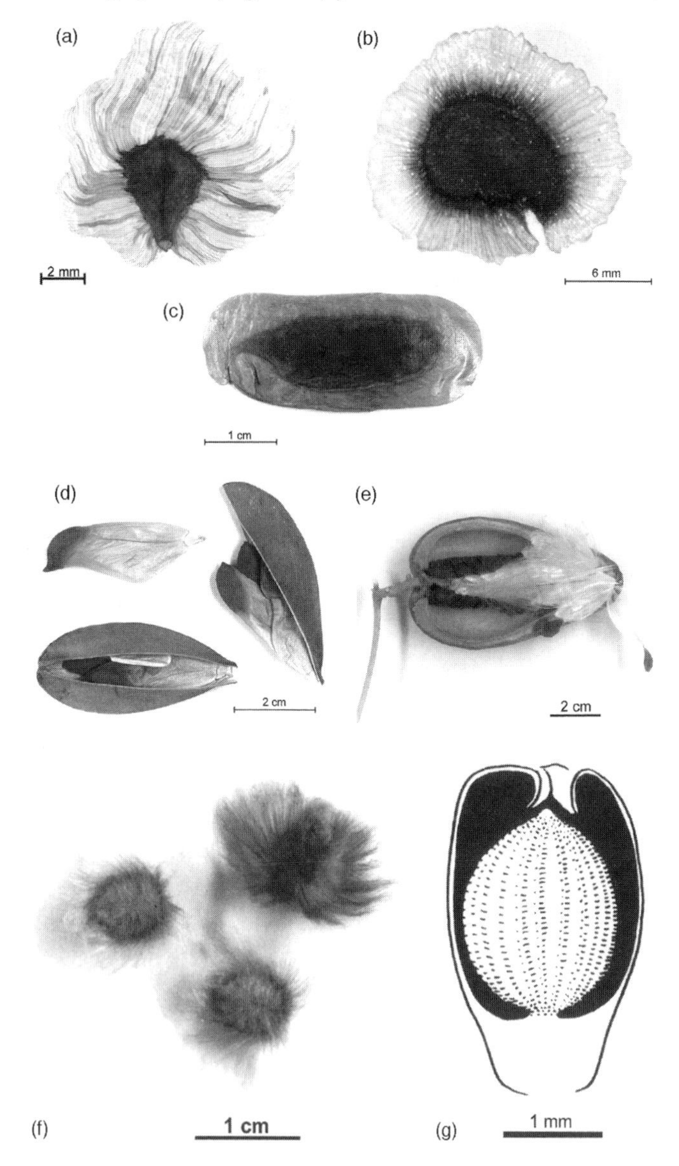

Fig. S.14. Testas aiding dispersal. (a) *Pandorea pandorana* (Bignoniaceae); seed with a peripheral wing. (b) *Spergularia media* (Caryophyllaceae); seed with a peripheral wing. (c) *Newtonia hildebrandtii* (Fabaceae); seed with a peripheral wing. (d) *Hippocratea parviflora* (Celastraceae); fruits with seeds bearing a wing on the raphal side. (e) *Marsdenia* sp. (Apocynaceae); seeds with a coma (i.e. a micropylar tuft of hairs). (f) *Ipomoea kituiensis* (Convolvulaceae); seed covered in hairs produced by the outer testa epidermis. (g) *Victoria amazonica* (Nymphaeaceae); seed with an aril developed from the funiculus that invests the whole seed and serves as an air float (aril cut away on one side). ((a–f) By Wolfgang Stuppy, copyright RBG Kew; (g) from Corner, E.J.H. (1976) *The Seeds of Dicotyledons.* Cambridge University Press, Cambridge, UK.)

2. Gymnosperms

Gymnospermous seeds display adaptations to a certain mode of dispersal. For example, the seeds of many Pinaceae (e.g. *Pinus*, *Cedrus*), Araucariaceae and *Welwitschia* (Fig. S.15) possess wings facilitating wind dispersal. To assist dispersal by animals, the mature seeds of Taxaceae and Podocarpaceae generally produce a bright red aril partly or entirely enveloping the seed. Similarly, the sarcotesta of the seeds of cycads, *Ginkgo* and *Gnetum* are conspicuously coloured to attract animal dispersers, e.g. bright red in *Encephalartos altensteinii*, pale yellow in *Encephalartos horridus*, orange-red in *Zamia floridana*, red in *Zamia lantifoliolata* and *Macrozamia fraseri* (Colour Plate 15F), salmon in *Microcycas* and white in *Dioon* and *Ceratozamia*. In *Ginkgo* the outer (orange-coloured) fleshy portion of the mature seed is (infamously) rich in butyric acid, emitting an odour like rancid butter. A thick and fleshy aril (which is of different origin than in the angiosperms) is present in Taxaceae and *Podocarpus*. (JD, WS)

Biswas, C. and Johri, B.M. (2001) Reproductive biology of gymnosperms. In: Johri, B.S. and Srivastava, P.S. (eds) *Reproductive Biology of Plants*. Springer, Berlin, Narosa Publishing House, New Delhi, India, pp. 215–236.

Bobrov, A.V.F.C., Melikian, A.P. and Yembaturova, E.Y. (1999) Seed morphology, anatomy and ultrastructure of *Phyllocladus* L.C. & A. Rich. ex Mirb. (Phyllocladaceae (Pilg.) Bessey) in connection with the generic system and phylogeny. *Annals of Botany* 83, 601–618.

Boesewinkel, F.D. and Bouman, F. (1984) The seed: structure. In: Johri, B.M. (ed.) *Embryology of Angiosperms*. Springer, Berlin, pp. 567–610.

Fig. S.15. Female cones of *Welwitschia mirabilis* (Welwitschiaceae), showing the papery winged seeds; though ovules are bitegmic the seedcoat is the outer integument (photo: Wolfgang Stuppy).

Boesewinkel, F.D. and Bouman, F. (1995) The seed: structure and function. In: Kigel, J. and Galili, G. (eds) *Seed Development and Germination.* Marcel Dekker, New York, USA, pp. 1–24.
Bresinsky, A. (1963) Bau, Entwicklungsgeschichte und Inhaltsstoffe der Elaiosomen. *Bibliotheca Botanica* 126, 1–54.
Corner, E.J.H. (1976) *The Seeds of Dicotyledons.* Cambridge University Press, Cambridge, UK.
Kapil, R.N., Bor, J. and Bouman, F. (1980) Seed appendages in angiosperms. I. Introduction. *Botanische Jahrbücher für Systematik* 101, 555–573.
Schnarf, K. (1937) *Anatomie der Gymnospermen-Samen* (*Handbuch der Pflanzenanatomie = Encyclopedia of Plant Anatomy*, Bd. 10, 1) . Borntraeger, Berlin.
Werker, E. (1997) *Seed Anatomy.* Gebrüder Borntraeger, Berlin, Stuttgart (*Handbuch der Pflanzenanatomie = Encyclopedia of Plant Anatomy*, Bd. 10, Teil 3: Spezieller Teil) .

Seedcoats – functions

The **integuments** and later the seedcoat fulfil a wide variety of functions, both during seed development and in the mature seed. The role played by the seedcoat in seed longevity is diverse and changes throughout the life-span of the seed. During seed development the integument(s), i.e. the developing seedcoat can:

- act as a pathway for transport and conversion (e.g. of sucrose, in some cases) of amino acids and sugars from the pericarp into the ovule for the embryo's as well as the endosperm's and seedcoat's own development (**see: Development of seeds – nutrient supply**);
- accumulate temporary reserve materials for later use by the coat cells themselves;
- assist in gas exchange;
- possibly supply growth substances inwards to the growing embryo and outward to maternal organs;
- possibly photosynthesize: chloroplasts which may act photosynthetically are present in seedcoats of some species but their functional significance is unclear;
- protect the embryo and endosperm against desiccation and mechanical injury.

In the mature seed, the seedcoat:

- provides protection for the mature embryo against physical and biological damage;
- sometimes develops special structures assisting dispersal (e.g. sarcotesta, wings, etc.);
- acts as a regulator of water uptake – in some species it maintains seed dormancy by preventing water absorption and/or gaseous exchange (**hardseededness**); contrarily, in other species with permeable seedcoats it permits rapid germination when water is available. In some cases, such as in the highly-pigmented coats of certain legumes, the coat prevents too rapid an influx of water to the embryo, which could cause cell damage and excessive leakage of intracellular substances during imbibition of the embryo (**see: Germination – influences of water; Germination – physical factors**).

As a general rule, seeds that have only weak mechanical protection lose their viability or vigour faster in the field than those that are heavily protected. Also, properties such as permeability to water (and gas, **see: Germination – influence of gases**) strongly influence the longevity and germinability of a seed, and seedcoat-imposed **dormancy** (an important aspect of physical dormancy) can be entirely due to a water-impermeable seedcoat (hardseededness; **see: Dormancy**). Water-impermeable seedcoats, as in many Fabaceae and Malvaceae, help to retain a low moisture content within the seeds and therefore help to maintain seed viability for much longer periods of time (**see: Viability**). The hard-coated seeds of *Ipomoea batatas* (Convolvulaceae), for example, can retain their germinability for more than 20 years. Seeds of *Canna* have even proved to remain viable for 600 years, and seeds of the sacred lotus, *Nelumbo nucifera*, have been reported to germinate after more than 1000 years. However, the impermeability of the seedcoat is not an absolute condition for seed longevity. Seeds of some Poaceae and Chenopodiaceae have permeable coats and can still have high longevity.

Apart from acting as a barrier against the uptake of water and the exchange of gas (**see: Germination – influences of gases**), the seedcoat can inhibit germination by presenting a mechanical restraint against radicle protrusion (**see: Germination – radicle emergence**), it can prevent the exit of germination inhibitors from the embryo, or it can supply germination inhibitors to the embryo. In many seeds of the Fabaceae, for example, interference with water uptake is probably the only factor involved. In many cases, however, the situation is more complex and germination is inhibited by more than one effect acting simultaneously. Several of the possible constraints exerted by the coat might act at the same time to keep the seed in a dormant state (**see: Dormancy – coat imposed**).

The water-impermeable properties of the seedcoat are reflected in its anatomical structure. Often, there are one or more layers of tightly packed cells with no pores, intercellular spaces or stomata between them, and water-repellent materials are present either in or on their cell walls.

In a number of families of both dicotyledons and monocotyledons, such as Cannaceae, Convolvulaceae, Convallariaceae, Fabaceae, Malvaceae and Rhamnaceae, a layer of palisade-like macrosclereids, either in the exotesta or the exotegmen (Malpighian cells), is often responsible for impermeability (**see: Seedcoats – structure**, Fig. S.17). Within the impermeable layer, various specific cell zones or substances may be responsible for impermeability of the seedcoat in various species, especially substances deposited within the Malpighian cells, such as:

- Waxy layer over the outermost cell layer.
- Cuticle.
- Mucilages. The mucilage stratum, or sub-cuticular layer, can become hard and water-resistant.
- The zone of light line of Malpighian cells. A light line (*linea lucida*) is found in the Malpighian layer of both permeable and impermeable seeds, and does not seem to play a role in water permeability. Its peculiar light diffraction is due to a high density of cellulose microfibrils, without interfibrillar spaces. (**See: Seedcoats – structure,** Fig. S.17)
- Changes of the micellar structure of cellulose, which occur during dehydration.
- Contraction of the palisade cells on desiccation.
- Water-repellent substances in the cell walls such as cutin, suberin, lignin, callose, phenols and quinones.

The colour and anatomy of the seedcoat can give clues as to whether or not a seed is fully mature. Phenolic substances such as tannins, which are mostly responsible for the brown colouring of many seeds, are only produced at the end of the maturation process. Therefore, a less dark colour can be an indication that a seed is not fully matured. The same is also reflected in an incomplete sclerification (usually by lignification of the cell walls) of the mechanical layer or layers, if present, in the seedcoat. (JD, WS)

(See: Rumination)

Boesewinkel, F.D. and Bouman, F. (1984) The seed: structure. In: Johri, B.M. (ed.) *Embryology of Angiosperms*. Springer, Berlin, pp. 567–610.

Boesewinkel, F.D. and Bouman, F. (1995) The seed: structure and function. In: Kigel, J. and Galili, G. (eds) *Seed Development and Germination*. Marcel Dekker, New York, USA, pp. 1–24.

Corner, E.J.H. (1976) *The Seeds of Dicotyledons*. Cambridge University Press, Cambridge, UK.

Netolitzky, F. (1926) *Anatomie der Angiospermen – Samen (Handbuch der Pflanzenanatomie = Encyclopedia of Plant Anatomy*, Bd. 10) . Borntraeger, Berlin. 1.

Werker, E. (1997) *Seed Anatomy*. Gebrüder Borntraeger, Berlin, Stuttgart *(Handbuch der Pflanzenanatomie = Encyclopedia of Plant Anatomy*, Bd. 10, Teil 3: Spezieller Teil) .

Seedcoats – structure

To protect their contents from mechanical or microbial damage, seeds (the next generation plus its supply of nutrients) are usually covered by an envelope composed of the seedcoat (or testa) (**see: Structure of seeds**, Fig. S.61), or the seedcoat plus **pericarp** (fruit coat). If seeds are borne in indehiscent fruits such as **nuts** or **drupes**, the mechanical protection of the seed is usually achieved by the fruit wall (pericarp) or its inner part (**endocarp**), respectively. If this is the case, the integument(s) often only produces a rudimentary seedcoat without differentiation of any mechanical tissues (e.g. Anacardiaceae, Apiaceae, Cornaceae, Fagaceae, Juglandaceae, Sapindaceae, Urticaceae). Undifferentiated or degenerated seedcoats are also present in small (dust-like) anemochorous (wind-dispersed) seeds of dehiscent fruits, e.g. in Orchidaceae (Fig. S.16), where the seedcoat consists of a single layer of thin-walled transparent cells originating from the exotesta.

The structure of fully developed seedcoats, however, displays an enormous diversity. The seedcoat is primarily formed by the integument(s) and to various extents by the **raphe** and the **chalaza**; only in **pachychalazal** seeds is the seedcoat formed almost exclusively by chalazal tissue (e.g. in Sapindaceae, Lauraceae, Meliaceae, Tropaeolaceae).

The mechanical layer, usually a one- or multi-layered **sclerenchyma**, can originate from any layer of the integuments (or chalaza in pachychalazal seeds) or, in rare cases, from the **funiculus** (Cactaceae: Opuntioideae).

1. Angiosperms

While the term testa is generally used synonymously with that of seedcoat (as is generally the case in this Encyclopedia), in his 1976 two-volume work *Seeds of the Dicotyledons* Corner suggested a terminology for the seedcoat that is explained below (and is used in articles in which specific details are discussed concerning this structure). The derivatives of the outer **integument** form the 'testa' and those of the inner integument form the 'tegmen'. The product of the single integument of unitegmic ovules he also called – somewhat inconsistently – 'testa'. Further to that, he called seeds with a characteristic testa, 'testal' and seeds with a characteristic tegmen, 'tegmic'. Corner's main conclusion was that the most distinctive character of the seedcoat lies in the position and structure of the main mechanical layer. This mechanical layer is usually composed of thick-walled cells, variously impregnated with lignin, suberin or cutin, and can be one or more cells thick. The shape of the cells varies from simple cuboid cells to radially elongate, palisade-like (Malpighian) cells (Figs S.17, S.18) through to horizontally elongate fibres, whereby all these cell types can have either evenly or unevenly thickened walls. Depending in which layer within the outer or inner integument the mechanical tissue develops, Corner distinguished the following categories within testal seeds:

(a) *exotestal* (outer epidermis of the outer integument, e.g. Begoniaceae (Fig. S.20), Fabaceae, Ranunculaceae, Rhamnaceae);

(b) *mesotestal* (middle layer(s) of the outer integument), e.g. Paeoniaceae, Myrtaceae, Rosaceae, Theaceae, Fig. S.21);

(c) *endotestal* (inner epidermis of the outer integument e.g. Brassicaceae, Grossulariaceae (Fig. S.22) Dilleniaceae, Magnoliaceae, Myristicaceae).

Where the whole or part of the outer integument differentiates into a fleshy tissue it is called a **sarcotesta**.

Tegmic seeds are distinguished into:

(a) *exotegmic seeds* (outer epidermis of the inner integument, e.g. Celastraceae, Clusiaceae, Euphorbiaceae (Fig. S.18), Linaceae, Malvaceae (Fig. S.23) , Phyllanthaceae (Fig. S.19));

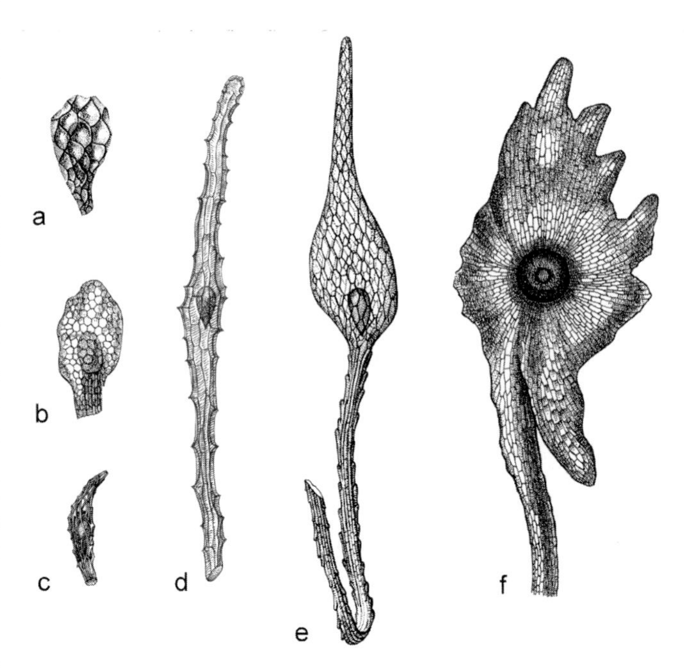

Fig. S.16. Seeds of Orchidaceae. (a) *Eulophia streptopetala*. (b) *Stanhopea aurea*. (c) *Cattleya loddegesii*. (d) *Acanthephippium bicolor*. (e) *Epidendrum cinnabarinum*. (f) *Cyrtosia lindleyana*. Taken from Takhtadzhyan (= Takhtajan), A. (ed.) (1985) *Anatomia Seminum Comparativa*, Vol. 1. Nauka, Leningrad (in Russian).

(b) *mesotegmic seeds* (middle layer(s) of the inner integument; this construction is rare and found in conjunction with an exotegmen or an endotegmen);

(c) *endotegmic seeds* (inner epidermis of the inner integument, e.g. Nandinaceae, now included in Berberidaceae, (Fig. S.24), Piperaceae, Saururaceae).

Most angiosperm families have one predominant mechanical layer and fall within one of these categories; but some seedcoats have two or three of these characteristics, e.g. those of Myristicaceae and Cucurbitaceae.

In the monocotyledons, the seedcoat structure is much less diverse than in dicotyledons; there are only a few exotestal and endotestal seeds and few, if any, tegmic ones.

Apart from thick-walled, mechanical tissues the seedcoat may provide additional protection through hardened compressed cell layers (seed surface in Fig. S.25) or by depositing toxic compounds (e.g. **phenolics**) in the cells to deter predators. A particular class of these phenolic substances is the dark-coloured tannins and their derivatives, such as quinones. Tannins are very common in seedcoats, especially in the walls and/or the lumen of the innermost epidermis, which they then stain conspicuously amber-brown. However, up until the last stage of development, these substances generally appear colourless and therefore can serve as an important indicator of seed maturity. Phenolics, like tannins, benefit the seed through the cross-linking of polysaccharides, which leads to increased wall rigidity and resistance to microorganisms.

As mentioned above, the seedcoat can be reduced in seeds that develop in indehiscent fruits. There are some rare and extreme cases, however, where the seedcoat might even be entirely absent. So-called ategmic ovules, for example, which never develop any integuments, or where they are much reduced, are present in five dicotyledonous families, four of which contain parasitic plants (Balanophoraceae, Loranthaceae, Santalaceae, Viscaceae). A second cause of a total loss of a seedcoat is that the integuments are totally absorbed during seed development, as happens in some Apocynaceae, Menispermaceae and Rubiaceae.

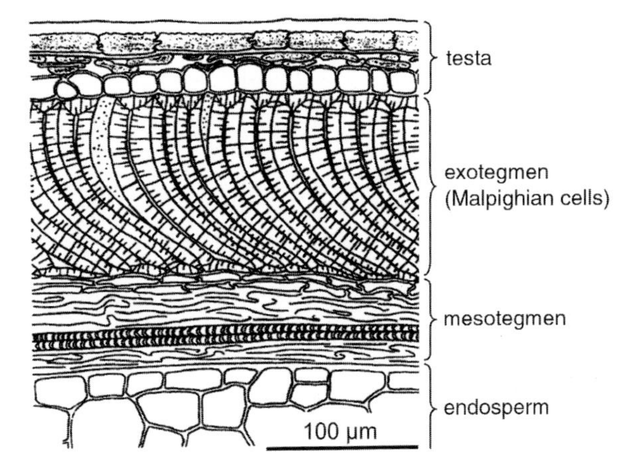

Fig. S.18. Longitudinal section of the exotegmic seedcoat of *Beyeria viscosa* (Euphorbiaceae). Shown is the typical exotegmic palisade of bent Malpighian cells without light line (*linea lucida*) as they are characteristic of the entire family (*sensu* Angiosperm Phylogeny Group (APG) (2003). An update of the Angiosperm Phylogeny Group classification for the orders and families of flowering plants: APG II. *Botanical Journal of the Linnean Society* 141, 399–436.), the basal end of the palisade generally points towards the micropyle. From Stuppy, W. (1996) Systematic morphology and anatomy of the seeds of the biovulate Euphorbiaceae. PhD Dissertation, Dept of Biology, University of Kaiserslautern, Germany (in German) .

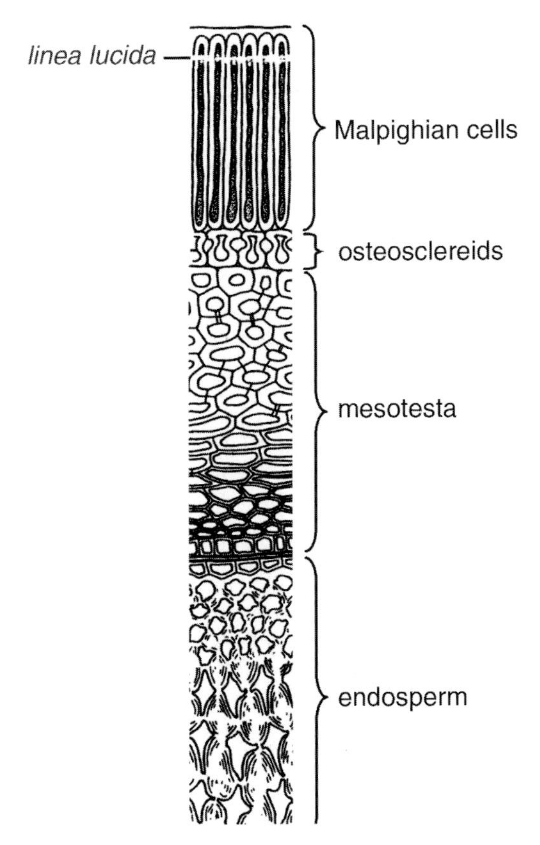

Fig. S.17. Cross section of the exotestal seedcoat of *Ceratonia siliqua*, carob (Fabaceae). Shown are the typical exotestal palisade layer of Malpighian cells with light line (*linea lucida*) and outer hypodermis of osteosclereids (bone-shaped cells) as they are characteristic of the entire family. From Fahn, A. (1990) *Plant Anatomy*, 4th edn. Pergamon Press, Oxford.

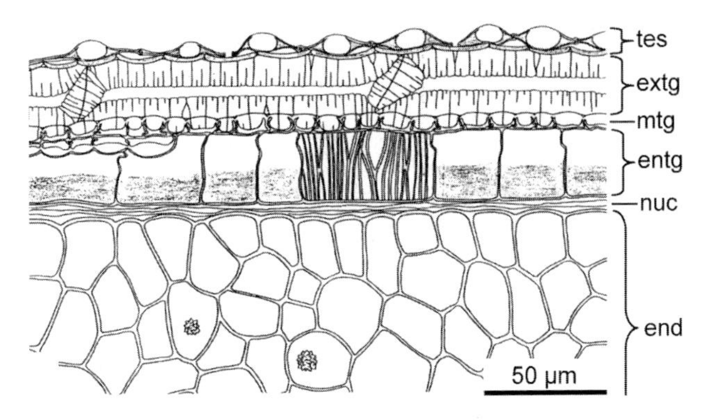

Fig. S.19. Longitudinal section of the exotegmic seedcoat of *Phyllanthus acidus* (Phyllanthaceae). The fibrous exotegmen is shown (tes = testa; extg = exotegmen; mtg = mesotegmen; entg = endotegmen; nuc = nucellus remains; end = endosperm). From Stuppy, W. (1996) *Systematic morphology and anatomy of the seeds of the biovulate Euphorbiaceae.* PhD Dissertation, Dept of Biology, University of Kaiserslautern, Germany (in German).

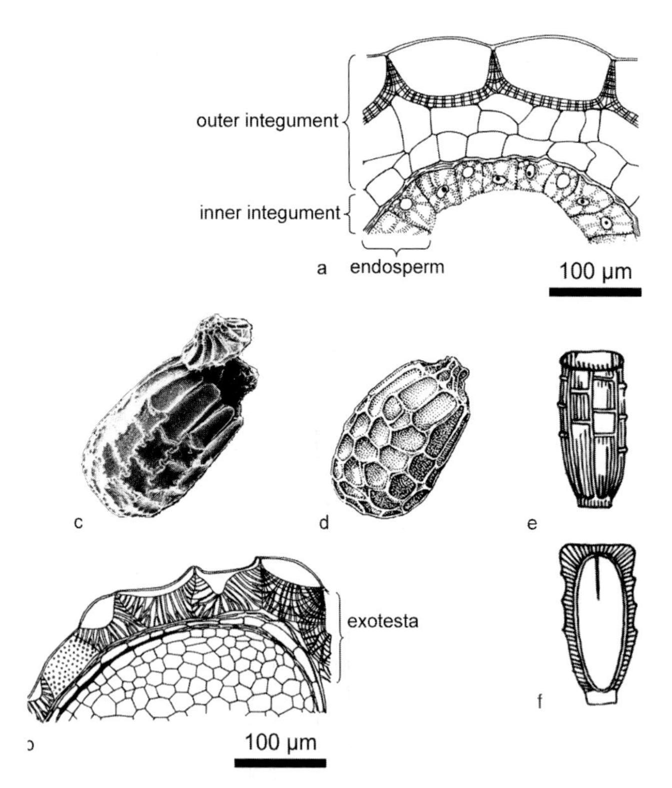

Fig. S.20. Seeds of Begoniaceae. (a) Cross section through developing and (b) mature exotestal seedcoat of *Begonia squamulosa* showing the three-layered outer integument; the mature seedcoat is exotestal with the radial and inner walls of the cells of the outer epidermis of the outer integument significantly thickened. (c) Germinating seed of *Begonia leptotricha* showing the lifting of the circumscissile lid (operculum) at the micropylar end. (d) Seed of *Begonia epipsila*; (e) and (f) Seed and longitudinal section of seed of *Begonia lobata*, the latter showing the straight storage-containing embryo (the endosperm is reduced to a single layer surrounding the embryo). (a, b) From Boesewinkel, F.D. and de Lange, A. (1983) Development of ovule and seed in *Begonia squamulata* Hook. f. *Acta Botanica Neerlandica* 32, 417–425. (c) From Boesewinkel, F.D. and Bouman, F. (1984) The seed: structure. In: Johri, B.M. (ed.) *Embryology of Angiosperms*. Springer, Berlin, pp. 567–610; (d) From Bouman, F. and de Lange, A. (1983) Structure, micromorphology of Begonia seeds. *The Begonian* 50, 70–78, 90–91. (e) From Irmscher, E. (1925) Begoniaceae. In: Engler, A. (ed.) *Die natürlichen Pflanzenfamilien*, Vol. 21, 2nd edn. Wilhelm Engelmann, Leipzig.

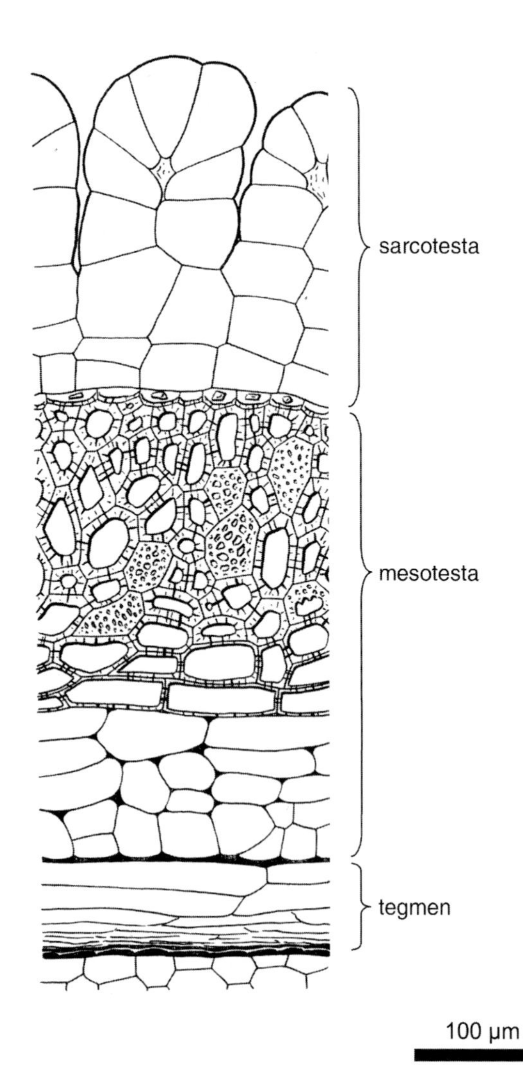

Fig. S.21. Transverse section through the mesotestal seedcoat of *Anneslea fragrans* (Theaceae). The outer testa epidermis is divided periclinally to form a 2–4-layered red sarcotesta; the mesotesta produces a thick outer sclerotic layer of lignified, pitted, isodiametric or shortly oblong cells (10–14 cells thick), the outermost cells often with a crystal and thin outer wall, and a thinner inner layer (3–7 cells thick) of thin-walled cells; the inner epidermis of the testa consists of small unspecialized cells; the tegmen remains unspecialized and is crushed in the mature seed. From Corner, E.J.H. (1976) *The Seeds of Dicotyledons*. Cambridge University Press, Cambridge, UK.

2. Gymnosperms

The typical gymnospermous seedcoat comprises three layers: an outer fleshy parenchymatous layer (sarcotesta), a middle sclerenchymatous (stony) layer (sclerotesta), and an innermost parenchymatous layer (**endotesta**), which generally collapses at maturity and forms a thin membranous layer (Fig. S.26). The thickness and number of cell layers can vary among these. For example, in some species of *Cycas* (*C. circinalis* and close relatives) the endotesta develops into a thick spongy layer that causes the seeds to float in water. The most important layer, however, is the sclerotesta, which provides the seed's mechanical protection. In many hard seeds the mature seedcoat consists almost exclusively of the sclerenchymatic layer (sclerotesta) while the other layers are desiccated and shrunk (e.g. in most Pinaceae the sarcotesta is only two or three cell layers thick). The sclerotesta itself is either rather homogeneous, or again composed of different layers. The latter behaviour is most prominently displayed by the Cycadaceae. Examples of seeds with homogeneous sclerotestae are certain *Pinus* species. The very thick sclerotesta of *Pinus cembra*, for example, uniformly consists of polygonous, isodiametric cells.

The epidermis covering the sarcotesta shows various modifications in the different families. In Pinaceae, the epidermis of the free part of the seed surface contributes towards the formation of the wing. The shape and arrangement of the epidermis cells can vary tremendously and stomata are found regularly. In Cycadaceae the cells of the outer epidermis are generally thick-walled and often form hairs which later, however, disappear during seed maturation.

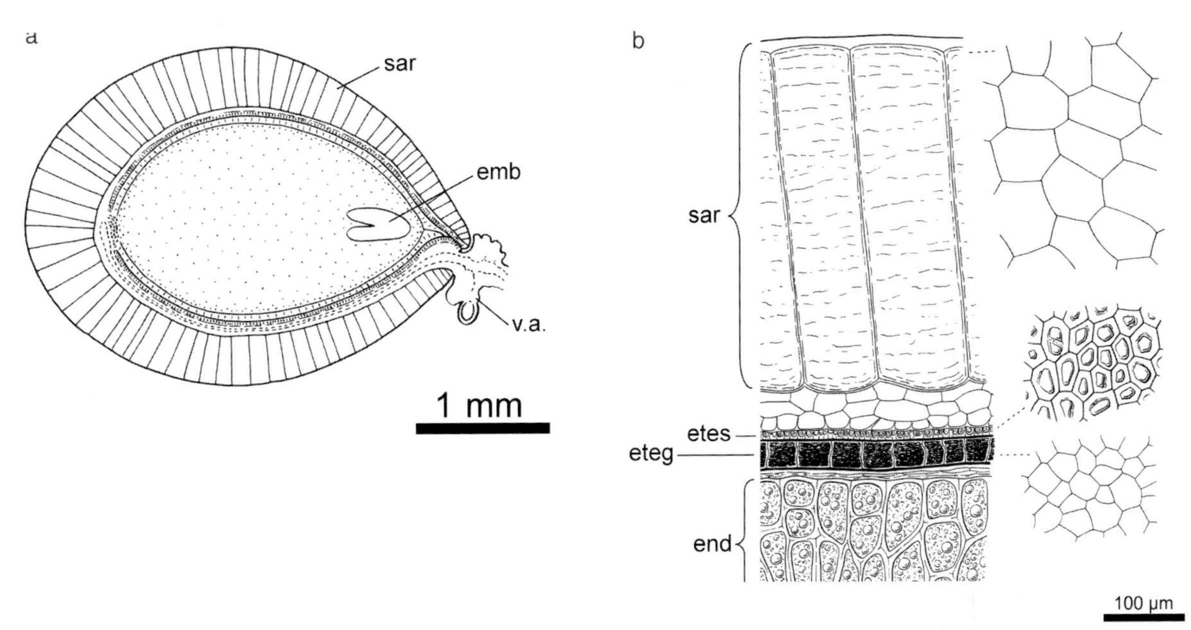

Fig. S.22. *Ribes sanguineum*, winter currant (Grossulariaceae). (a) Longitudinal section of seed containing a small embryo embedded in copious endosperm. (b) Transverse section of the endotestal seedcoat: the anatropous seeds of the Grossulariaceae are generally characterized by a sarcotesta mainly formed by the palisade-like cells of the outer testa epidermis, an endotesta consisting of a layer of small cuboid crystal-cells with thickened and lignified radial inner walls and a two-layered tegmen whose outer epidermis collapses while the inner epidermis develops into a layer of enlarged thin-walled cells with firm brown tanniniferous contents inside which is the endosperm; the drawings on the right show the respective layers in surface view (emb = embryo; v.a. = vestigial aril; sar = sarcotesta produced by the outer epidermis of the outer integument; etes = endotesta; eteg = endotegmen; end = endosperm). From Corner, E.J.H. (1976) *The Seeds of Dicotyledons*. Cambridge University Press, Cambridge, UK.

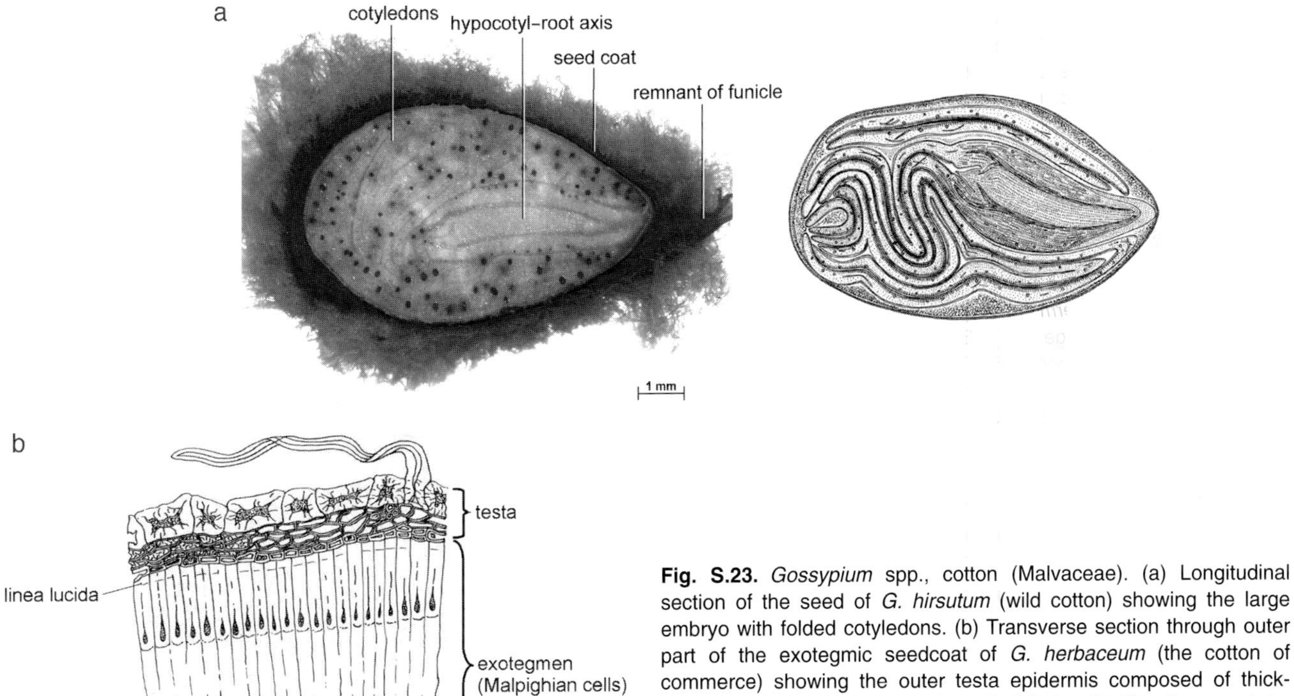

Fig. S.23. *Gossypium* spp., cotton (Malvaceae). (a) Longitudinal section of the seed of *G. hirsutum* (wild cotton) showing the large embryo with folded cotyledons. (b) Transverse section through outer part of the exotegmic seedcoat of *G. herbaceum* (the cotton of commerce) showing the outer testa epidermis composed of thick-walled cuboid cells with tannin and of hair-cells (forming aseptate, unbranched trichomes) and the exotegmic palisade layer of Malpighian cells with light line (*linea lucida*) as it is typical of the entire family. (a) A photograph by Wolfgang Stuppy, copyright RBG Kew, drawing from Takhtadzhyan (= Takhtajan), A. (ed.) (1992) *Anatomia Seminum Comparativa*, Vol. 4. Nauka, Leningrad (in Russian) ; (b) From Vaughan, J.G. (1970) *The Structure and Utilisation of Oil Seeds*. Chapman & Hall, London, UK.

The above covers those cases where only one integument is present. In the different groups of gymnosperms, however, ovules and seeds are covered by further envelopes whose significance for the formation of the seedcoat can be different. For example, in *Gnetum* all three integuments seem to contribute to the seedcoat, whereas in *Ephedra* and *Welwitschia* only the outer integument is involved. The wing of the seeds of *Welwitschia* originates from the outer integument.

The seedcoat is either formed by both chalaza and integument (Cupressaceae, *Gnetum*, *Ephedra*) or the seeds develop from pachychalazal ovules in which case the seedcoat develops mainly from the chalazal portion of the ovule. Pachychalazal ovules and seeds are quite frequently present in gymnosperms, e.g. in most cycads, the Pinaceae and the genus *Cephalotaxus*. (JD, WS)

(See: Gymnosperm seeds; Seedcoats – development; Seedcoats – dispersal aids; Seedcoats – functions; Tree seeds)

Angiosperm Phylogeny Group (APG) (2003) An update of the Angiosperm Phylogeny Group classification for the orders and families of flowering plants: APG II. *Botanical Journal of the Linnean Society* 141, 399–436.

Biswas, C. and Johri, B.M. (2001) Reproductive biology of gymnosperms. In: Johri, B.S. and Srivastava, P.S. (eds) *Reproductive Biology of Plants*. Springer, Berlin and Narosa Publishing House, New Delhi, India, pp. 215–236.

Boesewinkel, F.D. and Bouman, F. (1995) The seed: structure and function. In: Kigel, J. and Galili, G. (eds) *Seed Development and Germination*. Marcel Dekker, New York, USA, pp. 1–24.

Bresinsky, A. (1963) Bau, Entwicklungsgeschichte und Inhaltsstoffe der Elaiosomen. *Bibliotheca Botanica* 126, 1–54.

Corner, E.J.H. (1976) *The Seeds of Dicotyledons*. Cambridge University Press, Cambridge, UK.

Kapil, R.N., Bor, J. and Bouman, F. (1980) Seed appendages in angiosperms. I. Introduction. *Botanische Jahrbücher für Systematik* 101, 555–573.

Moise, J.A., Han, S., Gudynaite-Savitch, L., Johnson, D.A. and Miki, B.L. (2005) Seed coats: structure, development, composition and biotechnology. *In Vitro Cellular and Development Biology – Plant* 41, 620–644.

Netolitzky, F. (1926) Anatomie der Angiospermen – Samen (*Handbuch der Pflanzenanatomie* = *Encyclopedia of Plant Anatomy*, Bd. 10). Borntraeger, Berlin, Germany.

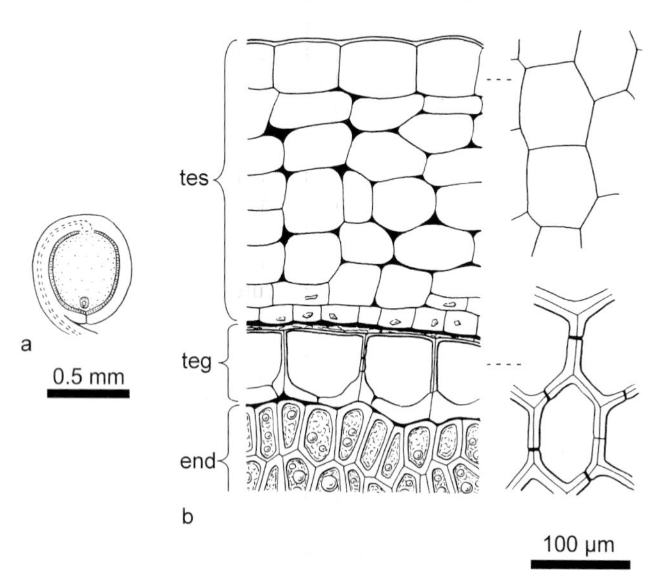

Fig. S.24. *Nandina domestica*, heavenly bamboo (Nandinaceae, now included in Berberidaceae). (a) Longitudinal section of seed showing the small embryo embedded in copious endosperm. (b) Transverse section of the endotegmic seedcoat which could be described as sarcotestal with a lignified endotegmen (the rest of the tegmen is crushed at maturity) but the sarcotesta collapses at maturity and does not function in seed dispersal; the drawings to the left of the transverse section show the outer testa epidermis and the endotegmen in surface view (tes = testa; teg = tegmen; end = endosperm). From Corner, E.J.H. (1976) *The Seeds of Dicotyledons*. Cambridge University Press, Cambridge, UK.

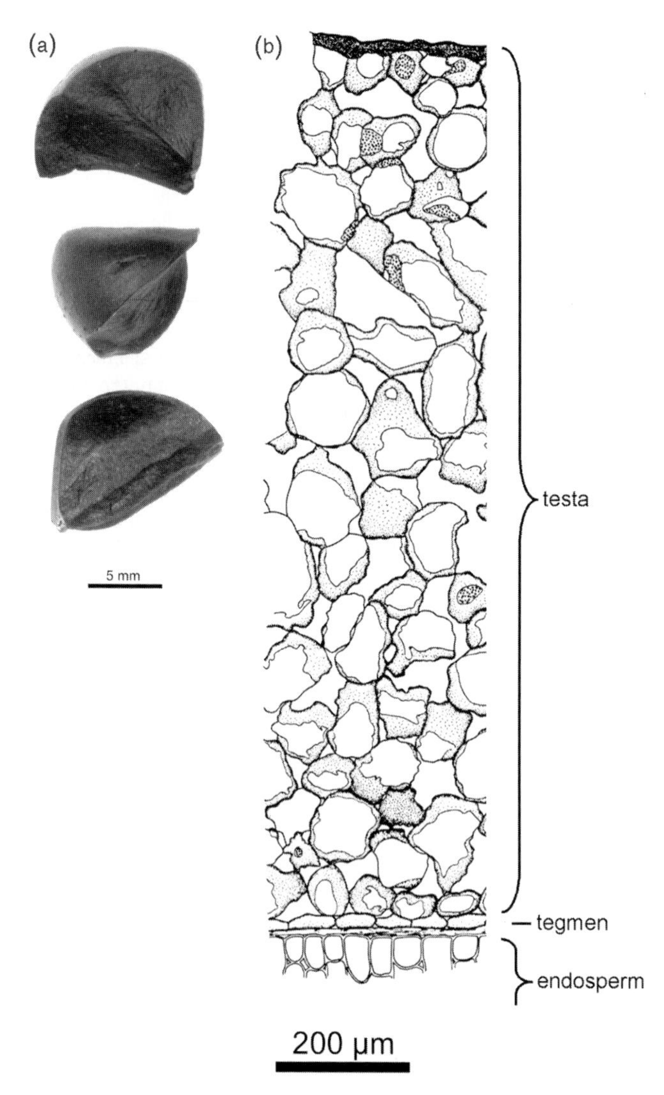

Fig. S.25. *Pancratium maritimum*, sea daffodil (Amaryllidaceae). (a) Seeds. (b) Transverse section of seedcoat; the thick protective outer layer is formed by several compressed cell layers. The walls of the cells of the entire outer integument, which constitutes the seedcoat, are impregnated with black quinones while the cell lumen is empty. (a) By Wolfgang Stuppy; (b) from Werker, E. and Fahn, A. (1975) Seed anatomy of *Pancratium* species from three different habitats. *Botanical Gazette* 136, 396–403.

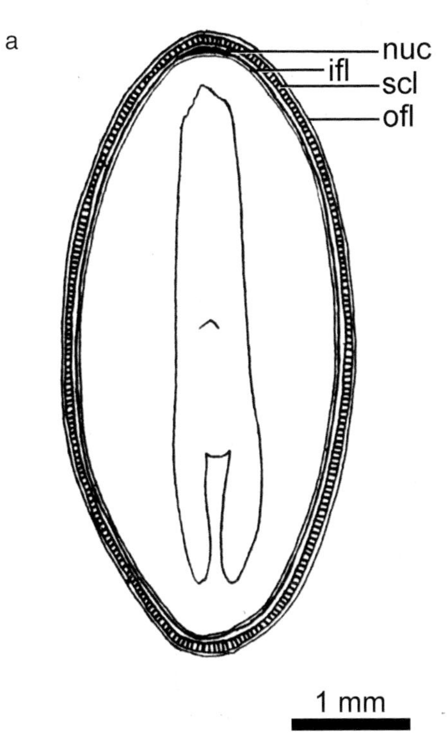

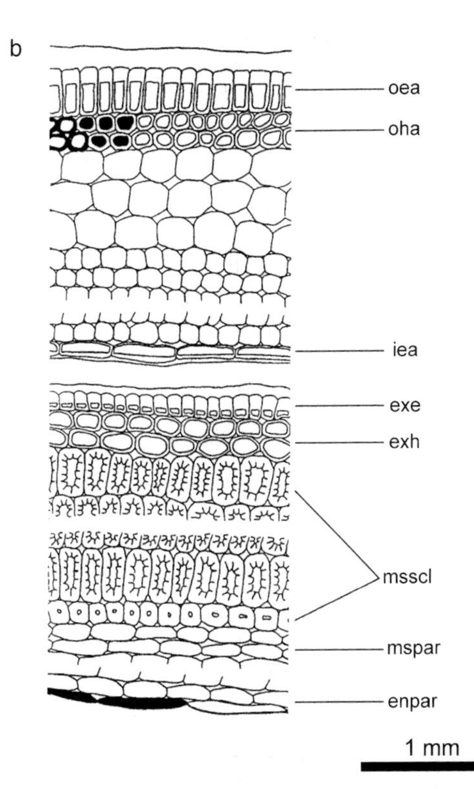

Fig. S.26. Seeds of the Coniferae. (a) Transmedian longitudinal section of the seed of *Pinus laricio* (Pinaceae) showing the three layers of the seedcoat (nuc = remains of the nucellus; ifl = remains of the inner fleshy (parenchymatous) layer (endotesta); scl = middle sclerenchymatous layer (sclerotesta); ofl = remains of the outer fleshy (parenchymatous) layer (sarcotesta)). (b) Transverse section of the aril and seedcoat of *Phyllocladus alpinus* (Phyllocladaceae) (oea = outer epidermis of the aril; oha = outer hypodermis of the aril; iea = inner epidermis of the aril; exe = exotesta; exh = exotesta: hypodermis of seedcoat; msscl = mesotesta sclerenchyma; mspar = mesotesta parenchyma; enpar = endotesta parenchyma). (a) From Coulter, J.M. and Chamberlain, C.J. (1901) *Morphology of Spermatophytes.* Appleton and Company, New York; (b) from Bobrov, A.V.F.C., Melikian, A.P. and Yembaturova, E.Y. (1999) Seed morphology, anatomy and ultrastructure of *Phyllocladus* L.C. & A. Rich. ex Mirb. (Phyllocladaceae (Pilg.) Bessey) in connection with the generic system and phylogeny. *Annals of Botany* 83, 601–618.

Schnarf, K. (1937) Anatomie der Gymnospermen-Samen (*Handbuch der Pflanzenanatomie = Encyclopedia of Plant Anatomy*, Bd. 10, 1) . Borntraeger, Berlin.

Stuppy, W. (2002) Seed characters and generic classification of Opuntioideae. *Succulent Plant Research* 6, 25–58.

Takhtadzhyan (= Takhtajan), A. (ed.) (1992) *Anatomia Seminum Comparativa*, Vol. 4. Nauka, Leningrad (in Russian) .

Werker, E. (1997) *Seed anatomy*. Gebrüder Borntraeger, Berlin, Stuttgart (*Handbuch der Pflanzenanatomie = Encyclopedia of Plant Anatomy*, Spezieller Teil, Bd. 10, Teil 3) .

Seeder

Device for sowing seed. (**See: Planting equipment**)

Seedless fruit

There is a strong consumer preference for soft fruit that lacks evident seeds, particularly where the seeds are naturally quite large, hard or indigestible stones, pips or pits. Such seedlessness is achieved using a range of breeding, selection and cultivation techniques.

Parthenocarpy (fruit development without fertilization) is a widespread trait of great antiquity in some clonal seedless crops, such as bananas and breadfruit, and the discovery of spontaneous mutants in citrus fruit led to the development of seedless varieties of oranges, such as the navel type. Pineapple is a **self-incompatible** parthenocarpic species and so seedless fruit is produced where only one cloned variety is grown. In practice many of the flowering clonal crops that have been selected for a vegetative or fruit product, such as *Rubus* fruits and strawberries, present often very acute flowering or fertility problems, and can only be bred from with difficulty, if at all. In bananas, for instance, many clones are so seed- and pollen-sterile that they cannot be bred by ordinary sexual means. Alternatively, parthenocarpic fruit set in some **cucurbits** can be induced by growth regulators, such as chlorflurenol (a morphactin growth inhibitor that interferes with **auxin** transport), in 'once-over' cucumber production for example. Indeed, experimentally, **auxins** or **gibberellins** can be used to induce parthenocarpy in several species.

Stenospermocarpy – the abortion of seed development due to mutant genes – occurs in vine grape, and has been used to develop a wide range of seedless varieties. In seedless table grapes, either the ovule or embryo aborts after pollination or fruit set. In these cases, the **embryo rescue** technique is widely used in breeding programmes, to obtain plants from crosses between lines. Some varieties are sprayed with **gibberellic acid** (GA_3) growth regulator to stimulate enlargement of the seedless berry clusters to an acceptable size.

Triploids, which cannot complete normal **meiosis** and so produce sterile seed, are used to produce seedless annual herbaceous crops. For example, in plants grown from F_1 hybrid triploid watermelon seed, which is produced by crossing a 'normal' **diploid** male parent with a tetraploid female parent obtained by doubling the **chromosome number**, the seeds normally abort shortly after pollination, leaving pale 'empty' **seedcoats** within the fruit.

A selection of vegetable crops, such as cucumbers, zucchini/courgettes, mangetout peas and 'baby corn' cobs, are cultivated to be superficially seedless by harvesting their fruits and pods whilst they are immature. (PH)

Rotino, G.L., Perri, E., Zottini, M., Sommer, H. and Spena, A. (1997) Genetic engineering of parthenocarpic plants. *Nature Biotechnology* 15, 1398–1401.

Seedling

The germinated seed develops into a seedling then a young plant. The stage at which a seedling becomes a young plant is rather arbitrary but it relates to the production of a few adult-type leaves. In agronomy, that developmental stage has been established by consensus, such as in maize, rice and wheat. Seedling growth may be **hypogeal** or **epigeal**. The timing of seedling **emergence** is a key factor in the establishment of cultivated crops (**see: Germination – field emergence models**). Some field crops (such as tobacco, true potato seed, and some rice, onion, cucurbits and horticultural brassicas) and most greenhouse plants, including those for **hydroponic production**, are established from seedlings raised in separate nurseries (**see: Transplants**).

Seedling disease complex

A set of soilborne diseases, caused by one or more of a group of soil-living fungal **pathogens** including the oomycte fungi, *Pythium*, *Phytophthora* and *Aphanomyces* spp., basidiomycetes such as *Rhizoctonia solani*, and hyphomycetes such as *Fusarium* spp. Symptoms, sometimes termed 'seedling **blight**', include:

- pre- and post-emergence seedling **damping-off** – **rotting** of the stem base near the surface of the soil, leading to pinching-off at the ground level or **wilting**;
- **wirestem** in older seedlings;
- soft, wet, tan-brown areas at or near the tips of young roots.

In severe infections very poor stands may require re-seeding.

These diseases can be controlled by use of non-systemic fungicide seed treatments, such as thiram and metalaxyl and high vigour seed, avoiding sowing in cold, wet soils (or warm moist soils in the cases of *Aphanomyces* and *Rhizoctonia*) and using seedlots with cracked seedcoats, in mechanically vulnerable species such as maize and sweetcorn. Few species are known to have natural resistance to *Pythium* and *Rhizoctonia solani*, though the latter is the target of breeding programmes. (**See: Disease; Pathogens**) (PH)

Seeds in space

Interest in germination and survival of seeds in space is part of the development of a biogenerative life support system (BLSS) which is needed to sustain long-term space exploration and colonization. Considerable technical difficulties have to be overcome to provide plants in space with an appropriate simulated environment in which to grow and reproduce. Limiting factors include the influence of microgravity (10^{-4}–10^{-6} *g*), lack of air movement, build-up of gases, e.g. ethylene, and high gamma radiation; these have resulted in poor plant growth and failure to produce viable seed.

Brassica rapa plants grown on the Mir space station in appropriately vented conditions underwent normal plant development, flowering, pollination and early seed development. But later during seed development they exhibited changes in the pattern of seed ripening compared to earth-bound controls. Specifically, ripening occurred from the tip of the **siliques** to the point of their attachment to the plant, whereas ripening of the seed was uniform throughout the siliques in the controls. This has been suggested to be the result of a metabolic gas gradient build-up inside the seeds in microgravity due to the absence of normal convection of the O_2 and CO_2 that occurs at 1 *g*. Even more striking was the difference in reserve deposition between the space-station-grown seeds and ground-control seeds. The latter synthesized storage protein and oil, whereas in the former substantial amounts of starch were additionally present in the mature seed. Protein bodies in the space-grown seeds were larger in number, oil bodies were fewer but larger, and distinct starch grains were present. The normal pattern of oil deposition in *Brassica* seeds is that it is formed after the synthesis of starch, and during its subsequent hydrolysis, using carbon from the sugars released from this polymer as it is degraded. (**See: Transitory starch**) Microgravity therefore upsets the pattern of some developmental events, which has implications for the vigour of seeds, particularly during early seedling growth, when there is a strong dependence upon stored reserves. Additionally, the nutritional quality of seeds grown in space could be compromised, resulting in seeds being an unpredictable source of nutrients in a BLSS.

Long-term survival of seeds in space has been tested in a Long Duration Exposure Facility (LDEF) during which they were exposed to microgravity for 6 years. Two million seeds, representing 106 species, were kept in sealed canisters or vented to expose the seeds to vacuum. Seeds in both conditions survived, but more in the sealed canisters. However, plants grown from maize seeds from an LDEF exhibited somatic mutations, including dwarfing and leaf discoloration, likely the result of long exposure of the seeds to higher gamma (cosmic) irradiation. Again, this is an undesirable consequence in respect of the development of long-term support systems, although in other LDEF experiments manipulating the storage conditions (e.g. shielding seeds from irradiation; regulating O_2 concentrations) has been shown to influence the subsequent germinability and vigour of seeds.

Research into seeds in space is still in its infancy, and since only relatively few experiments have been carried out, with seeds launched into space in a variety of conditions and facilities, it is not surprising that results have often been variable, and conclusions inconsistent. (JDB)

Musgrave, M.E. (2002) Seeds in space. *Seed Science Research* 12, 1–16.

Seed-to-seed production

Seed production in plots or fields established by planting with seed (in contrast to vegetative propagules, such as roots, bulbs and seedling transplants). (See: **Production for Sowing, I. Principles**)

Selection

Any process, whether in controlled breeding systems (Table S.6) or in nature (natural selection), which permits an increase in the proportion of certain genotypes of desired characteristics or landraces or groups in succeeding generations, by the preferential survival and reproduction or preferential elimination of individuals. Also, the resulting plant, line or strain produced by such selection process.

Selection, for example, is practised at all stages of pedigree breeding and at intervals in backcrossing (**see: Cross**), and is also used to extract variants from an existing cultivar or to eliminate off-types from an older cultivar. (PH)

Table S.6. Some major selection procedures used in plant breeding.

Mass selection	A large number of superior-looking plants are selected on the basis of appearance and harvested in bulk, and their seed is combined and replanted. Their progeny is further selected for the preferred characteristics, and the process is repeated for as many generations as is desired.
Progeny selection	Choice of breeding stock on the basis of the performance or testing of their offspring or descendants.
Pure-line selection	(1) numerous superior-appearing plants are selected from a genetically variable population; (2) progenies of the individual plant selections are grown and evaluated by simple observation for as long as possible, frequently for several years; and then (3) parameters are measured in extensive trials to determine whether the remaining selections are superior in yielding ability and other aspects of performance.
Bulk-population selection	In self-pollinated crops, populations are propagated as bulks until segregation has virtually ceased, before selection is initiated.
Pedigree selection	Pedigree methods are most useful to combine favourable traits from two or more parents. It requires growing a large F_2 population as this generation has the most variation. Selected plants produce seeds by inbreeding. From each selected plant, an F_3 family is grown for further selection. Inbreeding and selection may continue several more generations (up to eight) to develop true-breeding lines. The number of generations depends upon the differences between the original parents.
Marker-aided selection	Identification of one or more gene markers that are closely linked to genes associated with desirable traits, for which it is difficult or impossible to select directly, such as in the genetics of reduced **dormancy** or **preharvest sprouting**.

Simmonds, N.W. and Smartt, J. (1999) *Principles of Crop Improvement*, 2nd edn. Blackwell Science, Oxford, UK.

Self-incompatibility systems

Naturally occurring mechanisms found in angiosperms that inhibit the penetration or growth of the pollen tube into the stigma. Of the several kinds of SI systems that can be exploited for hybrid breeding and seed production, only the genetic homomorphic–sporophytic system has been used much in practice. The sporophytic SI system is widely used in the hybrid seed production of some diploid horticultural *Brassica* varieties, in the production of single-, double- and three-way-crosses, though it is not currently used for any of the major *Brassica* field crops, such as tetraploid **canola** rape.

Homomorphic SI systems (species with similar flowering structures on pollen-bearing and seed-bearing plants) depend on the existence of multiple **alleles** at the incompatibility locus, *S*, which determine the interactions of the style and the pollen grain. In essence, pollen or pollen tube growth is inhibited if the *S* alleles differ in the female and male parent, based on numerous and sometimes complex dominance patterns. Incompatibility is determined by a protein secretion over the surface of the stigma just prior to anthesis, which acts as a barrier to penetration by the germinating pollen grains. (In the alternative gametophytic system the hindrance to pollen tube growth lies in the style.) In the sporophytic SI systems, found in crucifers and composites, amongst other dicots, the pollen-producing plant dominates the interaction.

All SI systems used in hybrid seed production depend on at least one inbred line (for the female parent as a minimum) or several inbred lines, each of which are homozygous for their appropriate *S* allele. Double-cross systems, involving four *S* alleles, are used to produce large quantities of seed to overcome the yield limitations in single-cross systems.

Because self-incompatible parent lines are intrinsically infertile, they must be propagated by special, and expensive, methods that allow self-pollination. In the widely used 'bud pollination' (or 'bud selfing') method, the barrier is bypassed by placing pollen by hand on an immature stigma that has not yet developed the incompatibility reaction. However, this process, when repeated over generations, tends to select for self-fertility, and such lines eventually tend to produce in-line crosses (so-called '**sibs**') in sufficient proportions to cause problems with the quality of the resulting hybrid seed, particularly in varieties where a high degree of uniformity is essential, such as in Brussels sprouts. In some lines, seed set from selfing increases greatly if temperatures are high. (PH)

(**See: Brassica – horticultural**, Fig. B.5; **Production for sowing, III. Hybrids**)

Self-pollination and -fertilization

The process and outcome of pollination of the stigma by pollen produced on anthers within the same flower, or another flower on the same plant or within the same clone (also known as **autogamy**), usually resulting in self-fertilization or -seed setting ('selfing'). Self-pollinating plants therefore breed true-to-type so that any seed collected is similar to the parent. Examples of such species are shown in Table S.7.

Table S.7. Common crop plants which are normally and predominantly self-pollinated.

Barley	Bean	Chickpea	Cotton	Cowpea
Crambe	Flax (linseed)	Lentil	Lettuce	Millet (finger, foxtail)
Mungbean	Oat	Pea	Peanut (groundnut)	Pepper[a]
Potato	Rice	Sesame	Soybean	Tobacco
	Tomato	Triticale	Vetch	Wheat

[a]Also capable of appreciable cross-fertilization (allogamy).

Various mechanisms have evolved in Nature to ensure self-pollination in angiosperms. In grasses and small-grain cereals, for example, the flower is enclosed by bracts, and so is almost always pollinated from within. Most of the early group of domesticated plants are herbaceous annuals (with readily harvested and unusually large seeds, fruits or tubers) that are capable of selfing; their dependability would have been high for the first farmers because, as well as breeding true, they do not need a separate pollinator for successful reproduction. In most cases, therefore, the annual selfing habit in modern crop plants has been derived directly from their native progenitors. However, cultivated annual cotton evolved from wild perennial shrubs, and the self-pollinating cultivated soybean was converted from a wild outcrosser into a cultivated inbreeder. The production of hybrids in self-pollinated crops requires special genetic or manual techniques to ensure that out-crossing takes place. (PH)
(See: Compatibility; Cytoplasmic male sterility; Fertilization; Pollination; Production for sowing, III. Hybrids; Self-incompatibility system)

Hancock, J.E. (2003) *Plant Evolution and the Origin of Crop Species*, 2nd edn. CAB International, Wallingford, UK.

Seminal roots

Roots formed in the mature cereal or grass embryo, composed of the **radicle** (primary root) and lateral seminal roots, are collectively called seminal roots. Lateral seminal roots are formed from the tissue just above the scutellar node. Post-embryonic roots, also termed adventitious roots or crown roots, are produced after germination. In cereal plants, post-embryonic roots develop from the upper part of nodes, which mark the position on the stems where leaves are produced, including the base of the coleoptile. In maize, roots formed from nodes above ground are also called brace roots. Both seminal and post-embryonic roots develop branched structures, called lateral roots. These root systems differentiate to produce root hairs, which play a role in water and nutrition uptake. Often, primary root developed from the radicle diminishes early and does not contribute to the mature root system. In maize and rice, **mutants** have been isolated that lack lateral seminal and crown roots without effect on the formation of the radicle and primary roots. These mutants include *crown rootless 1* (*crl1*) in rice, *rootless* (*rtl*) and *rtcs* (*rootless for crown and seminal roots*) in maize. The isolation of such mutants indicates that the initiation of nodal roots requires genetic pathways distinct from those necessary for embryonic radicle development. **(See: Root meristem)** (O-AO, HS)

Senescence

A genetically programmed terminal developmental process which may involve the whole plant after a single reproductive cycle (monocarpic senescence). For example, many annual plants, including crop plants (e.g. wheat, soybeans), yellow and die during seed production even under optimal growth conditions. In the case of canola all leaves senesce prior to seed fill and the only remaining photosynthetic organs are the pod walls. Regulated degradation of endogenous protein for nitrogen recycling is a key component of senescence and involves expression of genes encoding for hydrolytic enzymes, such as proteases, ribonucleases and lipases. In soybean, the onset of seed filling triggers leaf senescence and remobilization of nitrogen compounds derived from metabolic degradation of protein accumulated in specialized leaf mesophyll cells. Seed removal delays this process. However, usually the onset of leaf senescence is controlled by the delivery of root-derived cytokinins to the leaves via the xylem. **(See: Programmed cell death)** (MT-H, CEO, JWP)

Separation distance

The spatial isolation required between a seed field and other sources of mechanical and genetic contamination, especially between cross-pollinated varieties. Also known as the isolation distance. **(See: Production for sowing, II. Agronomy, 4. Isolation; III. Hybrids, 2. Isolation)**

Separators

Equipment of various designs used in seed conditioning. Disc, cylinder or spiral separators exploit differences in one dimension of length, width or thickness where there is no appreciable difference in the other two. The vibratory separator separates rough or flat seeds from spherical ones on an inclined vibrating deck. Magnetic separators take advantage of the surface texture of seed, based on differential retention of finely ground iron powder. **(See: Conditioning, II. Cleaning)**

Serotiny

In some plant communities a seed bank is retained on the plant, a phenomenon usually called serotiny, which is strongly linked to fire, after which the seeds are released. Models suggest that serotiny is favoured by moderately frequent fires, so that the gap between fires does not exceed the life span of the canopy seed bank, and by low probabilities of recruitment between fires. It is also clear that serotiny is favoured by storage of seed in relatively massive structures (e.g. the cones of *Pinus* and Proteaceae) that protect the seeds from **predation** and from fire. In *Pinus halepensis*, a common Mediterranean tree, a substantial part of the annual seed crop is retained on the tree and many of these seeds can survive for up to 20 years. Canopy and soil seed banks often coexist. For example, in Greece both pines and *Cistus* spp. recruit after fires, the former by serotiny and the latter from a **soil seed bank**. Serotiny (in this context often called bradyspory) is common in desert species where seed release is linked to rainfall (**see: Deserts and arid lands; Dispersal**). (KT)

Sesame

1. The crop

Sesame (gingerlee) (*Sesamum indicum* (syn. *S. orientale*), 2n = 26, Pedaliaceae) is an annual plant grown in tropical and temperate zones within 40° of the equator. Major producers are India, China, Myanmar and Sudan, but many developing countries in Asia and Africa produce the crop for local consumption. The crop produces well under high temperatures and on stored moisture. However, low yields and the need to hand harvest, due to the absence of non-**shattering** varieties, limits the area of production. (**See: Crop Atlas Appendix, Map 2; Oilseeds – major,** Table O.1)

2. Origin

Archaeological evidence indicates that sesame was a prized oil crop at least 4000 years ago in present day Iraq, Syria and Pakistan. However, the distribution of wild related species suggests the centre of origin to be either India or Ethiopia.

3. The fruit and seed

The sesame **fruit** is a capsule (2.5–8.0 cm long by 0.5–2.0 cm in diameter), deeply grooved, usually flat sided and cylindrical in shape with a triangular beak. Each capsule contains 50–100 seeds and dehisces along the septa when mature. Seeds are small (2.5–3 mm long × 1.5 mm wide), pear-shaped, slightly flattened, surface smooth or reticulate weighing 2–4 g/1000 (Fig. S.27). Seed colour varies from white through yellow to black. Dark-coloured seeds have a higher oil content, but light-coloured seeds produce a desirable light-coloured oil and white seeds are preferred by the confectionery and baking trade. The seed contains about 47% oil and 25% high quality protein.

The ridged outer epidermis is a single layer of cells containing cap-shaped clusters of calcium oxalate crystals. The rest of the **testa** is made up of two layers of partially flattened cells containing single calcium oxalate crystals. A 'yellow membrane' links the testa to the two- to five-cell-thick **endosperm** made up of parenchyma containing **oil** and **protein bodies**. The **cotyledons** of the **embryo** have a single layer of palisade cells on their inner surface. The remainder of the cotyledons are made up of isodiametric cells containing oil and protein bodies with inclusions.

Fig. S.27. Sesame seeds (scale = mm) (photograph kindly produced by Ralph Underwood of AAFC, Saskatoon Research Centre, Canada).

4. Uses

Sesame seeds are primarily used as a source of high quality edible oil noted for its long shelf life and stability, due to the presence of the antioxidants, sesamol, sesamin and sesamolin (**see: Oilseeds – major,** Table O.2 for **fatty acid** composition). It is sometimes first heat-treated ('toasted') to confer a unique flavour. Low grade oil may be used in soap making. The oil is also used as a carrier for medicines and cosmetics. The meal, depending on the variety and oil extraction efficiency, may contain 35–50% **storage protein** that is high in methionine but low in lysine amino acid content. The meal may be processed into flour for human consumption but most is fed locally to animals. Whole seeds, often dehulled, are used in confectionery and baking operations. Ground seeds form the basis of the Middle Eastern and Mediterranean *tahini* and *halva*. (KD)

Ashri, A. (1989) Sesame. In: Robbelen, G., Downey, R.K. and Ashri, A. (eds) *Oil Crops of the World*. McGraw-Hill, New York, USA, pp. 375–387.

Deshpande, S.S., Deshpande, U.S. and Salunkhe, D.K. (1996) Sesame oil. In: Hui, Y.H. (ed.) *Bailey's Industrial Oil & Fat Products*, Vol. 2, 5th edn. John Wiley, New York, USA, pp. 457–496.

Vaughan, J.G. (1970) *The Structure and Utilization of Oil Seeds*. Chapman and Hall, London, UK.

Shattering

The loss of seed from the plant by the opening (dehiscence) or abscission of a fully mature fruit or seed-bearing organ (pod, ear, etc.): synonymous with 'shedding'. Many wild plants shatter readily at maturity, this being one of the major natural seed dispersal mechanisms. In cultivated cereal grains and podded crops such as the pea, non-shattering mutants were doubtless selected for during the earliest days of crop **domestication**, whether intentionally or not, and this remains a feature in some breeding programmes and a challenge in the domestication of novel crops.

Shattering is caused by the dissolution of cell walls in the abscission zones. The middle lamella of the cell wall has been shown in some species to be the major site of wall degradation, with disassembly of the primary and secondary cell walls varying according to the species. Several enzymes are involved in wall dissolution, e.g. cellulases, β-1,4-glucanases and polygalacturonases. In several cases it has been shown that these enzymes are up-regulated before abscission proceeds. **Hormones**, especially **ethylene** and **auxin**, have been implicated in the regulation of abscission, the latter hormone acting negatively, for example in *Brassica napus* silique shatter.

Pod shatter (fruit dehiscence) in *Arabidopsis* is controlled jointly by two genes *SHATTERPROOF* (*SHP1*) and *SHATTERPROOF2* (*SHP2*), which are *MADS box* genes. The two genes control differentiation of the dehiscence zone at the valve of the fruit (**a siliqua**) (thus allowing escape of the seeds) and promote lignification of adjacent cells. Shattering does not occur in the double mutant, *shp1 shp2* (Fig. S.28). Other genes have been identified which exert negative control over abscission and some which are involved in seed detachment from the fruit wall.

In some cultivated plants the shattering trait has been completely overcome as a consequence of selecting for gross

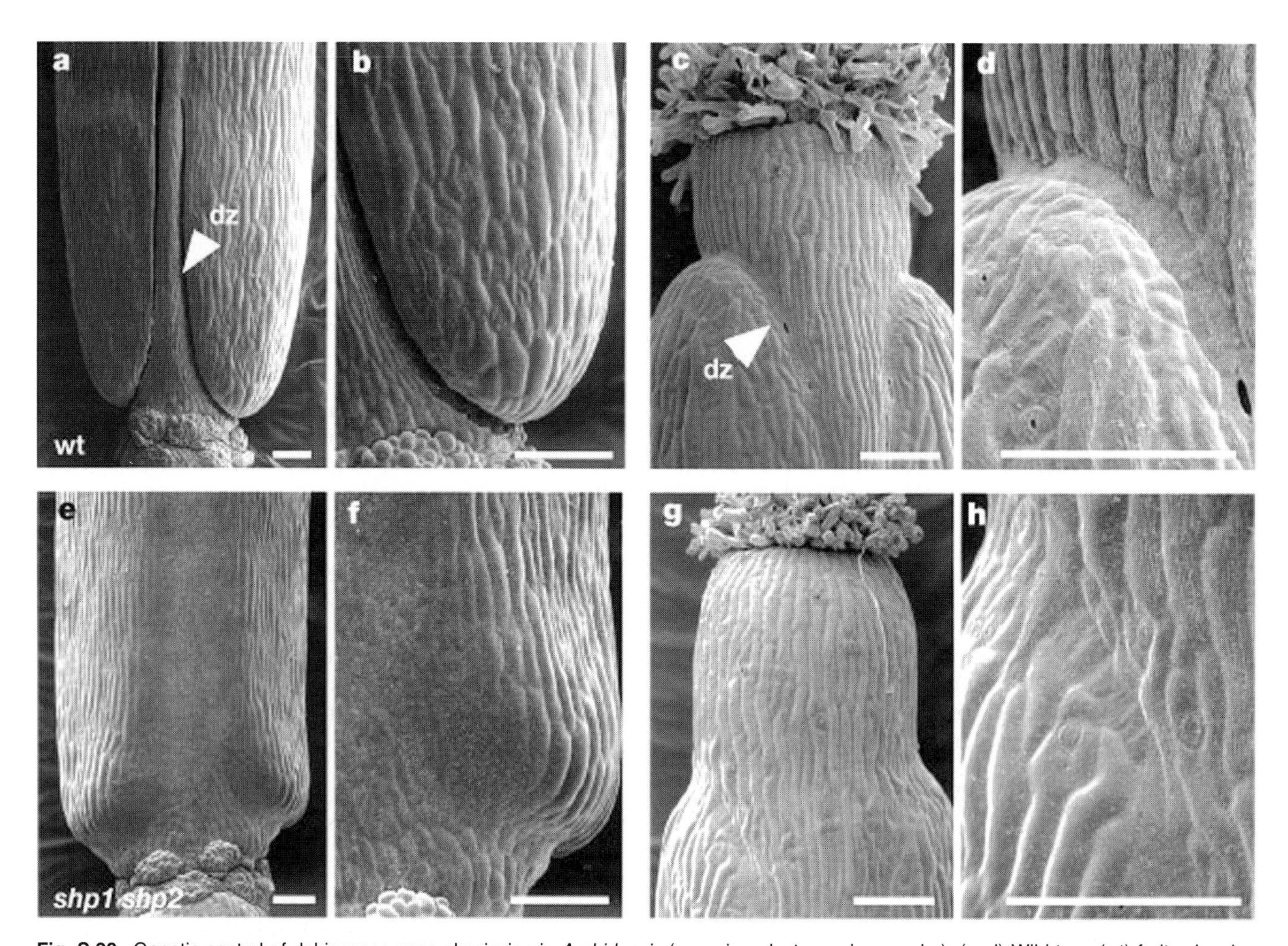

Fig. S.28. Genetic control of dehiscence zone abscission in *Arabidopsis* (scanning electron micrographs). (a–d) Wild-type (wt) fruits showing development of the dehiscence zones (dz); (a, b) separation of the valves from the replum is apparent at the fruit base. (e–h) *shp1 shp2* fruit: there is no development of dehiscence zones. Scale bars, 100 μm. From Liljegren, S.J., Ditta, G.S., Eshed, Y., Savidge, B., Bowman, J.L. and Yanofsky, M.F. (2000) *SHATTERPROOF* MADS-box genes control seed dispersal in *Arabidopsis*. *Nature* 404, 766–770. By permission of the Nature Publishing Group (www.nature.com) and the authors.

changes in morphology from the original wild type, most notably in the ear of maize, wheat and other cereals. In several modern crops, especially those with indeterminate flowering patterns where the floral structure contains seeds at different stages of maturity (such as oilseed rape and sunflower), a proportion of the seed-bearing structures on the plant or flower head is always dehiscent at harvest time. In these cases some shattering occurs readily in the field and during transportation, leading to seed loss and also the risk of establishing populations of 'volunteer' plants in the soil seed bank, which can cause contamination of subsequent crops. Losses, for example, can be considerable in oilseed rape/canola (often over 20%). For seed production purposes, shattering can be minimized by **windrowing** crops just as the seed reaches full maturity, and allowing it to cure for several days before harvesting. (MB, PH) **(See: Einkorn)**

Liljegren, S.J., Ditta, G.S., Eshed, Y., Savidge, B., Bowman, J.L. and Yanofsky, M.F. (2000) *SHATTERPROOF* MADS-box genes control seed dispersal in *Arabidopsis*. *Nature* 404, 766–770.

Pinopych, A., Ditta, G.S., Savidge, B., Liljegren, S.J., Baumann, E., Wiseman, E. and Yanofsky, M.F. (2003) Assessing the redundancy of MADS-box genes during carpel and ovule development. *Nature* 424, 85–88.

Rose, J.K.C., Catalá, C., Gonzalez-Carranza, Z.H. and Roberts, J.A. (2003) Cell-wall disassembly. In: Rose, J.K.C. (ed.) *The Plant Cell Wall*. Annual Plant Reviews, 8. Blackwell Publishing, Oxford, UK, pp. 264–324.

Shedding

See: Shattering

Shelling

In maize seed and grain production, the removal of seeds from the central **cob** of the dehusked ear – effectively precleaning. Also, generally in the food industry, the removal of a seed from its outer and often hard shell, hull, husk, cob or pod, such as **nuts** or drupes with woody or stony endocarp, but also legume pods and sunflower cypsels (**see: Fruit – types**).

Maize grain shelling is mechanized in the field using

specialized 'picker-sheller' or combine-harvester equipment. Alternatively, static shelling devices using burrs and lugs are used in many parts of world (or in some places the job is simply done by beating with a flexible stick); historically a great variety of hand- and motor-operated picker-wheel or spiked-disc type implements were used. For seed production purposes, maize shellers use a gentle rubbing/shearing action to minimize damage. (**See: Maize – cultivation**)

The removal of hard external shells involves the application of abrasion, crushing, cutting or high-impact shattering forces. For the fresh-consumption edible 'nut' market, the goal is usually to maximize the percentage of the higher-value whole **meats**. Further processing recovers saleable fragments, removes defects (such as 'stick-tight' adhering shell fragments) and foreign material, such as by oscillating screens and air separators (see: **Conditioning**) and **colour sorting**. By contrast, **palm** nuts require only crude shell breakage to extract the kernels for grinding into powder.

Very hard nuts, such as **walnut** and **macadamia**, require considerable mechanical force: machines use combinations of cutting blades or rollers with a fixed base plate. **Hazelnuts** for example are shelled using cones, rollers or traditional stone compressors, with sufficient clearance to avoid injuring the kernel; damage can be minimised by adjusting moisture content to 10%. Similarly, **Brazil nuts** are cracked after pre-treatment with high-pressure steam, and **chestnut** peel, made brittle by high temperature pre-treatment (brulage), is removed along with the pellicle by rubber-ended paddles moving against steel rods.

Cashew nuts are shelled manually, where labour costs are cheap: cutters, consisting of a pair of curved blades, score through the shell leaving the kernel untouched – a process which must be done with care, as the caustic oil contained in small pockets within the shell can cause skin irritation or severe burning in some people. In **pistachio**, where the endocarp splits naturally prior to maturity making the kernels easy to extract without mechanical cracking, the nut is largely marketed in-shell after removal of the mesocarp hull by abrasion.

Fresh vining peas and dry **peas** and **beans** are shelled from their pods by combine-harvesting in the field. Alternatively, dry pods gathered in **windrows** are shelled in hulling cylinders with abrasive surfaces, such as carborundum, or by crushing with revolving metal thrasher bars, combined with screens that allow the seed to pass through. Higher impact forces are used to shell **sunflower** seeds. (PH)
(**See: Scarification**)

Shoot meristem

A group of undifferentiated cells that become organized and proliferate to produce above-ground tissues and organs such as leaves, inflorescence and flowers. Based on functionality, the shoot meristem can be divided into three regions: central zone, which contains actively dividing undifferentiated cells; peripheral zone, derived from and surrounding the central zone, and serves as the region of differentiation into organ primordia; and rib zone, localized under the central zone, and contributes to the stem as well as acting as the regulatory region for maintaining proper meristem size. Meristem size and activity influence the speed and number of organs produced from shoots. In cereal embryos, the shoot meristem usually produces several leaf primordia before maturation. Despite functional similarity, shoot meristems appear to be controlled by a set of genes that are distinct from the ones establishing root meristem development. (O-AO, HS)
(**See: Embryonic axis; Plumule**)

Short-day plant

Species that initiate flowers best under short-day (long-night) regimes, although they may flower under long days after receiving the appropriate daylength.

In soybean, for example, a cultivar will not flower unless exposed to its characteristic long-night requirement; if grown in latitudes with shorter summer day-lengths, flowering is hastened, and yield and quality may be reduced. Hence a range of different 'maturity groups' are bred for sowing at different latitudes.

Shrinkage

Shrinkage is the loss of gross weight that occurs in the handling or treating of bulk grain, due to several factors, including physical loss of grain and **dust** during transportation and processing, combined with loss of moisture during handling, storage and drying. Resulting discrepancies (which are reportedly of the order of 0.2% at grain elevators in Canada, for example) can present a concern in trade. (**See: Storage management**)

SI system

See: **Self-incompatibility system**

Sib

In hybrid seed production: sibs are unwanted inbred seed contaminants arising from the self-pollination, instead of the desired crossing, of the parental lines in systems that rely on **self-incompatibility** (SI). The term is most notably used in the context of horticultural brassica F_1 hybrid production, where the parental SI lines are maintained by **bud pollination** (or bud selfing). Although these parental lines are intrinsically infertile, that process tends to select for a degree of self-fertility over time. As a consequence some self- or 'sister–brother' fertilization occurs almost invariably during the hybrid crossing stage, producing sib seed contaminants in a sufficient frequency to cause difficulty, notably in horticultural brassicas where the demand for uniformity is high. Sib plants are undesirable because they both represent a loss of genetic purity and lack hybrid vigour (**heterosis**). Sib seed impurities are very difficult to remove by conditioning and sorting. Their presence is commonly determined by greenhouse grow-out trials and isozyme analysis.

In a different, plant-breeding sense of the term, so-called half-sib and full-sib selection designs mate plants that have one or both parents in common, respectively; these inbreeding techniques are used both for selecting pure lines and for estimating genetic parameters within lines. (PH)

Signal transduction – an overview

When terminology is referred to in bold italics **see: Signal transduction – some terminology**.

Signal transduction represents the means by which internal and external signals are perceived then transmitted, both between cells and within the cell, and ultimately integrated to elicit a biological response. The transmission of signals among and within cells (that occurs in many **dormancy** and **germination** processes and in reserve mobilization) is frequently mediated by plant **hormones** such as **abscisic acid** (ABA), **auxin**, **gibberellin (GA)**, **ethylene (ethene)**, **brassinosteroids** and others. The presence and concentration of particular plant hormones or other primary signals are thought to be registered by specific *receptors*, which may be located at the cell surface in the plasma membrane, or within the cell itself. Light signals from the environment, which affect dormancy and germination, are also perceived by receptors (**see: Cryptochrome; Phytochrome**) and they eventually effect a biological change through transfer of information to the ultimate site of response. Sensitivity to hormones and to environmental factors, therefore, can depend on whether or not an appropriate receptor is present.

Once these primary signals have been perceived by the recipient cells, they must be converted into an altered physiological response. This action might ultimately involve changes in cell biochemistry, frequently involving changes in *gene expression*. Since the perception of the primary signal is generally spatially separated from the site of action (e.g. signal perception at the plasma membrane followed by changes in gene expression in the nucleus), signalling intermediates are required to relay information between these various compartments. Signalling intermediates come in many forms including proteins, *calcium* ions (Ca^{2+}), *inositol triphosphate*, *cyclic nucleotides* and others. The sum of the signalling intermediates that pass information from the site of perception to the site of action for a given stimulus is termed a signal transduction pathway. The term pathway is used to illustrate a series of sequential events by which the signal is relayed between intermediates each time the perceived stimulus is converted into a response. However, in plants, not all signal transduction pathways are linear.

An important general principle in signal transduction pathways is that sequential changes occur in the conformation of protein signalling intermediates. Direct interaction between the proteins themselves can result in conformation modifications, for example by the addition by an enzyme of a phosphate group (**phosphorylation**). Changes in the concentration of low molecular weight compounds referred to as *second messengers* (e.g. calcium Ca^{2+}, inositol triphosphate, cyclic nucleotides) can also affect the conformation of protein signalling intermediates. For example, Ca^{2+} ions act by binding to a specific protein named **calmodulin**, whose resultant change in conformation enables it to bind directly to other proteins and in turn change their conformation. The sequential changes in the conformation of the protein signalling intermediates represent the relay of the signal along a signal transduction pathway. A common example of the relay of information between proteins is mediated by enzymes called **protein kinases** which are responsible for the addition of a phosphate group (phosphorylation) on to another protein, as mentioned above. Protein phosphorylation (by **kinases**) and *dephosphorylation* (by *phosphatases*) are important ways of altering protein conformation and interactability. In a signalling cascade one kinase may phosphorylate and thereby activate a downstream kinase that in turn can phosphorylate another protein resulting in sequential changes in protein conformation. The interactions of proteins with small molecules and protein–protein interactions are not exclusive since signalling pathways often involve elements of both.

A biological signalling system must have two key properties: (i) specificity, such that distinct signals will initiate the signalling process; and (ii) sensitivity, so the perceived signal can be amplified in the relay or cascade and many downstream components (the targets of the signal transduction pathway) can be modulated. In plants, specific signalling pathways do not function in isolation. This is reflected by observations that different stimuli are capable of eliciting the same response in plant growth and development. In seeds, for example, the hormones gibberellin, ethylene and brassinosteroids are all individually capable of breaking seed dormancy or promoting germination in some species. The regulation of germination by multiple input variables has adaptive significance since many environmental variables must be assessed before the irreversible commitment to germinate. Protein signalling intermediates from different pathways are capable of interacting with each other to alter their activity, such that signals from different stimuli can converge to regulate common targets. The ability of one signalling pathway to influence another is termed *cross-talk*. Cross-talk does not occur exclusively in the form of protein–protein interactions, since the sensitivity to one hormone may be affected by another. For example, mutations in ethylene signalling pathway components lead to changes in abscisic acid signalling during germination, and the effects of both ethylene and abscisic acid are dependent on ambient sugar concentrations (**see: Sugar sensing**). Also, seeds unable to produce ABA are up to 100 times more sensitive to added GA than those capable of ABA synthesis. These complex biochemical interactions complicate our comprehension of signalling. As a consequence of cross-talk it can be difficult to analyse the specificity of signalling intermediates. For example, many different stimuli result in enhanced concentrations of intracellular calcium, yet the outcomes of the stimuli may differ. How does the cell distinguish between the second messenger arising from one signal or another? In the case of calcium, it is thought that intracellular Ca^{2+} oscillates in a signal-specific fashion. One aspect of specificity in signalling that is emerging involves the regulation of protein abundance (and thus activity) by specifically targeted degradation (enzymatic hydrolysis) by the **proteasome**; proteins thus destined for breakdown are phosphorylated by a kinase in the signalling cascade, and then tagged by a small protein called *ubiquitin*. This allows the tagged protein to be recognized by the proteasome complex that in turn degrades the protein. Protein degradation by the proteasome often acts to inhibit the response of the signalling pathway, an important element in the regulation of signal transduction pathways. The selective degradation of these response inhibitors, or negative regulators, by the proteasome represents the activation of the signalling pathway which is otherwise kept turned off, or inactive. These negative regulators therefore function as molecular switches in signalling pathways, keeping the pathway inactive until their targeted destruction by the proteasome.

Other important effects in signalling are exercised by sugars such as sucrose which, for example, interacts with hormonal signals and the 'cross talk' among hormones.

The ultimate outcome of signal transduction is often a change in gene expression. This is driven by the action of ***transcription factors***, proteins that bind specific sequences in the promoter regions of gene, and thus assist RNA polymerase in transcribing that particular gene. Activation or deactivation of specific transcription factors is frequently the end result of a signalling cascade and is therefore the ultimate mechanism by which expression of genes is signal-regulated. Transcription factors have been frequently isolated as components of signal transduction pathways.

Signalling intermediates in plants have been isolated following the identification of **mutants** with impaired responses, particularly in the model plant *Arabidopsis thaliana*. Seed germination assays have proven very useful in the identification of such *Arabidopsis* mutants, since this plant produces many small seeds of which hundreds may be assayed in a single Petri dish. Genetic, biochemical, physiological and cell biological studies with other seeds such as tobacco, barley and wild oats have also been very valuable. Consequently, many components of signal transduction have been identified that affect processes in seeds. (PM, GWB)

(See: Signal transduction – hormones; Signal transduction – light and temperature)

Gibson, S. (2004) Sugar and phytohormone response pathways: navigating a signalling network. *Journal of Experimental Botany* 55, 253–264.

Hancock, J. (1997) *Cell Signalling*. Longman, UK.

Hare, D., Seo, H.S., Yang, J.-Y. and Chua, N.-H. (2003) Modulation of sensitivity and selectivity in plant signalling by proteasomal destabilisation. *Current Opinion in Plant Biology* 6, 453–462.

Napier, R. (2004) Plant hormone binding sites. *Annals of Botany* 93, 227–233.

Trewavas, A. (2000) Signal perception and transduction. In: Buchanan, B.B., Gruissem, W. and Jones, R.L. (eds) *Biochemistry and Molecular Biology of Plants*. American Society of Plant Biology, Rockville, MD, USA, pp. 930–987.

Signal transduction - hormones

(See: Signal transduction – some terminology for terms in bold italics)

Numerous processes in seeds such as **development, maturation, dormancy, germination, reserve deposition** and **mobilization** involve the participation of hormones as regulatory factors. To effect a particular response the hormonal signal is perceived by a receptor, then transduced through a series of intermediates in the signal transduction pathway (**see: Signal transduction – an overview**). The signal transduction pathways for the hormones having action in seeds are discussed in this article. Knowledge of this subject in seeds is expanding at a rapid rate and as further details are revealed modifications are made to existing concepts.

1. Abscisic acid

Abscisic acid (ABA) is a plant hormone with a central role in the regulation of seed development and control of dormancy (both of which it promotes) and germination (which it inhibits). (**See: Desiccation tolerance – acquisition and loss; Dormancy; Dormancy – acquisition; Dormancy – coat-imposed; Dormancy – embryo (hypotheses); Dormancy – genes; Dormancy breaking; Germination – influences of hormones; Germination – molecular aspects; Germination – radicle emergence; Maturation – controlling factors**) Components of ABA signal transduction have been isolated following the identification of ABA-response **mutants** that are either insensitive to the ABA-mediated inhibition of germination (*ABA insensitive* alleles [*abi*], or hypersensitive to it – *enhanced response to ABA1* [*era1*]. ABA signal transduction in seeds is very complex, likely involving many pathways; this is reflected by the identification of at least ten *abi* alleles capable of circumventing ABA-mediated inhibition of germination.

The molecular biology of stomatal opening and closure has also provided much insight into ABA signal transduction, although it is probable that some of these mechanisms are unique to guard cells and hence do not apply to seeds. No ABA ***receptor*** has been identified, but there is evidence for receptor sites both on the cell surface and within the cell (Fig. S.29).

Downstream from these putative receptors, a primary response of guard cells to exogenous ABA is an elevation of intracellular ***calcium***, due to opening of plasma membrane calcium channels. The calcium channels are triggered in response to ABA-induced ***secondary messengers*** including ***inositol triphosphate***, ***cyclic ADP ribose***, hydrogen peroxide (H_2O_2) and nitric oxide (NO). Potassium channels are then activated in response to the changes in calcium, and changes in potassium concentrations are largely responsible for the regulation of stomatal opening via changes in osmotic potential of the guard cells. Many intermediates of ABA signal transduction related to stomatal regulation have been identified; however, only major regulators related to ABA action in seeds are discussed here.

Protein **phosphorylation** plays a central role in ABA signalling. Two genes *ABI1* and *ABI2* encode the proteins **ABI1** and **ABI2** – phosphatases which can ***dephosphorylate*** a putative ABA signalling element (e.g. ABI3), active only in the phosphorylated form, thus down-regulating the response to ABA. Two types of mutations in these two genes (*abi1* and *abi2*) disrupt many different ABA-regulated responses. Gain-of-function mutation (**see: Mutants**) (the dominant mutants *abi1-1*, *abi2-1*) results in insensitivity to ABA (because of greater phosphatase activity): loss-of-function (less or no phosphatase) mutants (recessive *abi1-2*, *abi2-2*) are hypersensitive to ABA. There is evidence for inactivation of ABI2 phosphatase by H_2O_2, one of the secondary messengers generated by ABA, which would therefore allow or enhance the ABA-dependent signalling process. In ERA1 is a farnesyltransferase enzyme responsible for the addition of a ***farnesyl*** group on to other proteins (e.g. downstream ABA ***transcription factors***). Like ABI1 and ABI2, ERA1 acts to inhibit ABA responses. ERA1 acts early in the ABA signal transduction pathway either in conjunction or immediately following ABI1 and ABI2, and functions in regulating the sensitivity of the germination response to ABA. The *era1* mutation confers an *e*nhanced *r*esponse to *A*BA.

Following the transduction of the ABA signal from the cytoplasm to the nucleus, ABA-induced changes in gene

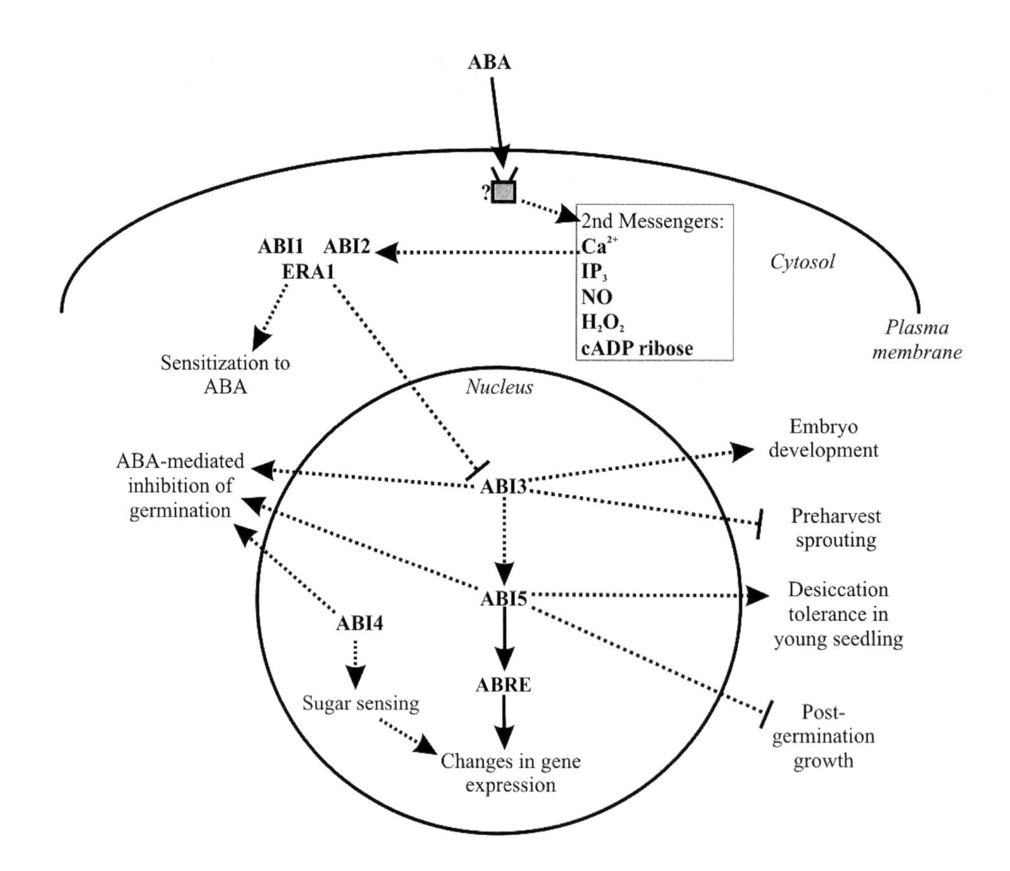

Fig. S.29. Schematic representation of ABA signal transduction. ABA is perceived by an unidentified receptor probably within the cytosol (grey box). Rapidly following the perception of ABA, concentrations of the second messengers – calcium ions (Ca^{2+}), inositol triphosphate (IP_3), nitric oxide (NO), hydrogen peroxide (H_2O_2) and cyclic ADP ribose – change. The activities of the cytosolic phosphatase proteins ABI1 and ABI2 and the farnesyltransferase ERA1 are altered, and the signal that ABA is present is transmitted into the nucleus. Within the nucleus the activity of the resident transcription factors ABI3, ABI4 and ABI5 are altered to mediate ABA-induced changes in gene expression. Specifically, ABI5 binds to the ABA-Response Element (ABRE) present in the promoters of genes responsive to ABA, thus regulating their expression. Lines with a flat end represent inhibition and solid arrows represent promotion of an event. Dotted lines represent interactions not elucidated, while solid arrows represent direct interactions.

expression are mediated by changes in abundance and activity of transcription factors such as ABI3, ABI4 and ABI5, which are expressed primarily in seeds. ABI3 maintains developing embryo identity by binding to the promoter of **storage protein** genes inducing their transcription, while suppressing precocious germination during seed development. Seeds from the maize *viviparous1* (*vp1*) mutant, an **orthologue** of *abi3*, germinate during development. The incorrect splicing (removal of the transcribed non-coding, intron regions from RNA) of the *ABI3*-like transcript in cereals has been linked to **preharvest sprouting** in wheat. The amount of expression of an *ABI3*-like gene is also correlated with the degree of dormancy in wheat. The ABI5 protein interacts with the ABI3 protein, and possibly they regulate gene expression while binding together to the DNA. ABI5 is also proposed to act later than ABI3 in the signal transduction pathway to regulate ABA-mediated post-germination growth arrest during conditions of insufficient moisture to support seedling growth. The abundance of the ABI5 protein is strongly correlated with **desiccation tolerance** in the growth-arrested seedlings. The ABI5 protein is degraded by the **proteasome**, and stabilized by ABA-induced phosphorylation. ABI4 is another major regulator of germination-related ABA signalling and is involved in the integration of sugar sensing in the regulation of germination.

Genes that are up-regulated by ABA have consensus sequences in their promoters, such as the ***ABRE*** (*AB*A *R*esponsive *E*lement) domain (C/TACGTGGC), to which ABA-induced transcription factors bind, such as ABI5. However these motifs are not exclusive to ABA-induced genes, and the sequence of flanking nucleotides is important for their function. Post-transcriptional regulation of messenger RNA transcribed from ABA-responsive genes by processing, and stability by RNA-binding proteins, are other facets of ABA signalling. (PM, GWB)

Himmelbach, A., Yang, Y. and Grill, E. (2003) Relay and control of abscisic acid signalling. *Current Opinion in Plant Biology* 6, 470–479.

Lovegrove, A. and Hooley, R. (2000) Gibberellin and abscisic acid signalling in aleurone. *Trends in Plant Science* 5, 102–110.

Nambara, E. and Marion-Poll, A. (2003) ABA action and interactions in seeds. *Trends in Plant Science* 8, 213–217.

Shen, Q., Gomez-Cadenas, A., Zhang, P., Walker-Simmons, M.K., Sheen, J. and Ho, T.H. (2001) Dissection of abscisic acid signal transduction pathways in barley aleurone layers. *Plant Molecular Biology* 47, 437–448.

2. Gibberellins

The **gibberellins** (GAs) are a complex family of terpenoid plant hormones. They are generally considered as promoters of germination and act to oppose the action of ABA (**see: Germination – influences of hormones**). GAs are involved in seed development, are required for the induction of enzymes involved in the degradation of reserves within storage tissues of cereals, and in the endosperm cell walls of certain dicot seeds, where they mediate the release from coat-imposed **dormancy**; they may be involved in the induction of embryo growth during germination (**see: Germination – molecular effects; Germination – radicle emergence; Mobilization of reserves – cereals**). Analysis of the GA transduction pathway has been carried out primarily in *Arabidopsis* GA response mutants and also by the study of the stimulation of the GA-induced hydrolytic enzyme α-amylase in the aleurone layer of cereal grains following germination, a property that has a commercial application in the brewing industry (**see: Malting; Mobilization of reserves – a monocot overview; Mobilization of reserves – cereals**).

A **receptor** for GA has not been identified; however, biochemical evidence suggests the existence of a plasma membrane-bound receptor (Fig. S.30) and evidence from a mutant rice for a putative soluble receptor has been reported.

Following perception of GA by the cell there are rapid increases in cytosolic calcium (Ca^{2+}) and *cyclic GMP* (inhibition of cyclic GMP synthesis inhibits the production of α-amylase in aleurone layer cells). Activation of ***G protein*** α-subunits is also involved in GA signalling. A protein named PHOTOPERIOD RESPONSIVE1 (PHOR1) is localized in the cytoplasm, and following the perception of GA it moves into the nucleus where it effects a positive action on the GA signal transduction pathway. PHOR1 is therefore capable of transducing the GA signal from the cytoplasm to the nucleus and is proposed to participate in the proteasome-mediated degradation of factors named ***DELLA*** proteins, which act to inhibit GA responses. DELLA proteins reside in the nucleus until the GA signal is perceived, at which point they are phosphorylated at a tyrosine residue by an unidentified **kinase**, then **ubiquitinated**, and promptly degraded by the **proteasome**, most likely still within the nucleus. Mutations that inactivate the GA sensitivity of a DELLA protein from *Arabidopsis*, named RGL2 (*R*EPRESSOR OF *G*A1-3-*L*ike*2*), are capable of conferring GA-independent germination. *DELLA* genes are the 'Green Revolution genes', since the dwarf wheat bred by Nobel laureate Norman Borlaug carries a mutation in the wheat *DELLA* gene resulting in reduced GA responses and the subsequent dwarf phenotype. SPINDLY (SPY) is another GA signalling intermediate, and mutants of it, like those of DELLA proteins, are capable of conferring GA-independent germination in *Arabidopsis* seeds. *SPY* encodes an ***N-acetyl glucosamine transferase***, an enzyme that will specifically transfer sugar moieties to proteins. SPY is also an inhibitor of GA signalling but its role remains obscure.

The *Arabidopsis SLEEPY1* (*SLY1*) gene and the rice orthologue *GIBBERELLIN INSENSITIVE DWARF2*

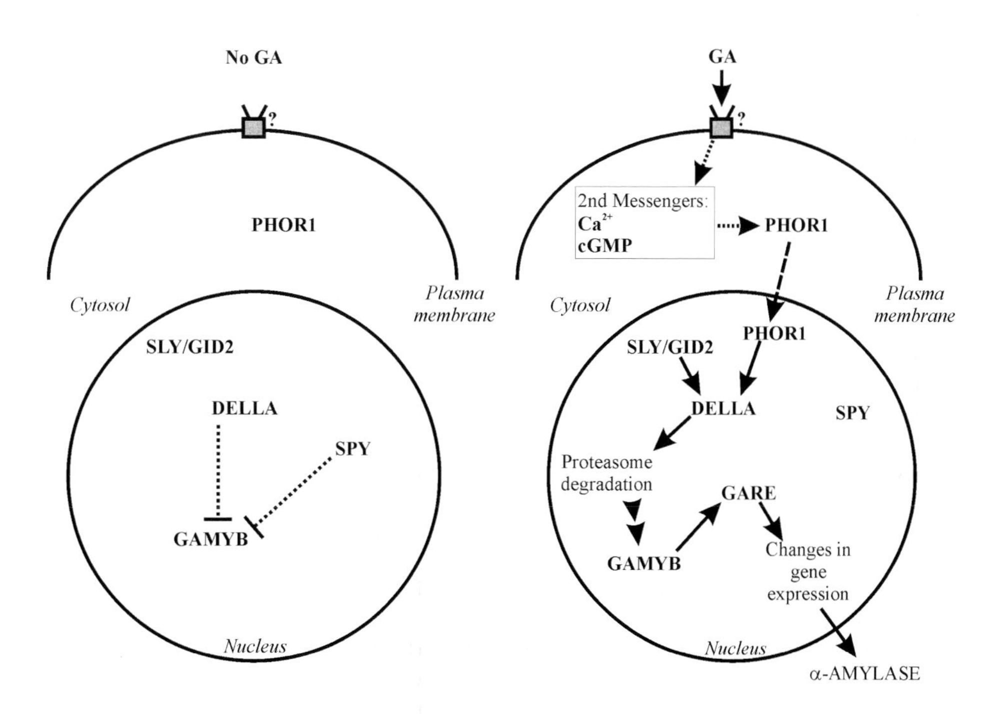

Fig. S.30. Schematic describing the sequence of events in GA signal transduction. In the absence of GA (left) the PHOR1 protein resides in the cytosol and DELLA and SPY proteins act to inhibit GA-inducible responses, such as the production of the GAMYB transcription factor. Following the perception of GA (right) on the plasma membrane by the unidentified receptor (grey box) the secondary messengers, calcium ions (Ca^{2+}) and cyclic GMP are employed. The PHOR1 protein then moves from the cytosol to the nucleus (dashed arrow) where, in conjunction with SLY/GID2, it is proposed to participate in the proteasome-mediated degradation of DELLA proteins. Following the degradation of DELLA, GA responses proceed (arrowheads), and GAMYB is produced which binds to the GA Response Element (GARE) within the promoter of genes responsive to GA, including the hydrolytic enzyme α-amylase. Lines with a flat end represent inhibition and solid arrows represent promotion of an event. Dotted lines represent interactions not elucidated, while solid arrows represent direct interactions.

(*GID2*) encode the ***F-box protein*** subunit of the proteasome, a protein that is responsible for the specific recognition of proteins destined for degradation by the proteasome. Both the *SLY1* and *GID2* proteins are involved in the proteasome-mediated degradation of the DELLA protein(s) in their respective species. *Arabidopsis* seeds with a *sly1* mutation are incapable of germinating even in the presence of GA since they are unable to execute GA responses in the absence of the degradation of their DELLA proteins. However, *gid2* mutant seeds are indeed capable of germinating, suggesting that a GA response is not required for the initiation of radicle elongation in this species.

DELLA proteins suppress the expression of a GA-induced MYB transcription factor (***GAMYB***) which is capable of binding directly to the ***GARE***, the *GA-R*esponsive *E*lement (TAACAAG/A) in the promoter of α-amylase in aleurone layer cells. GAMYB induces transcription of α-amylase, representing the downstream target of GA action in these cells.

GA exhibits ***cross-talk*** with ABA by countering the action of the latter and promoting seed germination, although in some species, such as lettuce, **cytokinins** are also required to neutralize the effects of applied ABA. Expression of some GA-regulated hydrolase enzymes such as α-amylase in the aleurone layers of cereal grains (**see: Mobilization of reserves – cereals**) and others, e.g. β-1,3-glucanase, partly responsible for the softening of micropylar tip endosperm cell walls in tomato and other seeds (**see: Germination – radicle emergence**), can be repressed by ABA. But surprisingly most of the GA-regulated, 'weakening' genes of tomato are less sensitive to ABA, and endosperm cell wall degradation in the micropylar tip of intact seeds therefore proceeds even at what would be expected to be inhibitory concentrations of the hormone. Several forms of cross-talk between the GA and ABA signal transduction pathways have been described. GA can induce ABA catabolism in *Arabidopsis* seeds and in secondary dormant lettuce seeds. GA may also be capable of inhibiting the ABA signalling pathway, for example by the rapid reduction of *ABI3* transcripts. Genes whose expression decreases in the presence of GA contain an ***ABRE*** (ABA response element) in their promoter, such that both GA and ABA signalling pathways act on the promoters of the same genes. Studies on the cereal aleurone layer demonstrate that the simultaneous treatment with GA and ABA fails to produce α-amylase; however, the DELLA protein is degraded. This demonstrates that the ABA signal to inhibit enzyme production is downstream, or after, the GA signal promotes both DELLA degradation and enzyme production. (GWB, PM)
(**See: Mobilization of reserves – cereals,** Fig. M.19)

Bassel, G.W., Zielinska, E., Mullen, R.T. and Bewley, J.D. (2004) Down-regulation of *DELLA* genes is not essential for germination of tomato, soybean and Arabidopsis seeds. *Plant Physiology* 136, 2782–2789.

Cao, D., Hussain, A., Cheng, H. and Peng, J. (2005) Loss of function of four DELLA genes leads to light- and gibberellin-independent seed germination in Arabidopsis. *Planta* 223, 105–13.

Gomi, K. and Matsuoka, M. (2003) Gibberellin signalling pathway. *Current Opinion in Plant Biology* 6, 489–493.

Itoh, H., Matsuoka, M. and Steber, C.M. (2003) A role for the ubiquitin-26S-proteasome pathway in gibberellin signalling. *Trends in Plant Science* 8, 492–497.

Lovegrove, A. and Hooley, R. (2000) Gibberellin and abscisic acid signalling in aleurone. *Trends in Plant Sciences* 5, 102–110.

Richards, D.E., King, K.E., Ait-ali, T. and Harberd, N.P. (2001) How gibberellin regulates plant growth and development: a molecular genetic analysis of gibberellin signalling. *Annual Review of Plant Physiology and Plant Molecular Biology* 52, 67–88.

Sun, T.-P. and Gubler, F. (2004) Molecular mechanism of gibberellin signalling in plants. *Annual Reviews of Plant Biology* 55, 197–223.

Ueguchi-Tanaka, M., Ashikari, M., Nakajima, M., Itoh, H., Katoh, E., Kobayashi, M., Chow, T., Hsing, Y.C., Kitano, H., Yamaguchi, I. and Matsuoka, M. (2005) *GIBBERELLIN INSENSITIVE DWARF1* encodes a soluble receptor for gibberellin. *Nature* 437, 693–698.

3. Ethylene (ethene)

This plant hormone is a simple olefin, which modulates many aspects of plant development, including germination (**see: Germination – influences of gases**). Our understanding of **ethylene (ethene)** signal transduction is more extensive than other hormone signalling pathways, and has been facilitated by the identification of *Arabidopsis* ethylene response mutants. Ethylene is perceived by the ***ETR/ERS*** family of membrane-borne ***histidine kinase*** receptors (five in *Arabidopsis*), thought to be located in the endoplasmic reticulum (Fig. S.31). These receptors act to negatively regulate the ethylene-signalling pathway; they repress the signalling pathway in the absence of ethylene and become inactive when ethylene is bound. The active receptors interact with and activate a downstream ***kinase***, CTR1, also a negative regulator of ethylene signalling. CTR1 is thought to be the first kinase of a ***MAP kinase*** (MAPK) signalling pathway, and probably downstream of this is a membrane-bound positive regulator ***EIN2*** whose function or location is not known. Downstream of this are the nuclear-localized EIN3 and EIN3-Like (EIL) transcription factors that bind to ethylene response elements (***EREs***) in the promoters of ethylene-responding genes.

Ethylene is proposed to promote seed germination through cross-talk with the ABA signal transduction pathway, acting negatively on the later pathway. For example, mutations in the ETR1 receptor and in the EIN2 transcription factor increase the sensitivity of *Arabidopsis* seed to ABA, suggesting these proteins act to overcome the ABA-mediated inhibition of germination. (PM, GWB)

Ghassemian, M., Nambara, E., Cutler, S., Kawaide, H., Kamiya, J. and McCourt, P. (2000) Regulation of abscisic acid signalling by the ethylene response pathway in Arabidopsis. *The Plant Cell* 12, 1117–1126.

Guo, H. and Ecker, J.R. (2004) The ethylene signalling pathway: new insights. *Current Opinion in Plant Biology* 7, 40–49.

4. Other hormones

Seed development, dormancy and germination are influenced by other classes of plant hormones in addition to those compounds discussed above (**see: Germination – influences of hormones**).

(a) **Brassinosteroids**. These are plant steroid hormones that play a role in the germination of *Arabidopsis* seed (**see: Brassinosteroids; Hormones**). Brassinosteroids (BR) are perceived by a plasma membrane-bound receptor ***kinase***, BRI1 (Fig. S.32). In the absence of BR the cytoplasmic kinase

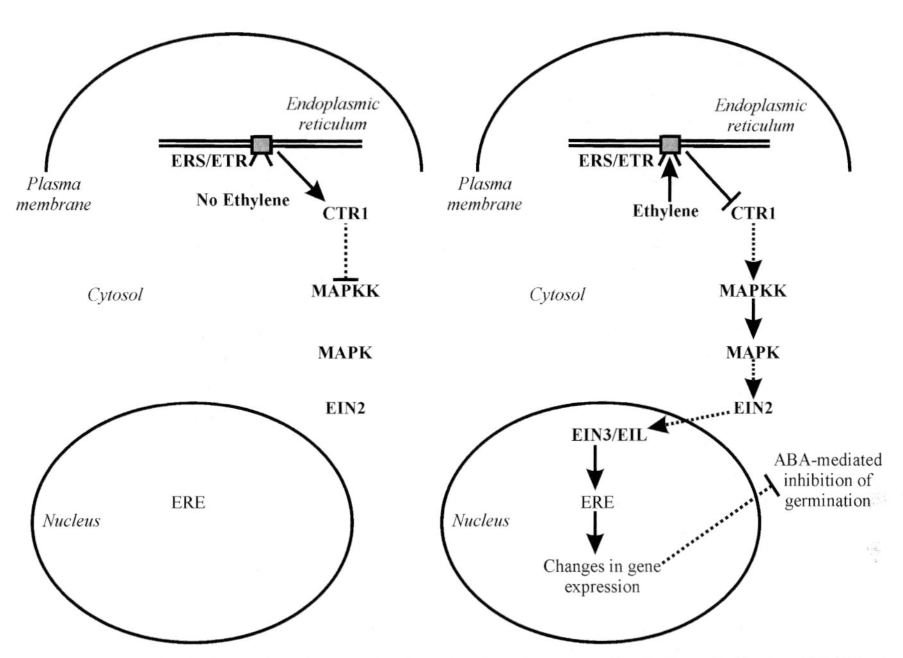

Fig. S.31. Schematic representation of ethylene signal transduction. In the absence of ethylene (left) the ETR/ERS receptors (grey box), located on the endoplasmic reticulum membrane, promote the activity of the kinase CTR1, which in turn inhibits ethylene responses. Following the perception of ethylene (right), CTR1 is inhibited and a MAP Kinase (MAPKK, MAPK) signalling cascade is initiated. This leads to the promotion of EIN2 activity, in turn promoting the activity of the nuclear-localized EIN3/EIL transcription factors. These proteins bind to the Ethylene Response Element (ERE) within the promoters of ethylene responsive genes that now likely alleviate the ABA-mediated inhibition of seed germination. Lines with the flat end represent inhibition and arrows represent promotion of an event. Dotted lines represent interactions not elucidated while solid arrows represent direct interactions.

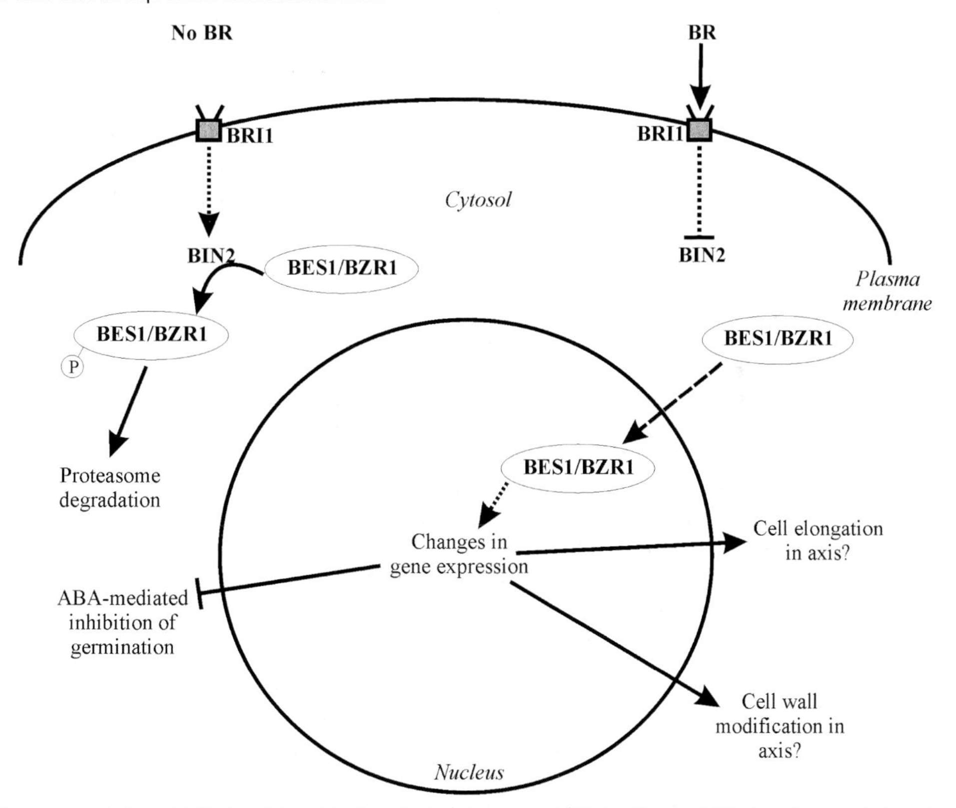

Fig. S.32. Schematic representation of BR signal transduction. In the absence of BR the kinase BIN2 is active, and phosphorylates BES1/BZR1 targeting these latter proteins for destruction by the proteasome. When BR is bound by the BRI1 receptor, BIN2 activity is inhibited and BES1/BZR1 proteins are translocated from the cytoplasm to the nucleus where they mediate changes in gene expression. Within seeds, BR responses are involved in overcoming the ABA-mediated inhibition of germination. Lines with solid arrows represent promotion of an event and those with a flat end represent inhibition of an event. Dotted lines represent interactions not elucidated, while solid lines represent direct interactions.

BIN2 is active and **phosphorylates** BES1 and BZR1, two closely related proteins. BES1 and BZR1 are then degraded by the proteasome in their phosphorylated state. Following the perception of BR by BRI1, the cytoplasmic kinase BIN2 is inactivated such that it no longer phosphorylates BES1 and BZR1, and these latter proteins are stabilized and re-localized from the cytoplasm to the nucleus where they mediate BR-induced changes in gene expression. BRs may be capable of enhancing germination by modifying the radicle cell wall and stimulating cell elongation, as has been demonstrated to occur in vegetative tissues. The signal transduction pathway employed by BR to stimulate germination is independent of the GA signal transduction pathway. (PM, GWB)

Nemhauser, J.L. and Chory, J. (2004) BRing it on: new insights into the mechanism of brassinosteroid action. *Journal of Experimental Botany* 55, 265–270.

Yin, Y., Wang, Z.-Y., Santiago, M.-C., Li, J., Yoshida, S., Asami, T. and Chory, J. (2002) BES1 accumulates in the nucleus in response to brassinosteroids to regulate gene expression and promote stem elongation. *Cell* 109, 181–191.

(b) **Cytokinins** regulate cell division in plants, including in the early stages of seed development. When supplied to seeds they may also promote germination (often atypical, i.e. cotyledon emergence precedes that of the radicle) (**see: Germination – influences of hormones**) but no role for endogenous cytokinins in germination is known. The cytokinin receptors are thought to be CK1, CRE1, AHK2 and AKH3, plasma membrane-located ***histidine kinases***. Following the perception of cytokinin, these kinases autophosphorylate and subsequently activate AHP proteins. Active AHP proteins move from the cytoplasm to the nucleus where it is postulated that they cause the ***transcription factor*** ARR proteins 1, 2 and 10 to be released from repressor proteins with which they are physically associated. Liberated ARR proteins then bind to promoter regions of cytokinin-activated genes and promote transcription. The role of cytokinin signalling in the regulation of seed germination remains to be elucidated. (PM, GWB)

Hutchinson, C.E. and Kieber, J.J. (2002) Cytokinin signalling in *Arabidopsis*. *The Plant Cell* 14, S47–S59.

Hwang, I. and Sheen, J. (2001) Two-component circuitry in Arabidopsis cytokinin signal transduction. *Nature* 413, 383–389.

(c) **Jasmonic acid** is a fatty acid-derived octadecanoide-based compound that is primarily associated with wound responses in plants, but also has many other physiological roles, amongst which is a negative effect on germination (**see: Germination – influences of hormones**). Although a considerable amount is known about stress signal transduction leading to its synthesis, little is known about jasmonic acid signalling. No jasmonate receptor has been identified. The *Arabidopsis* jasmonate-insensitive mutation *coi1* is in a gene encoding an ***F-box protein*** which targets other proteins, presumably negative regulators, for destruction by the **proteasome**. No substrates for COI1 have been identified. (PM, GWB)

Turner, J.G., Ellis, C. and Devoto, A. (2002) The jasmonate signal pathway. *The Plant Cell* Supplement 14, 153–164.

Signal transduction – light and temperature

(**See: Signal transduction – some terminology** for terms in bold italics)

Seed development, dormancy and germination are affected by light and temperature (**see: Dormancy; Dormancy breaking; Dormancy – ecology; Dormancy – ecophysiology;** entries under **Germination; Light, dormancy and germination**). Two different classes of photoreceptors have been identified, **phytochrome** and **cryptochrome**, both capable of sensing changes in light quality and initiating signal transduction. Phytochrome is a long-recognized regulator of seed dormancy and germination, and cryptochrome may be implicated in light inhibition of germination. Temperature is involved in dormancy breaking: chilling treatments break dormancy of hydrated seeds of many species, for example. There is no evidence for a single molecular receptor for temperature in seeds, nor do we know how the temperature signal is transduced. There is, however, information relating to temperature-signalling mechanisms in other plant parts, from which extrapolation to seeds may be tentatively made.

1. Phytochrome

Seeds use light as an environmental cue to tell them where they are positioned with respect to leaf canopy or soil cover, and also as a measure of daylength or season (**see: Light, dormancy and germination**). The protein of phytochrome (PHY), encoded for by a small gene family, is the prime receptor/effector for red and far-red light signals. Phytochrome **chromoproteins** detect light through a **tetrapyrrole chromophore** that is attached to the PHY protein which is thought to undergo a conformational change. PHY acts essentially as a molecular light switch: red light is absorbed by the chromoprotein which is converted to the active Pfr form of PHY and turns the switch on, whereas absorption of far-red light by the Pfr form returns PHY to the Pr form and turns the switch off (Fig. S.33). It is the balance between these two forms of PHY, as Pfr and Pr, that determines the physiological response. Two types of phytochrome, phyA and phyB, having slightly different PHY proteins, are thought to regulate germination in *Arabidopsis* at different light photon flux densities. PhyA controls the **very low fluence response**, in which seeds can detect extremely low levels of light (levels comparable to that produced by a fire-fly for example), and phyB controls the **low fluence response** (light levels comparable to a few seconds of shaded sunlight). In *Arabidopsis* and lettuce seeds, phyB responses (and possibly phyA also) are capable of reversibly inducing the expression of genes encoding GA 3-oxidases (sometimes called hydroxylases), enzymes responsible for the production of bioactive **gibberellins**, which likely break dormancy and promote germination (**see: Light, dormancy and germination; Phytochrome**).

Phytochrome signal transduction is a complex field in which there is evidence for many means by which signals can be transduced. There is strong biochemical evidence for the involvement of the ***second messengers***, ***calcium and cyclic GMP*** in the phytochrome signalling pathway (Fig. S.33).

Some forms of phytochrome are themselves light-regulated **protein kinases**. However, it is not clear if protein kinase

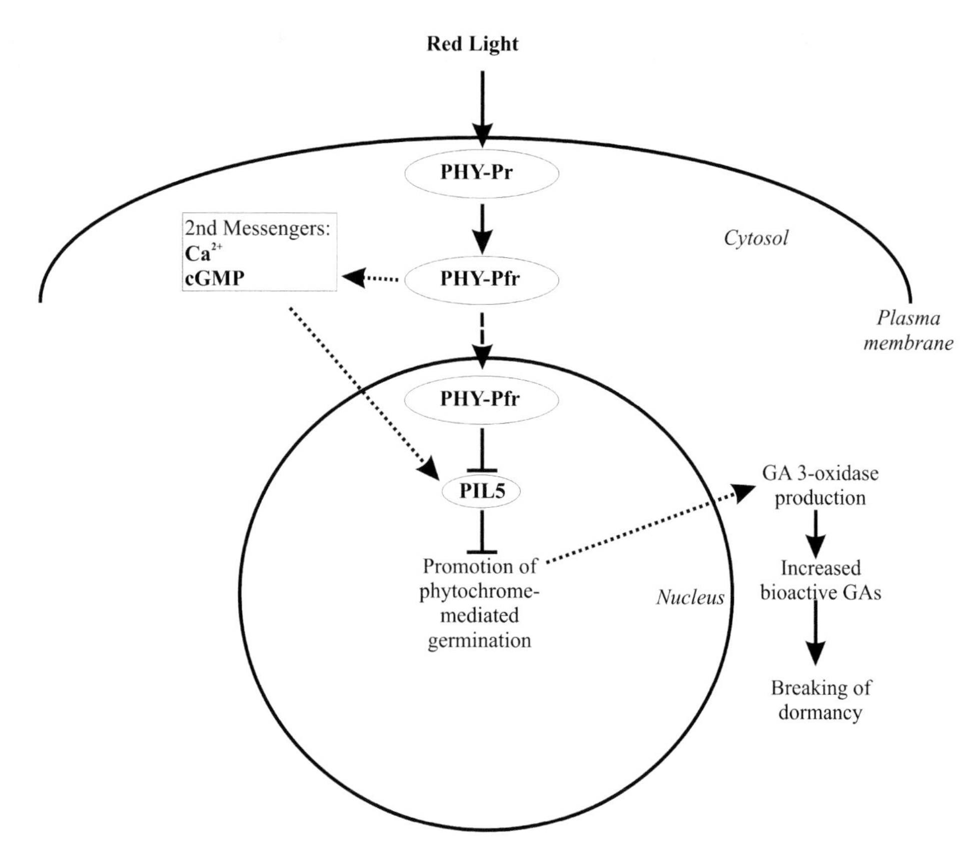

Fig. S.33. Schematic representation of the phytochrome signal transduction pathway. Red light is perceived by the Pr state of phytochrome in the cytosol changing it to Pfr (the protein part is indicated by PHY). This then translocates into the nucleus (represented by the bold arrow with a dashed line). This conversion of phytochrome also stimulates the production of the second messengers cytosolic calcium (Ca^{2+}) and cyclic GMP (cGMP) which transfer information to the nucleus by a mechanism not elucidated (indicated by dotted lines). Once in the nucleus the PHY-Pfr form of phytochrome interacts with and represses PIL5, a protein inhibiting the phytochrome-mediated promotion of seed germination. A gene possibly repressed by PIL5 is that for the enzyme GA_3-oxidase, repression that is released by Pfr. GA_3-oxidase activity produces bioactive GAs likely to be involved in the release from dormancy and the promotion of germination. Arrows indicate stimulation of a reaction or translocation of a compound, the bar indicates inhibition.

activity is a property of all classes of phytochrome. Active phytochrome (Pfr) is also translocated from the cytoplasm into the nucleus, where it can interact with and phosphorylate ***transcription factors*** such as PIF3 to regulate gene expression. A PIF3-like protein (PIL5) identified in *Arabidopsis* regulates the phytochrome-mediated promotion of seed germination in this species. The PIL5 protein resides in the nucleus where it inhibits seed germination, possibly by inhibiting expression of GA 3-oxidase. The active Pfr enters the nucleus, PIL5 is repressed and seed germination proceeds (Fig. S.33). PIL5 is the first identified regulator of phytochrome-mediated seed germination.

2. Cryptochrome

The red end of the spectrum is not the only informational quality of light. Blue light can also provide essential signals as to time and place, possibly in the case of seed germination by inhibiting early events in radicle growth. Blue light is primarily perceived by two receptors called **cryptochromes** 1 and 2 (CRY1 and 2). These proteins have two light-absorbing chromophores, **pterin** and **flavin**-derived FAD. The CRY proteins may be nuclear located; CRY1 is stable but CRY2 is **phosphorylated** and degraded when illuminated by blue light. The amino terminal end of the CRY proteins appears to be inhibitory since over-expression of the carboxy terminal alone results in constitutive photomorphogenesis. Blue light may act through CRY by increasing the stability of **transcription factors** that enhance the expression of blue-light regulated genes. Additionally, there is some evidence that CRY and PHY can interact directly.

3. Low temperature

Dormancy in some instances may be broken by a period of cold treatment of the imbibed seed, termed **chilling** or cold **stratification**. How the cold treatment breaks dormancy is little understood, although signal transduction of the cold response in the adult plant is beginning to be dissected. It is possible that calcium is a second messenger in the signalling of cold in seeds, since this is a common response to cold treatment in other cell types. Transcription of the *CBF/DREB1* transcription factor encoding genes is enhanced by cold, and this in turn promotes the transcription of genes such as those in the *RD* and *COR* families, associated with induced cold tolerance in vegetative tissues. *RD* genes encode hydrophilic proteins of unknown function; *COR* genes encode proteins thought to stabilize membranes against freeze-induced damage. Whether the stratification response in seeds is mediated by the *CBF/DREB1* transcription factors is not established.

There is evidence, for example in beech seeds, for the up-regulation by cold stratification of a phosphatase-type ABA signalling intermediate (of the ABI1, ABI2 type) that reduces seed sensitivity to ABA (**see: Signal transduction – hormones,** ***Abscisic acid***).

Arabidopsis seed stratification, like the phytochrome response, is capable of inducing the expression of a gene encoding GA 3-oxidases, enzymes responsible for the production of bioactive **gibberellins**. It has been proposed that elevated concentrations of bioactive GAs are responsible for the breaking of dormancy and the completion of germination.

There is a view that low temperature might act in seeds through effects upon cell membranes by causing phase changes and/or by acclimation adjustments of the membrane phospholipids. Such changes might possibly be part of the signalling processes. (PM, GWB)

Gyula, P., Schäfer, E. and Nagy, F. (2003) Light perception and signalling in higher plants. *Current Opinion in Plant Biology* 6, 446–452.

Kevei, E. and Nagy, F. (2003) Phytochrome-controlled signalling cascades in higher plants. *Physiologia Plantarum* 117, 305–313.

Oh, E., Kim, J., Park, E., Kim, J.I., Kang, C. and Choi, G. (2004) PIL5, a phytochrome-interacting basic helix-loop-helix protein, is a key negative regulator of seed germination in *Arabidopsis thaliana*. *The Plant Cell* 16, 3045–3058.

Shinozaki, K., Yamaguchi-Shinozaki, K. and Seki, M. (2003) Regulatory networks of gene expression in the drought and cold stress responses. *Current Opinion in Plant Biology* 6, 410–417.

Smith, H. (2000) Phytochromes and light signal perception by plants – an emerging synthesis. *Nature* 407, 585–591.

Sung, S. and Amasino, R.M. (2004) Vernalization and epigenetics: how plants remember winter. *Current Opinion in Plant Biology* 7, 4–10.

Toyomasu, T., Kawaide, H., Mitsuhashi, W., Inoue, Y. and Kamiya, Y. (1998) Phytochrome regulates gibberellin biosynthesis during germination of photoblastic lettuce seeds. *Plant Physiology* 118, 1517–1523.

Yamaguchi, Y., Ogawa, M., Kuwahara, A., Hanada, A., Kamiya, Y. and Yamaguchi, S. (2004) Activation of gibberellin biosynthesis and response by low temperature during imbibition of *Arabidopsis* seeds. *The Plant Cell* 16, 367–378.

Signal transduction – some terminology

Some relevant terms may also be found as separate entries in the encyclopedia.

ABRE
Abscisic Acid Response Element, a short DNA sequence that frequently occurs in the promoter of ABA-responsive genes with a consensus sequence of C/TACGTGGC. This element can be bound by ABA-induced transcription factors, such as ABI5, and thus help to confer ABA inducibility upon a gene.

cADP ribose
Cyclic adenosine diphosphoribose, a second messenger produced from adenosine diphosphoribose implicated in ABA signal transduction.

Calcium
A divalent cation (Ca^{2+}), which forms a common second messenger in cells. Intracellular concentrations are generally low, and an extracellular signal may prompt its release from intracellular Ca^{2+} stores, or influx of extracellular Ca^{2+} through specific ion channels.

Cross-talk
This describes the ability of one perceived stimulus initiating a signal transduction cascade of events to alter the activity of another signal transduction pathway.

Cyclic ADP ribose
See: cADP ribose

Cyclic GMP (cGMP)
Cyclic guanosine monophosphate. A second messenger produced from guanosine triphosphate (GTP) by the enzyme guanalyl cyclase, and in plants implicated in gibberellin (GA) and phytochrome signalling.

Cyclic nucleotide
A **nucleotide** in which the phosphate group is in diester linkage to two positions on the sugar residue.

DELLA
A class of proteins containing DELLA domains (a specific short amino acid sequence: D = aspartic acid, E = glutamic acid, L = leucine, A = alanine) that function as negative regulators of gibberellin (GA) signalling, and are deactivated in response to GA signals. (**See: Green revolution; Mobilization of reserves – cereals**)

Dephosphorylation
The enzymic action of a phosphatase, involving the removal of a phosphate group from a molecule (e.g. a protein).

EIN
A protein taking its name from the mutant form, ethylene insensitive.

EREs
Ethylene-Response Element, a short DNA sequence (AGCCGCC) present in the promoter of ethylene-induced genes. Transcription factors EIN3 and EIL bind to these sequences to regulate **gene expression**.

ETR/ERS
Ethylene Resistant/Ethylene Response Sensor, the names given to different members of the ethylene receptor gene family.

F-box protein
A component of the multiprotein proteasome complex. It is responsible for the specific recognition and recruitment of proteins destined for degradation by the proteasome.

Farnesyl group
Also called a prenyl group. A 15-carbon linear grouping of three isoprene units that are transferred to a specific cysteine residue on a target protein by farnesyl transferase (FTase) from farnesyl diphosphate (FPP). This is a post-translational

modification called farnesylation or prenation, and permits the protein to carry out its specific function; in many instances it is associated with the binding of proteins to membranes.

G proteins
A GDP-binding enzyme complex that is activated by a membrane-bound receptor, upon which GDP is exchanged for GTP. The GTP-binding form of the G protein is active and will in turn activate other proteins such as phospholipase D or potassium ion channels. The active G protein complex has intrinsic GTPase activity, which cleaves GTP to GDP, thus turning off the signal. These proteins may be involved in GA signal transduction

GAMYB
A gibberellin (GA)-regulated MYB-class transcription factor, identified as an activator of α-amylase gene expression in barley, thought to bind specifically to the GARE response element. MYBs are a class of transcription factors taking their name from the transcription factor involved in induction of myeloblastosis in animals and humans. (**See: Mobilization of reserves – cereals**)

GARE
Gibberellin (GA) Response Element, a short DNA sequence (TAACAA/GA) present in the promoter of GA-responsive genes such as those for α-amylase in the cereal aleurone layer. This sequence is thought to be bound by the GAMYB transcription factor. (**See: Mobilization of reserves – cereals**)

Gene expression
The active production of mRNA transcripts (**transcription**) and protein (**translation**) corresponding to a particular gene.

Histidine kinase
An enzyme that will transfer a phosphate group to a histidine residue of a protein. Ethylene perception in plants is through a family of autophosphorylating histidine kinases.

Inositol triphosphate
A secondary messenger involved in signalling, e.g. of abscisic acid. It consists of the 6-carbon cyclic inositol with phosphate groups at the 1, 4 and 5 positions.

Kinase
An enzyme that will transfer a phosphate group from adenosine triphosphate (ATP) on to an acceptor molecule (phosphorylation). There are different classes of kinase, with the most abundant class being that of the protein kinases, i.e. that phosphorylate proteins. (**See: Dormancy breaking – cellular and molecular aspects**)

K^+ channels
A species of membrane-bound protein whose function is to allow ingress or egress of a particular ion, in this case potassium, in response to a given signal. Important in the ABA response in guard cells, and possibly in other tissues.

MAP kinase
Mitogen Activated Protein kinase (so named from the properties of the mammalian version). A class of eukaryotic protein kinases, implicated in the intracellular signalling of stress (cold, drought, salt) and plant hormones (ethylene, ABA). MAP kinases form part of a signal transduction cascade in which a chain of protein kinases phosphorylate and activate the next downstream component of the cascade, thus amplifying the signal.

N-Acetyl glucosamine transferase
An enzyme that will catalyse the transfer (glycosylation) of a certain sugar moiety (N-acetyl glucosamine) to a specific sequence in a protein. The properties of that protein are subsequently changed. The addition of sugars to proteins may be analogous to their phosphorylation in that it leads to changes in their conformation and activity.

Negative regulator
A protein intermediate of a signal transduction pathway that acts to inhibit the signal from being passed along.

Phosphatase
An enzyme that removes a phosphate group from a molecule.

Positive regulator
A protein intermediate of a signal transduction pathway that acts to promote the signal being passed along.

Receptors
Proteins that will detect and bind a specific compound (signal) and then relay a signal further downstream. These proteins are frequently plasma membrane-bound but may be on intracellular membranes.

Second messenger
A signal-transmitting molecule synthesized or released in the cell in response to a signal from a primary signal, such as perception of a hormone. Examples of second messengers include **calcium**, hydrogen peroxide and nitric oxide (NO). (**See: Mobilization of reserves – cereals**)

Transcription factors
Proteins that regulate transcription (synthesis of mRNAs). Transcription factors may be general or specific to particular classes of genes. They bind to specific DNA sequences in the promoter region of genes.

Ubiquitination
The conjugation of the small protein ubiquitin to the lysine residues of proteins targeted for degradation by the proteasome. Ubiquitination is carried out by enzymes called ubiquitin ligases that have an important role in the regulation of signalling proteins. (PM, GWB)

Siliqua

A fruit: long, thin capsule with two generally multi-seeded compartments separated by a longitudinal, false septum; characteristic of the Cruciferae. A silicule is a siliqua which is as broad, or broader, than it is long. (**See: Brassica – horticultural; Brassica oilseeds; Brassica oilseeds – cultivation; Fruit – types**)

Silk-cut

A congenital disorder of some maize cultivars characterized by a horizontal cut or split in the pericarp around the kernel near to the embryo end. Breakage at this weak point may expose the starchy endosperm to infection by fungi. **(See: Disease)**

Single seed descent

One method of producing true-breeding pure-lines of a self-pollinating crop from individual plants. **(See: Line; Pure-line)**

Site-specific recombinases

Enzymes used in biotechnology that catalyse reciprocal recombination (breakage and rejoining) between both strands of a DNA molecule at specific recognition sequences. Bacterial and fungal site-specific recombinases, both in nature and in the laboratory, are used to generate specific chromosomal (or plasmid) rearrangements that alter the normal linear relationships of genes. Recombinases have site-specificity, i.e. they recognize specific sequences, or protein binding sites, that flank the target site for breakage and DNA strand exchange (rejoining). Modifying or mutating the recognition sites can make the excision or inversion event permanent. An example of their use is illustrated in the article on **GMO – patent protection technologies** where there is the directed removal of a blocking sequence containing a repressor protein gene, employing the CRE-LOX recombinase from bacteriophage P1. The recognition sites for this recombinase are small and have directionality; in the LOX site the recognition sequence is composed of 34 nucleotide pairs. Site-specific recombinases have also been successfully employed to remove antibiotic selectable marker genes in the production of transgenic crops for commercial use. (MJO)

Size of seeds

Seeds vary greatly in size among species. The extremes are the dust-like seeds of some **orchid** species (10^{-6} g) (**see:** Fig. S.69 in **Structure of seeds – identification characters**) and the double coconut palm (*Lodoicea maldivica*) (over 20 kg). Even within plant communities seed size can vary by five or six orders of magnitude. In general, larger plants produce larger seeds, so trees have larger seeds (global average 328 mg) than shrubs (69 mg), which have larger seeds than herbaceous plants (7 mg). However, after allowing for plant size, seed size is negatively related to seed number (Fig. S.34). For a given plant size and reproductive allocation (the amount of resources put into seed production) a plant can divide its resources into many small seeds, or fewer larger seeds. The seed size and number of a plant have consequences for the fate of seeds (discussed below), and evolutionary theory suggests the seed size exhibited by a plant should be governed by the relative benefits of having many small or few large seeds. The relative benefits are determined by the abiotic and biotic environment. **(See: Number of seeds – ecological perspective)**

The evolutionary history of a species therefore determines its average seed size. Some species have a constant seed size, most famously the **carob** (*Ceratonia siliqua*), the basis of the carat unit of weight used for gold (**see: Weights and measures**). However, while the average seed size of a species is relatively constant, most species exhibit wide variation about this average, typically fourfold. Furthermore, this variation occurs within individual plants rather than between individuals. This is explained by evolutionary theory, which suggests that the variety of conditions experienced by the seeds of a plant mean there is no single seed size which is ideal and a range of seed sizes provides the best solution to this variety. **(See: Evolutionary ecology)**

These explanations are based on two fundamental ecological consequences of having a particular seed size which are found when comparing seeds within, or between, species. First, seedling establishment is generally better for larger seeds. Seedlings from large seeds are more successful than those from smaller seeds when growing in shaded conditions under plant canopies and are more likely to win in competition with other seedlings. There is also some evidence that larger-seeded individuals have an advantage when establishing in conditions of low moisture or nutrient deprivation. They are able to emerge after burial at greater soil depths, and better tolerate defoliation by herbivores. Three mechanisms have been proposed to explain these advantages. Larger seeds: (i) produce larger seedlings which can better access resources such as light, moisture or nutrients; (ii) contain greater amounts of reserves which can sustain the seedlings for longer in conditions where external resources are limited (**see: Storage reserves**); and (iii) have been shown to have slower relative growth rates, which may allow longer survival under adverse conditions. Secondly, **dispersal** distances of seeds are often greater for smaller seeds because they fall more slowly and so travel further when dispersed by, for example, wind, water, or explosive pods. This relationship does not hold for animal-dispersed seeds. If more seeds are produced a greater absolute number will disperse long distances from the parent. So this leads to a more general consequence of the seed number-size trade-off on dispersal.

Thus, the seed size-number trade-off leads to a trade-off between the abilities to establish seedlings and to disperse seeds. This is linked to many other aspects of seed **evolution** and trade-offs in other seed characteristics. (JMB)

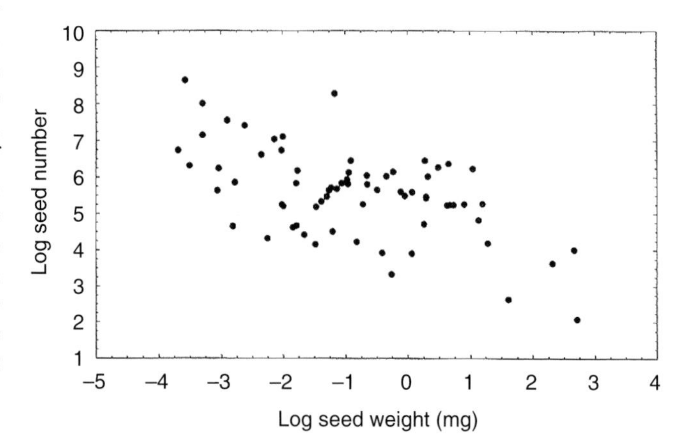

Fig. S.34. The negative relationship between seed number and size, from a study of 72 grassland plants. The correlation has a *r* value of 0.55. From Jakobsson, A. and Eriksson, O. (2000) A comparative study of seed number, seed size, seedling size and recruitment in grassland plants. *Oikos* 88, 494–502.

Leishman, M., Wright, I.J., Moles, A.T. and Westoby, M. (2000) The evolutionary ecology of seed size. In: Fenner, M. (ed.) *Seeds: The Ecology of Regeneration in Plant Communities*. CAB International, Wallingford, UK, pp. 31–57.

Moles, A.T. and Westoby, M. (2004) Seedling survival and seed size: a synthesis. *Journal of Ecology* 92, 372–383.

Moles, A.T., Ackerly, D.D., Webb, C.O., Tweddle, J.C., Dickie, J.B. and Westoby, M. (2005) A brief history of seed size. *Science* 307, 576–580.

For data on seed sizes see: Seed Information Database (SID) www.rbgkew.org.uk/data/sid/

Sizing

The process of separating material in a seedlot according to physical dimensions. (**See: Conditioning, II. Cleaning; Grain inspection**)

Smuts

A group of pathogenic fungi of the order Ustilaginales (Basidiomycetes) and, more informally, the disease symptoms that they cause. Smuts occur worldwide on a variety of plant species, but in cultivated crops are of the greatest economic concern in cereals such as wheat, barley, rye and maize where they affect grain quality and cause significant yield losses by infecting ovaries or developing kernels, as well as other parts of the plant. The name 'smut' derives from the characteristic dusty black or soot-coloured masses of spores that typically accompany the latter stages of disease development. The harvesting and threshing of smutted ears releases the spores and contaminates the healthy grain. Some diseases are interchangeably called 'stinking smuts' or common '**bunts**' because of the foul smell they impart to grain, such as in wheat (*Tilletia caries*).

In 'covered smuts', such as in barley (*Ustilago hordei*), the spore masses are held together by the intact grain seedcoat and glumes. By contrast, in 'loose smuts' generally the entire glumes and grain become completely transformed into powdery spore-masses, which may shatter off leaving a bare blackened spike at harvest, such as in wheat and barley (*U. tritici* and *U. nuda*) and rice smuts (*Tilletia barclayana*, *Neovossia horrida*, *Ustilaginoidea virens*). Common (blister) smut of sweetcorn and field maize (*U. zeae* or *U. maydis*) is recognized early in the season by the presence of silvery-white tumour-like galls (Colour Plate 10F), which later mature, rupture and release large masses of spores. (The galls are cultivated as a crunchy mushroom-flavoured food delicacy – known as '*huitlacoche*' in Mexico – which is harvested in the early summer.)

The disease cycle begins with the establishment of seedborne fungal mycelium in the embryo at flowering, which becomes dormant as the seed matures. When the infected seed germinates, or the seedling gets infected from the soil, the mycelium develops systemically, advancing with the growing point, invading the flower primordia when the panicle is being formed, and ultimately transforming individual mature grains into spore balls. Control measures include systemic seed treatment fungicides, planting of cleaned certified seed produced in smut-free fields and resistant varieties. (PH)

(**See: Disease; Pathogens**)

Soaking

See: Steeping

SOD

See: Superoxide dismutase

Sod seeding

Sowing directly into a sward without previous soil cultivation, also known as **direct drilling**. (**See: Tillage**)

Soil bacteria

See: Rhizosphere microorganisms

Soil fungi

See: Disease; Pathogens; Rhizosphere microorganisms; Seedling disease complex

Soil seed banks

Viable seeds on or in the soil form a soil seed bank. If these seeds persist for only a short time, such that successive seed crops do not overlap, the seed bank is described as transient. If at least some seeds survive for more than 1 year, the seed bank is described as persistent. In seasonal temperate climates, transient seed banks can be further divided into Type I, autumn-germinating species whose seeds are present during the summer only, and Type II, spring-germinating and present mainly during the winter (Fig. S.35). Seeds of both may be dormant when shed, this dormancy being broken by a period of high or low temperatures in Types I and II, respectively (**see: Dormancy breaking – temperature**). Persistent seed banks may be similarly divided into Types III and IV, depending on whether a smaller or larger fraction of the seed output enters the seed bank (Fig. S.35), but in fact this fraction may vary from year to year and from place to place in the same species. A more useful, but still arbitrary, distinction may be made between seeds that persist for only a few years (<5 years, short-term persistent) or a long time (>5 years, long-term persistent). Thompson *et al.* (1997) summarize the available information on seed bank types for the flora of northwest Europe.

1. Detecting and measuring seed banks

More often than not, seed banks are sampled by taking soil cores of varying depths, spreading the soil in trays under conditions suitable for germination and counting the seedlings. Sometimes seeds are extracted from the soil and counted, but extraction is not easy, and very small seeds are both easily lost and difficult to identify. In any event, extraction should always be combined with some attempt to determine whether the seeds found are alive, since this cannot be assumed. Physical extraction frequently reveals more species than does germination, suggesting that the latter method may fail to detect species with dormant seeds or stringent germination requirements. A useful compromise is to reduce the soil volume by sieving and then germinate the seeds remaining in the reduced soil volume. This has the advantages of both reducing the space required and accelerating germination.

Relatively few soil samples may be adequate to determine the species composition of a seed bank, while many more are usually necessary to determine seed densities of individual

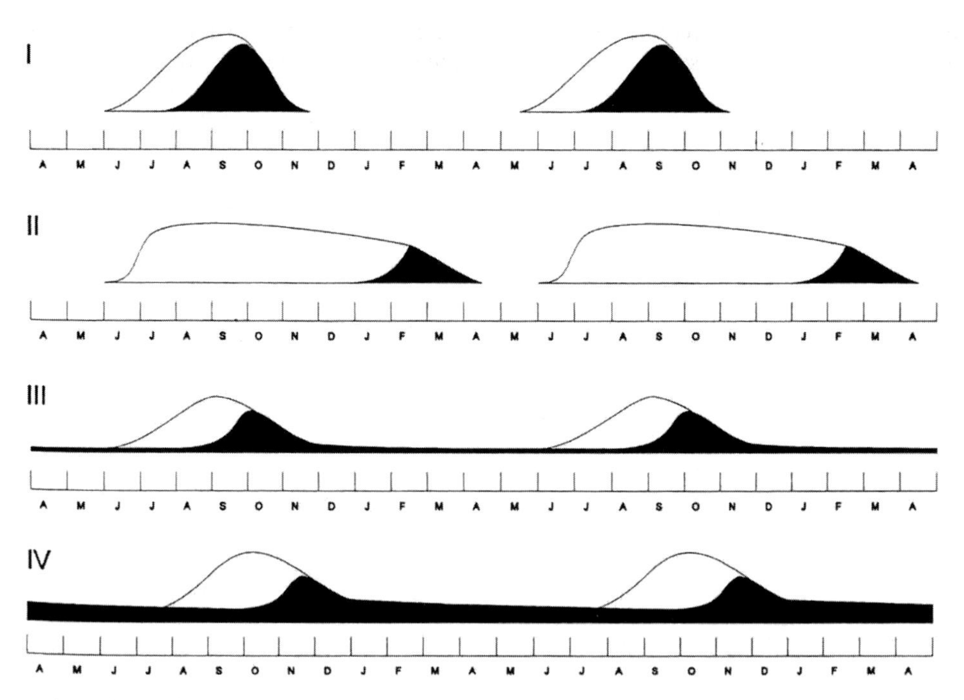

Fig. S.35. Four types of seed bank of common occurrence in temperate regions. See text for an explanation of the four types. Curves illustrate the seasonal abundance of immediately germinable (shaded) and viable but dormant (unshaded) seeds both in the soil and on the soil surface. Redrawn from Thompson, K. and Grime, J.P. (1979) Seasonal variation in the seed banks of herbaceous species in ten contrasting habitats. *Journal of Ecology* 67, 893–921.

species with acceptable precision. As a general rule, it is much better to take many small samples than fewer large ones. Seed densities are normally expressed as number per unit area. Since seed density in most soils declines rapidly with depth, and different studies often sample to widely varying depths, it is not helpful to express densities per unit volume.

Seed densities tend to be low beneath woodlands (tropical and temperate) and Arctic and alpine communities, and high beneath disturbed habitats such as arable fields, heathlands and some wetlands. Seed densities beneath some other communities, for example grasslands, are very variable (Table S.8). Since small seeds are produced in greater numbers than large ones, and small seeds are often more likely to persist in the soil, densities of individual species are strongly negatively related to seed size. The highest density recorded for any species is 488,708/m^2 for the very small seeds (*ca.* 0.05 mg) of *Spergularia marina*, in an inland salt marsh in Ohio, USA. **(See: Number of seeds – ecological perspective; Size of seeds)**

Table S.8. Ranges of seed bank densities in various habitats.

Habitat	Range of densities (per m^2)
Tundra	0–3,367
Temperate deciduous forest	0–13,650
Tropical forest	81–4,700
Tropical disturbed vegetation	48–18,900
Grassland	287–31,344
North American wetlands	191–171,830
Arable	5,025–130,300

Data from Leck, M.A., Parker, V.T. and Simpson, R.L. (eds) (1989) *Ecology of Soil Seed Banks*. Academic Press, London.

Direct evidence of minimum longevity comes from species that are no longer present in the community but are still present as viable seeds in the soil, for example weed seeds beneath formerly arable grasslands and seeds of light-demanding species beneath woodlands and plantations of known age. Nearly all records of extremely ancient seeds have been dated from circumstantial evidence, but the oldest reliably dated seeds are of the sacred lotus (*Nelumbo nucifera*) obtained from the dried bed of a former lake in northeast China. Radiocarbon dating showed that the oldest germinated seed was 1288 ± 250 years old. All apparent evidence of seeds of very great age in the soil should be treated with caution. **(See: Ancient seeds)**

Longevity can also be inferred from the vertical distribution of seeds in soil, since deeply buried seeds are usually older than those near the surface. Longevity can also be determined from burial experiments, in which seeds are deliberately buried and exhumed over a period of years or occasionally decades. The classic example of the genre is the burial of seeds of 21 species by W. J. Beal in 1879. They have been monitored at intervals (of 5, 10 and now 20 years) ever since and *Malva pusilla*, *Verbascum blattaria* and an unidentified (possibly hybrid) *Verbascum* were still alive after 120 years.

2. Predicting seed persistence

Dormancy and persistence in the soil are only tenuously related. Seeds in the soil may be in a variety of physiological states, many of them far from what physiologists would understand by 'dormant'; dormancy is therefore neither a necessary nor sufficient condition for seed persistence in the soil. For most seeds, the role of dormancy in seed persistence is confined to regulating the time of year at which seeds may respond to germination stimuli, or to preventing germination during the period immediately following seed shedding (**see: Dormancy – ecology; Dormancy – ecophysiology**).

There is a near-universal relationship between seed **size** and persistence; small seeds are more persistent than large ones. Since a first crucial step in the formation of a persistent soil seed bank is burial, it seems reasonable to assume that this relationship is connected to the higher probability of burial in small seeds. On the soil surface, seeds are very likely either to germinate or to be predated (**see: Predation**), while both are much less likely after burial. It has also been suggested that small seeds are inherently less likely to be eaten, but this remains controversial. Seeds that fail to persist in the soil, even for quite short periods, usually have rather flexible germination requirements; in particular, they are indifferent to light and therefore are likely to germinate even if buried. The major hazard facing persistent seeds appears to be attack by fungal or bacterial **pathogens**. There is some evidence that long-persistent seeds possess effective chemical defences against such attack.

Much of the work on seed persistence has been conducted in moist temperate habitats. Here seeds are at least intermittently imbibed. Given that longevity in artificial storage is increased by drying seeds to low moisture levels (**see: Gene banks; Seed banks**), it may seem surprising that imbibed seeds can persist for many decades in the soil. It seems that the genetic and membrane damage that reduces viability occurs in both dry and imbibed seeds, but in the latter it is rapidly repaired by the metabolically active seed (**see: Deterioration and longevity**). The principles underlying seed persistence may be quite different in arid habitats, where seeds are dry most of the time. Certainly persistent seeds with impermeable seedcoats ('hard' seeds) are much more frequent in arid habitats. (**See: Deserts and arid lands**)

3. Ecological significance of seed banks

Why do plants accumulate persistent seed banks? Because it is a very powerful way of averaging out the effects of environmental heterogeneity. An annual plant species with no seed bank will suffer severely the first time either reproduction or establishment fail completely, while one with a seed bank will not. In fact, simulations demonstrate that although a persistent seed bank may slow the maximum rate of population growth under consistently benign conditions, it also drastically reduces the probability of extinction in the face of environmental uncertainty. Investment in a seed bank is a direct function of the probability of complete failure of reproduction in any one year. Seed banks are most effective when heterogeneity is chiefly temporal, that is opportunities for establishment are unpredictable in time but relatively predictable in space. Dispersal may be a better option for dealing with spatial heterogeneity.

Theory, therefore, predicts that seed banks should be most advantageous in communities of annual plants occupying habitats that experience frequent and catastrophic, but relatively unpredictable, disturbances. At the opposite extreme, we would expect seed banks to be unimportant in stable habitats such as mature woodlands. Observations largely support these predictions. Seed persistence is greater in annuals (which generally produce more numerous, smaller and longer-lived seeds) than in related perennials. In northern England, woodlands and arable habitats represented the lower and upper extremes of longevity, respectively (Fig. S.36). A number of factors combine to produce the low seed persistence characteristic of woodlands. First, shade selects for increased seed size, which itself reduces the effects of environmental variability on regeneration from seed. Because large seeds (or at least, large seedlings) are more tolerant of shade and drought, they 'see' the environment as less variable, which reduces the selective value of persistence. Second, large seeds are more attractive to predators, if only because they are less likely to become buried. Finally, since mature woodlands are (by definition) relatively undisturbed, even the most long-lived buried seeds are unlikely to survive from one disturbance event to the next.

Thus, there are strong associations between habitat and seed persistence; most arable weeds have persistent seeds, most woodland plants do not. These strong associations are a consequence of the characteristic spatial and temporal patterns of disturbance normally experienced by those habitats. If the habitat-disturbance link is broken or weakened, the association between habitat and persistence may also be altered. If disturbance frequency in woodlands is increased by human harvesting, bringing the interval between successive disturbances within the normal compass of seed persistence, seed banks can come to be very important in **regeneration** in woodlands. Conversely, reducing the disturbance experienced by arable fields, for example by the introduction of a reduced cultivation regime, can diminish the importance of a persistent

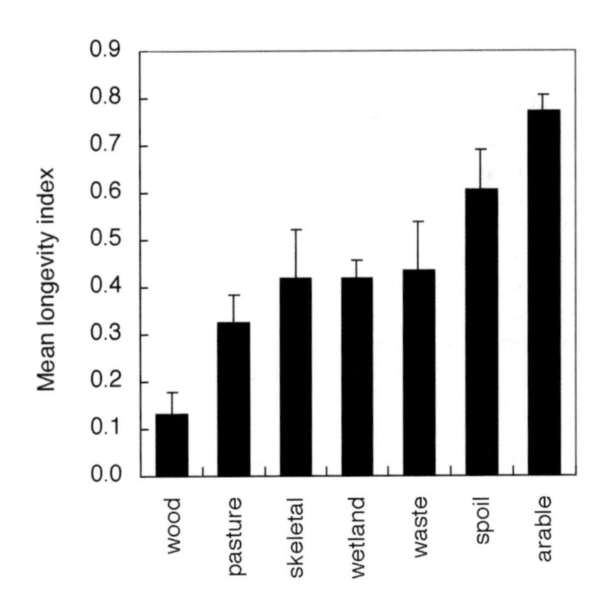

Fig. S.36. Mean seed longevity indices of habitat specialists from seven major habitat types in central England. Longevity index expresses seed persistence in soil on a scale of 0–1. Data from Thompson, K., Bakker, J.P. and Bekker, R.M. (1997) *The Soil Seed Banks of North West Europe: Methodology, Density and Longevity*. Cambridge University Press, Cambridge, UK. Bars are standard errors. Figure reproduced from Thompson, K., Bakker, J.P., Bekker, R.M. and Hodgson, J.G. (1998) Ecological correlates of seed persistence in soil in the NW European flora. *Journal of Ecology* 86, 163–169.

seed bank. Some habitat types are not conspicuously associated with either low or high seed persistence, largely because such plant communities are compatible with a wide range of disturbance regimes. Grasslands, for example, may be relatively undisturbed, or they may experience severe disturbance by grazing animals, drought and fire. Not surprisingly, therefore, many grasslands contain species with a wide spectrum of seed longevity. In 26 calcareous grassland fragments in Germany, the rate of local extinction for species with seed longevity >5 years was only half that of species with shorter-lived seeds.

Similarly, some wetlands are relatively undisturbed, while some shallow wetlands in dry climates may disappear entirely in times of severe drought. Both habitat types illustrate that, from the perspective of seed persistence, the regularity of disturbance is more important than its severity. For example, although Mediterranean grasslands are highly disturbed by drought, this seasonally recurrent disturbance does not select for a persistent seed bank. It is the irregularity of disturbance, rather than its severity, that selects for persistent seeds.

4. Leaving the seed bank

It is axiomatic that a seed bank only has any ecological significance if at least some of the seeds in it eventually germinate and establish successfully. In most plant communities with a closed canopy, this establishment usually requires at least some degree of disturbance to provide areas free of competition from established plants. Gaps that are created by any agency in vegetation can be considered 'competitor-free spaces' that provide opportunities for seedling establishment. The study of gaps and their role in promoting recruitment has an enormous literature, but here it is necessary simply to point out that sensitivity to gaps is an important point of difference between transient and persistent seed banks. Seeds that germinate soon after shedding generally have relatively undiscriminating germination requirements. Their germination is often timed to coincide with a reliable seasonal window of opportunity for seedling establishment. Seeds in a persistent seed bank tend to be far more discriminating; after all, there is little point in spending years or decades in the soil, only then to germinate in the wrong place or at the wrong time. Persistent seeds respond to one or more of a number of environmental variables, all of which have some value in indicating what sort of conditions the seedling is likely to face. These are chiefly variations in light, temperature and nutrients or other chemicals. For example: (i) seeds germinate after fires by responding to high temperatures or to the chemicals in smoke (**see: Germination – influences of smoke**); (ii) germination in response to light guarantees that the seed is at or near the soil surface, while a high red/far-red light ratio (which promotes germination, **see: Light – dormancy and germination; Phytochrome**) indicates light that is largely unfiltered by a plant canopy, i.e. a large gap; (iii) a positive response to temperature alternations restricts germination to near the soil surface or in gaps (surface soil and soil in large gaps experiences the largest temperature fluctuations); (iv) growing plants reduce soil nitrate concentrations, so seeds that are sensitive to elevated soil nitrate concentrations are more likely to germinate in gaps (**see: Germination – influences of chemicals**). (KT)

Fenner, M. and Thompson, K. (2005) *The Ecology of Seeds.* Cambridge University Press, Cambridge, UK.

Priestley, D.A. (1986) *Seed Aging: Implications for Seed Storage and Persistence in the Soil.* Cornell University Press, Ithaca, NY, USA.

Telewski, F.W. and Zeevaart, J.A.D. (2002) The 120-year period for Dr. Beal's seed viability experiment. *American Journal of Botany* 89, 1285–1288.

Thompson, K., Bakker, J.P. and Bekker, R.M. (1997) *The Soil Seed Banks of North West Europe: Methodology, Density and Longevity.* Cambridge University Press, Cambridge, UK.

Soil sickness

Seedling **blight** or **damping-off** symptoms of seedling **diseases**, caused by a variety of soilborne **pathogens**.

Soil types

Because of their physical and chemical properties soils exert a variety of effects on seed germination and seedling establishment. Physically, soil is made up of soil particles and rock fragments, organic matter, living organisms, and air spaces that may be partially or fully filled with water; its structure is determined by the arrangement of the particles relative to each other. Soil texture is determined by analysis using mechanical separation into the basic soil particle size categories of sand, silt and clay components (Table S.9). By common convention, soils are grouped into textural classes depending upon the relative proportions of these size components (Fig. S.37). (**See: Rhizosphere microorganisms; Seedbed environment**). For the variety of 'soilless' media used to plant seeds for crop cultivation, **see: Compost; Hydroponic production**.

Table S.9. Size classes of soil components (according to USDA classification).

	Particle size range (mm diameter)
Very coarse sand	1.0–2.0
Coarse sand	0.5–1.0
Medium sand	0.25–0.5
Fine sand	0.1–0.25
Very fine sand	0.05–0.1
Silt	0.002–0.05
Clay	less than 0.002

Solarization

A soil-sterilizing technique for the broad-spectrum control or delay of the action of soilborne disease pathogens such as the **seedling disease complex**, weeds, nematodes and insect pests. Seedbeds in the field or plots in the greenhouse are covered with transparent or black plastic sheets for several weeks before seed sowing or transplantation to raise the soil temperature, typically by about 3° to 8°C. The technique is used in horticultural and tobacco crop raising, and is well suited to tropical countries because high solar irradiation is required. (**See: Mulch; Transplantation**)

Solarization techniques are also used to a limited extent to control seedborne pathogens. (**See: Thermotherapy**) (PH)

Solid matrix priming

See: Matrix – priming

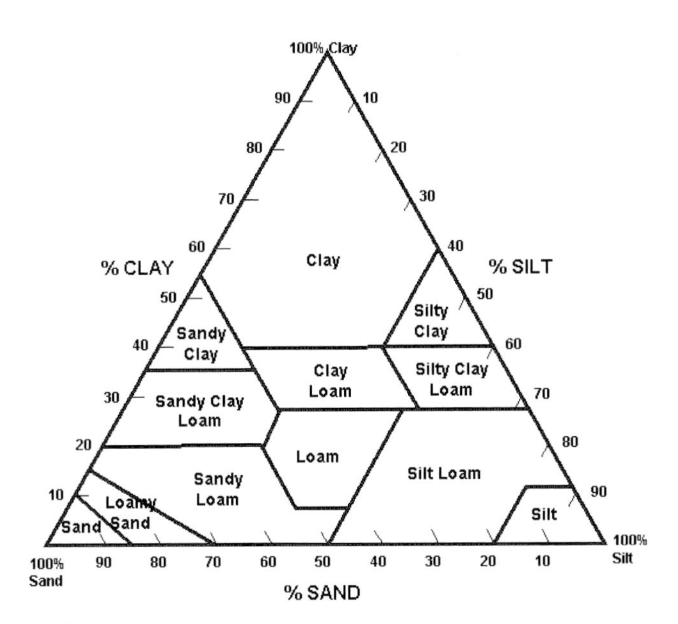

Fig. S.37. Soil texture classes in terms of their sand, silt and clay composition. The classes may be modified with suitable adjectives when rock fragments are present in substantial amounts (such as stony or chalky). The sand, loamy sand, and sandy loam classes may be further subdivided on the basis of the proportions of the various sand separates present (such as coarse, fine, very fine), or the presence of organic matter (such as **peat**). Redrawn, from Soil Science Society of America Glossary page www.soils.org/sssagloss/search.html

Somaclonal variation

Clones produced from any kind of **tissue culture** are called somaclones. They feature in the production of **somatic embryos** for **synthetic seeds**. Any variation of *in vitro* cloned plants is known as somaclonal variation. These variations are usually genetic in origin. Some types of tissue culture are more likely to show variation for a variety of reasons. The first source may be the explant that is to be induced; the parent plant may already exhibit mosaicism (cells of genetically different type) or polysomaty (polysomy, one or more chromosomes are repeated), having regions with different chromosome content. The result is induction of genetically variable tissue cultures. Generally, induction of organogenesis from young or **meristematic** tissues, such as shoot or root tips, or axillary buds (those in the axis of a bud or leaf) is less prone to such variation than adventitious (in a place other than the usual growing regions) induction from other older tissues.

Once induced, the increasing age of the culture may result in accumulation of genetic and physiological variants. The variation can be obvious – cultures lose their organogenic or embryogenic capacity, and closer analysis reveals chromosome abnormalities, such as endopolyploidy (multiple chromosome sets in some cells), **aneuploidy** (excess or reduction of particular chromosomes) or **polyploidy** (three or more chromosome sets in all cells). The basis for these chromosomal variations is mainly errors during mitosis. More cryptic variations in DNA methylation or in genes are detectable if powerful molecular tools are brought to bear, but this is generally a poorly researched area. A possible source of variation is likely to be found in any physiological conditions that diminish DNA repair mechanisms over time. Tissue cultures are fast growing, and errors quickly accumulate. Many tissue cultures must be reinitiated to avoid problems associated with increased variation, which must be eliminated to avoid sexual transmission to the offspring.

When organogenic or **somatic embryogenic** cultures are allowed to develop into plants there may be mosaicism within individual plants.

Some somaclonal variation is purely phenotypic (epigenetic), originating as physiological responses to *in vitro* conditions, and therefore is not transmissible to the offspring. Variation in size and number of parts, leaf arrangement (phyllotaxy) and a host of other morphological characteristics may be under very little genetic control in some species, with the result that organs and embryos developing *in vitro* show greater variation than *in vivo*.

To avoid problems in variation, explant selection must be carried out with care, and with some prior knowledge of variation in the source plant. Once cultures are initiated, it is a useful practice where possible to cryopreserve material so that reinitiation is more easily accomplished. Certitude concerning somaclonal variation is only obtained by vigilance and careful genetic testing. In theory, this somaclonal variation can be exploited if experiments are designed to select useful genotypes. In practice, tissue culturists spend much more effort avoiding somaclonal variation and examples of its effective exploitation are few. (PvA)

Somatic

Related to cells that are non-reproductive, i.e. those other than the **gametes**. **Somatic embryogenesis** is the production of embryos from vegetative plant cells, not from the zygote, which produces a zygotic embryo.

Somatic embryogenesis

The asexual embryo formation from somatic cells (Fig. S.38).

1. Principles

The diploid sporophytic (vegetative) generation of a plant's life cycle alternates with the haploid gametophytic (gamete forming) generation. In seed plants, male gametes are produced from pollen, and eggs are produced within female gametophytes (gymnosperms) or embryo sacs (angiosperms). Fertilization and fusion of gametes produces a diploid zygote, the first cell of the sporophyte generation; the process is known as zygotic embryogenesis. All cells of a sporophyte generation are considered to be somatic cells. If these cells produce embryos asexually, this process is called somatic embryogenesis.

In nature, somatic embryogenesis spontaneously occurs in a variety of organisms, from ferns to flowering plants. Embryos may appear on leaves, but more commonly asexual embryos develop from tissues within the ovule. Two different processes are responsible: **apomixis**, when somatic cells of either the **nucellus** or **embryo sac** produce embryos; or cleavage polyembryony which occurs later in seed development, e.g. zygotic embryos undergo polyembryogenesis, resulting in clonal multiplication of embryos. It is possible to have zygotic (sexual) and somatic (asexual) embryos in the same seed.

In common with all embryos, somatic embryos are generally considered to arise from single cells, but there are exceptions (cleavage polyembryony) and some researchers maintain that it is possible for a somatic embryo to be of multicellular origin.

Somatic embryogenesis *in vitro* is a form of vegetative propagation; it may occur either directly or indirectly. The difference in practical terms is that cells which are the originators of indirect embryogenesis must initially differentiate into **callus** and then further differentiate into embryonic cells. Directly produced somatic embryos do not have to go through a callus phase. It is possible that somatic embryogenesis is a continuous event, resulting in numerous identical embryos, since its induction may initiate a highly repetitive process of embryo formation.

Somatic embryogenesis provides a method for mass propagation of elite clones from breeding programmes and it can be used to produce larger numbers of genetically identical propagules than might otherwise be possible. These may serve as uniform stock for clonal plantations. In species in which somatic embryos are easily stored, a range of strategies for seedling deployment can be developed.

Somatic embryo development does not recapitulate all stages of zygotic embryo development (**see: Embryogenesis**). In zygotic embryo development, the first division of the zygote is polar, setting up an apical–basal gradient. Similar apico-basal divisions have been recorded in many somatic embryos, but it is apparent that much more variation is possible during the early developmental stages of somatic embryos than of zygotic embryos of the same species. The later stages, in contrast, are generally very similar morphologically and physiologically, although mature somatic embryos frequently do not resemble their zygotic counterparts (Fig. S.38).

Somatic embryogenesis is not the only form of *in vitro* embryogenesis. Embryogenesis from gametophytes or gametes is generally known as **haploid embryogenesis**. More specifically, embryos originating from cells of male gametophytes develop via androgenesis: embryos that originate from female gametophytes or **embryo sacs** develop via gynogenesis.

Fig. S.38. Mature somatic embryos of white spruce (*Picea glauca*). Each embryo has numerous cotyledons, but consists largely of **hypocotyl** and root cap tissue.

2. Techniques

Somatic embryogenesis has been induced in hundreds of species. An extensive list of species and techniques is available in George (1996).

Somatic embryos are induced from explants (dissected tissue). The type of induction (direct vs. indirect, Fig. S.39) depends on the predetermined state of the explant. Direct induction is easiest from plant tissues that have embryogenic developmental competence, such as parts of ovules and zygotic embryos. Somatic embryo formation from such tissue occurs in the absence of any applied **auxin**, although low amounts of auxin have sometimes been found to be effective. Indirect induction, in contrast, requires high concentrations of auxin, to cause cell proliferation and callus formation. When these tissues are subsequently placed in low auxin environments they are able to form embryos. Indirect embryogenesis is the much more popular method because the number of embryos that can be produced from the intervening callus is much greater than can be directly produced on an explant. Consequently, indirect embryogenesis is also favoured from the practical consideration of embryo production. Techniques by which direct and indirect embryogenesis are initiated to result in cultures of repetitively embryogenic tissues are similar. Once the embryonic state is induced, the initial differences of explant choice and treatment are of little importance. An important practical consideration when comparing these two methods of initiation is that direct embryogenesis is to be preferred as it results in much less **somaclonal variation** than indirect embryogenesis.

The role of auxin in induction is of central importance to somatic embryogenesis. Its use is dictated by a long empirical tradition, but there is no molecular mechanism known to explain auxin's effect on regulating cell division and differentiation. Various **polyamines** and **cytokinins** are also known to induce somatic embryogenesis.

(a) *Direct somatic embryogenesis* (Fig. S.39). The assumption in explant selection is that the tissue is already embryo-genically committed or competent. A variety of tissues has been used, including ovular tissues such as **nucellus**, **cotyledons** and **hypocotyls** of zygotic embryos, or parts of immature and mature plant organs, generally leaves or leaf parts such as petioles. **Protoplast** suspensions have also been induced to directly produce embryos. In many of these studies, no auxin is applied, or if present, it is at low concentrations in the range of 0.5–1.0 mg/l. The most commonly used auxin is 2,4-dichlorophenoxyacetic acid (2,4-D). A **cytokinin**, usually benzyl amino purine (BAP), is sometimes present. Embryos are produced that can be germinated and converted to plantlets.

(b) *Indirect somatic embryogenesis* (Fig. S.39). Embryogenesis begins with induction of an explant to form early embryo stages. Although it is possible to achieve induction and maturation of embryos in one step, it generally it takes two or more steps.

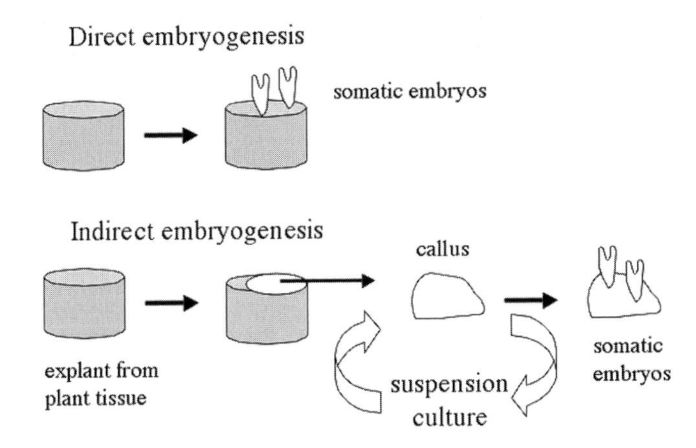

Fig. S.39. Schematic diagram comparing direct and indirect embryogenesis. Explants of plant tissue may produce somatic embryos directly, or they may initially produce a callus, which can be cultured and multiplied by suspension culture.

For some plant tissues, induction of embryogenic callus occurs in only one step. This callus may form embryos that are able to complete maturation directly without subsequent transfer to other media. It is more common, however, to find that following induction an embryogenic callus is produced that is able to form embryos that must be transferred either as individual embryos, or en masse with the embryogenic callus, to a low auxin medium before any further maturation can take place.

In some species, induction results in a non-embryogenic callus, and embryos are formed only if the tissue is transferred to a medium with no or low auxin concentration. The callus then acquires embryogenic competence. This is characteristic of tissues that require two or more steps to produce embryos.

During indirect embryogenic induction, auxin (usually 2,4-D) concentrations of 1–10 mg/l are used, often in combination with cytokinin, particularly BAP. The ratios of auxin to cytokinin can be quite important. For species in which multiple steps are required to achieve induction, the amount and type of auxin is also important. Auxins other than 2,4-D are often favoured in the later stages of multiple-step procedures.

The most effective type and amount of both auxin and cytokinin need to be determined species-by-species, if not genotype-by-genotype. Not only can the number of embryos formed depend on auxin concentration, but also genotypes may differ in their auxin response.

Explants that produce non-embryogenic callus are commonly found in tissue culture. These are also often a source material for adventitious shoot formation. Although both types of culture, indirect somatic embryogenesis and indirect shoot formation, require auxin and cytokinin to be present, the ratio of auxin to cytokinin is generally higher (around 3:1) for indirect somatic embryogenesis induction than for indirect shoot formation (around 2:1).

Other factors also influence indirect embryogenesis, and there are many different refinements based on alteration of media components and factors. Somatic embryogenesis can be influenced by altering the nitrogen concentration of a medium and by changing the form in which the nitrogen is available (organic, inorganic, reduced, or oxidized). Various amino acids, especially glutamine, sugar type and concentration also influence induction and embryo maturation. Improved embryo production can be achieved by alteration of other media factors including: overall salt concentration, micronutrient composition, osmotic potential, pH, redox potential and concentration of activated charcoal.

3. Commercial potential, successes

The degree of commercial success achieved depends on whether somatic embryogenesis is considered a propagation technique in its own right or as an integral part of genetic engineering.

Somatic embryogenesis can be viewed as a gateway technology, one which provides an entrance to novel breeding techniques. In this respect, it has been very successful commercially, as it has allowed transformation of plants, leading to new breeding stock that contains important genes for pest resistance (e.g. the insect resistance gene from *Bacillus thuringiensis*, *Bt*). Somatic embryos are aseptic cultures that are free of viruses, bacteria and fungi and therefore provide a platform for technologies that depend on uncontaminated tissues. They may serve as uniform stock for transformation experiments by providing a technically more accessible tissue; both *Agrobacterium* and particle bombardment technologies can be used, as well as protoplast-based micromanipulator-mediated transformations (see: **Transformation technology**).

Some aspects of somatic embryogenesis are commercially successful, as the technique is a well-established tool in the molecular biological arsenal. Most Canadian and US patents that involve somatic embryogenesis involve one of the following: (i) mass propagation of superior genotypes; (ii) provision of a vehicle for a transformation technique; (iii) *in vitro* production of valuable biochemicals (i.e. taxol); or (iv) acquisition of desirable physiological properties necessary for artificial seed technology (i.e. **desiccation tolerance**, cold resistance, encapsulation methods). Commercial production of novel plants via somatic embryogenesis, however, is extremely limited. (See: **Synthetic seed (synseed)**)

4. Limitations

There are limitations to the ability to induce somatic embryogenesis that are related to parental genotype, induction methodology, multiplication steps and developmental stage of the source material. There are also limitations on genetic uniformity and especially phenotypic uniformity.

(a) *Limits to initiation.* Problems with initiation are the biggest hurdle to the application of this technology. Studies to separate genetic from environmental effects have shown that induction may be strongly under genetic control; certain genotypes lend themselves to somatic embryogenesis. Another related problem is the frequency of initiation. Some genotypes produce somatic embryos less readily than others, requiring much larger numbers of explants.

Most initiation methods use plant material that has been grown in optimal conditions; using parent plants that are not as healthy as possible results is uncontrolled variation. Explants are placed on media that have been supplemented with plant growth regulating substances, and various combinations of carbohydrates and inorganic and organic nitrogen. Since any of these can be limiting, media considerations are critically important. Although many of the more commonly used media are variations of the high salt medium designed by Toshio Murashige and Folke Skoog ((1962) A revised medium for rapid growth and bioassays with tobacco tissue cultures. *Physiologia Plantarum* 15, 509–513), many species prosper on media that differ in salt concentration and composition, vitamin supplements, and micronutrient formulation. There are hundreds of media formulations in the scientific literature, from which it is clear that experimental protocols in which one or many components are altered can result in successful somatic embryogenesis.

Multiplication rates need to be high enough to justify the high cost of laboratory-intense protocols. Once the hurdle of induction is overcome, losses occur at each step of a protocol. The final number of embryos that turn into plants in the greenhouse (**conversion**) is not related to the number that are induced, but rather to the number of induced embryos that can be multiplied and, in turn, matured.

Induction of somatic embryogenesis is often difficult to achieve from mature plants. Unfortunately, mature plants are usually the stage at which breeders are able to conclude whether their predicted gains have been realized. To overcome this problem, breeders have taken parents with known gains and crossed them to create F_1 seed, which is then used to induce somatic embryos. Consequently, this is unproven material.

The long time that it takes to produce embryos from induction through to maturation is a disadvantage because cultures need to be maintained, which increases cost. As they age, cultures often lose their embryogenicity, either progressively or absolutely, they show an increase in somaclonal variation, and there is increased risk of contamination and loss.

(b) *Genetic variation of somatic embryo cultures*. Genetic variation due to somaclonal variation has been most commonly reported from chromosome studies. Mutation events that influence development are undesirable, whereas more neutral mutation events can be tolerated. There are a number of possible origins of somaclonal variation. Tissue explants used as source cultures for initiation of somatic embryos may show genetic mosaicism, which may be magnified or distorted by the somatic embryogenic process. Ideally, the starting material should be genetically uniform to avoid such problems, e.g. induction from early embryos. The uniformity of the resulting cultures is only as good as the care taken in selection of single embryos as source; failure to separate competing minor genetically unique embryos within an ovule will compromise the uniformity of the culture. If indirect embryogenesis is the only route available, it is possible to diminish the overall somaclonal variation by choosing cultures that produce their embryos after little subculture. If subculture is necessary then it should be frequent, since long periods between subcultures are deleterious. Diminishing the amounts of growth regulator during culture is considered desirable.

(c) *Phenotypic variation of somatic embryo cultures*. Phenotypic uniformity during embryogenesis is desirable if the final embryo phenotype is important. If developmental variation is not strictly controlled, then a variety of phenotypes will be present, not all of which will yield mature embryos. In such cases, selection of growth types within cultures becomes critically important. Phenotypic variation is much more evident than genotypic variation, and even though there are few parts to an embryo (radicle, hypocotyl, cotyledons and epicotyl), each can show variation *in vitro*. Pattern events during embryogenesis are not as uniformly under genetic control as generally supposed. In the few instances where this has been experimentally analysed, somatic embryos have been shown to vary in size and in number of parts and, for some species, they may even show greater phenotypic variation than in natural populations. (PvA)

Bonga, J.M. and von Aderkas, P. (1992) In Vitro *Culture of Trees*. Kluwer Academic, Dordrecht, The Netherlands.

Cyr, D. (2000) Seed substitutes from the laboratory. In: Black, M. and Bewley, J.D. (eds) *Seed Technology and its Biological Basis*. Sheffield Academic Press, Sheffield, UK, pp. 326–374.

George, E.F. (1993) *Plant Propagation by Tissue Culture: Part I. The Technology*. Exegetics Ltd, Edington, UK, pp. 1–574.

George, E.F. (1996) *Plant Propagation by Tissue Culture: Part II. In Practice*. Exegetics Ltd, Edington, UK, pp. 575–1361.

Peña, L. and Séquin, A. (2001) Recent advances in the genetic transformation of trees. *Trends in Biotechnology* 19, 500–506.

Sorghum

Sorghum is *Sorghum bicolor* (syns: *Andropogon sorghum, Holchus sorghum*) and *Sorghum vulgare*. Other names for sorghum are: *bachanta* (Ethiopia); *cholam* (India); *durra* (Sudan); great millet; guinea corn (West Africa); *jola* (India); *jawa* (India); Kafir corn (South Africa); milo (USA); *kaoliang* (China); sorgho (USA); and sweet sorghum (USA). It is often included with millets but its seeds are larger than those of the millets and do not store as well. It can be cultivated in arid or semi-arid regions. Sorghum, like maize, exhibits the more efficient **C4** mode of **photosynthesis**. (**See: Cereals; Millets**)

1. Distribution and economic importance

Sorghum is the world's fourth most important cereal, following wheat, maize and rice (**see: Cereals,** Tables C.8, C.9). Global production in order is: NAFTA (Mexico, USA, Canada, 33%), Africa (31%), Asia (22%), South America (8%), Australia (3%) and Europe (1%). Africa has the highest area under cultivation (52% of the global sorghum area). Approximately 55% of the African grain is produced in West Africa and ±30% in eastern and northeast Africa. The three highest African grain producers are Nigeria, Sudan and Ethiopia. In most African countries, annual yields are still ≤1 t/ha. Countries growing hybrids, with the exception of Sudan, have higher national yields, e.g. South Africa (1.9–2.7 t/ha). Hybrid yields under good management exceed 10 t/ha. Irrigation may also increase national yield, e.g. Egypt (4.7–5.8 t/ha).

Of the total world production in 2002 of 52 million t about 6.3 million t was exported at a trade value of about US$669 million (FAOSTAT). (**See: Crop Atlas Appendix, Maps 1, 9, 21, 25**)

2. Origin and domestication

Sorghum (*Sorghum bicolor*) was first cultivated over 5000 years ago in the region of Ethiopia or Chad, but like several other

African cereals it had spread to India by 4000 years ago, somewhat later to China, and to southern Africa by about 1500 years ago. Sorghums can be grouped into five broad groups (sometimes known as **races**). *Durra* is the most important in India and is widely grown in arid sub-Saharan Africa, as is *Caudatum*. *Guinea* sorghum is grown in humid West Africa, *Kafir* sorghums in southern Africa, and the low-yielding, rather primitive, *Bicolor* sorghums on a small scale in Africa and Asia. Sorghum is the most important cereal in sub-Saharan Africa, because of its resistance to dry, hot conditions. Sorghum has been grown in the Americas since the 18th century, mainly as animal feed. It is now found from 40°S to 45°N latitude, largely in drought-prone areas with low rainfall but also in high rainfall areas (>1000 mm/year) and mountainous regions (altitude >1000 m above sea level) in Ethiopia, Uganda and Lesotho.

3. Grain structure

In most sorghum varieties grains are near-spherical (diameter 2–4 mm) with a relatively large embryo (Colour Plate 9B) (**see: Cereals,** section 4). Seed weight varies according to cultivar and provenance but lies in the range 40–100 mg/seed. A single outer epidermis of the pericarp surrounds the grain. Pericarp **starch granules** are up to 6 µm diameter; smaller than those in the **endosperm**. A second unusual feature of sorghum anatomy is the absence of the testa in many types. The nucellar layer is, however, well developed; in fact it is the most conspicuous of all the bran layers and may be up to 50 µm thick and coloured yellow or brown. **Aleurone layer** cells are similar in size and appearance to those of maize as are the inner endosperm cells and the starch granules that they contain.

In some sorghums pigmentation occurs; all tissues may be coloured, but not all together. The six kernel appearance classes are: white, yellow, red, brown, buff and other. The grain is lustrous or not lustrous. The starchy endosperm may be colourless or yellow, the aleurone layer may contain pigmentation. The pericarp and nucellar layers may contain variable amounts of tannins – polyphenolic compounds responsible for red coloration. Types with large proportions of condensed tannins in the pericarp are known as 'bird proof' or 'bird repellent' because, it is assumed, of the unpalatability of the polyphenols (see sections 4, 5).

4. Composition and nutritional quality

Composition varies widely among cultivars and according to provenance. Mean values, however, have been reported as (dry weight basis): **storage protein** 11%, **starch** 69.5% (70–80% amylopectin), **oil** (triacylglycerol) 3.3%. The albumins and globulins (**see: Osborne fractions**) are relatively high and together comprise 17% of the total protein. High lysine **mutants** have been developed. The B vitamins and mineral elements (calcium, iron, phosphorus) are present as is **phytate**. Darkly coloured seeds contain tannins (polyphenols) which are anti-nutritional as they bind with protein, reducing its digestibility; they also impart a bitter taste.

5. Uses

Most of the sorghum is used in countries where it is produced but for different purposes. In North and Central America, South America and Oceania most is used for animal feed whereas in 'developing' countries, such as in Africa, most is for human consumption. The embryo makes the flour become rancid during storage, and to avoid having rancid flour, fresh grinding is done frequently. The grain can be prepared into products similar to sweetcorn or popcorn. Sorghum is made into a number of traditional food products (often after mixing with the flour of other cereals or cassava) such as: porridges; gruels; fermented, steamed and boiled dough products; rice-type products; wheatless pancakes, tortillas, cakes, cookies and breads, noodles and pastas; traditional alcoholic and non-alcoholic fermented beverages or beers (from sorghum malt), distilled beverages (by the Chinese). Fermentation, to make bread or beer, breaks down indigestible tannins and prolamin in the grain. Sorghum beer must contain at least 51% sorghum grain in the malt, and the product is an opaque beer that is traditionally popular is some regions of Africa. In South Africa, up until 1962, apartheid regulations allowed native Africans to only drink sorghum beer, while whites were permitted to drink the lighter European beers. Much African opaque beer is now brewed from SMM (sorghum/maize/millet) grain, and there is a so-called kaffir beer made from sorghum and wheat (**see: Malting**).

The starch can be made into biodegradable coatings, adhesives, thickening for pie fillings, gravies and many derivatives. Waxy endosperm types yield starch with properties similar to tapioca. The grain coat produces a natural red dye. White grains are preferred for food purposes (see section 4) while the darker seeds are used in brewing. The embryos produce an oil for cooking and salads.

(AE, MN, WAJdeM, MB, JDB)

For information on the particular contributions of the seed to seed science **see: Research seed species – contributions to seed science**.

Dendy, D.A.V. (ed.) (1994) *Sorghum and Millets: Chemistry and Technology*. American Association of Cereal Chemists, St Paul, MN, USA.

www.fao.org/DOCREP/T0818e/T0818E00.htm

Sorghum – cultivation

Sorghum, botanical name *Sorghum bicolor* (syns. and other vernacular names are also used, **see: Sorghum**) is an important world cereal, especially in Africa, Asia and North America. (**See: Crop Atlas Appendix, Maps 1, 9, 21**) Five races are distinguished (Bicolor, Guinea, Caudatum, Kafir, Durra), along with ten intermediate races that combine characteristics of these five. Sorghum is a member of the Poaceae family, subfamily Panicoideae, along with its wild relatives (in South Africa, e.g. *S. bicolor* ssp. *arundinaceum*, *S. bicolor* ssp. *drummondii*, *S. halepense*, and *S. versicolor*). *S. bicolor* has several duplicated loci; the chromosome base number is x = 5, with 2n = 20 chromosomes in *S. bicolor*, and 2n = 40 in *S. halepense*. It may be a paleopolyploid (an ancient **polyploid**); polyploidy in grasses is an ongoing widespread process and may reoccur from independent events. For some details concerning the geographic origins, world distribution and economic importance **see: Sorghum**. Also **see: Millets**.

Ninety per cent of the world's sorghum crop area is in Africa and Asia, mainly in drought-prone areas with low rainfall, though it will tolerate high rainfall areas (more than 1 m/year) and mountainous regions (above 1000 m), in Ethiopia, Uganda, and Lesotho. It becomes necrotic when temperatures drop below 0°C, but the crop regrowing after cutting (ratooned) can withstand cold and wetness better than pearl millet.

Unlike many other crops, sorghum can become dormant under adverse conditions, such as severe, early drought, and can resume growth once water becomes available again. Sorghum can resume tiller production and may produce good yields in relation to nutrients and water availability when planted out after generative development in small plugged trays, involving experimentally prolonged storage (>100 days). It also survives temporary waterlogging, but not sustained flooding.

Plant height varies from 0.8 to more than 4 m. Stalks can be dry, juicy, with sweet or insipid juice, and the stem and leaves may have a waxy bloom to varying degrees. There are six classes for leaf midrib colour and 13 classes are distinguished for inflorescence compactness and shape – ranging from very lax panicle through combinations of loose, drooping, erect, compact, broom and other types. Glume colour at maturity is highly variable – white, yellow, mahogany, orange-red, purple, black, grey and others.

1. Genetics and breeding

There are a considerable number of subsistence sorghum varieties. The first hybrids were made in the USA in the 1950s, based on **cytoplasmic male sterility**, and since then have became popular in Australia, China, India, Sudan and South Africa. Grain hybrids tend to be short (<2.5 m), photoperiod-insensitive, management responsive, and have a good **harvest index** (grain/straw ratio).

Sorghum could become a model for understanding the molecular genetics of the grasses, because it has a relatively small genome (equivalent to 30% of the maize genome, 25% of sugarcane and less than 5% of wheat) and **transgenics** can be made. Indeed, several laboratories already use the sorghum genome as a model for maize. In the near future, marker-assisted breeding and in the more distant future, gene transfer, will improve the rate, quality and quantity of improvements. Sorghum has the potential to produce a higher quality and quantity of food and feed, with fewer chemical inputs; in addition, plants may become bioproducers of **nutraceuticals**, chemicals and pharmaceuticals. The first transgenic sorghum was reported in 1993, but a drawback for the development of such varieties is that sorghum readily out-crosses with wild grasses, and thus transgenic genes can move into the wild ecosystem.

2. Development and production

Commonly, varieties take 65 to 150 days to mature, but some take longer and, in very dry areas of Botswana, Sudan and Eritrea, may take only 45 days. The period to maturity is determined by genotype, daylength and temperature; in the USA, for example, different varieties can be grown with a 15-h day and high temperatures, or at a 12-h day with low temperatures, or with graduations in between, over a range of latitudes. In West Africa, photoperiod-sensitive sorghums only flower towards the end of the rains, and thus escape grain damage from grain moulds and seed-sucking insects.

Sorghum is a short-day plant. Flowering, which starts about half the period to grain maturity, occurs from the top of the panicle downwards, and the full flowering period may extend over 6 to 15 days. **Spikelets** open at night (for about 2 h), and the stigmas and anthers emerge after about 10 min. Cross-pollination occurences may range from 0 to 40%, and average 6%.

At the seed stage, cultivars differ in susceptibility to **shattering** from very low to very high. Good exsertion (several cm) of the head out of the top leaf sheath or boot, and erect peduncle, facilitate harvesting, but the curved peduncles complicate mechanical harvesting.

Hybrid seed production requires isolation to avoid cross-pollination by wild relatives and cultivated sorghums; on the other hand the use of male sterility systems makes female parental lines very vulnerable to **ergot**.

During handling, extended exposure of the seed to the sun or the rain should be avoided. For proper storage, the moisture content of the seed should be close to 12% and remain between 6 and 14%. Relative humidity should be kept below 70% as excessive moisture may lead to grain moulds and insect infestation during **storage management**. However, moisture contents below 6% may cause irreversible damage.

4. Quality, vigour and dormancy

Sorghum vigour, though less satisfactory than in maize, is about the same as that in pearl millet and superior to that in finger millet.

Dormancy may last for up to several months after harvest and is associated with low sprouting during wet conditions after grain maturation. For a discussion of the genetics of sorghum dormancy, **see: Preharvest sprouting – genetics, 4. Sorghum**. Shattered seeds may thus survive winter and create a weed problem in the following crop.

5. Sowing

A fine, firm seedbed is required but cannot always be obtained, in particular in rain-fed, small-scale productions. The seed can be sown (seed rate ≤5 kg/ha), **broadcast** (seed rate up to 15 kg/ha), dry planted, or transplanted (at 3–4 weeks). It is grown in rows up to 5 cm deep with 50–100 cm between rows, 20–100 cm within rows (according to rainfall) on the flat, on hills and on ridges. It is a labour-intensive crop requiring seedbed preparation, weeding, bird scaring, harvesting and threshing. Drilling and planting in lines facilitates inter-row weeding. It can be a sole crop or in a mixed crop. With a high irrigation input, crop yields exceed 10 t/ha.

The time of planting affects yield and grain quality. Delayed planting usually lowers yield. The ripening of grain during the rains may lead to severe grain mould damage and yield reductions because of pathogens affecting the flowers and seed development.

Sorghum succeeds on a wide range of soils, including poor soils and in drought-prone areas, but may exhaust soil fertility. Its adaptation comes from low transpiration per kg plant tissue (83% of maize, 58% of wheat, 56% of potato and 34% of lucerne), small leaves, large absorption capacity of fibrous roots, and growth resumption when relieved of moisture stress. Consequently, on irrigated farms in drought-prone areas with high costs for irrigation water, it may be more economical to grow irrigated sorghum than maize.

6. Treatment

Sorghum is affected by a wide range of **pathogens** and **pests**, often each able to destroy the yield. Several important diseases (**smuts**, ergot, the grain mould complex) mainly affect the crop at or after flowering, though in-depth review of the seedborne and seed transmitted nature of the different pathogens is required. Disease resistance breeding is the major control strategy but, along with seed treatments, health testing and chemical control in crops provide, at best, modest control. One problem is that several pathogens are highly genetically diverse, e.g. *Colletotrichum graminicola*. Research on the integrated management of pests, including beneficial insects and diseases, is consequently important.

However, seed treatments, mainly against fungi, have been developed for sorghum and used in high input agriculture. In the USA, temporary (several decades) complete control was obtained for some diseases (e.g. downy mildew) in combination with seed treatments, and head smut (*Sporisorium reilianum*). In Africa, smut control is attempted with local traditional treatments as well as with chemical treatments. The recently developed resistance-inducing chemicals provide new opportunities for a wide spectrum, incomplete control in support of plant resistance and cultural management practices.

Since the mid-1990s, a start has been made with on-farm **priming** (soaking in water for 10 h) in Africa and India. Benefits include: faster emergence, better stand, a lower incidence of re-sowing, more vigorous plants, better drought tolerance, earlier flowering and harvest, and higher yield. **Transplant** production has similar benefits, producing even hardier plants; the extra costs of nurseries and trays are distinctly less than the returns, as yields may double.

(WAJdeM)

INTSORMIL, ICRISAT (1997) *Proceedings of the International Conference on Genetic Improvement of Sorghum and Pearl Millet*. 22–27 Sept. 1996, Lubbock, Texas, USA. Publication No. 97–5.

Leslie, J.F. (ed.) (2002) *Sorghum and Millets Diseases. World Agriculture Series*. Iowa State Press, Iowa, USA.

Mottram, A. and Young, E.M. (2002) *Transplanting Sorghum and Millet as a Means of Increasing Food Security in Semi-arid, Low Income Countries. A Project Update*. Centre for Arid Zone Studies, University of Wales, Bangor. www.bangor.ac.uk/transplanting/project.htm

Smith, C.W. and Frederiksen, R.A. (eds) (2000) *Sorghum: Origin, History, Technology, and Production*. John Wiley, New York, USA.

Sorption isotherms

All seeds are hygroscopic, including those with impermeable **seedcoats** once this physical barrier has been breached. Thus, seeds automatically absorb or desorb moisture by diffusion along a water potential gradient between the seed and the surrounding air until equilibrium is reached. The temperature-dependent relationship between the **equilibrium moisture content** of seeds and relative humidity is described by so-called sorption isotherms. Isotherms for **orthodox** seeds delineate three main **water binding** regions (important with respect to seed **longevity**):

I Water strongly bound at ionic sites (e.g. charged carboxyl or amino groups of proteins, lipids and cell walls).

II Water weakly bound at polar, non-ionic sites.

III Water loosely bound through the bridging of hydrophobic moieties.

Fig. S.40 provides a graphic illustration of the reverse sigmoidal relationship between the equilibrium moisture content of seeds and relative humidity at a given temperature.

(RJP, SHL)

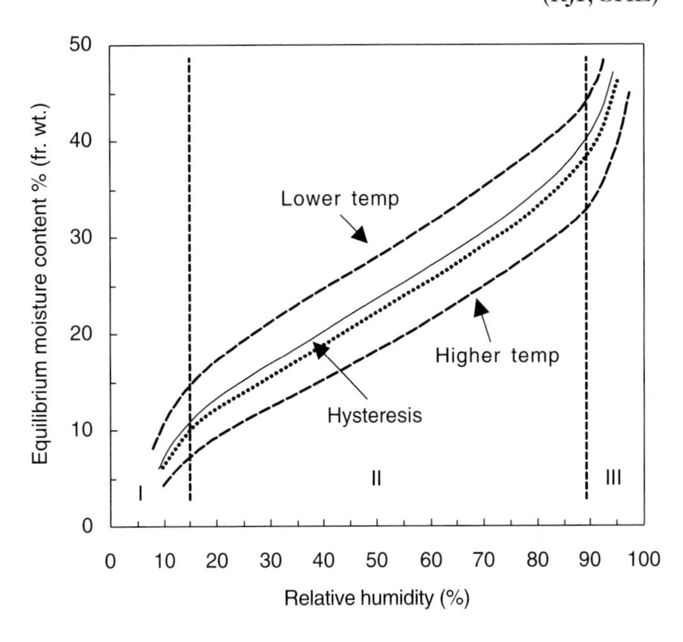

Fig. S.40. Illustration of sorption isotherms. The solid line depicts a typical desorption isotherm for non-oily seeds. Dashed lines show the effect of lowering or raising the temperature on moisture equilibria. The dotted line illustrates an absorption isotherm and the phenomenon of hysteresis; for a given relative humidity, seeds hold slightly less water on absorption than desorption. Vertical dashed lines delineate the three water-binding regions I, II and III.

Sowing

The dispersal or placement of seed on or in the ground or other planting medium for it to germinate. (The words 'sow' and 'seed' actually share the same etymological root.)

See: Planting equipment – agronomic requirements

Sowing methods – brief history

From ancient times, seeds have been sown, **broadcast**, sprinkled or dropped by hand into the ground, and covered in a separate operation, by raking, harrowing or simply treading in. Ancient sowing methods recorded in Chinese literature distinguished three sowing methods: broadcasting, sowing in rows and sowing individual seeds, and different methods were evidently employed according to the crop, size of field, soil or weather conditions. Seed droppers, consisting of a tube attached to a plough drawn by oxen, through which seed could be dropped by hand at regular intervals, are represented on Sumerian seals four millennia ago, which show a double-handled stilt with a single seed tube. A similar system is recorded in an first-century tomb in Shanxi, China, in which seed flow was apparently regulated by sieves, in combination with a rhythmic side-to-side movement to throw seeds alternately into each of two tubes set in a rectangular frame.

Such drills have survived to modern times in Iraq, Iran and parts of India. Indeed, simple hand-sowing practices are still widespread in rural areas across Asia and Africa where labour is plentiful, using techniques such as **dibbling** to drop seeds singly into holes and transplantation of **rice**.

The first recognizably modern type of mechanical seed drill was patented in 16th-century Venice. Jethro Tull, in early 18th-century England, designed one of the most famous, which has evolved into modern mass-flow meter types. Its key feature was metering through a notched rotating axle at the bottom of the seed box, through which seed flow was regulated by a brass cover and an adjustable spring. Subsequent meter designs consisted of wheels bearing small spoons that dipped into the seed hopper and guided it to the furrows in standard amounts. However, feeding mechanisms would often clog and perform erratically, despite improvements, until well into the 19th century, when mechanization of agriculture generally took hold in western Europe, its colonial countries and the New World. Modern tractor-pulled seed drills have a variety of metering systems and furrow openers, with devices dragged behind them to cover the seed and firm the **seedbed**. **(See: Planting equipment – agronomic requirements; Planting equipment – metering and delivery)** (PH)

Bray, F. (1984) *Joseph Needham: Science and Civilisation in China*. Vol. 6, *Biology and Biological Technology*: Part 2, *Agriculture*. Cambridge University Press, Cambridge, UK.

Soybean

1. World importance and distribution

Soybean (*Glycine max*) is the most highly used legume seed, sometimes called the miracle crop. The seeds are the world's foremost provider of **storage protein** and **oil** and an amazing diversity of products that contribute to the health and well-being of people all over the world. Soybean is a mainstay in the food processing, vegetable oil and animal feed industries. The seed supplies 30% of world vegetable oil and 60% of the vegetable protein. In 2002 there were 79.4 million ha of soybean planted worldwide of which approx. 37% were in the USA (FAOSTAT). Other important producers are Brazil (16 million ha), Argentina (11 million ha), China (9 million ha) and India (5.6 million ha). These five countries account for over 90% of world soybean production (see section 6). **(See: Crop Atlas Appendix, Map 2, 3, 13, 22; Legumes; Oilseeds – major,** Table O.1; **Soybean – cultivation)**

2. Origin

The soybean probably originated in northeast China and emerged as a domesticated species at least 3000 years ago. The cultivated soybean (*G. max*) is considered to have evolved from the wild species *G. soya*. It is an ancient food crop in China, Japan and Korea, having been taken out of China, it is thought, by Buddhist missionaries about 1500 years ago: it soon became a dietary staple. It was introduced to Europe in about 1691 (but it became known as a food plant only in the 18th century) and from Europe to the USA around 1804. Utilization of soybean in the western world rapidly expanded in the 20th century.

3. Seed structure and types

The soybean ovary initiates development of the fruit (pod) immediately following fertilization and it reaches nearly maximum length and width prior to the initiation of **ovule** (seed) development. Mature pods contain 2–4 small, hard, round or ovoid seeds 5–10 mm in diameter. The **endosperm** develops before the **embryo** and is almost completely absorbed by the two developing **cotyledons** prior to embryo maturity. A single row of cells, often called the aleurone layer, and flattened parenchyma between the **testa** and embryo are all that remain of the endosperm. See Fig. S.6.

Thus, the mature seed essentially consists of a thin seedcoat (**testa**) surrounding a large embryo. The seedcoat is marked with a **hilum** scar (black, brown, or yellow), which is the point of separation of the **funiculus** from the ovary. The mature, quiescent embryo consists of two large fleshy cotyledons, a **plumule** with two well-developed primary leaves enclosing one trifoliolate leaf primordium, and a **hypocotyl–radicle** axis that rests in a shallow depression formed by the cotyledons. The mature seed of most cultivated cultivars are yellow and near spherical in shape: however, there are genotypes with flattened and black, brown or green seeds (Fig. S.41). Seedling growth is **epigeal**.

4. Composition and nutritional value

Soybeans contain all three of the macro-nutrients required for good nutrition. **Protein** content is approx. 40% fw (about 90% of which is globulin, with a good balance of the **essential amino acids** needed for the human diet). There is about 20% fw insoluble carbohydrate 'fibre', about 11% soluble

Fig. S.41. Soybean seeds (see text for dimensions). These seeds, from the National Soybean Germplasm Collection housed at Urbana, Illinois, show a range of colours, sizes and shapes. Photo courtesy of Scott Bauer, ARS/USDA.

carbohydrate (of which 41–68% is sucrose, 12–35% stachyose, 5–16% **raffinose**) and a trace of **starch**. **Oil** (triacylglycerols) content is 15–20% fw in **oil bodies** of the cotyledons. **See: Oilseeds – major**, Table 2, for oil composition. Vitamins (B complex) are present, and minerals, including calcium and iron. Soybeans, especially the outer hull, are an excellent source of dietary fibre. The seeds also contain several important biologically active compounds such as the steroidal saponins isoflavones or phytoestrogens (**see: Pharmaceuticals and pharmacologically active compounds**).

Anti-nutritional factors. Soybeans have a relatively high content of inhibitors of trypsin and other proteases (**see: Protease inhibitors**) which are destroyed during cooking and fermentation and which are left in the supernatant liquid after curd and tofu preparation. Noxious haemagglutinins in the seeds are also inactivated by various preparative procedures: the agglutinin has an important clinical use in bone marrow transplantation procedures (**see: Soybean lectin**). Seeds are high in phytin, which can bind mineral ions (calcium, iron, magnesium) thus lowering the food value. Fermentation removes the phytin but much of it remains in tofu, bean curd and cooked beans.

5. Uses

Based on the differences in use, there are two distinct types of soybean: food and oil beans. In the Far East, soybean is made into various foods for human consumption, including tofu, soymilk, soy sprouts, *miso*, *natto* and *tempeh*, whereas in the west, most soybeans are crushed for oil for food and defatted, high-protein meal, primarily for animal feed. Food cultivars do not differ fundamentally from oil cultivars, except that they usually have a specific seed size requirement, lighter seedcoat, clear hilum, are higher in protein, lower in oil and demand a superior marketing grade.

There is an amazing diversity of soybean uses and products (about 180) for humans and livestock. Indeed, so numerous are its uses that soybean has been described as the 'miracle bean'. The two basic products of the soybean are protein meal and oil. In the USA, more than 90% of the oil is consumed as margarine, shortening, mayonnaise, salad oil and other edible products; the rest is used in industrial products such as paint, varnish, linoleum, and rubber fabrics and as a source of the natural emulsifier, lecithin. Soybean meal is the major source of the protein supplement used in livestock feeds, which utilize 98% of the total meal produced. In the protein-short areas of the world and elsewhere, soybean meal is finding increasing use in human food products. It was possibly around 2200 years ago when the Chinese learned to precipitate bean curd or tofu from a cooked puree of soybean by treatment with calcium or magnesium sulphate. These products are the major form in which processed soybean is consumed in Asia and increasingly in other parts of the world. Fermented soybean products are also prominent in the Asian diet. The beans are also used to make a milk substitute, soymilk. In addition to the provision of many food products soybean protein and oil are sources of numerous industrial products. Fig. S.42 summarizes the uses of soybean.

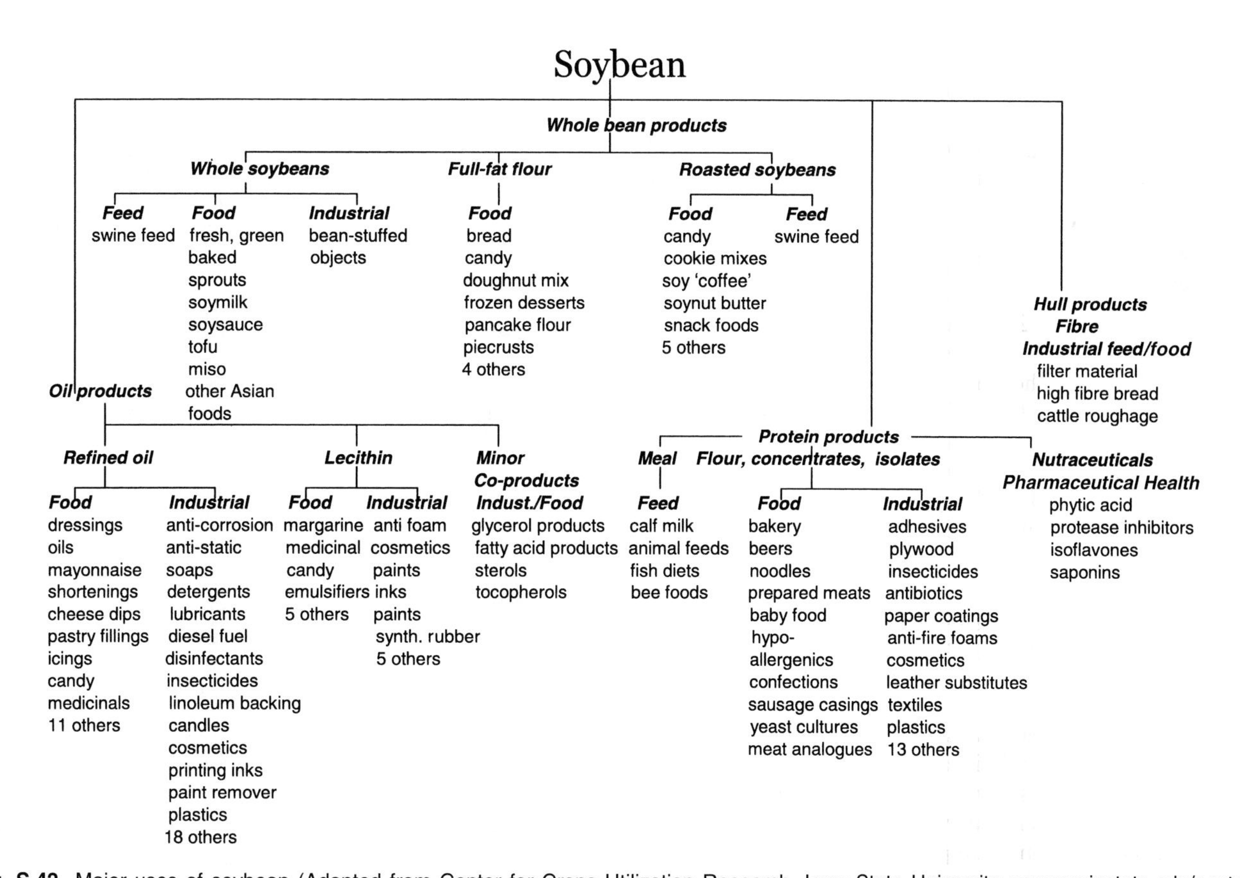

Fig. S.42. Major uses of soybean (Adapted from Center for Crops Utilization Research, Iowa State University, www.ag.iastate.edu/centers/ccur/soy).

(See: Oils and fats; Oils and fats – genetic modification; Oilseeds – major; Storage protein; Storage protein – processing for food)

6. Marketing and economics

Since the USA accounts for 35–40% of the world cultivated area and production of soybean, data from this country are a good illustration about marketing of soybean seeds. According to the statistics published by the United Soybean Board of the USA, in 2000 54% of the world's soybean trade originated from the USA. The European Union was the number one market for these soybeans though China was the largest single-country customer, with Japan second. The Philippines was the largest customer of USA soybean meal, followed by Canada. Mexico, China and Korea were prominent buyers of USA soybean oil. Information on world production and trade is given in Tables S.10 and S.11.

7. Plant types

The soybean is an erect, branched, hairy, annual plant about 0.6–1.5 m above ground level and 2 m below ground level. The original type appears to have been a trailing or semi-trailing plant, and from this it is supposed that the erect types have evolved. It has large trifoliolate leaves, small white or purple flowers and short, hairy pods with one to four seeds. At least two growth habits are recognized, determinate, erect plants, and indeterminate, trailing or semi-trailing plants. The reaction of soybean varieties and types to daylength (photoperiod) is one of the major developmental characteristics of the species. From a practical agricultural point of view in the USA soybeans are classified into nine groups according to their adaptation to a particular zone designated by their climatic conditions, mainly daylength.

Table S.10. Examples of soybean production and exports (2002).[a]

Country	Production (million t)	Value (million US$)
Argentina	30	1,119
Brazil	42	3,032
Canada	2	139
China	17	77
India	4	0.391
Paraguay	3	59
USA	75	5,623
World	181	10,545

[a]Excludes oil, curd, paste and sauce. Source: FAOSTAT.

Table S.11. Major soybean imports, 2002.[a]

Country	Amount (1000 t)	Value (million US$)
Belgium	1,752	378
China	13,848	3,019
Germany	4,346	908
Indonesia	1,365	299
Japan	5,039	1,223
Korea (South)	1,475	318
Mexico	4,383	925
Netherlands	5,602	1,164
Spain	3,352	696

[a]Excludes oil, curd, paste and sauce. Source: FAOSTAT.

Soybean is a major target species for genetic improvement by biotechnology, especially by the introduction of herbicide resistance. In 2002, more than 40% of the crop in the USA was of this type. (OV-V, MB, KD, DTK)

For information on the particular contributions of the seed to seed science **see: Research seed species – contributions to seed science.**

Brar, G.S. and Carter, T.E., Jr (1993) Soybean, *Glycine max* (L.) Merrill. In: Kalloo, G. and Bergh, B.O. (eds) *Genetic Improvement of Vegetable Crops*. Pergamon Press, Oxford, UK, pp. 427–463.

Caldwell, B.E. (1973) *Soybeans: Improvement, Production and Uses.* American Society of Agronomy, Madison, WI, USA.

Fehr, W.R. (1989) Soybean. In: Robbelen, G., Downey, R.K. and Ashri, A. (eds) *Oil Crops of the World*. McGraw-Hill, New York, USA, pp. 283–300.

Hymowitz, T. (1995) Soybean. In: Smartt, J. and Simmonds, N.W. (eds) *Evolution of Crop Plants*, 2nd edn. Longman, London, UK, pp. 326–332.

Liu, KeShun (1997) *Soybeans: Chemistry, Technology, and Utilization.* C.H.I.P.S., Texas, USA.

Nieuwenhuis, R. and Nieuwelink, J. (2002) *Cultivation of Soya and Other Leguminous Crops*. Agromisa, CTA, Wageningen, The Netherlands.

Shanmugasundaram, S. and Sulzberger, E.W. (eds) (1985) *Soybean in Tropical and Subtropical Cropping Systems*. AVRDC, Taiwan.

Sipos, E.F. and Szuhaj, B. (1996) Soybean oil. In: Hui, Y.H. (ed.) *Bailey's Industrial Oil & Fat Products*, Vol. 2, 5th edn. John Wiley, New York, USA, pp. 497–602.

Soybean Abstracts. CAB International, Wallingford, UK.

TeKrony, D.M., Egli, D.B. and White, G. (1987) Seed production and technology. In: Wilcox, J.R. (ed.) *Soybeans: Improvement, Production and Uses*, 2nd edn. American Society of Agronomy Monograph 16, Madison, Wisc, USA, pp. 295–353.

Vaughan, J.G. (1970) *The Structure and Utilization of Oil Seeds.* Chapman and Hall, London, UK.

Verma, D.P.S. and Shoemaker, R.C. (1997) *Soybean: Genetics, Molecular Biology and Biotechnology. Biotechnology in Agriculture Series*, No 14. Oxford University Press, Oxford, UK.

Soybean – cultivation

See: Soybean for information on the origins, the seed, composition, uses and economic importance

Soybean (*Glycine max*, Fabaceae) is a photoperiod-sensitive, **short-day plant**. Thus, latitude and summer daylength play a major role in cultivar development and adaptation to various geographic regions. On the American continent, for example, soybean cultivars have been divided into 12 **maturity** groups (MG) ranging from MG 000 adapted to high latitudes (Canada) to MG X adapted to Central America. If a cultivar is grown in latitudes with too short a summer daylength, flowering is hastened, and yield and quality may be reduced. Within the range of 10 to 30°C, increasing mean temperature hastens flowering.

1. Genetics and breeding

Improvements are needed through genetics and plant breeding in many soybean food traits such as beany flavour, flatus-producing ability (raffinose-series **oligosaccharides**), oxidative and flavour instability and the presence of anti-nutritional factors, but grain yield for oil use has received the greatest emphasis by plant breeders for many decades (**see: Soybean**). Progress has been made in selecting for yield, resistance to pathogens and nematodes and tolerance to production hazards, such as lodging and **shattering**. Recent estimates indicate that soybean yields are improving at the rate of ~23 kg/ha/yr due to improved genetics, improved production practices and higher CO_2 levels. As the demand for food beans has increased, breeders have placed an increased emphasis on developing cultivars specifically targeted for direct human consumption. Modifications of oil composition have resulted in lower saturated fat content and acceptable flavour stability without hydrogenation (**see: Oilseeds**). There is an expectation, therefore, that the manipulation of genes that determine seed composition will have a significant influence on the quality and utility of soybean for food uses in the future (such as higher lysine, methionine, stearic and oleic acid contents, and lower linolenic acid and stachyose). (**See: Oils and fats – genetic modification**)

Genetic improvements have been accomplished mainly through conventional plant breeding methods for self-pollinated crops, although molecular-based plant breeding techniques are assuming an increasingly important role. Breeding lines are selected to provide genetic variability for the targeted traits, including the use of techniques such as recurrent selection, crossing (**hybridization**), mutagenesis, molecular markers and genetic transformation; these populations are then advanced, with or without further selection, to produce relatively homozygous lines that are subject to yield and other trait evaluation. A given cycle may take 5 to 7 years (using winter nurseries) to release improved pure-line cultivars. The requirements for the commercial production of hybrids are not currently available however.

For many years plant breeders in the public sector were responsible for genetic improvement, but privately developed cultivars are now used extensively in many soybean production areas in North and South America. The private development of herbicide-tolerant soybean lines, using **genetic modification** (GM) technology, had a major impact on the soybean industry since their introduction in 1996. Initially the glyphosate-tolerance gene was transferred to cultivars by backcrossing, but more recently many breeding programmes have incorporated the gene directly into the majority of their lines. Producer uptake of herbicide-tolerance has been rapid: in 2003, more than 90% of the crop in the USA, and 55% of the global crop, was grown from such varieties. (**See** inset in: **Crop Atlas Appendix, Map 13**) However, this has exceeded consumer acceptance of GM soybean products in some areas of the world and has led to the implementation of **identity preservation** schemes to safeguard the quality of non-GM commodity grain crops.

2. Development and production

Reproductive development in soybean plants is induced by day-length causing axillary buds to develop into flower clusters (racemes) of two to 35 flowers each, in plants about 0.6–1.5 m tall. There are two types of stem growth habit and floral initiation: **determinate** erect plants and **indeterminate** trailing or semi-trailing annual plants, which are adapted to long and short growing seasons, respectively. The trailing types appear to be the original form from which the erect types evolved. Seed development begins with fertilization of highly self-fertile flowers (pollination typically occurs before the flowers open, especially in cool weather), and the soybean ovary immediately initiates development into the fruit (pod), which reaches nearly maximum length and width prior to the initiation of ovule (seed) development. The short hairy pod commonly contains two to three seeds. The endosperm develops before the embryo and is almost completely absorbed by the two developing cotyledons prior to embryo maturity; thus, the mature seed essentially consists of a thin **seedcoat** (**testa**) surrounding a large **embryo** (**See: Seedcoats – development**, Fig. S.6). Seed development can occur over a period from 20 to 70 days, but usually occurs in about 60 days, and is usually divided into three phases, as described in **Development of seeds – influence of external factors**.

Physiological maturity (PM) occurs at relatively high seed moisture concentrations (55%) as the colour of seeds and pods change from green to yellow, both of which are excellent indirect indicators of PM in individual seeds, and as predictors of PM of the soybean plant community. The onset of seed filling triggers leaf **senescence**. Plants harvested at reproductive growth stage R7 (defined as one pod on the main stem having reached its mature pod colour) had the same seed weight (yield) as plants harvested at full maturity (95% mature pods); thus, R7 represents an accurate indicator of PM for a plant community. Maximum seed viability occurs early in seed development (~35% final dry seed wt), but the ability to produce most **normal seedlings** and maximum seed vigour occurs at or slightly before PM. Seeds are too high in moisture for commercial harvest at PM, so harvest must be delayed until the seeds first dry to ≤14% moisture content (**harvest maturity**); the time from PM to harvest is usually 10 to 25 days, depending on environmental conditions.

Factors contributing to soybean grain quality for commercial processing are similar to those described above for seed quality for replanting (except that germination is not a concern): diseases or insects in the field or storage, and physical seed damage during seed handling operations. Soybean is considered to be mature and ready for harvest when the seed moisture declines to less than 15% in the field, and is above the 11% risk-threshold for physical damage.

Physical seed injury can occur at any time during seed harvesting, drying, conditioning or general handling, characterized by cracks or breaks in either the seedcoat or cotyledons or a broken hypocotyl–radicle axis. Effects may range from impaired primary root development to germination failure, or providing sites for infection after sowing. Physical damage is directly related to seed moisture and increases markedly as that decreases below 11%. Thus, if the field environment following PM is extremely hot and dry so that seed is harvested and handled at low moisture levels, physical injury and reductions in germination and vigour will occur. Physical injury can be reduced by: (i) harvesting before seed moistures decline below 11%; (ii) proper adjustment of the threshing mechanism of harvester or by using rotary threshing systems; (iii) selecting rubber-flight or belt conveyers and running conveyers at full capacity; (iv) not handling low-moisture seeds at freezing temperatures; and (v) conveying and conditioning seeds at moisture levels between 12 and 15%.

If grain moisture exceeds 13.5% after harvest, the seeds must be dried prior to storage to maintain quality and prevent growth of storage fungi (*Aspergillus* and *Penicillium* spp.) and bacteria (see: **Storage management**). Natural or sun drying is practised in developing countries; however, in developed countries continuous batch or in-bin dryers are commonly used. Rapid drying, which hardens the outer layers of the seed and seals moisture within the inner tissues, and excess heating (not to exceed 76°C) which can cause both discoloration and protein denaturation, should both be avoided.

Grades and standards for evaluating grain quality have been developed to promote fair-trading of soybean grain within and between countries. These **grain inspection** standards consider test weight per unit, damaged kernels (e.g. by heat, frost, mould or insects), foreign material, split grains, odours and colours other than yellow (or mottled seeds) to determine the soybean grade that is used by the buyer and seller in marketing.

3. Quality, vigour and dormancy

In contrast to other major grain crops (such as maize, wheat and rice), soybean seed is of variable physiological quality and short longevity, which means that nearly all seed for planting the crop in the spring must be produced the summer before. If seed is 'carried over' to the second planting season following production, it must be monitored carefully. Although there are many important components of soybean seed quality, the most chronic problem facing a seed producer and farmer is physiological germination and vigour. It is generally accepted that seedlots with an acceptable standard germination (≥ 80%) will produce acceptable seedling populations under near ideal field conditions. Unfortunately, soil conditions at planting are not always ideal and seed quality is sometimes low, which can cause reduced seedling emergence and low populations.

Although many factors influence physiological seed quality, the two major ones are physical seed injury (see previous section) and *Phomopsis* seed decay (**see: Pathogens**). Both are strongly influenced by environmental conditions during seed production, and several management and cultural practices that modify conditions are used during seed harvesting and handling to improve quality.

Phomopsis seed **infection**, stimulated by warm, moist conditions during seed development and maturation or delayed harvests, causes direct losses to both germination and vigour. Heavily infected seeds are badly shrivelled, elongated and mouldy, and usually cannot germinate, whereas lightly infected seeds are normal in size and appearance with disease primarily in the seedcoat. Seeds with latent infection, however, will germinate but often develop lesions on the cotyledons, seen as **abnormal seedlings** before the germination test is completed; increasing degrees of infection have a direct inverse effect on seed germination and vigour, which can frequently be reduced to unacceptable levels. Several management practices have been identified to assist producers in reducing seed infection: (i) crop rotation, since soybean residue provides the primary source of inoculum; (ii) organizing later crop maturity by delayed planting of early-maturing cultivars or, in the Americas, by production of cultivars at the cooler, drier northern (USA) or southern (South America) edges of their zones of adaptation; (iii) using foliar fungicides by monitoring rainfall, temperature and pod infection during seed development; and (iv) harvesting as soon as seed moisture declines to a harvestable level (≤14%, fresh weight basis).

Purple seed stain, another seedborne disease, causes conspicuous purple discoloration on seedcoats and other plant parts, affecting many cultivars; although not a major problem in terms of seed quality or yield, severe infections (>5%) can reduce the commercial grade of harvested grain.

The single, most recognized and accepted laboratory index of soybean seed quality is the standard ('warm') germination test, which is conducted under near ideal laboratory conditions. Soybean seedlots that have nearly identical and acceptable germination (80–90%) in the standard germination test may have quite different vigour potential, however. This difference between standard germination and vigour led to the development and standardization of several vigour testing protocols for soybean, including two stress tests, the **accelerated ageing** test and a **cold test** (some variation on the exposure of imbibing seed to 10°C for 7 days, often in the presence of non-sterile field soil, followed by a **grow-out test** period of, usually, 4 to 7 days) and an electrical **conductivity test**, measuring membrane integrity after imbibition. The accelerated ageing test predicts field performance under these stressful soil conditions more accurately than does the standard germination test and it has been accepted as a standardized vigour test for soybean seed. Physical injury is commonly measured by the topographical **tetrazolium test** or, for a rapid determination of seedcoat damage, the sodium hypochlorite soak test. (**See: Vigour testing**)

Seed **dormancy** is of little or no concern in the soybean cultivars that are used for commercial crop production. However, genotypes with dormant seed due to impermeable seedcoats (**'hardseededness'**) have been identified, and this trait has been incorporated into some cultivars to reduce seed infection by *Phomopsis longicolla*. Unfortunately, any advantage for higher seed quality has been outweighed by the disadvantages caused by irregular and delayed field emergence of the hard seeds leading to poor field populations.

4. Sowing and treatment

Soybean seedling emergence is **epigeal**, producing wide, deeply notched cotyledons. Some cultivars with short hypocotyls require shallower seeding depths than the usual 2.5 to 4 cm. The optimum temperature for hypocotyl elongation is 30°C, however germination begins at 10°C and seedling emergence occurs over a wide range of temperatures up to 35°C (**see: Germination – influences of temperature**). At lower temperatures, soybean seeds are susceptible to imbibitional **chilling injury**. In ideal soil conditions seedling emergence may occur in 5 days, whereas in cool temperatures emergence may be delayed 2 weeks or more. Conventionally tilled soils provide warmer soil temperatures and an ideal seedbed for seedling emergence, while reduced or no-tillage seedbeds generally have cooler soil temperatures, which may delay emergence.

When high-quality seed is used, adequate stands are generally achieved without the need for fungicide seed treatment, however under stressful soil conditions a seed-applied fungicide can aid stand establishment, and control potential disease problems. Fungicide seed treatment can increase the germination of seeds infected with *P. longicolla* to acceptable levels (≥80%), for example, provided the level of infection is not too severe. Specific seed treatments have also been shown to be effective in controlling three soilborne diseases (*Pythium* spp., *Phytophthora* spp. and *Rhizoctonia* spp.) that can reduce seedling emergence and hence yield.

Adequate supplies of *Bradyrhizobium japonicum* must be present to produce a crop without requiring added nitrogen fertilizer. There is little need to inoculate succeeding crops, providing a well-nodulated soybean crop has been grown within the past 2 years. Otherwise **rhizobial inoculation** is usually carried out on farm, using 'planter box' techniques to mix seed with liquid or solid formulations; preinoculation is also being developed, using **filmcoating** techniques to maintain rhizobia on dry seed for several weeks before sowing in sufficient numbers to nodulate seedlings. (DTK)

Carlson, J.B. and Larson, N.R. (1987) Reproductive morphology. In: Wilcox, J.R. (ed.) *Soybeans: Improvement, Production and Uses*, 2nd edn. American Society of Agronomy Monograph 16, Madison, WI, USA, pp. 95–133.

Egli, D.B. (1998) *Seed Biology and the Yield of Grain Crops.* CAB International, New York, USA.

McGee, D.C. (1992) *Soybean Diseases: A Reference Source for Seed Technologists.* APS Press, St Paul, MN, USA.

Orf, J.H., Diers, B.W. and Boerma, H.R. (2004) Genetic improvement: conventional and molecular-based strategies. In: Boerma, H.R. and Specht, J. (eds) *Soybeans: Improvement, Production and Uses*, 3rd edn. American Society of Agronomy Monograph, Madison, WI, USA.

TeKrony, D.M., Egli, D.B. and White, G. (1987) Seed production and technology. In: Wilcox, J.R. (ed.) *Soybeans: Improvement, Production and Uses*, 2nd edn. American Society of Agronomy Monograph 16, Madison, WI, USA, pp. 295–353.

Soybean lectin

Soybean seeds contain moderate quantities (<1% of the total seed protein) of a typical legume **lectin** that is called soybean agglutinin (SBA). This is a tetrameric protein (120 kDa) composed of two slightly different glycosylated subunits of approximately 30 kDa (253 amino acids). The lectin exhibits a preferential binding specificity towards N-acetylgalactosamine and **oligosaccharides** containing terminal N-acetylgalactosamine. SBA also binds galactose but with a much lower affinity. Although not cytotoxic and not considered to be a toxin, dietary SBA reduces the growth rate of young monogastric animals and induces hypertrophic (swelling) growth of the small intestine and pancreatic hypertrophy. Accordingly, the soybean lectin acts as a minor antinutrient. Possibly, the anti-nutritive effect of SBA is enhanced by the simultaneous presence of trypsin **(protease) inhibitors** (which are the major soybean seed antinutrients; **see: Soybean, composition and nutritional quality**). The SBA present in soybean seeds and soybean meal can readily be inactivated by heat treatment. Some soybean lines are SBA-negative (the lectin gene is not transcribed due to the insertion of a 3.4 kb DNA segment) but most, if not all, commercial cultivars probably contain it.

SBA is capable of discriminating T-cells from bone marrow stem cells and this particular property of the lectin is exploited in medicine for purging bone marrow of T-cells in order to reduce the risk of graft-versus-host disease. Therefore, SBA is of huge importance to the field of clinical bone marrow transplantation. (EVD, WJP)
(See: Lectins)

Van Damme, E.J.M., Peumans, W.J., Pusztai, A. and Bardocz, S. (1998) *Handbook of Plant Lectins: Properties and Biomedical Applications.* John Wiley, Chichester, UK.

Space planting

Any of several sowing practices by which single or multiple seeds are planted at each station in the soil, followed by thinning out after emergence to remove superfluous seedlings if required to achieve the ideal plant spacing. **(See: Check-row planting; Dibbling; Hill dropping; Precision drilling; Vacuum precision seeding)**

Spears

Instruments of several designs (such as the Dynamic Spear or Nobbe Trier) used in **sampling** of seedlots, to obtain representative sample for **quality testing** after insertion into the seed mass. **(See: Sampling, 1. Spears and Probe samplers)**

Spelt

Spelt (*Triticum aestivum* ssp. *spelta*) (Fig. S.43) is widely recognized as a progenitor of common hexaploid **wheat**. Spelt most likely occurred as a spontaneous outcross of a cultivated tetraploid **emmer** wheat with the wild diploid *T. tauschii*,

Fig. S.43. Ears of spelt wheat.

followed by spontaneous doubling of the chromosome number to produce a hexaploid with 42 chromosomes (AABBDD, **see: Wheat**, Table W.1 **and Einkorn**, Fig. E.1). Spelt has a relatively tough **rachis**, but it disarticulates upon threshing although the rachis segment remains attached to the **spikelets**. The grain of spelt is not free-threshing as the spikelets remain intact and enclose the grain. The **hull** must be removed in a postharvesting operation.

Spelt has the **gluten** protein that gives wheat its unique ability to produce a yeast-leavened loaf of bread and it is used for bread today in several parts of the world. Its primary use, however, because the hulls are difficult to remove from the grain, is as a feed grain (undehulled).

It also appears that common wheat arose from spelt relatively early by mutation of the **Q locus**. Spelt does not seem to have been widely cultivated in the Middle East, but the derived, free-threshing common wheat has been dominant in the region for many thousands of years. Spelt was better adapted to cooler regions and was much more extensively grown in central Europe in higher elevations. Most of the spelt grain produced today is grown organically where it is felt to be more competitive with weeds than common wheat.

There is considerable diversity among the spelts with a range of colours of **chaff** from black to brown to white, and some which are covered with short, fuzzy hairs. Some are awned and others awnless, most are tall and have weak straw. Most spelt is relatively susceptible to diseases. Both spring and winter types occur and the winter hardiness is similar to that of common wheat. Some types have higher amounts of gluten protein and are more suitable for bread than others. There has been relatively little breeding work done on spelt and most modern varieties are pure line selections from landraces. (DF)

Leonard, W.H. and Martin, J.H. (1963) *Cereal Crops*. Macmillan, New York, USA.

Simmonds, N.W. (ed.) (1976) *Evolution of Crop Plants*. Longman, New York, USA.

Stallknecht, G.F., Gilbertson, K.M. and Ranney, J.E. (1996) Alternative wheat cereals as food grains: einkorn, emmer, spelt, kamut, and triticale. In: Janick, J. (ed.) *Progress in New Crops*. ASHS Press, Alexandria, VA, USA, pp. 156–170.

Spermatophytes

(Greek: *sperma*, *spermatos* = seed; *phyton* = plant) Seed plants, a division of the plant kingdom, characterized by the female **gametophyte** being developed and retained within the **megasporangium** (i.e. the nucellus) which itself is covered by one or two **integument**(s), the entire structure producing a seed after fertilization of the egg cell in the embryo sac. **(See: Evolution of seeds; Reproductive structures; 1. Female)**

Spermosphere

The zone in the soil surrounding the seed where populations of microorganisms develop, which may be derived from those in the soil, or already present within the seed, or on its surface. **(See: Rhizosphere microorganisms)**

Spices and flavours

The *Oxford English Dictionary* defines spice as 'a strongly flavoured or aromatic substance of vegetable origin obtained from tropical plants commonly used as condiments'; and herbs as 'plants whose stems, leaves or both are used as a food flavouring, in medicines or for their scent'. According to Weiss (2002) 'aromatic or fragrant products from tropical plants

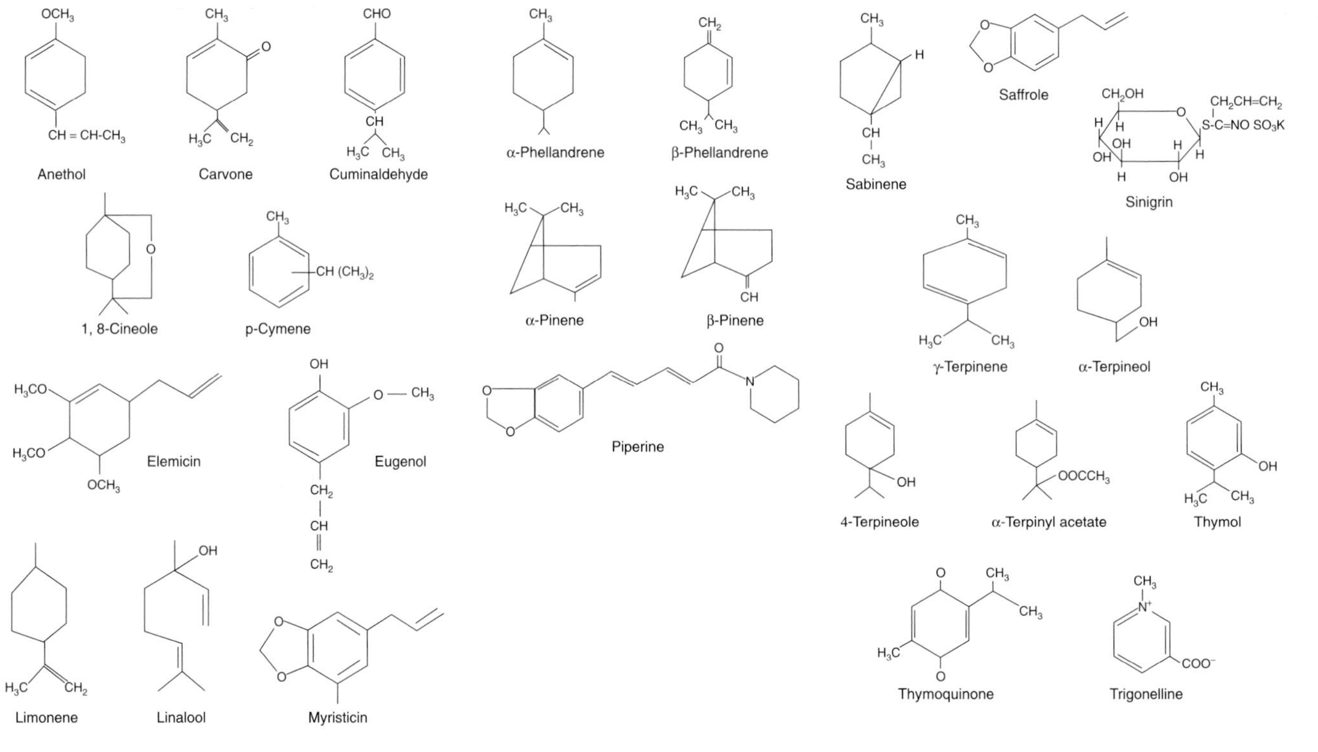

Fig. S.44. The main compounds responsible for aromas and flavours of spices.

used to flavour foods or beverages are considered a spice and those from temperate plants can be either a spice, usually a fruit or seed, or a herb when green parts are used'.

Spices, many of which are seeds or fruits, have played an important role in human history and civilization. They have been used for embalming, as incense, ointments, perfumes, poison antidotes, cosmetics, medicine, preservatives and condiments for flavouring food and confectionery. In ancient times spices were considered primarily as medicines rather than condiments. The Romans were mainly responsible for the popularization of spices as condiments and this use has spread to the rest of the world. It was the Arabs who controlled the earlier spice trade followed by the Europeans. The Portuguese, Dutch, Spanish and the British tried to exercise control over the Spice Islands (Indonesia): this led to voyages of discovery (to find alternative routes to the islands) and numerous local wars; and it laid the foundations for their colonial empires.

The major spices in which the fruits or seeds are used are shown in Table S.12. Certain families are particularly rich in spice seed species: the Apiaceae, for example, contains seven major spices – **ajowan, anise, caraway, coriander, cumin, dill** and **fennel**.

The relatively high content of a wide variety of terpenes, terpenoids, and in some, phenolics, in the **essential oils** of the seeds accounts for their distinct and characteristic aromas and flavours, together with the **oleoresins** (Fig. S.44). Almost all the spices are of tropical origin (centres of diversity) though the present-day major production regions may differ from the original locations (Table S.12). The mustards are the leading spice by volume produced and traded but black pepper is economically the most important. This spice accounts for 33% of the world trade followed by chilli pepper (capsicums) 22%, cardamom 4%, other seed spices 15% and vanilla 2%, though the ratios change from year to year. In 2001, the value of world spice exports was approximately US$413 million, India accounting for about 16%. Total world production in 2002 amounted to over 1 million t of which India was by far the major producer (Table S.13). In addition to their use in intact form or powders for culinary or medical purposes, several of the spices, such as anise, cardamom, fennel and mustard are valued as a source of fixed or essential oils. (KVP, NB)

See: Ajowan; Allspice; Anise; Annatto; Black cumin; Black pepper; Caraway; Cardamom (large and small); Coriander; Cumin; Dill; Fennel; Fenugreek; Mustard; Nutmeg and mace

Table S.13. Some examples of production and exports of spices.

Country	2002 Production (1000 t)	2001 Exports (US$1000)
Bangladesh	4.8	51
Burkina Faso	5.6	1
China	63	17,932
India	800	66,530
Nepal	15	23
Pakistan	45	7,444
Turkey	32	25,907
Spain	2.5	33,691
Sri Lanka	4.3	734
USA	NA	19,599
World	1,065	413,780

NA, not available.
From FAOSTAT: http://apps.fao.org/page/collections

Anonymous (1985) *The Wealth of India*, Vol. I. Publications and Informations Directorate, New Delhi, India, pp. 227–229.

Devon, T.K. and Scott, A.I. (1975) *A Hand Book of Naturally Occurring Compounds*, Vol. 1. Academic Press, New York, USA.

Guenther, E. (1978) *The Essential Oils*, Vol. 2. Robert E Krieger Publishing, New York, USA.

Peter, K.V. (ed.) (2001) *Handbook of Herbs and Spices*. Woodhead Publishing Co, Cambridge, UK.

Pickersgill, B. (2005) Spices. In: Prance, G. and Nesbitt, M. (eds) *Cultural History of Plants*. Routledge, New York, USA, pp. 153–172.

Table S.12. 'Seed' spices – origins and production.

Type	Origin (centre of diversity)	Present major production
Ajowan (*Trachyspermum ammi*)	Asia minor	India, Afghanistan, Iran, Pakistan
Allspice (*Pimenta officinalis*)	Caribbean	Jamaica, Central America
Anise (*Pimpenella anisum*)	Eastern Mediterranean	India, Iran, Italy, Spain, Turkey
Black cumin (*Nigella sativa*)	Mediterranean, North India	India, Middle East
Cardamom (small) (*Elettaria cardamomum*)	Indo-Malaya region	India, Guatemala, Sri Lanka
Cardamom (large) (*Amomum subulatum*)	Sikkim	India, Southeast Asia
Caraway (*Carum carvi*)	Mediterranean, West Asia	Europe
Coriander (*Coriandrum sativum*)	South Europe, Eastern Mediterranean	Europe, India, Morocco, Pakistan, USA
Cumin (*Cuminum cyminum*)	Eastern Mediterranean, Egypt	Egypt, India, Iran, Morocco, Turkey, China, Russia, Algeria, Japan
Dill (*Anethum graveolens*)	Mediterranean, West Asia	India, Pakistan
Fennel (*Foeniculum vulgare*)	South Europe, Mediterranean	Argentina, Europe, India, USA, former USSR, Rumania
Fenugreek (*Trigonella foenum-graecum*)	Southeast Europe, West Asia	Argentina, Egypt, India, Mediterranean
Mace (*Myristica fragrans*)	Moluccas islands	Grenada, Indonesia
Mustard		
White (*Brassica alba*)	Mediterranean	Europe, USA
Brown (*B. juncea*)	India	India
Black (*B. nigra*)	India	India, Asia
Nutmeg (*Myristica fragrans*)	Moluccas islands	Grenada, Indonesia
Pepper (black, green, white) (*Piper nigrum*)	Western Ghats (India)	Brazil, China, India, Indonesia, Sri Lanka, Vietnam

Purseglove, J.W., Brown, E.G., Green, C.L. and Robbins, S.R.J. (1981) *Spices*, Vol. 1. Tropical Agriculture Series, Longman, New York, USA.

Vaughan, J.G. and Geissler, C.A. (1997) *The New Oxford Book of Food Plants*. Oxford University Press, Oxford, UK.

Weiss, E.A. (2002) *Spice Crops*. CAB International, Wallingford, UK.

Windholz, M. (1983) *The Merck Index: An Encyclopedia of Chemicals, Drugs and Biologicals*, 10th edn. Merck, Rathway, NJ, USA.

Spike (spikelet)

A type of inflorescence with stalkless flowers on an elongated unbranched axis. The fruiting spike of cereals, especially maize. (**See: Floret**, Fig. F.7; **Infloresence**)

Spinach

'True' spinach (*Spinacia oleracea*, Chenopodiaceae) is cultivated as a popular direct-seeded annual leaf-vegetable: a total world crop area of 817 kha (compared to the 1020 kha of lettuce grown – FAOSTAT data, 2004), mainly in China (80%), Indonesia, the EU, Japan, Turkey and the USA. The crop thrives in cool, moist growing conditions, reaching edible maturity within about 1–1½ months from spring sowings; it is also cold-hardy, and autumn-sown, cold-hardy varieties, with different growing and **bolting** characteristics, are over-wintered in some areas. The two main types are distinguished by their foliage forms: smooth-leaved and crinkle-leaved 'Savoy', each with prostrate, semi-erect, and upright rosettes. Smooth and 'semi-Savoy' types are used for processing, and all types for salad and other fresh uses, with an increasing trend to densely sown 'baby-leaf' production for the bag-mix market. There are both open-pollinated and F_1 hybrid varieties, which predominate in many countries.

1. Origins

Cultivation originated in south-western Asia, and reached China by 7th century, but was not prevalent in Europe until the early 16th century and popular in North America until the early 20th century. The crop gradually took the place of many other similar though not closely related spinach-like vegetables, which are still cultivated in various parts of the world, such as **amaranth** (*Amaranthus* spp.), orach or mountain spinach (*Atriplex hortensis*), spinach-beet/Swiss chard (*Beta vulgaris* ssp. *cicla*) (**see: Sugarbeet**), New Zealand spinach (*Tetragonia tetragonioides*), and the vines Malabar/water (*Ipomoea aquatica*) and Indian/Ceylon spinach (*Basella* spp.).

2. Breeding and development

Although *S. oleracea* is mainly monoecious in nature (with separate male and female flowers on the same plant, **see: Reproductive structures**), the range of sex expression is highly variable; it is possible to produce highly male and female lines by selection, and from these single-cross and three-way cross **hybrids** are constructed.

Flowers develop in clusters of about 6-20 and are wind-pollinated. Seeds, which tend to ripen unevenly, are bladder-like one-seeded indehiscent fruits ('utricles'), with a mean size of 3.5 mm and a **thousand seed weight** of about 10 g; they consist at maturity of the pericarp, testa, annular embryo and perisperm (the endosperm is digested by the developing embryo). Depending on variety, seedcoats have different thicknesses and are either smooth/round or 'prickly/spiny' – hence 'spinach', from the Latin 'spina': types that were formerly known as 'summer' and 'winter', though these designations are now less accurate.

3. Production

Spinach seed is produced in Denmark mainly (about 85% of world production), and in France, the Netherlands and the USA (Washington and Florida).

Diseases, whose pathogens can be both seedborne and seed-transmitted, are a major consideration, and long (8- to 10-year or more) rotation periods are necessary to prevent their build-up, even when using resistant varieties. Major diseases include: the Leaf Spot Complex (*Cladosporium variabile*, *Stemphylium botryosum*, and other **pathogens**), which can result in substantial seed losses and high levels of seed infection; Verticillium Wilt (*Verticillium dahliae*), where infection of seedlots can restrict export to some markets; and Downy Mildew (*Peronospora effusa*/*P. farinosa* f. sp. *spinaciae*), whose oospores or mycelia can infest seed. Control measures include resistant varieties, removal or thorough decomposition of plant debris, and an array of crop fungicides. Fusarium wilt (*Fusarium oxysporum* f. sp. *spinaciae*) can also infect seed, as can *Phoma*.

Plants showing off-type atypical leaf characters or infection are removed by **roguing**, along with the male-parental plants in the case of hybrid seed production. The seed crop is combined in the field or, because it is very susceptible to **shattering**, is cut and dried in **windrows**, often on sheets to catch the seed.

4. Treatment

Seedborne pathogens can be eradicated with variable success by treatment with hot water (50°C, 25 min) or sodium hypochlorite; precisely controlled parameters are necessary so as not to damage germination.

Seeds are treated with fungicides (such as metalaxyl and thiram) directed against pre- and post-emergence damping **off** losses, and some seed with pyrethroid insecticides against insect pests. Seed is often **encrusted** or **filmcoated**.

In hybrid seed production, priming of the female parental line has been advocated to make germination faster and more uniform, and make **bolting** earlier and hence improve **nicking** efficiency. (PH)

Sporophyte

The diploid (2n) generation in a life cycle exhibiting alternation of generations; it is the predominant one in seed-bearing plants, since all vegetative (somatic) tissues are **diploid**. (**See: Evolution of seeds**)

Sporophytic-incompatibility system

See: Self-incompatibility systems (SI)

Sprouted seeds – food

Seeds of several species are germinated and as young seedlings are eaten by humans as 'health food'. The storage reserves are partly mobilized making them more readily available for digestion, and other beneficial changes occur in the content of various dietary nutrients such as vitamins, antioxidants etc. The following are the most common sprouted seeds in use: **adzuki bean**, **alfalfa**, broccoli, **buckwheat**, cabbage, **chickpea**, clover, **fenugreek**, green **pea**, **lentil**, **mung bean**, **mustard**, radish, **rye**, **sesame**, **soybean**, **triticale**, **wheat**.

Squash

The squashes are *Cucurbita pepo*, *C. maxima* and *C. moschata*. Squash seeds are consumed as snack food, similarly to **pumpkin**. The seed characteristics (composition, etc.) are close to those of pumpkin.

Staple length

A group of fibres with uniform properties, especially length. The staple length of cotton refers to the average length of a combed and straightened bundle of fibres when measured under defined conditions of temperature and relative humidity. (**See: Cotton; Fibres**)

Starch

A reserve carbohydrate material produced by plants, which serves as the primary component of most human and animal diets. Starches from **cereal** grains such as wheat, barley, rice, and maize and from tubers such as potato and cassava are major food sources for the human population worldwide. In developed countries, starch comprises at least 35% of the human daily caloric intake, but in under-developed regions, it can provide 80% or more of total calories. In addition to serving as food and animal feed, starch is an important industrial commodity and renewable raw material. Starches are utilized as additives in the food industry, serving as emulsifying or thickening agents. These qualities also are beneficial for the manufacture of adhesives, binding agents, strengthening agents and coating materials that are essential components for a range of commercial products.

Starch is produced in most tissues of a plant (**see: Starch – synthesis**). In leaves, **transitory starch** is synthesized during the day as the end product of photosynthesis, and degraded during the dark cycle to simple sugars to supply the energy needs of the growing plant. During this process (**see: Starch – mobilization**) sucrose is delivered to various tissues of the plant, including sink tissues such as seeds. In the seed endosperm, transported sucrose initiates the synthesis of storage starch, the primary energy reserve for the emerging seedling in the next generation. Storage starch from seeds also is an important agricultural commodity.

Starch utility stems largely from the chemical and structural features of the linear and branched polymers of glucose that constitute starch, **amylose** and **amylopectin**. Amylopectin is the primary component, comprising approximately 75% of most starches, and its complex branched-chain structure is responsible for starch crystallinity. Amylose accounts for the remaining ~25% of starch and is essentially a linear molecule, with only a small number of branches. Together, these molecules assemble in an organized manner to form insoluble, crystalline **starch granules**. Changes in the amylose:amylopectin ratio of granules or in the crystalline structure of amylopectin can alter the functional properties of starch and confer different food or industrial utilities.

Native starch as extracted from seeds are insoluble in cold water and most solvents. However, when starch granules are heated in water, they swell and lose their crystallinity, in a process termed gelatinization. The temperature at which starch gelatinizes is important in terms of identifying its functionality. For example, starch gel formation is critical for the production of puddings, salad dressings, paper coatings and adhesives. Gelatinized starches with high amylose percentages form firm gels and strong, tough films, whereas starches with little or no amylose produce nearly transparent, viscous pastes that do not gel. Both types have specific industrial utilities that are based on their distinct functional properties.

Maize is the dominant crop for starch production, with starch comprising 70–75% of the dry weight of the kernel. Normal, unmodified maize is widely used as livestock feed, and is also used for the industrial production of goods such as paper, pressed board, laundry starch and textiles. Starches with different structures or compositions also are available in several maize lines that have been modified genetically. For example, *waxy* **mutants** lack the enzyme necessary to make amylose, and thus have starch granules comprised solely of amylopectin, whereas other lines have been developed that produce high amylose starches. Also, maize *sugary2* mutants produce starch with 40% amylose as well as shorter amylopectin chains, which improves both digestibility and freeze–thaw stability. Another example of the utility of genetic mutants is found in the *sugary1* and *shrunken2* mutants of maize, which accumulate large amounts of soluble sugars in the endosperm and thus serve as standard sweetcorn varieties. In addition to genetic mutants, chemical modifications and mild heat treatments are also routinely used to alter the physical properties of starch, often with the aim of reducing starch viscosity or rendering starches more stable to processing conditions.

Starch is also an increasingly important resource for the production of chemical feedstock and fuel. Much of the starch from maize is hydrolysed in wet- or dry-milling operations (**see: Maize**) to glucose. Although glucose can potentially be converted to a range of organic chemicals, current industrial practices focus primarily on its conversion to two key products. One product is fructose, which is derived from the isomerization of glucose for the production of high fructose corn syrup, a sweetening agent for foods and beverages. The other conversion product is ethanol, which results from the large-scale fermentation of glucose in yeast cells. Expanded use of ethanol as an alternative fuel source indicates the potential for starch to serve as a renewable energy supply for the future. (MGJ)

Burrell, M.M. (2003) Starch: the need for improved quality or quantity – an overview. *Journal of Experimental Botany* 54, 451–456.

Johnson, L.A., Baumel, C.P., Hardy, C.L. and White, P.J. (1999) Identifying valuable corn quality traits for starch production. In: *A Project of the Iowa Grain Quality Initiative Traits Task Team.* Center for Crops Utilization, Iowa Agriculture and Home Economics Experiment Station, Iowa State University, Ames, IA, USA.

Starch granules

Small bodies of **starch** that provide packaging for reserve carbohydrates in plants. Sometimes these are inappropriately called starch grains. Starch granules consist almost entirely of

amylose and **amylopectin** along with some associated lipid and protein. Granules are synthesized and accumulate in **plastids**, primarily amyloplasts in seeds and chloroplasts in leaves. In seeds, storage of starch in granules is a central feature of plant evolution, as it enables the dense packing of large amounts of carbohydrate material in a dehydrated form. In this manner, desiccated seeds are able to maintain energy reserves for the next generation over extended periods of time. In leaves, starch in smaller granules is readily convertible to sugars by enzymatic digestion, and thus is available to meet the on-going energy needs of the plant for growth and development (**see: Starch – mobilization**). Digestion of raw starch granules by human, animal, fungal, or bacterial enzymes is also what gives starch its broad utility as food source or as industrial feedstock.

Starch granules from various sources contain differing amounts of **lipid** and attached phosphate, and also may vary in size and shape. Sizes range from the sub-micron level (in leaf chloroplasts) up to approximately 100 microns (in potato or canna tubers). Granules of the cereal starches are typically 5–20 microns in diameter. Starch granule shapes include ovals and spheres, cuboidal or polyhedral forms, discs, elongated cylinders, needles, and even filamentous forms, depending on the botanical origin. Furthermore, some granules are compound (e.g. oat), containing small particles of starch that appear to have developed separately, but simultaneously. The diversity of granule size, morphology and composition gives rise to diverse functional properties that result in many end-uses for starch.

Clues to the internal organization of starch granules have been gleaned from microscopic and X-ray diffraction analysis of intact and partially digested granules, and from structural analyses of the amylopectin component of starch. Observed under polarizing light microscopy, starch granules exhibit a characteristic dark cross ('maltese cross') that signifies birefringence. The intensity of birefringence depends on the degree of crystallinity of the starch, and can reveal the orientation of crystallites. Examination of partially digested starch granules by electron microscopy reveals an internal pattern of concentric 'growth rings' that represent alternating regions of amorphous and semi-crystalline material with distinct refractive indices and sensitivities to digestion. Growth ring formation is thought to be subject to circadian rhythms in tuber starches, and to diurnal fluctuations in substrate availability in cereal starches.

Another level of granule organization is described in terms of starch crystallinity. X-ray diffraction analysis of native starch granules reveals there are two main patterns of crystallinity, termed 'A' and 'B'. A-type crystallinity is typical of cereal grain starches such as maize, wheat and rice, and B-type crystallinity is typical of tuber and root starches. The differences between the A and B patterns relate to the packing of the helical chains in amylopectin (the branched polymer of starch), and the number of water molecules that stabilize these helices. A third crystallinity pattern, termed 'C', is a mixture of the A and B forms.

By itself, the structure of amylopectin, the glucosyl polymer that comprises ~75% of starch, is sufficient to determine the physical properties of starch granules, based on evidence that mutants lacking amylose produce near-normal granules. The fundamental architecture of amylopectin is based on an organized spatial arrangement of linear chains and branch linkages, described in the 'cluster model' of starch structure. According to this model, adjacent linear chains are paired as double helices and tightly packed in parallel arrays to form crystalline lamellae (Fig. S.45A). These linear arrays alternate with regions rich in branch linkages, termed amorphous lamellae. The periodic clustering of the branches permits a dense packing of the intervening linear chains, providing an efficient mechanism for nutrient storage. The 9–10 nm length of each repeating cluster unit, which comprises one amorphous and one crystalline lamella, is generally constant in starches from various botanical sources. Higher orders of amylopectin structure also are known, in which cluster units are grouped into discrete elongated structures of approximately 100 nm termed 'blocklets' that vary in size and shape and in their relative orientation to one another (Fig. S.45B).

In cereal grain **endosperms**, starch granule formation is developmentally regulated (e.g. beginning 8–10 days after fertilization in maize). The molecular mechanisms for granule initiation are not known, but spontaneous crystallization of a 'pro-starch coacervate' seemingly creates a nucleus or hilum for the developing starch granule. Granules then grow outward by apposition, taking on a characteristic shape and size. The shapes and sizes of starch granules seemingly are under genetic control, suggesting that different enzymatic activities are responsible for distinct granule properties. Increasingly, these enzymatic activities are being manipulated by genetic or transgenic approaches to alter starch granule properties for expanded industrial use. (MGJ)

(See: Cereals – composition and nutritional quality; Malting – barley)

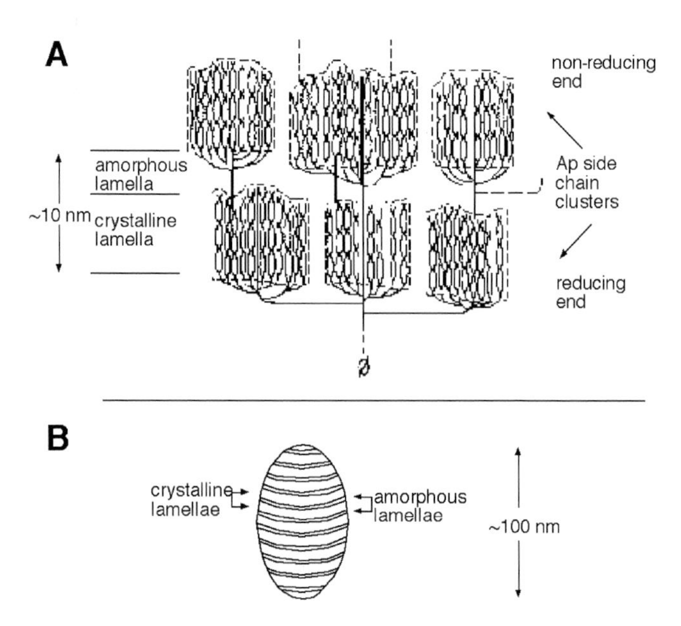

Fig. S.45. Diagrammatic representation of two levels of amylopectin structure. (A) Unit cluster level of amylopectin (Ap) structure. Solid lines indicate glucan chains, and intersections between them indicate branch linkages. Dotted lines indicate the boundaries of amylopectin side chain clusters in which primarily unbranched chains associate in tightly packed double helices. (B) Blocklet level of amylopectin structure, made up of amylopectin side chain clusters in a discrete unit.

James, M.G., Denyer, K. and Myers, A.M. (2003) Starch synthesis in the cereal endosperm. *Current Opinion in Plant Biology* 6, 215–222.

Myers, A.M., Morell, M.K., James, M.G. and Ball, S.G. (2000) Recent progress toward understanding the amylopectin crystal. *Plant Physiology* 122, 989–997.

Pilling, E. and Smith, A. (2003) Growth ring formation in the starch granules of potato tubers. *Plant Physiology* 132, 1–7.

Starch – mobilization

A degradative process that involves the enzymatic conversion of **starch** to simple sugars in order to meet the energy needs of the growing plant. Mobilization of transient starch occurs in leaves in accordance with the diurnal cycle. In the light phase, **starch granules** accumulate in leaf chloroplasts as the end products of photosynthesis. In the dark, this leaf starch is eventually broken down to glucose monomers by the coordinated actions of a number of degradative enzymes. Sucrose, the disaccharide sugar that is synthesized from glucose and fructose, is transported and distributed to tissues throughout the plant. Sucrose serves as the carbon source material for energy production in a number of metabolic pathways. In plants that store starch in non-photosynthetic tissues, such as seeds in cereals, potato tubers, or embryos of certain dicots (e.g. pea), the transported sucrose from the leaves fuels re-synthesis of large amounts of starch that persist for long time periods (i.e. storage starch) (**see: Starch synthesis**). Eventually, upon seed germination or tuber sprouting, storage starches are mobilized to meet the energy needs of the subsequent generation (**see: Mobilization of reserves – cereals**). In plant species that store oils in seeds rather than starch (e.g. rapeseed, *Brassica*, cereal embryos), starch accumulates transiently in developing seeds and declines as the carbon is further reduced to form **oils**. The processes governing the degradation of transient and storage starch granules are highly regulated.

Starches are mobilized by a combination of phosphorolytic and hydrolytic degradation (Fig. S.46). No single enzyme completely converts starch to simple sugars. Starch phosphorylases catalyse a reaction that inserts a phosphoryl group from inorganic pyrophosphate into an $\alpha(1\rightarrow4)$ glucoside bond, releasing glucose-1-phosphate. Although this reaction is potentially reversible, phosphorylases are thought to function primarily in starch depolymerization, releasing one glucose unit at a time. Starch degradation also involves the activities of $\alpha(1\rightarrow4)$-specific hydrolases of the α-amylase and β-amylase classes. The α-amylases are endo-acting enzymes that randomly cleave linear chains of glucan polymers internally, whereas β-amylases are exo-acting enzymes that cleave **maltose** units from the non-reducing ends of linear chains. The degradative activities of both amylase types are halted when they encounter a branch linkage, producing **limit dextrins**. This indicates that starch debranching enzymes (DBEs), which specifically hydrolyse $\alpha(1\rightarrow6)$ branch linkages, also are required for starch mobilization. Isoamylase-type DBEs may function in the hydrolysis of longer branch chains,

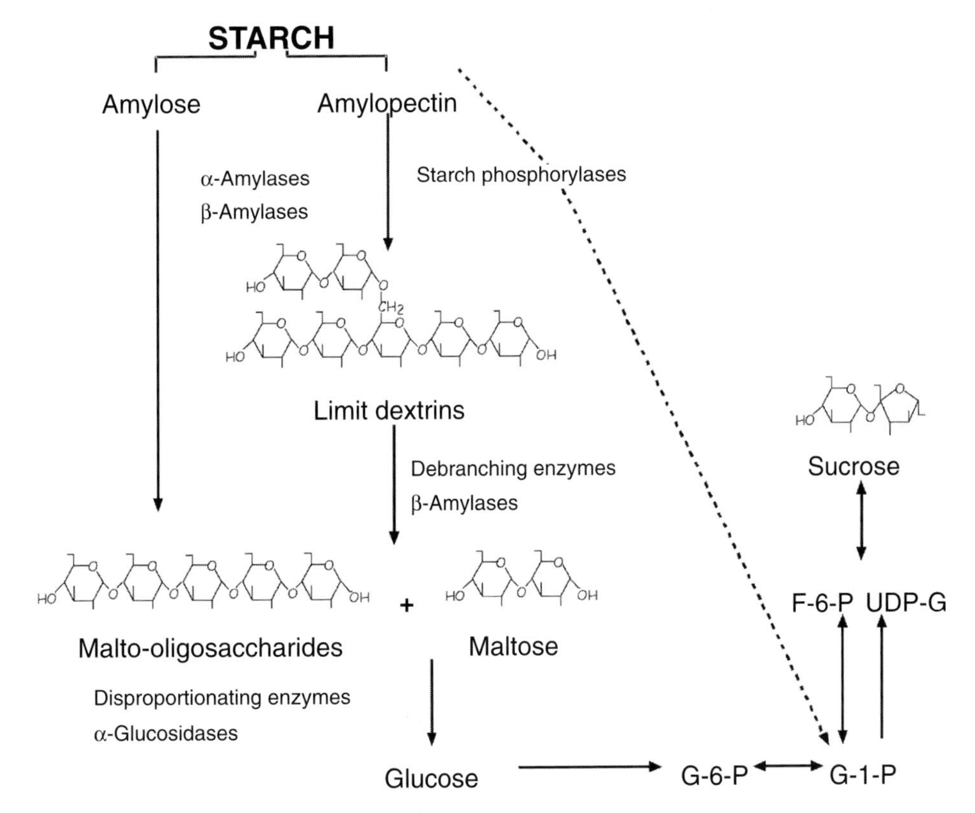

Fig. S.46. Enzymes and intermediates involved in the mobilization of starch. The final product of starch mobilization, sucrose, which is transported to the seedling, is produced by sucrose synthase, combining UDP-glucose (UDP-G) with fructose-6-phosphate (F-6-P). Starch phosphorylases act upon amylose and amylopectin.

and pullulanase-type DBEs may function to hydrolyse short branch chains, possibly those remaining after β-amylase digestion. In addition, α-glucosidases, which cleave α(1→4) or α(1→6) linkages to release single glucose units from the non-reducing ends of short malto-oligosaccharides, undoubtedly play a role in starch breakdown. Finally, disproportionating enzymes (DEs) also are thought to have a degradative function, because as they transfer short glucosyl lengths from one chain to another, a single glucose unit is released.

Although limit dextrins are products that result from hydrolysis of starch by amylases, they may not accumulate in the plant during starch degradation; they are most likely hydrolysed by the debranching enzymes, concurrently with α- and β-amylase activities (Fig. S.46). They do occur when starch is hydrolysed by the amylases during its *ex situ* processing.

Some measure of control of starch mobilization potentially occurs at the level of isoform specificity. Two starch phosphorylases, four DBEs, three α-amylases and up to nine β-amylases are known in plants. The reasons why plants have so many degradative enzymes is not clear, but it is possible that different enzyme isoforms function in separate tissues, or that they respond in different manners to plant-based or environmental signals. For example, analysis of 'starch excess' mutant plants that cannot completely mobilize leaf starch has shed light on the involvement of specific amylase isoforms in transient starch degradation. Furthermore, because one of these mutations lacks the activity of a kinase-type enzyme that adds phosphate groups to molecules, the direct phosphorylation of starch also may be required in order for starch mobilization to occur.

The mobilization of storage starch in cereal grains such as barley and wheat occurs following germination, and is initiated by degradative enzymes in response to the import of gibberellin into the **aleurone layer**. Many of these enzymes are produced *de novo* in aleurone layer cells in response to these signals. (**See: Malting – barley; Mobilization of reserves – cereals**) Together, they function to completely degrade starch to glucose, which then becomes a substrate for sucrose production. Sucrose provides much of the energy necessary for the growth and development of the young seedling until it achieves photosynthetic capacity. (MGJ)

Ritchie, S., Swanson, S.J. and Gilroy, S. (2000) Physiology of the aleurone layer and starchy endoperm during grain development and early seedling growth: new insights from cell and molecular biology. *Seed Science Research* 10, 193–212.

Starch – synthesis

Synthesis of both the **amylose** and **amylopectin** components of starch begins with the formation of an activated glucosyl donor, ADP-glucose (ADPG), in a reaction catalysed by ADPG pyrophosphorylase (AGPase). AGPase consists of two large subunits and two small subunits, each of which is encoded by distinct genes. In the cereal endosperm, the AGPase enzyme is extra-plastidial but it is plastidial in other cereal tissues and in all tissues of non-cereal plants (Fig. S.47). Thus, in the cereal endosperm, ADPG synthesized in the cytosol must be transported into plastids in order for starch synthesis to proceed; this may have functional importance for channelling large amounts of carbon into starch when sucrose is plentiful.

Within **amyloplasts**, ADPG is used to elongate a growing glucan chain in reactions catalysed by various starch synthases (SSs) that join glucosyl units via α-(1→4) linkages to build linear chains. Chain elongation in amylose exclusively utilizes a granule-bound SS, but elongation of amylopectin chains occurs via several different soluble or partially granule-bound SS isoforms, termed SSI, SSII, SSIII and SSIV. Furthermore, in cereals there are two SSII forms, SSIIa and SSIIb. Elucidation of the roles of the individual SSs is essential to an understanding of amylopectin biosynthesis, and hence, starch biosynthesis. SSI may be primarily responsible for the production of the shortest chains, those having ten glucosyl units or less. Further extension to produce longer chains is then catalysed by SSII or SSIII isoforms.

The structural organization of starch is based not only on non-random distributions of linear chains, but also on the frequency and positioning of branch linkages. Branches in amylopectin are introduced by branching enzymes (BEs, also called starch branching enzymes, SBEs) that catalyse a sequential, two-step reaction in which an α-(1→4) linkage in a linear chain is cleaved and the released reducing end is transferred to a C6 carbon, thus creating a new α-(1→6) linkage. Two classes of BEs are known, BEI and BEII, with the cereals having two BEII isoforms, BEIIa and BEIIb. Like the SSs, individual BE isoforms undoubtedly have distinct roles in starch synthesis. Current models propose that BEI transfers longer chains than either BEII form.

The involvement of starch debranching enzymes (DBEs), which hydrolyse some branch linkages in starch synthesis, is implicated by genetic evidence gained from mutations in a number of plant species. Mutations in DBE genes result in production of a water-soluble polysaccharide, **phytoglycogen**, at the expense of amylopectin. Phytoglycogen has twice the

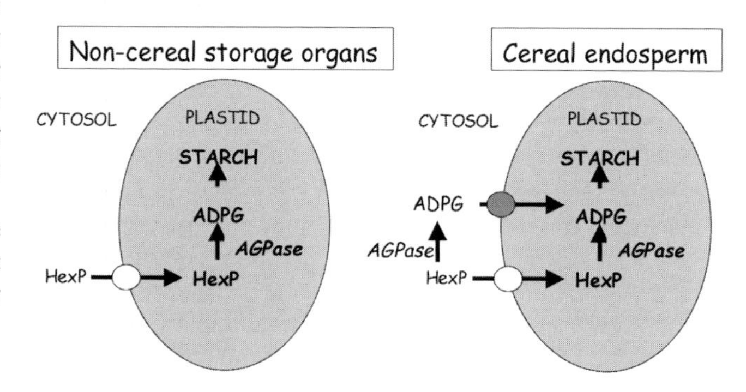

Fig. S.47. The synthesis of starch in the amyloplasts of cereal grains and non-cereal seeds. The substrate for starch synthesis, ADP-glucose (ADPG), is made by ADP-glucose pyrophosphorylase (AGPase). In the storage organs of most plants, AGPase exists entirely within the plastids and synthesizes ADPG from hexose phosphate (HexP, Glc-1-P or Glc-6-P) imported from the cytosol via a transporter in the plastid envelope (white circle). In cereal endosperms, 85–95% of the AGPase exists in the cytosol. ADPG is imported into the plastids via a specific ADPG transporter (dark grey circle). Some ADPG is also made from imported HexP within the plastids of cereal endosperm. Starch synthases add glucose units from ADPG to the growing amylose and amylopectin chains of starch.

number of branch linkages as amylopectin, a shorter and more uniform average chain length, and lacks branch clustering. These architectural differences suggest that DBEs influence amylopectin structural organization directly by providing a selective trimming function that removes inappropriately positioned branches. Alternatively, DBEs could affect amylopectin synthesis indirectly by preventing the build-up of alternative substrates such as phytoglycogen that might compete with amylopectin for enzymatic activity. Two DBE classes are conserved in plants, isoamylase-type and pullulanase-type enzymes, which differ in their substrate specificities. The isoamylase class has three distinct isoforms, ISO1, ISO2 and ISO3. Of these, ISO1 and ISO2 are thought to have biosynthetic functions, and ISO3 is suggested to function in starch degradation. There is evidence that the lone pullulanase-type enzyme functions in both starch synthesis and degradation.

Two additional classes of starch metabolizing enzymes may have biosynthetic functions, although their roles are largely thought to be in **starch mobilization**. These include starch phosphorylases, which theoretically are capable of elongating glucan chains, and disproportionating enzymes, which can transfer cleaved linear oligosaccharides from one linear chain to another.

Little is known about factors that regulate starch synthesis, although protein–protein interactions and protein modifications are suggested. AGPase, for example, may undergo post-translational modification, and be converted from an inactive dimerized state when two enzyme molecules are linked by a disulphide bond, to an activated state when these bonds are broken, in a reaction involving thioredoxin. In maize and barley, when one component of the starch pathway is altered genetically, secondary effects on other enzymes are noted. For example, *sugary1* mutations in maize that condition loss of ISO1 also affect the functions of the pullulanase-type DBE, BEIIa and SSIII. Also, in barley, the *sex6* mutation that results in loss of SSIIa also causes the release of several proteins from the granule, such as SSI and BEIIb. Such pleiotropic effects suggest that interactions occur among specific starch biosynthetic enzymes, possibly via direct physical associations. Protein phosphorylation is also implicated in the regulation of starch biosynthetic activities (e.g. possible activation by phosphorylation of SBEI and II and SSIIa), as is the binding of 14-3-3 proteins (proteins that bind to other proteins and modulate their activities) and/or molecular chaperones (proteins associated with the correct folding of other proteins). Increased understanding of starch synthesis will require more information on how protein modifications regulate enzymatic activity and protein complex formation, and discoveries of additional regulatory factors. (MGJ)

(See: Transitory starch)

Dinges, J.R., Colleoni, C., James, M.G. and Myers, A.M. (2003) Mutational analysis of the pullulanase-type debranching enzyme in maize indicates multiple functions in starch metabolism. *The Plant Cell* 15, 666–680.

James, M.G., Denyer, K. and Myers, A.M. (2003) Starch synthesis in the cereal endosperm. *Current Opinion in Plant Biology* 6, 215–222.

Morell, M.K., Kosar-Hashemi, B., Cmiel, M., Samuel, M.S., Chandler, P., Rahman, S., Buleon, A., Batey, I.L. and Li, Z. (2003) Barley *sex6* mutants lack starch synthase IIa activity and contain a starch with novel properties. *The Plant Journal* 34, 173–185.

Tetlow, I.J., Morell, M.K. and Eames, M.J. (2004) Recent developments in understanding the regulation of starch metabolism in higher plants. *Journal of Experimental Botany* 55, 2131–2145.

Starch – transitory

See: Transitory starch

Starter fertilizer

The placement of small quantities of nutrients in a concentrated zone close to the point of seed placement in the field at the time of planting: about 5 (± 2.5) cm away directly below, to the side, or to the side and below the seed in the **furrow**. Starter fertilization is considered to be distinct from application of pre-emergence fertilizer, whether broadcast or application in a band on the soil surface, because the nutrient placement is not strategically placed to be available to early seedling development and growth.

Applying starter fertilizers near to the seed is a well-accepted practice in several direct-seeded crops, such as forage legume and grass species, maize, grain-sorghum, soybean, carrot, tomato and onion – especially for sowing in cool, wet soils. Starter fertilizer solutions are also commonly applied during transplantation of young plants by injection into the soil: for example, while transplanting tomato, pepper, eggplant, lettuce, melons and cole crop (cabbage, cauliflower and broccoli) seedlings. The primary reason is to stimulate rapid early growth of the seedlings, which is associated with crop yield increases.

Starter fertilizers are typically rich in nitrogen and phosphorus (for example, ammonium polyphosphate or urea-ammonium nitrate dissolved in water, or dry di- or mono-ammonium phosphate); potassium or micronutrients are occasionally included as well. The amount that can be safely applied is limited, depending upon the salt content ('salt index') of the fertilizer, its distance from the seed and the soil texture. Positioning starter fertilizer in contact with the seed is another option, which requires a great amount of care to prevent germination or early seedling injury; many grass seeds, however, can tolerate seed-applied soluble phosphate fertilizer without injury, due to the protection afforded by the **lemma** and **palea**. (PH)

(See: Banding, band drilling)

Steckling

A type of plant-to-seed production used in biennial root crops, such as **sugarbeet**. Overwintered young plants of the parental line(s) are raised in so-called steckling beds and transplanted into seed production fields in the spring, at the stage when the roots are about 4 cm in diameter. **(See: Production for Sowing, I. Principles, 1. Starting material)**

Steeping

In seed technology, steeping (or soaking) procedures in excess water are used for both seed **enhancement** and **disinfection** treatments, or as a preliminary step in **priming** schemes.

Indeed, the term '**hydropriming**' in one of its current usages in the scientific literature is effectively synonymous with steeping. Since water availability is not limited during the wet stage, at least some seeds will eventually complete germination even though submerged, and the process must be arrested at a specific time to prevent the onset of Phase III of water uptake (**see:** Fig. P.31 in **Priming – technology**). In grain processing technology, steeping is also intrinsic to **malting** and other industrial processes, such as the wet milling of **maize**.

At its simplest, steeping is an agricultural practice of some antiquity, and can be seen as the forerunner of **pre-germination** and **hardening** procedures used in large-scale farming and mechanized sowing situations. Overnight steeping without drying is still practised in parts of the world today where circumstances allow, notably to start the sprouting (chitting) of rice seed before it is transplanted into dry-seedbed nurseries. Small quantities of seed are wrapped in cloth or sacks before immersion, and after draining may be covered to insulate the damp seed mass where ambient temperatures are low. On-farm steeping is advocated by researchers in parts of Africa and the Indian subcontinent as a pragmatic, low cost and risk method for enhancing crop establishment, such as for barley, chickpea, cowpea, groundnut (including bambara), lentil, linseed, maize, millets, mung bean, pigeon pea, upland rice, sorghum, soybean and wheat crops. The benefits variously include faster emergence, a better stand, a lower incidence of re-sowing, more vigorous plants, better drought tolerance, earlier flowering and harvest, and a higher yield. Rice is also steeped in mechanized systems, including increasing seed weight to aid sowing in **water-seeding**.

Submerged aerated hydration has been proposed to enhance the germination of horticultural brassicas, for example. Steeping also can remove residual amounts of water-leachable germination inhibitors from the seedcoat to speed subsequent germination, such as in sugarbeet and umbelliferous varieties (akin to the prewashing step before **germination testing** of certain species, and to one type of **dormancy breaking** in nature).

Steeping is also used to infiltrate chemical fungicides for the control of deep-seated seedborne diseases, or liquid formulations of other seed protectants or disinfectants. Treatment usually involves submersion or percolation of seed (at up to 30°C for several hours), followed by draining and drying back to the original seed moisture content (**see: Germination – physical factors, 2. Water uptake by the seed**). Short 'hot water' steeps (so-called **thermotherapy**), typically at about 50°C for about 30 min, are used to disinfect or eradicate certain seedborne fungal, bacterial or viral **pathogens**. Care needs to be taken in administering such types of treatment to avoid damaging seed quality.

Steeping is also involved in some **density sorting** procedures (including, for example, **IDS** used in tree seeds), and in vigour testing to measure solute **leakage**, such as from legume seed (**see: Vigour testing – biochemical, conductivity test**). Treatment with excess water is also intrinsic to **prechilling** and **stratification** treatments, to the wet-seed method of seed extraction, fermentation and washing, such as in tomato and cucurbit species where soft fruits are harvested (**see: Production for Sowing, V. Drying, 1. Pre-drying**), and in the slurrying step of **hydroseeding**. (PH)

Harris, D., Joshi, A., Khan, P.A., Gothkar, P. and Sodhi, P.S. (1999) On-farm seed priming in semi-arid agriculture: development and evaluation in maize, rice and chickpea in India using participatory methods. *Experimental Agriculture* 35, 15–29.

Maude, R.B. (1996) *Seedborne Diseases and their Control: Principles and Practices.* CAB International, Wallingford, UK.

Thornton, J.M. and Powell, A.A. (1992) Short-term aerated hydration for the improvement of seed quality in *Brassica oleracea* L. *Seed Science Research* 2, 41–49.

www.seedpriming.org

Sterility

In plant breeding and seed production: the failure or lack of capacity to complete fertilization and produce seed as a result of defective pollen, ovules or other causes. More generally: the failure to produce offspring, spores, etc. (**See: Cytoplasmic male sterility**)

Stick samplers

Instruments of several designs (such as Deep Bin Probe or Neate Sampler, the Multi Chamber type or Walking Stick Sampler, and Cargo Samplers) used in **sampling** of seedlots, all of which have apertures that can be opened and closed after insertion into the seed mass to take representative samples for **quality testing**. (**See: Sampling, 1. Spears and probe samplers**)

Stigma

The upper end of a carpel which is designed to receive the pollen grains, usually connected to the ovary by a style; the stigma is usually covered with papillae (short hair-like protruding cells) and secretes a sticky liquid to help retain the pollen grains. (**See: Pollination; Reproductive structures, 1. Female**)

Stimulants of germination

The seeds of certain **parasitic plants** such as witchweed (*Striga hermonthica*) and clover broomrape (*Orobanche minor*) need to be stimulated to germinate by compounds exuded by the roots of the host. Several such compounds have been identified. Strigol was the first to be discovered, and was found in the root exudate of a non-host plant, cotton. Nonetheless, it is active on *Striga* seeds at femtomolar concentration. Strigol and another active compound, sorgolactone, have since been found in the root exudates of host plants. Sorgolactone and the related alectrol are collectively called strigolactones. A hydroquinone exuded by sorghum roots, called 'sorghum xenognosin for *Striga* germination' (SXSg) and several very closely related compounds, are also active stimulants of germination. Structurally different compounds, also called xenognosins, play a different role as signals for the formation of the haustoria that absorb water and nutrients from the host. These stimulants have an obvious application in clearing **soil seed banks**. However, they are unstable in soil, so there is interest in finding other more stable compounds. One group that has been studied are certain fungal toxins such as cotylenin from *Cladosporium* spp. and fusicoccin from *Fusicoccum amygdali*, which also stimulate the germination of *Striga* and *Orobanche* (see: **Germination – influences of chemicals**; Fig. P.1 in **Parasitic plants,** for chemical details).

These compounds act upon *Striga* through the induction of **ethylene**, and inhibitors of ethylene synthesis and perception will block germination. This does not seem to be the case in *Orobanche*.

Many other compounds, including ethylene, **kinetin**, **zeatin**, **ABA**, inositol, methionine, scopoletin and sodium hypochlorite will stimulate the germination of *Striga* seeds but they are probably not active in field conditions. However, it is clear that root exudates contain a number of structurally different compounds that will stimulate germination. The absence of one may be compensated by the presence of others. It is also clear that the mechanisms of action are different in different parasitic plants. (GL, JR)

Stock seed

Usually refers to such as Foundation or Registered seed that will be used for further seed multiplication. (**See: Certification systems; Foundation seed**)

Storage fungi

Loss of grain during storage due to the action of storage fungi is a continuing major worldwide problem, particularly in tropical developing countries where losses in some locations are estimated to represent up to about a third of the annual harvest. Spoilage effects include discoloration (Fig. S.48), mustiness, fat or oil deterioration leading to increases in free fatty acid content accompanied by rancid odours and flavours, reduced seed germination, the production of fungal **mycotoxins** and disease (such as human aspergillosis, **see: Dust**), seed 'caking', and decay in the final stages of deterioration. Such key quality factors in seed intended for human consumption are scrutinized by **grain inspection** at trade trans-shipment points and by continual monitoring of seed bulks during storage. Visibly mouldy grain will be already tainted, although some species of mites feed on fungi and may mask evidence of fungal growth.

The specific composition of the fungal population on stored grain is highly dependent on seed water content and temperature; very small changes may change the amount and composition of microflora substantially. Mechanical damage, cracking and breakage of seed during harvest, drying and handling also increase susceptibility to subsequent invasion. **Pathogenic** or **saprophytic** field fungi dominate in the initial phase of development while relative humidity is high (above about 95%), such as *Alternaria*, *Cladosporium* and *Fusarium* spp. which have invaded during seed maturation or **swathing**, as kernel or ear rots or the like. As relative humidity falls to about 70–90%, the so-called storage fungi – species universally present on decaying plant debris or in soil surface layers – germinate from spores that became attached during harvest operations. Characteristically yeast and *Penicillium* dominate at first, followed by a succession of species of *Aspergillus*, the predominant genus involved in storage losses, each of which is active within a narrow moisture range. As species adapted to low moisture start the decay process, their

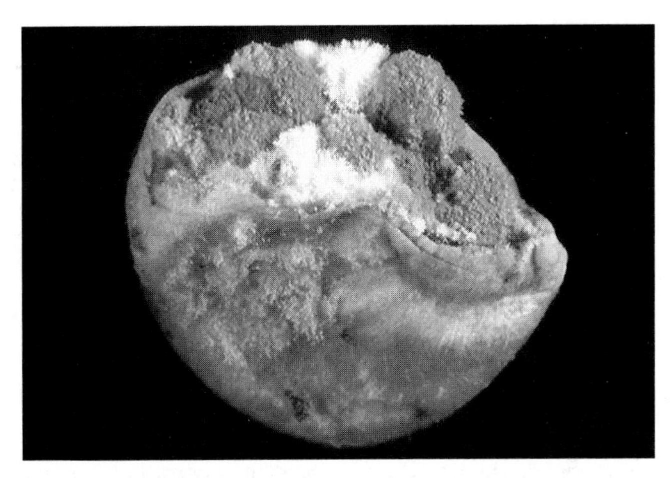

Fig. S.48. Storage mould on maize grain. Copyright Nigel Cattlin/Holt Studios International (from the CABI Compendium website).

metabolic activity produces heat and moisture that create conditions favourable to another species. **Storage pests** – insects such as grain weevils and mites – and condensation caused by **moisture migration** may also accelerate the process of deterioration by increasing seed temperature and moisture content and transmitting spores. Mycotoxin production is likely after a few weeks at above about 80% relative humidity. The later phases of seed deterioration can be accompanied by local grain heating in 'hot-spots' (up to about 60°C, such as in spout lines and the centres of storage bins), and the proliferation of thermotolerant fungi such as species of *Absidia* and *Mucor*. Further associated chemical reactions on rotting grain combined with the action of thermophilic bacteria can drive temperatures to about 75°C, and in time cause spontaneous combustion.

Storage losses are controlled by the proper design of seed storage structures, ventilation and handling procedures, to ensure seed is undamaged, efficiently cleaned from dust, and kept uniformly dry and cool (**see: Storage management**). The fungi continue growing slowly at near 0°C, so cooling alone is insufficient for long-term storage of damp grain. Propionic acid or ammonia treatment of seed is a long-standing method to control yeast and moulds, though these treatments render grain only fit for feeding to livestock; caustic soda treatment is used in a similar way, which causes grain swell and so makes silo storage impractical. (PH)

(**See: Cereals – storage; Disease; Pathogens**)

Agarwal, V.K. and Sinclair J.B. (1996) *Principles of Seed Pathology*, 2nd edn. Lewis Publishers/CRC Press, Boca Raton, FL, USA.

Neergard, P. (1977) *Seed Pathology*, Vols I and II. John Wiley, New York, USA.

Storage management

A considerable amount of seed is lost, or rendered less valuable or useless each year due to improper storage conditions after harvest, and associated fungal infection and insect and mite infestations that occur when grain is stored with high moisture content. The problem is most acute in parts of the developing world. Though few assessments are available, quantitative losses are conservatively estimated as at least 10%, and up to 40% recorded in some areas; losses due to deterioration of nutritive quality are more difficult to determine. Also important worldwide are the losses in germination quality and **vigour** in seed for sowing the next crop, or for malting.

After seed has been dried and cleaned, bulk postharvest storage management technology comprises targeted pest control, general good hygiene and aeration to obtain and maintain even, cool temperatures, and moisture contents to prevent local heating and drying, along with the regular observation and monitoring of conditions throughout the grain mass, by detector probes or direct seed **sampling**. During storage, loading and unloading, and trans-shipment, seed should be protected from the weather (water and, as far as possible, temperature fluctuations), contamination, the action of **storage fungi**, rodents, birds, insects and other predators (**see: Storage pests**), fire and contamination with other seed, dirt and debris. Storage structures should be designed and ventilated so as to avoid the development of cold or hot spots, and condensation. Because seed is a poor conductor of heat, such hazards are most acute in large storage structures. These effects, together with the discoloration, mustiness, decrease in germination, and production of **mycotoxins** all reduce quality grade and commercial value. Though the quantity of damaged grain is usually small in proportion to the whole bulk, some may be unavoidably mixed with undamaged grain during unloading.

1. Structures, containers and packages

The material and organizational resources to achieve effective postharvest storage management vary tremendously across the world, depending on the kind and volume of seed grain, the local climate and economic resources. In countries with the most developed distribution systems, cereal, oilseed and pulse grains are cleaned and stored in bulk at local farm, cooperative, merchant or governmental levels (at or after the point of official **grain inspection**, in countries operating such schemes), before shipping onward by road or rail to feed, flour or seed mills, to wholesalers, retailers or abroad. Storage structures range from small buildings to large warehouses and silos; in the former, seeds may be contained in bags, pallet boxes, bulk wood or metal bins, in bulk piles on the floor or platforms; in the latter, metal or concrete silos (also known as bins, sheds or grain **elevators**) have in-built conveying systems at their top and bottom, and in the largest facilities have vertical and horizontal storage elements to handle different bulks at the same time. In the major exporting and importing countries, commodity grain is conveyed into terminal silos or elevators; those for wheat at the main ports on the lower St Lawrence river in Canada, for example, have huge capacities – some holding at least 400,000 t of grain in total, in an array of silos to give the flexibility to preserve crop and batch identity. For practical handling purposes, local cylindrical-tower silos may hold up to about 1000–6000 t grain, depending on grain bulk density and moisture content, though they can be much larger at big grain handling facilities – where there can be up to about 50-m-diameter structures that hold more than 45,000 t. Flat-floor structures are also used to store larger grain lots. The largest ocean-going bulk carrier ships hold cargoes of up to 100,000 t of grain. Shipment of commodities such as malting barley and seeds in sealed bulk bags, sometimes referred to as envelopes, is becoming increasingly popular, as an alternative to the time-consuming expensive and not always adequate cleaning of special bulk containers. Separate from the bulk commodity distribution are **identity preservation** systems that segregate high-value crop products throughout the supply chain, such as coffee, cocoa nuts, and grain with unique product quality traits.

Much of the seed produced and stored each year for the purpose of sowing needs to be kept only until the next planting season, and may not require conditions other than normal air temperature and relative humidity. Except in humid climates, seeds are packed and stored in various bags or containers (made of burlap, cotton, paper, fibreboard or plastic/laminate films), soon after conditioning and treatment or at a later stage. These packaging materials are not physiologically ideal for seed storage purposes (judged by the principles adopted for the efficient **conservation of seed**, as are adopted in seed banks) except where the ambient humidity conditions are kept under good control, because they do not prevent fairly rapid seed moisture re-equilibration with the external atmosphere. But in practice they are usually sufficient to permit marketing without loss of seed viability, providing that seed is adequately dried in the first place. In humid areas and for high-value seeds moisture-barrier hermetic packages are used, such as lidded plastic tubs and buckets and metal cans.

In developing countries, most subsistence and small-scale farmers and village cooperatives retain seed for their own food, cattle feed and seed sowing purposes, and sell any surplus (such as rice) to traders or buyers immediately after harvest as they lack adequate storage capacity. (**See: Industry supply Systems, II. Informal**) On-farm storage, maybe only for a few days or weeks and seldom longer than to the next planting season, requires small constructions or simple containers, adapted to local conditions. Traditional storage granaries, such as aboveground chambers or bins, underground pits, inside houses or in open courtyards, are constructed with a wide variety of locally available materials (variously, wood, dung, earth, concrete, plastic, metal sheeting or recycled drums, sealed gourds or earthenware pots, jute or other bags, or woven baskets from straw, bamboo, reeds or plastic fibres). Recently, flexible durable structures are being developed: using butyl or PVC-plastic sheet to line temporary bunkers constructed on the open ground, or in bag storage systems suitable for small-scale farm use. These widely contrasting storage practices help explain the range of quantitative and qualitative storage losses in developing countries.

2. Maintenance of quality in bulk storage

Just-harvested grain in temperate climates is cooled to maintain quality and prevent spoilage during the initial periods of storage before seed is fully dried. This can be done by continuous blowing or drawing of ambient air through the seed mass at low rates (about 10 $m^3/h/t$), reaching 15°C within about 2 weeks. But keeping potentially damp grain in warm and humid climates is very hazardous, since metabolic processes as well as mould growth are accelerated at high temperatures; full use must be made of cool hours and days. Chilled aeration can be used on commodities that can justify the added expense of dehumidification and refrigeration.

Mould growth and mite reproduction stop at 65% relative humidity; grain is thus at risk of spoilage until dried to moisture contents of 14.5% for cereals and 7.5% for oilseeds (according to their respective **equilibrium relative humidity** relationships). Most storage pest insects breed rapidly at 25–33°C and not at all below 15°C (though grain weevils can proliferate above about 12°C); above 40°C most insects die within a day, and below 5°C they cannot feed and slowly die, although populations of mites (and fungi) can increase down to 5°C in moist grain, very slowly. Mycotoxin formation is most likely between 15 and 25°C.

In tropical countries, sun drying (**solarization**) is probably the most popular method of moisture reduction and pest control. In temperate climates artificial drying is necessary after harvest, using forced and usually heated air at 40°C or higher, preferably in relatively shallow layers of seed according to fan speed and the starting moisture content of the grain (**see: Production for sowing, V. Drying**). Drying may also be necessary at the bulk grain elevator stage. Hot air in a continuous dryer is likely to disinfest grain, but as grain cools naturally it becomes vulnerable to infestation. Except in the simplest batch dryer designs, grain is mixed and re-circulated during drying to give more uniform exposure to the air so that over-drying and heat damage are limited; in continuous dryers re-circulation is not necessary. Higher drying temperatures give higher throughput but excess heat can damage quality, especially protein functionality and germination: a maximum of 50°C is safe for malting grain, 65°C for wheat and barley at moisture contents of 20%, or much higher temperatures for the drying of feed grain. High-speed, high temperature drying may also cause injury in some seed species – for example, producing more stress-cracked maize kernels than after low-temperature drying, making them more liable to break during handling. Grain must be cooled immediately after drying to prevent storage problems, which can be done with ambient air providing it is about 6°C cooler than the seed mass. The overall process has been termed 'dryeration', and during the 'tempering' process, during which the moisture gradient that develops inside each kernel is allowed to equilibrate and internal stresses to dissipate, grain is held in a so-called 'steeping' bin.

Alternatively, bulk drying using ambient or slightly warmed air about 5°C warmer than the seed, sometimes after dehumidification, is carried out for longer periods of time, in relatively deep beds adjusted to the airflow resistance of the crop. Drying wheat from 20% moisture content requires airflow of at least 180 $m^3/h/t$ to reduce moisture by 0.5% a day, for example. Under these conditions, a 'drying zone' layer develops at the air inlet and then moves through the rest of the bulk, which in the meantime is cooled somewhat by the consequent ventilation. The immediate challenge is to complete drying before fungi and mites exceed acceptable levels ahead of the drying zone, where the grain may still be warm enough to produce ideal spoilage conditions. Under these conditions some moisture may condense on the upper surface of the grain mass, especially when there is cool, moist night air; grain stirrer augers can help mitigate this effect.

3. Cleaning and handling

It is important to minimize mechanical damage during all the processing and conveying operations from harvest, through drying to handling, especially in vulnerable heavy seeds (such as maize, bean, pea and soybean) that can be easily fractured by impact upon a hard or firm surface, the more so if they contain too much or too little moisture. Generally, seed may be injured if it is force-fed through restricted openings or pressure rollers (see also **conveyors**), or if augers are not run full. Damage from such causes is accumulative.

Storing grain with a minimum of fine material helps prevent many hard-to-manage bulk storage problems. Foreign matter (chaff, stalks, grain **dust**) presents a more suitable moist substrate for mould development than whole grains; likewise many insects that infest stored grains thrive on the presence of broken kernels and fine material; also extraneous plant debris, sand, earth, stones and the like impede the effective application of air to prevent the bulk from heating or to apply fumigant gases. Such factors can increase the temperatures within any accumulated pockets of fine material, and they lead to aeration and heating problems associated with the chain reaction of deterioration in the surrounding grain, and ultimately to possible spontaneous combustion (see next section). Cleaning grain before binning is the most effective technique, such as using rotary cleaners, scalpers and aspirators (**see: Production for sowing, V. Drying**).

The technique of silo/bin filling is also important to reduce the formation of concentrated pockets of fine material. Most commonly a grain spreader is installed at the top to distribute all material more or less homogeneously. Alternatively, in the core removal technique, grain is loaded in stages directly into the middle of a round silo/bin above its emptying spout, in such a way that fine material tends to accumulate in a central core; after each partial filling step, a fraction of the seed is removed from below, which effectively withdraws the fines from the bin.

4. Aeration

The major objective for aerating dried grain in bulk static storage is temperature management. Efficient aeration may also be necessary to deliver and purge fumigants efficiently throughout the bulk.

As the relative humidity of surrounding air changes over time, so can the moisture content of the grain in the store. In particular the moisture content of surface layers can rise or fall due to ambient conditions; condensation can be very marked in warm subtropical climates, in which the ambient temperature may drop considerably by night. In cool climates, as outdoor temperatures decline in autumn and winter, the insulating characteristics of grain prevent temperatures in the centre of the mass from falling as rapidly as those near the silo/bin walls. The resulting temperature differences produce convection currents and moisture migration. Cooled denser air settles towards the bin floor, moves across the floor towards the centre of the bin and upwards through the grain, where it warms and becomes less dense absorbing small amounts of moisture; as the rising air is cooled in the upper zones of the bin, moisture condenses from the air on to the cold grain, resulting in the wetting and crusting of surface grain, favouring the rapid development of mould spoilage fungi and sometimes bacteria, and sometimes enough to allow seed to germinate. (Similar problems can also occur in the holds of ships.) Surface moisture content may rise to 18% in wheat or more (i.e. about 82% RH at 15°C), encouraging mites and mycotoxin production, even where the bulk moisture content is low; under those conditions the 5 parts per billion maximum permissible level set in current EU regulations for ochratoxin A in cereals, produced by *Penicillium*, can be exceeded in 2 weeks. Where crusting occurs, the grain surface can be raked, stirred, or mixed in by moving bulk grain from one bin to another, or even removed.

Respiration due to grain and, mainly, microfloral metabolism in bulks with excessive moisture content may become very intensive, resulting in loss of dry weight, the creation of temperature gradients and moisture migration within the seed mass; in the extreme it can cause 'spontaneous heating' (up to about 60°C) and can reach the point of combustion (at about 75°C). High moisture and temperature also promotes the growth of such insect pests as the grain weevil, the lesser grain borer and the angoumois grain moth, and secondary pests.

Forced aeration upwards or downwards through grain mass, either intermittently or continuously, prevents moisture migration and reduces hot spots, equalizing temperatures throughout the bin. Aeration in the autumn lowers the temperature of the centre of the mass near to that of the outer portions, and in the spring and early summer rewarms the mass uniformly. Operational skills lie in the timing of initiating cooling or warming cycles according to changes in external temperatures, and in minimizing the extent of condensation areas where warm and cool grain meet.

5. Control of insect pests

Insecticides, fungicides, rodenticides and fumigants that limit the deterioration of stored grains supplement these storage practices. Some spray-chemicals are suitable for use in all parts of the store; others are restricted to surfaces that will not be in contact with grain. Dust formulations of diatomaceous earth, which act by desiccating insects, may be applied to dead spaces and structural surfaces. Cleaning alone will not eliminate all pests in empty stores, nor will pesticide treatment.

Protectant insecticides, which retain their insect-toxic properties for an extended time, are spray-applied on to grain as it enters storage. These active ingredients are not highly volatile, and have only limited penetration into infested kernels, so they may not destroy all life stages of some stored grain insects, such as eggs or internally feeding insects. There are also considerations about whether and how soon treated grain may be used for milling, baking and human or animal consumption. Examples that have been, or are being, used for these purposes include, cyfluthrin and chlorpyrifos-methyl (which are registered for use in the USA to control residual amounts of the lesser grain borer and other pests of small-grain cereals), synthetic pyrethroids, and methoprene (a synthetic analogue of an insect growth regulator). A recurrent problem with grain protectants is the development of resistance in the target pests.

Fumigation with toxic gases is effective in airtight structures and is an economical method carried out on a commercial scale (**see**: **Fumigants**). These chemicals rapidly kill all life stages, although not all individuals, and provide no residual activity against reinfestation. In Australia up to 80% of grain is currently fumigated with phosphine at some time during storage, for example.

By and large, fumigation technology is not yet applicable to insect prevention in storage at the farm level in many developing countries because storage structures are not airtight and are located inside or near residential areas where fumigation may be dangerous. The use of mercury is reported to be a local tradition in south Asia particularly in the Punjab provinces of Pakistan and India and nearby districts, despite its potential health hazards. Surveys indicate widespread use of **neem** (*Azardirachta indica*) by farmers in the Indo-Pakistan subcontinent and south Asia, where it grows abundantly; food grain stored in bags is sometimes mixed with whole leaves or extracts of them. (**See** also: **Storage pests, 5. Management strategies**)

6. Atmosphere modification

Storage in essentially sealed containers has long been recognized as an effective way to minimize losses due to pests and predation. Assorted pots, gourds and metal drums have long been adopted for this purpose for local storage in some tropical countries; for greatest effectiveness, containers should be sufficiently airtight and completely filled. For example, in parts of Africa such as in Botswana, it is sometimes the practice to mix millet in with stored beans to pack the intergranular spaces and hence physically impede infestation of the beans by braccate beetles.

In recent years renewed attention has been focused on technologies based on hermetic storage as an alternative or complement to reliance on chemical pesticides. Such techniques use the principle of self-regulated atmospheres: exploiting the metabolic activity of the insects and other aerobic organisms in the grain mass themselves to deplete O_2 and accumulate CO_2 concentrations below those permitting further insect development. Sealing of grain warehouses using plastic liners, for example, is part of a current integrated preservation regime in China that combines atmospheric manipulation with fumigation. The evacuation of sealed containers to low pressures, or flushing with CO_2 or N_2 gas, or pre-treating seed with heated air in spouted beds before filling, all offer other new potential approaches. (**See** also: **Hermetic storage**) (PH)

CSIRO Stored Grain Research Laboratory, Canberra, Australia. sgrl.csiro.au

Desai, B.B. (2004) *Seeds Handbook: Biology, Production, Processing and Storage* (2nd edn). Marcel Dekker, New York, USA.

FAO Information Network on Post-Harvest Operations (INPhO) www.fao.org/inpho/

Heaps, J.W. (ed.) (2006) *Insect Management for Food Storage and Processing* (2nd edn). AACC Press, St Paul, MN, USA.

Home-Grown Cereals Authority (UK) (2003) *The Grain Storage Guide*. Downloadable from www.hgca.com/research/grainstorage

Jayas, D.S., White, N.D. and Muir, W. E. (eds) (1995) *Stored Grain Ecosystems*. Marcel Dekker, New York, USA.

Navarro, S. and Noyes, R. (eds) (2001) *The Mechanics and Physics of Modern Grain Aeration Management*. CRC Press, Boca Raton, Florida, USA.

Reed, C. (2006) *Managing Stored Grain to Preserve Quality and Value.* American Association of Cereal Chemists, St Paul, MN, USA.

Sauer, D.B. (ed.) (1992) *Storage of Cereal Grains and Their Products.* 4th edn. American Association of Cereal Chemists, St Paul, MN, USA.

Subramanyam, B. and Hagstrum, D. (eds) (1996) *Integrated Management of Insects in Stored Products.* Marcel Dekker, New York, USA.

Storage pests

1. Damage caused by primary and secondary pests

Seeds in storage are a difficult food resource for pests to exploit. They are dry, hard and many contain toxic chemicals that protect them from being eaten. Nevertheless insects, including several beetles (Coleoptera) and one species of moth (Lepidoptera), are able to attack intact dry seeds directly. These are referred to as the primary storage pests, because they attack and destroy whole, undamaged grain. A rather larger group of secondary pests, mostly but not exclusively beetles and moths, feeds on seeds (and a range of other food products as well) once the **seedcoat** has been damaged by other primary pests or by mechanical injury. Besides insects, several species of mite (*Acarina*) are important secondary pests that also rely on prior damage. Both rodents (Rodentia) and birds (Aves), although not normally defined as either primary or secondary pests, may also be found consuming grain in stores that are inadequately screened against them. Scavengers that feed on the by-products of stored grain produced by storage pests, such as detritus and fungal moulds, are also usually not placed in the category of secondary pests.

Insect pests can be the cause of substantial losses of grain, particularly in the hot areas of tropical developing countries where food security may be at risk. In Africa, estimates made in the 1970s of grain weight losses due to infestation by insects such as weevils (*Sitophilus* spp.) showed that up to 5% loss was common. This is by no means insignificant as physical losses are accompanied by qualitative losses and, in the case of seed, significant losses of viability. In addition, the losses are experienced during the lean season before the new harvest so having an adverse effect on farming families at a particularly critical period. In the late 1970s, the arrival in Africa of a neotropical grain pest, the larger grain borer (*Prostephanus truncatus*), resulted in an increase to about 10% in weight losses in areas where maize is the staple and where the new pest was not adequately controlled.

In cooler areas of the world, generally less damage is caused by storage pests because seed storage systems are more sophisticated and/or because the climate is less favourable to their proliferation. However, although direct damage may be less severe there are still significant financial losses owing to the high costs of pest control operations to maintain seed stocks free of pests and, where pest management is inadequate, the loss of quality premium when grain consignments are downgraded due to the presence of live or dead insects. Even if these are killed, mechanical sieving will not remove the fragments of faecal material, frass (refuse left behind by boring insects) and hard body parts that can contaminate processed foods destined for human consumption.

Some seeds may be harvested with the pests already within them. This is commonly the case with primary pests, which may even start to cause damage in the field before harvest, and may not be completely removed during **conditioning**. In other cases, insect attack may be initiated in store. Grain intended as seed for sowing the next crop is usually stored for the entire period from harvest and is thus particularly exposed to the dangers presented by storage pests. For this reason, farmers need to take special care of their seed stocks. These are usually stored separately from food grain stocks and in much more secure containers, which may be insect proof, hermetically sealed. Nevertheless, insect infestation of seed for sowing is an important problem in some areas, particularly for large-volume arable crops, such as in India.

Although there are two or three hundred insect species that have been associated with grain and grain products, relatively few are significant pests. A summary of major species is presented in Table S.14 and a description of some of these is given below.

2. Selected primary pests of cereals

(a) *Grain weevils. Sitophilus* spp. including *S. zeamais* (Fig. S.49A), *S. oryzae* and *S. granarius* are important primary pests of cereals. Adult female weevils lay eggs singly in tiny holes that they gnaw in the grain and protect them by secreting a waxy 'egg-plug'. Upon hatching from the egg, the larva begins to feed in the **endosperm**, producing a cavity in the grain as the grub grows (Fig. S.49B). During the immature stages of its pre-adult life, *Sitophilus* is hidden inside the grain and is extremely difficult to detect. Eventually the fully grown larva pupates, and the adult emerges by biting its way out of the grain leaving a characteristically large, somewhat irregular emergence hole. Adult weevils feed especially on grain endosperm that has been exposed by breakage, or by entering emergence holes.

S. granarius is essentially a temperate pest and is not found in tropical countries except occasionally in cooler, upland areas.

S. zeamais and *S. oryzae* are commonly found throughout the world in tropical and subtropical regions especially where ambient humidities are fairly high. They thrive best at warm temperatures (around 28°C) and in grain in which the moisture content is not much less than 13%. Both species can infest any cereal grain that is large enough to support their development as well as other starchy foods such as dried cassava roots and even spaghetti. *S. zeamais* is particularly associated with maize and milled rice, but is often found on wheat and other small grains. It flies strongly, and often infests commodities in the field before harvest. *S. oryzae* is particularly associated with small grains, but is often found infesting maize and paddy rice. Certain rare strains of this species have been found breeding on lentils and split peas. Most strains are infrequent fliers, and are particularly common as pests of commerce in warehouses, transport vehicles and silos where they have been introduced by humans in transported grain. (**See: Cereals – storage**)

(b) *Grain borers. Rhyzopertha dominica* (Fig. S.50A) and *Prostephanus truncatus* are cylindrical brown beetles, which are able to thrive on cereal grain.

R. dominica is widespread throughout the tropics and subtropics and is most important as a pest of wheat and paddy

Table S.14. Major insect pests of seeds.

Species	Primary (1) or secondary (2) pest	Range as a significant pest	Food
Beetles (Coleoptera)			
Sitophilus granarius	1	Temperate regions	Whole cereal grains, mostly wheat and barley
Sitophilus oryzae	1	Tropics and subtropics	Whole cereals grains especially maize, sorghum and paddy rice, rarely pulses such as peas and grams
Sitophilus zeamais	1	Tropics and subtropics	Whole cereal grains especially maize, wheat, sorghum, also milled rice
Rhyzopertha dominica	1	Tropics and subtropics	Whole cereals, especially wheat, sorghum and paddy rice
Prostephanus truncatus	1	Tropics	Maize, especially on the cob
Acanthoscelides obtectus	1	Tropics and subtropics	Beans (*Phaseolus* spp.)
Callosobruchus analis	1	Tropical Asia	Cowpea and gram (*Vigna radiata*)
Callosobruchus chinensis	1	Tropics and subtropics	Cowpea, lentils, chickpea, adzuki bean
Callosobruchus maculatus	1	Tropics and subtropics	Cowpea, lentils and gram (*Vigna radiata*)
Callosobruchus rhodesianus	1	Africa	Cowpea
Callosobruchus subinnotatus	1	West Africa	Bambara groundnut
Caryedon serratus	1	West Africa	Groundnuts and tamarind seeds
Cryptolestes spp.	2	Temperate through to tropical climates	Damaged cereals, pulses, oilseeds, nuts and dried fruit
Oryzaephilus spp.	2	Temperate through to tropical climates	Damaged cereals, pulses, oilseeds, nuts and dried fruit
Trogoderma granarium	2	Tropics and subtropics, especially in areas with a hot dry climates	Oilseeds, damaged cereals and pulses
Tribolium castaneum	2	Tropics and subtropics	Damaged cereals, pulses, oilseeds, nuts and dried fruit
Tribolium confusum	2	Subtropics and temperate regions	Damaged cereals, pulses, oilseeds, nuts and dried fruit
Moths (Lepidoptera)			
Sitotroga cerealella	1	Tropics and subtropics, especially in areas with a hot dry climate	Whole cereal grains, especially sorghum, paddy rice and maize
Ephestia cautella	2	Tropics and subtropics	Damaged cereals, pulses, oilseeds, nuts and dried fruit
Plodia interpunctella	2	Subtropics and tropics	Damaged cereals, pulses, oilseeds, nuts and dried fruit
Corcyra cephalonica	2	Tropics and subtropics	Damaged cereals, pulses, oilseeds, nuts and dried fruit

rice, although it does also occur on other cereals and roots such as dried cassava.

P. truncatus is a sporadic but locally serious pest of maize stored on the cob in Central America. In the late 1970s, it became established in western Tanzania, where it became an extremely serious pest of farm-stored maize, roughly doubling average farm store losses from 5 to 10%; in individual cases, farmers might lose as much as 30%. Subsequently, it has spread to many countries in both East and West Africa. It is a serious pest in a wide range of environments but is particularly favoured by hot dry habitats. *P. truncatus* is now a quarantine pest in many countries.

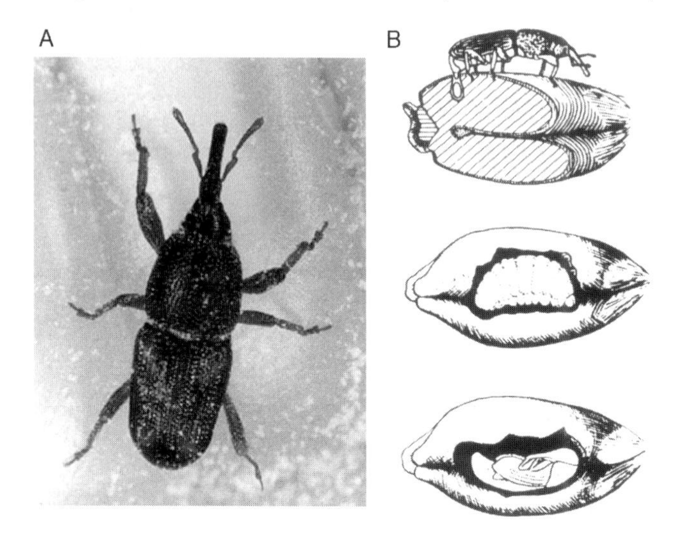

Fig. S.49. (A) *Sitophilus zeamais* (adult life size 2.5–4.5 mm). (B) Life cycle of *Sitophilus* sp. showing egg laying (top), larval development within a grain (middle) and pupation (bottom).

When infesting grain, female *R. dominica* and *P. truncatus* lay eggs at the end of tunnels they excavate in the seed. Subsequent development usually takes place within a grain, but unlike *Sitophilus*, larvae may bore out of one grain and into another (Fig. S.50B). After pupation the newly developed adult escapes from the grain by chewing its way out, but unlike *Sitophilus* spp. then continues to bore through grains. The adults feed throughout their lives, producing large quantities of dust and frass containing a high proportion of undigested fragments which can themselves support the development of larvae if sufficiently compacted. Similarly, ground produce or very small grains can be infested to a limited extent if compacted and stabilized.

R. dominica and *P. truncatus* are adapted to rather higher temperatures and lower moisture contents than *Sitophilus* spp. and they are therefore the dominant pest in areas which are hot, dry or both. In such areas sorghum is often grown, so *R. dominica* is frequently associated with this crop. *P. truncatus* is almost exclusively associated with maize and dried cassava chips, but very rarely with sorghum as the grain size is usually too small to support its development.

(c) *Angoumois grain moth.* The moth *Sitotroga cerealella* (Fig. S.51A) is an important primary pest of cereals and can

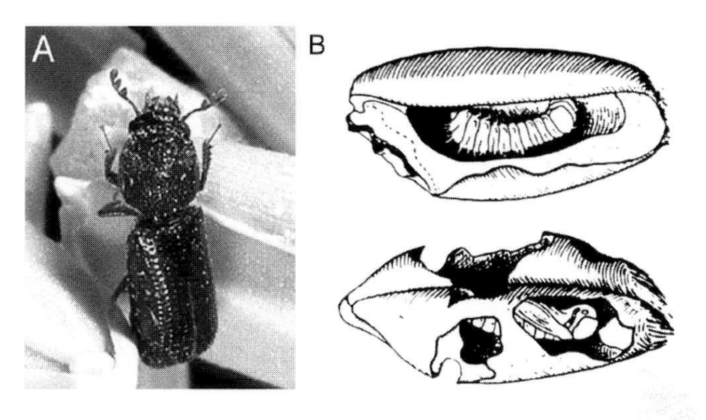

Fig. S.50. (A) *Rhyzopertha dominica* (life size 3–4.5 mm), and (B) its larva and pupa in a cereal grain.

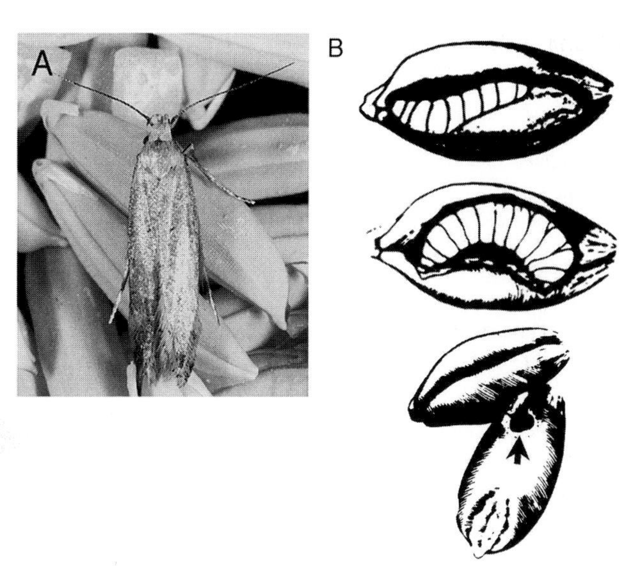

Fig. S.51. (A) *Sitotroga cerealella* adult (wing span 10–18 mm) and (B) its larva within a grain (middle right), pupa (top right) and grain showing 'trap door' (arrowed, right) for escape of the mature insect.

infest grain in the field before harvest, especially maize and sorghum.

Females lay eggs in masses on the commodity, and upon hatching the larvae bore into the grain. Subsequent development takes place within the grain (Fig. S.51B), but the larvae may leave one grain and enter another, especially if the grains are small. The larva prepares the point of eventual emergence by chewing a feeding chamber but leaving a thin area of grain seedcoat, known as a window, separating it from the exterior. After pupation, a rather characteristic hole is left behind, with a partial covering remaining at its edge in the form of a 'trap-door' (Fig. S.51B).

Sitotroga cerealella attacks any cereal with grains large enough to support larval development. It is widespread over tropical and subtropical parts of the world, sometimes entering warmer temperate areas.

3. Selected primary pests of pulses

(a) *Bean weevil. Acanthoscelides obtectus* is a common pest of *Phaseolus* beans. It sometimes attacks other legumes, but on these is seldom a serious pest.

The adult beetles are able to infest beans before or after harvest. Eggs are laid loosely in or around pods or beans, often under cracks in the **testa**. After hatching the larvae bore into the bean and spend their larval life feeding on the **cotyledons**, excavating a feeding chamber as they grow. The larvae pupate within the bean, but prepare the site of eventual escape by chewing their way close to the outside, leaving only the testa separating the pupation chamber from the exterior. The adult, which has relatively feeble mouthparts, is able to penetrate the testa and escape. The area of undermined testa is easily seen before adult emergence and is known as a window. The window itself is usually completely removed on emergence leaving a neat round hole in the bean.

A. obtectus develops most rapidly at around 30°C and 70% relative humidity, but is capable of tolerating quite low temperatures. This has resulted in it being able to spread to cool highland regions of the world and into some temperate areas.

(b) *Callosobruchus* spp. are important primary pests of a number of legumes including cowpeas, pigeon peas, chickpeas, adzuki beans, peas, grams and (occasionally) soybeans. They do not usually attack kidney beans or butter beans (*Phaseolus* spp.), however.

The adults (Fig. S.52A) are of the same general form as *A. obtectus*, but are usually a little smaller. The wing cases of some species are distinctly patterned, especially in the females, also in the female they do not cover the last abdominal segment. The life cycle of *Callosobruchus* spp. is similar to that of *A. obtectus*, except that the eggs are stuck firmly to the testa of the host seed or to the wall of a pod. Upon hatching, the larva bores through the floor of the egg, directly into the seed or the pod.

(c) *Zabrotes subfasciatus* is a common pest of kidney beans and butter beans, and seldom attacks other legumes. The life cycle of *Z. subfasciatus* is similar to that of *Callosobruchus* spp., including a stage where eggs are glued on to the testa. The pest originated in tropical America, but is now common in many tropical and subtropical regions, especially central and East Africa, Madagascar, the Mediterranean and India.

4. Selected secondary pests of cereals and pulses

(a) *Khapra beetle. Trogoderma granarium* adults are small oval beetles (Fig. S.52B) that are a very serious pest of cereal grains and oilseeds. Massive populations may develop and grain stocks can be almost completely destroyed. In addition, the cast larval skins may cover the surface of infested grain, where their abundant hairs present a health hazard to seed store workers and consumers. Attack occurs in large-scale stores; on the other hand the pest appears not to have been reported from farm stores. Many countries have specific quarantine regulations against possible importation of this pest on grain, and its presence results in an order to carry out expensive pest control measures or a rejection of the shipment. Although winged they are not known to fly and appear to rely on transport in old bags, etc. to get from one store to another.

T. granarium is very tolerant of high temperatures (up to 40°C) and low humidities (down to 2% RH). It is therefore a pest in hot, dry regions where other storage pests cannot survive. In addition, the larvae are able to enter diapause (a resting stage) when physical conditions are unfavourable. In this state they move very little, or not at all, and their metabolic rate is lowered

Fig. S.52. (A) *Callosobruchus maculatus* (life size 2.0–3.5 mm), (B) *Trogoderma granarium*, Kaphra beetle adult (life size 2.0–3.0 mm) and larva, (C) *Tribolium castaneum*, flour beetle adult (life size 2.5–4.5 mm), (D) *Ephestia cautella*, warehouse moth adult (wing span 11–28 mm).

so that they can survive several years of adverse conditions. In this state therefore they are very difficult to kill with residual insecticides or fumigants, although they would otherwise be susceptible to the usual storage insecticides and fumigants. They are usually hidden in cracks or crevices in the store, and thus protected against contact insecticides; their low metabolic activity also helps to reduce the rate of pesticide uptake.

T. granarium is widespread in the Indian subcontinent and adjacent areas and in many hot dry regions around the world. It is usually not found in humid regions.

(b) *Flour beetles. Tribolium* spp., of which the most common are *T. castaneum* (Fig. S.52C) and *T. confusum*, are brown slightly flattened beetles that feed on a range of commodities, especially cereals, but also groundnuts, nuts, spices, coffee, cocoa, dried fruit and occasionally pulses. They will also feed on animal tissues, including the bodies of dead insects, and they will attack and eat small or immobile stages of living insects, especially eggs and pupae. Under conditions of overcrowding cannibalism may take place.

Under optimum conditions (33–35°C at about 70% RH) adults live for many months. Throughout their lives females lay eggs loosely among their food and the larvae feed and complete their life cycle without necessarily leaving the food commodity. Development can be very quick (about 30 days) and population growth is very rapid. Heavy infestations can produce disagreeable odours and flavours in commodities due to the production of quinones from the abdominal and thoracic defence glands of the adults.

(c) *Warehouse moth. Ephestia cautella* is a common and important secondary pest of cereals, cereal products, cocoa, dried fruit, nuts and many other commodities (Fig. S.52D). Adult *E. cautella* are fairly short-lived (from 7 to 14 days typically) and do not feed. Newly emerged adults can mate within a few hours of emergence, and eggs are laid soon afterwards (usually within a day) loosely on the surface of the seed. The larvae move extensively as they feed and, as they move about, they spin copious quantities of silk, called webbing. The webbing from heavy infestations can mat together the commodity and render it unfit for consumption. Just prior to pupation, the larvae move out of the stored commodity and wander about freely until they find a suitable site – usually in cracks, crevices, and frequently the gaps between grain bags.

5. Management strategies to reduce insect infestations

(a) *Temperature.* Most granivorous insects are tropical or subtropical species and breed over a wide range of temperatures. Generally, at about 35°C reproduction is reduced, and at higher temperatures insects die; conversely, at about 15°C insects stop laying eggs and development stops, and at lower temperatures they die. Aeration immediately after harvest in large storage silos and bins is used to cool seed and remove temperature differences. Aeration may be combined with thermal disinfestations using either heat, such as during drying after harvest, or cold, such as in subsequent storage in latitudes where winter ambient temperatures are extremely low.

Subsequent prolonged changes in outside temperature can result in moisture movement and consequent condensation in the static storage mass, leading to fungal growth, heating and spoiled grain, all of which can favour insect build-up. It is good practice to monitor for temperature gradients and unnatural odours in the stored grain mass which, as well as being symptomatic of moisture management problems, can reflect the action of secondary grain pests. Because species' responses to these changes vary, proper seed sampling or trapping and identification of insect pests are all also important.

(b) *Insecticidal techniques.* Formulations of residual insecticides are sprayed on all interior surfaces in storage structures to eliminate insects a few weeks before the grain is brought in. Grain and seed may be admixed with persistent insecticidal materials of natural or synthetic origin. Grain protectants are sprayed or dusted on to grain as it is augered into the store, particularly where relatively long-term storage is intended, between cold and warm seasons. Although even distribution of treating the product during application is desirable, treatment of every kernel may not be necessary. In 'surface treatments', used to help prevent the establishment of migrating pests in the grain mass, the protectant is applied as a top-dressing on exposed surfaces of the grain mass, by raking in to a depth of about 0.1 to 0.3 m. Though grain protectants retain their insect-toxic properties for an extended time, they have only limited penetration into infested kernels and so may not destroy some internally feeding insects immediately after exposure, or all insect life stages, such as eggs and pupae. (**See: Treatments – pesticides, 1. Introduction**)

Alternatively, or in addition, grain fumigants are used to treat storage structures, or to control pests already present inside grain. They rapidly kill all life stages, but provide no residual activity against reinfestation. Active ingredients in commonly used fumigants include dichlorvos, methoprene (a synthetic analogue of an insect growth regulator that prevents the development of larvae into adults), and synthetic pyrethroids, such as cyfluthrin, phenothrin and deltamethrin. Development of resistance to some of the commonly used insecticides is a widespread major problem.

(c) *Diatomaceous earth.* These mineral earths are soft whitish powders formed from the fossils of algae that lived in salt and fresh water, consisting of almost pure silica. Forms of quarried diatomite are ground into abrasive inert desiccant dusts for application to dry grain as it is augered into the bin. The powders are also suited for application to the walls and floors of empty stores and farm equipment (for example, headers and augers), either as a dry dust or a water-based slurry. When beetles come in contact with the powder, the waxy covering on their exoskeleton is abraded and absorbed, leaving them prone to dehydration and death over about 6 weeks. Efficacy is reduced in humid environments, where the amount of water vapour is high enough to allow insect survival. Advantages compared to conventional insecticides include very low toxicity to humans and animals; but the dusts may increase wear in grain-handling machinery due to their abrasive characteristics, and alter grain flow characteristics.

(d) *Impact killing.* Some species of stored-product insects are vulnerable to the physical impact caused by moving grain. Pneumatic grain conveyors have a tendency to subject grain kernels to large forces because they operate at such high pressures. (**See** also: **Storage management, 5. Control of insect pests**) (RH)

CSIRO Stored Grain Research Laboratory (SGRL) sgrl.csiro.au

Gorham, J.R. (1987) *Insect and Mite Pests in Food: An Illustrated Key.* Vols I and II. US Department of Agriculture, Agriculture Handbook Number 655.

Haines, C.P. (ed.) (1991) *Insect and Arachnids of Stored Products: Their Biology and Identification*, 2nd edn. Natural Resources Institute, Chatham Maritime, Kent, UK.

Hodges, R.J. (2002) Pests of durable crops – insects and arachnids. In: Golob, P., Farrell, G. and Orchard, J.E. (eds) *Crop Post-harvest: Science and Technology.* Vol. 1. *Principles and Practice.* Blackwell, Oxford, UK, pp. 94–112.

Mound, L. (ed.) (1989) *Common Insect Pests of Stored Food Products, Economic Series* No. 15, 7th edn. British Museum (Natural History), London, UK.

Storage protein

Seed storage proteins can be defined as proteins whose primary role is as a store of carbon, nitrogen and sulphur, which are utilized to support germination to a limited extent, but mainly to support early seedling growth. They vary in total amount but often account for about half of the total seed proteins. In addition, the synthesis of storage proteins in the seed is regulated by the availability of nutrients to the parent plant, notably nitrogen and sulphur, with high nutrient levels resulting in disproportionate increases. For example, the proportion of hordein storage proteins in **barley** grain increases from about 35 to 50% of the total protein when the grain nitrogen availability increases from about 1 to 3%. Mild or severe sulphur deprivation leads to a substantial or an almost complete decline in S-containing legumin synthesis in pea, due to instability of its mRNA, whereas S-poor vicilin synthesis is unaffected, or even increases.

For details on the types of seed storage proteins and their location within the seed, **see: Osborne fractions; Storage protein – synthesis**).

The major types of storage proteins present in the seeds of important crops are summarized in Table S.15. Trivial names are often used for these proteins, which may be based on the species of origin (e.g. hordein for barley, *Hordeum vulgare*).

1. Albumins

These are present in seeds of many dicot plants, including major crops such as **sunflower**, oilseed **rape/canola**, **mustard**, **lupin**, **soybean** and **peanut**, but they have not been characterized from monocots (despite an initial report in seed of *Yucca*, a member of the Liliaceae). However, the presence of related proteins in species of fern clearly demonstrates an ancient origin before the separation of flowering plants.

These albumin storage proteins typically have sedimentation coefficients, S values (a measure of the protein mass, determined by sedimentation equilibrium ultracentrifugation) of about 2 Svedberg units (hence the common name 2S albumins) with molecular weights (MW) of about 10,000–15,000. The 2S albumins are not glycosylated and polymorphism arises from the presence of small multigene families. Four genes are present in *Arabidopsis* but one of these appears to account for the bulk of the protein that accumulates. At least 11 proteins have been characterized from ***Brassica*** spp. with 10–16 genes present in *B. napus*, while sunflowers and **Brazil nut** contain 11–13 and at least six albumin forms, respectively.

The three-dimensional structures for 2S albumins from *Brassica*, **castor bean** and sunflower show that the proteins are tightly folded and comprise five α-helices arranged in a right-handed superhelix. Despite having highly conserved three-dimensional structures the individual 2S albumins vary widely in their amino acid sequences. Of particular interest in relation to their nutritional quality is the presence in some species of albumins of exceptionally high proportions of the sulphur-containing amino acids, cysteine and methionine. All six albumins in Brazil nut are rich in methionine, with 14 or 15 of these being present in their large subunits. This results in the whole albumin fraction containing about 19 mol% methionine and 8 mol% cysteine. In contrast, only two of the 11–13 albumins in sunflower seeds are rich in methionine, with the best characterized of these (SFA8) containing 16 methionine and eight cysteine residues in a mature protein molecule of 103 residues. Other methionine-rich 2S albumins are present in cottonseed and *Amaranthus*, and cysteine-rich albumins (i.e. containing more than the eight cysteines typical of 2S albumins) in quinoa, a **pseudocereal**, and **pea**.

Wider comparisons demonstrate that 2S albumins belong to a larger group of proteins called the Prolamin Superfamily. (**See: Protein superfamilies; Storage protein – synthesis**)

Shewry, P.R. and Pandya, M.J. (1999) The 2S albumin storage proteins. In: Shewry, P.R. and Casey, R. (eds) *Seed Proteins.* Kluwer Academic, Dordrecht, The Netherlands, pp. 563–587.

Table S.15. The major types and common names of storage proteins present in seeds of important crops.

Family and species	Storage tissue	Storage protein
Dicotyledoneae		
Compositae		
Sunflower	Cotyledons (embryo)	11S helianthinin
(*Helianthus annuus*)		2S albumin
Leguminosae (Fabaceae)		
Soybean	Cotyledons (embryo)	7S conglycinin
(*Glycine max*)		11S glycinin
Broadbean (faba)	Cotyledons (embryo)	7S vicilin
(*Vicia faba*)		11S legumin
Pea	Cotyledons (embryo)	7S vicilin/convicilin
(*Pisum sativum*)		11S pea legumin
		2S albumin
Common, French or kidney bean (*Phaseolus vulgaris*)	Cotyledons (embryo)	7S phaseolin
Cruciferae		
Oilseed rape	Cotyledons (embryo)	11S cruciferin
(*Brassica napus*)		2S albumin (napin)
Malvaceae		
Cotton	Cotyledons (embryo)	11S gossypin
(*Gossypium hirsutum*)		7S congossypin
Monocotyledoneae		
Palmae		
Oil palm	Endosperm	7S globulin
(*Elais guineensis*)		
Gramineae (Poaceae)		
Wheat	Starchy endosperm	Prolamin (gliadin + glutenin)
(*Triticum aestivum*)	Embryo/aleurone layer	7S globulin
Barley	Starchy endosperm	Prolamin (hordein)
(*Hordeum vulgare*)	Embryo/aleurone layer	7S globulin
Oat	Starchy endosperm	Prolamin (avenin)
(*Avena sativa*)		11S globulin
	Embryo/aleurone layer	7S globulin
Rice	Starchy endosperm	11S globulin (glutelin)
(*Oryza sativa*)		Prolamin (oryzenin)
	Embryo	7S globulin
Maize	Starchy endosperm	Prolamin (zein)
(*Zea mays*)	Embryo	7S globulin

Taken from Shewry, P.R. (2000) Seed proteins. In: Black, M. and Bewley, J.D. (eds) *Seed Technology and its Biological Basis.* Sheffield Academic Press, Sheffield, UK, pp. 42–84.

2. Globulins

Globulins with sedimentation coefficients of 11–12 (11–12S) are widely distributed in seeds of monocot and dicot species and may ultimately prove to be universal in flowering plants. In addition, many species also contain 7–8S globulins. These two types of storage proteins are often called simply 7S and 11S globulins, or legumins and vicilins based on their characteristic occurrence in legumes and the legume tribe Viciae, respectively. Despite having little similarity in their amino acid sequences, they have similar structures, being based on subunits assembled to form trimers and hexamers.

The 7S globulin subunits typically have MWs (Mr) of about 40,000–60,000 with the mature trimeric protein having a MW of about 150,000–190,000. These subunits lack cysteine residues and hence the trimeric structure is stabilized by non-covalent interactions between the proteins. Despite this essentially simple structure the 7S globulins are highly polymorphic within and between species, the polymorphism arising in two ways. First, 7S globulins are encoded by small multigene families, leading to the presence of a population of 7S globulins comprising different combinations of subunits. For example, soybean β-conglycinin consists of at least seven isoforms based on combinations of three subunits: α, α′ and β. Secondly, in some species the 7S globulin subunits may undergo post-translational modification, either proteolysis and/or glycosylation. For example, in **pea**, proteolytic processing of Mr 50,000 subunits at one or two sites leads to polypeptide chains ranging in Mr from about 12,000 to the Mr 50,000 unprocessed subunits, with glycosylation occurring at a single position on some subunits only. In contrast, the 7S globulin subunits of *Phaseolus* (bean) are not proteolytically processed and glycosylation may take place at one or two sites. The processed polypeptides remain associated in subunits and are released when the 7S globulins are denatured with chaotropic solvents or detergents (**see: Storage protein – synthesis**).

11S globulins are assembled from subunits which typically have MWs of about 60,000. Proteolytic processing always occurs at a single site to give acidic or α- and basic or β-polypeptide chains that typically have masses of about 40,000 and 20,000, respectively. However, this proteolytic processing only occurs after the formation of a single inter-chain disulphide bond between cysteine residues in the basic and acidic chains. Glycosylation of 11S subunits rarely, if ever occurs. The 11S globulin subunits are initially assembled into trimers and then, after proteolytic processing, into mature hexamers with MWs of about 320,000–450,000. Polymorphism of 11S globulins again arises from the presence of multigene families.

The acidic and basic chains of the 11S globulin subunits appear to be related to the N- and C-terminal parts, respectively, of the 7S globulin subunits. Furthermore, the N- and C-terminal parts of each subunit form similar structures, indicating that they have evolved from a single ancestral protein which duplicated to form two domains corresponding to the 11S acidic chain/7S N-terminus and 11S basic chain/7S C-terminus (Fig. S.53).

The three-dimensional structures determined for several trimeric forms of 7S and 11S globulins show that the two domains of each subunit form core β-barrel structures with α-helical regions forming loops (Fig. S.54). The mature 7S globulin trimer has dimensions of about 80–90Å × 80–90Å × 35–40Å while the 11S globulin hexamer appears to be a 'dimer' of the two 'intermediate trimers'. In fact, the structures of the subunits and trimeric forms of the 7S and 11S proteins are so similar that they can be readily overlaid.

Other proteins from plants and microbes have similar three-dimensional structures, notably the presence of a core β-barrel structure. These include germins/oxalate oxidase and **auxin** (hormone)-binding binding proteins from plants. The name 'cupin' has been coined for this novel **protein superfamily** based on the Latin *cupa*, which means small barrel or cask.

Casey, R. (1999) Distribution and some properties of seed globulins. In: Shewry, P.R. and Casey, R. (eds) *Seed Proteins.* Kluwer Academic, Dordrecht, The Netherlands, pp.159–169.

3. Prolamins

Unlike the 2S albumins, 7S and 11S globulins, the importance of the prolamins relates not to their wide distribution (they are only present in seeds of grass species, e.g. cereals), but to the fact that cereals are the major crops harvested and utilized by humankind. Consequently, they have immense importance in food processing and in human and livestock nutrition.

The prolamins show extensive variation in their structure and properties but the following discussion will be confined to only two species, **wheat** and **maize**, which are typical of temperate small-grain cereals (which include **rye** and **barley**) and tropical panicoid cereals (including **millets** and **sorghum**), respectively.

Wheat prolamins comprise a mixture of at least 50 individual proteins that vary in MW from about 30,000 to 90,000. They are classically divided into two groups, the monomeric prolamins and polymeric glutenin subunits, which are subdivided further on the basis of electrophoretic mobility at low pH for gliadins (α-, β- γ- and ω-gliadins) and on the basis of subunit MW (high and low MW subunits) for glutenins. This classification is still widely used by cereal chemists mainly because the gliadins and glutenins have different functional properties in relation to the processing of wheat. However, comparisons of amino acid sequences demonstrate that it is more valid to classify wheat prolamins into three groups: high molecular weight (HMW), sulphur-rich (S-rich) and sulphur-poor (S-poor) prolamins.

A schematic comparison of the structures of typical HMW (HMW subunit 1Dx5), S-rich (γ-type gliadin) and S-poor (ω-gliadin) prolamins is shown in Fig. S.55. All have distinct domain structures, with one domain comprising repeated sequences based on one or more short peptide motifs. In the S-poor ω-gliadins these sequences comprise the majority of the protein with only short non-repetitive sequences at the N- and C-termini. In addition, the proteins lack cysteine residues and hence cannot form disulphide bonds.

The repetitive sequences present in the S-rich γ-gliadin are based on a similar consensus motif to those in the ω-gliadins, indicating a common origin for these two groups of protein. However, an extensive non-repetitive C-terminal domain is also present and contains eight cysteine residues that form four intra-chain disulphide bonds. Six of these cysteine residues are also present in other groups of S-rich prolamins, the α-gliadins and LMW subunits of glutenin, where they form the same pattern of disulphide bonds. The latter also

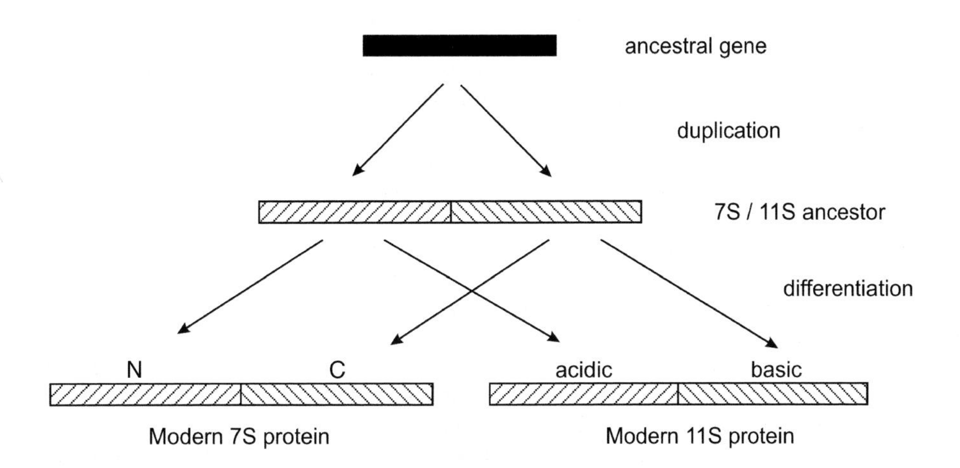

Fig. S.53. Schematic summary of the evolutionary relationships of 7S and 11S globulin subunits.

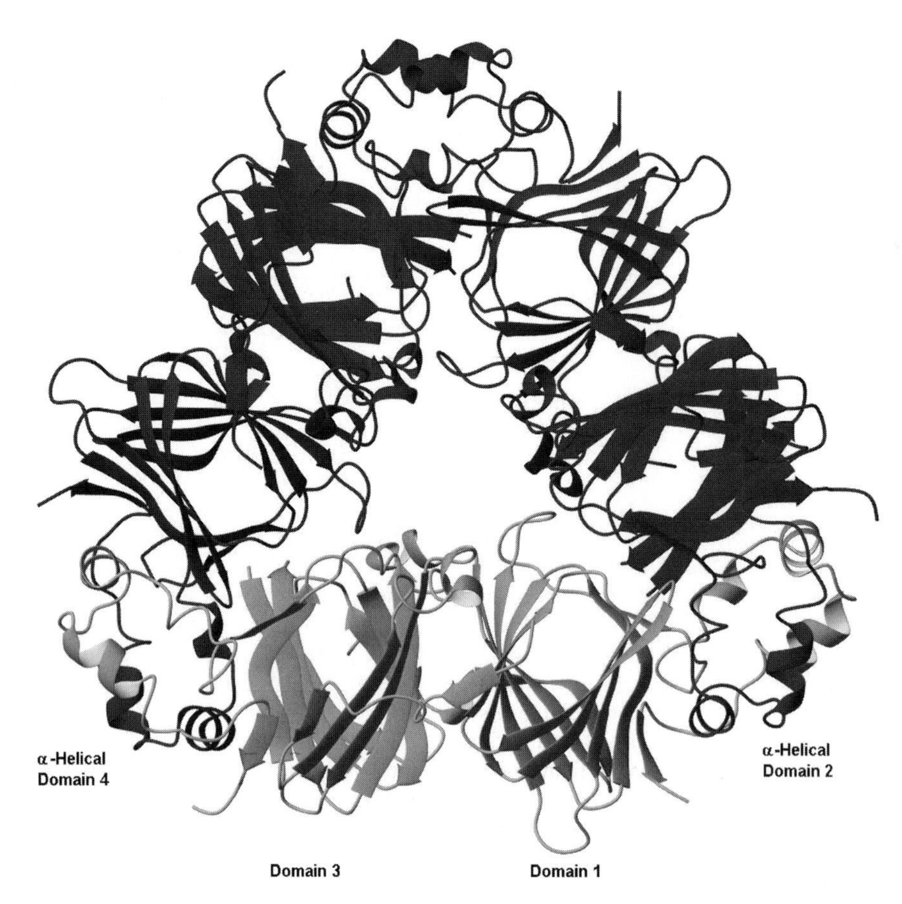

Fig. S.54. Three-dimensional structure of the polypeptide backbone of the trimeric 7S globulin form of storage protein.

contain one or two additional 'unpaired' cysteine residues that allow their incorporation into glutenin polymers by inter-chain disulphide bonds. The repeats present in the polymeric HMW subunits of glutenin are not related to those in the S-rich and S-poor prolamins, although the sequences of the N- and C-terminal non-repetitive domains of the HMW subunits are related to those present in the C-terminal domains of S-rich prolamins. The N- and C-terminal domains of the HMW subunits also contain up to six cysteine residues, which may form inter- and intra-chain disulphide bonds.

The presence of extensive repetitive sequences strongly influences the amino acid compositions of the whole proteins, with the ω-gliadins being rich in glutamine (40–50 mol%), proline (20–30 mol%) and phenylalanine (8–9 mol%), the S-rich prolamins in glutamine (30–40 mol%) and proline (15–20 mol%) and the HMW prolamins in glutamine (10–15 mol%) and glycine (15–20 mol%).

Wheat prolamins contain non-repetitive and repetitive domains that have quite different structures. The non-repetitive domains have compact globular structures and are rich in α-helix while the repetitive domains form extended structures, which comprise loose spirals, based on repeated β-reverse turns. These 'β-spirals' are similar to structures formed by fibrous and elastic proteins from animal tissues and may contribute to the elastomeric properties of wheat gluten.

The prolamins of maize, the zeins, are classified into four groups based on their apparent MWs shown by sodium dodecylsulphate polyacrylamide gel electrophoresis (SDS-PAGE) (Fig. S.56A) and their amino acid sequences. The major group, the α-zeins, is separated into two broad bands with apparent MWs of about 19,000 (Z19) and 21,000 (Z21) but their

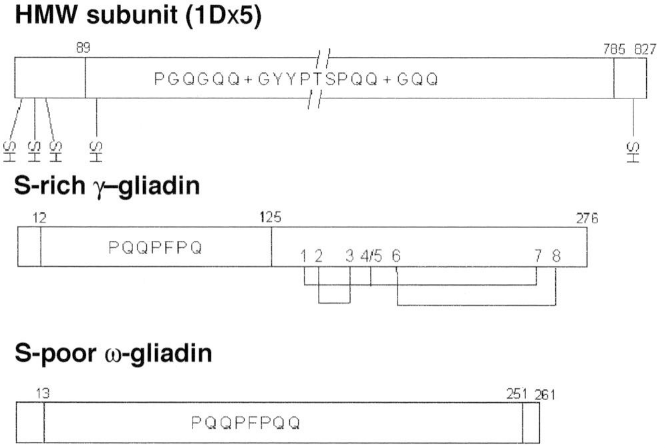

HMW subunit (1Dx5)

89 785 827

PGQGQQ + GYYPTSPQQ + GQQ

SH SH SH SH SH

S-rich γ-gliadin

12 125 276

PQQPFPQ

1 2 3 4/5 6 7 8

S-poor ω-gliadin

13 251 261

PQQPFPQQ

Fig. S.55. Schematic comparison of the amino acid sequences of typical HMW (HMW subunit 1Dx5), S-rich (γ-gliadin) and S-poor (ω-gliadin) prolamins of wheat. Note the presence of repeated sequences based on one or more short peptide motifs. Standard single letter abbreviations for amino acid are used: F, phenylalanine; G, glycine; P, proline; Q, glutamine; S serine; T, threonine; Y, tyrosine. Numbers 1–8 indicate the location of cysteine amino acids, which are linked by disulphide bonds. The numbers on top of the proteins indicate the locations of certain amino acids, at the C-terminus or delineating repeat sequences.

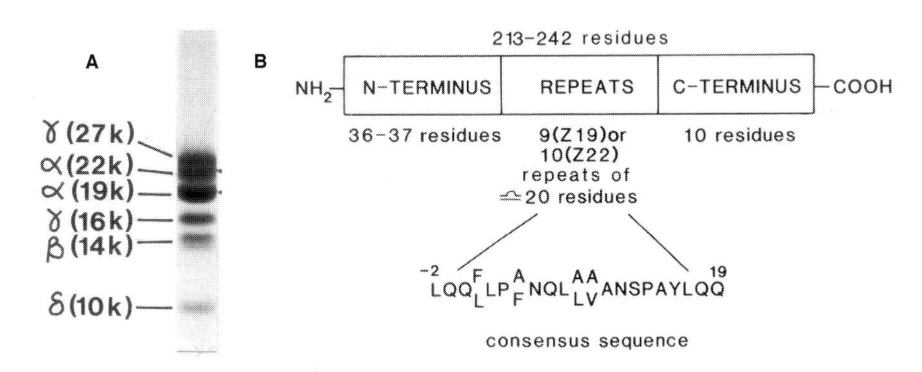

Fig. S.56. (A) SDS-PAGE of total prolamins (zeins) from maize showing the α-, β-, γ- and δ- groups of proteins. (B) Schematic summary of the sequence of α-zein showing the presence of highly degenerate repeats. Standard single letter abbreviations for amino acid are used: A, alanine; F, phenylalanine; L, leucine; N, asparagine; P, proline; Q, glutamine; S, serine; V, valine; Y, tyrosine.

true MWs are about 23,000–24,000 and 26,500–27,000, respectively. Each band contains a number of components that can be separated by isoelectric focusing and are encoded by a large multigene family (over 70–100 copies), which includes pseudogenes (genes in which the DNA has become altered slightly so that they are not expressed) as well as expressed genes. The proteins contain 9 (Z19) or 10 (Z21) degenerated repeats of about 20 amino acid residues (Fig. S.56B). However, these repeats are considered to from α-helical rather than β-spiral structures.

The α-zeins contain 0 or 1 cysteine residues and so form either monomers or dimers that are readily extracted in aqueous alcohols. In contrast, the β, γ and δ zeins all form alcohol-insoluble polymers. The γ-zeins also contain repeated sequences, either eight copies of a conserved hexapeptide (Pro Pro Pro Val) His Leu (28,000 MW γ-zein) or two copies of a degenerate form of this peptide (16,000 MW γ-zein). They also contain a short region of alternating proline residues called the Pro-X region. The reduced γ-zein subunits are water-soluble, unlike other prolamins. The β-zeins and δ-zeins contain about 11 mol% and 22 mol% methionine, respectively, these residues being concentrated in a specific region in β-zeins and clustered towards the centre of the δ-zeins, often as Met. Met doublets. The β-, γ- and δ-zeins are encoded by small gene families, probably comprising one, two to three and two genes, respectively. (PRS)

Coleman, C.E. and Larkins, B.A. (1999) The prolamins of maize. In: Shewry, P.R. and Casey, R. (eds) *Seed Proteins*. Kluwer Academic, Dordrecht, The Netherlands, pp. 109–139.

Shewry, P.R., Tatham, A.S. and Halford, N.G. (1999) The prolamins of the Tritceae. In: Shewry, P.R. and Casey, R. (eds) *Seed Proteins*. Kluwer Academic, Dordrecht, The Netherlands, pp. 35–78.

Storage protein – intolerance and allergies

The most widespread adverse human reaction to the consumption of seed protein is coeliac disease, also called 'gluten-sensitive enteropathy', which has been estimated to affect up to 0.5% of the population in Western Europe.

Coeliac disease is an intolerance rather than an allergic response (which is mediated by the antibody IgE, as discussed below) and is characterized by changes in the small intestine (duodenum and jejunum), often leading to loss of villi, flattening of the mucosa and hyper-proliferation of enterocytes in the crypts, resulting in malabsorption of nutrients and diarrhoea. However, the symptoms are highly variable in magnitude, meaning that the condition may not be diagnosed. The pathogenesis is incompletely understood, but is considered to be a T cell-mediated auto-immune response triggered by the ingestion of **wheat** (or related) proteins.

Dietary challenges show that withdrawal of wheat, **barley** and **rye** products results in a rapid recovery, including the development of a normal mucosal structure in the small intestine. Studies with cultured tissues obtained by biopsy and direct instillation of peptides into the gut of patients, have demonstrated that a short peptide corresponding to residues 31 to 43 of A (α-)-gliadins is particularly active. This has the sequence: leucine, glycine, glutamine, glutamine, glutamine, proline, and phenylalanine, proline, proline, glutamine, glutamine, proline, tyrosine. However, it is probable that other peptides with related sequences are also toxic, since all types of gliadins and related proteins of barley and rye also appear to be capable of triggering an adverse response. All types of wheat (including pasta wheat, **spelt**, **Kamut**, Polish wheat, and **einkorn**) should be considered as toxic.

A second type of intolerance associated with ingestion of wheat gluten is dermatitis herpetiformis, in which a skin rash is associated with granular deposits of IgA antibody in uninvolved skin.

Allergic, also called hypersensitive, reactions are the consequence of an inappropriate immune response to a foreign protein, resulting in the generation of a type of antibody called IgE. An individual can be exposed in various ways, notably by inhalation, skin contact, or ingestation. Seed proteins may result in allergic responses by all of these routes, but they are principally food allergens, and are responsible for four of the seven major food-related allergies: **cereals**, **peanut**, soybean and tree nuts.

Although the range of plant species that can result in allergic responses is wide, with little apparent relationship to the taxonomy of the species themselves, the proteins responsible fall into a small number of types. Many of them are small sulphur-rich seed proteins of the prolamin superfamily, notably 2S albumin storage proteins (which are dietary allergens in mustards, **Brazil nut**, **castor bean**, **sunflower**, **walnut**, **sesame**, **peanut** and **cotton**), non-specific lipid transfer proteins (dietary allergens in **maize**, barley and sunflower seeds), cereal α-amylase/trypsin inhibitors (dietary allergens in rice and wheat, respiratory allergens in baker's asthma to wheat) and soybean hydrophobic protein (respiratory allergy to soybean hulls). The major 7S and 11S storage globulins are also dietary allergens in soybean, peanut and walnut (a non-legume species) but not apparently in other species. In general the prolamin storage proteins of cereal grain are not major allergens, but wheat gluten has been associated with atopic dermatitis and exercise-induced anaphylaxis.

Other seed protein allergens are thiol and Kunitz trypsin inhibitor in soybean, **lectin** in peanut and peroxidases in cereals (wheat and barley).

Allergenic responses to soybean, peanut and hazelnut may also result from respiratory exposure to related proteins present in pollen, either profilins (an actin-binding protein, called Bet v 2 in birch pollen) or Bet v 1 (a major birch pollen antigen related to a group of pathogenesis-related proteins).

There is considerable interest in predicting the potential allergenicity of proteins present in novel (including genetically engineered) foods, as well as the removal of allergenic proteins or epitopes by genetic manipulation (e.g. hypoallergenic rice with lower α-amylase/trypsin inhibitor content), breeding or food processing (heating, protease treatment). (PRS)
(See: Allergen; Amylase inhibitors; Cereals – composition and nutritional quality; Lectins; Legumes; Phytohaemagglutinin; Protease inhibitors; Soybean lectin; Thionins)

Kasarda, D.D. (2001) Grains in relation to celiac disease. *Cereal Foods World* 46, 209–210.
Mills, E.N.C., Madsen, C., Shewry, P.R. and Wichers, H.J. (2000) Plant food protein allergens-the role of structure and function in allergenic potential. *Food Allergy and Intolerance* 2, 194–209.

Storage protein – mobilization

Proteolysis of storage proteins to amino acids is a multistep process that ultimately supplies the growing seedling with required nutrients, especially amino acids, until it can function as a photosynthetic autotrophe. The **proteases** present in germinating seeds and young seedlings have been studied primarily using exogenous, typically animal, protein substrates, although there are studies on the molecular changes to seed storage proteins during their mobilization, and on the action of purified seed proteases on endogenous storage proteins. In any given species of seed, a suite of proteases belonging to different mechanistic classes, and with different specificities, is involved. Furthermore, while some of these proteases are present already in the dry quiescent seed, others appear only after germination or growth have commenced. **(See: Mobilization of reserves – cereals; Mobilization of reserves – dicots)**

Some of the most complete studies have been carried out in legume seeds, including vetch (*Vicia sativa*), **soybean** (*Glycine max*), **mung bean** (*Vigna radiata*), and garden, kidney, French or common **bean** (*Phaseolus vulgaris*). In these seeds (as in many other dicots) the main protein reserves are stored in the membrane-bound **protein bodies** of the parenchyma cells of the cotyledons. Upon imbibition, the protein bodies are converted to lytic vacuoles, where hydrolysis of the storage proteins takes place. The mobilization (hydrolysis, degradation) of each storage protein must be initiated by a specific endopeptidase before it can be further degraded by other proteases (Fig. S.57). This initial hydrolysis is limited, cleaving a single or a limited small number of peptide bonds located at exposed surface positions on the storage protein molecule. The result is one or more specific, but transient, intermediates. This limited proteolysis is necessary for a conformational change in the storage protein, which now becomes susceptible to further hydrolysis by other less specific proteases, and possibly further hydrolysis by the initiating endopeptidase.

In a particular species of seed the identity of the initiating endopeptidase can vary with the type of storage protein, or indeed with the particular subunit of a multi-subunit protein. This is well illustrated in the soybean seed. Mobilization of the two larger α′ and α subunits of β-conglycinin, the soybean vicilin, is initiated by a serine endopeptidase, Protease C1. However, this enzyme has no affect on the smaller β subunit of β-conglycinin, whose hydrolysis is initiated by a cysteine endopeptidase, Protease C2, which also further degrades the products resulting from the action of Protease C1 on the α′ and α subunits. Degradation of the soybean Bowman-Birk and Kunitz type trypsin inhibitors (located in the same protein bodies as β-conglycinin) is initiated by another cysteine endopeptidase, Protease K1. **(See: Protease inhibitors)**

The mobilization of homologous proteins in different legume species often is initiated by homologous proteases. For example, the vicilins of vetch, garden bean and black gram (*Vigna mungo*), all of which contain β-subunits which resemble those of soybean β-conglycinin, are initially degraded by papain-like cysteine endopeptidases homologous to soybean Protease C2. There are seeds of legumes, however, in which the initiating protease may be quite different. While the degradation of the soybean Bowman-Birk trypsin inhibitor is initiated by the aforementioned cysteine endopeptidase Protease K1, the homologous mung bean Bowman-Birk type trypsin inhibitor is first hydrolysed by a serine endopeptidase, Proteinase F.

Following initial hydrolysis by the combined endopeptidase reactions, the products are further degraded by one or more serine carboxypeptidases, which are common components in the protein body/lytic vacuoles of legumes. The combined action of the multiple cysteine and serine endopeptidases and the serine carboxypeptidases on the storage proteins results in a mixture of free amino acids and small peptides. These are then transported (by an undefined transport system) out of the lytic vacuole (now forming due to the fusion of the emptying protein bodies) into the cytosol. The peptides are presumably further degraded to amino acids by the aminopeptidases and dipeptidases located in the cytosol (Fig. S.57). Finally, the amino acids are either used *in situ*, or exported via the vascular system to other parts of the seedling. While this scheme has been largely derived from work on legume seeds, it seems likely that it applies to dicots in general, as suggested by similar processes in the degradation of pumpkin (*Curcubita moschata*) seed storage globulins.

In the majority of monocot seeds that have been studied, most of the storage protein is present in the starchy **endosperm** (a non-living triploid tissue), with smaller

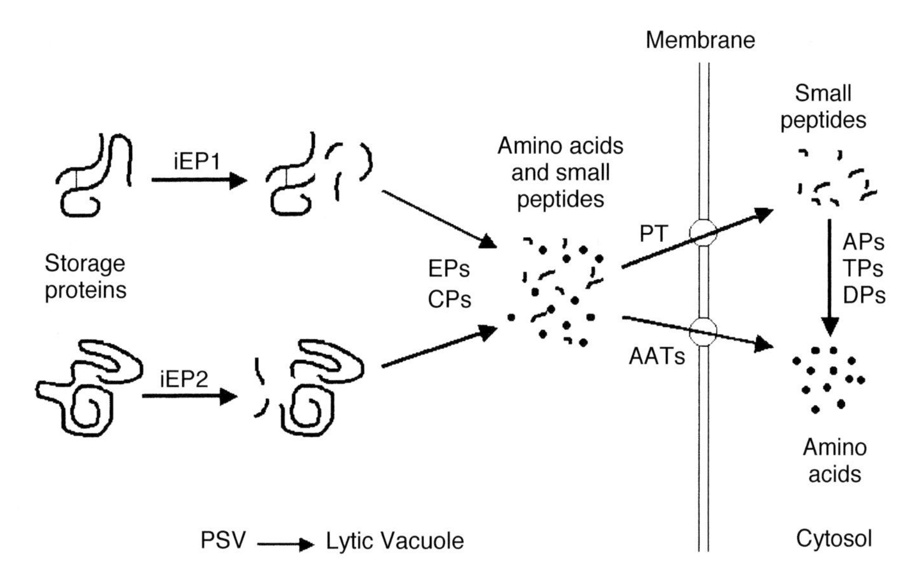

Fig. S.57. Generalized pathway for the degradation of storage proteins contained in protein storage vacuoles (PSV, protein bodies). After imbibition of the seed, the PSV becomes a lytic vacuole. For each storage protein the degradation is begun by an initiating endopeptidase (iEP1, iEP2) which inflicts a limited number of relatively specific cleavages upon the storage protein molecule. The partially degraded storage protein then becomes susceptible to profound degradation to amino acids and small peptides through the action of other endopeptidases (EPs, possibly including the iEP) and serine carboxypeptidases (CPs). The amino acids and small peptides are transported across the vacuolar membrane by amino acid (AATs) and peptide (PT) transporters in the membrane. Degradation of the small peptides is completed in the cytosol by a suite of aminopeptidases (APs), dipeptidases (DPs) and tripeptidases (TPs).

amounts in the embryo and the **aleurone layer**. Mobilization in the monocot seed has been especially well studied in barley (*Hordeum vulgare*). Upon imbibition of the seed, proteins stored in the protein bodies in the aleurone layer and in the embryo are first mobilized. This is accomplished in part by proteases already present in the protein bodies in the dry quiescent seed, most notably an aspartic endopeptidase, and also by proteases that appear during and after germination (especially the cysteine endopeptidases). Protein mobilization in the endosperm occurs somewhat later, with large increases in proteolytic activity occurring about day 3 after the start of imbibition. Most or all of this proteolysis is due to the synthesis and secretion of proteases by the aleurone layer, which also secretes malic acid, acidifying the endosperm to approximately pH 5, thus optimizing the conditions for protease activity. The combined action of endopeptidases and the abundant serine carboxypeptidases in the endosperm results in the conversion of the storage proteins to a mixture of free amino acids and short **oligopeptides** that cannot be further degraded in this tissue. The products are taken up by the growing seedling through the scutellum using amino acid and **peptide transporters** located in the plasma membrane of the scutellar epidermis. The peptides are subsequently hydrolysed to free amino acids by peptidases in the scutellum. (**See: Mobilization of reserves – a monocot overview; Mobilization of reserves – cereals**)

Examination of the barley seed protease complement by two-dimensional gel electrophoresis has identified 42 electrophoretically distinct proteases, composed of four aspartic endopeptidases, 27 cysteine endopeptidases, eight serine endopeptidases, and three metalloendopeptidases.

Proteolysis in the barley endosperm may also be aided by reduction of the storage proteins by **thioredoxin**, although this appears to be less important in this species compared to some other cereals such as wheat (*Triticum durum*).

For controlled accumulation and subsequent mobilization of storage proteins in seeds to occur the appearance and action of proteases must be closely regulated. In some seeds, storage proteins accumulate during seed development in the same compartment, the protein body, as a protease that will hydrolyse them during or following germination. This might occur because conditions within the protein bodies of the developing seed are not suitable for protease activity, e.g. because the pH is very different from that necessary for enzyme activity. The accumulation of the Bowman-Birk type trypsin inhibitor and Proteinase F in the same mung bean protein bodies may be explained this way. Proteinase F can hydrolyse the trypsin inhibitor over a narrow range of acidic pHs. During seed development, the pH of the protein bodies is too high for the Proteinase F to be active, but following maturation and germination proton pumps in the protein body membrane acidify the contents, allowing proteolysis. In other instances, the mature storage protein may be stored with a protease that cannot hydrolyse it unless the enzyme is first modified. For example, storage proteins are laid down along with carboxypeptidases in the protein bodies of many species (e.g. mung bean, barley) but they are only rendered susceptible to these enzymes after their initial hydrolysis by the endopeptidases that appear during and after germination. Another scenario is the joint accumulation of the storage protein and a catalytically inactive protease/protease inhibitor complex during seed development. A metalloendopeptidase that hydrolyses the major storage protein of buckwheat (*Fagopyrum esculentum*) accumulates in the protein body as an inactive complex with an inhibitor protein. This inhibitor apparently acts by binding to Zn^{2+} in the active site of the metalloendopeptidase. The inhibition is lifted following germination by the release of zinc ions (due to the hydrolysis of phytic acid salts), which bind to the inhibitor and inactivate it.

Hormonal signals from the cereal grain embryo play an important role in regulating proteolysis in the endosperm. **Gibberellin** released by the barley embryo following germination triggers the synthesis and release of proteases, in particular cysteine endopeptidases, from the aleurone layer, as well as stimulating acidification of the endosperm. Hormonal signals from the embryonic axis may play a similar role in regulating protease synthesis in some dicot seeds since detachment of the cotyledon from the axis disrupts the appearance of cysteine endopeptidases. There are reports of the normal pattern of protease appearance being re-established by the application of phytohormones to detached cotyledons, but these reports are not consistent with respect to the hormone involved, with gibberellic acid acting in some species (such as pea, *Pisum sativum*; and castor bean, *Ricinus communis*), cytokinins in other species (e.g. garden bean, *Phaseolus vulgaris*; lemon, *Citrus limon*; and squash, *Cucurbita maxima*); and ethylene in another (chickpea, *Cicer arietinum*). Regulation may depend upon a combination of these hormones, and the increase in proteases induced by the phytohormones is often much smaller than that occurring in the presence of the axis. (KAW)

(See: Peptide transport; Protease; Protease inhibitors)

Müntz, K. (1996) Proteases and proteolytic cleavage of storage proteins in developing and germinating dicotyledonous seeds. *Journal of Experimental Botany* 47, 605–622.

Shutov, A.D. and Vaintraub, I.A. (1987) Degradation of storage proteins in germinating seeds. *Phytochemistry* 26, 1557–1566.

Wilson, K.A. (1986) Role of proteolytic enzymes in the mobilization of protein reserves in the germinating dicot seed. In: Dalling, M.J. (ed.) *Plant Proteolytic Enzymes*, Vol. II. CRC Press, Boca Raton, Florida, USA, pp. 19–47.

Zhang, N. and Jones, B.L. (1995) Development of proteolytic activities during barley malting and their localization in the green malt kernel. *Journal of Cereal Science* 22, 147–155.

Storage protein – processing for food

The nutritional quality of seed proteins is important for animal feed and for human nutrition in some parts of the world where plant products account for a high proportion of the diet, but seeds are rarely consumed as a source of protein in developed countries. In these latter countries the major impact of seed proteins on the human diets is to confer specific functional properties that are required for the production of food. These include water absorption and binding, viscosity, foaming and emulsification, gelation, cohesion and adhesion, elasticity, and the absorption and binding of fats and flavours. Two groups of seed proteins are particularly important in this respect, the gluten proteins of **wheat**, and the 7S/11S globulin storage proteins of **soybean**.

The wheat gluten proteins correspond to the prolamin storage proteins, which are initially deposited in protein bodies in the starchy endosperm cells of the developing grain. These protein bodies coalesce during the later stages of grain maturation to give a continuous matrix in the mature starchy endosperm cells. On milling of the grain the starchy endosperm cells give white flour that can be hydrated and kneaded to give dough. When this occurs the gluten protein matrix present in the individual flour particles is brought together to give a continuous network in the dough. This network provides cohesive properties to the dough, but more importantly, it is also visco-elastic. These properties allow the dough to be formed into noodles or pasta but also allow the production of unleavened and leavened breads. In particular, the network is expanded during dough fermentation to give a light porous crumb structure, which is characteristic of leavened pan and hearth-baked breads. The precise balance of viscosity and elasticity varies with genotype and environment, which in turn affects the suitability of the dough for different end uses. In particular, highly elastic (strong) doughs are required for making bread, but weaker doughs for cakes and biscuits (cookies). Gluten can also be separated from wheat flour on an industrial scale and used to fortify weak flours or as an ingredient in food products. Much work has been directed towards understanding the molecular basis for gluten visco-elasticity (**see: Storage protein, 3. Prolamins**).

Whereas wheat proteins are usually used for food processing in the form of whole flour, soybean proteins are usually extracted and used as isolates. The value of soybean proteins is in conferring a wide range of properties to food (including most of those listed above), and also in textured form as meat substitutes. Also, of particular importance in Asia is their ability to form gels (i.e. tofu). Gelation is a phenomenon in which the protein forms an ordered network capable of holding water; gelation properties vary between different types of 7S and 11S globulins.

The important functional properties of wheat gluten and soybean proteins mean that they are remarkably pervasive in food systems, and it may be difficult to identify processed foods which do not contain one or both of them.

Kinsella, J.E. (1979) Functional properties of soy proteins. *Journal of the American Oil Chemists Society* 56, 242–258.

Shewry, P.R. (2000) Seed proteins. In: Black, M. and Bewley, J.D. (eds) *Seed Technology and its Biological Basis*. Sheffield Academic Press, Sheffield, UK, pp. 42–84.

Storage protein – synthesis

Seed storage proteins are deposited in the cell in discrete **protein bodies** (protein storage vacuoles) that originate from the endomembrane system of the cell. This system comprises the **endoplasmic reticulum** (ER) and structures derived from it, including the Golgi apparatus and various types of vesicle that transport components to specific cellular destinations (Fig. S.58) (**see: Cells and cell components**). This includes components that are secreted from the cell by fusion with the plasma membrane of vesicles derived from the Golgi apparatus. Consequently, the endomembrane system is sometimes called the secretory system and the proteins which are trafficked via the endomembrane system are called secretory proteins. They include the major seed **storage proteins** (prolamins, 7S/11S globulins, 2S albumins; **see: Osborne fractions**).

Seed storage proteins are synthesized from their mRNA templates on ribosomes attached to the endoplasmic reticulum

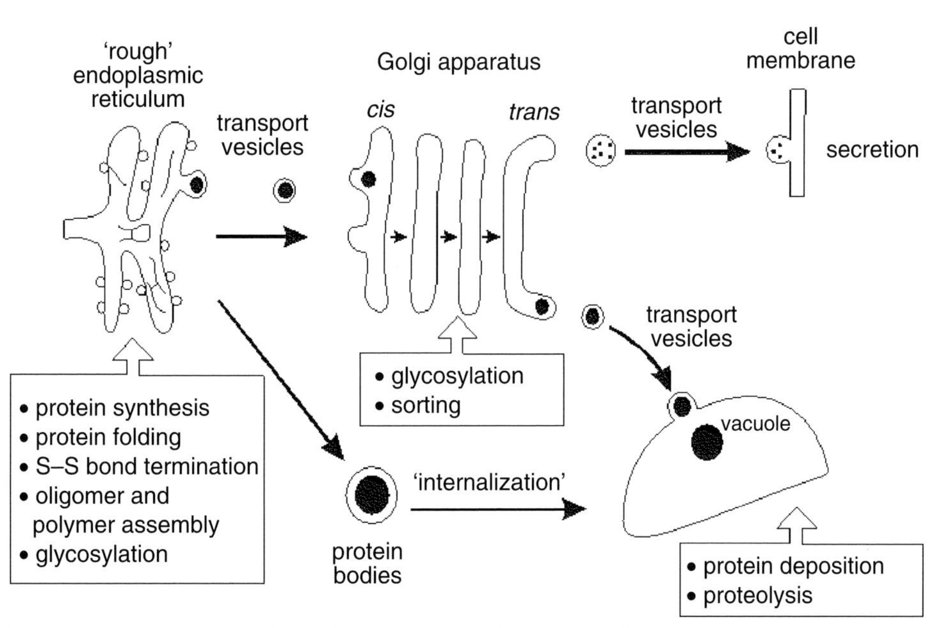

Fig. S.58. Summary of the major events occurring during the trafficking and deposition of seed storage proteins. Trimming of glycans on glycosylation proteins may also occur in the vacuole.

(**polysomes**). The new synthesized polypeptide differs from that deposited in the protein body by the presence of a short (approximately 20 amino acids) N-terminal signal sequence, which interacts with a complex of proteins in the ER membrane (the translocon) to lead the polypeptide through the membrane into the ER lumen, thus entering the endomembrane system. Once the polypeptide enters into the lumen the signal peptide is proteolytically removed by a specific signal peptidase enzyme. The nascent polypeptide also starts to fold into its mature three-dimensional structure. These three events, translocation, signal peptide cleavage and folding, all occur while the C-terminal part of the protein is still undergoing synthesis on the ribosome and hence are termed 'co-translational'. Protein folding is assisted by a complex of proteins and other factors, including the 'molecular chaperone' BiP (binding protein). These ensure that the protein attains the correct final structure (**see: Proteins and amino acids**). An integral part of the folding of proteins is the formation of disulphide bonds, either between cysteine residues in the same subunit (intra-chain) or, in the case of multisubunit proteins, between subunits (inter-chain). This may be assisted by a specific enzyme, protein disulphide isomerase (PDI), which catalyses the formation, exchange or reduction of disulphide bonds.

Albumin and globulin storage proteins are transported via the ER lumen and Golgi apparatus to specialized storage vacuoles where they form dense deposits. These vacuoles may subsequently divide to form protein bodies. However, the three types of protein differ in the modifications that occur during trafficking and deposition.

The 7S globulins do not contain disulphide bonds but their folding, assembly into trimers, and glycosylation (the addition of specific sugar residues to a protein) may all occur in the ER lumen. Not all 7S globulin subunits are glycosylated and those that are may be glycosylated on one or two asparagine residues (*N*-glycosylation). For example, one asparagine residue is usually glycosylated in phaseolin (the 7S globulin of *Phaseolus*) with a second site being less commonly glycosylated, while a single asparagine residue is glycosylated in some but not all vicilin subunits of pea. Furthermore, the glycosylation may occur in several stages, with a high mannose sugar residue (i.e. containing two *N*-acetylglucosamines, nine mannoses and three glucoses) being attached co-translationally in the ER and subsequently modified by enzymes in the ER lumen and Golgi apparatus, to give a more complex structure, and finally trimmed in the vacuole. The biological significance of these complex processes is not known but they may promote correct folding and assembly leading to greater protein stability.

The 7S globulin subunits may also be proteolytically processed in the vacuole in some species, e.g. proteolysis at one or two sites occurs in some 7S globulin subunits of pea, but not in others, e.g. *Phaseolus*.

The 11S globulins are not glycosylated but the subunits form a single intra-chain disulphide bond and assemble into 'intermediate' trimers within the ER lumen. Proteolysis at a single asparagine residue occurs within the vacuole to release the acidic and basic chains which remain associated by the single disulphide bond that is formed within the ER. This proteolytic processing appears to be required for the final assembly of the intermediate trimers into the mature hexamers.

The 2S albumins are synthesized as single polypeptide chains. Folding and formation of four intra-chain disulphide bonds occur in the ER and the protein is then transported to the vacuole without any further modification. Once within the vacuole, proteolysis occurs to release large and small subunits, which may involve the loss of short peptides from the N- and C-termini and from between the large and small subunit chains (a linker peptide). The four disulphide bonds present in the precursor protein form two intra-chain bonds in the large subunit and two inter-chain bonds. The processing of a typical 2S albumin (napin of oilseed rape) is summarized in Fig. S.59.

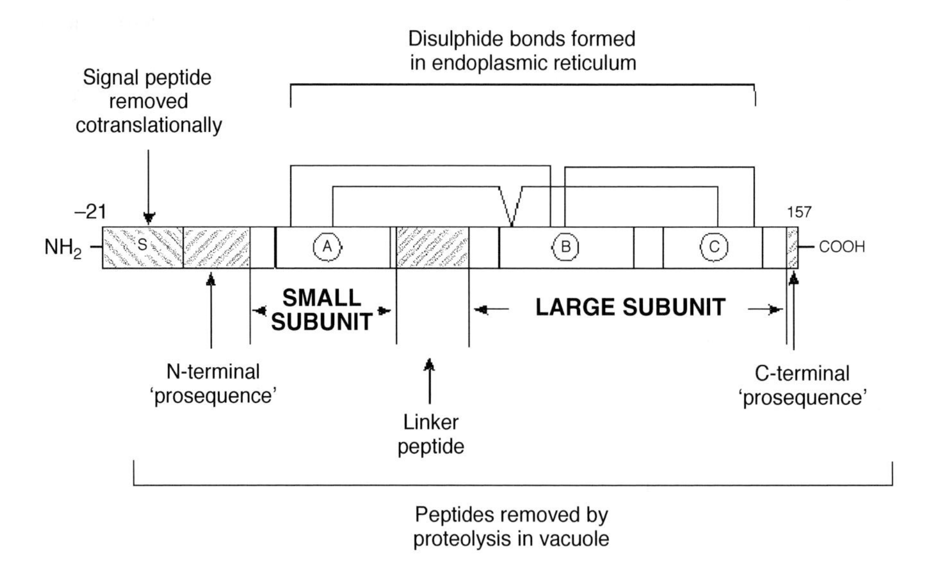

Fig. S.59. Summary of the processing of a typical 2S albumin, napin of oilseed rape, within the seed secretory system.

Prolamins are also synthesized on the rough ER and are co-translationally transported into the lumen with the cleavage of an N-terminal signal peptide. No further proteolytic processing or glycosylation of the proteins occurs, the sole modification to the covalent structure being disulphide bond formation. Most prolamins contain intra-chain disulphide bonds that are presumed to be formed co-translationally when the nascent polypeptides fold within the ER lumen. However, some prolamins are also assembled into polymers stabilized by inter-chain disulphide bonds. It is probable that these polymers are also formed within the lumen of the ER but it is not known whether this occurs concurrently with, or after, folding of the monomers. Similarly, it is not known whether the polymers remain stable in size and structure or undergo 'growth' or rearrangement during cereal grain development.

Whereas the albumin and globulin storage proteins are always deposited in storage vacuoles, the prolamins may be deposited in two types of protein body. In **rice** and **maize**, and probably also in **sorghum** and **millets**, prolamins accumulate directly within the lumen of the ER to form a population of protein bodies which are surrounded by an ER membrane and studded with ribosomes. In contrast, prolamins are deposited in protein bodies of both vacuolar and ER origin in **barley**, **wheat** and **rye**. It is possible that one group of storage proteins, the polymeric prolamins, is preferentially retained within the ER, whereas another group, the gliadins, is predominantly transported to the vacuoles. In wheat the two types of protein body (vacuolar and ER) coalesce during the later stages of grain development to give a continuous proteinaceous matrix in the endosperm cells of the dry mature grain.

The trafficking of all proteins through the endomembrane system is a highly regulated process with the destinations of individual proteins being determined by the presence, or absence, of specific inherent signals. A number of non-storage proteins are permanently resident in the ER lumen, including molecular chaperones (e.g. BiP) and protein disulphide isomerase (PDI). Their retention results from the presence of a specific retention sequence, which is a tetrapeptide located at the protein C-terminus. This is usually either KDEL (lysine, aspartic acid, glutamic acid, leucine) or HDEL (histidine, aspartic acid, glutamic acid, leucine). Such a retention sequence is not present in any of the storage prolamins that are retained within the ER lumen of cereal endosperm cells. BiP may play a role in the retention of prolamins in rice, by transiently binding to them to allow their assembly into protein bodies, but this mechanism has not been demonstrated for other species. An alternative explanation is that retention results from the physical properties of the prolamins, particularly their insolubility in aqueous media, which results in their aggregation to form protein deposits which can then be 'budded off' to form protein bodies. This aggregation may be a 'self assembly' process or may be facilitated by chaperones.

The precise sorting and trafficking mechanisms for storage proteins are not known. However, 2S albumins, 7S and 11S globulins, and prolamins may need to form electron-dense aggregates within the Golgi apparatus to assure that they are sorted into electron-dense vesicles for transport to the vacuoles. This aggregation is presumably determined by the structures of the individual proteins, with hydrophobic residues mediating protein–protein interactions.

Seeds often accumulate two or more different types of storage protein, which may be deposited in the same or different protein bodies. The 7S and 11S globulins have essentially similar structures and consequently they appear to be stored as a homogeneous mixture in the same protein bodies, e.g. in legume seeds. Similarly, in at least some species, the 2S albumins and 11S globulins also appear to be stored as a homogeneous mixture. However, protein bodies may also consist of separate phases with different components, a notable example being **castor bean** in which a matrix of 2S albumins and other proteins (e.g. **lectins**) contains crystalloid inclusions of 11S globulins and globoid inclusions of **phytin**. In rice the different routes taken by prolamins (ER) and 11S globulins (vacuole) result in two distinct populations of protein body called PBI (prolamins) and PBII (globulins). However, in wheat and **oats** these two populations of protein body fuse, resulting in the presence of globulins as inclusions in a prolamin matrix in wheat and vice versa in oats. Different types of zein (prolamins) are spatially concentrated within the ER-derived protein bodies of maize, with the β-zeins and γ-zeins being concentrated at the periphery and the α-zein in the centre.

The major processing events that occur to storage proteins within the secretory system are summarized in Fig. S.58 and Table S.16. (PRS)

(See: Storage reserves synthesis – regulation)

Kermode, A.R. and Bewley, J.D. (1999) Synthesis, processing and deposition of seed proteins: the pathway of protein synthesis and deposition in the cell. In: Shewry, P.R. and Casey, R. (eds) *Seed Proteins*. Kluwer Academic, Dordrecht, The Netherlands, pp. 807–841.

Storage reserves

Seeds of almost all species contain relatively large amounts of carbon- and nitrogen-containing compounds deposited during seed development that are later used by the nascent seedling until photosynthetic autotrophy is established (mobilization). The major reserves are insoluble, high-molecular mass compounds laid down in sub-cellular storage compartments – **protein bodies, oil bodies, starch granules** and **cell walls.** The principal storage organs are the **cotyledons, endosperm, perisperm** (rarely) and **megagametophytic** tissue. There are differences among species as to what organ is the major site of the reserves (e.g. endosperm in grasses, cereals; cotyledons in many legumes). Rarely, the embryonic axis itself is the location of the bulk reserves but frequently this organ contains quantitatively minor reserves, mostly soluble compounds of low molecular mass, such as sucrose and **raffinose-series oligosaccharides**: in some species, small amounts of **protein** and **oil** may also be present.

Most of our knowledge about the nature and amount of reserves comes from studies on food and crop seeds but the information reflects in principle the situation in non-crop species. Table S.17 shows the reserves in several species and the main storage tissue. The proportions of the different reserves in any one species is set between fairly fixed limits which are only slightly changed by selection and breeding. The controls operating to set these proportions are not understood. (MB)

Table S.16. Post-translational processing of storage proteins within the endomembrane system.

	ER	Golgi	Vacuole
7S globulins	signal peptide cleavage, glycosylation, trimer assembly	glycan modification	proteolytic processing, glycan trimming,
11S globulins	signal peptide cleavage, disulphide bond formation, assembly to form 'intermediate' trimers		proteolytic processing, hexamer assembly,
2S albumins	signal peptide cleavage, disulphide bond formation		proteolytic processing
prolamins	signal peptide cleavage, disulphide bond formation, polymer formation		

Table S.17. The major seed storage reserves in some crop and other species.[a]

	Protein	Oil[b]	Carbo-hydrate[c]	Major storage organ
Cereals				
Barley	12	3	76	Endosperm
Dent corn (maize)	10	5	80	Endosperm
Oats	13	8	66	Endosperm
Rye	12	2	76	Endosperm
Wheat	12	2	75	Endosperm
Legumes				
Broad bean	23	1	56	Cotyledons
Garden pea	25	6	52	Cotyledons
Groundnut	31	48	12	Cotyledons
Soybean	50	21	26	Cotyledons
Other				
Castor bean	18	64	Negligible	Endosperm
Oil palm	9	49	28	Endosperm
Pine	35	48	6	Megagametophyte
Rape (canola)	21	48	19	Cotyledons

[a]Percentage weight of harvested seed.
[b]In cereals oils are mostly in the embryo.
[c]Mostly starch.
From Bewley, J.D. and Black, M. (1994) *Seeds: Physiology of Development and Germination.* Plenum Press, New York, USA.

(See: Hemicellulose; Oils and fats; Oils and fats – mobilization; Oils and fats – synthesis; Phytin; Raffinose-series oligosaccharides; Starch; Starch – mobilization; Starch – synthesis; Storage protein; Storage protein – mobilization; Storage protein – synthesis; Storage reserves synthesis – regulation)

Bewley, J.D. and Black, M. (1978) *Physiology and Biochemistry of Seeds in Relation to Germination.* Vol. 1, *Development, Germination and Growth.* Springer, Berlin, Germany.
www.kew.org/data/sid/ for oil contents of several thousand species.

Storage reserves synthesis – regulation

Seeds store large amounts of **storage reserves** such as **proteins**, oils and polysaccharides (starch or hemicelluloses), which are utilized following germination for growth of the seedling before it becomes autotrophic. The accumulation of reserves in seed (seed filling) generally starts at mid-embryogenesis, following histodifferentiation, and ceases with the beginning of desiccation in late embryogenesis (**see: Development of seeds – an overview**). During this period, the cells in the embryo and endosperm are undergoing expansion.

The photoassimilate imported into seeds consists mainly of sucrose and amino acids, particularly glutamine and asparagine (**see: Development of seeds – nutrient supply**), all of which are taken up by the embryo/endosperm and used for the synthesis of the different types of seed reserves. Sucrose is hydrolysed into hexoses by an invertase-mediated unloading process while it is being imported into developing seeds. During the early stages of seed development, for example of pea and broad or faba bean (*Vicia faba*), this creates a high ratio of hexose to sucrose in the embryo, which promotes cell division. Later during seed development, a decrease in the hexose to sucrose ratio causes a transition from cell division to storage synthesis (starch), accompanied by a large increase in fresh weight. Therefore, changes in sugar concentration, which are determined by several enzymes including cell wall invertases, hexokinases, and also membrane sensors (**see: Sugar-sensing**), have been implicated in the control of the shift from mitotic activity to storage-product synthesis in developing seeds. SnRK1 (SNF_1-related protein kinase-1) is activated in response to high intracellular sucrose and/or low intracellular glucose. It is required for the derepression of some genes repressed by glucose, and it also regulates metabolic enzymes including acetyl-CoA carboxylase and glycogen synthase by modulating their phosphorylation state, suggestive of its role in metabolic signalling and carbon portioning in developing seeds. In oil-storing rapeseed, however, changing hexose:sucrose ratios do not influence these events.

Signals from **hormones** and other factors (e.g. temperature and nutrient availability, especially nitrogen and sulphur) also influence reserve synthesis during seed development. However, reserve synthesis should not be viewed in isolation because it overlays the developmental progression of embryogenesis and is controlled by integrated developmental and hormonal signals in the embryo.

1. Abscisic acid (ABA)

In developing seeds, ABA is associated with inhibition of precocious germination, acquisition of **desiccation tolerance** and regulation of the synthesis/accumulation of storage reserves. In intact, developing seeds of wheat, *Brassica* and soybean, the highest amounts of ABA are present during the most active phase of seed enlargement and deposition of storage reserves (**see: Development of seeds – hormone content**). In isolated embryos of rapeseed (*Brassica napus*), the synthesis of storage proteins (napin and cruciferin) requires the addition of ABA to the culture medium; if ABA is absent, the synthesis of napin mRNA ceases. Similar responses to ABA occur in the isolated embryos of wheat, with the promotion of synthesis of several storage proteins (a 7S globulin, wheat germ agglutinin and Em protein). The major storage proteins in the cereal endosperm do not appear to be influenced by ABA, however. In transgenic tobacco seed, ABA promotes a two- to threefold increase in the expression of the chimeric β-glucuronidase (GUS) gene attached to pea vicilin or *Brassica napus* napin storage-protein gene promoters. In addition to the major storage proteins, in some embryos there is an increase in inhibitory or maturation-related proteins (Table S.18).

Table S.18. Some proteins whose synthesis is promoted by ABA during seed development.

Species (tissue)	Protein
Wheat (embryo)	Wheat germ agglutinin Em protein (LEA-type protein) [a]Globulin
Maize (embryo)	[a]Globulin
Barley (aleurone layer)	Proteinase α-Amylase inhibitor
Soybean (cotyledon)	[a]β Subunit of β-conglycinin
Rapeseed (embryo)	[a]Napin [a]Cruciferin
Phaseolus bean (embryo)	[a]Storage protein
Sunflower (embryo)	[a]Helianthinin
Arabidopsis (embryo)	[a]2S protein [a]12S protein
Cotton (embryo)	LEA proteins

[a] Storage protein.

In other seeds there is no correlation between ABA availability and storage protein accumulation, e.g. the supply of ABA has no effect on the accumulation of protein reserves in excised embryos of *Pisum sativum* (deposition of legumin) and cotton. In developing somatic embryos, ABA may initially enhance storage protein synthesis, e.g. in those of white spruce, although the presence of osmoticum is required for synthesis to be maintained. During somatic embryogenesis of alfalfa (lucerne), only osmoticum and not ABA is stimulatory of storage protein synthesis.

ABA might also be involved in the synthesis of other storage reserves, although there are no reports that ABA induces any of the major enzymes associated with starch synthesis. However, microspore-derived *B. napus* embryos treated with ABA have a total fatty acid content 40% higher than controls, and the amounts of eicosenoic and erucic acids (the major fatty acid or oil reserves) are increased three- to fourfold after ABA treatment.

ABA promotes triacyglycerol synthesis in isolated wheat embryos. ABA also induces the synthesis of an **oleosin**, a protein associated with **oil bodies**, in isolated embryos of rapeseed in culture. Using **transgenic** tobacco, it is possible to create ABA-free seeds, by introducing an antibody gene encoding a protein that binds this hormone. These seeds accumulate less seed storage protein and oil bodies compared to wild-type ones.

Evidence for the involvement of ABA in storage reserve synthesis comes from ABA-deficient or -insensitive **mutants**. Seeds of ABA-deficient mutants of maize, tobacco, wheat, *Arabidopsis* and tomato exhibit reduced protein accumulation. In seeds of the *abi3* mutant (insensitive to ABA) of *Arabidopsis*, only two-thirds storage protein and about one-third of eicosenoic acid accumulate compared to that in the wild-type seeds. In the ABA-insensitive mutant *vp1* of maize, the embryo-specific globulins GLB1 and 2 are not accumulated in the embryo *in vitro*. In the ABA-deficient mutant of maize, *vp5*, embryos do not accumulate GLB1 and 2, but do so when ABA is supplied to embryos *in vitro*.

In developing wheat embryos in culture, inhibition of the synthesis of gibberellins (GAs) results in the synthesis of several proteins associated with maturation, including a storage globulin. This mimics the effects of applying ABA to the embryos; thus, GA and ABA may be antagonistic to each other with respect to the regulation of developmental events, and a changing hormone balance in the embryo may provide control over development and maturation. (**See: Storage protein – synthesis**)

2. Environmental factors

During seed development, environmental factors influence reserve deposition in seeds, thus affecting seed quality. For example, stress conditions such as water stress decrease starch deposition in developing wheat grains. In oilseed rape, oil (triacyglycerol) is deposited during the later stages of seed development. At maturity, in a cool environment and with an adequate water supply, the seeds are larger and have a high oil content. High temperature (30°C) reduces oil content but increases its oleic acid and decreases its linolenic acid content, the converse of what happens at low temperature (12°C) and under water stress. During seed development in evening primrose, temperature also affects the fatty acid composition of lipid such that the ratio of saturated/unsaturated fatty acids increases in the seeds of incubated fruits with rising temperatures. In immature, detached seeds, increasing incubation temperatures cause a decrease in the linolenic acid fraction of the oil, with an equivalent increase in the oleic acid. In mature seeds grown on plants subjected to increasing temperatures, the total saturated fatty acids including oleic acid increases, but the sum of the linoleic and linolenic acids decreases.

Nutrient availability during seed development also affects seed storage product deposition. For instance, sulphur (S) availability in the field affects seed S content and the relative abundance of conglycinin and glycinin storage proteins, both of which are S-containing, during soybean seed development. The amount of transcription of the storage protein genes is strongly influenced by the presence of available S. As availability increases from 12 to 62 mg S per plant, the amount of conglycinin varies from 15 to 40% of storage protein in the cotyledons. The glycinin fraction of storage protein decreases from the normal 60% to less than 30% when seed development occurs under S stress.

Although environmental factors regulate storage product synthesis in seeds, it is not understood how environmental changes are perceived, and what enzymes that are involved in storage product synthesis are affected by these changes.

3. Regulation of gene expression, and post-transcriptional control

Our understanding of the regulation of the genes related to seed storage reserve synthesis is based on studies of the expression patterns of transcripts of the relevant enzymes or proteins involved. For instance, the synthesis of globulin and prolamin reserve proteins is unique to the seed, and occurs only at specific times during its development. During this time, mRNAs for these storage protein genes are transcribed exclusively in the storage tissues (the cotyledons), by processes that are regulated both spatially and temporally. For instance, in many species of legumes, vicilins accumulate before legumins. Moreover, synthesis of the constituent subunits of vicilins may commence at different times. For example, the synthesis of α- and α′- subunits of β-conglycinin in soybean begins about 15–17 days after flowering, but that of the β subunit does not start until 5–7 days later.

Maturation drying of seeds results in the cessation of storage reserve synthesis and upon subsequent rehydration there is the promotion of germination- and post-germination-related gene expression. The promoter region of the phaseolin storage protein from *Phaseolus vulgaris* (**common**, French **bean**) is sensitive to desiccation, for when tobacco is transformed with the *phas* storage protein gene and its own promoter, it is expressed in the transgenic seed during its development, but not following maturation drying. (**See: Maturation – effects of drying**)

Studies have been carried out on transgenic plants expressing storage protein genes driven by their own promoters. Generally, these storage protein genes, even in a different (transgenic) seed, are still expressed only in the storage tissues, e.g. in transgenic tobacco, legume storage protein genes are expressed almost exclusively in the cotyledons and at the correct time during their development. Thus the promoter regions of different storage protein genes carry information that can be recognized by similar transcription factors in the nuclei of widely different species. Sequences regulating the temporal expression and tissue specificity have been identified for some storage proteins, and it is evident that more than one element in the **promoter** region is required for correct temporal and spatial gene expression. Part of the promoter functions to prevent expression in non-seed tissues.

Detailed analysis of the promoter region of the *phas* gene that encodes β-phaseolin, a vicilin synthesized in the developing French bean (*Phaseolus vulgaris*) seed, has revealed several target sequences (*cis*-elements) to which specific DNA-binding proteins (*trans*-acting factors) bind to activate or repress gene expression. This gene is highly expressed during the storage reserve synthesis period of embryogenesis, but it is silent throughout development of the vegetative plant. When the *phas* gene and its promoter are introduced into both *Arabidopsis* and tobacco, this pattern of expression is maintained. Two regions appear to affect temporal expression. A seed-specific enhancer region up-regulates expression late in seed development, whereas a minimal promoter drives low expression early in development. Some elements involved in temporal expression can be deleted without altering spatial expression. The major regulatory elements identified in the upstream regions of this and other storage protein genes are detailed in Table S.19. Certain *trans*-acting factors can bind to the DNA in several places because the regulatory region contains repeats of the target sequence that they recognize. The concomitant activation of different elements within the promoter region may be necessary to obtain full expression of a particular storage protein gene. The promoter sequences of these genes have common *cis* sequences that confer seed-specific expression (e.g. the prolamin box in cereal grains; legumin and vicilin boxes in their respective genes in dicots).

To illustrate this, in the promoter region of the *phas* vicilin gene, several *cis*-elements have been recognized, e.g. G-box, *CCAAATT* box, the E-site, RY elements and vicilin box. Specific DNA-binding proteins (the *trans*-acting factors) bind to these different regions within the promoter to activate or repress the expression of the storage protein gene. The G-box, *CCAATT* box and the E-site mediate high expression of the *phas* gene in the developing embryo; the RY elements have both a positive and negative regulatory role, and the vicilin box

Table S.19. The major regulatory (*cis*) regions in the DNA in the promoters of storage protein (SP) genes and the *trans*- (transcription factor) regulatory proteins that bind to them.

Regulatory element	Structural gene	Functions
CATGCATG (RY element)	Seed-specific genes Storage protein (SP) genes in legumes and some cereals	Essential for expression of some 11S and 7S SP genes, alone or in combination with other elements. Suppresses SP expression in vegetative tissues. *Trans*-acting factor identified (PvALF in *Phaseolus* bean).
TGTAAAG (Prolamin box)	Prolamin and legumin genes	Not essential for SP expression. May affect expression quantitatively. *Trans*-acting factor known (a zinc-finger protein in maize).
CACA (CA-rich box)	Legume and cereal SP genes. Seed and vegetative tissue genes.	General regulatory function, not essential for all SP genes. *Trans*-acting factor is CA-1 nuclear protein.
GCCAC(C/T)TC (Octanucleotide box or E-site)	7S,11S, and lectin SP protein	Confers seed-specific expression and suppresses transcription in stems and roots. May affect temporal expression. Putative *trans*-acting factor identified.
AA/G/CCCA	7S soybean SP gene not confirmed.	Role in seed SP gene expression *Trans*-acting factor known (SEF3 in soybean).
T/CACGTC/A/C (G-box) (ABRE)	Seed-specific and vegetative genes. Zein SP genes. ABA-responsive genes	Essential for gene expression. Binds transcription factor 02 (bZIP) in maize and CAN in *Phaseolus* bean.
Vicilin box I	Vicilin genes of legumes	May determine temporal or cell-specific patterns of expression. Binds *trans*-acting factors ROM1 and ROM2 in *Phaseolus* bean.
Vicilin box II	Most vicilin genes of legumes	Not essential for SP gene expression. Binds TEMP-1 *trans*-acting factor
Legumin box	Legumin genes of legumes	CATGCATG motif (RY element) in legumin box; important for seed-specific expression

Based on: Bewley, J.D., Hempel, F.D., McCormick, S. and Zambryski, P. (2000) Reproductive development. In: Buchanan, B.B., Gruissem, W. and Jones, R.L. (eds) *Biochemistry and Molecular Biology of Plants*. American Society of Plant Physiologists, Rockville, MD, USA, pp. 988–1043.

is a strongly negative regulatory element. *Trans*-acting factors bind to specific regions of the promoter to confer correct spatial expression in transgenic plants, and it is evident that interactions between the promoter regions are necessary for the correct expression of the gene.

As noted in **1. Abscisic acid**, in developing seeds of some species, the synthesis of several proteins, including storage proteins, is positively influenced by ABA. More than 20 *cis*-sequences specific for ABA-induced gene expression (ABA response elements: ABREs) have been identified in plants. Several (but not all) storage protein genes contain ABREs in their promoter region, which frequently includes a G-box motif that binds a specific type of *trans*-acting **transcription** factor (bZIP). The G-box ABRE is the major ABRE and E-site may be a coupling element. For instance, the G-box region of the *phas* promoter appears to act as a functional ABRE. **(See: Signal transduction – hormones)**

Expression of seed storage products is also regulated by post-transcriptional controls. For instance, two small S-rich albumins are both encoded by a single gene (*PA1*) in pea seeds, the product of which is cleaved post-transcriptionally to yield two different albumins (PA1a and PA1b). **Starch synthesis** in wheat endosperm is regulated by post-translational redox modification of ADP-Glc pyrophosphorylase (AGPase), in addition to its transcriptional and allosteric regulation. Dimerization of AGPase leads to inactivation of the enzyme as a result of a marked decrease of the substrate affinity and sensitivity to allosteric effectors. AGPase can be reactivated and its dimerization reversed *in vitro* by incubating the enzyme with a reductant such as dithiothreitol (Fig. S.60). (XG)

Chandrasekharan, M.B., Bishop, K.J. and Hall, T.C. (2003) Module-specific regulation of the β-phaseolin promoter during embryogenesis. *The Plant Journal* 33, 853–866.

DeLisle, A.J. and Crouch, M.L. (1989) Seed storage protein transcription and mRNA levels in *Brassica napus* during development and in response to exogenous abscisic acid. *Plant Physiology* 91, 617–623.

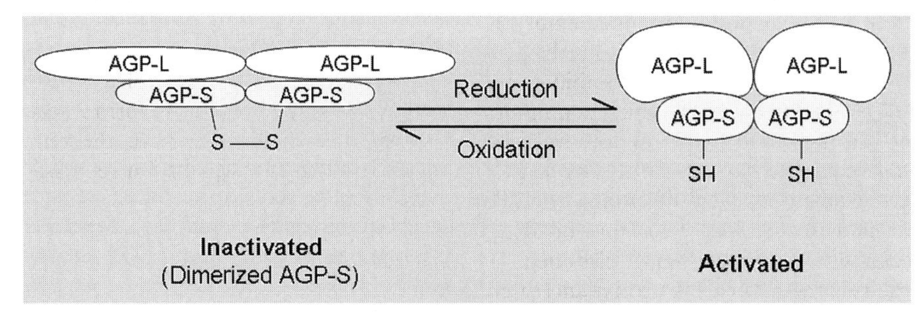

Fig. S.60. Model for redox modulation of AGPase activity. Activation of the enzyme is thought to occur by a reduction of an intermolecular disulphide bridge between the AGP-S subunits (small subunits of AGPase) at a conserved cysteine residue, in addition to a conformational change induced by the substrates. Reductive activation operates in response to cellular sucrose concentrations, in an unknown manner. AGP-L, large subunit of AGPase. From: Tetlow, I.J., Morell, M.K. and Emes, M.J. (2004) Recent developments in understanding the regulation of starch metabolism in higher plants. *Journal of Experimental Botany* 55, 2131–2145. With permission.

Sexton, P.J., Naeve, S.L., Paek, N.C. and Shibles, R. (1998) Sulfur availability, cotyledon nitrogen:sulfur ratio, and relative abundance of seed storage proteins of soybean. *Crop Science* 38, 983–986.

Srivastava, L.M. (2002) *Plant Growth and Development*. Academic Press, Amsterdam, The Netherlands, pp. 431–446.

Weber, H., Borisjuk, L. and Wobus, U. (1997) Sugar import and metabolism during seed development. *Trends in Plant Science* 2, 169–174.

Stored mRNA

Long-lived mRNA molecules that are synthesized during the latter stages of seed development, are stored in the dry seed, and are translated into proteins during germination. They are probably in association with proteins, as messenger ribonucleoprotein complexes (mRNPs). (**See: Germination – metabolism**)

Strain

A group of individuals from (or presumed to be from) a common origin, more narrowly defined than a **cultivar** in the case of plant breeding, and physiologically but not necessarily morphologically distinguishable in the case of an isolated fungus or bacterium. (**See: Race**)

Stratification

A term synonymous with the chilling or warm-temperature treatments used to break dormancy. The word comes from the old-established practice of arranging seeds in layers ('strata') in a wet substrate (e.g. sand) for the temperature treatment.

Though an old term it is still used as a term for dormancy breakage in certain species, including several tree species of economic importance (prechilling is the name for the more modern equivalent), by the natural or artificial burial of seeds in temperate climates over the winter months. The combination of moist seeds, temperature and time stimulates subsequent germination. For processing of seed, this was once generally carried out in a specially dug, vermin-proof stratification pit in alternating layers with either moist soil or sand, to promote germination the following spring. However, in present-day practice, it is less common to employ natural, winter conditions to overcome seed dormancy. Instead, moist, dormant seeds are incubated under controlled refrigeration, for species that require prechilling. Alternatively, warm stratification may be used as a moist **afterripening** treatment for overcoming dormancy caused by underdeveloped embryos, or for softening hard seedcoats to relieve mechanical dormancy. Dormancy treatments may be combined in sequence: ash (*Fraxinus*) seeds, for example, which commonly possess two types of dormancy due to underdeveloped embryos and prechilling requirements, are treated by warm-moist followed by cold-moist stratification. ('Dry, afterripening', a technique generally used for breaking dormancy in many cultivated crops, is not classified as stratification because the treatment does not involve increased moisture content.) (**See: Tree seeds**)

(a) *Moisture content*. There are very few studies on the amount of moisture necessary for different species to respond to the prechill process, but an emerging rule-of-thumb appears to indicate that a moisture content percentage a few per cent below full imbibition probably leads to the most rapid response. In the case of most temperate conifers this is about 30–40% moisture content (fresh weight basis), and for temperate broadleaved trees it is about 40–60%.

(b) *Duration*. The prechill duration required to overcome dormancy also varies widely between species. For example, the seeds of some species (e.g. *Fagus*, *Quercus*, *Pinus*, *Abies*, alpine eucalypts) may need only a few days or weeks whereas others require many months or years (e.g. holly, yew). But the geographical source, environmental conditions under which the seed developed and matured, the time of collection, and how the seed has been processed and stored also affect the depth of dormancy and hence length of prechill required.

(c) *Temperature*. The most effective prechill temperatures appear to be between 0 and 9°C with about 4°C usually being optimal.

The term stratification probably refers to the ancient European forestry and horticultural nursery practice of outdoor burial of dormant, woody plant seeds in layers (or strata) alternating with either soil or sand. For example, J. Evelyn (1664) described in *Sylva, or a discourse of forest trees and the propagation of timber* how seeds could be 'prepared for the vernal by being barrell'd or potted up in moist sand or earth stratum during the winter, at the expiration whereof you will find them sprouted'. And M. Cook (1679) in his treatise, *The manner of raising, ordering and improving forest and fruit trees* provides details of how to prepare ash seeds (so-called 'keys') before sowing – 'lay one laying of sand and a laying of keys .. any time in winter .. keep it all that year .. then about the latter end of January sow them.' (PG)

(**See: Dormancy breaking – temperature**)

(The term 'stratification' has another, completely different meaning in seed conditioning, which is noted here in passing only: as a term for mechanized procedures which arrange seed into layers – such as with light seed on top, heavy seed on bottom – based on their specific gravity. **See: Conditioning – cleaning**)

Stress tests

A range of **vigour testing** procedures in which seeds are subjected to one or more physiological stresses, such as mechanical impedance used in the Hiltner test, or cold or cool temperatures, to simulate environmental conditions in the field. (**See: Vigour tests – physiological, 3. Stress tests**)

Strip-till planting

An area 30 to 50 cm wide tilled sufficiently through living **mulch** or standing residue to form a **seedbed** for each row. At planting or at first cultivation, the remaining mulch in the middle row is cut loose, killed or retarded.

Strobilus

See: Cone

Strophiole

A localized, elongated aril in the region of the hilum, formed by a proliferation of the **raphe**. Used synonymously with the term **caruncle**. (**See: Seedcoats – dispersal aids,** Colour Plate 15B; **Seedcoats – structure**)

Structure of seeds

A seed is defined in the strict botanical sense as the fertilized **ovule** of an angiosperm or a gymnosperm, which together make up the Spermatophytes, or seed plants. A broader, and widely used definition also includes relatively small (usually), hard (indehiscent) fruits resembling seeds. In either case, the seed's major structural elements are three (Fig. S.61): an **embryo** (the new generation); a food store (usually); and a protective coat (almost always). These three reflect the functions of the seed, which are to ensure the dispersal and survival of embryos in more or less stressful environments, and thus maximize the chances of successful seedling establishment. Despite sharing this basic organization (*bauplan*), diverse evolutionary trajectories mean that the seeds of different species vary enormously in their size and structural complexity, from orchids (0.003–0.005 mg), Orobanchaceae (*ca.* 0.001 mg) (**see: Parasitic plants**) to Coco de Mer (up to 20 kg), with an equally large underlying anatomical variation in their component cells and tissues.

1. Embryo

The embryo develops by cell division and growth from the zygote (**embryogenesis**), which is the diploid product of the fusion of the egg nucleus and one of the pollen (generative) nuclei (both haploid) at fertilization. It represents the early stages of the sporophyte generation in the life cycle. In most, but not all species the final stage of seed development, maturation, is marked by a more or less protracted suspension of embryo growth (**quiescence**), often associated with its drying out, along with the rest of the seed, prior to **dispersal** (**see: Desiccation tolerance**; **Maturation** entries). At maturity, the absolute and relative size of the embryo varies considerably among species, as does its degree of differentiation. Structurally, the embryo generally consists of a radicle/hypocotyl axis bearing one (monocotyledons), two (dicotyledons) or more (polycotyledons, commonly in gymnosperms) **cotyledons**. An epicotyl, the region between insertion of the cotyledons and the first true leaves (the **plumule**), may also be present in some species. The seed becomes rehydrated upon **imbibition**, and **germination** leads to conversion of the embryo to a seedling and establishment of the young plant. In many species, radicle elongation and protrusion from the seed does not resume immediately on imbibition, despite it being fully rehydrated, since the seed may be dormant (**see: Dormancy**). During dormancy, physiological and/or morphological changes take place in the embryo, eventually leading to loss of dormancy and to completion of germination at a time when conditions are optimal for seedling survival and establishment. (**See: Embryo – structure**)

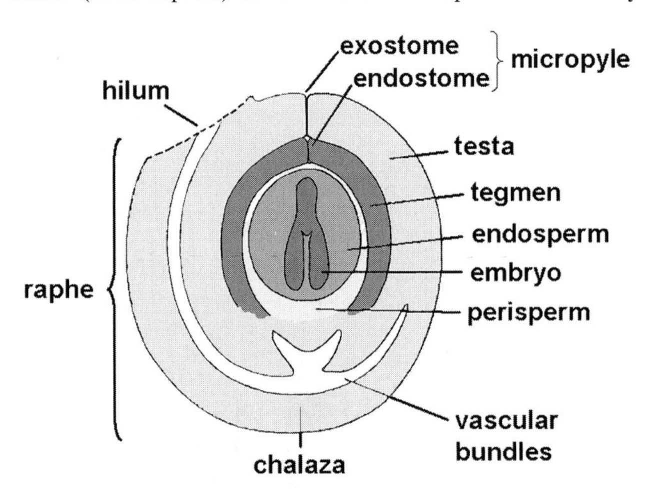

Fig. S.61. The general structure of a mature angiosperm seed. After the (double) fertilization of the ovule, the primary (triploid) endosperm nucleus starts dividing, forming the endosperm which may persist in the mature seed. The nucellus usually gets absorbed during seed development but in some families (e.g. Caryophyllaceae and relatives, Piperaceae, Zingiberaceae) it sometimes persists and develops into a storage tissue called the perisperm. The zygote develops into the embryo and represents the new sporophytic (next vegetative) generation. The integuments (or in unitegmic ovules, the single integument) together with the raphe and chalaza develop into the seedcoat in which the testa derives from the outer integument (as well as the descendants of the single integument of unitegmic ovules) and the tegmen from the inner integument. The exostome and endostome are openings in the outer and inner integuments of a bitegmetic ovule, through which the pollen tube passed during fertilization. The hilum is the point of attachment to the funiculus (stalk of the ovule), through which the vascular tissue passes from the parent plant to the seedcoat. Diagram modified after Boesewinkel, F.D. and Bouman, F. (1984) The seed: structure. In: Johri, B.M. (ed.) *Embryology of Angiosperms*. Springer, Berlin, pp. 567–610.

2. Food store

Embryos of the majority of species are dispersed with food reserves in varying amounts, either in the embryo itself or in structures which surround it. Thus, sufficient nutrients (**carbohydrates**, **oils**, storage protein, minerals) for germination and establishment are provided before the seedling becomes fully autotrophic. Classic exceptions include seeds of groups such as **orchids**, with extremely small, light, undeveloped seeds, lacking any food storage material. They are adapted to long-distance dispersal through the air ('dust seeds') (**see: Structure of seeds – identification characters**) and rely on the establishment of symbiotic associations (with soil fungi or other microrganisms) for germination and seedling growth.

A long established functional classification of the seeds of angiosperm species into two major types relies on the location of nutritive (reserve) material in the mature seed. In one type, the so-called endospermic (albuminous) seeds, the major reserve tissue is the **endosperm**. Unique to **angiosperms**, the true endosperm results from the triple fusion of one of the haploid pollen (male) sperm nuclei and two haploid (female) polar nuclei in the embryo sac. The resulting triploid endosperm (one paternal, two maternal genomes) grows and accumulates nutrients throughout seed development, which may be used by the developing embryo, or in the case of persistent endosperms, by the early growing seedling. At maturity, the relative size of the endosperm varies considerably, with the size of the embryo varying accordingly, depending on how much reserve material has been transferred to the embryo.

Included in the 'albuminous' category, though actually having little or no endosperm at maturity, are the so-called 'perispermic' seeds. The major reserve tissue in these is the **perisperm**, which is derived from the **nucellus** of the ovule. Perisperm is diploid maternal tissue, in contrast to the true endosperm. Perispermic seeds are relatively rare, being characteristic of just a few families, e.g. Chenopodiaceae and Cactaceae (Centrospermae). In the majority of species the nucellus or perisperm diminishes early in seed development. In the Cyanastraceae, for example, seeds of some species develop a chalazosperm, a nutritional tissue within the **chalaza**, which is the region where there is no differentiation between the **integuments** and the nucellus.

The other major category includes those seeds with no discernible endosperm at maturity, referred to as non-endospermic (exalbuminous) (**see: Non-endospermic seeds**). Here the endosperm and its food materials are mostly or completely used up during embryo development; the role of food storage falls to certain parts of the embryo, which in the majority of species is the specially adapted first seedling leaves, or **cotyledons**. There are seeds that have a very small, residual endosperm in the mature seed, which is one to a few layers thick. In some species, other organs fill this storage role, e.g. the swollen hypocotyl in *Bertholletia excelsa* (**Brazil nut**).

The division of seeds into endospermic and non-endospermic, though useful, is artificial. The reality is a continuum, from relatively small embryo with copious endosperm to well-developed embryo with storage organs and no endosperm. In many species, food storage is in a combination of embryonic and *extra*-embryonic tissues. On the basis of the occurrence of the two types of seed in extant and fossil taxa, botanists hypothesize that there has been transference of the nutrient storage function in mature seeds from endosperm (relatively primitive) to embryo (advanced) through evolution. The ontogeny of seeds of the non-endospermic (advanced) group, with their transient endosperm, may thus reflect their phylogeny.

The storage tissue of gymnosperms is sometimes erroneously referred to as 'endosperm', when in fact it is haploid tissue derived from the development of the **megagametophyte** (i.e. it is entirely maternal).

Interestingly, seeds consist of three distinct generations. The outer seedcoat is derived from maternal sporophyte (diploid) tissue. This surrounds the haploid megagametophyte food store in gymnosperms or the triploid endosperm in angiosperms, which in turn almost surrounds the (diploid) embryo, the new sporophyte generation. (**See: Embryo; Endosperm; Gymnosperm seeds**)

3. Covering structures – seedcoat

Seeds of the overwhelming majority of species have a protective coat of varying thickness and strength at maturity, consisting basically of dead cells, whose walls are thickened with lignin and other substances. As well as providing mechanical and chemical protection for the embryo and food store inside, the coats of many species are also structurally adapted to provide means of dispersal, e.g. wings, hairs, fleshy outgrowths (**see: Seedcoats – dispersal aids**). The seedcoat in some species also promotes dispersal in time through its involvement in certain types of dormancy (**see: Coat-imposed dormancy**). Seed covering structures may be impermeable to water and thus prevent or delay imbibition (**hardseededness**), or by their strength they may prevent or delay completion of germination (radicle emergence) in imbibed seeds (**see: Germination – radicle emergence; Seedcoats – functions**). In both angiosperms and gymnosperms the true seedcoat is derived only from the **integument**(s) of the ovule, following fertilization and development, and is thus wholly maternal. However, in many angiosperms the protective and dispersal functions of the true seedcoat have been transferred through evolution to the fruit wall (**pericarp**), as in the indehiscent dry fruit. This gives rise to the broader definition of seed alluded to in the opening paragraph and the term often used – dispersal unit. There are several distinct types of these dry indehiscent fruits containing (usually single) seeds (hence they are 'seeds' in the broad functional sense). These include **achenes** (e.g. some *Ranunculus* spp., *Helianthus*), **caryopses** (Poaceae) (**see: Cereals**), **cypselas** (Asteraceae, Dipsacaceae, Fig. S.62),

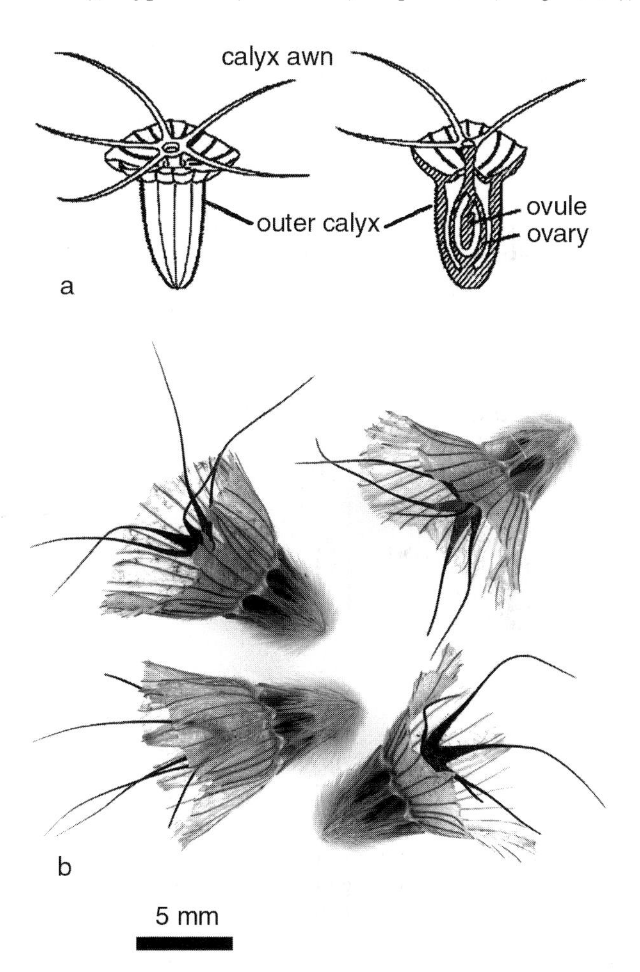

Fig. S.62. Cypsela of *Scabiosa stellata*, star flower (Dipsacaceae). The fruits of members of the Dipsacaceae possess a thin, air-containing, envelope around them. This envelope is not derived from a part of the flower, but represents an outer calyx formed by four laterally fused bracts below the inferior ovary. Surrounding the ovary, this outer calyx forms a kind of collar which represents an additional (besides the air-containing envelope) adaptation to wind dispersal. Together with the calyx awns that remain in the fruit, the outer calyx can also facilitate epizoochoral (attached to an animal) dispersal. (a) From Leins, P. (2000) *Blüte und Frucht*. Schweizerbart'sche Verlagsbuchhandlung. Stuttgart, Berlin, after Hegi (1918) *Illustrierte Flora von Mittel-Europa; mit besonderer Berücksichtigung von Deutschland, Oesterreich und der Schweiz*. J.F. Lehmanns, München. (b) Photograph by W. Stuppy, RBG Kew.

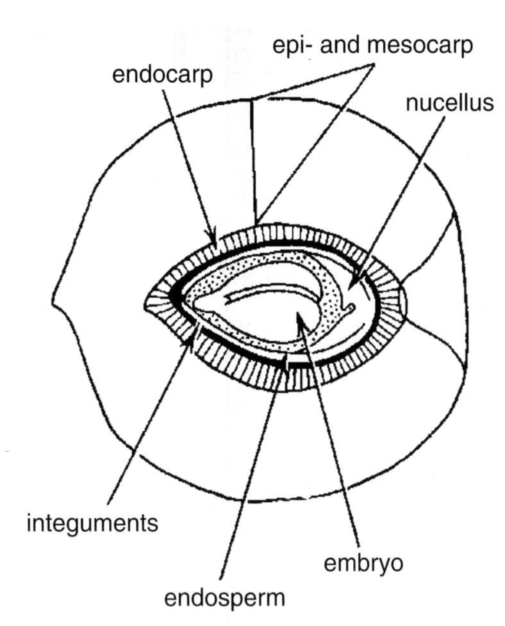

Fig. S.63. Longitudinal section of the drupe of *Prunus persica*, peach (Rosaceae). The epi-, meso- and endocarp are the outer, middle and inner regions of the fruit coat. The endocarp forms the 'stone' of the fruit. From Esau, K. (1965) *Plant Anatomy*, 2nd edn. John Wiley, New York.

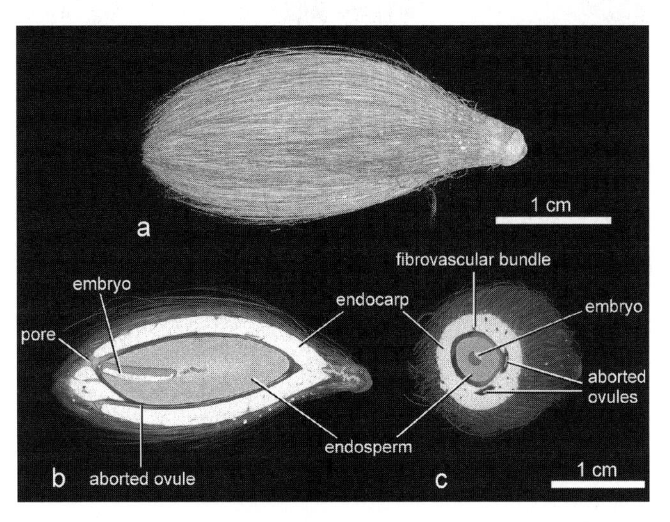

Fig. S.64. High-resolution X-ray computed tomography (HRCT) images of the fruit of the South American palm *Syagrus flexuosa*, Coco do Vaqueiro (Palmae (Arecaceae)). HRCT allows the three-dimensional analysis and reconstruction of an object from every conceivable angle. (a) Aspect of the stone of the fruit (epicarp and mesocarp have been removed). (b) Transmedian longitudinal section of the stone. (c) Transverse section of the stone. All morphologically relevant parts and organs such as the mesocarpic fibres, endocarp with fibrovascular bundles, endosperm and embryo can be distinguished in the HRCT images. Other significant morphological details include a central cavity in the endosperm (as a result of the nuclear endosperm formation), one of the three pores at the base of the fruit, which represent openings in the hard endocarp, and the two aborted ovules. (a, c) © The Digital Library of Morphology (www.digimorph.org), The University of Texas at Austin; (b) reprinted from Stuppy, W.H., Maisano, J.A., Colbert, M.W., Rudall, P.J. and Rowe, T.B. (2003) Three-dimensional analysis of plant structure using high-resolution X-ray computed tomography. *Trends in Plant Science* 8, 2–6, with permission from Elsevier.

mericarps (Apiaceae), **nuts** (e.g. *Corylus*) and **samaras** (e.g. *Fraxinus*). Fleshy fruits are largely beyond the scope of this account, except to note that in the case of **drupes** (e.g. *Prunus* – Fig. S.63) and other Rosaceae, *Cocos* and other Arecaceae (Fig. S.64) it is the endocarp (inner layer of the fruit wall) that becomes lignified to form the thick, hard outer protective shell of the stone or pyrene (dispersal unit, or 'seed' in the broad sense). Because they do not possess carpels, gymnosperms have not had the evolutionary option of replacing the functions of the seedcoat proper with pericarp structures. **(See: Rumination; Seedcoats – structure)** (JD, WS)

Boesewinkel, F.D. and Bouman, F. (1995) The seed: structure and function. In: Kigel, J. and Galili, G. (eds) *Seed Development and Germination*. Marcel Dekker, New York, USA, pp. 1–24.

Corner, E.J.H. (1976) *The Seeds of Dicotyledons*. Cambridge University Press, Cambridge, UK.

Fahn, A. (1990) *Plant Anatomy*, 4th edn. Butterworth-Heinemann, Oxford, UK.

Werker, E. (1997) *Seed anatomy*. Gebrüder Borntraeger, Berlin, Stuttgart, Germany (*Handbuch der Pflanzenanatomie = Encyclopedia of Plant Anatomy*, Spezieller Teil, Bd. 10, Teil 3).

Structure of seeds – identification characters

Within the **angiosperms**, the internal morphology of seeds as well as the anatomical structure and sculpture of the seedcoat vary tremendously and provide extremely valuable information with respect to identification and classification.

The variations of the internal morphology of seeds primarily concern the relative size, shape and location of the embryo (**see: Embryo,** Fig. E.7). The phylogenetic significance of the internal morphology of seeds is demonstrated by the facts that, for example, lateral embryos are characteristic of the Poaceae (grasses) and peripheral embryos are typical of almost the entire order Caryophyllales (carnation family and relatives). However, though internal seed characters are isolated from immediate environmental pressures because of their enclosure in the ovary, seed external structure is potentially influenced by conditions such as the type of the fruit or the density of the seeds within it.

A largely independent and greater variety of even more detailed characters is provided by the anatomy of the seedcoat. While more evolutionarily primitive angiosperms generally have medium- to large-size seeds with complex multilayered seedcoats (**see: Seedcoats – dispersal aids,** Fig. S.10), the most advanced flowering plants such as the Orchidaceae in the monocotyledons and the Asteridae in the dicotyledons have small, simplified seeds with few- or even just single-layered seedcoats. Because the structure of the seedcoat is genetically determined and, since the seeds are enclosed in the ovary (and hence not immediately exposed to environmental adaptation pressures), they have long been known to provide an excellent tool for identification and the natural (monophyletic) circumscription of families and even genera.

One practical and very important example of how seeds can serve as an identification tool is in forensic science. In 2003 the Anatomy Section of the Jodrell Laboratory of the Royal Botanic Gardens, Kew, in London was involved in the 'Torso in the Thames' murder enquiry. Hazel Wilkinson, a member of the

Section, identified segments of the seedcoat of the poisonous calabar bean (*Physostigma venenosum*) in the stomach of the boy whose dismembered body was found in the Thames near Shakespeare's Globe Theatre. In West Africa the calabar bean or ordeal bean is used for withcraft purposes indicating that the boy was a victim of a ritual killing. It contains the ordeal poison physostigmine, an alkaloid that inhibits acetylcholine esterase, thus interfering with nerve impulse transmission causing paralysis or even death. (**See: Archaeobotany; Pharmaceuticals and pharmacologically active compounds; Poisonous seeds**)

The most distinctive character of the seedcoat (testa) lies in the position and structure of the main mechanical layer, which is usually composed of thick-walled cells (**see: Seedcoats – structure**). This mechanical layer can be one or more cells thick and consists of radially elongate palisade cells, which are horizontally elongate fibres or cuboid cells, with either evenly or unevenly thickened walls. Such unevenly thickened walls, especially if they occur in the cells of the exotesta (the outer testa epidermis), are often the reason for the intricate and highly characteristic sculpture patterns of the seed surface, used for the classification of many groups such as Begoniaceae, Crassulaceae, Caryophyllales (e.g. Cactaceae, Fig. S.65), Caryophyllaceae (Fig. S.66), Gentianaceae and Orchidaceae. In some cases seed structure, especially sophisticated surface patterns, can even be species-specific.

Whilst surface patterns are the most important source of information in exotestal seeds with unilayered seedcoats (i.e. seedcoats in which the remaining layers of the seedcoat are either absent or undifferentiated and crushed, e.g. Cactaceae, Caryophyllaceae, Orchidaceae), it is the detailed anatomy of the mature integument(s) in taxa with thick and complex seedcoats that provides the most valuable clues. Most importantly, the principal structure of the seedcoat is often uniform within a natural family and in its detail it is often even species-specific. Leguminosae (Fabaceae) (**see: Seedcoats – structure,** Fig. S.17), for example, are characterized by an exotestal palisade layer of Malpighian cells and a hypodermis of hourglass cells, and members of the Euphorbiaceae are highly uniform in having an exotegmen composed of sabre-like palisade layer of Malpighian cells (Fig. S.67f).

The general independence of their anatomical structure from environmental conditions explains why a certain type of seedcoat anatomy is often characteristic for whole families, irrespective of the different environments inhabited by the plants. One such example of an extremely widespread and ecologically diversified group is the Fabaceae, in which, despite species inhabiting almost all climates and habitats, the principal seedcoat structure hardly shows any visible environmental adaptations. The members of this family are invariably (apart from those with indehiscent fruits, see below) characterized by an exotesta of Malpighian cells showing a **light line** (*linea lucida*) and an outer hypodermis of so-called hourglass cells or osteosclereids (**see: Seedcoats – structure,** Fig. S.17). The phylogenetic significance of seed characters is also demonstrated by the recent dismemberment of the long-established family Euphorbiaceae. Based on the differences he found in the anatomy of the seedcoats (Figs S.67 and S.68), Corner (1976) suggested that the family is probably polyphyletic (has more than one ancestor) and that the three uniovulate (single ovule) subfamilies (today the Euphorbiaceae *sensu stricto*) with their palisadal exotegmen (Fig. S.67) obviously represent a separate entity from the two biovulate subfamilies (now Phyllanthaceae and Picrodendraceae) with (usually) fibrous exotegmen (Fig. S.68). This observation has been supported phylogenetically by analysis of DNA sequences of members of the various families (Savolainen *et al.*, 2000).

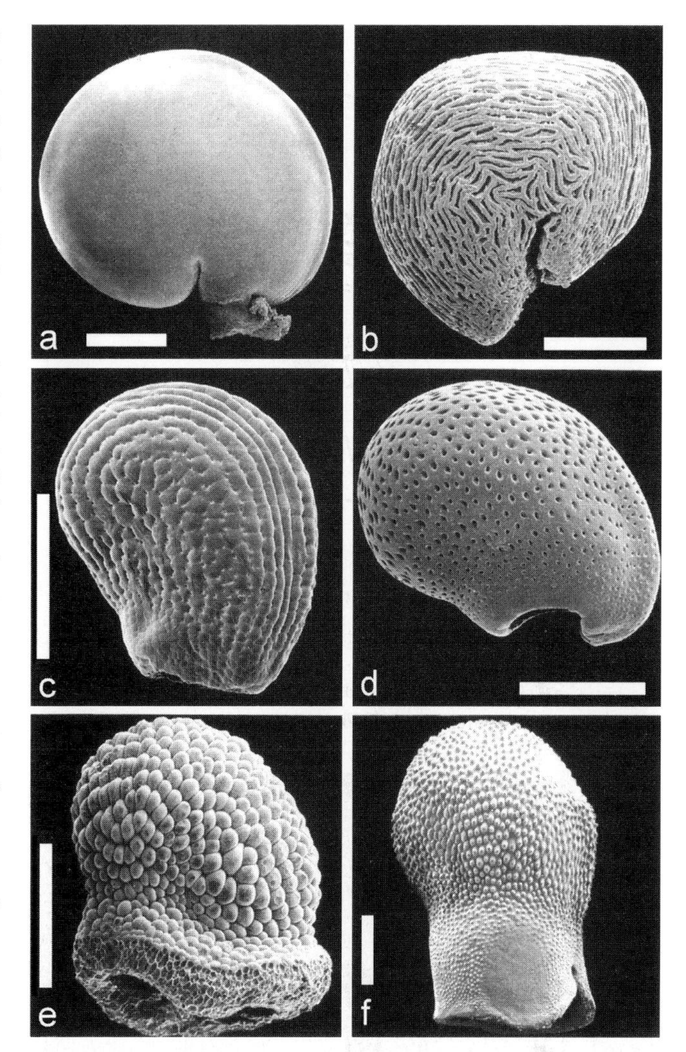

Fig. S.65. Campylotropous seeds in the Cactaceae. Scanning electron microscope (SEM) pictures (scale bars = 0.5 mm). (a) *Pereskia guamacho.* (b) *Pelecyphora aselliformis.* (c) *Cleistocactus buchtienii.* (d) *Ferocactus hamatacanthus* ssp. *sinuatus.* (e) *Gymnocalycium ritterianum.* (f) *Thelocactus bicolor.* From Bouman, F. and Boesewinkel, F.D. (1991) The campylotropous ovules and seeds, their structure and functions. *Botanische Jahrbücher für Systematik* 113, 255–270.

As already noted, the majority of seeds do not show any strong anatomical or morphological adaptations to the environment and can therefore be considered as having evolved in a strictly phylogenetic way; but nevertheless some significant structural adaptations do exist. Some relate to dispersal (**see: Seedcoats – dispersal aids**), such as **sarcotesta** and **aril** (as adaptations to zoochory, animal dispersal) or wings and hairs (as adaptations to anemochory,

Fig. S.66. Campylotropous seeds in the Caryophyllaceae (SEM pictures; scale bars = 0.5 mm): (a) *Arenaria serpyllifolia.* (b) *Saponaria officinalis.* (c) *Lychnis flos-cuculi* (d) *Moehringia trinervia.* (e) *Spergula arvensis.* (f) *Spergularia maritima.* From Bouman, F. and Boesewinkel, F.D. (1991) The campylotropous ovules and seeds, their structure and functions. *Botanische Jahrbücher für Systematik* 113, 255–270.

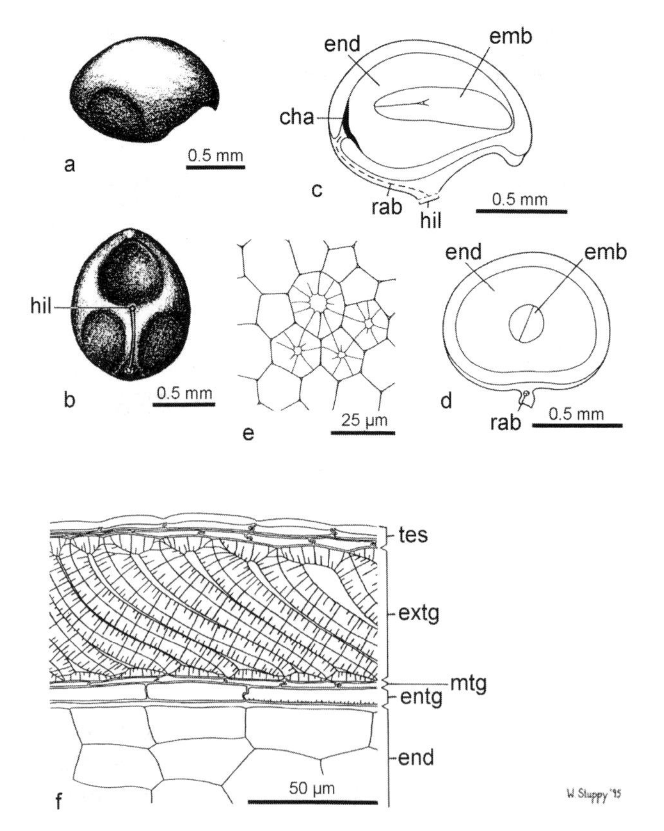

Fig. S.67. *Dysopsis glechomoides* (Euphorbiaceae). (a) Lateral view of seed. (b) Ventral view of seed. (c) Transmedian longitudinal section of seed. (d) Transverse section of seed. (e) Exotegmen cells in surface view. (f) Longitudinal section of seedcoat (end = endosperm; emb = embryo; cha= chalaza; hil = hilum; rab = raphe bundle; tes = testa; extg = exotegmen; mtg = mesotegmen; entg = endotegmen). From Stuppy, W. (1996) *Systematic morphology and anatomy of the seeds of the biovulate Euphorbiaceae.* PhD Dissertation, Dept of Biology, University of Kaiserslautern, Germany (in German with English summary).

wind dispersal). One interesting adaptation to dispersal is the syndrome of minute wind-dispersed seed, so-called dust- and (if there is an air space between the seedcoat and the embryo) balloon-seeds, generally smaller than 0.3 mm (Fig. S.69). This syndrome has arisen independently in monocotyledon and dicotyledon families with remarkable convergences. Without any obvious adaptive advantage, it is striking that such seeds are found in families in nutrient-poor environments (epiphytes, e.g. Orchidaceae, or helophytes, i.e. marsh plants of the Ericaceae, Pyrolaceae, Saxifragaceae), or otherwise nutritionally specialized plants (e.g. parasites and hemiparasites, e.g. Monotropaceae (now assigned to Ericaceae), Orobanchaceae, or carnivorous plants, e.g. Droseraceae, Lentibulariaceae).

Other obvious adaptations concern seed germination. Members of the palms and Zingiberales (ginger and relatives) often have seeds with plug-like structures in the micropylar area (called 'opercula') that become disjoined to permit radicle emergence by circumscissile dehiscence (Fig. S.70). The seeds of some legumes (e.g. *Albizia lophantha*, *Acacia* spp.) possess a specialized structure called 'imbibition lid' or 'strophiolar plug' which is a raised area below the hilum that erupts on heating (**see: Germination – influences of water**). This mechanism allows the previously impermeable seed to germinate after exposure to extreme heat (e.g. forest fires). Similar imbibition lids are known from *Sida spinosa* (Malvaceae).

Other adaptations such as hairs, hooks or **mucilage** on the seed surface help, for example, to position or fix it on to a substrate. The outer epidermis of the seeds of *Blepharis*, for example, whose cell walls consist of cellulose elementary fibrils that are embedded in amorphous mucilage, gives the wet seed a rather bizarre appearance (**see: Deserts and arid lands**). A similar example from the monocotyledons is the epiphytic orchid *Chiloschista lunifera*. At the micropylar end its seeds possess mucilaginous epidermal cells that contain helical wall thickenings which turn into hair-like threads (up to 4 mm long) on wetting, fixing the seeds to the underlying substrate (usually bark of tree branches).

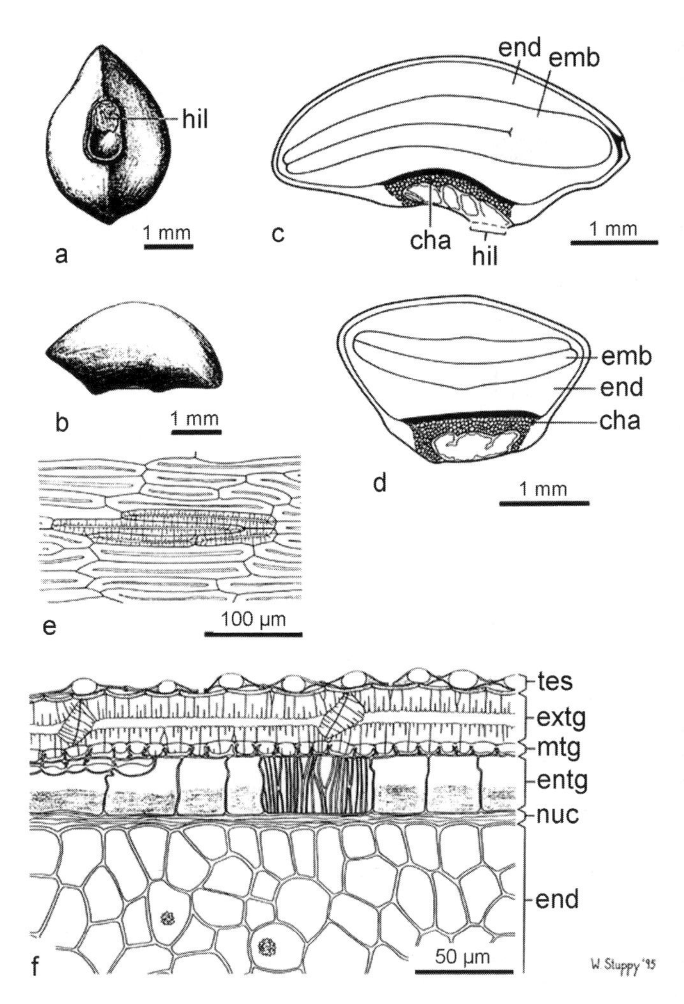

Fig. S.68. *Phyllanthus acidus*, Otaheite gooseberry (Phyllanthaceae). (a) Ventral view of seed. (b) Lateral view of seed. (c) Transmedian longitudinal section of seed. (d) Transverse section of seed. (e) Exotegmen cells in surface view. (f) Longitudinal section of seedcoat (end = endosperm; emb = embryo; cha = chalaza; hil = hilum; tes = testa; extg = exotegmen; mtg = mesotegmen; entg = endotegmen; nuc = nucellus). From Stuppy, W. (1996) *Systematic morphology and anatomy of the seeds of the biovulate Euphorbiaceae.* PhD Dissertation, Dept of Biology, University of Kaiserslautern, Germany (in German with English summary).

The strongest influence on the formation of the seedcoat, however, is exerted by an indehiscent pericarp (fruit coat), often developed by uniovulate ovaries, where, because there is only one seed, there is obviously no need for dehiscence. In such seeds (e.g. lettuce, sunflower) the true seedcoat is reduced, since it no longer serves as the mechanical protection of the seed. It then usually fails to develop the anatomical details that normally provide the phylogenetic clues.

With only 840 species of **gymnosperms** compared to 249,300 species of angiosperms, their seeds show considerably less diversity in their seed characters. Nevertheless, the differences between the four major groups (conifers, *Ginkgo*, Gnetales and cycads) are profound and even within these groups seed characters can provide taxonomically significant information, e.g. for *Cycas*, *Phyllocladus*, Podocarpaceae and Pinaceae. (JD, WS)

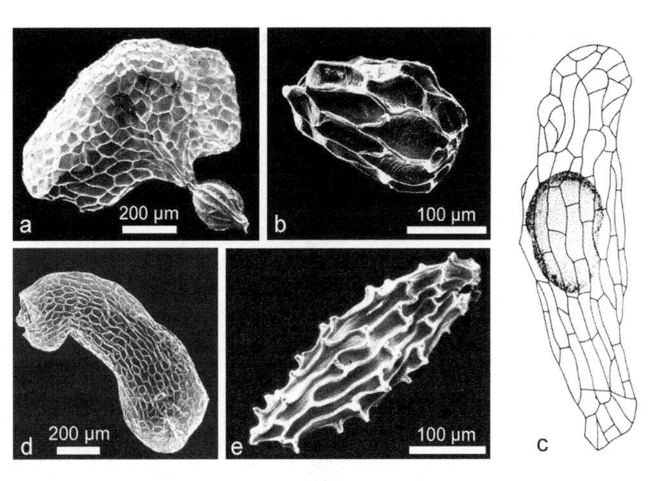

Fig. S.69. Dust and balloon seeds. (a) *Pterospora andromeda* (Ericaceae); dust seed covered by a tight seedcoat with elongate anticlinal pattern and a chalazal outgrowth forming a wing. (b) *Orobanche lutea*, yellow broomrape (Orobanchaceae); dust seed with reticulate anticlinal pattern. (c) *Parnassia palustris*, marsh grass of Parnassus (Parnassiaceae); balloon seed. (d) *Catasetum russelianum* (Orchidaceae); balloon seed. (e) *Orchis* sp. (Orchidaceae); balloon seed showing the reticulate exotesta forming a loose bag around the small, undifferentiated embryo. (a, b, d) From Rauh, W., Barthlott, W. and Ehler, N. (1975) Morphologie und Funktion der Testa staubförmiger Flugsamen. *Botanische Jahrbücher für Systematik* 96, 353–374; (c) from Boesewinkel, F.D. and Bouman, F. (1984) The seed: structure. In: Johri, B.M. (ed.) *Embryology of Angiosperms.* Springer, Berlin, pp. 567–610; (e) from Fahn, A. (1990) *Plant Anatomy*, 4th edn. Pergamon Press, Oxford.

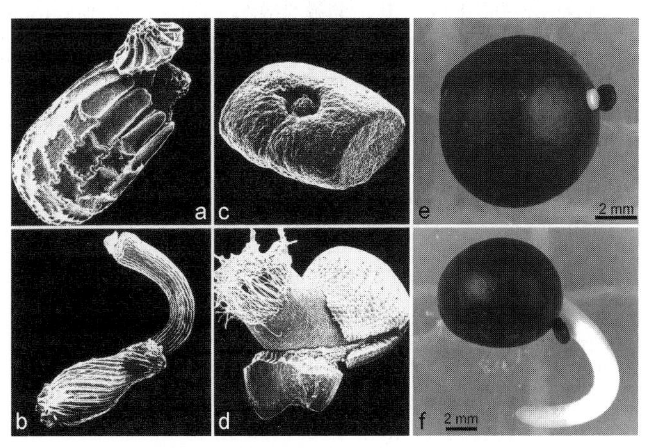

Fig. S.70. Operculate seeds. SEM (a–d) and light micrographs (e, f). (a) *Begonia leptotricha* (Begoniaceae). (b) *Philydrum lanuginosum* (Philydraceae). (c) *Stanfieldiella imperforata.* (d) *Cereus aethiops* (Cactaceae). (e) and (f) *Sabal minor* var. *louisiana* (Palmae). (a–d) From Boesewinkel, F.D. and Bouman, F. (1984) The seed: structure. In: Johri, B.M. (ed.) *Embryology of Angiosperms.* Springer, Berlin, pp. 567–610; (e, f) by Helen Vautier, RBG Kew.

Boesewinkel, F.D. and Bouman, F. (1984) The seed: structure. In: Johri, B.M. (ed.) *Embryology of Angiosperms*. Springer, Berlin, Germany, pp. 567–610.

Boesewinkel, F.D. and Bouman, F. (1995) The seed: structure and function. In: Kigel, J. and Galili, G. (eds) *Seed Development and Germination*. Marcel Dekker, New York, USA, pp. 1–24.

Corner, E.J.H. (1976) *The Seeds of Dicotyledons*. Cambridge University Press, Cambridge, UK.
Mabberley, D.J. (1997) *The Plant-book*, 2nd edn. Cambridge University Press, Cambridge, UK.
Netolitzky, F. (1926) *Anatomie der Angiospermen – Samen* (*Handbuch der Pflanzenanatomie = Encyclopedia of Plant Anatomy*, Bd. 10). Borntraeger, Berlin, Germany.
Savolainen, V., Fay, M.F., Albach, D.C., Backlund, A., van der Bank, M., Cameron, K.M., Johnson, S.A., Lledó, M.D., Pintaud, J.-C., Powell, M., Sheahan, M.C., Soltis, D.E., Soltis, P.S., Weston, P., Whitten, W.M., Wurdack, K.J. and Chase, M.W. (2000) Phylogeny of the eudicots: a nearly complete familial analysis based on rbcL gene sequences. *Kew Bulletin* 55, 257–309.
Schnarf, K. (1937) *Anatomie der Gymnospermen-Samen* (*Handbuch der Pflanzenanatomie = Encyclopedia of Plant Anatomy*, Bd. 10, 1). Borntraeger, Berlin, Germany.
Singh, H. and Johri, B.M. (1972) Development of gymnosperm seeds. In: Kozlowski, T.T. (ed.) *Seed Biology*, Vol. 1: *Importance, Development, and Germination*. Academic Press, New York, USA, pp. 21–75.
Stuppy, W. (1996) Systematic morphology and anatomy of the seeds of the biovulate Euphorbiaceae. PhD Dissertation, Dept of Biology, University of Kaiserslautern, Germany (in German with English summary).
Werker, E. (1997) *Seed Anatomy*. Gebrüder Borntraeger, Berlin, Stuttgart, Germany (*Handbuch der Pflanzenanatomie = Encyclopedia of Plant Anatomy*, Bd. 10, Teil 3: Spezieller Teil).

Sucrose synthesis

Sucrose, a disaccharide containing glucose and fructose, is the principal transport sugar within higher plants. It is the major product of photosynthesis and of the mobilization of starch, hemicelluloses and oils within seeds (**see: Hemicellulose – mobilization; Oils and fats – mobilization; Starch – mobilization**). It is also stored within the vegetative tissues of some plants, e.g. sugarcane, sugarbeet and carrots, but it is not stored in seeds in any appreciable quantities and where present it is almost all in the axis. However, in specially bred varieties such as 'super-**sweet corn**' (vegetable maize) and 'sugar snap' peas, sucrose is present in the main storage tissues.

During mobilization of **starch** and **hemicelluloses**, breakdown of the polymers is accompanied by conversion of the released sugars to glucose-1-phosphate. This is combined with the nucleotide phosphate UTP to form UDP-glucose (below). The UDP-glucose is then converted to sucrose-6-phosphate by its addition to fructose-6-phosphate, and the phosphate is finally removed to yield free sucrose.

Glucose-1-phosphate + UTP ↔ UDP-glucose + PPi
UDP glucose pyrophosphorylase

UDP-glucose + fructose-6-phosphate ↔ Sucrose-6-phosphate + UDP
Sucrose-phosphate synthase

Sucrose-6-phosphate + H_2O → Sucrose + Pi
Sucrose-phosphate phosphatase

The formation of sucrose by the process known as gluconeogenesis from the **fatty acids** released from triacylglycerols (TAGs) is more complex. The fatty acids undergo extensive conversion by β-oxidation and the **glyoxylate cycle** to form succinate, which in turn is converted to oxaloacetate (**see: Oils and fats – mobilization**). This is then converted by the enzyme phosphoenolpyruvate carboxykinase to phosphoenolpyruvate (PEP), which can be utilized to produce the phosphorylated form of fructose; this is combined with UDP-glucose to make sucrose.

Fatty acids → Acetyl CoA → Succinate → Oxaloacetate
β-oxidation *Glyoxylate cycle*
PEP carboxykinase ↓
Sucrose ← Glucose-1-phosphate ← PEP

(JDB)

Sugar sensing

The allocation of resources within a plant is dependent upon its ability to sense altered sugar concentrations, and thus tailor the demands of its sink tissues with the metabolism of source tissues. Sugars are key regulators of several aspects of metabolism within the plant, which affect gene expression during growth and development; thus they play a dual role as a nutrient and as a signalling molecule in signal transduction mechanisms.

There are multiple targets of sugar sensing, which include: (i) hexokinase (HXK)-linked sensing enzymes; (ii) membrane-bound sensors or transporters on the plasma membrane; (iii) intracellular membrane sensors (e.g. tonoplast and endoplasmic reticulum); and (iv) respiratory metabolites where HXK is an intermediary metabolite. These ultimately elicit a response in the cell by altering the transcription of the sugar-sensitive genes.

Sugar sensing is also dependent upon the cleavage of certain sugars, by invertase and sucrose synthase, enzymes which are pivotal in the determination of carbon resources and in the initiation of hexose-based sugar signals. Invertase-derived hexoses influence: (i) maize grain development at the stage of ovary development; soluble invertases are targets of drought stress which results in an increase in ovary abortion; (ii) carrot **somatic embryo** development; transformed embryos lacking vacuolar or cell-wall invertase develop abnormally; and (iii) cell division and reserve storage in developing large-seeded legumes; sucrose favours storage deposition and seed maturation, whereas hexoses promote cell division. (**See: Storage reserves synthesis – regulation**)

Sugar concentrations also affect germination. For example, low concentrations (10 mM) of mannose or deoxyglucose inhibit *Arabidopsis* germination, whereas glucose overcomes the inhibitory effect of **abscisic acid** (ABA). Here, exogenous glucose can modulate ABA concentration through influencing its synthesis and degradation, although there may be other sugar-sensitive events that are more influential in controlling germination such as modulation of hormonal signals. (JDB)

(**See: Dormancy breaking – cellular and molecular aspects, 3. Gene expression; Signal transduction; Signal transduction – hormones, 1. Abscisic acid**, Fig. S.29)

Andersen, M.N., Asch, F., Wu, Y., Jensen, C.R., Næsted, H., Mogensen, V.O. and Koch, K.E. (2002) Soluble invertase expression is an early target of drought stress during the critical, abortion-sensitive phase of young ovary development in maize. *Plant Physiology* 130, 591–604.

Gibson, S.I. (2005) Control of plant development and gene expression by sugar signaling. *Current Opinion in Plant Science* 8, 93–102.

Price, J., Li, T.-C., Kang, S.G. and Na, J.-C. (2003) Mechanisms of glucose signaling during germination of Arabidopsis. *Plant Physiology* 132, 1424–1438.

Sugarbeet

Sugarbeet (*Beta vulgaris*), a member of the Chenopodiacae, is grown in temperate or Mediterranean climates. The crop is almost entirely confined to the northern hemisphere, the exceptions being Chile and Uruguay. Modern varieties produce conical white roots that typically contain 15–20% sucrose (**see: Carbohydrates**) on a fresh weight basis at harvest time, deposited in concentric zones of parenchymatous tissue. In the present day, beet supplies about 37 million t of the approximately 125 million t total world annual consumption of raw sugar (being far outstripped by that produced from **sugarcane**). *B. vulgaris* also contains vegetable garden red or table beets (cultivated for their swollen taproots, in an assortment of shapes and colours, which quickly develop during the first growing season), and spinach beet and Swiss chard (cultivated for their leaves) and mangels or fodder beet (grown for animal feed). Sugarbeet is also a source of animal feed, which is compounded from the pressed dry pulp by-product remaining after the sugar extraction combined with the molasses, a mixture of residual sugar and impurities remaining after crystallization of the extracted sugar solution.

Sugarbeet is a biennial plant that is grown and harvested as a root crop at the end of its first season of growth. It is usually sown as early as possible in the spring, consistent with avoiding severe frosts (<−6°C) that would damage the young seedlings and any prolonged, cool periods that would **vernalize** the seedlings so that they bolt, flower and produce seed in the first year. **Bolting** detracts from root yield, and can also lead to the undesirable build up of 'volunteer' weed plants in the soil seed bank, which affect the cultivation of subsequent beet crops. In most locations the crop is harvested during autumn or winter, before freezing can damage the roots or the effects of vernalization cause the compressed stem to elongate. In Mediterranean areas (Egypt, Morocco, and parts of Spain and Italy) where the winter is so mild that **vernalization** is very slow, beet are sown in late summer and grown over winter to be harvested early the next year. Field **rots**, and the prevalence of severe leaf diseases, such as *Cercospora*, prevent beet from being grown at latitudes of less than about 30°.

1. History and origins

The leaf types of beet were recorded as vegetables in biblical times. All were derived from the subspecies *B. maritima*, which grows on Mediterranean and temperate seashores. The discovery in 1747 by Marggraf that the sugar crystallized from alcohol extracts of beet roots was the same as that from sugarcane led over the next half century to the development of the first white conical beet types and the first beet-sugar processing factory in Prussia. Napoleon in 1811, responding to the British blockade of European ports that cut off the supply of cane sugar from the Caribbean, promoted the industry in France and interest was kept alive, even after the resumption of cane sugar imports, particularly in those countries (Russia, Austria, and Hungary) that had no cane-growing colonies. Varieties were developed from Silesian fodder beet by selection for high sugar concentration, and later also for high root yield. By the 1880s sugar production throughout the world from beet exceeded that from cane, and by 1900 beet sugar production had spread across almost all of mainland Europe, from Sweden to Spain to the south, and Russia to the east. The industry also developed in North America, starting in California in 1870 and spread to the inter-mountain areas and to the northern plains and the Great Lakes states, where it is now centred.

2. Development, genetics and breeding

Beet is naturally a diploid plant, being mostly self-sterile and cross-pollinated by the wind. Its biennial life cycle requires a long, cool vernalization period to induce flowering. After the switch from the vegetative to the reproductive phase, the stem elongates (the plant 'bolts') and grows up to about 1.5 m high. Flowers without petals are borne in the axils of small leaves on the stem. Naturally these flowers occur in clusters, giving rise to clumps of fruits that usually each contain a single true seed. As many as four or five 'multigerm seeds' can be fused together, which are very difficult to separate by milling, but in modern commercial genetic monogerm varieties the flowers are borne singly. (**See Colour Plate 6D**)

The aim of modern sugar beet varieties is to drill-to-a-stand, that is to obtain a predetermined uniformly spaced plant population. Originally it was necessary for the farmer to remove the surplus seedlings by thinning of the excess seedlings (singling), leaving just one plant per position to give the desired population density. (Indeed, in areas of the world where manual labour is more plentiful, such as in Egypt, multigerm seed is still used to sow, and seedlings are thinned and weeds controlled by hand. Multigerm table red beet varieties are also still used in a similar way.) Natural **monogermicity** permits seed to be more accurately spaced in a precision drill, greatly increasing the chance of being presented with a single seedling when thinning the stand. 'Mechanical monogerm' seed – sometimes called 'precision seed' – was originally produced by rubbing, chopping or breaking the large multigerm clusters, followed by size grading (Colour Plate 6D).

A further major step forward was the discovery in 1942 of **cytoplasmic male sterility** and associated restorer ('O' type lines), which enabled traits to be introduced more quickly through hybrid breeding. The recessive genetic monogerm seed character (first isolated in Kiev in the 1930s) was introduced commercially in the USA and Europe from the late 1950s to the mid-1960s, and is present in nearly all modern cultivars, being introduced to the hybrid cross on the male-sterile maternal lines. Initially higher yielding triploid genetic monogerm varieties were produced from diploid × autotetraploid (**autopolyploid**, with 36 chromosomes) parent lines; but since the early 1990s diploid pollinators have again been widely used to produce diploid varieties. Any seed on the pollinators is removed from the field by **roguing** before seed harvest, and any multigerm by size grading and gravity separators during the **conditioning** stage.

Genetically modified varieties have been developed with tolerance to broad-spectrum herbicides (glyphosate or glufosinate-ammonium) and extensively tested, but at present have not been commercialized.

3. Production

Most genetic monogerm sugarbeet seed used for commercial beet production in Western Europe is produced either in southwest France or in the Po Valley of Italy, and for the USA in Oregon. Production locations preferably have dry conditions during seed development, to avoid the proliferation of seedborne disease, and warm, as any cold experienced by the ripening seed crop may cause seed to be vernalized while still on the plant and so increase the proportion of bolting plants in the subsequent crop.

Seed is produced by the root-to-seed method, and sometimes by the seed-to-seed method (**see: Production for Sowing, I. Principles**). In the former method, the parental lines are planted at the same time in nursery beds in late summer. Plants (**stecklings**) are harvested in the following early spring, and the pencil-shaped roots with attached trimmed foliage are transplanted into seed production fields, typically in a 2:6 row ratio of male:female plants. Wind-borne pollen can fertilize beet flowers over great distances, and long **isolation distances** (at least about 1 km, and often 10 km in practice) are prescribed to separate hybrids produced with different male pollinator lines. Male plants are usually removed after flowering in midsummer. Maternal plants are clipped before pollination to stimulate the production of side branches and thus more seed, to reduce plant height in order to avoid lodging, and to optimize the size and the quality of the seed produced. Harvesting is usually done by equipment which either lays the seed on the **swath**, from which it is combined after drying about 7 days later, or by direct combining after the application of a **desiccant** to reduce water content.

After local precleaning to remove field debris, seed is sent for further precleaning (**scalping** and **aspiration**) and conditioning. Rubbing, grading and polishing removes the substantial amount of the **pericarp** (the corky parenchymatous cortex) using specialized debearding equipment, and seed is size and density graded by screening and gravity tables (**see: Conditioning**). **X-ray analysis** is commonly used to monitor the quality of seed, for small, malformed, cracked **embryos** (**see:** Fig. X.1 for an illustration of a high-resolution version of the technique). The resultant processed seeds typically weigh about 10 mg, containing a germ of about 3 mg (in which the embryo surrounds a starchy **perisperm**) surrounded by a substantial amount of remaining pericarp, including the lid-like seed cap (**operculum**) which is forced open at the end of germination by the expanding **radicle**. (**See: Colour Plate 6D**)

4. Quality

Monogermicity is typically more than 99%, and viability more than 90%, and frequently more than 95%. Seed **vigour** is an important factor in crop establishment, and a variety of proprietary stress tests are used in the industry as an adjunct to the conditioning process.

The cortex contains water-soluble substances that inhibit germination, which have been identified as **phenolic** compounds and salts (Na and K) (**see: Germination inhibitors**). Sufficient inhibitors remain after conditioning to slow germination in laboratory conditions. For these reasons, standard germination testing procedures prescribe a preliminary rinsing, involving several changes of water, and drying of seed. The seeds are also sensitive to the amount of water used in tests, which is ascribed to the blockage of the basal pore, which impedes oxygen supply to the embryo. Soils that hold water, such as heavy clays, may therefore limit the rate of seed germination more easily than well-drained soils.

5. Sowing and treatment

The yields of beet crops are closely and positively related to the amount of solar energy intercepted by their foliage. To ensure adequate interception, the plant population density needs to be 75–100 thousand plants per hectare. To maximize yield it is valuable to extend the growing season for as long as possible, particularly by sowing early to capture more light in late spring and early summer. However, sowing earlier means that seeds may be exposed for long periods to temperatures only a few degrees in excess of the **base temperature** for germination (3°C). Sowing sugarbeet in **transplant** modules is practised in Japan; long 'paper pots' are used to ensure that the primary root grows straight, as any contortions will affect the shape of the mature root, and hence yield at harvest. Table beets are also transplanted.

Also the timing of sowing must be chosen to limit exposure of plants to vernalizing conditions (above freezing but below about 12°C) which could induce bolting, flowering and seed production in the first year. In environments where the temperatures are in this range for a long period (cool maritime regions), the need to produce varieties that are bolting resistant (needing a relatively long period of vernalization) has been an important plant-breeding objective.

The need for accurate seed placement by the drill (the practice of 'drilling to a stand'), combined with the poor flow characteristics of the discus shape of monogerm beet seed, led to the widespread use of seed **pelleting** in the European beet industry, along with **encrusting** in North America. Depending on the coating materials used, seed weights are commonly increased by up to a factor of about 1.5 to 3. On average more than 75% of seeds sown establish seedlings in the field.

Most of the monogerm seed supplied around the world is treated with one or two fungicides, such as thiram or metalaxyl to control seedborne fungi (such as deep-seated *Phoma betae*) and hymexazol to control soilborne **damping-off** fungi such as *Pythium* and *Aphanomyces*. Seeds may also be treated with one or two contact or systemic insecticides, such as a pyrethroid (tefluthrin or β-cyfluthrin) and a neonicotinoid (imidacloprid or clothianidin, currently) to control soil-inhabiting arthropod pests; the systemic and persistent activity of neonicotinoids also controls foliage-eating pests and early invasions of aphids, which carry viral pathogens to the established crop. (**See: Treatments – pesticides**)

Steeping seed to remove the water-soluble inhibitors is widely used to improve germination rate and, in some countries, **priming** is also used commercially to 'advance' germination and seedling emergence, and increase yield as a consequence. This treatment leads to an increase in the proportion of cells in the **root meristem** that are in the 4C stage of the **cell cycle**. (KJ, PH)

For information on the particular contributions of the seed to seed science **see: Research seed species – contributions to seed science**.

Cooke, D.A. and Scott, R.K. (eds) (1993) *The Sugar Beet Crop: Science into Practice*. Chapman & Hall, London, UK.

Coumans, M., Côme, P. and Gaspar, T. (1976) Stabilized dormancy in sugar beet fruits. 1. Seed coats as a physiochemical barrier to germination. *Botanical Gazette* 37, 274–278.

Durrant, M.J., Mash, S.J. and Jaggard, K.W. (1993) Effects of seed advancement and sowing data on the establishment, bolting and yield of sugarbeet. *Journal of Agricultural Science* 121, 333–341.

Perry, D.A. and Harrison, J.G. (1974) Studies on the sensitivity of monogerm sugar beet germination to water. *Annals of Applied Biology* 77, 51–60.

Sliwinska, E., Jing, H.-C., Job, C., Job, D., Bergervoet, J.H.W., Bino, R.J. and Groot, S.P.C. (1999) Effect of harvest time and soaking treatment on cell cycle activity in sugarbeet seeds. *Seed Science Research* 9, 91–99.

Whitney, E.D. and Duffus, J.E. (1995) *Compendium of Beet Diseases and Insects*. APS Press, Minnesota, USA.

Sugarcane

Cultivated sugarcane (*Saccharum* spp.) clones (cultivars) are propagated vegetatively through cane stem sections ('setts'), and flower rarely. All modern commercial cultivars are complex hybrid derivatives, with chromosomes introduced from wild forms (*S. spontaneum* and others) that have been **hybridized** and backcrossed on to 'noble canes' (*S. officinarum*). They have a high degree of **polyploidy** combined with much complex **aneuploidy** due to irregularities in **meiosis** (**chromosome numbers** generally vary between 2n = 100 and 2n = 130), and there is much variation in fertility and in the ability to produce seed. Flowering is daylength-sensitive and is negligible at very low (0°) or high (20–30°) latitudes. However, seed production can be a nuisance at intermediate latitudes, as it leads to rapid maturity and reduces total sugar yield, and the trait must be selected against in breeding for those regions. At the same time, breeding requires the facility to produce true seeds in order to obtain genetic recombination between parental lines. The **inflorescence structure** consists of an open-branched panicle ('arrow') bearing thousands of mainly cross-pollinated flowers in pairs of **spikelets** that after fertilization produce very small seeds (200–300/g), many of which are nonviable or maintain **viability** for only a short period under normal storage conditions. The seeds ('fuzz') have long silky hairs that facilitate wind-dispersal, and are usually collected by enclosing the tassel in a bag (**see: Seedcoats – dispersal aids**). (PH)

Summer annual

An annual plant that germinates in the spring and produces seeds in late summer. (**See: Dormancy – cycles; Dormancy – ecophysiology; Winter annual**)

Sunflower

1. The crop

Sunflower (*Helianthus annus*, 2n = 34, Compositae or Asteraceae) is the world's fourth most important vegetable oil source (**see: Oilseeds – major**, Table O.1). It is an annual plant unique among oil crops in having a single seed-bearing head (capitulum). The French and Spanish names for sunflower, *tournesol* and *girasol*, respectively, mean 'turn with the sun', which sunflower heads do until anthesis, after which they face east.

The sunflower is grown for its seed in many countries with the major producers, in order of importance, being the former USSR, Argentina, France, USA, China, India, Spain, the Balkan states, Hungary, Turkey and South Africa (**see: Crop Atlas Appendix, Maps 2, 17, 22**). In 2002 just over 24 million t of seeds were produced but only 2.3 million t were exported (FAOSTAT). Shipping whole seed is costly due to the bulk and low value of the fibrous hull. Thus, the extracted oil rather than the seed is traded internationally. Sunflower is one of the four major annual species grown for oil and feed meal, others being soybean, rapeseed (canola) and peanut.

2. Origin

Sunflower and its close relatives are all native to the Americas. The cultivated type evolved to a single-headed plant from the wild multi-branched form. Archaeological evidence suggests that sunflower cultivation began in the Arizona and New Mexico region as early as 3000 BC and may have been domesticated before maize. The distribution of the crop at the time of European exploration extended from the states of Arkansas to the Dakotas and eastward to Ontario, Canada and Pennsylvania. The sunflower was introduced into Europe in the 16th century and became established as an oil crop in Russia. Selection for high oil content, which Russian breeders initiated in 1860, increased the oil percentage from about 30% to between 40 and 50% (**see: Oilseeds – major**, Table O.2). This greatly increased the economics of crop production, oil extraction and transportation, allowing expanded production in many countries. The discovery and introduction of a **cytoplasmic male sterility**–nuclear fertility restorer system in the early 1970s made possible the production of today's high-yielding hybrid varieties. (**See: Sunflower – cultivation**)

3. Fruit and seed

The seed of sunflower is an **achene** or fruit made up of the kernel surrounded by an adhering **pericarp** or **hull**. Achenes vary in size from 7 to 25 mm long and 4 to 14 mm wide, pointed at the base and rounded at the top. Size and other attributes, such as number of seeds per head, vary considerably between large-seeded confectionery and small-seeded oil-seed types. For example, recorded seed weights in confectionery types lie in the range 91–125 mg per seed; oil-seed types are 43–86 mg. Oil and protein contents in confectionery types are, respectively, 29–31% and 16–18% (on a dry weight basis); in oil-seed types the respective values are 46–56% and 10–18%. The position of the seed on the head can affect both kernel:seed ratio and kernel oil concentration to a lesser degree within a given variety or type. Both attributes are often lower in seeds from the periphery compared to the centre of the capitulum, and some variation is also found in response to crop population density. Large achenes usually have thick hulls that are only partially filled, while small achenes usually have thin, tightly fitting hulls. The achenes mature from the periphery of the head to the centre. The colour of the achene may vary from white, through shades of brown and grey, with dark striping of various widths in confectionery types, to the characteristic solid black of most oilseed varieties. (**See: Colour Plates 12E–F**)

The pericarp or hull, which develops with or without fertilization, consists of an outer layer, four to six cells deep, arranged in radial rows with the outer layer of cells having thick walls and a thin cuticle coating. Twin short, straight hairs, normally fused together over most of their length, emerge at intervals from a basal cell that is slightly raised above the surrounding epidermal cells. The hairs are largely removed in the harvesting and cleaning process. The middle layer has a single layer of parenchyma cells that run in a radial direction at regular intervals, separating polygonal cells into bundles that run longitudinally. At maturity the outer cells of the middle layer become a disorganized dark mass, the **phytomelanin** layer, commonly called the 'armour layer', that provides protection against the sunflower moth (*Homoesoma electellum* Hulst.) and the European sunflower moth (*H. nebuella* Hbm.). Thin-walled, loosely packed parenchyma cells make up the inner layer.

The thin **testa** encloses the **endosperm** and **embryo** and is made up of an inner and outer layer of parenchyma and a middle layer of spongy parenchyma. The endosperm consists of one or two layers of **aleurone layer** cells containing **oil bodies** and very small **protein bodies**. The embryo is made up of two long cotyledons, containing almost all the seed's oil, and a pointed radicle/hypocotyl axis. The cotyledons are mostly made up of palisade parenchyma containing oil and protein bodies. (**See: Sunflower – cultivation** for information on seed growth)

4. Uses

Achenes of small, thin-hulled varieties are processed for oil while the large-seeded types are sold as bird feed or roasted and consumed as food snacks. In the oil extraction process, the hull is removed by decortication and the kernels processed for their 46+% oil and 40% protein meal. The oil is marketed as a premium salad oil due to its clear colour, low linolenic content and high level of nutritionally desirable linoleic acid (68%) (**see: Fatty acids; Oilseeds – major,** Table O.2, for composition). However, the oil lacks stability for commercial frying applications. To overcome this shortcoming, varieties producing a high oleic oil (80%) were developed but the National Sunflower Association in the USA has now endorsed the development of varieties producing oils containing 65% oleic and 26% linoleic acid that is being marketed as NuSun Oil. (**See: Oils and fats; Oils and fats – genetic modification**)

Dehulled sunflower meal is used as a protein supplement for swine and poultry but it is lower in lysine and higher in methionine than **soybean** meal (**see: Storage proteins**). Meal from partially dehulled sunflower can be successfully fed to ruminants. Sunflower hulls may be fed to animals as roughage, used as fuel, pressed into fireplace logs, or used to produce ethyl alcohol. (KD, AJH, AJdlV)

(**See: Sunflower – cultivation** for further information on seed properties, breeding and production)

Davidson, H.F., Campbell, E.J., Bell, R.J. and Pritchard, R.A. (1996) Sunflower oil. In: Hui, Y.H. (ed.) *Bailey's Industrial Oil & Fat Products*, Vol. 2, 5th edn. John Wiley, New York, USA, pp. 603–688.

Fick, G.N. (1989) Sunflower. In: Robbelen, G., Downey, R.K. and Ashri, A. (eds) *Oil Crops of the World*. McGraw-Hill, New York, USA, pp. 301–318.

Putt, E.D. (1997) Early history of sunflower. In: Seiler, G.T. (ed.) *Sunflower Technology and Production*. American Society of Agronomy, Madison, WI, USA, pp. 1–19.

Schneiter, A.A. (1997) Sunflower anatomy and morphology. In: Seiler, G.T. (ed.) *Sunflower Technology and Production*. American Society of Agronomy, Madison, WI, USA, pp. 67–111.

Vaughan, J.G. (1970) *The Structure and Utilization of Oil Seeds*. Chapman and Hall, London, UK.

Sunflower – cultivation

Sunflower is successfully grown over a widely scattered geographical area and is considered a crop adapted to a wide range of environmental conditions. (**See: Crop Atlas Appendix, Map 17.**) The 'seed' or grain is actually an **achene** (botanically a fruit) but we will use here the term, seed. The embryo and perisperm of the seed together are often referred to as a kernel, although this term is sometimes applied to the embryo alone. Further information on the seed, origins and economic importance, uses, etc. are found in **Sunflower**. (**See: Colour Plate 12E, F**)

1. Genetics and breeding

Sunflower is a cross-pollinating (allogamous) species, which shows significant hybrid vigour (**heterosis**) for both seed and oil yield. Although **open-pollinated** and **synthetic varieties** may still play a role in marginal growing regions, development of single-cross and three-way hybrids is the primary objective of most breeding programmes around the world. Commercial hybrid seed production in sunflower is based on **cytoplasmic male sterility** (CMS) and **nuclear fertility restoration**.

The breeding procedures for developing inbred lines and hybrids are similar to those utilized in **maize** and **sorghum**, with certain modifications required due to floral morphology and specific traits. Preharvest selection for agronomic traits and postharvest selection for achene characteristics (oil content, **fatty acid** composition, etc.), and seedling-disease reaction are all used. Over time, breeders have tended to select for smaller grain size in oil-seed hybrids.

The fatty acid composition of the oil is a key determinant of its quality (**see: Oilseeds – major**), although the content of minor components such as tocopherols is becoming increasingly important. Protein concentration of the embryo tends to vary inversely to oil content. Much of the sunflower grown for oil will produce high-linoleic and low-oleic oil, with other fatty acids found in small amounts. High oleic oil is less susceptible to oxidative changes during refining, storage and frying. The linoleic:oleic ratio shifts towards the latter in crops grown in warm environments, with night temperature playing an important controlling role. Since the development of the first high oleic cultivar via chemical mutagenesis in 1976, breeders have been able to generate hybrids that produce high oleic acid oil in both cool and warm environments. (**See: Sunflower**) Because of the low number of genes involved, their dominant nature and stability, and the embryo control of the trait that allows the use of the **half-seed technique**, the introgression of this character into breeding lines is relatively easy to manage. Work is now being conducted to obtain variants with high palmitic and stearic acid oils.

The industrial value of the seed is strongly determined by the proportion of **pericarp** (usually an unwanted by-product which dilutes the protein concentration of the meal obtained in the crushing process, with very little value for processors – it is often burnt for fuel), ease of hull removal during processing ('hullability') and the constitution of the kernel. Much breeding effort has gone into altering these attributes. As a result, there is much interest in processes that allow the hull to be removed before oil is extracted from the kernel, and in the determinants of hullability.

Backcross-converted inbred lines with immunity to all currently known races have been developed to downy **mildew** (*Plasmopara halstedii* – see below) and such resistance is being incorporated into hybrids. Most commercial hybrids are resistant to races 100, 300 and 330, but many new races have been reported in the main sunflower regions recently. This includes isolates showing physiological resistance to metalaxyl, one of the main fungicidal controls available (see below).

Genetic stocks derived from crossings with a wild population collected in Kansas, showing resistance to the imidazolinone herbicides, are being used to develop hybrids, which should significantly improve weed control in the crop. This technology also constitutes a complement for recent developments, such as pre-planting, intra-row cultivation and deep fertilization, aimed at adapting sunflower to non-till cultivation. (**See: Tillage**)

2. Development and production

Primitive and modern cultivated types are monocephalic, the stem terminating in a large inflorescence (capitulum) bearing attractive, large-petalled, sterile, infertile ray florets around the rim and numerous fertile, seed-producing, disc florets across the surface of the receptacle, which lack conspicuous petals and produce the seeds.

Quality for industrial purposes is largely determined by seed and kernel size/shape (especially for confectionery types), kernel:seed ratio, hullability, and oil concentration/fatty acid composition (especially for oil-seed types), although protein content and that of minor components such as tocopherols are also important. Variability in seed size is of interest both to seed producers and farmers (who have to deal with the effects of heterogeneity), and also to processors of seed for both oil and confectionery purposes, since both hullability and the value of product, respectively, are size-related.

The three components of crop yield, then, are seed number and size and, in oil-seed hybrids, oil concentration. These components are determined sequentially, with some overlapping, during the crop development cycle. Also sunflower has a remarkable capacity for compensation between seed number per plant and size in response to variations in crop density, so that yield is relatively constant over a wide range of plant populations.

(a) *Seed number* is the dominant component of yield, and the one that exhibits the strongest responses to environment and genotype. Seed number is determined over a broad developmental window, running from the floret differentiation phase (which is completed well before anthesis) to about 2 weeks after anthesis. During this interval the number is most sensitive to both intercepted radiation and temperature (to which it has a negative response above 22°C).

(b) *Seed size* can show important variations in response to genotype, environment (including plant population density), and their interactions, and to position on the head. Potential seed size is also dependent on crop population density.

The dynamics of the growth of the parts (hull, kernel, oil) of oil-seed and confectionery types are clearly distinct (Fig. S.71). Physiological maturity is achieved at about 650d°C (**day-degrees**, expressed on a base temperature, T_b, of 6°C), and the critical period for grain weight determination is from 250 to 450d°C after anthesis. Many of the effects on size derive from conditions to which the crop is exposed during seed filling, particularly water and nitrogen stresses, radiation and temperature; but there is also evidence that pre-anthesis conditions can condition seed size, perhaps by determining the maximum size achievable by the pericarp. The effects of post-anthesis conditions can be ascribed to changes in the rate and duration of the growth of the seed and its parts, particularly the embryo, although the growth of the pericarp may also be affected.

(c) *Oil concentration*. In oil-seed types, this has been importantly increased by breeding, attributable to changes in both kernel oil concentration and kernel:seed ratio. Genotypes with a white-pigmented achene hypodermis show lower kernel:seed ratio and lower oil concentration than those with an uncoloured hypodermis. Because the components shown in Fig. S.71 commence their growth sequentially, any stress affecting the crop after the start of seed filling (e.g. late drought, low light irradiation or high temperature stress), will reduce kernel size and oil content with respect to unstressed crops. This is of particular importance given the virtual absence of oil in the hull, which contributes substantially to seed weight. Effects of this type can interact with the pre-anthesis experience of the crop imposed by such factors as crop population density, which can alter potential seed size.

Time of sowing and environment can both also alter oil concentration. Interestingly, the response of oil to the time of sowing is dominated by changes in kernel oil concentration and has little to do with changes in the hull:kernel ratio. Numerous studies have shown that oil yield in sunflower is reduced when normal spring sowing dates are delayed, in both temperate and subtropical environments. Interactions have been found between genotype and sowing date for yield, which suggests the existence of genetic variability for traits related to specific adaptation to late plantings. The lower yields associated with late plantings have been variously hypothesized as being due to: (i) warm temperatures during the early growth period, which would promote an early excessive stem growth and reduce time to flowering; and (ii) cooler temperatures and reduced incident radiation post-anthesis, which would affect the dynamics of seed filling. The time of sowing effects on seed size and oil concentration do appear – in unstressed crops, and within limits – to be strongly and positively related to the radiation environment experienced during the critical period for seed weight determination.

While the broad picture of seed size and oil concentration responses to radiation and temperature during filling seem well established, it is not yet understood why oil concentration can be reduced in crops where filling occurs under very favourable environments or when photoassimilate source/sink

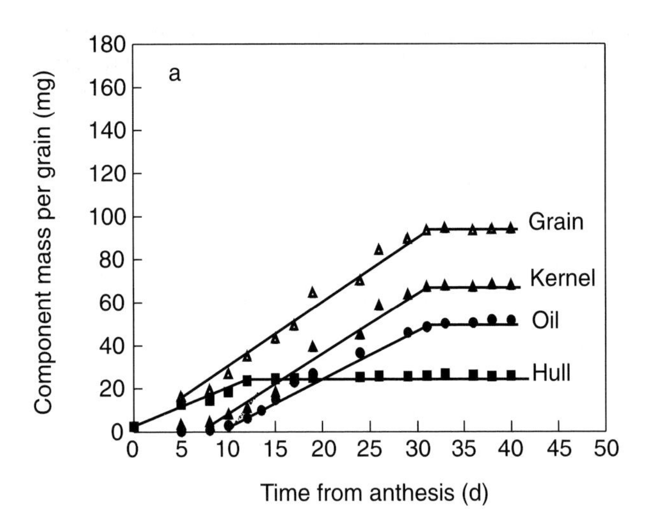

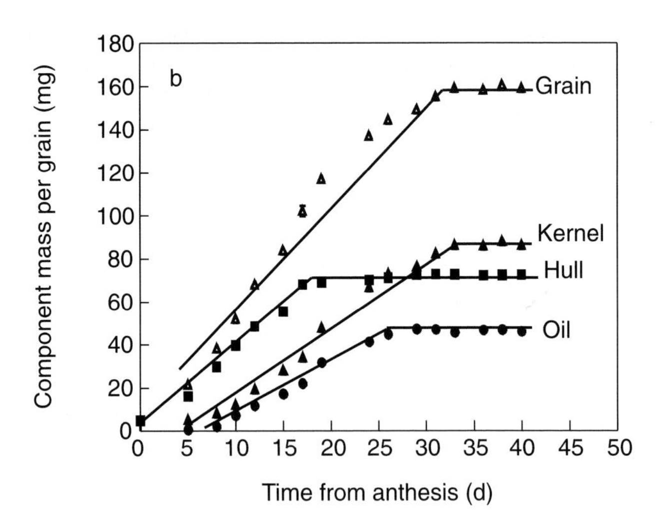

Fig. S.71. Dynamics of the growth of the whole seed (grain) and its components in oil-seed (a) and confectionery (b) sunflower. Protein tends to accumulate steadily in the grain once the growth of the kernel becomes established. By kind permission of A. Mantese.

ratios are high. Similarly, the physiological bases of the widely recognized effects of high nitrogen availability in depressing oil concentration have yet to be satisfactorily elucidated.

3. Production

Increases of the different lines are initially accomplished under bags in the breeding nursery and under screened cages, into which hives of honeybees are placed to ensure pollination of the lines to be crossed.

In hybrid seed production fields, the best compromise between maximum yield and desired seed size is normally reached at 50,000 to 60,000 plants/ha. Field increase of **Foundation** and commercial hybrid seed free of outcrossing requires isolation from other pollen sources, such as farmers' crops, wild and volunteer sunflowers. Adequate and timely pollen production is necessary for good pollen-stigma overlapping development between parental lines (**nicking**); pollination is ensured by placing beehives/ha generally in the seed production field at the beginning of anthesis.

Harvesting, drying, conditioning and storage of sunflower seeds have many peculiarities, associated with its physical and morphological attributes and differences with those of other major crops. Harvesting of seed should not start until the seed moisture has reached 11 to 13%. Chemical **desiccants** are used to hasten dry-down of sunflower after **physiological maturity** is reached. Excessively low harvest moisture (<9%), results in high levels of pericarp damage, which in turn can lead to an increased level of fungal infection of the seed. During **conditioning** air-screening may be carried out to remove non-seed material, and seeds are sized and usually precision graded according to thickness and width.

Seed that has been sized to fit certain types of planting equipment is evaluated on actual planters to ensure the seed will be planted properly under normal conditions. Because of the fairly large range in seed sizes due to effects of position on the inflorescence, seeds in some markets are sold in a range of size grades, and there may be some scope for using seed coatings to increase uniformity.

4. Quality, vigour and dormancy

Germination tests differ according to AOSA rules (20°C) and ISTA rules (20, 25, or alternating 20 to 30°C) for 7 days. Significant differences in emergence have been reported between **pure-live-seed**lots differing in vigour under low soil moisture conditions. **Accelerated ageing**, **cold tests**, and leachate **conductivity tests** are used for indirect evaluations of seed vigour, and have been found to provide better predictions on field emergence than conventional germination tests. Seed size also appears to be negatively correlated with vigour. Germination of sunflower seed after sowing is very sensitive to soluble salt in the soil and fertilizer applied in the row at seeding.

Freshly harvested sunflower seed normally expresses postharvest **dormancy**, for up to about 50 days depending on variety, an important feature for seed producers and farmers. Dormancy is attributable to each of the pericarp, perisperm and embryo-components that are lost at different rates during dry storage of the seed. **Gibberellin**, **abscisic acid** and **ethylene**, and tissue sensitivities to these hormones, are all considered to be involved in the control of dormancy and its breakage. Sunflower appears to be unusual among crop species in that dormancy is heightened when the mother plant experiences high temperatures during seed filling. Quality evaluation of dormant seed can be performed using the **tetrazolium test**. Alternatively, germination tests can be conducted by artificially breaking dormancy using ethylene or its precursors.

There has been little systematic work done on the biochemistry and control of sunflower seed germination, in spite of the fact that farmers regard germination and establishment of this species as less predictable than for other crops. Likewise, understanding of the control of seed longevity and ageing in sunflower is somewhat limited.

5. Diseases and treatment

(a) *Seeds and seedlings*. Sunflower is a host for over three dozen pathogenic organisms. Of these, fewer than a dozen cause serious economic losses on a regular basis and only two of them are major seedling diseases, both of fungal nature: downy **mildew** and *Alternaria* blight, both of which may be controlled by seed treatments.

Downy mildew is caused by soilborne, windborne and seedborne *Plasmopara halstedii* and is endemic wherever sunflower is cultivated. It produces systemic and localized infections, the first being the most relevant in terms of economic losses. Cool, waterlogged soil during a short period around germination and emergence strongly favours infection. Systemic root-initiated infection symptoms include **damping-off** and seedling blight. Infected seedlings that survive will produce stunted plants (shortened internodes) with chlorotic and puckered leaves and erect, platform heads with little, if any, seed. Chemical control relies on seed treatment with metalaxyl or oxadixyl, but the most effective way to control the disease is to incorporate resistance into the host.

At least four species of *Alternaria* are *bona fide* pathogens of sunflower, causing seedling blights, especially with contaminated seeds or when *Alternaria*-infected debris remains on the soil surface from the previous season. Genotypic variability for quantitative resistance to *Alternaria* spp. has been reported, but no sources of immunity have been identified. Chemical control of *Alternaria* seedling blight was effective by seed treatments using chlorothalonil, thiabendazole, benomyl, captan, mancozeb, thiram, iprodione and triadimenol. (**See: Pathogens**) (AJH, AJdlV)

Aguirrezábal, L.A.N., Lavaud, Y., Dosio, G.A.A., Izquierdo, N.G., Andrade, F.H. and González, L.M. (2003) Intercepted solar radiation effect during grain filling determines sunflower weight per seed and oil concentration. *Crop Science* 43, 152–161.

Chimenti, C.A., Hall, A.J. and López, M.S. (2001) Embryo growth rate and duration in sunflower as affected by temperature. *Field Crops Research* 69, 81–88.

de la Vega, A.J. and Hall, A.J. (2002) Effects of planting date, genotype, and their interaction on sunflower yield. II. Components of oil yield. *Crop Science* 42, 1202–1211.

Miller, J.F. (1987) Sunflower. In: Fehr, W.R. (ed.) *Principles of Cultivar Development*. Macmillan, New York, USA, pp. 626–668.

Santalla, E.M., Dosio, G.A.A., Nolasco, S.M. and Aguirrezábal, L.A.N. (2002) Effects of intercepted solar radiation on sunflower (*Helianthus annuus* L.) seed composition from different head positions. *Journal of the American Oil Chemists Society* 79, 69–74.

Saranga, Y., Levi, A., Horcicka, P. and Wolf, S. (1998) Large sunflower seeds are characterized by low embryo vigor. *Journal of the American Society for Horticultural Science* 3, 470–474.

Schneiter, A.A. (ed.) (1997) *Sunflower Technology and Production*. Crop Science Society of America, Madison, WI, USA.

Velasco, L., Pérez-Vich, B. and Fernández-Martínez, J.M. (2004) Grain quality in oil crops. In: Benech Arnold, R.L. and Sánchez, R.A. (eds) *Seed Physiology: Applications to Agriculture*. Food Products Press, New York, USA, pp. 427–452.

Superoxide dismutase (SOD)

A family of **antioxidant** enzymes thought to be involved in protection against desiccation damage and seed deterioration, that catalyse reactions that convert the superoxide radical to hydrogen peroxide. Superoxide radical is formed when an electron is added to molecular oxygen. Though the molecule has relatively weak oxidizing tendencies, it directly attacks enzymes in the Krebs Cycle and is the precursor to more reactive cytotoxic molecules, such as peroxynitrite or hydroxyl radical in the presence of transition metals. SOD enzymes are dimers or tetramers (MW ~30,000 to 40,000) that are distinguished by the metal groups at the active site: copper-zinc, manganese, or iron (in chloroplasts of plants; not present in animals). SODs are ubiquitous among organisms, with only a few aerobic bacteria lacking members of this enzyme family (but these organisms have high catalase activities). SOD is considered essential for aerobic life based on studies showing enhanced tolerance to O_2 toxicity in over-expressing organisms, and lethality in SOD-defective strains of animals. SOD activity in cells is induced by elevated O_2 and, in plants, SOD activity increases with acclimation to other stresses and is correlated with greater tolerance, suggesting that oxidative damage results from many stresses. Elevated SOD activity is considered symptomatic of repair processes following stress and the decreased activity of SOD and other antioxidant enzymes, measured in deteriorated seeds, may indicate a loss of repair capabilities. (CW)

(**See: Desiccation tolerance – protection against oxidative damage**)

Bailly, C. (2004) Active oxygen species and antioxidants in seed biology. *Seed Science Research* 14, 93–108.

Bailly, C., Benamar, A., Corbineau, F. and Côme, D. (1998) Free radical scavenging as affected by accelerated ageing and subsequent priming in sunflower seeds. *Physiologia Plantarum* 104, 646–652.

Halliwell, B. and Gutteridge, J.M.C. (1999) *Free Radicals in Biology and Medicine*, 3rd edn. Oxford University Press, Oxford, UK.

Suspension culture

Cells dispersed in liquid and allowed to grow freely. Cultures are derived from **callus** or tissues placed in suspension. The primary reason to use suspension cultures is to increase growth rate, which is much faster than in callus cultures or on semi-solid media. The initial explants (dissected tissues) may themselves dissociate or they may produce cells that dissociate. To maintain high levels of dispersion and prevent aggregation, **auxin** concentrations are increased, since this prevents middle lamella formation between cells. Suspensions can be used to rejuvenate lines that have slowed with age. Suspension cultures are either batch cultures in which cells are grown to an end point in a fixed volume of medium or continuous culture in which the medium is continually replenished. Suspension cultures are grown in many types of vessels, but large scale culturing is carried out in **bioreactors**. (PvA)

Suspensor

A structure ('embryo carrier') of a limited number of cells (often less than ten) that develops from the **zygote** into a stalk-like organ, to which the embryo is attached at the end away from the micropyle. The suspensor may play a role as the conduit of hormones and nutrients to the developing embryo. It is crushed as the embryo develops, and is absent from the mature seed, except vestigially in some gymnosperms and grass embryos. (**See: Development of embryos – dicots**)

Swath

Cut plant material (synonymous with **windrow**), which may include seed or grain, left lying in a band on the ground; a process used to ripen seed by post-maturation drying before threshing. (**See: Production for Sowing, IV. Harvesting**)

Swede

See: Brassica – horticultural

Sweetcorn

A popular vegetable in much of the world, as both a fresh and processed food. Commercial sweetcorn (*Zea mays*) varieties are based on one or more recessive genetic mutations that alter the carbohydrate content of the maize endosperm and allow sugars to accumulate. Sweetcorn quality is also determined by endosperm flavours and textures, tenderness of the pericarp, and kernel size and colour. From pre-Columbian times to the 1960s, the mutation that defined sweetcorn was *sugary1* (high content of the water-soluble polysaccharide (WSP), **phytoglycogen**), but mutations in other genes, such as *shrunken2*, produce a higher sugar content ('super-sweet', with high sucrose content) and are now used for many modern sweetcorn varieties. All sweetcorn is harvested at an immature stage of kernel development, about 18–20 days after pollination. Currently, nearly 100 sweetcorn varieties are available that differ in ear appearance, kernel colour, seed tenderness, aroma, and WSP content, as well as sugar amount. (MGJ)

(**See: Maize; Mutants; Phytoglycogen**)

Sword bean

Swordbean (*Canavalia gladiata*) seeds are similar to those of **jack bean**. Unlike the latter, the origins of the species are in the Old World. The plant is a climbing perennial. The long (15–40 cm) pods each contain 5–10 dark, reddish seeds. Confusingly, *C. ensiformis* (jack bean) can also be called sword bean.

Syncarp

A compound (aggregate) fleshy fruit derived from a flower with distinct carpels in the flowering stage that become fused (concrescent) together at maturity, e.g. *Annona squamosa* (sugar or custard apple). A synocarpous gynoecium is a multilocular ovary resulting from the fusion of two or more carpels (coenocarpous gynoecium) retaining their walls (septae).

Synergids

The two cells flanking the egg cell in the embryo sac of the angiosperms. (**See: Fertilization; Reproductive structures, 1. Female**)

Synthesis of reserves

See: Fatty acids – synthesis; Hemicellulose – synthesis; Oils and fats – synthesis; Starch – synthesis; Storage protein – synthesis; Storage reserves synthesis – regulation

Synthetic seed (Synseed)

Synthetic, or artificial seeds, are made by coating meristems, buds, or **somatic embryos** (**encapsulation**) with a matrix such as alginate. The synthetic seed can be stored in uniform conditions until required for outplanting. The matrix may contain substances to aid germination and hinder diseases. The potential benefit of synthetic seed resides in the added value of the genotype or in the novel breeding procedure that led to its creation (i.e. bioengineered). Costs generally outweigh benefits, and technical problems are often prohibitive, e.g. encapsulation of apical and axillary buds of kiwi fruit reduces the vigour and vegetative growth of these microcuttings. Clonal propagation of plants and encapsulation for planting is a target in some segments of the forestry industry, but **conversion** rates and instability of somatic embryos are issues precluding successful synthetic seed production on a commercial scale. Conifer somatic embryos are converted to plants and sold. Synthetic seed is not commercially available. (PvA)

Synthetic variety (or cultivar)

A mixture of strains, clones, inbreds, or inter-hybrid components, maintained individually and reconstituted at regular intervals to maintain the cultivar by open-pollination for a specified number of generations. The approach exploits **heterosis** by using genotypes of known ability to give superior hybrid performance when crossed in all combinations. Synthetic varieties are produced from self-incompatible species such as *B. rapa* (turnip rape), for example, by mixing two parental populations in the field; the resultant seed due to outcrossing is composed of ~50% hybrid seed. Other examples include sunflower (**see: Sunflower – cultivation**) and multiple-cross maize cultivars produced for use in parts of the world lacking the infrastructure for the production and marketing of **hybrids**, and some forage sunflower varieties.

Systemicity

A process or response that is not localized to one part of a plant (or animal) but is dispersed throughout the whole system of the organism.

In the context of plant pathology: systemic plant **disease** results by spread of **pathogen** throughout the host organism from a single infection, such as by seed-transmission of **bunts** and **smuts**. Systemic defence responses to a disease are expressed in many parts of the plant (such as in all the leaves), triggered by the first infection (the phenomenon of systemic acquired resistance, SAR).

In the context of applied plant protection: systemic pesticides applied to seed (or to soil or foliage) are absorbed and translocated by the vascular system to other parts of the plant and impart protection from insect or disease attacks, or induce systemic defence responses or, in the case of systemic herbicides, kill the entire plant. (**See: Treatments – pesticides**) (PH)

Tailings

Unwanted materials removed in the conditioning process. (See: **Conditioning**)

Take-all disease

See: **Pathogens, 15. Gaeumannomyces**

Tamarind

The tamarind (*Tamarindus indica*, syns. *T. occidentalis*; *T. officinalis*, Fabaceae) is best known because the internal pulp of the pod is used in cooking and food preparation, but the seeds also have some uses. They are ovate, about 1.6 cm long, borne in pods 6–20 cm in length. They contain approximately 63% **starch**, 14–18% **storage protein**, and 4.5–6.5% **oil**, rich in unsaturated **fatty acids**.

Tamarind seeds have limited uses as food. After they are roasted and soaked in water and the seedcoats removed, the seeds are boiled or fried, or ground to make a flour. Roasted, ground seeds are a substitute for coffee or are used as an adulterant in it. Seeds also yield a **pectin** substitute used in the manufacture of jellies, jams and marmalades, in preserved fruits and as a stabilizer in ice cream and mayonnaise.

The seeds or its products have various industrial uses, for example in sizing and finishing cotton and paper, textile colour printing, treating leather, and as a binder, thickener or glue.

The seed oil is used as an illuminant and as a varnish, and is also of culinary quality.

The medical applications are as a paste for drawing boils and, with certain additives, as a treatment for chronic diarrhoea and dysentery.

The Manilla tamarind or Guamachil, is a different species (*Pithecellobium dulce*) with edible fruits and seeds, the latter also yielding an oil. (MB)

Tapes

In these and related sowing formats, seed is impregnated on, or between, layers or strips of paper or plastic, which are laid in the ground or in seeding trays under cover. These formats share some of the advantages of mulching with plastic sheets: increasing soil temperatures; reducing soil drying, compaction or crusting; and suppressing competitive weed growth (**see: Mulch**). Nutrients may also be incorporated. (**See: Mats**)

Teff

Teff (*Eragrostis tef*) is an annual grass that is grown as a grain crop in the Horn of Africa (Ethiopia) at the higher altitudes with higher rainfall. It is a type of **millet**. The seed is quite small compared to other cereal grain crops. It is ground into flour and used mainly in certain types of flatbreads called *ingera*. Teff has not spread from its original site of domestication, unlike many other cereal grains. Whether this is due to lack of broad adaptation or lack of emigration of the people who were using it is not clear. Teff is used in some other Mediterranean climatic regions as a **forage** grass and for hay. (DF)

Dendy, D.A.V. (ed.) (1994) *Sorghum and Millets: Chemistry and Technology*. American Association of Cereal Chemists, St Paul, MN, USA.

Tegmen

Part of the **seedcoat** derived from the inner **integument**. (**See: Seedcoats – structure**)

Teleomorph

The sexual (perfect) stage of the life cycle of fungi; corresponding to the asexual **anamorph** stage. (**See: Diseases; Pathogens**)

Tepary bean

1. World importance and distribution

Tepary bean (*Phaseolus acutifolius*) (a **warm-season legume**) has been grown for a long time in Mesoamerica, mainly as a vegetable in desert zones or areas with a long dry period. Because they perform well under drought conditions, tepary beans are an excellent back-up crop when conditions are too dry for common beans. The cultivated form is a heliophyte, tolerating excessive sun. (**See: Legumes**)

2. Origin

Archaeological evidence shows that tepary bean was grown 5000 years ago in Puebla (Mexico), penetrating into the southeastern USA about 1200 years ago. Geographical distribution of the cultivated form extends from Arizona and New Mexico to Guanacaste, Costa Rica, on the dry subtropical slope of the Pacific.

3. Seed types

Basically two forms occur: one fairly small, rounded, white or black; and another a larger-sized angular, rhombohedric seed that may be white, greenish with grey, bay, dark yellow, mahogany, black, blue or purple mottling (Colour Plate 1F), or coffee-coloured. The average weight of tepary bean seed is between 10 and 20 g per 100 seeds.

4. Usage

Tepary beans are best suited for use as dry beans. They are eaten like other dry beans, first soaked, then boiled or baked. The American Indians of the southwest USA have developed various other uses for this bean such as in soups and stews and ground for meal.

5. Nutritive value

Tepary beans have a relatively high **storage protein** (23–25% fresh weight, fw) and carbohydrate (60% fw) content and they contain relatively less anti-nutritional factors than other legumes, for example less than half as much trypsin inhibitors (**see: Protease inhibitors**) than are present in soybeans. Tepary beans are high in **lectin**. But neither the trypsin inhibitors nor lectin pose a serious anti-nutritional threat as adequate cooking destroys most of these agents. The beans are rich in soluble fibre; they are a good source of iron and calcium.

6. Plant growth habits

There are two types of cultivated groups: the indeterminate shrubby varieties with short guide leaves and the indeterminate creepers with long guide leaves, which climb if they find support. (OV-V)

Miklas, P.N., Rosas, J.C., Beaver, J.S., Telek, L. and Freytag, G.P. (1994) Field performance of select tepary bean germplasm in the tropics. *Crop Science* 34, 1639–1644.

'Terminator' technology

A sobriquet used to describe patent protection technology by those opposed to it. (**See: GMO – patent protection technologies**)

Testa

Part of the seedcoat derived from the outer or single integument. Most frequently used in a more general sense as a term for the complete seedcoat. (**See: Seedcoats – structures**)

Tetrapyrrole

A biological molecule consisting of four linked pyrrole rings connected by methine bridges in a cyclic configuration. A pyrrole is an organic compound of the heterocyclic series characterized by a ring structure composed of four carbon atoms and one nitrogen atom (molecular formula C_4H_5N). A variety of side chains is often attached to a tetrapyrrole. These may be metalled, e.g. with iron to form heme.A or with magnesium to form chlorophyll; or non-metalled, e.g. the bilins in bile pigments or the **phytochrome** chromophore, phytochromobilin.

Tetrazolium Test

The Topographical Tetrazolium Test (TTZ or TZ test) is a rapid means of determining the **viability** of a seed sample using a 'vital stain', based on the principle that only living seed tissues contain active dehydrogenase enzymes that can chemically reduce the colourless tetrazolium salt to the red non-diffusible dye, formazan. The test is particularly useful in evaluating dormant seed at harvest, and seed prior to dressing and the application of appropriate chemical treatments. Dormant (or 'fresh') seeds and hard seeds react normally to TZ staining and so may be assessed as viable even though certain requirements must be fulfilled before they are able to germinate. There are difficulties with the test, in particular the necessity for experienced analysts and the subjective nature of the evaluation, which is not conducive to routine reproducible application between analysts and laboratories.

The staining responses of seed tissues in the Tetrazolium Test are:

Tissue condition	Tetrazolium produces
Live	Red staining
Deteriorated	Deep or mottled staining with flaccid texture
Dead	No staining

The TZ test reveals viability percentages based strictly on internal seed condition, manifested by the pattern of tissue staining; the **germination** test, on the other hand, reveals the combined performance of seed quality, manifested by the physiological ability to grow into a normal seedling in the defined test conditions. Two major areas of evaluation difficulties can occur when comparing the results of germination and TZ tests, though viability evaluations from the two approaches are often similar.

- The TZ test does not reveal potential damage to germination that can occur due to the presence of seedborne diseases. For example, the presence in wheat seed lots of diseases such as *Septoria nodorum* and *Microdochium nivale* will lead to lower germination results than would be anticipated from the TZ test results, unless an appropriate chemical fungicidal seed treatment is applied prior to germination testing.
- The TZ test cannot quantify damage due to chemicals and external treatment factors – for example, any phytotoxicity in response to applied agrochemical seed treatments (fungicides and insecticides) or any damage caused by the use of glyphosate as a preharvest desiccant.

The Tetrazolium Test can also be used to detect vigour differences. (**See: Vigour tests – biochemical, 3. Tetrazolium test**)

Technically, the TZ test is relatively straightforward to perform, and the means of evaluation are reasonably simple. Seeds are imbibed in water to activate enzyme systems and facilitate uniform dye absorption, for which they are physically prepared by methods according to the species involved. For example, cereals and large-seeded grasses are cut longitudinally through the centre of embryo to expose the **cotyledonary** leaves and root nodes, and fine-seeded grasses are punctured or cut across the endosperm immediately behind the **scutellum**, whereas seeds of legumes and other crops, which can readily absorb the dye through their **seedcoats**, are stained after imbibition without puncturing or cutting. Staining in the tetrazolium solution (0.1–1.0% 2,3,5-triphenyltetrazolium chloride (TTC) or bromide (TTB) in phosphate buffer, pH 6.5–7.5) is carried out in darkness at between 30 and 40°C for up to approximately 4 h for oats, wheat, barley and other cut seeds, and for from 6 to 24 h for peas, beans, fine-seeded grasses and other intact seeds. Excess dye is then washed away and the individual structures of each embryo are examined under magnification for the presence and extent of poorly stained areas in organs associated with the potential formation of normal or acceptable seedlings (Colour Plate 14A). Variations in colour patterns, texture, bruises, fractures, abnormal structures, damaged areas and insect infestation are all factors of importance that can be detected by this method. (RD)

ISTA (2003) *Working Sheets on Tetrazolium Testing. Volume I – Agricultural, Vegetable and Horticultural Species. Volume II – Tree and Shrub Species.* Edited by N. Leist and S. Krämer. Zurich, Switzerland.

Thaumatin

Thaumatin is an extremely sweet-tasting protein in the **arils** of *Thaumotococcus daniellii* seeds. It is one of the sweetest substances known: on a weight basis it is approximately 3000 times sweeter than sucrose and on a molar basis about 100,000 times. The thaumatin molecule has 207 amino acid residues, eight disulphide bridges (from the 16 cysteine residues) and a molecular mass of about 22 kDa. At least two slightly different forms of the polypeptide occur in the seed aril.

The sweetener is used in West Africa, for example to improve the palatability of over-fermented palm wine. The protein is of potential commercial interest as a sweetener for use in cooking, in flavouring, and in making confections. It has a slow-onset, lingering sweetness with a liquorice-like after-taste.

Thaumatin cDNA has been cloned, from which several transformants have been made, for example transgenic yeast which consequently produces thaumatin protein. Tomato and apple have also been transformed to contain the sweet protein in the fruits, demonstrating that transgenic expression of thaumatin might be useful for modifying fruit taste. The thaumatin gene has been isolated and sequenced, leading to the identification of a family of thaumatin-like genes in various plant parts (for example in wheat, maize, rice and barley grains, chestnuts, pollen, leaves, stigmas) that produce non-sweet proteins some of which have anti-fungal action, enzymic activity (e.g. as a hydrolase of β-1,3 glucans and of chitin), function in stress responses (e.g. osmotin-like), inhibit proteases (**see: Protease inhibitors**) and have **allergenic** and other properties. (MB)
(See: Thaumatococcus)

Bartoszewski, G., Niedziela, A., Szwacka, M. and Niemirowicz-Szczytt, K. (2003) Modification of tomato taste in transgenic plants carrying a thaumatin gene from *Thaumatococcus daniellii* Benth. *Plant Breeding* 122, 347–351.

Dudler, R., Mauch, F. and Reimmann, C. (1994) Thaumatin-like proteins. In: Witty, M. and Higginbotham, J.D. (eds) *Thaumatin.* CRC Press, Boca Raton, FL, USA, pp. 193–199.

Thaumatococcus

Thaumatococcus daniellii (Marantaceae) the sweet prayers plant or *katemfe*, occurs widely in the tropical rainforests and coastal zone of West Africa, as part of the forest undergrowth flora. Its seed is of interest as the source of **thaumatins**, intensely sweet-tasting proteins. The plant is a perennial, monocotyledonous herb with a shrubby growth habit. The fruits contain a pulp in which are embedded one to three brown-black seeds, each about 1 cm long, with a hard, thick coat. At one end of the seed is the relatively large, whitish **aril** where the thaumatin is located. The fruits are collected and sold for thaumatin extraction. (MB)

Thermal time (Thermotime)

A concept used in some mathematical models that describe the germination and emergence behaviour of seed populations. According to this model, an individual imbibed seed below its optimum germination temperature completes germination when it has accumulated sufficient heat units in relation to a **base-temperature** threshold (T) characteristic of its rank (g) in the seed lot. The concepts of thermal time and **hydrotime** have been combined in the development of **hydrothermal time** models of germination. (**See: Germination – field emergence models; Germination – influences of temperature; Germination rate**)

Thermodormancy

A form of **secondary dormancy** that is induced by elevated temperatures. After removal of the temperature stress, the seeds remain dormant and require a **dormancy-breaking** treatment, e.g. chilling, before they can germinate.

Thermogradient bar (table)

A rectangular or square plate of heat-conducting metal on the surface of which seeds are placed for germination. One side is heated and the opposite side is cooled electrically or by fluids circulating along the edges in order to obtain a continuous thermal gradient, the characteristics of which are defined by the temperatures of the edges (Fig. T.1A). The two temperatures

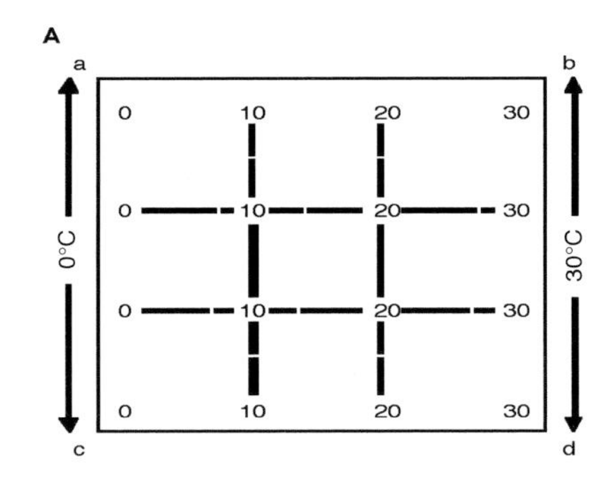

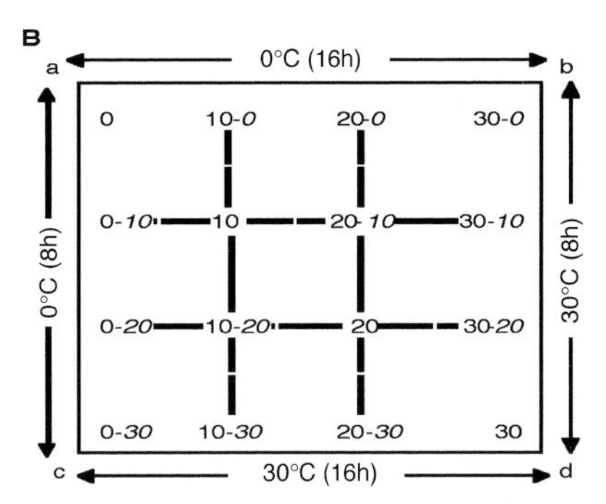

Fig. T.1. Principle of a thermogradient bar. Fluids at given temperatures circulate along the edges ab, cd, ac or bd of the plate and establish a thermal gradient between two opposite edges. (A) Temperatures obtained on the plate if a fluid at 0°C circulates continuously in ac and another fluid at 30°C in bd. (B) Temperatures obtained on the plate if the temperatures of the fluids are alternately 0°C in ab and 30°C in cd for 16 h, and then 0°C in ac and 30°C in bd for 8 h. The numbers in italics correspond to the temperatures the durations of which are the longest. The temperatures are constant on the diagonal, ad.

can be reversed at regular intervals to obtain alternate temperatures to simulate field conditions. Similar instruments with a double gradient, one in one direction and the second in the opposite direction, provide a large range of thermal combinations (Fig. T.1B). (DC, FC)

Fox, D.J.C. and Thompson, P.A. (1971) A thermo-gradient bar designed for use in biological studies. *Journal of Experimental Botany* 22, 741–748.

Murdoch, A.J., Roberts, E.H. and Goedert, C.O. (1989) A model for germination responses to alternating temperatures. *Annals of Botany* 63, 97–111.

Thermoinhibition

Inhibition of germination of a non-dormant seed solely by elevated temperatures. When the temperature stress is no longer present, the seed readily germinates. Contrast with **thermodormancy**. (**See: Germination – influences of temperature**)

Thermotherapy

Seed treatment by dry heat to control seedborne viruses, or with hot water or steam to control seedborne bacterial or fungal **pathogens**, sometimes in combination with chemicals, at temperature-times that are only slightly injurious to the host. Such remedial treatments are possible when the pathogen superficially contaminates the seed or is only slightly invasive, though **disinfestation** is not always completely successful because the conditions needed to destroy the pathogen generally impair seed viability. The disinfection of pathogens located 'deep-seatedly' in the embryo can be particularly difficult without harming the seed.

Short hot water treatments are used for certain small-seeded vegetable and flower seeds, typically by immersion at temperatures of about 50–60°C for up to about 30 min, followed by cooling in cold water and drying. Such treatments have been superseded in cereals from about the 1960s since the advent of efficient fungicides.

Very brief exposure to hot humid air ('aerated steam', or 'vapour-heat', treatment), immediately followed by rapid cooling, to give a selective heating of external layers of the seed where most of the important pathogens are located without the seeds reaching temperatures higher than about 60°C, is being developed for disinfestation of seedborne fungal pathogens from cereal and other seeds, especially because there is a demand for alternatives to chemical treatments. (**See: Organic seed**) Precise control of conditions is critical, and continual stirring to secure rapid penetration of heat to maintain a uniform transient temperature in the seed batches is important. The high temperatures can differently influence different species or seedlots; in each case, the optimal temperature must be found by pretesting.

Dry-heat protocols (1 or more days at 70 or 80°C, for example) have been tested or applied for the control of mosaic viruses, mainly those carried externally on seedcoats, including tomato mosaic virus (MV) on freshly extracted seeds, cucumber green mottle MV, capsicum MV, cowpea MV and lettuce MV. Short treatments with Na_3PO_4 or concentrated HCl solutions are also used to inactivate tomato MV, and orthophosphate steeps are also effective in eliminating capsicum MV from pepper seeds. Dry hot air has been used to disinfest stored grain of insects while it is being loaded into storage structures. (**See: Storage Pests, 5. Management strategies**)

A very simple, though imprecise, means of thermal treatment is **solarization**, which is sometimes applied on a small scale in warm countries, whereby seeds are heated by irradiation from the sun. Microwave irradiation has been tested, but mainly due to unreliable effects it has not been commercialized. (PH)

Forsberg, G. (2001) Heat sanitation of cereal seeds with a new, efficient cheap and environmental friendly method. In: Biddle, A.J. (ed.) *Seed Treatment – Challenges and Opportunities*. BCPC Symposium No. 76. BCPC, Farnham, UK, pp. 69–72.

Maude, R.B. (1996) *Seedborne Diseases and their Control: Principles and Practice*. CAB International, Wallingford, UK.

Rast, A.Th.B and Stijger, C.C.M.M. (1987) Disinfection of pepper seed infected with different strains of capsicum mosaic virus by trisodium phosphate and dry heat treatment. *Plant Pathology* 36, 583–588.

Thionins

A family of small proteins consisting of a single polypeptide chain of 45–47 amino acid residues. The first thionins were isolated from wheat seeds and called purothionins because of their high content of sulphur-containing amino acids. Though all thionins resemble the purothionins, there is some heterogeneity in primary structure and the number of intra-chain disulphide bonds. Basically, there are two subgroups: the eight-cysteine (four disulphide bridges) and the six-cysteine (three disulphide bridges) thionins. The amino acid sequences of the different thionins are highly divergent but all thionins are folded structurally in a very similar manner. Thionins are abundant proteins in the **endosperm** of wheat and barley (representing several per cent of the total seed protein) but are also expressed in the leaves of cereals. Other well-known thionins are the viscotoxins from *Viscum album*, the phoratoxins from *Phoradendron* sp. and crambin from *Crambe abyssinica*.

Thionins are **amphipathic** proteins (i.e. they contain polar and non-polar regions) that inhibit the growth of bacteria and fungi. Their mode of action most probably relies on their ability to form pores in membranes. Some thionins are cytotoxic for mammalian cells but since those in seeds of food and feed plants are not toxic upon oral administration they do probably not act as antinutrients. However, their high protease resistance reduces the overall conversion rate of the diet.

Most thionins presumably play a role in plant defence against pathogenic bacteria and fungi, and experiments with transgenic plants have confirmed that thionins are valuable candidates as resistance factors against these pathogens. (**See: Cereals – composition and nutritional quality, 1. Composition and quality; Proteins – non-storage**) (EVD, WJP)

Garcia-Olmedo, F. (1999) Thionins. In: Shewry, P.R. and Casey, R. (eds) *Seed Proteins*. Kluwer Academic, Dordrecht, The Netherlands, pp. 709–726.

Thioredoxin

A small (12 kDa) protein involved in **redox reactions** in the cell. It contains the conserved sequence –Trp-Cys-Gly/Ala-

Pro-Cys-, and in its oxidized form the two cysteine residues form a disulphide bond. Multiple forms of thioredoxin are present in plants. Type-m and type-f thioredoxins occur in chloroplasts and are involved in the reductive activation of photosynthetic enzymes. Type-h thioredoxins are probably located in the cytosol. A thioredoxin h is thought to play an important role in **storage protein mobilization** in the **endosperm** of cereal grains. Following imbibition of the wheat (*Triticum durum*) grain, thioredoxin h is converted to its reduced form by NADP-thioredoxin reductase, using **NADPH** generated by the oxidative pentose phosphate cycle in the endosperm. This reduced thioredoxin in turn reduces one or more disulphide bonds in the major storage proteins gliadin and glutenin, making them more susceptible to proteolysis by the endogenous proteases. Similar storage protein reduction occurs in rice (*Oryza sativa*) and barley (*Hordeum vulgare*) endosperms.

In addition to increasing the susceptibility of storage proteins to proteolytic degradation, thioredoxin is also involved in the activation of a calcium-dependent protease, thiocalsin. The activated thiocalsin then is capable of hydrolysing the reduced gliadins and glutenins. Thioredoxin is probably involved in the activation, or maintenance of activity, of the numerous cysteine endopeptidases in the endosperm. Thioredoxin also may contribute to the activation of other hydrolytic enzymes in the endosperm, in particular those involved in starch hydrolysis, by inactivating endogenous protein inhibitors of these hydrolases. For example, in barley, the **α-amylase**/subtilisin **inhibitor**, which inhibits the endogenous α-amylase, and an inhibitor of the barley starch debranching enzyme (pullulanase) are inactivated by reduced thioredoxin.

The possibility that thioredoxin plays a similar role in storage protein mobilization in dicots has not been demonstrated. However, thioredoxin does reduce a number of dicot seed storage proteins *in vitro*, including the Kunitz and Bowman-Birk type trypsin inhibitors of soybean (*Glycine max*), the 2S albumin of castor bean, and 2S globulins and a legumin in the peanut (*Arachis hypogaea*). (KAW)

(**See: Mobilization of reserves – a monocot overview; Storage protein – mobilization**)

Yano, H., Wong, J.H., Lee, Y.M., Cho, M. and Buchanan, B.B. (2001) A strategy for the identification of proteins targeted by thioredoxin. *Proceedings of the National Academy of Sciences USA* 98, 4794–4799.

Thornapple

Seeds of this species (*Datura stramonium*, Solanaceae) contain many different alkaloids and are sometimes used because of the psychoactive effects that they exert. (**See: Pharmaceuticals and pharmacologically active compounds; Psychoactive seeds**)

Thousand seed (or grain) weight

The weight of 1000 seeds, usually expressed in grams: a standard expression used in seed production, conditioning, treatment and **quality testing**.

Threshing

The initial separation of seed during harvest from parent-plant material (straw) either by hand or mechanically. Modern combine-harvesters are derived from 18th and 19th century threshers that used drums covered with wire screens and rapidly revolving iron beaters. Simple mechanical threshers, based on treadle-, water- or animal-operated revolving drums equipped with parallel rows of steel teeth or loops are still common in small-scale intensive farms throughout China. Historical manual threshing methods include the use of sticks, flails, rollers, combs, or the beating of sheaves on a mat, board or slab. Threshing floors are still used in parts of the world, and threshing tubs are used for rice in Asia, alongside the still widespread ancient technique of treading rice or millet by foot. Enterprising farmers in the Rajasthan region of India, for example, lay their crops on the road so that they are threshed by the tyres of passing vehicles (Fig. T.2). (**See: Production for Sowing, IV. Harvesting**)

Ti plasmid

Tumor-inducing plasmid of *Agrobacterium* that contains genes which encode for opines and cytokinins in its T-DNA that is transferred to infected plant cells to promote cell division and tumour formation. (**See: *Agrobacterium tumefaciens*; Transformation**)

Tillage

The use of implements to cultivate the soil, such as discs, harrows and firming devices. The most important reason for tillage is to prepare a **seedbed** for sowing by planter equipment. In the past, tillage has also been required for weed control and burial of residue from the previous crop. These factors have declined for many field crops with the widespread advent of chemical herbicides and the decreased importance of burying residue. However, some 'pre-emergence' fertilizers need to be incorporated into the soil and some weeds may need to be controlled by inter-row cultivation, such as in organic farming systems (**see: Organic seed**).

The choice of tillage and planting system requires matching the operations to the crop sequence, topography, soil type and weather conditions.

1. Conventional tillage (intensive tillage)

Any system that attempts to cover most of the residue, leaving less than 30% (often considerably less) of the soil surface covered with residue after planting. Ploughing, followed by secondary tillage operations such as discing and harrowing, has been the most common, traditional system, creating a well-tilled, residue-free seedbed.

Fig. T.2. Using passing vehicles as the tool for threshing of a crop (unidentified) spread on the road. Photo: M. Black.

2. Conservation and reduced tillage

Any tillage and planting system that covers 30% or more of the soil surface with protective crop residue cover (or 15% in the case of 'reduced tillage' systems), after planting, to reduce soil erosion by wind or water and water losses. In most systems, the soil is not tilled between the harvest of one crop and planting of the next. Many newer planters are designed to operate despite trash on the soil surface. With limited tillage, weed seeds are not turned up near to the soil surface, and those that do germinate do not have as much light or space to establish plants. Although effective in reducing erosion, these practices have increased concern about seed placement and general planter performance.

3. No-tillage system

Also referred to as zero-till or 'direct-seeding'. No-till systems disturb only a narrow strip of soil, leaving most of the residue on the soil surface, to provide the best erosion control. In most cases, planter-mounted devices till a narrow seedbed strip to assist in seed and fertilizer placement, using seeding openers that both physically prepare the seedbed and sow the seed. Three basic slot shapes are created, 'V', 'U' and inverted 'T'; alternatively, punch-type drills are also used (**see: Planting equipment – placement in soil**). Advantages of uniformly spread residues include increased water infiltration and reduced soil moisture evaporation but, in some poorly drained soils in the early spring, warming and drying, and hence germination and emergence, may be delayed.

4. Ridge tillage

Planting on raised ridges that were formed during the cultivation of the previous crop, with inter-row cultivation to reform ridges and control weeds. Ridge tillage is normally used in heavier soils, which dry sooner and warm faster.

5. Raised-bed tillage

Planting crops on top of a raised and shaped bed: a popular system in vegetable production to facilitate subsequent irrigation, crop husbandry and harvesting. Bed-forming machines (Fig. T.3) consist of mould-board ploughs that roll and then compress soil. Machinery can also apply **fumigant**, dry or liquid fertilizers, herbicides, or lay plastic **mulch** or drip-irrigation tapes as the bed is formed. Cross ditches may be cut across the bed to allow for sideways movement of run-off water in the field. (PH)

(See: Planting equipment – agronomic requirements, 2. Germination, emergence and establishment)

Fig. T.3. Raised bed tillage attachment to a tractor. From members.aol.com/kenncomfg/bed-info.htm (Kennco Manufacturing, FL, USA).

Baker, C.J., Saxton, K.E. and Ritchie, W.R. (1996) *No-Tillage Seeding: Science and Practice.* CAB International, Wallingford, UK.

Tissue culture

Plant tissue culture is the practice of growing cells and tissues on artificial media, e.g. organ cultures, meristem cultures, shoot cultures, node cultures, embryo cultures, root cultures, callus cultures, suspension cultures, **protoplast** cultures and anther cultures. The plant material is placed in controlled environments in which abiotic factors such as light, temperature, nutrients and growth matrix are controlled. The tissue is isolated not only from the parent tissue, but from other organisms that might reduce its growth, such as bacteria and fungi. Tissues and cells growing in aseptic isolation are valuable for many reasons, which gives this technology numerous practical applications, e.g. production of **somatic embryos**. Tissues can be grown free of various agriculturally important diseases. Such stock can be propagated to produce material that is certifiably free of pathogenic viruses, bacteria or fungi. To supplement conservation of genetic resources in the world of ecologically, socially, agriculturally or medically valuable species, tissues, cells or embryos can be induced *in vitro* and stored *in vitro*, or **cryostored** in liquid nitrogen. Tissue cultures provide a tool in novel breeding strategies that require uncontaminated cells and tissues to be used for genetic engineering. Tissue-cultured plants have many applications in clonal agriculture, horticulture and forestry. (PvA)

Tobacco

1. World importance, distribution and origin

Commercial cultivated tobacco, *Nicotiana tabacum*, is the only *Nicotiana* species grown on a wide scale. A limited amount of *N. rustica* is grown for snuff and for nicotine extraction, and several species are grown as ornamentals. The geographical limits for the tobacco crop are approximately latitudes 60°N to 45°S, but the crops grown at the extreme latitudes are minor amounts of local types. The most northerly tobacco-growing country is Poland (54°N), and the most southerly is Australia (37°S).

The greatest use of tobacco is for cigarettes, but it is also used for cigars, pipe tobacco, dry snuff, wet snuff and chewing tobacco. Tobacco has, over the years, been of enormous economic significance, both as a consumable commodity and as a revenue earner; in many countries (e.g. Zimbabwe) and US states (e.g. North Carolina and Kentucky) the industry is an important employer. However, the economic contribution of tobacco in many countries is slowly declining as consumption decreases with health concerns and pressure from the anti-smoking lobby.

Tobacco was first used by Native Americans, who chewed, sniffed and smoked it. Tobacco originated in South and Central America, as did the majority of *Nicotiana* species. Its susceptibility to frost, small seed, large leaves and low light photosynthetic requirement suggest that it grew in these regions in low to mid-altitude forest margins. Smoking was first reported by Christopher Columbus, and was introduced into England by Sir Walter Raleigh.

2. Classification and taxonomy

The genus *Nicotiana* (Solanaceae) is divided into three subgenera and 14 sections. *N. tabacum*, belongs to the *Tabacum* subgenus and the Genuinae section. It is one of 67 species (66 naturally occurring, one synthetic ornamental) in the *Nicotiana* genus, all of which produce alkaloids. Within the species, tobacco types are further classified by genotype, production and curing methodology, and also by use. The unique characteristics of each type limit the amount of germplasm that can be exchanged between them without adversely affecting quality.

N. tabacum (2n = 48) is a naturally occurring allotetraploid derived from the spontaneous hybridization of *N. sylvestris* (2n = 24) and *N. tomentosiformis* (2n = 24), with subsequent chromosome doubling.

Tobacco is a perennial, but is grown as an annual. The ovate or oblong-lanceolate leaves (generally sessile, but petiolate in Oriental types) are borne spirally on a single stem with a large terminal inflorescence (a panicle) of several hundred individual bisexual flowers. The flowers are tubular, about 50 mm long, five-lobed and usually pink, but range in colour from white to carmine.

3. Seed structure and types

Because tobacco is grown for its leaf, the seed has no economic value other than as a means of reproduction, and plant breeders therefore have no interest in modifying seed traits. Though leaf and plant types have been manipulated to such an extent that modern tobacco bears little resemblance to the original primitive tobacco, the seed morphology has remained unchanged.

Seeds are extremely small (there are about 11,000/g), with a flattened, ovoid shape, and a typical size of about 0.75 mm × 0.5 mm × 0.5 mm, although size does vary considerably. Few other crops have such small seeds or produce so many per plant: a single **capsule (ovary)** produces up to 3000 seeds, and a single plant can produce more than a million. The seeds are dark brown, with a finely reticulated surface caused by the thin outer walls infolding at maturity. There is a prominent **raphe** along one side, terminating in the projecting **hilum** at the point of the seed. The endosperm (three to five layers of thick-walled cells) makes up the bulk of the seed and serves as a food store by virtue of its **oil** (triacylglycerol) and **storage protein** content. The **micropylar endosperm** surrounds the radicle tip at the point of radicle protrusion. The slightly curved embryo is about 0.7 mm long, with two cotyledons, six cells thick, and with the radicle/hypocotyl oriented towards the hilum. There is a single layer of nucellar tissue, three subepidermal layers, and the testa, which consists of cutinized and lignified dead cells (see Fig. T.4).

The seed is used in studies of germination and dormancy (**see: Germination – molecular aspects; Germination – radicle emergence**)

5. Genetics and breeding

Tobacco is an inbred crop, and breeding methods have been almost exclusively the traditional pedigree and backcross

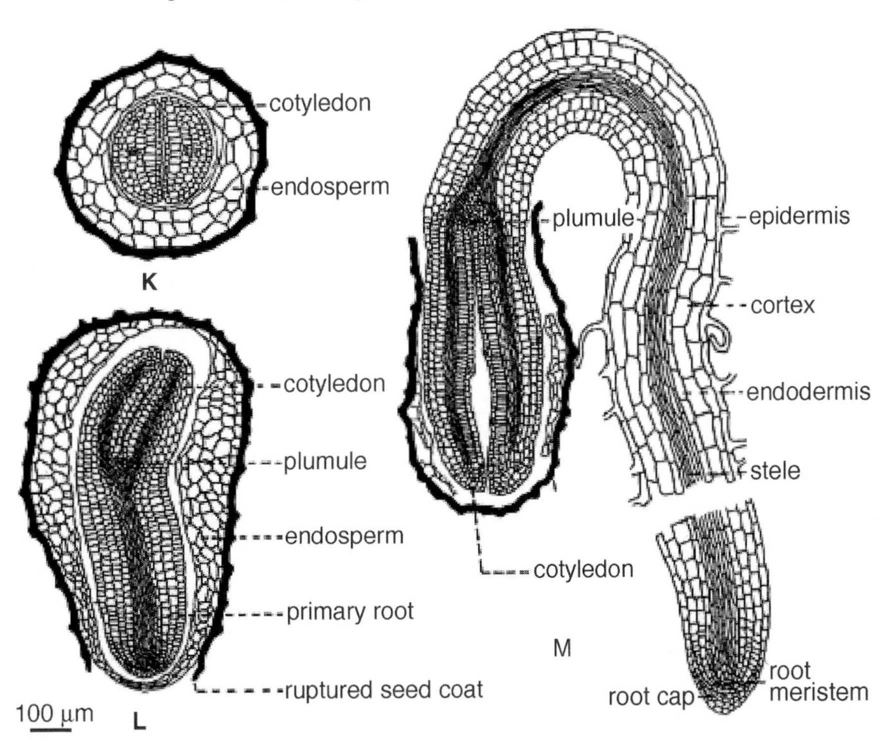

Fig. T.4. Tobacco seed structure. (K) Transverse section through quiescent seed at the cotyledonary level (× 70). (L) Longitudinal-median section through seed with radicle starting to extend, showing ruptured seedcoats (× 50). (M) Longitudinal-median section through seedling 9 days after the start of imbibition (× 50). From Avery, G.S., Jnr (1933) Structure and germination of tobacco seed and the developmental anatomy of the seedling plant. *American Journal of Botany* 20, 309–327. Reprinted with permission from the American Journal of Botany.

methods, with interspecific hybridization to introduce several major resistance genes. However, more and more new varieties are **hybrids**, and in some countries (e.g. USA, Zimbabwe) this is true of almost all new releases. The rationale for making hybrids is not the same as for outcrossing crops such as maize, for there is little **heterosis** in tobacco. Rather the benefits of hybrids in tobacco are: (i) several monogenic dominant resistance genes (e.g. to tobacco mosaic virus, and black shank fungus) are more acceptable in heterozygous than in homozygous form; (ii) a single inbred line can be used as the parent of several hybrids. Another major consideration in hybrid use is commercial control by the breeder of the seed source germplasm. Because it effectively contains two genomes, tobacco can withstand more chromosome imbalance and chromosome loss than can most plants – the full range of **monosomics** and **nullisomics** are viable, as are **haploids**. Dihaploids, both maternal (via *N. africana*) and anther-derived, have been used in breeding programmes.

The tobacco plant was the first to undergo genetic transformation with foreign genes, and is an eminently suitable subject for this process. Many **genetically modified** (GM) tobacco varieties have been produced, but few have been released and none is traded on the world market. Customer antipathy to GM tobacco is so strong that there are extensive programmes in place to ensure that the supply chain is free of GM tobacco.

6. Development

Tobacco is an almost entirely self-pollinated crop. The pollen is not windborne and cross-pollination by insects and hummingbirds is less than 5%. When the flower is ready for pollination, the stigma is moist and sticky and remains receptive for several days. The pollen tubes begin to grow down the style within a few hours of pollination and fertilization is generally complete within 4 days. Once fertilized, the ovules swell rapidly and the two-lobed ovary fills out. The developing seeds are initially white but turn brown as they approach maturity.

7. Production

The seed capsules are ready for harvest when the capsule is brown but the calyx is still yellow; this is generally about a month after pollination, although the seed is physiologically mature 21 to 24 days after pollination. If the capsules are left on the plant for too long after maturation, seed can be lost due to dehiscence and the quality of the seed can deteriorate, especially in wet weather.

Like all other seeds, tobacco seed requires drying before going into storage, particularly if the capsules are harvested in wet weather. To retain quality, seed should be stored at a moisture content of no more than 10%. Generally, seed is dried on racks with warm forced air, for which the temperature should not exceed 38°C. If stored correctly, tobacco seed can remain viable for many years. Four factors are important: the initial seed quality and moisture, and the temperature and relative humidity of the storage environment. Poor-quality seed will deteriorate much faster than good quality seed in storage. High humidity and high temperature interact to adversely affect seed quality in storage; ideal storage conditions are cool and dry. (**See: Deterioration kinetics; Drying of seed (for storage)**)

Seed is cleaned in two stages. The capsules are crushed and the debris sieved out; then the sieved seed is put through a pneumatic separator or a gravity table to remove inferior, light seed.

8. Quality, vigour and dormancy

The most obvious measure of seed quality is germination percentage, but other factors such as rate and uniformity of germination and **vigour** are becoming increasingly important with the advent of precision seeding. For seed certification purposes, raw (unpelleted) seed is tested for germination percentage, purity (percentage of inert matter) and weed seeds. Most countries specify a maximum of 1% inert matter and zero tolerance for weed seeds, but the germination standards can vary between countries (in 2002 in the USA it was 80%, in Zimbabwe 90%). Seed to be pelleted must meet these standards before **pelleting**, but there is no restriction on the amount of inert matter permitted in the final product.

Some seedlots have a short period of dormancy after harvest, which tends to diminish after storage, but the **afterripening** requirement is neither consistent nor universal. This dormancy is not variety specific, and is more marked when the maternal plant is subject to adverse conditions, especially drought. Dormancy is breakable by a short exposure to light (**see: Light – dormancy and germination**). In traditional seedbeds, the seed is sown on the surface and exposed to light, so dormancy is not a problem.

9. Sowing

Tobacco **transplants** can be produced in conventional plant beds or by **hydroponic** production methods. When producing transplants in the former, growers must pay attention to key factors such as site, fumigation, fertilization, seeding date and rate, cover management, water management, and insect and disease control. The plant bed site should be free from shade and contain fertile, well-drained soil that is high in organic matter. Weed control is typically obtained by fumigating the beds with methyl bromide prior to seeding. After a complete analysis, fertilizer is lightly incorporated into the soil at a rate of 25 g N/m^2, the bed is seeded at the rate of 55 mg of seed/m^2. The bed should be covered with a cotton, polypropylene, or nylon cover until approximately 1 week before transplanting, when it should be removed to harden the transplants. Insects and diseases must be controlled by timely applications of pesticides. The beds should be watered immediately after seeding, and as needed throughout the transplant production period.

The alternative hydroponic 'floating tray' production system of tobacco transplants, in which plants are grown in Styrofoam trays filled with soil-less potting mix, is widely utilized in most major tobacco production areas. This system provides easier management of transplant production, minimizes transplanting shock and eliminates plant pulling. Transplants are typically produced in greenhouses, but may be produced in outdoor waterbeds in temperate climates.

10. Treatments

Tobacco diseases carried in or on the seed or seed debris are: **anthracnose** (*Colletotrichum gloeosporioides*), the bacterial diseases wildfire and angular leaf spot (*Pseudomonas syringae* pv. *tabaci*), tobacco ringspot virus and TMV (tobacco mosaic virus). Seed treatment with silver nitrate is effective against the bacterial diseases, and this is compulsory in some countries where these diseases are endemic (e.g. Zimbabwe). Sodium hypochlorite solution is sometimes used as a seed surface sterilant, particularly against TMV. There is no seed treatment for anthracnose or tobacco ringspot virus.

The most common diseases of young seedlings are seedling **damping-off** (*Pythium* spp.), soreshin (*Rhizoctonia solani*), blackleg (*Erwinia carotovora* ssp. *carotovora*), collar rot (*Sclerotinia sclerotiorum*) and botrytis (*Botrytis cinerea*). The most common seedling pests are cutworm, aphids, slugs, armyworm, nematodes, ants, flea beetles and crickets; depending on the locality, some of these are only minor and occasional pests.

No pesticides or inoculants are currently applied to tobacco seed to control any of these problems. This is partly because the small size of tobacco seed has made it impractical to apply sufficient amount of pesticide to be effective. It is possible that there is potential for including pesticides and biological agents such as *Trichoderma* in tobacco seed pellets (**see: Rhizosphere microorganisms**), but this is not current practice.

Primed seed generally germinates faster and more uniformly than unprimed seed, and this difference is particularly marked at lower temperatures. However, the benefits of **priming** are variety-specific, and some varieties perform better unprimed. Priming considerably reduces the shelf life of seed.

Because they are small, seeds are very difficult to sow singly, but this has become necessary since the advent of float trays and precision seeding; although mass seeding was quite adequate for the traditional seedbeds, with the demise of methyl bromide **fumigant** this sowing practice is rapidly being phased out. For this reason, **pelleting** or coating with inert material is now widely used to increase seed size to the point where seeds can be handled individually. There are two types of pellets distinguished by the behaviour after imbibition; those that split, exposing the germinating seed, and those that disintegrate. Because of the requirements of **precision seeding**, seed merchants generally pellet only the highest quality seed. (AJ, RDM)

For information on the particular contributions of the seed to seed science **see: Research seed species – contributions to seed science**.

Hutchens, T.W. (1999) 4A: Tobacco seed. In: Davis, D.L. and Nielsen, M.T. (eds) *Tobacco Production, Chemistry and Technology*. Blackwell Science, Oxford, UK, pp. 66–69.

Smith, W.D. (1999) 4B: Seedling production. In: Davis, D.L. and Nielsen, M.T. (eds) *Tobacco Production, Chemistry and Technology*. Blackwell Science, Oxford, UK, pp. 70–75.

Tomato

The tomato in now one of the world's foremost vegetables, and is used both for fresh consumption and processing into a variety of food products. Two main types are cultivated. The determinate or bushy tomato, which produces flowers in a limited time period, is mainly used for processing. The indeterminate type, which produces flowers year-round in suitable climates, is largely used for the production of fresh fruits in gardens and in greenhouses where plants are normally trained to a single stem with side shoots or suckers removed, and suspended from high wires.

Tomato belongs to the Solanaceae, which includes other important food crops, notably **pepper and eggplant/** aubergine (in which the seeds have similar shape and internal structure to tomato) and potato, and also **tobacco**. Though the cultivated tomato is usually known as *Lycopersicon esculentum*, the synonym *Solanum lycopersicum* is sometimes used.

All wild species of tomato are native to the Andean region, and the first extensive domestication occurred in Mexico where the crop was already fairly well established. The crop was taken to Europe by the Spaniards, where it became cultivated in the south in the mid-16th century and widespread in the northwest by the late 18th century, and thence was spread to other parts of the world.

The tomato seed is a popular model experimental system used in basic research studies of seed development and germination, including the conceptual development and application of **hydrothermal time** models to germination, dormancy and priming. Many well-characterized **mutants** exist, including those affecting the production and response to **hormones**. Furthermore, the seed itself has the practical advantage that its fairly large size facilitates the study of the location of cytological and biochemical activities in different seed parts, which are easily dissected. Extensive research has addressed the interplay between the expansion force of the embryo against the mechanical restraint imposed by the endosperm, through which the radicle must penetrate at the end of germination. (**See: Germination – molecular aspects; Germination – physical factors; Germination – radicle emergence; Priming – physiological mechanisms**)

For information on the particular contributions of the seed to seed science **see: Research seed species – contributions to seed science**.

1. Genetics and breeding

Tomatoes have a more than 200-year history of genetic improvement, most recently in selection and breeding for greenhouse production purposes. **Hybridization** has been the most popular breeding technique, but pedigree selection, single-seed descent, backcross breeding (suitable for the transfer of monogenic qualitative characters) or a combination of these methods have also been used. **Hybrid** cultivars produced by artificial crossing have almost replaced standard open-pollinated cultivars in fresh-market production.

The *Lycopersicon* genus contains the cultivated tomato and seven closely related wild species (Table T.1), which are all native to the Andean region.

Many useful traits from these wild *Lycopersicon* species have been incorporated by breeders in modern tomato cultivars. Among the most important recent and future targets are:

1. Disease-resistance to fungal (e.g. *Fusarium*, *Clasdosporium*, *Verticillium*, *Pyrenochaeta*, *Altenaria*), bacterial (*Pseudomonas*) or viral (Tomato Yellow Leaf Curl Virus) pathogens or resistance to nematodes (mostly the *Meloidogyne* species).
2. Tolerance to high and low temperature, nutrient deficiency.

Table T.1. General features of the *Lycopersicon* species.

Species	Commercial use	Reproduction
Lycopersicon pimpinellifolium	Currant tomato	Self- and cross-pollinated
Lycopersicon cheesmanii	Wild species	Self-pollinated
Lycopersicon chmielewskii	Wild species	Cross-pollinated
Lycopersicon parviflorum	Wild species	Self-pollinated
Lycopersicon hirsutum	Wild species	Self-fertile, incompatible
Lycopersicon chilense	Wild species	Self-incompatible
Lycopersicon peruvianum	Wild species	Self-incompatible

Genome formula 2n = 2x = 24; the first five species belong to the *esculentum* complex, and the last two species belong to *peruvianum* complex.

3. **Parthenocarpy**, which allows for fruit production under low temperatures.
4. Improved fruit quality (such as shelf life, taste, improved appearance, **lycopene** content) and processing attributes (total soluble solids, dry matter, titratable pH).
5. Earliness, to extend the production season. (The inclusion of *Lycopersicon pimpinellifolium* as parental stock in combination with *Lycopersicon esculentum*, has resulted in progeny whose seeds germinate more rapidly than the latter.)
6. Growth habit (including the multiplicity of fresh product fruit types – 'on truss', cherry etc.)

Tomato was one of the first genetically modified crops accepted for commercialization ('Flavr Savr'®) with delayed fruit softening.

2. Development

The tomato seed attains its maximum dry weight between 35 and 50 days after pollination (DAP). During this period seeds become fully germinable and water content decreases to about 50% on a fresh weight basis. The milky endosperm becomes solid by 40 DAP. During maturation there is an accumulation of ABA and **storage proteins**, and **desiccation tolerance** and **primary dormancy** (where present) are established. Seed maturity is achieved when no further increase in dry weight occurs, though full germinability can be achieved later than 50 DAP, depending on the temperature.

The seed of a commercial tomato cultivar is a flattened ovoid, up to 5 mm long, 4 mm wide and 2 mm deep (Colour Plate 6E). The **embryo** has a spiral shape and is embedded in a thick-walled **endosperm** (a cap region enclosing the radicle tip) surrounded by a seedcoat (Fig. T.5). The endosperm provides nutrition for the initial growth of the embryo as the endosperm walls contain relatively large amounts of galactomannans. The hard testa, formed from the condensation of the inner layers of the integument, is covered with large soft hairs (which are embedded in **mucilage** within the fruit).

3. Production

Most processing tomatoes are produced from open-pollinated seed, whereas most fresh market tomato production is based on hybrid seed. Production of good quality, hybrid seeds requires not only labour-intensive hand emasculation of the female line during the late bud stage and pollination, but also requires optimal environmental conditions during seed development and careful attention to procedures. Cross-pollination is carried out three times a week either by dipping the stigma of flowers on the maternal parent into a pool of pollen, or transferring it on a fingertip, and hand removal of all naturally pollinated (i.e. non-hybrid) units from the female plants. Most hybrid seed is produced in the field, although some for the protected cultivation sector can be produced in greenhouses

(a) *Cultivation factors influencing seed production.* Seed production is influenced by environmental factors, such as light intensity, soil conditions and nutrients, as is the production of fruits. Stress conditions may have an adverse effect on the vigour of the seed. Thus, nutrient, water and soil pH management should be optimal to achieve good fruit and seed yield. To prevent the build up of pests and diseases, rotation with non-solanaceous crops is needed. Soil

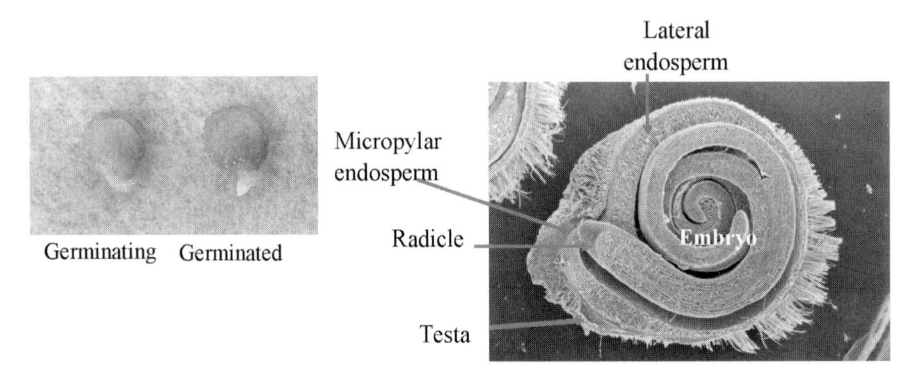

Fig. T.5. Tomato seeds. Germinated seeds (left) and longitudinal section (right). For dimensions see section 2 (above) (image by J.D. Bewley).

sterilization or fumigants are also used in greenhouses where rotation is more difficult. Humidity above 60% at the time of fruit maturity increases disease problems, which reduces seed yields.

(b) *Harvesting and extraction*. Harvesting and extracting seeds at the point of full maturity is necessary to avoid detriments to seed viability and germination. Fruits can be harvested mechanically in a single operation or by hand. Fruits from hybridization are always handpicked from the female lines to ensure that only those from cross-pollination are included and the hybrid seed is usually extracted by hand to maximize recovery. Seed quality is associated with fruit colour (red-firm at 70 DAP) rather than with the seed dry weight. Delayed harvest (80–90 DAP) may result in a decline in seed quality though this is not always the case (see below). Precocious germination can occur in overripe fruit.

Seed extraction can be completely mechanized (i.e. fruit is harvested automatically before passing into a crusher) resulting in a mixture of gelatinous seed, which is passed into a revolving cylindrical screen to separate seeds from debris. Seeds are then cleaned either by fermentation or with hydrochloric acid (0.7% HCl), which is the method adopted by commercial seed companies as it produces very bright clean seeds. Seeds can be optionally treated at this stage with sodium carbonate and HCl to control the seedborne tobacco mosaic virus (TMV), and with tri-sodium orthophosphate as an extra safeguard, immediately followed by washing and drying.

Seeds are then dried (3–4 days at 28–30°C), such as using rotary-paddle or hot air batch dryers. Drying needs to carried out slowly to avoid shrinkage of the seedcoat around the embryo, which reduces seed quality. In warm climates seeds can be air-dried (**solarization**), first in the shade, then in partial and finally in full sun. Subsequent seed conditioning includes **debearding** and sizing steps. Seeds can be readily stored for at least 3 to 5 years. In general, **viability** is retained between 18 and 24°C and a large range of relative humidities, though preferably moisture contents are kept at about 5.5%.

4. Quality, vigour and dormancy

Tomato seeds germinate best in the dark and, in some cultivars, prolonged light is inhibitory (**see: Light, dormancy and germination**). Optimum temperatures are between 20 and 25°C and a heat-sum requirement of 88 **day-degrees** above a base temperature (T_{min}) of 8.7°C is needed to achieve 50% germination. However, 13°C is generally the 'practical' minimum temperature for germination, with some variation among cultivars (**see: Germination – influences of temperature**). Several studies have shown that maximum quality (expressed in terms of germination capacity and seedling growth) is achieved several weeks after seed mass maturity. There are reports that quality is greater in seeds produced in fruits higher up the plant. Vigour can be assessed by germination tests before and after **accelerated ageing** and **controlled deterioration**. Priming treatments can improve vigour but not all cultivars respond favourably (see section 6 below). Both primary and secondary dormancy types are found in tomato seeds. Dormancy found mainly in freshly harvested seed batches can be relieved, for example, by cold **stratification**.

Some varieties suffer from the 'blind plant' syndrome, where in a proportion of seedlings the primary shoot **meristem** dies soon after seedling establishment (**see: Blind seed**).

Tomato can be affected by several fungal, bacterial (e.g. canker) and viral diseases (e.g. tomato mosaic virus). The seed **health testing** methods used include **ELISA**, bio-assay and liquid plating Bio-PCR.

5. Sowing

Tomatoes are direct-seeded or transplanted in the field or greenhouse, depending on location, environmental conditions and harvesting techniques. For example, direct sowing is limited in the humid tropics because of adverse growing conditions (excess soil water). Crops that are to be harvested mechanically, such as for processing, are usually directly seeded (500–1000 g seeds/ha) in raised or flat beds, in single or twin rows, and singly or in clusters. Where earliness is desired or expensive hybrids are involved, such as for fresh tomato greenhouse production, the use of **transplants** is preferable. Using transplants for field crops guarantees greater seedling uniformity and permits a better control of weeds, pest or diseases. In soil-less greenhouse cultivation the use of transplants is almost obligatory, and seeds are germinated in **peat** plugs or **rockwool** cubes according to choice. **Grafted seedlings** are sometimes used.

In general, tomato seeds will germinate at soil water potentials ranging from just above the wilting point to field capacity, though optimum germination is usually obtained at 50–75% field capacity. Reduced germination at high soil water content is probably the result of reduced oxygen availability to the seeds, since tomato shows dramatic reductions in germination under those conditions. High salt levels also decrease tomato seed germination, although less so than during subsequent growth.

6. Treatment

(a) *Pelleting and coating*. Tomato seeds are commonly **pelleted** or coated in many markets for space planting. These techniques are used to apply fungicide against seedborne and soilborne pathogens (see below).

(b) *Priming*. **Priming** treatments are applied to commercial production seed, and there is particular interest in their use to invigorate rootstock varieties. Priming has been widely reported in the research literature to hasten tomato seed germination and emergence and increases uniformity of germination, not only under optimal germination conditions but also under stress conditions, and also to permit germination over a wider range of temperatures. It has been reported for example that **osmopriming** increased loosening of the endosperm and testa which permitted further germination progress at sub-optimal temperatures. (**See: Priming – physiological mechanisms, 3. Endosperm cell-wall degradation**) Priming can have also a positive effect on the overall quality of lower quality seed lots, though negative effects on storage capability sometimes may accompany this.

(c) *Protection*. Various seed **treatments** are applied in tomato against fungal, bacterial and viral diseases and pests, in addition those mentioned above. Examples are the fungicides metalaxyl and related compounds, the antibiotic streptomycin against bacterial spot and chemical treatments such as trisodium phosphate and sodium hypochlorite against seed surface viruses (e.g. mosaic viruses). Hot water seed soaks are sometimes used against **anthracnose**, bacterial spot, bacterial canker and *Phoma* **rot**. Biopriming, for example with *Pseudomonas fluorescens*, is sometimes carried out. (EH, JMC)

Alvarado, A.D. and Bradford, K.J. (1988) Priming and storage of tomato *Lycopersicon lycopersicum* seeds. I. Effects of storage temperature on germination rate and viability. *Seed Science and Technology* 16, 601–612.

Atherton, J.G. and Rudich, J. (1987) *Tomato Crop: A Scientific Basis for Improvement*. Chapman & Hall, London.

Demir, I. and Ellis, R.H. (1992) Changes in seed quality during seed development and maturation in tomato. *Seed Science Research* 2, 81–87.

Dhaliwal, M.S., Kaur, P. and Singh, S. (2003) Genetic analysis of biochemical constituents in tomato (*Lycopersicum esculentum* Mill.). *Advances in Horticultural Science* 17, 37–41.

Heuvelink, E. (ed.) (2005) Tomatoes. *Crop Production Science in Horticulture*, No. 13. CAB International, Wallingford, UK.

Hilhorst, H.W.M., Groot, S.P.C. and Bino, R.J. (1998) The tomato seed as a model system to study seed development and germination. *Acta Botanica Neerlandica* 47, 169–183.

Mauromicale, G. and Cavallaro, V. (1995). Effects of seed osmopriming on germination of tomato at different water potential. *Seed Science and Technology* 23, 393–403.

Nevins, D.J. and Jones, R.A. (eds) (1987) *Tomato Biotechnology*. Alan Liss, New York, USA.

Valdes, V.M. and Gray, D. (1998) The influence of the stage of fruit maturation on seed quality in tomato (*Lycopersicum esculentum* L. Karsten). *Seed Science and Technology* 26, 309–319.

Yoder, J.I. (1993) *Molecular Biology of the Tomato: Fundamental Advances and Crop Improvements*. Technomic Publishing, Lancaster, PA, USA.

Topographical Tetrazolium test

See: Tetrazolium test (TTZ or TZ test)

Toxic seeds

See: Poisonous seeds

Toxins

See: Pharmaceuticals and pharmacologically active compounds; Poisonous seeds

TPS

True Potato Seed. (**See: Potato**)

Tramlining

A modular sowing system in mechanized agriculture that uses heavier and more efficient equipment for crop husbandry operations (tractors, boom sprays, seeders and harvesters) equipped with wider low-pressure tyres, to accommodate wheels whilst avoiding soil compaction, reducing rutting and row damage they sometimes cause. Wheel tracks are confined to specific lanes in the field. This system is commonly used for combinable arable crops in some parts of the world, and is assisted at sowing time in some locations by global position guidance systems. (**See: Planting equipment – placement in soil**)

Transcription

The production of RNA from the DNA template by RNA polymerases. Only about 1% of the DNA sequence is composed of functional genes, which are copied into messenger RNAs (mRNAs), which are exported from the nucleus to the sites of protein synthesis within the cell for **translation**. Transcription is promoted by transcriptional activators (transcription factors) which are proteins that bind to the promoter ('upstream', that is towards the 5′ deoxyribose end of the **nucleic acid** backbone chain) region of the gene. Many genes contain regions (introns) which are not represented in their mature mRNAs because they are removed (spliced out) during processing of the primary transcript. Hence the result is that only the exon regions of the genes are represented in the mature mRNA following transciption.

Transcriptomics

Analysis of the transcriptome (also called 'RNA' profiling) in a given sample yields a measurement of the relative amount of each mRNA. This reflects the balance between the **transcription** rate of a gene and the rate at which its specific transcript mRNA is being degraded. Several methods exist for determining the amounts of cellular mRNAs; most utilize nucleic acid 'probes' (restricted DNA, or RNA) covalently bound to glass slides or 'chips'. A major technology used is **cDNA microarray**, in which cDNAs (DNA copies of the mRNA population of a plant, that are typically 220–2000 bases long) are used as probes. Another predominant technology is Affymetrix chips on which cRNA oligonucleotides (25 bases long) are used attached to the chips as probes; the probes are designed specifically for known genes and chemically synthesized on the slide. Thousands of gene probes can be represented on a single chip. mRNA is isolated from tissue samples and used as a template to prepare a 'target', to be hybridized with the DNA on the chip, i.e. as cDNA (or cRNA for Affymetrix chips) that is labelled, usually with a fluorescent dye. The labelled target is hybridized to the probe on the microarray. Individual cDNAs/cRNAs from the target hybridize (bind) with the corresponding probe proportionally to their representation in a sample. A specialized scanner detects the hybridized molecules by fluorescence (Colour Plate 14B).

Microarray technology is expensive and time-intensive, both to set up and to use; however, the power of mRNA profiling is immense. It can be used to analyse changes in gene expression in seeds, for example, during development and germination, and in response to chemical and environmental treatments that break dormancy, and the response in different tissues can be compared. It can identify changes in gene activation in **mutants** compared to wild-type plants or between closely related cultivars. One particularly exciting application is gene discovery. Genes whose expression patterns mirror one another over time and are activated by the same stimuli are candidates for being involved in the same phenotypic response. Likewise, regulatory genes might be expected to be up-regulated or down-regulated just prior to the onset of expression of genes encoding enzymes of a particular metabolic pathway. Some examples of changes in mRNA accumulation during seed development using transcriptomics are to be found in the following reference. (AJR)

Girke, T., Todd, J., Ruuska, S., White, J., Benning, C. and Ohlrogge, J. (2000) Microarray analysis of developing *Arabidopsis* seeds. *Plant Physiology* 124, 1570–1581.

Transfer cells

See: Development of endosperm – cereals; Development of seeds – nutrient and water import

Transformation

The process of introducing isolated or cloned genetic material into the **genome** (DNA), either nuclear or organellar, of a plant cell in order to transform a plant from one genotype to another. In molecular genetic studies of seeds an important technique involves the introduction of genetic material to modify seed function. This may be carried out by transformation as described here.

The plant that is the recipient of the inserted DNA is referred to as a transgenic plant (or GMO, **genetically modified organism**; or more correctly GEO, genetically engineered organism), a product of genetic engineering. The goal of plant transformation is to facilitate a stable insertion of the isolated or cloned DNA such that it is transmitted through the germline to future generations in the same manner as the inherent genetic material. The isolated or cloned DNA usually consists of a gene of interest (the transgene), either engineered in a lab or isolated whole from a different plant species or organism, along with associated marker genes or DNA.

Plant transformation can be achieved in a variety of ways but all methods fall into one of two categories: those that require tissue culture, and those that deliver the DNA directly to an intact plant. Most transformation techniques fall into the first category where DNA is delivered to a plant cell or tissue from which a whole (transgenic) plant is regenerated in culture. The regeneration of plants from tissues or individual cells in culture occurs in one of two ways: somatic embryogenesis or organogenesis, both of which are controlled by plant **hormones** added to the culture medium. **Somatic embryogenesis** is the generation of embryos from non-germline diploid cells (somatic cells), e.g. callus, leaves or hypocotyls cells. The somatic embryos can then be 'germinated' to generate whole plants, as would embryos isolated from seeds. Organogenesis is the generation of organs from a tissue explant (a piece of tissue removed from a plant for use in culture), usually leaf or stem tissue. In most cases the organ produced from these methods is a shoot that can be excised, rooted and grown into a mature transgenic plant. Not all plants are easy to regenerate via either somatic embryogenesis or organogenesis, and in many cases the inability to regenerate plants from culture has limited the use of biotechnology for certain crops. Such limitations and the expense of tissue culture has led to considerable efforts to develop simple methods to introduce DNA into plants directly. Most of these efforts have been directed towards delivering the DNA into the normal gene transfer pathway during fertilization and zygote formation, i.e. through the flower during or post-pollination.

DNA can be delivered into a plant cell for integration into the plant chromosome(s) by several means, either as naked DNA, through a bacterial intermediary, or packaged to allow ease of transfer through the plant cellular membranes. The major task for all methods of DNA delivery is to bypass the formidable barriers of the rigid plant cell wall and the plasma membrane and to deposit the DNA inside the nucleus for integration into the plant genome. Bacterial DNA delivery systems package the DNA, penetrate the cellular barriers, and facilitate its incorporation into the plant chromosomes. However, it also appears that naked DNA, once placed within the cell, will naturally migrate into the nucleus of the cell for which it has an affinity. Once in the nucleus, presumably during a DNA replication cycle, it can become incorporated or stably integrated into the plant chromosomes. The following methods for plant transformation have been employed successfully to achieve the delivery of DNA to the nucleus of plant cells and produce transgenic plants for experimental or commercial use.

1. *Agrobacterium*-based transformation

This type of plant transformation was the first to be used to generate transgenic plants and takes advantage of the ability of a group of soil bacteria of the family *Agrobacterium* (primarily *Agrobacterium tumefaciens*) to transform plant cells. *Agrobacteria* are pathogenic bacteria that infect a plant host through a wound site and cause the plant to produce a mass of undifferentiated embryonic cells at the infection site that develops into a characteristic 'plant gall' or tumour. *Agrobacterium tumefaciens* is the causative agent for crown gall disease, the crown being the part of the plant that is situated at the stem-root junction in many plants. *Agrobacterium* causes the production of undifferentiated embryonic cells by transferring a section of its own DNA, carried on a tumour-inducing (Ti) **plasmid**, and inserting it into the nuclear genome of the plant host cells (transformation). The DNA fragment that is transferred from the bacterial plasmid to the plant genomic DNA is called Transferred DNA or T-DNA. T-DNA carries two important sets of genes. One set encodes enzymes that synthesize plant hormones and are responsible for the gall formation in the infected host, the other genes direct the synthesis of opines, amino acid derivatives, that are excreted by the crown gall cells to provide *Agrobacterium* with a source of carbon and nitrogen. The T-DNA is processed and transferred by the products (Vir-proteins) of the *Virulence* or *Vir*-genes that are also situated on the Ti plasmid (Fig. T.6).

Plant genetic engineers have taken advantage of this natural plant transformation method to introduce genes of interest into target plants. This is accomplished by developing engineered Ti plasmids in which the bacterial genes within the T-DNA are replaced by the genes to be introduced into the plant. This insertion occurs between what are known as the left (L) and right (R) borders of the T-DNA. Along with the inserted gene(s), selectable marker genes are often included in the transformation. Selectable marker genes encode proteins that can detoxify a harmful chemical that is added to the tissue culture medium during regeneration. Typically these chemicals are antibiotics that disrupt plant cells, e.g. kanamycin or hygromycin, or herbicides, e.g. gluphosinate (glufosinate, Basta, Liberty). The purpose of these chemicals is to prevent or inhibit the growth and development of plant cells that do not contain the inserted gene(s) of interest, thus making it easier to isolate those that are transformed.

Agrobacterium transformation is usually performed by inoculating (soaking in a bacterial suspension) sterile pieces of leaves, cotyledons, hypocotyls and stem segments, callus or

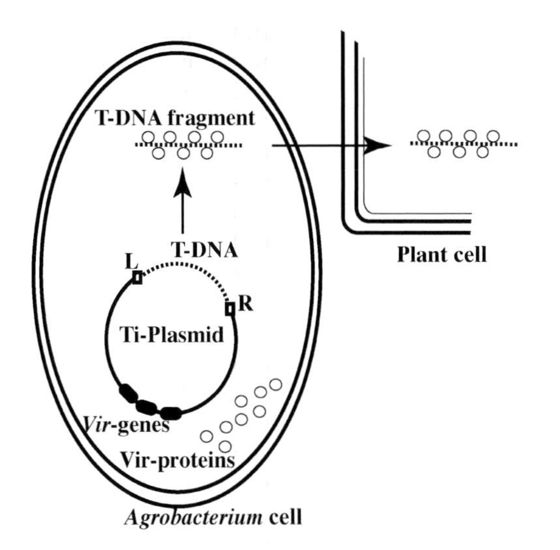

Fig. T.6. *Agrobacterium*-mediated delivery of T-DNA to plant cells (see text for explanation).

cell suspension cultures, or germinating seeds (Fig. T.7). Once inoculated, the tissues are given sufficient time for gene transfer to occur before the *Agrobacteria* are killed by the introduction of a chemical to which they alone are susceptible (e.g. carbenicillin) and the tissue placed into tissue culture for selection and regeneration of transgenic plants. Generally, these procedures require that the tissue is wounded prior to inoculation. The cutting of the tissue for inoculation is generally sufficient to provide a wound site, although bombarding the tissue with microscopic glass beads or ultrasound (sonication) can increase the effectiveness and efficiency of transformation. *Agrobacterium*-based transformation is very attractive because it is easy to perform and requires little in the way of specialized equipment. In addition, transgenic plants that are derived from *Agrobacterium* transformations often contain simple single copy insertions of the gene(s) of interest that makes downstream commercial breeding efforts less complex. On the negative side, *Agrobacterium*-based methods are often limited to one particular variety or cultivar of a crop (which requires extensive breeding efforts to transfer the transgene to an elite commercial genotype), and the plant host range of the particular *Agrobacterium* strains in use.

2. Protoplast transformation

With the ability of individual plant cells to regenerate into whole plants in tissue culture, via somatic embryogenesis or organogenesis, has come the possibility to develop transformation technologies where individual cells can be targeted for DNA delivery and transformation. Single cell transformation in plants has one major advantage: the barrier of the cell wall can be physically removed and, following DNA delivery, can be reconstituted. The plant cell lacking a cell wall is called a **protoplast** and the only barrier left for DNA delivery is the plasmalemma or cell membrane. The cell wall can be removed by either mechanical means or by digestion with cell-wall-degrading enzymes that are usually obtained from fungi. Protoplasts can be isolated from a variety of sources including directly from leaf tissue. However,

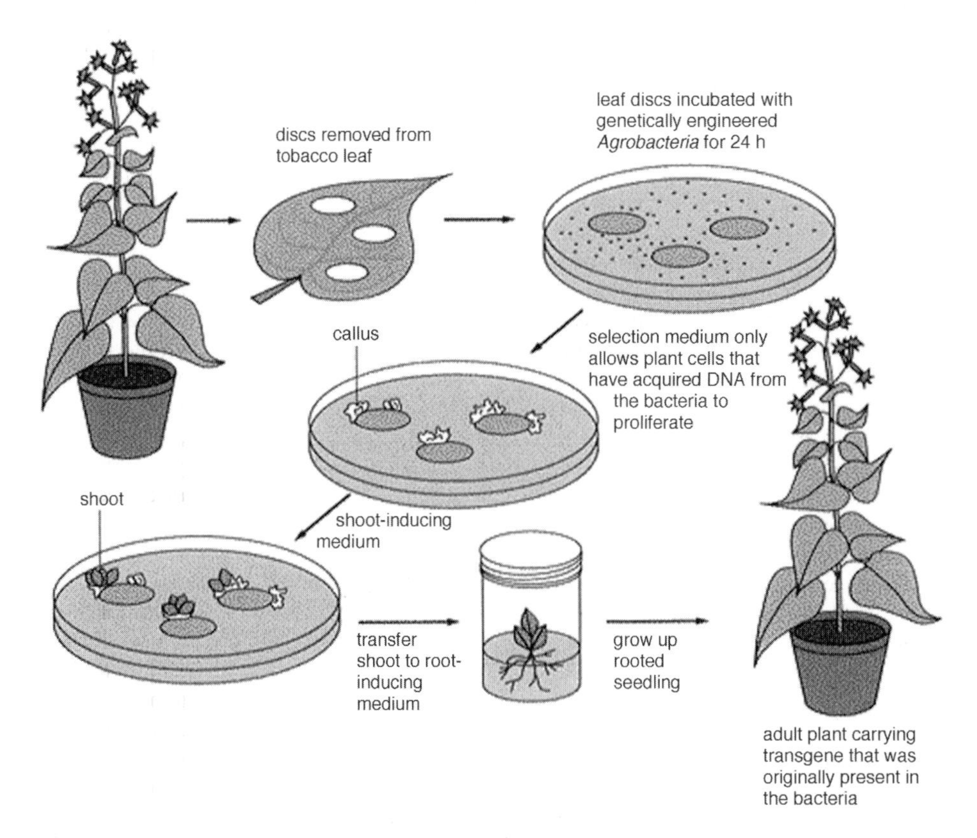

Fig. T.7. An overview of the steps involved in the transformation of tobacco plants using *Agrobacterium* to introduce new genes.

protoplasts are most commonly isolated from cell suspension cultures derived from embryogenic **callus**; they are undifferentiated masses of cells derived from such tissues as embryos, anthers, and other floral tissues. Once the cell wall is removed, DNA can be delivered to the cell nucleus by one of several methods including: (i) direct DNA uptake; (ii) the use of polyethylene glycol (PEG) to gently perforate the plasmalemma to increase the rate and extent of direct DNA uptake; (iii) the use of an electrical current to increase DNA movement across the plasma membrane (electroporation); and (iv) the encapsulation of DNA in small lipid vesicles (liposomes) which can fuse with the plasmalemma and deliver the DNA into the cell. *Agrobacterium* can also be used to deliver DNA to transform protoplasts. Protoplast transformation has been very successful for the transformation of monocot plants such as maize and rice, which have proved to be difficult to transform with *Agrobacterium*-based methods. Protoplast transformation techniques tend to be variety (genotype) independent and more broadly applicable than purely *Agrobacterium*-based transformation methods.

3. Biolistics

Biolistic transformation methods rely on a 'brute force' approach to delivering DNA through the cell wall and membrane barriers. The term biolistics is derived from the term ballistics, which is defined in *Webster's Dictionary* as 'the modern science dealing with the motion and impact of projectiles, especially those discharged from firearms'. In biolistics, the projectiles are microscopic particles, often gold or tungsten microspheres, coated with DNA, which, in early protocols, were discharged into plant tissues from a .22 caliber rifle casing in an apparatus appropriately known as the Gene Gun. Later protocols use gases, such as helium or carbon dioxide, or an electrical discharge to accelerate the microprojectiles into plant tissues or cells. The only limitation for plant transformation with this technology is the ability to regenerate fertile plants from the target tissue or cells. Almost any plant tissue can be used, including cell cultures, from apparently any variety or cultivar; this makes the technology extremely attractive to commercial ventures. The use of baffling screens to moderate particle acceleration, and the ever-decreasing size of the microprojectiles available for use in this technique, has vastly improved the efficiency of biolistic plant transformations. Nevertheless, there are problems with this approach. The transgenic plants obtained by biolistic transformation often reveal very complex patterns of transgene integration that complicates further breeding efforts and it is often difficult to deliver large DNA fragments intact into the plant cell. Coating the microprojectiles with *Agrobacterium*, carrying the desired transgenes, which protects the DNA and integrates it into the plant chromosomes in a simpler arrangement, has circumvented some of these problems.

4. *In planta* transformation

This type of plant transformation delivers transgenes, as naked DNA or via *Agrobacterium*, directly into an intact plant, negating the need for regeneration and tissue culture. Such methods have the obvious advantages of speed, from time of DNA delivery to a transgenic plant, and economy. Many of the *in planta* methods that have been developed target DNA delivery to the natural fertilization processes of plants, attempting to target the germinating pollen grain or the ovule during zygote formation. The most successful of the *in planta* transformation strategies has been the *Agrobacterium*-based floral dipping or infiltration method developed for *Arabidopsis*, a model plant for transgenic studies. *Arabidopsis* flowers are simply dipped into a suspension of actively growing *Agrobacteria* and allowed to continue their normal development. The seeds from the treated flowers are germinated on selective media and occasionally a transgenic seedling is obtained. The sheer number of seeds that can be collected for screening more than compensates for the very low efficiency of transformation that is displayed by this method. This procedure is now the standard method for those scientists who use transgenic studies to understand gene function in plants. However, the means by which the *Agrobacterium* gains entry into the developing zygote or seed to deliver the transgenes is unknown. Other methods of *in planta* transformation attempt to directly inject DNA into ovules during or immediately post-fertilization. Although successful, these procedures are difficult to perform and are highly irreproducible.

Delivery of transgenes to germinating pollen grains, prior to artificial fertilization, has also been moderately successful. Some methods culture the pollen in solutions or media containing DNA, for direct uptake into the pollen as it germinates; others germinate pollen in the presence of *Agrobacteria* containing the transgenes of interest. The biolistic bombardment of pollen suspensions has been successful in tobacco, and this approach could be more widely useful for producing transgenic crops since it circumvents some of the varietal barriers of many transformation strategies. (MJO)
(See: Biopharmaceuticals; Bioplastics; Genetic modification; Genetically-modified organism (GMO); GMO – patent protection technology)

Hansen, G. and Wright, M.S. (1999) Recent advances in the transformation of plants. *Trends in Plant Science* 4, 226–231.

Newell, C.A. (2000) Plant transformation technology: developments and applications. *Molecular Biotechnology* 16, 53–65.

Transgene

Common term for a gene that has been moved from one biological source to another, by artificial means. A transgenic seed (variety, crop) is one produced through a breeding programme involving genetic modification. It is a term used interchangeably with genetically modified seed (variety, crop) or, popularly, a 'biotech' variety, in contrast to a 'traditional' or 'conventional' seed (variety, crop). **(See: Genetic modification; Genetically-modified organism (GMO); Transformation)**

Transitory starch

A reserve carbohydrate that accumulates in leaves during photosynthesis. Transitory starch is synthesized during the light cycle, and mobilized (i.e. degraded to simple sugars) during the dark phase in order to meet the energy needs of the plant. Transitory starch production also occurs during apical **meristem** formation, during **seedcoat** development, prior to floral transition, prior to nectar production, and in the **cotyledons** of some **endospermic legumes** during mobilization of **mannan**-rich endosperm cell walls and during development of oil-storing seeds and cereal **embryos**. This is an indication that transient starch metabolism has a key role in many aspects of plant growth and development. Like storage starch in seeds or tubers, transitory starch granules are comprised of **amylopectin** and **amylose**, although the amylose content is lower. (MGJ)

Translation

The synthesis of a **protein** (polypeptide) on **polysomes** whose amino acid sequence reflects the nucleotide sequence of its encoding messenger RNA (mRNA) molecule. Amino acids are donated in order by tRNA molecules which recognize their appropriate codon sequence on the mRNA, and peptide bond synthesis is catalysed by the translation complex, which includes the ribosome and many cytoplasmic proteins, including initiation and elongation factors. Translation proceeds from the 5′ deoxyribose end of the **nucleic acid** backbone chain to the 3′ end, and the corresponding polypeptide synthesis proceeds from the N-Terminus to the C-terminus of the **protein**. (**See: Nucleic acids; Storage proteins – synthesis**)

Transplants

Many horticultural and floricultural crops are established from transplants (i.e. seeds already germinated (pregerminated) and grown into substantial seedlings) in field nursery beds or trays, either outdoors or under cover, for replanting in another location. This improves establishment and lengthens the growing period of the crop. It also reduces the need to thin-out surplus plants in the field, hastens crop maturity, gives a competitive advantage against weeds and other pests in the field and, generally, optimizes the germination and growth of seeds and seedlings. Although transplants can also be grown in outdoor plant beds and field nurseries, greenhouses provide the best control of environmental conditions and are preferred by many for transplant production. Greenhouse-grown transplants are now widely utilized for high-value vegetable and flower crops, in breeding programmes, in tree raising, in most major **tobacco** production areas of the world, and some paddy **rice** crop systems, as well as for crops that are grown entirely under cover up until harvest. Some seed production systems also use transplants (**see: Production for sowing**).

Transposons

Sequences of 'mobile' DNA that become inserted into genes thus affecting their expression. (**See: Mutants; Research seed species – contributions to seed science, 10. Maize**)

Treatment

In the context of seed technology: 'treatment' is a broad generic term to describe the range of materials, formulations, techniques, equipment and processes that may be intentionally applied to seed after the application of conventional seed **conditioning** technologies to improve seed physiological performance, 'planting value' and physical handling characteristics, and provide protection or improved growth to seeds, emerging seedlings and established plants after sowing. In the context of grain technology: treatment is the equivalent application of materials and practices to protect against the action of storage pests and diseases, or to reduce the degree of **dust** generated while handling seed to control explosion and health-and-safety risks.

The main categories of pre-sowing seed treatments used in modern developed agricultural systems can be classified by their objectives and basic technologies, as follows. Their inter-relationships with conditioning and seed testing technologies are illustrated schematically in Fig. T.8.

(1) Application, or **dressing**, of pesticides or other **crop protection** materials or processes to protect the seed and seedling by controlling or repelling the action of pests and diseases (fungal, viral, and bacterial) that are not controlled by resistant varieties or by biological, cultural, physical or sanitary means. Treatment with seed-applied fungicide and insecticide formulations is the traditional and main connotation of the term 'seed treatment', and comprises by far the largest economic market sector worldwide. Using the seed as a carrier is a well targeted way to deliver protective agents to crops, is convenient to the farmer, and has the environmental advantages of using greatly reduced chemical usage rates on the whole field compared to foliar spray or soil-applied alternatives. Protection is increasingly being extended further into the life of the established plant by the development of **systemic** agrochemicals. (**See: Treatments – pesticides**)

(2) Eradication of seedborne pathogens is also achieved using moist or dry heat (**thermotherapy**), by steeping in chemicals, or exposure to certain **electromagnetic enhancements,** using radiation, electrical or magnetic fields of various strengths (some of the latter are applied with the intention of improving seed physiological performance). The requirements of producing **organic seed** have led to a resurgence of non-pesticidal seed treatments.

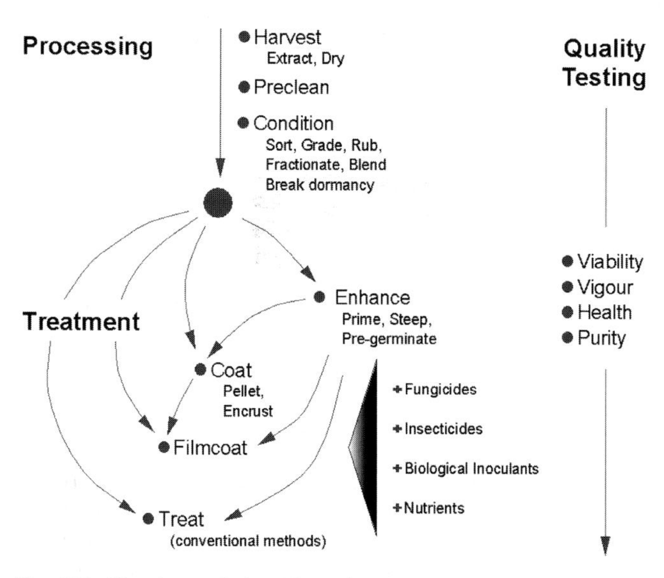

Fig. T.8. The inter-relationships of seed processing technologies used in agronomic practice to process and treat seed, after harvest, for growing the next crop. Processing involves various techniques, according to seed type. Optional treatments are used singly or in permutation, according to agronomic and market needs. Quality tests are conducted at stages throughout processing and treatment. From: Halmer, P. (2000) Commercial seed treatment technology. In: Black, M. and Bewley, J.D. (eds) *Seed Technology and its Biological Basis.* Sheffield Academic Press, UK, pp. 257–286.

(3) Application of nutrients or inoculants, such as symbiotic biological fertilizers as in the **rhizobial inoculation** of legume seed, or inoculation with beneficial antagonistic **rhizosphere microorganisms**.

(4) Coating to modify seed handling characteristics, or colour seeds, such as **encrusting** and **pelleting** to facilitate planting (**see: Planting equipment – agronomic requirements**). These techniques are also used, as is **filmcoating**, to securely apply other treatment materials such as the pesticides, micronutrients or inoculants.

(5) Seed physiological manipulations, such as hydration techniques in **hardening** or **priming** to increase the rate and uniformity of germination, **pregermination**, **soaking** or **steeping** to remove inhibitory compounds from the seed, **humidification** to protect against mechanical injury during seed handling and hydration damage during imbibition, or exposure to plant **hormone**, or various energy forms (e.g. radiation, heat, magnetism).

Some authors regard seed **conditioning** technologies as constituting another class of 'seed treatment' in themselves – that is, preparing seed for sowing by mechanical cleaning, purification, fractionation and other 'upgrading' techniques, exploiting seed physical properties such as size, shape, texture, density, buoyancy and colour.

It is also common terminology in both the seed industry and the scientific literature, to group all the non-pesticide treatments together under the heading of either 'functional seed treatments' or 'seed **enhancements**'. (PH)

Taylor, A.G. and Harman, G.E. (1990) Concepts and technologies of selected seed treatments. *Annual Review of Phytopathology* 28, 321–339.

Taylor, A.G., Allen, P.S., Bennett, M.A., Bradford, K.J., Burris, J.S. and Misra, M.K. (1998) Seed enhancements. *Seed Science Research* 8, 245–256.

Treatments - brief history

There are only scattered references to seed treatment and technology in the 'historic' period, before the start of the scientific age. These consist of written contemporaneous records of the Greek, Roman, Mediterranean and Near-Eastern Babylonian cultures and the Chinese and Hindu civilizations. Practices in pre-Columbian Central and South America, primarily the Aztec and Inca civilizations, were recorded very much later, soon after the Spanish conquest in the 16th century. Often, ancient authors uncritically report current agricultural practices and remedies, and may be reflecting more in the way of folklore, magic, mythology, superstition and undigested hearsay than firsthand experience. Knowledge of the medieval European feudal period comes largely from the compilations of monastic writers, who tended to harmonize older knowledge and practices with church teachings. (**See also: History of seed research; History of seeds in Mexico and Central America**)

1. Seed storage, control of pests, sowing enhancement and nutrition

The virtues of harvesting, selecting and preserving high quality seed for sowing must have been realized long before the start of recorded history. Clearly the great urban civilizations of Mesopotamia, northern India, Egypt and Rome depended upon successful grain storage technology – as illustrated by the Biblical story of the huge Egyptian granaries that successfully outlasted a prolonged famine (Genesis, chapter 41), and by the comments of Theophrastus (371–287 BC), the most important botanist of antiquity, on the death of seeds in storage.

There was awareness of techniques that we would recognize today as seed **enhancement**. The Roman author Columella (~60 CE), for example, mentions the use of water **flotation** to collect heavier, more viable seeds, and the use of brine or some other liquid with a higher specific gravity than water is a common practice throughout Asia. Beliefs in China in the ritual washing of seed-grain, perhaps stemming originally from the discovery of the advantages of **pre-germination**, were commonly associated with wet rice, and also applied to a number of other crops. A 6th-century treatise, for example, recommends pre-germinating both rice (both wet and dry) and hemp sown during the rainy season: for rice this involved soaking for 5 days, keeping warm and damp in a basket for another 3 days, and broadcast sowing when the seedlings had grown to 5 cm. Sunlight was held to be beneficial to seeds, and authors recommend drying seed in the sun after washing or soaking (a form of **solarization**). In the 18th century Chang Tsung-Fa (Szechwan) conveys knowledge of **hardening** techniques:

> to facilitate the germination of vegetables: dry the seed in the sun during the hottest month of the summer ... in the time that seed sun-dried one year grows one inch, seed dried two years running will grow two inches and seed dried for several years will grow several inches tall.

John Evelyn in 1664 notes the need for low-temperature **stratification** to promote germination of some **tree seeds**.

It has also apparently long been common for farmers to treat seed with simple fertilizing and insect-repellent substances. The Roman, Pliny the Second, writing in the 1st century, relates that when 'individual leek, kale, lettuce, celery, endive and watercress seeds were placed inside hollowed pellets of dung, they all come up wonderfully.' A 6th-century Chinese agricultural treatise describes **coating** seed grain with a porridge-like concoction of extracts of horse bones (collagen) and aconite (*Aconitum*, which produces an alkaloid that is highly toxic to mammals) followed by sun drying; according to a 17th-century treatise, peasants in one part of Szechwan were 'growing aconite for no other purpose than to treat seed grain'. Other treatments also recorded from this historic era include protecting seeds from insects and strengthening drought resistance by various decoctions of such things as silkworm droppings, eels' heads, cottonseed oil, arsenic, ashes, the root of *Sophora evansecens* mixed with lime.

2. Control of disease

Appreciating the state of knowledge in antiquity, and the actual practices and procedures used to protect cultivated plants from diseases, is a more complicated matter. Until the modern era, most **diseases** were variously held to be due to astrological causes or the weather, such as 'imbalances between nourishment and the surrounding air' or 'clinging moisture' – in contrast to the comprehension of the harmful roles of insect pests, whose sometimes spectacular effects were easy to recognize. Arguably the clearest exception was the recognition in pre-Columbian America of the association between a fungus and a seedborne disease in the common blister **smut** in maize (caused by *Ustilago zeae*), whose dramatic symptoms include the formation of mushroom-flavoured galls. Care must also be taken in interpreting descriptions of plant diseases in ancient and medieval writings. Familiar descriptions of seed diseases that are seemingly equivalent to the modern terms '**rust**', '**mildew**' and '**smut**' does not mean that they were always used to describe the same diseases that we recognize today.

Perhaps the earliest recorded seed treatment, noted by the natural philosopher Democritus (5th–4th century BC), was to use the juice of a succulent plant, the houseleek (*Sempervivum tectorum*). In ancient Mediterranean cultures, great reliance was evidently placed on the lees of olive oil ('amurca') as a universal disease and pest remedy. The Roman Virgil (1st century BC), for example, writes of medicating seed by a steep in nitre and amurca 'to obtain a fuller produce in the deceitful pods' (although he warns 'these measures can be in vain'), and of the use of amurca and soda as protection against weevils in stored grain and beans. His near-contemporary Varro writes of treating grain with wormwood and chalk. Pliny, in his compilation of 1st-century practices, also writes of steeping cereal seed in amurca and urine, whereas Palladius, in the 4th or 5th century, writes of steeping in ox gall. Ancient Indian culture (in the millennium starting from about the 5th century BC) drew a close relationship between medical knowledge and plant pathology, and common standby remedies for plant diseases included water, milk, honey and ghee. It is difficult to establish how effective farmers considered these procedures to be, in practical agronomic terms. But in general, the real nature and cause of plant disease did not become apparent until the start of the scientific era, with the rise of modern plant pathology in the 19th century. It was not until the 1850s, for example, that the true nature of **ergot** disease was understood, whose occurrence was so common that it had been thought to be part of the rye plant. (**See: Diseases; Pathogens**)

The foundations of modern seed treatment fungicides were the development of sulphur, copper-based formulations and mercury compounds. Remarkably the last of these have been abandoned only relatively recently (and the first two are still used widely and effectively, though in non-seed treatment applications). In probably the first recorded scientific seed treatment experiment, Tillet (1750) concluded that bunt-infected wheat could be greatly reduced by treatment with lime in combination with either salt or nitre, and soon after came the suggested uses of solutions of arsenic or mercury chloride (in 1755, by Aucante) and of copper sulphate (in 1761, by Schulthuss). Initially copper sulphate was applied by **steeping**, but large seed volumes came to be treated by a 'barn floor' method, in which seed was sprinkled with the solution. Copper treatment only became widely used in the mid-19th century, however, with suggestions (in 1873, by Dreich) of a further treatment step with lime to convert the copper to its insoluble oxycarbonate salt, which gave greater persistence and avoided toxic effects on the germination of any damaged seed. Later (in 1917, by Darnell-Smith) came the practice of dusting seed with basic copper carbonate powder instead. Formalin treatments were also introduced as a treatment to control cereal smuts in North America and Australia at the very end of the century, and later in Europe.

In 1914 Rheime introduced 'chlorophenyl mercury' as a less dangerous alternative to mercury chloride. This was the first of what became a wide family of broad-spectrum eradicant and protectant 'organomercurial' seed treatment fungicides (including alkyl, alkoxyl or aryl derivates) that were introduced throughout the 20th century as more efficient replacements to the copper and formalin alternatives for the control of most cereal smuts and some seedling blights, and were only phased out of use during the last two decades.

The origin of replacement non-mercurial synthetic organic seed treatment fungicides began in the early 1930s, and the range of chemicals available, the area and frequency of their use, and their effectiveness, have increased enormously since the mid-20th century. Several of the 'middle-aged' non-systemic protectant fungicide classes have been used steadily for well over 30 years, including the dithiocarbamates such as thiram, maneb and mancozeb, heterocyclic compounds such as the phthalimide, captan, and the aromatic compounds such as the chlorophenyls. The first **systemic** fungicides were introduced in the late 1960s and 1970s, beginning a new era in seed treatment. These included the oxathiin, carboxin, and the benzimidazole, benomyl, and the development of the pyrimidines and a succession of analogues of the triazoles and imidazoles and, latterly the phenylpyrroles and strobilurin – which are generally used in relatively small amounts, because of their more potent action against plant pathogens. The organochlorine insecticides were first introduced in the 1940s, followed by the organophosphates, carbamates and pyrethroids, and the systemic neonicotinoids were introduced in the mid-1990s. Some major historical milestones are shown in Table T.2. (**See: Sowing methods – brief history; Treatments – pesticides**) (PH)

Bray, F. (1984) *Joseph Needham. Science and Civilisation in China. Vol. 6. Biology and Biological Technology. Part 2. Agriculture.* Cambridge University Press, Cambridge, UK.

Brent, K.J. (1995) Fungicide resistance in crop pathogens: how can it be managed? *FRAC Monograph No. 1.* Global Crop Protection Federation, Brussels, Belgium.

Evenari, M. (1980/81) The history of germination research and the lesson it contains for today. *Israel Journal of Botany* 29, 4–21.

Jeffs, K.A. (1986) A brief history of seed treatments. In: Jeffs, K.A. (ed.) *Seed Treatment.* BCPC Publications, Thornton Heath, UK.

Maude, R.B. (1996) *Seedborne Diseases and their Control: Principles and Practice.* CAB International, Wallingford, UK.

Orlob, G.B. (1973) Ancient and medieval plant pathology. *Pflanzenschutz Nachrichten Bayer* 26, 65–294.

Treatments – micronutrients

Seed is sometimes used as a delivery vehicle for micronutrients known to be lacking from the soil. For example, sources of zinc are included in seed treatments for some rice, molybdate for some soybean (in the USA), and manganese for some cereals and legumes (in the UK). In appropriate locations, seed is treated with small amounts of major element fertilizers, including nitrate and phosphates. (**See: Starter fertilizer**)

Table T.2. Some major historical milestones in the evolution of modern chemical seed treatments.

	Introductions or withdrawals
1600s	Soaking in salt water to control cereal bunts, and with liming to dry grain
Mid–late 1700s	Introduction of arsenic, soaking in copper salts, or lime (with salt or nitre)
Late 1800s	Hot water treatment against seedborne pathogens. Formaldehyde soaking
1900s	First seed pelleting patents, to apply fertilizers to seeds
1910s	Organomercurial compounds and dry copper salts as eradicant fungicides against seedborne infections (e.g. cereal bunts)
1940s	First seed treatment insecticides: organochlorines (lindane) Broadening use of organomercurial fungicides First broad-spectrum seed treatments: organic sulphur (thiram, dithiocarbamates) and chlorinated aromatic fungicides (chlorophenyls)
1950s	Pelleting in USA and Europe for horticultural crops and sugar beet Organic sulphur phthalimide fungicide (captan) against soilborne damping-off fungi
1960s	First systemic fungicides for seedborne pathogens (oxathiin, carboxin) and against certain soil-borne fungal and aerial pathogens First broad-spectrum insecticides (carbofuran)
1970s	First systemic seed treatment fungicides for air-borne pathogens: benzimidazoles, triazoles (triadimenol), hydroxypyrimidine (ethirimol), phenylamide (metalaxyl)
1980s	Start of withdrawal of organomercurial fungicides Nonsystemic pyrethroid insecticides (tefluthrin) against soilborne insects Priming (initially osmoconditioning in horticultural crops)
1990s	Low-rate fungicides (fludioxonil, difenoconazole, tebuconazole, triticonazole) New classes of broad-spectrum insecticides: neonicotinoids (imidacloprid, thiamethoxam), fipronil, and the progressive banning of lindane

For details on the common names referred to in this table, **see: Treatments – pesticides**. Modified from Brandl, F. (2001) Seed treatment technologies: evolving to achieve crop genetic potential. In: Biddle, A. (ed.) *Seed Treatment: Challenges and Opportunities*. British Crop Protection Council, Farnham, UK, pp. 3–18.

Treatments – pesticides

Fungicides and insecticides are economically the two most important forms of seed treatment used in crop cultivation. Increasingly they are used in combination with each other. In the first instance, seed-applied fungicides and insecticides were targeted at the control of seedborne and soilborne pests and diseases that affect early plant growth or, in the case of **systemically** transmitted seedborne diseases, later growth stages. But since the 1980s fungicides and insecticides have been developed with systemic modes of action that can protect plants for several months into the life of the crop (**see: Treatments – brief history**).

In 2000, the worldwide market for seed treatment pesticides had a total value of around US$0.9 billion at the agrochemical manufacturer level, distributed between fungicides (58%), insecticides (26%) and mixtures of the two (16%). The major crops treated were small-grain cereals (40% – plus 7% in rice), maize (15%), oilseeds (12%), sugarbeet (6%), cotton (5%) and other crops (8% – plus 'seed' potatoes, 7%). In recent years this market sector has grown due to the introduction of agrochemicals that are capable of delivering high levels of efficacy at much-reduced usage rates compared to delivery by soil or early foliar application by sprays or granules. Also the advent of higher-value broadacre crop seed varieties, including **genetically modified** ones, has increased the importance of protection.

1. Introduction

(a) *Categories*. 'Pesticide', in its broadest scientific and legal senses, is a generic term for any substance(s) intended for preventing, destroying, repelling, reducing or mitigating the growth, detrimental action or effects of any pest populations. In this context 'pest-' has a wide meaning too – covering harmful insects and other arthropods, unwanted plants (weeds), pathogenic fungi, bacteria, viruses, rodents and 'other animals'. As a consequence there are several major groups of 'pesticide' usage; and several of them are directly relevant to seed technology:

- fungicides and insecticides, widely used as pre-sowing or storage treatments as already mentioned, and other antimicrobial agents, such as bactericides and viricides (see next four sections);
- **fumigants**, used in seed storage primarily to control pests, and to eliminate pests, diseases and weeds in seedling bed preparation;
- **defoliants** or **desiccants**, used to kill off crop plant foliage before seed harvest;
- nematicides, now being developed as seed treatments for the management of parasitic nematodes such as on cotton;
- herbicides, rodenticides and miticides.

Moreover, under the scope of registration laws and regulations (see section below), 'pesticides' are also understood to include:

- plant growth regulators, which are used for example in some seed enhancement treatments (such as **priming**);
- **safeners**, used for example to protect seeds against phytotoxic effects that would otherwise be caused by certain crop-applied herbicides;
- beneficial microorganisms or **'biopesticides'** including 'microbial pesticides' – living fungi, bacteria, viruses, protozoa and nematodes – and naturally occurring substances that control by non-toxic mechanisms;
- attractants or repellents that lure a pest to, or divert it from, a particular site.

However, excluded from the regulatory span of 'pesticides' are the following:

- 'biofertilizers' such as **rhizobia** and **mycorrhizae**;
- **rhizosphere microorganisms** that may be inoculated on to seeds or into growing media in seedling transplant regimes;
- the so-called 'plant-incorporated pesticides' – that is, plants that have been genetically engineered to contain insecticidal genes, such as those derived from *Bacillus thuringiensis* (*Bt*).

(b) *Modes of action*. The ending '-cide' (derived from the Latin, 'to kill') is often a misnomer: not all pesticides act by destroying or eradicating their target organisms. Many act by suppression – inhibiting their growth, feeding or reproduction – or by stimulating defence mechanisms in the protected plant. Specificity is achieved either through differential toxic action or through the manner in which the pesticide is used: its formulation, dosage, timing or placement.

Broad-spectrum pesticides are non-selective and may suppress a wide range of organisms; they are preferably used in situations where more than one pest is being targeted, but may also indiscriminately destroy or harm beneficial organisms.

Selective pesticides target a small number of individual pest species, while sparing most other fauna or flora, including beneficial species.

Protectant fungicides or insecticides provide local protection against invasion or infection from soilborne pathogens and pests for short periods of time. They may be released and effect their action in the surrounding soil. Many protectant fungicides are of this type, although some have **eradicant** activity, and tend to have broad-spectrum activity. Thiram, captan or fludioxonil, for example, are applied to seed to protect from soilborne fungi that cause pre- and post-emergence seedling **damping-off** and death during crop **establishment**, and also to eradicate or greatly reduce certain seedborne pathogens. Protectants have limited penetrative activity when applied to the seed surface and generally are not mobile within seed tissues, being non-systemic.

Systemic insecticides and fungicides can be absorbed by and move within the plant as it grows – entering the young seedling via the soil through the roots and being translocated through the plant accumulating in the cotyledons and young leaves – preventing for several weeks the action of early-season insect pests or diseases in the crop. Systemic fungicides, such as those to control loose **smut** on wheat, may have fairly narrow activity. Systemic seed treatments may also act non-systemically, directly against certain targets at the site of application, as well as penetrating the tissues of the emerging seedlings.

(c) *Nomenclature*. Several names are used to refer to substances with pesticide activity and to their formulations, particularly where synthetic chemicals are involved. These names are agreed according to the guidelines approved by regulatory authorities, international and national groups, and recognized committees on pesticide nomenclature (e.g. the International Organization for Standardization, ISO, Technical Committee 81). The full systematic 'chemical name' identifies the molecular structure; it often has a complex form, and may differ between countries for the same **active ingredient** (ai). For trade, registration and legislation purposes, therefore, and for use in popular and scientific publications, a short, distinctive, non-proprietary and widely accepted 'common (standard) name' is given to each ai, regardless of its manufacturer, formulator or trade name. Where the compound is a simple one, the common name may be the same as the chemical name, but more usually it is an abbreviated or simplified derivative, typically consisting of four to six syllables – such as 'imidacloprid' and 'metalaxyl'. Group names describe the chemical family to which the ai belongs, usually in terms of the functional chemical moiety – such as 'neonicotinoid' and 'phenylamide'. This generally indicates that the chemicals in the group share a common biochemical mode of action and cross-resistance potential. Lastly, proprietary brand, trade or product names are given to different formulations of one or more ai; occasionally this name is the same as the common name or resembles it, but often is unrelated and chosen for marketing purposes for specific target uses on specific crop groups. There can be different brand names of seed treatment formulations, for instance, for the same ai, from one country to another.

2. Treatment strategies

Decisions on the use of seed treatments are complex. Cost considerations aside, it may not be biologically necessary to treat for a given problem in all localities against seedborne and soilborne or airborne fungi and insects.

Treatment is often recommended as a routine practice where plant stands and yields are known to increase substantially and the cost is relatively inexpensive or when the crop has a high value and quality is paramount, and there is a lack of resistant varieties or adequate cultural control practices. Many soilborne diseases are more serious under cooler wetter conditions, and in temperate climates seed treatment is regarded as necessary in cereals (wheat, maize, rice), oilseeds, cotton and sugarbeet crops. Less than half of the varieties of cereal grain used today, for example, are highly resistant to **smuts**, and seed treatment is necessary to keep these diseases at a non-destructive level. Similarly, it is commonly held that there would be large losses in horticultural and ornamental crops without seed treatment.

At the same time, the concept that pesticides should be used on the principle of 'treatment according to need' has become widely accepted, and this notion has been extended to seed treatments as well. Many countries thus have general policies aimed at reducing the use of pesticides, and are increasingly focusing on integrated crop/pest management systems ('sustainable' approaches involving minimizing compatible cultural, chemical and biological inputs, to maintain pest populations at levels below those causing economic injury while minimizing environmental risks to users, non-target organisms and wildlife).

Policies vary on the use of prophylatic seed treatment – that is, on seed that is not known to be unhealthy, or is going to be sown across a range of soils where potential damage from pests or diseases cannot be predicted. In cases of risk of seedborne disease, it is highly inadvisable to sow untreated seed (whether **certified** or **farm-saved**) without testing for the pathogens (**see: Testing – health**). Treatments with fungicides may then be limited to infected lots, using threshold levels of infection determined through combination of research and empirical experience of disease expression and the risk of its spread into neighbouring and following crops; these thresholds may change due to the adoption of new plant varieties and treatment policies.

3. Fungicides

(a) *Targets.* Pre-sowing seed-applied fungicidal treatments are used predominantly for three reasons: (i) to control seedborne pathogens, carried as spores or other resting structures on the surface (disinfestation, or seed-surface **disinfection**) or deep inside the seed (disinfection), such as when introduced by **soaking** and **steeping** techniques, including the **pathogens** of economic concern that cause **rusts**, **smuts**, **mildews**, moulds later in the life of the plant, using ais with systemic modes of action; (ii) to protect the seed and young seedling from soilborne pathogens that cause such diseases as seedling blights and root rots, seed rots and pre- and post-emergence seedling damping-off that may prevent or delay seedling emergence, for sowing situations where there is an appreciable infection risk, such as cold or wet seedbeds; and (iii) to control early-season diseases, generally on aboveground plant parts (such as cereal powdery **mildews**), as an alternative to foliar spray application, again using ais with systemic modes of action.

Only occasional use is now made of controlling seedling disease by placement of a fungicide beside the seed at or shortly after planting time – such as carpropamid to control rice blast, *Pyricularia oryzae*, which can be applied either on the seed or in the seedbed. Certain fungicides are used to control yeast and mould fungi in stored grains and grain storage areas: a liquid formulation of propionic acid for spraying, for example. (**See: Storage fungi**)

(b) *Classification.* Fungicides can be classified into main primary chemical groups on the basis of modes of action: based on inhibition of the electron transport chain, nucleic acid synthesis, mitosis and cell division, protein synthesis, lipid and membrane synthesis, sterol biosynthesis or 'multi-site processes' (Table T.3). Some groups control a rather wide range of plant pathogens, whereas others have a limited spectrum of activity against one or two specific taxonomic groups; they also differ in terms of the relative ease with which fungi develop resistance to them.

Seed treatment fungicides are combined and co-formulated to complement each other – such as the ability to control *Pythium* spp. which cause damping-off under cold soil conditions, and other oomycete pathogens. The main diseases on major crops typically can be controlled by several different types of fungicide, which may all be registered for use in the country in question; by contrast there may be few if any registered treatments for certain 'minor uses'.

Modern cereal seed treatments, for example, are based on systemic fungicides representing several main chemical groups – including azoles, benzimidazoles, carboxamides, dicarboximides, guanidines, imidazole, phenylamides, pyrimidines and triazoles. Several of these groups require application at low rates, 1–5 g ai/100 kg seed. Latterly, fluquinconazole and, very recently, the development of the benzamide fungicide, silthiofam, have allowed control of take-all disease (*Gaeumannomyces graminis* var. *tritici*).

4. Insecticides

The most important and commonly used groups of synthetic insecticide seed treatments comprise the chlorinated hydrocarbons (organochlorines, such as lindane, which is now withdrawn in many markets), the organophosphates, the carbamates, the pyrethroids and the neonicotinoids (Table T.4). A major development in the 1990s has followed from the discovery of the last named class of compounds and has allowed, for the first time, the protection of seeds and young seedlings not only against a large number of soil-dwelling insects but also against a broad spectrum of early leaf-feeding and leaf-sucking insect pests, and has thereby expanded the scope of insecticidal seed treatment. After uptake by the germinating seed and young seedling, redistribution in the foliage provides control for up to 40 days after planting against aphids, thrips, whiteflies, leaf miners and various other pests.

Insecticides are also used as grain and seed protectants against damage by **storage pests**, or to protect empty bin surfaces. (**See: Storage management**) Though intended to retain their insect-toxic properties for an extended time, many are not highly volatile, and penetration into infested kernels is limited. Also **fumigants**, which form poisonous gases, may be applied to stored grain in gas-tight sealed silos or to soils for sterile seedbed preparation.

5. Bactericides

Bacterial pathogens may sometimes be controlled with bactericides or antibiotics (such as streptomycin), and have been successfully used as seed treatments to control if not completely eradicate seedborne bacteria (such as bacterial blight pathogens on dry bean seed) by reducing populations on the seed surface, mainly by short steeping or soaking treatments. However, their use as seed treatments is proscribed in many countries because of their priority for use in human medicine. Phytotoxicity is also a potential concern, such as affecting chloroplast development and causing reduced germination and stunted seedling development.

6. Viricides

There are no viricides as such, but seedborne viruses can be controlled by physical heat treatments (**see: Thermotherapy**). Many virus diseases are transmitted by insects, mites or nematodes, and control is sometimes possible by using insecticides, miticides and nematicides to destroy these vectors. For example, it is possible to control the first generation of leaf-feeding and leaf-sucking insects by systemic insecticidal seed treatment, and hence reduce the transmission of virus yellow diseases in sugarbeet (beet mild yellowing virus and beet yellows virus). (**See: 4. Insecticides**)

7. Formulations

In the pure ('technical-grade') state, most solid or liquid ais are in physically inconvenient forms for the purpose of application and biological effectiveness, and have unsatisfactory storage, handling and safety properties. For this reason, pesticides are commercially formulated to make the product more safe, effective and convenient to use, usually as mixtures of the ai with 'inert' ingredients that have no pesticide action. A single ai is often registered and sold in

Table T.3. Examples of major fungicidal active ingredients used as pre-sowing seed treatments (subject to the registration status in individual states and countries), and their biochemical mode of action on target pathogens.

Chemical group and mode of action	Chemical group name	Common name	Outline of seed treatment targets and crop uses
Benzimidazole Inhibition of mitotic spindle (β-tubulin) assembly	Benzimidazole		Systemic, protective and curative action against all pathogens, except oomycetes. All break down to carbendazim in the field
		benomyl	Seedborne *Phoma*, *Alternaria*, smuts and bunts (*Ustilago* spp.). Recently withdrawn
		carbendazim, fuberidazole, thiabendazole	Target pathogens variously include seedling damping-off complex, *Fusarium*
	Benzimidazole precursor fungicides (methyl benzimidazole carbamates)	thiophanate, thiophanate-methyl	
Dicarboximide NADH cytochrome c reductase in lipid peroxidation (proposed)	Dicarboximide	iprodione	Protectant and limited eradicant activity; control of *Alternaria*, *Fusarium*, *Helminthosporium*, *Rhizoctina* and *Sclerotinia* spp. and other pathogens
DMI (Demethylation inhibitor) – conazoles C14-demethylation in ergosterol biosynthesis	Imidazole	imazalil	Systemic and protectant activity, sometimes at very low application rates. Used in cereals to control *Fusarium*, *Helminthosporium* and barley leaf stripe and net blotch (*Pyrenophora graminea* and *P. teres*)
		perfurazoate	Prevention and cure of rice bakanae (*Fusarium moniliforme*), rice blast (*Pyricularia oryzae*) and leaf spot (*Cochliobolus miyabeanus*)
		prochloraz	Primarily used on cereals, flax/linseed; and as seed soak on rice in Asia
		triflumizole	Control of *Ustilago* spp., *Tilletia caries* and *Pyrenophora* in cereals and other crops
	Triazoles	bitertanol, cyproconazole, difenoconazole, flutriafol, hexaconazole, ipconazole, metconazole, tebuconazole, tetraconazole, triadimenol, triticonazole	Used, variously, to control seedborne cereal smuts and bunts (*Ustilago* spp.), and other seedborne and soilborne diseases, including *Fusarium* head blight and *Septoria* seedling blight, and powdery mildew in wheat (*Erysiphe graminis* f.sp. *tritici*)
		prothioconazole	Systemic and curative control of cereal pathogens including *Rhizoctonia* spp., *Pyrenophora* spp. and *Microdochium nivale*. Soilborne peanut pathogens such as *Sclerotinia* spp.
		fluquinconazole	As above, plus control of take-all *Gaeumannomyces graminis*

Table T.3. *Continued*

Chemical group and mode of action	Chemical group name	Common name	Outline of seed treatment targets and crop uses
Phenylamide Inhibition of RNA polymerase I	Anilide/Acylalanine	benalaxyl benalaxyl-R (the R-enantiomer of benalaxyl)	Control of *Plasmopara halstedii* in sunflower
		metalaxyl metalaxyl-m (mefenoxam: the R-enantiomer of metalaxyl)	Systemic and curative action to eliminate deep-seated infections. Used to control soilborne diseases, such as *Pythium* and *Phytophthora*, damping-off and downy mildew on a wide variety of crops
Oxathiin Inhibition of complex II in fungal respiration (succinate dehydrogenase)	Anilide/Carboxamides	carboxin	Systemic activity, particularly against basidiomycetes. Used to control early-season diseases on wide range of crops, such as seedborne cereal smuts and bunts (*Ustilago* spp.) of cereals
Hydroxypyrimidine Adenosine-deaminase	Pyrimidinol	ethirimol	Systemic activity against powdery mildew. Used, e.g. in cereal seed treatment combinations
Anilinopyrimidine Inhibition of methionine biosynthesis (proposed)	Anilinopyrimidine	cyprodinil pyrimethanil	Prevents and cures a broad range of ascomycetes and deuteromycetes; used in cereal seed treatment combinations
STAR ('Strobilurin Type Action and Resistance') (QOI 'Quinone Outside Inhibitors') Inhibition of respiration at complex III (ubiquinol oxidase, cytochrome bc1)	Strobilurines: Methoxyacrylate	azoxystrobin trifloxystrobin	Control of a wide range of pathogens; used, e.g. as seed treatment to control soilborne *Rhizoctonia* on cotton and soybean. Derived from *Strobilurus tenacellus*
		fluoxastrobin	Systemic control in cereals of leaf spot (*Rhynchosporium secalis*, *Pyrenophora teres*, *P. triticirepentis*), Septoria (*S. tritici*, *Leptosphaeria nodorum*), rust, and mildew
Phenylpyrroles Inhibition of MAP protein kinase in osmotic signal transduction	Phenylpyrrole	fenpiclonil fludioxonil	Fludioxonil is chemically related to the natural antibiotic pyrrolnitrin. Broad-spectrum activity against a wide range of seedborne and soilborne pathogens in maize, sorghum
Aromatic hydrocarbons Lipid peroxidation (proposed)	Chlorophenyl	chloroneb, tolclofos-methyl, quintozene (PCNB)	Control of seedling diseases
Melanin Biosynthesis Inhibitors (MBI-D, in cell wall) Inhibition of dehydratase		carpropamid	Systemic activity against rice blast (*Pyricularia oryzae*), either as seed treatment or in nursery boxes
Hindered Silyl amide ATP production (proposed)	Thiophene/Benzamide	silthiofam	Protective control of primary infection with soilborne take-all *Gaemannomyces graminis* var. *tritici* by release into the soil
Antibiotics Inhibition of protein synthesis		blasticidin kasugamycin streptomycin	Active against rice blast (*Pyricularia oryzae*). Some against bacterial diseases, e.g. fire blight
Antibiotics Inhibition of carbohydrate metabolism		validamycin	
Phenylurea Inhibition of cell division (proposed)		pencycuron	Non-systemic activity against soilborne *Rhizoctonia solani*

Continued

Table T.3. *Continued*

Chemical group and mode of action	Chemical group name	Common name	Outline of seed treatment targets and crop uses
Plant host defence inducers Induction of the salicylic acid pathway	Benzo-thiadiazole (BTH)	acibenzolar-S-methyl	All pathogen groups (including viruses and bacteria)
'Urea fungicide' Mechanism unknown	Cyano-acetamide oxime	cymoxanil	Curative and protective activity to control *Peronospora, Phytophthora, Plasmopara* spp.; limited uses, e.g. in pea
Inhibit DNA/RNA synthesis (proposed)	Isoxazole	hymexazol	Systemic control of damping-off diseases, including *Aphanomyces* in sugarbeet
Multi-site contact activity	Inorganics	copper salts	Active against oomycetes; includes 'Bordeaux mixture'
	Dithiocarbamates and relatives (include polymers, e.g. with Mn)	mancozeb, maneb, thiram	Very widely used as protective fungicides to control seedborne and soilborne pathogens of all groups, including smuts, bunts, seedling blight in cereals, damping-off, rots, blights caused by *Pythium* and *Fusarium* spp., *Phomopsis, Rhizoctonia*
	Phthalimides	captan	Protective and curative fungicides to control broad spectrum of diseases include those caused by *Fusarium, Pythium, Phoma, Phomopsis* and *Rhizoctonia* in maize, cotton, oilseed rape, vegetables
	Guanidines	guazatine	Primarily used against cereal seedborne diseases, to protect against bunts, *Fusarium* spp., *Septoria nodorum*
Mechanism unknown	Phosphonate	fosetyl-aluminium	Systemic control of *Phycomycetes*
	Benzotriazines	triazoxide	Non-systemic control of leaf stripe and net blotch in barley (*Pyrenophora graminea* and *P. teres*)

Active ingredients are frequently made available in combinations with up to three or more other fungicides to broaden the spectrum of disease control, according to the regional prevalence of pathogens and to manage resistance. Many are also used in foliar applications.
Extracted and expanded from FRAC (the specialist Technical Group of the Global Crop Protection Federation, GCPF, www.gcpf.org).

several different formulations between and even within countries, depending on the inclusion of other ais, user requirements and technical application needs. The common types of seed treatment formulations are shown in Table T.5.

In conventional seed treatment, formulations may be applied directly to seed without further modification, though stickers may be used to improve adhesion where high dose rates are used (**see: Filmcoating**). Dry powder formulations require the minimum equipment for application, such as drum mixers, with or without additional sticking agents added at the same time, but these treatment formulations are very vulnerable to loss by **dust-off** from the seed during handling, transport and planting. Dry formulations have now largely been replaced by slurried wettable or water-dispersible powders, or flowable concentrates (emulsion or micro-encapsulated formulations), or other liquid-based systems which are better suited to application of large doses of ais to seed, and for use in commercial-scale application facilities. Liquid applications help ensure much firmer adhesion and spreading and retention of formulations on seed surfaces, provide more accurate and thorough seed coverage between individual seeds, and are safer to operators.

Agents and adjuvants used to carry and improve the properties of ais in seed treatment (and other) formulations include:

- carriers and diluents (such as calcium carbonate, talc, kaolin, montmorillonite, attapulgite and synthetic silicas, with appropriate particle sizes according to the intended method of application);
- surfactants, spreading and wetting agents, to allow wettable powders to mix with water;
- spreaders and stickers, to allow the pesticide to cover and stay on the seed surface;
- suspension agents, for physical stability during storage;
- and defoamers and viscosity regulators for liquid formulations.

Most seed-treatment pesticide formulations now come from the manufacturer with a conspicuous dye or insoluble pigment added either as a convenience to the operator and farmer to

Table T.4. Examples of major insecticidal active ingredients used as pre-sowing seed treatments (subject to the registration status in individual states and countries).

Chemical group	Chemical name	Common name	Outline of seed treatment targets and crop uses
Carbamates		bendiocarb benfuracarb carbofuran carbosulfan furathiocarb methiocarb	Partly systemic; soil insects and some early leaf-feeding insects
		thiodicarb	Lepidoptera
Pyrethroids		bifenthrin (alpha)-cypermethrin beta-cyfluthrin permethrin tefluthrin	Non-systemic; soil insects
Organophosphates **Heterocyclic derivatives**		acephate chlorpyrifos chlorfenvinphos	Non-systemic. Many voluntarily cancelled and others lost registrations
Organochlorines		gamma HCH (lindane)	Non-systemic. Being withdrawn
Neonicotinoid	chloronicotinyl	imidacloprid acetamiprid thiacloprid	Soil and systemic (leaf-feeding and sucking) insects. Used on sugarbeet, wheat, barley, maize, soybean, sorghum, cotton, oilseed rape/canola and lettuce amongst others
	thianicotinyl	thiamethoxam clothianidin	Systemic. Used on cotton, canola, cereals, maize, oilseed rape, rice, sunflower, sorghum, peas, beans, sugarbeet, peanuts, several vegetables, amongst others
Phenylpyrazole		fipronil	Broad-spectrum neurotoxin, soil-living insect pests
Naturally occurring organisms	avermectin	abametin/emamectin	Derived from the soil bacterium *Streptomyces avermitilis*; nematicide, such as against parasitic root nematodes
	spinosyns	spinosad	Toxic metabolites derived from *Saccharopolyspora spinosa*; produced during fermentation in ratios according to the conditions of the fermentation process

Table T.5. Common seed treatment formulation types.

Type	Properties
Dry dust and powder-based treatments (usually designated as DS)	The first type of formulation developed for seed treatments; in decline as seed treatments, in favour of liquids, apart from in some developing countries
Wettable powders (WP), water-dispersible powders (WS), water soluble powder (SS)	Capable of being slurried or dissolved in water
Flowable concentrates (FS) and suspension concentrates (SC)	Stable aqueous suspensions of finely ground particles, for application to the seed usually directly, or after dilution with water
True solutions (LS)	Usually based on organic solvents, such as propylene glycol and glycol ethers, which are non water-miscible, and dry quickly
Emulsifiable concentrate (ES)	Usually a concentrated miscible oil solution of the ai that when added to water spontaneously disperses as fine droplets to form a stable emulsion
Encapsulated (CS)	Small solid particles, droplets of liquid, or dispersions of solids in liquids in non-volatile envelopes, in aqueous suspensions, using thin polymeric coatings such as of gelatin or polyvinyl-compounds; used for the controlled release of ais such as those with low vapour pressures, or otherwise to extend their periods of release by diffusion

identify treated seed, but also by regulation to deliberately give seed an unnatural appearance, and so prevent inadvertent use in the human food or animal feed chain. Some seed processors prefer to add additional dye or colorant with the pesticides at their treatment plants so that a desired colour is obtained. A wide range of ready-mixed coloured filmcoating polymers is commercially available in many countries for these uses. Care has to be taken to ensure that these colorant materials cause no injury to seed germination or danger to those who process or handle the seed; the USA Environment Protection Agency, for instance, lists dyes approved for this purpose.

For controlling insects in grain storage, formulations can be applied on to grain and storage structures as diluted sprays, usually emulsifiable concentrates and wettable powders. Also, certain volatile insecticides are formulated in slow-release polyvinyl strips for hanging in grain overspaces, to reduce adult moths during warm periods when they are expected to be laying eggs, for example. (**See: Fumigants**)

8. Registration and labelling

Pesticide registration is the stringent scientific, legal and administrative procedure through which every active ingredient and its formulation must be registered, or re-registered after a period of time, for its specific applications – in effect the granting of an official licence for sale and use. Many currently registered seed treatment formulations incorporate different amounts of an ai for each crop, or to suit different application systems. Some are mixtures of more than one ai to broaden the range of crop protection: for example, two or more fungicides and in combinations with insecticide(s). Pesticides are registered on the basis of specified formulations for use on designated crops, and may need to be registered again if a new or substantially changed formulation is brought forward.

Registrants submit dossiers of data to the designated national or state authorities detailing formulation, application rates, the means of application or delivery, efficacy against the target organisms on the crop on which it is to be used, worker safety, environmental acceptability and fate (short- or long-term toxicological effects on humans and non-target organisms) and storage and disposal practices. Seed treatment dossiers typically also require information about the effect on germination viability, immediately and after storage.

A number of active ingredients have moderate side effects on seed performance, by slowing germination – an aspect of reduced **vigour** – or producing seedling abnormalities by imposing phytotoxic stresses. Classic symptoms are a thick, contorted or stunted seedling shoot or root that count as **abnormal seedlings** in the approved **germination testing** conditions, though these effects may be less apparent if the test is carried out instead by planting in growing media, rather than on paper or sand. Though it is accepted that several classes of chemicals – such as the high-dose neonicotinoid insecticides – cause slow emergence in some crop species, this may be tolerated on balance, considering the protection benefits delivered to the crop. For example, the use of triazole-based seed treatments on barley varieties with inherently short **coleoptiles** is likely to exacerbate emergence problems. Seed **coating**, safening or **priming** techniques may be used to mitigate this response.

A successfully officially accepted data package results in the issuing of a 'pesticide label', a term that refers both: (i) to the information dossier placed with the authority (in which sense a registered pesticide is said to be 'labelled'); and (ii) to the specific form of information (chemical and common names, composition, directions for use, safety precautions, and so on) that must be physically incorporated on the outside of the pesticide container, along with any required accompanying supplemental information. In most countries, separate regulations prescribe the 'seed label' information required for identifying the treated seed, including such items as treatment names, handling health and safety, and how to dispose of any surplus.

Registration laws typically govern the physical application process and, in the case of seed treatments, not the sowing operation. Many countries thus currently do not restrict the sowing of imported seed that has already been treated elsewhere (providing it is on any relevant variety list), even though the treatments applied are not registered for application to seed within the importing country itself, as long as the treatment is stated on the seed label. Furthermore, some countries have seed treatment registrations for 'export only' purposes.

Agrochemical companies tend to focus marketing efforts on crops that will generate the revenue to justify the considerable costs of development and registration of new crop protection chemicals. Typically the registration process is time consuming and expensive, and thus often not feasible for minor use substances, such as for horticultural crops. To deal with this, some countries have so-called 'off label' registration systems for minor crop seed treatments, through which some elements of the required data package can be extrapolated from a major crop registration, and formulations may be used at the farmer's risk. (PH)

Allison, D. (2002) *Seed Treatments: Trends and Opportunities*. Asgrow Reports, PJB Publications Ltd, London and New York.

Brandl, F. (2001) Seed treatment technologies: evolving to achieve crop genetic potential. In: Biddle, A. (ed.) *Seed Treatment: Challenges and Opportunities*. British Crop Protection Council, Farnham, UK, 76, pp. 3–18.

Hewitt, H.G. (1998) *Fungicides in Crop Protection*. CAB International, Wallingford, UK.

Maude, R.B. (1996) *Seedborne Diseases and their Control: Principles and Practices*. CAB International, Wallingford, UK.

Tomlin, C.D.S. (2000) *The Pesticide Manual*, 12th edn. British Crop Protection Council, Farnham, UK.

Waller, J.M., Lenné, J.M. and Waller, S.J. (2002) *Plant Pathologist's Handbook*. CAB International, Wallingford, UK.

Ware, G. and Whitacre, D. (2004) *The Pesticide Book*, 6th edn. Meister Publications, Willoughby, Ohio, USA.

Compendium of Pesticide Common Names. www.hclrss.demon.co.uk

Treatments – pesticides application

The quality of seed pesticide treatment application relies upon efficient, accurate and uniform loading of active ingredients, in a way that is also safe to the operator and the environment. A wide range of equipment is capable of dealing with different seed species, formulation types and scales of operation, from small laboratory batches to custom installations capable of fully automated, continuous treatment integrated directly into seed **conditioning** production lines.

Newer classes of fungicides typically require very low rates of product application (in cereals, of the order of 1–5 g active ingredient/100 kg seed), whereas the insecticides that give prolonged systemic protection demand that relatively large amounts of chemical be applied (the neonicotinoid insecticide imidacloprid, for example, is applied at rates of 0.60 or 1.34 mg/kernel maize in the USA).

1. Quality standards

The main quality requirements for application of seed treatments are:

- *loading efficiency* – the correct target dose should be reliably applied to the seed bulk (expressed as active ingredient per weight or number of seed), and maintained on the seed up to the point of sowing;
- *uniformity* – materials should be distributed as uniformly as possible from seed to seed (though, in general, there seems to be no evidence that an even coverage of active ingredients over the surface of each seed is of any particular advantage to the biological efficacy of a pesticide treatment);
- *'drillability'* – materials should be securely attached to the seed surface, and non-tacky, so as not to impair flow through the metering and placement elements of seed **planting equipment** (some formulations used alone can inhibit seed flow, and may cause bridging or other problems, and small amounts of talc or graphite are included at the end of the treatment application to mitigate this effect);
- *seed quality* – causing minimal detrimental physiological (in terms of **viability** and **vigour**, especially as over-treatment with some active ingredients may injure the seed, and maintaining seed moisture at acceptable levels for seed storage) and physical (mechanical damage) effects of the active ingredients or the process of applying them;
- *health and environmental safety* – minimizing dustiness, and precautions to assure the safety of application operators and those who handle treated seed, to avoid the toxic or irritant properties of the chemicals;
- *appearance* – use of colorants and other materials to obtain attractive and uniform finishes (and as far as possible with continuous coverage over the seed surface), and the absence of visible dustiness – both important factors in the perception (though are not necessarily a true reflection) of loading distribution quality, especially when the pesticide adds substantially to the cost.

Seed loading (or pesticide recovery) analysis is now increasingly an important quality standard in the commercial seed treatment of many crops. Particularly with the advent of the high-value systemic insecticide treatments (based on neonicotinoids) it is becoming common practice for agro-chemical companies to monitor routine commercial seed production operations to check that loading rates are within acceptable limits, both to ensure effective performance and, in the case of expensive chemicals, that fair measure has been given. Recovery analysis is now routine in certain cereal and sugar beet treatment markets. Direct analytical techniques include chromatography (such as gas–liquid and, most commonly, high performance liquid chromatography), and absorbance and fluorescence spectrophotometry. In some cases a marker dye is used to give an indirect measure, which is suited for direct quality control purposes in treatment facilities.

2. Equipment

Machinery is designed both to meet these quality considerations and for practical criteria: correct rate application and coverage, ease of operation, the ability to handle a range of formulations and seed species at sufficient throughput rates, avoidance of mechanical damage to fragile seed such as maize, soybean and peanuts, minimization of any waste of formulation, and for easy cleaning to prevent cross-contamination between seed bulks. Operation requires frequent cleaning, calibration and adjustment of equipment.

Seed treatment application techniques fall into three basic categories: (i) Conventional treatment where the amount of applied liquid is relatively low and usually can be satisfactorily absorbed by the seed itself (by 'self-drying'). Drying of treated seed is often unnecessary where relatively little water is involved: for practical purposes, where the change in seed moisture content after treating is less than about 1% of the seed weight. If necessary, a separate drying stage can be included, or a so-called 'drying powder' (such as talc) added to sufficiently blot up the excess. The use of filmcoating binder techniques as an adjunct has become more prominent, particularly in cases where the retentive capacity of the seed is limited by the size of the seed; depending on the amount of materials used, this may or may not result in a conspicuously coated seed.

More elaborate coating technologies can be used where relatively large amounts of liquid are involved, such as (ii) specialized **filmcoating** equipment with integral drying, so that typically each seed goes through several cycles of spraying and drying within the coating equipment, and seed moisture content remains virtually unchanged throughout the treatment. (iii) **Pelleting** or encrusting are techniques used to alter seed dimensions or to add weight to facilitate flow through planting equipment; formulations can be applied concurrently. In all these coating techniques seed is effectively treated with several layers; in pelleting, it is therefore possible to apply different materials sequentially to help avoid interactions between chemicals, or mitigate phytotoxic effects on the seed; and filmcoating can be done with a 'topcoat layer' to cover the layer that contains the formulation.

Conventional seed treatment equipment includes 'dust' and liquid, slurry, mist treaters, which variously can handle dry or liquid flowables, true liquids, or emulsifiable concentrates or wettable powders that are dissolved, dispersed or suspended in water (**see: Treatments – pesticides, 7. Formulations**). A common feature is the controlled feeding of seed and formulations in the correct relative proportions, using pumps, volumetric and dispensers.

(a) *Dry or 'dust' treaters* were the first commercially available seed-treating equipment. Formulations are mechanically mixed with seed, in measured amounts of powdered pesticide, and are continuously applied to seed using a vibrating or auger-type feeder. Dust-treating equipment adds no moisture to the seed and is easy to clean and operate. The major disadvantages are poor adherence on to the seed surface causing lack of uniform dosage, sifting-off of treatments during handling, as well as worker safety and protection concerns (**see: Dust**) (Fig. T.9a).

(b) *Continuous-throughput systems* are used to treat high-volume crops or agricultural crops (small-grain cereals, maize, pulses, oilseeds, cotton, grasses and sugarbeet) at seed throughputs up to 60 t/h. Batch treaters are used for the low-volume or minor crops (chiefly horticultural vegetable and flower species), and also on a batch-continuous basis for the high-volume crops. The most advanced procedures and equipment are designed and operated to accurately apply measured quantities of a pesticide to a given volume, weight or number of seeds.

In these designs, the accuracy and uniformity of treatment application depend on the principles and precision of seed and liquid (slurry) metering, on how liquid is distributed on to the seed, and on the means and degree of secondary re-mixing. The more sophisticated machines have fail-safe controls to ensure that this metering occurs correctly, including safety interlocks and alarms designed to ensure operations occur in the correct sequence, and can place the various formulations and additives as layers on the seed.

- In older designs, seed is metered gravimetrically using counterbalanced weigher buckets, and chemical metering is determined by the size of cups. In a newer design, seed is metered by flow over a cone which is restricted by a variable regulating collar seated around it, while liquid flow is controlled through metering pumps or volumetric dosing jars – the operation of the two being electronically geared together (Fig. T.9b).
- Application to the seed itself is typically by rotary atomizers – often spinning discs – or less commonly though air or airless nozzles. 'Mist treaters', for example, use controlled-drop application on to cup-shaped spinning discs to 'atomize' the formulation into a droplet spray that is dispersed on to a curtain of seed falling over a dispersion cone.
- Almost all machines incorporate a remixing stage: a coating chamber into which seed and formulation are then passed for further blending and redistribution of material from seed to seed, before conveying the treated seed to the discharge point. Horizontal auger-type mixers may be equipped with internal fingers or solid flights, or brushes in the case of fragile seed such as maize, soybean and pulses. Drum-type coating chambers consist of an open drum rotating around a horizontal or inclined axis, and may have additional mixing elements inside, to tumble the seed and chemical together. Continuous-throughput or batch versions are both used, in some cases with a current of warm air to partially redry the seed.

Alternatively, horizontal drums rotating with a gentle mixing action and incorporating direct liquid spray systems are used to directly apply formulation to species that are susceptible to mechanical damage.

(c) *Rotary batch coaters* are now widely used, including in batch-continuous operation modes. Seed is dropped into a cylindrical mixing chamber whose floor is a rapidly rotating dish-shaped plate, which lifts the seed up in a toroidal (doughnut-shaped) mass that folds outward, inward and around the chamber sidewall. A spinning disc, mounted in the centre of the chamber above the plate, atomizes the treatment liquid on to the turning seed mass (Fig. T.9c). The dwell-time

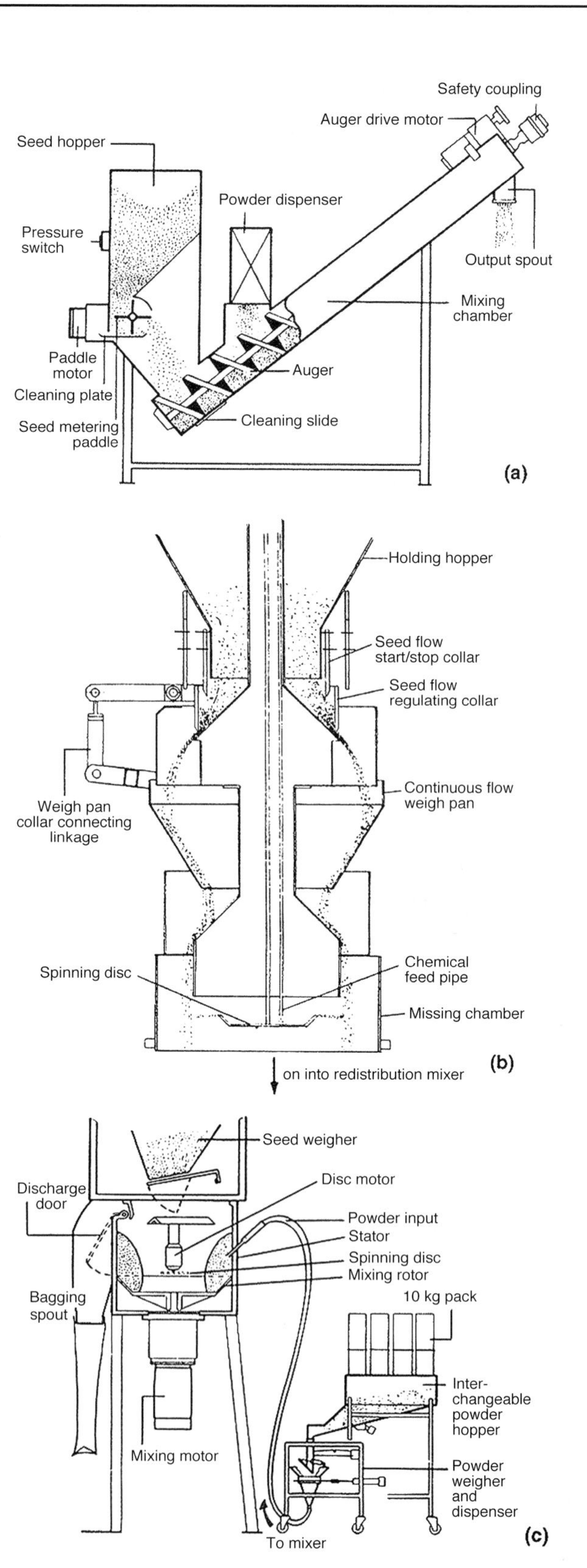

Fig. T.9. Sectional diagrams of agricultural seed treaters. (a) Plantector, (b) Mist-O-Matic®, (c) Rotostat®. a, c Reproduced from Pflanzenschutz-Nachrichten Bayer 34/1981, 3, with permission of Bayer PLC; b reproduced by courtesy of Dow Elanco. Maude, R. (ed.) *Seedborne Diseases and Their Control, Principles & Practice.* CAB International, Wallingford, UK.

of seed in the machines can be of the order of tens of seconds. Powdered dry formulations can also be applied effectively in this system at the same time, by direct augering of material on to the seed; indeed, by adding large amounts of material this technique can be used to produce encrusted or pelleted seed forms, for subsequent drying (**see: Pelleting**).

(d) *On-farm equipment* is much simpler in design. At the simplest, seed is mixed with dry powders or liquid formulations by stirring with a paddle or stick in the seed-box of the **planting equipment** (variously known as seed-, drill-, planter- or hopper-box application) or in a revolving barrel. Alternatively, auger- or drum-type machines treat seed with slurried or sprayed formulations as it is moved into the bulk seed planter. These methods are suited mostly for application of formulations that have a short shelf-life, or contain a bacteria or fungus as the active ingredient such as **rhizobial inoculation**, or in small-scale farming operations such as in developing countries.

(e) *Infiltration*. **Steeping** seed for various periods of time with fungicides in (partial) solution, such as thiram, is also used to penetrate seed tissues and control deep-seated seedborne pathogens: eradicating (or **disinfecting**), by removing or destroying established pathogens. Soaking with antibiotics, such as streptomycin, to control seedborne bacteria is possible, but commercial use is proscribed in many crop situations because of their importance as antibiotics in medicine.

3. Production organizational aspects

Markets and crops vary in terms of whether seed treatment is carried out by seed breeding companies and merchants, who typically treat high-value, low-volume seeds or high-volume hybrid seeds, or by local retailers or seed treaters, or on-farm which is typically where the low-value, high-volume seeds are treated; quantities of **farmer-saved seed** are treated either using the farmer's own equipment or by specialist mobile treating operations.

Treatment presents the formal seed industry with special challenges in production logistics and inventory management. Often the time available is short. Seed markets are characteristically seasonal and all seed for a crop must be processed and treated within a similar timeframe. The seed itself may be only recently produced. Much autumn sown cereal and oilseed rape in Europe, for example, is cleaned, processed, tested and treated all within about a 10-week period following harvest. But even where processed seed is available in good time, sales orders may not be finalized until a relatively late date. A range of varieties, some perhaps in relatively limited quantities, must be considered in possible combination with alternative seed treatments, perhaps with different application rates or formulations. Also there is the economic cost of the treatments, the volumes and value of the seed involved to bear in mind, as well as the potential decline in seed quality during storage and the potential cost of safe disposal of unsold treated seed, such as through food or feed distribution channels. For these reasons, seed companies typically choose not to carry over large amounts of unsold surplus seed in inventory to the following season if it can be avoided, and seed-applied crop protection treatments – especially those with higher value – tend in general to be produced to order, rather than speculatively. Treatment systems, therefore, must have the capacity and flexibility to be operated on a quick turnaround just-in-time basis. (PH)

Clayton, P.B. (1993) Seed treatment. In: Matthews, G.A. and Hislop, E.C. (eds) *Application Technology for Crop Production*. CAB International, Wallingford, UK, pp. 329–349.

Halmer, P. (2000) Commercial seed treatment technology. In: Black, M. and Bewley, J.D. (eds) *Seed Technology and its Biological Basis*. Sheffield Academic Press, UK, pp. 257–286.

Jeffs, K.A. (ed.) (1986) *Seed Treatment*. BCPC Publications, Thornton Heath, UK.

Tree seeds

1. Introduction

A 'tree' is considered to include any tall, usually woody plant, which forms a canopy that creates an empty, shaded space beneath. This somewhat vague description is deliberately intended to include the timber and pulp-producing **angiosperms** and **gymnosperms** that are traditionally thought of as trees, together with the many useful palms (e.g. bananas, oil and date **palms**). It excludes small, woody ornamental shrubs. A plant community that is made up of a significant proportion of trees is usually referred to as either a 'forest' or 'woodland'. These terms are often used synonymously, but sometimes 'forest' (or 'plantation forest') is used in a stricter sense to mean artificially established trees, whereas 'woodland' is reserved for natural tree cover. In 1990, the 'World Resources Institute' estimated that forests occupied about 30% of the world's land area, of which over 97% was 'natural' and only 3% had been artificially established. The forests of Siberia, about 6 Megahectares (Mha) in area, are almost twice the size of the Amazon rainforest and therefore constitute the largest single forest area on the planet. FAO (**Food and Agriculture Organization of the United Nations**) is the agency with a mandate to deal with global forestry issues and their role can be explored at www.fao.org/forestry.

There are many markedly different types of forest that have developed in response to the various climates in different regions of the world. Table T.6 provides a rough classification and worldwide distribution of some of these different types of forest and includes the names and origins of approximately 20 notable trees.

Most trees can take a decade or more to flower or fruit and perhaps even centuries to reach their final size and stature. By human standards, growing trees as a crop, waiting for natural woodland regeneration, or embarking on a tree improvement programme, therefore, are long-term endeavours.

2. History, uses, world distribution and economic importance

Throughout human history, the commonest fate of forests and woodlands has been land clearance for agriculture; and the commonest use for trees has been as firewood. In the developed world today, trees and forests are almost exclusively associated with commercial timber and paper pulp. Trees also directly provide numerous raw materials (such as cork, fibre, gums, resins, rubber, turpentine), and indirectly provide us with several others (such as honey, rattan, perfumes,

Table T.6. Summary of world forest types and example tree species of which they are composed.

	Climate	Locations	Notable trees
High latitude forests Boreal forest (taiga)	Long, severe winters (up to 6 months with mean < 0°C), and short summers (50–100 frost-free days). Annual rainfall: 37.5–50 cm, with little evaporation.	An almost continuous belt across much of North America and Eurasia	Pines, spruces, firs, larches, birches, alders and aspens
Mid-latitude forests – northern hemisphere Temperate, deciduous broadleaf (and conifer) forest	Warm, wet summers with cool/cold, wet winters.	Western and central Europe, eastern North America, and eastern Asia (including parts of China, Korea and Japan)	Oak, maple, beech, chestnut, hickory, walnut, elm, lime
Mid-latitude forests – southern hemisphere Temperate, deciduous broadleaf and conifer		Parts of South America, Australia, New Zealand and Africa	Araucarias, podocarps, southern beeches (*Nothofagus* spp.)
'Mediterranean' forests	Warm, dry summers, with cold, wet winters: a short, but predictable growing season. Rainfall: 40–100 cm per annum.	Around the Mediterranean Sea (Europe, North Africa, Asia Minor)	Oaks, pines, cedars, olive, carob
		California (Locally called 'chaparral' from the Spanish *chapa* for scrub oak)	Oaks, pines, *Ceanothus*
		Australia	*Eucalyptus* spp. (Locally called 'mallee')
		Chile (Locally called 'matorral' from the Spanish *mata* for shrub)	*Nothofagus*, Chilean palm (*Jubaea chilensis*)
		South Africa (locally called 'fynbos')	Proteas and cycads
Tropical rainforest Between 10°N and 10°S at elevations below ~900 m	Year-round warmth (mean ~17.5°C). Annual rainfall: ~250 cm.	Central and South America	Cocoa (*Theobroma cacao*); *Cordia alliodora*, jacaranda, rubber (*Hevea brasiliensis*)
		West and Central Africa, eastern Madagascar	Oil palm, Iroko (*Chlorophora excelsa*, *C. regia*)
		West coast of India, Southeast Asia, parts of Myanmar and Indonesia	Teak (*Tectona grandis*), mahogany (*Swietenia mahogoni*, *S. macrophylla*), mango (*Mangifera indica*), oranges and lemons (*Citrus* spp.)
Tropical dry forest	Year round warmth (mean ~17.5°C). Annual rainfall: 75–125 cm, with a distinct dry season for at least 5 months (< 10 cm rainfall a month).	Parts of Africa (Kenya, Tanzania, Zimbabwe, Botswana, South Africa, Namibia), South America (Brazil, Columbia, Venezuela), India and Australia	Palms, pines, acacias, eucalypts

medicines, dyes, polishes) from the plants and animals that could not exist in the absence of the forest. In addition, trees and forests are a vital part of the global ecosystem – they fix carbon dioxide, recycle oxygen, preserve water quality, influence weather patterns, stabilize soils, harbour wildlife – and are widely used for landscaping and in parks and gardens. Humans also exploit trees for their fruits and seeds. For example, they provide us with drinks (e.g. **coffee, cocoa**), food (e.g. edible **nuts**, citrus, apples, bananas, mangoes, etc.), **spices** (e.g. **nutmeg, mace**) and **oils** (e.g. **olives** and **palms**).

Finally, most trees are propagated from their seeds, with the notable exceptions of a few palms and fewer **conifers** (e.g. *Chamaecyparis* spp., *Cryptomeria japonica*, *Juniperus* spp., *Pinus radiata*) and broadleaf species (such as eucalypts, poplars and willows) that can be vegetatively propagated from 'offsets'.

Trees, and their fruits and seeds, therefore have considerable economic importance. Forest products are the third most valuable commodities in world trade after oil and gas. Nevertheless, human efforts to achieve sustainable forestry are inadequate.

3. Geographic origin

Trees are relatively long-lived individuals, and tree species which have colonized large geographic areas over periods of millennia often exhibit slightly different characteristics in the different places they grow. For example, Sitka spruce (*Picea sitchensis*) has a natural range that extends almost 3000 km from California to Alaska and, although individuals throughout the range are remarkably similar to each other morphologically, local populations have adapted to variations in latitude, day-length, soil and rainfall. Seed from these different populations, therefore, is usually better adapted to grow in a similar environment to that which its parent and its parents' ancestors have inhabited. Hence, if a particular species of tree is going to be grown elsewhere, then knowledge of seed source (and ecotype) is vital.

There are two terms used to define a tree seed source in forestry: 'provenance' is the place from where the seedlot has been collected; whereas 'origin' describes where the most remote ancestors can be traced to. Anyone aiming to grow trees with good vigour, health, stem form, crown habit, timber properties, etc. should make every effort to obtain their seed from an 'origin' where the environment is similar to that of the proposed planting site.

4. Seed development and collection

Seeds may be gathered directly from forests, or produced in specifically cultivated seed orchards.

(a) *Forests.* In some years trees flower well, and pollination, fertilization, seed development, maturation and ripening all appear to progress perfectly; the result can be an extremely large (or so-called '**mast**') crop of seeds. In other years there might be no crop at all, or only moderate yields.

(b) *Seed orchards.* The tree seed orchard industry accounts for a significant and growing amount of stock for reforestation in North America. Problems can occur when environmental conditions do not mimic those in natural stands, particularly factors such as elevation and temperature, which can have a negative impact on seed development. Tree seed orchards involved in the production of seeds of lodgepole pine (*Pinus contorta*), yellow cedar and other species often yield seeds that are shrivelled (indicative of poor seed fill) and exhibit poor germination rates and impaired seedling growth. Spraying with **gibberellin** (GA) to induce male and female flowers is a common practice in tree seed orchards and may contribute to resource limitation (inadequate allocation of carbon and nitrogen resources from the parent tree to developing seeds and ovules) and thereby reduce seed fill or viability.

Insects may attack the reproductive structures of conifers throughout their development from cone bud initiation to seed maturity, and even during cone collection and seed extraction. Forest insects that feed directly on conifer cones and/or seeds are termed conophytes; only about 100 of 50,000 species of insects known in Canada are conophytic and far fewer have an economic impact. In conifer seed orchards, which are normally managed to produce cone crops on a more regular basis than would occur in natural stands, conophytic insects tend to be more prevalent and cause more extensive damage.

(c) *Indicators of maturity.* Tree-seed collectors need to have a sound understanding of the complete sequence of development and maturation events for the species to be collected, especially of the ability to recognize the stage just prior to seed shed. Fruit and seed maturation is often accompanied by distinct changes in size, colour, surface texture, firmness, odour and taste, and experienced seed collectors use combinations of these changes to assess ripeness. For example, the cones of many pines change from green to brown on ripening, and many wind- (e.g. *Acer*, *Fraxinus*) and water-dispersed species (e.g. *Alnus* spp., *Cocos nucifera*) change from green to a straw or brown colour. In contrast, many edible fleshy fruits, which depend upon birds or animals for dispersal, ripen from green to much more striking colours such as white (*Symphoricarpos albus*), blue (some podocarps and junipers, *Prunus spinosa*), red (*Sorbus aucuparia*) and yellow/orange (several palms); at the same time the sour, bitter or astringent flesh may become more palatable, sweeter and edible to potential dispersal agents.

(d) *Harvesting.* A common practical difficulty in the collection of tree seed is how to retrieve seeds from such potentially tall plants. In fruit and nut orchards, the solution is to keep the trees short. But this is not an option in most other instances, such as agroforestry. If the trees produce large seeds that fall straight to the ground, one solution is to either sweep up the seeds from below or spread out nets or tarpaulins. (**See: Collection**, Fig. C.9.) But it is usually vital to do this on at least a daily basis to avoid animals thieving the crop. Climbing (using ladders, spurs or a 'tree bicycle'), mechanical shakers, hoists and even helicopters are occasionally used, but the commonest solution in forestry is simply to fell the fruiting trees or branches and pick or rake the seeds from them. It is hardly surprising that **collection** is generally the most expensive phase of tree seed production. (**See also: Gymnosperms – seed production; Seedcoats – dispersal aids; Seedcoats – structure**)

5. Drying, handling and processing

It is rarely possible to collect an entire tree-seedlot in less than a few days or weeks, so the accumulating crop needs very careful preliminary handling and temporary storage at the collection site. This is because freshly collected cones, fruits or seeds usually have a relatively high moisture content and are often at their most perishable. Protection from vermin and rough handling are always vital. But other conditions will vary with species. For example, some fruits will benefit from sun-drying, whereas others will be killed. Cones containing **orthodox** seeds are likely to benefit from dry, well-ventilated conditions, whereas **recalcitrant** seeds will need to be on a cool, dry surface, and covered with a waterproof tarpaulin to shelter them from wind, rain and sun. Damage done at this stage is likely to be irreversible. Seedlots therefore need daily inspections during acquisition.

Transportation between the collection site and main processing plant must also take account of seed characteristics. Recalcitrant seeds will need to be protected from drying, cold-sensitive seeds from low temperatures, and the cones and fruits of many orthodox seeds may actually benefit from the forced air circulation created by the travel.

A summary of cleaning, curing and extraction processes that can take place at the collection site or main processing plant are presented in Table T.7. Exact methods vary according to the fruit and seed type, their condition at harvest and whether the seed will be stored or not. Most of the more complicated seed processing takes place in the specialized equipment at the main processing plant. In alphabetical order, de-pulpers, de-wingers, kilns, macerators, shakers, threshers, tumblers and winnowers may all be used to prepare cleaned

Table T.7. Tree seed processing procedures after collection.

Process	Purpose
At or near the collection site (usually)	
Preliminary clean	Removes empty and decaying fruits; larger twigs, leaves, debris, etc. primarily to reduce the weight and volume needing transport.
Preliminary cure – **pre-curing**	Allows fruits and seeds which have been deliberately or inadvertently collected immature to ripen; or permits fruit/seed changes to take place which will ease subsequent extraction (e.g. pre-drying of **serotinous** cones).
At the processing plant	
Fruit/seed extraction	Removes cone/fruit parts which are not necessary to subsequent propagation. May include 'depulping' of fleshy fruits, 'dewinging' of winged fruits.
Cleaning	Removes seeds of other species, unproductive seeds (such as empty, dead and insect damaged) and inert matter (such as fruit/cone fragments, dust, soil, resin particles).

pure seeds with good physical quality and physiological performance which, if necessary, are suitable for storage.

6. Seed storage

The sporadic nature of seed production by the majority of tree species means that most collections are carried out with the intention of providing at least 3 to 5 years' worth of seed supply. In practice it is not uncommon for some tree species to be stored for 10–20 years, occasionally even longer (**see: Conservation of seed; Seed banks**). Before putting it into storage, seed is tested to ensure moisture content is suitably low, germination (or viability) percentage is sufficiently high, and often for species purity, lack of damage and infection with either insects or fungi.

(a) *Moisture content.* Most pine and spruce seeds should be dried to between 6 and 8% moisture content (mc), whereas firs and cedars are best stored at 10–12%, and most broadleaved trees at 10–15%. As a guideline, if all other storage conditions are kept constant, every 1% increase in moisture content above the minima quoted above will half the storage life. (**See: Harrington's Thumb Rules**)

A significant minority of tree species has recalcitrant fruits. Typically, recalcitrant tropical tree seeds should not be stored below 60% moisture content, whereas temperate tree seeds, which are slightly less susceptible to drying damage, should not be dried to less than 40%.

Once the moisture content of both orthodox and recalcitrant seeds has been adjusted for storage, every effort should be made to prevent the moisture content from changing. In the case of orthodox seeds at low moisture content, this is best achieved by packing the seeds into airtight containers such as sealed polythene bags, plastic drums or tins. In the case of recalcitrant seeds, sealed, closed containers must be avoided as sufficient gaseous exchange to permit respiration is essential, and a high humidity is important to retard drying. It is often beneficial to soak or spray recalcitrant seeds during storage.

(b) *Temperature.* This is probably the most important and commonly controlled environmental parameter during tree seed storage. Most orthodox pines and spruces (at 6–8% mc) store well for a decade or two at 0–2°C and even longer at sub-zero temperatures, whereas orthodox seeds which prefer slightly higher moisture contents (firs, cedars, broadleaved trees) typically require temperatures of −10 to −20°C, although again lower temperatures can be better. The deterioration of recalcitrant tree seeds can merely be retarded: it cannot be prevented. The best storage temperature range for recalcitrant, temperate tree seeds is about −3 to +5°C, to slow fungal growth as well as minimize seed deterioration; however, many recalcitrant, tropical tree seeds are killed by temperatures below +15°C.

7. Seed quality and performance

The quality and performance of commercially available tree seedlots lags considerably behind that of agricultural, horticultural vegetable and flower seeds, nearly all of which are supplied from dry-storage and ready for immediate sowing. Tree seeds on the other hand often have characteristics that can contribute to the failure of outwardly intact seeds to germinate: immature or dead embryos, empty seeds (lacking an embryo), **dormancy**, insect or physical damage and fungal infection.

(a) *Empty and damaged seeds.* The percentages of empty, insect- and physically damaged seeds in a seedlot are relatively easy to identify by cutting open a small sample. **X-ray analysis** is a relatively common, non-destructive alternative test (Fig. T.10).

Empty, and insect-damaged seeds are both relatively easy to remove from mixtures with good quality seeds using a variety of standard seed processing techniques such as **gravity tables**, aspirators, 'flotation' or equivalent (**see: Conditioning**). Physically-damaged seeds are less easy to remove, but a technique called **PREVAC** (Pressure-Vacuum) can be used to do this. Filled-dead seeds can be removed from filled-live seeds using **IDS** (Imbibition/Drying/Separation) techniques.

(b) *Fungal-infected seeds.* The cones, fruits and seeds of many tree species often carry microorganisms such as fungi, bacteria and viruses, either internally or on their surface. Fungi are generally held to be a fairly serious threat to tree seed and seedling health and ultimately plant production success. The frequency of seedborne *Fusarium* in British Columbia, Canada, for example, is equivalent regardless of the source of seeds (i.e. from tree seed orchards or natural stands). Seeds and cones harbouring this fungus can contaminate processing equipment, and spread to previously uncontaminated seeds. (**See: Storage fungi**)

Moist conditions (usually warm, but sometimes cold) generally maximize fungal growth. Hence, fungal damage is most prevalent during the following stages of the plant propagation cycle: (i) shortly after collection, when fruits, cones and seeds are still moist and need temporary storage at the forest; (ii) during transport to the seed extraction/

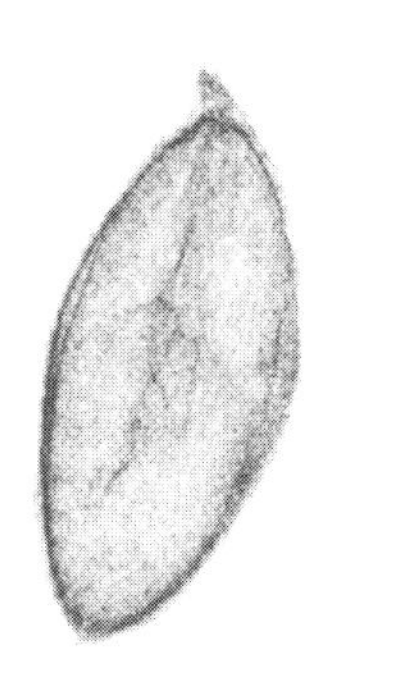

Fig. T.10. X-radiograph of Douglas fir (*Pseudotsuga menziesii*) seeds showing (left to right) empty, insect-damaged and filled seeds.

processing unit when fruits, cones and seeds are still moist; (iii) during the storage of recalcitrant seeds which must be maintained at high moisture contents to avoid **desiccation damage**; (iv) during the warm and/or cold moist pretreatment periods required to overcome the dormancy of many temperate seeds; and (v) under the moist conditions necessary to encourage germination in either laboratory tests, seedbeds or containers. Dry conditions minimize fungal growth, but have no adverse effect on fungal spores. Hence, warm/dry seed extraction from cones, or cold/dry storage of orthodox seeds have enormous potential for spreading the spores associated with fungal contamination.

Seed **treatments** to combat fungal diseases and their spread are available but little used. For contaminated seedlots, persistence of the fungus during imbibition and moist chilling is common; thus, in Canada, the implementation of a hydrogen peroxide treatment (e.g. 30% v/v for less than 1 h or 3% for 4–8 h) is usually warranted. A heat treatment (called '**thermotherapy**') has been reported for acorns but has proved difficult to apply to bulk quantities of seeds because an exact temperature of 41°C for 2.5 h is vital to success. Similarly, although many chemical treatments are effective at killing fungi, they can also be phytotoxic and substantially reduce expected germination percentage or seedling survival. Hence, fungicidal treatments are rarely used prophylactically; they tend to be used only after a fungal outbreak.

Some fungi may be beneficial to tree seed germination, because they are suspected of either breaking down hard seedcoats, or releasing gibberellins in the vicinity of dormant seeds, both of which may overcome dormancy and stimulate germination.

(c) *Dead seeds.* Many tree species habitually produce relatively large percentages of seeds that are outwardly intact, inwardly filled but nevertheless dead. Dead seeds tend to have very similar physical characteristics to live seeds and are therefore quite difficult to remove. Some success on a few species has been achieved with IDS, which induces density differences between the two categories of seed.

Table T.8. Classification and characteristics of seed dormancy as applied to tree species (**see: Dormancy**).

			Dormancy-breaking stimulus	
Dormancy type	Characteristics	Examples of occurrence	Natural	Seed handling
a. Immature embryo	Seeds are physiologically immature for germination	*Fraxinus excelsior*, *Ginkgo biloba*	Post-dispersal development	Afterripening
b. Mechanical dormancy	Embryo germination is physically restricted due to hard seed/fruit coat	*Pterocarpus*, some *Terminalia* spp., *Melia volkensii*	Gradual decomposition of hard structures, e.g. by termites	Mechanical cracking of restricting structure
c. Physical dormancy	Imbibition is impeded because of impermeable seedcoat or fruit	Mainly hard seed Leguminosae, plus some Myrtaceae and others	Abrasion by sand, high temperatures, temperature fluctuations, ingestion by animals, or other	Mechanical scarification (e.g. abrasion or burning), boiling water or acid pretreatment
d. Chemical dormancy	Fruit and seed contain chemical inhibitory compounds that prevent germination	Fleshy fruit such as berries, drupes and pomes, plus some dry seeds	Ingestion by frugivores, leaching by rain, gradual decomposition of fruit pulp	Removal of fruit pulp plus leaching with water
e. Requirement for light	Seeds fail to germinate unless exposed to appropriate light conditions/regime. Is operated by a phytochrome mechanism	Many temperate species, e.g. *Betula*. Humid tropical pioneer species e.g. *Spathodea* and some eucalypts	Exposure to light conditions likely to promote seedling survival viz. white light or light relatively rich in red light	Exposure to light, normally during germination, sometimes a distinct light–dark cycle of variable duration
f. Temperature-associated dormancy	Germination poor without pretreatment with appropriate temperatures	Most temperate species, e.g. *Fagus*, *Quercus*, *Pinus*	Exposure to low winter temperature	Stratification or chilling. High temperature, e.g. kiln or light burning.
		Dry zone tropical–subtropical pioneers, e.g.*Hakea*, *Pinus*, *Eucalyptus*, *Banksia*	Exposure to grass, bush or forest fires	Fluctuating temperature
		Humid tropical pioneers	Diurnal fluctuating temperature in gaps	

Adapted from Schmidt, L. (2000) *Guide to Handling of Tropical and Subtropical Forest Seed.* Danida Forest Tree Centre, Humlebaek, Denmark.

(d) *Immature seeds.* Tree seeds that are immature at dispersal are not uncommon. For example, Scots pine and Norway spruce seeds, which form on trees growing at the northernmost limits of their range, may simply not have enough time to complete development. In these cases a few days of **priming** or invigoration is all that is required to mature. Ash fruits on the other hand are usually shed while the embryo within is still very small. These fruits tend to need incubation for several months under warm, moist conditions in order to mature. Immature seeds are also very common amongst the ancient conifer group known as podocarps. In this group, immaturity is often so pronounced that in nature the fruits must become buried in a moist substrate almost immediately and remain there for years to complete development and acquire an ability to germinate.

8. Seed dormancy treatment

Tree seeds show some of the most pronounced and complicated forms of seed **dormancy** in the plant kingdom (Table T.8), which creates some of the most severe problems for propagators of woody plants. Some of these seeds are non-germinable because they have immature embryos or possess a layer of tissues that is impermeable to water. But many have forms of dormancy caused by chemicals that inhibit germination, or complex metabolic blocks to germination, or combinations of these mechanisms. In practice the raising of tree seedlings requires artificial, pre-sowing treatments ('pretreatments'), which mimic the natural dormancy breakage process, such as weakening of hard seedcoats (**see: Scarification**) and prechilling (**see: Stratification**) to stimulate germination. Some species possess two or more types of dormancy, and different treatments may need to be carried out in the correct sequence to break them both, and even through several cycles in some cases.

Fig. T.11 illustrates one aspect of tree seed dormancy: the effect of pretreatment on germination over a range of temperatures. This demonstrates some of the important distinctions between two sorts of dormancy in temperate tree seeds and also one exhibited in tropical tree seeds with hard seed- or fruit coats.

9. Establishment – seedbed and nursery sowing

There are generally three methods used to establish trees from seeds.

(a) *Nursery sowing and transplantation* is generally considered to be the quickest, most reliable and economic means of establishing trees on the widest range of sites. Seeds are sown into the controlled environment and only the best quality seedlings are transported to the final planting site. A wide range of container types is used, which though more expensive allows greater flexibility in planting time; in other cases plants are transported as 'bare-root' stock. Many leguminous trees are actinorhizal, depending on symbiotic associations with **rhizobia** and either **mycorrhiza** or *Frankia*, and must be inoculated with both at the seedling stage, using soil collected from under trees of the same species, crushed nodules, or from pure cultures.

(b) *Direct sowing at the final site of the proposed crop.* Because of the relatively low proportion of seeds that establish seedlings and the long time-scales involved, direct sowing is usually used on a small scale, and is most successful with cheap seed on new planting sites, where ground preparation,

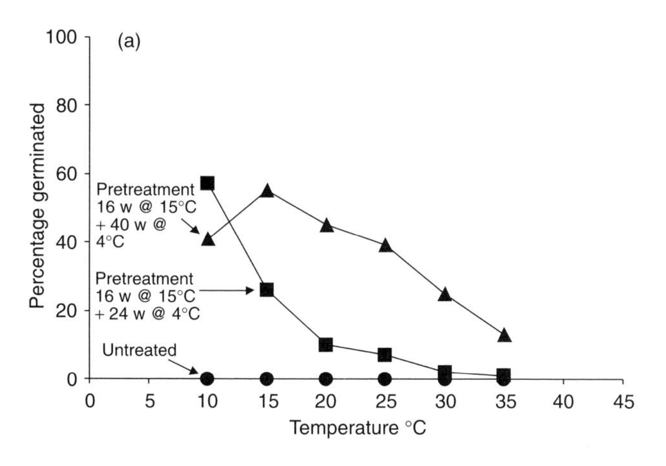

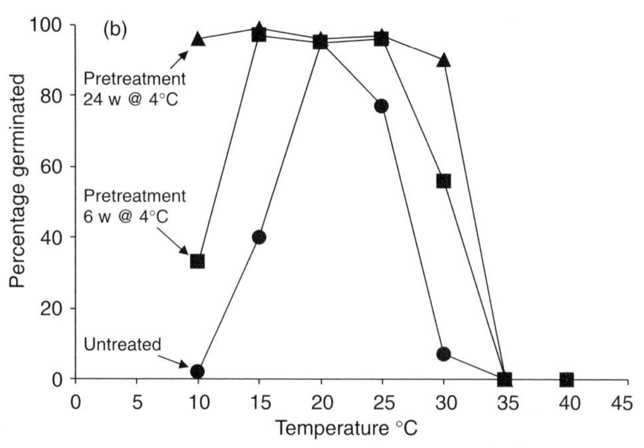

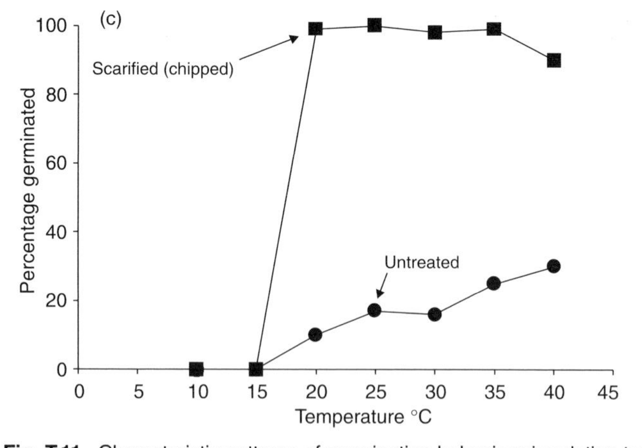

Fig. T.11. Characteristic patterns of germination behaviour in relation to temperature, seen in three major types of dormancy exhibited by tree seeds. (a) 'Deeply dormant' seeds: untreated, deeply dormant seeds do not germinate at any temperature (●); a relatively long pretreatment is required to stimulate any germination at all (■); even the longest and best pretreatment only stimulates about 50% germination at lower temperatures – there is a steady decrease in germination at higher temperatures (▲). (b) 'Shallowly dormant' seeds: untreated seeds show almost 100% germination at some temperatures, but no germination at all at others (●); a relatively short chilling pretreatment widens the range of temperatures over which seeds germinate (■); longer pretreatment stimulates almost 100% germination at most temperatures – but not the highest (▲). (c) 'Hard-seeded', tropical seeds: some untreated seeds germinate at all, especially the higher, temperatures (●); pretreatment by **scarification** stimulates almost 100% germination at most temperatures – but not below 15°C (■). w: weeks.

weed and pest control are easiest. It is least successful on reafforestation sites.

(c) *Natural regeneration* is authentically the unaided process by which trees and woodlands establish. In practical 'human-assisted natural regeneration', establishment from natural seed- or seedling-banks is artificially promoted by a number of pre-planned operations such as leaving 'seed-trees', applying carefully timed ground disturbance, minimizing seed **predation**, and protecting seedlings from weed competition and browsing. (**See: Regeneration**)

10. Genetics and improvement

In common with most plants the differentiated cells and tissues of trees can exhibit 'totipotency' – that is, the ability of a single cell, or group of cells to regenerate the whole plant. In theory, therefore, it should be possible to select a mature tree with exactly the characteristics required, take a small tissue sample from it and propagate numerous identical copies. Hence, individuals with bountiful flowering characteristics, improved fruit quality, faster timber production or whatever, should be very easy to select and multiply. The mass cloning of mature trees, selected for desirable characteristics should be achievable. (**See: Somatic embryogenesis**)

With the notable exceptions of some palms, which can be vegetatively propagated from 'offsets', and a few conifers (*Chamaecyparis* and *Juniperus* species and *Cryptomeria japonica*), most conifers and angiosperms are difficult or even impossible to propagate from cuttings taken from mature trees. Other trees often require such highly labour intensive techniques, such as grafting or micropropagation, that they are much too expensive to apply in forestry and can only be used to propagate higher-valued species such as rubber trees or palms. In yet other instances the vegetatively derived cells lose the superior characteristics they were selected for and show reduced growth rates, poorer flowering or fruiting characteristics or different branching patterns. As a consequence, vegetative propagation from mature trees is generally limited to poplars, willows, eucalypts and *Cryptomeria japonica* (for wood production), and some high value trees like palms, ornamental conifers (e.g. *Juniperus* species and *C. japonica*) and fruit trees (e.g. olive).

As a consequence of all of the above, tree breeding, in common with most other plant breeding programmes, relies on sexual reproduction: identification of suitable population(s) as sources of individuals with a sufficient level of phenotypic variation in the characteristic(s) chosen; selection and mating of so-called 'plus trees'; testing offspring in progeny trials; and propagation for commercial use. The difference between tree breeding and other plant breeding programmes is the significantly longer life cycle of most trees. The testing phase can take decades to carry out, and even then it will be necessary to return to the original parents of successful matings, if they still survive, and propagate larger numbers of the improved crosses. Noticeable progress can often take more than one human lifetime to accomplish. Marker-aided selection offers breeders the prospect of selecting for desirable traits that are associated with extremes of variation, e.g. very high wood density, or very good stem form or frost resistance. (PG)

Bonga, J.M. and von Aderkas, P. (1992) *In Vitro Culture of Trees.* Kluwer Academic, Dordrecht, The Netherlands.

Poulsen, K.M., Parratt, M.J. and Gosling, P.G. (eds) (1998) *ISTA Tropical and Sub-tropical Tree and Shrub Seed Handbook.* International Seed Testing Association, Zurich, Switzerland.

State of the World's Forests (2003) Food and Agricultural Organization of the United Nations, Rome, Italy.

World Resources 1990–91: A Guide to the Global Environment. Oxford University Press, New York, USA.

Triacylglycerols (TAGs)

Triacylglycerols (fats, oils) are non-polar lipids (neutral lipids) that are the main energy and carbon store in the cells of oilseeds and pollen. Seed triacylglycerol is a molecule with three fatty acids esterified with a glycerol backbone (Fig. T.12). Triacylglycerols are derived from glycerol-3-phosphate and fatty acyl-CoA molecules in the cytosol of oilseeds. (AJK) (**See: Fatty Acids; Oils and fats – synthesis**)

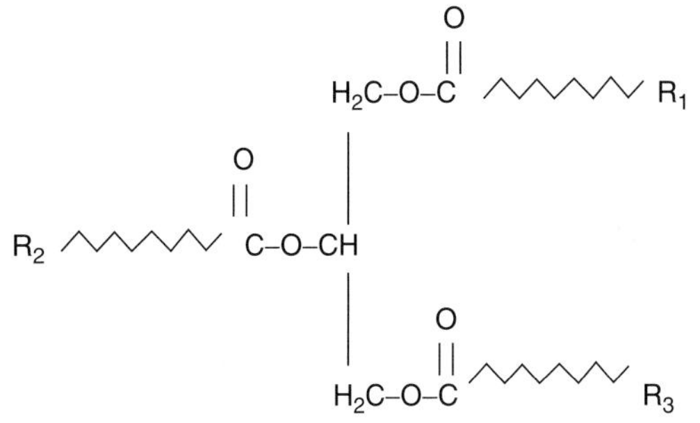

$H_2C{-}O{-}C({=}O) \sim\sim R_1$

$R_2 \sim\sim C({=}O){-}O{-}CH$

$H_2C{-}O{-}C({=}O) \sim\sim R_3$

Fig. T.12. A triacylglycerol molecule. Three fatty acyl chains (R_1–R_3) are esterified with a glycerol molecule.

Trichome

Fibres growing from the ovule epidermis or from the inner surfaces of seed pods. Cotton and milkweed fibre are examples of **ovule** epidermal trichomes, whereas kapok fibre originates from the inner surface and placenta of seed **pods**. (**See: Coir; Cotton; Fibres; Kapok; Milkweed**)

Triers

Perforated probes of several designs are used in **sampling** of seedlots, to obtain a representative sample for **quality testing** after insertion into the seed mass, with apertures that can be opened and closed after insertion into the seed mass. (**See: Sampling, 1. Spears and probe samplers**)

Triploid

Triploid plants possess three copies of the basic **chromosome number** (2n = 3x), and are usually **autopolyploids** that arise spontaneously in nature, or are constructed by breeders from the cross of a 2n = 4x (**tetraploid**) and a 2n = 2x (**diploid**). Triploids cannot complete normal **meiosis** because of uneven pairing of the chromosomes, and so characteristically produce sterility or seed abortion, which is exploited to produce **seedless fruit** varieties in crops such as watermelons, and is why the fruit of the widely available commercial banana clones is seedless. Triploid varieties are also bred in **sugarbeet**.

Also, the **endosperms** of seeds are composed of cells that are triploid (3n) because of the fusion of one male gamete with two female gametes during **fertilization**.

Triticale

Triticale (x*Triticosecale*) is a synthetic (i.e. produced by plant breeders, not in nature) **hybrid** usually of durum **wheat** and **rye** (AABBRR, **see: Wheat,** Table W.1). Both tetraploid wheats and hexaploid wheats can be crossed with rye, although there are some wheat genotypes that will cross much better than others. The 4x wheat crossed with rye is used to produce a hexaploid triticale (6x = 42) while a hexaploid wheat crossed with a rye produces an octoploid triticale (8x = 56). Triticale has several agronomic advantages over wheat such as lower water requirement, greater resistance to cold, higher yields under certain circumstances, and higher nutritional quality. In 2002 world production was 11 million t principally in Poland, France, ex-USSR, Australia, Portugal, USA, Brazil and Germany. Only about 311,000 t was traded at a value of nearly US$36 million (FAOSTAT).

The grains are more slender and pointed than wheat grains but they also have a crease and indeed share many of the features described for wheat (Colour Plate 9C). The bead-like appearance of the cell walls of the pericarp is less distinct than in wheat. Two populations of **starch granules** are present as in wheat.

The main compositional difference from wheat grains is the slightly higher **storage protein** content (10–13%) with a higher lysine value (3–4% lysine).

Triticale has been grown as a crop (mainly as animal feed) since the 1950s but the grains are also used as raw material for baking, beer and porridge. Its quality is not as good as that of wheat for making bread, although in Poland it is quite popular for making a light 'rye' bread. There is no market specifically for triticale grain, and so it has not developed into a large demand crop on its own. It has some popularity as a component of health foods. (AE, MN, DF)

Guedes-Pinto, H., Darvey, N. and Carnide, V.P. (eds) (1996) *Triticale: Today and Tomorrow. Developments in Plant Breeding. V.* Kluwer Academic, Dordrecht, The Netherlands.

True Potato Seed

Also commonly abbreviated to TPS. (**See: Potato**)

Trypsin inhibitor

See: Protease inhibitor

TTZ or TZ test

See: Tetrazolium test

Tundu disease

A seedborne disease of **wheat**, prevalent in northwestern India, also known as Tannau or yellow slime disease, caused by the combination of the bacterium *Rathayibacter* (also known as *Corynebacterium*) *tritici* and the nematode *Anguina tritici*. Symptoms include the rotting of developing seeds and the formation of excessive amounts of bacterial slime on various plant parts, which may dry to become a varnish-like coating, for instance, in and around the grain groove and other seed tissues. The nematode forms galls in the host, and appears to act as a vector in transmitting the bacterial pathogen to the growing apical meristem. (PH)

Tung

Tung oil is an industrial oil obtained from two deciduous tree species, *Aleurites fordii* and *A. montana* (2n = 22) (Euphorbiaceae), commonly called the tung oil tree and the China wood oil tree, respectively. *A. fordii* supplies about 90% of the traded product. The tung tree species are native to central and western China where plantings and use are traditional. China is indeed the main producer (86%) of tung oil with Paraguay (7.8%) and Argentina (4.5%) minor producers. Small amounts are also produced in Malawi, Madagascar and Brazil. (**See: Oilseeds – major**, Table O.1; **Oils and fats – genetic modification**)

The female flowers develop into hard roundish green or purplish-green fruits 5–8 cm in diameter with four or five carpels. The pericarp of *A. fordii* fruits is smooth while those of *A. montana* are strongly ribbed. Mechanical shelling of the **pericarp** normally yields four or five **nut-**like seeds. The seeds are roughly triangular in cross section and vary in length from 14 to 35 mm long and 13 to 25 mm wide. The **testa** is thick (1.5 mm) and hard. The **embryo** with two delicate **cotyledons** is surrounded by a thick **endosperm** that contains most of the oil. The arrangement and histology of the seed tissues are very similar to those of **castor bean**.

The seeds are expeller pressed to yield 33% oil containing 75 to 80% of a unique fatty acid, eleostearic (18:3), which is responsible for the oil's quick drying property, producing a tough waterproof glossy surface (**see: Oilseeds – major**, Table O.2 for fatty acid composition). The oil is used in paints, waterproofing products, in inks, resins and cleaning and polishing products. Tung meal contains 20 to 25% protein but also a poisonous saponin, so it can only be used as a fertilizer. The fruit shells can be used for mulching. (**See: Oils and fats**) (KD)

Duke, J.A. (1983) *Aleurites fordii* Hemsl. NewCROP: the New Crop Resource Online Program, Centre for New Crops and Plant Products, Purdue Univ. Last update 19 Dec. 1997. www.hort.purdue.edu/newcrop/duke_energy/Aleurites_ fordii.html

Vaughan, J.G. (1970) *The Structure and Utilization of Oil Seeds.* Chapman and Hall, London, UK.

Turgor

Turgor describes the hydrostatic pressure (P) generated in a cell as a result of the cell wall resisting the volume change caused by water movement into the cell down a **water potential** (ψ) gradient. Once water equilibrium is restored, water potentials inside (ψ_i) and outside (ψ_o) the cell are identical. Since seeds are hydraulically isolated from the parent plant, osmotic pressure (**see:** *osmotic* in **Development of seeds – nutrient and water transport**) of the apoplasmic sap (π_o) is the major factor influencing the outside water potential. Water potential inside the cell is a balance between cell turgor and osmotic (π_i) pressures. Thus at water equilibrium:

$$-\pi_o = P - \pi_i$$

and hence cell turgor of a seed cell is given by:

$$P = \pi_i - \pi_o$$

(MT-H, CEO, JWP)

Turnip

See: Brassica – horticultural

Ultra-dry storage

The use of ultra-dry conditions for seed storage was conceived as a means of maximizing the **longevity** of seeds in circumstances where facilities for low temperature storage were not available. The term 'ultra-dry' was first used to describe seeds dried below 5% moisture content (fresh weight basis) and was originally intended for oily seeds, recognizing that in several cases the optimum moisture content for storage was below the 5% level recommended for seed **gene banks**. Subsequently, 'ultra-dry' has been used to describe seeds dried to equilibrium at 10–12% RH and 20°C.

There is general agreement that it is possible to over-dry seeds to a point where their subsequent longevity in storage is compromised. There is also agreement that the optimum moisture content for storing oily seeds is less than non-oily seeds. However, there has been considerable debate as to whether the optimum moisture content depends on storage temperature. Essentially, the debate hinges on the shape of **sorption isotherms** at low temperatures and whether or not the **equilibrium relative humidity** (eRH) of seeds drops to sub-optimal levels when seed containers are moved from the temperature used for drying to the final storage temperature. It has been argued that the optimum eRH for seed storage probably lies between 15 and 20%. It follows that for seeds to be within this range when they are stored at the recommended temperature for seed gene banks (–18°C), the relative humidity of seed drying rooms, operating at 10–25°C, will need to be significantly higher than the 10–15% RH recommended by **FAO** and **IPGRI** in 1994. (**See: Deterioration in storage**)

(RJP, SHL)

Umbellifers

Previously known as the Umbelliferae and now known as the Apiaceae. They contain a range of edible vegetable, spice and herb crops, such as **ajowan**, **anise**, **caraway**, carrot, celeriac, celery, **coriander**, **cumin**, **dill**, **fennel**, parsley and parsnip, whose edible plant parts variously include leaves, roots, seeds, stems and stalks (Table U.1). Most umbellifers are cool-season crops with annual, biennial and perennial characteristics, and are generally produced in temperate climates or high elevations, with an optimal growing range between 15 and 21°C, though coriander and fennel have a greater adaptation to warmer temperatures than the rest of the umbellifer crops.

The majority of umbellifer vegetable and condiment crops are propagated from seed. This article focuses on the two crops of highest economic importance: carrot and celery. World carrot production is over 16 million t per year, with the Asian countries being the main producers, followed by Europe and the Americas. It is a good dietary source of sugars, vitamins A and C and fibre and an important source of nutrition, especially for infants. Celery production takes place mostly in North America and Europe.

1. History and origins

(a) *Carrot*, *Daucus carota* (2n = 2x = 18) includes 11 species, all found in the Mediterranean region. The genus *Daucus* is centred north and south of the Mediterranean, spreading to North Africa, Southwest Asia and Ethiopia, and the most widespread wild forms of this species belong to var. *carota*. The cultivated species belongs to var. *sativum*, and are of two main types.

- Eastern or Asiatic carrots are often called 'anthocyanin' carrots because of their purple roots, although some are yellow. Their pubescent leaves give them a grey-green colour, and they are readily susceptible to **bolting**. The greatest diversity of these carrots is found in Afghanistan, Russia, Iran and India. It is thought that these carrots originated some 5000 years ago in middle Asia around Afghanistan. They are still cultivated in Asia, but are being rapidly replaced by orange-rooted Western carrots. Because anthocyanin is an antioxidant, purple carrots are being promoted as a functional food in the USA.
- Western or 'carotene' carrots have orange, red or white roots. Most likely, western carrots were derived by selection among hybrid progenies of yellow eastern carrots, white carrots and wild subspecies grown in the Mediterranean region. The first two of these colours probably originated by mutation, possibly in Turkey. Orange carrots were first cultivated relatively recently, in the 16th or 17th century in the Netherlands, and our present cultivars probably originate from 'long orange' varieties developed there. Adaptation to northern latitudes has been accompanied by cultivar changes in photoperiodic response.

Cultivar types are classified by root shape and by their **bolting** ability, which can reduce marketable yield. There are temperate ('Chantenay', 'Danvers', 'Imperator', 'Nantes') and subtropical ('Brasilia', 'Kuroda', 'Tropical Nantes') types. In Europe 'Nantes' types are predominantly grown for fresh-market uses, whereas in North America the 'Imperator' types are the most popular. Processing carrot varieties grown in both Europe and North America are 'Flakee', 'Paris Market' and 'Oxheart'.

(b) *Celery* (*Apium graveolens*) (diploid species of 2n = 2x = 22) and its relatives are cultivated in three horticultural varieties: *dulce* celery (with succulent petioles), *rapaceum* celeriac (with enlarged root/hypocotyls) and *secalinum* smallage (for its leaves). There are two main cultivar types: 'self-blanching' or yellow, mostly grown in Europe, and Pascal or green celery, grown mostly in North America and Australia. Celery is rich in fibre

Table U.1. Edible species of the Apiaceae, listing their primary uses and plant portions utilized.

Botanical name	Common name	Uses[a]	Plant portion used[b]
Aegopodium podagraria	Bishop's elder, Bishop's weed	V	L, S
Anethum graveolens	Dill	C,V	S, L
Anethum sowa	Indian dill	C	S, L
Angelica archangelica	European/garden angelica	C,V	St, S, R, L
Angelica atropupurea	American/purple angelica	V	St, S, R
Angelica edulis	Japanese angelica	V	Stk, L
Angelica sylvestris	Wild parsnip	V,C	L, St, R, S
Anthriscus cerefolium	Chervil, French parsley, garden/salad chervil	V,C	L, S
Anthriscus sylvestris	Cow parsley	V	L
Apium graveolens var. *dulce*	Celery	V,C	L, S
Apium graveolens var. *rapaceum*	Celeriac, knob celery	V	R
Apium graveolens var. *secalinum*	Smallage	V,C	L, S
Arracacia xanthorrhiza	Arracacha	V	R, L, St
Bunium bulbocastanum	Great earthnut	V,C	T, L, F
Bunium persicum	Black cumin	C	S
Carum carvi	Caraway	C	S, R, L
Carum nigrum	Black caraway	C	S
Centella asiatica	Asiatic/Indian pennywort	V	L
Chaerophyllum bulbosum	Turnip-rooted chervil	V	R, L
Coriandrum sativum	Cilantro, Chinese/Mexican parsley	V	L
	Coriander	V,C	S, L
Crithmum maritimum	Rock samphire, sea fennel	V	L
Cryptotaenia canadensis	White/wild chervil	V	L, St, R, F
Cryptotaenia japonica	Japanese parsley/hornwort, Mitsuba	V	L, St, R
Cuminum cyminum	Cumin	C	S
Daucus carota var. *sativus*	Carrot	V	R, L
Eryngium foetidum	Culantro, Java coriander, stinkweed	V	L
Eryngium maritimum	Sea holly	V	R, Sh, L
Ferula assa-foetida	Asafetida, devil's dung, stinking gum (giant fennel)	C	R, Sh, (S)
Ferula communis	Common giant fennel	C,V	S, L
Foeniculum vulgare var. *azoricum*	Florence fennel, finocchio	V,C	L, S
Foeniculum vulgare var. *dulce*	Sweet anise/fennel (fennel)	C, (V)	S, L
Foeniculum vulgare var. *vulgare*	Bitter fennel	C	S
Heracleum lanatum	Cow parsnip	V,C	R, L, S
Heracleum sphondylium	Common cow parsnip	V	Sh, L
Hydrocotyle sibthorpioides	Lawn pennywort	V	L, St
Levisticum officinale	Lovage	C,V	L, S, R
Ligusticum scoticum	Scotch lovage	V	L, Sh, R
Myrrhis odorata	Sweet chervil/cicely, (garden) myrrh	C	L, S, R
Oenanthe javanica	Water celery/dropwort	V	L, Sh
Pastinaca sativa	Parsnip	V	R, L
Perideridia gairdneri	Squawroot, Yampah, Epos	V	R, S
Petroselinum crispum	Parsley	V,C	L
Petroselinum crispum var. *tuberosum*	Hamburg/turnip-rooted parsley	V,C	R, L
Pimpinella anisum	Anise	C	S, L
Pimpinella major	Greater burnet saxifrage	C	R, S, L
Pimpinella saxifraga	Burnet saxifrage	C	S, L, R
Sium sisarum	Skirret	V	R
Smyrnium olusatrum	Black lovage, horse parsely, Alexanders	V,C	L, Stk
Trachyspermum ammi	Ajwain, ajowan	C	S, L

[a]V, vegetable; C, condiment or flavouring.
[b]F, flowers; L, leaves; R, roots; S, seeds; Sh, shoots; St, stems; Stk, stalks; T, tubers.
From: Rubatzky, V.E., Quiros, C.F. and Simon, P.W. (1999) *Carrots and Related Vegetable Umbelliferae.* CAB International, Wallingford, UK.

and some chemical compounds that are health-promoting. Detrimental compounds include furanocoumarins, such as psoralens, and the **allergen** profiling; but high concentrations are not a problem in most varieties, though content varies depending on environment.

A. graveolens is a Mediterranean species, but is found from Sweden to India. Wild species of the genus *Apium*, of which there are approximately 20, are widely distributed throughout the world, including the Mediterranean, Australia, New Zealand, South Africa and South America (the southern zone of which is a rich centre of origin). China might have been an early centre of domestication; the plant was recorded in the writings of Confucius. Celery was first used as a medicinal plant and as a condiment (as seed extracts, variously, to cure headaches, as anti-flatulents, diuretic and aphrodisiacs) and for ornamentation and leafy herbs by the Greeks, Romans, Egyptians and Chinese. Celery was developed in Italy and France as a vegetable in the 16th century, resulting in selections with solid stems. (**See: Spices**)

2. Development

Though biennial, carrot and celery are cultivated as annuals. The compound inflorescences – umbels – comprise 50 or more umbellets, each bearing approximately 50 flowers, that are characteristically arranged in convex or flat-topped flower clusters, in which individual pedicels all arise from the same apex. (**See: Inflorescence**, Fig. I.1) Flowers are functionally male and female. They have five white to greenish-white petals and five stamens, which fall when the stigma becomes receptive. **Protandry** (in which anthers mature first), typical of the Apiaceae, promotes outcrossing by insects; the wind also promotes cross-pollination. When the flowers open, the two folded stigmata become erect and separate from each other, becoming receptive 4 to 5 days after anthesis. By that time most of the anthers have dropped from the flower: pollen shed and stigma receptivity are not synchronized and do not facilitate self-pollination within a flower. Each locule of the inferior, bilocular ovary contains a functional ovule. Nectaries attract a great variety of pollinating insects including flies, wasps and bees. The flowering period is characteristically protracted – ranging from 7 to 10 days for an individual umbel, and 30 to 50 days for all the umbels on the plant. The peripheral flowers of the umbel open first.

The fruit is composed of two dry, indehiscent, one-seeded **mericarps**, and the seeds are variously ornamented. The actual 'seeds' are constituted by the separated mericarps. The embryo constitutes only a few per cent of the seed weight, the majority being composed of the **endosperm** that develops prior to embryo growth during seed development. The primary carrot umbel (the 'king umbel') produces the highest seed quality. However, the majority of the seed yield is derived from the secondary umbels that develop from laterals on the primary flowering stem. Later umbels (tertiary or quaternary) generally have poorer quality seed. The staggered development of seeds in different umbel orders results in seed of different maturities, including embryo size, being formed on the same plant. This variation in seed maturity affects uniformity of the crop, because less-developed embryos take longer to germinate. However, it is difficult physically to grade seed into maturity classes to improve uniformity.

3. Genetics and breeding

Most global breeding work is directed to the temperate types because of their larger market share and crop value. These cultivars, however, are marked by lower seedling vigour and slower plant growth rate. There is a strong trend towards the use of F_1 carrot **hybrids** as replacements for open-pollinated cultivars, especially for fresh market production in Europe and North America, due to the increased demand for plant uniformity for precision planting equipment. Inbreeding results in severe depression, although it is still possible to produce inbred lines for hybrid production. Carrot hybrids are usually three-way crosses for increased seed productivity (**see: Production for sowing, III. Hybrids, Crosses**), though single-cross hybrids are on average more uniform. Care must be taken to isolate breeding plots from any wild carrots in the area, since cross-pollination occurs; in most cases, these hybrids will display deformed or forked roots. Unlike carrot, almost all celery varieties are open-pollinated.

There is no self-incompatibility in umbellifer crops, so emasculation is necessary, which is inefficient and difficult because the flowers are small and each produces only two seeds. However, because of protandry, emasculation is achievable by washing away the anthers and pollen from the flower with a stream of water before the stigmata have become receptive. Then the flowers are covered for 4 or 5 days and pollinated using a brush when the stigmata become receptive.

Cytoplasmic male sterility (CMS) is available for carrot hybrid production, the two main sources being: (i) *petaloid* and *carpeloid*, a system found in wild carrot in which the stamens are modified to petals or carpels and (ii) *brown anthers*, which dry and deform due to tissue degeneration, as found in 'Tendersweet' over 50 years ago. The petaloid CMS is mostly used for hybrid seed production in the USA because it is more stable than the brown anther system; however, in Europe and Asia the latter system is preferred because it is more stable there than petaloid CMS. Newer sources for both types of CMS have been found in recent years.

CMS in carrot is due to interaction of mitochondrial and nuclear genes. Carpeloid CMS seems to result from disrupted interaction between the products of nuclear and mitochondrial genes leading to reduced expression of some of the **MADS** box genes, which are known to be involved in floral development. The inheritance of CMS is not simple and has not been completely solved. Both types are dominant and may be determined by triplicated nuclear genes, one dominant (*M*) and two recessive (*t* and *l*), which interact with sterility-inducing cytoplasm (S) (as distinguished from (N), the normal fertility-inducing cytoplasm): Male sterile – (S)*MMlltt*; Restorer – (S)*MmLltt* or (S)*MmllTt*; and Maintainer – (N)*MMlltt*. Male-sterile lines produce less seed than fertile ones after pollination, perhaps because the flowers are less attractive to pollinators due to their reduced nectaries.

4. Production

Seed production is best in regions with dry summer weather. Biennial crops, such as carrot, parsnips and celeriac, must be vernalized to induce reproductive growth. In regions with mild, frost-free winters, the plants can be kept in the field for natural **vernalization**; otherwise this is done in cold chambers for approximately 8 weeks at 5 to 6°C.

There are two basic systems of seed production for root crops. In the seed-to-seed system, the plants are left in the field to overwinter and then the seed is harvested in the spring. In the root-to-seed system, roots are harvested, selected and stored in cold chambers and then replanted when the danger of frost is over; these plants will flower soon after planting. (**See: Production for sowing, I. Principles**) For foliage crops such as celery, selected plants for breeding purposes are lifted, trimmed and vernalized in cold chambers, and replanted in the field or in pots for controlled pollination in glasshouses.

Plant spacing and density in the field are critical to maximize yield, quality and uniformity. For hybrid carrot seed production the spacing of the transplants is normally from 60 to 90 cm between rows and 10 to 30 cm within rows. The female to male ratio is 4:1 or 8:2, where there are four two-row beds of female plants and one two-row bed of pollinators. In order to assure pollination, 15–20 beehives per hectare are distributed on the border of the fields. For celery, in the seed-to-seed scheme a typical plant spacing is 1 m apart. Typical seed yields for carrot and celery are 900 kg/ha, and good yields could reach close to 1200 kg/ha. Recommended isolation distance for these crops to avoid pollen contamination is 3 km. Normal-looking but **embryoless seeds** are often produced though the endosperm and seedcoat are undamaged. This injury is produced by various species of *Lygus* bugs, which feed preferentially on meristematic tissues but destroy the embryo at almost any development stage.

Umbels are ready for harvesting when they curl and become brittle, and seeds turn brown. For small fields or plots, hand harvesting is used, which enables workers to take only from umbels with ripe seed. The cut seed stalks are placed on a smooth surface or a canvas or hung in bunches for further drying in piles for 4–7 days under warm and dry conditions. The inflorescences are threshed by hand rubbing to release the seed, or passed through stationary **threshing** machines. For commercial-scale seed production, mobile combine harvesters are used for the harvesting and threshing and separation operations, after cutting the crop and leaving it in **windrows** to dry, and further debris separation is accomplished by airflow aspiration cleaning equipment. Spiny seeds, like those of carrots (Colour Plate 6F), dill and caraway must be processed further to remove the spines by milling or gently rubbing using specialized equipment, by **debearding**.

After cleaning, the seeds are dried to moisture ranges of 7 to 10% (which is fairly representative for all umbellifer seeds) by warm air at 40 to 50°C for 1 to several hours. Seed viability varies with species, but when stored under favourable conditions (such as 10°C, with low moisture content, in moisture-proof containers) most umbellifer seeds will last at least 3 years, with the exception of parsley and parsnip seeds which do not store well beyond a year. For longer preservation, storage at sub-zero temperature is necessary.

5. Quality, vigour and dormancy

Seed germination is slow and variable within seed lots, sometimes taking 2 weeks for seedlings to emerge. Celery germination is inhibited at temperatures of approximately 30°C. Exposure of seed to light or low temperatures is a pre-requisite in some species to break **dormancy**. In general, seedling growth is slow until the seedlings develop true leaves.

Vigour and germination capacity are influenced by environment, nutritional status and health of the seed plant and, of course, by timing of harvesting. Seeds from primary and secondary umbels are physiologically more mature and are therefore larger and heavier due to a larger endosperm to embryo ratio. However, embryo length seems to be the main determinant of seed quality, as measured by rate of germination, seedling emergence and weight. Delaying harvest results in longer embryos at the risk of seed lost due to **shattering**.

6. Treatments

Seed enhancements are used to improve germination and emergence of both carrot and celery.

Seed conditioning by **priming** is sometimes used, in aerated solutions of polyethylene glycol or mineral salts such as K_3PO_4 or KNO_3 for 1 to 3 weeks with continuous aeration. Priming increases the percentage of seedling emergence and increases germination rate; although their ageing is accelerated, seeds can be stored for several months to a year, depending upon the conditions. In addition to light and cold temperature treatments, dormancy can be overcome by soaking the seed in a mixture of **gibberellin** $GA_{4/7}$ and ethephon at 5°C. **Pre-germination** or **chitting** is an older seed-conditioning procedure where seeds are imbibed and then held at low temperatures (typically 1°C) in aerated water to delay radicle protrusion: the seed is sown just before radicle emergence by **fluid drilling** using a gel medium.

Pelleting is the method of choice for direct seeding of high-value carrot and celery, and also for celery **transplant production** in some countries.

Seed treatments for disease control include hot-water treatment, bleach and fungicides for eradication of seedborne **diseases** such as celery late **blight** (*Septoria apiicola*), carrot seedling blight and black root **rot** (*Alternaria dauci*, *A. radicina*), Phoma rot (*Phoma* spp.) and bacterial blight (*Xanthomonas campestris* pv. *carotae*) (**see: Pathogens**). In some countries, the insecticide tefluthrin is registered as a seed treatment to control attack by first generation carrot fly.

(CFQ)

George, R.A.T. (1999) *Vegetable Seed Production*, 2nd edn. CAB International, Wallingford, UK, pp. 238–264.

Heywood, V.H. (1983) Relationship and evolution in the *Daucus carota* complex. *Israel Journal of Botany* 32, 51–65.

Linke, B., Nothnagel, T. and Borner, T. (2003) Flower development in carrot CMS plants: mitochondria affect the expression of MADS box genes homologous to GLOBOSA and DEFICIENS. *The Plant Journal* 34, 27–37.

Peterson, C.E. and Simon, P.W. (1986) Carrot breeding. In: Bassett, M.J. (ed.) *Breeding Vegetable Crops*. AVI Publishing, Westport, Connecticut, USA, pp. 321–356.

Quiros, C.F. (1994) Celery. In: Kalloo, G. and Bergh, B.O. (eds) *Genetic Improvement of Vegetable Crops*. Pergamon Press, New York, USA, pp. 523–534.

Uniformity

In seed technology, 'uniformity' (in its familiar meaning in the sense of 'being alike' or homogeneous) describes populations of seeds that have closely similar sizes or shapes to each other as a result of **grading** or **pelleting**, or that are uniformly blended and uniformly conveyed in **storage management** systems. In addition, 'uniformity of germination' describes the behaviour of a population of seeds which more or less all complete germination (**emergence** of the **radicle**) over a limited period of time; that is, they have a 'steep' **germination curve**. Uniformity of germination, and hence seedling emergence, is generally a desirable agronomic trait in cultivated crops, whether seeds are being sown in the field or raised as **transplants**. It is characteristically associated with a high degree of seed **vigour** and to the uniformity of planting depth and the absence of shallow **dormancy**, and is increased in response to **enhancement** treatments such as **priming** and is reduced as a result of seed **ageing**. Also, in crops such as sunflower, maize, grain sorghum and sugarbeet, **yield potential** can be improved by increased uniformity of **space planting** within the optimum population range. A combination of uniformity of emergence and spacing is advantageous in these and other crops, notably in many horticultural and ornamental species, because it leads to uniformity of seedling development, to uniformity of crop development and ultimately contributes to greater uniformity of the harvested heads, roots, bulbs or whole plants. (**See: Conditioning; Dormancy – importance in agriculture; Planting equipment – agronomic requirements; Quality testing; Size of seeds; Vigour testing**) (PH)

Units

Apart from its familiar common uses in the sense of 'single item' (such as an individual seed, propagule), various 'seed units' are accepted measures of quantity of a seedlot of a crop in many markets, defined either by weight or number or, sometimes, volume and usually approximate to the quantity needed to sow a convenient unit area (hectare, acre), or a convenient size sack for manual handling (such as the 50 lb sack that has been used for maize and soybean in the USA). Likewise, in crop protection seed treatment, formulation registered application rates are defined in terms of a weight or volume of formulation per unit weight or number of seeds.

Number-based seed units are commonly used for crops that are precision space-planted. For example, sugarbeet seedlots are widely quantified in terms of statutory units of 100,000 seeds. In many horticultural seed markets, seed is packed and sold in 'unit packs', which are adopted by a seed company or by consensus (rather than by statute) within a given market. For example, onion seed is sold in units of 250,000 seeds in some markets. Maize in North America is increasingly sown in standard 80,000-seed unit sacks (instead of the 50 lb sacks, which can vary in weight from about 14 to 32 kg because of the differences in seed size).

The 'unit weight' is thus the weight of the stated number of seeds, which may be the units of sale, as above. But this term is also sometimes used as a synonym for the **thousand-grain weight** (TGW).

A 'sampling unit' is that which is defined in official protocols as the required maximum weight of the seedlot to be sampled, where seedlots are packed in small containers such as the tins, cartons or packets used in the retail trade. Regulations dictate the number of small containers from which samples are to be taken and combined.

Conversely, throughout recorded history, some seed species have been used to define units of weight, volume and length, because they were perceived to be reliably uniform (**see: Weights and measures**). (PH)

Upgrading

Improvement of seed quality through removal of contaminating material and rotten, cracked, broken or low quality seed. (**See: Conditioning; Enhancement**)

UPOV

International Union for the Protection of New Varieties of Plants. The organization that provides the internationally recognized system that harmonizes **variety** protection (intellectual property) rights, including the specification of methods used to conduct **Distinctness, Uniformity and Stability** (**DUS**) trials. (**See: Laws and regulations**)

Ureides

Ureides are nitrogen compounds (allontoin and allontic acid) exported from nitrogen-fixing nodules in the xylem. These nodules are spherical with determinate internal meristematic activity present in roots of common bean and soybean. In contrast, the major nitrogen compounds exported in the xylem from nitrogen-fixing nodules of most other legume species are the amides, glutamine and asparagine. In this case, the nodules are elongate with indeterminate apical meristematic activity. Ureides are present in negligible quantities in the developing embryos of legume seeds; they are converted to glutamine and asparagine in the pod and seed coat before import. (**See: Development of seeds – nutrient and water import**)

(MT-H, CEO, JWP)

V

Vacuum precision seeding

A sowing system for producing seedling **transplants** for field crops or bedding plants, in which seeds or pellets (in the case of irregularly shaped seeds) are held by suction against countersunk holes in a rectangular flat plate, or an array of nozzles, until the pressure is released. A wide selection of seeder plate sizes and configurations are readily available to fit chosen trays; industry suppliers offer, for example, arrays of between about 25 and 500 holes, in a choice of at least seven different hole diameters, to grip the seed and provide the highest 'singulation' accuracy for a wide range of small flower and vegetable seed species. Similar devices are commonly used in seed testing laboratories, to set seeds on flat paper blotters or towels at the start of **germination testing**.

The 'vacuum disc' metering principle is also intrinsic to a major type of field precision drill, which is widely used in horticultural production and is becoming increasingly popular for sowing field crops. (PH)

(See: **Planting equipment – metering and delivery**)

Value for Cultivation and Use (VCU)

A term covering the concept that, to be registered and be placed on a National Variety List, a new plant **variety** must undergo performance testing to ensure that it is superior, or at least acceptably equivalent, to existing alternative varieties under local growing conditions.

The object of VCU trials is to quantify the characteristics of a new variety that are of interest to the grower, which are assessed in multi-site agronomic performance trials, typical of the region for which the variety is intended. The most important criteria of VCU assessment are: yield, and its stability; quality characteristics of harvested crop; and disease and pest resistance. Since these criteria vary from location to location in the same year, and from year to year at the same location, the results are expressed relative to a 'control' established variety, the field characteristics of which are relatively well known and relatively stable. Positive VCU and **Distinctness, Uniformity and Stability (DUS)** test results together constitute the basis for varietal listing in a country (National Listing) and the granting of ownership rights (**see: Laws and Regulations**). (RD)

VAM

See: **Arbuscular mycorrhiza**

Vanilla

Authentic vanilla flavouring comes from the so-called bean or pod of the **orchid**, *Vanilla fragrans*. The 'bean' is actually a fruit – a type of capsule (**see: Fruit structure**), 10–25 cm long – containing pulp in which a vast number of tiny seeds are embedded. The full flavour develops when the beans are 'cured', a lengthy process involving several fermentative changes, a number of drying stages and exposure to sunlight, during which the fruit becomes black. Although the higher proportion of the flavouring compounds (vanillin and other components) is provided by the fruit wall, the inner pulp and seeds are also flavoursome. Evidence that genuine vanilla has been used rather than synthetic flavouring, for example in ice cream, can often be seen from the presence of tiny black specks, the seeds. Several recipes call specifically for the use of the seeds, scraped from inside the fruit, such as in vanilla sugar and for various cakes and desserts. (MB)

Variety

A clearly distinguishable group of bred cultivated plants which, when reproduced under controlled conditions, whether by seed or vegetatively, retain their clear distinguishing morphological, physiological, cytological, chemical or other characteristics. A **hybrid** variety has, in addition, a particular specified 'formula of hybridization' that is intrinsic to its maintenance. The terms variety and **cultivar** ('*culti*vated *var*iety') are synonymous.

According to the **UPOV** 1991 definition, a variety is a plant grouping within a single botanical taxon of the lowest known rank, which grouping can be:

- defined by the expression of the characteristics resulting from a genotype or combination of genotypes;
- distinguished from any other plant grouping by at least one of the said characteristics; and
- considered as a unit with regard to its suitability for being propagated unchanged.

A variety must be different from similar already known varieties by:

- one characteristic that is important, precise and subject to little fluctuation; or
- several characteristics the combination of which is such as to give it the status of a new variety.

Certain types of variety are distinguished in **laws and regulations**. For example: a '**synthetic variety**' is one produced by open-pollination from several specified elements so that, although not homozygous, it is at genetic equilibrium; a 'composite variety' is the first generation produced by random mating of a large number of specified parents; and the term 'local variety' describes a group of plants originating from a defined region without being the result of explicit breeding work, but that has been shown by official tests to have sufficient uniformity, stability and distinctness to warrant recognition.

A National Variety List is produced by countries that operate variety registration and testing schemes, and is a regularly updated list of plant varieties that are authorized for sale in a particular region or geographical area. Some schemes make listing mandatory before seed is sold; others operate on a recommended basis, to give guidance alone. Listing is based on the outcome of **Distinctness, Uniformity and Stability** (DUS) and **Value for Cultivation and Use** (VCU) trial results.

(PH)

(See: Laws and regulations; Purity testing)

VCU

See: Value for Cultivation and Use

Vermiculite

Vermiculites are used as seed encapsulants, as soil conditioners, in 'soilless' seed sowing and seedling growing media, and as substrates in **hydroponic production** systems. They provide both aeration and drainage, can retain and hold substantial amounts of water and later release it as needed, are sterile and free from diseases, and have a fairly neutral pH.

Technically, vermiculite is the mineralogical name given to hydrated laminar magnesium-aluminium-ironsilicate, which resembles mica in appearance. When its ores are subjected to heat, vermiculite exfoliates or expands, typically by a factor of eight to 12 times, into worm-like pieces (Latin *vermiculare* – to breed worms) by the rapid conversion of contained water to steam. This characteristic of mechanical separation of the layers is the basis for the wide range of industrial uses of the mineral. **(See: Compost; Perlite)** (PH)

Vernalization

A process to cold-treat seeds, bulbs or seedlings of a plant to shorten its vegetative period and thereby induce earlier flowering and fruiting. Vernalization, with a cold spell followed by warmer lengthening days, is a process necessary to induce flowering in biennial crops, and thus is an important consideration in the production of seed of **sugarbeet**. Vernalization is effective in some cereals (e.g. winter barley, rye, wheat). **(See: Wheat – cultivation)**

Very low fluence response

A response triggered by a very small amount of light. The VLFR can trigger germination in imbibed seeds and opening of cotyledons in dark-grown seedlings. It is mediated by phytochrome A but unlike some other **phytochrome** responses, the VLFR is irreversible – both red and far-red light can activate this response. **(See: Light – germination and dormancy)**

Vesicular arbuscular mycorrhiza (VAM)

See: Arbuscular mycorrhiza

Vetch

The true vetches are species of *Vicia*. Some produce edible seeds. The **broad bean** (*V. faba*) strictly is a vetch. Whole pods of *V. villosa* (hairy vetch) are used as green manure. But some leguminous species with edible seeds that are not *Vicia* are also named vetches, e.g *Lathyrus sativus*, *L. cicera* – chickling vetch (**see: Chickling pea**), *L. latifolius* – bitter vetch and *Astragulus diffisus* – milk vetch.

Viability

The potential for a seed or seed population to germinate. Thus a viable seed is one that has sufficient living tissue to be capable of germination in the absence of **dormancy** and if environmental conditions such as water, temperature and light are suitable. Conversely, a non-viable seed will be incapable of germination in such circumstances but will not necessarily be completely dead. Loss of viability may be caused by ageing, especially at relatively high temperatures and humidities, but could also be due to damage sustained during drying if the seeds were not mature or if they were desiccation intolerant (**see: Recalcitrant seeds**). A seed could also lose viability as a result of insect **infestation** if the embryo is consumed. These non-viable seeds may contain living tissue but they will not be capable of germination.

The term 'dead' should only be used to describe seeds that have lost viability. It should not be used to describe seeds that were never viable, such as empty seeds or seeds with poorly developed or aborted embryos. Such seeds are clearly non-viable but since they were not derived from viable seeds they should not be called dead.

The viability of a seed population is usually, and correctly, expressed as a percentage of individuals that germinate when tested (assuming the absence of dormant individuals in the non-germinating fraction) or, a percentage of individuals that stain positively in an alternative viability test such as the **tetrazolium test**. Some organizations, however, restrict the use of 'viability test' to these alternative tests thus drawing a distinction between tests that actually measure germination and those that aim to measure the potential to germinate (**see: Viability equations**). (RJP, SHL)

Viability equations

In the 1960s, it was recognized that in a population of **orthodox** seeds the distribution of individual seed life spans was normally distributed. Thus when a seed population loses **viability**, at a given seed **moisture content** and temperature, the relationship between cumulative percentage germination, as a measure of viability, and storage period is described by a sigmoidal curve. When percentage germination is converted to **probits** these sigmoidal survival curves become linearized and the relationship between probit germination and storage period is described by the basic viability equation:

$$v = K_i - P/\sigma \quad (1)$$

where v is germination percentage (viability) in probits, K_i is the intercept on the y axis, corresponding to initial viability, P is the storage period and σ, the reciprocal of the slope of the transformed seed survival curve, is the standard deviation of the distribution of individual seed life spans.

The slope of transformed seed survival curves, in effect a measure of the rate of loss in viability, is affected by changes in seed moisture content and/or temperature. There is a linear relationship between the logarithm of seed moisture content and the logarithm of the time taken for viability to fall by one probit (σ), a convenient measure of seed **longevity**, described by the equation:

$$\text{Log}\sigma = K - C_W \log m \quad (2)$$

where K is the intercept on the y axis, CW is the slope describing the relative effect of change in moisture content (m) on longevity. Although previously thought to be constant at the species level, the value of CW can vary between ecotypes.

The relative effect of temperature on longevity is described by a quadratic equation:

$$\text{Log}\sigma = K - C_H t - C_Q t^2 \quad (3)$$

where K is the intercept on the y axis and C_H and C_Q are constants describing the relative effect of temperature (t) on longevity. The latter effect has been shown to be more or less constant across a diverse range of species giving rise to the so-called 'universal temperature constants' whose values are: $C_H = 0.0329$; $C_Q = 0.000478$.

Because the effects of seed moisture content and temperature are independent, equations (2) and (3) can be combined:

$$\text{Log}\sigma = K_E - C_W \log m - C_H t - C_Q t^2 \quad (4)$$

and equation (4) can be substituted into equation (1) giving the so-called 'improved viability equation':

$$v = K_i - p / 10^{K_E - C_W \log m - C_H t - C_Q t^2} \quad (5)$$

where K_E is an intercept constant which describes the value of log σ at 0°C and 1% moisture content. The value of K_E, also previously thought to be constant at the species level, has been shown to vary during seed maturation.

The improved viability equation, after the experimental determination of the viability constants K_E and C_W (accepting the universal temperature constants), can be used to predict seed longevity over a wide range of seed moisture and temperatures relevant to conservation of seed and commercial seed storage. Because estimates of seed viability constants have error margins attached, predictions of seed longevity using the viability equations must be treated as a rough guide only.

Viability constants have been determined for over 50 species including domesticated and wild plant species (see the *Seed Information Database*, www.rbgkew.org.uk/data/sid). Major gaps in knowledge still exist, with little or no published data available for several orders of plant families. (RJP, SHL)

Ellis, R.H. and Roberts, E.H. (1980) Improved equations for the prediction of seed longevity. *Annals of Botany* 45, 31–37.

Viability testing

Seed viability is a measure of a seed's capacity to produce a normal seedling and can be measured directly in **germination testing**, or indirectly in the **excised embryo test** or biochemical tests, notably the topographical **tetrazolium test**.

Vicilin

A storage protein (7S globulin) that is present in the seeds of many plants. (**See: Osborne fractions; Storage protein, 2. Globulins**)

Vigour

The vigour of a seed or a population of seeds is, in effect, a measure of the extent of damage that has accumulated as viability declines. When a seed is stored, damage accumulates inside the cells, at a rate mainly determined by the moisture content and temperature of storage, until it loses the ability to germinate and eventually dies. In a population of seeds the lifespans of individuals are normally distributed such that some seeds die quickly, some die slowly, but most die around the mean time to death of the population (**see: Longevity; Viability**).

No seeds in a population are immune from the effects of ageing. Thus after a given period of storage when even the shortest lived individuals in a population are still capable of germination, all seeds will have lost 'vigour' and the tell-tale signs of accumulated damage will be present. Often one of the clearest signs of this damage is that the time taken for individual seeds to germinate will have increased. However, in a germination test of viability, conducted under optimum conditions, it may be difficult to detect this decline in quality. Thus a seed**lot** showing few signs of ageing in a laboratory test under optimum conditions might perform very poorly under more stressful field conditions. Vigour tests are designed to expose these subtle signs of damage so that test results are a more reliable indicator of field performance. Vigour tests usually employ sub-optimal temperatures. (RJP, SHL)
(**See: Vigour testing**)

Vigour testing

Vigour tests are conducted in order to obtain information on seed quality in addition to that provided by the germination test, within a relatively short period of time. Vigour tests are variously used by seed companies and researchers to select seed lots for sowing, evaluate storage potential and the degree of deterioration of carryover seed, to assess the effect of injury, fungicide treatment and other pre- and postharvest adverse factors, as an aid during seed conditioning and at selection steps during plant breeding.

The concept of seed **vigour** arose due to the failure of the standard germination test to predict the differences in field emergence observed amongst highly germinable seed lots, particularly in poor field conditions. This suggested that there is a further physiological aspect to the quality of a seed lot – beyond the simple ability of a proportion of seeds to germinate eventually under the favourable conditions used in standard **germination testing**, and to produce the seedling morphological structures essential for a normal plant to develop. This additional physiological facet of seed quality has come to be referred to as seed vigour. Seed lots having high germination but poor emergence are referred to as 'low vigour' seeds, whereas those also giving good emergence are termed 'high vigour' seeds. Differences in vigour are also reflected in the rate of germination and seedling growth, in both favourable and unfavourable conditions for germination and emergence (**see: Seedbed environment**). Furthermore, a high vigour seedlot has good storage potential and retains high germination during storage, whereas low vigour seed lots show poor storage potential and may show a rapid decline in germination in store. The characteristics of high and low vigour seeds are summarized in Table V.1.

Table V.1. Comparison of the characteristics of high and low vigour seed lots.

	Vigour level	
	High	Low
Mean rate of germination	Fast	Slow
Synchrony of germination	Good	Poor
Mean seedling size	Large, uniform	Small, variable
Emergence potential	Good in most soil conditions	Poor in less than optimum soil conditions
Storage potential	Good	Poor

1. Definition

The definition of seed vigour has been modified many times due to difficulties in agreeing a description that incorporates all its characteristics. However, clear definitions have now been approved by the two organizations that are responsible for the development and standardization of seed testing.

The definition adopted by the **Association of Seed Analysts** (AOSA) states: 'Seed vigor comprises those seed properties that determine the potential for rapid, uniform emergence, and development of normal seedlings under a wide range of field conditions'.

The **International Seed Testing Association** (ISTA) definition, which was accepted into the ISTA Rules in June 2001, is broader in context and specifies some of the characteristics encompassed within the term vigour:

> Seed vigour is a sum of those properties that determine the activity and level of performance of seed lots of acceptable germination in a wide range of environments.
>
> Seed vigour is not a single measurable property, but is a concept describing several characteristics associated with the following aspects of seedlot performance.
>
> - Rate and uniformity of seed germination and seedling growth.
> - Emergence ability of seeds under unfavourable environmental conditions.
> - Performance after storage, particularly the retention of the ability to germinate.
>
> A vigorous seedlot is one that is potentially able to perform well even under environmental conditions that are not optimal for the species.

The major cause of differences in vigour is seed ageing, which involves the accumulation of degenerative changes until the ability to germinate is lost. Differences in vigour therefore arise due to the position of a seedloton the seed survival curve (**see: Vigour testing – ageing**, Fig. V.1a). All seed lots that have high germination are found in the initial phase of the survival curve, where there is only a small gradual decline in germination. A high vigour lot (e.g. lot A) is placed early in this phase, has therefore undergone little ageing and is physiologically young. In contrast a low vigour lot (e.g. lot C) has undergone a period of ageing and is therefore placed further along the initial phase and is physiologically older. (**See: Deterioration kinetics; Deterioration reactions**)

Other sources of vigour differences lie in conditions that arise during seed production, differences between seed **physiological maturity** and **harvest maturity**, and the timing of harvest, and in the genetics of a race or **variety** – notably hybrid vigour (**heterosis**).

2. Test principles and strategies

The essential characteristics of a vigour test have been identified on many occasions. A test should first of all have a good theoretical background, which ensures that its outcome can be explained and unexpected results can be more easily identified and examined further. A vigour test should be simple to complete, and hence applicable in a wide range of seed testing environments, and rapid, to enable prompt reporting of results. The results of the test should correlate with a practical expression of seed vigour, such as emergence in the field or glasshouse, or seed **longevity** in store. The test, preferably, should also provide a quantitative method of assessment to avoid subjective judgements and hence make standardization easier. Furthermore, if the test is to be used in many laboratories, it should be repeatable both within and between them, making results comparable. Finally – and a critical aspect in the running of any seed testing laboratory – the test should be economically practical.

The theoretical background of most official vigour tests lies in seed **deterioration** and ageing, and its consequences in clearly identifiable physiological and biochemical outcomes. A wide range of quantifiable parameters have been proposed in the research literature, but only some of these have attained the status of official seed vigour tests. Vigour tests can be divided into three broad categories that reflect different measurable aspects of the ageing process, namely: physiological tests, biochemical tests, and tests that apply ageing processes to the whole seed, before physiological or biochemical assessment. In addition, vigour tests can also be described as being either 'direct' or 'indirect'.

3. Direct tests

Direct tests measure differences in terms of an aspect of germination or growth, often under stress conditions. One advantage of these types of test is that in effect they evaluate all factors that influence vigour at once, such as ageing, mechanical injuries and morphological factors. Another is that they also often apply environmental conditions similar to those in the field, or measure time courses, and hence appear easier to interpret.

(a) *Physiological tests* measure the impact of ageing on an aspect of germination or early seedling growth. These tests are essentially modifications of the germination test, designed to reveal the reduced rate and uniformity of germination that is found in aged seeds, which, in turn, leads to the production of smaller seedlings, when assessed at a particular time after sowing.

- Most simply, vigour can be assessed as the rate of germination, based on an 'early' (or 'first') count at a set time during the time-course of the standard germination test; this approach is often used. Alternatively, the rate of germination may also be assessed by calculation of the **mean germination time, mean germination value** and **peak germination value**. Seedling growth following germination is also used to assess vigour in the 'seedling growth rate test' and the 'seedling growth test'. (**See: Germination – field emergence models**)

- Stress tests apply, in controlled conditions, one or more non-optimal environmental parameters that may be encountered in the field. Low vigour seeds are revealed by the reduced tolerance to these stresses, measured usually in terms of final germination or seedling emergence. Officially-recognized tests include the cold- and cool-germination tests, and the Hiltner test. Seed researchers and companies use variations on these tests for in-house vigour testing. (**See: Vigour testing – physiological**)

(b) *Whole-seed ageing tests* are based on the manipulation of the rate of seed ageing by holding seeds for a specified period of time at a raised temperature and moisture content, or in an atmosphere of high relative humidity. The seeds therefore age rapidly, so that reduction in germination, normally seen over a period of months or years, occurs within days or even hours, and can be measured in a standard germination test. Ageing tests include the accelerated ageing and controlled deterioration tests. (**See: Vigour testing – ageing**)

4. Indirect tests

These tests focus on one aspect of vigour, rather than evaluating the effect of several or all vigour factors together on the germination or growth process of the intact seed. Because they usually involve more controlled test conditions these tests can be more reproducible than direct tests, are generally less time-consuming and may use less equipment.

'Biochemical tests' are indirect tests that largely measure one of the consequences of ageing. Official tests of this type include measurement of the leakage of solutes from the seed, as in the conductivity test, and enzyme activity, as in the tetrazolium test and GADA test. The research literature contains a range of other proposed 'metabolic markers' that may have use for biochemical vigour testing, and in future **genomics, transcriptomics, proteomics** and **metabolomics** may offer useful measurable vigour parameters. (**See: Vigour testing – biochemical**) (AAP)

Anon. (2003) *Seed Vigor and Vigor Tests*. The Ohio State University, Department of Horticulture and Crop Science. www.ag.ohio-state.edu/~seedsci/

AOSA (2002) In: Spears, J. (ed.) *Seed Vigor Testing Handbook*. AOSA, USA.

Ferguson, J.M. (1993) AOSA perspective of seed vigor testing. *Journal of Seed Technology* 17, 101–104.

Hampton, J.G. (1992) Vigour testing within laboratories of the International Seed Testing Association: a survey. *Seed Science and Technology* 20, 199–203.

ISTA (1995) In: Hampton, J.G. and TeKrony, D.M. (eds) *Handbook of Vigour Test Methods*, 3rd edn. ISTA, Zurich, Switzerland.

ISTA (2001) Rules amendments. *Seed Science and Technology* 29, supplement 2.

Matthews, S. (1985) Physiology of seed ageing. *Outlook on Agriculture* 14, 89–94.

Vigour testing – ageing

Two related 'direct' tests, used officially in vigour testing, exploit in practice the theoretical principles that underlie the seed **viability equations**. These principles hold that: (i) in a seed population stored at a given moisture content and temperature, the relationship between cumulative percentage germination and the duration of storage is a measure of the initial **viability** of the seedlot; and (ii) the relationship can be used to predict both seedlot **longevity** over the range of seed moisture and temperature conditions relevant to commercial seed storage and the existing ability of the seedlot to perform under more stressful field conditions. Even when individual seeds in an ageing population remain capable of germination, they will already have lost 'vigour' so that, for example, the time taken for them to complete germination increases (**see: Deterioration kinetics**). This decline in quality may not be detected by a germination test of viability, conducted under optimum conditions, but will be revealed if the seedlot is artificially aged before the same test is carried out.

1. Accelerated ageing test

This test was initially developed to assess the storage potential of a number of species. However, the results have subsequently been shown to reflect vigour in more general terms and hence reveal both the field emergence and storage potential of a seed lot.

In the accelerated ageing (AA) test, seeds are held at high relative humidity (close to 100%) at raised temperature (41–45°C) for 48–72 h, the exact conditions depending on the species. The seeds absorb moisture from the humid atmosphere, reaching moisture contents ranging from 28 to 45% depending on the species. At this high moisture content and the raised temperature, the seeds age rapidly. After the period of ageing, the seeds are used in a standard germination test. The germination after accelerated ageing is compared with the standard germination of the seedlot to provide an assessment of vigour. Where the AA germination is similar to the standard germination, the seedlot has high vigour. If the AA germination is less than the standard germination, the lot has medium or low vigour. This vigour assessment enables seed lots to be ranked in terms of their emergence and storage potential.

Evaluation of the effects of the variables shown to influence the results of an AA test, in particular the sample size, temperature, seed moisture content and timing of the test has led to the standardization of the AA method for soybean. The AA test for soybean is therefore recommended by the **AOSA** and appears in the **ISTA Rules** as one of the two validated vigour tests. Both the AOSA and ISTA also publish details of suggested AA test conditions for a range of other species. However, there has not been extensive comparative testing to provide data on the repeatability of the test as applied to these species or to confirm the relationship between the test results and an expression of vigour.

2. Controlled deterioration test

The controlled deterioration (CD) test was developed to detect differences in the vigour of seed lots of small-seeded vegetable species. The aim of the CD test is to expose each seedlot to exactly the same amount of deterioration or ageing.

Thus, in this test, the seed **moisture content** of all seed lots to be tested is accurately raised to the same moisture content at the start of the test. This is achieved through imbibition of seeds of known moisture content on moist germination papers and frequent weighing to determine when the required moisture content has been reached. Seeds are then sealed in a foil packet and the moisture allowed to equilibrate throughout the seed overnight at 5–10°C before being placed in a water bath at 45°C for 24 h. After the period of deterioration, seed germination is tested.

The precise period of deterioration during CD effectively moves each seedlot along the seed survival curve by the same amount (Fig. V.1). Thus a physiologically young seed (high vigour) will retain a high germination after CD, whilst a physiologically older seed (low vigour) will show reduced germination. The differences in germination after CD allow seed lots to be ranked in terms of their vigour. CD test results have been shown to relate to emergence both in the field and glasshouse, and to seed storage potential under commercial conditions. (AAP)

ISTA (1995) In: Hampton, J.G. and TeKrony D.M. (eds) *Handbook of Vigour Test Methods*, 3rd edn. ISTA Zurich, Switzerland.

ISTA (2001) Rules amendments. *Seed Science and Technology* 29, supplement 2.

Powell, A.A. (1995) The controlled deterioration test. In: van de Venter, H.A. (ed.) *Seed Vigour Testing Seminar*. ISTA, Zurich, Switzerland, pp. 73–87.

Powell, A.A. and Matthews, S. (1984) Prediction of the storage potential of onion seed under commercial storage conditions. *Seed Science and Technology* 12, 641–647.

TeKrony, D.M. (1993) Accelerated ageing test. *Journal of Seed Technology* 17, 110–120.

Vigour testing – biochemical

An assortment of 'indirect' tests is used officially in vigour testing, based on various measures of cellular or metabolic integrity that change as a consequence of seed **deterioration**.

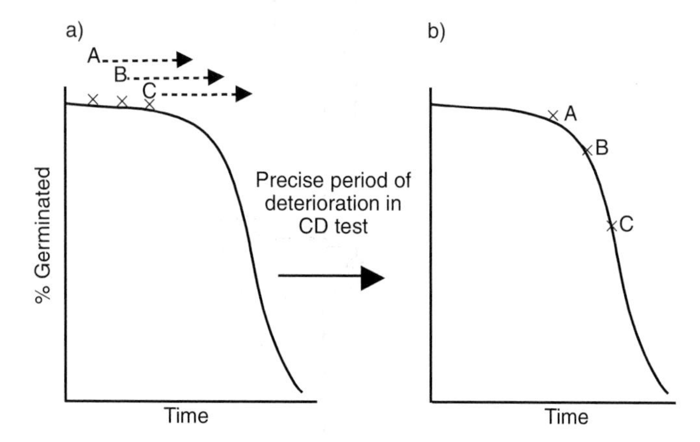

Fig. V.1. Seed survival curve (a) with seed lots A, B and C having similar high percentage of germination, but differing in the extent that they have undergone ageing and hence in their vigour level. Deterioration of these lots at high moisture content and temperature in the CD test reveals clear differences in germination (b) that are indicative of seed vigour.

1. Conductivity test

The conductivity test measures the solute leakage, particularly potassium ions, into water. Hence an increase in the electrical conductivity of seed soak water indicates high solute leakage. The test has been well developed for and is commonly used in the assessment of the quality of seed peas (*Pisum sativum*). It also applies to a wide range of grain legumes (e.g. soybean, *Glycine max*; common or French bean, *Phaseolus vulgaris*; mung bean, *Vigna radiata*; field bean, *Vicia faba*) as well as some other species. However, the methods used for other species have not been standardized and the relationships with field emergence not always consistently proven.

Loss of solutes from seeds reflects the integrity of cell membranes and/or the presence of dead tissues within the seed. Membrane integrity and the incidence of dead tissue are both influenced by the two major factors determining seed vigour in grain legumes, namely seed ageing and the incidence of **imbibition** damage (or soaking injury), and also by the greater sensitivity of aged seed to **membrane damage**. Thus the measurement of conductivity assesses the incidence of the major causes of reduced vigour in grain legumes.

The method for the conductivity test for garden peas is standardized and is reproducible and repeatable. Four replicates of 50 seeds are soaked in 250 ml deionized or distilled water for 24 h at 20°C, after which the electrical conductivity of the seed soak water is measured. Seed lots that show high amounts of solute leakage, i.e. high conductivity, are low vigour lots, high vigour seed lots exhibit low leakage and low conductivity. The correlation between conductivity test results and the field emergence of peas has been proven many times. Preferably, results are expressed in terms of seed weight.

2. Single-seed conductivity measurements

The bulk conductivity test provides an average evaluation of the conductivity of several replicates of bulk samples of seeds. Wide variations between replicates could occur, however, as a result of high leakage from individual seeds that show extensive physical or physiological deterioration. In this case, an evaluation of the leakage from individual seeds may be desirable. Studies with pea, soybean, cotton, field bean, maize and small-seeded crops have demonstrated that analysis of the single-seed leachate conductivity, using the instruments described above, successfully evaluated the seed vigour.

Commercial instruments are available to measure the leachate conductivity from individual seeds. Seeds are soaked singly in each compartment of a 100-cell tray, into which electrodes are lowered; the same basic procedure and precautions are followed as for the bulk conductivity test. Conductivity is measured automatically in µS/cm (S is a unit of measure, Siemens) but it is recommended that each seed be weighed prior to testing so readings also can be recorded per gram individual seed weight.

The measurement of the leachate conductivity of single seeds offers the opportunity to determine the extent of variability in solute leakage within a seed lot. It can identify whether high leakage results from the presence of a few highly damaged seeds or whether there is an overall high amount of leakage within the seed population. In particular, it can be used to determine the incidence of seedcoat damage that is not discernible visibly, but which influences the rate of water uptake and hence imbibition damage, and the incidence of hard, or slowly imbibing seeds, that show little leakage.

Measurement of single seed conductivity has also been proposed as a means of predicting the germination of a seed sample. In this case, any seed having a leachate conductivity above a so-called partition value is said to be non-germinable. However, a common partition value cannot be applied to all seeds within a species due to differences in leakage that arise due to genotype and seed size. In addition, areas of dead tissue within the seed that contribute to high leakage may not be located in areas of the seed that prevent germination.

Hepburn, H.A., Powell, A.A. and Matthews, S. (1984) Problems associated with the routine application of electrical conductivity measurements of individual seeds in the germination testing of peas and soyabeans. *Seed Science and Technology* 12, 403–413.

ISTA (1995) In: Hampton, J.G. and TeKrony D.M. (eds) *Handbook of Vigour Test Methods*, 3rd edn. ISTA, Zurich, Switzerland.

ISTA (2001) Rules amendments. *Seed Science and Technology* 29, supplement 2.

Powell, A.A., Matthews, S. and Oliveira, M. de A. (1984) Seed quality in grain legumes. *Advances in Applied Biology* 10, 217–285.

3. Tetrazolium test

The topographical tetrazolium test (TTZ), commonly used in the viability testing of seed based upon evaluating the presence and location of living tissue within the seed, can also be used to detect vigour differences. The same procedures are used, except that vigour is explicitly assessed by appraising the pattern and intensity of red staining in the context of subsequent seedling development. This form of the test has been applied to detect vigour differences in cereals and, in the USA, for vigour assessment in crops including soybean, cotton, pea and clover. Its limitations lie in the requirements for experienced analysts and the subjective nature of the evaluation, which make routine reproducible application difficult between analysts and laboratories. For details of the test, **see: Tetrazolium test**.

4. Glutamic acid decarboxylase activity (GADA) test

Many of the observations of reduced enzyme activity associated with seed ageing accompany the stage of decline in germination in the population, rather than vigour. However, the activity of glutamic acid decarboxylase has been clearly shown to decrease in maize before germination begins to fall, that is as the vigour of germinable seeds declines. But though GADA was correlated with the storage potential of maize, no relationship has been established between GADA and field emergence. Furthermore, no clear link between GADA and vigour has been found in soybean, beans or peanut. There has been little or no evaluation of the repeatability of GADA as a vigour test. The relatively complex nature of enzyme extraction and assay suggests that GADA is not particularly suited to application as a routine test, but is more likely to be used as a means of evaluating vigour within research programmes. (AAP)

AOSA (2002) In: Spears, J. (ed.) *Seed Vigor Testing Handbook*. Association of Official Seed Analysts, USA.

ISTA (1995) In: Hampton, J.G. and TeKrony, D.M. (eds) *Handbook of Vigour Test Methods*, 3rd edn. ISTA, Zurich, Switzerland.

Vigour testing – physiological

An assortment of so-called 'direct' tests are used in vigour testing, variously based on expressions of the germination population kinetics, measures of seedling growth rate, or on germination performance under mechanical or temperature stress. Some of these tests have official status under ISTA and AOSA Rules.

1. Germination rate and related parameters

Various expressions of rate (or, strictly speaking, their mathematical reciprocals – **see: Germination rate**) are used as measures of seed vigour. One often used index is calculated as the **mean germination time**:

$$\text{Mean germination time (MGT)} = \Sigma f_x \, . \, x \, / \, \Sigma f$$

where f_x = number of newly germinated seeds on day x and x = the number of days since the beginning of the germination test. A lower MGT value indicates more rapid germination.

Peak germination value. A value describing germination rate that is used to indicate seed vigour, particularly of tree seeds.

Peak value = cumulative number of normal seedlings/days of final count.

Mean germination value. An assessment of the seed vigour of some tree species, by measuring the relative rate of germination. A high germination value is said to reflect high seed vigour.

Germination value = peak value × total number of normal seedlings/days of final count.

2. Seedling growth tests

(a) *Seedling dry weight*. This test is applied to soybean and maize. It is based on the standard germination test, but in this case the moisture content of the paper towels is more clearly defined (approximately 30 g water per towel) in order to give more standardized growth conditions. At the end of the test, normal, abnormal and dead seeds are counted and the shoots and roots of normal seedlings are cut free from the rest of the seed tissues (cotyledons or endosperm) and dried at 80°C for 24 h. The mean dry weight per seedling is then calculated. Seedling dry weight, measured in this way at a specific time after the seeds started to imbibe, reflects the time taken to germinate, since the subsequent seedling growth rate (within a given variety) is not appreciably influenced by seed ageing. A high mean dry weight reflects rapid germination and high vigour, whereas a low dry weight indicates slower germination and a lower vigour level. Repeatability of this test can only be achieved where there is accurate temperature control of incubators/germination rooms. Moreover, the dry weight measurements are time-consuming to complete, and growth may be different between cultivars.

(b) *Shoot elongation*. This test has been applied to the cereals, barley, wheat, oats and maize, and to ryegrass and fescue grass. In a modification of the rolled towel germination test, a series of parallel lines is drawn, 1 cm apart, from the centre of the towel upwards, and the seeds are positioned along the length of the centre baseline. After completion of the germination test the numbers of normal and abnormal seedlings, and dead seeds are counted. In addition, the number of shoots from normal seedlings between each pair of parallel lines are counted and a vigour rating calculated as:

$$\text{Vigour rating} = \frac{(n \times 1) + (n \times 2) \ldots + (n \times 15)}{N}$$

where n = number of shoot tips within the pair of parallel lines, of which the uppermost line is 1, 2 … … 15 (cm) from the baseline; N = total number of seeds originally set to germinate. The vigour rating is therefore influenced by the percentage of germinable seeds and, as in the dry weight growth rate test, the time that individual seeds take to germinate: a higher vigour rating is produced from rapidly emerging seeds that have a longer period for growth within the test period.

Problems of repeatability in this test, due to variations in temperature, can be overcome by using a high vigour reference seedlot on each test occasion. Thus the growth of the reference lot should be adjusted to 10 and that of other lots scaled on a proportional basis

i.e. $$L = \frac{Y \times 10}{C}$$

where L = adjusted mean plumule length, Y = value of test sample and C = value of the reference lot.

Image capture systems, linked with image analysis computer algorithms to calculate shoot lengths, have been developed to calculate indexes.

3. Stress tests

(a) *Hiltner test*. The principle of the Hiltner test is to impose a physical impedance stress during germination. Initially developed to detect seedborne infection by *Fusarium* spp., the test was subsequently shown to reflect other defects that prevented normal seedling growth. Seeds are sown within layers of sterile brick grit (crushed stone) or coarse sand with a particle size of 2–3 mm. Seed vigour status is indicated by the degree of emergence of normal seedlings that successfully penetrate these mechanical barriers. The test has been applied in Europe for vigour testing of cereals but is little used elsewhere. Whilst it is reasonably reproducible, the relationship with field emergence has been variable. Furthermore, the test is expensive, time consuming and requires a great deal of space. There may also be difficulties in obtaining supplies of brick grit and in the washing and drying of the grit.

(b) *Cold test*. This is probably the best known of the physiological vigour tests that are based on the application of stress during germination: according to a 2001 ISTA survey for instance, it was the vigour test conducted frequently by ISTA seed laboratories and most frequently by the majority in North America. The conditions of the test simulate the adverse conditions of early spring sowings and the ability of seeds to produce normal seedlings, and its results are commonly thought to express the interaction between genotype, soilborne pathogens, seed quality and seed treatment. A close association between its results and field emergence has been shown in maize and this relationship has also been evaluated for a number of other species. Maize seedlots show differing sensitivity to physiological **chilling injury**, which influence their susceptibility to fungal pathogens, and hence emergence potential, in the field. Vulnerability varies between different genotypes and may be further increased due to seed ageing, or reduced by seed treatments with fungicides.

In the most widely used test for maize in North America and Europe, seeds are set to germinate, often in paper towels or shallow trays, at low temperature (most commonly 10°C, for 7 days – because chilling sensitivity is maximal 2–3 days after the start of **imbibition**), followed by a further period of 4–7 days at 25°C. Both the **AOSA** and **ISTA Rules** suggest covering seeds with a thin layer of a 1:1 sand:field-soil mixture, although the necessity for this is not agreed. The cold test method has not been standardized, and individual laboratories implement their own in-house methods. There are closely corresponding tests for soybean, such as imbibing seed at 10°C for 7 days, often in the presence of non-sterile field soil, followed by a grow-out test of, usually, 4 to 7 days.

(c) *Cool-temperature germination test*. This was developed specifically for cotton, and involves the assessment of normal germination at 18°C in the dark, in rolled towels in which the moisture content is controlled. Though cotton can germinate at temperatures from 12 to 40°C, the optimum is 25–34°C, and the test in effect evaluates the rate of germination. (AAP)

AOSA (2002) In: Spears, J. (ed.) *Seed Vigor Testing Handbook*. Association of Official Seed Analysts, USA.

ISTA (1995) In: Hampton, J.G. and TeKrony, D.M. (eds) *Handbook of Vigour Test Methods*, 3rd edn. ISTA, Zurich, Switzerland.

Nichols, M.A. and Heydecker, W. (1968) Two approaches to the study of germination data. *Proceedings of the International Seed Testing Association* 33, 531–540.

Nijenstein, J.H. and Kruse, M. (2000) The potential for standardisation in cold testing of maize (*Zea mays* L.). *Seed Science and Technology* 28, 837–851.

Sako, Y., McDonald, M.B., Fujimura, K., Evans, A.F. and Bennett, M.A. (2001) A system for automated seed vigor assessment. *Seed Science and Technology* 29, 625–636.

Visible germination

The first indication of **radicle** production from a germinated seed, although in some species another part of the embryo is the first to emerge. (**See: Germination**)

Vitrification

A process by which the viscosity of a liquid is rapidly raised so that the liquid is solidified, not into a crystalline phase but into an amorphous **glass**. It occurs in seeds as they dry.

Vivipary

Also known as viviparous germination. The germination of the developing seed while still on the parent plant: absence of the maturation phase of seed development. This is generally regarded as being different from **preharvest sprouting**, which is the germination of the mature seed on the parent plant when the environment is very humid. The phenomenon is likely to be caused by the absence of ABA action, because of either its abnormally low content or a low sensitivity to it. There are numerous **mutants**, especially of *Arabidopsis* and maize, that show vivipary. In the latter, for example, the viviparous *vp1* mutant is ABA insensitive while the *vp5* mutant is ABA deficient. There is also evidence that endogenous gibberellin promotes vivipary. (See: **Abscisic acid; Maturation – control factors; Precocious germination; Signal transduction – hormones; *Vp1* gene**)

White, C.N., Proebsting, W.M., Hedden, P. and Rivin, C.J. (2000) Gibberellins and seed development in maize I. Evidence that gibberellin/abscisic acid balance governs germination versus maturation pathways. *Plant Physiology* 122, 1081–1088.

VLFR

See: **Very low fluence response**

Voacanga africana

Voacanga africana (Apocynaceae), a native of West Africa, is an evergreen tree that grows up to 6–10 m. Embedded in the pulp of each fruit are numerous dark brown, ellipsoid seeds. These have psychoactive effects by virtue of their indole alkaloid content, including voacamine, voacangine and many related compounds, which can reach 10% of the fresh weight.

In parts of West Africa shamans are said to ingest the seeds for visionary purposes. The alkaloids cause a mild to strong stimulation lasting several hours, certain doses having a strong hallucinogenic effect. The seeds also have medicinal uses, for example as an analgesic and for treating mental disorders (MB)

(See: **Pharmaceuticals and pharmacologically active compounds; Psychoactive seeds**)

Volunteer

A usually undesirable plant derived from seed left from a previous crop, deposited into the soil seed **bank** after **bolting**, **shattering** or spillage during grain harvesting. Volunteer plants may be removed by **tillage**, herbicide or **roguing** practices.

Vp1 gene

Gene symbol, deriving its name from the *VIVIPAROUS* gene. Homologues of *Vp1* have been identified in wild oat, maize and wheat. The gene is **orthologous** with *ABI3* and, hence, encodes for a **transcription** factor that is involved in the suppression of viviparous germination: hence, when the mutant form of the gene is present (*vp1*) **vivipary** is exhibited. (See: **Dormancy genes; Preharvest sprouting – mechanisms; Signal transduction – hormones; Vivipary**)

Walnut

There are about 15 types of walnut (*Juglans* spp., Juglandaceae), all temperate climate nut trees. The principal walnut of commerce is *Juglans regia*, the common, English or Persian walnut (native from the Carpathian mountains of central Europe to eastern Asia) although there is some limited production of the American black walnuts (*J. nigra*), which are primarily gathered from native trees. *J. cinerea* is the white walnut or butternut of North America.

Nearly 1.5 million t of walnuts were harvested in 2003, the leading producers being China (25% of world production), the USA (20%) and Iran (11%). (**See: Crop Atlas Appendix, Map 4**)

The walnut is a drupe-like fruit that has a green, fibrous outer husk enclosing the hard shell of the nut (Colour Plate 5F) – a true nut in the botanical sense (**see: Nut**). The husk dehisces irregularly at maturity. *J. regia* nuts have a hard, thin shell that separates readily from the seed (**kernel**), consisting of an **embryo** with deeply-divided **cotyledons** and a thin **testa**. The nuts are 3–4 cm long and 2.5–3 cm wide: the seeds are slightly smaller (Fig. W.1).

Walnuts are used as snack or dessert food, in confectionery and ice cream. The **oil** (triacylglycerol), about 70% fw, is extracted for use in paints or for culinary purposes. (**See: Nut**, Table N.2) (VP, MB)

Vaughan, J.G. and Geissler, C.A. (1997) *The New Oxford Book of Food Plants*. Oxford University Press, Oxford, UK, pp. 32–33.

Warm-season grain legumes

The warm-season grain legumes are characterized by **epigeal seedling growth**, a period of rapid vegetative growth, followed by flowering when daylengths become progressively shorter during the growing season. The most important species in this group are the **common bean** (*Phaseolus vulgaris* L.) on the American continent, **cowpea** (*Vigna unguiculata*) in Africa and **adzuki bean** (*Vigna radiata*) in Asia. (**See: Legumes**)

Summerfield, R.J. and Roberts, E.H. (eds) (1985) *Grain Legume Crops*. William Collins, London, UK.

Water activity

See: Relative humidity (RH)

Water binding

The attraction of water molecules to surfaces through electrostatic, hydrophilic or van der Waals interactions. Also known as bound water, water of hydration, vicinal water or interfacial water, the intermolecular associations of water and the surface affect the thermodynamic and kinetic properties of

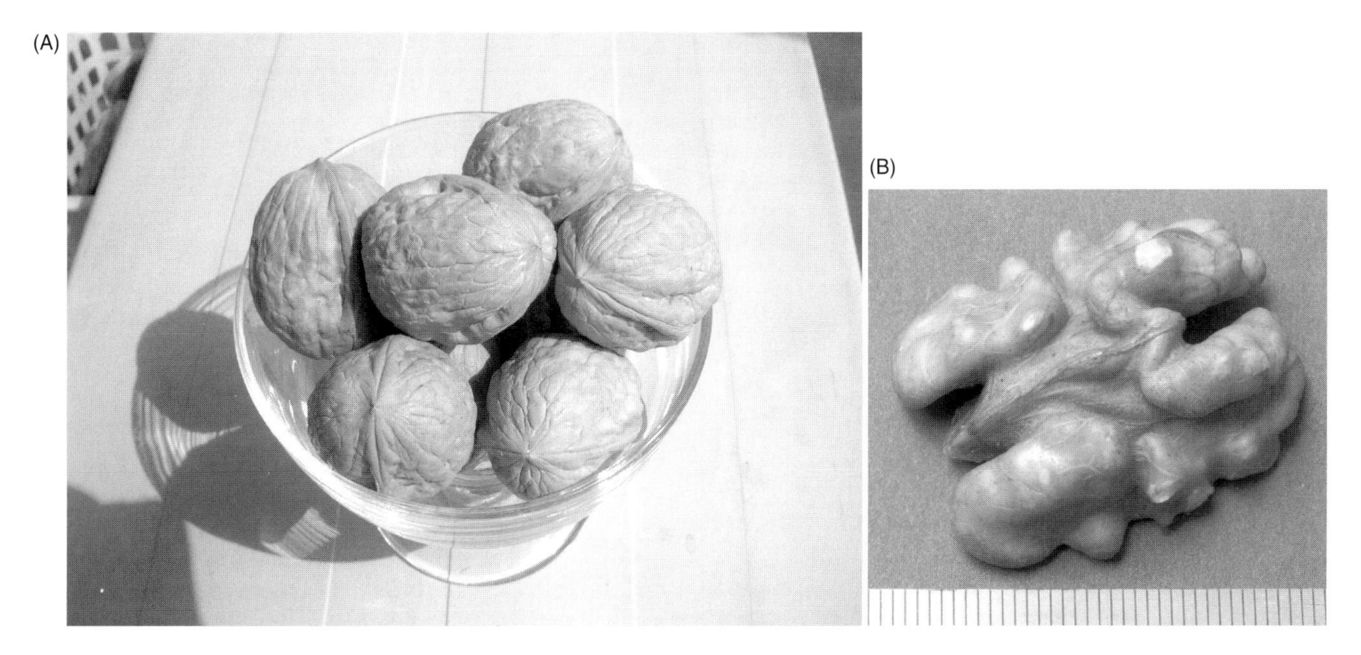

Fig. W.1. Walnut. (A) Intact nuts (image by Mariko Sakamoto). (B) A cotyledon with radicle (image by Mike Amphlett, CABI).

water, distinguishing them from properties of dilute solution water, which is sometimes called bulk or free water. Bound water can also be described functionally by the work required to remove it from macromolecules either through freezing, evaporation, or mechanical expulsion. The properties used to identify water binding are similar to those used to define phase changes, and **phase** behaviour is sometimes used as a conceptual model to explain water binding. Using phase behaviour concepts, the transition of bound water to solution water occurs at a critical water content (analogous to transition temperature), which is a function of the amount of attracting surfaces and the proximity of perturbed water molecules to the surface. Contrary to the name, bound water molecules are not anchored to the surface; they are sufficiently mobile to exchange, albeit more slowly, with other water molecules in liquid and vapour phases. Often, the terms bound water and 'unfreezable' water are used interchangeably, though water with weak surface interactions can freeze and extremely viscous, but not bound, water cannot freeze within a specified time. (It is inadvisable to use the term unfreezable water because it connotes a thermodynamic improbability rather than the more likely mechanism of kinetic restrictions.) 'Structural' water is a special case of bound water required to maintain the conformation of macromolecules. Protein structure is unstable and polar lipids undergo liquid crystalline to gel phase transitions when dried below critical water contents. The nature of surface–water interactions of structural water is still conjectural. (**See: Critical water content; Desiccation damage; Hydration force explanation; Water replacement hypothesis**). (CW)

D'Arcy, R.L. and Watt, I.C. (1970) Analysis of sorption isotherms of non-homogeneous sorbents. *Transactions of the Faraday Society* 66, 1236–1245.

Vertucci, C.W. and Leopold, A.C. (1987) Water binding in legume seeds. *Plant Physiology* 85, 224–231.

Water content

In the context of seeds, this is a measure of the concentration of water within the seed, expressed on either an absolute scale or on a relative scale. Water content (moisture content) can be expressed as a percentage, calculated using the equation:

seed w.c. % = ((**fresh weight** – **dry weight**)/fresh weight) × 100

It can also be related to the dry mass as units water per unit dry mass, e.g. g water/g dry mass.

Gravimetric methods for determining water content mostly indicate only the average content over the whole seeds but not any unequal distribution of water within the seeds. The application of nuclear magnetic resonance spectroscopy (NMR) techniques does, however, reveal patterns of water distribution especially in seeds that are not fully hydrated. This technique has shown, for example, the redistribution of water in immature seeds in response to ethylene, water in non-viable and germinable seeds and localized water in dry, afterripening seeds.

Fountain, D.W., Forde, L.C., Smith, E.E., Owens, K.R., Bailey, D.G. and Callaghan, P.T. (1998) Seed development in *Phaseolus vulgaris* L. cv. Seminole. 3. NMR imaging of embryos during ethylene-induced precocious germination. *Seed Science Research* 8, 357–365.

Manz, B., Müller, K., Kucera, B., Volke, F. and Leubner-Metzger, G. (2005) Water uptake and distribution in germinating tobacco seeds investigated *in vivo* by nuclear magnetic resonance imaging. *Plant Physiology* 138, 1538–1551.

Watermelon

Watermelon (*Citrullus lanatus* syn. *C. vulgaris*) seeds are produced in abundance in the large, nearly spherical fruits that can be 25 cm in diameter. The roughly ovoid seeds are very variable in size according to the variety (approx. 6000–60,000 seeds/kg) but typically are 8–10 mm long, with relatively hard red, brown, black, yellow or white coats. They contain approx. 40% (fresh weight) **storage protein**, 40% **oil** and 7% sucrose. In China, parts of Africa, India and south Asia the kernels (embryos) are eaten, raw or roasted, and may be ground into an edible flour, or used in stews, soups, etc. The extracted oil is edible. (**See: Cucurbits**) (MB)

Water potential

Water potential (ψ_w) is an expression of the thermodynamic state of water. It corresponds to the work required to raise the bound water to the potential level of pure water, and is expressed in terms of energy per unit mass (J/kg, Joules.kg), convertible most frequently into pressure (MPa, MegaPascals) via the relation -1 MPa = -10 bar = 1 J/g. The water potential of a medium results from the sum of four factors: (i) solute (osmotic) potential (ψ_π), which depends on solute concentration; (ii) pressure potential (ψ_p), which corresponds to the physical forces exerted on water by its environment, by the cell wall for example; (iii) matric potential (ψ_m), corresponding to the binding forces between solid matrices and water; and (iv) gravitational potential (ψ_g):

$$\psi_w = \psi_\pi + \psi_p + \psi_m + \psi_g.$$

In the case of seeds, ψ_g can be disregarded. Seed water potential is very low (between -350 and -50 MPa) when seeds are 'dry', almost entirely because of the extremely low matric potential. Pure water has the highest potential, assigned as 0 (zero) value. Water diffusion occurs down an energy gradient from high to low potentials. The initial inward flow of water during seed **imbibition** phase (Phase I of germination) (**see: Germination; Germination – physical factors**) is a consequence of ψ_m. As imbibition advances, the difference in water potential between the moist medium (pure water in Petri dishes or soil solution) and the seed decreases and the rate of water absorption by the seed decreases. After this initial imbibition phase, ψ_m is high (i.e. close to zero), as is ψ_π, and there is almost no water entry into the seed (Phase II). Later on, ψ_m plays a minor role in water absorption by the germinating and germinated seed. The increase in water uptake following **radicle** growth (Phase III) results mainly from the products of **mobilization** of **storage reserves** (proteins, oils and carbohydrates) to solute molecules which are transported to the growing regions of the seedling and help decrease ψ_π of the cells. (**See: Seedbed environment**) (DC, FC)

Bradford, K.J. (1995) Water relations in seed germination. In: Kigel, J. and Galili, G. (eds) *Seed Development and Germination*. Marcel Dekker, New York, USA, pp. 351–396.

Hadas, A. (1982) Seed–soil contact and germination. In: Khan, A.A. (ed.) *The Physiology and Biochemistry of Seed Development, Dormancy and Germination*. Elsevier Biomedical Press, Amsterdam, The Netherlands, pp. 507–527.

Water replacement hypothesis

A proposed explanation for how small molecules such as trehalose and sucrose protect the structure of macromolecules during drying, e.g. of seeds. The process is considered to be important in protecting seeds from **desiccation damage** and **deterioration** (**see: Desiccation tolerance – protection by stabilization of macromolecules**). The premise for this hypothesis is that drying removes structural water from molecules (**see: Phase transition; Water binding**) but that hydrophilic solutes can stabilize molecular structures in the absence of water by providing the necessary hydrophilic interactions or spacing. The substitution of water by solutes prevents molecules from aggregating, thereby reducing the likelihood of phase transitions in **membrane** lipids or precipitation of proteins. The model uses optimal geometrical positioning of solutes on the macromolecular surface to explain the protective capacity of specific solutes. An alternative, though not mutually exclusive, explanation of macromolecular stabilization by hydrophilic solutes is given by the **Hydration force explanation**. (CW)

Clegg, J.S. (1986) The physical properties and metabolic status of *Artemia* cysts at low water contents: The Water Replacement Hypothesis. In: Leopold, A.C. (ed.) *Membranes, Metabolism and Dry Organisms*. Cornell University Press, Ithaca, NY, USA, pp. 169–187.

Crowe, L. (2002) Lessons from nature: the role of sugars in anhydrobiosis. *Comparative Biochemistry and Physiology Part A* 131, 505–513.

Water-seeding

A method of direct broadcast sowing of dry or presoaked rice either manually or by dropping seed into flooded fields from an airplane. Seed is submerged in water for approximately 1 day and allowed to drain for about half a day to initiate the germination process. About a day after the aerial sowing operation, the field is drained to allow for seedling establishment. Water-seeded culture, as an alternative to conventional dryland planting, is especially valuable for the suppression of serious infestations of red rice weeds (which are unable to germinate in the anaerobic conditions) and/or to permit continuous culture of rice in the same field. (**See: Rice – cultivation, 5. Sowing**)

Wattle

Wattle comprises several species of *Acacia*, including *A. victoriae*, *A. murrayana* and *A. pycnantha* (Fabaceae). The seeds have long been a food for Australian indigenous peoples and there is now a steadily increasing demand for it in the popular market as a 'bushfood'. The potential for commercial production in Australia is under consideration. The seeds are nutritious, having about 22% (fresh weight) **storage protein**, about 35% **starch**, trace amounts of sugars, approx. 4% **oil** and a mineral element content typical of **legume** seeds. Seeds are eaten roasted and ground, and incorporated into jams, biscuits, cakes and mustards. (MB)

Maslin, B.R., Thomson, L.A.J., McDonald, M.W and Hamilton-Brown, S. (1998) *Edible Wattle Seeds of Southern Australia*. CSIRO Publishing, Australia.

Wax – storage

The only seed known to contain liquid wax instead of **triacylglycerols** as an energy store is the desert shrub jojoba (*Simmondsia chinensis*). Over 60% of its seed dry weight consists of linear wax esters of ω9 monounsaturated C20, C22 and C24 **fatty acids** and alcohols. Jojoba wax is widely used in cosmetics and medical applications. It has been possible to successfully replace most of the storage triacylglycerol of transgenic oilseed plants with liquid wax. There are three enzymes necessary and sufficient for this seed wax formation, a β-ketoacyl-CoA synthase, a fatty acyl reductase and a wax synthase. Genes encoding these enzymes were cloned from jojoba and expressed in the model oilseed *Arabidopsis*. Large quantities of short chain liquid wax accumulated in the transgenic seeds in place of triacylglycerol, representing up to 70% of the total seed oil. (AJK)

Waxy endosperm

A **mutant** that affects the synthesis of **starch** in cereal **endosperms**. The *wx* mutation results in reduced activity of the starch biosynthetic enzyme, granule-bound starch synthase (GBSS), resulting in severely decreased **amylose** synthesis, and proportionately more **amylopectin** in the **starch granule**. Starch that is high in amylopectin is valuable in the food industry because it improves the texture and freeze-thaw stability of frozen foods; it is also of importance in providing strength to the products of the paper and adhesive industries. (**See: Starch – synthesis**)

Weevil

The term weevil is often used to describe any insect found infesting grain or grain products. Strictly however, grain weevils are beetles belonging to the Curculionidae, which includes species that pollinate flowers, eat ferns and other plants, feed on decomposing vegetation, and destroy crops and foodstuffs. The weevil pests of stored products belong to the genera *Sitophilus*, most importantly *S. oryzae* the rice weevil, *S. zeamais* the maize weevil, and *S. granarius* the granary weevil. (**See: Storage pests**)

Here follows one graphic account of a devastating weevil attack:

> During the height of the plague, when walking through the lanes between the infested stacks, there was a regular hissing noise to be heard, caused by the movement of weevils in bags. At certain times, the *Calandra* [=*Sitophilus*] would swarm out of the bags and leave the wheat, and, in an endeavour to minimise the damage caused by them as much as possible, the insects falling to the ground would pile up 4 to 5 inches high alongside the stacks, were swept up every day and destroyed by burning and scalding. … When the plague was at its worst, up to one ton of weevils per day (about 10^9) were gathered up … and this cleaning and destroying went on for some months.
>
> Extract from Winterbottom, D.C. (1922) *Weevil in Wheat and Storage of Grain in Bags: a Record of Australian Experience During the War Period, 1915–1919*. Adelaide, Australia. (PH)

Weights and measures

Because they were perceived to be of reliably uniform weight and size, seeds were used in many parts of the world as standards to fix weight, volume and length. Several examples are considered here.

1. Weight (mass)

One of the earliest recorded weights is the *shekel* of Middle Eastern countries which is a weight equivalent to 180 barley or wheat grains. This formed the basis of larger units, for example in Babylonia and Sumeria, where the *mina* and *talent* were, respectively, composed of 60 and 3600 *shekels*. It is thought that the pound weight is derived from the *mina*. The *shekel*, because it was used to quantify precious metal in coinage, became associated with a type of currency.

Cereal grains – wheat and barley – were prominent in early metrological systems, especially as standards for weighing the precious metals, gold and silver. An early unit of mass was the 'grain', the smallest unit in the apothecary, troy and avoirdupois systems of Europe. Apothecaries' weights were used in Europe from about 1270 for pharmaceutical ingredients, and this system, based on the grain (now 65 mg) survived in Britain, for example, until at least the middle of the 20th century. The grain unit was also of much wider use. According to some authorities the wheat grain was used to standardize coinage while barley was employed for other materials. Henry III of England stipulated that every penny coin should weigh the equivalent of 32 grains of wheat. But by Elizabethan times the pennyweight was equivalent to the weight of 24 wheat grains and the troy ounce was made up of 20 pennyweights: twelve ounces (i.e. 5760 grains) constituted the troy pound. Troy weights were used for precious metals and stones while the avoirdupois system (one pound = 7000 grains) was of more general application.

Seeds of other species were also used as standards for precious metals and gems. In India, mustard seed was a standard for gold, and in the Middle East and Mediterranean regions, seeds of liquorice and especially **carob** were chosen. The Arabic and Greek words commonly used for carob, *quirrat* and *keration*, respectively, gave us the carat unit for gemstones and the karat for gold. In eastern Asia, seeds of *Abrus precatorius* were used as a standard weight for precious metals and gemstones. These highly attractive seeds, also employed as **jewellery**, are extremely poisonous (**see: Poisonous seeds**).

Calibration systems using seeds were developed. For example, in certain north Indian writings from the second century BC, three small poppy seeds were held to equal in mass one black mustard seed, three of which equal one white mustard seed. Six of these were equivalent to one barley grain, three of which are equal in weight to one *rati* (*Abrus precatorius*) – a standard for gold and jewels. The *dirhem* of the Islamic world (also a silver coin) was defined in mediaeval Egypt as the weight of 60 'average', huskless barley grains, while another Islamic unit, the *miskal*, was the weight of 6000 mustard seeds. In the 13th century AD, in what is now Sri Lanka, a table of weights was described thus: '3 **sesame** seeds = 1 *amu** seed; 3 *amu* seeds = 1 paddy rice seed; 8 paddy rice seeds = 1 *madatiya** seed; 20 *madatiyas* = 1 *kalan* (a unit of weight – ed.)'. And at about the same time, the renowned Italian mathematician Leonardo of Pisa (Fibonacci), in his treatise *Liber Abaci* (on financial mathematics), described the complex system in use at the time: 'Pisan hundredweights ... have in themselves one hundred parts each of which is called a roll, and each roll contains 12 ounces and each of which weighs one half of 39 pennyweights: and each pennyweight contains six carobs, and a carob is four grains of corn (i.e. barley or wheat – ed.)' (from Sigler, L.E. (2002) *Fibonacci's Liber Abaci: A Translation into Modern English of Leonardo Pisano's Book of Calculation*. Springer, New York, USA).

*amu = *Adenanthera pavonina* (Fabaceae), red bead tree or red sandal wood or Circassian seeds; *madatiya* = *Paspalum scrobiculatum* (Poaceae), kodo or ditch millet. Seeds of both of these species were also used as currency.

2. Volume

It is thought that to compare the capacities of different containers these were filled with seeds which were later counted to provide an assessment of relative volumes. Seed number was later used as a standard actually to define volume. In 13th-century England, for example, the gallon was defined as the volume equivalent of eight troy pounds of wheat, i.e. 5760 grains.

3. Length

There are early records that seed size was used to define linear dimension. For example, in China in 1000 BC the grain of **foxtail millet** served as a standard for length, including lengths of strings in various musical instruments. Much later, the barley grain ('barleycorn') was a standard length in Anglo-Saxon England, which was retained after the Norman invasion to become, in 1305, the defined inch of three grains laid end to end, from which the length of the foot of 12 inches was derived. (MB)

Chapman, C.R. (1996) *Weights, Money and Other Measures Used by Our Ancestors*. Genealogical Publishing, Baltimore, MD, USA.

Klein, H.A. (1988) *The Science of Measurement. A Historical Survey*. Constable, New York, USA.

Zupko, R.E. (1977) *British Weights and Measures; a History from Antiquity to the Seventeenth Century*. University of Wisconsin Press, Madison, WI, USA.

Wheat

There are about 16 cultivated species of wheat but the two most important ones are bread wheat (*Triticum aestivum*) and durum (or macaroni) wheat (*T. durum*) of both of which there are thousands of **cultivars**. Wheat is a crop of temperate climate but it is cultivated in the higher lands of the tropics and subtropics. (**See: Wheat – cultivation**)

1. Distribution and economic importance

Worldwide wheat production has now moved into third place behind maize and rice (**see: Cereals,** Table C.8) but with the latter it qualifies as a foremost cereal for human food. Major producers are listed in that table: other significant producers are Argentina, Australia, Canada, France, Germany, Turkey and Ukraine. There is a large international trade in wheat: in 2002 over 121 million t were exported (more than one-fifth of total production) at a value of about US$15 billion (FAOSTAT). Bread wheat dominates world production while durum wheat accounts for 5%. (**See: Crop Atlas Appendix, Maps 1, 5, 21, 23, 25**)

2. Origin

Both wheat (*Triticum*) and **barley** (*Hordeum*) were first domesticated in the Fertile Crescent of the Near East, 10,000–11,000 years ago. The earliest cultivated wheats were two hulled species, **einkorn**, *T. monococcum* (chromosome complement AA, Table W.1) and **emmer**, *T. turgidum* ssp. *dicoccum* (AABB) both with grains tightly enclosed by tough husks (**glumes**). Although both species were to spread to Europe from about 9000 years ago, they are now rare, cultivated in mountainous areas of central and eastern Europe. At about 10,000 years ago, free-threshing or naked forms of wheat evolved, including tetraploid durum or macaroni wheat and hexaploid bread wheat. Bread wheat (*T. aestivum*, AABBDD) formed through the accidental hybridization of emmer and a wild wheat, *T. tauschii*. Of numerous minor species of wheat, the most important is **spelt** *T. aestivum*, var. *spelta* or *T. spelta* (AABBDD), a hulled species closely related to bread wheat, first widely cultivated in central Europe from 4000 years ago. Wild emmer (*Triticum dicoccoides*, AABB) probably arose first, from a relative of wild einkorn (*Triticum urartu*, AA) and wild spelt (*Aegilops speltoides* BB). Through hybridization with a further wild wheat (*Triticum tauschii*, DD), spelt and bread wheat finally arose. Some early relatives of modern wheat are shown in Fig. W.2.
(See: **Domestication**)

3. Grain structure

In bread (or common) wheat and durum wheat the grain is naked and threshed free of the lemma and palea at harvest. Much rarer cultivated wheats such as spelt, emmer and einkorn can be hulled, i.e. they retain **spikelet** structures including **glume, lemma, palea** and **rachis**.

Grain dimensions vary greatly depending on growing conditions, variety and position within the ear but typically lengths range between 3 and 10 mm, maximum diameters between 3 and 5 mm and grain weights between 25 and 65 mg.

Table W.1. Genomic constitution of wheats and some relatives.

Species	Genome constitution	Common name[a]
Diploid (2n = 14)		
Triticum urartu	AA	Two grain einkorn
T. monococcum	AA	**Einkorn** (cultivated)
Aegilops speltoides	BB	Wild **spelt**
T. tauschii	DD	Wild **wheat**
Secale cereale L.	RR	**Rye** (cultivated)
Tetraploid (2n = 28)		
Triticum dicoccoides	AABB	**Emmer** (wild)
T. turgidum ssp. *dicoccum*	AABB	Emmer (cultivated)
T. turgidum (*T. durum*)	AABB	Durum
T. turgidum, ssp. *turanicum*	AABB	**Kamut**
Hexaploid (2n = 42)		
Triticum spelta	AABBDD	Spelt (cultivated)
T. aestivum L.	AABBDD	Common wheat (cultivated)
x *Triticosecale*	AABBRR	**Triticale**

[a]Types in bold lettering have separate entries.

Fig. W.2. Some early relatives of modern wheat (from www.hort.purdue.edu/. ../V3–156.html).

Grain colour ranges from yellow to reddish brown (Colour Plate 9D) but grains are usually classed as red or white.

They exhibit few differences from the generalized structure (**see: Cereals, 4. Basic grain anatomy**). The most striking morphological characteristic is a re-entrant 'crease' parallel to its long axis, and on the ventral side – that is the opposite side to the embryo. Wheat **aleurone layer** tissue is one cell thick, the cells being approximately cubic (Fig. W.3).

4. Composition and nutritional quality

By mass, stored whole wheat typically comprises moisture (9–18%), **starch** (60–68%), **storage protein** (8–15%),

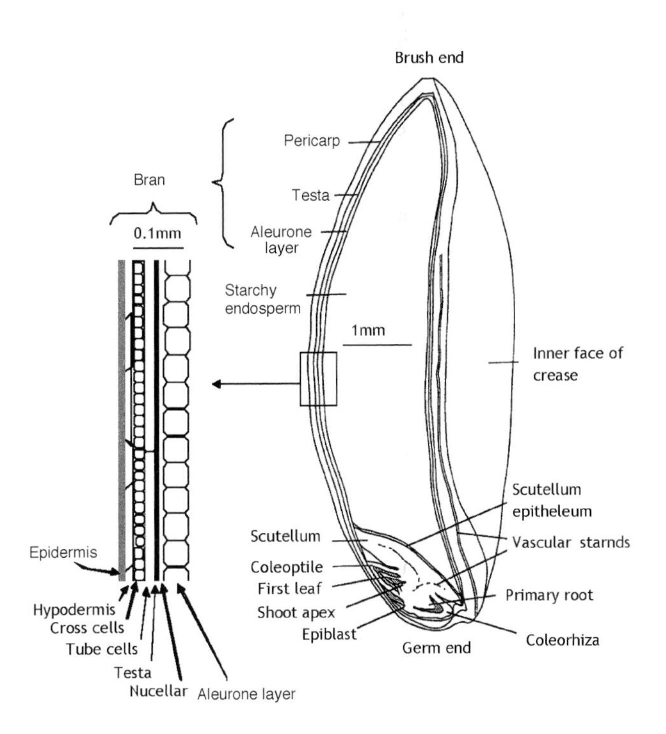

Fig. W.3. Wheat grain structure. The drawing on the right shows a half-grain cut away to reveal the crease, embryo and endosperm. The drawing (left) is a representation of the bran structure (from MJG).

cellulose (2–2.5%), **oil** (1.5–2.0%), sugars (2–3%) and mineral matter (1.5–2%). Internally, the grain is dominated by a starchy **endosperm**, accounting for between 85 and 65% of the grain dry matter, depending on whether the grain is well filled or shrivelled. The endosperm is extracted as white flour when milled and, at typical moisture contents of 13–15.5%, contains the starch (65–70%), but less protein (7–13%), cellulose (trace–0.2%), oil (triacylglycerol) (0.8–1.5%), sugar (1.5–2.0%) and mineral matter (0.3–0.6%), compared to the whole grain.

Two distinct populations of **starch granules** are present, two-thirds of the starch mass being contributed by large lenticular granules between 8 and 30 μm (A granules) and one-third by near-spherical granules of less than 8 μm diameter (B granules). **Amylose** and **amylopectin** content of the starch is typically 20–30% and 70–80%, respectively.

Protein concentration in the endosperm decreases towards the centre as the starch concentration and cell size increases though protein content per cell is relatively constant. The principal storage proteins of the grain are in the endosperm. About 5% of the total seed protein is globulin ('triticins'), but the major storage proteins are prolamins (**see: Osborne fractions**). These comprise about 50 different components that appear to have been derived from a common ancestral protein (**see: Storage protein**). They possess relatively high amounts of proline and glutamine but low contents of lysine. Historically, wheat prolamins have been divided between monomers (gliadins) and polymers (glutenins). The prolamins constitute the **gluten**, which gives bread wheat dough its physical and architectural properties. Gluten has allergenic properties, causing coeliac disease. (**See: Allergens; Storage proteins – intolerance and allergies**)

The endosperm is surrounded for the most part by the aleurone layer, testa (**seedcoat**) and **pericarp** (fruit wall); these are separated from the endosperm as bran in milling which accounts for about 15% of the dry matter, 20% of the protein, the majority of the fibre, and several important minerals and vitamins. The remaining dry matter, which includes B vitamins and protein, are accounted for by the **embryo** and **scutellum** that are also removed during milling. (**See: Cereals – composition and nutritional quality; Storage protein – processing for food**)

5. Processing and uses

Some wheat is destined for animal feed but most is used for human food, for which it is first processed by milling. Wheat beers (of German origin) are brewed from 40 to 75% raw or malted wheat plus the malted barley. They are usually unfiltered and keg- or bottle-fermented for a hefeweizen (hazy wheat), but it is not uncommon to encounter kristallklar (crystal clear) or filtered wheat beers as well. Unmalted wheat is also added as an adjunct to barley malt in the brewing of beers to enhance head retention and foam stability (**see: Malting**).

(a) **Crushing**, *stone grinding and hammer milling*. These and several domestic processes carried out on small batches convert, in one stage, whole grains into powders (Fig. W.4). The powders may comprise particles with a relatively wide range of sizes because the different tissues present in the grain respond differently to the forces supplied. Starchy endosperm is relatively friable while pericarp and seedcoats become flattened into flakes. This provides the opportunity for separating these tissues by sieving and this may or may not be carried out according to the intended end-use. Sieving is integral in more sophisticated multi-stage processes such as flour milling of bread wheat in which sieving alternates with passages of intermediate products (collectively known as 'stocks') through pairs of rollers. In this treatment, grains are first passed through a gap between pairs of ridged, fluted rollers, each of which is independently driven, allowing for one to rotate more quickly than the other (the 'differential'). This first passage (the first break) opens the grain, releasing some endosperm as large chunks and flattening the bran.

Stocks produced during flour milling are defined by their particle size but, because the grain components behave differently during grinding, the composition is consistent within a size fraction and different from other fractions. Having passed through the first break and the subsequent sieving stage, stocks are streamed for further treatment according to particle size and composition. The coarsest fraction, consisting mainly of outer coats (pericarp and seedcoats, collectively called bran) with some endosperm remaining attached, is passed through rollers with finer fluting to separate the two components by scraping the endosperm from the bran. After sieving, further scrapings achieve the best possible separation.

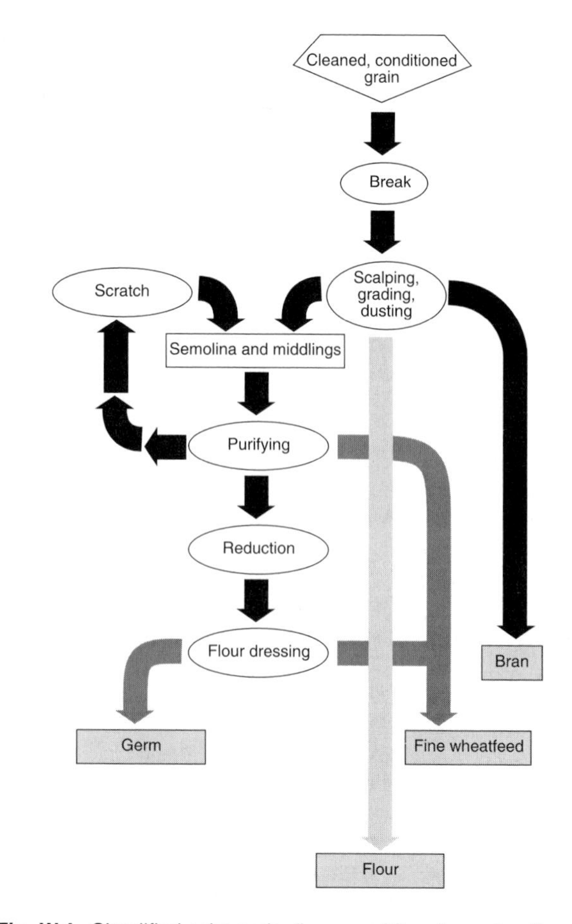

Fig. W.4. Simplified schematic diagram of flow through a flour mill to produce various fractions used in the food and feed industry.

Chunks of endosperm that constitute the finer fraction are subjected to further crushing between a series of pairs of smooth or frosted rollers (the reduction system) to reduce their particle size to that of flour. Remaining fragments of bran are removed by sieving and by 'purifying', a process that separates particles on the basis of specific gravity and terminal velocity of particles in an upward air current. At each grinding stage some flour (particles passing through the finest sieve) is produced. Flour streams from all passages are combined to give a 'straight-run' flour. The composition of all flour streams is mainly starchy endosperm but the quantity of minor inclusions varies so specialist products can be produced by combining selected streams. Other products of wheat milling are particles whose main constituent is bran. These are generally combined as 'wheatfeed', which is used as a component of livestock feeds. It is also possible to concentrate fragments of the embryonic axis as 'wheat germ'.

The debranning or pearling technique used in rice milling has been introduced as a pretreatment to wheat grains. Because the crease is deeper in a wheat grain than in a rice grain, some bran is inaccessible to surface abrasion and has to be separated by the normal wheat milling processes that follow debranning. At the present time it is applied only in a small proportion of flour mills.

Bread wheat flour is used for making a variety of different breads, cakes, biscuits and pastry. The selection of the type of flour depends partially on its protein content, which reflects the protein content of the original endosperm. Flours containing relatively high amounts of protein are described as 'strong', and are used for bread. 'Weak' flours with less protein are used for biscuits, cakes, etc. Durum (macaroni) wheat is mainly but not exclusively used for pasta and couscous. Milling of this wheat, which has a very hard endosperm (see section 6), is similar to the above but efforts are made to reduce production of particles of flour fineness. Grinding pressures are lower than those used in conventional flour milling and greater use is made of purifiers to clean up the semolina stocks. As in flour production, semolina produced at all machine passages is blended to produce the raw material for pasta production. In recent years, techniques for producing pasta from flour have led to the use of conventional flour milling techniques for treating durum wheat.

6. Major varietal types

Seed characteristics and components are used to classify wheat varieties for trade, use and breeding. These characteristics include seedcoat colour (red or white), endosperm texture (hard or soft), and properties that influence dough rheology (weak to strong). The intensity of red pigmentation (phlobaphene) in the seedcoat is principally determined by three functionally equivalent genes at homeologous loci (R/r) on the long arms of chromosomes 3A, 3B and 3D. Red varieties carry one or more of the red (dominant) alleles and intensity of pigmentation increases as 'gene dosage' increases to three. Red varieties exhibit more dormancy than white varieties so are favoured in climates conducive to **preharvest sprouting**. White wheats are more suited to areas that are dry during ripening and harvest, and are favoured for the manufacture of certain types of flat bread and noodles.

The hardness of a variety or seed relates to the resistance encountered when it is milled, i.e. the harder the wheat the greater the force required (see section 5). When milled, hard wheats break down along the outlines of endosperm cell walls, or through the cells across the starch grains and proteins alike, to yield coarse, smooth-sided granules that flow easily over surfaces and through sieves. Conversely, milling soft wheat produces a mass of fine, angular shaped particles of cell debris. Hardness is inherited comparatively simply and is mostly under the control of a major gene encoded on the short arm of chromosome 5D. Friabilin is a surface protein complex occurring on the surface of starch granules, abundant in soft wheats, scarce in hard wheats and absent from durum wheat. Two of the main components of friabilin are the isoform polypeptides puroindoline-a and puroindoline-b. The presence of both isoforms appears to be necessary for the expression of softness, and mutation in either can be associated with hardness. Hard wheats are favoured for breadmaking, not only because they are easier to mill, but because the higher grinding pressures that can be tolerated causes more damage to starch granules which, in turn, leads to a greater potential water adsorption. Soft wheats are suitable for products where dryness is a favoured characteristic, i.e. biscuits (cookies).

Leavened bread production is largely limited to the genomes coding for the necessary proteins to generate an elastic, strong dough suitable for the capture of gas in bubbles during fermentation, thus allowing the dough to rise. Precise dough rheological properties are the result of interactions between all major constituents of flour (i.e. starch, lipid and protein) and water. However, the unique elastic properties of doughs made from wheat flours are largely due to the type and amount of **gluten** present, containing the prolamins. Varieties with a high gliadin:glutenin ratio tend to have viscous, extensible doughs often suitable for biscuit making. Those having a low gliadin:glutenin ratio have better elasticity and strength, more suitable for breadmaking. Allelic variation in the high molecular weight (HMW) subunits is closely linked with variation in breadmaking quality and dough resistance. These subunits are composed of 580–730 amino acid residues and are controlled by genes at loci located on the long arms of chromosomes 1A, 1B and 1D notated as Glu-A1, Glu-B1 and Glu-D1. Varieties of bread wheat contain between three and five major HMW glutenin subunits. Two of these are coded by genes at Glu-D1, one or two by Glu-B1, and one or none by Glu-A1. Over 50% of the variation in baking potential and/or dough rheology within wheat collections of several countries has been explained on the basis of HMW-glutenin subunit composition. (**See: Storage protein; Storage protein – processing for food**) (AE, MN, MJG)

For information on the particular contributions of the seed to seed science **see: Research seed species – contributions to seed science.**

Gooding, M.J. and Davies, W.P. (1997) *Wheat Production and Utilization*. CAB International, Wallingford, UK.

Jones, G. (2001) *The Millers*. Carnegie Publishing, Lancaster, UK.

Kent, N.L. and Evers, A.D. (1994) *Technology of Cereals*, 4th edn. Pergamon Press, Oxford, UK.

Pomeranz, Y. (ed.) (1998) *Wheat Chemistry and Technology*. American Association of Cereal Chemists, St Paul, MN, USA.

Posner, E.S. and Hibbs, A.N. *Wheat Flour Milling*. American Association of Cereal Chemists, St Paul, MN, USA.

Wheat - cultivation

Wheat (*Triticum* spp.) is a cool season grass belonging to the family Poaceae and tribe Triticeae. The modern cultivated wheats are allopolyploids; the two most important species are bread wheat (*Triticum aestivum*, hexaploid, 2n = 42) and durum (or macaroni) wheat (*T. durum*, tetraploid, 2n = 28). Other species, largely of historical or minor commercial interest are *T. monococcum* (**einkorn**), *T. dicoccoides* and *T. turgidum* (wild and cultivated **emmer**), *T. spelta* (**spelt**) and *T. turgidum*, ssp. *turanicum*. For the geographic origins, evolution, history, world distribution and economic importance **see: Wheat**.

1. Genetics and breeding

Wheat is generally self-pollinated but some cross-pollination is possible. Hybrids (F_1) have been commercially developed in certain countries since the mid-1970s, either by exploiting **cytoplasmic male sterility** or nuclear male sterility, or by spraying chemical hybridizing agents (CHAs). However, hybrid varieties do not occupy more than a very small percentage of any national wheat area, which compares poorly with other cereals, most notably the hybrids of **maize**, **rice** and **rye**. Hybrid vigour (**heterosis**) for yield in wheat is often too small to confer sufficient advantage over the highest yielding elite inbred lines that would justify the extra costs and difficulties of hybrid production. There are several reasons for this: wheat is a poor pollinator for cross-fertilization and seed set can be heavily dependent on weather conditions and cultivar; candidate cultivars with good potential for cross-fertilization are severely limited in number; and combining parents for greatest heterotic effect on yield often compromises end-use quality. In addition, heterosis is thought to be relatively low in wheat because much of the effect derives from dominance, that is, by the masking of unfavourable genes rather than by the presence of more than one allele at a locus.

As wheat is an allopolyploid it should exhibit some introgenomic heterosis, that is, it could be regarded as a fixed heterozygote. It is, for example, possible to develop inbred lines with equal or superior yield to the F_1 hybrids from which they were selected. The heterotic effect of some hybrids may, therefore, only represent a temporary improvement over the best available inbred lines, and this will be further compromised if the choice of parents is constrained by considerations of cross-fertilizing potential.

Despite these difficulties, some evidence of over-dominance and/or favoured intermediate phenotypes that cannot be achieved in inbred lines has been demonstrated for some wheat hybrids. Additionally, the development of improved CHAs in the 1990s has led to some local successes with, for example, hybrid wheats in France contributing an estimated 5% of total wheat production in 1998.

The genetic transformation of wheat has lagged behind that of other important crops. Particle bombardment and *Agrobacterium*-mediated **transformation** systems have contributed to the development of glyphosate herbicide-tolerant varieties, whose approval is currently being sought in a number of countries. Experimentally, transformed wheats have been produced with modified **storage proteins** and increased disease resistance.

2. Development

The rate of development and maturation of a wheat genotype largely depends on temperature, vernalization and photoperiod; cultivars vary in their response to these factors, the extent to which they interact, and the relative sensitivity to them at different growth stages. The development of winter wheats is hastened following **vernalization** (typically 3 to 10°C, for 6 to 8 weeks), and also by exposure to long days, although short days can sometimes substitute for vernalization. This variation contributes to the wide adaptation and distribution of wheat in world agriculture. Several genes or loci have been mapped in wheat that are associated with life-cycle duration including those for vernalization, photoperiod and development rate or 'earliness'.

Wheat development is **determinate**. The **spike**, or ear, has a main axis (**rachis**) consisting of short internodes bearing sessile spikelets alternately on opposite sides, culminating in a terminal spikelet positioned in a plane at a right angle to the other spikelets. Each spikelet is subtended by two basal bracts (**glumes**) and has a central axis (rachilla) bearing two rows of alternate **florets**, totalling between two and nine per spikelet (though not all florets develop green anthers, reach anthesis and set grains). Each perfect floret has a pistil, three stamens and two lodicules, all enclosed in two flowering glumes (the inner **palea** and the outer **lemma**). Ears, spikelets and mature grains are shown in Figs W.5 and W.6.

Anthesis (pollen release) commences typically between 3 and 8 days after ear emergence, depending on temperature and cultivar. Flowering starts in the basal florets of the central spikelets (halfway up the ear) and proceeds basipetally and acropetally, and acropetally within the spikelet. Though flowering within an individual spike is usually complete within 2 to 3 days, the process over the whole plant and crop may extend over 5 to 10 days, due to variations in tiller maturity. The optimum temperature for fertilization is 18–24°C. Sterility can be caused by various factors: by heat and cold stress during ear emergence; anthesis nutrient deficiency, particularly of boron; droughts; waterlogging, and extremes of humidity and alkalinity. For most of these factors there are strong interactions with cultivar.

The stages through which the grain passes, as the embryo, endosperm and other tissues develop after fertilization, are conventionally named according to colour and texture, as the reserve precursors accumulate and the final reserve composition is established – the so-called watery, milk, soft dough, hard dough, kernel hard and harvest ripe stages. For a detailed and illustrated account, **see: Development of embryos – cereals; Development of endosperm – cereals;** and www.wheatbp.net.

3. Production, drying and storage

Crops for wheat seed are produced in broadly similar ways to those produced for human and livestock grain consumption, but with special attention to weed and disease control and seed quality. Seed rates are similar except for early multiplicative generations when seed is scarce. Production fields should have been free of wheat for at least 2 years, and of other small-grained cereals for at least 1 year.

Wheat is self-fertilizing and, therefore, only needs to be isolated from other wheat crops by a hedge, ditch and/or a

Fig. W.5. A near-mature awn-less wheat ear. Ears are held vertically within the canopy during dry-down. Individual spikelets dry at different rates; the last to dry down will be at the base of the ear (these retain a slight green colour). From www.wheatbp.net.

2–4 m non-cropped strip, to prevent physical mixing at harvest. Reduced late nitrogen application avoids lodging and its accompanying grain shrivelling, green tillers, **preharvest sprouting** and harvesting difficulties. In temperate, moist and well-fertilized conditions, such as in western Europe, fungicides are applied to delay the death of the canopy and help maintain grain filling and increase mean grain weight. Hand weeding may be worthwhile if the weed species is easy to locate by height and specified in certification standards – wild oats in the UK, for example, at low to moderate populations (one plant in every 20–30 m^2).

Seed crops are usually combine-harvested, though in smaller-scale production they may be **swathed** (cut) and later transported to a stationary thresher. Preharvest rainfall can, for example stimulate grain sprouting, encourage grain diseases, and increase the need for intensive drying to achieve reliable storage. Harvesting of seed crops can occur over a wide range of moisture contents (8–20% for combining; 30% for swathing) depending on region, but threshing at extremes increases the risk of bruising and cracking that can reduce viability. **Moisture contents** of 13 to 15% are often ideal for combine-harvesting, and are also usually adequate for short-term storage; seed at 14% moisture can be safely stored at

Fig. W.6. An individual spikelet 40 days after pollination. Floret 2 has been cut away and its component parts are separated. The grain has started to shrink and the colour has changed to a light brown. The glumes, lemmas and palea, which have protected the grain, are now dry and brittle. From www.wheatbp.net.

15°C for up to 35 weeks, for example, with respect to grain viability and the development of fungal growth. (**See: Cereals – storage; Storage fungi**)

Grain harvested at too high moisture contents can be dried using forced air at near-ambient temperatures in the bulk store, paying careful attention to avoid mould formation, because several days are often required for the process. Alternatively, high temperature (about 45–55°C, or more) continuous-flow, batch- or storage-dryers, are used to dry grain sufficiently quickly to avoid significant deterioration by microorganisms, though viability can be reduced at excessive temperatures and drying durations, or with high moisture-content grain. Dried grain must be cooled, usually by ventilating with ambient air.

Grain precleaning starts while combine harvesting, as loose chaff is blown away. The most sophisticated techniques are reserved for seed crops and a wide variety of **conditioning** machines may be used, including indented cylinder separators. Roll mill separators catch the awns of wild oat on the rotating velvet-like surface. Specific gravity **separators** classify grain particles of differing density, and remove impurities such as ergot from seed.

4. Quality, vigour and dormancy

Minimum standards of purity and germinability vary between and within countries and in different generations and grades of certified seed. Typically, total purity needs to be above 98% and there are often specific maximum standards of weeds (such as wild oats, couch and certain bromes) and other wheat varieties and cereals. Similarly, depending on region and the availability of seed treatment, there will be specific standards for seedborne **pathogens** such as **ergot** (*Claviceps purpurea*), loose **smut** (*Ustilago nuda*), **bunts** (*Tilletia* spp.) and *Fusarium* spp. Minimum viability, as assessed in standardized germination tests, is typically either 85 or 90%. Large grain size is often desired because of its association with vigour (i.e. rapid, successful and uniform emergence from depth), and to limit losses when removing weed seeds and other impurities during cleaning. Seed vigour can be influenced by pre- and postharvest conditions such as weathering (e.g. wet/dry rainfall cycles, temperature) mechanical (threshing and handling), and various other types of damage for example by seed drying. Both pre- and postharvest dormancy can vary according to environmental factors operating during grain development. Relatively low temperatures tend to favour the onset of dormancy while higher temperatures, especially close to maturation, reduce it, partially by encouraging **afterripening** (**see: Dormancy breaking**). The dormancy status of grain partially determines its propensity to undergo preharvest sprouting, which severely affects grain quality especially for milling and baking purposes. For a discussion of the genetics of wheat dormancy, **see: Preharvest sprouting – genetics, 2. Wheat**.

5. Sowing

Wheat can be established in soil prepared in diverse ways ranging from ploughing cultivation to direct drilling (zero tillage). Sowing depths commonly range between 25 and 100 mm, with the optimum for temperate and moist conditions about 30 mm, but deeper planting in drier conditions or where predation pressure is high. Drill sowing gives greater control of depth and row spacing, but **broadcast seeding** followed by light harrowing is quicker and allows sowing on wetter soils or on rough terrain.

For wheat in moist and fertile conditions, when there is good control of disease and minimal lodging, the responses of both dry matter and grain yield per unit area to plant population and seed rate is approximately asymptotic. Harvest index is not greatly affected by seed rate in such conditions. The crop compensates for low plant population mostly by: increasing the production and survival of tillers to form ears; less consistently by increasing the number of grains per ear; and more rarely still, by increasing mean grain weight. Optimal plant populations can, therefore, be lower when tillering is encouraged, for example by sowing winter wheat earlier in the autumn, and greater when the time for compensatory growth is reduced, by sowing later or in the spring.

Sowing rate to achieve a plant population relies greatly on the quality of the seed and the seedbed. After **probit** transformation, percentage seedling field emergence and percentage germination in standard laboratory viability tests are linearly related to each other; for untransformed data therefore, the difference between the two increases rapidly both as laboratory germination decreases, and as field conditions deteriorate. For example, a model derived from experimental data predicts field emergence values in sandy-loam soil under poor (5–6°C, 15.5% soil moisture content) and good (10–11°C, 12% soil moisture content) conditions, as follows: where seed viability is 98%, emergence is 67 and 93%, respectively; and where the viability is 85% emergence is 29 and 67%. (**See: Germination – field emergence models**) Seed size is often positively correlated with early seedling development and biomass. In good growing conditions the effects of such differences are comparatively small and disappear as development progresses. Rapid establishment and early season vigour, however, are useful traits for improving yields in more marginal and drought-stressed environments. In average UK conditions, for instance, with certified seed the optimum sowing rate is often around 250 seeds/m^2 or 130–150 kg/ha for seed of mean weight of 50–60 mg; farmers may sow more than this to reduce the risk of poor establishment in adverse circumstances. Optimum seed rates decline in less fertile conditions, particularly where nitrogen and moisture availability is low, and for many wheat-producing areas of the world where production is less intensive, sowing rates are usually less than 100 kg/ha.

The sowing date is largely governed by climate and the requirements of the crop rotation. In general, the higher the latitude, the cooler the summer temperatures, the longer the potential cropping period, and the earlier the drilling of autumn sown crops. In hotter climates cropping is restricted to the cooler winter months. In extremes, such as the Sudan, sowing date is critical to make sure that the reproductive phase coincides with the coldest part of the season, and the total growing season is restricted to just 90–100 days. Similarly, in Southeast Asia, sowing is often rotated with rice so that wheat can be planted in the cooler, drier season when land would otherwise be fallow.

6. Treatments

Wheat seed is commonly treated prior to sowing with seed treatment fungicide(s) and/or insecticide. Cheaper, protectant fungicides are commonly used to 'disinfect' the surface of the seed against peripheral pathogens such as **bunt** (*Tilletia tritici*) and *Fusarium* spp. In addition to controlling these **pathogens**, fungicides show varying degrees of systemicity (from the carboxamide, triazole, cyanopyrole and guanidine groups), and can also be used to control the seedborne pathogens that cause loose **smut** (*Ustilago nuda*) and/or protect the young seedling from foliar pathogens such as *Septoria* spp. and powdery **mildew** (*Erysiphe graminis*). Latterly, the development of a benzamide fungicide (silthofam) has allowed useful control of take-all disease (*Gaeumannomyces graminis* var. *tritici*) by application as a seed treatment. Neonicotinoid insecticides (such as imidacloprid) can reduce damage from wireworms and virus transmission by aphids. (**See: Treatments – pesticides**) **(MJG)**

Evans, L.T., Wardlaw, I.F. and Fischer, R.A. (1975) Wheat. In: Evans, L.T. (ed.) *Crop Physiology*. Cambridge University Press, Cambridge, UK, pp. 101–150.

Gooding, M.J. and Davies, W.P. (1997) *Wheat Production and Utilization: Systems Quality and the Environment*. CAB International, Wallingford, UK.

Hu, T., Metz, S., Chay, C., Zhou, H.P., Biest, N., Chen, G., Cheng, M., Feng, X., Radionenko, M., Lu, F. and Fry, J. (2003) *Agrobacterium*-mediated large-scale transformation of wheat (*Triticum aestivum* L.) using glyphosate selection. *Plant Cell Reports* 21, 1010–1019.

Khah, E.M., Ellis, R.H. and Roberts, E.H. (1986) Effects of laboratory germination, soil temperature and moisture content on the emergence of spring wheat. *Journal of Agricultural Science* 107, 431–438.

Pinthus, M.J. (1985) Tritici. In: Halvey, A.H. (ed.) *CRC Handbook of Flowering*. Vol. IV. CRC Press, Boca Raton, Florida, USA, pp. 418–443.

Snape, J.W., Butterworth, K., Whitechurch, E. and Worland, A.J. (2001) Waiting for fine times: genetics of flowering time in wheat. *Euphytica* 119, 185–190.

Wild oat

The seeds of wild oat (*Avena fatua*) show degrees of **dormancy**, depending on the strain, in some cases a deep **embryo dormancy**. Because of the dormancy, wild oat is a pernicious weed among cereals, especially wheat (**see: Dormancy – importance in agriculture**). Seeds of this species have been much used in studies of the genetics of dormancy (**see: Dormancy – genetics**). They have also featured in studies of the cell biology of gibberellin action on the aleurone layer (**see: Mobilization of reserves – cereals; Signal transduction – hormones**).
(**See: Dormoat**)

Wild rice

Wild rice (*Zizania palustris*) is an aquatic grass, native to North America that grows predominantly in the Great Lakes area. It has been consumed by indigenous populations since prehistoric times, and was probably introduced to Europeans via the fur traders. Native Ojibway, Menomini and Cree tribes called it *Manomio*, but European settlers called it wild rice or Indian rice because of its resemblance to the oriental rice (*Oryza sativa*); both are members of the Poaceae but in different genera. Other names include: water oats (the French settlers saw a resemblance of the inflorescences to those of oats), Canadian rice, squaw rice, Indian wild rice, blackbird oats and marsh oats.

Wild rice in its natural habitat is harvested by hand, using canoes into which the grains are shaken. Its tendency to **shatter** means that harvesting is a hit and miss process, and losses are large. Efforts to grow wild rice as a field crop began in the 1950s, and breeding to produce shatter-resistant lines has been successful. Commercial production occurs in Canada, some northern states of the USA, and particularly in California, the largest growing area. Commercial seeding and mechanical harvesting have increased the number of hectares under cultivation. Annual productivity is very variable, due to fluctuations in weather and water levels.

Kernels of the mature grain are approx. 1–2 cm long and 3–4 mm wide, and the grains are dark grey to black (Colour Plate 9E). Nutritionally, wild rice is higher in **storage protein** (12–14%) and vitamin B than white rice, has about the same carbohydrate (mostly **starch**) content, but is much lower in **oil**. Its protein also has a higher amount of lysine and methionine than most cereals, and is similar to oats in amino acid balance. (JDB)

Oelke, E.A. (1993) Wild rice: domestication of a native North American genus. In: Janick, J. and Simon, J.E. (eds) *New Crops*. Wiley, New York, USA, pp. 235–243.

Wilting point

The approximate soil water content at which a plant is unable to extract sufficient water from the soil to meet its needs. In other words, water is being held by the soil particles with greater tension than the plant can overcome. (**See: Seedbed environment**)

Wilts

Characteristic symptom and description of a range of **diseases** caused mainly by infections of the fungi *Fusarium oxysporum* and *Verticillium* spp., which may attack plants from the seedling stage onwards, developing conidia that move passively upward in the xylem vessels. Plants respond by forming plugs (tyloses) in the plant vascular tissues, which block the movement of water. Most species are persistent soilborne **pathogens** favoured by high soil temperatures, but in some circumstances they may also be seedborne, carried either on the seedcoat or more deep-seatedly. Bacterial wilts, such as seedborne *Clavibacter michiganensis* (bacterial canker of tomato), similarly cause lysigenous cavities in xylem vessels, which fill with viscous granular deposits and bacterial masses. In maize, *Erwinia stewartii* which can be seedborne or transmitted by flea beetles, causes seedling wilt (and leaf blight, with early similar symptoms when infected after tassels emerge). (PH)

Windrow

A row of cut or pulled plant material (synonymous with **swath**), which may include seed or grain, left loosely on the ground during harvesting for post-maturation drying 'by the wind' before threshing. (**See: Production for sowing, IV. Harvesting, 3. Harvest methods**)

Wing

Structure assisting wind dispersal (**anemochory**), which can be formed by the **ovary** wall (e.g. samara, samarium), floral elements (pseudosamara) or the seedcoat itself (so-called alate seeds). (**See: Seedcoats – dispersal aids**)

Winged bean

Also known as the Goa bean, manila bean, four-angled bean and sometimes the asparagus pea or bean, the winged bean (*Psophocarpus tetragonolobus*) is a tropical, leguminous, vine-like plant, closely related to **cowpea**. It is thought to be native to India and Southeast Asia but the species grows in the wild in East Africa and so it has been suggested that Africa is its origin. On the other hand, the plants may have been introduced from Asia and then escaped. It is now cultivated in all of the tropics including South America and the Caribbean. The pods, which can reach 50 cm in length, have four longitudinal projections, hence the name winged or four-angled bean. The seeds, similar in appearance to **soybean**, are rich in **storage protein** (30–40%), **oil** (15–18%) and carbohydrates, mainly **starch** (24–42%), thiamine, calcium, phosphorus and iron. The immature pods are eaten, prepared like snap or green beans and resemble asparagus in taste. The seeds are roasted, baked or steamed and can be used to make curd, 'milk', a flour or a coffee-like drink. A cooking oil is extracted from the oil-rich seeds. The winged bean is not grown on a large scale anywhere. Further exploitation is limited by the plant's climbing habits, the fact that the beans do not ripen simultaneously, and the tannin concentration in the beans which gives them an unpleasant taste and makes them somewhat indigestible. (MB)
(**See: Legumes**)

Kalloo, G. (1993) Winged bean, *Psophocarpus tetragonolobus* (L.) DC. In: Kalloo, G. and Bergh, B.O. (eds) *Genetic Improvement of Vegetable Crops*. Pergamon Press, Oxford, UK, pp. 465–469.

Winnow

Manual cleaning process of seed (or other substances) such as after **threshing** and husking, that removes only light, loose chaffy material from heavier grain by the action of wind or air movement. At its simplest, baskets and sieves are rhythmically flicked up in the air to remove the chaff, or grain is tossed with a fork or shovel in breezy weather, or slowly poured from a certain height into a pile. An arrangement using a sieve suspended from a tripod is still common today in China and Southeast Asia.

Mechanical devices incorporating fans were developed in China, and this style of winnowing fan is still used in many developing countries. Winnowing machines with adjustable air streams are used, for example, for cleaning tree seed as it falls or is shaken on an undulating surface. The physical principle of air current separation is also incorporated into such cleaning equipment as air-screen machines and aspirations. (**See: Conditioning, II. Cleaning**) (PH)

Winter annual

An annual plant that germinates before winter and produces seeds in the following year. (**See: Dormancy – cycles; Dormancy – ecophysiology; Summer annual**)

Wirestem

A common name of diseases (part of the soilborne **seedling disease complex**) affecting older seedlings, whose symptoms include rotting of the stem base (also typical of **damping-off**), so that the part below the ground and the upper taproot become constricted and the seedling collapses.

Witloof chicory

See: ***Cichorium* (endive and chicory)**

Wrinkled pea

Early experiments with wrinkled and smooth peas provide the basis for classical Mendelian genetics. In the mid-19th century, Gregor Mendel crossed true-breeding strains of common garden peas that had either smooth-coated (round) or wrinkled seeds. First-generation progeny of these crosses (F_1) were allowed to self-pollinate, producing the F_2 generation. From these experiments, Mendel established the principle of dominance, because the F_1 seeds were not a blend of the two traits, but were round like one parent. Thus, the trait that appeared in the F_1 generation became the 'dominant' trait, and the trait that failed to appear (i.e. wrinkled seeds) the 'recessive' trait. When one-quarter of the F_2 progeny were found to have wrinkled seeds, Mendel established his 'Rule of Segregation', which holds that the factors responsible for these traits (alleles) segregate from each other during gamete formation.

A mutation in the *rugosus* gene (insertion of a **transposon**) of pea produces the wrinkled pea phenotype. This gene codes for a starch branching enzyme, termed BEI (SBEI) in pea, which introduces branches to the growing **amylopectin** molecule in early **starch synthesis**. Pea BEI is analogous to the BEIIb isoform in cereal plants. Loss of this particular BE activity during pea embryo development results in significant reductions of both amylopectin and total starch, and the formation of deeply fissured granules. Together with increased water content, these alterations cause pea seeds to become 'wrinkled' when they undergo **maturation drying**. (**See: Pea**) (MGJ)

X

Xenia

The effect of the pollen genes on the development of the fruit or the seeds (**see: Development of seeds – cereals; Imprinting**). Specifically, in the developing maize kernel, it is the joint effect of the genotypes of the pollen and seed parent on endosperm characters. Depending on the inheritance of dominant or recessive genes between pollen and mother plant, and the mixture of pollen to which it is exposed in the field, different xenia effects may result in different individual kernels on the same cob. For example, the triploid starchy endosperm may have different yellow-to-white colours (see Table X.1); other endosperm characters that exhibit xenia are purple versus colourless **aleurone layer**, nonshrunken versus shrunken, and non-waxy versus *waxy* starch. (PH)

Table X.1. Effect of endosperm colour genes in maize: an example of xenia.

Mother-plant polar nuclei	Pollen sperm	Resulting endosperm	
YY	Y	YYY	Deep yellow
YY	y	YYy	Medium yellow
yy	Y	Yyy	Light yellow
yy	y	yyy	White

X-ray analysis

A technique for inspecting seed internal structures and anatomy. It is particularly useful for sugarbeet seed (Fig. X.1), native species, trees, shrubs and **nuts** which can be observed internally without breaking the shell, and is used as an adjunct in seed processing to determine fractures on otherwise damaged internal structures or immature embryos, and to assess the quality of accessions acquired in scientific seed **collection**. (**See: Tree seeds, 7. Seed quality and performance**, Fig. T.10) For

(A)

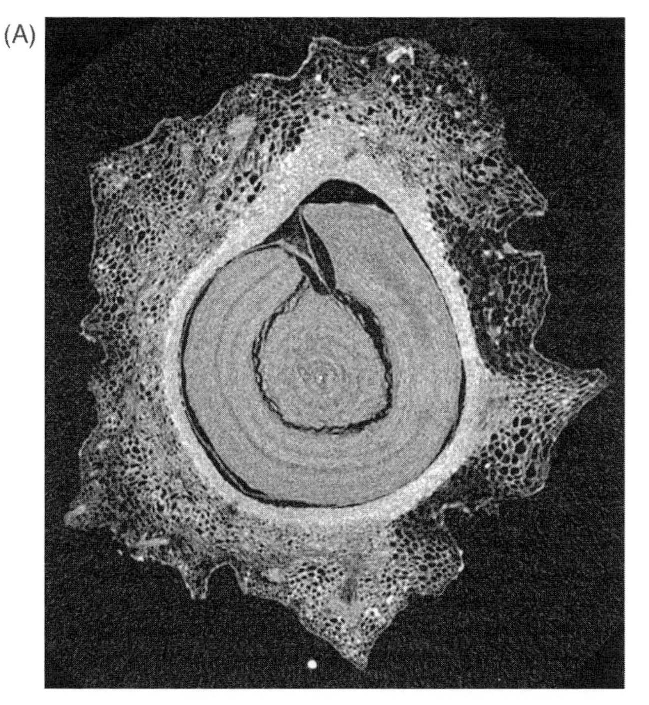

(B)

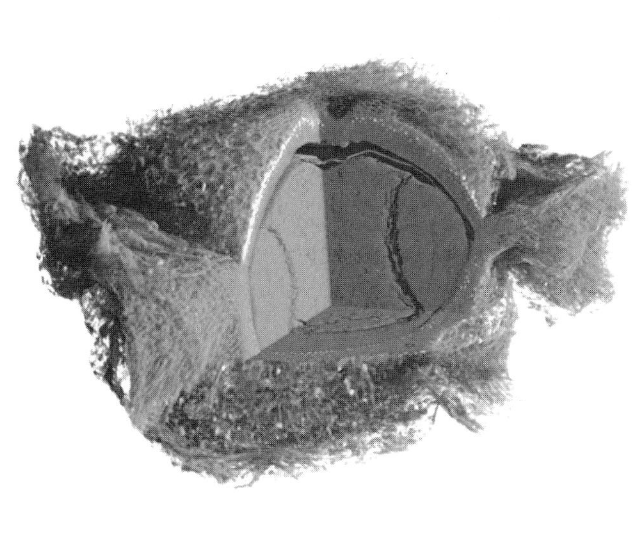

Fig. X.1. Reconstructed high-resolution X-ray pictures of a sugarbeet seed: (A) a 2-dimensional cross-section; (B) a 3-dimensional 'cut-away' visualization. The pictures are computer-generated tomographs using X-ray images obtained from 400 different angles within a 2-dimensional plane. The lightest part of the image shows the greatest density of plant material. The structure and proportions of the outer pericarp, the inner embryo and central starchy perisperm and other minor tissues can be analysed in great detail. Copyright Strube-Dieckmann GmbH & Co/Fraunhofer Gesellschaft, Development Centre for X-ray Technology (EZRT).

illustrations of the high resolution X-ray tomography technique, see Fig. X.1; and **see: Structure of seeds**, Fig. S.64)

X-ray diffraction is used to analyse the crystallinity of biological structures, such as seed **starch granules** and **fibres**. X-ray spectrometry is used to analyse cationic composition, such as of **phytin** deposits in seeds. The ionizing effect of X-rays is also used as a mutagenic principle to induce plant **mutants**. (PH)

Xyloglucan

A polymeric carbohydrate (**hemicellulose** polysaccharide) with a backbone of β-1,4-linked glucose residues and single α-1,6-linked xylose side chains (and can contain low amounts of galactose and arabinose). A cross-linking xylan in primary cell walls, it also occurs in some seeds as a storage carbohydrate, particularly in the cotyledons of non-endospermic legumes, *Impatiens*, *Tropaeolum*, *Cyclamen* and mustard. These polysaccharides are also called **amyloids** because of their staining reaction with the starch dye, iodine/potassium iodide. The xyloglucan-rich seed flour of the African leguminous plant, *Detarium senegalese*, is traditionally used in Nigeria as a thickening agent in food, suggestive of a use for this hemicellulose in the food industry.

Mobilization of xyloglucans, which accounts for 33% by weight of the reserves stored in the cotyledons of nasturtium (*Tropaeolum majus*) seeds, occurs following germination; likewise in the seeds of the tropical tree *Hymenaea courbaril* (kerosene tree, coubaril, West Indian locust), where this cell wall hemicellulose breakdown appears to be under the control of auxins which increase in the cotyledons at the time of expansion of the first leaves. The enzymes involved are endo-(1,4)-β-glucanase, α-xylosidase and β-galactosidase, which increase in activity as the xyloglucans are hydrolysed. (JDB)

dos Santos, H.P., Purgatto, E., Mercier, H. and Buckeridge, M.S. (2004) The control of storage xyloglucan mobilization in cotyledons of *Hymenaea cuorbaril*. *Plant Physiology* 135, 287–299.

Yield

The product resulting from the growth or cultivation of a plant or crop. Yield used without a qualifier generally refers to economic yield, i.e. that portion of the plant harvested for its economic value. Economic yield encompasses a variety of plant parts or organs including whole shoots (e.g. forage crops), roots and tubers ((e.g. potato (*Solanum tuberosum*), cassava (*Manihot esulenta*), beet (*Beta vulgaris*)), stems (e.g. sugarcane (*Saccharum officinarum*)), leaves (e.g. lettuce (*Lactuca sativa*), tea (*Camellia sinensis*)), or mature fruits or seeds (e.g. cereal and legume crops such as maize (*Zea mays*), wheat (*Triticum* spp.), rice (*Oryza sativa*), soybean (*Glycine max*)).

Yield may be expressed on a per plant (g/plant) or a per area basis (g/m^2 or kg/ha), but the appropriate expression depends upon the method of production. Yield per plant is most meaningful if the crop is produced as a series of isolated plants. Yield per unit area, however, is most meaningful for crops produced in plant communities (most agronomic crops). Yield per plant in plant communities is usually inversely related to plant population, so yield per plant changes with population, but yield per area may be constant. Focusing on yield per plant in these situations tends to obscure real changes in plant productivity.

Yield is the weight of plant material harvested, but it may be expressed as **dry weight**, **fresh weight**, or weight at some specified water concentration. Dry weight best characterizes the productivity of the plant or plant community, but the other expressions may be better related to marketing the crop (e.g. grain yield at a moisture content relating to market standards) and therefore have more practical utility.

Scientists may also measure total yield, which refers to all plant material, or more frequently, all above-ground material (roots are very difficult to harvest) and includes the economic yield plus all other plant parts. Total yield is a measure of the total biological productivity of the plant or plant community whereas economic yield in most crops represents only a portion of the total productivity. Thus, economic yield can increase without an increase in total biological productivity.

Yield of agronomic crops has increased steadily over the past 50 years as a result of the development of higher yielding cultivars and improved management practices. Theoretically, there must be a biological limit to yield, but where this limit lies relative to today's yields is a matter of fierce debate among crop physiologists. (DBE)

Loomis, R.S. and Conner, D.J. (1992) *Crop Ecology: Productivity and Management in Agricultural Systems*. Cambridge University Press, Cambridge, UK.

(In the completely different, *biomechanical sense* of the term 'yield' – for an account of the yielding of the cell wall during the process of embryo cell expansion, **see: Germination – physical factors**.)

Yield components

The study of crop yield can be simplified if yield (**see: Yield; Yield potential**) is divided into its individual parts or components. The primary yield components of a grain crop are the number of seeds per unit area or per plant (**see: Charles-Edwards model**), and the average weight per seed (seed size) (**see: Development of seeds – influence of external factors**), but more complex representations are possible, e.g. yield = (plants/area) (pods/plant) (seeds/pod) (weight/seed). Such complex representations are species specific, depending upon the growth habit and the characteristics of the seed-bearing structures, but the primary components (seed number and seed size) apply to all grain crops. Yield components are more commonly used with grain crops where seeds are the yield, although theoretically, the approach could be applied to any crop.

Yield components are useful and necessary for research purposes because the physiological processes that determine yield act on the yield components, not on yield *per se*. Yield components also emphasize the sequential nature of yield production, i.e. seed number is determined before seed size, which must be considered to evaluate environmental effects on yield.

Yield-component compensation occurs when a change in one component results in an opposite compensating change in another component, with yield remaining the same. Yield component compensation has frustrated plant breeders trying to increase yield by selecting for individual components, but compensation does not negate the value of focusing on yield components when studying environmental effects or the physiological processes underlying yield. (DBE)

Yield determination

1. Source

A nutrient source is an organ or tissue that exhibits a net export of a specified nutrient species.

2. Sink

A nutrient sink is an organ or tissue that exhibits a net import of a specified nutrient.

3. Sink strength

The potential capacity of a sink to import a nutrient species independent of external influences affecting nutrient supply.

Sink strength is distinguished from sink mobilizing ability. The latter measures actual nutrient import (absolute growth rate) and is an integrated outcome of all influences on nutrient transport within and beyond the sink boundary (**see: Development of seeds – nutrients and water import**).

4. Harvest index (HI)
Harvest index provides a measure of crop productivity as the ratio of economic yield to biological yield. Since biomass is primarily derived from photoassimilates (carbon, hydrogen and oxygen account for 90% or more of plant biomass), HI measures the cumulative outcome of photoassimilate partitioning across crop development at harvest. Thus HI includes current and remobilized biomass accumulated prior to seed development. For grain crops, HI is an estimate of the proportionate amount of biomass partitioned into developing seeds. Increases in potential economic yield of grain crops over the millennia have resulted largely from changes in HI rather than biological yield. The largest variation in total seed yield per plant is contributed by differences in seed number. The latter parameter is set by the degree of floret abortion that is sensitive to nutrient supply at floret and seed set. Variation in final seed size is a secondary contributor and this character is governed by cell number set during the pre-storage phase of seed development (**see: Charles-Edward model; Development of seeds – nutrients and water import; Potential seed number**).

5. Fill parameters
Final seed biomass is a product of fill rate and duration. These two determinants of yield are related inversely. In most circumstances, fill rate (nutrient import) is the most important regulator of final seed biomass. Photoassimilate (carbon compounds, mainly sucrose and amino nitrogen compounds) import rates are limited by the seed's capacity to accumulate photoassimilates. Increases or decreases in photoassimilate production have little impact on rates of seed fill (Table Y.1). Hence import of nutrients by seeds and their sink strength are identical. Seed sink strength for photoassimilates is a function of seed cell number and the activities of seed physiological processes including that of membrane sucrose transporters. Sink strength is an inherent genetic property of seed filial tissues (endosperm/cotyledons), but genotypic differences are also reflected by sink activity and sink properties of maternal seed tissues. Potential photoassimilate sink activities and sizes are set during the pre-storage phase of seed development. Duration of seed fill is under both genetic and environmental control. Alterations in seed fill or duration have little impact on final seed biomass as one parameter compensates for the other (**see: Charles-Edwards model; Development of seeds – nutrients and water import**).

6. Source/sink limitation
Manipulating production of, and inter-sink competition for, photoassimilates has demonstrated that the site of nutrient limitation depends upon the phase of seed development. Seed set and cell division (pre-storage) phases are acutely sensitive to alterations in supplies of photoassimilates and amino nitrogen compounds (Table Y.1). In contrast, seed filling is less dependent upon current photoassimilate supplies (Table Y.1). During seed fill, any shortfall in photoassimilate supplies is buffered by remobilization of stored reserves. Cessation of seed growth appears to result from anatomical or metabolic changes in seeds rather than from a lack of nutrient supply (**see: Development of seeds – nutrient and water import**).

(MT-H, CEO, JWP)

Jenner, C.F., Ugalde, T.D. and Aspinall, D. (1991) The physiology of starch and protein deposition in the endosperm of wheat. *Australian Journal of Plant Physiology* 18, 211–216.

Yield penalty

Lower yields associated with the use of specific cultivars, perhaps with unique characteristics, or possibly the adoption of certain management practices, constitute a yield penalty. Thus, yields are not as high as they could have been if other cultivars or management practices had been used. Yield penalty is currently widely used to refer to lower yields of genetically engineered cultivars with herbicide, disease or insect resistance in the absence of weeds or pests. Cultivars with high seed oil (**see: Oilseeds – major**) or protein concentrations, or other unique seed traits may also exhibit a yield penalty. Lower yields of soybean (*Glycine max*) planted late, after a wheat (*Triticum aestivum*) crop is harvested (i.e. double cropped), are an example of a yield penalty associated with a management system. A cultivar or management system that involves a yield penalty may still have an economic advantage over a cultivar or system that maximizes yield, leading to its use by producers. The producer must decide if the specific trait is more valuable than the loss of yield.

Table Y.1. Summary of experimental observations used to deduce whether seed growth is source- or sink-limited for photoassimilates and amino nitrogen compounds at key stages of development.

Seed development phase	Limitation	Experimental evidence	
Seed set/cell division	Photoassimilate and amino nitrogen supply	Negative:	Shading Defoliation
		Positive:	CO_2 fertilization Sink removal *Rht* dwarfing genes
Seed filling	Sink-limited for photoassimilates	Photoassimilate supply independent. *In vivo* and *in vitro* growth rates similar.	
	Source-limited for amino nitrogen	Increased *in vivo* or *in vitro* supplies enhance seed storage protein content	

Yield penalties are probably more common in the first cultivars released from breeding programmes for pest resistance, or in the development of genetically engineered cultivars. Yield penalties were a major concern with early genetically engineered cultivars, but continued breeding has virtually eliminated the penalty in many crops. (DBE)

Evans, L.T. (1993) *Crop Evolution, Adaptation and Yield.* Cambridge University Press, Cambridge, UK.

Yield potential

The yield of a cultivar of a crop species grown in an environment where water, nutrients, pests, diseases, weeds and other stresses are not limiting (**see: Yield**). Yield potential represents the ability of the plant to exploit a specific environment when not limited by stress, technology, or economic concerns. Potential yield is higher than attainable yield (yields obtained by skilful use of best technology, yields obtained by the best producers), which is, in turn, higher than actual yield (average yield of a large area, such as a district, state or country, represents the average use of technology and average skill of producers). There are more factors limiting actual yield than attainable yield, while at the potential level, yield is determined only by the plant and the unmanageable aspects of the environment – principally solar radiation and temperature. Yield potential is, therefore, a characteristic of a location as determined by solar radiation, temperature and the suitability of the plant for the specific environment.

Yield potential is hard to measure accurately because it is difficult to be sure that no stresses are limiting plant growth, but as a theoretical concept, it sets the upper limit to yield and provides a guide to evaluate yield improvement. Valid questions are: has yield potential increased over time or has technology eliminated more of the stresses limiting yield, thus moving attainable yield closer to potential? (DBE)

Evans, L.T. and Fischer, R.A. (1999) Yield potential: its definition, measurement, and significance. *Crop Science* 39, 1544–1551.

Yopo

Known also as cebil, this South American species (*Anadenanthera colubrina*, Fabaceae) has seeds which contain tryptamine-derived alkaloids (**see: Pharmaceuticals and pharmacologically active compounds**). The seeds are toasted, pulverized and used by Indians in the Orinoco basin in Colombia, Venezuela and possibly in southern parts of the Brazilian Amazon either as a psychedelic snuff or smoked or in an enema. (**See: Psychoactive seeds**)

Z

Zeatin

The first natural **cytokinin** to be discovered and one of the major active forms in most plants, zeatin gets its name because it was isolated from developing seeds of maize (*Zea mays*). (See: Fig. C.30 in **Cytokinin**)

Zein

The major storage **prolamin** present in the **endosperm** of **maize** kernels. There are classes of zeins, (α, β, γ, δ), of which the α type constitutes about 70% of the total. (**See: Storage proteins**)

Zoochory

The dispersal of the **diaspores** (seeds and fruits) of a plant by animals, e.g. by fish: ichthyochory, by mammals: mammaliochory, by ants: myrmecochory, by birds: ornithochory, by reptiles: saurochory, and in the mouth by birds and ants: stomatochory, synzoochory. (**See: Dispersal**)

Zygote

The fertilized egg, which results from the fusion of the female gamete, the egg and the male gamete – in seed plants a sperm nucleus from the pollen. The zygote develops to form the embryo of the seed. (**See: Embryogenesis; Fertilization**)

Zygotic embryo

An embryo that is produced from the fertilized egg, with a diploid (2n) complement of chromosomes that result from the fusion of a male pollen nucleus (n) with the egg cell (n). Contrast with: (i) **somatic embryo** that is formed from diploid somatic cells only, and has an identical chromosome complement to the plant from which it is derived; and (ii) embryos produced by **apogamy** and **apomixis**. (**See: Development of embryos – cereals; Development of embryos – dicots; Embryogenesis; Reproductive structures**)

Crop Atlas Appendix

I. World Crop Areas Harvested in 2002

A. World regional areas of major crop groups

Data for North and Central America, South America, Africa, Europe, West Asia, Southern Asia, East and Southeast Asia, China and Oceania, as shown.

Map 1. Major Cereals.
Map 2. Major Oilseeds (including delinted cottonseed and oil palm kernels).
Map 3. Grain Legumes (including pulses, soybean and groundnut in shell).
Map 4. Tree Nuts.

B. National areas of individual crops

Map 5. Wheat.
Map 6. Rice.
Map 7. Maize.
Map 8. Barley.
Map 9. Sorghum.
Map 10. Millets.
Map 11. Oat.
Map 12. Rye.
Map 13. Soybean.
Map 14. Cottonseed (de-linted).
Map 15. Groundnut (in shell).
Map 16. Rapeseed.
Map 17. Sunflower.
Map 18. Oil Palm and Coconut.
Map 19. Total Pulses.
Map 20. Coffee and Cacao.

An area is counted each time it is sown or planted (such as twice for double-sown rice).

Crop areas are drawn at the same scale between the individual crop maps, except for maps 11, 12 and 20.

For the individual crop maps: insets at bottom left – the distribution of crop areas and yields between continental regions; insets at top left, except for maps 18 and 20 – world crop area and yields at 5-year intervals between 1962 and 2002; plus, for maps 7, 13, 14 and 16, insets at top left, the annual world crop area grown with genetically modified (GM) varieties between 1996 and 2003.

II. Annual World Commodity Balances

Mean of 5-year production and consumption data (1998–2002) within world regions as shown (North and Central America, South America, Africa, Europe, West Asia, Northwest Asia, Southern Asia, East and Southeast Asia, China, Japan and Oceania).

'Production' includes domestic agricultural and non-commercial production but excludes harvesting losses; 'consumption' includes domestic uses for human food, food processing, animal feed, non-food manufacturing, crop sowing and wastage during storage and transportation. The difference between production and consumption equates to the balance between net imports, exports and internal stock changes.

Map 21. Major Cereal Grains.
Map 22. Major Oilseeds.

III. Major Routes of the International Grain Trade

The 5-year mean annual volumes (1998–2002) of grain transported intact: that is, excluding processed grain products, and grain movements within a country or the EU trade group.

Map 23. Wheat (from the major exporting countries and trade groups: Canada, the USA, Argentina, the EU, the Russian Federation, Ukraine, Kazakhstan and Australia).
Map 24. Maize (from the major exporting countries: the USA, Argentina, Brazil and China).

IV. Seeds in Human Nutrition

Map 25. Contribution of combined major grains – cereals, pulses and oilseeds – to: (i) calorific intake in relations to national population; and (ii) the mean dietary calorific and protein consumption in nine world regions (North and Central America, South America, Africa, Europe, West Asia, Southern Asia, East and Southeast Asia, China and Oceania).

All maps drawn for this Encyclopedia by PH, using data from FAOSTAT (faostat.fao.org) and ISAAA (International Service for the Acquisition of Agri-Biotech Applications, www.isaaa.org)

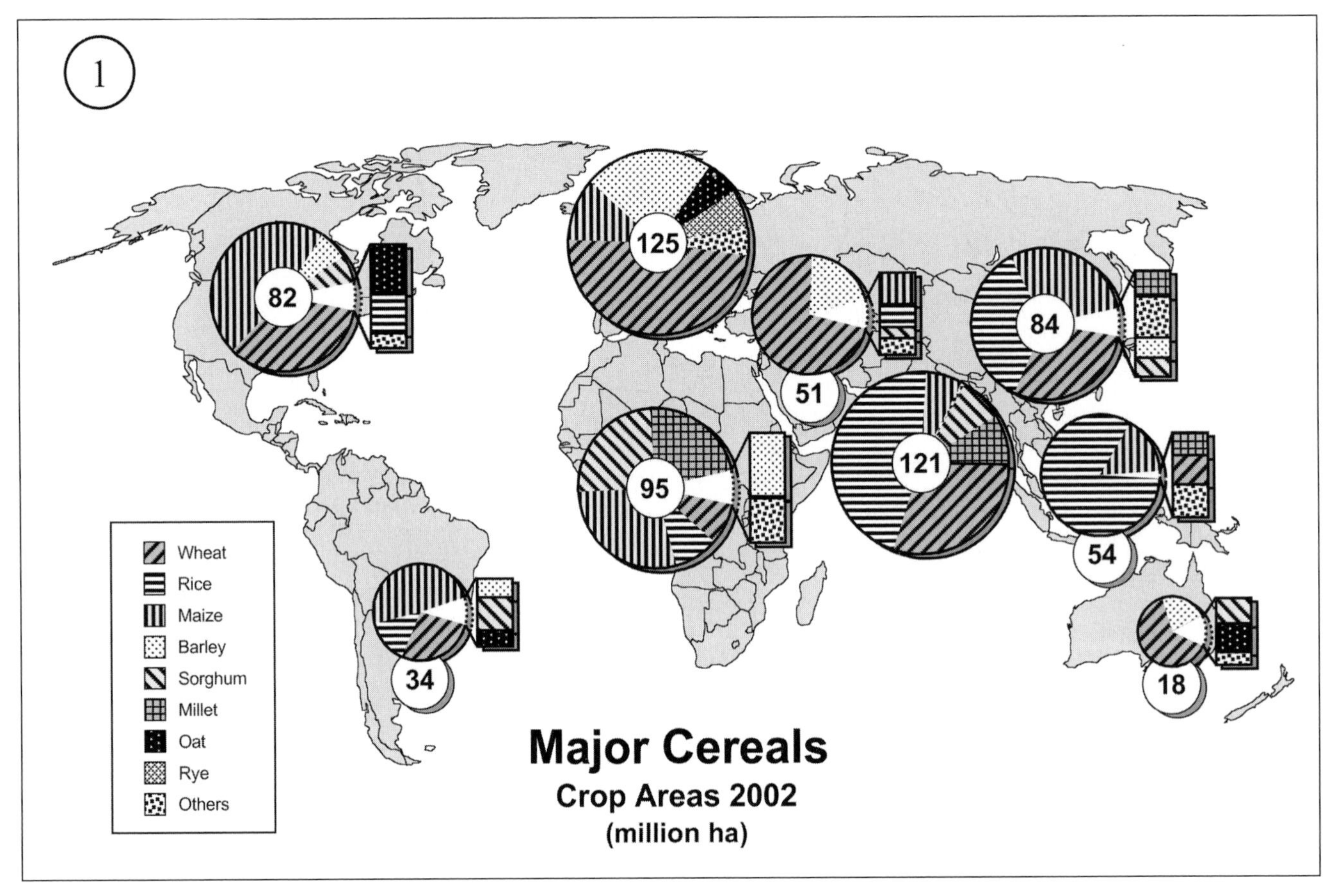
1
82
125
51
84
121
95
54
34
18
Wheat
Rice
Maize
Barley
Sorghum
Millet
Oat
Rye
Others
Major Cereals
Crop Areas 2002
(million ha)

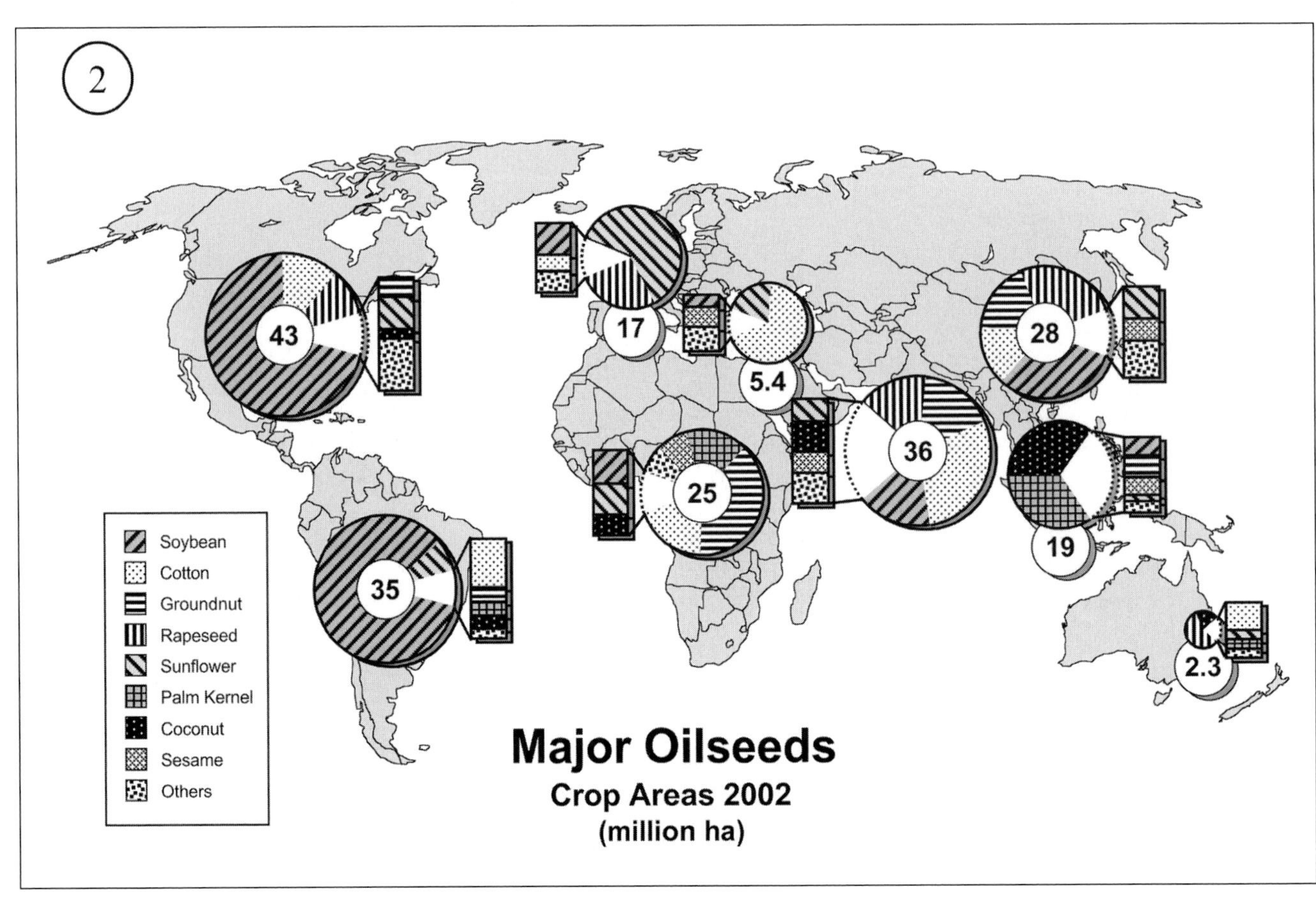
2
43
17
5.4
28
36
25
19
35
2.3
Soybean
Cotton
Groundnut
Rapeseed
Sunflower
Palm Kernel
Coconut
Sesame
Others
Major Oilseeds
Crop Areas 2002
(million ha)

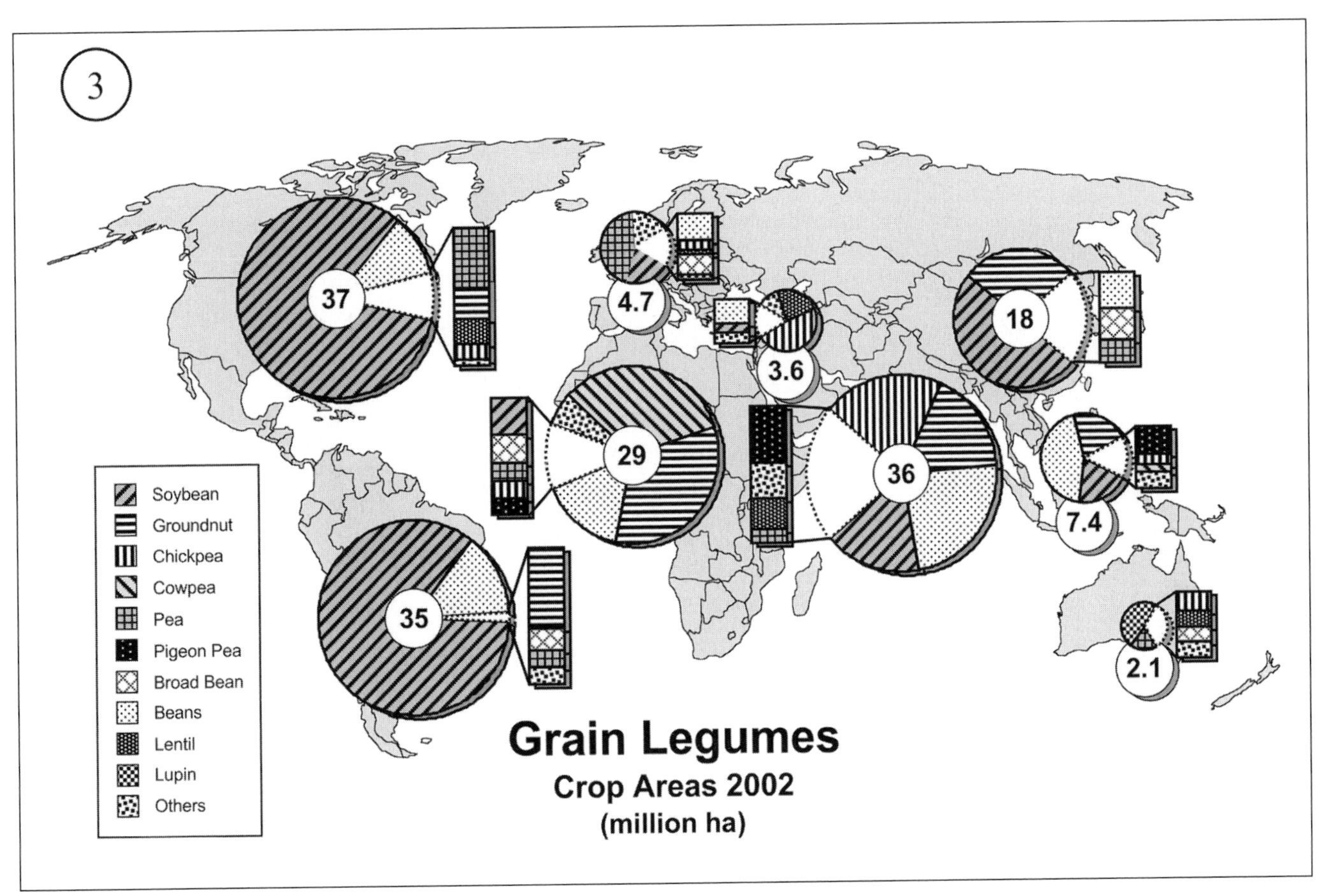
3
37
4.7
3.6
18
29
36
7.4
35
2.1
Soybean
Groundnut
Chickpea
Cowpea
Pea
Pigeon Pea
Broad Bean
Beans
Lentil
Lupin
Others
Grain Legumes
Crop Areas 2002
(million ha)

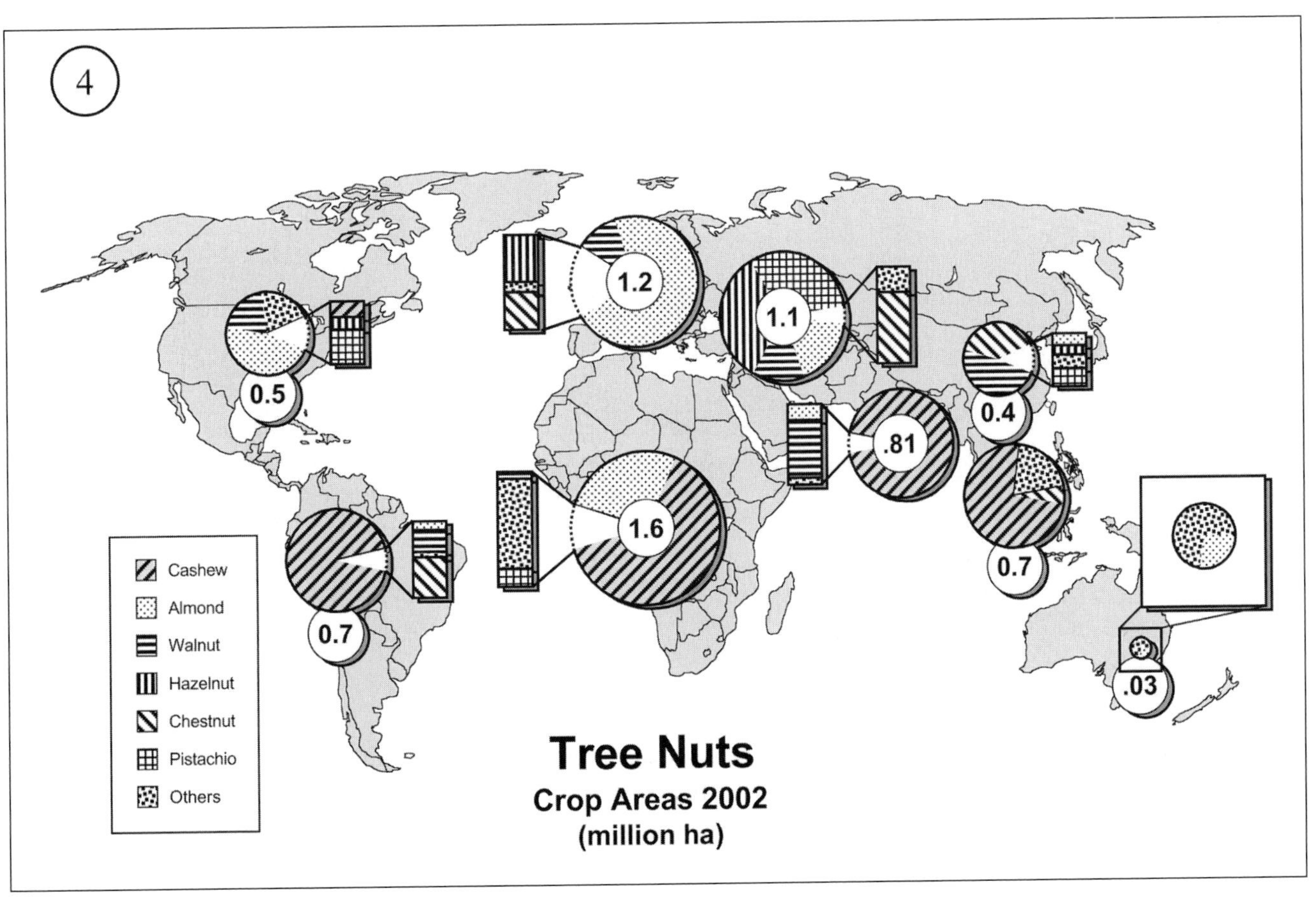
4
1.2
1.1
0.5
0.4
.81
1.6
0.7
0.7
.03
Cashew
Almond
Walnut
Hazelnut
Chestnut
Pistachio
Others
Tree Nuts
Crop Areas 2002
(million ha)

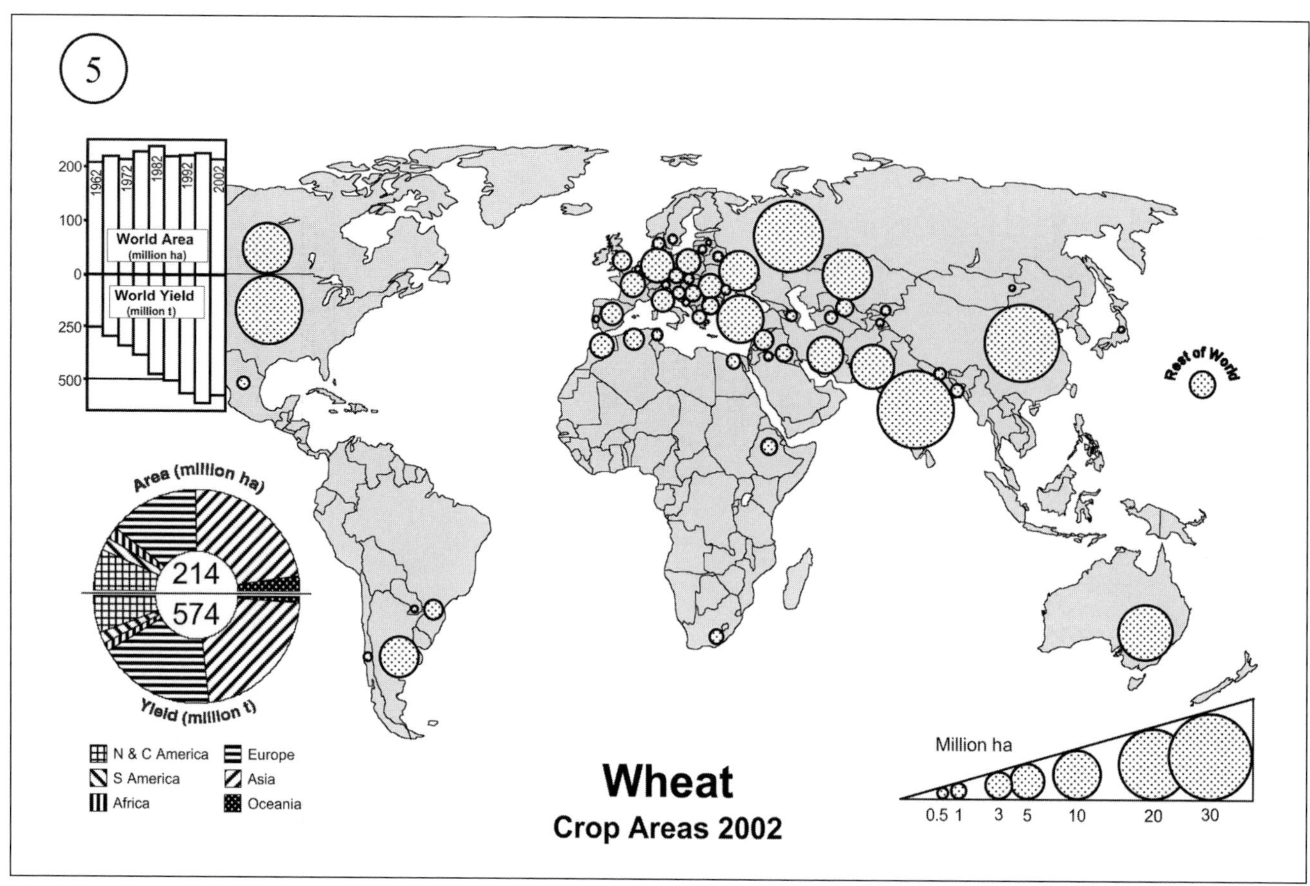
5
200
100
0
250
500
1962
1972
1982
1992
2002
World Area
(million ha)
World Yield
(million t)
Area (million ha)
214
574
Yield (million t)
N & C America
S America
Africa
Europe
Asia
Oceania
Rest of World
Million ha
0.5 1 3 5 10 20 30
Wheat
Crop Areas 2002

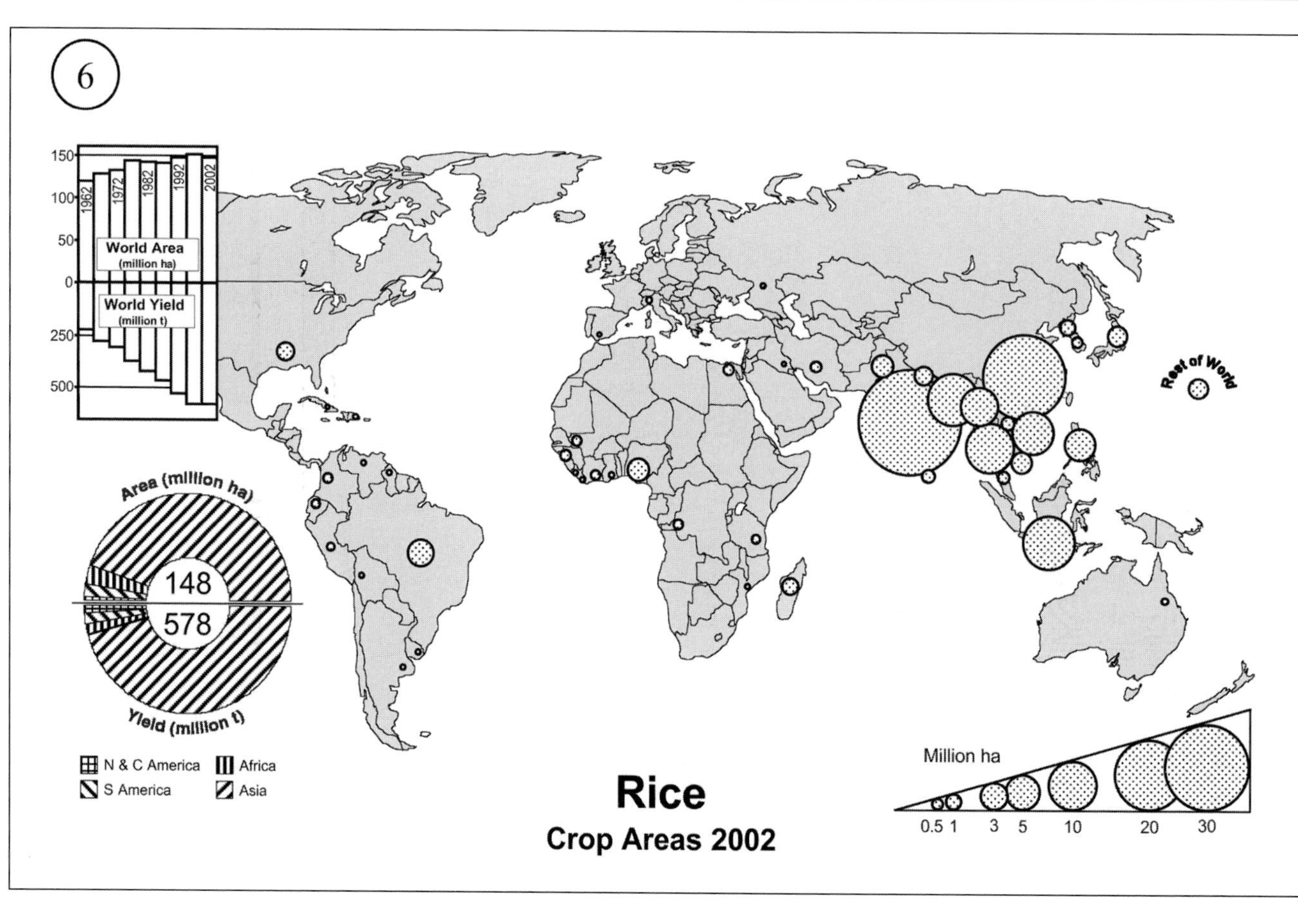
6
150
100
50
0
250
500
1962
1972
1982
1992
2002
World Area
(million ha)
World Yield
(million t)
Area (million ha)
148
578
Yield (million t)
N & C America
S America
Africa
Asia
Rest of World
Million ha
0.5 1 3 5 10 20 30
Rice
Crop Areas 2002

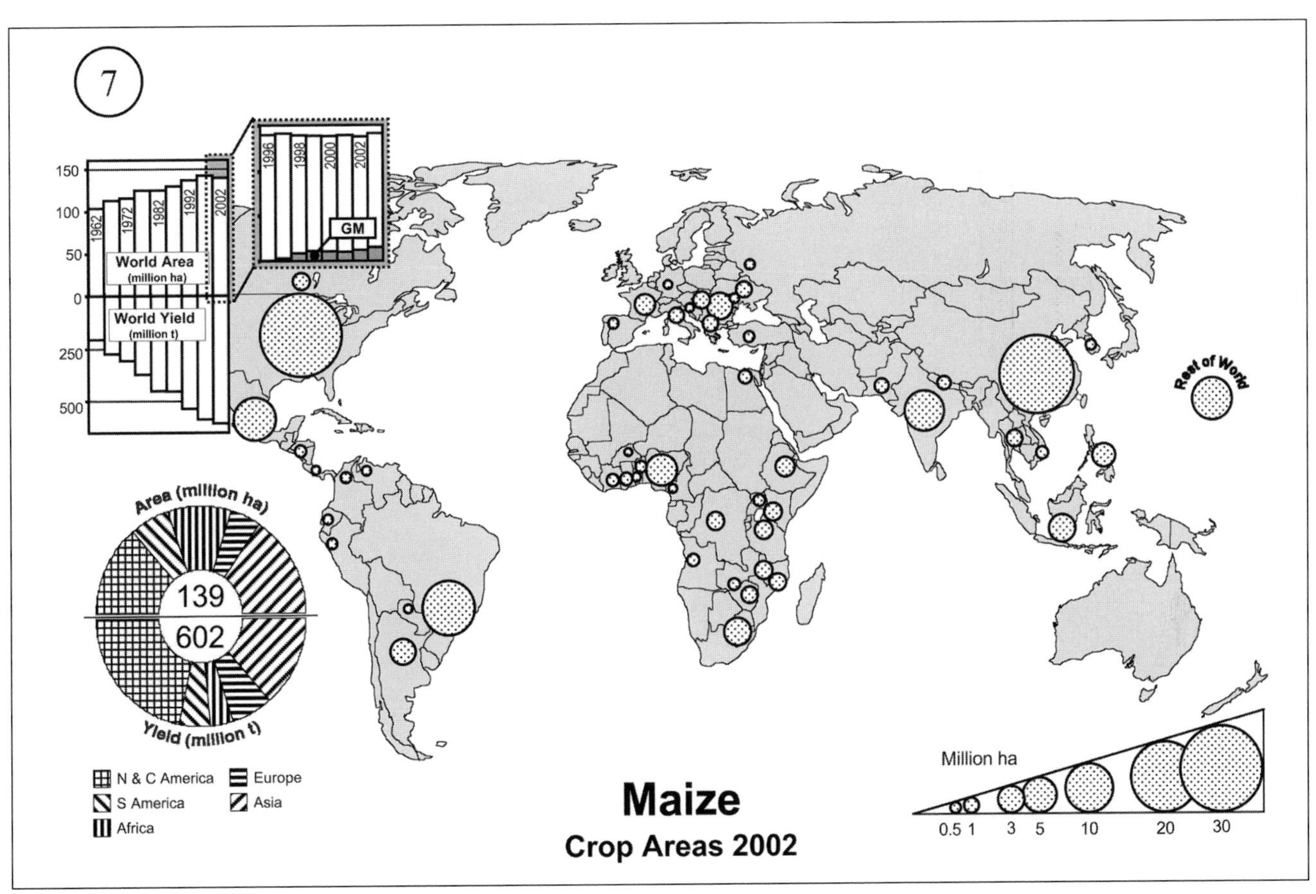

Maize
Crop Areas 2002

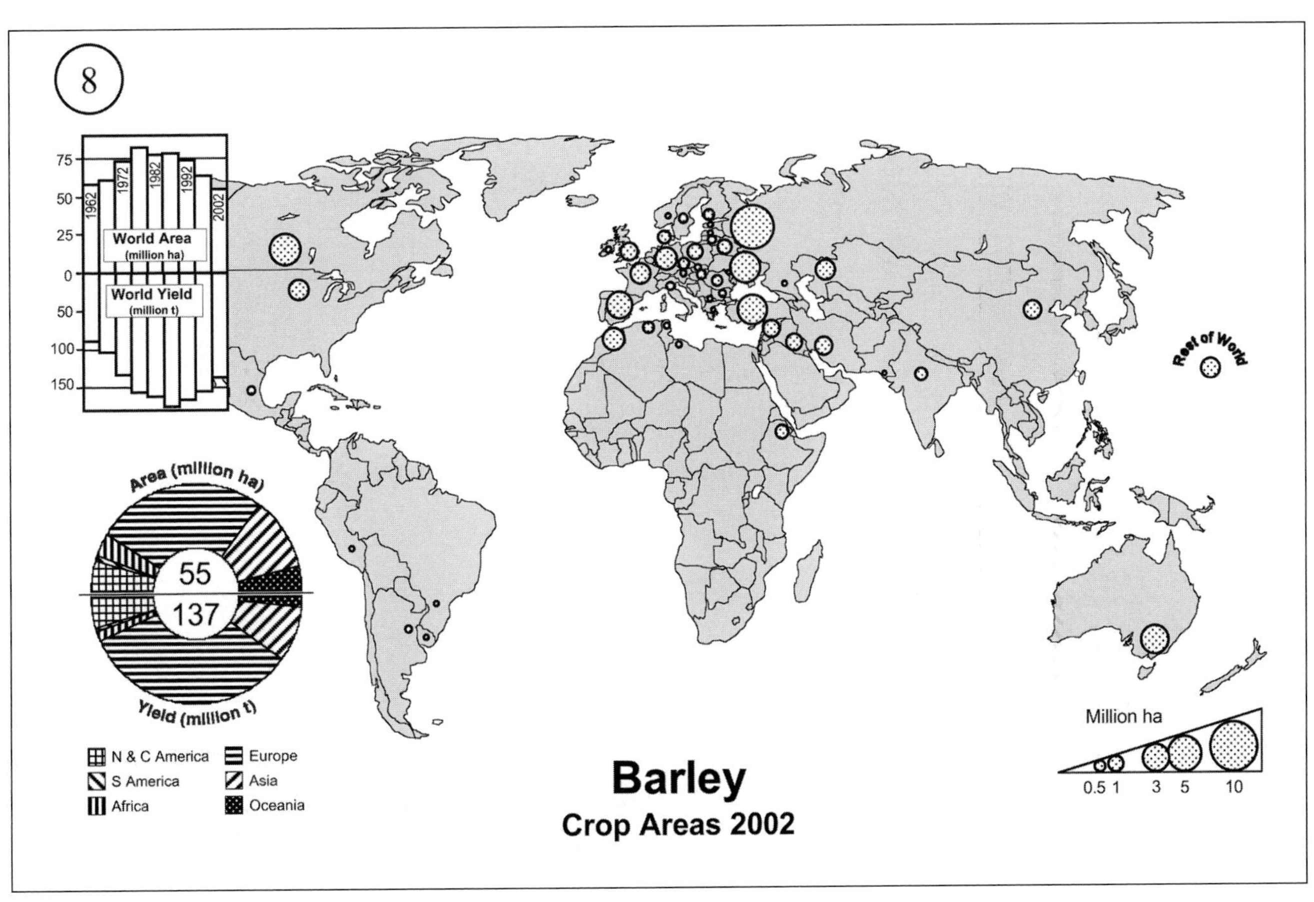

Barley
Crop Areas 2002

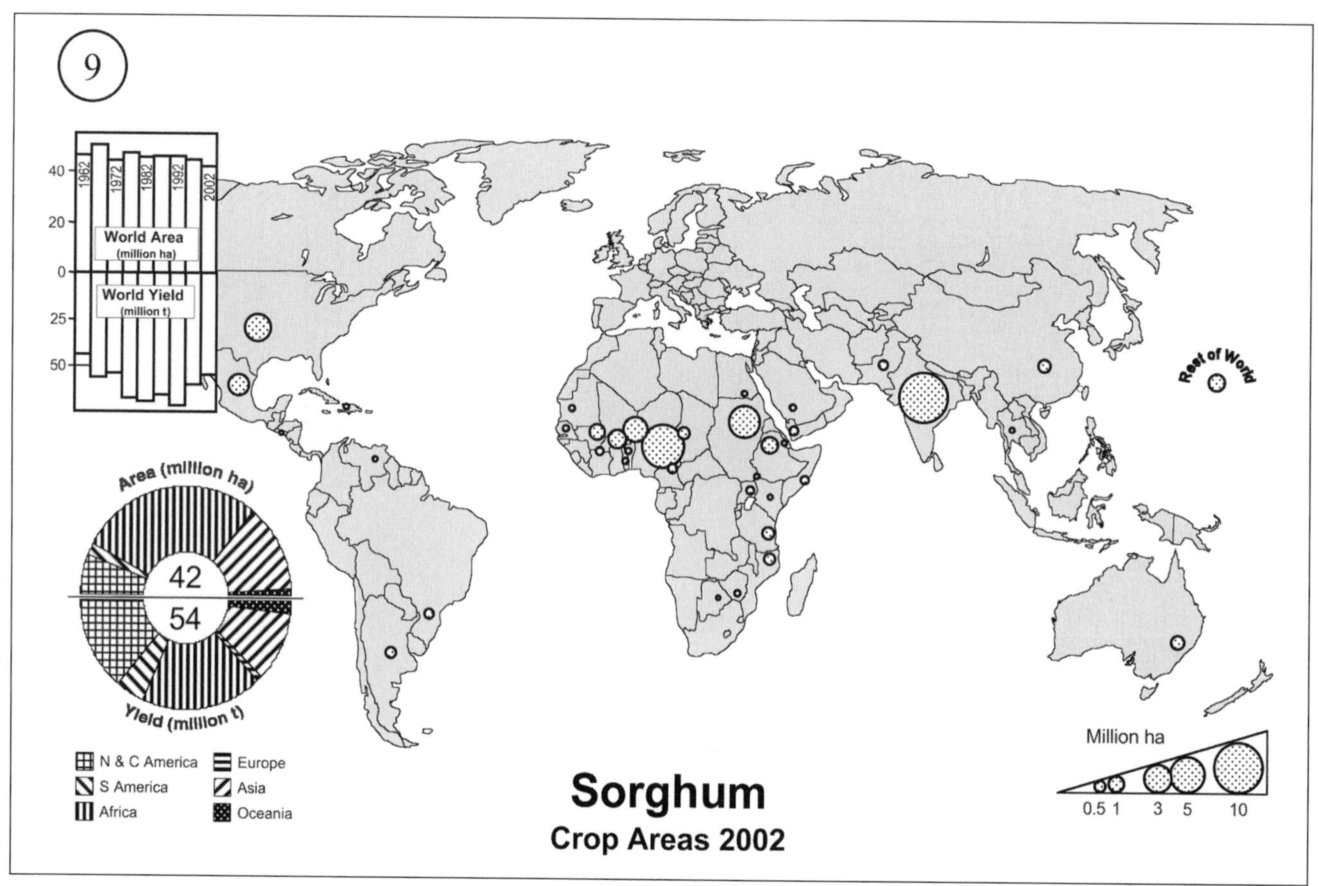

9
40
20
0
25
50
1962
1972
1982
1992
2002
World Area
(million ha)
World Yield
(million t)
Area (million ha)
42
54
Yield (million t)
N & C America
S America
Africa
Europe
Asia
Oceania
Rest of World
Million ha
0.5 1 3 5 10
Sorghum
Crop Areas 2002

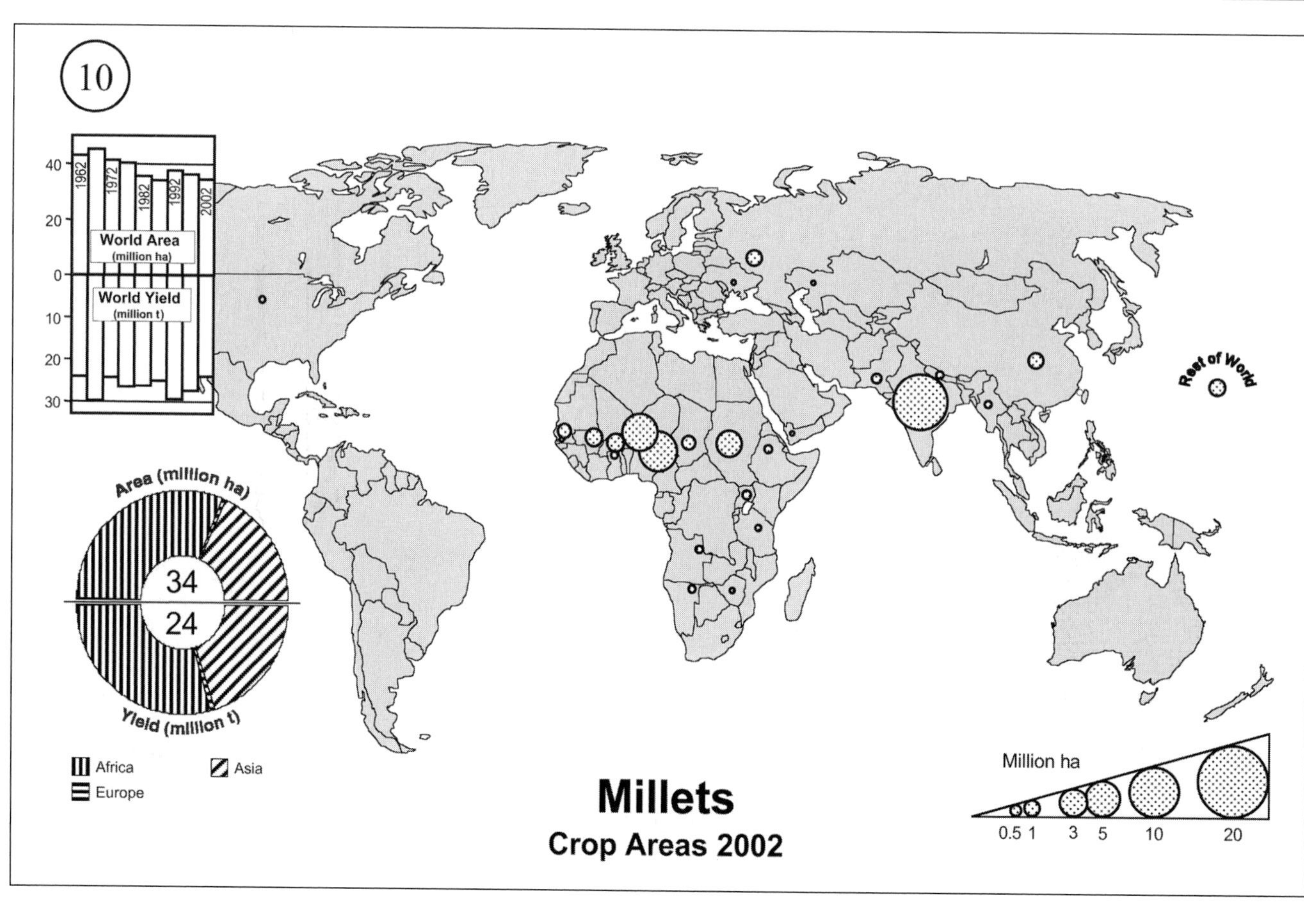

10
40
20
0
10
20
30
1962
1972
1982
1992
2002
World Area
(million ha)
World Yield
(million t)
Area (million ha)
34
24
Yield (million t)
Africa
Europe
Asia
Rest of World
Million ha
0.5 1 3 5 10 20
Millets
Crop Areas 2002

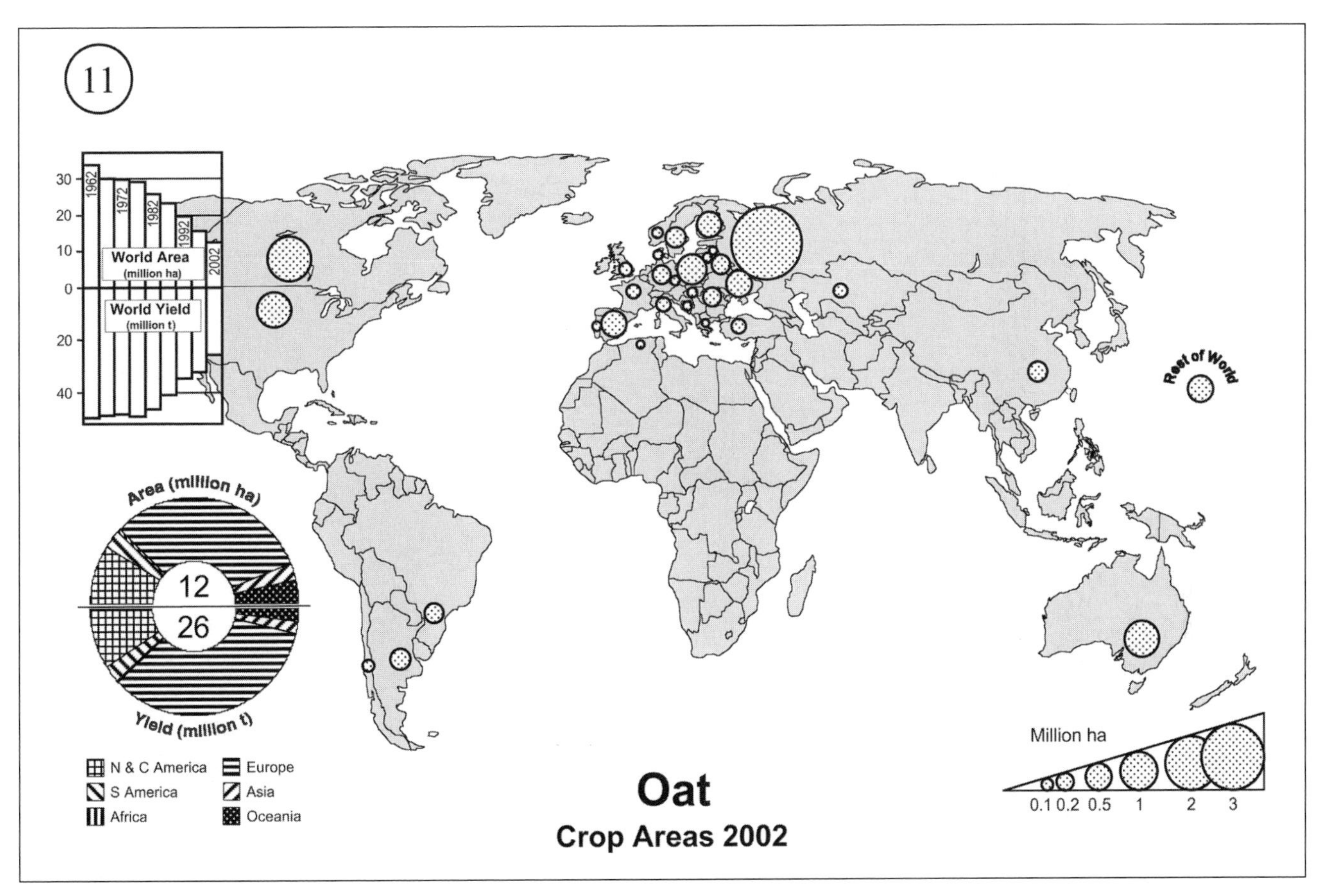
11
30
20
10
0
20
40
1962
1972
1982
1992
2002
World Area
(million ha)
World Yield
(million t)
Area (million ha)
12
26
Yield (million t)
N & C America
S America
Africa
Europe
Asia
Oceania
Rest of World
Million ha
0.1 0.2 0.5 1 2 3
Oat
Crop Areas 2002

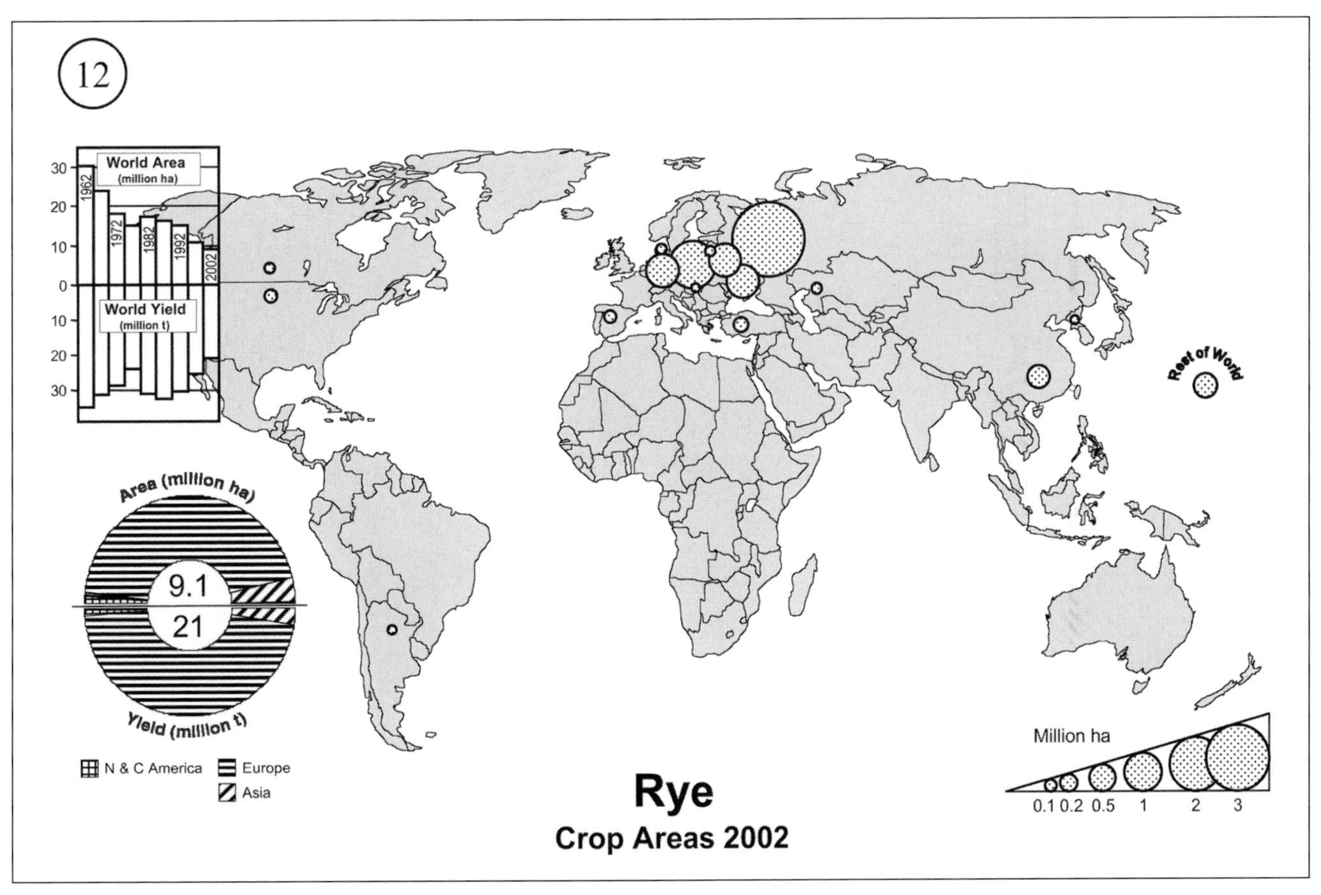
12
30
20
10
0
10
20
30
1962
1972
1982
1992
2002
World Area
(million ha)
World Yield
(million t)
Area (million ha)
9.1
21
Yield (million t)
N & C America
Europe
Asia
Rest of World
Million ha
0.1 0.2 0.5 1 2 3
Rye
Crop Areas 2002

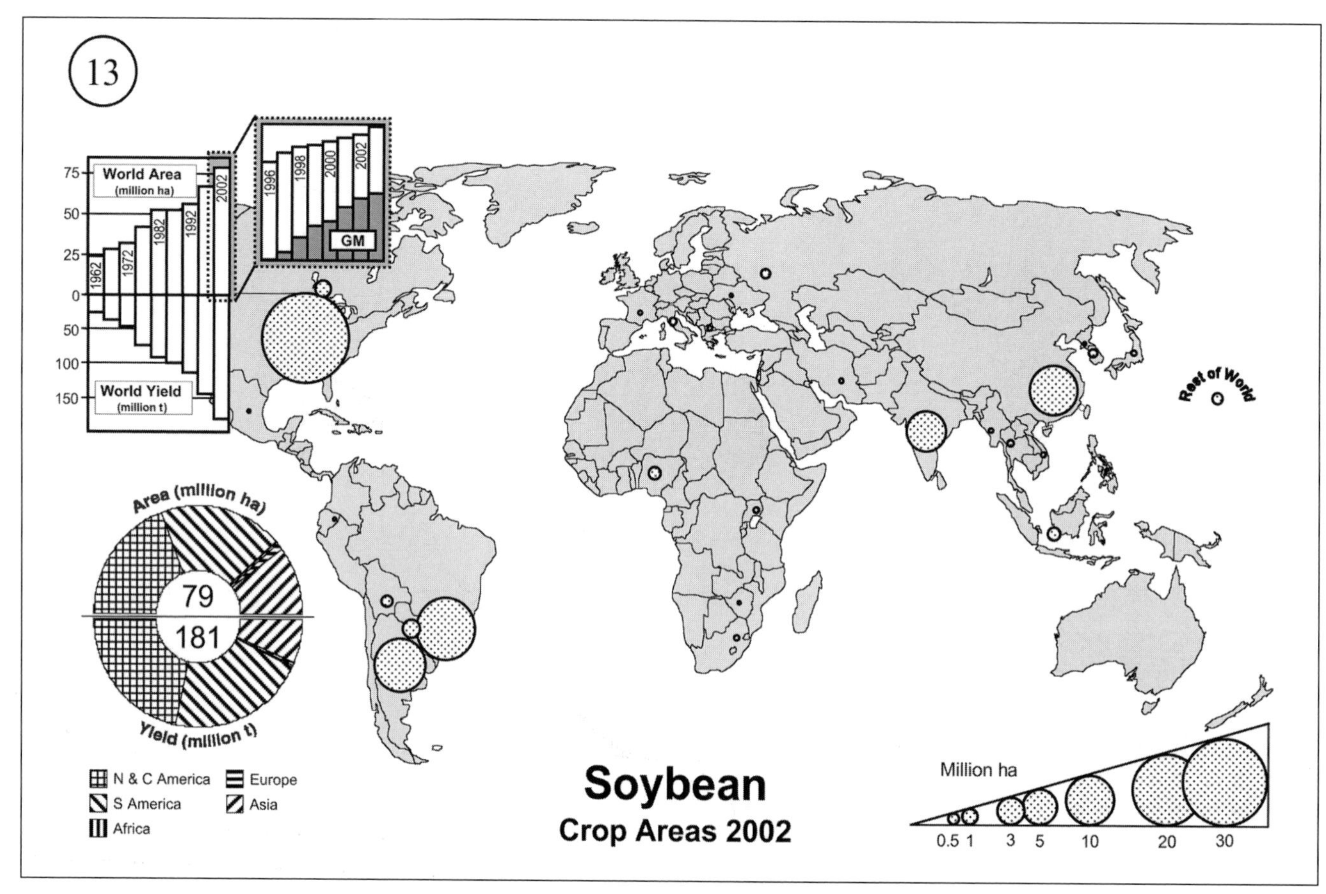
13
World Area
(million ha)
World Yield
(million t)
75
50
25
0
50
100
150
1962
1972
1982
1992
2002
1996
1998
2000
2002
GM
Area (million ha)
79
181
Yield (million t)
N & C America
S America
Africa
Europe
Asia
Rest of World
Soybean
Crop Areas 2002
Million ha
0.5 1 3 5 10 20 30

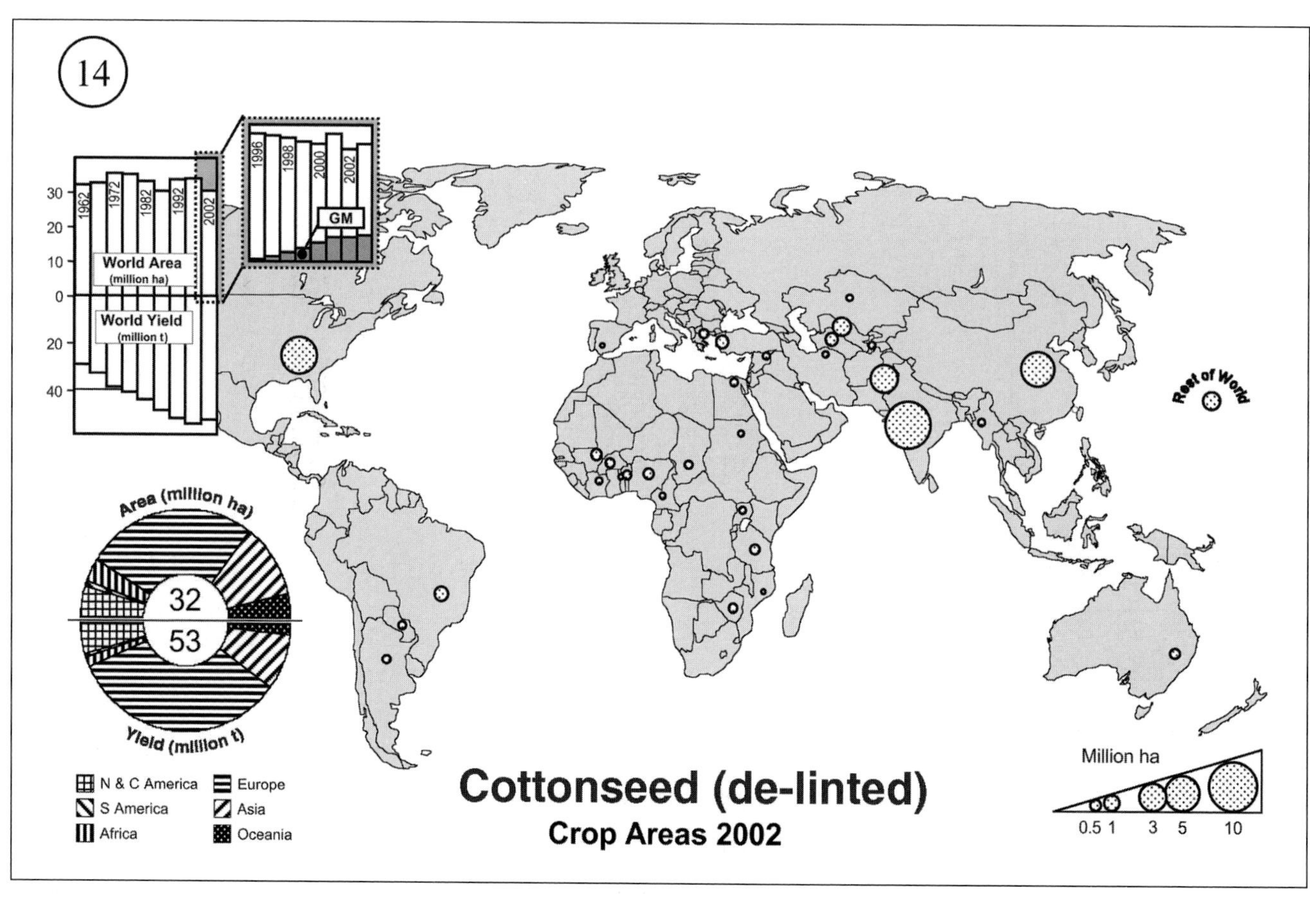
14
World Area
(million ha)
World Yield
(million t)
30
20
10
0
20
40
1962
1972
1982
1992
2002
1996
1998
2000
2002
GM
Area (million ha)
32
53
Yield (million t)
N & C America
S America
Africa
Europe
Asia
Oceania
Rest of World
Cottonseed (de-linted)
Crop Areas 2002
Million ha
0.5 1 3 5 10

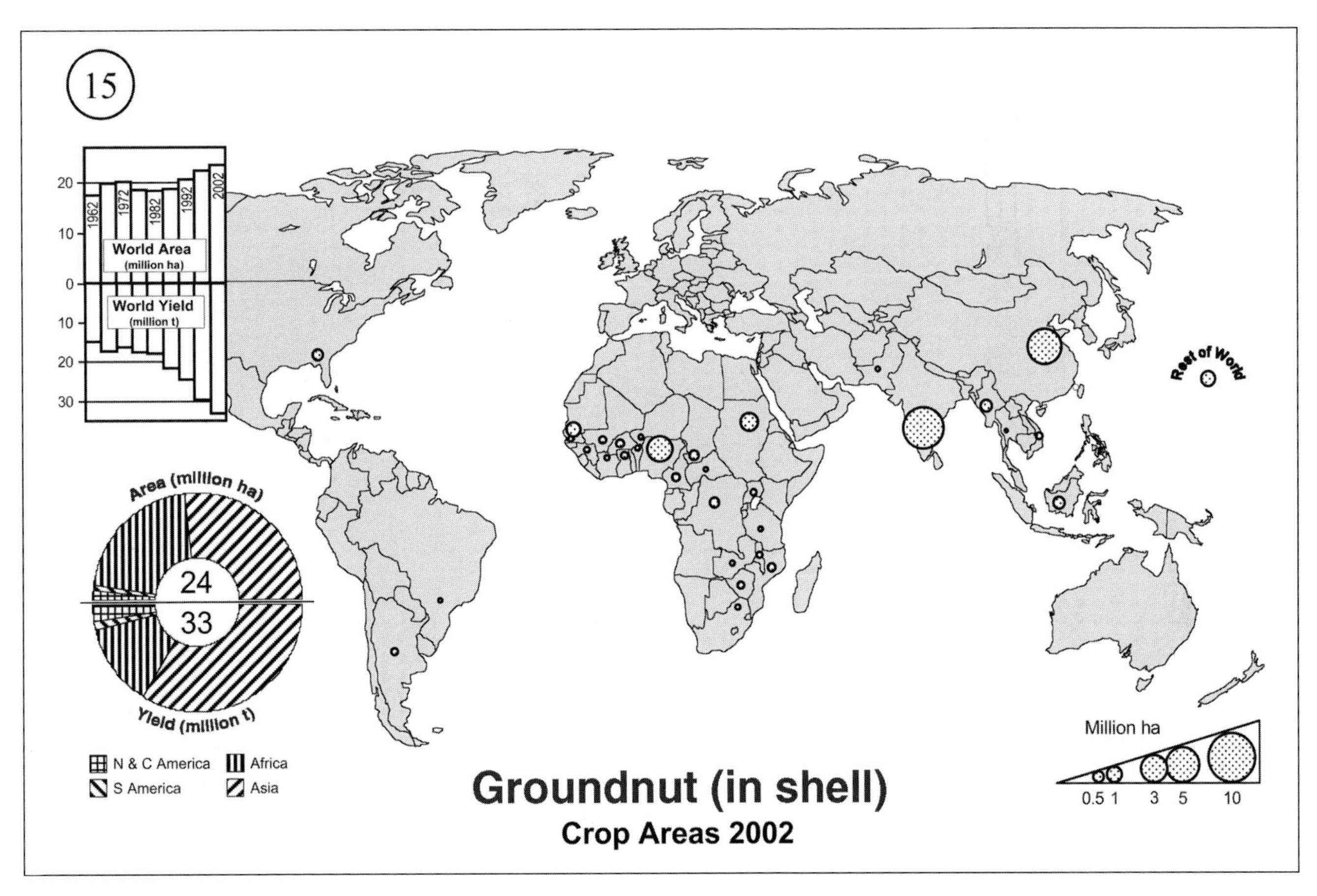
15
20
10
0
10
20
30
1962
1972
1982
1992
2002
World Area
(million ha)
World Yield
(million t)
Area (million ha)
24
33
Yield (million t)
N & C America
S America
Africa
Asia
Rest of World
Million ha
0.5 1 3 5 10
Groundnut (in shell)
Crop Areas 2002

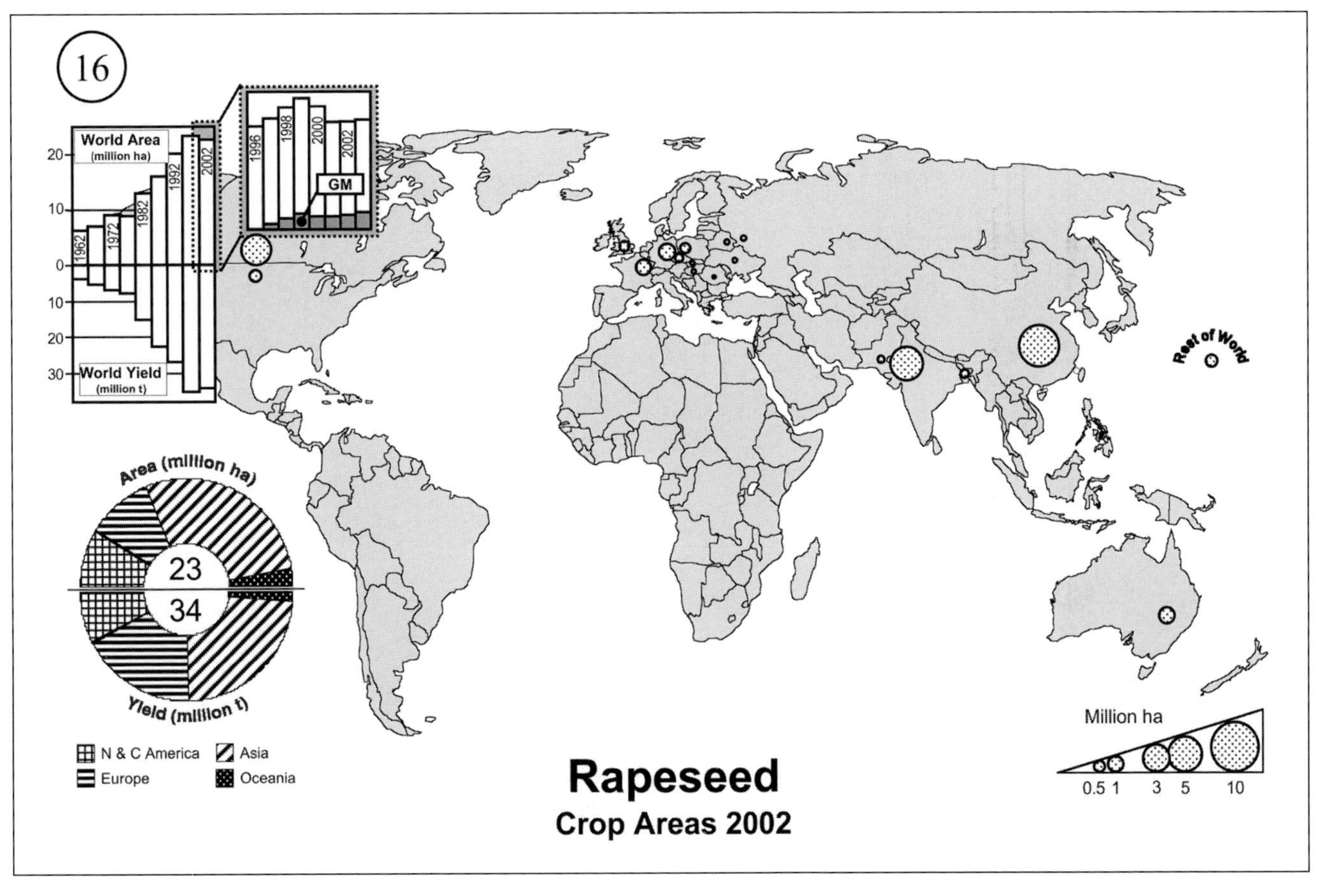
16
World Area
(million ha)
20
10
0
10
20
30
1962
1972
1982
1992
2002
1996
1998
2000
2002
GM
World Yield
(million t)
Area (million ha)
23
34
Yield (million t)
N & C America
Europe
Asia
Oceania
Rest of World
Million ha
0.5 1 3 5 10
Rapeseed
Crop Areas 2002

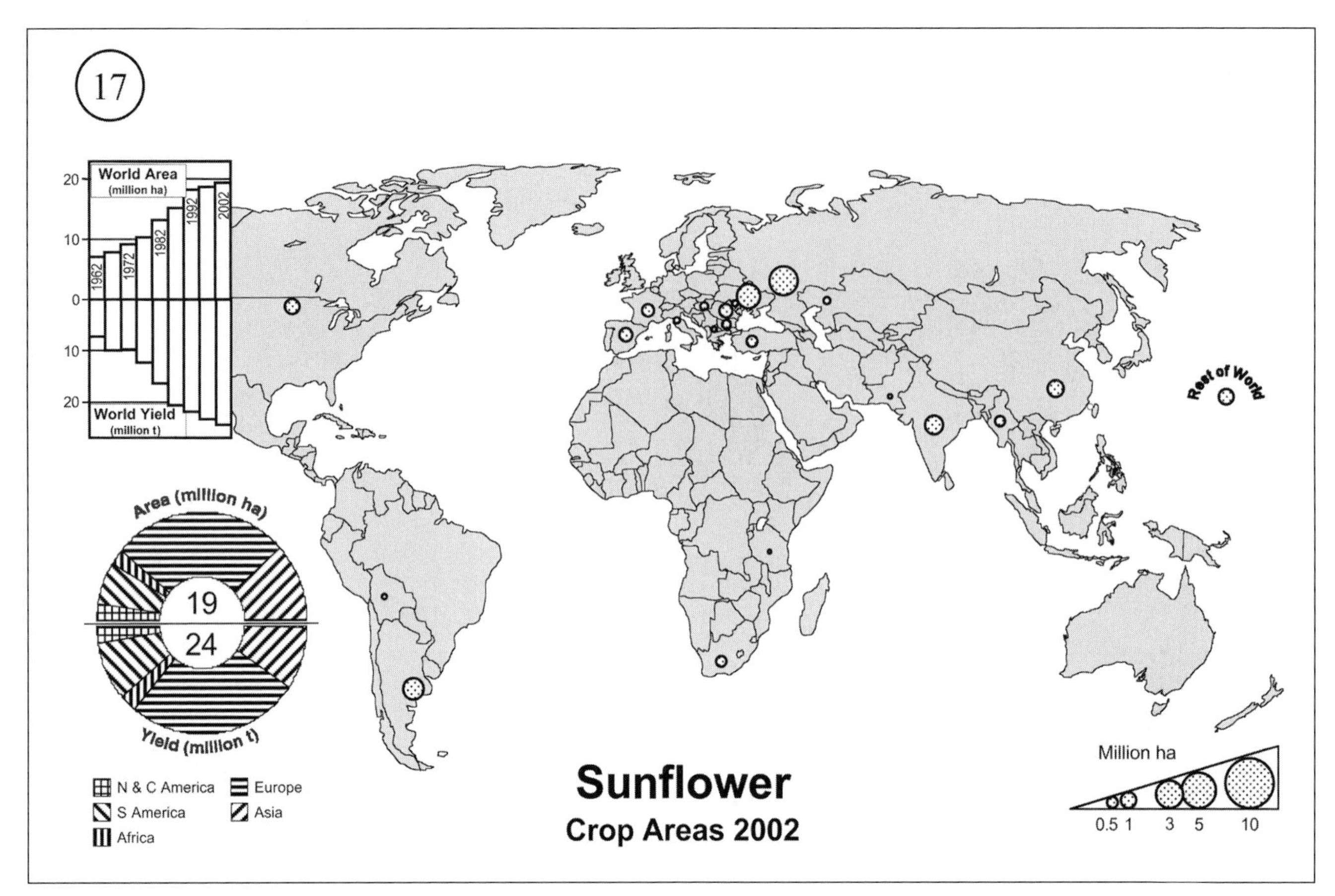
17
World Area (million ha)
World Yield (million t)
1962
1972
1982
1992
2002
20
10
0
10
20
Area (million ha)
19
24
Yield (million t)
Rest of World
N & C America
S America
Africa
Europe
Asia
Million ha
0.5 1 3 5 10
Sunflower
Crop Areas 2002

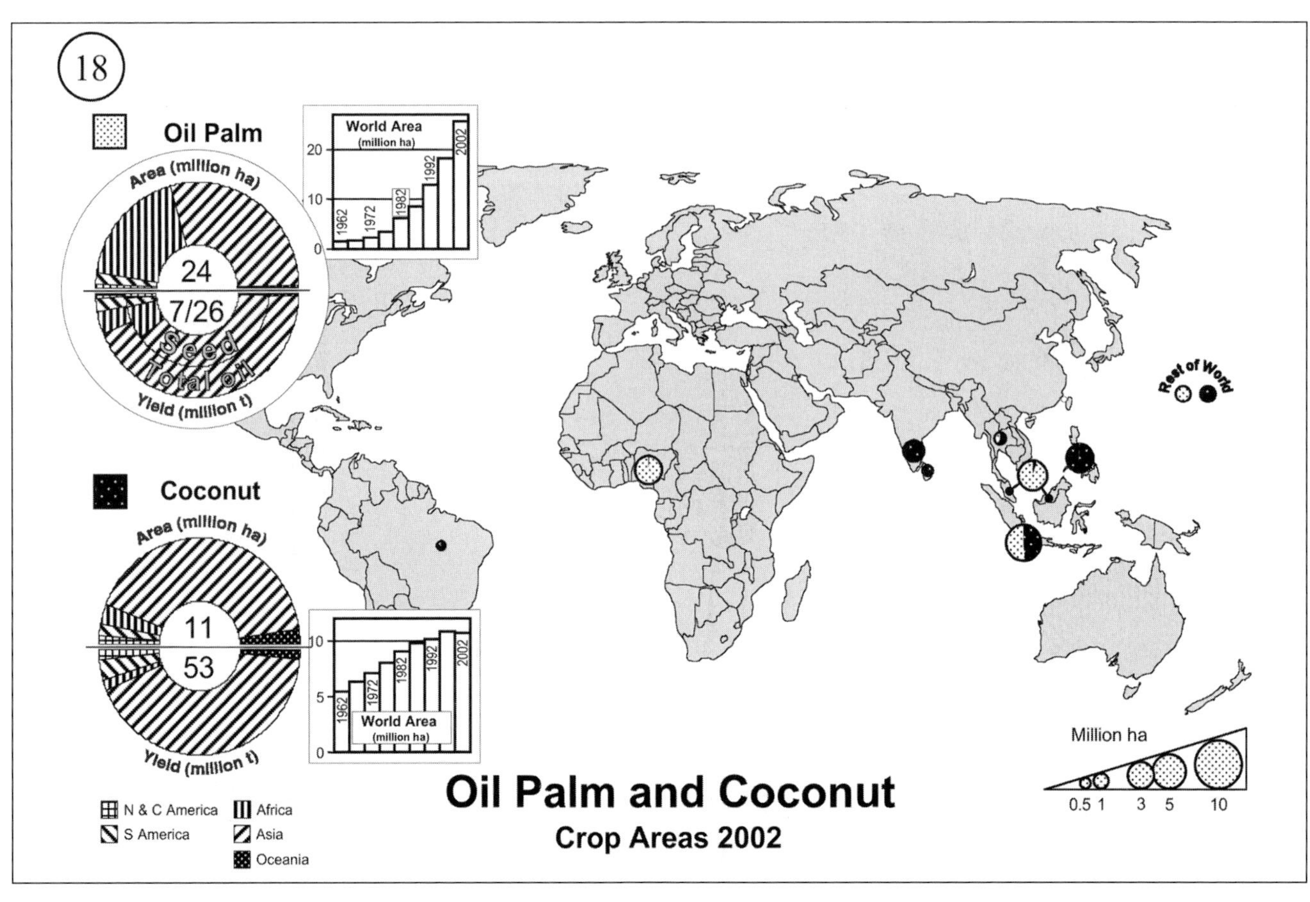
18
Oil Palm
World Area (million ha)
1962
1972
1982
1992
2002
20
10
0
Area (million ha)
24
7/26
Seed
Total oil
Yield (million t)
Coconut
Area (million ha)
11
53
Yield (million t)
World Area (million ha)
10
5
0
Rest of World
N & C America
S America
Africa
Asia
Oceania
Million ha
0.5 1 3 5 10
Oil Palm and Coconut
Crop Areas 2002

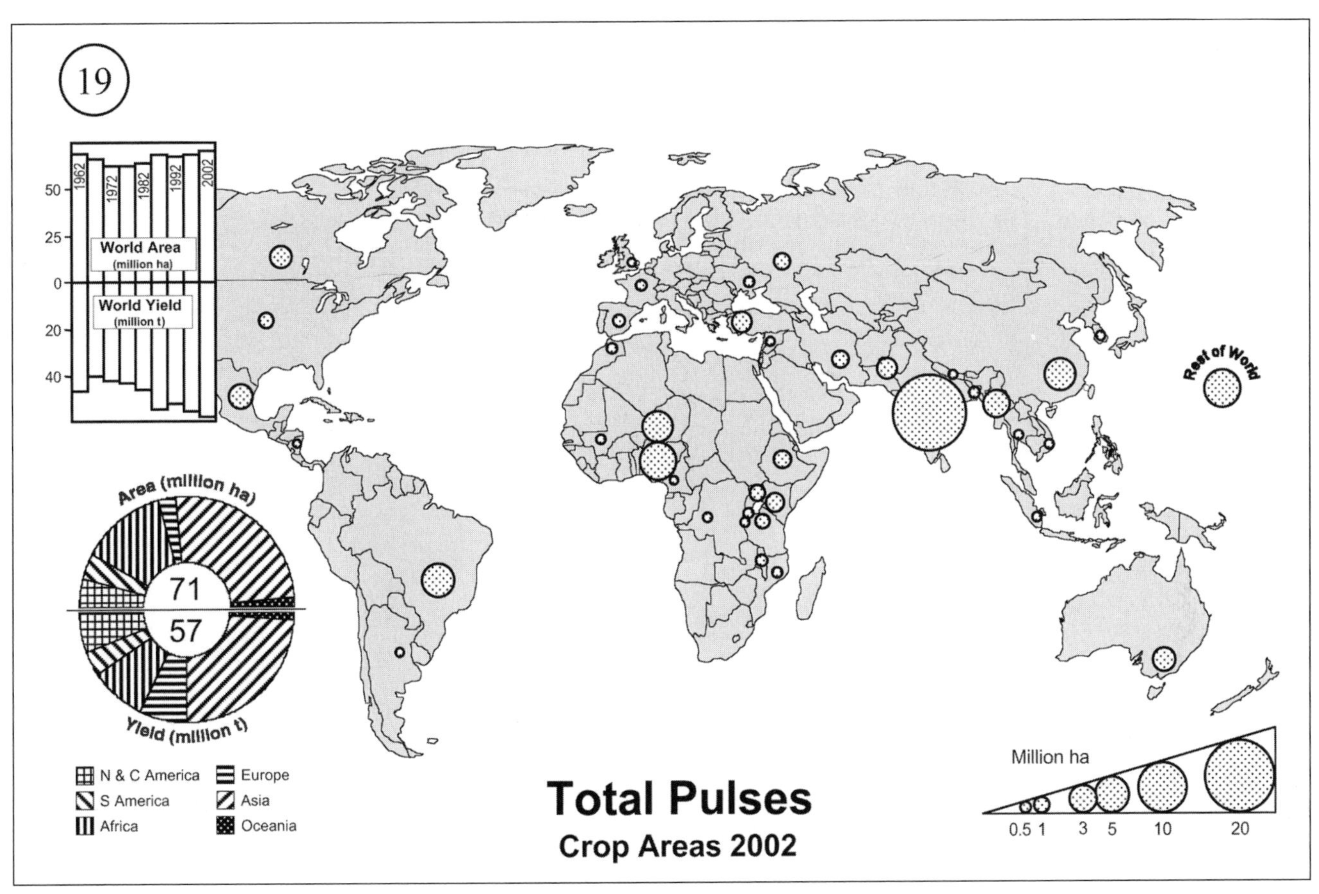
19
1962
1972
1982
1992
2002
50
25
0
20
40
World Area (million ha)
World Yield (million t)
Rest of World
Area (million ha)
71
57
Yield (million t)
N & C America
S America
Africa
Europe
Asia
Oceania
Million ha
0.5 1 3 5 10 20
Total Pulses
Crop Areas 2002

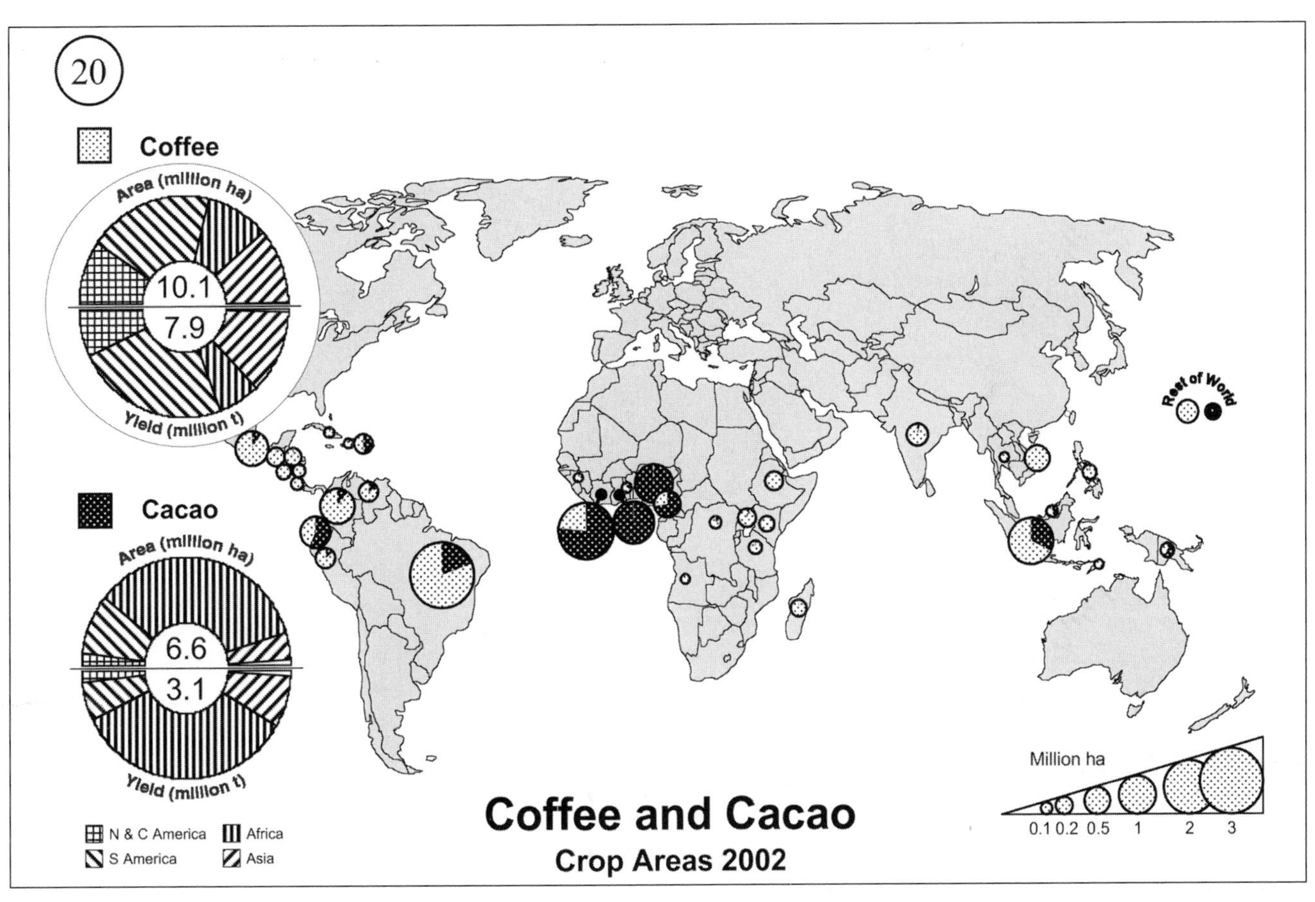
20
Coffee
Area (million ha)
10.1
7.9
Yield (million t)
Cacao
Area (million ha)
6.6
3.1
Yield (million t)
N & C America
S America
Africa
Asia
Rest of World
Million ha
0.1 0.2 0.5 1 2 3
Coffee and Cacao
Crop Areas 2002

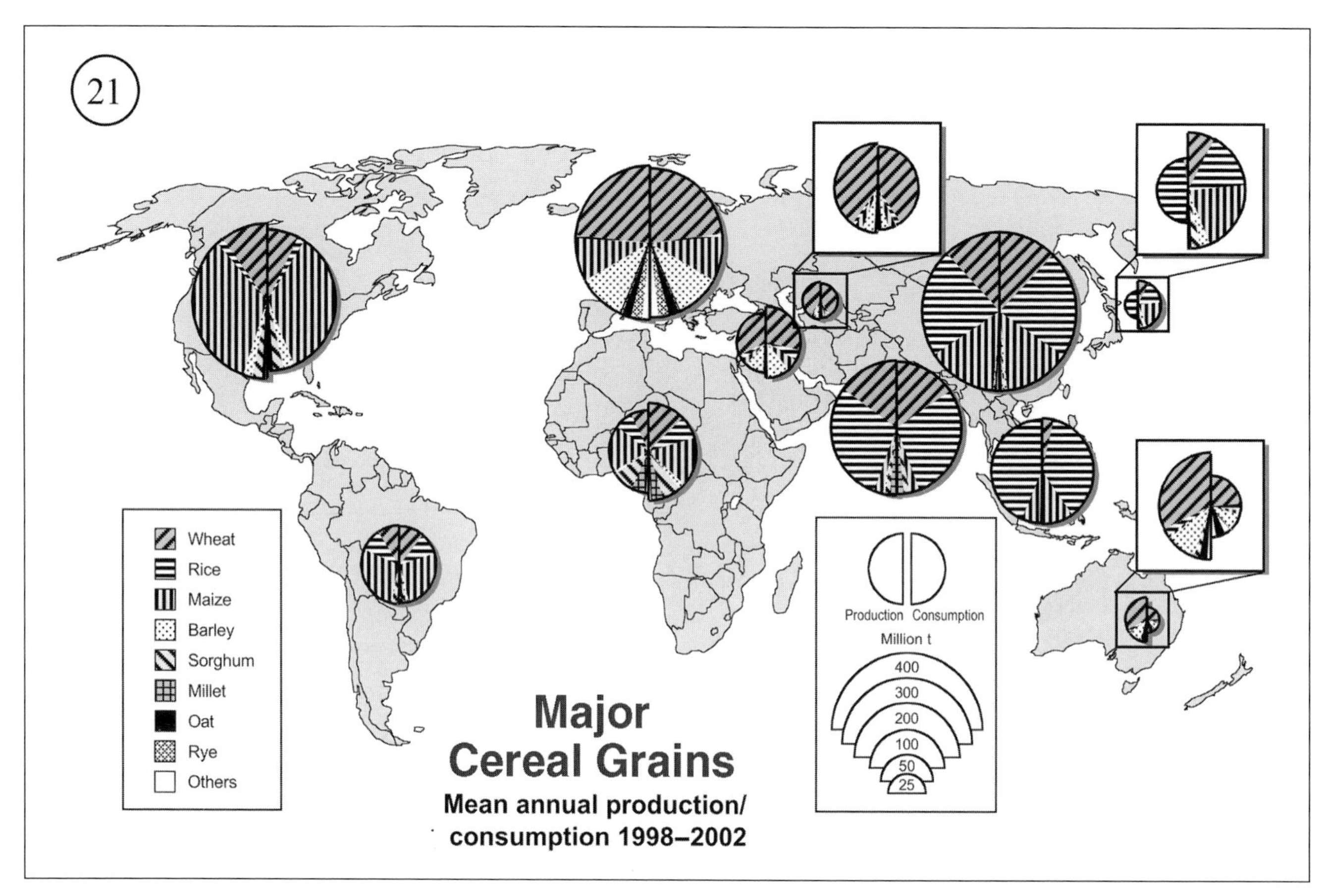
21
Wheat
Rice
Maize
Barley
Sorghum
Millet
Oat
Rye
Others
Production Consumption
Million t
400
300
200
100
50
25
Major
Cereal Grains
Mean annual production/
consumption 1998–2002

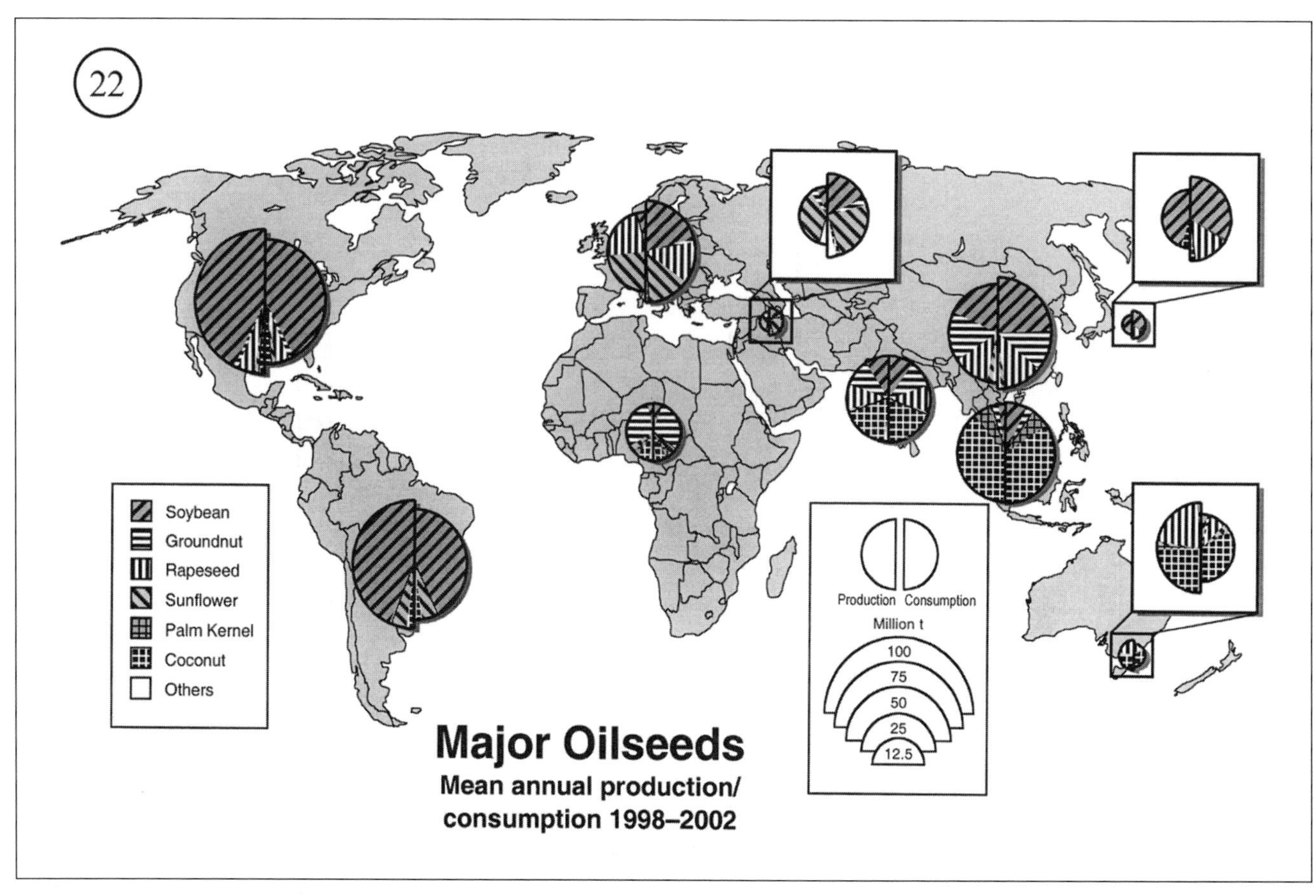
22
Soybean
Groundnut
Rapeseed
Sunflower
Palm Kernel
Coconut
Others
Production Consumption
Million t
100
75
50
25
12.5
Major Oilseeds
Mean annual production/
consumption 1998–2002

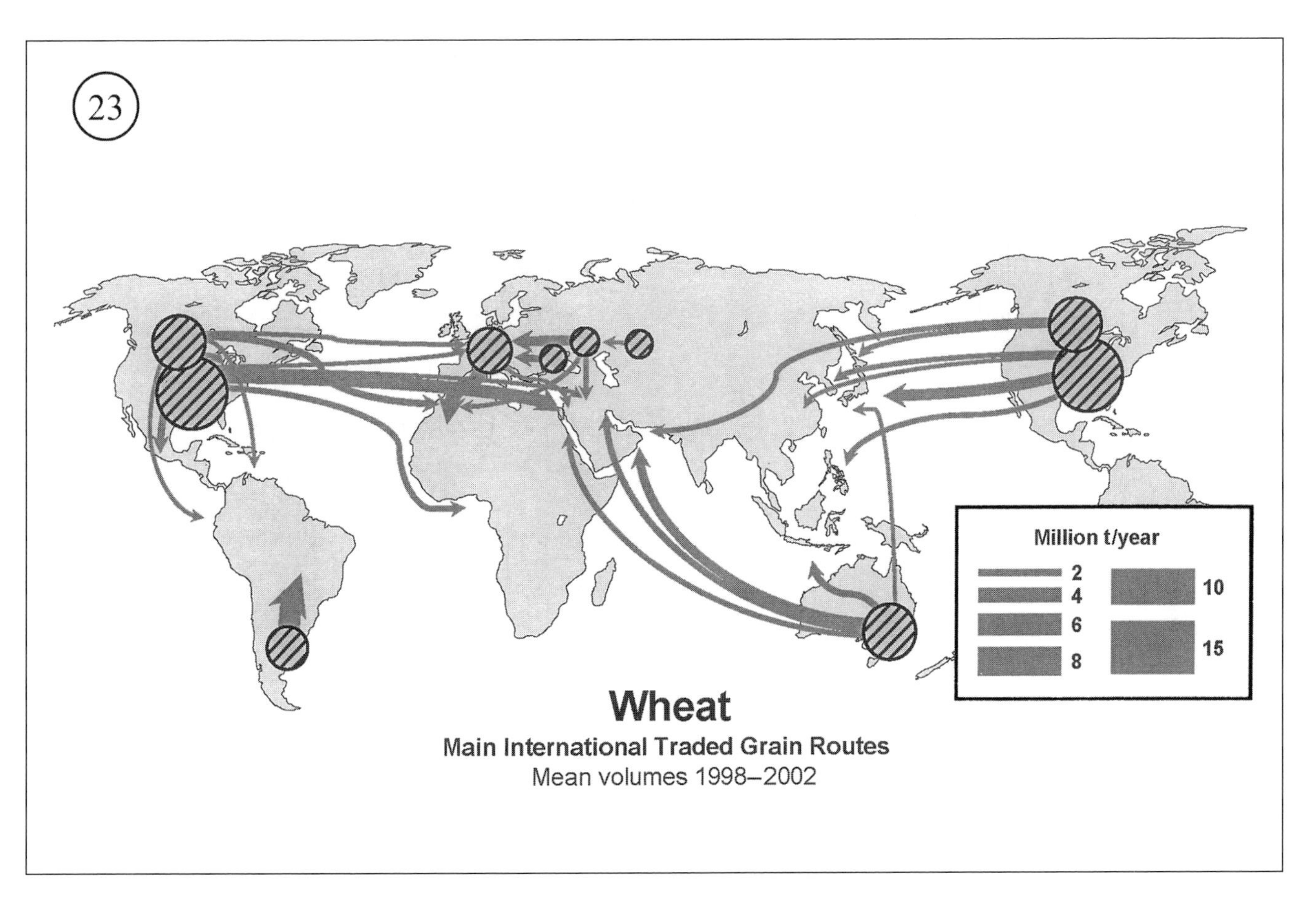

Wheat
Main International Traded Grain Routes
Mean volumes 1998–2002

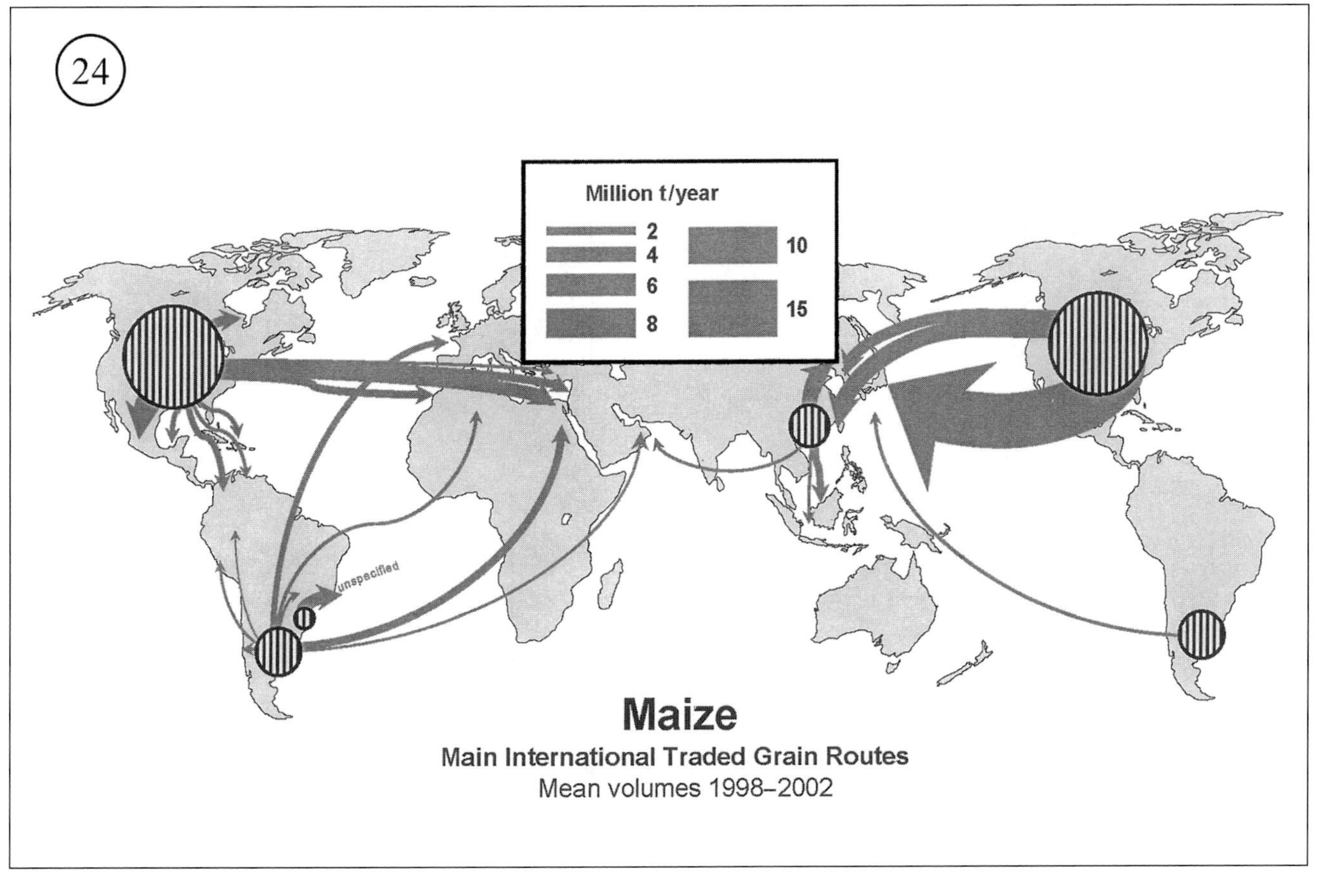

Maize
Main International Traded Grain Routes
Mean volumes 1998–2002

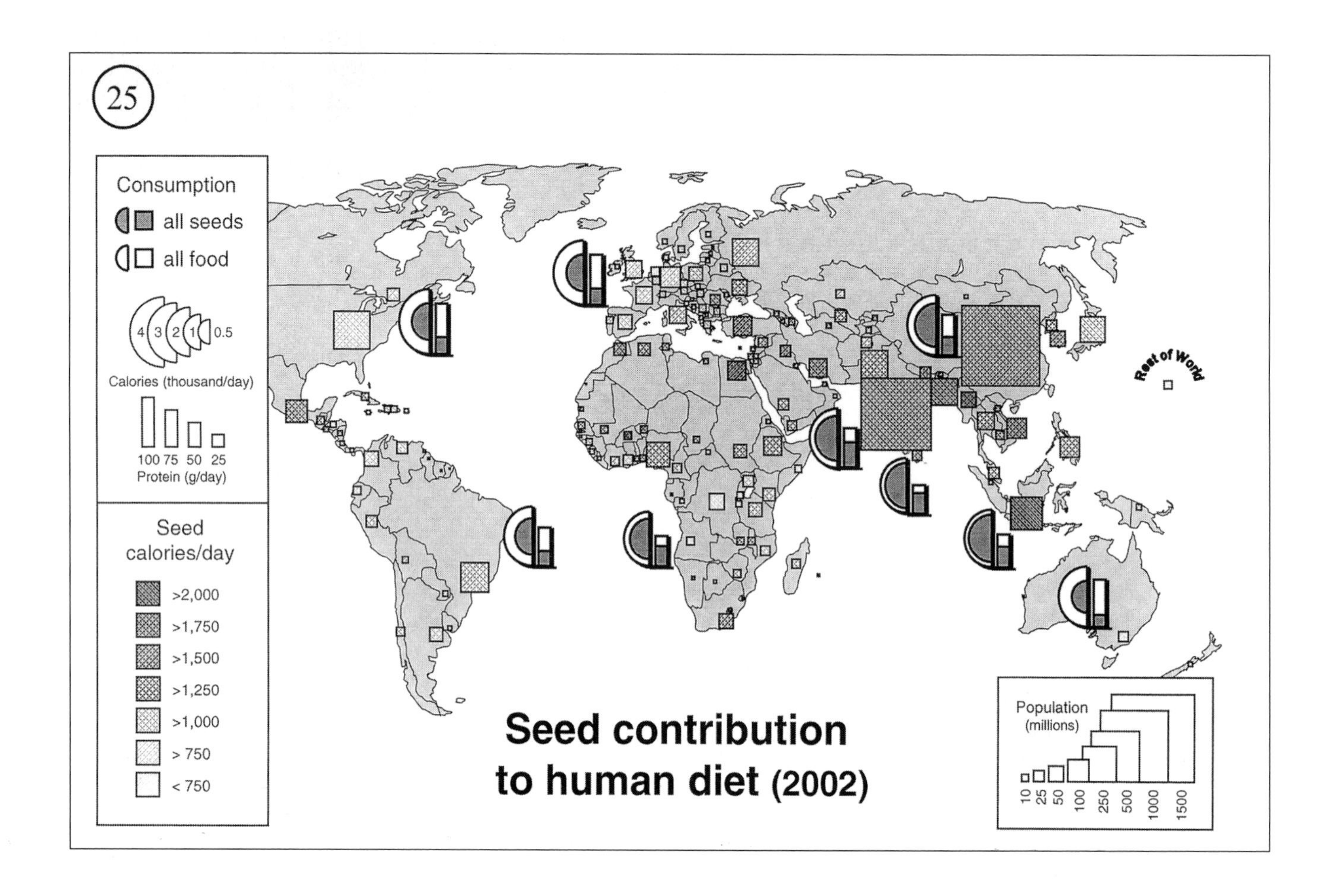
25
Consumption
all seeds
all food
4 3 2 1 0.5
Calories (thousand/day)
100 75 50 25
Protein (g/day)
Seed
calories/day
>2,000
>1,750
>1,500
>1,250
>1,000
> 750
< 750
Rest of World
Seed contribution
to human diet (2002)
Population
(millions)
10
25
50
100
250
500
1000
1500

Index of Species

Index

BOCA RATON PUBLIC LIBRARY, FLORIDA
3 3656 0460525 2